Proper and Improper Forcing
Second Edition

Since their inception, the Perspectives in Logic and Lecture Notes in Logic series have published seminal works by leading logicians. Many of the original books in the series have been unavailable for years, but they are now in print once again.

This volume, the 5th publication in the Perspectives in Logic series, studies set-theoretic independence results (independence from the usual set-theoretic ZFC axioms), in particular for problems on the continuum. The author gives a complete presentation of the theory of proper forcing and its relatives, starting from the beginning and avoiding the metamathematical considerations. No prior knowledge of forcing is required. The book will enable a researcher interested in an independence result of the appropriate kind to have much of the work done for them, thereby allowing them to quote general results.

Saharon Shelah works in the Institute of Mathematics at The Hebrew University of Jerusalem and in the Department of Mathematics at Rutgers University, New Jersey.

PERSPECTIVES IN LOGIC

The *Perspectives in Logic* series publishes substantial, high-quality books whose central theme lies in any area or aspect of logic. Books that present new material not now available in book form are particularly welcome. The series ranges from introductory texts suitable for beginning graduate courses to specialized monographs at the frontiers of research. Each book offers an illuminating perspective for its intended audience.

The series has its origins in the old *Perspectives in Mathematical Logic* series edited by the Ω-Group for "Mathematische Logik" of the Heidelberger Akademie der Wissenschaften, whose beginnings date back to the 1960s. The Association for Symbolic Logic has assumed editorial responsibility for the series and changed its name to reflect its interest in books that span the full range of disciplines in which logic plays an important role.

Arnold Beckmann, Managing Editor
Department of Computer Science, Swansea University

Editorial Board:

Michael Benedikt
Department of Computing Science, University of Oxford

Elisabeth Bouscaren
CNRS, Département de Mathématiques, Université Paris-Sud

Steven A. Cook
Computer Science Department, University of Toronto

Michael Glanzberg
Department of Philosophy, University of California Davis

Antonio Montalban
Department of Mathematics, University of Chicago

Simon Thomas
Department of Mathematics, Rutgers University

For more information, see www.aslonline.org/books_perspectives.html

PERSPECTIVES IN LOGIC

Proper and Improper Forcing

SAHARON SHELAH
Hebrew University of Jerusalem

ASSOCIATION FOR SYMBOLIC LOGIC

CAMBRIDGE
UNIVERSITY PRESS

CAMBRIDGE
UNIVERSITY PRESS

University Printing House, Cambridge CB2 8BS, United Kingdom

One Liberty Plaza, 20th Floor, New York, NY 10006, USA

477 Williamstown Road, Port Melbourne, VIC 3207, Australia

4843/24, 2nd Floor, Ansari Road, Daryaganj, Delhi – 110002, India

79 Anson Road, #06–04/06, Singapore 079906

Cambridge University Press is part of the University of Cambridge.

It furthers the University's mission by disseminating knowledge in the pursuit of education, learning, and research at the highest international levels of excellence.

www.cambridge.org
Information on this title: www.cambridge.org/9781107168367
10.1017/9781316717233

Association for Symbolic Logic
Richard A. Shore, Publisher
Department of Mathematics, Cornell University, Ithaca, NY 14853
http://www.aslonline.org

First published 2017

Printed in the United Kingdom by Clays, St Ives plc

A catalogue record for this publication is available from the British Library.

ISBN 978-1-107-16836-7 Hardback

Perspectives in Mathematical Logic

This series was founded in 1969 by the Omega Group consisting of R. O. Gandy, H. Hermes, A. Levy, G. H. Müller, G. E. Sacks and D. S. Scott. Initially sponsored by a grant from the Stiftung Volkswagenwerk, the series appeared under the auspices of the Heidelberger Akademie der Wissenschaften. Since 1986, *Perspectives in Mathematical Logic* is published under the auspices of the Association for Symbolic Logic.

Mathematical Logic is a subject which is both rich and varied. Its origins lie in philosophy and the foundations of mathematics. But during the last half century it has formed deep links with algebra, geometry, analysis and other branches of mathematics. More recently it has become a central theme in theoretical computer science, and its influence in linguistics is growing fast.
The books in the series differ in level. Some are introductory texts suitable for final year undergraduate or first year graduate courses, while others are specialized monographs. Some are expositions of well-established material, some are at the frontiers of research. Each offers an illuminating perspective for its intended audience.

Dedicated to My Beloved Son Omri

מקדש לבני האהוב עמרי

Table of Contents

Introduction

Even though the title of this book is related to the old Lecture Notes "Proper Forcing", it is a new book: not only did it have six new chapters and some sections moved in (and the thirteenth chapter moved out to the author's book on cardinal arithmetic), but the old material and also the new have been revised for clarification and corrected several times.

Now, twenty years after its discovery, I feel that perhaps "proper forcing" is a household concept in set theory, and the reader probably knows the basic facts about forcing and proper forcing. However we demand no prerequisites except some knowledge of naive set theory (including stationary sets, Fodor Lemma, strongly inaccessible, Mahlo, weakly compact etc.; occasionally we mention some large cardinals in their combinatorial definitions (measurable, supercompact), things like $0^{\#}$ and complementary theorems showing some large cardinals are necessary, but ignorance in those directions will not hamper the reader), and the book aims at giving a complete presentation of the theory of proper and improper forcing from its beginning avoiding the metamathematical considerations; in particular no previous knowledge of forcing is demanded (though the forcing theorem is stated and explained, not proved). This is the main reason for not just publishing the additional material in a shorter book. Another reason is the complaints about shortcomings of the Proper Forcing Lecture Notes.

Forcing was founded by Cohen's proof of the independence of the continuum hypothesis; Solovay and many others developed the theory (works prior to 1977). Particularly relevant to this book are Solovay and Tennenbaum [ST] and Martin and Solovay [MS], Jensen (see [DeJo]), Silver [Si67], Mitchell [Mi1], Baumgartner [B1], Laver [L1], Abraham, Devlin and Shelah [ADS:81].

We do not elaborate here on the history of the subject (but of course there are credits and references to it in each chapter or in its sections, e.g. on Baumgartner's axiom A see VII §4). And in the first two chapters we review classical material.

Our aim is to try to develop a theory of iterated forcing for the continuum. In addition to particular consistency results that are hopefully of some interest, I try to give to the reader methods which he could use for such independence results. Many of the results are presented in an "axiomatic" framework for this reason.

The main aim of the book is thus to enable a researcher interested in an independence result of the appropriate kind, to have much of the work done for him, thus allowing him to quote general results.

We know how for any partial order P (which we call here a forcing notion) we can find an extension V^P of the universe V of set theory (e.g. thinking of V as being countable); this is explained in Chapter I. So for $G \subseteq P$ generic over V (i.e. not disjoint from any dense subset of P from V), $V[G]$ is an extension of V, a model of set theory with the same ordinals, consisting of sets constructible from V and G. By fine tuning P we can get universes of set theory with various desired properties, we are particularly interested in those of the form "for every x there is y such that ...".

So, e.g., for every Souslin tree there is a c.c.c. forcing notion P changing the universe so that it is no longer a Souslin tree, but to prove the consistency of "there is no Souslin tree" we need to repeat it till we "catch our tail". For this an iteration $P * \underset{\sim}{Q}$ is defined, and $(V^P)^{\underset{\sim}{Q}} = V^{P * \underset{\sim}{Q}}$ is proved (i.e. two successive generic extensions can be conceived as one). More delicate is the limit case, and for this case "finite support iteration" $\langle P_i, \underset{\sim}{Q}_j : i \leq \alpha, j < \alpha \rangle$ works; by natural bookkeeping we can consider all Souslin trees and even all Souslin trees in V^{P_i} for $i < \alpha$, and as the c.c.c. is preserved if $\mathrm{cf}(\alpha) > \aleph_1$, since a Souslin tree can be coded as $A \subseteq \omega_1$, we "catch our tail".

In other cases countable support iteration can serve for "catching our tail"; some chain condition is then needed.

The main issue is thus *preservation*, i.e. if each $\underset{\sim}{Q}_i$ has (in V^{P_i}) a property, does P_α, the limit of an iteration, have it (this depends of course on the kind of the iteration, that is on the support). The most basic property here is not collapsing $\aleph_1$, which is treated for several kinds of iteration, i.e. supports.

Classically, a natural condition guaranteeing that $\aleph_1$ is not collapsed is (the countable chain condition) c.c.c. under finite support iterations; it is preserved (see Chapter II). It is natural to ask that the stationarity of subsets of ω_1 be preserved as well (as if $\langle S_n : n < \omega \rangle$ is a partition of ω_1 to stationary sets, we can force closed unbounded subsets E_n of ω_1 without collapsing $\aleph_1$ (see Baumgartner, Harrington and Kleinberg [BHK]) but in the limit necessarily $\aleph_1$ is collapsed). So it is natural to strengthen this somewhat (to preserve the stationarity of subsets of $\mathcal{S}_{\leq\aleph_0}(\lambda)$ for every λ), and we get the notion of proper forcing, which is preserved under the countable support iteration (see Chapter III, see alternative proof in XII §1 - and for $\aleph_1$ - free iteration in IX). This is a major notion here and the proof of its preservation serve as paradigm here.

However some forcing notions preserving $\aleph_1$ are not proper: Prikry forcing, Namba forcing. We have to use such forcing (i.e. non-proper) when during the iteration, in some intermediate stage we have to change the cofinality of some uncountable regular cardinal (e.g. $\aleph_2$) to $\aleph_1$. To deal with them we prove the preservation of semiproperness under revised countable support (see Chapter X) and show that Prikry forcing is (always) semiproper and Namba forcing is "often" semiproper. But there may be not semiproper non-proper forcing notions, in particular Namba forcing may be not semiproper (see XII §2), so we introduced properties like the S-condition (guaranteeing that no real is added (see Chapter XI)) and UP($\mathbb{I}$) (see Chapter XV), and another one (for continuum $> \aleph_2$) see Chapter XIV.

But of course we really would like to have a general framework for preserving other properties. This is dealt with in Chapter VI §1, §2; a prototypical property is preserving "every new $f \in {}^\omega\omega$ is dominated by an old $g \in {}^\omega\omega$" which is done in V §4 (but this is done only under the assumption that the forcing does not collapse $\aleph_1$ by some of the earlier versions), but includes many examples like "(f, g)-bounding for a closed enough family of pairs (f, g)". Later we deal with more general theorems (XIII §3, with a forerunner VI §3, which deals e.g. with "no new $f \in {}^\omega\omega$ dominates F").

However the case of "no new reals" is somewhat harder and it is treated in several places: in Chapter V §3 (for S-complete forcing, $S \subseteq S_{\leq\aleph_0}(\lambda)$ stationary), in Chapter V §6, §7 (for $(< \omega_1)$-proper $\mathbb{D}$-complete (e.g. $\mathbb{D}$ is a simple $\aleph_1$-completeness system)), Chapter VIII §4 (for a generalization, in particular $\mathbb{D}$ is a simple 2-completeness system) and Chapter XVIII §2 (for essentially strongly proper forcing notions). Some other properties do not fall (or are not presented) under any of those cases: strongly proper (see IX, VI §6, IX §4), not collapsing $\aleph_2$ for $\aleph_1$-complete forcing when we use mixed support (VIII §1).

In all iterations somewhere we need to prove a suitable chain condition in order to catch our tails; for finite support iteration this is done directly (II), for countable support iteration see III §4, and more VII §1, VIII §2, and for revised countable support later in XVII §4 this plays a central role.

Every iteration theorem + chain condition gives the consistency of an axiom, see VII §3, VIII §4, XVII §1, §2, §3. This stresses the problem of having a general iteration theorem for "continuum $\geq \aleph_3$".

Of course much of the book deals with specific problems which serve both as an illustration of the methods and for their interest per se (see the annotated content and the separate introductions to the chapters).

The mathematical work was done between 1977 and 1989; the author approaches other aspects of our subject in Judah and Shelah [JdSh:292], [Sh:630] (on nicely defined forcing, e.g. Borel), Rosłanowski and Shelah [RoSh:470] (on even more nicely defined forcing, quite explicitly in fact), [Sh:176] §7, §8, Goldstern and Shelah [GoSh:295] (on amalgamation and projective sets in the final model) and [Sh:311] (continuing Chapter XV), [Sh:587], [Sh:F259], [Sh:655] (on replacing $\aleph_0, \aleph_1$ by λ, λ^+) and [Sh:592] (more on FS iteration of c.c.c. forcing notions).

Prerequisites: we assume that the reader has some knowledge of naive set theory (including stationary sets, Fodor lemma, etc.). The metamathematical side is avoided by stating the forcing theorem without proof in Chapter I. If you have read Jech [J] or Kunen [Ku83] you should be well prepared. Several places present some preservation theorems with less generality and they may be of some help to the reader (and they can be treated as good preparation for this book too). Let us list some of them: Baumgartner [B3], Abraham [Ab], Bartoszyński and Judah [BaJu95], Goldstern [Go], Jech [J86], Judah and Repicky [JuRe].

* * *

There are many people who gave indispensable help for the book in various stages.

Concerning the old "Proper Forcing" I heard from Baumgartner the idea of avoiding the metamathematical side, Azriel Levy, who has a much better name than the author in such matters, made notes from the lectures in the Hebrew University, rewrote them, and they appear as Chapters I, II, and part of III. These chapters were somewhat corrected and expanded by Rami Grossberg and the author. Most of XI §1–5 were lectured on and (the first version) written up by Shai Ben David. And most of all Rami Grossberg has taken care of it in all respects and Danit Sharon typed it.

Concerning the present work, Azriel Levy has helped in transforming the book from Troff to TEX. Menachem Kojman in addition wrote up first version of section I §7 from my lectures in the eighties.

Various parts of the book benefitted a lot from proofreading and pointing gaps by my students, postdocs and co-workers. All of them contributed in various ways and I am very grateful for their help, though I failed to make a complete list of the contributors.

Most of all Martin Goldstern has helped in some very distinct capacities. Not only did he proofread and revise several chapters but also with his magic touch in various TEX-nical aspects he made the appearance of this book possible.

Last but not least I thank Hagit Levy and Gosia Rosłanowska for typing and retyping and retyping ... the book.

Thank you all.

Saharon Shelah

Institute of Mathematics
The Hebrew University of Jerusalem
91904 Jerusalem, ISRAEL

Department of Mathematics
Rutgers University
Hill Center – Busch Campus
New Brunswick, NJ 08854, USA

shelah@math.huji.ac.il
http://www.math.rutgers.edu/~shelah

The author thanks "The Israel Science Foundation" administered by The Israel Academy of Sciences and Humanities, and the United States-Israel Binational Science Foundation, and National Science Foundation USA for partially supporting this research[†] (the old and the new one) in various times.

† This footnote was omitted by mistake in [Sh:b].

Notation

Natural numbers are denoted by k, ℓ, m, n.

Ordinals are denoted by $i, j, \alpha, \beta, \gamma, \delta, \xi, \zeta$, θ where δ is reserved for limit ordinals.

Cardinals (usually infinite) are denoted by $\lambda, \mu, \kappa, \chi$. Let $\aleph_\alpha$ be the α'th infinite cardinal, $\omega_\alpha = \aleph_\alpha$, $\omega = \omega_0$. Let ${}^\beta\alpha = \{f : f$ a function from β to $\alpha\}$, ${}^{\beta>}\alpha = \bigcup_{\gamma<\beta} {}^\gamma\alpha$. For sequences of ordinals (i.e., members of some ${}^\beta\alpha$), $\ell g(\eta) = \mathrm{Dom}\eta$, $\eta(i)$ the i'th ordinal so $\eta = \langle \eta(i) : i < \ell g(\eta)\rangle$. We write also $\langle \eta(0), \ldots, \eta(n)\rangle$ or $\eta(0), \ldots, \eta(n)$ as seems fit. We denote sequences, usually of ordinals, by η, ν, ρ and also τ. The concatanation of η, ν is denoted by $\eta\hat{\ }\nu$ Let c.l.u.b. or club mean closed unbounded.

Let $|A|$ denote the cardinality of the set A, $\mathcal{P}(A)$ denote the power set of A, and $\mathrm{cf}(\alpha)$ the cofinality of α.

"A real" means here a subset of ω, or its characteristic functions. The quantifiers "$\exists^* n$", "$\forall^* n$" (sometimes written $\exists^\infty$ and $\forall^\infty$, respectively) are abbreviations for "for infinitely many $n \in \omega$" and "for all but finitely many $n \in \omega$", respectively.

Let φ, ψ denote first order formulas. Let $\varphi(x_0, \ldots, x_{n-1})$ means every free variable of φ appears in $\{x_0, \ldots, x_{n-1}\}$.

Let P (and also Q and R) denote a partially ordered set or even a quasi ordered set (i.e., $p \leq q \leq p$ does not necessarily imply $p = q$). We call such P a forcing notion, and assume $\emptyset$ ot $\emptyset_P$is a minimal member of P. We use G for a generic subset of P, (usually G_α or G_{P_α} for a generic subset of P_α), (for a definition of generic see I §1). Let p, q, r denote members of forcing notions, we say p, q are incompatible in P if they have no common upper bound in P.

We do not distinguish strictly between a model M, or a forcing notion P, and their universe. $\bigtimes_{i\in I} A_i$ is the cartesian product sometimes also denoted by $\prod_{i\in I} A_i$. Distinguish (in principle) it from $\prod_{i\in I} \mu(i)$ (multiplication of cardinals). We shall not distinguish notationally between multiplication of cardinals or ordinals. For an uncountable cardinal λ of uncountable cofinality, $\mathcal{D}_\lambda$ stands for the filter generated by the closed unbounded subsets of λ; when $S \subseteq \lambda$ is stationary (also denoted by $S \not\equiv 0 \bmod \mathcal{D}_\lambda$) $\mathcal{D}_\lambda + S$ is the filter generated by $\mathcal{D}_\lambda \cup \{S\}$.

For ordinals α, β such that $\beta < \alpha$ we let $S^\alpha_\beta = \{\gamma < \aleph_\alpha : \mathrm{cf}\gamma = \aleph_\beta\}$ but when μ, λ are (infinite) cardinals such that $\mu < \lambda$ then $S^\lambda_\mu = \{\gamma < \lambda : \mathrm{cf}\lambda = \mu\}$ when no confusion arise.

We made a special effort to uniformize the notation we use but still there may be some exceptions in the chapters, for example in Chapter III, $\mathcal{S}_{\aleph_0}(A)$ is used to denote the family of countable subsets of A (see Definition III 1.2) but the same family in Chapter V is denoted by $\mathcal{S}_{<\aleph_1}(A)$ (see Definition V 3.3). So since some of the notions are redefined the reader is advised to check the nearest definition rather than the first.

Models are denoted by letters M, N perhaps with an index. We shall not always distinguish between the model and its universe, but always $|M|$ will denote the universe of M and $\|M\|$ its cardinality.

Content by Subject

1. Results Outside Set Theory

In the Appendix, Sect. 2 there are results on the power of Ext $(G, \mathbb{Z})$ assuming various weak diamonds.

2. Results in Naive Set Theory

In the Appendix §1 we investigate weak variants of the diamond (continuing Devlin and Shelah [DvSh:65].)

In the Appendix §3 we prove that CH implies some kind of weak diamond to S_1^2 (and generalizations).

In IX §3 we discuss various specializations of Aronszajn trees.

In XIII §4 a sufficient condition for the existence of Ulam filters is given.

3. Basics of Forcing

In Chapter I we explain how to use forcing and discuss some basic forcing. In Chapter II we explain iteration with $< \kappa$-support, (in particular finite support) deal for example with Martin Axiom, and more examples.

4. Specific Independence Results on Trees

In III 5.4 we present a proof that every $\aleph_1$-Aronszajn tree can be specialized by a c.c.c. forcing notion (and generally in III §5 deal with κ-trees).

In III §6 we present a proof of the consistency of " ZFC$+2^{\aleph_0} = \aleph_2+$ there is no $\aleph_2$-Aronszajn trees".

In V §6, §7, we present a proof of CON (ZFC+G.C.H.+ every Aronszajn tree is special) which seems to us more adaptable to further needs (e.g. $2^{\aleph_1}$ large).

In V §8 we present a proof of the consistency of the Kurepa hypothesis.

In VII §3 we prove consistency results on Aronszajn trees strengthening the previous ones, motivated by general topology. This implies the consistency of G.C.H. + there is a countably paracompact regular space which is not normal (see VII 3.25).

In VII §3 we present a proof of the consistency of a strengthening of CON (ZFC+G.C.H. + there is no $\aleph_1$-Kurepa tree.)

In VIII §3 we prove the consistency of ZFC+CH+SH+$2^{\aleph_1} > \aleph_2$.

In IX§4 we prove that SH$\not\Rightarrow$ " every Aronszajn tree is special" and variations.

5. Theorems on $\aleph_1$-Complete Forcing

In VIII §1 we prove that we can iterate $\aleph_2$-complee forcing and $\aleph_1$-complete forcing satisfying a strong $\aleph_2$-chain condition, without collapsing $\aleph_1$ and $\aleph_2$.

In VIII 2.7A we remark on another strong $\aleph_2$-chain condition preserved by CS iteration.

6. Chain Conditions

The c.c.c. and κ-c.c. are presented in Chapter II and the preservation of the c.c.c. by FS is proved there.

In III 4.1 we prove that a CS iteration of proper forcing notions of power $< \kappa$, κ regular $(\forall \mu < \kappa)\ \mu^{\aleph_0} < \kappa$, of length κ satisfies the κ-c.c. (by proving that if the length is $< \kappa$, it has a dense subset $< \kappa$).

Of course IV is dedicated to oracle-c.c.

Lemma V 1.5 proves the $\aleph_2$-c.c. of a CS iteration of E-complete forcing notions each of power $\aleph_1$, (assuming CH).

VII §1 deals with a strong κ-chain (e.c.c.), such that if we have a CS iteration of length κ with the condition we use for not adding reals (in V§7, VIII§4) then the forcing satisfies the κ-c.c. So this helps to get consistency results with ZFC+CH.

In VII §2 we deal with κ-pic (= κ-properness isomorphism conditions). If P satisfies it, then P satisfies the κ-c.c., and adds $< \kappa$ reals, and an iteration of length κ still satisfies the κ-c.c. This helps to get consistency results with ZFC+$2^{\aleph_0} = \aleph_2 + 2^{\aleph_1} = \lambda$ (so we start with $V \models$ "$CH + 2^{\aleph_1} = \kappa$" and this holds for the intermediate stages). Application is starting with $V \models$ "CH+$2^{\aleph_1} = \kappa$" and use a CS iteration $\bar{Q} = \langle P_i, Q_i : i < \omega_2 \rangle$ to specialize all Aronszajn trees without adding reals.

On the κ-c.c. for RCS iteration of length κ see X §5 (5.3, 5.4) and XI 6.3(2).

On λ-c.c. for κ-RS iterations see XIV.

7. Preservation of Properness and Variants

As the book was written in the generic way, i.e. as the author advance, there are several such proofs. In III §3 the preservation of properness by CS iteration is proved. In IX §2 the preservation of properness under $\aleph_1$-free limit is proved. In X §2 the preservation of properness and semi-properness under RCS is proved. (remember that semi-proper forcing may change the cofinality of some regular $\lambda > \aleph_1$ to ω). In XII §2 the preservation of properness under CS iteration is reproved, using the definition of properness by games (similarly for semiproperness) .

We deal with preservation of α-properness and $(\omega, 1)$- properness in V §2, §3, X §7.

8. Consistency Results on the Uniformization Properties and Variants

In II §4 we prove the consistency of "some family $\mathcal{P}$ of $\aleph_1$ subsets of ω has k-uniformization property" i.e. if $f_A : A \to k$ for $A \in \mathcal{P}$ then for some $f : \omega \to k$ we have

$$A \in \mathcal{P} \Rightarrow (\forall^* n \in A)(f(n) = f_A(n)).$$

In V§1, we prove the consistency with G.C.H. of "for some stationary $S \subseteq \omega_1$, $\langle A_\delta : \delta < \omega_1 \rangle$ have the $\aleph_0$-uniformization property if each A_δ is a set of order type ω, $\sup(A_\delta) = \delta$".

In VIII §4 we prove the consistency with G.C.H. of $\neg \Phi^3_{\aleph_2}$. Moreover, e.g.: for $\langle A_\delta : \delta < \omega_1 \text{ limit} \rangle$ as above, $n_\delta < 3$ then for some $f : \omega_1 \to 3$, for every $\delta < \omega_1$ limit, for some $m_\delta \in \{0, 1, 2\} \setminus \{n_\delta\}$, we have $(\forall^* \alpha \in A_\delta)(f(\alpha 0 = m_\delta)$.

9. Consistency of "Large Ideals"

In XIII we get consistency results on ω_1, in XVI use smaller ones, in XIV with larger continuum.

In XVII §4 we deal with properties consistent with $\neg 0^{\#}$ (on ω_1 being an α-th function from ω_1 to ω_1 even $\alpha = 2^{\aleph_1}$).

10. Other Consistency Results

In III §7 we deal with " for a family of $\aleph_1$ countable subsets A_i of $\omega_1 (i < \omega_1)$, order type $(A_i) < \mathrm{Sup} A_i$, there is a club $C \subseteq \omega_1$ $\bigwedge_i \mathrm{Sup}(C \cap A_i) < \mathrm{Sup} A_i$.

In VI §4 we prove the consistency of " there is no P-point".

In VI §5 we prove the consistency of " there is a unique Ramsey ultrafilter (on ω)".

In X, XI we prove various independence results on ω_2, read XI §1, X 8.4, and XI §7's theorem.

In VII §3, §4, many applications are listed.

In XVIII §4 we prove the consistency of "there is a unique P-point (on ω)" (which necessarily is a Ramsey ultrafilter).

11. Other Preservations of Generalizations of Properness

In V §3 we prove the preservation of ω-properness+ the ω^ω-bounding property (properness suffices - see VI).

In VI §1, §2, §3, XVIII §3 we deal with general preservation theorems, e.g. of covering models, hence of various specific properties which can be formulated this way (F-bounding properties, Sacks property, Laver property, PP-property, D generates a Ramsey ultrafilter, D generates a P-point (ultrafilter)). Now VI §2 relies on VI §1 to give specific results (using covering models); a different approach, guaranteeing only preservation continue to hold in limit stages, is VI §3, which apply e.g. to "$F \subseteq {}^\omega\omega$ is not dominated by $\cdot$"; this is further developed in XVIII §3.

In IX §4 we deal with preservation of "(T^*, S)-preserving" which means T^* looks like Souslin trees at levels $\delta \notin S$, and in the end we comment on possible generalizations.

Not adding reals is dealt with in V §1, §2 and mainly V §7, VIII §4 and X §7 and in XVIII §2.

In V §1 and X §3 we deal with preservation of generalizations of $\aleph_1$-completeness.

In XI §5, §6 we deal with the preservation of the S-condition (always satisfied by Nm which change the cofinality of $\aleph_2$ to $\aleph_0$ but guarantees reals are not added).

In XV §3 we deal with a generalization, i.e. proving preserving of a condition implying $\aleph_1$ not collapsed, which is satisfied by all proper forcing notions and all forcing satisfying the S-condition.

12. Forcing Axioms

On consistency, see III §4, VII §2, VIII §3, X 2.6, XIII, XIV, XVI §2. Also XVII §3 (SPFA$\nvdash$ weak Chang conjecture), XVII §1 (SPFA$\equiv$ MM), XVII §2 (SPFA$\nvdash$ PFA$^+$)

13. Counterexamples

In VII §5 we build an iteration $\langle P_n, \underset{\sim}{Q}_n : n < \omega\rangle$ such that each $\underset{\sim}{Q}_n$ does not collapse any stationary subset of ω_1; but any limit we take collapses $\aleph_1$.

In III §4 we build (in ZFC) a forcing notion of power $\aleph_1$, not collapsing $\aleph_1$ but also not preserving the stationarity of some $S \subseteq \omega_1$.

Examples of iteration $\langle P_n, \underset{\sim}{Q}_n : n < \omega\rangle$ each $\underset{\sim}{Q}_n$ is α-proper for each countable α not adding reals, but any limit we take that collapses $\aleph_1$ is presented in V 5.1 using Appendix §1 (really the previous result of Devlin and Shelah [DvSh:65]).

In XII §2 we show that many forcing notions satisfy the conditions from XI but are not semiproper. In fact if there is an $\{\aleph_1\}$-semiproper forcing notion changing cofinaltity of $\aleph_2$ to $\aleph_0$ (e.g. if Nm is $\{\aleph_1\}$-semiproper then Chang conjecture holds.

Examples of iteration $\langle P_n, \underset{\sim}{Q}_n : n < \omega\rangle$, each Q_n is a α-proper and is a $\mathbb{D}$-complete for some simple $\aleph_1$-complete completeness system but any limit adds reals is presented in XVIII §1.

Annotated Content

I. FORCING, BASIC FACTS

§0. Introduction

§1. Introducing Forcing

We define generic sets, names for a forcing notion, and formulate the forcing theorems.

§2. The Consistency of CH

Our aim is to construct by forcing a model of ZFC were CH holds. First we explain the problem of not collapsing cardinals, and second prove that $\aleph_1$-complete forcing notion does not add reals.

§3. On the Consistency of the Failure of CH

We construct a model of ZFC in which the Continuum Hypothesis fails; define the c.c.c., prove that forcing with c.c.c. forcing preserves cardinalities and cofinalities, and prove also the Δ-system lemma for finite sets.

§4. More on the Cardinality $2^{\aleph_0}$ and Cohen Reals

We construct for every cardinal λ in V which satisfies $\lambda^{\aleph_0} = \lambda$ a model $V[G]$ such that $V[G] \models 2^{\aleph_0} = \lambda$. Also Cohen reals are defined.

§5. Equivalence of Forcing Notions, and Canonical Names

We define when two forcing notions are equivalent, introduce canonical names and prove that for every P-name $\underset{\sim}{\tau}$ there is a canonical P-name $\underset{\sim}{\sigma}$ such that $\Vdash_P$ "$\underset{\sim}{\tau} = \underset{\sim}{\sigma}$".

§6. Random Reals, Collapsing Cardinals and Diamonds

We introduce random reals and the Levy collapse, and prove that for regular λ Levy$(\aleph_0, < \lambda)$ satisfies the λ-c.c. For every uncountable regular λ and a stationary $S \subseteq \lambda$ define a forcing notion P which preserve the regularity of λ and stationarity of S add no bounded subsets to λ and such that $V^P \models \diamondsuit_S$.

§7. $\clubsuit$ Does Not Imply Diamond

$\clubsuit$ is a weak relative of diamond, for $S \subseteq \lambda$ stationary $\clubsuit(S)$ says that we can find $\langle A_\delta : \delta \in S\rangle$ $A_\delta \subseteq \delta = \sup(A_\delta)$, and for every unbounded $A \subseteq \lambda$ for some ($\equiv$ stationarily many) $\delta \in S$, $A_\delta \subseteq A$. If CH, $\clubsuit(\aleph_1) \equiv \diamondsuit_{\aleph_1}$, so we prove the consistency of $\clubsuit(\aleph_1) + \neg CH$, but forcing three times.
Start with $V \models GCH$, for $\diamondsuit_s$ for $S = \{\delta < \aleph_2 : \mathrm{cf}(\delta) = \aleph_0\}$, using it construct a "very good" $\clubsuit(S)$ sequence $\langle A_\delta : \delta \in S\rangle$. Then force by adding $> \aleph_2$ subsets of $\aleph_1$ by countable conditions, $\bar{A}$ was constructed to withstood it, lastly collapse $\aleph_1$ to $\aleph_0$ by Levy$(\aleph_0, \aleph_1)$. As any new unbounded subset of $\aleph_2$ contains an old one, $\bar{A}$ still witness $\clubsuit(S)$ but now S is a stationary subset of $\aleph_1$ of the last universe.

II. ITERATION OF FORCING

§0. Introduction

§1. The Composition of Two Forcing Notions

Composition of two notions and state the associativity lemma are defined.

§2. Iterated Forcing

We define iterated forcing, and prove that the c.c.c is preserved by FS (finite support) iteration.

§3. Martin's Axiom and Few Applications

We prove that ZFC $+2^{\aleph_0} > \aleph_1$+MA is consistent. Use MA to prove many simple uniformization properties.

§4. The Uniformization Property

Here we deal with more general uniformization properties some of which contradict MA. We strenghten the demand of almost disjointness to being a kind of tree and prove the consistency of a version contradicting MA.

§5. Maximal Almost Disjoint Families of Subsets of ω

A maximal almost disjoint (MAD) subset of $\mathcal{P}(\omega)$ is a family of infinite subsets of ω such that the intersection of any two members is finite and maximal with this property. We prove using MA that every MAD set has cardinality $2^{\aleph_0}$. Also the other direction: for every $\aleph_1 \leq \lambda < 2^{\aleph_0}$ there exists a generic extension of V by c.c.c. forcing such that in it there exist mad set of power λ.

III. PROPER FORCING

§0. Introduction

§1. Introducing Properness

We define "P is a proper forcing notion ", prove some definitions are equivalent (and deal with the closed unbounded filter $\mathcal{D}_{\aleph_0}(\lambda)$).

§2. More on Properness

We define "p is (N, P)-generic " and deal more with equivalent definitions of properness.

§3. Preservation of Properness Under CS Iteration

We prove the theorem mentioned in the title, CS is countable support.

§4. Martin's Axiom Revisited

We discuss the popularity of the c.c.c., whether we can replace it by a more natural and weaker condition. We give a sufficient condition for a countable support iteration of length κ to satisfy the κ-c.c. We prove the consistency (assuming existence of an inaccessible cardinal) of "ZFC + $2^{\aleph_0} = \aleph_1$+ Ax[for forcing notions not destroying stationary subsets of ω_1 of cardinality $\aleph_1$]". We show that the last demand cannot be replaced by "not collapsing cardinalities or cofinalities".

§5. On Aronszajn Trees

We define κ-Aronszajn and κ-Souslin trees. We then present existence theorems (for λ^+ when $\lambda = \lambda^{<\lambda}$) and prove that under MA every Aronszajn tree is special.

§6. Maybe There Is No $\aleph_2$-Aronszajn Tree

We prove the consistency of ZFC $+2^{\aleph_0} = \aleph_2$+ there is no $\aleph_2$-Aronszajn tree, the method being collapsing successively all $\lambda, \aleph_1 < \lambda < \kappa$ (κ a weakly compact cardinal) and treating every potential initial segment of an $\aleph_2$-Aronszajn tree to ensure it will not actually become an initial segment of such a tree.

§7. Closed Unbounded Subsets of ω_1 Can Run Away from Many Sets

We prove the consistency of ZFC $+2^{\aleph_0} = \aleph_2$ with if for $i < \omega_1$, $A_i \subseteq \omega_1$ has order type $<$ SupA_i, then for some closed unbounded $C \subseteq \omega_1, (\forall i)[\mathrm{Sup}(C \cap A_i) < \mathrm{Sup} A_i]$.

IV. ON ORACLE-C.C., THE LIFTING PROBLEM OF THE MEASURE ALGEBRA, AND "$\mathcal{P}(\omega)$/FINITE HAS NO NON-TRIVIAL AUTOMORPHISM"

§0. Introduction

The oracle-c.c. method enables us to start with $V \models \diamondsuit_{\aleph_1}$ and extend the set of reals ω_2-times (by iterated forcing), in the intermediate stages $\diamondsuit_{\aleph_1}$ holds, and we omit types of power $\aleph_1$ along the way, i.e. promise that some intersections of $\aleph_1$ Borel sets remain empty.

§1. On Oracle Chain Condition

One way to build forcing notions satisfying the $\aleph_1$-c.c., is by successive countable approximations including promises to maintain the predensity of countably many subset, many times using the diamond. We formalize a corresponding property ($\bar{M}$-c.c., $\bar{M}$ an oracle) and prove the equivalence of some variants of the definition.

§2. The Omitting Type Theorem

We prove that if the intersection of $\aleph_1$ Borel sets is empty and even if we add a Cohen real it remains empty, (and $\diamondsuit_{\aleph_1}$) *then* for some oracle $\bar{M}$, for every forcing notion P satisfying the $\bar{M}$-c.c., in V^P the intersection of the Borel sets (reinterpreted) is still empty.

§3. Iterations of $\bar{M}$-c.c. Forcings

We show that for Finite Support iteration $\bar{Q} = \langle P_i, \underset{\sim}{Q}_i : i < \alpha \leq \omega_2\rangle$, if $\bar{M}_i \in V^{P_i}$ is an $\aleph_1$-oracle large enough for $\langle\langle \bar{M}_j, P_j, \bar{Q}_j\rangle : j < i\rangle$, and Q_i satisfies the $\bar{M}_i$-c.c. *then* $P_\alpha = \mathrm{Lim}\bar{Q}$ satisfies the $\bar{M}_0$-c.c. The first three sections give the exact formulation of the aim stated in the introduction and prove that it works.

§4. The Lifting Problem of the Measure Algebra

We show how to apply the method described in §1 - §3 in order to get a model in which the natural homomorphism $\mathcal{B} \to \mathcal{B}/$ (measure zero sets) does not lift. Where $\mathcal{B}$ is the Boolean algebra of Borel sets of reals.

§5. Automorphisms of $\mathcal{P}(\omega)$/finite

We use our method to prove that the Boolean algebra "power set of ω divided by the ideal of finite sets" can have only trivial automorphisms, where those are defined as the ones induced by permutations of ω or "almost" permutation of ω. This is equivalent to having the topological space $\beta(\mathbb{N}) \setminus \mathbb{N}$ having only trivial autohomeomorphisms. However a main lemma is delayed to the next section.

§6. Proof of Main Lemma 5.6

The point missing in §5 is: if F is an automorphism of $\mathcal{P}(\omega)$/finite, $\bar{M}$ an $\aleph_1$-oracle, then there is forcing notion P satisfying the $\bar{M}$-c.c., and a P-name $\underset{\sim}{X}$ of a real such that in V^P, for no $Y \subseteq \omega$, $(\forall A, B \in \mathcal{P}(\omega)^V)[X \cap A =_{ae} B \Rightarrow \underset{\sim}{Y} \cap F(A) =_{ae} F(B)]$, moreover even a Cohen forcing does not introduce such a Y, where $=_{ae}$ means equal modulo finite. We try to build such P, $\underset{\sim}{X}$ and prove that if we always fail F is trivial.

V. α-PROPERNESS AND NOT ADDING REALS

§0. Introduction

§1. $\mathcal{E}$-Completeness – a Sufficient Condition for Not Adding Reals

We define what it means to be $\mathcal{E}$-complete e.g., if $P \subseteq ({}^{\omega_1 >}2, \triangleleft)$, $\mathcal{E} \subseteq \omega_1$ stationary and $f_n \subseteq f_{n+1} \in P$, $\mathrm{Sup}(\mathrm{Dom} f_n) \in \mathcal{E}$. We show that properness + $\mathcal{E}$-completeness are preserved by CS iteration and get corresponding Axiom. We also introduce a forcing axiom which is consistent with CH use it to prove a uniformization property which implies existence of a non free Whitehead group. In this way we get universe in which ω_1 is "schizophrenic": for two disjoint stationary subsets S_1, S_2 we have $\diamondsuit$ on S_1 but on S_2 the situation is as with MA.

§2. Generalizations of Properness

We introduce various variants of properness. We find it interesting and beneficial to have properties like properness dealing with sequences of countable models defining in particular α-properness.

§3. α-Properness and $(\mathcal{E}, \alpha)$-Properness Revisited

We repeat the previous section in more detail.

§4. Preservation of "ω-Properness + the ${}^{\omega}\omega$ Bounding Property"

P satisfies the ${}^{\omega}\omega$-bounding property if $[\forall f \in ({}^{\omega}\omega)^{V^P}][\exists g \in ({}^{\omega})^V]$ $(\wedge_n f(n) < g(n))$ i.e.: every new function $f : \omega \to \omega$ is dominated by an old one. We prove in great detail the theorem stated in the title as it serve a prototype of having preservation of "properness + X", dealt with later and is a case of a central theme to this book.

§5. Which Forcings Can We Iterate Without Adding Reals

We explain why "not adding reals" is not preserved by any kind of iteration, and suggest a remedy - $\mathbb{D}$-completeness. More elaborately the weak diamond tell us we have to exclude some forcing notion and $\mathbb{D}$-completeness seems a simple way to exclude then.

§6. Specializing an Aronszajn Tree Without Adding Reals

We prove that every Aronszajn tree can be specialized by a "nice" forcing: α-proper for every $\alpha < \omega_1$ and $\mathbb{D}$-complete for some $\aleph_1$-completeness system $\mathbb{D}$. Together with the next section this gives a proof of Con(ZFC + $\exists\kappa[\kappa$ inaccessible]) $\to$ Con(ZFC + GCH + SH) and with Chapter VIII a new proof of Jensen's Con(ZFC + GCH + SH) where SH is the Souslin Hypothesis.

§7. Iteration of $(E, \mathbb{D})$-Complete Forcing Notions

We prove that the limit of a CS iteration of Q_i each is α-proper for every $\alpha < \omega_1$, and $\mathbb{D}$-complete for some simple $\aleph_1$-completeness from V does not add reals.

§8. The Consistency of SH + CH + There Are No Kurepa Trees

VI. PRESERVATION OF ADDITIONAL PROPERTIES AND APPLICATIONS

§0. Introduction

§1. A General Preservation Theorem

We present a way to prove preservation of "properness $+\varphi$" for properties φ restricting our set of reals. Our hope is that this framework is easy to be applied to many properties. In the end we present a version, more general in one respect, less general in most respects (for readability).

§2. Examples

We specify more the framework in §1 to capture more of the common properties. Then we prove that the ${}^\omega\omega$-bounding property, the Sacks property, the Laver property, the PP-property, some (f, g)-bounding properties and some others come under the framework of §1; the "Sacks" and "Laver" properties appear first and most characteristically in the forcing notions bearing the respective names.

§3. Preservation of Unboundedness

We prove a presentation theorem suitable to prove the preservation of: no $g \in {}^\omega\omega$ satisfies $\bigwedge_{f \in F} f \leq^* g$ for a fixed $F \subseteq {}^\omega\omega$. We then look at other examples and prove the consistency of "ZFC + $\mathfrak{s} > \mathfrak{b}$" i.e. there is a non-dominated $F \subseteq {}^\omega\omega$, $|F| = \aleph_1$, but for every $\mathcal{B} \subseteq \mathcal{P}(\omega)$/finite of cardinality $\aleph_1$, some infinite $A \subseteq \omega$ induce an ultrafilter $\{B \in \mathcal{B} : A \subseteq_{ae} B\}$, but relaying on the existence of a forcing notion presented in §6.

§4. There May Be No P-Point

We present a proof of this theorem, using the preservation of the PP-property. This may serve as a preliminary test, whether our general machinery simplifies and clarifies proofs.

§5. There May Exist a Unique Ramsey Ultrafilter

The main result is the consistency of ZFC + $2^{\aleph_0} = \aleph_2+$ "there is a unique Ramsey ultrafilter on ω up to permutations of ω". For this we have to prove that "D generates a Ramsey ultrafilter " is preserved - by another application of §1, and of course mainly to work on each iterand.

§6. On the Splitting Number $\mathfrak{s}$ and Domination Number $\mathfrak{b}$ and on $\mathfrak{a}$

In §3 we have proved the consistency of $\mathfrak{s} > \mathfrak{b}$, modulo the existence of a suitable forcing notion: the one that adds an infinite $A \subseteq \omega$ including an ultrafilter on the "old" $\mathcal{P}(\omega)$ (helping to prove $\mathfrak{s}$ large), is proper of course and is almost ${}^{\omega}\omega$-bounding (helping to prove $\mathfrak{b}$ small). We define suitable creatures, and then the forcing notion build from them, and prove that it has the desired properties. We then prove parallel lemmas on a similar forcing notion "respecting" an ideal (usually one coming from a MAD family enough indistructible). This helps to get a universe with $\mathfrak{a}$ large.

§7. On $\mathfrak{s} > \mathfrak{b} = \mathfrak{a}$

We prove the consistency of the statement in the title. The main point is that if we force as for "CON(ZFC + $\mathfrak{s} > \mathfrak{b}$)" and we build in ground model a "bad enough MAD family", a suitable preservation theorem shows that it remains MAD.

§8. On $\mathfrak{h} < \mathfrak{s} = \mathfrak{b}$

We prove the consistency of the statement in the title; for this we use an iteration in which we add many Cohen reals making $\mathfrak{s}$ large (as in §6) and dominating real. We use the Cohen reals to construct a family witnessing $\mathfrak{h}$ is small (i.e. $\aleph_1$). To show that this works use our iterating only "nicely definable" forcing notions.

VII. AXIOMS AND THEIR APPLICATION

§0. Introduction

§1. On the κ-Chain Condition, When Reals Are Not Added

When we iterate $\aleph_2$ times forcings not adding real, (but not necessarily $\aleph_1$-complete) we suggest a condition called $\aleph_2$-e.c.c. so that if each Q_i satisfies the $\aleph_2$-e.c.c., then P_{ω_2} satisfies the $\aleph_2$-c.c.

§2. The Axioms

We suggest some axioms whose consistency follows from the theorems on preservation under iteration of various properties.

§3. Applications of Axiom II

We prove several applications of an axiom consistent with G.C.H.

§4. Applications of Axiom I

We prove some applications and mention others of an axiom consistent with $2^{\aleph_0} = \aleph_2$.

§5. A Counterexample Connected to Preservation

An example is given of a countable support iteration of length ω of forcing not collapsing stationary subsets of ω_1, but the limit collapse $\aleph_1$.

VIII. κ-PIC AND NOT ADDING REALS

§0. Introduction

§1. Mixed Iteration

We prove that we can iterate $\aleph_2$-complete forcings and $\aleph_1$-complete forcings satisfying a strong $\aleph_2$-chain condition, without collapsing $\aleph_1$ and $\aleph_2$.

§2. Chain Conditions Revisited

We suggest another condition, κ - pic, to ensure the limit of the iteration P_κ satisfies the κ-c.c. The aim is e.g., to start with $V \models$ "$2^{\aleph_0} = \aleph_1 \wedge 2^{\aleph_1} > \aleph_2$", and use CS iteration $\bar{Q}$ of length ω_2 , each time dealing with "all problems" (there are $2^{\aleph_1}$) at once.

§3. The Axioms Revisited

We discuss what axioms we can get according to the four possibilities of the truth of $2^{\aleph_0} = \aleph_1$, $2^{\aleph_1} = \aleph_2$ but assuming always $2^{\aleph_0} \leq \aleph_2$.

§4. More on Forcings Not Adding ω-Sequences and on the Diagonal Argument

We prove e.g., that CH does not imply $\Phi^3_{\aleph_1}$ (a kind of uniformization) by dealing with completeness systems which are 2-complete. Our main results are of the form: Suitable CS iteration does not add reals. So we continued the proof in Ch.V on CS iterations not adding reals, weakening the demand on the comleteness system from getting $\aleph_1$-complete filter to "the intersection of any two members is non empty, and we try to get properties preserved by the iterations.

IX. SOUSLIN HYPOTHESIS DOES NOT IMPLY "EVERY ARONSZAJN TREE IS SPECIAL"

§0. Introduction

§1. Free Limits

We look at Boolean algebras generated by a set of sentences in infinitary propositional calculus (mainly $L_{\omega_1,\omega}$). This enables us to define free limit.

§2. Preservation by Free Limit

We prove that an iteration in which we use $L_{\omega_1,\omega}$-free limit at limit stages, preserve properness, in some sense this gives a more natural proof of the preservation of properness.

§3. Aronszajn Trees: Various Ways to Specialize

We introduce some new ways to specialize Aronszajn trees, and present the old ones, as well as some connection between those properties.

§4. Independence Results

Here are the main results of the chapter. We use an iterated forcing $S -$ st-specializing any Aronszajn tree. The problem is to make sure that some fixed tree T^* will remain not special. In fact we get that on one stationary S all Aronszajn trees are special but some trees are "Souslin" on $\omega_1 \setminus S$, a strong way to be not special. We introduce such a property of forcing "(T^*, S)-preserving" and show that is is preserved in iteration. There is a discussion of the problem and our strategy in the beginning of the section and a discussion of open problems and how the preservation theorem can be generalized.

X. ON SEMI-PROPER FORCING

§0. Introduction

We would like to deal with forcing notions not preserving e.g. "δ has uncountable cofinality".

§1. Iterated Forcing with RCS (Revised Countable Support)

The standard countable support iteration cannot be applied when some uncountable cofinalities are changed to ω, we introduce the revised version suitable for this case. Though harder to define, this iteration conforms better with our intuition concerning iterations.

§2. Proper Forcing Revisited

We define semi-properness, and prove that it is strongly preserved by RCS iteration.

§3. Pseudo-Completeness

We prove that a weakening of $\aleph_1$-completeness (compatible with changing some cofinalities to $\aleph_0$) is strongly preserved by RCS iteration.

§4. Specific Forcings

We deal with Prikry forcing, Namba forcing and generalizations which are semi-proper when we use "large" filters which may exist even on small cardinals.

§5. Chain Conditions and Abraham's Problem

We prove that under reasonable conditions when we iterate up to κ the κ-c.c. holds and get the first application: a universe V in which for every $A \subseteq \omega_1$ there is a countable subset of ω_2^V which does not belong to $L(A)$.

§6. Reflection Properties of S_0^2. Refining Abraham's Problem and Precipitous Ideals

For some large cardinal κ, by iteration we find a forcing notion P, such that $V^P \models$ "$\kappa = \aleph_2$ and $A = \{\delta < \kappa : \mathrm{cf}\delta = \aleph_0, \delta$ regular in $V\}$ is stationary ". So we may make A large in some sense, as mentioned in the title.

§7. Friedman's Problem

We collapse some large κ, by iterated forcing, which sometimes collapses $(2^{\aleph_2})^+$ to $\aleph_1$, sometimes changes the cofinality of $\aleph_2$ to $\aleph_0$, and sometimes adds a closed unbounded $C \subseteq S$ of order type ω_1, where $S \subseteq S_0^2$ is stationary. We get a model V in which every stationary $S \subseteq S_0^2 = \{\delta < \aleph_2 : \mathrm{cf}(\delta) = \aleph_0\}$ contains a closed copy of ω_1. By stronger hypothesis we get it for every stationary $S \subseteq S_0^\alpha$, $\mathrm{cf}(\aleph_\alpha) > \aleph_0$.

XI. CHANGING COFINALITIES: EQUI-CONSISTENCY RESULTS

§0. Introduction

We try here to weaken semiproperness.

§1. The Theorems

Here we describe what kind of a condition on forcing notions we want (i.e. satisfied by some specific notions and preserved by suitable iterations). Then we proceed to get consistency results. The proof uses RCS-iteration of length κ, κ a strongly inaccessible cardinal. In each step, we allow Namba forcing. The consistency results are mostly from Chapter X but here we use the minimal large cardinals required.

§2. The Condition

We describe here the condition, called the S-condition or $\mathbb{I}$-condition for S a set of regular cardinals $> \aleph_1$, $\mathbb{I}$ a set of $\aleph_2$-complete ideals, and some helping definitions and conventions.

§3. The Preservation Properties Guaranteed by the S-Condition

We prove that, assuming a forcing notion P satisfies our condition, forcing with P implies $\aleph_1$ is not collapsed, and (assuming CH) no real is added; and for this we need partitions theorems on trees.

§4. Forcing Notions Satisfying the S-Condition

We show that Namba forcing, Nm satisfies the $\{\aleph_2\}$-condition that Nm and Nm$'$ are really different forcing notions; that Nm, Nm$'$ may satisfy the $\aleph_4$-c.c. (while $2^{\aleph_0} = \aleph_1, 2^{\aleph_1}$ is large) We also prove $\aleph_1$-complete forcing and a forcing notion shooting a closed unbounded subset of order type ω_1 through a stationary $S \subseteq S^2_0$ satisfies our condition.

§5. Finite Composition

We prove that under suitable hypothesis, a composition of forcing satisfying an S-condition satisfies it. For this we prove a combinatorial theorem on trees.

§6. Preservation of the $\mathbb{I}$-Condition by Iteration

Here we prove that if we iterate forcing notions satisfying our conditions, but enough times collapse the present $2^{|P|}$ to $\aleph_1$, the composite forcing satisfies the condition. So usually we have large segments of cardinals which we have to collapse by $\aleph_1$-complete forcings, but for strongly inaccessible we can use Nm straight away (by 6.5).

§7. Further Independence Results

We prove the equiconsistency of "ZFC + κ is Mahlo" and " ZFC +$\aleph_2$ has the Friedman property", and a further result using weakly compact cardinal. We also prove the equiconsistency of "ZFC + κ is 2-Mahlo " and of "ZFC+ there is the club of $\aleph_2$ consisting of regular cardinals of L".

§8. Relativising to a Stationary Set

XII. IMPROPER FORCING

§0. Introduction

§1. Games and Properness

Equivalent definitions of variants of properness by games are given, and it is exemplified how the proofs of the preservation theorems in this context look like.

§2. When Is Namba Forcing Semiproper, Chang's Conjecture and Games

We prove e.g., that if some $\{\aleph_1\}$-semi proper forcing changes the cofinality of $\aleph_2$ to ω then Namba forcing is semi-proper, and Chang's Conjecture holds hence $0^\# \in V$. So without some fairly large cardinals (in some inner model) all semi proper forcing notions are proper, so actually Chapter X cannot do what Chapter XI does.

XIII. LARGE IDEALS ON ω_1

§0. Introduction

§1. Semi-Stationarity

We define and prove the basic properties of semi-stationarity of subsets of $\mathcal{S}_{\leq\aleph_0}(\lambda)$, and the connection with semiproper iterations up to measurables.

§2. S-Suitable Iterations and Sealing Forcing

S-suitable iterations, for $S \subseteq \omega_1$ stationary, are semi-proper iterations in which we "promise" for some subalgebras of $\mathfrak{B}{\restriction}S = \mathcal{P}(\omega_1)/(\mathcal{D}_{\omega_1} + S)$ in some V^{P_i}, that they remain $\lessdot\ \mathfrak{B}^{V^{P_j}}$ for $j > i$ ($\lessdot$ means a complete subalgebra - see 0.1(4)); usually the subalgebras are $\mathfrak{B}{\restriction}S$ in V^{P_i}. They are reasonable if we try to get a universe in which not only does $\mathcal{P}(\omega_1)/(\mathcal{D}_{\omega_1} + S)$ satisfies the $\aleph_2$-c.c. but also has more specific structure. We prove various lemmas on how we can continue such iterations, mainly using sealing forcing, which "seal" various antichains.

§3. On $\mathfrak{B} = \mathcal{P}(\omega_1)/\mathcal{D}_{\omega_1}$ Being Layered or the Levy Algebra

We prove, starting from enough supercompacts the consistency of G.C.H. + $(\mathfrak{B}\restriction S) \stackrel{\text{def}}{=} \mathcal{P}(\omega_1)/(\mathcal{D}_{\omega_1} + S)$ is layered, which means that ($S \subseteq \omega_1$ is stationary and) "almost" every $\mathcal{A} \subseteq \mathfrak{B}\restriction S$ of power $\aleph_1$, satisfies $\mathcal{A} \lessdot \mathfrak{B}$ (almost means a club of cofinality $\aleph_1$); we can have $S = \omega_1$, if we omit CH. Moreover we can do this without adding reals and have some forcing axiom. We then get a stronger condition of this form: $\mathfrak{B}\restriction S$ is (the completion of) the Levy($\aleph_0, < \aleph_2$) algebra; in fact gives two proofs of this with extra things as above.

§4. On $\mathcal{P}(\omega_1)/(\mathcal{D}_{\omega_1} + S)$ is Reflective or Ulam

We prove, again starting from supercompacts, that in some generic extensions not adding reals, a positive answer to the Ulam question holds: there are $\aleph_1$ measures on ω_1 (really $\{0, 1\}$-measures) with *countable* additivity such that every set $\subseteq \omega_1$ is measurable with respect to at least one of them. For this we define when a filter is Ulam, and give a combinatorial sufficient condition for it (which take some space to show). Then we have forcing doing it, with extra conclusions as in the solution for layerness. Note that here we have, for $\delta < \kappa$ of cofinality $\aleph_0$, a demand dual to the one when we want to get the Levy algebra: decreasing sequences of members of $\bigcup_{i<\delta} \mathfrak{B}^{P_i}$ has a positive intersection if this is reasonable. We also prove the consistency of a weaker statement - among any $\aleph_2$ stationary subsets of S, there are $\aleph_2$, the intersection of any countably many is stationary. Before all this we deal with reflective Boolean algebra.

XIV. ITERATED FORCING WITH UNCOUNTABLE SUPPORT

§0. Introduction

We try to deal with "P does not change the cofinality of every $\mu \leq \kappa$".

§1. κ-Revised Support Iteration

We define and investigate κ-RS (κ-revised support), intended for iterating forcing not collapsing cardinals $\leq \kappa$, but possibly changing some cardinals with cofinality $> \kappa$ to ordinals with cofinality $< \kappa$.

§2. Pseudo-Completeness

We do not know to generalize the theorem on proper forcing (III) and semi proper forcing (Chapter X) but we generalize properties like pseudo completeness (see X §3). We deal with such properties, some do not add bounded subsets to κ some may add even reals. For this we use a forcing notion which has also a notion of pure extensions $[\leq_0]$. We define properties like (S,γ)-Pr_1^+ saying we have pure decidability and $\leq_0$ is γ-complete; this is less restrictive than it may look by 2.4. For such forcing notions we define an iteration with some kind of mixed support and prove the appropriate preservation theorems and theorems on chain conditions.

§3. Axioms

We deal with the relevant forcing axioms, and prove that it applies to some forcing notions.

§4. On Sacks Forcing

We prove that we can apply our condition to Sacks forcing.

§5. Abraham's Second Problem – Iterating Changing Cofinality to ω

We here apply the previous theorem to solve Abraham's second problem (solved for $\aleph_2$ in Chapter X). For this we use a forcing notion based on an initial segment of a play of a game (using a fixed winning strategy) rather than positive sets for a fixed ideal.

XV. A MORE GENERAL ITERABLE CONDITION ENSURING $\aleph_1$ IS NOT COLLAPSED

§0. Introduction

A drawback of Chapter IX is that it implies "the forcing notion does not add real"; our aim is to overcome it.

§1. Preliminaries

§2. Trees of Models and UP

We define a property of a forcing notion, $UP(\mathbb{I}, \mathbf{S}, \mathbf{W})$ where $\mathbb{I}$ is a set of ideals, $\mathbf{S}$ a set of regular cardinals (most popular value $\{\aleph_1\}$), $\mathbf{W} \subseteq \omega_1$ stationary (e.g. ω_1); this condition generalizes the one in chapter XI, gives an equivalent definition for the interesting cases there. We prove some basic properties: it does not destroy the stationarity of subsets of $\mathbf{W}$, and in suitable cases it is preserved by composition. For this we again have to deal with the partition theorems on trees.

§3. Preservation of the $UP(\mathbb{I}, \mathbf{S}, \mathbf{W})$ by Iteration

We prove that if we iterate forcing notions with such a property, each time collapsing soon enough, then the limit satisfies a similar condition. We first define (3.1), deal with limit of length ω (3.2), ω_1 (3.3), note continuity and chain condition (3.4), iterates up to an inaccessible (3.5), do one more step (3.6), prove that we have enough freedom in reorganizing the forcings such that the previous cases suffice (3.7, see 1.7) and conclude enough such iterations exists (3.8).

§4. Families of Ideals and Families of Partial Orders

A bothersome point in §2, §3 is that for $\underset{\sim}{Q}$, a P-name of a forcing notion satisfying some $UP(\underset{\sim}{\mathbb{I}}_1)$, we demand that the members of $\underset{\sim}{\mathbb{I}}_1$ were ideals in V, not V^P. We show here that we can replace a general $\underset{\sim}{\mathbb{I}}$, a P-name of set of ideals in V^P by a family of ideals $\mathbb{I}^1$ in V (not changing much the relevant properties). For this we find it better to replace the ideals by partial orders, revising the definition of UP. We then restate our iteration theorem, and have a theorem on not adding reals parallel to $\mathbf{W}$-complete.

XVI. LARGE IDEALS ON ω_1 FROM SMALLER CARDINALS

§0. Introduction

§1. Bigness of Stationary $T \subseteq \mathcal{S}_{\leq \aleph_0}(\lambda)$

We consider various bigness properties of subsets of $\mathcal{S}_{\leq \aleph_0}(\lambda)$, connected reflection properties and large cardinal properties and interrelation.

§2. Getting Large Ideal on ω_1

Here we deal with the appropriate lemmas on iterated forcing and sealing maximal antichains of $\mathcal{P}(\omega_1)/\mathcal{D}_{\omega_1}$, and then get consistency results of the form: (CH)+$\mathcal{P}(\omega_1)/(\mathcal{D}_{\omega_1}+S)$ is $\aleph_2$-saturated or even $\mathcal{P}(\omega_1/\mathcal{D}_{\omega_1}+S)$ is layered (hence we can get non regular ultrafilter) or even in the Levy algebra.

XVII. FORCING AXIOMS

§0. Introduction

§1. Semiproper Forcing Axiom Implies Martin's Maximum

We prove the SPFA=MM. Also we prove that SPFA$\vee Ax_1[\aleph_1$- complete] implies that a forcing notion is semi proper iff it does not destroy the stationarity of subsets of ω_1. We then show that SPFA implies $P(\omega_1/\mathcal{D}_{\omega_1})$ is $\aleph_2$-saturated.

§2. SPFA Does Not Imply PFA$^+$

We prove that SPFA does not imply Ax_1 [semi proper]=SPFA$^+$, and even AX_1[proper]=PFA$^+$. We first prove this consistency starting with a supercompact limit of a supercompact and then show that one supercompact suffices. We then show that properness is (provably in ZFC) not productive, and that SPFA implies Ax_1 [$\aleph_1$-complete] or even Ax_1[c.c.c. $*$ $\aleph_1$-complete].

§3. Canonical Functions for ω_1

We start with $g : \omega_1 \to \omega_1$, such that for no stationary $A \subseteq \omega_2$ and $\alpha < \omega_2$ is $g{\restriction}A$ equal to an α-th function, and κ supercompact and find a κ-c.c. semi proper forcing notion forcing $\kappa = \aleph_2$, PFA$^+$, and g retains its property hence Chang conjecture fails. For this we define "Q is a g-small proper forcing notion". We then prove for $\alpha < \omega_1$ additively indecomposable, Ax[α-proper] is consistent with the existence of $\langle c_\delta : \delta < \omega_1,\ \alpha \text{ divides } \delta\rangle$, $c_\delta \subseteq \delta$ closed unbounded of order type α and for every club E of ω_1 for stationarily many $\delta < \omega_1$, $c_\delta \subseteq E$. So for $\alpha' < \alpha'' < \omega_1$ additively indecomposable $\mathrm{Ax}_{\omega_1}[\alpha'$-proper] $\not\Rightarrow$Ax[α'-proper]. Note that CS iteration of $\aleph_1$-complete and $\aleph_1$-c.c. forcing notion, is α-proper for every $\alpha < \omega_1$.

§4. A Largeness of $\mathcal{D}_{\omega_1}$ in Forcing Extensions of L and Canonical Functions

We define a property $(*)^1_\lambda$ of a cardinal λ, such that if it holds in V it holds in L (and appropriate Erdös cardinals satisfies it, and assuming $0^\#$ we show that $L^{\mathrm{Levy}(\aleph_0,<\kappa)} \models (*)^1_\kappa$ holds when $\lambda > \kappa > \aleph_0$).

Now we can iterate (CS) first adding λ functions from ω_1 to ω_1 and then kill all stationary subsets of ω_1 which contradict "f_α is an α-th function for $\alpha < \lambda$ and for every $g \in {}^{\omega_1}\omega_1$, $\{\mathrm{eq}(g, f_\alpha) : \alpha < \lambda\}$ is predense in $P(\omega_1)/\mathcal{D}_{\omega_1}$ (we get a more explicit version: for every $x \in H(\chi)$ there is a countable N, $x \in N \prec (H(\chi), \in, <^*_\chi)$ such that $[g \in {}^{\omega_1}\omega_1 \cap N \Rightarrow \bigvee_{\alpha \in \lambda \cap N} g(N \cap \omega_1) = f_\alpha(N \cap \omega_1)$. We then play with cardinal arithmetic.

XVIII. MORE ON PROPER FORCING

§0. Introduction

§1. No New Reals: A Counterexample and New Questions

We prove that the weak diamond (see AP§1) is not the only obstacle to "the limit of a CS iteration of proper forcing not adding real does not add reals." This points to some quite simple forcing notions, such that if we iterate them with CS the theorem in Ch V (and the weak diamond) does not tell us whether reals may be added in limits or not.

§2. Not Adding Reals

We give here a quite weak condition guaranteeing that CS iterations of forcing notions satisfying those conditions do not add reals. Those conditions are particularly strong for forcing notions of cardinality $\aleph_1$ and answer the questions raised by the first section.

§3. Other Preservations

We deal with preservations of the properties (by CS iteration of proper forcing). This was done in VI, here our framework is weaker so more general. We then carry various examples, including preservation of the P-point.

§4. There May Be a Unique P-Point

We prove the consistency of " there is a unique P-point"; it is (necessarily) a Ramsey ultrafilter but there is no other P-point. In VI§5 we have proved " there is a unique Ramsey ultrafilter". The missing part proved here is: given $D_0 \leq_{\rm RK} D_1$, D_0 a Ramsey ultrafilter, D_1 a P-point, $D_1 \not\leq_{\rm RK} D_0$, find a forcing notion Q, $\Vdash_Q$ "D_0 is not a P-point nor can it be completed to a P-point by further forcing notions with all the required properties".

APPENDIX: ON WEAK DIAMONDS AND THE POWER OF EXT

§0. Introduction

§1. Unif: a Strong Negation of the Weak Diamond

Introduce a generalization of the negation of the weak diamond (i.e., $\Phi^2_{\aleph_1}$) and prove cases of this principle from an appropriate replacement of $2^{\aleph_0} < 2^{\aleph_1}$.

§2. On the Power of Ext and Whitehead's Problem

We show that for "every non free abelian group G is not Whitehead, moreover EXT$(G, \mathbb{Z})$ large", much less than $V = L$ is needed; this exemplifies the use of §1.

§3. Weak Diamond for $\aleph_2$ Assuming CH

We prove that every ladder system $\bar{\eta} = \langle \eta_\delta : \delta \in S^2_1 \rangle$ when η_δ is continuous cannot be uniformized assuming $2^{\aleph_0} = \aleph_1$. This shows some hopeful theorem on cases of "an CS iteration of $\aleph_2$-c.c. forcing notion not adding reals gives an $\aleph_2$-c.c. forcing notion not adding reals", similarly for "($\leq \aleph_1$)-support iteration of $\aleph_1$-complete forcing notions not collapsing $\aleph_2$ gives a forcing notion not collapsing $\aleph_2$.

REFERENCES

I. Forcing, Basic Facts

§0. Introduction

In this chapter we start by introducing forcing and state the most important theorems on it (done in §1); we do not prove them as we want to put the stress on applying them. Then we give two basic proofs:
in §2, we show why CH (the continuum hypothesis) is consistent with ZFC, and in §3 why it is independent of ZFC. For this the $\aleph_1$-completeness and c.c.c.(=countable chain conditions) are used, both implying the forcing does not collapse $\aleph_1$ the later implying the forcing collapse no cardinal. In §4 we compute exactly $2^{\aleph_0}$ in the forcing from §3 (in §3 we prove just $V[G] \models$ "$2^{\aleph_0} \geq \lambda$"; we also explain what is a "Cohen real"). In §5 we explain canonical names.

Lastly in §6 we give more basic examples of forcing: random reals, forcing diamonds. The content of this chapter is classical, see on history e.g. [J]. (Except §7, 7.3 is A. Ostaszewski [Os] and 7.4 is from [Sh:98, §5], note that later Baumgartner has found a proof without collapsing and further works are:

P. Komjáth [Ko1], continuing the proof in [Sh:98] proved it consistent to have MA for countable partial orderings $+\neg$CH, and $\clubsuit$. Then S. Fuchino, S. Shelah and L. Soukup [FShS:544] proved the same, without collapsing $\aleph_1$ and M.Džamonja and S.Shelah [DjSh:604] prove that $\clubsuit$ is consistent with SH (no Souslin tree, hence $\neg$CH).)

§1. Introducing Forcing

1.1 Discussion. Our basic assumption is that the set theory ZFC is consistent. By Godel's completeness theorem it has a countable model. We make the following further assumptions about this model.

(a) The membership relation of the model is the real membership relation; and therefore the model is of the form $(V, \in)$.

(b) The universe V of the model is a transitive set, i.e., $x \in y \in V \to x \in V$.

Assumptions (a) and (b) are not essential but it is customary to assume them, and they simplify the presentation. So "V a model of ZFC", will mean "a countable model of ZFC satisfying (a) and (b)", and the letter V is used exclusively for such models.

Cohen's forcing method is a method of extending V to another model $V^\dagger$ of ZFC. It is obvious that whatever holds in the model $V^\dagger$ cannot be refuted by a proof from the axioms of ZFC, and therefore it is compatible with ZFC. If we show that a statement and its negation are both compatible with ZFC then we know that the statement is undecidable in ZFC.

Why do we look at extensions of V and not at submodels of V? After all, looking at subsets is easier since their members are already at hand: To answer this question we have to mention Godel's constructibility. The constructible sets are the sets which must be in a universe of set theory once the ordinals of that universe are there. Godel showed that the class L of the constructible sets is a model of ZFC and that one cannot prove in ZFC that there are any sets which are not constructible. Therefore, for all we know, V may contain only sets which are constructible and in this case every transitive subclass $V^\dagger$ of V which contains all ordinals of V and which is a model of ZFC must coincide with V, and therefore it gives us nothing new.

1.2 Discussion. Now we come to the concept of forcing. A forcing notion $P \in V$ is just a partially ordered set (not empty of course). Usually a partial order is required to satisfy $p \leq q \,\&\, q \leq p \Rightarrow p = q$, but we shall not (this is just a technicality), this is usually called pre-partial order or quasi order. It is also called a forcing notion. We normally assume that P as a minimal element denoted by $\emptyset_P$, i.e.

$$(\forall q \in P)(P \models \emptyset_P \leq q),$$

really from Chapter II on, we do not lose generality as by adding such a member we get an equivalent forcing notion, see §5. We want to add to V a subset G of P as follows.

(1) G is directed (i.e., every two members of G have an upper bound in G) and downward closed (i.e., if $x \leq y \in G$ then also $x \in G$).

Trivial examples of a set G which satisfies (1) is the empty set $\emptyset$ and $\{x : x \leq p\}$ for $p \in P$.

The following should be taken as a declaration of intent rather than an exactly formulated requirement.

(2) We want that $G \notin V$ and moreover G is "general" or "random" or "without any special property".

We aim at constructing a (transitive) set $V[G]$ which is a model of ZFC with the same ordinals as V, such that $V \subseteq V[G]$ and $G \in V[G]$, and which is minimal among the sets which satisfy these requirements.

So we can look at P as a set of approximations to G, each $p \in P$ giving some information on G, and $p \leq q$ means q gives more information; this view is helpful in constructing suitable forcing notions.

Where does the main problem in constructing such a set $V[G]$ lie? In the universe of set theory the ordinals of V are countable ordinals since V itself is countable. But an ordinal of V may be uncountable from the point of view of V (since V is a model of ZFC and the existence of uncountable ordinals is provable in ZFC). Since for each ordinal $\alpha \in V$ the information that α is countable is available outside V, G may contain i.e. code that information for each $\alpha \in V$. In this case every ordinal of V (and hence of $V[G]$) is countable in $V[G]$ and thus $V[G]$ cannot be a model of ZFC. How do we avoid this danger?

By choosing G to be "random" we make sure that it does not contain all that information.

While we choose a "random" G we do not aim for a random $V[G]$, but we want to construct a $V[G]$ with very definite properties. Therefore we can regard p as the assertion that $p \in G$ and as such p provides some information about G. All the members of G, taken together, give the complete information about G.

Now we come back to the second requirement on G and we want to replace the nebulous requirement above by a strict mathematical requirement.

1.3 Definition. (1) A subset $\mathcal{I}$ of P is said to be a *dense* subset of P if it satisfies

$$(\forall p \in P)(\exists q \in P)(p \leq q \,\&\, q \in \mathcal{I})$$

(2) Call $\mathcal{I} \subseteq P$ open (or upward closed) if for every $p, q \in P$

$$p \geq q \,\&\, q \in \mathcal{I} \Rightarrow p \in \mathcal{I}$$

1.4. Discussion. Since we want G to contain as many members of P as possible without contradicting the requirement that it be directed, we require:

$(2)'$ $G \cap \mathcal{I} \neq \emptyset$ for every dense open subset $\mathcal{I}$ of P which is in V.

1.4A Definition. A subset G of P which satisfies requirements (1) and $(2)'$ is called *generic* over V (we usually omit V), where this adjective means that G satisfies no special conditions in addition to those it has to satisfy.
The forcing theorem will assert that for a generic $G, V[G]$ is as we intended it to be.

Does $(2)'$ imply that $G \notin V$? Not without a further assumption, since if P consists of a single member p then $G = \{p\}$ satisfies (1) and $(2)'$ and $G \in V$. However if we assume that P has no trivial branch, in the sense that above every member of P there are two incompatible members, then indeed $G \notin V$ (incompatible means having no common upper bound). To prove this notice that if $G \in V$, then $P \setminus G$ is a dense open subset of P in V, remember that G is downward closed, and by (2)$'$ we would have $G \cap (P \setminus G) \neq \emptyset$, which is a contradiction.

1.5 The Forcing Theorem, Version A. (1) If G is a generic subset of P over V, then there is a transitive set $V[G]$ which is a model of ZFC, $V \subseteq V[G]$, $G \in V[G]$ and V and $V[G]$ have the same ordinals and we can allow V as a class of $V[G]$ (i.e. in the axioms guaranting (first order) definiable sets exists "$x \in V$" is allowed as a predicate).
(2) P has a generic subset G, moreover for every $p \in P$ there is a $G \subseteq P$ generic over V, $p \in G$. $\square_{1.5}$

1.6 Discussion. We shall not prove 1.5(1), but we shall prove 1.5(2). Since V is countable, P has at most $\aleph_0$ dense subsets in V; let us denote them with $\mathcal{I}_0, \mathcal{I}_1, \mathcal{I}_2, \ldots$ we shall construct by induction a sequence p_n. We take an arbitrary p_0. We choose p_{n+1} so that $p_n \leq p_{n+1} \in \mathcal{I}_n$; this is possible since $\mathcal{I}_n$ is dense. We take $G = \{q \in P : \exists n (q \leq p_n)\}$. It is easy to check that this G is generic.

Since we want to prove theorems about $V[G]$ we want to know what are the members of $V[G]$. We cannot have in V full knowledge on all the members of $V[G]$ since this would cause these sets to belong to V. So we have to agree that we do not know the set G, but, as we want as much knowledge on $V[G]$ as possible, we require that except for that we have in V full knowledge of all members of $V[G]$, more specifically V contains a prescription for building that member out of G. We shall call these prescriptions "names". We shall be guided in the construction of the names by the idea that $V[G]$ contains only those members that it has to.

Remember:

1.7 Definition. We define the rank of any $a \in V$:

$$\text{rk}(a) \text{ is } \bigcup \{\text{rk}(b) + 1 : b \in a\}$$

(note if $a = \emptyset, \mathrm{rk}(a) = 0$), the union of a set of ordinals is an ordinal, hence $\mathrm{rk}(a)$ is an ordinal if defined, and by the axiom of regularly $\mathrm{rk}(a)$ is defined for every a. So

1.8 Definition. We define what is a P-name (or name for P or a name in P) τ of rank $\leq \alpha$, and what is its interpretation $\tau[G]$. If P is clear we omit it.

This is done by induction on α. τ is a name of rank $\leq \alpha$ if it has the form $\tau = \{(p_i, \tau_i) : i < i_0\}$, $p_i \in P$ and each τ_i is a name of some rank $< \alpha$.

The interpretation $\tau[G]$ of τ is $\{\tau_i[G] : p_i \in G, i < i_0\}$

1.9 Definition.

(1) Let $\mathrm{rk}_n(\tau) = \alpha$ if τ is a name (for some P) of rank $\leq \alpha$ but not a name of rank $\leq \beta$ for any $\beta < \alpha$.

(2) For $a \in V$ and forcing notion P, $\dot{a}$ is a P-name defined by induction on $\mathrm{rk}(a)$;

$$\dot{a} = \{(p, \dot{b}) : p \in P, b \in a\}$$

(3) $\underset{\sim}{G} = \{(p, \dot{p}) : p \in P\}$ (when necessary we denote it by ($\underset{\sim}{G}_P$).

(4) $\mathrm{rk}_r(\tau)$, the revised rank of a P-name τ is defined as follows: $\mathrm{rk}_r(\tau) = 0$ iff $\tau = \dot{a}$ for some $a \in V$

Otherwise

$$\mathrm{rk}_r(\tau) = \bigcup\{\mathrm{rk}_r(\sigma) + 1 : (p, \sigma) \in \tau \text{ for some } p\}$$

1.9A Remark. 1) Usually, we use $\underset{\sim}{\tau}, \underset{\sim}{f}, \underset{\sim}{a}$ etc. to denote P-names not necessarily of this form.

Eventually we lapse to denoting $\dot{a}$ (the P-name of a) by a, abusing our notation, in fact, no confusion arrives.

1.10 Claim. Given a forcing notion P, and $G \subseteq P$ generic over V, we have:

(1) $\mathrm{rk}_r(\tau) \leq \mathrm{rk}_n(\tau)$ and $\mathrm{rk}(\tau[G]) \leq \mathrm{rk}_n(\tau)$ for any P-name τ.

(2) for $a \in V$, $\dot{a}[G] = a$

(3) $\underset{\sim}{G}[G] = G$.

(4) $\mathrm{rk}_r(\tau), \mathrm{rk}_n(\tau)$ are well defined ordinals, for any P-name τ.

Proof. Trivial. $\square_{1.10}$

1.11 Discussion. Notice that while every name belongs to V, the values of the names are not necessarily in V since the definition of the interpretation of a name cannot be carried out in V. It turns out that these names are sufficient in the sense that the set of their values is a set $V[G]$ as required:

1.12 The Forcing Theorem (strengthened), Version B. In version A, in addition $V[G] = \{\tau[G] : \tau \in V$, and τ is a P-name $\}$. $\square_{1.12}$

We want to know which properties hold in $V[G]$. The properties we are interested in are the first order properties of $V[G]$, i.e., the properties given by formulas of the predicate calculus. We shall refer to the members of $V[G]$ by their names so we shall substitute the names in the formulas.

1.13 Definition. If $\tau_1 \ldots, \tau_n$ are names, for the forcing notion P, $\varphi(x_1, \ldots, x_n)$ a first-order formula of the language of set theory with an additional unary predicate for V, then we write $p \Vdash_P$ "$\varphi(\tau_1 \ldots, \tau_n)$" ($p$ forces $\varphi(\tau_1 \ldots, \tau_n)$ for the forcing P) *if* for every generic subset G of P which contains p we have:

$\varphi(\tau_1[G] \ldots, \tau_n[G])$ is satisfied (=is true) in $V[G]$,

in symbols $V[G] \models$ "$\varphi(\tau_1[G], \ldots, \tau_n[G])$".

1.14 The Forcing Theorem, Version C. If G is a generic subset of P *then* (in addition to the demands in versions A and B we have:) for every $\varphi(\tau_1 \ldots, \tau_n)$ as above there is a $p \in G$ such that $p \Vdash_P$ "$\neg\varphi(\tau_1 \ldots, \tau_n)$" or $p \Vdash_P$ "$\varphi(\tau_1 \ldots, \tau_n)$". Therefore $V[G] \models$ "$\varphi(\tau_1[G] \ldots, \tau_n[G])$" iff for some $p \in G$ $p \Vdash_P$ "$\varphi(\tau_1 \ldots, \tau_n)$". Moreover $\Vdash$ (as a relation) is definable in V. $\square_{1.14}$

This is finally the version we shall actually use, but we shall not prove this theorem either.

The *forcing relation* $\Vdash_P$ clearly depends on P. If we deal with a fixed P we can drop the subscript P. We refer to P as the *forcing notion.*

The rest of the section is devoted to technical lemmas which will help to use the forcing theorem.

1.15 Definition. For $p, q \in P$ we say that p and q are *compatible* if they have an upper bound. $\mathcal{I} \subseteq P$ is an *antichain* if every two members of $\mathcal{I}$ are incompatible. $\mathcal{I} \subseteq P$ is a *maximal antichain* if $\mathcal{I}$ is an antichain and there is no antichain $\mathcal{J} \subseteq P$ which properly includes $\mathcal{I}$. We say $\mathcal{I} \subseteq P$ is pre-dense (above $p \in P$) if for every $q \in P$ $(q \geq p)$ some $q^\dagger \in \mathcal{I}$ is compatible with q. We say $\mathcal{I} \subseteq P$ is dense above $p \in P$ if for every $q \in P$ such that $q \geq p$ there is r, $q \leq r \in \mathcal{I}$; we may omit "above p". We define "$\mathcal{I} \subseteq P$ is pre-dense above $p \in P$" similarly.

1.16 Lemma. Let G be a downward closed subset of P. Then: G is generic (over V) *iff* for every maximal antichain $\mathcal{I} \in V$ of P we have $|G \cap \mathcal{I}| = 1$.

Proof. Suppose G is generic. Since G is directed it cannot contain two incompatible members and hence $|G \cap \mathcal{I}| \leq 1$. Given $\mathcal{I} \in V$, a subset of P, let $\mathcal{J} = \{p \in P : (\exists q \in \mathcal{I}) p \geq q\} \in V$, i.e., $\mathcal{J}$ is the upward closure of $\mathcal{I}$. So $\mathcal{J}$ is obviously upward closed i.e. an open subset, we shall now show that if $\mathcal{I}$ is a maximal antichain of P, then $\mathcal{J}$ is dense. For any $r \in P$ clearly r is compatible with some member q of $\mathcal{I}$ (otherwise $\mathcal{I} \bigcup \{r\}$ would be an antichain properly including the maximal antichain $\mathcal{I}$), let $p \geq r, q$. Then, by the definition of $\mathcal{J}$, $p \in \mathcal{J}$ and we have proved the density of $\mathcal{J}$.
Since $\mathcal{J}$ is dense and open by Definition 1.4A we know $G \cap \mathcal{J} \neq \emptyset$, let $p \in G \cap \mathcal{J}$. Since $p \in \mathcal{J}$, there is a $q \in \mathcal{I}$ such that $q \leq p$, and since $p \in G$ and G is generic, $q \in G$ and so $q \in G \cap \mathcal{I}$, hence $|G \cap \mathcal{I}| \geq 1$. So (assuming $G \subseteq P$ is generic over V) we have proved: for every maximal antichain $\mathcal{I} \in V$ of P, $|G \cap \mathcal{I}| = 1$, thus proving the only if part of the lemma.

Now assume that for every maximal antichain $\mathcal{I} \in V$ we have $|G \cap \mathcal{I}| = 1$. First let $\mathcal{J} \in V$ be a dense subset of P and we shall prove $G \cap \mathcal{J} \neq \emptyset$. By Zorn's lemma there is an antichain $\mathcal{I} \subseteq \mathcal{J}$ which is maximal among the antichains in $\mathcal{J}$, i.e. the antichains of P which are subsets of $\mathcal{J}$. We claim that $\mathcal{I}$ is a maximal antichain. Let $r \in P$, we have to prove that r is compatible with some member of $\mathcal{I}$ (and hence $\mathcal{I}$ cannot be properly extended to an antichain). Since $\mathcal{J}$ is dense there is a $p \in \mathcal{J}$ such that $p \geq r$. Since $p \in \mathcal{J}$ and $\mathcal{I}$ is an antichain maximal in $\mathcal{J}$ necessarily p is compatible with some member q of $\mathcal{I}$, hence r is also compatible with q; so we have finished proving "$\mathcal{I}$ is a maximal antichain of P". So by our present assumption $|G \cap \mathcal{I}| = 1$ hence $G \cap \mathcal{J} \supseteq G \cap \mathcal{I} \neq \emptyset$.

Secondly to see that G is directed let $q, r \in G$ and let $\mathcal{J} = \{p \in P : p \geq q, r$ or p is incompatible with q or p is incompatible with $r\}$. Clearly $\mathcal{J} \in V$, to prove that $\mathcal{J}$ is dense let $s \in P$. If s is incompatible with q then $s \in \mathcal{J}$. Otherwise there is a $t \in P$ such that $s, q \leq t$. If t is incompatible with r then $t \in \mathcal{J}$, and we know that $t \geq s$. Otherwise there is a $w \in p$ such that $w \geq t, r$. Since $t \geq s, q$ we have $w \geq q, r$ and hence $w \in \mathcal{J}$. Since $w \geq t \geq s$ we know $\mathcal{J}$ is dense. By what we have shown above, $G \cap \mathcal{J} \neq \emptyset$. Let $p \in G \cap \mathcal{J}$. We shall see that p cannot be incompatible with q or with r, therefore, since $p \in \mathcal{J}$, $p \geq q, r$. We still have to prove that no two members of G, such as p and q, are incompatible. Suppose $p, q \in G$ and p and q are incompatible. We extend the antichain $\{p, q\}$, by Zorn's lemma to a maximal antichain $\mathcal{I} \in V$. We have $\mathcal{I} \cap G \supseteq \{p, q\}$, contradicting $|\mathcal{I} \cap G| = 1$.

As part of the assumption of 1.16 is "$G \subseteq P$ is downward closed", and we have proved G is directed, and $[\mathcal{J} \in V$ is a dense subset of $P \Rightarrow G \cap \mathcal{J} \neq \emptyset]$, we have proved that G is a generic subset of P over V (see Definition 1.4A). Hence we have finished proving also the if part of the lemma. $\square_{1.16}$

1.17 Lemma. If $\mathcal{J}$ is a pre-dense subset of P in V and G is a generic subset of P *then* $G \cap \mathcal{J} \neq \emptyset$.

Proof. Let $\mathcal{J}^{\dagger} = \{p \in P : (\exists q \in \mathcal{J}) p \geq q\}$. Let us prove that $\mathcal{J}^{\dagger}$ is a dense open subset of P. Now $\mathcal{J}^{\dagger}$ is obviously upward-closed. Let $r \in P$. Since $\mathcal{J}$ is pre-dense there is a $q \in \mathcal{J}$ such that q is compatible with r. Therefore, there is a $p \in P$ such that $p \geq q, r$. By the definition of $\mathcal{J}^{\dagger}$ we have $p \in \mathcal{J}^{\dagger}$. Thus we have proved that for every $r \in P$ there is a $p \in \mathcal{J}^\dagger$ such that $p \geq r$, and so $\mathcal{J}^\dagger$ is dense. Since $\mathcal{J} \in V$ and $\mathcal{J}^\dagger$ is constructed from $\mathcal{J}$ in V we have $\mathcal{J}^\dagger \in V$. Since G is generic over V we have $G \cap \mathcal{J}^\dagger \neq \emptyset$. Let $p \in G \cap \mathcal{J}^\dagger$. By the definition of $\mathcal{J}^\dagger$ there is a $q \in \mathcal{J}^\dagger$ such that $q \leq p$. Since G is downward closed we have $q \in G$ and hence $q \in G \cap \mathcal{J} \neq \emptyset$, which is what we had to prove. $\square_{1.17}$

1.18 Lemma. Let $q \in P$, and let $\mathcal{I}$ be a subset of P in V which is pre-dense above q. For every generic subset G of P if $q \in G$ then $G \cap \mathcal{I} \neq 0$.

Proof. Let $\mathcal{I}^\dagger = \mathcal{I} \bigcup \{p \in P : p \text{ is incompatible with } q\}$. Since $\mathcal{I} \in V$ also $\mathcal{I}^\dagger \in V$. Let us prove that $\mathcal{I}^\dagger$ is a pre-dense subset of P. Let $r \in P$. If r is incompatible with q then $r \in \mathcal{I}^\dagger$. If r is compatible with q then there is an $s \in P$ such that $s \geq r, q$. Since $\mathcal{I}$ is pre-dense above q, necessarily s is compatible with some member of $\mathcal{I}$, and hence r is compatible with the same member of $\mathcal{I}$ which neccessarily is also in $\mathcal{I}^\dagger$. Thus we have shown that $\mathcal{I}^\dagger$ is pre-dense. Let G be a generic subset of P such that $q \in G$. Since $\mathcal{I}^\dagger$ is pre-dense and $\mathcal{I}^\dagger \in V$ we have $G \cap \mathcal{I}^\dagger \neq \emptyset$. Let $t \in G \cap \mathcal{I}^\dagger$. Since $t, q \in G$, t is compatible with q, hence by the definition of $\mathcal{I}^\dagger$ we must have $t \in \mathcal{I}$ and thus $t \in G \cap \mathcal{I} \neq \emptyset$. $\square_{1.18}$

1.19 Lemma. Let $\mathcal{I} = \{p_i : i < i_0\}$ be an antichain in P and $\{\tau_i : i < i_0\}$ a corresponding indexed family of P-names (in V). *Then* there is a name τ such that: for every $i < i_0$ and for every generic G, if $p_i \in G$ then $\tau[G] = \tau_i[G]$ (and $\tau[G] = \emptyset$ if $G \cap \{p_i : i < i_0\} = \emptyset$). (We recall that a generic G contains at most one member of $\mathcal{I}$ and if $\mathcal{I}$ is a maximal antichain of P then G contains exactly one member of $\mathcal{I}$).

1.19A Remark. This means we can define a name by cases.

Proof. Suppose $\tau_i = \{\langle p_{i,j}, \tau_{i,j}\rangle : j < j_i\}$, (of course $j_i = 0$ is possible) and let $\tau = \{\langle r, \tau_{i,j}\rangle : j < j_i,\ i < i_0, r \geq p_{i,j} \text{ and } r \geq p_i\}$. $\square_{1.19}$

We note also:

1.20 Claim. Let G be a downward closed directed subset of P. The following are equivalent:

(a) G is generic.

(b) $G \cap \mathcal{I} \neq \emptyset$ for every dense open subset $\mathcal{I}$ of P.

(c) $G \cap \mathcal{I} \neq \emptyset$ for every dense subset $\mathcal{I}$ of P.

(d) $G \cap \mathcal{I} \neq \emptyset$ for every pre-dense subset $\mathcal{I}$ of P.

(e) $G \cap \mathcal{I} \neq \emptyset$ for every maximal antichain $\mathcal{I}$ of P.

1.20A Remark. Clearly for $\mathcal{I} \subseteq P$,

(1) $\mathcal{I}$ is dense open $\Rightarrow \mathcal{I}$ is dense $\Rightarrow \mathcal{I}$ is pre-dense,

(2) $\mathcal{I}$ is a maximal antichain of $P \Rightarrow \mathcal{I}$ is pre-dense.

Proof. By Remark 1.20A(1) clearly (d)$\Rightarrow$(c)$\Rightarrow$(b), by 1.20A(2) clearly (d)$\Rightarrow$(e) by 1.16 (e)$\Rightarrow$(a), trivially (a)$\Rightarrow$(b); by the closing up of subsets of P clearly (b)$\Rightarrow$(c)$\Rightarrow$(d); together we have finished. $\square_{1.20}$

§2. The Consistency of CH (The Continuum Hypothesis)

Usually the consistency of CH, i.e., of $2^{\aleph_0} = \aleph_1$, is proved by showing that it holds in L (the class of constructible sets) but we do not want to go in this way. So

2.1 Theorem. Model A. There is a model of ZFC in which $2^{\aleph_0} = \aleph_1$.

Proof. Let us first review the main points of the construction of the model. We start with a countable transitive model V of ZFC in which $2^{\aleph_0}$ is either $\aleph_1$ or greater. We shall extend it to a model $V[G]$ in which G will essentially be a counting of length $\aleph_1$ of all sets of natural numbers, i.e. a function g from ω_1 onto the family of sets of natural numbers. Each condition, i.e., each member of P, is an approximation of the generic object G and therefore will consist of partial information about the counting. Since every two members of G are compatible, the members of G, taken together, yield a counting of subsets of ω. The three things about which we have to worry in the proof are the following:

a) How do we know that every subset a of ω in V will occur in the counting given by G?

This will be answered by showing that the set of all partial countings in which a occurs is a dense subset of P, and hence G contains such a partial counting.

b) How will the new subsets of ω, i.e., the subset of ω which are in $V[G]$ but not in V, be counted when the members of P, being in V, can count only subsets of ω which are in V?

Here we shall make sure that $V[G]$ has no new subsets of ω.

c) Is $\aleph_1$ of V, which we have mapped on the set of all subsets of ω in $V[G]$ also the $\aleph_1$ of $V[G]$?

Here the answer is easily positive because V and $V[G]$ have the same sets of natural numbers.

In $V[G]$ we want to obtain a function g from $\aleph_1^{V[G]}$ (where $\aleph_1^{V[G]}$ denotes the ordinal which is the $\aleph_1$ of $V[G]$, i.e., the least ordinal α such that $V[G]$ does not contain a mapping of ω onto α) onto $\mathcal{P}(\omega)^{V[G]}$ (where $\mathcal{P}(\omega)$ is the power set of ω and the superscript $V[G]$ means that $\mathcal{P}(\omega)^{V[G]}$ is the power set of ω in $V[G]$). It will turn out that $\aleph_1^{V[G]} = \aleph_1^V$, and the only subsets of ω available in $V[G]$ are the members of $\mathcal{P}(\omega)^V$. By easy considerations for every countable ordinal α, the function $g \restriction \alpha$ has to belong to V (as it can be coded by a set of natural numbers). Therefore partial information about g is given by functions from countable ordinals into $\mathcal{P}(\omega)$ in V. Thus it is natural to define

$$P = \{f : f \text{ is a function from a countable ordinal into } \mathcal{P}(\omega)\}$$

where the definition is inside V: with $f \leq g$ *iff* $f \subseteq g$.
A member f of P is understood to "claim" that g is like f on the domain of f. When does a member $f^\dagger$ of P give us more information that f? When $f \subseteq f^\dagger$. Therefore we take the partial order $<$ on P to be proper inclusion.

Let G be a generic subset of P. By the definition of the concept of a generic object every two members of G are compatible, hence $\bigcup G = \bigcup\{f : f \in G\}$ is a function, we shall denote it with f_G. The domain of f_G is the union of the domains of the members of G and hence it is a union of ordinals $< \aleph_1^V$ and therefore the domain of f_G is an ordinal $\leq \aleph_1^V$. If A is a countable set in V then in V there is a function r from ω onto A. Since V is a transitive set, V contains already all the objects which are members of A in the universe and hence A has no new members in $V[G]$. Now $V[G]$ has the same set ω of natural numbers as V and therefore r maps ω on A also in $V[G]$ and A is countable in $V[G]$. In particular every ordinal which is countable in V is also countable in $V[G]$. Therefore, if $\aleph_1^V$ is uncountable in $V[G]$ then $\aleph_1^V$ is the least ordinal which is uncountable in $V[G]$, i.e., $\aleph_1^V = \aleph_1^{V[G]}$. If $\aleph_1^V$ is countable in $V[G]$ then $\aleph_1^V < \aleph_1^{V[G]}$, hence in either case $\aleph_1^V \leq \aleph_1^{V[G]}$. Therefore the domain of f_G is an ordinal $\leq \aleph_1^{V[G]}$ (since it is $\leq \aleph_1^V$). We have in fact deal with

2.2 Definition. 1) A first order formula $\varphi(x)$ (in the language of set theory) is upward absolute *if* when $V \subseteq V^\dagger$ are models of ZFC with the same ordinals (so by our conventions both are transitive sets) and $a \in V$, then: $V \models \varphi[a] \Rightarrow V^\dagger \models \varphi[a]$. Note that properties and function are interpreted by first order formulas.
2) We say φ is absolute if both φ and $\neg\varphi$ are upward absolute (i.e. we have "iff" above).

Obviously (or see [J]):

2.3 Lemma. 1) The following are upward absolute: "α an ordinal", "α not an ordinal" "α is (not) a natural number", "$\alpha = n$", "$\alpha \neq n$", "$\alpha = \omega$", "$\alpha \neq \omega$", "α is not a cardinal", "α is not regular", "A has cardinality $\leq \alpha$", "$A \subseteq B$", "$A \supseteq B$", "f is (not) a (one-to-one) function from A to (onto) B".
2) In fact any relation (function, property) defined by a $\sum_1$ - formula, i.e., by $(\exists y)\varphi(x_0 \ldots, x_{n-1}, x_n, y)$ when φ has only bounded quantifiers i.e., of the form $(\forall z_1 \in z_2), (\exists z_1 \in z_2)$ is upward absolute. $\square_{2.3}$

We have now to prove the following facts.

2.4 Fact. Every $A \in \mathcal{P}(\omega)^V$ is in the range of f_G.

2.5 Fact. $\mathcal{P}(\omega)^{V[G]} = \mathcal{P}(\omega)^V$.

Once these facts are proved we know that the range of f_G is $\mathcal{P}(\omega)^{V[G]}$ and in $V[G]$ the function f_G maps an ordinal $\leq \aleph_1^{V[G]}$ onto $\mathcal{P}(\omega)$ hence in $V[G]$ the set $\mathcal{P}(\omega)$ has at most $\aleph_1$ subsets, which establishes $2^{\aleph_0} = \aleph_1$ in $V[G]$.

Remember that a subset $\mathcal{I}$ of P is said to be *pre-dense* in P if every member p of P is compatible with some member of $\mathcal{I}$. In particular $\mathcal{I}$ is pre-dense if $\mathcal{I}$ is a maximal antichain or if $\mathcal{I}$ is such that for every p in P there is a $q \in \mathcal{I}$ such that $q \geq p$.

Proof of Fact 2.4. Let $A \in \mathcal{P}(\omega)^V$; we want to prove that $A \in \mathrm{Rang}(f_G)$, i.e., for some $f \in G$ we have $A \in \mathrm{Rang}(f)$, or, in other words $G \cap \{f \in P : A \in \mathrm{Rang}(f)\} \neq \emptyset$. For this purpose it suffices to show that the set $\mathcal{I} = \{f \in P : A \in \mathrm{Rang}(f)\}$ is dense, since this set is obviously in V. Let $p \in P$ then $p : \alpha \to \mathcal{P}(\omega)$ in V, where α is a countable ordinal. Let f be the function with domain $\alpha + 1$, which is also a countable ordinal in V, into $\mathcal{P}(\omega)^V$ given by $f(\xi) = p(\xi)$ for $\xi < \alpha$ and $f(\alpha) = A$. Clearly $f \in P, f \geq p$ and $f \in \mathcal{I}$. Thus $\mathcal{I}$ is dense and Fact 2.4 is established. $\square_{2.4}$

Now Fact 2.5 will follow from 2.7, 2.8 below, thus completing the proof of Theorem 2.1.

2.6 Definition. 1) A forcing notion P is said to be $\aleph_1$-*complete*, or *countably complete* if every increasing (by $\leq$) sequence $\langle p_n : n < \omega \rangle$ of members of P, i.e., every sequence $\langle p_n : n < \omega \rangle$ such that $p_0 \leq p_1 \leq p_2 \leq \ldots$, has an upper bound p in P (i.e., $p \geq p_n$ for every $n < \omega$).
2) A partial order P is λ-complete if for any $\gamma < \lambda$ and increasing (by $\leq = \leq^P$) sequence $\langle p_i : i < \gamma \rangle$ of members of P, the sequence has an upper bound in P.

2.7 Lemma. Our present set P is countably complete in V.

Proof. Let $\langle p_n : n < \omega \rangle$ be a nondescending sequence of members of P in V. Since the p_n's are pairwise compatible $p = \bigcup_{n<\omega} p_n$ is a function. The domain of p is the union of the domains of the p_n's. Thus the domain of p is a countable union of countable ordinals, and is therefore a countable ordinal. The range of p is the union of the ranges of the p_n's and it consists therefore of members of $\mathcal{P}(\omega)^V$. Thus $p \in P$, and obviously $p \geq p_n$ for every $n < \omega$. $\square_{2.7}$

2.8 Theorem. 1) For every countably complete forcing notion P in V and every generic subset G of P, $V[G]$ contains no new ω-sequence of members of V, i.e., if $\langle a_n : n < \omega \rangle \in V[G]$ and $a_n \in V$ for $n < \omega$ then also $\langle a_n : n < \omega \rangle \in V$. In particular if $a \subseteq \omega$ and $a \in V[G]$ then also $a \in V$.
2) If the forcing notion P is λ-complete, $G \subseteq P$ is generic over V *then* $V[G]$ has no new (i.e.$\notin V$) bounded subsets of λ, not even new sequences of length $< \lambda$ of members of V.

Proof. 1) Let $\langle a_n : n < \omega \rangle \in V[G]$, then $\langle a_n : n < \omega \rangle$ has a name $\underaccent{\sim}{\tau}$. By the forcing theorem (i.e. 1.14) there is a $q \in G$ such that $q \Vdash$ "$\underaccent{\sim}{\tau}$ is an ω-sequence of members of V". We shall prove that the subset $\{p \in P : p \Vdash \text{"}\underaccent{\sim}{\tau} \in V\text{"}\}$ of P is pre-dense above q and, by 1.18, therefore G contains a p such that $p \Vdash$ "$\underaccent{\sim}{\tau} \in V$" and therefore $\underaccent{\sim}{\tau}[G] \in V$, i.e., $\langle a_n : n < \omega \rangle \in V$ is true in $V[G]$. Here we use the fact that $\{p \in P : p \Vdash \text{"}\underaccent{\sim}{\tau} \in V\text{"}\}$ is in V; this is the case since forcing is definable in V.

Let us prove now that $\mathcal{I} = \{p \in: p \Vdash \text{"}\underaccent{\sim}{\tau} \in V\text{"}\}$ is pre-dense above q. Let $q^\dagger \geq q$; it "knows" (i.e. forces) that every a_n is in V, since already q forces this statement, but even $q^\dagger$ does not necessarily "know" the identity of a_n. We shall see that we can extend $q^\dagger$ to a condition which "knows" a_0(=force a value), then to a condition which "knows" a_1, and so on, and as a consequence of the countably completeness of P, $q^\dagger$ can be finally extended to a condition p which "knows" all the a_n's. This will imply p "knows" that $\langle a_n : n < \omega \rangle$ is some particular member of V, and $p \Vdash$ "$\underaccent{\sim}{\tau} \in V$", which establishes the pre-density of $\mathcal{I}$ above q.

We define a sequence $\langle p_n : n < \omega\rangle$ of conditions and a sequence $\langle a_n : n < \omega\rangle$ of members of V as follows. Let us mention now that the forthcoming definition is carried out entirely within V and therefore the obtained sequences are members of V. We set $p_0 = q^\dagger$. For $n \geq 0$ we choose p_{n+1} and a_n so that $p_{n+1} \geq p_n$ and $p_{n+1} \Vdash$ "$\underset{\sim}{\tau}(n) = \dot{a}_n$", where $\underset{\sim}{\tau}(n) = \dot{a}_n$ is an abbreviation of "the n-th term of the sequence $\underset{\sim}{\tau}$ is a_n" and $\dot{a}_n$ is the P-name of a_n (see Definition 1.9(2) and 1.9A). Do such p_{n+1} and a_n exist? To prove their existence we go out of V, but this does not matter since once we know they exist the definition proceeds entirely within V. Let $G^\dagger$ be any generic subset which contains p_n.

In $V[G^\dagger]$ we have $\underset{\sim}{\tau}[G^\dagger]$ is an ω-sequence of members of V, since $q \leq p_n \in G^\dagger$ and $q \Vdash$ "τ is an ω-sequence of members of V". Let a_n be the n-th term of the sequence $\underset{\sim}{\tau}[G^\dagger]$, then $a_n \in V$ and $\underset{\sim}{\tau}[G^\dagger](n) = a_n$ is true in $V[G^\dagger]$. By the forcing theorem there is an $r \in G^\dagger$ such that $r \Vdash$ "$\underset{\sim}{\tau}(n) = \dot{a}_n$". Since $r, p_n \in G^\dagger$ they are compatible. Choose $p_{n+1} \geq r, p_n$ then also $p_{n+1} \Vdash$ "$\underset{\sim}{\tau}(n) = \dot{a}_n$", and p_{n+1} and a_n are as required. In order to choose a definite p_{n+1} in P we assume that we have some fixed well ordering of P in V and p_{n+1} is chosen to be the least member of P in that well-ordering for which there exists an a_n so that $p_{n+1} \geq p_n$ and $p_{n+1} \Vdash$ "$\underset{\sim}{\tau}(n) = \dot{a}_n$". Note that a_n is uniquely determined by p_{n+1} since if also for some $b \neq a_n$ we have $p_{n+1} \Vdash$ "$\underset{\sim}{\tau}(n) = \dot{b}$" then for every generic $G^\dagger$ which contains p_{n+1} we have that the n-th term of $\underset{\sim}{\tau}[G^\dagger]$ is both a_n and b, which is impossible.

Since P is countably complete there is a $p \in P$ such that $p \geq p_n$ for all $n < \omega$. We have, obviously $p \geq p_0 = q^\dagger$ and for every $n < \omega$ we know $p \geq p_{n+1}$ and hence $p \Vdash$ "$\underset{\sim}{\tau}(n) = \dot{a}_n$". Thus for every generic subset $G^\dagger$ of P which contains p we have: $\underset{\sim}{\tau}[G^\dagger]$ is an ω-sequence and $\underset{\sim}{\tau}[G^\dagger](n) = a_n$ for every $n < \omega$, hence $\underset{\sim}{\tau}[G^\dagger] = \langle a_n : n < \omega\rangle \in V$ (note $\langle a_n : n < \omega\rangle \in V$ since this sequence was defined in V). By the definition of the forcing relation we have $p \Vdash$ "$\underset{\sim}{\tau} \in V$", which is what we had to prove.

If $a \in V[G]$ and $a \subseteq \omega$ then let $\langle a_n : n < \omega\rangle$ be the characteristic function of a. Since each a_n is 0 or 1 we have $\langle a_n : n < \omega\rangle$ is a sequence of members of V and hence, by the present theorem $\langle a_n : n < \omega\rangle \in V$ and a can be easily obtained from $\langle a_n : n < \omega\rangle$ within V.

2) Left to the reader. $\square_{2.8,2.5,2.1}$

§3. On the Consistency of the Failure of CH

We first prove a technical lemma, and then prove that $2^{\aleph_0} = \aleph_\alpha$ is possible for almost any α.

3.1 The Existential Completeness Lemma. 1) If $p_0 \Vdash_P$ "$(\exists x)\varphi(x)$" *then* there is a name $\underset{\sim}{\tau}$ such that $p_0 \Vdash_P \varphi(\underset{\sim}{\tau})$, where $\varphi(x)$ is a formula which may mention names.

2) Moreover for every formula $\varphi(x)$ as above for some P-name $\underset{\sim}{\tau}$

$\Vdash_P$ "$(\exists x)[\varphi(x)] \rightarrow \varphi(\underset{\sim}{\tau})$" and $\Vdash_P$ "$\neg(\exists x)[\varphi(x)] \rightarrow \underset{\sim}{\tau} = \emptyset$".

Proof. 1) The idea of the proof is as follows. The condition p_0, "knows" that $(\exists x)\varphi(x)$ but this does not tell us directly that p_0 knows a particular name of a set x which satisfies $\varphi(x)$. However with more information than that in p_0 we know names of sets which satisfy $\varphi(x)$. What we have to do is to combine the various names to a single name which equals each of those names just when the name satisfies $\varphi(x)$.

Let

$$\mathcal{J} \stackrel{\text{def}}{=} \{q : q \Vdash_P \text{"}\neg(\exists x)\varphi(x)\text{" or for some name } \underset{\sim}{\tau} \text{ we have } q \Vdash_P \text{"}\varphi(\underset{\sim}{\tau})\text{"}\}.$$

$\mathcal{J}$ is defined in V, hence $\mathcal{J} \in V$. We shall now see that $\mathcal{J}$ is dense. Let $r \in P$, but r does not force $\neg(\exists x)\varphi(x)$. Then there is a generic $G \subseteq P$ such that $r \in G$, and $V[G] \models$ "$(\exists x)\varphi(x)$", by r's choice. Since every member of $V[G]$ is the value of some name we have $V[G] \models$ "$\varphi(\underset{\sim}{\tau}[G])$" for some P-name $\underset{\sim}{\tau}$. By the forcing theorem there is an $r^\dagger \in G$ such that $r^\dagger \Vdash_P$ "$\varphi(\underset{\sim}{\tau})$". Since G is directed and $r \in G$ without loss of generality $r^\dagger \geq r$ and by definition, $r^\dagger \in \mathcal{J}$. Now if $r \Vdash_P$ "$\neg(\exists x)\varphi(x)$", trivially $r \in \mathcal{J}$. Thus we have shown that $\mathcal{J}$ is dense. Let $\mathcal{I}$ be a maximal subset of $\mathcal{J}$ of pairwise incompatible members. We shall see that also $\mathcal{I}$ is pre-dense. Let $q \in P$ and suppose, in order to obtain a contradiction, that q is incompatible with every member of $\mathcal{I}$. By what we proved about $\mathcal{J}$ there is a $q^\dagger \in \mathcal{J}$ such that $q^\dagger \geq q$. Also $q^\dagger$ is, clearly, incompatible with every member of $\mathcal{I}$. Now $\mathcal{I} \bigcup \{q^\dagger\}$ is a subset of $\mathcal{J}$ of pairwise incompatible members which properly includes $\mathcal{I}$, which contradicts our choice of $\mathcal{I}$.

Let $\mathcal{I} = \{q_i : i < \alpha\}$, since $\mathcal{I} \subseteq \mathcal{J}$ there is for every $i < \alpha$ a name $\underset{\sim}{\tau}_i$ such that $q_i \Vdash_P$ "$\varphi(\underset{\sim}{\tau}_i)$" or $q_i \Vdash_P$ "$\neg(\exists x)\varphi(x)$". Let $\underset{\sim}{\tau}$ be the name

$$\underset{\sim}{\tau} = \begin{cases} \underset{\sim}{\tau}_i & \text{if } q_i \Vdash_P \text{ "}\varphi(\underset{\sim}{\tau}_i)\text{"} \\ \emptyset & \text{otherwise} \end{cases}$$

which we have proved to exist (1.19) for pairwise incompatible q_i's. We claim that $p_0 \Vdash$ "$\varphi(\underset{\sim}{\tau})$". To prove that let G be a generic subset of P and $p_0 \in G$. Since $\mathcal{I}$ is pre-dense we have $G \cap \mathcal{I} \neq \emptyset$ and hence for some $i < \alpha$ we have $q_i \in G$. If $\underset{\sim}{\tau}_i$ is not defined $q_i \Vdash_P$ "$\neg(\exists x)\varphi(x)$", but then q_i, p_0 are incompatible, but both are in G, contradiction. So $q_i \Vdash$ "$\varphi(\underset{\sim}{\tau}_i)$" hence we have $V[G] \models \varphi(\underset{\sim}{\tau}_i[G])$. Also since $q_i \in G$ we have, by the definition of $\underset{\sim}{\tau}$, $\underset{\sim}{\tau}[G] = \underset{\sim}{\tau}_i[G]$, hence $V[G] \models$ "$\varphi(\underset{\sim}{\tau}[G])$", which establishes $p_0 \Vdash$ "$\varphi(\underset{\sim}{\tau})$".

2) The second part in the lemma was really proved too. $\square_{3.1}$

3.2 Theorem. Model B. There is a model in which the continuum hypothesis fails. Moreover, for every cardinal λ there is a forcing notion P such that $\Vdash_P$ "$2^{\aleph_0} \geq \lambda$ and every cardinal of V is a cardinal (of $V[G]$))".

Convention. We use the word "real" as meaning a subset of ω or its characteristic function.

Proof. We want to add to V λ real numbers (i.e., functions from ω into 2). Each condition gives us some information about them, so we have to make sure that the information contained in a single condition will not suffice to compute one of the reals since in this case this real will be already in V. Therefore we shall define the conditions so that each one will contain only a finite amount of information, and therefore each condition is clearly insufficient for computing any of the reals. We shall regard the λ reals, each of which is an ω-sequence of

0's and 1's, as written in a long sequence one after the other to form a sequence of length λ of 0's and 1's. The members of P will be finite approximations to this member of ${}^{\lambda}2$. Therefore we take $P = \{f : f$ is a finite function from λ into $\{0,1\}\}$ where by " a finite function from λ" we mean a function from a finite subset of λ. For the partial order on P we take proper inclusion, i.e., $f < g$ if $f \subset g$. This forcing is called "adding λ Cohen reals."

3.3 Lemma. $\Vdash_P$ "there are at least λ reals".

Proof of the Lemma. We shall prove the existence of the reals by giving them names. Since for every generic G there is a function g which satisfies $g = \bigcup_{f \in G} f$ we have $\Vdash_P$ "$(\exists x)(x = \bigcup_{f \in \underset{\sim}{G}_P} f)$" and by the first lemma in this section, 3.1 there is a name $\underset{\sim}{g}$ such that $\Vdash_P$ "$\underset{\sim}{g} = \bigcup_{f \in \underset{\sim}{G}_P} f$". Using again the same method we get, for every $i < \lambda$ a name $\underset{\sim}{a}_i$ defined by $\underset{\sim}{a}_i(n) = \underset{\sim}{g}(i+n)$. Now a_i is forced to be a name of a real number provided g is a name of a function on λ into $\{0,1\}$. We shall prove it in the next sublemma.

3.4 Sublemma. g is a function from λ into $\{0,1\}$ (i.e., this is forced).

Proof of the Sublemma. g is a function since G is a directed set of functions, hence its union g is a function. Also g is into $\{0,1\}$ since each member of G is into $\{0,1\}$. Next $\mathrm{Dom}(g) \subseteq \lambda$ since for every $f \in G$ we have $f \in P$ and hence $\mathrm{Dom}(f) \subseteq \lambda$; we still have to prove that $\mathrm{Dom}(g) = \lambda$. Let $i < \lambda$; it suffices to prove that the set $\mathcal{J}_i = \{f \in P : i \in \mathrm{Dom}(f)\}$, which is clearly is V, is pre-dense. Let $f \in P$, if $i \in \mathrm{Dom}(f)$ then $f \in \mathcal{J}_i$, otherwise let $f^\dagger = f \bigcup \{\langle i, 0\rangle\}$ then $f^\dagger \in P$ and $f^\dagger > f$ and $f^\dagger \in \mathcal{J}_i$. $\square_{3.4}$

We return to the Lemma 3.3. Note that since we have proved that $\Vdash_P$ "$\underset{\sim}{g}$ is a function from λ into $\{0,1\}$", it is enough to prove for each $i \neq j < \lambda$ that $\Vdash_P$ "$\underset{\sim}{a}_j \neq \underset{\sim}{a}_j$". For this it suffices to prove that for every $p \in P$ there is an $r \geq p \in P$ and an $n < \omega$ such that $r \Vdash_P$ "$\underset{\sim}{a}_i(n) \neq \underset{\sim}{a}_j(n)$", since this proves that the set of all $p \in P$ such that $p \Vdash_P$ "$\underset{\sim}{a}_i \neq \underset{\sim}{a}_j$" is pre-dense.

Let $p \in P$. Since $\mathrm{Dom}(p)$ is finite there is an n_0 such that for every $n \geq n_0$ $i+n \notin \mathrm{Dom}(p)$ and an n_1 such that for every $n \geq n_1$ we have $j+n \notin \mathrm{Dom}(p)$. We set $r = p \bigcup \{\langle i+k, 0\rangle, \langle j+k, 1\rangle$, where $k \geq n_0, n_1$. Clearly r is a function since $i+k, j+k \notin \mathrm{Dom}(p)$ and $i+k \neq j+k$ since $i \neq j$. Obviously $p < r$ and (r forces that) $\underset{\sim}{a}_i(k) = 0$, $\underset{\sim}{a}_j(k) = 1$ hence $\underset{\sim}{a}_i(k) \neq \underset{\sim}{a}_j(j)$. $\square_{3.3}$

Continuation of the Proof of 3.2. We started with λ, which is a cardinal of V and we proved that in $V[G]$ there are at least λ real numbers, but is λ in $V[G]$ the "same" cardinal as it was in V? As the matter stands now we do not even know whether the continuum hypothesis fails in $V[G]$ since even though λ may be a large cardinal in V it may be countable or $\aleph_1$ in $V[G]$. We shall now prove that all the cardinals of V are still cardinals in $V[G]$ so for example if $\lambda = \aleph_2^V$ then λ is still the third infinite cardinal in $V[G]$ and thus $\lambda = \aleph_2^{V[G]}$. We shall prove that the cardinals of V are not collapsed in $V[G]$ for forcing notions P which satisfy the countable chain condition and that P satisfies this condition. I.e. the cardinals of V are still cardinals in $V[G]$, and as V, $V[G]$ have the same ordinals and no non-cardinals of V are cardinals of $V[G]$ we have: V, $V[G]$ have the same cardinals; this is done in 3.6, 3.8 below. This is important general theorem which we shall use a lot. This will finish the proof of 3.2.

3.5 Definition.

(1) A forcing notion Q satisfies the countable chain condition (c.c.c.) if Q has no uncountable subset of pairwise incompatible members, i.e., if every uncountable subset of Q contains two compatible members.

(2) A forcing notion Q satisfies the λ-chain condition (λ-c.c.) if there are no λ pairwise incompatible members of P.

3.6 Lemma. If a forcing notion Q satisfies the c.c.c. then

(i) forcing with Q does not collapse cardinals and cofinalities, (i.e., $\Vdash_Q$ "every cardinal of V is a cardinal (of $V[G]$)), and the cofinality is preserved".

(ii) For every ordinal α and every Q-name $\underset{\sim}{\tau}$ there is, in V, a function F from α (to V) such that for every $\beta < \alpha$ we have $|F(\beta)| \leq \aleph_0$ and $\Vdash_Q$ "if $\underset{\sim}{\tau}$ is a function from α into V then $(\forall \beta < \alpha)[\underset{\sim}{\tau}(\beta) \in F(\beta)]$".

Proof of Lemma 3.6. Proof of (ii): We define the function F on α by $F(\beta) = \{a \in V : (\exists q \in Q)(q \Vdash_Q$ "$\underset{\sim}{\tau}$ is a function from α and $\underset{\sim}{\tau}(\beta) = \dot{a}$")$\}$. We have to prove that the right hand side is a set, and not a proper class of V and moreover is countable. We shall assume it now and prove it later. The right hand side is the set of all possible values of $\underset{\sim}{\tau}(\beta)$ in all the $V[G]$'s in which $\underset{\sim}{\tau}[G]$ is a function from α into V. To see this suppose G is a generic set such that $\underset{\sim}{\tau}[G]$ is a function from α into V. Then for some $a \in V$ $V[G] \vDash$ "$\underset{\sim}{\tau}$ is a function from α and $\underset{\sim}{\tau}(\beta) = \dot{a}$". By the forcing theorem there is a $q \in Q$ such that $q \Vdash_Q$ "$\underset{\sim}{\tau}$ is a function from α and $\underset{\sim}{\tau}(\beta) = \alpha$". Hence $a \in F(\beta)$ and therefore $V[G] \vDash$ "$(\forall \beta < \alpha)[\underset{\sim}{\tau}(\beta) \in F(\beta)]$". Since this is the case for every generic G we have what is claimed in (ii).

Now we shall prove not only that the class $\{a \in V : (\exists q \in Q)(q \Vdash_Q \underset{\sim}{\tau}$ is a function on α and $\underset{\sim}{\tau}(\beta) = \dot{a}$")$\}$ is a set but even that it is a countable set, and thus $|F(\beta)| \leq \aleph_0$. Suppose $\{a_i : i < \omega_1\}$ is a subset of this class. For each such a_i there is a $q_i \in Q$ such that $q_i \Vdash_Q$ "$\underset{\sim}{\tau}$ is a function on α and $\underset{\sim}{\tau}(\beta) = a_i$". Since Q satisfies the c.c.c. there must be some $i \neq j$ such that q_i and q_j are compatible. Let $q \geq q_i, q_j$ and let $G \subseteq Q$ be a generic subset of Q which contains q. We have $q_i, q_j \in G$ and hence, since $q_i \Vdash_Q$ "$\underset{\sim}{\tau}(\beta) = a_i$" and $q_j \Vdash_Q$ "$\underset{\sim}{\tau}(\beta) = a_j$", we have $V[G] \vDash$ "$a_i = \underset{\sim}{\tau}(\beta) = a_j$", hence $a_i = a_j$. Thus we have shown that at least two of the a_i's must be equal therefore the class $\{a \in V : \exists q \in Q(q \Vdash_Q$ "$\underset{\sim}{\tau}$ is a function from α and $\underset{\sim}{\tau}(\beta) = a$")$\}$ must be countable. The proof that $F(\beta)$ is a set is similar.

Proof of (i). Let λ be an uncountable cardinal of V and suppose λ is not a cardinal in $V[G]$. Then there is an ordinal $\alpha < \lambda$ and a function $f \in V[G]$ which maps α onto λ. Let $\underset{\sim}{\tau}$ be a name for f in $V[G]$. By part (ii) of our lemma (already proved) there is a function F from α in V such that $f(\beta) \in F(\beta)$ and $|F(\beta)| \leq \aleph_0$ for every $\beta < \alpha$. We have therefore $\lambda = \text{Rang}(f) \subseteq \bigcup_{\beta<\alpha} F(\alpha)$. But in V we have $|\bigcup_{\beta<\alpha} F(\beta)| \leq |\alpha|\aleph_0 < \lambda$, since λ is an uncountable cardinal

of V and $\alpha < \lambda$, and therefore also $|\alpha| < \lambda$. Thus we have proved that the uncountable cardinals are not collapsed in $V[G]$. Also $\aleph_0$ is not collapsed since the finite ordinals and ω of V are also finite ordinals and ω in $V[G]$.
The preservation of the cofinality is proved similarly. $\square_{3.6}$

Similarly one can prove for uncountable λ.

3.7 Claim. If Q satisfies the λ-c.c. then

(i) forcing by Q preserve cardinals and cofinalities which are $\geq \lambda$.

(ii) for every Q-name and ordinal α there is a function F with domain α, ($F \in V$) such that $\Vdash_Q$ " if $\underset{\sim}{\tau}$ is a function from α to V then $\underset{\sim}{\tau}(\beta) \in F(\beta)$ for every $\beta < \alpha$", and $|F(\beta)| < \lambda$ for $\beta < \alpha$.

3.8 Lemma. The forcing notion P which we use here for Model B satisfies the c.c.c.

3.8A Remark. Once we prove this lemma we know that all the cardinals in V here are cardinals also in $V[G]$ and therefore λ is a cardinal also in $V[G]$, and if λ is the α-th infinite cardinal $\aleph_\alpha$ in V it is also the α-th infinite cardinal in $V[G]$.

Proof. Suppose $\{f_i : i < \aleph_1\} \subseteq P$, in order to prove Lemma 3.8, it is enough to prove that two of the functions are compatible. By 3.10 below for some uncountable $A \subseteq \omega_1$ and finite w for every $i \neq j$ from A, we have $\mathrm{Dom}(f_i) \cap \mathrm{Dom}(f_j) = w$. The number of possible $f_i \restriction w$ is finite (i.e. $\leq 2^{|w|}$), so without loss of generality $f_i \restriction w = f^*$ for every $i \in A$. Now for any $i \neq j \in A$, we know $f_i \cup f_j \in P$ is a common upper bound of f_i, f_j. $\square_{3.8}$
We have promised the so called Δ-system lemma.

3.9 Definition. A family F of finite sets is called a Δ-system if there is a set w such that for any A, B in F we have $A \cap B = w$.
In this formulation our problem reduces to:

3.10 Lemma. Given an indexed family F of finite sets, $|F| = \aleph_1$ there is $F^\dagger \subseteq F; |F^\dagger| = \aleph_1$ such that $F^\dagger$ is a Δ-system.

Proof. The cardinality of each A in F is a member of ω but there are $\aleph_1$ elements in F, so by the pigeonhole principle some n is obtained uncountably many times. In similar cases in the future we will just say: w.l.o.g. $|A| = n$ for any A in F (since all we need is a family of $\aleph_1$ finite sets.)

Now we proceed by induction on n:

For $n = 1$: this is the pigeonhole principle for $\aleph_1$.

For $n > 1$ we distinguish two cases.

1) There is $a \in \bigcup\{A : A \in F\}$ such that there are uncountably many B in F such that $a \in B$, then you have an easy induction step (you take a "out" and put it back after using induction hypothesis).
2) If there is no such a then we build a sequence $\langle A_\alpha : \alpha < \omega_1\rangle$ such that $\alpha \neq \beta \Rightarrow A_\alpha \cap A_\beta = \emptyset$ $(A_\alpha \in F)$; suppose A_β is defined for $\beta < \alpha$. Now $\bigcup_{\beta<\alpha} A_\beta$ is countable. The subfamily of F of members which contain an element of this union is clearly countable so there is $A \in F$, such that $A \cap \bigcup_{\beta<\alpha} A_\beta = \emptyset$, let $A_\alpha = A$ and we are done (in this case $w = \emptyset$). $\square_{3.10,3.2}$

§4. More on the Cardinality $2^{\aleph_0}$ and Cohen Reals

Now we want to find what is exactly the power of the continuum in $V[G]$ for the model from Theorem 3.2.

4.1 Theorem. Let $P = \{f : f$ is a finite function from λ to $\{0,1\}\}$. We have already shown that $\Vdash_P$ "$2^{\aleph_0} \geq \lambda$", we shall show now that if $\lambda^{\aleph_0} = \lambda$ then $\Vdash_P$ "$2^{\aleph_0} = \lambda$".

Proof. The idea is to construct a family of $\lambda^{\aleph_0}$ canonical names for the real numbers and then prove that for every name $\underset{\sim}{\tau}$ there is a canonical name $\underset{\sim}{\tau}'$ of a real such that $\Vdash_P$ " if $\underset{\sim}{\tau}$ is a real then $\underset{\sim}{\tau} = \underset{\sim}{\tau}'$". Since a real is a function from ω into $\{0,1\}$, a real r is given by telling for each $n < \omega$ whether $r(n) = 0$ or $r(n) = 1$. For a P-name $\underset{\sim}{\tau}$ of a real the answer whether $\underset{\sim}{\tau}(n) = 0$ or $\underset{\sim}{\tau}(n) = 1$ depends on which condition is in G therefore we shall have a maximal antichain $\langle p_{n,i} : i \le \alpha_n \rangle$ in P, where $\alpha_n \le \omega$ (because of the c.c.c.), and a function f_n on α_n which tells us that if $p_{n,i} \in G$ then $\underset{\sim}{\tau}(n) = f_n(i)$. Since each generic $G \subseteq P$ contains exactly one of the $p_{n,i}$, for $i \le \alpha_n$, this will give, for any given G, the value of $\underset{\sim}{\tau}(n)$ in a unique way. For any pair of sequences $\langle \langle p_{n,i} : i < \alpha_n \rangle : n < \omega \rangle$, $\langle f_n : n < \omega \rangle$ where for every $n < \omega$, f_n is a function from α_n into $\{0,1\}$, and $\langle p_{n,i} : i < \alpha_n \rangle$ a maximal antichain of P we can construct a P-name, which we shall denote with $\underset{\sim}{r}(\langle f_n : n < \omega \rangle, \langle \langle p_{n,i} : i < \alpha_n \rangle : n < \omega \rangle)$ such that $\underset{\sim}{r}$ is a name of a function from ω such that for every $n < \omega$ if $p_{n,i} \in G$ then $\underset{\sim}{r}(n) = f_n(i)$. There is no difficulty in obtaining such a name by the methods we described above (or by 3.1). A name of this form is called, in this proof, canonical.

Let us estimate now, in V, the number of the canonical names. For a fixed n, and a given α_n, there are $2^{|\alpha_n|} \le 2^{\aleph_0}$ different suitable f_n's. Therefore there are $\le \Pi_{n<\omega} 2^{\aleph_0} = 2^{\aleph_0}$ possibilities for $\langle f_n : n < \omega \rangle$. P is included in the set of all finite subsets of $\lambda \times 2$ and $|\lambda \times 2| = \lambda$ and so, obviously, also $|P| = \lambda$. The number of countable sequences from P is therefore $\lambda^{\aleph_0}$, thus the number of possibilities for $\langle p_{n,i} : i < \alpha_n \rangle$ is at most $\lambda^{\aleph_0}$, and for sequence $\langle \langle p_{n,i} : i < \alpha_n \rangle : n < \omega \rangle$ the number is at most $(\lambda^{\aleph_0})^{\aleph_0} = \lambda^{\aleph_0}$. For each sequence $\langle \langle p_{n,i} : i < \alpha_n \rangle : n < \omega \rangle$ we have at most $2^{\aleph_0}$ corresponding sequences $\langle f_n : n < \omega \rangle$, thus the total number of possibilities is at most $2^{\aleph_0} \cdot \lambda^{\aleph_0} = \lambda^{\aleph_0}$. Since $\lambda^{\aleph_0} = \lambda$ there are at most λ canonical names.

4.2 Lemma. For every P-name $\underset{\sim}{\tau}$ there is a canonical name $\underset{\sim}{r}$ (as defined in the beginning of the proof of 4.1 above) such that $\Vdash_P$ " if $\underset{\sim}{\tau}$ is a real then $\underset{\sim}{\tau} = \underset{\sim}{r}$".

Proof of the Lemma. For every n let $\mathcal{J}_n = \{p \in P : p \Vdash_P$ "$\underset{\sim}{\tau}$ is not a real " *or* $p \Vdash_P$ "$\underset{\sim}{\tau}$ is a real and $\underset{\sim}{\tau}(n) = \ell$" for some $\ell \in \{0,1\}\}$. For each n the set $\mathcal{J}_n$ is a dense subset of P, since every $q \in P$ can be extended to a $p \in \mathcal{J}_n$. (If

G is a generic subset of P which contains q then in $V[G]$ either $\underset{\sim}{\tau}[G]$ is not a real or $\underset{\sim}{\tau}[G]$ is a real in which case $\underset{\sim}{\tau}[G](n) = 0$ or $\underset{\sim}{\tau}[G](n) = 1$. By the forcing theorem some member p of G forces the statements mentioned above which hold in $V[G]$, and, without loss of generality, we can assume $p \geq q$. This p is in $\mathcal{J}_n$). Let $\mathcal{I}_n$ be a maximal antichain contained in $\mathcal{J}_n$. Since P satisfies the c.c.c. we have $|\mathcal{I}| \leq \aleph_0$ and we take $\alpha_n = |\mathcal{I}_n|$, $\mathcal{I}_n = \{p_{n,i} : i < \alpha_n\}$. We define f_n on α_n by:

$$f_n(i) = \begin{cases} 0 & \text{if } p_{n,i} \Vdash_P \text{“}\underset{\sim}{\tau}(n) = 0\text{”} \\ 1 & \text{otherwise} \end{cases}$$

Let $\underset{\sim}{\tau}^* = \underset{\sim}{\tau}(\langle f_n : n < \omega\rangle, \langle\langle p_{n,i} : i < \alpha_n\rangle : n < \omega\rangle)$, and we shall prove that for every generic $G \subseteq P$ we have $V[G] \vDash$ “ if $\underset{\sim}{\tau}$ is a real then $\underset{\sim}{\tau} = \underset{\sim}{\tau}^*[G]$”. Assume that $\underset{\sim}{\tau}[G]$ is indeed a real. Since τ^* is a name of a real, by its construction $\underset{\sim}{\tau}[G] = \underset{\sim}{\tau}^*[G]$ will be established once we prove that for every $n < \omega$ we have $\underset{\sim}{\tau}[G](n) = \underset{\sim}{\tau}^*[G](n)$. Since $\{p_{n,i} : i < \alpha_n\}$ is a maximal antichain in P (in V) G contains $p_{n,i}$ for a unique i. Since $p_{n,i} \in \mathcal{J}_n$ and since $p_{n,i}$ cannot force that $\underset{\sim}{\tau}$ is not a real (as $\underset{\sim}{\tau}[G]$ is a real in $V[G]$), we have $p_{n,i} \Vdash_P$ “$\underset{\sim}{\tau}(n) = \ell$”, for some $\ell \in \{0, 1\}$. By the definition of $f_n(i)$ we have $f_n(i) = \ell$ for the same ℓ. By the definition of $\underset{\sim}{\tau}^*$ we have $\underset{\sim}{\tau}^*[G](n) = f_n(i)$ if $p_{n,i} \in G$. Since $p_{n,i} \Vdash_P$ “$\underset{\sim}{\tau}(n) = \ell$” we have $V[G] \vDash$ “$\underset{\sim}{\tau}[G](n) = \ell = f_n(i) = \underset{\sim}{\tau}[G](n)$” which is what we have to prove.

$\square_{4.2}$

Continuation of the proof of 4.1. To prove $\Vdash_P$ “$2^{\aleph_0} \leq \lambda$” we shall show that for every generic $G \subseteq P$ we have $V[G] \vDash$ “$2^{\aleph_0} \leq \lambda$”. Suppose $V[G] \vDash$ “$2^{\aleph_0} \geq \lambda^+$”, then in $V[G]$ there is a one-to-one function f from λ^+ into the set of reals. Let $\underset{\sim}{\sigma}$ be a name of this function. For every $i < \lambda^+$ there is a canonical name $\underset{\sim}{\tau}_i^*$ of a real such that $\Vdash_P$ “if $\underset{\sim}{\sigma}(i)$ is a real then $\underset{\sim}{\sigma}(i) = \underset{\sim}{\tau}_i^*$” (since we can either regard $\underset{\sim}{\sigma}(i)$ itself as a name and use the lemma above, or else use an earlier lemma, 3.1, which establishes the existence of a name $\underset{\sim}{\tau}$ such that $\Vdash_P$ “$\underset{\sim}{\sigma}(i) = \underset{\sim}{\tau}$” whenever $\Vdash_P$ “$(\exists x)(\underset{\sim}{\sigma}(i) = x)$” and then use the lemma to obtain $\Vdash_P$ “$\underset{\sim}{\tau} = \underset{\sim}{\tau}_i^*$”.) Since there are at most λ canonical names, and λ^+ of $V[G]$ is also $\geq \lambda^+$ of V there is a $j \neq i$ such that $\underset{\sim}{\tau}_j^* = \underset{\sim}{\tau}_i^*$. Now in $V[G]$ we have $f(i) = \underset{\sim}{\sigma}[G](i) = \underset{\sim}{\tau}_i^*[G] = \underset{\sim}{\tau}_j^*[G] = \underset{\sim}{\sigma}[G](j) = f(j)$, contradicting the fact that f is one-to-one. So we have proved 4.1. $\square_{4.1}$.

4.3 Definition. Cohen Generic Reals. Let us take $P = \{f : f$ is a finite function from ω into $\{0,1\}\}$, with $p \leq q$ being defined as $p \subseteq q$. This is called Cohen forcing. It is easily seen that if G is a generic subset of P then $\bigcup G$ is a real. Let us see now that we can reconstruct G from $\bigcup G$ by taking G to be the set of all finite subsets of $\bigcup G$. If $p \in G$ then p is obviously a finite subset of $\bigcup G$. If p is a finite subset of $\bigcup G$ then $p = \{\langle k_i, \ell_i \rangle : i < n\}$ for some $n < \omega$. Since $\langle k_i, \ell_i \rangle \in G$ there is a $p_i \in G$ such that $\langle k_i, \ell_i \rangle \in p_i$. Since G is directed there is a $q \in G$ such that $q \geq p_i$ for all $i < n$. Obviously $p = \{\langle k_i, \ell_i \rangle : i < n\} \leq q$, and since G is downward-closed we have also $p \in G$. Since G and $\bigcup G$ can be constructed from each other we can identify them and speak of a generic real $\bigcup G$. Given V, we shall call a real number g a Cohen real over V if for some $G \subseteq P$ which is generic over V we have $g = \bigcup G$.

4.3A Discussion. When we talk here about reals we talk about members of the cantor set ${}^{\omega}2$, given an open subset A of ${}^{\omega}2$ in V this A is also a set of reals in the (true) universe, but we are interested not in A itself in the universe but in the subset of ${}^{\omega}2$ in the universe that has there the same description that A has in V. For example, suppose A is $\{r \in {}^{\omega}2 : V \models r \supseteq p\}$ for some $p \in P$, then we are interested in the set $A^* = \{r \in {}^{\omega}2 : r \supseteq p\}$.

Obviously $A = A^* \cap V$ and A^* contains reals which are not in V (since, in the universe, $|A^*| = 2^{\aleph_0}$ while $|V| = \aleph_0$) and hence not in A. The analogous situation for the real line $\mathbb{R}$ is when we look at a rational interval $(a,b)^V = \{x \in \mathbb{R}^V : a < x < b\}$, and the corresponding set in the universe is $(a,b) = \{x \in \mathbb{R} : a < x < b\}$. Obviously $(a,b)^V = (a,b) \cap V$ and (a,b) contains reals which are not in V (since $|(a,b)| = 2^{\aleph_0}$ and $|V| = \aleph_0$) and hence not in $(a,b)^V$. Let us denote with B_p the basic open set $\{r \in {}^{\omega}2 : r \supseteq p\}$ in the Cantor space, and with B_p^V the corresponding set $\{r \in {}^{\omega}2 : V \models r \supseteq p\}$ in V. Obviously $B_p^V = B_p \cap V$. Given an open set A^V in V we define $A^* = \bigcup\{B_p : p \in P, B_p^V \subseteq A^V\}$. A^* is obviously an open subset of ${}^{\omega}2$ and $A^V = A^* \cap V$. For a closed set

C^V we take $A = {}^\omega 2 \cap V \setminus C^V$ and $C^* = {}^\omega 2 \setminus A^*$, we can write it also as $C^* = \{r \in {}^\omega 2 : (\forall n < \omega)\ [B^V_{r\restriction n} \cap C^V \neq \emptyset]\}$. One can easily see that if A^V is a clopen (i.e. closed and open) set then A^* is the same set whether we regard A^V as an open set or as a closed set.

We shall use A^V only when $A^V \in V$; in fact for every Borel set $A \in V$ there is a unique Borel set A^* in the universe such that they have a common definition (in V) and then $A^* \cap V = A$.

The close connection between Cohen forcing and Cohen (generic) real is a quite universal phenomena for the forcing in use.

4.4 Theorem. A real number r is a Cohen real (or Cohen generic real) over V *iff* r belongs to no C^* where C^V is a closed nowhere dense subset of ${}^\omega 2$ in V, or in other words, if r belongs to every set A^* for every set $A \in V$ which is a dense open set of ${}^\omega 2$ in V.

Proof. Let r be a Cohen real and let A^V be a dense open set in V. We have $r = \bigcup G$ for some generic subset G of P where P is from 4.3. Let $\mathcal{I} = \{p \in P : B^V_p \subseteq A^V\}$. Since A^V is a dense open set, clearly $\mathcal{I}$ is a dense subset of P, and hence $G \cap \mathcal{I} \neq \emptyset$. Let $p \in G \cap \mathcal{I}$. Since $p \in G$ clearly $r = \bigcup G \supseteq p$, i.e., $r \in B_p$. Since $p \in \mathcal{I}$ we have $B^V_p \subseteq A^V$ hence $B_p \subseteq A^*$ and $r \in A^*$.

Now assume that for every dense open set $A^V, r \in A^*$. Let G be the set of all finite subsets of r, then $r = \bigcup G$. We shall prove that G is a generic subset of P over V and hence r is a Cohen real (over V). Let $\mathcal{I}$ be a dense subset of P. Then $A^V \stackrel{\text{def}}{=} \bigcup\{B^V_p : p \in \mathcal{I}\}$ is a dense open subset of $({}^\omega 2)^V$, and therefore $r \in A^*$, hence $r \in B^V_q$ for some $q \in P$ such that $B^V_q \subseteq A^V$. Since B^V_q is a closed set in the compact space ${}^\omega 2 \cap V$ and $\{B^V_p : p \in \mathcal{I}\}$ is an open cover of A^V and hence of B^V_q there is a finite subset $\{B^V_{p_i} : 1 \leq i \leq n\}$, with $\{p_i : 1 \leq i \leq n\} \subseteq \mathcal{I}$ which covers B^V_q. Let m be such that m includes the domains of all p_i, $1 \leq i \leq n$, and q. Let $r^\dagger$ be such that $r^\dagger(j) = r(j)$ for $j < m$ and $r^\dagger(j) = 0$ for $j > m$. Since $r \in B_q$ also $r^\dagger \in B^V_q$ and since $r^\dagger \in V$ also $r^\dagger \in B^V_q$. Therefore for some $1 \leq i \leq n$, $r^\dagger \in B^V_{p_i}$ and $r^\dagger\restriction\mathrm{Dom}(p_i) = p_i$. But $r^\dagger$

coincides with r on m which includes $\mathrm{Dom}(p_i)$, hence $r\restriction\mathrm{Dom}(p_i) = p_i$. Lastly $r\restriction\mathrm{Dom}(p_i) \in G$ since it is a finite subset of r, hence $p_i \in G$. Since $p_i \in \mathcal{I}$ we have $G \cap \mathcal{I} \neq 0$ and G is generic. $\square_{4.4}$

4.5 Corollary. The set of all Cohen reals over a model V is a comeager subset of ${}^\omega 2$.

Proof. The reals which are not Cohen reals are exactly the reals which do not belong to some set A^* where A^V is a dense open subset of ${}^\omega 2 \cap V$ in V. Let $\mathcal{I} = \{p \in P : B_p^V \subseteq A^V\}$. Since A^V is a dense open subset of ${}^\omega 2 \cap V$ in V, clearly $\mathcal{I}$ is a dense subset of P and $A^* = \bigcup\{B_p : p \in \mathcal{I}\}$ is a dense open subset of ${}^\omega 2$. The set of the reals which are not Cohen reals is $\bigcup\{{}^\omega 2 \setminus A^* : A^V \in V, A^V$ is dense open in $({}^\omega 2)^V\}$. This is the union of $\aleph_0$ nowhere dense sets ($2^\omega \setminus A^*$ is nowhere dense since A^* is dense open, the union is countable since V is) and is thus a meager set. Therefore the set of all Cohen reals is comeager. $\square_{4.5}$

Remark. Does the Cohen forcing collapse cardinals? No, since P is countable and hence, obviously it satisfies the c.c.c. See Lemma 3.6 (i).

§5. Equivalence of Forcings Notions, and Canonical Names

We deal with forcing with a subset of P.

5.1 Lemma. Let $(P, \leq)$ be a forcing notion in V and let Q be a dense subset of P.

(a) If G is a generic subset of P over V *then* $H \stackrel{\text{def}}{=} G \cap Q$ is a generic subset of Q over V and $G = \{p \in P : (\exists q \in H) p \leq q\}$.

(b) If H is a generic subset of Q over V *then* $G \stackrel{\text{def}}{=} \{p \in P : (\exists q \in H) p \leq q\}$ is a generic subset of P over V and $H = G \cap Q$.

(c) The assertion (a) and (b) above establish a one-one correspondence between the generic subsets of P and those of Q. If G and H are corresponding generic sets then $V[G] = V[H]$ and therefore the same statements are forced by the forcing notions P and Q (they are equivalent - see Definition 5.2 below).

(d) Any Q-name is a P-name, and if $\underset{\sim}{\tau}_1, \dots, \underset{\sim}{\tau}_n$ are Q-names, $\varphi(x_1, \dots, x_n)$ a first order formula then $\Vdash_P \varphi(\underset{\sim}{\tau}_1, \dots, \underset{\sim}{\tau}_n)$ iff $\Vdash_Q \varphi(\underset{\sim}{\tau}_1, \dots, \underset{\sim}{\tau}_n)$.

(e) For any P-name $\underset{\sim}{\tau}$ there is a Q-name $\underset{\sim}{\sigma}$ such that $\Vdash_P$ "$\underset{\sim}{\tau} = \underset{\sim}{\sigma}$".

Proof of the Lemma. (a) It is obvious that $G \cap Q$ is downward closed in Q. Let $p, p^\dagger \in G \cap Q$. Since G is directed there is a $p'' \in G$ such that $p'' \geq p, p^\dagger$. Now Q is dense, and hence $\{q \in Q : q \geq p''\}$ is dense above p''. Since $p'' \in G$ clearly $G \cap \{q \in Q : q \geq p''\} \neq \emptyset$; let q be such that $p'' \leq q \in G \cap Q$, so $q \geq p'' \geq p, p^\dagger$. Thus we have seen that $G \cap Q$ is directed. Let $\mathcal{I} \in V$ be dense in Q, then, as easily seen, $\mathcal{I}$ is also dense in P. Therefore we have $\mathcal{I} \cap (Q \cap G) = (\mathcal{I} \cap Q) \cap G = \mathcal{I} \cap G \neq \emptyset$. We can conclude by 1.20 that $G \cap Q$ is a generic subset of Q (over V).

Obviously $\{p \in P : (\exists q \in G \cap Q) p \leq q\} \subseteq G$. In the other direction, if $r \in G$ then, as we have seen for p'' above, there is a $q \in G \cap Q$ such that $q \geq p$. Therefore $r \in \{p \in P : (\exists q \in G \cap Q) p \leq q\}$ and so

$$\{p \in P : (\exists q \in G \cap Q) p \leq q\} = G.$$

(b) G is obviously downward closed and directed. Let $\mathcal{I}$ be a dense upward closed subset of P, then $\mathcal{I} \cap Q$ is obviously a dense subset of Q. Then $\emptyset \neq H \cap (\mathcal{I} \cap Q) = H \cap \mathcal{I} \subseteq G \cap \mathcal{I}$. Lastly $H = G \cap Q$ is obvious by the definition of G as H is downward closed.

(c) Since we have $G \in V[H]$, because G can be easily computed from H, we can evaluate all the P-names in $V[H]$ and we get therefore $V[G] \subseteq V[H]$. Similarly $H \in V[G]$ implies $V[H] \subseteq V[G]$, hence $V[H] = V[G]$.

(d),(e) are left to the reader. $\square_{5.1}$

In fact in 5.1 we have proved that P, Q are equivalent where

5.2 Definition. The forcing notions P, Q are equivalent if there are $\underset{\sim}{\tau}, \underset{\sim}{\sigma}$, a P-name and a Q-name respectively such that:

(1) $\Vdash_P$ "$\underset{\sim}{\tau}$ is a generic (over V) subset of Q".

(2) $\Vdash_Q$ "$\underset{\sim}{\sigma}$ is a generic subset of P"

(3) for $G \subseteq P$ generic, $G = \underset{\sim}{\sigma}[\underset{\sim}{\tau}[G]]$

(4) for $G \subseteq Q$ generic, $G = \underset{\sim}{\tau}[\underset{\sim}{\sigma}[G]]$

Clearly equivalence of forcing notion is an equivalence relation.

5.3 Definition. (1) A function f from P into Q is called a complete embedding if: for any maximal antichain $\mathcal{I} \subseteq P$, $f(\mathcal{I}) = \{f(p) : p \in \mathcal{I}\}$ is a maximal antichain of Q and $f{\restriction}\mathcal{I}$ is one to one of course and $P \models$ "$p \leq q$" $\Rightarrow Q \models$ "$f(p) \leq f(q)$".

(2) We write $P \lessdot Q$ if $P \subseteq Q$ (which mean: $p \in P \Rightarrow p \in Q$, for p, $q \in P$ we have $P \models$ "$p \leq q$" $\Leftrightarrow Q \models$ "$p \leq q$"), and the identity mapping is a complete embedding of P into Q.

5.4 Lemma. 1) If f is a complete embedding of P into Q, *then* there is a Q-name $\underset{\sim}{\sigma}$, $\Vdash_Q$ "$\underset{\sim}{\sigma}$ is a generic subset of P". If in addition the range of f is a dense subset of Q, then P, Q are equivalent.

2) If $P \lessdot Q$, and p_1, $p_2 \in P$ then: p_1, p_2 are compatible in P iff p_1, p_2 are compatible in Q.

3) Assume $P \subseteq Q$. Then $P \lessdot Q$ iff every pre-dense $\mathcal{I} \subseteq P$ is pre-dense in Q too.

Proof. Easy. $\square_{1.5}$

5.5 Definition. (1) For a forcing notion P, and $p, q \in P$, we say $p \approx q$ (p, q are equivalent) if any $r \in P$ is compatible with p iff it is compatible with q. Clearly $\approx$ is an equivalence relation on P.

(2) We define $P/\approx$ as follows: the members are $p/\approx$ (for $p \in P$) and we define a partial order: $(p/\approx) \le (q/\approx)$ *iff* there are $p^\dagger \in p/\approx$ and $q^\dagger \in q/\approx$ such that every $r \in P$ compatible with $q^\dagger$ is compatible with $p^\dagger$.

5.6 Claim. (1) In Definition 5.5 we have:

(a) $(p/\approx) \le (q/\approx)$ *iff* $q \Vdash_P$ "$p \in \underset{\sim}{G}_P$".

(b) $\approx$ is an equivalence relation.

(c) in 5.5 (2), "there are $p^\dagger \in (p/\approx), q^\dagger \in (q/\approx)$" can be replaced by "for every $p^\dagger \in (p/\approx)$ and $q^\dagger \in (q/\approx)$"

(d) $(p/\approx) = (q/\approx)$ iff $(p/\approx) \le (q/\approx) \,\&\, (q/\approx) \le (p/\approx)$.

(e) $P/\approx$ is a partial order.

(2) The function $p \to p/\approx$ is a complete embedding of P into $P/\approx$ with a dense range ; hence $P, P/\approx$ are equivalent. The function preserves "$p \le q$", "p, q compatible", "p, q incompatible" (though not necessarily "$\neg p \le q$").

Proof. 1)(a) If the right side fails, let $G \subseteq P$ be generic over V such that $q \in G$ but $p \notin G$. Choose a maximal antichain $\mathcal{I}$ such that $p \in \mathcal{I} \in V$, then for some r we have $G \cap \mathcal{I} = \{r\}$ and q, r are compatible (because they are in G) while p, r are not compatible (because $r \ne p$ as one is in G the other not and $\{r, p\} \subseteq \mathcal{I}$). By clause (c) of 5.6(1) this implies the failure of the left side, so we have proved the only if direction. The other direction in the "iff" is proved similarly.

(b), (c) Easy.

(d), (e) Left to the reader.

2) Easy. $\square_{5.6}$

It is interesting to note that in each equivalence class of equivalent forcing we can choose canonically a representative (unique up to isomorphism), which is essentially a complete Boolean algebra (without the one).

5.7 Definition. For any forcing notion P let:

(1) A set $A \subseteq P$ is called open-regular if it is open and for every $p \in P \setminus A$ there is $q \ge p$ incompatible with A, i.e., $(\forall r \in P)(r \ge q \to r \notin A)$.

(2) $RO(P)$ is the partial order whose set of elements is the nonempty open regular subsets of P, and $A \leq B$ (for $A, B \in RO(P)$), if $B \subseteq A$, equivalently if every $p \in P$ incompatible with A (i.e. incompatible with every $q \in A$) is incompatible with B.

(3) For $p \in P$ let $ro(p) = \{q \in P :$ there is no $r \geq q$ incompatible with $p\}$.

5.8 Theorem. (1) The mapping $p \mapsto ro(p)$ is a complete embedding of P into $RO(P)$;

(2) $RO(P)/{\approx}$, if we add to it a maximal element 1, becomes a complete Boolean algebra;

(3) If B is a complete Boolean algebra, and $P = B \setminus \{1_B\}$, (with the usual ordering of a Boolean algebra) *then* $RO(P)/{\approx}$ is isomorphic to P, moreover, we can use the isomorphism $p \mapsto ro(p)$.

Proof. Well known. $\square_{5.8}$

5.9 Theorem. The forcing notions P, Q are equivalent *iff*

$RO(P)/{\approx}$ and $RO(Q)/{\approx}$ are isomorphic.

Proof. The proof is easy, using:

5.10 Claim. Suppose $A \in RO(P)$ and $\mathcal{I}$ a maximal antichain such that for every $p \in \mathcal{I}$ either $p \in A$ or p is incompatible with A. *Then* A, $\bigcup_{p \in A \cap \mathcal{I}} ro(p)$ are equal (in $RO(P)$, union as in a complete Boolean algebra). $\square_{5.10,5.9}$

5.11 Definition. For any $a \in V^{\dagger}$ and $V \subseteq V^{\dagger}$ we define $\mathrm{rk}_V(a)$: $\mathrm{rk}_V(a) = 0$ iff $a \in V$, and $\mathrm{rk}_V(a) = \bigcup\{\mathrm{rk}_V(b) + 1 : b \in a\}$ otherwise.

5.12 Definition. We define when a P-name $\underset{\sim}{\tau}$ is canonical by induction on its rank α: if $\underset{\sim}{\tau} = \{(p_i, \underset{\sim}{\tau}_i) : i < i_0\}$, then τ is canonical if:

(1) If $\Vdash_P$ "$\mathrm{rk}(\underset{\sim}{\tau}) \leq \beta$", then $\beta \geq \alpha$, and if $\Vdash_P$ "$\mathrm{rk}_V(\underset{\sim}{\tau}) \leq \beta$" then $\beta \geq 1 + \mathrm{rk}_r(\underset{\sim}{\tau})$

(2) if $\Vdash_P$ " $\underset{\sim}{\tau}$ has power $< \lambda$", λ is a regular cardinal and P satisfies the λ-c.c. *then* $i_0 < \lambda$"

(3) each τ_i is canonical, moreover: if $p_i \Vdash$ "$\mathrm{rk}(\underset{\sim}{\tau}_i) \leq \beta$" then $\mathrm{rk}_n(\underset{\sim}{\tau}_i) \leq \beta$, and if $p_i \Vdash$ "$\mathrm{rk}_V(\underset{\sim}{\tau}_i) \leq \beta$" then $\beta \geq 1 + \mathrm{rk}_r(\underset{\sim}{\tau}_i)$ and if $\Vdash_P$ "$\underset{\sim}{\tau}_i$ has power $< \lambda$", λ a regular cardinal and $P\restriction\{p : p \geq p_i\}$ satisfies the λ-c.c. then $|\underset{\sim}{\tau}_i| < \lambda$.

5.13 Theorem. For every P-name $\underset{\sim}{\tau}$ there is a canonical P-name $\underset{\sim}{\sigma}$ such that $\Vdash_P$ "$\underset{\sim}{\tau} = \underset{\sim}{\sigma}$".

Proof. We prove by induction on the rank of $\underset{\sim}{\tau}$, that

$(*)$ if $r \in P$, $r \Vdash$ "$\mathrm{rk}_V(\underset{\sim}{\tau}[\underset{\sim}{G}]) \leq \alpha$ and $\mathrm{rk}_V(\underset{\sim}{\tau}[\underset{\sim}{G}]) \leq \beta$, and $\underset{\sim}{\tau}[\underset{\sim}{G}]$ has power $< \lambda$", and λ is a regular cardinal, and $P\restriction\{r' : r \leq r' \in P\}$ satisfies the λ-c.c., *then* we can find a canonical P-name $\underset{\sim}{\sigma}$ such that: $r \Vdash$ "$\underset{\sim}{\tau} = \underset{\sim}{\sigma}$" and $\mathrm{rk}_n(\underset{\sim}{\sigma}) \leq \alpha$ and $\mathrm{rk}_r(\underset{\sim}{\sigma}) \leq 1 + \beta$, and if $\underset{\sim}{\sigma} = \{(q_i, \underset{\sim}{\sigma}_i) : i < i_0\}$ then $i_0 < \lambda$.

If $\mathrm{rk}_n(\underset{\sim}{\tau}) = 0$, let $\underset{\sim}{\sigma} = \underset{\sim}{\tau}$ and there are no problems. If not, but $\Vdash_P$ "$\underset{\sim}{\tau} \in V$" let $\mathcal{I}$ be a maximal antichain of P above r such that for every $p \in \mathcal{I}$ for some $a_p \in V$, $p \Vdash_P$ "$\underset{\sim}{\tau} = \underset{\sim}{a}_p$". We then let $\sigma = \{(p, \dot{b}) : p \in \mathcal{I}, b \in a_p\}$. So assuming neither occurs, we can find a maximal antichain $\mathcal{I}$ of P above r, so that for each $p \in \mathcal{I}$, for some ordinals α_p and β_p and λ_p we have $p \Vdash$ "$\mathrm{rk}(\underset{\sim}{\tau}[\underset{\sim}{G}]) = \alpha_p$ and $\mathrm{rk}_V(\underset{\sim}{\tau}[\underset{\sim}{G}]) = \beta_p$, and the power of $\underset{\sim}{\tau}[\underset{\sim}{G}]$ is λ_p, (which is a cardinal in $V[\underset{\sim}{G}]$)". Let $\underset{\sim}{f}_p$ be a P-name such that $p \Vdash$ "$\underset{\sim}{f}_p$ a function from λ_p onto $\underset{\sim}{\tau}$", for each $p \in \mathcal{I}$ (use Lemma 3.1). Let $\underset{\sim}{\tau} = \{(p_i, \underset{\sim}{\tau}_i) : i < i_0\}$, and let $\mathcal{J}_\gamma$ for $\gamma < \sup\{\lambda_p : p \in \mathcal{I}\}$) be a maximal antichain of P above r, so that each $q \in \mathcal{J}_\gamma$ is above some member p of $\mathcal{I}$, and if $\gamma < \lambda_p$, for some $i = i(q, \gamma)$, $q \geq p_i$, $q \Vdash$ "$\underset{\sim}{f}_p(\gamma) = \underset{\sim}{\tau}_i$" and $q \Vdash_P$ "$\mathrm{rk}(\underset{\sim}{\tau}_i[\underset{\sim}{G}]) = \alpha_q^{i,\gamma}$, $\mathrm{rk}_V(\underset{\sim}{\tau}[\underset{\sim}{G}]) = \beta_q^{i,\gamma}$ and the power of $\underset{\sim}{\tau}_i[G]$ is $\lambda_q^{i,\gamma}$ (which is a cardinal in $V[\underset{\sim}{G}]$)".

Now apply $(*)$ by the induction hypothesis for $\underset{\sim}{\tau}_{i(q,\gamma)}, q$ for any $q \in \mathcal{J}_\gamma$ and get canonical $\underset{\sim}{\sigma}_{\gamma,q}$. Let

$$\underset{\sim}{\sigma} = \{(q, \underset{\sim}{\sigma}_{\gamma,q}) : \gamma < \sup\{\lambda_p : p \in \mathcal{I}\} \text{ and } q \in \mathcal{J}_\gamma\}$$

As $P\restriction\{r' : r \le r' \in P\}$ satisfies the λ-c.c., $|\mathcal{I}| < \lambda$, and as $r \Vdash$ "$\underset{\sim}{\tau}[\underset{\sim}{G}]$ has power $< \lambda$", each λ_p ($p \in \mathcal{I}$) is $< \lambda$. As λ is regular, $\bigcup_{p\in\mathcal{I}} \lambda_p$ is $< \lambda$, so we can finish. $\square_{5.13}$

5.14 Theorem. If P_1, P_2 are equivalent and disjoint, then there is Q, $P_\ell \lessdot Q$, P_ℓ a dense subset of Q, (for $\ell = 1, 2$). $\square_{5.14}$

5.15 Claim. The λ-c.c. property is preserved by equivalence. $\square_{5.15}$

5.16 Definition. (1) $Tc(x)$ (the transitive closure of x) is defined by induction on the rank of x,

$$Tc(x) = \{x\} \cup \bigcup_{y\in x} Tc(y)$$

and it is the minimal transitive set to which x belongs
(2) $H(\lambda) = \{x : |Tc(x)| < \lambda\}$

5.17 Claim. Let λ be an uncountable regular cardinal.

(1) If $\underset{\sim}{\tau}$ is a canonical P-name, $P \in H(\lambda)$ and $V \models \mathrm{cf}(\lambda) > |P|$ *then:*

$$\underset{\sim}{\tau} \in H(\lambda) \textit{ iff } \Vdash_P \text{“}\underset{\sim}{\tau} \in H(\lambda)^{V[\underset{\sim}{G}]}\text{”}$$

(2) $(H(\lambda), \in)$ satisfies all axioms of ZFC except possibly the power set axiom.

Proof. (1) If $\underset{\sim}{\tau} \in H(\lambda)$ we can prove by induction on $\mathrm{rk}_n(\underset{\sim}{\tau})$ that $\underset{\sim}{\tau}[G] \in H(\lambda)^{V[G]}$ hence obviously $\Vdash_P$ "$\underset{\sim}{\tau} \in H(\lambda)^{V[\underset{\sim}{G}]}$". Now P satisfies the $|P|^+$-chain condition and $H(\lambda) = \bigcup\{H(\mu^+) : \mu < \lambda\}$ so w.l.o.g. λ is a successor cardinal; clearly $|P| < \lambda$ (by the definition of $H(\lambda)$). By Claim 3.7, λ is a cardinal also in $V[G]$ so Definition 5.12(1) gives the other direction.
(2) Is obvious. Furthermore if λ is strongly inaccessible $(H(\lambda), \in)$ is a model of all ZFC. $\square_{5.17}$

5.18 Claim. 1) If P is a forcing notion satisfying the c.c.c. *then* the number of canonical P-names of an ordinal $< \mu$ is $\leq (|P| + \mu + \aleph_0)^{\aleph_0}$.
2) In (1) the number of canonical P-names of a function from μ to λ, is $\leq (|P| + \mu + \lambda + \aleph_0)^{\mu}$.
3) If P satisfies the κ-c.c., κ regular, the numbers in (1) and (2) should be $(|P| + \mu + \aleph_0)^{<\kappa}$, $(|P| + \mu + \lambda + \aleph_0)^{<(\kappa+\mu^+)}$ respectively.

Proof. Included, essentially, in the proof of 4.1. $\square_{5.18}$

§6. Random Reals, Collapsing Cardinals and Diamonds

Random Reals.

6.1 Definition. Let $\mathcal{C}$ be the set of all closed subsets of the real line, $\mathcal{B}$ the set of all Borel subsets of the real line and $\mathcal{M}$ the set of all (Lebesgue) measurable subsets of the real line. For $X \in \{\mathcal{C}, \mathcal{B}, \mathcal{M}\}$ we take $P_X = \{A \in X : A$ is of positive Lebesgue measure $\}$, and $p \leq q$ if $q \subseteq p$. We use here the real line, but we could have used also the Cantor space ${}^{\omega}2$ as the real line. For $A \in \mathcal{M}$, let $\mathrm{Leb}(A)$ be the Lebesgue measure of A.

6.1A Discussion. Every measurable set of positive measure includes a closed set of positive measure, and thus $P_{\mathcal{C}}$ is dense in $P_{\mathcal{M}}$. Since $\mathcal{C} \subseteq \mathcal{B} \subseteq \mathcal{M}$, the three notion of forcing $P_{\mathcal{M}}$, $P_{\mathcal{B}}$, and $P_{\mathcal{C}}$ are interchangeable (in fact equivalent see 5.1(c) and Definition 5.2). We shall work mostly with $P_{\mathcal{C}}$ and when we do not care with which of these notions we work we shall write just P.

First we shall prove that $P_{\mathcal{M}}$ satisfies the c.c.c. Let p_i, $i < \aleph_1$ be pairwise incompatible conditions. For each $i < \aleph_1$ we know $\mathrm{Leb}(p_i) > 0$, where Leb denotes Lebesgue measure. Therefore there is a positive integer n_i such that $\mathrm{Leb}(p_i \cap (-n_i, n_i)) \geq \frac{1}{n_i}$. Since there are $\aleph_1$ i's and $\aleph_0$ positive integers there is an uncountable subset S of $\aleph_1$ and an $n < \omega$ such that $n_i = n$ for all $i \in S$. Since for $i, j < \aleph_1$, $i \neq j$, p_i and p_j are incompatible we have $\mathrm{Leb}(p_i \cap p_j) = 0$. Let $T \subseteq S$ contain $2n^2 + 1$ members, then $\mathrm{Leb}((-n, n)) \geq \mathrm{Leb}(\bigcup_{i \in T} p_i \cap$

$(-n,n)) = \sum_{i\in T}\mathrm{Leb}(p_i\cap(-n,n)) \geq \sum_{i\in T}\frac{1}{n} = (2n^2+1)\frac{1}{n} = 2n+\frac{1}{n}$, which is a contradiction.

For every positive real ε in V the set $\{[n\varepsilon,(n+1)\varepsilon]\colon n \text{ is an integer }\}$ is obviously a maximal antichain, therefore G contains a closed interval of length ε. For $p\in P_{\mathcal{C}}$ let $p^{V[G]}$ denote the closed set with the same description in $V[G]$ as p has in V. If $p_i\in G\subseteq P_{\mathcal{C}}$ for $i=1,\dots,n$ we have $\mathrm{Leb}(\bigcap_{i=1}^n p_i^V)\neq\emptyset$. Since $p_i^{V[G]}\supseteq p_i^V$ we have $\bigcap_{i=1}^n p_i^{V[G]}\neq\emptyset$. Let us consider the set $\{p^{V[G]} : p^V\in G\subseteq P_{\mathcal{C}}\}$. The intersection of every finite subset of it is nonvoid and its members are closed sets, some of them bounded, hence $\cap_{p\in G}p^{V[G]}\neq\emptyset$ by compactness. Moreover, since for every positive integer m the set $\{p^{V[G]} : p^V\in G\}$ contains an interval of length $\frac{1}{m}$ (since if the length of p^V is $\frac{1}{m}$ so is the length of $p^{V[G]}$) necessarily the set $\cap_{p\in G}p^{V[G]}$ consists of a single real g. A real obtained in this way is called a *random real* (over V). We shall see that we can reconstruct G from g. Now let us mention that since $P_{\mathcal{C}}$ is dense in $P_{\mathcal{B}}$, we have, for every $p\in P_{\mathcal{B}}\cap G$ a condition $q\geq p$ such that $q\in P_{\mathcal{C}}\cap G$, i.e., $q\subseteq p$. Since the generic real g belongs to $q^{V[G]}$, and since $q\subseteq p$ implies $q^{V[G]}\subseteq p^{V[G]}$ also $g\in p^{V[G]}$. Thus for every Borel set p^V in G we have $g\in p^{V[G]}$.

To reconstruct G from g it suffices to tell which closed sets belong to G, since if we deal with $P_{\mathcal{B}}$ or $P_{\mathcal{M}}$, G will consist of all members, of $P_{\mathcal{B}}$ or $P_{\mathcal{M}}$ which are $\leq$ these closed sets. Now we shall see that in $V[G_{P_{\mathcal{C}}}]$, $G=\{A^V\in P_{\mathcal{C}} : g\in A^{V[G]}\}$. We have already seen that if $A^V\in G$ then $g\in A^{V[G]}$. Now we assume that $g\in A^{V[G]}$ and prove that $A^V\in G$. Assume $A^V\notin G$ and extend $\{A^V\}$ to an antichain S of $P_{\mathcal{C}}$ which is maximal among the antichains which consist of pairwise disjoint sets. We shall see that S is a maximal antichain of $P_{\mathcal{C}}$. If this is not the case there is an $E\in P_{\mathcal{C}}$ such that E is incompatible with every member of S, i.e., $\mathrm{Leb}(E\cap B)=0$ for every $B\in S$. By the c.c.c. S is countable, hence $\mathrm{Leb}(E\setminus\bigcup S)=\mathrm{Leb}(E)\setminus\mathrm{Leb}(E\cap\bigcup S)$, but $\mathrm{Leb}(E\cap\bigcup S)\leq\sum_{B\in S}0=0$ hence $\mathrm{Leb}(E\setminus S)=\mathrm{Leb}(E)>0$.

Now $E\setminus\bigcup S$ is a Borel set of positive measure, hence it includes a closed set W of positive measure. For every $B\in S$ we know that $W\cap B\subseteq W\cap\bigcup S=\emptyset$, contradicting the maximality of S. Now that we have proved that S is a maximal antichain we know that G contains some member B^V of S, which is different from A^V (since $A^V \notin G$) and therefore $B^V \cap A^V = \emptyset$. Since $B^V \in G$ we have $g \in B^{V[G]}$, hence $g \in B^{V[G]} \cap A^{V[G]}$. Let $[n, n+1]^{V[G]}$ be an interval which contains g, then $B^{V[G]} \cap A^{V[G]} \cap [n, n+1]^{V[G]} \neq \emptyset$.

On the other hand, since $(B^V \cap [n, n+1]^V) \cap [A^V \cap [n, n+1]^V) = \emptyset$ the distance between $B^V \cap [n, n+1]^V$ and $A^V \cap [n, n+1]^V$ is some $d > 0$. Let e be a rational number $< d$. There is a finite number of intervals of length e which separate $V_B^V \cap [n, n+1]^V$ from $A^V \cap [n, n+1]^V$. The "same" intervals separate also $B^{V[G]} \cap [n, n+1]^{V[G]}$ from $A^{V[G]} \cap [n, n+1]^{V[G]}$ contradicting $B^{V[G]} \cap A^{V[G]} \cap [n, n+1]^{V[G]} \neq \emptyset$.

6.2 Theorem. A real number r is random over V *iff* for every Borel set $A^V \in V$ such that $\mathrm{Leb}(A^V) = 0$ we have $r \notin A^*$ *iff* this holds for every G_σ set A^V (where A^* is the Borel set in the universe with the same description as A.)

Proof. Assume that r is random over V, $r \in \cap\{B^{V[G]} : B^V \in G \subseteq P_{\mathcal{B}}\}$. Let A^V be a Borel set with $\mathrm{Leb}(A^V) = 0$. As easily seen the complement $\mathbb{R}^V \setminus A^V$, where $\mathbb{R}^V$ is the set of all reals in V, is such that $\{\mathbb{R}^V \setminus A^V\}$ is a maximal antichain in $P_{\mathcal{B}}$ and hence $\mathbb{R}^V \setminus A^V \in G$. Therefore $r \in \mathbb{R}^{V[G]} \setminus A^{V[G]} \subseteq \mathbb{R} \setminus A^*$, i.e., $r \notin A^*$.

Now assume that $r \notin A^*$ for every G_σ - set A^V (i.e. countable intersection of open sets) such that $\mathrm{Leb}(A^V) = 0$. Define $G \subseteq P_{\mathcal{C}}$ by $B^V \in G$ iff $r \in B^*$. If $A^V \supseteq B^V \in G$ then also $A^* \supseteq B^*$ and $r \in B^V$, hence $r \in A^*$ and $A^V \in G$. If $A^V, B^V \in G$ then $r \in A^* \cap B^*$. If $\mathrm{Leb}(A^V \cap B^V) = 0$ then since $A^V \cap B^V$ is closed and therefore is a G_σ-set, we would get $r \notin A^* \cap B^*$. Therefore $\mathrm{Leb}(A^V \cap B^V) > 0$, and since $r \in A^* \cap B^*$, we get $A^V \cap B^V \in G$. Finally, let S be a maximal antichain in $P_{\mathcal{C}}$, we have to prove $G \cap S \neq \emptyset$ in order to prove that G is generic. Let $E^V = \mathbb{R}^V \setminus \bigcup\{A : A \in S\}$, so E is obviously a G_σ - set. Since S is maximal $\mathrm{Leb}(E^V) = 0$, therefore $r \notin E^*$. Since $E^V = \mathbb{R}^V \setminus \bigcup\{A : A \in S\}$, so we have $E^* = \mathbb{R} \setminus \bigcup_{B \in S} B^*$. Since $r \notin E^*$ we have $r \in B^*$ for some $B \in S$, and therefore $B \in G$ and $G \cap S \neq 0$. $\square_{6.2}$

The Levy Collapse.

6.3 Definition. 1) $\text{Levy}(\aleph_0, \lambda) = \{f : f \text{ is a finite function from } \omega \text{ into } \lambda\}$, where λ is an uncountable cardinal. For the partial order on $\text{Levy}(\aleph_0, \lambda)$ we choose inclusion.

2) $\text{Levy}(\kappa, \lambda) = \{f : f \text{ is a partial function from } \kappa \text{ to } \lambda \text{ such that } |\text{Dom}(f)| < \kappa\}$ ordered by inclusion.

6.4 Discussion. In $V[G]$, where G is a generic subset of $\text{Levy}(\aleph_0, \lambda)$ it is easily seen that $g = \bigcup G$ is a function on ω onto λ. Therefore $|\lambda|^{V[G]} = \aleph_0$.

Since in $V[G]$ there are no cardinals between λ and λ^+ all the ordinals $< \lambda^+$ are countable in $V[G]$ so $(\lambda^+)^V = \aleph_1^{V[G]}$ provided λ^+ is a cardinal in $V[G]$, thus we must check what is the fate of the chain condition for $\text{Levy}(\aleph_0, \lambda)$. $\text{Levy}(\aleph_0, \lambda)$ has λ pairwise incompatible members, for example $\{\langle 0, \alpha\rangle : \alpha < \lambda\}$. However, it is easy to see that $|\text{Levy}(\aleph_0, \lambda)| = \lambda$ and hence $\text{Levy}(\aleph_0, \lambda)$ satisfies the λ^+-chain condition and the cardinal λ^+ is not collapsed by forcing with $\text{Levy}(\aleph_0, \lambda)$.

We define

6.5 Definition. 1) $\text{Levy}(\aleph_0, < \lambda) = \{f : f \text{ is a finite function from } \lambda \times \omega \text{ into } \lambda \text{ such that } f(0, n) = 0 \text{ and for } \alpha \neq 0 \text{ we have } f(\alpha, n) < \alpha\}$. The partial order on $\text{Levy}(\aleph_0, < \lambda)$ is inclusion.

2) $\text{Levy}(\kappa, < \lambda) = \{f : f \text{ is a partial function from } \lambda \times \kappa \text{ into } \lambda \text{ such that } |\text{Dom}(f)| < \kappa \text{ and } f(0, \alpha) = 0 \text{ and } \alpha > 0 \Rightarrow f(\alpha, i) < \alpha\}$ ordered by inclusion.

6.6 Discussion. Let G be a generic subset of $\text{Levy}(\aleph_0, < \lambda)$ over V, and let $f_G = \bigcup G$. Obviously for every $0 < \alpha < \lambda$ the function $f_G(\alpha, -)$ is a mapping of ω onto α, and hence α is countable in $V[G]$. What about λ, is it, too, countable in $V[G]$ or else is it $\aleph_1^{V[G]}$? If λ is singular in V it stays singular also in $V[G]$, hence it cannot be $\aleph_1^{V[G]}$ and it is countable.

6.7 Theorem. If λ is regular then $\text{Levy}(\aleph_0, < \lambda)$ satisfies the λ-chain condition, and hence $\lambda = \aleph_1^{V[G]}$ in $V[G]$.

We shall prove the following more general version.

6.8 Theorem. Let λ be a regular uncountable cardinal, $|W| = \lambda$, let $\langle A_x : x \in W\rangle$ be such that $|A_x| < \lambda$ for $x \in W$ and let $P = \{f : f$ is a finite function on W such that $f(x) \in A_x$ for each $x \in \mathrm{Dom}(f)\}$ be a forcing notion such that: if $f, g \in P$ agree on $\mathrm{Dom}(f) \cap \mathrm{Dom}(g)$ then f, g are compatible. *Then* P *satisfies the* λ*-chain condition.*

Proof. Without loss of generality we can assume that $W = \lambda$ and $A_x \subseteq \lambda$ for each $x \in W$. Let $\langle f_\alpha : \alpha < \lambda\rangle$ be a sequence of members of P. For $\alpha < \lambda$ we define $h(\alpha) = \mathrm{Max}[\{0\} \bigcup (\mathrm{Dom}(f_\alpha) \cap \alpha)]$. Since f_α is a finite function we have $h(\alpha) < \alpha$ for $\alpha > 0$. Thus h is a regressive function on $\lambda \setminus \{0\}$ and therefore it has a fixed value, γ_0 on a stationary subset S of λ by Fodor's Lemma. Let $P^\dagger$ be the set of all members of P whose domain is included in $\gamma_0 + 1$. For every finite subset u of $\gamma_0 + 1$

$$|\{f \in P^\dagger : \mathrm{Dom}(f) = u\} = |\{f \in P : \mathrm{Dom}(f) = u\}| =$$

$$= \Pi_{\alpha \in u} |A_\alpha| \leq \aleph_0 + \mathrm{Max}_{\alpha \in u} |A_\alpha| < \lambda.$$

The number of finite subsets u of $\gamma_0 + 1$ is $\leq |\gamma_0| + \aleph_0 < \lambda$ hence, since λ is regular $|P^\dagger| = \sum_{u \subseteq \gamma_0 + 1,\, u \text{ is finite}} |\{f \in P : \mathrm{Dom}(f) = u\}| < \lambda$. For each $g \in P^\dagger$ let $S_g = \{\alpha \in S : f_\alpha \restriction (\gamma_0 + 1) = g\}$. Clearly $\bigcup_{g \in P^\dagger} S_g = S$ since for every $\alpha \in S$ we have $\alpha \in S_{f_\alpha \restriction (\gamma_0 + 1)}$. Since $|P^\dagger| < \lambda$ one of the S_g's say S_{g_0} must be stationary (since the union of $< \lambda$ nonstationary sets is nonstationary). Let $C = \{\delta < \lambda : \delta$ a limit ordinal satisfying $(\forall \alpha < \delta)[\mathrm{Dom}(f_\alpha) \subseteq \delta]\}$, clearly C is a closed unbounded subset of λ, hence $S' = S_{g_0} \cap C$ is a stationary subset of λ. Let $\alpha \in S_{g_0}$ and let $\beta \in S_{g_0}$ be such that $\alpha < \beta \in S$ hence β is a strict upper bound of $\mathrm{Dom}(f_\alpha)$, we shall see that f_α and f_β are compatible. Since $\beta \in S_{g_0} \subseteq S$, clearly $\mathrm{Max}(\mathrm{Dom}(f_\beta) \cap \beta) = \gamma_0$ and since $\mathrm{Dom}(f_\alpha) \subseteq \beta$ we have $\mathrm{Dom}(f_\alpha) \cap \mathrm{Dom}(f_\beta) \subseteq \gamma_0 + 1$. Since $\alpha, \beta \in S_{g_0}$ we have $f_\alpha \restriction (\gamma_0 + 1) = f_\beta \restriction (\gamma_0 + 1) = g_0$, hence f_α and f_β are compatible. $\square_{6.8,6.7}$

On Diamonds.

6.9 Theorem. If λ is regular uncountable cardinal, $S \subseteq \lambda$ stationary, then for some forcing notion P:

(1) $\Vdash_P$ "$\Diamond_S$ holds " (see below)

(2) forcing with P, preserve the cardinals $\mu \leq \lambda$ (in fact it is λ-complete and hence add no new α-sequences of ordinals for $\alpha < \lambda$.

(3) $|P| \leq \sum_{\mu<\lambda} 2^\mu$, so if $\lambda^{<\lambda} = \lambda$ forcing with P does not change cardinalities and cofinalities,

where: "$\Diamond_S$ holds " means: there is a sequence $\langle A_\alpha : \alpha \in S\rangle$, $A_\alpha \subseteq \alpha$ such that for any $A \subseteq \lambda$ the set $\{\alpha \in S : A \cap \alpha = A_\alpha\}$ is stationary.

Proof. Now the idea is as usual to construct a generic object, by approximations inside V.

So we define: $P = \{\langle A_i : i \in S \cap \alpha\rangle : \alpha < \lambda$ and $A_i \subseteq i$ for every $i \in S \cap \alpha\}$, ordered by: being an intial segment.

Part (3) is easy by the definition of P.

Part (2) holds as P is obviously λ-complete (take union) and the inequality stated in (3) is straigthforward - [the fact that no new α-sequences are added follows from regularity of λ]. Now what we are left with is:

(*) if G is P-generic then $V[G] \models$ "$\Diamond_S$ holds"

Let A be included in λ, and let C be closed and unbounded subset of λ, both in $V[G]$, then C, A have names $\underset{\sim}{C}, \underset{\sim}{A}$ repectively and for some f in G, $f \Vdash$ "$\underset{\sim}{C}$ is a club of λ and $\underset{\sim}{A}$ is a subset of λ". Let $f \leq p \in P$. All we have to do is find $q > p$ such that $q = \langle A_i : i \in S \cap \alpha\rangle$, $\alpha < \lambda$, $A_i \subseteq i$ and $q \Vdash$ "$i \in \underset{\sim}{C} \,\&\, \underset{\sim}{A} \cap i = A_i$" for some $i \in S \cap \alpha$; this will prove that the set of such q's is dense above f hence that G contains one of them so $V[G] \models$ "$\{\alpha \in S : A \cap \alpha = A_\alpha$ and $\alpha \in \underset{\sim}{C}\} \neq \emptyset$" and as this holds for any club $C \in V[G]$ clearly $V[G] \models$ "$\{\alpha \in S : A \cap \alpha = A_\alpha\}$ is stationary" i.e., $V[G] \models$ "$\Diamond_S$ holds as exemplified by $\underset{\sim}{\bar{A}}$" where $\underset{\sim}{\bar{A}} = \bigcup\{f : f \in G\}$.

So let us find q, we define by induction on $\zeta < \lambda$, $\alpha_\zeta < \lambda$, $p_\zeta = \langle A_i : i \in S \cap \alpha_\zeta\rangle$, B_ζ and β_ζ such that

1) for $\xi < \zeta$, $\alpha_\xi < \alpha_\zeta$, $p_\xi < p_\zeta$, and $p_0 = p$

2) $p_{\zeta+1} \Vdash$ "$\beta_\zeta \in \underset{\sim}{C}$"

3) $p_{\zeta+1} \Vdash$ "$\underset{\sim}{A} \cap \alpha_\zeta = B_\zeta$" for some $B_\zeta \in V$ such that $B_\zeta \subseteq \alpha_\zeta$.

4) for limit ζ, we have $p_\zeta = \bigcup_{\xi<\zeta} p_\xi = \langle A_i : i \in S \cap \alpha_\zeta \rangle$.

5) $\alpha_\zeta < \lambda$ is (strictly) increasing continuous and β_ζ is strictly increasing continous.

6) $\alpha_\zeta < \beta_{\zeta+1} < \alpha_{\zeta+1}$.

The definition is easy, remember for ζ successor $f \Vdash$ "$\underset{\sim}{C}$ is an unbounded subset of λ", so there is $\beta \in \underset{\sim}{C}$, $\beta > \alpha_\zeta$. For ζ limit remember $f \Vdash$ "$\underset{\sim}{C}$ is closed ".

For (3) remember that P is λ-complete hence does not add new bounded subset of λ.

Note that for ζ limit $\alpha_\zeta = \beta_\zeta$ (by clauses 5),6)) and that $\xi < \zeta \Rightarrow B_\xi = B_\zeta \cap \alpha_\xi$. In the end $\{\beta_\zeta : \zeta < \lambda\}$ is a club in V, but S is a stationary subset of λ (in V) hence for some limit ζ, $\beta_\zeta \in S$, but then $p_\zeta \Vdash$ "for $\varepsilon < \zeta$, $\underset{\sim}{A} \cap \alpha_\varepsilon = B_\varepsilon$ and $\beta_\varepsilon \in \underset{\sim}{C}$ hence $\underset{\sim}{A} \cap \beta_\zeta = B_\zeta = \bigcup_{\xi<\zeta} B_\xi$ and $\beta_\zeta \in \underset{\sim}{C}$". Let $q = \langle A'_i : i \in S \cap (\beta_\zeta + 1) \rangle$, where A'_i is: A_i if $i < \beta_\zeta$ and B_ζ if $i = \beta_\zeta$. Easily $p \leq q$, and q is as required.

Note that we also have proved that S remains stationary. $\square_{6.9}$

§7. ♣ Does Not Imply ◇

Note: ♣ is a weak version of the diamond.

7.1 Definition. For a regular uncountable cardinal λ and stationary $S \subseteq \lambda$ set of limit ordinals let us state the combinatorial principle ♣:

♣(S) = "there exists a witness i.e. a sequence $\langle A_\alpha : \alpha \in S \rangle$ such that for every $\alpha \in S$ we have $A_\alpha \subseteq \alpha$ and $\sup A_\alpha = \alpha$ and for every unbounded subset X of λ there exists an $\alpha \in S$ such that $A_\alpha \subseteq X$".

When (S) is omitted, it is $(\aleph_1)$.

7.2 Observation. This form of stating ♣ implies the apparantly stonger form of ♣:

"$\langle A_\alpha : \alpha \in S\rangle$ is as above, and for every unbounded X in λ there are stationarily many points $\alpha \in S$ such that $A_\alpha \subseteq X$".

Proof. Assume the first form holds, let X be unbounded and let C be an arbitrary club in λ. We must show that there is a point $\alpha \in C \cap S$ such that $A_\alpha \subseteq X$. Define by induction an increasing sequence $\langle \beta_i : i < \lambda\rangle$ of ordinals in X as follows: γ_i is the first member of C greater than all $\langle \beta_i : j < i\rangle$, and β_i is the first member of X greater than γ_i. As both C and X are unbounded, this is well defined. Now $X' \stackrel{\text{def}}{=} \langle \beta_i : i < \lambda\rangle$ is unbounded, therefore there exists an $\alpha \in S$ such that $A_\alpha \subseteq X'$. But the demand $\sup A_\alpha = \alpha$ implies that α is also a limit of members of C - as between two consecutive members of X' there is a member of C. By closeness of C, $\alpha \in C$. $\square_{7.2}$

Next we prove

7.3 Fact. if $CH + ♣_S$ hold, then also $\diamondsuit_S$ holds, where S is a stationary subset of $\aleph_1$.

Proof. suppose that $\langle A_\alpha : \alpha \in S\rangle$ is a witness that $♣_S$ holds. Using CH let $\langle B_i : i < \aleph_1\rangle$ be a list in which every bounded subset of $\aleph_1$ appears $\aleph_1$ times, and such that $\sup(B_i) \leq i$. To get such a list start with a function g from $\aleph_1$ onto $S_{\leq\aleph_0}(\aleph_1) \times \aleph_1$ (this is where CH comes in, where $S_{\leq\aleph_1}(A)$ is the family of countable subsets of A). Define $i : \aleph_1 \to S_{\leq\aleph_0}(\aleph_1)$ by cases as follows: suppose $g(i) = (B_\alpha, \beta)$. Set B_i to be B_α if $\sup(B_\alpha) \leq i$, and $\emptyset$ otherwise. So it is easy to check that i is a listing of all bounded subsets of $\aleph_1$ which satisfies our requirements. Now define a sequence $\langle D_\alpha : \alpha \in S\rangle$ as follows: $x \in D_\alpha$ iff there is an $i \in A_\alpha$ such that $x \in B_i$. Let us verify now that this sequence demonstrates $\diamondsuit_S$. Let X be a subset of $\aleph_1$. If X is bounded, let X' be the set of indices of X in our list, namely all i such that $B_i = X$. From our assumption that each bounded set appears $\aleph_1$ times in the list, X' is unbounded, and so there are stationarily many points $\alpha \in S$ such that $A_\alpha \subseteq X'$. This implies that D_α is X. Now suppose X is unbounded, and define by induction a function

$j : \aleph_1 \to \aleph_1$ as follows: $j(\alpha)$ = minimal j greater than all the ordinals in $\{j(\beta) : \beta < \alpha\}$ such that B_i is the bounded set $X \cap \sup\{j(\beta) : \beta < \alpha\}$. Now as j is strictly increasing, for every α we have $j(\alpha) \geq \alpha$ (the first counterexample yields an immediate contradiction). So there is a club C whose members are closed under j. Set X' as the range of j. This is clearly an unbounded set by the monotonicity of j. So there are stationarily many $\delta \in S$ for which $A_\delta \subseteq X'$. Hence there are stationary many $\delta \in S \cap C$ for which $A_\delta \subseteq X'$. For each such δ, let us prove that D_δ is exactly $X \cap \delta$: As A_δ contains only members of X', which are in particular indices of bounded subsets of the form $X \cap \alpha$, it is clear that D_δ - which is the union of these sets - is contained in $X \cap \delta$. To see equality, fix an arbitrary $\beta < \delta$. It suffice to show that in the union D_δ appears an initial segment of the form $X \cap \alpha$ with $\alpha \geq \beta$. As $\delta \in C$, clearly δ is closed under the function j so $j(\beta) < \delta$, and since A_δ is cofinal in δ, there is an ordinal γ such that $\delta > \gamma > j(\beta)$ and $\gamma \in A_\delta$. But $A_\delta \subseteq X'$ hence $\gamma = j(\alpha)$ for some α, now by the monotonicity of j, $\alpha > \beta$. So by definition of j, $B_{j(\alpha)}$ is an initial segment of X which is obtained by cutting X somewhere higher than $j(\beta)$, but the latter is greater or equal to β. So D_δ includs $X \cap \beta$ for arbitrarily large $\beta < \delta$, hence D_δ include $X \cap \delta$; hence equality follows. $\square_{7.3}$

7.4 Theorem. ♣ does not imply ◇.

Discussion.

It is clear that the principle ◇ implies ♣, and under CH, we have seen that ♣ implies ◇. It was asked whether ♣ → ◇. As we shall now see, the answer is negative. We shall build a model of ZFC in which ♣ holds, but CH fails. As trivially ◇ → CH, this necessarily implies that ◇ also fails.

Proof. Out intention is to begin with a ground model satisfying GCH (or just $2^{\aleph_0} = \aleph_1$, $2^{\aleph_1} = \aleph_2$; for example a model of $V = L$) which has a ◇-sequence on $\aleph_2$. Using the ◇ we will define a ♣-sequence on $\aleph_2$ which is immune to $\aleph_1$-complete forcing, i.e. is still a ♣-sequence in the universe after forcing with an $\aleph_1$-complete forcing notion. The next step is adding $\aleph_3$ subsets to $\aleph_1$ using an $\aleph_1$-complete forcing notion. The last stage will be collapsing $\aleph_1$, thus making of the cardinal that previously was $\aleph_2$ our present $\aleph_1$. Although this will not be done by an $\aleph_2$-complete forcing, our $\clubsuit$-sequence on $\aleph_2$, which has has now dropped to $\aleph_1$, will remain a $\clubsuit$-sequence, while there will be $\aleph_2$ subsets of $\aleph_0$ (which were formerly $\aleph_3$ subsets of $\aleph_1$). So we change our universe three times. We will present each stage of our plan in detail:

Stage A: defining a $\clubsuit$-sequence on $\aleph_2$ using $\diamondsuit(\aleph_2)$

We first note here that a $\diamondsuit$-sequence can guess not only ordinary subsets of $\aleph_2$, but also elementary substructures of any structure over $\aleph_2$. Suppose a structure M over $\aleph_2$ (in a countable language) is given. For each relation symbol or a function symbol R fix a subset of A_R $\aleph_2$ of size $\aleph_2$ and a 1-1 function $F_R : \aleph_2^{lg(R)} \to A_R$ such that two different A_R's are disjoint where we let $\ell g(R) = n$ if R is an n-place relation symbol, and $\ell g(R) = n + 1$ if R is an n-place function symbol. Thus the set $F_R(R^M)$ codes the relation R^M. Let A be the union of all $F_R(R^M)$. So A codes the structure M. If $\langle A_i : i \in S\rangle$ is a $\diamondsuit$-sequence, then A is guessed stationarily often, namely for stationarily many $i \in S$ we have $A \cap i = A_i$. Clearly, the set $C = \{i < \aleph_2$: for all R we have $F_R(R^{M\restriction i}) \subseteq i\}$ is closed unbounded in $\aleph_2$. Moreover, the set of α such that $M\restriction\alpha$ is an elementary substructure of M is a club by the Skolem-Löwenhein theorem and the continuity of elementary chains. So the intersection of the two clubs with our stationary set is the stationary set S^M of all α such that $M\restriction\alpha$ is an elementary substructure of M and for all R we have $F_R(R^{M\restriction\alpha}) \subseteq \alpha$ and $A_\alpha = \bigcup_R F_R(R^{M\restriction\alpha})$ (that is, $M\restriction\alpha$ contains its own coding and A_α equals this coding). In short we say that for every α in S^M, A_α guess the elementary substructure $M\restriction\alpha$.

Let S be the subset of $\aleph_2$ of all ordinals having cofinality $\aleph_0$. We wish to make use of $\diamondsuit_S$. Why does this principle hold in our ground model? Because, e.g. we could have forced it easily by a preliminary forcing which appeares in 6.9.

Now we come to defining the $\clubsuit$-sequence. Let us choose coding for a language with two relation signs, $<^*$ and a two place relation $R(\ ,\ )$, and let $\langle M_\alpha = (\alpha, <^*_\alpha, R_\alpha) : \alpha \in S\rangle$ be a diamond sequence for such models. So for every structure M with universe $\aleph_2$ for this language, for stationarily many $\delta \in S$ the substructures $M \restriction \delta$ of it are guessed by our $\diamondsuit$-sequence i.e. $M \restriction \delta = M_\delta$. Restrict attention now only to those places in which the guessed substructure satisfies the following sentences (the set of such $\delta \in S$ will be called S'):

(i) $<^*$ is a partial order.

(ii) if $\beta <^* \gamma$ then $R(\beta, x) \to R(\gamma, x)$

(iii) $(\forall \alpha)(\forall \beta)(\exists \gamma > \alpha)(\exists \xi > \beta)\ [R(\gamma, \xi)]$

Note that there are stationarily many such places (by guessing one such structure on $\aleph_2$, for example). In each such place δ we define now a subset of δ of order type ω which we call D_δ : let δ be fixed, and let $\langle \beta_n^\delta : n < \omega \rangle$ be a cofinal increasing sequence in δ; we define by induction on $n < \omega$ a sequence $\langle \gamma_n^\delta, \xi_n^\delta : n < \omega \rangle$ such that

1. $\gamma_n^\delta <_\delta^* \gamma_{n+1}^\delta$
2. $\xi_n^\delta > \beta_n^\delta$
3. $R_\delta(\gamma_n^\delta, \xi_n^\delta)$

So let γ_0^δ, $\xi_0^\delta < \delta$ be such that $R(\gamma_0^\delta, \xi_0^\delta)$ & $\beta_0^\delta < \xi_0^\delta$, exists by clause (iii) above, and the induction step i.e. choosing γ_{n+1}^δ, ξ_{n+1}^δ such that $\gamma_n^\delta <_\delta^* \gamma_{n+1}^\delta$ & $\beta_{n+1}^\delta < \xi_{n+1}^\delta$ & $R(\gamma_{n+1}^\delta, \xi_{n+1}^\delta)$ is handled using (iii). Set $D_\delta = \{\xi_n^\delta : n < \omega\}$.

Our ♣-sequence will be the sequence $\langle D_\delta : \delta \in S' \rangle$.

Claim B. $\langle D_\delta : \delta \in S' \rangle$ is a ♣-sequence.

Proof. Suppose X is unbounded in $\aleph_2$. Define a structure $(\aleph_2, <^*, R)$ with $<^*$ be the natural order on $\aleph_2$, and $R(y, x)$ iff $x \in X$. So for stationarily many δ in S' the model is guessed by A_δ, namely $\langle \delta, A_\delta \rangle$ is an elementary substructure of our structure. So each $\xi_n \in X$ - which implies that $D_\delta \subseteq X$.

Stage C: adding $\aleph_3$ subsets to $\aleph_1$.

We will force now with $P_1 = \{$countable functions from $\aleph_3$ to $2\}$, ordered by inclusion. The advantage of our ♣-sequence to other ♣-sequences is that it is preserved under the forcing notions we are about to apply, and this is what we are about to check now. We must check first, however, the weaker condition, that this forcing preserves stationarity of stationary subsets of $\aleph_2$ namely that a subset S of $\aleph_2$ which intersects every "old" club ("old" meaning "in the universe before the forcing") intersects also every "new" club. This follows from the following couple of claims:

Claim D. P_1 satisfies the $\aleph_2$-.c.c.

Claim E. Every P that satisfies the $\aleph_2$-.c.c preserves the stationarity of subsets of $\aleph_2$.

Proof of Claim D. let χ be any regular cardinal which is large enough to have the power set of P_1 in $H(\chi)$. Suppose that $\mathcal{I}$ is a maximal antichain of P_1 whose cardinality is greater than $\aleph_1$. Now define an increasing sequence of elementary submodels of $(H(\chi), \in)$ of length $\aleph_1$ called $\langle N_i : i < \aleph_1 \rangle$, as follows:

(i) $P_1, \mathcal{I}$ are members of N_0, and N_0 is countable, has cardinality $\aleph_1$.

(ii) every countable subset of N_i is an element of N_{i+1}.

(iii) $j < i \Rightarrow N_j \prec N_i$.

There is no problem to carry out the construction, because at each stage in the construction the cardinality of the model at hand is at most $\aleph_1$, therefore it may have at most $\aleph_1$ countable subsets (we have CH in the ground model). So close these $\aleph_1$ elements together with everything you already have in an elementary submodel of cardinality $\aleph_1$ using the Skolem-Löwenheim method. At limits take unions, for example. Let us denote the union of the increasing chain as N. So N is an elementary submodel of $(H(\chi), \in)$ of cardinality $\aleph_1$. Furthermore, every countable set of elements of N is a subset of one of the models along the construction (say N_i) by the regularity of $\aleph_1$, therefore it is an element of N_{i+1}, and therefore of N. In short we say that "N is closed under countable subsets".

We recall that $P_1, \mathcal{I} \in N$. As $\mathcal{I}$ is greater in cardinality than N, there must be an element p of $\mathcal{I}$ which is not a member of N. (Remember that p is a countable function from $\aleph_3$ to 2.) Look at p', which we define as $p \cap N$. From the countability of p, p' is clearly countable, and as N is closed under countable sets and p' is certainly a countable subset of N, we have $p' \in N$. Now $(H(\chi), \in)$ satisfies the sentence "there is a member of $\mathcal{I}$ which extends p'" (it is p). As N is an elementary submodel of it, and contains as members all the constants mentioned in this sentence - therefore there must be an element $q \in \mathcal{I} \cap N$ which according to N extends p'. We lack only one more detail to derive a contradiction: the domain of q is countable, therefore there is an enumeration of it *in* N. This implies that $\mathrm{Dom}(q)$ is contained in N. So inspect the union $q \cup p$. Clearly q extends the part of p which is in N, and contradicts nothing that is outside of N, as we just saw. Therefore p and q are compatible members of $\mathcal{I}$.

Proof of Claim E. Let S be a stationary subset of $\aleph_2$ in V, and suppose $\underset{\sim}{C}$ is a P_1-name for a club of $\aleph_2$ with the condition p in the generic set forcing $p \Vdash$ "$\underset{\sim}{C}$ is a club in $\aleph_2$". Attach to each ordinal $\alpha < \aleph_2$ a maximal antichain $\mathcal{I}_\alpha$ of extentions of p such that for each $q \in \mathcal{I}_\alpha$ there is a $\beta(q)$ such that $q \Vdash$ "$\beta(q)$ is the minimal member of $\underset{\sim}{C}$ above α". Define as B_α the set of all ordinals γ for which some member of $\mathcal{I}_\alpha$ forces that γ is the first member of $\underset{\sim}{C}$ above α. Clearly B_α has cardinality $\leq |\mathcal{I}_\alpha|$ which is $< \aleph_2$ hence B_α is bounded in $\aleph_2$. Using our standard argument, we see that $C^* := \{\delta < \aleph_2 :$ for all $\alpha < \delta$ the set B_α is included in $\delta\}$ is a club of $\aleph_2$. But this club, being defined in V, is an old club! Therefore it meets S, let us say in δ. As for each $\mathcal{I}_\alpha$, $\alpha < \delta$ the generic set must choose a condition from $\mathcal{I}_\alpha$, and B_α is contained in δ, δ is a limit of the realization of $\underset{\sim}{C}$, therefore in it.

Stage F. So we know now that our S' from the definition of the ♣-sequence is still a stationary subset of $\aleph_2$ after forcing with P_1. But is the sequence still a ♣-sequence? We verify this now.

Let $\underset{\sim}{X}$ be a P_1-name, $p \in P_1$ and $p \Vdash_{P_1}$ "$\underset{\sim}{X} \subseteq \aleph_2$ is unbounded and $\underset{\sim}{C}$ is a club of $\aleph_2$". We show that the set of conditions which force "there exists $\delta \in S \cap \underset{\sim}{C}$ such that $D_\delta \subseteq \underset{\sim}{X}$" is dense above p. Fix an elementary submodel

N of $(H(\chi), \in)$, $|N| = \aleph_2$, N closed under subsets of size $\leq \aleph_1$, with p, P_1, $\underset{\sim}{X}$ and $\underset{\sim}{C}$ members of N. Enumerate in the sequence $\langle p_i : i < \aleph_2 \rangle$ the member of $P_1 \cap M$ which are $\geq p$ and define a structure M with the same language as above with universe $\aleph_2$ with: $i <^* j$ iff $P_1 \models p_i \leq p_j$, and $R(i, x)$ iff as $p_i \Vdash$ "$x \in \underset{\sim}{X}$". From the definition of our diamond sequence $\langle M_\alpha : \alpha \in S' \rangle$, stationarily many elementary substructures of this structure are guessed by it i.e. $S'' = \{\alpha \in S' : M_\alpha = N \restriction \alpha\}$ is stationary. Let α be one such coordinate so $D_\alpha = \{\xi_n^\alpha : n < \omega\}$ where $\xi_n^\alpha < \xi_{n+1}^\alpha < \alpha = \bigcup_{n<\omega} \xi_n^\alpha$, $N \models \gamma_n^\alpha <_\alpha^* \gamma_{n+1}^\alpha$, and $N \models R(\gamma_n^\alpha, \xi_n^\alpha)$. Then we have at hand now an increasing sequence in P_1 of conditions $\langle q_n^\alpha : n < \omega \rangle$ such that $q_n \Vdash$ "$\xi_n^\alpha \in \underset{\sim}{X}$", just let $q_n^\alpha = p_{\gamma_n^\alpha}$. From elementaricity, this is really an increasing sequence of conditions, and let us bound it by q, which existed by $\aleph_1$-completeness. So $q \Vdash$ "$D_\alpha \subseteq \underset{\sim}{X}$" and q is certainly an extention of p.

So we have added $\aleph_3$ subsets to $\aleph_1$ forcing with P_1 without destroying out $\clubsuit$-sequence nor collapsing cardinals.

Stage G: collapsing $\aleph_1$.

We collapse now $\aleph_1, P_2$ being the finite functions from $\aleph_0$ to $\aleph_1$. This is a forcing notion of size $\aleph_1$ and therefore leaves the cardinals above $\aleph_1$ unchanged. To show that our sequence, which was formerly on $\aleph_2$ is now on $\aleph_1$ still a $\clubsuit$-sequence, it will suffice to prove the following.

Fact H. In the world after the collapse every new unbounded set of (the new) $\aleph_1$ contains an old unbounded set of (the old) $\aleph_2$.

Proof. Suppose p is a condition and $p \Vdash$ "$\underset{\sim}{X}$ is an unbounded subset of $\aleph_2$". So for every $i < \aleph_2$ there is an ordinal $\tau(i)$ and an extention $q(i)$ of p such that $q(i) \Vdash$ "$\aleph_2 > \tau(i) > i$ and $\tau(i) \in \underset{\sim}{X}$". By the pigeon hole principle there are $\aleph_2$ coordintes i with a fixed $q(*) = q(i)$. Set $X \stackrel{\text{def}}{=} \{\tau(i) : q(*) = q(i)\}$ - then surely X is unbounded, so $q(*)$ is an extention of p which forces that X is an unbounded subset of $\underset{\sim}{X}$.

From the lemma we easily deduce both the preservation of stationary sets (in a new club there is an old unbounded set, the closure of which is an old club; alternatively use Claim E) and the preservation of our ♣-sequence, for instead of guessing a new set, it is enough to guess an old subset. $\square_{7.4}$

II. Iteration of Forcing

§0. Introduction

Suppose $V_{\ell+1}$ is a generic extension of V_ℓ, for $\ell = 0, 1$. Is V_2 a generic extension of V_0? In §1 we present the possible answer, in fact if $V_{\ell+1} = V_\ell[G_\ell]$, G_0 is a subset of P generic over V_0, G_1 is a subset of $\underset{\sim}{Q}[G_0]$ generic over V_1, we can get V_2 by some subset G of $P * \underset{\sim}{Q}$ generic over V_0, and there are natural mappings between the family of possible pairs $[G_0, G_1]$ and the family of possible G's. In §2 we deal with iterations $\langle P_\ell, \underset{\sim}{Q}_\ell : \ell < \alpha\rangle$ of length an ordinal α.

This seems suitable to deal with proving the consistency of "for every x there is y such that ..." each Q_α producing a y_α for some $x_\alpha \in V^{P_\alpha}$. However V^{P_α} is not $\bigcup_{i<\alpha} V^{P_i}$, still if we speak of, say, $x \in H(\lambda)$ and $\mathrm{cf}(\alpha) \geq \lambda$ and $P_\alpha = \bigcup_{i<\alpha} P_i$, and P_α satisfies the c.c.c. (or less), then no "new" x appear in V^{P_α}, so we can "catch our tail."

An important point is what we do for limit ordinals δ. We choose $P_\delta = \bigcup_{i<\delta} P_i$ (direct limit), this is the meaning of FS (finite support iteration). An important property is (see 2.8): if each $\underset{\sim}{Q}_i$ satisfies the c.c.c. then so does P_δ.

In §3 we present MA (Martin's axiom) and prove its consistency. The axiom says inside the universe, for any c.c.c. forcing notion P we can find directed $G \subseteq P$ which are "quite generic", say not disjoint to $\mathcal{I}_i$ for $i < i^*$ if $\mathcal{I}_i \subseteq P$ is dense and $i^* < 2^{\aleph_0}$. The proof of its consistency (3.4) is by iterations as in §2 of c.c.c. forcing notions, the point being the right bookkeeping and the "catching of your tail." We then give some applications; if $A_i \subseteq \omega$ $(i < \lambda < 2^{\aleph_0})$ are infinite almost disjoint, $S \subseteq \lambda$ then for some $f : \omega \rightarrow \{0,1\}$ $f{\restriction}A_i =_{ae} 1_{A_i}$ iff $i \in S$ (this is 3.7), and we can omit "$A_i \subseteq \omega$" (by 3.5). Can we, for $\langle f_i \in {}^{A_i}\{0,1\} : i < \lambda \rangle$, $\lambda > \aleph_0$ find $f : {}^{\omega}\{0,1\}$ such that $f{\restriction}A_i =_{ae} f_i$? (we say $\langle A_i : i < \lambda \rangle$ has uniformization). If the A_i's are like branches of trees and the f_i are constant then yes (by 3.9).

In §4 we continue with this question. *If* $A_i \subseteq \omega$ $(i < \lambda)$ and the A_i has splitting ≤ 2 (if not branches of a tree this mean: if we know the first n members of A_i, there are ≤ 2 possibilities for the nth member) *then* uniformization fail (see 4.2 and 4.4). Moreover if MA holds, then $\langle A_i : i < \lambda \rangle$ has a subsequence as above so the answer is no. Still a positive answer is consistent (see 4.6) where we fix $\langle A_i : i < \lambda \rangle$ and preserve a strong negation of: for no uncountable $S \subseteq \lambda$, $\langle A_i : i \in S \rangle$ has spliting ≤ 2. For this we demand each Q_i to satisfy a strong version of the c.c.c.

Lastly §5 deals with the existence of mad (=maximal almost disjoint) families of subsets of ω, showing the consistency of the existence.

Also this chapter presents old material. The material is mostly from Solovay, Tennenbaum [ST] (consistency of Souslin hypothesis using FS iteration of c.c.c. forcing notions) and Martin, Solovay [MS] (on applications of MA), but note that [ST] use Boolean algebras. But the iteration like here was shortly later known, and MA was formulated and proved consistent independently by Martin and Rowbottom.

The material of §4 is from [Sh:98, §4, §1 (mainly 1.1(3), 1.2)] (where we phrase the iteration theorem more generally). See more (for higher cardinals) in Mekler and Shelah [MkSh:274].

§1. The Composition of Two Forcing Notions

1.1 Discussion. We shall now ask the following question. Suppose we start with V, extend it once by means of the forcing notion P to $V[G]$, where G is a generic subset of P, then we take a forcing notion Q in $V[G]$ and extend

$V[G]$ to $V[G][H]$, where H is a subset of Q generic over $V[G]$; can $V[G][H]$ be obtained from V by a single forcing extension? The answer is positive as we shall now see.

Since $Q \in V[G], Q$ does not necessarily belong to V but it has a P-name $\underset{\sim}{Q}$ in V. Since Q is a forcing notion we have, by the Forcing Theorem that $p \Vdash_P$ "$\underset{\sim}{Q}$ is a forcing notion (i.e. a partial order)", for some $p \in G$. We shall, however, make the stronger assumption, which suffices for all our needs, that $\Vdash_P$ "$\underset{\sim}{Q}$ is a forcing notion." If this stronger assumption would fail to hold we can define

$$\underset{\sim}{Q}^\dagger = \begin{cases} \underset{\sim}{Q} & \text{if } p \in G \\ \{0\} & \text{otherwise} \end{cases}$$

and then $\Vdash_P$ "$\underset{\sim}{Q}^\dagger$ is a P-name for Q in $V[G]$". To prove that "forcing twice" is like forcing once we want to define a forcing notion $P * \underset{\sim}{Q}$, i.e. the set and the order.

1.2 Definition. 1) $P * \underset{\sim}{Q} = \{\langle p, \underset{\sim}{q}\rangle : p \in P, \underset{\sim}{q}$ is a canonical P-name of a potential member of $\underset{\sim}{Q}$, i.e., $\Vdash_P$ "$\underset{\sim}{q} \in \underset{\sim}{Q}$", $\underset{\sim}{q}$ a canonical P-name"$\}$.

[What are the canonical names of a potential member of $\underset{\sim}{Q}$ and why do we use this concept? A member of $V[G]$ may have a proper class of names in V so $P * \underset{\sim}{Q}$ would be a proper class if we would not restrict the names to a certain representative set. What is important is that every member of $\underset{\sim}{Q}[G]$ for every generic subset G of P has a name in this set, e.g., has a canonical name. By I 5.13 this is true. Also there is a class of P-names $\underset{\sim}{q}$ such that: for some $p \in P$, we have $p \Vdash_P$ "$\underset{\sim}{q} \in Q$"; again demanding $\Vdash$ "$\underset{\sim}{q} \in Q$" suffice as for every P-name $\underset{\sim}{q}^0$ there is a P-name $\underset{\sim}{q}^1$ such that $\Vdash_P$ "$\underset{\sim}{q}^1 \in Q$" and $\Vdash_P$ "if $\underset{\sim}{q}^0 \in Q$ then $\underset{\sim}{q}^0 = \underset{\sim}{q}^1$" (why? by I 1.19, definition by cases as we assume $\underset{\sim}{Q} \neq \emptyset$).]

2) We define on $P * \underset{\sim}{Q}$ a (pre) partial order as follows:

$\langle p_1, \underset{\sim}{q}_1\rangle \leq \langle p_2, \underset{\sim}{q}_2\rangle$ *if* $p_1 \leq p_2$ (i.e. $P \models p_1 \leq p_2$) and $p_2 \Vdash_P$ "$\underset{\sim}{q}_1 \leq \underset{\sim}{q}_2$ in the partial order $\underset{\sim}{Q}$".

1.3 Claim. If $P \in V$ is a forcing notion, $\Vdash_P$ "$\underset{\sim}{Q}$ is a forcing notion" *then* $P * \underset{\sim}{Q}$ is a forcing notion.

Proof. First $\langle p, \underset{\sim}{q}\rangle \le \langle p, \underset{\sim}{q}\rangle$ since $p \Vdash_P$ "$\underset{\sim}{q} \le \underset{\sim}{q}$" because $\langle p, \underset{\sim}{q}\rangle \in P * \underset{\sim}{Q}$ implies $p \Vdash_P$ "$\underset{\sim}{q} \in \underset{\sim}{Q}$" and since $\Vdash_P$ "$(\underset{\sim}{Q}, \le)$ is a forcing notion". Now assume $\langle p_1, \underset{\sim}{q}_1\rangle \le \langle p_2, \underset{\sim}{q}_2\rangle \le \langle p_3, \underset{\sim}{q}_3\rangle$ then, since $p_3 \ge p_2 \ge p_1$ we have $p_3 \ge p_1$ and $p_3 \Vdash_P$ "$\underset{\sim}{q}_1 \le \underset{\sim}{q}_2 \,\&\, \underset{\sim}{q}_2 \le \underset{\sim}{q}_3$". Since also $\Vdash_P$ "$(\underset{\sim}{Q}, \le)$ is a forcing notion", clearly $p_3 \Vdash_P$ "$\underset{\sim}{q}_1 \le \underset{\sim}{q}_3$" and so $\langle p_1, \underset{\sim}{q}_1\rangle \le \langle p_3, \underset{\sim}{q}_3\rangle$. We shall use $P * \underset{\sim}{Q}$ as a forcing notion. $\square_{1.3}$

1.4 Theorem. This theorem asserts, essentially, that forcing by $P * \underset{\sim}{Q}$ is equivalent to forcing first by P and then by $\underset{\sim}{Q}[G]$.

(1) Let $G_{P*\underset{\sim}{Q}} \subseteq P * \underset{\sim}{Q}$ be generic over V then

a. $G_P \overset{\text{def}}{=} \{p \in P : (\exists p^\dagger)(\exists \underset{\sim}{q})(p \le p^\dagger \,\&\, \langle p^\dagger, \underset{\sim}{q}\rangle \in G_{P*\underset{\sim}{Q}})\} \subseteq P$ is generic over V.

b. $G_{P*\underset{\sim}{Q}}/G_P \overset{\text{def}}{=} \{\underset{\sim}{q}[G_P] : (\exists p)[\langle p, \underset{\sim}{q}\rangle \in G_{P*\underset{\sim}{Q}}]\}$ is a generic subset of $\underset{\sim}{Q}[G_P]$ over $V[G_P]$.

(2) If G_P is a generic subset of P over V and $H \subseteq \underset{\sim}{Q}[G_P]$ is a generic subset of $\underset{\sim}{Q}[G_P]$ over $V[G]$ *then* $G_P * H \overset{\text{def}}{=} \{\langle p, \underset{\sim}{q}\rangle : \langle p, \underset{\sim}{q}\rangle \in P * \underset{\sim}{Q} \,\&\, p \in G_P \,\&\, \underset{\sim}{q}[G] \in H\} \subseteq P * \underset{\sim}{Q}$ is a generic subset of $P * \underset{\sim}{Q}$ over V.

(3) The operations in (1) and (2) are one inverse of the other.

Proof. (1) a. G_P is downward closed by its definition. G_P is directed since $G_{P*\underset{\sim}{Q}}$ is. Now let $\mathcal{I}$ be a dense subset of P. Define $\mathcal{I}^+ = \{\langle p, \underset{\sim}{q}\rangle \in P * \underset{\sim}{Q} : p \in \mathcal{I}\}$. We shall see that $\mathcal{I}^+$ is a dense subset of $P * \underset{\sim}{Q}$. Let $\langle p, \underset{\sim}{q}\rangle \in P * \underset{\sim}{Q}$, then there is a $p^\dagger \ge p$ such that $p^\dagger \in \mathcal{I}$. Now $\langle p^\dagger, \underset{\sim}{q}\rangle \in P * \underset{\sim}{Q}$. Also $\langle p^\dagger, \underset{\sim}{q}\rangle \ge \langle p, \underset{\sim}{q}\rangle$ (since $p^\dagger \Vdash$ "$\underset{\sim}{q} \le \underset{\sim}{q}$", as $\Vdash_P$ "$(\underset{\sim}{Q}, \le)$ is a forcing notion"). Since $\langle p^\dagger, \underset{\sim}{q}\rangle \in \mathcal{I}^+$ this shows that $\mathcal{I}^+$ is dense. Therefore $\mathcal{I}^+ \cap G_{P*\underset{\sim}{Q}} \ne \emptyset$, let $\langle p, \underset{\sim}{q}\rangle \in \mathcal{I}^+ \cap G_{P*\underset{\sim}{Q}}$, then $p \in \mathcal{I} \cap G_P$, hence $\mathcal{I} \cap G_P \ne \emptyset$. As we prove it for every $\mathcal{I}$, G_P is a generic subset of P over V.

(1)b. (i) We want first to prove that $G_{P*\underset{\sim}{Q}}/G_P$ is downward closed, so let $\underset{\sim}{Q}[G_P] \models$ "$q^\dagger \le \underset{\sim}{q}[G_P]$" in $V[G_P]$ and assume $\underset{\sim}{q}[G_P] \in G_{P*\underset{\sim}{Q}}/G_P$. Then by the Forcing Theorem (and I 5.13) there is a canonical name $\underset{\sim}{q}^\dagger$ such that $q^\dagger = \underset{\sim}{q}^\dagger[G_P]$, and for some $p^\dagger \in G_P$ we have $p^\dagger \Vdash$ "$\underset{\sim}{q}^\dagger \le \underset{\sim}{q}$ and $\underset{\sim}{q}^\dagger \in \underset{\sim}{Q}$". By the "definition by cases" w.l.o.g. $\Vdash_P$ "$\underset{\sim}{q}^\dagger \in \underset{\sim}{Q}$" (we will usually forget to mention this explicity). As we assume $\underset{\sim}{q}[G_P] \in G_{P*\underset{\sim}{Q}}/G_P$ clearly for some $p \in G_P$ we have $\langle p, \underset{\sim}{q}\rangle \in G_{P*\underset{\sim}{Q}}$. Since $p^\dagger \in G_P$ there are $p'', \underset{\sim}{q}''$ such that $\langle p'', \underset{\sim}{q}''\rangle \in G_{P*\underset{\sim}{Q}}$ and $p^\dagger \le p''$ (by the definition of G_P.) Since $G_{P*\underset{\sim}{Q}}$ is directed there is $\langle p^*, \underset{\sim}{q}^*\rangle \in G_{P*\underset{\sim}{Q}}$ such that $\langle p^*, \underset{\sim}{q}^*\rangle \ge \langle p, \underset{\sim}{q}\rangle$, $\langle p'', \underset{\sim}{q}''\rangle$. We claim that $\langle p^*, \underset{\sim}{q}^\dagger\rangle \le \langle p^*, \underset{\sim}{q}^*\rangle$. Since $\langle p^*, \underset{\sim}{q}^*\rangle \ge \langle p, \underset{\sim}{q}\rangle$, we have $p^* \Vdash$ "$\underset{\sim}{q}^* \ge \underset{\sim}{q}$". Since also $p^* \ge p'' \ge p^\dagger$ and $p^\dagger \Vdash$ "$\underset{\sim}{q}^\dagger \le \underset{\sim}{q}$" we have $p^* \Vdash$ "$\underset{\sim}{q}^\dagger \le \underset{\sim}{q}$", so together with the previous sentence $p^* \Vdash$ "$\underset{\sim}{q}^\dagger \le \underset{\sim}{q}^*$", hence $\langle p^*, \underset{\sim}{q}^*\rangle \ge \langle p^*, \underset{\sim}{q}^\dagger\rangle$. Since $\langle p^*, \underset{\sim}{q}^*\rangle \in G_{P*\underset{\sim}{Q}}$ also $\langle p^*, \underset{\sim}{q}^\dagger\rangle \in G_{P*\underset{\sim}{Q}}$ and $q^\dagger = \underset{\sim}{q}^\dagger[G_P] \in G_{P*\underset{\sim}{Q}}/G_P$. Thus $G_{P*\underset{\sim}{Q}}/G_P$ is downward closed.

To see that $G_{P*\underset{\sim}{Q}}/G_P$ is directed let $q, q^\dagger$ be in $G_{P*\underset{\sim}{Q}}/G_P$. Then $q = \underset{\sim}{q}[G_P], q^\dagger = \underset{\sim}{q}^\dagger[G_P]$ for some $\underset{\sim}{q}, \underset{\sim}{q}^\dagger$ such that there are $p, p^\dagger$ satisfying $\langle p, \underset{\sim}{q}\rangle$, $\langle p^\dagger, \underset{\sim}{q}^\dagger\rangle \in G_{P*\underset{\sim}{Q}}$. Since $G_{P*\underset{\sim}{Q}}$ is directed there is a $\langle p'', \underset{\sim}{q}''\rangle \in G_{P*\underset{\sim}{Q}}$ such that $\langle p'', \underset{\sim}{q}''\rangle \ge \langle p, \underset{\sim}{q}\rangle, \langle p^\dagger, \underset{\sim}{q}^\dagger\rangle$. We have $p'' \Vdash$ "$\underset{\sim}{q}, \underset{\sim}{q}^\dagger \le \underset{\sim}{q}''$" and since $p'' \in G_P$ we have $\underset{\sim}{q}[G_P], \underset{\sim}{q}^\dagger[G_P] \le \underset{\sim}{q}''[G_P]$ and obviously $q''[G_P] \in G_{P*\underset{\sim}{Q}}/G_P$.

(ii) Let $\mathcal{I}$ be a dense subset of $\underset{\sim}{Q}[G_P]$ in $V[G_P]$, and we shall show that $\mathcal{I} \cap G_{P*\underset{\sim}{Q}}/G_P \ne \emptyset$. Clearly $\mathcal{I}$ has a P-name $\underset{\sim}{\mathcal{I}}$ and for some $p_0 \in G_P$ we have $p_0 \Vdash_P$ "$\underset{\sim}{\mathcal{I}}$ is a dense subset of $\underset{\sim}{Q}$".

Let $\mathcal{I}^+ = \{\langle p, \underset{\sim}{q}\rangle \in P * \underset{\sim}{Q} : p \Vdash_P \text{"}\underset{\sim}{q} \in \underset{\sim}{\mathcal{I}} \,\&\, p \ge p_0\text{"}\}$. Since $p_0 \in G_P$ and, as we can replace p_0 by any p', $p_0 \le p' \in G_P$, w.l.o.g. there is a $\underset{\sim}{q}_0$ such that $\langle p_0, \underset{\sim}{q}_0\rangle \in G_{P*\underset{\sim}{Q}}$. We claim $\mathcal{I}^+$ is dense above $\langle p_0, \underset{\sim}{q}_0\rangle$ in $P * \underset{\sim}{Q}$. Let $\langle p^\dagger, \underset{\sim}{q}^\dagger\rangle \ge \langle p_0, \underset{\sim}{q}_0\rangle$ and let $G^\dagger$ be any generic subset of P such that $p^\dagger \in G^\dagger$. In $V[G^\dagger]$ we know that $\underset{\sim}{\mathcal{I}}[G^\dagger]$ is a dense subset of $\underset{\sim}{Q}[G^\dagger]$, hence (by I 3.1, I 5.13) there is a canonical name $\underset{\sim}{q}''$ of a member of $\underset{\sim}{Q}$ such that $\underset{\sim}{q}''[G^\dagger] \ge \underset{\sim}{q}^\dagger[G^\dagger]$ and $q''[G^\dagger] \in \underset{\sim}{\mathcal{I}}[G^\dagger]$. Let $p'' \in G^\dagger$ be such that $p'' \Vdash_P$ "$\underset{\sim}{q}'' \in \underset{\sim}{\mathcal{I}}$" and $p'' \Vdash_P$ "$\underset{\sim}{q}'' \ge \underset{\sim}{q}^\dagger$"

and (as $G^\dagger$ is directed) $p'' \geq p^\dagger$. Then clearly $\langle p'', \underset{\sim}{q}'' \rangle \geq \langle p^\dagger, \underset{\sim}{q}^\dagger \rangle$ and $\langle p'', \underset{\sim}{q}'' \rangle \in \mathcal{I}^+$. So we have proved that $\mathcal{I}^+$ is dense in $P * \underset{\sim}{Q}$, hence $\mathcal{I}^+ \cap G_{P*Q} \neq \emptyset$. Let $\langle p_1, \underset{\sim}{q}_1 \rangle \in \mathcal{I}^+ \cap G_{P*Q}$, then $\underset{\sim}{q}_1[G_P] \in \underset{\sim}{\mathcal{I}}[G_P] = \mathcal{I}$. So really in $V[G_P]$ we have $\mathcal{I} \cap (G_{P*\underset{\sim}{Q}}/G_P)$ is not empty, as required.

(2) (i) To see that $G_P * H$ is downward closed let $\langle p, \underset{\sim}{q} \rangle \in G_P * H$ and $\langle p^\dagger, \underset{\sim}{q}^\dagger \rangle \leq \langle p, \underset{\sim}{q} \rangle$. Then $p \in G_P, p^\dagger \leq p$ hence also $p^\dagger \in G_P$. Now $\underset{\sim}{q}[G_P] \in H$ and $q^\dagger[G_P] \leq q[G_P]$ (since $p \in G_P$) hence $\underset{\sim}{q}^\dagger[G_P] \in H$, therefore $\langle p^\dagger, \underset{\sim}{q}^\dagger \rangle \in G_P * H$; so we have proved that $G_P * H$ is downward closed.
To see that $G_P * H$ is directed let $\langle p, \underset{\sim}{q} \rangle, \langle p^\dagger, \underset{\sim}{q}^\dagger \rangle \in G_P * H$. Since $p, p^\dagger \in G_P$ there is a $p'' \geq p, p^\dagger$ in G_P. Since $\underset{\sim}{q}[G_P], \underset{\sim}{q}^\dagger[G_P] \in H$ there is a canonical name $\underset{\sim}{q}''$ of a potential member of $\underset{\sim}{Q}[G_P]$ such that $\underset{\sim}{q}''[G_P] \geq \underset{\sim}{q}[G_P]$, $q^\dagger[G_P]$ and $\underset{\sim}{q}''[G_P] \in H$. Since G_P is directed we can assume, without loss of generality that $p'' \Vdash_P$ "$\underset{\sim}{q}'' \in \underset{\sim}{Q} \,\&\, \underset{\sim}{q}'' \geq \underset{\sim}{q} \,\&\, \underset{\sim}{q}'' \geq \underset{\sim}{q}^\dagger$". Thus $\langle p'', \underset{\sim}{q}'' \rangle \in P * \underset{\sim}{Q}$ and $\langle p'', \underset{\sim}{q}'' \rangle \geq \langle p, \underset{\sim}{q} \rangle, \langle p^\dagger, \underset{\sim}{q}^\dagger \rangle$.

(ii) Let $\mathcal{I}$ be a dense open subset of $P * \underset{\sim}{Q}$. Let $\mathcal{I}/G_P = \{\underset{\sim}{q}[G_P] : (\exists p \in G_P) [\langle p, \underset{\sim}{q} \rangle \in \mathcal{I}]\}$. We shall see that $\mathcal{I}/G_P$ is a dense subset of $\underset{\sim}{Q}[G_P]$ in $V[G_P]$. Let $q_0 \in \underset{\sim}{Q}[G_P]$, then for some canonical name $\underset{\sim}{q}_0$ of a member of $\underset{\sim}{Q}$ we have $\underset{\sim}{q}_0[G_P] = q_0$, and let $p_0 \in G_P$, so $p_0 \Vdash_P$ "$\underset{\sim}{q}_0 \in \underset{\sim}{Q}$". Then $\langle p_0, \underset{\sim}{q}_0 \rangle \in P * \underset{\sim}{Q}$. Let $\mathcal{I}_{q_0} = \{p \in P : p \geq p_0 \,\&\, (\exists \underset{\sim}{q}^\dagger)[p \Vdash_P \text{"}\underset{\sim}{q}^\dagger \geq \underset{\sim}{q}_0\text{"}$ and $\langle p, \underset{\sim}{q}^\dagger \rangle \in \mathcal{I}]\} \in V$. We shall see that $\mathcal{I}_{q_0}$ is dense in P above p_0. Let $p_1 \geq p_0$, then $\langle p_1, \underset{\sim}{q}_0 \rangle \in P * \underset{\sim}{Q}$. Since $\mathcal{I}$ is dense in $P * \underset{\sim}{Q}$ there is a $\langle p^\dagger, \underset{\sim}{q}^\dagger \rangle \in \mathcal{I}$ such that $\langle p^\dagger, \underset{\sim}{q}^\dagger \rangle \geq \langle p_1, \underset{\sim}{q}_0 \rangle$. We have also $p^\dagger \Vdash_P$ "$\underset{\sim}{q}^\dagger \geq \underset{\sim}{q}_0$" hence $p^\dagger \in \mathcal{I}_{q_0}$, $p^\dagger \geq p_1$ and so $\mathcal{I}_{q_0}$ is dense in P above p_0. Since $p_0 \in G_P$ by I 1.18 there is p_2 such that $p_0 \leq p_2 \in G_P$ and $p_2 \in \mathcal{I}_{q_0}$, hence for some $\underset{\sim}{q}^\dagger$ we have $p_2 \Vdash_P$ "$\underset{\sim}{q}^\dagger \geq \underset{\sim}{q}_0$" and $\langle p_2, \underset{\sim}{q}^\dagger \rangle \in \mathcal{I}$. Since $p_2 \in G_P$ we have $\underset{\sim}{q}^\dagger[G_P] \geq \underset{\sim}{q}[G_P] = q$ and $q^\dagger[G_P] \in \mathcal{I}/G_P$ (as $\langle p_2, \underset{\sim}{q}^\dagger \rangle \in \mathcal{I}$). So really $\mathcal{I}/G_P$ is a dense subset of $\underset{\sim}{Q}[G_P]$ (in $V[G_P]$). Since H is a generic subset of $\underset{\sim}{Q}[G_P]$ over $V[G_P]$ we have $(\mathcal{I}/G_P) \cap H \neq \emptyset$, let $\underset{\sim}{q}[G_P] \in (\mathcal{I}/G_P) \cap H$. Since $\underset{\sim}{q}[G_P] \in \mathcal{I}/G_P$ there is a $p \in G_P$ such that $\langle p, \underset{\sim}{q} \rangle \in \mathcal{I}$. Therefore $\langle p, \underset{\sim}{q} \rangle \in \mathcal{I} \cap (G_P * H)$, which establishes the genericity of $G_P * H$.
(3) Left to the reader. $\square_{1.4}$

1.5 Lemma. For every $P * \underset{\sim}{Q}$ -name $\underset{\sim}{\tau}$ there is a P-name $\underset{\sim}{\tau}^*$ such that for every generic subset G_P of P, $\underset{\sim}{\tau}^*[G_P]$ is a $\underset{\sim}{Q}[G_P]$-name in $V[G_P]$ and for every subset G_Q of Q which is generic over $V[G_P]$ we have $(\underset{\sim}{\tau}^*[G_P])[G_Q] = \underset{\sim}{\tau}[G_P * G_Q]$. We use the notation $\underset{\sim}{\tau}^*[G_P] = \underset{\sim}{\tau}/G_P$.

Proof. By induction on the rank α of the $(P * \underset{\sim}{Q})$-name $\underset{\sim}{\tau}$. Suppose $\underset{\sim}{\tau} = \{\langle\langle p_i, \underset{\sim}{q}_i\rangle, \underset{\sim}{\tau}_i\rangle : i < i_0\}$ where $\langle p_i, \underset{\sim}{q}_i\rangle \in P * \underset{\sim}{Q}$ and $\underset{\sim}{\tau}_i$ is a $(P * \underset{\sim}{Q})$-name of rank $< \alpha$. By the induction hypothesis each $\underset{\sim}{\tau}_i$ has a translation $\underset{\sim}{\tau}_i^*$ to a P-name such that $\underset{\sim}{\tau}_i^*[G_P]$ is a $\underset{\sim}{Q}[G_P]$-name and $\underset{\sim}{\tau}_i^*[G_P][G_Q] = \underset{\sim}{\tau}_i[G_P * G_Q]$. Let $\underset{\sim}{\sigma}_i$ be the P-name of $\langle \underset{\sim}{q}_i, \underset{\sim}{\tau}_i^*\rangle$ and let $\underset{\sim}{\tau}^* = \{\langle p_i, \underset{\sim}{\sigma}_i\rangle; i < i_0\}$; this is clearly suitable. (See I 1.8). $\square_{1.5}$

1.6 Definition. If $P \lessdot Q$ (see Definition I 5.3(2)), and $G \subseteq P$ is generic over V, *then* let $Q/G \in V[G]$ be the following forcing notion:

(1) its set of elements is $\{q \in Q : q \text{ is compatible in } Q \text{ with every } p \in G\}$,

(2) its order is inherited from Q.

Sometimes we write Q/P instead $Q/\underset{\sim}{G}_P$, (so it is a P-name) and if h is a complete embedding of P into Q (or even to $RO(Q)$) we write $Q/(P, h)$.

1.7 Lemma. (1) $P \lessdot P * \underset{\sim}{Q}$ (when P a forcing notion, $\Vdash_P$ "$\underset{\sim}{Q}$ a forcing notion" and we identify $p \in P$ with $\langle p, \emptyset_{\underset{\sim}{Q}}\rangle$)

(2) The forcing notions $(P * \underset{\sim}{Q})/P$ and $\underset{\sim}{Q}$ are equivalent (i.e., this is forced by P). Moreover for any generic $G \subseteq P$, the function f, $f(\langle p, \underset{\sim}{q}\rangle/\approx) = \underset{\sim}{q}[G]/\approx$ (for $\langle p, \underset{\sim}{q}\rangle \in P * \underset{\sim}{Q}/G$, equivalently $p \in G$) is an isomorphism from $(P * \underset{\sim}{Q})[G]/\approx$, onto $\underset{\sim}{Q}[G]/\approx$ (where $\approx$ denotes the relation defined in I 5.5).

(3) If $P \lessdot Q$ (both forcing notions in V) *then* Q is equivalent to $P * (Q/P)$.

$\square_{1.7}$

It is not hard to see that

1.8 The Associative Law Lemma. If P is a forcing notion, $\Vdash_P$ "$\underset{\sim}{Q}$ a forcing notion", $\Vdash_{P*Q}$ "$\underset{\sim}{R}$ a forcing notion", then $(P * \underset{\sim}{Q}) * \underset{\sim}{R}$, and $P * (\underset{\sim}{Q} * \underset{\sim}{R})$ are equivalent. $\square_{1.8}$

§2. Iterated Forcing

2.1 Discussion. We saw already that two successive extensions of V by forcing are equivalent to an appropriate single extension. We want to ask now the same question about an infinite number of extensions by forcing. The need for this arose in the following case.

By a classical theorem, if a linearly ordered set is dense and complete, has no first or last member, and has a dense countable subset then it is order-isomorphic to the real numbers. Souslin raised the question whether one can replace the last requirement, that there is a dense countable subset, by the requirement that every set of pairwise disjoint intervals is at most countable. The statement that these two additional requirements are equivalent is called *Souslin's hypothesis*, and an ordered set which is a counterexample to Souslin's hypothesis is called a *Souslin's continuum.* Jech and Tennenbaum proved the independence of Souslin's hypothesis of the axioms of ZFC by using forcing to obtain a universe in which there is a Souslin continuum. Later Jensen proved that in the constructible universe there is a Souslin continuum. To prove the consistency of Souslin's hypothesis Solovay and Tennenbaum, [ST] proceeded as follows: Given V and a Souslin continuum C, one can construct a generic extension $V[G]$ of V in which C is no more a Souslin continuum by a generic introduction of an uncountable set of pairwise disjoint intervals in C. In this case we say that we have "destroyed" C. Using the same method we can go on destroying more and more Souslin continua. It can be shown that a Souslin's continuum has exactly $2^{\aleph_0}$ points, hence there are at most $2^{2^{\aleph_0}}$ Souslin continua (up to isomorphism). However, when we destroy one Souslin continuum, new ones will be created and we must be sure we have destroyed them all. Since we have infinite time at our disposal this may be possible if we can "catch out tail", but for this we need that in the end no new Souslin continua arise which look doubtful by the above estimate. However we can show that it is enough to deal with subsets of Souslin continuum of power $\aleph_1$, so there are only $2^{\aleph_1}$ such orders. We shall iterate λ times, so that in the new universe, $2^{\aleph_0} = 2^{\aleph_1} = \lambda$, so if the cofinality of the length of the iteration is $> \aleph_1$ we have a chance to catch our tail.

2.2. Definition. We shall call $\bar{Q} = \langle \underset{\sim}{Q}_j : j < \alpha \rangle$ (or $\bar{Q} = \langle P_j, \underset{\sim}{Q}_j : j < \alpha \rangle$ or $\bar{Q} = \langle P_i, \underset{\sim}{Q}_j : j < \alpha \text{ and } i \leq \alpha \rangle$), for some ordinal α, *a system of FS (finite support) iterated forcing* (or FS iteration) if each $\underset{\sim}{Q}_j, j < \alpha$ is a name, for the forcing notion P_j, of a forcing notion (=quasi-order), i.e., $\Vdash_{P_j}$ "$\underset{\sim}{Q}_j$ is a forcing notion", where P_j for $j \leq \alpha$ is defined by recursion as the set of all finite functions f with domain included in j such that for all $i \in \mathrm{Dom}(f)$ we have $f(i)$ is a canonical name for the forcing notion P_i of a potential member of $\underset{\sim}{Q}_i$ and we call P_α the direct limit of $\bar{Q}$ and denote it $\lim_{<\aleph_0}(\bar{Q})$. So $i < j \leq \alpha$, $p \in P_j \Rightarrow p{\restriction}i \in P_i$ and $P_i \subseteq P_j$. We use freely I 5.13.

This is called iterated forcing *with finite support* since the functions f we use in the P_j's are finite functions. The P_j's are sets since we restrict the choice of the $f(i)$'s to be canonical names of members of $\underset{\sim}{Q}_i$. We can replace "finite support" by CS ("countable support") or "$<\kappa$-support", see Chapter III.

The partial order on P_j for $j \leq \alpha$ is defined as follows: $f \leq g \Leftrightarrow \mathrm{Dom}(f) \subseteq \mathrm{Dom}(g)$ & $(\forall i \in \mathrm{Dom}(f))$ $[g{\restriction}i \Vdash_{P_i}$ "$f(i) \leq_{\underset{\sim}{Q}_i} g(i)$"$]$.

2.2A Fact. In Definition 2.2:

(1) If $i < j \leq \alpha$ then $P_i \subseteq P_j$ as sets and even as partial orders.

(2) If $i < j \leq \alpha$ and $p \in P_j$ then $p{\restriction}i \in P_i$; moreover $P_j \models$ "$p{\restriction}i \leq p$" and if $p{\restriction}i \leq q \in P_i$ then $r \stackrel{\text{def}}{=} q \cup p{\restriction}(j \setminus i)$ belong to P_j and is the least upper bound of q, p in P_j (actually as we are dealing with quasi order we should say a least upper bound).

(3) If $i < j \leq \alpha$ then $P_i \lessdot P_j$ and $q \in P_i, p \in P_j \Rightarrow P_j \models q \leq p \leftrightarrow P_i \models q \leq p{\restriction}i$.

(4) If $j \leq \alpha$ is a limit ordinal then $P_j = \bigcup_{i<j} P_i$.

(5) The sequence $\langle \underset{\sim}{Q}_j : j < \alpha \rangle$ uniquely determines the sequence $\langle P_j, \underset{\sim}{Q}_j : j < \alpha \rangle$ and vice versa and similarly for $\langle P_i, \underset{\sim}{Q}_j : j < \alpha$, and $i \leq \alpha \rangle$.

(6) If $\underset{\sim}{Q}'_i$ is a P_i-name, such that $\Vdash$ "$\underset{\sim}{Q}'_i$ is a dense subset of $\underset{\sim}{Q}_i$" *then* $P'_i = \{f \in P_i$: for every $j \in \mathrm{Dom}(f)$ we have: $\Vdash_{P_i}$ "$f(j) \in \underset{\sim}{Q}'_i$"$\}$ is a dense subset of P_i. Moreover we can define and prove by induction on $i \leq \alpha$, that $P''_i = \{f \in P_i$: for every $j \in \mathrm{Dom}(f)$ we have: $f(j)$ is a P''_i-canonical name of a member of $\underset{\sim}{Q}'_i\}$ is a dense subset of P_i and $\underset{\sim}{Q}''_i$ is a canonical P''_i-name satisfying $\Vdash_{P_i}$ "$\underset{\sim}{Q}''_i = \underset{\sim}{Q}'_i$" and $\langle P''_{j_0}, \underset{\sim}{Q}''_{j_1} : j_0 \leq i, j_1 < i \rangle$ is a FS iteration.

(6A) Assume Q'_i is a set of canonical P_i-names of member of $\underset{\sim}{Q}_i$ such that for every P_i-name $\underset{\sim}{p}$ for some $\underset{\sim}{q} \in Q'_i$ we have $\Vdash_{P_i}$ "if $\underset{\sim}{p} \in \underset{\sim}{Q}_i$ then $\underset{\sim}{Q}_i \models$ "$\underset{\sim}{p} \leq \underset{\sim}{q}$"". Then $P'_i = \{f \in P_i$: for every $j \in \mathrm{Dom}(f)$ we have $f(j) \in Q'_i\}$ is a dense subset of P'_i and $\langle P'_i, \underset{\sim}{Q}_j : i \leq \alpha, j < \alpha \rangle$ satisfies (1) — (4) above.

(7) If λ is regular uncountable, $\alpha < \lambda$ and $\Vdash_{P_i}$ " the density of $\underset{\sim}{Q}_i$ is $< \lambda$" *then* the density of P_α is $< \lambda$.

(8) If for $i < \alpha$, $\Vdash_{P_i}$ "$\underset{\sim}{Q}_i \in H(\lambda_i)$", $\langle \lambda_i : i \leq \alpha \rangle$ is an increasing sequence of regulars, $2^{\lambda_i} < \lambda_{i+1}$ and for limit $\delta \leq \alpha$, $\sum_{i<\delta} \lambda_i < \lambda_\delta$, *then* $\bar{Q} \in H(\lambda_\alpha)$.

2.3 Definition. Let $\langle P_i, \underset{\sim}{Q}_j : i \leq \alpha, j < \alpha \rangle$ be a FS iteration. For $\beta \leq \gamma \leq \alpha$ we define $P_{\beta,\gamma}$ and if $\gamma < \alpha$, $\underset{\sim}{Q}^{[\beta]}_\gamma$ by recursion on γ as follows: $P_{\beta,\gamma}$, $\underset{\sim}{Q}^{[\beta]}_\gamma$ are P_β-names, $P_{\beta,\gamma}$ is the set of all finite functions f from $\gamma \setminus \beta$ such that for $i \in \mathrm{Dom}(f)$, $f(i)$ is a canonical P_β-name for a canonical $P_{\beta,i}$-name of a potential member of $\underset{\sim}{Q}^{[\beta]}_i$. Now if $\gamma < \alpha$ then essentially $P_\gamma = P_\beta * \underset{\sim}{P}_{\beta,\gamma}$ (see 2.4(a)) and let $\underset{\sim}{Q}^{[\beta]}_\gamma$ be $\underset{\sim}{Q}_\gamma / \underset{\sim}{G}_{P_\beta}$ (see Definition 1.6).

The next theorem is given here without proof.

2.4 The General Associativity Theorem.

(a) $P_\gamma \approx P_\beta * \underset{\sim}{P}_{\beta,\gamma}$ for $\beta \leq \gamma \leq \alpha$, where $\underset{\sim}{P}_{\beta,\gamma}$ is a name in the forcing notion P_β for the forcing notion which is $P_{\beta,\gamma}$ in $V[G]$, where G is a generic subset of P_β, and where the Q_j's, $\beta \le j < \alpha$ are translated to names for the forcing notions $P_{\beta,j}$, $\approx$ means that one of the two forcing notions are isomorphic to a dense subset of the other, so that they represent essentialy the same forcing notion (see I §5).

(b) $P_{\beta,\beta+1} \approx \underset{\sim}{Q}_\beta$ over $V[G_\beta]$ where G_β is a generic subset of P_β.

(c) If $\langle \beta_i : i \le \gamma \rangle$ is an increasing and continuous sequence such that $\beta_0 = 0$ and $\beta_\gamma = \alpha$, then $\langle P_{\beta_i,\beta_{i+1}} : i < \gamma \rangle$ is an iterated forcing equivalent (in the $\approx$ sense) to $\langle \underset{\sim}{Q}_i : i < \alpha \rangle$. This is, in some sense, a general associative rule. $\square_{2.4}$

2.5 The Definition by Induction Theorem. (One can construct $\underset{\sim}{Q}_i$'s by a given recursive recipe.) If F is a function and α is an ordinal then there is a unique FS iterated forcing $\langle \underset{\sim}{Q}_j : j < \alpha_0 \rangle$ such that for all $j < \alpha_0$, $\underset{\sim}{Q}_j = F(\langle \underset{\sim}{Q}_i : i < j \rangle)$ and either $\alpha_0 = \alpha$ or else $F(\langle \underset{\sim}{Q}_i : i < \alpha_0 \rangle)$ is not suitable for $\underset{\sim}{Q}_{\alpha_0}$, i.e., it is not a name of a forcing notion in the forcing notion P_{α_0}.

Proof. This theorem is an obvious consequence of the standard definition-by-recursion theorem. $\square_{2.5}$

2.6 Theorem. If P and $\underset{\sim}{Q}$ are as in the definition of $P * \underset{\sim}{Q}$ and P and $\underset{\sim}{Q}$ satisfy the c.c.c., where by "$\underset{\sim}{Q}$ satisfying the c.c.c.", we mean $\Vdash_P$ "$\underset{\sim}{Q}$ satisfies the c.c.c.", *then* $P * \underset{\sim}{Q}$ satisfies the c.c.c.

Proof. Let $\{\langle p_i, \underset{\sim}{q}_i \rangle : i < \aleph_1\}$ be a sequence of conditions in $P * \underset{\sim}{Q}$. We claim first that there is a $p \in P$ such that $p \Vdash_P$ "$|\{i : p_i \in \underset{\sim}{G}_P\}| = \aleph_1$", where G_P is the generic subset of P. Suppose this is not the case, then $\Vdash_P$ "$|\{i : p_i \in G_P\}| = \aleph_0$". Let $B = \{\zeta : (\exists r \in P)\ [r \Vdash \text{"}\sup\{i : p_i \in \underset{\sim}{G}_P\} = \zeta\text{"}]\}$. Since each i for which p_i is defined is a countable ordinal and $\{i : p_i \in G_P\}$ is a countable set in $V[G_P]$ also each $\zeta \in B$ is a countable ordinal in $V[G_P]$ and hence in V (since P satisfies the c.c.c., by I 3.6 we know $\aleph_1^{V[G_P]} = \aleph_1^V$). Since P satisfies the c.c.c again by I 3.6 we know B is countable, so let $\xi \overset{\text{def}}{=} \sup(B)$, we have $\xi < \aleph_1$.

Obviously $p_{\xi+1} \Vdash$ "$\xi+1 \in \{i : p_i \in G_P\}$" hence $p_{\xi+1} \Vdash$ "$\sup\{i : p_i \in G_P\} > \xi$" on the other hand we have $\Vdash_P$ "$\sup\{i : p_i \in G_P\} \le \xi$", which is a contradiction.

Hence now we know that for some $p \in P$ we have $p \Vdash_P$ "$|\{i : p_i \in G_P\}| = \aleph_1$"; let G_P be a generic subset of P such that $p \in G_P$. Let $A = \{i : p_i \in G_P\}$, then the set $\{\underset{\sim}{q}_i[G_P] : i \in A\}$ is an uncountable subset of $\underset{\sim}{Q}[G_P]$, in $V[G_P]$ of course. Since $\underset{\sim}{Q}[G_P]$ satisfies the c.c.c. there are $i, j \in A$ such that $\underset{\sim}{q}_i[G_P]$ and $\underset{\sim}{q}_j[G_P]$ are compatible, hence there is a $\underset{\sim}{q}[G_P] \in \underset{\sim}{Q}[G_P]$ such that $\underset{\sim}{q}[G_P] \ge \underset{\sim}{q}_i[G_P], \underset{\sim}{q}_j[G_P]$. Since $p_i, p_j \in G_P$ and G_P contains conditions which force $\underset{\sim}{q} \ge \underset{\sim}{q}_i$ and $\underset{\sim}{q} \ge \underset{\sim}{q}_j$, because G_P is directed *there exists* a $p^* \in G_P$ such that $p^* \ge p_i, p^* \ge p_j$, $p^* \ge p$ and $p^* \Vdash$ "$\underset{\sim}{q} \ge \underset{\sim}{q}_i$" and $p^* \Vdash$ "$\underset{\sim}{q} \ge \underset{\sim}{q}_j$", thus $\langle p^*, \underset{\sim}{q}\rangle \in P * \underset{\sim}{Q}$ and $\langle p^*, \underset{\sim}{q}\rangle \ge \langle p_i, \underset{\sim}{q}_i\rangle$, $\langle p_j, \underset{\sim}{q}_j\rangle$ and so $\langle p_i, \underset{\sim}{q}_i\rangle$ $\langle p_j, \underset{\sim}{q}_j\rangle$ are compatible (in $P * \underset{\sim}{Q}$). $\square_{2.6}$

Within the proof of this theorem we have proved the following:

2.7 Observation. Let P be a forcing notion which satisfies the c.c.c. and let $\{p_i : i < \omega_1\}$ be a sequence of members of P. *Then* for some $p \in P$ we have $p \Vdash_P$ "$|\{i < \omega_1 : p_i \in \underset{\sim}{G}_P\}| = \aleph_1$", where $\underset{\sim}{G}_P$ denotes the generic subset of P (in fact for every $\xi < \omega_1$ large enough, $p = p_\xi$ satisfies the conclusion). $\square_{2.7}$

2.8 Theorem. If $\langle P_i, \underset{\sim}{Q}_i : i < \alpha\rangle$ is a system of FS iterated forcing and for each $i < \alpha$ the forcing $\underset{\sim}{Q}_i$ satisfies the c.c.c. (i.e., $\Vdash_{P_i}$ "$\underset{\sim}{Q}$ satisfies the c.c.c.") *then* P_α satisfies the c.c.c.

Proof. We proceed by induction on α. For $\alpha = 0$, P_0 consists of the null function only, hence it satisfies, trivially, the c.c.c. If α is a successor let $\alpha = \beta + 1$ then by 2.4 we know $P_\alpha = P_{\beta+1} \approx P_\beta * P_{\beta,\beta+1} \approx P_\beta * \underset{\sim}{Q}_\beta$. By the induction hypothesis P_β satisfies the c.c.c. and since also $\underset{\sim}{Q}_\beta$ satisfies the c.c.c., theorem 2.6 establishes that also $P_\beta * \underset{\sim}{Q}_\beta$ satisfies the c.c.c. The relation $\approx$ obviously preserves the c.c.c. (see I 5.15), hence P_α too satisfies the c.c.c.

So assume α is a limit ordinal. Let $\{p_i : i < \omega_1\} \subseteq P_\alpha$. Now $\{\mathrm{Dom}(p_i) : i < \omega_1\}$ is an uncountable family of finite subsets of α.

If $\mathrm{cf}(\alpha) > \aleph_1$ then for some $\xi < \alpha$ we have $\bigcup_{i<\omega_1} \mathrm{Dom}(p_i) \subseteq \xi$, so $p_i \in P_\xi$ for $i < \omega_1$ and we can apply the induction hypothesis.

If $\mathrm{cf}(\alpha) = \aleph_0$, let $\alpha = \bigcup_{n<\omega} \alpha_n$, $\alpha_n < \alpha_{n+1} < \alpha$. So for each $i < \omega_1$ for some $n(i) < \omega$ we have $\mathrm{Dom}(p_i) \subseteq \alpha_{n(i)}$. So for some $n(*) < \omega$ the set $A \overset{\mathrm{def}}{=} \{i < \omega_1 : n(i) = n^*\}$ is uncountable. So $\{p_i : i \in A\}$ is an uncountable subset of $P_{\alpha_{n(*)}}$ so by the induction hypothesis for some $i \neq j$ from A, p_i, p_j are compatible in $P_{\alpha_{n(*)}}$ hence in P_α as required.

Lastly, assume $\mathrm{cf}(\alpha) = \aleph_1$, so let $\langle \alpha_i : i < \aleph_1 \rangle$ be a (strictly) increasing continuous sequence of ordinals with limit α. Clearly for every $i < \omega_1$, $\mathrm{Dom}(p_i)$ being a finite subset of α is included in $\alpha_{g(i)}$, for some $g(i) < \omega_1$ and is disjoint to $[\alpha_{f(i)}, \alpha_i)$ for some countable ordinal $f(i) < i$ when i is limit ordinal (remember α_i is increasing continuous).

So clearly $E = \{i < \aleph_1$: i is a limit ordinal and for every $j < i$ we have $g(j) < i\}$ is a club of $\aleph_1$, and by Fodor lemma for some $j(*)$ the set $S = \{i < \aleph_1 : f(i) = j(*)\}$ is stationary and let $\xi = \alpha_{j(*)}$. Now $\{p_i \restriction \xi : i \in E \cap S\}$ is an uncountable subset of P_ξ, hence by induction hypothesis there are in it two compatible numbers $p_i \restriction \xi$, $p_j \restriction \xi$, i.e., there is $q \in P_\xi$ such that $q \geq p_i \restriction \xi$, $p_j \restriction \xi$, but then clearly p_i, p_j are compatible in P_α. e.g. $q \bigcup (p_i \restriction [\xi, \alpha) \bigcup (p_j \restriction [\xi, \alpha))$ is a common bound. $\square_{2.8}$

2.9 Lemma. Assume $\langle P_i, \underset{\sim}{Q} : i < \alpha \rangle$ is a FS iteration of c.c.c. forcing notions, $\Vdash_{P_i}$ "$|\underset{\sim}{Q}_i| \leq \lambda$" (forcing) and $\lambda^{\aleph_0} = \lambda$ and $|\alpha| \leq \lambda$.

1) If $\Vdash_{P_i}$ " the set of elements of $\underset{\sim}{Q}_i$ is $\subseteq V$" (for each $i < \alpha$) *then* $|P_\alpha| \leq \lambda$.

2) Without this extra assumption, P_α has a dense subset of cardinality $\leq \lambda$.

3) In (1) if $Y \subseteq V$, $|Y| \leq \lambda$ then the number of canonical P_i-names of members of Y is $\leq \lambda$.

Proof. 1) We prove it by induction on α. For each $i < \alpha$, by the induction hypothesis, $|P_i| \leq \lambda$, by 2.8 the forcing notion P_i satisfies the c.c.c. so by I 3.6 there is a set $Y_i \in V$ of cardinality $\leq \lambda$ (in V) such that $\Vdash_{P_i}$ " every member of $\underset{\sim}{Q}_i$ belongs to Y_i". As in I §4 the number of canonical P_i-names of members of Y_i (hence of members of $\underset{\sim}{Q}_i$) is at most λ. Let $\hat{Q}_i$ be the set of canonical P_i-names of members of $\underset{\sim}{Q}_i$. So P_α is the set of functions f, with domain a finite subset of α, and $i \in \mathrm{Dom}(f) \Rightarrow f(i) \in \hat{Q}_i$. Clearly

$$|P_\alpha| \leq \sum_n |\alpha|^n \cdot (\sup_i |\hat{Q}_i|)^n \leq \lambda.$$

2) Let $\underset{\sim}{f}_\beta$ be a P_β-name for a one to one function for some ordinal ($\leq \lambda$) onto $\underset{\sim}{Q}_\beta$. Now $P'_\alpha = \{p \in P_\alpha$: for each $\beta \in \mathrm{Dom}(p)$ for some canonical P_β-name $\underset{\sim}{\tau}$ of an ordinal, $p(\beta) = \underset{\sim}{f}_\beta(\underset{\sim}{\tau})\}$.

We prove by induction on α that P'_α is a dense subset α of cardinality $\leq \lambda$.

3) Left to the reader. $\square_{2.9}$

§3. Martin's Axiom and Few Applications

What is the meaning of MA (Martin's Axiom, discovered by Martin and Rowbottom independently). It says that we can find quite generic sets inside our universe. As we have noted before (see I 1.4), if P has no trivial generic subset (i.e., above any $p \in P$ there are two incompatible members of P) then we cannot find a generic subset of P over V. But we may well find such $G \subseteq P$ generic over some $V^\dagger \subseteq V$. So it is plausible that for any family of $< \kappa$ dense subsets of P there is a directed $G \subseteq P$ not disjoint to any of them. How can we build a model V satisfying such a requirement? We extend and re-extend the universe, in stage α we extend the universe we have got V_α, to $V_{\alpha+1} = V_\alpha[G_\alpha]$ by forcing by some forcing notion Q_α. The hope is that in the end $V_\lambda = \bigcup_{\alpha<\lambda} V_\alpha$ is as required, as if $R \in V_\lambda$ is a suitable forcing notion (satisfying the c.c.c. with elements from V and cardinality $< \lambda$ in our case) and $\mathcal{I}_i \subseteq R$ a dense subset for $i < i_0$, then R and $\langle \mathcal{I}_i : i < i_0 \rangle$ belongs to some V_i, and for some j, $i \leq j < \lambda$, $Q_j = R$. So the generic object $G_{j+1} \subseteq P_{j+1}$ will give a generic subset of $\underset{\sim}{Q}_j[G_{j+1} \cap P_j]$ as required: essentially $G_{j+1}/(G_{j+1} \cap P_j)$; so this construction is similar in some sense, to the consturction of λ-saturated models in model theory. Now $V_\lambda = \bigcup_{\alpha<\lambda} V_\alpha$ is impossible when $V_\alpha = V[G_\alpha]$, $G_\alpha \subseteq P_\alpha$ is generic over V, but $H(\lambda)^{V_\lambda} = \bigcup_{\alpha<\lambda} H(\lambda)^{V_\alpha}$ is reasonable and is enough.

This is carried out by iterated forcing.

3.1 Definition. Martins's Axiom for κ. MA_κ. If P is a forcing notion satisfying the c.c.c. and $\mathcal{I}_i \subseteq P$ is dense in P for $i < \kappa$ *then* there is a directed subset G of P such that for every $i < \kappa$ we have $G \cap \mathcal{I}_i \neq \emptyset$.

3.2 Observation. In order to have MA_κ it suffices to require that the definition of MA_κ hold only for forcing notions P such that $|P| \leq \kappa$.

Proof. To prove this let P be a forcing notion satisfying the c.c.c. and for $i < \kappa$ let $\mathcal{I}_i \subseteq P$ be a dense subset of P. For $i < \kappa$ let f_i be a function with domain P such that for $p \in P$ we have $f_i(p) \geq p$ and $f_i(p) \in \mathcal{I}_i$. Let g be a function with domain $P \times P$ such that for $p, q \in P$ if p and q are compatible then $g(p,q) \geq p, q$. Let p_0 be any member of P and let Q be the closure of $\{p_0\}$ under the functions f_i for $i < \kappa$ and g (i.e. this is its set of elements, the order is inherited.) We have $|Q| \leq \kappa$. Let us see that $\mathcal{I}_i \cap Q$ is dense in Q. Let $p \in Q$ then $f_i(p) \geq p$, $f_i(p) \in \mathcal{I}_i$, and $f_i(p) \in Q$ since Q is closed under f_i. Now let us prove that Q satisfies the c.c.c. Let A be an antichain in Q; we claim that A is also an antichain in P and hence $|A| \leq \aleph_0$. Let $p, q \in A$ be distinct, then p, q are incompatible in P since if p, q were compatible in P we would have $g(p,q) \geq p, q$, but Q is closed under g hence also $g(p,q) \in Q$ and p, q are compatible also in Q, contradicting the assumption that A is an antichain in Q. By the version of MA_κ for $|P| \leq \kappa$ there is a directed subset G of Q such that $G \cap \mathcal{I}_i \cap Q \neq \emptyset$ for all $i < \kappa$. This G is as required. $\square_{3.2}$

3.3 Definition. Martin's Axiom MA. $(\forall \kappa < 2^{\aleph_0})\mathrm{MA}_\kappa$.

$\mathrm{MA}_{\aleph_0}$ is true since if for $i < \omega$ the set $\mathcal{I}_i$ is a dense subset of P then let us pick by induction a sequence $\langle p_n : n < \omega \rangle$ such that $p_n \in \mathcal{I}_n$ and $p_{n+1} \geq p_n$, then choose $G = \{p_n : n < \omega\}$.

Therefore we have that the CH implies MA. However, usually one means by MA, MA with the negation of the CH.

3.4 Theorem. If $\aleph_0 < \lambda = \lambda^{<\lambda}$ then there is a forcing notion P, $|P| = \lambda$ which satisfies the c.c.c. and such that $\Vdash_P$ "$2^{\aleph_0} = \lambda$ & MA".

Proof. First we shall show that if $\lambda = \lambda^{<\lambda}$, $|P| \leq \lambda$ and P satisfies the c.c.c. then $\Vdash_P$ "$\lambda = \lambda^{<\lambda}$". Since P satisfies the c.c.c. and $\lambda > \aleph_0$, by I 3.6(i) the ordinal λ is an uncountable cardinal also in $V[G]$.

To prove the theorem we shall give a canonical name to every function from μ to λ, where μ is a cardinal $< \lambda$. The canonical names (for this context) will be as follows: such a name has the form

$$\{\langle p_{i,n}, \langle i, j_{i,n}\rangle\rangle : i < \mu \text{ and } n < \omega\}$$

where $j_{i,n} < \lambda$, $p_{i,n} \in P$. For each P-name $\underset{\sim}{\tau}$, $\Vdash_P$ "$\underset{\sim}{\tau}$ a function from μ to λ" choose for each $i < \mu$, a maximal antichain $\{p_{i,n} : n < \omega\} \subseteq P$ (possibly with repetitions) such that $p_{i,n} \Vdash_P$ "$\underset{\sim}{\tau}(i) = j_{i,n}$" for some $j_{i,n} < \lambda$ (this is possible as $\mathcal{I}_i = \{p : p \Vdash_P$ "$\underset{\sim}{\tau}(i) = j$" for some $j < \lambda\}$ is a dense open subset of P, as P satisfies the c.c.c.)(if the maximal antichain is finite we can change notation or use "possibly with repetitions"). Let

$$\underset{\sim}{\tau}^\dagger = \{\langle p_{i,n}, \langle i, j_{i,n}\rangle\rangle : i < \mu, n < \omega\}$$

Then clearly $\Vdash_P$ "$\underset{\sim}{\tau}^\dagger = \underset{\sim}{\tau}$". So we can consider only canonical $\underset{\sim}{\tau}$. What is their number? For each $i < \mu$ we choose two ω-sequences, one from P and one from λ, so we have $\leq \lambda^{\aleph_0}|P|^{\aleph_0}$; and so the numbers of such names is $\leq (\lambda^{\aleph_0}|P|^{\aleph_0})^\mu \leq \lambda^\mu \leq \lambda^{<\lambda} = \lambda$. Hence clearly in V^P, $\lambda^\mu = \lambda$ for $\mu < \lambda$. Now we return to the proof.

Let $\bar{S} = \langle S_{\gamma,\mu} : \gamma < \lambda,\ \mu < \lambda, \mu$ is a cardinal $\rangle$ be a partition of λ to disjoint sets each of cardinality λ such that $i \in S_{\gamma,\mu} \Rightarrow i \geq \gamma$. We shall define $\underset{\sim}{Q}_i$ by induction such that $\Vdash_{P_i}$ "the members of $\underset{\sim}{Q}_i$ are from V" and $|P_i| \leq \lambda$. Assuming we have arrived to α, we know P_α satisfies the c.c.c. by 2.8 (we carry

the definition by 2.5). At stage α all the $\underset{\sim}{Q}_\beta$'s, for $\beta < \alpha$, are defined, and hence P_α is defined. Let $\langle \underset{\sim}{\leq}_\xi : \xi \in S_{\alpha,\mu} \rangle$ be a list of the canonical names for the forcing P_α of quasi-orders of μ. We shall use below $\underset{\sim}{\leq}_\xi$ in the ξ-th stage of the construction but since for $\xi \in S_{\alpha,\mu}$ we have $\xi \geq \alpha$ this does not spoil the induction. But can we find such a list, i.e. is $S_{\alpha,\mu}$ large enough? So how many such canonical names are there? By the induction hypothesis and 2.9, $|P_\alpha| \leq \lambda$. A quasi-order on μ is a function on $\mu \times \mu$ into $\{0, 1\}$, and as we saw in I 4.2 the number of canonical names in the forcing P_α is $\leq \lambda^{<\lambda} = \lambda$, since $|P_\alpha| \leq \lambda$.
Now by the choice of $\bar{S}$ there are unique $\gamma_\alpha < \lambda$ and $\mu_\alpha < \lambda$ (a cardinal) such that $\alpha \in S_{\gamma_\alpha,\mu_\alpha}$. By the demand above this implies $\alpha \geq \gamma_\alpha$, hence $\langle \underset{\sim}{\leq}_\xi : \xi \in S_{\gamma_\alpha,\mu_\alpha} \rangle$ is already well defined. So in particular $\underset{\sim}{\leq}_\alpha$ is a P_{γ_α}-name of a partial order on μ_α. As $\gamma_\alpha \leq \alpha$ we know $P_{\gamma_\alpha} \lessdot P_\alpha$ hence $\underset{\sim}{\leq}_\alpha$ is also a P_α-name of partial order on μ_α.

We define now

$$\underset{\sim}{Q}_\alpha = \begin{cases} \langle \mu_\alpha, \underset{\sim}{\leq}_\alpha \rangle & \text{if } \Vdash_{P_\alpha} \text{"}\langle \mu_\alpha, \underset{\sim}{\leq}_\alpha \rangle \text{ satisfies the c.c.c."} \\ \langle 1, \{\langle 1, 1 \rangle\} \rangle & \text{otherwise} \end{cases}$$

It is now obvious that $\Vdash_{P_\alpha}$ "$\underset{\sim}{Q}_\alpha$ is a forcing notion, it satisfies the c.c.c. and its elements are ordinals " since if this does not hold for $\langle \mu_\alpha, \underset{\sim}{\leq}_\alpha \rangle$ then we have chosen Q_α as $\langle 1, \{\langle 1, 1 \rangle\} \rangle$ which obviously satisfies the c.c.c. So we have carried the induction. Therefore $P \overset{\text{def}}{=} P_\lambda$ also satisfies the c.c.c. by 2.8. Our argument above that $|P_\alpha| \leq \lambda$ works also for $\alpha = \lambda$ hence $|P| \leq \lambda$. It is true also that $2^{\aleph_0} \geq \lambda$ since, as we shall see in the next lemma $\mathrm{MA}_\mu \Rightarrow 2^{\aleph_0} > \mu$ and of course $2^{\aleph_0} \leq \lambda$ so equality holds.

Let $G \subseteq P$ $(= P_\lambda)$ be generic over V, we should prove $V[G] \models$ "MA_μ" for a given $\mu < \lambda$. Let $\mu < \lambda$, and let R be a c.c.c. forcing notion in $V[G]$ and let $\mathcal{I}_i$, for $i < \mu$, be dense subsets of R in $V[G]$. As we saw by 3.2 we can assume without loss of generality that the set of members of R is μ (if $|R| < \mu$ we can introduce many "copies" of a single member and setting each of them $\leq$ than all others on μ). Let $\mathcal{I} = \{\langle i, j \rangle : i \in \mathcal{I}_j\} \subseteq \mu \times \mu$. So for some P-names $\underset{\sim}{R}$, $\mathcal{I}$, $\mathcal{I}_i$ we have $R = \underset{\sim}{R}[G]$ and $\mathcal{I} = \underset{\sim}{\mathcal{I}}[G]$, $\mathcal{I}_i = \underset{\sim}{\mathcal{I}}_i[G]$, w.l.o.g. $\Vdash_P$ "$\underset{\sim}{R}$ is a quasi order on μ and $\underset{\sim}{\mathcal{I}}$ is a subset of $\mu \times \mu$ and $\mathcal{I}_i = \{j : \langle i, j\rangle \in \underset{\sim}{\mathcal{I}}\}$ is a dense subset of $\underset{\sim}{R}$" (we could have add "and $\underset{\sim}{R}$ satisfies the c.c.c." and slightly save later). For each pair $\langle i, j\rangle \in \mu \times \mu$ there is a maximal antichain $\mathcal{I}_{i,j}$ in P which determines the truth value of $\underset{\sim}{R} \models i \leq j$ and a maximal antichain $\mathcal{J}_{i,j}$ in P which determines the membership of $\langle i, j\rangle$ in $\mathcal{I}$. Now $\bigcup\{\mathrm{Dom}(p) : p \in \bigcup_{i,j<\mu} \mathcal{I}_{i,j} \cup \mathcal{J}_{i,j}\}$ is a subset of λ in V of cardinality $\leq \mu < \lambda$. Now $\lambda = \lambda^{<\lambda}$ hence λ is regular (since $\lambda^{\mathrm{cf}\lambda} > \lambda$) and therefore $\gamma \stackrel{\mathrm{def}}{=} \sup \bigcup\{\mathrm{Dom}(p) : p \in \bigcup_{i,j<\mu} \mathcal{I}_{i,j} \cup \mathcal{J}_{i,j}\} + 1 < \lambda$, and $\mathcal{I}_{i,j}, \mathcal{J}_{i,j} \subseteq P_\gamma$ for $i, j < \mu$, and so the P-names $\underset{\sim}{R}$ and $\underset{\sim}{\mathcal{I}}$ of R and $\mathcal{I}$ are names for the forcing notion P_γ. For the generic subset G of P let $G_\gamma = \{p\restriction\gamma : p \in G\}$. So G_γ is a generic subset of P_γ. This can be shown in any one of the following two ways. One way is to use the fact that $P_\lambda \approx P_\gamma * \underset{\sim}{P}_{\gamma,\lambda}$ (see 2.4) and then G_γ is the first component of $G \subseteq P_\gamma * \underset{\sim}{P}_{\gamma,\lambda}$ and we have already proved that it is generic in P_γ. Another way is to prove directly that G_γ is a generic subset of P_γ using 2.2A. Since in computing $\underset{\sim}{R}[G] = R$ and $\underset{\sim}{\mathcal{I}}[G] = \mathcal{I}$ only G_γ is used we have $\underset{\sim}{R}[G_\gamma] = R$, $\underset{\sim}{\mathcal{I}}[G_\gamma] = \mathcal{I}$ and

$$V[G_\gamma] \models \text{“}\underset{\sim}{R} \text{ is a quasi-order with set of elements } \mu \,\&\, \underset{\sim}{\mathcal{I}}_i \text{ for } i < \mu \text{ is dense in } \underset{\sim}{R} \text{ and } \underset{\sim}{R} \text{ satisfies the c.c.c. ”}.$$

Hence there is a $p \in G_\gamma$ such that $p \Vdash_{P_\gamma}$ "$\underset{\sim}{R}$ is a quasi-order $\mu \,\&\, R$ satisfies the c.c.c. ".

$$\text{Let } \underset{\sim}{\leq}^* \stackrel{\mathrm{def}}{=} \begin{cases} \underset{\sim}{R} & \text{if } p \in G_\gamma \\ \in \restriction\mu & \text{otherwise} \end{cases}$$

then $\Vdash_{P_\gamma}$ "$\underset{\sim}{\leq}^*$ is a quasi-order with set of elements μ satisfying the c.c.c. ". Therefore there is a $\xi \in S_{\gamma,\mu}$ such that $\underset{\sim}{\leq}_\xi = \underset{\sim}{\leq}^*$. Since $\xi \in S_{\gamma,\mu}$ we know $\xi \geq \gamma$ hence $G_\xi \supseteq G_\gamma$ and since $p \in G_\gamma$ we have $\underset{\sim}{Q}_\xi[G_\xi] = \underset{\sim}{\leq}_\xi[G_\xi] = R$. $P_{\xi+1} = P_\xi * \underset{\sim}{Q}_\xi$ hence $G^* = G_{\xi+1}/G_\xi$ is a generic subset of $\underset{\sim}{Q}_\xi[G_\xi] = R$ over $V[G_\xi]$ (provided that $(\mu, \underset{\sim}{\leq}_\xi)$ satisfies the c.c.c. in $V[G_\xi]$, but this follows from $V[G] \models$ "R satisfies the c.c.c."). For $i < \mu$ we know $\mathcal{I}_i \in V[G_\gamma] \subseteq V[G_\xi]$ and $\mathcal{I}_i$ is dense in $\underset{\sim}{Q}_\xi[G_\xi] = R$ hence $G^* \cap \mathcal{I}_i \neq \emptyset$, and G^* is a directed subset of R.

3.4A. Lemma. $\mathrm{MA}_\mu \to \mu < 2^{\aleph_0}$.

Proof. Assume $\mu \geq 2^{\aleph_0}$. Let P be the forcing notion of all finite functions from ω into $\{0,1\}$, with proper inclusion as the partial order (i.e. the Cohen forcing). For each $\eta \in {}^{\omega}\{0,1\}$ let $\mathcal{I}_\eta = \{p \in P : p \not\subseteq \eta\}$, and for each $n < \omega$ let $\mathcal{I}_n = \{p \in P : n \in \mathrm{Dom}(p)\}$. Obviously each $\mathcal{I}_\eta$ and $\mathcal{I}_n$ is dense (and open) and there are $2^{\aleph_0} \leq \mu$ such sets. By MA_μ there is a directed subset G of P such that $G \cap \mathcal{I}_\eta \neq \emptyset$ for each $\eta \in {}^{\omega}\{0,1\}$ and $G \cap \mathcal{I}_n \neq 0$ for each $n < \omega$. Since G is directed $g = \bigcup G$ is a function, since $G \cap \mathcal{I}_n \neq 0$ we know $n \in \mathrm{Dom}(g)$ for each $n < \omega$, hence $g \in {}^{\omega}\{0,1\}$. Now $G \cap \mathcal{I}_g = \emptyset$ since for every $p \in G$ we know $p \subseteq \bigcup G = g$ hence $p \notin \mathcal{I}_g$, but this contradicts $G \cap \mathcal{I}_g \neq \emptyset$ which we get by MA_μ. $\square_{3.4A,3.4}$

Some Applications of $\mathrm{MA}+2^{\aleph_0} > \aleph_1$.

3.5 Theorem. Assume MA and let λ be a cardinal $\aleph_0 \leq \lambda < 2^{\aleph_0}$ and let $\langle A_i : i < \lambda \rangle$ be a family of infinite pairwise almost disjoint subsets of ω (i.e., if $i \neq j$ and $i, j < \lambda$ then $A_i \cap A_j$ is finite), and let S be a subset of λ. There is a function f on ω into $\{0,1\}$ such that for all $i < \lambda$ we have: $f{\restriction}A_i =_{ae} 1_{A_i}$ iff $i \in S$, where 1_{A_i} is the function on A_i with the fixed value 1 and $=_{ae}$ denotes that two functions have the same values for all elements of their domain except (possibly) finitely many.

Proof. Let

$P = \{f : f$ is a function such that $\mathrm{Dom}(f) = A_{i_1} \cup \cdots \cup A_{i_n} \cup w$ for some $i_1, \ldots, i_n \in S$ and a finite $w \subseteq \omega$ and for $1 \leq \ell \leq n$ we have $f{\restriction}A_{i_\ell} =_{ae} 1_{A_{i_\ell}}\}$.

The partial order on P is inclusion.

Since, for each $f \in P, f^{-1}[\{0\}]$ is finite we can take for each $f \in P$ a finite $w_f \supseteq f^{-1}[\{0\}]$ to play the role of w in the definition of P.

To see that P satisfies the c.c.c. let $\langle f_i : i < \aleph_1 \rangle$ be a sequence of members of P. Since all the $f{\restriction}w_f$ belong to the countable set of all finite functions from ω into $\{0,1\}$ we have $i \neq j$, $i, j < \aleph_1$ such that $f_i{\restriction}w_{f_i} = f_j{\restriction}w_{f_j}$. Obviously $f_i \bigcup f_j$ is a function and a member of P and above f_i and f_j, hence f_i and f_j are compatible.

We shall specify below λ dense subsets of P (called $\mathcal{I}^*_n, \mathcal{I}_i, \mathcal{I}_{n,i}$). By MA there is a directed $G \subseteq P$ such that $\mathcal{I} \cap G \neq \emptyset$ for each one of the specified dense sets $\mathcal{I}$. Let $g = \bigcup G$. Since G is directed every two members of G are compatible and g is a function from a subset of ω into $\{0,1\}$. We establish now the following properties of G.

1. $\mathrm{Dom}(g) = \omega$. For $n < \omega$ let $\mathcal{I}^*_n = \{f \in P : n \in \mathrm{Dom}(f)\}$. Now $\mathcal{I}^*_n$ is dense since for $f \in P$ if $f \notin \mathcal{I}^*_n$ then $f \bigcup \{\langle n, 1\rangle\} \in \mathcal{I}^*_n$. Since $G \cap \mathcal{I}^*_n \neq \emptyset$ there is an $f \in G$ such that $n \in \mathrm{Dom}(f)$, hence $n \in \mathrm{Dom}(f) \subseteq \mathrm{Dom}(g)$.

2. If $i \in S$ then $g{\restriction}A_i =_{ae} 1_{A_i}$. For $i \in S$ let $\mathcal{I}_i = \{f \in P : \mathrm{Dom}(f) \supseteq A_i\}$. To see that $\mathcal{I}_i$ is dense in P let $f \in P$ then $\mathrm{Dom}(f) = A_{i_1} \cup \cdots \cup A_{i_n} \cup w$ where $i_1, \ldots, i_n \in S$ and w is finite. If $i \in \{i_1, \ldots, i_n\}$ then $f \in \mathcal{I}_i$. If $i \notin \{i_1, \ldots, i_n\}$ then $A_i \cap \mathrm{Dom}(f) \subseteq \bigcup_{1 \le \ell \le n} (A_i \cap A_{i_\ell}) \bigcup w$ and each set participating in this union is finite. It follows immediately now that $f \cup [(A_i \setminus \mathrm{Dom}(f)) \times \{1\}] \in P$ and this member of P obviously is above f and belongs to $\mathcal{I}_i$. Hence we have now $G \cap \mathcal{I}_i \neq \emptyset$. Let $f \in G \cap \mathcal{I}_i$, $f{\restriction}A_i =_{ae} 1_{A_i}$ and since $g \supseteq f$ we have $g{\restriction}A_i =_{ae} 1_{A_i}$.

3. If $i \notin S$ then $g{\restriction}A_i$ obtains the value 0 for infinitely many members of A_i. Let $\mathcal{I}_{i,n} = \{f \in P : (\exists m \in A_i)\ (m \ge n \,\&\, f(m) = 0)\}$. To see that $\mathcal{I}_{i,n}$ is dense let $f \in P$, and let f be as in the definition of P. Since $i \notin S$, neccessarily $i \notin \{i_1, \ldots, i_n\}$, therefore, as we saw above, $\mathrm{Dom}(f) \cap A_i$ is finite. Since A_i is infinite there is an $m \in A_i \setminus \mathrm{Dom}(f)$ such that $m \ge n$. Now $f \cup \{\langle m, 0\rangle\} \in P$ hence $f \cup \{\langle m, 0\rangle\} \in \mathcal{I}_{n,i}$ also $f \le f \cup \{\langle m, 0\rangle\}$; hence we have shown $\mathcal{I}_{n,i}$ is dense. Since $\mathcal{I}_{n,i} \cap G \neq \emptyset$ there is an $f \in G$ such that there is an $m \in A_i$ satisfying $m \ge n$ and $f(m) = 0$, hence $g(m) = 0$. Thus $g(m) = 0$ for arbitrarily large $m \in A_i$. $\square_{3.5}$

3.5A. Conclusion. MA_λ implies $2^\lambda = 2^{\aleph_0}$. $\square_{3.5A}$

It is natural to ask

3.6 Question. Under the assumptions of Theorem 3.5, is there an $f : \omega \to \{0,1\}$ such that $f{\restriction}A_i =_{ae} 1_{A_i}$ for $i \in S$ and $f{\restriction}A_i =_{ae} 0_{A_i}$ for $i \notin S$?
We shall return to this later. Note however

3.7 Theorem. In Theorem 3.5 we can omit the requirement "$A_i \subseteq \omega$", requiring only $|A_i| = \aleph_0$.

Proof. We let $P = \{f : f$ is a function whose domain is $A_{i_1} \cup \cdots \cup A_{i_n}$ and whose range is $\subseteq \{0,1\}$ where for some $n < \omega$ we have $i_1 \in S, \ldots, i_n \in S$, $f^{-1}[\{0\}]$ finite $\}$, ordered by inclusion.

Let $p_i \in P$ (for $i < \aleph_1$) be $\aleph_1$ conditions, $\mathrm{Dom}(p_i) = \bigcup_{\alpha \in u_i} A_\alpha$ with $u_i < \lambda$ is finite, so w.l.o.g. $i \neq j \Rightarrow u_i \cap u_j = u^*$. By the definition of P, there are only countably many possible $p_i \restriction \bigcup_{\alpha \in u^*} A_\alpha$, so w.l.o.g. $p_i \restriction \bigcup_{\alpha \in u^*} A_\alpha = f$ for every $i < \aleph_1$. Let $w_i \overset{\mathrm{def}}{=} p_i^{-1}[\{0\}]$, it is a finite set so w.l.o.g. $i \neq j \rightarrow w_i \cap w_i = w^*$ & $|w_i| = \ell(*)$. So for each i the cardinality of $\{j : \mathrm{Dom}(p_i) \cap w_j \setminus w^* \neq \emptyset\}$ is at most the cardinality of $\mathrm{Dom}(p_i)$ hence this set is countable, so w.l.o.g. $i < j \Rightarrow \mathrm{Dom}(p_i) \cap w_j \subseteq w^*$.

Now if $i < j$ and p_i, p_j are incompatible, then there is $x \in \mathrm{Dom}(f_i) \cap \mathrm{Dom}(f_j)$ such that $f_i(x) \neq f_j(x)$, so $0 \in \{f_i(x), f_j(x)\}$ hence $x \in w_i \cup w_j$, in fact $x \in (w_i \setminus w^*) \cup (w_j \setminus w^*)$; but by the previous sentence $x \notin w_j \setminus w^*$ so $x \in w_i \setminus w^*$; also $x \notin \mathrm{Dom}(f)$ (as $p_i \restriction \mathrm{Dom}(f) = p_j \restriction \mathrm{Dom}(f)$) so $x \in \bigcup_{\alpha \in u_j \setminus u^*} A_\alpha$, hence $(w_i \setminus w^*) \cap (\bigcup_{\alpha \in u_j \setminus u^*} A_\alpha) \neq \emptyset$. Let $w_i \setminus w^* = \{x_{i,\ell} : \ell < \ell(*)\}$; so rephrasing if $i < j < \aleph_1$ and p_i, p_j are incompatible then for some $\alpha(i,j) \in u_j \setminus u^*$ and $\ell(i,j) < \ell(*)$ we have $x_{i,\ell(i,j)} \in A_{\alpha(i,j)}$. Let D be a nonprincipal ultrafilter on ω, so for each $j \in [\omega, \omega_1)$ for some $\ell_j < \ell(*)$ we have $\{i < \omega : \ell(i,j) = \ell\} \in D$ and as u_j is finite, for some $\alpha_j \in u_j \setminus u^*$ we have $\{i < \omega : \alpha(i,j) = \alpha_j\} \in D$. The number of possible ℓ_j is $\ell(*)$ so for some $j(1) \neq j(2) \in [\omega, \omega_1)$ we have $\ell_{j(1)} = \ell_{j(2)}$. Hence

$$A = \{i < \omega : \ell(i, j(1)) = \ell_{j(1)} = \ell_{j(2)} = \ell(i, j(2)) \text{ and}$$

$$\alpha(i, j(1)) = \alpha_{j(1)} \text{ and } \alpha(i, j(2)) = \alpha_{j(2)}\}$$

belongs to D, so $A_{\alpha_{j(1)}} \cap A_{\alpha_{j(2)}}$ includes $\{x_{i,\ell_{j(1)}} : i \in A\}$ which is infinite, hence we have $\alpha_{j(1)} = \alpha_{j(2)}$ and $(u_{j(1)} \setminus u^*) \cap (u_{j(2)} \setminus u^*) \neq \emptyset$ contradicting the choice of u^*. So P satisfies the c.c.c.

Now if $i < \lambda$, $i \notin S$ and $A_i \cap (\bigcup_{j \in S} A_j)$ is infinite, then $\mathcal{I}_{i,n} \stackrel{\text{def}}{=} \{p \in P : A_i \cap p^{-1}(\{0\})$ has cardinality $\geq n\}$ is a dense subset of P. Also for each $x \in \bigcup_{i<\lambda} A_i$ we have $\mathcal{J}_x \stackrel{\text{def}}{=} \{p \in P : x \in \mathrm{Dom}(p)\}$ is a dense subset of P. Lastly for $i \in S$ we have $\mathcal{I}_i = \{p \in P : A_i \subseteq \mathrm{Dom}(p)\}$ is a dense subset of P.

So there is a directed $G \subseteq P$ such that $G \cap \mathcal{I}_{i,n} \neq \emptyset$ for $i < \lambda$, $i \notin S$, $n < \omega$ and $G \cap \mathcal{I}_i \neq \emptyset$ for $i \in S$, and $G \cap \mathcal{J}_x \neq \emptyset$ for $x \in \bigcup_{i<\lambda} A_i$. Let f^* be $\bigcup_{f \in G} f$ and f^{**} be the function extending f^* with domain $\bigcup_{i<\lambda} A_i$ and being constantly zero on $\bigcup_{i<\lambda} A_i \setminus \bigcup_{i \in S} A_i$. It is easy to check f^{**} is as required. $\square_{3.7}$

A partial answer to the Question 3.6 is

3.8 Definition. A sequence $\langle A_i : i < \lambda \rangle$ of infinite pairwise almost disjoint subset of ω is called a *tree* if for any $i, j < \lambda$ if $n \in A_i \cap A_j$, then $A_i \cap n = A_j \cap n$.

An example of a tree of $2^{\aleph_0}$ subsets of the set of all finite sequences of 0's and 1's is the set $\{T_f : f \in {}^\omega 2\}$ where $T_f = \{f \restriction n : n < \omega\}$.

3.9 Theorem. Assume MA, let λ be a cardinal such that $\aleph_0 \leq \lambda < 2^{\aleph_0}$ and let $\langle A_i : i < \lambda \rangle$ be a family of infinite pairwise almost disjoint subsets of ω which is a tree. Let $S \subseteq \lambda$ then there is an $f : \omega \to \{0, 1\}$ such that for all $i \in S$ we have $f \restriction A_i =_{ae} 1_{A_i}$ and for all $i \in \lambda \setminus S$ we have $f \restriction A_i =_{ae} 0_{A_i}$.

Proof. Let

$P = \{f : f$ is a function and for some $i_1, \ldots, i_n < \lambda$ and a finite subset w of ω we have $\mathrm{Dom}(f) = A_{i_1} \cup \cdots \cup A_{i_n} \cup w$, and for $\ell = 1, \ldots, n$: $f \restriction A_{i_\ell} =_{ae} 1_{A_{i_\ell}}$ if $i_\ell \in S$ and $f \restriction A_{i_\ell} =_{ae} 0_{A_{i_\ell}}$ if $i_\ell \notin S\}$.

The partial order on P is inclusion.

Let us prove now that P satisfies the c.c.c. Let $\langle f_i : i < \aleph_1 \rangle$ be a sequence of members of P. For f_i let $\mathrm{Dom}(f_i) = A_{\alpha(i,1)} \cup \cdots \cup A_{\alpha(i,n(i))} \cup w_i$, where $\alpha(i, 1), \alpha(i, 2), \ldots < \lambda$. Let $k_i < \omega$ be such that k_i is a strict upper bound of w_i and for each ℓ, $1 \leq \ell \leq n(i)$, $A_{\alpha(i,\ell)} \cap k_i$ contains a number m such that $f_i(m') = f_i(m)$ for every number $m' \in A_{\alpha(i,\ell)}$ which is $> m$. There is such a k_i since each $f_i \restriction A_{\alpha(i,\ell)}$ has the value 0 almost everywhere or the value 1 almost everywhere. We demand in addition that for $1 \leq \ell, m \leq n(i)$, $\ell \neq m$,

$A_{\alpha(i,\ell)} \cap k_i \not\subseteq A_{\alpha(i,m)} \cap k_i$ and $A_{\alpha(i,\ell)} \cap A_{\alpha(i,m)} \subseteq k_i$; without loss of generality we can assume that for all $i < \aleph_1$ $n(i)$ is fixed, i.e., $n(i) = n$ for $i < \aleph_1$, k_i is fixed, i.e., $k_i = k^*$ for $i < \aleph_1$ and for $1 \leq \ell \leq n$, the set $A_{\alpha(i,\ell)} \cap k^*$ is fixed (for each ℓ) and $f_i \restriction k^*$ is fixed and w_i is fixed. We shall now prove that for $i \neq j$ $i, j < \aleph_1$ the conditions f_i and f_j are compatible. Consider the union $f_i \cup f_j$. If this union is a function it obviously belongs to P and is above f_i and f_j (thus completing the proof of c.c.c.). Suppose it is not a function. Since $f_i \restriction k^* = f_j \restriction k^*$ there is a $k \geq k^* (k < \omega)$ such that $f_i(k) \neq f_j(k)$. Since $w_i = w_j \subseteq k^*$ clearly $k \notin w_i = w_j$, hence for some $1 \leq \ell_0, \ell_1 \leq n$, $k \in A_{\alpha(i,\ell_0)}$ and $k \in A_{\alpha(j,\ell_1)}$. Let $m_0 = \mathrm{Max}\ A_{\alpha(i,\ell_0)} \cap k^*$, since $k \cap A_{\alpha(i,\ell_0)}$ contains a number m such that we have $(\forall m')[m \leq m' \in A_{\alpha(i,\ell_0)}, f_i(m') = f_i(m)]$, clearly $f_i(m_0) = f_i(k)$. Note $\ell_0 \neq \ell_1 \Rightarrow A_{\alpha(i,\ell_0)} \cap k^* \neq A_{\alpha(i,\ell_1)} \cap k^* = A_{\alpha(j,\ell_1)} \cap k^*$. Since $\langle A_i : i < \aleph_1 \rangle$ is a tree and $k \in A_{\alpha(i,\ell_0)} \cap A_{\alpha(j,\ell_1)}$, $k \geq k^*$, we have $\ell_0 = \ell_1$ and therefore $m_0 = \mathrm{Max}(A_{\alpha(j,\ell_1)} \cap k^*)$ and, as for i, $f_j(k) = f_j(m_0)$. Since $m_0 < k^*$ and $f_i \restriction k^* = f_j \restriction k^*$ we have $f_i(m_0) = f_j(m_0)$ hence $f_i(k) = f_j(k)$, contradicting our choice of k.

Looking at the proof of Theorem 3.5 it is clear how to choose dense subsets $\mathcal{I}$ of P such that if G is a directed subset of P which intersects each one of them then $g = \bigcup G$ has the following properties: $\mathrm{Dom}(g) = \omega$, for $i \in S$ we have $g \restriction A_i =_{ae} 1_{A_i}$ and for $i \in \lambda \setminus S$ we have $g \restriction A_i =_{ae} 0_{A_i}$. $\square_{3.9}$

3.10 Theorem. In 3.9 "$A_i \subseteq \omega$" is not required, provided $|A_i| \leq \aleph_0$.

Proof. The proof is analogous to the proof of Theorem 3.7. $\square_{3.10}$

§4. The Uniformization Property

Let us present now the setting with which we shall deal with the problem 3.6.

4.1 Definition. Let $\bar{A} = \langle A_i : i < \alpha \rangle$ be a sequence of sets (or we use a family of sets), each of which is (usually countable and always) infinite and h a function from $\bigcup_{i<\alpha} A_i$ to the class of nonzero ordinals. We say that $(\bar{A}, h)$ has the *uniformization property* if for every family of functions $\langle f_i : i < \alpha \rangle$ such that $f_i : A_i \to \mathrm{Ord}$ and $f_i(a) < h(a)$, there is a function $f : \bigcup_{i<\alpha} A_i \to \mathrm{Ord}$ such that for each $i < \alpha$ the function f_i is almost included in f (i.e., included except for $< |\mathrm{Dom}(f_i)|$ members a of $\mathrm{Dom}(f_i)$; which usually means except for finitely many $a \in \mathrm{Dom}(f)$). We shall denote $\bigcup_{i<\alpha} A_i$ with $D(\bar{A})$, so $\mathrm{Dom}(h) = D(\bar{A})$. If h is constantly λ we may write $(\bar{A}, \lambda)$ and if $\lambda = 2$ we may write $\bar{A}$. We shall say that $\bar{A}$ is a tree if there is a partial order $<$ on $D(\bar{A})$ such that:

a) $(D(\bar{A}), <)$ is a tree such that for all $x \in D(\bar{A})$ the set $\{y : y < x\}$ is a finite set linearly ordered by $<$.

b) For each i the set A_i is a branch in $D(\bar{A})$, i.e., if $x \in A_i$ and $y < x$ then also $y \in A_i$ and any two members of A_i are $<$-comparable. Since each A_i is infinite and since by clause (a) the tree $(D(A), <)$ is of height $\leq \omega$, each A_i is a maximal branch, i.e., it goes all the way up.

We shall denote $(D(\bar{A}), <)$ with T and T_n will denote the n-th level of T, that is $T_n = \{x : |\{y : y < x| = n\}$. For a branch A we denote with $(A)_{[n]}$ the only member of A in the level n, i.e., the only member of $A \cap T_n$.

4.2 Claim. If $\bar{A} = \langle A_i : i < \alpha \rangle$, $\alpha \geq \omega_1$, is a tree in the above sense and $D(\bar{A}) \subseteq {}^{\omega>}2$ and the tree order is $\lhd$ *then* $\bar{A}$ does not have the uniformization property.

Proof. So let $\eta_i \in {}^{\omega}2$ be such that $A_i = \{\eta_i{\restriction}n : n < \omega\}$ so $(A_i)_{[n]} = \eta_i{\restriction}n$. Let us define the function f_i on A_i as follows. $f_i(\eta_i{\restriction}n) = \eta_i(n)$. The value of f_i for this member of $D(\bar{A})$ tells us which way the branch goes at $\eta_i{\restriction}n$, right or left, since

$$\eta_i{\restriction}(n+1) = \eta_i{}^\frown\langle \eta_i(n) \rangle$$

Assume now that there is a function f uniformizing all f_j's, or even all f_i's for some uncountable set $I \subseteq \alpha$. For each $i \in I$ let $n(i)$ be such that $f_i(x) = f(x)$ for every $x \in A_i$ of level $\geq n(i)$. Since I is uncountable there is an $n^* < \omega$ such that $\{i \in I : n(i) = n^*\}$ is uncountable, we denote this set with I_1. Since $D(\bar{A})$ has at most 2^{n^*} members of level n^* there is a $\rho \in {}^{n^*}2$ such that $I_2 = \{i \in I_1 : \eta_i{\restriction}n^* = \rho\}$ is uncountable; we denote this set with I_2. For all $i, j \in I_2$ we have $\eta_i{\restriction}n^* = \rho = \eta_j{\restriction}n^*$ and since A_i and A_j are branches we have $\eta_i(k) = \eta_j(k)$ for all $k < n^*$. For $i \neq j$ such that $i, j \in I_2$ we know that A_i and A_j are different branches hence there is a least ℓ such that $\eta_i(\ell) \neq \eta_j(\ell)$, since $\ell < n^* \Rightarrow n(i) = n(j)$ clearly $\ell \geq n^*$ hence $f_i(\eta_i{\restriction}\ell) = f(\eta_i{\restriction}\ell)$ and $f_j(\eta_j{\restriction}\ell) = f(\eta_j{\restriction}\ell)$. By the definition of f_i and f_j we have $\eta_i{\restriction}(\ell+1) = (\eta_i{\restriction}\ell)\hat{}\langle\eta_i(\ell)\rangle = (\eta_i{\restriction}\ell)\hat{}\langle f(\eta_i{\restriction}\ell)\rangle = (\eta_j{\restriction}\ell)\hat{}\langle f(\eta_j{\restriction}\ell)\rangle$ (since $\eta_i{\restriction}\ell = \eta_j{\restriction}\ell$ by the minimality of ℓ) $= (\eta_j{\restriction}\ell)\hat{}\langle\eta_j(\ell)\rangle = \eta_j{\restriction}(\ell+1)$, contradicting the choice of ℓ. $\square_{4.2}$

4.3 Theorem. Assume MA. Suppose $\bar{A} = \langle A_i : i < \lambda\rangle$ is a family of pairwise almost disjoint countable sets and $\lambda < 2^{\aleph_0}$. *Then* $(\bar{A}, \aleph_0)$ has the uniformization property provided that:

(*) for every countable A, $\{i < \lambda : A \cap A_i$ is infinite $\}$ is countable.

Proof. Let $f_i : A_i \to \omega$ for $i < \lambda$, and we shall find f, almost extending each f_i. Let $P = \{f :$ there are $i_1, \dots, i_n < \lambda$ such that $\mathrm{Dom}(f) = A_{i_1} \cup \cdots \cup A_{i_n}$, $\mathrm{Range}(f) \subseteq \omega$ and f almost extend f_i for $\ell = 1, \dots, n\}$ ordered by inclusion.

It suffices to prove that P satisfies the c.c.c., and proceed as in e.g. 3.5.; but this is not hard (proof similar to 3.9). $\square_{4.3}$

4.4 Discussion. Theorem 4.2 asserts that if $D(\bar{A})$ is a tree which at each node branches into at most two branches then $\bar{A}$ does not have the uniformization property (the fact that in the theorem $D(\bar{A})$ was actually a subset of ${}^{\omega}2$ is obviously irrelevant). The assumption in the theorem that $D(\bar{A})$ is a tree can be replaced by the following weaker assumption. For each $i < \alpha$ let A_i be given as a sequence of length ω enumerating its elements with no repetitions, which we write as $\langle a_{i,0}, a_{i,1}, \dots\rangle$ so $\{a_{i,n} : n < \omega\} = A_i$. The weaker hypothesis is that $\bar{A}$ is uncountable and for each sequence $\langle b_0, \dots, b_{n-1}\rangle$ of members of $D(\bar{A})$ the set $\{a_{i,n} : \langle a_{i,0}, \dots, a_{i,n-1}\rangle = \langle b_0, \dots, b_{n-1}\rangle\}$ has at most two members, i.e., for all A_i given $a_{i,0}, \dots, a_{i,n-1}$ there are at most two possibilities for $a_{i,n}$. (This assumption obviously holds if $D(\bar{A})$ is a tree with branching as above and each sequence $\langle a_{i,0}, a_{i,1}, \ldots\rangle$ is such that $a_{i,n} = (A_i)_{[n]}$ for all $n < \omega$). We shall see that also under this weaker assumption $\bar{A}$ does not have the uniformization property.

Let $\bar{A}$ be as above. For each $i < \alpha$ let $B_i = \{\langle a_{i,0}, \ldots, a_{i,n-1}\rangle : n < \omega\}$, $\bar{B} = \langle B_i : i < \alpha\rangle$. Now $\bar{B}$ is obviously a tree which at each node branches to at most two branches, and for $i \neq j$ we have $B_i \neq B_j$ since $A_i \neq A_j$. We can prove by induction on n that $\{a_{i,n} : i < \alpha\}$ has $\leq 2^n$ elements, hence $\bigcup_{i<\alpha} B_i$ is countable. Let $\langle f_i : i < \alpha\rangle$ where $f_i : B_i \to \{0,1\}$ be a family of functions which cannot be uniformized, by claim 4.2. We define a family $\langle g_i : i < \omega_1\rangle$ where $g_i : A_i \to \{0,1\}$ as follows. Let $g_i(a_{i,n}) = f_i(\langle a_{i,0}, \ldots, a_{i,n}\rangle)$. We shall see that this family too cannot be uniformized. Suppose $g : D(\bar{A}) \to \{0,1\}$ uniformizes this family. Define f on $D(\bar{B})$ as follows: $f(\langle a_{i,0}, \ldots, a_{i,n-1}\rangle) = g(a_{i,n-1})$ if $n > 0$ and $f(\langle\rangle)$ can be given any value. We shall see that f is an uniformization of $\langle f_i : i < \alpha\rangle$. Given $i < \alpha$ we know that for some $k < \omega$, for all $n > k$ we have $g_i(a_{i,n}) = g(a_{i,n})$ hence, by the definitions of g_i and f we have $f_i(\langle a_{i,0}, \ldots, a_{i,n}\rangle) = f(\langle a_{i,0}, \ldots, a_{i,n}\rangle)$.

Let us remark now that the theorem is true also for a tree $D(\bar{A})$ which has a bounded branching at each node (and hence also in the more general case mentioned above). We shall show it here for the case where $D(\bar{A}) \subseteq {}^{\omega>}4$, which is sufficiently general to exhibit the proof of the general case. For a natural number k let $d_\ell(k)$ denote the coefficient of 2^ℓ in the binary expansion of k, $d_\ell(k) \in \{0,1\}$. We define now on the branch A_i of ${}^{\omega>}4$ two functions f_i^0 and f_i^1 by setting $f_i^\ell((A_i)_{[n]}) = d_\ell((A_i)_{[n+1]})$, for $\ell = 0,1$. Let f^ℓ be the function uniformizing the functions $\langle f_i^\ell : i < \alpha\rangle$ for $\ell = 0,1$. Let n^*, z and I be such that $I \subseteq \alpha$, $|I| > \aleph_0$, $(A_i)_{[n^*]} = z$ for $i \in I$ and $f_i^\ell(n) = f^\ell(n)$ for $n \geq n^*$ (as in the proof of the theorem). We get now a contradiction from the fact that for $i, j \in I$ $i \neq j$ there is a least level where A_i and A_j go to their separate ways, while at each node (from level n^* and above) f^0 and f^1 determine completely the way the branch goes in the next level so A_i and A_j must go the same way.

We deal here exclusively with trees $\bar{A}$ and other systems where $|D(\bar{A})| = \aleph_0$. In 3.8 we dealt with a different definition of a tree, namely we called a family w of subsets of ω a tree if for all $x, y \in w$ if $n \in x \cap y$ then $x \cap n = y \cap n$.

In this case let us define $<^*$ on ω by $k <^* \ell$ if $k < \ell$ and for every $A \in w$ if $\ell \in A$ then also $k \in A$. It is easily seen that $<^*$ is a partial order; the set $\{k : k <^* n\}$ is linearly ordered by $<^*$ which coincides with $<$ on this set and that each $A \in w$ is a branch in the tree $\langle \omega, <^* \rangle$. Thus a tree in the sense of 3.8 is also a tree in the present sense.

Now let $\bar{A}$ be a tree as defined here with $|D(\bar{A})| = \aleph_0$ with $<^*$ as the partial order relation of the tree. Identify the members of $D(\bar{A})$ with the natural numbers in such a way that if $k <^* \ell$ then $k < \ell$. Now suppose $n \in A_i \cap A_j$ and $k < n$, $k \in A_i$. Since A_i is a branch of the tree we have $k <^* n$ or $n <^* k$; $n <^* k$ would imply $n < k$, contradiction; hence $k <^* n$. Since also A_j is a branch, $n \in A_j$ and $k <^* n$ we get also $k \in A_j$. Since this works in both directions we have $A_i \cap n = A_j \cap n$. Thus a tree in the present sense is isomorphic to a tree in the sense of Definition 3.8.

4.5 Theorem. Assuming MA+$2^{\aleph_0} > \aleph_1$. Let $\bar{A} = \langle A_i : i < \aleph_1 \rangle$, $A_i = \{a_{i,n} : n < \omega\}$ be a set of $\aleph_1$ countably infinite pairwise distinct sets such that $|D(\bar{A})| = \aleph_0$. *Then* there is an uncountable subset $W \subseteq \aleph_1$ which has the property that for every sequence $\langle b_0, \ldots, b_{n-1} \rangle$ the set $\{a_{i,n} : \langle a_{i,0}, \ldots, a_{i,n-1} \rangle = \langle b_0, \ldots, b_{n-1} \rangle$, and $i \in W\}$ has at most two members. As a consequence $\bar{A}$ does not have the uniformization property, since $\bar{A} \restriction W$ does not have it.

Proof. Remember A_i is $\{a_{i,0}, a_{i,1}, \ldots\}$ (with no repetitions). Let $P \stackrel{\text{def}}{=} \{w \subseteq \aleph_1 : w$ is finite and for all $\langle b_0, \ldots, b_{n-1} \rangle$ we have

$$|\{a_{n,i} : i \in w \text{ and } \langle a_{i,0}, \ldots, a_{i,n-1} \rangle = \langle b_0, \ldots, b_{n-1} \rangle\}| \leq 2\}$$

and P is partially ordered by inclusion. By deleting at most $\aleph_0$ members of $\bar{A}$ we can make sure that for all $i < \aleph_1$ and $n < \omega$:

$$|\{j < \aleph_1 : \langle a_{j,0}, \ldots, a_{j,n-1} \rangle = \langle a_{i,0}, \ldots, a_{i,n-1} \rangle\}| = \aleph_1$$

[For every finite sequence $\langle b_0, \ldots, b_{n-1} \rangle$ of members of $D(\bar{A})$, and there are $\aleph_0$ such sequences, if $0 < |\{j < \aleph_1 : \langle a_{j,0}, \ldots, a_{j,n-1} \rangle = \langle b_0, \ldots, b_{n-1} \rangle\}| \leq \aleph_0$ we delete all the A_j's with j in this set, so altogether we deleted countably many sets.]

Let us see now that P satisfies the c.c.c. Let $\langle w_t : t < \aleph_1\rangle$ be a sequence of members of P. Let us write w_t as $\{i_1^t, i_2^t, \ldots, i_{n(t)}^t\}$. Let $k(t)$ be the least number k such that the sequences $\langle a_{i_r^t,0}, a_{i_r^t,1}, \ldots, a_{i_r^t,k-1}\rangle$ for $r = 1, \ldots, n(t)$ are pairwise distinct. Without loss of generality we can assume that for all $t < \aleph_1$ we have $n(t) = n$, $k(t) = k$ and for each $r = 1, \ldots, n$ the sequence $\langle a_{i_r^t,0}, a_{i_r^t,1}, \ldots, a_{i_r^t,k-1}\rangle$ is fixed for all $t < \aleph_1$, we shall now see that for $s, t < \aleph_1$ we have $w_s \cup w_t \in P$ and hence we have w_s and w_t are compatible. Proving $w_s \cup w_t \in P$ is equivalent to showing that for all sequences $\langle b_0, \ldots, b_{m-1}\rangle$ we have $|\{a_{i,m} : i \in w_s \cup w_t \text{ and } \langle a_{i,0}, \ldots, a_{i,m-1}\rangle = \langle b_0, \ldots, b_{m-1}\rangle\}| \leq 2$. If $m < k$ then the initial $(m+1)$- tuples $\langle a_{i,0}, \ldots, a_{i,m-1}, a_{i,m}\rangle$ of the A_i's for $i \in w_s \bigcup w_t$ are exactly these $(m+1)$ - tuples for $i \in w_s$ since for $m = 1, \ldots, n$ we have $\langle a_{i_m^s,0}, \ldots, a_{i_m^s,k-1}\rangle = \langle a_{i_m^t,0}, \ldots, a_{i_m^t,k-1}\rangle$, hence $\{a_{i,m} : i \in w_s \bigcup w_t$ and $\langle a_{i,0}, \ldots, a_{i,m-1}\rangle = \langle b_0, \ldots, b_{m-1}\rangle\} = \{a_{i,m} : i \in w_s$ and $\langle a_{i,0}, \ldots, a_{i,m-1}\rangle = \langle b_0, \ldots, b_{m-1}\rangle\}$, and the latter set has at most two members since $w_s \in P$. If $m \geq k$ then since the initial sequences of length k of all A_i for $i \in w_s$ are all distinct, and the same holds for all $i \in w_t$ so $\langle b_0, \ldots, b_{m-1}\rangle$ can be an initial sequence of A_i for at most one $i \in w_s$ and at most one $i \in w_t$ so that $|\{i \in w_s \bigcup w_t : \langle a_{i,0}, \ldots, a_{i,m-1}\rangle = \langle b_0, \ldots, b_{m-1}\rangle\}| \leq 2$ hence $|\{a_{i,m} : i \in w_s \bigcup w_t,$ and $\langle a_{i,0}, \ldots, a_{i,m-1}\rangle = \langle b_0, \ldots, b_{m-1}\rangle\}| \leq 2$. So P satisfies the c.c.c.

For $\varepsilon < \aleph_1$ let $\mathcal{I}_\varepsilon$ be the subset of P, $\mathcal{I}_\varepsilon = \{w \in P : (\exists j > \varepsilon)[j \in w]\}$. Let us prove that $\mathcal{I}_\varepsilon$ is a dense subset of P. Let $w \in P$ and let k be such that the sequences $\langle a_{i,0}, a_{i,1}, \ldots, a_{i,k-1}\rangle$ are all distinct for different $i \in w$. Take a fixed $i \in w$ (the case where $w = \emptyset$ is trivial). By our assumption there are $\aleph_1$ ordinals j such that $\langle a_{j,0}, \ldots, a_{j,k-1}\rangle = \langle a_{i,0}, \ldots, a_{i,k-1}\rangle$, hence there is such a $j >$ max(w), ε. We shall see that $w \cup \{j\} \in P$, to prove that we have to show that for all $\langle b_0, \ldots, b_{m-1}\rangle$

$$(*) \qquad |\{a_{\gamma,m} : \gamma \in w \bigcup \{j\} \text{ and } \langle a_{\gamma,0}, \ldots, a_{\gamma,m-1}\rangle = \langle b_0, \ldots, b_{m-1}\rangle\}| \leq 2.$$

If $m < k$ then since the first k members of A_j are the first k members of A_i, where $i \in w$, the left side of $(*)$ remains unchanged if $w \cup \{j\}$ is replaced by w and since the inequality holds for w, as $w \in P$, it holds also for $w \cup \{j\}$. If $m \geq k$

then, since the sequences $\langle a_{\gamma,0}, a_{\gamma,1}, \ldots, a_{\gamma,k-1}\rangle$ are all distinct for different $\gamma \in w$ there is at most one $\gamma \in w$ such that $\langle a_{\gamma,0}, \ldots, a_{\gamma,n-1}\rangle = \langle b_0, \ldots, b_{m-1}\rangle$, hence there are at most two $\gamma \in w \cup \{j\}$ which satisfy this equality and $(*)$ follows immediately.

By MA there is a directed subset G of P which intersects each $\mathcal{I}_\varepsilon$. Therefore $W = \bigcup G$ is a cofinal subset of $\aleph_1$. We take now $\bar{A}^\dagger = \langle A_i : i \in W\rangle$. We still have to prove that for all $\langle b_0, \ldots, b_{m-1}\rangle$ we have $|\{a_{i,m} : i \in W$ and $\langle a_{i,0}, \ldots, a_{i,m-1}\rangle = \langle b_0, \ldots, b_{m-1}\rangle\}| \leq 2$. Assume that for some $\langle b_0, \ldots, b_{m-1}\rangle$ this set has three members $a_{i_1,m}, a_{i_2,m}, a_{i_3,m}$ where $i_1, i_2, i_3 \in W$. Since $W = \bigcup G$ and G is directed we have $i_1, i_2, i_3 \in w$ for some $w \in G \subseteq P$. Then we have $|\{a_{i,m} : i \in w$ and $\langle a_{i,0}, \ldots, a_{i,m-1}\rangle = \langle b_0, \ldots, b_{m-1}\rangle\}| \geq 3$, contradicting $w \in P$. $\square_{4.5}$

4.6 Theorem. It is consistent that there is a tree $\bar{A} = \langle A_\alpha : \alpha < \omega_1\rangle$ such that $|D(\bar{A})| = \aleph_0$ and $\bar{A}$ has the uniformization property.

4.6A Remark. We can phrase a condition on forcing notions preserved by FS iteration (it depends on an $\bar{A}$ with the relevant property) and by this prove the consistency of an axiom; see [Sh:98].

Proof. Let T be a tree with ω levels, $|T| = \aleph_0$, T_n (the n'th level) is finite for $n < \omega$ and for all $x \in T_n$ we have $|\{y \in T_{n+1} : y > x\}| \geq 2^{n^2}$. We shall obtain a generic extension of V in which there is a tree $\bar{A}$ of length $\aleph_1$ which has the uniformization property. We saw above (in 4.4) that if the branching of a tree at each node is bounded the tree does not have the uniformization property. Here we shall see that if at level n each node branches to 2^{n^2} branches the tree may have the uniformization property. There is a gap here which one should narrow or even eliminate.

First one introduces $\aleph_1$ generic branches of T by the following set Q_0 of conditions

$Q_0 = \{f : f$ is a finite function on $\omega_1 \times \omega$ such that:

$$\langle \alpha, n\rangle \in \mathrm{Dom}(f) \Rightarrow f(\alpha, n) \in T_n,$$

$$\langle \alpha, n+1\rangle \in \mathrm{Dom}(f) \Rightarrow \langle \alpha, n\rangle \in \mathrm{Dom}(f),$$
$$f(\alpha, n) < f(\alpha, n+1) \text{ (when defined)}\}$$

The partial order relation on Q_0 is inclusion. If G is a generic subset of Q_0 let $F = \bigcup G$, $A_\alpha = \{F(\alpha, n) : n < \omega\}$ and $\bar{A} = \langle A_\alpha : \alpha < \omega_1\rangle$. Now $\bar{A}$ will be the tree which has the uniformization property (it is obvious that for $\alpha \neq \beta$, $\alpha, \beta < \omega_1$ we have $A_\alpha \neq A_\beta$, since the subset $\{f \in Q_0 : \exists n[\langle \alpha, n\rangle, \langle \beta, n\rangle \in \mathrm{Dom}(f) \,\&\, f(\alpha, n) \neq f(\beta, n)]\}$ of Q_0 is dense in Q_0). '

The next step is to carry out an iterated forcing so that at each step a different system $\bar{g} = \langle g_i : i < \omega_1\rangle$ where $g_i : A_i \to \{0,1\}$ gets uniformized. Eventually we obtain that $\bar{A}$ has the uniformization property if all appropriate $\bar{g}$'s appear. We shall describe first only a single step of this iterated forcing. The conditions we use now are the following. Let $\bar{g}$ be a name of a system of functions as above in the forcing Q_0. We define $Q(\bar{g}) = \{h : h$ is a finite function from $\aleph_1$ to ω and for every $\alpha, \beta \in \mathrm{Dom}(h)$ the functions $g_\alpha \restriction \{x \in A_\alpha : \mathrm{height}(x) \geq h(\alpha)\}$ and $g_\beta \restriction \{x \in A_\beta : \mathrm{height}(x) \geq h(\beta)\}$ are compatible $\}$.

The partial order relation on $Q(\bar{g})$ is inclusion. This forcing is done over $V[G_0]$, where G_0 is a generic subset of Q_0. Let G be a generic subset of $Q(\bar{g})$ and $H = \bigcup G$. We can now define the uniformizing function $g : T \to \{0,1\}$ as follows. For $x \in T$ if $x \in A_\alpha$ and heigth$(x) \geq H(\alpha)$ then $g(x) = g_\alpha(x)$; if x belongs to no A_α such that $H(\alpha) \leq \mathrm{height}(x)$ we define $g(x)$ arbitrarily. If $x \in A_\alpha$, $\mathrm{height}(x) \geq H(\alpha)$ and also, $\beta \neq \alpha$, $x \in A_\beta$ and $\mathrm{height}(x) \geq H(\beta)$ then since G is directed and $H = \bigcup G$ there is an $h \in G$ such that $\alpha, \beta \in \mathrm{Dom}(h)$, $h(\alpha) = H(\alpha)$ and $h(\beta) = H(\beta)$. Since $h \in Q(\bar{g})$, $h(\alpha), h(\beta) \leq \mathrm{height}(x)$ we, by the definition of $Q(\bar{g})$, have $g_\alpha(x) = g_\beta(x)$, so that $g(x)$ is well defined. It is obvious from the definition of g that it uniformizes all g_α's for $\alpha < \aleph_1$.

Let us consider now the set $Q_0 * \underset{\sim}{Q}(\bar{g})$ and prove that it satisfies the c.c.c. Let W be an uncountable subset of $Q_0 * \underset{\sim}{Q}(\bar{g})$ and we shall prove that there are two distinct members which are compatible. Each $\langle p, \underset{\sim}{q}\rangle \in W \subseteq Q_0 * \underset{\sim}{Q}(\bar{g})$ is first extended as we shall tell. If two of the extended conditions will turn out to be compatible then the corresponding original conditions are compatible too. First we extend p to p_1 which decides exactly the value of $\underset{\sim}{q}$ ($\underset{\sim}{q}$ is the name of a finite subset of $\omega_1 \times \omega$), so we shall now regard q as a finite function from ω_1 to ω and not as a name. Then we extend p_1 to p_2 so that the $\mathrm{Dom}^\dagger(p_2) \stackrel{\mathrm{def}}{=} \{\alpha : (\exists n)[\langle\alpha, n\rangle \in \mathrm{Dom}(p_2)]\}$ (which is the set of branches on which p "speaks") will include the domain of q. Let u_p denote $\mathrm{Dom}^\dagger(p)$ i.e., the set of indices of the branches about which p contains information. For a sufficiently large $c < \omega$ we can extend p_2 to p_3 so that $c > |u_{p_2}|$, $\mathrm{Dom}(p_3) = u_{p_2} \times \{0, \ldots, c\}$ and $p_3(i, c) \neq p_3(j, c)$ for $i, j \in u_{p_2}$ such that $i \neq j$ (i.e., different $i \in u_{p_2}$ are indices of branches of $\bar{A}$ which branch off from each other at a level up to level c), and $c > q(i)$ for $i \in u_p$. ¿From now on we shall assume that all the members of W have the properties of $\langle p_3, q\rangle$.

Using the standard technique we can delete members from W so as to obtain a set W with the following properties. There are a $c < \omega$, a finite subset $v \subseteq \aleph_1$; finite sequences $v^\alpha = \langle j_1^\alpha, \ldots, j_n^\alpha\rangle$ for $\alpha < \aleph_1$ of n different members such that the sets $\{j_1^\alpha, \ldots, j_n^\alpha\}$ (which we shall also call v^α) are disjoint from each other for different α's and are disjoint from v; a function p with domain $v \times \{0, \ldots, c\}$, a function ρ with domain $\{1, \ldots, n\} \times \{0, \ldots, c\}$, a partial function q with domain v and a partial function h on $\{1, \ldots, n\}$ such that $W = \{\langle p_\alpha, q_\alpha\rangle : \alpha < \aleph_1\}$ and $\mathrm{Dom}(p_\alpha) = (v \cup v_\alpha) \times \{0, \ldots, c\}$, $p = p_\alpha \restriction (v \times \{0, \ldots, c\})$, $p_\alpha \restriction (v_\alpha \times \{0, \ldots, c\}) = \{\langle\langle j_k^\alpha, m\rangle, \rho(k, m)\rangle : 1 \leq k \leq n\}$ and $q \leq q_\alpha$ and $q_\alpha \setminus q = \{\langle j_k^\alpha, h(k)\rangle : 1 \leq k \leq n\}$. The members $\rho(1, c), \rho(2, c), \ldots, \rho(n, c)$ and $p(\alpha, c)$ (for $\alpha \in V$) of T_c are pairwise distinct (by the properties of $\langle p, q\rangle$ obtained above). Above each one of them there are at least 2^{c^2} different members of T_{c+1}. Let $\langle t_{k,\ell} : 1 \leq \ell \leq 2^{c^2}\rangle$ be a sequence of 2^{c^2} different members of T_{c+1} which are above $\rho(k, c)$. For $\alpha, \ell = 1, \ldots, 2^{c^2}$ let $p^*_{\alpha,\ell} = p_\alpha \bigcup \{\langle\langle j_k^\alpha, c+1\rangle, t_{k,\ell}\rangle : 1 \leq k \leq n\}$. Notice that for different α's the initial parts of the branches with indices $j_1^\alpha, \ldots, j_n^\alpha$ were the same, but this is not the case for the $p^*_{\alpha,\ell}$'s since we introduced different branchings at level $c+1$. Let $p^*_\ell = p^*_{\ell,\ell}$ for $\ell = 1, \ldots, 2^{c^2}$. Since all the p^*_ℓ's behave the same way on v and have otherwise disjoint domains also $\bigcup_{1 \leq \ell \leq 2^{c^2}} p^*_\ell \in Q_0$. Extend this condition to a condition p^* which determines the values of $g_{j_m^\alpha}(\rho(m, k))$ for all $1 \leq \alpha \leq 2^{c^2}$, $1 \leq m \leq n$ and $1 \leq k \leq c$. If we keep α fixed and let k, m vary we have here 2^{c^2} functions on a set with at most $n \times c < c^2$ members into $\{0,1\}$, hence there must be two different $\beta \neq \gamma$, $1 \leq \beta, \gamma \leq 2^{c^2}$ such that $\underset{\sim}{g}_{j_k^\beta}(\rho(k,m)) = \underset{\sim}{g}_{j_k^\gamma}(\rho(k,m))$ for all $1 \leq k \leq n$, $1 \leq m \leq c$ (they are Q_0-names but we mean the values p^* force for them). We claim that $\langle p^*, q_\beta \cup q_\gamma \rangle \in Q_0 * Q(\underset{\sim}{\bar{g}})$, hence $\langle p_\beta, q_\beta \rangle$ and $\langle p_\gamma, q_\gamma \rangle$ are compatible. What may prevent $\langle p^*, q_\beta \cup q_\gamma \rangle$ from being a condition when each of $\langle p^*, q_\beta \rangle$ and $\langle p^*, q_\gamma \rangle$ is (since $\langle p_\beta, q_\beta \rangle$ and $\langle p_\gamma, q_\gamma \rangle$ are conditions and $p_\beta \leq p^*$, $p_\gamma \leq p^*$ by the choice of p^*)? There may be k_1, m, k_2, such that $m \geq h(k_1)$, $m \geq h(k_2)$, $\underset{\sim}{F}(j_{k_1}^\beta, m) = \underset{\sim}{F}(j_{k_2}^\gamma, m)$ such that $\underset{\sim}{g}_{j_{k_1}^\beta}(\rho(k_1,m)) \neq \underset{\sim}{g}_{j_{k_2}^\gamma}(\rho(k_2,m))$ (where $\underset{\sim}{F} = \bigcup\{f : f \in \underset{\sim}{G}_{Q_0}\}$, see the choice of the A_j's). But by the choice of p^* this can occur only for $m \leq c$, so we get $\rho(k_1,m) = \rho(k_2,m)$. This is the case where the corresponding functions g_j give different values to the same member $\rho(k_1,m) = \rho(k_2,m)$ of T and this member is above the place in the two branches with indices $j_{k_1}^\beta$ and $j_{k_2}^\gamma$ where the uniformization is supposed to occur. By our choice of β and γ we have $g_{j_{k_2}^\beta}(\rho(k_2,m)) = g_{j_{k_2}^\gamma}(\rho(k_2,m))$, hence $g_{j_{k_1}^\beta}(\rho(k_1,m)) \neq g_{j_{k_2}^\beta}(\rho(k_2,m))$ while $\rho(k_1,m) = \rho(k_2,m)$, but this shows that $\langle p^*, q_\beta \rangle$ is not a condition in $Q_0 * Q(\gamma)$, which is a contradiction.

This of course does not yet prove Theorem 4.6. If we want to carry on the iteration, somehow imitating the scheme of Martin's axiom, we should do two things: first isolate some property of $\langle A_i : i < \omega_1 \rangle$ in V^{Q_0}, a property which is the "reason" why $Q(\bar{g})$ is c.c.c. Then we should formulate a property of forcing notion which would ensure the property of $\langle A_i : i < \omega_1 \rangle$ is preserved, at last we should show by induction on $\alpha \leq \omega_2$ that the iteration of $Q(\bar{g})$ over all names for candidates has the desired property. Where by candidate we mean a $\bar{g} = \langle g_i : i < \omega_1 \rangle, g_i : A_i \to \{0,1\}$.

Assume T was chosen such that:

$$(\forall m)(\exists n > m)(\forall x \in T_n)[|\{y \in T_{n+1} : y > x\}| \geq 2^{|\{y \in T : (y \leq x)\}|^m}]$$

What we get from the genericity of the $\aleph_1$-branches is:

⊗ for every $k < \omega$ and distinct A_i^ℓ for $i < \omega_1, \ell < k$ from $\bar{A}$ there are $n < \omega$, and pairwise distinct $a_1, \ldots, a_k \in T_n$ and $w \subseteq \omega_1$ such that:

(i) $a_i = (A^\ell_i)_{[n]}$ for every $i \in w$

(ii) $i \neq j \in w \Rightarrow (A^\ell_i)_{[n+1]} \neq (A^\ell_j)_{[n+1]}$

(iii) $|w| > \prod_{\ell<k} 2^{|\{y|y<a_\ell\}|} = i(a_0, \dots, a_{k-1})$ (the i is for notational convenience)

For $\bar{g} = \langle g_\alpha : \alpha < \omega_1 \rangle$, $g_\alpha : A_\alpha \to \{0,1\}$ the forcing notion $Q(\bar{g})$ is defined above, let $\langle P_\alpha, \underset{\sim}{Q}_\alpha : \alpha < \alpha^* \rangle$ denote a Finite Support iteration of length α^* such that Q_0 is as above and for $\alpha > 0$, $\underset{\sim}{Q}_\alpha = Q(\underset{\sim}{\bar{g}}{}_\alpha)$ uniformizes a candidate $\underset{\sim}{\bar{g}}{}_\alpha$ where $\bar{g}_\alpha = \langle g^\alpha_\xi : \xi < \omega_1 \rangle$ such that $\Vdash_{P_\alpha}$ "$\bar{g}^\alpha$ is a candidate ". We prove by induction on $\alpha \le \alpha^*$ the following condition.

$(*)_\alpha$: if $k < \omega$ and $p_i \in P_\alpha$, $A^\ell_i \in \bar{A}$ for $i < \omega_1$, $\ell < k$, then $\forall n_1 < \omega \exists n < \omega$, $n > n_1$ $\exists a_0, \dots, a_{k-1} \in T$ $\exists w \subseteq \omega_1$ such that

(i) $a_\ell = (A^\ell_i)_{[n]}$ for $\ell < k, i \in w$ and $a_{\ell_1} = a_{\ell_2} \Leftrightarrow (\forall i \in w)(A^{\ell_1}_i = A^{\ell_2}_i)$.

(ii) for $i \neq j \in w$ we have $(A^\ell_i)_{[n+1]} \neq (A^\ell_j)_{[n+1]}$ or $A^\ell_i = A^\ell_j$.

(iii) $|w| > i(a_0, \dots, a_{k-1})$

(iv) $\exists q \in P_\alpha$ such that $p_i \le q$ for each $i \in w$

Note that $(*)_\alpha \Rightarrow P_\alpha$ satisfies the c.c.c. So if we succeed to prove that, the rest of the proof is like the proof of 3.4. Also note that proving $(*)_\alpha$ w.l.o.g. for each i the sequence $\langle A^\ell_i : \ell < k \rangle$ is without repetitions. So let us carry the induction.

Case (1) $\alpha = 0$ nothing to prove.

Case (2) $\mathrm{cf}(\alpha) \ge \aleph_2$. Then for some $\beta < \alpha$ we have: $i < \omega_1 \Rightarrow p_i \in P_\beta$ so $(*)_\beta$ gives the conclusion.

Case (3) α limit, $\mathrm{cf}(\alpha) = \omega$.

Let $\alpha = \bigcup_{n<\omega} \alpha_n$ then for each $i < \omega_1$ there is $n(i) < \omega$ such that: $p_i \in P_{\alpha_{n(i)}}$ so for some n we have $|\{i : n(i) = n\}| = \aleph_1$ then $(*)_{\alpha_{n(i)}}$ gives the conclusion.

Case (4) α limit, $\mathrm{cf}(\alpha) = \aleph_1$ so $\alpha = \bigcup_{i<\omega_1} \alpha_i$ and $\langle \alpha_i : i < \omega_1 \rangle$ is increasing and continuous; for each i we let $h(i) = \mathrm{Min}\{j : \mathrm{Dom}(p_i \restriction \alpha_i) \subseteq \alpha_j\}$ this is a pressing down function on the set of limit ordinals $< \omega_1$, so by Fodor's Lemma for some i_0 we have: $S \overset{\mathrm{def}}{=} \{i : h(i) \le \alpha_{i_0}\}$ is stationary. W.l.o.g. $p_i \in P_{\alpha_{i+1}}$, let $p'_i = p_i \restriction \alpha_i$, so $p'_i \in P_{\alpha_{i_0}}$. Now for any finite $w \subseteq \omega_1$, if $\{p'_i : i \in w\}$ has an upper bound p^* in P_{i_0}, then $\{p_i : i \in w\}$ has a common upper bound p in P_α: $p^* \cup \bigcup_{i \in w} (p_i{\restriction}[\alpha_i, \alpha_{i+1}))$. Hence by $(*)_{\alpha_{i_0}}$ (which holds by the induction hypothesis) we can complete the proof of $(*)_\alpha$.

Case (5) $\alpha = \beta + 1$.

If $\alpha + 1$ this is clear, so assume $\beta > 0$. Remember that $Q(\underset{\sim}{\bar{g}}_\beta)$ is the set of functions f such that : $\mathrm{Dom}(f)$ is a finite subset of ω_1 and for $\xi \in \mathrm{Dom}(f)$, $f(\xi) < \omega$ and

$$(\forall \zeta, \xi \in \mathrm{Dom}(f))(\forall n)\Big[f(\zeta) \leq n \wedge f(\xi) \leq n$$

$$\& \, (A_\zeta)_{[n]} = (A_\xi)_{[n]} \Rightarrow g_\beta((A_\zeta)_{[n]}) = g_\beta((A_\xi)_{[n]})\Big]$$

W.l.o.g. for each $i < \omega_1$ we have $\beta \in \mathrm{Dom}(p_i)$, so $p_i(\beta)$ is (the name of) a finite function from ω_1. W.l.o.g. $p_i(\beta)$ is an actual function, and $|\mathrm{Dom}(p_i(\beta))|$ is constant so let $\mathrm{Dom}(p_i(\beta)) = \{A_i^\ell : k \leq \ell < k(0)\}$ (where k comes from the case of $(*)_\alpha$ we are trying to prove) w.l.o.g. $p_i(\beta)(A_i^\ell)$ depends on ℓ only. We apply $(*)_\beta$ to $k(0)$, $p_i{\restriction}\beta \in P_\beta$ and $A_i^\ell (i < \omega_1, \ell < k(0))$; we thus get n, $a_\ell(\ell < k(0))$ w_0 and q_0. Clearly we can find $q_1 \in P_\beta$, $q_0 \leq q_1$ such that for each $i \in w_0$, $k \leq \ell < k(0)$ and $m \leq n$ we have $q_1 \Vdash_{P_\beta}$ "$\underset{\sim}{g}_\gamma(m) = c_i(\ell, m)$" where $A_i^\ell = A_\gamma$; of course $c_i(\ell, m) \in \{0, 1\}$. The number of possible functions is $i(a_k, \ldots, a_{k(0)-1})$ as $|w_0| > i(a_0, \ldots, a_{k(0)-1})$ so for some c, $w = \{i \in w_0 : c_i = c\}$ has cardinality $> i(a_0, \ldots, a_{k-1})$. Now $q = q_1 \cup \{\langle \beta, \bigcup_{i \in w_0} p_i(\beta)\rangle\} \in P_\alpha$ and q, $a_\ell(\ell < k)$, w exemplify $(*)_\alpha$. $\square_{4.6}$

§5. Maximal Almost Disjoint Families of Subsets of ω

5.0 Definition. By a mad (maximally almost disjoint) subset of $\mathcal{P}(\omega)$ we mean an infinite subset F of $\mathcal{P}(\omega)$ such that for all $x, y \in F$ we have $|x| = \aleph_0$ but $|x \cap y| < \aleph_0$ and for every $z \in \mathcal{P}(\omega)$ there exists $x \in F$ such that $|z \cap x| = \aleph_0$.

5.1 Claim. There is a mad subset of $\mathcal{P}(\omega)$ of cardinality $2^{\aleph_0}$.

Proof. Replace ω by the set of all nodes of the full binary tree, i.e., with the set of all finite sequences of 0's and 1's. This tree has $2^{\aleph_0}$ branches and every two of them are almost disjoint. Extend the set of all branches to a mad set by Zorn's lemma. $\square_{5.1}$

5.2 Observation. No countable subset of $\mathcal{P}(\omega)$ is mad.

Proof. Let $\langle a_i : i < \omega \rangle$ be a sequence of infinite pairwise almost disjoint subsets of ω. For each $i < \omega$ we have $a_i \setminus \bigcup_{j<i} a_j = a_i \setminus \bigcup_{j<i} a_j \cap a_i$, and since each set $a_j \cap a_i$ is finite and a_i is infinite there is an $x_i \in a_i \setminus \bigcup_{j<i} a_j$. For $j < i$ we have $x_j \in a_j$ while $x_i \notin a_j$ hence $x_j \neq x_i$, therefore the set $b = \{x_i : i < \omega\}$ is an infinite set. Since for $j < i$, $x_i \notin a_j$ we have $b \cap a_j \subseteq \{x_0, \dots, x_j\}$. Thus the intersection of b with each a_j is finite and therefore $\{a_i : i < \omega\}$ is not mad. $\square_{5.2}$

So if the continuum hypothesis holds then there are mad sets of cardinality $2^{\aleph_0} = \aleph_1$ and of no other cardinality. If the continuum hypothesis does not hold we are faced with the question whether there are mad sets of cardinalities $\geq \aleph_1$ but $< 2^{\aleph_0}$.

5.3 Theorem. Martin's axiom implies that every mad set is of cardinality $2^{\aleph_0}$.

Proof. Let $\mathcal{A}$ be an infinite set of infinite pairwise almost disjoint subsets of ω, $|\mathcal{A}| < 2^{\aleph_0}$. Let $P_{\mathcal{A}} = \{\langle a, t \rangle : a$ is a finite subset of ω and t is a finite subset of $\mathcal{A}\}$. For $p = \langle a, t \rangle$ and $q = \langle b, s \rangle$ in $P_{\mathcal{A}}$ we define: $p \leq q$ if: $a \subseteq b$, $s \subseteq t$ and for each $u \in t$ we have $b \cap u = a \cap u$. Now $\leq$ is easily seen to be a partial order. The meaning of the condition $\langle a, t \rangle$ is that the set w which we are constructing will include a, but for every $u \in t$ we will have $w \cap u = a \cap u$, so no additional members of u will be in w (other than those already in a).

If $\langle a, t \rangle$, $\langle a, s \rangle$ are in $P_{\mathcal{A}}$ then, obviously, also $\langle a, t \bigcup s \rangle \in P_{\mathcal{A}}$ hence $\langle a, t \rangle$ and $\langle a, s \rangle$ are compatible. Therefore incompatible members of $P_{\mathcal{A}}$ must have different first components. Since there are only $\aleph_0$ finite subsets of ω every subset of $P_{\mathcal{A}}$ of pairwise incompatible members must be countable, and the c.c.c. holds.

For every $u \in \mathcal{A}$ let $\mathcal{I}_u = \{\langle a, t\rangle \in P_{\mathcal{A}} : u \in t\}$. Now $\mathcal{I}_u$ is a dense subset of $P_{\mathcal{A}}$ since for every $\langle b, s\rangle \in P_{\mathcal{A}}$ we have $\langle b, s \cup \{u\}\rangle \in \mathcal{I}_u$. For $n < \omega$ let $\mathcal{I}_n = \{\langle a, t\rangle \in P_{\mathcal{A}} : (\exists k \geq n) k \in a\}$. Let us prove that $\mathcal{I}_n$ is a dense subset of $P_{\mathcal{A}}$. Let $\langle b, s\rangle \in P_{\mathcal{A}}$. Since s is finite and $\mathcal{A}$ infinite there is a $v \in \mathcal{A} \setminus s$. Now $v \cap (\bigcup s) = v \cap \bigcup\{v \cap u : u \in s\}$ is finite hence $v \setminus \bigcup s$ is infinite. Let $k \in v \setminus \bigcup s$, $k \geq n$, we shall see that $\langle b \cup \{k\}, s\rangle \geq \langle b, s\rangle$. To prove that let $u \in s$; now $(b \cup \{k\}) \cap u = b \cap u$ since $k \notin \bigcup s$ hence $k \notin u$.

By Martin's axiom there is a directed subset G of $P_{\mathcal{A}}$ which intersects each $\mathcal{I}_u$ and each $\mathcal{I}_n$. Let A be the union of the first components of G. Let $u \in \mathcal{A}$, let $\langle a, s\rangle \in \mathcal{I}_u \cap G$, then $u \in s$. We shall see that $A \cap u = a \cap u$ and hence $A \cap u$ is finite. We know $\langle a, s\rangle \in G$ so $a \subseteq A$ hence $a \cap u \subseteq A \cap u$; assume that $m \in A$, then $m \in b$ for some $\langle b, t\rangle \in G$. Since G is directed there is an $\langle c, r\rangle \in G$ such that $\langle c, r\rangle \geq \langle a, s\rangle$, $\langle b, t\rangle$ so $m \in b \subseteq c$ but $m \in u$ by its choice so $m \in c \cap u$. Since $u \in s$ and $\langle c, r\rangle \geq \langle a, s\rangle$ we have $u \cap c = u \cap a$, hence $m \in u \cap a$. So we have proved $m \in A \cap u \Rightarrow m \in a \cap u$; hence $A \cap u \subseteq a \cap u$; so by a previous sentence $A \cap u = a \cap u$, hence $A \cap u$ is finite. A is infinite since for every n we know G contains a member $\langle a, t\rangle$ of $\mathcal{I}_n$ hence A contains a number $\geq n$.

Thus we have constructed an infinite set A which is almost disjoint from every member of $\mathcal{A}$, and hence $\mathcal{A}$ is not mad. $\square_{5.3}$

5.4 Theorem. If V does not satisfy the continuum hypothesis and λ is a cardinal such that $\aleph_1 \leq \lambda \leq 2^{\aleph_0}$ in V *then* V can be generically extended by a c.c.c. extension which preserves also $2^{\aleph_0}$ (automatically it still is a cardinal) to an extension $V[G]$ where there is a mad subset of $\mathcal{P}(\omega)$ of cardinality λ.

Proof. We start in V with a sequence $\langle A_\alpha : \alpha < \lambda\rangle$ of almost disjoint subsets of ω (there is such by claim 5.1). Now we proceed with a system of FS iterated forcing of length ω_1 as follows. At step α we assume that we have constructed a sequence $\langle A_i : i < \lambda + \alpha\rangle$ of pairwise almost disjoint subsets of ω. We take $F_\alpha = \{A_i : i < \lambda + \alpha\}$ and we use at this step the forcing notion P_{F_α}, where

$P_{\mathcal{A}}$ is as in the proof of Theorem 5.3, and we introduce by means of it a new infinite subset $A_{\lambda+\alpha}$ of ω almost disjoint with each member of F_α. (This follows immediately from the proof of Theorem 5.3.) Since P_{F_α} is a c.c.c. forcing, by 2.8 our iterated forcing is c.c.c. For P_{F_α} we can use, instead of pairs $\langle a, t\rangle$ where a is a finite subset of ω and t is a finite subset of F_α, such pairs where t is a finite subset of $\lambda + \alpha$ and each $i \in t$ stands for the corresponding A_i. Thus each P_{F_α} will consist of elements from V (while its set of elements is not necessarily in V), and the cardinality of P_{F_α} is therefore $\mathrm{Max}(\alpha, \lambda) = \lambda$. For the iterated forcing we can use only standard names in the set P of conditions, hence $|P| = \lambda$. Since P is a c.c.c. forcing, standard arguments (as in 3.4) show that $|\mathcal{P}(\omega)|^{(V[G])} = [(\lambda)^{\aleph_0}]^V \leq (\lambda^{\aleph_0})^{V[G]} \leq [(2^{\aleph_0})^{\aleph_0}]^V = [2^{\aleph_0}]^V$, hence in $V[G]$ we know $2^{\aleph_0}$ is the same as in V.

Finally let us prove that $\{A_\alpha : \alpha < \lambda + \omega_1\}$ is a mad subset of $\mathcal{P}(\omega)$. The "almost disjoint" is trivial. Let $A \subseteq \omega$, $A \in V[G]$. For each $n \in \omega$ let $\mathcal{I}_n$ be a maximal antichain in P of conditions which decide $\underset{\sim}{n} \in \underset{\sim}{A}$, where $\underset{\sim}{A}$ is a name of A. Since P satisfies the c.c.c. we know $|\mathcal{I}_n| = \aleph_0$. Let $\mathcal{I} = \bigcup_{n<\omega} \mathcal{I}_n$, so $|\mathcal{I}| = \aleph_0$. For each $q \in \mathcal{I}$ the set $\mathrm{Dom}(q)$ is a finite subset of ω_1, hence there is an $\alpha < \omega_1$ such that for all $q \in \mathcal{I}$ we have $\mathrm{Dom}(q) \subseteq \alpha$. If G_α is the component of G in the iterated forcing up to and not including α, we have $A \in V[G_\alpha]$. We shall see that $A \cap A_\beta$ is infinite for some $\beta \leq \lambda + \alpha$ which establishes that $\{A_i : i < \lambda + \omega_1\}$ is mad. Moreover, we shall show that if $A \cap A_\beta$ is finite for every $\beta < \lambda + \alpha$ then $A \cap A_\alpha$ is infinite. This follows from the following lemma, which ends the proof. $\square_{5.4}$

5.5 Lemma. Let F be a set of pairwise almost disjoint subsets of ω, and let $B \subseteq \omega$ be such that for every finite subset $F^\dagger$ of F the set $B \setminus \bigcup F^\dagger$ is infinite. Let G be a generic subset of P_F over V and let A be the union of the first components of the members of G, *then* $A \cap B$ is infinite.

Proof. For every n let

$$\mathcal{I}_n = \{\langle a, t\rangle \in P_F : (\exists m \geq n)[m \in B \cap a]\}.$$

We shall see that $\mathcal{I}_n$ is dense. Let $\langle b, t\rangle \in P_F$. By our assumption $B \setminus \bigcup t$ is infinite hence it contains an $m \geq n$. Now $\langle a \cup \{m\}, t\rangle \in \mathcal{I}_n$ and since $m \notin \bigcup t$ $\langle a \cup \{m\}, t\rangle \leq \langle a, t\rangle$. Since $\mathcal{I}_n$ is dense, G contains some $\langle a, t\rangle \in \mathcal{I}_n$, hence $A \supseteq a$ and so A contains an $m \in B$ such that $m \geq n$. Since this holds for every $n < \omega$ we have $A \cap B$ is infinite. $\square_{5.5}$

Discussion. We shall prove that essentially all countable forcing notions are equivalent. If one carries out the proof of the last theorem for the case where $\alpha = \aleph_1$, but one starts with a sequence $\langle A_i : i < \omega\rangle$ of pairwise almost disjoint subsets of ω then each P_F is a countable forcing and therefore equivalent to the addition of a Cohen real. Thus the FS iterated addition of $\aleph_1$ Cohen reals yields an extension of V with a mad set of cardinality $\aleph_1$. But the iterated extension of $\aleph_1$ Cohen reals is the same as the simultaneous addition of $\aleph_1$ Cohen reals (as in the proof of I 3.2).

5.6 Theorem. Every two countable forcing notions where above every condition there are two incompatible conditions are equivalent.

Proof. Let P be the set of all finite sequences of natural numbers ordered by proper inclusion and let Q be a countable forcing as in the theorem. We can assume that Q has a minimal element $\emptyset_Q$ (or change the proof slightly). We shall show that P and Q are equivalent by constructing an isomorphism F from P onto a dense subset of Q. We shall define $F(\eta)$, for $\eta \in P$, by induction on the length of η. Let $F(\langle\rangle) = \emptyset_Q$, where $\emptyset_Q$ is the minimal member of Q. Let $Q = \{q_n : n < \omega\}$. We shall take care to satisfy the following in the definition.

(i) For every $\eta \in P$ the set $\{F(\eta\hat{\ }\langle i\rangle) : i < \omega\}$ is a maximal set of incompatible members of Q greater than $F(\eta)$.

(ii) For every n there is an $\eta \in P$ of length $n+1$ such that $F(\eta) \geq q_n$.

It is easy to check that this suffice. We assume, as an induction hypothesis, that $\{F(\eta) : \ell g(\eta) = n\}$ is pre-dense in P. Since, by clause (i), for every $\eta \in \bigcup_{\ell<n} {}^{\ell}\omega$ we have $\{F(\eta\hat{\ }\langle i\rangle) : i < \omega\}$ is a maximal set of incompatible members of

Q above $F(\eta)$, the set $\{F(\nu) : \ell g(\nu) = n\}$ is a maximal antichain of Q. To define $F(\eta\hat{}\langle i\rangle)$, for all $i < \omega$ and $\eta \in {}^n\omega$, define a set $\mathcal{I}^*_\eta$ of pairwise incompatible members of Q above $F(\eta)$ as follows. If $\ell g(\eta) = n$ and $F(\eta)$ is compatible with q_n let $s_0 \geq F(\eta), q_n$ (in Q) otherwise let $s_0 \geq F(\eta)$. Let t_1, s_1 be two incompatible members of Q greater than s_0 and, by induction, let t_{n+1}, s_{n+1} be two incompatible members of Q greater than s_n. Take $\mathcal{I}_\eta = \{t_n : 1 \leq n < \omega\}$, $\mathcal{I}_\eta$ is a set of pairwise incompatible members, and if $F(\eta)$ is compatible with q_n then each member of $\mathcal{I}_\eta$ is $\geq q_n$. Let $\mathcal{I}^*_\eta$ be a maximal set of pairwise incompatible members above $F(\eta)$ which includes $\mathcal{I}_\eta$. Now $\mathcal{I}^*_\eta$ is countable, since Q is, and we define the $F(\eta\hat{}\langle i\rangle)$'s so that $\{F(\eta\hat{}\langle i\rangle) : i < \omega\} = \mathcal{I}^*_\eta$. Since (ii) holds $\{F(\eta) : \eta \in P\}$ is a dense subset of Q. $\square_{5.6}$

5.7 Discussion. Using the method of Theorem 5.4 we can extend V so that in the extension for every λ such that $\aleph_0 < \lambda \leq 2^{\aleph_0}$ there is a mad subset of cardinality λ but $2^{\aleph_0}$ is preserved. To do this let $\langle \lambda_\alpha : \alpha < \mu\rangle$ be the cardinals $\aleph_0 < \lambda < 2^{\aleph_0}$ in increasing order. For $\alpha < \mu$ we construct a mad subset $\{A^\alpha_i : i < \lambda_\alpha + \omega_1\}$ of $\mathcal{P}(\omega)$ as follows. $\{A^\alpha_i : i < \lambda_\alpha\} \in V$ is a family of pairwise almost disjoint subsets of ω. We extend V by iterating $\mu \times \omega_1$ times and at the step of ordinal $\mu \times i + \alpha$ we add the set $A^\alpha_{\lambda+i} \subseteq \omega$ as an almost disjoint set of $\{A^\alpha_j : j < \lambda_\alpha + i\}$ as in the proof 5.5: forcing by $P_{\{A^\alpha_j : j < \lambda_\alpha + i\}}$.

III. Proper Forcing

§0. Introduction

In Sect. 1 we introduce the property "proper" of forcing notions: preserving stationarity not only of subsets of ω_1 but even of any $S \subseteq \mathcal{S}_{\leq\aleph_0}(\lambda)$. We then prove its equivalence to another formulation.

In Sect. 2 we give more equivalent formulations of properness, and show that c.c.c. forcing notions and $\aleph_1$-complete ones are proper.

In Sect. 3 we prove that countable support iteration preserves properness (another proof, for a related iteration, found about the same time is given in IX 2.1; others are given in X §2 (with revised support) and XII §1 (by games)). Also we give a proof by Martin Goldstern (in §3).

In Sect. 4 it is proved that starting with V with one inaccessible κ, for some forcing notion P: P is proper of cardinality κ, do satisfy the κ-c.c. and $\Vdash_P$ "*if* Q is a forcing notion of cardinality $\aleph_1$, not destroying stationarity of subsets of ω_1 and $\mathcal{I}_i \subseteq Q$ is dense for $i < \omega_1$, *then* for some directed $G \subseteq Q$, $\bigwedge_{i<\omega_1} G \cap \mathcal{I}_i \neq \emptyset$". For this we need to give a sufficient condition for Lim$\bar{Q}$ to satisfy the κ-c.c. (where $\bar{Q} = \langle P_i, \underset{\sim}{Q}_i : i < \kappa\rangle$ is a CS iteration of proper forcing such that for each $i < \kappa$ we have $\Vdash_{P_i}$ "$|\underset{\sim}{Q}_i| < \kappa$"). For this we show that the family of hereditarily countable conditions is dense in each P, so $i < \kappa \Rightarrow P_i$ has density $< \kappa$.

In sections 5, 6 we present known theorems on speciality of Aronszajn trees.

In Sect. 7 we prove: for V satisfying CH there is an $\aleph_2$-c.c. proper P such that $\Vdash_P$ " *if* for $i < \omega_1$, the set $A_i \subseteq \omega_1$ is countable with no last element and $\mathrm{otp}(A_i) < \sup(A_i)$, *then* for some club C of ω_1 we have $i < \omega_1 \Rightarrow \sup(C \cap A_i) < \sup(A_i)$".

In Sect. 8 we prove the consistency of the Kurepa hypothesis (first proved by Silver [Si67] and see more Devlin). This is a proof from the author's lecture in 1987.

§1. Introducing Properness

1.1 Discussion. When we iterate we are faced with the problem of obtaining for the iteration the good properties of the single steps of iteration. Usually, in our context, the worst possible vice of a forcing notion is that it collapses $\aleph_1$. The virtue of not collapsing $\aleph_1$ is not inherited by the iteration from its single components. As we saw, the virtue of the c.c.c. is inherited by the FS iteration from its components. However in many cases the c.c.c. is too strong a requirement. We shall look for a weaker requirement which is more naturally connected to the property of not collapsing $\aleph_1$, and which is inherited by suitable iterations.

We shall now study a certain generalization of the concepts of a closed unbounded and a stationary subset of ω_1. They were introduced and investigated by Jech and Kueker.

1.2 Definition. For A uncountable let $\mathcal{S}_{\aleph_0}(A) = \{s : s \subseteq A, |s| \leq \aleph_0\}$. Let $W \subseteq \mathcal{S}_{\aleph_0}(A)$ be called closed if it is closed under unions of increasing (by $\subseteq$ of course) ω-sequences. $W \subseteq \mathcal{S}_{\aleph_0}(A)$ is called unbounded (in $\mathcal{S}_{\aleph_0}(A)$) if for every $s \in \mathcal{S}_{\aleph_0}(A)$ there is a $t \in W$ such that $t \supseteq s$. If $W \subseteq \mathcal{S}_{\aleph_0}(A)$, the closure of W is $\mathrm{cl}(W) = \{\bigcup_{n<\omega} s_n : s_n \in W \text{ and } s_n \subseteq s_{n+1} \text{ for } n < \omega\}$, (clearly $W \subseteq \mathrm{cl}(W)$ and $\mathrm{cl}(W)$ is closed).

1.3 Lemma. The intersection of $\aleph_0$ closed unbounded subsets W_i, $i < \omega$, of $\mathcal{S}_{\aleph_0}(A)$ is a closed unbounded subset of $\mathcal{S}_{\aleph_0}(A)$.

Proof. Since each set W_i is closed, the intersection $\bigcap_{i<\omega} W_i$ is obviously closed too. Let us prove now that $\bigcap_{i<\omega} W_i$ is unbounded too. Let $s \in \mathcal{S}_{\aleph_0}(A)$; we have to prove the existence of a set $t \supseteq s$ such that $t \in \bigcap_{i<\omega} W_i$. We shall define a sequence $\langle s_\alpha : \alpha < \omega^2 \rangle$ of members of $\mathcal{S}_{\aleph_0}(A)$ as follows. Let $s_0 = s$, for $\alpha > 0$, $\alpha = \omega \cdot k + \ell$ choose s_α as an arbitrary member of W_ℓ which includes $\bigcup_{\beta<\alpha} s_\beta \in \mathcal{S}_{\aleph_0}(A)$; it exists as W_ℓ is unbounded. We take now $t = \bigcup_{\alpha<\omega^2} s_\alpha$. Obviously $t \supseteq s_0 = s$. For a fixed $i < \omega$ and every $\alpha < \omega^2$, let $\alpha = \omega \cdot k + \ell$ then $\alpha < \omega(k+1) + i$, hence, by the definition of $s_{\omega(k+1)+i}$ we have $s_\alpha \subseteq s_{\omega(k+1)+i}$. Therefore $t = \bigcup_{\alpha<\omega^2} s_\alpha = \bigcup_{k<\omega} s_{\omega \cdot k+i}$. The sequence $\langle s_{\omega \cdot k+i} : k < \omega \rangle$ is a $\subseteq$-increasing ω-sequence of members of W_i, and since W_i is closed also its union t is in W_i. Thus $t \in \bigcap_{i<\omega} W_i$, which is what we had to prove. $\square_{1.3}$

1.4 Definition. By the last lemma we know that the closed unbounded subset of $\mathcal{S}_{\aleph_0}(A)$ generate an $\aleph_1$-complete filter, namely the filter of all subsets of $\mathcal{S}_{\aleph_0}(A)$ which include a closed unbounded set. We denote this filter with $\mathcal{D}_{\aleph_0}(A)$ or $\mathcal{D}_{<\aleph_1}(A)$ or $\mathcal{D}(A)$. A subset of $\mathcal{S}_{\aleph_0}(A)$ is called *stationary* if it meets every closed unbounded subset of $\mathcal{S}_{\aleph_0}(A)$, i.e., if it meets every member of $\mathcal{D}_{\aleph_0}(A)$.

We shall now present the lemma which says that for $|A| = \aleph_1$, $\mathcal{S}_{\aleph_0}(A)$ and $\mathcal{D}_{\aleph_0}(A)$ do not differ significantly from ω_1 and the filter $\mathcal{D}_{\omega_1}$ generated by the closed unbounded subsets of ω_1.

1.5 Lemma. A subset of ω_1 is a closed unbounded subset of ω_1 (in the usual sense of a closed unbounded subset of an ordinal) iff it is a closed unbounded subset of $\mathcal{S}_{\aleph_0}(\omega_1)$.

Proof. Easy. $\square_{1.5}$

We shall now introduce a more restricted set of generators for $\mathcal{D}(A)$.

1.6 Definition. M will denote an algebra, with universe A, and with countably many functions. Let

$Sm(M) = \{s : s \subseteq A, |s| \leq \aleph_0, s \text{ is closed under the operations of } M\}$, i.e., $Sm(M)$ is the set of countable subalgebras of M. Now $Sm(M)$ is obviously a closed unbounded subset of $\mathcal{S}_{\aleph_0}(A)$ (even if M is a partial algebra).

A subset of $\mathcal{S}_{\leq\aleph_0}(A)$ of the form $Sm(M)$, is called an Sm-generator of $\mathcal{D}(A)$.

1.7 Lemma. For every closed unbounded subset W of $\mathcal{S}_{\aleph_0}(A)$ there is an algebra M on A such that $Sm(M) \subseteq W$.

Proof. We shall define, for every finite sequence $\bar{a} = \langle a_0, \ldots, a_{n-1}\rangle$ of members of A, by induction on the length n, a set $s(\bar{a}) \in W$ such that $s(\bar{a}) \supseteq \{a_0, \ldots, a_{n-1}\}$ and $s(\bar{a}) \supseteq s(\langle a_0, \ldots, a_{n-2}\rangle)$ when $n \geq 1$ (of course if $n = 1$ $\langle a_0, \ldots, a_{n-2}\rangle$ is the empty sequence). This is obviously possible because W is unbounded. We define now n-place functions $F^n_\ell, \ell < \omega$, for all $n < \omega$ such that $s(\langle a_0, \ldots, a_{n-1}\rangle) = \{F^n_\ell(a_0, \ldots, a_{n-1}) : \ell < \omega\}$. Let $M = (A, F^n_\ell)_{n<\omega, \ell<\omega}$. Let $s = \{a_0, a_1, \ldots\}$ be a subalgebra of M. Denote $s_n = s(\langle a_0, \ldots, a_{n-1}\rangle) = \{F^n_\ell(a_0, \ldots, a_{n-1}) : \ell < \omega\}$. We have:

a) $s_n \subseteq s$, since s is a subalgebra.
b) $s_n \subseteq s_{n+1}$, by definition of $s(\langle a_0, \ldots, a_n\rangle)$.
c) $a_n \in s_{n+1}$, also the choice of $s(\langle a_0, \ldots, a_{n-1}\rangle)$
d) $s_n \in W$.

By (a) and (c) we have $s = \bigcup_{n<\omega} s_n$; by (b) and (d) we get $s \in W$. Thus we have shown $Sm(M) \subseteq W$. $\square_{1.7}$

We have now seen that the filter $\mathcal{D}_{\aleph_0}(A)$ is generated by the family of sets $Sm(M)$ where M is an algebra on A as above. We shall now see one use of this fact.

1.8 Theorem. Let $P \in V$ be a forcing notion which satisfies the c.c.c., let λ be an uncountable cardinal, and let G be a generic subset of P over V. Every closed unbounded subset B of $\mathcal{S}_{\aleph_0}(\lambda)^{(V[G])}$ in $V[G]$ includes the closure of a set which is a closed unbounded subset of $\mathcal{S}_{\aleph_0}(\lambda)^{(V)}$ in V. In fact $\mathcal{D}_{\aleph_0}(\lambda)^{(V[G])}$ is generated by the closures of the Sm-generators of $\mathcal{D}_{\aleph_0}(\lambda)^{(V)}$ (for any Sm-generator $(Sm(M)^V$ of $\mathcal{D}(\lambda)^{(V)}$ in V, its closure in $V[G]$, is an Sm-generator of $\mathcal{D}(\lambda)^{(V[G])}$ in $V[G]$, as it is $(Sm(M))^{(V[G])})$.

Proof. By what we have proved above we have the following in $V[G]$. There is an algebra $M = (\lambda, F^n_\ell)_{n,\ell<\omega}$ such that $Sm(M) \subseteq B$. In V the function F^n_ℓ has a name $\underset{\sim}{F}^n_\ell$ (moreover we have in V the sequence $\langle \underset{\sim}{F}^n_\ell : n < \omega, \ell < \omega \rangle$). W.l.o.g. $\Vdash$ "$\underset{\sim}{F}^n_\ell$ is an n-place function from λ to λ". Because of the c.c.c., by Lemma I. 3.6 (ii) for all $\alpha_0, \dots, \alpha_{n-1} < \lambda$ we know in V that the set of possible values of $\underset{\sim}{F}^n_\ell(\alpha_0, \dots, \alpha_{n-1})$ is countable and not empty. We define the functions $F^n_{\ell,k}$ for $k < \omega$ so that these $\leq \aleph_0$ values are $\{F^n_{\ell,k}(\alpha_0, \dots, \alpha_{n-1}) : k < \omega\}$. So we know, for all $n, \ell < \omega$ and $\alpha_0, \dots, \alpha_{n-1} < \lambda$ that in $V[G]$ we have $F^n_\ell(\alpha_0, \dots, \alpha_{n-1}) \in \{F^n_{\ell,k}(\alpha_0, \dots, \alpha_{n-1}) : k < \omega\}$. So $N = (\lambda, F^n_{\ell,k})_{n,\ell,k<\omega}$ is an algebra in V and (in $V[G]$) every subalgebra of N is clearly a subalgebra of M. We have $Sm(N)^{(V)} \subseteq Sm(N)^{V[G]} \subseteq Sm(M)^{(V[G])} \subseteq B$ and $Sm(N)^{(V)}$ is a closed unbounded subset of $\mathcal{S}_{\aleph_0}(\lambda)^{(V)}$ in V, and the closure $Sm(N)^V$ in $V[G]$ is $Sm(N)^{V[G]} \subseteq B$. $\square_{1.8}$

A consequence of this theorem is that in a c.c.c. extension $V[G]$ of V every stationary subset of $\mathcal{S}_{\aleph_0}(\lambda)^{(V)}$ in V is also a stationary subset of $\mathcal{S}_{\aleph_0}(\lambda)^{(V[G])}$ in $V[G]$; in short, the extension does not destroy the stationarity of stationary subsets of $\mathcal{S}_{\aleph_0}(\lambda)$. We shall use this property to define the concept of proper forcing. While it is a consequence of the fact that $\mathcal{D}_{\aleph_0}(\lambda)^{V[G]}$ is generated by the closures of the members of $\mathcal{D}_{\aleph_0}(\lambda)^V$ is does not seem to require as much.

1.9 Definition. A forcing notion P is called *proper* if for every (uncountable) cardinal λ, forcing with P preserves stationarity modulo $\mathcal{D}_{\aleph_0}(\lambda)$. We shall denote this condition for λ with $Con_1(\lambda)$ (more exactly $\mathrm{Con}_1(\lambda, P)$ but we omit P when, as usual, P is clear from the context). Note that $Con_1(\aleph_0)$ is meaningless, or trivially true if you like.

Note that properness is preserved by equivalence of forcing notions.

1.10 Theorem. 1) P is proper *iff* the following condition holds for each cardinal λ.

$Con_2(\lambda) = Con_2(\lambda, P)$: Assume that $\{p_{i,j} : j < \alpha_i, i < \alpha\} \subseteq P$ is such that $\alpha \leq \lambda$ and $\alpha_i \leq \lambda$ for $i < \alpha$, and such that for all $i < \alpha$ the set $\{p_{i,j} : j < \alpha_i\}$ is pre-dense in P. Then for all $p \in P$:

$$\{s \in \mathcal{S}_{\aleph_0}(\lambda) : (\exists q \in P)[q \geq p \text{ and } \{p_{i,j} : j < \alpha_i, j \in s\} \text{ is pre-dense above } q \text{ for any } i \in s, i < \alpha]\} \in \mathcal{D}_{\aleph_0}(\lambda).$$

2) Moreover, for any $\lambda \geq |P| + \aleph_1, Con_1(\lambda)$ is equivalent to $Con_2(\lambda)$; and if P is a complete Boolean algebra without 1 then $Con_1(\lambda)$ is equivalent to $Con_2(\lambda)$ for every uncountable λ.

Proof. We assume first $\neg Con_2(\lambda)$, i.e., there are $\alpha \leq \lambda$, $\alpha_i \leq \lambda$ for $i < \alpha$, $\{p_{i,j} : j < \alpha_i\}$ which is pre-dense in P and $p \in P$ such that the set $T \stackrel{\text{def}}{=} \{s \in \mathcal{S}_{\aleph_0}(\lambda)$: for no $q \in P$ do we have: $q \geq p$ and $\{p_{i,j} : j < \alpha_i, j \in s\}$ is pre-dense above q for any $i \in s, i < \alpha\}$ is stationary.
Let $G \subseteq P$ be a generic subset of P such that $p \in G$. Now G meets every pre-dense set hence there is in $V[G]$ a function f on α such that $p_{i,f(i)} \in G$ for all $i < \alpha$. For the algebra (λ, f) we have $Sm((\lambda, f)) = Sm((\lambda, f))^{V[G]} \in \mathcal{D}_{\aleph_0}(\lambda)^{(V[G])}$. We shall show that $T \cap Sm((\lambda, f)) = \emptyset$ thus T which is stationary in V is no longer stationary in $V[G]$. Assume $s \in T \cap Sm((\lambda, f))$, then, as $T \in V$, clearly $s \in V$. Since $s \in Sm((\lambda, f))$ clearly s is closed under f hence $V[G] \models (\forall i \in s)$ $(i < \alpha \rightarrow (\exists j \in s)(j < \alpha_i \,\&\, p_{i,j} \in G))$, hence some $r \in G$ forces this statement. Since G is directed and $p \in G$ there is a $q \in G$ such that $q \geq p, r$ hence $q \Vdash_P$ "$(\forall i \in s)(i < \alpha \rightarrow (\exists j \in s)\ (j < \alpha_i \,\&\, p_{i,j} \in \underset{\sim}{G}_P))$". Therefore for every $i \in s$, such that $i < \alpha$ we know $\{p_{i,j} : j < \alpha_i \,\&\, j \in s\}$ is pre-dense above q (if this were not the case then for some $q^* \geq q$, q^* is incompatible with each member of $\{p_{i,j} : j < \alpha_i \,\&\, j \in s\}$ for some $i \in s$, $i < \alpha$, and for a generic G which contains q^* we cannot have $(\exists j \in s)(j < \alpha_i \,\&\, p_{i,j} \in G)$). Thus s satisfies exactly the condition of not belonging to T, which is a contradiction.

We have proved that $\neg Con_2(\lambda)$ implies $\neg Con_1(\lambda)$. We shall prove that $Con_2(\lambda)$ implies $Con_1(\lambda)$ for $\lambda \geq |P| + \aleph_1$, or for all $\lambda \geq \aleph_1$ if P is a complete

Boolean algebra without 1. This suffices to finish the proof of part (2) (of 1.10). It also suffices for part (1) i.e. for proving that $(\forall\lambda)Con_2(\lambda)$ implies $(\forall\lambda)Con_1(\lambda)$ since, as we shall now see if $\lambda > \mu \geq \aleph_1, Con_1(\lambda)$ implies $Con_1(\mu)$ (see 1.13, note that by 2.1 if $\lambda > \mu$, $Con_2(\lambda)$ implies $Con_2(\mu)$). For this purpose we shall prove the following lemmas (1.11, 1.12, 1.13 and then return to the proof of 1.10).

1.11 Lemma. For any sets D, E we denote by $D\bar{\cup}E$ the set $\{x\cup y : x \in D \,\&\, y \in E\}$. For all disjoint uncountable sets A, B we have: W is a closed unbounded (or stationary) subset of $\mathcal{S}_{\aleph_0}(A)$ iff $W\bar{\cup}\mathcal{S}_{\aleph_0}(B)$ is a closed unbounded (or stationary) subset of $\mathcal{S}_{\aleph_0}(A \cup B)$.

Proof. We can deal separately with the case any of the sets is empty, easily, so w.l.o.g. they are not empty. The proof that if W is closed unbounded in $\mathcal{S}_{\aleph_0}(A)$ then $W\bar{\cup}\mathcal{S}_{\aleph_0}(B)$ is closed unbounded in $\mathcal{S}_{\aleph_0}(A \cup B)$ is trivial. Now assume that W is stationary in $\mathcal{S}_{\aleph_0}(A)$, and suppose $W\bar{\cup}\mathcal{S}_{\aleph_0}(B)$ is not stationary in $\mathcal{S}_{\aleph_0}(A \cup B)$. Then there is a model $M = (A \cup B, F^n_\ell)_{n,\ell<\omega}$ such that $(W\bar{\cup}\mathcal{S}_{\aleph_0}(B)) \cap Sm(M) = \emptyset$. We can assume, without loss of generality, that the set of functions $\{F^n_\ell : n, \ell < \omega\}$ is closed under substitution. We define a function $\hat{F}^n_\ell$ for n-tuples of members of A as follows:

$$\hat{F}^n_\ell(a_0, \ldots, a_{n-1}) = \begin{cases} F^n_\ell(a_0, \ldots, a_{n-1}) & \text{if } F^n_\ell(a_0, \ldots, a_{n-1}) \in A \\ \text{any member of } A & \text{otherwise} \end{cases}$$

Let $\hat{M} = (A, \hat{F}^n_\ell)_{n,\ell<\omega}$. We shall see that if $s \in Sm(\hat{M})$ then for some $t \in \mathcal{S}_{\aleph_0}(B)$, $s \cup t \in Sm(M)$. Let t be the subalgebra of M generated by s; we have to prove that $t \setminus s \subseteq B$. Let $b \in t \setminus s$, then since the set $\{F^n_\ell : n, \ell < \omega\}$ is closed under substitution, $b = F^n_\ell(a_0, \ldots, a_{n-1})$ for some $n < \omega$, $\ell < \omega$ and $a_0, \ldots, a_{n-1} \in s$. If $b \in A$ then by the definition of $\hat{F}^n_\ell$ we know that $\hat{F}^n_\ell(a_0, \ldots, a_{n-1}) = b$, and since $s \in Sm(M)$, s is closed under $\hat{F}^n_\ell$, clearly $b \in s$, which cannot be the case since $b \in t \setminus s$. Therefore $b \notin A$, hence $b \in B$ and we have proved $t \setminus s \subseteq B$. We claim that $W \cap Sm(\hat{M}) = \emptyset$, contradicting our assumption that W is a stationary subset of $\mathcal{S}_{\aleph_0}(A)$. Suppose $s \in W \cap Sm(\hat{M})$, then, as we have shown, for some $t \in \mathcal{S}_{\aleph_0}(B)$ we have $s \cup t \in Sm(M)$. However $s \cup t \in W \bar{\cup} \mathcal{S}_{\aleph_0}(B)$ contradicting $(W \bar{\cup} \mathcal{S}_{\aleph_0}(B)) \cap Sm(M) = \emptyset$.

Thus we have proved the "only if" part. The "if" part can be proved similarly or by applying the "only if" part to $W_1 = \mathcal{S}_{\aleph_0}(A) \setminus W$. $\square_{1.11}$

1.12 Claim. (1) If f is a one-to-one function from A into B, then for $X \subseteq \mathcal{S}_{\aleph_0}(A)$: X is a stationary subset of $\mathcal{S}_{\aleph_0}(A)$ iff $\{a \in \mathcal{S}_{\aleph_0}(B) : f^{-1}(a) \in X\}$ is a stationary subset of $\mathcal{S}_{\aleph_0}(B)$.
(2) If $f : A \to B$ is one-to-one onto, *then* f induces a mapping from $\mathcal{P}(\mathcal{S}_{\aleph_0}(B))$ onto $\mathcal{P}(\mathcal{S}_{\aleph_0}(A))$ preserving Boolean operations and stationarity.
(3) If $V \subseteq V^\dagger$ are models of ZFC, $A, B \in V$, $V \models$ "$|A| = |B|$", then the stationarity of some $X \subseteq \mathcal{S}_{\aleph_0}(B)^V$ is destroyed in $V^\dagger$ iff the stationarity of some $X \subseteq \mathcal{S}_{\aleph_0}(A)^V$ is destroyed in $V^\dagger$ (where $X \in V$ of course)

Proof. Note that $\{a \subseteq \mathcal{S}_{\aleph_0}(A \cup B)$: if $y = f(x)$ then $x \in a \Leftrightarrow y \in a\} \in \mathcal{D}_{\aleph_0}(A \cup B)$.

The proof is left to the reader. $\square_{1.12}$

1.13 Claim. If $\lambda > \mu \geq \aleph_1$ then $Con_1(\lambda)$ implies $Con_1(\mu)$.

Proof. Let W be a stationary subset of $\mathcal{S}_{\aleph_0}(\mu)^V$. Then, as we have proved in 1.11, $\mathcal{S}_{\aleph_0}(\lambda \setminus \mu)^V \bar{\cup} W$ is a stationary subset of $\mathcal{S}_{\aleph_0}(\lambda)^V$. Since $Con_1(\lambda)$ holds $\mathcal{S}_{\aleph_0}(\lambda \setminus \mu)^V \bar{\cup} W$ is also a stationary subset of $\mathcal{S}_{\aleph_0}(\lambda)^{V[G]}$ in $V[G]$. We claim that W is a stationary subset of $\mathcal{S}_{\aleph_0}(\mu)^{V[G]}$ in $V[G]$. If this is not the case then there is a closed unbounded subset C of $\mathcal{S}_{\aleph_0}(\mu)^{V[G]}$ in $V[G]$ such that $C \cap W = \emptyset$. By Lemma 1.11 $\mathcal{S}_{\aleph_0}(\lambda \setminus \mu)^{V[G]} \bar{\cup} C$ is a closed unbounded subset of $\mathcal{S}_{\aleph_0}(\lambda)^{V[G]}$ in $V[G]$. Since $C \cap W = \emptyset$ we have $(\mathcal{S}_{\aleph_0}(\lambda \setminus \mu)^{V[G]} \bar{\cup} C) \cap (\mathcal{S}_{\aleph_0}(\lambda \setminus \mu)^V \bar{\cup} W) = \emptyset$ contradicting what we got that $\mathcal{S}_{\aleph_0}(\lambda \setminus \mu)^V \bar{\cup} W$ is a stationary subset of
$\mathcal{S}_{\aleph_0}(\lambda)^{V[G]}$ in $V[G]$. $\square_{1.13}$

Continuation of the Proof of 1.10. We return now to the proof that $Con_2(\lambda)$ implies $Con_1(\lambda)$ for any uncountable $\lambda \geq |P| + \aleph_1$ or for all $\lambda \geq \aleph_1$ if P is a complete Boolean algebra without 1. Let T be a stationary subset of $\mathcal{S}_{\aleph_0}(\lambda)$ in V. To prove that T is also a stationary subset of $\mathcal{S}_{\aleph_0}(\lambda)$ in $V[G]$ we have to prove that for every P-name $\underset{\sim}{M} = (\lambda, \underset{\sim}{F}^n_\ell)_{\ell,n<\omega}$ of an algebra, $\emptyset \Vdash_P$ "$T \cap Sm(\underset{\sim}{M}) \neq \emptyset$". Let $p \in P$, we shall prove that there is a $q \geq p$ such that $q \Vdash_P$ "$T \cap Sm(\underset{\sim}{M}) \neq \emptyset$". Let $h : {}^{\omega>}\lambda \to \lambda$ be a one-to-one function. We denote the restriction of h to n-tuples with h^n. Let h_ℓ, for $\ell < \omega$ be a function such that for $n > \ell$ we have $h_\ell(h^n(\beta_0, \ldots, \beta_{n-1})) = \beta_\ell$. For each $i < \lambda$ if $i = h(n, \ell, \beta_0, \ldots, \beta_{n-1})$ let $\mathcal{I}_i$ be a maximal antichain of P of conditions which force definite values for $\underset{\sim}{F}^n_\ell(\beta_0, \ldots, \beta_{n-1})$. If $|P| \leq \lambda$ then clearly $|\mathcal{I}_i| \leq \lambda$. If P is a Boolean algebra without 1 (and the order inherited from the Boolean algebra, not the inverse, so "x, y incompatible means $\models x \cup y = 1$), then for each $\beta < \lambda$ we can put in $\mathcal{I}_i$ the minimal condition (equivalently the lub in the Boolean algebra of all conditions) which force $\underset{\sim}{F}^n_\ell(\beta_0, \ldots, \beta_{n-1}) = \beta$, if there are such conditions and then $\mathcal{I}_i$ will be a maximal set of conditions which force definite values on $\underset{\sim}{F}^n_\ell(\beta_0, \ldots, \beta_{n-1})$ and $|\mathcal{I}_i| \leq \lambda$ and $\mathcal{I}_i$ is a maximal antichain. We define for $i < \lambda$ the ordinal α_i and the set $\{p_{i,j} : j < \alpha_i\}$ so that $\{p_{i,j} : j < \alpha_i\} = \mathcal{I}_i$. Let $\gamma(i,j)$ be such that $p_{i,j} \Vdash$ "$\underset{\sim}{F}^n_\ell(\beta_0, \ldots, \beta_{n-1}) = \gamma(i,j)$". Let $M^* = (\lambda, h^n, h_n, \gamma, n)_{n<\omega}$, then $Sm(M^*) \in \mathcal{D}_{\aleph_0}(\lambda)$. Let $W \stackrel{\text{def}}{=} \{s \in \mathcal{S}_{\aleph_0}(\lambda) : (\exists q \geq p)\ [\{p_{i,j} : j \in \alpha_i \cap s\}$ is pre-dense over q for all $i \in s]\}$; we know that $W \in \mathcal{D}_{\aleph_0}(\lambda)$ by the assumption $Con_2(\lambda)$. Since T is a stationary subset of $\mathcal{S}_{\aleph_0}(\lambda)$ there is an $s \in T \cap Sm(M^*) \cap W$. Since $s \in W$ let $q \geq p$ be such that $\{p_{i,j} : j \in \alpha_i \cap s\}$ is pre-dense above q for each $i \in s$. Assume $\beta_0, \ldots, \beta_{n-1} \in s$. By definition of M^* and since $s \in Sm(M^*)$, $\omega \subseteq s$. Thus $n, \ell \in s$ and since s is closed under h^{n+2} also $i = h^{n+2}(n, \ell, \beta_0, \ldots, \beta_{n-1}) \in s$. Since $\{p_{i,j} : j \in \alpha_i \cap s\}$ is pre-dense over q, every generic filter G which contains q contains some $p_{i,j}$ for $j \in \alpha_i \cap s$, and therefore $F^n_\ell(\beta_0, \ldots, \beta_{n-1}) = \gamma(i,j)$ in $V[G]$. Since $i, j \in s$ and since s is closed under the function γ we have $F^n_\ell(\beta_0, \ldots, \beta_{n-1}) \in s$. Thus we have in $V[G]$ that $\forall n \forall \ell (\forall \beta_0, \ldots, \beta_{n-1} \in s)\ [F^n_\ell(\beta_0, \ldots, \beta_{n-1}) \in s]$, hence $s \in Sm(\underset{\sim}{M})$. Since this holds for every G which contains q we have $q \Vdash$ "$s \in Sm(\underset{\sim}{M})$", i.e., $q \Vdash$ "$T \cap Sm(\underset{\sim}{M}) \neq \emptyset$" (since $s \in T$). So T is still stationary in V^P, as required.

$\square_{1.10}$

1.15 Observations. It can be seen, by means of $Con_2(\lambda)$ that in order that the forcing P be proper it suffices to require $Con_2(\lambda)$, or $Con_1(\lambda)$ for some $\lambda \geq 2^{|P|}$ (see Lemma 2.2). We can also replace, equivalently, $\alpha, \alpha_i \leq \lambda$ in $Con_2(\lambda)$ by $\alpha = \alpha_i = \lambda$, and we can replace the pre-dense sets by maximal antichains and $2^{|P|}$ by the number of the maximal antichains (see Lemma 2.2 in the next section). $\square_{1.15}$

1.16 Lemma. If P is a proper forcing *then* in $V[G]$ every countable set of ordinals is included in a countable set of ordinals of V (and hence $\aleph_1^V$ is uncountable in $V[G]$).

Proof. Let a be a countable set of ordinals in $V[G]$, then for some cardinal λ we have $a \in \mathcal{S}_{\aleph_0}(\lambda)^{V[G]}$; now the set $\{s \in \mathcal{S}_{\aleph_0}(\lambda)^{V[G]} : s \supseteq a\}$ is obviously a closed unbounded subset of $\mathcal{S}_{\aleph_0}(\lambda)^{V[G]}$ in $V[G]$. But $\mathcal{S}_{\aleph_0}(\lambda)^V$ is a stationary subset of $\mathcal{S}_{\aleph_0}(\lambda)^V$ in V, hence, since the forcing P is proper, it is also a stationary subset of $\mathcal{S}_{\aleph_0}(\lambda)^{V[G]}$ in $V[G]$. Thus $\mathcal{S}_{\aleph_0}(\lambda)^V \cap \{s \in \mathcal{S}_{\aleph_0}(\lambda)^{V[G]} : s \supseteq a\} \neq \emptyset$ and λ has a subset countable in V which includes a.

An alternative proof is that if $s \in \mathcal{S}_{\aleph_0}(\lambda)^{V[G]}$, then $W_1 = \{t : s \subseteq t \in \mathcal{S}_{\aleph_0}(\lambda)\} \in \mathcal{D}_{\aleph_0}(\lambda)$ in $V[G]$, but in V we have $W_0 = \mathcal{S}_{\aleph_0}(\lambda)^V \in \mathcal{D}_{\aleph_0}(\lambda)$, so in V W_0 is stationary, hence it is stationary in $V[G]$ hence $W_0 \cap W_1 \neq 0$ which is just what we need. $\square_{1.16}$

As a consequence, if α is an ordinal such that $\mathrm{cf}(\alpha) > \aleph_0$ in V, we have also $\mathrm{cf}(\alpha) > \aleph_0$ in $V[G]$.

§2. More on Properness

Discussion. It is worth noticing that one can use for a set of generators of $\mathcal{D}_{\aleph_0}(A)$ not only the set $\{Sm(M) : M$ is a model, the universe of M is A and M is an algebra with $\aleph_0$ operations $\}$ but also a somewhat wider set $\{Sm(M)$: M is a model, the universe of M is A and M is a partial algebra with $\aleph_0$ operations $\}$. This can be done since if M is a partial algebra, i.e., an algebra whose operations are not necessarily defined for all arguments, then $Sm(M)$ is also a closed unbounded subset of $\mathcal{S}_{\aleph_0}(A)$.

2.1 Claim. For $\mu < \lambda$, $Con_2(\lambda, P) \Rightarrow Con_2(\mu, P)$.

Proof. To see that let $\{p_{i,j} : j < \alpha_i, i < \alpha\}$ be as required by $Con_2(\mu)$, i.e., $\alpha \leq \mu$ and $\alpha_i \leq \mu$ for $i < \alpha$. Since $\mu < \lambda$ we can apply $Con_2(\lambda)$ and obtain $D \stackrel{\text{def}}{=} \{s \in \mathcal{S}_{\aleph_0}(\lambda) : \exists q[p \leq q \in P \,\&\, (\forall i \in s \cap \alpha)\ [\{p_{i,j} : j \in s \cap \alpha_i\}$ is pre-dense above $q]]\} \in \mathcal{D}_{\aleph_0}(\lambda)$. Since $\alpha_i \leq \mu$ for $i < \alpha$ and $\alpha \leq \mu$ we have $D = (D \cap \mathcal{S}_{\aleph_0}(\mu))\bar{\cup}\mathcal{S}_{\aleph_0}(\lambda \setminus \mu)$ (where $A\bar{\cup}B = \{x \cup y : x \in A \,\&\, y \in B\}$). By Lemma 1.11 for $T \subseteq \mathcal{S}_{\aleph_0}(A)$, if $T\bar{\cup}\mathcal{S}_{\aleph_0}(B) \in \mathcal{D}_{\aleph_0}(A \cup B)$ then $T \in \mathcal{D}_{\aleph_0}(A)$. Therefore $D \cap \mathcal{S}_{\aleph_0}(\mu) \in \mathcal{D}_{\aleph_0}(\mu)$ which establishes $Con_2(\mu)$ since $D \cap \mathcal{S}_{\aleph_0}(\mu)$ is exactly the set required for $Con_2(\mu)$. $\square_{2.1}$

2.2 Lemma. $Con_2(2^{|P|}) \Rightarrow (\forall\lambda \geq \aleph_0)Con_2(\lambda)$, and hence, since $\mu < \lambda$ and $Con_2(\lambda) \Rightarrow Con_2(\mu)$, therefore $(\exists\sigma \geq 2^{|P|})(Con_2(\sigma)) \Leftrightarrow (\forall\lambda)Con_2(\lambda)$.

Proof. It clearly suffices (see 2.1) to prove that for $\lambda > 2^{|P|}$ we have $Con_2(2^{|P|}) \Rightarrow Con_2(\lambda)$. Let p, $\langle p_{i,j} : j < \alpha_i, i < \alpha\rangle$ be as in $Con_2(\lambda)$. Let $\mathcal{I}_i$ denote the subset $\{p_{i,j} : j < \alpha_i\}$ of P. Let $\langle \mathcal{J}_i : i < 2^{|P|}\rangle$, be a listing possibly with repetitions of all pre-dense subsets of P. Let $\langle q_{i,j} : j < \beta_i\rangle$ be a listing of the members of $\mathcal{J}_i$, then we can have $\beta_i \leq |P|$. We define a partial function $F : \lambda \to 2^{|P|}$ by $F(i) =$ the first γ such that $\mathcal{J}_\gamma = \mathcal{I}_i$, for $i < \alpha$. We define also two partial functions G and H on $\lambda \times \lambda$, into λ by $G(i,j) =$ the γ such that $p_{i,j} = q_{F(i),\gamma}$, for $i < \alpha$, $j < \alpha_i$, and $H(i,j) =$ the least γ such that $p_{i,\gamma} = q_{F(i),j}$, for $i < \alpha$, $j < \beta_{F(i)}$. Since $Con_2(2^{|P|})$ holds the set $A \stackrel{\text{def}}{=} \{s \in \mathcal{S}_{\aleph_0}(2^{|P|}) : (\exists q \geq p)(\forall i \in s)(\{q_{i,j} : j \in s \cap \beta_i\}$ is pre-dense above $q)\}$ is in $\mathcal{D}_{\aleph_0}(2^{|P|})$. Therefore there is a partial algebra M with universe $2^{|P|}$ such that $A \supseteq Sm(M)$. Let N be the partial algebra whose universe is λ and whose partial operations are those of M together with F, G and H (which were defined above). We shall show that for every $s \in Sm(N)$ there is a $q \geq p$ such that for all $i \in s \cap \alpha$, $\{p_{i,j} : j \in s \cap \alpha_i\}$ is pre-dense above q. This will establish

$Con_2(\lambda)$ since the set which is required by $Con_2(\lambda)$ to be in $\mathcal{D}_{\aleph_0}(\lambda)$ has been shown to include $Sm(N)$ which is in $\mathcal{D}_{\aleph_0}(\lambda)$.

Let $s \in Sm(N)$; since N contains all the partial operations of M we have $s \cap 2^{|P|} \in Sm(M)$. Since $Sm(M) \subseteq A$ we have $s \cap 2^{|P|} \in A$; therefore there is a $q \geq p$ such that

$$\otimes\ (\forall i \in s \cap 2^{|P|})(\{q_{i,j} : j \in s \cap \beta_i\} \text{ is pre-dense above } q).$$

We shall show that for this q we have $(\forall i \in s \cap \alpha)(\{p_{i,j} : j \in s \cap \alpha_i\}$ is pre-dense above q), which is all what is left to prove. Let $i \in s \cap \alpha$, since s is closed under F also $F(i) \in s \cap 2^{|P|}$ (since $\text{Rang}(F) \subseteq 2^{|P|}$) hence, by $\otimes$, $\{q_{F(i),j} : j \in s \cap \beta_i\}$ is pre-dense above q. We shall see that $\{q_{F(i),j} : j \in s \cap \beta_i\} = \{p_{i,j} : j \in s \cap \alpha_i\}$ and this will establish that $\{p_{i,j} : j \in s \cap \alpha_i\}$ is pre-dense above q. For $j \in s \cap \beta_i$ we know $q_{F(i),j} = p_{i,H(i,j)}$ by the definition of H. Since $i, j \in s$ also $H(i,j) \in s$ and $H(i,j) < \alpha_i$ by the definition of H. Thus $q_{F(i),j} = p_{i,H(i,j)} \in \{p_{i,j} : j \in s \cap \alpha_i\}$. In the other direction, for $j \in s \cap \alpha_i$ $p_{i,j} = q_{F(i),G(i,j)} \in \{q_{F(i)} : j \in s \cap \beta_i\}$, since s is closed under G. $\square_{2.2}$

2.3 Theorem. Let $M = (|M|, \ldots,)$ be a model with countably many relations and functions, if M is uncountable then:

$$\{|N| \in \mathcal{S}_{\aleph_0}(|M|) : N \prec M\} \in \mathcal{D}_{\aleph_0}(|M|).$$

Proof. Let $M^\dagger$ be an algebra with universe M and with Skolem functions of M as operations (their choice is not unique, but is immaterial; we can e.g. expand M by a well ordering $<^*$ of its universe, and use all functions definable in $(M, <^*)$). Then, as is well known, $Sm(M^\dagger) \subseteq \{N \in \mathcal{S}_{\aleph_0}(|M|) : N \prec M\}$. Since $Sm(M^\dagger) \in \mathcal{D}_{\aleph_0}(|M|)$ also $\{|N| \in \mathcal{S}_{\aleph_0}(|M|) : N \prec M\} \in \mathcal{D}_{\aleph_0}(|M|)$. $\square_{2.3}$

2.4 Definition. For a cardinal λ we denote with $H(\lambda)$ the set of all sets whose transitive closure is of cardinality $< \lambda$. For a regular uncountable λ we know that $(H(\lambda), \in)$ is a model for all axioms of ZFC except maybe for the power set axiom. If not said otherwise we assume λ is like that, for simplicity.

2.5 Definition. Let N be an elementary substructure of $(H(\lambda), \in)$ and let $P \in N$ be a forcing notion. For $q \in P$ we say that q is (N, P)-generic, (or N-generic if it is clear which P we are dealing with), *if* for every subset $\mathcal{I}$ of P which is pre-dense and is in N the set $\mathcal{I} \cap N$ is pre-dense above q.

2.6 Lemma. A condition q is (N, P)-generic *iff* for every $\underset{\sim}{\tau}$ which is a name of an ordinal in the forcing notion P, if $\underset{\sim}{\tau} \in N$ then $q \Vdash$ "$\underset{\sim}{\tau} \in N$" (i.e., if the name is in N then q forces the value to be in N) *iff* for every P-name $\underset{\sim}{\tau} \in N$, $q \Vdash$ " if $\underset{\sim}{\tau} \in V$ then $\underset{\sim}{\tau} \in N$".

Proof. We prove only the first "iff", the second has the same proof. Assume that q is N-generic and let $\underset{\sim}{\tau} \in N$ be a name of an ordinal. Let $\mathcal{I} = \{r \in P : r \Vdash$ "$\underset{\sim}{\tau} = \alpha$", for some ordinal $\alpha\}$. $\mathcal{I}$ is obviously pre-dense. $\mathcal{I}$ is definable from P and $\underset{\sim}{\tau}$ in $(H(\lambda), \in)$, hence $\mathcal{I} \in N$. Since q is N-generic, $\mathcal{I} \cap N$ is pre-dense above q. Let f be the function on $\mathcal{I}$ defined by $f(r) =$ that α for which $r \Vdash$ "$\underset{\sim}{\tau} = \alpha$", then f is definable in $(H(\lambda), \in)$ from $\underset{\sim}{\tau}$, hence $f \in N$. Since $\mathcal{I} \cap N$ is pre-dense above q, $q \Vdash$ "$\underset{\sim}{G}_P \cap (\mathcal{I} \cap N) \neq \emptyset$", i.e., $q \Vdash$ "$(\exists r \in \mathcal{I} \cap N) r \in \underset{\sim}{G}_P$". Therefore if G is a subset of P generic over V and $q \in G$ then $\underset{\sim}{\tau}[G] = f(r)$ holds in $V[G]$, where $r \in (\mathcal{I} \cap N) \cap G$. Since $r \in N$, also $f(r) \in N$ (as $f \in N$) thus $\underset{\sim}{\tau}[G] \in N$ in $V[G]$. Therefore $q \Vdash$ "$\underset{\sim}{\tau} \in N$".

Now assume that for every P-name $\underset{\sim}{\tau}$ of an ordinal, if $\underset{\sim}{\tau} \in N$ then $q \Vdash$ "$\underset{\sim}{\tau} \in N$". Let $\mathcal{I} \in N$ be pre-dense in P. There is an $f \in H(\lambda)$ which maps $|\mathcal{I}|$ onto $\mathcal{I}$, hence there is such an f in N. We take $\underset{\sim}{\tau} = \text{Min}\{i : f(i) \in \underset{\sim}{G}_P\}$. Since f, $P \in N$ and $\underset{\sim}{\tau}$ is definable from f and P in $(H(\lambda), \in)$, also $\underset{\sim}{\tau} \in N$, and $\underset{\sim}{\tau}$ is obviously a P-name of an ordinal. By our assumption $q \Vdash$ "$\underset{\sim}{\tau} \in N$", hence $q \Vdash$ "$(\exists i \in N)(f(i) \in \underset{\sim}{G}_P)$". Since f maps the members of $N \cap |\mathcal{I}|$ to members of $N \cap \mathcal{I}$, being in N, we have $q \Vdash$ "$(\exists r \in \mathcal{I} \cap N)(r \in \underset{\sim}{G}_P)$". Therefore $\mathcal{I} \cap N$ is pre-dense above q, which is what we had to prove. $\square_{2.6}$

2.7 Remark. For $\lambda \geq |P|$ it does not matter in $Con_2(\lambda)$ whether we require that for each $i < \alpha$ the set $\{p_{i,j} : j < \alpha_i\}$ is pre-dense or whether this set is pre-dense above p. Why? on the face of it the version where we require the set

$\{p_{i,j} : j < \alpha_i\}$ to be pre-dense is weaker since it makes a stronger assumption, but we now prove from it the stronger version. Suppose each $\{p_{i,j} : j < \alpha_i\}$ is pre-dense above p. Blow each such set by adding to it all members of P incompatible with p, to get the set $\{p_{i,j} : j < \beta_i\}$. Since $|P| \leq \lambda$ we know $|\beta_i| \leq \lambda$ so we can apply the weak version of $Con_2(\lambda)$ (β_i may be $> \lambda$ but since only the cardinality figures here it is O.K. as long as $\beta_i < \lambda^+$). We obtain a set A in $\mathcal{D}_{\aleph_0}(\lambda)$ such that for $s \in A$ we have a $q \geq p$ for which for each $i \in s \cap \alpha$ the set $\{p_{i,j} : j \in s \cap \beta_i\}$ is pre-dense above q. For $\alpha_i \leq j < \beta_i$, $p_{i,j}$ is incompatible with p and hence also with q, therefore also the set $\{p_{i,j} : j \in s \cap \alpha_i\}$ is pre-dense above q which establishes the stronger version of $Con_2(\lambda)$. For P being a complete Boolean algebra with 1 omitted it suffice to add -p for $p \in P$, p not minimal, so $\lambda \geq \aleph_0$ suffice.

2.8 Theorem. (1) Let $\lambda > 2^{|P|}$, λ regular and assume $P \in H(\lambda)$ (this adds little since $H(\lambda)$ contains an isomorphic copy of P). P is a proper forcing notion iff for every countable elementary substructure N of $(H(\lambda), \in)$ satisfying P, $p \in N$ there is a condition q, $p \leq q \in P$ such that q is (N, P)-generic.
(2) For $\lambda \geq 2^{|P|}$, $P \in H(\lambda)$, P is proper iff[†] $\{N : N \prec (H(\lambda), \in)$ is countable and there is an (N, P)-generic $q \geq p\ \} \in \mathcal{D}_{\aleph_0}(H(\lambda))$ for every $p \in P$.

Proof. (1) We first prove that "if" part, i.e. suppose the condition of the theorem holds, and we shall prove $Con_2(2^{|P|})$ (suffice by 1.10 and 2.2). Let $\langle p_{i,j} : i < \alpha, j < \alpha_i \rangle$ be as in $Con_2(2^{|P|})$. Let $N \prec (H(\lambda), \in)$ be such that $P, p, \langle p_{i,j} : i < \alpha, j < \alpha_i \rangle \in N$. For $i \in N$, $\mathcal{I}_i \stackrel{\text{def}}{=} \{p_{i,j} : j < \alpha_i\} \in N$ since it is definable in $(H(\lambda), \in)$ from $\langle p_{i,j} : i < \alpha, j < \alpha_i \rangle$ and i. Let $q \geq p$ be (N, P)-generic; since $\{p_{i,j} : j < \alpha_i\} \in N$ and it is pre-dense we have that $\mathcal{I}_i \cap N = \{p_{i,j} : j \in N \cap \alpha_i\}$ is pre-dense above q. Therefore to establish $Con_2(2^{|P|})$ it suffices to prove that the set $A = \{N \cap 2^{|P|} : N \prec (H(\lambda) \in)$, and $p, P, \langle p_{i,j} : i < \alpha, j < \alpha_i \rangle \in N$, $|N| \leq \aleph_0\}$ is in $\mathcal{D}_{\aleph_0}(2^{|P|})$. The set $A^\dagger = \{N \in \mathcal{S}_{\aleph_0}(H(\lambda)) : N \prec (H(\lambda), \in)$ and $p, P, \langle p_{i,j} : i < \alpha, j < \alpha_i \rangle \in N\}$

† We do not strictly distinguish between the set of N's and the set of their universes.

is in $\mathcal{D}_{\aleph_0}(H(\lambda))$ by Theorem 2.3. This implies that also $A \in \mathcal{D}_{\aleph_0}(2^{|P|})$, by the technique of using a model with operations closed under composition which we have already used several times, or more exactly by 1.12(1).

Now we prove the other direction of the theorem, so let $p \in P$, $\{p, P\} \subseteq N \prec (H(\lambda), \in)$, N countable and we shall find q as required. Assume that P is a proper forcing, i.e., $Con_2(2^{|P|})$. Since $\lambda > 2^{|P|}$ in $H(\lambda)$ there is a sequence $\langle p_{i,j} : i < \alpha, j < \alpha_i \rangle$ where $\alpha_i \leq |P|$, $\alpha \leq 2^{|P|}$ such that for $i < \alpha$ the set $\mathcal{I}_i \stackrel{\text{def}}{=} \{p_{i,j} : j < \alpha_i\}$ varies over all the pre-dense subset of P. By $Con_2(2^{|P|})$ there is a partial algebra with universe $2^{|P|}$ such that $Sm(M) \subseteq \{s \in \mathcal{S}_{\aleph_0}(2^{|P|}) : (\exists q \geq p)\ (\forall i \in s \cap \alpha)[\{p_{i,j} : j \in s \cap \alpha_i\}$ is pre-dense over $q]\}$; necessarily $M \in H(\lambda)$. Since there are such $\langle p_{i,j} : i < \alpha, j < \alpha_i \rangle$ and M in $H(\lambda)$ there are such also in N. N obviously contains all the natural numbers. Since M is given as a mapping of ω on all partial operations of M, all these operations belong to N and hence N is closed under them (if you prefer to see M as $(|M|, F)$, F a function with domain the vocabulary of M, which is countable, it works as well). Therefore $N \cap 2^{|P|} \in Sm(M)$ and therefore there is a $q \geq p$ such that $(\forall i \in N \cap \alpha)(\{p_{i,j} : j \in N \cap \alpha_i\}$ is pre-dense over $q)$. Let $\mathcal{I}$ be any subset of P in N which is pre-dense in P. Since in $(H(\lambda), \in)$ it is true that "for every pre-dense subset of P there is an $i < \alpha$ such that $\mathcal{I} = \mathcal{I}_i$ " (since this is the way we get $\langle p_{i,j} : i < \alpha, j < \alpha_i \rangle$ in $(H(\lambda), \in))$ and this sequence belongs to N; clearly this is true in N. Therefore $\mathcal{I} = \mathcal{I}_i$ for some $i \in N \cap \alpha$. For this i, if $j \in N \cap \alpha_i$ then also $p_{i,j} \in N$ (since $\langle p_{i,j} : j \in N \cap \alpha_i \rangle$ can be taken to be one-to-one, if $p_{i,j} \in N$ also $j \in N \cap \alpha_i$). Thus $\{p_{i,j} : j \in N \cap \alpha_i\} = \{p_{i,j} : j < \alpha_i\} \cap N = \mathcal{I}_i \cap N = \mathcal{I} \cap N$ and we know that this set is pre-dense above q (since $i \in N \cap \alpha$). Thus we have shown $q \geq p$ to be (N, P)-generic.

(2) Left to the reader. $\square_{2.8}$

2.9 Discussion. We shall now present another proof of the fact that if P satisfies c.c.c. then it is proper. We shall prove that if $\underset{\sim}{\tau}$ is a name of an ordinal, $N \prec (H(\lambda), \in)$ and $\underset{\sim}{\tau} \in N$ then $\emptyset \Vdash$ "$\underset{\sim}{\tau} \in N$". There is a maximal antichain $\mathcal{I}$ of P such that for each $p \in \mathcal{I}$, $p \Vdash$ "$\underset{\sim}{\tau} = \alpha$" for a unique α. Because of the c.c.c. $|\mathcal{I}| \leq \aleph_0$ so we can take $\mathcal{I} = \{p_i : i < \alpha\}$, $\alpha \leq \omega$. The sequence $\langle p_i : i < \alpha \rangle$ is in $H(\lambda)$ and its properties can be formulated in $(H(\lambda), \in)$. Therefore there is such a sequence in N. Since $\omega \subseteq N$ we have $p_i \in N$ for every $i < \omega$, and if α_i is the ordinal such that $p_i \Vdash$ "$\underset{\sim}{\tau} = \alpha_i$", then α_i is defined in $(H(\lambda), \in)$ from P, $\underset{\sim}{\tau}$ and p_i hence $\alpha_i \in N$. As $\{p_i : i < \alpha\}$ is a maximal antichain in P we conclude $\emptyset \Vdash$ "$\underset{\sim}{\tau} \in \{\alpha_i : i \in \alpha\}$", but $\{\alpha_i : i < \alpha\} \subseteq N$ (see above) so $\emptyset \Vdash$ "$\underset{\sim}{\tau} \in N$".

So for P satisfying the c.c.c., for λ, p, N as in 2.8(1), we have: p is (N, P)-generic.

2.10 Theorem. If the forcing notion P is $\aleph_1$-complete then it is proper.

Proof. Let λ be large enough, i.e., λ regular and $\lambda > 2^{|P|}$, let $p, P \in N \prec (H(\lambda), \in)$, $|N| = \aleph_0$. Let $\langle \mathcal{I}_i : i < \omega \rangle$ be a list of all pre-dense sets which are in N. We define the sequence $\langle p_n : n < \omega \rangle$ of members of $N \cap P$ by induction on n: $p_0 = p$ and $p_{n+1} \geq p_n, r_n$ for some $r_n \in \mathcal{I}_n \cap N$. There is such a $p_{n+1} \in N$ since "p_n is compatible with some members of $\mathcal{I}_n$"; $\mathcal{I}_n$ being pre-dense in N. By the $\aleph_1$- completeness of P there is a q such that $q \geq p_n$ for all $n < \omega$. Now q is (N, P)-generic since for every pre-dense subset $\mathcal{I}_n$ of P in N, $\mathcal{I}_n \cap N$ is pre-dense above q since $q \geq r_n$, $r_n \in \mathcal{I}_n \cap N$. (Remember a set Q is pre-dense above q if for every $p \geq q$ there is a member of Q which is compatible with p, but does not have to be $\geq q$). $\square_{2.10}$

Though the following is simple it has misled some.

2.11 Theorem. Let $P \in N \prec (H(\lambda), \in)$, and let G be a generic subset of P (over V). Let $N[G] = \{\underset{\sim}{\tau}[G] : \underset{\sim}{\tau}$ is a name $\& \underset{\sim}{\tau} \in N\}$. *Then* we have $N[G] = (N[G], \in) \prec (H^{(V[G])}(\lambda), \in)$ (and $N \subseteq N[G]$ of course).

Proof. By repeating the Forcing theorems for N and $H(\lambda)$, Claim I 5.17 implies $N[G] \subseteq H(\lambda)^{V[G]} = H(\lambda)[G]$. Let $\varphi(x, y_1, \dots, y_n)$ be a first order formula. We shall prove that if $(H(\lambda), \in)^{V[G]} \models (\exists x)\varphi(x, y_1, \dots, y_n)$ for some $y_1, \dots, y_n \in N[G]$ then there is an $x \in N[G]$ such that

$$(H(\lambda), \in)^{V[G]} \models \varphi(x, y_1, \dots, y_n).$$

Then, by the Tarski-Vaught criterion we shall have $N[G] \prec (H(\lambda), \in)^{V[G]}$. Since $y_1, \ldots, y_n \in N[G]$, let $\underset{\sim}{\tau}_1, \ldots, \underset{\sim}{\tau}_n \in N$ be P-names such that $y_1 = \underset{\sim}{\tau}_1[G], \ldots, y_n = \underset{\sim}{\tau}_n[G]$. So

$$V[G] \models \text{“}(H(\lambda), \in)^{V[G]} \models (\exists x)\varphi(x, \underset{\sim}{\tau}_1, \ldots, \underset{\sim}{\tau}_n)\text{”}.$$

By the “existential completeness” of the forcing names (see I 3.1) there is a P-name $\underset{\sim}{\sigma}$ such that

$$\emptyset \Vdash \text{“}\underset{\sim}{\sigma} \in H(\lambda) \text{ and } (H(\lambda), \in)^{V[\underset{\sim}{G}]} \models (\exists x)\varphi(x, \underset{\sim}{\tau}_1, \ldots, \underset{\sim}{\tau}_n) \rightarrow \varphi(\underset{\sim}{\sigma}, \underset{\sim}{\tau}_1, \ldots, \underset{\sim}{\tau}_n)\text{”}.$$

By I 5.17 and I 5.13, there is a name $\underset{\sim}{\tau} \in H(\lambda)$ such that $\emptyset \Vdash$ “$\underset{\sim}{\sigma} = \underset{\sim}{\tau}$”, therefore $\emptyset \Vdash$ “$(H(\lambda), \in)^{V[\underset{\sim}{G}]} \models (\exists x)\ \varphi(x, \underset{\sim}{\tau}_1, \ldots, \underset{\sim}{\tau}_2) \rightarrow \varphi(\underset{\sim}{\tau}, \underset{\sim}{\tau}_1, \ldots, \underset{\sim}{\tau}_n)$”, where $\underset{\sim}{\tau} \in H(\lambda)$. Forcing statements relativized to $H^{V[G]}(\lambda)$ can be defined in $(H(\lambda), \in)$, hence $(H(\lambda), \in) \models (\exists \text{ a } P\text{-name } \underset{\sim}{\tau}) \Big[\emptyset \Vdash \text{“}(H(\lambda), \in)^{V[\underset{\sim}{G}]} \models [(\exists x\varphi(x, \underset{\sim}{\tau}_1, \ldots, \underset{\sim}{\tau}_n)) \rightarrow \varphi(\underset{\sim}{\tau}, \underset{\sim}{\tau}_1, \ldots, \underset{\sim}{\tau}_n)]\text{”}\Big]$. By the Tarski-Vaught criterion for $N \prec (H(\lambda), \in)$ there is such a name $\underset{\sim}{\tau} \in N$. Thus $V[G]$ satisfies:

$$(H(\lambda), \in) \models \text{“}(\exists x)\varphi(x, \underset{\sim}{\tau}_1[G], \ldots, \underset{\sim}{\tau}_n[G]) \rightarrow \varphi(\underset{\sim}{\tau}[G], \tau_1[G], \ldots, \underset{\sim}{\tau}_n[G])\text{”}.$$

We finish as $\underset{\sim}{\tau}_\ell[G] = y_\ell$ for $\ell = 1, \ldots, n$ and $\underset{\sim}{\tau}[G] \in N[G]$. $\square_{2.11}$

2.11A Remark.

Do we have, in 2.11, also $(N[G], N, \in) \prec (H^{V[G]}(\lambda), H(\lambda), \in)$? This holds iff $N[G] \cap H(\lambda) = N$.

2.12 Theorem. Under the assumptions of the last theorem, the following three conditions are equivalent.

(a) $G \cap N$ is N-generic, i.e., for every $\mathcal{I} \in N$ which is pre-dense in P, $\mathcal{I} \cap N \cap G \neq \emptyset$.

(b) $N[G] \cap \text{Ord} = N \cap \text{Ord}$

(c) $N[G] \cap V = N \cap V$

(d) replace in (a) pre-dense by dense

(e) replace in (a) pre-dense by maximal antichain

Proof. (a) $\Rightarrow$ (c). Let $x \in N[G] \cap V$. We shall prove $x \in N$. Since $x \in N[G]$, $x = \underset{\sim}{\tau}[G]$ for some $\underset{\sim}{\tau} \in N$. Let $\mathcal{I} \stackrel{\text{def}}{=} \{p \in P : (\exists y)(p \Vdash \text{"}\underset{\sim}{\tau} = y \text{ (i.e } \dot{y})\text{"}) \vee p \Vdash \text{"}\tau \notin V\text{"}\} = \{p \in P : (\exists y \in H(\lambda))[p \Vdash \text{"}\tau = \dot{y}\text{"}) \vee p \Vdash \text{"}\tau \notin H(\lambda)\text{"}]\}$.

(Remember that by Claim I 5.17, if $\underset{\sim}{\tau} \in H(\lambda)$ and $p \Vdash \text{"}\underset{\sim}{\tau} = y\text{"}$, then $y \in H(\lambda)$.) $\mathcal{I}$ is obviously pre-dense in P. Since $\mathcal{I}$ is definable in $(H(\lambda), \in)$ from $\underset{\sim}{\tau}$, P and $\underset{\sim}{\tau}$, $P \in N$ also $\mathcal{I} \in N$. By (a) there is a $p \in \mathcal{I} \cap N \cap G$. Since $V[G] \models \text{"}\underset{\sim}{\tau}[G] \in V\text{"}$ we cannot have $p \Vdash \text{"}\underset{\sim}{\tau} \notin H(\lambda)\text{"}$, hence for some $y \in H(\lambda)$, $p \Vdash \text{"}\underset{\sim}{\tau} = y\text{"}$ and as $p, \underset{\sim}{\tau}$ are in N and y is definable from them, necessarily $y \in N$, hence $x = \underset{\sim}{\tau}[G] = y \in N \cap H(\lambda) \subseteq N \cap V$.

(c) $\Rightarrow$ (b) is obvious.

(b) $\Rightarrow$ (a). Let $\mathcal{I} \in N$ be pre-dense in P. Let $\mathcal{J} = \{q \in P\text{: for some } p \in \mathcal{I}$ we have $p \leq q\}$. Let $\mathcal{I}^\dagger$ be a subset of $\mathcal{J}$, an antichain in P and maximal under those two conditions (for $\subseteq$). As $\mathcal{I}$ is pre-dense in P clearly $\mathcal{J}$ is dense and open hence $\mathcal{I}^\dagger$ is a maximal antichain of P. By the definition of $\mathcal{J}$, as $\mathcal{I}^\dagger \subseteq \mathcal{J}$ there is a function f from $\mathcal{I}^\dagger$ to $\mathcal{I}$ such that for every $p \in \mathcal{I}^\dagger$, $p \geq f(p)$ and $f(p) \in \mathcal{I}$. Since $\mathcal{I} \in N \prec (H(\lambda), \in)$, there is such an $\mathcal{I}^\dagger \in N$ and so w.l.o.g. ($\mathcal{I}^\dagger \in N$ and) $f \in N$, since in $H(\lambda)$ there is a sequence $\langle q_\beta : \beta < \alpha \rangle$ listing the members of $\mathcal{I}^\dagger$, there is such a sequence in N. Let $\underset{\sim}{\tau}$ be the canonical P-name such that for $\beta < \alpha$ we have: $\underset{\sim}{\tau}[G] = \beta$ if $q_\beta \in G$ (since $\mathcal{I}^\dagger$ is a maximal antichain of P there is one and only one such β). So $\underset{\sim}{\tau}$ is a P-name of an ordinal and $\underset{\sim}{\tau} \in N$. By (b) we have $\underset{\sim}{\tau}[G] \in N$. So $\gamma \stackrel{\text{def}}{=} \underset{\sim}{\tau}[G] \in N$. Since $\langle q_\beta : \beta < \alpha \rangle \in N$ also $q_\gamma \in N$. Since $\underset{\sim}{\tau}[G] = \gamma$, $q_\gamma \in G$ but $q_\gamma \in \mathcal{I}^\dagger$ so $f(q_\gamma) \in \mathcal{I}$, hence $f(q_\gamma) \in N$ (as q_γ, $f \in N$) and $f(q_\gamma) \in G$ (as $q_\gamma \geq f(q_\gamma)$ & $q_\gamma \in G$) hence clearly $f(q_\gamma) \in \mathcal{I} \cap N \cap G$ and so $\mathcal{I} \cap N \cap G \neq \emptyset$.

(e) $\Rightarrow$ (a) $\Rightarrow$ (d) Left to the reader. $\square_{2.12}$

2.13 Corollary. Assume P is a forcing notion and $P \in N \prec (H(\lambda), \in)$ and $q \in P$, then the following are equivalent:

(a) q in (N, P)-generic.

(b) $q \Vdash$ "$N[\underset{\sim}{G}_P] \cap \mathrm{Ord} = N \cap \mathrm{Ord}$".

(c) $q \Vdash$ "$N[\underset{\sim}{G}_P] \cap V = N \cap V$".

(d) for every maximal antichain $\mathcal{I}$ of P which belongs to N we have $q \Vdash$ "$N \cap \mathcal{I} \cap \underset{\sim}{G}_P \neq \emptyset$".

(e) for every dense open subset $\mathcal{I}$ of P which belongs to N we have: $q \Vdash$ "$N \cap \mathcal{I} \cap \underset{\sim}{G}_P \neq \emptyset$".

(f) $q \Vdash$ "$(N[\underset{\sim}{G}_P], N, \in) \prec (H(\lambda)^{V[\underset{\sim}{G}_P]}, H(\lambda), \in)$".

Proof. Each of the present (a) - (e) is equivalent to the statement that the corresponding condition in the last theorem holds for all generic subsets G of P which contain q. $\square_{2.13}$

§3. Preservation of Properness Under Countable Support Iteration

3.1 Definition. We call $\bar{Q} = \langle P_i, \underset{\sim}{Q}_i : i < \alpha \rangle$ (or $\langle \underset{\sim}{Q}_i : i < \alpha \rangle$) a system of countable support iterated forcing (or a CS iterated forcing system or a CS iteration) *if* the following holds (on canonical names see Definition I 5.12, Theorem I 5.13):

$$P_i = \{f : \mathrm{Dom}(f) \text{ is a countable subset of } i \text{ and } (\forall j \in \mathrm{Dom}(f))[f(j) \text{ is a canonical } P_j\text{-name and } \Vdash_{P_j} \text{“} f(j) \in \underset{\sim}{Q}_j\text{”}]\}.$$

$\underset{\sim}{Q}_i$ is a P_i-name of a forcing notion.

The partial order $\leq_i$ on P_i is defined by

$$f \leq_i g \Leftrightarrow (\forall j \in \mathrm{Dom}(f))[g{\restriction}j \Vdash \text{“} f(j) \leq_{\underset{\sim}{Q}_j} g(j)\text{”}].$$

For every $j \notin \mathrm{Dom}(f)$ we take $f(j)$ to be a name $\emptyset_j = \emptyset_{Q_j}$ of the minimal member of Q_j. Let P_α be defined like P_i above. We say: the forcing notion defined by this system is the (partial order) P_α. We say $P_\alpha = \mathrm{Lim}_{<\aleph_1} \langle \underset{\sim}{Q}_j : j < \alpha \rangle$ or $P_\alpha = \mathrm{Lim}_{<\aleph_1} |\bar{Q}|$. We may omit the "$< \aleph_1$".

Instead "$f(i)$ is a canonical P_i-name" we can use other variants.

3.1A Fact. For Definition 3.1 the parallel of II 2.2A hold (only in part (7) $(\forall \alpha < \lambda)[|\alpha|^{\aleph_0} < \lambda]$ is needed), i.e. in Definition 3.1:

(1) If $i < j \leq \alpha$ then $P_i \subseteq P_j$ as sets and even as partial orders.

(2) If $i < j \leq \alpha$ and $p \in P_j$ then $p \restriction i \in P_i$; moreover $P_j \models$ "$p \restriction i \leq p$" and if $p \restriction i \leq q \in P_i$ then $r \stackrel{\text{def}}{=} q \cup p \restriction (j \setminus i)$ belong to P_j and is the least upper bound of q, p in P_j (actually a least upper bound).

(3) If $i < j \leq \alpha$ then $P_i \lessdot P_j$ and $q \in P_i, p \in P_j \Rightarrow P_j \models q \leq p \Leftrightarrow P_i \models q \leq p \restriction i$.

(4) If $j \leq \alpha$ is a limit ordinal of uncountable cofinality then $P_j = \bigcup_{i<j} P_i$.

(5) The sequence $\langle \underset{\sim}{Q}_j : j < \alpha \rangle$ uniquely determines the sequence $\langle P_j, \underset{\sim}{Q}_j : j < \alpha \rangle$ and vice versa and similarly for $\langle P_i, \underset{\sim}{Q}_j : j < \alpha$, and $i \leq \alpha \rangle$.

(6) If $\underset{\sim}{Q}'_i$ is a P_i-name, such that $\Vdash$ "$\underset{\sim}{Q}'_i$ is a dense subset of $\underset{\sim}{Q}_i$" *then* $P'_i = \{f \in P_i$: for every $j \in \mathrm{Dom}(f)$ we have: $\Vdash_{P_i}$ "$f(j) \in \underset{\sim}{Q}'_i$"$\}$ is a dense subset of P_i. Moreover we can define and prove by induction on $i \leq \alpha$, $P''_i = \{f \in P_i$: for every $j \in \mathrm{Dom}(f)$ we have: $f(j)$ is a P''_i-canonical name of a member of $Q'_i\}$ is a dense subset of P_i and $\underset{\sim}{Q}''_i$ is a canonical P''_i-name satisfying $\Vdash_{P_i}$ "$\underset{\sim}{Q}''_i = \underset{\sim}{Q}'_i$" and $\langle P''_{j_0}, \underset{\sim}{Q}''_{j_1} : j_0 \leq i, j_1 < i \rangle$ is a FS iteration.

(6A) Assume Q'_i is a set of canonical P_i-names, such that for every P_i-name $\underset{\sim}{p}$ for some $\underset{\sim}{q} \in Q'_i$, $\Vdash_{P_i}$ "if $\underset{\sim}{p} \in Q_i$ then $Q_i \models \underset{\sim}{p} \leq \underset{\sim}{q}$". Then $P'_i = \{f \in P_i$: for every $j \in \mathrm{Dom}(f)$ we have: $f(i) \in Q'_i\}$ is a dense subset of P_i and $\langle P'_i, \underset{\sim}{Q}_i : i < \alpha \rangle$ satisfies (1) - (4) above.

(7) Moreover we can define and prove by induction on $i < \alpha$, F_i, P''_i such that

$$P''_i = \{f \in P'_i : \text{ for every } j \in \mathrm{Dom}(f), f(j) \in \mathrm{Rang}(F_j)\}$$

is a dense subset of P_i and F_i is a function with domain the P''_i-names of members of Q'_i, satisfying: $\underset{\sim}{p} \in Q'_i \Rightarrow F_i(\underset{\sim}{p})$ is a canonical P''_i-name forced to be equal to $\underset{\sim}{p}$. So letting $\underset{\sim}{Q}''_i = \{\underset{\sim}{p} : \underset{\sim}{p}$ a P''_i-name of a member of $Q_i\}$, we have: $\langle P''_i, \underset{\sim}{Q}''_i : i < \alpha \rangle$ is a countable support iteration, $P''_i \subseteq P_i$ is a dense subset.

(8) If $\bar{Q} = \langle P_i, \underset{\sim}{Q}_j : i \leq \alpha, j < \alpha \rangle$ is a CS iteration, and for $i < \alpha$, $\Vdash_{P_i}$ "$\underset{\sim}{Q}_i \in H(\lambda_i)$" and $\langle \lambda_i : i \leq \alpha \rangle$ is an increasing sequence of regulars satisfying $2^{\lambda_i} < \lambda_{i+1}$ and for limit $\delta \leq \alpha \Rightarrow (\sum_{i<\delta} \lambda_i)^{\aleph_0} < \lambda_\delta$ *then* $\bar{Q} \in H(\lambda_\alpha)$

3.1B The Definition by Induction Theorem. (one can construct $\underset{\sim}{Q}_i$'s by a given recursive recipe). If F is a function and α is an ordinal then there is a unique CS iterated forcing $\langle \underset{\sim}{Q}_j : j < \alpha_0 \rangle$ such that for all $j < \alpha_0$ $\underset{\sim}{Q}_j = F(\langle \underset{\sim}{Q}_i : i < j \rangle)$ and either $\alpha_0 = \alpha$ or else $F(\langle \underset{\sim}{Q}_i : i < \alpha_0 \rangle)$ is not suitable for $\underset{\sim}{Q}_{\alpha_0}$, i.e., it is not a name of a forcing notion in the forcing notion P_{α_0}.

Proof. This theorem is an obvious consequence of the standard definition-by-recursion theorem. $\square_{3.1B}$

3.2 Theorem. If $\langle P_i, \underset{\sim}{Q}_i : i < \alpha \rangle$ is a countable-support iterated forcing system and for each $i < \alpha$, $\Vdash_{P_i}$ "$\underset{\sim}{Q}_i$ is proper" then P_α is proper.

Remark. The reader may look at the alternative proofs presented in the book: IX §2 (for one using alternative iteration) XII §1 (the one using games) and another proof later in this section.

Proof. In Theorem 2.8(1) we showed that P is a proper forcing iff for some $\lambda > 2^{|P|}$ every countable elementary substructure N of $(H(\lambda), \in)$ such that $P, p \in N$ has a q , $p \leq q \in P$ such that q is N-generic. As easily seen from the proof or by 2.8(2) it suffices to require this only for all such N which contain some fixed member y of $H(\lambda)$.

For our present proof we choose a regular cardinal λ which is large enough with respect to $|P_\alpha|$, and the definition of the iteration and we shall show that P_α is proper by showing that for every countable elementary substructure N of $(H(\lambda), \in)$ such that $\langle P_i, \underset{\sim}{Q}_i : i < \alpha \rangle \in N$, $P_\alpha \in N$ and for all $p \in P_\alpha \cap N$ there is a q, $p \leq q \in P_\alpha$ which is N-generic. We shall show, by induction on $j \leq \alpha$ such that $j \in N$, a somewhat stronger property:

$(*)$ For all $i < j$, $i \in N$ and for all $p \in N \cap P_j$, and $q \in P_i$ if q is (N, P_i)-generic and $q \geq p{\restriction}i$ then there is an $r \in P_j$ such that r is (N, P_j)-generic, $r \geq p$ and $r \geq q$ and $r{\restriction}i = q$ (we could add $\mathrm{Dom}(r) \cap [i, j) = N \cap [i, j)$).

For $j = 0$ the statement $(*)$ is vacuously true. Now we assume $(*)$ for j and prove it for $j+1$. Since $j+1 \in N$ also $j \in N$. Therefore, since $(*)$ holds for j we may assume, without loss of generality that $i = j$. Let G_j be a generic subset of P_j which contains q. By Theorem 2.11 we have

$$N[G_j] \prec (H(\lambda), \in)[G_j] = (H(\lambda), \in)^{V[G_j]},$$

since $P_j \in N$ (because j, $\langle P_i, \underset{\sim}{Q}_i : i < \alpha\rangle \in N$ and P_j is definable in $(H(\lambda), \in)$ from j and $\langle \underset{\sim}{Q}_i : i < \alpha\rangle$) and $\underset{\sim}{Q}_j \in N$ and hence $\underset{\sim}{Q}_j[G_j] \in N[G_j]$. Remember that $\underset{\sim}{Q}_j[G_j]$ is a proper forcing in $V[G_j]$. Since $p, j \in N$ also $p(j) = p_j \in N$ and $p_j[G_j] \in \underset{\sim}{Q}_j[G_j] \cap N[G_j]$; since λ is still sufficiently large and $\underset{\sim}{Q}_j[G_j]$ is proper there is an $r_j \in \underset{\sim}{Q}_j[G_j]$ such that $r_j \geq p_j[G_j]$ and r_j is $(N[G_j], \underset{\sim}{Q}_j[G_j])$-generic. Since the only requirement we had about the generic subset G_j of P_j was that is contains q, q forces the existence of an r_j as above. By the existential completeness lemma I 3.1 there is a name $\underset{\sim}{r}_j$ such that $q \Vdash_j$ "$\underset{\sim}{r}_j \in \underset{\sim}{Q}_j \,\&\, \underset{\sim}{r}_j \geq p_j \,\&\, \underset{\sim}{r}_j$ is $(N[\underset{\sim}{G}_j], \underset{\sim}{Q}_j)$-generic" and w.l.o.g. $\Vdash_{P_j}$ "$\underset{\sim}{r}_j \in \underset{\sim}{Q}_j$". We set now $r = q \cup \{\langle j, \underset{\sim}{r}_j\rangle\}$. Obviously $r \in P_{j+1}$ and $r \restriction j = q$. Also since $q \geq p\restriction j$ and $q \Vdash$ "$\underset{\sim}{r}_j \geq p_j$" we have $r \geq p$. We still have to prove that r is (N, P_{j+1})-generic.

By the corollary in 2.13 in order to prove that r is (N, P_{j+1})-generic it suffices to prove that for all generic subsets G of P_{j+1} which contain r, $N[G] \cap \mathrm{Ord} = N \cap \mathrm{Ord}$. Let G_j be the part of G up to j i.e. $G \cap P_j$. Since $r \in G$ clearly $q \in G_j$. Since q is (N, P_j)-generic we have $N[G_j] \cap \mathrm{Ord} = N \cap \mathrm{Ord}$. Let $G^* \subseteq \underset{\sim}{Q}_j[G_j]$, be the "$j$-th component" of G. Since $r \in G$ clearly $\underset{\sim}{r}_j[G] \in G^*$. Since $\underset{\sim}{r}_j[G_j]$ is $(N[G_j], \underset{\sim}{Q}_j[G_j])$-generic we have $N[G_j][G^*] \cap \mathrm{Ord} = N[G_j] \cap \mathrm{Ord}$, and using the equality above we get $N[G_j][G^*] \cap \mathrm{Ord} = N \cap \mathrm{Ord}$. We have to observe that $N[G] \subseteq N[G_j][G^*]$, then we have $N[G] \cap \mathrm{Ord} = N \cap \mathrm{Ord}$. For every P_{j+1}-name $\underset{\sim}{\tau}$ in N there is a name $\underset{\sim}{\tau}^* \in N$ as in Lemma II 1.5 ($\underset{\sim}{\tau}^*$ is definable from $\underset{\sim}{\tau}$ and P_{j+1}, and is hence in N). By Lemma II 1.5, $\underset{\sim}{\tau}[G] = \underset{\sim}{\tau}^*[G_j][G^*] \in N[G_j][G^*]$.

Now we come to deal with the case where j is a limit ordinal. Let $\langle \underset{\sim}{\tau}_n : n < \omega\rangle$ be a sequence of all P_j-names of ordinals which are in N. Note that $N \cap j$

is a countable set with no last element, so let $i = i_0 < i_1 < i_n < \cdots, (n < \omega)$, be a sequence cofinal in it.

For $p' \in P_j$ and $q' \in P_{i_n}$ such that $q' \geq p' \restriction i_n$ let $q' \bar{\cup} p'$ denote $q' \cup (p' \restriction (j \setminus i_n))$. Since $q' \geq p' \restriction i_n$, $q' \bar{\cup} p' \in P_j$ and $q' \bar{\cup} p' \geq p_1$. For $p_1, p_2 \in P_j$ we write $p_1 \approx p_2$ for $p_1 \leq p_2 \,\&\, p_2 \leq p_1$.

We define now two sequences $\langle q_n : n < \omega \rangle$ and $\langle p_n : n < \omega \rangle$ such that $q_0 = q, p_0 = p$ and for all $n < \omega$:

(1) $q_n \in P_{i_n}$ and q_n is (N, P_{i_n})-generic

(2) $q_{n+1} \restriction i_n = q_n$

(3) $p_n \in P_j$ and $\mathrm{Dom}(p_n) \subseteq N \cap j$

(4) $q_n \geq p_n \restriction i_n$

(5) $p_{n+1} \restriction i_n = p_n \restriction i_n$ and $q_n \bar{\cup} p_{n+1} \geq p_n$

(6) $q_n \Vdash_{P_{i_n}}$ "$(\exists s \in P_j \cap N)\ (\exists q^\dagger \geq q_n)(s \restriction i_n \leq q^\dagger \ \&\, q^\dagger \in \underset{\sim}{G}_{P_{i_n}}$
$\&\, q^\dagger \bar{\cup} s \cong q^\dagger \bar{\cup} p_n)$"

(7) $q_n \bar{\cup} p_{n+1} \Vdash_{P_j}$ "$\underset{\sim}{\tau}_n \in N$"

Let us assume that q_n and p_n are defined and that they satisfy (1) - (7). We shall now define p_{n+1} and q_{n+1}. Let G be a generic subset of P_{i_n} such that $q_n \in G$. We shall see that there are $q^\dagger \in G$, s and s^* such that

(a) $q^\dagger \geq q_n$ and $s \in P_j \cap N$, $s \restriction i_n \leq q^\dagger$, $q^\dagger \bar{\cup} p_n \cong q^\dagger \bar{\cup} s$.

(b) $s^* \restriction i_n \leq q^\dagger$, $s^* \in P_j \cap N$, $s^* \geq s$ and s^* decides the value of τ_n.

By (6) there is an $s \in P_j \cap N$ and a $q^\dagger \geq q_n$, $s \restriction i_n \leq q^\dagger$ such that $q^\dagger \in G$ and $q^\dagger \bar{\cup} s \approx q^\dagger \bar{\cup} p_n$. The set $\mathcal{I}_0 = \{s^* : s^* \geq s$ and s^* decides the value of $\tau_n\}$ is obviously a pre-dense subset of P_j above s. This set belongs to N since it is definable from the parameters s and τ_n which are in N. Let $\mathcal{I}$ denote the set which consists of the restrictions of the members of $\mathcal{I}_0$ to i_n, since $\mathcal{I}_0 \in N$ and $s \in N \cap P_j$ clearly $\mathcal{I} \in N$ and $\mathcal{I}$ is a pre-dense subset of P_{i_n} above $s \restriction i_n$. Since q_n is (N, P_{i_n})-generic and $s \restriction i_n \leq q_n$ we have $\mathcal{I} \cap N$ is pre-dense above q_n. Therefore $\mathcal{I} \cap N \cap G \neq \emptyset$. Let $r \in \mathcal{I} \cap N \cap G$, then "$r$ is a restriction to i_n of a $s^* \in P_j$ such that $s^* \geq s$ and s^* decides $\underset{\sim}{\tau}_n$ and $s^* \restriction i_n \in G$" is true in $(H(\lambda)^{V[G]}, \in)$. By Tarski-Vaught's criterion there is such an s^* in $N[G]$, but $s^* \in P_j \subseteq V$ and by 2.13(c) (as $q_n \in G$, q_n is (N, P_{i_n})-generic) we know $N[G] \cap V = N$, together $s^* \in N$. Thus $r = s^* \restriction i_n \in G$, and we can take $q^\dagger$ to be $\geq s^* \restriction i_n$. Thus

$q^\dagger, s^*$ are as required by (a) and (b). Therefore $V[G] \models$ "$(\exists s^*)(\exists q^\dagger \in G)(s^*$ and $q^\dagger$ are as in (a) and (b), and s^* is the first such element in some fixed well ordering of $P_j)$". By the existential completeness lemma there is a P_{i_n}-name $\underset{\sim}{s}^*$ such that $q_n \Vdash$ "$(\exists q^\dagger \in \underset{\sim}{G})$ $[\underset{\sim}{s}^*, q^\dagger$ are as in (a) and (b) and $\underset{\sim}{s}^*$ is the least such]". Since each possible $\underset{\sim}{s}^*[G]$ is in N and it satisfies that $|\mathrm{Dom}(\underset{\sim}{s}^*)| \le \aleph_0$ in $(H(\lambda), \in)$, it satisfies this also in N, hence $\mathrm{Dom}(s^*) \subseteq N$ (since an enumeration of $\mathrm{Dom}(s^*)$ is in N). We define p_{n+1} as follows. Let $p_{n+1}\restriction i_n = p_n \restriction i_n$. For $\gamma \in j \cap N \setminus i_n$ let $p_{n+1}(\gamma)$ be the P_γ-name of the member of $\underset{\sim}{Q}_\gamma$ determined by $\underset{\sim}{s}^*$ (i.e., if $\underset{\sim}{s}^*$ is a set of pairs of members of P_{i_n} and members of $P_j \cap N$ then $p_{n+1}(\gamma) = \{\langle r, t\rangle : (\exists r^\dagger \le r)(\exists s)\ (\langle r^\dagger, s\rangle \in \underset{\sim}{s}^* \,\&\, (\bar{\exists} r'' \le r)(\langle r'', t\rangle \in s(\gamma)))\})$.

Now let us define q_{n+1}. For each $s \in P_j \cap N$ such that $s \restriction i_n \le q_n$ there is, by the induction hypothesis an $q_{n+1}(s) \in P_{i_{n+1}}$ such that $q_{n+1}(s)\restriction i_n = q_n$, $q_{n+1}(s) \ge s\restriction i_{n+1}$ and $q_{n+1}(s)$ is $(N, P_{i_{n+1}})$-generic. We define q_{n+1} as follows. The domain of q_{n+1} is the union of all the domains of the $q_{n+1}(s)$'s for $s \in N$ as above, and since N is countable the domain of q_{n+1} is countable. Let $q_{n+1}\restriction i_n = q_n$. For $i_n \le \gamma < i_{n+1}$ such that $\gamma \in \mathrm{Dom}(q_{n+1})$ if $q_n \in G$, and G is a generic subset of P_{i_n}, then $V[G] \models (\exists u)(\exists s \in P_j \cap N)\ ([q_n \bar{\cup} s \approx q_n \bar{\cup} p_{n+1}] \,\&\, u = q_{n+1}(s))$. By the existential completeness lemma there is a P_{i_n}-name $\underset{\sim}{u}$ of a $P_{i_{n+1}}$-condition such that $q_n \Vdash_{P_{i_n}}$ "$(\exists s \in P_j \cap N)$ $(q_n \bar{\cup} s \approx q_n \bar{\cup} p_{n+1} \,\&\, \underset{\sim}{u} = q_{n+1}(s))$". Now $\underset{\sim}{u}$ determines canonically a P_γ-name of a Q_γ-condition: $\underset{\sim}{u}(\gamma)$, which is taken to be the value of $q_{n+1}(\gamma)$.

We shall not present here the proof that p_{n+1} and q_{n+1} thus defined satisfy (1) - (7).

Now we define $r = \cup_{n<\omega} q_n$. Clearly r belongs to P_j. We claim that for every $n, r \ge p_n$. To prove that we have to show that for every $\gamma \in \mathrm{Dom}(p_n)$, $r\restriction\gamma \Vdash$ "$p_n(\gamma) \le r(\gamma)$". Since $\gamma \in \mathrm{Dom}(p_n)$ we have, by (3), $\gamma \in N$. Let k be minimal such that $\gamma < i_k$ then, by clause (4) we have $q_k \ge p_k \restriction i_k$, hence $q_k\restriction\gamma \Vdash$ "$p_k(\gamma) \le q_k(\gamma)$". By the definition of r, $q_k\restriction\gamma = r\restriction\gamma, q_k(\gamma) = r(\gamma)$. Also, by (5) if $n \ge k$ then $p_k(\gamma) = p_n(\gamma)$, hence $r\restriction\gamma \Vdash_{P_\gamma}$ "$p_n(\gamma) \le r(\gamma)$". So assume $n < k$ (hence $k > 0$); now for $\ell < k$ we have: $q_\ell \bar{\cup} p_{\ell+1}\restriction\gamma \Vdash_{P_\gamma}$ "$p_\ell(\gamma) \le_{Q_\gamma} p_{\ell+1}(\gamma)$" (by clause (5)) but $q_k\restriction\gamma \ge q_\ell \bar{\cup} p_{\ell+1}\restriction\gamma$ for $\ell < k$ (by clauses (4) and (5)) hence $q_k\restriction\gamma \Vdash_{P_\gamma}$ "$p_0(\gamma) \le p_1(\gamma) \le \ldots \le p_k(\gamma)$", hence $q_k\restriction\gamma \Vdash_{P_\gamma}$ "$p_n(\gamma) \le p_k(\gamma)$", but we have proved above $q_k{\restriction}\gamma \Vdash_{P_\gamma}$ "$p_k(\gamma) \le q_k(\gamma)$", hence $q_k{\restriction}\gamma \Vdash_{P_k}$ "$p_n(\gamma) \le q_k(\gamma)$". However $r{\restriction}\gamma = q_k{\restriction}\gamma$ and $q_k(\gamma) = r(\gamma)$ (as $\gamma < i_k$) hence this means $r{\restriction}\gamma \Vdash_{P_\gamma}$ "$p_n(\gamma) \le r(\gamma)$" as required. So we have really proved $P_j \models$ "$p_n \le r$". Thus, by (7), $r \Vdash$ "$\underset{\sim}{\tau}_n \in N$" and therefore r is (N, P_j)-generic which finishes our proof. $\square_{3.2}$

3.3 Alternative proof of 3.2.

3.3A Advice to the reader. There are situations where it is enough to understand and believe the statement of a theorem (as opposed to its proof). For example, we took this attitude in Chapter 1 when we discussed the fundamental theorem of forcing.

However, this approach should not be used here. Not only is the preceding theorem basic for the theory to be developed in the rest of the book, it is (in the author's opinion) also essential for the reader to understand the proof, since variations and extensions of this proof will appear throughout the book.

To help the reader understand the proof of the Theorem 3.2 better we now give a reformulation of this proof which is due to Goldstern [Go]. This version emphasizes the fact that the conditions p_n are in N by constructing the whole sequence $(p_n : n < \omega)$ *before* constructing the generic conditions q_n.

N is an elementary submodel of some $H(\chi)$ for some large χ containing $\langle P_\alpha, Q_\alpha : \alpha < \varepsilon\rangle$.

3.3B Fact. If $\beta > \alpha$, $q \in P_\alpha$, $p \in P_\beta$, $q \ge^* p{\restriction}\alpha$, then $Q^+ \stackrel{\text{def}}{=} q \cup p{\restriction}[\alpha, \beta)$ is in P_β, and $q^+ \ge^* p$ (i.e., $q^+ \Vdash p \in G$).

3.3C "Existential Completeness Lemma". For any forcing P, and any condition $p \in P$, any formula $\varphi(x)$:

$$p \Vdash \exists\, x\ \varphi(x) \quad \textit{iff} \quad \text{there is a name } \underset{\sim}{\tau} \text{ such that } p \Vdash \varphi(\underset{\sim}{\tau}).$$

Proof. By I 3.1.

3.3D Preliminary Lemma. (This lemma does not require properness.) Assume $\alpha_1 \leq \alpha_2 \leq \beta$, $\underset{\sim}{p}_1$ is a P_{α_1}-name for a condition in P_β. Let $\mathcal{I}$ be a dense open set of P_β. Then $\emptyset_{P_{\alpha_2}} \Vdash_{P_{\alpha_2}} \exists p_2 \varphi(p_2)$, where $\varphi(p_2)$ is the conjunction of the following clauses:

(1) $p_2 \in P_\beta$, $p_2 \geq^* \underset{\sim}{p}_1$.

(2) $p_2 \in \mathcal{I}$.

(3) If $p_1 \restriction \alpha_2 \in G_{\alpha_2}$, then $p_2 \restriction \alpha_2 \in G_{\alpha_2}$.

3.3E Remark. By the existential completeness lemma there is an α_2-name $\underset{\sim}{p}_2$ for a condition in P_β such that $\Vdash_{P_{\alpha_2}} \varphi(\underset{\sim}{p}_2)$.

3.3F Remark. The P_{α_1}-name $\underset{\sim}{p}_1$ corresponds naturally to a P_{α_2}-name, which we also call $\underset{\sim}{p}_1$.

Proof. Assume not, then there exists a condition $r \in P_{\alpha_2}$ such that

$$r \Vdash \text{"there is no } p_2 \text{ satisfying (1)–(3)".}$$

We may assume that r decides what $\underset{\sim}{p}_1$ is, (i.e. $r \Vdash \underset{\sim}{p}_1 = p_1$ for some $p_1 \in V$), and r also decides whether $p_1 \restriction \alpha_2 \in G_{\alpha_2}$.

Case 1. $r \Vdash p_1 \restriction \alpha_2 \notin G_{\alpha_2}$:

But then (3) is true for any p_2, so

$$r \Vdash \text{"there is no } p_2 \text{ satisfying (1)–(2)"}$$

which is a contradiction since $\mathcal{I}$ is a dense open.

Case 2. $r \Vdash p_1 \restriction \alpha_2 \in G_{\alpha_2}$, i.e. $r \geq^* p_1 \restriction \alpha_2$. Now let $r' = r \cup p_1 \restriction [\alpha_2, \beta) \geq^* p_1$, and find $r'' \in D$, $r'' \geq r'$. Then

$$r'' \restriction \alpha_2 \Vdash r'' \text{ satisfies (1)–(3),}$$

again a contradiction, because $r'' \restriction \alpha_2 \geq r$.

3.3G "Composition Fact". $q \in P_{\alpha+1}$ is $(P_{\alpha+1}, N)$-generic iff:

$$q{\restriction}\alpha \text{ is } (P_\alpha, N)\text{-generic, and } q{\restriction}\alpha \Vdash \text{"}q(\alpha) \text{ is } (Q_\alpha, N[G_\alpha \cap N])\text{-generic."}$$

Proof. See §2.

3.3H Induction Lemma. For all $\beta \in N \cap \varepsilon$, for all $\alpha \in N \cap \varepsilon$, all $\underset{\sim}{p} \in N$ assume $\underset{\sim}{p}$ is a P_α-name for a condition in P_β, and

(a) $q \in P_\alpha$

(b) q is (P_α, N)-generic.

(c) $q \Vdash_{P_\alpha}$ "$\underset{\sim}{p}{\restriction}\alpha \in G_\alpha \cap N$".

Then there is a condition q^+:

(a)$^+$ $q^+ \in P_\beta$, $q^+{\restriction}\alpha = q$

(b)$^+$ q^+ is N-generic

(c)$^+$ $q^+ \Vdash_{P_\beta}$ "$\underset{\sim}{p} \in G_\beta \cap N$".

(Note that "q is N-generic" implies already "$q \Vdash \underset{\sim}{p} \in N$", so the main point of (c) is to say that $q \Vdash p{\restriction}\alpha \in G_\alpha$)

(For $\alpha = 0$ this shows that P_β is proper.)

Proof.

The proof is by induction on β.

Successor step.

Let $\beta = \beta' + 1$. Since we can first use the induction hypothesis on α, β'' to extend q to a condition $q' \in P_{\beta''}$ satisfying the appropriate version of (a)–(c), we may simplify the notation by assuming $\beta = \alpha + 1$.

Clearly $q \Vdash_{P_\alpha}$ "$N[G_\alpha] \prec H(\chi)^{V[G_\alpha]}$", $q \Vdash_{P_\alpha}$ "there is a $(Q_\alpha, N[G_\alpha])$-generic condition $\geq \underset{\sim}{p}(\alpha)$". By "existential completeness", there is a P_α-name $q^+(\alpha)$ for it. By the "composition fact", we are done.

Limit step.

Let $\beta \in N$ be a limit ordinal, $\beta \in N = \bigcup \alpha_n$, $\alpha_0 = \alpha$, $\alpha_n \in N$. Let $\langle \mathcal{I}_n : n < \omega \rangle$ enumerate all dense subsets of P_β that are in N.

First we will define a sequence $\langle \underset{\sim}{p}_n : n < \omega \rangle$, $\underset{\sim}{p}_n \in N$ such that *in* N the following will hold:

(0) $\underset{\sim}{p}_n$ is a P_{α_n}-name for a condition in P_β

(1) $\Vdash_{P_{\alpha_{n+1}}} \underset{\sim}{p}_{n+1} \geq^* \underset{\sim}{p}_n$

(2) $\Vdash_{P_{\alpha_{n+1}}} \underset{\sim}{p}_{n+1} \in \mathcal{I}_n$

(3) $\Vdash_{P_{\alpha_{n+1}}}$ "If $\underset{\sim}{p}_n \restriction \alpha_{n+1} \in G_{\alpha_{n+1}}$ then $\underset{\sim}{p}_{n+1} \restriction \alpha_{n+1} \in G_{\alpha_{n+1}}$".

For each n we thus get a name $\underset{\sim}{p}_n$ that is in N. For each n we can use the "preliminary lemma" (and Remark 3.3E before its proof) in N to obtain $\underset{\sim}{p}_{n+1}$. Now we define a sequence $\langle q_n : n < \omega \rangle$, $q_n \in P_{\alpha_n}$, and q_n satisfies (a), (b), (c) (if we write q_n for q, p_n for p, and α_n for α).
$q_{n+1} = q_n^+$ can be obtained by the induction hypothesis, applied to α_n, α_{n+1}, and $\underset{\sim}{p}_n \restriction \alpha_{n+1}$. By (c)$^+$ we know

$$q_n^+ \Vdash_{P_{\alpha_n}} \text{“}(\underset{\sim}{p}_n[G_{\alpha_n}]) \restriction \alpha_{n+1} \in G_{\alpha_{n+1}}\text{”}.$$

Hence, by (3) and the genericity of q_{n+1} we have

$$q_{n+1} \Vdash_{P_{\alpha_{n+1}}} \text{“}(\underset{\sim}{p}_{n+1}[G_{\alpha_{n+1}}]) \restriction \alpha_{n+1} \in G_{\alpha_{n+1}} \cap N\text{”}.$$

Since $q_{n+1} \restriction \alpha_n = q_n$, $q = \lim q_n$ exists and is $\geq q_n$ for all n.

We have to show that $q \Vdash p \in G_\beta \cap N$ and that q is generic. Let G_β be a generic filter containing q. We will write p_n for $\underset{\sim}{p}_n[G_{\alpha_n}]$. (Note that $p_n \in N$, because q_n was N-generic and $q_n \in G_{\alpha_n}$.) Since $q_n \in G_\beta$, we have $p_n \restriction \alpha \in G_{\alpha_n} \cap N$ and $N \models p_n \geq^* p_{n-1} \geq^* \ldots \geq^* p_0$. Hence $p \restriction \alpha_n \in G_{\alpha_n} \cap N$ for all n, and therefore $p \in G_\beta \cap N$. Similarly, $p_n \in G_\beta$ for all n.
Consider a dense set $\mathcal{I}_n \subseteq P_\beta$. Since $q_{n+1} \Vdash \underset{\sim}{p}_{n+1} \in \mathcal{I}_n$, we have $p_{n+1} \in G_\beta \cap \mathcal{I}_n \cap N$. Hence q is generic. $\square_{3.3}$

More advice to the reader. It may also be helpful to look at the proof in Chapter XII, §1, which uses games. (Chapter XII, §1 can be read independently of chapters IV to XI.)

3.4 The General Associativity Theorem. Suppose $\langle P_i, \underset{\sim}{Q}_i : i < \alpha \rangle$ is a CS iterated forcing system each $\underset{\sim}{Q}_i$ proper then the parallel to II 2.4 holds.

Proof. Left to the reader. $\square_{3.4}$

3.5 Theorem. Suppose $\langle \underset{\sim}{Q}_j : j < \alpha \rangle$ is a $(< \kappa)$-support iterated forcing, $P_j = \mathrm{Lim}\langle \underset{\sim}{Q}_i : i < j \rangle$.

If $\Vdash_{P_j}$ "$\underset{\sim}{Q}_j \lessdot \underset{\sim}{Q}_j^\dagger, \underset{\sim}{Q}_j$ a dense subset of $\underset{\sim}{Q}_j^\dagger$" and $P_j^\dagger = \mathrm{Lim}\langle \underset{\sim}{Q}_i^\dagger : i < j \rangle$, *then* $P_j \lessdot P_j^\dagger$ is a dense subset of $P_j^\dagger$.

Remark. By Lemma I 5.1 (a), we can replace $\underset{\sim}{Q}_j$ by any equivalent $\underset{\sim}{Q}_j^\dagger$ (just use 3.5 a few times).

Proof. Left to the reader. $\square_{3.5}$

§4. Martin's Axiom Revisited

Why is c.c.c. forcing so popular? I think the main reason is that such forcing notions preserve cardinalities and cofinalities, so why shall we not be interested in the property "P does not collapse cardinals " instead "P satisfies the c.c.c. ". In particular Magidor and Stavi had wondered on the role of the c.c.c. mainly in MA and asked:

"Is it consistent that for any forcing notion P of power $\aleph_1$ not collapsing cardinals (i.e., $\aleph_1$) and dense $\mathcal{I}_i \subseteq P$ (for $i < \aleph_1$) there is a directed $G \subseteq P$, such that $G \cap \mathcal{I}_i \neq \emptyset$ for $i < \aleph_1$?"

In particular Baumgartner, Harrington and Kleinberg [BHK] proved that if $S \subseteq \omega_1$ is stationary co-stationary, and CH holds, then there is a forcing notion $P_S = \{C : C \text{ a countable closed subset of } S\}$ with the order $C_1 \leq C_2$ iff $C_1 = C_2 \cap (\mathrm{Sup} C_1 + 1)$ which does not change cardinalities and cofinalities and which collapse S (i.e., collapse its stationarity, i.e., $\Vdash_{P_S}$ "$S \subseteq \omega_1$ is not stationary ".

So why not include such forcing in MA? Because we can find pairwise disjoint stationary sets $S_n \subseteq \omega_1, \omega_1 = \cup_{n<\omega} S_n$. If we make each S_n in turn not stationary, ω_1 must be collapsed. More exactly, if we try to iterate the forcings P_{S_n}, after ω steps $\aleph_1$ collapses, no matter how the limit is taken. It does not matter if we look at the desired version of MA, in some V and let $\mathcal{I}^n_\alpha = \{C \in P_{S_n} : \sup(C) \geq \alpha\}$. Thus if $G \cap \mathcal{I}^n_\alpha \neq \emptyset$ for $n < \omega, \alpha < \omega_1$, then in V, each S_n is not stationary.

You can still argue that CH is the cause of the problem but we shall prove in Theorem 4.4 that even $2^{\aleph_0} > \aleph_1, S \subseteq \omega_1$ stationary co-stationary there is a forcing notion P of power $\aleph_1$, not changing cardinalities and cofinalities but still collapsing S.

So it is natural to change the question to "P of power $\aleph_1$, not collapsing stationary subsets of $\aleph_1$", and we shall answer it positively, assuming there is a model V of ZFC with a strongly inaccessible cardinal.

The natural scheme is to iterate (by CS iteration) proper forcing of power $\aleph_1$, in an iteration of length ω_2. However to prove the consistency of almost anything by iterating proper forcing we usually have to prove the κ-chain condition is satisfied, where κ will be the new $\aleph_2$ and the length of the iteration. We have a problem even if $|P| = \aleph_1, \Vdash_P$ "$|\underset{\sim}{Q}| = \aleph_1$", $P * \underset{\sim}{Q}$ have a large power because of the many names. We can overcome this either by using κ strongly inaccessible, or showing that the set of names which are essentially hereditarily countable is dense.

Another problem is that "not destroying stationary subsets of ω_1" is not the same as "proper". However we shall prove that if P is not proper, then $\Vdash_{\mathrm{Levy}(\aleph_1, 2^{|P|})}$ "P destroys a stationary subset of ω_1". So instead of "honestly" dealing with a candidate P i.e., a forcing notion which does not destroy stationary subsets of ω_1, but is not proper we cheat and make it to destroy a stationary subset of ω_1.

4.1 Theorem. Suppose $\bar{Q} = \langle P_i, \underset{\sim}{Q}_i : i < \kappa \rangle$ is a CS iteration $\Vdash_{P_i}$ "$\underset{\sim}{Q}_i$ is a proper forcing notion which has power $< \kappa$", κ is regular and $(\forall \mu < \kappa)\mu^{\aleph_0} < \kappa$.

Then $P_\kappa = \lim\bar{Q}$ *satisfies the* κ*-c.c., and each* $P_i (i < \kappa)$ *even has a dense subset of power* $< \kappa$. *Hence for* $i < \kappa, \Vdash_{P_i}$ "$2^{\aleph_0} < \kappa$".

Proof. Easily (twice use 3.4), w.l.o.g. the set of elements of $\underset{\sim}{Q}_i$ is a cardinal $\underset{\sim}{\mu}_i < \kappa$; (i.e., $\underset{\sim}{\mu}_i$ is a P_i-name of a cardinal $< \kappa$).

4.1A Definition. For a forcing notion P and P-name $\underset{\sim}{Q}$ of a forcing notion with set of elements $\underset{\sim}{\mu}$ (a P-name of cardinal) with minimal element $\emptyset_{\underset{\sim}{Q}}$ (can demand it to be 0) we define a hereditary countable P-name of a member of $\underset{\sim}{Q}$: it is the closure of the set of ordinals $< \underset{\sim}{\mu}$ (see (*) below) by the two operations (a) and (b) (see below):

(*) the names $\dot{\alpha}$ for $\alpha < \underset{\sim}{\mu}$ or more exactly for an ordinal α the P-name $\underset{\sim}{\tau}_\alpha$ is such that $\underset{\sim}{\tau}_\alpha[G] = \alpha$ if $\alpha < \underset{\sim}{\mu}[G_P]$ and $\underset{\sim}{\tau}_\alpha[G] = \emptyset_Q[G_P]$ if $\alpha \geq \underset{\sim}{\mu}[G_P]$. Of course we can restrict to $\underset{\sim}{\tau}_\alpha$ such that $\nVdash_P \alpha \geq \underset{\sim}{\mu}$. Also if $\underset{\sim}{\mu} = \mu$ we can use just $\dot{\alpha}$, $\alpha < \mu$

(a) if $\underset{\sim}{\tau}_n (n < \omega)$ are such names, and $p_n \in P$ (for $n < \omega$) *then* let $\underset{\sim}{\tau}$ be the $\underset{\sim}{\tau}_n$ for the least n satisfying $p_n \in \underset{\sim}{G}_P$, and $\emptyset_{\underset{\sim}{Q}}$ if there is no such n.

(b) if $\underset{\sim}{\tau}_{n,m} (n < \omega, m < \omega)$ are such names, let $\underset{\sim}{\tau}$ be the least ordinal $\alpha < \underset{\sim}{\mu}$ such that for every n, $\{\underset{\sim}{\tau}_{n,m} : m < \omega\}$ is pre-dense over τ (in Q); and $\emptyset_{\underset{\sim}{Q}}$ if there is no such α. (Remember: the members of $\underset{\sim}{Q}$ are ordinals $< \underset{\sim}{\mu}$).

We shall prove by induction on $\xi \leq \kappa$ that P_ξ satisfies the κ-chain condition.

Suppose this holds for every $\zeta < \xi$, so for $\zeta < \xi$ by Claim I 3.7 clause (ii) we have $< \kappa$ possible values for $\underset{\sim}{\mu}_\zeta$, each is $< \kappa$ so

$$\mu_\zeta \overset{\text{def}}{=} \sup\{\mu : \nVdash_{P_\zeta} \text{“}\mu \neq \underset{\sim}{\mu}_\zeta\text{”}\}$$

is $< \kappa$ (as κ is regular). So for $\zeta < \xi$ w.l.o.g. $\underset{\sim}{\mu}_\zeta = \mu_\zeta$ (as we can add to $\underset{\sim}{Q}_\zeta$ the ordinals i, $\underset{\sim}{\mu}_\zeta \leq i < \mu_\zeta$ such that i is $\approx$ to the minimal element $\emptyset_{Q_i}$). Let us define by induction on ζ, $P^\dagger_\zeta = \{f : f$ a function with domain a countable subset of ζ, $f(i)$ is a hereditarily countable $P_i^\dagger$-name of an ordinal $< \mu_i\}$. Let $P_\xi^\dagger \subseteq P_\xi$ inherit its order. We now can prove by induction on $\zeta \le \xi$, that $P_\zeta^\dagger$ is a dense subset of P_ζ, using the proof that properness is preserved by CS iteration. It is clear that $|P_\xi^\dagger| \le |\xi|^{\aleph_0}$, so for $\xi < \kappa$, P_ξ has a dense subset of power $< \kappa$. So we finish.

For $\xi = \kappa$, if $p_i \in P_\kappa$ for $i < \kappa$, clearly $S \overset{\text{def}}{=} \{i < \kappa : \text{cf}(i) = \aleph_1\}$ is stationary, $f(i) = \text{Sup}[i \cap \text{Dom}(p_i)] < i$ is a pressing down function, hence by Fodor Lemma on some stationary $S_1 \subseteq S, h$ has a constant value γ. There is a closed unbounded $C \subseteq \kappa$ such that: if $\beta \in C, \alpha < \beta$, then $\text{Dom}(p_\alpha) \subseteq \beta$. So $S_1 \cap C$ is still stationary, hence has power κ and for $\alpha, \beta \in C \cap S_1, p_\alpha, p_\beta$ are compatible iff $p_\alpha \restriction \gamma, p_\beta \restriction \gamma$ are compatible (in P_γ or P_κ, does not matter). But we have proved that P_γ satisfies the κ-chain condition, so we finish. $\square_{4.1}$

We have proved

4.1B Claim. For $\bar{Q} = \langle P_i, \underset{\sim}{Q}_i : i < \alpha \rangle$, a *CS* iteration of proper forcing, such that for each $i < \alpha$ it is forced that $\underset{\sim}{Q}_i$ is with set of elements $\subseteq$ Ord, we have, for $i < \alpha$:

1) $P_i' = \{f \in P_i$: for $j \in \text{Dom}(f)$, $f(j)$ is a hereditarily countable P_j'-name$\}$ is a dense subset of P_i, and $i < j \le \alpha$ and $f \in P_j' \Rightarrow f \restriction i \in P_i'$
2) If $\underset{\sim}{f}$ is a $\underset{\sim}{P}_\alpha$-name of a function from ω to Ord and $\text{cf}(\alpha) > \aleph_0$ *then* for a dense open set of $q \in P_\alpha$, for some $\beta < \alpha$ and P_β-name $\underset{\sim}{g}$ of a function from ω to Ord, $q \Vdash_{P_\alpha}$ "$\underset{\sim}{f} = \underset{\sim}{g}$".

4.2 Theorem. Suppose P is not proper, then there is an $\aleph_1$-complete forcing notion Q, in fact $Q = \text{Levy}(\aleph_1, 2^{|P|})$ will do, such that $|Q| \le 2^{|P|}$ and $\Vdash_Q$ "P collapses some stationary $S \subseteq \omega_1$".

Proof. As P is not proper, there is a stationary $S \subseteq \mathcal{S}_{\aleph_0}(\mu)$ which P destroys, $\aleph_0 < \mu \le 2^{|P|}$. So there are P-names $\underset{\sim}{F}_n^\ell$ of n-place functions from μ to μ, such that $\Vdash_P$ "$Sm((\mu, \underset{\sim}{F}_0^\ell, \dots)) \cap S = \emptyset$". Let $Q = \text{Levy}(\aleph_1, \mu) = \{f : f$ a function from some $\alpha < \omega_1$ into $\mu\}$.

Fact A. $\Vdash_Q$ "S is a stationary subset of $\mathcal{S}_{\aleph_0}(\mu)$".

This is because Q is $\aleph_1$ -complete hence, by 2.10, proper.

Fact B. The statement $\Vdash_P$ "$S \subseteq \mathcal{S}_{\aleph_0}(\mu)$ is not stationary " is absolute, i.e., if it holds in V it holds in V^Q.

We just have to check that the P-names $\underset{\sim}{F}^{\ell}_{n}$ continue to satisfy the suitable requirement (and Q adds no new member to P and no new member to S).

Fact C. $\Vdash_Q$ " the ordinal μ has power $\aleph_1$".

This is trivial.

Fact D. If forcing by P destroys a stationary subset of $\mathcal{S}_{\aleph_0}(A)(A = \mu$ in our case), A of power $\aleph_1$, *then* forcing by P destroys some stationary subset of ω_1. (follows from Lemma 1.5 and 1.12(3)). $\square_{4.2}$

4.3 Theorem. Suppose ZFC has a model with a strongly inaccessible cardinal κ. Then ZFC has a model in which $2^{\aleph_0} = \aleph_2$ and

$(*)$ If P is a forcing notion of power $\aleph_1$ not destroying stationary subsets of ω_1, and $\mathcal{I}_i \subseteq P$ is dense for $i < \omega_1$ *then* there is a directed $G \subseteq P$ satisfying $G \cap \mathcal{I}_i \neq \emptyset$ for $i < \omega_1$.

Proof. Notice that *if* $V \models$ "$\kappa > \aleph_0 \,\&\, \kappa^{<\kappa} = \kappa \,\&\, |P| \leq \kappa \,\&\, (\exists\lambda \leq \kappa)[P$ has the λ-c.c.]" *then* $V^P \models$ "$\kappa^{<\kappa} = \kappa \,\&\, \kappa > \aleph_0$".

This is proved exactly as the parallel fact in Theorem II 3.4. Now let $\{S_\alpha : \alpha < \kappa\}$ be a partition of κ to κ sets such that $\beta \in S_\alpha \Rightarrow \beta \geq \alpha$, and $|S_\alpha| = \kappa$. Define by induction on $i < \kappa$ a CS iterated forcing system $\langle P_i, \underset{\sim}{Q}_i : i < \kappa\rangle$. Let $\langle \underset{\sim}{\leq}_\xi : \xi \in S_\alpha\rangle$ be a list of the canonical P_α-names of partial orders on ω_1. The induction hypothesis for $i < \kappa$ is:

(1) Q_j is proper for $j < i$.

(2) the density of P_i is $< \kappa$ (i.e. it has a dense subset of cardinality $< \kappa$).

Assuming $\langle P_j, \underset{\sim}{Q}_j : j < i\rangle$ is already defined, let

$$\underset{\sim}{Q}_i = \begin{cases} \langle \omega_1, \underset{\sim}{\leq}_i \rangle & \text{if } \Vdash_{P_i} \text{“}(\omega_1, \underset{\sim}{\leq}_i) \text{ is proper”} \\ \text{Levy}(\aleph_1, 2^{\aleph_1}) & \text{otherwise} \end{cases}$$

where Levy$(\aleph_1, 2^{\aleph_1})$ means “$2^{\aleph_1}$ and the Levy collapse are interpreted in V^{P_i}”, so are P_i-names. Clearly P_i is proper (by Theorem 3.2 and remembering that Levy$(\aleph_1, 2^{\aleph_1})$ is $\aleph_1$-complete hence by Theorem 2.10 proper). We still have to check that density$(P_i) < \kappa$ but it is easy, note that we use Theorem 4.1. Finally also P_κ is proper by 3.2 and (again by Theorem 4.1) satisfy the κ-c.c. which makes it possible to prove $V^{P_\kappa} \models (*)$ exactly as in the proof of Theorem II 3.4, but using 4.2 above. $\square_{4.3}$

Note that in view of Theorem 4.1 we have a parallel of MA for proper forcing without assuming an inaccessible. We now return to a promise.

4.4 Theorem. Suppose $S \subseteq \omega_1$ is stationary co-stationary (i.e., also $\omega_1 \setminus S$ is stationary too). *Then* there is a forcing notion P_S which shoots a closed unbounded $C \subseteq S$ (i.e., add such a set) without collapsing cardinals (or changing cofinalities).

Remark. So we cannot answer Magidor, Stavi's question positively in the original version.

Remark. Assuming CH this was done by Baumgartner, Harrington, Kleinberg [BHK]. Without CH, Abraham [A] and Baumgartner [B3] introduce forcing notions which add a new closed unbounded subset of ω_1 (for different purposes). We can adapt each for proving 4.4., and will use a forcing similar to Abraham's.

Proof. Let $P = \text{Levy}(\aleph_0, < \aleph_1)$. So P is essentially adding $\aleph_1$ Cohen reals. If $G_P \subseteq P$ is (directed and) generic over V, then G_P is also generic over L (the constructible universe) as $P \in L$ and also over $L[S]$. By Theorem I 6.7 the forcing P satisfies the countable chain condition and $\aleph_1^{L[G_P]} = \aleph_1^V = \aleph_1^{V[G_P]}$ and $V, V[G_P]$ have the same cardinals and cofinalities. Let

$Q = \{C : C$ a closed bounded subset of S which belongs to $L[S, G_P]\}$

$C_1 \leq C_2$ iff $C_1 = C_2 \cap (\mathrm{Max}(C_1) + 1)$

Clearly Q is a forcing notion of power $\aleph_1$, so it cannot collapse cardinals or regularity of cardinals except possibly $\aleph_1$ (all finite subsets of S belong to Q). So we shall prove that $P * \underset{\sim}{Q}$ does not collapse $\aleph_1$. So let (in V) $N \prec (H(\lambda), \in)$, N countable, $(p, \underset{\sim}{q}) \in P * \underset{\sim}{Q} \in N, (p, \underset{\sim}{q}) \in N$, $\delta \stackrel{\text{def}}{=} N \cap \omega_1 \in S$ and suppose $G_P \subseteq P$ is generic over V and $p \in G_P$. Note that as S is a stationary subset of ω_1, there is such N (in V). So it is enough to find $(q', \underset{\sim}{p}') \geq (q, \underset{\sim}{p})$ which is $(N, P * \underset{\sim}{Q})$-generic. As P satisfies the countable chain condition, p is (N, P)-generic (by 2.9), hence $N[G_P] \cap V = N$. Clearly $Q[G_P] \in L[S, G_P] \subseteq V[G_P]$, now $N[G_P]$ does not necessarily belongs to $L[S, G_P]$, but $N[G_P] \cap L[S, G_P]$ is $N[G_\gamma] \cap L_\delta[S, G_P] = L_\delta[S, G_P] \in L[S, G_P]$ and is a countable set in $L[S, G_P]$. In $L[S, G_P]$ we have an enumeration of $\underset{\sim}{Q}[G_P] \cap N[G_P]$ (of length ω), say $\langle q_n : n < \omega \rangle$ (but not of the set of dense subsets); in fact we have it even in $L[S, G_P \restriction (\delta + 1)]$ (and as we use $\mathrm{Levy}(\aleph_0, < \aleph_1)$ not $\mathrm{Levy}(\aleph_0, < \kappa)$ even in $L[S, G_P \restriction \delta])$. Now in $L[S, G_P]$ there is a Cohen generic real over $V[G_P \restriction (\delta + 1)]$ say $r^* \in {}^\omega\omega$ and we use it to construct a sequence $\bar{C} = \langle C_n : n < \omega \rangle$ such that $C_n \in \underset{\sim}{Q}[G_P]$ in $L[S, G_P]$ i.e. $\bar{C} \in L[S, G_P]$, e.g. we choose C_n by induction on n; we let $C_0 = \underset{\sim}{q}[G_P]$ and we let C_{n+1} be $q_{m(n)}$ where $m(n)$ is the first natural number m such that: $m \geq r^*(n)$ and $\underset{\sim}{Q}[G_P] \models$ "$C_n \leq q_m$". Let $G \stackrel{\text{def}}{=} \{q : q \in \underset{\sim}{Q}[G_P]$, and $q \in N[G_P]$ and for some n, $\underset{\sim}{Q}[G_P] \models q \leq C_n\}$. Clearly $G \subseteq N[G_P] \cap \underset{\sim}{Q}[G_P]$ is generic over $V[G_P \restriction (\delta + 1)]$. So $q^\dagger = \bigcup_n C_n \cup \{\delta\} \in \underset{\sim}{Q}[G_P]$ is $(N[G_P], \underset{\sim}{Q}[G_P])$-generic. Going to names we finish. $\square_{4.4}$

Remark. Also N in $V[G_P]$ is O.K. as it still belongs to some $V[G_P \restriction \alpha]$ for some $\alpha < \omega_1$.

§5. On Aronszajn Trees

5.1 Definition. 1) A cardinal κ is said to have the *tree* property if every tree of height κ in which every level has $< \kappa$ members has a branch of length κ. A tree which is a counterexample to the tree property of κ is called a *κ-Aronszajn tree.* By the König infinity lemma $\aleph_0$ has the tree property.

2) A κ-Aronszajn tree in which every antichain is of cardinality $< \kappa$ is called a *κ-Souslin tree.* An $\aleph_1$-Aronszajn tree, and an $\aleph_1$-Souslin tree are called an Aronszajn and a Souslin tree, respectively. A λ^+-Aronszajn tree is said to be *special* if it is the union of λ antichains, if $\lambda = \aleph_0$ we may omit it. A special Aronszajn tree cannot be Souslin, since in a Souslin tree every antichain is countable, hence the tree, being uncountable, cannot be the union of $\aleph_0$ antichains. A λ-wide Aronszajn tree is a tree with ω_1 levels, λ nodes and no ω_1-branch.

5.1A Remark. It is easy to show that an Aronszajn tree T is special iff there is a function $f : T \to \mathbb{Q}$ which is order preserving. A λ^+-tree which is special is a λ^+-Aronszajn tree. The following was proved by Aronszajn, [Ku35] (and 5.3 is a well known generalization).

5.2 Theorem. There is a special Aronszajn tree T.

Proof. The members of the α-th level T_α of T will be increasing bounded sequences of rational numbers (of length α) with $\trianglelefteq$ as the tree relation. When we come to define T_α we assume that for all $\beta < \gamma < \alpha$ and for all $x \in T_\beta$ and every rational $q > \mathrm{Sup}(\mathrm{Rang}(x))$ there is a $y \in T_\gamma$ such that $x \triangleleft y$ and $\mathrm{Sup}(\mathrm{Rang}(y)) = q$, and $|T_\beta| = \aleph_0$. If $\alpha = 0$ take $T_\alpha = \{<>\}$. If $\alpha = \gamma + 1$ take $T_\alpha = \{z\hat{\ } < q >: z \in T_\gamma \,\&\, q \in \mathbb{Q} \,\&\, q > \mathrm{Sup}(\mathrm{Rang}(z))\}$, where $\mathbb{Q}$ is the set of all rational numbers. Obviously $|T_\alpha| = |T_\beta| \cdot \aleph_0 = \aleph_0$. The induction hypothesis holds also for $\beta < \alpha$, as easily seen. If α is a limit ordinal then for every $x \in \bigcup_{\beta<\alpha} T_\beta$ and every $q > \mathrm{Sup}(\mathrm{Rang}(x))$ we shall construct a sequence y of length α which extends x such that $\mathrm{Sup}(\mathrm{Rang}(y)) = q$. Let $\langle \beta_n : n < \omega \rangle$ be a (strictly) increasing sequence such that $x \in T_{\beta_0}$ and $\mathrm{Sup}\{\beta_n : n < \omega\} = \alpha$. Let $\langle q_n : n < \omega \rangle$ be an increasing sequence of rationals such that $q_0 \geq \mathrm{Sup}(\mathrm{Rang}(x))$ and $\mathrm{Sup}_{n<\omega} q_n = q$. We define now a member $x_n \in T_{\beta_n}$ such that $\mathrm{Sup}(\mathrm{Rang}(x_n)) = q_n$ as follows: $x_0 = x$. Assume $x_n \in T_{\beta_n}$ is defined; $\operatorname{Sup}(\operatorname{Rang}(x_n)) = q_n < q_{n+1}$ then by the induction hypothesis there is an $x_{n+1} \in T_{\beta_{n+1}}$ such that $x_n \lhd x_{n+1}$ and $\operatorname{Sup}(\operatorname{Rang}(x_{n+1})) = q_{n+1}$. Take $y = \bigcup_{n<\omega} x_n$, then the length of y is $\cup_{n<\omega}\beta_n = \alpha$ and $\operatorname{Sup}(\operatorname{Rang}(y)) = \operatorname{Sup}(\operatorname{Rang}(x_n)) = \operatorname{Sup}_{n<\omega} q_n = q$. As y was chosen for x and q we let $y = y_{x,q}$. Lastly let $T_\alpha = \{y_{x,q}: x \in \bigcup_{\beta<\alpha} T_\alpha \text{ and } \operatorname{Sup}(\operatorname{Rang}(x)) < q \in \mathbb{Q}\}$. Since we introduced one such y for each $x \in \cup_{\beta<\alpha} T_\beta$ and $q > \operatorname{Sup}(\operatorname{Rang}(x))$ and there are only $\aleph_0$ such pairs clearly $|T_\alpha| \leq \aleph_0$.

T has no branch of length ω_1 since if S is such a branch then $\cup S$ is an increasing sequence of rationals of length ω_1, which is impossible.

By our construction of T, for every $x \in T$ we know $\operatorname{Sup}(\operatorname{Rang}(x))$ is a rational number. Therefore $T = \bigcup_{q\in\mathbb{Q}}\{x \in T : \operatorname{Sup}(\operatorname{Rang}(x)) = q\}$, and each set $\{x \in T : \operatorname{Sup}(\operatorname{Rang}(x)) = q\}$ is clearly an antichain. Thus the tree T is a special Aronszajn tree. $\square_{5.2}$

When we want to construct a κ^+-Aronszajn tree we use, instead of the rationals, the set $\mathbb{Q}_\kappa$ of all sequences of ordinals $< \kappa$ of length κ which are eventually 0, ordered lexicographically. We can proceed as in the construction of the $\aleph_1$-Aronszajn tree, but when we construct T_α, for a limit ordinal α such that $\operatorname{cf}(\alpha) < \kappa$, we have to put in T_α all the increasing sequences y of members of $\mathbb{Q}_\kappa$ of length α such that $y{\restriction}\beta \in T_\beta$ for every $\beta < \alpha$. Otherwise we have no assurance that we can carry out the construction of T_α for a limit ordinal α such that $\operatorname{cf}(\alpha) = \kappa$. In order to be sure that $|T_\alpha| \leq \kappa$, for every $\alpha < \kappa^+$ we need that $\kappa^{<\kappa} = \sum_{\mu<\kappa} \kappa^\mu = \kappa$, since this will enable us to prove that if for a limit ordinal α with $\operatorname{cf}(\alpha) < \kappa$ we construct T_α as mentioned above we still have $|T_\alpha| \leq \kappa$.

So we have presented a proof of the well known:

5.3 Theorem. If $\kappa = \kappa^{<\kappa}$ then there is a κ^+-Aronszajn tree.

If the continuum hypothesis holds then $\aleph_1^{\aleph_0} = \aleph_1$ and therefore there is an $\aleph_2$-Aronszajn tree. Therefore, if we look for a model with no $\aleph_2$-Aronszajn tree, the continuum hypothesis should fail to hold in such a model. There is a theorem which says that in such a model $\aleph_2$ is a weakly compact cardinal in L, hence the consistency of the inexistence of $\aleph_2$-Aronszajn trees is at least as strong as the consistency of the existence of a weakly compact cardinal; we shall see that these two consistency assumptions are equivalent. Mitchell had proved this theorem, and Baumgartner gave a simpler proof by proper forcing.

The following theorem is due to Baumgartner, Malitz and Reinhart [BMR].

5.4 Theorem. For every tree T of height ω_1 with no branch of length ω_1 (no restrictions on its cardinality) there is a c.c.c. forcing notion P such that in the generic extension of V by P the tree T is special. If $|T| < 2^{\aleph_0}$ *then* by Martin's axiom it follows that T is special.

As a consequence, if we assume Martin's axiom and $2^{\aleph_0} > \aleph_1$, then all Aronszajn trees are special and hence there are no Souslin trees.

Proof. Let P be the set of all finite functions p from T into ω such that if $p(x) = p(y)$ then x and y are incomparable. For every $x \in T$ the set $\mathcal{I}_x$ of all members of P whose domain contains x is obviously dense in P, hence if G is a generic subset of $P, F = \cup G$ is defined on all of T and if $F(x) = F(y)$ then x and y are incomparable. If we have Martin's axiom and $|T| < 2^{\aleph_0}$ then there are $< 2^{\aleph_0}$ dense sets $\mathcal{I}_x$ and the directed set G can be taken to intersect all of them, and $F = \cup G$ is as above, i.e., it specializes T since $T = \cup_n \{x : F(x) = n\}$.

We still have to prove that P satisfies the c.c.c. Suppose there is an uncountable subset W of P whose members are pairwise incompatible. Without loss of generality we can assume that all members of W have the same cardinality, that their domains form a Δ-system with the heart s and that for all $p \in W, p{\restriction}s$ is the same function. Denote $W = \{p_\alpha : \alpha < \omega_1\}$ and let $\mathrm{Dom}(p_\alpha) \setminus s = \{x_{\alpha,1}, \ldots, x_{\alpha,n}\}$. Let $\alpha, \beta < \omega_1$, p_α and p_β are incompatible, hence $p_\alpha \cup p_\beta \notin P$. Since p_α and p_β coincide on s and the rest of their domains are disjoint we must have for some $1 \leq k, \ell \leq n$, $p_\alpha(x_{\alpha,k}) = p_\beta(x_{\beta,\ell})$ while $x_{\alpha,k}$ and $x_{\beta,\ell}$ are comparable. Let $Y_{\alpha,k,\ell} = \{\beta < \omega_1 : \beta \neq \alpha, p_\alpha(x_{\alpha,k}) = p_\beta(x_{\beta,\ell})$ and $x_{\alpha,k}$ and $x_{\beta,\ell}$ are comparable $\}$. As we saw $\bigcup_{1\leq k,\ell \leq n} Y_{\alpha,k,\ell} = \omega_1 \setminus \{\alpha\}$. Let E be a uniform ultrafilter on ω_1, then for every α there are k and ℓ such that $Y_{\alpha,k,\ell} \in E$, let $k(\alpha)$ and $\ell(\alpha)$ be such. Therefore for an uncountable subset A of $\omega_1, k(\alpha) = k$ and $\ell(\alpha) = \ell$ for $\alpha \in A$. Let $\alpha, \beta \in A$ then $Y_{\alpha,k,\ell}$,

$Y_{\beta,k,\ell} \in E$ hence $Y_{\alpha,k,\ell} \cap Y_{\beta,k,\ell} \in E$ and therefore $|Y_{\alpha,k,\ell} \cap Y_{\beta,k,\ell}| = \aleph_1$. Let $\gamma \in Y_{\alpha,k,\ell} \cap Y_{\beta,k,\ell}$ then $x_{\alpha,k}$ and $x_{\beta,k}$ are comparable with $x_{\gamma,\ell}$. Now $x_{\gamma,\ell}$'s with different $\gamma \in Y_{\alpha,k,\ell} \cap Y_{\beta,k,\ell}$ are different, and since there are only countably many members of T below $x_{\alpha,k}$ or below $x_{\beta,\ell}$ (in T's sense) there must be some $\gamma \in Y_{\alpha,k,\ell} \cap Y_{\beta,k,\ell}$ such that $x_{\gamma,\ell}$ is greater than both $x_{\alpha,k}$ and $x_{\beta,k}$ (in T's sense) and since T is a tree, $x_{\alpha,k}$ is comparable with $x_{\beta,k}$. This holds for all $\alpha, \beta \in A$ hence T has a linearly ordered subset of cardinality $\aleph_1$: $\{x_{\alpha,k} : \alpha \in A\}$, and therefore a branch of length ω_1, contradicting our assumption. $\square_{5.4}$

§6. Maybe There Is No $\aleph_2$-Aronszajn Tree

Toward this we mention (see for history, 6.2 below):

6.1 Lemma. 1) Assume $V \models$ "$2^{\aleph_0} > \aleph_1$ & T is an $\aleph_2$-Aronszajn tree." Let P be an $\aleph_1$-complete forcing notion. Then $V[P] \models$ "T has no cofinal branches". ($\aleph_2$ may become of cardinality $\aleph_1$ in $V[P]$ so it does not have to stay an $\aleph_2$-Aronszajn tree.)

2) Assume:

(a) T is a tree with δ^* levels such that $\mathrm{cf}(\delta^*) > \aleph_0$

(b) for no limit $\delta < \delta^*$ of cofinality $\aleph_0$ can we find pairwise distinct $x_\eta \in T_\delta$ for $\eta \in {}^\omega 2$ such that: $[\alpha < \delta \Rightarrow \{x_\eta \restriction \alpha : \eta \in {}^\omega 2\}$ is finite$]$ ($x_\eta \restriction \alpha$ is the unique $y <_T x_\eta$, $y \in T_\alpha$)

(c) P is an $\aleph_1$-complete.

Then forcing by P add no new δ^*-branch to T.

Proof. 1) Assume that $p_0 \Vdash$ "$\underset{\sim}{B}$ is a cofinal branch in T". We shall define in V two functions $F : {}^{\omega>}2 \to T\restriction\alpha$ for some $\alpha < \omega_2$ and $S : {}^{\omega>}2 \to P$ such that:

(i) $F(<>) =$ the root of T, $S(<>) = p_0$

(ii) for all $x \in 2^{<\omega}$ we have $S(x) \Vdash$ "$F(x) \in \underset{\sim}{B}$".

(iii) $x \lhd y \Rightarrow S(x) <_P S(y)$, $F(x) <_T F(y)$, and

(iv) $F(x\char`^ <0>)$ and $F(x\char`^ <1>)$ are incomparable in T.

$F(\eta)$ and $S(\eta)$ are defined by induction on the length of η. Assume $S(\eta)$, $F(\eta)$ are defined we shall define $S(\eta\hat{}\langle\ell\rangle)$, $F(\eta\hat{}\langle\ell\rangle)$ for $\ell = 0, 1$. Since $S(\eta) \geq_P p_0$, $S(\eta)$ has, for every $\beta < \omega_2$, an extension which forces some member of T_β (i.e., the set of vertices of height β in the tree) to be in $\underset{\sim}{B}$. If $\{t : F(\eta) <_T t$ and there is $p \geq_P S(\eta)$ such that $p \Vdash$ "$t \in \underset{\sim}{B}$"$\}$ was a set of pairwise comparable members of T then they would be a branch of T in V, contradicting our hypothesis. Therefore there are two incomparable t's in this set, take one to be $F(\eta\hat{} < 0 >)$ and the other to be $F(\eta\hat{} < 1 >)$ and choose $S(\eta\hat{} < 0 >)$ and $S(\eta\hat{} < 1 >)$ as conditions $\geq S(\eta)$ such that $S(\eta\hat{} < \ell >) \Vdash$ "$F(\eta\hat{} < \ell >) \in \underset{\sim}{B}$" for $\ell \in \{0, 1\}$. Since the range of F is countable it is included in some $T{\restriction}\alpha$ for some $\alpha < \omega_2$. Since P is $\aleph_1$ -complete, for every $\eta \in 2^\omega$, P contains a condition p_η which is an upper bound of $\{S(\eta{\restriction}n) : n < \omega\}$. Since $p_\eta \geq p_0$ there is a $q_\eta \geq p_\eta$ and a $t_\eta \in T_\alpha$ such that $q_\eta \Vdash$ "$t_\eta \in \underset{\sim}{B}$". Let $\nu \neq \eta$, and $\nu, \eta \in 2^\omega$. Let n be the least such that $\nu{\restriction}n \neq \eta{\restriction}n$, then by requirement (iv) above we have that $F(\nu{\restriction}n)$ and $F(\eta{\restriction}n)$ are incomparable in T. Now $P \models$ "$q_\eta \geq p_\eta \geq S(\eta{\restriction}n)$", hence also $q_\eta \Vdash$ "$F(\eta{\restriction}n) \in \underset{\sim}{B}$". Since q_η forces that $\underset{\sim}{B}$ is a branch of T and that $t_\eta, F(\eta{\restriction}n) \in \underset{\sim}{B}$ clearly t_η and $F(\eta{\restriction}n)$ are comparable in T. Since the height of $F(\eta{\restriction}n)$ is $< \alpha$ and the height of t_η is α we have $F(\eta{\restriction}n) <_T t_\eta$. Similarly also $F(\nu{\restriction}n) <_T t_\nu$ and since $F(\eta{\restriction}n)$ and $F(\nu{\restriction}n)$ are $<_T$-incomparable also t_η and t_ν are $<_T$-incomparable and hence different. Thus T_α contains $2^{\aleph_0} > \aleph_1$ different members t_η, contradicting the assumption that T is an $\aleph_2$-Aronszajn tree.

2) Similar proof. $\square_{6.1}$

6.2 Theorem. If ZFC is consistent with the existence of a weakly compact cardinal then ZFC is consistent with $2^{\aleph_0} = \aleph_2$ and the non-existence of $\aleph_2$-Aronszajn trees.

Remark. By what was mentioned in the last section we have "iff" in this theorem. Mitchell had proved the theorem and Baumgartner [B3] gave a simpler proof by proper forcing.

Proof. Let κ be a weakly compact cardinal. We shall use a system $\langle P_i, \underset{\sim}{Q}_i : i < \kappa\rangle$ of iterated forcing with countable support. Q_i will be the composition of two forcing notions $Q_{i,0}$ and $Q_{i,1}$. Now $Q_{i,0}$ will be the forcing notion of countable functions from ω_1 into ω_2 (in $V[P_i]$) which collapses $\aleph_2$ i.e. Levy$(\aleph_1, \aleph_2)$. Now $Q_{i,0}$ is obviously $\aleph_1$-complete. $V[P_i][Q_{i,0}]$ contains wide trees of cardinality $\aleph_1$ and ω_1 levels with no ω_1-branches (e.g., $\{\langle\alpha,\beta\rangle : \beta < \alpha < \omega_1\}$ with $\langle\alpha,\beta\rangle <_T \langle\alpha^\dagger,\beta^\dagger\rangle$ iff $\alpha = \alpha^\dagger \,\&\, \beta < \beta^\dagger$). Let W be the disjoint union of all such trees (up to isomorphisms), as a single tree with at most $2^{\aleph_1}$ roots (we take all the trees to be $\omega_1 \times \{i\}$ with some partial orderings). The tree have ω_1 levels, $\leq 2^{\aleph_1}$ nodes and no ω_1-branch. By Theorem 5.4 there is a c.c.c. forcing $Q_{i,1}$ which makes this tree special, and hence makes every $\aleph_1$-wide tree of cardinality $\aleph_1$ special, provided it has no ω_1- branch.

Note that in V^{P_i}, $Q_{i,0}$ has cardinality $\leq 2^{\aleph_1}$, and $Q_{i,1}$ in $V^{P_i * \underset{\sim}{Q}_{i,0}}$ has cardinality $\leq 2^{\aleph_1}$. Let us notice that these descriptions of $Q_{i,0}$ in $V[P_i]$ and $Q_{i,1}$ in $V[P_i][Q_{i,0}] = V[P_i * Q_{i,0}]$ really yield corresponding names $\underset{\sim}{Q}_{i,0}$ and $\underset{\sim}{Q}_{i,1}$ by the Lemma of the existential completeness which we proved.

Since $Q_{i,0}$ is $\aleph_1$-complete over $V[P_i]$ and since $Q_{i,1}$ satisfies the c.c.c. over $V[P_i][Q_{i,0}]$, both are proper, hence $Q_{i,0} * Q_{i,1}$ is proper and therefore also each $P_i, i \leq \kappa$, is proper. Thus $\aleph_1$ is not collapsed even in $V[P_\kappa]$ (by 3.2). Let λ be an inaccessible cardinal, $\lambda \leq \kappa$. Our construction of P_i and Q_i are such that for $i < \lambda$ we have $|Q_i| < \lambda$, hence each P_i has a dense subset of cardinality $< \lambda$. Therefore, as we proved in 4.1 also P_λ satisfies the λ-c.c. and therefore λ is not collapsed in $V[P_\lambda]$ and thus $V[P_\lambda] \vDash$ "$\lambda \geq \aleph_2$" (since $\aleph_1$ too is not collapsed). Before finishing we prove two lemmas.

6.3 Lemma. $V[P_\lambda] \vDash$ " there are at least λ real numbers " for every $\lambda \leq \kappa$.

Proof. It suffices to prove that for every $i < \lambda$ there is a real in $V[P_{i+1}] \setminus V[P_i][Q_{i,0}]$. We shall see that a forcing notion such as $Q_{i,1}$ introduces a Cohen real over the previous universe. Let us simplify the notation by writing $V^\dagger$ for $V[P_i][Q_{i,0}]$, and Q for $Q_{i,1}$ and T for the tree in $V^\dagger$ which Q makes special. Let $\langle a_j : j < \omega\rangle$ be an ascending sequence in T in $V^\dagger$ (i.e., $j < k < \omega \rightarrow$

$a_j <_T a_k$). Now Q introduces a function $\underset{\sim}{F}$ on T into ω such that if $a, b \in T$ and $\underset{\sim}{F}(a) = \underset{\sim}{F}(b)$ then a and b are $<_T$-incomparable. For $j < \omega$ let $\underset{\sim}{t}_j = 0$ if $\underset{\sim}{F}(a_j)$ is even and $\underset{\sim}{t}_j = 1$ if $\underset{\sim}{F}(a_j)$ is odd. We shall see that $\underset{\sim}{t} = \langle \underset{\sim}{t}_j : j < \omega\rangle$ is a Cohen real over $V^\dagger$, i.e., for every dense subset $\mathcal{I}$ of ${}^{\omega>}2$ in $V^\dagger, t\restriction n \in \mathcal{I}$ for some $n < \omega$. For $p \in Q$ let $p^* = \{\langle j, s\rangle : j < \omega$ and $a_j \in \mathrm{Dom}(p)$ and $([s = 0 \,\&\, p(a_j)$ is even$]$ or $[s = 1 \,\&\, p(a_j)$ is odd$]\,)\}$. Let $Q_{\mathcal{I}} = \{p \in Q : p^* \in \mathcal{I}\}$; we shall see that $Q_{\mathcal{I}}$ is a dense subset of Q in $V^\dagger$. Clearly $Q_{\mathcal{I}} \in V^\dagger$ since it is defined in $V^\dagger$. For $q \in Q$ let $r \geq q^*, r \in \mathcal{I}$; there is such an r since $\mathcal{I}$ is dense in ${}^{\omega>}2$. Let n be a strict upper bound of the range of q. Let $p = q \cup \{\langle a_j, 2n + 2j + r(j)\rangle : j \in \mathrm{Dom}(r) \,\&\, a_j \notin \mathrm{Dom}(q)\}$. Obviously $p \in Q$, and $p^* = r \in \mathcal{I}$ hence $p \in Q_{\mathcal{I}}$. Since $p \geq q$ we know $Q_{\mathcal{I}}$ is dense. Let G be the generic subset of Q then there is a $p \in G \cap Q_{\mathcal{I}}$ such that $p^* \in \mathcal{I}$, and for some $n < \omega, p^* \in {}^n2$ (as $p^* \in Q = {}^{\omega>}2$). Since $p \subseteq \underset{\sim}{F}[G]$ we have $\underset{\sim}{t}[G]\restriction n = p^*$, hence $\underset{\sim}{t}[G]\restriction n \in \mathcal{I}$, which establishes that t is a Cohen real. $\square_{6.3}$

6.4 Lemma. For every inaccessible $\lambda \leq \kappa, V[P_\lambda] \models$ "$\lambda = \aleph_2$".

Proof. Let G be a generic subset of P_λ. We saw already that $V[G] \models$ "$\lambda \geq \aleph_2$". Suppose now that $V[G] \models$ "$\aleph_2 = \mu$", where $\mu < \lambda$. Let $F \in V[G]$ be a function on $\mu \times \omega_1$ such that for all $0 < \alpha < \mu$ we have $\{F(\alpha, \beta) : \beta < \omega_1\} = \alpha$, i.e., $F(\alpha, -)$ is a mapping of ω_1 on α. Let $\underset{\sim}{F}$ be a name of F and let $p_0 \in G$ force that F is as we described. For each $\alpha < \mu$ and $\beta < \omega_1$ let $\mathcal{I}_{\alpha,\beta}$ be a maximal antichain of members of P_λ which are $\geq p_0$ and which give definite values to $\underset{\sim}{F}(\alpha, \beta)$. Since as we saw, P_λ satisfies the λ-c.c. condition, clearly $|\mathcal{I}_{\alpha,\beta}| < \lambda$. Let $\mathcal{I} = \bigcup_{\alpha<\mu,\beta<\omega_1} \mathcal{I}_{\alpha,\beta}$. Since λ is regular $|\mathcal{I}| < \lambda$. Since each member of $\mathcal{I} \cup \{p_0\}$ is a countable function on λ there is a $\gamma < \lambda$ such that $\mathcal{I} \cup \{p_0\} \subseteq P_\gamma$. Let $G_\gamma = G \cap P_\gamma$, then clearly $F \in V[G_\gamma]$ (since we define $F(\alpha, \beta)$ in $V[G_\gamma]$ to be that γ for which there is a $q \in \mathcal{I}_{\alpha,\beta} \cap G_\gamma$ such that $q \Vdash$ "$F(\alpha, \beta) = \gamma$"). Then $V[G_\gamma] \models \mu \leq \aleph_2$, but since $V[G] \models$ "$\mu = \aleph_2$" we have $V[G_\gamma] \models$ "$\mu = \aleph_2$". But since we force above $V[G_\gamma]$ with $Q_{\gamma,0}[G_\gamma]$ which collapses the $\aleph_2$ of $V[G_\gamma]$, we have $V[G_\gamma][G_{\gamma,0}] \models$ "$\mu < \aleph_2$" hence $V[G] \models$ "$\mu < \aleph_2$", which is a contradiction.

$\square_{6.4}$

Continuation of the Proof of Theorem 6.2. Now let us go on with the proof of the theorem. Assume that there is a $p_0 \in P_\kappa$ such that $p_0 \Vdash_{P_\kappa}$ " there is an $\aleph_2$-Aronszajn tree," i.e., $p_0 \Vdash_{P_\kappa}$ " there is a κ-Aronszajn tree T on κ and a function F on κ such that for $\alpha < \kappa, F(\alpha)$ is the rank of α in T, (since by the lemma $V[P_\kappa] \models$ "$\aleph_2 = \kappa$"), i.e. $p_0 \Vdash_{P_\kappa}$ " there is a transitive relation T on κ such that for all $\alpha, \beta, \gamma \in \kappa$ we have: $[\alpha T \gamma \,\&\, \beta T \gamma \Rightarrow \alpha T \gamma]$ and there is a function F from κ into κ such that for all $\alpha, \beta < \kappa$: if $\alpha T \beta$ then $F(\alpha) < F(\beta)$, for all $\alpha \in \kappa$ and $\gamma < F(\alpha)$ there is a $\beta T \alpha$ such that $F(\beta) = \gamma$, and for all $\gamma \in \kappa$ there is a $\beta \in \kappa$ such that for all $\alpha \in \kappa$ if $F(\alpha) = \gamma$ then $\alpha \leq \beta$, and for all $B \subsetneq \kappa$ there are $\alpha, \beta \in B$ such that $\alpha \neq \beta \wedge \neg \alpha T \beta \wedge \neg \beta T \alpha$, or else there is a $\beta < \kappa$ such that $B \subseteq \beta$". This implies, by the existential completeness that there are canonical names of $\underset{\sim}{T}$ and $\underset{\sim}{F}$ of relations on κ such that:

(∗) $p_0 \Vdash$ "$\underset{\sim}{T}$ is a transitive relation on κ and $(\forall \alpha, \beta, \gamma < \kappa)(\alpha \underset{\sim}{T} \gamma \,\&\, \beta \underset{\sim}{T} \gamma \rightarrow \alpha \underset{\sim}{T} \gamma)$ and $(\forall \alpha, \beta, \gamma < \kappa)$ $[\alpha = \beta \vee \alpha \underset{\sim}{T} \beta \vee \beta \underset{\sim}{T} \alpha]$ and $(\exists \underset{\sim}{F} : \kappa \rightarrow \kappa)(\forall \alpha, \beta < \kappa)$ $[(\alpha \underset{\sim}{T} \beta \rightarrow F(\alpha) < F(\beta)) \,\&\, \ (\beta < \underset{\sim}{F}(\alpha) \rightarrow (\exists \gamma < \kappa)(\gamma \underset{\sim}{T} \alpha \wedge \underset{\sim}{F}(\gamma) = \beta))]$ and $(\forall \gamma < \kappa)(\exists \beta < \kappa)(\forall \alpha < \kappa)(\underset{\sim}{F}(\alpha) = \gamma \rightarrow \alpha < \beta)$"

and for every canonical name $\underset{\sim}{B}$ of a subset of κ

(∗∗) $p_0 \Vdash_\kappa$ "$(\exists \alpha, \beta \in \underset{\sim}{B})(\alpha \neq \beta \wedge \neg \alpha \underset{\sim}{T} \beta \wedge \neg \beta \underset{\sim}{T} \alpha) \vee (\exists \beta \in \kappa)(\underset{\sim}{B} \subseteq \beta)$".

Now a name $\underset{\sim}{X}$ of a subset of $\kappa \times \kappa$ (like $\underset{\sim}{T}$) or κ (like $\underset{\sim}{B}$) or even a subset of $H(\kappa)^V$, we can assume the name is canonical (see Definition I 5.12 and Theorem I 5.13). So $\underset{\sim}{X}$ is a subset of $\{(p, x) : p \in P_\kappa$ and $x \in H(\kappa)\}$, so a subset of $H(\kappa)$ and even assume that for each $x \in H(\kappa)$ the set $\mathcal{I}_{\underset{\sim}{X}, x} \overset{\text{def}}{=} \{p : (p, x) \in \underset{\sim}{X}\}$ is an antichain of P_κ. But P_κ satisfies the κ.c.c. hence $x \in H(\kappa) \Rightarrow |\mathcal{I}_{\underset{\sim}{X}, x}| < \kappa$, hence $E = \{\mu < \kappa : \mu$ strong limit singular, and $[j < \mu \Rightarrow \bar{Q} \restriction j \in H(\mu)]$ and $x \in H(\mu) \Rightarrow \mathcal{I}_{\underset{\sim}{X}, x} \in H(\mu)\}$ is a club of κ. So for $\mu \in E$, $\underset{\sim}{X} \cap H(\mu)$ is a P_μ-name.

Consider now the structure $(H(\kappa), \in, \underset{\sim}{T}, \underset{\sim}{F})$. The statement (∗) is a first order statement about this structure, and that (∗∗) holds for every $\underset{\sim}{B}$ as mentioned is a Π^1_1 statement about this structure i.e. a statement of the form: for every subset X of the model some first model sentence holds. We now use one of the equivalent forms of the definition of weakly compact (can be read from the proof, or see e.g. [J]). Since κ is weakly compact and therefore Π^1_1-

indescribable there is an inaccessible cardinal $\lambda < \kappa$, from E such that $(H(\lambda), \in, \underset{\sim}{T} \cap \lambda \times \lambda, \underset{\sim}{F} \cap \lambda \times \lambda)$ is an elementary substructure of $(H(\kappa), \in, \underset{\sim}{T}, \underset{\sim}{F})$ and satisfies the Π^1_1 statement mentioned above. P_κ and $\Vdash_{P_\kappa}$ are definable in $(H(\kappa), \in, \underset{\sim}{T}, \underset{\sim}{F})$ and the same definitions give P_λ and $\Vdash_{P_\lambda}$ in $(H(\lambda), \in, \underset{\sim}{T} \cap \lambda \times \lambda, \underset{\sim}{F} \cap \lambda \times \lambda)$. Therefore $(\underset{\sim}{T} \cap \lambda \times \lambda)[G_\lambda]$ is a λ-Aronszajn tree in $V[G_\lambda]$ (where G is the generic subset of P_κ over V and $G_\lambda = G \cap P_\lambda$). We claim that $(\underset{\sim}{T} \cap \lambda \times \lambda)[G_\lambda]$ is the part of the tree $\underset{\sim}{T}[G]$ up to level λ. Let $\langle \alpha, \beta \rangle \in (\underset{\sim}{T} \cap \lambda \times \lambda)[G_\lambda]$ then $\alpha, \beta < \lambda$ and for some $p \in G_\lambda$, $p \Vdash_{P_\lambda} \langle \underset{\sim}{\alpha}, \underset{\sim}{\beta} \rangle \in (T \cap \lambda \times \lambda)$, hence $p \Vdash_{P_\kappa}$ "$\langle \alpha, \beta \rangle \in \underset{\sim}{T}$" (by the relation between the two above mentioned structures), hence, since $p \in G, \langle \alpha, \beta \rangle \in T[G]$, shows that $(\underset{\sim}{T} \cap \lambda \times \lambda)[G_\lambda]$ is included in the part of $\underset{\sim}{T}[G]$ up to level λ. The proof of the equality of these two trees will be completed once we show that in $\underset{\sim}{T}[G]$ all the ordinals in the levels below λ are $< \lambda$. Let $\mu < \lambda$ then, since $(*)$ holds for the structure $(H(\lambda), \ldots)$ there is a $p \in G_\lambda$ and an ordinal $\beta < \lambda$ such that $p \Vdash_{P_\lambda} (\forall \alpha \in \lambda)\ (\underset{\sim}{F} \cap \lambda \times \lambda)(\alpha) = \mu \rightarrow \alpha < \beta)$, therefore $p \Vdash_{P_\kappa} (\forall \alpha \in \kappa)\ (\underset{\sim}{F}(\alpha) = \mu \rightarrow \alpha < \beta)$, and since $p \in G$ we have in $V[G]$ that all the ordinals in the level μ are $< \beta < \lambda$.

Thus in $V[P_\lambda]$ we know $\underset{\sim}{T} \cap (\lambda \times \lambda)$ is a λ-Aronszajn tree, i.e., an $\aleph_2$-Aronszajn tree. Now we saw that in $V[P_\lambda]$ there are at least λ real numbers, i.e., $2^{\aleph_0} \geq \aleph_2$ in $V[P_\lambda]$ and we know $Q_{\lambda,0}$ is an $\aleph_1$-complete forcing notion in $V[P_\lambda]$, therefore, by Lemma 6.1 $T \cap (\lambda \times \lambda)$ still has no cofinal branch in $V[P_\lambda][Q_{\lambda,0}]$. Since in $V[P_\lambda][Q_{\lambda,0}]$ we have $|\lambda| = \aleph_1$ there is a subset a of λ of order-type ω_1. Let b be the set of all the ordinals in $T \cap \lambda$ whose level is in a, so $T \cap b$ is an $\aleph_1$-tree with no cofinal branch in $V[P_\lambda][Q_{\lambda,0}]$. Now $Q_{\lambda,1}$ makes this tree special (dealing with a tree isomorphic to it). Thus the tree $T \restriction \{t : F(t) \in a\}$ is a special $\aleph_1$-tree in $V[P_{\lambda+1}]$ and therefore it stays so also in $V[P_\kappa]$ since the function which makes it special is in $V[P_{\lambda+1}]$ and hence also in $V[P_\kappa]$. Thus in $V[P_\kappa]$ we have $T \restriction \lambda$ is a tree which has no cofinal branch and $\{t \in T : F(t) \in a\}$ is (in $V^{P_{i+1}}$) an $\aleph_1$-wide Aronszajn tree, but this is a contradiction since $T \cap (\lambda \times \lambda)$ is the part up to level λ of the κ-tree T and as such it must have branches of length λ. $\square_{6.2}$

We used a weakly compact κ to obtain a generic extension in which there are no $\aleph_2$-Aronszajn trees. If all we want is to obtain an extension of V in which there are no special $\aleph_2$-Aronszajn trees (i.e., trees which are the union of $\aleph_1$ antichains) then it suffices to use a Mahlo cardinal κ (see [B3] on this).

§7. Closed Unbounded Subsets of ω_1 Can Run Away from Many Sets

Baumgartner [B3] has proved the consistency of the following with $ZFC+2^{\aleph_0} = \aleph_2$: if $A_i \subseteq \omega_1$, for $i < \omega_1$, is infinite countable then there is a closed unbounded $C \subseteq \omega_1$ such that $A_i \not\subseteq C$ for every $i < \omega_1$. We prove a somewhat stronger assertion.

7.1 Theorem. $ZFC + 2^{\aleph_0} = \aleph_2$ is consistent with:

(*) if $A_i \subseteq \omega_1$ has no last element and is nonempty and has order type $< \mathrm{Sup} A_i$, for $i < \omega_1$, *then* there is a closed unbounded subset C of ω_1, such that $C \cap A_i$ is bounded in A_i for every i.

Remark. Abraham improved Baumgartner's result to:

$ZFC + 2^{\aleph_0} =$ anything $+$

(**) there are $\aleph_2$ closed unbounded subsets of $\aleph_1$, the intersection of any $\aleph_1$ of them is finite.

Galvin proved previously that CH implies (*) fail. Our proof is similar to [Sh80 §4].

Proof. We start with a model V satisfying CH, and use CS iterated forcing of length ω_2, such that in the intermediate stages CH still holds. So by 4.1, it suffices to prove.

7.2 Lemma. Suppose V satisfies CH. There is a proper forcing notion of power $\aleph_1$ which adds a closed unbounded subset of C of ω_1 and $\mathrm{Sup}[C \cap A] < \mathrm{Sup} A$ for any infinite $A \subseteq \omega_1$ with no last element and order type $< \mathrm{Sup}(A)$, which belongs to V.

Discussion. A plausible forcing to exemplify 7.2 is:

$$Q = \{C : C \text{ a closed countable subset of } \omega_1 \text{ so that if } A \in V \text{ is a set of ordinals } < \omega_1,\ \delta = \mathrm{Sup}(A) \text{ is a limit ordinal, } A \text{ has order type } < \delta, \text{ then } \mathrm{Sup}[A \cap C] < \delta\}$$

(so if $\delta > \sup(C)$ this holds trivially), with the order

$$C_1 < C_2 \text{ iff } C_2 \cap [(\mathrm{Max} C_1) + 1] = C_1.$$

So the elements of Q are approximations to the required C. It is clear that a generic subset of Q gives a C as required, provided that $\aleph_1$ is not collapsed; hence the main point is to prove the properness of Q. Unfortunately it seems Q is not proper, in fact has no infinite members. However if we want to add a $C \in Q$ which is (N, Q)-generic for some $N \prec (H(\lambda), \in)$ a c.c.c., forcing is enough. So we could first force with some P, $|P| = \aleph_2, P$ satisfies the c.c.c. such that

$$\Vdash_P \text{“}2^{\aleph_0} = \aleph_2 \text{ and MA holds”}.$$

Then we define $\underset{\sim}{Q}$ in $V[G_P]$ for $G_P \subseteq P$ generic over V; similar to the definition above but members of the forcing notion are from $V[G_P]$ whereas the A for which we demand $\sup(C) \cap A < \sup(A)$ are from V. So

$Q = \{C \in V[G]$: C a closed countable subset of ω_1, and if $A \in V$ is infinite with no last element and (order type A) $<$ Sup$A \leq$ Max$C$, then $\mathrm{Sup}(C \cap A) < \mathrm{Sup} A\}$.

Now Q is proper, and adds a C as required, so $P * \underset{\sim}{Q}$ adds a C as required. Unfortunately $P * \underset{\sim}{Q}$ also collapses $\aleph_2$, so if we are willing to use some strongly inaccessible $\kappa > \aleph_0$, there is no problem. Otherwise, we use a restricted version of MA, which is consistent with $2^{\aleph_0} = \aleph_1$, so 7.3, 7.4, 7.5 below prove Lemma 7.2 hence Theorem 7.1. Throughout we use the order $<$ on Q defined above.

7.3 Claim. Suppose

(a) V satisfies CH, P is a forcing notion of power $\aleph_1$ satisfying the c.c.c. and $G \subseteq P$ is generic over V.

(b) $\delta < \omega_1$ is a limit ordinal, $(\forall \alpha, \beta < \delta)[\alpha + \beta < \delta]$, $R \in V[G]$ is a countable family of closed bounded subsets of δ, ordered by $C_1 \leq C_2$ iff $C_1 = C_2 \cap (\text{Max} C_1 + 1)$

(c) Define (in $V[G]$):
$Q = Q_R \stackrel{\text{def}}{=} \{(C, \{(A_i, \alpha_i) : i < n\}) : C \in R, n < \omega$, for each $i < n$, $A_i \in V$ is a subset of δ of order type $< \delta$, and $C \cap A_i \subseteq \alpha_i$ and α_i is an ordinal $< \delta\}$,
the order is
$(C^1, \{(A^1_i, \alpha^1_i) : i < n^1\}) \leq (C^2, \{(A^2_i, \alpha^2_i) : i < n^2\})$ *iff* $C^1 < C^2, n^1 \leq n^2$ and for every $i < n^1$ for some $j < n^2$ we have $(A^1_i, \alpha^1_i) = (A^2_j, \alpha^2_j)$ and $C^2 \setminus C^1$ is disjoint to A^1_i.

Then

1) Q_R satisfies the c.c.c.

2) if for every $C \in R$, $\{\beta < \delta : C \cup \{\beta\} \in R\}$ has order type δ then $\Vdash_Q$ "$\bigcup_{C \in \underset{\sim}{G}_Q} C$ is an unbounded subset of δ".

Proof. 1) Trivial, as any two conditions with the same first coordinate are compatible, and there are only countably many possibilities for the first coordinate.

(2) Trivial, because if $A_1, \ldots, A_n$ are subsets of δ of order type $< \delta$, their union has order type $< \delta$ by Dushnik, Miller [DM]. So for every $p = (C, \{(A_\ell, \alpha_\ell) : \ell < n\}) \in Q$ the set

$$B_p = \{\beta < \alpha : p \leq (C \cup \{\beta\}, \{(A_\ell, \alpha_\ell) : \ell < n\}) \in Q \text{ and } \beta > \max(C)\}$$

has order type δ because

$$B_p = \{\beta : C \cup \{\beta\} \in R\} \setminus (\bigcup_{\ell < n} A_\ell \cup (\max C + 1))$$

and the first set has order type δ (by an assumption) whereas the second has order type $< \delta$ (by the previous sentence as $\text{otp}(A_\ell) < \delta$ by the definition of Q, and $\text{otp}([0, \max C + 1)) < \delta$ by the assumption on R). Now $B_p \subseteq \delta$ being of order type δ, necessarily is unbounded in δ and we are done. $\square_{7.3}$

7.4 Claim. Suppose V satisfies CH. There is a forcing notion P of power $\aleph_1$ satisfying the c.c.c., such that the following statement is forced:

$(*)$ Suppose $\delta < \omega_1$ is limit, $(\forall \alpha, \beta < \delta)[\alpha + \beta < \delta]$ (equivalently δ is an ordinal power of ω) and R is a countable family of closed bounded subsets of δ such that $(\forall C \in R)(\forall \beta)$ $(\text{Max}C < \beta < \delta \rightarrow C \cup \{\beta\} \in R)$ and for $n < \omega$ we have: $\mathcal{I}_n$ is a dense open subset of R such that $\mathcal{I}_n^{\dagger} = \{(C, \emptyset) : C \in \mathcal{I}_n\} \subseteq Q_R$ is pre-dense in Q_R (where Q_R is from 7.3 clause (c)). *Then* there are $C_n \in R$, such that $C_n < C_{n+1}$, $\text{Sup}_n \text{Max}C_n = \delta, C_{n+1} \in \mathcal{I}_n$ and for every $A \in V, A \subseteq \delta$ of order type $< \delta$, $\text{Sup}[A \cap (\cup_n C_n)] < \delta$; moreover we can choose $C_0 \in R$ arbitrarily.

Proof. We can use an FS iteration $\langle P_i, \underset{\sim}{Q}_i : i < \omega_1 \rangle$ of forcing notions satisfying the c.c.c., such that for every possible R and δ (which are in V or appear in V^{P_i} for some $i < \omega$) for uncountably many $j < \omega_1$ in V^{P_j} we have $Q_j = Q_R$. $\square_{7.4}$

7.5 Claim. Suppose V satisfies CH, P is as in 7.4, and $\underset{\sim}{Q} = \{C : C \in V^P$ a closed bounded subset of ω_1, such that for every infinite countable $A \in V$, $A \subseteq \omega_1$ with no last element, and (order type of A) $<$ sup$A$, if Sup$A \leq$ Max$C$ then $\text{Sup}(A \cap C) < \text{Sup}A\}$ ordered by: $C_1 \leq C_2$ iff $C_1 = C_2 \cap (\text{ max}C + 1)$. *Then* $\underset{\sim}{Q}$ is proper (i.e., $\Vdash_P$ "$\underset{\sim}{Q}$ is proper ") and has cardinality $\leq \aleph_1$.

Proof. Let $G \subseteq P$ be generic over V, λ be regular big enough and let in $V[G]$

$S \stackrel{\text{def}}{=} \{N : N \prec (H(\lambda), \in), P \in N,\ H(\aleph_2)^V \in N,\ G \in N,$ and there is a sequence $\langle N_i : i < \delta \rangle$ such that $N_i \prec N$, $\langle N_j : j \leq i \rangle \in N_{i+1}$, $N = \cup_{i<\delta} N_i$ and $\delta = N \cap \omega_1$ (automatically $\underset{\sim}{Q}[G] \in N$)$\}$

It is easy to check the following facts, and by 2.8 they imply Q is proper, so we finish the proof of 7.2, hence 7.1.

Fact A. $S \in \mathcal{D}_{\aleph_0}((H(\lambda))$ (in $V[G]$, remember we do not distinguish strictly between N and its set of elements).

Fact B. In $V[G]$: if $N \in S, \delta = N \cap \omega_1$, $R \stackrel{\text{def}}{=} \underset{\sim}{Q}[G] \cap N$, and $\mathcal{I} \in N$ is a dense open subset of $\underset{\sim}{Q}[G]$, *then*

(a) $\{(C, \{(A_i, \alpha_i) : i < n\}) \in Q_R : C \in \mathcal{I} \cap R\}$ is a dense subset of Q_R and $\{(C, \emptyset) : C \in \mathcal{I} \cap R\}$ is a pre-dense subset of Q_R.

(b) For every $C \in \underset{\sim}{Q}[G] \cap N$ there is C^*, $C \leq C^* \in \underset{\sim}{Q}[G]$ and C^* is $(\underset{\sim}{Q}[G], N)$-generic.

Proof. Let $\langle N_i : i < \delta \rangle$ be as in the definition of S. Let $p_0 = (C_0, \{(A_\ell, \alpha_\ell) : \ell < n\}) \in Q_R$ and let $\mathcal{I} \in N$ be a dense open subset of $\underset{\sim}{Q}[G]$, so for some $i < \delta$ we have $\underset{\sim}{\mathcal{I}}$, $p \in N_i$. Let δ_j be $N_j \cap \omega_1$. Let $A'_\ell = \{j < \delta : A_i \cap [\delta_j, \delta_{j+1}) \neq \emptyset\}$, so $\text{otp}(A'_\ell) \leq \text{otp}(A_\ell) < \delta$. hence as in the proof of 7.3 there is $j \in (i, \delta)$ which does not belong to A'_ℓ for $\ell < \omega$. Now $p_1 = (C_0 \cup \{\delta_j\}, \{(A_\ell, \alpha_\ell) : \ell < n\})$ belongs to R and is $\geq p$. There is $C_2 \in \underset{\sim}{Q}[G] \cap \underset{\sim}{\mathcal{I}}[G]$ such that $C \cup \{\delta_j\} \leq C_2$ (in $\underset{\sim}{Q}[G]$), hence there is such C_2 in N_{j+1} (as relevant parameters belong to it). Now $p_2 = (C_2, \{(A_\ell, \alpha_\ell) : \ell < n\})$ belongs to $\underset{\sim}{Q}[G] \cap N$ and is $\geq p_1 \geq p_0$. This proves clause (a).

Let $\langle \mathcal{I}^0_n : n < \omega \rangle$ list the dense open subsets of $\underset{\sim}{Q}[G]$ which belong to N and $C \in \underset{\sim}{Q}[G] \cap N$, and let $\mathcal{I}_n = \mathcal{I}^0_n \cap N$, and let $R = \underset{\sim}{Q}[G] \cap N$ with the inherited order. Trivially $C \in R$ & $\max(C) < \beta < N \cap \omega_1 \Rightarrow C \leq_R C \cup \{\beta\}$, and the other assumption in $(*)$ of 7.4 holds by the previous paragraph, so there is $\langle C_n : n < \omega \rangle$ as guaranteed there (with $C \leq C_0$).

Now $C^* = \bigcup_{n<\omega} C_n \cup \{N \cap \omega_1\}$ belongs to $\underset{\sim}{Q}[G]$, $C_n \leq C^*$ and $C_{n+1} \in \mathcal{I}_n \subseteq \mathcal{I}^0_n$, so as $\mathcal{I}_n$ was open, $C^* \in \mathcal{I}_n$. So we have proved clause (b) too. $\square_{7.5,7.2,7.1}$

§8. The Consistency of SH + CH + There Are No Kurepa Trees

8.1 Definition. For any regular κ, a κ-Kurepa tree is a κ-tree such that the number of its κ-branches is $> \kappa$. Let the κ-Kurepa Hypothesis (in short κ–KH) be the statement "there exists κ-Kurepa tree". We may write "KH" instead of ω_1-KH. (Be careful: KH says "there *are* Kurepa trees", but SH says "there are *no* Souslin trees"!)

Solovay proved that Kurepa trees exist if $V = L$, more generally Jensen [Jn] proved the existence of κ-Kurepa's trees follows from Jensen's $\diamondsuit^+$, which holds in L for every regular uncountable κ which is not "too large". But $\neg$KH is consistent with of ZFC + GCH, which was first shown by Silver in [Si67], starting from a strongly inaccessible κ. The method of his proof is as follows: collapse every λ, $\omega_1 < \lambda < \kappa$ using Levy's collapse $\text{Levy}(\aleph_1, < \kappa) = \{p : |p| \leq \aleph_1 \text{ \& } p \text{ is a function with } \text{Dom}(p) \subseteq \kappa \times \omega_1 \wedge \forall \langle \alpha, \xi \rangle \in \text{Dom}(p)(p(\alpha, \xi) \in \alpha)\}$. Now $\text{Levy}(\aleph_1, < \kappa)$ can be viewed as an iteration of length κ, and satisfied the κ-c.c. on the one hand, and $\aleph_1$-completeness on the other hand. Therefore $\aleph_1$ does not get collapsed, as well as any cardinal $\aleph_\alpha \geq \kappa$. Suppose now that $T \in V^P$ is an ω_1-tree. So it has appeared already at an earlier stage along the iteration, say $T \in V^{P'}$, where V^P is obtained from $V^{P'}$ by an $\aleph_1$-complete forcing. In $V^{P'}$ the tree T has at most $2^{\aleph_1}$ branches, and this is less than κ. Note that by 6.1(2) the tree T can have no new ω_1-branches in V^P. So T is not a Kurepa tree in V^P.

Devlin in [De1] and [De2] has shown, starting from a strongly inaccessible, the consistency of GCH + SH + $\neg$KH. For a proof by iteration see Baumgartner [B3].

8.2 Remark. In both proofs the inaccessible cardinal is necessary, for $\neg$KH implies that $\aleph_2$ is an inaccessible cardinal of L.

The main point in Silver's proof, is the fact that $\aleph_1$-complete forcing notions do not add new branches to ω_1-trees. In this section we prove that the property of not adding branches is preserved under CS iterations and use this to give another proof of CON(SH + $\neg$KH) from the consistency of "$(\exists \kappa)\ \kappa$ inaccessible". This serve as a prelude and motivation to Chapter V (and even more Chapter VI), which deals with preservation of such properties. In chapter V we will show that moreover the iteration we construct here does not add reals, so (since we start from a model of CH) we will get a model of "CH + SH + $\neg$KH".

8.3 Definition. A forcing notion Q is *good* for an ω_1-tree T (so the α-th level of T is T_α etc.), if for any countable elementary submodel $N \prec H(\chi, \in)$, for χ large enough, with $T, Q \in N$, and every condition $p \in N \cap Q$, there exists an (N, Q)−generic condition $q \geq p$ such that if $\underset{\sim}{\tau} \in N$ is a name, $q \Vdash$ "either $\underset{\sim}{\tau}[\underset{\sim}{G}_P]$ is an old branch of T or $\underset{\sim}{\tau}[\underset{\sim}{G}_P] \cap T_{<\delta_N}$ is not a branch of $T_{<\delta_N}$ with a bound $x \in T_{\delta_N}$", where δ_N denotes $N \cap \omega_1 = sup(N \cap \omega_1)$.

8.4 Fact. Q is good for an ω_1-tree T iff Q is proper and Q does not add a new branch to T.

Proof. $\Rightarrow$: Suppose Q is good for T. The properness of Q follows trivially. Let $p \Vdash_Q$ "$\underset{\sim}{\tau}$ is a new branch of T", and we shall derive a contradiction; let $\{T, p, Q\} \in N \prec (H(\chi), \in)$, χ large enough and N countable. So let $q \geq p$ be as in the definition of *good.*

If $\underset{\sim}{\tau}[G]$ is an old branch — we are done. If not, $\underset{\sim}{\tau}[G] \cap T_{<\delta_N} \neq B_x = \{y : y <_T x\}$ for all $x \in T_{\delta_N}$. But this implies that $\underset{\sim}{\tau}[G]$ being linearly ordered by $<_T$ has no member of level $\geq \delta_N$, so it cannot be a ω_1-branch of T.

Conversely, suppose that Q is proper and does not add a new ω_1-branch to T. Let $\underset{\sim}{\tau}, p \in N$ be as in the definition, and pick $q \geq p$ which is (N, Q)-generic, and a generic subset G of P over V with $q \in G$. So $\underset{\sim}{\tau}[G] \in N[G] \prec H(\chi, \in)[G]$, and $\underset{\sim}{\tau}[G]$ is either an old ω_1-branch, or is not an ω_1-branch at all. In the first case we are done. Now if $\underset{\sim}{\tau}[G]$ is not an ω_1-branch, then either $(\exists \alpha)\underset{\sim}{\tau}[G] \cap T_\alpha = \emptyset$ or $\exists x, y \in \underset{\sim}{\tau}[G]$ such that x, y are not comparable in T. By elementaricity of $N[G]$, such an α or such x, y exist also in $N[G]$. So q forces what is required by the definition. $\square_{8.4}$

8.5 Theorem. If T is an ω_1-tree and $\bar{Q} = \langle P_i, \underset{\sim}{Q}_i : i < \alpha \rangle$ is a countable support iteration such that for all i the forcing notion $\underset{\sim}{Q}_i$ (is forced to be) good for T, *then* also $P_\alpha = \mathrm{Lim}(\bar{Q})$ is good for T.

Proof. We break here the proof into two parts. The first part is nothing more than another proof of the preservation of properness under countable support iteration. It is meant to help those readers who find the proof in III 3.2 hard to follow. In the second part we show how to extend the first part in order to get a full proof of the theorem.

Let $\bar{Q} = \langle P_i, \underset{\sim}{Q}_j : i \leq \alpha, j < \alpha \rangle$ be a countable support iteration such that $\Vdash_{P_j}$ "$\underset{\sim}{Q}_j$ is proper" for all $j < \alpha$. We fix some regular χ which is large enough for what we need. As in III 3.2 we prove by induction on $j \leq \alpha$ the condition $(*)_j$, which is stronger than the properness of P_j:

$(*)_j$ P_j is good for T, and

(a) forcing with P_j add no ω_1-branches to T and:

(b) *for all* $i < j$ and countable $N \prec (H(\chi), \in)$ such that $i, j, \underset{\sim}{Q} \in N$ and $p \in P_j \cap N$ and an (N, P_i)- generic $q \in P_i$ which satisfies $q \geq p{\restriction}i$ *there is* r such that:

(i) $r \in P_j$

(ii) $r{\restriction}i = q$

(iii) r is (N, P_j)-generic

(iv) $p \leq r$

(v) $\mathrm{Dom}(q) \cap [i, j) = N \cap [i, j)$.

The proof is split to cases. Note that though in the statement $(*)_j$(b) we say "for $i < j$" it holds for $i = j$ too.

Case 1. $j = 0$

Trivial.

Case 2. j a successor ordinal

Let $j = j_1 + 1$, now (a) of $(*)_j$ holds as (a) of $(*)_{j_1}$ holds and $\underset{\sim}{Q}_i$ is good for T, so we shall deal with (b) of $(*)_j$. So by the induction hypothesis applied to j_1 and i (see remark above) w.l.o.g. $i = j_1$ and continue as in the proof of III 3.2.

Case 3. j a limit ordinal

We first look at a case of clause (b) of $(*)_j$ and/or of clause (b) of $(*)_j$ (by 8.4 this suffice) so i, N, p, q are given as there.

As N is countable, we can pick a sequence of ordinals of order type ω, $\langle i_n : n < \omega \rangle$ which is cofinal in $N \cap j$ and such that $i_0 = i$ and $i_n \in N$ for all n. Let $\langle \mathcal{I}_n : n < \omega \rangle$, enumerate all the dense sets of P_j in N. Let $\langle (\underset{\sim}{\mathcal{I}}_n, \underset{\sim}{x}_n) : n < \omega \rangle$

enumerate the pairs $(\underset{\sim}{\tau}, x)$ where $\underset{\sim}{\tau}$ is a P_j-name from N and $x \in T_{\delta_N}$ where $\delta_N \stackrel{\text{def}}{=} N \cap \omega_1$.

We define by induction on n a condition q_n and a P_{i_n}-name of conditions $\underset{\sim}{p}_n$ such that:

(a) $q_n \in P_{i_n}$ is (N, P_{i_n})-generic.

(b) $q_0 = q$

(c) $q_{n+1} \restriction i_n = q_n$, $\mathrm{Dom}(q_n) = \mathrm{Dom}(q_0) \cup ([i_0, i_n) \cap N)$

(d) $\underset{\sim}{p}_0 = p$ and $\underset{\sim}{p}_n$ is a P_{i_n}-name of a member of P_j

(e) $q_n \Vdash_{P_{i_n}}$ "$\underset{\sim}{p}_n \in P_j \cap \mathcal{I}_n \cap N$"

(f) $q_n \Vdash_{P_{i_n}}$ "$\underset{\sim}{p} \restriction i_n \in \underset{\sim}{G}_{P_{i_n}}$"

(g) $q_n \Vdash_{P_{i_n}}$ "$\underset{\sim}{p}_n \leq \underset{\sim}{p}_{n+1}$"

(h) if $G_j \subseteq P_j$ is generic over V and $q_n \in G_j \cap P_{i_n}$ and $\underset{\sim}{p}_n[G_j \cap P_{i_n}] \in G_{P_j}$ then either $(\exists \alpha < \delta_N)(\underset{\sim}{\tau}_{i_n}[G_j] \cap T_\alpha \not\subseteq \{t : t <_T x_n\}$ or $T \cap \underset{\sim}{\tau}_{i_n}[G_j]$ is an ω_1-branch of T from V.

Let us carry the induction. For $n = 0$ there is no problem.

Suppose now that we have defined $q_n, \underset{\sim}{p}_n$ and let us define $\underset{\sim}{p}_{n+1}, q_{n+1}$. Pick a generic subset $G_{P_{i_n}}$ of P_{i_n} such that $q_n \in G_{i_n}$. So by clause (e) we have $\underset{\sim}{p}_n[G_{i_n}] \in P_j \cap N$, let $p_n^* \stackrel{\text{def}}{=} \underset{\sim}{p}_n[G_{i_n}]$. Define, in V, the set $\mathcal{J}_n = \{u \in P_{i_n} : (\exists r)[p_n^* \leq r \;\&\; r \in \mathcal{I}_n \;\&\; u = r \restriction i_n]\} \in V$. As $\underset{\sim}{p}_n[G_{i_n}]$ belongs to N, so does $\mathcal{J}_n$. Clearly, $\mathcal{J}_n$ is dense above $\underset{\sim}{p}_n[G_{i_n}] \restriction i_n$. Define $\hat{\mathcal{J}}_n = \mathcal{J}_n \cup \{u : u$ is incompatible with $p_n^* \restriction i_n\} \in V$. So $\hat{\mathcal{J}}_n \in N$ is a dense subset of P_{i_n}. By the genericity of q_n, the set $\hat{\mathcal{J}}_n \cap N$ is predense above q_n, and as a consequence there exists a condition $u_0 \in \hat{\mathcal{J}}_n \cap N \cap G_{i_n}$. As by clause (f) we have $p_n^* \restriction i_n = \underset{\sim}{p}_n[G_{i_n}] \restriction i_n \in G_{i_n}$, clearly u_0 cannot be incompatible with it, but $u_0 \in \hat{\mathcal{J}}_n$ so by the definition of $\hat{\mathcal{J}}_n$ necessarily $u_0 \in \mathcal{J}_n$. There is, therefore, a condition $r_0 \in \mathcal{I}_n$ such that $u_0 = r_0 \restriction i_n$. By elementaricity of N, we can assume that $r_0 \in N$.

In $V[G_{i_n}]$, for any $p \in P_j/G_{i_n} = \{p \in P_j : p \restriction i_n \in G_{i_n}\}$ let $B^n_p = B^n_p[G_{i_n}] = \{t \in T : p \nVdash_{P_j/G_{i_n}} \text{“}t \notin \underset{\sim}{T}_n\text{”}\}$, equivalently $\{t \in T :$ for some p satisfying $p \leq p' \in P_j/G_{i_n}$ we have $p' \Vdash \text{“}t \in \underset{\sim}{T}_n\text{”}\}$. We now choose p^0_{n+1}, α_n such that:

(i) $p^0_{n+1} \in P_j/G_{i_n}$

(ii) $r_0 \leq_{P_j} p^0_{n+1} \in N[G_{i_n}]$

(iii) one of the following occurs

(a) $B^n_p[G_{i_n}] \cap T_{\alpha_n} \not\subseteq \{t : t <_T x_n\}$

(b) $p^0_{n+1} \Vdash_{P_j/G_{i_n}} \text{“}T \cap \underset{\sim}{T}_n$ is an ω_1-branch of T”.

Why is this possible? If for some r, $r_0 \leq r \in P_j/G_{i_n}$ and $B^n_r[G_{i_n}]$ is disjoint from some T_α then there are such r, $\alpha_n \in N[G_{i_n}]$ and $p^0_{n+1} = r$ is as required. If for some $\alpha < \omega_1$, $B^n_{r_0}[G_{i_n}] \cap T_\alpha$ has at least two members, then there is such $\alpha_n < \omega_1$, and so there is $t_n \in B^n_{r_0}[G_{i_n}] \cap T_{\alpha_n} \subseteq N$ such that $\neg(t_n <_T x_n)$. By the definition of $B^n_{r_0}[G_{i_n}]$ there is p^0_{n+1} satisfying $r_0 \leq p^0_{n+1} \in P_j/G_{i_n}$ such that $p^0_{n+1} \Vdash_{P_j} \text{“}t_n \in \underset{\sim}{T}_n\text{”}$, and p^0_{n+1}, α_n, t_n are as required. Again by elementaricity w.l.o.g. $p^0_{n+1} \in N[G_{i_n}]$ so $p^0_{n+1} \in N$ (as $N[G_{i_n}] \cap V = N$).

Define now $\underset{\sim}{p}_{n+1}$, a P_{i_n}-name by cases. Let $\underset{\sim}{p}_{n+1}[G_{i_n}]$ be $\underset{\sim}{p}_n[G_{i_n}]$ if q_n is not in the generic set G_{i_n}, and equals p^0_{n+1} as described above otherwise. For the definition of q_{n+1} we utilize the induction hypotheses $(*)_{i_n}$. We have just given a prescription, i.e. a name, for $\underset{\sim}{p}_{n+1}$. We can choose a maximal antichain $\mathcal{J} = \{u^n_\zeta : \zeta < \zeta_n(*)\}$ of P_{i_n} of conditions which decide this name, namely $u^n_\zeta \Vdash \text{“}\underset{\sim}{p}_{n+1} = p^{n+1}_\zeta\text{”}$ for some p^{n+1}_ζ, and $u^n_\zeta \geq q_{i_n}$ or u_ζ is incompatible with q_{i_n}.

For each $\zeta < \zeta(*)$ we can apply the induction hypothesis $(*)_{i_{n+1}}$ holds so apply $(*)_{i_{n+1}}$ clause (b) with i_n, i_{n+1}, N, u_ζ, p^{n+1}_ζ here standing for i, j, N, q, p there and get $q^{n+1}_\zeta \in P_{i_n+1}$ as guaranteed there. Define q_{n+1} as follows: $\mathrm{Dom}(q_{n+1}) = (\mathrm{Dom}(q_n)) \cup (N \cap [i_n, i_{n+1}))$, for $\gamma \in \mathrm{Dom}(q_{n+1})$: if $\gamma \in \mathrm{Dom}(q_n)$ then $q_{n+1}(\gamma) = q_n(\gamma)$; if $\gamma \in N \cap [i_n, i_{n+1})$ then $q_{n+1}(\gamma)$ is a P_γ-name: if $\zeta < \zeta_n(*)$ and $u^n_\zeta \in \underset{\sim}{G}_{P_\gamma}$ then it is $p^{n+1}_\zeta(\gamma)$. Check that q_{n+1} is as required.

So we have succeed to carry the induction. Let $q = \bigcup_{n<\omega} q_n$, clearly $q \in P_j$ and $q \restriction i_n = q_n$. As in the proof of III 3.2 we can show that:

$(*)$ if $G_j \subseteq P_j$ is generic over V and $q \in G_j$ then $\underset{\sim}{p}_n[G_j \cap P_{i_n}] \in G_j$.

So q is (N, P_j)-generic and as

$$q_n \Vdash_{P_{i_n}} \text{“}p \restriction i_{n+1} = p_0 \restriction i_{n+1} \leq_{P_{i_n}} p_1 \restriction i_{n+1} \leq \dots \leq \underset{\sim}{p}_{i_{n+1}} \restriction i_{n+1} \leq q_{n+1}\text{”}$$

clearly q is above p. This show that q is as required in clause (b) of $(*)_j$. But by the choice of p^0_{n+1} (and the list $\langle(\underset{\sim}{\tau}_n, x_n) : n < \omega\rangle$) necessarily $q \Vdash$ “for every $\tau \in N[\underset{\sim}{G}_j]$, if τ is not an old ω_1-branch of T then $\tau \cap N[\underset{\sim}{G}_j]$ is not of the form $\{t : t <_T x\}$, for $x \in T_{\delta_N}$”. So q is as required in clause (b) of $(*)_j$ and in Definition 8.3. $\square_{8.5}$

8.6 Theorem. If CON(ZFC + κ is inaccessible) then CON(ZFC + GCH +SH + $\neg$KH).

Proof. Described in 8.1, using 8.5. For CH we need to use the results from chapter V, sections 6 and 7.

IV. On Oracle-c.c., the Lifting Problem of the Measure Algebra, and "$\mathcal{P}(\omega)$/finite Has No Non-trivial Automorphism"

§0. Introduction

We present here the oracle chain condition and two applications: the lifting problem for the measure algebra, and the automorphism group of $\mathcal{P}(\omega)$/finite.

Let $\mathcal{B}$ be the family of the Borel subsets of $(0, 1)$ (i.e. sets of reals which are > 0 but < 1). Let I_{mz} be the family of $A \in \mathcal{B}$ which have Lebesgue measure zero. Clearly I_{mz} is an ideal. The lifting problem is: "Can the natural homomorphism from $\mathcal{B}$ to $\mathcal{B}/I_{mz}$ be lifted ($\equiv$ split), i.e. does it have a right inverse? Equivalently, define on $\mathcal{B}$ an equivalence relation: $A_1, A_2 \in \mathcal{B}$ are equivalent if $(A_1 \setminus A_2) \cup (A_2 \setminus A_1)$ has Lebesgue measure zero: is there a set of representatives which forms a Boolean algebra? If CH holds the answer is positive (see Oxtoby [Ox]). (This holds for any $\aleph_1$-complete ideal). We will show, in §4, that a negative answer is also consistent with ZFC.

Since the problem of splitting the measure algebra is simpler, we will consider it first, but in the introduction we use the second problem to describe the main idea of our technique.

It is well known that if CH holds then $\mathcal{P}(\omega)$/finite is a saturated (model-theoretically) atomless Boolean algebra of power $\aleph_1$, hence has $2^{2^{\aleph_0}}$-many automorphisms, as any isomorphism from one countable subalgebra to another can be extended to two different automorphisms. It was not clear what the situation is if CH fails.

On the other hand any one-to-one function f from one co-finite subset of ω onto another co-finite subset of ω induces an automorphism of $\mathcal{P}(\omega)/\text{finite}$: A/finite is mapped to $f(A)/\text{finite}$. We call such an automorphism trivial. A priori it is not clear whether $\mathcal{P}(\omega)/\text{finite}$ has nontrivial automorphisms. Our main conclusion is that possibly all automorphisms are trivial (i.e. this holds in some model of ZFC); In fact in some generic extension of V. Hence there is e.g.-no Borel definition of such automorphism.

In this chapter our main aim is to present "oracle chain condition forcing". Iterating forcing satisfying the $\aleph_1$-chain condition, introduced by Solovay and Tennenbaum [ST], is well known (see Ch II). We use mainly the same framework: we start, e.g., with the constructible universe $V = L$, and use (finite support) iteration of length ω_2 of forcing notions, $\langle P_i, \underset{\sim}{Q}_i : i < \omega_2 \rangle$, each Q_i satisfies the $\aleph_1$-c.c. and is of power $\aleph_1$. At stage α we guess an automorphism f_α of $\mathcal{P}(\omega)/\text{finite}$ in V^{P_α}, more exactly a P_α-name of such an automorphism, such that for any P_{ω_2}-name $\underset{\sim}{f}$ of an automorphism of $\mathcal{P}(\omega)/\text{finite}$, there are stationarily many $\alpha < \aleph_2$, for which we have: $\text{cf}(\alpha) = \aleph_1$, and $\underset{\sim}{f}_\alpha \subseteq \underset{\sim}{f}$ (this is possible if we have a diamond sequence on $\{\delta < \omega_2 : \text{cf}(\delta) = \aleph_1\}$ and CH holds, by some simple considerations). Our means of killing the automorphism, i.e., guaranteeing f_α cannot be extended to an automorphism of $\mathcal{P}(\omega)/\text{finite}$ in $V^{P_{\omega_2}}$, is to add a set $X \subseteq \omega$ so that for no $Y \subseteq \omega$:

$$(*)\ X \cap A \text{ is finite} \Leftrightarrow Y \cap f_\alpha(A) \text{ is finite for every } A \in [\mathcal{P}(\omega)]^{V^{P_\alpha}}.$$

The demand $(*)$ helps us since if $f \supseteq f_\alpha$ is an automorphism of the Boolean algebra $[\mathcal{P}(\omega)]^{V^{P_{\omega_2}}}/\text{finite}$ then $Y/\text{finite} = f(X/\text{finite})$ satisfies $(*)$.

This looks very reasonable, but even if we succeed to show that when we add X (by forcing Q_α) we do not add a Y satisfying $(*)$, how can we know that such a Y is not added later on during the iteration?

This is the role of the oracle chain condition. The best way we can explain the construction is as follows. We can look at iterated forcing as a construction of the continuum in $\aleph_2$ steps, i.e., at each successor step we add more reals, and even at limit steps of cofinality $\aleph_0$ we do so (but not at limit steps of cofinality $\aleph_1$). Now promising not to add a Y as in $(*)$ is an omitting type obligation, and building a model of power $\aleph_2$ by a chain of ω_2 approximations, promising to omit types along the way, is a widely used method in model theory (mainly for $\aleph_1$ instead of $\aleph_2$). See e.g., Keisler's work on $L(Q)$, for $\aleph_1$; the $\aleph_2$ case like ours is somewhat more difficult, see [Sh:82], [Sh:107], [HLSh:162].

The oracle chain condition is an effective version of the $\aleph_1$-c.c. Assume that we want to construct an $\aleph_1$-c.c. forcing notion P with an underlying set ω_1. If $\bar{M} = \langle M_\delta : \delta < \omega_1 \rangle$ (the $\aleph_1$-oracle) is a $\diamondsuit$-sequence on ω_1; i.e. for each $A \subseteq \omega_1$ the set $\{\delta : A \cap \delta \in M_\delta\}$ is stationary, and we demand:

(†) If $A \in M_\delta$ is pre-dense in $P{\restriction}\delta$, then A is pre-dense in P, then, as we will see in 1.6(1), P satisfies $\aleph_1$-c.c.

We call (a variant of) this the $\bar{M}$-c.c., (assuming each M_δ is closed enough). The connection between the $\bar{M}$-c.c. and the omitting type argument discussed above is the following.

Suppose B_i are Borel sets of reals ($i < \aleph_1$) such that their intersection is empty even if we add a Cohen real. (Note: finite support iteration always adds Cohen reals at limit stages of cofinality $\aleph_0$. Hence we have to deal with Cohen extensions.) If $\diamondsuit$ holds then, as we will see in §2, there is an $\aleph_1$-oracle $\bar{M}$ such that in any generic extension of V by a forcing satisfying the $\bar{M}$-chain condition no real is in all the B_i's (note that in a bigger universe we reinterpret each B_i using the same definition).

Clearly in order to apply oracle chain condition forcings we have to prove relevant lemmas on composition and preservation by direct limit. It is also very helpful to replace a sequence $\langle \bar{M}_i : i < \aleph_1 \rangle$ of oracles by a single oracle $\bar{M}$ such that any P satisfying the $\bar{M}$-c.c. will satisfy all $\bar{M}_i$-c.c.'s. For this we have to choose the right variant of the definition. Altogether the situation is that at step α we are given an $\aleph_1$-oracle which Q_α has to satisfy, and are allowed to demand, for $\aleph_1$ Borel types, that $V^{P_{\omega_2}}$ will omit them provided that not only our universe V^{P_α} omits them, but even forcing by Cohen forcing does not change this. This is done in §1, 2, 3, and of course does not depend on any understanding of the automorphism of $\mathcal{P}(\omega)$/finite or of homomorphism from $\mathcal{B}/I_{mz}$ into $\mathcal{B}$. So it is sufficient to read §1-3 if you want to apply the oracle chain condition; though you may want to read 4.6 (which is the proof of Theorem 4.3) to see how all the threads are put together (modulo specific lemma, in that case 4.5), and it is commonly believed that seeing examples helps to understand a method.

Of course, to apply this we have to look at the specific application at some point; this is done in the induction step, i.e., working in V^{P_α} and being given an automorphism f_α, and an ω_1-oracle $\bar{M}$ we have to find $\underset{\sim}{Q}_\alpha$ adding a real $\underset{\sim}{X}$ and satisfying the $\bar{M}$-c.c. so that $(*)$ holds for no $\underset{\sim}{Y}$ even if we add a Cohen real.

Of course if f_α is trivial this cannot be done. We shall construct Q_α, assume that there is a ($Q_\alpha\times$Cohen)-name $\underset{\sim}{Y}$ forced to satisfy $(*)$, and analyzing this situation we shall eventually prove f_α is trivial. We have to note that if $\underset{\sim}{f}$ is a P_{ω_2}-name of a nontrivial automorphism of $\mathcal{P}(\omega)$/finite, then for a closed unbounded $C \subseteq \omega_2$, for every $\delta \in C$ with $\mathrm{cf}(\delta) = \omega_1$, $\underset{\sim}{f}\restriction[\mathcal{P}(\omega)/\text{finite}]^{V^{P_\delta}}$ has a P_δ-name forced (for P_δ) to be a nontrivial automorphism of $\mathcal{P}(\omega)$/finite in V^{P_δ}, hence a diamond on $\{\delta < \omega_2 : \mathrm{cf}(\delta) = \omega_1\}$ can guess it stationarily often, hence we have "killed" it somewhere in the process. We can have the results of §4 and §5 together.

A natural question is:

Question. Find a parallel to $\bar{M}$-c.c., replacing Cohen forcing (in the assumptions of the omitting type theorem) by random real forcing, or any other reasonable forcing adding a real.
In [Sh-b, IV], only the case "every automorphism of $\mathcal{P}(\omega)$/finite is trivial" appear. The other is from [Sh:185]. On the first use of the method see [Sh:100]. On further works see [BuSh:437], [Ju92], [Ve86], [Ve93], [ShSr:296], [ShSr:315], [ShSr:427].

Note that if $P \subseteq Q$ then $P \lessdot Q$ iff every pre-dense subset of P is pre-dense in Q (to see that this is a necessary condition, assume $\mathcal{I} \subseteq P$ is pre-dense; let $\mathcal{I}^* = \{p \in P : (\exists q \in \mathcal{I}) p \geq q\}$ and let $\mathcal{I}^{**}$ be a maximal set of pairwise incompatible conditions contained in $\mathcal{I}^*$; then $\mathcal{I}^{**}$ is a maximal antichain of P, hence of Q, therefore it is pre-dense in Q and so is $\mathcal{I}$, see I 5.4(3)).

§1. On Oracle Chain Conditions

1.1 Definition. An $\aleph_1$-*oracle* is a sequence $\bar{M} = \langle M_\delta : \delta$ is a limit ordinal $< \omega_1\rangle$, where M_δ is a countable transitive model of ZFC^- (or a large enough portion of ZFC) such that $\delta + 1 \subseteq M_\delta$, $M_\delta \models$ "δ is countable" and $\bar{M}$ satisfies: $(\forall A \subseteq \omega_1)[\{\delta : A \cap \delta \in M_\delta\}$ is stationary]. (ZFC$^-$ is ZFC without the power set axiom).

Remark. Note that the existence of an $\aleph_1$-oracle is equivalent to the holding of $\diamondsuit_{\aleph_1}$ (by a theorem of Kunen, we shall use only the trivial implication: $\diamondsuit_{\aleph_1}$ implies the existence of an $\aleph_1$-oracle), and that $\diamondsuit_{\aleph_1}$ implies CH.

To help the reader to understand our intention to associate a condition on forcing notions with each $\aleph_1$-oracle $\bar{M}$, we give here a tentative definition (which we will modify later):

1.2 First try. A forcing P with universe ω_1 satisfies the $\bar{M}$-*c.c.* if for every $\delta < \omega_1$:

(†) $(\forall A \in M_\delta)$ $[A \subseteq \delta \& A$ is pre-dense in $P{\restriction}\delta \Rightarrow A$ is pre-dense in $P]$.

But if we adopted this definition, it could happen that we have two isomorphic forcing notions, one satisfying the definition and the other not. The solution will be to require the property (†) not for all δ, but only for a large enough set of δ's.

1.3 Definition. With each $\aleph_1$-oracle $\bar{M}$ we associate the filter $D_{\bar{M}}$ over ω_1 generated by the sets $I_{\bar{M}}(A) = \{\delta < \omega_1 : A \cap \delta \in M_\delta\}$ for $A \subseteq \omega_1$. Let $\mathcal{P}(\delta)$ be $\{A : A \subseteq \delta\}$.

1.3A Remark. Recall that $\langle S_\delta : \delta < \omega_1 \rangle$ is a $\diamondsuit^*$-sequence if $|S_\delta| \leq \aleph_0$ and for every $A \subseteq \omega_1$, $\{\delta : A \cap \delta \in S_\delta\}$ contains a closed unbounded subset of ω_1; and if $V = L$, then a $\diamondsuit^*$-sequence exists (by Jensen's work).

1.4 Claim. 1) If $\langle M_\delta \cap \mathcal{P}(\delta) : \delta < \omega_1 \rangle$ is a $\diamondsuit^*$-sequence (see 1.3A) then every generator of $D_{\bar{M}}$, $I_{\bar{M}}(A)$, contains a closed unbounded set;
2) For every $A, B \subseteq \omega_1$ there is $C \subseteq \omega_1$ such that $I_{\bar{M}}(C) = I_{\bar{M}}(A) \cap I_{\bar{M}}(B)$. So: for a set $Z \subseteq \omega_1$, $Z \in D_{\bar{M}} \Leftrightarrow (\exists A)[I_{\bar{M}}(A) \subseteq Z]$.
3) $D_{\bar{M}}$ is a proper normal filter, containing every closed unbounded set of limit ordinals $< \omega_1$.
4) For every $A \subseteq \omega_1 \times \omega_1$ or even $A \subseteq H(\aleph_1)$ the set $I_{\bar{M}}(A)$ belongs to $D_{\bar{M}}$.

Proof. 1) By the definitions of a $\diamondsuit^*$-sequence and of $I_{\bar{M}}(A)$.
2) Let $g : \omega_1 \to \omega_1$ be the map defined by $g(\alpha) = 2\alpha$, and let $f : \omega_1 \to \omega_1$ be the map defined by $f(\alpha) = 2\alpha + 1$. If $\delta < \omega_1$ is a limit ordinal, then δ is closed under g, f, and $g \restriction \delta, f \restriction \delta \in M_\delta$.

Now, taking $C = g(A) \cup f(B)$ we obtain what we need (remembering M_δ is a model of ZFC^-).

3) a) $D_{\bar{M}}$ *contains every closed unbounded set of limit ordinals* $< \omega_1$.

For this, we have to construct, for a given $\bar{M}$ and an increasing continuous sequence $\langle \delta_i : i < \omega_1 \rangle$ of limit ordinals, a subset A of ω_1 such that if $\delta < \omega_1$, $\delta \neq \delta_i$ for all $i < \omega_1$, then $A \cap \delta \notin M_\delta$. We construct such A piece by piece, namely determining $A \cap [\delta_i, \delta_i + \omega)$ by induction on i (we begin with $i = -1$ taking $\delta_{-1} = 0$; outside these intervals all ordinals are in A). Having determined $A \cap \delta_i$, we have $2^{\aleph_0}$ possibilities for $A \cap (\delta_i + \omega)$; we choose one which does not belong to the countable set $\{B \cap (\delta_i + \omega) : B \in M_\delta,\ \delta_i < \delta < \delta_{i+1}\}$. It is easy to check that A satisfies our requirement.

b) $D_{\bar{M}}$ *is a proper filter.*

By 2) every set in $D_{\bar{M}}$ contains some $I_{\bar{M}}(A)$ which is stationary by the assumption "$\bar{M}$ is an $\aleph_1$-oracle", hence it is nonempty.

c) $D_{\bar{M}}$ *is normal.*

It suffices to show that if $A_i \subseteq \omega_1 (i < \omega_1)$ then there is $A \subseteq \omega_1$ such that $A \cap \delta \in M_\delta$ implies $A_i \cap \delta \in M_\delta$ for all $i < \delta$. By a) and 2) it suffices if the implication holds for a closed unbounded set of δ's.

Let $\langle -, - \rangle$ be a nicely defined pairing function from ordinals to ordinals, (preserving countability), and let $C = \{\delta < \omega_1 : \delta \text{ is closed under } \langle -, - \rangle\}$. For $\delta \in C$, the restriction of $\langle -, - \rangle$ to $\delta \times \delta$ belongs to M_δ because the nice definition of $\langle -, - \rangle$ is absolute for M_δ, hence C is closed unbounded in ω_1.

Let $A = \{\langle i, \alpha \rangle : \alpha \in A_i \text{ and } i < \omega_1\}$, assume $\delta \in C$ and $A \cap \delta \in M_\delta$, and let $i < \delta$. Then also $A_i \cap \delta \in M_\delta$, which completes the proof.

4) Similar to the proof of 3). $\square_{1.4}$

1.5 Definition (the real one). We define when a forcing notion P *satisfies the $\bar{M}$-c.c.* by cases

a) If $|P| \leq \aleph_0$, always.

b) If $|P| = \aleph_1$ and for some (every) $f : P \to \omega_1$ which is one-to-one,

$$\{\delta < \omega_1 : \text{for } A \in M_\delta, A \subseteq \delta, f^{-1}(A) \text{ is pre}-\text{dense in } f^{-1}\{i : i < \delta\}$$
$$\text{implies } f^{-1}(A) \text{ is pre}-\text{dense in } P\} \in D_{\bar{M}}$$

To see the equivalence of the "some" and "every" versions let $f : P \to \omega_1$ witness the "some" version and let $g : P \to \omega_1$ be another one-to-one function. Then the set

$$B \stackrel{\text{def}}{=} \{\delta < \omega_1 : f^{-1}(\{i : i < \delta\}) = g^{-1}(\{i : i < \delta\})$$
$$\text{and for every } A \in M_\delta$$
$$[f^{-1}(A) \text{ is pre-dense in } f^{-1}(\{i : i < \delta\}) \Rightarrow f^{-1}(A) \text{ is pre-dense in } P]$$
$$\text{and } (f \circ g^{-1}) \restriction A \in M_\delta\}$$

belongs to $D_{\bar{M}}$ and witnesses that g also satisfies clause b).

c) If $|P| > \aleph_1$ and for every $P^\dagger \subseteq P$: if $|P^\dagger| \leq \aleph_1$ then there are P'' such that $|P''| \leq \aleph_1$, $P^\dagger \subseteq P'' \subseteq P$ and $f : P'' \to \omega_1$, as in b) such that: $P'' \subseteq_{ic} P$, which means, if $p, q \in P''$ then: $P \models p \leq q \Leftrightarrow P'' \models p \leq q$, and: p, q are compatible in P *iff* p, q are compatible in P''.

Notation. $P \models \bar{M}$-c.c. denotes that P satisfies $\bar{M}$-c.c.

Remark. In this chapter we use iterations of length ω_2 starting with $V = L$ and every iterand is of size ω_1, so we work explicitly only with the case $|P| = \aleph_1$.

1.6 Claim. 0) If P_1, P_2 are isomorphic forcing notions, then: P_1 satisfies the $\bar{M}$-c.c. if P_2 satisfies the $\bar{M}$-c.c.

1) If P satisfies the $\bar{M}$-c.c. for some $\aleph_1$-oracle $\bar{M}$, then P satisfies the $\aleph_1$-c.c.

2) Let $\bar{M}$ be an $\aleph_1$-oracle, $P \lessdot Q$. If Q satisfies the $\bar{M}$-c.c., then P satisfies the $\bar{M}$-c.c.

3) In Definition 1.5 for the case $|P| > \aleph_1$, we can demand $P'' \lessdot P$ and get an equivalent definition (remember we are assuming CH as we are assuming $\diamondsuit_{\aleph_1}$).

Proof. 0) Trivial (by the equivalence proved in Definition 1.5 clause (b)).

1) Clearly, it suffices to prove it for P with universe ω_1. So, assume that $\mathcal{J}$ is a maximal antichain in P of power $\aleph_1$. Then there is a closed unbounded subset C of ω_1 such that: if $\delta \in C$, $q < \delta$ then there is $p \in \mathcal{J} \cap \delta$ compatible with q, and also if $p, q < \delta$ are compatible then they have a common upper bound in $P{\restriction}\delta$.

As P satisfies the $\bar{M}$-c.c., there is $\delta \in C \cap I_{\bar{M}}(\mathcal{J})$ with the property: $[A \in M_\delta$ & A is a pre-dense subset of $P{\restriction}\delta] \Rightarrow [A$ is pre-dense in $P]$ (actually, the set of these δ's is in $D_{\bar{M}}$). But $\mathcal{J} \cap \delta$ is a counterexample as: $\mathcal{J} \cap \delta \in M_\delta$ (because $\delta \in I_{\bar{M}}(\mathcal{J})$), and $\mathcal{J} \cap \delta$ is a pre-dense subset of $P{\restriction}\delta$ (because $\delta \in C$ and the definition of C) hence (by the previous-sentence) any $p \in \mathcal{J} \setminus \delta$ is compatible with some $q \in \mathcal{J} \cap \delta$; contradiction.

2) Assume first that $|P| = |Q| = \aleph_1$. W.l.o.g. the universe of Q is ω_1. We have to show that for a set in $D_{\bar{M}}$ of δ's, if A is pre-dense in $P{\restriction}\delta$ and $A \in M_\delta$ then A is pre-dense in P; since $P \lessdot Q$ we know $P \subseteq_{ic} Q$ hence it suffices to

show that A is pre-dense in Q. Since Q satisfies the $\bar{M}$-c.c., it suffices to show that A is pre-dense in $Q{\restriction}\delta$.

For $q \in Q$ define $\mathcal{I}_q \overset{\text{def}}{=} \{r \in P : r$ is incompatible with q, or $(\forall r^\dagger \geq r)(r^\dagger \in P \Rightarrow r^\dagger$ is compatible with $q)\}$.

Let $\mathcal{J}_q$ be a maximal set of pairwise incompatible elements in $\mathcal{I}_q$. As clearly $\mathcal{I}_q$ is dense in P, $\mathcal{J}_q$ is a maximal antichain in P, hence in Q (since $P \lessdot Q$). Pick $r_q \in \mathcal{J}_q$ such that $(\forall r^\dagger \geq r_q)(r^\dagger \in P \Rightarrow r^\dagger$ is compatible with $q)$; such r_q exists since otherwise q would contradict the fact that $\mathcal{J}_q$ is a maximal antichain in Q. We may assume that δ is such that: $q < \delta$ implies $r_q < \delta$, and any $p_1, p_2 \in P{\restriction}\delta$ compatible in P are compatible in $P{\restriction}\delta$, and the same holds for Q.

Now let A be pre-dense in $P{\restriction}\delta$ and let $q \in Q{\restriction}\delta$. Since $r_q \in P{\restriction}\delta$, there is $p \in A$ compatible with r_q. Let $r^\dagger \in P{\restriction}\delta$ be above p and r_q. By the choice of r_q we know that $r^\dagger$ is compatible with q, hence p is compatible with q, so A is pre-dense in $Q{\restriction}\delta$. Hence we have finished the case $|P| = |Q| = \aleph_1$.

Before proving the claim for all P, Q we present the following simple fact on forcing notions.

1.6A Fact. (CH). If $P_1 \subseteq P_2$, $|P_1| \leq \aleph_1$, and P_2 satisfies the $\aleph_1$-c.c., *then* there is P_3 such that $|P_3| \leq \aleph_1$ and $P_1 \subseteq P_3 \lessdot P_2$.

Proof of Fact. We define, by induction, an increasing continuous sequence $\langle P^{(\alpha)} : \alpha < \omega_1 \rangle$ of subsets of P of cardinality $\leq \aleph_1$ as follows:

$P^{(0)} = P_1$,

$P^{(\alpha)} = \cup_{\beta<\alpha} P^{(\beta)}$ for limit α.

For $\alpha = \beta + 1$, for every countable subset $\mathcal{I}$ of $P^{(\beta)}$ which is not pre-dense in P_2, pick $p_{\mathcal{I}} \in P_2$ exemplifying this. Let $P_0^{(\alpha)}$ be the set obtained by adding the $p_{\mathcal{I}}$'s to $P^{(\beta)}$. Let $P^{(\alpha)}$ be the set obtained from $P_0^{(\alpha)}$ by adding a common upper bound for every pair in $P_0^{(\alpha)}$ having one in P_2.

Now, $P_3 = \cup_{\alpha<\omega_1} P^{(\alpha)}$ is as required, proving the Fact. $\square_{1.6A}$

Continuation of the proof of 1.6

Returning to our claim, we distinguish the following cases:

i) $|P| \leq \aleph_0$. Then trivially P satisfies the $\bar{M}$-c.c.

ii) $|P| = \aleph_1$. Then since $Q \models \bar{M}$-c.c. there is $P^\dagger \subseteq_{\rm ic} Q$, $P \subseteq P^\dagger$, such that $P^\dagger$ satisfies the $\bar{M}$-c.c. and $|P^\dagger| = \aleph_1$. Since $P \lessdot Q$, clearly $P \lessdot P^\dagger$, and now by what we have already proved P satisfies the $\bar{M}$-c.c.

iii) $|P| > \aleph_1$. Let $P^\dagger \subseteq P$, $|P^\dagger| \leq \aleph_1$. Using the Fact 1.6A, we obtain P'', such that $|P''| \leq \aleph_1$ and $P^\dagger \subseteq P'' \lessdot P$. Since $P \lessdot Q$ it follows that $P'' \lessdot Q$, so by i) or ii) we have: P'' satisfies the $\bar{M}$-c.c. Hence P satisfies the $\bar{M}$-c.c.

3) Let P satisfy the $\bar{M}$-c.c. (by Definition 1.5), $|P| > \aleph_1$, and let $P^\dagger \subseteq P$, $|P^\dagger| \leq \aleph_1$ and (by 1.6A) choose P'', $|P''| \leq \aleph_1$ such that $P^\dagger \subseteq P'' \lessdot P$. From 2) we know that P'' satisfies the $\bar{M}$-c.c., hence is as required. $\square_{1.6}$

1.7 Fact.

1) If $P \lessdot R$, $P \subseteq Q \subseteq_{ic} R$ then $P \lessdot Q$

2) P_1, $P_2 \lessdot Q$, $P_1 \subseteq P_2$ then $P_1 \lessdot P_2$; if P_1, $P_2 \subseteq_{ic} Q$, $P_1 \subseteq P_2$ then $P_1 \subseteq_{ic} P_2$

3) $\subseteq_{ic}$ is a partial order. $\square_{1.7}$

§2. The Omitting Type Theorem

2.1 Lemma. ($\diamondsuit_{\aleph_1}$). Suppose $\psi_i(x)(i < \omega_1)$ are Π^1_2 (i.e., $\forall$ real $\exists$ real ...) formulas on reals (with a real parameter, possibly). Suppose further that there is no solution to $\bigwedge_{i<\omega_1} \psi_i(x)$ in V; and moreover even if we add a Cohen real to V there will be none. *Then* there is an $\aleph_1$-oracle $\bar{M}$ such that:

$(*)$ $P \vDash \bar{M}$-c.c. implies: in V^P there is no solution to $\bigwedge_i \psi_i(x)$.

Proof. Let $n < \omega$ be large enough so that the forcing theorems can be proved from $\sum_n$ sentences in ZFC, and the assumption of the lemma can be formulated as a $\sum_n$ statement.

Now, for a given countable forcing notion P and a given P-name for a real $\underset{\sim}{\tau}$ (w.l.o.g. canonical), let $M(P, \underset{\sim}{\tau})$ be a countable $\sum_n$-elementary submodel of V containing P, $\underset{\sim}{\tau}$ and $\langle \psi_i : i < \omega_1 \rangle$. Let $\mathcal{I}(P, \underset{\sim}{\tau})$ be the collection of all pre-dense subsets of P lying in $M(P, \underset{\sim}{\tau})$.

2.1A Claim.

1) If $P \subseteq_{ic} P^\dagger$ and every $A \in \mathcal{I}(P, \underset{\sim}{\tau})$ is pre-dense in $P^\dagger$, then

$$V^{P^\dagger} \models \neg \bigwedge_i \psi_i(\underset{\sim}{\tau}[G])$$

(note that $\underset{\sim}{\tau}$ is also a $P^\dagger$-name; we can restrict ourselves to $i \in M(P, \underset{\sim}{\tau})$).

2) Assume that P is a countable forcing, $\underset{\sim}{\tau}$ a P-name of a real, a canonical one, φ a Π^1_1-formula, $p \Vdash_P$ "$\varphi(\underset{\sim}{\tau})$". Then for some pre-dense $\mathcal{I}_n \subseteq P$ for $n < \omega$, *if* $P \subseteq_{ic} P^\dagger$, each $\mathcal{I}_n$ is pre-dense in $P^\dagger$ *then* $p \Vdash_{P^\dagger}$ "$\varphi(\underset{\sim}{\tau})$".

Proof of Claim. 1) Let G be $P^\dagger$-generic over V. Since every $A \in \mathcal{I}(P, \underset{\sim}{\tau})$ is pre-dense in $P^\dagger$ and therefore intersects G, $G \cap P$ is P-generic over $M(P, \underset{\sim}{\tau})$. (Why is $G \cap P$ directed? If $p, q \in G \cap P$ then $\mathcal{J} = \{r \in P : r$ incompatible with p (in P or equivalently in $P^\dagger$), *or* r incompatible with q *or* $p \leq r$ & $q \leq r\} \in M(P, \underset{\sim}{\tau})$ is predense in P hence in $P^\dagger$, hence not disjoint to $G \cap P$.) Also $\underset{\sim}{\tau}[G] = \underset{\sim}{\tau}[G \cap P]$. Since $M(P, \underset{\sim}{\tau})$ is a $\sum_n$-elementary submodel of V, it satisfies the assumption of the lemma, hence $M(P, \underset{\sim}{\tau})[G \cap P] \models$"$\neg\psi_i(\underset{\sim}{\tau}[G])$ for some $i < \aleph_1$". So for some $i < \aleph_1$, $i \in M(P, \underset{\sim}{\tau})$ we have $M(P, \underset{\sim}{\tau})[G \cap P] \models \neg\psi_i(\underset{\sim}{\tau}[G])$ (remember that P is countable (even in the sense of M) and therefore adding a generic subset of P is equivalent to adding a Cohen real). Since $\sum^1_2$ statements are upward absolute from $M(P, \underset{\sim}{\tau})[G \cap P]$ to $V[G]$ by Schoenfields's absoluteness theorem, it follows that $V[G] \models$ "$\neg\psi_i(\underset{\sim}{\tau}[G])$", proving the claim.

2) Proved above. $\square_{2.1A}$

Continuation of the proof of 2.1: Now, using $\diamondsuit_{\aleph_1}$ and a closing process, we can obtain an $\aleph_1$-oracle $\bar{M} = \langle M_\delta : \delta < \omega_1 \rangle$ satisfying: if P is a forcing notion, has universe $\delta < \omega_1$ and $P, \underset{\sim}{\tau} \in M_\delta$ then $\mathcal{I}(P, \underset{\sim}{\tau}) \subseteq M_\delta$ (remember that $P, \mathcal{I}(P, \underset{\sim}{\tau})$ are countable). We will show that $\bar{M}$ satisfies the requirement of the lemma.

Assume, on the contrary, that $P^\dagger \models \bar{M}$-c.c. but $\underset{\sim}{\tau}$ is a $P^\dagger$-name of a real so that $V^{P^\dagger} \models \bigwedge_i \psi_i(\underset{\sim}{\tau})$. We may assume that $|P^\dagger| = \aleph_1$ because:

a) $|P^\dagger| \le \aleph_0$ is impossible by the assumption of the lemma.

b) if $|P^\dagger| > \aleph_1$ then by 1.6(1) and 1.6(3) we can find P'' such that $\underset{\sim}{\tau}$ is a P''-name, $|P''| = \aleph_1$, $P'' \lessdot P^\dagger$ and $P'' \models \bar{M}$-c.c. Since $P'' \lessdot P^\dagger$, $V^{P''} \models \neg\psi_i(\underset{\sim}{\tau})$ would imply $V^{P^\dagger} \models \neg\psi_i(\underset{\sim}{\tau})$(again by an absoluteness argument), hence $V^{P''} \models \bigwedge_i \psi_i(\underset{\sim}{\tau})$. So P'' can replace $P^\dagger$ in the sequel.

W.l.o.g. the universe of $P^\dagger$ is ω_1. We can find $\delta < \omega_1$ such that letting $P = P^\dagger \restriction \delta$ the following will hold:

i) $\underset{\sim}{\tau}$ is a P-name.

ii) $P, \underset{\sim}{\tau} \in M_\delta$.

iii) $P \subseteq_{ic} P^\dagger$.

iv) $A \in M_\delta, A$ is a pre-dense subset of $P \Rightarrow A$ is pre-dense in $P^\dagger$.

From these facts, the claim and the construction of $\bar{M}$, it follows that $V^{P^\dagger} \models \neg \bigwedge_i \psi_i(\underset{\sim}{\tau})$, a contradiction. $\square_{2.1}$

2.2 Example. If $V \models [\Diamond_{\aleph_1} \& A \subseteq \mathbb{R}\ \&\ A$ is of the second category] *then* there is an ω_1-oracle $\bar{M}$ such that:

$P \models \bar{M}$-c.c. $\Rightarrow V^P \models$ "A is of the second category".

Proof. Let $A = \{r_i : i < \omega_1\}$, $\psi_i(x) = (r_i \in B_x)$ where $B_x = \cup_{n<\omega} B^n_x, B^n_x$ closed nowhere dense, where B^n_x is "simply" described by x such that every set of the first category is a subset of B_x for some x and absolutely, for every real x and $n < \omega$, B^n_x is nowhere dense. Each ψ_i is trivially Π^1_2. We still have to show:

$(*)$ if P is countable, the following fails: $\Vdash_P$ " for some real x, for every $i < \omega_1$ we have $r_i \in B_x$".

Suppose, on the contrary, that there are $r \in P$ and a P-name $\underset{\sim}{x}$, so that $r \Vdash_P$ "$\underset{\sim}{x}$ is a counterexample". Then for every i, for some $p \in P$ and n we have $p \Vdash_P$ "$r_i \in B^n_{\underset{\sim}{x}}$". Let $A^n_p = \{r_i : p \Vdash_P$ "$r_i \in B^n_{\underset{\sim}{x}}$"$\}$; we have shown $A \subseteq \cup_{n,p} A^n_p$. We shall show that every A^n_p is nowhere dense and get a contradiction.

Let $n < \omega$, $p \in P$, (c_1, c_2) a rational interval. It suffices to find a rational subinterval disjoint from A^n_p. But $\Vdash_P$ "$B^n_{\underset{\sim}{x}}$ is nowhere dense". So there are $p^\dagger \geq p$, $(c^\dagger_1, c^\dagger_2) \subseteq (c_1, c_2)$ such that $p^\dagger \Vdash_P$ "$B^n_{\underset{\sim}{x}} \cap (c^\dagger_1, c^\dagger_2) = \emptyset$". Clearly:

$$r_i \in (c^\dagger_1, c^\dagger_2) \Rightarrow p^\dagger \Vdash_P \text{“} r_i \notin B^n_{\underset{\sim}{x}} \text{”} \Rightarrow$$

$$\Rightarrow p \nVdash_P \text{“} r_i \in B^n_{\underset{\sim}{x}} \text{”} \Rightarrow r_i \notin A^n_p.$$

$\square_{2.2}$

The arguments above give the following fact, too:

2.2A Fact. If $A \subseteq \mathbb{R}$ is of the second category and P is the forcing notion of adding α Cohen reals to V, *then* $V^P \models$ "A is of the second category". $\square_{2.2}$

It simplifies to note:

2.3 Fact. Let P be a countable forcing notion, Q Cohen forcing

1) If $\mathcal{J}$ is a pre-dense subset of $P \times Q$, then we can find pre-dense subsets $\mathcal{I}_n$ of P, for $n < \omega$ such that:

(*) if P' is a forcing notion, $P \subseteq P'$ and each $\mathcal{I}_n$ is a pre-dense subset of P' too, *then* $\mathcal{J}$ is a pre-dense subset of $P' \times Q$.

2) The same applies to $P \times \underset{\sim}{Q}$, $P' \times \underset{\sim}{Q}$ as for any forcing R, $R \times Q$ is a dense subset of $R \times \underset{\sim}{Q}$.

Proof. 1) Let $\{q_n : n < \omega\}$ list the members of Q and let $\mathcal{I}_n = \{p \in P :$ for some q we have: $q_n \leq q \in Q$ and (p, q) is above some member of $\mathcal{J}\}$.

2) Should be clear. $\square_{2.3}$

§3. Iterations of $\bar{M}$-c.c. Forcings

3.1 Claim. If $\bar{M}_i$ (for $i < \omega_1$) are $\aleph_1$-oracles then there is an $\aleph_1$-oracle $\bar{M}$ such that:

$$(P \models \bar{M}\text{-c.c.}) \Rightarrow (\bigwedge_{i<\omega_1} P \models \bar{M}_i\text{-c.c.}).$$

Proof. Let $\bar{M}_i = \langle M^i_\delta : \delta < \omega_1 \rangle$.

Choose M_δ such that $\cup_{i<\delta} M^i_\delta \subseteq M_\delta$, M_δ a countable transitive model of ZFC$^-$. Clearly $\bar{M} = \langle M_\delta : \delta < \omega_1 \rangle$ is an $\aleph_1$-oracle. It suffices to prove that:

$(*)$ if $\bar{M}_1, \bar{M}_2$ are $\aleph_1$-oracles, $\{\delta : M^1_\delta \subseteq M^2_\delta\}$ is co-bounded (i.e., with a bounded complement) then $\bar{M}_2$-c.c. $\Rightarrow$ $\bar{M}_1$-c.c.

Since for each $A \subseteq \omega_1$, every sufficiently large $\delta \in I_{\bar{M}_1}(A)$ is in $I_{\bar{M}_2}(A)$, it suffices to show that if δ has the required property with respect to $\bar{M}_2$ (and δ is sufficiently large) then it has the required property with respect to $\bar{M}_1$. But this follows trivially from $M^1_\delta \subseteq M^2_\delta$. (Of course in $(*)$ we can require e.g. just "$\{\delta : M^1_\delta \not\subseteq M^2_\delta\}$ is not stationary or just $= \emptyset$ mod $D_{\bar{M}^2}$"). $\square_{3.1}$

3.2 Claim. If $P_i (i < \alpha)$ is the result of a finite support iteration, each P_i satisfying the $\bar{M}$-c.c. and $P = \cup_{i<\alpha} P_i$, *then* P satisfies the $\bar{M}$-c.c.

Proof. We demand superficially less on the P_i's: $P_i \lessdot P_j$ for $i < j$, and $P_\delta = \cup_{i<\delta} P_i$ for limit $\delta < \alpha$, and each P_i (for $i < \alpha$) satisfies the $\bar{M}$-c.c.

Case I: α successor.

Clearly $P \in \{P_i : i < \alpha\}$ (it is $P_{\alpha-1}$) so by the assumption P satisfies the $\bar{M}$-c.c.

Case II: $\mathrm{cf}(\alpha) > \aleph_1$.

If $P^\dagger \subseteq P$, $|P^\dagger| = \aleph_1$, then there is $i < \alpha$ with $P^\dagger \subseteq P_i$ because $\mathrm{cf}(\alpha) > \aleph_1$. Since $P_i \models \bar{M}$-c.c. we can find an $\bar{M}$-c.c. $P'' \subseteq_{\mathrm{ic}} P_i$ of size $\aleph_1$ such that $P^\dagger \subseteq P''$. So $P \models \bar{M}$-c.c.

Case III: $\mathrm{cf}(\alpha) = \aleph_1$.

Obviously it is sufficient to deal with the case $\alpha = \omega_1$. We will now show that it is sufficient to deal with the case $|P| = \aleph_1$.

Indeed, if $|P| < \aleph_1$ then the conclusion of our claim holds. So assume $|P| > \aleph_1$, and P does not satisfy the $\bar{M}$-c.c. Then there is $P^\dagger \subseteq P$ of power $\aleph_1$ so that $P^\dagger \subseteq P'' \subseteq_{\mathrm{ic}} P$, $|P''| = \aleph_1$ imply that P'' does not satisfy the $\bar{M}$-c.c. We note that P satisfies the $\aleph_1$-c.c. (See Claim 1.6(1) and the proof of Lemma II 2.8).

Let N be an elementary submodel of $(H(\chi), \in, <^*_\chi)$ for a large enough regular χ such that $P^\dagger$, $\langle P_i : i \leq \omega_1 \rangle \in N$, $N^\omega \subseteq N$ and $\|N\| = \aleph_1$. Take $P''_i \stackrel{\text{def}}{=} P_i \cap N$. Then $P''_i \lessdot P_i$ (as in the proof of 1.6A) and $\langle P''_i : i \leq \omega_1 \rangle$ is increasing continuous and $P^\dagger \subseteq P''_{\omega_1}$. But $|P''_{\omega_1}| = \aleph_1$ and it does not satisfy the $\bar{M}$-c.c., so we have reduced our counterexample to one of power $\aleph_1$.

We assume now, w.l.o.g., that $P_i \setminus \cup_{j<i} P_j \subseteq \omega_1 \times \{i\}$. We let $A_i \stackrel{\text{def}}{=} \{\delta :$ if $C \in M_\delta$ is a pre-dense subset of $P_i \restriction (\delta \times \delta)$ then C is pre-dense in $P_i\}$.

For each $i < \omega_1, A_i \in D_{\bar{M}}$ (as $P_i \models \bar{M}$-c.c.). Since $P_i \lessdot P$ and $P_i \models \aleph_1$-c.c., by the proof of claim 1.6(2), we can find a function pr_i from $P \times \omega$ into P_i such that: $p \in P_i$ is compatible with $q \in P$ in P iff for some n, $pr_i(q, n), p$ are compatible in P_i. We let $B_i = \{\delta : pr''_i((\delta \times \delta) \times \omega) \subseteq \delta \times \delta$ and $pr_i \cap (\delta \times \delta) = pr_i \restriction ((\delta \times \delta) \times \omega) \in M_\delta\}$.

For each $i < \omega_1$, $B_i \in D_{\bar{M}}$.

We let $A \stackrel{\text{def}}{=} \nabla_i A_i = \{\delta :$ for every $i < \delta, \delta \in A_i\}$ and $B = \nabla_i B_i \stackrel{\text{def}}{=} \{\delta :$ for every $i < \delta, \delta \in B_i\}$. As $D_{\bar{M}}$ is a normal filter, clearly A, B, $A \cap B \in D_{\bar{M}}$. Assume now that $\delta \in A \cap B$, $Y \in M_\delta$ is a pre-dense subset of $P \restriction (\delta \times \delta)$. Notice that

$(*)$ $p \in P_i$ is compatible with some $q \in Y$ iff it is compatible with some $r \in pr_i(Y \times \omega)$.

We know that for all $i < \delta$ we have $pr_i(Y \times \omega) \in M_\delta$ (as $\delta \in B_i$) and it is a pre-dense subset of $P_i \restriction (\delta \times \delta)$ by $(*)$. Therefore, as $\delta \in A_i$, $pr_i(Y \times \omega)$ is pre-dense in P_i. Using $(*)$ again, (as $P_\delta = \cup_{i<\delta} P_i, P \restriction (\delta \times \delta) \subseteq P_\delta$), Y is pre-dense in P_δ, and since $P_\delta \lessdot P$, Y is pre-dense in P. Thus we have proved that P satisfies the $\bar{M}$-c.c. (Note that for some/any bijection $f : \omega_1 \times \omega_1 \to \omega_1$, $\{\delta : f''(\delta \times \delta) = \delta\}$ is closed unbounded).

Case IV: $\mathrm{cf}(\alpha) = \aleph_0$.

One can deal with this case as with Case III (the diagonal intersections are replaced by countable intersections). $\square_{3.2}$

We shall deal now with the "real" successor case:

3.3 Claim. Assume $V \models \diamondsuit_{\aleph_1}$, $\bar{M}$ is an $\aleph_1$-oracle, P is a forcing notion of power $\aleph_1$ satisfying the $\bar{M}$-c.c. *Then* in V^P there is an $\aleph_1$-oracle $\bar{M}^*$ such that: if $\underset{\sim}{Q} \in V^P$ satisfies the $\bar{M}^*$-c.c. (in V^P) *then* $P * \underset{\sim}{Q}$ (which is in V) satisfies the $\bar{M}$-c.c.

Proof. First we observe that if $\bar{M}^*$ in V^P is good for all Q of power $\leq \aleph_1$ then it is good for all Q. Hence we shall treat $\underset{\sim}{Q}$ as a P-name of a binary relation on ω_1 (but we shall define $\bar{M}^*$ without depending on a given $\underset{\sim}{Q}$). We also assume that P has universe ω_1, and we fix for the rest of the proof a generic $G \subseteq P$ (which has a canonical P-name in V). Note that $\omega_1 \times \omega_1$ is a dense subset of $P * \underset{\sim}{Q}$, so we can use it as the set of members.

For $A \subseteq \omega_1 \times \omega_1$ and $G \subseteq P$ we define $A[G] = A^G = \{\alpha : (\exists p \in G)[(p, \alpha) \in A]\}$.

Let $\delta < \omega_1$ be a limit ordinal. In $V[G]$ we let M^*_δ be a countable transitive model of ZFC$^-$ containing δ, M_δ and $\{A^G : A \subseteq \delta \times \delta, A \in M_\delta\}$. We will first show that $\bar{M}^*$ is a $\aleph_1$-oracle in $V[G]$.

The following fact implies this statement.

3.3A Fact. $(\forall A \in D^{V[G]}_{\bar{M}^*})(\exists B \in D_{\bar{M}})[B \subseteq A]$.

Proof of the Fact. Let $\underset{\sim}{S}$ be a P-name of a subset of ω_1. Let $T = \{(p, \alpha) : p \in P$ and $p \Vdash_P$ "$\alpha \in \underset{\sim}{S}$"$\}$. Clearly $\underset{\sim}{S}[G] = T[G]$. Since P satisfies $\aleph_1$-c.c. and so $(\forall \alpha < \omega_1)(\exists \beta < \omega_1)\left[T[G] \cap \alpha = T[G \cap \beta] \cap \alpha\right]$, we have in V a club $C \subseteq \omega_1$ such that $\Vdash_P$ "$T[\underset{\sim}{G}] \cap \delta = T[\underset{\sim}{G} \cap \delta]$" for each $\delta \in C$. So $\delta \in C$ and $A = T \cap (\delta \times \delta) \in M_\delta$ implies $A^G = T[G \cap \delta] \cap \delta = T[G] \cap \delta \in M^*_\delta$. But these restrictions give an element of $D_{\bar{M}}$ by 1.4(3). So we have proved that for every P-name $\underset{\sim}{S} \subseteq \omega_1$, for some $T \subseteq \omega_1 \times \omega_1$ we have $\Vdash$ "$I_{\bar{M}^*}(\underset{\sim}{S})^{V^P} \supseteq I_{\bar{M}}(T)^V$" (well, $T \subseteq \omega_1 \times \omega_1$, and not $T \subseteq \omega_1$, so use a pairing function, see 1.4(4)).

$\square_{3.3A}$

Thus we have proved that $\bar{M}^*$ is an $\aleph_1$-oracle in $V[G]$.

We now assume that $A \subseteq \delta \times \delta$, $A \in M_\delta$ and A is pre-dense in $(P * \underset{\sim}{Q}) \restriction (\delta \times \delta)$. We want to show that A is pre-dense in $P * \underset{\sim}{Q}$. The proof will be broken into three steps, and as we proceed we will impose restrictions on δ, the reader can easily verify that these restrictions are allowable (in the sense of $D_{\bar{M}}$) and do not depend on A and G.

First Step. If $G \subseteq P$ is generic over V, then $A^G \subseteq \delta$, $A^G \in M^*_\delta$ and A^G is pre-dense in $\underset{\sim}{Q}[G]\restriction\delta$.

Proof. The first two facts are immediate.

Let $(p,q),(\alpha,\beta) \in P * \underset{\sim}{Q}$. We define $f_{(p,q)}(\alpha,\beta)$, whenever (p,q) and (α,β) are compatible, as the least ordinal which is the possible P-part of a condition above (p,q) and (α,β) in $P * \underset{\sim}{Q}$. We may assume that for each p, $q < \delta$ $f''_{(p,q)}(\delta \times \delta) \subseteq \delta$ and for each $q < \delta$ the mapping $p \mapsto f_{(p,q)}\restriction(\delta\times\delta)$ for $p < \delta$ belongs to M_δ.

Let $q \in \underset{\sim}{Q}[G]\restriction\delta$, and define $\mathcal{I}_q = \bigcup_{p<\delta} f_{(p,q)}(A)$. Then $\mathcal{I}_q \subseteq \delta$, $\mathcal{I}_q \in M_\delta$ and $\mathcal{I}_q$ is dense in $P\restriction\delta$ (as A is pre-dense in $(P * \underset{\sim}{Q})\restriction(\delta\times\delta)$ and by the definition of $f_{(p,q)}$). Hence we may assume that $\mathcal{I}_q$ is pre-dense in P (by our assumption that P satisfies the $\bar{M}$-c.c.).

Let $r \in \mathcal{I}_q$. Then for some $p < \delta$, $s \in Q$ and $(\alpha,\beta) \in A$ we have in $P * \underset{\sim}{Q}$ $(r,s) \geq (p,q),(\alpha,\beta)$. As $P \models$ "$r \geq \alpha$", $r \Vdash_P$ "$\alpha \in \underset{\sim}{G}_P$", so $r \Vdash_P$ "$\beta \in A^{\underset{\sim}{G}_P}$"; as $r \Vdash_P$ "$\beta \leq_{\underset{\sim}{Q}} s$ and $q \leq_{\underset{\sim}{Q}} s$" it follows that $r \Vdash_P$ "q is compatible with some element of A^G". As this holds for all $r \in \mathcal{I}_q$ and $\mathcal{I}_q$ is pre-dense in P, it is true in $V[G]$, and as this holds in $V[G]$ for all $q \in \underset{\sim}{Q}[G]\restriction\delta$, we obtain that A^G is pre-dense in $\underset{\sim}{Q}[G]\restriction\delta$.

Second Step. If $G \subseteq P$ is generic over V, then A^G is pre-dense in $\underset{\sim}{Q}[G]$.

Proof. This follows from the assumption that in $V[G]$, the forcing notion $\underset{\sim}{Q}[G]$ satisfies the $\bar{M}^*$-c.c. We only have to observe here that the restriction to a set in $D_{\bar{M}^*}$ can be replaced by a restriction to a set in $D_{\bar{M}}$, by Fact 3.3A.

Third Step. A is pre-dense in $P * \underset{\sim}{Q}$.

Proof. Let $(p, q) \in P * \underset{\sim}{Q}$. Since for any $G \subseteq P$ generic over V, in $V[G]$ the set A^G is pre-dense in $\underset{\sim}{Q}[G]$, there exist $r \geq p$, $(\alpha, \beta) \in A$ and $s \in Q$ such that $r \Vdash_P$ "$\alpha \in \underset{\sim}{G}_P$ and $\beta \leq_{\underset{\sim}{Q}} s$, and $q \leq_{\underset{\sim}{Q}} s$". As $r \Vdash_P$ "$\alpha \in \underset{\sim}{G}_P$", $r \geq \alpha$, so $(r, s) \geq (p, q), (\alpha, \beta)$, in $P * \underset{\sim}{Q}$, proving what we need.

$\square_{3.3}$

§4. The Lifting Problem of the Measure Algebra

4.1 Notation. Let $\mathcal{B}$ be the family of Borel subsets of $(0,1)$. Every Borel set $\subseteq (0,1)$ has a definition φ (in the propositional calculus $L_{\omega_1,\omega}$), i.e. let φ be a sentence in the $L_{\omega_1,\omega}$ propositional calculus, with vocabulary $\{t_q : q \in \mathbb{Q}\}$, where $\mathbb{Q}$ denotes the rational numbers, (t_q stands for the statement $r < q$). We let $A = \text{Bo}[\varphi]$ be the Borel set correspondending to this definition, i.e. $A = \{r : r \in (0,1)$, and if we assign to the propositional variables t_q the truth value of "$r < q$", the sentence φ becomes true$\}$. Notice that the answer to "$r \in \text{Bo}[\varphi]$" is absolute.

If $B \in \mathcal{B}$ in V, and $V[G]$ is a generic extension of V, then let $B^{V[G]}$ be the unique B_1 such that for some φ, $V \models$"$B = \text{Bo}(\varphi)$", and $V[G] \models$"$B_1 = \text{Bo}[\varphi]$". Note that the choice of φ is immaterial.

Let I_{mz} be the family of $A \in \mathcal{B}$ of measure zero and I_{fc} be the family of $A \in \mathcal{B}$ which are of the first category. If a, b are reals, $0 \leq a, b \leq 1$, let

$$(a, b) = \{x : a < x < b \ \text{ or } \ b < x < a\}$$

$[a, b]$ is defined similarly. Again this is absolute. We can use $\mathbb{R}$ or $(0,1)$, does not matter.

4.2 Definition. If B is a Boolean algebra, I is an ideal, we say that B/I splits if there is a homomorphism $h : B/I \to B$ such that $h(x/I)/I = x/I$.

Equivalently there is a homomorphism $h : B \to B$ with kernel I such that $h(x) = x \bmod I$.

4.3 Theorem. It is consistent with ZFC that $\mathcal{B}/\mathcal{B} \cap I_{mz}$ does not split (if ZFC is consistent).

4.4 Discussion. (1) If CH holds then $\mathcal{B}/I_{mz}$ splits (see Oxtoby [Ox]); in fact this holds for any $\aleph_1$-complete ideal.

(2) Note that $\mathfrak{B}/I_{mz}$ has a natural set of representatives:

$$h^m(X) = \{a \in R : 1 = \lim_{\varepsilon \to 0}[\text{Leb}(X \cap (a - \varepsilon, a + \varepsilon))/2|\varepsilon|]\}$$

where Leb(X) is the Lebesgue measure (for a set of reals). Unfortunately h^m is not a homomorphism.

Now for $X \in \mathcal{B}$, $h^m(X) \in \mathcal{B}$ and $h^m(X) = X \bmod I_{mz}$ (see Oxtoby [Ox]).

(3) Note also that $(\mathcal{B} + I^{\oplus}_{mz})/I^{\oplus}_{mz}$ splits where $\mathcal{B} + I^{\oplus}_{mz}$ is the Boolean algebra of subsets of $(0,1)$ generated by $\mathcal{B}$ and $I^{\oplus}_{mz}$ and $I^{\oplus}_{mz} = \{A : A$ is a subset of some member of $I_{mz}\}$. A function exemplifying this can be defined as follows: for each real r let E_r be an ultrafilter on $(0,1)$ such that if $A \subseteq (0,1)$ and $1 = \lim_{\varepsilon \to 0}[\text{Leb}(A \cap (r - \varepsilon, r + \varepsilon))/2|\varepsilon|]$, then $A \in E_r$ and for every $X \in \mathcal{B} + I_{mz}$ let

$$h(X) = \{r : X \in E_r, r \in (0,1)\}$$

Clearly $h(X) = h^m(X) \bmod I_{mz}$ hence $h(X) = X \bmod I_{mz}$.

(4) The dual problem, replacing "measure zero" by "meager" can be solved too. I.e.

($*$) it is consistent with ZFC that $\mathcal{B}/\mathcal{B} \cap I_{\text{fc}}$ does not split (if ZFC is consistent).

The proof is like the proof of 4.3, replacing "measure zero" by "meager", it is done explicitly in [Sh:185] [were also Theorem 4.3 was proved].

(5) The author was asked this problem by Fremlin and Talagrand during the Kent Conference, summer 79, who said it is very important for measure theory; Talagrand even promised me flowers on my grave from measure theorists. I did not check that yet ...

4.5 Main Lemma. Let $\bar{M}$ be an $\aleph_1$-oracle (so CH and even $\diamondsuit_{\aleph_1}$ hold) and h be a homomorphism from $\mathcal{B}$ to $\mathcal{B}$ with kernel I_{mz}, such that $h(X) = X \mod I_{mz}$ for every $X \in \mathcal{B}$.

Then there is a forcing notion P of cardinality $\aleph_1$ satisfying the $\bar{M}$-chain condition, and a P-name $\underset{\sim}{X}$ (of a Borel set) such that for every generic $G \subseteq P \times Q$ over V (where Q is the Cohen forcing) there is no Borel set A in $V[G]$ satisfying:

α) $A = \underset{\sim}{X}[G] \mod I_{mz}^{V[G]}$

β) for every $B \in \mathcal{B}^V$, if $B^{V[G]} \subseteq \underset{\sim}{X}[G] \mod I_{mz}$ then $(h(B))^{V[G]} \subseteq A$,

γ) for every $B \in \mathcal{B}^V$, if $B^{V[G]} \cap \underset{\sim}{X}[G] = \emptyset \mod I_{mz}$ then $(h(B))^{V[G]} \cap A = \emptyset$.

4.6. *Proof of the Theorem from the Main Lemma.* For simplicity our ground model is a model of $V = L$. We intend to define a finite support iteration $\langle P_i, Q_i : i \leq \omega_2 \rangle$ such that $|P_i| < \aleph_2$ and $P_i \models \aleph_1$-c.c. for $i < \omega_2$.

We also want to define a sequence $\langle F_i : i < \omega_2 \rangle$, $F_i \in V^{P_i}$, $F_i \subseteq [\mathcal{B} \times \mathcal{B}]^{V^{P_i}}$ (more exactly a P_i-name of such set), such that for any $F \subseteq [\mathcal{B} \times \mathcal{B}]^{V^{P_{\omega_2}}}$, the set

$$\{i < \omega_2 : cf(i) = \aleph_1, F \restriction [\mathcal{B}]^{V^{P_i}} = F_i\}$$

is stationary in ω_2.

This is possible, using $\diamondsuit_{\{\delta < \omega_2 : cf(\delta) = \aleph_1\}}$.

(F_i is a P_i-name. So it becomes meaningful after we have defined P_i, but using $\diamondsuit_{\{\delta < \omega_2 : cf(\delta) = \omega_1\}}$ we obtain a schematic definition of the F_i-s which does not depend on the actual construction of the P_i's or just guess $P \restriction i$ and $\underset{\sim}{F}_i \restriction i$. See more detatils for the second case in Claim 5.3).

So how do we define P_i for $i \leq \omega_2$? As we are going to iterate with finite support, $P_i (i \leq \omega_2)$ will be determined as soon as we define $\langle \underset{\sim}{Q}_i : i < \omega_2 \rangle$ by the relations $P_0 = \{\emptyset\}$, $P_{i+1} = P_i * \underset{\sim}{Q}_i$ and $P_\delta = \cup_{i<\delta} P_i$ for limit δ and $P_{j,i} = P_i / P_j$. Actually, we will define by induction $\underset{\sim}{Q}_i$, $P_{j,i} \in V^{P_j}$ and $\aleph_1$-oracles $\bar{M}_i \in V^{P_{i+1}}$ (actually a P_{i+1}-name) and $\bar{M}_i^j \in V^{P_i}$ for $j < i < \omega_2$ (remember that having defined $\underset{\sim}{Q}_i$ we have determined P_{i+1} so we can go about our next task). We define by induction demanding that:

$(*)_i$ for each $j < i$ in the universe $V^{P_{j+1}}$, the forcing notion $P_{j+1,i}$ satisfies the $\bar{M}_j$-c.c (if $j+1 = i$ this is an empty demand).

For $i = 0$ we have nothing to check. For i limit use the claims on FS iteration from II and Claim 3.2. So to carry the induction assume that $i < \omega_2$ and $\underset{\sim}{Q}_j$ $(j < i)$ and $\bar{M}_j \in V^{P_{j+1}}$ (for $j < i$) have been defined (so also P_j for $j \leq i$, and $(*)_i$ holds). For each $j < i$, let $P_{j,i} \in V^{P_j}$ be the (P_j-name of the) forcing notion (of power $\aleph_1$) such that $P_i = P_j * \underset{\sim}{P}_{j,i}$ i.e. $P_i/\underset{\sim}{G}_{P_j}$. We assume that $F_i \in V^{P_i}$ is a splitting homomorphism; otherwise we let $\underset{\sim}{Q}_i$ be a P_i-name of the Cohen forcing and $\bar{M}_i$ be any $\aleph_1$-oracle in $V^{P_{i+1}}$ (remember that $\diamondsuit_{\aleph_1}$ holds in the ground model and the P_i's satisfy the $\aleph_1$-c.c. and have power $< \aleph_2$ and Cohen forcing satisfies every $\aleph_1$-oracle). Let $\bar{M}_i^j \in V^{P_i}$ (for $j < i$) be an $\aleph_1$-oracle with the property: if $Q \models \bar{M}_i^j$-c.c. in V^{P_i} *then* $P_{j+1,i} * \underset{\sim}{Q} \models \bar{M}_j$-c.c. in $V^{P_{j+1}}$ (note that $\bar{M}_i^j$ exists by Claim 3.3., by the induction hypothesis). Now, let $\bar{M}^* \in V^{P_i}$ be an $\aleph_1$-oracle such that $\bar{M}^*$-c.c. implies $\bar{M}_i^j$-c.c. for $j < i$ (existing by Claim 3.1). Apply now the main lemma 4.5 in V^{P_i} (for $h = F_i$) to obtain $P = Q_i \in V^{P_i}$ and $\underset{\sim}{X}_i$ (a Q_i-name so actually a P_{i+1}-name) satisfying the conclusions of 4.5, so by the choice of $\bar{M}^*$, $j < i$ implies in $V^{P_{j+1}}$, the forcing notion $P_{j+1,i+1} = P_{j+1,i} * \underset{\sim}{Q}_i$ satisfies the $\bar{M}_j$-c.c. Finally apply Lemma 2.1 in $V^{P_{i+1}} = V^{P_i * \underset{\sim}{Q}_i}$ for the type appearing in $(\alpha) - (\gamma)$ of 4.5, to obtain an $\aleph_1$-oracle $\bar{M}_i \in V^{P_{i+1}}$ such that the above mentioned type is omitted in any generic extensions of $V^{P_{i+1}}$ with forcing notions satisfying the $\bar{M}_i$-c.c. and such that: if $\underset{\sim}{R}$ is a (P-name of a) forcing notion satisfying the $\bar{M}_i$-c.c. in $V^{P_{i+1}}$ then $Q_i * \underset{\sim}{R}$ satisfies the $\bar{M}^*$-c.c. in V^{P_i}. Now we can check $(*)_{i+1}$.

As P_{ω_2} satisfies the $\aleph_1$-c.c. (by 1.6(1)) and has power $\aleph_2$, there are at most $\aleph_2^{\aleph_0} = \aleph_2$ P_{ω_2}-names of reals; on the other hand we have added reals $\aleph_2$ times, so $V^{P_{\omega_2}} \models$"$2^{\aleph_0} = \aleph_2$".

To conclude the proof of 4.3, assume that $F \in V^{P_{\omega_2}}$ is a splitting homomorphism. Then there is some $i < \omega_2$ such that $F \restriction [\mathcal{B}]^{V^{P_i}} = F_i$ is a splitting homomorphism in V^{P_i} (and $\mathrm{cf}(i) = \aleph_1$). So at stage i we had $\underset{\sim}{Q}_i$ introducing $\underset{\sim}{X} \in \mathcal{B}$ and an $\aleph_1$-oracle $\bar{M}_i \in V^{P_{i+1}}$ such that $F(\underset{\sim}{X})$ cannot exist in generic extensions of $V^{P_{i+1}}$ with forcing notions satisfying the $\bar{M}_i$-c.c. Examining our construction and remembering Claim 3.2 it is easy to prove by induction on $\alpha \leq \omega_2$ that for $\alpha > i$ we have $V^{P_{i+1}} \models$ "$P_{i+1,\alpha}$ satisfies the $\bar{M}_i$-c.c.". Hence $V^{P_{\omega_2}} = (V^{P_{i+1}})^{P_{i+1,\omega_2}}$ is such an extension of $V^{P_{i+1}}$, a contradiction. $\square_{4.3}$

4.7. *Proof of the Main Lemma 4.5.* In this proof let Se denote the set of sequences $\bar{a} = \langle a_i : i \leq \omega \rangle$ such that the sequence is monotone, $a_i \neq a_{i+1}$, for $i < \omega, a_i$ is rational, but a_ω is irrational, and $\langle a_i : i < \omega \rangle$ converges to a_ω (and they are from the interval $(0,1)$).

4.7A Definition. Let $P = P(\langle \bar{a}^\alpha : \alpha < \beta \rangle)$ where $\beta \leq \omega_1, \bar{a}^\alpha \in Se$ and a^α_ω for $\alpha < \beta$ are pairwise distinct, denote the following forcing notion: $p \in P$ iff the following three conditions hold:

(a) $p = (U_p, f_p)$, where U_p is an open subset of $(0,1)$, $\mathrm{cl}(U_p)$ is of measure $< 1/2$, and f_p is a function from U_p to $\{0,1\}$;

(b) there are n and b_l, (for $l \leq n$), I_l (for $l < n$) such that $0 = b_0 < b_1 < \cdots < b_{n-1} < b_n = 1$ and $U_p = \bigcup_{l=0}^{n-1} I_l$, I_l is an open subset of (b_l, b_{l+1}) and even $\mathrm{cl}(I_l) \subseteq (b_l, b_{l+1})$.

(c) I_l is either a rational interval, $f_p \restriction I_l$ constant, *or* for some $\alpha < \beta$ and $k_l < \omega$, $I_l = \bigcup_{k_l \leq m < \omega} (a^\alpha_{2m}, a^\alpha_{2m+1})$, $f_p \restriction (a^\alpha_{4m+2i}, a^\alpha_{4m+2i+1})$ is constantly i when $k_l \leq 2m$, $m < \omega$, $i \in \{0,1\}$.

The order on P is: $p \leq q$ iff $U_p \subseteq U_q, f_p \subseteq f_q$, and $\mathrm{cl}(U_p) \cap U_q = U_p$.

Finally we let $\underset{\sim}{X}_P = \bigcup\{f_p^{-1}\{0\} : p \in G_P\}$.

4.7B Stage 1. We will define here a statement, in the next stage prove that it suffices for proving the main lemma, and later prove it.

(St) Let $P_\delta = P(\langle \bar{a}^\alpha : \alpha < \delta \rangle), \delta < \omega_1$ be given, $\delta \geq \omega$ as well as a countable model M^*_δ of ZFC$^-$ such that $P_\delta \in M^*_\delta$, a condition $(p^*, r^*) \in P_\delta \times Q$ and a $(P_\delta \times Q)$-name $\underset{\sim}{\varphi}$ of a definition of a Borel set (this is a candidate for $h(\underset{\sim}{X}_P)$ i.e. the A in 4.5; Q is of course Cohen forcing).

Then we can find $\bar{a}^\delta \in Se$ such that $(\forall \alpha < \delta)[a^\delta_\omega \neq a^\alpha_\omega]$ and letting $P_{\delta+1} = P(\langle \bar{a}^\alpha : \alpha \leq \delta \rangle)$, the following conditions hold:

(A) Every pre-dense subset of P_δ which belongs to M_δ is a pre-dense subset of $P_{\delta+1}$ (note that if M_δ is quite closed this implies the same for $(P_\delta \times Q, P_{\delta+1} \times Q)$, see 2.3(1)).

(B) There is $(p',r') \in P_{\delta+1} \times Q$, such that $(p^*,r^*) \leq (p',r')$ and one of the following holds, for some n:

(B1) $(p',r') \Vdash_{P_{\delta+1}\times Q}$ "$a^\delta_\omega \in \mathrm{Bo}[\underset{\sim}{\varphi}]$, and $\bigcup_{n<m<\omega}(a^\delta_{4m+2}, a^\delta_{4m+3}) \cap \underset{\sim}{X}_{P_{\delta+1}} = \emptyset$" and "$a^\delta_\omega \in h(\bigcup_{n<m<\omega}(a^\delta_{4m+2}, a^\delta_{4m+3}))$";

or

(B2) $(p',r') \Vdash_{P_{\delta+1}\times Q}$"$a^\delta_\omega \notin \mathrm{Bo}(\underset{\sim}{\varphi})$ and $\bigcup_{n<m<\omega}(a^\delta_{4m}, a^\delta_{4m+1}) \subseteq \underset{\sim}{X}_{P_{\delta+1}}$" and "$a^\delta_\omega \in h(\bigcup_{n<m<\omega}(a^\delta_{4m}, a^\delta_{4m+1}))$".

4.7C Stage 2. It is enough to prove the statement (St).

We choose M^α and $\bar{a}^\alpha \in Se$ by induction on $\alpha < \omega_1$ such that $\beta < \alpha \Rightarrow a^\alpha_\omega \neq a^\beta_\omega$, and we also choose pre-dense subsets $\mathcal{I}_\gamma$ of $P(\langle \bar{a}^\beta : \beta < \alpha\rangle)$ for $\gamma < \omega\alpha + \omega$ such that: $\gamma < \omega\alpha_1 + \omega$, $\alpha_1 < \alpha_2$ implies $\mathcal{I}_\gamma$ is a pre-dense subset of $P(\langle \bar{a}^\beta : \beta < \alpha_2\rangle)$ (of course $\alpha_1 < \alpha_2$ implies $P(\langle \bar{a}^\alpha : \alpha < \alpha_1\rangle) \subseteq P(\langle \bar{a}^\alpha : \alpha < \alpha_2\rangle)$, moreover $P(\langle \bar{a}^\alpha : \alpha < \alpha_1\rangle) \subseteq_{\mathrm{ic}} P(\langle \bar{a}^\alpha : \alpha < \alpha_2\rangle)$, read Definition 4.7A). For $\alpha \leq \omega$ choose any $\bar{a}^\alpha \in Se$ with $\mathcal{I}_\gamma = \{(\emptyset, \emptyset)\}$. Generally make sure $\{\mathcal{I}_{\omega\alpha+n} : n < \omega\}$ include $\{\mathcal{I} : \mathcal{I} \in \bigcup_{\beta\leq\alpha} M_\beta, \mathcal{I}$ a pre-dense subset of $P(\langle \bar{a}^\beta : \beta < \alpha\rangle)\}$. Now in stage $\alpha \geq \omega$ a bookkeeping gives us a $(P_\delta \times Q)$-name $\underset{\sim}{\varphi}_\alpha$ as in (St). Choose M^α including $\bigcup_{\gamma\leq\alpha} M_\gamma$, $\underset{\sim}{\varphi}_\alpha$, $\langle \bar{a}^\beta : \beta < \alpha\rangle$, choose $\bar{a}^\alpha$ as in (St) and choose $\{\mathcal{I}_{\omega\alpha+n} : n < \omega\}$ such that all forcing statement mentioned in clause (B) of (St) continue to hold if we replace $P_{\delta+1}$ by any P such that $P_{\delta+1} \subseteq_{ic} P$ and each $\mathcal{I}_{\omega\alpha+n}$ is pre-dense in P for $n < \omega$ and each predense subset of P_δ which belongs to M_δ appears in $\{\mathcal{I}_{\omega\alpha+n} : n < \omega\}$ (possible by 2.1(A)(2)). Clearly $P = P(\langle \bar{a}^\alpha : \alpha < \omega_1\rangle)$ satisfies the $\bar{M}$-c.c. For the conclusion of 4.5 let $\underset{\sim}{X} = \underset{\sim}{X}_P$ and $(p^*,r^*) \in P \times Q$ force $\underset{\sim}{A}$ is a counterexample, (so $(\alpha), (\beta), (\gamma)$ there hold) so for some $\delta \in [\omega, \omega_1]$, $\underset{\sim}{A} = \underset{\sim}{\varphi}_\delta$.

Remember that if $(p',r') \Vdash_{P_{\delta+1}\times Q}$ "$a^\delta \in \mathrm{Bo}(\underset{\sim}{\varphi})$" then this remains valid if we replace $P_{\delta+1}$ by any forcing notion $P, P_{\delta+1} \subseteq_{\mathrm{ic}} P$ provided that certain (countably many) maximal antichains of $P_{\delta+1}$ remain maximal antichains of P

which we have guaranteed.

If clause (B1) holds then $B = \bigcup_{n<m<\omega}(a^\delta_{4m+2}, a^\delta_{4m+3})$ contradicts clause (γ) of 4.5 (for $\underset{\sim}{A}$ =Bo($\underset{\sim}{\varphi}$)) and if clause (B2) holds then $B = \bigcup_{n<m<\omega}(a^\delta_{4m}, a^\delta_{4m+1})$ contradicts clause (β) of 4.5 (for $\underset{\sim}{A}$ =Bo($\underset{\sim}{\varphi}$)). We get a contradiction, so P, $\underset{\sim}{X}$ from above proves the main lemma 4.5 (which suffices for proving theorem 4.3).

So from now on we concentrate on the proof of (St).

4.7D Stage 3. Choosing $\bar{a}^\delta$. So let P_δ, $\langle \bar{a}^\alpha : \alpha < \delta \rangle$, $\underset{\sim}{\varphi}$, M^*_δ and (p^*, r^*) be given, as in the assumption of (St), choose λ big enough (i.e. $\lambda = \beth^+_8$), N a countable elementary submodel of $(H(\lambda), \in)$ containing $P_\delta, \langle \bar{a}^\alpha : a < \delta \rangle, \underset{\sim}{\varphi}, M^*_\delta, h$.

Choose a real a^δ_ω, which belongs to $(0,1) \setminus \text{cl}(U_{p^*})$ but does not belong to any Borel set of measure zero which belongs to N. This is possible as by demand (a) in the definition of $P(\langle \bar{a}^\alpha : \alpha < \delta \rangle), \text{cl}(U_{p^*})$ has measure $< 1/2$. So $(0,1) \setminus \text{cl}(U_{p^*})$ has positive measure, whereas the union of all measure zero Borel sets in N is a countable union hence has measure zero. So a^δ_ω is a random real over N and $N[a^\delta_\omega]$ is a model of enough set theory: ZFC$^-$+ "$\mathcal{P}^8(\omega)$ exists" (where $\mathcal{P}(A)$ is the power set, $\mathcal{P}^{n+1}(A) = \mathcal{P}(\mathcal{P}^n(A)))$ (all those facts are well known; see e.g. Jech [J]). Clearly $a^\delta_\omega \in h((0, a^\delta_\omega))$ or $a^\delta_\omega \in h((a^\delta_\omega, 1))$, so w.l.o.g. the former occurs. It is also clear that for every $\varepsilon > 0$, $a^\delta_\omega \in h((a^\delta_\omega - \varepsilon, a^\delta_\omega))$. [Otherwise, choose a rational b, $a^\delta_\omega - \varepsilon < b < a^\delta_\omega$, then $a^\delta_\omega \notin h((a^\delta_\omega - \varepsilon, a^\delta_\omega)) \Rightarrow a^\delta_\omega \in h((0,b))$. Hence $a^\delta_\omega \in h((0,b)) \setminus (0,b)$, but this set has measure zero (by the property of h) and obviously belongs to N].

Now let $\langle b_n : n < \omega \rangle \in N[a^\delta_\omega]$ be a strictly increasing sequence of rationals converging to a^δ_ω. Next in $N[a^\delta_\omega]$ we define a forcing notion R (the well known dominating function forcing):

$R = \{(f, g) : f$ a function from some $n < \omega$ to ω, satisfying $(\forall i < n) f(i) > i$, and g is a function from ω to $\omega\}$.

The order is

$$(f,g) \le (f',g') \text{ iff } f \subseteq f', (\forall l) g(l) \le g'(l) \text{ and}$$
$$(\forall i)[i \in \mathrm{Dom}(f') \& i \notin \mathrm{Dom}(f) \Rightarrow g(i) \le f'(i)].$$

Choose a subset G of R (remember that $R \in N[a^\delta_\omega]$) generic over $N[a^\delta_\omega]$, next define a function $f^* = \underset{\sim}{f}^*[G] = \cup\{f : (f,g) \in G\} \in (N[a^\delta_\omega])[G]$. So it is known that $(N[a^\delta_\omega])[G] = (N[a^\delta_\omega])[f^*]$ and a finite change in f^* preserve genericity, i.e. if $f' : \omega \to \omega$, $(\forall i < \omega) f(i) > i$ and $\{n : f^*(n) \neq f'(n)\}$ is finite then for some subset G' of R generic over $N[a^\delta_\omega]$ we have $f' = \underset{\sim}{f}[G']$ and $(N[a^\delta_\omega])[G] = (N[a^\delta_\omega])[f^*] = (N[a^\delta_\omega])[f'] = (N[a^\delta_\omega])[G']$). We shall work for a while with the model $N[a^\delta_\omega][f^*]$. We define (in this model) a sequence of natural numbers $\langle n(l) : l < \omega\rangle$, defining $n(l)$ by induction on l. Let $n(0) = 0$, and $n(l+1) = f^*(n(l))$. Now we define for $m < 4$ and $k < \omega$ a set A^k_m, $A^k_m = \bigcup_{k \le l < \omega} (b_{n(4l+m)}, b_{n(4l+m+1)})$.

So $A^0_m (m = 0,1,2,3)$ is a partition of (b_0, a^δ_ω) modulo measure zero as $\{b_n : n < \omega\}$ has measure zero, but remember $a^\delta_\omega \in h((b_0, a^\delta_\omega))$ hence for some unique $m(*)$, $a^\delta_\omega \in h(A^0_{m(*)})$. Note that $A^k_m \in N[a^\delta_\omega][f^*]$, but $h \restriction N[a^\delta_\omega][f^*]$ does not necessarily belong to this model, so we determine $m(*)$ in V. As we could have made a finite change in f^* (replacing $f^*(0)$ by $f^*(n(m(*)))$, legal as $f^*(n(m(*))) > n(m(*)) > m(*) \ge 0$) we can assume $a^\delta_\omega \in h(A^0_0)$.

As $a^\delta_\omega \in h((a^\delta_\omega - \varepsilon, a^\delta_\omega))$ for every ε, and as h is a homomorphism, $a^\delta_\omega \in h(A^k_0)$ for every k.

First let us try to choose $\bar{a}^\delta = \langle b_{n(l)} : l < \omega\rangle {}^\frown \langle a^\delta_\omega\rangle$.

4.7E Stage 4. Condition (A) of (St) holds.

This means:

4.7E1 Subclaim. *Every pre-dense subset $\mathcal{J}$ of P_δ which belongs to M_δ is pre-dense in $P_{\delta+1}$.*

Proof of the Subclaim. As $M_\delta \in N$ clearly $\mathcal{J} \in N$. Let $p \in P_{\delta+1}$, $p \notin P_\delta$, so by the definition of $P_{\delta+1}$ there are $q \in P_\delta$ and rational numbers c_0, c_1 and a natural number $l(0)$ such that

$$0 < c_0 < a^\delta_\omega < c_1 < 1, \quad c_0 < b_{n(4l(0))}, \quad c_0 > b_{n(4l(0))-1},$$
$$\mathrm{cl}(U_q) \cap [c_0, c_1] = \emptyset, \quad U_p = U_q \cup A^{l(0)}_0 \cup A^{l(0)}_2, \quad f_p = f_q \cup 0_{A^{l(0)}_0} \cup 1_{A^{l(0)}_2}.$$

(0_A is the function with domain A which has constant value 0; similarly 1_A.)

Before we continue we prove:

4.7E2 Fact. Let $r \in P_\delta, \mathcal{J} \subseteq P_\delta$ be dense, $(c_0, c_1) \subseteq (0,1)$ be an open interval disjoint from U_r. Then

$$C = \{x \in (c_0, c_1) : \text{ there is } r_1 \in \mathcal{J} \text{ such that } r_1 \geq r \text{ and } x \notin \mathrm{cl}(U_{r_1}\}$$

has measure $|c_1 - c_0|$ (subtraction as reals).

Proof of the Fact. The condition is equivalent to "$(c_0, c_1) \setminus C$ has measure zero", so we can partition (c_0, c_1) into finitely many intervals and prove the conclusion for each of them. So w.l.o.g. the measure of (c_0, c_1) is $< 1/2$. Now for every $\varepsilon > 0$ we can find r_0, $r \leq r_0 \in P_\delta$ such that $U_{r_0} \cap (c_0, c_1) = \emptyset$ and U_{r_0} has measure $\geq 1/2 - \varepsilon$ (but of course $< 1/2$). As $\mathcal{J} \subseteq P_\delta$ is dense, there is $r_2 \in \mathcal{J}, r_0 \leq r_2 \in P_\delta$. So (ignoring the sets $\mathrm{cl}(U_{r_l}) \setminus U_{r_l}$, $l = 0, 2$ which have measure zero):

(i) $(c_0, c_1) \setminus U_{r_2} \subseteq C$ (by C's definition);

(ii) $(c_0, c_1) \cap U_{r_2} \subseteq U_{r_2} \setminus U_{r_0}$.

Hence

(iii) $\mathrm{Leb}((c_0, c_1) \setminus C) \leq \mathrm{Leb}((c_0, c_1) \cap U_{r_2}) \leq \mathrm{Leb}(U_{r_2} \setminus U_{r_0}) \leq \mathrm{Leb}(U_{r_2}) - \mathrm{Leb}(U_{r_0})$
$\leq 1/2 - (1/2 - \varepsilon) = \varepsilon$.

As this holds for every ε we finished the proof of the fact. $\square_{4.7E2}$

4.7E3. *Continuation of the proof of the subclaim.* Let $\mathcal{J}_1 = \{r \in P_\delta : (\exists q_1 \in \mathcal{J})(q_1 \leq r)\}$, so $\mathcal{J}_1$ is a dense open subset of P_δ and $\mathcal{J}_1 \in N$. Now for every $k \geq n(4l(0))$ let

$T_k = \{t : t \in P_\delta, U_t$ is the union of finitely many intervals whose endpoints are from $\{b_l : n(4l(0)) \leq l < k\}$ and $\mathrm{Leb}(U_q \cup U_t) < 1/2\}$.

So T_k is finite, and for every $t \in T_k$, $q \leq q \cup t \in P_\delta$, and $a^\delta_\omega \notin \mathrm{cl}(U_t)$ (where $q \cup t = (U_q \cup U_t,\ f_q \cup f_t)$, of course; on q see beginning of the proof of 4.7E).

Now in the model N (as $M_\delta \in N$ hence $\mathcal{J}_1 \in N$) we can define, for each k and $t \in T_k$:

$D_t \stackrel{\text{def}}{=} \{x \in (c_0, c_1) : \text{there is } r \in \mathcal{J}_1, r \geq q \cup t, x \notin \text{cl}(U_r)\}$.

By the fact 4.7E2 we have proved, we know that $(c_0, c_1) \setminus D_t \setminus U_t$ has measure zero (note, $\text{cl}(U_t) \setminus U_t$ has measure zero). As $a^\delta_\omega \in (0,1) \setminus U_t$ and a^δ_ω does not belong to any Borel set of measure zero which belongs to N, clearly,

(*) for every $k \geq n(4l(0))$ and $t \in T_k$ we have $a^\delta_\omega \in D_t$.

So for each $k \geq n(4l(0))$ and $t \in T_k$, there is $r_t \in P_\delta, r_t \in \mathcal{J}_1$, $t \cup q \leq r_t$ such that $a^\delta_\omega \notin \text{cl}(U_{r_t})$. Hence for some $g(t) < \omega$, $[b_{g(t)}, a^\delta_\omega] \cap \text{cl}(U_{r_t}) = \emptyset$ and $(a^\delta_\omega - b_{g(t)}) < 1/2 - \text{Leb}(U_{r_t})$. As T_k is finite, we can define a function $g : \omega \to \omega$, by $g(k) = \max\{g(t) : t \in T_k\}$. Clearly $g \in N[a^\delta_\omega]$, because, in defining g we have used $\mathcal{J}_1$, $n(4l(0))$, $\langle b_l : l < \omega\rangle$, a^δ_ω, but not $\langle b_{n(l)} : l < \omega\rangle$, $\langle n(l) : i < \omega\rangle$ or f^*. Hence for every large enough $l, g(l) < f^*(l)$, because f^* dominates $N[a^\delta_\omega] \cap {}^\omega\omega$. So by the choice of the $n(l)$'s, for every large enough l, $g(n(l)) < n(l+1)$. Choose a large enough l, let $k = n(4l)+1, U_t = U_p \restriction [b_{n(4l(0))}, b_k], f_t = f_p \restriction U_t, t = (U_t, f_t)$ and it belongs to T_k. Now r_t and p are compatible, (the "$(a^\delta_\omega - b_{g(t)}) < 1/2 - \text{Leb}(U_{r_t})$" guarantees that $\text{Leb}(U_p \cup U_{r_t}) < 1/2$: other parts should be clear too). So we finish the proof of the subclaim 4.7E1. $\square_{4.7E1}$

This really proves part (A) of (St).

4.7F Stage 5. Condition (B) of (St). Remember $\underset{\sim}{\varphi}$ is a $P_\delta \times Q$-name of a (code of a) Borel set (and it belongs to N, hence the $\aleph_0$ maximal antichains involved in its definition are, by 4.7E1 maximal antichains of $P_{\delta+1} \times Q$), so it is also a $(P_{\delta+1} \times Q)$-name. Pick a large enough k such that:

$p^*_1 = (U_{p^*} \cup A^k_0 \cup A^k_2, f_{p^*} \cup 0_{A^k_0} \cup 1_{A^k_2})$ belongs to $P_{\delta+1}$.

So $p^*_1 \in N[a^\delta_\omega][f^*]$ and $(p^*_1, q^*) \geq (p^*, q^*)$ in $P_{\delta+1} \times Q$, so there is $(p', r') \geq (p^*_1, q^*)$ forcing an answer to "$a^\delta_\omega \in \text{Bo}[\underset{\sim}{\varphi}]$", i.e. $(p', r') \Vdash_{P_{\delta+1} \times Q}$ "$a^\delta_\omega \in \text{Bo}[\underset{\sim}{\varphi}]$" or $(p', r') \Vdash_{P_{\delta+1} \times Q}$ "$a^\delta_\omega \notin \text{Bo}[\underset{\sim}{\varphi}]$". If the second possibility holds, then (B2) holds, so condition (B) of (St) holds (remember we have made $a^\delta_\omega \in h(A^k_0)$ for every k in stage 3).

So suppose $(p', r') \Vdash_{P_{\delta+1} \times Q}$ "$\underset{\sim}{a}^{\delta}_{\omega} \in \mathrm{Bo}[\underset{\sim}{\varphi}]$". Observe that the truth value of such a statement can be computed in $N[a^{\delta}_{\omega}][f^*]$ (i.e., we get the same result in the universe and in this countable model). But f^* is R-generic over $N[a^{\delta}_{\omega}]$. So if something holds, then there is $(f_0, g_0) \in G$ is such that:

(α) $(f_0, g_0) \Vdash_R$ "$(p', r') \Vdash_{P_{\delta+1} \times Q}$ "$\underset{\sim}{a}^{\delta}_{\omega} \in \mathrm{Bo}[\underset{\sim}{\varphi}]$" ",

(β) $f_0 \subseteq f^*, (\forall i < \omega)[i \notin \mathrm{Dom}(f_0) \Rightarrow g_0(i) \leq f^*(i)]$.

(Note that the definition of $P_{\delta+1}$ depends on f^*.) So if we change f^* in finitely many places, maintaining (β), it will still be true that $(p', r') \Vdash_{P_{\delta+1} \times Q}$ "$\underset{\sim}{a}^{\delta}_{\omega} \in \mathrm{Bo}[\underset{\sim}{\varphi}]$". But we can do it in such a way that only finitely many $n(l)$'s are changed and for some n the old A^k_0 becomes A^n_2, and the old A^k_2 becomes A^{n+1}_0, so now clause (B1) of (St) will hold.

In any case (B) of (St) holds, hence we have finished proving (St) hence by 4.7C we have finished the proof of the main lemma and by 4.6 we have finished the proof of the theorem. $\square_{4.5}$

$\square_{4.3}$

§5. Automorphisms of $\mathcal{P}(\omega)$/finite

General topologists have been interested in what $AUT(\mathcal{P}(\omega)/\text{finite})$ can be. ($\mathcal{P}(\omega)$ is viewed as a Boolean algebra of sets, $\mathcal{P}(\omega)$/finite is the quotient when dividing by the ideal of finite subsets of ω). Or equivalently: what can be the group of autohomeomorphisms of $\beta(\mathbb{N}) \setminus \mathbb{N}$?

Note that $AUT(\mathcal{P}(\omega))$ is isomorphic to the set of permutations of ω ($Per(\omega)$). It is well known that:

5.1 Claim. (CH) $\mathcal{P}(\omega)$/finite is a saturated atomless BA (= Boolean algebra) of power $\aleph_1$, so it has $2^{\aleph_1}$ automorphisms (saturated in the model theoretic sense). $\square_{5.1}$

5.1A Remark. This claim will not be used here.

It is also easy to note that:

(*) If $f \in Per(\omega)$ the set of permutations of ω, then $A/\text{finite} \to f(A)/\text{finite}$ is an automorphism of $\mathcal{P}(\omega)/\text{finite}$.

Moreover:

5.2 Claim. If f is a one-to-one function, $\text{Dom}(f), \text{Rang}(f)$ are cofinite subsets of ω then $A/\text{finite} \to f(A)/\text{finite}$ is an automorphism of $\mathcal{P}(\omega)/\text{finite}$. (Such an f is called an almost permutation of ω, and the induced automorphism is called trivial, and we say it is induced by f). $\square_{5.2}$

5.3 Claim. If $V = L$, or just $V \models$ "$2^{\aleph_0} = \aleph_1$, $2^{\aleph_1} = \aleph_2$ and $\diamondsuit_{S^2_1}$ holds where $S^2_1 = \{\delta < \aleph_2 : \text{cf}(\delta) = \aleph_1\}$" *then* we can define for $i \in S^2_1$, P^*_i, F_i such that:

(a) P^*_i is a c.c.c. forcing of cardinality $\aleph_1$ and for simplicity the set of elements of P^*_i is i.

(b) $\underset{\sim}{F}_i$ is a P^*_i-name of a subset of $[\mathcal{P}(\omega) \times \mathcal{P}(\omega)]^{V^{P^*_i}}$

(c) if $\langle P_i : i < \omega_2 \rangle$ is $\lessdot$-increasing continuous sequence of c.c.c. forcing notions of cardinality $\aleph_1$, the set of elements of P_i is $\subseteq \omega_2$ and $P_{\omega_2} = \bigcup_{i<\omega_2} P_i$, $\underset{\sim}{F}$ is a P_{ω_2}-name and $\underset{\sim}{F} \subseteq [\mathcal{P}(\omega) \times \mathcal{P}(\omega)]^{V^{P_{\omega_2}}}$, i.e. this is forced *then* the set $\{i : S^2_1 : P^*_i = P^*_i \text{ and } \underset{\sim}{F}\restriction[\mathcal{P}(\omega)]^{V^{P_i}} = \underset{\sim}{F}_i\}$ is stationary in ω_2.

Proof. Translate everything to P_{ω_2}-names and apply $\diamondsuit_{\{\delta<\omega_2:\text{cf}(\delta)=\aleph_1\}}$ (remember the $\aleph_1$-chain condition). $\square_{5.3}$

5.4 Claim.

Suppose P_i $(i \leq \omega_2)$ are as in Claim 5.3, $F \in [AUT(\mathcal{P}(\omega)/\text{finite})]^{V^{P_{\omega_2}}}$ and

$\{i < \omega_2 : \text{cf}(i) = \aleph_1$ and $F\restriction[\mathcal{P}(\omega)/\text{finite}]^{V^{P_i}}$ is

(i) not in V^{P_i} or

(ii) not an automorphism of $[\mathcal{P}(\omega)/\text{finite}]^{V^{P_i}}$ or

(iii) induced by some almost permutation f of ω

($f \in V^{P_i}$ i.e. a P_i-name, of course)$\}$

is stationary.

Then F is trivial.

Proof. Since the sets defined by (i) and (ii) are not stationary,

$$S = \{i < \omega_2 : \text{cf}(i) = \aleph_1, F\restriction[\mathcal{P}(\omega)/\text{finite}]^{V^{P_i}} \text{ is in } V^{P_i}, \text{ and is a trivial automorphism }\}$$

is stationary.

For $i \in S$, let $f_i \in V^{P_i}$ be an almost permutation of ω inducing $F\restriction[P(\omega)/\text{finite}]^{V^{P_i}}$. By a suitable version of Fodor's lemma we can find $f^* \in V^{P_{\omega_2}}$, $S^* \subseteq S$ stationary, such that $(\forall i \in S^*)$ $[f_i = f^*]$. Clearly f^* induces $F\restriction[P(\omega)/\text{finite}]^{V^{P_{\omega_2}}}$. $\square_{5.4}$

We want to show:

5.5 Main Theorem. CON(ZFC) implies CON(ZFC $+2^{\aleph_0} = \aleph_2+$ every automorphism of $\mathcal{P}(\omega)/\text{finite}$ is trivial). In fact, if $\diamondsuit_{\aleph_1}$ and $\diamondsuit_{S_1^2}$, where $S_1^2 = \{\delta < \omega_2 : \text{cf}(\delta) = \aleph_1\}$, then for some c.c.c. forcing notion P of cardinality $\aleph_2$, the conclusion holds in V^P.

We postpone the work concerning specifically automorphisms of $\mathcal{P}(\omega)/\text{finite}$ to the next section. To be precise we will prove there the following.

5.6 Main Lemma. Suppose

$V \models$ "$2^{\aleph_0} = \aleph_1$, $\bar{M}^*$ is an $\aleph_1$-oracle, $F \in AUT(\mathcal{P}(\omega)/\text{finite})$ is not trivial".

Then there is a forcing notion P such that

1) $|P| = \aleph_1$,
2) P satisfies the $\bar{M}^*$-c.c.
3) P introduces a subset X of ω, such that if Q is the Cohen forcing, then in $V^{P*\underset{\sim}{Q}}$ there is no $Y \subseteq \omega$ satisfying $\{Y \cap F(C) =_{ae} F(B) \Leftrightarrow X \cap C =_{ae} B : C, B \in [\mathcal{P}(\omega)]^V\}$ (where $A =_{ae} A^\dagger$ means $A \subseteq_{ae} A^\dagger$ and $A^\dagger \subseteq_{ae} A$, where $A \subseteq_{ae} A^\dagger$ iff $A \setminus A^\dagger$ is finite) or even just satisfying $\{Y \cap F(C) =_{ae} F(B) : X \cap C =_{ae} B \text{ where } B, C \in [\mathcal{P}(\omega)]^V\}$.

5.6A Remark. Note we really replace F by some $F^\dagger$ from $\mathcal{P}(\omega)$ to $\mathcal{P}(\omega)$ such that $F(A/\text{finite}) = F^\dagger(A)/\text{finite}$; but we do not distinguish.

Proof of 5.5 (assuming 5.6). The proof of 5.5 from 5.6 is very similar to the proof of 4.6 (just replace "splitting homomorphism" by "nontrivial automorphism". Note that 5.6(3) corresponds to 4.5 (α)–(γ)). $\square_{5.5}$

5.7 Claim. In our scheme (of oracle c.c., as used in 4.6) we can demand:

1) for any second category $A \subseteq [\mathcal{P}(\omega)]^{V^{P_{\omega_2}}}$ we have $\{\alpha < \omega_2 : \text{cf}(\alpha) = \aleph_1 \Rightarrow A \cap [\mathcal{P}(\omega)]^{V^{P_i}} \in V^{P_i}$ and is of 2nd category$\}$ contains a closed unbounded set.

2) if $A \subseteq [\mathcal{P}(\omega)]^{V^{P_i}}$, $A \in V^{P_i}$, A is of 2nd category, *then* there is an oracle $\bar{M}$ (in V^{P_i}) such that for any forcing notion $P \in V^{P_i}$ satisfying the $\bar{M}$-c.c., in $(V^{P_i})^P$ the set A remains of the 2nd category.

Proof. For 1) It follows by 2). For 2) see 2.2.

5.8 Conclusion. 1) We can add in 5.5 that in V^P any second category set contains such a subset of cardinality $\aleph_1$.

2) We can find $\aleph_1$-dense sets of reals, some of the first category, some of the second category.

3) Baumgartner's construction [B4] cannot be carried out here as his conclusion fails.

Proof. 1), 2) Left to the reader.

3) Baumgartner [B4] proved the consistency of: if for $\ell = 1, 2$, $A_\ell \subseteq \mathbb{R}$, $|A_\ell \cap (c, d)| = \aleph_1$ for any reals $c < d$ then $(A_1, <) \cong (A_2, <)$; but this fails in V^P by part (2). His proof was: for given A_1, A_2 in a universe satisfying CH, builds a c.c.c. forcing notion P, such that $\Vdash_P$ "$(A_1, <) \cong (A_2, <)$". So for an $\aleph_1$-oracle $\bar{M}$, and A_1, A_2 from V (where $V \models \diamondsuit_{\aleph_1}$) there is no such P satisfying the $\bar{M}$-c.c.

§6. Proof of Main Lemma 5.6

Since the proof is quite long it will be divided into stages.

Stage A. Preliminaries. For a sequence $\bar{A} = \langle A_i : i < \alpha \rangle$ $(\alpha \leq \omega_1)$ of almost disjoint infinite subsets of ω, define

$$P(\bar{A}) = \{f : f \text{ is a partial function from } \omega \text{ to } 2 = \{0,1\},$$
$$f \text{ is a finite union of functions of the form:}$$
$$(1) \text{ finite functions,}$$
$$(2)\ 1_{A_i} \setminus \text{ finite,}$$
$$(3)\ 0_{A_i} \setminus \text{ finite}\}$$

where ℓ_A $(l \in \{0,1\})$ is the function $f(\alpha) = \ell$ if $\alpha \in A$, undefined otherwise, and $f \setminus$finite means $f \restriction (\mathrm{Dom}(f) \setminus A)$ for some finite A.

The idea of the proof is as follows: we try to define $\bar{A} = \langle A_i : i < \omega_1 \rangle$ such that: $P(\bar{A})$ will satisfy the requirements of the lemma for X whose characteristic function is $\cup\{p : p \in G_{P(\bar{A})}\}$. We will have a general pattern for defining $\bar{A}$. Our work will be to show that if all instances of this pattern fail then F must be trivial.

Our first worry is how to make $P(\bar{A})$ satisfy the $\bar{M}^*$-c.c. The following $(*1)$ takes care of this while leaving us with a sufficient degree of freedom in the construction of $\bar{A}$.

$(*1)$ If $\bar{A} = \langle A_j : j < \alpha \rangle$, $\alpha < \omega_1$, M_α is a countable collection of pre-dense subsets of $P(\bar{A})$, $B \subseteq \omega$ is almost disjoint from each $A_j (j < \alpha)$ and is infinite, *then* we can find disjoint, infinite C_1, C_2, $C_1 \cup C_2 = B$ such that for every $\ell \in \{1,2\}$ and every $A \subseteq C_\ell$, if $|A| = |C_\ell \setminus A| = \aleph_0$, then every $\mathcal{I} \in M_\alpha$ is pre-dense in $P(\bar{A}\char`^\langle A, C_\ell \setminus A \rangle)$.

Remark. Note that by 2.3, the same statement holds if we replace $P(\bar{A})$ by $P(\bar{A}) \times Q$, where Q denotes the Cohen forcing.

Proof of $(*1)$. We define by induction on $n < \omega$ a number $k_n < \omega$ and disjoint finite subsets C^1_n, C^2_n of B such that $C^1_n \cup C^2_n = B \cap k_n$ and

$$m < n \Rightarrow C^1_m = C^1_n \cap k_n \ \& \ C^2_m = C^2_n \cap k_n.$$

In the end we will let $C_l = \bigcup_{n<\omega} C^l_n$.

In each stage we deal with an obligation of the following kind:

$\mathcal{I} \in M_\alpha$ is a pre-dense subset of $P(\bar{A})$, τ is a term for a member of $P(\bar{A}\char`^\langle A, C_1 \setminus A\rangle)$ (or C_2 in the other case); note: A, C_1, are not known yet, still $A \subseteq C_1$; our task is to ensure τ will be compatible with some member of $\mathcal{I}$ no matter how A and C_1 are defined as long as $C_1 \subseteq B$, $C^1_n \subseteq C_1$ and $C^2_n \cap C_1 = \emptyset$.

τ has the form $f \cup i_{A\setminus k} \cup j_{C_1\setminus A\setminus k}$ where $i, j \in \{0, 1\}$, $k < \omega$, $f \in P(\bar{A})$ (if this union is not a function, then we don't need to do anything.) We don't know what $A \cap C^1_n$ is. There are $2^{|C^1_n|}$ subsets of C^1_n : $A_1, \ldots, A_\ell, \ldots$, for $1 \leq \ell \leq 2^{|C^1_n|}$; look at $f_\ell \stackrel{\text{def}}{=} f \cup i_{A_\ell\setminus k} \cup j_{C^1_n\setminus A_\ell\setminus k}$. Suppose this is a function, then it belongs to $P(\bar{A})$, hence is compatible with some $g_\ell \in \mathcal{I}$. Now $\mathrm{Dom}(g_\ell) \cap B$ is finite (for B being almost disjoint from every A_j and $g_\ell \in \mathcal{I} \subseteq P(\bar{A})$).

Let : $k_{n,\ell} = \mathrm{Max}(\mathrm{Dom}(g_\ell) \cap B)$ for $1 \leq \ell \leq 2^{|C^1_n|}$, and $k_{n,0} = k_n$;

$$k_{n+1} \stackrel{\text{def}}{=} \mathrm{Max}\{(k_{n,\ell} + 1) : 0 \leq \ell \leq 2^{|C^1_n|}\};$$

$$C^1_{n+1} = C^1_n;$$

$$C^2_{n+1} = C^2_n \cup [(k_{n+1} \setminus C^1_n) \cap B].$$

Taking additional care to make $C_1 = \cup_{n<\omega} C^1_n$ and $C_2 = \cup_{n<\omega} C^2_n$ infinite, we obtain what we need. $\square_{(*1)}$

In view of $(*1)$, it is convenient to introduce the following notation:

Q will denote Cohen forcing. For $\alpha \leq \omega_1$, P_α will denote the forcing notion corresponding to $\bar{A} = \langle A_j : j < \alpha\rangle$ (we delete α from our further notation when α is fixed).

$P_\ell[A]$ will denote $P(\bar{A}\char`^\langle A, C_\ell \setminus A\rangle)$ for $\ell \in \{1, 2\}$, $A \subseteq C_\ell$ as in $(*1)$.

$P^Q_\ell[A]$ will denote $P_\ell[A] \times Q$.

If $p, q \in P_{\omega_1} \times Q$ then we let $p = ((p)_0, (p)_1)$ and $q = ((q)_0, (q)_1)$, we will write $p \cup q$ for $((p)_0 \cup (q)_0, (p)_1 \cup (q)_1)$. For $p \in P_{\omega_1}$, $q \in P_{\omega_1} \times Q$, $p \cup q$ means $(p \cup (q)_0, (q)_1)$, etc. We identify P_{ω_1} with $P_{\omega_1} \times \{\emptyset\} \subseteq P_{\omega_1} \times Q$.

For $A \subseteq B$, Ch_B^A will denote $1_A \cup 0_{B \setminus A}$.

Stage B: The outline of the proof. We say that F is trivial on $A \subseteq \omega$ iff there are $A' \equiv_{\mathrm{ae}} A$ and $f : A' \xrightarrow{1-1} \omega$ such that $F(B) \equiv_{\mathrm{ae}} f(B)$ for each $B \subseteq A'$. Let $J = \{A \subseteq \omega : F \text{ is trivial on } A\}$.

We should distinguish two cases:

Case 1: J is dense, i.e. for every $A \in [\omega]^{\aleph_0}$ we have $J \cap [A]^{\aleph_0} \neq \emptyset$

Case 2: there is $A^+ \in [\omega]^{\aleph_0}$ such that $[A^+]^{\aleph_0} \cap J = \emptyset$.

We will construct the sequence $\bar{A} = \langle A_\alpha : \alpha < \omega_1 \rangle$ by induction. During the induction we will require $\oplus_1$, where

$\oplus_1$ $A_\alpha \in J$ in Case 1 for each $\alpha < \omega_1$ and $\omega \setminus A_0 \subseteq A^+$ in Case 2.

Note that in Case 2 $A_\alpha \subseteq A^+$ guarantees $A_\alpha \notin J$ for $0 < \alpha < \omega_1$.

In stage C below we shall describe the inductive construction of $\langle A_\alpha : \alpha < \omega_1 \rangle$ and $\langle M_\alpha : \alpha < \omega_1 \rangle$ and we show that either $P(\langle A_\alpha : \alpha < \omega_1 \rangle)$ satisfies our requirements or in some step α the statement $(*2)$ defined in stage C holds.

In stage D we shall show that if $(*2)$ holds, then F is trivial on any B provided that B is almost disjoint from $\{A_i : i < \alpha\}$, so $B \in J$, hence Case 1 holds.

In stage E we shall construct a function g which generates F on ω, which proves that F is trivial. This concludes the proof of the main Lemma.

Stage C: Note that $(*1)$ enables us to construct $\bar{A}$ as follows: We let $\langle A_n : n < \omega \rangle$ be any pairwise disjoint infinite subsets of ω satisfying $\oplus_1$ from Stage B. At each later stage, we can use $(*1)$ for any infinite B which is almost disjoint from each earlier set (we can always find such B: arrange the earlier sets in an ω-sequence $\langle B_n : n < \omega \rangle$, and let $B = \{b_n : n < \omega\}$ where $b_n \in B_n \setminus \cup_{m<n} B_m$). Then we choose by induction on $\alpha \in [\omega, \omega_1)$ the set A_α.

What about M_α? It includes M_α^* (in order to prepare the satisfaction of the $\bar{M}^*$-c.c., $\bar{M}^*$ is given in the assumption of 5.6), as well as additional pre-dense subsets which will be specified in the sequel, when their necessity becomes apparent; and of course M_α increases with α and is a model of ZFC$^-$. Sometimes, these will be subsets of $P_\alpha \times Q$ which are to remain pre-dense in $P_\ell^Q[A]$, but we can deal with such subsets in the framework of $(*1)$ without difficulty by 2.3(1). Also, we will preserve pre-density above a given condition, but all these pre-dense sets are in M_α provided M_α is closed enough.

Altogether our induction hypothesis on $\alpha \in [\omega, \omega_1)$ are $\otimes_1$ and

$\oplus_2$ $\langle A_\beta : \beta < \alpha \rangle$ is a sequence of infinite, pairwise almost disjoint subsets of ω and $P_\alpha \stackrel{\text{def}}{=} P(\langle A_i : i < \alpha \rangle)$ (and P_{ω_1} will be $P(\langle A_i : i < \omega_1 \rangle)$).

$\oplus_3$ N_α is a countable set of pre-dense subsets of $P(\langle A_\beta : \beta < \alpha \rangle)$ increasing (but not continuous) in α, $M_\alpha^* \subseteq M_\alpha$, and N_α is equal to $M_\alpha \cap \{\mathcal{I} : \mathcal{I}$ a pre-dense subset of $P(\langle A_\beta : \beta < \alpha \rangle)\}$ where M_α is the Skolem hull of N_α in $(H(\beth_8^+), \in, <^*_{\beth_8^+})$.

Now, in using $(*1)$ we can choose between various candidates for B, and for each of them we can choose between various continuations of $\bar{A}$ (namely, we can choose $\ell \in \{1, 2\}$ and $A \subseteq C_\ell$). How do we choose?

In the verification of 3) of the lemma, we have to consider all possible $\underset{\sim}{Y}$ which are $P_{\omega_1} \times Q$-names of subsets of ω. By CH and the $\aleph_1$-c.c., we can find $P_\alpha \times Q$-names $\underset{\sim}{Y}_\alpha$, which are determined schematically before the actual construction of $\bar{A}$, so that the sequence

$$\langle \underset{\sim}{Y}_\alpha : \omega \leq \alpha < \omega_1, \alpha \text{ an even ordinal } \rangle$$

will contain all possible $\underset{\sim}{Y}$, each appearing $\aleph_1$-times. [More explicitly, we can identify a $P_{\omega_1} \times Q$-name of a $\underset{\sim}{Y} \subseteq \omega$ with $\langle q_{i,j}, \mathbf{t}_{i,j} : i, j < \omega \rangle$, $q_{i,j} \in P_{\omega_1} \times Q$, $\mathbf{t}_{i,j}$ a truth value, for each $i < \omega$ the sequence $\langle q_{i,j} : j < \omega \rangle$ is a maximal antichain, $q_{i,j} \Vdash$ "$[i \in \underset{\sim}{Y}] \equiv \mathbf{t}_{i,j}$". We consider $\underset{\sim}{Y}_\alpha$ only if M_α already contains each $\langle q_{i,j} : j < \omega \rangle$, so they remain maximal antichains. Of course we do not know apriori P_{ω_1} but we can make a list of length ω_1 including all possible $P_i (i < \omega_1)$ and $\underset{\sim}{Y}_\alpha$. As there are $\leq \aleph_1$ candidates, and each candidate appears $\aleph_1$ times everything is clear]. Our choice of B (hence C_1, C_2 by $(*1)$), ℓ, A will be as follows: whenever possible, we choose them so that $\oplus_1$ from Stage B holds and

$$[Ch^A_{C_\ell} \Vdash_{P^Q_\ell[A]} \text{“}\underset{\sim}{Y}_\alpha \cap F(C_\ell) =_{ae} F(A)\text{”}] \text{ is false .}$$

Assume now that P_{ω_1} was obtained by such a procedure fails to satisfy the requirements of the lemma. As 5.6(1) is clearly satisfied and we ensured the satisfaction of 5.6(2), only 5.6(3) may be false. We will show that this is exemplified by $\underset{\sim}{X} = \cup\{f^{-1}(\{1\}) : f \in \underset{\sim}{G}\}$ where $\underset{\sim}{G} \subseteq P_{\omega_1}$ is the name of the generic subset of P_{ω_1}. If the existence of an appropriate Y is not forced already by $(\emptyset, \emptyset)$ the minimal condition in $P_{\omega_1} \times Q$, then some $r \in P_{\omega_1} \times Q$ forces that there is such Y and in this case $P_{\omega_1} \restriction \{p : p \geq (r)_0\}$ satisfies the requirements of the lemma (as the M_α's take care of the satisfaction of the $\bar{M}^*$-c.c. for such restrictions of P_{ω_1} as well). Hence we may assume that $(\emptyset, \emptyset)$ forces the existence of such $\underset{\sim}{Y}$. So, there is $\underset{\sim}{Y} = \underset{\sim}{Y}_\alpha (\omega \leq \alpha < \omega_1, \alpha \text{ even})$ which realizes the relevant type, and w.l.o.g. α is large relative to $\underset{\sim}{Y}$ (remember each $\underset{\sim}{Y}$ appears $\aleph_1$ times.) Since $Ch^A_{C_\ell} \Vdash_{P_{\omega_1} \times Q}$ "$\underset{\sim}{X} \cap C_\ell = A$" (referring to the C_ℓ and A with which we actually choose at stage α), it follows that $Ch^A_{C_\ell} \Vdash_{P_{\omega_1} \times Q}$ "$\underset{\sim}{Y} \cap F(C_\ell) =_{ae} F(A)$". ($F$ from the assumption of 5.6.)

Now, when we are at stage α we know $\underset{\sim}{Y}_\alpha, P_\alpha$ but not P_{ω_1}. So we want to show that this is already true if we consider forcing in $P^Q_\ell[A]$. As the forcing relation for each formula $n \in \underset{\sim}{Y}$ is not affected by this shift, our only problem arises from the fact that equality is forced to hold only *ae*. This difficulty is overcome as follows: if, on the contrary, there is $p \geq Ch^A_{C_\ell}$ in $P^Q_\ell[A]$ forcing "$\underset{\sim}{Y} \cap F(C_\ell) \neq_{ae} F(A)$", then for each $k < \omega$ the set of those conditions "knowing" some counterexample to the equality above k is pre-dense above p in $P^Q_\ell[A]$; so, as we remarked above, we can assume this pre-density is still true in $P_{\omega_1} \times Q$, but (as $p \geq Ch^A_{C_\ell}$) $p \Vdash_{P_{\omega_1} \times Q}$ "$\underset{\sim}{Y} \cap F(C_\ell) =_{ae} F(A)$", a contradiction (i.e. we can discard this case instead taking care of it).

So, $Ch^A_{C_\ell} \Vdash_{P^Q_\ell[A]}$ "$\underset{\sim}{Y} \cap F(C_\ell) =_{ae} F(A)$". But we chose B, ℓ and A doing our best to avoid this, so we must have had no better choice; thus:

Fact. If P_{ω_1} does not satisfy our requirements *then* there is $\alpha < \omega_1$, $\alpha \geq \omega$ such that (∗2) below holds:

(∗2) Assume $A_j (j < \alpha)$, $M_i (i < \alpha)$ were determined as well as a P_α-name $\underset{\sim}{Y} = \underset{\sim}{Y}_\alpha$ of a subset of ω. Let B be any infinite subset of ω almost disjoint from each $A_j (j < \alpha)$ such that $B \in J$ if Case 1 holds, (in Stage B), and let $B = C_1 \cup C_2$ be some partition obtained as (∗1). *Then* for every $\ell \in \{1, 2\}$ and every $A \subseteq C_\ell$ such that $|A| = |C_\ell \setminus A| = \aleph_0$, we have:

$$Ch^A_{C_\ell} \Vdash_{P^Q_\ell[A]} \text{“}\underset{\sim}{Y} \cap F(C_\ell) =_{ae} F(A)\text{”}.$$

Note that (∗2) is a kind of definition of F, not so nice though; so our aim is to show that if F is definable in this weak sense then it is very nicely definable: is trivial.

Stage D: Under (∗2) we have F is trivial on B (i.e. for some one to one $g : B \to \omega$, for every $B' \subseteq B$, $F(B') =_{\text{ae}} \{g(m) : m \in B'\}$.

We will eventually show that (∗2) implies that F is trivial but first we want to prove that $F \restriction \mathcal{P}(B)$ is trivial (this will be accomplished in (∗13) below; however those proofs will be used later, too).

We now assume B is as in (∗2) and concentrate on C_1; C_2 can be treated similarly, putting together we can get the result on B. So now we shall use $P^Q_1[A]$ for $A \subseteq C_1$.

Let $P_\alpha \times Q = \{p_\ell : \ell < \omega\}$, and let for $n < \omega$, $\mathcal{I}_n$ be a dense subset of $P_\alpha \times Q$, $\mathcal{I}_n = \{q^n_i : i < \omega\}$, such that $q^n_i \Vdash_{P_\alpha \times Q}$ “$[n \in \underset{\sim}{Y}] \equiv \mathbf{t}^n_i$” ($\mathbf{t}^n_i$ denotes a truth value, 1 is true, 0 is false).

We want to show that $F \restriction \mathcal{P}(C_1)$ is trivial (i.e. for some $f : C_1 \to \omega$ for every $C' \subseteq C_1$, $F(C') =_{\text{ae}} \{f(n) : n \in C_1\}$ (so $\text{Rang}(f) =_{\text{ae}} F(C_1)$). To do this we partition C_1 into three pieces, $C_1 = A^*_0 \cup A^*_1 \cup A^*_2$ and we shall prove (step by step):

(a) $F \restriction A^*_i$ is nicely defined by (∗6) on a set of 2nd category.

(b) $F \restriction A^*_i$ is continuous on a set of 2nd category.

(c) $F \restriction A^*_i$ is trivial apart from a set of first category.

(d) F is trivial on A^*_i.

We shall now define by induction on $k < \omega$ a natural number n_k and a function $f_k : a_k \to \{0,1\}$ where $a_k = [n_k, n_{k+1}) \cap C_1$. We will take $A^*_i \stackrel{\text{def}}{=} \bigcup_{m<\omega} a_{3m+i}$ for $i < 3$. We want our n_k, f_k to satisfy (∗3) and (∗4) below:

Let us denote by $IF(k)$ the set of the functions f, $\mathrm{Dom}(f)$ a proper initial segment of $C_1 \setminus [0, n_k), \mathrm{Rang}(f) \subseteq \{0,1\}$.

(∗3) (a) $n_0 = 0$

(b) $n_{k+1} > n_k, \mathrm{Rang}(f_k) = \{0,1\}$

(c) For any $\ell < k, m < n_k$, a function f from $C_1 \cap [0, n_k)$ to $\{0,1\}$ and $\mathbf{t}$ denoting a truth value, the following holds:

if for some i, $\mathbf{t}^m_i = \mathbf{t}$ and $f \cup f_k \cup p_\ell \cup q^m_i \in P_\alpha \times Q$

then for some j, $\mathbf{t}^m_j = \mathbf{t}$ and for every $h \in IF(k+1)$,

$f \cup f_k \cup h \cup p_\ell \cup q^m_j \in P_\alpha \times Q$.

(∗4) For every $\ell < k$ and g_0, g_1 functions from $[0, n_k) \cap C_1$ to $\{0,1\}$, one of the following holds: (F is from the hypothesis of the Main Lemma 5.6):

Case (d_α): for *no* $m \in F(C_1) \setminus [0, n_{k+1})$ are there $i(0) < \omega$, $i(1) < \omega$, $\mathbf{t}^m_{i(0)} \neq \mathbf{t}^m_{i(1)}$, and $h \in IF(k+1)$, *such that* for every $h^\dagger$, $h \subseteq h^\dagger \in IF(k+1)$ implies $g_0 \cup f_k \cup h^\dagger \cup p_\ell \cup q^m_{i(0)}$ and $g_1 \cup f_k \cup h^\dagger \cup p_\ell \cup q^m_{i(1)}$ are both in $P_\alpha \times Q$.

Case (d_β): for every $k(*) > k$ and function h from $[n_{k+1}, n_{k(*)}) \cap C_1$ to $\{0,1\}$ there are $m \in F(C_1) \cap [n_{k(*)}, n_{k(*)+1})$, $i(0) < \omega$, $i(1) < \omega$ *such that* $\mathbf{t}^m_{i(0)} \neq \mathbf{t}^m_{i(1)}$, and for every $h^\dagger \in IF(k(*)+1)$, $g_0 \cup f_k \cup h \cup f_{k(*)} \cup h^\dagger \cup p_\ell \cup q^m_{i(0)}$ and $g_1 \cup f_k \cup h \cup f_{k(*)} \cup h^\dagger \cup p_\ell \cup q^m_{i(1)}$ are both in $P_\alpha \times Q$.

We have now to convince the reader that we can define n_k, f_k that satisfy (∗3), (∗4). Assume that we have defined $n_0, \ldots, n_{k-1}, n_k$ and $f_0, \ldots, f_{k-1}$ and we want to define n_{k+1}, f_k. We have finitely many tasks, of three types:

(i) instances of (∗3), (c) namely: given $\ell < k$, $m < n_k$, $f : C_1 \cap [0, n_k) \to \{0,1\}$ and $\mathbf{t}$, we have to satisfy "*if* for some $i \ldots$"

(ii) instances of $(*4)$ (d_α), namely: given $\ell < k$, $g_0, g_1 \in \{0,1\}^{[0,n_k)\cap C_1}$, we try to satisfy "for no $m\ldots$" (but sometimes we fail – then we will remember this failure in (iii), at all later stages).

(iii) failures of $(*4)(d_\alpha)$ in the past: given a task of type (ii) which we did not fulfill at a previous step, we have to satisfy (d_β) for $k(*) =$ our present k.

We will define f_k by taking approximations of it in $IF(k)$, each one of them intends to fulfill one task and contains its predecessor – this ensures that once we have constructed an approximation fulfilling a given task, the f_k that we will eventually obtain will also fulfill it. Our initial approximation is an arbitrary element of $IF(k)$ assuming the values 0 and 1.

So now we deal separately with each type:

(i) Assume such i exists (with f_k interpreted as our current approximation of it). The addition of h can create a contradiction only inside C_1 (by the definition of $IF(k+1)$), and since C_1 is almost disjoint from the domain of any condition in P_α (in this case $(p_\ell \cup q_i^m)_0$), by defining f_k in the dangerous finite portion $C_1 \cap \mathrm{Dom}(p_\ell \cup q_i^m)_0$, as $(p_\ell \cup q_i^m)\restriction C_1$ (and extending for making its domain a (proper) initial segment of $C_1 \setminus [0, n_k)$) we fulfill our task.

(ii) Let $IF_c(k)$ denote the subset of $IF(k)$ consisting of those functions containing our current approximation of f_k. Assume that:

$\otimes_1$ There exists $f \in IF_c(k)$ such that for every $m \in F(C_1)$, $m > \mathrm{Sup}(\mathrm{Dom}(f))$, and for every $f^\dagger \in IF_c(k)$, $f^\dagger \supseteq f$, and for every $i(0), i(1) < \omega$, *if* for every $f'' \in IF_c(k)$, $f'' \supseteq f^\dagger$ both $g_0 \cup f'' \cup p_\ell \cup q^m_{i(0)}$ and $g_1 \cup f'' \cup p_\ell \cup q^m_{i(1)}$ are in $P_\alpha \times Q$ *then* $\mathbf{t}^m_{i(0)} = \mathbf{t}^m_{i(1)}$.

Assuming this, we take such f as our next approximation of f_k, which will satisfy this instance of (d_α). So if we fail, we put burden on the future cases of (iii) but we know that:

$\otimes_2$ For every $f \in IF_c(k)$ there are $m \in F(C_1)$, $m > \mathrm{Sup}(\mathrm{Dom}(f))$ and $f^\dagger \in IF_c(k)$, $f^\dagger \supseteq f$ and $i(0), i(1) < \omega$ *such that* $\mathbf{t}^m_{i(0)} \neq t^m_{i(1)}$ and for every $f'' \in IF_c(k)$, $f'' \supseteq f^\dagger$ both $g_0 \cup f'' \cup p_\ell \cup q^m_{i(0)}$ and $g_1 \cup f'' \cup p_\ell \cup q^m_{i(1)}$ are in $P_\alpha \times Q$.

(iii) Adapting our notation to that of (ii), we denote our present k as $k(*)$ and consider some $k < k(*)$ and assume that the property $\otimes_2$ (written at the end of (ii)) held for $\ell < k$ and $g_0, g_1 \in {}^{[0,n_k)\cap C_1}\{0,1\}$. (This we have found in stage k, note there are for each k only finitely many triples (ℓ, g_0, g_1).)

By the formulation of case (d_β) we are given $h \in {}^{[n_{k+1}, n_{k(*)})\cap C_1}\{0,1\}$; let $f^0_{k(*)}$ be our current approximation of $f_{k(*)}$ and use the property $\otimes_2$ for $f = f_k \cup h \cup f^0_{k(*)}$ to obtain $m, f^\dagger, i(0), i(1)$ as described there. Letting $f^1_{k(*)} = f^\dagger \setminus (f_k \cup h) = f^\dagger \restriction (\mathrm{Dom}(f^\dagger) \setminus [n_k, n_{k(*)}])$ and extending it if necessary to ensure that $m < n_{k(*)+1}$, we obtain the next approximation to $f_{k(*)}$ fulfilling our task.

So the definition of n_k, f_k can be carried out, so we have gotten $(*3)$ and $(*4)$.

$\square_{(*3),(*4)}$

Our treatment from here on concentrates on $A^* = \cup_{k<\omega} a_{3k+1}$ (remember $a_k = [n_k, n_{k+1}) \cap C_1$). The other "two thirds" of C_1 can be treated similarly. Let $f^* = \cup_{k<\omega}(f_{3k} \cup f_{3k+2})$, $A_1 = f^{*-1}(\{1\})$. Notice that by $(*3)$ (b), both A_1 and its complement in $\mathrm{Dom}(f^*)$ are infinite.

Thus, from $(*2)$ it follows that:

$(*5)$ For every $A \subseteq A^*$, $Ch^A_{A^*} \cup f^* \Vdash_{P_1^Q[A\cup A_1]}$ "$\underset{\sim}{Y} \cap F(C_1) =_{ae} F(A \cup A_1)$".
Hence, for every $A \subseteq A^*, Ch^A_{A^*} \cup f^*$ can be extended to a condition (in $P_1^Q[A \cup A_1]$) forcing equality above a certain n and in particular deciding for each $m \geq n$ in $F(C_1)$ whether it belongs to $\underset{\sim}{Y}$ (as no other names are involved in the equality). Looking at $\mathcal{P}(A^*)$ as a topological space, we conclude (as $|P_\alpha \times Q| = \aleph_0$):

$(*6)$ There exists a condition $p^* \in P_\alpha \times Q$, a finite function h^* satisfying $\mathrm{Dom}((p^*)_0) \cap A^* \subseteq \mathrm{Dom}(h^*) \subseteq A^*$, and $n^* < \omega$ such that

$$\mathcal{A} \overset{\mathrm{def}}{=} \{A \subseteq A^* : h^* \subseteq Ch^A_{A^*} \text{ and for each } m \geq n^* \text{ in } F(C_1) \text{ we have: } [p^* \cup Ch^A_{A^*} \cup f^* \Vdash_{P_1^Q[A\cup A_1]} \text{“}m \in \underset{\sim}{Y}\text{”} \text{ or } p^* \cup Ch^A_{A^*} \cup f^* \Vdash_{P_1^Q[A\cup A_1]} \text{“}m \notin \underset{\sim}{Y}\text{”}]\}$$

is of the second category everywhere "above" h^* (i.e., its intersection with any open set determined by a finite extension of h^* is of the second category).

We now define, for $A \in \mathcal{A}$,

$$F_1(A) = \{m \in F(C_1) : p^* \cup Ch^A_{A^*} \cup f^* \Vdash_{P^Q_1[A \cup A_1]} \text{“}m \in \underset{\sim}{Y}\text{”}\}\ .$$

It follows from (*5) that

(*6A) $F(A \cup A_1) =_{ae} F_1(A)$ for all $A \in \mathcal{A}$.

We will verify next that $\mathcal{A}$ is a Borel set. The demand $h^* \subseteq Ch^A_{A^*}$ defines an open set, and then we have a conjunction over all $m \geq n^*$ in $F(C_1)$ of conditions each requiring that $p^* \cup Ch^A_{A^*} \cup f^*$ decide whether m belongs to $\underset{\sim}{Y}$. Denoting $\mathcal{A}_q = \{A \subseteq A^* : q$ is incompatible with $p^* \cup Ch^A_{A^*} \cup f^*\}$ for $q \in P_\alpha \times Q$, and remembering that $\mathcal{I}_m = \{q^m_i : i < \omega\}$ is dense in $P_\alpha \times Q$ and we can preserve its pre-density in $P^Q_1[A \cup A_1]$, we see that $A \subseteq A^*$ satisfies the requirement for m iff it belongs to $[\cap\{\mathcal{A}_{q^m_i} : i < \omega, \mathbf{t}^m_i$ is true $\}] \cup [\cap\{\mathcal{A}_{q^m_i} : i < \omega, \mathbf{t}^m_i$ is false$\}]$. As each $\mathcal{A}_q$ is open, we have shown that $\mathcal{A}$ is Borel.

As $\mathcal{A}$ is a Borel set, we know by a theorem of Baire that (by extending h^*) we may assume that $\{A \subseteq A^* : A \notin \mathcal{A}$, and $h^* \subseteq Ch^A_{A^*}\}$ is of the first category.

Let k be large enough so that $n_{3k} \geq n^*$, maxDom(h^*) and $\ell < 3k - 1$ where $p^* = p_\ell$.

For $A \in \mathcal{A}$, let $\mathbf{t}^m_A$ denote the truth value of $m \in F_1(A)$. Then for every $m \in [n_{3k}, n_{3k+3}) \cap F(C_1)$ we know $p^* \cup Ch^A_{A^*} \cup f^*$ is compatible with some q^m_i but is not compatible with any q^m_i such that $\mathbf{t}^m_i \neq \mathbf{t}^m_A$ (remember the definitions of $F_1(A)$ and $\mathcal{A}$ and observe that $m \geq n^*$, $m \in F(C_1)$).

Hence by (c) of (*3), $\mathbf{t}^m_A$ is determined, for $A \in \mathcal{A}$ and $m \in [n_{3k}, n_{3k+3}) \cap F(C_1)$, by $Ch^A_{A^*}{\restriction}[0, n_{3k+4}) = Ch^A_{A^*}{\restriction}[0, n_{3k+2})$; that is: $\mathbf{t}^m_A = \mathbf{t}$ iff for some i, $\mathbf{t}^m_i = \mathbf{t}$ and for every $h \in IF(3k+4)$ we have

$$Ch^A_{A^*}{\restriction}[0, n_{3k+2}) \cup f^*{\restriction}[0, n_{3k+4}) \cup h \cup p^* \cup q^m_i \in P_\alpha \times Q.$$

Now we want to show that $\mathbf{t}^m_A$ is determined (for A, m, k as above) by $Ch^A_{A^*}\restriction a_{3k+1}$ alone. So let us assume that $A^{(0)}, A^{(1)} \in \mathcal{A}$ and $Ch^{A^{(0)}}_{A^*}\restriction a_{3k+1} = Ch^{A^{(1)}}_{A^*}\restriction a_{3k+1}$, and for some $m \in [n_{3k}, n_{3k+3}) \cap F(C_1)$ we have $\mathbf{t}^m_{A^{(0)}} \neq \mathbf{t}^m_{A^{(1)}}$. Let for $t = 0, 1$, $g_t = (Ch^{A^{(t)}}_{A^*} \cup f^*)\restriction([0, n_{3k-1}) \cap C_1)$. From the way $\mathbf{t}^m_{A^{(0)}}, \mathbf{t}^m_{A^{(1)}}$ are determined (see end of previous paragraph) and from $\mathbf{t}^m_{A^{(0)}} \neq \mathbf{t}^m_{A^{(1)}}$ it follows that at stage $3k-1$ of (*4), for the above g_0, g_1 and $p^* = p_\ell$, clause (d_α) fails (take $h = f^*\restriction(a_{3k} \cup a_{3k+2} \cup a_{3k+3}) \cup Ch^{A^{(t)}}_{A^*}\restriction a_{3k+1}$, which does not depend on t). Hence (d_β) holds in this case. Using this, we want to construct $B^{(0)}, B^{(1)} \in \mathcal{A}$ such that $A^{(i)} \cap n_{3k-1} = B^{(i)} \cap n_{3k-1}$ for $i = 0, 1$, and $B^{(0)} \setminus n_{3k-1} = B^{(1)} \setminus n_{3k-1}$ and we shall choose them in such a way that for infinitely many $m \in F(C_1)$, $Ch^{B^{(0)}}_{A^*}$ and $Ch^{B^{(1)}}_{A^*}$ determine $\mathbf{t}^m_{B^{(0)}}, \mathbf{t}^m_{B^{(1)}}$, respectively, to be distinct. Since $B^{(0)} =_{ae} B^{(1)}$, also by (*6A) we know $F_1(B^{(0)}) =_{ae} F(B^{(0)} \cup A_1) =_{ae} F(B^{(1)} \cup A_1) =_{ae} F_1(B^{(1)})$, so this will be a contradiction. Thus, it remains to show how to carry out the construction of $B^{(0)}, B^{(1)}$. We do it by an ω-sequence of finite approximations, the set of finite approximations is $\{(e_1, e_2) :$ for some n, e_1, e_2 are finite functions from $A^* \cap [0, n)$ to $\{0, 1\}$ e_i, includes $Ch^{A(i) \cap n_{3k+1}}_{A^* \cap n_{3k+1}}$ and $e_1\restriction[n_{3k+1}, n) = e_2\restriction[n_{3k+1}, n)\}$.

We start with the respective characteristic functions of $A^{(0)}, A^{(1)}$ up to n_{3k-1}. As $A^{(0)}, A^{(1)} \in \mathcal{A}$, both characteristic functions contain h^*.

As $\{A \subseteq A^* : A \notin \mathcal{A}$ and $h^* \subseteq Ch^A_{A^*}\}$ is of the first category, we have a countable family of nowhere dense subsets of $\mathcal{P}(A^*)$ which are to be avoided by $B^{(0)}, B^{(1)}$, to ensure $B^{(0)}, B^{(1)} \in \mathcal{A}$. So we arrange all these tasks (each time dealing with either $B^{(0)}$ or $B^{(1)}$ and one nowhere dense subset) in an ω sequence, and by the definition of "nowhere dense" we obtain each time a finite extension of our former approximations which ensures the implementation of our task. Say $\langle \mathcal{A}_n : n < \omega \rangle$ is this sequence of nowhere dense sets, and we choose (e^i_1, e^i_2) for $i < \omega$ approximations such that for $i = 3j$, (e^{i+1}_1, e^{i+1}_2) ensure $B^{(0)} \notin \mathcal{A}_j$ and $\max\mathrm{Dom}(e^i_1) > j$, for $i = 3j+1$ (e^{i+1}_1, e^{i+1}_2) ensure $B^{(1)} \notin \mathcal{A}_j$, and for $i = 3j+2$ (e^{i+1}_1, e^{i+1}_2) ensure that for some $k < j$ and $m \in [n_k, n_{k+1})$ is as required (i.e. $Ch^{B^{(0)}}_{A^*}$, $Ch^{(1)}_{A^*}$ determine $\mathbf{t}^m_{B^{(0)}}, \mathbf{t}^m_{B^{(1)}}$ are distinct). If we succeed, $B^{(0)} = \bigcup_{i<\omega} e^i_1$, $B^{(1)} = \bigcup_{i<\omega} e^i_2$ are as required. Why can we carry the construction? The least trivial case is the last.

We will show that our construction is all right by proving: if $k(*) > 3k-1$ and $k(*) \equiv 1 \pmod 3$ then there is $m \in F(C_1) \cap [n_{k(*)}, n_{k(*)+1})$ such that for $t = 0, 1$ there are $i(t) < \omega$ such that $\mathbf{t}^m_{i(0)} \neq \mathbf{t}^m_{i(1)}$ and for every $h \in IF(k(*)+3)$ we have $Ch^{B(t)}_{A^*} \restriction [0, n_{k(*)}) \cup f_{k(*)} \cup f^* \restriction [0, n_{k(*)+3}) \cup h \cup p^* \cup q^m_{i(t)} \in P_\alpha \times Q$. To prove this, use (d_β) for such $k(*)$ and $h = (Ch^{B(t)}_{A^*} \cup f^*) \restriction [n_{3k}, n_{k(*)})$. We thus get the desired contradiction.

So, by the last four paragraphs and (*6A), as F is an automorphism and A_1 is fixed, we have proven (for the last equality: we can just subtract $F(A_1)$):

(*7) There are functions $G'_k : \mathcal{P}(a_{3k+1}) \to \mathcal{P}(F(A^*) \cap [n_{3k}, n_{3k+3}))$ such that for every $A \in \mathcal{A}$ we have $F(A) =_{ae} F(A \cup A_1) \setminus F(A_1) =_{ae} F_1(A) \setminus F(A_1) =_{ae} \cup_{k<\omega} G'_k(A \cap a_{3k+1})$.

Now, (*7) would be good enough (as we shall see later) if it were not restricted to $\mathcal{A}$. So we want to achieve a similar result without this restriction.

We define by induction a sequence of pairs $\langle u_\ell, g_\ell \rangle$ $(\ell < \omega)$ such that $\langle u_\ell : \ell < \omega \rangle$ is a partition of $A^* \setminus \mathrm{Dom}(h^*)$ into finite subsets and $g_\ell : u_\ell \to \{0, 1\}$, as follows: for $r < \omega$, let $\mathcal{A}_r$ be the r-th element in a sequence of nowhere dense subsets of $\mathcal{P}(A^*)$ showing that $\{A \subseteq A^* : A \notin \mathcal{A}, h^* \subseteq Ch^A_{A^*}\}$ is of the first category. Each $\langle u_\ell, g_\ell \rangle$ is defined so that:

(i) u_ℓ is disjoint from u_m for all $m < \ell$.

(ii) if $\ell \in (A^* \setminus \mathrm{Dom}(h^*)) \setminus \cup_{m<\ell} u_m$ then $\ell \in u_\ell$.

(iii) if $A \subseteq A^*$, $g_\ell \subseteq Ch^A_{A^*}$ and $\ell = 2k$ or $\ell = 2k+1$ then $A \notin \mathcal{A}_k$.

To see that this is possible, use the fact that $\mathcal{A}_\ell$ is nowhere dense taking into account one after another all the elements of ${}^v\{0,1\}$ where $v = (\cup_{m<\ell} u_m) \cup \mathrm{Dom}(h^*)$.

Now, for $t = 0, 1$ let $B^*_t = \cup_{r<\omega} u_{2r+t}$ and $g^*_t = \cup_{r<\omega} g_{2r+t}$, $B_t = g^{*-1}_t(\{1\})$. Then $\mathrm{Dom}(g^*_t) = B^*_t$ and $\langle B^*_0, B^*_1, \mathrm{Dom}(h^*) \rangle$ is a partition of A^*. From (iii) it follows that if $A \subseteq A^*$, $h^* \cup g^*_t \subseteq Ch^A_{A^*}$ then $A \in \mathcal{A}$.

From $(*7)$ we can thus derive a definition for $F(A \cap B_0^*)$ working uniformly for all $A \subseteq A^*$ extending G'_k to G''_k

$$G''_k(A \cap B_0^* \cap a_{3k+1}) = G'_k([(A \cap B_0^*) \cup B_1] \cap a_{3k+1}) \setminus F(B_1)$$

(where $F(B_1) \in \mathcal{P}(\omega)$ denotes one fixed representative of the equivalence class $F(B_1)$). Similarly we can derive a definition for $F(A \cap B_1^*)$. From these two definitions (we can use $F(\bigcup_{n<\omega} u_{2n+1})$, $F(\bigcup_{n<\omega} u_{2n})$,) remembering that $\mathrm{Dom}(h^*)$ is finite, we obtain:

$(*8)$ There are functions $G_k : \mathcal{P}(a_{3k+1}) \to \mathcal{P}(F(A^*) \cap [n_{3k}, n_{3k+3}))$ such that for every $A \subseteq A^*$ we have $F(A) =_{ae} \cup_{k<\omega} G_k(A \cap a_{3k+1})$.

With this result in hand, we have good control of $F \restriction \mathcal{P}(A^*)$. We prove first:

$(*9)$ For every large enough k and disjoint $b_1, b_2 \subseteq a_{3k+1}$,

$$G_k(b_1) \cap G_k(b_2) = \emptyset.$$

Otherwise there is an infinite $S \subseteq \omega$, and $b_1^k, b_2^k \subseteq a_{3k+1}$ for $k \in S$, $b_1^k \cap b_2^k = \emptyset$, $G_k(b_1^k) \cap G_k(b_2^k) \neq \emptyset$. Then let $A_1 = \cup_{k \in S} b_1^k$, $A_2 = \cup_{k \in S} b_2^k$. Clearly $A_1 \cap A_2 = \emptyset$ and $A_1, A_2 \subseteq A^*$, and $F(A_1) \cap F(A_2) \neq_{ae} \emptyset$. This contradicts that F/finite commutes with $\cap$.

As F commutes with $\cup$ we can get similarly:

$(*10)$ For every large enough k and disjoint $b_1, b_2 \subseteq a_{3k+1}$,

$$G_k(b_1 \cup b_2) = G_k(b_1) \cup G_k(b_2),$$

and as F is onto (and monotonic) we have:

$(*11)$ For every large enough k, for any singleton $b \subseteq a_{3k+1}$ we have: $G_k(b)$ is a singleton (and also G_k is onto $F(A^*) \cap [n_{3k}, n_{3k+3})$.

Define g, $\mathrm{Dom}(g) = A^*$, $G_k(\{\ell\}) = \{g(\ell)\}$ for $\ell \in a_{3k+1}$, k large enough. So clearly by (∗8)–(∗11):

(∗12) For every $A \subseteq A^*$, $F(A) =_{ae} g(A) = \{g(i) : i \in A\}$ and g is one-to-one.

Remember that A^* is one third of C_1. Doing this separately for each third and then the same for C_2, we obtain for every B which is almost disjoint from each $A_j (j < \alpha)$: (that we can get a one-to-one g is proved like (∗9)):

(∗13) There is a one-to-one function g, $\mathrm{Dom}(g) =_{ae} B$, such that $(\forall A \subseteq B)$ $[g(A) =_{ae} F(A)]$.

Stage E: F is trivial on ω.

Let I be the ideal generated by $\{A_j : j < \alpha\}$ and the finite subsets of ω. Let $I_1 \overset{\text{def}}{=} \{B \subseteq \omega : (\forall A \in I)|B \cap A| < \aleph_0\}$. Clearly, I_1 is also an ideal.

Remember $J = \{A \subseteq \omega : F$ is trivial on $A\}$. In stage D we have shown that F is trivial on sets in I_1 (if (∗2) holds). If Case 2 from stage B holds then there is $A^+ \in [\omega]^{\aleph_0}$ such that $[A^+]^{\aleph_0} \cap J = \emptyset$, hence $\omega \setminus A_0 \subseteq A^+$ but there are B's as in stage B and by it (∗2) holds hence by stage D we know that $B \in J$ and obviously $B \cap A_0$ is finite. So B gives contradiction. Hence Case 1 of Stage B holds i.e. J is dense. We now outline the rest of the proof, and then shall give it in details. Using the knowledge we gathered on F in the previous stage we construct countably many functions $\{g_p : p \in P_\alpha \times Q\}$ such that for each $B \in I_1 \cap J$ there is $p \in P_\alpha \times Q$ such that $g_p \restriction B$ induces F on B.

Next we shall put these functions together to get a single function g such that g induces F on every $A \in I_1$. Finally we shall show that in this case g defines F.

Notice:

(∗14) Every A which is not in I contains an infinite subset in I_1.

To see this, let $\langle B_n : n < \omega\rangle$ be an ω-enumeration of the generators of I (i.e. $\{A_j : j < \alpha\}$). We try to define a sequence $\langle a_n : n < \omega\rangle$ such that $a_m \neq a_n$ for $m < n$ and $a_n \in A \setminus \cup_{m<n} B_n$. If we succeed we obtain a subset as required.

If we cannot define a_n, then $A \subseteq (\cup_{m<n} B_m) \cup \{a_m : m < n\}$, hence $A \in I$, contrary to our assumption.

Let $B \in J \cap I_1$ be infinite, $B = C_1 \cup C_2$ as in (*1).

(*15) Let g induce $F \restriction \mathcal{P}(B)$ (in the sense of (*13)), and let $\ell \in \{1, 2\}$. Then for every $p \in P_\alpha \times Q$ there are q, $p \leq q \in P_\alpha \times Q$ and n such that:

$(\forall m \in C_\ell)(m \geq n \Rightarrow [q \cup \{\langle m, 0\rangle\} \Vdash_{P_\alpha \times Q} \text{“}g(m) \notin \underset{\sim}{Y}\text{”}])$

and:

$(\forall m \in C_\ell)(m \geq n \Rightarrow [q \cup \{\langle m, 1\rangle\} \Vdash_{P_\alpha \times Q} \text{“}g(m) \in \underset{\sim}{Y}\text{”}])$

Proof of (*15). It is enough to prove it for $m \in C_1$, so $\ell = 1$. Suppose it fails for g, p. Let $\{q_\ell : \ell < \omega\}$ be the set of all conditions in $P_\alpha \times Q$ above p. Let $n \mapsto ((n)_0, (n)_1)$ be a mapping of ω onto $\omega \times \omega$.

We will define an increasing sequence of finite functions functions $h_n (n < \omega)$, $\mathrm{Dom}(h_n) \subseteq C_1$, $\mathrm{Rang}(h_{n+1} \setminus h_n) = \{0, 1\}$, and if $n \in C_1$ then $n \in \mathrm{Dom}(h_{n+1})$; we will also define $m_n \in C_1 (n < \omega)$, as follows:

We let $h_0 = (p)_0 \restriction (\mathrm{Dom}((p)_0) \cap C_1)$.

Assume that we have defined h_n, and we will see how we define m_n and h_{n+1}. If h_n is not compatible with $q_{(n)_0}$ then m_n is chosen arbitrarily and $h_{n+1} \supseteq h_n$ is chosen in accordance with our requirements above. If h_n is compatible with $q_{(n)_0}$, we use our assumption for $q = h_n \cup q_{(n)_0}$. Notice that if m is large enough then $m \notin \mathrm{Dom}((q)_0) \cap C_1$ and $[m \in C_1 \Rightarrow g(m)$ is defined, $g(m) \in F(C_1)$, $g(m) \geq (n)_1]$. So we can find $m = m_n \in C_1$ satisfying all these, and $k_n \in \{0, 1\}$ such that $h_n \cup q_{(n)_0} \cup \{\langle m_n, k_n\rangle\} \nVdash_{P_\alpha \times Q} \text{“}g(m_n) \notin \underset{\sim}{Y} \text{ iff } k_n = 0\text{”}$. Note: $k_n = 0$ is a failure of the first conclusion of (*15), $k_n = 1$ is a failure of the second conclusion of (*15) . We will choose $k_n = 0$ if possible.

Let $q_n^\dagger \geq h_n \cup q_{(n)_0} \cup \{\langle m_n, k_n\rangle\}$ force “$g(m_n) \in \underset{\sim}{Y}$ iff $k_n = 0$”. Let $h_{n+1}^\dagger = (q_n^\dagger)_0 \restriction (\mathrm{Dom}((q_n^\dagger)_0) \cap C_1)$, and let $h_{n+1} \supseteq h_{n+1}^\dagger$ satisfy $\mathrm{Rang}(h_{n+1} \setminus h_n) = \{0, 1\}$ and $n \in \mathrm{Dom}(h_{n+1})$ if $n \in C_1 \setminus \bigcup_{m \leq n} \mathrm{Dom}(h_m)$.

Now, $\mathrm{Dom}(\cup_{n<\omega} h_n) = C_1$. Let $A \subseteq C_1$ be the subset with characteristic function $\cup_{n<\omega} h_n$. So A and $C_1 \setminus A$ are infinite, so by (*2) we know $Ch^A_{C_1} \Vdash_{P_1^Q[A]}$ “$\underset{\sim}{Y} \cap F(C_1) =_{ae} F(A)$”. By our choice of h_0, $Ch^A_{C_1} = \cup_{n<\omega} h_n$ is compatible with p, so some extension of $p \cup Ch^A_{C_1}$ in $P_1^Q[A]$ forces the equality above some integer. So there is n such that:

$$q_{(n)_0} \cup Ch^A_{C_1} \Vdash_{P_1^Q[A]} \text{“}\underset{\sim}{Y} \cap F(C_1) \setminus (n)_1 = F(A) \setminus (n)_1\text{”}$$

and $F(A) \setminus (n)_1 = g(A) \setminus (n)_1$.

We refer now to the definition of m_n, h_{n+1}, k_n, suppose that $k_n = 0$, the other case is similar. Since $h_n \subseteq Ch^A_{C_1}$ is compatible with $q_{(n)_0}$, we had $q^\dagger_n \Vdash_{P_\alpha \times Q}$ “$g(m_n) \in \underset{\sim}{Y}$” where $q^\dagger_n$ was above $h_n \cup q_{(n)_0} \cup \{\langle m_n, 0\rangle\}$. Now, $m_n \notin A$ since h_{n+1} says so, hence $g(m_n) \notin g(A)$. As $g(m_n) \geq (n)_1$, $g(m_n) \notin F(A)$; but $g(m_n) \in F(C_1)$, hence $q_{(n)_0} \cup Ch^A_{C_1} \Vdash_{P_1^Q[A]}$ “$g(m_n) \notin \underset{\sim}{Y}$”. But $q^\dagger_n$ forces the opposite (we know it in $P_\alpha \times Q$, but it does not matter), and they are compatible (because $q_{(n)_0} \subseteq q^\dagger_n$ and $Ch^A_{C_1}$ and $q^\dagger_n$ are compatible by our choice of $h^\dagger_{n+1}$), a contradiction. So we have finished the proof of (*15).

We define now, for every $p \in P_\alpha \times Q$, a partial a function from ω to ω as follows: $g_p(m) = k$ iff $[p \cup \{\langle m, 0\rangle\} \Vdash_{P_\alpha \times Q}$ “$k \notin \underset{\sim}{Y}$” and $p \cup \{\langle m, 1\rangle\} \Vdash_{P_\alpha \times Q}$ “$k \in \underset{\sim}{Y}$” and k is the only one satisfying this$]$.

Our intention in defining g_p is to have countably many functions which induce $F \restriction \mathcal{P}(B)$ uniformly for all $B \in I_1$.

(*16) For every $B \in J \cap I_1$ and every $p \in P_\alpha \times Q$ there is q, $p \leq q \in P_\alpha \times Q$, such that $B \subseteq_{ae} \mathrm{Dom}(g_q)$ and $(\forall A \subseteq B)[F(A) =_{ae} g_q(A)]$.

Proof of (*16). If B is finite, then there is nothing to prove, so we assume that B is infinite. Let g exemplify $B \in J$ (remember $I_1 \subseteq J$).

Let us call k a candidate to be $g_p(m)$, if it satisfies the first two parts of the definition (but maybe not the uniqueness requirement). We notice that if k is a candidate to be $g_p(m)$ and $k^\dagger$ is a candidate to be $g_p(m^\dagger)$ and $m \neq m^\dagger$, then $k \neq k^\dagger$ (consider $p \cup \{\langle m, 0\rangle, \langle m^\dagger, 1\rangle\}$, it forces both $k \notin \underset{\sim}{Y}$ and $k^\dagger \in \underset{\sim}{Y}$). Notice also that in (*15) we have shown that for every large enough $m \in C_\ell$ (and appropriate q) $g(m)$ is a candidate to be $g_q(m)$.

Thus, given B and p, we apply (*15) first for $\ell = 1$ and p, and then for $\ell = 2$ and the condition obtained for $\ell = 1$, obtaining $q \geq p$ such that for every large enough $m \in B$, $g(m)$ is a candidate to be $g_q(m)$. Now, it suffices to show that $B \subseteq_{ae} \mathrm{Dom}(g_q)$, since whenever $g_q(m)$ is defined there is only one candidate, but $g(m)$ is one, so for almost all $m \in B$ we have $g_q(m) = g(m)$ and we know that g induces $F{\restriction}\mathcal{P}(B)$.

Assume that $B \not\subseteq_{ae} \mathrm{Dom}(g_q)$. Then w.l.o.g. B is disjoint from $\mathrm{Dom}(g_q)$ and we have k_m $(m \in B)$ such that k_m and $g(m)$ are distinct candidates to be $g_q(m)$. Let $C \stackrel{\mathrm{def}}{=} \{k_m : m \in B\}$. Then C is infinite and disjoint from $g(B)$ (as no k is a candidate for two m's). We let $D' = F^{-1}(C)$; more exactly $F(D') =_{\mathrm{ae}} C$ (possible as F is onto). Choose an infinite $D \subseteq D'$ such that $D \in J$. Then as $g(B) =_{ae} F(B)$, D' is almost disjoint from B.

Case I: $D \in I_1$.

In this case, we apply (*15) for D and q twice (as we did above for B and p) to obtain $q^\dagger \geq q$ such that for large enough $m^\dagger \in D$, $g^\dagger(m^\dagger)$ is a candidate to be $g_{q^\dagger}(m^\dagger)$ (where $g^\dagger$ induces $F{\restriction}\mathcal{P}(D)$). But for large enough $m^\dagger \in D$, $m^\dagger \notin B$ and $g^\dagger(m^\dagger)$ is a candidate to be $g_{q^\dagger}(m)$ for some $m \in B$ (as $g^\dagger(D) =_{ae} F(D)$, $F(D) \subseteq_{\mathrm{ae}} \{k_m : m \in B\}$, and $q \leq q^\dagger \in P_\alpha \times Q$), a contradiction (as in the beginning of the proof of (*16) we observe that no k is a candidate to be $g_{q^\dagger}(n)$ for two n's).

Case II: $D \notin I_1$.

Then w.l.o.g. $D \subseteq A_j$ for some $j < \alpha$. By strengthening q we can assume w.l.o.g. that $1_{A_j}\backslash\text{finite} \leq q$ or $0_{A_j}\backslash\text{finite} \leq q$. Assume that the former is the case (the second case is dealt with similarly). Then $q \Vdash_{P_{\omega_1} \times Q}$ "$A_j \subseteq_{ae} \underset{\sim}{X}$", hence $q \Vdash_{P_{\omega_1} \times Q}$ "$F(A_j) \subseteq_{ae} \underset{\sim}{Y}$". As we have seen in the proof of (*2), by preserving the pre-density of (countably many) appropriate subsets above corresponding conditions, we can conclude that $q \Vdash_{P_\alpha \times Q}$ "$F(A_j) \subseteq_{ae} \underset{\sim}{Y}$". Hence there are n and $q^\dagger \geq q$ in $P_\alpha \times Q$ such that $q^\dagger \Vdash_{P_\alpha \times Q}$ "$n^\dagger \in \underset{\sim}{Y}$" for all $n^\dagger \in F(A_j)$, $n^\dagger \geq n$. But $F(D) \subseteq_{ae} C \cap F(A_j)$, so taking $n^\dagger = k_m$ large enough we know that $n^\dagger$ is a candidate to be $g_{q^\dagger}(m)$ – a contradiction.

So we have finished the proof of (*16).

(*17) For $p_1, p_2 \in P_\alpha \times Q$, $B = \{n : g_{p_1}(n), g_{p_2}(n) \text{ are defined and distinct}\} \in I$.

Proof of (*17): If not, then by (*14) there exists an infinite $B' \subseteq B$ which belongs to I_1, but as we are in Case 1 (from stage B, as said in the begining of stage D) there is an infinite $B_1 \subseteq B'$ which belongs to J. Together $B_1 \in I_1 \cap J$

is such that g_{p_1}, g_{p_2} are defined everywhere in B_1 but they never agree there. By $(*16)$ we can find for $i = 1, 2$ condition $q_i \geq p_i$ such that $B_1 \subseteq_{ae} \mathrm{Dom}(g_{q_i})$ and $(\forall A \subseteq B_1) F(A) =_{ae} g_{q_i}(A)$. It follows that $g_{q_1} \restriction B_1 =_{ae} g_{q_2} \restriction B_1$; otherwise, we have an infinite $B_2 \subseteq B_1$ on which the functions never agree, and we can divide B_2 into three parts each having disjoint images under the two functions (decide inductively for $n \in B_2$ to which part is belongs, a good decision always exists as the functions are one-to-one); one part at least is infinite, call it B_3, then $F(B_3)$ is almost equal to both $g_{q_1}(B_3)$ and $g_{q_2}(B_3)$ which are disjoint. But, whenever g_{p_1}, g_{q_1} are both defined they must agree (since $q_1 \geq p_1$), and the same holds for p_2, q_2, so we obtain that $g_{p_1} \restriction B_1 =_{ae} g_{p_2} \restriction B_1$, contradicting our assumption. So $(*17)$ holds.

By arranging the conditions in $P_\alpha \times Q$ in an ω-sequence and using $(*17)$ we obtain that there is a partial function g^0 from ω to ω such that for every $p \in P_\alpha \times Q$ we have

$$\{n : g_p(n) \text{ is defined but } g^0(n) \text{ is not defined or is } \neq g_p(n)\} \in I.$$

Hence, by $(*16)$, for every $B \in J \cap I_1$, $B \subseteq_{ae} \mathrm{Dom}(g^0)$ and $(\forall A \subseteq B) F(A) =_{ae} g^0(A)$.

We have almost achieved our goal of inducing $F \restriction \mathcal{P}(B)$ uniformly for all $B \in J \cap I_1$ – the missing point is that we want g^0 to be one-to-one. But this must be true after discarding from $\mathrm{Dom}(g^0)$ a set from I, because otherwise we can construct $\langle b_n : n < \omega \rangle$ with $b_n \neq b_m$ for $n \neq m$, $g^0(b_{2k}) = g^0(b_{2k+1})$ and $B = \{b_n : n < \omega\} \in J \cap I_1$ (see the proof of $(*14)$ noting $J \cap I_1$ is an ideal); then for large enough n and an appropriate g_q (inducing $F \restriction \mathcal{P}(B)$) $g^0(b_n) = g_q(b_n)$, so g_q is not one-to-one, a contradiction. Since discarding a set in I does not affect the other properties of g^0, we have a one-to-one g^0 inducing $F \restriction \mathcal{P}(B)$ uniformly for all $B \in J \cap I_1$. But any element of I_1 contains an element of $J \cap I_1$, so g^0 induces $F \restriction \mathcal{P}(B)$ for all $B \in I_1$. [Why? Assume $B \in I_1$, $\neg[F(B) =_{ae} g^0(B)]$, then one of the following occurs:

(a) $F(B) \setminus g^0(B)$ is infinite, so as F is onto for some infinite $B_1 \subseteq B$, $F(B_1) \cap g^0(B) = \emptyset$ and we can find $B_2 \subseteq B_1$ which belongs to J, so $F(B_2) =_{ae}$

$g^0(B_2)$ but $g^0(B_2) \subseteq g^0(B)$ whereas $F(B_2) \subseteq_{ae} F(B_1) \leq_{ae} \omega \setminus g^0(B)$ contradiction,

(b) $g^0(B) \setminus F(B)$ is infinite so there is an infinite $B_1 \subseteq B$ such that $g^0(B_1) \subseteq g^0(B) \setminus F(B)$ and we get a similar contradiction.]

As we are in Case 1 (from Stage B, see beginning of Stage E) $(\forall j < \alpha) A_j \in J$, and they are almost disjoint, there is g^1, one-to-one partial function from ω to ω, so that for every $B \in I$, $F(B) =_{ae} g^1(B)$ and $B \subseteq_{ae} \mathrm{Dom}(g^1)$. Clearly $\mathrm{Dom}(g^1) \cup \mathrm{Dom}(g^0)$ is co-finite (use $(*14)$).

Let I^* be the ideal that $I \cup I_1$ generates. Let $D \stackrel{\mathrm{def}}{=} \{n : g^1(n) \neq g^0(n)$ and both are defined$\}$.

If $D \notin I^*$ we can find (as we found $B_3 \subseteq B_2$ in the proof of $(*17)$) $D_1 \subseteq D$ such that $D_1 \notin I^*$ and $g^1(D_1) \cap g^0(D_1) = \emptyset$.

As F is onto, for some $D_1^\dagger \subseteq D_1$, $F(D_1^\dagger) =_{ae} F(D_1) \cap g^0(D_1)$. Then $D_1^\dagger \in I_1$, otherwise it has an infinite subset $D_1'' \in I$, so $F(D_1'') =_{ae} g^1(D_1'')$ hence $F(D_1'') \subseteq_{ae} g^1(D_1)$, but $F(D_1'') \subseteq_{ae} F(D_1^\dagger) \subseteq_{ae} g^0(D_1)$, a contradiction. Similarly, $D_1 \setminus D_1^\dagger \in I$; so $D_1 \in I^*$, a contradiction.

So $D \in I^*$, hence by trivial changes in g^0, g^1 we get $D = \emptyset$. Let $g = g^1 \cup g^0$. For every $A \in I^*$, $F(A) =_{ae} g(A)$, so as I^* is dense, F is an automorphism, this holds for any A.

As $F(\omega) =_{ae} g(\mathrm{Dom}(g))$, $\mathrm{Rang}(g)$ is co-finite.

Why can we assume g is one-to-one? No integer can have an infinite origin set A, since then $F(A)$ is infinite while $g(A)$ is a singleton. Only finitely many integers can have a non-singleton origin set, otherwise we would have two disjoint infinite sets with the same infinite image under g. So we can throw out the problematic finite part of g.

Thus F is trivial. $\square_{5.6,5.5}$

V. α-Properness and Not Adding Reals

§0. Introduction

Next to not collapsing $\aleph_1$, not adding reals seems the most natural requirement on a forcing notion. There are many works deducing various assertions from CH and many others which do it from diamond of $\aleph_1$. If we want to show that the use of diamond is necessary, we usually have to build a model of ZFC in which CH holds but the assertion fails, by iterating a suitable forcing. A crucial part in such a proof is showing that the forcing notions do not add reals even when we iterate them. So we want a reasonable condition on $\underset{\sim}{Q}_i$ (in V^{P_i}) which ensures that forcing with P_α does not add reals when $\langle P_i, \underset{\sim}{Q}_i : i < \alpha \rangle$ is a CS iterated forcing system. Another representation of the problem is "find a parallel of MA consistent with G.C.H.".

The specific question which drew my attention to the above was whether there may be a non-free Whitehead group of power $\aleph_1$ (from [Sh:44] we know that there is no such group if $V = L$ or even if $\diamondsuit_S$ holds for every stationary $S \subseteq \omega_1$, and that there is such a group if MA $+2^{\aleph_0} > \aleph_1$ holds). This is essentially equivalent to: "Is there a stationary $S \subseteq \omega_1$, and for each $\delta \in S$ an unbounded subset A_δ of order-type ω, such that $\bar{A} = \langle A_\delta : \delta \in S \rangle$ has the uniformization property" (see II 4.1, i.e. if $\bar{h} = \langle h_\delta : \delta \in S \rangle$, h_δ a function from A_δ to $2 = \{0, 1\}$ then for some $h : \bigcup_{\delta \in S} A_\delta \to 2$ for every δ, $h_\delta \subseteq^* h$ i.e. $\{\alpha \in A_\delta : h_\delta(\alpha) \neq h(\alpha)\}$ is finite). It is easy to see that $\diamondsuit_S$ implies $\langle A_i : i \in S \rangle$

does not have the uniformization property (and in II 4.3, we proved, from ZFC + MA $+2^{\aleph_0} > \aleph_1$, that $\bar{A}$ has the uniformization property).

The solution was surprising. By Devlin and Shelah [DvSh:65] (see AP §1 here), there is a weak form $\Phi^2_{\omega_1}$ of $\diamondsuit_{\aleph_1}$ which follows from CH; in fact is equivalent to $2^{\aleph_0} < 2^{\aleph_1}$. This statement implies many consequences of the diamond (see on it in Appendix §1; see [Sh:87a], [Sh:87b], [Sh:88] and a systematic development in Abraham and Shelah [AbSh:114] and [Sh:192] and lately [Sh:576], [Sh:600]). In particular $\langle A_\delta : \delta \in S\rangle$ does not have the uniformization property when $S \in \mathcal{D}_{\omega_1}$ i.e. S contains a closed unbounded subset of ω_1. This still leaves open the question for S a stationary costationary subset of ω_1. Now for such sets it was proved in [Sh:64] that the uniformization property may hold for a fixed stationary costationary subset S of ω_1, i.e. for all $\langle A_\delta : \delta \in S\rangle$, $A_\delta \subseteq \delta$ unbounded of order type ω. However $\diamondsuit_{\omega_1 \setminus S}$ (and even $\diamondsuit^*_{\omega_1 \setminus S}$) may still hold. On the situation for $\lambda > \aleph_1$ see [Sh:186], Mekler and Shelah [MkSh:274] and [Sh:587]. More information on the connection between unifomization and group theoretic questions see [Sh:98] or see the book [EM] and lately Eklof, Mekler, Shelah [EMSh:441], [EMSh:442], Eklof, Shelah [EkSh:505].

We also deal with "when does a CS iteration of proper forcing add no new reals?" For this we need two properties. One is $\mathbb{D}$-completeness (see §5) which is a way to exclude the impossible cases, and another is α-proper for $\alpha < \omega_1$, where we replace a countable elementary submodel by tower of height α of such models (in §3, and in §2 for more general case). The iteration theorem is proved in §7, but to apply it to the classical problem of SH we need "good forcing notion", this is done in §6; Jensen's original proof use a different forcing. Lastly, in §8 we deal with KH giving a proof in our context to results of Silver and Devlin. We also start investigating preservation of additional property: in §4 we deal with ${}^\omega\omega$-bounding.

Notation. In this chapter λ, μ will stand for uncountable cardinals, if not explicitly stated otherwise.

§1. $\mathcal{E}$-Completeness – a Sufficient Condition for Not Adding Reals

1.1 Definition. Let $\mathcal{E}$ be a family of subsets of $\mathcal{S}_{\aleph_0}(\mu)$ (we assume always $\mathcal{S}_{\aleph_0}(\mu) \in \mathcal{E}$, so μ is reconstructible from $\mathcal{E}$ but using specific $\mathcal{E}$ we may forget to write $\mathcal{S}_{\aleph_0}(\mu)$). In an abuse of notation, instead of a singleton $\{E\}$, $E \subseteq \mathcal{S}_{\aleph_0}(\mu)$, we write E. When $\mathcal{E} = \{a \in \mathcal{S}_{\aleph_0}(\mu) : a \cap \omega_1 \in S\}$ we write $\{S\}$ or S (here $S \subseteq \mathcal{S}_{\aleph_0}(\omega_1)$ or just S a subset of ω_1), similarly if $S \subseteq \mathcal{S}_{\aleph_0}(\mu_1)$, we may interpret it as $\{a \in \mathcal{S}_{\aleph_0}(\mu) : a \cap \mu_1 \in S\}$. Remember $\mathcal{D}_{\aleph_0}(A)$ is the filter $\{A \subseteq \mathcal{S}_{\aleph_0}(A) : A$ include some club of $\mathcal{S}_{\aleph_0}(A)\}$ (see Definition III 1.4), it is $\aleph_1$-complete, fine (i.e. $x \in A \Rightarrow \{a : x \in a\} \in \mathcal{D}_{\aleph_0}(A)$) and normal (i.e. $A_x \in \mathcal{D}_{\aleph_0}(A)$ for $x \in A$ implies $\{a : (\forall x \in a)(a \in A_x)\} \in \mathcal{D}_{\aleph_0}(A)$).

(1) We say that $\mathcal{E}$ is nontrivial if for every λ large enough, there is a countable $N \prec (H(\lambda), \in)$ such that $\mathcal{E} \in N$ and $N \cap \mu \in A$ for every $A \in \mathcal{E} \cap N$. We say in such cases that N is *suitable* for $\mathcal{E}$.

(2) We say, for a nontrivial $\mathcal{E}$, that a forcing notion P is $\mathcal{E}$-complete if for every λ large enough, and $N \prec (H(\lambda), \in)$ countable, suitable for $\mathcal{E}$, to which P belongs, the pair (N, P) is complete (see below).

(3) The pair (N, P) is complete if every generic sequence $\langle p_n : n < \omega \rangle$ for (N, P) has an upper bound in P, where:

(4) $\langle p_n : n < \omega \rangle$ is a generic sequence for (N, P) if $p_n \in P \cap N$, $P \models p_n \leq p_{n+1}$, and for every dense open subset $\mathcal{I}$ of P which belongs to N, $\mathcal{I} \cap \{p_n : n < \omega\} \neq \emptyset$.

1.2 Claim.

1) If $\mathcal{E}$ is nontrivial and $\mathcal{E} \in H(\lambda)$, *then* the set of suitable N's is unbounded in $\mathcal{S}_{\aleph_0}(H(\lambda))$. Moreover $\mathcal{E}$ is nontrivial $\subseteq \mathcal{P}(\mathcal{S}_{\aleph_0}(\mu))$ iff the fine normal filter on $\mathcal{S}_{\aleph_0}(\mu)$ it generates is a proper filter. So if $\mathcal{E} = \{E\}$, we can add "iff E is a stationary subset of $\mathcal{S}_{\aleph_0}(\mu)$".

2) In the definition, in (1) we get the same answer for all λ for which $\mathcal{E} \in H(\lambda)$; if we replace "$\mathcal{E} \in N \prec (H(\lambda), \in)$" by "$N \prec (H(\lambda), \in, \mathcal{E})$" we get the same

answer for all λ for which $\mathcal{S}_{\aleph_0}(\mu) \in H(\lambda)$, so we may replace the universal quantifier on λ by an existential.

3) If $\mathcal{E}$ is nontrivial, it has the finite (in fact, countable) intersection property.
4) If $S \subseteq \omega_1$ is stationary, then $\mathcal{E} = \{S\}$ is nontrivial (for $\mu = \omega_1$).
5) If P is $\mathcal{E}$-complete for some nontrivial $\mathcal{E}$, then P does not add reals.
6) If $N \prec (H(\lambda), \in)$ is suitable for $\mathcal{E}$, $P \in N$ a forcing notion, $q \in P$ is generic for (N, P), then $q \Vdash_P$ "$N[\underset{\sim}{G}_P]$ is suitable for $\mathcal{E}$".
7) If $\mathcal{E} \subseteq \mathcal{P}(\mathcal{S}_{\aleph_0}(\mu))$, $\mu_1 = |\mathcal{E}| + \mu$, $\mathcal{E} = \{X_i : i < |\mathcal{E}|\}$ and $E^* = \{a \in \mathcal{S}_{\aleph_0}(\mu_1)$: if $i \in a$ and $i < |\mathcal{E}|$ then $a \cap \mu \in X_i\}$, *then*: $\mathcal{E}$ is nontrivial iff $\{E^*\}$ is nontrivial; also for any forcing notion P, P is $\mathcal{E}$-complete iff P is E^*-complete.
8) If P, Q are $\mathcal{E}$-complete, then $P \times Q$ is $\mathcal{E}$-complete.
9) If $N \prec (H(\lambda), \in)$ and μ, $E \in N$ and $E \subseteq \mathcal{S}_{\leq\aleph_0}(\mu)$ then: N is suitable for $\{E\}$ *iff* $N \cap \mu \in E$.

Proof. 1) Fix λ and a countable $a \in H(\lambda)$. We want to find a suitable N such that $a \in N$ (then also $a \subseteq N$). Assume

$(*)$ there is no suitable $N \prec (H(\lambda), \in)$, $a \in N$.

Then for some $\lambda_1 > \lambda$ we have $(H(\lambda_1), \in) \models$ "$(\exists a \in H(\lambda))[(*)]$" and $H(\lambda) \in H(\lambda_1)$ of course, and λ_1 is as required in Definition 1.1(1). Let $N_1 \prec (H(\lambda_1), \in)$ be suitable for $\mathcal{E}$. Then for some $a \in N_1 \cap H(\lambda)$, $N_1 \models$ "$(*)$". Now consider $N = N_1 \cap H(\lambda)$ and get a contradiction. The other two sentences are easy too; on the normal fine filter on $\mathcal{S}_{\aleph_0}(\mu)$ which $\mathcal{E}$ generates see 1.4.

2) Easy, by an argument similar to III 2.2.

3)-9): Easy, (for (6) see 1.3(1)). $\square_{1.2}$

1.2A Explanation of 1.1(4). λ is large enough to ensure everything about P, forcing, etc., is expressible in $H(\lambda)$, now as N is an elementary submodel, it is legitimate to ask what goes on when you force with P starting in N, of course a generic sequence is not far from being a generic subset, so what (4) says is that for any generic extension $N[G]$, there is $p \in P$ which knows everything about it (so $G \subseteq P \cap N$ is generic over N).

1.3 Theorem.

(1) If P is $\mathcal{E}$-complete (so $\mathcal{E}$ nontrivial in V), *then* $\Vdash_P$ "$\mathcal{E}$ is nontrivial".

(2) If $\bar{Q} = \langle P_i, \underset{\sim}{Q}_i : i < \alpha \rangle$ is a countable support iteration, $\Vdash_{P_i}$ "$\underset{\sim}{Q}_i$ is $\mathcal{E}$-complete", *then* $P_\alpha = \lim \bar{Q}$ is $\mathcal{E}$-complete.

1.3A Remark. So in (2) it is enough to assume $\mathcal{E}$ is not trivial in V and $\Vdash_{P_i}$ " if $\mathcal{E}$ is not trivial then $\underset{\sim}{Q}_i$ is $\mathcal{E}$-complete."

Proof.

(1) Note: $\mathcal{S}_{\aleph_0}(\mu)^V = \mathcal{S}_{\aleph_0}(\mu)^{V^P}$. Let λ be large enough and $p \in P$. Let $N \prec (H(\lambda), \in)$ be suitable for $\mathcal{E}$, $P \in N$, $p \in N$, hence (N, P) is complete (see Definition 1.1(2)). Choose $\langle p_n : n < \omega \rangle$, a generic sequence for (N, P), $p_0 = p$ and choose $p^* \geq p_n$ for all $n < \omega$. Since p^* is (N, P)-generic by Corollary III 2.13 (see clauses (a), (f)) we have, $p^* \Vdash$ "$(N[G], \in) \prec (H(\lambda)^{V[G]}, \in)$; $\mathcal{E} \in N[G]$ and $N[G] \cap \mu = N \cap \mu$ and $H(\lambda)^V \cap N = H(\lambda)^V \cap N[G]$" and as $\mathcal{E} \in V$ also $N[G] \cap \mathcal{E} = N \cap \mathcal{E}$, hence $\mathcal{S}_{\aleph_0}(\mu) \cap N = \mathcal{S}_{\aleph_0}(\mu) \cap N[G]$ (as forcing by P adds no new countable subsets of μ) hence

$$N[G] \cap \mu = N \cap \mu \in \bigcap_{\substack{A \in \mathcal{E} \\ A \in N}} A = \bigcap_{\substack{A \in \mathcal{E} \\ A \in N[G]}} A$$

So $p^* \geq p$ forces $N[G]$ to exemplify that $\mathcal{E}$ is not trivial.

(2) Let λ be large enough, $N \prec (H(\lambda), \in)$ suitable for $\mathcal{E}$. Let $\langle p_n : n < \omega \rangle$ be a generic sequence for (N, P_α), (note $p_n \in N$, $\mathrm{Dom}(p_n)$ countable hence $\mathrm{Dom}(p_n) \subseteq N$). Define $p^* \in P_\alpha$: its domain is $N \cap \alpha$, and for $i \in N \cap \alpha$, $p^*(i)$ is a member of $\underset{\sim}{Q}_i$ which is an upper bound for $\{p_n(i) : n < \omega \text{ and } i \in \mathrm{Dom}(p_n)\}$ if there is such upper bound (in $\underset{\sim}{Q}_i$, say first such upper bound in some well ordering $\underset{\sim}{\leq}_i$ of $\underset{\sim}{Q}_i$ (a P_i-name)). We now prove by induction on $i \in N \cap \alpha$ that $p^* \restriction i \geq p_n \restriction i$ for every n (note $i \in \mathrm{Dom}(p_n)$ for every n large enough as $\{p \in P_\alpha : i \in \mathrm{Dom}(p)\}$ is a dense open subset of P_α which belongs to N). There are no special problems.

$\square_{1.3}$

1.4 Claim.

(1) The minimal normal fine filter on $\mathcal{S}_{\aleph_0}(\mu)$ which includes $\mathcal{E}$ is $\mathcal{D} = \mathcal{D}(\mathcal{E})$ which is defined by:

$A \in \mathcal{D}$ if and only if there is $C \in \mathcal{D}_{\aleph_0}(\mu)$ and $A_i \in \mathcal{E} \cup \{\mathcal{S}_{\aleph_0}(\mu)\}$ for $i < \mu$, such that $\{a \in C : (\forall i \in a) a \in A_i\} \subseteq A$.

(2) $\mathcal{E}$ is nontrivial if and only if $\emptyset \notin \mathcal{D}(\mathcal{E})$.

(3) P is $\mathcal{E}$-complete if and only if P is $\mathcal{D}(\mathcal{E})$-complete.

Proof. Easy. $\square_{1.4}$

1.5 Lemma. Assume $2^{\aleph_0} = \aleph_1$. If $\bar{Q} = \langle P_i, \underset{\sim}{Q}_i : i < \alpha \rangle$ is a countable support iteration, $\Vdash_{P_i}$ "$|\underset{\sim}{Q}_i| = \aleph_1$", $\mathcal{E}$ a family of subsets of $\mathcal{S}_{\aleph_0}(\mu)$ which is nontrivial, and each $\underset{\sim}{Q}_i$ is $\mathcal{E}$-complete. Then $P_\alpha = \operatorname{Lim} \bar{Q}$ satisfies the $\aleph_2$-chain condition.

Proof. Let $\{p_i \in P_\alpha : i < \aleph_2\}$ be given. We shall find two compatible conditions among them. Pick λ regular large enough, for every $i < \aleph_2$, let $N_i \prec (H(\lambda), \in)$ be countable such that $\{\bar{Q}, p_i, i, \mathcal{E}\} \subseteq N_i$ and N_i is suitable for $\mathcal{E}$.

1.5A Fact. We can find $i < j < \omega_2$ and an isomorphism $h : N_i \to N_j$ (onto N_j) such that $h(p_i) = p_j$ and $h \restriction (N_i \cap N_j) = id_{N_i \cap N_j}$.

Proof of the Fact. Denote $S^2_1 = \{\gamma < \aleph_2 : \operatorname{cf}(\gamma) = \aleph_1\}$, clearly it is a stationary set; define $f(\gamma) = \operatorname{Min}\{\beta : N_\gamma \cap (\cup_{i<\gamma} N_i) = N_\gamma \cap (\cup_{i<\beta} N_i)\}$. Since $||N_\gamma|| = \aleph_0$ and $[\gamma \in S^2_1 \Rightarrow \operatorname{cf}(\gamma) = \aleph_1]$ clearly f is a regressive function on S^2_1, hence by Fodor's lemma there exists $S \subseteq S^2_1$ stationary and $\beta < \aleph_2$ such that $f[S] = \{\beta\}$. The number of countable subsets of $\cup_{i<\beta} N_i$ is

$$|| \cup_{i<\beta} N_i||^{\aleph_0} \leq (|\beta| \cdot \aleph_0)^{\aleph_0} = \aleph_1^{\aleph_0} = (2^{\aleph_0})^{\aleph_0} = 2^{\aleph_0} = \aleph_1 \,,$$

therefore we may choose $T \subseteq S$ of cardinality $\aleph_2$ and a set B^* such that $(\forall \gamma \in T)[N_\gamma \cap (\cup_{i<\beta} N_i) = B^*]$. For every $\gamma \in T$ define $N^\dagger_\gamma = \langle N_\gamma, p_\gamma, c \rangle_{c \in B^*}$. The number of isomorphism types is $\leq 2^{\aleph_0} = \aleph_1$ hence we may choose $S^\dagger \subseteq T$, $|S^\dagger| = \aleph_2$ such that $i \neq j \in S \Rightarrow N^\dagger_i \cong N^\dagger_j$. Pick such i, j from $S^\dagger$. Let

$h : N_i \to N_j$ be the isomorphism. As each $c \in B^*$ is an individual constant, h is an isomorphism over B^* (i.e., it is the identity on B^*) and similarly $h(p_i) = p_j$. This is the isomorphism we promised in the Fact so we have proved the fact.

$\square_{1.5A}$

Continuation of the proof of 1.5: Let $i < j$ and h be as in Fact 1.5A. Now choose $\{p_i^n \in P_\alpha \cap N_i : n < \omega\}$ such that $p_i = p_i^0 \leq p_i^1 \leq p_i^2 \leq \cdots$ and for every dense subset $\mathcal{I} \in N_i$ of P_α there exists n such that $p_i^n \in \mathcal{I}$ and let $\{p_j^n \in P_\alpha \cap N_j : n < \omega\}$ be defined by $p_j^n = h(p_i^n)$. Define a condition r as follows: $\mathrm{Dom}(r) = (\alpha \cap N_i) \cup (\alpha \cap N_j)$, for $\xi \in \alpha \cap N_i \setminus N_j$, $r(\xi)$ will be a P_ξ-name of an upper bound of $\{p_i^n(\xi) : n < \omega\}$ if there is such a bound, and otherwise $\emptyset = \emptyset_{P_\xi}$. For $\xi \in \alpha \cap N_j$, $r(\xi)$ will be a name of an upper bound of $\{p_j^n(\xi) : n < \omega\}$ if there is such an element, and otherwise $\emptyset = \emptyset_{P_\xi}$. It suffices to prove that for every $n < \omega$ we have $p_i^n \leq r$ and $p_j^n \leq r$. We shall prove by induction on $\gamma \leq \alpha$ that for every $n < \omega$, $p_i^n{\restriction}\gamma, p_j^n{\restriction}\gamma \leq r{\restriction}\gamma$. This suffices as for $\gamma = \alpha$ we get that r is a common upper bound of $p_i{\restriction}\alpha = p_i$ and $p_j{\restriction}\alpha = p_j$ (in P_α). For $\gamma = 0$ this is trivial.

For γ limit, it follows from the induction hypothesis (and the definition of the order).

For $\gamma = \xi + 1$, notice that $\mathrm{Dom}(p_i^n) \subseteq N_i \cap \alpha$, $\mathrm{Dom}(p_j^n) \subseteq N_j \cap \alpha$, and divide to 4 cases:

1. $\xi \notin N_i$ and $\xi \notin N_j$; trivial.

2. $\xi \in N_i \setminus N_j$, it suffices to prove

(*) $r{\restriction}\xi \Vdash_{P_\xi}$ "$p_i^n(\xi) \leq r(\xi)$".

If $\{p_i^n(\xi) : n < \omega\}$ has an upper bound in $\underset{\sim}{Q}_\xi$ then this is true by construction (i.e. the choice of $r(\xi)$).

By the choice of $\langle p_i^n : n < \omega\rangle$ as $P_\xi \lessdot P_\alpha$ clearly $\langle p_i^n{\restriction}\xi : n < \omega\rangle$ is a generic sequence for (N_i, P_ξ), hence by the induction hypothesis $r{\restriction}\xi$ is (N_i, P_ξ)-generic. So by 1.2(6) and the definition of the order of P_α we have $r{\restriction}\xi \Vdash_{P_\xi}$ "$N_i[\underset{\sim}{G}_\xi]$ is

$\mathcal{E}$-suitable and $\langle p_i^n(\xi) : n < \omega \rangle$ is a generic sequence for $(N_i[\underset{\sim}{G}_\xi], \underset{\sim}{Q}_\xi)$". Hence $r{\restriction}\xi \Vdash_{P_\xi}$ "$\langle p_i^n(\xi) : n < \omega \rangle$ has an upper bound in $\underset{\sim}{Q}_\xi$", so we finish.

3. $\xi \in N_j \setminus N_i$, symmetric proof to 2 (using the choice of h).

4. $\xi \in N_i \cap N_j$; remember that w.l.o.g. $\underset{\sim}{Q}_\xi$ (set of elements) is ω_1 and as above by the induction hypothesis $r{\restriction}\xi$ is (N_i, P_ξ)-generic and (N_j, P_ξ)-generic. Since $N_i \cap \omega_1$ and $N_j \cap \omega_1$ are initial segments of ω_1 and $N_i \cong N_j$ (and $\omega_1 \in N_i \cap N_j$) clearly $N_i \cap \omega_1 = N_j \cap \omega_1$. Also $r{\restriction}\xi$ determines $G_{P_\xi} \cap N_i$ and $G_{P_\xi} \cap N_j$ hence for every m there is an $n \in \omega$, and $\alpha^m \in N_i \cap \omega_1$, such that $p_i^n{\restriction}\xi \Vdash_{P_\xi}$ "$p_i^m(\xi) = \alpha^m$". To this relation in N_i we can apply h, which yields $p_j^n{\restriction}\xi \Vdash_{P_\xi}$ "$p_j^m(\xi) = \alpha^m$" (since $\alpha^m \in N_i \cap N_j$). Hence $r{\restriction}\xi \Vdash_{P_\xi}$ "$p_i^m(\xi) = p_j^m(\xi)$" for all $m < \omega$. Now continue as in the previous case 2.

$\square_{1.5}$

1.6 Theorem. Suppose that CH holds in V, and $\mathcal{E}$ is a nontrivial family of subsets of $\mathcal{S}_{\aleph_0}(\mu)$, $\chi^{\aleph_1} = \chi = \mathrm{cf}(\chi)$. *Then* V has a generic extension V_1 by proper forcing in which:

(*) (a) CH holds, $\mathcal{E}$ is not trivial, $2^{\aleph_1} = \chi$, and

(b) *If* P is an $\mathcal{E}$-complete proper forcing notion, $|P| = \aleph_1$ and $\mathcal{I}_i \subseteq P$ is dense for $i < i_0 < \mathrm{cf}(\chi)$, *then* there is a directed $G \subseteq P$ such that $G \cap \mathcal{I}_i \neq \emptyset$ for $i < i_0$.

1.6A Remark.

(1) Properness is not essential in the proof of the theorem (except for having it in the conclusion), its use will appear in 1.7.

(2) Also the reader should be aware of the fact that $\mathcal{E}$-completeness and properness is *more* than properness alone, otherwise 1.6 would say:

$$\mathrm{MA}_{\aleph_1} \,\&\, \mathrm{CH},$$

which is of course impossible.

Proof. We use countable support iterated forcing systems $\langle P_i, \underset{\sim}{Q}_i : i < \alpha \rangle$ such that

(∗) $\alpha < \chi^+$ and $\Vdash_{P_i}$ "$|\underset{\sim}{Q}_i| = \aleph_1, \underset{\sim}{Q}_i$ is proper and is $\mathcal{E}$-complete".

For any such system $\bar{Q}$, $\operatorname{Lim} \bar{Q}$ is an $\mathcal{E}$-complete forcing notion which satisfies the $\aleph_2$-chain condition (by 1.3(2) and 1.5 respectively). Also by III §3 $\operatorname{Lim}\bar{Q}$ is proper.

By usual bookkeeping it is enough to prove the subfact below (note that for $R = \{f : \text{for some } \alpha < \omega_1, f : \alpha \to \{0,1\}\}$ ordered by inclusion, always (a) of (iii) below holds).

1.6B Subfact. If $\bar{Q}^1$ satisfies (∗), $P^1 = \operatorname{Lim}\bar{Q}^1$ and $\underset{\sim}{R}$ is a P^1-name of a forcing notion, *then* there is a $\bar{Q}^2$ such that:

(i) $\bar{Q}^2$ satisfies (∗).

(ii) $\bar{Q}^1$ is an initial segment of $\bar{Q}^2$.

(iii) for some maximal antichain $\mathcal{I}$ of P^2, for every $p \in \mathcal{I}$ (where $P^2 = \operatorname{Lim}\bar{Q}^2$):

either (a) $p \Vdash_{P^2}$ " there is a directed subset of $\underset{\sim}{R}$, generic over V^{P_1}" (in fact, it is the generic subset of some $\underset{\sim}{Q}_\beta$, $\beta \in [\ell g(\bar{Q}^1), \ell g(\bar{Q}^2)]$),

or (b) for no $\bar{Q}$, and q do we have: $\bar{Q}$ satisfies (∗) and $\bar{Q}^2$ is as initial segment of $\bar{Q}$ and $p \le q \in \operatorname{Lim}\bar{Q}$, and $q \Vdash_{\operatorname{Lim}\bar{Q}}$ "$\underset{\sim}{R}$ is a proper $\mathcal{E}$-complete forcing with universe ω_1".

Proof. Immediate. $\square_{1.6B,1.6}$

1.6C Remark. 1) This is different from the situation of II 3.4, where we had "c.c.c." instead of "$\mathcal{E}'$-complete for some suitable $\mathcal{E}'$".

2) So the Schemma of the proof of 1.6 is more general than the one in II 3.4.

3) Assume that P, R are forcing notions in V, $\mathcal{E} \subseteq \mathcal{S}_{\le\aleph_0}(\mu)$ is nontrivial:

(a) if R is not proper in V, P is proper, then R is not proper in V^P (use the equivalent definition in III 1.10(1) and for simplicity the set of members of Q is an ordinal): similarly for $\mathcal{E}$-proper (see Definition 2.2(5) below). The proof is included in in the proof of III 4.2.

b) If R is not $\mathcal{E}$-complete in V, P is e.g. $\mathcal{E}$-complete, *then* R is not $\mathcal{E}$-complete in V^P .

c) Without "P is proper", clause (a) is not necessarily true.

4) By (3)(a) (of the Remark 1.6C), as $[\beta < \alpha \Rightarrow P_\beta/P_\alpha$ is proper$]$, we can use in the proof of 1.6 the older Schemma.

5) We can omit $\chi = \mathrm{cf}(\chi)$ in 1.6 and replace $i < i_0 < \mathrm{cf}(\chi)$ by $i < \chi_1$ where $\chi_1 \leq \chi$ is regular. (And use iterated forcing of length $\delta < \chi^+$, $\mathrm{cf}(\delta) = \chi_1$. Instead χ^+ we can use an inaccessible.)

1.7 Conclusion. In the model from 1.6, if $\mathcal{E} = \{\omega_1 \setminus S\}$, $S \subseteq \omega_1$, stationary costationary the following holds:

a) for any $\langle A_j : \delta \in S\rangle$, such that $A_\delta \subseteq \delta$ unbounded of order type ω for $\delta \in S$, we have: $(\langle A_\delta : \delta \in S\rangle, \aleph_0)$ has the uniformization property.

b) S is still stationary (after the forcing, by properness).

Remark. Remember, we say a family $\mathcal{P} = \{A_\alpha : \alpha \in S\}$ of sets has the κ-uniformization property (or $(\mathcal{P}, \kappa)$ has the uniformization property) *if* for every family $\{f_\alpha : \alpha \in S\}$, f_α a function from A_α to κ, there is $f : \bigcup_{\alpha \in S} A_\alpha \to \kappa$ such that $\bigwedge_\alpha f_\alpha =_{ae} f{\restriction}A_\alpha$, where $f =_{ae} g$ if $|\{\alpha : f(\alpha) \neq g(\alpha), \alpha \in \mathrm{Dom}(f)\}| < |\mathrm{Dom}(f)|$ (note: this is symmetric and transitive only if we demand $\mathrm{Dom}(f) = \mathrm{Dom}(g)$).

Proof. Let $\langle A_\delta : \delta \in S\rangle$ be as above, $f_\delta : A_\delta \to \omega$, and $Q \overset{\mathrm{def}}{=} \{f : f$ a function from some $\alpha < \omega_1$ to ω, such that for every limit $\delta \leq \alpha, \delta \in S$ we have $f{\restriction}A_\delta =_{\mathrm{ae}} f_\delta\}$, ordered by $\subseteq$.

We have to check the following four facts:

Fact A. If $p \in Q, \mathrm{Dom}(p) \leq \alpha < \omega_1$ *then*, there is q, $p \leq q \in Q$, $\mathrm{Dom}(q) = \alpha$.

Fact B. Q is $\mathcal{E}$-complete.

Fact C. If $p \in Q$, $A \subseteq \omega_1 \setminus \mathrm{Dom}(p)$ is finite, f a function from A to ω, *then* there is q, $p \leq q \in Q$, $f \subseteq q$.

Fact D. Q is proper.

Proof of Fact A. Let $\{\delta_\ell : \ell < j \le \omega\}$ be a list of all limit ordinals $\delta \in S$, such that $\mathrm{Dom}(p) < \delta \le \alpha$. As A_{δ_ℓ} has order type ω, and $\mathrm{Sup}(A_{\delta_\ell}) = \delta_\ell$, clearly $A_{\delta_\ell} \cap A_{\delta_m}$ is finite for $m < \ell$, hence we can define by induction on $\ell < \omega$, $\beta_\ell < \delta_\ell$ such that $\beta_\ell \ge \mathrm{Dom}(p)$ and $\beta_\ell > \mathrm{Max}(A_{\delta_\ell} \cap A_{\delta_m})$ for $m < \ell$. Now define q, a function from α to ω:

$$q(i) = \begin{cases} p(i) & i \in \mathrm{Dom}(p) \\ f_{\delta_\ell}(i) & i \in (A_{\delta_\ell} \setminus \beta_\ell) \\ 0 & \text{otherwise (but } i < \alpha) \end{cases}$$

Now q is well defined as the β_ℓ were defined such that the $A_{\delta_\ell} \setminus \beta_\ell$ are pairwise disjoint and disjoint to $\mathrm{Dom}(p)$. It is trivial to check $p \le q \in Q$.

Proof of Fact B. Trivial. (Note that if $\langle p_n : n < \omega \rangle$ is an increasing sequence of members of Q, then $\bigcup_{n<\omega} p_n$ satisfies almost all the requirements, the problematic one is: if $\delta \in S$ is $\bigcup_{n<\omega} \mathrm{Dom}(p_n)$ (i.e. the supremum of the domain) then $(\bigcup_{n<\omega} p_n){\restriction}A_\delta =_{ae} f_\delta$. But by Fact A the set $\bigcup_{n<\omega} \mathrm{Dom}(p_n)$ is $N \cap \omega_1$ if $\langle p_n : n < \omega \rangle$ is a generaic sequence for (N, Q), so if $N \cap \omega_1 \in \omega_1 \setminus S$ the sequence has an upper bound, and this holds for N suitable for $\mathcal{E}$.)

Proof of Fact C. Let $A = \{\alpha_\ell : \ell < m\}$ increasing with ℓ and we define by induction $p_\ell \in Q$ (for $\ell \le 2m$), $p_0 = p$, $\mathrm{Dom}(p_{2\ell+1}) = \alpha_\ell$, $\mathrm{Dom}(p_{2\ell+2}) = \alpha_\ell + 1$, $p_{2\ell+2}(\alpha_\ell) = f(\alpha_\ell), p_\ell \le p_{\ell+1}$. Now the existence of $p_{2\ell+1}$ follows by Fact A, and $p_{2\ell+2}$ belongs to Q as for every limit δ, $[\delta \le \mathrm{Dom}(p_{2\ell+1}) \Leftrightarrow \delta \le \mathrm{Dom}(p_{2\ell+2})]$. Now $q = p_{2m}$ is as required.

Proof of Fact D. Let λ be large enough, $\mu > (2^\lambda)^+$, $<^*_\mu$ a well ordering of $H(\mu)$, for which λ is the first element and let $<^*_\lambda = <^*_\mu {\restriction} H(\lambda)$. It suffices to prove that for any given countable $N \prec (H(\mu), \in, <^*_\mu)$, $Q \in N$, $p \in N \cap Q$ there is $q \in P$ which is (N, Q)-generic, $p \le q$. So let $\delta \overset{\mathrm{def}}{=} N \cap \omega_1$ and choose $\alpha_n < \delta$, such that $\alpha_n < \alpha_{n+1}$, $\delta = \cup_{n<\omega} \alpha_n$. Let $\{b_\ell : \ell < \omega\}$ be a list of all members of $N \cap H(\lambda)$ and N_k be the Skolem hull of $\{Q, p\} \cup \{b_\ell : \ell < k\} \cup \{i : i < \alpha_k\}$ in the model $N^\dagger \overset{\mathrm{def}}{=} N {\restriction} H(\lambda) \prec (H(\lambda), \in, <^*_\lambda)$. So clearly $N_k \in N$ (as λ is definable in $(H(\mu), \in, <^*_\lambda)$, being the first, hence $(H(\lambda), \in, <^*_\lambda)$ belongs to N).

It is also clear that $\cup_{k<\omega} N_k = N^\dagger$, so every pre-dense $\mathcal{I} \subseteq Q, \mathcal{I} \in N$ belongs to $N^\dagger$ hence to some N_k. Now, define by induction on n, p_n such that:

a) $p_0 = p$, $p_n \leq p_{n+1}$

b) $p_n \in N_n \cap Q$

c) if $\delta \in S$, then p_n, f_δ agree on $A_\delta \cap N_n \setminus \mathrm{Dom}(p_0)$

d) if $\ell < n$, b_ℓ an open dense subset of Q then $p_n \in b_\ell$

e) $\alpha_n \in \mathrm{Dom}(p_{n+1})$

For $n = 0, p_0 = p$ satisfies all the requirements.

If p_n is defined and satisfies the requirements, first note that $A_\delta \cap N_{n+1}$ is finite as: $N_{n+1} \in N$, $\mathrm{Sup}(N_{n+1} \cap \omega_1) \in N$ hence $\mathrm{Sup}(N_{n+1} \cap \omega_1) < \delta$, whereas A_δ has order type ω with $\mathrm{Sup}(A_\delta) = \delta$. By Fact C there is $p_n^\dagger \geq p_n$, $p_n^\dagger \in Q$, $p_n^\dagger \supseteq f_\delta \restriction (N_{n+1} \cap A_\delta \setminus \mathrm{Dom}(p_n))$ and by Fact A w.l.o.g. $\alpha_n \subseteq \mathrm{Dom}(p_n^\dagger)$ (if $\delta \notin S$, we use only fact A). As $N_{n+1} \prec (H(\lambda), \in, <^*_\lambda)$, $\{p_n, Q, \alpha_n, f_\delta \restriction (N_{n+1} \cap A)\} \in N_{n+1}$, we can find such $p_n^\dagger \in N_{n+1}$. Now $p_n^\dagger$ satisfies all the requirements on p_{n+1} (for c) use the induction hypothesis) except maybe d) for $\ell = n$. So if b_n is an open dense subset of Q, we choose $p_{n+1} \in b_n$, $p_{n+1} \geq p_n^\dagger, p_{n+1} \in N_{n+1}$, and if b_n is not an open dense subset of Q, we choose $p_{n+1} = p_n^\dagger$.

So, we have completed the definition by induction of the p_n's. Now $q = \cup_{n<\omega} p_n$ is a member of Q because: by a)(and b)), q is a function from an ordinal to ω; by b) we have $\mathrm{Dom}(p_n) \subseteq \delta$, and by e) we have $\alpha_n \subseteq \mathrm{Dom}(p_n)$ hence $\mathrm{Dom}(q) = \delta$; for $\delta_1 < \delta$ we have $q \restriction A_{\delta_1} =_{ae} f_{\delta_1}$ as $q \restriction A_{\delta_1} \subseteq p_n \in Q$ for some n and if $\delta \in S$ then $q \restriction A_\delta =_{ae} f_\delta$ by c). Also q belongs to every b_ℓ which is an open dense subset of Q (by d) as $p_{\ell+1} \leq q$). But, every open dense subset of Q which belongs to N, belongs to $H(\lambda)$ (as λ is large enough) hence to $N^\dagger$ hence is b_ℓ for some ℓ, so q is (N, Q)-generic. $\square_{1.7}$

* * *

We can easily get similarly (more exactly, combining 1.7 and [Sh:44], [Sh:64]):

1.8 Conclusion. In the model from 1.6, if $\mathcal{E} = \{\omega_1 \setminus S\}$, $S \subseteq \omega_1$ stationary costationary, the following holds: If G is an abelian group, $G = \cup_{i<\omega_1} G_i$, G_i

increasing, countable, free, and G_i/G_j is free when $j \notin S$, $j < i$, *then* G is a Whitehead group. $\square_{1.8}$

Also note that if $\mathcal{E}$ is a normal filter on ω_1, it is well know that by $\mathcal{E}$-complete forcings we can "shoot" through all $A \in \mathcal{E}$ closed unbounded subsets of ω_1. We can in 1.7 (hence 1.8) replace $\{\omega_1 \setminus S\}$, by a normal nontrivial ideal on ω_1; which we can assume is dense in the sense that every stationary $S \subseteq \omega_1$ contains a subset in the ideal.
The reason for 1.8 is

1.8A Fact. For $\bar{G} = \langle G_i : i < \omega_1 \rangle$, $S \subseteq \omega_1$ as in 1.8, H an abelian group, h a homomorphism from H onto G with kernel $\mathbb{Z}$, letting $H_i = h^{-1}(G_i)$, the following forcing notion is proper and $(\omega_1 \setminus S)$-complete: $P = \{g : g$ a homomorphism from G_i into H_i such that $(h{\restriction}H_i) \circ g = \mathrm{id}_{G_i}\}$. $\square_{1.8A}$

Note

1.9 Claim. If P is $\{S\}$-complete, S a stationary subset of ω_1 and in V we have $\diamondsuit_S$ *then* in V^P we also have $\diamondsuit_S$.

Proof. Straightforward (as in IV, we use $\diamondsuit_S$ to given the isomorphism type of a countable elementary submodel and a name of a subset of ω_1). $\square_{1.9}$

§2. Generalizations of Properness

We shall repeat most of this section in the next one, with more details and less generality.

2.1 Definition. For an uncountable cardinal λ, countable ordinal α and $\ell < \omega$:

(1) Let $SQS^{\ell}_{\alpha}(\lambda)$ be the set of sequences $\langle N_i : i \leq \alpha \rangle$ such that:

a) N_i a countable submodel of $(H(\lambda), \in)$.

b) $i \in N_i$ and $\langle N_j : j \leq i\rangle \in N_{i+1+\ell}$ or, at least, $\langle N_j : j \leq i\rangle$ is definable in $N_{i+1+\ell}$.

c) If φ is first order, $\bar{a} \in N_i$ and $(H(\lambda), \in) \models$ "$\exists x\ \varphi(x, \bar{a})$", *then* for some $b \in N_{i+\ell}$, $(H(\lambda), \in) \models \varphi(b, \bar{a})$ (so for limit $\delta \leq \alpha$, $N_\delta \prec (H(\lambda), \in)$).

d) N_i $(i \leq \alpha)$ is increasing and continuous.

(2) A forcing notion P is (α, ℓ)-proper, if (for λ large enough): for every $\bar{N} = \langle N_i : i \leq \alpha\rangle \in SQS^0_\alpha(\lambda)$ (the zero is intended), such that $P \in N_0$, and for every $p \in N_0$, $p \in P$, there is an $r \in P, r \geq p$ which is $(\bar{N}, P, \ell)$-generic (or (P, ℓ)-generic for $\bar{N}$, or ℓ-generic for $\bar{N}$), which means: for every i, $r \Vdash_P$ "$N_i[\underset{\sim}{G}_P] \cap V \subseteq N_{i+\ell}$", where:

(3) If $P \in N \subseteq H(\lambda)$, $G \subseteq P$ generic, *then* $N[G] = \{\underset{\sim}{\tau}[G] : \underset{\sim}{\tau}$ a P-name first order definable from parameters from $N\}$. We define $\bar{N}[G]$ similarly, for $\bar{N} = \langle N_i : i \leq \alpha\rangle$.

2.1A Remark.

1) Note that for $\ell = 0$, $N_i \prec (H(\lambda), \in)$.

2) Note that by Lemma 2.5 it follows that $(\mathcal{E}, \alpha, \ell)$-properness is equivalent to $(\mathcal{E}, \alpha, k)$-properness for $k, \ell > 0$. See 2.5A(0).

2.2 Definition.

(1) $\mathcal{S}^\alpha_{\aleph_0}(\mu) = \{\langle a_i : i \leq \alpha\rangle : a_i \in \mathcal{S}_{\aleph_0}(\mu)$ and the sequence is increasing continuous$\}$.

(2) We call $\mathcal{E} \subseteq \cup_{i \leq \alpha} \mathcal{P}(\mathcal{S}^i_{\aleph_0}(\mu))$, (α, ℓ)-nontrivial, if for every large enough λ, $SQS^\ell_\alpha(\lambda, \mathcal{E}) \neq \emptyset$, where $SQS^\ell_\alpha(\lambda, \mathcal{E})$ is the set of $\bar{N} = \langle N_i : i \leq \alpha\rangle \in SQS^\ell_\alpha(\lambda)$ such that: $\mathcal{E} \in N_0$ (and $\mu \in N_0$) and *if* $A \in \mathcal{E} \cap N_0 \cap \mathcal{P}(\mathcal{S}^\gamma_{\aleph_0}(\mu))$ and $\gamma \leq \alpha$, *then* $\langle N_i \cap \mu : i \leq \gamma\rangle \in A$. In such a case, we call $\bar{N}$ suitable for $\mathcal{E}$.

(3) P is $(\mathcal{E}, \alpha, \ell)$-proper means: *if* λ is large enough, and $\bar{N} \in SQS^0_\alpha(\lambda, \mathcal{E})$, $\beta < \gamma \leq \alpha$, γ a limit ordinal, $P \in N_\beta$ and $p \in P \cap N_\beta$, *then* there is q, $p \leq q \in P$ such that q is $(\bar{N}{\restriction}[\beta, \gamma], P, \ell)$-generic.

(4) In Definition 2.1, 2.2 we may suppress ℓ when it is zero.

(5) A forcing notion which is $(\mathcal{E}, 0, 0)$-proper will be called $\mathcal{E}$-proper.

(6) A forcing notion will be called (α, ℓ)-proper if it is $(\{\mathcal{S}_{\aleph_0}(\mu)\}, \alpha, \ell)$-proper.

2.3 Theorem. Assume $\mathcal{E}$ is (α, ℓ)-nontrivial. Countable support iteration preserves $(\mathcal{E}, \alpha, \ell)$-properness, provided that $\ell = 0$ or α is a limit ordinal.

Before proving we show 2.4, 2.5 below.

2.3A Remark. There are examples that the notions are distinct, proper is $(\{\mathcal{S}_{\aleph_0}(\mu)\}, 0, 0)$-proper.

2.4 Claim.

1) If $P \in N \prec (H(\lambda), \in, <^*_\lambda)$, N countable, $G \subseteq P$ generic over V, *then* $N[G] = \{\underset{\sim}{\tau}[G] : \underset{\sim}{\tau} \in N \text{ a } P\text{-name}\}$.

2) If $\bar{N} \in SQS^0_\alpha(\lambda), P \in N_0, G \subseteq P$ is generic, *then* $V^P \models$ "$\bar{N}[G] \in SQS^0_\alpha(\lambda)$".

3) If $\bar{N} \in SQS^0_\alpha(\lambda)$, $\bar{Q} = \langle P_\ell, \underset{\sim}{Q}_\ell : \ell < n\rangle \in N_0$ is iterated forcing, $\underset{\sim}{r}_\ell$ a P_ℓ-name of a member of $\underset{\sim}{Q}_\ell$, is $(\bar{N}[\underset{\sim}{G}_\ell], \underset{\sim}{Q}_\ell, k)$-generic ($\underset{\sim}{G}_\ell \subseteq P_\ell$ the generic set), *then* $r = \langle r_0, \underset{\sim}{r}_1, \ldots, \underset{\sim}{r}_{n-1}\rangle$ is $(\bar{N}, P_n, nk)$-generic.

4) If P is $(\alpha, 1)$-proper and $\alpha = \omega\beta$, *then* P is $(\beta, 0)$-proper.

5) If P is (α_1, ℓ_1)-proper, $\alpha_1 \geq \alpha_2$, $\ell_1 \leq \ell_2$, then P is (α_2, ℓ_2)-proper. Also P is $(0, 0)$-proper iff P is proper.

Proof. 1) Straightforward.

2) like 1.3(1).

3) Left to the reader.

4), 5) Check. $\square_{2.4}$

2.5 Lemma. Consider the following properties of a forcing notion P, $p \in P$, countable limit ordinal α, λ regular large enough and $<^*_\lambda$ a well ordering of $H(\lambda)$, $\bar{\gamma}$ is an α-sequence of ordinals, strictly increasing, $\gamma(i) < i + \omega$, $\bar{k}$ an α-sequence of natural numbers, $k(i) \geq 3$, are equivalent:

(1) There is a function F, $\mathrm{Rang}(F) \subseteq \mathcal{S}_{<\aleph_1}(\mathcal{P}(P))$, $\mathrm{Dom}(F) = \mathcal{S}^{<\alpha}_{<\aleph_1}(\mathcal{P}(P))$ such that (but for $\mathcal{I} \subseteq P$, $A \subseteq \mathcal{P}(P)$ we let $\mathcal{I} \cap A = \{p \in \mathcal{I} : \{p\} \in A\}$): for every increasing continuous sequence of countable subsets of $\mathcal{P}(P)$, $A_i (i < \alpha)$ satisfying: $F(\langle A_j : j \leq i \rangle) \subseteq A_{i+1}$, $p \in P \cap A_0$ *there is* a $q \in P$, $q \geq p$, such that for every $i < \alpha$ and $\mathcal{I} \in A_i$ a maximal antichain of P, $\mathcal{I} \cap F(\langle A_j : j \leq i \rangle)$ is pre-dense above q.

(2) For every $N_i \subseteq (H(\lambda), \in, <^*_\lambda)$ for $i < \alpha$ satisfying (a), (b), (c) listed below and $p \in N_0 \cap P$ there is a $q \in P$, $q \geq p$ such that

(∗) for every $i < \alpha$ and P-name $\underset{\sim}{\beta}$ of an ordinal, $\underset{\sim}{\beta} \in N_i$, we have

$$q \Vdash_P \text{“}\underset{\sim}{\beta} \in N_{i+1}\text{”}$$

where

(a) $\langle N_j : j \leq i \rangle \in N_{i+1}$, N_i continuously increasing, N_i is countable and $P \in N_0$, $\alpha \in N_0$.

(b) For every (first-order) formula $\varphi(x, \bar{a}), \bar{a} \in N_i$, $i < \alpha$, $(H(\lambda), \in, <^*_\lambda) \models (\exists x)\varphi(x, \bar{a})$ implies $(H(\lambda), \in, <^*_\lambda) \models (\exists x \in N_{i+1})\ \varphi(x, \bar{a})$

(c) In b) we can allow $\bar{a} \subseteq N_i \cup \{\langle N_j : j \leq i \rangle\}$.

$(3)_{\bar{\gamma}}$ The same as (2) omitting (c), replacing N_{i+1} by $N_{\gamma(i)+3}$ in (∗).

$(4)_{\bar{k}}$ The same as in (2), replacing N_{i+1} by $N_{i+k(i)}$ in (∗), omitting (c).

(5) Like (4) for $\bar{k}$ constantly 3.

2.5A Remark. (0) The point of this lemma is to show that some natural variants of Definition 2.1(2) are equivalent.

(1) Note that clause (2) is just a case of $(\alpha, 1)$-properness.

(2) “λ large enough” just means $\mathcal{P}(P) \in H(\lambda)$; we can replace $\mathcal{P}(P)$ by the family of maximal antichains of P.

(3) As (1) of 2.5 does not depend on λ, we get the equivalence of the others for all suitable λ, similarly concerning $\bar{\gamma}$ and $\bar{k}$.

(4) We can replace $\alpha \in N_0$ by “$i + 1 \subseteq N_i$”.

Proof. <u>(1) ⇒ (5):</u> We can assume that the F exemplifying (1) is definable in $(H(\lambda), \in, <_\lambda)$ (by a formula with the parameters P and α only); just take the $<_\lambda$-first F satisfying (1). Moreover we can assume there is an $\bar{f} = \langle f_0, f_1, \ldots \rangle_{n<\omega} \in H(\lambda)$ similarly definable such that $F(\langle A_j : j \leq i \rangle) \subseteq \{f_n(\langle A_j : j \leq i \rangle) : n < \omega\}$. Clearly for every $i < \alpha$, $n < \omega$ and $\langle A_j : j \leq i \rangle$ we can find some first order $\varphi_n(x, P, p, \langle A_j : j \leq i \rangle)$ such that the unique $x \in H(\lambda)$ satisfying it in $(H(\lambda), \in, <^*_\lambda)$ is $f_n(\langle A_j : j \leq i \rangle)$. Hence, if $N_i (i < \alpha)$ are as in $(4)_{\bar{k}}$, then for every $i < \alpha$, $n < \omega$, $f_n(\langle N_{2j} \cap \mathcal{P}(P) : j \leq i \rangle)$ is definable with parameters from N_{2i+1}, hence is an element of N_{2i+2}. So $F(\langle N_{2j} \cap \mathcal{P}(P) : j \leq i \rangle) \subseteq N_{2i+2}$. So, if q exemplifies the satisfaction of (1) for $\langle N_{2j} \cap \mathcal{P}(P) : j < \alpha \rangle$, then it exemplifies the satisfaction of $(4)_{\bar{k}}$ for $\langle N_j : j < \alpha \rangle$ where $k_i = 3$.

<u>(1) $\Rightarrow$ (2):</u> Similar proof, but we use $F(\langle N_j \cap \mathcal{P}(P) : j \leq i \rangle)$, made possible by use of (c) from (2).

<u>(2)$\Rightarrow$ (1):</u> To define $F(\langle A_j : j \leq i \rangle)$, we define a sequence $\langle N_\zeta(\langle A_j : j \leq i \rangle) : \zeta \leq 2i+2 \rangle$ as follows: $N_\zeta = \bigcup_{\gamma<\zeta} N_\gamma$ for limit ζ, and $N_\zeta = N_\zeta(\langle A_j : j \leq i \rangle) =$ the Skolem hull of $\{\langle N_\gamma : \gamma \leq \beta \rangle : \beta < \zeta\} \cup \{A_j : 2j < \zeta\}$ in the model $(H(\lambda), \in, <^*_\chi)$ for a successor ζ and $\zeta = 0$. Note that $N_\zeta(\langle A_j : j \leq i_0 \rangle) = N_\zeta(\langle A_j : j \leq i_1 \rangle)$, if $\zeta \leq 2i_0+2, \zeta \leq 2i_1+2$. Let $F(\langle A_j : j \leq i \rangle) = \mathcal{P}(P) \cap N_{2i+2}$. If $\langle A_i : i < \alpha \rangle$ obeys F, let $N_\zeta = N_\zeta(\langle A_j : j \leq i \rangle)$ for some (or all) i such that $\zeta \leq 2i+2$. Then $\langle N_\zeta : \zeta \leq \alpha \rangle$ satisfies (a), (b),(c). For $p \in A_0 = \mathcal{P}(P) \cap N_0$, find q as guaranteed in (2). Let $\mathcal{I} \in A_i$. Then $\mathcal{I} \in N_{2i+1}$, so, by (2), $\mathcal{I} \cap N_{2i+2}$ is pre-dense above q and it includes $\mathcal{I} \cap F(\langle A_j : j \leq i \rangle)$, just as required.

<u>(5) $\Rightarrow$ for some $\bar{k}$, $(4)_{\bar{k}}$:</u> Trivial.

<u>$(4)_{\bar{k}}$ $\Rightarrow$ for some $\bar{\gamma}$ $(3)_{\bar{\gamma}}$:</u> (i.e. $\bar{\gamma}$ depends on $\bar{k}$): Given $\bar{k}$ define $\gamma(0) = k(0)$, $\gamma(i) = \cup_{j<i}\gamma(j) + k(\cup_{j<i}\gamma(j)) + 8$.

<u>$(3)_{\bar{\gamma}} \Rightarrow (1)$:</u> Similar to the proof of (2) $\Rightarrow$ (1), only we define $N_\zeta(\langle A_j : j \leq i \rangle)$ for $\zeta \leq \gamma(i) + 8 \times (\gamma(i) + 1 - \sup\{\delta : \delta \text{ limit} \leq \gamma(i)\})$. Note that as $\{P, \alpha\} \in N_\delta$ $\bar{\gamma}$ belongs to P or at least some.

Putting together all the implications, we have finished. $\square_{2.5}$

Proof of 2.3. We seperate the proof to the two natural cases. Let $\langle P_\zeta, \underset{\sim}{Q}_\xi : \zeta \le \zeta^*, \xi < \zeta^* \rangle$ be a countable support iteration and let λ be large enough.

Case A: $\ell = 0$.

We prove by induction on $\zeta \le \zeta^*$ and then by induction on $\gamma \le \alpha$ the following

$(*)_{\zeta,\gamma}$ *if* $\xi < \zeta$, $\bar N \in SQS^0_\alpha(\lambda, \mathcal{E}), \bar Q \in N_0$, $\beta < \gamma \le \alpha$, $\{\zeta, \xi\} \in N_\beta$ and $q \in P_\xi$ is $(\bar N \restriction [\beta,\gamma], P_\xi)$-generic and $p \in P_\zeta \cap N_\beta$ satisfies: $p \restriction \xi \le q$, or just $\underset{\sim}{p}$ is a P_ξ-name, $\Vdash_{P_\xi}$ "$\underset{\sim}{p} \in N_\beta \cap P_\zeta$, $\underset{\sim}{p} \restriction \xi \in G_{P_\xi}$", *then* there is $r \in P_\zeta$ such that: r is $(\bar N \restriction [\beta,\gamma], P_\zeta])$-generic, $r \restriction \xi = q$ and $p \le q$ and $\mathrm{Dom}(r) \setminus \xi = N_\gamma \cap \zeta \setminus \xi$.

As case B is more involved we do it in more details.

Case B: $\ell = 1$

Note that by 2.4, each $\underset{\sim}{Q}_i$ is proper. We prove by induction on $\zeta \le \zeta^*$ and then by induction on $\gamma \le \alpha$ and $\zeta \le \zeta^*$ the following:

$(*)_{\zeta,\gamma}$ *if* $\xi < \zeta \le \zeta^*$, $\bar N \in SQS^0_\alpha(\lambda, \mathcal{E})$, $\bar Q \in N_0$, $\beta \le \gamma \le \alpha$, γ a limit ordinal and β is a non limit ordinal, $\{\zeta, \xi\} \in N_\beta$ and $q \in P_\xi$ and $q \Vdash_{P_\xi}$ "if $\beta \le \beta_1 < \gamma$ then $N_{\beta_1}[\underset{\sim}{G}_{P_\xi}] \cap V \subseteq N_{\beta_1 + n}$ for some $n < \omega$" and $p \in P_\zeta \cap N_\beta$ satisfies $p \restriction \xi \le q$ or just p is a P_ξ-name of a member of $P_\zeta \cap N_\beta$ such that $p \restriction \xi \in G_{P_\xi}$ and $\mathrm{Dom}(q) = N_\gamma \cap \xi$ and $u \subseteq [\beta, \gamma)$ is a finite set of non limit ordinals *then* there is $r \in P_\zeta$ such that

(a) $r \Vdash_{P_\zeta}$ " if $\beta \le \beta_1 < \gamma$ then $N_{\beta_1}[\underset{\sim}{G}_{P_\zeta}] \cap V \subseteq N_{\beta_1+n}$ for some $n < \omega$"

(b) $r \Vdash_{P_\zeta}$ "if $\beta_1 \in u$ so $\beta \le \beta_1 < \gamma$, then $N_{\beta_1}[\underset{\sim}{G}_{P_\zeta}] \cap V^{P_\xi} = N_\beta[\underset{\sim}{G}_{P_\xi}]$"

(c) $p \le r$

(d) $r \restriction \xi = q$.

Note that when $\xi = \zeta$ the assertion is trivial.

case 1: $\zeta = 0$. There is nothing to prove.

case 2: $\zeta = \zeta_1 + 1$.

So $\xi \le \zeta_1$, so by the induction hypothesis (and the form of what we are trying to prove) w.l.o.g. $\xi = \zeta_1$ and $\beta \in u$ and $(\forall \beta')[\beta' + 1 \in u \ \& \ \beta' \ge \beta \rightarrow \beta' \in u]$, and $\beta \in u$. Let $u = \{\beta_0, \beta_1 + 1, \ldots, \beta_n - 1\}$, $\beta = \beta_0 \le \beta_1 < \ldots < \beta_{n-1}$ and let $\beta_n = \gamma$. Let $q \in G_\xi \subseteq P_\xi$, G_ξ generic over V and let $N'_i = N_i[G_\xi]$

for $i \in [\beta,\gamma]$, so $\langle N'_i : i \in [\beta,\gamma]\rangle \in SQS^0_{(\gamma-\beta)+1}(\lambda,\mathcal{E})$ in $V[G_\xi]$. If $\gamma = \alpha$ let $N'_{\alpha+1} \prec (H(\chi)[G_\xi],\in)$ be countable such that $\{\bar{N}, G_\xi, p, q, \bar{Q}, \xi, \zeta\} \in N'_{\alpha+1}$. Clearly $p(\xi)[G_\xi] \in \underset{\sim}{Q}_\xi[G_\xi] \cap N'_{\beta_0}$, so there is $p_0 \in N'_{\beta_0+1}$ which is $(N'_{\beta_0}, \underset{\sim}{Q}_\xi[G_\xi])$-generic (because as noted above, $\Vdash_{P_\xi}$ "$\underset{\sim}{Q}_\xi[\underset{\sim}{G}_\xi]$ is a proper forcing").

We now choose by induction on ℓ, $p_\ell \in N'_{\beta_\ell+1}$ such that:

$(*)_1$ if $\beta \le j \le \beta_\ell$, j is a limit ordinal *then* $p_\ell \Vdash N'_j[\underset{\sim}{G}_{Q_\xi}] \cap V[G_\xi] = N'_j$

$(*)_2$ if $\beta \le j \le \beta_\ell$ *then* $p_\ell \Vdash$ "$N'_j[\underset{\sim}{G}_{Q_\xi}] \cap V[G_\xi] \subseteq N'_{\min\{j+n,\beta_\ell\}}$ for some $n < \omega$".

$(*)_3$ $\underset{\sim}{Q}_\xi[G_\xi] \models p(\xi)[G_\xi] \le p_\ell \le p_{\ell+1}$

For $\ell = 0$ this was done above, for $\ell = n$ this complete the proof for the present case so let us choose $p_{\ell+1}$ assuming we have already chosen p_ℓ.

Now if $\beta_{\ell+1} = \beta_\ell + 1$ we just use $\Vdash_{P_\xi}$ "$\underset{\sim}{Q}_\xi$ is proper", so assume $\beta_{\ell+1} > \beta_\ell + 1$, so by a demand on u we know that $\beta_{\ell+1}$ is a limit ordinal. So first choose $p'_\ell \in \underset{\sim}{Q}_\xi[G_\xi] \cap N'_{\beta_\ell+2}$ which is above p_ℓ and is $(N_{\beta_\ell+1}, \underset{\sim}{Q}_\xi[G_\xi])$-generic (using again properness) and then choose $p_{\ell+1} \in \underset{\sim}{Q}_\xi[G_\xi] \cap N_{\beta_{\ell+1}+1}$ above p'_ℓ and satisfying $(*)_1 + (*)_2$, which is possible by the induction hypothesis on γ (and $\beta_{\ell+1}$ being a limit ordinal), so we have finished the induction step on ℓ hence the present case.

case 3: ζ a limit ordinal.

First as in the proof of the previous case, w.l.o.g. $u = \emptyset$. Now use diagonalization as usual. $\square_{2.3}$

§3. α-Properness and $(\mathcal{E},\alpha)$-Properness Revisited

In §1 we gave some solution to "which forcings do not add reals". What occurs is that we may have a small stationary subset of ω_1, on which e.g. uniformization properties hold. But we want e.g. to be able to prove the consistency of CH + SH, which is impossible by §1's method, because it is possible that the model V_1 from Theorem 1.6 satisfies also $\diamondsuit_{\omega_1}$, and even $\diamondsuit^*_{\omega_1 \setminus S}$ (see 1.9 or [Sh:64]).

Here we make an investment for this goal by developing α-properness (and $(\mathcal{E}, \alpha)$-properness) which is a generalization of properness, when the genericity is obtained for some tower of models simultaneously. In almost all cases the proof that properness holds gives α-properness. The point is that for some properties X, for "X + α-properness" it is easier to prove preservation by CS iteration.

To a large degree we redo here §2, with more explanation and, for notational simplicity, only for $\ell = 0$.

3.1 Definition. For $\alpha < \omega_1$ the forcing notion P is said to be *α-proper* if for every sufficiently large λ and for every sequence $\langle N_i : i \leq \alpha \rangle$ such that N_i is a countable set, $N_i \prec (H(\lambda), \in)$, if the sequence $\langle N_i : i \leq \alpha \rangle$ is continuously increasing, $i \in N_i$, $\langle N_j : j \leq i \rangle \in N_{i+1}, P \in N_0$ and $p \in P \cap N_0$, *then* there is a q, $p \leq q \in P$ which is (N_i, P)-generic for every $i \leq \alpha$.

3.2 Remarks.

(1) Obviously, a forcing notion P is 0-proper if and only if it is proper.

(2) It is also obvious that if $\beta < \alpha$ and P is α-proper then P is also β-proper (every sequence $\langle N_i : i \leq \beta \rangle$ which satisfies the above conditions can be extended to a sequence $\langle N_i : i \leq \alpha \rangle$ which satisfies these conditions and, since P is α-proper there is a $p \leq q \in P$ which is (N_i, P)-generic for every $i \leq \alpha$). Therefore, in particular, every α-proper P is proper.

(3) If P is α-proper it is also $(\alpha + 1 + \alpha)$-proper. To see this let $\langle N_i : i \leq \alpha + 1 + \alpha \rangle$, p be as required. Since P is α-proper there is a q_0, $p \leq q_0 \in P$ which is (N_i, P)-generic for every $i \leq \alpha$. Since $N_{\alpha+1} \prec (H(\lambda), \in)$ and p, P, $\langle N_i : i \leq \alpha \rangle \in N_{\alpha+1}$ there is such a $q_0 \in N_{\alpha+1}$. Since P is α-proper there is q_1, $q_0 \leq q_1 \in P$ which is $(N_{\alpha+1+i}, P)$-generic for every $i \leq \alpha$. Since $q_1 \geq q_0, q_0$ is also (N_i, P)-generic for every $i \leq \alpha + 1 + \alpha$.

(4) Note that if α is limit, $\langle N_i : i < \alpha \rangle$ increasing and continuous, p is (N_i, P)-generic for $i < \alpha$ (and $P \in N_0$), then p is (N, P)-generic where $N = \cup_{i<\alpha} N_i$. As a consequence of this and (3), if P is proper it is n-proper for all $n < \omega$ and if P is ω-proper it is α-proper for all $\omega \leq \alpha < \omega^2$. And:

P is γ_0-proper iff P is γ_1-proper when $\gamma_0\omega = \gamma_1\omega$. Hence it is enough to deal with additively indecomposable γ (i.e. $(\forall\beta < \gamma)(\beta + \beta < \gamma)$).

(5) For $\langle N_i : i \leq \alpha\rangle$ as in 3.1, α additively indecomposable, as $\alpha \in N_\alpha$, for some $\beta < \alpha$, $\alpha \in N_\beta$; now $\alpha = \beta + \alpha$, so with easy manipulations this definition is equivalent to the one with $\alpha \in N_0$.

3.3 Definition. $\mathcal{S}^\alpha_{\aleph_0}(A) = \mathcal{S}^\alpha_{<\aleph_1}(A) = \{\langle a_i : i \leq \alpha\rangle : a_i \in \mathcal{S}_{\aleph_0}(A)$ for all $i \leq \alpha$ and $\langle a_i : i \leq \alpha\rangle$ is continuously increasing$\}$. Let F be a function from $\bigcup_{\beta\leq\alpha} \mathcal{S}^\beta_{<\aleph_1}(A)$ into $\mathcal{S}_{<\aleph_1}(A)$ and let $G(F) = \{\langle a_i : i \leq \alpha\rangle \in \mathcal{S}^\alpha_{<\aleph_1}(A) : (\forall i < \alpha)$ $(\forall$ finite $b \subseteq a_{i+1})$ $F(\langle a_j : j \leq i\rangle, b) \subseteq a_{i+1}) \wedge (\forall$ finite $b \subseteq a_0)$ $F(b) \subseteq a_0\}$ where we write $F(b)$ instead $F(\langle b\rangle)$ and $F(\bar{a}, b)$ instead $F(\bar{a}\hat{\ }\langle b\rangle)$. Let F_n, $n < \omega$, be functions into $\mathcal{S}_{<\aleph_1}(A)$ and let F be given by $F(x) = \bigcup_{n<\omega} F_n(x)$, then $G(F) \subseteq \bigcap_{n<\omega} G(F_n)$, hence the set of all $G(F)$'s generates an $\aleph_1$-complete filter $\mathcal{D}^\alpha_{<\aleph_1}(A)$ on $\mathcal{S}^\alpha_{<\aleph_1}(A)$.

3.4 Theorem. The forcing notion P is α-proper if and only if it preserves the property of being a stationary subset of $\mathcal{S}^\alpha_{<\aleph_1}(A)$ (i.e. being a set of positive measure) with respect to the filter $\mathcal{D}^\alpha_{<\aleph_1}(A)$ for every uncountable A.

Proof. Similar to the proof of the corresponding fact for proper forcing. $\square_{3.4}$

3.5 Theorem. For each $\alpha < \omega_1$, α-properness is preserved by countable support iterations.

Proof. Again the proof is similar to the one on properness, or see 2.3(1). $\square_{3.5}$

Now we add $\mathcal{E}$ as a parameter, where $\mathcal{E}$ is similar to what we did in §1.

3.6 Definition. A family $\mathcal{E}$ of subsets of $\bigcup_{\gamma<\omega_1} \mathcal{S}^\gamma_{\aleph_0}(\mu)$ is α-nontrivial if: For every λ large enough, there is a continuous sequence $\bar{N} = \langle N_i : i \leq \alpha\rangle$ of countable elementary submodels of $(H(\lambda), \in)$, $\langle N_j : j \leq i\rangle \in N_{i+1}$, $\mathcal{E} \in N_0$ such that: $\langle N_i \cap \mu : i \leq \alpha\rangle$ belongs to $\bigcap\{Y : Y \in \mathcal{E} \cap N_0\}$. In this case we call $\bar{N}$ suitable for $\mathcal{E}$ and for $(\mathcal{E}, \alpha)$. Let $\mathcal{D}_\alpha(\mathcal{E}) = \{S \subseteq \mathcal{S}^\alpha_{\aleph_0}(\mu) : \mathcal{E} \cup \{\mathcal{S}^\alpha_{\aleph_0}(\mu) \setminus S\}$ is α-trivial$\}$.

3.6A Remark. 1) If $\alpha < \omega_1$, $\mathcal{E}_{\beta,\gamma} \subseteq \mathcal{P}(S^{\gamma-\beta}_{\aleph_0}(\mu))$ for $\beta < \gamma \leq \alpha$, β not limit, *then* we can find $\mathcal{E} \subseteq \mathcal{P}(S^{\alpha}_{\aleph_0}(2^\mu))$ such that: if $\lambda > 2^\mu$, $\langle N_i : i \leq \alpha\rangle \in SQS^0_\alpha(\lambda)$ and $\mu \in N_0$, *then* (a)⇔(b) where:

(a) $\langle N_i \cap \mu : \beta \leq i \leq \gamma\rangle \in \mathcal{E}_{\beta,\gamma}$ when $\mathcal{E}_{\beta,\gamma}$ is defined.

(b) $\langle N_i \cap 2^\mu : i \leq \alpha\rangle \in \mathcal{E}$.

2) We also use in this section the following stronger demand than 3.6:

$$\text{if } \beta < \alpha, \text{ then } \langle N_{\beta+\gamma} \cap \mu : \gamma \leq \alpha - \beta\rangle \in \bigcap\{Y : Y \in \mathcal{E} \cap N_\beta\}.$$

3) The point of 3.6A(1) (and its parallel for 3.6A(2)) is the variation in Definition 3.6 do not give a really new notion.

3.7 Definition. A forcing notion P is $(\mathcal{E},\alpha)$-proper ($\mathcal{E}$ as above, α-nontrivial) *if* for every $\bar{N}$ which is suitable for $(\mathcal{E},\alpha)$ and $p \in N_0$, $p \in N_0 \cap P$ there is $q \geq p$ (in P) such that q is (N_i, P)-generic for every $i \leq \alpha$.

The following repeats 2.3.

3.8 Theorem. Suppose $\mathcal{E}$ is α-nontrivial, $\mathcal{E} \subseteq \bigcup_{\gamma<\omega_1}\mathcal{P}(S^{\gamma}_{\aleph_0}(\mu))$.

(1) If P is $(\mathcal{E},\alpha)$-proper, then $\Vdash_P$ "$\mathcal{E}$ is α-nontrivial".

(2) If $\bar{Q} = \langle P_i, \underset{\sim}{Q}_i : i < \beta\rangle$ is a countable support iteration, $\Vdash_{P_i}$ "$\underset{\sim}{Q}_i$ is $(\mathcal{E},\alpha)$-proper", $P_\beta = \operatorname{Lim}\bar{Q}$, *then* P_β is $(\mathcal{E},\alpha)$-proper.

(3) If $P \in N_0$, $\bar{N}$ is $\mathcal{E}$-suitable, $p \in P$ is (N_i, P)-generic for every $i \leq \ell g(\bar{N})$, *then* $p \Vdash_P$ "$\bar{N}[G]$ is $\mathcal{E}$-suitable".

3.8A Remark. So in (2) (by (1)) it suffices to assume $\Vdash_{P_i}$ "if $\mathcal{E}$ is α-nontrivial (in V^{P_i}) then $\underset{\sim}{Q}_i$ is $(\mathcal{E},\alpha)$-proper".

Proof. No new point. $\square_{3.8}$

3.9 Theorem. If P is $(\mathcal{E},\alpha)$-proper and Q is not $(\mathcal{E},\alpha)$-proper, *then* $\Vdash_P$ "Q is not $(\mathcal{E},\alpha)$-proper".

Proof. Easy. $\square_{3.9}$

§4. Preservation of ω-Properness + the ${}^{\omega}\omega$-Bounding Property

4.1 Definition. A forcing notion P has the ${}^{\omega}\omega$-bounding property if: for any $f \in ({}^{\omega}\omega)^{V[G]}$ ($G \subseteq P$ generic) there is $g \in ({}^{\omega}\omega)^{V}$ such that $f \leq g$ (i.e. $(\forall n < \omega) f(n) \leq g(n)$).

4.2 Discussion. Clearly the ${}^{\omega}\omega$-bounding property can be considered as an approximation to the property "not adding reals". Also this property, and similar properties play crucial parts in many independence proofs. That is, many times we want on one hand to add many reals, but on the other hand to preserve something. e.g. to preserve: the set of old (or constructible) reals is of the second category or does not have measure zero, or every new real belongs to an old Borel set of special kinds, etc. In the next chapter we shall deal with various such properties. But here we choose to deal with ${}^{\omega}\omega$-bounding, as it is very natural, and as the proof of its preservation is a prototype for many other such proofs. To be more exact we do not prove that it is preserved, only that together with ω-properness it is preserved. (This will be eliminated in the next chapter). The proof also serve as introduction to the proof of preservation of "no new reals" in §7 and to VI. Of course VI §2 gives an alternative proof of the theorem 4.3.

4.3 Theorem. The property "ω-properness + the ${}^{\omega}\omega$-bounding property" is preserved by countable support iteration.

Proof. Let $\langle P_i, \underset{\sim}{Q}_i : i < \alpha \rangle$ be a CS iterated forcing system. We prove that it has the ${}^{\omega}\omega$-bounding property by induction on α (the preservation of ω-properness follows from Theorem 3.5). For $\alpha = 0$ there is nothing to prove. For $\alpha + 1$ we have $V^{P_{\alpha+1}} = (V^{P_\alpha})^{Q_\alpha}$. If $f \in V^{P_{\alpha+1}} = (V^{P_\alpha})^{Q_\alpha}$, then there is a function

$g \in V^{P_\alpha}$ such that $g \geq f$ (since Q_α has the ${}^{\omega}\omega$-bounding property) and, by the induction hypothesis, there is an $h \in ({}^{\omega}\omega)^V$ such that $h \geq g$, so $h \geq f$.

If α is a limit ordinal and $\mathrm{cf}(\alpha) > \aleph_0$, then every $f \in V^{P_\alpha}$ already appears in some V^{P_i}, $i < \alpha$ (by III 4.1B(2)), so we can apply the induction hypothesis.

So we are left with the case α is a limit ordinal of cofinality $\aleph_0$.

The following lemma is the main point.

4.4 Lemma. Suppose

a) P a proper forcing notion, $\underset{\sim}{Q}$ a P-name of a proper forcing notion, $(p, \underset{\sim}{q}) \in P * \underset{\sim}{Q}$ and $P, \underset{\sim}{Q}$ have the ${}^{\omega}\omega$-bounding property, λ large enough, $N_0 \prec N_1 \prec N_2 \prec (H(\lambda), \in)$, and $P, \underset{\sim}{Q}, P * \underset{\sim}{Q}$ and $(p, \underset{\sim}{q})$ belong to N_0.

b) $N_0 \in N_1, N_1 \in N_2$, and each N_ℓ is countable.

c) $r \in P$ is (N_ℓ, P)-generic for $\ell = 0, 1, 2$ and $r \geq p$.

d) $\langle \mathcal{I}_\ell : \ell < \omega \rangle$ is a list of all maximal antichains of P which belong to N_0, $\mathcal{I}^*_\ell \subseteq \mathcal{I}_\ell \cap N_0$ is finite, $\mathcal{I}^*_\ell$ pre-dense above r and $\langle \mathcal{I}_\ell : \ell < \omega \rangle \in N_1$, $\langle \mathcal{I}^*_\ell : \ell < \omega \rangle \in N_1$.

e) $\langle \mathcal{J}_\ell : \ell < \omega \rangle \in N_1$ is a list of the maximal antichains of $P * \underset{\sim}{Q}$ which belong to N_0.

Then there is a $\underset{\sim}{q}_1 \in \underset{\sim}{Q} \cap N_2$, $\mathcal{J}^*_\ell \subseteq \mathcal{J}_\ell \cap N_0$ finite for $\ell < \omega$, such that: $(r, \underset{\sim}{q}_1) \geq (p, \underset{\sim}{q})$, and each $\mathcal{J}^*_\ell$ is pre-dense above $(r, \underset{\sim}{q}_1)$ (hence $(r, \underset{\sim}{q}_1)$ is $(N_0, P * \underset{\sim}{Q})$- generic) and $\langle \mathcal{J}^*_\ell : \ell < \omega \rangle \in N_2$.

4.4A Remarks.

(1) Instead of a maximal antichain, we can look at a name of an ordinal, or dense subsets.

(2) The situation for P, N_0, N_1, r in the assumption is similar to the situation of $P * \underset{\sim}{Q}, N_0, N_2, (r, \underset{\sim}{q}^1)$ in the conclusion when $\underset{\sim}{q}^1 \geq \underset{\sim}{q}_1$ is $(N_2, \underset{\sim}{Q})$ generic. So we preserve the situation while not increasing the condition in P. So, every time we advance one step in the iteration, we lose genericity for one of the models (N_1). This will give us the induction step in the proof of 4.3 for $\mathrm{cf}(\alpha) = \aleph_0$.

Proof of Lemma 4.4. For helping us in understanding let $G = G_P \subseteq P$ be generic over V, and we shall work sometimes in $V[G]$, sometimes in V. Note that if $r \in G$ (which is the interesting case for us) then for $\ell = 0, 1, 2$ we have $N_\ell[G] \cap H(\lambda)^V = N_\ell$ and $N_\ell[G] \prec (H(\lambda)[G], \in)$ and even $(N_\ell[G], N_\ell, \in) \prec (H(\lambda)[G], H(\lambda), \in)$ and $N_\ell[G] \in N_{\ell+1}[G]$. Alternatively, we could rewrite statements of the form $V[G] \models \ldots$ as $r \Vdash \ldots$

First try:
As $\mathcal{J}_0$ is a maximal antichain in $P * \underset{\sim}{Q}$, the set $\{\underset{\sim}{q}^0[G] : (p^0, \underset{\sim}{q}^0) \in \mathcal{J}_0,\ p^0 \in G\}$ is a maximal antichain of $\underset{\sim}{Q}[G]$. Hence $\underset{\sim}{q}[G]$ is compatible with some such $\underset{\sim}{q}^0[G]$. Let $\underset{\sim}{p}^0, \underset{\sim}{q}^0$ be P-names such that:
$\Vdash_P$ "$\underset{\sim}{p}^0 \in \underset{\sim}{G}_P, (\underset{\sim}{p}^0, \underset{\sim}{q}^0) \in \mathcal{J}_0$ and, $\underset{\sim}{q}, \underset{\sim}{q}^0$ are compatible in $\underset{\sim}{Q}$". Let $\mathcal{I}_0^\dagger = \{p_\eta : \eta \in T_0\}$, where $T_0 \subseteq {}^1\mu$ for some μ codes a maximal antichain in P deciding which element of $\mathcal{J}_0$, $(\underset{\sim}{p}^0, \underset{\sim}{q}^0)[G]$ will be, i.e, $p_\eta \Vdash$ "$(\underset{\sim}{p}^0, \underset{\sim}{q}^0) = (p_\eta^0, \underset{\sim}{q}_\eta^0)$", where $(p_\eta^0, \underset{\sim}{q}_\eta^0) \in \mathcal{J}_0$ iff $\eta \in T_0$. Then $p_\eta \Vdash$ "$p_\eta^0 \in G$", so without loss of generality $p_\eta \geq p_\eta^0$, and $p_\eta \Vdash$ "$\underset{\sim}{q}$ and $\underset{\sim}{q}_\eta^0$ are compatible in $\underset{\sim}{Q}$".

Similarly for each $\eta \in T_0$; if $p_\eta \in G_P$ and there are $p^1 \geq p_\eta$, and $(p_1, \underset{\sim}{q}_1) \in \mathcal{J}_1$ such that $p_1 \leq p^1$, $p^1 \in G_P$ and $p^1 \Vdash_P$ "$\underset{\sim}{q}, \underset{\sim}{q}_\eta^0, \underset{\sim}{q}_1$ are compatible". So, there is a T_1, $T_1 \subseteq {}^2\mu$ for some μ, $\eta \in T_1 \Rightarrow \eta \restriction 1 \in T_0$ and for every $\eta_0 \in T_0$ for some $\eta_1 \in T_1$, $\eta_0 = \eta_1 \restriction 1$, and $\mathcal{I}_1^\dagger = \{p_\eta : \eta \in T_1\}$ is a maximal antichain of P, $p_\eta \geq p_{\eta \restriction 1}$, $p_\eta \geq p_\eta^1$, $(p_\eta^1, \underset{\sim}{q}_\eta^1) \in \mathcal{J}_1$ and $p_\eta \Vdash$ "$\underset{\sim}{q}, \underset{\sim}{q}_{\eta \restriction 1}^0, \underset{\sim}{q}_\eta^1$ are compatible in $\underset{\sim}{Q}$".

So, we can easily define inductively on n, T_n, $p_\eta (\eta \in T_n), \mathcal{I}_n^\dagger$ and $(p_\eta^n, \underset{\sim}{q}_\eta^n) \in \mathcal{J}_n$ (for $\eta \in T_n$).

Looking at the way we have defined this, clearly we can assume T_n, $\langle p_\eta : \eta \in T_n \rangle \in N_0$ (i.e. T_n and the function $\eta \mapsto p_\eta$ in N_0) and $\langle (p_\eta^n, \underset{\sim}{q}_\eta^n) : \eta \in T_n \rangle \in N_0$, But as $\langle \mathcal{I}_n : n < \omega \rangle$ does not necessarily belong to N_0 (in fact it cannot), we do not try to claim $\langle T_n : n < \omega \rangle \in N_0$, etc., but we can assume that $\langle \langle T_n, \langle p_\eta : \eta \in T_n \rangle, \langle (p_\eta^n, \underset{\sim}{q}_\eta^n) : \eta \in T_n \rangle \rangle : 0 \leq n < \omega \rangle$ belongs to N_1.

Now as each $\mathcal{I}_\ell^\dagger \in N_0$ is a maximal antichain of P, for some $n(\ell) < \omega$, $\mathcal{I}_\ell^\dagger = \mathcal{I}_{n(\ell)}$, hence $\mathcal{I}_{n(\ell)}^* \subseteq \mathcal{I}_{n(\ell)} \cap N_0 = \mathcal{I}_\ell^\dagger \cap N_0$ is pre-dense above r and is finite. Let $T_\ell^* = \{\eta \in T_\ell : p_\eta \in \mathcal{I}_{n(\ell)}^*\}$. So, it is natural to look for $q_1 \in \underset{\sim}{Q}[G]$

(where $r \in G$) such that for each $\ell < \omega$, $\{\underset{\sim}{q}^\ell_\eta[G] : \eta \in T^*_\ell\}$ is pre-dense above q_1. This will be sufficient – it implies (in $V[G]$) that q_1 is $(N_0[G], \underset{\sim}{Q}[G])$-generic, $q_1 \geq \underset{\sim}{q}[G]$, and in V, for some P-name $\underset{\sim}{q}_1$ we have $\mathcal{J}^*_\ell \overset{\text{def}}{=} \{(p^\ell_\eta, \underset{\sim}{q}^\ell_\eta) : \eta \in T^*_\ell\}$ is a finite subset of $\mathcal{J}_\ell \cap N_0$ pre-dense above $(r, \underset{\sim}{q}_1)$; moreover clearly we can choose $q_1 \in N_2$, in fact $q_1 \in N_1$.

Unfortunately, there is no reason to asssume q_1 exists. Look at the extreme case $T^*_n = \{\eta_n\}$ (e.g. when P is $\aleph_1$-complete and r determines $\underset{\sim}{G}_P \cap N_0$). So in $V[G]$ we know q, $q^\ell_{\eta_\ell} (\ell < \omega)$ and we know $\{q, q^\ell_{\eta_\ell} : \ell < \ell_0\}$ is compatible for every $\ell_0 < \omega$; this is not a good reason to assume $\{q, q_{\eta_\ell} : \ell < \omega\}$ is compatible, except when Q is $\aleph_1$-complete and any two compatible members have a least upper bound.

Second Try:

Let $\mathcal{J}_\ell \cap N_0 = \{(p^\ell_m, \underset{\sim}{q}^\ell_m) : m < \omega\}$ ($\mathcal{J}_\ell$ from (e) of 4.4) and as $N_0 \in N_1$, $\langle \mathcal{J}_\ell : \ell < \omega\rangle \in N_1$, we can assume that $\langle\langle(p^\ell_m, \underset{\sim}{q}^\ell_m) : m < \omega\rangle : \ell < \omega\rangle \in N_1$. Let $\underset{\sim}{S}_\ell = \{m < \omega : p^\ell_m \in \underset{\sim}{G}_P\}$. This is a P-name, $\underset{\sim}{S}_\ell \in N_1$ and even $\langle \underset{\sim}{S}_\ell : \ell < \omega\rangle \in N_1$. If $N_0[G] \cap V = N_0$ then in $V[G]$ there is a function $f : \omega \to \omega$ such that

$$q_1 = \underset{\sim}{q}[G] \wedge \bigwedge_{\ell<\omega} \bigvee_{\substack{m \leq f(\ell) \\ m \in \underset{\sim}{S}_\ell[G]}} \underset{\sim}{q}^\ell_m[G]$$

is consistent (because $\underset{\sim}{Q}$ has the ${}^{\omega}\omega$-bounding property and $N_0[G] \cap V \subseteq N_0$). More formally, this means that in $\underset{\sim}{Q}[G]$ there is a $q_1 \geq \underset{\sim}{q}[G]$ such that for every $\ell < \omega$, $\{\underset{\sim}{q}^\ell_m[G] : m \leq f(\ell)$ and $m \in \underset{\sim}{S}_\ell[G]\}$ is pre-dense above q_1 in $\underset{\sim}{Q}[G]$. And also, equivalently, there is $q_1 \in \underset{\sim}{Q}[G]$ such that $q_1 \Vdash_{Q[G]}$ "$\underset{\sim}{q} \in \underset{\sim}{G}_Q$ and for every $\ell < \omega$ for some $m \leq f(\ell)$ we have $p^\ell_m \in G$, $\underset{\sim}{q}^\ell_m[G] \in \underset{\sim}{G}_Q$" (anyhow, the expression q_1 has intuitive meaning, formally see later in this section).

But as P has the ${}^{\omega}\omega$-bounding property, we can assume that $f \in V$. Also as $N_1[G] \prec (H(\lambda)[G], \in)$, and the parameters appearing in the requirements on f belong to $N_1[G]$, we can assume $f \in V \cap N_1[G] = N_1$.

Now in V we have a P-name of it, $\underset{\sim}{f} \in N_1$ such that $\Vdash_P$ "$N_0[\underset{\sim}{G}_P] \cap V = N_0$ implies that $\underset{\sim}{f}$ is as above; also in any case $\underset{\sim}{f} \in ({}^\omega\omega)^V$"; so in particular r forces $\underset{\sim}{f}$ to be as above. As r is (N_1, P)-generic we have just countably many candidates for $f \in ({}^\omega\omega)^V \cap N_1$ i.e. for $\underset{\sim}{f}[G]$; and clearly there is in V a function $f^* \in {}^\omega\omega$ such that for every $g \in N_1 \cap {}^\omega\omega$ we have, $g \leq_{ae} f^*$ (i.e. $\{n < \omega :$ not $g(n) \leq f^*(n)\}$ is finite) and $f^* \in N_2$. So it is reasonable to try

$$q_1 = \underset{\sim}{q}[G] \wedge \bigwedge_{\ell} \bigvee_{\substack{m \leq f^*(\ell) \\ m \in S_\ell[G]}} \underset{\sim}{q}^\ell_m[G],$$

(i.e. it is consistent; see in the beginning of the second try or end of the section concerning an exact definition.) The $(N_0, P * \underset{\sim}{Q})$-genericity of (r, q_1) and $\langle \mathcal{J}^*_\ell : \ell < \omega \rangle \in N_2$ should be clear. So the question is whether

$$r \Vdash_P \text{“}\underset{\sim}{q}[\underset{\sim}{G}_P] \wedge \bigwedge_{\ell} \bigvee_{\substack{m \leq f^*(\ell) \\ m \in \underset{\sim}{S}_\ell[\underset{\sim}{G}_P]}} \underset{\sim}{q}^\ell_m[\underset{\sim}{G}_P] \text{ is consistent in } \underset{\sim}{Q}[\underset{\sim}{G}_P]\text{”}.$$

(If so we can use a suitable P-name for q_1; let $\underset{\sim}{q}_1 \in N_2$ be the above expression (or just a condition forcing it) if it exists and $\underset{\sim}{q}$ otherwise).

Unfortunately, though f^* is a very plausible candidate, the fact is that if $G \subseteq P$ is generic over V, $r \in G$, the relation $\underset{\sim}{f}[G](n) \leq f^*(n)$ may fail for some n, though necessarily only for finitely many n's.

Third try:

The second try almost succeeded, except that the function f^* did not work on a finite set. So we try to take care of all finite sets that could occur, using the first try. Remember $\mathcal{I}^\dagger_\ell = \mathcal{I}_{n(\ell)}$, and $\mathcal{I}^*_{n(\ell)}$ is a finite subset of $\mathcal{I}_{n(\ell)} \cap N_0$ (pre-dense above r). Let $\mathcal{I}^*_{n(\ell)} = \{p_\eta : \eta \in T^*_\ell\}$, T^*_ℓ a finite subset of T_ℓ and so for some $k(\ell) < \omega$, for every $\eta \in T^*_\ell$ we have $(p^\ell_\eta, \underset{\sim}{q}^\ell_\eta) \in \{(p^\ell_m, \underset{\sim}{q}^\ell_m) : m < k(\ell)\}$ when $(p^\ell_m, \underset{\sim}{q}^\ell_m)$ were chosen in the second try.

Clearly $\langle k(\ell) : \ell < \omega \rangle \in N_1$, (as it can be computed from $\langle n(\ell) : \ell < \omega \rangle$ and $\langle \mathcal{I}^*_n : n < \omega \rangle$ and $\langle \langle (p^\ell_m, \underset{\sim}{q}^\ell_m) : m < \omega \rangle : \ell < \omega \rangle$ all of which belong to N_1) and so in the second try w.l.o.g. $k(\ell) \leq f^*(\ell)$ for every $\ell < \omega$.

Now return to the beginning of the argument in the second try. We know that for every $q^\dagger \in \underset{\sim}{Q}[G] \cap N_0[G]$ there is a $f \in ({}^\omega\omega)^V[G]$ such that

$$q^\dagger \wedge \bigwedge_{\ell<\omega} \bigvee_{\substack{m \leq f(\ell) \\ m \in \underset{\sim}{S}_\ell[G]}} \underset{\sim}{q}^\ell_m[G]$$

is consistent (i.e., as said above, some member of $\underset{\sim}{Q}$ forcs all those pieces of information). In particular for every $\eta = \langle m_0, \ldots, m_{n-1}\rangle (m_i < \omega)$ there is an $f = f_\eta \in ({}^\omega\omega)^{V[G]}$ such that: if $\{\underset{\sim}{q}[G], \underset{\sim}{q}^0_{m_0}[G], \underset{\sim}{q}^1_{m_1}[G], \ldots, \underset{\sim}{q}^{n-1}_{m_{n-1}}[G]\}$ is compatible (in $\underset{\sim}{Q}[G]$) then

$$(*)_\eta \qquad \underset{\sim}{q}[G] \wedge \bigwedge_{\ell<n} \underset{\sim}{q}^\ell_{m_\ell}[G] \wedge \bigwedge_{n \leq \ell < \omega} \bigvee_{\substack{m \leq f(\ell) \\ m \in \underset{\sim}{S}_\ell[G]}} \underset{\sim}{q}^\ell_m[G]$$

is consistent (i.e. some member of $\underset{\sim}{Q}[G]$ force this). Without loss of generality $f_\eta \in ({}^\omega\omega)^V$. Let, for $i < \omega$, $f^+(i) = \mathrm{Max}\{f_\eta(i) : \eta = \langle m_0, \ldots, m_{n-1}\rangle, n \leq i$ and $m_0 \leq k(0), \ldots, m_{n-1} \leq k(n-1)\}$. The maximum is taken over a finite set, hence, it is a well defined natural number, so $f^+ \in ({}^\omega\omega)^{V[G]}$. So there is in V a function $f^\dagger \in ({}^\omega\omega)^V$ such that $f^+ \leq f^\dagger$.

Now we work in V. For each $\eta \in {}^{\omega>}\omega$ there is a P-name $\underset{\sim}{f}_\eta$ of a function from ω to ω which belongs to V, such that if f_η as above exists, then $\underset{\sim}{f}_\eta$ is such a function; w.l.o.g. $\langle \underset{\sim}{f}_\eta : \eta \in {}^{\omega>}\omega\rangle \in N_1$, and remember that $\langle k(\ell) : \ell < \omega\rangle \in N_1$. Hence $\underset{\sim}{f}^+$ (which is defined from them as above) belongs to N_1; as well as $\underset{\sim}{f}^\dagger$. Note that $r \Vdash$ "the $\underset{\sim}{f}_\eta$'s and $\underset{\sim}{f}^+$, $\underset{\sim}{f}^\dagger$ are as above". Let $f^* \in N_2$ be as in the second try be such that $k_\ell < f^*(\ell)$, so we know $\underset{\sim}{f}^\dagger \leq_{\mathrm{ae}} f^*$.

Now we shall prove that

$$r \Vdash_P \text{"}\underset{\sim}{q} \wedge \bigwedge_{\ell<\omega} \bigvee_{\substack{m \leq f^*(\ell) \\ m \in \underset{\sim}{S}_\ell[\underset{\sim}{G}_P]}} \underset{\sim}{q}^\ell_m \text{ is consistent (in } \underset{\sim}{Q}[\underset{\sim}{G}_P])\text{"}.$$

As remarked in the end of the second try this suffices. So let $G \subseteq P$ be a subset of P generic over V, $r \in G$. So $f^\dagger = \underset{\sim}{f}^\dagger[G] \in N_1 \cap ({}^\omega\omega)^V$ $(\subseteq V)$ hence

$f^\dagger \leq_{ae} f^*$. So in $V[G]$ for some $i < \omega$, for every j, $i \leq j < \omega \Rightarrow f^\dagger(j) \leq f^*(j)$. Also there is a unique ω-sequence η of ordinals such that $\eta\restriction\ell \in T_\ell$, $p^\ell_{\eta\restriction\ell} \in G$ for $1 \leq \ell < \omega$ (from the first try; remember $\{p_\eta : \eta \in T^*_\ell\}$ is a maximal antichain of P and $p_{\eta\restriction m} \leq p_\eta$). So, $(p^\ell_{\eta\restriction\ell}, \underset{\sim}{q}^\ell_{\eta\restriction\ell}) = (p^\ell_{m_\ell}, \underset{\sim}{q}^\ell_{m_\ell})$ for some $m_\ell < \omega$. By the definition of $k(\ell)$, and m_ℓ we have $m_\ell < k(\ell) \leq f^*(\ell)$.

Let $\eta = \langle m_0, \ldots, m_{i-1}\rangle$ (where i was chosen above). Then by $(*)_\eta$,

$$\underset{\sim}{q}[G] \wedge \bigwedge_{\ell<i} \underset{\sim}{q}^\ell_{m_\ell}[G] \wedge \bigwedge_{\ell\geq i} \bigvee_{\substack{m\leq \underset{\sim}{f}_\eta[G](\ell) \\ m\in \underset{\sim}{S}_\ell[G]}} \underset{\sim}{q}^\ell_m[G]$$

is consistent, so the result follows, since $l \geq i \Rightarrow f_\eta(\ell) \leq f^+(\ell) \leq f^\dagger(\ell) \leq f^*(\ell)$.

$\square_{4.4}$

Continuation of the proof of the Theorem 4.3:
We were proving by induction on α that if $\Vdash_{P_i}$ "$\underset{\sim}{Q}_i$ is ω-proper and has the ${}^\omega\omega$-bounding property", $\bar{Q} = \langle P_i, \underset{\sim}{Q}_i : i < \alpha\rangle$ is a CS iteration then $P_\alpha = \operatorname{Lim}\bar{Q}$ is ω-proper and has the ${}^\omega\omega$-bounding property. The ω-properness follows by 3.5, and for the ${}^\omega\omega$-bounding property only the case of $\operatorname{cf}(\alpha) = \aleph_0$ was left. Now by III 3.3 w.l.o.g. $\alpha = \omega$. Let $\underset{\sim}{f}$ be a P_α-name, $p \in P_\alpha$, $p \Vdash$ "$\underset{\sim}{f} \in {}^\omega\omega$", and we have to find $g \in ({}^\omega\omega)^V$ and q satisfying $p \leq q \in P_\alpha$ such that $q \Vdash$ "$\underset{\sim}{f} \leq g$". We can assume w.l.o.g. that

$$(*)\ \underset{\sim}{f}(n) \text{ is a } P_n\text{-name}$$

(this follows from the proof that P_ω is proper, see III 3.2).

Let $N_\ell \prec (H(\lambda), \in)$ (λ large enough) be an increasing chain such that, p, $\langle P_n, \underset{\sim}{Q}_n : n < \omega\rangle \in N_0$, $N_\ell \in N_{\ell+1}$ each N_ℓ countable (Note that $N_\ell \prec N_{\ell+1}$ follows from $N_\ell \in N_{\ell+1}$, N_ℓ countable, and $N_\ell, N_{\ell+1} \prec (H(\lambda), \in)$).

We want to find $q \in P_\omega$, $q \geq p$, $q \Vdash_{P_\omega}$ "$\underset{\sim}{f} \leq g$" for some $g \in ({}^\omega\omega)^V$. For this we now define by induction on n a sequence $\langle q_n : n \in \omega\rangle$, where each q_n is in P_n such that the following will hold:

1) $q_{n+1}\restriction n = q_n$, $p\restriction n \leq q_n$

2) q_n is (N_k, P_n)-generic for $k = 0$, and $n+1 \leq k < \omega$

3) there is a function $F_n \in N_{n+1}$, whose domain is the set of maximal antichains of P_n which belong to N_0, and for every $\mathcal{I} \in \mathrm{Dom}(F_n), F_n(\mathcal{I})$ is a finite subset of $\mathcal{I} \cap N_0$ pre-dense above q_n.

Clearly, if we succeed then $q = \cup_{n<\omega} q_n \in P_\omega$ is as required as then we can define $g(n)$ as the minimal $g(n)$, $q_n \Vdash$ "$\underset{\sim}{f}(n) \leq g(n)$", $g(n)$ exists by 3) and $(*)$.

For $n = 0$ use the ω-properness of $\underset{\sim}{Q}_0$, and for $n+1$ we use first the lemma 4.4 and then ω-properness. $\square_{4.3}$

4.5 Definition. 1) For a forcing notion Q, let Q^+ be the following forcing notion, first defining Q_0^+:

(a) the set of members of Q_0^+ is the closure of Q under the operation $p \wedge q$, $p \vee q$, $\neg p$, $\bigwedge_{n<\omega} p_n$, $\bigvee_{n<\omega} p_n$ (assuming no accidental equality)

(b) $\underset{\sim}{G}_Q^+$ is the P-name of the following subset of Q_0^+:

for $r \in Q$, $r \in \underset{\sim}{G}_Q^+$ *iff* $r \in \underset{\sim}{G}_Q$

for $r = p \wedge q$, $r \in \underset{\sim}{G}_Q^+$ *iff* $p \in \underset{\sim}{G}_Q^+$ and $q \in \underset{\sim}{G}_0^+$

for $r = p \vee q$, $r \in \underset{\sim}{G}_Q^+$ *iff* $p \in \underset{\sim}{G}_Q^+$ or $q \in \underset{\sim}{G}_Q^+$

for $r = \neg p$, $r \in \underset{\sim}{G}_Q^+$ iff $p \notin \underset{\sim}{G}_Q^+$

for $r = \bigwedge_{n<\omega} p_n$, $r \in \underset{\sim}{G}_Q^+$ *iff* $p_n \in \underset{\sim}{G}_Q^+$ for every $n < \omega$

for $r = \bigvee_{n<\omega} p_n$, $r \in \underset{\sim}{G}_Q^+$ iff $p_n \in \underset{\sim}{G}_Q^+$ for some $n < \omega$

(c) for $r_1, r_2 \in Q_0^+$, we define $r_1 \leq^{Q^+} r_2$ *iff* $\Vdash_Q$ "if $r_2 \in \underset{\sim}{G}_Q^+$ then $r_1 \in \underset{\sim}{G}_Q^+$"

(d) $Q^+ = \{q \in Q^+ :$ for some $r \in Q, r \Vdash$ "$q \in \underset{\sim}{G}_Q^+$"$\}$.

4.6 Fact. 1) $\Vdash_Q$"$\underset{\sim}{G}_Q^+$ is a generic subset of Q^+ (or V) and $\underset{\sim}{G}_Q^+ \cap Q = \underset{\sim}{G}_Q$"

2) Q is a dense subset of Q^+

3) essentially $Q = Q^+ \restriction Q$ i.e. for $p, q \in Q$, $Q^+ \models$ "$p \leq q \Leftrightarrow \neg(\exists r)(r \in Q \,\&\, p \leq r \,\&\, [q, r$ incompatible$])$". $\square_{4.6}$

4.6A Remark. We can continue and do iteration in this context, see X §1.

§5. Which Forcings Can We Iterate Without Adding Reals

In Sect. 1 we have proved that we can iterate forcing notions of special kind ($\mathcal{E}$-complete) without adding reals. As a result we get a parallel of MA for such forcings and get the consistency of some uniformization property (see more in Chapters VII, VIII). However this axiom, quite strong in some respects, is consistent with diamond on $\aleph_1$: (see [Sh:64], [Sh:98] or 1.9 here). On some stationary subsets of ω_1 it can say much, but on others nothing.

So we shall try here to find another property of forcing notions, so that forcing with $\operatorname{Lim} \bar{Q}, \bar{Q} = \langle P_i, \underset{\sim}{Q}_i : i < \alpha \rangle$ a CS iteration of such forcing, does not add reals.

5.1 Example. Assume $2^{\aleph_0} = \aleph_1$ (or even $2^{\aleph_0} < 2^{\aleph_1}$ suffices).

Let $A_\delta \subseteq \delta$ be unbounded of order type ω, for $\delta < \omega_1$ limit, so by [DvSh:65] (or see AP §1), $\langle A_\delta : \delta < \omega_1 \rangle$ does not have the uniformization property, hence there are $f_\delta : A_\delta \to \{0, 1\}$ such that for no $f : \omega_1 \to \{0, 1\}$, is $f \restriction A_\delta =_{ae} f_\delta$ for every δ. Let $\bar{f} = \langle f_\delta : \delta < \omega_1 \rangle$, $P_{\bar{f}} = \{f : \operatorname{Dom}(f)$ is an ordinal $\alpha < \omega_1$, $\delta \le \alpha \Rightarrow [f \restriction A_\delta =_{ae} f_\delta]\}$, ordered by inclusion. Consider the dense sets.

$$\mathcal{I}_i = \{f : i \le \operatorname{Dom}(f) \text{ and } f \in P_{\bar{f}}\}$$

So clearly there is no directed $G \subseteq P_{\bar{f}}$ such that $G \cap \mathcal{I}_i \ne \emptyset$ for every $i < \omega_1$.

5.1A Remark. Previously Jensen (see Devlin and Johnsbraten [DeJo]) showed, that though forcing with Souslin trees does not add reals, starting with $V = L$ (at least with $V \models \diamondsuit_{\aleph_1}$) there is a CS iteration of such forcing of length ω, such that forcing by the limit adds reals. This, however, does not exclude a suitable MA for the example above, because MA for this forcing implies $\neg$CH.

Now, $P_{\bar{f}}$ is a very nice forcing – e.g. it is α-proper for every $\alpha < \omega_1$, but our desired property should exclude it. The following is a try to exclude this case by a reasonable condition.

We shall return to this subject in VIII, §4 (going deeper but also having presentational variations of the definitions).

5.2 Definition.

(1) We call $\mathbb{D}$ a completeness system if for some μ, $\mathbb{D}$ is a function defined on the set of triples $\langle N, P, p\rangle$, $p \in N \cap P$, $P \in N$, $N \prec (H(\mu), \in)$, N countable such that (P is meant here as a predicate on N, i.e., $P \cap N$): $\mathbb{D}_{\langle N,P,p\rangle} = \mathbb{D}(N, P, p)$ is a filter, or even a family of nonempty subsets of $\mathrm{Gen}(N, P) = \{G : G \subseteq N \cap P,\ G$ directed and $G \cap \mathcal{I} \neq \emptyset$ for any dense subset $\mathcal{I}$ of P which belongs to $N\}$ such that if $G \in \mathrm{Gen}(N, P)$ belongs to any member of $\mathbb{D}_{\langle N,P,p\rangle}$, then $p \in G$.

(2) We call $\mathbb{D}$ a λ-completeness (λ may also be finite or $\aleph_0$ or $\aleph_1$) system if each family $\mathbb{D}_{\langle N,P,p\rangle}$ has the property that the intersection of any i elements is nonempty for $i < 1 + \lambda$ (so for $\lambda \geq \aleph_0$, $\mathbb{D}_{\langle N,P,p\rangle}$ generates a filter). Now, such $\mathbb{D}$ can be naturally extended to include $N \prec (H(\mu^\dagger), \in)$, $\mu \in N$, $\mu < \mu^\dagger$ by $\mathbb{D}(N, P, p) = \mathbb{D}(N \cap H(\mu), P, p)$. We do not distinguish strictly.

(3) We say $\mathbb{D}$ is on μ. We not always distinguish strictly between $\mathbb{D}$ and its definition.

5.3 Definition.

(1) Suppose P is a forcing notion, $\mathcal{E}$ a nontrivial family of subsets of $\mathcal{S}_{\aleph_0}(\mu)$ and $\mathbb{D}$ a completeness system on μ.
We say P is $(\mathcal{E}, \mathbb{D})$-complete if for every large enough λ, *if* $P, \mathcal{E}, \mathbb{D} \in N$, $p \in P \cap N$, $N \prec (H(\lambda), \in)$, N countable, $A \in \mathcal{E} \cap N \Rightarrow N \cap \mu \in A$, *then* the following set contains some member of $\mathbb{D}_{\langle N,P,p\rangle}$ (i.e., $\mathbb{D}_{\langle N\cap H(\mu),P,p\rangle}$):

$$\mathrm{Gen}^+(N, P) = \{G \in \mathrm{Gen}(N, P) : p \in G \text{ and there is}$$
$$\text{an upper bound for } G \text{ in } P\}$$

(2) If $\mathcal{E} = \{\mathcal{S}_{\aleph_0}(\mu)\}$ we write just $\mathbb{D}$-complete.

5.4 Remark.

(1) We can think of $\mathbb{D}_{\langle N,P,p\rangle}$ as a filter on the family of directed subsets G of $P \cap N$ generic over N, to which p belongs. The demand "$(\mathcal{E}, \mathbb{D})$-complete" means that (for $\mathcal{D}(\mathcal{E})$-majority of such N's) the "majority" of such G's have an upper bound in P hence the name $(\mathcal{E}, \mathbb{D})$-completeness.

(2) In some sense the definitions above are trivial: if P is $\mathcal{E}$-proper and does not add reals, then there is a κ-completeness system $\mathbb{D}$ such that P is $(\mathcal{E}, \mathbb{D})$-complete for all κ simultaneously. Because, given $\langle N, P, p\rangle$, we extend p to $q \in P$ which is (N, P)-generic. If $\{\mathcal{I}_n : n < \omega\}$ is a list of the dense subsets of P which belong to N, $\mathcal{I}_n \cap N = \{p_{n,k} : 0 < k < \omega\}$, we can define a P-name $\underset{\sim}{x}$:

$$\underset{\sim}{x} = \{\langle n, k\rangle : n < \omega \text{ and } k \text{ is minimal such that } p_{n,k} \in \underset{\sim}{G}_P\}$$

Clearly $q \Vdash_P$ "$\underset{\sim}{x} \in {}^\omega\omega$", and since P does not add reals there is an $x^* \in ({}^\omega\omega)^V$, and r, $q \le r \in P$, $r \Vdash_P$ "$x^* = \underset{\sim}{x}$". Let $G_r = \{p^\dagger \in P \cap N : p^\dagger \le r\}$. Clearly $G_r \in \mathrm{Gen}(N, P)$ and let

$$\mathbb{D}_{\langle N,P,p\rangle} = \{\{G_r\}\}$$

So what is the point of such a definition? We shall use almost always completeness systems restricted in some sense: $\mathbb{D}_{\langle N,P,p\rangle}$ is defined in a reasonably simple way. The point is that usually when we want to decide whether some $G \in \mathrm{Gen}(N, P)$ has an upper bound, we do not need to know the whole P, but rather some subset of N, e.g. a function f from N to itself. Check the example we discussed before: if $\delta = N \cap \omega_1$, then we just need to know $f{\restriction}\delta$. But two $f{\restriction}\delta$'s may give incompatible demands, so for it the system is only a 1-completeness system. So if we deal with $\aleph_0$-completeness system, we exclude it (in fact later we shall discuss even 2-completeness system).

An explication of "defined in a reasonably simple way" is:

5.5 Definition.

(1) A completeness system $\mathbb{D}$ is called simple if there is a first order formula ψ such that:

$$\mathbb{D}(N,P,p) = \{A_x : x \text{ a finitary relation on } N, \text{ i.e. } x \subseteq N^k, \text{ for some } k \in \omega\}$$

where

$$A_x = \{G \in \mathrm{Gen}(N,P) : (N \cup \mathcal{P}(N), \in, p, P, N) \models \psi[G,x]\}$$

(2) A completeness system $\mathbb{D}$ is called almost simple over V_0 (V_0 a class, usually a subuniverse) *if* there is a first order formula ψ such that:

$$\mathbb{D}(N,P,p) = \{A_{x,z} : x \text{ a relation on } N, z \in V_0\}$$

where

$$A_{x,z} = \{G \in \mathrm{Gen}(N,P) : \langle V_0 \cup N \cup \mathcal{P}(N), \in^{V_0}, \in^{N\cup\mathcal{P}(N)}, p, P, V_0, N\rangle \models \psi[G,x,z]\}$$

where $\in^A = \{(x,y) : x \in A, y \in A, x \in y\}$.

(3) If in (2) we omit z we call $\mathbb{D}$ simple over V_0.

5.6 Claim.

(1) A λ-completeness system (see Definition 5.2(2)) is a λ^*-completeness system for every $\lambda^* \leq \lambda$.

(2) P is $(\mathcal{E},\mathbb{D})$-complete for some $\mathbb{D}$ if and only if P is $\mathcal{E}$-proper and (forcing with P) does not add new reals.

Proof. (1) Trivial.

(2) The direction $\Leftarrow$ (i.e. "if") was proved in Remark 5.4(2) above. So, let us prove the "only if" part. So P is $(\mathcal{E},\mathbb{D})$-complete.

Suppose $N \prec (H(\lambda),\in)$, $p \in P$ and $\{p,P,\mathcal{E},\mathbb{D},\mu\} \in N$, N countable and $N \cap \mu \in \bigcap_{A \in \mathcal{E} \cap N} A$. So $B = \{G \in \mathrm{Gen}(N,P) : G$ has an upper bound and

$p \in G\} \in \mathbb{D}_{\langle N,P,p\rangle}$, hence $B \neq \emptyset$ (by Definition 5.2) and let $G \in B$. So G has an upper bound q (by the definition of $(\mathcal{E},\mathbb{D})$-completeness), $G \in \mathrm{Gen}(N,P)$, and by Definition 5.3, $p \in G$. So $q \geq p$ is (N,P)-generic. If $p \Vdash_P$ "$\underset{\sim}{f} \in {}^{\omega}\omega$", $\underset{\sim}{f} \in N$, then for every n, $\mathcal{I}_n = \{r \in P : r \Vdash$ "$\underset{\sim}{f}(n) = k$" for some $k < \omega\}$ is a dense subset of P which belongs to N, hence $\mathcal{I}_n \cap G \neq \emptyset$. Hence q determines the value of $\underset{\sim}{f}(n)$. So q determines $\underset{\sim}{f}(n)$ for every n. Hence it determines $\underset{\sim}{f}$, i.e. $\underset{\sim}{f}$ is not a new real. Now if there were a new real, some p would without loss of generality force $\underset{\sim}{f}$ is such a real. Choosing N as above we get a contradiction.

$\square_{5.6}$

5.7 Example. Forcing with a Souslin tree T is not $\mathbb{D}$-complete for any simple 2-completeness system $\mathbb{D}$.

Let $N \prec (H(\lambda), \in)$ be countable, $T \in N$, $\delta = N \cap \omega_1$. Note that $\mathrm{Gen}(N,P)$ consists of all branches of $T \cap N$, and $\mathrm{Gen}^+(N,P)$ consists of the branches of $T \cap N$ which have an upper bound, i.e. $A_x = \{y \in T : y < x\}$, where $x \in T_\delta =$ the δ-th level of T. Now N "does not know" what is the set of such branches of $T \cap N$, and two disjoint sets are possible.

The above is an argument, not a proof. To be exact, we can, assuming diamond of $\aleph_1$, build a Souslin tree, such that no first order formula ψ defines a simple 2-completeness system for which T is $\mathbb{D}$-complete.

§6. Specializing an Aronszajn Tree Without Adding Reals

The traditional test for generalizing MA has been Souslin Hypothesis. Jensen has proved the consistency of the Souslin Hypothesis with G.C.H. (see Devlin and Johnsbraten [DeJo]). He iterates forcing notions of Souslin trees, in limit points of cofinality $\aleph_0$ he uses diamond to refine the inverse limit of the trees, in limit points of cofinality $\aleph_1$ he uses the square on $\aleph_2$ (and preparatory measures in previous steps). In successor stages he specializes a specific tree, by first forcing a closed unbounded set and then building a Souslin tree using $\diamondsuit^*_{\aleph_1}$ (more precisely he adds $\aleph_2$ closed unbounded subsets in the beginning).

We shall prove that there is a $\mathbb{D}$-complete forcing notion P_T specializing an Aronszajn tree T, for $\mathbb{D}$ a simple $\aleph_1$-completeness system. The proof is close to Jensen's successor stage. We feel that the ideas of the proof are applicable to related problems, see [AbSh:114], [AbSh:403], [DjSh:604].

Notation. For an $\aleph_1$-tree T, T_i is the i-th level, $T{\restriction}i = \cup_{j<i} T_j$, and for $x \in T_\beta$, $\alpha \leq \beta$, $x{\restriction}\alpha$ is the unique $y \in T_\alpha$, $y \leq x$.

6.1 Theorem. There is a simple $\aleph_1$-completeness system $\mathbb{D}$, such that for every Aronszajn tree T, there is a $\mathbb{D}$-complete forcing notion P_T, specializing it, i.e. $\Vdash_{P_T}$ "T is a special Aronszajn tree", also P_T is α-proper for every countable ordinal α.

Proof.

First Approximation:

Let

$$P^0_T = \{f : f \text{ a function from } T{\restriction}(\alpha+1) \text{ to } \mathbb{Q} \text{ (set of rational numbers)} \\ \text{such that } \alpha < \omega_1, x < y \Rightarrow f(x) < f(y)\}.$$

The order is inclusion. If $f \in P^0_T$, $\mathrm{Dom}(f) = T{\restriction}(\alpha+1)$ we say f has height α, $\mathrm{ht}\,(f) = \alpha$.

Clearly P^0_T specializes T, but we have to prove that it is proper and does not add reals and more. Let $N \prec (H(\lambda), \in)$ with T, $P^0_T \in N$ and $N \cap \omega_1 = \delta$, (N countable). Let (the δ-th level of T be) $T_\delta = \{x_n : n < \omega\}$. It is trivial that we can extend any condition to a condition of arbitrarily large height. So we have to define an increasing sequence of conditions $p_n \in P^0_T \cap N$, which will be generic for N (hence their heights converge to δ) and has an upper bound. Now in order that $\{p_n : n < \omega\}$ has an upper bound, it is necessary that for each $\ell < \omega$, the sequence (of rationals) $\langle p_n(x_\ell{\restriction}\mathrm{ht}\,(p_n)) : n < \omega\rangle$ is bounded. So a natural condition to ensure it is e.g.

$(*)$ for $\ell < n$ we have $p_n(x_\ell{\restriction}\mathrm{ht}(p_n)) + 1/2^n > p_{n+1}(x_\ell{\restriction}\mathrm{ht}(p_{n+1}))$

This is not difficult by itself, but we have also to ensure the genericity of $\langle p_n : n < \omega\rangle$. So it clearly suffices to prove, for each n

$(**)_n$ if $p \in P^0_T \cap N$, $n < \omega$, $\mathcal{I}$ an open dense subset of P^0_T which belongs to N, $x_0, \dots, x_{n-1} \in T_{N\cap\omega_1}$ and $\varepsilon > 0$ (a rational or real), *then* there is a $q \in P^0_T \cap N$, $p \le q$, $q \in \mathcal{I}$ and

$$p(x_\ell{\restriction}\mathrm{ht}(p)) + \varepsilon > q(x_\ell{\restriction}\mathrm{ht}(q)) \text{ for } \ell < n$$

Unfortunately we see no reason for $(**)_n$ to hold.

In fact, it is false, and for every natural number n, $\mathcal{I}_n = \{p \in P^0_T :$ for every $x \in \mathrm{ht}(p)$, we have $p(x) \ge n\}$ is dense.

Second Approximation:

We can remedy this by using $P^1_T = \{f : f \in P^0_T$, and: *if* $\beta < \mathrm{ht}(f), x \in T_\beta$ and $\varepsilon > 0$ and T is a Souslin tree, *then* for some y we have $x < y \in T_{ht(f)}$ and $f(x) < f(y) < f(x) + \varepsilon\}$.

Now for $n = 1$, $(**)_n$ is true; more generally for any $n < \omega$, $(**)_n$ is true if:

$(***)_n$ $\{(y_0, \dots, y_{n-1}) : \bigwedge_{\ell<n} y_\ell \in T \,\&\, y_\ell < x_\ell\}$ is generic for $(N, T^n)\}$.

(T^n - the n-th power of T i.e. the set of elements is in $\bigcup_{i<\omega_1} {}^n(T_i)$, and $\bar{x} \le \bar{y}$ iff $\bigwedge_{\ell<n} \bar{x}(\ell) <_T \bar{y}(\ell)$.)

Why? Though it is not used we shall explain. For a given p and rational ε let $R = R_{p,\mathcal{I},\varepsilon} = \{(y_0, \dots, y_{n-1}) :$ for some $\alpha < \delta, \alpha > \mathrm{ht}(p) \,\&\, y_\ell \in T_\alpha$, and for some $q \in P^1_T, p \le q \in \mathcal{I}$, $\mathrm{ht}(q) = \alpha, q(y_\ell) < q(x_\ell{\restriction}\mathrm{ht}(p)) + \varepsilon\}$. Now R is a dense subset of $\{(y_0, \dots y_{n-1}) :$ for some γ, for each $\ell, x_\ell{\restriction}\mathrm{ht}(p) \le y_\ell \in T_\gamma\}$. [Why? as given $(y_0, \dots, y_{n-1}) \in {}^n(T_\gamma)$ we can find r, $p \le r \in P_T$, $\mathrm{ht}(r) = \gamma$, $r(y_\ell) < r(y_\ell{\restriction}\mathrm{ht}(p)) + \varepsilon$ (by a density argument), let $\varepsilon_1 = \mathrm{Min}\{r(y_\ell) - r(y_\ell{\restriction}\mathrm{ht}(p)) : \ell < n\}$, and let us choose q, $r \le q \in \mathcal{I}$, without loss of generality $q \in N$. Now by the definition of P^1_T we can find y'_ℓ, $y_\ell \le y'_\ell \in T_{\mathrm{ht}(q)}$ such that

$q(y'_\ell) < q(y_\ell) + \varepsilon_1$. Clearly q exemplifies $(y'_\ell : \ell < n) \in R$ as required]. Hence there is a $(y_0, \dots y_{n-1}) \in R, \bigwedge_\ell y_\ell \leq x_\ell$.

However, there may be Souslin trees which do not satisfy $(***)_n$ for $n > 1$.

6.1A Explanation. So we shall change P^0_T somewhat by adding "promises" such that if (the parallel to) $(**)$ fails, then we can add one more promise to p guaranteeing that p has no extension in $\mathcal{I}$, a contradiction to $\mathcal{I}$ being open and dense.

The Actual Proof.

6.2 Definition. We call Γ a promise (more exactly a T-promise) if there are a closed unbounded subset C of ω_1 and $n < \omega$ (denoted by $C(\Gamma)$, $n(\Gamma)$ respectively) such that:

a) the members of Γ are n-tuples $\langle x_0, \dots, x_{n-1} \rangle$ of distinct elements from T_α where $\alpha \in C$. We say $\langle x_0, \dots, x_{n-1} \rangle \leq \langle y_0, \dots, y_{n-1} \rangle$ if $x_0 \leq y_0, \dots, x_{n-1} \leq y_{n-1}$,

b) if $\alpha < \beta$ are in C, $\bar{x} \in \Gamma \cap {}^n(T_\alpha)$, *then* there are infinitely many $\bar{y}$'s, $\bar{x} < \bar{y} \in \Gamma \cap {}^n(T_\beta)$ which are pairwise disjoint (i.e. the ranges of the sequences are disjoint),

c) $\Gamma \cap {}^n(T_{\min C(\Gamma)})$ is not empty.

6.3 Definition. We let $P_1 = \{(f, C) : C$ is a characteristic function of a closed subset of some successor ordinal $\alpha + 1 < \omega_1$, with the last element $\alpha = \ell t(C)$, and f is a monotonically increasing function from $\bigcup_{i \in C} T_i$ to $\mathbb{Q}\}$. Let $(f_1, C_1) \leq (f_2, C_2)$ if and only if $C_1 \subseteq C_2$ (equivalently $C_1 = C_2 \restriction (\ell t(C_1) + 1))$ and $f_1 \subseteq f_2$.

6.4 Definition. We say that $(f, C) \in P_1$ fulfills or satisfies a promise Γ if: $\ell t(C) \in C(\Gamma)$ and $C(\Gamma) \supseteq C \setminus \mathrm{Min} C(\Gamma)$ and for every $\alpha < \beta$ in $C(\Gamma) \cap C$ and $\bar{x} \in \Gamma \cap {}^n(T_\alpha)$ (where $n = n(\Gamma)$) the following holds:

$\oplus$ for every $\varepsilon > 0$ there are infinitely many pairwise disjoint $\bar{y} \in \Gamma\cap^n(T_\beta)$ such that $f(x_\ell) < f(y_\ell) < f(x_\ell) + \varepsilon$ for $\ell < n$ and $\bar{x} < \bar{y}$.

6.5 The Main Definition. $P = P_T = \{(f, C, \Psi) : (f, C) \in P_1$, and Ψ is a countable set of promises which (f, C) fulfills $\}$

$$(f_1, C_1, \Psi_1) \le (f_2, C_2, \Psi_2) \text{ if:}$$

$(f_1, C_1) \le (f_2, C_2)$ (in P_1) and $\Psi_1 \subseteq \Psi_2$ and: $\alpha \in C_2 \backslash C_1$ implies $\alpha \in \bigcap_{\Gamma\in\Psi_1} C(\Gamma)$ (actually follows).

6.5A Notation. If $p = (f, C, \Psi)$ we write $f = f_p$, $C = C_p, \Psi = \Psi_p$, $\ell t_p = \ell t(C_p)$.

6.6 Fact. If $p \in P$, $\beta < \omega_1$, then

(1) there is a $q \in P$, $q \ge p$, and $\ell t_q \ge \beta$,

(2) moreover, if $\beta \in \bigcap_{\Gamma\in\Psi_p} C(\Gamma)$ and $\beta > \ell t_p$, then we can have $\ell t_q = \beta$,

(3) moreover, if $m < \omega$, $y_0, \dots, y_{m-1} \in T_\beta$, $\varepsilon > 0$ we can in addition to (2) demand $f_p(y_i{\restriction}\ell t_p) < f_q(y_i) < f_p(y_i{\restriction}\ell t_p) + \varepsilon$ for $i < m$.

Proof. (1) Clearly $\bigcap_{\Gamma\in\Psi_p} C(\Gamma)$ is a closed unbounded subset of ω_1 (as Ψ is countable and each $C(\Gamma)$ is a closed unbounded subset of ω_1). Hence there is an ordinal $\beta^\dagger, \beta^\dagger > \beta, \beta^\dagger > \ell t_p$ and $\beta^\dagger \in \bigcap_{\Gamma\in\Psi_p} C(\Gamma)$, and apply (2).

(2) Let $\alpha = \ell t_p$. We define $C_q = C_p \cup \{\beta\}$, $\Psi_q = \Psi_p$, so we still have to define f_q, but as we want to have $f_p \subseteq f_q$, we have to define just $f_q{\restriction}T_\beta$. We have two demands on it, in order that $q \in P$:

(i) monotonicity: $f_p(x{\restriction}\alpha) = f_q(x{\restriction}\alpha) < f_q(x) \in \mathbb{Q}$ for $x \in T_\beta$

(ii) $\oplus$ from Definition 6.4 for $\alpha_1 < \alpha_2$ in $C_q \setminus \mathrm{Min}(C(\Gamma))$ (hence in $C(\Gamma)$, $\Gamma \in \Psi_p = \Psi_q$, $\bar{x} \in \Gamma\cap^{n(\Gamma)}(T_{\alpha_1})$ when $\alpha_2 = \beta$ (for $\alpha_2 < \beta$ use $p \in P$)).

If we succeed to define $f_q{\restriction}T_\beta$ such that it satisfies (i) and (ii), then q is well defined, and trivially belongs to P and is $\ge p$.

Now, (ii) consists of countably many demands on the existence of infinitely many $\bar{y} \in {}^n(T_\beta)$.

Let $\{(\Gamma_m, \gamma_m, \bar{x}^m) : m < \omega\}$ be a list of the triples $(\Gamma, \gamma, \bar{x})$, $\Gamma \in \Psi_p, \bar{x} \in \Gamma \cap {}^{n(\Gamma)}(T_\gamma)$, $\gamma < \beta$, $\gamma \in C_p \cap C(\Gamma)$, each appearing infinitely often (if this family is empty, we have no work at all).

We now define by induction on m, a function f_m such that:

a) f_m is a function from a finite subset of T_β to $\mathbb{Q}$ such that $f_p(x \restriction \alpha) < f_m(x)$ for $x \in \mathrm{Dom}(f_m)$.

b) $f_m \subseteq f_{m+1}$

c) There is a $\bar{y}^m \subseteq \mathrm{Dom}(f_{m+1}) \setminus \mathrm{Dom}(f_m)$, $\bar{y}^m \in \Gamma_m$, $\bar{x}^m < \bar{y}^m$ and for every $\ell < n(\Gamma_m)$ (which is the length of $\bar{x}^m$) $f_p(x^m_\ell) < f_m(y^m_\ell) < f_p(x^m_\ell) + 1/m$.

This will be enough, as any triple appears infinitely often and the $\bar{y}^m$'s are pairwise disjoint and $1/m$ converges to zero, so any completion of $\cup_m f_m$ to a function from T_β to $\mathbb{Q}$ satisfying (i) is as required.

We let f_0 be arbitrary satisfying (a), e.g. the empty function.

If f_m is defined, consider $\Gamma = \Gamma_m$. Let $n = n(\Gamma)$, if $\gamma_m = \alpha$ we know that Γ is a promise, $\gamma_m \in C(\Gamma)$, (part of requirements of $(\Gamma_m, \gamma_m, \bar{x}^m)$) and $\beta \in C(\Gamma)$ (by the hypothesis of Fact 6.6(2)). Hence (by the definition of a promise) there are infinitely many pairwise disjoint $\bar{y}$'s, $\bar{x}^m < \bar{y}$, $\bar{y} \in \Gamma \cap {}^n(T_\beta)$. As the domain of f_m is finite there is such a $\bar{y}$ disjoint from $\mathrm{Dom}(f_m)$. So we let:

$$\mathrm{Dom}(f_{m+1}) = \mathrm{Dom}(f_m) \cup \{y_0, \ldots, y_{n-1}\}$$

$$f_{m+1}(y_\ell) = f_p(y_\ell \restriction \alpha) + 1/(2m)$$

If $\gamma_m < \alpha$, we use the fact that $(f_p, C_p) \in P_1$, satisfies the promise Γ, $\gamma_m \in C(\Gamma)$ and $\alpha \in C(\Gamma)$ (by Definition 6.4 $C(\Gamma) \supseteq C_p \setminus \mathrm{Min}C(\Gamma)$ and $\alpha \in C_p, \alpha > \gamma_m \geq \mathrm{Min}C(\Gamma)$). So there is a $\bar{z} \in \Gamma \cap {}^n(T_\alpha)$ such that $\bar{x}^m < \bar{z}$, and $f_p(z_\ell) < f_p(x^m_\ell) + 1/3m$. Now we apply the argument above, replacing $\bar{x}^m$ by $\bar{z}$.

(3) The same proof as that of (2), using our freedom to choose f_0.

So we finish the proof of Fact 6.6. $\square_{6.6}$

Now we shall prove the crux of the matter: the parallel of (**).

6.7 Fact.

(1) If $N \prec (H(\lambda), \in)$ (λ large enough) $P, p \in N$, $p \in P$, N countable, $N \cap \omega_1 = \delta$, $\varepsilon > 0$ and $x_0, \dots, x_{n-1} \in T_\delta$ (are distinct) and $\mathcal{I} \in N$ is an open dense subset of P, *then* there is a $q \in \mathcal{I} \cap N$, $q \geq p$, $\ell t_q = \delta$ and $f_q(x_\ell \restriction \ell t_q) < f_p(x_\ell \restriction \ell t_p) + \varepsilon$.

(2) In (1), we can instead of $x_0, \dots, x_{n-1}$ have $B_0, \dots, B_{n-1}, \delta$-branches of $T_\delta \cap N$ (i.e. $B_\ell = \{x_i^\ell : i < \delta\}$, $x_i^\ell \in T_i$, $x_i^\ell < x_j^\ell$ for $i < j$). Define $B_\ell \restriction \alpha$ as the unique $x \in B_\ell \cap T_\alpha$, and replace the conclusion of (1) by $f_q(B_\ell \restriction \ell t_q) < f_p(B_\ell \restriction \ell t_p) + \varepsilon$.

Proof. (1) By (2), using $B_\ell = \{y \in T : y < x_\ell\}$.

(2) Suppose $p, N, \varepsilon, B_0, \dots, B_{n-1}$ form a counterexample, for simplicity ε rational and let $\alpha = \ell t_p$ and $x_\ell = B_\ell \restriction \alpha$, $\bar{x} = \langle x_0, \dots, x_{n-1} \rangle$. Let

$\Gamma_1 = \{\bar{y} : \bar{y} \in {}^n(T_\beta)$ for some $\beta \geq \alpha$, $\bar{x} \leq \bar{y}$, and there is no $(q, \gamma, \bar{z})$ such that: $\bar{z} \leq \bar{y}, \gamma \geq \alpha, \bar{z} \in {}^n(T_\gamma), q \geq p, \ell t_q = \gamma, q \in \mathcal{I}$ and $\forall \ell < n[f_q(z_\ell) < f_q(x_\ell) + \varepsilon]\}$.

So Γ_1 is, in a sense, the set of "bad" $\bar{y}$'s; the places to which we cannot extend p suitably. More explicitly:

6.7A. Subfact. If $\beta \in \bigcap_{\Gamma \in \Psi_p} C(\Gamma) \setminus (\ell t_p + 1)$, $\beta < \delta$, then

$$\langle B_0 \restriction \beta, \dots, B_{n-1} \restriction \beta \rangle \in \Gamma_1.$$

Otherwise, for some $\alpha \leq \gamma \leq \beta$ there exists a $(q, \gamma, \bar{z})$ witnessing

$$\langle B_0 \restriction \beta, \dots, B_{n-1} \restriction \beta \rangle \notin \Gamma_1.$$

But then q exemplifies p, N, ε, $B_0, \dots, B_{n-1}$ do not form a counterexample, except that maybe $q \notin N$.

Clearly Γ_1 is definable in $H(\lambda)$ using parameters which are in N, hence $\Gamma_1 \in N$. Now, the requirements on q are also first order with parameters in N, so w.l.o.g. $q \in N$. So subfact 6.7A holds. $\square_{6.7A}$

Let $C_1 = \bigcap_{\Gamma \in \Psi_p} C(\Gamma) \setminus \ell t_p$, so again $C_1 \in N$. By the subfact 6.7A, $N \models$ " for every $\gamma \in C_1$ there is a $\bar{y} \in \Gamma_1 \cap {}^n(T_\gamma)$" but $N \prec (H(\lambda), \in)$, hence also the universe V satisfies the statement.

Our plan is to get a promise $\Gamma \subseteq \Gamma_1$ in N, and show that $r \overset{\text{def}}{=} (f_p, C_p, \Psi_p \cup \{\Gamma\}) \in P$, $p \leq r$, and above r there is no member of $\mathcal{I}$, thus getting a contradiction to "$\mathcal{I}$ is an open dense subset of P".

Let $\Gamma_2 = \{\bar{y} \in \Gamma_1$: there are uncountably many $\bar{z} \in \Gamma_1, \bar{y} < \bar{z}\}$. By the above, $\bar{x} \in \Gamma_2$. We shall prove later:

6.7B. Subfact. There is a closed unbounded $C^* \subseteq \omega_1$, $\alpha = \mathrm{Min} C^*$, $C^* \subseteq C_1$, such that $\Gamma = \{\bar{y} \in \Gamma_2$: for some $i \in C^*$, $\bar{y} \in {}^n(T_i)\}$ is a promise.

Let us show that this will be enough to prove 6.7, hence Theorem 6.1 except checking simplicity.

As before, we can assume $C^* \in N$; and as $\mathrm{Min} C(\Gamma) = \mathrm{Min} C^* = \alpha = \ell t_p$, clearly $p^\dagger = (f_p, C_p, \Psi_p \cup \{\Gamma\}) \in P \cap N$ and $p \leq p^\dagger$. As $\mathcal{I}$ is an open and dense subset of P there is a $q \geq p^\dagger$ in $\mathcal{I}$. As $q \in P$, (f_q, C_q) satisfies the promise Γ, – so as $\alpha \in C(\Gamma) \cap C_q$, also $\beta = \ell t_q \in C(\Gamma) \cap C_q$. Hence by the definition of "fulfilling a promise" and as $\bar{x} \in \Gamma$ (see above), there is a $\bar{y} \in {}^n(T_\beta) \cap \Gamma$ such that $\bar{x} \leq \bar{y}$ and $f_q(y_\ell) < f_p(x_\ell) + \varepsilon$ for each $\ell < n$. So by Γ_1's definition, $\bar{y} \notin \Gamma_1$ (as $q \in \mathcal{I}$) but $\bar{y} \in \Gamma \subseteq \Gamma_2 \subseteq \Gamma_1$. We arrive at a contradiction thus proving Fact 6.7, except that we need:

Proof of Subfact 6.7B. Note that

a) if $\bar{z} \in \bigcup_{i \geq \alpha} {}^n(T_i)$, $\bar{z} < \bar{y} \in \Gamma_1$, then $\bar{z} \in \Gamma_1$; so clearly Γ_2 has this property too.

Next note that

b) for any $\bar{y} \in \Gamma_2$ the set $\{\bar{z} \in \Gamma_2 : \bar{y} \leq \bar{z}\}$ is uncountable.

(Why? If not, for some γ such that $\delta < \gamma < \omega_1$ there is no $\bar{z} \in \Gamma_2$, $\bar{y} \leq \bar{z}$, $\bar{z} \notin \bigcup_{i<\gamma} {}^n(T_i)$. But there are distinct $\bar{z}^i \in \Gamma_1$, $\bar{z} \leq \bar{z}^i$ for $i < \aleph_1$, so w.l.o.g. $\bar{z}^i \in {}^n(T_{\gamma(i)})$, $\gamma(i) \geq i$. Now, there are just countably many possible $\langle z^i_0 \restriction \gamma, \ldots, z^i_{n-1} \restriction \gamma \rangle$ (for $\gamma \leq i < \omega_1$), (as T_γ is countable), hence for some

$\bar{z} \in {}^n(T\gamma)$ the set $\{i : \gamma \le i < \omega_1, z^i_\ell \restriction \gamma = z_\ell \text{ for } \ell < n\}$ is uncountable, hence $\bar{z} \in \Gamma_2$).

c) for any $\bar{y} \in \Gamma_2 \cap {}^n(T_i)$, $i < j < \omega_1$ there is a $\bar{z} \in {}^n(T_j) \cap \Gamma_2$, $\bar{y} < \bar{z}$.

This is just a combination of a) and b).

d) for any $\bar{y} \in \Gamma_2 \cap {}^n(T_i)$ there is a j such that $i < j < \omega_1$ and disjoint $\bar{z}^1, \bar{z}^2 \in \Gamma_2 \cap {}^n(T_j)$, $\bar{y} < \bar{z}^1, \bar{y} < \bar{z}^2$.

Otherwise for $i < j < \omega_1$, let $\bar{z}^j \in \Gamma_2 \cap {}^n(T_j)$, $\bar{z} \le \bar{z}^j$ (by c)). So for $i < \xi < \zeta < \omega_1$, for some ℓ and k, $z^\xi_\ell < z^\zeta_k$ (otherwise use (c) on $\bar{z}^\xi$ with ξ, ζ here standing for i, j there to get a contradiction). This contradicts T being Aronszajn by the proof of Theorem III 5.4.

e) for any $\bar{y} \in \Gamma_2 \cap {}^n(T_i)$ and $m < \omega$ there are $j < \omega_1$, $j > i$ and pairwise disjoint $\bar{z}^1, \dots, \bar{z}^m \in \Gamma_2 \cap {}^n(T_j)$ $\bar{y} < \bar{z}^\ell$ for $\ell = 1, \dots, m$.

Just by induction on n, using d) and c).

Now we prove the subfact itself. For any $\bar{y} \in \Gamma_2$ there are (by (e)) $j_m(\bar{y})$ ($m < \omega$) such that there are m pairwise disjoint members of $\Gamma_2 \cap {}^n(T_{j_m(\bar{y})})$ which are $> \bar{y}$. By c) this holds for any $j \ge j_m(\bar{y})$. Now, if $j \ge \bigcup_{m<\omega} j_m(\bar{y})$, then we can find for every m, m pairwise disjoint members of $\Gamma_2 \cup {}^n(T_j)$ which are $> \bar{y}$. Let $\{\bar{z}^i : i < i_0\}$ be a maximal subset of $\{\bar{z} \in \Gamma_2 \cap {}^n(T_j) : \bar{y} \le \bar{z}, j \in \omega_1\}$ whose members are pairwise disjoint. If $i_0 < \omega$, choose another such set $\{\bar{y}^\ell : \ell < ni_0 + 1\}$ (exists as $j \ge \bigcup_m j_m(\bar{y})$). Now, at least one $\bar{y}^\ell$ should be disjoint from all $\bar{z}^i$'s, a contradiction to the maximality of $\{\bar{z}^i : i < i_0\}$. Hence i_0 is infinite. Let $C^* = \{i : i \text{ is } \alpha \text{ or } i > \alpha,\ i \in C_1 \text{ and } \bar{y} \in \Gamma_2 \cup \bigcup_{j<i} {}^n(T_j) \text{ implies that } i > \bigcup_{m<\omega} j_m(\bar{y})\}$.

It is easy to check C^* is as required in 6.7B. So we finish the proof of 6.7. $\square_{6.7B, 7}$

Continuation of the Proof of 6.1. The only point left is to prove the existence of the appropriate simple $\aleph_1$-completeness system $\mathbb{D}$. This is trivial. (It is easy to see that if $x : T \cap N \to \omega$ codes the branches of $T \cap N$ (use x which is eventually constant on each δ-branch of $T \cap N$ where $\delta \stackrel{\text{def}}{=} N \cap \omega_1$), then $\text{Gen}^+(N, P)$ can be written as A_x with some suitable ψ as in 5.5. The point is that $\bigcap_{i<\omega} A_{x_\ell} \ne \emptyset$ because we prove not only 6.7(1) but also 6.7(2).) $\square_{6.1}$

§7. Iteration of $(\mathcal{E}, \mathbb{D})$-Complete Forcing Notions

The discussion in the two previous sections lacked the crucial point that we can iterate such forcing notions without adding reals. In order to get a reasonable form of MA we need to iterate up to some regular $\kappa > \aleph_1$ and have the κ-c.c. For $\kappa = \aleph_2$, Lemma 1.5 does not suffice as $|P_T| = \aleph_2$ (P_T from the proof of Theorem 6.2; not to speak of the lack of $\mathcal{E}$-completeness) but meanwhile κ strongly inaccessible will suffice (see VII §1, or VIII §2 for eliminating this). An aesthetic drawback of the proof is that we do not prove that the forcing we get by the iteration enjoys the same property we require from the individual forcing notions but see VIII§4, which contains more detailed proofs of stronger theorems.

7.1 Theorem. Let $\bar{Q} = \langle P_i, \underset{\sim}{Q}_i : i < \alpha \rangle$ be a countable support iteration, $P_\alpha = \operatorname{Lim} \bar{Q}$; $\mathcal{E}$ a nontrivial family of subsets of $\mathcal{S}_{\aleph_0}(\mu)$.

(1) If each $\underset{\sim}{Q}_i$ is β-proper for every $\beta < \omega_1$, and $(\mathcal{E}, \mathbb{D}_i)$-complete for some simple $\aleph_1$-completeness system $\mathbb{D}_i$ (so $\mathbb{D}_i$ is a P_i-name), *then* P_α does not add reals.

(2) We can replace in (1) "simple" by "almost simple over V" (note: V and not V^{P_i}).

Combining the ideas of the proofs of 7.1 and of 4.3 we can prove

7.2 Theorem. In Theorem 7.1 we can weaken "$\aleph_1$-completeness system" to "$\aleph_0$-completeness system".

However we shall not prove it now (see VIII §4 for more).

Proof of Theorem 7.1. Note: $\mathbb{D}$ is a function with domain α, $\mathbb{D}_i$ is a P_i-name (of an $\aleph_1$-completeness system or more acurately a definition of such a system). For clarity of presentation we first deal with the case $\alpha = \omega$ (for $\alpha < \omega$ there is nothing to prove).

Let $N_i \prec (H(\lambda), \in)$ be countable (for $i < \omega$), $\mathbb{D}, \bar{Q} \in N_0$, $N_i \in N_{i+1}$ (hence $N_i \prec N_{i+1}$) each N_i is suitable for $\mathcal{E}$ (remember Definition 1.1(1), really just N_0 suitable suffice) and $p \in P_\omega \cap N_0$, and $\underset{\sim}{f} \in N_0$ be a P_ω-name of a real.

Now we shall define by induction on $n < \omega$ conditions r_n, p_n such that:

(A) (1) $r_n \in P_n$, $r_n = r_{n+1} \restriction n$

(2) r_n is (N_i, P_n)-generic for $i = 0$ and $n + 1 \leq i < \omega$.

(B) (1) There is a $G^*_n \in \mathrm{Gen}(N_0, P_n)$ which is bounded by r_n and belongs to N_{n+1}.

(2) $p_n \leq p_{n+1}$, $p_n \in P_\omega \cap N_0$, $p_0 = p$ and $p_n \restriction n \leq r_n$ (equivalently, $p_n \restriction n \in G^*_n$).

(3) Let $\{\mathcal{I}_n : n < \omega\}$ be a list of the open dense subsets of P_ω which belong to N_0; then $p_{n+1} \in \mathcal{I}_n$.

Finally, let r be such that $\forall n[r \restriction n = r_n]$. Then $r \geq p_n$, so r decides all values of $\underset{\sim}{f}(k)$ (as for each k for some n we have $\mathcal{I}_n = \{q \in P_\omega : q$ force a value to $\underset{\sim}{f}(k)\}$.

Let us carry out the induction.

$n = 0$: Trivial (Note $P_0 = \{\emptyset\}$).

$n + 1$: We shall first define p_{n+1}, then G^*_{n+1}, and finally r_{n+1}.

First step. We want to find $p_{n+1} \in P_\omega \cap N_0$, $p_{n+1} \geq p_n$, $p_{n+1} \restriction n \in G^*_n$ and $p_{n+1} \in \mathcal{I}_n$. As $\mathcal{I}_n \subseteq P_\omega$ is dense and $\mathcal{I}_n \in N_0$, above every $q \in P_\omega \cap N_0$ there is $r \in P_\omega \cap N_0 \cap \mathcal{I}_n$, $r \geq q$. Let $\mathcal{J}_n = \{r \in P_n$: there is an $r^* \in P_\omega$, $r^* \geq p_n$, $r^* \in \mathcal{I}_n$, $r^* \restriction n = r\}$, clearly it is dense above $p_n \restriction n$ (in P_n). But $p_n \restriction n \in G^*_n \in \mathrm{Gen}(N_0, P_n)$, and $\mathcal{J}_n \in N_0$, hence there is $r \in G^*_n \cap \mathcal{J}_n$, and so there is an $r^* \in P_\omega$, $r^* \geq p_n$, $r^* \in \mathcal{I}_n$, $r^* \restriction n = r$. So, clearly, r^* can be chosen as p_{n+1} we need.

Second step. Let $G_n \subseteq P_n$ be generic over V, $r_n \in G_n$ (hence $G^*_n \subseteq G_n$). We shall try to see what are the demands on G^*_{n+1}. Really we want in $V[G_n]$, to find a member of $\mathrm{Gen}(N_0[G_n], \underset{\sim}{Q}_n[G_n])$ which has an upper bound in $\underset{\sim}{Q}[G_n]$ and $p_n(n)$ belongs to it.

Note that $\mathbb{D}_n$ is a P_n-name which belongs to N_0. So there is also a P_n-name $\underset{\sim}{\psi}_n$ for the formula ψ appearing in the definition of simplicity (or almost simplicity), and it belongs to N_0. As we "know" $G_n \cap N_0 = G_n^*$, we "know" $\psi_n = \underset{\sim}{\psi}_n[G_n]$, i.e. some member of G_n^* force $\underset{\sim}{\psi}_n = \psi_n$. So we know that for some $A_{x,y}$ (x a relation on N_0, $y \in V$, see Definition 5.5) every $G \in A_{x,y}$ has an upper bound (in $\underset{\sim}{Q}_n[G_n]$), where

$$A_{x,y} = \{G \in \mathrm{Gen}(N_0[G_n], \underset{\sim}{Q}_n[G_n]) : (V \cup N_0[G_n] \cup \mathcal{P}(N_0[G_n])^{V[G_n]},$$

$$\in^V, \in^{N_0[G_n] \cup \mathcal{P}(N_0[G_n])}, p_{n+1}(n)[G_n], \underset{\sim}{Q}_n[G_n], V, N_0[G_n]) \models \psi[G, x, y]\}.$$

So we have P_n-names $\underset{\sim}{x}, \underset{\sim}{y}$ for x and y. Now $\underset{\sim}{x}, \underset{\sim}{y}$ are quite unlikely to be in N_0 (as their definitions used N_0 as a parameter) but they can be chosen in $N_{n+1}[G_n]$ (remember $N_{n+1}[G_n] \prec (H(\lambda)[G_n], \in)$ as $r_n \in G_n$, r_n is (N_{n+1}, P_n)-generic, and $N_0[G_n] \in N_{n+1}[G_n]$).

Moreover, though we need to know G_n to be able to find their exact values, we know that they are in V (remember $\mathcal{P}(N[G_n]) \in V$ and P_n does not add reals and even ω-seqeunces of a member of V); well formally $N_{n+1}[G_n]$ cannot be in V as it has members like G_n, but the isomorphism type of $(N_{n+1}[G_n], N_{n+1}, G_n, c)_{c \in N_{n+1}}$ does, and so does the isomorphism type of the model appearing in the definition of $A_{x,y}$. If you are still confused see VIII §4, where essentially we make the set of members of $\underset{\sim}{Q}_n[G_n]$ and $N_n[G_n]$ to be a set of ordinals.

So as r_n is (N_{n+1}, P_n)-generic $r_n \Vdash$ "$\underset{\sim}{x}, \underset{\sim}{y} \in N_{n+1}$", so we have just countably many possible pairs (those in N_{n+1}). Now N_{n+1} "thinks" there are uncountably many possibilities, but as $N_{n+1} \in N_{n+2}$, $N_{n+2} \models$ "N_{n+1} is countable", in N_{n+2} we "know" that $\bigcap\{A_{x,y} : x, y \in N_{n+1}\}$ is nonempty (remember $G_n^* \in N_{n+1}$ hence $N_0[G_n^*] \in N_{n+1}$). So, in N_{n+2} there is a $G^n \subseteq \underset{\sim}{Q}_n[G_n] \cap N_0[G_n^*]$ which belongs to the intersection. So though we do not know exactly what $\underset{\sim}{x}$ and $\underset{\sim}{y}$ will be, we know (as long as $r_n \in \underset{\sim}{G}_n$)) that $G^n \in \underset{\sim}{A}_{x,y}$. From G^n, G_n^* we can easily compute $G_{n+1}^* \in \mathrm{Gen}^+(N_0, P_{n+1})$, $G_{n+1}^* \cap P_n = G_n^*$, $G_{n+1}^* = \{\langle \underset{\sim}{q}_0, \underset{\sim}{q}_1, \dots, \underset{\sim}{q}_n\rangle : \langle \underset{\sim}{q}_0, \underset{\sim}{q}_1, \dots, \underset{\sim}{q}_{n-1}\rangle \in G_n^*, \underset{\sim}{q}_n \in N_0 \text{ and } \underset{\sim}{q}_n[G_n] \in G^n\}$.

Third step. We now have to define $r_{n+1} \in P_{n+1}$, so as we require $r_{n+1}\restriction n = r_n$, we just have to define $r_{n+1}(n)$. What are the requirements on it? Looking at the induction demand, just:

$r_n \Vdash_{P_n}$ "$r_{n+1}(n)$ is above each $q(n)[\underset{\sim}{G}_n]$ for members q of G^*_{n+1} and is $(N_i[\underset{\sim}{G}_{P_n}], \underset{\sim}{Q}_n[\underset{\sim}{G}_{P_n}])$- generic for $n+2 \leq i < \omega$",

and there is no problem in this. We have finished the proof of 7.1 for $\alpha = \omega$.

Now we have to turn to the general case i.e. with no restriction on α.

Let $p, \bar{Q} \in N_0 \prec (H(\lambda), \in), N_0$ countable, N_0 suitable for $\mathcal{E}$ and $p \in P_\alpha$. Let $N_0 \cap (\alpha+1) = \{\beta_i : i \leq \gamma\}$, $i < j \Rightarrow \beta_i < \beta_j$ (so $N_0 \cap (\alpha+1)$ has order type $\gamma+1$). Now we can find $N_i \prec (H(\lambda), \in)$ for $i \leq \gamma$, N_i countable, $\langle N_i : j \leq i\rangle \in N_{i+1}$ (just define by induction on i) and let $N_\delta = \cup_{i<\delta} N_i$ for limit $\delta \leq \gamma$.

As $N_0 \in N_1$, $\gamma + 1 \subseteq N_1$, hence $i \in N_i$.

7.3 Definition.
A pair (r, G^*) is called an (i,ζ)-th approximation if ($i < \zeta \leq \gamma$ and) :

a) $r \in P_{\beta_i}$ and r is (N_j, P_{β_i}) generic for $j = 0$ and $i+1 \leq j \leq \zeta$.

b) $G^* \in \mathrm{Gen}(N_0, P_{\beta_i})$, and G^* is bounded by r and $G^* \in N_{i+1}$.

Now it suffices to prove

7.4 Claim. If $0 \leq i < j \leq \zeta \leq \gamma$ and (r, G^*) in an (i,ζ)-th approximation, $p \in P_{\beta_j} \cap N_0$, $p\restriction\beta_i \in G^*$, *then* there is a (j,ζ)-th approximation $(r^\dagger, G^\dagger)$ such that $p \in G^\dagger$, $r^\dagger\restriction\beta_i = r$ and $G^\dagger \cap P_{\beta_i} = G^*$ (actually $G^\dagger$ depends just on G^* and $\langle N_\beta : \beta \leq \gamma\rangle$, but not on r).

Why is the claim sufficient? Just use $i = 0$, $j = \zeta = \gamma$ (so $\beta_\gamma = \alpha$), and we get what we need.

Proof of 7.4. Now, the proof of the claim is by induction on j (for all i, ζ). Then for successor this is just like the induction step for $\alpha = \omega$, and for limit j we diagonalize using the induction hypothesis (also taking care of clause (a) of Definition 7.3). $\square_{7.4,7.1}$

§8. The Consistency of SH + CH + There Are No Kurepa Trees

We wish now to give yet another application of the technique of not adding reals in iterations.

8.1 Definition. For any regular κ, a κ-Kurepa tree is a κ-tree such that the number of its κ-branches is $> \kappa$. Let the κ-Kurepa Hypothesis (in short $\kappa-$KH) be the statement "there exists κ-Kurepa tree". We may write "KH" instead of ω_1-KH. (Be careful: KH says "there *are* Kurepa trees", but SH says "there are *no* Souslin trees"!)

Solovay proved that Kurepa trees exist if $V = L$, more generally Jensen [Jn] proved the existence of κ-Kurepa's trees follows from Jensen's $\diamondsuit^+$, which holds in L for every regular uncountable κ which is not "too large". But $\neg$KH is consistent with of ZFC + GCH, which was first shown by Silver in [Si67], starting from a strongly inaccessible κ. The method of his proof is as follows: collapse every λ, $\omega_1 < \lambda < \kappa$ using Levy's collapse $\text{Levy}(\aleph_1, < \kappa) = \{p : |p| \leq \aleph_1 \;\&\; p \text{ is a function with } \text{Dom}(p) \subseteq \kappa \times \omega_1 \wedge \forall\langle\alpha, \xi\rangle \in \text{Dom}(p)(p(\alpha, \xi) \in \alpha)\}$. Now $\text{Levy}(\aleph_1, < \kappa)$ can be viewed as an iteration of length κ, and satisfied the κ-c.c. on the one hand, and $\aleph_1$-completeness on the other hand. Therefore $\aleph_1$ does not get collapsed, as well as any cardinal $\aleph_\alpha \geq \kappa$. Suppose now that $T \in V^P$ is an ω_1-tree. So it has appeared already at an earlier stage along the iteration, say $T \in V^{P'}$, where V^P is obtained from $V^{P'}$ by an $\aleph_1$-complete forcing. In $V^{P'}$ the tree T has at most $2^{\aleph_1}$ branches, and this is less than κ. Note that by 6.1(2) the tree T can have no new ω_1-branches in V^P. So T is not a Kurepa tree in V^P.

Devlin in [De1] and [De2] has shown, starting from a strongly inaccessible, the consistency of GCH + SH + $\neg$KH. For a proof by iteration see Baumgartner [B3].

8.1A Remark. In both proofs the inaccessible cardinal is necessary, for $\neg$KH implies that $\aleph_2$ is an inaccessible cardinal of L.

8.2 Definition. Suppose T is an ω_1-tree, and that Q is a forcing notion. We say that Q is *good for* T if for every $p \in Q$ and a countable elementary submodel $N \prec (H(\chi), \in)$, for χ large enough such that T, Q, $p \in N$, there is an (N, Q)-generic condition $q \geq p$ such that for every name $\underset{\sim}{\tau} \in N$ of a branch of T, either $q \Vdash_Q$ "$\underset{\sim}{\tau}[\underset{\sim}{G}_Q] \in N$ and is an old cofinal branch of T" or $q \Vdash_Q$ "$\underset{\sim}{\tau}[G] \cap T(\delta(N))$ is not $b(a) = \{x \in T(\delta(N)) : s < a\}$ for any $a \in T_{\delta(N)}$".

8.2A Fact. A forcing notion Q is good for an ω_1-tree T iff Q is proper and in V^Q there are no new cofinal branches of T.

Proof. $\Rightarrow$: Suppose Q is good for T. The properness of Q follows trivially. Let $p \Vdash_Q$ "$\underset{\sim}{\tau}$ is a new branch of T", and we shall derive a contradiction; let $\{T, p, Q\} \in N \prec (H(\chi), \in)$, χ large enough and N countable. So let $q \geq p$ be as in the definition of *good.*

If $\underset{\sim}{\tau}[G]$ is an old branch — we are done. If not, $\underset{\sim}{\tau}[G] \cap T_{<\delta_N} \neq B_x = \{y : y <_T x\}$ for all $x \in T_{\delta_N}$. But this implies that $\underset{\sim}{\tau}[G]$ being linearly ordered by $<_T$ has no member of level $\geq \delta_N$, so it cannot be a ω_1-branch of T.

Conversely, suppose that Q is proper and does not add a new ω_1-branch to T. Let $\underset{\sim}{\tau}, p \in N$ be as in the definition, and pick $q \geq p$ which is (N, Q)-generic, and a generic subset G of P over V with $q \in G$. So $\underset{\sim}{\tau}[G] \in N[G] \prec H(\chi, \in)[G]$, and $\underset{\sim}{\tau}[G]$ is either an old ω_1-branch, or is not an ω_1-branch at all. In the first case we are done. Now if $\underset{\sim}{\tau}[G]$ is not an ω_1-branch, then either $(\exists \alpha) \underset{\sim}{\tau}[G] \cap T_\alpha = \emptyset$ or $\exists x, y \in \underset{\sim}{\tau}[G]$ such that x, y are not comparable in T. By elementaricity of $N[G]$, such an α or such x, y exist also in $N[G]$. So q forces what is required by the definition. $\square_{8.4}$

8.3 Lemma. If Q is an $\aleph_1$-complete forcing notion and T is an ω_1-tree, then Q adds no new cofinal branches to T.

Proof. It is enough to show that Q is good for T. Suppose that $N \prec (H(\chi), \in)$ is a countable elementary submodel and that T, $Q \in N$. Let $\langle \mathcal{I}_n : n < \omega \rangle$ be a list of all dense open subsets of Q which belong to N. Let $p \in Q \cap N$. Let

$\langle(x_n, \underset{\sim}{\tau}_n) : n < \omega\rangle$ be a list of all pairs (x_n, τ_n) such that $x_n \in T_{\delta(N)}$ and $\underset{\sim}{\tau}_n \in N$ is a Q-name of a branch in of T.

By induction on n we construct a sequence of conditions p_n such that:

(1) $p_0 = p$

(2) $p_n \leq p_{n+1} \in Q \cap N \cap \mathcal{I}_n$.

(3) $p_{n+1} \Vdash_Q$ "τ_n is an old cofinal branch of T" or there is some $x \in T \cap N$ such that $x \not\leq_T x_n$ and $p_{n+1} \Vdash_Q$ "$x \in \underset{\sim}{\tau}$".

Suppose first that the construction is carried out. Let $q \in Q$ extend all p_n (q exists by $\aleph_1$-completeness). Clearly, q is (N, Q)-generic. For every $\underset{\sim}{\tau} \in N$, a Q-name of a branch of T, either $q \Vdash$ "$\underset{\sim}{\tau}$ is an old branch of T", of $q \Vdash (\forall n)[\underset{\sim}{\tau} \cap T \cap N \neq \{x : x <_T x_n\}]$. In the first case, as q is generic, $q \Vdash \underset{\sim}{\tau} \in N$". In the second, clearly q forces that $\underset{\sim}{\tau}$ is not a cofinal branch of T. Therefore Q is good for T.

We still have to show that the construction can be carried out. Suppose p_n is picked. First find an extention $p' \geq p_n$ that decides whether $\underset{\sim}{\tau}_n$ is an old branch or a new branch. If $p' \Vdash$ "$\underset{\sim}{\tau}$ is old", define $p_{n+1} = p''$ for some $p' \leq p'' \in \mathcal{I}_n \cap N$. Else, $p' \Vdash$ "$\underset{\sim}{\tau}$ is new". Let $B = \{x \in T : \exists p'' \geq p', p'' \Vdash$ "$x \in \underset{\sim}{\tau}$"$\}$. Clearly, B is downward closed, and $B \in N$. If B were a cofinal branch, this would contradict $p' \Vdash$ "$\underset{\sim}{\tau}$ is new". Therefore there are two incomparable elements in B. By elementarity, there are two such elements in N. Therefore we can pick in N a condition $p_{n+1} \geq p_n$ such that $p_{n+1} \Vdash$ "$\tau \cap T_\delta = x$" for some x such that $x \not\leq_T x_n$. $\square_{8.3}$

8.4 Theorem. Suppose T is an ω_1-tree, $\langle P_i, Q_i : i < \alpha\rangle$ is a countable support iteration, and no Q_i adds new cofinal branches to T, then also P_α does not add cofinal branches to T.

Proof. By induction on α. If $\alpha = 0$ or α is a successor ordinal, there is not much to prove. Suppose that α is limit. Let N be a countable elementary submodel as usual and suppose that $p \in P_\alpha \cap N$. Pick a sequence $\langle \alpha : n < \omega\rangle$ of ordinals such that $\alpha_n \in N \cap \alpha$, $\alpha_{n+1} > \alpha_n$ and $\bigcup_{n<\omega} \alpha_n = \sup(\alpha \cap N)$.

Let $\langle \mathcal{I}_n : n < \omega \rangle$ be an enumerations of all dense open subsets of P which belong to N. Let $\langle (x_n, \underset{\sim}{\tau}_n) : n < \omega \rangle$ be a enumeration of $T_{\delta(N)} \times \{\underset{\sim}{\tau} \in N : \underset{\sim}{\tau}$ is a P_α-name of a branch of $T\}$.

By induction on n we pick p_n, q_n as in the proof of preservation of properness under countable support iteration in III, §3. In addition to the conditions there we demand:

(*) $q_n \Vdash$ "$p_{n+1} \Vdash (\exists \alpha)\tau{\restriction}\alpha \neq \{x \in T(\alpha) \ \wedge \ x <_T x_n\}$" or

$q_n \Vdash$ ["$p_{n+1} \Vdash \tau$ is old"].

We show how to pick p_{n+1}. let $G \subseteq P_{\alpha_n}$ be generic such that $q \in G$. Then there is a $p \in \mathcal{I}_n \cap N$ such that $p{\restriction}\alpha_n \in G$. There is some extention $p' \geq p$ which decides whether $\underset{\sim}{\tau}_n$ is old or new. The rest is as in the proof of the previous Lemma. $\square_{8.4}$

We shall utilize now the preservation Theorem we just proved to reprove Devlin's result:

8.5 Theorem. If CON(ZFC + κ is inaccessible) then CON(ZFC + GCH +SH + $\neg$KH).

Proof. Let κ be strongly inaccessible. we define a countable support iteration of lengh κ of proper forcing notions, $\langle P_i, \underset{\sim}{Q}_j : i \leq \kappa, j < \kappa \rangle$. When i is odd, $\underset{\sim}{Q}_i = \text{Levy}(\aleph_1, \aleph_2)^{V[P_i]}$. When i is even, $\underset{\sim}{Q}_i = Q(\underset{\sim}{T}_i)$ is the forcing notion defined in §6 which specializes some given tree $\underset{\sim}{T}_i$ (a P_i-name) without adding reals and every such names appear.

8.6 Claim.

(a) P_κ has the $\aleph_2$-cc.

(b) P_κ does not add reals.

(c) $V^{P_\kappa} \models$ SH.

(d) $V^{P_\kappa} \models \neg$KH.

Proof. Clause (a) is easy by III 4.1. For (b) see §6 and §7. Clause (c) is clear, as every ω_1-tree gets specialized along the way by suitable bookkeeping. Suppose now that T is some ω_1-tree in V^{P_κ}. There is some intermediate stage P_i for

$i < \kappa$ such that $T \in V^{P_i}$. In V^{P_i} the tree T has at most $2^{\aleph_1}$ (of V^{P_i}) branches. As κ is inaccessible in V^{P_i}, the number of branches of T is $< \kappa$. So there is some $j < \kappa$ such that in V^{P_j} the tree T has at most $\aleph_1$ *old* branches. What is left to see is that the rest of the forcing does not introduce new branches to T. By theorem 8.4 it is enough to show that no Q_j adds branches (for $j \geq i$). In case Q_j is Levy$(\aleph_1, \aleph_2)$, this is known from Lemma 8.3, because the collapse is $\aleph_1$-complete. The remaining case is that $Q_i = Q(T_i)$ is the forcing notion defined in §6 for specializing a given Aronszajn tree T_i. So our proof is finished once we know that this forcing notion does not add branches to an ω_1-tree T.

8.7 Claim. Suppose T is an $\aleph_1$-Aronszajn tree, and T^* is any $\aleph_1$-tree. Then $Q = Q(T)$ is good for T, where $Q(T)$ is the forcing for specializing T defined in §6.

Proof. Suppose that $N \prec (H(\chi), \in)$ is a countable elementary submodel such that Q, T, $T^* \in N$ and $p \in Q \cap N$. We shall find a condition $q \geq p$ such that q is (Q, N)-generic and such that for every Q-name $\underset{\sim}{\tau} \in N$ of a branch of T^*, $q \Vdash$ "$\underset{\sim}{\tau}$ is an old branch of T^* or $\underset{\sim}{\tau} \cap T^* \cap N \neq b(a)$ for all $a \in T^*_{\delta(N)}$".

In the proof we shall follow the proof of Theorem 6.1, in which the properness of Q was shown.

8.8 Definition. Suppose that $p \in Q$ and $\delta > \ell t(p)$. For $\bar{x} = (x_0, \ldots, x_{n-1}) \in {}^n(T_\delta)$ and $\varepsilon > 0$ we say that $q \geq p$ with $\ell t(q) < \delta$ *respects* $\bar{x}$ *by* ε iff $f_q(x_\ell \restriction \ell t(q)) < f_p(x_\ell \restriction \ell t(p)) + \varepsilon$ for all $\ell < n$.

The main point is the following claim which is the parallel of 6.7.

8.9 Claim. Suppose that $N \prec (H(\chi), \in)$ is as usual, and $p \in N$ is a condition and $\underset{\sim}{\tau} \in N$ is a Q-name of an ω_1-branch of T^*. Let $\delta = N \cap \omega_1$, $\bar{x} \in {}^n(T^*_\delta)$, $a \in T^*_\delta$ and $\varepsilon > 0$. *Then* there is a condition $q \in N$, $q \geq p$ and q respects $\bar{x}$ by ε and such that:

(i) $q \Vdash$ "$\underset{\sim}{\tau} \restriction \delta \neq b(a) = \{y : y <_{T^*} a_\ell\}$ or

(ii) $q \Vdash$ "$\underset{\sim}{\tau}$ is an old branch of T".

Proof. Suppose there is no $q \geq p$ in N as required and which satisfies (i). We shall see that there is one which satisfies (ii). Define the set of "bad" $\bar{y}$ as follows: Γ_1 is the collection of all $\bar{y} \in {}^n(T^*_\beta)$ such that $\beta > \alpha$ and

(a) For every $\gamma < \beta$ and $\bar{z} \in {}^m(T^*_\beta)$, $m < \omega$, there are $\gamma' \in [\gamma, \beta)$ and an extension of p of height $\gamma' < \beta$ which respects $\langle z_\ell \restriction \gamma' : \ell < \ell g \bar{x} \rangle$ by ε and which determines the value of $\underset{\sim}{\tau} \restriction \gamma$,

(b) There are no two extensions, q_0 and q_1, of p such that q_k respects $\bar{y}$ by ε (for $k = 1, 2$) and q_1 and q_0 force contradictory information about $\underset{\sim}{\tau}$.

Let $B(\Gamma_1)$ be the set of levels β for which there is $\bar{y} \in \Gamma_1$ of height β. Observe that $\bar{x} \in \Gamma_1$. Why? Clause (a) follows from 6.7; if (b) does not hold then there is a condition q such that $q \Vdash$ "$\underset{\sim}{\tau}$ is not $b(a)$", contrary to our assumptions, remembering a is constant in 8.9. Observe also that by 6.7, there is a club E of ω_1 such that if $\bar{z} \in \Gamma_1$ and $\bar{y} < \bar{z}$ is of height $\beta \in E$, then $\bar{y} \in \Gamma_1$. W.l.o.g. $E \in N$. As $\bar{x} \in \Gamma_1$, $\delta \in B(\Gamma_1) \cap E$. Therefore $B(\Gamma_1) \cap E$ is unbounded, and clearly it is closed.

Let $\Gamma_2 = \{\bar{y} \in \Gamma_1$: there are uncountably many $\bar{z} \in \Gamma_1$ such that $\bar{y} < \bar{z}\}$. For each $\bar{y} \in \Gamma_2$ of height β, define $t(\bar{y})$ to be the set of nodes $t \in T$ such that some extension of p of height β, which respects $\bar{y}$ by ε forces that $t \in \underset{\sim}{\tau}$. Then $t(\bar{y})$ is linearly ordered and contains in fact a branch of height β. If $\bar{y}_1$ and $\bar{y}_2$ are any n-tuples in Γ_2, then $t(\bar{y}_1)$ and $t(\bar{y}_2)$ are compatible. (Why? By clause (b).) Thus the function t defines a cofinal branch of T and the intersection of this branch with T^*_δ is just $b(a)$. Since Γ_2 is definable in N, this branch is an old branch in N. Now, as before, we get a promise Γ out of Γ_2 and we add this promise to p in order to obtain the desired q. It follows now that $q \Vdash$ "$\underset{\sim}{\tau}$ is this old branch".

$\square_{8.9,8.6,8.5}$

VI. Preservation of Additional Properties, and Applications

This chapter contains results from three levels of generality: some are specific consistency results; some are preservation theorems for properties like "properness + ${}^{\omega}\omega$-bounding", and some are general preservation theorems, with the intention that the reader will be able to plug in suitable parameters to get the preservation theorem he needs. We do not deal here with "not adding reals" - we shall return to it later (in VIII §4 and XVIII §1,§2).

Results of the first kind appear in 3.23, §4, §5, §6, §7, §8. In §4 we prove the consistency of "there is no P-point (a kind of ultrafilter on ω)". We do this by CS iteration, each time destroying one P-point; but why can't the filter be completed later to a P-point? (If we add enough Cohen reals it will be possible.) For this we use the preservation of a property stronger than ${}^{\omega}\omega$-bounding, enjoyed by each iterand.

More delicate is the result of §5 "there is a Ramsey ultrafilter (on ω) but it is unique, moreover any P-point is above it" (continued in XVIII §4). Here we need in addition to preserve "D continues to generate an ultrafilter in each V^{P_α}".

In 3.23 we prove the consistency of $\mathfrak{s} > \mathfrak{b} = \aleph_1$; i.e. for every subalgebra $\mathbb{B}$ of $\mathcal{P}(\omega)/\text{finite}$ of cardinality $\aleph_1$, there is $A \subseteq \omega$ which induce on $\mathbb{B}$ an ultrafilter $\{B/\text{finite}: B \in \mathbb{B} \text{ and } A \subseteq^* B\}$; but there is $F \subseteq {}^{\omega}\omega, |F| = \aleph_1$ with no $g \in {}^{\omega}\omega$ dominating every $f \in F$. We use a forcing Q providing a "witness" A for $\mathbb{B} = (\mathcal{P}(\omega)/\text{finite})^V$; not adding g dominating $({}^{\omega}\omega)^V$; we iterate it (CS). After ω_2 steps the first property is O.K., but we need a preservation lemma to show the second is preserved. The definition of this Q and the proof of its relevant properties are delayed to §6. In §7 (i.e. 7.1) we prove the consistency of $\mathfrak{a} > \mathfrak{b}$. Lastly in §8 (i.e. in 8.2) we prove the consistency of $\mathfrak{h} < \mathfrak{b} = \mathfrak{a}$. On history concerning §6, §7, §8 see introduction to §6. See relevant references in the section.

* * *

We now review most of the preservation theorems appearing here for countable support iteration of proper forcing; actually this is done for more general iterations (including RCS, a pure finite/pure countable, FS-finite support), see 0.1 and we can weaken "proper". You can read it being interested only in CS iteration of proper forcing, ignoring all adjectives "pure" and the properties "has pure (θ_1, θ_2)-decidability" (or feeble pure (θ_1, θ_2)-decidability), so letting $\leq_{\rm pr} = \leq$.

0.A Theorem. For any CS iteration $\langle P_i, \underset{\sim}{Q}_j : i \leq \delta, j < \delta \rangle$ if for each $i < \alpha$ we have $\Vdash_{P_i}$ "Q_i satisfies X" *then* P_δ satisfies X; for each of the following cases:

1) X = "Q is proper and ${}^\omega\omega$-bounding" [Why? By 2.8D, i.e. by 2.3 + 2.8B + 2.8C].

2) Let $f, g : \omega \to \omega + 1 \setminus \{0, 1\}$ be functions diverging to infinity [i.e. $(\forall n < \omega)(\exists k < \omega)(\forall m)(k < m < \omega \Rightarrow f(m) > n \,\&\, g(m) > n)$] and:
X = "Q is proper and *for every* $\ell < \omega$ and $\eta \in (\prod_n f(n)^{[g(n)^\ell]})^{V^Q}$ there is a sequence $\langle u_n : n < \omega \rangle \in V$ such that $\bigwedge_n \eta(n) \in u_n$ and $|u_n| > 1 \Rightarrow |u_n| \leq g(n)^{1/\ell}$. [Why? By 2.11F.]

3) X = "Q proper and every dense open $A \subseteq {}^{\omega>}\omega$ includes an old such set". [Why? See 2.15D; or see 2.15B(2) for an equivalent formulation, then by 2.15C, 2.3(5) we can apply 2.3(2)].

Remark. Particular cases of $0.A(2)$ are the Sacks property (f constantly ω, all g's), and the Laver property (f, g vary on all legal members of ${}^\omega\omega$), the names were chosen for the most natural forcing notions with these properties. Other pairs $f, g \in {}^\omega\omega$ were introduced in and important for [Sh:326 §2]. Concerning the PP-property and the strong PP-property see 2.12, 3.25-6.

For some other properties we can prove that in limit stages, violation does not arise; but leave to the specific iteration the burden for the successor stages. We say "X is preserved in limit".

0.B Theorem. For CS iteration of proper forcing, $\bar{Q} = \langle P_\alpha, \underset{\sim}{Q}_\beta : \alpha \leq \delta, \beta < \delta \rangle$, δ a limit ordinal.

1) If for $\alpha < \delta$, in V^{P_α} there is no new $f \in {}^\omega\omega$ dominating all $h \in ({}^\omega\omega)^V$ *then* this holds for V^{P_δ} [see 3.17(1)],

2) If for $\alpha < \delta$, in V^{P_α} there is no new $f \in {}^\omega\omega$ dominating all $h \in ({}^\omega\omega)^V$ and no real which is Cohen over V *then* this holds for V^{P_δ} [see 2.13D(2); more on Cohen see 2.17].

3) If for $\alpha < \delta$ in V^{P_α} there is no random real over V *then* this holds for V^{P_δ} [see 3.18].

* * *

We now turn to the third kind of results.

In §1 we present a general context suitable for something like: for every $\eta \in ({}^\omega\omega)^{V^Q}$ there is a "small" tree $T \subseteq {}^{\omega>}\omega$ from V such that $\eta \in \lim(T)$; so we assume that the family of small trees has some closure properties. In 2.1 - 2.7 we more specify our context, so that we can get preservation in successor stages too. In 1.16, 1.17 we deal with a generalization where we have several kinds of $\eta \in {}^\omega\omega$ (but for simplifying the presentation, we restrict generality in other directions). A reader who feels our level of generality is too high (or goes over to this view while reading 2.1-2.8) can prefer a simplified version (which is [Sh:326, A2 pp 387-399]), so read only 1.16, 1.17 for the case $k^* = 1$ and then look at any of 2.9 - 2.17 (each dedicated to a specific property being preserved) ignoring the undefined notions.

In 3.1 - 3.13 we give another context (tailored for "there is no dominating reals"). Here for successor stages we use a stronger property (like almost ${}^\omega\omega$-bounding). In XVIII §3 we give another such general theorem.

The reader is tuned now to countable support iteration of proper forcing but we shall later consider other contexts (semiproperness in Chapter X; forcing with additional "partial order $\leq_{\rm pr}$" (pr for pure) plus some substitute of properness in Chapter XIV, XV). To save repetition, in 0.1 below we describe the various contexts. The subscript θ has a role only when $\leq_{\mathrm{pr}}$ is present (cases D-F below) and its meaning is described in 0.1(3). Note that also FS iteration of c.c.c. forcing is a particular case: $\leq_{\mathrm{pr}}$ is equality and $\theta = \aleph_1$ (the relevant results will be presented in §3). Let θ missing mean $\theta \equiv 1$. We may write e.g. $0.1_{\theta=\aleph_0}$ rather than $0.1_{\aleph_0}$ to stress this.

0.1_θ *Iteration Context:*

1) We shall use iteration $\bar{Q} = \langle P_j, \underset{\sim}{Q}_i : j \leq \alpha, i < \alpha \rangle$ of one of the following forms:

(A) Countable support iteration of proper forcing (see III). In this case $\leq_{\mathrm{pr}}$ is the usual order, 1.11 is just III 1.7; "purely" can be omitted; similarly for (B) (C).

(B) Like (1) but for $\delta < \alpha$ limit we weaken "$\underset{\sim}{Q}_\delta$ is proper" to "for arbitrarily large $i < \delta, P_{\delta+1}/P_{i+1}$ is proper or even just E-proper" where $E \subseteq \mathcal{S}_{\leq\aleph_0}(\mu)$ is a fixed stationary set (we can use similar variants of the other cases).

(C) RCS iteration which is a semiproper iteration (see Chapter X).

(D) Each forcing notion $\underset{\sim}{Q}_i$ has also a partial order $\leq_{\mathrm{pr}}$, $[p \leq_{\mathrm{pr}} q \Rightarrow p \leq q]$; a minimal element $\emptyset_Q$ and is purely proper (i.e. *if* $p \in Q \cap N, Q \in N$, N countable and $N \prec (H(\chi), \in, <^*_\chi)$, *then* there is a (N, Q)-generic $q, p \leq_{\mathrm{pr}} q \in P$). The iteration is defined as $P_i = \{p : p$ a function with domain a countable subset of i, for $j \in \mathrm{Dom}(p)$ we have: $\Vdash_{P_j}$ "$p(j) \in Q_j$" and $\{j :$ not $\Vdash_{P_j}$ "$\emptyset_{Q_j} \leq_{\mathrm{pr}} p(j)$" $\}$ is finite$\}$.
A particular case is FS iteration of c.c.c forcing. This (i.e. clause (D)) is a particular case of Chapter XIV.

(E) The iterations $\overline{Q}$ which are GRCS as in XV §1 (and see 0.3), such that: for each $\alpha < \ell g(\overline{Q})$ for some n we have $\Vdash_{P_{\alpha+n}}$ "$(2^{\aleph_1}) + |P_\alpha|$ is collapsed to $\aleph_1$" and each $\underset{\sim}{Q}_\alpha$ is purely semiproper.

(F) The GRCS iterations as in XV §3 (so each Q_i satisfies $UP(\mathbb{I}, \mathbf{W})$, where $\mathbf{W} \subseteq \omega_1$ is stationary.

(G) The GRCS iteration as in XV §4.

2) We say "P purely adds no f such that $(\forall x \in V)\varphi(x, f)$" *if* for every $p \in P$ and P-name $\underset{\sim}{f}$, for some $q \in P$ and $x \in V$: $p \leq_{\rm pr} q$ and $q \Vdash$ "$\underset{\sim}{f}$ does not satisfy $\varphi(x, \underset{\sim}{f})$".

3) $\theta \in \{1, 2, \aleph_0, \aleph_1\}$ and: $\theta = 1$ means no demand, $\theta \geq \aleph_0$ means each Q_α (or each $P_\alpha, P_\alpha/P_{\beta+1}$) has pure $(\theta, 2)$-decidability (see Definition 1.9) and $\theta = 2$ means they have pure $(2, 2)$-decidability (see Definition 1.9).

Remark. We shall concentrate on case F in 0.1(1) as it is the hardest.

0.2 Definition.

1) We say W is absolute if it is a *definition* (possibly with parameters) of a set so that if $V^1 \subseteq V^2$ are extensions of V (but still models of ZFC with the same ordinals) and $x \in V^1$ then: $V^2 \models$ "$x \in W$" iff $V^1 \models$ "$x \in W$". Note that a relation is a particular case of a set. It is well known that Π^1_2 relations on reals and generally κ-Souslin relations are absolute.

2) We say that a player absolutely wins a game if the definition of legal move, the outcomes and the strategy (which need not be a function with a unique outcome) are absolute and its being a winning strategy is preserved by extensions of V.

3) We can relativize absoluteness to a family of extensions, e.g. for a given universe V and family K of forcing notions we can look only at $\{V^Q : Q \in K\}$; so for V^{Q_0} we consider only the extensions $\{V^Q : Q_0 \lessdot Q \in K\}$, or even demand Q/Q_0 has a specified property. We do not care to state this all the time.

Though Case D is covered by Chapter XIV, (and XV) we may note:

0.3 Theorem. 1) The iteration in case (D) preserves "purely proper".

2) X§2 is generalized to "purely semiproper is preserved" by GRCS iterations.

§1. A General Preservation Theorem

An important part of many independence proofs using iterated forcing, is to show that some property X is preserved (if satisfied by each iterand). We have dealt with such problems in Chapter V (preserving e.g. "ω-properness + the ${}^{\omega}\omega$-bounding property"), [Sh:b] Chapter VI (general context and many examples), [Sh:207], [Sh:177] (replacing the weak form of ω-proper by proper), Blass and Shelah [BsSh:242] (preserving ultrafilters which are P-points), [Sh:326]; in [Sh:b] Chapter X §7 we have dealt with semiproperness. Here we redo [Sh:b] Chapter VI §1, giving a general context which serves for many examples replacing proper by the weaker condition semiproper and even UP and "CS iteration" by "GRCS iterations" i.e. revised countable/finite support with purity (and correcting it). You may read this section replacing everywhere: UP by proper, RCS iteration by countable support iteration, $\leq_{\mathrm{pr}}$ by the usual order, $\mathbf{S}$ by the class of regular cardinals, $\mathbf{W} = \omega_1$, semi-generic by generic, omit $\mathbb{I}$-suitable, then 1.9, 1.10 are not necessary.

In fact there is more in common between the examples discussed later even than expressed by the stricter context suggested here (fine covering model) (i.e., the use of trees $T, T \cap {}^{n}\omega$ finite and absoluteness in the definitions of covering models) but the saving will not be so large; we shall return to this in §2.

Unfortunately "adding no reals" will require special treatment (as is the case even if we assume properness). We have dealt with it separately in Chapter V and will return to it in VIII §4, XVIII §1, §2.

For applications it suffices to read Definitions 1.1 - 1.5 (the fine covering models and preservation of them); also 1.9 and Theorem 1.12 (on more general preservation theorems). Another general way to get such preservation theorems is presented in XVIII §3. A simpler version of the theorem is presented in 1.16, 1.17 here (and see 1.3(10); earlier see [Sh:326, Appendix A2 pp. 387-399] (but also for a finite sequence of covering models)).

1.1 Definition. We call (D, R) a *weak covering model* (in V) if:

a) D a set, R a two place relation on D, xRT implies that T is a closed subtree of ${}^{\omega>}\omega$ (i.e., $\langle\rangle \in T$, T is closed under initial segments, and above any $\eta \in T$ there are arbitrarily long members of T),

b) (D, R) covers, i.e. for every $\eta \in {}^{\omega}\omega$ and $x \in \mathrm{Dom}(R)(= \{x : (\exists T)xRT\})$ there is $T \in D$ such that xRT and $\eta \in \lim T$, where

$$\lim T = \{\eta \in {}^{\omega}\omega : \eta \restriction k \in T \text{ for every } k < \omega\}$$

1.1A Remark. The intuitive meaning is: xRT means T is a closed tree of "size" at most x. In Definition 1.2, which exploits more of our intuition, we have an order on the set of possible x's, $x \leq y$, with the intuitive meaning "x is a smaller size than y". So it would be natural to demand:

$$xRT, x < y \Rightarrow yRT \text{ and } xRT, T^{\dagger} \subseteq T \Rightarrow xRT^{\dagger}$$

However, no need arises. Note also that sometimes x appears trivially (e.g. see the ${}^{\omega}\omega$-bounding model in 2.8).

1.2 Definition. (1) A *fine covering model* is $(D, R, <)$ such that:

(α) (D, R) is a weak covering model

(β) $<$ is a partial order on $\mathrm{Dom}(R)$, such that

(i) $(\forall y \in \mathrm{Dom}(R))(\exists x \in \mathrm{Dom}(R))(x < y)$

(ii) $(\forall y, x \in \mathrm{Dom}(R))(\exists z \in \mathrm{Dom}(R))(x < y \rightarrow x < z < y)$

(iii) if $y < x, yRT$ then for some $T^* \in D$, $T \subseteq T^*$ and xRT^*

(iv) if $y < x$ and for $l = 1, 2$ yRT_l *then* there is $T \in D$ such that: xRT, $T_1 \subseteq T$ and for some n, $[\nu \in T_2 \mathbin{\&} \nu \restriction n \in T_1 \Rightarrow \nu \in T]$

(γ) (a) If $x > x^{\dagger} > y_{n+1} > y_n$ for $n < \omega$ and $T_n \in D, y_nRT_n$ (for $n < \omega$) *then* there is $T^* \in D, xRT^*$ and an infinite set $w \subseteq \omega$ such that:

$$\lim T^* \supseteq \{\eta : \eta \text{ is in } {}^{\omega}\omega \text{ and for every } i \in w, \eta \restriction \min(w \setminus (i+1)) \in \bigcup_{\substack{j<i \\ j\in w}} T_j \cup T_0\}$$

(b) if $\eta, \eta_n \in {}^\omega\omega$, $\eta \restriction n = \eta_n \restriction n$ for each $n < \omega$ and $x \in \mathrm{Dom}(R)$ *then* for some $T \in D$, xRT, $\eta \in \lim T$ and $\eta_n \in \lim T$ for infinitely many n.

(δ) condition (γ) continues to hold in any generic extension in which (α) holds.

(2) For a property X of forcing notions, $(D, R, <)$ is a fine covering model for X-forcing if Definition 1.2(1) holds when we restrict ourselves in (δ) to X-forcing notions only.

(3) We say $(D, R, <)$ is a temporarily fine covering model if it satisfies $(\alpha), (\beta), (\gamma)$ i.e. is a fine covering model for trivial forcing.

1.3 Remark. 1) In an abuse of notation we do not always distinguish between $(D, R, <)$ and (D, R).

2) Look carefully at (δ), it is in a sense, meta-mathematical.

3) So if $(D, R, <)$ is a fine covering model and P is a (D, R)-preserving forcing notion (see Definition 1.5 below) *then* in V^P the model $(D, R, <)$ is still a fine covering model. [Why? In Definition 1.2(1) clause (α) holds as P is (D, R)-preserving, clause (β) holds as it is absolute, clause (γ) holds as in V, $(D, R, <)$ is a fine covering model by clause (δ) of Definition 1.2(1) and clause (δ) by its transitive nature.]

4) In $(\gamma)(a)$ of 1.2(1), we can replace "$y_n RT_n$" by $x^\dagger RT_n$ (by (β) (ii) (iii)).

5) We write in 1.2(1)(β) (iv)$^+$ if $n = 0$.

6) If we assume 1.2(1)(β)(iv)$^+$, then in 1.2(1)(γ)(a) w.l.o.g. $T_n \subseteq T_{n+1}$ hence the conclusion in $(\gamma)(a)$ is:

$$\lim T^* \supseteq \{\eta \in {}^\omega\omega : \text{ for every } i \in w,\ \eta \restriction i \in T_{\max[(w \cap i) \cup \{0\}]}\}.$$

7) We can in (γ) add "and $0 \in w$".

8) A condition stronger than $(\gamma) = (\gamma)_0$ of 1.2(1) is:

$(\gamma)_1 = (\gamma)^+$ if $x > x^\dagger > y_{n+1} > y_n$ for $n < \omega$ and $T_n \in D, y_n RT_n$ (for $n < \omega$) *then* there is $T^* \in D, xRT^*$ and an infinite set $w \subseteq \omega$ such that:

$$\lim T^* \supseteq \{\eta : \eta \text{ is in } {}^\omega\omega \text{ and for every } i \in w, \eta \restriction i \in \bigcup_{\substack{j \le i \\ j \in w}} T_j\}$$

(I.e. it implies both (a) and (b) of 1.2(γ) (when (D,R) covers, of course).)
If we assume $(\beta)(\text{iv})^+$, then in 1.2(1)(γ) w.l.o.g. $T_n \subseteq T_{n+1}$ hence the demand in $(\gamma)^+$ is $\lim T^* \supseteq \{\eta \in {}^\omega\omega : \text{for every } i \in w,\ \eta \restriction i \in T_i\}$.
Why? Let y'_n be: $y'_0 = y_0$, $y'_{n+1} = y_{n+2}$. We choose by induction on n, T'_n such that $y_n R T'_n$ and $T'_0 = T_0$, and $T'_n \subseteq T'_{n+1}$ and for some k_n we have $\eta\restriction k_n \in T'_n \ \&\ \eta \in \bigcup_{m \le n+1} T_m \Rightarrow \eta \in T'_{n+1}$. Now by clause $(\gamma)^+$ there are an infinite $w' \subseteq w$ and T^* such that xRT^* and $\lim(T^*) \supseteq \{\eta \in {}^\omega\omega :$ for every $i \in w'$ we have $\eta\restriction i \in \bigcup_{\substack{j \le i\\ j \in w}} T'_j\}$. Let $w' = \{n_i : i < \omega\}$ with $n_i < n_{i+1}$. Let $j(\ell)$ $(\ell < \omega)$ be increasing fast enough, i.e. $n_{j(\ell+1)} > k_{n_{j(\ell)-1}}$, $w \overset{\text{def}}{=} \{n_{j(\ell)} : \ell < w\}$. It is enough to prove that w and T^* are as required in clause (γ). So assume $\eta \in {}^\omega\omega$ belongs to the set on the right hand side of the inclusion in clause (γ), and we shall prove $\eta \in \lim T^*$. So we are assuming that for every $\ell < \omega$ we have $\eta\restriction n_{i(\ell+1)} \in \bigcup_{j<\ell} T_{n_j} \cup T_0$. So it is enough to prove that η appears in the right side of the inclusion in (γ) for w', $\langle T'_i : i < \omega\rangle$. So let $i < \omega$ and we should prove that $\eta\restriction n_i \in \bigcup_{\ell \le i} T'_{n_\ell}$ (as $w' = \{n_i : i < \omega\}$, n_i increasing with i).
Let ℓ be such that $j(\ell) \le i < j(\ell+1)$, so by the assumption on η we have $\eta\restriction n_i \lhd \eta\restriction n_{j(\ell+1)} \in \bigcup_{m<\ell} T_{j(m)} \cup T_0$. We prove this by induction on i.
Case 1: $\eta\restriction n_{j(\ell+1)} \in T_0$

So $\eta\restriction n_i \in T_0$, but $T - 0 = T'_0 \subseteq T'_{n(i)}$ hence $\eta\restriction n_i \in T_{n(i)} \subseteq \bigcup_{\ell \le i} T'_{n(\ell)}$ as required.

Case 2: There is $m < \ell$ such that $\eta\restriction n_{j(\ell+1)} \in T_{n_{j(m)}}$

Necessarily $i \ge j(\ell) > j(m)$ so by the induction hypothesis on i we have $\eta\restriction n_{j(\ell)-1} \in \bigcup_{k \le j(\ell)-1} T'_{n_k}$ but $T'_n \subseteq T'_{n+1}$ so $\eta\restriction n_{j(\ell)-1} \in T'_{n_{j(\ell)-1}}$ as by assumption $\eta\restriction n_{j(\ell+1)} \in T_{n_{j(m)}}$, $m < \ell$, by the choice of $T'_{n_{j(\ell)}}$ as $j(\ell+1) > k_{n_{j(\ell)-1}}$ necessarily $\eta\restriction n_{j(\ell+1)} \in T'_{n_{j(\ell)}}$ but $T'_{n_{j(\ell)}} \subseteq T'_{n_i}$ and $n_i \le n_{j(\ell+1)}$ hence $\eta\restriction n_i \in T'_{n_i} \le \bigcup_{\ell \le i} T'_{n_\ell}$ as required.
8A) In clause (γ) w.l.o.g. $T_n \subseteq T_{n+1}$ (i.e. this weaker version implies the original version using (α), (β) of course).
[Why? By 2.4D (note 2.4A, 2.4B, 2.4C, 2.4D do not depend on the intermediate material).]

9) Note in 1.2(1)(γ)(a), that any infinite $w' \subseteq w$ is o.k.

10) In some circumstances clause (b) of 1.2(1)(γ) is a too strong demand, e.g. preservation of P-points. We can overcome this by letting $R = \bigvee_{\ell<k} R_\ell$ (k is finite) and demanding 1.2(γ)(a) for each R_ℓ whereas instead 1.2(γ)(b) we demand

(b)′ if $\eta_n, \eta \in {}^\omega\omega$, $\eta_n \restriction n = \eta \restriction n$ for $n < \omega$, and for some $m < k$ we have

$$(\forall x \in \mathrm{Dom}(R_m))(\exists T)(x R_m T \ \&\ \eta \in \lim T) \quad \text{and}$$

$$(\forall x \in \mathrm{Dom}(R_m)) \bigwedge_n (\exists T)(x R_m T \ \&\ \eta_n \in \lim T)$$

then for every $x \in \mathrm{Dom}(R_m)$ for some T we have: $x R_m T$ and for infinitely many $n < \omega$, $\eta_n \in \lim(T)$.

See more on this in 1.16, 1.17 and §5.

Proof. E.g.

9) Assume $\eta \in {}^\omega\omega$ and

$(*)_0$ $i \in w' \Rightarrow \eta \restriction \min(w' \setminus (i+1)) \in \bigcup_{j \in w', j<i} T_j \cup T_0$;

we have to prove $\eta \in \lim(T^*)$. For this it suffices to prove:

$(*)_1$ $i \in w \Rightarrow \eta \restriction (\min(w \setminus i+1)) \in \bigcup_{j \in w, j<i} T_j \cup T_0$.

Let $i \in w$, define $i_1 = i$, $j_1 = \min(w' \setminus i)$, $i_2 = \min(w \setminus (i_1+1))$, $j_2 = \min(w' \setminus (j_1+1))$; so in particular $i_1 \le j_1 \in w'$, $i_1 < i_2 \le j_2$, $j_1 < j_2$. As $(*)_0$ holds apply it to j_1 and get $\eta \restriction j_2 \in \bigcup_{j \in w', j<j_1} T_j \cup T_0$, hence for some j_0, $j_0 = 0 \vee (j_0 < j_1 \ \&\ j_0 \in w')$ and we have $\eta \restriction j_2 \in T_{j_0}$. As $i_2 \le j_2$ clearly $\eta \restriction i_2 \in T_{j_0}$. As $j_1 = \min(w' \setminus i_1)$ we know that $i_1 \le j_1$ and $[i_1, j_1) \cap w' = \emptyset$ and thus $j_0 = 0 \vee (j_0 < i_1 \ \&\ j_0 \in w')$. Hence $j_0 = 0 \vee (j_0 < i_1 \ \&\ j_0 \in w)$. So $\eta \restriction i_2 \in T_{j_0} \subseteq \bigcup_{j \in w, j<i} T_j \cup T_0$, as required. (See more 2.4D.) $\square_{1.3}$

1.4 Convention. If the order $<$ is not specified then $< = <_{\mathrm{dis}}$ (see below). Let $<_0$ be such that:

$$x <_0 y \ \text{ iff } \ x, y \in {}^\omega\omega \ \&\ x(0) <_0' y(0)$$

where $(\omega, <'_0)$ is isomorphic to $(\mathbb{Q}, <)$ (i.e. the rationals). Let $<_{\rm dis}$ be:

$$x <_{\rm dis} y \text{ iff } x, y \in {}^\omega\omega, 1 \leq x(n) \leq y(n) \text{ for every } n$$
$$\text{and } y(n)/x(n), x(n) \text{ diverge to } \infty.$$

Let $<^*_{\rm dis}$ be: $x <^*_{\rm dis} y$ iff $y <_{\rm dis} x$. (Note: in $\mathrm{DP}({}^\omega\omega) \stackrel{\rm def}{=} \{x \in {}^\omega\omega : x(n) \geq 1$, $\langle x(n) : n < \omega\rangle$ diverges to infinity$\}$, $<_0$, $<_{\rm dis}$ and $<^*_{\rm dis}$ satisfy clauses (β)(i), (ii), (iii) of Definition 1.2(1)).

1.5 Definition. Let (D, R) be a weak covering model. We say that a forcing notion P *preserves* (D, R) or is (D, R)*-preserving* if $\Vdash_P$ "(D, R) is a weak covering model". We add "purely" if: for every $p \in P$ and $\underset{\sim}{f}$ such that $p \Vdash$ "$\underset{\sim}{f} \in {}^\omega\omega$" and $x \in \mathrm{Dom}(R)$, for some q, T we have $p \leq_{\rm pr} q \in P, xRT$ and $q \Vdash$ "$\underset{\sim}{f} \in \lim T$".

1.6 Definition. 1) For a weak covering model (D, R) and $y \in \mathrm{Dom}(R)$, $(D, R) \models$ "$\varphi_{\rm dis}(y)$" if:

for every $\eta^* \in {}^\omega\omega$ and function F from $D \times \omega$ to $\mathrm{Rang}(R) = \{T\colon (\exists x \in D)xRT\}$ such that $(\forall n)(\forall z \in \mathrm{Dom}(R))[zRF(z,n)$ and $\eta^* {\restriction} n \in F(z,n)]$.
there are T^*, yRT^* and an infinite set w of natural numbers, and $z_\ell \in \mathrm{Dom}(R)$ for $\ell \in w$ such that:

$$T^* \supseteq \{\eta \in {}^{\omega>}\omega : \text{there is } \ell \in w \text{ such that } \eta \restriction \ell \lhd \eta^*, \text{ and } \eta \in F(z_\ell, \ell)\}.$$

Note that the truth value of $(D, R) \models$ "$\varphi_{\rm dis}(y)$" depends on V (remember $\lhd$ means initial segment).

1A) For a weak covering model (D, R) and $y \in \mathrm{Dom}(R)$ we write $(D, R) \models$ "$\varphi^*_{\rm dis}(y)$" if:
for every $\eta^* \in {}^\omega\omega$, and a function $F : D \times \omega \to \mathrm{Rang}(R)\}$ such that $(\forall n)(\forall z \in \mathrm{Dom}(R))zRF(z,n)$ and a function H from $D \times \omega$ into ${}^\omega\omega$ such that $\eta^* \restriction n \lhd H(z,n) \in \lim F(z,n)$ (so $\eta^* \restriction n \in F(z,n)$)
there are T^*, yRT^* and an infinite $w \subseteq \omega$ and $n_\ell < \omega$, and $z_\ell \in \mathrm{Dom}(R)$ for

$\ell \in w$ such that:

$$T^* \supseteq \{\eta \in {}^{\omega>}\omega : \text{there is } \ell \in w \text{ such that } \eta \restriction \ell \lhd \eta^*,$$
$$\text{and } \eta \restriction n_\ell \lhd H(z_\ell, \ell) \text{ and } \eta \in F(z_\ell, \ell)\}.$$

(If $n_\ell \geq \ell$, then "$\eta \restriction \ell \lhd \eta^*$" is not necessary and if $n_\ell \leq \ell$, then $\eta \restriction n_\ell \lhd H(z_\ell, \ell)$ is not necessary). Note that the truth value of $(D, R) \models \varphi^*_{\text{dis}}(y)$ depends on V and $\varphi_{\text{dis}}(y) \Rightarrow \varphi^*_{\text{dis}}(y)$ (as in φ^*_{dis} the set of $\eta \in {}^{\omega>}\omega$ which we demand to be in T^* is smaller than for φ_{dis}).

2) We call (D, R) a *covering model* if it is a weak covering model and

(c) for every $y \in \text{Dom}(R)$, $(D, R) \models \varphi_{\text{dis}}(y)$ or at least $(D, R) \models \varphi^*_{\text{dis}}(y)$.

3) For a weak covering model (D, R) and $\bar{x} = \langle x_n : n < \omega \rangle$ and z, where $\{x_n : n < \omega\} \cup \{z\} \subseteq \text{Dom}(R)$ we say that $(D, R) \models \psi_{\text{dis}}(\bar{x}, z)$ if:

$(*)$ *for every* $\eta^* \in {}^\omega\omega$ and a set $\{T_{n,j} : n, j < \omega\}$ such that $x_n R T_{n,j}$ for $n, j < \omega$ *there are* $\langle T^\alpha : \alpha \leq \omega \rangle$ such that:

(i) $T^n \subseteq T^{n+1}$ and $T^0 \subseteq T^\omega$

(ii) zRT^ω (so $T^\omega \in D$)

(iii) $\eta^* \in \lim T^0$

(iv) if $n, j < \omega$ and $\nu \in (\lim T_{n,j}) \cap (\lim T^n) \cap (\lim T^\omega)$, *then* for some k:

$$(\forall \rho)[\nu \restriction k \trianglelefteq \rho \in T_{n,j} \Rightarrow \rho \in T^{n+1} \cap T^\omega]$$

4) (D, R) is a strong covering model if it is a covering model and

(d) For every $z \in \text{Dom}(R)$ there are $x_n (n < \omega)$ such that:

$$(D, R) \models \psi_{\text{dis}}(\langle x_0, x_1, \ldots \rangle, z)$$

1.7 Definition. 1) Let K be a property of weak covering models. We say that a forcing notion P is K-preserving if:

for any $(D, R) \in V$ satisfying K, P preserves (D, R). We add "purely" if for any (D, R) satisfying K, P purely preserves (D, R).

2) We call a covering model (D, R) *smooth* if:

for any (D, R)-preserving forcing notion P, $\Vdash_P$ "(D, R) is a covering model".

3) We call a strong covering model (D, R) *strongly smooth* if:
for any (D, R)-preserving forcing notion P we have $\Vdash_P$ "(D, R) is a strong covering model".

1.8 Claim. 1) If $(D, R, <)$ is a fine covering model *then* (D, R) is a strongly smooth strong covering model.
2) The following is a sufficient condition for $(D, R) \models \psi_{\mathrm{dis}}(\langle x_n : n < \omega\rangle, z)$:
$(*)$ for some $y_n \in \mathrm{Dom}(R)$ (for $n < \omega$) :
$(a)_n$ *if* $n < \omega$, $x_n R T_j$ for $j < \omega$ and $y_n R T$ *then* for some $\langle n_j : j < \omega\rangle$ and T^*:

(i) $y_{n+1} R T^*$

(ii) $T \subseteq T^*$

(iii) $\eta \in T_j$ & $\eta \restriction n_j \in T \Rightarrow \eta \in T^*$

(b) if $y_n R T^n$ for $n < \omega, T^n \subseteq T^{n+1}$ *then* for some T^*, $z R T^*$ and $\bigwedge_n T^n \subseteq T^*$.
3) For a weak covering model (D, R) we have: if (D, R) is a strong covering model then it is a covering model, and strongly smooth implies smooth.

Proof: 1) By 1.2(1)(α) we have: (D, R) is a weak covering model. Now we show that it is a strong covering model. So by 1.6(2), (4) we have to check conditions (c), (d) of Definition 1.6.
Proof of (c) We are going to prove that for $y \in \mathrm{Dom}(R)$ we have $(D, R) \models \varphi^*_{\mathrm{dis}}(y)$.
So suppose $\eta^* \in {}^\omega\omega$, F is a function from $D \times \omega$ to $\mathrm{Rang}(R)$, and H is a function from $D \times \omega$ to ${}^\omega\omega$ such that:

$$(\forall n)(\forall x \in \mathrm{Dom}(R))[x R F(x, n) \ \&\ \eta^* \restriction n \lhd H(x, n) \in \lim F(x, n)]$$

First we use a stronger assumption.

Proof of (c) assuming $(\gamma)^+$ of 1.3(8): So there exist, by (β)(i), (ii) of Definition 1.2, $y^\dagger$, x_n (for $n < \omega$) such that $y > y^\dagger > x_{n+1} > x_n > \ldots > x_0$ (choose $y^\dagger$ and then, inductively on n, x_n). Let $z_\ell \stackrel{\mathrm{def}}{=} x_\ell$ and $n_\ell \stackrel{\mathrm{def}}{=} \ell$ and let $T_n \stackrel{\mathrm{def}}{=} F(x_n, n)$. Apply condition $(\gamma)^+$ of Definition 1.2(1) (i.e. 1.3(8)) to get T^* and an infinite

$w \subseteq \omega$ such that $\lim T^* \supseteq \{\eta \in {}^\omega\omega : \text{for every } i \in w, \eta \upharpoonright i \in \bigcup_{\substack{j\in w\\ j\le i}} T_j\}$ and yRT^*. Remember $n_\ell \ge \ell$.

We shall show that T^* and w and $\langle n_\ell : \ell < \omega\rangle$ are as required in 1.6(1A). We have to prove that (for each $\ell \in w$ and $\eta \in {}^{\omega>}\omega$):

$(*)$ $\eta \upharpoonright n_\ell \trianglelefteq H(z_\ell, \ell)$ & $\eta \in F(z_\ell, \ell) \Rightarrow \eta \in T^*$.

(Note: $\eta \upharpoonright \ell \triangleleft \eta^*$ follows from $\eta \upharpoonright \ell = \eta \upharpoonright n_\ell \trianglelefteq H(z_\ell, \ell)$ because $\eta^* \upharpoonright \ell \triangleleft H(z_\ell, \ell)$ by the assumptions on H in 1.6(1A).)

So assume $\eta \upharpoonright n_\ell \trianglelefteq H(z_\ell, \ell)$ and $\eta \in F(z_\ell, \ell)$ (so $\eta \in T_\ell$), and we have to prove $\eta \in T^*$. We can choose ν, $\eta \triangleleft \nu \in \lim F(z_\ell, \ell)$, so it suffices to show that for any $i \in w$ we have $\nu \upharpoonright i \in \bigcup_{\substack{j\le i\\ j\in w}} T_j$. If $i \ge \ell$, then: $\nu \upharpoonright i \in F(z_\ell, \ell) = T_\ell \subseteq \bigcup_{\substack{j\le i\\ j\in w}} T_j$ and if $i < \ell$ then: $\nu{\upharpoonright}i = \eta \upharpoonright i = \eta^* \upharpoonright i \triangleleft H(z_i, i) \in \lim F(z_i, i)$, hence $\nu \upharpoonright i \in F(z_i, i) = T_i \subseteq \bigcup_{\substack{j\le i\\ j\in w}} T_j$ (remember $i \in w$). So by the conclusion of $(\gamma)^+$ (in 1.3(8) which we have applied) $\nu \in \lim T^*$ hence $\eta \in T^*$ is as required; so we have proved condition (c).

The full proof of (c): Let $y^\dagger, x_n$ be as above. Now we prove (c) using (γ) of 1.2(1) only. So we are given η^*, F and H as in the assumptions of 1.6(1A). Apply condition (γ)(b) of Definition 1.2(1) with $x_0, \eta^*, H(x_n, n)$ (for $n < \omega$) here standing for x, η, η_n (for $n < \omega$) there, and get an infinite $w_0 \subseteq \omega$ and $T_0 \in \text{Rang}(R)$ such that $x_0 R T_0$ and $\bigwedge_{n\in w_0} H(x_n, n) \in \lim T_0$, hence $\eta^* \in \lim T_0$.

Let $w_0 = \{k_\ell : \ell < \omega\}$, k_ℓ increasing with ℓ, of course w.l.o.g. $k_\ell + 1 < k_\ell$ (hence $\ell + 1 < k_\ell$). Applying (β)(iv) (of 1.2(1)) choose $T_{\ell+1}$ such that $x_{k_\ell+1} R T_{\ell+1}$, $T_0 \subseteq T_{\ell+1}$, even $T_\ell \subseteq T_{\ell+1}$, and for some $m_\ell < \omega$ we have: $[\rho{\upharpoonright}m_\ell \in T_0$ & $\rho \in F(x_{k_\ell}, k_\ell) \Rightarrow \rho \in T_{\ell+1}]$. So by (γ)(a) (of 1.2(1)) for some T^* and infinite $w_1 \subseteq \omega$ we have: yRT^* and for every $\eta \in {}^\omega\omega$ we have $[\bigwedge_{i\in w_1} \eta{\upharpoonright}\min(w_1 \setminus (i+1)) \in \bigcup_{\substack{j<i\\ j\in w_1}} T_j \cup T_0] \Rightarrow \eta \in T^*$. We define $w = \{k_\ell : \ell + 1 \in w_1\}$ and for $\ell + 1 \in w_1$ let us define n_{k_ℓ} as the first natural number $n = n_{k_\ell}$ such that $n_{k_\ell} > \ell$, in the interval (ℓ, n_{k_ℓ}) there are at least two members of w_1, and $n_{k_\ell} > m_\ell$.

We are going to prove that w, $\langle n_j : j \in w\rangle$ are as required (in 1.6(1A)). Remembering that the general members of w have the form k_ℓ with $\ell+1 \in w_1$, it suffices to prove that (the replacement of $\eta \in {}^{\omega>}\omega$ by $\nu \in {}^{\omega}\omega$ is as in the proof above of clause (c) from $(\gamma)^+$ of 1.3(8)):

(*) for any ℓ and ν we have $\otimes \Rightarrow \oplus$ where

$\otimes$ (A) $k_\ell \in w$ (i.e. $\ell + 1 \in w_1$)

(B) $\nu \in {}^{\omega}\omega$

(C) $\nu{\restriction}k_\ell \triangleleft \eta^*$

(D) $\nu{\restriction}n_{k_\ell} \triangleleft H(z_{k_\ell}, k_\ell)$

(E) $\nu \in \lim(F(z_{k_\ell}, k_\ell))$

$\oplus$ $\nu \in \lim(T^*)$

By the choice of T^*, for getting $\oplus$ it suffices to prove

$\otimes_1$ if $i = i_1 \in w_1$, and $i_2 = \min(w_1 \setminus (i_1 + 1))$ then $\nu{\restriction}i_2 \in \bigcup_{j \in w_1, j < i_1} T_j \cup T_0$.

Note that $k_\ell \in w_0$ (see above before choice of the T_ℓ's). We split the proof of $\otimes_1$ accordingly to how large i is.

Case 1: $\neg(\exists j)[\ell < j \in w_1 \cap i_1]$

By the choice of n_{k_ℓ} we know that in interval (ℓ, n_{k_ℓ}) there are at least two members of w_1, but $i_1 \leq \min(w_1 \setminus (\ell + 1))$ and $i_2 = \min(w_1 \setminus (i_1 + 1))$ so necessarily $i_2 < n_{k_\ell}$. Hence (by the previous sentence, by $\otimes$(D), by the choice of T_0, and trivially respectively) we have

$$\nu{\restriction}i_2 \triangleleft \nu{\restriction}n_{k_\ell} \triangleleft H(z_{k_\ell}, k_\ell) \in \lim(T_0) \text{ and thus } \nu{\restriction}i_2 \in \bigcup_{j \in w_1, j < i_1} T_j \cup T_0$$

as required (in $\otimes_1$).

Case 2: $(\exists j)[\ell < j \in w_1 \cap i_1]$

Let $i_0 = \max(w_1 \cap i_1)$, so by the assumption of the case not only i_0 is well defined but also it is $> \ell$. Looking at the desired conclusion of $\otimes_1$ and the definition of i_0 it suffices to prove that $\nu{\restriction}i_2 \in T_{i_0}$. But we know that $[n < \omega \Rightarrow T_n \subseteq T_{n+1}]$ and (by the previous sentence) $\ell < i_0$, hence $T_{\ell+1} \subseteq T_{i_0}$, so it suffices to prove $\nu{\restriction}i_2 \in T_{\ell+1}$. For this by the choice of $T_{\ell+1}$ it suffices to show the following:

$\otimes_2$ (A) $\nu{\restriction}m_\ell \in T_0$

(B) $\nu{\restriction}i_2 \in F(x_{k_\ell}, k_\ell)$

As for clause $\otimes_2$(B), by the assumption $\otimes$(E) it holds. As for clause $\otimes_2$(A), we know

$$\nu{\restriction}m_\ell \lhd \nu{\restriction}n_{k_\ell} \lhd H(z_{k_\ell}, k_\ell) \in \lim(T_0).$$

[Why? first $\lhd$ holds as $m_\ell < n_{k_\ell}$ by the choice of n_{k_ℓ}, second $\lhd$ holds by assumption $\otimes$(D), and the last "$\in$" holds as $k_\ell \in w_0$ and the choice of T_0, w_0.]

So both clauses of $\otimes_2$ hold hence $\otimes_1$ holds in case 2 hence in general, hence we have proved $(*)$. Thus we have finished proving clause (c) in the general case. Having proved condition (c) we shall now prove condition (d).

Proof of (d). Choose $x^\dagger$ and then by induction on $n < \omega, x_n$ such that $x_n < x_{n+1} < \ldots < x^\dagger < z$ (they exist by (β) of Definition 1.2(1)).

So it suffices to prove that $(D, R) \models \psi_{\mathrm{dis}}(\langle x_0, x_1, \ldots \rangle, z)$. Let $\eta^* \in {}^\omega\omega$ and $\langle T_{n,j} : n, j < \omega \rangle$ be as in $(*)$ of Definition 1.6(3).

For each $n < \omega$, by applying ω times Def 1.2(1)(β)(ii), we can find $x_{n,j}$ (for $j \le \omega$) such that $x_n < x_{n,0} < x_{n,1} < \ldots < x_{n,\omega} < x_{n+1}$ (first choose $x_{n,\omega}$ and then $x_{n,0}, x_{n,1}, \ldots$). We now define by induction on n, T^*_n such that $x_n R T^*_n$ and $T^*_n \subseteq T^*_{n+1}$. First let T^*_0 be such that $x_0 R T^*_0, \eta^* \in \lim T^*_0$ (possible by 1.1(b) and 1.2(1)(α)). Second, assuming T^*_n was defined, we can choose by induction on j trees $T'_{n,j}$ satisfying: $T'_{n,j} \subseteq T'_{n,j+1}$, $T'_{n,0} = T^*_n$, $x_{n,j} R T'_{n,j}, \eta^* \in \lim T'_{n,j}$ and such that for some $m = m(n,j)$ we have

$$(\forall \rho)[\rho \in T_{n,j} \& \rho \restriction m \in T'_{n,j} \Rightarrow \rho \in T'_{n,j+1}]$$

(possible by 1.2(1)(β)(iv)). Now by (γ)(a) of Def 1.2(1) we can find $w(n) \subseteq \omega$ infinite and T^*_{n+1} such that $x_{n+1} R T^*_{n+1}$ and

$$T^*_{n+1} \supseteq \{\eta : \text{ for every } i \in w(n),\ \eta \restriction \min(w(n) \setminus (i+1)) \in \bigcup_{\substack{j<i \\ j \in w(n)}} T'_{n,j} \cup T^*_n\}.$$

Necessarily $T^*_n \subseteq T^*_{n+1}$.

Then applying (γ)(a) of Definition 1.2(1) we can find $w \subseteq \omega$ infinite and T^*_ω such that zRT^*_ω and:

$$T^*_\omega \supseteq \{\eta : \text{ for every } i \in w,\ \eta \restriction \min(w \setminus (i+1)) \in \bigcup_{\substack{j<i \\ j \in w}} T^*_j \cup T^*_0\}.$$

As said above $T^*_n \subseteq T^*_{n+1}$ for each n, clearly from the condition above zRT^*_ω and $T^*_0 \subseteq T^*_\omega$ and in particular $\eta^* \in \lim T^*_0 \subseteq \lim T^*_\omega$. So in 1.6(3) (*), (with T^*_α here for T^α there) conditions (i), (ii), (iii) are satisfied. As for condition (iv), let $\nu \in (\lim T_{n,j}) \cap (\lim T^*_n) \cap (\lim T^*_\omega)$. Then any $k < \omega$ such that: $k > \min\{i \in w : |i \cap w \setminus (n+1)| > 1\}$ and $k > \min\{i : |i \cap w(n) \setminus (j+1)| > 1\}$ and $k > m(n,j)$ is as required. So T^*_ω is as required in 1.6(3), i.e. we have proved (d) from 1.6(4).

Now why is (D,R) strongly smooth? By remark 1.3(3). Suppose P is (D,R)-preserving then in V^P still (D,R) is a weak covering model as P is (D,R)-preserving, hence (α) of Definition 1.2(1) holds in V^P, (β) is trivial, and $(\gamma),(\delta)$ hold by (δ). So $(D,R,<)$ is a fine covering model in V^P hence, by what we already proved it is temporarily a strong covering model. As this holds for every P we finish.

2) Similar proof.

3) Read the definitions. $\square_{1.8}$

1.9 Definition. A forcing notion Q has pure (θ_1,θ_2)-decidability if: for every $p \in Q$ and Q-name $\underset{\sim}{t} < \theta_1$, there are $a \subseteq \theta_1, |a| < \theta_2$ (but $|a| > 0$) and $r \in Q$ such that $p \leq_{\text{pr}} r$, and $r \Vdash_Q$ "$\underset{\sim}{t} \in a$" (for $\theta_1 = 2$, alternatively, $\underset{\sim}{t}$ is a truth value), [if $\theta = \theta_1 = \theta_2$ we write just θ].

1.9A Remark. 1) If $\aleph_0 > \theta_2 > 2$, pure $(\theta_2, 2)$-decidability is equivalent to pure $(2,2)$-decidability.

2) Q purely semiproper implies Q has $(\aleph_1,\aleph_1)$-decidability.

3) If Q is purely proper *then* Q has $(\lambda,\aleph_1)$-decidability for every λ.

4) If $\leq_{\text{pr}} = \leq$ and Q is proper or Q has the c.c.c. (and we let $\leq_{\text{pr}}$ be equality if not defined) *then* Q is purely proper (see 0.1 case D).

1.10 Lemma. For $(\theta_1, \theta_2) \in \{(2,2), (\aleph_0, 2)\}$ the property "Q has pure (θ_1, θ_2)-decidability" is preserved by GRCS iteration as 0.1.

Proof: In quoting we refer to case F. We prove it by induction on the length of the iteration (for all $q, \underset{\sim}{t}$ and generic extension of V). By the distributivity of the iteration (in case F claim XV 1.7) it suffices to deal with the following five cases:

Case 1. $\underline{\alpha \leq 1}$ Trivial.

Case 2. $\underline{\alpha = 2}$ Easy.

case 3. $\underline{\alpha = \omega_1}$ If there is q_1, $q \leq_{\rm pr} q_1 \in P_\alpha$ such that for some $\beta < \alpha$, and P_β-name $\underset{\sim}{t}_1$,: $q_1 \in P_\beta$ and $q_1 \Vdash_{P_\alpha}$ "$\underset{\sim}{t} = \underset{\sim}{t}_1$", then we can use the induction hypothesis. By XV 3.3 this holds

Case 4. $\underline{\alpha \text{ strongly inaccessible}, \alpha > |P_i| \text{ for } i < \alpha}$: Even easier than the case $\alpha = \omega_1$.

case 5. $\underline{\alpha = \omega}$: So $\theta_2 = 2$, and w.l.o.g. $p = \{p_n : n < \omega\}, p_n$ a P_n-name of a member of $\underset{\sim}{Q}_n$. We define q_n such that:

(i) q_n a P_n-name of a member of $\underset{\sim}{Q}_n$

(ii) $\Vdash_{P_n}$ "$p_n \leq_{\rm pr} q_n$"

(iii) in V^{P_n}, q_n decides $\underset{\sim}{s}_n$, where:

for $G_{n+1} \subseteq P_{n+1}$ generic over V, $\underset{\sim}{s}_n[G_{n+1}]$ is $k+1$ *iff* there is $r \in P_\omega/G_{n+1}$ such that $\mathrm{Dom}(r) = [n+1, \omega)$, $P_\omega/G_{n+1} \models p \restriction [n+1, \omega) \leq_{\rm pr} r$ and $r \Vdash_{P_\omega/G_{n+1}}$ "$\underset{\sim}{t} = k$", with k minimal under those conditions; otherwise (i.e. if there is no such k) $\underset{\sim}{s}_n = 0$. (Actually q_n is a P_n-name of a member of $\underset{\sim}{Q}_n[\underset{\sim}{G}_n]$.) (If $\theta_1 = \aleph_0$ - clear, if $\theta_1 = 2$ - use Definition 1.9 twice, see 1.9A(1)).

Now $q = \{q_n : n < \omega\} \in P_\omega, p \leq_{\rm pr} q$; clearly there is r, $q \leq r \in P_\omega$ and $\ell < \theta_1$ such that $r \Vdash$ "$\underset{\sim}{t} = \ell$". Also w.l.o.g. for some $n(*)$, $[n(*) \leq n < \omega \Rightarrow r \restriction \{n\}$ is pure]; hence $r \restriction n(*) \Vdash_{P_{n(*)}}$ "$P_\omega / P_{n(*)} \models p \restriction [n(*), \omega) \leq_{\rm pr} r \restriction [n(*), \omega)$".

We can prove by downward induction on $m \leq n(*)$ that for some $\ell > 0$ we have $(r \restriction m) \cup \{q_m\} \Vdash$ "$\underset{\sim}{s}_m = \ell$".

For $m = 0$ we easily finish (by the definition of $\underset{\sim}{s}_m$). $\square_{1.10}$

1.10A Claim. Assume that Q is (D,R)-preserving, (D,R) a is weak covering model, Q has pure (θ_1,θ_2)-decidability and for some λ and stationary $S \subseteq S_{\leq\aleph_0}(\lambda)$, the forcing notion Q is purely S-proper (or the parallel for semiproper and $|D| = \aleph_1$, follows from 0.1(1) in all cases there).

(a) If $(\theta_1,\theta_2) = (\aleph_0, 2)$ *then* Q is purely (D,R)-preserving.

(b) If $(\theta_1,\theta_2) = (\aleph_0,\aleph_0)$ and for every $x \in \mathrm{Dom}(R)$ there is $y \in \mathrm{Dom}(R)$ such that for each $n < \omega$:

$$(\forall T_1,\dots,T_n \in \mathrm{Rang}(R))(\exists T \in \mathrm{Rang}(R))[\bigwedge_{\ell=1}^{n} yRT_\ell \rightarrow xRT \ \& \ \bigwedge_{\ell=1}^{n} T_\ell \subseteq T]$$

then Q is purely (D,R)-preserving.

Proof. Straight. $\square_{1.10A}$

1.11 Claim. 1) Assume $\bar{Q} = \langle P_n, \underset{\sim}{Q}_n : n < \omega\rangle$ a GRCS iteration with $\underset{\sim}{Q}_n$ having pure $(\aleph_0, 2)$ decidability, as XV 3.1. Then for every $p \in P_\omega$, $p \Vdash$ "$\underset{\sim}{f} \in {}^\omega\omega$" there is $q, p \leq_{\mathrm{pr}} q \in P_\omega$, such that $q \Vdash$ "$\underset{\sim}{f}(n) = \underset{\sim}{k}_n$" where $\underset{\sim}{k}_n$ is a P_n-name.

2) If we assume in addition: $p \Vdash_{P_\omega}$ "$\underset{\sim}{f} \leq g$", $g \in {}^\omega\omega$" (and $g \in V$) *then* we can replace "having pure $(\aleph_0, 2)$-decidability" by "having pure $(2,2)$-decidability".

Proof. Straightforward. $\square_{1.11}$

1.12 Theorem. Suppose (D,R) is a smooth strong covering model, $\bar{Q} = \langle P_i, \underset{\sim}{Q}_i : i < \delta\rangle$ a GRCS iteration as in 0.1, e.g. satisfying $\langle \mathbb{I}_{i,j}, \lambda_{i,j}, \mu_{i,j}, \mathbf{S}_{i,j}, \mathbf{W} : \langle i,j\rangle \in W\rangle$ (as in XV 3.1), $\mathbb{I} = \cup\{\mathbb{I}_{i,j} : \langle i,j\rangle \in W\}$, and $S \subseteq \mathbf{S}_{i,j}$ for $\langle i,j\rangle \in W$, each $\underset{\sim}{Q}_i$ with pure (θ_1,θ_2)-decidability and

$(*)$ $(\theta_1, \theta_2) \in \{(\aleph_0, 2), (2,2)\}$ and† if $(\theta_1,\theta_2) = (2,2)$ then for each T, x, k there is $F \in {}^\omega\omega$ such that

$$\forall\eta[xRT \ \& \ \eta \in {}^{\omega>}\omega \ \& \ (\exists k)(\eta(k) \geq F(k)) \ \& \ \eta \restriction k \in T) \Rightarrow \eta \in T]$$

† The meaning of this is like 1.11(2), i.e. we are not interested in all ${}^\omega\omega$ just in $\{\eta : \eta \in {}^\omega\omega$ and $\eta \leq g$ (i.e. $(\forall n)(\eta(n) \leq g(n))\}$.

and

(**) $|D| \leq \aleph_1$ or at least every regular uncountable $\kappa \leq |D|$ belongs to **S**.

If P_i is purely (D, R)-preserving for each $i < \delta$ (δ a limit ordinal) *then* P_δ is purely (D, R)-preserving.

1.12A Remark. If $\leq_{\text{pr}}^{Q_i} = \leq^{Q_i}$, the pure (θ_1, θ_2)-decidability is always trivially true for $(\aleph_0, 2)$ and even $(\infty, 2)$.

Proof: By XV 1.7 it is enough to consider only the cases $\delta = \omega, \delta = \omega_1, \delta$ strongly inaccessible $\bigwedge_{i<\delta} \delta > |P_i|$. In the last two cases $\mathbb{R}^{V^{P_\delta}} = \bigcup_{i<\delta} \mathbb{R}^{V^{P_i}}$ so w.l.o.g. $\delta = \omega$.

Suppose $p \in P_\omega, \underset{\sim}{f}$ a P_ω-name, $p \Vdash_{P_\omega}$ "$\underset{\sim}{f} \in {}^\omega\omega$" and $z \in \text{Dom}(R)$.

By 1.11 above (using part (1) if $(\theta_1, \theta_2) = (\aleph_0, 2)$ and using part (2) and (*) if $(\theta_1, \theta_2) = (2, 2)$) w.l.o.g $\underset{\sim}{f}(k)$ is a P_k-name of a natural number.

For notational simplicity we shall write the members of P_n as $\langle \underset{\sim}{q}_\ell : \ell < n \rangle, \Vdash_{P_\ell}$ "$\underset{\sim}{q}_\ell \in \underset{\sim}{Q}_\ell$" and similarly for P_ω. Let $p = \langle \underset{\sim}{q}_\ell : \ell < \omega \rangle$, and let for $m < n$ (in V^{P_m}) $P_n / P_m = \underset{\sim}{Q}_m * \underset{\sim}{Q}_{m+1} * \dots * \underset{\sim}{Q}_{n-1}$.

Now we define by induction on $n < \omega$, a condition $p^n \in P_n$ such that $p^n = \langle \underset{\sim}{q}_0^n, \dots, \underset{\sim}{q}_{n-1}^n \rangle$, and for each $m \leq n$ a P_m-name $\underset{\sim}{t}_{n,m}$ such that:

α) $p \restriction n \leq_{\text{pr}} p_n$, and $p_n \leq_{\text{pr}} p_{n+1} \restriction n$, moreover

$$\Vdash_{P_\ell} \text{"}\underset{\sim}{q}_\ell \leq_{\text{pr}} \underset{\sim}{q}_\ell^n \leq_{\text{pr}} \underset{\sim}{q}_\ell^{n+1}\text{"}.$$

β) If $G_m \subseteq P_m$ is generic (over V) $m \leq n$, then in $V[G_m]$ we have

$$\langle \underset{\sim}{q}_m^n, \dots, \underset{\sim}{q}_{n-1}^n \rangle \Vdash_{P_n/P_m} \text{"}\underset{\sim}{f}(n) = \underset{\sim}{t}_{n,m}[G_m]\text{"};$$

so $\underset{\sim}{t}_{n,n} = \underset{\sim}{f}(n)$. Equivalently, $\langle \emptyset, \emptyset, \dots, \emptyset, \underset{\sim}{q}_m^n, \dots, \underset{\sim}{q}_{n-1}^n \rangle \Vdash_{P_n}$ "$\underset{\sim}{f}(n) = \underset{\sim}{t}_{n,m}$".

γ) $\Vdash_{P_m} \left[\underset{\sim}{q}_m^n \Vdash_{\underset{\sim}{Q}_m} \text{"}\underset{\sim}{t}_{n,m+1} = \underset{\sim}{t}_{n,m}\text{"} \right]$

This is easily done: define $\langle \underset{\sim}{q}_\ell^n : \ell < n < \omega \rangle$ by induction on n, for each n let $\underset{\sim}{t}_{n,n} = \underset{\sim}{f}(n)$ and define $\underset{\sim}{q}_\ell^n, \underset{\sim}{t}_{\ell,n}$ by downward induction on $\ell < n$.

Let $\underset{\sim}{f}_m$ be the P_m-name of a function from ω to ω defined by: for $n \geq m$ we have: $\underset{\sim}{f}_m(n) = \underset{\sim}{t}_{m,n}$ and for $n < m$ we have: $\underset{\sim}{f}_m(n) = \underset{\sim}{f}(n)$. So clearly we have:

$\delta)$ $\langle \emptyset, \ldots, \emptyset, q^n_m \rangle \Vdash_{P_{m+1}}$ "$\underset{\sim}{f}_m \restriction n = \underset{\sim}{f}_{m+1} \restriction n$".

$\varepsilon)$ $\Vdash_{P_n}$ "$\underset{\sim}{f}_n \restriction n = \underset{\sim}{f} \restriction n$".

By Definition 1.6(4)(d) there are $x_n (n < \omega)$ such that

$$(D, R) \models \psi_{dis}(\langle x_0, \ldots \rangle, z).$$

Now let χ be large enough, and we split our requirement according to the kind of iteration. (The cases are from 0.1, cases A,B of 0.1 are covered by the later cases).

Let N be countable (the cases listed cover all possibilities):

Case D or C: $N \prec (H(\chi), \in, <^*_\chi)$, N is countable such that $(*)$ below holds.

Case E,F,G: Let $\langle N_\eta : \eta \in (T, \mathbf{I}) \rangle$ be an $(\mathbb{I}, \mathbf{W})$-suitable tree of models, $N = N_{<>}$ such that

$(*)$ $\langle P_\ell, \underset{\sim}{Q}_\ell : \ell < \omega \rangle$, $P \in N$, and also $(D, R), \underset{\sim}{f}, \langle q^n_\ell : \ell < n < \omega \rangle$, $\langle \underset{\sim}{t}_{n,m} : m \leq n < \omega \rangle$ belong to N and $N \cap \omega_1 \in \mathbf{W}$ in cases E, F, G.

Let $\langle T_{n,j} : j < \omega \rangle$ enumerate $\{T \in D \cap N : x_n R T\}$ and η^* be f_0 (which is a P_0-name, i.e. a function in V). Now let $\langle T^\alpha : \alpha \leq \omega \rangle$ be as guaranteed in $(*)$ of 1.6(3).

We now define (in V!) by induction on n conditions $r^n = \langle r_0, \ldots, r_{n-1} \rangle \in P_n$ (so trivially $r^n = r^{n+1} \restriction n$) such that:

a) $p \restriction n \leq_{\mathrm{pr}} r^n$,

b) $r^n \Vdash$ "$\underset{\sim}{f}_n \in (\lim T^n) \cap (\lim T^\omega)$",

c) *Case D*: r^n is generic for (N, P_n).

Case C: r^n is semi-generic for (N, P_n).

Case E,F,G: for some P_n-name $\underset{\sim}{\eta}_n$, letting $N_{\underset{\sim}{\eta}_n} = \bigcup_{k<\omega} N_{\underset{\sim}{\eta}_n \restriction k}$

we have: r^n is semi-generic for $(N_{\underset{\sim}{\eta}_n}, P_n)$

and $r^n \Vdash_{P_n}$ "$N_{\underset{\sim}{\eta}_n}[G_{P_n}]$ is $(\bigcup_{l \geq n} \mathbb{I}_\ell)$-suitable model for χ"

If we succeed then we easily finish; clearly $r^\dagger = \langle r_0, r_1, \ldots, r_n, \ldots\rangle$ satisfies $p \leq_{\rm pr} r^\dagger$; also for $n < \omega$:

$$(r^\dagger \restriction n) \Vdash_{P_n} \text{“}\underset{\sim}{f} \restriction n = \underset{\sim}{f}_n \restriction n\text{”}.$$

Hence $(r^\dagger \restriction n) \Vdash_{P_n}$ “$\underset{\sim}{f} \restriction n \in T^n \cap T^\omega$” and therefore $r^\dagger \Vdash_{P_\omega}$ “$\underset{\sim}{f} \in \lim T^\omega$ ”. As zRT^ω (by 1.6(3)(*)(ii)) clearly $r^\dagger, T^\omega$ are as required.

So we have just to carry out the induction. There is no problem for $n = 0$ (by the choice of η^*). So we have to do the induction step. Assume r^n is defined, and we shall define r^{n+1}.

Note. as P_{n+1} purely preserves (D, R) we can deduce:

⊗ $\underset{\sim}{Q}_n$ (in V^{P_n}) purely preserves (D, R).

Let $G_n \subseteq P_n$ be generic over $V, r^n \in G_n$, so $\underset{\sim}{f}_{n+1}$ becomes a $\underset{\sim}{Q}_n[G_n]$-name $\underset{\sim}{f}_{n+1}/G_n$ of a member of ${}^\omega\omega$. But $(D, R, <)$ is purely preserved by P_{n+1}, hence for every $q \in \underset{\sim}{Q}_n[G_n]$, and $y \in \mathrm{Dom}(R)$ there is a condition $q^\dagger$, $q \leq_{\rm pr} q^\dagger$ (where $q^\dagger \in \underset{\sim}{Q}_n[G_n]$), such that $q^\dagger \Vdash_{\underset{\sim}{Q}_n[G_n]}$ “$\underset{\sim}{f}_{n+1}/G_n \in \lim T^\dagger$ ” for some $T^\dagger \in D$ satisfying $yRT^\dagger$. Also there are $\langle q^\dagger_\ell : \ell < \omega\rangle$, and ν such that:

$\nu \in \lim T^\dagger$, in $\underset{\sim}{Q}_n[G_n]$ we have $q^\dagger \leq_{\rm pr} q^\dagger_0 \leq_{\rm pr} q^\dagger_1 \leq_{\rm pr} \ldots$ and $q^\dagger_\ell \Vdash_{\underset{\sim}{Q}_n[G_n]}$ “$\underset{\sim}{f}_{n+1} \restriction \ell = \nu \restriction \ell$”. [Why? $\nu = \underset{\sim}{f}[G_n]$ can serve.]

We can use choice functions, so let $\nu = F_1(q, z)$ and $q^\dagger_\ell = F_{2,\ell}(q, z)$ and $T^\dagger = F_0(q, z)$, and $q^\dagger = F_2(q, z)$. By our hypotheses (smoothness) in $V[G_n]$ we know that (D, R) is still a covering model. Note also that w.l.o.g. $F_0, F_1, F_{2,\ell}$ belong to $N[G_n]$. Remember (by (a)) that in Case F $N[G_n]$ is an $(\bigcup_{\ell \geq r} \mathbb{I}_\ell)$-suitable model for χ.

So now we apply condition c) of Definition 1.6(2) (the definition of a covering model) and get that in $V[G_n]$ the statement $\varphi^*_{\rm dis}(x_{n+1})$ holds. Look at the definition of $\varphi^*_{\rm dis}$ (1.6(1A)) and apply it to $\eta^* \stackrel{\rm def}{=} \underset{\sim}{f}_n[G_n]$ (which is an actual member of ${}^\omega\omega$ in $N[G_n]$), the function H with domain $D\times\omega$, $H(z, m) \stackrel{\rm def}{=} F_1(q^m_n, z)$ and the function F : $\mathrm{Dom}(R) \times \omega \to D$ defined by $F(z, m) \stackrel{\rm def}{=} F_0(q^m_n, z)$. So we get a tree $T^*_n \in D$, and an infinite set w_n as described there.

However note: $D \subseteq V$, so though T^*_n is defined in $V[G_n]$ it is an element of V. Working in V we have P_n-names $\underset{\sim}{T}^*_n, \underset{\sim}{w}_n$.

In fact without loss of generality $\underset{\sim}{T}^*_n \in N$, hence (by assumption (**) of 1.12 and condition (c) on r^n) we have $\langle r_0, \ldots, r_{n-1}\rangle \Vdash_{P_n}$ "$\underset{\sim}{T}^*_n \in D \cap N$" so for some P_n-name $\underset{\sim}{j} = \underset{\sim}{j}(n)$ (of a natural number) we have $\langle r_0, \ldots, r_{n-1}\rangle \Vdash_{P_n}$ "$\underset{\sim}{T}^*_n = T_{n,\underset{\sim}{j}}$". Now $\langle r_0, \ldots, r_{n-1}\rangle$ forces $\underset{\sim}{f}_n \in (\lim T^n) \cap (\lim T^\omega) \cap (\lim \underset{\sim}{T}^*_n) = (\lim T^n) \cap (\lim T^\omega) \cap (\lim T_{n,\underset{\sim}{j}(n)})$.

Hence, working in $V[G_n]$, by the choice of $\langle T^\alpha : \alpha \leq \omega\rangle$ (see 1.6(3)(iv)) there is $k < \omega$ (which depends on $\underset{\sim}{f}_n[G_n]$) such that:

(A) $\underset{\sim}{f}_n[G_n] \restriction k \lhd \rho \in T_{n,\underset{\sim}{j}(n)[G_n]} \cap T^\omega \Rightarrow \rho \in T^{n+1} \cap T^\omega$.

Now w.l.o.g. we can increase k, so w.l.o.g. $k \in \underset{\sim}{w}_n[G_n]$ (and $k > n$); (k was defined in $V[G_n]$). By the choice of q^k_n and the $\underset{\sim}{f}_\ell$'s:

(B) $q^k_n \Vdash_{Q_n[G_n]}$ "$\underset{\sim}{f}_n[G_n] \restriction k \lhd \underset{\sim}{f}_{n+1}$",

also by the choice of $F_0, F_1, F_2, F_{2,\ell}$:

(C) $F_{2,\ell}(q^k_n, x_{n+1}) \Vdash_{\underset{\sim}{Q}_n[G_n]}$ "$\underset{\sim}{f}_{n+1} \restriction \ell \lhd F_1(q^k_n, x_{n+1})$" and

(D) $q^k_n \leq F_{2,\ell}(q^k_n, x_{n+1}) \in \underset{\sim}{Q}_n[G_n] \cap N[G_n]$, and

(E) $H(x_{n+1}, k) = F_1(q^k_n, x_{n+1})$ and $F(x_n, n) = F_0(q^k_n, x_n)$.

Now by the choice of $T^*_n = T^*_{n,\underset{\sim}{j}[G_n]}$ for some ℓ

(F) $H(x_{n+1}, k) \restriction l \trianglelefteq \rho \in F(x_n, n) \Rightarrow \rho \in T_{n,\underset{\sim}{j}[G_n]}$

So together.

(G) $F_{2,l}(q^k_n, x_{n+1})$ is a member of $\underset{\sim}{Q}_n[G_n] \cap N[G_n]$, it is a pure extension of p_n and it forces $\underset{\sim}{f}_{n+1}$ (really $\underset{\sim}{f}_{n+1}[G_n]$) to belong to $\lim T^{n+1} \cap \lim T^\omega$.

Now we can choose $r_n, F_{2,l}(q^k_n, x_{n+1}) \leq_{\rm pr} r_n \in Q_n[G_n]$ to satisfy (c) thus finishing the induction and the proof. $\square_{1.12}$

So e.g.

1.13 Corollary. Suppose

(α) $\langle P_i, \underset{\sim}{Q}_j : i \leq \alpha, j < \alpha\rangle$ is a GRCS iteration as in XV 3.1. (i.e. 0.1F)

(β) (D, R) is a fine covering model,

(γ) $\Vdash_{P_i}$ "$\underset{\sim}{Q}_i$ is purely (D, R)-preserving"

(δ) D has cardinality $\aleph_1$ (or just $(**)$ of 1.12)

(ε) each $\underset{\sim}{Q}_i$ has pure $(\aleph_0, 2)$-decidability.

Then P_α is purely (D, R)-preserving.

Proof: We prove by induction on α that $\Vdash_{P_\alpha}$ "(D, R) is a smooth strong covering model" and P_α is purely (D, R)-preserving.

Case 1. $\alpha = 0$ By 1.8(1) we know (D, R) is a smooth strong covering model.

Case 2. $\alpha = \beta + 1$ By the induction hypothesis $V^{P_\beta} \models$ "(D, R) is a smooth strong covering model", as $\underset{\sim}{Q}_\beta$ is (D, R)-preserving $V^{P_\beta} \models [\Vdash_{Q_\beta}$ "(D, R) is a weak covering model"$]$, hence (see Definition 1.7(3)):

$V^{P_\beta} \models [\Vdash_{Q_\beta}$ "(D, R) is a strong covering model"$]$.

Let $\underset{\sim}{R}$ be a P_α-name of a (D, R)-preserving forcing notion; easily $\Vdash_{P_\beta}$ "$\underset{\sim}{Q} * \underset{\sim}{R}$ is (D, R)-preserving" so as above $\Vdash_{P_\beta}$ "$[\Vdash$ "$_{\underset{\sim}{Q}_\beta * \underset{\sim}{R}}$ (D, R) is a strong covering model"$]$" .

So in $V^{P_\alpha} = (V^{P_\beta})^{\underset{\sim}{Q}_\beta}$, for every (D, R)-preserving forcing notion $\underset{\sim}{R}$, $\Vdash_{\underset{\sim}{R}}$ "(D, R) is a strong covering model".

So in V^{P_α}, (D, R) is a smooth strong covering model. As for "P_α is purely (D, R)-preserving", by 1.10A it follows by the previous sentence and clause (α).

Case 3. α limit The real case, done in 1.12. $\square_{1.13}$

1.13A Corollary. Suppose:

(α) $\bar{Q}$ is a countable support iteration of proper forcing

(β) (D, R) is a fine covering model

(γ) $\Vdash_{P_i}$ "$\underset{\sim}{Q}_i$ is (D, R)-preserving "

Then P_α is (D, R)-preserving.

1.13B Remark. 1) We have parallel conclusions to 1.13 weakening (ε) to

(ε)′ "$\underset{\sim}{Q}_i$ has (2,2)-decidability "

if we add the requirement from 1.12($*$) for $(\theta_1, \theta_2) = (2, 2)$.

2) We can have parallel conclusions to 1.13 weakening (ε) to

$(\varepsilon)''$ "$\underset{\sim}{Q}_i$ has $(\aleph_0, \aleph_0)$-decidability"

if we add

(ζ) each $\underset{\sim}{Q}_i$ is purely ${}^\omega\omega$-bounding.

1.14 Definition. 1) A class ($\equiv$ property) K of objects $(D, R, <)$ is a fine class of covering models if:

(i) each member satisfies $(\alpha), (\beta), (\gamma)$ of Definition 1.2.

(ii) *if* Q is a forcing notion, K-preserving (i.e. each $(D, R, <) \in K^V$ is a weak covering model even in V^Q) *then* in V^Q: each $(D, R, <) \in K^V$ is in K^{V^Q} and satisfies (γ) of Definition 1.2(1); note that clauses (α), (β) of 1.2(1) follows.

2) "K is a (smooth) (strong) class of covering models" are defined similarly.

1.15 Theorem. In 1.12, 1.13 (and 1.13B) we can replace the covering model by a class of covering models.

* * *

1.16 Definition.

1) $(\overline{D}, \overline{R})$ is a weak covering k^*-model if: $\overline{D} = \langle D_k : k < k^* \rangle$, $\overline{R} = \langle R_k : k < k^* \rangle$, $k^* < \omega$ and

(a) for each $k < k^*$, D_k is a set, R_k is a two place relation on D_k, xR_kT implies T is a closed subtree of ${}^{\omega>}\omega$.

(b) $(\overline{D}, \overline{R})$ covers, i.e. for every $\eta \in {}^\omega\omega$, for some $k < k^*$, η is of the k-th kind which means: for every $x \in \mathrm{Dom}(R_k) = \{x : (\exists T) x R_k T\}$ there is $T \in D_k$ such that xR_kT and $\eta \in \lim(T)$.

2) $(\overline{D}, \overline{R}, \overline{<})$ is a fine covering k^*-model if

(α) $(\overline{D}, \overline{R})$ is a weak covering k^*-model

(β) $\overline{<} = \langle <_k : k < k^* \rangle$, $<_k$ is a partial order on $\mathrm{Dom}(R_k)$ such that

(i) $(\forall y \in \mathrm{Dom}(R_k))(\exists x \in \mathrm{Dom}(R_k))(x <_k y)$

(ii) $(\forall y, x \in \mathrm{Dom}(R_k))(\exists z \in \mathrm{Dom}(R_k))(x <_k y \rightarrow x <_k z <_k y)$

(iii) if $y <_k x, yR_kT$ then for some $T^* \in D_k$, $T \subseteq T^*$ and xR_kT^*

(iv) if $y <_k x$ and for $\ell = 1, 2$ we have yR_kT_ℓ *then* there is $T \in D_k$ such that:

xR_kT, $T_1 \subseteq T$ and for some n, $[\nu \in T_2 \ \&\ \nu \restriction n \in T_1 \Rightarrow \nu \in T]$

(γ) (a) for each $k < k^*$ the following holds. If $x >_k x^\dagger >_k y_{n+1} >_k y_n$ for $n < \omega$ and $T_n \in D_k, y_nR_kT_n$ (for $n < \omega$) *then* there is $T^* \in D_k, xR_kT^*$ and an infinite set $w \subseteq \omega$ such that:

$$\lim T^* \supseteq \{\eta \in {}^\omega\omega: \text{ for every } i \in w, \eta \restriction \min(w \setminus (i+1)) \in \bigcup_{\substack{j<i \\ j \in w}} T_j \cup T_0\}$$

(b) if $k < k^*$, $\{\eta\} \cup \{\eta_n : n < \omega\} \subseteq {}^\omega\omega$, $\eta \restriction n = \eta_n \restriction n$ and $x <_k y$, and η, η_n are of the k-kind (see below), *then* for some $T \in D_k$ we have yR_kT & $\eta \in \lim(T)$ and for infinitely many n, $\eta_n \in \lim(T)$.

(δ) condition (γ) continues to hold in any generic extension in which (α) holds.

3) For a property X of forcing notions, $(\overline{D}, \overline{R}, \overline{<})$ is a fine covering k^*-model for X-forcing if Definition 1.16(2) holds when in (δ) we restrict ourselves to X-forcing notions only.

4) We say $(\overline{D}, \overline{R}, \overline{<})$ is a temporarily fine covering k^*-model if it satisfies $(\alpha), (\beta), (\gamma)$ i.e. it is a fine covering k^*-model for trivial forcing.

5) We say $\eta \in {}^\omega\omega$ is of (k, x)-kind (or just the x-th kind when $\langle \mathrm{Dom}(R_k) : k < k^* \rangle$ are pairwise disjoint) if there is T such that $\eta \in \lim(T)$ and xR_kT (note: $(\bar{D}, \bar{R})$ covers iff for any $\eta \in {}^\omega\omega$ and $\bar{x} = \langle x_k : k < k^* \rangle \in \prod_{k<k^*} \mathrm{Dom}(R_k)$ for some k, the sequence η is of the (k, x_k)-kind). We say η is of the k-th kind if it is of the (k, x)-kind for every $x \in \mathrm{Dom}(R_k)$.

For simplicity we restrict ourselves to the fine case (and not the parallel of smooth strong covering).

1.17 Theorem. Assume $(\overline{D}, \overline{R}, \overline{<})$ is a fine covering k^*-model.

1) If $\overline{Q} = \langle P_i, \underset{\sim}{Q}_j : i \leq \delta, j < \delta \rangle$ is a CS iteration, each $\underset{\sim}{Q}_j$ preserves $(\overline{D}, \overline{R}, \overline{<})$ then so does P_δ

2) Similarly for other iterations as in 0.1 (with pure preserving).

Proof. For simplicity $\mathrm{Dom}(R_k)$ are pairwise disjoint so let $<= \bigcup_{k<k^*} <_k$. We concentrate on part 1). By V 4.4, if δ is of uncountable cofinality then there is no problem, as all new reals are added at some earlier point. So we may suppose that $\mathrm{cf}(\delta) = \aleph_0$ hence by associativity of CS iterations of proper forcing (III) without loss of generality $\delta = \omega$.

We claim that $\Vdash_{P_\omega}$ "$(\overline{D}, \overline{R}, \overline{\leq})$ covers." (Note that this suffices for the proof of the theorem.)

So let p^* be a member of P_ω and $\underset{\sim}{f}$ a P_ω-name such that $p^* \Vdash_{P_\omega}$ "$\underset{\sim}{f} \in {}^\omega\omega$", and $x_k^* \in \mathrm{Dom}(R_k)$ for $k < k^*$. It suffices to prove that for some k, T, and p we have: $p^* \leq p \in P_\omega$, $x_k R_k T_k$ (so $T_k \in D_k$) and $p \Vdash_{P_\omega}$ "$\underset{\sim}{f} \in \lim(T_k)$". As we can increase p^* w.l.o.g. above p^*, for every n, $\underset{\sim}{f}(n)$ is a P_n-name. Let χ be large enough and let N be a countable elementary submodel of $(H(\chi), \in, <^*_\chi)$ to which $\{x_0, \ldots, x_{k^*-1}, p^*, \underset{\sim}{f}, \overline{Q}\}$ belongs.

For clarity think that our universe V is countable in the true universe or at least $\beth_3(|P_\omega|)^V$ is. We let $K = \{(n,p,G) : n < \omega,\ p \in P_\omega$ is above p^*, $G \subseteq P_n$ is generic over V and $p{\restriction}n \in G\}$. On K there is a natural order: $(n,p,G) \leq (n',p',G')$ if $n \leq n'$, $P_\omega \models p \leq p'$ and $G \subseteq G'$. Also for $(n,p,G) \in K$ and $n' \in (n,\omega)$ there is G' such that $(n,p,G) \leq (n',p,G')$ as $\bar{Q}$ is an iteration of proper forcing notions. Also if $(n,p,G) \in K$ and $p \leq p' \in P_\omega/G$ (i.e. $p' \in P_\omega$ and $p'{\restriction}n \in G$) then $(n,p,G) \leq (n,p',G)$. For $(n,p,G) \in K$ let $L_{(n,p,G)} = \{g : g \in ({}^\omega\omega)^{V[G]}$ and there is an increasing sequence $\langle p_\ell : \ell < \omega\rangle$ in $V[G]$ of conditions in P_ω/G, $p \leq p_0$, such that $p_\ell \Vdash \underset{\sim}{f}{\restriction}\ell = g{\restriction}\ell\}$. So:

$(*)_1$ $K \neq \emptyset$

$(*)_2$ $(n,p,G) \in K \Rightarrow L_{(n,p,G)} \neq \emptyset$

$(*)_3$ $g \in L_{(n,p,G)} \Rightarrow (\underset{\sim}{f}{\restriction}n)[G] = g{\restriction}n$.

Note also

$(*)_4$ $L_{(n,p,\underset{\sim}{G}_n)}$ is a P_n-name.

$(*)_5$ if $(n,p,G) \leq (n',p',G')$ then $L_{(n',p',G')} \cap V \subseteq L_{(n,p,G)}$

1.17A Fact. There are $k < k^*$ and $(n,p,G) \in K$ such that if $(n,p,G) \leq (n',p',G') \in K$ *then* for some $(n'',p'',G'') \in K$, $(n',p',G') \leq (n'',p'',G'')$ there is $g \in L_{(n'',p'',G'')}$ which is of the k'th kind.

[Why? otherwise choose by induction (n^ℓ, p^ℓ, G^ℓ) for $\ell \le k^*$, in K, increasing such that: $L_{(n^{\ell+1}, p^{\ell+1}, G^{\ell+1})}$ has no members of the ℓ'-kind for $\ell' \le \ell$. So $L_{(n^{k^*}, p^{k^*}, G^{k^*})} = \emptyset$, a contradiction.]

So choose k and $(n^\otimes, p^\otimes, G^\otimes) \in K$ as in the fact, w.l.o.g. $n^\otimes = 0$. Remember that $\underset{\sim}{f}(n)$ is a P_n-name for each n.

1.17B Fact. If $(n^\otimes, p^\otimes, G^\otimes) \le (n, p, G) \in K$ and $x \in \mathrm{Dom}(R_k)$ then there is $g \in L_{(n,p,G)}$ which is of the (k, x)-kind.

Proof. By the choice of $(n^\otimes, p^\otimes, G^\otimes)$ there is $(n', p', G') \in K$ such that $(n, p, G) \le (n', p', G')$ and $L_{(n', p', G')}$ has a member g of the k-th kind. So there are $T \in \mathrm{Rang}(R_k)$ and $\langle p'_\ell : \ell < \omega \rangle$ such that $g \in \lim(T)$, $x R_k T$, $p'_0 = p'$, $p'_\ell \le p'_{\ell+1}$, $p'_\ell \in P_\omega / G'$, $p'_\ell \Vdash_{P_\omega/G'}$ "$\underset{\sim}{f} \restriction \ell = g \restriction \ell$". Note that $T \in V$. From the point of view of $V[G]$, all this is just forced by some $q \in G'$, so q forces that $\langle \underset{\sim}{p}'_\ell : \ell < \omega \rangle$, T, $\underset{\sim}{g}$ are as above. So we can find $\langle q_\ell : \ell < \omega \rangle$, $q_\ell \in P_{n'}/G$, q_ℓ increasing, $q \le q_\ell$ and q_ℓ forces a value to $\underset{\sim}{p}'_\ell$, say p''_ℓ and to $\underset{\sim}{g} \restriction \ell$ and is above $p''_\ell \restriction n'$.

And we are done.

1.17C Fact. If $\mathcal{T} \subseteq \mathrm{Rang}(R_k)$ is countable and $x <_k y$, and $(\forall T \in \mathcal{T})(\exists z \le_k x)(z R_k T)$ and $T^0 \in \mathcal{T}$ then for some $T^1 \in \mathrm{Rang}(R_k)$ we have $y R_k T^1$, $T^0 \subseteq T^1$ and for each $T \in \mathcal{T}$ for some m we have:

$$(\forall \nu)(\nu \in T \ \&\ \nu \restriction m \in T^0 \Rightarrow \nu \in T^1).$$

Proof. Let $\langle T_n : n < \omega \rangle$ list $\mathcal{T}$ (possibly with repetitions) such that $T_0 = T^0$. Let $x <_k x' <_k y$, choose inductively x_n, $x <_k x_n <_k x_{n+1} <_k x'$ (possible by clause (β)(ii) of Definition 1.16(2)). Choose inductively $T'_n \in \mathrm{rang}(R_k)$ such that $T'_0 = T_0 = T^0$ and $x_n R_k T'_n$, $T'_n \subseteq T'_{n+1}$ and for some $k_n < \omega$ we have: $\nu \in T_n$, $\nu \restriction k_n \in T'_n \Rightarrow \nu \in T'_{n+1}$ (possible by clause (β)(iv) of Definition 1.16(2)). Choose, for each n, $T''_n \in \mathrm{Rang}(R_k)$ such that $x' R_k T''_n$, $T'_n \subseteq T''_n$ (possible by clause (β)(iii) of Definition 1.16(2)). Next use 1.16(1)(γ)(a) to find an infinite $w \subseteq \omega$ and $T^1 \in \mathrm{Rang}(R_k)$ such that yRT^1, $T_0'' \subseteq T^1$ and

$$i \in w \ \&\ \nu{\restriction}\min(w \setminus (i+1)) \in \bigcup_{j<i, j \in w} T_j'' \cup T_0'' \Rightarrow \nu \in T^1.$$

Check that T^1 is as required. $\square_{1.17C}$

Continuation of the proof of 1.17.

Choose $x' < x_k^*$ and then inductively on n choose x_n such that $x_n <_k x'$ $x_n <_k x_{n+1}$, and choose a countable $N \prec (H(\chi), \in)$ (with $\chi = \mathrm{cf}(\chi) > \beth_\omega(|P_\omega|)$) such that all the elements $\langle x_n : n < \omega \rangle$, $\langle P_n, \underset{\sim}{Q}_n : n < \omega \rangle$, $\underset{\sim}{f}$, $p^\otimes$ belong to N. Now, working in V, we choose by induction on n sequences $\langle p_\eta : \eta \in {}^n\omega \rangle$, $\langle \underset{\sim}{f}_\eta : \eta \in {}^n\omega \rangle$, $\langle q_\eta : \eta \in {}^n\omega \rangle$, and T_n such that

(A) p_η is a $P_{\ell g(\eta)}$-name of a member of $P_\omega \cap N$, $p_{\langle\rangle} = p^\otimes$, $p_\eta \le p_{\eta \frown \langle \ell \rangle}$, $p_\eta{\restriction}n \le q_{\eta{\restriction}n}$.

(B) q_η is $(N, P_{\ell g(\eta)})$-generic, $q_\eta \in P_{\ell g(\eta)}$ and $[\ell < \ell g(\eta) \Rightarrow q_\eta{\restriction}\ell = q_{\eta{\restriction}\ell}]$.

(C) $\underset{\sim}{f}_\eta$ is a $P_{\ell g(\eta)}$-name of a member of ${}^\omega\omega$ and $q_\eta \Vdash_{P_{\ell g(\eta)}}$ "$\underset{\sim}{f}_\eta \in \lim(T_n) \cap N[G_{P_{\ell g(\eta)}}]$ is of the (k, x_{3n})-kind and belongs to $L_{(n, p_\eta[\underset{\sim}{G}_{P_{\ell g(\eta)}}], \underset{\sim}{G}_{P_{\ell g(\eta)}})}$" when $\eta \in {}^n\omega$.

(D) $x_{3n} R_k T_n$ and $T_n \subseteq T_{n+1}$.

(E) $p_{\eta \frown \langle \ell \rangle} \Vdash_{P_\omega}$ "$\underset{\sim}{f}_\eta{\restriction}\ell = \underset{\sim}{f}_{\eta \frown \langle \ell \rangle}{\restriction}\ell = \underset{\sim}{f}{\restriction}\ell$".

Suppose we succeed in this endeavour. By (β)(iii) of 1.16(1) we can find T_n' such that $T_n \subseteq T_n'$, $x' R_k T_n'$ (as $x_{3n} <_k x'$). Let w and T^* be as guaranteed by clause (γ)(a) of Definition 1.16(1) (for $\langle T_n' : n < \omega \rangle$, x', x) and let $\langle n_i : i < \omega \rangle$ be the increasing enumeration of w. So $x R_k T^*$ and: if $\eta{\restriction}n_{i+1} \subseteq \bigcup_{j<i} T_{n_j}' \cup T_0'$ for each $i < \omega$ then $\eta \in T^*$.

Let $g(i) \overset{\mathrm{def}}{=} n_i$. Let $\nu = \langle n_{2j+1} : j < \omega \rangle$.

So it is enough to prove that for some $q \in P_\omega$ which is above $p^\otimes$, we have $q \Vdash_{P_\omega}$ "$\underset{\sim}{f} \in \lim(T^*)$". We choose $q \in P_\omega$ by $q{\restriction}i = q_{\nu{\restriction}i}$, by clause (B) we have: $q \in P_\omega$ is well defined and above each $q_{\nu{\restriction}i}$ and above each $p^\otimes{\restriction}i$ hence above $p^\otimes$.

We just have to prove: $q \Vdash$ "$\underset{\sim}{f}{\restriction}n_{i+1} \in \bigcup_{j<i} T_{n_j}' \cup T_0'$". As $q{\restriction}(i+1) = q_{\nu{\restriction}(i+1)}$, by clause (A) we have $p_{\nu{\restriction}(i+1)} \le q{\restriction}(i+1)$; by clause (E) letting $\eta = \nu{\restriction}i$, $\ell = \nu(i)$

we have $q \Vdash_{P_\omega}$ "$\underset{\sim}{f}_\eta \restriction \ell = \underset{\sim}{f}_{\eta \frown \langle \ell \rangle} \restriction \ell = \underset{\sim}{f} \restriction \ell$", but $\ell = \nu(i) = n_{2i+1}$, so we have $q \Vdash$ "$\underset{\sim}{f} \restriction n_{2i+1} = \underset{\sim}{f}_{\nu \restriction i} \restriction n_{2i+1}$"; now by clause (C) applied to $\eta = \nu \restriction i$ remembering $T_n \subseteq T'_n$ we have $q \Vdash$ "$\underset{\sim}{f}_{\nu \restriction i} \in \lim(T'_i)$" hence by the last two statements $q \Vdash$ "$(\underset{\sim}{f} \restriction n_{2i+1}) \in T''_i$". So, as $n_j < n_{j+1}$, for $i = 0$ we have $q \Vdash$ "$\underset{\sim}{f} \restriction n_i \in T''_0$", and for $i > 0$ we have $q \Vdash$ "$\underset{\sim}{f} \restriction n_{2i+1} \in T'_i \subseteq T'_{2i-1}$" and for $i \geq 0$ we have $q \Vdash$ "$\underset{\sim}{f} \restriction n_{2i+2} \trianglelefteq \underset{\sim}{f} \restriction n_{2i+3} \in T'_{i+1} \subseteq T'_{2i}$" so $q \Vdash$ "$\underset{\sim}{f} \restriction n_{i+1} \in \bigcup_{j<i} T'_{n_j} \cup T''_0$" holds (check by cases).

Hence we have finished proving $\Vdash_{P_\omega}$ "$(\overline{D}, \overline{R})$ covers ${}^\omega\omega$". So it suffices to carry out the induction.

There is no problem for $n = 0$.

Let us deal with $n + 1$. By fact 1.17C (above) there are $T_{n,i} \in \mathrm{Rang}(R_k)$ for $i < \omega$ such that

(*) (i) $T_{n,0} = T_n$

(ii) $T_{n,i} \subseteq T_{n,i+1}$

(iii) $x_{3n+i} R_k T_{n,i}$

(iv) if $T \in (\mathrm{Rang}(R_k)) \cap N$ and $(\exists z)(z \leq_k x_{3n+i}$ & $z R_k T)$ then for some $m = m_T < \omega$ we have $\nu \in T$ & $\nu \restriction m \in T_{n,i} \Rightarrow \nu \in T_{n,i+1}$.

Let $T_{n+1} = T_{n,3}$.

Next we define $p_{\eta \frown \langle \ell \rangle}$, $\underset{\sim}{f}_{\eta \frown \langle \ell \rangle}$, $q_{\eta \frown \langle \ell \rangle}$ for $\eta \in {}^n\omega$, $\ell < \omega$. It is enough to define then in $N[G_{P_n}]$ where G_{P_n} is any generic subset of P_n to which q_η belongs (note that e.g. $p_{\eta \frown \langle \ell \rangle}$ is a P_{n+1}-name, and if $q_\eta \notin G_{P_n}$ the requirements on it are trivial to satisfy).

Let $\eta \in {}^n\omega$, and let G_{P_n} be a subset of P_n generic over V such that $q_\eta \in G_{P_n}$. So now p_η is in $(P_\omega / G_{P_n}) \cap N[G_{P_n}]$, and $f_\eta \stackrel{\mathrm{def}}{=} \underset{\sim}{f}_\eta[G_{P_n}]$ is a member of ${}^\omega\omega$ of the (k, x_{3n})-kind which belongs to $L_{(n, p_\eta, G_{P_n})}$, moreover $f_\eta \in N[G_{P_n}]$. So in $N[G_{P_n}]$ there is an increasing sequence $\langle p^0_{\eta \frown \langle \ell \rangle} : \ell < k \rangle$ of members of P_ω / G_{P_n}, $p_\eta = p^0_{\eta \frown \langle 0 \rangle}$, $p^0_{\eta \frown \langle \ell \rangle} \Vdash_{P_\omega / G_{P_n}}$ "$\underset{\sim}{f} \restriction \ell = f_\eta \restriction \ell$" w.l.o.g. $p^0_{\eta \frown \langle \ell \rangle} \restriction n = p_\eta \restriction n$. If $G_{P_{n+1}} \subseteq P_{n+1}$ is generic over V extending G_{P_n} and $p^0_{\eta \frown \langle \ell \rangle} \restriction (n+1) \in G_{P_{n+1}}$ then $(n+1, p^0_{\eta \frown \langle \ell \rangle}, G_{P_{n+1}}) \geq (n^\otimes, p^\otimes, G^\otimes)$ is from K, so by Fact 1.17B there are $f_{\eta,\ell} \in ({}^\omega\omega)^{V[G_{P_{n+1}}]}$ and an increasing sequence $\langle p^1_{\eta \frown \langle \ell \rangle, j} : j < \omega \rangle$ of conditions from $P_\omega/G_{P_{n+1}}$ starting with $p^0_{\eta\char`^\langle\ell\rangle}$ such that $p^1_{\eta\char`^\langle\ell\rangle,j} \Vdash$ "$\underset{\sim}{f}\restriction j = f_{\eta,\ell}\restriction j$" and $f_{\eta,\ell}$ is of the (k,x_{3n})-kind, say $f_{\eta,\ell} \in \lim(T^0_{\eta,\ell})$, $x_{3n}R_kT^0_{\eta,\ell}$.

Letting $Q_n = \underset{\sim}{Q}_n[G_{P_n}]$ we have Q_n-names for these objects so $\langle \underset{\sim}{f}_{\eta,\ell} : \ell < \omega\rangle$ is a Q_n-name of an ω-sequence of members of ${}^\omega\omega$ of the (k,x_{3n})-kind and $\underset{\sim}{T}^0_\eta$ and $\langle \underset{\sim}{p}^1_{\eta\char`^\langle\ell\rangle,j} : j<\omega\rangle$ are $\underset{\sim}{Q}_n$-names as above.

W.l.o.g. $\langle(\underset{\sim}{f}_{\eta,\ell}, \underset{\sim}{T}^0_{\eta,\ell}, \langle \underset{\sim}{p}^1_{\eta\char`^\langle\ell\rangle,j} : j<\omega\rangle) : \ell<\omega\rangle \in N[G_{P_n}]$.

So we can find $\langle (p^1_{\eta,\ell}, T^1_{\eta,\ell}) : \ell<\omega\rangle$ such that:

$$Q_n \vDash \text{“}p^0_{\eta,\ell} \le p^1_{\eta,\ell}(n)\text{”},$$

$$p^1_{\eta,\ell} \Vdash_{Q_n} \text{“}\underset{\sim}{T}^0_\eta = T^1_{\eta,\ell} \text{ hence } \underset{\sim}{f}_{\eta,\ell} \in \lim(T^1_{\eta,\ell})\text{”}$$

$$\text{and } x_{3n}R_kT^1_{\eta,\ell}.$$

Also we can find $g^1_{\eta,\ell}$, $p^1_{\eta,\ell,j}$ ($\ell<\omega$, $j<\omega$) such that $p^1_{\eta,\ell,0} = p^1_{\eta,\ell}$, $p^1_{\eta,\ell,j} \le p^1_{\eta,\ell,j+1}$ and $p^1_{\eta,\ell,j} \Vdash_{P_\omega/G_{P_n}}$ "$\underset{\sim}{f}_{\eta,\ell}\restriction j = g_{\eta,\ell}\restriction j$" where $g_{\eta,\ell} \in {}^\omega\omega$ necessarily $g_{\eta,\ell}\restriction j \in T^1_{\eta,\ell}$ hence $g_{\eta,\ell} \in \lim(T^1_{\eta,\ell})$. W.l.o.g. $\langle p^1_{\eta,\ell,j} : \ell<\omega, j<\omega\rangle$, $\langle g_{\eta,\ell} : \ell<\omega\rangle$ belongs to $N[G_{P_n}]$. So $g_{\eta,\ell}$, $f_\eta \in {}^\omega\omega$ are of the (k,x_{3n})-kind and $g_{\eta,\ell}\restriction\ell = f_\eta\restriction\ell$, so by clause (γ)(b) of Definition 1.16(2), there is $T^2_\eta \in \mathrm{Rang}(R_k)$ such that $x_{3n}R_kT_\eta$ and $B^\eta_0 = \{\ell<\omega : g_{\eta,\ell} \in \lim(T^2_\eta)\}$ is infinite.

Now as $f_\eta \in \lim(T_n)$, $x_{3n}R_kT^2_\eta$ and $T^2_\eta \in N[G_{P_n}] \cap \mathrm{Rang}(R_k)$ clearly, by $(*)$ above, for some $m_\eta<\omega$, $f_\eta\restriction m_\eta \trianglelefteq \nu \in T^2_\eta \Rightarrow \nu \in T_{n+1}$. Hence

$$\ell \in B^\eta_0 \;\&\; \ell \ge m_\eta \;\Rightarrow\; g_{\eta,\ell} \in \lim(T^2_\eta) \;\&\; g_{\eta,\ell}\restriction m_\eta = f_\eta\restriction m_\eta \;\Rightarrow\; g_{\eta,\ell} \in \lim(T_{n,1}).$$

As $x_{3n}R_kT^1_{\eta,\ell}$, $T^1_{\eta,\ell} \in N[G_{P_n}] \cap \mathrm{Rang}(R_k)$ clearly for some $m_{\eta,\ell} \in (m_\eta,\omega)$ we have $g_{\eta,\ell}\restriction m_{\eta,\ell} \triangleleft \nu \in T^1_{\eta,\ell} \Rightarrow \nu \in T_{\eta,2}$ and hence

$$p_{\eta,\ell,m_{\eta,\ell}} \Vdash_{Q_n} \text{“}\underset{\sim}{f}_{\eta,\ell}\restriction m_{\eta,\ell} = g_{\eta,\ell}\restriction m_{\eta,\ell} \text{ and } \underset{\sim}{f}_{\eta,\ell} \in \lim(T^1_{\eta,\ell})\text{”}.$$

Thus $p_{\eta,\ell,m_{\eta,\ell}} \Vdash_{Q_n}$ "$\underset{\sim}{f}_{\eta,\ell} \in \lim(T_{n,2})$". (Note that $\underset{\sim}{B}^\eta_0$, $\underset{\sim}{T}^1_{\eta,\ell}$, $\underset{\sim}{m}_n$, $\underset{\sim}{m}_{\eta,\ell}$ are P_n-names.)

Now, at last, we define $p_{\eta\,\hat{}\,\langle i\rangle}$ for $i < \omega$. So $p_{\eta\,\hat{}\,\langle i\rangle}\restriction n = p_\eta$, and we define $p_{\eta\,\hat{}\,\langle i\rangle}(n)$ in $V[G_{P_n}]$ where $q_\eta \in G_{P_n}$ (justified above). Let $\ell(i)$ be the ℓ-th member of $B^\eta_o \setminus m_\eta$, and $p_{\eta\,\hat{}\,\langle i\rangle}(n) = p_{\eta,\ell(i),m_{\eta,\ell(i)}}$ and $\underset{\sim}{f}_{\eta\,\hat{}\,\langle i\rangle}$ be $\underset{\sim}{f}_{\eta,\ell(i)})$.

Lastly let $q_{\eta\,\hat{}\,\langle i\rangle} \in P_{n+1}$ be such that $q_{\eta\,\hat{}\,\langle i\rangle}\restriction = q_\eta$, $q_{\eta\,\hat{}\,\langle i\rangle}$ above $p_{\eta\,\hat{}\,\langle i\rangle}$ and is (N, P_{n+1})-generic (possible as in the proof of preservation of properness by iteration. $\square_{1.17}$

§2. Examples

In this section we use the machinery from the previous section. First (2.1–2.7) we try to restate the results in a way easier to apply by putting more of the common part of the examples in the general results, but you can deal directly with the examples i.e. you can essentially ignore 2.4-2.5, start with 2.7, and use 1.15 (instead 2.1 - 2.5) but have to check somewhat more. Then we deal with several properties which we call: ${}^\omega\omega$-bounding property, Sacks property, Laver property, (f, g)-bounding and more. Several have been used (explicitly or implicitly) and we show that their preservation by countable support iteration follows from 1.13A (so actually from 1.12; really we use 1.15). We usually present the "classical" examples of such forcing.

Names (Sacks, Laver) come from the forcing which seems to be "the example" of a forcing with this property. However as Judah comments, maybe "Sacks property" is confusing as Sacks's forcing satisfies a stronger condition. For simplicity:

2.0 Convention. Forcing notions are from the first case of 0.1 (e.g. proper) and $V^\dagger$ subuniverse of V means, if not said otherwise, $V = (V^\dagger)^Q$, Q as above.

2.1 General Discussion and Scheme.

For usual notions we have two variants of the preservation theorem. We first define a family K of candidates for covering models, usually they have all the same definition, φ but applied in *some* subuniverse $V^\dagger$ (with the same $\aleph_1$) and we get $\varphi(V^\dagger)$, and demand that it is a weak covering model (or a family of covering models; this restricts the family of $V^\dagger$; we can further restrict ourselves to the case $V = V^\dagger[G]$ where G is a subset of some forcing notion $P \in V^\dagger$ generic over V). We then write $K = K_\varphi$ (φ - the definition, possibly with parameters). Then we prove:

(A) any model from K_φ is actually a temporary fine covering model.

So

(B) if $(D, R, <) \in (K_\varphi)^V$ still covers in V^P *then* it is (in V^P) still a temporary fine covering model.

This implies that

(C) if $\bar{Q} = \langle P_j, Q_i : j \leq \alpha, i < \alpha \rangle$ is an iteration as in 0.1, α a limit ordinal, $(D, R, <) \in K_\varphi$ in V and for every $\beta < \alpha$ we have $\Vdash_{P_\beta}$ "$(D, R, <)$ still covers, so it is a weak covering model" *then* $(D, R, <)$ covers in V^{P_α}.

But we may want a nicer preservation theorem in particular dealing with the composition of two.

2.1A Definition. 1) For a formula $\varphi = \varphi_x$ (possibly with a free parameter x) defining for any universe $V^\dagger$ which satisfies $x \in V^\dagger$ a weak covering model $\varphi_x[V^\dagger]$ (the definition in $V^\dagger$) and a property Pr of forcing notions, we do the following. Let

$$K_\varphi^{Pr} = \varphi^{Pr}(V) = \{\varphi_x[V^\dagger] : V^\dagger \text{ a subuniverse of } V,\ V = (V^\dagger)^Q \text{ for some forcing notion } Q \text{ satisfying } \mathrm{Pr}, x \in V^\dagger, \varphi_x[V^\dagger] \text{ covers in } V, \text{ so } Q \text{ is } \varphi_x[V^\dagger]\text{-preserving}\},$$

so $\varphi_x[V]$ is a member of $\varphi^{Pr}(V)$. We omit Pr if Q fits into the appropriate case of 0.1 (see 2.0); for simplicity we concentrate on this case†.

2) A forcing notion P is K_φ^{Pr}-preserving or φ-preserving *if* it preserves each $(D, R, <) \in K_\varphi^{Pr}$. We may add "purely" to all of them.

3) Writing $D^\varphi, R^\varphi, <^\varphi$ we mean $\varphi[V] = (D^\varphi, R^\varphi, <^\varphi)$; if φ has a free parameter x and a fixed parameter t we write $\varphi_t[V; x]$, or $\varphi_{t,x}[V]$.

---

† it is reasonable to deal only with Pr preserved by the relevant iterations, and everything is similar.

2.2 Restatement of Definition.

1) φ is a temporarily definition of weak covering models if (each instance satisfies):

   (α) (a) 1.1(a)

   (b) 1.1(b) (i.e. $\varphi[V^\dagger]$ covers in the $V^\dagger$ in which we define)

1A) φ is a temporary fine definition of covering models if (α) (above) and in addition:

   (β) 1.2(1)(β)

   (γ) 1.2(1)(γ) i.e. $\varphi[V^\dagger]$ satisfies it in $V^\dagger$

2) φ is a fine definition of covering models if in addition:

   (δ) if $Q \in V$ is $\varphi(V)$-preserving (i.e. each member of $\varphi(V)$ covers in V^Q i.e. $(\alpha)(b)$ holds also in V^Q) *then* in V^Q still each member of $\varphi(V^Q)$ satisfies 1.2(1)(γ).

3) φ is a finer definition of covering models (for simplicity with no free parameter) *if* in addition:

   $(\alpha)(c)$ $<^\varphi$ is absolute for φ-preserving extensions i.e. if V^1 is a class of $V^2, \varphi(V^1)$ covers in V^2 (remember 2.1A(3) and 2.0), $x, y \in V^1$ then: $V^1 \models x <^\varphi y$ iff $V^2 \models x <^\varphi y$. Similarly for D^φ, R^φ.

   (ε) if Q is $\varphi(V)$-preserving, $\varphi[V] \models$ "$y < x$", and $T^* \in V^Q$ and $\varphi[V^Q] \models yRT^*$ *then* for some $T^{**} \in V$ we have: $T^* \subseteq T^{**}$ and $\varphi[V] \models xRT^{**}$ moreover

   $(\varepsilon)^+$ like (ε) above but Q is demanded only to be $\varphi[V]$-preserving.

4) φ is a finest definition of covering models if in addition:

   (ζ) if Q is $\varphi(V)$-preserving, and $x \in \mathrm{Dom}(R^\varphi[V^Q])$ *then* there is a $y \in \mathrm{Dom}(R^\varphi[V])$, such that $\varphi[V^Q] \models y < x$.

5) the φ-covering model is $\varphi[V]$; a φ-covering model is a $\varphi[V^\dagger]$ for an appropriate subuniverse $V^\dagger$ so it belongs to $\varphi(V)$.

6) $(D, R, <)$ is 2-directed when: if $y < x, yRT_1, yRT_2$ (so $x, y, T_1, T_2 \in D$) *then* for some T, xRT and $T_1 \cup T_2 \subseteq T$. We say φ is 2-directed if every $\varphi[V]$ is (see 1.2(1)(β)(iv) and 1.3(5)).

2.3 Restatement of Theorems.

1) If φ is a fine definition of covering models, $\bar{Q} = \langle P_i, \underset{\sim}{Q}_j : i \le \alpha, j < \alpha \rangle$ is an iteration as in $0.1_{\theta=\aleph_0}$ and $\underset{\sim}{Q}_j$ is purely φ-preserving for $j < \alpha$ *then* P_α is purely φ-preserving, hence: $\varphi[V]$ covers and φ is a fine definition of covering models, even in V^{P_α}.
2) If φ is a finer definition of covering models, $\bar{Q} = \langle P_i, \underset{\sim}{Q}_j : i \le \alpha, j < \alpha \rangle$ is as in $0.1_{\theta=\aleph_0}$ and
   (∗) each $\underset{\sim}{Q}_j$ is purely $\varphi[V^{P_j}]$-preserving (and (see 0.1) $\underset{\sim}{Q}_j$ has pure $(\aleph_0, 2)$-decidability)

   then P_α is purely $\varphi[V]$-preserving.
3) In (2) we can weaken (∗) to
   $(*)^-$ for $i \le j < \alpha, i$ non limit we have P_{j+1}/P_i is purely $\varphi[V^{P_i}]$-preserving
4) fine $\Leftarrow$ finer.
5) finer $\Leftarrow$ finest.
6) If φ is a finer definition of covering models, and Q is $\varphi[V]$-preserving *then* Q is $\varphi(V)$-preserving.
7) We can replace pure $(\aleph_0, 2)$-decidability by "pure (2,2)-decidability" if each $\varphi(V')$ is as in 1.12(∗).

Proof: Straightforward. E.g.

6) Suppose $\varphi[V'] = (D', R', <') \in \varphi(V)$, so $V = (V')^{Q'}$, Q' as in 0.1 and $(D', R', <')$ covers in V too. Suppose further that $p \in Q$ and $p \Vdash$ "$\underset{\sim}{f} \in {}^\omega\omega$" and $x \in \mathrm{Dom}(D')$; choose $y \in \mathrm{Dom}(R')$, $y <' x$.
By clause (α)(c) (see 2.2(3)) $\varphi[V] \models$ "$y < x$" (and $x, y \in \mathrm{Dom}(R^{\varphi[V]})$. As Q is purely $\varphi[V]$-preserving there are q and T_1 such that: $p \le_{\mathrm{pr}} q \in Q$, $T_1 \in \mathrm{Dom}(R^{\varphi[V]})$, $\varphi[V] \models$ "yRT_1" and $q \Vdash$ "$\underset{\sim}{f} \in \lim(T_1)$". By clause $(\varepsilon)^+$ (see 2.2(3)) there is $T_0 \in \mathrm{Rang}(R')$ such that $xR'T_0$, $T_1 \subseteq T_0$. So q, T_0 are as required. $\square_{2.3}$

We can save somewhat using: (we shall usually use 2.4(2))

2.4 Claim.

1) Suppose (i) $(D, R, <)$ is a temporarily fine covering model in V, and:

(ii) V is a subuniverse of $V^\dagger$ and $(D, R, <)$ covers in $V^\dagger$, or just $V^\dagger = V^Q$, Q is $(D, R, <)$-preserving,

(iii) every countable $a \subseteq D$ from $V^\dagger$ is a subset of some countable $b \in V$ (e.g., Q is proper or: Q preserves $\aleph_1$, $V \models$ "$|D| \leq \aleph_1$"),

(iv)* there are one-to-one functions $h_n : \omega \to \omega$ such that $h_n \restriction n = h_{n+1} \restriction n$, and $(\mathrm{Rang}(h_n)) \cap (\mathrm{Rang}(h_m)) \subseteq \mathrm{Rang}(h_n \restriction \mathrm{Min}\{n, m\})$ and: for every $x \in D$ for some $y \in D$, for every T_1 such that yRT_1 there is T_0, xRT_0 such that: $\eta \in \lim T_1$ implies $\langle \eta(h_n(\ell)) : \ell < \omega \rangle \in \lim T_0$ for infinitely many n. In fact $\langle h_n : n < \omega \rangle$ may depend on x.

Then $(D, R, <)$ is a temporarily fine covering model in $V^\dagger$.

2) We can replace (iv)* by

(iv)** there are an infinite $w \subseteq \omega$ and functions $h_n : \omega \to \omega$ and a sequence $\langle (g_k, \bar{f}^k) : k < \omega \rangle$ such that

(α) $h_n \restriction n = h_{n+1} \restriction n$

(β) for $k < \omega$ the set $v_k \stackrel{\text{def}}{=} \{(n, \ell) : \ell \geq n - 1, n < \omega \text{ and } h_n(\ell) = k\}$ is finite, g_k is a function from ${}^{(v_k)}\omega$ to ω, $\bar{f}^k = \langle f^k_{(n,\ell)} : (n, \ell) \in v_k \rangle$, $f^k_{(n,\ell)} : \omega \to \omega$ such that $f^k_{(n_0,\ell_0)}(g_k(\ldots, m_{(n,\ell)}, \ldots)_{(n,\ell) \in v_k}) = m_{(n_0,\ell_0)}$

(γ) for every $x \in \mathrm{Dom}R$ for some $y \in \mathrm{Dom}R$ we have: if yRT_1 then there is T_0 satisfying xRT_0 and

$$(\forall \eta \in \lim T_1)(\exists^\infty n)[\langle f^{h_n(\ell)}_{(n,\ell)}(\eta(h_n(\ell))) : \ell < \omega \rangle \in \lim T_0]$$

3) We can replace (iv)* by

(iv)*** for every $x \in \mathrm{Dom}R$ for some Borel function $\mathbf{B}$ from $\{\langle \eta_\alpha : \alpha \leq \omega \rangle : \eta_\alpha \in {}^\omega\omega \text{ and } \eta_\omega \restriction n = \eta_n \restriction n\}$ into ${}^\omega\omega$, *there is* $y \in \mathrm{Dom}R$ such that: for every T_1 satisfying yRT_1 there is T_0 satisfying xRT_0 such that

$$\langle \eta_\alpha : \alpha \leq \omega \rangle \in \mathrm{Dom}(\mathbf{B}) \; \& \; \mathbf{B}(\langle \eta_\alpha : \alpha \leq \omega \rangle) \in \lim T_1 \Rightarrow (\exists^\infty n)(\eta_\alpha \in \lim T_0).$$

Remark: Applying 2.4 we may wonder if (iii) is a burden. At first glance, if $V^\dagger = V^Q$, Q not proper, this may be so. But actually we need it only for the limit cases, and there in the cases of iteration of non-proper forcing notions, we usually assume that in some earlier stage the cardinality of D becomes $\aleph_1$.

Before proving 2.4, we make some observations of some interest, among them a proof.

2.4A Observation. If $\bigwedge_n T_n \subseteq T_{n+1}$ and T_n, T are perfect subtrees of ${}^{\omega>}\omega$ and w is a witness for $(*)^1_{\langle T_n:n<\omega\rangle,T}$ *then* u is a witness for $(*)^1_{\langle T_n:n<\omega\rangle,T}$ if $\otimes$ holds, where

$(*)^1_{\langle T_n:n<\omega\rangle,T}$ T_n, T perfect trees $\subseteq {}^{\omega>}\omega$, $T_0 \subseteq T$ and for some $w = \{n_0, n_1, \ldots\}$ (strictly increasing called a witness): $\eta \in {}^{\omega>}\omega$, $\bigwedge_i[\eta{\restriction}n_{i+1} \in \bigcup_{j<i} T_{n_j} \cup T_0] \Rightarrow \eta \in T$,

$\otimes$ $u \subseteq \omega$ is infinite and: if $i_0 < i_1 < i_2$ are successive members of w *then* $|u \cap (i_0, i_2)| \leq 1$ and the second member of w is smaller than the second member of u.

Proof. Let $w = \{n_i : i < \omega\}$, and $u = \{m_i : i < \omega\}$, both in increasing order. Assume $\eta \in {}^{\omega>}\omega$ and $\bigwedge_i \eta{\restriction}m_{i+1} \in \bigcup_{j<i} T_{m_j} \cup T_0$ and it suffices to prove that $\bigwedge_i \eta{\restriction}n_{i+1} \in \bigcup_{j<i} T_{n_j} \cup T_0$. As each T_j is perfect without loss of generality $\ell g(\eta) = n_{i(*)}$ for some $i(*) > 0$, and we shall prove by induction on $i < i(*)$ that $\eta{\restriction}n_{i+1} \in \bigcup_{j<i} T_{n_j} \cup T_0$. For $i = i(*) - 1$ we will get the desired conclusion by the choice of w. For $i = 0$ we have $\bigcup_{j<i} T_{m_j} \cup T_0 = T_0 = \bigcup_{j<i} T_{n_j} \cup T_0$ so as $m_1 \geq n_1$ the conclusion should be clear.

For $i + 1 > 1$, as $|u \cap (n_{i-1}, n_{i+1})| \leq 1$ (holds by $\otimes$), if $u \cap [0, n_{i-1}] = \emptyset$ then by $u \cap [0, n_{i+1})$ has at most one member hence $m_1 \geq n_{i+1}$ and we do as above. So there is $j < \omega$ such that $m_{j+1} \geq n_{i+1}$, $m_{j-1} \leq n_{i-1}$. Now we know $\eta{\restriction}m_{j+1} \in \bigcup_{\varepsilon<j} T_{m_\varepsilon} \cup T_0$, so if $\eta{\restriction}m_{j+1} \in T_0$ then $\eta{\restriction}n_{i+1} \trianglelefteq \eta{\restriction}m_{j+1} \in T_0 \subseteq \bigcup_{\varepsilon<i} T_{n_\varepsilon} \cup T_0$ and we are done. So for some $\varepsilon < j$, $\eta{\restriction}m_{j+1} \in T_{m_\varepsilon}$ hence $\eta{\restriction}n_{i+1} \trianglelefteq \eta{\restriction}m_{j+1} \in T_{m_\varepsilon} \subseteq T_{m_{j-1}} \subseteq T_{n_{i-1}} \subseteq \bigcup_{\zeta<i} T_{n_\zeta} \cup T_0$ as required. $\square_{2.4A}$

2.4B Observation. Suppose $h : \omega \to \omega$ is one to one (or just finite to one), T_n, S_n, T are perfect subtrees of ${}^{\omega>}\omega$, $\bigwedge_n S_n \subseteq S_{n+1}, T_0 \subseteq S_0$ and for each n for some $m \in [n, \omega)$ we have $(*)^2_{T_{h(n)},S_n,S_{m+1}}$ holds (see below).

Then $(*)^1_{\langle S_n:n<\omega\rangle,T} \Rightarrow (*)^1_{\langle T_n:n<\omega\rangle,T}$ where:

$(*)^1_{\langle T_n:n<\omega\rangle,T}$ T_n, T perfect trees $\subseteq {}^{\omega>}\omega$, $T_0 \subseteq T$ and for some $w = \{n_0, n_1, \ldots\}$ (strictly increasing): $\eta \in {}^{\omega>}\omega$, $\bigwedge_i [\eta\restriction n_{i+1} \in \bigcup_{j<i} T_{n_j} \cup T_0] \Rightarrow \eta \in T$,

$(*)^2_{T_1,T_2,T_3}$ for some $k < \omega$ (the witness) $\rho \in T_1$ & $\rho\restriction k \in T_2 \;\Rightarrow\; \rho \in T_3$.

Remark. Note $(*)^2_{T_1^*,T_2^*,T_3^*}$ & $T_1' \subseteq T_1^*$ & $T_2' \subseteq T_2^*$ & $T_3^* \subseteq T_3' \Rightarrow (*)^2_{T_1',T_2',T_3'}$

Proof. We want to prove $(*)^1_{\langle T_n:n<\omega\rangle,T}$, so we have to find an appropriate w. Let $w_1 = \{n_i : i < \omega\}$ (the increasing enumeration) witness $(*)^1_{\langle S_n:n<\omega\rangle,T}$ and for $j < \omega$ let k_j be such that it witnesses $(*)^2_{T_{h(j)},S_j,S_m}$ (for the first possible $m > n$, note that $(*)^2_{T_{h(j)},S_j,S_{m+1}}$ is preserved by increasing m as $S_m \subseteq S_{m+1}$). By 2.4A above without loss of generality

$\oplus$ $\bigwedge_i n_i \in \mathrm{Rang}(h)$, and $k_{n_i} < n_{i+1}$ and $(\forall k)(h(k) \leq n_i \Rightarrow k < n_{i+1})$, hence $n_i < h(n_{i+1}) < n_{i+2}$, and also for some $m \in (n_i, n_{i+1})$ we have $(*)^2_{T_{h(n_i)},S_{n_i},S_m}$ we get $(*)^2_{T_{h(n_i)},S_{n_i},S_{n_{i+1}}}$).

Choose $m_i = h(n_{4i+4})$. Now we shall prove that $w \stackrel{\text{def}}{=} \{m_i : i < \omega\}$ is a witness to $(*)^1_{\langle T_n:n<\omega\rangle,T}$ thus finishing the proof of 2.4B. So we assume $\eta \in {}^{\omega>}\omega$, $\bigwedge_i \eta\restriction m_{i+1} \in \bigcup_{j<i} T_{m_j} \cup T_0$ and we have to prove that $\eta \in T$. As $w_1 = \{n_i : i < \omega\}$ witnesses $(*)^1_{\langle S_n:n<\omega\rangle,T}$, it suffices to prove: for each $i < \omega$ we have $\eta\restriction n_{i+1} \in \bigcup_{j<i} S_{n_j} \cup S_0$.

We prove it by induction on i. If $n_{i+1} \leq m_1$ then as $\eta\restriction m_1 \in T_0$, $T_0 \subseteq S_0$, T_0 is perfect, clearly $\eta\restriction n_{i+1} \in S_0 \subseteq \bigcup_{j<i} S_j \cup S_0$. But $n_{i+1} \leq m_1$ holds if $n_{i+1} \leq h(n_8)$ what implies $i < 9$. So we assume $i \geq 9$. Let $4i(*) + 2 \leq i < 4(i(*) + 1) + 2$ (so $i(*) \geq 1$). So by the assumption $\otimes$, we have $\eta\restriction n_{i+1} \triangleleft \eta\restriction h(n_{4(i(*)+1)+4}) = \eta\restriction m_{i(*)+1} \in \bigcup_{j<i(*)} T_{m_j} \cup T_0$. Stipulating $m_{-1} = 0$, for some $j(*) \in \{-1, 0, \ldots, i(*) - 1\}$ we have $\eta\restriction m_{i(*)+1} \in T_{m_{j(*)}}$. If $j(*) = -1$, then $\eta\restriction n_{i+1} \triangleleft \eta\restriction m_{i(*)+1} \in T_0 \subseteq S_0 \subseteq \bigcup_{j<i} S_{n_j} \cup S_0$ as required. So assume $j(*) \in \{0, \ldots, i(*) - 1\}$. But we know that $(*)^2_{T_{m_{j(*)}},S_{n_{4j(*)+4}},S_{n_{4j(*)+5}}}$ [Why? As by the definition $m_{j(*)} = h(n_{4j(*)+4})$, and by $\oplus$ above], and we want to apply it to $\rho \stackrel{\text{def}}{=} \eta\restriction m_{i(*)+1}$. The first assumption of $(*)^2_{T_{m_{j(*)}},S_{n_{4j(*)+4}},S_{4j(*)+5}}$ was deduced above: $\rho = \eta\restriction m_{i(*)+1} \in T_{m_{j(*)}}$. The second assumption there is

$\rho\restriction k_{n_{4j(*)+4}} \in S_{n_{4j(*)+4}}$ (by the choice of $k_{n_{4j(*)+4}}$), now we know $j(*) < i(*)$ and $\rho\restriction k_{n_{4j(*)+4}} = \eta\restriction k_{n_{4j(*)+4}} \triangleleft \eta\restriction n_{4j(*)+5} \in S_{n_{4j(*)+4}}$

[Why? First, the equality holds as:

(a) $\rho = \eta\restriction m_{i(*)+1}$

(b) $k_{n_{4j(*)+4}} \leq m_{i(*)+1}$, because $m_{i(*)+1} = h(n_{4(i(*)+1)+4}) > n_{4(i(*)+1)+3}$ $= n_{4i(*)+7} \geq n_{4(j(*)+1)+7} > n_{4j(*)+6} > k_{n_{4j(*)+4}}$ (why? by definition of $m_{i(*)+1}$, by $\oplus$, arithmetic, as $i(*) \geq j(*)$, arithmetic and $\oplus$ respectively).

Secondly, the $\triangleleft$ holds as $k_{n_{4j(*)+4}} \leq n_{4j(*)+5}$ by $\oplus$.

Finally, the membership holds - by the induction hypothesis on i, and $\langle S_n : n < \omega\rangle$ being increasing, note the induction hypothesis can be applied as $j(*) < i(*)$ hence $4j(*) + 5 \leq 4(i(*) - 1) + 5 = 4i(*) + 1 < i$].

So we can actually apply $(*)^2_{T_{m_{j(*)}}, S_{n_{4j(*)+4}}, S_{n_{4j(*)+5}}}$ and get $\rho = \eta\restriction m_{i(*)+1}$ belongs to $S_{n_{4j(*)+5}}$. As $\eta\restriction n_{i+1} \triangleleft \eta\restriction m_{i(*)+1} = \rho$ (see above) and $4j(*) + 5 \leq 4(i(*) - 1) + 5 < 4i(*) + 2 \leq i$, really $\eta\restriction n_{i+1} \in \bigcup_{j<i} S_{n_j} \cup S_0$ as required, thus we have finished. $\square_{2.4B}$

2.4C Observation. If $(*)^1_{\langle T_n : n<\omega\rangle, T}$ holds as witnessed by w, and $\bigwedge_n T_n \subseteq T_{n+1}$ and $h : \omega \to \omega$ is such that $\mathrm{Rang}(h)$ is infinite, $h(0) = 0$ and we let $T^1_n \stackrel{\mathrm{def}}{=} T_{h(n)}$ *then* $(*)^1_{\langle T^1_n : n<\omega\rangle, T}$.

Proof. Let $u \subseteq \omega$ be infinite such that $h\restriction u$ is one to one (possible as $\mathrm{Rang}(h)$ is infinite), $h\restriction u$ is strictly increasing and for $i < j$ in u, $h(i) < j$ & $i < h(j)$, moreover, $|(h(i), j) \cap w| \geq 2$. Now u is as required by 2.4A above. $\square_{2.4C}$

2.4D Observation. In 1.2(1)(γ)(a) we can add the assumption $T_n \subseteq T_{n+1}$ and get an equivalent condition (assuming 1.2(1)(α), (β) of course).

Proof. Of course we only need to assume this apparently weaker version and prove the original version. Let $x_0 < \ldots < x_n < x_{n+1} < \ldots y^+ < y$, $x_n R T^0_n$ be given. We define by induction on n, T^1_n such that: $T^1_n \subseteq T^1_{n+1}$, $T^1_0 = T^0_0$, $x_{n+1} R T^1_n$ and $(*)^2_{T^0_n, T^1_n, T^1_{n+1}}$ (possible by 1.2(1)(β)(iv) which says: if $y < x$, $y R T_\ell$ then $(\exists T)[T_1 \subseteq T$ & $(*)^2_{T_2, T_1, T}]$). So $T^1_n \subseteq T^1_{n+1}$ and (as we are assuming the weaker version of 1.2(1)(γ)(a)) $(*)^1_{\langle T^1_n : n<\omega\rangle, T}$ holds for some T such that yRT. By 2.4B, (with $S_n \stackrel{\text{def}}{=} T^1_n$, $T_n \stackrel{\text{def}}{=} T^0_n$ and $h(n) = n$) we get $(*)^1_{\langle T^0_n : n<\omega\rangle, T}$ as required. $\square_{2.4D}$

2.4E Observation. If V, $V^\dagger$, $(D, R, <)$ satisfy conditions (i), (ii), (iii) of claim 2.4(1) *then* $(D, R, <)$ satisfies $(\gamma)(a)$ of 1.2(1) also in $V^\dagger$.

Proof: Let $x > x^\dagger > y_{n+1} > y_n$ for $n < \omega$ and $T_n \in V$ be such that $y_n R T_n$ (but the sequence $\langle T_n : n < \omega\rangle$ may be from $V^\dagger$. Let b be a countable set from V such that $\{T_n : n < \omega\} \subseteq b \subseteq V$. Let $\langle S^0_n : n < \omega\rangle \in V$ enumerate $\{T \in b : (\exists y)(y < x^\dagger \& yRT)\}$, so $\{T_n : n < \omega\} \subseteq \{S^0_n : n < \omega\}$. Without loss of generality $S^0_0 = T_0$ and for each n for infinitely many m we have $S^0_m = S^0_n$. By 1.2(1)(β) we can find in V a sequence $\langle z_n, S^1_n, k_n : n < \omega\rangle, z^\dagger$ such that $x^\dagger < z_n < z_{n+1} < z^\dagger < x$ for $n < \omega$ (of course $S^1_n \in V$), $k_n < \omega$ such that $z_n R S^1_n$, and $S^1_0 = T_0, S^1_n \subseteq S^1_{n+1}$ and $[\rho \in S^0_n \ \& \ \rho\restriction k_n \in S^1_n \Rightarrow \rho \in S^1_{n+1}]$ (choose them inductively). So $(*)^2_{S^0_n, S^1_n, S^1_{n+1}}$; now $(*)^2_{-,-,-}$ has obvious monotonicity properties in its variables (see 2.4B), hence $n_0 \le n < n_1 \Rightarrow (*)^2_{S^0_n, S^1_{n_0}, S^1_{n_1}}$. Choose by induction on n, $h(n)$ as

$$\min\{m : T_n = S^0_m \text{ and } m > n \text{ and } m > \sup\{h(k) : k < n\}\},$$

well defined by the choice of $\langle S^0_m : m < \omega\rangle$. So we know $(*)^2_{S^0_{h(n)}, S^1_n, S^1_{h(n)+1}}$.

We want to apply 2.4B with $\langle S^1_n : n < \omega\rangle$, $\langle T_n : n < \omega\rangle$, h here standing for $\langle S_n : n < \omega\rangle$, $\langle T_n : n < \omega\rangle$, h there; we have here almost all the assumptions (including h is one to one (even strictly increasing) and $\bigwedge_n \bigvee_{m \ge n} (*)^2_{T_{h(n)}, S^1_n, S^1_{m+1}}$) but still need to choose T^* and prove that xRT^* and $(*)^1_{\langle S^1_n : n<\omega\rangle, T^*}$.

Apply (γ)(a) of 1.2(1) in V (which holds by (i)) with $\langle S^1_n : n < \omega\rangle, \langle z_n : n < \omega\rangle, z^\dagger, x$ here standing for $\langle T_n : n < \omega\rangle, \langle y_n : n < \omega\rangle, x^\dagger$, x there, and get T^* (in V!) such that $(*)^1_{\langle S^1_n : n<\omega\rangle, T^*}$ holds and xRT^*. So we can really apply 2.4B hence get that $(*)^1_{\langle T_n : n<\omega\rangle, T^*}$ holds, as required. $\square_{2.4E}$.

2.4F Proof of 2.4(1). From definition 2.2(1A), part (α) and (β) should be clear. By 2.4E we know that (γ)(a) of 1.2(1) holds, so it suffices to prove (γ)(b) of 1.2(1).

Given $x \in \mathrm{Dom}(R)$ and η, $\eta_n \in {}^{\omega}\omega$ such that $\eta{\restriction}n = \eta_n{\restriction}n$, let y and h_n $(n < \omega)$ be as in (iv)*. We can find $\nu \in {}^{\omega}\omega$ such that: for each $n < \omega$ we have $\eta_n(k) = \nu(h_n(k))$ (note that there is such $\nu \in {}^{\omega}\omega$ because if $\ell = h_{n_1}(k_1) = h_{n_2}(k_2)$ then $\ell \in \mathrm{Rang}(h_{n_1}) \cap \mathrm{Rang}(h_{n_2})$ hence $k_1, k_2 < \min\{n_1, n_2\}$, so $h_{n_2}(k_2) = h_{n_1}(k_1) = h_{n_2}(k_1)$, but h_{n_2} is one to one so $k_1 = k_2$). As $(D, R, <)$ covers we can find T_1 such that yRT_1 and $\nu \in \lim(T_1)$. Now let T_0 be as guaranteed by (iv)* (of 2.4). $\square_{2.4}$

2.4G Proof of 2.4(2), (3). Similar.

2.5 Claim.

(1) We can get the conclusion of 2.4 and even strengthen it by " in $V^\dagger$ the model $(D, R, <)$ still satisfies $(\gamma)_1$ (see 1.3(8))" if we replace (iv)* by:

 (iv) every $f \in ({}^{\omega}\omega)^{V^\dagger}$ is dominated by some $g \in ({}^{\omega}\omega)^V$,

 (v) $(D, R, <)$ satisfies $(\gamma)_1$ of 1.3(8),

 (vi)′ $(D, R, <)$ is 2-directed (see 2.2(6)).

(2) In 2.4(1) we can replace (iv)$^+$ by (iv)$^-$, (v)′ below and (vi)′ above, where

 (iv)$^-$ no $f \in ({}^{\omega}\omega)^{V^\dagger}$ dominates $({}^{\omega}\omega)^V$

(v)′$(\gamma)_2$ if $y, x, y^\dagger, x_n, T_n \in D_\varphi[V]$, $x_0 < x_1 < \ldots < y^\dagger < y$, $x_n R T_n$, $T_n \subseteq T_{n+1}$ (for each $n < \omega$) *then* for some $k < \omega$ and $\langle T^\ell : \ell < k\rangle$, $\langle B_\ell : \ell < n\rangle$ we have:

 (a) $\omega = \bigcup_{\ell<k} B_\ell$

 (b) if $n \in B_\ell$, $\eta \in T_n$, $\eta{\restriction}n \in T_0$ then $\eta \in T^\ell$

 (c) yRT^ℓ.

(3) Assuming (α), (β) of 2.11 we have $(\gamma)_3 \Rightarrow (\gamma)$, $(\gamma)_3 \Rightarrow (\gamma_2)$ where

 $(\gamma)_3$ like $(\gamma)_2$ replacing (b) by

 (b)$^+$ if $\ell < k$, and $(\forall n)(\eta{\restriction}n \in \bigcup\{T_m : m \leq n \text{ and } m \in B_\ell\})$ implies $\eta \in \lim T^\ell$.

Remark. Can we phrase a maximal $(\gamma)_n$? Like $(\gamma)_2$ but without T_n.

Proof. 1) As in the proof in 2.4F, we have (α), (β), (γ)(a) of 1.2(1) and it suffices to prove (γ)(b) and $(\gamma)_1$ of 1.3(8), but the latter implies the former. So let $V^\dagger$, $\langle x_n : n < \omega \rangle$, $x^\dagger$, x and $\langle T_n : n < \omega \rangle$ be given as there. So $x_n R T_n$, $\{x^\dagger, x\} \subseteq \mathrm{Dom}(R)$, $x_n < x_{n+1} < x^\dagger < x$. As in the proof of 2.4E we can find $\langle (z_\ell, S^0_\ell) : \ell < \omega \rangle \in V$ such that $\{(x_n, T_n) : n < \omega\} \subseteq \{(z_n, S^0_n) : n < \omega\}$, and w.l.o.g. $n < \omega \Rightarrow z_n R S^0_n$ & $z_n < x^\dagger$. By the 2-directness we can find a sequence $\langle (y_n, S^1_n) : n < \omega \rangle \in V$ such that $y_n < y_{n+1} < x^\dagger$ and $S^1_n = T_0$ and $y_n R S^1_n$ and $S^0_n \cup S^1_n \subseteq S^1_{n+1}$ (possible by (vi)$'$ which). Define $h \in ({}^\omega\omega)^{V^\dagger}$ by $h(n) = \min\{m : T_n = S^0_m\}$ and choose a strictly increasing function $g \in ({}^\omega\omega)^V$ such that $[n < \omega \Rightarrow h(n) < g(n)$ & $n < g(n)]$. By $(\gamma)_1$ of 1.3(8) applied to $\langle S^1_{g(i)} : i < \omega \rangle$ in V there are $T^* \in \mathrm{Rang}(R)$ and infinite $w_1 \subseteq \omega$ such that

$(*)_1$ xRT

$(*)_2$ $\lim(T^*) \supseteq \{\eta : \eta \in {}^\omega\omega \text{ and } i \in w_1 \Rightarrow \eta{\restriction}i \in \bigcup_{j \in w_1, j \le i} S^1_{g(j)}\}$

Let us prove that T^* and w are as required. So we assume

$(*)_3$ $\eta \in {}^\omega\omega$ and

$$i \in w_1 \Rightarrow \eta{\restriction}i \in \bigcup_{j \in w_1, j \le i} T_j.$$

We should prove that $\eta \in \lim(T^*)$, but by $(*)_2$ it suffice to prove:

$(*)_4$ $i \in w_1 \Rightarrow \eta{\restriction}i \in \bigcup_{j \in w_1, j \le i} S^1_{g(j)}$

As $T_j \subseteq S^1_{g(j)}$ this is immediate.

2) As in the proof of 2.4(1) we can deal with conditions (α), (β), (γ)(a) (the first two: trivially, the last one by 2.4E). For (γ)(b) let η, $\eta_n \in ({}^\omega\omega)^{V^\dagger}$, $\eta_n{\restriction}n = \eta{\restriction}n$ and $y \in \mathrm{Dom}(R)$ be given and choose $x^\dagger$; $\langle x_n : n < \omega \rangle$, $\langle T_n : n < \omega \rangle$, $(x = y)$ $\langle S^1_n : n < \omega \rangle$, x_n, $y^\dagger$ as in the proof of 2.5(1), so in particular $\eta \in S^1_0$ and $h(n) = \min\{m : m > n \text{ and } \eta_n \in \lim(S^1_n)\}$ are well defined; note $S^1_n \subseteq S^1_{n+1}$. Let $g \in {}^\omega\omega$ be strictly increasing, $g(0) = 0$ such that $A = \{n : h(n) < g(n)\}$ is infinite. We can find such g by clause (iv)$^-$ of the assumption. Now apply $(\gamma)_2$ to y_n $(n < \omega)$, $x^\dagger$, y, $\langle S^1_{g(n)} : n < \omega \rangle$, and get $k < \omega$, $\langle B_\ell : \ell < k \rangle$, $\langle T^\ell : \ell < k \rangle$ as there (in particular yRT^ℓ). Now for each $n \in A$ for some $\ell(n) < k$, we have $(\nu \in {}^{\omega >}\omega)$ & $\nu{\restriction}n \in S^1_{g(0)}$ & $\nu \in S^1_{g(n)} \Rightarrow \nu \in T^{\ell(n)}$. So, if $n \in A$, $\eta_n{\restriction}n = \eta{\restriction}n \in S^1_0 = S^1_{g(0)}$, $\eta_n \in S^1_{h(n)} \subseteq S^1_{g(n)}$ hence $\eta_n \in T^{\ell(n)}$. So for some $\ell < k$, $\{n \in A : \ell(n) = \ell\}$ is infinite and we are done. $\square_{2.5}$

* * *

2.6 Definition. $TTR = \{T \cap {}^{m\geq}\omega : m < \omega,\ T \subseteq {}^{\omega>}\omega$ a closed tree and for every $n < \omega$ we have $T \cap {}^{n\geq}\omega$ finite $\}$, where "T is a closed tree" means, as usual: $T \neq \emptyset$, $[\eta \in T \ \&\ \nu \triangleleft \eta \Rightarrow \nu \in T]$, $[\eta \in T \Rightarrow \bigvee_{i<\omega} \eta\hat{\ }\langle i\rangle \in T]$. Note that TTR has a natural tree structure: $t < s$ if $t = s \cap {}^{n\geq}\omega$ for some n. For $t \in TTR$ let $\mathrm{ht}\,(t) = \min\{n : t \subseteq {}^{n\geq}\omega\}$ and $TTR_n = \{t \in TTR : \mathrm{ht}\,(t) = n\}$.

2.6A Notation. $DP({}^{\omega}\omega) = \{x \in {}^{\omega}\omega : x(n) \geq 1$ for every n and $\langle x(n) : n < \omega\rangle$ diverges to infinity, i.e. for every $m < \omega$ for some $k < \omega$, for every $n \geq k, x(n) \geq m.\}$

2.6B Remark. Usually we can replace x by x', $x'(n) = \mathrm{Min}\{x(m) : n \leq m < \omega\}$, hence without loss of generality x is nondecreasing.

2.7 Fact. Each closed tree $T \subseteq {}^{\omega\geq}\omega$ such that $(\forall n)[|T \cap {}^{n\geq}\omega| < \aleph_0]$ induces a branch $\{T \cap {}^{n\geq}\omega : n < \omega\}$ (in the tree TTR) and is its union. Now TTR is isomorphic to ${}^{\omega>}\omega$.

* * *

Now we deal with some examples: we do not state the aim - the preservation theorems by combining with 2.1-2.7 - for each φ_i separately but usually we mention the case of CS iteration of proper forcing.

2.8A Definition. [${}^{\omega}\omega$-bounding]: 1) We define $\varphi = \varphi_1^{cm}$ (a definition of covering models) by letting $\varphi[V] = (D, R)$ if:

a) $D = H(\aleph_1)^V$

b) xRT iff $x, T \in D$, $x \in DP({}^{\omega}\omega)$, T is a closed tree and $(\forall n)(T \cap {}^{n}\omega$ is finite) (so x has really no role)

c) $<=<_0$ (see 1.4)

2) A forcing notion P is ${}^{\omega}\omega$-bounding (in V) if it is φ_1^{cm}-preserving (see 2.8C-equivalent to the definition from V).

2.8B Claim. $\varphi = \varphi_1^{cm}$ is a finest definition of covering models for proper forcing, and it is 2-directed.

Remark.

1) Instead of "proper" we can use "forcing Q such that every countable subset of $\min\{(2^{\aleph_0})^{V^1} : \varphi^{cm}[V^1] \text{ covers } V\}$ from V^Q is included in one from V".
2) We just "forget" to mention the pure version.

Proof: Let us check the conditions in Definition 2.2, the 2-directed should be clear.

(α) (a) Trivial by a), b) of Definition 2.8A.

(α) (b) Trivial (a tree with one branch).

(α) (c) and (β) are trivial.

(γ), $(\gamma)^+$ Let $x > y, yRT_n$ (remember 1.3(4)).

Put $w \stackrel{\text{def}}{=} \omega$, $T^* \stackrel{\text{def}}{=} \{\eta$: for every $k < \omega, \eta \restriction k \in \bigcup_{j \leq k} T_j\}$ and check that T^* is as required.

(δ) By 2.4 for (γ), 2.5(1) for $(\gamma)^+$.

$(\varepsilon)^+$ Now we use Fact 2.7 applied to T^*, (i.e. to the branch of TTR which T^* induced). So there is a closed tree $C \subseteq TTR, C \in D, xRC$ and for every $n, T^* \cap {}^{n>}\omega \in C$. Let $T^{**} = \{\eta \in {}^{\omega>}\omega$: for some $t \in C, \eta \in t\}$. Clearly $T^{**} \in D$ (as $C \in D, D = H(\aleph_1), T^{**}$ is a closed tree $\subseteq {}^{\omega>}\omega$), and $T^* \subseteq T^{**}$. In addition, for every n

$$T^{**} \cap {}^n\omega = \bigcup\{t \cap {}^n\omega : t \in C \cap TTR_{n+1}\},$$

so, being a finite union of finite sets, $T^{**} \cap {}^n\omega$ is finite.

(ζ) Easy. $\square_{2.8B}$

2.8C Fact. If $V \subseteq V^\dagger$ then $\varphi_1^{cm}[V]$ covers in $V^\dagger$ *if and only if*

$$(\forall f \in ({}^\omega\omega)^{V^\dagger})(\exists g)(f <^* g \in ({}^\omega\omega)^V)$$

if and only if

$$(\forall y \in \mathrm{DP}\,({}^\omega\omega)^{V^\dagger})(\exists x \in \mathrm{DP}\,({}^\omega\omega)^V)[y <^*_{\text{dis}} x]$$

if and only if

$$(\forall y \in \mathrm{DP}\,({}^{\omega}\omega)^{V^{\dagger}})(\exists x \in \mathrm{DP}\,({}^{\omega}\omega)^{V})[y <_{\mathrm{dis}} x].$$

2.8D Conclusion. For a CS iteration $\bar{Q} = \langle P_i, \underset{\sim}{Q}_j : i \leq \alpha, j < \alpha\rangle$ if $\Vdash_{P_j}$ "$\underset{\sim}{Q}_j$ is proper and ${}^{\omega}\omega$-bounded" for each i *then* P_α is proper and ${}^{\omega}\omega$-bounding.

Proof. By 2.3(2) and 2.8B + 2.8C.

* * *

2.9A Definition. [The Sacks property] Define $\varphi = \varphi_2^{cm}$ (a definition of a covering model) by letting $\varphi[V] = (D, R)$ be

a) $D = H(\aleph_1)$

b) xRT iff $x, T \in D$ and $x \in DP({}^{\omega}\omega), T \subseteq {}^{\omega>}\omega$ and for every $n < \omega$, $T \cap {}^{n}\omega$ has at most $x(n)$ elements.

c) $<=<_{\mathrm{dis}}$ (see 1.4).

2.9B Claim. $\varphi = \varphi_2^{cm}$ is a finest definition of covering models.

Proof. Let us check the conditions in Definition 2.2.

(α) (a) Trivial by a), b) of Definition 2.9A.

(α) (b) Trivial.

(α) (c) Trivial.

(β) Trivial, by the definition of the partial order (1.4).

$(\gamma)^+$ Let $y_n R T_n$ and $y_n <_{\mathrm{dis}} y_{n+1} <_{\mathrm{dis}} x^{\dagger} <_{\mathrm{dis}} x$ (for $n < \omega$). Define n_k inductively as the first $n < \omega$ such that $\ell < k \Rightarrow n_\ell < n$ and for every ℓ, $n \leq \ell < \omega$, we have $(k+2) \cdot x^{+}(\ell) \leq x(\ell)$. Let $w = \{n_k : k < \omega\}$ and

$$T^* = \{\eta : n \in w \Rightarrow \eta \restriction n \in \bigcup_{\substack{m \leq n \\ m \in w}} T_n\}.$$

(δ) Immediate by 2.4(1).

$(\varepsilon)^+$ There is by 2.7 a closed tree $C \subseteq TTR$ in D, $T^* \cap {}^n\omega \in C$, zRC where $z(m) = x(m)/y(m)$. Let $C_1 = \{t \in C$: for every $n, |t \cap {}^n\omega| \le y(n)\}$ and let C_2 be the maximal closed tree $\subseteq C_1$. Clearly $C_2 \in D$ and $T^* \cap {}^n\omega \in C_2$ for every n, now $T^{**} = \bigcup\{t : t \in C_2\} \in D$ is as required.

(ζ) Easy by 2.8C. $\square_{2.9B}$

2.9C Claim.

1) If $V \subseteq V^\dagger$, $\varphi_2^{cm}[V]$ covers in $V^\dagger$ *iff* for every $\eta \in ({}^\omega\omega)^{V^\dagger}$ and $y \in DP({}^\omega\omega)^V$ there is $\langle a_\ell : \ell < \omega \rangle \in V$, $a_\ell \subseteq \omega$, $|a_\ell| \le y(\ell)$ and $\bigwedge_\ell \eta(\ell) \in a_\ell$.

2) If $V \subseteq V^\dagger$ and $\varphi_2^{cm}[V]$ covers in $V^\dagger$ *then* $\varphi_1^{cm}[V]$ covers in $V^\dagger$.

Proof. Straight.

2.9D Conclusion. For a CS iteration $\bar{Q} = \langle P_i, \underset{\sim}{Q}_j : i \le \alpha, j < \alpha \rangle$ if $\Vdash_{P_j}$ "$\underset{\sim}{Q}_j$ is proper and has the Sacks property" *then* P_α is proper and has the Sacks property.

Proof. By 2.3(2) and 3.9B + 2.9C.

* * *

2.10A Definition. [The Laver Property] 1) We define $\varphi = \varphi_3^{cm}$ by letting $\varphi[V] = (D, R, <)$ (the Laver model) be

a) $D = H(\aleph_1)^V$.

b) xRT iff ($x, T \in D$ and) $x \in DP({}^\omega\omega)$, $T \subseteq {}^{\omega >}\omega$ a closed tree and: $(\forall n)$ [the set $\{\eta(n) : \eta \in T, \ell g(\eta) = n+1, (\forall i \le n)\eta(i) < x(2i)\}$ has power $\le x(2n+1)$].

c) $x < y$ iff $\langle x(2n+1) : n < \omega \rangle <_{\rm dis} \langle y(2n+1) : n < \omega \rangle$ (see 1.4B) and $\langle x(2n) : n < \omega \rangle = \langle y(2n) : n < \omega \rangle$.

2.10B Claim. $\varphi = \varphi_3^{cm}$ is a finest definition of a covering model.

Proof. It can be proved very similarly to the proof of 2.9B. The proof of (α), (β) is totally trivial and (δ) follows from (γ) by 2.4(1), so we shall prove $(\gamma)^+$.

Let $x > y > y_{n+1} > y_n$, T_n, be given and $y_n R T_n$.

We can choose $n_0 < n_1 < n_2 < \dots$ (by induction) such that: for $k \geq n_\ell$, $(\ell + 2) \times y(2k+1) < x(2k+1)$, and let $w = \{n_\ell : \ell < w\}$,

$T^* = T_0 \cup \{\eta : \text{for every } i \in w, \eta \restriction i \in \bigcup_{\substack{j \leq i \\ j \in w}} T_j\}$

$T^0 = \{\eta \in {}^{\omega>}\omega : \text{for every } i < \ell g(\eta), \eta(i) < x(2i)\}$

Clearly $T^* \subseteq {}^{\omega>}\omega$ is a closed tree, and for any k, $|T^0 \cap T^* \cap {}^k\omega| \leq x(2k+1)$, because, letting $n_\ell \leq k < n_{\ell+1}$, $\{\eta(k) : \ell g\eta > k, \eta \in T^0 \cap T^*\}| \leq |\{\eta(k) : \ell g\eta > k, \eta \in T^0 \cap (\bigcup_{j \leq \ell+1} T_{n_j})\}| \leq \sum_{j \leq \ell+1} |\{\eta(k) : \ell g(\eta) > k, \eta \in T^0 \cap T_{n_j}\}| \leq \sum_{j \leq \ell+1} y(2k+1) < x^\dagger(2k+1)$.

So T^* the definition of xRT^* is satisfied and T^* is as required.

$(\varepsilon)^+$, (ζ) left to the reader - similar to the proof of (δ). $\square_{2.10B}$

2.10C Claim. 1) If $V \subseteq V^\dagger$, $\varphi_3^{cm}[V]$ covers in $V^\dagger$ *iff* for every $\eta \in (\prod_{n<\omega}(n+1))^{V^\dagger}$ and $y \in DP({}^\omega\omega)^V$ there is $\langle a_\ell : \ell < \omega \rangle \in V$, $a_\ell \subseteq \omega$, $|a_\ell| \leq y(\ell)$ and $\bigwedge_\ell \eta(\ell) \in a_\ell$ *iff* for every $f \in (DP({}^\omega\omega))^V$ for every $y \in (DP({}^\omega\omega))^{V^+}$ and $\eta \in (\prod_{n<\omega} f(n))^{V^+}$ there is $\langle a_\ell : \ell < \omega \rangle \in V$, $|a_\ell| \leq y(\ell)$ and $\bigwedge_{\ell<\omega} \eta(\ell) \in a_\ell$ *iff* similarly for some f.

2) A forcing notion P has the Sacks property (i.e. is φ_2^{cm}-preserving) iff it has the ${}^\omega\omega$-bounding property (i.e. is φ_1^{cm}-preserving), and the Laver property (i.e. is φ_3^{cm}-preserving).

Proof. Easy.

2.10D Conclusion. For a CS iteration $\bar{Q} = \langle P_i, \underset{\sim}{Q}_j : i \leq \alpha, j < \alpha \rangle$, if $\Vdash_{P_i}$ "$\underset{\sim}{Q}_i$ is proper and has the Laver property" *then* P_α is proper and has the Laver property.

Proof. By 2.3(2) and 2.10B + 2.10C.

* * *

The next example deal with trying to have: every new $\eta \in \prod_{n<\omega} f(n)$ belong to some old $\prod_{n<\omega} a_n$, $\langle |a_n| : n < \omega\rangle$ quite small, where $f(n)$ can be finite. Below we could have used $Y = \{\mathrm{id}\}$, but in applying it is more convenient to have Y. See more on this in [Sh:326] and much more in Roslanowski, Shelah [RoSh:470].

2.11A Definition. Let f denote a one place function, $\mathrm{Dom}(f) = \omega, 1 \le f(n) \le \omega$, f diverges to ∞ and g denote a two place function from ω to $\{\alpha : 1 \le \alpha \le \omega\}$; both nondecreasing, for clarity. Let $Y \subseteq DP({}^\omega\omega)$ have absolute definition and $<=<_Y$ be an absolute dense order on Y with no minimal member, and those properties are absolute (so Y may be countable, if $Y = DP({}^\omega\omega)$ we omit it). Finally, let H denote a family of such pairs (f,g). If $H = \{(f,g)\}$ we write f, g.

We define $\varphi = \varphi^{cm}_{4,Y;H}$, but if $Y = DP({}^\omega\omega)$ we may omit it, by letting for a universe $V, \varphi[V]$ be $(D, R, <)$ where

a) $D = H(\aleph_1)$,

b) $\mathrm{Dom}(R)$ is the set of triples (z, f, g) for $z \in Y$, $(f,g) \in H$; more formally member $x \in DP({}^\omega\omega)$ such that $\langle x(3\ell + i) : \ell < \omega\rangle$ codes z when $i = 0$, f when $i = 1$ and g when $i = 2$; we write $x = (z^x, f^x, g^x)$. We define: xRT *iff* $x, T \in D$, $x \in \mathrm{Dom}(R)$, $T \subseteq {}^{\omega>}\omega$ is a closed tree and for each n the set $\{\eta(n) : \eta \in T,\ \ell g(\eta) = n+1, (\forall i \le n)\eta(i) < f^x(i)\}$ has cardinality $< 1 + g^x(n, z^x(n))$. (So for $g^x(n, z^*(n)) = \omega$ this means "finite".)

c) $<_Y$ is the dense order of Y (e.g. $<_0$ or $<_{\mathrm{dis}}$) and $\langle z^1, f^1, g^1\rangle < \langle z^2, f^2, g^2\rangle$ iff $f^1 = f^2$, $g^1 = g^2$ and $z^1 < z^2$.

We may use also g with positive real (not integer) values, but still algebraic.

Let us note that φ^{cm}_1 (${}^\omega\omega$-bounding), φ^{cm}_2 (Sacks), φ^{cm}_3 (Laver) are particular cases of $\varphi^{cm}_{4,Y;H}$:

2.11B Claim. 1) Let $f = \omega$ (i.e. the function with constant value ω), $g(n,i) = \omega$ and $Y = DP({}^\omega\omega)$. Then for universes $V \subseteq V^\dagger, \varphi^{cm}_1[V]$ covers in $V^\dagger$ iff $\varphi^{cm}_{4,f,g}[V]$ covers in $V^\dagger$ (hence a forcing notion Q is φ^{cm}_1-preserving iff it is $\varphi^{cm}_{4,f,g}$-preserving).

2) Let g be $g(n,i) = 1+i$. For universes $V \subseteq V^\dagger, \varphi_3^{cm}[V]$ covers in $V^\dagger$ *iff* for every $f \in DP({}^\omega\omega)^V$, $\varphi_{4,f,g}^{cm}[V]$ covers in $V^\dagger$ (hence a forcing notion Q is φ_3^{cm}-preserving iff it is $\varphi_{4,f,g}^{cm}$-preserving for every $f \in DP({}^\omega\omega)$).

Proof. Check.

2.11C Claim. 1) Assume

(i) H is a family of pairs (f,g) and $Y \subseteq DP({}^\omega\omega)$ (an absolute definition, dense with no minimal element),

(ii) each $(f,g) \in H$ is as in 2.11A and $x <_Y y \Rightarrow \langle g(n,y(n))/g(n,x(n)) : n < \omega\rangle$ diverges to ∞,

(iii) *for every* $(f,g) \in H$ and $y \in Y$ *there are* $x \in Y$ and $(f',g') \in H$ and $h_n : \omega \to \omega$ one to one, $h_n{\restriction}n = h_{n+1}{\restriction}n$, $[n < m \Rightarrow \mathrm{Rang}(h_n) \cap \mathrm{Rang}(h_m) = \mathrm{Rang}(h_n{\restriction}n)]$ and $g'(h_n(\ell), x(h_n(\ell))) \le g(\ell, y(\ell))$ and $f'(h_n(\ell)) \ge f(\ell)$.

Then $\varphi_{4,Y,H}^{cm}$ is a fine definition of covering models.

2) In part (1) we can replace clause (iii) by

(iii)$^-$ *for every* $(f,g) \in H$ and $y \in Y$ *there are* $x \in Y$ and $(f',g') \in H$ and $h_n : \omega \to \omega$ such that:

(α) $h_n{\restriction}n = h_{n+1}{\restriction}n$

(β) for every $k < \omega$, letting $w_k = \{(n,\ell) : \ell \ge n-1 \text{ and } h_n(\ell) = k\}$ we have $\prod\limits_{(n,\ell)\in w_k} f(\ell) \le f'(k)$

(γ) $g(n,\ell) \ge g'(n, h_n(\ell))$

3) Assume we replace (iii) by

(iii)* *for* $(f,g) \in H, x_1 < y$ in Y *there are* $x_2 \in DP({}^\omega\omega), (f',g') \in H$ such that: for every n large enough $f'(n) \ge f(n)^{g(n,x_1(n))}$ and $g(n,y(n)) \ge g(n,x_1(n)) \times g'(n, x_2(n))$.

Then $\varphi_{4,Y,H}^{cm}$ is a finest definition of a family of covering models.

Proof. 1) Let us check the conditions in Definition 2.2. Let $(f,g) \in H$ and we deal with each $\varphi_{4,Y,f,g}^{cm}[V]$ separately (this is enough).

(α) (a) Trivial by definition 2.11A.

(α) (b) Trivial.

(α) (c) Trivial.

(β) Check.

$(\gamma)^+$ Let $y < x$, yRT_n for $n < \omega$ (remember 1.3(4) letting $y = x^\dagger$).

Choose n_k by induction on k such that: $\bigwedge_{\ell<k} n_\ell < n_k$ and $n_k < \omega$ and $[n_k \leq \ell < \omega \Rightarrow k \times g(\ell, y(\ell)) < g(\ell, x(\ell))]$ (possible by assumption (ii)). Now $w = \{n_k : k < \omega\}$ and $T^* = \{\eta \in {}^{\omega>}\omega : \text{for every } n \in w, \eta \restriction n \in \bigcup_{\substack{\ell \leq n \\ \ell \in w}} T_\ell\}$ are as required.

(δ) By 2.4(1) (for (iv)* use the assumption (iii) of 2.11C(1)) and 2.7.

2) The proof is similar to the proof of part (1) using 2.4(2) instead 2.4(1) in proving clause (δ).

3) Note that (iii)* $\Rightarrow$ (iii)$^-$ easily, so demands (α), (β), (γ), $(\gamma)^+$, (δ) hold.

$(\varepsilon)^+$ Straightforward (use a tree T, x_2RT, to "catch" the T in a narrow tree $\subseteq TTR$).

(ζ) Check. $\square_{2.11C}$

2.11D Conclusion. For Y, H satisfying (i), (ii), (iii)* of 2.11C, for any CS iteration $\bar{Q} = \langle P_i, \underset{\sim}{Q}_j : i \leq \alpha, j < \alpha \rangle$, if $\Vdash_{P_i}$ "$\underset{\sim}{Q}_i$ is proper and $\varphi^{cm}_{4,Y;H}[V^{P_i}]$-preserving" *then* P_α is proper, $\varphi^{cm}_{4,Y;H}[V]$-preserving.

2.11D Definition. We say that a forcing notion Q is (f, g)-bounding (where $f, g \in {}^\omega(\omega + 1 \setminus \{0, 1\})$) if for every $\eta \in (\prod_{n<\omega} f(n))^{V^Q}$ there is $\langle a_n : n < \omega \rangle \in V$ such that $|a_n| \leq g(n)$ and $\eta \in \prod_{n<\omega} a_n$.

2.11F Conclusion. Assume

$(*)$ $f, g \in {}^\omega(\omega + 1 \setminus \{0, 1\})$ are diverging to infinity.

If $\bar{Q} = \langle P_i, \underset{\sim}{Q}_j : i \leq \alpha, j < \alpha \rangle$ is a CS iteration such that Q_i is $(f^{g^\ell}, [g]^{1/\ell})$-bounding in V^{P_i} for every $\ell < \omega$ *then* P_α is proper and (f^{g^ℓ}, g)-bounding for every $\ell < \omega$.

Proof. We use 2.11C(3) (and 2.11A, 2.11B and 2.3). We let $Y = \{x \in {}^\omega\omega : x$ constant $\}$, so we can identify x with $x(0)$, let $\{a_n : n < \omega\}$ list the positive rationals, define $x < y \Leftrightarrow a_{x(0)} < a_{y(0)}$ and let $g_\ell(n, x(n)) = [g(n)^{a_{x(n)}/\ell}]$ (=the integer part, note: x is constant so $x(n) = x(0)$) and $f_\ell(n) = f(n)^{[g(n)^\ell]}$.

Lastly let $H = \{(f_\ell, g_\ell) : \ell < \omega\}$, so $\varphi^{cm}_{4,Y;H}$ is well defined.

Now we show that $\varphi^{cm}_{4,Y;H}$ is a finest definition of covering model, to get this we would like to apply 2.11C(3). Among the three assumptions there, clause (i) holds by the choice of Y and H. Also the first phrase in clause (ii) holds, as for the second, if $x < y$ and $(f, g) \in H$, then for some ℓ, $(f, g) = (f_\ell, g_\ell)$, hence

$$g(n, y(n))/g(n, x(n)) = [g(n)^{a_{y(0)}/\ell}]/[g(n)^{a_{x(0)}/\ell}]$$

$$\geq (g(n)^{a_{y(0)}/\ell} - 1)/(g(n)^{a_{x(0)}/\ell}) = g(n)^{(a_{y(0)} - a_{x(0)})/\ell} - g(n)^{-a_{x(0)}/\ell}$$

as $a_{y(0)} > a_{x(0)} > 0$, this clearly diverges to infinity.

Lastly for clause (iii)*, let $(f, g) \in H$ (so for some ℓ, $(f, g) = (f_\ell, g_\ell)$) and let $x_1 < y$ in Y. Now choose $x_2 \in Y$ such that $\varepsilon + x_2(0) < y(0) - x_1(0)$ for some $\varepsilon > 0$ and choose m such that $a_{x_1(0)}/\ell < m$ and let $(f', g') = (f_{\ell+m}, g_{\ell+m})$. Let us check:

$$f'(n) = f_{\ell+m}(n) = f(n)^{g(n)^{\ell+m}} = (f(n)^{g(n)^\ell})^{g(n)^m} = f_\ell(n)^{g(n)^m}$$

$$\geq f_\ell(n)^{g_\ell(n, x_1(n))}$$

(the last inequality because $g_\ell(n, x_1(n)) = [g(n)^{a_{x_1(n)}/\ell}]$ and $a_{x_1(0)}/\ell < m$)

$$g(n, y(n)) = g_\ell(n, y(n)) =$$

$$[g(n)^{a_{y(n)}/\ell}] \geq g(n)^{a_{y(n)}/\ell} - 1 \geq (g(n)^{a_{x_1(n)}/\ell})(g(n)^{a_{x_2(n)}/\ell})(g(n)^{\varepsilon/\ell}) - 1$$

$$\geq g(n, x_1(n))g(n, x_2(n))g(n)^{\varepsilon/\ell} - 1 > g(n, x_1(n))g(n, x_2(n))$$

(the last inequality: for n large enough).

So really (iii)* of 2.11C(3) holds hence 2.11C(3) applies and $\varphi^{cm}_{4,Y;H}$ is a finest definition of covering models, so 2.3(2) applies.

Lastly, we can check that by monotonicity

⊗ Q is $\varphi^{cm}_{4,Y;H}$-preserving iff Q is $(f^{g^\ell}, [g^{1/\ell}])$-bounding for every $\ell < \omega$.

So by the last two sentences we are done. $\square_{2.11E}$

* * *

2.12A Definition. [The *PP* property] 1) We define $\varphi = \varphi_5^{cm}$ (a definition of a covering model) by letting $\varphi[V] = (D, R, <)$ (the *PP* model) where:

a) $D = H(\aleph_1)$

b) xRT iff $x, T \in D, x \in {}^\omega\omega$ is strictly increasing, $T \subseteq {}^{\omega>}\omega$ is a closed subtree and $T \cap {}^n\omega$ is finite for every n and:

(*) for arbitrarily large n there are k, and $n < i(0) < j(0) < i(1) < j(1) \leq \ldots < i(k) < j(k) < \omega$ and for each $\ell \leq k$, there are $m(\ell) < \omega$ and $\eta^{\ell,0}, \ldots, \eta^{\ell,m(\ell)} \in T \cap {}^{j(\ell)}\omega$, such that: $j(\ell) > x(i(\ell) + m(\ell))$ and

$$(\forall \eta \in T \cap {}^{j(k)}\omega) \bigvee_{\ell,m} \eta^{\ell,m} \trianglelefteq \eta.$$

c) $<$ is $<^*_{\text{dis}}$

Remark: concerning the *PP* property, there is a strong version ("strong *PP*-property") proved in 4.4 and 5.6 for the forcing notion there and a weak version ("weak *PP*-property") derived in 2.12D below and used in 4.7 and 5.8 (though in the statement the "*PP*-property" appears). See Definition 2.12E.

2.12B Claim. 1) If the forcing notion P is φ_5^{cm}-preserving then it has the ${}^\omega\omega$-bounding property; if P has the Sacks property (i.e. is φ_2^{cm}-preserving) *then* it is φ_5^{cm}-preserving.

2) If $(D, R, <)$ is a Sacks model (i.e. $\varphi_2^{cm}[V]$) *then*
$(\forall \eta \in {}^\omega\omega)(\forall x)(\exists T \in D)\ [x \in (\text{Dom}(R^{\varphi_5^{cm}})) \cap D \Rightarrow xR^{\varphi_5^{cm}}T \ \&\ \eta \in \lim T]$

3) If $(D, R, <)$ is a *PP*-model (see 2.2(5)) *then*
$(\forall \eta \in {}^\omega\omega)(\forall x)(\exists T \in D)\ [x \in (\text{Dom}(R^{\varphi_1^{cm}})) \cap D \Rightarrow xR^{\varphi_1^{cm}}T \ \&\ \eta \in \lim T]$.

Proof. Easy.

2.12C Claim. $\varphi = \varphi_5^{cm}$ is a finest definition of a covering model, 2-directed.

Proof. Let us check the conditions in Definition 2.2.

(α) (a) Trivial by a), b) of Definition 2.12A.

(α) (b) Check.

(α) (c) Check.

(β) Trivial.

(γ)(a) So let $(D, R, <)$ be $\varphi[V]$. Let $x > y$, and yRT_n. Let $h_m : \omega \to \omega$ be such that for any n there are $i(0) < j(0) < \ldots < j(k), \eta_{\ell,i}$ (for $i \leq m(\ell), \ell \leq k$) witnessing $(*)$ of Definition 2.12A(1)(b) for yRT_m and n (so $n < i(0)$) such that $j(k) < h_m(n)$. Now we define n_i by induction on $i, n_0 = 0$ and n_{i+1} is such that: choose $\ell_0, \ell_1, \ldots, \ell_{i+1}$ as follows: $\ell_0 = n_i, \ell_{j+1} = h_{n_j}(\ell_j) + 1$, and $n_{i+1} = \ell_{i+1}$. Let $T^* = \{\eta :$ for every $i, \eta \restriction n_{i+1} \in \bigcup_{j<i} T_{n_j}\}$ i.e. we choose $w = \{n_i : i < \omega\}$.

So clearly $xR^{\varphi_5^{cm}}T^*$ is as required.

(γ)(b) Easy.

(δ) We use 2.4(2) with $h_n(\ell) = \ell$. So let $x \in \mathrm{Dom}(R)$, and we choose $y = x$. Now $w_\ell = \{(n, \ell) : \ell \geq n\}$, g_n is any one to one function from ${}^n\omega$ onto ω, $f^n_{(n,\ell)}$ is thus determined. Now check.

$(\varepsilon)^+$ So we know Q is $\varphi_5^{cm}[V]$-preserving, $\varphi_5^{cm}[V] \models y < x$ and $T^* \in V^Q$, and $\varphi[V^Q] \models yRT^*$. We should find $T^{**} \in V$ such that: $T^* \subseteq T^{**}$ and $\varphi[V] \models xRT^{**}$. We work in $V^+ = V^Q$ but $(D, R, <) = \varphi_5^{cm}[V]$. The proof is straight but still we elaborate. Let $h^* : \omega \to \omega$ be defined for T^* as h_m was defined for T_m in the proof of clause (γ)(a). So by 2.12B(3) there is $h^{**} \in (D \cap {}^\omega\omega)^V$ such that h^{**} is strictly increasing and $(\forall n)[h^*(n) \leq h^{**}(n)]$.

We now choose z such that for every n, there are $n = m^n_0 < m^n_1 < \ldots < m^n_{n+1}, m^n_{\ell+1} = h^{**}(m^n_\ell) + m^n_\ell + 1$, and let $z(n) = m^n_{n+1}$. Clearly $z \in D^V$.

So remembering 2.7, we can apply the "covering property" of (D, R) to T^* (i.e., the branch T^* induces in TTR). Apply it for z and we get an appropriate closed subtree $C \in D = H(\aleph_1)^V$ of TTR, (so $T^* \cap {}^{n\geq}\omega \in C$ for every n). Clearly $T^{**} = \bigcup_{t \in C} t$ is a closed subtree of ${}^{\omega>}\omega$, it belongs to D, and there is no problem to prove $T^* \subseteq T^{**}$. The only point left is why xRT^{**}.

Let $C^\dagger$ be the set of $t \in C$ such that if $n < \omega, h^{**}(n) \leq ht(t)$ then for some $k < ht(t)$ and $n < i(0) < j(1) < \ldots < i(k) < j(k) \leq ht(t)$ the statement in

$(*)$ of Definition 2.12 A clause (b) holds (for t and y). Let C'' be the maximal closed tree $\subseteq C^\dagger$. It is easy to check that $C'' \in D$, and that T^* induces a branch $\subseteq C''$, so without loss of generality $C = C^\dagger = C''$.

Now for arbitrarily large n, there are $k < \omega$, and $n < i(0) < j(0) < i(1) < j(1) < \ldots < i(k) < j(k) < \omega$, and for each $\ell < k$ there are $m(\ell) < \omega$, $t_{\ell,0}, \ldots, t_{\ell,m(\ell)} \in C \cap TTR_{j(\ell)}$ such that $j(\ell) > z(i(\ell) + m(\ell))$ and

$$(\forall t \in C \cap TTR_{j(k)})[\bigvee_{l,n} t_{l,n} \leq t].$$

By the definition of z, there are $\xi(\ell, 0) < \ldots < \xi(\ell, m(\ell) + 1)$ such that $i(\ell)+m(\ell)+1 < \xi(\ell,0)$ and $\xi(\ell, m(\ell)+1) < j(\ell)$, and $h^{**}(\xi(\ell,m)) < \xi(\ell,m+1)$. So by the assumption on $C (= C^\dagger)$ for each such $\ell < k$, $m < m(\ell)$, there are $k_{\ell,m}$, $\xi(\ell,m) < i(0,\ell,m) < j(0,\ell,m) < i(1,\ell,m) < j(1,\ell,m) < \ldots < i(k_{\ell,m},\ell,m) < j(k_{\ell,m},\ell,m) < \xi(\ell,m+1)$ and $n(\alpha,\ell,m)$ (for $\alpha < k_{\ell,m}$) such that $j(\alpha,\ell,m) > x(i(\alpha,\ell,m) + n(\alpha,\ell,m))$ and $\eta_{\alpha,\beta,\ell,m} \in ({}^{j(\alpha,\ell,m)}\omega) \cap t_{\ell,m}$ (for $\beta < n(\alpha,\ell,m)$) and $(\forall \nu \in t_{\ell,m} \cap {}^{\xi(\ell,m+1)}\omega)[\bigvee_{\alpha,\beta} \eta_{\alpha,\beta} \lhd \nu]$.
Now the set of $i(\alpha,\ell,m), j(\alpha,\ell,m), n(\alpha,\ell,m)$ and $\eta_{\alpha,\ell,m,\beta}$ for $\beta < n(\alpha,\ell,m)$ supplies the required witnesses.

(ζ) Easy (by 2.8 C). $\square_{2.12C}$

2.12D Claim. Assume $V \subseteq V'$ and $\varphi_5^{\mathrm{cm}}(V)$ covers in V'. *Then* for every $\eta \in ({}^\omega 2)^{V'}$ there is an infinite $w \subseteq \omega$ from V and $\langle k_n, \langle i_n(\ell), j_n(\ell) : \ell \leq k_n \rangle : n \in w \rangle$ from V such that:

(a) $n < i_n(0) < j_n(0) < i_n(1) < j_n(1) < \ldots < i_n(k_n) < j_n(k_n) < \min(w \setminus (n+1))$.

(b) for every $n \in w$ for some $\ell \leq k_n$ we have $\eta(i_n(\ell)) = \eta(j_n(\ell))$.

Remark. Only the x defined by $x(\ell) = 2^\ell$ suffices.

Proof. Easy.

2.12E Definition. 1) A forcing notion Q has the PP-property *iff* it is φ_5^{cm}-preserving.

2) A forcing notion Q has the weak PP-property if V, V^Q satisfies the conclusion of 2.12D.

3) A forcing notion Q has the strong PP-property if changing φ_5^{cm} to $\varphi_{5.5}^{\mathrm{cm}}$ in Definition 2.12A by demanding $k = 0$ in $(*)$, we have: $\varphi[V]$ covers in V^Q.

2.12F Claim. For a forcing notion Q:

1) the strong PP-property implies the PP-property.

2) The PP-property implies the weak PP-property.

2.12G Conclusion. For a CS iteration $\bar{Q} = \langle P_i, \underset{\sim}{Q}_j : i \leq \alpha, j < \alpha \rangle$, if $\Vdash_{P_i}$ "$\underset{\sim}{Q}_i$ is proper with the PP-property" *then* P_α is proper with the PP-property.

Proof. By 2.3(2), by 2.12C.

* * *

The following deals with "no Cohen real + no real dominates F ($\subseteq {}^{\omega>}\omega$, see §3)".

2.13A Definition. We define $\varphi = \varphi_6^{cm}$ (a definition of a covering model) by letting $\varphi[V] = (D, R, <)$ where

a) $D = H(\aleph_1)$

b) xRT *iff* T is a perfect nowhere dense tree, $x \in \mathrm{DP}\,({}^\omega\omega)$.

c) $< = <_0$

2.13B Observation. Q is φ_6^{cm} preserving iff Q adds no Cohen real.

2.13C Claim. φ_6^{cm} is a 2-directed, fine definition of a covering model for forcing which are φ_1^{cm}-preserving† (= ${}^\omega\omega$-bounding) *or* even just not adding a dominating real†† and are proper (or satisfy UP) (caution: not preserved under composition).

---

† Of course as ${}^\omega\omega$-bounded forcing necessarily add no Cohen reals.

†† On this see 3.17(2),(3).

Proof. (α) (a), (b), (c) Trivial

(β) Trivial

(γ) Check, even $(\gamma)_1$ (see 1.3(8)) and $(\gamma)_2$ (of 2.5(2)) hold. E.g. concerning $(\gamma)_2$, given $\langle T_n : n < \omega \rangle$, nowhere dense trees, choose by induction on $i < \omega$, $n_i < \omega$ as follows: $n_0 = 0$, n_{i+1} is minimal n such that $n \in (n_i, \omega)$ and for every $\eta \in {}^{n_i \geq}(n_i + 1)$ there is ν: $\eta \triangleleft \nu \in {}^{n>}n$ such that $\nu \notin \bigcup_{j \leq n_i} T_j$. Let for $\ell < 2$, $T^\ell \stackrel{\text{def}}{=} \{\eta \in {}^{\omega>}\omega : \text{for some } i = \ell \mod 2, \text{ and } n \in [n_i, n_{i+1}) \text{ we have } \eta{\restriction}n \in T_0,$ and $\eta \in T_n\}$.

(δ) By 2.5(2). $\square_{2.13C}$

2.13D Conclusion. 1) For a CS iteration $\bar{Q} = \langle P_i, \underset{\sim}{Q}_i : i \leq \alpha, j < \alpha \rangle$, if $\Vdash_{P_i}$ "$\underset{\sim}{Q}_i$ is proper not adding a Cohen real (over V^{P_i})" and P_α adds no dominating real over V *then* P_α is proper and adds no Cohen real over V.

2) The property "P purely does not add a Cohen real nor an $\underset{\sim}{\eta} \in {}^\omega\omega$ dominating F" where $F \subseteq {}^\omega\omega$ is fixed not dominated in the old universe, is preserved in limit of iterations as in $0.1_{\theta=\aleph_0}$.

Remark. 1) Note we have $(|D|, \aleph_1)$-covering in §1.

2) What if in 0.1 we use (D), $\leq_{\text{pr}}$ is $=$ (so we use FS iterations satisfying the c.c.c.)? In the limit we add a Cohen real, necessarily the family is empty. We cannot apply it as for P a c.c.c. forcing, $\leq_{\text{pr}}$ is equality so "P purely preserves φ_0^{cm}" always fails.

3) Of course we can interchange using/not using F in parts (1) and (2) of 2.13D.

Proof. 1) By 2.3(1) and 2.13C applied to φ_6^{cm}.

2) Usually using in addition 3.17. $\square_{2.13D}$

* * *

The following deals with "every new real belongs to some old closed set of Lebesgue measure zero".

2.14A Definition. We define φ_7^{cm} (a definition of a covering model) by letting $\varphi[V] = (D, R, <)$ where

(a) $D = H(\aleph_1)$.

(b) xRT means T is a perfect tree, with $\lim T$ having Lebesgue measure zero.

(c) $<=<_0$

So if Q is φ_7^{cm}-preserving then Q adds no random real but not inversely.

2.14B Claim. φ_7^{cm} is a 2-directed fine definition of a covering model for forcing which are purely φ_1^{cm}-preserving (= ${}^\omega\omega$-bounding property) or even just not adding a dominating real (caution! not preserved under composition.)

Proof.

(α) Trivial

(β) Trivial

$(\gamma)^+$ Check

(δ) By 2.5(1). $\square_{2.17D}$

2.14C Conclusion. 1) The property "P purely does not add any real, which does not belong to any old closed measure zero set from V and is purely ${}^\omega\omega$-bounding and has pure $(2,2)$-decidability" is preserved by limit (for iterations as in $0.1_{\theta=2}$) [but not necessarily composition].

2) The property "P purely does not add any real not belonging to any closed old set of measure zero from V and adds no real dominating F" is preserved in limits for iterations as in 0.1, where $F \subseteq {}^\omega\omega$ is a fixed undominated family.

Proof. Like the proof of 2.13C. $\square_{2.14C}$

* * *

The following deals with "every new dense open subset of ${}^{\omega>}\omega$ is included in some old one".

2.15A Definition. Let $\langle \rho^*_\ell : \ell < \omega \rangle$ enumerate ${}^{\omega>}\omega$. Let $T^* \subseteq {}^{\omega>}\omega$ be a perfect tree such that for every $\nu \in \lim T^*$, $A_\nu \stackrel{\text{def}}{=} \{\rho^*_\ell : \ell < \omega$ and $\nu(2\ell) = 1\}$ is open, and $\rho^*_\ell {}^\wedge \rho^*_{\nu(2\ell+1)} \in A_\nu$ (hence A_ν is dense)$\}$, and such that for every dense open subset A of ${}^{\omega>}\omega$ there is $\nu \in \lim T^*$ such that $A = A_\nu$.

We define φ^{cm}_8 (a definition of a covering model) by letting

$\varphi^{\text{cm}}_8(V) = (D, R, <)$:

(a) $D = H(\aleph_1)$

(b) xRT means that $x \in \text{DP}\,({}^\omega\omega)$ and $T \subseteq T^*$ is perfect satisfying: $\bigcap\{A_\nu : \nu \in \lim T\}$ is dense open.

(c) $< = <_0$ (see 1.4)

2.15B Claim. 1) For $A \subseteq {}^{\omega>}\omega$ there is a closed $T = T_A \subseteq T^*$ such that: if $\eta \in \lim T^*$ then A_η (which is dense open) include A iff $\eta \in \lim T_A$. So $T \in \text{Rang}(R)$ iff for some dense open $A \subseteq {}^{\omega>}\omega$ we have $T \subseteq T_A$

2) A forcing notion Q is φ^{cm}_8 (V)-preserving *iff* every open dense subset of ${}^{\omega>}\omega$ in V^Q include a dense open subset of ${}^{\omega>}\omega$ from V *iff* for some (every) subuniverse $V^\dagger$ of V such that $\varphi^{\text{cm}}_8(V^\dagger)$ covers in V, Q is $\varphi^{\text{cm}}_8(V^\dagger)$ preserving.

3) If Q_0 is $\varphi^{\text{cm}}_8(V)$-preserving and $\Vdash_{Q_0}$ "$\underset{\sim}{Q}_1$ is $\varphi^{\text{cm}}_8(V^{Q_0})$-preserving" *then* $Q_0 * \underset{\sim}{Q}_1$
is $\varphi^{\text{cm}}_8(V)$-preserving.

4) If Q is φ^{cm}_8-preserving *then* Q is ${}^\omega\omega$-bounding.

Proof. 1) - 3) Check.

4) For $h \in ({}^\omega\omega)^{V^Q}$ let $A_h = \{\eta \in {}^{\omega>}\omega :$ for some n, $\bigwedge_{\ell<n} \eta(\ell) = 0$, $\eta(n) \neq 0$ and $\ell g(\eta) \geq n + h(n) + 1\}$. So $A_h \subseteq {}^{\omega>}\omega$ is dense open, so there is $A \subseteq {}^{\omega>}\omega$, dense open, $A \in V, A \subseteq A_h$. Define $g : \omega \to \omega$ in V by: $g(n) = \min\{\ell g(\eta) : \eta \in A, n = \min\{\ell : \eta(\ell) > 0\}\}$. Then $g : \omega \to \omega$, $g \in V$ and $(\forall n) h(n) < g(n)$. $\square_{2.15B}$

2.15C Claim. φ^{cm}_8 is a finest definition of a covering model which is 2-directed.

Proof. $(\alpha), (\beta)$, 2-directed: easy, now we prove more than (γ) (see 2.5(3)).

$(\gamma)_3$ Assume yRT_n, so for each n there is a dense open $A_n \subseteq {}^{\omega>}\omega$ such that $T_n \subseteq T_{A_n}$. We choose by induction on $n < \omega$, $k_n < \omega$ such that $\bigwedge_{\ell<n} k_\ell < k_n$, and if $\ell \leq n$ there is $\rho \in \bigcap_{m \leq n} A_m$ such that $\rho^*_\ell \lhd \rho$, $\rho \in \{\rho^*_m : m < k_n\}$. Now let $k = 2$, and for $i < k$ let

$$B_i = \{n : \text{ for some } \ell = i \mod 2 \text{ we have } k_\ell \leq n < k_{\ell+1}\}$$

and let

$$T_i = \{\nu \in T^* : \text{ if } n \leq \ell g(\nu) \text{ then } \nu \restriction n \in \bigcup \{T_{A_m} : m < n \text{ and } m \in B_i\}.$$

(δ) by 2.5(2) (remember that $(\gamma)_3 \Rightarrow (\gamma)_2$ by 2.5(3)).

$(\varepsilon)^+$, (ζ) left to the reader. $\square_{2.15C}$

Remark. alternatively, instead of 2.15 B,C, look in XVIII §3.

2.15D Conclusion. For CS iteration $\langle P_i, \underset{\sim}{Q}_j : i \leq \alpha, j < \alpha\rangle$, if $\Vdash_{P_i}$ "Q_i is proper and any new open dense subset of ${}^{\omega>}\omega$ includes an old one" *then* P_α is proper and any dense open subset $A \in V^P$ of ${}^{\omega>}\omega$ includes a dense open subset $A \in V$ of ${}^{\omega>}\omega$.

Proof. By 2.3(2) and 2.16C (and 2.16B(2)).

* * *

2.16 Conclusion. For $\theta = \aleph_0$ the property " Q is purely $\varphi_\ell[V]$-preserving with pure $(\theta, 2)$ decidability" is preserved by iteration as in 0.1_θ for $\ell = 1, \ldots, 8$ (e.g. by CS iterations of proper forcing). This is true for $\theta = 2$, $\ell = 1, \ldots, 8$ when $(*)$ of 1.12 holds (recheck).

* * *

The following is an addition from early nineties, inspired by the interest in "adding no Cohen real". It is dual to 2.13, see 2.17 C(1) below.

2.17A Definition. For $Y \subseteq DP({}^\omega\omega)$ with an absolute definition we define $\varphi^{\rm cm}_{9,Y}$, a definition of a covering model. For a universe V we let $\varphi[V] = (D, R, <)$ be:

(a) $D = H(\aleph_1)$.

(b) xRT iff

(i) $x = \langle x^{[0]}, x^{[1]}, x^{[2]}\rangle$, say $x(3n+i)$ codes $x^{[i]}(n)$, $x^{[i]} \in Y$, $x^{[1]}(n+1) > x^{[1]}(n)$ and the difference is a power of 2, $x^{[2]}[n] \leq \log_2[x^{[1]}(n+1) - x^{[1]}(n)]$,

(ii) $T \in {}^{\omega>}\omega$ closed subtree,

$$\eta \in {}^{\omega>}\omega \ \&\ \eta\restriction n \in T \ \&\ \eta(n) \geq x^{[0]}(n) \ \Rightarrow\ \eta \in T.$$

(iii) for each n the following holds:

$(*)_n$ for any $m < (x^{[1]}(n+1) - x^{[1]}(n))/2^{x^{[2]}(n)}$ there is a function $g = g_{n,m}$ with domain the interval $[x^{[1]}(n) + m\cdot 2^{x^{[2]}(n)}, x^{[1]}(n) + (m+1)\cdot 2^{x^{[2]}(n)})$, $(\forall \ell)[g(\ell) < x^{[0]}(\ell)]$ and $[\eta \in T \ \&\ \ell g(\eta) \geq x^{[1]}(n+1) \Rightarrow g_{n,m} \not\subseteq \eta]$. So $g \stackrel{\rm def}{=} \bigcup_{n,m} g_{n,m}$ belongs to $\prod_{\ell<\omega} x^{[0]}(\ell)$, we call g a witness.

(c) $x < y$ iff $x^{[0]} = y^{[0]}$, $x^{[1]} = y^{[1]}$ and $x^{[2]} <_{\rm dis} y^{[2]}$ (see 1.4).

Explanation. So what is the meaning of xRT? The interesting part of T is $T' \stackrel{\rm def}{=} \{\eta\restriction\ell : \eta \in (\lim T) \cap \prod_{n<\omega} x^{[0]}(n)\}$ and T is in a way "explicitly nowhere dense" i.e. for some $g \in \prod_{n<\omega} x^{[0]}(n)$, for every $\eta \in \lim(T')$ for every n for many subintervals I of $[x^{[1]}(n), x^{[1]}(n+1))$ we have $g\restriction I \neq \eta\restriction I$.

2.17B Claim. Assume $V \subseteq V^+$. Then $(\gamma) \Rightarrow (\beta) \Rightarrow (\alpha)$, where:

(α) "for every $f \in DP({}^\omega\omega)^V$ and $g \in (\prod_{\ell<\omega} f(\ell))^{V^+}$ there is $h \in (\prod_{\ell<\omega} f(\ell))^V$ such that $\{\ell : h(\ell) = g(\ell)\}$ is finite.

(β) $\varphi^{\rm cm}_{9,Y}[V]$ covers in V^+ for $Y = DP({}^\omega\omega)$.

(γ) every covering model from $\varphi^{\rm cm}_{9,DP({}^\omega\omega)}(V)$ covers in V^+.

2.17C Remark. 1) It is well known that: condition (α) implies there is no Cohen real over V in V^+; and if $V_0 \subseteq V_1 \subseteq V_2$, in V_1 there is a real f in ${}^\omega\omega$ dominating $({}^\omega\omega)^{V^0}$; and in V_2 there is $g \in \prod_{\ell<\omega} f(\ell)$ contradicting (α) *then* in V_2 there is a Cohen real over V_0.

The preservation theorem below implies a Cohen real is not added in limits.

2) Note that making $x^{[0]}$ smaller makes being in $\lim T$, xRT, harder.

3) The absoluteness requirements can be restricted as usual.

4) What we deduce below is complimentary in a sense to 2.13 A-C.

5) Why in 2.17A the $2^{x^{(2)}(n)}$? Of course a more general notion will use norms (see [RoSh:470]).

6) If Y is closed enough then in 2.17B we have $(\gamma) \Leftrightarrow (\beta)$.

2.17D Claim. 1) Assume

(i) Y is a subset of $DP({}^\omega\omega)$,

(ii) for every $x \in Y$ there is $y \in Y$ and there are $\langle \ell^*_n : n < \omega \rangle$ such that:

  (a) $\lim_n \ell^*_n = \infty$

  (b) $1 \leq \ell^*_n \leq \ell^*_{n+1}$, ℓ^*_n a power of 2 (for technical reasons)

  (c) $x^{[1]} = y^{[1]}$

  (d) $x^{[0]} = y^{[0]}$

  (e) $y^{[2]}(n) = x^{[2]}(n) - \log_2(\ell^*_n) \geq 0$

(iii) for $<$ from 2.17A(c) $(Y, <)$ is dense with no minimal member.

Then $\varphi^{\text{cm}}_{9,Y}[V]$ is a fine covering model.

2) Assume (i) Y is an absolute definition of a subset of $DP({}^\omega\omega)$,

(ii) clause (1)(ii) holds absolutely,

(iii) clause (1)(iii) holds absolutely.

Then $\varphi^{\text{cm}}_{9,Y}$ is a fine definition of a covering model.

Proof. We check the condition in definition 2.2.

(α) (a)(b)(c) Check.

(β) Check.

$(\gamma)^+$ We check on $\varphi^{\text{cm}}_{9,Y}[V]$.

So let $x_n < x_{n+1} < x^\dagger < y$ be given, $x_n R T_n$. Now any thin enough infinite $w \subseteq \omega$ will work as:

$\otimes$ $\bigcup_{\ell<\ell(*)} T_\ell$ satisfies $(*)_n$ from (b)(iii) of Definition 2.17A for y if $\bigwedge_\ell xRT_\ell$, $x^{[0]} = y^{[0]}$, $x^{[1]} = y^{[1]}$ and $2^{x^{[2]}(n)/y^{[2]}(n)}$ is $\geq \ell(*)$.

(δ) We use 2.4(3) (and for checking the demand (iv)*** there we use assumption (ii)).

Let $x \in \mathrm{Dom}(R)$ be given and we shall define y and $\mathbf{B}$ as required in clause (iv)*** of 2.4(3). So let y be as defined by clause (ii) of 2.17D(1). Now we define the Borel function $\mathbf{B}$; we let $\mathbf{B}(\langle \eta_\alpha : \alpha \leq \omega \rangle) = \nu$ (where $\eta_\alpha \in {}^\omega\omega$, $\eta_\alpha \restriction \alpha = \eta_\omega \restriction \alpha$, $\nu \in {}^\omega\omega$) if:

($\oplus_1$) Let $n < \omega$, m, $(x^{[1]}(n+1) - x^{[1]}(n))/2^{x^{[2]}(n)}$

(a) $\langle \eta_\omega(x^{[1]}(n) + m \cdot 2^{x^{[2]}(n)} + i) : i < 2^{x^{[2]}(n)} \rangle$ is equal to $\langle \nu(y^{[1]}(n) + m \cdot 2^{x^{[2]}(n) \cdot \ell^*_n} + i) : i < 2^{x^{[2]}(n)} \rangle$

(b) if $k < \ell^*_n - 1$ *then* $\langle \eta_k(x^{[1]}(n) + m \cdot 2^{x^{[2]}(n)} + i) : i < 2^{x^{[2]}(n)} \rangle$ is equal to $\langle \nu(y^{[1]}(n) + m \cdot 2^{x^{[2]}(n) \cdot \ell^*_n} + (k+1) \cdot 2^{x^{[2]}(n)} + i) : i < 2^{x^{[2]}(n)} \rangle$.

So assume T_1 satisfying yRT_1 is given and we should define appropriate T_1. As yRT_1, there are functions $g^1_{n,m}$ for $n < \omega$, $m < (y^{[1]}(n+1) - y^{[1]}(n))/2^{y^{[2]}(n)}$ witnessing it. For $n < \omega$, $m < (y^{[1]}(n+1) - y^{[1]}(n))/2^{x^{[2]}(n)}$ we define a function $g^0_{n,m}$ as follows. Its domain is of course $(x^{[1]}(n) + m \cdot 2^{x^{2]}(n)}, x^{[1]}(n) + (m+1) \cdot 2^{x^{[2]}(n)})$ and for each $k < \ell^*_n$, we have $g^1_{n,\ell^*_n \cdot m + \ell} \subseteq g^0_{n,m}$.
The checking is straight. $\square_{2.17D}$

Remark. Note that for many pairs (x_0, x_1) from T_1, $x_1 R T_1$ we can produce T_0, $x_0 R T_0$, where T_1 in a way codes T_0.

2.17E Conclusion. For $\varphi^{\mathrm{cm}}_{9,Y}$ as in 2.17D(1), for CS iteration $\bar{Q} = \langle P_i, \underset{\sim}{Q}_j : i \leq \delta, j < \delta \rangle$, if $\Vdash_{P_i}$ "Q_i is proper" and P_i preserve $\varphi^{cm}_{9,Y}[V]$ for $i < \delta$ *then* P_δ preserve $\varphi^{cm}_{9,Y}[V]$.

§3. Preservation of Unboundedness

3.1 Notation. 1) ψ may denote an absolute definition of a two-place relation on ${}^\omega\omega$ which we denote $R^\psi[V]$ (so when extending the universe, we reinterpret R, but we know that the interpretations are compatible). We write xRy instead of $R(x,y)$. Sometimes ψ is an absolute definition of a three-place relation R on ${}^\omega\omega$ and then we write $xR^z y$ instead of $R(x,y,z)$.

Let $\bar{R}$ denote $\langle R_n : n < \omega\rangle$ (each R_n as above) so $\bar{R}^m = \langle R_n^m : n < \omega\rangle$. We identify $\langle R : n < \omega\rangle$ with R.

Remember $\mathcal{S}_{<\kappa}(A) = \{B \subseteq A : |B| < \kappa\}$ and if κ is regular uncountable then $\mathcal{D}_{<\kappa}(A)$ is the filter on $\mathcal{S}_{<\kappa}(A)$ generated by the sets $G(M) = \{|N| : N \prec M, ||N|| < \kappa$ and $N \cap \kappa$ is an ordinal$\}$ for M a model with universe A and $< \kappa$ relations.

3.2 Definition. 1) For $F \subseteq {}^\omega\omega$ and a two place relation R on ${}^\omega\omega$, we say that F is R-bounding if $(\forall g \in {}^\omega\omega)(\exists f \in F)[gRf]$.

2) $F \subseteq {}^\omega\omega$ is $\bar{R}$-bounding if it is R_n-bounding for each n (where $\bar{R} = \langle R_n : n < \omega\rangle$).

3) For $F \subseteq {}^\omega\omega$, $\bar{R}$ (each R_n two place) and $S \subseteq \mathcal{S}_{<\aleph_1}(F)$ the pair $(F, \bar{R})$ is S-*nice* if:

α) F is $\bar{R}$-bounding.

β) For any $N \in S$, for some $g \in F$, for every $n_0, m_0 < \omega$ player II has a winning strategy for the following game and, moreover, the strategy is absolute. The game is defined for each countable set N (but only $N \cap F$ is needed) and it lasts ω moves.

In the kth move: *player I* chooses $f_k \in {}^\omega\omega, g_k \in F \cap N$, such that $f_k \restriction m_{\ell+1} = f_\ell \restriction m_{\ell+1}$ for $0 \leq \ell < k$ and $f_k R_{n_k} g_k$ and then *player* II chooses $m_{k+1} > m_k$ and $n_{k+1} > n_k$.

In the end player II wins if $\left(\bigcup_{k<\omega} f_k \restriction m_k\right) R_{n_0} g$, (or if player I can't choose in the k'th move he lose).

4) We say $(F, \bar{R})$ is $S/\mathcal{D}_{\aleph_0}(F)$-nice if: for some $C \in \mathcal{D}_{\aleph_0}(F)$, we have: $(F, \bar{R})$ is $(S \cap C)$-nice.

5) We omit S when this holds for some $S \in \mathcal{D}_{\aleph_0}(F)$.

3.3 Notation. $<^*$ is the partial order on ${}^\omega\omega$ defined as: $f <^* g$ iff for all but finitely many $n < \omega, f(n) < g(n)$. In this case we say that g dominates f. We say that g dominates a family $F \subseteq {}^\omega\omega$ if g dominates every $f \in F$.

3.4 Definition. 1) A family $F \subseteq {}^\omega\omega$ is dominating if every $g \in {}^\omega\omega$ is dominated by some $f \in F$.
2) A family $F \subseteq {}^\omega\omega$ is unbounded (or undominated) if no $g \in {}^\omega\omega$ dominates it.

3.5 Definition. 1) A forcing notion P is almost ${}^\omega\omega$-bounding if: *for every* P-name $\underset{\sim}{f}$ of a function from ω to ω and $p \in P$ *for some* $g : \omega \to \omega$ (from V!) *for every* infinite $A \subseteq \omega$ (again A from V) there is $p', p \leq p' \in P$ such that:

$$p' \Vdash_P \text{"for infinitely many } n \in A, \underset{\sim}{f}(n) < g(n)\text{"}$$

2) A forcing notion P is weakly bounding (or F-weakly bounding, where $F \subseteq ({}^\omega\omega)^V)$) if $({}^\omega\omega)^V$ (or F) is an unbounded family in V^P.

3.6 Claim.

1) If a forcing notion P is weakly bounding, and $\underset{\sim}{Q}$ $(\in V^P)$ is almost ${}^\omega\omega$-bounding, *then* their composition $P * \underset{\sim}{Q}$ is weakly bounding.
2) If Q is almost ${}^\omega\omega$-bounding, $F \subseteq {}^\omega\omega$ an unbounded family (from V) *then* F is still an unbounded family in V^Q.
3) If Q is adding λ Cohens (i.e. $Q \stackrel{\text{def}}{=} \{f : f$ a partial finite function from λ to $\{0,1\}\}$ ordered by inclusion) *then* Q is almost ${}^\omega\omega$-bounding.

Proof. 1) By part (2) (apply it in V^P to $F = ({}^\omega\omega)^V$ and the forcing notion Q).
2) Assume $p \in Q$ forces that $\underset{\sim}{f}$ dominates F and we shall get a contradiction. Let $g \in ({}^\omega\omega)^V$ be as in Definition 3.5(1). As in V, F is unbounded, for some $f^* \in F$ we have $\{n < \omega : g(n) < f^*(n)\}$ $(\in V)$ is infinite, so choose this set as A, so by Definition 3.5(1) we know that for some p':

(a) $p \leq p' \in Q$

(b) $p' \Vdash_Q$ "for infinitely many $n \in A$, $\underset{\sim}{f}(n) < g(n)$ (hence by A's definition $\underset{\sim}{f}(n) < g(n) < f^*(n))$"

and this contradicts $p \Vdash_Q$ "$\underset{\sim}{f}$ dominates F".

3) Easy. $\square_{3.6}$

3.7 Definition. $R^{\psi_0} = \psi_0(V)$ is: fRg iff $\{n : f(n) \leq g(n)\}$ is infinite.

3.8 Claim. A forcing notion P (in V) is weakly bounding ($\equiv$ adds no dominating real) *iff* $\Vdash_P$ "F is R-bounding" where $F = ({}^\omega\omega)^V$, $R = R^{\psi_0}$. $\square_{3.8}$

3.9 Claim. Let $R = R^{\psi_0}$ and $F \subseteq {}^\omega\omega$ be an R-bounding set, such that $(\forall f_0, \dots, f_n, \dots \in F)(\exists g \in F)[\bigwedge_{n<\omega} f_n <^* g]$. *Then* (F, R) is nice.

Proof. We have to describe g and an absolute winning strategy for N (and n_0, m_0). Choose $g \in F$ such that $(\forall f \in N)[f \in F \Rightarrow f <^* g]$. As for the strategy, n_ℓ is irrelevant, we just choose $m_{k+1} = \min\{m :$ there are at least k numbers $i < m$ such that $g(i) > f_k(i)\}$. $\square_{3.9}$

3.10 Claim. Suppose that $\mathcal{P} \subseteq [\omega]^{\aleph_0}$ is a P-filter (i.e. it is a filter containing the cobounded subsets and for any $A_n \in \mathcal{P}$ $(n < \omega)$ for some $A^* \in \mathcal{P}$ we have $(\forall n)[A^* \subseteq_{ae} A_n])$ and $\mathcal{P}$ has no intersection (i.e. there is no $X \in [\omega]^{\aleph_0}$ such that $X \subseteq_{ae} A$ for every $A \in \mathcal{P}$; recall that $X \subseteq_{ae} A$ means "$X \setminus A$ is finite"). Let R be:

$$xRy \text{ iff } x \notin [\omega]^{\aleph_0} \text{ or } y \notin [\omega]^{\aleph_0} \text{ or } x \not\subseteq_{ae} y.$$

(We identify $x \subseteq \omega$ with its characteristic function. The case "$y \notin [\omega]^{\aleph_0}$" will be irrelevant.)

Then

1) $(\mathcal{P}, R)$ is nice.

2) Let Q be a proper forcing notion. $\mathcal{P}$ is R–bounding in V^Q iff $\Vdash_Q$ " the filter $\mathcal{P}$ generates is a P-filter with no intersection " (i.e. every $q \in Q$ forces one statement iff it forces the other).

Proof. 1) Clause α) of Definition 3.2(3) is obvious as "$\mathcal{P}$ has no intersection" (see above). In (β) choose $g = A^* \in \mathcal{P}$ such that

$$(\forall A \in N)[A \in \mathcal{P} \Rightarrow A^* \subseteq_{ae} A].$$

Again, the least obvious point is the winning strategy; again n_k is irrelevant and player II chooses $m_k = \min\{m : f_k \cap m \setminus g \text{ has power} > k\}$.

2) Left to the reader. $\square_{3.10}$

3.10A Remark. We can use ${}^{\omega}\lambda$ instead ${}^{\omega}\omega$.

Sometimes we need a more general framework (but the reader may skip it, later replacing H_z, R^z_n by F, R_n).

3.11 Notation. If H is a set of (ordered) pairs, let $\text{Rang}(H) = \{y : (\exists x)[\langle x, y\rangle \in H]\}$ and $\text{Dom}(H) = \{x : (\exists y)[\langle x, y\rangle \in H]\}, H_x = \{y : \langle x, y\rangle \in H\}$.

We shall treat a set F (from e.g. Definition 3.2) as the following set of pairs: $\{\langle 0_\omega, x\rangle : x \in F\}$ where 0_ω is the function with domain ω and constant value 0 (so e.g. 3.13 applies to 3.2 too).

3.12 Definition. 1) For a set $H \subseteq {}^{\omega}\omega \times {}^{\omega}\omega$, and $\bar{R}$ (an ω-sequence of three place relations written as $xR^z y$) and $S \subseteq \mathcal{S}_{<\aleph_1}(H)$ we say that $(H, \bar{R})$ is S-nice if:

α) H is $\bar{R}$-bounding which means: for every $z \in \text{Dom}(H)$, H_z is $\bar{R}^z$-bounding, i.e. $(\forall n)(\forall f \in {}^{\omega}\omega)(\exists g \in H_z)[f R^z_n g]$ letting $\bar{R}^z = \langle R^z_n : n < \omega\rangle$.

β) For any $N \in S$, $z \in \text{Dom}(H \cap N)$ and for every $n_0, m_0 < \omega$ for some $g \in H_z$ and $z_0 \in \text{Dom}(H) \cap N$ player II absolutely wins the following game which lasts ω moves.

In the kth move: *player* I chooses $f_k \in {}^{\omega}\omega, g_k \in \text{Rang}(H \cap N)$ such that $f_k \restriction m_{\ell+1} = f_\ell \restriction m_{\ell+1}$ for $0 \le \ell < k$ and $f_k R^{z_k}_{n_k} g_k$; then *player* II chooses $m_{k+1} > m_k, n_{k+1} > n_k$ and $z_{k+1} \in \text{Dom}(H \cap N)$.

At the end of play, player II wins iff $(\bigcup_k f_k \restriction m_{k+1}) R^z_{n_0} g$.

2) $(H, \bar{R})$ is $S/\mathcal{D}_{\aleph_0}(H)$-nice if for some $C \in \mathcal{D}_{\aleph_0}(H)$ we have: $(H, \bar{R})$ is $(S \cap C)$-nice.

3.13 Lemma.

1) Suppose

(i) $\bar{Q} = \langle P_j, \underset{\sim}{Q}_i : i < \delta, j \leq \delta \rangle$ is an iteration as in 0.1, for[†] $\mathbb{I}$

(ii) $S \subseteq \mathcal{S}_{<\aleph_1}(H)$ is stationary in V, and if we are in one of the cases (C), (E), (F) of 0.1, then $|H| = \aleph_1$

(iii) $(H, \bar{R})$ is $S/\mathcal{D}_{<\aleph_1}(H)$-nice

(iv) for every $i < \delta$, in V^{P_i} we have: H is $\bar{R}$-bounding

(v) all Q_i have pure $(\aleph_0, 2)$-decidability (see Definition 1.9)

(vi) $|H| = \aleph_1$ or at least

$$(\forall a \in [V^{P_\delta}]) \left[|a| \leq \aleph_0 \ \&\ a \subseteq H \Rightarrow \left(\exists b \in (\mathcal{S}_{\aleph_0}(H))^V\right)[a \subseteq b]\right].$$

Then in V^{P_δ}, H is $\bar{R}$-bounding.

2) We can weaken (v) to

(v)$^-$ all Q_i has pure $(2,2)$-decidability

provided that for some fixed $f^* \in {}^\omega\omega$, $\left[(\exists i)[f(i) > f^*(i)] \Rightarrow f R^z_n g\right]$ for any $z \in \mathrm{Dom}(H)$ and $g \in H_z$.

3) Assume $\bar{R} = \langle R_n : n < \omega \rangle$ a decreasing sequence of (absolute definitions of) three place relations on ${}^\omega\omega$, $F \subseteq {}^\omega\omega$ is $\bar{R}$ bounding (i.e. we are in the context of Definition 3.2 not 3.11, 3.12). Assume further (i), (ii), (iii), (iv) and (vi) from (1), replacing H by F and

(v)$_f$ all Q_i have pure feeble $(\aleph_0, 2)$-decidability (see Definition 3.14 below).

Then in V^{P_δ}, F is $\bar{R}$-bounding.

4) Assume, as in (2) that for some fixed $f^* \in {}^\omega\omega$, $[(\exists i)f(i) > f^*(i) \Rightarrow f R_n g]$ for any f, $g \in {}^\omega\omega$. The results of (3) holds if we replace (v)$_f$ by (v)$_f^-$ meaning replacing there $(\aleph_0, 2)$ by $(2,2)$.

---

† So the reader may think on CS of proper forcing so $\leq_{\mathrm{pr}} = \leq$. The $\mathbb{I}$ is from clause (F) there, so can be ignored for the cases (A)-(E), e.g. the two cases just mentioned.

3.13A Remark.

1) You can read the proof with $n_0 = 0, F$ instead H, R instead $R_n^{z_n}$ (see 3.11).
2) The proof gives somewhat more than the lemma, i.e. it applies to more cases. "H is $\bar{R}$-bounding" means that (α) of 3.12 holds.
3) We can weaken 3.12(1)(β) to " in no generic extension of V, no strategy of player I is a winning strategy"(and 3.13 still holds). The proof is similar, only we choose the G_k in $V^{\mathrm{Levy}(\aleph_0, 2^{|P_\omega|})}$.
4) Part (3) (or (4)) of 3.13 is suitable for FS iteration of c.c.c. forcing by 3.16(4) below.

3.14 Definition. 1) A forcing notion Q has pure feeble (θ_1, θ_2)-decidability *if*: for every $p \in Q$ and Q-name $\underset{\sim}{\tau}$ satisfying $p \Vdash_Q$ "$\underset{\sim}{\tau} < \theta_1$" there are $a \subseteq \theta_1, |a| < \theta_2$ and $q, p \leq_{\mathrm{pr}} q \in Q$ such that q weakly decides $\underset{\sim}{\tau} \in a$; where
2) $q \in Q$ weakly decides $\underset{\sim}{\tau} \in a$ (or any other statement) if no pure extension of q decides this is false.
3) A forcing notion Q has pure weak (θ_1, θ_2)-decidability if for each $p \in Q$ in the following game, player II has a winning strategy.

In the n'th move player I chooses $\underset{\sim}{\tau}_n$, a Q-name of an ordinal $< \theta_1$ and player II chooses $a_n, a_n \subseteq \theta_1$, $|a_n| < \theta_2$. In the end player II wins the play if for every $n < \omega$ there is q_n, $p \leq_{\mathrm{pr}} q_n \in Q$, q_n weakly decides $\bigwedge_{\ell<n} \underset{\sim}{\tau}_\ell \in a_\ell$.

3.15 Proof of 3.13 (1). We speak mainly on cases (A) and (F) of 0.1(1). W.l.o.g. $\mathrm{cf}(\delta) = \aleph_0$ or for every $i < \delta$ we have $\Vdash_{P_i}$ "$\mathrm{cf}(\delta) > \aleph_0$" (by 3.16 below we have associativity; use a maximal antichain of conditions deciding and restrict yourselves above one member; then if necessary use renaming.)

If $\mathrm{cf}(\delta) > \aleph_0$, then any real in V^{P_δ} belongs to V^{P_j} for some $j < \delta$ (see III 4.1B(2), (or X or XIV or XV); hence there is nothing to prove, so we shall assume $\mathrm{cf}(\delta) = \omega$. By III, 3.3 or XV 1.7, w.l.o.g. $\delta = \omega$.

Suppose $p \in P_\omega$, $z \in \mathrm{Dom}(H)$, $n_0 < \omega$ and $\Vdash_{P_\omega}$ "$\underset{\sim}{f} \in {}^\omega\omega$" ; we shall find r, $p \leq_{\mathrm{pr}} r \in P_\omega$ and $g \in H_z$ such that $r \Vdash_{P_\omega}$ "$\underset{\sim}{f} R^z_{n_0} g$". Let $m_0 < \omega$. Let N be a countable elementary submodel of $(H(\lambda), \in)$ (λ regular large enough) to which

$\langle P_j, \underset{\sim}{Q}_i : i < \omega, j \leq \omega\rangle, p, \underset{\sim}{f}, z, S, H$ belong as well as the parameters involved the definitions of the R_n's. The set of such N belongs to $\mathcal{D}_{<\aleph_1}(H(\lambda))$, hence for some such N, $N \cap H \in S$ (and N is $\mathbb{I}$-suitable for case (F) of 0.1).

By 1.11 w.l.o.g. for each $n < \omega, \underset{\sim}{f}(n)$ is a P_n-name, and we let $p = \langle p_n^0 : n < \omega\rangle$ where $\Vdash_{P_n}$ "$p_n^0 \in \underset{\sim}{Q}_n$". Let $g \in H_z$ and $z_0 \in N \cap \mathrm{Dom}(H)$ be as in clause (β) of Definition 3.12 (for $N \cap H$ and z, n_0, m_0).

We shall now, by induction on $k < \omega$, define $q_k, \underset{\sim}{p}_k, \underset{\sim}{g}_k, \underset{\sim}{z}_k, \underset{\sim}{m}_k, \underset{\sim}{n}_k$ such that

- (a) $q_k \in P_k$ is (N, P_k)-generic (for (A) of 0.1(1)) or (N, P_k)-semi-generic (for (F) of 0.1(1)) and $q_k \Vdash_{P_k}$ "$N[\underset{\sim}{G}_{P_k}] \cap H = N \cap H$"
- (b) $q_k \restriction n = q_n$ for $n < k$
- (c) $\underset{\sim}{p}_k \in P_\omega$, in fact is a P_k-name of a member of P_ω
- (d) $\underset{\sim}{p}_k \restriction k \leq_{\mathrm{pr}} q_k$
- (e) $\underset{\sim}{p}_{k+1} \restriction k = \underset{\sim}{p}_k \restriction k$ and $\underset{\sim}{p}_k \leq_{\mathrm{pr}} \underset{\sim}{p}_{k+1}$
- (f) $q_k \Vdash_{P_k}$ "$\underset{\sim}{p}_k \in N[\underset{\sim}{G}_{P_k}]$" i.e. $\underset{\sim}{p}_k$ is a P_k-name of a member of $N[\underset{\sim}{G}_{P_k}] \cap (P_\omega/\underset{\sim}{G}_{P_k})$
- (g) $\underset{\sim}{z}_k$ is a P_k-name of a member of $\mathrm{Dom}(H) \cap N$
- (h) $\underset{\sim}{m}_k < \underset{\sim}{m}_{k+1}$ and $\underset{\sim}{n}_k < \underset{\sim}{n}_{k+1}$
- (i) $\underset{\sim}{m}_k$, $\underset{\sim}{n}_k$ are P_k-names of natural numbers

Note that (a) implies that $N \cap H$ belongs to the club of $\mathcal{S}_{<\aleph_1}(H)$ involving "$(H, \bar{R})$ is $S/\mathcal{D}_{<\aleph_1}(H)$-nice".

For $k = 0$ we let $q_0 = \emptyset, p_0 = p$.

For $k+1$, we work in $V[G_k]$, G_k a generic subset of P_k satisfying $q_k \in G_k$. So $p_k = \underset{\sim}{p}_k[G_k] \in N[G_k]$ and $p_k \restriction k \in G_k$. In $N[G_k]$ we can find an $\leq_{\mathrm{pr}}$-increasing sequence of conditions $p_{k,i} \in P_\omega/G_k$ for $i < \omega$, such that $p_{k,0} = \underset{\sim}{p}_k[G_k]$ and $p_{k,i} \in N[G_k]$, moreover even $\langle p_{k,i} : i < \omega\rangle \in N[G_k]$ and $p_{k,i}$ forces values for $\underset{\sim}{f}(j)$ for $j \leq i$. So for some function $f_k \in N[G_k]$ we have $f_k \in {}^\omega\omega$ and $p_{k,i} \Vdash_{P_\omega/P_k}$ "$\underset{\sim}{f} \restriction i = f_k \restriction i$". As $N[G_k] \prec (H(\lambda)[G_k], \in)$ (see III 2.11), for some $g_k \in N[G_k] \cap H_{z_k} = N \cap H_{z_k}$, we have $N[G_k] \models$ "$f_k R^{z_k}_{n_k} g_k$"†. Now we use the absolute strategy (from Definition 3.6(2) for $N \cap H$) to choose

† Really n_k, g_k are P_k-names so we should have written $\underset{\sim}{n}_k[G_k]$ but ignore this.

$z_{k+1}, n_{k+1}, m_{k+1}$ (the strategy's parameters may not be in N, but the result is) and we want to have $p_{k+1} = p_{k,m_{k+1}}$. However all this was done in $V[G_k]$, so we have only suitable P_k-names which is O.K. In the end, let $r \in P_\omega$ be defined by $r \restriction k = q_k \restriction k$ for each k; by requirement (b) we know that r is well defined and belongs to P_ω. Suppose $r \in G_\omega \subseteq P_\omega, G_\omega$ generic over V. As in the proof of the preservation of properness we can prove by induction on k that $\underset{\sim}{p}_k \leq_{\text{pr}} r$ for each k. Then in $V[G_\omega]$ we have made a play of the game from Definition 3.12(1)(β), player II using his winning strategy so $((\bigcup_k \underset{\sim}{f}_k \restriction k)[G_\omega])R^z_{n_0} g$ holds in $V[G_\omega]$, but clearly $p_{k,n_k} \leq_{\text{pr}} \underset{\sim}{p}_{k+1} \leq_{\text{pr}} r$ hence $p_{k,n_k} \in G_\omega$ hence $(\underset{\sim}{f} \restriction m_k)[G_\omega] = (\underset{\sim}{f}_k \restriction m_k)[G_\omega]$. Consequently $\underset{\sim}{f}[G_\omega] = (\bigcup_k \underset{\sim}{f}_k \restriction k)[G_\omega]$ and $\underset{\sim}{f}[G_\omega]R^z_{n_0} g$ holds in $V[G_\omega]$ and r forces the required information .

Proof of 3.13(2): Similar.

Proof of 3.13(3): Like 3.13(1). We use freely 3.16 below, but note that no harm is caused if player II increases m_k, n_k (not z_k!). A play (or an initial segment of the play) in which player II do this is said to weakly follow the strategies. Now the strategy in use is to weakly follow all possible subplays. I.e. above (in the proof of 3.13(1)) we, by induction on $k < \omega$, choose q_k, $\underset{\sim}{p}_k$, $\underset{\sim}{g}_k$, $\langle(\underset{\sim}{z}_v, \underset{\sim}{m}_v, \underset{\sim}{n}_v) : k \in v \subseteq k+1\rangle$: and $\underset{\sim}{m}_k$, $\underset{\sim}{n}_k$ such that:

(a) - (f) and (h) as before

(g)′ $\underset{\sim}{z}_v$ is a P_k-name of a member of $\mathrm{Dom}(H) \cap N$

(i) $\underset{\sim}{m}_v$, $\underset{\sim}{n}_v$ are P_k-names of natural numbers, and

$$k = \max(v) \Rightarrow \underset{\sim}{m}_v < \underset{\sim}{m}_{v\setminus\{k\}} \ \&\ \underset{\sim}{n}_v < \underset{\sim}{n}_{v\setminus\{k\}}$$

(j) $\underset{\sim}{m}_k = \max\{m_v + k : v \subseteq k+1\}$, $\underset{\sim}{n}_k = \max\{\underset{\sim}{n}_v + k : v \subseteq k+1\}$.

In the induction step, $p_{k,i}$ $(i < \omega)$, f_k are chosen such that: $p_k \leq_{\text{pr}} p_{k,0}$, $p_{k,i} \leq_{\text{pr}} p_{k,i+1}$ and no pure extension of $p_{k,i}$ in P_ω/G_k forces $\underset{\sim}{f}{\restriction} i \neq f_k{\restriction} i$. Now for each v such that $k \in v \subseteq k+1$ we pretend that the play so far involve only player I choosing $\langle(f_\ell, g_\ell) : \ell \in v\rangle$ and player II choosing $\langle(\underset{\sim}{m}_{v\cap(\ell+1)}, \underset{\sim}{n}_{v\cap(\ell+1)}, \underset{\sim}{z}_{v\cap(\ell+1)}) : \ell \in v \setminus \{k\}\rangle$ and player's II given winning strategy dictates $(\underset{\sim}{m}_v, \underset{\sim}{n}_v, \underset{\sim}{z}_v)$. Lastly $\underset{\sim}{m}_{k+1}$, $\underset{\sim}{n}_{k+1}$ are computed by clause (j).

We have defined a name for a strategy; we can show that it is forced that unboundedly often we have made the right move, so moving to the appropriate subplay we are done.

Proof of 3.13(4): Similar. $\square_{3.13}$

3.16 Claim. 1) For $(\theta_1, \theta_2) \in \{(2,2), (\aleph_0, 2)\}$ the property "Q has pure feeble (θ_1, θ_2)-decidability" is preserved by iteration as in 0.1.

2) Similarly[†] for "pure weak".

3) Q has pure feeble (θ_1, θ_2)-decidability if Q has pure weak decidability.

4) If Q has feeble pure $(\theta^*, 2)$-decidability and θ^* is uncountable and $\leq_{\rm pr}$ is equality (as we do for FS iteration of c.c.c. forcing) or $\theta^* \geq 2$ and $\leq_{\rm pr}$ is $\leq^Q$ (as for CS iteration of proper forcing) *then* Q has pure feeble $(\theta, 2)$-decidability for every θ.

5) For $(\theta_1, \theta_2) \in \{(n, 2) : 2 \leq n < \omega\}$, every Q has pure feeble (θ_1, θ_2)-decidability.

Proof. 1) We copy the proof of 1.10, changing (iii) (in the proof of case 5 $(\alpha = \omega)$) to

(iii)$'$ first for $n < \omega$ we define a P_{n+1}-name $\underset{\sim}{s}_n$: for $G_{n+1} \subseteq P_{n+1}$ generic over $V, \underset{\sim}{s}_n[G_{n+1}]$ is $k+1$ if there is $r \in P_\omega/G_{n+1}$ such that $\mathrm{Dom}(r) = [n+1, \omega)$, $P_\omega/G_{n+1} \models$ "$p \restriction [n+1, \omega] \leq_{\rm pr} r$" and r weakly decides $\underset{\sim}{t} = k$, i.e. for no $r', r \leq_{\rm pr} r' \in P_\omega/G_{n+1}$ does $p' \Vdash_{P_\omega/G_{n+1}}$ "$\underset{\sim}{t} \neq k$"; if there is no such r, then $\underset{\sim}{s}_n[G_{n+1}] = 0$.

Second let $q_n \in \underset{\sim}{Q}_n[G_n]$ be such that $p_n \leq_{\rm pr} q_n$ and q_n weakly decides the value of $\underset{\sim}{s}_n$, (i.e. of $\underset{\sim}{s}_n/G_n$) (if $\theta_1 = 2$, use Definition 3.13A twice).

Also in the end we prove by downward induction on $m \leq n(*)$ that $(r \restriction m) \cup \{q_m\}$ weakly decides $\underset{\sim}{s}_m = \ell$.

2) Similar proof (using 3.16(1)).

3) Read the definitions.

4) Straight.

[†] Alternatively use XIV §2.

5) Easy. $\square_{3.16}$

* * *

We now give some applications. Concerning 3.17 if you want also "no Cohen", see 2.13.

3.17 Conclusion. 1) The property "P is weakly bounding" i.e. "P does not add a dominating real over V" is preserved in limit (for iterations as in $0.1_{\theta=2}$, see 0.1(3)) provided that $\mathfrak{b}^V = \aleph_1$ in the non-proper case.

2) If $F \subseteq {}^\omega\omega$ is not $<^*$-bounded then "P does not add a $<^*$-bound to F" is preserved in limit (for iterations as in $0.1_{\theta=2}$) provided that e.g. $|F| = \aleph_1$ in the non proper-cases.

3) In parts (1)+(2) we can use iterations as in 0.1 with pure feeble $(\aleph_0, 2)$-decidability.

Proof. 1), 2) Let $\bar{Q}$ be such an iteration, $F = ({}^\omega\omega)^V$ for 3.17(1), given for 3.17(2) and R is defined by ψ_0 (see Definition 3.7). By 3.9 (F, R) is a nice pair in V. Even for every $i < \ell g(\bar{Q})$, in V^{P_i} the set F is still unbounded and every countable subset of F in V^{P_i} is included in a countable subset of F from V; hence by 3.9 (F, R) is a nice pair even in V^{P_i}. By 3.13(3) this is true also in V^{P_δ}(where $\delta = \ell g(\bar{Q})$.

3) Similar proof (to that of 3.13(1)) or by 3.13(3)). $\square_{3.17}$

3.18 Lemma. The property "P_α purely adds no random real over V" is preserved under limits for iterations as in $0.1_{\theta=2}$ or just by iteration as in 0.1 with every $\underset{\sim}{Q}_i$ having pure feeble $(\aleph_0, 2)$-decidability (see 3.16(4)).

Remark. Concerning the successor case see XVIII 3.20(i). Before we prove 3.18 we need some definitions and claims. Now for $T \subseteq {}^{\omega>}\omega$, and $\eta \in {}^{\omega>}2$ we let $T^{\langle\eta\rangle} = \{\nu : \eta\hat{\ }\nu{\restriction}[\ell g(\eta), \omega) \in T\}$. Note that Lemma 3.18 includes the case of FS iterations.

3.19 Definition. 1) We let ψ_1 be as follows:
$xR^{\psi_1}y$ *iff* y is a perfect subtree of ${}^{\omega>}2$ with positive Lebesgue measure, $x \in {}^{\omega}2$ and $(\forall n < \omega)(\forall \rho \in {}^{n}2)[\rho\hat{\ }(x\restriction[n,\omega)) \notin \lim y]$.
2) Let H_1^V be $\{\langle y_1, y_2\rangle : y_1, y_2$ are perfect subtrees of ${}^{\omega>}2$ with positive Lebesgue measure such that: $\lim y_2 \subseteq \{\eta \in {}^{\omega}2$: for some $n < \omega$ and $\rho \in {}^{n}2$ we have $\rho\hat{\ }(\eta\restriction[n,\omega)) \in \lim y_1\}\ \}$

3.20 Claim. 1) H_1^V is an $\aleph_1$-directed partial order.
2) Suppose $V \subseteq V_1$, and for any countable $a \subseteq H_1^V$ from V_1 there is a countable $b \subseteq H_1^V, a \subseteq b \in V_1$ (and $R = R^{\psi_1}$). The following are equivalent:

(i) no real in V_1 is random over V

(ii) $\mathrm{Dom}(H_1^V)$ is R-bounding in V_1

(iii) $(\mathrm{Dom}(H_1^V), R)$ is nice in V_1 (here Definition 3.2(3) is the relevant one, with $\mathrm{Dom}(H_1^V)$ here having the role of F there).

Proof. 1) Easy
2) (i) $\Rightarrow$ (ii): Let $x \in ({}^{\omega}2)^{V_1}$. As x is not random over V there is a Borel set $B \in V$ of Lebesgue measure 0 such that $x \in B$ (i.e. x belongs to the V_1-interpretation of B). Without loss of generality B is closed under $=^*$ (i.e. if $\eta_1, \eta_2 \in {}^{\omega}2$ and $\eta_1 =^* \eta_2 (\equiv \bigvee_{n<\omega} \eta_1\restriction(n,\omega) = \eta_2\restriction[n,\omega))$ then $\eta_1 \in B \equiv \eta_2 \in B)$. There is $T \subseteq {}^{\omega>}2$ perfect, $T \in V$ such that $\lim T$ has positive measure and $\lim(T) \cap B = \emptyset$. So it is enough to prove that xRT, i.e. $(\forall n < \omega)[x \notin \lim T^{\langle n\rangle}]$ where $T^{\langle n\rangle} \overset{\mathrm{def}}{=} \{\eta$: for some $\rho \in T$ we have $\ell g(\rho) = \ell g(\eta)$ and $(\forall \ell)(n \le \ell < \ell g\rho \rightarrow \rho(\ell) = \eta(\ell))\}$ i.e. $x \in {}^{\omega}2 \setminus \bigcup_{n<\omega} \lim(T^{\langle n\rangle})$, but this follows from $x \in B$.

(ii) $\Rightarrow$ (iii): Condition (α) of Definition 3.2(3) is clear. For condition (β) let $N \prec (H(\chi), \in)$ be countable, $z_0 \in N \cap \mathrm{Dom}(H_1)$, so for some a we have $N \cap H_1^V \subseteq a \subseteq H_1^V$, $a \in V$, $V \models |a| = \aleph_0$. So there is $T \in \mathrm{Dom}(H_1^V)$, such that ${}^{\omega}2 \setminus \bigcup_{n<\omega} T^{\langle n\rangle}$ contains all $2^{\omega} \setminus \bigcup_{n<\omega} \lim(T_1^{\langle n\rangle})$ for $T_1 \in N \cap \mathrm{Dom}(H_1^V)$, hence it contains all Borel measure zero sets from V which are in N.

We have to give the winning strategy for player II.

In stage k, f_k, g_k are given $f_k R g_k$, $g_k \in N \cap \mathrm{Dom}(H_1^V)$, so g_k is a perfect subtree of ${}^{\omega>}2$ of positive Lebesgue measures. Then $\rho\hat{\ }(f_k\restriction[n,\omega) \notin \lim g_k$ for $\rho \in {}^{n}2$,

$n < \omega$ and $(g_k, T) \in H_1$; together with the choice of T we know that for $n < \omega$ and $\rho \in {}^n2$ we have $\rho\hat{\ }(f_k{\restriction}[n,\omega)) \notin \lim T$.

Choose $m_{k+1} > m_k$ large enough such that: for every $n \leq m_k$, $\rho \in {}^n2$ we have: $\rho\hat{\ } f_k \restriction [n, m_{k+1}) \notin T$.

(iii) $\Rightarrow$ (i): Immediate. $\square_{3.20}$

3.21 *Proof of lemma 3.18*

Let $F \stackrel{\text{def}}{=} \mathrm{Dom}(H_1^V)$. Let $\bar{Q}$ be an iteration as in 0.1, $\ell g(\bar{Q}) = \delta$, and in no $V^{P_i}(i < \delta)$ is there a real random over V. So by the claim 3.20(2) we know that $(F, \bar{R})$ is nice in V^{P_i}. Hence by 3.13(3) it is nice in V^{P_δ}, hence by claim 3.20(2) in V^{P_δ} there is no real random over V. $\square_{3.18}$

We now give an application of 3.17, taken from [Sh:207], Lemma 3.22 is proved in §6 (see 6.13). On history see introduction to §6.

3.22 Lemma. There is a forcing notion Q such that

(a) Q is proper

(b) Q is almost ${}^\omega\omega$-bounding.

(c) $|Q| = 2^{\aleph_0}$

(d) In V^Q there is an infinite set $A^* \subseteq \omega$ such that for every infinite $B \subseteq \omega$ from V we have $A^* \cap B$ is finite or $A^* \setminus B$ is finite.

3.22A Remark. For 3.23 it is enough to prove 3.22 assuming CH.

3.23 Theorem. Assume $V \models$ CH.

1) For some forcing notion P^*, P^* is proper, satisfies the $\aleph_2$-c.c., and

(*) In V^{P^*}, $2^{\aleph_0} = \aleph_2$, there is an unbounded family of power $\aleph_1$, but no splitting family (see below) of power $\aleph_1$.

2) We can also demand that in V^{P^*} there is no MAD of power $\aleph_1$ (see Definition 3.24(2)).

3.24 Definition. 1) $\mathcal{P}$ is a splitting family if $\mathcal{P} \subseteq [\omega]^{\aleph_0}$ (= the family of infinite subsets of ω) and for every $A \in [\omega]^{\aleph_0}$ for some $B \in \mathcal{P}$ we have: $|A \cap B| = |A \setminus B| = \aleph_0$.

2) A family $\mathcal{A}$ is MAD (maximal almost disjoint) if:

(a) $\mathcal{A}$ is a subset of $[\omega]^{\aleph_0}$

(b) for any distinct $A, B \in \mathcal{A}$ the intersection $A \cap B$ is finite

(c) $\mathcal{A}$ is maximal under (a) + (b).

3) Let $\mathfrak{b} = \min\{|F| : F \subseteq {}^\omega\omega$ is not dominated$\}$ where "F not dominated" means that for every $g \in {}^\omega\omega$ for some $f \in F$ we have $\neg f \leq^* g$. Let $\mathfrak{d} = \min\{|F| : F \subseteq {}^\omega\omega$ is dominating$\}$ where "F is dominating" means that for every $g \in {}^\omega\omega$ for some $f \in F$ we have $g <^* f$. Let $\mathfrak{s} = \min\{|\mathcal{P}| : \mathcal{P} \subseteq [\omega]^{\aleph_0}$ is a splitting family (see above)$\}$

Proof of 3.23. 1) We define a countable support iteration of length $\aleph_2$: $\langle P_\alpha, \underset{\sim}{Q}_\alpha : \alpha < \omega_2 \rangle$ with (direct) limit $P^* = P_{\omega_2}$. Now each $\underset{\sim}{Q}_\alpha$ is the Q from 3.22 for V^{P_α}, so $V^{P_\alpha} \models$ "$|\underset{\sim}{Q}_\alpha| = 2^{\aleph_0}$". As $V \models$ CH we can prove by induction on $\alpha < \omega_2$ that $\Vdash_{P_\alpha}$ "CH" (see III, Theorem 4.1). We also know that P^* satisfies the $\aleph_2$-c.c. (see III, Theorem 4.1). If $\mathcal{P}$ is a family of subsets of ω of power $\leq \aleph_1$ in V^{P^*} then for some $\alpha < \omega_2, \mathcal{P} \in V^{P_\alpha}$, and forcing by Q_α gives a set A^*_α exemplifying $\mathcal{P}$ is not a splitting family by clause (d) of 3.22. So from all the conclusions of 3.23 only the existence of an undominated family of power $\aleph_1$ remains. Now we shall prove that $F = ({}^\omega\omega)^V$ is as required. By 3.8 it is enough to show

(*) $\Vdash_{P_{\omega_2}}$ "F is R^{ψ_0}-bounding" (see Definition 3.7).

Now note: F has power $\aleph_1$ as $V \models$ CH. We prove that F is R^{ψ_0}-bounding in V^{P_α} by induction on $\alpha \leq \omega_2$. For $\alpha = 0$ this is trivial; $\alpha = \beta + 1$: as $\underset{\sim}{Q}_\beta$ is almost ${}^\omega\omega$-bounding (see 3.22 clause (b)) and by Fact 3.6(1); if $\mathrm{cf}(\alpha) \geq \aleph_0$ by Conclusion 3.17(1).

2) Similar. We use a countable support iteration $\langle P_j, \underset{\sim}{Q}_i : i < \omega_2, j \leq \omega_2 \rangle$ such that:

(a) for every $i < \omega_2$, and MAD $\langle A_\alpha : \alpha < \omega_1 \rangle \in V^{P_i}$, for some $j > i$, *either* $\underset{\sim}{Q}_{2j}$ = adding $\aleph_1$ Cohen reals, and $\underset{\sim}{Q}_{2j+1} = \{p \in \underset{\sim}{Q}^{V^{P_{2j+1}}} :$

$p \geq p_{2j+1}\}$ where in $V^{P_{2j+1}}$ we have $p_{2j+1} \Vdash_{\underset{\sim}{Q}}$ "$\langle A_\alpha : \alpha < \omega_1 \rangle$ is not a MAD family"

or $\underset{\sim}{Q}_{2j}$ = adding $\aleph_1$-Cohen reals, $\underset{\sim}{Q}_{2j+1} = Q[I_{2j+1}]$ where I_{2j+1} is the ideal (of subsets of ω) which $\langle A_\alpha : \alpha < \omega_1 \rangle$ and the cofinite sets generate (on $Q[I]$ see Definition 6.10).

(b) For j even $\underset{\sim}{Q}_j$ is adding $\aleph_1$ Cohen reals.

(c) For j odd, $\underset{\sim}{Q}_j$ is $\underset{\sim}{Q}$, or $\{p \in \underset{\sim}{Q} : p \geq p_j\}$, or it is $Q[I_{\underset{\sim}{A}}]$ where $\underset{\sim}{A}$ is a P_{j-1}-name of a MAD family of cardinality $\aleph_1$ and $V^{P_j} \models [\Vdash_{\underset{\sim}{Q}}$ "$\mathcal{A}_j$ is MAD"]

(d) for $\aleph_2$ ordinals j, Q_{2j+1} is $Q^{V^{P_{2j+1}}}$.

There is no problem to carry out the definition. Each $\underset{\sim}{Q}_j$ is almost ${}^\omega\omega$-bounding by 3.22 (i.e. see 6.13, when $\underset{\sim}{Q}_j = Q^{V^{P_j}}$), by 6.22 (when $\underset{\sim}{Q}_j = Q[I_{\mathcal{A}_j}]$ we can apply it to $V^{P_{j-1}}$ as $\underset{\sim}{Q}_{j-1}$ is adding $\aleph_1$ Cohens so 6.15 applies and the second possibility in 6.22 fails by clause (c) above) and 3.6(3) (if j is even i.e. Q_j is adding $\aleph_1$ Cohens). So as in part (1), P_{ω_2} preserves "$({}^\omega\omega)^V$ unbounded". Also $\mathfrak{s} = \aleph_2$ and $2^{\aleph_0} = \aleph_2$ are proved as part (1) by clause (d). Lastly assume $\mathcal{A} \subseteq \mathcal{P}(\omega)$ is a MAD family, $|\mathcal{A}| = \aleph_1$, so for some i, $\mathcal{A} \in V^{P_i}$. So there is j as in clause (a). Work over $V_0 = V^{P_{2j}}$ so $\underset{\sim}{Q}_{2j}$ is adding $\aleph_1$ Cohens. If $p \Vdash_{\underset{\sim}{Q}_{2j} * \underset{\sim}{Q}}$ "$\mathcal{A}$ is not MAD" for some $p \in \underset{\sim}{Q}_{2j} * \underset{\sim}{Q}$ *then* w.l.o.g. $p \in \underset{\sim}{Q}$ (as $\underset{\sim}{Q}_{2j}$ is homogeneous) and we use the second possibility in clause (c). If not, we use the third possibility of clause (c). $\square_{3.23}$

* * *

We add the following in Summer'92 after a question of U. Abraham. In the proof of the consistency of "there is no P-point" below (§4) we use the "PP-property" (see 2.12A-F). We actually prove a stronger property called "the strong PP-property" which implies the "PP-property" which we have proved is preserved, so Abraham asked whether it itself is preserved. The following variants would have sufficed for the purpose of §4 which was the reason of existence.

3.25 Lemma. Assume that $\mathbf{f} : \omega \to \omega + 1 \setminus \{0,1\}$, $\mathbf{h} : \omega \to \omega \setminus \{0,1\}$ and $F \subseteq \prod_{\mathbf{f},\mathbf{h}} \stackrel{\text{def}}{=} \{f : \mathrm{Dom}(f) = \omega$ and $f(n)$ a subset of $\mathbf{f}(n)$ of cardinality $< \mathbf{h}(n)$, $\lim_{n\to\infty} (|f(n)|/\mathbf{h}(n)) = 0\}$ are such that:

(∗) for any countable $A \subseteq F$ there is $f \in F$ such that $\bigwedge_{g\in A} (\forall^* n)[g(n) \subseteq f(n)]$.

Let R be defined as: gRf *iff*

(a) $g, f \in \prod_{\mathbf{f},\mathbf{h}}$ (i.e. we consider a member of ${}^\omega\omega$ as coding such sequences)

(b) $(\exists^* n) g(n) \subseteq f(n)$.

Let $S \subseteq \mathcal{S}_{<\aleph_2}(F)$ be stationary and assume F is R-bounding.

Then (F, R) is S-nice. (Hence, we have a preservation theorem for a limit).

Proof. Check Definition 3.2(3). Part (α), F is R-bounding, should be clear. For clause (β), given $N \in S$, let $f_N \in F$ be such that $f \in N \cap F \Rightarrow (\forall^* n)[f(n) \subseteq f_N(n)]$ (it exists by the assumption (∗)). The winning strategy is clear: choose m_{k+1} such that $\{i < m_{k+1} : f_k(i) \subseteq g_k(i) \subseteq f_N(i)\}$ has at least k members. $\square_{3.25}$

But of course it is nicer to have also preservation for composition of two forcing notions.

3.26 Lemma. 1) Let $\mathbf{f} : \omega \to \omega + 1 \setminus \{0,1\}$, $\mathbf{h}^t : \omega \to \omega \setminus \{0\}$ for $t \in \mathbb{Q}$ be such that for $s < t$ (from $\mathbb{Q}$) we have $0 = \lim_{n\to\infty} (\mathbf{h}^s(n)/\mathbf{h}^t(n))$ and for each $t \in \mathbb{Q}$ the set $\prod_{\mathbf{f},\mathbf{h}^t}$ satisfies (∗) of 3.25. The following property is preserved by iterations as in $0.1_{\theta=2}$ and as in 0.1 with each Q_i having pure feeble $(2,2)$-decidability:

$(*)_1$ (a) Q is purely ${}^\omega\omega$-bounding.

(b) *for every* $s < t$ from $\mathbb{Q}$, $f \in V^Q$ such that $f \in \prod_{\mathbf{f},\mathbf{h}^s}$ and $k_0 < k_1 < \dots$ (so $\langle k_i : i < \omega\rangle \in V$) *for some* $g \in V$ such that $g \in \prod_{\mathbf{f},\mathbf{h}^t}$ we have $(\exists^* i)\, [\bigwedge_{\ell\in[k_i,k_{i+1})} f(\ell) \subseteq g(\ell)]$.

1A) So e.g. in (1), if $\bar{Q} = \langle P_i, \underset{\sim}{Q}_j : i \leq \alpha, j < \alpha\rangle$ is CS iteration, $\Vdash_{P_i}$ "Q_i is proper satisfying $(*)_1$" then P_α is proper satisfying $(*)_1$.

2) We can replace $(*)_1$, by

$(*)_2$ (a) Q is purely ${}^\omega\omega$-bounding

(b) *for every* $s < t$ from $\mathbb{Q}$ and $f \in V^Q$, such that $f \in \prod_{\mathbf{f},\mathbf{h}^s}$ *for some* $g \in V$, $g \in \prod_{\mathbf{f},\mathbf{h}^t}$ and for every infinite $A \in ([\omega])^{\aleph_0})^V$, for infinitely may $i \in A$, $f(i) \subseteq g(i)$.

3) In (1) we can assume $F = ({}^\omega\omega)^{V'}$ (for some $V' \subseteq V$, a just reasonably closed) is unbounded $\aleph_1$-directed by $<^*$ and replace $(*)_1$ by

$(*)_3$ (a) Q purely preserves "F is unbounded"

(b) like (b) of $(*)_1$ for $\langle k_i : i < \omega\rangle \in F$.

Proof. Similarly to the previous Lemma one may deal with limit cases using 3.13 for the respective variant of clause (b) (and by 2.8 for $(*)_1$(a), $(*)_2$(a), 3.17 for $(*)_3$(a)). So now it suffices to prove this for iteration of length two: $P_2 = Q_0 * \underset{\sim}{Q}_1$ (so $P_1 = Q_0$, P_0 is trivial). First we prove part (2). Let $f \in (\prod_{\mathbf{f},\mathbf{h}^s})^{V^{P_2}}$, $s < t$ from $\mathbb{Q}$. Choose $t' \in (s,t)_{\mathbb{Q}}$. Applying $(*)_2$ to f, s, t', V^{P_1}, V^{P_2} we can find $f' \in [\prod_{\mathbf{f},\mathbf{h}^{t'}}]^{V^{P_1}}$ satisfying the requirements there on g. Next we apply $(*)_2$ on f', t', t, V, V^{P_1} and get $g \in [\prod_{\mathbf{f},\mathbf{h}^t}]^V$.

Now for any infinite $A \subseteq \omega$, $A \in V$, by the choice of g we know that $A' =: \{i \in A : f'(i) \subseteq g(i)\} \in V^{Q_0}$ is infinite. Hence by the choice of f' we know that $A'' \stackrel{\text{def}}{=} \{i \in A' : f(i) \subseteq f'(i)\}$ is infinite and clearly it belongs to V^{P_2}. Putting together $A'' = \{i \in A : f(i) \subseteq f'(i) \subseteq g(i)\}$ is infinite. So g is as required.

Now we prove part (1), so we are given $s < t$ from $\mathbb{Q}$ and $\langle k_i : i < \omega\rangle \in V$ (strictly increasing) and $f \in (\prod_{\mathbf{f},\mathbf{h}^s})^{V^{P_2}}$. Let $t' \in (s,t)_{\mathbb{Q}}$. Applying $(*)_1$, to f, s, t', $V^{P_1}, V^{P_2}, \langle k_i : i < \omega\rangle$ we get $f' \in (\prod_{\mathbf{f},\mathbf{h}^{t'}})^{V^{P_1}}$ satisfying the requirements on g in $(*)_1$. So $A =: \{i :$ for every $\ell \in [k_i, k_{i+1})$ we have $f(\ell) \subseteq f'(\ell)\}$ is infinite. As Q_0 is ${}^\omega\omega$-bounding there is a sequence $\ell(0) < \ell(1) < \dots$ (in V) such that $A \cap [\ell(i), \ell(i+1)) \neq \emptyset$ for every $i < \omega$. Let $k'_i = k_{\ell(i)}$ and apply $(*)_1$ to $f', t', t, V, V', \langle k'_i : i < \omega\rangle$ and get g, which is as required.

The proof of part (3) is similar. $\square_{3.26}$

3.26A Remark. In 3.25 we have the requirement $F \cap \prod_{\mathbf{f},\mathbf{h}}$ satisfies $(*)$ of 3.25. We can work as 3.26(2) and weaken it to:

$(*)^-$ for any $s < t$ and countable $A \subseteq \prod_{\mathbf{f},\mathbf{h}^t}$ there is $f \in \prod_{\mathbf{f},\mathbf{h}^t}$ such that $\bigwedge_{g \in A} (\forall^* n)[g(n) \subseteq f(n)]$.

§4. There May Be No P-Point

We define the forcing notion $P(F)$ (introduced by Gregorief) which, for an ultrafilter F, adds a set A such that $\omega \setminus A$, $A \neq \emptyset \mod F$, see definition 4.1. If F is a P-point (see definition 4.2A) this forcing is α-proper for every $\alpha < \omega_1$, and has the PP-property. Our point is that $P(F)^\omega$ enjoys all these properties and in addition $\Vdash_{P(F)^\omega}$ "F cannot be completed to a P-point ". We will argue in the following way: as we use $P(F)^\omega$, we can define a new subset A_n of ω such that $\Vdash_{P(F)^\omega}$ "$A_n \in \underset{\sim}{E}$", where $\underset{\sim}{E}$ is an extension of F to an ultrafilter in the generic extension, but for each $g \in {}^\omega\omega \cap V$ we have $\Vdash_{P(F)^\omega}$ "$\bigcap_{n<\omega}(A_n \cup g(n)) \equiv \emptyset \mod F$".

We originally (see the presentation in Wimmer [Wi]) use the stronger version of the PP-property, but there were problems with the preservation theorem i.e., in that version the essential forcing was not an iteration.

Note that, we continue to add reals after forcing with $P(F)^\omega$, so in fact we prove the above described argument works with Q instead of $P(F)^\omega$ provided $P(F)^\omega \lessdot Q$, Q has the PP-property. So the importance of proving that this property of Q is preserved is clear. The iteration in the end is standard.

The proof presented in [Wi] uses not exactly $P(F)^\omega$. Rather we note that if Q satisfies the c.c.c. then for any P-point F_0 in V^P there is $F_1 \stackrel{\text{def}}{=} \{A \subseteq \omega : A \in V, \Vdash_Q \text{"}A \in \underset{\sim}{F}_0\text{"}\}$ which is a filter enjoying some of the properties of $F_0 : \mathcal{P}(\omega)/F_1$ (in V) is a Boolean algebra satisfying the c.c.c. and (if Q has the ${}^\omega\omega$-bounding property), for every $A_n \in F_1$ there is $A \in F_1$, $\bigwedge_{n<\omega} A \subseteq_{ae} A_n$. Let $\{F^i : i < \aleph_2\}$ (assuming G.C.H.) list all such filters in V. Now the product P with countable support of all the $P(F^i)^\omega$ satisfies: in V^P, no F^i can be extended to a P-point by an argument as mentioned above. However to close the proof we need "P satisfies the c.c.c. ", which fails. But we replace P by a subset which satisfies it and still has the desirable other properties. I expect that the proof can be modified to have $2^{\aleph_0} > \aleph_2$ (but this was not carefully checked), whereas for the present proof we do not know how to do this.

4.1 Definition. For a filter D on a set I (we always assume all co-finite subsets of I are in D), we define the following forcing notions ordered by inclusion:

1) $P(D) = \{f : f$ is a function from B to $\{0,1\}$ for some $B = \emptyset \bmod D$, i.e. $B \subseteq I, I \setminus B \in D\}$,
2) $P^\dagger(D) = \{f \in P(D) : f^{-1}(\{1\})$ is finite $\}$,
   $P'(D) = \{f : f$ a function from B to $\{0,1\}$, $B \neq I \bmod D\}$,
   $P''(D) = \{f \in P'(D) : f^{-1}(\{1\})$ is finite $\}$.

4.1A Remark. Mathias [Mt3] used $P''(D)$ for the filter D of co-finite subsets of ω; Silver used $P^\dagger(D)$ for the filter D of cofinite subsets of ω and for an ultrafilter D, Gregorief used $P(D)$ for an ultrafilter D, and proved that it collapses $\aleph_1$ iff it is not a P-point.

4.2 Lemma. If F is a P-point (see below) *then* $Q = P(F)^\omega$ is proper (in fact α-proper for every $\alpha < \omega_1$) and has the PP-property.

Proof. It will follow from 4.3 and 4.4. $\square_{4.2}$

4.2A Definition.

1) A filter F on I is called a P-filter or P-point filter *if* (it contains all co-finite subsets of I and) for every $A_n \in F$ (for $n < \omega$) there is $A \in F$ such that $A \subseteq_{ae} A_n$ for every n. Just "a P-point" means an ultrafilter.
2) We call F fat if for every family of finite pairwise disjoint $w_n \subseteq I$ (for $n < \omega$) there is an infinite $S \subseteq \omega$ such that $\bigcup_{n \in S} w_n = \emptyset \bmod F$. (Clearly every P-point is fat.)
3) F is a Ramsey ultrafilter (on I) if for every $h : I \to \omega$ there is $A \in F$ such that $h{\restriction}A$ is a constant or $1-1$ (and F contains all co-finite subsets of I). Note that a Ramsey ultrafilter is a P-filter.

4.3 Fact. Assume F is a fat P-filter on ω. Let F^* be $\{A \subseteq \omega \times \omega$: for every n for some $B \in F$, $A \cap (\{n\} \times \omega) = \{n\} \times B\}$. *Then* F^* is a fat P-filter on $\omega \times \omega$, and the forcing notion $P(F)^\omega$ is isomorphic to $P(F^*)$.

Proof of the Fact.

First condition: F^* is a filter on $\omega \times \omega$ including all co-finite sets. Check.

Second condition: F^* is a P-filter. Let $A_k \in F^*$, then $A_k \cap (\{n\} \times \omega) = \{n\} \times B_{k,n}$ for some $B_{k,n} \in F$. As F is a P-filter there is $B^* \in F$, $B^* \subseteq_{ae} B_{k,n}$ for every k, n. Let $B_n^* \overset{\text{def}}{=} B^* \cap \bigcap_{k<n} B_{k,n}$ and $A^* \overset{\text{def}}{=} \bigcup_{n<\omega}(\{n\} \times B_n^*)$. Clearly $B_n^* \in F$ (as we have assumed that F is a filter), hence $A^* \in F^*$. Now $A^* \setminus A_k \subseteq \bigcup_{n<\omega}[\{n\} \times (B_n^* \setminus B_{k,n})] \subseteq \bigcup_{n\leq k}[\{n\} \times (B_n^* \setminus B_{k,n})]$, but $B_n^* \setminus B_{k,n} \subseteq B^* \setminus B_{k,n}$, hence it is finite. Therefore $A^* \setminus A_k$ is finite and hence $A^* \subseteq_{ae} A_k$. But k is arbitrary, so A^* is as required.

Third condition: F^* is fat. Let $w_n \subseteq \omega \times \omega$ be finite and pairwise disjoint for $n < \omega$. We define by induction on n infinite sets $S_n \subseteq \omega$, $S_{n+1} \subseteq S_n$ such that $\{i < \omega : \langle n, i\rangle \in \bigcup_{\ell \in S_n} w_\ell\} = \emptyset \bmod F$. We can do this with no problem, and let $k(0) = \mathrm{Min}(S_0)$, $k(n) = \mathrm{Min}(S_n \setminus \{k(0) \dots k(n-1)\})$. As every cofinite subset of ω belongs to F, it is easy to check $\bigcup_{n<\omega} w_{k(n)} = \emptyset \bmod F^*$.
So we have established the first conclusion of 4.3.

The isomorphism of $P(F)^\omega$ and $P(F^*)$ is trivial, for $p = \langle f_0, f_1, f_2 \dots\rangle \in P(F)^\omega$, let $H(p) \in P(F^*)$ be $H(p)(\langle n, k\rangle) = f_n(k)$. $\square_{4.3}$

So it suffices (for proving Lemma 4.2) to prove:

4.4 Lemma. If F is a fat P-filter on a countable set *then* $P(F)$ is proper (in fact α-proper for every $\alpha < \omega_1$) and has the PP-property.

4.4A Remark. We will really prove the strong PP-property, see remark to 2.12A and Def 2.12E(1),(3).

Proof of 4.4. W.l.o.g. F is a filter on ω. So let $p_0 \in P(F)$, $\{p_0, F\} \in N \prec (H(\lambda), \in)$, N is countable and λ is large enough. Now, before proving properness, we prove:

4.5 Crucial Fact. For every $p \in P(F)$ and every $P(F)$-name $\underset{\sim}{t}$ of an ordinal and $n < \omega$, there is $q \in P(F)$, $p \leq q$, such that $q{\restriction}n = p{\restriction}n$ and for every $g : n \to 2$, there is an ordinal α_g such that $(q{\restriction}[n, \infty)) \cup g \Vdash_{P(F)}$ "$\underset{\sim}{t} = \alpha_g$".

Proof. Let g_i (for $i < 2^n$) be a list of all functions $g : n \to 2$. We shall define by induction, an increasing sequence of conditions $p_i \in P(F)$ (so $p_i \leq p_{i+1}$), for $i \leq 2^n$. Let $p_0 = p$, and if p_i is defined let

$$p_i^\dagger = (p_i{\restriction}[n, \infty)) \cup g_i.$$

Clearly $p_i^\dagger \in P(F)$ hence there are α_{g_i} and $p_i'' \in P(F), p_i^\dagger \leq p_i''$ such that $p_i'' \Vdash_{P(F)}$ "$\underset{\sim}{t} = \alpha_{g_i}$". Let $p_{i+1} = p_i \cup p_i''{\restriction}[n, \omega)$. Clearly $p_i \leq p_{i+1} \in P(F)$, and $(p_{i+1}{\restriction}[n, \omega)) \cup g_i \Vdash$ "$\underset{\sim}{t} = \alpha_{g_i}$". So $p_{(2^n)}$ is as required from q. So we have proved Fact 4.5. $\square_{4.5}$

Before we prove 4.4 we also note

4.6. Fact. Assume F is a fat P-filter (on ω). If $p_n \leq p_{n+1}$ (for $n < \omega$), $p_n \in P(F)$ *then* there is $q \in P(F)$, $q \geq p_0$ such that $q \geq p_n{\restriction}[n, \infty)$ for infinitely many n.

Proof of Fact 4.6. Let A_n be the domain of p_n, so $A_n = \emptyset \bmod F$ hence $\omega \setminus A_n \in F$. As F is a P-filter there is A, $\omega \setminus A \in F$, such that for every n, $(\omega \setminus A) \subseteq_{ae} (\omega \setminus A_n)$ i.e., $A_n \subseteq_{ae} A$. Hence there are $k_n < \omega$ such that $(A_n \setminus [0, k_n)) \subseteq A$. w.l.o.g. $A_0 = \mathrm{Dom}(p_0) \subseteq A$ and $k_n > n$.

Now we shall choose natural numbers $\ell(0) < \ell(1) < \ell(2) < \ldots$, and want to choose them such that

$$q = p_0 \cup \bigcup_n (p_{\ell(n)}{\restriction}[\ell(n), \omega)) \in P(F)$$

So as q is a function from a subset of ω to 2, (because $p_n \le p_{n+1}$) we only have to take care of the demand $\mathrm{Dom}(q) = \emptyset \bmod F$. Note that $\mathrm{Dom}(q) \setminus A = \bigcup_{n<\omega}(\mathrm{Dom}(p_{\ell(n)}) \setminus A \setminus [0, \ell(n)) = \bigcup_{n<\omega}(A_{\ell(n)} \setminus A \setminus [0, \ell(n))) = \bigcup_n (A_{\ell(n)} \cap [\ell(n), k_{\ell(n)}) \setminus A)$ (remember $A_n \setminus [0, k_n) \subseteq A, A_0 \subseteq A$).

Now let $w_\ell = [\ell, k_\ell)$, (for $\ell < \omega$) which is a finite set. As $\mathrm{Min}(w_\ell) = \ell$, there is an infinite $S \subseteq \omega$ such that $\{w_\ell : \ell \in S\}$ is a family of pairwise disjoint sets. Since F is fat there is an infinite $S_1 \subseteq S$ such that $\bigcup\{w_\ell : \ell \in S_1\} = \emptyset \bmod F$. So let $\{\ell(n) : n < \omega\}$ be a list of the members of S_1, $\ell(n) < \ell(n+1)$. Then $q \in P(F)$ which proves Fact 4.6. $\square_{4.6}$

Continuation of the Proof of 4.4. Let $\{\underset{\sim}{\tau}_i : i < \omega\}$ be a list of all $P(F)$-names of ordinals which belong to N. Using the crucial fact 4.5 we can define by induction on n, $p_n \in P(F) \cap N$, $p_n \le p_{n+1}$ (p_0 is already defined) such that:

(*) if $g : n \to 2 = \{0, 1\}$ and $\ell < n$ then for some ordinal $\alpha(g, \ell)$

$$(p_n \restriction [n, \omega)) \cup g \Vdash_{P(F)} \text{“}\underset{\sim}{\tau}_\ell = \alpha(g, \ell)\text{”}.$$

Applying the Fact 4.6 for the sequence $\langle p_n : n < \omega \rangle$ constructed above with the property (*) we obtain a $q \in P(F)$ such that $q \ge p_0$ and $q \ge p_n \restriction [n, \omega)$ for infinitely many n. For such an n we have $(p_n \restriction [n, \omega)) \cup q \restriction [0, n) \le q$, hence $q \Vdash$ “for $\ell < n$, $\underset{\sim}{\tau}_\ell = \alpha(g, \ell)$ for some function $g : n \to 2$ extending $q \restriction [0, n)$”. So q is $(N, P(F))$-generic, and as $q \ge p_0$, we have proved that $P(F)$ is proper. In fact not only q is $(N, P(F))$-generic, but even $q \restriction [n, \omega)$ for any $n < \omega$ is, as every $g \cup q \restriction [n, \omega)$ is, for $g : n \to \{0, 1\}$.

Let us prove $P(F)$ is α-proper for any countable α, by induction on α. Let $\langle N_i : i \le \alpha \rangle$ be as in V.3.1, $p \in P(F)$ and $\{p, F\} \in N_0$; we shall prove that not only is there $q \ge p$, $(N_i, P(F))$-generic for every $i \le \alpha$, but also $q \restriction [n, \omega)$ is $(N_i, P(F))$-generic (for $i \le \alpha, n < \omega$). If $\alpha = 0$ we have proved this, if α is a successor use the induction hypothesis. So assume α is limit and let $\alpha = \bigcup_{n<\omega} \alpha_n$, $\alpha_n < \alpha_{n+1}$. By the induction hypothesis we can define $q_n \in N_{\alpha_n+1}$, such that for every $i \le \alpha_n$ and $k < \omega$ we have $q_n \restriction [k, \infty)$ is $(N_i, P(F))$-generic and $q_0 \ge p, q_{n+1} \ge q_n$.

Apply Fact 4.6 to $p_0, q_0, q_1, \ldots$ and get q as required, remember that as q is $(N_i, P(F)))$-generic for every $i < \alpha$, it is also $(N_\alpha, P(F))$-generic.

¿From Lemma 4.4 only the strong PP-property remains to be shown (see on it 2.12, particularly 2.12E(3); this suffices by 2.12F(1)). So let $x \in {}^\omega\omega$ diverge to infinity, and let N be as above, $\underset{\sim}{f} \in N$ be a $P(F)$-name of a member of ${}^\omega\omega$, and w.l.o.g. we can assume that for $n < \omega$, $\underset{\sim}{\tau}_{2n} = \underset{\sim}{f}(n)$ where $\{\underset{\sim}{\tau}_n : n < \omega\}$ was a list of the names of ordinals used in the proof of the properness of $P(F)$. When we define the p_n, by induction on n, make one change in $(*)$ above (in this proof): instead of considering $\ell < n$, we consider the ℓ such that $\ell \leq 2x(n+1+2^{n+1})+2$. So we have:

$(*)'$ if $g : n \to 2$ and $\ell \leq 2(x(n+1+2^{n+1})+1)$ then for some ordinal $\alpha(g, \ell)$, $p{\restriction}[n, \omega) \cup g \Vdash_{P(F)}$ "$\underset{\sim}{t}_\ell = \alpha(g, \ell)$".

Now we let $k_n = 0$, $m_n(0) = 2^n$, $i_n(0) = n+1$, $j_n(0) = x(n+1+2^n)+1$. Then, by $(*)'$, we have $p_n{\restriction}[n, \omega) \Vdash_{P(F)}$ "$\underset{\sim}{f}{\restriction}j_n \in \{h_{g,j_n} : g : n \to 2\}$", where $h_{g,j_n}(\ell) = \alpha(g, 2\ell)$. So $p_n{\restriction}[n, \omega)$ allows $\underset{\sim}{f}{\restriction}j_n(0)$ at most 2^n possibilities which is $m_n(0)$. As $q \geq p_n{\restriction}[n, \omega)$ for infinitely many n, and for each such n, q "allows" $\underset{\sim}{f}{\restriction}j_n$ less than $m_n(0)+1$ possibilities, clearly $k_n = 0, m_n(0), i_n(0), j_n(0)$ witness n is as required in 2.12E(3) (i.e. 2.12A(b)$(*)$ with $k = 0$), so we have finished.

$\square_{4.4,4.2}$

4.7 Lemma. Suppose F is a P-point and $P(F)^\omega \lessdot P$ and P has the PP-property (or just it is ${}^\omega\omega$-bounding and has the weak PP-property, see Definition 2.12E).

Then in V^P, F cannot be extended to a P-point.

Proof. Suppose $p \in P$ forces that $\underset{\sim}{E}$ is an extension of F to a P-point (in V^P). Let $\langle \underset{\sim}{r}_n : n < \omega \rangle$ be the sequence of reals which $P(F)^\omega$ introduces (i.e. $\underset{\sim}{r}_n(i) = \ell$ iff for some $\langle f_0, f_1, \ldots \rangle \in \underset{\sim}{G}_{P(F)^\omega}$ we have $f_n(i) = \ell$). Define a P-name:

$\underset{\sim}{h}(n)$ is 1 if $\{i < \omega : \underset{\sim}{r}_n(i) = 1\} \in \underset{\sim}{E}$ and

$\underset{\sim}{h}(n)$ is 0 if $\{i < \omega : \underset{\sim}{r}_n(i) = 0\} \in \underset{\sim}{E}$.

So $p \Vdash_P$ "$\underset{\sim}{h} \in {}^\omega 2$". Now as P have the PP-property, by 2.12D (and see 2.12E), there is $p_1 \geq p$, $(p_1 \in P)$, and for each $n < \omega$ there are $k(n) < \omega$, $i_n(0) < j_n(0) < i_n(1) < j_n(1) < \ldots < i_n(k(n)) < j_n(k(n))$, and $j_n(k(n)) < i_{n+1}(0)$ such that:

$$p_1 \Vdash_P \text{" for every } n < \omega \text{ for some } \ell \leq k(n) \text{ we have } \underset{\sim}{h}(i_n(\ell)) = \underset{\sim}{h}(j_n(\ell))\text{"}.$$

Now define the following P-names:

$$\underset{\sim}{A}_n = \{m < \omega : \text{ for some } \ell \leq k(n), \underset{\sim}{r}_{i_n(\ell)}(m) = \underset{\sim}{r}_{j_n(\ell)}(m)\}.$$

4.7A Fact. $p_1 \Vdash_P$ "$\underset{\sim}{A}_n \in \underset{\sim}{E}$".

This is true because p_1 forces that for some $\ell \leq k(n)$ we have $\underset{\sim}{h}(i_n(\ell)) = \underset{\sim}{h}(j_n(\ell))$ and by the definition of $\underset{\sim}{h}$ we know:

$$p \Vdash_P \text{"}\{m < \omega : \underset{\sim}{r}_{i_n(\ell)}(m) = \underset{\sim}{h}(i_n(\ell))\} \in \underset{\sim}{E}\text{"}$$
$$p \Vdash_P \text{"}\{m < \omega : \underset{\sim}{r}_{j_n(\ell)}(m) = \underset{\sim}{h}(j_n(\ell))\} \in \underset{\sim}{E}\text{"}.$$

Putting together these three things (and $p \leq p_1$) we get $p_1 \Vdash_P$ "$\{m < \omega :$ for some $\ell \leq k(n)$ we have $r_{i_n(\ell)}(m) = \underset{\sim}{h}(i_n(\ell)) = \underset{\sim}{h}_j(j_n(\ell)) = r_{j_n(\ell)}(m)\} \in \underset{\sim}{E}$" but this set is included in $\underset{\sim}{A}_n$, hence $p_1 \Vdash$ "$\underset{\sim}{A}_n \in \underset{\sim}{E}$". So Fact 4.7A holds. $\square_{4.7A}$

So $p_1 \Vdash$ "$\{\underset{\sim}{A}_n : n < \omega\} \subseteq \underset{\sim}{E}$", but as $p_1 \Vdash$ "$\underset{\sim}{E}$ is a P-point" for some $\underset{\sim}{g}$ we also have $p_1 \Vdash_P$ "$\underset{\sim}{g} \in {}^\omega\omega$ and $\bigcap_{n<\omega}(\underset{\sim}{A}_n \cup [0, \underset{\sim}{g}(n)]) \in \underset{\sim}{E}$". Now as P has the PP-property by 2.12B(1) (or 2.12F(4)), it has the ${}^\omega\omega$-bounding property, hence there is p_2, $p_1 \leq p_2 \in P$ and $g \in {}^\omega\omega$ (in V) such that $p_2 \Vdash_P$ "$\underset{\sim}{g}(n) \leq g(n)$ for every n". Hence

$$p_2 \Vdash_P \text{"} \bigcap_{n<\omega} (\underset{\sim}{A}_n \cup [0, g(n)]) \in \underset{\sim}{E}\text{"}$$

and therefore

$(*)$ $\qquad p_2 \Vdash_P$ "$\bigcap_{n<\omega}(\underset{\sim}{A}_n \cup [0, g(n)]) \neq \emptyset \bmod F$".

As $p_2 \in P$ and $P(F)^\omega \lessdot P$, there is $q = \langle f_0, f_1, \ldots\rangle \in P(F)^\omega$ such that p_2 is compatible (in P) with any $q^\dagger, q \leq q^\dagger \in P(F)^\omega$. As F is a P-point, there is $A^* \subseteq \omega$ (in V) such that $A^* = \emptyset \mod F$ and $\mathrm{Dom}(f_i) \subseteq_{ae} A^*$ for every i. Choose, by induction on $n < \omega$, $\alpha_n < \alpha_{n+1} < \omega$, such that $\alpha_n > g(n)$ and for $\ell \leq k(n)$:

$$(\mathrm{Dom}(f_{i_n(\ell)}) \cup \mathrm{Dom}(f_{j_n(\ell)})) \setminus [0, \alpha_n) \subseteq A^*.$$

Now we shall define $q^\dagger = \langle f_\ell^\dagger : \ell < \omega \rangle$, $q \leq q^\dagger \in P(F)^\omega$. Let: $f^\dagger_{i_n(\ell)} = f_{i_n(\ell)} \cup 0_{[\alpha_n, \alpha_{n+1}) \setminus A^*}$ and $f^\dagger_{j_n(\ell)} = f_{j_n(\ell)} \cup 1_{[\alpha_n, \alpha_{n+1}) \setminus A^*}$ (where 0_B is the function with domain B and constant value 0, 1_B defined similarly.) Otherwise $f^\dagger_m = f_m$.

Plainly, $f^\dagger_m$ is a function (by the definition of α_n), its domain is the same as that of f_m plus a finite subset of ω, hence $\mathrm{Dom}(f^\dagger_m) \subseteq \omega$, $\mathrm{Dom}(f^\dagger_m) = \emptyset \mod F$. Also $f_m \subseteq f^\dagger_m$, hence $q \leq q^\dagger = \langle f^\dagger_0, f^\dagger_1, \ldots \rangle \in P(F)^\omega$. Clearly $q^\dagger \Vdash_P$ "$\underset{\sim}{A}_n$ is disjoint to of $[\alpha_n, \alpha_{n+1}) \setminus A^*$" (by the definition $\underset{\sim}{A}_n$ and $f^\dagger_{i_n(\ell)}, f^\dagger_{j_n(\ell)}$).

Also $g(n) < \alpha_n$, hence

$$q^\dagger \Vdash_P \text{“}\underset{\sim}{A}_n \cup [0, g(n)] \setminus A^* \text{ is disjoint to } [\alpha_n, \alpha_{n+1})\text{”}$$

and thus

$$q^\dagger \Vdash_P \text{“} \bigcap_{n<\omega} (\underset{\sim}{A}_n \cup [0, g(n)] \setminus A^*) \text{ is disjoint to } \bigcup_n [\alpha_n, \alpha_{n+1}) = [\alpha_0, \omega)\text{”}.$$

Consequently (as $A^* = \emptyset \mod F$ and $[0, \alpha_0]$ is finite)

$$q^\dagger \Vdash_P \text{“} \bigcap_{n<\omega} (\underset{\sim}{A}_n \cup [0, g(n)]) = \emptyset \mod F\text{”}.$$

By the choice of q we know that p_2, $q^\dagger$ are compatible in P so let $p_3 \in P$ be a common upper bound of p_2, $q^\dagger$, hence $p_3 \Vdash_P$ "$\bigcap_{n<\omega} (\underset{\sim}{A}_n \cup [0, g(n)]) = \emptyset \mod F$" which contradicts $(*)$. $\square_{4.7}$

4.8 Theorem. It is consistent with ZFC $+2^{\aleph_0} = \aleph_2$ that there is no P-point.

Proof. It is left to the reader, or see the 5.13, where a similar proof is carried out. $\square_{4.8}$

4.9 Claim. Assume iteration $\overline{Q} = \langle P_i, \underset{\sim}{Q}_j : i \leq \delta, j < \delta \rangle$ is as in 0.1, E is a non principal ultafilter in V which is a P-point (i.e. if $A_n \in E$ for $n < \omega$ then for some $A \in E$ we have $A \setminus A_n$ finite for each $n < \omega$) and if P_δ is not proper, E is generated by $\leq \aleph_1$ sets. If E is (pedantically generates) an ultrafilter in V^{P_i} for each $i < \delta$ (and δ is a limit ordinal) *then* E is a P-point in V^{P_δ}.

Remark. We weaken the assumption on E (in V) to

$(*)_0$ $E \subseteq \mathcal{P}(\omega)$ and $\mathrm{fil}(E) = \{A \subseteq \omega : (\exists B \in E) A \supseteq B\}$ is a non principal ultrafilter on ω, which is a P-point.

Proof.

We shall use 1.17 (see Definition 1.16, for our family of forcing notions). Let $k^* = 2$, $D_\ell = H(\chi)^V$, and

xR_0T *iff* ($x \in ({}^\omega\omega) \cap D$ and) for some $A = A^0_T \in E$ for every $\eta \in \lim(T)$ we have $\{n < \omega : \eta(n) = 1\} \supseteq A$

xR_1T *iff* for some $A = A^1_T \in E$ for every $\eta \in \lim(T)$ we have $\{n < \omega : \eta(n) = 1\} \cap A = \emptyset$

(so the x's are not important).

Then clearly $(\overline{D}, \overline{R}) = (\langle D_0, D_1 \rangle, \langle R_0, R_1 \rangle)$ is a weak covering 2-model, in particular it covers in V; use the standard $<_i$. Note

$(*)_1$ $(\overline{D}, \overline{R})$ covers in V^P *iff* E generates an ultrafilter in V^P

$(*)_2$ if in V^P the family E generates an ultrafilter then E generates a P-point in V^P (of course provided that P is proper or P preserves $\aleph_1$ and $|E| = \aleph_1$ or $\Vdash_P$ "$\mathcal{S}_{<\aleph_1}(|E|)^V$ is cofinal in $\mathcal{S}_{<\aleph_1}(|E|)^{V^P}$").

We next prove that $(\overline{D}, \overline{R}, \overline{<})$ is a fine covering 2-model. We check Definition 1.16(2).

Clauses (α), (β) are trivial.

Clause (γ)(a): Let $k < 2$, xR_kT_n. By symmetry let $k = 0$ so xR_0T_n. So let $A_n = A^k_{T_n}$, so $A_n \in E$. We can find $B \in E$ such that $B \subseteq A_0$, $B \subseteq_{ae} A_n$ for each n. Let $T^* = \{\eta \in {}^\omega\omega : (\forall i \in B)\eta(i) = 1\}$ so the choice $A^k_{T^*} = B$ exemplifies xR_kT^* (for any x), so it suffices to prove the inclusion from (γ)(a). Toward this let for $i < \omega$ let $m_i \in (i, \omega)$ be such that $B \setminus m_i \subseteq A_i$, and let $w \subseteq \omega$ be infinite such that if $i < \omega$, $j \in w$, $i \leq \max(w \cap j)$ then $m_i < j$. Now suppose $\eta \in {}^\omega\omega$ and $i \in w \Rightarrow \eta{\restriction}(\min(w \setminus (i+1)) \in \bigcup_{j<i, j\in w} T_j \cup T_0$, and we are going to prove $\eta \in \lim(T^*)$. This means that we should prove $A' = \{n < \omega : \eta(n) = 1\} \supseteq B$; so let $n \in B$; if n is smaller than the second member of w, then applying the assumption for $i =$ the first member of w we get $\eta{\restriction}(n+1) \in T_0$ which implies the desired conclusion. If not let $i_0 < i_1 \leq n < i_2$, where i_0, i_1, i_2 are successive members of w. So by the assumption $\eta{\restriction}i_2 \in \bigcup_{j\leq i_0} T_j \cup T_0$, now if $\eta{\restriction}i_2 \in T_0$ we are done, so assume $\eta{\restriction}i_2 \in T_j$, $j \leq i_0$, hence $m_j \leq i_1 \leq n$. Then $A^k_{T_j} \cap [m_j, \infty) \supseteq B \cap [m_j, \infty) \supseteq \{n\}$, hence as $\eta \in T_j$ we get $\eta(n) = 1$ as required.

Clause (γ)(b): We can find $B_i \in E$ such that $\{n < \omega : \eta_i(n) = k\} \supseteq B_i$, then we can find $B \in E$ such that $B \setminus B_i$ is finite say $\subseteq [0, m_i)$ for $i < \omega$. Choose $\langle n_j : j < \omega \rangle$ as in the proof of (γ)(a); by symmetry w.l.o.g. $\bigcup_j [n_{2j}, n_{2j+1})$ belongs to the ultrafilter which E generated. So $B \cap \bigcup_j [n_{2j}, n_{2j+1}) \cap \{i : \eta(i) = k\}$ belongs to this ultrafilter, hence it includes some $B' \in E$. Consequently, $\{i < \omega : \eta_{n_{2j}}(i) = k\} \supseteq B'$ for each $j < \omega$ and $\{i < \omega : \eta(i) = k\} \supseteq B'$, as required.

Clause (δ): Straight by $(*)_1$, $(*)_2$. $\square_{4.9}$

4.10 Remark. 1) Mekler [Mk84] considers the generalizations to finitely additive measures $\mu : \mathcal{P}(\omega) \to [0,1]_{\mathbb{R}}$, generalizing this proof to prove the consistency of "the parallels of P-points do not existence". Though there the PP-property fails, he showed that ${}^\omega\omega$-bounding suffices. Still we felt the PP-property is inherently interesting.

2) Baumgartner [B6] was interested in ultrafilters with properties which weaken "being P-points". Answering his question we prove that if in the iteration above we use unboundedly often random real then there is no P-point (see above), and there is a measure zero ultrafilter (see [B6]).

3) The question of whether there are always NWD-ultrafilters (see van Douwen [vD81], and [B6]) is answered negatively in [Sh:594], generalizing the proof here (and continuing the "use of E" from [Sh:407]). There the PP-property is used.

§5. There May Exist a Unique Ramsey Ultrafilter

Usually it is significantly harder to prove that there is a unique object than to prove there is none. The proof is similar to the one in the previous section, but here we are destroying other Ramsey ultrafilter (in fact "almost" all other P-points) while preserving our precious Ramsey ultrafilter. By a similar proof we can construct a forcing notion P such that e.g. in V^P there are exactly two Ramsey ultrafilters (in both cases up to the equivalence induced by the Rudin Keisler order) or any other number.

More exactly we shall prove the consistency of "there is a unique Ramsey ultrafilter F_0 on ω, up to permutation of ω, moreover for every P-point F, $F_0 \leq_{RK} F$".

Note that if there is a unique P-point it should be Ramsey; however, concerning the question of the existence of a unique P-point we return to it in XVIII §4.

Our scheme is to start with a universe with a fixed Ramsey ultrafilter F_0, to preserve its being an ultrafilter and even a Ramsey ultrafilter. Our ultrafilter will be generated by $\aleph_1$ sets. Now in each stage we shall try to destroy a given P-point F such that $F_0 \not\leq_{RK} F$. The forcing from §4 does not work, but if we use a version of it in the direction of Sacks forcing it will work.

5.1 Claim.

1) If F is a P-point in V, P is a proper forcing notion and $\Vdash_P$ "F generates an ultrafilter " *then* it (more exactly the one it generates) is a P-point in V^P.

2) If the ultrafilter F is Ramsey in V, and P is ${}^\omega\omega$-bounding, proper and $\Vdash_P$ "F generates an ultrafilter", *then* in V^P, F still generates a Ramsey ultrafilter.

Proof.

1) As for being a P-filter, let $p \Vdash_P$ "$\{\underset{\sim}{A}_n : n < \omega\}$ is included in the ultrafilter

which F generates ". So w.l.o.g. $p \Vdash_P$ "$\underset{\sim}{A}_n \in F$", and by properness for some q, $p \le q \in P$, and $A_{n,m} \in F$ (for $n, m < \omega$) we have $q \Vdash_P$ " for each n, $\underset{\sim}{A}_n \in \{A_{n,m} : m < \omega\}$". As F is a P-point in V and $\{A_{n,m} : n, m < \omega\} \subseteq F$ belong to V, there is $A \in F$ which is almost included in every $A_{n,m}$ hence in each $\underset{\sim}{A}_n$; (note: e.g., if F is generated by $\aleph_1$ sets, then "P does not collapse $\aleph_1$" is sufficient instead "P is proper").

2) As F generates a P-point in V^P, the following will suffice: let $0 = \underset{\sim}{n}_0 < \underset{\sim}{n}_1 < \underset{\sim}{n}_2 \dots$ and $p \in P$; then we can find $A \in F$ and $q \ge p$ such that $q \Vdash$ "$A \cap [\underset{\sim}{n}_i, \underset{\sim}{n}_{i+1})$ has at most one element for each i" (i.e. F is a so called Q-point). Remember P has the ${}^\omega\omega$-bounding property. So there are $h \in {}^\omega\omega \cap V$, and $q \ge p$ such that $q \Vdash_P$ " $(\forall i) \underset{\sim}{n}_i < h(i)$". W.l.o.g. h is strictly increasing.

Define n_i^* (in V by induction on i): $n_0^* = 0, n_{i+1}^* = h(n_i^* + 1) + 1$. Now for no i, j we have $\underset{\sim}{n}_i[G] \le n_j^* < n_{j+1}^* < \underset{\sim}{n}_{i+1}[G]$. [Why? Assume this holds and, of course, $i < j$; as $\underset{\sim}{n}_\ell < \underset{\sim}{n}_{\ell+1}$, clearly $\ell \le \underset{\sim}{n}_\ell[G]$ hence

$$n_{j+1}^* > h(n_j^* + 1) \ge h(\underset{\sim}{n}_i[G] + 1) \ge h(i+1) \ge \underset{\sim}{n}_{i+1}[G]$$

(remember h is strictly increasing), a contradiction]. Also F is an ultrafilter in $V[G]$, by the assumption. As in V, F is a Ramsey ultrafilter and $\langle n_i^* : i < \omega \rangle \in V$, there is $A \in F$ such that $A \cap [n_i^*, n_{i+1}^*)$ has at most one element for each i. Let $G \subseteq P$ be generic over V be such that $q \in G$. Checking carefully in $V[G]$ we see that for every i we have $A \cap [\underset{\sim}{n}_i[G], \underset{\sim}{n}_{i+1}[G])$ has at most two elements and in this case they are necessarily successive members of A. Let $A_0 = \{k \in A : |A \cap k| \text{ is even}\}$, so either A_0 or $A \setminus A_0$ belong to the ultrafilter which F generates, and both are as required. $\square_{5.1}$

5.2 Lemma.

1) "F generates an ultrafilter in V^Q which is a P-point, Q is proper" is preserved by countable support iteration for F a P-point.

2) "F generates an ultrafilter in V^Q which is Ramsey + Q is ${}^\omega\omega$-bounding + Q is proper " is preserved by countable support iteration for F a Ramsey ultrafilter.

Proof. 1) By 4.9 and see 5.1(1).

2) Combine (1), 5.1(2) and 2.8. $\square_{5.2}$

5.3 Definition. For F a filter on ω, let $\mathrm{SP}(F)$ be $\{T : T$ is a perfect tree $\subseteq {}^{\omega>}2$ and for some $A \in F$, for every $n \in A$, $\eta \in T \cap {}^n2$ implies $\eta\char`^\langle 0\rangle \in T$ & $\eta\char`^\langle 1\rangle \in T\}$. The order is the inverse inclusion. We denote the maximal such A by $\mathrm{spt}(T)$.

5.3A Remark.

1) So $\mathrm{SP}(F)$ is a "mixture" of $P(F)$ and Sacks forcing and $\mathrm{SP}^*(F)$ (defined below) is half way between $\mathrm{SP}(F)$ and $\mathrm{SP}(F)^\omega$.
2) Remember $T_{[\eta]} \overset{\mathrm{def}}{=} \{\nu \in T : \nu \trianglelefteq \eta$ or $\eta \trianglelefteq \nu\}$ for any $\eta \in T$ and $T^{[n]} \overset{\mathrm{def}}{=} \{\eta \in T : \ell g(\eta) = n\}$ for any $n < \omega$.

5.4 Definition. Let $T_n^{\otimes} \overset{\mathrm{def}}{=} \bigtimes_{\ell<n}({}^\ell 2)$, $T^{\otimes} \overset{\mathrm{def}}{=} \bigcup_{n<\omega} T_n^{\otimes}$ ordered by the being initial segment, i.e. for $f \in T_n^{\otimes}$ and $g \in T_m^{\otimes}$ we set $f \triangleleft g$ iff $f(i) = g(i)$ for each $i < n$. (Note $f(i) \in {}^i2$). For a filter F on ω, let $\mathrm{SP}^*(F)$ be

$$\{T : T \text{ is a perfect tree } \subseteq T^{\otimes} \text{ and for every } k < \omega \text{ we have } \mathrm{spt}_k(T) \in F\},$$

where

$$\mathrm{spt}_k(T) = \{n < \omega : \text{for every } \eta \in T^{[n]}(= T \cap T_n^{\otimes}) \text{ and } \nu \in {}^k2 \text{ there is } \rho \in {}^n2, \eta\char`^\langle\rho\rangle \in T_{n+1}^{\otimes} \cap T \text{ such that } \rho{\restriction}k = \nu\}.$$

The order is the inverse inclusion.

5.5 Claim. Let F be a filter on ω and Q be $\mathrm{SP}(F)$ or $\mathrm{SP}^*(F)$.

1) If $T \in Q, T^{[n]} = \{\eta_1, \dots, \eta_k\}$ (with no repetition), $T_\ell = T_{[\eta_\ell]}, T^\dagger_\ell \in Q$, $T_\ell \le T^\dagger_\ell$ (i.e. $T^\dagger_\ell \subseteq T_\ell$) *then* $T \le T^\dagger \stackrel{\text{def}}{=} \bigcup_{\ell=1}^k T^\dagger_\ell \in Q$ and $T^\dagger \Vdash$ "for some $\ell \in \{1, \dots, k\}$ we have $T^\dagger_\ell \in \underset{\sim}{G}_Q$".
2) If $\underset{\sim}{t}$ is a P-name of an ordinal $T \in Q$ and $n < \omega$ *then* there are $T^\dagger$, $T \le T^\dagger \in Q$ and A such that $T^\dagger \Vdash_Q$ "$\underset{\sim}{t} \in A$" and $|A| \le |T^{[n]}|$, and $\bigcup_{\ell \le n} T^{[\ell]} \subseteq T^\dagger$. Moreover for each $\eta \in T^{[n]}, T^\dagger_{[\eta]}$ determines $\underset{\sim}{t}$.

Proof. 1) Observe that $\mathrm{spt}_j(T^\dagger) \supseteq \bigcap_{1 \le \ell \le k} \mathrm{spt}_j(T_\ell) \setminus (n+1)$.
2) For each $\eta \in T^{[n]}$ there is T^η, $T_{[\eta]} \le T^\eta$ such that T^η decides the value $\underset{\sim}{t}$. Now amalgamate the T^η together by applying part 1). $\square_{5.5}$

5.6 Lemma. Let F be a P-point ultrafilter on ω. *Then*

1) SP(F) is proper, in fact α-proper for every $\alpha < \omega_1$, and has the strong PP-property; and so is SP$(F)^\omega$
2) SP$^*(F)$ is also proper, α-proper for every $\alpha < \omega_1$ and has the strong PP-property.

Proof. Similar to the proof of 4.4. For its proof we shall use the following theorem, of Galvin and McKenzie, (but later we shall prove a similar theorem in detail (5.11)); note that we use only the "only if" direction.

5.7 Theorem. Let F be an ultrafilter on ω. Then F is a P-point [Ramsey ultrafilter] iff in the following game player I has no winning strategy:

in the n-th move:
player I chooses $A_n \in F$
player II choose $w_n \subseteq A_n, w_n$ is finite [a singleton].
In the end player II wins if $\bigcup_{n<\omega} w_n \in F$.

Proof of 5.6 from 5.7. We just have to define a strategy for player I, (in the game from 5.7): playing on the side with the conditions in the forcing. From the two forcing listed in the lemma we concentrate on proving only the properness of SP$^*(F)$ (the other have similar proofs and this is the only one we shall use). Let $N \prec (H(\chi), \in, <^*_\chi)$ be countable with $F \in N$, so SP$^*(F) \in N$; and let

$T \in \mathrm{SP}^*(F) \cap N$ and let $\langle \mathcal{I}_n : n < \omega \rangle$ be a list of the dense subsets of $\mathrm{SP}^*(F)$ which belong to N. We shall define now a strategy for player I. In the n'th move player I chooses "on the side" condition $T_n \in \mathrm{SP}^*(F) \cap N$ in addition to choosing $A_n \in F$ and player II chooses finite $w_n \subseteq A_n$. For $n = 0$, player I chooses $T_0 = T$ and $A_0 = \omega$.

For $n > 0$, for the n'th step player I, using 5.5, chooses $T_n \in \mathrm{SP}^*(F) \cap N$, such that $T_{n-1} \leq T_n$, $T_{n-1}^{[k_n]} = T_n^{[k_n]}$, where $k_n \stackrel{\text{def}}{=} \max[\bigcup\{w_{n'} : n' < n\} \cup \{n\}] + n + 1$ and $(\forall \eta \in T_n^{[k_n]})\ ((T_n)_{[\eta]} \in \mathcal{I})$. Then player I plays $A_n = \mathrm{spt}_n(T_n)$. Note that whatever are the choices of player II, we have $T_n \in N$ and we can let player I choose T_n as the first one which is as required by the well ordering $<^*_\chi$. As F is a P-point, by 5.7 there is a play in which he uses the strategy described above and player II wins the play; this will give us the desired sequence of conditions. Indeed, $T = \bigcap_{n<\omega} T_n \in \mathrm{SP}^*(F)$ satisfies $\mathrm{spt}_n(T) \supseteq \bigcup\{w_k : k \in [n, \omega)\}$ (for each $n < \omega$) and hence T belongs to $\mathrm{SP}^*(F)$. $\square_{5.6}$

Similar argument is carried out in more detail in the proof of 5.12.

5.8 Lemma. 1) If F is a P-point ultrafilter, $\mathrm{SP}(F)^\omega \lessdot Q$, and Q has the PP-property *then* in V^Q, F cannot be extended to a P-point ultrafilter.
2) If F is a P-point ultrafilter, $SP^*(F) \lessdot Q$, Q has the PP-property *then* in V^Q, F cannot be extended to a P-point ultrafilter.

Proof. The proof is almost identical with the proof of 4.7, so we do not carry out it in detail. (In fact we get the variant with weaker assumption as proved in 4.7).

This is particularly true for part (1). For part (2) copy the proof of 4.7, replacing $P(F)$ by $\mathrm{SP}^*(F)$ and defining $\underset{\sim}{r}_n$ as:

$\underset{\sim}{r}_n(i) = \ell$ *iff* $i \leq n \Rightarrow \ell = 0$ and

$$i > n \Rightarrow (\exists T \in \underset{\sim}{G}_{\mathrm{SP}^*(F)})(\exists \eta \in T_{i+1}^{\otimes})[T = T_{[\eta]}\ \&\ (\eta(i))(n) = \ell].$$

This is done up to and including the choice of p_2 (i.e. $(*)$ in the proof of 4.7).

As $p_2 \in P$ and $\mathrm{SP}^*(F) \lessdot P$ clearly there is $q \in \mathrm{SP}^*(F)$ such that p_2 is compatible in P with any q' satisfying $q \le q' \in \mathrm{SP}^*(F)$. For $k < \omega$, as $q \in \mathrm{SP}^*(F)$ by Definition 5.4 we know that $\mathrm{spt}_k(q) \in F$, so as F is a P-point there is $B^* \in F$ such that $B^* \setminus \mathrm{spt}_k(q)$ is finite for every $k < \omega$. Choose by induction on $n < \omega$, $\alpha_n < \omega$ such that $\alpha_n < \alpha_{n+1}$, $\alpha_n > g(n)$ and $\alpha_n > j_n(k(n))$ and $B^* \setminus \mathrm{spt}_{j_n(k(n))+1}(q) \subseteq [0, \alpha_n)$. Define $q' \stackrel{\mathrm{def}}{=} \{\eta : \eta \in q$ and for every $m < \omega$ we have: if $\alpha_n \le m < \ell g(\eta)$, $m < \alpha_{n+1}$ and $m \in \mathrm{spt}_{j_n(k(n))+1}(q)$ then for each $\ell \le k(n)$ we have $(\eta(m))(i_n(\ell)) = 0$ and $(\eta(m))(j_n(\ell)) = 1\}$.

Now

(a) $q' \subseteq T^{\otimes}$ is closed under initial segments and $\langle\rangle \in q'$

[Why? Read the definition of q']

(b) q' has no $\lhd$-maximal element

[Why? Assume $\eta \in q' \cap T^{\otimes}_m$. If $m < \alpha_0$ then any $\nu \in \mathrm{Suc}_q(\eta)$ belongs to q'. So let $\alpha_n \le m < \alpha_{n+1}$; if $m \notin \mathrm{spt}_{j_n(k(n))+1}(q)$ again any $\nu \in \mathrm{Suc}_q(\eta)$ belongs to q', so assume $m \in \mathrm{spt}_{j_n(k(n))+1}(q)$, which means

$$(\forall \eta' \in q \cap T^{\otimes}_m)(\forall \rho \in {}^{j_n(k(n))+1}2)(\exists \nu)[\eta'{}^\frown\langle\nu\rangle \in q \;\&\; \nu{\restriction}j_n(k(n)) + 1 = \rho].$$

Apply this for η' and for the $\rho^* \in {}^{j_n((k(n))+1}2$ defined by $\{\ell < j_n(k(n)) + 1 : \rho^*(\ell) = 1\} = \{j_n(\ell) : \ell \le k(n)\}$, and find ν satisfying $\rho^* \trianglelefteq \nu$ and such that $\eta{}^\frown\langle\nu\rangle \in \mathrm{Suc}_q(\eta)$ and even $\eta{}^\frown\langle\nu\rangle \in \mathrm{Suc}_{q'}(\eta)$.]

(c) If $\alpha_n \le m < \alpha_{n+1}$, $m \in \mathrm{spt}_{j_n(k(n))+1}(q)$ then $m \in \mathrm{spt}_{i_n(0)}(q')$.

[Why? Same proof as of clause (b) noting that for any $\rho_1 \in {}^{i_n(0)}2$ we can find ρ^* such that $\rho_1 \lhd \rho^* \in {}^{j_n(k(n))+1}2$, such that for $m \in [i_n(0), j_n(k(n)) + 1)$, we have $\rho^*(m) = 1 \Leftrightarrow m \in \{j_n(\ell) : \ell \le k(n)\}$]

(d) Let $k < \omega$, then $\mathrm{spt}_k(q') \in F$.

[Why? Choose $n(*)$ such that $k < i_{n(*)}(0)$. Now if $m \in B^* \setminus \alpha_{n(*)}$ then for some n, $n(*) \le n < \omega$ and $\alpha_n \le m < \alpha_{n+1}$ hence $m \in \mathrm{spt}_{j_n(k(n))+1}(q)$ and so by clause (c) we have $m \in \mathrm{spt}_{i_n(0)}(q')$. But $\mathrm{spt}_\ell(q')$ decreases with ℓ and $k < i_{n(*)}(0) \le i_n(0)$, so $m \in \mathrm{spt}_k(q')$. Together $B^* \setminus \alpha_{n(*)} \subseteq \mathrm{spt}_k(q')$, but the former belongs to F.]

(e) $q' \Vdash_{\mathrm{SP}^*(F)}$ "$\bigcap\limits_{n<\omega} (\underset{\sim}{A}_n \cup [0, g(n)))$ is disjoint to $B^* \setminus \alpha_0$"

[Why? Because if $\alpha_n \leq m < \alpha_{n+1}$ and $m \in B^*$ then: by the definitions of $\underset{\sim}{r}_{i_n(\ell)}, \underset{\sim}{r}_{j_n(\ell)}$ $(\ell \leq k(n))$ and $\underset{\sim}{A}_n$ (which is $\{\alpha < \omega$: for some $\ell \leq k(n)$, $r_{i_n(\ell)}(\alpha) = r_{j_n(\ell)}(\alpha)\}$) we know $m \notin \underset{\sim}{A}_n$, also $m \geq \alpha_n > g(n)$, together this suffices.]

Now q', p_2 are compatible members of P (see the choice of q and remember $q \leq q' \in SP^*(F)$), so let $p_3 \in P$ be such that $p_2 \leq p_3$, $q' \leq p_3$. So by clause (e) the condition p_3, being above q', forces that $\bigcap_{n<\omega} (\underset{\sim}{A}_n \cup [0, g(n)))$ is disjoint to a member of F. So as $p_2 \leq p_3$ clearly p_2 cannot force $\bigcap_{n<\omega} (\underset{\sim}{A}_n \cup [0, g(n))) \neq \emptyset \mod F$. But this contradicts the choice of p_2. $\square_{5.8}$

We now state some well known basic facts on the Rudin-Keisler order on ultrafilters.

5.9 Definition.

1) Let F_1, F_2 be ultrafilters on I_1, I_2, respectively. We say $F_1 \leq_{RK} F_2$ *if*: there is a function f from I_2 to I_1 such that $f(I_2) = \{f(i) : i \in I_2\} \in F_1$ and: $A \in F_1$ iff $f^{-1}(A) \in F_2$

2) In this case we say $F_1 = f(F_2)$, if $|I_1| \leq |I_2|$ we can assume w.l.o.g. f is onto I_1.

5.9A Remark. We shall use only ultrafilters on ω, which are not principal, i.e. in $\beta(\omega) \setminus \omega$ in topological notation.

It is known (see e.g. [J])

5.10 Theorem.

1) $\leq_{RK}$ is a quasi-order.

2) An ultrafilter F on ω is minimal *iff* it is Ramsey (minimal means $F^\dagger \leq_{RK} F \Rightarrow F \leq_{RK} F^\dagger$ (see part (4)).

3) If F is a P-point, $F^\dagger \leq_{RK} F$ *then* $F^\dagger$ is a P-point.

4) If $F^1 \leq_{RK} F^2 \leq_{RK} F^1$, *then* there is a permutation f of ω such that $F_2 = f(F_1)$.

Proof. Well known. $\square_{5.10}$

5.11 Lemma. Suppose F_0, F_1 are ultrafilters on ω. *Then* the condition (A) and condition (B) below are equivalent.

(A) F_1 is a P- point, F_0 is a Ramsey ultrafilter, and not $F_0 \leq_{RK} F_1$.

(B) in the following game player I has no winning strategy:

in the n-th move, n even:

player I chooses $A_n \in F_0$

player II chooses $k_n \in A_n$

in the n-th move, n odd:

player I chooses $A_n \in F_1$

player II chooses a finite set $w_n \subseteq A_n$

In the end player II wins if

$$\{k_n : n < \omega\text{even}\} \in F_0 \text{ and } \bigcup\{w_n : n < \omega \text{ odd }\} \in F_1.$$

Proof. $\neg$(A)$\Rightarrow$ $\neg$(B): If F_1 is not a P-point or F_0 is not Ramsey then player I can win by 5.7. (I.e., if F_1 is not a P-point, then are $B_n \in F_1$ for $n < \omega$ such that for no $B \in F_1$ do we have $B \setminus B_n$ is finite for every n, now player I has a strategy guaranteeing: for n odd, $A_n = \bigcap_{\ell \leq (n-1)/2} B_\ell \setminus (\sup \bigcup\{w_\ell : \ell < n \text{ odd}) + 1)$, this is a winning strategy. If F_0 is not a Ramsey ultrafilter there are $B_n \in F_0$ for $n < \omega$ such that for no $k_n \in B_n$ (for $n < \omega$) do we have $\{k_n : n < \omega\} \in F_0$, now player I has a strategy guaranteeing $A_{2n} = B_n$, this is a winning strategy.) So we can assume F_1 is a P-point and F_0 is Ramsey, so by $\neg$(A) necessarily $F_0 \leq_{RK} F_1$, hence some $h : \omega \to \omega$ witnesses $F_0 \leq_{\mathrm{RK}} F_1$. Then player I can play such that $\bigcup\{h^{-1}(k_n) : n \in \omega\}$ and $\bigcup\{w_n : n \in \omega\}$ will be disjoint. So one of them is not in F_1, thus player I wins.

(A)$\Rightarrow$(B): Suppose H is a wining strategy of player I. Let λ be big enough, $N \prec (H(\lambda), \in), \{F_0, F_1, H\} \in N$ and N is countable. As F_ℓ is a P-point there is $A^*_\ell \in F_\ell$ such that $A^*_\ell \subseteq_{ae} B$ for every $B \in F_\ell \cap N$.

Now we can find an increasing sequence $\langle M_n : n < \omega\rangle$ of finite subsets of $N, N = \bigcup_{n<\omega} M_n$ such that it increases rapidly enough; more exactly:

α) $H, F_0, F_1 \in M_0, M_n \in M_{n+1}$; also can demand $x \in M_n$ & x finite $\Rightarrow x \subseteq M_n$; also $M_n \cap \omega$ is an initial segment of ω,

β) if $\varphi(x, a_0, \ldots)$ is a formula of length $\leq 1000 + |M_n|$ with parameters from $M_n \cup \{M_n\}$ satisfied by some $x \in N$, then it is satisfied by some $x \in M_{n+1}$,

γ) for $\ell = 0, 1$ if $B \in F_\ell \cap N$, $B \in M_n$ then $B \cup M_{n+1} \supseteq A^*_\ell$,

δ) $M_0 \cap \omega = \emptyset$.

Let $u_{n+1} = (M_{n+1} \setminus M_n) \cap \omega$. So $\langle u_n : n < \omega\rangle$ forms a partition of ω. As F_ℓ is an ultrafilter, there are $S_\ell \subseteq \omega$ such that $\bigcup\{u_n : n \in S_\ell\} \in F_\ell$, and $n < m$ & $\{n, m\} \subseteq S_\ell \Rightarrow m - n \geq 10$.

Can we demand also $n \in S_0$, $m \in S_1$ implies the absolute value of $n - m$ is ≥ 5? For the S_0, S_1 we have, for each $n \in S_0$ there is at most one $m \in S_1$ such that $|n - m| \leq 4$ and vice versa. So in the bad case there are $S^\dagger_\ell \subseteq S_\ell$, $f : S^\dagger_0 \to S^\dagger_1$ one to one and onto, $n - 4 \leq f(n) \leq n + 4$, $\bigcup\{u_n : n \in S^\dagger_\ell\} \in F_\ell$ for $\ell = 0, 1$; moreover, for any $S^*_\ell \subseteq S^\dagger_\ell$,

$$\bigcup\{u_n : n \in S^*_0\} \in F_0 \quad \text{iff} \quad \bigcup\{u_n : n \in S^*_1\} \in F_1$$

provided that $S^*_1 = f(S^*_0)$. Also as F_0 is a Ramsey ultrafilter, there are $k_n \in u_n$ (for $n \in S^\dagger_0$) such that $\{k_n : n \in S^\dagger_0\} \in F_0$. So the function $f^* : \omega \to \omega$ defined by $f^*(\ell) = k_n$ for $\ell \in u_{f(n)}$, $n \in S^\dagger_0$, and $f^*(\ell) = 0$ otherwise, exemplifies $F_0 \leq_{RK} F_1$, contradiction.
So without loss of generality

$(*)$ for $n \in S_0, m \in S_1$ we have $n - m$ has absolute value ≥ 5,

$(**)$ there are $k^*_n \in u_n \cap A^*_0$ (for $n \in S_0$) such that $\{k^*_n : n \in S_0\} \in F_0$ (because F_0 is Ramsey.)
It is also clear that by (γ) above, as $A^* \in F_1$:

$(***)$ For $n \in S_1$ let $v_n \stackrel{\text{def}}{=} u_n \cap \bigcap\{A : A \in F_1 \cap M_{n-2}\}$. Then

$$\bigcup\{v_n : n \in S_1\} \in F_1,$$

also $h^*_\ell \in \bigcap\{A : A \in F_0 \cap M_{n_2}\}$.
[Simply note $u_n \cap A^* \subseteq v_n$ and w.l.o.g. $\min(S_\ell) > 2$].

Now there is no problem to define by induction on $\ell < \omega$, $n_\ell < \omega$ and an initial segment $\bar{t}^\ell$ of length ℓ of a play of the game (both increasing) such that: the initial segment belong to M_{n_ℓ}; and every k^*_n will appear among the k's which player II have chosen in the play if $n \leq n_\ell$, $n \in S_0$; and every v_n will appear among the w's player II have chosen in the play if $n \leq n_\ell$, $n \in S_1$; and n_ℓ has the form $n^* + 2$ with $n^* \in S_0 \cup S_1$; and player I uses his strategy. But in the play we produce player II wins, contradiction. $\square_{5.11}$

5.12 Main Lemma. Suppose F_0 is a Ramsey ultrafilter (on ω), F is a P-point, and $Q = \mathrm{SP}^*(F)$, and $\Vdash_Q$ "F_0 is not an ultrafilter" *then* $F_0 \leq_{RK} F$.

Proof. Let $T_0 \in Q$, $\underset{\sim}{A}$ be a Q-name, $T_0 \Vdash_Q$ "$\underset{\sim}{A} \subseteq \omega$ and $\omega \setminus \underset{\sim}{A}, \underset{\sim}{A} \neq \emptyset$ mod F_0", and w.l.o.g. $\Vdash_Q$ "$\underset{\sim}{A} \subseteq \omega$", (such $T_0, \underset{\sim}{A}$ exists as after forcing with Q, F_0 will no longer generate an ultrafilter). Note that by the choice of T_0, $\underset{\sim}{A}$ for any $T \geq T_0$:

$$\{n : \text{ for some } T^\dagger \geq T, T^\dagger \Vdash_Q \text{“} n \in \underset{\sim}{A}\text{” and for some } T^\dagger \geq T, T^\dagger \Vdash_Q \text{“} n \notin \underset{\sim}{A}\text{”}\}$$

belongs to F_0.

Now we use the game defined in Lemma 5.11. We shall describe a winning strategy for player I. During the play, player I in his moves defines also $T_n \in Q$ preserving the following:

$(*)$ (a) $T_{n+1} \geq T_n$
 (b) $T_n \Vdash_Q$ "$k_\ell \in \underset{\sim}{A}$" for ℓ even, $\ell < n$
 (c) $T_{n+1}^{[m(n)]} = T_n^{[m(n)]}$ where $m(n) = 1 + \max[\bigcup\{w_\ell : \ell \text{ odd}, \ell < n\} \cup \{n\}]$
 (d) for $\ell < n$ odd we have: $w_\ell \subseteq \mathrm{spt}_\ell(T_n)$ (see Definition 5.4)

(e) for n even, for the play from 5.7 player I chooses

$$A_n \subseteq \{k : T_n \nVdash \text{“}k \notin \underset{\sim}{A}\text{”}\}$$

(f) for n odd, for the play from 5.7 player I chooses $A_n = \mathrm{spt}_{m(n)}(T_n)$.

More exactly, player I chooses T_{n+1} in the n-th move after player II's move (see below more).

This is enough, as if in the end $\bigcup\{w_\ell : \ell < \omega \text{ odd }\} \in F$, then $T \stackrel{\mathrm{def}}{=} \bigcap_n T_n \in Q$, because for each $\ell < \omega$, we have $n > \ell \Rightarrow \mathrm{spt}_\ell(T_{n+1}) \subseteq \mathrm{spt}_\ell(T_n)$ and $\mathrm{spt}_{\ell+1}(T_n) \subseteq \mathrm{spt}_\ell(T_n)$ so by clauses (c)+(d)

(∗) $\ell < m \leq k \Rightarrow w_k \subseteq \mathrm{spt}_\ell(T_m)$.

Hence $\mathrm{spt}_\ell(T) \supseteq \bigcap_{m>\ell} \mathrm{spt}_\ell(T_m) \supseteq \bigcup_{m\geq\ell} w_m \in F$ (as all cofinite subsets of ω belong to F). Now T forces $\{k_\ell : \ell < \omega \text{ even }\} \subseteq \underset{\sim}{A}$ (remember clause (b)), so $\{k_\ell : \ell < \omega \text{ even }\} \notin F_0$ by the hypothesis on $T_0, \underset{\sim}{A}$ (as $\{k_\ell : \ell < \omega\} \in V$, and $T_0 \leq T$, $T \Vdash_P$ “$\{k_\ell : \ell < \omega\} \subseteq \underset{\sim}{A}$” so $\{k_\ell : \ell < \omega\} \in F_0$ implies: $T \Vdash_Q$ “$\omega \setminus \underset{\sim}{A} = \emptyset \mod F$”, a contradiction). So the strategy defined above is a winning strategy for player I hence by Lemma 5.11, $F_0 \leq_{RK} F$. So it remains to show that player I can carry out the strategy i.e. can preserve (∗). Note that T_0 is defined.

Case 1: n even > 0: Player I lets $m(n) < \omega$ be $\max[\bigcup\{w_\ell : \ell < n \text{ odd}\} \cup \{n\}] + 1$, and let $T_n^{[m(n)]} = \{\eta_0, \ldots, \eta_{s(n)}\}$ with no repetition. For each η_ℓ ($\ell \leq s(n)$) clearly $(T_n)_{[\eta_\ell]}$ is $\geq T_0$ and belongs to Q, hence

$A_\ell^n = \{k < \omega$: there are $T'_{\ell,k}, T''_{\ell,k} \geq (T_n)_{[\eta_\ell]}$, such that $T'_{\ell,k} \Vdash_Q$ “$k \in \underset{\sim}{A}$”, and $T''_{\ell,k} \Vdash_Q$ “$k \notin \underset{\sim}{A}$”$\}$

belong to F_0.

Now: player I plays $A_n = \bigcap_{\ell \leq s(n)} A_\ell^n$ which is clearly a legal move.

Player II chooses some $k_n \in A_n$.

Player I (“on the side”) lets $T_{n+1} = \bigcup\limits_{\ell \leq s(n)} T'_{\ell,k_n}$ (it is as required in (∗)).

Case 2: n odd: Player I lets $A_n = \mathrm{spt}_{m(n)}(T_n)$ (note $Q = \mathrm{SP}^*(F)$). Note T_n has just been chosen.

Player II chooses a finite $w_n \subseteq A_n$ and player I lets on the side $T_{n+1} = T_n$. $\square_{5.12}$

5.13 Theorem. It is consistent with ZFC $+2^{\aleph_0} = \aleph_2$ that, up to a permutation on ω, there is a unique Ramsey ultrafilter on ω. Moreover any P-point is above it (in the Rudin-Keisler order).

Proof. We start with a universe satisfying $2^{\aleph_0} = \aleph_1 + 2^{\aleph_1} = \aleph_2$ and $\diamondsuit_{\{\delta<\aleph_2:\mathrm{cf}(\delta)=\aleph_1\}}$. There is a Ramsey ultrafilter F in V. We shall use a CS iterated forcing $\langle P_i, \underset{\sim}{Q}_i : i < \omega_2\rangle$ such that each Q_i is proper, has the PP-property (hence is ${}^\omega\omega$-bounding), has cardinality continuum and forces that F still generates an ultrafilter. So by 5.1, 5.2, F remains a Ramsey ultrafilter in V^{P_i} for $i \leq \omega_2$ and also we can show by induction on $i < \omega_2$, that in V^{P_i}, CH holds and P_i has cardinality $\aleph_1$; so by VIII §2 below P_{ω_2} satisfies the $\aleph_2$-chain condition. If $F_1 \in V[G_{\omega_2}]$ ($G \subseteq P_{\omega_2}$ generic) is a P-point, not above F, then there is a $p \in P_{\omega_2}$ forcing $\underset{\sim}{F}_1$ is a name of such ultrafilter, and for a closed unbounded set of $\delta < \aleph_2$, $\mathrm{cf}(\delta) = \aleph_1$ implies that $\underset{\sim}{F}^1_\delta \stackrel{\text{def}}{=} \underset{\sim}{F}_1 \cap \mathcal{P}(\omega)^{V^{P_\delta}} \in V^{P_\delta}$ and p forces that $\underset{\sim}{F}^1_\delta$ is a P-point not above F (in V^{P_δ}).

Now, by the diamond $\diamondsuit_{\{\delta<\aleph_2:\mathrm{cf}(\delta)=\aleph_1\}}$ we can assume that for some such δ, $\underset{\sim}{Q}_\delta = \mathrm{SP}^*(\underset{\sim}{F}^1_\delta)$.

Now by 5.12 forcing with Q_δ (over V^{P_δ}) preserves "F (generates) an ultrafilter", by 5.6(2) Q_δ has the PP-property hence (by 2.12B) Q_δ is ${}^\omega\omega$-bounding and trivially Q_δ has cardinality continuum; so Q_i is as required. Now as each Q_j ($i < j < \omega_2$) has the PP-property, P_{ω_2}/P_δ has the PP-property (by 2.12C+2.3). So by lemma 5.12 we know $\underset{\sim}{F}^1_\delta$ cannot be completed to a P-point in $V^{P_{\omega_2}}$. $\square_{5.13}$

§6. On the Splitting Number $\mathfrak{s}$ and Domination Number $\mathfrak{b}$ and on $\mathfrak{a}$

For a survey on this area, see van Douwen [D] and Balcar and Simon [BS].

Nyikos has asked us whether there may be (in our terms) an undominated family $\subseteq {}^\omega\omega$ of power $\aleph_1$ while there is no splitting family $\subseteq [\omega]^{\aleph_0}$ of power $\aleph_1$. He observed that it seems necessary to prove, assuming CH, the existence of a P-point without a Ramsey ultrafilter below it (in the Rudin-Keisler order).

In the third section we have proved a preservation lemma for countable support iterations whose first motivation is that no new $f \in {}^\omega\omega$ dominates all old ones, and prove (3.23(1)) the consistency of ZFC+$2^{\aleph_0} = \aleph_2 + \mathfrak{d} = \mathfrak{s} > \mathfrak{b}$ where $\mathfrak{d}$ is the minimal power of a dominating subfamily of ${}^\omega\omega$ (see 3.24(3)), and $\mathfrak{s}$ is the minimal power of splitting subfamily of $[\omega]^{\aleph_0}$ (see Def 3.24(1)) and $\mathfrak{b}$ is the minimal power of an undominated subfamily of ${}^\omega\omega$ (see Definition 3.24(2)).

However one point was left out in Sect. 3: the definition of the forcing we iterate, and the proof of its relevant properties: that it adds a subset $\underset{\sim}{r}$ of ω such that $\{A \in V : A \subseteq \omega, \underset{\sim}{r} \subseteq^* A\}$ is an ultrafilter of the Boolean algebra $\mathcal{P}(\omega)^V$; but in a strong sense it does not add a function $\underset{\sim}{f} \in {}^\omega\omega$ dominating all old members of ${}^\omega\omega$; this was promised in 3.22. Note that Mathias forcing adds a subset $\underset{\sim}{r}$ of ω as required above, but also adds an undesirable $\underset{\sim}{f}$. This is done here; its definition takes some space. This forcing notion makes the "old" $[\omega]^{\aleph_0}$ an unsplitting family. The proof of this is quite easy, but we have more trouble proving the "old" ${}^\omega\omega$ is not dominated. ¿From the forcing notion (and, in fact, using a simpler version), we can construct a P-point as above.

Then A. Miller told us he is more interested in having in this model "no MAD of power $\leq \aleph_1$" (MAD stands for "a maximal almost disjoint family of infinite subsets of ω") (i.e. $\mathfrak{s}$, $\mathfrak{a} > \aleph_1 = \mathfrak{b}$). A variant of our forcing can "kill" a MAD family and the forcing has the desired properties if we first add $\aleph_1$ Cohen reals (see 3.23(2), 6.16). We also like to prove the consistency of ZFC+$2^{\aleph_0} = 2^{\aleph_1} = \aleph_2 + \aleph_2 = \mathfrak{s} > \mathfrak{a} = \mathfrak{b} = \aleph_1$, where $\mathfrak{a} = \min\{|\mathcal{A}| : \mathcal{A}$ a maximal family of almost disjoint subsets of $\omega\}$ (see Definition 3.24(2)). In the seventh section we show that in the model we have constructed (in the proof of 3.23(1)) there is a MAD (maximal family of pairwise disjoint infinite subsets of ω) of power $\aleph_1$ (hence $\mathfrak{a} = \aleph_1$). This answers a question of Balcar and Simon:

they defined

$$\mathfrak{a}_s = \min\{|\mathcal{A}| : \mathcal{A} \text{ is a maximal family of almost disjoint subsets of } \omega \times \omega,$$
$$\text{which are graphs of partial functions from } \omega \text{ to } \omega\}.$$

They have proved $\mathfrak{s} \leq \mathfrak{a}_s$ and $\mathfrak{a} \leq \mathfrak{a}_s \leq 2^{\aleph_0}$, so our result implies that $\mathfrak{a} < \mathfrak{a}_S$ is consistent.

In the eighth section we prove the consistency (with ZFC $+2^{\aleph_0} = 2^{\aleph_1} = \aleph_2$) of $\aleph_1 = \mathfrak{h} < \mathfrak{a} = \mathfrak{b} = \aleph_2$ (where $\mathfrak{h}$ is the minimal cardinal κ for which $\mathcal{P}(\omega)$/finite is a $(\kappa, 2^{\aleph_0})$-distributive Boolean algebra).

The relations between the cardinals above are described by the following diagram.

$$\begin{array}{ccccccccc} & & \mathfrak{s} & \longrightarrow & \mathfrak{d} & & \longrightarrow & & 2^{\aleph_0} \\ & & \uparrow & & \uparrow & & & & \uparrow \\ \aleph_1 & \longrightarrow & \mathfrak{h} & \longrightarrow & \mathfrak{b} & \rightarrow & \mathfrak{a} & \rightarrow & \mathfrak{a}_s \end{array}$$

(where arrow means "$\leq$ is provable is ZFC") (see [D] and [Sh:207] for results not mentioned above, and two other cardinal invariants); sections 6, 7, 8 represent material from [Sh:207] (revised).

* * *

Now we turn to the definition of the forcing we iterate and the proof of its relevant properties: that it adds a subset $\underset{\sim}{r}$ of ω such that $\{A \in V : A \subseteq \omega, \underset{\sim}{r} \subseteq_{ae} A\}$ is an ultrafilter in the Boolean algebra $\mathcal{P}(\omega)^V$; but in a strong sense (that is, almost ${}^\omega\omega$-bounding) it does not add a function $\underset{\sim}{f} \in {}^\omega\omega$ dominating all old members of ${}^\omega\omega$.

More on such forcing notions see [RoSh:470].

6.1 Definition. 1) Let K_n be the family of pairs (s, h), s a finite set, h a partial function from $\mathcal{P}(s)$ (you can think of $h(t)$ when not defined as -1) to $n+1$ such that:

(a) $h(s) = n$

(b) if $h(t) = \ell + 1$ (so $t \subseteq s$), $t = t_1 \cup t_2$ then $h(t_1) \geq \ell$ or $h(t_2) \geq \ell$ and $|t| > 1$.

We may add

(c) if $t_1 \subseteq t_2$ are in $\mathrm{Dom}(h)$ then $h(t_1) \leq h(t_2)$.

2) $K_{\geq n}, K_{\leq n}, K_{n,m}$ are defined similarly, and $K = \bigcup_{n<\omega} K_n$.

We call s the domain of (s,h) and write $a \in (s,h)$ instead of $a \in s$. We call (s,h) standard if s is a finite subset of the family of hereditarily finite sets. We use the letter t to denote such pairs. We call (s,h) simple if $h(t) = [\log_2(|t|)]$ for $t \subseteq s$. If $t = (s,h) \in K$, let $\mathrm{lev}(t) = \mathrm{lev}(s,h)$ be the unique $n < \omega$ such that $t \in K_n$.

6.2 Definition. 1) Suppose $(s_\ell, h_\ell) \in K_{s(\ell)}$ for $\ell \in \{0,1\}$. We say $(s_0,h_0) \leq^d (s_1,h_1)$ (or (s_1,h_1) refines (s_0,h_0)) if:
$s_0 = s_1$ and $[t_1 \subseteq t_2 \subseteq s_0 \ \& \ h_1(t_1) \leq h_1(t_2) \Rightarrow h_1(t_1) \leq h_0(t_1) \leq h_0(t_2)]$ (so $\mathrm{lev}(s_0,h_0) \geq \mathrm{lev}(s_1,h_1)$ and $\mathrm{Dom}(h_1) \subseteq \mathrm{Dom}(h_0)$).
2) We say $(s_0,h_0) \leq^e (s_1,h_1)$ if for some $s'_0 \in \mathrm{Dom}(h_0)$, $(s'_0, h_0 \restriction \mathcal{P}(s'_0)) = (s_1,h_1)$
3) We say $(s_0,h_0) \leq (s_1,h_1)$ if for some (s',h'), $(s_0,h_0) \leq^e (s',h') \leq^d (s_1,h_1)$.

6.3 Fact. The relations $\leq^d$, $\leq^e$, $\leq$ are partial orders of K. $\square_{6.3}$

6.4 Definition.

1) Let L_n be the family of pairs (S,H) such that:
   a) S is a finite tree with a root called $\mathrm{root}(S)$.
   b) H is a function whose domain is $\mathrm{in}(S)$ =the set of non-maximal points of S and with values H_x for $x \in \mathrm{in}(S)$.
   c) For $x \in \mathrm{in}(S)$, $(\mathrm{Suc}_S(x), H_x) \in K_{\geq n}$, where $\mathrm{Suc}_S(x)$ is the set of immediate successors of x in S, so $H_x(\mathrm{Suc}_S(x)) \geq n$.
2) We say $(S^0,H^0) \leq (S^1,H^1)$ if $S^0 \supseteq S^1$, they have the same root, $\mathrm{in}(S^1) = S^1 \cap \mathrm{in}(S^0)$ and for every $x \in \mathrm{in}(S^1)$, $(\mathrm{Suc}_{S^0}(x), H^0_x) \leq (\mathrm{Suc}_{S^1}(x), H^1_x)$ and of course $\mathrm{Suc}_{S^1}(x) = \mathrm{Suc}_{S^0}(x) \cap S^1$.
3) Let $\mathrm{int}(S) \stackrel{\mathrm{def}}{=} S \setminus \mathrm{in}(S)$, $\mathrm{lev}(S,H) = \max\{n : (S,H) \in L_n\}$, $x \in (S,H)$ means $x \in S$. A member of L_n is standard if $\mathrm{int}(S) \subseteq \omega$ and $\mathrm{in}(S)$ consists of hereditarily finite sets not in ω. Let for $x \in S$, $(S,H)^{[x]} = (S^{[x]}, H \restriction S^{[x]})$ where $S^{[x]}$ is $S \restriction \{y \in S : S \models x \leq_S y\}$.

4) For $\mathbf{t} \in L_n$ let $\mathbf{t} = (S^{\mathbf{t}}, H^{\mathbf{t}})$ and let $\mathrm{lev}(\mathbf{t}) = \max\{n : \mathbf{t} \in L_n\}$
5) We say $\mathbf{t}^1, \mathbf{t}^2 \in \bigcup_{n<\omega} L_n$ are disjoint *if*: $S^{\mathbf{t}^1} \cap S^{\mathbf{t}^2} = \emptyset$.
6) Let $\mathrm{int}(\mathbf{t}) = \mathrm{int}(S^{\mathbf{t}})$.
7) Let $L = \bigcup\limits_{n<\omega} L_n$

6.5 Fact. The relation $\leq$ is a partial order of $L = \bigcup_n L_n$. $\square_{6.5}$

6.6 Fact. If $(S, H) \in L_n$ then $(S', H') \stackrel{\mathrm{def}}{=} \mathrm{half}(S, H)$ belongs to $L_{[(n+1)/2]}$ and $(S, H) \leq (S', H')$ where $S' = S$, $H'_x(A) = [H_x(A) - \mathrm{lev}(S, H)/2]$ where $[x]$ is the largest integer $\leq x$ and $\mathrm{Dom}(H'_x) = \{A : H_x(A) \geq \mathrm{lev}(S, H)/2\}$. $\square_{6.6}$

6.7 Fact. If $(S, H) \in L_{n+1}$, $\mathrm{int}(S) = A_0 \cup A_1$ *then* there is $(S^1, H^1) \geq (S, H)$, $(S^1, H^1) \in L_n$ such that $[\mathrm{int}(S^1) \subseteq A_0$ or $\mathrm{int}(S^1) \subseteq A_1]$.

Proof. Easy by induction on the height of the tree (using clause (b) of Def 6.1(1)). $\square_{6.7}$

6.8 Definition. We define the forcing notion Q:

1) $p \in Q$ *if* $p = (w, T)$ where w is a finite subset of ω, T is a countable (infinite) set of pairwise disjoint standard members of L and $T \cap L_n$ is finite for each n, moreover for simplicity the convex hulls of the $\mathrm{int}(\mathbf{t})$ for $\mathbf{t} \in T$ are pairwise disjoint; let $\mathrm{cnt}(T)$ and $\mathrm{cnt}(p)$ mean $\bigcup_{(H,S)\in T} \mathrm{int}(S, H)$. Writing $T = \{\mathbf{t}_n : n < \omega\}$ we mean $\langle \min(\mathrm{int}(\mathbf{t}_n)) : n < \omega\rangle$ strictly increasing.
2) Given $\mathbf{t}_1 = (S_1, H_1), \ldots, \mathbf{t}_k = (S_k, H_k)$ all from L such that $S_i \cap S_j = \emptyset$ $(i \neq j)$, and given $\mathbf{t} = (S, H)$ from L, we say $\mathbf{t}$ is *built* from $\mathbf{t}_1, \ldots, \mathbf{t}_k$ if: there are incomparable nodes $a_1, \ldots, a_k$ of S such that every node of S is comparable with some a_i, and such that, letting $S(a_i) = \{b \in S : b \geq_S a_i\}$ we have $(S_i, H_i) = (S(a_i), H \restriction S(a_i))$.
3) $(w^0, T^0) \leq (w^1, T^1)$ iff: $w^0 \subseteq w^1 \subseteq w^0 \cup \mathrm{cnt}(T^0)$, and, letting $T^0 = \{\mathbf{t}^0_0, \mathbf{t}^0_1, \ldots\}$, $T^1 = \{\mathbf{t}^1_0, \mathbf{t}^1_1 \ldots\}$, there are finite, nonempty pairwise disjoint subsets of ω, B_0, $B_1, \ldots$, and there are $\hat{\mathbf{t}}_i \geq \mathbf{t}^0_i$ for all $i \in \bigcup_j B_j$ such that

for each n only finitely many of the $\hat{\mathbf{t}}_i$ are inside L_n and such that for each j, letting $B_j = \{i_1, \ldots, i_k\}$, $\mathbf{t}^1_j$ is built from $\hat{\mathbf{t}}_{i_1}, \ldots, \hat{\mathbf{t}}_{i_k}$.

4) We call (w, T) standard if $T = \{\mathbf{t}_n : n < \omega\}$, $\max(w) < \min[\mathrm{int}(\mathbf{t}_n)]$, $\max[\mathrm{int}(\mathbf{t}_n)] < \min[\mathrm{int}(\mathbf{t}_{n+1})]$ and $\mathrm{lev}(\mathbf{t}_n)$ is strictly increasing (and writing $T = \{\mathbf{t}_n : n < \omega\}$ we mean this).

6.9 Definition. For $p = (w, T)$ we write $w = w^p$, $T = T^p$. We say q is a pure extension of p ($p \leq_{\mathrm{pr}} q$) if $q \geq p$, $w^q = w^p$. We say p is pure if $w^p = \emptyset$, and $p \leq^* q$ means omitting finitely many members of T^q makes $q \geq p$.

The following generalization will be used later.

6.10 Definition. 1) For an ideal I of $\mathcal{P}(\omega)$ (which includes all finite sets) let $Q[I]$ be the set of $p \in Q$ such that for every $A \in I$, for infinitely many $\mathbf{t} \in T^p$, $\mathrm{int}(\mathbf{t}) \cap A = \emptyset$. The main case is $I =$ family of finite subsets of ω (then $Q[I] = Q$).

2) Let $Q'[I]$ be $\{p \in Q :$ there is q such that $Q \models p \leq q$ and $q \in Q[I]\}$ (so $Q[I]$, $Q'[I]$ are equivalent).

6.10A Remark. 1) So if $p = (w, \{\mathbf{t}_n : n < \omega\}) \in Q[I]$ then $p \leq (w, \{\mathrm{half}(\mathbf{t}_n) : n < \omega\}) \in Q[I]$.

2) More generally if $p = (w, \{\mathbf{t}_n : n < \omega\}) \in Q[I]$ and $h : \omega \to \omega$ is a function from ω to ω going to ∞ (i.e. $\lim \inf_{n<\omega} h(n) = \infty$) and $\mathbf{t}'_n \geq \mathbf{t}_n$, or even $\mathbf{t}'_n \geq \mathbf{t}_n^{[x_n]}$ and $\mathrm{lev}(\mathbf{t}'_n) \geq h(\mathrm{lev}(\mathbf{t}_n))$ then $(w, \{\mathbf{t}'_n : n < \omega\}) \in Q[I]$.

6.11 Fact. 0) Q is a partial order.

1) If $p \in Q$ and $\underset{\sim}{\tau}_n$ (for $n < \omega$) are Q-names of ordinals, then there is a pure standard extension q of p such that: letting $T^q = \{\mathbf{t}_\ell : \ell < \omega\}$, for every $n < \omega$ and $w \subseteq \max[\mathrm{int}(\mathbf{t}_n)] + 1$, if we let $q^n_w = (w, \{\mathbf{t}_\ell : \ell > n\})$, then for $k \leq n$:

q^n_w forces a value on $\underset{\sim}{\tau}_k$ *iff* some pure extension of q^n_w forces a value on $\underset{\sim}{\tau}_k$.

Moreover if $T^p = \{\mathbf{t}^0_n : n < \omega\}$, we can demand $\bigwedge_{\ell<n^*} \mathbf{t}_\ell = \mathbf{t}^0_\ell$ but then the demand on n, w above is for $n \geq n^* - 1$ only.

2) Q is proper (in fact α-proper for every $\alpha < \omega_1$).

3) $\Vdash_Q$ "$\{n : (\exists p \in \underset{\sim}{G}_Q)[n \in w^p]\}$ is an infinite subset of ω which $\mathcal{P}(\omega)^V$ does not split."

Proof. Easy (for (3) use 6.7, see more in 6.16(3)). $\square_{6.11}$

6.12 Lemma. Let $q, \underset{\sim}{\tau}_n$ be as in 6.11(1). Then for some pure standard extension r of q, letting $T^r = \{\mathbf{t}'_n : n < \omega\}$, (lev$(\mathbf{t}'_n)$ strictly increasing, of course and) the following holds.

(*) *For every* $n < \omega$, $w \subseteq [\max(\text{int}(\mathbf{t}'_{n-1})) + 1]$, and $\mathbf{t}''_n \geq \mathbf{t}'_n$ (so we ask only lev$(\mathbf{t}''_n) \geq 0$) *there is* $w' \subseteq \text{int}(\mathbf{t}''_n)$, such that $(w \cup w', \{\mathbf{t}'_\ell : \ell > n\})$ forces a value on $\underset{\sim}{\tau}_m$ for $m \leq n$ (we let $\max\text{int}(\mathbf{t}'_{-1})$ be $\max(w^q \cup \{-1\})$).

This lemma follows easily from claim 6.14 (see below) (choose by it the $\mathbf{t}'_n$ by induction on n) and is enough for a proof of Lemma 3.22, which we now present.

6.13 Proof of Lemma 3.22. By 6.11(2), clause (a) (of 3.22 i.e. Q is proper) holds (more fully use the last clause of 6.11(1) to get a sequence of conditions as needed); and by 6.11(3) clause (d) (of 3.22 i.e. inducing an ultrafilter on the old $\mathcal{P}(\omega)$) holds; and clause (c) (of 3.22 i.e. $|Q| = 2^{\aleph_0}$) is trivial. For proving clause (b) (i.e. Q is almost ${}^\omega\omega$-bounding, see Definition 3.5(1)) let $\underset{\sim}{f} \in {}^\omega\omega$ and $p \in Q$ be given. Let $\underset{\sim}{\tau}_n = \underset{\sim}{f}(n)$, apply 6.11(1) to get q and then apply (on q, $\underset{\sim}{\tau}_n$ $(n < \omega)$) 6.12 getting $r = (w^p, \{\mathbf{t}'_n : n < \omega\}) \geq q$. We have to define $g \in {}^\omega\omega$ (as required in Definition 3.5(1)). Let $g(n) = \max\{k + 1 :$ for some $w \subseteq [\max(\text{int}(\mathbf{t}'_n)) + 1]$ we have $(w, \{\mathbf{t}'_\ell : \ell > n) \Vdash$ "$\underset{\sim}{f}(n) = k$"$\}$. Let A be any infinite subset of ω, and we define $p' = (w^p, \{\mathbf{t}'_n : n \in A\})$, so $p' \geq r \geq p$. We have to show that $p' \Vdash_Q$ "for infinitely many $n \in A$, $\underset{\sim}{f}(n) < g(n)$". So it is enough, given $n_0 < \omega$ and p^2, $p' \leq p^2 \in Q$ to find $n \in A \setminus n_0$ and p^3 such that $p^2 \leq p^3 \in Q$ and $p^3 \Vdash_Q$ "$\underset{\sim}{f}(n) < g(n)$". So assume that $n_0 < \omega$ and

$p' \le p^2 \in Q$, and $p^2 = (w^2, T^2)$, $T^2 = \{\mathbf{t}^2_n : n < \omega\}$, and w.l.o.g. for some $i(*) > n_0$ for every n we have $\min[\mathrm{int}(\mathbf{t}^2_n)] > \max[\mathrm{int}(\mathbf{t}'_{i(*)})] > \sup(w^2)$. As $p^2 \ge p'$, we can find $k < \omega$, $i_1 < \cdots < i_k$ from A, and $\mathbf{t}^*_{i_\ell} \ge \mathbf{t}'_{i_\ell}$ such that $\mathbf{t}^2_0$ is built from $\mathbf{t}^*_{i_1}, \ldots, \mathbf{t}^*_{i_k}$; by the previous sentence $i_1 > i(*)$. By $(*)$ from 6.12 (as $w^2 \subseteq \max[\mathrm{int}(\mathbf{t}'_{i(*)})] + 1$, and $i(*) < i_1$ and r from 6.12 is standard), there is $w'' \subseteq \mathrm{int}(\mathbf{t}^*_{i_k})$ (hence $w'' \subseteq \mathrm{int}(\mathbf{t}^2_0)$) such that $p^3 = (w^2 \cup w'', \{\mathbf{t}'_j : j \in (i_k, \omega)\})$ forces a value, say m to $\underset{\sim}{f}(i_k)$, so by the definition of g clearly $m < g(i_k)$. But clearly p^2, p^3 have a common upper bound: $p^4 = (w^2 \cup w'', \{\mathbf{t}^2_n : n \in (n_4, \omega)\})$ for every $n_4 < \omega$ large enough (really $n_4 = 0$ is O.K.!). So we are done. $\square_{3.22}$

6.14 Claim. Let $(\emptyset, T)$ be a pure condition, and let W be a family of finite subsets of $\mathrm{cnt}(T)$ so that

$(*)$ for every $(\emptyset, T') \ge (\emptyset, T)$, there is a $w \subseteq \mathrm{cnt}(T')$ such that $w \in W$.

Let $k < \omega$. Then there is $\mathbf{t} \in L_k$ appearing in some $(\emptyset, T') \ge (\emptyset, T)$ such that:

$$\mathbf{t}' \ge \mathbf{t} \quad \Rightarrow \quad (\exists w \in W)[w \subseteq \mathrm{int}(\mathbf{t}')].$$

Proof. Let T^* be arbitrary such that $(\emptyset, T) \le (\emptyset, T^*) \in Q$, and $T^* = \{\mathbf{t}_n : n < \omega\}$. For notational simplicity, without loss of generality let W be closed upward.

Stage A: There is n such that for every $\mathbf{t}'_\ell \ge \mathrm{half}(\mathbf{t}_\ell)$ (for $\ell < n$) we have $\bigcup_{\ell<n} \mathrm{int}(\mathbf{t}'_\ell) \in W$. This is because the family of $\langle \mathbf{t}'_\ell : \ell < n\rangle$, $n < \omega$, $\mathbf{t}'_\ell \ge \mathrm{half}(\mathbf{t}_\ell)$ form an ω-tree with finite branching and for every infinite branch $\langle \mathbf{t}'_\ell : \ell < \omega\rangle$ by $(*)$ there is an initial segment $\langle \mathbf{t}'_\ell : \ell < n\rangle$ with $\bigcup_{\ell<n} \mathrm{int}(\mathbf{t}'_\ell) \in W$. [Why? Define $(S^\ell, H^\ell) \in L$ such that $S^\ell = S^{\mathbf{t}'_\ell}$ and $H^\ell_x(A) = H^{\mathbf{t}_\ell}_x(A)$ (and not $H^{\mathbf{t}'_\ell}_x(A)$!) when $x \in \mathrm{in}(S^\ell)$, $A \subseteq \mathrm{Suc}_{S_\ell}(x)$, so letting $T' =: \{(S^\ell, H^\ell) : \ell < \omega\}$ we have: $\mathrm{lev}(S^\ell, H^\ell) \ge \mathrm{lev}(\mathbf{t}_\ell)/2^\ell - 1/2$ and $(\emptyset, T^*) \le (\emptyset, T')$. Now apply $(*)$ remembering W is upward closed.] By König's lemma we finish.

Stage B: There are $n(0) < n(1) < n(2) < \ldots$ such that for every m and $\mathbf{t}'_\ell \ge \mathrm{half}(\mathbf{t}_\ell)$ for $n(m) \le \ell < n(m+1)$, the set $\bigcup\{\mathrm{int}(\mathbf{t}'_\ell) : n(m) \le \ell < n(m+1)\} \in W$. The proof is by repeating stage A (changing T^*).

Stage C: There are $m(0) < m(1) < \ldots$ such that: if $i < \omega$, for every function h with domain $[m(i), m(i+1))$ such that $h(j) \in [n(j), n(j+1))$ and $\mathbf{t}'_\ell \geq \mathrm{half}(\mathbf{t}_\ell)$ for all relevant ℓ then $\bigcup\{\mathbf{t}'_{h(j)} : j \in [m(i), m(i+1))\}$ belongs to W.
The proof is parallel to that of stage B; as there it is enough, assuming $m(i^*)$ was chosen, to find appropriate $m(i^*+1) > m(i^*)$. The set of branches corresponds to $\{\langle \mathbf{t}'_\ell : \ell \in [m(i^*), \omega)\rangle$: for some function $h \in \prod_{\ell \in [m(i^*), \omega)} [n(\ell), n(\ell+1))$ for every $\ell \in [m(i^*), \omega)$, $\mathbf{t}'_\ell \geq \mathrm{half}(\mathbf{t}_{h(\ell)})\}$. So if the conclusion fails i.e. for every $m > m(i^*)$ if we assign $m(i^*+1) = m$, for some function h_m with domain $[m(i^*), m)$, $h(\ell) \in [n(\ell), n(\ell+1))$ and $\langle \mathbf{t}^m_\ell : \ell \in [m(i^*), m)\rangle$, where $\mathbf{t}^m_\ell \geq \mathrm{half}(\mathbf{t}_{h_m(\ell)})$ the desired conclusion fails. So by König's lemma we can find $h \in \prod_{\ell \in [m(i^*), \omega)} [n(\ell), n(\ell+1))$, $\langle \mathbf{t}'_\ell : \ell \in [m(i^*), \omega)\rangle$ such that for every $m' \in [m(i^*), \omega)$ for infinitely many $m \in [m', \omega)$ we have

$$\ell \in [m(i^*), m') \Rightarrow h_m(\ell) = h(\ell) \ \&\ \mathbf{t}'_\ell = \mathbf{t}^m_\ell.$$

As before using $\langle \mathbf{t}'_{h(\ell)} : \ell < \omega\rangle$ we can contradict the assumption $(*)$.

Stage D: We define a partial function H from finite subsets of ω to ω: let $H(u) \geq 0$ if for every $\mathbf{t}'_\ell \geq \mathrm{half}(\mathbf{t}_\ell)$ (for $\ell \in u$) we have $(\bigcup_{\ell \in u} \mathrm{int}(\mathbf{t}'_\ell)) \in W$ and let $H(u) \geq m+1$ if $[u = u_1 \cup u_2 \Rightarrow H(u_1) \geq m \vee H(u_2) \geq m]$.

We have shown that $H([n(i), n(i+1))) \geq 0$, and $H([n(m(i)), n(m(i+1)))) \geq 1$, (for the later, assuming $u = [n(m(i)), n(m(i+1))) = u_1 \cup u_2$ we have that: either u_1 contains an interval $[n(j), n(j+1))$ for some $j \in [m(i), m(i+1))$ *or* u_2 has a member in each such interval so it contains $\{h(j) : j \in [m(i), m(i+1))\}$ for some $h \in \prod_{\ell=m(i)}^{m(i+1)-1} [n(\ell), n(\ell+1))$; now apply stage B to show that in the first case $H(u_1) \geq 0$ and Stage C to show that in the second case $H(u_2) \geq 0$).

It clearly suffices to find u, $H(u) \geq k$. [We then define $\mathbf{t} = (S, H)$ as follows: $S = \bigcup_{\ell \in u} S^{\mathbf{t}_\ell} \cup \{u\}$, u is the root with set of immediate successors being $\{\mathrm{root}(\mathbf{t}_\ell) : \ell \in u\}$; and the order restricted to $S^{\mathbf{t}_\ell}$ is as in $\mathbf{t}_\ell$; and for $x \in S^{\mathbf{t}_\ell}$ we have $H^{\mathbf{t}}_x = H^{\mathrm{half}(\mathbf{t}_\ell)}_x$ and $H^{\mathbf{t}}_u(A) \overset{\mathrm{def}}{=} H(\{\ell : \mathrm{root}(S^{\mathbf{t}_\ell}) \in A\})$.] We prove the existence of such u by induction on k, (e.g. simultaneously for all T',

$(\emptyset, T') \geq (\emptyset, T))$. This is done by repeating the proof above (alternatively, we just repeat 2^k times getting an explicit member of K_k in the root). $\square_{6.14}$

The rest of this section deals with $Q[I]$. Note that by 6.21(2) below in the interesting case the set of standard $p \in Q[I]$ is dense. For the rest of this section:

6.15 Notation. 1) Let Q^0 be the forcing of adding $\aleph_1$ Cohen reals $\langle r_i : i < \omega_1 \rangle$, $r_i \in {}^\omega\omega$. We usually work in $V_1 = V^{Q^0}$.

2) Let $\mathcal{A} = \{A_i : i < \alpha^*\}$ denote an infinite family of infinite subsets of ω (usually the members are pairwise almost disjoint).

3) Let $I = I_{\mathcal{A}}$ be the ideal of $\mathcal{P}(\omega)$, including all finite subsets of ω but $\omega \notin I$ and generated by $\mathcal{A} \cup \{[0, n) : n < \omega\}$. So $I_{\mathcal{A}}$ depends on the universe (the interesting case here is $\mathcal{A}$ a MAD family in V, of the form $\{A_i : i < \omega_1\}$, $Q[I_{\mathcal{A}}]$ means in $V_1 = V^{Q^0}$). If not said otherwise we assume $\emptyset \notin I_{\mathcal{A}}$.

6.16 Claim. Assume $\mathcal{A} \in V$ is a family of subsets of ω (not necessarily MAD), and we work in $V_1 = V^{Q^0}$, and $I = I_{\mathcal{A}}$ so $Q[I]$ is from V^{Q^0}:

1) If $p \in Q[I]$ and $\underset{\sim}{\tau}_n$ $(n < \omega)$ are $Q[I]$-names of ordinals *then* there is a pure standard extension q of p such that: $q \in Q[I]$, and letting $T^q = \{\mathbf{t}_n : n < \omega\}$, for every $n < \omega$ and $w \subseteq [\max \operatorname{int}(\mathbf{t}_n) + 1]$ let $q^n_w = (w, \{\mathbf{t}_\ell : n < \ell < \omega\})$, then ($q^n_w \in Q[I]$, of course, and) for every $k \leq n$ we have: q^n_w forces a value on $\underset{\sim}{\tau}_k$ iff some pure extension of q^n_w in $Q[I]$ forces a value on $\underset{\sim}{\tau}_K$.
2) $Q[I]$ is proper, (moreover α-proper for every $\alpha < \omega_1$ (not used)).
3) $\Vdash_{Q[I]}$ "$\{n : (\exists p \in G_{Q[I]}) n \in w^p\}$ is an infinite subset of ω which is almost disjoint from every $A \in I$ (equivalently $A \in \mathcal{A}$)."

Proof. 1) Let λ be regular large enough, N a countable elementary submodel of $(H(\lambda), \in, V \cap H(\lambda))$ to which I, $\langle r_i : i < \omega_1 \rangle$, $Q[I]$, p and $\langle \underset{\sim}{\tau}_n : n < \omega \rangle$ belong and $N' = N \cap V \in V$ (remember we are working in V_1). Let $\delta = N \cap \omega_1$ (so $\delta \notin N$). So $N = N'[\langle r_i : i < \delta \rangle]$ belongs to $V[\langle r_i : i < \delta \rangle]$.

We define by induction on $n < \omega$, $q^n \in Q[I] \cap N$, $\mathbf{t}_n$ and $k_n < \omega$ such that:

a) each q^n is a pure extension of p.

b) $q^n \geq q^\ell$ for $\ell < n$ and if $w \subseteq k_n$, $m < n+1$ and some pure extension of (w, T^{q^n}) forces a value on $\underset{\sim}{\tau}_m$, then (w, T^{q^n}) does it.

c) $k_n > k_\ell$ and $k_n >$ max int$(\mathbf{t}_\ell)$ for $\ell < n$.

d) every $\ell \in \text{cnt}(q^{n+1})$ is $> k_n$ i.e. $\mathbf{t} \in T^{q^{n+1}} \Rightarrow \min[\text{int}(\mathbf{t})] > k_n$.

e) $\mathbf{t}_n \geq \mathbf{t}'_n$ for some $\mathbf{t}'_n \in T^{q^n}$ and $\text{lev}(\mathbf{t}_n) > n$ and $\min[\text{int}(\mathbf{t}_n)]$ is $> k_n$.

There is no problem in doing this: in stage n, we first choose k_n, then q^n and at last $\mathbf{t}_n$. We want in the end to let $T^q = \{\mathbf{t}_n : n < \omega\}$ (and $w^q = w^p$). One point is missing. Why does $q = (w^p, T^q)$ belong to $Q[I]$ (not just to Q)? But we can use some function in $V[\langle r_i : i < \delta \rangle]$ to choose k_n, q^n and then let $\mathbf{t}_n$ be the $r_\delta(n)$-th member of T^{q^n} which satisfies the requirement (in some fixed well ordering from V of the hereditarily finite sets). As $\mathcal{A} \in V$ and $r_\delta \in {}^\omega\omega$ is Cohen generic over $V[\langle r_i : i < \delta \rangle]$, this should be clear.

2) Easy by part (1).

3) Use Definition 6.10 and Fact 6.7. $\square_{6.16}$

6.17 Claim. Assume $\mathcal{A} = \{A_i : i < \alpha^*\} \in V$ is a MAD family, and in V_1 we have that $\Vdash_Q$"$\{A_i : i < \alpha^*\}$ is a MAD family". In V_1, let I be the ideal generated by $\{A_i : i < \alpha^*\}$ and the finite subsets of ω. Then: $(w, \{\mathbf{t}_n : n < \omega\})$ is a [standard] condition in $Q'[I]$ *iff*

it is a [standard] condition in Q and there are finite (non empty) pairwise disjoint $u_\ell \subseteq \alpha^*$ (for $\ell < \omega$) such that for each ℓ, for every k for some $n < \omega$, for some $\mathbf{t}'_n$, $\mathbf{t}'_n \geq \mathbf{t}_n$, $\text{lev}(\mathbf{t}'_n) \geq k$ and $\text{int}(\mathbf{t}'_n) \subseteq \bigcup_{i \in u_\ell} A_i$ *iff*

as before but there are singletons u_ℓ as above.

6.17A Remark. Note: if $\mathcal{A} \in V$, by 6.17 the standard $q \in Q[I]$ are dense in $Q[I]$, but otherwise we do not know. In the proof it does not matter.

Proof. The third condition implies trivially the second. We shall prove [second $\Rightarrow$ first] and then [first$\Rightarrow$third]. Suppose there are u_ℓ $(\ell < \omega)$ as in the second condition above and we shall prove the first one. So for each $\ell < \omega$ we can find $\langle \mathbf{t}'_n : n \in B_\ell \rangle$, $B_\ell \subseteq \omega$ is infinite, $\mathbf{t}'_n \geq \mathbf{t}_n$, $\text{lev}(\mathbf{t}'_n) \geq |B_\ell \cap n|$ and

$\text{int}(\mathbf{t}'_n) \subseteq \bigcup_{i \in u_\ell} A_i$. Wlog $\langle B_\ell : \ell < \omega \rangle$ are pairwise disjoint, so $p \leq p' \stackrel{\text{def}}{=} (w, \{\mathbf{t}'_n : n \in \bigcup_{\ell<\omega} B_\ell\})$ and $p' \in Q$, so it suffices to show $p' \in Q[I]$. Now every $B \in I$ is included in $\bigcup_{i \in u} A_i \cup \{0, \ldots, n^* - 1\}$ for some finite $u \subseteq \omega_1$ and $n < \omega$. But for some ℓ, u_ℓ is disjoint from u, hence $B \cap (\bigcup_{i \in u_\ell} A_i)$ is finite. We know that for infinitely many $n \in B_\ell$, $\text{int}(\mathbf{t}'_n) \subseteq \bigcup_{i \in u_\ell} A_i$ and the $\text{int}(\mathbf{t}'_n)$ $(n < \omega)$ are pairwise disjoint, hence for the infinitely many $n < \omega$, $\text{int}(\mathbf{t}_n) \cap B = \emptyset$, as required in the first condition.

Lastly assume the first condition and we shall prove the third one. Suppose $p = (w, \{\mathbf{t}_n : n < \omega\}) \in Q'[I]$, see Definition 6.10(2), w.l.o.g. $p \in Q[I]$. We choose by induction on m a finite $u_m \subseteq \alpha^*$, disjoint from $\bigcup_{\ell<m} u_\ell$ such that

$$B_m = \{n < \omega : \text{for some } \mathbf{t}'_n \geq \mathbf{t}_n \text{ we have } \text{lev}(\mathbf{t}'_n) \geq \text{lev}(\mathbf{t}_n)/2 - 1 \text{ and } \text{int}(\mathbf{t}_n) \subseteq \bigcup_{i \in u_m} A_i\}$$

are infinite and moreover u_m is a singleton.

Assume we have arrived to stage m. Let $B \stackrel{\text{def}}{=} \{n : \text{int}(\mathbf{t}_n)$ is disjoint to $\bigcup\{A_i : i \in \bigcup_{\ell<m} u_\ell\}\}$, so B is necessarily infinite (by the Definition of $Q[I]$), moreover $p^0 \stackrel{\text{def}}{=} (w, \{\text{half}(\mathbf{t}_n) : n \in B\})$ belongs to $Q[I]$ and is above p. Now clearly $Q[I] \subseteq Q$, hence $p^0 \in Q$. By an assumption of 6.17, we know $\mathcal{A} = \{A_i : i < \alpha^*\}$ is a MAD family even after forcing by Q, so there are $p^1 = (w', \{\mathbf{t}'_n : n < \omega\}) \in Q$, $p^0 \leq p^1$ and $i_0 < \alpha^*$ such that

(∗) $\quad p^1 \Vdash$ "$\{n : (\exists q \in \underset{\sim}{G}_Q)[n \in w^q]\} \cap A_{i_0}$ is infinite".

Let n^* be $> \sup(A_{i_0} \cap \bigcup\{A_i : i \in \bigcup_{\ell<m} u_\ell\})$. By 6.7 (more exactly, as in the proof of 6.16(3)), without loss of generality, $\bigcup_{n<\omega} \text{cnt}(\mathbf{t}'_n) \subseteq A_{i_0} \setminus n^*$ or $\bigcup_{n<\omega} \text{cnt}(\mathbf{t}'_n) \cap A_{i_0} = \emptyset$, but the second possibility contradicts (∗) so the first holds.

But $p^1 \geq p^0$ (in Q) so for each $n < \omega$ for some $k < \omega$ and $j_{n,0} < \cdots < j_{n,k-1}$ from B, $\mathbf{t}'_n$ is built from $\text{half}(\mathbf{t}_{j_{n,0}}), \ldots, \text{half}(\mathbf{t}_{j_{n,k-1}})$. So for some $y \in \mathbf{t}'_n$ we have $(\mathbf{t}'_n)^{[y]} \geq \text{half}(\mathbf{t}_{j_{n,0}})$, hence clearly (or see 6.20 below) there is $\mathbf{t}''_{j_{n,0}} \geq \mathbf{t}_{j_{n,0}}$, such that $\text{int}(\mathbf{t}''_{j_{n,0}}) \subseteq A_{i_0} \setminus n^*$ and $\text{lev}(\mathbf{t}''_{j_{n,0}}) \geq \text{lev}(\mathbf{t}_{j_{n,0}})/2$. Lastly let $u_m = \{i_0\}$ (i.e. all depend on m). $\square_{6.17}$

6.18 Claim. Let V_1, $\mathcal{A}$, $I = I_{\mathcal{A}}$ be as in 6.16 + 6.17 (so $\mathcal{A}$ is a MAD family in V, V_1 and V^Q). Assume we are given $k^* < \omega$, $(\emptyset, T) = (\emptyset, \{\mathbf{t}_n : n < \omega\}) \in Q[I]$, and a family W of finite subsets of $\mathrm{cnt}(T)$ such that
$(*)$ if $(\emptyset, T) \leq (\emptyset, T') \in Q[I]$ *then* there is $w \subseteq \mathrm{cnt}(T')$ such that $w \in W$.
Then there is $\mathbf{t} \in L_{k^*}$ appearing in some $(T', \emptyset) \geq (T, \emptyset)$ such that:

$$\mathbf{t}' \geq \mathbf{t} \Rightarrow (\exists w \in W)[w \subseteq \mathrm{int}(\mathbf{t}')].$$

Proof. Without loss of generality T is standard (by 6.16(1)) and W upward closed (check). Moreover we may assume that $\mathrm{lev}(\mathbf{t}_n) \geq 2k^*$ for each $n < \omega$.

We know, by 6.17 above, that there is $T^* = \{\mathbf{t}_n : n < \omega\}$, such that $(\emptyset, T^*) \in Q$ is standard, $(\emptyset, T) \leq (\emptyset, T^*)$ and for some sequence $\langle j_m : m < \omega \rangle$ of pairwise distinct ordinals $< \alpha^*$ and partition $\langle B_m : m < \omega \rangle$ of ω to infinite sets we have:

$$n \in B_m \Rightarrow \mathrm{int}(\mathbf{t}_n) \subseteq A_{j_m}.$$

For every finite $u \subseteq \omega$ define

$$\begin{aligned} \mathrm{nor}(u) = \max\{m : & \text{ for every cover } \langle u_\ell : \ell < 2^m \rangle \text{ of } u \\ & \quad (\text{i.e. } u_\ell \subseteq u \text{ and } \bigcup_{\ell < 2^m} u_\ell = u), \\ & \text{for some } \ell < 2^m \text{ and for every } \mathbf{t}'_i \geq \mathrm{half}(\mathbf{t}_i) \\ & \text{for } i \in u_\ell \text{ we have: } [\bigcup_{i \in u_\ell} \mathrm{int}(\mathbf{t}'_i)] \in W\}. \end{aligned}$$

If for some finite $u \subseteq \omega$, $\mathrm{nor}(u) \geq k^*$ we can finish. Why? Just as in the end of the proof of 6.14 we define $\mathbf{t} = (S, H)$ as follows: $S = \bigcup\{S^{\mathbf{t}_\ell} : \ell \in u\} \cup \{u\}$, u is the root, its set of (immediate) successor is $\{\mathrm{root}(S^{\mathbf{t}_\ell}) : \ell \in u\}$, the order is defined by: restricted to $S^{\mathbf{t}_\ell}$ is as in $\mathbf{t}_\ell$, for $x \in S^{\mathbf{t}_\ell}$ we let $H^{\mathbf{t}}_x = H^{\mathrm{half}(\mathbf{t}_\ell)}_x$ and $H^{\mathbf{t}}_u$ is defined by: for $v \subseteq u$ let $H^{\mathbf{t}}_u(\{\mathrm{root}(S^{\mathbf{t}_\ell}) : \ell \in v\}) = \mathrm{nor}(v)$. We know $H_u(\{\mathrm{root}(S^{\mathbf{t}_\ell}) : \ell \in u\}) \geq k^*$. This suffices as $\oplus_1$ below holds. Clearly by the definition of nor we have

$\oplus_0$ $\quad \mathbf{t} \geq \mathbf{t}' \Rightarrow (\exists w \in W)[w \subseteq \mathrm{int}(\mathbf{t}')]$

Now we have to prove

$\oplus_1$ if $v = v_1 \cup v_2$, $\mathrm{nor}(v) \geq m+1$ then: $\mathrm{nor}(v_1) \geq m$ or $\mathrm{nor}(v_2) \geq m$.

Proof of $\oplus_1$. If $\mathrm{nor}(v_1) \not\geq m$, then there is a cover $\langle v^1_\ell : \ell < 2^m \rangle$ of v_1 such that:

$(*)_1$ for every $\ell < 2^m$ for some $\mathbf{t}'_m \geq \mathrm{half}(\mathbf{t}_m)$ (for $m \in v^1_\ell$) we have $[\bigcup_{m \in v^1_\ell} \mathrm{int}(\mathbf{t}'_m))] \notin W$.

Similarly, if $\mathrm{nor}(v_2) \not\geq m$ then there is a cover $\langle v^2_\ell : \ell < 2^m \rangle$ of v_2 such that:

$(*)_2$ for every $\ell < 2^m$ for some $\mathbf{t}'_m \geq \mathrm{half}(\mathbf{t}_m)$ (for $m \in v^2_\ell$) we have $\bigcup\limits_{m \in v^2_\ell} \mathrm{int}(\mathbf{t}'_m) \notin W$

Define for $i < 2^{m+1}$:

$$v_i = \begin{cases} v^1_i & \textit{if } i < 2^m \\ v^2_{i-2^m} & \textit{if } i \in [2^m, 2^{m+1}). \end{cases}$$

So if the conclusion fails then $\langle v_i : i < 2^{m+1} \rangle$ exemplifies $\mathrm{nor}(v) \not\geq m+1$, a contradiction.

We can conclude from all this that, toward contradiction we can assume that

$\otimes$ $u \subseteq \omega$ finite $\Rightarrow \mathrm{nor}(u) \not\geq k^*$.

So

$\otimes_1$ for every $n = \{0, \ldots, n-1\}$, $\mathrm{nor}(n) \not\geq k^*$ so there is a cover $\langle v^n_\ell : \ell < 2^{k^*} \rangle$ of n such that:

$\oplus$ for every ℓ for some $\mathbf{t}'_i \geq \mathrm{half}(\mathbf{t}_i)$ (for $i \in v^n_\ell$) we have $[\bigcup_{i \in v^n_\ell} \mathrm{int}(\mathbf{t}'_i)] \notin W$.

By König's lemma there is a sequence $\langle v_\ell : \ell < 2^{k^*} \rangle$ of subsets of ω such that for every $m < \omega$ for some $n = n(m) > m$ we have $v_\ell \cap m = v^n_\ell \cap m$.

Now for some $\ell = \ell(*) < 2^{k^*}$, for infinitely many $m < \omega$ for infinitely many $n \in B_m$ we have $n \in v_\ell$ (on the B_m's, see beginning of the proof of 6.18), so by 6.17 we know that $(\emptyset, \{\mathbf{t}_i : i \in v_{\ell(*)}\}) \in Q[I]$ (and of course is $\geq (\emptyset, T)$). (Alternatively, for some $\ell < 2^{k^*}$, for every $A \in I_A$, for infinitely many $n \in v_\ell$ we have $\mathrm{int}(\mathbf{t}_n) \subseteq \omega \setminus A$. If not then for each ℓ some $A_\ell \in I_A$ fails it. So let $A = \bigcup\limits_{\ell < 2^{k^*}} A_\ell \in I_A$ and we get contradiction to $(\emptyset, \{\mathbf{t}_i : i < \omega\}) \in Q[I]$.) Now for every k letting $n = n(k)$ be such that $v_{\ell(*)} \cap k = v^n_{\ell(*)} \cap k$, we apply $\oplus$. So there are $\mathbf{t}'_i \geq \mathrm{half}(\mathbf{t}_i)$ (for $i \in v^n_{\ell(*)})$) such that $\bigcup_{i \in v^n_{\ell(*)}} \mathrm{int}(\mathbf{t}'_i) \notin W$, and by monotonicity $\bigcup_{i \in v_{\ell(*)} \cap k} \mathrm{int}(\mathbf{t}'_i) \notin W$. By König's lemma (as W is upward closed) there is $\langle \mathbf{t}^*_i : i \in v_{\ell(*)} \rangle$, $\mathbf{t}^*_i \geq \mathrm{half}(\mathbf{t}_i)$ such that for every n we have

$\bigcup_{i \in n \cap v_{\ell(*)}} \mathrm{int}(\mathbf{t}_i^*) \notin W$. So (again as in the proof of 6.14, see 6.20 below) choose $(S^\ell, H^\ell) \in L$ such that $\mathrm{int}(S^\ell) = \mathrm{int}(\mathbf{t}_\ell^*)$ and $\mathrm{lev}(S^\ell, H^\ell) \geq \mathrm{lev}(\mathbf{t}_\ell)/2$, i.e. $S^\ell = S^{\mathbf{t}_\ell^*}$, $H_x^\ell(v) = \min\{[\mathrm{lev}(\mathbf{t})/2], H_x^{\mathbf{t}_\ell}(v)\}$ (when $v \subseteq \mathrm{Suc}_{S^\ell}(x)$). So clearly $(\emptyset, T^*) \leq (\emptyset, \{(S^\ell, H^\ell) : \ell < \omega\}) \in Q[I]$ (see 6.10A(2)) and we apply $(*)$ from the assumption and we get a contradiction, so finishing the proof of 6.18. $\square_{6.18}$

6.19 Claim. Let $\mathcal{A}$, $I = I_{\mathcal{A}}$ be as in 6.18. Let q, $\underset{\sim}{\tau}_n$ be as in 6.16(1). Then for some pure standard extension $r \in Q[I]$ of q, letting $T^r = \{\mathbf{t}'_n : n < \omega\}$, (standard (see Definition 6.8(4)) so $\mathrm{lev}(\mathbf{t}'_n)$ strictly increasing, of course) the following holds:

$(*)$ *For every* $n < \omega$, $w \subseteq [\max(\mathrm{int}(\mathbf{t}'_{n-1})) + 1]$, and $\mathbf{t}''_n \geq \mathbf{t}'_n$ (so we ask only $\mathrm{lev}(\mathbf{t}''_n) \geq 0$) *there is* $w' \subseteq \mathrm{int}(\mathbf{t}''_n)$, such that the condition $(w \cup w'$, $\{\mathbf{t}_\ell : \ell > n\})$ forces a value on $\underset{\sim}{\tau}_m$ for $m \leq n$ (we let $\max \mathrm{int}(\mathbf{t}'_{-1})$ be $\max(w^q \cup \{-1\})$).

Proof. Like the proof of 6.16(1) but using as the induction step claim 6.18. $\square_{6.19}$

6.20 Fact. If $\mathbf{t}_1 \geq \mathrm{half}(\mathbf{t}_0)$, *then* for some $\mathbf{t}_2 \geq \mathbf{t}_0$ we have $\mathrm{int}(\mathbf{t}_2) = \mathrm{int}(\mathbf{t}_1)$, $\mathrm{lev}(\mathbf{t}_2) \geq \mathrm{lev}(\mathbf{t}_0)/2$.

Proof. Included in earlier proof: 6.14. $\square_{6.20}$

6.21 Conclusion. Let V_1, $\mathcal{A}$, $I = I_{\mathcal{A}}$ be as in 6.18.

1) If $p \in Q[I]$ and $\omega = \bigcup_{\ell<k} A_\ell$ where $k < \omega$ *then* for some p', $p \leq_{\mathrm{pr}} p' \in Q[I]$ and for some $\ell < k$ we have $\mathrm{cnt}(T^p) \subseteq A_\ell$.

2) The set of standard $p \in Q[I]$ is dense, in fact for any $p \in Q[I]$ there is a standard q, $p \leq q \in Q[I]$, $w^q = w^p$ and $T^q \subseteq T^p$.

Proof. 1) By repeated use w.l.o.g. $k = 2$. Let $p \in Q[I]$ and $T^p = \{\mathbf{t}_n : n < \omega\}$. For each n apply 6.7 to find $\mathbf{t}'_n \geq \mathbf{t}_n$ such that $\mathrm{int}(\mathbf{t}'_n) \subseteq A$ or $\mathrm{int}(\mathbf{t}'_n) \subseteq \omega \setminus A$ and $\mathrm{lev}(\mathbf{t}'_n) \geq \mathrm{lev}(\mathbf{t}_n) - 1$. Let $Y_0 = \{n : \mathrm{int}(\mathbf{t}'_n) \subseteq A\}$, $Y_1 = \omega \setminus Y_0$, so for some

$\ell \in \{0,1\}$ we have: for every $X \in I$ the set $\{n \in Y_\ell : \mathrm{int}(\mathbf{t}'_n) \cap X = 0\}$ is infinite. [Why? If X_ℓ contradict the demand for $\ell = \{0,1\}$ then $X_0 \cup X_1 \in I$ contradict $p \in Q[I]$ by Definition 6.10.] So $(w^p, \{\mathbf{t}'_n : n \in Y_\ell\}) \in Q[I]$ is above p, and it forces $\bigcup\{w^r : r \in \underset{\sim}{G}_{Q[I]}\} \setminus w^p$ is included in A_ℓ.

We give also an alternative proof, which can be applied for more general question. Let $p = (w, \{\mathbf{t}_n : n < \omega\}) \in Q[I]$. By the proof of [first condition $\Rightarrow$ third condition] in 6.17, there are pairwise distinct $j_m < \omega_1$ (for $m < \omega$) such that for each m the set

$$B_m \stackrel{\text{def}}{=} \{n < \omega : \text{ there is } \mathbf{t}'_n \geq \mathrm{half}(\mathbf{t}_n) \text{ such that } \mathrm{int}(\mathbf{t}'_n) \subseteq A_{j_m}\}$$

is infinite. So we can find $B'_m \subseteq B_m$ for $m < \omega$ such that: $\langle B'_m : m < \omega \rangle$ is a sequence of infinite pairwise disjoint sets. For each $m < \omega$, $n \in B_m$ choose $\mathbf{t}'_n \geq \mathrm{half}(\mathbf{t}_n)$ such that $\mathrm{int}(\mathbf{t}'_n) \subseteq A_{j_m}$. Let $\mathbf{t}''_n \geq \mathbf{t}_n$ be such that $\mathrm{int}(\mathbf{t}''_n) = \mathrm{int}(\mathbf{t}'_n)$ and $\mathrm{lev}(\mathbf{t}''_n) \geq \mathrm{lev}(\mathbf{t}_n)/2$.

If $\mathrm{lev}(\mathbf{t}''_n) > k$, let $\mathbf{t}^3_n \geq \mathbf{t}''_n$ be such that $\mathrm{lev}(\mathbf{t}^3_n) \geq \mathrm{lev}(\mathbf{t}''_n) - k$ (really $\geq \mathrm{lev}(\mathbf{t}''_n) - [1 + \log_2(k)]$ suffices) and for some $\ell = \ell(n)$, $\mathrm{int}(\mathbf{t}^3_n) \subseteq A_{\ell(n)}$. For each $m < \omega$, for some ℓ_m the set $B''_m = \{n \in B_m : \mathrm{lev}(\mathbf{t}^3_n) > k, \ell(n) = \ell_m\}$ is infinite and for some $\ell(*) < k$ the set $\{m < \omega : \ell_m = \ell(*)\}$ is infinite. Now $p \stackrel{\text{def}}{=} (w^p, \{\mathbf{t}^3_n :$ for some m we have $\ell_m = \ell(*)$, and $n \in B''_m\})$ is as required.

2) Left to the reader (or see 6.16(1)). $\square_{6.21}$

Now we pay a debt needed for the proof of 3.23(2).

6.22 Claim. Assume V_1, $\mathcal{A}$, $I = I_{\mathcal{A}}$ are as in 6.18. Then $Q[I]$ is almost ${}^\omega\omega$-bounding *or* for some $p \in Q$ we have $p \Vdash_Q$ "$\{A_i : i < \aleph^*\}$ is not a MAD."

Proof. Assume the second possibility fails. So let $p \in Q[I]$ and $\underset{\sim}{f}$ be a $Q[I]$-name of a function from ω to ω. Let $\underset{\sim}{\tau}_n = \underset{\sim}{f}(n)$, and apply 6.16(1) and get q as there. Next apply 6.19 to those q, $\underset{\sim}{\tau}_n$ and get r which satisfies $(*)$ from 6.19.

By 6.17, 6.21(2) and we can find $r_1 = (w^p, \{\mathbf{t}'_n : n < \omega\})$, a standard member of $Q[I]$ such that $r \leq r_1$ and for some pairwise distinct $j_m < \omega_1$ the sets

$B_m \stackrel{\text{def}}{=} \{n < \omega : \text{int}(\mathbf{t}'_n) \subseteq A_{j_m}\}$ are infinite. Clearly also r_1 satisfies $(*)$ of 6.19. Choose pairwise distinct $n(m, \ell)$ for $m < \ell < \omega$ such that $n(m, \ell) \in B_m \setminus \{0\}$ and $\min \text{int}(\mathbf{t}'_{n(m,\ell)}) > \ell$. Now we define a function $g : \omega \to \omega$ (in V_1) by

$$\begin{aligned} g(\ell) = \ \max\{\{\ell+1\} \cup \{k : & \text{ for some } m < \ell \text{ and } w \subseteq [0, \ \max \text{int}(\mathbf{t}'_{n(m,\ell)-1})] \\ & \text{and } w_1 \subseteq \text{int}(\mathbf{t}'_{n(m,\ell)}) \text{ we have} \\ & (w^p \cup w_1, \{\mathbf{t}'_n : n > n(m,\ell)\}) \Vdash_{Q[I]} \text{“}\underset{\sim}{f}(\ell) = k\text{”}\}\}. \end{aligned}$$

So $g \in ({}^\omega\omega)^{V_1}$, and let $A \subseteq \omega$ ($A \in V_1$) be infinite, and let $p_A = (w^p, \{\mathbf{t}'_{n(m,\ell)} : m < \ell \text{ and } \ell \in A\})$. Now clearly $r_1 \leq p_A \in Q$, p_A standard and even $p_A \in Q[I]$ because still for each $m < \omega$ the set $\{n : \mathbf{t}'_n \in T^{p_A} \text{ and } \text{int}(\mathbf{t}'_n) \subseteq A_{j_m}\}$ is infinite: it includes $\{n : n = n(m, \ell) \text{ for some } \ell \in A \setminus (m+1)\}$. Now one can easily finish the proof. $\square_{6.16}$

A trivial remark is

6.23 Fact. Cohen forcing and even the forcing for adding λ Cohen reals (by finite information) is almost ${}^\omega\omega$-bounding.

§7. On $\mathfrak{s} > \mathfrak{b} = \mathfrak{a}$

See background in §6.

7.1 Theorem. Assume $V \models CH$. Then for some forcing notion P^*, P^* is proper, satisfies the $\aleph_2$-c.c., is weakly bounding and:

$(*)$ In V^{P^*} we have $2^{\aleph_0} = \aleph_2$, there is an unbounded family of ${}^\omega\omega$ of power $\aleph_1$ (i.e. $\mathfrak{b} = \aleph_1$) and also a MAD family of power $\aleph_1$ i.e. $\mathfrak{a} = \aleph_1$, but there is no splitting family of power $\aleph_1$ i.e. $\mathfrak{s} > \aleph_1$ (so $\mathfrak{s} = \aleph_2$).

Proof. The forcing $\langle P_\alpha, \underset{\sim}{Q}_\alpha : \alpha < \omega_2\rangle$, $P^* = P_{\omega_2}$ are as in the proof of 3.23(1). So the only new point is the construction of a MAD of power $\aleph_1$. This will be done in V; the proof of its being MAD will be done directly rather than through a preservation theorem (though the proof is similar).

Let $\{\langle B^i_n : n < \omega\rangle : i < \aleph_1\}$ enumerate (in V) all sequences $\langle B_n : n < \omega\rangle$ of finite nonempty subsets of ω (remember CH holds in V). Next choose a MAD family $\langle A_\alpha : \alpha < \aleph_1\rangle$ such that

(∗∗) for each infinite ordinal $\alpha < \omega_1$ and $i < \alpha$: *if* for every $k < \omega$, $\alpha_1, \dots, \alpha_k < \alpha$ for every m for some (equivalently infinitely many) $n < \omega$, $\min(B^i_n) > m$ and $B^i_n \cap (A_{\alpha_1} \cup \ldots \cup A_{\alpha_k}) = \emptyset$

then

(a) for infinitely many $n < \omega$, $B^i_n \subseteq A_\alpha$

(b) for any $k < \omega$ and $\alpha_1, \dots, \alpha_k \le \alpha$ for infinitely many $n < \omega$ we have $B^i_n \cap (\bigcup_{\ell=1}^{k} A_{\alpha_\ell}) = \emptyset$.

[How? let $A_n = \{k^2 + n : k \in (n, \omega)\}$, and then choose A_α for $\alpha \in [\omega, \omega_1)$ by induction on α as required in (∗∗).]

Let λ be a regular large enough cardinal, $\alpha \le \omega_2$. For a generic $G_\alpha \subseteq P_\alpha$, a model $N \prec (H(\lambda)[G_\alpha], \in)$ is called *good* if it is countable, G_α, $\langle P_j, \underset{\sim}{Q}_i : i < \alpha, j \le \alpha\rangle$, $\langle A_i : i < \omega_1\rangle$, $\langle\langle B^i_n : n < \omega\rangle : i < \omega_1\rangle \in N$ and for every set $\{B_n : n < \omega\} \in N$ of finite nonempty subsets of ω, letting $\delta = N \cap \omega_1$ we have *if*

$\otimes_1$ $(\forall m, k < \omega)(\forall \alpha_1, \dots, \alpha_k < \delta)(\exists^* n < \omega)[B_n \cap (A_{\alpha_1} \cup \ldots \cup A_{\alpha_k}) = \emptyset \ \&\ \min(B_n) > m]$

then $(\exists^* n)[B_n \subseteq A_\delta]$ (remember $\exists^* n$ stands for "for infinitely many").

Note that in the definition of goodness, we have that $\otimes_1$ is equivalent to

$\otimes_2$ $(\forall m, k < \omega)(\forall \alpha_1, \dots, \alpha_k < \omega_1)(\exists^* n < \omega)[B_n \cap (A_{\alpha_1} \cdots \cup A_{\alpha_k}) = \emptyset \ \&\ \min(B_n) > m]$

(as $N \prec (H(\lambda)[G_\alpha], \in)$).

We shall prove by induction on $\alpha \le \omega_2$, that:

$(\circledast)_\alpha$ for every $\beta < \alpha$, a countable $N \prec (H(\lambda), \in)$ to which $\langle P_j, \underset{\sim}{Q}_i : i < \alpha, j \le \alpha\rangle$, and α, β belong and generic $G_\beta \subseteq P_\beta$ *if* $N[G_\beta] \cap \omega_1 = N \cap \omega_1$ (so $N[G_\beta] \cap V = N$), $N[G_\beta]$ is good (in $V[G_\beta]$ of course) and

$p \in N[G_\beta] \cap P_\alpha/G_\beta$ *then* there is $q \in P_\alpha/G_\beta$, $q \geq p$, $\mathrm{Dom}(q) \setminus \alpha = N \cap \beta \setminus \alpha$, q is $(N[G_\beta], P_\alpha/G_\beta)$-generic and: if $G_\alpha \subseteq P_\alpha$ is generic, $G_\beta \subseteq G_\alpha$, $q \in G_\alpha$, then $N[G_\alpha]$ is good.

This is proved by induction. The case $\alpha = \omega_2$, $\beta = 0$ gives the desired conclusion.

[Why? If not for some $p \in P^* = P_{\omega_2}$ and a P_{ω_2}-name $\underset{\sim}{B} = \{\underset{\sim}{k}_n : n < \omega\}$ we have

$$p \Vdash_{P_{\omega_2}} \text{“}\underset{\sim}{B} \text{ is an infinite subset of } \omega, \text{ moreover } \underset{\sim}{k}_n < \underset{\sim}{k}_{n+1} < \omega \text{ for } n < \omega,$$
$$\text{and } \underset{\sim}{B} \cap A_\alpha \text{ is finite for every } \alpha < \omega_1\text{”}.$$

Let $N \prec (H(\lambda), \in)$ be countable such that $\langle P_\alpha, \underset{\sim}{Q}_\alpha : \alpha < \omega_2 \rangle$, $P^* = P_{\omega_2}$, p, $\underset{\sim}{B}$, $\langle \underset{\sim}{k}_n : n < \omega \rangle$ belong to N, and let $\delta \stackrel{\text{def}}{=} N \cap \omega_1$. Clearly $N \cap \{\langle B^i_n : n < \omega \rangle : i < \omega_1\} = \{\langle B^i_n : n < \omega \rangle : i < \delta\}$, so by the choice of the A_α's (see (**) above), N is good (in $V = V^{P_0}$). Hence there is $q \in P_{\omega_2}$ such that $p \leq q$, q is (N, P_{ω_2})-generic and $q \Vdash_{P_{\omega_2}}$ “$N[\underset{\sim}{G}_{\omega_2}]$ is good”. Let $G \subseteq P_{\omega_2}$ be generic over V, $q \in G$, (hence $p \in G$) so $N[G]$ is good, $N[G] \cap \omega_1 = \delta$ and $\langle \{\underset{\sim}{k}_n[G]\} : n < \omega \rangle$ belongs to $N[G]$. Hence by the definition of good, $(\exists^* m)[\underset{\sim}{k}_m[G] \in A_\delta]$, but this means $A_\delta \cap \underset{\sim}{B}[G]$ is infinite, contradicting the choice of p (as $p \in G$).]

The case $\alpha = 0$ is trivial (saying nothing) and the case α limit is similar to the proof of 3.13. In the case α successor, by using the induction hypothesis we can assume $\alpha = \beta + 1$.

By renaming $V[G_\beta]$, $N[G_\beta]$ as V, N we see that it is enough to prove that for any good N and $p \in Q \cap N$ (remember $Q_\beta = Q^{V[G_\beta]}$) *there is* $q \geq p$ which is (N, Q)-generic and $q \Vdash_Q$ “$N[G]$ is good”.

Let $\delta = N \cap \omega_1$, and let $\delta = \{\gamma(\ell) : \ell < \omega\}$. Let $\{\underset{\sim}{\tau}_\ell : \ell < \omega\}$ be a list of all Q-names of ordinals which belong to N, and $\{\langle \underset{\sim}{B}^\ell_n : n < \omega \rangle : \ell < \omega\}$ be a list of all Q-names of ω-sequences of nonempty finite subsets of ω which belong to N, and which are forced to satisfy $\otimes_2$, each appearing infinitely often. For notational simplicity only, assume p is pure. We shall define by induction on $\ell < \omega$ pure $p_\ell = (\emptyset, T^{p_\ell}) = (\emptyset, \{\mathbf{t}^\ell_n : n < \omega\})$ such that:

a) $p_\ell \in N$, p_ℓ standard (so $\max \mathrm{int}(\mathbf{t}^\ell_n) < \min \mathrm{int}(\mathbf{t}^\ell_{n+1})$),

b) $p_0 = p$, $p_{\ell+1} \geq p_\ell$,

c) $\mathbf{t}^\ell_n = \mathbf{t}^{\ell+1}_n$ for $n \leq \ell$ and $\mathrm{lev}(\mathbf{t}^\ell_\ell) \geq \ell$,

d) for any finite $w \subseteq \omega$ and finite $T' \subseteq T^{p_{\ell+1}}$ we have $(w, T^{p_{\ell+1}} \setminus T') \Vdash_Q$ "$\underset{\sim}{\tau}_\ell \in C^*_\ell$" for some countable set of ordinals C^*_ℓ which belongs to N,

e) for every $w_0 \subseteq (\max[\mathrm{int}(\mathbf{t}^\ell_\ell)] + 1)$, $m < \ell$, and $\mathbf{t} \geq \mathbf{t}^{\ell+1}_{\ell+1}$ there is $w_1 \subseteq \mathrm{int}(\mathbf{t})$ such that the condition $(w_0 \cup w_1, \{\mathbf{t}^{\ell+1}_i : \ell+1 < i < \omega\})$ forces that

$$\text{“}(\exists j)[\min(\underset{\sim}{B}^m_j) > \ell \text{ and } \underset{\sim}{B}^m_j \subseteq A_\delta]\text{”}.$$

Below we shall let $p^m_\ell = (\emptyset, \{\mathbf{t}^{\ell,m}_n : n < \omega\})$. Let $p_0 = p$.

Suppose p_ℓ is defined. By 6.12 there is a pure $p^0_\ell \geq p_\ell$ in N such that $\mathbf{t}^{\ell,0}_i = \mathbf{t}^\ell_i$ for $i \leq \ell$, and for any finite $w \subseteq \omega$ and finite $T' \subseteq T^{p^0_\ell}$ we have $(w, T^{p^0_\ell} \setminus T') = p^0_\ell \Vdash$ "$\tau_\ell \in C^*_\ell$" for some countable set of ordinals C^*_ℓ from N [why? read $(*)$ of 6.12].

Given p^0_ℓ we define:

$\mathbb{B} = \{B : B \subseteq \omega$ is finite, $\min(B) > \ell$, and there is standard

$p^* = (\emptyset, \{\mathbf{t}^*_n : n < \omega\}) \geq p^0_\ell$ such that $\bigwedge_{i \leq \ell} \mathbf{t}^*_i = \mathbf{t}^\ell_i$

(so $\mathrm{lev}(\mathbf{t}^*_{\ell+1}) \geq \ell + 1$) and:

for every $w_0 \subseteq \max \mathrm{int}(\mathbf{t}^\ell_\ell) + 1$ and $\mathbf{t} \geq \mathbf{t}^*_{\ell+1}$ and $m < \ell$, for some

$w_1 \subseteq \mathrm{int}(\mathbf{t})$, the condition $(w_0 \cup w_1, \{\mathbf{t}^*_i : i > \ell + 1 \text{ and } i < \omega\})$

forces " for some $j < \omega$ we have $\underset{\sim}{B}^m_j \subseteq B$"$\}$.

Clearly

(A) $\mathbb{B} \in N$

(B) $\mathbb{B}$ satisfies $(\otimes_1)$ from the definition of good.

[Why? Let $k < \omega$, $\alpha_1, \ldots, \alpha_k < \delta$. By the assumption, $\Vdash_Q$ "for each $m < \ell$, the sequence $\langle \underset{\sim}{B}^m_\ell : j < \omega \rangle$ satisfies $\otimes_2$" hence $\Vdash$ " for every $m < \ell$ for some $n = n(m)$ we have $\min(\underset{\sim}{B}^m_n) > \ell$ and $\underset{\sim}{B}^m_n \cap (A_{\alpha_1} \cup \ldots A_{\alpha_k}) = \emptyset$. Hence there is standard $q = (\emptyset, \{\mathbf{s}_{\ell+1}, \mathbf{s}_{\ell+2}, \ldots\}) \in Q$, $q \geq (\emptyset, \{\mathbf{t}^{\ell,0}_{\ell+1}, \mathbf{t}^{\ell,0}_{\ell+2}, \ldots\})$ such that $\mathrm{lev}(\mathbf{s}_{\ell+1}) \geq \ell + 1$ and:

$\oplus$ if $w_0 \subseteq \max(\mathrm{int}(\mathbf{t}^\ell_\ell)) + 1$, $m < \ell$, $\mathbf{t} \geq \mathbf{s}_{\ell+1}$ then for some $w_1 \subseteq \mathrm{int}(\mathbf{t})$ and $n^m_{w_0,w_1} < \omega$ and $C^m_{w_0,w_1}$ we have

(α) $(w_0 \cup w_1, \{\mathbf{s}_{\ell(1)+1}, \mathbf{s}_{\ell(1)+2}, \ldots\}) \Vdash$ "$\underset{\sim}{B}^m_{n^m_{w_0,w_1}} = C^m_{w_0,w_1}$".

(β) $C^m_{w_0,w_1} \subseteq \omega \backslash \ell$, and $C^m_{w_0,w_1}$ is nonempty finite disjoint to $A_{\alpha_0} \cup \ldots \cup A_{\alpha_k}$.

So necessarily $\bigcup\{C^m_{w_0,w_1} : w_0 \subseteq \max(\mathrm{int}(\mathbf{t}^\ell_\ell)) + 1$ and $\mathbf{t} \geq \mathbf{s}_{\ell+1}$, and $w_1 \subseteq \mathrm{int}(\mathbf{t})$ and $C^m_{w_0,w_1}$ is well defined and $m < \ell\} \in \mathbb{B}$ is as required finishing the proof of clause (B). We could have demand in $\oplus$ above for one w_1 to be O.K. for all $m < \ell$.]

(C) We can define $p_{\ell+1}$.

[Why? As $\mathbb{B} \in N$ satisfies $(\otimes_1)$ and N is good necessarily there is $B \in \mathbb{B}$, $B \subseteq A_\delta$. For this B there is p^* as in the definition of $\mathbb{B}$. Let $\mathbf{t}^{\ell+1}_n = \mathbf{t}^\ell_n$ for $n \leq \ell$, $\mathbf{t}^{\ell+1}_n = \mathbf{t}^*_n$ for $n > \ell$. So $p_{\ell+1} = (\emptyset, \{\mathbf{t}^{\ell+1}_n : n < \omega\})$ is defined.]

So we have defined $p_{\ell+1}$ satisfying (a)-(e). So we can define p_ℓ for $\ell < \omega$ and now $q \overset{\mathrm{def}}{=} (\emptyset, \{\mathbf{t}^n_n : n < \omega\})$ is as required. $\square_{7.1}$

§8. On $\mathfrak{h} < \mathfrak{s} = \mathfrak{b}$

See background in §6. We first recall well known definitions.

8.1 Definition. 1) Let $\mathfrak{h}$ be the minimal cardinal λ such that there is a tree T with λ levels (not normal!) and $A_t \in [\omega]^{\aleph_0}$ for $t \in T$ such that $[t < s \Rightarrow A_s \subseteq_{ae} A_t]$ and $(\forall B \in [\omega]^{\aleph_0})(\exists t \in T)[A_t \subseteq_{ae} B]$ and if $t, s \in T$ are $<_T$-incomparable then $A_s \cap A_t$ is finite. See Balcar, Pelant and Simon [BPS] on it (and in particular why it exists which was for long an open problem).

2) Let $Q^d = \{(n, f) : n < \omega, f \in {}^\omega\omega\}$ with the order defined by

$(n_1, f_1) \leq (n_2, f_2)$ if and only if

$$n_1 \leq n_2,\ f_1 \restriction n_1 = f_2 \restriction n_1 \text{ and } f_1 \leq f_2 (\text{ i.e. } \bigwedge_\ell f_1(\ell) \leq f_2(\ell)).$$

This forcing adds a dominating real and it satisfies c.c.c. This is called Hechler forcing or dominating real forcing.

8.2 Theorem. Assume $V \models CH$.
For some proper forcing P of power $\aleph_2$ satisfying the $\aleph_2$-c.c., in V^P, $\mathfrak{h} = \aleph_1$, $\mathfrak{b} = \mathfrak{s} = \aleph_2$ (and $2^{\aleph_0} = 2^{\aleph_1} = \aleph_2$).

Proof. We shall use the direct limit P of the CS iteration $\langle P_i, \underset{\sim}{Q}_i : i < \omega_2\rangle$ where:

A) letting $i = (\omega_1)^3\gamma + j, j < (\omega_1)^3$, if $j \neq \omega_1, \omega_1 + 1$ then $\underset{\sim}{Q}_i$ is Cohen forcing; if $j = \omega_1$ then $\underset{\sim}{Q}_i$ is Q from Definition 6.8 (in V^{P_j}) and if $j = \omega_1 + 1$ then $\underset{\sim}{Q}_i$ is Q^d (see Definition 8.1(2), also other nicely definable forcing notions are O.K.).

B) We use the presentation of countable support defined in III, proof of Theorem 4.1, i.e. using only hereditarily countable names. We let $\underset{\sim}{r}_i$ be the generic real of $\underset{\sim}{Q}_i$.

Clearly $|P| = \aleph_2$, P satisfies the $\aleph_2$-c.c. and is proper (see III §3, §4), hence forcing by P preserves cardinals. Clearly in V^P, $\mathfrak{s} \geq \aleph_2$ (because for unboundedly many $i < \aleph_2$, $\underset{\sim}{Q}_i = Q$ (from Definition 6.6, and 6.11(3)) and $\mathfrak{b} \geq \aleph_2$ (because for unboundedly many $i < \aleph_2$, $\underset{\sim}{Q}_i = Q^d$) and $2^{\aleph_0} = \aleph_2$. Hence in V^P we have $\mathfrak{s} = \mathfrak{b} = \aleph_2$ (so $\mathfrak{d} = \aleph_2$) and always $\mathfrak{h} \geq \aleph_1$. So the only point left is $V^P \models$ "$\mathfrak{h} \leq \aleph_1$".

We define by induction on $i < \omega_2$ (an ordinal $\alpha(i)$ and) $P_{\alpha(i)}$-names $\underset{\sim}{\eta}_i, \underset{\sim}{A}_i$ such that

(a) $\alpha(i) = (\omega_1)^3(i+1)$,

(b) $\underset{\sim}{\eta}_i \in \bigcup_{\beta<\omega_1} {}^{\beta+1}(\omega_2) \setminus \{\underset{\sim}{\eta}_j : j < i\}$ and for every successor $\beta < \ell g(\underset{\sim}{\eta}_i)$ we have $\underset{\sim}{\eta}_i \restriction \beta \in \{\underset{\sim}{\eta}_j : j < i\}$) (i.e. those things are forced),

(c) $\underset{\sim}{\eta}_j \lhd \underset{\sim}{\eta}_i \Rightarrow \underset{\sim}{A}_i \subseteq_{ae} \underset{\sim}{A}_j$ (for $j < i$) and $\underset{\sim}{A}_i$ is an infinite subset of ω,

(d) if $\underset{\sim}{A} \subseteq \omega$ is infinite and $\underset{\sim}{A} \in V^{P_j}$ then for some $i < j + \omega_1$, $\underset{\sim}{A} \subseteq \underset{\sim}{A}_j$,

(e) $\underset{\sim}{A}_i$ includes no infinite set from $V^{P_{\alpha(j)}}$ when $j < i$, and moreover is a subset of the generic real of $Q_{\omega_1^3 i+3}$,

(f) if $\underset{\sim}{\eta}_i$, $\underset{\sim}{\eta}_j$ are $\lhd$-incomparable then $\underset{\sim}{A}_i \cap \underset{\sim}{A}_j$ is finite (i.e. this is forced).

There is no problem to do this if you know the known way to build trees exemplifying the definition of $\mathfrak{h}$ (by Balcar, Pelant and Simon [BPS]), provided that no ω_1-branch has an intersection. I.e. for no $\eta \in {}^{\omega_1}(\omega_2)$ and $B \in [\omega]^{\aleph_0}$ (in $V^{P_{\omega_2}}$) do we have $B \subseteq_{ae} A_{i_\alpha}$ where $\eta\restriction(\alpha+1) = \eta_{i_\alpha}$ for $\alpha < \omega_1$; by clause (e) above necessarily i_α is strictly increasing. Let $i(*) = \bigcup_{\gamma<\omega_1} i_\gamma$ and $\alpha(*) = \bigcup_{\gamma<\omega_1} \alpha(i_\gamma)$, in $V^{P_{\alpha(*)}}$ there is no intersection by clause (e) (even in the case $\eta \notin V^{P_{\alpha(*)}}$). So it is enough to prove this for a fixed $i(*)$ hence also $\alpha(*)$.

We can look, in $V^{P_{\alpha(*)}}$, at the iteration $\bar{Q}' = \langle P'_\beta, \underset{\sim}{Q}_\gamma : \alpha(*) \le \gamma < \omega_2, \alpha(*) \le \beta \le \omega_2\rangle$, where $P'_\beta \overset{\text{def}}{=} P_\beta/P_{\alpha(*)}$. Let $G_1 \subseteq P_{\alpha(*)}$ be generic, $V_1 = V[G_1]$. Note that every element of P'_{ω_2} can be represented by a countable function from ordinals ($< \omega_2$) to hereditarily countable sets (built from ordinals $< \omega_2$). The set of elements of P'_{ω_2} as well as its partial order are definable from ordinal parameters only (all this in $V[G_1]$). Suppose $p \in P'_{\omega_2}$ forces $\underset{\sim}{B}$ (a P'_{ω_2}-name of a subset of ω) and $\underset{\sim}{i}_\gamma$ (for $\gamma < \omega_1$) to be as above (so with limit $i(*)$). W.l.o.g. for each $n < \omega$ there is an antichain $\langle q_{n,\ell} : \ell < \omega\rangle$ which is predense above p, such that $q_{n,\ell} \Vdash$ "$n \in \underset{\sim}{B}$ *iff* $\mathbf{t}_{n,\ell}$", $\mathbf{t}_{n,\ell}$ a truth value. So for some $j(*) < \alpha(*)$ we have p, $\langle\langle q_{n,\ell} : \ell < \omega\rangle : n < \omega\rangle \in V[G_1 \cap P_{j(*)}]$.

There is p_1, $p \le p_1 \in P'_{\omega_2}$ such that $p_1 \Vdash$ "$\underset{\sim}{i}_\gamma = i$" for some γ, i such that $j(*) < (\omega_1)^3 i < \alpha(*)$ so $p_1 \Vdash$ "$\underset{\sim}{B} \subseteq \underset{\sim}{r}_{\omega_1^3 i+3}$" where $\underset{\sim}{r}_{\omega_1^3 i+3}$ is the generic real that the set $G_1 \cap Q_{\omega_1^3 i+3}$ gives (see the end of clause (e)). Now using automorphisms of the forcing $P_{\alpha(*)}/P_{j(*)}$ we see that there is p_2, $p \le p_2 \in P'_{\omega_2}$ such that $p_2 \Vdash$ "$\underset{\sim}{B}$ is almost disjoint from $\underset{\sim}{r}_{\omega_1^3 i+3}$". From this we can conclude that $p \Vdash$ "$\bigcup_{\gamma<\omega_1} \underset{\sim}{\eta}_{i_\gamma} \notin V[G_1]$" (otherwise some $p_0 \ge p$ forces a particular value and repeat the argument above for p_0). Hence it suffices to prove by induction on $\beta \in [\alpha(*), \omega_2]$ that forcing with P'_β adds no new ω_1-branches to the tree $T \in V_1$ where $T = \{\underset{\sim}{\eta}_i[G_1] : i < i(*)\}$, ordered by $\lhd$, (i.e. all are on $V^{P_{\alpha(*)}}$). Let $\eta_i = \underset{\sim}{\eta}_i[G_1]$ for $i < i(*)$.

We prove by induction on $\beta \in [\alpha(*), \omega_2]$ that

$(*)^1_\beta$ P'_β adds no new ω_1-branch to $T \in V_1$.

So assume $p_0 \in P'_\beta$ is such that $p_0 \Vdash$ "$\langle \underset{\sim}{\nu}_\gamma : \gamma < \omega_1 \rangle$ is a new ω_1-branch of $\{\eta_i : i < i(*)\} \in V_1$".

In V_1 choose a sequence $\langle N_m : m < \omega \rangle$ of countable elementary submodels of $(H(\chi), \in, <^*_\chi)$ such that β, $\bar{Q}'$, P, $\underset{\sim}{B} \in N_0$, $N_m \in N_{m+1}$, $N_\omega = \bigcup_{m<\omega} N_m$. Let $\delta_m = N_m \cap \omega_1$, and let

$$A_m = \{\bar{\nu} = \langle \nu_\gamma : \gamma < \delta_m \rangle : \bar{\nu} \in N_{m+1} \text{ and for every } \gamma < \delta_m, \bar{\nu} \restriction \gamma \in N_m\}.$$

So $\langle A_m : m < \omega \rangle \in V_1$, and we can list $A_m = \{\bar{\nu}^{m,\ell} : \ell < \omega\}$, $\langle \langle \bar{\nu}^{m,\ell} : \ell < \omega \rangle : m < \omega \rangle \in V_1$. The real $\underset{\sim}{r}_{i(*)}$ is a Cohen real over V_1 (as $\langle i_\gamma : \gamma < \omega_1 \rangle$ is strictly increasing with limit $i(*)$), and we can interpret $Q_{i(*)}$ as ${}^{\omega>}\omega$, so let $\underset{\sim}{r}_{i(*)} = \langle \underset{\sim}{\ell}_m : m < \omega \rangle$.

Clearly for proving $(*)^1_\beta$ it is enough to find q such that:

$(*)_2$ $q \in P'_\beta$, $p_0 \le q$, q is (N_m, P'_β)-generic for each $m < \omega$ and $q \Vdash_{P'_\beta}$ "$\langle \underset{\sim}{\nu}_\gamma : \gamma < \delta_m \rangle \ne \langle \nu_\gamma^{m,\underset{\sim}{\ell}_m} : \gamma < \delta_m \rangle$" for each m.

The proof splits to cases, the first four cases give $(*)^1_\beta$ directly, the last three do it through $(*)_2$.

Case 1: For $\beta = \omega_2$ no new branches appear (by $\aleph_2$-c.c.).

Case 2: For $\beta = i(*)$ trivial.

Case 3: For $\beta = \alpha + 1$, Q_α Cohen: use "Q_α is the union of $\aleph_0$ directed sets" (and such forcing notions do not add a new ω_1-branch to any old tree).

Case 4: For $\beta = \alpha + 1$, $Q_\alpha = Q^d$: similarly, as

$$Q^d = \bigcup_n \{\{(n, f) : f \in {}^\omega\omega \ \&\ f \restriction n = \eta\} : n < \omega, \eta \in {}^n\omega\}.$$

Case 5: For $\beta = \alpha + 1$, $Q_\alpha = Q$: so for some γ we have $\alpha = \omega_1^3\gamma + \omega_1$, shortly we shall work in the universe $V^{P'_{\omega_1^3\gamma}}$. Let $q' \in P'_\alpha$ be (N_m, P'_α)-generic for each $m < \omega$, $p_0 \restriction \alpha \le q'$ (such q' exists as all those forcing notions are ω-proper and ω-properness is preserved by CS iteration by V §3). Let $G'_\alpha \subseteq P'_\alpha$ be generic over V_1 such that $q' \in G'_\alpha$, and we work in $V_2 = V_1[G'_\alpha]$. Let $N'_m = N_m[G'_\alpha]$, $w = w^{p_0(\alpha)}$ (a finite subset of ω, actually it is $w^{p_0(\alpha)[G'_\alpha]}$).

We choose by induction on $m < \omega$, $q_m = (w, T_m) \in Q_\alpha$, $T_m = \langle \mathbf{t}^m_n : n < \omega \rangle$ such that:

(a) $q_m \in N'_m \cap Q_\alpha$, $q_m \le q_{m+1}$, $[n < m \Rightarrow \mathbf{t}^m_n = \mathbf{t}^{m+1}_n]$, $p_0(\alpha) \le q_0$,

(b) q_{m+1} is (N'_m, Q_α)-generic

(c) $q_{m+1} \Vdash$ "$\langle \underset{\sim}{\nu}_\gamma : \gamma < \delta_m \rangle \ne \langle \nu^{m,\ell_m}_\gamma : \gamma < \delta_m \rangle$".

This clearly suffices and for the induction step, clauses (a), (b) are possible by the proof of "Q_α is proper" (in §6), and reflecting on the proof there also clause (c) [in more details given q_m, let $\langle \underset{\sim}{\tau}_{m,k} : k < \omega \rangle$ list the Q-names of ordinals which belong to N_m, now we choose by induction on $k < \omega$, $q_{m,k} = (w, T_{m,k})$ and $\gamma_{m,k}$, $T_{m,k} = \{\mathbf{t}^{m,k}_n : n < \omega\}$ standard, $\{q_{m,k}, \gamma_{m,k}\} \in N_m$, $q_m = q_{m,0}$, $q_{m,k} \le q_{m,k+1}$, $[n \le m + k \Rightarrow \mathbf{t}^{m,k}_n = \mathbf{t}^{m,k}_{n+1}]$, and for every $w_0 \subseteq \max[\mathrm{int}(\mathbf{t}^{m,k}_{m+k})] + 1$ and $\mathbf{t} \ge \mathbf{t}^{m,k}_{m+k+1}$, for some $w_1 \subseteq \mathrm{int}(\mathbf{t})$ we have $(w_0 \cup w_1, \{\mathbf{t}^{m,k}_{m+k+2}, \mathbf{t}^{m,k}_{m+k+3}, \ldots\})$ force $\underset{\sim}{\nu}_{\gamma_{m,k}} = \rho$, $\rho \ne \nu^{m,\ell_m}_{\gamma_{m,k}}$ and forces a value to $\underset{\sim}{\tau}_{i,k}$; for the induction step get a candidate for all $\gamma < \omega_1$ and use Δ-system (and "the branch is new" i.e. not from $V^{P'_\alpha}$).]

Case 6: $\beta > \alpha(*)$ is a limit ordinal; $\mathrm{cf}(\beta) = \aleph_0$

Quite straightforward as in the proof of the preservation of ω-properness (of course we could work in V rather than in V_1 and use the induction hypothesis). Choose $\langle \beta_n : n < \omega \rangle$ such that $\beta = \bigcup_{n<\omega} \beta_n$, $i(*) = \beta_0$, $\beta_n < \beta_{n+1} < \beta$, and $\beta_n \in N_0$. We choose by induction on n, q_n, p_n such that:

(a) $q_n \in P'_{\beta_n}$, $\mathrm{Dom}(q_n) = (\bigcup_{k<\omega} N_k) \cap [\alpha(*), \beta_n)$

(b) q_n is (N_i, P_{β_n})-generic for each $i \le \omega$

(c) p_n is a P_{β_n}-name of a member of $P_\beta \cap N_n$

(d) $p_n \restriction \beta_n \le q_n$, $q_{n+1} \restriction \beta_n = q_n$

(e) $p_n \le p_{n+1}$

(f) p_{n+1} is (N_n, P_β)-generic (i.e. forced to be)

(g) $q_n \cup (p_n \restriction [\beta_n, \beta)) \Vdash_{P_\beta}$ "$\langle \underset{\sim}{\nu}_\gamma : \gamma < \delta_0 \rangle \ne \langle \nu^{n,\ell_n}_\gamma : \gamma < \delta_0 \rangle$".

The induction should be clear and $q \overset{\mathrm{def}}{=} \bigcup_n q_n = \bigcup_n (q_{n+1} \restriction [\beta_n, \beta_{n+1}))$ is as required.

Case 7: $\beta > \alpha(*)$ a limit ordinal, $\mathrm{cf}(\beta) > \aleph_0$

Like case 6, but $\beta_n = \sup(N_n \cap \beta)$. $\square_{8.2}$

Concluding Remark. The proof of "no new ω_1-branch" has little to do with the specific problem. More on definable forcing notions see [Sh:630].

VII. Axioms and Their Application

§0. Introduction

In the first section we introduce the κ-e.c.c. (κ-extra chain condition). We prove that if we have an iteration of length $\leq \kappa$ of $(< \omega_1)$-proper forcing notions which do not add reals, and if, moreover each forcing used is $\mathbb{D}$-complete for some simple $\aleph_1$-completeness system $\mathbb{D}$, *then* the limit satisfies the κ-c.c. . This helps us e.g. in iterations of length ω_2 of forcings among which none add reals, but each adds many subsets of $\aleph_1$.

In the second section we deal with forcing axioms; essentially our knowledge is good when we want $2^{\aleph_0} = 2^{\aleph_1} = \aleph_2$ and reasonable when we want $2^{\aleph_0} = \aleph_1$ and even $2^{\aleph_0} = \aleph_2$. In the third section we discuss applications of the forcing axiom which is consistent with CH as just mentioned. In the fourth section we discuss the forcing axiom which is consistent with $2^{\aleph_0} = 2^{\aleph_1} = \aleph_2$, and in the fifth section we give an example of a CS iteration collapsing $\aleph_1$ only in the limit. See relevant references in the sections.

§1. On the κ-Chain Condition, When Reals Are Not Added

When we prove various consistency results by iterating proper forcings, we often have to check that the $\aleph_2$-chain condition holds.

Remember, we deal with a CS iteration $\bar{Q} = \langle P_i, \underset{\sim}{Q}_i : i < \alpha \rangle$ where $\underset{\sim}{Q}_i$ is a P_i-name of a proper forcing (in V^{P_i}), $P_\alpha = \{f : \mathrm{Dom}(f)$ is a countable subset of α and $i \in \mathrm{Dom}(f)$ implies $\emptyset \Vdash_{P_i}$ "$f(i) \in \underset{\sim}{Q}_i$", i.e. $f(i)$ is a P_i-name of a member of $\underset{\sim}{Q}_i\}$. So, P_α is proper by III 3.2. Here, we concentrate on the case when no real is added, in fact when we have a sufficient condition for it. The case without this restriction will be discussed again in VIII §2 .

Remark. Note that even if $\Vdash_{P^i}$ "$|\underset{\sim}{Q}^1| = \aleph_1$", still may be $|P_2| = 2^{\aleph_1}$, as there may be many P_1-names of elements of Q_1.

1.1. Lemma. If κ is regular, $(\forall \mu < \kappa)\mu^{\aleph_0} < \kappa$ and $V^{P_\alpha} \models$ "$|Q_\alpha| < \kappa$" *then* P_κ satisfies the κ-chain condition.

Proof. See III 4.1.

1.2. Definition. P satisfies the κ-e.c.c. (κ-extra chain condition) provided that there is a two place relation R on P (usually pRq is intended to mean that "p and q have a least upper bound") such that:

A) for any $p_i \in P$ (for $i < \kappa$) there are pressing down functions $f_n : \kappa \to \kappa$ (i.e. $(\forall \alpha) f_n(\alpha) < 1 + \alpha$) for $n < \omega$ such that: $0 < i, j < \kappa$ and $\bigwedge_{n<\omega}(f_n(i) = f_n(j))$ imply $p_i R p_j$.

B) if in P we have $p_0 \leq p_1 \leq p_2 \leq \ldots \leq p_n \leq p_{n+1} \leq \ldots p_\omega$ and $q_0 \leq q_1 \leq q_2 \leq \ldots \leq q_n \leq q_{n+1} \leq \ldots q_\omega$ and $\bigwedge_n p_n R q_n$, *then* there is an r such that $\bigwedge_n r \geq p_n$ & $\bigwedge_n r \geq q_n$.

1.2A Remark. This is very similar to the condition used in [Sh:80] (and similar to a work of Baumgartner, see VIII 1.1, 1.1A(1)). The real difference is the absence of $\aleph_1$-completeness. The fact that there (in [Sh:80], clause (C) there is a parallel to clause (A) here) we use only one function and closed unbounded C, and demand $i, j \in C$, $\mathrm{cf}(i) = \mathrm{cf}(j) = \aleph_1$, is just a variant form which was more convenient to represent there. The role of p_ω, q_ω is just to show that $\{p_n : n < \omega\}$ and $\{q_n : n < \omega\}$, each has an upper bound (so for a $\aleph_1$-complete forcing they are not needed, hence, also $\ell g(\bar{Q}) \leq \kappa$ is not needed). Even closer is Stanley and Shelah [ShSt:154], [ShSt:154a]. We can ask in (A) only that there are p'_i (for $i < \kappa$) such that $P \models p_i \leq p'_i$ and $[0 < i, j < \kappa, \bigwedge_n (f_n(i) = f_n(j)) \Rightarrow p'_i R p'_j]$.

1.3. Lemma. Suppose $V \models 2^{\aleph_0} = \aleph_1$ while $(\forall \mu < \kappa)\mu^{\aleph_0} < \kappa$ and κ is regular. Suppose further that $\bar{Q} = \langle P_i, \underset{\sim}{Q}_i : i < \alpha_0 \leq \kappa \rangle$ is a CS iteration.

In addition:

(a) Each Q_α is $(< \omega_1)$-proper ($= \alpha$-proper for each $\alpha < \omega_1$);

(b) Each Q_α is $\mathbb{D}$-complete for some $\aleph_1$-completeness system *from* V (see definition in V §7) and, for simplicity, the set of elements of Q_α is a subset of $\lambda = 2^{|P_{\alpha_0}|}$ and $\lambda > \kappa$;

(c) Each Q_α satisfies the κ-e.c.c.

Then P_{α_0} satisfies the κ-c.c. ($= \kappa$-chain condition).

1.3A Remark.

1) Compare with Theorem 7.1 from V. We add part (c) to the hypothesis (and $\alpha_0 \leq \kappa$), and get the κ-c.c. of P_{α_0}. In fact, to prove 1.3, we shall repeat the proof of V7.1, after an appropriate preparatory step.

2) We can weaken Definition 1.2 so that the proof of 1.3 still works, e.g. by strengthening the hypothesis on the p_n, q_n in clause (B). For example, we could have demanded that $q_{n+1} \in \mathcal{I}[q_n]$, $p_{n+1} \in \mathcal{I}[p_n]$ where for $r \in P$ we have that $\mathcal{I}[r] \subseteq P$ is a dense subset of P (or even $q_{n+1} \in \mathcal{I}[q_0, q_1, \ldots q_n]$, $p_{n+1} \in \mathcal{I}[p_0, \ldots, p_n]$).

3) Note that when P is $\aleph_1$-complete, 1.2(B) is satisfied for $R =$ "having a least upper bound". Also 1.2(A) can be weakened by e.g. demanding the conclusion to be true only for $i, j \in A$, for some A in some appropriate filter on $\kappa \times \kappa$, *or* for some A which is not in an appropriate ideal which is precipituous.

Proof. As in the proof of Lemma III 4.1 we can conclude that if the Lemma holds for each $\alpha_0 < \kappa$, then it also holds for $\alpha_0 = \kappa$. So, w.l.o.g. $\alpha_0 < \kappa$. Let $\underset{\sim}{R}_i$ be (a P_i-name of a 2-place relation) exemplifying Definition 1.2 for $\underset{\sim}{Q}_i$. Let $p_\alpha \in P_{\alpha_0}$ $(\alpha < \kappa)$ be given. We now define, by induction on $n < \omega$, countable models N^n_α (for all $\alpha < \kappa$ simultaneously) such that:

i) $N^n_\alpha \prec (H(\lambda), \in), P_{\alpha_0} \in N^n_\alpha, \bar{Q} \in N^n_\alpha, p_\alpha \in N^0_\alpha, ||N^n_\alpha|| = \aleph_0$, and $\alpha \in N^0_\alpha$.

ii) $N^n_\alpha \prec N^{n+1}_\alpha$, and the additional conditions below are satisfied.

For $n = 0$, choose any N^0_α satisfying (i).

If we have defined N^n_α for n, let $N^n_\alpha \cap P_{\alpha_0} = \{p^n_{\alpha,\ell} : \ell < \omega\}$, $p^n_{\alpha,0} = p_\alpha$.

For $i < \alpha_0$ and $\ell < \omega$, consider the sequence $\langle p^n_{\alpha,\ell}(i) : \alpha < \kappa\rangle$ (if $i \neq \mathrm{Dom}(p^n_{\alpha,\ell})$ then we are stipulating $p^n_{\alpha,\ell}(i) = \emptyset_{\underset{\sim}{Q}_i}$). In V^{P_i} it is a sequence of length κ of elements of Q_i. But Q_i (in V^{P_i}) satisfies the κ-e.c.c., so there are ω-sequences of functions $\bar{f}^n_{i,\ell} = \langle f^n_{i,\ell,k} : k < \omega\rangle$ exemplifying that condition (A) from Definition 1.2 holds. In V we have P_i-names $\underset{\sim}{\bar{f}}^n_{i,\ell}$ for $\bar{f}^n_{i,\ell}$.

We now define N^{n+1}_α such that $N^n_\alpha \prec N^{n+1}_\alpha \prec (H(\lambda), \in)$, and

iii) $\langle\langle \underset{\sim}{\bar{f}}^n_{i,\ell} : i < \alpha_0\rangle, \langle p^n_{\gamma,\ell} : \gamma < \kappa\rangle : \ell < \omega\rangle \in N^{n+1}_\alpha, N^n_\alpha \in N^{n+1}_\alpha$ and $\langle N^n_\beta : \beta < \kappa\rangle \in N^{n+1}_\alpha$.

Let $N^\omega_i = \bigcup_{n<\omega} N^n_i$.

Now, using that $(\forall \gamma < \kappa)\gamma^{\aleph_0} < \kappa$ and that κ is regular, we can easily find $i < j < \kappa$, and h such that:

a) h is an isomorphism from N^ω_i onto N^ω_j such that $h(i) = j$.

b) h is the identity on $N^\omega_i \cap N^\omega_j$ (so it maps to themselves: $\langle p_\alpha : \alpha < \kappa\rangle$, $\langle p^n_{\alpha,\ell} : \alpha < \kappa\rangle, \langle\langle \underset{\sim}{\bar{f}}^n_{i,\ell} : i < \alpha_0\rangle : \ell < \omega\rangle$).

c) $N^\omega_i \cap \kappa \subseteq j$.

d) $N^\omega_i \cap (\alpha_0 + 1) = N^\omega_j \cap (\alpha_0 + 1) = \{\alpha(\xi) : \xi < \xi_0 < \omega_1\}, \alpha(\xi)$ increasing. Also $N^\omega_j \cap j = N^\omega_i \cap i = N^\omega_j \cap i$.

Now we choose countable $N_\zeta \prec (H(\lambda), \in)$ for $\zeta \leq \xi_0$ such that $h \in N_0$, and $N^\omega_i, N^\omega_j \in N_0$, $\langle N_\alpha : \alpha \leq \zeta\rangle \in N_{\zeta+1}$.

We now repeat the proof of V 7.1, more exactly:

1.4. Claim. Suppose that $0 \leq \xi < \zeta \leq \xi_0$, $N_i (i \leq \xi_0)$, $N_i^\omega \in N_0$, $N_j^\omega \in N_0$, are as above, $r \in P_{\alpha(\xi)}$, $G^* \subseteq (N_i^\omega \cup N_j^\omega) \cap P_{\alpha(\xi)}$, $G^* \cap N_i^\omega$ is generic for $(N_i^\omega, P_{\alpha(\xi)})$ (i.e. $G^* \cap N_i^\omega$ is directed and if $\mathcal{I} \in N_i^\omega$ and $\mathcal{I}$ is pre-dense in $P_{\alpha(\xi)}$, then $\mathcal{I} \cap G^* \cap N_i^\omega \neq \emptyset$), and h maps $G^* \cap N_i^\omega$ onto $G^* \cap N_j^\omega$. In addition, every element of G^* is $\leq r$, r is (N_ε, P_ξ)-generic for any ε satisfying $\varepsilon = 0$ or $\xi \leq \varepsilon < \xi_0, G^* \in N_\xi$ and $p_i \restriction \alpha(\xi)$, $p_j \restriction \alpha(\xi) \in G^*$ and $G^* \in N_{\xi+1}$.

Then there is a $G \subseteq (N_i^\omega \cup N_j^\omega) \cap P_{\alpha(\zeta)}$ with $G^* \subseteq G, G \in N_{\zeta+1}$, $G \cap N_i^\omega$ is generic for $(N_i^\omega, P_{\alpha(\zeta)})$ such that h maps $G \cap N_i^\omega$ onto $G \cap N_j^\omega$ and $r \wedge \bigwedge_{q \in G} q \neq \emptyset$ (Boolean intersection). In other words: $r \Vdash_{P_{\alpha(\xi)}}$ "G has an upper bound "in $P_{\alpha(\zeta)}/\underset{\sim}{G}_{\alpha(\xi)}$".

Proof. The proof is as in V 7.1. The only difference is the case $\zeta = \xi + 1$, here we use clause (B) of Definition 1.2: necessarily G^* "tells" us the functions have the same values, as they are pressing down. $\square_{1.4}$

Continuation of the proof of 1.3 From Claim 1.4, it follows that p_i and p_j are compatible in P_{α_0}, for i and j that we fixed earlier.

$\square_{1.3}$

Remark. Note in Lemma 1.1, if the iteration is defined such that we have a support of power $\leq \mu$, and $(\forall \chi < \kappa) \chi^\mu < \kappa, \kappa$ regular, still P_κ satisfies the κ-c.c. (On free limits, see IX §1, 2)

1.6. Lemma. We can replace (in 1.3) "$\aleph_1$-completeness system" by "$\aleph_0$-completeness system".

Proof. Using V 7.2 instead of V 7.1.

1.7. Remark. We can even replace "$\aleph_1$-completeness system" by "2-completeness system", using VIII 4.5, 4.13.

§2. The Axioms

AXIOM I.

1) $2^{\aleph_0} = 2^{\aleph_1} = \aleph_2$ and:
2) if $|P| = \aleph_2, P$ proper, $\mathcal{I}_i \subseteq P$ pre-dense (for $i < \omega_1$), *then* there is a directed $G \subseteq P$, $\bigwedge_{i<\omega_1} G \cap \mathcal{I}_i \neq \emptyset$.

2′) Moreover if $|P| = \aleph_2, P$ proper, $\mathcal{I}_i \subseteq P$ pre-dense (for $i < \omega_2$), $P = \bigcup_{i<\omega_2} P_i$ where P_i are increasing and $|P_i| \leq \aleph_1$, *then* there is an $\alpha < \omega_2$, $\text{cf}(\alpha) = \omega_1$ and a directed $G \subseteq P_\alpha$ such that:

$$\bigwedge_{i<\alpha} (G \cap \mathcal{I}_i \neq \emptyset).$$

3) If $|P| = \aleph_1, P$ is proper iff forcing with P does not destroy stationary subsets of ω_1.

AXIOM II.

1) $2^{\aleph_0} = \aleph_1$ and $2^{\aleph_1} = \aleph_2$.
2) If $|P| \leq 2^{\aleph_1}$, P is α-proper for every $\alpha < \omega_1$, and P is $\mathbb{D}$-complete for some simple $\aleph_0$-completeness system $\mathbb{D}$, $\mathcal{I}_i \subseteq P$ (for $i < \alpha_0 < 2^{\aleph_1}$) and each $\mathcal{I}_i$ is pre-dense, *then* there is a directed $G \subseteq P$, such that, $\bigwedge_{i<\alpha_0} G \cap \mathcal{I}_i \neq \emptyset$ (we can also define 2′ like we did in Axiom I.).

AXIOM II′ [S]. ($S \subseteq \omega_1$ stationary, costationary)
Similar to AXIOM II, but $\mathbb{D}$-completeness refers only to those N for which $N \cap \omega_1 \notin S$ (i.e. we have $(\mathbb{D}, S)$-completeness); also in the definition of $(< \omega_1)$-properness we can demand $N_i \cap \omega_1 \notin S$ provided we add properness to the set of hypotheses.

2.1. Theorem. Suppose CON(ZFC + "κ is 2^κ-supercompact"). *Then* CON(ZFC+ Axiom I).

Proof. We start as in III 4.3, defining $\langle P_i, \underset{\sim}{Q}_i : i < \kappa \rangle$. Given P_i, define $\underset{\sim}{Q}_i$ by induction on $i < \kappa$ (κ is 2^κ- supercompact in V), as usual.

Case I. If i is not strongly inaccessible, Q_i is the Levy collapse of $(2^{|P_i|})^{V^{P_i}}$ to $\aleph_1$: $Q_i = \{f : |\mathrm{Dom}(f)| = \aleph_0,\ \mathrm{Dom}(f) \subseteq \omega_1,\ \mathrm{Rang}(f) \subseteq (2^{|P_i|})^{V^{P_i}}\}$.

Case II. If i is strongly inaccessible and, in V^{P_i}, there is a proper forcing P, with universe i and $\mathcal{I}_j \subseteq i$ (for $j < i$), each $\mathcal{I}_j$ pre-dense, and $\{j :$ there is a $G \subseteq j$, directed by $\leq$ and $\bigwedge_{\xi<j} G \cap \mathcal{I}_\xi \neq 0\}$ is not stationary (subset of i), *then* $Q_i = P$ i.e. Q_i is one of those P's.

Case III. not I nor II – proceed as in Case I.

Now P_κ is proper, has density $\leq \kappa$ and satisfies the κ-c.c. (by Lemma 1.1), so in V^{P_κ}, $2^{\aleph_0} = \kappa = \aleph_2$ (Why? $\kappa \leq \aleph_2$ by case I, $2^{\aleph_0} \leq \kappa$ as P_κ satisfies the κ-c.c., has density κ and κ is strongly inaccessible, $\aleph_2 \leq 2^{\aleph_0}$ as clause (2) of the axiom I holds, as is proved below). So clauses (1), (3) of Axiom I hold – as in III 4.5 and clause 2) follows from clause 2′), so we are left with proving clause 2′) of Axiom I.

Suppose in the end that $\underset{\sim}{R}$, $\underset{\sim}{\mathcal{I}}_\alpha(\alpha < \kappa)$ are P_κ-names of a counterexample to 2′).

Let E be an ultrafilter over $\mathcal{S}_{<\kappa}(2^\kappa)$ exemplifying that κ is 2^κ-supercompact; and we code P_κ, $\underset{\sim}{R}$, $\underset{\sim}{\mathcal{I}}_\alpha(\alpha < \kappa)$ on κ. Let $S_0 = \{A \in \mathcal{S}_{<\kappa}(2^\kappa) : \mathcal{P}(A \cap \kappa) \subseteq A$, A is closed under reasonable operation, and $A \cap \kappa$ is a strongly inaccessible cardinal$\}$. Clearly $S_0 \in E$.

Then, for $A \in S_0$, the forcing notions $P_\kappa \cap A = P_{A\cap\kappa}$ and $\underset{\sim}{R}{\restriction}A$ are proper (see the definition of proper for a discussion of this: in $2^{|P|}$ we can get a witness for properness etc). So, we have proved that $i = A \cap \kappa$, $P_i = P_{A\cap\kappa}$ and, $\underset{\sim}{R} \cap A, \langle \underset{\sim}{\mathcal{I}}_\alpha \cap A : \alpha \in A \cap \kappa\rangle$ are candidates for the case II in the definition of Q_i. So, if $A \in S_0$, then $i = A \cap \kappa$ is inaccessible and there are some $P^i, \mathcal{I}^i_j(j < i)$ which we have actually chosen.

By the properties of E, there are such $i(0) < i(1), P^{i(0)} \subseteq P^{i(1)}, \mathcal{I}^{i(0)}_j = \mathcal{I}^{i(1)}_j \cap i(0)$ for $j < i(0)$ and we almost get contradiction to the choice of $P^{i(1)}$

Using such $i(\xi)$, $\xi \leq \lambda, i(\lambda) = \lambda < \kappa$ which form a stationary subset of λ we get a contradiction. $\square_{2.1}$

Remark. We essentially use the proof that $\diamondsuit_\kappa$ holds for κ measurable, which is well known.

2.2. Theorem.

1) CON(ZFC + κ is 2^κ-supercompact) implies CON(ZFC + Axiom II + G.C.H.).
2) In both cases (2.1, 2.2(1)) we can relativize to S ($S \subseteq \omega_1$ stationary, costationary). If in Axiom II we assume $|P| < 2^{\aleph_1}$, no large cardinality is needed.

Proof.

1) Similar.
2) Just note that the same iteration works. $\square_{2.2}$

2.3. Discussion. In almost all the applications we need a weaker version of the axioms for whose consistency we do not need a large cardinal.

Usually our task is to show that for every $A \subseteq H(\aleph_1)$ with $|A| = \aleph_1$, there is $B \subseteq H(\aleph_1)$ such that

$$H(\aleph_1) \models \varphi(A, B)$$

where for Axiom II, φ is any first order (or L_{ω_1,ω_1}) sentence, and for Axiom I, φ should have quantifications on A, B only.

So we iterate ω_2 times only, each time forcing a B for a given A, till we catch our tail in ω_2 steps (we can have $|A| = \aleph_2$ and visit $A \cap V^{P_\alpha}$ for stationarily many $\alpha < \omega_2, \mathrm{cf}(\alpha) = \omega_1$).

So, only Consis(ZFC) is needed.

For Axiom I part (3), κ inaccessible suffices (see III 4.1,2,3), whether it is necessary is still not clear. We can get Axiom I without inaccessible in the cases above, when we are able, provably, to find Q_A in an intermediate V. For P_{ω_2} to satisfy the $\aleph_2$-c.c. we need $|Q_A| = \aleph_1$. With κ inaccessible, Q_A should just have a cardinality smaller than κ.

For $|P| = \aleph_1$, if $P \in V^{P_{\omega_2}}$, P is proper in $V^{P_{\omega_2}}$, and for arbitrarily large $\beta < \alpha$, $\underset{\sim}{Q}_\beta = \mathrm{Levy}(\aleph_1, 2^{\aleph_1})$, then P is proper in V^{P_α} for every α large enough, (by III 4.2 though not vice versa). So Axiom I 2) for $|P| = \aleph_1$ comes under the previous discussion.

There may be applications where we really have to use information on $V^{P_\alpha}(\alpha < \omega_2)$ to build Q_α, mainly like in Axiom I, when we want to build a forcing giving B for a given A (see above) and we want to use CH or $\diamondsuit_{\aleph_1}$ for the building. But with Axioms I, II (or the versions we can get with an inaccessible) we can use the axiom: collapse $2^{\aleph_0}$, building a forcing and look at the composition (see §3, Application G).

Sometimes we use properties of P_α (like ${}^\omega\omega$-boundedness) which we usually demand from each Q_i, and from P in the axiom, and we have to prove that P_α satisfies it, (see Chapter VI).

However still Axioms I,II look like a reasonable choice. We shall use them, and can remark, for suitable applications, that only CON(ZFC) is needed.

As we mentioned (see III 4.3) CON(ZFC+ κ inaccessible) implies the consistency of

AXIOM I_a*:*
If $|P| = \aleph_1, P$ does not destroy stationary subsets of ω_1 and $\mathcal{I}_i \subseteq P$ pre-dense (for $i < \omega_1$), *then* there is a directed $G \subseteq P$, such that

$$\bigwedge_{i<\omega_1} (\mathcal{I}_i \cap G \neq \emptyset).$$

We can ask whether we can get something like Axiom I for $2^{\aleph_0} = \aleph_3$. Roitman (see [B]) proved that this is difficult, by proving that:

2.4. Theorem. (Roitman)

1) If $\bar{Q} = \langle P_n, \underset{\sim}{Q}_n : n < \omega\rangle$ is a CS iteration, $\underset{\sim}{Q}_n$ nontrivial (i.e. above every element there are two incompatible ones) *then* Lim$\bar{Q}$ does not satisfy the $2^{\aleph_0}$-chain condition.
2) If $\bar{Q} = \langle P_i, \underset{\sim}{Q}_i : i < \omega_1\rangle$ is a CS iteration, $\underset{\sim}{Q}_i$ nontrivial, $2^{\aleph_0} > \aleph_1$, then Lim$\bar{Q}$ collapses $2^{\aleph_0}$ to $\aleph_1$.

2.5. Question.

1) What kind of axioms can we get with:
   A) $2^{\aleph_0} = \aleph_3$?
   B) with $\aleph_1 < 2^{\aleph_0} < 2^{\aleph_1}$?
2) Can we define properness so that it works for higher cardinals (e.g, for SH.)?
3) Similar to (1), but we ask about iterations.

   For 1), a solution for particular problems appears in [AbSh:114].

2.6. More Discussion. In connection with the beginning of the previous discussion, there is no problem if for $A \subseteq H(\aleph_1)$, $|A| = \aleph_1$, we need to force by some Q_A of large cardinality to get some $B \subseteq \mathrm{Ord}$. We iterate up to the first larger cardinal. Baumgartner and Mekler have improved Theorem 2.1 as follows.

2.7 Theorem.

1) (Mekler) In 2.2(1) we can weaken the hypothesis on κ to π^2_2-indescribability (the advantage of this property is that if κ satisfies it in V, it satisfies it in L).
2) (Baumgartner) If κ is supercompact in some forcing extension $V^\dagger$ of V, $\aleph_1^{V^\dagger} = \aleph_1^V, \aleph_2^{V^\dagger} = \kappa$, $2^{\aleph_0} = \aleph_2$ and if

   (∗) P is a proper forcing, and $\underset{\sim}{S}_i (i < \omega_1)$ are P-names such that $\Vdash_P$ "$\underset{\sim}{S}_i$ is a stationary subset of ω_1",

   then there is a directed $G \subseteq P$, and stationary $S_i \subseteq \omega_1$ $(i < \omega_1)$ such that for every j, $i < \omega_1$, for some $p \in G, p \Vdash$ "$j \in \underset{\sim}{S}_i$" and $j \in S_i$, or $p \Vdash$ "$j \notin \underset{\sim}{S}_i$" and $j \notin S_i$.

Proof.

1) For simplicity assume $V = L$.

We define the CS iteration $\langle P_i, \underset{\sim}{Q}_i : i < \kappa \rangle$, such that $|P_i| < \kappa$, and each $\underset{\sim}{Q}_i$ is proper, and $\underset{\sim}{Q}_i$ is the first P_i-name of a counterexample for "$V^{P_i} \models$ "$\underset{\sim}{Q}_i$ satisfies the axiom"". If P_κ is not as required look at $\{ \langle P, \underset{\sim}{Q}, \langle \underset{\sim}{\mathcal{I}}_i : i < \omega_1 \rangle \rangle :$

$p \in P_\kappa, \underset{\sim}{Q}, \underset{\sim}{\mathcal{I}}_i$ are P_κ names, $p \Vdash \underset{\sim}{Q}, \underset{\sim}{\mathcal{I}}_i$ (for $i < \omega_1$) form a counterexample$\}$. Choose the first (by the canonical order of L) member $\langle p, \underset{\sim}{Q}, \bar{\underset{\sim}{I}}\rangle$. Now use indescribability.

2) By Laver [L], w.l.o.g. κ remains supercompact if we force by any κ-complete forcing. Force by $\{\bar{Q} : \bar{Q}$ is a CS iteration of proper forcing of power $< \kappa$, $\ell g(\bar{Q}) < \kappa\}$, ordered by being an initial segment. The generic object is, essentially such $\bar{Q}$ of length κ, so force by $\lim \bar{Q}$. $\square_{2.7}$

Probably better is the following:

2.8 Definition. Let κ be a supercompact cardinal. We call $f : \kappa \to H(\kappa)$ a *Laver diamond* if for every cardinal λ and $x \in H(\lambda)$, there is a normal fine ultrafilter D on $\mathcal{S}_{<\kappa}(H(\lambda))$ such that the set

$$A_D(x) \overset{\text{def}}{=} \{a \in \mathcal{S}_{<\kappa}(H(\lambda)) : x \in a, a \cap \kappa \in \kappa, \text{ and in the Mostowski collapse } MC_a \text{ of } a,\ x \text{ is mapped to } f(a \cap \kappa)\}$$

is in D.

By Laver $[L]$, if κ is a supercompact cardinal, we can assume that a Laver diamond for it exists.

2.9 Lemma. Suppose κ is a supercompact and f^* is a Laver diamond for it. Define $\underset{\sim}{Q}_i$ by induction on $i < \kappa$, as follows:

If $f^*(i)$ *is a* P_i*-name,* $\Vdash_{P_i}$ "$f^*(i)$ *proper*", *i limit,* then $\underset{\sim}{Q}_i = f^*(i)$.

Otherwise $\underset{\sim}{Q}_i = \text{Levy}(\aleph_1, 2^{2^{\aleph_1}})$.

Then $\Vdash_{P_\kappa}$ "(*) of 2.7(2) , i.e. Ax_{ω_1}[proper] " (see Definition 2.10 below).

Proof. By the properness iteration lemma, P_κ is proper, and also it satisfies the κ-c.c. Let $\underset{\sim}{Q}$ be a P_κ-name for a proper forcing, and λ a regular cardinal such that $\underset{\sim}{Q} \in H(\lambda)$; without loss of generality $\Vdash_{P_\kappa}$ "$2^{|\underset{\sim}{Q}|} < \lambda$". We use the formulation of Definition 2.10: Ax_{ω_1}[proper]. Let $\mathcal{I}_i (i < \omega_1)$ and $\underset{\sim}{S}_\beta (\beta \leq \omega_1)$ be given as in Definition 2.10 (i.e., they are P_κ-names of such objects). Apply

Definition 2.8 to $x = \underset{\sim}{Q}$ and λ such that $\underset{\sim}{Q} \in H(\lambda)$ and even $2^{|P_\kappa * \underset{\sim}{Q}|} < \lambda$, and get D as there. Choose $a \in A_D(x)$ such that $\langle \underset{\sim}{\mathcal{I}}_i : i < \omega_1 \rangle$ and $\langle \underset{\sim}{S}_\beta : \beta \leq \omega_1 \rangle$ belong to a, $(a, \in)$ is isomorphic to some $(H(\chi), \in)$ and letting MC_a be the Mostowski Collapse of a (i.e. the unique isomorphism from $(a, \in)$ onto $(H(\chi), \in)$ and $\mu \stackrel{\text{def}}{=} a \cap \kappa \in \kappa$, we have $f(\mu) = MC_a(\underset{\sim}{Q})$. Note $MC_a(\underset{\sim}{Q})$ is a P_μ-name of a proper forcing etc. Easily, $\underset{\sim}{Q}_\mu = f^*(\mu)$ in V^{P_μ}, and $\underset{\sim}{Q}_\mu$ is isomorphic to $a \cap \underset{\sim}{Q}$, so we can finish.

$\square_{2.9}$

2.10 Definition. (1) Let $\alpha \leq \omega_1, \varphi$ a property of forcing notion, λ a cardinal. Then $Ax_{\alpha,\beta}[\varphi, \lambda]$ means:

if

(i) P is a forcing notion satisfying φ and $P \in H(\lambda)$.

(ii) $\mathcal{I}_i$ is a pre-dense subset of P for $i < i^* \leq \beta$.

(iii) $\underset{\sim}{S}_i$ is a P_i-name of a stationary subset of ω_1, for $i \leq \alpha$.

then there is a G such that:

G is a directed subset of P,

G is not disjoint to $\mathcal{I}_i$ for $i < i^*$

$\underset{\sim}{S}_i[G] = \{\zeta < \omega_1 : \text{for some } p \in G \text{ we have } p \Vdash_P \text{"}\zeta \text{ in } \underset{\sim}{S}_i\text{"}\}$,

is a stationary subset of ω_1 for each $i < \alpha$.

(2) If $\lambda = \aleph_2$ we omit it. If $\beta = \omega_1$ we may omit it. If $\alpha = 0$, we omit it. $Ax^+[\varphi, \lambda]$ is $Ax_1[\varphi, \lambda]$.

§3. Applications of Axiom II (so CH Holds)

3.1 Application A. Axiom II implies SH($= \mathrm{SH}_{\aleph_1}$), in fact – every Aronszajn tree is special. (See V. 6.1.) Alternatively, see Application F.

3.2 Application B. On isomorphisms of Aronszajn trees on a closed unbounded set of levels, etc, see U. Abraham and S. Shelah [AbSh:114].

3.3 Application C: Uniformization. Axiom II implies: If η_δ is an increasing ω-sequence converging to δ, for any limit $\delta < \omega_1$ *then* for every $F : \omega_1 \to \omega$ there is a $g : \omega_1 \to \omega$ such that: $(\forall \delta < \omega_1)[\delta \text{ limit} \to (\exists n)[n \geq F(\delta) \ \& \ n = g(\eta_\delta(k))$ for all but finitely many $k < \omega]]$

Proof. Let η_δ for limit $\delta < \omega_1$ be given. Let $P_F = \{f : \mathrm{Dom}(f)$ is an ordinal $< \omega_1, \mathrm{Rang}(f) \subseteq \omega$, and for every limit $\delta \leq \mathrm{Dom}(f)$ the condition above holds$\}$. The order on P_F is the extension of functions. Now, α-properness is very easy. We prove $\mathbb{D}$-completeness for a simple $\aleph_0$-completeness system $\mathbb{D}$.

We have to note that if N is a countable elementary submodel of $(H(\lambda), \in)$, $P_F \in N$, $N \cap \omega_1 = \delta$, $p \in N$ and $\eta^1_\delta, F^1(\delta), \ldots, \eta^n_\delta, F^n(\delta)$ are n "candidates" for η_δ, $F(\delta)$, *then* we can choose $\alpha_\ell < \delta$ (for $\ell = 1, \ldots, n$) and $m > \mathrm{Max}_{\ell=1,n} F^\ell(\delta)$, such that $A = \bigcup_\ell (\mathrm{Rang}(\eta^\ell_\delta) \setminus \alpha_\ell)$ has order type ω and is disjoint to $\mathrm{Dom}(f)$, and find a $q \geq p$, q is (N, P_F)-generic, $q{\restriction}A$ is constantly m.

More formally, (see V. 5.2, 5.3), we shall define $\mathbb{D}_{\langle N, P\cap N, p\rangle}$, as a filter on $A_0 = \{G \subseteq P \cap N : p \in G, G$ is directed not disjoint to any $\mathcal{I} \in N, \mathcal{I} \subseteq P, \mathcal{I}$ pre-dense$\}$ such that it depends only on the isomorphism type of (N, P, p). Let $N \cap \omega_1 = \delta$.

The filter will be generated by $A^n_{\eta,p}$, where for $n < \omega, \eta$ an ω-sequence converging to δ, $A^n_{\eta,p} = \{G \subseteq A_0 :$ for some $k < \omega$, and $k \geq n$, for every $q \in G$ and $\ell < \omega$, if $\eta(\ell) \in (\mathrm{Dom}(q) \setminus \mathrm{Dom}(p))$, then $q(\eta(\ell)) = k\}$. $\square_{3.3}$

A conclusion is the following:

3.4 Application D. G.C.H. $\not\to \Phi^{\aleph_0}_{\aleph_1}$. (For a definition - see below).
But we know $(2^{\aleph_0} < 2^{\aleph_1}) \Rightarrow \Phi^2_{\aleph_1}$ (Devlin and Shelah [DvSh:65], or here AP §1).

3.4A Definition. The statement Φ^κ_λ is defined as: For every $G : {}^{\lambda>}2 \to \kappa$ there is an $F : \lambda \to \kappa$ such that for every $g \in {}^\lambda 2$ we have $\{i < \lambda : G(g{\restriction}i) = F(i)\}$ is stationary.

Question. Does G.C.H. $\rightarrow \Phi^3_{\aleph_1}$? (See the next chapter.)

3.5 Application D*. Axiom II implies: If $\langle \eta_\delta : \delta < \omega_1, \text{ limit } \rangle$ is as above, and $c_\delta \in {}^\omega\omega$, *then* there is an $f : \omega_1 \rightarrow \omega$ such that $(\forall \delta)(\exists^\infty n)[f(\eta_\delta(n)) = c_\delta(n)]$ (this was proved in U. Abraham, K. Devlin and S. Shelah [ADSh:81]).

There an application of this to a problem of Hajnal and Mate on the coloring number of graphs is given.

Proof. Easy by now. $\square_{3.5}$

3.6 Application E. Fleissner showed:

$\Phi^{\aleph_0}_{\aleph_1} \Rightarrow$ (topological statement A) $\Rightarrow$ not $\Big[$there is a tree $\Phi = \langle \eta_\delta : \delta < \omega_1 \rangle$, with $\mathrm{Dom}(\eta_\delta) = \omega$, for each δ, and $\eta_\delta(n)$ (for $n < \omega$) are increasing with $\delta = \bigcup_n \eta_\delta(n)$, such that:

$$(\forall h : \omega_1 \rightarrow \omega)(\exists f : \omega_1 \rightarrow \omega)(\exists h' : \omega_1 \rightarrow \omega)(\forall \ \text{limit} \ \ \delta < \omega_1)$$

$$(\exists \ m_\delta < \omega)(\forall n > m_\delta)[h(\delta) \leq f(\eta_\delta(n)) \leq h'(\delta)].\Big].$$

For a definition of the topological statement A see 3.20(3), 3.25A, 3.25B.

Clearly we can choose any Φ and then use application C to see that this Φ satisfies the conclusion above:

So Axiom II $\Rightarrow$ not (topological statement A) and Axiom II is consistent with ZFC + G.C.H, so

3.7 Conclusion. G.C.H. $\not\Rightarrow$ (topological statement A) (again CON(ZFC) suffice).

3.8 Application F. Fleissner asks about the consistency of the following with G.C.H. $(*_E)$ there is a special Aronszajn tree T such that letting $T^\dagger \stackrel{\text{def}}{=} \{t \in T : \mathrm{ht}\,(t) \text{ limit}\}$ (where $\mathrm{ht}\,(t)$ is the level or height of t) we have (we may assume that $T = \langle \omega_1, <_T \rangle$ with i-th level $[\omega i, \omega i + \omega)$, and we may write $x < y$ instead $x <_T y$):

$$\otimes_T \ (\forall f : T \to \omega)\Big[(\forall t \in T^\dagger)(\exists x < t)(\forall y)[x < y < t \to f(y) \le f(t)] \Rightarrow (\exists g : T \to \omega)(\forall t \in T^\dagger)(\exists x < t)(\forall y)[x < y \le t \to f(y) \le g(y) = g(x)]\Big].$$

(This is sufficient for the existence of some examples in general topology, see the end of the application and 3.25 below; we can add $(\forall y)(f(y) \le g(y))$.) Again CON(ZFC) suffices (as Claim 3.19 F10 holds), i.e. we prove:

Claim. [AxII] Every Aronszajn tree T satisfies $\otimes_T$ (remembering that AxII is consistent with G.C.H. and implies that every Aronszajn tree is special, we get the desired answer).

The proof is quite similar to that of Application A, see V 6.1. However, here we have to do more; an incidental point is that here we have to find a $g : T \to \omega$, not from a club of levels but from all of them.

Let T be an Aronszajn tree and $f^* : T \to \omega$ be such that:

$$(\forall t \in T^\dagger)(\exists x < t)(\forall y)(x < y < t \to f^*(y) \le f^*(t)).$$

Let $F = \{(g, C)$: for some countable ordinal i we have $\mathrm{Dom}(g) = T_{\le i} \stackrel{\mathrm{def}}{=} \bigcup_{\alpha \le i} T_\alpha$, $\mathrm{Rang}(g) \subseteq \omega$ (and, if you like, $(\forall x \in \mathrm{Dom}(g))[f^*(x) \le g(x)]$), $C \subseteq (i+1)$, C closed, $i \in C$, and

$$(\forall t \in T^\dagger \cap T_{\le i})(\exists x < t)(\forall y)[x < y \le t \to f^*(y) \le g(y) = g(x)]\}.$$

For $(g, C) \in F$, let $i(g)$ be the unique i such that $\mathrm{Dom}(g) = \bigcup_{\alpha \le i} T_\alpha$. We order F by

$$(g_1, C_1) \le (g_2, C_2) \ \text{ iff } \ g_1 \subseteq g_2, C_1 = C_2 \cap (i(g_1) + 1).$$

A generic subset of F gives g as required, but F not only is not necessarily $\aleph_1$-complete but may collapse $\aleph_1$, or add reals.

So what do we do? We add obligations.

3.9 F1 Definition. I (more formally (I, C, m)) is called an *obligation for* (T, f^*) if $C = C(I)$ is a closed unbounded subset of ω_1, $m = m(I) < \omega$ and:

a) $I \subseteq \bigcup\{(T_\alpha)^{m(I)} : \alpha \geq \text{Min}(C)\}$ and

$$(a_1, \ldots, a_{m(I)}) \in I \Rightarrow \bigwedge_{k \in [1, m(I)]} \bigwedge_{\ell < k} (a_\ell \neq a_k),$$

b) if $\text{Min}(C) \leq \alpha < \beta$, and $(a_1, \ldots, a_{m(I)}) \in (T_\beta)^{m(I)} \cap I$, then

$$(a_1 \restriction \alpha, \ldots, a_{m(I)} \restriction \alpha) \in (T_\alpha)^{m(I)} \cap I$$

(of course for $a \in T_\beta$, $a \restriction \alpha$ is the unique $b \in T_\alpha$ such that $b <_T a$),

c) if $\alpha = \text{Min}(C)$, then $(T_\alpha)^{m(I)} \cap I$ has $\aleph_0$ pairwise disjoint members,

d) if $\alpha \in C, (a_1^*, \ldots, a_{m(I)}^*) \in (T_\alpha)^{m(I)} \cap I$ and $\beta \in C, \alpha < \beta$, then

$$\{(a_1, \ldots, a_{m(I)}) \in (T_\beta)^{m(I)} \cap I : \bigwedge_{\ell=1}^{m(I)} a_\ell \restriction \alpha = a_\ell^*\}$$

contains $\aleph_0$ pairwise disjoint members,

e) there are $n_\ell(I) < \omega$ (for $\ell = 1, \ldots, m(I)$) such that: *if* $\bar{a} \in I \cap (T_\alpha)^{m(I)}$, $\bar{b} \in I \cap (T_\beta)^{m(I)}$ and $\alpha, \beta \in C, \alpha < \beta$ and $\bar{a} \leq \bar{b}$ (see below), *then* $n_\ell(I) \geq \text{Max}\{f^*(t) : a_\ell \leq t < b_\ell\} < \omega$.

Notation. If $\bar{a} \in (T_\alpha)^{m(I)}$, then $\alpha(\bar{a}) \stackrel{\text{def}}{=} \alpha$.

3.10 F2 Definition. For $\bar{a}, \bar{b} \in I, \bar{a} \leq \bar{b}$ holds *if* $\bar{a} \in (T_{\alpha(\bar{a})})^{m(I)}$, $\bar{b} \in (T_{\alpha(\bar{b})})^{m(I)}$, $\alpha(\bar{a}) \leq \alpha(\bar{b})$, and

$$\bigwedge_{\ell=1}^{m(I)} [a_\ell = b_\ell \restriction \alpha(\bar{a})].$$

Then we say $\bar{b}$ extends $\bar{a}$ or $\bar{b}$ is an extension of $\bar{a}$.

Let $\alpha_\xi(I)$ be the ξ-th element of $C(I)$, in the increasing enumeration of $C(I)$.

3.11 F3 Definition. (g, C) fulfills the obligation I if a), b) and c) below hold, where:

a) $i(g) \in C(I)$, $i(g) \geq \alpha_1(I)$ and $C \setminus \alpha_0(I) \subseteq C(I)$.

Subdefinition (F3i). We say that g is I-good for $\bar{a} \in I$ *if* $\alpha(\bar{a}) \leq i(g)$ and $(\forall \ell)(\forall y)[1 \leq \ell \leq m(I) \ \&\ a_\ell \restriction \alpha_0(I) \leq y \leq a_\ell \rightarrow f(y) \leq g(y) = g(a_\ell \restriction \alpha_0(I))]$,
moreover $(\forall \ell)(\forall y)[1 \leq \ell \leq m(I) \ \&\ a_\ell \restriction \alpha_0(I) \leq y \leq a_\ell \rightarrow g(y) = n_\ell(I)]$.

b) There are $\bar{a}_{k,\ell} \in I \cap (T_{\alpha_1(I)})^{m(I)}, \bar{a}_k \in I \cap (T_{\alpha_0(I)})^{m(I)}, (k < \omega, \ell < \omega)$ such that: $\bar{a}_k \leq \bar{a}_{k,\ell}$; $\{\bar{a}_k : k < \omega\}$ are pairwise disjoint and for each k we have: $\{\bar{a}_{k,\ell} : \ell < \omega\}$ are pairwise disjoint and g is I-good for each $\bar{a}_{k,\ell}$.

Subdefinition (F3ii). We say (g, C) is I-very good for $\bar{a}$ *if* $\alpha(\bar{a}) \in C \cap C(I)$, and for any β, $\alpha(\bar{a}) < \beta \in C \cap (i(g) + 1)$, $\bar{a}$ has $\aleph_0$ pairwise disjoint extensions in $I \cap (T_\beta)^m$ for which g is I-good.

c) If $\alpha < \beta < \gamma$ are in $C \cap C(I)$, $\bar{a} \in I \cap (T_\alpha)^{m(I)}$ and, (g, C) is I-very good for $\bar{a}$ *then* $\bar{a}$ has $\aleph_0$ pairwise disjoint extensions in $I \cap (T_\beta)^{m(I)}$ for which (g, C) is I-very good (γ appear just to make β not of maximal level as then I-very good is meaningless).

3.12 F4 Definition. $P_{(T,f^*)} = \{(g, C, B) : (g, C) \in F, B$ a countable family of obligations for (T, f^*) which (g, C) satisfies$\}$.

For short, we may write P_T for $P_{(T,f^*)}$.

3.13 (F5) Claim. $\mathcal{I}_i \stackrel{\text{def}}{=} \{(g, C, B) \in P_{(T,f^*)} : i \leq i(g)\}$ is a dense subset of $P_{(T,f^*)}$ (for each $i < \omega_1$).

Proof. Let $(g, C, B) \in P_{(T,f^*)}$ and $i < \omega_1$.

As B is countable, $\cap_{I \in B} C(I)$ is a closed unbounded subset of ω_1. So we can find a ξ such that: $\xi > i$, $i(g)$ and $\xi \in C(I)$ for every $I \in B$. We let $C^\dagger = C \cup \{\xi\}$, $B^\dagger = B$.

Now we shall define g' (such that $(g', C^\dagger, B^\dagger) \geq (g, C, B)$ and $(g', C^\dagger, B^\dagger) \in \mathcal{I}_i$ will exemplify the desired conclusion).

The nontrivial part in defining g' is to satisfy Definition (F3)(c). There, the nontrivial case (and it implies the others) is $\beta = i(g), \gamma = \xi$. So, for each $\alpha < \beta$ $(= i(g))$ and $I \in B$, $\alpha \in C(I) \cap C$ and $\bar{a} \in (T_\alpha)^{m(I)} \cap I$ such that (g, C) is I-very good for $\bar{a}$, we have to provide $\aleph_0$ pairwise disjoint $\bar{b} \geq \bar{a}$ such that $\bar{b} \in (T_\beta)^{m(I)} \cap I$ and $(g', C^\dagger)$ is I-very good for $\bar{b}$. (Hence $(g', C^\dagger)$ will be I-very good for $\bar{a}$.)

Let $\{\langle I_k, \alpha_k, \bar{a}_k\rangle : k < \omega\}$ be a list of all triples as above. As (g, C) is I_k-very good for $\bar{a}_k$, there is a set of pairwise disjoint sequences $\{\bar{b}_{k,\ell} : \ell < \omega\} \subseteq (T_\beta)^{m(I_k)} \cap I_k$ such that g is I_k-good for $\bar{b}_{k,\ell}$ and $\bar{a}_k < \bar{b}_{k,\ell}$.

Now we can easily find an infinite $S_k \subseteq \omega$ (for $k < \omega$) such that $\ell_1 \in S_{k_1}$, $\ell_2 \in S_{k_2}$, $k_1 \neq k_2 \Rightarrow \bar{b}_{k_1,\ell_1} \cap \bar{b}_{k_2,\ell_2} = \emptyset$ (more exactly, the intersection of their ranges is empty). Also for $k, \ell < \omega$ we can choose $\bar{c}^m_{k,\ell} \in (T_\gamma)^{m(I_k)} \cap I_k$ for $m < \omega$ such that: $\bar{b}_{k,\ell} < \bar{c}^m_{k,\ell}$ and $m(1) \neq m(2) \Rightarrow \mathrm{Rang}(\bar{c}^{m(1)}_{k,\ell}) \cap \mathrm{Rang}(\bar{c}^{m(2)}_{k,\ell}) = \emptyset$; remember $\gamma = \xi$.

Let $\bar{b}_{k,\ell} = \langle b_{k,\ell,e} : 1 \leq e \leq m(I_k)\rangle$ and $\bar{c}^m_{k,\ell} = \langle c^m_{k,\ell,e} : 1 \leq e \leq m(I_k)\rangle$. Let $\gamma^{m(1),m(2)}_{k,\ell,e} \stackrel{\text{def}}{=} \mathrm{Max}\{\zeta : c^{m(1)}_{k,\ell,e} \restriction \zeta = c^{m(2)}_{k,\ell,e} \restriction \zeta\}$, so by Ramsey theorem and basic properties of trees for any $k, \ell < \omega$, $e \in [1, m(I_k)]$, either $(\exists \gamma)(\forall m)[\gamma^m_{k,\ell,e} = \gamma]$ or $\gamma^{m(1),m(2)}_{k,\ell,e}$ does not depend on $m(2)$ and is strictly increasing in $m(1)$.

Now we have to define $g' \restriction (\bigcup\{T_\zeta : i(g) < \zeta \leq \xi\})$. If $t \in \bigcup\{T_\zeta : i(g) < \zeta \leq \xi\}$, $b_{k,\ell,e} < t \leq c^m_{k,\ell,e}$, then note that by clause (e) of 3.9

$$\mathrm{Max}\{f(s) : b_{k,\ell,e} \leq s \leq t\} \leq n_e(I_k),$$

in which case we let $g'(t) = n_e(I_k)$ (by the above choice of $\bar{b}_{k,\ell}$ there is no contradiction, as $b_{k,\ell,e} = b_{k_1,\ell_1,e_1} \Rightarrow k = k_1 \,\&\, \ell = \ell_1 \,\&\, e = e_1$).
By the assumption on the $\bar{c}^m_{k,\ell,e}$'s, if $t \in \bigcup\{T_\zeta : i(g) < \zeta \leq \xi\}$ and $g'(t)$ is not yet defined *then:* $g'(y)$ is not yet defined for any large enough $y < t$ or $t \notin T^\dagger$.

Let $\{t_n : n < \omega\}$ be a list of all $t \in \bigcup\{T_\zeta : i(g) < \zeta \leq \xi\}$, such that for arbitrarily large $s < t$ we have that $g'(s)$ is still undefined and the height of t is a limit ordinal. We can easily define by induction on n, sets $A_n \subseteq \bigcup\{T_\zeta : i(g) < \zeta \leq \xi\}$, such that: on every $t \in A_n$, $g'(t)$ is undefined, A_n are pairwise disjoint, each A_n is linearly ordered, each A_n is convex (i.e,

$x, y \in A_n \wedge x < z < y \Rightarrow z \in A_n$), and for every t_n, for some $x_n < t_n$, and $n' \leq n$ we have $\{y : x_n \leq y \leq t_n\} \subseteq A_{n'}$, and $\mathrm{Sup}_{x \in A_n} f^*(x) < \omega$ (see the hypothesis on f^*; A_n may be empty).

Define $g'(t)$ for $t \in A_n$ as $\mathrm{Max}_{x \in A_n} f^*(x)$. Complete g' to its required domain $\bigcup_{\alpha \leq \xi} T_\alpha$ by $g'(t) = f^*(t)$ if $g'(t)$ not already defined. It is easy to check $(g', C^\dagger) \in F$, $(g', C^\dagger, B^\dagger) \in P_{(T,f^*)}$ and $(g, C, B) \leq (g', C^\dagger, B^\dagger)$. $\square_{3.13}$

3.14 F6 Claim. If $(g, C, B) \in P_{(T,f^*)}$, $\xi > i(g)$, $\xi \in \bigcap_{I \in B} C(I)$ is a limit ordinal and $\mathbf{t}_1, \ldots, \mathbf{t}_n$ are branches of $\bigcup_{\alpha < \xi} T_\alpha$, and $y_1, \ldots y_n$ are $<_T$-incomparable, $y_\ell \in \mathbf{t}_\ell$, $y_\ell \notin \mathrm{Dom}(g)$, $n_\ell \geq \mathrm{Max}\{f^*(x) : x = z_\ell \text{ or } y_\ell \leq x \in \mathbf{t}_\ell\}$, *then* there is a $(g^\dagger, C^\dagger, B^\dagger) \in P_T$ such that $i(g^\dagger) = \xi, (g, C, B) \leq (g^\dagger, C^\dagger, B)$ and $\bigwedge_{\ell=1}^{n} (\forall x \in \mathbf{t}_\ell)(x \notin \mathrm{Dom}(g) \;\&\; y_\ell \leq x \rightarrow g'(x) = n_\ell)$.

3.15 Remarks. By the demand on f^* we know that n_ℓ, y_ℓ always exist, if t_ℓ have distinct an upper bound in T (in particular the Max is well defined).

Proof. Same proof, assuring $b_{k,\ell,e} \notin t_\ell$. $\square_{3.14}$

3.16 F7 Claim. If λ is large enough ($2^{\aleph_1}$ should be o.k.), $\{P_{(T,f^*)}, T, f^*\} \in N \prec (H(\lambda), \in)$, $\|N\| = \aleph_0, \delta \stackrel{\text{def}}{=} \omega_1 \cap N$ and $\mathbf{t}_1, \ldots, \mathbf{t}_n$ are (distinct) δ-branches of $N \cap T = \bigcup_{\alpha < \delta} T_\alpha$, $(g, C, B) \in P_T \cap N, \mathcal{J} \in N$ a maximal antichain of P_T, $y_\ell \in \mathbf{t}_\ell$, $y_1, \ldots, y_n$ are pairwise $<_T$-incomparable, $y_\ell \in \bigcup_{i \leq i(g)} T_i$ and for every ℓ and z, $y_\ell \leq z \in \mathbf{t}_\ell$ we have

$$n_\ell \geq \mathrm{Max}\{f^*(z) : y_\ell \leq z \in \mathbf{t}_\ell, \mathrm{ht}\,(z) > i(g)\}$$

and

$$n_\ell = g(y_\ell) = g(z) \text{ when } y_\ell \leq z \in \mathbf{t}_\ell, \mathrm{ht}\,(z) \leq i(g)$$

(remember, $\mathrm{ht}\,(z)$ is the height of z in the tree), *then* there is a $(g', C^\dagger, B^\dagger) \in P_T \cap N$ such that $(g', C^\dagger, B^\dagger) \geq (g, C, B)$ and $(g', C^\dagger, B^\dagger)$ is above some member of $\mathcal{J}$ and $(\forall x \in \mathrm{Dom}(g'))[(x \in \mathbf{t}_\ell \setminus \mathrm{Dom}(g)) \Rightarrow g'(x) = n_\ell]$.

Proof. Suppose not. Define

$$\begin{aligned}
I_0 \stackrel{\text{def}}{=} \{(a_1,\dots,a_m) :\ & \text{for some } \beta<\omega_1, \beta>i(g), a_1,\dots,a_m \in T_\beta,\\
& a_\ell \geq \mathbf{t}_\ell{\restriction} i(g) = \text{ the unique } x\in \mathbf{t}_\ell\cap T_{i(g)},\\
& (\text{so necessarily } a_1,\dots,a_m \text{ are distinct}), \beta \in \bigcap_{I\in B} C(I),\\
& \bigwedge_\ell (\forall x)(x\leq a_\ell \wedge x\notin \mathrm{Dom}(g) \Rightarrow f^*(x)\leq n_\ell)\\
& \text{and there is no } (g',C^\dagger,B^\dagger)\geq (g,C,B) \text{ such that:}\\
& \quad \text{a) } i(g')\leq\beta\\
& \quad \text{b) } (g',C^\dagger,B^\dagger) \text{ is above some member of } \mathcal{J}\\
& \quad \text{c) } \bigwedge_{\ell=1}^{m} (\forall x)[x\leq a_\ell \,\&\, \mathrm{ht}\,(x)>i(g) \,\&\, \mathrm{ht}\,(x)\leq i(g')\\
& \qquad\qquad \Rightarrow g'(x)=n_\ell]\}.
\end{aligned}$$

Clearly $I_0\in N$, and if $a_\ell\in \mathbf{t}_\ell\cap T_\beta$ (for $\ell=1,\dots,m$) are distinct (and for $\beta\in[i(g),\delta)$ there are such a_ℓ) (a_ℓ is determined by $\mathbf{t}_\ell,\beta$), then $(a_1,\dots,a_m)\in I_0$, provided that $i(g)<\beta\in\bigcap_{I\in B}C(I)$. Note: $\bar a\in I_0\Rightarrow \alpha(\bar a)\in \bigcap_{I\in B} C[I]$

But in N, the set $\bigcap_{I\in B}C(I)\in N$ is unbounded below δ.

So $N\models$ "for arbitrarily large $\beta<\omega_1$ there is an $\bar a\in I_0\cap (T_\beta)^m$". So by N's choice this really holds.

We now define, by induction on $\varepsilon<(2^{\aleph_1})^+$, a set I_ε. I_0 was already defined. For limit ε, we set $I_\varepsilon=\bigcap_{\zeta<\varepsilon}I_\zeta$, and if $\varepsilon=\zeta+1$ then $\bar a\in I_\varepsilon$ iff $\bar a\in I_\zeta$ and for arbitrarily large $\gamma<\omega_1$ we have: there is $\bar b\in (T_\gamma)^m$ such that $\bar a<\bar b\in I_\zeta$. Clearly, if $\bar a<\bar b$ are in I and $\bar b\in I_\varepsilon$ then $\bar a\in I_\varepsilon$; also for some $\varepsilon(*)<(2^{\aleph_1})^+$ we have $\varepsilon(*)\leq\varepsilon<(2^{\aleph_1})^+\Rightarrow I_\varepsilon=I_{\varepsilon(*)}$. For $\bar a\in I_0$ let $\varepsilon(\bar a)=\mathrm{Min}\{\varepsilon:\varepsilon=\varepsilon(*)$, or $\bar a\notin I_{\varepsilon+1}\}$. Returning to $\mathbf{t}_\ell$'s, easily $\varepsilon(\langle \mathbf{t}_\ell{\restriction}\beta:\ell=1,\dots,n\rangle)$ is nonincreasing with β for $\beta\in(i(g),\,N\cap\omega_1)$, hence eventually constant, hence is constantly $\varepsilon(*)$. So $\langle \mathbf{t}_\ell{\restriction}\beta:\ell=1,\dots,n\rangle$ is in $I_{\varepsilon(*)}$ for $\beta\in(i(g),\delta)$, hence $I_{\varepsilon(*)}\neq\emptyset$.

By the proof of Theorem III 5.4 (as in V §6), there is a closed unbounded $C^*\subseteq\bigcap_{I\in B}C(I)\setminus(i(g)+1)$ such that there are $\aleph_0$ pairwise disjoint members of $I_{\varepsilon(*)}$ in $(T_\delta)^m$, moreover there are $\aleph_0$ pairwise disjoint members of $I_{\varepsilon(*)}$ in $(T_\delta)^m$ which are above $\bar a$, if $\bar a\in(T_\beta)^m$, $\bar a\in I_0$, $\beta<\delta$, $\delta\in C^*$ and $\bar a\in (T_\beta)^m\cap I_{\varepsilon(*)}$.

Define $I \stackrel{\text{def}}{=} \{\bar a :$ for some γ we have $\gamma > \beta \ge \mathrm{Min}(C^*)$, $\bar a \in (T_\beta)^m, \bar b \in I_{\varepsilon(*)} \cap (T_\gamma)^m$, $\gamma \in C^*$, $\bar a < \bar b\}$.

Then I (more formally (I, C^*, m)) is an obligation for (T, f^*). By a variant of 3.14 F6 we can find a $(g', C^\dagger, B^\dagger)$, where $g' \ge g, i(g) = \alpha_1(I) =$ second element of C^*, $C^\dagger = C \bigcup \{\alpha_0(I), \alpha_1(I)\}$, $B^\dagger = B \cup \{I\}$, such that for infinitely many pairwise disjoint $\bar a \in I \cap (T_{\alpha_0(I)})^m$, for infinitely many pairwise disjoint $\bar b$ we have $\bar a < \bar b \in I \cap (T_{\alpha_1(I)})^m$ and $\bigwedge_{\ell=1}^m (\forall x)[x \le b_\ell \wedge \mathrm{ht}\,(x) > i(g) \to g'(x) = n_\ell]$. So $(g, C, B) \le_{P_{(T,f^*)}} (g', C^\dagger, B^\dagger)$.

So, there is a $(g'', C'', B'') \ge (g', C^\dagger, B^\dagger)$ in $P_{(T,f^*)}$ which is above a member of $\mathcal{J}$. As $I \in B^\dagger$, clearly $i(g'') \in C^* = C(I)$, and there is an $\bar a \in I \cap (T_{i(g'')})^m$ for which g'' is I-good. This contradicts the definition of I_0, I.

$\square_{3.16}$

3.17 F8 Claim.

1) P_T is proper

2) P_T is α-proper for every $\alpha < \omega_1$.

Proof.

1) If N is as in 3.16 F7 and $\delta \stackrel{\text{def}}{=} N \cap \omega_1$, while $(g_0, C_0, B_0) \in P_{(T,f^*)} \cap N$, let $\langle \mathcal{J}_n : n < \omega \rangle$ be a list of maximal antichains of P_T which belong to N. We define $(g_n, C_n, B_n) \in P_{(T,f^*)} \cap N$ which are increasing, (g_n, C_n, B_n) is above a member of $\mathcal{J}_{n-1}$ (when $n > 0$) and "on the side" we all the time have more comitments of the form that appear in 3.16F7. More specifically, together with $(g_n,\ C_n,\ B_n)$ we have $\mathbf{t}_\ell$, y_ℓ, m_ℓ, for $\ell = 1, \dots, k_n$, such that $\mathbf{t}_\ell$ is a branch of $T \cap N$ with an upper bound in T_δ, $y_\ell \in \mathbf{t}_\ell$, $y_\ell \in \bigcup_{i \le i(g)} T_i$, $\langle \mathbf{t}_\ell \restriction i(g_n) : \ell = 1, \dots, k_n \rangle$ are pairwise distinct, $y_\ell \le y \in \mathbf{t}_\ell \cap \bigcup_{i \le i(g)} T_i \Rightarrow g_n(y) \le n_\ell$ and $n_\ell \ge \mathrm{Max}\{f^*(s) : y_\ell \le s \in \mathbf{t}_\ell\}$ and if $t_\ell \in T_\delta$ is the upper bound of $\mathbf{t}_\ell$ then $n_\ell \ge f^*(t_\ell)$. We can continue by 3.16 F7 in order that in the end we get a condition. We have two kinds of tasks:

A) for every $t \in T_\delta$, there is an $x \in \bigcup_{j<\delta} T_j, x < t$ such that

$$(\forall y)(x < y < t \to f^*(y) \le g(y) = g(x) \ \&\ f^*(t) \le g(x))$$

If our promise until now is $\mathbf{t}_1, n_1, y_1 \dots, \mathbf{t}_\ell, n_\ell, y_\ell$ we let $\mathbf{t}_{\ell+1} = \{x : x <_T t\}$, $n_{\ell+1} = \mathrm{Min}\{n :$ for some $z \in \mathbf{t}_{\ell+1}$, for every y, $z \le y \le t \Rightarrow f(y) \le n\}$ and then choose $y_{\ell+1}$ appropriately.

B) for every $I \in B_n$, $\alpha \in C_n \cap C(I), \bar{a} \in (T_\alpha)^m \cap I, (g_n, C_n)$ is I-very good for $\bar{a}$, and $k < \omega$, we want that there will be k pairwise disjoint $\bar{b}$'s in $(T_\delta)^m \cap I$, so that $(\cup g_n, \cup C_n)$ will be I-good for $\bar{b}$.

(As we do this eventually for every k, we can get $\aleph_0$ pairwise disjoint $\bar{b}$'s.) This is again quite easy.

We have only $\aleph_0$ tasks, so there is no problem. Having defined $\langle (g_n, C_n, B_n) : n < \omega \rangle$ it is straightforward to find an upper bound $(g^*, C^*, B^*) \in P_{(T,f^*)}$, with $i(g^*) = \delta$, which is $(N, P_{(T,f^*)})$-generic and is above (g_0, C_0, B_0).

2) A similar proof. $\square_{3.17}$

3.18 F9 Claim. P_T is $\mathbb{D}$-complete for some simple $\aleph_1$-completeness system $\mathbb{D}$.

Proof. As in the proof of 3.17 F8 above, we have to prove that *if* for one N (with $\delta \stackrel{\text{def}}{=} N \cap \omega_1$) we are given countably many possible pairs

$\langle T_\delta, \langle I \cap (T_\delta)^m : I \in N$ an obligation for $(T, f^*)\rangle\rangle$

then we can define a sequence $\langle (g_n, C_n, B_n) : n < \omega \rangle$ which is appropriate for all of them at once. This is trivial as in the proof of 3.17 F8 we do not actually need to assume that $\mathbf{t}_\ell$ has an upper bound in T_δ $(\ell = 1, \dots, k_n)$, just that it is well defined as a δ-branch of T_δ with $\sup\{f^*(\mathbf{t}_\ell \restriction \alpha) : \alpha < \delta\} < \omega$. $\square_{3.18}$

The following is not needed for applying AxII, but is needed if we want to use the weaker variant equiconsistent with ZFC.

3.19 F10 Claim. P_T satisfies the $\aleph_2$-e.c.c. (see §1).

Proof. Trivial: define $h : P_T \to \omega_1$ such that

$$h(g, C, B) = h(g', C^\dagger, B^\dagger) \textit{ iff } g = g' \text{ and } C = C^\dagger$$

(this is possible as $2^{\aleph_0} = \aleph_1$). Let $(g, C, B)R(g', C^\dagger, B^\dagger)$ mean $(g, C) = (g', C^\dagger)$ and in Definition 1.2(B) let

$$f(i) = h(p_i).$$

Remember that (g, C, B), $(g, C, B^\dagger)$ have a lub: $(g, C, B \bigcup B^\dagger)$. $\square_{3.19,3.8}$

3.20 Discussion.

1) The proof here is appropriate for Application A; the small gain is that we directly find a function specializing T rather than finding one specializing a closed unbounded set of levels, and then using a theorem saying this is equivalent.
2) In applications A and F, compared to Jensen's CON(ZFC + G.C.H. + SH), we can get also CON(ZFC + CH + SH + $2^{\aleph_1}$=anything of cofinality $> \aleph_1$) using Lemma 1.3, and iterating ω_2 times, each time specializing all Aronszajn trees. The same proof works for all the relevant cases.
3) *Fleissner's Question.* It is unknown whether ZFC $\vdash$ " there is a countably paracompact non normal Moore space". (Equivalently, there is a countably paracompact not hereditarily countably paracompact Moore space).
   Such spaces can be constructed by Wager's technique from normal non-metrizable Moore spaces.
   Application E gives the first example of such a space not constructed in this way.
   Application F shows it can even be a Jones road space – a more traditional space than the space constructed in §3 E; see more in 3.25.

3.21 Application G. Ax II implies: There are no Kurepa trees. Moreover, every $\aleph_1$-tree (a tree of height ω_1 with all levels countable) is essentially specialized, i.e. there is an $f : T \to \mathbb{Q}$ (rationals) such that: $t \leq s \Rightarrow f(t) \leq f(s)$, and $t \leq s_1, t < s_2, f(t) = f(s_1) = f(s_2) \Rightarrow (s_1 \leq s_2 \text{ or } s_2 \leq s_1)$. (Why does this imply that T is not a Kurepa tree? On any ω_1-branch B, f is eventually constant, so choose the minimal $x \in B$, $f{\restriction}\{y \in B : y \geq x\}$ is constant, call it $x(B)$. Then $B_1 \neq B_2$ ω_1-branches $\Rightarrow x(B_1) \neq x(B_2)$, so T has $\leq \aleph_1$ branches.)

3.22 Remark. Here Con(ZFC $+\exists$ inaccessible) is sufficient (and necessary).

Proof. First, let P_0 be a Levy collapse of $\aleph_2$ to $\aleph_1$ (which is $\aleph_1$-complete). In V^P, by Silver (or see III 6.1), T has at most $\aleph_1$ many ω_1-branches. Let $\{B_i : i < \omega_1\}$ be a list of the ω_1-branches of T, w.l.o.g. they are pairwise disjoint (choose $B_i^\dagger \subseteq B_i$ pairwise disjoint end segments by induction).

Now Q is a forcing (in V^P) which specializes T as in Application A, *but* on each B_ℓ the function is constant. The proof is the same.

So $P * \underset{\sim}{Q}$ essentially specializes T and so guarantees that T has $\leq \aleph_1$ branches. A directed $G \subseteq P * \underset{\sim}{Q}$ defines the function f which essentially specializes T if it meets the following $\aleph_1$ dense sets:

$\mathcal{I}_t = \{p : p \Vdash \text{“}f(t) = q\text{”} \text{ for some } q \in \mathbb{Q}\}$ for $t \in T$.

So by Axiom II there is a G as required, provided that $P * \underset{\sim}{Q}$ is $(< \omega_1)$-proper and $\mathbb{D}$-complete for a simple $\aleph_1$-completeness system $\mathbb{D}$.

For $(< \omega_1)$-properness: P obviously is, $\underset{\sim}{Q}$ – as in Application A, (in V^P) and so by III 3.2. applied to $(< \omega_1)$-properness, $P * \underset{\sim}{Q}$ is $(< \omega_1)$-proper. The task of checking for $\mathbb{D}$-completeness is left to the reader. $\square_{3.21}$

3.23 History for G. Baumgartner Malitz Reinhart [BMR] prove MA $+\neg$ CH $\Rightarrow$ every Aronszajn tree is special.

Silver proves that by Levy collapsing κ=first (strongly) inaccessible to $\aleph_2$ one obtains that there are no Kurepa trees (this includes the lemma we quote). Devlin proved: CON(ZFC + MA + $2^{\aleph_0} = \aleph_2$+ no Kurepa trees).

Shelah [Sh:73] used essential specialization function as the function above and showed that we can use $\aleph_1$-c.c. forcings to essentially specializes $\aleph_1$-trees with few branches (the proof uses more particularly unnecessary information).

Baumgartner [B3], using proper forcing, defines essential specialization and strengthens Devlin's result to: CON(ZFC + MA + every tree of power $\aleph_1$ and height $\aleph_1$ is essentially special). Independently, Todorcevic proved these results too.

It is well known that for such a consistency result an inaccessible cardinal is necessary.

3.24 Application H. Assume Axiom II′[S]. For $\delta \in S$ let η_δ be an ω-sequence converging to δ. Then $\langle \eta_\delta : \delta \in S \rangle$ has the $\aleph_0$-uniformization property. See [Sh:64], and [Sh:98]; this result should be an exercise to a reader who arrives here (but you may want more refined results as in [Sh:98], then proofs there are still of interest).

3.25 On Countable Paracompactness. Some general topologists consider suspiciously application F. So, let us give the derivation of the solution of the original problem.

3.25A Problem. Is the existence of a countably paracompact regular space which is not normal consistent with G.C.H.?

3.25B Definition. A topological space X is countably paracompact if *for every* family of open sets U_n of X (for $n < \omega$), which forms a cover (i.e. satisfies $X = \bigcup_{n<\omega} U_n$), *there are* open U'_n $(n < \omega)$ which refine U_n $(n < \omega)$ (i.e. $\bigwedge_n \bigvee_m U'_n \subseteq U_m$) and form a cover of X (i.e. $X = \bigcup_{n<\omega} U'_n$) which is locally finite (i.e. for every $x \in X$, $\{n : x \in U'_n\}$ is finite).

3.25C Definition. We shall consider a tree T as a topological space as follows: the set of points of the space is the set of nodes of T, for $t \in T$ its neighbourhood basis is:

$$\Big\{\{t\}\Big\} \text{ if } \mathrm{ht}\,(t) \text{ non limit.}$$

$$\Big\{\{y : x <_T y \leq_T t\} : x <_T t\Big\} \text{ if } \mathrm{ht}\,(t) \text{ is a limit ordinal.}$$

3.25D Fact. For a ω_1-tree T we have:
(∗) as a topological space, T is Hausdorff (use the normality of the tree) and even regular. $\square_{3.25D}$

3.25E Claim. If T is an ω_1-tree satisfying $\otimes_T$ (the conclusion of 3.8(=Application F)) above, *then*, as a topological space, T is countably paracompact.

Proof. Let $U_n \subseteq T$ be open with $T = \bigcup_{n<\omega} U_n$. Define a function f from T to ω by: $f(t) = \mathrm{Min}\{n : t \in U_n\}$.

First check that f satisfies the antecedent of $\otimes_T$, i.e. $(\forall t \in T^\dagger)(\exists x < t)(\forall y)[x < y \leq t \Rightarrow f(y) \leq f(t)]$ (the order is of the tree).

So let $t \in T^\dagger$, i.e. $t \in T$, $t \in T_\delta$, δ a limit ordinal. For some n, $n = f(t)$ so $t \in U_n$, hence for some $x < t$, $\{y : x < y \leq t\} \subseteq U_n$, hence $x < y \leq t \Rightarrow f(y) \leq n = f(t)$, as required. So by $\otimes_T$ there is a function $g : T \to \omega$ satisfying the conclusion of $\otimes_T$, i.e. $(\forall t \in T^\dagger)(\exists x < t)\forall y[x < y \leq t \to f(y) \leq g(y) = g(x)]$. Now, without loss of generality as asid above we can force it, still we derive it. [Why?

$(*)$ $f(t) \leq g(t)$ for every $t \in T$.

Let $t \in T^\dagger$, so for some $x_1 < t$ we have:

$(*)_1$ $x_1 < y < t \Rightarrow f(y) \leq f(t)$

(this is true, as we have verified the antecedent of $\otimes_T$)

and for some $x_2 < t$

$(*)_2$ $x_2 < y \leq t \Rightarrow f(y) \leq g(y) = g(x_2)$

(this is possible by the choice of g).

Now if $y \in A_t \overset{\text{def}}{=} \{y : y < t, x_1 < y, x_2 < y\}$, then $f(y) \leq f(t)$ (by $(*)_1$), $f(t) \leq g(t)$ (by $(*_2)$, by the "$f(y) \leq g(y)$" there) and $g(t) = g(x_2)$ (by $(*)_2$, by the "$g(y) = g(x_2)$" there) and $g(x_2) = g(y)$ (by $(*)_2$, by the "$g(y) = g(x_2)$" there). Together, $y \in A_t \Rightarrow f(y) \leq g(y)$. So, $\{t : f(t) > g(t)\}$ is necessarily a set of isolated points with no accumulation point. Hence we can change the values of g on it while not harming the conclusion of $\otimes_T$].

Define $U'_{n,\ell} = \{x \in U_n : g(x) = \ell\}$ (for $n \leq \ell < \omega$). First, clearly $U'_{n,\ell} \subseteq U_n$. Second, each $U'_{n,\ell}$ is also an open set: if $t \in T^\dagger \cap U'_{n,\ell}$, let $x_2 < t$ be such that $x_2 < y \leq t \Rightarrow f(y) \leq g(y) = g(x_2) = g(t)$, so $x_2 < y \leq t \Rightarrow g(y) = \ell = g(x_2)$ (there is such an x_2 by the choice of g). Let $x_1 < t$ be such that $x_1 < y \leq t \Rightarrow y \in U_n$ (there is such an x as we have verified the antecedent of $\otimes_T$) and choose an $x < t$, $x > x_1, x > x_2$, clearly $\{y : x \leq y \leq t\} \subseteq U_{n,\ell}$. If $t \in T \setminus T^\dagger$, obviously $\{t\} \subseteq U'_{n,\ell}$.

Third, $T = \bigcup_{n \leq \ell < \omega} U'_{n,\ell}$ because if $t \in T$, then for some n, we have that $t \in U_n \setminus \bigcup_{m<n} U_m$, hence by the choice of f, $f(t) = n$ and for some ℓ we have $g(t) = \ell$, so $t \in U'_{n,\ell}$ and $n \leq \ell$ as $n = f(t) \leq g(t)$.

Fourth, $\{U_{n,\ell} : n \le \ell < \omega\}$ is locally finite: if $t \in T$, $\ell^* = g(t)$ then $\{U_{n,\ell} : n \le \ell < \omega, t \in U_{n,\ell}\} \subseteq \{U_{n,\ell} : n \le \ell = \ell^*\}$ which has $\ell^* + 1$ members. $\square_{3.25E}$

It was proved in [DvSh:85] (using the weak diamond) that

3.25F Claim. (CH) No special Aronszajn-tree is normal (as a topological space, in the topology we considered). $\square_{3.25F}$

So we can solve Watson's problem:

3.26 Conclusion. AxII (which is consistent with G.C.H.) implies that every Aronszajn ω_1-tree T is special and $\otimes_T$ holds. Hence, we have a countably paracompact, non normal, regular topological space which is an Aronszajn tree. In fact, it suffices to use a weaker version of AxII, for whose consistency (even with GCH), CON(ZFC) suffices. $\square_{3.26}$

§4. Applications of Axiom I

4.1. Claim. P is proper and even α-proper for every $\alpha < \omega_1$ if at least one of the following holds:

1. P satisfies the $\aleph_1$-c.c.

2. P is $\aleph_1$-complete (then P is even strongly proper and ${}^\omega\omega$-bounding)

3. P is Sacks forcing, or Silver forcing, or Gregorief forcing, or a product with countable supports of such forcings (then P is even strongly proper and ${}^\omega\omega$-bounding) for definitions see Lemma VI 2.14(2); Remark VI 4.1A; Definition VI 4.1(1) and VI 4.1A, Definition IX 2.6, Definition V 4.1 respectively).

4. P is Laver forcing ($P = \{T : T \subseteq {}^{\omega>}\omega$, T non empty closed under initial segments, no XX-minimal element and for some $\eta \in T$ such that: $T \cap {}^{\ell g(\eta)}\omega = \{\eta\}$ and $\eta \trianglelefteq \nu \in T \Rightarrow (\exists^* n)(\nu{}^\frown\langle n\rangle \in T)\}$ ordered by inverse inclusion).

Proof. An exercise. $\square_{4.1}$

4.2. Discussion. Baumgartner [B3], independently of the author's work on proper forcing, and at about the same time, introduced Axiom A forcing defined below. It covers a large part of the application of proper forcing, but to many it seems easier to handle.

P satisfies Axiom A if there are partial orders $\leq_n$ on P such that:

i) $\leq_0$ is the usual order $\leq$,

ii) $x \leq_{n+1} y \Rightarrow x \leq_n y$,

iii) if $\bigwedge_n x_n \leq_n x_{n+1}$ then $\{x_n : n < \omega\}$ has an upper bound in $\leq$,

iv) if $\underset{\sim}{\tau}$ is a name of an ordinal, $p \in P, n < \omega$, then there are $q \in P, p \leq_n q$ and a countable $A \subseteq \mathrm{Ord}, A \in V$ such that $q \Vdash_P$ "$\underset{\sim}{\tau} \in A$"

Baumgartner proved "P satisfies *Axiom* $A \Rightarrow P$ is proper"; in fact "P satisfies $A \Rightarrow P$ is $(< \omega_1)$-proper". Which forcing notions are equivalent to ones satisfying Axiom A? See XIV 2.4.

4.3. Discussion. Baumgartner [B3] found many applications for proper forcing: new ones and simplified proofs for the old ones (see proofs there). It is a matter of taste whether to deduce them from Axiom I, or build a forcing doing it. Some of them are:

4.3A. P_a, the forcing giving finite information on the enumeration of a closed unbounded subset of ω_1, is proper.

Remark. But P_a is not ω-proper, so this shows proper $\neq$ ω-proper.

This club does not include any old infinite sets. So if we iterate ω_2 times (by Axiom I or any variant) we obtain that for any $\aleph_1$ infinite subsets of ω_1 there is a club which does not include any of them. So this statement is consistent with $2^{\aleph_0} = \aleph_2$. In fact $2^{\aleph_0}$ can be anything, since if P_{ω_2} is the forcing for doing the above, its product with adding κ Cohen generic is also O.K.

U. Abraham proved the consistency of a similar assertion for $\aleph_2$ (with CH).

4.3B. Every tree of height $\aleph_1$ and power $\aleph_1$ is essentially special.

See §3 Application G. Only before forcing with P, add a Cohen real. Here the forcing specializing T consists of finite information, but the first part is the same (see history there, more exactly in 3.23).

4.3C. CON(ZFC + no $\aleph_2$-Aronszajn trees) (originally proved by Mitchell and Silver). We use κ weakly compact.

4.3D. Laver's consistency of Borel's conjecture: Baumgartner iterates $\aleph_2$ times the forcing $P = \{A : A \subseteq \omega \text{ is infinite}\}$, ordered by $A <_P B$ iff $B \leq_{ae} A$ for adding a Ramsey ultrafilter, and Mathias forcing for those ultrafilters (note that we can use the preservation of the Laver property (see Definition VI 2.9 and Conclusion 2.12)).

4.3E. CON(ZFC + "there are no $\aleph_2$ subsets of $\aleph_1$ of power $\aleph_1$, with pairwise countable intersection"), originally proved by Baumgartner.

4.4. On isomorphism of Aronszajn trees see Abraham and Shelah [AbSh:114].

§5. A Counterexample Connected to Preservation

5.1 Example. If we iterate forcing which does not destroy stationary subsets of ω_1 we may destroy $\aleph_1$.

Proof. For every $\alpha < \omega_2$, let $\alpha = \bigcup_{i<\omega_1} A_i^\alpha$ where A_i^α are countable, increasing and continuous in i and let $h_\alpha(i) =$ the order type of A_i^α.

Let $\mathcal{D} = \mathcal{D}_{\omega_1}$ (the filter of closed unbounded subsets of ω_1). Then $g_1 <_\mathcal{D} g_2$ means $\{\alpha < \omega_1 : g_1(\alpha) < g_2(\alpha)\} \in \mathcal{D}$.

Then $\alpha < \beta \Rightarrow h_\alpha <_\mathcal{D} h_\beta$. Suppose

(∗) $(\forall \alpha < \omega_2)[h_\alpha <_\mathcal{D} g]$ for some $g : \omega_1 \to \omega_1$ (g exists e.g., if $V = L$).

Define $P_g = \Big\{(h, s, F, T) : h$ is a function from some $i < \omega_1$ to ω_1, i is a successor ordinal, s is a characteristic function of a closed subset of i, and $h <_s g$, which means $s(j) = 1 \Rightarrow h(j) < g(j)$, F is a countable subset of $\{h_\alpha : \alpha < \omega_2\}$, T is a function from F to the set of closed subsets of i such that $h_\alpha \in F \,\&\, j \in T(h_\alpha) \Rightarrow h_\alpha(j) < h(j)\Big\}$.
We call i the domain of the condition.

The partial ordering is obvious. P_g is not necessarily $\aleph_1$-closed.

5.1A Fact. If CH, then P_g satisfies the $\aleph_2$-chain condition (moreover there is a function $\mathbf{h} : P_g \to \omega_1$ such that if $\mathbf{h}(p_0) = \mathbf{h}(p_1)$ then p_0, p_1 and a lub).

Proof. For the first phrase find a Δ-system then take the union.
(For the second let $\mathbf{h}((h, s, F, T))$ code (h, s).) $\square_{5.1A}$

5.1B Fact. Forcing with P_g does not destroy stationary subsets of ω_1 and does not add reals.

Proof. Let $S \subseteq \omega_1$ be stationary, $p \in P_G$ and $\underset{\sim}{C}$ a P_g-name of a closed unbounded subset such that $p \Vdash_{P_g}$ "$\underset{\sim}{C}$ is disjoint to S".

First take $N \prec (H(\aleph_7), \in, P, \underset{\sim}{C}, S, p)$, $||N|| = \aleph_1$ such that $\omega_1 + 1 \subseteq N$.

Let $N \cap \omega_2 = \varepsilon$, so $\varepsilon = \{\zeta(i) : i < \omega_1\}$, and without loss of generality N is such that $N = \bigcup_{i<\omega_1} N_i$, $N_i \prec N$, N_i are increasing continuous, $||N_i|| = \aleph_0$, and $\langle N_j : j \leq i\rangle \in N_{i+1}$.

Now $h_\varepsilon <_{\mathcal{D}} g$ by the assumption on g. So let C_0 be a closed unbounded set such that $i \in C_0 \Rightarrow h_\varepsilon(i) < g(i)$.

For each $i < \omega_1, \zeta(i) < \epsilon$ so $h_{\zeta(i)} <_{\mathcal{D}} h_\varepsilon$. So let C^i be a closed unbounded subset of ω_1 such that $j \in C^i \Rightarrow h_{\zeta(i)}(j) < h_\varepsilon(j)$.

Let $C_1 \subseteq \omega_1$ be closed unbounded such that

$$i \in C_1 \Rightarrow N_i \cap \varepsilon = \{\zeta(j) : j < i\}.$$

Now

$$C \stackrel{\text{def}}{=} \{i \in C_0 \cap C_1 : (\forall j < i) i \in C^j\}$$

is known to be closed unbounded subset of ω_1. Choose a $\delta \in C$ and let $\delta = \bigcup_n j_n, j_n < \delta$.

Now, by induction on $n < \omega$, define $p_n \in P_g \cap N_\delta, p = p_0, p_n \leq p_{n+1}$ such that $j_n \subseteq \mathrm{Dom}(h^{p_n}), p_{n+1} \Vdash$ "$\gamma_n \in \underset{\sim}{C}$" for some γ_n, $j_n < \gamma_n < \delta$.

Now $\bigcup_{n<\omega} p_n$ can be extended to a condition $p^* \in P_g$ by adding δ to the domain because $\delta \in C$ and $h_\varepsilon(\delta)$ can serve as the value of $h(\delta)$ i.e. $p^* = (h^{p^*}, s^{p^*}, F^{p^*}, T^{p^*})$ is defined by: $\mathrm{Dom}(h^{p^*}) = \delta + 1$, $h^*\restriction\delta = \bigcup_{n<\omega} h^{p_n}$ (well defined as $j_n \subseteq \mathrm{Dom}(h^{p_n})$), and $h^{p^*}(\delta) = h_\varepsilon(\delta)$, s^{p^*} is a function with domain $\delta+1$, $s^{p^*}\restriction\delta = \bigcup_{n<\omega} s^{p_n}$ and $s^{p^*}(\delta) = 1$, $F^{p^*} = \bigcup_{n<\omega} F^{p_n}$ (which is $\subseteq N$ as each p_n belongs to N) and lastly if $h_\alpha \in F^{p^*}$ let $n(\alpha) = \mathrm{Min}\{n : h_\alpha \in F^{p_n}\}$ and let $T^{p^*}(h_\alpha) = (\bigcup_{n\in[n(\alpha),\omega)} T^{p_n}(h_\alpha)) \cup \{\delta\}$. Why $p^* \in P_g$? As $\delta \in C_0$, we have $h_\varepsilon(\delta) < g(\delta)$ and $\zeta \in N \cap \omega_2 \Rightarrow \zeta = \zeta(i)$, for some $i < \delta$, hence $h_\zeta(\delta) < h_\varepsilon(\delta)$ (as $\delta \in C$) but $h_\varepsilon(\delta) < g(\delta)$ so $h_\zeta(\delta) < g(\delta)$. Now $p^* \Vdash$ "$\delta \in \underset{\sim}{C}$" as $\underset{\sim}{C}$ is closed and $\delta = \bigcup_n j_n = \bigcup_n \gamma_n$ and $p_{n+1} \Vdash$ "$\gamma_n \in \underset{\sim}{C}$". But $\delta \in S$, so p cannot force that $\underset{\sim}{C}$ is disjoint to S. Also, if $\underset{\sim}{r} \in N_0$ is a P_g-name of a real then we can arrange that p_{n+1} forces a value to $\underset{\sim}{r}(n)$ hence $p^* \Vdash$ "$\underset{\sim}{r} = r$" for some old real r. $\square_{5.1B}$

P_g in general is not proper. Now if $G \subseteq P_g$ is generic, we can find a generic function h such that

$$(\forall \alpha < \omega_2) h_\alpha <_{\mathcal{D}} h <_{\mathcal{D}} g\,.$$

So P_h is well defined.

If we iterate ω-times (taking any kind of limit) we necessary destroy ω_1. Why?

We get $h_0, h_1 \ldots$ functions from ω_1 to ω and $h_{n+1} <_{\mathcal{D}} h_n$ (any closed unbounded subset of ω_1 remains closed unbounded), contradiction. (Note: the kind of limit we take at ω is irrelevant.) $\square_{5.1}$

VIII. κ-pic and Not Adding Reals

§0. Introduction

In the first section we show that we can iterate $\aleph_2$-complete forcing, and $\aleph_1$-complete forcing which satisfy the $\aleph_2$-c.c. in a strong sense.

In the second section we deal with a strong version of the $\aleph_2$-c.c. called $\aleph_2$-pic. It is useful for proving that for CS iteration of length ω_2 of proper forcing notions, the limit still satisfies the $\aleph_2$-c.c. This in turn will be used in order to get universes with $2^{\aleph_1} > 2^{\aleph_0} = \aleph_2$.

In the third section we deal again with the axioms; starting with a model of ZFC (not assuming the existence of large cardinals) we phrase the axioms we can get. There are four cases according to whether $2^{\aleph_0}$ is $\aleph_1$ or $\aleph_2$, and $2^{\aleph_1}$ is $\aleph_2$ or larger [our knowledge on the case $2^{\aleph_0} \geq \aleph_3$ is slim].

In the fourth section we return to the problem of when a CS iteration of proper forcing preserves "not adding reals". We weaken "each $\underset{\sim}{Q}_i$ (a P_i-name) is $\mathbb{D}$-complete for some $\mathbb{D}$ a $(\lambda, 1, \kappa)$-system", by replacing "each $\mathbb{D}_x$ is an $\aleph_1$-complete filter" or even just "each $\mathbb{D}_x$ is a filter" by "each $\mathbb{D}_x$ is a family of sets, the intersection of e.g. any two is nonempty". So we can deduce ZFC+CH $\nvdash \Phi^3_{\aleph_1}$. We also try to formulate the property preserved by iteration weaker than this completeness. See references in the relevant sections.

§1. Mixed Iteration – $\aleph_2$-c.c., $\aleph_2$-Complete

1.1 Lemma. Suppose P_α $(\alpha \le \alpha_0)$ are forcing notions in V and $V \models$ CH and $\underset{\sim}{Q}_\alpha$ are such that:

1) $\underset{\sim}{Q}_\alpha$ is a P_α-name of a forcing in V^{P_α},
2) $P_\alpha = \{f : |\{2\beta : 2\beta < \alpha \text{ and } 2\beta \in \mathrm{Dom}(f)\}| \le \aleph_0 \text{ and } |\{2\beta+1 : 2\beta+1 < \alpha$ and $2\beta + 1 \in \mathrm{Dom}(f)\}| \le \aleph_1$ and $\emptyset \Vdash_{P_i}$ "$f(i) \in \underset{\sim}{Q}_i$" for all $i \in \mathrm{Dom}(f)\}$,
3) $\underset{\sim}{Q}_{2\beta+1}$ is $\aleph_2$-complete (i.e. in $V^{P_{2\beta+1}}$, if $q_i \in \underset{\sim}{Q}_{2\beta+1} (i < \delta < \omega_2)$ are increasing, then $(\exists q_\delta \in \underset{\sim}{Q}_{2\beta+1}) \bigwedge_{i<\delta} q_i \le q_\delta)$,
4) for $\alpha = 2\beta$, in V^{P_α} there is an $\underset{\sim}{h}_\alpha : \underset{\sim}{Q}_\alpha \to \omega_1$, such that: $\underset{\sim}{h}_\alpha(p) = \underset{\sim}{h}_\alpha(q) \Rightarrow$ (p, q has a least upper bound $p \wedge q$),
5) $\underset{\sim}{Q}_{2\beta}$ is $\aleph_1$-complete,
6) $V \models CH$.
7) The order on P_α is as usual: $P_\alpha \models$ "$p \le q$" *iff* for every $\beta \in \mathrm{Dom}(p)$ we have $q \in \mathrm{Dom}(p)$ and $q{\restriction}\beta \Vdash_{P_\beta}$ "$p(\beta) \le q(\beta)$"

Then: P_{α_0} is $\aleph_1$-complete and does not collapse $\aleph_2$.

1.1A Remark.

1) Condition 4) was introduced by Baumgartner for getting a weak MA for a $\aleph_1$-complete forcing.

2) For simplicity, we assume that $\emptyset \in Q_\beta$ is minimal, $h_{2\beta}(\emptyset) = 0$, and adopt the convention: if $f \in P_\alpha$ and $f(\beta)$ is not defined otherwise, then $f(\beta) = \emptyset$.

3) Of course the decision to use odd and even ordinals for the two different cases is arbitrary, since any other iteration along two disjoint sets of ordinals for the two different cases can be translated into such an iteration.

4) For a better theorem — see Chapter XIV.

5) We can replace $\aleph_1$ by κ if κ is regular, $\kappa = \kappa^{<\kappa}$ (so "countable" is replaced by "of cardinality $< \kappa$".)

Proof. Let λ be large enough, $N \prec (H(\lambda), \in)$, $P_{\alpha_0} \in N$, $||N|| = \aleph_1$, and every countable subset of N belongs to N. Let $p \in P_{\alpha_0} \cap N$, and $\langle \mathcal{I}_\alpha : \alpha < \omega_1 \rangle$ be a list of all maximal antichains of P_{α_0} which belong to N (so $\langle \mathcal{I}_\alpha : \alpha < \omega_1 \rangle \notin N$ but for each $\alpha^* < \omega_1$ we have that $\langle \mathcal{I}_\alpha : \alpha < \alpha^* \rangle \in N$, by the choice of N). It is trivial that P_{α_0} is $\aleph_1$-complete. It then suffices to prove the existence of a $p^* \in P_{\alpha_0}$, $p \leq p^*$ such that p^* is (N, P_{α_0})-generic, so proving that P_{α_0} is "somewhat proper".

By CH, we can let $\{\langle \alpha^\xi, A^\xi, \langle \gamma^\xi_{\beta,n} : \beta \in A^\xi, n < \omega \rangle \rangle : \xi < \omega_1\}$ be a list of all triples of the form $\langle \alpha, A, \langle \gamma_{\beta,n} : \beta \in A, n < \omega \rangle \rangle$ such that $\alpha < \omega_1$, $A \subseteq \{2\beta : 2\beta \in \alpha_0 \cap N\}, |A| \leq \aleph_0$ and $\gamma_{\beta,n} < \omega_1$.

We now inductively define conditions $p_\xi \in N \cap P_{\alpha_0}$ (for $\xi < \omega_1$) which are increasing, with $p_0 = p$, such that:

A) $\mathrm{Dom}(p_\xi) \cap \{2\beta : 2\beta < \alpha_0\} \subseteq \mathrm{Dom}(p)$

B) $2\beta \in \mathrm{Dom}(p_\xi) \Rightarrow p_\xi(2\beta) = p(2\beta)$

C) *if* there are $p^n \in P_{\alpha_0} \cap N$ (for $n < \omega$) such that

(i) $p_\xi \leq p^0 \leq p^1 \leq p^2 \ldots$,

(ii) $2\beta \in A^\xi \Rightarrow (p^{n+1} \restriction 2\beta) \Vdash_{P_{2\beta}}$ "$\underset{\sim}{h}_{2\beta}(p^n(2\beta)) = \gamma^\xi_{2\beta,n}$", and

(iii) for some $q \in \mathcal{I}_{\alpha^\xi}$ we have $q \leq p^0$

then there are such $p^0, p^1 \ldots$ such that:

$[2\beta + 1 \in \bigcup_{n<\omega} \mathrm{Dom}(p^n)] \Rightarrow \bigwedge_{n<\omega} [\Vdash_{P_{2\beta+1}}$ " if $p^0(2\beta+1) \leq p^1(2\beta+1) \leq \ldots \leq p^n(2\beta+1)$(in $\underset{\sim}{Q}_{2\beta+1}$) then $p^n(2\beta+1) \leq p_{\xi+1}(2\beta+1)$"$]$.

We let $p^n_\xi = p^n$ and $q_\xi = q$.

There is no problem in the definition; we can assume w.l.o.g. that

$$\Vdash_{P_{2\beta+1}} \text{"}p^n_\xi(2\beta+1) \leq p^{n+1}_\xi(2\beta+1)\text{"}$$

for every n, as we can replace p^n_ξ by r^n where $r^n(2\beta) = p^n_\xi(2\beta)$ and $r^n(2\beta+1) = p^\ell_\xi(2\beta+1)$ for the maximal $\ell \leq n$ such that $\langle p^m_\xi(2\beta+1) : m \leq \ell \rangle$ is increasing (this is of course a $P_{2\beta+1}$-name).

Now we define p^* : if $2\beta \in \mathrm{Dom}(p)$, $p^*(2\beta) = p(2\beta)$, and if $2\beta + 1 \in \bigcup_{\xi<\omega_1} \mathrm{Dom}(p_\xi)$ then $p^*(2\beta + 1)$ is (a $P_{2\beta+1}$-name of) an upper bound (in $\underset{\sim}{Q}_{2\beta+1}$) of $\{p_\xi(2\beta + 1) : \xi < \omega_1\}$ if it exists, and $p(2\beta + 1)$ otherwise. So $p = p_0 \leq p^* \in P_{\alpha_0}$. So if $(2\beta+1) \in N \cap \alpha_0$ then $p^*{\restriction}(2\beta+1) \Vdash_{P_{2\beta+1}}$ "$p^*(2\beta+1)$ is an upper bound of $\{p_\zeta(2\beta + 1) : \zeta < \omega_1$ (and $2\beta + 1 \in \mathrm{Dom}(p_\zeta))\}$".

Now we prove that for each $\alpha < \omega_1$, $\mathcal{I}_\alpha \cap N$ is pre-dense above p^*. Clearly, there are p^0_* and $q_* \in P_{\alpha_0}$ such that $p^* \leq p^0_*$, $q_* \leq p^0_*$, $q_* \in \mathcal{I}_\alpha$.

Let $A_0 = \mathrm{Dom}(p^0_*) \cap \{2\beta : 2\beta < \alpha_0\} \cap N$. Now, by the $\aleph_1$-completeness of P_{α_0}, there is a $p^1_* \in P_{\aleph_0}$, $p^0_* \leq p^1_*$, such that for every $2\beta \in A_0$,

$$(p^1_*{\restriction}2\beta) \Vdash_{P_{2\beta}} \text{“}\underset{\sim}{h}_{2\beta}(p^0_*(2\beta)) = \gamma_{2\beta,0}\text{”}$$

for some $\gamma_{2\beta,0}$. We continue to define $p^{n+1}_* \geq p^n_*$ such that

$$(p^{n+1}_*{\restriction}2\beta) \Vdash_{P_{2\beta}} \text{“}\underset{\sim}{h}_{2\beta}(p^n_*(2\beta)) = \gamma_{2\beta,n}\text{”}$$

for every $2\beta \in A_n \overset{\mathrm{def}}{=} \mathrm{Dom}(p^n_*) \cap \{2\beta : 2\beta < \alpha_0\} \cap N$. We now define $A \overset{\mathrm{def}}{=} (\bigcup_{n<\omega} A_n)$, $\gamma_{2\beta,n} = 0$ for $2\beta \in A \setminus A_n$. So, for some ξ we have $\alpha_\xi = \alpha$, $A^\xi = A$, $\langle \gamma_{2\beta,n} : 2\beta \in A, n < \omega\rangle = \langle \gamma^\xi_{2\beta,n} : 2\beta \in A, n < \omega\rangle$. As these objects are countable subsets of N, by the choice of N, they belong to N.

As $N \prec (H(\lambda), \in)$, for this ξ there are q, p^n as mentioned in clause (C) above, $p^0 \geq q \in \mathcal{I}_\alpha$. Again, without loss of generality, $p^n \in P_{\alpha_0} \cap N$ and $q \in \mathcal{I}_\alpha \cap N$, so $q_\xi, \langle p^n_\xi : n < \omega\rangle$, as in (C), are well defined. Now by the properties of $h_{2\beta}$, we can prove, by induction on $\gamma \leq \alpha_0$, that

$(*)$ *for every* $\zeta < \gamma$, and $r \in P_\alpha$ such that $p^n_*{\restriction}\zeta \leq r$ and $p^n_\xi{\restriction}\zeta \leq r$ for $n < \omega$, there is an $r^* \in P_\gamma$ such that $p^n_*{\restriction}\gamma \leq r^*$, $p^n_\xi{\restriction}\gamma \leq r^*$ for $n < \omega$, and $r^*{\restriction}\zeta = r$, and $\mathrm{Dom}(r^*) \cap [\zeta, \gamma) \subseteq \bigcup_{n<\omega} \mathrm{Dom}(p^n_\xi) \cup \bigcup_{n<\omega} \mathrm{Dom}(p^n_*)$.

For γ a limit there are no problems: use the induction hypothesis and the "bound" on the domain of r^*. For $\gamma = 2\beta + 2$, w.l.o.g. $\zeta = 2\beta + 1$ (by the induction hypothesis for "$\gamma^\dagger = 2\beta + 1$"). By clause (C) and the induction hypothesis we know:

$$r{\restriction}(2\beta+1) \Vdash_{P_{2\beta+1}} \text{``}p^n_*(2\beta+1) \geq p^*(2\beta+1) \geq p_{\xi+1}(2\beta+1) \geq p^n_\xi(2\beta+1)\text{''},$$

so by the $\aleph_1$-completeness of $\underset{\sim}{Q}_{2\beta+1}$, r exists. Lastly, for $\gamma = 2\beta+1$, the nontrivial case is that 2β belongs to $\bigcup_{n<\omega} \mathrm{Dom}(p^n_\xi)$ and also to $\bigcup_{n<\omega} \mathrm{Dom}(p^n_*)$ hence $2\beta \in N \cap \alpha_0$. Again w.l.o.g. $\zeta = 2\beta$, and by the hypothesis on r and $p^n, p^n_*(n<\omega)$:

$$r{\restriction}(2\beta) \Vdash_{P_{2\beta}} \text{``}\underset{\sim}{h}_{2\beta}(p^n_\xi(2\beta)) = \underset{\sim}{h}_{2\beta}(p^n_*(2\beta))\text{''},$$

$$r{\restriction}(2\beta) \Vdash_{P_{2\beta}} \text{``}p^n_\xi(2\beta) \leq p^{n+1}_\xi(2\beta)\text{''},$$

$$r{\restriction}(2\beta) \Vdash_{P_{2\beta}} \text{``}p^n_*(2\beta) \leq p^{n+1}_*(2\beta)\text{''}.$$

So r forces that $p^n_\xi(2\beta), p^n_*(2\beta)$ have a least upper bound $p^n_\xi(2\beta) \wedge p^n_*(2\beta)$; and as $p^n_\xi(2\beta) \leq p^{n+1}_\xi(2\beta)$, and $p^n_*(2\beta) \leq p^{n+1}_*(2\beta)$ (i.e. r forces this), also $p^n_\xi(2\beta) \wedge p^n_*(2\beta) \leq p^{n+1}_\xi(2\beta) \wedge p^{n+1}_*(2\beta)$, hence by the $\aleph_1$-completeness of $\underset{\sim}{Q}_{2\beta}$, there is a $q(2\beta)$ such that $r{\restriction}(2\beta) \Vdash_{P_{2\beta}}$ "$\bigwedge_{n<\omega}(p^n_\xi(2\beta) \wedge p^n_*(2\beta)) \leq r^*(2\beta)$" and $\emptyset \Vdash_{P_{2\beta}}$ "$r^*(2\beta) \in \underset{\sim}{Q}_{2\beta}$", so we are done.

Taking $\zeta = 0, \gamma = \alpha_0$ in $(*)$, we see that the set $\{p^n_\xi,\ p^n_* : n < \omega\}$ has an upper bound which necessarily is a common upper bound to q_ξ, q_*, so as $\mathcal{I}_\alpha$ is an antichain, $q_* = q_\xi \in N$, so $\mathcal{I}_\alpha \cap N$ is pre-dense above p^*, and we finish. $\square_{1.1}$

1.2 Remark. The reason for including this is as follows. It was a consequence of the work on proper forcing that we can iterate $\aleph_1$-complete and $\aleph_1$-c.c. forcings together. So it was natural to ask the parallel for $\aleph_2$-complete and $\aleph_1$-complete with the $\aleph_2$-c.c. But we do not know how to iterate the second kind alone (and in general this is impossible since $\aleph_2$ will collapse). So it is reasonable to replace $\aleph_2$-c.c. by something stronger (here – clause (4) of the lemma). (Remember $p^n_*(2\beta)$ is $\emptyset$ when $2\beta \notin \mathrm{Dom}(p^n_*)$, and $h_{2\beta}(\emptyset) = 0$). Of course, much better would be to find one condition unifying the two conditions - see Chapter XIV.

However as the interaction has no applications now, we shall not discuss it further (there are other tries at $\aleph_2$-c.c., see [Sh 80]).

Note also that the analogous lemma for $\aleph_1$-complete, $\aleph_1$-c.c. forcing holds, but now it has no application.

1.3 Claim. If $\diamondsuit_{\aleph_1}$ holds, then in 1.1 we can change the iteration to the usual $(<\aleph_2)$-support iteration and the conclusion still holds.

Proof. We let $N = \bigcup_{\xi<\omega_1} N_\xi$, $N_\xi \prec N$ where N_ξ are countable, increasing and continuous. By $\diamondsuit_{\aleph_1}$ there are for $\xi < \omega_1$, 2-place functions f_ξ from

$$Y_\xi \stackrel{\text{def}}{=} \{\langle \alpha, \beta \rangle : \beta, \alpha \in N_\xi, \alpha, \beta \text{ ordinals}\}$$

into ω_1, such that for every 2-place $f : \bigcup_{\xi<\omega_1} Y_\xi \to \omega_1$ the set $\{\xi : f{\restriction}Y_\xi = f_\xi\}$ is stationary. Repeat the proof of 1.1 but in the definition of the p_ξ's, we replace A), B), C) by:

A) if $\xi < \zeta < \omega_1$ and $2\beta \in N_\xi \cap \alpha_0$, then $p_\zeta(2\beta) = p_\xi(2\beta)$

B) $p_\xi \leq p_\zeta \in N \cap P_{\alpha_0}$ for $\xi \leq \zeta < \omega_1$

C) *If* $\xi < \omega_1$, and there are q and $p^i (i < \omega)$ such that:

(i) $q \leq p^0$, $q \in \mathcal{I}_\alpha$, and $p_\xi \leq p^0$

(ii) for $i < j < \omega, p^i \leq p^j$, moreover $\Vdash_{P_\gamma}$ "$p^i(\gamma) \leq p^j(\gamma)$" for each $\gamma \in \text{Dom}(p^i)$

(iii) $p^{i+1}_\xi{\restriction}(2\beta) \Vdash_{P_{2\beta}}$ "$\underset{\sim}{h}_{2\beta}(p^i(2\beta)) = f_\xi(i, 2\beta)$" for $2\beta \in N_\xi \cap \alpha_0$,

then there are q, $p^i_\xi (i < \omega)$ satisfying (i), (ii), (iii), such that:

for $i < \omega$ and $\gamma \in \text{Dom}(p^i_\xi) \setminus \{2\beta : \beta \in N_\xi \text{ and } 2\beta < \alpha_0\}$, we have $\Vdash_{P_\gamma}$ "$p^i_\xi(\gamma) \leq p_{\xi+1}(\gamma)$"

In the end we define p^*, $\text{Dom}(p^*) = \alpha_0 \cap N, p^*(\gamma)$ is $p_\xi(\gamma)$ for any even γ, and for any ξ such that $\gamma \in N_\xi$, and it is any upper bound of $\{p_\xi(\gamma) : \xi < \omega_1\}$ for γ odd. $\square_{1.3}$

1.4 Remark. See more in [Sh:186], [Sh:587].

§2. Chain Conditions Revisited

We here deal again with problems like those of §1 from VII, but allowing the continuum to increase somewhat. Here, κ is a fixed cardinal.

2.1 Definition. P satisfies the κ-p.i.c. (κ-properness isomorphism condition) provided the following holds, for λ large enough:

Suppose $i < j < \kappa$, $\kappa \in N_i \prec (H(\lambda), \in, <_\lambda)$, ($<_\lambda$ is a well ordering of $H(\lambda)$) and $\kappa \in N_j \prec (H(\lambda), \in, <_\lambda)$, $||N_i|| = ||N_j|| = \aleph_0$, $P \in N_i \cap N_j, i \in N_i, j \in N_j$, $N_i \cap \kappa \subseteq j, N_i \cap i = N_j \cap j$, $p \in P \cap N_i$, h an isomorphism from N_i onto N_j, $h{\restriction}(N_i \cap N_j)=$ the identity and $h(i) = j$.

Then there is a $q \in P$, such that:

(a) $p, h(p) \leq q$, and for every maximal antichain $\mathcal{I} \subseteq P$, $\mathcal{I} \in N_i$ we have that $\mathcal{I} \cap N_i$ is pre-dense above q, and similarly for $\mathcal{I} \in N_j$ (but clause (b) below implies that this follows from the rest of (a))

(b) for every $r \in N_i \cap P$ and $q^\dagger$ such that $q \leq q^\dagger \in P$ there is a $q'', q^\dagger \leq q'' \in P$ such that $[r \leq q''$ iff $h(r) \leq q'']$; equivalently;

(a$'$+b$'$) letting $\underset{\sim}{G}$ be the P-name of the generic set

$$q \Vdash_P \text{“}(\forall r \in N_i \cap P)(r \in \underset{\sim}{G} \text{ iff } h(r) \in \underset{\sim}{G})\text{”},$$

$$q \Vdash_P \text{“}p \in \underset{\sim}{G}\text{”},$$

$$\text{and } q \text{ is } (N_i, P) - \text{generic.}$$

2.2 Claim.

1) If Definition 2.1 holds for P, $H(\lambda), <_\lambda$, *then* it holds for any $\lambda_1 > 2^\lambda$ and well ordering $<_1$ of $H(\lambda_1)$ (in fact, we can omit the well ordering).

2) If Definition 2.1 holds for $P, H(\lambda), <_\lambda$ *then* it holds for some $\lambda_1, <_1$ such that $\lambda_1 \leq (\mu + |P|)^+$, where μ is the number of maximal antichains of P (w.l.o.g. $P \in H(|P|^+)$).

Proof. Similar to the proof in III §1,2. $\square_{2.2}$

2.3 Lemma. Suppose $(\forall \mu < \kappa)\mu^{\aleph_0} < \kappa$ where κ is regular and P satisfies the κ-p.i.c. *Then* P satisfies the κ-chain condition.

Proof. Let $p_i \in P$ for $i < \kappa$ be given. Let $(H(\lambda), \in, <_\lambda)$ be as in Definition 2.1. Find, for $i < \kappa$, models N_i such that $i, p_i \in N_i \prec (H(\lambda), \in, <_\lambda)$, $||N_i|| = \aleph_0$. Define $f(i) \overset{\text{def}}{=} \text{Sup}(N_i \cap i)$, so $\text{cf}(i) > \aleph_0 \Rightarrow f(i) < i$.
By Fodor's Lemma, for some γ the set $\{i : f(i) = \gamma\}$ is stationary. As $(\forall \mu < \kappa)$ $\mu^{\aleph_0} < \kappa$ and κ is regular, for some $A \subseteq \gamma$, $S = \{i : N_i \cap i = A\}$ is stationary. Similarly, we can assume that for some B, $i \neq j \in S \Rightarrow N_i \cap N_j = B$ (see the proof of V1.5A). Also $C = \{\delta < \kappa : (\forall i < \delta)(N_i \cap \kappa \subseteq \delta)\}$ is closed unbounded, so $S_1 = S \cap C$ is stationary. Now there are κ models $(N_i, p_i, i, a)_{a \in B}$ where p_i, i, a $(a \in B)$ are individual constants and $\kappa > 2^{\aleph_0}$, and the number of isomorphism types of such models is $2^{\aleph_0}$, so for some $i < j$ there is an isomorphism $h : N_i \to N_j$ (onto), $h(p_i) = p_j, h{\restriction}B =$ the identity. Now apply Definition 2.1 for N_i, N_j, h, p_i. $\square_{2.3}$

2.4 Lemma. Suppose $\overline{Q} = \langle P_i, \underset{\sim}{Q}_j : i \leq \alpha_0, j < \alpha_0 \rangle$ is an iteration with countable support. Suppose further

(*) Q_α satisfies the κ-pic (for each $\alpha < \alpha_0$) and κ is regular.

Then 1) If $\alpha_0 < \kappa$, P_{α_0} satisfies the κ-p.i.c.

2) If $\alpha_0 \leq \kappa$, P_{α_0} satisfies the κ-chain condition, provided that

$$(\forall \mu < \kappa)(\mu^{\aleph_0} < \kappa).$$

Proof. 1)Let $(H(\lambda), \in, <_\lambda)$, h, i, j be as required for Definition 2.1. Let $\underset{\sim}{G}_\beta$ be the P_β-name of the generic subset of P_β. Without loss of generality $\overline{Q} \in N_i \cap N_j$ (because $P_{\alpha_0} \in N_i \cap N_j$), hence $\alpha_0 \in N_i \cap N_j$; as $\alpha_0 < \kappa$, $N_i \cap \alpha_0 = N_j \cap \alpha_0$, (read Definition 2.1) and the proof is now similar to the proof of properness.

We prove by induction on $\xi \leq \alpha_0$ that:

(**) *for every* $\zeta < \xi$, $\zeta \in N_i \cap \alpha_0$, $\xi \in N_i \cap \alpha_0$ and $p \in N_i \cap P_{\alpha_0}$ or even just a P_ξ-name $\underset{\sim}{p}$ of such a condition, $q_\zeta \in P_\zeta$, $q_\zeta \geq p{\restriction}\zeta$, $q_\zeta \geq h(p){\restriction}\zeta$, such that q_ζ is (N_i, P_ζ)-generic and (N_j, P_ζ)-generic and $q_\zeta \Vdash_{P_\zeta}$ "$(\forall r \in N_i \cap P_\zeta)$ $[r \in \underset{\sim}{G}_\zeta$ iff

$h(r) \in \underset{\sim}{G}_\zeta]$", *there is a* $q_\xi \in P_\xi$ such that $q_\xi{\restriction}\zeta = q_\zeta$, $q_\xi \geq p{\restriction}\xi$, $q_\xi \geq h(p){\restriction}\xi$; q_ξ is (N_i, P_ξ)-generic and (N_j, P_ξ)-generic and $q_\xi \Vdash_{P_\xi}$ "$(\forall r \in N_i \cap P_\xi)$ $[r \in \underset{\sim}{G}_\xi$ iff $h(r) \in \underset{\sim}{G}_\xi]$".

Note that $(**)$ with p an element implies the apparently more general version with p being $\underset{\sim}{p}$, a P_ξ-name of a member of $N \cap P_{\alpha_0}$, such that $q_\xi \Vdash_{P_\xi}$ " for some $p \in N_i \cap P_{\alpha_0}$, $\underset{\sim}{p}[\underset{\sim}{G}_{P_\xi}] = p$, and $p{\restriction}\zeta \leq q$ and $h(p){\restriction}\zeta \leq q$". (Used in the inductive proof for ξ limit.)

For ξ a successor, we first, by the induction hypothesis, define $q_{\xi-1}$ as required (necessarily $\xi - 1 \in N_i \cap N_j$); then notice that, by the induction hypothesis, if we force with $P_{\xi-1}$ and get a generic $G_{\xi-1} \subseteq P_{\xi-1}$ and $q_{\xi-1}$ is in this generic set, then h is still an isomorphism etc, so we can use the κ-p.i.c. on $\underset{\sim}{Q}_{\xi-1}[G_{\xi-1}]$.

For $\mathrm{cf}(\xi) = \aleph_0$, we work as in the proof of properness (III 3.2), using the induction hypothesis. Noticing that $q_\xi \Vdash_{P_\xi}$ "$(\forall r \in P_\xi)$ $(r \in \underset{\sim}{G}_\xi$ iff $h(r) \in \underset{\sim}{G}_\xi)$" makes no problem in the limit, we do not have to take special care. For cf $(\xi) \geq \aleph_1$ the proof is similar but easier.

2) Trivial by 1) and 2.3 and the proof of III 4.1. $\square_{2.4}$

2.5 Lemma. If P is proper and $\kappa > |P|$ *then* P satisfies the κ-p.i.c.

Proof. We start with $i, j, N_i, N_j, (H(\lambda), \in, <_\lambda)$, p and κ as in 2.1. Remember that $P \in N_i \cap N_j$. In $(H(\lambda), \in, <_\lambda)$ there is a $<_\lambda$-first one-to-one function g from $|P|$ onto P, so $g \in N_i \cap N_j$. Also as $|P| \in N_i \cap N_j$, by the assumption on N_i, N_j, κ from Definition 2.1, $|P| < i, j$ and hence $N_i \cap |P| = N_j \cap |P|$. Hence (using the function g), $N_i \cap P = N_j \cap P$ and so h is the identity on $P \cap N_i$. Now it may well be that there is an $\mathcal{I} \in N_i$, a pre-dense subset of P, which does not belong to N_j. But if q is (N_j, P)-generic then for any $\mathcal{I} \in N_i$ a pre-dense subset of P, $h(\mathcal{I}) \in N_j$ is a pre-dense subset of P and $\mathcal{I} \cap N_i = \mathcal{I} \cap (P \cap N_i) = h(\mathcal{I}) \cap (P \cap N_j)$, hence $\mathcal{I} \cap N_i = h(\mathcal{I}) \cap N_j$ is pre-dense above q. So q is (N_i, P)-generic too and we can use the properness of P to define q as required (with clause (b) of Definition 2.1 being trivial). $\square_{2.5}$

2.6 Definition. The κ-p.i.c* is defined similarly to κ-p.i.c, but we add one assumption:
for any $a \in N_i$ there is a sequence $\langle a_\alpha : \alpha < \kappa \rangle$ in $N_i \cap N_j$ such that $a_i = a$ (this implies the corresponding condition on N_j); equivalently N_i is the Skolem hull of $(N_i \cap N_j) \bigcup \{i\}$, and N_j is the Skolem Hull of $(N_i \cap N_j) \bigcup \{j\}$.

2.7 Lemma.

1) The κ-p.i.c. implies the κ-p.i.c*.

2) Lemmas 2.2–2.5 hold for κ-p.i.c*, (and we call them 2.2*,..., respectively).

3) P satisfies the κ-p.i.c* if P satisfies the conditions from [Sh:80] which are:

a) P is $\aleph_1$-complete.

b) for any $p_i \in P$ $(i < \kappa)$ there are $p_i^\dagger \in P, p_i \leq p_i^\dagger$ and pressing down functions $F_n : \kappa \setminus \{0\} \to \kappa$, (i.e., $F_n(\alpha) < \alpha$) for $n < \omega$, such that: if $i < j$ and $\bigwedge_n F_n(i) = F_n(j)$, *then* $p_i^\dagger, p_j^\dagger$ have a least upper bound in P, called $p_i^\dagger \wedge p_j^\dagger$.

2.7A Remark. So 2.7(2), (3), 2.4*, 2.4* give an alternative proof of [Sh:80], for the case $\alpha_0 \leq \kappa$. In fact, 2.4* holds for α_0 not necessarily $< \kappa$, when each $\underset{\sim}{Q}_i$ is $\aleph_1$-complete, and this gives an alternative axiom for [Sh:80].

Proof. 1) Trivial.

2) The least trivial part is 2.3. Here the extra assumption is the least obvious. So, by induction, we define $N_i^k (k < \omega)$. For $k = 0$ we choose N_i^0 such that: $\{p_i, i\} \in N_i^0 \prec (H(\lambda), \in, <_\lambda), ||N_i^0|| = \aleph_0$. Suppose N_i^k (for each i) has been defined and let $\{a_{i,e}^k : e < \omega\}$ enumerate the members of N_i^k. We choose N_i^{k+1} such that $N_i^k \in N_i^{k+1} \prec (H(\lambda), \in, <_\lambda)$, $||N_i^{k+1}|| = \aleph_0$, and $\langle \langle a_{j,e}^k : j < \kappa \rangle : e < \omega \rangle \in N_i^{k+1}$.

Now let $N_i = \bigcup_{k<\omega} N_i^k$ and proceed as in 2.3.

3) Let N_i, N_j, h be as in the definition of κ-p.i.c*, $p \in N_i$. Let $\langle \mathcal{I}_n : n < \omega \rangle$ be a list of all maximal antichains of P which belong to N_i. For every $a \in N_i$ let $\mathrm{seq}_a \in N_i \cap N_j$ be a sequence of length κ such that $a = \mathrm{seq}_a(i)$. We define,

by induction, conditions p_n:

$p_0 = p$,

if p_{2n} is defined, choose $p_{2n+1} \geq p_{2n}$, such that $p_{2n+1} \in P \cap N_i$, $p_{2n+1} \geq$ (some $q_n \in \mathcal{I}_n$),

if p_{2n+1} is defined, consider $\text{seq}_{p_{n+1}} = \langle r_{\alpha,n} : \alpha < \kappa \rangle$. We can assume w.l.o.g. $(\forall \alpha < \kappa)\ r_{\alpha,n} \in P$. So there are $\langle F_n : n < \omega \rangle$, $\langle r^{\dagger}_{\alpha,n} : \alpha < \kappa \rangle$ as mentioned in 2.7(3)(b). As $\langle r_{\alpha,n} : \alpha < \kappa \rangle \in N_i \cap N_j$ and $<_\lambda$ is a well ordering, we can assume that $\langle F_m : m < \omega \rangle$, $\langle r^{\dagger}_{\alpha,n} : \alpha < \kappa \rangle \in N_i \cap N_j$. Let $p_{2n+2} = r^{\dagger}_{i,n}$ (remember $p_{2n+1} = r_{i,n} \leq r^{\dagger}_{i,n}$). Notice that $h(r_{i,n}) = r_{j,n}$, and by the choice of N_i, N_j we have $\bigwedge_m F_m(i) = F_m(j)$. So $r^{\dagger}_{i,n} = p_{2n+2}$ and $r^{\dagger}_{j,n} = h(p_{2n+2})$ have a least upper bound $q_{2n+2} \stackrel{\text{def}}{=} p_{2n+2} \wedge h(p_{2n+2})$. In the end:

$$p_2 \leq p_4 \leq p_6 \leq \dots$$

$$h(p_2) \leq h(p_4) \leq h(p_6) \leq \dots$$

Now $q_{2n+2} \leq q_{2n+4}$, as they are *least* upper bounds. So by $\aleph_1$-completeness there is a q, $\bigwedge_n q \geq q_n$. Now q is as required. $\square_{2.7}$

2.8 Lemma. 1) All forcings used in VII §3 (= applications of Axiom II) satisfy the $\aleph_2$-p.i.c*, (but of course Levy($\aleph_1, < \kappa$) if $\kappa > \aleph_2$)

2) Moreover for each application we can find a forcing notion doing all the assigments of this kind present in the current universe and satisfies the $\aleph_2$-p.i.c*, and in fact all are $(< \omega_1)$-proper and $\mathbb{D}$-complete for a simple $\aleph_0$-completeness system $\mathbb{D}$.

Proof. We elaborate two of them leaving rest to the reader.

application F: Let $\overline{T} = \langle (T^\alpha, f^*) : \alpha < \alpha^* \rangle$ be a sequence of pairs (T, f^*), T an Aronszajn $\aleph_1$-tree, $f^* : T \to \omega$ satisfies the antecendent of $\otimes$ in VII 3.8 (in the main case: listing all of such pairs). We define a forcing notion $P_{\overline{T}}$. A member p of $P_{\overline{T}}$ has the form $p = (i, w, \overline{g}, \bar{C}, B)$, where:

(i) $w \subseteq \alpha^*$ is countable,

(ii) $i < \omega_1$, $\bar{C} = \langle C_\alpha : \alpha \in w \rangle$, C_α the characteristic function of a closed subset of $i+1$ to which i belongs

(iii) $\bar{g} = \langle g_\alpha : \alpha \in w \rangle$, g_α a function from $T^\alpha_{\leq i} = (T^\alpha)_{\leq i}$ to ω such that (g_α, C_α) is as in VII 3.11.

Notation: For finite $u \subseteq \alpha^*$ we letting T^u be the disjoint union of $\{T^\alpha : \alpha \in u\}$ (i.e. make them disjoint).

(iv) B is a countable family, for each member I for some finite $u = u(I) \subseteq \omega$, $(g^{[u]}, \bigcap_{\alpha \in u} C_\alpha, \{I\}) \in P_{(T^u, f^*_u)}$

application C: Use product with countable support. $\square_{2.8}$

2.8A Remark. We do not investigate the connection between κ-p.i.c, and κ-e.c.c. However, κ-e.c.c. was introduced to deal with the case in which we iterate forcings which are $\mathbb{D}$-complete for some $\mathbb{D}$. We introduce the κ-p.i.c. to deal with the case in which we want to get $V \models$ "$\aleph_2 = 2^{\aleph_0} < 2^{\aleph_1}$". So we use an iteration of length ω_2 where each iterand does not add reals. On the other hand, κ-p.i.c.* seems to replace κ-p.i.c. totally.

Note that the property of being κ-p.i.c. *essentially* (but seemingly not formally) implies properness.

2.9 Claim. If $\alpha_0 < \kappa$, $\langle P_\alpha, \underset{\sim}{Q}_\alpha : \alpha < \alpha_0 \rangle$ is a CS iteration, $\alpha_0 < \kappa$, each Q_α satisfies the κ-p.i.c* and $(\forall \mu < \kappa)\mu^{\aleph_0} < \kappa$; *then,* in $V^{P_{\alpha_0}}$ we have $(\forall \mu < \kappa)\mu^{\aleph_0} < \kappa$ and $2^{\aleph_0} < \kappa$.

Proof. Trivial.

§3. The Axioms Revisited

3.1 Thesis. Proper forcing is efficient for getting models in which $2^{\aleph_0} \leq \aleph_2$ and important things it gets are such universes of set theory which in addition satisfy conditions of the form "for every $A \subseteq \omega_1 \ldots$". The reason is that we, at present, can iterate only ω_2 times without collapsing $\aleph_2$ (of course, if we are interested in c.c.c. forcing, we can increase this to $2^{\aleph_0}$). So we have a division to four main cases we can reasonably handle:

I) $2^{\aleph_0} = 2^{\aleph_1} = \aleph_2$,

II) $2^{\aleph_0} = \aleph_1, 2^{\aleph_1} = \aleph_2$,

III) $2^{\aleph_0} = \aleph_2 < 2^{\aleph_1}$,

IV) $2^{\aleph_0} = \aleph_1, \aleph_2 < 2^{\aleph_1}$.

3.2 Discussion. We have dealt in VII §2 with I), II), and have the appropriate axioms. Also in previous works we dealt with mainly I), II); sometimes we get more for free: e.g. in Laver [L1] (consistency of Borel conjecture), the value of $2^{\aleph_1}$ was immaterial.

For getting such models with some extra properties we iterate ω_2 times. At some stages we increase $2^{\aleph_1}$, or add "a few" reals (and preserve CH meanwhile (see 2.9) if $\aleph_2$ in the end is a given inaccessible κ, "few" can be interpreted as $< \kappa$) (according to the case – for I, III each time we add a few reals, for III, IV we start by adding many $A \subseteq \omega_1$). In other stages (for I and II) we consider $A \subseteq \omega_1$ and force "for" "it" some B. If we want III or IV, we consider all $A \subseteq \omega_1$ of a certain kind and simultaneously add for each such A an appropriate B. Sometimes we want the forcing to preserve something (e.g. ${}^{\omega}\omega$-boundedness, or a Ramsey ultrafilter etc.) but we shall not deal with those things here, for the number of axioms arising is not bounded.

For other possibilities of $2^{\aleph_0}, 2^{\aleph_1}$ the situation is not clear. On some consistency results see Abraham and Shelah [AbSh:114]. We get there results with $2^{\aleph_0} > \aleph_2$ but the results are on $A \subseteq \omega_1$. Resolving the problematic cases, first of all $2^{\aleph_0} \geq \aleph_3$ seems to me a major problem; we shall discuss this later.
Note: in case IV for example, we are restricted by our iterations being of length ω_2.

Generally, for getting the consistency of stronger axioms we have to assume the consistency of ZFC+ some large cardinal. Here we concentrate on assuming the consistency of ZFC only.

3.3 Notation. φ, ψ are first order sentences (in the language with $=, \in$ and one predicate P).

M is a model with universe ω_1 and language of cardinality $\leq \aleph_0$, $M = (|M|, \dots, R_i \dots)_{i<i_0\leq\omega}$. Let N denote an expansion of M, again with $\leq \aleph_0$ relations and φ a first order sentence in N's language.

3.4 Lemma. If ZFC is consistent, then so are ZFC + each one of the following axiom schema (separately):

1) $2^{\aleph_0} = 2^{\aleph_1} = \aleph_2 +$

Axiom Schema. I_b: For each (ψ, φ), for every M with universe $\subseteq H(\aleph_1)$ such that $(H(\aleph_1), \in, M) \models \psi$, there is an expansion N of M such that $N \models \varphi$, provided that

$(*)_{I_b}$: the following is provable from ZFC + G.C.H.: *if* $(H(\aleph_1), \in, M) \models \psi$ *then* for some proper forcing notion P satisfying the $\aleph_2$-p.i.c.*, $|P| \leq \aleph_2$, $\Vdash_P$ "there is an expansion N of M, N satisfying φ and $2^{\aleph_0} = \aleph_1$".

2) $2^{\aleph_0} = \aleph_1 + 2^{\aleph_1} = \aleph_2$ (+ G.C.H. if you want) +

Axiom Schema. II_b: For each (ψ, φ), for every M with universe $\subseteq H(\aleph_1)$ such that $(H(\aleph_1), \in, M) \models \psi$, there is an expansion N of M such that $N \models \varphi$, provided that

$(*)_{II_b}$ the following is provable from ZFC + G.C.H.: *if* $(H(\aleph_1), \in, M) \models \psi$ *then* for some $(< \omega_1)$-proper P, $\mathbb{D}$-complete for some simple $\aleph_0$-completeness system $\mathbb{D}$, satisfying the $\aleph_2$-p.i.c.*, $|P| \leq \aleph_2$ and $\Vdash_P$ " there is an expansion N of M satisfying φ". We can use here $H(\aleph_2)$ instead of $H(\aleph_1)$.

3) $2^{\aleph_0} = \aleph_1 + 2^{\aleph_1} =$ any cardinality of cofinality $\geq \aleph_2 +$

Axiom Schema. IV_b: For each (ψ, φ), for every M with universe $\subseteq H(\aleph_1)$ such that $(H(\aleph_1), \in, M) \models \psi$, there is an expansion N of M such that $N \models \varphi$, provided that

$(*)_{IV_b}$ the following is provable in ZFC + CH: there is a $(< \omega_1)$-proper forcing notion P, $\mathbb{D}$-complete for some simple $\aleph_0$-completeness system $\mathbb{D}$, satisfying the

$\aleph_2$-p.i.c.*, $|P| \leq 2^{\aleph_1}$, such that $\Vdash_P$ "for every $M \in V$, if $(H(\aleph_1)^V, \in, M) \models \psi$, then there is (in V^P) an expansion N of M satisfying φ".

4) $2^{\aleph_0} = \aleph_2 < 2^{\aleph_1} =$ anything of cofinality $\geq \aleph_2$+

Axiom Schema. III_b: For each (ψ, φ), for every M with universe $\subseteq H(\aleph_1)$ such that $(H(\aleph_1), \in, M) \models \psi$, there is an expansion N of M satisfying φ, provided that

$(*)_{III_b}$: the following is provable from ZFC + CH: there is a proper forcing P satisfying the $\aleph_2$-p.i.c*., $|P| \leq 2^{\aleph_1}$ such that:

$\Vdash_P$ "CH and for every $M \in V$ with universe $\subseteq H(\aleph_1)$,
if $(H(\aleph_1)^V, \in, M) \models \psi$, then
there is (in V^P) an expansion N of M satisfying φ".

Proof. Straightforward by now, when we use the relevant theorems on forcing. $\square_{3.4}$

3.4A Remarks on 3.4(3).

A) Notice that we use CH instead of G.C.H, and we here put first $\exists P$ and then $\forall M$. We can also assume that P is an iteration with countable support satisfying the above conditions.

B) If $2^{\aleph_1}$ is such that $(\forall \mu < 2^{\aleph_1})$ $[\mu^{\aleph_0} < 2^{\aleph_1}]$, we can replace $H(\aleph_1)$, by $H(2^{\aleph_1})$.

Of course, as we use "larger" cardinals κ to be collapsed to $\aleph_2$, we can get stronger axioms:

3.5 Lemma. If "ZFC + $\exists$ an inaccessible cardinal" is consistent, then so are "ZFC + each of the following" (separately).

1) $2^{\aleph_0} = 2^{\aleph_1} = \aleph_2$+

Axiom Schema. I_a: like I_b, but we replace (in $(*)_{I_b}$ the demand) "$\ldots P$, satisfying the $\aleph_2$-p.i.c.*", by "$\ldots P$, $|P| <$ first strongly inaccessible".

2) G.C.H. +

Axiom Schema. II_a: like II_b, but we replace "$\ldots, P$ satisfies the $\aleph_2$-p.i.c.*, $|P| \leq \aleph_2$" by "$\ldots, |P| <$ first strongly inaccessible".

3) $2^{\aleph_0} = \aleph_1 + 2^{\aleph_1} =$ first inaccessible +

Axiom Schema III_a. Like III_b but we replace "$\ldots P$, P satisfies the $\aleph_2$-p.i.c.*, $|P| \leq 2^{\aleph_1}$" by "$\ldots P$, $|P| <$ first inaccessible".

4) $2^{\aleph_0} = \aleph_2 = 2^{\aleph_1} =$ first inaccessible +

Axiom Schema IV_a. like IV_b but we replace "$\ldots, P$ satisfying the $\aleph_2$-p.i.c.*" by "$\ldots P, |P| <$ first inaccessible".

3.5A Remark. 1) Our use here of I_a is not the same as in VII §2, but we can take their union as I_a.

2) We can replace in this section "$\mathbb{D}$ simple $\aleph_0$-completeness system" by $\mathbb{D} \in V^{P_0}$ is a 2-completeness system (see §4).

3) We can replace in 3.5 "$|P| < \kappa$" by "$P \models \kappa$-p.i.c.*".

Proof. Again easy.

Remark. We can of course try more axioms, but those mentioned above seem to suffice.

§4. More on Forcing Not Adding ω-Sequences and on the Diagonal Argument

4.1 Discussion. We have proved in VII §3, application D, that CH $\not\to \Phi_{\aleph_1}^{\aleph_0}$ whereas in Devlin and Shelah [DvSh:65] (or see somewhat more [Appendix §1]) it is shown that CH $\to \Phi_{\aleph_1}^2$. But Axiom II does not prove the consistency of "not $\Phi_{\aleph_1}^3$" with CH. More generally, we can ask whether we can make the condition on the Q_α (in V 7.1, 7.2) weaker.

We saw no point in trying to weaken the assumption "α-proper for every $\alpha < \omega_1$" to e.g. ω-proper, as it seemed to us that every natural example of forcing will satisfy it (truly, sometimes we want to destroy some stationary subsets of ω_1 and under reasonable conditions we can succeed, for example, the proof the consistency of "the closed unbounded filter on ω_1 is precipitous" (see Jech, Magidor, Mitchell and Prikry [JMMP]), but we can amalgamate such a proof with our constructions). The hard part seems to be the $\mathbb{D}$-completeness, where $\mathbb{D} \in V$ or $\mathbb{D}$ is simple, again, do not seem to be a serious obstacle to anything; but the requirement that any finitely many possibilities are compatible (i.e. $\mathbb{D}_{\langle N,P,p\rangle}$ generates a filter) seemed to be an obstacle – e.g. to the natural forcing for making $\Phi^3_{\aleph_1}$ false. Remember, $\mathbb{D}_{\langle N,\dots\rangle}$ was a family of subsets of $\mathcal{P}(N)$ with the finite intersection property. We shall try to replace this requirement by the requirement that the intersection of any two is nonempty. As an application we get consistency with CH of variants of $\Phi^3_{\aleph_1}$ (we hope there will be more). Note that we replace here $\mathbb{D}_{\langle N,P,p\rangle}$ by another equivalent formulation.

Note another drawback, which at present is only aesthetical, the $\mathbb{D}$-completeness is not preserved; i.e. we have not stated a natural condition, preserved by CS iterations, that implying that no ω-sequence of ordinals is added. Note that we do not use the full generality of Definition 4.2. We treat it in 4.14—4.22. A minor difference with Chapter V is that we use countable subsets of some λ instead of $H(\lambda)$, but this is just a matter of presentation.

For another point of view and more results see [Sh:177] or better yet XVIII §1, §2 .

4.2 Definition. 1) Let λ be a cardinal. $\mathbb{D}$ is called a $(\lambda, 1, k)$-system (or completeness system) if:

(i) $\mathbb{D}$ is a function (where $\mathbb{D}(x)$ may be written as $\mathbb{D}_x$),

(ii) $\mathbb{D}_{\langle a,<^*,x,p\rangle}$ is well defined iff $a \in \mathcal{S}_{\aleph_0}(\lambda)$, $<^*$ is a partial order of a, $x \subseteq a \times a$ and $p \in a$,

(iii) $\mathbb{D}_{\langle a,<^*,x,p\rangle}$ is a family of subsets of $\mathcal{P}(a)$, the intersection of any i of them is nonempty when $i < 1 + k$.

2) We say that $\mathbb{D}$ is a k-completeness system if for some λ it is $(\lambda, 1, k)$-completeness system.

4.3 Definition. A forcing notion P is called $(\mathcal{D}, \mathbb{D})$-complete, where $\mathcal{D}$ is a filter over $\mathcal{S}_{\aleph_0}(\lambda)$ and $\mathbb{D}$ a $(\lambda, 1, k)$-system, if $\lambda \geq 2^{|P|}$, P is isomorphic to $P^* = (P^*, \leq^*)$, $P^* \subseteq \lambda$, and for every $p \in P^*$ and $\mathcal{I}_\alpha = \{p_i^\alpha : i < i_\alpha \leq \lambda\}$ pre-dense subsets of P^* (for $\alpha < \lambda$), for some $x \subseteq \lambda \times \lambda$, the family of all $a \in \mathcal{S}_{\aleph_0}(\lambda)$ that satisfy the following, is in $\mathcal{D}$:

$(*)$ the following contains, as a subset, a member of $\mathbb{D}_{\langle a, <^* \restriction a,\ x \restriction a \times a,\ p \rangle}$:

$\{G \subseteq a$: 1) for every $\alpha \in a$, for some $i \in a \cap i_\alpha$, $p_i^\alpha \in G$
2) $(\exists q \in P^*)(\forall r \in G)(r \leq^* q)$
3) $p \in G\}$.

If $\mathcal{D} = \mathcal{D}_{<\aleph_1}(\lambda)$ (see V §2) we may omit it and write $\mathbb{D}$ instead of $(\mathcal{D}, \mathbb{D})$. If $\mathcal{D} = \mathcal{D}_{<\aleph_1}(\lambda) + S$, where $S \subseteq \mathcal{S}_{<\aleph_1}(\lambda)$ is stationary, then we may write S instead of $\mathcal{D}$.

4.3A Remark.

1) In Definitions 4.2, 4.3 we can replace λ by any set of this cardinality or by any larger cardinality.
2) We omit $\mathcal{D}$ in 4.5 below from laziness only.
3) In 4.4 and 4.3 of [Sh:b] we use $\bar{a} \in \mathcal{S}_{\aleph_0}^\alpha(\lambda)$ for some fixed α, but we use only the case $\alpha = 1$ in Theorem 4.5; to help the reader we delay this generality to a later part of this section.
4) If $\mathcal{D}_{\lambda_0}^1$ is a fine normal filter on $\mathcal{S}_{<\aleph_1}(\lambda_0)$, and $\lambda_1 \geq \lambda_0$, we let $\mathcal{D}_{\lambda_1}^1$ be the fine normal filter on $\mathcal{S}_{<\aleph_1}(\lambda_1)$ generated by

$$\left\{\{a \in \mathcal{S}_{<\aleph_1}(\lambda_1) : a \cap \lambda_0 \in X\} : X \in \mathcal{D}_{\lambda_0}^0\right\}.$$

5) If $S \subseteq \mathcal{S}_{<\aleph_1}(\lambda)$ is stationary we may write S instead of $\mathcal{D}_{<\aleph_1}(\lambda) + S$.
6) Instead of $\mathcal{I}_\alpha$ for $\alpha < \lambda$ we can use $\mathcal{I}_\alpha$ for $\alpha < \alpha^*$, $\alpha^* \leq \lambda$.

4.3B Fact. 1) In Definition 4.3, any choice of $(P^*, \leq^*)$ gives an equivalent definition. Also we can increase λ in a natural way.

2) Definitions 4.3 and 4.4 (simplicity) are compatible with the definition of completeness systems from V 5.2, 5.3, 5.5.

Specifically:

(A) For a forcing notion P, $k \leq \aleph_1$ and a family $\mathcal{E}$ of subsets of $\mathcal{S}_{<\aleph_1}(\mu)$, the following are equivalent:

(i) For some $\lambda \geq \mu$, P is $(\mathcal{D}_{<\aleph_1}(\lambda) + \mathcal{E}, \mathbb{D})$-complete for some simple $(\lambda, k, 1)$-completeness system $\mathbb{D}$ in the sense of 4.3, 4.4 below, where $\mathcal{D}_{<\aleph_1}(\lambda) + \mathcal{E}$ is the fine normal filter on $\mathcal{S}_{<\aleph_1}(\lambda)$ generated by

$$\{\{a \in \mathcal{S}_{<\aleph_1}(\lambda) : a \cap \mu \in X\} : X \in \mathcal{E}\}.$$

(ii) P is $(\mathcal{E}, \mathbb{D})$-complete for some simple k-completeness system in the sense of V §5.

(B) For a forcing notion P, $k \leq \aleph_1$, $\mathcal{E}$ a family of subsets of $\mathcal{S}_{<\aleph_1}(\mu)$, and a subuniverse V_0 (V_0 a transitive sub-class of V containing all ordinals and being a model of ZFC) such that $\mathcal{S} \in V_0$, the following are equivalent:

(i) For some $\lambda \geq \mu$ and $(\lambda, k, 1)$-completeness system $\mathbb{D} \in V_0$, the forcing notion P is $(\mathcal{D}_{<\aleph_1}(\lambda) + \mathcal{E}, \mathbb{D})$-complete in the sense of V §5.

(ii) For some $\lambda \geq \mu$ and k-complete system $\mathbb{D}$ which is almost simple over V_0, the forcing notion P is $(\mathcal{E}, \mathbb{D})$-complete in the sense of V §5.

Proof. Straightforward. $\square_{4.3B}$

4.4 Definition. We call $\mathbb{D}$ simple if for some first order formula $\psi(v, u)$, $\mathbb{D}_{\langle a, <^*, x, p\rangle} = \{\{G \subseteq a : p \in G, \langle a \bigcup \mathcal{P}(a \times a), \in \restriction a, <^*, x, p\rangle \models \psi(G, u)\} : u \subseteq a\}$, ($\psi$ can have a countable sequence of ordinals as a parameter).

4.5 Theorem. If $\langle P_\alpha, \underset{\sim}{Q}_\alpha : \alpha < \alpha_0\rangle$ is an iteration with countable support and each $\underset{\sim}{Q}_\alpha$ is β-proper, for every $\beta < \omega_1$ and $\mathbb{D}^\alpha$-complete for some $(\lambda_\alpha, 1, 2)$-system, $\mathbb{D}^\alpha \in V$ (possibly $\mathbb{D}^\alpha, \lambda_\alpha$ are actually P_α-names $\underset{\sim}{\mathbb{D}}^\alpha, \underset{\sim}{\lambda}_\alpha$ but it does not matter), *then* forcing with P_{α_0} does not add any new ω-sequences of ordinals.

Note:

4.6 Claim. Any simple system in V^{P_α} (where P_α is a forcing notion adding no new ω-sequences) is in V.

Note:

4.6A Remark. Every $\underset{\sim}{Q}_\alpha$ is ${}^\omega\omega$-bounding because it is $\mathbb{D}^\alpha$-complete, which implies it does not add (to V^{P_α}) reals and even does not add ω-sequences of ordinals.

Proof of 4.5. We prove some claims and then the theorem becomes obvious (Claim 4.10 is the heart of the matter).

4.7 Definition. Let $\overline{A} = \langle A_i : i \leq \beta \rangle$, β countable, each A_i is a countable set of ordinals, $A_i (i \leq \beta)$ is (strictly) increasing and continuous. For $\xi \leq \zeta, \xi \in A_0, \zeta \in A_0, \overline{A}$ as above, we define when $\overline{A}$ *is long for* (ξ, ζ), by induction on ζ:

Case (i). $\zeta = \xi$: $\overline{A}$ is long for (ξ, ζ) (under the assumptions above) if $\beta > 0$.

Case (ii). ζ a successor, $\zeta > \xi$: $\overline{A}$ is long for (ξ, ζ) if for some $\beta^\dagger < \beta$ we have that $\langle A_i : i \leq \beta^\dagger \rangle$ is long for $(\xi, \zeta - 1)$.

Case (iii). ζ a limit: $\overline{A}$ is long for (ξ, ζ) if there are $\beta_i (i \leq \omega^2)$ (the ordinal square of ω) such that: $i < j \leq \omega^2 \Rightarrow \beta_i < \beta_j$; $\beta_{\omega^2} + \omega + 1 < \beta$; and for every i and (ξ_1, ζ_1) we have: $\xi_1 \in A_{\beta_i}, \zeta_1 \in A_{\beta_i}$, $\xi \leq \xi_1 \leq \zeta_1 < \zeta$ and $i < \omega^2$, implies that $\langle A_j : \beta_i + 2 \leq j \leq \beta_{i+1} \rangle$ is long for (ξ_1, ζ_1).

4.8 Claim.

1) If $\xi < \zeta \in A_0$, $\langle A_i : i \leq \beta \rangle$ is as in the assumptions of Definition 4.7, $\beta_0 < \beta_1 \leq \beta$, and $\langle A_i : \beta_0 \leq i \leq \beta_1 \rangle$ is long for (ξ, ζ) *then* $\langle A_i : i < \beta \rangle$ is long for (ξ, ζ).
2) $\langle A_i : i \leq \beta \rangle$ is long for (ξ, ζ) iff $\langle A_i \cap (\zeta + 1) \setminus \xi : i \leq \beta \rangle$ is long for (ξ, ζ).

Proof. 1) By induction on ζ, and there are no problems.

2) Easy. $\square_{4.8}$

4.9 Claim. Let $\lambda \geq \omega_1$, $\beta < \omega_1$ and for $i < \beta$ we have $N_i \prec (H(\lambda), \in)$, and N_i are countable increasing continuous and $\langle N_j : j \leq i \rangle \in N_{i+1}$ and $\xi < \zeta \in N_0$. *Then* we can find an α such that $\beta \leq \alpha < \omega_1$, and a countable N_i for $\beta \leq i \leq \alpha$ such that $N_i \prec (H(\lambda), \in)$, $\langle N_j : j \leq i \rangle \in N_{i+1}$ for $i < \alpha$, N_i are countable increasing continuous in i and $\langle N_i \cap \lambda : i < \alpha \rangle$ is long for (ξ, ζ).

Proof. Again by induction on ζ.

4.10 Claim. Suppose $\underset{\sim}{Q}_i, P_i, \alpha_0$ are as in Theorem 4.5, λ is large enough and for $i \leq \beta (< \omega_1)$ $N_i \prec (H(\lambda), \in, <^*_\lambda)$ is countable, increasing, continuous and

$$\langle N_j : j \leq i \rangle \in N_{i+1},\ \xi \leq \zeta \in N_0 \cap (\alpha_0 + 1),\ \langle P_i, \underset{\sim}{Q}_i : i < \alpha_0 \rangle \in N_0.$$

Suppose further that $\langle N_i \cap \alpha_0 : i \leq \beta \rangle$ is long for (ξ, ζ), $G_e (e = 0, 1)$ are directed subsets of $P_\xi \cap N_\beta$, $r_e \in P_\xi$ (for $e = 0, 1$), $(\forall q \in G_e) q \leq r_e$, $G_0 \cap N_0 = G_1 \cap N_0$ and $i \leq \beta \Rightarrow \mathcal{I} \cap G_e \cap N_i \neq \emptyset$ for every pre-dense $\mathcal{I} \subseteq P_\xi$ with $\mathcal{I} \in N_i$ and $e = 0, 1$. Suppose also $p \in P_\zeta \cap N_0$ and $p \restriction \xi \in G_0$.

Then there is a directed $G^* \subseteq P_\zeta \cap N_0$ such that $G_0 \cap N_0 \subseteq G^*$, G^* not disjoint to any pre-dense $\mathcal{I} \subseteq P_\zeta, \mathcal{I} \in N_0, p \in G^*$, and $r_e \Vdash_{P_\xi}$ "$\{q \restriction [\xi, \zeta) : q \in G^*\}$ has an upper bound in P_ζ / P_ξ" for $e = 0, 1$.

Proof. By induction on ζ (for all $\xi \leq \zeta$).

Note that the assertion (for $\xi = 0$) implies that forcing by P_ζ does not add ω-sequences of ordinals.

Also note if $p \leq q$ are in $P_\xi \cap N_0$, $q \in G_0 \cap N_0$ then $p \in G_0 \cap N_0$ (as $\{r \in P_\xi : r \geq p$ or r, p incompatible in $P_\xi\}$ is pre-dense in P_ξ and belongs to N_0); similarly for N_β instead N_0 and/or for G_1 instead of G_0. Also, for $e = 0, 1$ and $i < \beta$, we have $G_e \cap N_i \in N_{i+1}$. Why?

This is easy, as the set $\mathcal{I} \subseteq P_\xi$, defined below, is pre-dense and belongs to N_{i+1}, hence is not disjoint from $G_e \cap N_{i+1}$. So there is an $r^\dagger \in \mathcal{I} \cap N_{i+1}$ with

$r^\dagger \in G_e \cap N_{i+1}$; but by the assumption $(\forall q)[q \in G_e \cap N_i \to q \leq r_e]$ and $r^\dagger \leq r_e$. Hence, by the definition of $\mathcal{I}^0$ below, we know that $r^\dagger \notin \mathcal{I}^0$, so $r^\dagger \in \mathcal{I}^1$ (see below), and necessarily $\{q \in N_i \cap P_\xi : q \leq r^\dagger\} = G_e \cap N_i$. So $G_e \cap N_i \in N_{i+1}$ as it is defined from parameters $(N_i, P_\xi, r^\dagger)$ in it. Here is the definition of $\mathcal{I}$:

$$\mathcal{I} = \mathcal{I}^0 \bigcup \mathcal{I}^1, \text{ where}$$

$$\begin{aligned}\mathcal{I}^0 = \{q \in P_\xi :\ & \text{there are no } r \in P_\xi, r \geq q \text{ and } G \subseteq N_i \cap P_\xi \text{ such that} \\ & (\forall p' \in G) \quad p' \leq r \text{ and} \\ & (\forall \mathcal{J} \in N_0)\ [\mathcal{J} \text{ pre-dense in } P_\xi \to \mathcal{J} \cap G \neq \emptyset]\} \\ \mathcal{I}^1 = \{q \in P_\xi :\ & \text{there is a } G \subseteq N_0 \cap P_\xi \text{ such that } (\forall p' \in G) \quad p' \leq q \text{ and} \\ & (\forall \mathcal{J} \in N_0)[\mathcal{J} \text{ pre-dense in } P_\xi \to \mathcal{J} \cap G \neq \emptyset]\}.\end{aligned}$$

Case (i): $\zeta = \xi$.

Trivial. Just let $G^* = G_0 \cap N_0$.

Case (ii): ζ a successor ordinal $> \xi$.

So by Definition 4.7, for some $\gamma < \beta$, $\langle N_i \cap \lambda : i \leq \gamma \rangle$ is long for $(\xi, \zeta - 1)$. For $e = 0, 1$, we can find $r^a_e \in G_e \cap N_{\gamma+1}$ such that r^a_e is above every member of $G_e \cap N_\gamma$ (see the proof above). Hence, by the induction hypothesis, there is a $G^* \subseteq N_0 \cap P_{\zeta-1}$ such that:

(a) for every pre-dense $\mathcal{I} \subseteq P_{\zeta-1}, \mathcal{I} \in N_0$, the intersection $G^* \cap \mathcal{I}$ is not empty,

(b) $r^a_e \Vdash_{P_\xi}$ "G^* has an upper bound in $P_{\zeta-1}/P_\xi$",

(c) $G_e \cap N_0 = G^* \cap P_\xi$.

Now without loss of generality G^* is definable (in $(H(\lambda), \in, <^*_\lambda)$) from the parameters $\overline{Q}$, $\langle N_i : i \leq \gamma \rangle$, r^a_0, r^a_1, ζ, ξ, hence

(d) $G^* \in N_{\gamma+1}$.

Now we want to define $G^{**} \subseteq P_\zeta \cap N_0$ as in the conclusion of the claim. G^* determines $\mathbb{D}^{\zeta-1} \in V$, as $\zeta \in N_0$ (i.e. some member of G^* forces $(\Vdash_{P_{\zeta-1}})$ $\underset{\sim}{\mathbb{D}}^{\zeta-1}$ to be some $\mathbb{D}^{\zeta-1}$). Let $G^* \subseteq G^f_{\zeta-1} \subseteq P_{\zeta-1}$ where $G^f_{\zeta-1}$ is generic (over V). Clearly, some members of $G^f_{\zeta-1}$ force $N_0 \cap \lambda_{\zeta-1}$ ($\lambda_{\zeta-1}$ is from the definition of $\mathbb{D}^{\zeta-1}$) to be in the appropriate closed unbounded subset of $\mathcal{S}_{\aleph_0}(\lambda_{\zeta-1})$, so we "know" $\mathbb{D} = \mathbb{D}^{\zeta-1}$ and it belongs to $N_{\gamma+1}[G^f_{\zeta-1}]$ (and to V; why not to N_1? we need G^*; "know" means independently of the particular choice of $G^f_{\zeta-1}$). In $V^{P_{\zeta-1}}$, let $(Q^*_\zeta, <^*_\zeta) \overset{h}{\cong} (Q_\zeta, <)$, $Q^*_\zeta \subseteq \lambda$, $x \subseteq \lambda \times \lambda$ code a list $\langle \mathcal{J}_\alpha = \langle p^\alpha_i : i < i_\alpha \rangle : \alpha < \alpha^* \rangle$ of the pre-dense subsets of $(Q^*_{\zeta-1}, <^*_{\zeta-1})$ from Definition 4.3 and x is as in Definition 4.3. They may be $P_{\zeta-1}$-names, but without loss of generality $\langle \underset{\sim}{\mathcal{J}}_\alpha = \langle \underset{\sim}{p}^\alpha_i : i < \underset{\sim}{i}_\alpha \rangle : \alpha < \underset{\sim}{\alpha}^* \rangle$, $(\underset{\sim}{Q}^*_{\zeta-1}, \underset{\sim}{\leq}^*_{\zeta-1}), \underset{\sim}{h}$ and $\underset{\sim}{x}$ belong to $N_0[G^f_{\zeta-1}]$. [Why? By III 2.13.] And we can compute $\underset{\sim}{x} \restriction (N_0 \cap \lambda_{\zeta-1})$, $(\underset{\sim}{Q}^*_\zeta, \underset{\sim}{\leq}^*_{\zeta-1}) \restriction (N_0 \cap \lambda_{\zeta-1})$ so that $N_{\gamma+1}[G^f_{\zeta-1}] \prec (H(\lambda), \in, <^*_\lambda)$. Finally, let $y = \langle N_0 \cap \lambda_{\zeta-1}, \underset{\sim}{\leq}^*_{\zeta-1} \restriction (N_0 \cap \lambda_{\zeta-1}), \underset{\sim}{x} \restriction ((N_0 \cap \lambda_{\zeta-1}) \times (N_0 \cap \lambda_{\zeta-1})), \underset{\sim}{h}(p) \rangle$. There is a set $A \in \mathbb{D}_y$ such that (in $V[G^f_{\zeta-1}]$) for any $G \in A$, we have that $G \subseteq N_0 \cap \underset{\sim}{Q}^*_{\zeta-1}[G^f_{\zeta-1}]$ is not disjoint to any pre-dense $\mathcal{I} \subseteq Q^*_{\zeta-1}$ which belongs to $N_0[G^f_{\zeta-1}]$ and G has an upper bound in $\underset{\sim}{Q}^*_{\zeta-1}[G^f_{\zeta-1}]$. Note that $G^f_{\zeta-1} \subseteq P_{\zeta-1}$ is an arbitrary generic set which includes G^*, so there is a $P_{\zeta-1}$-name $\underset{\sim}{\tau}$ such that $\Vdash_{P_{\zeta-1}}$ " if there is an A as above then $\underset{\sim}{\tau}$ names it ". Again without loss of generality $\underset{\sim}{\tau} \in N_{\gamma+1}[G^f_{\zeta-1}]$.

Now $\mathcal{I}^* = \{q \in P_{\zeta-1} : q \text{ forces } \underset{\sim}{\tau} \text{ to be some specific } A \in \mathbb{D}_y\}$ is pre-dense in $P_{\zeta-1}$ and belongs to $N_{\gamma+1} \subseteq N_\beta$. So, by the assumptions on G_e, there are r^*_e for $e = 0, 1$, such that $r^*_e \in P_{\zeta-1} \cap N_\beta$, $r^*_e \restriction \xi \in G_e$, $(\forall q \in G^*)(q \leq_{P_{\zeta-1}} r^*_e)$ and $r^*_e \Vdash_{P_{\zeta-1}}$ "$\underset{\sim}{\tau} = A_e$", where $A_e \in \mathbb{D}_y, A_e \in N_\beta$. By the hypothesis on $\mathbb{D}^{\zeta-1}$, $A_0 \cap A_1 \neq \emptyset$ so as A_0, $A_1 \in N_\beta$ there is a $G^\otimes \in A_0 \cap A_1 \cap N_\beta$.

Then

$$G^{**} = \{q : q \in N_0 \cap P_\zeta \text{ and } q \restriction (\zeta - 1) \in G^* \text{ and for some } r \in G^\otimes, \\ r \Vdash_{P_{\zeta-1}} \text{"}h(q(\zeta-1)) = r\text{"}\},$$

is as required (it is well defined as though $\underset{\sim}{h}$ is a $P_{\zeta-1}$-name, it belongs to N_0 hence $\underset{\sim}{h} \restriction N_0$ can be computed from G^*).

Case (iii). ζ a limit.

So there are $\beta_i (i \leq \omega^2)$ such that $i < j \leq \omega^2 \to \beta_i < \beta_j$, $\beta_{\omega^2} + \omega + 1 < \beta$ and for any (ξ_1, ζ_1) in N_{β_i} if $\xi \leq \xi_1 \leq \zeta_1 < \zeta$ then $\langle N_j \cap \zeta : \beta_i + 2 \leq j \leq \beta_{i+1} \rangle$ is long for (ξ_1, ζ_1) and $\langle \beta_i : i \leq \omega^2 \rangle$ is increasing continuous (we can assume this by 4.8(1)).

Let $\beta^* = \beta_{\omega^2}$ and w.l.o.g. we can assume $\langle \beta_i : i \leq \omega^2 \rangle \in N_{\beta^*+1}$ (as there is such a sequence in N_{β^*+1}, because it exists in V). Similarly $i < \omega^2 \Rightarrow \langle \beta_j : j \leq i \rangle \in N_{\beta_i+1}$. Choose $\zeta_n \in N_0 \cap \zeta$ such that $\xi = \zeta_0 < \zeta_1 < \dots \zeta_n < \zeta_{n+1} \dots < \zeta$ and $[\gamma \in N_0 \;\&\; \gamma < \zeta \Rightarrow \bigvee_n \gamma < \zeta_n]$.

Now we define by induction on $n < \omega$, $k_n = k(n) < \omega$, G^n_η (for $\eta \in {}^{(k(n))}2$), r^*_n, $\underset{\sim}{\eta}_n$ such that:

1) r^*_n is a member of $P_{\zeta_n} \cap N_{\beta^*+\omega+1}$ with domain $\zeta_n \cap N_{\beta^*+\omega}$, $r^*_0 \leq r_0, r^*_0 \leq r_1$ (see below for formal problems or let $r^*_0 = \emptyset$ (so $\mathrm{Dom}(r^*_0) = \emptyset$) but then we use $r^*_n \cup r_e$ to force anything),
2) $r^*_{n+1} \restriction \zeta_n = r^*_n$, (or you can say that r^*_n is a P_ζ-name of such a condition with r_0, r_1 deciding the value but see (8) or the beginning of the proof below)
3) if $n + 2 \leq m < \omega$, r^*_n is $(N_{\beta^*+m}, P_{\zeta_n})$-generic,
4) if $\mathcal{I} \subseteq P_{\zeta_n}$ is a maximal antichain, $\mathcal{I} \in N_{\beta^*+1}$ *then* for some finite $\mathcal{J} \subseteq \mathcal{I} \cap N_{\beta^*+1}$, $\mathcal{J}$ is pre-dense above r^*_n (note that, as described in the beginning of the proof of 4.10, this implies that the function giving $\mathcal{J}$ (from $\mathcal{I}$) belongs to N_{β^*+n+2}),
5) $k_0 = 1$, $G^0_{\langle e \rangle} = G_e \cap N_{\beta^*}$,
6) $G^n_\eta \subseteq P_{\zeta_n} \cap N_{\beta^*}, G^n_\eta \in N_{\beta^*+1}$ for $\eta \in {}^{k(n)}2$,
7) if $m < n$ and $\eta \in {}^{k(n)}2$, then $G^m_{\eta \restriction k(m)} = G^n_\eta \cap P_{\zeta_m}$,
8) $\underset{\sim}{\eta}_n$ is a P_{ζ_n}-name which belongs to N_{β^*+1},

$$r^*_n \Vdash \text{“}\underset{\sim}{\eta}_n \in {}^{k(n)}2\text{”},$$

$$r^*_{n+1} \Vdash \text{“}\underset{\sim}{\eta}_{n+1} \restriction k_n = \underset{\sim}{\eta}_n\text{”},$$

$$r^*_n \Vdash \text{“}G^n_{\underset{\sim}{\eta}_n} \text{ is included in the generic subset of } P_{\zeta_n}\text{”},$$

9) if $j \leq \beta^*$, *but* for no $k \leq k(n)$ is $\beta_{\omega k} + 1 < j \leq \beta_{\omega(k+1)}$ *then* for every pre-dense $\mathcal{I} \subseteq P_{\zeta_n}$, $\mathcal{I} \in N_j$, and $\eta \in {}^{k(n)}2$ we have $N_j \cap \mathcal{I} \cap G^n_\eta \neq \emptyset$,
10) $\eta, \nu \in {}^{k(n)}2$, $\eta \restriction k = \nu \restriction k$, $k < k(n)$ implies $G^n_\eta \cap N_{\beta_{\omega k}+1} = G^n_\nu \cap N_{\beta_{\omega k}+1}$ and we denote both by $G^n_{\eta \restriction k}$,
11) $(\forall q \in N_{\beta^*} \cap P_{\zeta_n})\ [(\exists q^\dagger \in G^n_\eta)(q \leq q^\dagger) \rightarrow q \in G^n_\eta]$ for $\eta \in {}^{k(n)}2$.

There is no problem for $n = 0$ (how do we define r_0^*? we can assume that $\underset{\sim}{Q}_\gamma$ is closed under disjunction. Let $\varepsilon \in N_\beta \cap (\zeta + 1)$ be maximal such that $(\forall r \in N_{\beta+\omega} \cap P_\varepsilon)(r \leq r_0 \equiv r \leq r_1)$, by the definition of CS iteration it is well defined. As before we can find $r_e' \in N_{\beta^*+\omega+1}$ which is below r_e and $(\forall r \in N_{\beta+\omega} \cap P_\zeta)(r < r_e' \equiv r < r_e)$, $r_0' \restriction \varepsilon = r_1' \restriction \varepsilon$, $\varepsilon < \zeta \Rightarrow r_0' \restriction \varepsilon \Vdash_{P_\varepsilon}$ "$r_0'(\varepsilon), r_1'(\varepsilon)$ are incompatible in $\underset{\sim}{Q}_\varepsilon$" and define r_0^* as follows: $\mathrm{Dom}(r_0^*) = \mathrm{Dom}(r_0') \cup \mathrm{Dom}(r_1')$, and

$$\begin{aligned} r_0^*(\gamma) &= r_0'(\gamma) \text{ if } \gamma < \varepsilon, \\ r_0^*(\gamma) &= r_0'(\gamma) \vee r_1'(\varepsilon) \text{ if } \gamma = \varepsilon, \\ r_0^*(\gamma) &= r_1'(\gamma) \text{ if } r_1' \restriction (\varepsilon + 1) \in \underset{\sim}{G}_{P_{\varepsilon+1}}, \\ r_0^*(\gamma) &= r_0'(\gamma) \text{ otherwise.)} \end{aligned}$$

So assume we have defined for n and we shall define for $n + 1$.

First we define, by induction on $\ell \leq k(n)$, for every $\eta \in {}^{\ell}2$, the sets G_η^{n+1} (see (10) above), and we have to satisfy (9),(7), and r_n^* should force G_η^{n+1} is (bounded by) a condition, if $\eta = \underset{\sim}{\eta}_n \restriction \ell$.

This makes no problem, using the induction hypothesis on ζ and $(< \omega_1)$-properness.

Second we want to define $k(n+1), G_\eta^{n+1}$ (for $\eta \in {}^{k(n+1)}2$). Let $G_{\zeta_n} \subseteq P_{\zeta_n}$ be generic, $r_n^* \in G_{\zeta_n}$ and work for a while in $V[G_{\zeta_n}]$.

For each $p' \in N_{\beta^*}[G_{\zeta_n}] \cap (P_{\zeta_{n+1}}/G_{\zeta_n})$, there is a $G \subseteq P_{\zeta_{n+1}}/G_{\zeta_n}$, $p' \in G$, G has an upper bound, and *if* $p' \in N_j, j \leq \beta^*$ is as in (9) then $G \cap N_j$ is generic for $(N_j, P_{\zeta_{n+1}}/G_{\zeta_n})$ [equivalently, if $\mathcal{I} \subseteq P_{\zeta_{n+1}}$ is pre-dense, $\mathcal{I} \in N_j$, *then* the set $\mathcal{I} \cap N_j \cap G$ is not empty]. So there is a function F giving such a G with $\mathrm{Dom}(F) = (P_{\zeta_{n+1}}/G_{\zeta_n}) \cap N_{\beta^*}[G_{\zeta_n}]$. So, in V, we have a P_{ζ_n}-name $\underset{\sim}{F}$ for it. As its domain is countable, $\Vdash_{P_{\zeta_n}}$ "$\underset{\sim}{F} \in V$" (the domain is essentially $\subseteq N_{\beta^*}$). Also it is clear that, without loss of generality, $\underset{\sim}{F} \in N_{\beta^*+1}$ as $\langle N_i : i \leq \beta^* \rangle \in N_{\beta^*+1}$, $r_n^* \in G_{\zeta_n}$ and condition (9), (4).

So, by condition (4), there are $F_1, \ldots, F_m \in N_{\beta^*+1}$ such that $r_n^* \Vdash$ "$\underset{\sim}{F} \in \{F_1, \ldots, F_n\}$", $F_1 \ldots \in V$ (note that so their domain is computed by $G_{\underset{\sim}{\eta}_n} \ldots$).

By renaming, choose $k(n+1)$, F_η (for $\eta \in {}^{k(n+1)}2$) such that for every $\eta \in {}^{k(n)}2$ we have $\{F_1, \ldots, F_n\} = \{F_\nu : \nu \in {}^{k(n+1)}2, \nu \restriction k(n) = \eta\}$. Now for any

$\eta \in {}^{k(n+1)}2$ we can first define $G^{n+1}_\eta \cap N_{\omega k(n+1)+2}$ so that it depends on $\eta \restriction k(n)$ and not η, and then let $G_\eta = F_\eta(\bigwedge(G^{n+1}_\eta \cap N_{\beta_{\omega k_{n+1}}+2}))$.

Third we define r^*_{n+1} by V 4.5. $\square_{4.10}$

Proof of 4.5. Immediate by 4.10. $\square_{4.5}$

4.10A Remark.

1) So now everywhere we can use 4.5 instead of V 7.1 and strengthen Axiom II, II_a, II_b etc.
2) We could use shorter sequences, e.g. $\beta = \beta^* + 2$ is o.k. (see implicitly VI §1, and explicitly XVIII 2.10).
3) By easy manipulations, it does not matter in Theorem 4.3, whether $\mathbb{D}^\alpha$ is a P_α-name of a member of V, or simply a member of V (i.e. the function $\alpha \mapsto \mathbb{D}^\alpha$ is in V).

4.11 Conclusion. It is consistent with ZFC + G.C.H. that:

($*$) If $k < \omega$ and η_δ is an ω-sequence converging to δ for any limit $\delta < \omega_1$ and $\langle A_\delta : \delta < \omega_1 \rangle$ is such that $A_\delta \subseteq k$, $|A_\delta| < k/2$
then there is an $h : \omega_1 \to k$ such that for every limit $\delta < \omega_1$, $\langle h(\eta_\delta(n)) : n < \omega \rangle$ is eventually constant, and its constant value $\notin A_\delta$.

4.12 Conclusion. G.C.H. $\not\Rightarrow \Phi^3_{\aleph_1}$.

4.13 Lemma.

1) The demand on each $\underset{\sim}{Q}_\alpha$ in 4.5 follows from: $\underset{\sim}{Q}_\alpha$ is γ-proper for every $\gamma < \omega_1$, and $\mathbb{D}_\alpha$-complete for some simple 2-completeness system $\mathbb{D}_\alpha$.
2) We can demand in 4.5 that each $\underset{\sim}{Q}_\alpha$ is γ-proper for every $\gamma < \omega_1$, and $\mathbb{D}_\alpha$-complete for some almost simple 2-completeness system over V.

Remark. See Definition V 5.5.

Proof. Straightforward. $\square_{4.13}$

* * *

Finally, we indicate how to rephrase the proof in the form of a condition which is preserved by iteration.

4.14 Definition. We[†] call $E \subseteq \mathcal{S}^{<\omega_1}_{\le\aleph_0}(\lambda)$ *stationary* if for every $\chi > \lambda$ (equivalently, some $\chi > \lambda^{\aleph_0}$) and for every $x \in H(\chi)$ there is a sequence $\overline{N} = \langle N_i : i \le \alpha\rangle$ such that

(a) $\overline{N} = \langle N_i : i \le \alpha\rangle$ is an increasing continuous sequence of countable elementary submodels[††] of $(H(\chi), \in, <^*_\chi)$, $\alpha < \omega_1$ and $\overline{N}\restriction(i+1) \in N_{i+1}$ for $i < \ell g(\overline{N})$,

(b) $x \in N_0$,

(c) $\langle N_i \cap \lambda : i \le \alpha\rangle \in E$.

4.15 Definition. Let λ be a cardinal, $E \subseteq \mathcal{S}^{<\omega_1}_{\le\aleph_0}(\lambda)$ stationary, and κ a cardinal (may be finite). We call $\mathbb{D}$ a (λ, E, κ)-system if:

(A) $\mathbb{D}$ is a function (written $\mathbb{D}_x$).

(B) $\mathbb{D}_{\langle\bar{a},<^*,x,p\rangle}$ is defined iff $\bar{a} = \langle a_i : i \le \alpha\rangle \in E$, $<^*$ a partial order of a_α, x is a binary relation on a_α and $p \in a_0$. If $\mathbb{D}_{\langle\bar{a},<^*,x,p\rangle}$ is well defined then it is a family of subsets of $\mathcal{P}(a_0)$.

(C) *If* $i < 1+\kappa$ and $\langle\bar{a}, \le^*_j, x_j, p\rangle \in \mathrm{Dom}(\mathbb{D})$ for $j < i$ (note: same $\bar{a}$ and p and possibly distinct $\le^*_j, x_j$) and $\le^*_j \restriction a_0 = \le^*_0 \restriction a_0$, $x_j\restriction a_0 = x_i\restriction a_0$ and $A_j \in \mathbb{D}_{\langle\bar{a},\le^*,x_j,p\rangle}$ for $j < i$, *then* $\bigcap_{j<i} A_j \ne \emptyset$.

4.16 Definition. A forcing notion P is called $(\mathcal{D}, \mathbb{D})$-complete, where $\mathcal{D}$ is a (fine normal) filter on $\mathcal{S}_{\le\aleph_0}(\lambda)$ and $\mathbb{D}$ is a (λ, E, κ)-system (so $E \subseteq \mathcal{S}^{<\omega_1}_{\le\aleph_0}(\lambda)$ is stationary) *if*:

(A) $\lambda > 2^{|P|}$, P is isomorphic to $(P^*, <^*)$, $P^* \subseteq \lambda$,

(B) for any $p \in P^*$ and pre-dense subsets $\mathcal{I}_\alpha$ of P^*, $\mathcal{I}_\alpha = \{p^\alpha_i : i < i_\alpha \le \lambda\}$ (for $\alpha < \lambda$), for some $x \subseteq \lambda \times \lambda$ and a nonstationary $Y \subseteq \mathcal{S}^{<\omega_1}_{\le\aleph_0}(\lambda)$, we have:

[†] Remember $\mathcal{S}^{<\omega_1}_{\le\aleph_0}(\lambda) \stackrel{\text{def}}{=} \{\bar{A} : \bar{A}$ an increasing continuous sequence of countable subsets of λ of length $< \omega_1\}$.

[††] Of course we can omit $<^*$ and still get an equivalent formulation.

(*) If $\bar{a} = \langle a_i : i \le \alpha\rangle \in E \setminus Y$ for $i \le \alpha$, *then* the following set includes a member of $\mathbb{D}_{\langle \bar{a}, <^* \restriction a_0, x \restriction a_0, p\rangle}$:

$$\Big\{G \subseteq a_0\text{: 1) for every } \alpha \in a_0 \text{ for some } i \in a_0,\ p_i^\alpha \in G.\ \ 2)(\exists q \in P^*)(\forall r \in G)[r \le q]\ \ 3) p \in G\Big\},$$

(C) for every $Y \in \mathcal{D}$ we have $\{\bar{a} \in E : a_0 \notin Y\}$ is not stationary (as a subset of $\mathcal{S}^{<\omega_1}_{\le\aleph_0}(\lambda)$).

(D) if $Y \subseteq \mathcal{S}_{\le\aleph_0}(\lambda)$, $Y \ne \emptyset \bmod \mathcal{D}$ then $\{\bar{a} \in E : a_0 \in Y\}$ is stationary.

4.17 Convention. If $\mathcal{D} = \mathcal{D}_{\le\aleph_0}(\lambda) + F_0$, $F_0 \ne \emptyset \bmod \mathcal{D}_{\le\aleph_0}(\lambda)$ *then* we shall write F_0 instead of $\mathcal{D}$. If $\mathcal{D} = \mathcal{D}_{\le\aleph_0}(\lambda)$ we shall write $\mathbb{D}$ instead of $(\mathcal{D}, \mathbb{D})$. We do not always distinguish between $\mathcal{D}$, a fine normal filter on $\mathcal{S}_{\le\aleph_0}(\lambda)$ in V, and the fine normal filter it generates in a generic extension of V.

4.18 Claim. (1) If P is $(\mathcal{D}, \mathbb{D})$-complete, *then* forcing with P adds no new ω-sequence of ordinals (in particular, no reals).

(2) Suppose P is $(\mathcal{D}, \mathbb{D})$-complete for a (λ, E, κ)-completeness system $\mathbb{D}$, and $\mu \ge \lambda$.

Then for some $\mathbb{D}'$, E', $\mathcal{D}'$ we have :

(a) $E' \subseteq \mathcal{S}^{<\omega_1}_{\le\aleph_0}(\mu)$ is stationary and for $\mathbb{D}'$ a (μ, E', κ)-completeness system P is $(\mathcal{D}', \mathbb{D}')$-complete,

(b) $\mathcal{D}'$ is the normal fine filter on $\mathcal{S}_{\le\aleph_0}(\mu)$ generated by

$$\Big\{\{a : a \cap \lambda \in y\} : y \in \mathcal{D}\Big\}.$$

(c) $E' = \{\bar{a} : \langle a_i \cap \lambda : i \le \ell g(\bar{a})\rangle \in E \text{ and } \bar{a} \in \mathcal{S}^{<\omega_1}_{\le\aleph_0}(\mu) \text{ and } a_0 \cap \lambda \in \mathcal{D}\}$.

(d) $\mathbb{D}'$ is defined naturally.

(3) If $V \models 2^\lambda \le \mu$ then (in the first part) $\mathcal{D}'$ has the form $\mathcal{D}_{\le\aleph_0}(\lambda) + F$ for some stationary $F \subseteq \mathcal{S}_{\le\aleph_0}(\lambda)$.

4.19 Claim. 1) Assume $E \subseteq \mathcal{S}^{<\omega_1}_{\leq\aleph_0}(\lambda)$ is stationary and $\mathcal{D}, E$ satisfy condition (C) of 4.16. If $Y \in \mathcal{D}$ and H is a function from Y to λ, *then* $\{\bar{a} \in E : H(a_0) \notin a_1\}$ is not stationary.

2) In definition 4.14 the value of χ is immaterial and, if $\chi > 2^\lambda$, we can omit "$x \in N_0$".

3) Assume $\bar{Q} = \langle P_\xi, \underset{\sim}{Q}_\zeta : \xi \leq \zeta^* \text{ and } \zeta < \zeta^*\rangle$ is a CS iteration of proper forcings. Assume also that $\mathcal{D}$ is a (fine) normal filter on $\mathcal{S}_{\leq\aleph_0}(\lambda)$, and for $\zeta < \xi \leq \zeta^* < \lambda$, $E_{\zeta,\xi} \subseteq \mathcal{S}^{<\omega_1}_{\leq\aleph_0}(\lambda)$ is stationary and each quadruple $(\mathcal{D}, E, P_\zeta/P_\xi)$ satisfies in V^{P_ζ} conditions (C), (D) of Definition 4.16. *Then*

$$\begin{aligned} E \overset{\text{def}}{=} \{\bar{a} : \;& \bar{a} \in \mathcal{S}^{<\omega_1}_{\leq\aleph_0}(\lambda) \text{ and there is } \langle \varepsilon_\zeta : \zeta \in a_0 \cap (\zeta^*+1)\rangle \text{ increasing} \\ & \text{continuous, } 0 < \varepsilon_0,\ \varepsilon_{\zeta^*} < \ell g(\bar{a}), \text{ for each } \zeta \in a_0 \cap (\zeta^*+1) \text{ we have} \\ & \langle a_0\rangle \hat{}\, \langle a_\epsilon : \varepsilon_\zeta + 2 \leq \epsilon \leq \varepsilon_{\zeta+1}\rangle \in E_{\varepsilon_\zeta, \varepsilon_{\zeta+1}} \text{ and} \\ & a_0,\ a_{\ell g(\bar{a})-1},\ a_{\varepsilon_\zeta+1} \text{ belong to } \{b_0 : \bar{b} \in E_0\}\} \end{aligned}$$

is stationary and satisfies clauses (C), (D) of Definition 4.16.

(4) If $F \subseteq \mathcal{S}_{\leq\aleph_0}(\lambda)$ is stationary and in (3) we add $[\bar{a} \in E_\zeta \& \zeta < \zeta^* \Rightarrow \bigwedge_{i \leq \ell g(\bar{a})} a_i \in F]$ *then* we can replace E by $E' = \{\bar{a} \in E : \bigwedge_{i \leq \ell g(\bar{a})} a_i \in F\}$.

Proof. Straightforward. $\square_{4.19}$

4.19A Remark. If $\lambda < \chi, N_i \prec (H(\chi), \in, <^*_\chi)$ for $i \leq \alpha$ is countable, increasing, continuous (in i) and $\langle N_j : j \leq i\rangle \in N_{i+1}$ *then* we can find a limit $\beta > \alpha$ and N_i (for $i \in (\alpha, \beta]$) such that:

(a) $\langle N_i : i \leq \beta\rangle$ is increasing and continuous, $N_i \prec (H(\chi), \in, <^*_\chi)$ is countable, and $\langle N_j : j \leq i\rangle \in N_{i+1}$.

(b) $\lambda \in N_{\alpha+1}$ and *if* $E \in N_\beta$, $E \subseteq \mathcal{S}_{\leq\aleph_0}(\lambda)$ is stationary *then* for some i, j, we have $\alpha < i < j < \beta$, $E \in N_i$ and $\langle N_\gamma \cap \lambda : i \leq \gamma \leq j\rangle \in E$.

4.20 Theorem. Assume

(a) $\overline{Q} = \langle P_\alpha, \underset{\sim}{Q}_\alpha : \alpha < \alpha^*\rangle$ is a CS iteration,

(b) each $\underset{\sim}{Q}_\alpha$ is $(<\omega_1)$-proper (in V^{P_α}),

(c) $\kappa \geq 2, \lambda \geq \omega_1$ a cardinal, $\alpha^* \leq \lambda, \lambda = \lambda^{\aleph_0}$, and P_{α^*} has a dense subset of cardinality $\leq \lambda$,

(d) λ is a cardinal, $\mathcal{D}$ is a normal filter on $\mathcal{S}_{\leq\aleph_0}(\lambda)$ (or $\mathcal{D} = \mathcal{D}_{\leq\aleph_0}(\lambda) + F$),

(e) for every α, $\Vdash_{P_\alpha}$ "$\underset{\sim}{Q}_\alpha$ is $(\mathcal{D}, \underset{\sim}{\mathbb{D}}^\alpha)$-complete", $\underset{\sim}{\mathbb{D}}^\alpha$ a P_α-name of a $(\lambda_\alpha, \underset{\sim}{E}_\alpha, \kappa)$-completeness system from V, $\underset{\sim}{E}_\alpha$ is a P_α -name of a member of V which is a stationary subset of $\mathcal{S}^{<\omega_1}_{\leq\aleph_0}(\lambda)$ (in V) satisfying (C), (D) of 4.18.

Then:

(1) P_{α^*} is $(\mathcal{D}, \mathbb{D})$-complete for some (λ, E, κ)-completeness system $\mathbb{D} \in V$, for some stationary $E \subseteq \mathcal{S}^{<\omega_1}_{\leq\aleph_0}(\lambda)$, satisfying:

$$\bigcup_{\bar{a}\in E} \operatorname{Rang}(\bar{a}) \subseteq \{a_i : \nVdash_{P_\alpha} \text{"}\bar{a} \notin E^\alpha \text{ for some } \alpha < \alpha^*\text{" and } i \leq \ell g(\bar{a})\},$$

(2) moreover, if $\alpha < \beta \leq \alpha^*$, then in V^{P_α}, P_β/P_α is $(\mathcal{D}, \mathbb{D})$-complete for some $(\lambda, E^{\alpha,\beta}, \kappa)$-completeness system $\mathbb{D} \in V$, for some $\underset{\sim}{E}^{\alpha,\beta} \in V$ as above.

4.20A Remark. Why can we assume that λ is constant? see 4.18(2).

Proof. This is proved by induction on α^*. Fix a one to one function H^* from $\mathcal{S}^{<\omega_1}_{\leq\aleph_0}(\lambda)$ onto λ and let $\langle(\mathbb{D}^\zeta, E^\zeta) : \zeta < \lambda\rangle$ list, in V, all pairs $(\mathbb{D}, E)$, $\mathbb{D}$ a (λ, E, κ)-completeness system such that for some $\alpha < \beta < \alpha^*, \Vdash_{P_\alpha}$ "$(\mathbb{D}, E)$ is $<^*_{\lambda^+}$-first for which P_β/P_α is $(\mathcal{D}, \mathbb{D})$-complete, $\mathbb{D}$ is a (λ, E, κ)-completeness system and $\mathbb{D}, E \in V$". There is such a list of length λ, as $\alpha^* \leq \lambda$ and P_{α^*} has a dense subset cardinality $\leq \lambda$ (see assumption (a) of 4.20).

By the following subclaim, without loss of generality $(\underset{\sim}{\mathbb{D}}^\alpha, \underset{\sim}{E}^\alpha)$, for $\alpha < \alpha^*$, are really $(\mathbb{D}^\alpha, E^\alpha)$ (i.e. members of V rather than names of such members) and $E^\alpha \cap \mathcal{S}^{\leq 2}_{\leq\aleph_0}(\lambda) = \emptyset$ and similarly $(\mathbb{D}^{\alpha,\beta}, E^{\alpha,\beta})$, for $\alpha < \beta < \alpha^*$. (We can alternatively redefine the iteration, inserting many trivial forcings, i.e. replace Q_i by $Q'_{\lambda\times i+\zeta}(\zeta < \lambda)$ which is $(\mathcal{D}, \mathbb{D}^\zeta)$-complete (one of them is Q_i, the others are trivial).)

4.21 Subclaim. Under the assumption in 4.20, for each $\alpha < \alpha^*$, for some $(\mathbb{D}, E) \in V$, $\mathbb{D}$ is a (λ, E, κ)-completeness system in V, $E \in \mathcal{S}^{<\omega_1}_{\leq\aleph_0}(\lambda)$ is stationary and $\Vdash_{P_\alpha}$ "$\underset{\sim}{Q}_\alpha$ is $(\mathcal{D}, \mathbb{D})$-complete".

Proof. Let

$$E^*_\alpha = \{\bar a : \bar a \in \mathcal{S}^{<\omega_1}_{\leq\aleph_0}(\lambda), \ell g(\bar a) > 2 \text{ and for some } \bar\varepsilon = \langle(\varepsilon^\zeta_0, \varepsilon^\zeta_1) : \zeta \in a_0 \cap \alpha\rangle$$
$$\text{for every } \zeta \in a_0, \varepsilon^\zeta_0 < \varepsilon^\zeta_1 < \ell g(\bar a) \text{ and } \bar\varepsilon\restriction\zeta \in N_{\varepsilon^\zeta_0} \text{ and}$$
$$\langle a_i : i = 0 \text{ or } \varepsilon^\zeta_0 \leq i \leq \varepsilon^\zeta_1\rangle \in E^\zeta\}.$$

By 4.19(3) we can show that E^* is of the right kind. $\square_{4.21}$

Continuation of the proof of 4.20: By the associativity law for CS iterations of proper forcing, the following cases suffice. Also, clause (1) of the conclusion is a special case of clause (2) and if $\beta < \alpha^*$ then the statement has already been proved (by the induction hypothesis).

First Case: $\alpha = 0,\ \beta = \alpha^* = 1$.

Trivial.

Second Case: $\alpha = 1,\ \beta = \alpha^* = 3$.

$$\text{Let } E^{1,3} = \{\bar a : \bar a \in \mathcal{S}^{\omega_1}_{\leq\aleph_0}(\lambda), \bar a = \langle a_i : i \leq i^*\rangle \text{ and for some } j^*_1, j^*_2 \text{ we have :}$$
$$0 < j^*_1 < j^*_2 < i^*, j^*_2 + 5 \leq i^* \text{ and } \bar a\restriction(j^*_1 + 1) \in E^1$$
$$\text{and } \langle a_0\rangle{}^\frown\bar a\restriction(j^*_1 + 3, j^*_2) \in E^2\}$$

Third Case: $\alpha = 0, \beta = \alpha^* = 2$.

Similar to the second case, but easier.

Fourth Case: $\alpha = 1, \beta = \alpha^* = \omega$.

$E^{\alpha,\beta} = \{\bar a$: for some $\langle\beta_j : j \leq \omega^2\rangle$ increasing and continuous, $\beta_0 = 0$, $\ell g(\bar a) = \beta_{\omega^2} + \omega + 1$, and for each n, $m < \omega$, $\langle a_{\beta_{\omega n}}\rangle{}^\frown\langle a_\gamma : \beta_{\omega m+n} + 2 \leq \gamma \leq \beta_{\omega m+n+1}\rangle$ belongs to $E^n\}$.

If $\kappa \geq \aleph_0$, $\langle\beta_j : j \leq \omega\rangle$ suffices.

Fifth Case: $\alpha = 1,\ \beta = \alpha^* = \mathrm{cf}(\alpha^*) > \aleph_0$.

$E^{\alpha,\beta} = \{\bar a \in \mathcal{S}^{<\omega_1}_{\leq\aleph_0}(\lambda) : \beta \in a_0$ and for some $\langle\beta_j : j \leq \omega^2\rangle$ increasing continuous $\beta_0 = 0, \ell g(\bar a) = \beta_{\omega^2} + \omega + 1$, and we can choose $\langle\alpha(n) : n < \omega\rangle$ such that: $\alpha(n) \in a_0 \cap \beta, \alpha(n) < \alpha(n+1), \alpha(0) = \alpha(= 1)$ and $\sup(\beta \cap a_0) = \bigcup_{n<\omega} \alpha(n)$, and for each $n, m < \omega$ we have $\langle a_{\beta_{\omega n}} \rangle \hat{\ } \langle a_\gamma : \beta_{\omega m+n} + 2 \leq \gamma \leq \beta_{\omega m+n+1} \rangle$ belongs to $E^{\alpha(n),\alpha(n+1)}\}$. $\square_{4.20}$

4.24 Concluding Remarks. 1) There is not much difference between using $(\mathcal{D}, \mathbb{D})$-completeness and $(F, \mathbb{D})$-completeness (where F is a stationary subset of $\mathcal{S}_{\leq\aleph_0}(\lambda)$ for some λ), as long as we do not mind increasing λ (see 4.18).
2) In Theorem 4.20 ($< \omega_1$)-proper can be weakened by restricting ourselves to e.g., E^α_W-proper for every α, where $E^\alpha_W = \{\langle a_i : i \leq \alpha \rangle \in \mathcal{S}^\alpha_{\leq\aleph_0}(\lambda) : a_i \cap \omega_1 \in W\}$ for a stationary subset $W \subseteq \omega_1$; demanding that each E_β is a subset of $\bigcup_{\alpha<\omega_1} E^\alpha_W$, then also in the definition of "long " (4.17) we restrict ourselves to the case $\bigwedge_i N_i \cap \omega_1 \in W$.
3) We could have replaced E^α by $\mathrm{Dom}(\mathbb{D}^\alpha)$, etc.
4) We may note the following generalization. Call $E \subseteq \mathcal{S}^{<\omega_1}_{\leq\aleph_o}(\lambda)$ unambiguous if for no $\bar{a}, \bar{b} \in E$ is $\bar{a}$ a proper initial segment of $\bar{b}$. Then for $\bar{c} \in \mathcal{S}^{<\omega_1}_{\leq\aleph_0}(\lambda)$, we let $\zeta_E(\bar{c})$ be the unique $\zeta \leq \ell g(\bar{c})$ such that $\bar{c}{\restriction}\zeta \in E$ (maybe $\zeta_E(\bar{c})$ is not defined.)

Now in Definition 4.15 we will have also $E^* \subseteq \mathcal{S}^{<\omega_1}_{\leq\aleph_0}(\lambda)$ stationary such that $E \subseteq \mathrm{Dom}(\zeta_{E^*})$ and now we call $\mathbb{D}$ a $(\lambda, E, E^*, \kappa)$-system and note that only now (in (B)) $\mathbb{D}_{\langle \bar{a}, <^*, \lambda, p\rangle}$ is a family of subsets of $\mathcal{P}(a_{\zeta_{E^*}(\bar{a})})$. Now, in Definition in 4.16, the family is:

$\{G :$ 1) for every $\varepsilon \leq \zeta(\bar{a}), \alpha \in a_\varepsilon$ for some $i \in a_\varepsilon$ we have $p^i_\alpha \in G$

2) $(\exists q \in P^*)(\forall r \in G)(r \leq q)$ (i.e. G has an upper bound in $(P^*, <^*)$)

3) $p \in G\}$.

(This generalization adds some indices to the proofs but no essential changes. This was the point of the original version of 4.2, 4.3.)
5) We can view (4) as a particular case of, more generally, putting an induction hypothesis on $G \cap N_0$.
6) See more in X §7 and XVIII §1, §2.

4.22 Definition. We say $E \subseteq \mathcal{S}^{<\omega_1}_{\leq\aleph_0}(\lambda)$ is simple if letting $H(\aleph_1, \lambda)$ be the closure of λ under taking countable subsets, E is first order definable in $(H(\aleph_1, \lambda), \in, <^*)$.

4.23 Claim. In 4.20 condition (e), instead of "$\mathbb{D}, E$ are from V", it is enough to demand that they are simple.

IX. Souslin Hypothesis Does Not Imply "Every Aronszajn Tree Is Special"

§0. Introduction

We prove that the Souslin Hypothesis does not imply "every Aronszajn tree is special"; solving an old problem of Baumgartner, Malitz and Reinhardt. For this end we introduce variants of the notion "special Aronszajn tree" and discuss them (this is §3, see references there). We also introduce a limit of forcings bigger than the inverse limit, and prove it preserves properness and related notions not less than inverse limit, and the proof is easier in some respects, and was done already in 78; see §1, §2. We can get away without using it for the present theorems, but we want to represent it somewhere. The Aronszajn trees are addressed in §4; we choose a costationary $S \subseteq \omega_1$ and make all $\aleph_1$-trees S-st-special, while on "$\omega_1 \setminus S$ the tree remains Souslin". If $S = \emptyset$ this means that every $\aleph_1$-tree is special when restricted to some unbounded set of levels, in fact while there is no antichains whose set of levels is stationary. See more in 4.9.

§1. Free Limits

1.1 Discussion and Definitions. For A a set of propositional variables, λ a regular cardinal, let: $L_\lambda(A)$ be the set of propositional sentences generated from A, by negation and conjunction and disjunctions on sets of power $< \lambda$.

Let $L_\mu(A) = \bigcup_{\lambda<\mu} L_\lambda(A)$ for μ a limit cardinal ($> \aleph_0$) or ∞. Let φ, ψ, θ denote sentences; Φ, Ψ set of sentences.

We define (in $L_\infty(A)$) $\vdash \psi$, or $\Phi \vdash \psi$ as usual (the rules of the finite case, and $\Phi \vdash \bigwedge \Phi$, and from $\Phi \vdash \varphi_i$ for $i \in I$ deduce $\Phi \vdash \bigwedge_{i \in I} \varphi_i$,) and let $\bigvee_i \varphi_i = \neg \bigwedge_i \neg\varphi_i$.

Always $\vdash$ means in $L_\infty(A)$ even if we deal with $L_\lambda(A)$.

The following is well known.

1.2 Theorem. The following are equivalent for Φ, φ:

(1) $\Phi \vdash \varphi$;

(2) there is no model of $\Phi \bigcup \{\neg\varphi\}$ with truth values in a complete Boolean algebra;

(3) if λ is such that $|\Phi|$, and the power of any set on which we make conjunction inside some sentence $\theta \in \Phi \bigcup \{\varphi\}$ are $\leq \lambda$ and $P = \mathrm{Levy}(\aleph_0, \lambda)$ i.e. the collapsing of λ to ω by finite functions, *then*

$$\Vdash_P \text{" there is no model of } \Phi \bigcup \{\neg\varphi\}\text{"}.$$

1.2A Remark. This can be proven by a small fragment of ZFC, admissibility axioms, at least when we prove only (1) $\Leftrightarrow$ (3). Hence (by proving not (1) implies not (3)):

1.3 Conclusion. If A is a transitive admissible set, $\Phi, \varphi \in A$ then "$\Phi \vdash \varphi$" has the same truth value in V and in A.

1.4 Definition. For given A and $\theta \in L_\infty(A)$, let $FF_\lambda(\theta)$ be $\{\psi : \psi \in L_\lambda(A), \theta \nvdash \neg\psi\}$ partially ordered by $\psi_1 \leq \psi_2$ if $\theta \wedge \psi_2 \vdash \psi_1$. ($FF$ denotes free-forcing; we can identify φ, ψ if $\varphi \leq \psi \leq \varphi$.)

Reversing the definition of $\leq$ and adding a minimal element, we get a Boolean algebra in which every set of $< \lambda$ elements has a least upper bound provided that we identify ψ_1, ψ_2 when $\theta \vdash \psi_1 \equiv \psi_2$.

1.5 Definition. For any forcing notion P let $\theta[P]$ be the following sentence: $\bigwedge\{(c \to \neg d)\wedge(b \to a) : a, b \in P, a \leq b, c, d \in P, c, d \text{ incompatible}\} \wedge \bigwedge\{\bigvee_{a\in\mathcal{I}} a : \mathcal{I} \subseteq P$ a maximal set of pairwise incompatible elements$\}$.

1.6 Definition. Let $P_i (i < \delta)$ be $\lessdot$-increasing, δ an ordinal (λ an infinite regular cardinal). Then their λ-free limit $(\mathrm{Flim}^{\lambda}_{i<\delta} P_i)$ is $FF_\lambda(\bigwedge_{i<\delta}\theta[P_i])$ (where the set of propositional variables is $\bigcup_{i<\delta} P_i$). If we omit λ we mean $\lambda = \aleph_1$.

1.7 Claim. $P \lessdot Q$ implies $\theta[Q] \vdash \theta[P]$, and $P \lessdot FF_\lambda(\theta[P])$.

Proof. The first statement is trivial, for the second see the proof of Claim 1.8.

$\square_{1.7}$

1.7A Remark. Our notation may be confusing, as for conditions $p, q \in P, p\wedge q$ is "p and q", i.e., both are in the generic set so in our order $p \wedge q$ is above p and above q as it give more information.

1.8 Claim. If as in Definition 1.6, P_δ is the λ-free limit of $P_i (i < \delta)$ *then* $P_i \lessdot P_\delta$ for $i < \delta$.

Proof. Let us check the conditions.

proof of clause (b) Let $\mathcal{I} \subseteq P_i$ be a maximal set of pairwise incompatible elements of P_i. Suppose $\varphi \in \mathrm{Flim}^{\lambda}_{i<\delta} P_i$ is incompatible with each $a \in \mathcal{I}$. As $\varphi \in \mathrm{Flim}^{\lambda}_{i<\delta} P_i$, by definition $\bigwedge_{i<\delta}\theta[P_i] \nvdash \neg\varphi$. So by 1.2, after some forcing there is a model of φ, $\bigwedge_{j<\delta}\theta[P_j]$. But $\bigvee_{a\in\mathcal{I}} a$ is a conjunct of the second sentence, so in the model some $q \in \mathcal{I}$ is true. So after some forcing, there is a model of $\varphi \wedge q, \bigwedge_{j<\delta}\theta[P_j]$, so by 1.2, $\bigwedge_{j<\delta}\theta[P_j] \nvdash \neg(\varphi \wedge q)$, so $\varphi \wedge q \in FF_\lambda(\bigwedge_{j<\delta}\theta[P_j])$; so φ, q are compatible in $FF_\lambda(\bigwedge_{j<\delta}\theta[P_j]) = \mathrm{Flim}^{\lambda}_{i<\delta} P_i$.

proof of clause (a) Let $a, b \in P_i$, if they are compatible in P_i, for some $c \in P_i$, $a \leq c, b \leq c$, and this clearly holds in P_δ by its definition.

If they are incompatible in P_i then $a \to \neg b$ appears as a conjunct in $\theta[P_i]$ and we can finish. Similarly for $a, b \in P_i, a \leq b$ in P_i implies $a \leq b$ in P_δ. $\square_{1.8}$

1.9 Remark. We can change the definition of $FF_\lambda(\theta)$ (hence of $F\lim$) by changing $\vdash$. The natural way is to let K be a class of complete Boolean algebras, and $\Phi \vdash_K \varphi$ iff any Boolean valued model of Φ is a Boolean valued model of φ provided that the complete Boolean algebra is from K. So $FF^K_\lambda(\theta) = \{\varphi \in L_\lambda : \theta \nvdash_K \neg\varphi\}$.

The most interesting K's seems

$K_1 = \{B :$ forcing with $B \setminus \{0\}$ satisfies $X\}$

where $X =$ does not add reals, does not collapse $\aleph_1$, does not collapse stationary subsets of $\aleph_1$, the $UP_\ell(S)$ condition (Min$S \geq \aleph_2, \ell = 0, 2$, see XV §3).

§2. Preservation by Free Limit

2.1 Theorem. If each P_i is a forcing notion; and P_i $(i < \delta)$ is $\lessdot$-increasing, each P_i is proper as well as $P_j/P_i (i < j < \delta)$ and for $\alpha < \delta, \mathrm{cf}(\alpha) = \aleph_0 \Rightarrow P_\alpha = \mathrm{Flim}^{\aleph_1}_{i<\alpha} P_i$ *then* their $\aleph_1$-free limit $P = \mathrm{Flim}^{\aleph_1}_{i<\delta} P_i$ is proper. Also P/P_i is proper for any $i < \delta$.

2.1A Remark. Similarly for μ-proper by [Sh:100] terminology if we take μ^+-free limit. We can restrict ourselves to non-limit i, j.

Proof. Let $N \prec (H(\chi), \in)$ be countable such that $\langle P_i : i < \delta\rangle \in N, p \in P \cap N$ and χ big enough (see III). Let $\{\mathcal{I}_n : n < \omega\}$ be a list of all pre-dense subsets of P which belong to N. Let $\delta(*) = \sup[\delta \cap N]$, and let $P_\delta = P$, so $N \cap P = N \cap P_\delta \subseteq N \cap P_{\delta(*)}$ and if $\delta(*) < \delta$ then $N \cap P_{\delta(*)} = \bigcup_{i<\delta(*)} N \cap P_i$. As necessarily $\mathrm{cf}(\delta(*)) = \aleph_0$ clearly $P_{\delta(*)} = \mathrm{Flim}^{\aleph_1}_{i<\delta(*)} P_i$. So it is enough to prove $p \wedge \bigwedge_n (\bigvee_{a \in N \cap \mathcal{I}_n} a) \in P_{\delta(*)}$ (in Boolean algebras terms: is not zero), for p being any member of $P \cap N$.

Now assume w.l.o.g. that everything is in some countable transitive model M (e.g. work in $V' = V^{\mathrm{Levy}(\aleph_0,\mu)}$, χ strong limit such that $\langle P_i : i < \delta\rangle \in H(\mu)$, $\mu < \chi$, let $M = H(\mu)^V$ and remember 1.3). We can find $\alpha_n < \alpha_{n+1}, \alpha_n \in N \cap \delta$, $\sup[\delta(*) \cap N] = \bigcup_n \alpha_n$.

Let $\langle \Phi_n : n < \omega \rangle$ be a list of all countable (in M) subsets of $P_{\delta(*)}$ which belongs to N.

We now define by induction on n, in V, G_n, p_n such that:

(1) $G_n \subseteq P_{\alpha_n}$, $G_n \subseteq G_{n+1}$,

(2) G_n is P_{α_n}-generic for M and $G_n \cap N$ is $(P_{\alpha_n} \cap N)$-generic for N,

(3) $p_n \leq p_{n+1}$, $p = p_0$, $p_n \in N \cap P$,

(4) p_n is compatible (in P) with every member of G_n,

(5) p_{2n+1} is $\geq q_n$ for some $q_n \in \mathcal{I}_n \cap N$,

(6) either $p_{2n+2} \vdash \wedge \Phi_n$ or $p_{2n+2} \vdash \neg r_n$ for some $r_n \in \Phi_n$.

The proof is trivial (provided you know about the composition of forcings and completeness theorem for the propositional calculus $L_{\omega_1,\omega}$).

In the end $G = \bigcup_n G_n$ gives us a model of $\bigwedge_{j<\delta(*)} \theta[P_j]$ (by: members of G are true, members of $\bigcup_{j<\delta} P_j \setminus G$ are false). This holds by clause (2).

For $r \in P \cap N, r$ is true in the model iff $p_n \geq r$ for some n (this is proved by induction on the complexity of r, (see conditions (4) and (6)). In the model p_n is true (for each $n < \omega$), hence $\bigvee_{a \in \mathcal{I}_n \cap N} a$ is true (for each $n < \omega$) hence $p \wedge \bigwedge_n (\bigvee_{a \in \mathcal{I}_n \cap N} a)$ is true there (p true as $p_0 = p$).

So in V there is a model of $\bigwedge_{j<\delta(*)} \theta[P_j]$, $p \wedge \bigwedge_n (\bigvee_{a \in \mathcal{I}_n \cap N} a)$ so $p \wedge \bigwedge_n (\bigvee_{a \in \mathcal{I}_n \cap N} a) \in P_{\delta(*)}$ as required. $\square_{2.1}$

2.1B Remark. Part of the proof is essentially a repetition of the completeness theorem for $L_{\omega_1,\omega}$ (propositional calculus). But note that in this proof there was no need (as in the ones for inverse limit) to use names. Also, almost all previous theorems on preservation hold for free iterations.

2.2 Definition. Let $\bar{Q} = \langle P_i, \underset{\sim}{Q}_i : i \leq i_0 \rangle$ be an ω_1-free iteration if :

(a) P_i is $\lessdot$-increasing,

(b) $P_{i+1} = P_i * \underset{\sim}{Q}_i = \{\langle p, q \rangle : p \in P_i, \Vdash_{P_i}$ "$q \in \underset{\sim}{Q}_i$"$\}$, with the order $\langle p, q \rangle \leq \langle p^\dagger, q^\dagger \rangle \Leftrightarrow p \leq p^\dagger \wedge [p^\dagger \Vdash_P q \leq q^\dagger]$; and we identify $p \in P_i$ with $\langle p, \emptyset \rangle$,

(c) for limit δ, P_δ is the $\aleph_1$-free limit of $\langle P_i : i < \delta \rangle$.

2.3 Definition. For an $\aleph_1$-free iteration $\bar{Q} = \langle P_i, \underset{\sim}{Q}_i : i < i_0 \rangle$ let $\text{Flim}^{\aleph_1} \bar{Q}$ be $P_\alpha * \underset{\sim}{Q}_\alpha$ if $i_0 = \alpha + 1$ and $\text{Flim}^{\aleph_1}_{i<i_0} P_i$ if i_0 is a limit ordinal, and we let $P_{i_0} = \text{Flim}^{\aleph_1} \bar{Q}$.

2.4 Definition. We say that $\aleph_1$-free iteration preserves a property if whenever each Q_i (in V^{P_i}) has it, then so does P_i.

2.5 Theorem. Properness is preserved by $\aleph_1$-free iteration.

Proof. See 2.1; and prove by induction on β that for $\alpha < \beta$,

(*) If $\langle P_i : i < i_{i_0} \rangle \in N \prec (H(\lambda), \in), ||N|| = \aleph_0, \alpha \le \beta \le i_0, \alpha \in N, \beta \in N, p \in P_\beta \cap N, q \in P_\beta$, and for every pre-dense $\mathcal{I} \subseteq P_\alpha$, $[\mathcal{I} \in N, \Rightarrow \mathcal{I} \cap N$ is pre-dense above q,$]$ and every $q^\dagger, q \le q^\dagger \in P_\alpha$ is compatible with p *then* for some $r \in P_\beta$, for every pre-dense $\mathcal{I} \subseteq P_\beta$ we have $[\mathcal{I} \in N \Rightarrow \mathcal{I} \cap N$ pre-dense above $r]$, $p \le r$ and for every $q^\dagger : q \le q^\dagger \in P_\beta \Rightarrow q^\dagger$ compatible with r. $\square_{2.5}$

The following Definition and Theorem are not really necessary for the rest of the chapter, but will help in understanding §4.

2.6 Definition.

1) P is strongly proper when: *if* λ is large enough (i.e. $\lambda > (2^{|P|})^+$), $P \in N \prec (H(\lambda), \in)$, $||N|| = \aleph_0, p \in P \cap N$ and $\mathcal{I}_n \subseteq N$ pre-dense in $N \cap P$ (but we do not ask $\mathcal{I}_n \in N$), *then* for some $q, p \le q \in P$, each $\mathcal{I}_n$ is pre-dense above q.
2) P is strongly α-proper *if* for large enough λ, $P \in N_i \prec (H(\lambda), \in)$, $||N_i|| = \aleph_0$, $\langle N_j : j \le i \rangle \in N_{i+1}$ for $i < \alpha$, N_i increasing continuous, $p \in P \cap N_0$, $P, i \in N_i$, $\mathcal{I}^i_n \subseteq P \cap N_i$ is a pre-dense subset of $P \cap N_i$ (for $n < \omega$) and $\langle \mathcal{I}^j_n : j \le i, n < \omega \rangle \in N_{i+1}$ (for $i < \alpha$) *then* there is a $q \in P, p \le q$, $\mathcal{I}^n_i$ pre-dense above q (in P, for each $n < \omega, i \le \alpha$). Note that we can replace $(H(\lambda), \in)$ by $(H(\lambda), \in, <^*_\lambda)$.

2.7 Theorem. Strong properness is preserved by $\aleph_1$-free iteration.

2.7A Remark. A similar theorem holds for strong α-properness and for CS iteration.

Proof. Let $\langle P_i, \underset{\sim}{Q}_j : i \le i_0, j < i_0 \rangle$ be an $\aleph_1$-free iteration. We prove by induction on $\alpha \le i_0$ that for any $\beta < \alpha$:

$(*)_{\beta,\alpha}$ Let $\langle P_i, \underset{\sim}{Q}_j : i \le i_0, j < i_0 \rangle \in N \prec (H(\lambda), \in)$ (where $\lambda > (2^{|P|})^+$), $||N|| = \aleph_0$, C a countable family of $\aleph_0$ pre-dense subsets of $P_\alpha \cap N$, closed under the operations listed below. Suppose $\beta < \alpha, p \in P_\alpha \cap N, \alpha \in N, \beta \in N, q \in P_\beta$, no $q^\dagger, q \le q^\dagger \in P_\beta$ is incompatible with p, and $[\mathcal{I} \subseteq P_\beta \& \mathcal{I} \in C \Rightarrow \mathcal{I}$ pre-dense above $q]$. *Then* there is $q_\alpha, p \le q_\alpha \in P_\alpha$, $q \le q_\alpha$, no $q \le q^\dagger \in P_\beta$ is incompatible with q_α and $[\mathcal{I} \in C \Rightarrow \mathcal{I}$ pre-dense above $q_\alpha]$.

The family of operations under which C is closed is (for $p \in N \cap P_{i_0}, \gamma \in N \cap i_0$ and $\mathcal{I} \in N$ a pre-dense subset of P_{i_0}):

(Op 1) $Op_1(\mathcal{I}, \gamma, p) = \{r : r \in P_\gamma \cap N$ and *either* for some $r^* \in P_{i_0}$ and $r_1 \in \mathcal{I}$ we have $r_1 \le r^*$ and $p \le r^*$ but no $r^\dagger, r \le r^\dagger \in P_\gamma \cap N$ is incompatible with r^* *or* r is incompatible with $p\}$ for $\gamma \in N, \mathcal{I} \in C$, $p \in P_\alpha$. (Note that for $p = \emptyset$ the last phrase is vacuous.)

For $\alpha = 0$. Totally trivial. For $\alpha = \gamma + 1$. So $\beta \le \gamma$, also as $\alpha \in N$ clearly $\gamma \in N$ and by the induction hypothesis for γ, $(*)_{\beta,\gamma}$ holds, so w.l.o.g. $\beta = \gamma$. So we want to use the hypothesis $P_\alpha = P_\gamma * \underset{\sim}{Q}_\gamma$, $\underset{\sim}{Q}_\gamma$ is strongly proper; then we use $\langle q, \underset{\sim}{r} \rangle$, $\underset{\sim}{r} \in \underset{\sim}{Q}_\gamma$ a name of an appropriate element of $\underset{\sim}{Q}_\gamma$. We have to prove that the appropriate subsets of $N[\underset{\sim}{G}_{P_\gamma}] \cap \underset{\sim}{Q}_\gamma[\underset{\sim}{G}_{P_\gamma}]$ are pre-dense subsets. But as C is closed under (Op 1) this is easy.

For α limit. Let $\alpha_n \in N$, $\bigcup_{n<\omega} \alpha_n$ is α or at least $[\bigcup_n \alpha_n, \alpha) \cap N = \emptyset$ (note $[\alpha^\dagger, \alpha)$ is interval of ordinals).

We work as in 2.1 using the induction hypothesis. $\square_{2.7}$

2.8 Claim. 1) If we iterate ω-proper, ${}^\omega\omega$-bounding forcings it does not matter whether we use $\aleph_1$-free iteration or countable support one (in the latter we get a dense subset of the first).

2) We can replace "ω-proper" by proper.

Proof. Left to the reader. $\square_{2.8}$

2.8A Remark. For the proof, see V §3 (and for (2) see XVIII §2). The parallel of 2.7 for countable support was noted by Harrington and the author.

By the way we note that unlike $\aleph_1$-c.c. forcing:

2.9 Example. There are proper forcing P, Q such that $P \lessdot Q$ but Q/P is not proper.

Proof. We let P_0 = adding a subset $\underset{\sim}{r}$ of ω_1 with a condition being a countable characteristic function.

Let $Q_0 \in V^{P_0}$, $Q_0 = \{f : \mathrm{Dom}(f) = \alpha < \omega_1, \mathrm{Rang}(f) = \{0,1\}, f^{-1}(\{0\})$ is a closed set of ordinals (not just closed subset of α!) included in $r\}$. (r denotes the generic subset of ω_1 which P_0 produces.)

Now $P_0, P_0 * Q_0$ are proper but in V^{P_0}, $Q_0 \cong P_0 * Q_0/P_0$ is not proper as it destroys the stationarity of $\omega_1 \setminus r_0$.

§3. Aronszajn Trees: Various Ways to Specialize

We introduce new variants of the notion "special Aronszajn tree", define some old ones (special, r-special) and prove some known theorems and some easy ones. See Kurepa [Ku35], Baumgartner, Malitz and Rienhard [BMR], Baumgartner [B] and also Devlin and Shelah [DvSh:65]. Recall

3.1 Definition.

(1) An ω_1-tree $T = (|T|, <_T)$ is a partially ordered set, such that (when no confusion arises, we write $<$ instead of $<_T$ and T instead of $|T|$):

(a) for every $x \in T, \{y \in T : y < x\}$ is well-ordered, and its order type which is denoted by $\mathrm{rk}(x) = \mathrm{rk}_T(x)$, is countable,

(b) $T_\alpha = \{x \in T : \text{rk}(x) = \alpha\}$ is countable, $\neq \emptyset$,

(c) if $\text{rk}(x) = \text{rk}(y)$ is a limit ordinal *then* $x = y \Leftrightarrow \{z : z < x\} = \{z : z < y\}$,

(d) if $x \in T_\alpha, \alpha < \beta$, *then* for some $y \in T_\beta, x < y$, in fact there are at least two distinct such y's.

If we wave (c) and (d) we call it an almost ω_1-tree; similarly for the other definitions.

(2) A set $B \subseteq T$ is a *branch* if it is totally (i.e. linearly) ordered (hence well ordered) and maximal; it is an α-branch if it has order type α.

(3) An Aronszajn tree is an ω_1-tree with no ω_1-branch.

(4) An ω_1-tree is Souslin or ω_1-Souslin tree if there is no uncountable antichain (= set of pairwise incomparable elements).

3.1A Remark. Condition (1)(d) is not essential, except to make every Souslin tree an Aronszajn tree. So except this implication all the definitions and results in this section hold for almost ω_1-trees.

3.2 Definition.

(1) For a set $S \subseteq \omega_1$ which is unbounded, we call an ω_1-tree S-special if there is a monotonic increasing function f from $\bigcup_{\alpha \in S} T_\alpha$ to $\mathbb{Q}$ (the rationals), i.e., $x < y \Rightarrow f(x) < f(y)$.

(2) A special ω_1-tree is an ω_1-special ω_1-tree (this is the classical notion).

(3) r-special, $S - r$-special are defined similarly when the function maps T into $\mathbb{R}$ (the reals).

(4) We say f specializes (S-specialize, etc.) T. We can replace S by a function h, $\text{Dom}(h) = \omega_1$, $\text{Rang}(h) = S, h$ increasing.

3.3 Definition. For a stationary $S \subseteq \omega_1$ we call an ω_1-tree S-st-special if there is a function f, $\text{Dom}(f) = \bigcup_{\alpha \in S \setminus \{0\}} T_\alpha$, and $x \in T_\alpha \Rightarrow f(x) \in \alpha \times \omega$ (cartesian product) such that $x < y \Rightarrow f(x) \neq f(y)$ when defined. If S is a set of limit ordinals we can assume $x \in T_\alpha \Rightarrow f(x) < \alpha$. So if $S \subseteq \omega_1$ is not stationary this says nothing.

3.4 Claim.

(1) If T is S-special or $S - r$-special ($S \subseteq \omega_1$ unbounded) or S-st-special ($S \subseteq \omega_1$ stationary) ω_1-tree *then* T is an Aronszajn tree but not Souslin. Any ω_1-Souslin tree is an Aronszajn tree.

(2) The following implications among properties of ω_1-trees hold (where $S_2 \subseteq S_1 \subseteq \omega_1$, S_2 unbounded in ω_1, $S_1 = \{\alpha(i) : i < \omega_1\}$, α_i increasing with i)

(a) S_1-special $\Rightarrow S_2$-special,

S_1-special $\Rightarrow S_1 - r$-special,

$S_1 - r$-special $\Rightarrow S_2 - r$-special,

$S_1 - r$-special $\Rightarrow S_1 \cap \{\alpha(i+1) : i < \omega_1\}$-special,

(b) for S_1 stationary: S_1-special $\Rightarrow S_1 - st$-special,

(c) $S_1 - st$-special $\Rightarrow S_2 - st$-special, (if S_1, S_1 are stationary subsets of ω_1).

(d) for $C \subseteq \omega_1$ closed unbounded: $S_1 \cap C - st$-special $\Leftrightarrow S_1 - st$-special,

(e) if $(\forall i) h_1(i) \leq h_2(i)$ and T is h_1-special then T is h_2-special.

Proof. Trivial:

(1) for S-special, and $S - r$-special – well known, for $S - st$-special by the Fodor lemma.

(2) Trivial – check. E.g. the last phrase in (a), if f is $S_1 - r$-specialize T, define $f^* : \bigcup_{i<\omega_1} T_{\alpha(i+1)} \to \mathbb{Q}$ by: if $x \in T_{\alpha(i+1)}$ then $f^*(x)$ is a rational $< f^*(x)$ but $> f(y)$ where $y \in T_{\alpha(i)}, y < x$. $\square_{3.4}$

Remark. By 3.4(2)(d) dealing with $S - st$-special we can assume all members of S are limit, and so $\mathrm{Rang}(f) \subseteq \omega_1$ in the Definition.

3.5 Claim.

(1) T is S-special *iff* $S \subseteq \omega_1$ is unbounded and there is $f : \bigcup_{\alpha \in S} T_\alpha \to \omega$, $[x < y \,\&\, \mathrm{rk}(x) \in S \,\&\, \mathrm{rk}(y) \in S \Rightarrow f(x) \neq f(y)]$.

(2) T is $\omega_1 - st$-special *iff* T is special.

Remark. See Claim 3.11.

Proof. (1) Well known (the "only if" part is trivial; for the "if" part, with given $f : \bigcup_{\alpha \in S} T_\alpha \to \omega$ we define $f_n : \{x : x \in \bigcup_{\alpha \in S} T_\alpha$ and $f(x) \leq n\} \to \mathbb{Q}$ by induction on n such that $f_n \subseteq f_{n+1}$, f_n satisfies the requirement (see Definition 3.2) and $\text{Rang}(f_n)$ is finite. Now $\bigcup_{n<\omega} f_n$ is as required).

(2) The "if" part is trivial.

So suppose f $\omega_1 - st$-specialize T. For every $x \in T$, let $K_x = \{t \in (rk(x)+1) \times \omega$: for no $y \leq x$ is $f(y) = t\}$. We now define by induction on $\alpha < \omega_1, g_\alpha$ and $A_{x,t}$ (for $t \in K_x, x \in \bigcup_{\beta<\alpha} T_\beta$) such that:

(a) g_α is a function from $T_{<\alpha} \stackrel{\text{def}}{=} \bigcup_{\beta<\alpha} T_\beta$ to ω,

(b) $x < y, x \in T_{<\alpha}, y \in T_{<\alpha} \Rightarrow g_\alpha(x) \neq g_\alpha(y)$,

(c) $\beta < \alpha \Rightarrow g_\beta \subseteq g_\alpha$,

(d) $A_{x,t}$(for $t \in K_x, x \in T_{<\alpha}$) is an infinite subset of ω,

(e) for every $x \in T_{<\alpha}, t \neq s \in K_x \Rightarrow A_{x,t} \cap A_{x,s} = \emptyset$,

(f) $t \in K_x, x \in T_{<\alpha} \Rightarrow A_{x,t} \cap \{g_\alpha(y) : y \leq x\} = \emptyset$,

(g) if $x < y \,\&\, x \in T_{<\alpha} \,\&\, y \in T_{<\alpha}, t \in K_x \cap K_y$ then $A_{x,t} = A_{y,t}$.

For $\alpha = 0, 1, \alpha$ limit – no problem.

For $\alpha + 1 > 1$ – let $x \in T_\alpha \subseteq T_{<(\alpha+1)}$, and $s = f(x)$, so by K's definition for some $y = y_x < x$, $s \in K_y$. We choose $g_{\alpha+1}(x) \in A_{y,s}$ ($= A_{z,s}$ for every z satisfying $y \leq z < x$) and let $g_{\alpha+1} \restriction T_{<\alpha} = g_\alpha$.

For $t \in K_x \setminus \bigcup_{z<x} K_z$ (there are $\aleph_0$ such t's) we choose $A_{x,t} \subseteq A_{y,s} \setminus \{g_{\alpha+1}(x)\}$ infinite pairwise disjoint. If $t \in K_z \setminus \{f(z') : z' \leq x\}$ for some $z < x$ we let $A_{x,t} = A_{z,t}$.

Now by 3.5(1) clearly $g = \bigcup_{\alpha<\omega_1} g_\alpha$ shows T is special. $\square_{3.5}$

3.6 Claim. Let $S \subseteq \omega_1$ be unbounded.

(1) If every Aronszajn tree is S-special *then* every Aronszajn tree is special.

(2) If every Aronszajn tree is $S-r$-special *then* every Aronszajn tree is special.

Proof. (1) Let T be an Aronszajn tree, $S = \{\alpha(i) : i < \omega_1\}$, $\alpha(i)$ increasing. Define T^* (a partial order): The set of elements is $\{\langle x, \gamma \rangle : x \in T, \gamma \leq \alpha(\text{rk}_T(x))$ and $y < x \Rightarrow \alpha(\text{rk}_T(y)) < \gamma\}$; the order in T^* is: $\langle x, \gamma \rangle <_{T^*} \langle x^\dagger, \gamma^\dagger \rangle$ if $x < x^\dagger$ or $x = x^\dagger, \gamma < \gamma^\dagger$.

Now T^* is almost an Aronszajn tree; the only missing part is in Definition 3.1, part (d) ("in fact there are at least two distinct such y's") the problem is e.g. when i is a limit ordinal, $\alpha(i) > \bigcup_{j<i} \alpha(j)$, in level $\bigcup_{j<i} \alpha(j)$. We can add more elements and find an Aronszajn tree, T^{**} such that $T^* \subseteq T^{**}$, and if g^{**} S-specializes the tree T^{**}, we let $g : T \to \mathbb{Q}$ be defined by $g(x) \stackrel{\text{def}}{=} g^{**}(\langle x, \alpha(\mathrm{rk}_T(x))\rangle)$, it specializes T.

(2) By 3.6(1) and 3.4(2) (a), last clause. $\square_{3.6}$

3.7 Lemma.

(1) ($\diamondsuit_{\omega_1}$) There is an r-special Aronszajn tree which is not special.

(2) Moreover (in (1)) there is no antichain $\mathcal{I}$ such that $\mathrm{rk}(\mathcal{I}) = \{\mathrm{rk}(x) : x \in \mathcal{I}\}$ contains a closed unbounded subset of ω_1.

(3) ($\diamondsuit^*_{\omega_1}$) There is an r-special Aronszajn tree, such that for no antichain $\mathcal{I} \subseteq T$ is $\mathrm{rk}(\mathcal{I}) = \{\mathrm{rk}(x) : x \in \mathcal{I}\}$ stationary.

Remark. Part (1) was proved by Baumgartner [B1].

Proof. We define by induction on $\alpha < \omega_1$ the tree $(T_{<\alpha}, <_T \restriction T_{<\alpha})$ and $f : T_{<\alpha} \to \mathbb{R}$ satisfying $x < y \Rightarrow f(x) < f(y)$ such that if $\beta < \gamma < \alpha$, $x \in T_\beta$, ε a real positive number (> 0), then for some $y, x < y \in T_\gamma$, $f(y) < f(x) + \varepsilon$; and $x \in T_{\alpha+1} \Leftrightarrow f(x) \in \mathbb{Q}$ and if $\delta < \omega_1$ is a limit ordinal, $x \in T_\delta$ then $f(x) = \sup\{f(y) : y < x\} \in \mathbb{R}$.

For $\alpha = 0$, α-successor of successor or α limit, no problem.

For $\alpha + 1$, α limit, we are given antichains $\mathcal{I}^\alpha_n \subseteq T_{<\alpha}$ for $(n < \omega)$ (by $\diamondsuit_{\aleph_1}$ or $\diamondsuit^*_{\aleph_1}$) and we can define $T_{<\alpha+1}$ (and hence $f \restriction T_{<\alpha+1}$) such that

$(*)$ if $x \in T_\alpha, n < \omega$ and $\{y \in T_{<\alpha} : y < x\} \cap \mathcal{I}^\alpha_n = \emptyset$ *then* for some $y < x$, and $\varepsilon > 0$, $f(y) < f(x) < f(y) + \varepsilon$, and there is no z, $z \in \mathcal{I}^\alpha_n, y < z \in T_{<\alpha}$ and $f(y) < f(z) < f(y) + \varepsilon$.

Now $T = \bigcup_{\alpha<\omega_1} T_{<\alpha}$ and f are defined in the end. Suppose $\mathcal{I} \subseteq T$ is an antichain Now $C = \{\alpha < \omega_1 : \alpha$ limit, and if $x \in T_{<\alpha}, \varepsilon > 0$, and there is $y \in \mathcal{I}, x < y, f(x) < f(y) < f(x) + \varepsilon$ then there is such $y \in T_{<\alpha}\}$ is closed unbounded (note that it suffices to consider $\varepsilon \in \{1/n : n$ positive natural number $\})$.

Now if $\alpha \in C, \mathcal{I} \cap T_{<\alpha} = \mathcal{I}^\alpha_{n(0)} \in \{\mathcal{I}^\alpha_n : n < \omega\}$, $\alpha \in \mathrm{rk}(\mathcal{I})$ then by $(*)$ we get $\mathcal{I} \cap T_\alpha = \emptyset$ (if $y \in \mathcal{I}, y \in T_\alpha$, by $(*)$ we know $\{z : z < y\} \cap \mathcal{I}^\alpha_{n(0)} \neq \emptyset$; let z be in it, then $z < y$ both in $\mathcal{I}$, but $\mathcal{I}$ is an antichain).

Now by defining $\mathcal{I}^\alpha_n$ using $\diamondsuit_{\omega_1}$ or $\diamondsuit^*_{\omega_1}$ we get (1), (2) and (3). $\square_{3.7}$

3.8 Claim. ($\diamondsuit^*_{\aleph_1}$) Let h be a function from ω_1 to ω_1. There is a tree T which is h_1-special iff $\{i : h(i) < h_1(i)\}$ contains a closed unbounded subset of ω_1 (see Definition 3.2(4)).

Proof. Similar to the proof of 3.7. $\square_{3.8}$

3.9 Lemma. Let $S \subseteq \omega_1$ be stationary, and assume $\diamondsuit^*_{\omega_1 \setminus S}$ hold. There is an $S - st$-special tree which is $S_1 - st$-special iff $S_1 \setminus S$ is not stationary; moreover there is no antichain $\mathcal{I}$, such that $\mathrm{rk}(\mathcal{I}) \setminus S$ is stationary. (If $S = \omega_1$ we do not need any hypothesis, $\diamondsuit^*_\emptyset$ is meaningless anyhow and this is the classical theorem on the existence of special Aronszajn trees of Aronszajn himself.) Also we can make the tree such that it is not h-special for any h.

Proof. We define by induction on $\alpha < \omega_1$, $(T_{<\alpha}, <_T \restriction T_{<\alpha})$, and $(f \restriction T_{<\alpha}) : T_{<\alpha} \to \alpha \times \omega_1$; such that $x \in T_{<\alpha} \setminus T_0 \,\&\, \mathrm{rk}(x) \in S \Rightarrow f(x) \in \mathrm{rk}(x) \times \omega$; $x \in T_0 \Rightarrow f(x) \in \{0\} \times \omega = 1 \times \omega$; $\mathrm{rk}(x) < \omega_1 \setminus S \Rightarrow f(x) \in (\mathrm{rk}(x) + 1) \times \omega$ and $x < y \Rightarrow f(x) \neq f(y)$, such that

(a) $\beta \in S$, $x \in T_\beta \Rightarrow |\beta \times \omega \setminus \{f(y) : y < x\}| = \aleph_0$, (if β is a non-limit ordinal this holds trivially.)

(b) if $x \in T_\beta, \beta < \gamma < \alpha$, $\{\langle \xi, n \rangle\} \bigcup A \subseteq ((\beta + 1) \times \omega \setminus \{f(z) : z \leq x\})$, $\langle \xi, n \rangle \notin A$ and A is finite, *then* there is $y \in T_\gamma$ such that $x < y$ and $\{f(z) : z \leq y\} \cap A = \emptyset$ but $\langle \xi, n \rangle \in \{f(z) : z \leq y\}$.

We can demand

(c) if α is limit, $\alpha \notin S$ and $x \in T_\alpha$ *then* $f(x) \notin \alpha \times \omega$

and note that we have

(d) if α is limit, $\alpha \in S$ and $x \in T_\alpha$ *then* $f(x) \in (\alpha \times \omega \setminus \{f(z) : z \leq x\})$.

We let $T_\alpha = [\alpha\omega, (\alpha+1)\omega)$ (so T has infinitely many minimal elements; we can add a root). Let $\mathcal{P}_\alpha \subseteq \mathcal{P}(\alpha)$ for $\alpha < \omega_1 \setminus S$ be countable such that $\langle \mathcal{P}_\alpha : \alpha < \omega_1 \setminus S \rangle$ a witness for $\diamondsuit^*_{\omega_1 \setminus S}$. Let us carry the definition.

Case 1: $\alpha = 1$

$\leq_T \restriction T_{<1}$ is the equality.

Let $f \restriction T_0$ be any function from T_0 to $1 \times \omega$.

Case 2: α limit: trivial.

Case 3: $\alpha = \beta + 2$.

Let $\langle A_x : x \in T_\beta \rangle$ be a partition of $T_{\beta+1} = [(\beta+1)\omega, (\beta+2)\omega)$ to (pairwise disjoint) infinite sets. Define the order on $T_{<\alpha}$ by:

for $x, y \in T_{<\alpha} : x <_T y$ iff

$$x <_{T_{<(\beta+1)}} y \text{ or } (\exists z \in T_\beta)(x \leq_{T_{<(\beta+1)}} z \;\&\; y \in A_z).$$

Case 4: $\alpha = \delta + 1, \delta \in S$ (δ limit).

As in the construction of special Aronszajn trees using (b) (and taking care of it).

Case 5: $\alpha = \delta + 1, \delta \notin S$ (δ limit).

Let $\{B^\delta_n : n < \omega\}$ be a list of maximal antichains of $T_{<\delta}$ including all maximal antichains which belongs to $\mathcal{P}_\delta$. Let $\{(\gamma_n, k_n) : n < \omega\}$ enumerate $\delta \times \omega$.

Choose $\langle \beta_n : n < \omega \rangle$ such that $\beta_0 = \beta, \beta_n < \beta_{n+1} < \delta$ and $\bigcup_{n<\omega} \beta_n = \delta$. For each $\beta < \delta, x \in T_\beta$ and finite $A \subseteq ((\beta+1) \times \omega \setminus \{f(z) : z \leq x\})$, we choose by induction on $n, y_n = y^\delta_n[A, x]$, and $\varepsilon_n = \varepsilon_n[A, x]$ such that:

(i) $y_0 = x, \quad y_n \in T_{\varepsilon_n}, y_n \leq_{T_{<\delta}} y_{n+1}, \delta > \varepsilon_n \geq \beta_n$

(ii) A is disjoint to $\{f(z) : z \leq y_n\}$

(iii) let $\ell(n) < \omega$ be minimal such that:

$$\langle \gamma_{\ell(n)}, k_{\ell(n)} \rangle \notin \{f(z) : z \leq y_n\} \cup A,$$

then $(\gamma_{\ell(n)}, k_{\ell(n)}) \in \{f(z) : z < y_{n+1}\}$.

(iv) either $\{z : z < y_{n+1}\} \cap B^\delta_n \neq \emptyset$, or there is no $y \in T_{<\delta}$ satisfying $y_n < y$, $\{f(z) : z \leq y\} \cap A = \emptyset$, $\{z : z < y\} \cap B^\delta_n \neq \emptyset$.

The induction step is by part (b) of the induction hypothesis.

Let $\langle (x^\delta_i, \beta^\delta_i, A^\delta_i) : i < \omega \rangle$ list all triplets (x, β, A) as above (i.e. $x^\delta_i \in T_{\beta^\delta_i}, \beta^\delta_i < \delta, A^\delta_i$ finite $\subseteq ((\beta^\delta_i + 1) \times \omega \setminus \{f(z) : z \leq x^\delta_i\})$.

Now we define

$$y_1 \leq_{T_{<\alpha}} y_2$$

iff for some $\beta < \delta$, $y_1 \leq_{T_{<\beta}} y_2$, or for some $i < \omega$ we have $y_2 = \delta + i$, $\bigvee_{n<\omega} y_1 \leq y^\delta_n[A^\delta_i, x^\delta_i]$. Let $f(\delta + i) = (\delta, 0)$. It is easy to see that we finish this case, too. So we have carried the inductive construction.

Now let us see that $T = T_{<\omega_1} = \bigcup_{\alpha<\omega_1} T_{<\alpha}$ is as required (clearly $f : T \to \omega_1 \times \omega$). Being S-st special is by the requirements

$$\mathrm{rk}(x) \in S \Rightarrow f(x) \in \mathrm{rk}(x) \times \omega$$

$$x < y \Rightarrow f(x) \neq f(y).$$

Now suppose $\mathcal{I} \subseteq T$ is an antichain, $S_1 \overset{\mathrm{def}}{=} \mathrm{rk}(\mathcal{I}) \setminus S$ stationary.
For each $\alpha \in S_1$, let $x_\alpha \in \mathcal{I}$, $\mathrm{rk}(x_\alpha) = \alpha$. Note: $f(x_\alpha) = (\alpha, 0)$. Now, by the definition of $\diamondsuit^*_{\omega_1 \setminus S}$, for some club C of ω_1:

$$\delta \in C \cap S_1 \Rightarrow \{x_\alpha : \alpha < \delta \cap S_1\} \in \mathcal{P}_\delta.$$

Choose $\delta(*) \in C \cap S_1$ such that $M_{\delta(*)} \prec M_{\omega_1}$ where for $\delta \leq \omega_1$ we define

$$M_\delta = \Big(T_{<\delta}, <_{T_{<\delta}}, f{\restriction}T_{<\delta}, \{x_\alpha : \alpha \in S_1 \cap \delta\}\Big).$$

So $\{x_\alpha : \alpha < \delta(*) \cap S_1\} \in \mathcal{P}_{\delta(*)}$ hence for some n we have $\{x_\alpha : \alpha < \delta(*) \cap S_1\} = B_n^{\delta(*)}$ and for some $i < \omega$ we have $x_{\delta(*)} = \delta(*) + i$. Now $x_{\delta(*)}$ (which is well defined as $\delta(*) \in S_1$) is in $T_{\delta(*)}$ and $y_m^{\delta(*)}[A_i^{\delta(*)}, x_i^{\delta(*)}] <_T x_{\delta(*)}$.

As $\alpha \in S_1 \Rightarrow f(x_\alpha) = \langle \alpha, 0 \rangle$, by the choice of $y_{n+1}^{\delta(*)}[A_i^{\delta(*)}, x_i^{\delta(*)}]$ (clause (iv)) and by the choice of C

$$\{z : z < y_{n+1}^{\delta(*)}[A_i^{\delta(*)}, x_i^{\delta(*)}]\} \cap \{x_\alpha : \alpha < \delta(*)\} \neq \emptyset$$

contradicting $\{x_\alpha : \alpha \in S_1\}$ being an antichain. $\square_{3.9}$

3.10 Lemma. ($\Diamond_{\omega_1}$) There is a special Aronszajn tree T, such that for no antichain $\mathcal{I} \subseteq T$ is $\mathrm{rk}(\mathcal{I})$ closed unbounded. (For stationary: there is necessarily: this is mentioned in Devlin and Shelah [DvSh:65] p. 25).

Remark. E.g., MA + $2^{\aleph_0} > \aleph_1$ implies that this fails.

Proof. We define by induction on α, $(T_{<\alpha}, <_T \restriction T_{<\alpha})$ and $f : T_{<\alpha} \to \mathbb{Q}$ monotonic, so that $\beta < \gamma < \alpha$, $x \in T_\beta$, $\varepsilon > 0$ implies that for some $y \in T_\gamma$ we have $x < y \,\&\, f(x) < f(y) < f(x) + \varepsilon$. For limit $\delta < \omega_1$ we are given an antichain $\mathcal{I}^\alpha \subseteq T_{<\alpha}$ (by $\Diamond_{\aleph_1}$) and demand that for $x \in T_\alpha$, either

$$(\exists y \in \mathcal{I}^\alpha) y < x$$

or $(\exists y < x)$ [there is no z, $y < z \in \mathcal{I}^\alpha, f(z) \leq f(x) \in \mathbb{Q}$].
[How? For each $y \in T_{<\delta}$ and rational $\varepsilon > 0$, choose if possible, $z = z_{y,\varepsilon}$ such that $y \leq_{T_{<\delta}} z \in \mathcal{I}_\delta$, $f(z) < f(y) + \varepsilon$, if not let $z_{y,\varepsilon} = y$. Let $q_{y,\varepsilon} = f(z_{y,\varepsilon}) + (f(y) + \varepsilon - f(z_{y,\varepsilon}))/2$. So we can demand that for every $x \in T_\delta$ for some $y \in T_{<\delta}$ and rational $\varepsilon > 0$ we have $z_{y,\varepsilon} <_{T_{<(\delta+1)}} x$ and $f(x) = q_{y,\varepsilon}$.]
The checking is easy. See the last two paragraphs of the proof of 3.7. $\square_{3.10}$

3.11 Lemma. T is $\{\alpha + 1 : \alpha < \omega_1\}$-special iff T is r-special.

Remark. Proved by Baumgartner [B1].

Proof. The direction $\Leftarrow$ already appears.

For $\Rightarrow$ let f $\{\alpha+1 : \alpha < \omega_1\}$-specialize T.

Let $g : \mathbb{Q} \to \mathbb{Q}$ and $\varepsilon : \mathbb{Q} \to \{1/n : n > 0 \text{ natural}\}$ be such that the intervals $[g(q) - \varepsilon(q), g(q) + \varepsilon(q)]$ are pairwise disjoint and g is order preserving (possible: let $\mathbb{Q} = \{q_n : n < \omega\}$ and define $f(q_n)$, $\varepsilon(q_n)$ by induction on n).

Now define f^* as follows: $x \in T_{\alpha+1} \Rightarrow f^*(x) = g(f(x))$

$$x \in T_\alpha, \alpha \text{ limit} \Rightarrow f^*(x) = \text{Sup}\{g(f(y)) : y < x, y \in T_{\beta+1}, \beta < \alpha\}\,.$$

Now f^* r-specializes T; the only point to check is:

$$x \in T_{\alpha+1}, \alpha \text{ limit} \Rightarrow g(f(x)) > \text{Sup}\{g(f(y)) : y < x, y \in T_{\beta+1}, \beta+1 < \alpha\}$$

which follows by g's definition (the sup is $\leq g(f(x)) - \varepsilon(f(x))$ as for every $y < x$, $g(f(x))$ is smaller than it). $\square_{3.11}$

§4. Independence Results

It is well known that

4.1 Claim. If T is an $\aleph_1$-Souslin tree, $\lambda > \aleph_1$, $N \prec (H(\lambda), \in)$, $||N|| = \aleph_0$ $T \in N, x \in T_\delta$, $\delta = \omega_1 \cap N$ *then* $B_T(x) = \{y \in T_{<\beta} : y < x\}$ is generic for (T, N), i.e., for every $\mathcal{I} \in N, \mathcal{I} \subseteq T$ which is pre-dense in T, we have:

$$\mathcal{I} \cap B_T(x) = \mathcal{I} \cap N \cap B_T(x) \neq \emptyset.$$

4.2 Definition. 1) For an Aronszajn tree T,
$Q(T) = \{(h, f) : h$ is a partial function from ω_1 to ω_1;

$$\alpha < \beta \;\&\; \{\alpha, \beta\} \in \text{Dom}(h) \Rightarrow 0 < \alpha \leq h(\alpha) < \beta \leq h(\beta);$$

f is a finite function,

$$\mathrm{Dom}(f) \subseteq \bigcup_{\alpha\in \mathrm{Dom}(h)} T_{h(\alpha)} ;$$
$$x \in T_{h(\alpha)} \Rightarrow f(x) \in \alpha \times \omega;$$

$$x < y \,\&\, [x, y \in \mathrm{Dom}(f)] \Rightarrow f(x) \neq f(y)\}$$

The order on $Q(T)$ is defined by $(h, f) \leq (h^\dagger, f^\dagger)$ if $h \subseteq h^\dagger, f \subseteq f^\dagger$;

We let $(h, f) \bigcup (h^\dagger, f^\dagger) \stackrel{\text{def}}{=} (h \bigcup h^\dagger, f \bigcup f^\dagger)$, $(h, f) \bigcup h^\dagger \stackrel{\text{def}}{=} (h \bigcup h^\dagger, f)$, $(h, f) \bigcup f^\dagger \stackrel{\text{def}}{=} (h, f \bigcup f^\dagger)$.

2) We say that T is (S, h)-st-specialized by f if: S is a stationary subset of ω_1, and h is a function from S to ω_1 satisfying $(\forall \alpha \in S)(\alpha \leq h(\alpha))$ and h is a function with domain $\bigcup\limits_{\alpha\in S} T_{h(\alpha)}$ satisfying

$$\alpha \in S \;\&\; x \in T_\alpha \Rightarrow f(x) \in \alpha \times \omega \quad \text{and}$$

$$\alpha < \beta \;\&\; x \in T_{h(\alpha)} \;\&\; y \in T_{h(\beta)} \;\&\; x <_T y \Rightarrow F(x) \neq f(y).$$

So T is (S, h)-st-special if some f (S, h)-st-specialized it.

(Easily implies T is not Souslin.)

4.3 Definition. For an Aronszajn tree T and stationary set S, $Q(T, S) = \{(h, f) : (h, f) \in Q(T)$, and $\alpha \in (\mathrm{Dom}(h)) \cap (S \setminus \{0\})$ implies $h(\alpha) = \alpha\}$, order – as before.

Explanation. Our aim is to get a universe in which SH (Souslin Hypothesis) holds (i.e., there is no Souslin tree) but not every Aronszajn tree is special. The question was raised by Baumgartner Malitz and Reinhardt [BMR], and later independently by U. Abraham, and is natural as, until now, the consistency of SH was proved by making every Aronszajn tree special; see the proof of Solovay and Tennenbaum [ST], Martin and Solovay [MS], Baumgartner, Malitz and Reinhard [BMR] without CH, and Jensen proof in Devlin and Johnsbraten [DeJo] with CH (and Laver and Shelah [LvSh:104] for $\aleph_2$-Souslin tree). For this aim we have introduced in §3 various notions of specializations (each implying the tree is not Souslin). So the program is to make every tree special in some weaker than the usual sense. The notion r-special which had been introduced by Kurepa [Ku35] is not suitable, as if every Aronszajn tree is r-special then every Aronszajn tree is special (see 3.6(2)). Similarly "h-special" for any fixed increasing $h : \omega_1 \to \omega$ is not suitable by 3.6(1) (see Definition 3.2(4)).

So a natural candidate is "h-special for some h" (i.e., for every tree there is an h for which it is h-special). Forcing with $Q(T)$ does the job for T – we take generic h and f. (It would be more natural to let f go to $\mathbb{Q}$ and be monotonically increasing, but by 3.5(2) the forcing $Q(T)$ makes T h-special for some h, and this way we have more uniformity with Definition 4.3.) So we should iterate such forcings, but retain some T as not special.

A second way is to make each T $S - st$-special for some fixed stationary S; for this $Q(T, S)$ is tailored. (Note that the f we get from a generic subset of $Q(T, S)$ has domain $\bigcup_{\alpha \in S_1} T_\alpha$ where $S \setminus S_1$ non-stationary.) For $S = \emptyset$ we get the previous case, so we shall ignore $Q(T)$.

This leads to a secondary problem: Can every Aronszajn tree be $S_1 - st$-special, but some Aronszajn trees are not $S_2 - st$-special ($S_2 \setminus S_1$ stationary, of course)? We answer positively.

4.4 Claim. 1) For T an Aronszajn tree, $S \subseteq \omega_1$, $Q(T, S)$ is proper.
2) For T an Aronszajn $\aleph_1$-tree, and $S \subseteq \omega_1$ we have:

$$\Vdash_{Q(T,S)} \text{"}T \text{ is not Souslin tree, in fact for some function } \underset{\sim}{h} \text{ is } (\omega_1, \underset{\sim}{h})\text{-st-special".}$$

3) In part (2), if S is stationary then $\Vdash_{Q(T,S)}$ "T is S-st-special".

Proof. We can assume w.l.o.g. $|T_0| = \aleph_0$. Let $\lambda > (2^{\aleph_1})$, $N \prec (H(\lambda), \in)$ be countable, $T, S \in N$, $p_0 = (h, f) \in Q(T, S) \cap N$, and let $\delta = N \cap \omega_1$.

Then $p_1 = (h \bigcup \{\langle \delta, \delta \rangle\}, f) \in Q(T, S)$ exemplifies what is required.

For checking, we really repeat the proof of Baumgartner Malitz and Reinhardt [BMR] that the standard forcing (now) for specializing an Aronszajn tree satisfies the $\aleph_1$-c.c. (or read the proof of demand (iii) in the proof of 4.6 – it is just harder).

2) Let $\underset{\sim}{h} = \bigcup \{h : (h, f) \in \underset{\sim}{G}_{Q(T,S)}\}$, and $\underset{\sim}{C} = \mathrm{Dom}(\underset{\sim}{h})$ and $\underset{\sim}{f} = \bigcup \{f : (h, f) \in$

$\underset{\sim}{G}_{Q(T,S)}\}$. We know (see III) that it is forced that $\underset{\sim}{C}$ is a club of ω_1, so T becomes $(\underset{\sim}{C}, \underset{\sim}{h})$-st-special.

3) Should be clear. $\square_{4.4}$

4.5 Definition. We call a forcing notion P, (T^*, S)-preserving (do you have a better name?), where T^* is an Aronszajn tree, $S \subseteq \omega_1$, if: for every $\lambda > (2^{|P|+\aleph_1})^+$, $\langle P, T^*, S\rangle \in N \prec (H(\lambda), \in), N$ countable, $\delta \overset{\text{def}}{=} N \cap \omega_1 \notin S$ and $p \in N \cap P$, there is p_1 which is preserving for (p, N, P, T^*, S); i.e.,

(i) $p \leq p_1 \in P$,

(ii) p_1 is (N, P)-generic,

(iii) for every $x \in T^*_\delta$, *if*

(∗) $x \in A \to (\exists y < x)(y \in A)$ holds for every $A \subseteq T^*, A \in N$,

then

(∗∗) for every P-name $\underset{\sim}{A}, \underset{\sim}{A} \in N$ such that $\Vdash_P$ "$\underset{\sim}{A} \subseteq T^*$"

the following holds:

$$p_1 \Vdash \text{“}x \in \underset{\sim}{A} \to (\exists y < x) y \in \underset{\sim}{A}\text{”}.$$

4.6 Lemma. If T^*, T are Aronszajn trees, $S \subseteq \omega_1$, *then* $Q(T, S)$ is (T^*, S)-preserving.

Remark. If T^* is Souslin tree then (∗) from Definition 4.5 is satisfied by every countable $N \prec (H(\lambda), \in)$ and $x \in T^*_\delta$ when $N \cap \omega_1 = \delta$ (this follows by 4.1).

Proof. Let $P \overset{\text{def}}{=} Q(T, S)$. Let $N \prec (H(\lambda), \in)$, $\delta \overset{\text{def}}{=} N \cap \omega_1 \notin S$, $||N|| = \aleph_0$, $\langle T^*, T, S\rangle \in N$ hence $P \in N$, $p = (h_0, f_0) \in P \cap N$ (as in Definition 4.5.), and (remembering $\delta = N \cap \omega_1$) let $\delta^* = \sup\{f(\delta) + 1 : f \in N, f(\delta)$ is an ordinal $< \omega_1\}$.

Define $p_1 = (h_0 \bigcup \{\langle \delta, \delta^*\rangle\}, f_0)$, demands (i) of Definition 4.5 is trivial, and demand (ii): its proof is easier than that of demand (iii), so let us check condition (iii). So suppose $x \in T^*_\delta$ and

(∗) if $A \subseteq T^*, A \in N, x \in A$ then $(\exists y)(y <_{T^*} x \,\&\, y \in A)$.

Let $\underset{\sim}{A}$ be a $Q(T,S)$-name of a subset of T^*, and $\underset{\sim}{A} \in N$. We shall prove that for every $p_2, p_1 \le p_2 \in Q(T,S)$, for some $p_3, p_2 \le p_3 \in Q(T,S)$, and $p_3 \Vdash$ "$x \notin \underset{\sim}{A}$" or $p_3 \Vdash$ "$y \in \underset{\sim}{A}$ for some $y <_{T^*} x$".

Let $p_2 = (h_2, f_2)$, if $p_2 \Vdash_P$ "$x \notin \underset{\sim}{A}$", then we can choose $p_3 = p_2$. Otherwise there is $p_2^\dagger \in P$, such that

$$p_2 \le p_2^\dagger \text{ and } p_2^\dagger \Vdash_P \text{“} x \in \underset{\sim}{A} \text{”}.$$

Let $p_2^\dagger = (h_2^\dagger, f_2^\dagger)$, $p_2^\dagger = p_2^a \bigcup p_2^b$, where $p_2^a = (h_2^a, f_2^a)$, $p_2^b = (h_2^b, f_2^b)$ where $h_2^a = h_2^\dagger \restriction \delta$, $h_2^b = h_2^\dagger \restriction [\delta, \omega_1)$ (closed open interval) and

$$f_2^a = f_2^\dagger \restriction T_{<\delta}, \quad f_2^b = f_2^\dagger \restriction (T \setminus T_{<\delta}).$$

Note that by the definition of $Q(T,S)$:

4.6A Fact.

(1) $p_2^a \in P \cap N$,

(2) $z \in \mathrm{Dom}(f_2^b) \Rightarrow \mathrm{rk}_T(z) \ge \delta^*$.

Now let $\alpha_0 = \mathrm{SupRang}(h_2^a)$ (which is $< \delta$) and we define a function F as follows:

$\mathrm{Dom}(F) = \{y \in T^* : \mathrm{rk}(y) >_{T^*} \alpha_0\}$,

$F(y) = \mathrm{Sup}\{\alpha^* < \omega_1 :$ there is $({}^*h_2^b, {}^*f_2^b)$ (in $Q(T,S)$) such that:

(a) $\mathrm{Min}(\mathrm{Dom}({}^*h_2^b)) = \mathrm{rk}_{T^*}(y)$,

(b) ${}^*h_2^b(\mathrm{rk}_{T^*}(y)) = \alpha^*$,

(c) $(h_2^a \bigcup {}^*h_2^b, f_2^a \bigcup {}^*f_2^b) \Vdash_{Q(T,S)}$ "$y \in \underset{\sim}{A}$"$\}$

(so we demand also that $(h_2^a \bigcup {}^*h_2^b, f_2^a \bigcup {}^*f_2^b)$ is in $Q(T,S)$).

Now clearly $F \in N$ (as it is defined by a (first-order) formula in $(H(\lambda), \in)$ whose parameters are in N). Clearly $F(y) \le \omega_1$ (for $y \in T^* \setminus T^*_{\le \alpha_0}$). Let $A^* = \{y \in T^* : \mathrm{rk}_{T^*}(y) > \alpha_0,\ F(y) = \omega_1\}$. (Note that $A^* \subseteq T^*$ is a set, not a P-name of a set.).

Let F^* be a function from ω_1 to ω_1 defined by:

$$F^*(\alpha) = \mathrm{Sup}\{F(y)+1 : y \in T^*_{\leq\alpha}, \mathrm{rk}_{T^*}(y) > \alpha_0, y \notin A^*, \text{ i.e., } F(y) < \omega_1\}.$$

As $|T^*_{\leq\alpha}| \leq \aleph_0$, we have $F^* : \omega_1 \to \omega_1$, and clearly $F^* \in N$ (same reason).

By the definition of δ^*, $F^*(\delta) < \delta^*$. But (h_2^b, f_2^b) exemplify $F(x) \geq \delta^*$, so necessarily $F(x) = \omega_1$. So by the definition of A^* above $x \in A^*$. Hence by the hypothesis $(*)$ there is $y <_{T^*} x$ such that $y \in A^*$. So (in $H(\lambda)$, hence in N) we can define a sequence $\bar{p} = \langle (h_2^{b,i}, f_2^{b,i}) : i < \omega_1 \rangle$ such that:

(a)′ $\mathrm{Min}(\mathrm{Dom}(h_2^{b,i})) = \mathrm{rk}_{T^*}(y) > \alpha_0$,

(b)′ $h_2^{b,i}(\mathrm{rk}(y)) \geq \alpha_0 + i$,

(c)′ $(h_2^a \bigcup h_2^{b,i}, f_2^a \bigcup f_2^{b,i}) \Vdash_{Q(T,S)}$ "$y \in \underset{\sim}{A}$".

For $i < \delta$ let $p_3^i = (h_2^a \bigcup h_2^b \bigcup h_2^{b,i}, f_2^a \bigcup f_2^b \bigcup f_2^{b,i})$. If $p_3^i \in Q(T,S)$ then by clause (c)′ and as $y <_T x$, this condition is as required.

Why can p_3^i be not in $Q(T,S)$? The first coordinate $(h_2^a \bigcup h_2 \bigcup h_2^{b,i})$ is O.K., as $h_2^a \subseteq h_2^{b,i} \in N$.

What about the second? Note that $f_2^a \bigcup f_2^b, f_2^a \bigcup f_2^{b,i}$ are O.K. as $p_2 \in Q(T,S)$ and by (c)′ above correspondingly. Hence the only danger is that there are $z_1 \in \mathrm{Dom}(f_2^b), z_2 \in \mathrm{Dom}(f_2^{b,i})$, $z_2 <_T z_1$ (as $f_2^{b,i} \in N$, $rk(z_1) \geq h_2^b(\delta) = \delta^*$ this is the only bad possibility).

But remember that in $H(\lambda)$ we have $z \in \mathrm{Dom}(f_2^{b,i}) \Rightarrow rk(z) \geq i$, so by a lemma on Aronszajn trees due to Baumgartner, Malitz and Reinhart (in their proof of $MA \vdash$ "every Aronszajn tree is special") which appears in the proof of III 5.4, there is a sequence $\langle i_n : i < \omega \rangle (i_n < \omega_1)$ such that

$$m \neq n \;\&\; z_1 \in \mathrm{Dom}(f_2)^{b,i_m} \;\&\; z_2 \in \mathrm{Dom}(f_2^{b,i_n}) \Rightarrow \begin{pmatrix} z_1 \not<_T z_2 \\ z_2 \not<_T z_1 \end{pmatrix}.$$

So again there is such a sequence in N, and all but at most $|\mathrm{Dom}(f_2^b)|$ are O.K., i.e., $p_3^i \in Q(T,S)$. So we finish. $\square_{4.6}$

4.7 Theorem. Let T^* be a Souslin tree. Suppose $P_\alpha(\alpha \leq \alpha_0)$, $\underset{\sim}{Q}_\alpha(\alpha < \alpha_0)$ form an $\aleph_1$-free iteration (i.e., $P_{\alpha+1} = P_\alpha * \underset{\sim}{Q}_\alpha$, $P_\delta = \mathrm{Flim}^{\aleph_1}_{\alpha<\alpha_0} P_\alpha$) and for every α at least one of the following holds:

(α) $\underset{\sim}{Q}_\alpha$ is (in V^{P_α}) (T^*, S)-preserving,

(β) there is a P_α-name $\underset{\sim}{\mathcal{I}}_\alpha$ of an antichain of T^* (in V^{P_α}), $\underset{\sim}{S}_\alpha \stackrel{\text{def}}{=} \text{rk}(\underset{\sim}{\mathcal{I}}_\alpha) \subseteq \omega_1 \backslash S$ where $\text{rk}(\underset{\sim}{\mathcal{I}}_\alpha) = \{\text{rk}(x) : x \in \underset{\sim}{\mathcal{I}}_\alpha\}$, and in V^{P_α}:

$\underset{\sim}{Q}_\alpha = Q_{club}(\omega_1 \setminus S_\alpha) = \{g : \text{for some } i < \omega_1, \text{Dom}(g) = i+1, \text{Rang}(g) = \{0,1\},$ $\{j \leq i : g(i) = 1\}$ is closed and is $\subseteq \omega_1 \setminus S_\alpha\}$,
Then P_{α_0} is (T^*, S)-preserving.

4.7A Remark. 1)We can amalgamate conditions (α) and (β) but it has no use.
2) See on such theorems in XVIII§3.
3) Note that the forcing notion $Q(T, S)$ (T is an Aronszajn tree, $S \subseteq \omega_1$ co-stationary) adds an antichain $\mathcal{I}$ of T such that $\text{rk}(\mathcal{I}) \setminus S$ is stationary. This is because by the proof of 4.4 $\Vdash_{Q(T,S)}$ "$\{\delta < \omega_1$: for some $(h, f) \in G_{Q(T,S)}$, we have $\delta \in \text{Dom}(h)$, $h(\delta) = \delta \in S\}$ is stationary" (together with Fodor's lemma).

Proof. We prove by induction on $\alpha \leq \alpha_0$ the following:

$\oplus_\alpha$ Suppose $\beta < \alpha \leq \alpha_0$, $N \prec (H(\lambda), \in)$, $\beta \in N, \alpha \in N$, $\langle P_i : i \leq \alpha\rangle \in N$, $\delta = N \cap \omega_1 \notin S$, $p \in P_\alpha \cap N$, $q_1 \in P_\beta$, and

- (i) (a) $p{\restriction}\beta \leq q_1$ (natural meaning: no $q^\dagger$, $q_1 \leq q^\dagger \in P_\beta$ is incompatible with p; if we deal with complete BA, $p{\restriction}\beta$ is the projection).
  (b) Moreover, if $p{\restriction}\beta \leq p^\dagger \in P_\beta \cap N$, then $q_1, p^\dagger$ are compatible;
- (ii) q_1 is (N, P_β)-generic,
- (iii) if $x \in T^*_\delta$ and $(\forall A \subseteq T^*)(A \in N \& x \in A \rightarrow (\exists y < x) y \in A)$ then for very P_β-name $\underset{\sim}{A} \in N$, we have $q_1 \Vdash_{P_\beta}$ "$x \in \underset{\sim}{A} \rightarrow (\exists y <_{T^*} x) y \in \underset{\sim}{A}$".
  Then there is $p_1 \in P_\alpha$ such that
- (i)′ (a) $p_1{\restriction}\beta = q_1$, $p \leq p_1$ (natural meaning).
  (b) if $p{\restriction}\alpha \leq p^\dagger \in P_\alpha \cap N$ then $p_1, p^\dagger$ are compatible,
  moreover (this implies (a)+(b)): *if* $p{\restriction}\alpha \leq p^\dagger \in P_\alpha \cap N$, $q_1 \leq q^\dagger \in P_\beta$, $p^\dagger{\restriction}\beta \leq q^\dagger$ *then* $p^\dagger, p_1, q^\dagger$ are compatible (= have an upper bound),

(ii)′ p_1 is (N, P_α)-generic,

(iii)′ the parallel of (iii) with $\beta \mapsto \alpha, q_1 \mapsto p_1$.

Case 1. $\alpha = 0$. Trivial.

Case 2. $\alpha + 1$. By the similarity between the assumptions on q_1 and the conclusion on p_1, we can assume w.l.o.g. $\beta = \alpha$. Let $G \subseteq P_\alpha$ be generic over V, $q_1 \in G$. Then $N[G] \prec (H(\lambda), \in)$ (see III 2.11).

Now in $V[G]$ (hence in $H(\lambda)[G]$) we can find in $\underset{\sim}{Q}_\alpha[G_{P_\alpha}]$ a condition $p_1^\dagger \geq p(\alpha)$, which is $(N[G], Q_\alpha[G])$-generic, as in Definition 4.5. Why?

Note that we can ignore (i)′(b), as we can take a disjunction over countably many possibilities one for each $r \in N[G]$, $P_{\alpha+1}/G$, $r \geq p$. More accurately, maybe in Q_α it does not exist, but we can make a trivial change in Q_α to ensure it, without affecting the iteration (in fact, the forcing notion we actually use has such a condition anyway).

Now our proof splits according to which of the conditions (α) or (β) from the theorem, $\underset{\sim}{Q}_\alpha[G]$ satisfies.

(α) Straightforward, by Definition 4.5.

(β) By the choice of T^* (a Souslin tree), by Claim 4.1 we have: $[x \in A \in N$ and $x \in T^*_\delta \Rightarrow (\exists y < x) y \in A]$. So by the assumption on q_1 for every $A \in V[G], A \subseteq T^*$, $A \in N[G]$, of course there is a P_α-name $\underset{\sim}{A} \in N$, $\underset{\sim}{A}[G] = A$; now we know $q_1 \Vdash_{P_\alpha}$ "$x \in \underset{\sim}{A} \rightarrow (\exists y < x) y \in \underset{\sim}{A}$", hence in $V[G], x \in A \cap T^*_\delta \Rightarrow (\exists y < x) y \in A$.

In particular, we can take $A = \underset{\sim}{\mathcal{I}}_\alpha[G] \in N[G]$ (remember $\underset{\sim}{\mathcal{I}}_\alpha[G] \in V^{P_\alpha}$ and as $\langle P_i : i \leq \alpha \rangle \in N$ hence w.l.o.g. $\langle \underset{\sim}{I}_j : j < \alpha$ and if $\underset{\sim}{Q}_j$ satisfies clause (β), $\underset{\sim}{I}_j$ is as there, otherwise $\underset{\sim}{I}_j = \emptyset \rangle$). So clearly if $x \in T^*_\delta \cap \underset{\sim}{\mathcal{I}}_\alpha[G]$ then $x \in A$ implies $\underset{\sim}{\mathcal{I}}_\alpha[G]$ is not an antichain, contradiction. So $T^*_\delta \cap \underset{\sim}{\mathcal{I}}_\alpha[G] = \emptyset$, so $\delta \notin \underset{\sim}{S}_\alpha[G] = \mathrm{rk}(\underset{\sim}{\mathcal{I}}_\alpha)[G]$, and then the desired conclusion is quite easy (remember Q_α's definition).

So we have p_1 as required. Now p_1 is in V^{P_α}, so in V we have a P_α-name $\underset{\sim}{p}_1$ for it, and let $p_1 = (q_1, \underset{\sim}{p}_1^\dagger) \in P_\alpha * \underset{\sim}{Q}_\alpha$ which by the usual thing for composition of forcing, is as required.

Case 3. α limit.

Choose α_n for $n < \omega$ such that $\beta = \alpha_1 < \dots \alpha_n < \alpha_{n+1} < \dots; \alpha_n \in N$ and $\alpha(*) \stackrel{\text{def}}{=} \sup(\alpha \cap N) = \bigcup_{0<n<\omega} \alpha_n$.

We define by induction on $n < \omega, n \geq 1$, $q_n \in P_{\alpha_n}$, $q_{n+1} \restriction \alpha_n = q_n$, q_1 is given and q_{n+1} is obtained from q_n by the induction hypothesis, with $p, q_n, q_{n+1}, \alpha_n, \alpha_{n+1}$ here standing for $p, q_1, p_1, \alpha, \beta$ there.

Let $\langle (\underset{\sim}{A}_n, x_n) : n < \omega \rangle$ be a list of all pairs $(\underset{\sim}{A}, x)$, where $\underset{\sim}{A}$ is a P_α-name of a subset of T^*, $x \in T^*_\delta$ and $\underset{\sim}{A}$ is in N; let $\langle \mathcal{J}_n : n < \omega \rangle$ be a list of all pre-dense subsets of P_α which belong to N. Let

$p_1 \stackrel{\text{def}}{=} p \wedge \bigwedge_n q_n \wedge \bigwedge_n (\bigvee_{r \in \mathcal{J}_n \cap N} r) \wedge \bigwedge_{n<\omega} [\bigvee \{p \in P_\alpha \cap N : p \Vdash_{P_\alpha} \text{"} y \in \underset{\sim}{A}_n \text{"}$ for some $y <_{T^*} x_n\} \vee \bigvee \{q_n \wedge \bigwedge_{r \in \mathcal{J}} r : \mathcal{J} \subseteq N \cap P_\alpha, \mathcal{J}$ is definable in $(N, \{y : y < x\})$ and $q_n \wedge \bigwedge_{r \in \mathcal{J}} r \Vdash_{P_\alpha} \text{"} x_n \notin \underset{\sim}{A}_n \text{"}\}]$.

There are two facts on p_1 we have to prove:

(A) $p_1 \in P_\alpha = \text{Flim}^{\aleph_1}_{i<\alpha(*)} P_i$, i.e., $\bigwedge_{i<\alpha(*)} \theta[P_i] \nvdash p_1$ (as clearly p_1 has the right form),

(B) (i)$'$, (ii)$'$, (iii)$'$ (of $\oplus_\alpha$ above) hold.

For proving both facts we do the following. We assume everything is in some countable transitive model M (or $M \Rightarrow V, V \Rightarrow V^*$, in V^* we have $|H(\lambda)^V|$ is countable which is easy by forcing).

Let $p^\dagger, q^\dagger$ be as in (i)$'$ (the "moreover" version).

We let $G_{\alpha_1} \subseteq P_{\alpha_1} = P_\beta$ be generic (i.e., M-generic) such that $p^\dagger \restriction \beta, q^\dagger \in G_{\alpha_1}$.

We shall find $G_{\alpha(*)} \subseteq P_{\alpha(*)}$ such that for each n the set $G_{\alpha(*)} \cap P_{\alpha_n}$ is generic (for (M, P_{α_n})), and the truth values it gives to all $p \in \bigcup_{n<\omega} P_{\alpha_n}$ make $p_1 \wedge p^\dagger$ true (so we have, in V, a model exemplifying $\bigwedge_{i<\alpha(*)} \theta[P_i] \nvdash \neg(p_1 \wedge p^\dagger)$ (= fact (A)), and $G_{\alpha_1} \subseteq G_{\alpha(*)}$.

As for fact (B), clause (B) (ii)$'$ holds trivially by the definition of p_1 (i.e., $\bigwedge_n (\bigvee_{r \in \mathcal{J}_n} r)$). Similarly the last conjunct takes care of (B) (iii)$'$.

The "moreover" phrase of (B) (i)$'$ holds by the free choice of $p^\dagger, q^\dagger$ (and the way $G_{\alpha_1}, G_{\alpha(*)}$ are chosen), hence $p_1 \restriction \beta = q_1$; the other inequality follows by p_1's definition. So it is enough to find $G_{\alpha(*)}$.

We define by induction G_{α_n}, p_n such that (as in the proof of 2.2):

(1) $G_{\alpha_n} \subseteq P_{\alpha_n}, G_{\alpha_n} \subseteq G_{\alpha_{n+1}}$,

(2) G_{α_n} is P_{α_n}-generic over M,

(3) $p_n \leq p_{n+1}, p_0 = p^\dagger, p_n \in P_\alpha \cap N$,

(4) p_n is compatible (in P_α) with every member of $G_{\alpha_n}, q_n \in G_{\alpha_n}$,

(5) p_{3n+1} is $\geq q_n^\dagger$ for some $q_n^\dagger \in \mathcal{J}_n \cap N$,

(6) $p_{3n+2} \vdash \wedge\Phi_n$ or $p_{3n+2} \vdash \neg r_n$ for some $r_n \in \Phi_n$, where $\langle \Phi_n : n < \omega\rangle$ is a list of all countable $\Phi \subseteq P_\alpha, \Phi \in N$,

(7) in $M[G_{\alpha_n}]$ for every $A \in N[G_{\alpha_n}]$, $A \subseteq T^*$ we have $[x \in T^*_\delta \,\&x \in A \rightarrow (\exists y < x) y \in A]$ holds ($q_n \in G_{\alpha_n}$ do the job),

(8) $p_{3n+3} \Vdash_{P_\alpha}$ "$(\exists y < x_n) y \in \underset{\sim}{A}_n$" or $p_{3n+3} \wedge \bigwedge_{r \in \mathcal{J}} r \Vdash_{P_\alpha}$ "$x \notin \underset{\sim}{A}_n$", for some $\mathcal{J}$, such that $\mathcal{J} \subseteq G_{\alpha_{n-1}}$, and $\mathcal{J}$ is definable in $(N, \{y : y <_{T^*} x_n\})$ (remember $\{(\underset{\sim}{A}_n, x_n) : n < \omega\}$ list the pairs $(\underset{\sim}{A}, x)$, $\underset{\sim}{A} \in N$ a P_α-name of a subset of T^*, and $x \in T^*_\delta$.)

As in the proof of 2.2, this suffices [for $\mathcal{J}$ as in (8), use the conjunct corresponding to $\mathcal{J} \cup \{p_{3n+3}\}$ in p_1]. The only nontrivial part in the definition is taking care of (8). So let $n = 3k+2, p_n, G_{\alpha_n}$, be defined, and we shall define $p_{n+1}, G_{\alpha_{n+1}}$. We define:

$\underset{\sim}{A}_k^\dagger = \{y \in T^* :$ there is $r \in P_\alpha, r \geq p_n$, which is compatible with every member of $\underset{\sim}{G}_{\alpha_n}$(= the name of the generic subset of P_{α_n}) such that $r \Vdash_{P_\alpha}$ "$y \in \underset{\sim}{A}_k$"$\}$.

Clearly $\underset{\sim}{A}_k^\dagger$ is a P_{α_n}-name (as we use $\underset{\sim}{G}_{\alpha_n}$ in the definition but not $\underset{\sim}{G}_{\alpha_n+1}$) and if $p_n \restriction \alpha_n \leq r \in P_{\alpha_n}$ then:

(*) $r \Vdash_{P_{\alpha_n}}$ "$y \notin \underset{\sim}{A}_k^\dagger$" implies $r \wedge p_n \Vdash_{P_{\alpha_n}}$ "$y \notin \underset{\sim}{A}_k$".

However the inverse implication does not follow. Now if we can choose p_{n+1}, such that $p_n \leq p_{n+1} \in P_\alpha \cap N, p_{n+1}$ compatible with every member of G_{α_n} (equivalently of $G_{\alpha_n} \cap N$) such that $p_{n+1} \Vdash_{P_\alpha}$ "$y \in \underset{\sim}{A}_k$" for some $y <_{T^*} x_k$, then we can proceed to define $G_{\alpha_{n+1}}$ with no problem.

So we assume that there is no such p_{n+1} and let $p_{n+1} = p_n$. Let

$$\mathcal{J} = \{\neg r : r \in P_{\alpha_n} \cap N \, , r \Vdash_{P_{\alpha_n}} \text{“} y \in \underset{\sim}{A}^\dagger_k \text{” for some } y <_{T^*} x_k\}.$$

Clearly $\mathcal{J}$ is definable in $(N, \{y : y <_{T^*} x_k\})$, $\mathcal{J} \subseteq P_{\alpha_n} \cap N$, and $\mathcal{J} \subseteq G_{\alpha_n}$ (as if $(\neg r) \in \mathcal{J}$, $(\neg r) \notin G_{\alpha_n}$ then $r \in G_{\alpha_n}$ so we would not have arrive here), and $p_n \leq p_{n+1} \in P_\alpha \cap N$ and $p_{n+1} \in P_\alpha / G_{\alpha_n}$ so it is enough to prove

(**) $p_n \wedge q_n \wedge \bigwedge_{r \in \mathcal{J}} r \Vdash_{P_\alpha}$ “$x_k \notin \underset{\sim}{A}_k$”.

Now $\underset{\sim}{A}^\dagger_k$ is a P_{α_n}-name of a subset of T^* (and it belongs to N), so by the choice of q_n:

$$q_n \Vdash_{P_{\alpha_n}} \text{“} x_k \in \underset{\sim}{A}^\dagger_k \rightarrow (\exists y <_{T^*} x_k) y \in \underset{\sim}{A}^\dagger_k \text{”}.$$

However for each $y <_{T^*} x_k$,

$$\mathcal{J}_y = \{r \in P_{\alpha_n} : r \Vdash_{P_{\alpha_n}} \text{“} r \in \underset{\sim}{A}^\dagger_k \text{” or } r \Vdash_{P_{\alpha_n}} \text{“} y \notin \underset{\sim}{A}^\dagger_k \text{”}\}$$

is a dense subset of P_{α_n} which belongs to N hence $\mathcal{J}_y \cap N$ is pre-dense above q_n (in P_{α_n}) (as $y \in N$). So q_n forces that if $y \in \underset{\sim}{A}^\dagger_k (y <_{T^*} x_k)$ then some $r \in \mathcal{J}_y \cap N$ is in the generic subset of P_{α_n}, and $r \Vdash_{P_{\alpha_n}}$ “$y \in \underset{\sim}{A}^\dagger_k$”. Hence $q_n \wedge p_n \in P_\alpha$ forces that: if $x_k \in \underset{\sim}{A}_k$, then necessarily $x_k \in \underset{\sim}{A}^\dagger_k$ (see (*)) hence some $y <_{T^*} x_k$ is in $\underset{\sim}{A}^\dagger_k$. Hence some $r \in \mathcal{J}_y \cap N$ for which $r \Vdash_{P_{\alpha_n}}$ “$y \in \underset{\sim}{A}^\dagger_k$” is in the generic set , and clearly $\neg r \in \mathcal{J}$. So clearly (as $p_n \in \mathcal{J}$) $q_n \wedge p_n \wedge \bigwedge_{r \in \mathcal{J}} r$ forces that: $x_k \in \underset{\sim}{A}_k$ leads to a contradiction (as r and $\neg r$ are incompatible) so it forces $x_k \notin \underset{\sim}{A}_k$ i.e. (**) holds as promised, so we have succeeded to define $P_{n+1} = P_{3k+3}$ as required. There is no problem to define $G_{\alpha_{n+1}}$, so we finish proving (8) hence the theorem. $\square_{4.7}$

4.8 Conclusion. Assume $S \subseteq \omega_1$ is co-stationary. For some forcing notion P, not collapsing $\aleph_1$, in V^P we have: every Aronszajn tree is $S - st$-special, but some Aronszajn tree T^* is not $S^* - st$-special for any $S^* \subseteq \omega_1 \setminus S$ stationary, moreover for every antichain $\mathcal{I}$ of T^*, $\mathrm{rk}(\mathcal{I}) \setminus S$ is not stationary. Also there is no Souslin tree. Assuming $2^{\aleph_0} = 2^{\aleph_1}$, $2^{\aleph_1} = \aleph_2$ we have: P is proper $\aleph_2$-c.c. of cardinality $\aleph_2$.

Remark. S is co-stationary – otherwise it is not interesting, but there is no other restriction e.g. S may be empty. See more 4.9(2).

Proof. Trivial by the previous Theorems 4.6, 4.7, but note that for ensuring in a transparent way that T^* remains an Aronszajn tree we would start the iterated forcing by $Q(T^*, S)$. As for the $\aleph_2$-chain condition, see VIII §2. Remember also that our forcings are proper and proper forcings preserves stationarity of subset of ω_1 (see III). In more details, by some preliminary forcing without loss of generality $V \models$ "$\diamondsuit_{\aleph_1} + 2^{\aleph_1} = \aleph_2$" and let T^* be a Souslin tree. We can define an $\aleph_1$-free iteration $\langle P_i, \underset{\sim}{Q}_j : i \le \omega_2, j < \omega_2\rangle$ as in 4.7, such that:

(a) $Q_0 = Q(T^*, S)$

(b) each Q_α satisfies one of the following:

(α) Q_α is proper and (T^*, S)-preserving of cardinality $\aleph_1$.

(β) for some P_α-name of an antichain $\underset{\sim}{\mathcal{I}}_\alpha$ of T^*, $\mathrm{rk}(\underset{\sim}{\mathcal{I}}_\alpha) \cap S = \emptyset$ and $\underset{\sim}{Q}_\alpha = Q_{\mathrm{club}}(\omega_1 \setminus \mathrm{rk}(\underset{\sim}{\mathcal{I}}_\alpha)) = \{g :$ for some $i < \omega_1$ g is a function from $i+1$ to $\{0,1\}$, $g^{-1}(\{1\})$ closed and included in $\omega_1 \setminus \mathrm{rk}(\underset{\sim}{\mathcal{I}}_\alpha)\}$.

(c) for every $\gamma < \omega_2$ and P_γ-name $\underset{\sim}{\mathcal{I}}$ of an antichain of T^* such that $\Vdash_{P_\gamma}$ "$\mathrm{rk}(\underset{\sim}{\mathcal{I}}_\alpha) \cap S = \emptyset$", *for some* $\beta < \omega_2$, $\underset{\sim}{\mathcal{I}}_\beta = \underset{\sim}{\mathcal{I}}$ (and $\gamma \in (\beta, \omega_2)$ and $\Vdash_{P_\beta}$ "$\underset{\sim}{Q}_\beta = Q_{\mathrm{club}}(\omega_1 \setminus \mathrm{rk}(\underset{\sim}{\mathcal{I}})$").

(d) for every $\gamma < \omega_2$ and P_γ-name $\underset{\sim}{T}$ of an ω_1-tree for some $\beta < \omega_2$ we have $\Vdash_{P_{\beta_2}}$ "$Q_\beta = Q(\underset{\sim}{T}_\beta, S)$, $\underset{\sim}{T}_\beta$ is an Aronszajn tree, and if $\underset{\sim}{T}$ is an Aronszajn tree (in V^{P_β}) then $\underset{\sim}{T}_\beta = \underset{\sim}{T}$".

Now:

(i) P_α (and P_α/P_β for $\beta < \alpha$) is S-proper [Why? as in both cases in (b), $\underset{\sim}{Q}_\alpha$ is S-proper]

(ii) P_α is (T^*, S)-preserving [Why? By 4.7].

(iii) P_α does not collapse $\aleph_1$ [Why? By (ii) as $S \subseteq \omega_1$ is co-stationary.]

(iv) in V^{P_α} (if $\alpha > 0$) T^* is an Aronszajn tree [Why? Q_0 ensures it: if $h^* = \cup\{h : (\exists f)[(h,f) \in \underset{\sim}{G}_{Q_0}]\}$, $f^* = \cup\{f : (\exists h)[(h,f) \in \underset{\sim}{G}_{Q_0}]\}$, $\mathrm{Dom}(h^*) = \{\alpha(j) : j < \omega_1\}$ is increasing and continuous (by density) we have: f^* is a function with domain $\bigcup_{i<\omega_1} T^*_{h^*(\alpha(i))}$ and $x \in T^*_{h^*(\alpha(i))} \Rightarrow h(x) \in \alpha(i) \times \omega$,

$[x \in \mathrm{Dom}(f^*) \,\&\, y \in \mathrm{Dom}(f^*) \,\&\, x <_{T^*} y \Rightarrow f^*(x) \neq f^*(y)]$ and $i \in S \Rightarrow \alpha(i) = i$ on S. So by Fodor's lemma, T^* has no uncountable antichains).]

(v) We can define $\langle P_i, \underset{\sim}{Q}_j : i \leq \omega_2, j < \omega_2 \rangle$ to satisfy condition (a), (b), (c), (d). [The least trivial point is to ensure an instance of condition (d), given by the bookkeeping, which is fine as for Aronszajn tree T, $Q(T, S)$ is proper (by 4.4.) and (T^*, S)-preserving by (4.6). We succeed in having the bookkeeping as $2^{\aleph_1} = \aleph_2$.]

(vi) P_{ω_2} satisfies the $\aleph_2$-c.c. [Why? As in III, using "(T^*, S)-preservance" here similarly to the way we use "S-properness" there. Remember we have assume $V \models \diamondsuit_{\aleph_1}$ hence $2^{\aleph_0} = \aleph_1$.]

(vii) P_{ω_2} collapses no cardinal and changes no cofinality.

(viii) in $V^{P_{\omega_2}}$, T^* is S-st-special [Why? use Q_0].

(ix) in $V^{P_{\omega_2}}$, for every antichain $\mathcal{I}$ of T^*, $\mathrm{rk}(\mathcal{I}) \setminus S$ is non-stationary (remember: $\mathrm{rk}(\mathcal{I}) = \{\mathrm{rk}(x) : x \in \mathcal{I}\}$) [Why? by condition (c)]).

(x) in $V^{P_{\omega_2}}$, every ω_1-Aronszajn is S-st-special [Why? By clause (d) and the definition of $Q(T, S)$]

(xi) in $V^{P_{\omega_1}}$ there is no ω_1-Souslin tree provided that [Why? By 3.4(1) and clause (x) above when S is stationary or 4.4(2).]

(xii) P_α preserve stationarity of subsets of ω_1, moreover is proper. [Why? It is $(\omega_1 \setminus S)$-proper by its being (T^*, S)-preserving clause (ii), and it is S-proper by clause (i)]

Putting together (i)—(xii) we have clearly finished. $\square_{4.8}$

4.9 Concluding Remarks.

(1) We can ask: can we do it with G.C.H. and can we get independence of other variants of "every Aronszajn tree is non-Souslin, special, etc." but we have not tried. For G.C.H. it is natural to use a variant of the forcing used in V §6 for the consistency of G.C.H. + SH with ZFC.

(2) By the definition of the forcing $Q(T, S)$; and by 3.5(2) (applied to an almost subtree), in 4.8 we get that every Aronszajn tree is $S^\dagger$-special for some $S^\dagger$ (the range of the generic h). So for S empty, we get: every Aronszajn tree is $S^\dagger$-

special for some $S^\dagger$ (equivalently h-special for some $h : \omega_1 \to \omega_1$) but some tree is not $S^* - st$-special for any stationary $S^* \subseteq \omega_1$.

(3) Note that case (β) in 4.7, is needed for the part of conclusion of 4.8 saying: for no antichain $\mathcal{I} \subseteq T^*$ is $\mathrm{rk}(\mathcal{I}) \setminus S$ stationary (we are adding a closed unbounded subset of ω_1 disjoint to any such $\mathrm{rk}(\mathcal{I}) \setminus S$). Waving this we can omit (β) in 4.7.

(4) Abraham noted that "T is h-special for some h" is equivalent to "T is $S-r$-special for some closed unbounded $S \subseteq \omega_1$". Note that we can define $S - P$-special for every partial order P, and if $\alpha_i \in P(i < \omega_1)$ implies $(\exists i < j < \omega_1)$ $\alpha_i \leq \alpha_j$ then "T $S - P$-special" implies "T is not Souslin". Note also that "$S - r$-special for some closed unbounded S" implies $\omega_1 - \mathbb{R} \times \mathbb{Q}$-special [$\mathbb{R}$-reals, $\mathbb{Q}$-rationals, the order-lexicographic]. So we have proved, e.g., "every Aronszajn tree is $\omega_1 - \mathbb{R} \times \mathbb{Q}$ special" does not imply "every Aronszajn tree is special".

(5) We can also try to get a model of ZFC where, e.g,

(A) (for some stationary co-stationary $S \subseteq \omega_1$) every Aronszajn tree is $S - st$-special, but some Aronszajn tree T^* is not h-special for any h; or

(B) there is no Souslin tree but some Aronszajn tree is not h-special for any h.

For (A) it is natural to define $Q^\dagger(T, S) = \{(h, f) : (h, f) \in Q(T, S), \mathrm{Dom}(f) \subseteq \bigcup_{h(\alpha)=\alpha} T_\alpha\}$. But T is the union of $\aleph_0$ disjoint copies of T^*, so $Q(T, S)$ cause "T^* is h-special for some h".

(6) We can generalize Definition 4.5. Let $\bar{M} = \langle M_\alpha : \alpha < \omega_1 \rangle$ be a sequence of countable models, the universe of M_α is γ_α, $\gamma_\alpha(\alpha < \omega_1)$ increasing continuous, and let $\varphi(x, y, U)$ be a quantifier free formula, where x, y are individual variables and U is a monadic predicate. We call a forcing notion $(\bar{M}, \varphi)$-preserving if for λ large enough and N a countable elementary submodel of $(H(\lambda), \in)$, $P \in N, p \in N$, there is $q \geq p$ which is (N, P)-generic and, letting $\delta = N \cap \omega_1$:

if $x \in M_{\delta+1}$ and $(\forall A \in N)(\exists y \in M_\delta)\varphi(x, y, A)$

then

$q \Vdash_P$ "$(\forall A \in N[G])(\exists y \in M_\delta)\varphi(x, y, A)$.

(7) Note that *if* $\alpha(i) < \omega_1$ is (strictly) increasing continuous in i, T is an ω_1-tree, h is a function, $\mathrm{Dom}(h) = \bigcup_{i<\omega} T_{\alpha(i+1)}$, $[x \in T_{\alpha(i+1)} \Rightarrow h(x) < \alpha(i) \times \omega]$ and $h(x) = h(y) \Rightarrow \neg(x <_T y)$, *then* there is $h^* : \bigcup_{i<\omega_1} T_{\alpha(i+1)} \to \mathbb{Q}$ such that $(*)$ $x, y \in \mathrm{Dom}(h^*) \,\&\, x <_T y \Rightarrow h^*(x) \neq h^*(y)$ and even $(*)^+$ $x, y \in \mathrm{Dom}(h^*) \,\&\, x <_T y \Rightarrow h^*(x) < h^*(y)$.

The proof with $(*)$ is similar to the proof of 3.5(2). So to derive $(*)^+$ first prove with $(*)$ and then use 3.5(1).

X. On Semi-Proper Forcing

§0. Introduction

We weaken the notion of proper to semiproper, so that some important properties (the most important is not collapsing $\aleph_1$, being preserved by some iterations) still hold for this weaker notion. But the class of semiproper forcing will also include some forcings which change the cofinality of a regular cardinal $> \aleph_1$ to $\aleph_0$. We will also describe how to iterate such forcings preserving semiproperness. So, using the right iterations, we can iterate such forcings without collapsing $\aleph_1$. As a result, we solve the following problems of Friedman, Magidor and Abraham respectively, by proving (modulo suitable large cardinals) the consistency of the following with G.C.H.:

(1) for every $S \subseteq \aleph_2, S$ or $\aleph_2 \setminus S$ contains a closed copy of ω_1,

(2) there is a normal precipitous filter D on $\aleph_2$, $\{\delta < \aleph_2 : \mathrm{cf}(\delta) = \aleph_0\} \in D$,

(3) for every $A \subseteq \aleph_2$, $\{\delta < \aleph_2 : \mathrm{cf}(\delta) = \aleph_0, \delta \text{ is regular in } L[\delta \cap A]\}$ is stationary.

However, the countable support iteration does not work, so we introduce the revised countable support. Though it is harder to define, it satisfies more of the properties we intuitively assume iterations satisfy and is applicable for the purpose of this chapter.

Notation.

Ord is the class of ordinals, Car the class of cardinals, ICar the class of infinite cardinals, UCar $=$ ICar $\setminus \{\aleph_0\}$ and RCar the class of infinite regular cardinals, SCar $=$ RCar $\cup \{2\}$, RUCar $=$ RCar $\cap$ UCar, and we let

$$S^\alpha_\beta = \{\delta < \aleph_\alpha : \mathrm{cf}\delta = \aleph_\beta\}.$$

§1. Iterated Forcing with RCS (Revised Countable Support)

Iterated forcing with countable support is widely used since Laver [L1]. One of its definitions is that at the limit stage with cofinality $\aleph_0$ we take the inverse limit, and at the limit stage with cofinality $> \aleph_0$ we take the direct limit. Another formulation is given in Definition III 3.1. However, the applications, as far as we remember, are for forcing notions which preserve the property "the cofinality of δ is uncountable", and in fact are E-proper, for some E which is a stationary subset of $\mathcal{S}_{\leq\aleph_0}(\cup E)$.

However, in our case we are interested just in forcing notions which do change some cofinality to $\aleph_0$. In these cases, we cannot break the iterated forcing into an initial segment and the rest (i.e., break $\langle P_i, \underset{\sim}{Q}_i : i < \alpha\rangle$ into $\langle P_i, \underset{\sim}{Q}_i : i < \beta\rangle$) and $\langle P_i/P_\beta, \underset{\sim}{Q}_i : \beta \leq i < \alpha\rangle$). The reason is that maybe the first forcing changes the cofinality of some $\delta, \beta < \delta < \alpha$ to $\aleph_0$; but then P_δ/P_β is not the inverse limit of $\langle P_i/P_\beta, \underset{\sim}{Q}_i : \beta \leq i < \delta\rangle$, and $\Vdash_{P_\beta}$ "$\langle P_i/P_\beta, \, , \underset{\sim}{Q}_i : \beta \leq i < \alpha\rangle$ is not a CS iteration". In fact, as every $p \in P_\delta$ has domain a bounded subset of δ, if $\Vdash_{P_\beta}$ "$\underset{\sim}{\alpha}_n \in (\beta, \delta), \underset{\sim}{\alpha}_n < \underset{\sim}{\alpha}_{n+1}, \delta = \bigcup_{n<\omega} \underset{\sim}{\alpha}_n$, and $\langle \underset{\sim}{p}_{n,i} : i < \lambda\rangle$ is a sequence of pairwise incompatible conditions in $Q_{\underset{\sim}{\alpha}_n}$ or just in $P_{\underset{\sim}{\alpha}_{n+1}}/P_{\underset{\sim}{\alpha}_n}$ i.e $P_{\underset{\sim}{\alpha}_n}$-names of members of $P_{\underset{\sim}{\alpha}_{n+1}}/\underset{\sim}{G}_{P_{\underset{\sim}{\alpha}_n}}$" and we let $\underset{\sim}{\tau} : \omega \to \lambda$ be $\underset{\sim}{\tau}(n) = i$ *if* $\underset{\sim}{p}_{n,i}[\underset{\sim}{G}_{P_\delta} \cap P_\beta]$ belongs to $\underset{\sim}{G}_{P_\delta}$ or there is no such i and we let $\underset{\sim}{\tau}(n) = 0$, *then* $\Vdash_{P_\delta}$ "$\underset{\sim}{\tau}$ is a function from ω onto $\lambda + 1$". So if each $\underset{\sim}{Q}_i$ has two incompatible members and δ is divisible by ω^2, then P_δ will collapse $\aleph_1$ and even $(2^{\aleph_0})^{V^{P_\beta}}$ for $\beta < \delta$.

Hence we suggest another iteration, RCS (revised countable support), which seems to be the reasonable solution to this dilemma.

The essence of the solution is that a name of a condition is really a condition. More exactly, in countable support iteration a condition may be $\{(\beta, \underset{\sim}{q})\}$ such that $\underset{\sim}{q}$ is a P_β-name of a member of $\underset{\sim}{Q}_\beta$, so $\underset{\sim}{q}$ is a name but β is a "real" ordinal. But now we allow $\underset{\sim}{\beta}$ to be a name. But a name with respect to which forcing notion? We would like to use P_α-names, but then we get a vicious circle, defining what is a condition of P_α using P_α-names. So we can allow P_γ-names $\underset{\sim}{\beta}$ for some $\gamma < \alpha$, such that $\Vdash_{P_\gamma}$ "$\gamma \leq \underset{\sim}{\beta} < \alpha$", and then allow a P_γ-name of condition as above etc (this is the successor case in clause (B) of Definition 1.2(1), and shall use it freely in later sections). The exact definition appears below; though it has a somewhat cumbersome definition, it seems to conform better to our intuitive idea of iteration. A first version of it can found in [Sh:119]. For other realizations of this (and alternatives to §1 here) see [Sh:250], which is redone here in Chapter XIV. In XIV §1 we deal with κ-RS. There, all the induction on γ disappears as $\kappa > \aleph_1$ makes it unavailable. An alternative way is XIV 2.6=[Sh:250, 2.6] where we simplify matters by demanding, e.g., for $\bar{Q}$-named ordinal $\underset{\sim}{\zeta}$ that: $q \Vdash$ "$\underset{\sim}{\zeta} = \xi$" $\Rightarrow q{\restriction}\xi \Vdash$ "$\underset{\sim}{\zeta} = \xi$", the price is the loss of the associativity law (see 1.1A(1)), this makes the treatment later less elegant, but does not cause real damage as far as we know: i.e. we cannot restrict inductive proofs to the cases the length δ of the iteration being 1, 2, ω, ω_1, κ inaccessible, but rather have 1, successor, for some $\alpha < \delta$, $p{\restriction}\alpha \Vdash \mathrm{cf}(\delta) = \aleph_0$ (where we are interested in the forcing above p) etc. As things are, we need to consider in e.g. 1.1(B), not only $r \in P_\xi$ but also $r \in P_{\xi+1}$ except when $\xi + 1 = \alpha$ (to avoid vicious circle), hence we have $\gamma = \beta + 1 < \alpha$ or $\gamma = \beta = \alpha - 1$ there. Compared to the previous (i.e. [Sh:b]) version, for smoothness we essentially complete the $\underset{\sim}{Q}_i$'s and we also give (for completeness) the equivalent outside definition of $\bar{Q}$-named ordinals (and conditions (1.3(2))).

1.0 Remark.

(1) If $P_1 = P_0 * \underset{\sim}{Q}_0$, $\underset{\sim}{x}$ a P_1-name, $G_0 \subseteq P_0$ generic, then in $V[G_0]$, $\underset{\sim}{x}$ can be naturally interpreted as a Q_0-name, called $\underset{\sim}{x}/G_0$, which has a P_0-name $\underset{\sim}{x}/\underset{\sim}{G}_0$ or $\underset{\sim}{x}/P_0$; but usually we do not care to make those fine distinctions.

(2) Using $\bar{Q} = \langle P_i, \underset{\sim}{Q}_i : i < \alpha \rangle$, P_α will mean Rlim $\bar{Q}$ (see Definition 1.1).

(3) If D is a filter on a set $J, D \in V$, $V \subseteq V^\dagger$ (e.g., $V^\dagger = V[G]$) then in an abuse of notation, D will denote also the filter it generates (on J) in $V^\dagger$.

(4) Formally, if $\Vdash_{P_0}$ "$\underset{\sim}{Q}_0$ is a forcing notion " then $P_0 * \underset{\sim}{Q}_0$ is a class, but this is for superficial reasons. We can demand that the set of members of $\underset{\sim}{Q}_0$ (in V^{P_0}) is a cardinal, and use only "canonical" P_0-names (as in 1.1 (B)), or restrict ourselves to members of some $H(\chi)$. In the iteration in this section (see 1.1), writing $|P|$, we mean $|P/\approx|$ (see I 5.5). We may use instead $d(P)$, the density character, which is defined as $\mathrm{Min}\{|P'| : P' \subseteq P, \forall p \in P\ \exists p' \in P'[p \leq p']\}$ or the essential density $d'(P) = \mathrm{Min}\{|P'| :$ for some P'', $P \lessdot P''$, P dense in P'' and $P' \subseteq P''$ and $(\forall p \in P)(\exists p' \in P')[p' \Vdash_{P''}$ "$p \in G_{P''}$"$]\}$ (we say P' is essentially dense in P; this means it is dense in the Boolean completion of P). The change does not make much difference.

(5) $\mathcal{D}_\kappa$ is the closed unbounded filter on κ.

(6) For a forcing notion Q, an almost member q of Q is $\{(p_i, q_i) : i < i^*\}$ such that $[p_i, q_i \in Q]$ &$[p_i, p_j$ compatible $\Rightarrow q_i = q_j]$, *and* for $r \in Q$, $q \leq r$ means $r \Vdash_Q$ "for every $i < i^*$ if $p_i \in \underset{\sim}{G}_Q$ then $q_i \in \underset{\sim}{G}_Q$"; if q', q'' are almost members of Q we define: $q' \leq q''$ iff $(\forall r \in Q)[q'' \leq r \Rightarrow q' \leq r]$. If, as we normally agree, $\emptyset_Q \in Q$ is minimal in Q then we can identify $r \in Q$ and the almost member $\{(\emptyset_Q, r)\}$. The set of almost members of Q will be denoted by $\hat{Q}$ (this is in fact just the completion of Q but if $p, q \in Q$ are equivalent (i.e. $\Vdash$ "$p \in \underset{\sim}{G}_Q \leftrightarrow q \in \underset{\sim}{G}_Q$" then in $\hat{Q}$, $p \leq q \leq p$ so they can be identified).

(7) Note that an almost member of $\hat{Q}$ is equivalent to a member of $\hat{Q}$, but is not a real almost member, but we usually ignore the distinction.

(8) See more on why the iteration is good in XI §1.

1.1 Definition. We define and prove the following (A), (B), (C), (D), Def. 1.2 and claims 1.3(1), 1.4, by *simultaneous induction on* α (also for generic extensions of V):

(A) $\bar{Q} = \langle P_i, \underset{\sim}{Q}_i : i < \alpha \rangle$ is an RCS iteration (RCS stands for revised countable support).

(B) a $\bar{Q}$-named ordinal (or $[j,\alpha)$-ordinal), (above a condition r).

(C) a $\bar{Q}$-named condition (or $[j,\alpha)$-condition), and we define $\underset{\sim}{q}\restriction\xi$, $\underset{\sim}{q}\restriction\{\xi\}$ for a $\bar{Q}$-named $[j,\alpha)$-condition $\underset{\sim}{q}$ and ordinal ξ and they are a member of P_ξ and a P_ξ-name of a member of $\hat{\underset{\sim}{Q}}_\xi$ respectively; of course $\xi \in [j,\alpha]$ (and $\xi \in [j,\alpha)$ respectively).

(D) the RCS-limit of $\bar{Q}$, $\mathrm{Rlim}\,\bar{Q}$ which satisfies $P_i \lessdot \mathrm{Rlim}\,\bar{Q}$ for every $i < \alpha$ and $p\restriction\xi$, $p\restriction\{\xi\}$ for $\xi < \alpha$, $p \in \mathrm{Rlim}\,\bar{Q}$.

(A) We define "$\bar{Q}$ is an RCS iteration"

$\alpha = 0$: no condition.

α is limit: $\bar{Q} = \langle P_i, \underset{\sim}{Q}_i : i < \alpha \rangle$ is an RCS iteration iff for every $\beta < \alpha$, $\bar{Q}\restriction\beta$ is one.

$\alpha = \beta + 1$: $\bar{Q}$ is an RCS iteration iff $\bar{Q}\restriction\beta$ is one, $P_\beta = \mathrm{Rlim}\,(\bar{Q}\restriction\beta)$ and $\underset{\sim}{Q}_\beta$ is a P_β-name of a forcing notion.

(B) We define "$\underset{\sim}{\zeta}$ is a $\bar{Q}$-named $[j,\alpha)$-ordinal of depth Υ above r" by induction on the ordinal Υ (and $\alpha = \ell g\bar{Q}$).

The intended meaning is an $(\mathrm{Rlim}\,\bar{Q})$-name of an ordinal of a special kind, however $\mathrm{Rlim}\,\bar{Q}$ is still not defined. So we use the part already known.

For $\Upsilon = 0$: "$\underset{\sim}{\zeta}$ is a $\bar{Q}$-named $[j,\alpha)$-ordinal of depth Υ above r" means $\underset{\sim}{\zeta}$ is a (plain) ordinal in $[j,\alpha)$, i.e., $j \le \underset{\sim}{\zeta} < \alpha, r \in P_{\zeta+1}$; but if $\underset{\sim}{\zeta} + 1 = \alpha$ then $r \in P_\zeta$.

For $\Upsilon > 0$: "$\underset{\sim}{\zeta}$ is a $\bar{Q}$-named $[j,\alpha)$-ordinal of depth Υ above r" means that for some $\beta < \alpha$, (letting $\gamma = \beta + 1$ if $\beta + 1 < \alpha$ and $\gamma = \beta$ otherwise) $r \in P_\gamma$, and for some antichain $\mathcal{I}$ of P_γ, pre-dense above r, $\mathcal{I} = \{p_i : i < i_0\} \subseteq P_\gamma$, $\{\Upsilon_i : i < i_0\}$ and $\{\underset{\sim}{\zeta}_i : i < i_0\}$, we have $P_\gamma \vDash$ "$(r\restriction\gamma) \le p_i$" (for simplicity), $\underset{\sim}{\zeta}_i$ is a $\bar{Q}$-named $[\,\max\{j,\beta\},\alpha)$-ordinal of depth Υ_i above $p_i, \Upsilon_i < \Upsilon$, and $\underset{\sim}{\zeta}$ is $\underset{\sim}{\zeta}_i$ if

p_i and r (i.e., if p_i, r will be in the generic set then $\underset{\sim}{\zeta}$ will be ζ_i; this is informal but clear, see formal version in 1.2(1)).

Without Υ : We say $\underset{\sim}{\zeta}$ is a $\bar{Q}$-named $[j, \alpha)$-ordinal above r, if it is such for some depth.

Without $r : r = \emptyset$.

Similarly, we omit "$[j, \alpha) -$ " when $j = 0$.

(C) We define "$\underset{\sim}{q}$ is a $\bar{Q}$-named $[j, \alpha)$-condition of depth Υ above r" and also $\underset{\sim}{q}\restriction\{\xi\}, \underset{\sim}{q}\restriction\xi$ and the $\bar{Q}$-named $[j, \alpha)$-ordinal $\zeta(\underset{\sim}{q})$ associated with $\underset{\sim}{q}$.

The definition is similar to (B).

For $\Upsilon = 0$: We say "$\underset{\sim}{q}$ is a $\bar{Q}$-named $[j, \alpha)$-condition of depth Υ above r" if for some ordinal $\zeta, j \le \zeta < \alpha$ and $\underset{\sim}{q}$ is a P_ζ-name of a member of $\hat{\underset{\sim}{Q}}_\zeta$ (see 1.0(6)), $r \in P_{\zeta+1}$ but if $\zeta + 1 = \alpha$ then $r \in P_\zeta$ and for simplicity $\underset{\sim}{q}$ is above $r\restriction\{\zeta\}$ i.e. if $\zeta + 1 < \alpha$ then $r\restriction\zeta \Vdash_{P_\zeta}$ " in $\hat{\underset{\sim}{Q}}_\zeta$, $r\restriction\{\zeta\} \le \underset{\sim}{q}$" (note: $r\restriction\zeta \in P_\zeta$, $r\restriction\{\zeta\}$ is a member of $\hat{\underset{\sim}{Q}}_\zeta$). We let

$$\underset{\sim}{q}\restriction\xi = \begin{cases} \underset{\sim}{q} & \text{if } \xi > \zeta + 1 \\ \underset{\sim}{q} & \text{if } \xi = \zeta + 1 \\ \emptyset_{P_\xi} & \text{if } \xi \le \zeta \end{cases}$$

$$\underset{\sim}{q}\restriction\{\xi\} = \begin{cases} \underset{\sim}{q} & \text{if } \xi = \zeta, \\ \emptyset_{\underset{\sim}{Q}_\xi} & \text{if } \xi \ne \zeta. \end{cases}$$

notes: $\emptyset \in P_0$ and remember 1.0(7). Finally we let $\zeta(\underset{\sim}{q}) = \zeta$. [What if we wave "$\underset{\sim}{q}$ above $r\restriction\{\zeta\}$"? Then $\xi = \zeta + 1$ need special attention as in $\bar{Q}\restriction\xi$, r may not be in P_ζ so we have to transfer the information of $\underset{\sim}{q}$ to "allowable" form, so $\underset{\sim}{q}\restriction\xi$ depend also on r; so $\underset{\sim}{q}$ should also tell us who is r or require $r\restriction\zeta \Vdash [\hat{\underset{\sim}{Q}}_\zeta \models r\restriction\{\zeta\} \le \underset{\sim}{q}$" or we should write $\underset{\sim}{q}\restriction_r\xi$, $\underset{\sim}{q}\restriction_r\{\xi\}$.]

For $\Upsilon > 0$: We say $\underset{\sim}{q}$ is a $\bar{Q}$-named $[j, \alpha)$-condition of depth Υ above r, if for some $\beta < \alpha$ (letting $\gamma = \beta + 1$ if $\beta + 1 < \alpha$ and $\gamma = \beta$ otherwise) for some $\bar{Q}$-named $[j, \alpha)$-ordinal of depth Υ above $r, \underset{\sim}{\zeta}$, defined by β, γ, $\{p_i : i < i_0\} \subseteq P_\gamma, \{\Upsilon_i : i < i_0\}$, $\{\zeta_i : i < i_0\}$, we have for each $i < i_0$ a $\bar{Q}$-named $[\max\{\beta, j\}, \alpha)$-condition $\underset{\sim}{q}_i$ of depth Υ_i above $r \bigcup p_i$ (see clause (c) in (D) below), so informally $\zeta(\underset{\sim}{q}_i) = \underset{\sim}{\zeta}_i$, and $\underset{\sim}{q}$ is $\underset{\sim}{q}_i$ if p_i and r are in the generic set of P_γ).

We then let $\zeta(\underset{\sim}{q}) = \underset{\sim}{\zeta}$.

Now we define $\underset{\sim}{q}\restriction\xi$ and $\underset{\sim}{q}\restriction\{\xi\}$; [really, we can just replace $\underset{\sim}{q}_i$ by $\underset{\sim}{q}_i\restriction\xi$, $\underset{\sim}{q}_i\restriction\{\xi\}$ respectively. In order to be pedantic, we need the following]. We define $\underset{\sim}{q}\restriction\xi$ as follows (below we ask $r \in \bigcup_{\varepsilon<\xi} P_{\varepsilon+1}$, because if ξ is a successor, $r \in P_\xi$ is a reasonable situation, if ξ a limit ordinal - not). *If* $r \in \bigcup_{\varepsilon<\xi} P_{\varepsilon+1}$ and $\beta+1<\xi$, then $\underset{\sim}{q}\restriction\xi$ is defined like $\underset{\sim}{q}$ replacing $\underset{\sim}{q}_i$ by $\underset{\sim}{q}_i\restriction\xi$. If $r \in \bigcup_{\varepsilon<\xi} P_{\varepsilon+1}$, $\beta+1=\xi=\alpha$, then $\underset{\sim}{q}\restriction\xi$ is $\underset{\sim}{q}$. If $r \in \bigcup_{\varepsilon<\xi} P_{\varepsilon+1}$, $\beta+1=\xi<\alpha$ then $\underset{\sim}{q}\restriction\xi$ is the following P_β-name of a member of $\underset{\sim}{\hat{Q}}_\beta$:

if $r\restriction\beta \in \underset{\sim}{G}_{P_\beta}$ then $\underset{\sim}{q}\restriction\xi$ is $\{(p_i\restriction\{\beta\}, \underset{\sim}{q}_i) : p_i\restriction\beta \in \underset{\sim}{G}_{P_\beta}, i<i_0\} \in \underset{\sim}{\hat{Q}}$.

If $r \in \bigcup_{\varepsilon<\xi} P_{\varepsilon+1}$, $\beta+1>\xi$ or $r \notin \bigcup_{\varepsilon<\xi} P_{\varepsilon+1}$ *then*: $\underset{\sim}{q}\restriction\xi$ is $\emptyset$ (or not defined).

Similarly for $\underset{\sim}{q}\restriction\{\xi\}$. If $r \in P_{\xi+1}$ (or $r \in P_\xi$), $\gamma \le \xi$ then $\underset{\sim}{q}\restriction\{\xi\}$ is defined like $\underset{\sim}{q}$ replacing $\underset{\sim}{q}_i$ by $\underset{\sim}{q}_i\restriction\{\xi\}$. If $r \in P_{\xi+1}$, $\beta<\gamma=\xi+1$ (hence $\beta=\xi<\alpha$) then $\underset{\sim}{q}\restriction\{\xi\}$ is the following P_β-name of a member of $\underset{\sim}{\hat{Q}}_\beta$: $\{(r\restriction\{\beta\} \cup \underset{\sim}{p}_i\restriction\{\beta\}, \underset{\sim}{q}_i\restriction\{\beta\}) : p_i\restriction\beta \in \underset{\sim}{G}_{P_\beta}$ and $r\restriction\beta \in P_\beta$ and $i<i_0\}$. If $r \in P_{\xi+1}$, $\beta=\gamma=\xi+1$ (actually is ruled out) or $\gamma>\xi+1$ then $\underset{\sim}{q}\restriction\{\xi\}$ is $\emptyset$. If $r \notin P_{\xi+1}$, then $\underset{\sim}{q}\restriction\{\xi\}$ is $\emptyset$ (or not defined).

[The definitions of $\underset{\sim}{\zeta}(q\restriction\xi), \underset{\sim}{\zeta}(q\restriction\{\xi\})$ are left to the reader].

We omit Υ and/or "$[j,\alpha)-$" if this holds for some ordinal Υ and/or $j=0$. We omit r when $r=\emptyset(=\emptyset_{P_0})$. We leave the definition of $\underset{\sim}{q}\restriction[\zeta,\xi)$ to the reader.

(D) We define Rlim$\bar{Q}$ as follows:

if $\alpha=0$: Rlim$\bar{Q}$ is trivial forcing with just one condition: $\emptyset=\emptyset_{P_0}$;

if $\alpha>0$: we call $\underset{\sim}{q}$ an atomic condition of *Rlim*$\bar{Q}$, if it is a $\bar{Q}$-named condition.

The set of conditions in $P_\alpha = $ Rlim$\bar{Q}$ is

$\{p : p$ a countable set of atomic conditions; and for every $\beta<\alpha$, $p\restriction\beta \overset{\text{def}}{=} \{r\restriction\beta : r \in p\} \in P_\beta$, and $p\restriction\beta \Vdash_{P_\beta}$ "$p\restriction\{\beta\} \overset{\text{def}}{=} \{r\restriction\{\beta\} : r \in p\}$ has an upper bound in $\underset{\sim}{\hat{Q}}_\beta$"$\}$.

The order is inclusion, (but in later sections we sometimes ignore the difference between $p \le q$ and $p \Vdash$ "$q \in \underset{\sim}{G}$")

Now we have to show:

(a) $P_\beta \lessdot \text{Rlim}\bar{Q}$ (for $\beta < \alpha$). [By 1.4(1) below.]

(b) For $\beta < \alpha$, any $(\bar{Q}\restriction\beta)$-named $[j,\beta)$-ordinal (or condition) above r is a $\bar{Q}$-named $[j,\alpha)$-ordinal (or condition) above r. [Why? Obvious.]

(c) If $\xi < \alpha$, $\underset{\sim}{q}$ is a $\bar{Q}$-named (atomic) condition above r, $r \in \bigcup_{\varepsilon<\xi} P_\varepsilon$, then $\underset{\sim}{q}\restriction\xi$ is a $(\bar{Q}\restriction\xi)$-named (atomic) condition above r. [Why? Obvious.]

(d) If $\beta_1 < \beta_2 < \alpha$, $p \in P_{\beta_2} \setminus P_{\beta_1}$, $p \le q$ in P_{β_2} then $q \notin P_{\beta_1}$ (though it may be equivalent to one).

(e) If $\xi < \alpha$, $\underset{\sim}{q}$ a $\bar{Q}$-named atomic condition above r, $r \in \bigcup_{\varepsilon<\xi} P_\xi$ *then* $\Vdash_P$ "$\underset{\sim}{q}\restriction\{\xi\}$ is a member of $\hat{\underset{\sim}{Q}}_\xi$ ".

1.1A Explanation. 1)What will occur if we simplify by letting in 1.1(B), for $\Upsilon > 0$, $\gamma = \beta$ always? Nothing happens, except that 1.5(3) is no longer true; though this is used later, we can manage without it too, though less esthetically; for variety, XIV 2.6 = [Sh:250, 2.6] is developed in this way (for a generalization called κ-RS, our case is $\kappa = \aleph_1$). For the case which interests us the two definitions are equivalent - by the proof of 2.6 (here).

2) So why in 1.1(B), for $\Upsilon > 0$, we do not let $\gamma = \beta + 1$ always? If $\beta + 1 = \alpha$, we fall into a vicious circle; defining $P_{\beta+1}$ using conditions in $P_{\beta+1}$; alternatively see XIV §1.

* * *

1.1B Remark. We can obviously define $\bar{Q}$-named sets; but for conditions (and ordinals for them) we want to avoid the vicious circle of using names which are interpreted only after forcing with them below.

1.2 Definition.

(1) Suppose $\underset{\sim}{\zeta}$ is a $\bar{Q}$-named $[j,\alpha)$-ordinal above r, $r \in G \subseteq \bigcup_{i<\alpha} P_i$ and $G \cap P_i$ generic over V (whenever $i < \alpha$) (say G is in some generic extension of V).

We define $\underset{\sim}{\zeta}[G]$ by induction on the depth: if the depth of $\underset{\sim}{\zeta}$ is 0, it is $\underset{\sim}{\zeta}$, if the depth of $\underset{\sim}{\zeta}$ is > 0, and it is defined by β, γ, $\{p_i : i < i_0\}$, $\{\underset{\sim}{\zeta}_i : i < i_0\}$, $\{\Upsilon_i : i < i_0\}$ as in Definition 1.1(B) *then* for a unique $i < i_0, p_i \in G$ and we let $\underset{\sim}{\zeta}[G] = \underset{\sim}{\zeta}_i[G]$ (remember $\Upsilon_i < \Upsilon$). (If there is no such i, it is not defined but as we demand $\{p_i : i < i_0\}$ is a predense above $r\restriction\gamma$ in P_γ above r and $\gamma < \alpha$ and $r \in G$, it will be defined).

If $r \notin G$ then $\underset{\sim}{\zeta}[G]$ is undefined, or we can give it a default value, like ∞.

For a $\bar{Q}$-named $[j, \alpha)$-condition $\underset{\sim}{q}$ above r, we define $\underset{\sim}{q}[G]$ similarly (with default value $\emptyset$).

(2) For $\underset{\sim}{\zeta}$ a $\bar{Q}$-named $[j, \alpha)$-ordinal above r, and $q \in \bigcup_{i<\alpha} P_i$ let $q \Vdash_{\bar{Q}}$ "$\underset{\sim}{\zeta} = \xi$" if for every $G \subseteq \bigcup_{i<\alpha} P_i$, such that each $G \cap P_i$ $(i < \alpha)$ is generic over V, $q \in G \Rightarrow \underset{\sim}{\zeta}[G] = \xi$, (similarly $q \Vdash_{\bar{Q}}$ "$\underset{\sim}{\zeta}$ undefined".)

1.3 Claim.

(1) Suppose $\underset{\sim}{\zeta}$ is a $\bar{Q}$-named ordinal [above r], ($\bar{Q}$ an RCS iteration, $\alpha = \ell g(\bar{Q})$). If $G \subseteq \bigcup_{i<\alpha} P_i$ [and $r \in G$] and each $G \cap P_i$ (where $i < \alpha$) is a generic subset of P_i over V, *then* for some ξ, $\underset{\sim}{\zeta}[G] = \xi, j \leq \xi < \alpha$. Moreover for some $q \in P_{\xi+1} \cap G$ we have $q \Vdash_{\bar{Q}}$ "$\underset{\sim}{\zeta} = \xi$" and $[\xi + 1 = \alpha \Rightarrow q \in P_\xi]$.

(2) Suppose $\bar{Q}$ is an RCS iteration of length α, $j < \alpha$, $\varphi(x, y)$ a definition with parameters in V and $r \in \bigcup_{i<\alpha} P_i$ such that:

(i) *If* G^* is generic over V for some forcing notion, in $V[G^*]$ we have $G \subseteq \bigcup_{i<\alpha} P_i$ is directed, for each $i < \alpha$ the set $G \cap P_i$ is generic over V and $r \in G$ *then* $V[G] \models (\exists! x)\varphi(x, G)$ and we call this unique x, $x_\varphi[G]$.

Suppose further that for such G^*, G we have $x_\varphi[G]$ is an ordinal $\zeta_\varphi = \zeta_\varphi[G] \in [j, \alpha)$ (or it is a pair $(\zeta_x, q_x) = (\zeta_\varphi[G], q_\varphi[G])$, with $\zeta_\varphi[G]$ an ordinal $\in [j, \alpha)$, $q_\varphi[G] \in \underset{\sim}{\hat{Q}}_{\zeta_\varphi[G]}[G \cap P_{\zeta_\varphi[G]}])$ and $r \in P_{\zeta_\varphi[G]+1}$.

(ii) If G^*, $G, x = x_\varphi[G]$ are as in (i), *then* for some $q \in G \cap P_{\zeta_\varphi[G]+1} \cap (\bigcup_{i<\alpha} P_i)$ we have:

$(*)^q_x$ *if* G^{**}, $G' \in V[G^{**}]$ satisfy the requirements on G^*, G and $q \in G'$ *then* $x_\varphi[G'] = x (= x_\varphi[G])$; note $\zeta_x = \alpha - 1 \Rightarrow q \in P_{\alpha-1}$ follows,

(iii) if $\delta < \alpha$ is limit, $r \in P_\beta$, $\beta < \delta$, and G^* generic over V and $G \in V[G^*]$ and $r \in G \subseteq \bigcup_{\varepsilon<\delta} P_\varepsilon$ and $G \cap P_\varepsilon$ generic over V for $\varepsilon < \delta$, *then*

either for some $q \in G$, and x, $(*)^q_x$ above holds, (so $\zeta_x < \delta$)

or for some $\beta_1 \in (\beta, \delta)$ and r^*, $r \leq r^* \in P_{\beta_1} \cap G$ we have:

for any $\beta' \in (\beta_1, \delta)$ for any r', x':

$$r^* \leq r' \in P_{\beta'} \& (*)^{r'}_{x'} \Rightarrow \zeta_{x'} \geq \delta$$

Then there is a $\bar{Q}$-named $[j, \alpha)$-ordinal above r, $\underset{\sim}{\zeta}$ [or $\bar{Q}$-named $[j, \alpha)$-condition $\underset{\sim}{q}$] such that:

If G^* is generic over V for some forcing notion, in $V[G^*]$, $G \subseteq \bigcup_{i<\alpha} P_i$ directed, for each $i < \alpha$ the set $G \cap P_i$ is generic over V and $r \in G$ *then* $x_\varphi[G] = \underset{\sim}{\zeta}[G]$ [or $x_\varphi[G] = \underset{\sim}{q}[G]$ (i.e. equivalent members of $\hat{Q}_{\underset{\sim}{\zeta}[\underset{\sim}{q}][G]}[G \cap P_{\underset{\sim}{\zeta}[\underset{\sim}{q}][G]})$].

1.3A Remark. 1) Concerning 1.3(2), of course every $\bar{Q}$-named ordinal (or condition) [above r] satisfies these conditions.

Proof. (1) The proof is by induction on the depth of $\underset{\sim}{\zeta}$.

(2) The proof is straightforward. For notational simplicity we deal with the case of $\bar{Q}$-named $[j, \alpha)$-ordinals only; but for easing the induction we define in Definition 1.1 clause (B) also "extended $\bar{Q}$-named ordinals" by just allowing ζ also values $\geq \alpha$ (but still $j < \alpha$ and now in $(*)^q_x$ we have $\zeta_x \geq \alpha - 1 \Rightarrow q \in P_{\alpha-1}$ (and we stipulate for α not successor, $\alpha - 1 = \alpha$)), and so similarly in 1.3(2)(i); clearly it suffices to prove 1.3(2) for this extension. Let β^* be minimal such that $r \in P_{\beta^*}$; we know $\beta^* < \alpha$. Let $\mathcal{I}$ be the set of $r^* \in \bigcup_{i<\alpha} P_i$ such that:

$(*)[r^*]$ for some β, γ we have: $r \leq r^* \in P_\gamma, j \leq \beta < \alpha$, $\beta \leq \gamma < \alpha, \gamma \leq \beta + 1$ and there is an extended $\bar{Q}$-named $[\beta, \infty)$-ordinal $\underset{\sim}{\zeta}$ such that:

if G^* is generic over V for some forcing notion, $G \in V[G^*]$, $G \subseteq \bigcup_{i<\alpha} P_i$, $G \cap P_i$ is generic over V for $i < \alpha$ and $r, r^* \in G$

then $x_\varphi[G] = \underset{\sim}{\zeta}[G]$.

Let $\mathcal{J} = \{p \in \mathcal{I} :$ for some $\gamma < \alpha$ we have $p \in P_\gamma \setminus \bigcup_{\varepsilon<\gamma} P_\varepsilon$ and for no γ', $j \leq \gamma' < \gamma$ is there $p' \in \mathcal{I} \cap P_{\gamma'}$, p' compatible with p (say, in P_γ)$\}$. It is enough to prove $r \in \mathcal{I}$, so assume that this fails. Choose χ large enough such that $\bar{Q} \in H(\chi)$, G^* be such that in $V[G^*]$ the cardinal 2^χ becomes a countable ordinal.

Now

$(*)_0$ If β, γ, r^* are as in $(*)[r^*]$ and $r^* \leq r^{**} \in P_\gamma$ then $r^{**} \in \mathcal{I}$
[this is trivial].

$(*)_1$ If $r \leq r^* \in P_\beta$, $\beta^* \leq \beta < \alpha$, $\mathcal{I} \cap P_\beta \setminus \bigcup_{\gamma<\beta} P_\gamma$ is pre-dense above r^* in P_β then $r^* \in \mathcal{I}$.
[Why? Straightforward by the inductive step in (B) of Definition 1.1].

For $\beta' < \alpha$, $r \in G \subseteq P_{\beta'}$, G generic over V, we define $\mathcal{I}^{[G]} = \{p \in \bigcup_{i<\alpha} P_i : p \in \bigcup_{\beta' \leq \varepsilon < \alpha} P_\varepsilon / G$ and for some $r' \in G$ we have $p \cup r' \in \mathcal{I}\}$.

$(*)_2$ Assume $r \in G \subseteq P_{\beta'}$, G is generic over V, $p \in \bigcup_{\beta' \leq \varepsilon < \alpha} P_\varepsilon / G$ and for some extended $\bar{Q}$-named $[j, \infty)$-ordinal $\underset{\sim}{\zeta}'$ above p we have: $G \subseteq G' \subseteq \bigcup_{\varepsilon<\alpha} P_\varepsilon \;\&\; p \in G' \;\&$ [for $\varepsilon < \alpha$, $G' \cap P_\varepsilon$ is generic over V] $\Rightarrow x_\varphi[G'] = \underset{\sim}{\zeta}'[G']$. *Then* $p \in \mathcal{I}^{[G]}$. [Why? Check, using the successor case in clause (B) of Definition 1.1.]

We shall prove by induction on $\beta \in [\beta^*, \bigcup_{\varepsilon<\alpha} \varepsilon]$ that

$\otimes$ if $\beta^* \leq \beta(0) < \beta$, $G_{\beta(0)} \subseteq P_{\beta(0)}$ is generic over V, $r \in G_{\beta(0)}$, $G_{\beta(0)} \cap \mathcal{I} = \emptyset$ *then* there is G_β such that $[\beta < \alpha \Rightarrow G_\beta \subseteq P_\beta$ is generic over $V]$, $[\beta = \alpha \Rightarrow G_\beta \subseteq \bigcup_{i<\alpha} P_i \& \bigwedge_{i<\alpha} G_\beta \cap P_i$ is generic over $V]$, $G_{\beta(0)} \subseteq G_\beta$ and $G_\beta \cap \mathcal{I} = \emptyset$.

It suffice to prove $\otimes$, as from $\otimes$ for $\beta = \bigcup_{\varepsilon<\alpha} \varepsilon$ we get 1.3(2); why? there is $G_{\beta^*} \subseteq P_{\beta^*}$ generic over V, such that $r \in G_{\beta^*}$ and $\mathcal{I} \cap G_{\beta^*} = \emptyset$ (otherwise by $(*)_1$ applied to $r^* = r$, $\beta^* = \beta$ we get $r \in \mathcal{I}$). Now use $\otimes$ with $\beta(0) = \beta^*$, $\beta = \bigcup_{\varepsilon<\alpha} \varepsilon$, and $G_{\beta(0)} = G_{\beta^*}$ and get G_β; contradiction to the assumption (ii) of 1.3(2), thus finishing the proof of 1.3(2).

Note that as $G_{\beta(0)} \cap \mathcal{I} = \emptyset$ also $G \cap \mathcal{I}^{[G]} = \emptyset$.

First case: $\underline{\beta = \beta^*}$. Empty.

Second case: $\underline{\beta = \beta_1 + 1 > \beta^*}$. So by the induction hypothesis without loss of generality $\beta(0) = \beta_1$. Clearly, $\beta < \alpha$ (otherwise we are done). As $G_{\beta(0)} \subseteq P_{\beta(0)}$ is generic over V (and $r \in G_{\beta(0)}$), there is $r^* \in G_{\beta(0)}$ such that $r \leq r^*$ and $r^* \Vdash$ "$\mathcal{I} \cap \underset{\sim}{G}_{\beta(0)} = \emptyset$". So there is no r', $r^* \leq r' \in P_{\beta(0)} \cap \mathcal{I}$. Is there $r' \in \underset{\sim}{Q}_{\beta(0)}[G_{\beta(0)}]$ incompatible with every $\{p \restriction \{\beta_1\} : p \in P_{\beta(0)+1} \cap \mathcal{I},\ p \restriction \beta_1 \in G_{(\beta(0))}\}$? (Note $(*)_0$ and remember $\beta_1 = \beta(0)$.) If so, no problem to find G_β as required; otherwise, without loss of generality, r^* forces this and by $(*)_1$, $r^* \in \mathcal{I}$, contradiction.

Third case: $\underline{\beta = \alpha \text{ is limit}}$. Without loss of generality in $V[G^*]$, P_α (and α) are countable. Let in $V[G^*]$, $\langle \beta_n : n < \omega \rangle$ be increasing with limit β, $\beta_0 = \beta(0)$. We define by induction on $m < \omega$, $G_{\beta_m} \subseteq P_{\beta_m}$ generic over V, increasing in n such that: $G_{\beta_m} \cap \mathcal{I} = \emptyset$. Let $n(0) = 0$, $G_{\beta_0} = G_{\beta(0)}$. For $m+1$, use the induction hypothesis. Now $\bigcup_{m<\omega} G_{\beta_m}$ is as required.

Fourth case: $\underline{\beta = \delta < \alpha \text{ is limit.}}$ Let $\alpha^* > \alpha$ be an ordinal never of the form $\zeta_\varphi[G]$. We shall define $\varphi'(x, y)$ such that for $\bar{Q}' = \bar{Q} \restriction \delta$, $r' = r, j' = j$ the assumption of 1.3(2) holds: if $r \in G \subseteq \bigcup_{\varepsilon<\delta} P_\varepsilon$ and $G \cap P_\varepsilon$ is generic over V for $\varepsilon < \delta$ *then:*

(a) if for some $q \in G$, $q \geq r$ and x, the statement $(*)^q_x$ holds then $x_\varphi[G] = x$.

(b) otherwise, $x_\varphi[G] = \alpha^*$.

Now to see that assumption (i) of 1.3(2) holds we use assumption (iii) of 1.3(2) and also the other assumption holds. So by the induction hypothesis on α, an extended $\bar{Q}'$-named $[j, \infty)$-ordinal $\underset{\sim}{\zeta}'$ exists, say of depth Υ. Looking at 1.1(B) there is a set T of strictly decreasing finite sequences of ordinals closed under initial segments and $\langle \underset{\sim}{\zeta}_\eta, \Upsilon_\eta, p_\eta, \beta_\eta : \eta \in T \rangle$, where

(α) $\underset{\sim}{\zeta}_{\langle\rangle} = \underset{\sim}{\zeta}'$, $p_{\langle\rangle} = r$, $\Upsilon_{\langle\rangle}$ the depth of $\underset{\sim}{\zeta}_{\langle\rangle}$, $r \in P_{(\beta_{\langle\rangle}+1)}$

(β) if η is maximal in T then $\Upsilon_\eta = 0$, $\beta_\eta < \delta$, $\underset{\sim}{\zeta}_\eta$ an ordinal $\underset{\sim}{\zeta}_\eta \geq \beta_\eta$, $p_\eta \in P_{(\beta_\eta+1)}$

(γ) if $\eta \in T$ is not maximal in T then $\nu \in \mathrm{Suc}_T(\eta) \Rightarrow p_\eta \leq p_\nu \in P_{(\beta_\eta+1)}$ & $\beta_\eta \leq$

β_ν, $\langle p_\nu : \nu \in \mathrm{Suc}_T(\eta)\rangle$ is a maximal antichain in $P_{(\beta_\eta+1)}$ above p_η, $\underset{\sim}{\zeta}_\eta$ is the following extended $\bar{Q}$-named $[\beta_\eta, \infty)$-ordinal above p_η: if p_ν then it is $\underset{\sim}{\zeta}_\nu$.

Suppose first: $[\eta$ maximal in T & $p_\eta{\restriction}\beta(0) \in G_{\beta(0)}$ & $\zeta_\eta = \alpha^* \Rightarrow p_\eta \in \mathcal{I}^{[G_{\beta(0)}]}]$. Let $T' = \{\eta \in T : p_\eta{\restriction}\beta(0) \in G_{\beta(0)}\}$; we define $\underset{\sim}{\zeta}$. Just for every maximal $\eta \in T'$ such that $\zeta_\eta = \alpha^*$, "plant" a witness to $p_\eta \in \mathcal{I}^{[G_{\beta(0)}]}$. In details, we prove that for every $\eta \in T$, there is a $\bar{Q}$-named $[j,\alpha)$-ordinal $\underset{\sim}{\zeta}^*_\eta$ above p_η such that: if $G_{\beta(0)} \subseteq G' \subseteq \bigcup_{\varepsilon<\alpha} P_\varepsilon$ & $p_\eta \in G'$ & $(\forall\varepsilon < \alpha)$ ($P_\varepsilon \cap G'$ is generic over V) then $x_\varphi[G'] = \underset{\sim}{\zeta}^*_\eta[G']$. This is shown by $\triangleleft$-downward induction on $\eta \in T$. In the case η is maximal in T, then: if $p_\eta{\restriction}\beta(0) \notin G_{\beta(0)}$ the demand is quite vacuous, if $\zeta_\eta \neq \alpha^*$ we can use a $\bar{Q}$-name of depth 0 and in the remaining case we know that $p_\eta \in \mathcal{I}^{[G_{\beta(0)}]}$ and this give the required conclusion. The remaining (=second) case is $\eta \in T$ not $\triangleleft$-maximal, and so use the induction hypothesis (and as in $(*)_1$, the successor case of clause (B), Definition 1.1). So we have gotten a name of the right kind in $V[G_{\beta(0)}]$, so by $(*)_2$ we get a contradiction. So for some maximal $\eta \in T$, $p_\eta{\restriction}\beta(0) \in G_{\beta(0)}, \zeta_\eta = \alpha^*$ and $p_\eta \notin \mathcal{I}^{[G_{\beta(0)}]}$. If for any such η, $\{q \in P_\delta : p_\eta \le q \in \mathcal{I}\}$ is pre-dense in $P_\delta/G_{\beta(0)}$ above p_η, we again can get a witness to $p_\eta \in \mathcal{I}^{[G_{\beta(0)}]}$ (reread clause (iii) of 1.3(2)), again contradiction. So some $q^* \in P_\delta$ is $\geq p_\eta$ and is incompatible with any $q \in \mathcal{I} \cap P_\delta$ in $P_\delta/G_{\beta(0)}$. Any $G_\delta \subseteq P_\delta$ generic over V which include $G_{\beta(0)} \cup \{q^*\}$ is as required. $\square_{1.3}$

1.4 Claim. Let $\bar{Q} = \langle P_i, \underset{\sim}{Q}_i : i < \alpha\rangle$ be an RCS iteration, $P_\alpha = \mathrm{Rlim}\bar{Q}$.

(1) If $\beta < \alpha$, then not only $P_\beta \lessdot P_\alpha$, but if $q \in P_\beta$, $p \in P_\alpha$, then q, p are compatible in P_α iff $q, p{\restriction}\beta$ are compatible in P_β. Moreover if $q \in P_\beta, p \in P_\alpha, P_\beta \models$ "$p{\restriction}\beta \le q$" then $p \cup q$ is a common upper bound of p, q in P_α (even a lub, and in particular $P_\beta \vDash$ "$q{\restriction}\alpha \le q$").

(2) If $\underset{\sim}{\beta}, \underset{\sim}{\gamma}$ are $\bar{Q}$-named $[j, \ell g(\bar{Q}))$-ordinals, then $\mathrm{Max}\{\underset{\sim}{\beta}, \underset{\sim}{\gamma}\}$ (defined naturally) is a $\bar{Q}$-named $[j, \ell g(\bar{Q}))$-ordinal.

(3) If $\alpha = \beta_0 + 1$, in Definition 1.1, part (D), in defining the set of elements of P_α we can restrict ourselves to $\beta = \beta_0$. Also in such a case, $P_\alpha = P_{\beta_0} * \underset{\sim}{Q}_{\beta_0}$ (essentially). More exactly, $\{p \bigcup \{\underset{\sim}{q}\} : p \in P_{\beta_0}$, $\underset{\sim}{q}$ a P_{β_0}-name of a member

of $\underset{\sim}{Q}_{\beta_0}\}$ is a dense subset of P_α, and the order $p_1 \bigcup\{\underset{\sim}{q}_1\} \leq_1 p_2 \bigcup\{\underset{\sim}{q}_2\}$ iff $[p_1 \leq p_2$ (in P_{β_0}) and $p_2 \Vdash_{P_{p_0}}$ "$\underset{\sim}{q}_1 \leq \underset{\sim}{q}_1$ in $\underset{\sim}{Q}_{\beta_0}$"] is equivalent to that of P_α, i.e., we get the same completion to a Boolean Algebra.

(4) The following set is dense in P_α : $P'_\alpha \stackrel{\text{def}}{=} \{p \in P_\alpha$: for every $\beta < \alpha$, if $r_1, r_2 \in p$, then $\Vdash_{P_\beta}$ "if $r_1 \restriction \{\beta\} \neq \emptyset$, $r_2 \restriction \{\beta\} \neq \emptyset$ then they are equal"$\}$.

(5) $|P_\alpha| \leq (\sum_{i<\alpha} 2^{|P_i|})^{|\alpha|}$, for limit α (i.e. we count conditions only up to equivalence).

(6) If $\Vdash_{P_i}$ "$|\underset{\sim}{Q}_i| \leq \kappa$", κ a cardinal, then $|P_{i+1}| \leq 2^{|P_i|} + \kappa$ (i.e. identifying equivalent names).

(7) If $\Vdash_{P_i}$ "$d(Q_i) \leq \kappa$" then $d(P_{i+1}) \leq d(P_i) + \kappa$ (where d is density).

(8) For α limit $d(P_\alpha) \leq 2^{\Sigma_{i<\alpha} d(P_i)}$.

Proof. Easy.

1.5 The Iteration Lemma.

(1) Suppose F is a function, *then* for every ordinal α there is RCS-iteration $\bar{Q} = \langle P_i, \underset{\sim}{Q}_i : i < \alpha^\dagger\rangle$, such that:

(a) for every $i, \underset{\sim}{Q}_i = F(\underset{\sim}{Q} \restriction i)$,

(b) $\alpha^\dagger \leq \alpha$,

(c) either $\alpha^\dagger = \alpha$ or $F(\bar{Q})$ is not an (Rlim $\bar{Q}$)-name of a forcing notion.

(2) Suppose $\beta < \alpha$, $G_\beta \subseteq P_\beta$ is generic over V, *then* in $V[G_\beta]$, $\bar{Q}/G_\beta = \langle P_i/G_\beta, \underset{\sim}{Q}_i : \beta \leq i < \alpha\rangle$ is an RCS-iteration and Rlim $(\bar{Q}) = P_\beta *$ (Rlim $\bar{Q}/\underset{\sim}{G}_\beta$) (essentially).

(3) The Associative Law.

If $\alpha_\xi(\xi \leq \xi(0))$ is increasing and continuous, $\alpha_0 = 0$, $\bar{Q} = \langle P_i, \underset{\sim}{Q}_i : i < \alpha_{\xi(0)}\rangle$ is an RCS-iteration, $P_{\xi(0)} =$ Rlim $\bar{Q}$, *then* so are

$$\langle P_{\alpha(\xi)}, P_{\alpha(\xi+1)}/P_{\alpha(\xi)} : \xi < \xi(0)\rangle \text{ and } \langle P_i/P_{\alpha(\xi)}, \underset{\sim}{Q}_i : \alpha(\xi) \leq i < \alpha(\xi+1)\rangle;$$

and vice versa.

(4) If $\bar{Q}$ is an RCS iteration, $p \in \mathrm{Rlim}\bar{Q}$, $P'_i = \{q \in P_i : q \geq p\restriction i\}, \underset{\sim}{Q}'_i = \{p \in \underset{\sim}{Q}_i : p \geq p\restriction\{i\}\}$ *then* $\bar{Q} = \langle P'_i, Q'_i : i < \ell g\bar{Q}\rangle$ is (essentially) an RCS iteration (and $\mathrm{Rlim}\bar{Q}'$ is $P'_{\ell g\bar{Q}}$).

Proof. (1) Easy.

(2) Pedantically, we should formalize the assertion as follows:

(∗) There is a function $F = F_0$ (= a definable class), such that for every RCS-iteration $\bar{Q}$, and $\ell g(\bar{Q}) = \alpha$, and $\beta < \alpha$, $F_0(\bar{Q}, \beta)$ is a P_β-name of $\underset{\sim}{\bar{Q}}^\dagger$ such that:

a) $\Vdash_{P_\beta}$ "$\underset{\sim}{\bar{Q}}^\dagger$ is a RCS-iteration of length $\alpha - \beta$".

b) $P_\beta * (\mathrm{Rlim}\underset{\sim}{\bar{Q}}^\dagger)$ is equivalent to $P_\alpha = \mathrm{Rlim}\bar{Q}$, by $F_1(\bar{Q}, \beta)$ (i.e., $F_1(\bar{Q}, \beta)$ is an isomorphism between the corresponding completions to Boolean algebras).

c) if $\beta \leq \gamma \leq \alpha \Vdash_{P_\beta}$ "$F_0(\bar{Q}\restriction\gamma, \beta) = F(\bar{Q}, \beta)\restriction(\gamma - \beta)$" and $F_1(\bar{Q}, \beta)$ extends $F_1(\bar{Q}\restriction\gamma, \beta)$ and $F_1(\bar{Q}\restriction\gamma, \beta)$ transfer the P_γ-name $\underset{\sim}{Q}_\gamma$ to a P_β-name of a $(\mathrm{Rlim}(\underset{\sim}{\bar{Q}}^\dagger\restriction(\gamma - \beta))$-name of $\underset{\sim}{Q}^\dagger_{\gamma-\beta}$ (when $\underset{\sim}{\bar{Q}}^\dagger = \langle \underset{\sim}{Q}^\dagger_i : i < \gamma - \beta\rangle$).

The proof is by induction on α, and there are no special problems.

(3) Again, pedantically the formulation is

(∗∗) For $\bar{Q}$ is an RCS-iteration, $\ell g(\bar{Q}) = \alpha_{\xi(0)}$, $\bar{\alpha} = \langle \alpha_\xi : \xi \leq \xi(0)\rangle$ increasing continuous, $F_3(\bar{Q}, \bar{\alpha})$ is an RCS-iteration $\bar{Q}^\dagger$ of length $\alpha_{\xi(0)}$ such that:

a) $F_4(\bar{Q}, \bar{\alpha})$ is an equivalence of the forcing notions $\mathrm{Rlim}\bar{Q}, \mathrm{Rlim}\bar{Q}^\dagger$.

b) $F_3(\bar{Q}\restriction\alpha_\zeta, \bar{\alpha}\restriction(\zeta + 1)) = F_3(\bar{Q}, \bar{\alpha})\restriction\zeta$.

c) $\underset{\sim}{Q}^\dagger_\xi$ is the image by $F_4(\bar{Q}\restriction\alpha_\xi, \bar{\alpha}\restriction(\xi + 1))$ of the $P_{\alpha_\xi} = \mathrm{Rlim}(\bar{Q}\restriction\alpha_\xi)$-name $F_0(\bar{Q}\restriction\alpha_{\xi+1}, \alpha_\xi)$

The proof again poses no special problems.

(4) Left to the reader. $\square_{1.5}$

1.6 Claim. If κ is regular, and $|P_i| < \kappa$ (or just $d(P_i) < \kappa$) for every $i < \kappa$, and $\bar{Q} = \langle P_i, \underset{\sim}{Q}_i : i < \kappa\rangle$ is an RCS-iteration, then:

(1) every $\bar{Q}$-named ordinal is in fact a $(\bar{Q}\restriction i)$-named ordinal for some $i < \alpha$,

(2) like (1) for $\bar{Q}$-named conditions,

(3) $P_\kappa = \bigcup_{i<\kappa} P_i$.

Proof. Easy.

1.7 Claim. Suppose $\bar{Q} = \langle P_i, \underset{\sim}{Q}_i : i < \delta \rangle$ is an RCS-iteration, δ limit and $p \in P_\delta$, and $\underset{\sim}{\zeta}$ is a $\bar{Q}$-named ordinal. *Then* there are $i < \delta$, and $p^\dagger \in P_{i+1}, p{\restriction}(i+1) \le p^\dagger$ such that $p^\dagger \Vdash_{\bar{Q}}$ "$\underset{\sim}{\zeta} = i$" (or $p^\dagger \Vdash$ "$\underset{\sim}{\zeta}[G]$ undefined " if we allow this). The same holds for $\bar{Q}$-named conditions (if $\underset{\sim}{Q}_i \subseteq V$).

Proof. Easy. By 1.3(1).

§2. Proper Forcing Revisited

2.1 Discussion. Properness is a property of forcing notions which implies that $\aleph_1$ is not collapsed by forcing with P, and is preserved by countable-support iteration (and also $\aleph_1$-free iteration, see IX.). This property was introduced in chapter III, and (see VII §3, 4) many examples of forcing not collapsing $\aleph_1$ were shown to be proper ($\aleph_1$-complete, c.c.c., Sacks forcing, Laver forcing and more). It was argued that proper forcing is essentially the most general property implying $\aleph_1$ is not collapsed and preserved under iteration. So the forcing of shooting a closed unbounded set through a stationary subset S of $\aleph_1$ (see Baumgartner, Harrington and Kleinberg [BHK], and III 4.4), though not collapsing $\aleph_1$, is excluded as if $\aleph_1 = \bigcup_{n<\omega} S_n, S_n$ pairwise disjoint stationary subsets of $\aleph_1$ and we shoot a closed unbounded subset through each $\omega_1 \setminus S_n$, in the limit $\aleph_1$ is collapsed. Of course we can "kill" stationary sets in a fixed normal ideal of $\aleph_1$ (see e.g. [JMMP]) and properness really demands somewhat more than not destroying stationary subsets of $\aleph_1$ (also stationary subsets of $\mathcal{S}_{\le\aleph_0}(\lambda) = \{A \subseteq \lambda : |A| {\le} \aleph_0$ should not be destroyed); but those seemed technical points.

However, in Chapters III–IX we were mainly interested in forcings of cardinality $\aleph_1$, so another restriction of properness was ignored: if P is proper, any countable set of ordinals in V^P is included in a countable set of V. So forcing changing the cofinality of some λ, $\mathrm{cf}\lambda > \aleph_1$, to $\aleph_0$, are not included. In fact, there are such forcings which do not collapse $\aleph_1$, and moreover, do not add reals: Prikry forcing [Pr] (which changes the cofinality of a measurable cardinal to $\aleph_0$) and Namba [Nm] which changes the cofinality of $\aleph_2$ to $\aleph_0$ (and do not add reals when CH holds).

We suggest here a property of forcing, called semiproperness, such that most theorems proved for proper forcing hold (when we use RCS-iteration) and it includes Prikry forcing. We did not know whether there is a forcing changing the cofinality of $\aleph_2$ to $\aleph_0$ which is semiproper (i.e., provably from ZFC), but we shall have an approximation to this, (but see XII §2).

So in this section we introduce the notion, and prove the preservation under RCS-iteration. In this we weaken a little the assumptions: for limit δ, Q_δ is not necessarily semiproper, only $P_{\delta+1}/P_{i+1}(i < \delta)$ is semiproper. This change does not influence the proof, but is useful, as we can exploit the fact that δ was a large cardinal in V. Note that the useful result is Corollary 2.8.

2.2 Definition. A forcing notion P is $\underset{\sim}{S}$-semiproper ($\underset{\sim}{S}$ a P-name of a class of uncountable cardinals of V) if for any large enough regular λ, and well-ordering $<^*$ of $H(\lambda)$, and countable $N \prec (H(\lambda), \in, <^*)$, such that $P \in N$, $\underset{\sim}{S} \in N$, and for every $p \in P \cap N$ there is q, $p \leq q \in P$ such that: for every cardinal $\kappa \in N$ and P-name $\underset{\sim}{\beta} \in N$ of an element of κ,

$q \Vdash_P$ "if $\kappa \in \underset{\sim}{S}$ then there is $A \in N, |A|^V < \kappa, \underset{\sim}{\beta} \in A$"

Equivalently, if $\underset{\sim}{S}$ consists of regular cardinals of V, $q \Vdash_P$ "if $\kappa \in \underset{\sim}{S}$ then $\mathrm{Sup}(N \cap \kappa) = \mathrm{Sup}(N[G] \cap \kappa)$"; or even $q \Vdash$ " if $\mathrm{cf}(\kappa)^V \in \underset{\sim}{S}$, then $\mathrm{Sup}(N \cap \kappa) = \mathrm{Sup}(N[G] \cap \kappa)$"; the case "$\underset{\sim}{S} = \{\aleph_1\}$ is the main case.

(Note that we write A and not $\underset{\sim}{A}$, i.e., A is in V; also when κ is regular in V, without loss of generality $A = \gamma$ for some $\gamma < \kappa$; this is the main case.)

We call q, under such circumstances, $\underset{\sim}{S}$-semi-(N, P)-generic. "Semiproper" means "$\{\aleph_1\}$-semiproper", and "semi-generic" means "$\{\aleph_1\}$-semi-generic" (we change the conventions of [Sh:b] where they mean $\underset{\sim}{\text{U}}\text{RCar}^{V^P}$- semiproper, $\underset{\sim}{\text{U}}\text{RCar}^{V^P}$-semi-generic respectively (see below)).

2.2A Remark. We could here change the definition to:

$q \Vdash_P$ " if $\kappa \in \underset{\sim}{S} \cap N[G]$ then, letting N' = the Skolem Hull of $N \cup \{\kappa\}$, we have $\mathrm{Sup}(N' \cap \kappa) = \mathrm{Sup}(N[G] \cap \kappa)$"

(in this case every $\kappa \in S$ is regular $> \aleph_0$). We have not looked into this variant.

2.2B Remark. When we write "P is $\text{U}\underset{\sim}{\text{C}}\text{ar}$-semiproper" or "$P$ is UCar-semiproper", UCar means $\{\delta : \delta = \aleph_1^V \text{ or } \mathrm{cf}^{V^P}(\delta) > \aleph_0\}$ so it is a P-name. Similarly for SCar, RUCar instead of RCar (and also $\aleph_1$) etc. But e.g. RUCar^V-semiproper means the regular uncountable cardinals of V.

2.3 Claim.

(1) If P is UCar^V-semiproper, or even S-semiproper, $S = \{\lambda : \mathrm{cf}\lambda > \aleph_0$ and λ a cardinal, in $V\}$, or even RUCar^V-semiproper, *then* P is proper, and vice versa. Moreover, in this case, q in Definition 2.2 is (N, P)-generic which means: if $\underset{\sim}{\beta} \in N$ is a P-name of an ordinal then $q \Vdash_P$ "$\underset{\sim}{\beta} \in N$".

(2) P is $\underset{\sim}{S}$-semiproper iff the condition of Definition 2.2 holds for some $\lambda > 2^{|P|}$, and well-ordering $<^*$ *iff* it holds for $\lambda = (2^{|P|})^+$ (provided that $P \in H(\lambda)$). Also, the well ordering $<^*$ is convenient but not really necessary.

(3) P is $\underset{\sim}{S}$-semiproper iff $(B^P \setminus \{0\}, \geq)$ is, where B^P is the complete Boolean algebra corresponding to P.

(4) In Definition 2.2, for $\kappa > \aleph_0$, and $\kappa > |P|$, the condition is trivially satisfied by any q, so only $\underset{\sim}{S} \cap \{\kappa : \aleph_0 < \kappa \leq |P|\}$ is relevant.

(5) If $P \lessdot Q$, $\underset{\sim}{S}$ a P-name and Q is $\underset{\sim}{S}$-semiproper *then* P is $\underset{\sim}{S}$-semiproper.

(6) If P is $\underset{\sim}{S}$-semiproper, $\Vdash_P$ "$\kappa \in \underset{\sim}{S}$", $\mathrm{cf}(\kappa) > \aleph_0$, *then* $\Vdash_P$ "$\mathrm{cf}(\kappa) > \aleph_0$". In particular, if $\aleph_1^V \in S$ then $\aleph_1^V = \aleph_1^{V^P}$.

(7) If $\Vdash_P$ "$\underset{\sim}{S}^1 \subseteq \underset{\sim}{S}^2$", P is $\underset{\sim}{S}^2$-semiproper then P is $\underset{\sim}{S}^1$-semiproper (similarly for semi generic).

Proof. Easy. $\square_{2.3}$

2.4 Definition.

(1) A property is preserved by RCS-iteration, provided that for any RCS-iteration $\bar{Q} = \langle P_i, \underset{\sim}{Q}_i : i < \alpha \rangle$, if $\underset{\sim}{Q}_i$ has the property (in V^{P_i}) for each i, then Rlim $\bar{Q}$ has the property.

(2) A property is strongly preserved by RCS-iteration provided that, for $\bar{Q} = \langle P_i, \underset{\sim}{Q}_i : i < \alpha \rangle$ an RCS-iteration, we have

(a) *if* for every $\gamma \leq \beta < \alpha$ such that γ not a limit ordinal, $P_{\beta+1}/P_\gamma$ has the property *then* Rlim$\bar{Q}$ has the property and

(b) if $\alpha = \beta + 1 > \gamma$, P_β/P_γ and $\underset{\sim}{Q}_\beta$ have the property, then P_α/P_γ has the property.

(3) We can replace RCS-iteration by any other kind of iteration in this definition.

2.4A Remark. In VI 1.6, 1.7, many properties were shown to be preserved by CS iteration. In fact we have proved they are strongly preserved for CS iteration – see VI 0.1(B) and even RCS iterations.

2.5 Claim.

(1) In Definition 2.4(1), (2) it suffices to consider the two-step iteration and the case where α is a regular cardinal and: $\gamma < \beta < \alpha$ implies P_β/P_γ has the property (where for 2.4(2) γ is zero or a successor ordinal).

(2) If a property is strongly preserved by RCS-iteration *then* the property is preserved by RCS-iteration.

(3) In (1), for α regular, we can add: [$\beta < \alpha \Rightarrow \Vdash_{P_\beta}$ "α is a regular cardinal"] provided that: P_α has the property *iff* $\{p \in P_\alpha : P_\alpha \restriction \{q : q \geq p\}$ has the property $\}$ is dense.

Proof. Easy, by induction on α; for (1) use the associative law 1.5(3). For (3) use 1.5(4). $\square_{2.5}$

2.6 The Semi-Properness Iteration Lemma.

(1) "Q is $\underset{\sim}{S}^Q$-semiproper" is strongly preserved by RCS-iteration for

$$\underset{\sim}{S}^Q = \{\aleph_1^V\} \cup \{\kappa : \text{ in } V^Q \text{ we have "}\kappa = \mathrm{cf}\kappa > \aleph_0\text{"}\},$$

so it is a Q-name.

(2) Suppose $\bar{Q} = \langle P_i, \underset{\sim}{Q}_i : i < \alpha \rangle$ is an RCS-iteration, for successor $j \leq \alpha$ for arbitrarily large non limit $i < j$, P_j/P_i is $\underset{\sim}{S}_{i,j}$-semiproper (and $\underset{\sim}{S}_{i,j}$ is defined, $\underset{\sim}{S}_{i,j}$ is a P_j-name). Let ($\underset{\sim}{S}$ is a P_α-name):

$\underset{\sim}{S} = \{\lambda : \lambda$ an uncountable regular cardinal, and for every i non-limit we have: $\mathrm{cf}(\lambda)^{V^{P_i}} \in \underset{\sim}{S}_{i,j}$ for every $j \in [i, \alpha)$, for which $\underset{\sim}{S}_{i,j}$ is well defined $\}$.

Then $P_\alpha = \mathrm{Rlim}\bar{Q}$ is $\underset{\sim}{S}$-semiproper provided that:

(C1) for every limit $\delta \leq \alpha$ there is $\xi < \delta$, such that

$$\Vdash_{P_\xi} \text{"}[\mathrm{cf}(\delta) = \aleph_0 \textit{ or } \text{ for every } \xi \leq i < j < \delta : \text{ if } \underset{\sim}{S}_{i,j} \text{ is defined}$$
$$\text{then } \Vdash_{P_j/P_i} \text{"}\mathrm{cf}(\delta)^{V^{P_i}} \in \underset{\sim}{S}_{i,j}]\text{"}.$$

(3) In (2) we can weaken (C1) by replacing ξ by a $(\bar{Q}\restriction\delta)$- named $[0, \delta)$-ordinal $\underset{\sim}{\xi}$ i.e. if $p \in P_{\xi+1}$, $p \Vdash$ "$\underset{\sim}{\xi} = \xi$" then, for $\xi \leq i < j \leq \delta$, i non-limit we have, $p\restriction\xi \Vdash_{P_\xi}$ " $[\mathrm{cf}\delta = \aleph_0$ *or* $p\restriction[\xi, j) \Vdash_{P_j/P_\xi}$ "$(\mathrm{cf}\delta)^{V^{P_i}} \in \underset{\sim}{S}_{i,j}$"]", and replace S by $\underset{\sim}{S} = \{\lambda$: for every non-limit $i < \alpha$ and $j \in [i, \alpha)$ (such that $\underset{\sim}{S}_{i,j}[\underset{\sim}{G}_{i,j}]$ well defined), the cofinality of λ as computed in V^{P_i} is $> \aleph_0$ and belongs to $\underset{\sim}{S}_{i,j}[\underset{\sim}{G}_{P_j}]\}$.

(4) In part (2) we can omit the condition (C1) and replace "for arbitrarily large non-limit $i < j$" by "for every $i_0 < j$ there is a $\bar{Q}$-named $[i_0, j)$-ordinal $\underset{\sim}{i}$ forced to satisfy the demand on i".

Remark.

(1) For $i < \alpha$ non-limit clearly $\underset{\sim}{S}_{i,i+1}$ is defined, so $\underset{\sim}{Q}_i$ is $\underset{\sim}{S}_{i,i+1}$-semiproper.

(2) In 2.6(2) and (3), in (C1) we can replace "for every" by "for arbitrarily large" assuming $\underset{\sim}{S}_{i,j}$ decreases with j.

(3) See XII§1 for an alternative proof, using games.

Proof. (1) Follows from (2).
(2) We prove the theorem by induction on α, for all $\bar{Q}$'s and even for $\bar{Q}$'s in forcing extensions of V.

Let $T = \{(i,j) : \underset{\sim}{S}_{i,j}$ is defined $\}$ (here $T \in V$).

Note that for any $\beta \leq \gamma \leq \alpha$,$\beta$ non-limit, $\bar{Q}{\restriction}[\beta,\gamma) = \langle P_i/P_\beta, \underset{\sim}{Q}_i : \beta \leq i < \gamma\rangle$ satisfies the hypothesis on $\bar{Q}$. Let λ be big enough, $<^*$ a well-ordering of $H(\lambda)$, $\bar{Q} \in H(\lambda)$, $N \prec (H(\lambda), \in, <^*)$, N countable, $\underset{\sim}{S} \in N$, $P_\alpha \in N$ hence w.l.o.g. $\bar{Q} \in N$ [because $(H(\lambda), \in, <^*) \models$ "there is $\bar{Q}$, an RCS-iteration as in 2.6(2) such that $P_\alpha = \text{Rlim}\,\bar{Q}$", as $P_\alpha \in N \prec (H(\lambda), \in, <^*)$ there is such a $\bar{Q}$ in N]. Similarly w.l.o.g. $\langle \underset{\sim}{S}_{i,j} : (i,j) \in T\rangle$ belongs to N. Furthermore, let $p \in P_\alpha \cap N$.

Case A. α non-limit.

The cases $\alpha = 0$, $\alpha = 1$ are too trivial to consider. For $\alpha > 1$ by the induction hypothesis on α and 1.5(3) we can assume $\alpha = 2$.

So by 2.3(3)+1.4(3) w.l.o.g. $P_2 = Q_0 * \underset{\sim}{Q}_1$, and let $p = (p_0, \underset{\sim}{p}_1) \in P_1 \cap N$. As clearly $Q_0 \in N$, there is $q_0 \in Q_0$, $p_0 \leq q_0$, which is $\underset{\sim}{S}_{0,1}$-semi (N,P)-generic. To help us in understanding let $G_0 \subseteq Q_0$ be generic, $q_0 \in G_0$. As $<^*$ is a well-ordering of $H(\lambda)$, $(H(\lambda)[G_0],\ H(\lambda), \in, <^*)$ has definable Skolem functions, and a definable well-ordering (and note: $H(\lambda)[G_0]$ is $H(\lambda)$ of the universe $V[G_0]$ as we know that any member of $H(\lambda)[G_0]$ has a name in $H(\lambda)$).

Now $N[G_0]$ is the Skolem Hull of N in $(H(\lambda)[G_0], \in, <^*)$. So: as $\underset{\sim}{p}_1[G_0] \in N[G_0]$ (because $\underset{\sim}{p}_1, G_0 \in N[G_0]$), $Q_1 = P_1/G_0$ is $\underset{\sim}{S}_{1,2}$-semiproper (i.e. $\underset{\sim}{S}_{1,2}[G_0]$-semiproper), and Q_1, $\underset{\sim}{p}_1[G_0] \in N[G_0] \prec (H(\lambda)[G_0], \in, <^*)$, *there is* $q_1 \in Q_1$ which is $\underset{\sim}{S}_{1,2}$-semi $(N[G_0], Q_1)$-generic and $q_1 \geq \underset{\sim}{p}_1[G_0]$. Let $G_1 \subseteq Q_1$ be generic, $q_1 \in G_1$. Note that $\underset{\sim}{S} \subseteq \underset{\sim}{S}_{0,1} \cap \underset{\sim}{S}_{1,2}$.

So if $\kappa \in N$ and $\text{cf}(\kappa) \in \underset{\sim}{S}_{0,1}[G_0]$ then as q_0 is $\underset{\sim}{S}_{0,1}$-semi (N, Q_0)-generic and $q_0 \in Q_0$ clearly $\text{Sup}(N \cap \kappa) = \text{Sup}(N[G_0] \cap \kappa)$; and similarly if $\kappa \in N$ and $\text{cf}^{V[G_0]}(\kappa) \in \underset{\sim}{S}_{1,2}[G_0, G_1]$ then $\text{Sup}(N[G_0] \cap \kappa) = \text{Sup}(N[G_0, G_1] \cap \kappa)$. We have described q_1 knowing G_0, hence there is an appropriate Q_0-name $\underset{\sim}{q}_1$ such that $q_0 \Vdash_{Q_0}$ "$\underset{\sim}{q}_1$ is as described above".

As $\underset{\sim}{S} \subseteq \underset{\sim}{S}_{0,1} \cap \underset{\sim}{S}_{1,2}$ and as G_0, G_1 were arbitrary except that $q_0 \in G_0$, $\underset{\sim}{q}_1 \in G_1$, clearly $(q_0, \underset{\sim}{q}_1)$ is $\underset{\sim}{S}$-semi (N, P_2)-generic.

Case B. α a limit ordinal and there are $\beta < \alpha$ and $p^\dagger$ such that $p{\restriction}\beta \leq p^\dagger \in P_\beta$ and $p^\dagger \Vdash_{P_\beta}$ "$\mathrm{cf}(\alpha) = \aleph_0$".

As $N \prec \langle H(\lambda), \in\rangle$, $\bar{Q} \in N$ and $\beta \in N$, $p \in N$, we can assume β is a successor ordinal and $p^\dagger \in N$, hence by 1.4(1) without loss of generality $p{\restriction}\beta = p^\dagger$. Moreover by Case A it suffices to prove that $P_\alpha/P_\beta, P_\beta$ are $\underset{\sim}{S}$-semiproper (for P_β, more exactly $\{\kappa :$ for no $q \in \underset{\sim}{G}_{P_\beta}, q \Vdash_{P_\alpha}$ "$\kappa \notin \underset{\sim}{S}$"$\}|$. By the induction hypothesis this holds for P_β; for P_α/P_β (we are working in $V[G_\beta]$, $G_\beta \subseteq P_\beta$ generic over V, $p^\dagger \in G_\beta$ by 1.5(1)) w.l.o.g. $\beta = 0$ so $\mathrm{cf}\alpha = \aleph_0$, and as $\bar{Q} \in N$, $\alpha \in N$, clearly there are $\alpha_n < \alpha, \alpha_n < \alpha_{n+1}, \alpha = \bigcup_{n<\omega} \alpha_n$, and w.l.o.g. each α_n is a successor ordinal or 0 and $\alpha_n \in N$, $\alpha_0 = \beta$ and $(\alpha_n, \alpha_{n+1}) \in T$.

Now let $\{(\underset{\sim}{\beta}_n, \kappa_n) : n < \omega\}$ be a list of the pairs $(\underset{\sim}{\beta}, \kappa)$, where $\kappa \in N$ and $\underset{\sim}{\beta}$ a P_α-name of an ordinal $< \kappa$, $\underset{\sim}{\beta} \in N$. We define by induction on $n < \omega$ p_n, q_n such that:

(1) p_n is a P_{α_n}-name of a member of $N \cap P_\alpha, p_0 = p^\dagger$.

(2) $q_n \in P_{\alpha_n}$, $q_{n+1}{\restriction}\alpha_n = q_n, q_n$ is $(\bigcap_{k<n} \underset{\sim}{S}_{\alpha_k, \alpha_{k+1}})$ -semi (N, P_{α_n})-generic,

(3) $p_n{\restriction}\alpha_n \leq q_n$, (i.e. this is forced)

(4) $p_{n+1} \Vdash_{P_\alpha}$ "$\underset{\sim}{\beta}_n < \underset{\sim}{\gamma}_n$ for some $\underset{\sim}{\gamma}_n$ a P_{α_n}-name of an ordinal $< \kappa_n$, $\underset{\sim}{\gamma}_n \in N$".

(5) $q_n \Vdash_{P_{\alpha_n}}$ "$p_n \leq p_{n+1}$ (in P_α)"

This is easy ($q_{n+1}{\restriction}[\alpha_n, \alpha_{n+1})$ can be constructed like $\underset{\sim}{q}_1$ in case A). Of course the point is that a P_{α_n}-name of a condition in $P_{\alpha_n+1}/P_{\alpha_n}$ is essentially a condition in $P_{\alpha_{n+1}}$. Now $\bigcup_{n<\omega} q_n$ is as required.

Case C. α a limit ordinal and for no $\beta < \alpha$, $p^\dagger \in P_\beta$, $p{\restriction}\beta \leq p^\dagger$ does $p^\dagger \Vdash_{P_\beta}$ "$\mathrm{cf}(\alpha) = \aleph_0$".

Let $\xi = \xi^*$ be as guaranteed by condition (C1) from the hypothesis. By case A without loss of generality $\xi = 0$. Let $\alpha_n \in N$, $\alpha_n < \alpha_{n+1}$, $\bigcup_{n<\omega} \alpha_n = \mathrm{Sup}(N \cap \alpha)$ (exists, as $\alpha \in N$), and $\alpha_0 = 0$, α_n non-limit; and repeat the previous proof getting $\langle q_n : n < \omega\rangle$, adding

(6) if $r \in p_n$ (so r is a $\bar{Q}$-named atomic condition) *then* for some m and P_{α_n}-name $\underset{\sim}{\xi}_m < \alpha$ we have

$$p_{m+1}\restriction(\underset{\sim}{\xi}_m + 1) \Vdash \text{“}\underset{\sim}{\zeta}(r) = \xi_m\text{”},$$

in other words for n, $k < \omega$ for some $m > n$ and P_{α_m}-name $\underset{\sim}{\xi}_m$ we have:

for every $G_{\alpha_m} \subseteq P_{\alpha_m}$ generic over V to which q_m belongs, letting r be the k-th member of p_n in the canonical well ordering of p_n of order type ω, we have:

either for $\xi = \underset{\sim}{\xi}_m[G_{\alpha_m}] < \alpha_n$, and some $p' \in G_{\alpha_m} \cap P_{\xi_m+1}$, $p' \Vdash_{P_{\xi+1}} \underset{\sim}{\zeta}(r) = \xi$” or for some $\xi \in [\alpha_n, \alpha)(\cap N[G_{\alpha_n}])$, we have $p_m[G_{\alpha_m}]\restriction(\xi+1) \Vdash_{P_\alpha/G_{\alpha_n}}$ “$\underset{\sim}{\zeta}(r) = \xi$”.

By condition (C1) from the hypothesis and as $\xi^* = 0$, we have $q_n \Vdash_{P_{\alpha_n}}$ “$N \cap \alpha$ is unbounded in $N[\underset{\sim}{G}_{\alpha_n}] \cap \alpha$, i.e. $\{\alpha_n : m < \omega\}$ is an unbounded subset of $N[\underset{\sim}{G}_{\alpha_n}] \cap \alpha$”. Let $q \in P_{\sup(N\cap\alpha)}$, $q\restriction\alpha_n = q_n$. The new point is that condition (3) above does not immediately give $p_n \leq q$, only yields $(*)$ $p_n\restriction(\bigcup_{i<\omega} \alpha_i) \leq \bigcup_{i<\omega} q_i$. But if $q \in G \subseteq P_\alpha$, G generic over V, then $p'_n \overset{\text{def}}{=} \underset{\sim}{p}_n[G \cap P_{\alpha_n}]$ is a member of $N[G_{\alpha_n}] \cap P_\alpha$, and for every $\bar{Q}$-named condition $r \in p'_n$ we know by (6) above that for some m, $\underset{\sim}{\xi}_m$ is a P_{α_m}-name and letting $\xi = \underset{\sim}{\xi}_m[G \cap P_{\alpha_m}]$ we have $p'_{m+1}\restriction\xi \Vdash_{P_\xi}$ “$\underset{\sim}{\zeta}(r) = \xi$”. But $\underset{\sim}{\zeta}(r) \in N[G_{\alpha_n}] \subseteq N[G]$ and q_n is $\{\mathrm{cf}\alpha\}$-semi $(N[G_{\alpha_n}], P_\alpha/P_{\alpha_n})$-generic hence $\underset{\sim}{\xi}_n[G] < \sup(N \cap \alpha)$ hence by $(*)$ we know $\{r\} \in G$. This insures that: *if* $q \in G \subseteq P_{\sup(N\cap\alpha)}$, G generic over V and $r \in p_n[G_{\alpha_n}]$ $(n < \omega)$ (so r is a $\bar{Q}$-named condition) *then* $\underset{\sim}{\zeta}(r)[G] < \sup(N \cap \alpha)$.

As this holds for every $r \in p'_n$ we necessarily have $p'_n \in G$ [as some $q^* \in G$ forces $(\forall r)(r \in p'_n \Rightarrow r \in G)$. Why? As this hold; assume toward contradiction that $q^* \nVdash$ “$p'_n \in \underset{\sim}{G}$” so, w.l.o.g. it force the negation, but you can check that $p'_n \cup q^* \in P_\alpha/G_{\alpha_n}$, contradiction].

As this holds for every appropriate G, we have $q \Vdash_{P_\alpha}$ “$p_n \in \underset{\sim}{G}_{P_\alpha}$” which is enough.

(3) A similar proof (only we increase p to determine ξ).

(4) The proof is like the proof of part (2), but in the case α is a limit ordinal (i.e. cases B, C), we use $\underset{\sim}{\alpha}_n$ a $\bar{Q}$-named ordinal, so conditions (1)-(5) (see case B) should be revised accordingly and if $n < \omega$, $\underset{\sim}{\xi}$ is a $\bar{Q}$-named ordinal in the Skolem-hull of $N \cup \{p_n\}$ then for some m, $\Vdash_{P_\alpha}$ "$\underset{\sim}{\xi} \leq \underset{\sim}{\alpha}_m$". $\square_{2.6}$

As we do not actually need 2.6(4) we have not elaborate. In fact, essentially we have proved above also the following, which will be useful e.g. for chain conditions:

2.7 Lemma. *If* $\bar{Q} = \langle P_i, \underset{\sim}{Q}_i : i < \delta \rangle$ is an RCS-iteration as in 2.6(2) or 2.6(3), δ a limit ordinal, and $\emptyset \Vdash_{P_i}$ "cf$\delta > \aleph_0$" (and of course $\Vdash$ "$\aleph_1 \in \underset{\sim}{S}$", really (C1) of 2.6(2) is needed for δ only), for every $i < \delta$, *then* $\emptyset \Vdash_{P_\delta}$ "cf$(\delta) > \aleph_0$". Moreover, in this case, $\bigcup_{i<\delta} P_i$ is a dense subset of P_δ more exactly essentially dense (i.e. for every $p \in P_\delta$ for some $q \in \bigcup_{i<\delta} P_i$ we have $q \Vdash$ "$p \in \underset{\sim}{G}_{P_\delta}$").

Proof. Let $p \in P_\delta$. Let χ be large enough, $<^*$ a well ordering of $H(\chi)$, $N \prec (H(\chi), \in, <^*)$ is countable, $\{\bar{Q}, \underset{\sim}{S}, p\} \subseteq N$. In the proof of 2.6, for $\alpha = \delta$, necessarily case C occurs. Now $q \in P_{\sup(N \cap \alpha)} \subseteq \bigcup_{\beta<\alpha} P_\beta$ is above p_0 which is p. Now in (C1) the second possibility always holds, so if $\underset{\sim}{\tau} : \omega \to \delta$ is a P_δ-name from N, then q forces each $\underset{\sim}{\tau}(n)$ to be equal to some $P_{\alpha_{k(n)}}$-name of an ordinal $< \delta$ from N, which q forces to be $< \sup(N \cap \delta)$. Together we finish. $\square_{2.7}$

Also note that the most useful case of 2.6 is

2.8 Corollary. Suppose $\langle P_i, \underset{\sim}{Q}_i : i < \delta \rangle$ is an RCS-iteration, and for every $j < \delta$ for arbitrarily large non-limit $i < j+1$, P_{j+1}/P_i is $\{\aleph_1\}$-semiproper, and for every $i < \delta$, $\Vdash_{P_{i+n}}$ " the power of P_i is $\aleph_1$" for some $n < \omega$. *Then* P_δ is $\{\aleph_1\}$-semiproper. *If* in addition $|P_i| < |\delta|$, for $i < \delta$ and δ is inaccessible *then* P_δ is $\underset{\sim}{S}$-semiproper, for $\underset{\sim}{S} = \{\aleph_1^V\} \cup \{\kappa : \Vdash_{P_\delta}$ "κ is a cardinal, $\kappa = \text{cf}(\kappa) > \aleph_0$"$\}$. If in addition cf$(\delta) = \aleph_1$ then $\bigcup_{i<\delta} P_i$ is a dense subset of P_δ more exactly essentially dense (i.e. for every $p \in P_\delta$ for some $q \in \bigcup_{i<\delta} P_i$ we have $q \Vdash$ "$p \in \underset{\sim}{G}_{P_\delta}$").

2.9 Remark. For iteration of proper forcings, there is really no difference between CS and RCS-iterations (see III 1.16), i.e. for $\bar{Q}$ an *RCS* iteration of proper forcing, $\{p \in \mathrm{RLim}\,\bar{Q}$: the set $\{\alpha$: for some $r \in p$ and $q \in P_{\beta+1}$ we have $q \Vdash_{P_{\beta+1}}$ "$\underset{\sim}{\zeta}(r) = \beta$"$\}$ is countable$\}$ is a dense subset (in a weak sense) of $\mathrm{Rlim}\,\bar{Q}$. In fact E-properness (for some stationary $E \subseteq \mathcal{S}_{\leq\aleph_0}(\cup E)$) suffices.

2.10 Conclusion. Suppose κ is supercompact (without loss of generality, with Laver indestructibility). *Then* for some κ-c.c. semiproper forcing notion P of power κ, $\Vdash_P$ "SPFA" and even $\Vdash_P$ "SPFA$^\alpha$" for all $\alpha \leq \omega_1$, where SPFA $=$ SPFA^0 and SPFA$^\alpha$ is the assertion Ax_α [semiproper], i.e.:
If Q is a semiproper forcing notion, $\langle \underset{\sim}{\tau}_i : i < \omega_1\rangle$ a sequence of Q-names of members of V, $\langle \underset{\sim}{S}_\beta : \beta < \alpha\rangle$ a sequence of Q-names of stationary subsets of ω_1, *then* for some directed $G \subseteq Q$:
(a) for every $i < \omega_1$, for some $q \in G$, q forces a value to $\underset{\sim}{\tau}_i$.
(b) for every $\beta < \alpha$, $\{\zeta < \omega_1 : \exists q \in G, q \Vdash$ "$\zeta \in \underset{\sim}{S}_\beta\}$ is a stationary subset of ω_1.
Proof. Same as PFA-see VII, 2.7(2) or VII 2.9. We use iteration as in 2.8, e.g. require $\underset{\sim}{Q}_{2i+1}$ is $\mathrm{Levy}(\aleph_1, 2^{|P_{2i+1}|})$. $\square_{2.10}$

§3. Pseudo-Completeness

A widely used family (or property) of forcing is $\aleph_1$-completeness, i.e., if $p_n \leq p_{n+1} \in P$, then there is $p \in P$, $p_n \leq p$ for every n. This is the simplest family of forcing which does not add reals, nor new ω-sequences of ordinals. In our perspective we want a condition parallel to this, including, e.g., Prikry forcing.

3.1 Definition. For a forcing notion P, a P-name $\underset{\sim}{S}$ of a set of cardinals of V, an ordinal δ (always a limit ordinal) and condition p we define a game $\partial^\delta_{\underset{\sim}{S}}(p, P)$: in the i-th move, player I chooses a cardinal (in V) λ_i and a P-name $\underset{\sim}{\beta}_i$ of an ordinal $< \lambda_i$, and player II has to find a condition p_i, and a set $A_i \subseteq \lambda_i$, $|A_i| < \lambda_i$, $A_i \in V$ such that:
(A) $p_i \Vdash$ "$\underset{\sim}{\beta}_i \in A_i$ or $\lambda_i \notin \underset{\sim}{S}$"; and
(B) $p_i \geq p$, $p_i \geq p_j$ for $j < i$.

The play continues for δ moves.

In a specific play, player II wins iff $\{p\} \bigcup \{p_i : i < \delta\}$ has an upper bound (and loses otherwise). If a player has no legal move (this can occur to player II only) then he loses instantly.

We say that a player wins the game if he has a winning strategy.

3.2 Claim.

(1) At most one player can win the game $\Game^\delta_{\underset{\sim}{S}}(p, P)$.

(2) If for every $\lambda_i \in \underset{\sim}{S}$ and $\mu \in \mathrm{SCar}$, $\mu \le \lambda_i \Rightarrow \mu \in \underset{\sim}{S}$, then in the definition of the game, it does not matter if we demand $|A_i| = 1$ (i.e., if one side has a winning strategy iff he has a winning strategy in the revised game).

(3) If μ_1 is regular, $\mu_1 < \mu_0$, δ divisible by μ_1 (and if $\mu_1 = \mu_2^+$ "δ divisible by $(\mu_2)^2$" suffice) and for every cardinal μ, $[\mu_1 \le \mathrm{cf}\mu \le \mu \le \mu_0 \Rightarrow \mu \in \underset{\sim}{S}]$ *then* in the definition of the game, it does not matter if we demand, when $\lambda_i = \mu_0$, that $|A_i| < \mu_1$.

(4) Also we can replace λ_i by any set $B \in V$, $|B| = \lambda_i$. If λ_i is regular (even if only in V) we can demand $A_i \in \lambda_i$ (i.e., it is a proper initial segment).

(5) If for every regular μ satisfying $\aleph_0 \le \mu \le \lambda$ we have $\mu \in \underset{\sim}{S}$ and there is $n \in \underset{\sim}{S}$, $1 < n < \aleph_0$ and for every $p \in P$, player II does not lose in the game $\Game^\delta_{\underset{\sim}{S}}(p, P)$, *then* forcing by P does not introduce new δ-sequences from λ. (Usually $n = 2$; for $n > 2$ we have to work somewhat more in the proof.)

(6) If $n \in \underset{\sim}{S}$, $n < \omega$, adding $\{m : n < m \le \aleph_0\}$ to $\underset{\sim}{S}$ does not change anything; also if $\mathrm{cf}(\lambda) \in \underset{\sim}{S}$ adding λ does not change anything.

(7) In Definition 3.1, if $\mathrm{cf}^V(\lambda) \in \underset{\sim}{S}$ we can add λ to $\underset{\sim}{S}$ with nothing being changed.

Proof. E.g.(3), player II can find a response in the revised game by playing $< \mu_1$ many moves in the original game, each time having a family $\mathcal{P}$ of $< \mu_1$ candidates, and for each $A \in \mathcal{P}$, if $\mathrm{cf}(|A|) \in [\mu_1, \mu_0]$ we replace it by a subset of smaller cardinal by one more, and if $\mathrm{cf}(|A|) < \mu_1$, we represent it as the union of $< \mu_1$ sets each of cardinality $< |A|$. In (6) (as well as in (2), (5)) just let player II use several moves to "answer" one question (if $m = \omega$ it is still finitely many though without an a priory bound). $\square_{3.2}$

3.3 Definition. The forcing P is $(\underset{\sim}{S}, \delta)$-complete if player II wins in the game $\Game^{\delta}_{\underset{\sim}{S}}(p, P)$ for every $p \in P$.

We define "P is $(\underset{\sim}{S}, < \beta)$-complete" similarly. P is pseudo κ-complete if it is $(\kappa^+ \cap \mathrm{SCar}^V, \mu)$-complete for every (cardinal) $\mu < \kappa$.

3.4 Lemma.

(1) If P is $|\delta|^+$-complete *then* it is (Car^V, δ)-complete.

(2) If P is $(\lambda^+ \cap \mathrm{SCar}^V, \delta)$-complete, $\delta \leq \lambda$, *then* forcing by P does not change the cofinality of any μ, $\aleph_0 < \mu \leq |\delta|$, and forcing by P does not add new δ-sequences from λ .

(3) In particular if P is $(\{2\}, \omega)$-complete (or even $(\{n\}, \omega)$-complete) *then* forcing by P does not add reals.

(4) If P is $(\underset{\sim}{S}, \omega)$-complete *then* P is $\underset{\sim}{S}$-semiproper.

(5) If P is (S_1, δ_1)-complete, *then* it is (S_2, δ_2)-complete provided that $(\forall \gamma \in S_2)(\exists \beta \in S_1)$ $[\mathrm{cf}(\gamma) = \beta$ or $\gamma = \beta]$ and $\delta_2 \leq \delta_1$.

(6) P is $(\underset{\sim}{S}, \delta)$-complete implies $(B^P \setminus \{0\}, \geq)$ is $(\underset{\sim}{S}, \delta)$-complete, ($B^P$ is the complete Boolean algebra corresponding to P). (See also 3.8.)

Proof. Easy. $\square_{3.4}$

3.5 Theorem.

(1) RCS-iteration strongly preserves $(\mathrm{S\underset{\sim}{C}ar}, \omega)$-completeness, and $(\mathrm{R\underset{\sim}{C}ar}, \omega)$--completeness and $(\mathrm{RU\underset{\sim}{C}ar}, \omega)$ -completeness. Moreover, if the assumption holds for the iteration $\bar{Q}$, $\bar{Q}$ has limit length, and λ is in the sets of cardinals mentioned above in each V^{P_i}, $i < \ell g\bar{Q}$, then it is so in V^{P_δ}.

(2) RCS-iteration strongly preserves (S, ω)-completeness for $S \subseteq \{2, \aleph_0, \aleph_1\}$, if we restrict ourselves to $\bar{Q}$'s satisfying $(\forall i < \ell g(\bar{Q}))[(\exists n) \Vdash_{P_{i+n}}$ "$|P_i| \leq \aleph_1$"$]$.

(3) The strong preservation in (2) holds even without the extra assumption.

3.5A Remark. Actually, we demanded in 3.1 that $\underset{\sim}{S}$ is a set of cardinals but, for example, $\mathrm{S\underset{\sim}{C}ar}$ is essentially $|P|^+ \cap \mathrm{S\underset{\sim}{C}ar}$.

3.5B Remark.

We can also imitate 2.6, and vice versa.

Proof. (1) We use Claim 2.5(1), so have to deal only with iteration $\bar{Q} = \langle P_i, \underset{\sim}{Q}_i : i < \alpha \rangle$ where $\alpha = 2$ or $\alpha = \lambda$ a regular cardinal.

Let $\underset{\sim}{S}$ be any one of those three classes of cardinals, (remember, the meaning of our $\underset{\sim}{S}$ depends on which forcing it applies to say $\underset{\sim}{S} = \underset{\sim}{S}^P$, which we know by the game we are using) so $\partial^\omega_{\underset{\sim}{S}}(p, P)$ means $\partial^\omega_{\underset{\sim}{S}^P}(p, P)$.

Case A. $\alpha = 2$.

Let $p = (p_0, \underset{\sim}{p}_1) \in Q_0 * \underset{\sim}{Q}_1$, and let $F_0, \underset{\sim}{F}_1$ be the winning strategies of player II in $\partial^\omega_{\underset{\sim}{S}}(p_0, Q_0)$, $\partial^\omega_{\underset{\sim}{S}}(\underset{\sim}{p}_1, \underset{\sim}{Q}_1)$ respectively. By 3.2(4), we can assume $\underset{\sim}{F}_1$ gives us an ordinal or a member of $\{0, 1\}$ if the corresponding λ is regular or 2 respectively. The idea of the proof is that the output $\underset{\sim}{F}_1$ gives us, a Q_0-name for an ordinal, can be used as input for F_0.

Let in the i-th move player I choose λ_i and a P_2-name $\underset{\sim}{\beta}_i$ of an ordinal $< \lambda_i$, and player II choose $(p_{0,i}, \underset{\sim}{p}_{1,i}) \in P_2$, a P_1-name $\underset{\sim}{A}_{1,i}$, and a set $A_{0,i} \subseteq \lambda_i$ (Officially player II plays $(p_{0,1}, \underset{\sim}{p}_{1,i})$, $A_{0,i}$, and chooses $\underset{\sim}{A}_{1,i}$ for himself). Player II preserves the following property:

(*)(a) $p_{0,i} \Vdash_{Q_0}$ "the following is an initial segment of the play of $\partial^\omega_{\underset{\sim}{S}}(\underset{\sim}{p}_1, \underset{\sim}{Q}_1)$ in which player II uses the strategy $\underset{\sim}{F}_1 : \langle \ldots, \langle \lambda_j, \underset{\sim}{\beta}_j \rangle, \langle \underset{\sim}{p}_{1,j}, \underset{\sim}{A}_{1,j} \rangle, \ldots \rangle_{j \le i}$".

(b) $p_{0,i} \Vdash_{Q_0}$ "$\underset{\sim}{A}_{1,i}$ is an ordinal $\underset{\sim}{\alpha}_i < \lambda_i$ if $\lambda_i \geq \aleph_0$ and a singleton $\{\underset{\sim}{\alpha}_i\} \subseteq \lambda_i$ if $\lambda_i = 2$ and $\underset{\sim}{A}_{1,i} \subseteq A_{0,i}$".

(c) $A_{0,i}$ is an ordinal $< \lambda_i$ if $\lambda_i \geq \aleph_0$ and a singleton $\subseteq 2$ if $\lambda_i = 2$.

(d) The following is an initial segment of a play of the game $\partial^\omega_{\underset{\sim}{S}}(p_0, Q_0)$ in which player II uses his winning strategy F_0: in the j-th move player I chooses

λ_j, $\underset{\sim}{\alpha}_j$ such that: $[\lambda_i \geq \aleph_0 \Rightarrow \underset{\sim}{A}_{1,i} = \alpha_i]$ and $[\lambda_i = 2 \Rightarrow A_{1,i} = \{\underset{\sim}{\alpha}_i\}]$ and player II chooses $p_{0,j}, A_{0,j}$.

It is easy to see that player II can do this and that it is a winning strategy.

Case B. $\alpha = \lambda$ a regular cardinal and $p \in P_\lambda$ and there are $\beta < \lambda, p^\dagger \in P_\beta$, $p{\restriction}\beta \leq p^\dagger$ such that $p^\dagger \Vdash_{P_\beta}$ "$\mathrm{cf}(\lambda) = \aleph_0$".

By the previous case, it suffices to prove that $P_\alpha/P_{\beta+1}$ is $(\underset{\sim}{S}, \omega)$-complete, so w.l.o.g. $\mathrm{cf}(\lambda) = \aleph_0$ and in fact $\lambda = \aleph_0$, and there are no problems. We leave the details as an exercise to the reader.

Case C. $\alpha = \lambda$ is regular and for every $\beta < \alpha$, $p{\restriction}\beta \Vdash_{P_\beta}$ "$\mathrm{cf}(\lambda) > \aleph_0$".

We will first give an informal sketch of II's strategy. We will also choose ξ_n, $0 = \xi_0 < \xi_1 < \ldots < \xi_n < \lambda$. After each move $(\lambda_n, \underset{\sim}{\alpha}_n)$ of player I, player II starts a new game $\eth_n = \eth^\omega_{\underset{\sim}{S}}\ [p_n{\restriction}[\xi_n, \xi_{n+1}), P_{\xi_{n+1}}/P_{\xi_n}]$, where $p_n{\restriction}[\xi_n, \alpha)$ is chosen such that it decides $\underset{\sim}{\alpha}_n$ up to a P_{ξ_n}-name $\underset{\sim}{\alpha}_{n,n}$. He then plays one step in each of the games $\eth_m$ $(m = n-1, \ldots, 0)$, simulating for I_m (i.e. first player in $\eth_m$) the move $\langle \lambda_n, \underset{\sim}{\alpha}_{n,m+1} \rangle$ and II_m answer $\langle p_n{\restriction}[\xi_m, \xi_{m+1}), \underset{\sim}{\alpha}_{n,m} \rangle$ (where $\underset{\sim}{\alpha}_{n,m}$ is a P_{ξ_m}-name) where we choose a constant winning strategy for II_m (it is a P_{ξ_m}-name) and player II answers in the true game in $\alpha_{n,0}$. The ξ_ℓ's must be big enough such that all the p_n's are eventually forced to be essentially in $\bigcup_{\ell<\omega} P_{\xi_\ell}$ (i.e. equivalent to a member). We only have to deal with countably many $\bar{Q}$-named ordinals, so we can take care of finitely many at each step n.

We now describe more formally the winning strategy of player II. By a hypothesis, for every non-limit $\beta < \gamma < \alpha$, and $r \in P_{\beta,\gamma} (= P_\gamma/P_\beta)$ player II has a winning strategy $\underset{\sim}{F}_{\beta,\gamma}(r)$ (a P_β-name) for winning the game $\eth^\omega_{\underset{\sim}{S}}(r, P_\gamma/P_\beta)$. We can change a little the rules of the game $\eth^\omega_{\underset{\sim}{S}}(r, P_\gamma/P_\beta)$, letting in stage n player I choose $k < \omega$ and a finite sequence $\langle \lambda^n_1, \underset{\sim}{\beta}^n_1, \ldots, \lambda^n_k, \underset{\sim}{\beta}^n_k \rangle$ ($\underset{\sim}{\beta}^n_\ell$ a P_γ/P_β-name of an ordinal $< \lambda^n_\ell$) and player II will choose $\alpha^n_1, \ldots, \alpha^n_k \in V^{P_\beta}$, and a condition $p_n \in P_\gamma/P_\beta$ satisfying $V^{P_\beta} \models$ "$\alpha^n_\ell < \lambda^n_\ell$", $p_n \Vdash_{P_\gamma/P_\beta}$ "*if* $\lambda^n_\ell \in \underset{\sim}{S}$ *then* $\underset{\sim}{\beta}^n_\ell < \alpha^n_\ell < \lambda^n_\ell$ when $\lambda^n_\ell \geq \omega$ and $\underset{\sim}{\beta}^n_\ell = \alpha^n_\ell$ when $\lambda^n_\ell = 2$" and $p_n \geq p, p_n \geq p_{n-1}$, (remember here $\underset{\sim}{S}$ is really a P_γ-name). Note: if player II wins the usual game he will win also the revised one.

Let for every $p \in P_\alpha$, $p = \{p^{[\ell]} : \ell < \omega\}, p^{[\ell]}$ a $\bar{Q}$-named condition.

Now player II's winning strategy uses some auxiliary games which he plays on the side. In stage n, player I chooses $\lambda_n, \underset{\sim}{\alpha}_n$ (a P_α-name of an ordinal $< \lambda_n$), but player II chooses not only p_n, A_n, but also a non-limit ordinal $\xi_n < \lambda$, and for $\ell \le n$ P_{ξ_ℓ}-names $\underset{\sim}{\alpha}_{n,\ell}$ of ordinals $< \lambda_n$ and for $k \le n$ also $\underset{\sim}{\beta}^n_k$ which is a P_{ξ_k}-name of ordinal $< \lambda$ such that:

a) $p \le p_n, p_{n-1} \le p_n$, $\xi_0 = 0$, $\xi_{n+1} = \mathrm{Max}\{\xi_n + 1, \beta^\ell_0 + 1 : \ell \le n\}$

b) $p_n \restriction [\xi_n, \alpha) \Vdash_{P_\alpha}$ "$\underset{\sim}{\alpha}_n = \underset{\sim}{\alpha}_{n,n}$, $\zeta(p^{[\ell]}_k)$ is $\le \underset{\sim}{\beta}^n_n < \lambda (= \alpha)$ (or undefined) for $\ell, k < n$ such that $\ell = n-1 \vee k = n-1$" where $\underset{\sim}{\alpha}_{n,n}$, $\underset{\sim}{\beta}^n_n$ are P_{ξ_n}-names.

c) for each $m < n$, the following is an initial segment of a play in the game $\underset{\sim}{G}^\omega_{\underset{\sim}{S}}(p_m \restriction [\xi_m, \xi_{m+1}), P_{\xi_{m+1}}/P_{\xi_m})$ in which player II uses his winning strategy $\underset{\sim}{F}_{\xi_m, \xi_{m+1}}(p_m \restriction [\xi_m, \xi_{m+1}))$:

$$\Big\langle \langle \lambda_{m+1}, \underset{\sim}{\alpha}_{m+1,m+1}, \lambda, \underset{\sim}{\beta}^{m+1}_{m+1} \rangle, \langle p_{m+1} \restriction [\xi_m, \xi_{m+1}), \underset{\sim}{\alpha}_{m+1,m}, \underset{\sim}{\beta}^{m+1}_m \rangle, \ldots, \\ \langle \lambda_n, \underset{\sim}{\alpha}_{n,m+1}, \lambda, \underset{\sim}{\beta}^n_{m+1} \rangle, \langle p_n \restriction [\xi_m, \xi_{m+1}), \underset{\sim}{\alpha}_{n,n}, \underset{\sim}{\beta}^n_m \rangle \Big\rangle$$

i.e.

$$\Big\langle (\langle \lambda_{m+i}, \underset{\sim}{\alpha}_{m+i,m+1}, \lambda, \underset{\sim}{\beta}^{m+i}_{m+1} \rangle, \langle p_{m+i} \restriction [\xi_m, \xi_{m+1}), \underset{\sim}{\alpha}_{m+i,m}, \underset{\sim}{\beta}^{m+i}_m \rangle) : \\ 1 \le i \le n-m \Big\rangle.$$

d) Player II choice of A_n, is $A_n = \alpha_{n,0}$ if $\lambda_n \ge \omega$ and $A_n = \{\alpha_{n,0}\}$ if $\lambda_n = 2$.

Player II can carry out his strategy easily, defining in stage n, first ξ_n, second $p_n \restriction [\xi_n, \alpha)$, $\underset{\sim}{\alpha}_{n,n}$ and $\underset{\sim}{\beta}^n_n$, third he defined by downward induction on $m < n$, $p_n \restriction [\xi_m, \xi_{m+1}]$, $\alpha_{n,m}$, $\underset{\sim}{\beta}^n_m$, and fourth play as in d).

2) The proof is left to the reader. (Compare 2.8).

3) Combining the proofs of 2.6(4) and part (1). $\square_{3.5}$

3.6 Definition.

For a forcing notion P, a P-name $\underset{\sim}{S}$ of a set of cardinals, an ordinal δ and a condition p we define the games $\mathrm{E}\partial^\delta_{\underset{\sim}{S}}(p, P)$, $\mathrm{R}\partial^\delta_{\underset{\sim}{S}}(p, P)$

(or $\mathrm{E}\partial^\delta(p, P, \underset{\sim}{S}), \mathrm{R}\partial^\delta(p, P, \underset{\sim}{S})$ respectively; E stands for essentially, R for really).

(1) In a play of the game $\mathrm{E}\eth^{\delta}_{\underset{\sim}{S}}(p, P)$ in the i-th move, player I chooses a cardinal λ_i and a P-name $\underset{\sim}{\beta}_i$ of an ordinal $< \lambda_i$ and player II has to find a set $A_i \subseteq \lambda_i$, $|A_i| < \lambda_i$, $(A_i \in V)$.

The play continues for δ moves. In the end player II wins if he can find a condition $p^\dagger \in P$, $p \leq p^\dagger$ such that for every $i < \delta$, $p^\dagger \Vdash_P$ "$\underset{\sim}{\beta}_i \in A_i$, or $\lambda_i \notin \underset{\sim}{S}$".

(2) In a play of the game $\mathrm{R}\eth^{\delta}_{\underset{\sim}{S}}(p, P)$ in the i-th move, player I chooses a condition q_i, $q_i \geq p_j$ for every $j < i$ and $q_i \geq p$, and a cardinal λ_i and a P-name $\underset{\sim}{\beta}_i$ of an ordinal $< \lambda_i$ and player II has to find a condition p_i and a set $A_i \subseteq \lambda_i$, $|A_i| < \lambda_i$, $(A_i \in V)$ such that

(A) $p_i \Vdash_P$ "$\underset{\sim}{\beta}_i \in A_i$ or $\lambda_i \notin \underset{\sim}{S}$",

(B) $p_i \geq q_i$.

The play continues for δ moves, and player II wins if $\{p\} \bigcup \{p_i : i < \delta\}$ has an upper bound.

Note: 3.6(1) is close to 3.1, 3.6(2) is stronger. Comparing Definition 3.6(2) with XIV Definition 2.1, the definition here is stronger when $\delta > \omega$.

3.7 Definition. The forcing P is essentially $(\underset{\sim}{S}, \delta)$-complete [really $(\underset{\sim}{S}, \alpha)$-complete] if player II wins in the game $\mathrm{E}\eth^{\delta}_{\underset{\sim}{S}}(p, P)$ [$\mathrm{R}\eth^{\delta}_{\underset{\sim}{S}}(p, P)$] for every $p \in P$.

3.8 Lemma.

(1) The parallels of 3.2, 3.4 hold.

(2) Let P be a forcing, B the corresponding Boolean algebra. Then P is essentially $(\underset{\sim}{S}, \alpha)$-complete iff $(B^P \setminus \{0\}, \geq)$ is $(\underset{\sim}{S}, \alpha)$-complete; and if $\alpha \geq \omega$, this implies P is $\underset{\sim}{S}$-semiproper. If P is complete (i.e. for any $\mathcal{I} \subseteq P$ there is p such that every $G \subseteq P$ generic over V: $p \in G$ iff $\mathcal{I} \cap G \neq \emptyset$ and $(\forall q \in \mathcal{I})(q \leq p)$) then P is $(\underset{\sim}{S}, \alpha)$-complete iff P is essentially $(\underset{\sim}{S}, \alpha)$-complete. If P is really $(\underset{\sim}{S}, \alpha)$-complete then P is $(\underset{\sim}{S}, \alpha)$-complete which implies essentially $(\underset{\sim}{S}, \alpha)$-complete.

Proof. Easy. $\square_{3.8}$

3.9 Theorem. (1) RCS-iteration strongly preserves the notions "essential (S,ω)-completeness" for $S \in \{\underset{\sim}{\mathrm{SCar}}, \underset{\sim}{\mathrm{RCar}}, \underset{\sim}{\mathrm{RUCar}}\}$. Similarly for "real (S,ω)-completeness.

(2) Moreover, if the assumption holds for the iteration $\bar{Q}$, $\bar{Q}$ has limit length, and the cofinality of λ is in the set of cardinals mentioned above in each V^{P_i}, $i < \ell g\bar{Q}$, then it is in V^{P_δ}.

(3) RCS-iteration strongly preserves essential (S,ω)-completeness for $S \subseteq \{2, \aleph_0, \aleph_1\}$, if we restrict ourselves to $\bar{Q}$'s satisfying

$$(\forall i < \ell g(\bar{Q}))[(\exists n) \Vdash_{P_{i+n}} \text{“}|P_i| \leq \aleph_1\text{”}]$$

(or even without it).

Proof. Similar to previous ones. $\square_{3.9}$

3.10 Definition. For $\mathbf{W} \subseteq \omega_1$ we call a forcing notion pseudo $(*, \mathbf{W})$-complete if for each $p \in P$ in the following game player I has a winning strategy. The play lasts ω moves. *In the n'th move:* player I chooses an ordinal $\alpha_n < \omega_1$ such that $\bigwedge_{\ell<n} \beta_\ell < \alpha_n$ and a P-name $\underset{\sim}{\tau}_n$ of a countable ordinal. Player II chooses ordinals $\beta_n, \gamma_n < \omega_1$ such that $\alpha_n < \beta_n$, $\bigwedge_{l<n} \beta_l < \beta_n$. In the end player II wins the play iff (a) or (b) where

(a) $\bigcup_{n<\omega} \alpha_n \notin \mathbf{W}$.

(b) there is $q \in P$ satisfying: $p \leq q$ and $q \Vdash_P$ "$\underset{\sim}{\tau}_n = \underset{\sim}{\gamma}_n$ for $n < \omega$".

3.10A Remark. We can define games and completeness variations of the earlier notions in this section with length of game ω with a stationary $\mathbf{W} \subseteq \omega_1$ as a parameter as we have done to $(\{2\},\omega)$ -completeness in 3.10 and the parallel theorems hold.

3.11 Claim. (1) Pseudo $(*, \mathbf{W})$-completeness is strongly preserved by RCS-iteration.

(2) If $\mathbf{W}$ is stationary (subset of ω_1) and P is $(*, \mathbf{W})$-complete *then* forcing with P preserves stationarity of subsets of $\mathbf{W}$ and adds no real.

3) If $\omega_1 \setminus \mathbf{W}$ is not stationary, P is pseudo $(*, \mathbf{W})$-complete *then* P is essentially $(\{\aleph_1\}, \omega)$-complete.

Proof. Left to the reader. $\square_{3.11}$

§4. Specific Forcings

We prove here for various forcings that they are semiproper and even (S, δ)-complete; of course, otherwise our previous framework will be empty. See [J], chapters 5-6, for a discussion of some of the large cardinals we use (which are standard).

Prikry forcing (adding an unbounded ω-sequence to a measurable cardinal without adding bounded subsets) satisfies all we can expect. But for our purposes, more important are forcings which change the cofinality of $\aleph_2$ to $\aleph_0$, without adding reals (or at least not collapsing $\aleph_1$). Namba [Nm] has found such a forcing, when CH holds.

However we do not know the answer to:

Problem. Is Namba forcing $\{\aleph_1\}$-semiproper? (But see XII §2).

However, Namba forcing is not necessarily $(\{2\}, \omega)$-complete; this is equivalent to "$\mathcal{D}^{\mathrm{cb}}_{\aleph_2}$ is Galvin" (see below).

We deal with a variant of Namba forcing, (for the original see XI 4.1), $\mathrm{Nm}'(\mathfrak{D})$ ($\mathfrak{D}$ a system of filters on sets of power $\aleph_2$, see below), and prove the relevant assertion (4.7). Then we prove that if each filter in $\mathfrak{D}$ has the $(\{2, \aleph_0, \aleph_1, \}, \omega)$-Galvin property (see 4.9, 4.9A), *then* $\mathrm{Nm}'(\mathfrak{D})$ is semiproper, moreover is $(\{\aleph_0, \aleph_1, 2\}, \omega)$-complete. The point is that when a large cardinal is collapsed to $\aleph_2$, if D was originally a normal ultrafilter, then after the collapse it may well have some largeness property like the Galvin property.

4.1 Definition. If D is a complete normal ultrafilter on κ, then the D-Prikry forcing, $PF(D)$, is:

$\{(f, A) : f$ a function, with domain $n < \omega, f$ is increasing, $(\forall i < n) f(i) < \kappa$, and A belongs to $D\}$.

$(f_1, A_1) \leq (f_2, A_2)$ iff $f_1 \subseteq f_2$, $A_1 \supseteq A_2$, and for $i \in \mathrm{Dom}(f_2) \setminus \mathrm{Dom}(f_1)$, $f_2(i) \in A_1$.

Prikry defined this notion and proved [Pr] in fact that:

4.2 Theorem. For any normal ultrafilter D over κ, $P = PF(D)$ is $(\mathrm{RCar}^P, \lambda)$-complete for every $\lambda < \kappa$, and changes the cofinality of only one cardinal, κ (to $\aleph_0$). (So remembering the notation introduced before 3.2, $(\mathrm{RCar}^P, \lambda)$-complete really means $(\mathrm{Car} \setminus \{\kappa\}, \lambda)$-complete).

4.3 Definition. (1) A filter-tagged tree is a pair $(T, \mathfrak{D})$ such that:

(a) T is a nonempty set of finite sequences of ordinals, closed under taking initial segments, and there is some maximal $\eta_0 \in T$ for which $[\nu \in T$, $\ell g(\nu) \leq \ell g(\eta_0) \Rightarrow \nu = \eta_0 \restriction \ell g(\nu)]$; we call η_0 the trunk of T, $\eta_0 = tr(T)$.

(b) $\mathfrak{D}$ is a function such that for every $\eta \in T$, $\mathfrak{D}_\eta = \mathfrak{D}(\eta)$ is a filter on some set $\subseteq \{\eta\char`^\langle\alpha\rangle : \alpha$ an ordinal$\}$ and if $\mathrm{tr}(T) \trianglelefteq \eta \in T$ *then* $\mathrm{Suc}_T(\eta) \stackrel{\mathrm{def}}{=} \{\nu \in T : \ell g(\nu) = \ell g(\eta) + 1, \nu \restriction \ell g(\eta) = \eta\} \neq \emptyset \bmod \mathfrak{D}_\eta$.

(2) We call $(T, \mathfrak{D})$ normal if $\mathrm{Dom}(\mathfrak{D}) = \{\eta \in T : \mathrm{tr}(T) \trianglelefteq \eta\}$ and for every such η, $\mathfrak{D}_\eta$ is a filter over $\mathrm{Suc}_T(\eta)$ (see below). For $\eta \in T$, $(T, \mathfrak{D})_{[\eta]} = (T_{[\eta]}, \mathfrak{D}) \stackrel{\mathrm{def}}{=} (\{\nu \in T : \nu \trianglelefteq \eta \text{ or } \eta \trianglelefteq \nu\}, \mathfrak{D})$.

(3) We call $(T, \mathfrak{D})$ λ-complete if each $\mathfrak{D}_\eta (\eta \in T)$ is λ -complete.

4.4 Definition. For filter-tagged trees $(T_1, \mathfrak{D}_1)$, $(T_2, \mathfrak{D}_2)$:

(1) We define: $(T_1, \mathfrak{D}_1) \leq (T_2, \mathfrak{D}_2)$ iff

(a) $T_2 \subseteq T_1$,

(b) For every $\eta \in T_2$, if $\eta \trianglerighteq \mathrm{tr}(T_2)$ then $\mathrm{Suc}_{T_2}(\eta) \neq \emptyset \bmod \mathfrak{D}_1(\eta)$ and $\mathfrak{D}_1(\eta) \restriction \mathrm{Suc}_{T_2}(\eta) = \mathfrak{D}_2(\eta) \restriction \mathrm{Suc}_{T_2}(\eta)$ where for a filter D over I, and $J \subseteq I$, $J \neq \emptyset \bmod D$ we let:

$$D \restriction J = \{A \cap J : A \in D\}.$$

(2) We define: $(T_1, \mathfrak{D}_1) \leq_{pr} (T_2, \mathfrak{D}_2)$ ("pure extension") if in addition $\mathrm{tr}(T_1) = \mathrm{tr}(T_2)$.

(3) We define: $(T_1, \mathfrak{D}_1) \leq_n (T_2, \mathfrak{D}_2)$ if in addition (to (2)) for η of length $\leq n$, $\eta \in T_1 \Leftrightarrow \eta \in T_2$.

(4) $\mathrm{Nm}'(T^*, \mathfrak{D}^*) \stackrel{\mathrm{def}}{=} \{(T, \mathfrak{D}) : (T^*, \mathfrak{D}^*) \leq (T, \mathfrak{D})\}$ ordered by $\leq$. We write $\eta \in (T, \mathfrak{D})$ for $\eta \in T$. If $p = (T^*, \mathfrak{D}^*)$ we write T_p for T^*, $\mathfrak{D}_p$ for $\mathfrak{D}^*$. Instead of T_p, $T_{p'}$, T_{p_1}, T_{p^k}, etc, we usually write just T, T', T_1, T^k, etc.

4.4A Remark. For every filter-tagged tree $(T, \mathfrak{D})$ for a unique normal $(T, \mathfrak{D}^\dagger)$ we have $(T, \mathfrak{D}) \leq (T, \mathfrak{D}^\dagger) \leq (T, \mathfrak{D})$.
2)So we can restrict ourselves to normal members of $\mathrm{Nm}'(T^*, \mathfrak{D}^*)$.

4.5 Claim. 1) If $\eta \in T \Rightarrow |\mathrm{Suc}_T(\eta)| \leq \aleph_2$ then $\mathrm{Nm}'(T, \mathfrak{D})$ is $(\{\lambda : \lambda = \mathrm{cf}\lambda > \aleph_2\}, \omega)$-complete.
2) Moreover, in the cases where we shall prove that Nm' is (S, ω)-complete, $S \subseteq \{2, \aleph_0, \aleph_1\}$, we could prove it is $(S \cup \{\lambda : \lambda = \mathrm{cf}\lambda > \aleph_2\}, \omega)$-complete (see 4.12).

Proof of 1). It is enough to prove:
$(*)$ if $p \in P = \mathrm{Nm}'(T, \mathfrak{D})$, $n < \omega$, $\underset{\sim}{\tau}$ a P-name of an ordinal, *then* there is $q \in P$, $p \leq_n q$ and a set A of ordinals, $|A| \leq \aleph_2$, $q \Vdash$ "$\underset{\sim}{\tau} \in A$".

Proof of $(*)$: Let
$T_0^* = \{\eta \in p : \ell g(\eta) \geq n, \mathrm{tr}(p) \trianglelefteq \eta$, and $(p)_{[\eta]}$ has a pure extension deciding the value of $\underset{\sim}{\tau}\}$,
$T_1^* = \{\eta \in T_0^*$: there is no $\nu \triangleleft \eta$, $\nu \in T_0^*\}$.

4.5A Subfact. T_1^* is a front of r for some r satisfying $p \leq_n r$; i.e every ω-branch of r contains one and only one element of T_1^*.

Proof of the Subfact. Clearly without loss of generality $\mathrm{tr}(p)$ has the length $\geq n$. By a partition theorem in [RuSh:117] (or see here XI 3.5 or XV 2.6B(2), and if CH see 4.6 below) there is $r \in \mathrm{Nm}'(T, \mathfrak{D})$, $p \leq_{pr} r$, such that:

either (a) for every $\eta \in \lim(r)$, $(\exists n)$ $[\eta{\restriction}n \in T^*_1]$

or (b) for no $\eta \in \lim(r)$, $(\exists n)$ $[\eta{\restriction}n \in T^*_1]$.

If (b) holds, then we can find p' and γ such that: $r \le p'$ and $p' \Vdash$ "$\underset{\sim}{\tau} = \gamma$". But then let $\nu \in p'$, $\ell g \nu \ge n$, $\ell g(\mathrm{tr}(p'))$. Then $\nu \in T^*_0$ (witnessed by $(p')_{[\nu]}$) hence for some $k \le n$, $\nu{\restriction}k \in T^*_1$. But this is a contradiction to (b), as $\nu \in r$. Hence (a) holds, hence T^*_1 is a front of r, and $p \le_n r$ because $p \le_{\mathrm{pr}} r$ and $\ell g(\mathrm{tr}(p)) \ge n$. $\square_{4.5A}$

Continuation of the proof of 4.5: Let, for $\nu \in T^*_1$, q^ν be a pure extension of $(r)_{[\nu]}$ satisfying

$$q^\nu \Vdash \text{“}\underset{\sim}{\tau} = \gamma^\nu\text{”}.$$

Then $q = \cup\{q^\nu : \nu \in T^*_1\}$ is a condition (i.e. $T_q = \bigcup\limits_{\nu \in T^*_1} T_{q^\nu}$ and $\mathfrak{D}_q = \mathfrak{D}_p$) such that $p \le_n r \le_n q$ and

$$q \Vdash \text{“}\underset{\sim}{\tau} \in \{\gamma^\nu : \nu \in T^*_1\}.$$

So $(*)$ is proved.

2) Check the proof of part (1). $\square_{4.5}$

4.5B Remark. 1) In 4.5(1), (2) we can replace $\aleph_2$ by any $\mu > \aleph_2$.

2) As in the proof of 4.5(1) we prove $(*)$ we can (e.g. in 5.5) use the preservation of RUCar-properness (3.5(1)) instead of 3.5(2).

4.6 Lemma. If $(T, \mathfrak{D})$ is a filter-tagged tree, which is λ^+-complete (i.e., each $\mathfrak{D}_\eta$ is a λ^+-complete filter) and $H : T \to \lambda$ and $\lambda^{\aleph_0} = \lambda$, *then* there is $(T^\dagger, \mathfrak{D}^\dagger)$, $(T, \mathfrak{D}) \le_{\mathrm{pr}} (T^\dagger, \mathfrak{D})$ such that $H(\eta)$ depends only on $\ell g(\eta)$, for $\eta \in T^\dagger$.

Remark. See Rubin and Shelah [RuSh:117] p. 47 – 48 on the history of this and such theorems there.

Proof. For any sequence $\bar\alpha = \langle \alpha_n : n < \omega \rangle$, $\alpha_n < \lambda$, we define a game $\partial_{\bar\alpha}$:

Let η_0 be the trunk of T.

In move 0 player I chooses $A_1 \subseteq \mathrm{Suc}_T(\eta_0)$, $A_1 = \emptyset \bmod \mathfrak{D}_{\eta_0}$, and player II chooses $\eta_1 \in \mathrm{Suc}_T(\eta_0) \setminus A_1$.

In move n, player I chooses $A_{n+1} \subseteq \mathrm{Suc}_T(\eta_n)$, $A_{n+1} = \emptyset \bmod \mathfrak{D}_{\eta_n}$ and player II chooses $\eta_{n+1} \in \mathrm{Suc}_T(\eta_n) \setminus A_{n+1}$.

In the end, player II wins the play if for every n we have $H(\eta_n) = \alpha_n$.

Now we prove:

(*) For some $\bar{\alpha} = \langle \alpha_n : n < \omega \rangle$, $\alpha_n < \lambda$, player II wins the game (i.e., has a winning strategy).

Clearly the game is closed, hence it suffices to prove that for some $\bar{\alpha}$, player I does not have a winning strategy. So assume that for every $\bar{\alpha}$ player I has a winning strategy $F_{\bar{\alpha}}$ in the game $\eth_{\bar{\alpha}}$, and we shall get a contradiction. A winning strategy is a function which, given the previous moves of the opponent ($\eta_1, \ldots, \eta_{n-1}$ in our case), gives a move to the player, so that in any play in which he uses the strategy he wins the play.

Now define by induction on n, $\eta_n \in T$ such that $\ell g(\eta_n) = \ell g(\mathrm{tr} T) + n$ and $\eta_{n+1} \restriction n = \eta_n$:

η_0 is the trunk of T

$\eta_{n+1} \in \mathrm{Suc}_T(\eta_n) \setminus \bigcup_{\bar{\alpha}} F_{\bar{\alpha}}(\langle \eta_1, \ldots, \eta_n \rangle)$.

Why does η_{n+1} exist? For every $\bar{\alpha}$, $F_{\bar{\alpha}}(\langle \eta_1, \ldots, \eta_n \rangle) = \emptyset \bmod \mathfrak{D}_{\eta_n}$, $\mathfrak{D}_{\eta_n}$ is λ^+-complete and the number of $\bar{\alpha}$'s is $\lambda^{\aleph_0} = \lambda < \lambda^+$. So $\bigcup_{\bar{\alpha}} F_{\bar{\alpha}}(\langle \eta_1, \ldots, \eta_n \rangle) = \emptyset \bmod \mathfrak{D}_{\eta_n}$, and so η_{n+1} exists as $\mathrm{Suc}_T(\eta_n) \neq \bmod \mathfrak{D}_{\eta_n}$ by Definition 4.3(1) clause (b).

But let $\alpha_n^* \stackrel{\mathrm{def}}{=} H(\eta_n)$ and $\bar{\alpha}^* = \langle \alpha_n^* : n < \omega \rangle$, so

$$F_{\bar{\alpha}^*}(<>), \eta_1, \ldots, F_{\bar{\alpha}^*}(\langle \eta_1, \ldots, \eta_n \rangle), \eta_{n+1}, \ldots$$

is a play of $\eth_{\bar{\alpha}^*}$ in which player I uses his strategy $F_{\bar{\alpha}^*}$, but he lost: contradiction, hence (*) holds.

Proof of the Lemma from (*). Let $\langle \alpha_n : n < \omega \rangle$ be as in (*), and W be the winning strategy of player II.

Let $T_0 = \{\eta \in T$: for some n, we have: $\ell g(\eta) = \ell g(\eta_0) + n$, and for some $A_1, \ldots, A_n$, for every $0 < \ell \le n$ we have $\eta \restriction (\ell g \eta_0 + \ell) = W(\langle A_1, \ldots, A_\ell \rangle)\} \cup \{\eta_0 \restriction \ell : \ell \le \ell g \eta_0\}$.

It is clear that T_0 is closed under initial segments. Now if $\eta \in T_0$, $\eta_0 \trianglelefteq \eta \in T_0$ then $\mathrm{Suc}_{T_0}(\eta) \ne \bmod \mathfrak{D}_\eta$, for otherwise if $n = \ell g(\eta) - \ell g(\eta_0)$, and $A_1, \ldots, A_n$ are "witnesses for $\eta \in T_0$", then player I could have chosen $A_{n+1} = \mathrm{Suc}_{T_0}(\eta)$, and then by definition $W(A_1, \ldots, A_{n+1}) \in T_0$ and also $W(A_1, \ldots, A_{n+1}) \notin \mathrm{Suc}_{T_0}(\eta) = A_{n+1}$ but $W(A_1, \ldots, A_{n+1}) \in \mathrm{Suc}_T(\eta)$ and $\mathrm{Suc}_{T_0}(\eta) = T_0 \cap \mathrm{Suc}_T(\eta)$, contradiction.

So $(T_0, \mathfrak{D}) \le_{pr} (T, \mathfrak{D})$ and $(T, \mathfrak{D})$ is as required. $\square_{4.6}$

4.7 Theorem. Suppose $(T^*, \mathfrak{D}^*)$ is an $\aleph_2$-complete filter-tagged tree. Let $P = \mathrm{Nm}'(T^*, \mathfrak{D}^*)$ *then*

(1) (CH) P does not add reals.

(2) If for every $(T, \mathfrak{D}^*) \in P$ for some $\eta \in T, \mathrm{tr}(T) \trianglelefteq \eta$ and for some $A \subseteq \mathrm{Suc}_\eta(T)$ and function $F : A \to \lambda$ we have $(\forall \alpha < \lambda)[F^{-1}(\{i : i < \alpha\}) \equiv \emptyset \bmod \mathfrak{D}^*_\eta]$ and $A \ne \emptyset \bmod \mathfrak{D}^*_\eta$ *then* $\Vdash_P$ "$\mathrm{cf}(\lambda) = \aleph_0$".

(3) P does not collapse $\aleph_1$ (and if $\mathfrak{D}^*$ is λ^+-complete, $\mathrm{cf}(\lambda) > \aleph_0$ then $\Vdash_P$ "$\mathrm{cf}(\lambda) > \aleph_0$").

4.7A Remark. If we waive CH, P may add reals but it does not collapse $\aleph_1$; sometimes it satisfies the $\aleph_4$-c.c. even though $2^{\aleph_1} > \aleph_4$ (see XI 4.3).

4.7B Notation. If $\mathrm{Dom}(\mathfrak{D}^*) = T$ let $\mathrm{Nm}'(\mathfrak{D}^*) = \mathrm{Nm}'(T, \mathfrak{D}^*)$, and if $T = {}^{\omega >}(\omega_2)$, $\mathfrak{D}^*(\eta) = \{\{\eta\hat{\ } < \alpha >: \alpha \in A\} : A \in D\}$, we let $\mathrm{Nm}'(T, D) = \mathrm{Nm}'(D) = \mathrm{Nm}'(\mathfrak{D}^*)$.

4.7C Remark. So if $P = \mathrm{Nm}'(D^*)$, $D^* \supseteq \mathcal{D}^{cb}_{\aleph_2} (\overset{\mathrm{def}}{=} \{A \subseteq \aleph_2 : A$ co - bounded$\})$, $G \subseteq P$ is generic, *then* $\bigcup\{\eta : \eta \in (T, D^*)$ for every $(T, D^*) \in G\}$ is a member of ${}^\omega(\omega_2)$ (in $V[G]$) and as $D^* \supseteq \mathcal{D}^{cb}_{\aleph_0}$, it is unbounded in ω_2 so $\Vdash_{\mathrm{Nm}'(D^*)}$ "$\mathrm{cf}(\aleph_2^V) = \aleph_0$".

Proof of 4.7.

(1) Now suppose $\underset{\sim}{\tau}$ is a name of an ω-sequence from ω_1, and let $(T, \mathfrak{D}^*) \in P$. It is easy to define by induction $(T_n, \mathfrak{D}^*)$ such that:

(a) $(T_0, \mathfrak{D}^*) = (T, \mathfrak{D}^*)$,

(b) $(T_n, \mathfrak{D}^*) \leq_n (T_{n+1}, \mathfrak{D}^*)$ and $(T_n, \mathfrak{D}^*) \leq_{pr} (T_{n+1}, \mathfrak{D}^*)$,

(c) for every $\eta \in T_{n+1}$, if $\ell g(\eta) = n+1$, then for some $\bar{\alpha}_\eta$ and $\ell \leq n$ we have: $(T_{n+1}, \mathfrak{D}^*)_{[\eta]} \Vdash_P$ "$\underset{\sim}{\tau} \restriction \ell = \bar{\alpha}_\eta$", and ℓ is maximal, i.e., either $\ell = n$, or there are no $T^\dagger, \alpha$ such that $\alpha < \omega_1$ and $(T^\dagger, \mathfrak{D}^*) \Vdash$ "$\underset{\sim}{\tau}(\ell) = \alpha$" and $(T_{n+1}, \mathfrak{D}^*)_{[\eta]} \leq_{pr} (T^\dagger, \mathfrak{D}^*)$.

Clearly $(\bigcap_{n<\omega} T_n, \mathfrak{D}^*) \in P$ and $(T_n, \mathfrak{D}^*) \leq (\bigcap_{n<\omega} T_n, \mathfrak{D}^*)$.

Now use Lemma 4.6 on $(\bigcap_{n<\omega} T_n, \mathfrak{D}^*)$, and $H, H(\eta) = \bar{\alpha}_\eta$ and get $(T^\dagger, \mathfrak{D}^*)$, $(\cap_{n<\omega} T_n, \mathfrak{D}^*) \leq_{pr} (T^\dagger, \mathfrak{D}^*)$, $H(\eta) = \bar{\alpha}^n$ for $\eta \in T^*$, $\ell g(\eta) = n+1$. Now for each ℓ, there is $(T'', \mathfrak{D}^*)$, $(T^\dagger, \mathfrak{D}^*) \leq (T'', \mathfrak{D}^*)$ and $\bar{\alpha}$ such that $(T'', \mathfrak{D}^*) \Vdash_P$ "$\underset{\sim}{\tau} \restriction \ell = \bar{\alpha}$", and let $\eta_0 \in T''$ be the trunk of T''; w.l.o.g. $\ell + 1 < \ell g(\eta_0)$. By the choice of $\bar{\alpha}_{\eta_0}, \ell \leq \ell g(\bar{\alpha}_{\eta_0})$ hence $\bar{\alpha} = \bar{\alpha}^k \restriction \ell$ for $k = \ell g(\eta_0)$, hence for every $\eta \in T^\dagger$, $\ell g(\eta) = \ell g(\eta_0)$ implies $\bar{\alpha}_\eta \restriction \ell = \bar{\alpha}_{\eta_0} \restriction \ell$, hence $(T^\dagger, \mathfrak{D}^*)_{[\eta]} \Vdash$ "$\underset{\sim}{\tau} \restriction \ell = \bar{\alpha}_{\eta_0} \restriction \ell = \bar{\alpha}^\ell$". But $(T^\dagger, \mathfrak{D}^*) \Vdash$ " for some $\eta \in T^\dagger$, $\ell g(\eta) = \ell$ and $(T^\dagger, \mathfrak{D}^*)_{[\eta]}$ belongs to $\underset{\sim}{G}_P$ (the generic subset of P)." So clearly $(T^\dagger, \mathfrak{D}^*) \Vdash$ "$\underset{\sim}{\tau} \restriction \ell = \bar{\alpha}^{\ell g(\eta_0)} \restriction \ell$", and as this holds for every ℓ we have $(T^\dagger, \mathfrak{D}^*) \Vdash$ "$\underset{\sim}{\tau} = \langle \bar{\alpha}^{m(n)}(n) : n < \omega \rangle$" when we choose the numbers $m(n)$ large enough, i.e., such that $n < \ell g(\bar{\alpha}^{m(n)})$.

(2) Clearly the following is a dense open subset of P, $\mathcal{I}_0 = \{(T, \mathfrak{D}) : (T, \mathfrak{D}) \in P$ and for every $\eta \in T$, *if* there are $A_i \subseteq \mathrm{Suc}_T(\eta)$, for $i < \lambda$, $\bigcup_{j<i} A_j = \emptyset \bmod \mathfrak{D}_\eta$, $\bigcup_{i<\lambda} A_i \neq \emptyset \bmod \mathfrak{D}_\eta$ *then* there is $F_\eta : \mathrm{Suc}_T(\eta) \to \lambda$ such that $\bigwedge_{\alpha<\lambda} \{\nu : \nu \in \mathrm{Suc}_T(\eta), F(\nu) < \alpha\} = \emptyset \bmod \mathfrak{D}_\eta\}$. Now for each $(T, \mathfrak{D}) \in \mathcal{I}_0$ let $B(T, \mathfrak{D}) = \{\eta \in T :$ there is F_η as above$\}$. Note: $(T_1, \mathfrak{D}) \leq (T_2, \mathfrak{D}) \Rightarrow B(T_1, \mathfrak{D}) \cap T_2 = B(T_2, \mathfrak{D})$; by [RuSh:117] or XI 3.5 or XV 2.6B(2) here, for every $(T, \mathfrak{D}) \in \mathcal{I}_0$ there is $(T', \mathfrak{D})$, $(T, \mathfrak{D}) \leq_{\mathrm{pr}} (T', \mathfrak{D})$ such that:

(a) $(\forall \eta \in \lim T')(\exists^{\aleph_0} n)[\eta \restriction n \in B(T, \mathfrak{D})]$, or

(b) $(\forall \eta \in \lim T')(\exists^{<\aleph_0} n)[\eta \restriction n \in B(T, \mathfrak{D})]$,

In the second case, applying again the partition theorem mentioned above we get a constant bound n to $\{\ell g(\eta) : \eta \in B(T, \mathfrak{D})\}$, and increasing the trunk contradict the hypothesis (of 4.7(2)). So $\mathcal{I}_1 = \{(T, \mathfrak{D}) \in \mathcal{I}_0$ and (a) holds$\}$ is dense open subset of P. Fix $(T, \mathfrak{D}) \in \mathcal{I}_1$. For $\eta \in B(T, \mathfrak{D})$, let F_η be as required above; and let $\underset{\sim}{\tau}$ be the unique ω-sequence such that for every $p \in \underset{\sim}{G}_P$, and $n < \omega$, $\underset{\sim}{\tau} \restriction n \in p$. Then $\lambda \cap \{F_{\underset{\sim}{\tau} \restriction n}(\underset{\sim}{\tau}(n)) : n < \omega$ (and $F_{\underset{\sim}{\tau} \restriction n}$ is well defined) $\}$ is a countable unbounded subset of λ.

3) Similar to part (1) using XI 3.7 instead of 4.6 (but not used here). $\square_{4.7}$

4.8 Problem. Is the forcing semiproper? (See XII.)

4.9 Definition. For a filter D on a set I, and a set S of cardinals, we call D an (S, α)-Galvin filter (and the dual ideal a Galvin ideal) if player II has a winning strategy in the following game, for every $J \subseteq I$, $J \neq \emptyset \bmod D$ (we call the game the (S, α)-Galvin game for (D, J)):

In the ith move player I chooses a function F_i from I to some $\lambda \in S$ and player II chooses $A_i \subseteq J \cap \bigcap_{j<i} A_j$ such that $|F_i(A_i)| < \lambda$. Player II wins if $\bigcap_{i<\alpha} A_i \neq \emptyset \bmod D$. For simplicity we can say J was chosen by player I in his first move.

4.9A Remark. Galvin suggests this game for $D^{cb}_{\omega_2}$ = the co-bounded subset of κ for a cardinal κ, $\alpha = \omega$ and $S = \{2\}$. So for $\alpha = \omega$, $S = \{2\}$ we omit (S, α). Note that only $S \cap (|I| + 1)$ is relevant for the game.

4.10 Definition. A filter D on κ has the Laver (or $\aleph_1$-Laver) property, if there is a family W of subsets of κ, $A \in W \Rightarrow A \neq \emptyset \bmod D$, W is dense [i.e. $\forall A \subseteq \kappa$, $A \neq \emptyset \bmod D \rightarrow (\exists B \in W)(B \subseteq A \bmod D)$], and W is closed with respect to countable intersections of descending chains.

Related to this property is the following game:

In the n'th move, player I chooses a set A_n and player II chooses a set B_n, such that for all n $A_n \supseteq B_n \supseteq A_{n+1} \neq \emptyset \bmod D$, II wins iff $\bigcap_{n<\omega} A_n \neq \emptyset \bmod D$.

Clearly, if D has the Laver property, then player II wins.
Galvin, Jech and Magidor [GJM] and Laver independently proved the following.

4.11 Theorem. If we start with a universe V, $V \models$ "$G.C.H. + \kappa$ is measurable" and use Levy collapsing of κ to $\aleph_2$ (so every λ, $\aleph_1 \le \lambda < \kappa$ now will have cardinality $\aleph_1$) *then* in the new universe $V[G]$, $\mathcal{D}^{cb}_{\omega_2}$ is a Galvin filter, in fact $(\mathrm{Car} \setminus \{\aleph_2\}, \omega)$-Galvin filter. Moreover if $D \in V$ was a normal ultrafilter on κ, then in $V[G]$ the filter D has the Laver property. [We identify here D with the filter it generates in $V[G]$ which is normal.]

More exactly, [GJM] proved that player II has a winning strategy in the play above for D a normal filter on λ, Laver proved the $\aleph_1$-Laver property in that context, but the difference is not essential in our context.
(We shall not prove it here.)

The relevance of this is:

4.12 Theorem. Let $S \subseteq \mathrm{SCar}$.

(1) If P is $\mathrm{Nm}'(T^\dagger, \mathfrak{D}^*)$ (see 4.4(4)), each $\mathfrak{D}^*_\eta$ is an (S,ω)-Galvin, $\aleph_2$-complete filter *then* P is S-semiproper and even (S,ω)-complete (and we can add all λ, $\mathrm{cf}(\lambda) > |T^\dagger|$ to S).

(2) We can strengthen the hypothesis in (1) by "$\mathfrak{D}^*_\eta$ is $|\alpha|^+$-Laver" and then get even "(S,α)-complete for pure extensions" (see XIV).

Proof. (1) Also easy, but we shall do it. By 3.4(4) it suffices to prove (S,ω)-completeness. Let $p^* \in P$ and we shall prove that the second player wins in $\Game^\omega_S(p,P)$. For every $\eta \in T^* \setminus \{\nu : \nu \triangleleft \mathrm{tr}(T^*)\}$, let H_η be a winning strategy of player II in the (S,ω)-Galvin game for $(\mathfrak{D}_\eta, \mathrm{Suc}_{T^*}(\eta))$.

We first prove

4.13 Fact. Suppose $p \in P$, ($P = \mathrm{Nm}'(T^\dagger, \mathfrak{D}^*)$, $\mathfrak{D}^*$ is $\aleph_2$-complete), $\lambda \in S$, and $\underset{\sim}{\alpha}$ is a P-name of an ordinal, $p \Vdash$ "$\underset{\sim}{\alpha} < \lambda$" and

(∗) $X \in \mathfrak{D}^+_\eta$, $F : X \to \lambda \Rightarrow (\exists\alpha)[\{\nu \in X : F(\nu) < \alpha\} \neq \emptyset \mod \mathfrak{D}]$ (this follows from $\mathfrak{D}_\eta$ being (S,ω)-Galvin).

Then there are $p^\dagger$ and $\alpha < \lambda$ such that $p \leq_{pr} p^\dagger \in P$ and $p^\dagger \Vdash$ "$[\lambda \geq \aleph_0 \Rightarrow \underset{\sim}{\alpha} < \alpha]$ and $[\lambda = 2 \Rightarrow \underset{\sim}{\alpha} = \alpha]$".

4.13A Remark. In the proof of the fact we do not use the Galvin property assumption; also the $\aleph_2$-completeness can be waived, see 4.14 below.

Proof of the fact 4.13. For notational simplicity only we assume, $2 \notin S$. Easily we can find $p_1 = (T_1, \mathfrak{D}^*)$, $p \leq_{pr} p_1$, such that for every $\eta \in T_1$, if there are $\beta < \lambda$, and q, $(T_1, \mathfrak{D}^*)_{[\eta]} \leq_{pr} q$, $q \Vdash$ "$\underset{\sim}{\alpha} < \beta$", then $(T_1, \mathfrak{D}^*)_{[\eta]} \Vdash$ "$\underset{\sim}{\alpha} < \beta$" for some β. For each $\eta \in T_1$, let β_η be such that $(p_1)_{[\eta]} \Vdash$ "$\underset{\sim}{\alpha} < \beta_\eta$", β_η may be undefined for some η but if $\eta \lhd \nu \in T_1$, β_η defined, then β_ν is defined and equal to β_η. So for every $\eta \in \lim T_1$, $\{\ell : \beta_{\eta \restriction \ell}$ not defined$\}$ is an initial segment of ω. By the $\aleph_2$-completeness and 4.6 if CH and XI 3.5 in general, there is T_2, $(T_1, \mathfrak{D}^*) \leq_{pr} (T_2, \mathfrak{D}^*)$ and a set $A \subseteq \omega$ such that $\forall \eta \in T_2$ β_η is defined iff $\ell g(\eta) \in A$ (A is an endsegment or the empty set (so there are only countably many possibilities, this is why XI 3.5 can be applied)). But $A = \emptyset$ is impossible by density. So for some n β_η is defined for every $\eta \in T_2$, $\ell g(\eta) = n$. We can (by induction on n using (∗) in the assumption, see below for a similar argument or again by XI 3.5) define p_3, $(T_2, \mathfrak{D}^*) \leq_{pr} p_3$, and $\beta < \lambda$ such that $[\eta \in p_3$ & β_η defined $\Rightarrow \beta_\eta < \beta]$ this implies $p_3 \Vdash$ "$\underset{\sim}{\alpha} < \beta$", so the Fact holds. $\square_{4.13}$

Continuation of the proof of 4.12(1): Remember $p^* = (T^*, \mathfrak{D}^*)$ is given; w.l.o.g. the trunk of T^* is $\langle\rangle$.

In the first move player I chooses $\lambda_0 \in S$ and a P-name $\underset{\sim}{\beta}_0$ of an ordinal $< \lambda$.

Player II chooses $\beta_0 < \lambda$ and $p_0 \in P$ such that $p^* \leq_{pr} p_0$, $p_0 \Vdash_P$ "$\underset{\sim}{\beta}_0 < \beta$" (possible by the Fact 4.13 above).

However if player II continues to play like this, he may loose as maybe $\bigcap_n T_n$ (where $p_n = (T_n, \mathfrak{D}^*)$) will be $\{\langle\rangle\}$.

So he is thinking how to make $\mathrm{Suc}_{\bigcap_{n<\omega} T_n}(<>) \neq \emptyset \bmod \mathfrak{D}^*_{<>}$. If he, on the other hand, will demand $p_0 \leq_1 p_{n+1}$, he will have $\mathrm{Suc}_{\bigcap_{n<\omega} T_n}(<>) \notin \emptyset \bmod \mathfrak{D}^*$, but it will be hard (and in fact impossible) to do what is required when, e.g., $\lambda_n = \aleph_1$. So what he will do is to decrease $\mathrm{Suc}_{T_n}(<>)$, but do it using his winning strategy $H_{<>}$ for the (S,ω)-Galvin game for $\mathfrak{D}_{<>}$. So in the second move player I chooses a cardinal $\lambda_1 \in S$ and P-name $\underset{\sim}{\beta}_1$ of an ordinal $< \lambda_1$. Player II, first for each $\eta \in p_0$, $\ell g(\eta) = 1$, chooses $p_1^\eta = (T_1^\eta, \mathfrak{D}^*)$ such that $(p_0)_{[\eta]} \leq_{pr} p_1^\eta$ and $p_1^\eta \Vdash_P$ "$\underset{\sim}{\beta}_1 \leq \beta_\eta$", this is possible by 4.13. This defines a function from $\mathrm{Suc}_{T_0}(<>)$ to λ_1, so player II consults the winning strategy $H_{<>}$, gets $A^0_{<>} \subseteq \lambda_1$, $|A^0_{<>}| < \lambda_1$, and lets $T_1 = \bigcup\{T_1^\eta : \beta_\eta \in A^0_{<>}\}$. Now at last player II actually plays: the condition $(T_1, \mathfrak{D}^*)$ and the ordinal $\sup A^0_{\langle\rangle}$.

In the third move, player II tries also to insure that $\{\eta \in \bigcap_n T_n : \ell g(\eta) = 2\}$ will be as required. Now player I chooses $\lambda_2 \in S$ and a P-name $\underset{\sim}{\beta}_2$. Player II chooses for every $\eta \in T_1$, $\ell g(\eta) = 2$ a condition p_2^η such that $(p_1)_{[\eta]} \leq_{pr} p_2^\eta$ and $p_2^\eta \Vdash_P$ "$\underset{\sim}{\beta}_2 \leq \beta_\eta$". So for every $\eta \in T_1$, $\ell g(\eta) = 1$, we have a function from $\mathrm{Suc}_\eta(T_1)$ to λ_2, so consulting the strategy H_η, player II chooses $A^1_\eta \subseteq \lambda_2$, $|A^1_\eta| < \lambda$. We can assume that each A^1_η is a proper initial segment (i.e., an ordinal) and for $\lambda = 2$, a singleton. So the number of possible A^1_η is λ. So now the function $\eta \mapsto A^1_\eta$ ($\eta \in \mathrm{Suc}_{T_1}(<>)$) is a function whose domain is $\mathrm{Suc}_{T_1}(<>)$. So player II can consult again the strategy $H_{<>}$, and find $A^2_{<>}$, and let $T_2 = \bigcup\{T_2^\eta : \ell g(\eta) = 2,\ \eta \in T_1,\ \beta_\eta \in A^1_{\eta\restriction 1}$ and $A^1_{\eta\restriction 1} \subseteq A^2_{\eta\restriction 0} = A^2_{<>}\}$. Now at last player II plays: the condition $(T_2, \mathfrak{D}^*)$ and the ordinal $\sup A^2_{\langle\rangle}$.

The rest should be clear (compare with the proof of 6.2).

(2) By 4.13 it should be clear $\square_{4.12}$

4.14 Remark. Really in 4.12(1) we can replace $\aleph_2$-completeness by $\aleph_1$-completeness by using XI 3.5 instead of 4.6. In fact even this can be waived. We use $\aleph_2$-completeness only in the proof of Fact 4.13; but we now give a proof which eliminate it. Instead of choosing T_2, we let $H : \lim T^\dagger \to 2$ be defined by $H(\eta) = 0$ iff $(\exists n)[\beta_{\eta\restriction n}$ is defined], and so there is T' such that $(T^\dagger, \mathfrak{D}^*) \leq_{pr} (T', \mathfrak{D}^*)$ and H is constant on $\lim T'$ (by XI 3.5 which does not need any completeness). Now on $B = \{\eta \in T' : (\forall \ell < \ell g(\eta))\ [\beta_{\eta \restriction \ell}$ is not defined$]\}$, we can define a rank:

$$\mathrm{rk}(\eta) = \bigcup\{\gamma + 1 : \{\nu : \nu \in B \cap \mathrm{Suc}_{T'}(\eta) \text{ and } \mathrm{rk}(\nu) \geq 0\} \neq \emptyset \bmod \mathfrak{D}_\eta\}.$$

If for some η, $\mathrm{rk}(\eta) = \infty$ we let $T'' \stackrel{\mathrm{def}}{=} \{\nu \in T' : \nu \trianglelefteq \eta$ or $\eta \triangleleft \nu \,\&\, \bigwedge_{\ell g\eta \leq \ell \leq \ell g\nu} \mathrm{rk}(\nu\restriction\ell) = \infty\}$; we get T'', $(T', \mathfrak{D}) \leq (T'', \mathfrak{D}), T'' \subseteq T'$, contradiction to the choice of T'. Otherwise (i.e. $\eta \in T' \Rightarrow \mathrm{rk}(\eta) < \infty$) we can prove by induction on its $\mathrm{rk}(\mathrm{tr}T')$ that we can find $p^\dagger$ as required.

(Note we are not assuming CH).

§5. Chain Conditions and Abraham's Problem

Chain conditions are very essential for iterated forcing. In Solovay and Tannenbaum [ST] this is the point, but even when other conditions are involved, we have to finish the iteration and exhaust all possibilities, so some chain condition is necessary to "catch our tail." In our main line we want to collapse some large κ to $\aleph_2$, in an iterated forcing of length (and power) κ, each P_i of power $< \kappa$. So we want that κ stays a regular cardinal, and the obvious way to do this is by the κ-chain condition. We prove it by the traditional method of the Δ-system. For general RCS iteration, we have to assume κ is Mahlo (i.e., $\{\lambda < \kappa$ strongly inaccessible$\}$ is stationary) and for iteration of semiproper forcing we ask for less.

Now we are able to answer the following problem of U. Abraham:

Problem. Suppose G.C.H. holds in V. Is there a set $A \subseteq \aleph_1$ so that every ω-sequence from $\aleph_2$, belongs to $L[A]$?

To construct a model where the answer is "no" we shall collapse some inaccessible κ, which is the limit of measurable cardinals, changing the cofinalities of arbitrarily large measurables $< \kappa$ to $\aleph_0$.

5.1 Definition.

(1) For any iteration $\bar{Q} = \langle P_i, \underset{\sim}{Q}_i : i < \alpha \rangle$ and a set $S \subseteq \alpha$ we call $\bar{p} = \langle p_i : i \in S \rangle$ a Δ-system if, for $i < j$ in $S, p_i \restriction i = p_j \restriction j$ and $p_i \in P_j$. We call $p_i \restriction i$ the heart of the Δ-system, $hr(\bar{p})$.

(2) For a forcing P, we call $\bar{p} = \langle p_i : i \in S \rangle$ a μ-weak Δ-system if $p_i \in P$, $\bigcup_{i \in S} i$ is a regular cardinal κ, and there is a condition $q = hr(\bar{p})$ (the heart of $\bar{p}$) such that for every r, $q \leq r \in P$ there is $\alpha < \kappa$ satisfying : if $\alpha < \alpha_i \in S$ for $i < \mu_1 < \mu$ then $\{r\} \bigcup \{p_{\alpha_i} : i < \mu_1\}$ has an upper bound in P.

5.2 Claim. Any Δ-system in an RCS iteration as in Definition 5.1 (1), with $\alpha = \sup S$ and $P_\alpha = \bigcup_{i<\alpha} P_i$ is an $\aleph_1$-weak Δ-system.

Proof. Easy. $\square_{5.2}$

5.3 The Chain Condition Lemma.

(1) Suppose $\bar{Q} = \langle P_i, \underset{\sim}{Q}_i : i < \kappa \rangle$ is an RCS iteration, κ regular, $|P_i| < \kappa$ for $i < \kappa$ and let $A = \{\lambda < \kappa : \lambda$ strongly inaccessible$\}$. *Then* for every sequence $\bar{p} = \langle p_j : j \in B \subseteq A \rangle$, we can find a closed unbounded $C \subseteq \kappa$ and a pressing down function h on $C \cap B$ (i.e., $h(j) < j$) such that for any α, $\langle p_j : j \in B \cap C, h(j) = \alpha \rangle$ is a Δ-system. (So in the non trivial case κ is strongly inaccessible Mahlo cardinal.)

(2) Assume $A^\dagger = \kappa$ and: $\underset{\sim}{Q}_i$ is RṴCar-semiproper (for all i) *or* semiproper, $\Vdash_{P_i + n_i}$ " $(2^{|P_i|})^V$ has cardinality $\aleph_1$" (for all i) *or* even $\bar{Q}$ as in 2.6(3), $\bar{Q}$ as in 2.6(4) and $A^\dagger = \{i : \emptyset \Vdash_{P_i}$ "$\bigcup_{j<i} P_j$ is dense in P_i"$\}$. *Then* in (1), we can replace A by $A^\dagger$ (We know that if each Q_i is semiproper (or just P_j/P_{i+1}) then $[\text{cf}^V(i) = \aleph_1 \Rightarrow i \in A^\dagger]$ and also:$[i$ limit, 2.6(2) or 2.6(3) apply to $\bar{Q} \restriction i$ and $\Vdash_{P_j}$ "cf$i > \aleph_0$" for every $j < i\} \Rightarrow i \in A^\dagger])$.

(3) If we agree to weaken the conclusion to "$\aleph_1$-weak Δ-system", we can replace "$|P_i| < \kappa$ for $i < \kappa$" by "$d(P_i) < \kappa$ for $i < \kappa$" or even, for any $A \subseteq \kappa$, "each $P_i (i < \kappa)$ satisfies the conclusion of (1) for A". In (2) we can assume just each Q_i is semiproper.

Before we prove the lemma note:

5.4 Corollary.

(1) If in 5.3(1), A is stationary or in 5.3(2), $A^\dagger$ is stationary, *then* $P_\kappa = \text{Rlim}\bar{Q}$ satisfies the κ-chain condition.

(2) If D is a normal ultrafilter on κ, $B \in D$, $B \subseteq A$ then (in 5.3(1)) for some $B^\dagger \in D$, $\langle p_j : j \in B^\dagger \rangle$ is a Δ-system.

Proof of 5.3. (1) If B is not stationary (as a subset of κ), the conclusion is trivial, so suppose B is stationary. Necessarily κ is strongly inaccessible (as κ is regular, every member of A is strongly inaccessible and $B \subseteq A$), hence by 1.6, $P_\kappa \stackrel{def}{=} \text{Rlim}\bar{Q} = \bigcup_{i<\kappa} P_i$. As $|P_i| < \kappa$ for every $i < \kappa$, there is a one to one function H from P_κ onto κ. Again as $|P_i| < \kappa$ for $i < \kappa$, clearly

$$C = \{i : H \text{ maps } \bigcup_{j<i} P_j \text{ onto } i \text{ and for } j < i\text{: if } j \in B \text{ then } p_j \in P_i\}.$$

is a closed unbounded subset of κ. We now define the function h with domain $B \cap C$: $h(i) = H(p_i \restriction i)$.

We first prove that h is pressing down. Clearly $p_i \restriction i \in P_i$, and if $i \in B \cap C$ then i is strongly inaccessible and $(\forall j < i)[|P_j| < i]$ hence by 1.6, $P_i = \bigcup_{j<i} P_j$, hence $p_i \restriction i \in \bigcup_{j<i} P_j$, so $h(i) < i$. Now looking at the definitions of h and C we see that $\langle p_j : j \in B \cap C, h(j) = \alpha \rangle$ is a Δ-system, for any α.

(2) The proof is similar, using 2.7 instead of 1.6.

(3) Left to the reader. $\square_{5.3}$

5.5 Theorem. Suppose Con (ZFC + "there is an inaccessible cardinal κ which is the limit of measurable cardinals"). *Then* the following theory is consistent: ZFC + G.C.H. $+(\forall A \subseteq \aleph_1)(\exists \bar{\alpha})$ ($\bar{\alpha}$ an ω-sequence of ordinals $< \aleph_2$, $\bar{\alpha} \notin L[A]$).

Proof. We start with a model V of ZFC + G.C.H + "κ is strongly inaccessible, and limit of measurables". We define an RCS iterated forcing $\langle P_i, \underset{\sim}{Q}_i : i < \kappa \rangle$, such that $|P_i| < \kappa$. We do it by induction on i, and clearly (see 1.4(6) for i limit) the induction hypothesis $|P_i| < \kappa$ continues to hold. If $\bar{Q}^i = \langle P_j, \underset{\sim}{Q}_j : j < i \rangle$ is defined, let κ_i be the first measurable cardinal $> |P_i|$, where $P_i = \text{Rlim}\bar{Q}_i$. It is known (see e.g. [J]) that κ_i is measurable in V^{P_i}, and any normal ultrafilter on it from V is an ultrafilter (and normal) in V^{P_i}, too. As $|P_i| < \kappa$, by hypothesis

$\kappa_i < \kappa$. So let $Q_{i,0}$ be $PF(D_i)$ (see 4.1) where $D_i \in V$ is any normal ultrafilter on κ_i, and let $\underset{\sim}{Q}_{i,1}$ be $P_i * \underset{\sim}{Q}_{i,0}$-name of the Levy collapse of κ_i^+ to $\aleph_1$ (i.e. $Q_{i,1} = \{f : \mathrm{Dom}(f)$ is an ordinal $< \aleph_1$, and $\mathrm{Rang}(f) \subseteq \kappa_i^+\}$, with inclusion as order). We let $Q_i = Q_{i,0} * \underset{\sim}{Q}_{i,1}$.

Now by 4.2, $Q_{i,0} = PF(D_i)$ is $(\mathrm{Car}^V \setminus \{\kappa_i\}, \omega)$-complete, $Q_{i,1}$ is (Car^V, ω)-complete trivially (by 3.4(1)) hence by 3.5 Q_i is $(\mathrm{Car}^V \setminus \{\kappa_i\}, \omega)$-complete.

Hence by 3.5(2), $P_\kappa = \mathrm{Rlim}\langle P_i, \underset{\sim}{Q}_i : i < \kappa\rangle$ is $(\{2, \aleph_0, \aleph_1\}, \omega)$-complete, hence it does not add reals and does not change the cofinality of $\aleph_1$. By 3.4(4) P_κ is semiproper. By 5.3(2) P_κ satisfies the κ-chain condition, so clearly if $G_\kappa \subseteq P_\kappa$ is generic then $\aleph_1^{V[G_\kappa]} = \aleph_1^V, \aleph_2^{V[G_\kappa]} = \kappa$, $V[G_\kappa]$ have the same reals as V, and $V[G_\kappa]$ satisfies the G.C.H.

Now if $A \subseteq \omega_1$, then as P_κ satisfies the κ-chain condition, A is determined by $G_i = G_\kappa \cap P_i$ for some $i < \kappa$. By 1.1(D), G_i is generic for P_i, so $L[A] \subseteq V[G_i]$, but in $V[G_i]$ an ω-sequence from $\aleph_2^{V[G_\kappa]}$ is missing: the Prikry sequence we shot through κ_{i+1} which was measurable in $V[G_i]$. $\square_{5.5}$

§6. Reflection Properties of S^2_0: Refining Abraham's Problem and Precipitous Ideals

In the previous section we have collapsed a large cardinal κ to $\aleph_2$, such that to "many" measurable cardinals $< \kappa$ we add an unbounded ω-sequence. However, "many" was interpreted as "unbounded set". This is very weak and we often desire for more, e.g. in 6.4, we would like to change cofinalities on a stationary set.

Notice that it is known that if we collapse a large cardinal by $\aleph_1$-complete forcing then $S^2_1 \overset{\mathrm{def}}{=} \{\delta < \aleph_2 : \mathrm{cf}(\delta) = \aleph_1\}$ has reflection and bigness properties, e.g., those from Definition 4.10. However, for S^2_0, we get nothing as it is equal to $\{\delta < \aleph_2 :$ in the universe before the collapse, $\mathrm{cf}(\delta) = \aleph_0\}$ and it is known, e.g., that on such a set there was no normal ultrafilter.

So we can ask whether S^2_0 can have some "large cardinal properties". The natural property to consider is precipitous normal filters D on $\aleph_2$ such that $S^2_0 \in D$. Such filters were introduced in Jech and Prikry [JP1] and studied in Jech and Prikry [JP2], Jech, Magidor, Mitchell and Prikry [JMMP].

Their important property is that if we force by $PP(D)$ (which is $\{A \subseteq \kappa : A \neq \emptyset \operatorname{mod} D\}$ ordered by an inverse inclusion), G is generic, the domain of D is I, and in $V[G]$, $E \supseteq D$ is the ultrafilter G generates (on old sets) then V^I/E (taking only old $f : I \to V$) is well-founded. Jech, Magidor, Mitchell and Prikry [JMMP] proved that the existence of a precipitous filter on $\aleph_1$ is equiconsistent with the existence of a measurable cardinal, and also proved the consistency of "$\mathcal{D}_{\aleph_1}$ (= the filter of closed unbounded sets) is precipitous". (Notice that the Laver property is stronger). Magidor asked:

Problem I. Is ZFC + G.C.H.+ there is a normal precipitous filter D on $\aleph_2$, $S^2_0 \in D$ consistent?

We answer positively, by collapsing suitably some κ to $\aleph_2$. Letting D be a normal ultrafilter on κ in V, provided that $A = \{\lambda < \kappa$: in the old universe λ is measurable $\} \in D$. We will force that in the new universe, D generates a normal precipitous filter (which we also call D) such that S^2_0 belongs to it.

This was proved previously and independently, using supercompact cardinals, by Gitik.

We can also consider the following strengthening of Abraham's problem:

Problem II. If V satisfies G.C.H., does there exist $A \subseteq \aleph_2$ such that, for every $\delta < \aleph_2$, every ω-sequence from δ belongs to $L[A \cap \delta]$?

Again we have to change the cofinality on a stationary set, and to iterate forcing such that stationarily often we change the cofinality of $\aleph_2$ to $\aleph_0$.

When we do this the first time, in stage λ for example, the forcing so far P_λ is just Levy's collapse Levy$(\aleph_1, < \lambda)$ so by 4.11, 4.12 we have a $(\text{Car}^V \setminus \{\aleph_2\}, \omega)$-complete forcing Q_λ doing this; but later the collapse P_λ is not even $\aleph_1$-complete. We have two ways to cope with this. One way is to look again at theorem 3.5 on iterated (S, ω)-complete forcing (for various S), from which we see that less is needed. If $P_\lambda = \mathrm{Rlim}\,\bar{Q}, \bar{Q} = \langle P_i, \underset{\sim}{Q}_i : i < \lambda \rangle$ collapses λ to $\aleph_2$, it suffices that $(\mathrm{Rlim}\bar{Q}/P_{i+1}) * \underset{\sim}{Q}_\lambda$ is $(\{2, \aleph_0, \aleph_1\}, \omega)$-complete. We will show that we can achieve this by using Namba forcing as Q_i and our induction hypothesis there. The second possibility is to demand e.g. each Q_i is quite pseudo-complete and prove that in V^{P_λ} we get a large ideal in λ. We use the first approach (but see 6.1A). For clarity of exposition, we first prove a weaker lemma.

6.1 Lemma. Suppose D is a normal ultrafilter on λ, $\bar{Q} = \langle P_i, \underset{\sim}{Q}_i : i < \lambda \rangle$ an RCS iteration and for all $i < \lambda$, $|P_i| < \lambda$. Suppose further $P_\lambda = \mathrm{Rlim}\bar{Q}$ is $(\{2, \aleph_0, \aleph_1\}, \omega)$-complete and collapses λ to $\aleph_2$. Consider the following game $\partial(p_0, \underset{\sim}{A}_0) = \partial^\omega(p_0, \underset{\sim}{A}_0, P_\lambda, D)$, for $p_0 \in P_\lambda$, $\underset{\sim}{A}_0$ a P_λ-name of a subset of λ such that $p_0 \Vdash_{P_\lambda}$ "$\underset{\sim}{A}_0 \neq \emptyset \bmod D$". (The game is played in V.)

In the first move:

Player I chooses P_λ-names $\underset{\sim}{\beta}_1$ (of an ordinal $< \aleph_1$) and $\underset{\sim}{F}_1$ (a function from λ to $\aleph_1$).

Player II has to choose $p_1 \in P_\lambda$, $p_0 \leq p_1$ and $\gamma_1 < \omega_1$ and $\beta_1 < \omega_1$ such that $p_1 \Vdash_{P_\lambda}$ "$\underset{\sim}{A}_1 = \underset{\sim}{A}_0 \cap \underset{\sim}{F}_1^{-1}(\{\gamma_1\}) \neq \emptyset \bmod D$, and $\underset{\sim}{\beta}_1 = \beta_1$".

In the n-th move, player I chooses P_λ-names $\underset{\sim}{\beta}_n < \omega_1, \underset{\sim}{F}_n : \lambda \to \aleph_1$, and player II chooses p_n, $p_{n-1} \leq p_n$ and $\gamma_n < \omega_1$ and $\beta_n < \omega_1$ such that $p_n \Vdash_{P_\lambda}$ "$\underset{\sim}{A}_n = \underset{\sim}{A}_{n-1} \cap \underset{\sim}{F}_n^{-1}(\{\gamma_n\}) \neq \emptyset \bmod D$, and $\underset{\sim}{\beta}_n = \beta_n$".

In the end, player II wins if $\{p_n : n < \omega\}$ has an upper bound $p \in P_\lambda$ such that $p \Vdash_{P_\lambda}$ "$\bigcap_{n<\omega} \underset{\sim}{A}_n \neq \emptyset \bmod D$".

Our conclusion is that player II wins the game.

6.1A Remark. *If* $\{i : P_\lambda/P_i$ is $\{(\{2, \aleph_0, \aleph_1\}, \omega)$-complete $\} \in D$ (i.e. for almost all i, for every $G_i \subseteq P_i$ generic over V, in $V[G_i]$, P_λ/G_i is $(\{2, \aleph_0, \aleph_1\}, \omega)$-complete) *then* in V^{P_λ}, D is a $\{2, \aleph_0, \aleph_1\}$-Galvin filter. Similarly for 6.2 (see XIII 1.9).

Proof. Let $p_0 \in P_\lambda$, $\underset{\sim}{A}_0$ a P_λ-name, $p_0 \Vdash_{P_\lambda}$ "$\underset{\sim}{A}_0 \neq \emptyset \bmod D$". We shall describe the winning strategy of player II in the game $\partial(p_0, \underset{\sim}{A}_0)$. Let the winning strategy of player II in $\Game^{\omega}_{\{2,\aleph_0,\aleph_1\}}(p, P_\lambda)$ be $H[p]$. By 3.2(2), we can assume that player II really determined the value of the P_λ-names of countable ordinals given to him. We can also assume player II is given by player I a pair of names of countable ordinals (instead of one).

Let $B_0 = \{i < \lambda$: there is $p \geq p_0$, $p \Vdash_{P_\lambda}$ "$i \in \underset{\sim}{A}_0$"$\}$. Now $B_0 \in D$ because otherwise, as D is an ultrafilter in V we have $B_0 = \emptyset \bmod D$, but since $p_0 \Vdash_{P_\lambda}$ "$\underset{\sim}{A}_0 \subseteq B_0$" (by B_0's definition) we have $p_0 \Vdash_{P_\lambda}$ "$\underset{\sim}{A}_0 = \emptyset \bmod D$", contradiction.

Now for every $i \in B_0$, there is $p_{0,i} \in P_\lambda$, $p_0 \leq p_{0,i}$ such that $p_{0,i} \Vdash_{P_\lambda}$ "$i \in \underset{\sim}{A}_0$".

So let player I's first move in $\Game(p_0, \underset{\sim}{A}_0)$ be choosing $\underset{\sim}{\beta}_1$ (a P_λ-name of an ordinal $< \aleph_1$), and $\underset{\sim}{F}_1 : \lambda \to \aleph_1$, $\underset{\sim}{F}_1$ a P_λ-name. Now for each $i \in B_0$, player II simulates a play of the game $\Game_i = \Game^{\omega}_{\{2,\aleph_0,\aleph_1\}}(p_{0,i}, P_\lambda)$. He plays $(\underset{\sim}{\beta}_1, \underset{\sim}{F}_1(i))$ (i.e., a pair of names of ordinals $< \aleph_1$) for player I_i, and by the strategy $H[p_{0,i}]$ gets a move for player II_i: $p_{1,i} \in P_\lambda$, $p_{0,i} \leq p_{1,i}$, and $\alpha_{1,i} < \aleph_1$, $\varepsilon_{1,i} < \aleph_1$ such that $p_{1,i} \Vdash_{P_\lambda}$ "$\underset{\sim}{\beta}_1 = \alpha_{1,i}$ and $\underset{\sim}{F}_1(i) = \varepsilon_{1,i}$". Now for some $B_1 \subseteq B_0$, $B_1 \in D$, and $\langle p_{1,i} : i \in B_1\rangle$ is a Δ-system with heart p_1 (see 5.4(2)), and we can also make $\langle \alpha_{1,i}, \varepsilon_{1,i} \in B_1\rangle$ constantly $(\alpha_1, \varepsilon_1)$ (for $i \in B_1$) since there are only $\aleph_1$ many possibilities.

Now player II can make his move in $\Game(p_0, \underset{\sim}{A}_0)$: he chooses p_1, α_1 and ε_1. It is easy to check that this is a legitimate move. (Use 5.2 to show $p_1 \Vdash$ "$\underset{\sim}{\beta}_1 = \alpha_1$".)

So player II continues to play such that after the n-th move:

$(*)_n$ there are $B_n \subseteq B_{n-1} \subseteq \ldots \subseteq B_1 \subseteq B_0$ all in D, $p_{\ell,i} \in P_\lambda$, for $0 \leq \ell \leq n$, $i \in B_\ell$, $p_{0,i} \leq p_{1,i} \leq \ldots \leq p_{\ell,i}$, $\langle p_{\ell,i} : i \in B_\ell\rangle$ is a Δ-system with heart p_ℓ (for $0 < \ell \leq n$) $p_0 \leq p_1 \leq \ldots \leq p_n$, and at the ℓ-th move player I chooses $\underset{\sim}{\beta}_\ell, \underset{\sim}{F}_\ell$, and player II chooses $p_\ell, \alpha_\ell, \varepsilon_\ell$ and (for $1 \leq \ell \leq n$ and $i \in B_\ell$) $p_{\ell,i} \Vdash$ "$\alpha_\ell = \underset{\sim}{\beta}_\ell$ and $\underset{\sim}{F}_\ell(i) = \varepsilon_\ell$". Also for each $\ell \leq n$, $\ell > 0$ and each $i \in B_\ell$, the following is an initial segment of a play of a game $\Game^{\omega}_{\{2,\aleph_0,\aleph_1\}}(p_{0,i}, P_\lambda)$, in which player II_i uses the winning strategy $H[p_{0,i}]$:

$$\langle \underset{\sim}{\beta}_1, \underset{\sim}{F}_1(i)\rangle, \langle p_{1,i}, \alpha_1, \varepsilon_1\rangle, \langle \underset{\sim}{\beta}_2, \underset{\sim}{F}_2(i)\rangle, \langle p_{2,i}, \alpha_2, \varepsilon_2\rangle, \ldots, \langle \underset{\sim}{\beta}_\ell, \underset{\sim}{F}_\ell(i)\rangle, \langle p_{\ell,i}, \alpha_\ell, \varepsilon_\ell\rangle.$$

It is easy to check that player II can use this strategy; moreover, by the choice of $H[p_{0,i}]$, for every $i \in \bigcap_n B_n$ the set $\{p_{n,i} : n < \omega\} \subseteq P_\lambda$ has an upper bound, say q_i; as $B_n \in D$, $\bigcap_{n<\omega} B_n \in D$ and clearly, by 5.4(2), for some $B_\omega \in D$, $B_\omega \subseteq \bigcap_{n<\omega} B_n$ and $\langle q_i : i \in \bigcap_n B_n \rangle$ is a Δ-system with heart p, clearly $p_n \leq p$ for each n, and so by 5.2 $p \Vdash_{P_\lambda}$ " $\{i \in \bigcap_n B_n$: $q_i \in \underset{\sim}{G}_{P_\lambda}$ hence for every ℓ $\underset{\sim}{\beta}_\ell = \alpha_\ell$ & $\underset{\sim}{F}_\ell(i) = \varepsilon_\ell\} \neq \emptyset \bmod D$". So clearly player II has won the play, hence the game.

$\square_{6.1}$

6.1B Remark. We could have used any S, $S \subseteq \{2, \aleph_0, \aleph_1\}$ instead of $\{2, \aleph_0, \aleph_1\}$ and would have obtained a parallel result. The same holds for 6.2 and 7.2 . Also in both we can replace completeness by essential completeness.

6.2 Lemma. Suppose λ is measurable, D a normal ultrafilter over λ, $\bar{Q} = \langle P_i, \underset{\sim}{Q}_i : i < \lambda \rangle$ an RCS iteration, each P_j, P_j/P_{i+1} is $(\{2, \aleph_0, \aleph_1\}, \omega)$-complete and $|P_i| < \lambda$ for $i < j < \lambda$.
Then, letting $\underset{\sim}{Q}_\lambda = \mathrm{Nm}'(D)$ in the universe $V^{P_\lambda * \underset{\sim}{Q}}$ the forcing notion $P_\lambda * \underset{\sim}{Q}_\lambda$ is $(\{2, \aleph_0, \aleph_1\}, \omega)$-complete.

Proof. Just combine the proofs of 6.1 and 4.12(1) (so now we will have a tree of conditions instead $p_{\ell,i}, i \in B_\ell$)). Let us give the details. We will only prove *essential* $(\{2, \aleph_0, \aleph_1\}, \omega)$- completeness (which is enough for all practical purposes) and indicate modifications for $(\{2, \aleph_0, \aleph_1\}, \omega)$-completeness (if we like to use only the essential version, naturally we should then also assume only that P_j, P_j/P_{i+1} are essentially $(\{2, \aleph_0, \aleph_1\}, \omega)$-complete; remember the implications from 3.8(2)). So let $S = \{2, \aleph_0, \aleph_1\}$, $r^* = (p^*, \ q^*) \in P_\lambda * \underset{\sim}{Q}_\lambda$ and we shall describe a winning strategy for player II in the game $\mathrm{E}\eth_S^\omega(r^*, P_\lambda * \underset{\sim}{Q}_\lambda)$ (see 3.6). As $S = \{2, \aleph_0, \aleph_1\}$, by 3.2(2) without loss of generality player II has to give actual values.

Without loss of generality $p^* \Vdash_{P_\lambda}$ "$\mathrm{tr}(q^*) = \eta^*$", $\eta^* \in {}^{\omega>}\lambda$. For notational convenience only (or considering $\underset{\sim}{Q}'_\lambda = \{T \in \underset{\sim}{Q}_\lambda : \mathrm{tr}(T) \trianglerighteq \eta^*\} \simeq \underset{\sim}{Q}_\lambda$) we may assume $\eta^* =<>$. In the n'th move ($n \geq 0$) player I will choose a $P_\lambda * \underset{\sim}{Q}_\lambda$-name

$\underset{\sim}{\beta}^n$ of a countable ordinal, and player II will choose a countable ordinal γ_n [and a condition for the "real" game $\Game^{\omega}_{S}(r^*, P_\lambda * \underset{\sim}{Q}_\lambda)$].

To make his choice, player II plays on the side also trees $T_n \subseteq {}^{n\geq}\lambda$, ordinals $\langle i_\eta : \eta \in T_n \cap {}^{n}\lambda\rangle$, and P_λ-names of $\underset{\sim}{Q}_\lambda$-conditions $\langle \underset{\sim}{q}^n_\eta : \eta \in T_n \cap {}^{n}\lambda\rangle$ and conditions $\langle p^n_\eta : \eta \in T_n\rangle$ in certain forcings appearing in the iteration $\bar{Q}$ and names of ordinals $\langle \underset{\sim}{\beta}^n_\eta : \eta \in T_n \cup \{\langle\rangle^-\}\rangle$, preserving the following:

(A)(1) $T_n \subseteq {}^{n\geq}\lambda$, each $\eta \in T_n$ is strictly increasing.

(2) If $\eta \in T_n$, $\ell < \ell g(\eta)$, then $\eta{\restriction}\ell \in T_n$.

(3) If $\eta \in T_n$, $\ell g(\eta) < n$, then $\{i > \lambda : \eta\hat{\ }\langle i\rangle \in T_n\} \in D$.

(4) $T_{n+1} \cap {}^{n\geq}\lambda \subseteq T_n$.

(B)(1) If $n \geq 1$, $\eta \in {}^{n}\lambda$, then we let $\eta^- = \eta{\restriction}(n-1)$.

(2) If $\eta \in T_n \cap {}^{n}\lambda$, then $i_\eta < \lambda$, i_η a successor ordinal $> \eta(n-1)$.

(3) If $\eta \in T_n \cap {}^{n}\lambda$, $n \geq 1$ then $i_\eta > i_{\eta^-}$

(4) $<>^-$ is not really defined, but we let $i_{<>^-} = 0$,

so $P_{i_{<>}}/P_{i_{<>^-}} = P_{i_{<>}}/P_0 = P_{i_{<>}}$.

(C)(1) For $\eta \in T_n$, $p^n_\eta \in P_{i_\eta}/P_{i_{\eta^-}}$. (So for $\eta = <>$, $p^n_{<>} \in P_{i_{<>}}$).

(2) For $\eta \in T_n \cap T_{n+1}$ we have $p^n_\eta \leq p^{n+1}_\eta$ (this is actually implied by (3) below).

(3) For $\eta \in T_n$, $\langle \underset{\sim}{\beta}^\ell_\eta, p^\ell_\eta, \underset{\sim}{\beta}^\ell_{\eta^-} : \ell g(\eta) < \ell \leq n\rangle$ is an initial segment of a play of $\Game^{\omega}_{S}[p^{\ell g(\eta)}_\eta, P_{i_\eta}/P_{i_{\eta^-}}]$ in which player II uses his winning strategy. So $\underset{\sim}{\beta}^\ell_\eta$ is a P_{i_η}-name for a countable ordinal and $\beta^\ell_{<>^-}$ is a real ordinal. Player II lets (in the actual play) $\gamma_\ell = \beta^\ell_{<>}$ (for the "purely" essential version we should just have p^ℓ_η be in the completion of $P_{i_\eta}/P_{i_{\eta^-}}$).

(4) For $p \in P_\alpha$, $p\prime \in P_\lambda/P_\alpha$, $p \cup p\prime$ is the element of P_λ corresponding to $(p, p\prime) \in P_\alpha * (P_\lambda/P_\alpha)$. We let $\bar{p}^n_\eta \stackrel{\text{def}}{=} \bigcup_{\ell \leq \ell g(\eta)} p^n_{\eta{\restriction}\ell} \in P_{i_\eta}$ (see C1).

(D)(1) $\underset{\sim}{q}^n_\eta$ is a P_λ-name for an element of $\underset{\sim}{Q}_\lambda = \mathrm{Nm}'(D)$ with trunk η (remember $\eta \in T_n \cap {}^{n}\lambda$).

(2) $(\emptyset, \underset{\sim}{Q}^n_\eta) \Vdash_{P_\lambda * Q_\lambda}$ "$\underset{\sim}{\beta}^n = \underset{\sim}{\beta}^n_\eta$" when $\eta \in T_n \cap {}^{n}\lambda$.

(3) For $\eta \in T_n \cap {}^{n}\lambda$, $\bar{p}^n_\eta \Vdash_{P_\lambda}$ "$[Q_\lambda \models q^{n-1}_{\eta^-} \leq q^n_\eta$ and $Q_\lambda \models [q^{n-1}_{\eta^-}]_{[\eta]} \leq_{\mathrm{pr}} q^n_\eta$"].

For the following, note that $\Vdash_{P_\lambda}$ "D is an $\aleph_2$-complete filter", so by 4.13 we have $\Vdash_{P_\lambda}$ " for every $q \in \underset{\sim}{Q}_\lambda$, every $\underset{\sim}{Q}_\lambda$-name $\underset{\sim}{\beta}$ of an ordinal $< \omega_1$, there is $\gamma < \omega_1$, and q', $q \leq_{\mathrm{pr}} q'$, $q \Vdash \underset{\sim}{\beta} = \gamma$". In move number 0, player I plays $\underset{\sim}{\beta}^0$. Player II finds $\underset{\sim}{q}^0_{<>} \in \underset{\sim}{Q}_\lambda$ and a P_λ-name $\underset{\sim}{\beta}^0_{<>}$ such that:

$\Vdash_{P_\lambda}$ "$q^* \leq_{\mathrm{pr}} q^0_{<>}$, so $\mathrm{tr}(q^0_{<>}) =<>$, and $q^0_{<>} \Vdash_{Q_\lambda}$ "$\underset{\sim}{\beta}^0 = \underset{\sim}{\beta}^0_{<>}$"".

But as $P_\lambda \models \lambda$-c.c., $\beta^0_{<>}$ is really a $P_{i_{<>}}$-name (for some successor ordinal $i_{<>} < \lambda$), so player II can find $p^0_{<>} \in P_{i_{<>}}$ such that $P_\lambda \models$ "$p^* \leq p^0_{\langle\rangle}$" and $p^0_{<>} \Vdash$ "$\underset{\sim}{\beta}^0_{<>} = \gamma_0$", for some $\gamma_0 < \omega_1$. Then player II lets $T_0 = \{<>\}$ and play γ_0.

In move $n+1$, player I plays a $P_\lambda * \underset{\sim}{Q}_\lambda$-name $\underset{\sim}{\beta}^{n+1}$. For each $\eta \in T_n \cap {}^n\lambda$, let (in V):

$B_\eta = \{\alpha < \lambda : \forall \ell < \ell g(\eta)[\alpha > \eta(\ell)]$ and $\exists p \in P_\lambda/P_{i_\eta}$, $\bar{p}^n_\eta \cup p \Vdash_{P_\lambda}$ "$\eta\hat{\ }\langle\alpha\rangle \in q^n_\eta$"$\}$.

6.2A Claim. *$B_\eta \in D$.*

Proof of the Claim. For each $\alpha < \lambda$ let $\underset{\sim}{t}^\alpha_\eta$ be the following P_{i_η}-name:

$\underset{\sim}{t}^\alpha_\eta = 0$ if $\Vdash_{P_\lambda/\underset{\sim}{G}_{P_{i_\eta}}}$ "$\eta\hat{\ }\langle\alpha\rangle \notin q^n_\eta$" and $\underset{\sim}{t}^\alpha_\eta = 1$ otherwise.

As there are (essentially) $< \lambda$ many possible such $P^{n+1}_{i_\eta}$-names [as $|P_{i_\eta}| < \lambda$, so $2^{|P_{i_\eta}|} < \lambda$], for some $A_\eta \in D$ and $\underset{\sim}{t}$, $\forall \alpha \in A_\eta : \underset{\sim}{t}^\alpha_\eta = \underset{\sim}{t}$. If $\bar{p}^n_\eta \Vdash$ "$\underset{\sim}{t} = 1$", then $A_\eta \subseteq B_\eta$ and we are done. Otherwise, there is $p' \geq \bar{p}^n_\eta$, $p' \in P_{i_\eta}$, $p' \Vdash$ "$\underset{\sim}{t} = 0$". But $\bar{p}^n_\eta \Vdash$ "$\{\alpha < \lambda : \eta\hat{\ }\langle\alpha\rangle \in q^n_\eta\} \neq \emptyset \bmod D$" and $p' \Vdash$ "$\forall \alpha \in A_\eta$, $\eta\hat{\ }\langle\alpha\rangle \notin q^n_\eta$", a contradiction (as $A_\eta \in D$). This ends the proof of the claim. $\square_{6.2A}$

Continuation of the proof of 6.2: For $\alpha \in B_\eta$, let $p^n_{\eta\hat{\ }\langle\alpha\rangle}$ be a p as in the definition of B_η. Then let $q^{n+1}_{\eta\hat{\ }\langle\alpha\rangle}$ be (a P_λ-name of a member of $\underset{\sim}{Q}_\lambda$) such that:

$$\bar{p}^{n+1}_{\eta\hat{\ }\langle\alpha\rangle} \Vdash_{P_\lambda} \text{“}\Big[[q^n_\eta]_{[\eta\hat{\ }\langle\alpha\rangle]} \leq_{\mathrm{pr}} q^{n+1}_{\eta\hat{\ }\langle\alpha\rangle} \text{ and } q^{n+1}_{\eta\hat{\ }\langle\alpha\rangle} \Vdash_{\underset{\sim}{Q}_\lambda} \text{“}\underset{\sim}{\beta}^{n+1} = \underset{\sim}{\beta}^{n+1}_{\eta\hat{\ }\langle\alpha\rangle}\text{”}\Big]\text{”},$$

where $\underset{\sim}{\beta}^{n+1}_{\eta\hat{\ }\langle\alpha\rangle}$ is a P_λ-name. Again by λ-c.c., for some large enough successor ordinal $i_{\eta\hat{\ }\langle\alpha\rangle} < \lambda$, $\underset{\sim}{\beta}^{n+1}_{\eta\hat{\ }\langle\alpha\rangle}$ is a $P_{i_{\eta\hat{\ }\langle\alpha\rangle}}$-name and $p^{n+1}_{\eta\hat{\ }\langle\alpha\rangle} \in P_{i_{\eta\hat{\ }\langle\alpha\rangle}}/P_{i_\eta}$ (and $i_\eta < i_{\eta\hat{\ }\langle\alpha\rangle}$) and $\alpha < i_{\eta\hat{\ }\langle\alpha\rangle}$. We can increase $p^{n+1}_{\eta\hat{\ }\langle\alpha\rangle}$ and $i_{\eta\hat{\ }\langle\alpha\rangle}$ such that

$\bar{p}^n_\eta \Vdash_{P_{i_\eta}}$ "$p^{n+1}_{\eta\hat{}\langle\alpha\rangle} \Vdash_{P_{i_\eta\hat{}\langle\alpha\rangle}/P_{i_\eta}}$ "$\underset{\sim}{\beta}^{n+1}_{\eta\hat{}\langle\alpha\rangle} = \underset{\sim}{\beta}^{n+1}_{\eta,\alpha}$"" for some P_{i_η}-name $\underset{\sim}{\beta}^{n+1}_{\eta,\alpha}$. As there are only $< \lambda$ many such names of countable ordinals, we can find $S^{n+1}_\eta \in D$, $S^{n+1}_\eta \subseteq B_\eta$, and a name $\underset{\sim}{\beta}^{n+1}_\eta$ such that for all $\alpha \in S^{n+1}_\eta$, $\underset{\sim}{\beta}^{n+1}_{\eta,\alpha} = \underset{\sim}{\beta}^{n+1}_\eta$. Now we can, for each $\eta = \nu\hat{}\langle\alpha\rangle \in T_n \cap {}^n\lambda$, play a step $[\underset{\sim}{\beta}^{n+1}_\eta,\ p^{n+1}_\eta,\ \underset{\sim}{\beta}^{n+1}_{\eta^-,\alpha}]$ in the game $\partial^\omega_S[p^n_\eta, P_{i_\eta}/P_{i_{\eta^-}}]$, to get $P_{i_{\eta^-}}$-name $\underset{\sim}{\beta}^{n+1}_{\eta^-,\alpha}$ (i.e. we play for player I_η the name $\underset{\sim}{\beta}^{n+1}_\eta$, and the winning strategy of player II_η gives us $p^{n+1}_\eta, \underset{\sim}{\beta}^{n+1}_{\eta^-,\alpha}$; α was the last element in η). Again, for each $\nu \in T_n$ of length $n-1$ there is a set $S^{n+1}_\nu \in D$, $S^{n+1}_\nu \subseteq \{\alpha : \nu\hat{}\langle\alpha\rangle \in T_n\}$, and for all $\alpha \in S^{n+1}_\nu$, $\bar{\beta}^{n+1}_{\nu\hat{}\langle\alpha\rangle} = \beta^{n+1}_\nu$.

We continue by downward induction (in each step k, $n \geq k \geq 0$ defining P_ν-names $\underset{\sim}{\beta}^{n+1}_{\nu,\alpha}$ for $\nu\hat{}\langle\alpha\rangle \in T_n \cap {}^k\lambda$, satisfying demand (C3) and then "uniformizing" using a set $S^{n+1}_\nu \in D$ as before). Finally, player II plays $\gamma_{n+1} = \beta^{n+1}_{<>}$, and define $T_{n+1} \subseteq {}^{n+1\geq}\lambda$ by:

(1) $T_{n+1} \cap {}^0\lambda = \{<>\}$

(2) for $\eta \in T_{n+1} \cap {}^k\lambda$, $\mathrm{Suc}_{T_{n+1}}(\eta) = \{\eta\hat{}\langle\alpha\rangle : \alpha \in S^{n+1}_\eta\}$, for $k \leq n$.

This completes the description of player II's strategy. Finally, define T as $T = \bigcup_{\ell<\omega} \bigcap_{n\geq\ell} (T_n \cap {}^\ell\lambda)$. Clearly $({}^{\omega>}\lambda, \mathfrak{D}) \leq_{\mathrm{pr}} (T, \mathfrak{D})$ (where $\mathfrak{D}_\eta = D$). For each $\eta \in T$ let $p_\eta \in P_{i_\eta}/P_{i_{\eta^-}}$ be $\geq p^n_\eta$ for every $n \geq \ell g(\eta)$ (i.e. this is $\Vdash_{P_{i_{\eta^-}}}$); (exists as we have used a winning strategy in $\partial^\omega_{\{2,\aleph_0,\aleph_1\}}[p^{\ell g(\eta)}_\eta,\ P_{i_\eta}/P_{i_{\eta^-}}]$). Let $\bar{p}_\eta = \bigcup_{\ell \leq \ell g(\eta)} p_{\eta\restriction\ell} \in P_{i_\eta}$. By repeated use of 5.4(2) we can find T', $(T,\mathfrak{D}) \leq_{\mathrm{pr}} (T', D)$ and $\langle p^+_\eta : \eta \in T\rangle$ such that for each $\eta \in T'$ we have $\langle p_{\eta\hat{}\langle\alpha\rangle} : \alpha < \lambda, \eta\hat{}\langle\alpha\rangle \in T\rangle$ is a Δ-system with heart $p^+_\eta \in P_\lambda/P_{i_\eta}$. Note: $\eta^- = \nu$, $\eta \in T' \Rightarrow p^+_\nu \leq p_\eta$.

It is easy to see that if $\langle p_\alpha : \alpha < \lambda\rangle$ is a Δ-system with heart p^+, then $p^+ \Vdash_{P_\lambda}$ "$\{\alpha : p_\alpha \in G_{P_\lambda}\} \neq \emptyset \bmod D$ in V^{P_λ}". Using this fact we can show that $p_{<>} \cup p^+_{<>} \Vdash$ "$\underset{\sim}{q} \overset{\mathrm{def}}{=} \{\eta \in T' : \bar{p}_\eta \cup p^+_\eta \in G_{P_\lambda}\} \in \underset{\sim}{Q}_\lambda = \mathrm{Nm}'(D)$ and $q^* \leq_{\mathrm{pr}} \underset{\sim}{q}$".

To finish the proof it is enough to show that $(p_{<>} \cup p^+_{<>}, \underset{\sim}{q}) \Vdash$ "$(\forall n)\underset{\sim}{\beta}^n = \gamma_n$". Assume that this is false, then there is a witness, i.e. $(p', \underset{\sim}{q}') \geq (p_{<>} \cup p^+_{<>}, \underset{\sim}{q})$, $n \in \omega$ and $\alpha^* < \omega_1$, such that $(p', \underset{\sim}{q}') \Vdash$ "$\underset{\sim}{\beta}^n = \alpha^*$", but $\alpha^* \neq \gamma_n$. Without loss of generality $\underset{\sim}{q}'$ has a trunk of length $> n$, and also there is $\eta \in$

${}^{\omega>}\lambda$ such that $p' \Vdash$ "$\mathrm{tr}(\underset{\sim}{q}') = \eta$" and $\ell g(\eta) = m > n$. As $p' \Vdash$ "$\eta \in \underset{\sim}{q}$", without loss of generality $\forall \ell \leq \ell g(\eta) : \bar{p}_{\eta\restriction\ell} \cup p^+_{\eta\restriction\ell} \leq p'$. By (D), $p' \Vdash$ "$\underset{\sim}{q}' \geq \underset{\sim}{q} \geq \underset{\sim}{q}^\ell_{\eta\restriction\ell}$" for all $\ell \leq \ell g(\eta)$. So for $\ell = n$ we get $(p', \underset{\sim}{q}') \geq (\bar{p}^n_{\eta\restriction n}, \underset{\sim}{q}^n_{\eta\restriction n})$, so $(p', \underset{\sim}{q}') \Vdash$ "$\underset{\sim}{\beta}^n = \gamma_n$" and we are done.

If we want to play $\Game^\omega_{\{2,\,\aleph_0,\,\aleph_1\}}$ and not only $\mathrm{E}\Game^\omega_{\{2,\,\aleph_0\,,\aleph_1\}}$, we also have to give conditions r_n forcing $\underset{\sim}{\beta}^n = \gamma_n$ at each step n.

Without loss of generality we may assume that for each $\eta \in T_n$, of length $< n$ we have $\langle p^n_{\eta\hat{}\langle\alpha\rangle} : \eta\hat{}\langle\alpha\rangle \in T_n\rangle$ forms a Δ-system with heart $\hat{p}^n_\eta$. Let $p_n = p^n_{<>} \cup \hat{p}^n_{<>}$, $q_n = \{\eta \in {}^{\omega>}\lambda : \ell g(\eta) \leq n \Rightarrow \bar{p}^n_\eta \cup \hat{p}^n_\eta \in G_{P_\lambda}, \ell g(\eta) \geq n \Rightarrow \eta \in \underset{\sim}{q}^n_{\eta\restriction n}\}$.

Then as above we can prove $(p_n, \underset{\sim}{q}_n) \Vdash$ "$\underset{\sim}{\beta}^n = \gamma_n$", then we have to show $(p_n, \underset{\sim}{q}_n) \leq (p_{n+1}, q_{n+1})$, and finally that $(p_{<>} \cup p^+_{<>}, \underset{\sim}{q})$ (from the end of the proof for $\mathrm{E}\Game^\omega_{\{2,\,\aleph_0,\,\aleph_1\}}$) is $\geq (p_n, \underset{\sim}{q}_n)$ for all n (in $P_\lambda * \underset{\sim}{Q}_\lambda$, or at least in $P_\lambda * \underset{\sim}{Q}_\lambda / \simeq$). These details are left to the reader. $\square_{6.2}$

6.3 Definition. A filter D on a set I (in a universe V) is called precipitous if the following holds:

$\Vdash_{PP(D)}$ "there are no $f_n : I \to$ ordinals, $f_n \in V$, such that $f_{n+1} <_{\underset{\sim}{E}} f_n$ for each n", where

(i) $PP(D) = \{A \subseteq I : A \neq \emptyset \bmod D\}$ ordered by reverse inclusion.

(ii) $\underset{\sim}{E}$ is the filter generated by the generic set of $PP(D)$,

(iii) $f <_{\underset{\sim}{E}} g$ means $\{\alpha \in I : f(\alpha) < g(\alpha)\} \in \underset{\sim}{E}$.

6.3A Remark. The following is an equivalent definition: a filter D over I is precipitous if player I does not have a winning strategy in the following game $\Game_{\mathrm{prec}}(D)$.

First move

player I chooses $A_1 \subseteq I, A_1 \neq \emptyset \bmod D$,

player II chooses $B_1 \subseteq A_1, B_1 \neq \emptyset \bmod D$;

n-th move

player I chooses $A_n \subseteq B_{n-1}, A_n \neq \emptyset \bmod D$,

player II chooses $B_n \subseteq A_n, B_n \neq \emptyset \bmod D$.

Player II wins if $\bigcap_{n<\omega} A_n$ (which is $= \bigcap_{n<\omega} B_n$) is nonempty (not necessarily $\neq \emptyset \bmod D$).

See Jech and Prikry [JP2], and Jech, Magidor, Mitchell and Prikry [JMMP].

6.4 Theorem. Suppose "ZFC + G.C.H. + κ is strongly inaccessible and $A = \{\lambda < \kappa : \lambda$ measurable $\}$ is stationary" is consistent. *Then:*

(1) The following statement is consistent with ZFC + G.C.H.: for every $B \subseteq \aleph_2$ for some $\delta < \aleph_2$ (in fact for some club C of λ for every $\delta \in A \cap C \cap (S^2_0)^{V[G]}$), $\mathrm{cf}(\delta) = \aleph_0$, but in $L[B \cap \delta]$, δ is a regular cardinal $> \aleph_1$ (in $V[G]$, $\lambda = \aleph_2$).

(2) If in the hypothesis $A \in D$, D is a normal ultrafilter on κ, then there is a normal precipitous filter on $\aleph_2$ to which S^2_0 belongs.

Proof. So let V be a model of ZFC +G.C.H., and let κ be a strongly inaccessible cardinal, such that $A = \{\lambda < \kappa : \lambda \text{ measurable}\}$ is stationary.

We now define by induction in $i < \kappa$ forcing notions $P_i \in V, \underset{\sim}{Q}_i \in V^{P_i}$, such that $|P_i| < \kappa$, $\langle P_j, \underset{\sim}{Q}_j : j < \kappa\rangle$ is an RCS iteration. So by 1.5(1) it suffices to define $\underset{\sim}{Q}_i$ for a given $\langle P_j, \underset{\sim}{Q}_j : j < i\rangle$.

Case A. $i = \lambda$ is a measurable cardinal, such that for every $j < \lambda$, $|P_j| < \lambda$. In this case let D_λ be a normal ultrafilter over λ (in V), and $Q_\lambda = \mathrm{Nm}'(D_\lambda)$. (In V^{P_λ}, D_λ is not an ultrafilter any more, since we may have $\Vdash_{P_\lambda} \lambda = \aleph_2$, but it will still be "large", see 6.1, 6.2).

Case B. Not case A.

In this case let Q_i be the Levy collapse of $(2^{|P_i|} + |i|^+)^V$ to $\aleph_1$, i.e., $\{f \in V^{P_i} : f$ a countable function from ω_1 to $2^{|P_i|} + |i|^+\}$.

Now by 3.5 and 6.2 it is easy to see that $P_\kappa = \mathrm{Rlim}\langle P_i, \underset{\sim}{Q}_i : i < \kappa\rangle$ is $(\{2, \aleph_0, \aleph_1\}, \omega)$-complete (note: if for i Case A occurs, then for every $j < i$, in $V^{P_{j+1}}$, D is still a normal ultrafilter), and by 5.4 it satisfies the κ-chain condition.

So clearly in V^{P_κ} G.C.H. holds, every real is from V, and $\aleph_1 = \aleph_1^V$, $\aleph_2 = \kappa$. Also if $\lambda \in A$, then $(\forall i < \lambda)\ |P_i| < \lambda$ (prove by induction on i for each λ). Let $G \subseteq P_\lambda$, be generic, and we shall prove that $V[G]$ satisfies the requirements.

Part 1. So let $B \subseteq \aleph_2$, and let $\underset{\sim}{B} \in V$ be a P_κ-name for it. Then $C_0 = \{\delta :$ for every $i < \delta$ we have $\underset{\sim}{B} \cap \{i\}$ has a P_j-name for some $j < \delta\}$ is a closed unbounded subset of κ, because P_κ satisfies the κ-chain condition (and $P_j \lessdot P_\kappa = \bigcup_{i<\kappa} P_i$ for $j < \kappa$) and obviously $C_0 \in V$.

Now if $\lambda \in C_0 \cap A$, then we can check that $|P_i| < \lambda$ for $i < \lambda$, so case A holds hence $Q_\lambda = \mathrm{Nm}'(D_\lambda)$, hence in $V[G]$, $\mathrm{cf}(\lambda) = \aleph_0$ by 4.7(2). On the other hand, clearly $G \cap P_\lambda$ is a generic subset of P_λ (as $P_\lambda \lessdot P_\kappa$), by 5.4 P_λ satisfies the λ-chain condition, so $\Vdash_{P_\lambda}$ "$\mathrm{cf}(\lambda) = \lambda$". Hence in $V[G \cap P_\lambda]$, $A \cap \lambda$ is present, but λ is a regular cardinal $> \aleph_1$. So also in $L[A \cap \lambda], \lambda$ is a regular cardinal $> \aleph_1$. Lastly as P_κ satisfies the κ-c.c. also in V^{P_λ}, A is a stationary subset of $\kappa = \aleph_2^{V^{P_\kappa}}$. Together we finish.

Part 2. The following implies the desired conclusion; it is essentially the same proof as [JMMP] who do it for the Levy collapse; and it suffices for (2) of the theorem. It follows from Magidor [Mg80] Theorem 2.1, and is included for completeness only. (By construction, $V^P \models$ "$A \subseteq S_2^0$ and $A \in D$").

6.5 Lemma. Suppose κ is measurable, D a normal ultrafilter over κ, $\bar{Q} = \langle P_i, \underset{\sim}{Q}_i : i < \kappa\rangle$ an RCS iteration, $|P_i| < \kappa$ for $i < \kappa$, $P = P_\kappa = \mathrm{Rlim}\bar{Q}$.

Then in V^P, D is a precipitous filter.

Proof. If not, in V^P there is $A_0 \in PP(D)$, $A_0 \Vdash_{PP(D)}$ "$\langle \underset{\sim}{f}_n : n < \omega\rangle$ is an ω-sequence of functions from κ to ordinals which belong to V^P which is decreasing mod $\underset{\sim}{E}, \underset{\sim}{f}_n \in V^P$" where $\underset{\sim}{E}$ is as in clause (ii) of Definition 6.3.

So there is $p \in P$, a P-name $\underset{\sim}{A}_0$, and $P * PP(D)$-names $\underset{\sim}{f}^\dagger_n$ of the $\underset{\sim}{f}_n$ such that $p \Vdash_{P_\lambda}$ "$\underset{\sim}{A}_0$, $\underset{\sim}{f}^\dagger_n$ are as above".

Let $B_0 = \{\lambda < \kappa : \lambda$ is strongly inaccessible and for some $p^\dagger \geq p$, $p^\dagger \in P$, and $p^\dagger \Vdash_{P_\lambda}$ "$\lambda \in \underset{\sim}{A}_0$"$\}$.

Because D is normal, κ measurable, $\{\lambda < \kappa : \lambda$ strongly inaccessible$\} \in D$, hence $B_0 \in D$. For each $\lambda \in B_0$ choose $p_{\lambda,0}$, $p \leq p_{\lambda,0} \in P$, $p_{\lambda,0} \Vdash$ "$\lambda \in \underset{\sim}{A}_0$". By 5.3 there is $B^\dagger_0 \subseteq B_0$, $B^\dagger_0 \in D$ such that $\langle p_{\lambda,0} : \lambda \in B^\dagger_0 \rangle$ is a Δ-system with heart $p^\dagger$.

Now we define by induction on $n < \omega$, $p_{\lambda,n}, p_n, p^\dagger_n, B_n, B^\dagger_n, \underset{\sim}{A}_n, \underset{\sim}{A}^\dagger_n, \underset{\sim}{g}_n, \alpha_{\lambda,n}$ (for $\lambda \in B^\dagger_n$) ($\underset{\sim}{A}_n, \underset{\sim}{A}^\dagger_n, \underset{\sim}{g}_n$ are P-names) such that:

(1) $\langle p_{\lambda,n} : \lambda \in B^\dagger_n \rangle$ is a Δ-system of members of P with heart $p^\dagger_n$.

(2) $B_{n+1} \subseteq B^\dagger_n \subseteq B_n$, $B_{n+1} \in D$,

(3) $p_{n+1} \geq p^\dagger_n \geq p_n$ all in P,

(4) $p_{\lambda,n+1} \geq p_{\lambda,n}$ both in P, $\underset{\sim}{g}_n$ a P-name of a function from κ to Ord,

(5) $p_{\lambda,n} \Vdash_P$ "$\lambda \in \underset{\sim}{A}_n$ and $\underset{\sim}{g}_{n-1}(\lambda) = \alpha_{\lambda,n-1}$", $\alpha_{\lambda,n} < \alpha_{\lambda,n-1}$ for $n > 0$,

(6) $\underset{\sim}{A}^\dagger_n = \{\lambda \in B^\dagger_n : p_{\lambda,n}$ is in the generic set of $P\}$,

(7) $p_{n+1} \Vdash_P$ "$\underset{\sim}{A}_{n+1} \in PP(D)$ and $\underset{\sim}{A}_{n+1} \subseteq \underset{\sim}{A}^\dagger_n$ and $[\underset{\sim}{A}_{n+1} \Vdash_{PP(D)}$ "$\underset{\sim}{f}_n/\underset{\sim}{E} = \underset{\sim}{g}_n/\underset{\sim}{E}$"$]$ and $\underset{\sim}{A}_{n+1} \subseteq \{i < \kappa : \underset{\sim}{g}_n(i) < \underset{\sim}{g}_{n-1}(i)\}$",

(8) $B_{n+1} = \{\lambda \in B^\dagger_n$: there is $p^\dagger \geq p_{\lambda,n}, p^\dagger \geq p_{n+1}$, such that for some α, $p^\dagger \Vdash$ "$\lambda \in \underset{\sim}{A}_{n+1}$ and $\underset{\sim}{g}_n(\lambda) = \alpha$"$\}$.

The definition is easy: for $n > 0$, we first we define p_n, $\underset{\sim}{A}_n$ and $\underset{\sim}{g}_{n-1}$ (by 7), then B_n and $p_{\lambda,n}$ (by 5 and 8), then $B^\dagger_n$ and $p^\dagger_n$ (by (1), using 5.4), finally $\underset{\sim}{A}^\dagger_n$ by (6).

Now as $B^\dagger_n \in D$, $\bigcap_{n<\omega} B^\dagger_n \neq \emptyset$, and if λ belongs to the intersection, $\langle \alpha_{\lambda,n} : n < \omega \rangle$ is strictly decreasing sequence of ordinals, contradiction. $\square_{6.5,6.4}$

§7. Friedman's Problem

Friedman [Fr] asked the following.

7.0 Problem. Is there for every $S \subseteq S_0^\alpha$ $(= \{i < \aleph_\alpha : \operatorname{cf}(i) = \aleph_0\})$, a closed set of order type ω_1, included in S or in $S_0^\alpha \setminus S$? We call this statement $\operatorname{Fr}(\aleph_\alpha)$. Let $\operatorname{Fr}^+(\aleph_\alpha)$ means that every stationary $S \subseteq S_0^\alpha$ includes a closed set of order type ω_1.

Van Liere proved that $\operatorname{Fr}(\aleph_2)$ implies $\aleph_2$ is a Mahlo strongly inaccessible cardinal in L; and $\operatorname{Fr}(\aleph_\alpha)$+ not $\operatorname{Fr}(\aleph_2)$ ($\aleph_\alpha$ regular $> \aleph_2$) implies $0^\#$ exists (using squares). We prove the consistency of $\operatorname{Fr}(\aleph_2)$+G.C.H. with ZFC, modulo the consistency of a measurable cardinal of order 1 . We recall the well known:

7.1 Definition. We define by induction on n what are a measurable cardinal of order n and a normal ultrafilter of order n. For $n = 0$ those are just a measurable cardinal and a normal ultrafilter. For $n+1, D$ is a normal ultrafilter of order $n+1$ on κ if $\{\lambda < \kappa : \lambda$ is measurable of order $n\} \in D$ and it is a normal ultrafilter. We call κ measurable of order $n+1$ if there is an ultrafilter of order $n+1$ on it.

7.2 Lemma. Suppose D is a normal ultrafilter on κ, $\bar{Q} = \langle P_i, \underset{\sim}{Q}_i : i < \kappa \rangle$ an RCS iteration and $|P_i| < \kappa$ for every $i < \kappa$.

Suppose further that $G \subseteq P_\kappa$ is generic, $S \subseteq (S_0^\kappa)^{V[G]}$, $S \in V[G]$ stationary and even $\neq \emptyset \bmod D$, and (in $V[G]$) let

$Q_\kappa = \{f :$ the domain of f is some successor ordinal $\alpha < \aleph_1$, f is into S and it is increasing and continuous $\}$

So let $\underset{\sim}{S}, \underset{\sim}{Q}_\kappa$ be P_κ-names for them and $\Vdash_{P_\kappa}$ "$\underset{\sim}{S} \neq \emptyset \bmod D$ and $S \subseteq S_0^\kappa$". We then conclude:

(1) If P_κ is $\{\aleph_1\}$-semiproper, *then* so is $P_\kappa * \underset{\sim}{Q}_\kappa$,

(2) If P_κ is essentially $(\{2, \aleph_0, \aleph_1\}, \omega)$-complete, then so is $P_\kappa * \underset{\sim}{Q}_\kappa$.

Proof. (1) The problem is that Q_κ may destroy a stationary subset (of ω_2), so it is not proper, though it obviously does not add ω-sequences. So let $\underset{\sim}{S}, \underset{\sim}{Q}_\kappa$ be P_κ-names for S, Q_κ. Let $A = \{\alpha < \kappa : \alpha$ a strongly inaccessible cardinal and $(\forall i < \alpha)(|P_i| < \alpha)$.

Let λ be regular, big enough, $\bar{Q}, \underset{\sim}{Q}_\kappa, \underset{\sim}{S} \in H(\lambda)$, let $<^*$ be a well ordering of $H(\lambda)$ and let $N \prec (H(\lambda), \in, <^*)$ be countable, $p, q, \bar{Q}, \underset{\sim}{S}, \underset{\sim}{Q}_\kappa \in N$, $(p, \underset{\sim}{q}) \in P_\kappa * \underset{\sim}{Q}_\kappa$, and we shall prove the existence of an $\{\aleph_1\}$-semi $(N, P_\kappa * \underset{\sim}{Q}_\kappa)$- generic condition $\geq (p, \underset{\sim}{q})$. In V (hence in $H(\lambda)$), we let

$$S_0 = \{\lambda \in A: \text{ there is } p^\dagger \in P_\kappa \text{ such that } p \leq p^\dagger \text{ and } p^\dagger \Vdash \text{“}\lambda \in \underset{\sim}{S}\text{”}\}.$$

As in previous proofs $S_0 \in D$, and for each $\lambda \in S_0$ let $p_{\lambda,0} \in P_\kappa$, $p_{\lambda,0} \geq p$, $p_{\lambda,0} \Vdash$ “$\lambda \in \underset{\sim}{S}$” and for some $S_1 \subseteq S_0$, $S_1 \in D$, and $\langle p_{\lambda,0} : \lambda \in S_1\rangle$ is a Δ-system (see 5.4(2)). As N was an elementary submodel we can assume S_0, S_1, $\langle p_{\lambda,0} : \lambda \in S_1\rangle$ and its heart p_0 belongs to N (but of course not all included in N). Let $S_2 = S_1 \cap \bigcap\{S^\dagger : S^\dagger \in D \text{ and } S^\dagger \in N\}$, so clearly $S_2 = \{\alpha_i : i < \kappa\} \subseteq S_1$ is an indiscernible sequence over $N \bigcup \omega_1$ in the model $(H(\lambda), \in, <^*)$ (but does not belong to N). Note that for a formulae $\varphi = \varphi(x_1, \ldots, x_\kappa; y_1, \ldots, y_n)$ with n parameters $y_1, \ldots, y_n$ from ω_1, and k parameters $x_1 < \ldots < x_k$, from κ, the corresponding function $f : [\kappa]^k \rightarrow \{true, \quad false\}^{\aleph_1^n}$ is in N and it is constant on S_2. (The function f is: for $\alpha_1 < \ldots < \alpha_\kappa < \kappa$ let $f(\alpha_1, \ldots, \alpha_\kappa) = \{(\beta_1, \ldots, \beta_n, \mathbf{t}) : \beta_1, \ldots, \beta_n < \omega_1$ the $\mathbf{t}$ is the truth value of $\varphi(\alpha_1, \ldots, \alpha_\kappa, \beta_1, \ldots, \beta_n)\}$). Clearly $p \leq p_0$.

Let $N \cap P_\kappa \subseteq P_\mu$, $S_3 = S_2 \setminus (\mu + 1)$ (with $\mu < \kappa$, of course).

Clearly $S_3 \in D$ hence $S_3 \neq \emptyset$. Let $\chi \in S_3$ be such that $\chi = \sup(S_3 \cap \chi)$, and N^* be the Skolem Hull (in $(H(\lambda), \in, <^*)$) of $N \bigcup\{\chi\}$, by the choice of S_2 (and Rowbotton theorem), clearly $\delta \stackrel{\text{def}}{=} N^* \cap \omega_1 = N \cap \omega_1$ and $\kappa\cap$(Skolem hull of $N \cup \chi) = \chi$ (or you can choose such χ).

Also $P_\chi \in N^*$ (as $\langle P_i, \underset{\sim}{Q}_i : i < \kappa\rangle \in N^*, \chi \in N^*$) and clearly $P_\chi \lessdot P_\kappa$. Now P_κ is $\{\aleph_1\}$-semiproper and $\langle p_{\lambda,0} : \lambda \in S_1\rangle \in N^*$ and $\chi \in N^*$, hence $p_{\chi,0} \in N^*$ and $p_0 \leq p_{\chi,0}$ and there is $p_1 \in P_\kappa, p_1 \geq p_{\chi,0}$, which is $\{\aleph_1\}$-semi (N^*, P_κ)-generic. As $N^* \cap \omega_1 = N \cap \omega_1$, p_1 is also $\{\aleph_1\}$-semi (N, P_κ)-generic.

Let $G \subseteq P_\kappa$ be generic, $p_1 \in G$, and we shall find $f \in \underset{\sim}{Q}_\kappa[G]$ which is $\{\aleph_1\}$-semi $(N[G], \underset{\sim}{Q}_\kappa[G])$-generic, this obviously suffices. We have $\delta = N^*[G] \cap \omega_1$. In $V[G]$, χ has cofinality $\aleph_0$, (as $p_{\chi,0} \Vdash$ “$\chi \in \underset{\sim}{S}$” and $\underset{\sim}{S} \subseteq (S_0^\chi)^{V^{P_\chi}}$). So there are $\alpha_0 < \ldots < \alpha_n < \alpha_{n+1} < \ldots, \bigcup_{n<\omega} \alpha_n = \chi$, $\alpha_n \in N^*[G]$, and let $\{\underset{\sim}{\gamma}_n : n < \omega\}$ be a list of all $\underset{\sim}{Q}_\kappa[G]$-names of countable ordinals which belong to $N[G]$ (not $N^*[G]$!). We let $f_0 = \underset{\sim}{q}[G] \in N[G]$, N_n be the Skolem Hull of $N \bigcup \{\alpha_0, \ldots, \alpha_n\}$ in $(H(\lambda), \in, <^*)$ and define by induction $f_n \in N_n[G]$ such that $\underset{\sim}{Q}_\kappa[G] \models$ "$f_{n+1} \geq f_n$", $\chi > \operatorname{Sup}\operatorname{Rang}(f_n) \geq \alpha_n$ and $f_n \Vdash_{\underset{\sim}{Q}_\kappa[G]}$ "$\underset{\sim}{\gamma}_n = \beta_n$" for some β_n. This will suffice because $\bigcup_{n<\omega} f_n \bigcup \{\langle \delta, \chi \rangle\} \in \underset{\sim}{Q}_\kappa[G]$ and is $\{\aleph_1\}$-semi $(N[G], \underset{\sim}{Q}_\kappa[G])$-generic because $N_n \cap \omega_1 \subseteq N^* \cap \omega_1 = N \cap \omega_1$. Defining f_{n+1}, the only nontrivial point is $\chi > \operatorname{Sup}\operatorname{Rang}(f_{n+1})$, but $f_{n+1} \in N_{n+1}[G]$, and $N_{n+1}[G] \cap \kappa \subseteq \chi$, (as by a version of Rowbottom's partition theorem on normal ultrafilters $S_3 \setminus \chi$ is indiscernible, in $(H(\lambda), \in, <^*)$, over $N \bigcup \chi$). Now, for every P_κ-name $\underset{\sim}{\beta} \in N_n$ of an ordinal $< \kappa$, for some $\beta \in N_n \cap \kappa, \Vdash_{P_\kappa}$ "$\underset{\sim}{\beta} < \beta$" (as P_κ satisfies the κ-c.c., see 5.3) hence $\operatorname{Sup}(N_n[G] \cap \kappa) = \operatorname{Sup}(N_n \cap \kappa) \leq \chi$. So we can define f_{n+1}, hence all the f_n's hence, as said above, we finish.

(2) By 3.8(2) the complete Boolean algebra $P = RO(P_\lambda)$ is (S, ω)-complete, where S will be $\{2, \aleph_0, \aleph_1\}$; let $\underset{\sim}{Q} = RO(\underset{\sim}{Q}_\kappa)$.

Let $(p, \underset{\sim}{q}) \in P * \underset{\sim}{Q}$. Clearly it is enough to describe the winning strategy of player II in $\mathrm{E}\eth^\omega_S\Big((p,\ \underset{\sim}{q}),\ P * \underset{\sim}{Q}\Big)$.

Suppose in the n-th move, player I chooses the $P * \underset{\sim}{Q}$-name $\underset{\sim}{\beta}_n$ of an ordinal $< \aleph_1$, and player II will choose β_n. Player II will do the following: after the n-th move he will have $(p, \underset{\sim}{q}_\eta) \in P * \underset{\sim}{Q}$ and $\underset{\sim}{\beta}'_\eta, \beta_\eta$ for every increasing sequence η of ordinals $< \kappa$ of length $\leq n$ such that:

(1) $(p, q_{<>}) = (p, \underset{\sim}{q})$,

(2) $(p, q_{\eta \restriction \ell}) \leq (p, \underset{\sim}{q}_\eta)$,

(3) $(p, \underset{\sim}{q}_\eta) \Vdash$ "$\underset{\sim}{\beta}_{\ell g(\eta)} = \underset{\sim}{\beta}'_\eta$", $\underset{\sim}{\beta}'_\eta$ a P-name (of an ordinal $< \aleph_1$).

(4) for some $A_n \in D$, for every increasing $\eta \subseteq A_n$ of length $\leq n$ we have $\beta_\eta = \beta_{\ell g(\eta)}$,

(5) $p \Vdash_P$ "$\operatorname{Sup}\operatorname{Rang}(\underset{\sim}{q}_\eta) > \operatorname{Max}\operatorname{Rang}(\eta)$",

(6) For $\eta \in {}^n\kappa$ increasing, $\langle \underset{\sim}{\beta}'_{\eta\restriction 0}, \beta_{\eta\restriction 0}, \underset{\sim}{\beta}'_{\eta\restriction 1}, \beta_{\eta\restriction 1}, \ldots, \underset{\sim}{\beta}'_{\eta\restriction n}, \beta_{\eta\restriction n}\rangle$ is an initial segment of a play of $\mathrm{E}\eth^\omega_S(p, P)$ in which player II uses his winning strategy.

Clearly Player II can do the above and it gives him a strategy. (i.e. the zero-th move is easy. In the $(n+1)$'th move; first for every increasing $\nu \in {}^n\kappa$ for every $\eta = \nu\hat{}\langle\alpha\rangle$ first choose q_η to satisfy (2), (5) and force $\underset{\sim}{\beta}_{\ell g(\eta)}$ to be equal to a P-name $\underset{\sim}{\beta}'_\eta$; then choose β_η to satisfy (6). Finally he chooses A_n and β_n to satisfy (4)). We have to prove that he wins by this strategy. So let $A = \bigcap_n A_n$, and for $\eta \in {}^\omega A$ increasing, we know that for some (by clause 6) $p_\eta \in P_\kappa$, $p_\eta \Vdash$ "$\underset{\sim}{\beta}'_{\eta\restriction\ell} = \beta_{\eta\restriction\ell}$" for $\ell < \omega$.

Let $K = \{T : T$ a tree of finite increasing sequences from A, closed under initial segments, $\langle\rangle \in T$ and for every $\eta \in T$, $\{i \in A : \eta\hat{}\langle i\rangle \in T\} \in D\}$ (we can replace D by $\mathcal{D}_\kappa + A$ or $\mathcal{D}^{cb}_\kappa + A$ in this context since we only need κ-completeness). Remember $\lim T = \{\eta : \ell g(\eta) = \omega,\ \eta\restriction k \in T$ for every $k < \omega\}$. So K is closed under intersection of $< \kappa$ elements. For each $T \in K, \eta \in T$, let x^T_η be, in $RO(P)$, $\mathrm{Sup}_P\{p_\nu : \nu \in \lim T$ and $\nu \trianglerighteq \eta\}$ (Remember, we replaced P_κ by a complete Boolean algebra B^P or see 1.4(9)). Clearly x^T_η decreases with T, so as P satisfies the κ-chain condition, for some $T, x^T_{\langle\rangle}$ is minimal (i.e., $T^\dagger \subseteq T, T^\dagger \in K$ implies $x^{T^\dagger}_{\langle\rangle} = x^T_{\langle\rangle}$), and similarly for every $\eta \in T$.

Obviously,

(1) $x^T_\eta = \mathrm{Sup}\{x^T_{\eta\hat{}\langle i\rangle} : \eta\hat{} < i > \in T\}$: (this holds for any tree),

(2) $RO(P) \models 0 < b \leq x^T_\eta$ implies $\{i : RO(P) \models b \cap x^T_{\eta\hat{}<i>} \neq 0\} \neq \emptyset \bmod D$ (by T's minimality).

Let $\underset{\sim}{T}^* = \{\eta : x^T_\eta$ belongs to the generic set of $P\}$. Hence $x^T_{\langle\rangle} \Vdash_P$ "$T^* \neq \emptyset$, in fact $\langle\rangle \in T^*$", and

(3) $x^T_{\langle\rangle} \Vdash_P$ " for any $\eta \in \underset{\sim}{T}^*$ for κ many i's we have $\eta\hat{}\langle i\rangle \in T^*$".

Now if $G \subseteq P$ is generic, $x^T_{\langle\rangle} \in G$ then $\underset{\sim}{S}[G]$ is a stationary subset of S^2_0, and $C = \{\delta$: if $\eta \in {}^{\omega>}\delta$, then $\mathrm{Rang}(\underset{\sim}{q}_\eta[G]) \subseteq \delta\}$ is closed unbounded. Hence for some η, δ with $\eta \in {}^\omega\delta$ the following holds: $\delta \in \underset{\sim}{S}[G] \cap C$, $(\forall k)\eta\restriction k \in \underset{\sim}{T}^*[G]$, and $\bigcup_{\ell<\omega} \eta(\ell) = \delta$. Let $q^* = \bigcup_{\ell<\omega} q_{\eta\restriction\ell} \bigcup\{\langle \mathrm{Sup} \bigcup_{\ell<\omega} \mathrm{Dom}(q_{\eta\restriction\ell}), \delta\rangle\} \in Q$. Let $\underset{\sim}{q}^*$ be the P-name of such a q^*. It is easy to check $(x^T_{\langle\rangle}, \underset{\sim}{q}^*)$ is as required. $\square_{7.2}$

7.2A Remark. In 7.2, we can weaken the assumption allowing $S = \emptyset \bmod D$ when e.g. $P_j, P_j/P_{i+1}$ are semiproper. (A complete proof of a better theorem appeared in XI.)

7.3 Theorem. If "ZFC + G.C.H. + there is a measurable of order 1" is consistent, then so is "ZFC + G.C.H.+ for every subset of S^2_0, either it or its complement, contains a closed copy of ω_1".

Remark. We do not try to get the weakest hypothesis. It will be interesting to find an equi-consistency result. (See XI.)

Proof. So let V satisfy G.C.H., $B \subseteq \kappa$ the set of measurables of order 0, not 1, and for every $\mu \in B$, let D_μ be a normal ultrafilter on μ; we know (see below why) that $\diamondsuit_B$ holds, and let $\bar{S} = \langle S_\mu : S_\mu \subseteq H(\mu), \mu \in B \rangle$, exemplify it. Moreover, if $S \subseteq H(\kappa)$, φ a Π^1_1 sentence, $(H(\kappa), \in, S) \models \varphi$ then $\{\mu \in B : S \cap H(\mu) = S_\mu, (H(\mu), \in, S_\mu) \models \varphi\}$ is a stationary subset of κ. It is well known that there are such S_μ. [Why $\bar{S}$ exists? Choose inductively $S_\mu \subseteq H(\mu)$ for $\mu \in B$ such that if possible $\{\mu' : \mu' \in \mu \cap B$ and $S_\mu \cap H(\mu') \neq S_{\mu'}\}$ is not a stationary subset of μ].

We define an RCS iterated forcing $\langle P_i, \underset{\sim}{Q}_i : i < \kappa \rangle$ by induction on i, such that $|P_i| < \kappa$, and for every measurable $\mu < \kappa$, $i < \mu \Rightarrow |P_i| < \mu$.

When we have defined $\underset{\sim}{Q}_j$ for $j < i$ then $P_j (j \leq i)$ are defined. If $i \in \kappa \setminus B$, $\underset{\sim}{Q}_i$ is $\{f : f$ a countable function from $\aleph_1$ to $|P_i|^+ + 2^{\aleph_2}\}$ ($2^{\aleph_2}$ of V^{P_i}).

If $i \in B$, $S_i = \langle p, \underset{\sim}{S} \rangle, p \in P_i$, $\underset{\sim}{S}$ a P_i-name, $p \Vdash_{P_i}$ "$\underset{\sim}{S}$ is a subset of S^i_0 and $\underset{\sim}{S} \neq \emptyset \bmod D$ for some normal ultrafilter $D \in V$ on i", then we let $\underset{\sim}{Q}_i$ be as in 7.2 if p is in the generic set, and trivial otherwise. We can finish as in previous proofs. $\square_{7.3}$

7.3A Remark. 1) We leave the checking that the forcing works, to the reader. For the normal ultrafilters D', D'' on κ if, for $B' \subseteq \kappa$ we have $B' \in D' \Leftrightarrow \{\lambda \in B : B' \cap \lambda \in D_\lambda\} \in D''$, then we can get in V^{P_κ} every $A' \in (D')^+$ contains a closed copy of ω_1.

2) Note that P_κ will be $\{\aleph_1\}$-semiproper and even $\{2, \aleph_0, \aleph_1\}$-complete (see 3.5).

3) In fact we could have gotten that every stationary $S \subseteq S^2_0$ contains a closed copy of ω_1, i.e. $Fr^+(\aleph_2)$ if we use 7.2A.

4) The forcing in the proof of 7.3 preserve "$\mathrm{cf}(\delta) > \aleph_0$". [Why? Use 2.7 and simple properties of the $\underset{\sim}{Q}_i$'s.]

7.4 Theorem. Suppose "ZFC + there are two supercompact cardinals" is consistent. Then so is ZFC + G.C.H. + "$Fr^+(\aleph_\alpha)$ for every regular $\aleph_\alpha (\alpha > 1)$".

Remark. Slightly better is XI 7.6. We can also get result like XI 7.2(c) i.e.

(∗) for every $\theta = \mathrm{cf}(\theta) > \aleph_4$ and stationary $W_i \subseteq \{\delta < \theta : \mathrm{cf}(\delta) = \aleph_0\}$ we can find an increasing continuous $h : \omega_1 \to S$ such that $h(i) \in W_i$

Proof. Let $V \models$ "$2^\mu = \mu^+$ for $\mu \geq \lambda$" and $\kappa < \lambda$ and κ, λ are supercompact.

By a theorem of Laver [L] we can assume no κ-complete forcing will destroy the supercompactness of κ. The following is known:

$(*)_0$ If $\aleph_\alpha \geq \lambda$ is regular, $S \subseteq S^\alpha_0$ is stationary, then for some $\mu, \kappa < \mu < \lambda$ and $\delta < \aleph_\alpha$, we have $\mathrm{cf}(\delta) = \mu$ and $S \cap \delta$ is stationary.

Let P be the Levy collapse of λ to κ^+. By Baumgartner [B2], in V^P,

$(*)_1$ for every stationary $S \subseteq \lambda \cap S^\infty_0$, for some $\delta < \lambda$, $\mathrm{cf}(\delta) = \kappa$ (in V^P), $S \cap \delta$ is stationary.

Moreover,

$(*)_2$ If in V^P, $\aleph_\beta > \lambda$, $\aleph_\beta$ regular, $S \subseteq S^\beta_0$ stationary, then for some $\delta < \aleph_\beta$, $\mathrm{cf}(\delta) = \kappa$, and $S \cap \delta$ is stationary in δ.

(why? as $|P| = \lambda < \aleph_\beta = \mathrm{cf}(\aleph_\beta)$, S is the union of λ sets from V, so at least one of them is stationary (subset of λ in V), so without loss of generality $S \in V$. Now by $(*)_0$ above we can find δ as there. But P is κ-complete and collapses δ to size κ, so $\mathrm{cf}(\delta)^{V^P} = \kappa$, and $S \cap \delta$ is stationary in V^P. We want to deduce $\Vdash_P$ "$S \cap \delta$ is stationary in δ", as $\delta \in S \Rightarrow \mathrm{cf}(\delta) = \aleph_0$ this is easy).

We can conclude

(∗) in V^P, if $\mu > \kappa$ is regular, $S \subseteq \mu \cap S_0^\infty$ is stationary then for some $\delta < \mu, \mathrm{cf}(\delta) = \kappa$ and $S \cap \delta$ is stationary.

Let Q be the forcing from 7.3 A(3) (or the proof of XI 7.1), we shall show that $V^{P * \underset{\sim}{Q}}$ is as required.

Note: $\aleph_1^{V^{P * \underset{\sim}{Q}}} = \aleph_1^V$, $\aleph_2^{V^{P * \underset{\sim}{Q}}} = \kappa$, $\aleph_3^{V^{P * \underset{\sim}{Q}}} = \lambda$ and every cardinal $\mu > \lambda$ of V remains a cardinal in $V^{P * \underset{\sim}{Q}}$ and the properties "δ a limit ordinal" "$\mathrm{cf}(\delta) = \aleph_0$", $\mathrm{cf}(\delta) > \aleph_0$ are preserved by P (being $\aleph_1$-complete) and Q (by 7.3A(4)) so S_0^∞ has the same interpretation in V, V^P and $V^{P * \underset{\sim}{Q}}$.

Let, in $V^{P * \underset{\sim}{Q}}$, μ be a regular cardinal $> \aleph_1$ and $S \subseteq \mu \cap S_0^\infty$ be stationary. If $\mu \leq \kappa$, apply the proof of 7.3, 7.3A(3). If $\mu > \kappa$ then, as $V^P \models$ "$|Q| = \kappa$", S is the union of κ subsets which belong to V^P, so at least one is stationary, so w.l.o.g. $S \in V^P$.

So in V^P, for some $\delta, \mathrm{cf}(\delta) = \kappa, \delta \cap S$ is stationary; as $\underset{\sim}{Q}$ satisfies the κ-chain condition, $S \cap \delta$ is still stationary in $V^{P * \underset{\sim}{Q}}$, as required. $\square_{7.4}$

XI. Changing Cofinalities; Equi-Consistency Results

§0. Introduction

We formulate a condition which is (strongly) preserved by revised countable support iteration, implies $\aleph_1$ is not collapsed, no real is added and is satisfied e.g. by Namba forcing, and any $\aleph_1$-complete forcing. So we can iterate forcing notions collapsing $\aleph_2$ but preserving $\aleph_1$ up to some large cardinal.

Our aim is to improve the results of chapter of X to equi-consistency results. If you want to add reals, look at Chapter XV. To prove the preservation we use partition theorems and Δ-system theorems on tagged trees (3.5, 3.5A, 3.7 (and 4.3A)). Some of them are from Rubin and Shelah [RuSh:117], see detailed history there on pages 47, 48 and more on mathematics see [RuSh:117], [Sh:136] 2.4, 2.5 (pages 111 – 113).

§1. The Theorems

1.1 Discussion. In this chapter we list the demands that we would like our condition to satisfy, and show how, having a condition satisfying these demands we can prove our theorems. Then, in the following sections we will formulate the condition and prove it satisfies all our demands. Lastly we shall prove some more complicated theorems applying the condition.

The Demands. We will have a condition for forcing notions such that:

(i) If P satisfies the condition *then* forcing with P does not collapse $\aleph_1$ and, moreover, (when CH holds) it does not add reals.

(ii) If $P = \mathrm{Rlim}\,\bar{Q}$, where $\bar{Q}$ is an RCS iteration of forcing notions such that each of them satisfies the condition *then* P satisfies it as well.

(RCS iteration was defined in X §1. In 1.9 we will recall its basic properties).

Really we do not get (ii) but a slightly different version (ii)$'$, which is as good for our purpose:

(ii)$'$ Assume V satisfies: if $\bar{Q} = \langle P_i, \underset{\sim}{Q}_i : i < \delta \rangle$ is an RCS iteration, $\underset{\sim}{Q}_{2i+1}$ is Levy collapse of $2^{|P_{2i+1}|+|i|}$ to $\aleph_1$ (by countable conditions), each $\underset{\sim}{Q}_{2i}$ satisfies the condition. *Then* P_δ, the revised limit of $\bar{Q}$, satisfies the condition (see also 6.2A).

(iii) If $\bar{Q} = \langle P_i, \underset{\sim}{Q}_i : i < \kappa \rangle$ is an RCS iteration as in (ii)$'$, κ is a strongly inaccessible cardinal and $|P_i| < \kappa$ for $i < \kappa$ *then* P_κ, the revised limit of $\bar{Q}$, satisfies the κ-c.c.

(iv) The condition is satisfied by the following forcing notions:

a. Namba forcing. (See 4.1, it adds a cofinal countable subset to ω_2 without collapsing ω_1.) We denote this forcing notion by Nm.

b. Any $\aleph_1$-closed forcing notion.

c. $P[S]$, where S is a stationary subset of ω_2 such that $\alpha \in S \Rightarrow \mathrm{cf}(\alpha) = \omega$, and $P[S] = \{f : f$ is an increasing and continuous function from $\alpha + 1$ into S, for some $\alpha < \omega_1\}$. Note that $P[S]$ shoots a closed copy of ω_1 into S hence collapses $\aleph_2$.

Remark. The condition on P is, by the terminology we shall use, essentially the $\{\lambda \le |P| : \lambda \text{ regular} > \aleph_1\}$-condition; more exactly, the definition of such a notion is in 6.7, where (ii)$'$ is proved. Now (iv)$_a$ holds by 4.4, (iv)$_b$ by 4.5, (iv)$_c$ holds by 4.6, (i) by 3.2 and (iii) automatically follows from 6.3A(1) as in X 5.3, see 1.13.

Remark. The preservation theorem in this chapter is in a sense orthogonal to the one of Chapter X, since here we are not interested in semiproperness of forcing notions (e.g. Namba forcing may fail to be semiproper, but it always satisfies the condition in this chapter). In chapter XV we will present a generalization of the S-condition which also generalizes semiproperness.

Assume we have a condition satisfying all of these demands and let us get to the proofs of our theorems.

1.2 Theorem. If "ZFC + G.C.H. + there is a measurable cardinal" is consistent then so is "ZFC + G.C.H. + there is a normal precipitous filter D on ω_2 such that $S^2_0 \in D$".

Remark.

(1) S^2_0 is $\{\alpha < \omega_2 : \mathrm{cf}(\alpha) = \aleph_0\}$.

(2) By [JMMP] the converse of this theorem is also true, so we have an equiconsistency result.

(3) In fact if "ZFC + there is a measurable cardinal" is consistent then so is "ZFC + G.C.H. + there is a measurable cardinal", so we can delete "G.C.H." from the hypothesis of our theorem.

Proof. We start with a model of ZFC + G.C.H. with a measurable cardinal κ. We iterate, by the RCS iteration, forcing with Nm κ many times. More exactly let $\bar{Q} = \langle P_i, Q_i : i < \kappa \rangle$ be an RCS iteration, $\underset{\sim}{Q}_{2i}$ is Nm (see (iv)$_a$ above), $\underset{\sim}{Q}_{2i+1} = \mathrm{Levy}(\aleph_1, 2^{|P_{2i+1}|})$. Let V denote our ground model, and P denote $\mathrm{Rlim}\,\bar{Q}$. We can prove by induction $|P_i| < \kappa$; moreover if $\lambda \le \kappa$ is strongly inaccessible then $i < \lambda \Rightarrow |P_i| < \lambda$, and P_λ has power λ.

By 1.1 (ii)$'$, P satisfies the condition (remembering (iv)$_a$+(iv)$_b$) hence by 1.1(i), forcing by P does not collapse $\aleph_1$ nor add reals and so $V^P \models CH$. On the other hand clearly $|P_i| \ge i$, hence $\underset{\sim}{Q}_{2i+1}$ collapses $|i|$ to $\aleph_1$, hence all λ, $\aleph_1 < \lambda < \kappa$ are collapsed by P. By 1.1(iii) (or X 5.3(1)) P satisfies the κ-chain condition hence κ is not collapsed. So clearly $\aleph_1^{V^P} = \aleph_1^V$, $\aleph_2^{V^P} = \kappa$ and V^P satisfies G.C.H.

Let F be a normal κ-complete ultrafilter over κ (in V), then by (iii) and X.6.5 (see references there), F generates a normal precipitous filter on κ in V^P. Let A be $\{\lambda < \kappa : \lambda \text{ is inaccessible}\}$ (in V) then $A \in F$ so we are done with the proof once we show that $\lambda \in A$ implies λ has cofinality ω in $V[P]$. As Nm satisfies our condition (by demand $(\mathrm{iv})_{\mathrm{a}}$) and λ is inaccessible in V we know that the iteration up to stage λ satisfies the λ-c.c. (by demand (iii), or by using X5.3(1) provided that we restrict A to Mahlo cardinals). Hence after forcing with P_λ we have $\lambda = \aleph_2$ and at the next step in the iteration Nm shoots a cofinal ω-sequence into λ, a sequence that exemplifies $\mathrm{cf}(\lambda) = \aleph_0$ in V^P, see 4.1A. $\square_{1.2}$

1.3 Theorem. If "ZFC + G.C.H. + there is a Mahlo cardinal" is consistent *then* so is "ZFC + G.C.H. + for every stationary $S \subseteq S_0^2$ there is a closed copy of ω_1 included in it".

Remark. Earlier Van-Liere has shown the converse and is a variant of a problem of Friedman, see on this X 7.0.

For the clarity of the exposition we prove here a weaker theorem and postpone the proof of the theorem as stated above to Sect. 7 of this chapter.

1.4 Theorem. If "ZFC + G.C.H. + there is a weakly compact cardinal" is consistent *then* so is "ZFC + G.C.H. + for every stationary $S \subseteq S_0^2$ there is a closed copy of ω_1 included in it".

Proof. The proof is very much like the proof of Theorem 7.3 of X; the only difference is that now we do not have to demand that there will be measurable cardinals below the weakly compact cardinal. We give here only an outline of the proof. Let κ be weakly compact, w.l.o.g. $V = L$, so by Jensen's work there is $\langle \bar{A}_\alpha : \alpha < \kappa,\ \alpha \text{ inaccessible}\rangle$, $\bar{A}_\alpha = \langle A_{\alpha,e} : e < n_\alpha\rangle$, a diamond sequence satisfying: $A_{\alpha,e} \subseteq H(\alpha)$, and for every finite sequence $\bar{A}$ of subsets of $H(\kappa)$, and Π_1^1 sentence ψ such that $(H(\kappa), \in, \bar{A}) \models \psi$ there is some inaccessible λ such that $\bar{A}{\restriction}H(\lambda) = \bar{A}_\lambda$ (i.e. $n_\lambda = \ell g(\bar{A})$ and $A_{\lambda,e} = A_e \cap H(\lambda)$) and

$(H(\lambda), \in, \bar{A}_\lambda) \models \psi$. Now we define an RCS iteration $\bar{Q} = \langle P_i, \underset{\sim}{Q}_i : i < \kappa \rangle$. Let $\underset{\sim}{Q}_\alpha = P[\underset{\sim}{S}_\alpha]$ (as it was defined in 1.1 (iv)$_c$) whenever $\bar{A}_\alpha = \langle P_\alpha, p_\alpha, \underset{\sim}{S}_\alpha \rangle$, α strongly inaccessible and $p_\alpha \in P_\alpha$ and $p_\alpha \Vdash_{P_\alpha}$ "$\underset{\sim}{S}_\alpha$ is a stationary subset of S_0^λ $(= \{\delta < \lambda : \mathrm{cf}(\delta) = \aleph_0\})$", and in all other cases we force with the usual Levy $\aleph_1$-closed conditions for collapsing $2^{|P_\alpha| + |\alpha|}$.

In the model we get after the forcing $\kappa = \aleph_2$ and every stationary subset of S_0^2 includes a closed copy of ω_1. (For checking the details note that our forcing notion, and any initial segment of it, satisfies our condition thus no reals are added, $\aleph_1$ is not collapsed and in any λ-stage for inaccessible λ, the initial segment of the forcing satisfies the λ-c.c. so at that stage $\lambda = \omega_2$, when we use $P[S_\lambda]$ we are forcing with $P[S]$ for S which is a stationary subset of S_0^2). $\square_{1.3}$

1.4A Remark. If κ is only a Mahlo cardinal then this proof suffices if we just want "for every $S \subseteq S_0^2$, S or $S_0^2 \setminus S$ contains a closed copy of ω_1 " or even if we want "if h is a pressing-down function on S_0^2, then for some α, $h^{-1}(\{\alpha\})$ contains a closed copy of ω_1". See more in 7.2.

1.5 Theorem. If "ZFC + G.C.H. + there is an inaccessible cardinal" is consistent *then* so is "ZFC + G.C.H. + there is no subset of $\aleph_1$ such that all ω-sequences of $\aleph_2$ are constructible from it".

Remark.

(1) This theorem answers a question of Uri Abraham who has also proved its converse.

(2) Again we can omit G.C.H. from the hypothesis.

Proof. Let κ be inaccessible and let $\bar{Q} = \langle P_i, \underset{\sim}{Q}_i : i < \kappa \rangle$ be an RCS iteration, $\underset{\sim}{Q}_{2i} = (\mathrm{Nm})^{V^{P_{2i}}}$, $\underset{\sim}{Q}_{2i+1} = \mathrm{Levy}(\aleph_2, 2^{|P_{2i}| + |i|})$. In the resulting model $\kappa = \aleph_2$ and as the forcing satisfies the κ-c.c. any subset of $\aleph_1$ is a member of a model obtained by some proper initial segment of our iteration, but the ω-sequence added to ω_2 by the next Nm forcing does not belong to this model so it is not constructible from this subset of $\aleph_1$. $\square_{1.5}$

1.6 Theorem. If the existence of a Mahlo cardinal is consistent with ZFC *then* so is "G.C.H. + for every subset A of $\aleph_2$ there is some ordinal δ such that $\mathrm{cf}(\delta) = \aleph_0$ but δ is a regular cardinal in $L[A \cap \delta]$".

Remark. Again this is an answer to a question of Uri Abraham and again he has shown that the converse of the theorem is true as well by using the square on λ.

Proof. Let κ be Mahlo (in a model V of ZFC + G.C.H.), w.l.o.g. $V = L$ and iterate as in the proof of 1.5. Let A be a subset of $\aleph_2$ in the resulting model. As the forcing notion satisfies the κ-c.c., we can find a closed and unbounded $C \subseteq \aleph_2$ such that for $\delta \in C$ we have $A \cap \delta \in V[P_\delta]$ where P_δ is the Rlim of δ'th initial segment of our iteration. As κ is Mahlo, $\{\lambda < \kappa : \lambda \text{ is inaccessible}\}$ is stationary in it so there is some inaccessible λ in C. Such λ exemplifies our claim. P_λ satisfies the λ-c.c., so in $V[P_\lambda]$ we have $\lambda = \aleph_2$ hence Nm at the λ-step of the iteration adds a cofinal ω sequence into λ, so in $V[P_\kappa]$, which is our model, $\mathrm{cf}(\lambda) = \aleph_0$. But $L[A] \subseteq V[P_\lambda]$ as $\lambda \in C$ (and $P_\delta \lessdot P = P_\kappa$). $\square_{1.6}$

One more answer to a question from Uri Abraham's dissertation is to get V such that if $A \subseteq \omega_2$ and $\aleph_2^{L[A]} = \aleph_2$ then $L[A]$ has $\geq \aleph_2$ reals. We had noted that for the statement to hold in V, it is enough to have: $L[\{\delta < \aleph_2^V : \mathrm{cf}^V \delta = \aleph_0\}]$ has at least $\aleph_2^V$ reals; more explicitly, it is enough to produce a model in which there are $\aleph_2$ distinct reals r, such that for some $\lambda \in \mathrm{Car}^L$ we have $r(\ell) = 0$ iff $\mathrm{cf}((\lambda^{+\ell})^L) = \aleph_0$ (i.e., the cofinality is in V, $\lambda^{+\ell}$ is computed in L). Then the answer below was obtained by Shai Ben-David using the same method as of 1.5, 1.6:

1.7 Theorem. [Ben David] The consistency of "ZFC + there exists an inaccessible cardinal" is equivalent with the consistency with ZFC of the statement: "There is no cardinal preserving extension of the universe in which there is a set $A \subseteq \aleph_2$ such that $L[A]$ satisfies C.H. and $\aleph_2^{L[A]} = \aleph_2$".

However, this proof relies on a preliminary version of this chapter in which forcing notions adding reals were permitted, which unfortunately seems doubtful and was abandoned. The framework given in 1.1, is not enough since in (iv) no forcing notions adding reals appear, but we can use XV §3 instead (i.e. for unboundedly many $i, \underset{\sim}{Q}_i$ is Cohen forcing)

Remark. In fact there is no class of V which is a model of ZFC, having the same $\aleph_1$ and $\aleph_2$ and satisfying C.H.

1.8 Remark. The partition theorems presented later can be slightly generalized to monotone families (instead of ideals) as done in the first version of this book. But this is irrelevant to our main purpose.

We now recall the main properties of RCS iterations. Whenever it is convenient, we will assume that all partial orders under consideration are complete Boolean algebras i.e. are $(B \setminus \{0\}, \geq)$.

1.9 Definition. We say that a sequence $\langle P_\alpha, \underset{\sim}{Q}_\beta, \mathbf{i}_\beta : \alpha \leq \delta, \beta < \delta \rangle$ is an RCS iteration of length δ (δ not necessarily limit ordinal), if:

(1) For all β, $\mathbf{i}_\beta$ is a dense embedding from $P_\beta * \underset{\sim}{Q}_\beta$ into $P_{\beta+1}$, or into the complete Boolean algebra generated by $P_{\beta+1}$. [We usually do not mention $\mathbf{i}_\beta$ and identify $P_\beta * \underset{\sim}{Q}_\beta$ with $P_{\beta+1}$].

(2) For all $\alpha < \beta \leq \delta$, $P_\alpha \lessdot P_\beta$. [We assume that for all $p \in P_\beta$, the projection of p to P_α exists, and we write it as $p{\restriction}\alpha$. We write $P_{\alpha,\beta}$ for the quotient forcing P_β/P_α or P_β/G_α in V^{P_α} or $V[G_\alpha]$, and let $p \mapsto p{\restriction}[\alpha, \beta)$ be the obvious map in V^{P_α}.]

(3) Whenever $\alpha \leq \delta$ is a limit ordinal, then $P_\alpha = \mathrm{Rlim}\langle P_\beta : \beta < \alpha \rangle$.

Also, if we write $\bar{Q} = \langle P_\alpha, \underset{\sim}{Q}_\alpha : \alpha < \delta \rangle$, we automatically define $P_\delta \overset{\mathrm{def}}{=} \mathrm{Rlim}\, \bar{Q}$ (if δ is a limit) or $P_\delta \overset{\mathrm{def}}{=} P_{\delta-1} * \underset{\sim}{Q}_{\delta-1}$ (if δ is a successor), respectively.

We say that $\langle P_\alpha : \alpha < \delta \rangle$ is an RCS iteration iff $\langle P_\alpha, \underset{\sim}{Q}_\alpha : \alpha < \delta \rangle$ is one, with $\underset{\sim}{Q}_\alpha \overset{\mathrm{def}}{=} P_{\alpha+1}/P_\alpha$.

We will not define here what Rlim $\bar{Q}$ actually is. A possible definition is in chapter X. Here we will only collect some properties of RCS iterations which we will use. First, we need a definition:

1.10 Definition. If $\underset{\sim}{\alpha}$ is a P_δ-name we say that $\underset{\sim}{\alpha}$ is *prompt*, if $\Vdash$ "$\underset{\sim}{\alpha} \leq \delta$", and for all ordinals $\xi \leq \delta$, all conditions $q \in P_\delta$:

$$\text{whenever } \ q \Vdash \text{“}\underset{\sim}{\alpha} = \xi\text{”}, \ \text{ then already } \ q{\restriction}(\xi + 1) \Vdash \text{“}\underset{\sim}{\alpha} = \xi\text{”}.$$

(where $q{\restriction}(\delta + 1) = q$)

Note that

1.10A Observation. 1) For a P_δ-name $\underset{\sim}{\alpha}$ we have:

$\underset{\sim}{\alpha}$ is prompt iff $\Vdash$ "$\underset{\sim}{\alpha} \leq \delta$", and for all $\xi \leq \delta$ and all $p \in P_\delta$:

$$\text{if } p \Vdash \text{“}\underset{\sim}{\alpha} \leq \xi\text{”}, \text{ then } p{\restriction}(\xi + 1) \Vdash \text{“}\underset{\sim}{\alpha} \leq \xi\text{”}.$$

Also the inverse implication holds, of course.

2) If S is a set of prompt P_δ-names, *then* also Sup(S) is a prompt P_δ-name and min(S) is a prompt P_δ-name.

Proof. Easy.

1.11 Definition. If $\underset{\sim}{\alpha}$ is a prompt P_δ-name, *then*

(a) $P_{\underset{\sim}{\alpha}} \overset{\text{def}}{=} \{p \in P_\delta : (\forall q \geq p) \ [\text{if } q \Vdash \text{“}\underset{\sim}{\alpha} = \xi\text{”}, \text{ then } q{\restriction}(\xi + 1) \Vdash_{P_\delta} \text{“}p \in \underset{\sim}{G}_\delta\text{”}]\}$

(b) for an atomic $\bar{Q}$-condition p, $p{\restriction}\underset{\sim}{\alpha}$ is naturally defined: for $G_\delta \subseteq P_\delta$ generic over V, $(p{\restriction}\underset{\sim}{\alpha})[G_\delta]$ is $p[G_\delta]$ if $\underset{\sim}{\zeta}_p[G_\delta] \leq \underset{\sim}{\alpha}[G_\delta]$ and $\emptyset$ otherwise

(c) for $p \in P_\delta$, let $p{\restriction}\underset{\sim}{\alpha} = \{r{\restriction}\underset{\sim}{\alpha} : r \in p\}$

It may be more illustrative to consider the following dense subset:

$$P^*_{\underset{\sim}{\alpha}} \overset{\text{def}}{=} \bigcup_{\xi \le \delta} \{p \in P_\xi : p \Vdash_{P_\delta} \text{“}\underset{\sim}{\alpha} = \xi\text{”}\}$$

1.11A Remark.

(a) For any prompt P_δ-name $\underset{\sim}{\alpha}$ we have $P_{\underset{\sim}{\alpha}} \lessdot P_\delta$. Moreover, if $\Vdash_{P_\delta}$ “$\underset{\sim}{\alpha}_1 \le \underset{\sim}{\alpha}_2$”, then $P_{\underset{\sim}{\alpha}_1} \lessdot P_{\underset{\sim}{\alpha}_2}$.

(b) If $\dot{\alpha}$ is the canonical name of α, then $P_\alpha = P^*_{\dot{\alpha}}$.

1.12 Properties of RCS iterations. Let $\bar{Q} = \langle P_\alpha, \underset{\sim}{Q}_\alpha : \alpha < \delta\rangle$ be an RCS iteration with RCS limit P_δ. *Then*

(0) Assume $\alpha(*) < \delta$, $\underset{\sim}{\alpha}$ is a prompt $P_{\alpha(*)+1}$-name of an ordinal $\ge \alpha(*)$, $\mathcal{I}$ is an antichain of $P_{\alpha(*)+1}$ such that $p \in \mathcal{I} \Rightarrow p \Vdash_{P_\delta}$ “$\underset{\sim}{\alpha} = \alpha(*)$” and for each $p \in \mathcal{I}$, $\underset{\sim}{\beta}_p$ is a prompt P_δ-name of an ordinal $\ge \alpha(*)$. *Then* for some prompt P_δ-name $\underset{\sim}{\gamma}$ we have $\Vdash_{P_\delta}$ “ if $p \in \mathcal{I} \cap \underset{\sim}{G}_{P_\delta}$ then $\underset{\sim}{\gamma} = \underset{\sim}{\beta}_p$ and if $\mathcal{I} \cap \underset{\sim}{G}_{P_\delta} = \emptyset$ then $\underset{\sim}{\gamma} = \underset{\sim}{\alpha}$”.

(1) Whenever $\langle \underset{\sim}{\alpha}_n : n \le \omega\rangle$ is a sequence of prompt P_δ-names, satisfying $\Vdash_{P_\delta}$ “$\underset{\sim}{\alpha}_n < \underset{\sim}{\alpha}_{n+1}$” for all n, and $\Vdash_{P_\delta}$ “$\underset{\sim}{\alpha}_\omega = \sup_n \underset{\sim}{\alpha}_n$”, then $P_{\underset{\sim}{\alpha}_\omega}$ is the inverse limit of $\langle P_{\underset{\sim}{\alpha}_n} : n < \omega\rangle$. So in particular: whenever $\langle p_n : n < \omega\rangle$ is a sequence of conditions in P_δ, and $p_n \in P_{\underset{\sim}{\alpha}_n}$, and $p_{n+1}\restriction\underset{\sim}{\alpha}_n = p_n$ for all n, then there is $p \in P_{\underset{\sim}{\alpha}_\omega}$ such that for all n, $p\restriction\underset{\sim}{\alpha}_n = p_n$. Moreover if $p_0 \in P_{\underset{\sim}{\alpha}_0}$ and $\underset{\sim}{p}_{n+1}$ is a $P_{\underset{\sim}{\alpha}_n}$-name of a member of $P_{\underset{\sim}{\alpha}_{n+1}}$ such that $\underset{\sim}{p}_{n+1}\restriction\underset{\sim}{\alpha}_n = \underset{\sim}{p}_n$ then there is p as above.

(2) Let $\alpha(*) < \delta$ be non-limit, $G_{\alpha(*)} \subseteq P_{\alpha(*)}$ generic over V, and $\langle \alpha_\zeta : \zeta \le \beta\rangle$ an increasing continuous sequence of ordinals in $V[G_{\alpha(*)}]$, $\alpha_0 = \alpha(*)$, $\alpha_\beta = \delta$, each $\alpha_{\zeta+1}$ a successor ordinal.
In $V[G_{\alpha(*)}]$, we define $P'_\zeta = P_{\alpha_\zeta}/G_{\alpha(*)}$, $\underset{\sim}{Q}'_\zeta = P_{\alpha_{\zeta+1}}/\underset{\sim}{G}_{\alpha_\zeta}$ (where $\underset{\sim}{G}_{\alpha_\zeta}$ is the P_{α_ζ}-name of the generic subset of P'_ζ, which essentially means a generic subset of P_{α_ζ} over V extending $G_{\alpha(*)}$ ($\underset{\sim}{Q}'_\zeta$ is still a P'_ζ-name) $\bar{Q}' = \langle P'_\zeta, \underset{\sim}{Q}'_\zeta : \zeta < \beta\rangle$. *Then* in $V[G_{\alpha(*)}]$, $\bar{Q}'$ is an RCS iteration with limit P'_β.

(3) If δ is limit, *then* for all $p \in P_\delta$,

$$\Vdash_{P_\delta} \text{“}p \in \underset{\sim}{G}_\delta \text{ iff } \forall \alpha < \delta\ [p{\restriction}\alpha \in \underset{\sim}{G}_\delta]\text{”}$$

(4) If δ is a limit, *then* for all $p \in P_\delta$ we have a countable set $\{\underset{\sim}{\zeta}^k(p) : k < \omega\}$ of prompt names with $\Vdash$ “$\underset{\sim}{\zeta}^k(p) < \delta$” for all k such that letting $\underset{\sim}{\zeta}^* = \sup\{\underset{\sim}{\zeta}^k(p) : k < \omega\}$, we have $\underset{\sim}{\zeta}^*$ is an almost prompt P_δ-name and $p \in P_{\underset{\sim}{\zeta}^*}$.

(5) If δ is a strongly inaccessible cardinal and for every $\alpha < \delta$ we have $|P_\alpha| < \delta$ *then* P_δ is the direct limit of $\bar{Q}$ that is $\bigcup\limits_{\alpha<\delta} P_\alpha$ is a dense sunset of P_δ.

1.13 More properties of RCS iterations. As corollaries of (4) and (5) above we get:

(1) Let $\bar{Q}$ be an RCS iteration as above, and assume

$$\Vdash_{P_\delta} \text{“cf}(\delta) > \aleph_0\text{”}$$

Then

(a) $\bigcup_{\alpha<\delta} P_\alpha$ is (essentially) a dense subset of P_δ that is for every $p \in P_\delta$ for some $q \in \bigcup\limits_{\alpha<\delta} P_\alpha$ we have $q \Vdash$ “$p \in \underset{\sim}{G}_{P_\delta}$” (so P_δ is the direct limit of $\langle P_\alpha : \alpha < \delta\rangle$)

(b) No new ω-sequences of ordinals are added in stage δ, i.e., whenever $p \in P_\delta$, $\underset{\sim}{\tau}$ a P_δ-name and $p \Vdash_{P_\delta}$ “$\underset{\sim}{\tau} : \omega \to \text{Ord}$”, *then* there is an $\alpha < \delta$, a P_α-name $\underset{\sim}{\tau}^*$ and a condition $q \geq p$ such that $q \Vdash$ “$\underset{\sim}{\tau} = \underset{\sim}{\tau}^*$”.

(2) *If* $\langle P_\alpha, \underset{\sim}{Q}_\alpha : \alpha < \kappa\rangle$ is an RCS iteration, κ a strongly inaccessible cardinal, P_κ does not collapse $\aleph_1$, and for all $\alpha < \kappa$ we have $|P_\alpha| < \kappa$, *then* P_κ satisfies the κ-chain condition.

Proof. (1a) Let $p \in P_\delta$, $\underset{\sim}{\zeta}^*$ as in 1.12(4), and let $q \geq p$ decide the value of $\underset{\sim}{\zeta}^*$, say $q \Vdash_{P_\delta}$ “$\underset{\sim}{\zeta}^* = \xi$”. Then $q{\restriction}(\xi+1) \in P_{\xi+1}$ is essentially stronger than p.

(1b) Not hard.

(2) Easy, since we take direct limits on the stationary set $S^\kappa_1 = \{\delta < \lambda : \text{cf}(\delta) = \aleph_1\}$, by (1)(a). $\square_{1.13}$

§2. The Condition

In this section we get to the heart of this chapter, the definition of our condition for forcing notions. We need some preliminary definitions.

2.1 Definition. A tagged tree (or an ideal tagged tree) is a pair $(T, \mathbf{I})$ such that:

(A) T is a tree i.e., a nonempty set of finite sequences of ordinals such that if $\eta \in T$ then any initial segment of η belongs to T; here with no maximal nodes if not said otherwise. T is partially ordered by initial segments, i.e., $\eta \triangleleft \nu$ iff η is an initial segment of ν.

(B) $\mathbf{I}$ is a function with domain including T such that for every $\eta \in T : \mathbf{I}(\eta)$ $(\overset{def}{=} \mathbf{I}_\eta)$ is an ideal of subsets of some set called the domain of $\mathbf{I}_\eta$, and $\mathrm{Suc}_T(\eta) \overset{def}{=} \{\nu : \nu \text{ is an immediate successor of } \eta \text{ in } T\} \subseteq \mathrm{Dom}(\mathbf{I}_\eta)$.

(C) For every $\eta \in T$ we have $\mathrm{Suc}_T(\eta) \neq \emptyset$ and above each $\eta \in T$ there is some $\nu \in T$ such that $\mathrm{Suc}_T(\nu) \notin \mathbf{I}_\nu$.

2.1A Convention. For any tagged tree $(T, \mathbf{I})$ we can define $\mathbf{I}^\dagger$, $\mathbf{I}^\dagger_\eta = \{\{\alpha : \eta\hat{\ }\langle\alpha\rangle \in A\} : A \in \mathbf{I}_\eta\}$; we sometimes, in an abuse of notation, do not distinguish between $\mathbf{I}$ and $\mathbf{I}^\dagger$; e.g. if $\mathbf{I}^\dagger_\eta$ is constantly I^*, we write I^* instead of $\mathbf{I}$. Sometimes we also write $\mathrm{Suc}_T(\eta)$ for $\{\alpha : \eta\hat{\ }\langle\alpha\rangle \in T\}$.

2.2 Definition. η will be called a splitting point of $(T, \mathbf{I})$ if $\mathrm{Suc}_T(\eta) \notin \mathbf{I}_\eta$ (just like ν in (3) above). Let $\mathrm{sp}(T, \mathbf{I})$ be the set of splitting points of $(T, \mathbf{I})$.

We call $(T, \mathbf{I})$ normal if $\eta \in T \setminus \mathrm{sp}(T, \mathbf{I}) \Rightarrow |\mathrm{Suc}_T(\eta)| = 1$ (we may forget to demand this).

2.3 Definition. We now define orders between tagged trees:

(a) $(T_2, \mathbf{I}_2) \leq (T_1, \mathbf{I}_1)$ if $T_1 \subseteq T_2$ and whenever $\eta \in T_1$ is a splitting point of T_1 then $\mathrm{Suc}_{T_1}(\eta) \notin \mathbf{I}_2(\eta)$ and $\mathbf{I}_1(\eta)\restriction\mathrm{Suc}_{T_1}(\eta) = \mathbf{I}_2(\eta)\restriction\mathrm{Suc}_{T_1}(\eta)$ (where $I\restriction A = \{B : B \subseteq A \text{ and } B \in I\}$) and $\mathrm{Dom}(I\restriction A) = A$.

(b) $(T_2, \mathbf{I}_2) \leq^* (T_1, \mathbf{I}_1)$ iff $(T_2, \mathbf{I}_2) \leq (T_1, \mathbf{I}_1)$ and every $\eta \in T_1$ which is a splitting point of T_2 is a splitting point of T_1 as well.

2.3A Notation. We omit $\mathbf{I}_1$ and denote a tagged tree by T_1 whenever $\mathbf{I}_\eta = \{A \subseteq \mathrm{Suc}_T(\eta) : |A| < |\mathrm{Suc}_T(\eta)|$ if $|\mathrm{Suc}_T(\eta)| > \aleph_0$, and $A = \emptyset$ if $|\mathrm{Suc}_T(\eta)| \leq \aleph_0\}$ for every $\eta \in T$.

2.4 Definition.

(1) For a set $\mathbb{I}$ of ideals, a tagged tree $(T, \mathbf{I})$ is an $\mathbb{I}$-tree if for every $\eta \in T, \mathbf{I}_\eta \in \mathbb{I}$ (up to an isomorphism) or $|\mathrm{Suc}_T(\eta)| = 1$.

(2) For a set S of regular cardinals, T is called an S-tree if for some $\mathbf{I}$, $(T, \mathbf{I})$ is an $\mathbb{I}_S$-tagged tree where $\mathbb{I}_S = \{\{A \subseteq \lambda : |A| < \lambda\} : \lambda \in S\}$

2.5 Definition.

(1) For a tree T, lim$T$ is the set of all ω-sequences of ordinals, such that every finite initial segment of them is a member of T. The set $\lim T$ is also called the set of "branches" of T.

(2) A subset J of a tree T is a front if $\eta, \nu \in J$ implies none of them is an initial segment of the other, and every $\eta \in$ lim$T$ has an initial segment which is a member of J.

2.6 Main Definition. Let S be a set of regular cardinals; we say that a forcing notion P satisfies the S-condition if there is a function F with values of the "right" forms, so that for every S-tree T:

if f is a function $f : T \to P$ satisfying

(a) $\nu \lhd \eta$ implies $f(\nu) \leq_P f(\eta)$ and

(b) there are fronts $J_n (n < \omega)$ (of T) such that $\bigcup_{n<\omega} J_n = \mathrm{sp}(T, I)$, every member of J_{n+1} has a proper initial segment belonging to J_n and $\eta \in J_n$ implies

$$\langle \mathrm{Suc}_T(\eta), \langle f(\nu) : \nu \in \mathrm{Suc}_T(\eta) \rangle \rangle = F(\eta, w[\eta], \langle f(\nu) : \nu \trianglelefteq \eta \rangle)$$

($w[\eta]$ is defined below) and $\mathrm{Suc}_T(\eta) = \{\eta\hat{}\langle\alpha\rangle : \alpha < \lambda\}$ for some $\lambda \in S$ (for simplicity).

then for every $T^\dagger$, $T \leq^* T^\dagger$ there is some $p \in P$ such that $p \Vdash_P$ "$\exists \eta \in \lim T^\dagger$ such that $\forall k < \omega$ we have $f(\eta \restriction k) \in \underset{\sim}{G}_P$" where $\underset{\sim}{G}_P$ is the P-name of the generic subset of P; note that in general η is not from V, i.e. it may be a branch which forcing by P adds.

2.6A Explanation. First for the notation: $w[\eta] = \{k < \ell g(\eta) : \eta\restriction k \in \bigcup_{\ell<\omega} J_\ell\}$.
Now for the meaning: One can regard the situation as a kind of a *game*. There are two players. In ω many steps they define a tree T and an increasing function $f : T \to P$. In the n'th move, player I defines an initial segment T_n of the tree T (so T_n will be the set of nodes up to some member of the front J_n) and a function $f_n : T_n \to P$ which is increasing such that $m < n \Rightarrow f'_m \subseteq f_n$ (see below). Player II end-extends the tree T_n to a tree T'_n by adding successors to each leaf (=node without successor) in T_n and extends f_n to a function f'_n on T'_n. Then player I plays T_{n+1} (an end extension of T'_n with no infinite branches), and a function f_{n+1} ($\supseteq f'_n$), etc. Finally, $T = \bigcup_n T_n$, $f = \bigcup_n f_n$. Player II wins a play if for all $T^\dagger$: if $T \leq^* T^\dagger$, then there is $p \in P$ such that $p \Vdash_P$ "$(\exists \eta \in \lim T^\dagger)(\forall \kappa < \omega) f(\eta\restriction k) \in \underset{\sim}{G}_P$". P satisfies the S-condition if there is a winning strategy F which at each point η depends only on what happened so far on the nodes below η.

However F, the "winning strategy" of player II, has only partial memory.

Remark. It does not matter if we require $\bigcup_n J_n = \mathrm{sp}(T, I)$ or $\bigcup_n J_n \subseteq \mathrm{sp}(T, I)$, or equivalently whether we allow player I to play any end extension T_n of T'_{n-1} or only end extensions with no new splitting points.

Remark. (1) If P is a dense subset of Q, then P has the $\mathbb{I}$-condition (see 2.7 below) iff Q has it.
(2) If $P \lessdot Q$, and Q has the $\mathbb{I}$-condition, then also P has the $\mathbb{I}$-condition.

The proof of (2) uses the fact that if: $f : T \to P$, then the existence of a branch in $\{\eta \in T : f(\eta) \in G_p\}$ is absolute between the universes V^P and V^Q.

2.6B Convention.

(1) In Definition 2.6, the value F gives to $\mathrm{Suc}_T(\eta)$ is w.l.o.g. $\{\eta\char`\^\langle\alpha\rangle : \alpha < \lambda\}$ for some λ, and we do not strictly distinguish between λ and $\mathrm{Suc}_T(\eta)$.

(2) The domain of F consists of triples of the form $\langle\eta, w, f\rangle$, where η is a finite sequence of ordinals, $w \subseteq \mathrm{Dom}(\eta)$, and f is an increasing function from $\{\eta\restriction k : k \leq \ell g(\eta)\}$ into P. The value $F(\eta, w, f)$ has two components: The first is of the form $\{\eta\char`\^\langle i\rangle : i \in A\}$ for some set A of ordinals (by (1), without loss of generality $A = |A|$) and the second component is a family of elements of P above $f(\eta)$, indexed by the first component.

When we define such a function F, we usually call the first component "$\mathrm{Suc}_T(\eta)$" (here "T" is just a label, not an actual variable), and we write the second component as $f\restriction\mathrm{Suc}_T(\eta)$ or $\langle f(\nu) : \nu \in \mathrm{Suc}_T(\eta)\rangle$ (i.e. we use the same variable "f" that appears in the input of F).

2.7 Definition. For a set $\mathbb{I}$ of ideals we define similarly when does a forcing notion P satisfies the $\mathbb{I}$-condition (the only difference is dealing with $\mathbb{I}$-trees instead of S-trees), so now

$$\langle \mathrm{Suc}_T(\eta), \mathbf{I}_\eta, \langle f(\nu) : \nu \in \mathrm{Suc}_T(\eta)\rangle\rangle = F(\eta, w[n], f\restriction\{\nu : \nu \trianglelefteq \eta\})$$

and $\mathrm{Suc}_T(\eta) = \mathrm{Dom}(\mathbf{I}_\eta)$. We allow ourselves to omit $\mathrm{Suc}_T(\eta)$ when it is well understood. (We can let the function depend on $\mathbf{I}_\nu(\nu \triangleleft \eta)$ too).

2.7A Remark. If I is restriction closed (i.e. $I \in \mathbb{I}$, $A \subseteq \mathrm{Dom}(I)$, $A \notin I$ then $I\restriction A \in \mathbb{I}$ at least for some $B \subseteq A$, $B \notin I^+$ and $J \in \mathbb{I}$ we have $I\restriction B \cong J$) *then* we can weaken the demand to

$$\mathrm{Suc}_T(\eta) \subseteq \mathrm{Dom}(\mathbf{I}_\eta), \quad \mathrm{Suc}_T(\eta) \notin I_\eta).$$

§3. The Preservation Properties Guaranteed by the S-Condition

3.1 Definition.

(1) An ideal I is λ-complete if any union of less than λ members of I is still a member of I.

(2) A tagged tree $(T, \mathbf{I})$ is λ-complete if for each $\eta \in T$ the ideal $\mathbf{I}_\eta$ is λ-complete.

(3) A family $\mathbb{I}$ of ideals is λ-complete if each $I \in \mathbb{I}$ is λ-complete.

3.2 Theorem. (CH) If P is a forcing notion satisfying the $\mathbb{I}$-condition for an $\aleph_2$-complete $\mathbb{I}$ *then* forcing with P does not add reals.

As an immediate conclusion we get:

3.3 Theorem. (CH) If P is a forcing notion satisfying the S-condition for a set S of regular cardinals greater than $\aleph_1$ *then* forcing with P does not add reals.

The main tool for the proof of the theorem is the combinatorial Lemma 3.5 from [RuSh:117], for which we need a preliminary definition. More on such theorems and history see Rubin and Shelah [RuSh:117].

3.4 Definition. We define a topology on $\lim T$ (for any tree T) by defining for each $\eta \in T$ the set $T_{[\eta]} = \{\nu : \eta \trianglelefteq \nu \text{ or } \nu \trianglelefteq \eta\}$ and letting $\{\lim T_{[\eta]} : \eta \in T\}$ generate the family of open subsets of $\lim T$ (so each such set $\lim(T_{[\eta]})$ is also closed and is called basic open, and an open subset is an arbitrary union of basic open sets). The family of Borel sets is the σ-algebra generated by the open sets.

3.5 Lemma. 1) If $(T, \mathbf{I})$ is a λ^+-complete tree and H is a function from $\lim T$ to λ such that for every $\alpha < \lambda$ the set $H^{-1}(\{\alpha\})$ is a Borel subset of $\lim T$ (in the topology that was defined in Definition 3.4) *then* there is a tagged subtree $(T^\dagger, \mathbf{I})$, $(T, \mathbf{I}) \leq^* (T^\dagger, \mathbf{I})$ (see 2.3(b)) such that H is constant on $\lim T^\dagger$.

2) In part (1) we can let H be multivalued, i.e. assume $\lim(T)$ is $\bigcup_{\alpha<\lambda} H_\alpha$, each H_α is a Borel subset of $\lim(T)$. If $(T, \mathbf{I})$ is λ^+-complete *then* there is $(T^\dagger, \mathbf{I})$ such that $(T, \mathbf{I}) \leq^* (T^\dagger, \mathcal{I})$ and for some α we have $\lim(T^\dagger) \subseteq H_\alpha$.

Proof. 1) First note that *if* $T_1 \subseteq T$ is such that: $\langle\rangle \in T_1$; for every $\eta \in T_1$ if η is a splitting point of $(T, \mathbf{I})$ then $\mathrm{Suc}_{T_1}(\eta) = \mathrm{Suc}_T(\eta)$ and if η is not a splitting point of T then $|\mathrm{Suc}_{T_1}(\eta)| = 1$, *then* $(T, \mathbf{I}) \leq^* (T_1, \mathbf{I}\restriction T_1)$, so w.l.o.g. we can assume that in T every point is either a splitting point or it has only one immediate extension.

For each $\alpha < \lambda$ let us define a game $\eth_\alpha$: in the first move player I chooses the node η_0 in the tree such that $\ell g(\eta_0) = 0$, player II responds by choosing a proper subset A_0 of $\mathrm{Suc}_T(\eta_0)$ such that $A_0 \in \mathbf{I}_{\eta_0}$, in the n-th move player I chooses an immediate extension of η_{n-1}, η_n such that $\eta_n \notin A_{n-1}$ or η_{n-1} is not a splitting point of $(T, \mathbf{I})$, and player II responds by choosing $A_n \in \mathbf{I}_{\eta_n}$.

Player I wins if for the infinite branch η defined by $\eta_0, \eta_1, \eta_2, \ldots$ we have $H(\eta) = \alpha$. By the assumption of the lemma this is a Borel game so by Martin's Theorem, [Mr75] one of the players has a winning strategy. We claim that there is some $\alpha < \lambda$ for which player I has a winning strategy in the game $\eth_\alpha$. Assume otherwise, i.e., for every $\alpha < \lambda$ player II has a winning strategy F_α. We construct an infinite branch inductively: let $\eta_0 = \langle\rangle$, $\eta_0 \in T$. At stage n let A_n be $\bigcup_{\alpha<\lambda} F_\alpha(\eta_0, \eta_1, \ldots, \eta_{n-1})$; now if η_{n-1} is a splitting point (of $(T, \mathbf{I})$) then $\mathbf{I}_{\eta_{n-1}}$ is λ^+-complete and each $F_\alpha(\eta_0, \ldots, \eta_{n-1})$ is a member of it, hence $A_n \in \mathbf{I}_{\eta_{n-1}}$, so clearly $\mathrm{Suc}_T(\eta_{n-1}) \nsubseteq A_n$.

If η_{n-1} is not a splitting point it has only one immediate successor and let it by η_n, otherwise since $\mathrm{Suc}(\eta_{n-1}) \notin I_{\eta_{n-1}}$, $A_n \in I_{\eta_{n-1}}$, we have $(\mathrm{Suc}(\eta_{n-1}) \setminus A_n) \neq \emptyset$ so we choose $\eta_n \in (\mathrm{Suc}_T(\eta_{n-1}) \setminus A_n)$. Let $\eta = \bigcup_{n<\omega} \eta_n$ be the infinite branch that we define by our construction and let $\alpha(*) = H(\eta)$. Now in the game $\eth_{\alpha(*)}$ if player I will choose η_n at stage n (for all n) and player II will play by $F_{\alpha(*)}$, player I will win although player II has used his winning strategy $F_{\alpha(*)}$, contradiction.

So there must be $\alpha(*)$ such that player I has a winning strategy $H_{\alpha(*)}$ for $\eth_{\alpha(*)}$ and let $T^\dagger$ be the subtree of T defined by $\{\eta : \langle \eta\restriction 0, \ldots, \eta\restriction(n-1)\rangle$ are the first n moves of player I in a play in which he plays according to $H_{\alpha(*)}\}$. Now for $\eta \in T^\dagger \cap \mathrm{sp}(T)$, let $A = \mathrm{Suc}_{T^\dagger}(\eta)$. Then $A \notin I_\eta$, otherwise player II could have played it as A_n. So $T \leq^* T^\dagger$, and $T^\dagger$ is as required.

2) Same proof replacing $H^{-1}(\{\alpha\})$ by H_α so $H(\eta) = \alpha$ by $\eta \in H_\alpha$. $\square_{3.5}$

3.5A Corollary. If $(T, \mathbf{I})$ is a λ^+-complete tree, and g is a function from T into λ, and $\lambda^{\aleph_0} = \lambda$, *then* there is a tagged subtree $(T^\dagger, I)$, $(T, I) \leq^* (T^\dagger, I)$ such that $g{\restriction}T^\dagger$ depends only on the length of its argument, i.e. for some function $g^\dagger : \omega \to \lambda$, for all $\eta \in T^\dagger$, $g(\eta) = g^\dagger(\ell g(\eta))$. $\square_{3.5A}$

Proof of Theorem 3.2. Let $\underset{\sim}{\tau}$ be a name of a real in $V[P]$ and $p_0 \in P$ and we will find a condition $p \in P$ forcing $\underset{\sim}{\tau}$ to be equal to a real from V and $p_0 \leq p$. Let f, $\langle T, \mathbf{I}\rangle$ be such that $\mathrm{Rang}(\mathbf{I}) \subseteq \mathbb{I}$, $f : T \to P$ and be defined as follows: we define by induction on k, for a sequence η of ordinals of length k, the truth value of $\eta \in T$, $f(\eta)$, and then $\mathbf{I}_\eta$. We let $\langle\rangle \in T, f(\langle\rangle) = p_0$. For $\eta \in T$ of even length $2k$, we use F from the definition of the $\mathbb{I}$-condition, to define $\mathrm{Suc}_T(\eta), \mathbf{I}_\eta, f{\restriction}\mathrm{Suc}_T(\eta)$. For $\eta \in T$ of length $2k+1$, we let $\mathrm{Suc}_\eta(T) = \{\eta\hat{}\langle 0\rangle\}$, and we define $f(\eta\hat{}\langle 0\rangle)$ such that it will be an extension of the value of f on its predecessor and such that $f(\eta)$ forces a value for $\underset{\sim}{\tau}(k)$ (the k'th place of the real that $\underset{\sim}{\tau}$ names).

We continue by defining $H : \lim T \to \mathbb{R}^V$ (as we assume C.H. clearly $|\mathbb{R}^V| = \aleph_1$, so it is just like a function from T to ω_1) by letting $H(\eta)(k) =$ the value forced by $f(\eta{\restriction}(2k+1))$ for $\underset{\sim}{\tau}(k)$. By Lemma 3.5 there is $(T^\dagger, \mathbf{I})$, $(T, \mathbf{I}) \leq^* (T^\dagger, \mathbf{I})$ on which H is constant, now let p be the forcing condition that by Definition 2.6 forces "$\exists \eta \in \lim T^\dagger$ such that $\forall k[f(\eta{\restriction}k) \in \underset{\sim}{G}_P]$". This p forces $\underset{\sim}{\tau}$ to equal the constant value of H on $T^\dagger$ which is a member of V, and $p \Vdash$ "$p_0 \in \underset{\sim}{G}_P$". $\square_{3.2}$

3.6 Theorem. If P is a forcing notion satisfying the S-condition for a set of regular cardinals S and $\aleph_1 \notin S$ *then* forcing by P does not collapse $\aleph_1$.

Remark. 1) Note that this is stronger than 3.3, as we do not assume C.H. and that we allow $\aleph_0 \in S$. The proof is quite similar to the proof of Theorem 3.2 but here we use a somewhat different combinatorial lemma. Note also that we shall not use this theorem;
2) We can generalize 3.2 and 3.6 to $\mathbb{I}$-condition when $\mathbb{I}$ is $\aleph_2$-complete, striaght-forwardly, see 3.8.

3.7 Lemma. Let $(T, \mathbf{I})$ be such that for some regular uncountable λ, for every $\eta \in T$ either $\mathbf{I}_\eta$ is λ^+-complete or $|\mathrm{Suc}_T(\eta)| < \lambda$, *then* for every $H : T \to \lambda$ satisfying $\{\eta \in \lim T : H(\eta) < \alpha\}$ is a Borel subset of $\lim T$ for any successor $\alpha < \lambda$, there is $\alpha < \lambda$ and $(T', \mathbf{I})$, $(T, \mathbf{I}) \leq^* (T', \mathbf{I})$ such that for all $\eta \in T'$ we have $H(\eta) < \alpha$, and for all η in T', if $|\mathrm{Suc}_T(\eta)| < \lambda$, then $\mathrm{Suc}_{T'}(\eta) = \mathrm{Suc}_T(\eta)$.

Proof of the lemma. We define for each successor $\alpha < \lambda$ a game ∂_α very much like the way we did it for proving Lemma 3.5, the only difference being that if $|\mathrm{Suc}_T(\eta_n)| < \lambda$ player II chooses A_n such that $|\mathrm{Suc}_T(\eta_n) \setminus A_n| = 1$ (otherwise player II chooses $A_n \in \mathbf{I}_{\eta_n}$ just like in 3.3); player I wins if for every $n < \omega$ $H(\eta_n) < \alpha$. Here again the game ∂_α is determined for every α (here simply because if player II wins a play he does so at some finite stage). Again we claim that there should be at least one α for which player I has a winning strategy. Assume the contrary and let F_α be player's II winning strategy for each $\alpha < \lambda$. We construct a subtree T^* deciding by induction on the height of the members of T which of them are the members of T^*. For η that is already in T^*, if $|\mathrm{Suc}_T(\eta)| < \lambda$ we include all the members of $\mathrm{Suc}_T(\eta)$ in T^*; otherwise $\mathbf{I}_\eta$ is λ^+-complete so $\mathrm{Suc}_T(\eta) \setminus \bigcup_{\alpha<\lambda} F_\alpha(\eta{\restriction}0, \eta{\restriction}1, \ldots, \eta)$ is not empty, so we pick one extension of η from this set and the rest of $\mathrm{Suc}_T(\eta)$ will not be in T^*. Now T^* is a tree of height ω branching to less than λ successors at each point, so as λ is regular uncountable $|T^*| < \lambda$ and there is some $\alpha^* < \lambda$ such that $\eta \in T^*$ implies $H(\eta) < \alpha^*$. Regarding the game ∂_{α^*}, there is a play of it in which player I chooses all along the way members of T^* and player II plays according to F_{α^*}, of course player I wins this game contradicting the assumption that F_{α^*} is a winning strategy for player II.

We define T' just like we did in the proof of Lemma 3.5, collecting all the initial segments of plays of player I in the game $\Game_{\alpha^*}$ when he plays according to his winning strategy H_{α^*}. $\square_{3.7}$

Proof of Theorem 3.6. Just like in the proof of 3.2, having a name $\underset{\sim}{\tau}$ of a function in $V[P]$ mapping ω into ω_1 we take an S-tree T, define a function $h : T \to P$ using the F in odd stages and in even stages forcing more and more values for $\underset{\sim}{\tau}$. Using 3.7 with $\lambda = \aleph_1$, we get a condition $p \in P$ forcing the function that $\underset{\sim}{\tau}$ names to be bounded below ω_1, so we are done. $\square_{3.6}$

Similarly we can prove:

3.8 Theorem. If P satisfies the $\mathbb{I}$-condition, and λ is regular uncountable and $(\forall I \in \mathbb{I})$ $[|\bigcup I| < \lambda$ or I is λ^+-complete] *then* $\Vdash_P$ "$\mathrm{cf}(\lambda) > \aleph_0$". If $(\forall I \in \mathbb{I})$ [I is λ^+-complete] and $\lambda = \lambda^{\aleph_0}$ *then* P adds no new ω-sequence from λ.

3.8 Warning. The statement "in $V^P, \underset{\sim}{Q}$ satisfies the S-condition" may be interpreted as "in $V^P, \underset{\sim}{Q}$ satisfies the $\mathbb{I}$-condition" in two ways:

(a) $\mathbb{I} = \{\{A \in V^P : A \subseteq \lambda, V^P \models |A| < \lambda\} : \lambda \in S\}$

(b) $\mathbb{I} = \{\{A \in V : A \subseteq \lambda, V \models |A| < \lambda\} : \lambda \in S\}$

Note that $I \in \mathbb{I}$ is identified with the ideal it generates.

However the two interpretations are equivalent if P satisfies the λ-chain condition (or is λ^+-complete) for each $\lambda \in S$ (and even weaker conditions) [and this will be the case in all our applications.]

§4. Forcing Notions Satisfying the S-Condition

4.1 Definition. Namba forcing Nm is the set $\{T : T$ is a tagged $\{\aleph_2\}$-tree, such that for every $\eta \in T$, for some ν, $\eta \trianglelefteq \nu \in T$ and $|\mathrm{Suc}_T(\nu)| = \aleph_2\}$ with the order $T \leq T'$ iff $T \supseteq T'$ (see 2.4); so smaller trees carry more information and we identify T and $(T, I^{bd}_{\omega_2})$, $I^{bd}_{\omega_2}$ is the ideal of bounded subsets of ω_2. We will write $\underset{\sim}{h}$ or $\underset{\sim}{h}_{Nm}$ for the generic branch added by Nm, i.e. $\Vdash$"$\underset{\sim}{h} : \omega \to \omega_2^V$", and for every p in the generic filter, $\mathrm{tr}(p) \stackrel{\mathrm{def}}{=} \mathrm{trunk}(p) \subseteq h$", where the trunk of T is $\eta \in T$ of maximal length such that $\nu \in T$ & $\ell g(\nu) \leq \ell g(\eta) \Rightarrow \nu \trianglelefteq \eta$. We can restrict ourselves to normal members: T such that $\eta \in T \Rightarrow |\mathrm{Suc}_T(\eta)| \in \{1, \aleph_2\}$. For I an ideal of ω_2, $\mathrm{Nm}(I)$ is defined similarly (not used in this chapter).

4.1A Claim. Nm changes the cofinality of $\aleph_2$ to $\aleph_0$ ($\underset{\sim}{h}_{Nm}$ exemplifies this).

Remark. In X 4.4 (4) the variant of Namba forcing Nm′ is the set of all trees of height ω such that each tree has a node, the trunk such that below its level the tree-order is linear and above it each point has $\aleph_2$ many immediate successors, the order is inversed inclusion. Namba introduces Nm in [Nm]. Both forcing notions add a cofinal ω-sequence to ω_2 (Nm by 4.1A, Nm′ by X 4.7(2)) without collapsing $\aleph_1$, and (if CH holds) neither of them adds reals (Nm by 4.4, Nm′ by 4.7(1), (3)), but they are not the same.

4.2 Claim. (Magidor and Shelah). Assume CH. If $h[h']$ is a Namba sequence [Namba′-sequence] *then* in $V[h]$ we cannot find a Namba′-sequence over V, nor can we in $V[h']$ find a Namba-sequence over V.

Proof. Trivially we can in Nm and Nm′ restrict ourselves to conditions which are trees consisting of strictly increasing finite sequences of ordinals. First we look in $V[h']$, let $\underset{\sim}{h}'$ be the Nm′-name of the Namba′ sequence, and let $\underset{\sim}{f}$ be a Nm′-name of an increasing function from ω to ω_2. Let $T^0 \in \mathrm{Nm}'$ and suppose $T^0 \Vdash_{Nm'}$ "$\mathrm{Sup\,Rang}(\underset{\sim}{f}) = \omega_2^V$". Now it is easy to find T^1 in Nm′, $T^1 \geq T^0$, such that for each $m, n < \omega$ the truth values of "$\underset{\sim}{f}(n) = \underset{\sim}{h}'(m)$", and "$\underset{\sim}{f}(n) < \underset{\sim}{h}'(m)$" are determined by T^1 (i.e., forced), (possible by X 4.7(1), as forcing by Nm′ does not add reals.)

Let $A = \{k < \omega$: for some m, T^1 forces that for every $i < k$ we have $\underset{\sim}{f}(i) < \underset{\sim}{h}'(m) \leq \underset{\sim}{f}(k)\}$, so A is an infinite subset of ω, in V, and let $A = \{k_\ell : \ell < \omega\}$, $k_0 < k_1 < k_2 < \dots$ and there are $\langle m_\ell : \ell < \omega\rangle$ such that $m_\ell < m_{\ell+1} < \omega$ and $T^1 \Vdash$ "$\underset{\sim}{f}(i) < \underset{\sim}{h}'(m_\ell) \leq \underset{\sim}{f}(k_\ell)$ for $i < k_\ell$ and $\ell < \omega$". Now

(*) $T^1 \Vdash_{\mathrm{Nm}'}$ " for every $F \in V$, an increasing function from ω_2 to ω_2, there is $\ell_0 < \omega$ such that for every $\ell > \ell_0 + 3$, $\underset{\sim}{f}(k_\ell) > F(\underset{\sim}{f}(k_{\ell-2}))$".

Why? This is because for every $F \in V$ (as above, without loss of generality strictly increasing) and $T^2 \geq T^1$ (in Nm′) if ℓ_0 is the length of the trunk of T^2 then for some $T^3 \geq T^2$;

(**) $T^3 \Vdash_{\mathrm{Nm}'}$ " if $\ell > \ell_0 + 3$ then $\underset{\sim}{f}(k_\ell) > F(\underset{\sim}{f}(k_{\ell-2}))$"

Why? Simply choose $T^3 = \{\eta \in T^2 : \text{if } \ell g(\eta) \geq \ell + 1 > \ell_0 \text{ then } \eta(\ell) > F(\eta(\ell - 1))\}$, clearly $T^3 \in \mathrm{Nm}', T^3 \geq T^2$ and T^3 satisfies (**).

So (*) holds, but it exemplifies $\underset{\sim}{f}$ is not a Nm-sequence, i.e., $T^1 \Vdash_{\mathrm{Nm}'}$ "$\underset{\sim}{f}$ is not a Namba-sequence". [Why? Because $\Vdash_{\mathrm{Nm}}$ " for some function $F \in V$ from ω_2 to ω_2 for arbitrarily large $\ell < \omega$ we have $\underset{\sim}{h}(k_\ell) < F(\underset{\sim}{h}(k_{\ell-2}))$" as if $T \in \mathrm{Nm}$ and for simplicity each $\eta \in T$ is strictly increasing we let $F : \omega_2 \to \omega_2$ be such that $F(\alpha) = \min\{\delta : \text{if } \eta \in T \cap {}^{\omega>}\delta \text{ then for some } \nu, \eta \triangleleft \nu \in T \cap {}^{\omega>}\delta\}$, and let $T' = \{\eta \in T$: if $\ell < \ell g(\eta)$ and $\eta{\restriction}\ell \in \mathrm{sp}(T)$ and $|\{m < \ell : \eta{\restriction}m \in \mathrm{sp}(T)\}| \in \bigcup\{[k_{10i}, k_{10i+5}) : i < \omega\}$ then $\eta(\ell) < F(\eta(\ell - 1))\}$, and T' forces the failure.] So we have proved one half of 4.2.

Now let us prove the second assertion in the claim, i.e., let $\underset{\sim}{f}$ be a Nm-name of an increasing function from ω to ω_2, and we shall prove that it is forced, not to be generic for Nm′ so assume $\Vdash_{\mathrm{Nm}}$ " $\bigcup_{m<\omega} \underset{\sim}{f}(n) = \omega_2$". Clearly this is enough.

Let $T \in \mathrm{Nm}$, then we can find $T^0 \geq T, T^0 \in \mathrm{Nm}$ (normal) such that for every splitting point η of T_0 and $\nu = \eta\char`^\langle\alpha\rangle \in T^0$:

1) for some n_ν, $T^0_{[\nu]} \Vdash_{Nm}$ "$n_\nu = \mathrm{Min}\{\ell : \underset{\sim}{f}(\ell) > \mathrm{Max\,Rang}(\nu)\}$,
2) for some γ_ν, $T^0_{[\nu]} \Vdash$ "$\underset{\sim}{f}(n_\nu) = \gamma_\nu$"
3) if ρ_ν is the trunk of $T^0_{[\nu]}$ then $(\forall\beta)[\rho_\nu\char`^\langle\beta\rangle \in T^0 \to \beta > \gamma_\nu]$.

If n_ν is not defined let $n_\nu = \omega$ (this occurs if $\nu \notin \mathrm{sp}(T)$.

Now by 3.5A there is T^1, $T^0 \leq^* T^1$ (in Nm) and $n_\ell (\ell < \omega)$ such that $n_\eta = n_{\ell g(\eta)}$ for every $\eta \in T^1$. Let $\{\ell_i : i < \omega\}$ be a list of $\{\ell < \omega : n_{\ell+1} \neq \omega\}$, such that $\ell_i < \ell_{i+1}$, so $\eta \in T^1$, $\ell g(\eta) = \ell_i$ implies η is a splitting point of T^1. Note that if $\eta \in T^1$, $\ell_{i+1} \in \mathrm{Dom}(\eta)$, then $T^1_{[\eta]} \Vdash$ "$\eta(\ell_i) < f(n_{\ell_i}) < \eta(\ell_{i+1})$". Let $T^2 = \{\eta \in T^1 : \text{if } \ell_{2i} < \ell g(\eta), \text{ then } \eta(\ell_{2i}) = \mathrm{Min}\{\alpha : (\eta{\restriction}\ell_{2i})\char`^\langle\alpha\rangle \in T^1\}$, $F(\alpha) = \mathrm{Min}\{\gamma : (\forall k < \omega)\ (\forall \nu \in T^2 \cap {}^{\omega>}\alpha)(\exists\rho \in {}^k\gamma)(\nu\char`^\rho \in T^2)\}$. So F is nondecreasing and $T^2_{[\eta]} \Vdash$ "$\underset{\sim}{f}(n_{\ell_{2i}}) \leq \eta(\ell_{2i+1}) < F(\eta(\ell_{2i-1})) \leq F(\underset{\sim}{f}(n_{\ell_{2i-1}}))$" for $i \in (0, \omega)$.

Let $A \stackrel{\text{def}}{=} \{n_{\ell_{2i}} : 0 < i < \omega\}$.

Then $T^2 \in \text{Nm}, T^2 \geq T^1$, and $F \in V$ is a function from ω_2 to $\omega_2, A \in V$ an infinite subset of ω and

$T^2 \Vdash_{Nm}$ "for every $n \in A, \underset{\sim}{f}(n) < F(\underset{\sim}{f}(n-1))$".

This shows that (T^2 forces) $\underset{\sim}{f}$ is not a Nm′-sequence. $\square_{4.2}$

4.3 Claim.

(1) Nm′, Nm do not satisfy the $2^{\aleph_0}$-chain condition.

(2) It is consistent with ZFC that $2^{\aleph_0} = \aleph_1, 2^{\aleph_1}$ arbitrarily large and Nm, Nm′ satisfies the $\aleph_4$-c.c.

4.3A Remark. The proof of (2) is inspired by the proof of Baumgartner of the consistency of: ZFC + $2^{\aleph_0}$ arbitrarily large + "there is no set of $\aleph_3$ subsets of $\aleph_1$ with pairwise countable intersection". Thinking a minute the close connection between the problems should be apparent. The other ingredient is the Δ-system theorem on trees from Rubin and Shelah (again see [RuSh:117]).

Note that Nm, Nm′ necessarily colapse $\aleph_3$ (see [Sh:g, VII 4.9]) so 4.3(2) is best possible.

Proof. (1) For every real η (i.e. $\eta \in {}^{\omega}2$), let $T_\eta = \{\nu : \nu$ a finite sequence of ordinals $< \omega_2$, and $n < \ell g(\nu) \Rightarrow \nu(n) + \eta(n)$ is an even ordinal$\}$.

Clearly $T_\eta \in$ Nm and $T_\eta \in$ Nm′, and the T_η's are pairwise incompatible (in Nm and in Nm′) and there are $2^{\aleph_0}$ such T_η's.

(2) Let V satisfy G.C.H. $\kappa > \aleph_2$ and $P = \{f : f$ a countable function from κ to $\{0,1\}\}$ ordered by inclusion. Suppose in $V^P, \underset{\sim}{Q}$ is Nm or Nm′, and it does not satisfy the $\aleph_4$-chain condition. So there is $p_0 \in P$ and P-names $\underset{\sim}{T}_i (i < \aleph_4)$ such that $p_0 \Vdash_P$ "each $\underset{\sim}{T}_i$ belongs to $\underset{\sim}{Q}$ (for $i < \aleph_4$) and they are pairwise incompatible in $\underset{\sim}{Q}$, $\underset{\sim}{Q}$ is Nm or Nm′". Without loss of generality $p_0 \Vdash$ " if $\underset{\sim}{Q} =$ Nm′, then every T_i has trunk $\langle\rangle$".

For each i we can now find a tree of conditions p^i_η deciding higher and higher splitting points of $\underset{\sim}{T}_i$. Specifically, we will define $A^i \subseteq {}^{\omega>}(\omega_2)$, p^i_η, ν^i_η for $\eta \in A^i$ such that

(a) $\langle\rangle \in A^i, p^i_{\langle\rangle} \geq p_0$

(b) $p^i_\eta \Vdash_P$ "$\eta \in \underset{\sim}{T}_i$" (and $p^i_\eta \in P$ of course).

(c) $p^i_\eta \Vdash_P$ "ν^i_η is a splitting point of $\underset{\sim}{T}_i, \eta \trianglelefteq \nu^i_\eta$, and: if $\rho \triangleleft \nu^i_\eta$ is a splitting point of T_i then for some $\ell < \ell g(\eta)$ we have $\rho = \nu^i_{\eta \restriction \ell}$".

(d) $\eta \in A^i, \rho \in A^i, \eta \triangleleft \rho$ implies $\nu^i_\eta \triangleleft \nu^i_\rho$

(e) $\nu^i_\eta {}^\frown \langle\alpha\rangle \in A^i$ iff for some $q \in P, p^i_\eta \leq q$ and $q \Vdash_P$ "$\nu^i_\eta {}^\frown \langle\alpha\rangle \in \underset{\sim}{T}_i$".

(f) if $\rho = \nu^i_\eta {}^\frown \langle\alpha\rangle \in A^i$ then $p^i_\eta \leq p^i_\rho$, and $p^i_\rho \Vdash_P$ "$\rho \in \underset{\sim}{T}_i$" [this actually follows from (b) and (d) and $\eta \trianglelefteq \nu^i_\eta$].

(g) if $\underset{\sim}{Q}$ is Nm$'$ then for every i and $\eta \in A^i$, $\nu^i_\eta = \eta$.

This is easily done, and let $T^0_i = \{\eta \restriction \ell : \ell \leq \ell g(\eta), \eta \in A^i\}$, and let $p^i_\eta (\eta \in T^0_i)$ be $p^i_\nu, \nu \in A_i$, where $\eta \trianglelefteq \nu$, and $(\forall \rho)[\eta \trianglelefteq \rho \triangleleft \nu \rightarrow \rho \notin A^i]$. By the Δ-system theorem on trees from [RuSh:117, Th.4.12, p.76] there is T^1_i satisfying $T^0_i \leq^* T^1_i$, and $q^i_\eta (\eta \in T^1_i)$ such that:

(α) $p^i_\eta \leq q^i_\eta$ hence $p_0 \leq q^i_\eta$ (and $q^i_\eta \in P$).

(β) if η is a splitting point of T^1_i, then $\eta {}^\frown \langle\alpha\rangle, \eta {}^\frown \langle\beta\rangle \in T^1_i$ & $\alpha \neq \beta$ implies $\mathrm{Dom}(q^i_{\eta {}^\frown \langle\alpha\rangle}) \cap \mathrm{Dom}(q^i_{\eta {}^\frown \langle\beta\rangle}) = \mathrm{Dom}(q^i_\eta)$ and $q^i_\eta \leq q^i_{\eta {}^\frown \langle\alpha\rangle}$, $q^i_\eta \leq q^i_{\eta {}^\frown \langle\beta\rangle}$

(γ) if η is not a splitting point of T^1_i, $(\eta \in T^1_i)$ then for the unique α such that $\eta {}^\frown \langle\alpha\rangle \in T^1_i$, we have $q^i_{\eta {}^\frown \langle\alpha\rangle} = q^i_\eta$.

Now by the usual Δ-system theorem there are $i < j < \aleph_4$ such that $T^1_i = T^1_j$ and for every $\eta \in T^1_i, q^i_\eta, q^j_\eta$ are compatible. Let

$$q^* = q^i_{\langle\rangle} \cup q^j_{\langle\rangle} \in P$$

$$\underset{\sim}{T} = \{\eta \in T^1_i : q^i_\eta \in \underset{\sim}{G}_P \ \& \ q^j_\eta \in \underset{\sim}{G}_P\}$$

Clearly $\underset{\sim}{T}$ is a P-name of a subset of ${}^{\omega>}(\omega_2)$, closed under initial segments, $\underset{\sim}{T} \subseteq \underset{\sim}{T}_i, \underset{\sim}{T}_j$, so it suffices to prove

$$q^* \Vdash \text{“}\underset{\sim}{T}_i \leq^* \underset{\sim}{T} \ \& \ \underset{\sim}{T}_j \leq^* \underset{\sim}{T} \ \& \ \underset{\sim}{T} \in \underset{\sim}{Q}\text{”}$$

which is easy. $\square_{4.3}$

4.4 Lemma. Nm satisfies the S-condition for any S such that $\aleph_2 \in S$.

Proof. To show our claim holds we have to describe F and then show that F does its work. At a point η where we use F, F has to determine $\mathrm{Suc}_T(\eta)$, and $f(\eta')$ for any immediate successor η' of η (see 2.6B for the notation). At such a point $f(\eta)$ is already known and is a condition in Nm. Let ν_η be a point of minimal height in $f(\eta)$ such that ν_η has $\aleph_2$ many immediate successors (in $f(\eta)$). Let $\mathrm{Suc}_T(\eta)$ be $\{\eta\char`^\langle\alpha\rangle : \nu_\eta\char`^\langle\alpha\rangle \in f(\eta)\}$ and for each $\eta\char`^\langle\alpha\rangle$ in $\mathrm{Suc}_T(\eta)$ let $f(\eta\char`^\langle\alpha\rangle)$ be the subtree of $f(\eta)$ consisting of members of $f(\eta)$ which are comparable with $\nu_\eta\char`^\langle\alpha\rangle$ (in the tree order of $f(\eta)$). When we want to check that our F does the work; we are given an S-tree T, fronts J_n and a function $f : T \to \mathrm{Nm}$ as above in 2.6 and we are given a subtree T', $T \leq^* T'$. We have to find a condition $r \in \mathrm{Nm}$ so that $r \Vdash$ “there exists an infinite η such that for every $n < \omega, \eta{\restriction}n \in T'$ and $f(\eta{\restriction}n) \in \underset{\sim}{G}$”. We produce r by passing from T' to a subtree $T'' \geq T'$ such that every point in T'' either belongs to some front J_n (and thus fits the demands of F and in particular has $\aleph_2$ many successors) and is a splitting point, or it has exactly one immediate successor. Now r is the tree of all the initial segments of trunks of $f(\eta)$ for some $\eta \in T'' \cap (\bigcup_{n<\omega} J_n)$; that is:

$$r = \{\rho : \exists \eta \in (\bigcup_{n<\omega} J_n) \cap T'' \text{ such that } \rho \trianglelefteq \nu_\eta\}$$

where ν_η is from the definition of $f(\eta)$ according to F. By the construction, if η_1, η_2 are $\triangleleft$-incomparable, then so are $\nu_{\eta_1}, \nu_{\eta_2}$, hence by the definition of Nm, r is a member of Nm. As any $p \in \mathrm{Nm}$ forces that “$\exists \eta \in \lim(p)$ such that for all n the subtree defined by $\eta{\restriction}n$ belongs to $\underset{\sim}{G}$”, it is not hard to see that r is as required. $\square_{4.4}$

4.4A Claim. If I^* is an $\aleph_2$-complete ideal on ω_2 to which every singleton belongs and $I^* \in \mathbb{I}$ then Nm satisfies the $\mathbb{I}$-condition.

Proof. Same, and really follows (see more generally in XV §4). $\square_{4.4A}$

4.5 Lemma. Any ω_1-closed forcing notion satisfies any S-condition.

Proof. This is trivial with no "real" demands on F, when we are in the relevant situation with an S-tree T', $T \leq^* T'$ and $f : T \to P$ we just pick $r \in P$ such that $r \geq \bigcup_{n<\omega} f(\eta{\restriction}n)$ for some $\eta \in \lim T'$, such r exists by the completeness of P and it forces that any smaller condition is a member of $\underset{\sim}{G}_P$, so we are done. $\square_{4.5}$

4.5A Remark. The same is true for strategically $\aleph_1$-closed forcing notions (games of length $\omega + 1$ suffice).

4.6 Lemma. Let W be a stationary subset of $S_0^2 = \{\alpha < \omega_2 : \mathrm{cf}(\alpha) = \omega\}$ and let $P[W] = \{h : h$ is an increasing and continuous function from $\alpha + 1$ into W for some $\alpha < \omega_1\}$ ordered by inclusion, *then* $P[W]$ satisfies the S-condition for any S such that $\aleph_2 \in S$.

Proof. We define the F and then show why it works. Each $F(\eta)$ will determine $\mathrm{Suc}_T(\eta)$ to be $\{\eta\hat{\ }\langle\alpha\rangle : \alpha < \omega_2\}$ and $f(\eta\hat{\ }\langle\alpha\rangle)$ a condition above $f(\eta)$ such that $\mathrm{Max}(\mathrm{Rang}(f(\eta\hat{\ }\langle\alpha\rangle))) > \alpha$ (note that by the definition of $P[W]$ each function h which belongs to $P[W]$ attains its maximum: $\max(\mathrm{Rang}(h)) = h(\max(\mathrm{Dom}(h)))$. Let us denote $\mathrm{Max}(\mathrm{Rang}(f(\eta)))$ by α_η. For proving that F works, assume T is an S-tree, J_n fronts, $f : T \to P[W]$ meets our requirements for F (see 2.6) and $T \leq^* T'$. Let C_1 be a closed unbounded subset of ω_2 such that if $\delta \in C_1$ and $\eta \in T'$ and $\eta \in {}^{\omega>}\delta$ then $\alpha_\eta < \delta$. Let C_2 be a closed unbounded subset of ω_2 such that for $\delta \in C_2, \eta \in T' \cap (\bigcup_{n<\omega} J_n)$ satisfying $\eta \in {}^{\omega>}\delta$ and $\alpha < \delta$ there is always some β such that $\alpha < \beta < \delta$ and $\eta\hat{\ }\langle\beta\rangle \in T'$. Now for some $\eta \in T'$ we pick $\delta \in C_1 \cap C_2 \cap W$ such that $\alpha_\eta < \delta$ and construct an $\triangleleft$-increasing sequence $\langle\eta_n : n < \omega\rangle$ in T' such that $\lim_{n\to\omega}\alpha_{\eta_n} = \delta$ (this is possible as $\mathrm{cf}(\delta) = \omega$ using the definitions of C_1 and

C_2). Let $r = \bigcup_{n<\omega} f(\eta_n) \cup \{\langle \mathrm{Sup}_{n<\omega} \mathrm{Dom}(f(\eta_n)),\ \delta\ \rangle\}$, clearly $r \in P[W]$ and forces each $f(\eta_n)$ to belong to the generic G so we are done. $\square_{4.6}$

4.6A Lemma. Let $\bar{W} = \langle W_i : i < \omega_1 \rangle$ be a sequence of stationary subsets of $S^2_0 = \{\alpha < \omega_2 : \mathrm{cf}(\alpha) = \omega\}$ and let the forcing notion $P[\bar{W}]$ be defined by

$$P[\bar{W}] \stackrel{\mathrm{def}}{=} \{f : f \text{ is an increasing and continuous function from } \alpha + 1 \text{ into } W_0 \text{ for some } \alpha < \omega_1, \text{and } h \text{ satisfies } \forall i \leq \alpha : h(i) \in W_i\}$$

(ordered by inclusion), *then* $P[\bar{W}]$ satisfies the S-condition for any S such that $\aleph_2 \in S$.

Proof. We define the F as in the previous lemma: Each $F(\eta)$ will determine $\mathrm{Suc}_T(\eta)$ to be $\{\eta\hat{\ }\langle\alpha\rangle : \alpha < \omega_2\}$ and $f(\eta\hat{\ }\langle\alpha\rangle)$ a condition above $f(\eta)$ such that $\mathrm{Max}(\mathrm{Rang}(f(\eta\hat{\ }\langle\alpha\rangle))) > \alpha$. Let us write α_η for $\mathrm{Max}(\mathrm{Dom} f(\eta))$ and δ_η for $\mathrm{Max}(\mathrm{Rang}(f(\eta)))$.

Now assume T is an S-tree, and $f : T \to P[\bar{W}]$ obeys F, and let $T \leq^* T'$. By Lemma 3.5 with $\lambda = \aleph_1$ (not $\lambda = 2^{\aleph_0}$) we can find a subtree T'', $T' \leq^* T''$ and an $\alpha < \omega_1$ such that whenever $\eta_0 \lhd \eta_1 \ldots$ are elements of T'', then $\lim_{n\to\omega} \alpha_{\eta_n} = \alpha$. Now as in the proof of 4.6 let $\delta \in W_\alpha$ be such that

$$(\forall \eta \in {}^{<\omega}\delta \cap T'')(\forall i < \delta)(\exists j < \delta)[\alpha_\eta < \delta,\ \eta\hat{\ }\langle j\rangle \in T'' \text{ and } i < \alpha_{\eta\hat{\ }\langle j\rangle} < \delta].$$

Again we can construct a sequence $\eta_0 \lhd \eta_1 \lhd \ldots$ in T'' such that

$$\lim \mathrm{SupRang}(\eta_n) = \delta.$$

Let $r = \bigcup_{n<\omega} f(\eta_n) \cup \{\langle \alpha, \delta\rangle\}$, then $r \in P[\bar{W}]$ and r forces each $f(\eta_n)$ to belong to the generic G. $\square_{4.6A}$

4.7 Lemma. Suppose

(a) P satisfies the $\mathbb{I}_0$-condition.

(b) For every $I_0 \in \mathbb{I}_0$ there is $I_1 \in \mathbb{I}_1$ such that $\bigcup_{A\in I_1} A \subseteq \bigcup_{A\in I_0} A$ and $(\forall B \in I_0)[B \subseteq \bigcup_{A\in I_1} A \rightarrow B \in I_1]$

Then P *satisfies the* $\mathbb{I}_1$*-condition.*

Proof. Trivial. $\square_{4.7}$

§5. Finite Composition

5.1 Theorem. Let Q_0 satisfies the $\mathbb{I}_0$-condition and let $\underset{\sim}{Q}_1$ be a Q_0-name of a forcing notion such that the weakest condition of Q_0 forces it to satisfy the $\mathbb{I}_1$-condition. Let

(a) μ be the first regular cardinal strictly greater than the cardinality of the domain of each member of $\mathbb{I}_0$

(b) λ be such that $\lambda = \lambda^{<\mu} \geq |Q_0|$

(c) assume $\Vdash_{Q_0}$ "$\mathbb{I}_1$ is λ^+-complete"

(d) let $\mathbb{I}$ be $\mathbb{I}_0 \cup \mathbb{I}_1$

Then $P = Q_0 * \underset{\sim}{Q}_1$ satisfies the $\mathbb{I}$-condition.

Remark. Note that $\mathbb{I}_1 \in V$ (we will not gain much by letting $\mathbb{I}_1 \in V^{Q_0}$.)

Proof. Once again we have to define the function F and then prove it does its work. We will need a combinatorial lemma and its proof will conclude the proof of the theorem. For $f(\eta) \in P$ we denote by $f^0(\eta)$ its Q_0-part and by $f^1(\eta)$ the $\underset{\sim}{Q}_1$-part (it is a Q_0-name of a condition in $\underset{\sim}{Q}_1$), let F^0 be the function exemplifying Q_0 satisfies the $\mathbb{I}_0$-condition and $\underset{\sim}{F}^1$ be the Q_0-name of the function exemplifying $\underset{\sim}{Q}_1$ satisfies the $\mathbb{I}_1$-condition.

We divide the definition of the F to even and odd stages. In even stages i.e., when $|w|$ is even, we will refer to the Q_0 part of P and use F^0. More precisely, let $\langle B, \mathbf{I}_\eta, \langle r_\nu : \nu \in B\rangle\rangle = F^0_{w_1}(\eta, \langle f^0(\eta{\restriction}\ell) : \ell \leq \ell g(\eta)\rangle)$ where $w_1 = \{\ell \in w : |\ell \cap w|$ even$\}$. Now let $\mathrm{Suc}_T(\eta) = B$ and for $\nu \in B$, $f^1(\nu) = f^1(\eta)$, $f^0(\nu) = r_\nu$. In odd stages we essentially do the same for the $\underset{\sim}{Q}_1$-part but we need a little modification; $\underset{\sim}{F}^1$ is just a $\underset{\sim}{Q}_0$-name of a function and we may not even know the domain of the ideal $\mathbf{I}_\eta$ it give (= the $\mathrm{Suc}_T(\eta)$), so first we extend $f^0(\eta)$ to a condition q_0' satisfying $f^0(\eta) \leq q_0' \in Q_0$ and forcing a specific value for $\mathbf{I}_\eta$ (hence for $\mathrm{Suc}_T(\eta)$) as defined by $\underset{\sim}{F}^1$, and then proceed like in the Q_0 part (of course we change each $f^0(\eta)$ there to $f^1(\eta)$ and so on) and we let the Q_0 part $f^0(\eta')$ for each $\eta' \in \mathrm{Suc}_T(\eta)$ be the q_0' we have picked (the $\underset{\sim}{Q}_1$ part will be defined by $\underset{\sim}{F}^1(\eta, \langle f^1(\eta \restriction \ell : \ell < \ell g(\eta))\rangle)$ (if we want to allow F to have just $\mathrm{Suc}_T(\eta) \neq \emptyset \bmod \mathbf{I}_\eta$, act as in the proof 6.2).

Before we can show that this definition works we need a definition and a combinatorial lemma.

5.2 Definition. For a subset A of T we define by induction on the length of η, $\mathrm{res}_T(\eta, A)$ for each $\eta \in T$. Let $\mathrm{res}_T(\langle\rangle, A) = \langle\rangle$. Assume $\mathrm{res}_T(\eta, A)$ is already defined and we define $\mathrm{res}_T(\eta\char`^\langle\alpha\rangle, A)$ for all members $\eta\char`^\langle\alpha\rangle$ of $\mathrm{Suc}_T(\eta)$. If $\eta \in A$ then $\mathrm{res}_T(\eta\char`^\langle\alpha\rangle, A) = \mathrm{res}_T(\eta, A)\char`^\langle\alpha\rangle$ and if $\eta \notin A$ then $\mathrm{res}_T(\eta\char`^\langle\alpha\rangle, A) = \mathrm{res}_T(\eta, A)\char`^\langle 0\rangle$. Thus $\mathrm{res}(T, A) \stackrel{\mathrm{def}}{=} \{\mathrm{res}_T(\eta, A) : \eta \in T\}$ is a tree obtained by projecting, i.e., gluing together all members of $\mathrm{Suc}_T(\eta)$ whenever $\eta \notin A$.

5.3 Lemma. Let λ, μ be uncountable cardinals satisfying $\lambda^{<\mu} = \lambda$ and let $(T, \mathbf{I})$ be a tree in which for each $\eta \in T$ either $|\mathrm{Suc}_T(\eta)| < \mu$ or $\mathbf{I}(\eta)$ is λ^+-complete. *Then* for every function $H : T \to \lambda$ there exist $T', (T, \mathbf{I}) \leq^* (T', \mathbf{I})$ such that (letting $A = \{\eta \in T : |\mathrm{Suc}_T(\eta)| < \mu\}$) for $\eta, \eta' \in T'$: $\mathrm{res}_T(\eta, A) = \mathrm{res}_T(\eta', A)$ implies: $H(\eta) = H(\eta')$ and $\eta \in A$ iff $\eta' \in A$, and if $\eta \in T' \cap A$, then $\mathrm{Suc}_T(\eta) = \mathrm{Suc}_{T'}(\eta)$. (Note that the lemma is also true for $\lambda = \mu = \aleph_0$).

5.4 Continuation of the proof of 5.1. Using the lemma let us prove the theorem. So we are given $(T, \mathbf{I}), f, J_n$ for $n < \omega$ as in 2.6 for our F, and consider $f^0 : T \to Q_0$ as a function to λ, (remember $|Q_0| \leq \lambda$).
We let $A = \{\eta : |\mathrm{Suc}_T(\eta)| < \mu\}$. By the lemma for every $(T', \mathbf{I})$ satisfying $(T, \mathbf{I}) \leq^* (T', \mathbf{I})$ there is a subtree $T'', (T', \mathbf{I}) \leq^* (T'', \mathbf{I})$ such that for every $\eta, \eta' \in T''$ we have: $f^0(\eta) = f^0(\eta')$, and $\eta \in A$ iff $\eta' \in A$ whenever $\mathrm{res}_T(\eta, A) =$

$\mathrm{res}_T(\eta', A)$. Let $T^* = \{\mathrm{res}_T(\eta, A) : \eta \in T''\}$, it is an $\mathbb{I}_0$-tree (since the "even" fronts of the original tree now become splitting points) and f^0 induces a function $\hat{f}^0$ from it to Q_0 i.e., $\nu = \mathrm{res}_T(\eta, A)$ implies $\hat{f}^0(\nu) = f^0(\eta)$ (by the conclusion of 5.3, $\hat{f}^0$ is well defined).

By the definition of F for even $|w|$'s and the assumption that F^0 exemplify the $\mathbb{I}_0$-condition we can find an $r_0 \in Q_0$ and a Q_0-name $\underset{\sim}{\eta}$ of a member of $\lim T^*$ such that $r_0 \Vdash_{Q_0}$ "for every $k < \omega$ we have $\hat{f}^0(\underset{\sim}{\eta}\restriction k) \in \underset{\sim}{G}_0$" where $\underset{\sim}{G}_0$ is the Q_0-name for the generic subset of Q_0". Let $\underset{\sim}{T}^+ \stackrel{\text{def}}{=} \{\rho \in T'' : \mathrm{res}(\rho, A) = \mathrm{res}(\underset{\sim}{\eta}, A)\}$; this is a Q_0-name of an $\mathbb{I}_1$-tree and by the definition of F in the odd stages (i.e. F_w when $|w|$ is odd) there are a Q_0-name $\underset{\sim}{r}_1$ of a member of Q_1 and a $Q_0 * \underset{\sim}{Q}_1$-name $\underset{\sim}{\nu}$ of an ω-branch of $\underset{\sim}{T}^+$ such that $r_0 \Vdash_{Q_0} [\underset{\sim}{r}_1 \Vdash_{Q_1}$ " $\underset{\sim}{\nu} \in \lim \underset{\sim}{T}^+$ is such that $f^1(\underset{\sim}{\nu}\restriction k) \in \underset{\sim}{G}_1$ for every $k < \omega$"] where $\underset{\sim}{G}_1$ is the name of the generic set for $\underset{\sim}{Q}_1$ and $\underset{\sim}{\nu}$ is forced to be a name in $\underset{\sim}{Q}_1$ of a member of $\lim \underset{\sim}{T}^+$. The condition in $P = Q_0 * \underset{\sim}{Q}_1$ which witnesses that the $\mathbb{I}$-condition holds is of course $\langle r_0, \underset{\sim}{r}_1\rangle$, since $\langle r_0, \underset{\sim}{r}_1\rangle \Vdash$ "$\underset{\sim}{\nu} \in \lim T$, and for all $k \in \omega$, $\underset{\sim}{f}^0(\underset{\sim}{\nu}\restriction k) = \hat{\underset{\sim}{f}}^0(\mathrm{res}(\nu\restriction k), A) \in \underset{\sim}{G}_0$, and $f(\nu\restriction k) \in G$. $\square_{5.1}$

We now pay our debt and prove Lemma 5.3; the proof is in the spirit of the proofs of the previous combinatorial Lemmas 3.3 and 3.5.

5.5. *Proof of Lemma 5.3* Without loss of generality $\eta\char`^\langle\alpha\rangle \in T \Rightarrow \alpha < |\mathrm{Suc}_T(\eta)|$. We will prove the lemma by induction on μ. We start with a successor μ, in such cases there is a cardinal κ such that $\mu = \kappa^+$ and for each $\eta \in T$ we have $res_T(\eta, A) \in {}^{\omega>}\kappa$. Let $\{\langle h_\alpha, g_\alpha\rangle : \alpha < \lambda\}$ be a list of all the pairs (h, g) such that g is a function from ${}^{\omega>}\kappa$ to $\{0, 1\}$ and h is in a function from ${}^{\omega>}\kappa$ to λ (by the assumption $\lambda^{<\mu} = \lambda$, hence there are at most λ many such pairs). For each $\alpha < \lambda$ we define a game ∂_α just like in the proof of 3.7, except that now player I wins if for the $\eta \in \lim T$ that they constructed along the play we have: $\eta\restriction k \in A$ iff $g_\alpha(\mathrm{res}_T(\eta\restriction k, A)) = 0$ and $H(\eta\restriction k) = h_\alpha(\mathrm{res}_T(\eta\restriction k, A))$ for every $k < \omega$.

If for every $\alpha < \lambda$ player II had a winning strategy we could build a subtree T^* by induction on the height of $\eta \in T$ taking into T^* all members of $\mathrm{Suc}_T(\eta)$

when $\eta \in A$, and otherwise picking as the only member of $\mathrm{Suc}_{T^*}(\eta)$ an element of $\mathrm{Suc}_T(\eta)$ that is not in any of the A_α's that are defined for player II at that stage by his winning strategy for ∂_α (this is possible as we assume that $\eta \notin A$ implies I_η is λ^+-complete).

The map $\eta \mapsto \mathrm{res}(\eta, A)$ is 1-1 on T^*, so there is a pair $\langle h_{\alpha_0}, g_{\alpha_0}\rangle$ in our list such that for each $\eta \in T^*$ we have $H(\eta) = h_{\alpha_0}(\mathrm{res}_T(\eta, A))$ and $g_{\alpha_0}(\eta) = 0$ iff $\eta \in A$. Now we define a play in the game ∂_{α_0}: player I plays choosing only members of T^* while II plays according to his winning strategy for ∂_{α_0}, but in such a play, player I surely wins and we get the desired contradiction.

So there exists some $\beta < \lambda$ for which player II has no winning strategy in ∂_β, but the game is determined hence player I has a winning strategy for ∂_β. Now let T' be the tree of all sequences η that can appear in a play where player I used this strategy. T' satisfies the requirements (similar to 3.5). This finishes the case where μ is a successor.

If μ is singular, then $\lambda \leq \lambda^{<\mu^+} = \lambda^\mu \leq (\lambda^{<\mu})^{\mathrm{cf}\mu} = \lambda^{\mathrm{cf}(\mu)} \leq \lambda^{<\mu} = \lambda$, so we can without loss of generality replace μ by μ^+. If μ is a regular limit cardinal (or just $\aleph_0 < \mathrm{cf}(\mu) < \mu$), then we first use Lemma 3.7 to find $T', (T, \mathbf{I}) \leq^* (T', \mathbf{I})$, and $\mu' < \mu$ such that for every $\eta \in T'$: $\mathbf{I}_\eta$ is λ^+-complete or $|\mathrm{Suc}_T(\eta)| \leq \mu'$, and then use the induction hypothesis on μ'. $\square_{5.3}$

5.6 Corollary to 5.3. Assume that $1 = \lambda_0 < \mu_0 < \lambda_1 < \mu_1 < \ldots$ are cardinals satisfying $\lambda_{k+1}^{<\mu_k} = \lambda_{k+1}$ for all k. Let $(T, \mathbf{I})$ be a tagged tree, and assume $T = \bigcup_k A_k$ where for all $\eta \in A_k$:

$$|\bigcup \mathbf{I}_\eta| < \mu_k \text{ and } I_\eta \text{ is } \lambda_k^+ - \text{complete.}$$

Let $f_k : T \to \lambda_k$, for $k < \omega$.

Then there is a tree T^* such that $(T, \mathbf{I}) \leq^* (T^*, \mathbf{I})$ and for all k and all η, ν in T^*:

$$(*) \qquad \text{if } \mathrm{res}(\eta, \bigcup_{i<k} A_i) = \mathrm{res}(\nu, \bigcup_{i<k} A_i), \text{ then} f_k(\eta) = f_k(\nu)$$

Proof. As $\lambda_0 = 1$ clearly $T_0 = T$ will satisfy the condition for $k = 0$. We apply Lemma 5.3 to T_0 (with $\lambda = \lambda_1$, $\mu = \mu_0$, $f = f_1$, $A = A_0$) to get a subtree T_1 satisfying the condition also for $k = 1$. We continue by induction. In the k-th step, we apply Lemma 5.3 to T_k with $\lambda = \lambda_{k+1}$, $\mu = \mu_k$, $f = f_{k+1}$, $A = \bigcup_{\ell<k+1} A_\ell$.

Finally, let $T^* = \bigcap_k T_k$. Clearly T^* satisfies $(*)$. Note that $\langle\rangle \in T^*$, and if $\eta \in T^* \cap A_k$, then by the conclusion of lemma 5.3,

$$\mathrm{Suc}_{T_k}(\eta) = \mathrm{Suc}_{T_{k+1}}(\eta) = \dots$$

so $\mathrm{Suc}_{T^*}(\eta) = \mathrm{Suc}_{T_k}(\eta)$. Hence T^* is a tree and $(T_0, \mathbf{I}) \leq^* (T^*, \mathbf{I})$. $\square_{5.6}$

§6. Preservation of the $\mathbb{I}$-Condition by Iteration

6.1 Definition. We say that $\bar{Q} = \langle P_i, \underset{\sim}{Q}_i : i < \alpha\rangle$ is *suitable* for $\langle \mathbb{I}_{i,j}, \lambda_{i,j}, \mu_{i,j} : \langle i,j\rangle \in W^*\rangle$ provided that the following hold:

(0) $W^* \subseteq \{\langle i,j\rangle : i < j \leq \alpha, i \text{ is not strongly inaccessible } \}$ and $\{\langle i+1, j\rangle : i+1 < j < \bigcup_{\beta\leq\alpha} \beta+1\} \subseteq W^*$ (we can use some variants, but there is no need)

(1) $\bar{Q}$ is a RCS iteration.

(2) $P_{i,j} = P_j/P_i$ satisfies the $\mathbb{I}_{i,j}$-condition for $\langle i,j\rangle \in W^*$.

(3) for every $I \in \mathbb{I}_{i,j}$ the set $\bigcup I$ is a uncountable cardinal, I is $\lambda^+_{i,j}$-complete, $\lambda_{i,j} < \bigcup I < \mu_{i,j}$, $\mu_{i,j}$ regular, and $|P_i| \leq \lambda_{i,j}$, and $\lambda^+_{i,j} \geq \aleph_2$ (note that $\mathbb{I}_{i,j}$ is from V and not V^{P_i}, and $i \leq \lambda_{i,j} < \mu_{i,j}$).

(4) if $i(0) < i(1) < i(2) \leq \alpha$, $\langle i(0), i(1)\rangle \in W^*, \langle i(1), i(2)\rangle \in W^*$ then $\lambda_{i(1),i(2)}{}^{<\mu_{i(0),i(1)}} = \lambda_{i(1),i(2)}$.

(5) for every $I \in \mathbb{I}_{i(2),i(3)}$ and $i(0) < i(1) \leq i(2) < i(3), I$ is $\lambda^+_{i(0),i(1)}$-complete.

6.2 Lemma. If $\bar{Q} = \langle P_n, \underset{\sim}{Q}_n : n < \omega\rangle$ is *suitable* for $\langle \mathbb{I}_{i,j}, \lambda_{i,j}, \mu_{i,j} : i < j < \omega\rangle$, and $\mathbb{I} = \bigcup_{n<\omega} \mathbb{I}_{n,n+1}$ *then* $P_\omega = \mathrm{Rlim}\bar{Q}$ satisfies the $\mathbb{I}$-condition.

Proof. Let $\mathbb{I}_i = \mathbb{I}_{i,i+1}$, $\lambda_i = \lambda_{i,i+1}$, $\mu_i = \mu_{i,i+1}$, note that $P_{i,i+1} = \underset{\sim}{Q}_i$, so $\underset{\sim}{Q}_i$ satisfies the $\mathbb{I}_i$-condition, $|P_i| \leq \lambda_i, \mu_i \leq \lambda_{i+1} = \lambda_{i+1}{}^{<\mu_i} < \mu_{i+1}$.

For each $i < \omega$, let $\underset{\sim}{F}_i$ be a P_i-name of a function witnessing that $\underset{\sim}{Q}_i$ satisfies the $\mathbb{I}_i$-condition. We will act as in the proof of Theorem 5.1, but now we have countably many $\underset{\sim}{F}_i$'s rather than two. We can a priori partition the tasks, so let $\omega = \bigcup_{i<\omega} B_i$, the B_i pairwise disjoint, each B_i infinite.

Now we shall define the function F which exemplifies "P_ω satisfies the $\mathbb{I}$-condition". So we have to define $F(\eta, w, f\restriction\{\nu : \nu \trianglelefteq \eta\})$, (see Definition 2.6). Let i be the unique $i < \omega$ such that $|w| \in A_i$, let $w^* = \{\ell \in w : |\ell \cap w| \in A_i\}$, and let $\langle \underset{\sim}{B}, \underset{\sim}{I}, \langle \underset{\sim}{r}_\nu : \nu \in \underset{\sim}{B}\rangle\rangle = \underset{\sim}{F}_i(\eta, w^*, \langle f(\eta\restriction\ell)(i) : \ell \leq \ell g(\eta)\rangle)$, (so $\Vdash_{P_i}$ "$\underset{\sim}{B} = \mathrm{Dom}(\underset{\sim}{I})$").

We choose $q_\eta \in P_i$, such that $(f(\eta)\restriction i) \leq q_\eta$ and for some λ_η, $q_\eta \Vdash_{P_i}$ "$|\underset{\sim}{B}| = \underset{\sim}{B} = \lambda_\eta$" and $q_\eta \Vdash_{P_i}$ "$\underset{\sim}{I}$ is $\mathbf{I}_\eta$ which belongs to $\mathbb{I}$, in fact to $\mathbb{I}_i = \mathbb{I}_{i,i+1}$ (by the natural isomorphism $\underset{\sim}{f}_\eta$)", (see 2.6B). Let $p_\eta = f(\eta)\restriction(i+1, \omega)$. We choose $Suc_T(\eta) = \{\eta\hat{\ }\langle\alpha\rangle : \alpha < \bigcup \mathbf{I}_\eta\}$, and define: $F(\eta, w, \langle f(\nu\restriction\ell) : \ell \leq \ell g(\eta)\rangle) = \langle \lambda_\eta, \mathbf{I}_\eta, \langle \underset{\sim}{r}_\nu \cup p_\eta \cup q_\eta : \nu \in Suc(\eta)\rangle\rangle$, [really we should replace $\underset{\sim}{r}_\nu$ by the function $\{\langle i, \underset{\sim}{r}_\nu\rangle\}$, and λ_η by $\{\eta\hat{\ }\langle i\rangle : i < \lambda_\eta\}$ but we shall ignore such problems].

We now have to prove that P_ω, $\mathbb{I}$, and F satisfy Definition 2.6. So let $(T, \mathbf{I})$, $J_k (k < \omega)$ and $f : T \to P$ be as in Definition 2.6 and $(T, \mathbf{I}) \leq^* (T^0, \mathbf{I})$ and we have to find a $p \in P$, such that $p \Vdash$ "$\exists \eta \in \lim T^0$ such that $(\forall k < \omega) f(\eta\restriction k) \in \underset{\sim}{G}_{P_\omega}$".

First define $f_k : T^0 \to P_k$ by $f_k(\eta) = f(\eta)\restriction k$. Let $A_k = \{\eta \in T^0 : \mathbf{I}_\eta \in \mathbb{I}_k\}$. By 5.6 we can find a tree T^*, $(T^0, \mathbf{I}) \leq^* (T^*, \mathbf{I})$, such that whenever $\eta, \nu \in T^*$ and $\mathrm{res}(\eta, \bigcup_{i<k} A_i) = \mathrm{res}(\nu, \bigcup_{i<k} A_i)$, then $f_k(\eta) = f_k(\nu)$. Let $T_k^* = \mathrm{res}(T^*, \bigcup_{i<k+1} A_i)$. Define $\underset{\sim}{f}_k^* : T_k^* \to \underset{\sim}{Q}_k^*$ by $f_k^*(\mathrm{res}(\eta, \bigcup_{i<k+1} A_i)) = f_k(\eta)(k)$.

By induction on $i = 0, 1, 2\ldots$ we can now define P_{i+1}-names $\underset{\sim}{\eta}_i$ and conditions $p(i) \in \underset{\sim}{Q}_i$ such that $\langle p(0), \ldots, p(i)\rangle \Vdash_{P_{i+1}}$ "$\underset{\sim}{\eta}_i \in \lim T_i^*$ and $(\forall \ell < \omega)\ f_i^*(\underset{\sim}{\eta}_i\restriction\ell) \in \underset{\sim}{G}_{Q_i}$" and for all $i < j$,

$$\Vdash_{P_{j+1}} \text{"}(\forall \ell)\ [\underset{\sim}{\eta}_i\restriction\ell = \mathrm{res}(\underset{\sim}{\eta}_j\restriction\ell, \bigcup_{k<i+1} A_k)]\text{"}.$$

Finally we can find a P_ω-name $\underset{\sim}{\eta}$ such that for all $\ell \Vdash_{P_\omega}$ "for all large enough i $\underset{\sim}{\eta}\restriction\ell = \underset{\sim}{\eta}_i\restriction\ell$". It is now clear that $\Vdash_{P_\omega}$ "$\underset{\sim}{\eta} \in \lim T^*$, and $\forall \ell\ f(\eta\restriction\ell) \in G_{P_\omega}$". $\square_{6.2}$

6.2A Remark. Note that here as well in the next theorem we need that the $\mathbb{I}_{i,j}$'s are well separated (compare 6.1(4), (5)) i.e. some have small underlying sets, others have large completeness coefficients (e.g. in the previous theorem we required that ideals $\mathbb{I}_i$ are on sets $\subseteq \mu_i$, and ideals in $\mathbb{I}_{i+1}$ had to be λ_{i+1}-complete, $\lambda_{i+1} \geq \mu_i$). To satisfy these requirements we will in applications only work with iterations in which in every odd step some large enough cardinal is collapsed, see 1.1(ii)$'$.

6.3 Lemma.

(1) If $\bar{Q} = \langle P_\alpha, \underset{\sim}{Q}_\alpha : \alpha < \omega_1 \rangle$ is suitable for $\langle \mathbb{I}_{\alpha,\beta}, \lambda_{\alpha,\beta}, \mu_{\alpha,\beta} : \alpha < \beta < \omega_1, \alpha$ non-limit$\rangle$, and $\mathbb{I} = \bigcup \{\mathbb{I}_{\alpha,\beta} : \alpha < \beta < \omega_1$ and α non-limit$\}$ *then* $P = P_{\omega_1} = \lim \bar{Q}$ satisfies the $\mathbb{I}$-condition.

(2) We can replace ω_1 by any δ such that $\aleph_0 < \mathrm{cf}^V \delta < \mathrm{Min}\{\lambda_{\alpha,\beta} : \langle \alpha, \beta \rangle \in W^*\}$.

Proof. 1) We will first prove this assuming CH (which is enough for all applications in this chapter), and then indicate how we can get rid of this extra assumption. The proof consists of two parts: In part A we define the function F, and in part B we show that it satisfies the requirements from definition 2.6.

Part A: To each $p \in P$ we have associated a countable set $\{\underset{\sim}{\zeta}^k(p) : k < \omega\}$ of prompt names, such that letting $\underset{\sim}{\zeta}^*(p) = \sup\{\underset{\sim}{\zeta}^k(p) : k < \omega\}$, we have $p \in P_{\underset{\sim}{\zeta}^*}$ (see 1.12(4)). Let $\bigcup_{i<\omega} B_i$ be the set of odd natural numbers > 2, the B_i infinite pairwise disjoint, be such that $(\forall \ell \in B_i)(i + 1 < \ell)$.

Let $\underset{\sim}{F}_{\alpha,\beta}$ be P_α-name of a function exemplifying "$P_{\alpha,\beta}$ satisfies the $\mathbb{I}_{\alpha,\beta}$-condition."

Let us explain our strategy; we cannot deal with all pairs (α, β) along a branch as the branch is countable, and α, β range over an uncountable set. So along each branch η we try to determine the $\bar{Q}$-named ordinals, $\underset{\sim}{\zeta}^m(f(\eta\restriction\ell))$, so we get a potential bound α^* to larger and larger parts of each $f(\eta\restriction\ell)$ and we shall use the functions $F_{\alpha^*_n,\alpha^*_{n+1}}$, where $\alpha^* = \bigcup \alpha^*_n$.

We shall define now the function F which exemplifies "P satisfies the $\mathbb{I}$-condition," so we have to define $F(\eta, w, \langle f(\eta\restriction\ell) : \ell \leq \ell g(\eta) \rangle)$. If $|w| \notin \bigcup_i B_i$

define it as any $\langle I, \langle f(\eta\hat{\ }\langle\xi\rangle) : \xi \in B\rangle\rangle$ such that $f(\eta\hat{\ }\langle\xi\rangle)$ forces a value to $\underset{\sim}{\zeta}^m(f(\eta\restriction\ell))$ for $m, \ell \leq \ell g(\eta)$.

Now let $|w| \in \bigcup_i B_i$.

Naturally we shall use one of the $\underset{\sim}{F}_{\alpha,\beta}$, but we have to determine which one. By the way we are defining F, we can assume that for $k < \ell g(\eta)$, $f(\eta\restriction(k+1))$ determines (i.e. forces a value to) $\underset{\sim}{\zeta}^m(f(\eta\restriction\ell))$ for $\ell, m \leq k$, so we can define the following:

Let $\alpha_0(\eta) = 0$, and for $0 < k < \ell g(\eta)$ let

$$\alpha_k(\eta) = \mathrm{Max}\{\underset{\sim}{\zeta}^m(f(\eta\restriction\ell)) + k : \ell, m \leq k\}$$

Note that for any finite or infinite sequence ν: if $\eta \triangleleft \nu$, $k < \ell g(\eta)$, then $\alpha_k(\eta) = \alpha_k(\nu)$.

Let i be such that $|w| \in B_i$. Then $i + 1 < |w| \leq \ell g(\eta)$, so

$$(*) \qquad \alpha \overset{\text{def}}{=} \alpha_i(\eta) \qquad \beta \overset{\text{def}}{=} \alpha_{i+1}(\eta)$$

are well defined.

Let $w^*_\eta = \{k \in w : |w \cap k| \in B_i\}$, and let

$$\langle \underset{\sim}{B}, \underset{\sim}{\mathbf{I}}, \langle \underset{\sim}{r}_\nu : \nu \in \underset{\sim}{B}\rangle\rangle = F_{\alpha,\beta}\left(\eta, w^*_\eta, \langle f(\eta\restriction\ell)\restriction[\alpha,\beta) : \ell \in w^*_\eta \cup \{\ell g(\eta)\}\rangle\right)$$

(recall α and β should have subscripts η and w, which we suppress for notational simplicity).

Now choose $q_\eta \geq f(\eta)\restriction\alpha$ such that $q_\eta \in P_\alpha$ and such that $q_\eta \Vdash_{P_\alpha}$ "$\underset{\sim}{\mathbf{I}}$ is isomorphic to $\mathbf{I}_\eta$, $\mathbf{I}_\eta \in \mathbb{I}_{\alpha,\beta}$", and let $F(\eta, w, \langle f(\nu\restriction\ell) : \ell \leq \ell g(\eta)\rangle) = \langle \mathbf{I}_\eta, \langle q_\eta \bigcup \underset{\sim}{r}_\nu : \nu = \eta\hat{\ }\langle\alpha\rangle \text{ and } \alpha < \mathrm{Dom}(\mathbf{I}_\eta)\rangle\rangle$.

Part B. Now we have to prove that $P, \mathbb{I}$ and F satisfy Definition 2.6. So let $(T, \mathbf{I})$, J_k $(k < \omega)$ and f be as in Definition 2.6 for the F chosen above, and $(T, \mathbf{I}) \leq^* (T^\dagger, \mathbf{I})$ and we have to find the required p. To each branch η of $T^\dagger$ we have associated a sequence $\langle \alpha_k(\eta) : k < \omega\rangle$ of countable ordinals. Since we assume CH we also have $\aleph_1^{\aleph_0} = \aleph_1$, so by Lemma 3.5A we can find T'' such that $(T^\dagger, \mathbf{I}) \leq^* (T'', \mathbf{I})$, and for some fixed sequence $\langle \alpha^*(k) : k < \omega \rangle$ we have $\alpha^*(k) = \alpha_k(\eta)$ for all $\eta \in \lim(T'')$.

Now we continue as in the proof of 6.2. We let $A_\ell = \{\eta : I_\eta \in \mathbb{I}_{\alpha^*(\ell),\alpha^*(\ell+1)}\}$ and $f_\ell(\eta) = f(\eta){\restriction}\alpha^*(\ell)$. By 5.6 we can find a tree T^*, $(T'', \mathbf{I}) \leq^* (T^*, \mathbf{I})$ such that for all η, ν in T^*:

$$\text{If } \mathrm{res}(\eta, \textstyle\bigcup_{\ell<k} A_\ell) = \mathrm{res}(\nu, \bigcup_{\ell<k} A_\ell), \text{ then } f(\eta){\restriction}\alpha^*(k) = f(\nu){\restriction}\alpha^*(k).$$

We let T^*_k be $\{\mathrm{res}_{T^*}(\eta, \bigcup_{\ell\leq k} A_\ell) : \eta \in T^*\}$ and $f^*_k : T^*_k \to P_{\alpha^*(k),\alpha^*(k+1)}$ is defined by

$$f^*_k(\mathrm{res}_{T^*}(\eta, \bigcup_{\ell\leq k} A_\ell)) = p \text{ iff } f(\eta){\restriction}[\alpha^*(k), \alpha^*(k+1)) = p$$

Now note that $\langle J_n : n \in B_0 \rangle$ is a system of fronts as in 2.6, and at each $\eta \in J_n$, if $n \in B_0$, then the function $F(\eta, \{k : \eta{\restriction}k \in \bigcup_m J_m\}, \langle f(\nu) : \nu \trianglelefteq \eta \rangle)$ used the function $\underset{\sim}{F}_{\alpha^*(0),\alpha^*(1)}(\eta, \{k : \eta{\restriction}k \in \bigcup_{m\in B_0} J_m\}, \langle f(\nu){\restriction}[\alpha^*(0), \alpha^*(1))\rangle)$ (but $\underset{\sim}{F}_{\alpha^*(0),\alpha^*(1)}$ is $F_{\alpha^*(0),\alpha^*(1)}$ as $\alpha^*(0) = 0$), so we can find $p_1 \in P_{\alpha^*(1)}$, such that for some $P_{\alpha^*(1)}$-name $\underset{\sim}{\eta}_0$

$$p_1 \Vdash_{P_{\alpha^*(1)}} \text{“}\underset{\sim}{\eta}_0 \in \lim T^*_0 \text{ and } (\forall \ell < \omega)\ f^*_0(\underset{\sim}{\eta}_0{\restriction}\ell) \in \underset{\sim}{G}_{P_{\alpha^*(1)}}\text{”}.$$

Continuing by induction, we define $p_n \in P_{\alpha^*(n+1)}$ satisfying $p_n{\restriction}\alpha^*(m + 1) = p_m$ for $m < n$, such that for some $P_{\alpha^*(n+1)}$-name $\underset{\sim}{\eta}_n$

$p_n \Vdash_{P_{\alpha^*(n+1)}}$ “$\underset{\sim}{\eta}_n \in \lim T^*_n$ and $f^*_n(\underset{\sim}{\eta}_n{\restriction}\ell) \in \underset{\sim}{G}_{P_{\alpha^*(n+1)}}$ for every $\ell < \omega$ and for $m < n, \ell < \omega$, $\eta_m{\restriction}\ell = \mathrm{res}_{T^*}\left(\underset{\sim}{\eta}_n{\restriction}\ell, \bigcup_{\ell\leq m} A_\ell\right)$”.

So $p = \bigcup_{n<\omega} p_n \in P_{\alpha^*}$, where $\alpha^* = \bigcup_n \alpha^*(n)$, and there is a P_{α^*}-name $\underset{\sim}{\eta}$ such that $p \Vdash_{P_{\alpha^*}}$ “$\underset{\sim}{\eta} \in \lim T^* \subseteq \lim T''$, and for every $n, m < \omega$, $\mathrm{res}_{T^*}(\underset{\sim}{\eta}{\restriction}n, \bigcup_{\ell\leq n} A_\ell) = \underset{\sim}{\eta}_m{\restriction}m$” (this determines $\underset{\sim}{\eta}$ uniquely as $T^* \subseteq T''$). So $p \Vdash_{P_{\alpha^*}}$ “for every $m, \ell < \omega$, $f(\underset{\sim}{\eta}{\restriction}\ell){\restriction}\alpha^*(m) \in \underset{\sim}{G}_{P_{\alpha^*}}$” hence, by the definition of RCS, as α^* is limit: $p \Vdash_{P_{\alpha^*}}$ “for every $\ell < \omega$, $f(\underset{\sim}{\eta}{\restriction}\ell){\restriction}\alpha^* \in \underset{\sim}{G}_{P_{\alpha^*}}$”.

We have here a problem: A priori, $f(\underset{\sim}{\eta}{\restriction}\ell)$ is not necessarily in P_{α^*}, (only in P_{ω_1}) so $f(\underset{\sim}{\eta}{\restriction}\ell){\restriction}\alpha^* \in \underset{\sim}{G}_{P_{\alpha^*}}$ seems to be weaker than the required “$f(\underset{\sim}{\eta}{\restriction}\ell) \in$

$\underset{\sim}{G}_{P_{\omega_1}}$". However, we have $f\left(\underset{\sim}{\eta}\restriction(m+\ell+1)\right)\restriction\gamma \Vdash_{P_\gamma}$ "$\underset{\sim}{\zeta}^m(f(\underset{\sim}{\eta}\restriction\ell)) = \gamma$" for some γ, so $\gamma < \alpha_{m+\ell+2}(\eta) = \alpha^*(m+\ell+2) < \alpha^*$

Since also $p \Vdash_{P_{\alpha^*}}$ "$f(\underset{\sim}{\eta}\restriction(m+\ell+2))\restriction\gamma \in \underset{\sim}{G}_{P_{\alpha^*}}$", we conclude that

$$p \Vdash \text{"}\underset{\sim}{\zeta}^m(f(\underset{\sim}{\eta}\restriction\ell)) < \alpha^*$$

for every $\ell, m < \omega$.

So p essentially forces $\forall\ell\, f(\underset{\sim}{\eta}\restriction\ell) \in P_{\alpha^*}$. Hence clearly $p \Vdash_{P_{\omega_1}}$ "$f(\underset{\sim}{\eta}\restriction\ell) \in \underset{\sim}{G}_{P_{\omega_1}}$".

If we do not have CH, we modify the proof as follows: let $g : \omega \to \omega$ be such that $g(m) \le m$ and $(\forall n)(\exists^\infty m)[g(m) = n]$. Next, for each $\alpha < \omega_1$ let $\langle \rho^\alpha_\ell : \ell < \omega\rangle$ list all finite sequences of the form $\langle (i_k, \beta_k) : k \le k^*\rangle$, $\beta_0 = 0$, $\beta_k < \beta_{k+1} \le \alpha$, $i_k < i_{k+1} < \omega$ such that if $\rho^\alpha_{\ell_1} \lhd \rho^\alpha_{\ell_2}$ then $\ell_1 < \ell_2$. Let $\rho^\alpha_\ell = \langle (i_k(\alpha,\ell), \beta_k(\alpha,\ell)) : k \le k^*(\alpha,\ell)\rangle$, second, we write the odd natural numbers > 2 as a doubly indexed union $\bigcup_{i,m} B_{i,m}$ of infinite disjoint sets (instead of $\bigcup_i B_i$). Then, instead of $(*)$, we define $\rho[w,\eta]$ as $\rho^{\alpha_i(\eta)}_\ell$ when $|w| \in B_{i,\ell}$ where $\alpha_i(\eta)$ was defined in Part A and so define $i = i_w$, $\ell = \ell_w$. Next, we define by induction on $|w|$ when (w,η) is nice: it is nice when $k^*(\alpha_i(\eta),\ell) = 0$ or for $k < k^*(\alpha_i(\eta),\ell)$ we have $i_k(\alpha_i(\eta),\ell) \in w$, and $\rho[w \cap i_k(\alpha,\ell), \eta\restriction i_k(\alpha,\ell)] = \rho^{\alpha_i(\eta)}_\ell\restriction(k+1)$.

Now if (w,η) is not nice we do nothing, if it is nice, we let $k = k[w,\eta]$ be the $k = g(k^*(\alpha_i(\eta),\ell))$. We let

$$\alpha = \alpha[w,\eta] = \beta_{i_k(\alpha_i(\eta),\ell)}(\alpha_i(\eta),\ell)$$

$$\beta = \beta[w,\eta] = \beta_{i_{k+1}[w,\eta]}(\alpha_i(\eta),\ell)$$

$w^* = w^*[w,\eta] = \{i : (\alpha[w\cap i, \eta\restriction i], \beta[w\cap i, \eta\restriction i]) = (\alpha,\beta)$ and for some $m < k^*$, $i = i_m(\alpha_i(\eta),\ell)\}$. Then we define the function F as before.

In part B, when we check that this construction works, we can only find a tree T'' with the property that for some α^*, for all branches η in T'', $\lim_{k\to\omega}\alpha_k(\eta) = \alpha^*$(using 3.5). Let $\langle \alpha^*(k) : k < \omega\rangle$ be a sequence of ordinals converging to α^*. Now we can shrink T'', so as to use only $F_{\alpha^*(n),\alpha^*(n+1)}$ $(n < \omega)$, i.e. let us define by induction on n (stipulating $J'_{-1} = \{\langle\rangle\}$ $J'_n =$

$\{\eta : \eta \in \bigcup_{n<\omega} J_m$, and letting $w = \{\ell < \ell g(\eta) : \eta{\restriction}\ell \in \bigcup_{m<\omega} J_m\}$ for some k^* and $i_0 < \dots i_{k^*-1}$ from w, $\bigwedge_{\ell<k^*} i_\ell \in J'_{\ell-1}$ and $\rho[w,\eta] = \langle (i_\ell, \alpha^*(\ell)) : \ell < k^*\rangle\char`\^\langle(\ell g\eta, \alpha^*(k^*))\rangle\}$ (it is a system of fronts, i.e., every branch of T'' meets each J_n infinitely often) and let T^0, $(T'', \mathsf{I}) \le (T^0, \mathsf{I})$ be such that:

if $\eta \in T^0, \eta \in \bigcup_n J_n$ then $(\forall\alpha)(\eta\char`\^\langle\alpha\rangle \in T'' \Rightarrow \eta\char`\^\langle\alpha\rangle \in T^0)$

if $\eta \in T^0, \eta \notin \bigcup_n J_n$ then $(\exists!\alpha)(\eta\char`\^\langle\alpha\rangle \in T^0)$.

Now we continue as before.

(2) Left to the reader (essentially the same proof). $\square_{6.3}$

6.3A Corollary. (1) For $P = \text{Rlim}\,\bar{Q}$ as in the previous lemma, $\bigcup_{\alpha<\omega_1} P_\alpha$ is (essentially) a dense subset of P i.e. for every $p \in P_{\omega_1}$ there are q and α such that $p{\restriction}\alpha \le q \in P_\alpha$, $q \Vdash$ "$p \in \underset{\sim}{G}_{P_{\omega_1}}$" (in fact $r \in p \Rightarrow q \Vdash$ "$\underset{\sim}{\zeta}(r) < \alpha$").

(2) For $\bar{Q} = \langle P_i, \underset{\sim}{Q}_i : i < \delta\rangle$ as in the previous lemma (so $\delta = \omega_1$ or just $\text{cf}(\delta) = \omega_1$), *if* $\underset{\sim}{\bar{\alpha}}$ is a P_δ-name of an ω-sequence of ordinals ($P_\delta = \text{Rlim}\,\bar{Q}$, of course) $p \in P_\delta$ *then* for some $i < \delta, q \in P_i$, and $\underset{\sim}{\bar{\beta}}$ a P_i-name of an ω-sequence of ordinals, $P_\delta \vDash$ "$p \le q$", and $q \Vdash_{P_\delta}$ "$\underset{\sim}{\bar{\alpha}} = \underset{\sim}{\bar{\beta}}$".

Proof: By 1.13 (or directly from the proof of 6.3). $\square_{6.3A}$

6.4 Lemma. Suppose $\bar{Q} = \langle P_\alpha, \underset{\sim}{Q}_\alpha : \alpha < \kappa\rangle$ is suitable for $\langle \mathbb{I}_{\alpha,\beta}, \lambda_{\alpha,\beta}, \mu_{\alpha,\beta} : \langle i,j\rangle \in W^*, \rangle$, κ is strongly inaccessible $|P_i| + \lambda_{i,j} + \mu_{i,j} + |\bigcup I| < \kappa$ for every $\langle\alpha,\beta\rangle \in W^*$, $I \in \mathbb{I}_{\alpha,\beta}$ and let $\mathbb{I} = \bigcup_{\alpha,\beta} \mathbb{I}_{\alpha,\beta}$. *Then* $P_\kappa = \text{Rlim}\,(\bar{Q})$ satisfies the $\mathbb{I}$-condition.

Proof. This is quite easy, because $P_\kappa = \bigcup_{\alpha<\kappa} P_\alpha$. So let $\underset{\sim}{F}_{\alpha,\beta}$ be a P_α-name of a witness to "$P_{\alpha,\beta} \stackrel{\text{def}}{=} P_\beta/P_\alpha$ satisfies the $\mathbb{I}_{\alpha,\beta}$-condition", for $\alpha < \beta < \kappa$, α non-limit, and let $\omega = \bigcup_{\alpha<\omega} A_i$, the A_i's are infinite, pairwise disjoint and $n \in A_{i+1} \Rightarrow n > 1 + i$ (so $\beta(k)$ is an ordinal $< \kappa$, not just a name). Now we shall define the function F, so we should define

$F(\eta, w, \langle f(\eta{\restriction}\ell) : \ell \le \ell g(\eta)\rangle)$ (See Definition 2.6). Let i be such that $|w| \in A_i$, $w^* = \{\ell \in w : |w \cap \ell| \in A_i\}$ and let $\beta(0) = 0$, $\beta(1) = 1$ and for $k > 1$, $k \le \ell g(\eta)$, let $\beta(k) = \text{Min}\{\gamma + k : \ell \le k \Rightarrow f(\eta{\restriction}\ell) \in P_\gamma\}$. Now we shall use $\underset{\sim}{F}_{\beta(i),\beta(i+1)}$, so let $\underset{\sim}{F}_{\beta(i),\beta(i+1)}(\eta, w^*, \langle f(\eta{\restriction}\ell){\restriction}[\beta(i), \beta(i+1) : \ell \leq \ell g(\eta)\rangle)$ be $\langle \underset{\sim}{I}, \langle \underset{\sim}{r}_\nu : \nu \in \bigcup \underset{\sim}{I} \rangle\rangle$, and choose $q^0_\eta \in P_{\beta(i)}$ such that $q^0_\eta \Vdash_{P_{\beta(1)}}$ "$\underset{\sim}{I} = \mathbf{I}_\eta$", and $P_{\beta(i)} \Vdash$"$f(\eta){\restriction}\beta(i) \leq q^0_\eta$". Now we can define $F(\eta, w, \langle f(\eta{\restriction}\ell) : \ell \leq \ell g(\eta)\rangle) = \langle \mathbf{I}_\eta, \langle f(\nu) : \nu \in Suc_T(\eta)\rangle\rangle$ and $Suc_T(\eta) = \{\eta\hat{\ }\langle\alpha\rangle : \alpha < \bigcup \mathbf{I}_\eta\}$, and $f(\nu) = q^0_\eta \cup \underset{\sim}{r}_\nu$. Note that $\beta(k)$ depends on $\eta{\restriction}k$, so we should have written $\beta(\eta{\restriction}k)$, see below.

Now suppose we are given $(T, \mathbf{I}), J_n, f$ as in Definition 2.6. (for $P = \text{Rlim}\,\bar{Q}$ and $\mathbb{I} = \bigcup_{i<j} \mathbb{I}_{i,j}$) and $(T, \mathbf{I}) \leq^* (T', \mathbf{I})$ and we have to find p as required. Let for every $\eta \in T$, $\beta(\eta)$ be 0 if $\ell g(\eta) = 0$, 1 if $\ell g(\eta) = 1$ and $\text{Min}\{\gamma + \ell g(\eta) : f(\eta) \in P_\gamma$ otherwise; so $\nu \triangleleft \eta \Rightarrow \beta(\nu) < \beta(\eta)$ and $\beta(\eta)$ is never a limit ordinal. So by a repeated use of Lemma 5.3 we can get T^*, $(T^\dagger, \mathbf{I}) \leq^* (T^*, \mathbf{I})$ such that:

$(*)$ for every $\eta \in T^*$ and $\eta \triangleleft \nu_\ell \in T^*$, $\text{res}(\nu_1, A_\eta) = \text{res}(\nu_2, A_\eta)$ then $\langle f_{\nu_1{\restriction}\ell}{\restriction}\beta(\eta)\rangle : \ell \leq \ell g\nu_1\rangle = \langle \rho_{\nu_2{\restriction}\ell}(\beta(\eta)) : \ell \leq \ell g\nu_2\rangle$ where $A_\eta = \{\nu \in T : |\text{Suc}_T(\nu)| < \mu_{\theta,\beta(\eta)}\}$

By induction on n we will now define prompt names $\underset{\sim}{\beta}_n$ of ordinals, conditions $p_n \in P_{\underset{\sim}{\beta}_n}$ and $P_{\underset{\sim}{\beta}_n}$-names $\underset{\sim}{\eta}_n$ and $\underset{\sim}{\nu}_n$ such that p_n forces the following

(1) $\underset{\sim}{\eta}_n \in T^*, \ell g(\underset{\sim}{\eta}_n) = n$

(2) $\underset{\sim}{\eta}_n \triangleleft \underset{\sim}{\nu}_n \in \lim(\text{res}(T^*, A_{\eta_n}))$

(3) $\forall \rho \in T^* \forall l < \omega$: if $\nu_n{\restriction}\ell = \text{res}(\rho, A_{\eta_n})$, then $f(\rho){\restriction}\underset{\sim}{\beta}_n \in G_{\underset{\sim}{\beta}_n}$ and $\underset{\sim}{\beta}_n = \beta(\underset{\sim}{\eta}_n)$.

For $n = 0$ there is nothing to do: ($\eta_0 = \langle\rangle$, $\beta_0 = 0$).
In stage $n+1$ we will work in $V[G_{P_{\beta_n}}]$, where $G_{P_{\beta_n}}$ is a generic filter on $P_{\underset{\sim}{\beta}_n}$ containing p_n, more formally, we have $\beta_n < \kappa$ and $\underset{\sim}{G}_{P_{\underset{\sim}{\beta}_n}} \subseteq P_{\beta_n}$ generic over V such that $\Vdash_{P_\kappa/G_{P_{\beta_n}}}$ "$\underset{\sim}{\beta}_n = \beta_n$". We let $\eta_{n+1} = \nu_n{\restriction}n+1$, $\beta_{n+1} = \beta(\eta_{n+1})$. Since we have used $F_{\beta(\eta_n),\beta(\eta_{n+1})}$ we can find a condition $p_{n,n+1}$ in $P_{\beta_n,\beta_{n+1}}$ and a $P_{\beta_n,\beta_{n+1}}$-name $\underset{\sim}{\nu}_{n+1}$ such that

$$\Vdash_{P_{\beta_{n+1}}/P_{\beta_n}} \text{“}\underset{\sim}{\nu}_{n+1}{\restriction}\ell = \text{res}_{T^*}(\rho, A_{\eta_{n+1}}) \Rightarrow f(\rho){\restriction}[\beta_n, \beta_{n+1}) \in \underset{\sim}{G}_{P_{\underset{\sim}{\beta}_n,\underset{\sim}{\beta}_{n+1}}}\text{”}$$

Now we can return to V and translate everything back to $P_{\underset{\sim}{\beta}_n}$-names, and get a condition p_{n+1} from p_n and $p_{n,n+1}$.

Since we are using RCS iteration see (1.12)(1), letting $\beta^* = \sup_n \underset{\sim}{\beta}_n$, we can after ω many steps find a condition $p \in P_{\beta^*}$ which is stronger than every p_n, and a P_{β^*}-name $\underset{\sim}{\nu}$ of a branch in T^*, defined by $\underset{\sim}{\nu} = \bigcup_n \underset{\sim}{\eta}_n$. This is as required.

$\square_{6.4}$

6.5 Lemma. Suppose $\bar{Q} = \langle P_\alpha, Q_\alpha : \alpha < \kappa \rangle$ is suitable for $\langle \mathbb{I}_{\alpha,\beta}, \lambda_{\alpha,\beta}, \mu_{\alpha,\beta} : \langle \alpha, \beta \rangle \in W^* \rangle$, κ is strongly inaccessible, $|\bigcup I| + \lambda_{\alpha,\beta} + \mu_{\alpha,\beta} < \kappa$ for any $\langle \alpha, \beta \rangle \in W^*$, and $I \in \mathbb{I}_{\alpha,\beta}$, $\underset{\sim}{Q}_\kappa$ is a P_κ-name of a forcing notion satisfying the $\mathbb{I}_\kappa$-condition and let $\mathbb{I}_0 = \bigcup_{\alpha,\beta} \mathbb{I}_{\alpha,\beta}, \mathbb{I} = \mathbb{I}_0 \cup \mathbb{I}_\kappa$. Let

$$A^* = \{\alpha < \kappa : \text{for every } i < \alpha, \Vdash_{P_i} \text{“cf}(\alpha) > \aleph_0\text{” and for every } I \in \mathbb{I}_0,$$
$$V \models \text{“}|\bigcup I| \geq \alpha \Rightarrow I \text{ is } |\alpha|^+\text{-complete” and}$$
$$V \models \text{“}|\bigcup I| \geq \text{cf}\alpha \Rightarrow I \text{ is } |\text{cf}\alpha^+|\text{-complete}\}$$

and assume:

(a) for every $I \in \mathbb{I}_\kappa$: either I is κ^+-complete or I is κ-complete and normal and $\kappa \setminus A^* \in I$.

(b) for some $I^* \in \mathbb{I}_\kappa$ and $\bigcup I^* = \kappa$ and all singletons are in I^*.

Then $P_\kappa * \underset{\sim}{Q}_\kappa$ satisfies the $\mathbb{I}$-condition.

Remark: In Gitik Shelah [GiSh:191], $(a) + (b)$ were weakened to: each $I \in \mathbb{I}$ is κ-complete (or see XV §3).

Proof. Let $\underset{\sim}{F}_{\alpha,\beta}$ witness “$P_{\alpha,\beta} = P_\beta / P_\alpha$ satisfies the $\mathbb{I}_{\alpha,\beta}$-condition” and $\underset{\sim}{F}_\kappa$ (a P_κ-name) witness “$\underset{\sim}{Q}_\kappa$ satisfies the $\mathbb{I}_\kappa$-condition” and $\bigcup_{i,j,k,m<\omega} A_{i,j,k,m} = \{3n+2 : n < \omega\}$, the $A_{i,j,k,m}$'s infinite pairwise disjoint and $n \in A_{i,j,k,m} \Rightarrow i,j,k,m < n$.

Now we shall define the function F, so we should define $\langle \mathbf{I}_\eta, \langle f(\nu) : \nu \in \text{Suc}_T(\eta) \rangle \rangle = F(\eta, w, \langle f(\eta \restriction \ell) : \ell < \ell g(\eta) \rangle)$ (see Definition 2.6). If $p \in P_\kappa * \underset{\sim}{Q}_k$ we will write $p \restriction \kappa$ for the P_κ-component of p and $p(\kappa)$ for the Q_κ-component.

Case i. $|w|$ is divisible by 3.

We let $\mathbf{I}_\eta = I^*$ and for $\nu \in \mathrm{Suc}_T(\eta)$ we have $f(\nu) \geq f(\eta)$ be such that: if $\ell < \ell g(\eta)$, $\eta(\ell) \in A^*$, then for some $\alpha < \eta(\ell), f(\nu)\restriction\eta(\ell) \in P_\alpha$ (possible by 6.3(2)). We denote the minimal such α by $\alpha_{\nu,\ell}$. Thus, $\alpha_{\nu,\ell} < \nu(\ell)$ and $f(\nu)\restriction\nu(\ell) \in P_{\alpha_{\nu,\ell}}$.

Case ii. $|w| + 2$ is divisible by 3. We act as in the proof of 5.1, i.e., we use our winning strategy for the game on $\underset{\sim}{Q}_\kappa$: let $w^* = \{\ell \in w : |w \cap \ell| + 2$ is divisible by 3$\}$, and we let $\langle \underset{\sim}{I}, \langle \underset{\sim}{r}_\nu : \nu \in \underset{\sim}{B}\rangle\rangle = \underset{\sim}{F}_\kappa(\eta, w^*, \langle f(\eta\restriction\ell)(\kappa) : \ell \leq \ell g(\eta)\rangle)$.

Choose $q_\eta \in P_\kappa$, $q_\eta \geq f(\eta)\restriction\kappa$, $q_\eta \Vdash_{P_\kappa}$ "$\underset{\sim}{I} = \mathbf{I}_\eta$" for some $\mathbf{I}_\eta$, and let $F(\eta, w, \langle f(\eta\restriction\ell) : \ell \leq \ell g(\eta)\rangle) = \langle \mathbf{I}_\eta, (q_\eta, r_\nu) : \nu = \eta\hat{\ }\langle\alpha\rangle, \alpha < \bigcup \mathbf{I}_\eta\rangle\rangle$

Case iii. $|w| + 1$ is divisible by 3.

So for a unique quadruple $\langle i, j, k, m\rangle, |w|$ belongs to $A_{i,j,k,m}$, (hence $i, j, k, m < |w| \leq \ell g(\eta)$) and let $w^* = \{\ell \in w : |w \cap \ell| \in A_{i,j,k,m}\}$.

Now we shall use $\underset{\sim}{F}_{\xi,\zeta}(\eta, w^*, \langle f(\eta\restriction\ell)\restriction[\xi,\zeta) : \ell < \ell g(\eta)\rangle)$ were $\xi < \zeta < \kappa$ are chosen as follows

if $\eta(i) < \eta(j), \eta(i) \in A^*, \eta(j) \in A^*, i < k, j < k, k < m$, $|w \cap k|$ and $|w \cap m|$ are divisible by 3, (so $\alpha_{\eta\restriction k,i}, \alpha_{\eta\restriction m,i}$ are well defined) *then* let $\xi = \alpha_{\eta\restriction k,i} + k + 1, \zeta = \alpha_{\eta\restriction m,i} + m + 1$

if $\eta(i) < \eta(j), \eta(i) \in A^*$, $\eta(j) \in A^*, i < k, j < k, k = m$, $|w \cap k| - 1$ is divisible by 3 and $\eta(i) < \alpha_{\eta\restriction k,i}$ *then* let $\xi = \eta(i), \zeta = \alpha_{\eta\restriction k,i} + k + 1$

if $\eta(i) \geq \eta(j), \eta(i) \in A^*$, $i < k < m$, $|w \cap k|$ and $|w \cap m|$ are divisible by 3 *then* let $\xi = \alpha_{\eta\restriction k,i} + k + 1$, $\zeta = \alpha_{\eta\restriction m,i} + m + 1$

if $\eta(i) \geq \eta(j), \eta(i) \in A^*$, $i < k = m$, $|w \cap k|$ is divisible by 3 *then* let $\xi = 0, \zeta = \alpha_{\eta\restriction k,i} + k + 1$

if none of the above occurs *then* let $\xi = 0, \zeta = 1$.

So let $(T, \mathbf{I}), J_n, f$ be as in Definition 2.6 and $(T, \mathbf{I})^* \leq^* (T', \mathbf{I})$, w.l.o.g. $\bigcup_{n<\omega} J_n$ is the set of splitting points of $(T, \mathbf{I})$, (shrink T considering T'), and for notational simplicity we assume $J_n = \{\eta \in T : \ell g(\eta) = n\}$. So for η we have used $\omega = \{\ell : \ell < \ell g(\eta)\}$. Let $\sigma_\eta \stackrel{\text{def}}{=} \eta(\ell g(\eta) - 1)$.

We have to find p as required there.

Assume for simplicity CH. So by Lemma 3.5A, we can find $T^1, (T^\dagger, \mathbf{I}) \leq^* (T^1, \mathbf{I})$ such that

(a) For some $B_0 \subseteq \omega \times \omega$, for every $\eta \in \lim T^1$, $\eta(i) < \eta(j)$ *iff* $\langle i, j\rangle \in B_0$

(b) For $\eta \in T^1 : \ell g(\eta) \in B_1$ *iff* $\mathbf{I}_\eta$ is κ^+-complete; also $\ell g(\eta) \in B_1$ implies and $\ell g(\eta) + 2$ is divisible by 3

(c) If $\ell g(\eta) \in B_2 \stackrel{\text{def}}{=} \{n < \omega : n + 2 \text{ is divisible by 3}, n \notin B_1\}$ then $\kappa \setminus A^* \in \mathbf{I}_\eta$ and $\mathbf{I}_\eta$ is a normal ideal on κ

(d) If $\ell g(\eta) \in B_3 \stackrel{\text{def}}{=} \{n : n + 1 \text{ is divisible by 3 }\}$ then $\bigcup \mathbf{I}_\eta$ has cardinality $< \kappa$.

Let $A^{**} = \{\alpha \in A^* : \alpha$ is strong limit and for every $\langle i, j\rangle \in W^*$, $I \in \mathbb{I}_{i,j}$, if $i < j < \alpha$ then $|\bigcup I| + \lambda_{i,j} + \mu_{i,j} + |P_i| < \alpha\}$

Clearly if $I \in \mathbb{I}_\kappa$ is not κ^+-complete then $\kappa \setminus A^{**} \in I$ (since I is normal)

(e) if $(\ell g(\eta) - 1) \in B_4 \stackrel{\text{def}}{=} \{n : n \text{ divisible by } 3\}$ and $\sigma_{\eta\restriction(m+1)} \in A^{**}$ then (being normal) $f(\eta)\restriction\sigma_{\eta\restriction(m+1)}$ is (equivalent to) a member of $\bigcup\{P_\gamma : \gamma < \sigma_{\eta\restriction(m+1)}\}$

say to some member of $P_{\alpha_{\eta,m}}$, where $\alpha_{\eta,m} < \eta(m)$ (see case (i)).

We can conclude (by (c) and (e) above) that without loss of generality

(f) If $\eta \in T^1, \ell g(\eta) - 1 \in B_2$ then $\sigma_\eta \in A^{**}$. Also $\eta(\ell) < \kappa \Rightarrow \eta(\ell) < \sigma_\eta$, and $[|\bigcup \mathbf{I}_{\eta\restriction m}| < \kappa \Rightarrow |\bigcup \mathbf{I}_{\eta\restriction m}| < \sigma_\eta]$ for every $\ell < \ell g(\eta) - 1$, $m < \ell g(\eta)$; and if $\eta\char`^\langle\alpha\rangle \in T^1, \mathbf{I}_\eta$ is κ^+-complete then $\alpha > \kappa$.

For $\eta \in T^1$ if $\ell g(\eta) - 1 \in B_2$ let

$A_\eta = \{\nu \in T^1 : \nu \trianglelefteq \eta \text{ or: } \eta \triangleleft \nu \text{ and } |\bigcup \mathbf{I}_\eta|^V < \sigma_\eta\}$

$A^\dagger_\eta = \{\nu \in T^1 : \nu \trianglelefteq \eta \text{ or: } \eta \triangleleft \nu \text{ and } |\bigcup \mathbf{I}_\eta|^V < \mathrm{cf}^V(\sigma_\eta)\}$

By a repeated use of 5.3 (starting at $\langle\rangle$ and going up in T^1) we can find T^2, $(T^1, \mathbf{I}) \leq^* (T^2, \mathbf{I})$ such that

(g) If $\eta \in T^2, \ell g(\eta) - 1 \in B_2$, $\ell g(\eta) \leq m < \omega$, m divisible by 3 and for $\ell = 1, 2$, $\nu_\ell \in T^2$, $\ell g(\nu_\ell) = m + 1$, $\mathrm{res}(\nu_1, A_\eta) = \mathrm{res}(\nu_2, A_\eta)$ then $\alpha_{\nu_1, \ell g(\eta)-1} = \alpha_{\nu_2, \ell g(\eta)-1}$ (notice that $\alpha_{\nu, \ell g(\eta)-1} < \sigma_\eta < \kappa$, so κ-completeness suffices).

(h) If $\eta \in T^2, \ell g(\eta) - 1 \in B_2$, $\ell g(\eta) \leq m < \omega$, m divisible by 3, then there is $\gamma_{\eta,m} < \sigma_\eta$ such that if $\nu \in T^2$, $\ell g(\nu) = m + 1$ then $\alpha_{\nu, \ell g(\eta)-1} \leq \gamma_{\eta,m}$.

Note

(i) If $\eta \in T^2, \ell g(\eta) - 1 \in B_2$, then σ_η has cofinality $> \aleph_0$ hence $\gamma_\eta = \bigcup\{\gamma_{\eta,m} : m \geq \ell g(\eta), m$ divisible by $3\}$ is $< \sigma_\eta$.

Now if $\eta \in T^2, \ell g(\eta) \in B_2$ then the function $\alpha \mapsto \gamma_{\eta\hat{}\langle\alpha\rangle}$ is a regressive function on a subset of κ which do not belong to $\mathbf{I}_\eta$. Hence for some γ, $\{\nu : \nu = \eta\hat{}\langle\alpha\rangle, \gamma_\nu = \gamma\}$ is not in $\mathbf{I}_\eta$.

So without loss of generality

(j) If $\eta \in T^2, \ell g(\eta) \in B_2$ then for some $\gamma_\eta < \kappa$:
$(\forall\alpha)[\eta\hat{}\langle\alpha\rangle \in T^2 \Rightarrow \gamma_{\eta\hat{}\langle\alpha\rangle} = \gamma_\eta]$
So w.l.o.g.

(k) If $\eta \in T^2, \ell g(\eta) \in B_2$, $\eta \lhd \nu \in T^2$ then $|\mathbf{I}_\nu| \leq \gamma_\eta$ or $\mathbf{I}_\nu$ is $|\sigma_{\nu\restriction(\ell g(\eta)+1)}|^+$-complete.
If $\eta \in T^2, \ell g(\eta) \in B_2$ let

$$T^2_\eta = \{\mathrm{res}(\nu, A_\eta) : \nu \in T^2\}.$$

So w.l.o.g.

(l) If $\eta \in T^2, \ell g(\eta) \in B_2$, and for $\ell = 1, 2$ $\nu_\ell \in T^2$, $\eta \lhd \nu_\ell$, $\mathrm{res}(\nu_1, A_\eta) = \mathrm{res}(\nu_2, A_\eta)$ then $p_{\nu_1}\restriction\gamma_\eta = p_{\nu_2}\restriction\gamma_\eta$ and $|\cup\mathbf{I}_{\nu_1}| < \gamma_\eta \Leftrightarrow |\bigcup\mathbf{I}_{\nu_2}| < \gamma_\eta \Rightarrow \mathbf{I}_{\nu_1} = \mathbf{I}_{\nu_2},]$.

Let $\ell_0 = 0$ and $\{\ell_m : 1 \leq m < \omega\}$ be an enumeration in increasing order of $\{\ell < \omega : \ell - 1 \in B_2\}$. For any $\eta \in T^2, \ell g(\eta) = \ell_m$ we define β_η as $\mathrm{Sup}\{\gamma_\nu : \eta \lhd \nu, \gamma_\nu$ defined and $\ell g(\nu) < \ell_{m+1}\}$. Remembering that if $\ell g(\nu) \notin B_2$ then $|\bigcup\mathbf{I}_\nu| < \kappa$ or $\mathbf{I}_\nu$ is κ^+-complete it is clear that w.l.o.g.

(m) If $\eta \in T^2, \ell g(\eta) = \ell_m$ then $\beta_\eta < \kappa$; and $\eta \lhd \nu \in T^2$, $\ell g(\nu) = \ell_{m+1}$ implies $\beta_\eta < \sigma_\nu$. Let, for $\eta \in T^2, \ell g(\eta) = \ell_m$,

$$A^*_\eta \stackrel{\mathrm{def}}{=} \{\nu \in T^2 : \nu \trianglelefteq \eta \text{ or } \eta \trianglelefteq \nu,\ |\bigcup\mathbf{I}_\nu| < \gamma_\eta\}.$$

For every $\eta \in T^2, \ell g(\eta) = \ell_m$ we define $T^2_\eta \stackrel{\mathrm{def}}{=} \{\mathrm{res}(\nu, A^*_\eta) : \nu \trianglelefteq \eta$ or $\eta \lhd \nu\}$ and we define f_η and $\mathbf{I}^\eta$ (functions with domain $\subseteq T^2_\eta$) by: $f_\eta(\rho) = f(\nu)\restriction\beta_\eta$ and $\mathbf{I}^\eta_\rho = \mathbf{I}_\eta$ if $\eta \trianglelefteq \nu \in T_2$, $\rho = \mathrm{res}(\nu, A^*_\eta)$, is except that $\mathbf{I}^\eta_\rho$ is defined only if $|\bigcup\mathbf{I}_\eta| < \gamma_\eta$.

Now look at $T^2_{\langle\rangle}, f_{\langle\rangle}$ and $p\restriction\alpha_{\langle\rangle}$ (w.l.o.g. $p\restriction\alpha_{\langle\rangle} = p\restriction\kappa$). As in case (iii) of the definition of F we work hard enough (repeating previous proofs) there is a $P_{\beta_{\langle\rangle}}$-name $\underset{\sim}{\eta}_{\langle\rangle}$ and $q_{\langle\rangle} \in P_{\beta_{\langle\rangle}}$ such that :

$q_{\langle\rangle} \Vdash_{P_{\beta_{\langle\rangle}}}$ "$f_{\langle\rangle}(\underset{\sim}{\eta}_{\langle\rangle}\restriction\ell) \in \underset{\sim}{G}_{\beta_{\langle\rangle}}$ for every ℓ" as our strategy will often have used $F_{\alpha_{\langle\rangle},\beta_{\langle\rangle}}$.

Now comes a crucial observation: if $G_{\beta_{\langle\rangle}} \subseteq P_{\beta_{\langle\rangle}}$ is generic (over V) $\eta \in T_2, \ell g(\eta) = \ell_1$ and $\mathrm{res}(\eta, A^*_0) \lhd \eta_{\langle\rangle}[G_{\beta_{\langle\rangle}}]$, $\nu = \eta\hat{\ } < \alpha > \in T^2$ then we can (as in 5.1 and above) find $q_\nu \in P_{\beta_\nu}$, and P_{β_ν} -name $\underset{\sim}{\eta}_\nu$ such that: q_ν is compatible with every member of $G_{\beta_{\langle\rangle}}$ and

$q_\nu \Vdash_{\beta_{\langle\rangle}}$ "$f_\nu(\underset{\sim}{\eta}_\nu\restriction\ell) \in \underset{\sim}{G}_{\beta_\nu}$ for every ℓ, and for every $\eta \in T^2$, $[\mathrm{res}(\eta, A^*_\nu) \lhd \underset{\sim}{\eta}_\nu \Rightarrow \mathrm{res}(\eta_{\langle\rangle}, A^*_{\langle\rangle}) \trianglelefteq \eta_{\langle\rangle}]$". Note that $q_{\langle\rangle} \Vdash$ "$\{\alpha : q_{\eta\hat{\ }\langle\alpha\rangle} \in G_{P_\kappa}\} \neq \emptyset \bmod \mathsf{I}_\eta$". Hence each q_η can play the role of $q_{\langle\rangle}$ in the next step:

We can continue and define $q_\nu, \underset{\sim}{\eta}_\nu$ for every $\nu \in T^2, \bigvee_{m<\omega} \ell g(\nu) = \ell_m$ with the obvious properties:

(1) $q_\nu \in P_{\gamma_\nu}$,

(2) $q_\nu\restriction[\sigma_\nu, \gamma_\nu) = q_\nu$ when $\ell g(\nu) > 0$,

(3) $q_\nu \Vdash_{P_{\beta_\nu}}$ "$\underset{\sim}{\eta}_\nu \in \lim T^2_\nu$"

(4) $q_\nu \Vdash_{P_{\beta_\nu}}$ "if for every $\ell_m < \ell g(\nu)$, $\mathrm{res}(\nu, A^*_{\nu\restriction\ell_m}) \lhd \underset{\sim}{\eta}_{\nu\restriction\ell_m}$ and

$$f_{\nu\restriction\ell_m}(\underset{\sim}{\eta}_\nu\restriction k) \in \underset{\sim}{G}_{\nu\restriction\ell_m} \text{ for every } k$$

then for every $k < \omega$ and $\ell_m < \ell g(\nu)$,

$$\mathrm{res}(\underset{\sim}{\eta}_\nu\restriction k,\ A^*_{\nu\restriction\ell_m}) \lhd \underset{\sim}{\eta}_{\nu\restriction\ell_m} \text{ and } f_\nu(\underset{\sim}{\eta}_\nu\restriction k) \in \underset{\sim}{G}_{\beta_\nu} \text{ and } \nu \trianglelefteq \underset{\sim}{\eta}_\nu\text{"}.$$

Now we define a P_κ-name of a subtree of $T^2 : \underset{\sim}{T}^3 = \{\nu\restriction n : n < \ell g(\nu)$, $\ell g(\nu) = \ell_m$ for some $m, q_\nu \in \underset{\sim}{G}_{P_\kappa}$ and for $i < m$ $q_{\nu\restriction\ell_i} \in \underset{\sim}{G}_{P_\kappa}$ and $\mathrm{res}(\nu, A^*_{\nu\restriction\ell_i}) \lhd \underset{\sim}{\eta}_{\nu\restriction\ell_i}\}$. Clearly,

$q_{\langle\rangle} \Vdash_{P_\kappa}$ "$\underset{\sim}{T}^3 \subseteq T^2, T_3$ is closed under initial segments,

$\mathrm{Suc}_{\underset{\sim}{T}^3}(\eta) \neq \emptyset \bmod \mathsf{I}_\eta$ for $\eta \in \underset{\sim}{T}^3$ and

if $\eta \in \underset{\sim}{T}^3, \ell g(\eta) \in B_1 \cup B_2$ then $\mathrm{Suc}_{\underset{\sim}{T}^3}(\eta) = Suc_{T^2}(\eta)$"

$$q_{\langle\rangle} \Vdash_{P_\kappa} \text{" for every } \eta \in \underset{\sim}{T}^3, f(\eta)\restriction[0,\kappa) \text{ belongs to } \underset{\sim}{G}_\kappa\text{"}.$$

Now we can use the hypothesis "$\underset{\sim}{F}_\kappa$ exemplifies that $\underset{\sim}{Q}_\kappa$ satisfies the $\mathbb{I}_\kappa$-condition" and case (ii) in the definition of F to finish. $\square_{6.5}$

6.6 Lemma. Suppose $\bar{Q} = \langle P_i, \underset{\sim}{Q}_j : i \le \alpha, j < \alpha\rangle$ is suitable for $\langle \mathbb{I}_{i,j}, \lambda_{i,j}, \mu_{i,j} : i < j \le \alpha$, i is non-limit$\rangle$, $i(*) < \alpha$ is non-limit, $G_{i(*)} \subseteq P_{i(*)}$ generic over V, and $\langle i_\zeta : \zeta \le \beta\rangle$ is an increasing continuous sequence of ordinals in $V[G_{i(*)}]$, $i_0 = i(*)$, $i_\beta = \alpha$, each $i_{\zeta+1}$ a successor ordinal .
In $V[G_{i(*)}]$ we define $P'_\zeta = P_{i_\zeta}/G_{i(*)}$, $\underset{\sim}{Q}'_\zeta = \underset{\sim}{Q}_{i_\zeta}/[G_{i(*)}]$, (still a P_ζ-name) $\bar{Q}' = \langle P'_\zeta, \underset{\sim}{Q}'_\xi : \zeta \le \beta, \xi < \beta\rangle$, *then* in $V[G_{i(*)}]$, $\bar{Q}'$ is suitable for $\langle \mathbb{I}_{i_\zeta,i_\xi}, \lambda_{i_\zeta,i_\xi}, \mu_{i_\zeta,i_\xi} : \zeta < \xi \le \beta$, ζ non-limit $\rangle$.

Remark. Had we allowed $\mathbb{I}_{i,j}$, $\lambda_{i,j}$, $\mu_{i,j}$ to be suitable names we would have obtained here a stronger theorem.

Proof. Straightforward. $\square_{6.6}$

6.7 Conclusion. Suppose

(a) $\bar{Q} = \langle P_i, Q_i : i < \alpha\rangle$ is an RCS iteration.

(b) $\underset{\sim}{Q}_i$-satisfies the $\mathbb{I}_i$-condition, and $\mathbb{I}_i$ is $\aleph_2$-complete (in V^{P_i}, but $\mathbb{I}_i \in V$)

(c) if $\mathrm{cf}(i) < i \vee (\exists j < i)|P_j| \ge i$, then for some λ, μ we have $\bigcup_{j \ge i} \mathbb{I}_j$ is λ^+-complete and $(\forall I \in \bigcup_{j<i} \mathbb{I}_j)(|\bigcup I| < \mu)$ and $|P_i| \le \lambda = \lambda^{<\mu}$

(d) if $\mathrm{cf}(i) = i$ & $(\forall j < i)|P_j| < i$ then every $I \in \mathbb{I}_i$ is i^+-complete *or* normal, and e.g. $A^*_i = \{\alpha < i : \mathrm{cf}(\alpha) \ne \aleph_1\} \in I$.

Then $\mathrm{Rlim}\,\bar{Q}$ satisfies the $(\bigcup_{i<\alpha} \mathbb{I}_i)$-condition; if in addition κ is strongly inaccessible and $\bigwedge_{i<\kappa} |P_i| < \kappa$ then $\mathrm{Rlim}\,\bar{Q}$ satisfies the κ-c.c.

Proof. We should prove by induction on $i \le \alpha$ that for every $j < i$, P_i/P_j satisfies the $\bigcup\{\mathbb{I}_\gamma : j \le \gamma < i\}$-condition using 5.1, 6.2, 6.3, 6.4, 6.5 (and 5.6). $\square_{6.7}$

6.8 Conclusion. We can satisfy the demands of 1.1 by "P satisfies $\{\lambda : \aleph_2 \leq \lambda \leq |P|, \lambda$ regular $\}$"-condition.

Concluding Remark. We could have strengthened somewhat the result 6.6, but with no apparent application by using a larger A_i^*.

§7. Further Independence Results

In this section we complete some independence results.

7.1 Theorem. The following are equiconsistent.

(a) ZFC + there is a Mahlo cardinal.

(b) ZFC + G.C.H.+ $Fr^+(\aleph_2)$ (where $Fr^+(\aleph_2)$ means that every stationary $S \subseteq S_0^2 = \{\delta < \aleph_2 : \mathrm{cf}(\delta) = \aleph_0\}$ contains a closed copy of ω_1).

Remark. Our proof will use, in addition to the ideas of the proof of Theorem 1.4 also ideas of the proof of Harrington and Shelah [HrSh:99], but, for making the iteration work, we build a quite generic object rather than force it (as in [Sh:82]).

2) In b) we can also contradict G.C.H. (using XV §3 for a) $\Rightarrow$ b)).

Proof. The implication b) $\Rightarrow$ a) was proved by Van Lere, using the well known fact that if in L there is no Mahlo cardinal, then the square principle holds for $\aleph_2$. So the point is to prove a) $\Rightarrow$ b). As any Mahlo cardinal in V is a Mahlo cardinal in L, we can assume $V = L$, κ a strongly inaccessible Mahlo cardinal.

We shall define a revised countable support iteration $\bar{Q} = \langle P_i, \underset{\sim}{Q}_i : i < \kappa \rangle$, $|P_i| \leq \aleph_{i+1}$. If i is not a strongly inaccessible cardinal $\underset{\sim}{Q}_i$ is the Levy collapse of $2^{\aleph_1}$ to $\aleph_1$ by countable conditions (in V^{P_i}). If i is strongly inaccessible then $\underset{\sim}{Q}_i$ is $P[\underset{\sim}{S}_i]$ (see 4.6), $\underset{\sim}{S}_i$ is a P_i-name of a stationary subset of $S_0^i = \{\delta < i : \mathrm{cf}(\delta) = \aleph_0$ in $V\}$ (note $\Vdash_{P_i}$ "$i = \aleph_2$" by 1.1(3)), where $\underset{\sim}{S}_i$ will be carefully chosen as described below. Note that $\aleph_1^V = \aleph_1^{V^{P_\kappa}}$, $\kappa = \aleph_2^{V^{P_\kappa}}$ (again we use 1.1(3)) so P_κ collapse no cardinal $\geq \kappa$ and in V^{P_κ} the GCH holds.

We let $P_\kappa = \mathrm{Rlim}\,\bar{Q}$. In V^{P_κ} we define an iterated forcing $\bar{Q}^* = \langle P_i^*, \underset{\sim}{Q}_i^* : i < \kappa^+\rangle$, with support of power $\aleph_1$, such that (in V^{P_κ}) for each i, $\underset{\sim}{S}_i^*$ is a P_i^*-name of a subset of S_0^2 which does not contain a closed copy of ω_1, and:

$\underset{\sim}{Q}_i^* = \{f \in (V^{P_\kappa})^{P_i^*} : f$ an increasing continuous function from some $\alpha < \omega_2$ into $\omega_2 \setminus \underset{\sim}{S}_i^*$ and if α is a limit ordinal, $\bigcup_{i<\alpha} f(i) \notin \underset{\sim}{S}_i^*\}$ (so $\underset{\sim}{Q}_i$ makes S_i^* nonstationary).

We shall prove that (if the $\underset{\sim}{S}_i$'s were chosen suitably then):

(*) For every $\alpha < \kappa^+$, forcing by P_α^* does not add new ω_1-sequences (to V^{P_κ}) and P_α^* contains a dense subset of power $\leq \aleph_2$ (everything in V^{P_κ}).

This implies that $P_{\kappa^+}^*$ satisfies the κ^+-chain condition, so by a suitable bookkeeping every $P_{\kappa^+}^*$-name of $S \subseteq S_0^2$ which does not contain a closed copy of ω_1 is $\underset{\sim}{S}_i^*$ for some i. So easily we can conclude that it is enough to prove (*).

So let $\alpha^* < \kappa^+, p \in P_{\alpha^*}^*$, and $\underset{\sim}{\tau}$ be a $P_{\alpha^*}^*$-name of a function from ω_1 to ordinals.

For all those things we have P_κ-names, (but α^* is an actual ordinal in V, as P_κ satisfies the κ-c.c.). Now in V we can define an increasing continuous sequence $N_i^*(i < \kappa)$ of elementary submodels of $H(\kappa^{+++})$ of cardinality $< \kappa$ such that P_κ, $\bar{Q}$ and all the names involved belong to N_0^*, $\langle N_i^* : i \leq j\rangle \in N_{j+1}^*, N_{j+1}^*$ is closed under sequence of length $\leq |N_j^*|$.

Now in V, as $V = L$, $\diamondsuit_{\{\lambda<\kappa:\lambda \text{ is strongly inaccessible}\}}$ holds, so we have guessed $\langle N_i^* : i \leq \lambda\rangle$, $\underset{\sim}{\tau}, p, \ldots$ in some stage λ, $(\bigcup_{i<\lambda} N_i^*) \cap \kappa = N_\lambda^* \cap \kappa = \lambda$. Really we are only guessing subsets of κ, so we can only guess the isomorphism type of N_λ, etc., or equivalently, its Mostowski collapse. I.e. let f be a one to one function from κ onto $\bigcup_{i<\kappa} N_i^*$, let h be a one to one function from κ^3 onto κ, and $h_\ell : \kappa \to \kappa$ for $\ell < 3$ be such that $\varepsilon = h(\alpha,\beta,\gamma) \Leftrightarrow \alpha = h_0(\alpha)$ & $\beta = h_1(\beta)$ & $\gamma = h_2(\gamma)$. Let

$$A = \{h(0,\alpha,\beta) : f(\alpha) \in f(\beta) \text{ and } \alpha,\beta < \kappa\} \cup \{h(1,i,\alpha) : f(\alpha) \in N_i\}$$

$$\cup\{h(2,0,\alpha) : f(\alpha) = \langle \bar{Q}, \bar{Q}^*, \underset{\sim}{T}, p, \alpha^*\rangle\}.$$

Let $\langle A_\lambda : \lambda < \kappa \text{ inaccessible}\rangle$ be a diamond sequence. Let N^λ be the unique transitive subset of $H(\kappa)$ isomorphic to $(\lambda, \in^\lambda)$ where $\in^\lambda = \{(\alpha,\beta) : \alpha < \lambda, \beta < \lambda, h(0,\alpha,\beta) \in A_\lambda\}$, if there is one. Let $g^\lambda : \lambda \to N^\lambda$ be the isomorphism. Let for $i < \lambda$, $N_i^\lambda = N^\lambda \restriction \{g^\lambda(\alpha) : h(1,i,\alpha) \in A_\lambda\}$ and $x^\lambda = (\bar{Q}^\lambda, \bar{Q}^{*,\lambda}, \underset{\sim}{T}^\lambda, p^\lambda, \alpha^{*,\lambda})$ be such that $x^\lambda = g^\lambda(\alpha)$, $\alpha = \min\{\beta : h(2,0,\beta) \in A_\lambda\}$ (if there are such α). Now necessarily $W = \{\lambda < \kappa : \lambda \text{ inaccessible, and } A \cap \lambda = A_\lambda, \text{ and } f \text{ maps } \lambda \text{ onto } N_\lambda^*\}$ is stationary and for $\lambda \in W$, N^λ is well defined and isomorphic to N_λ^* say by $g^* : N^\lambda \to N_\lambda^*$ and $g^*(N_i^\lambda) = N_i^*$ (for $i < \lambda$), $\bar{Q}^\lambda = \bar{Q}\restriction\lambda$, $g^*(\bar{Q}^{*,\lambda}) = \bar{Q}^*$, $g^*(\underset{\sim}{T}^\lambda) = \underset{\sim}{T}$, $g^*(p^\lambda) = p$, $g(\alpha^{*,\lambda}) = \alpha^*$.

So now we will explain what we did in stage λ to take care of this situation.

First we will give an overview of how to get the sets S_λ; In stage λ (an inaccessible below κ, so work in V^{P_α}) we use $\diamondsuit$, i.e. A_λ to obtain a continuous increasing sequence $\langle N_i^\lambda : i \leq \lambda\rangle$ of quite close models (which guesses (the isomorphic type of a) a sequence $\langle N_i^* : i \leq \kappa\rangle$ as above). We also guess an ordinal $\alpha = \alpha^{*,\lambda} \in N^\lambda$ (so actually we are only guessing $\text{otp}(\alpha^* \cap N_\lambda^*)$) and x^λ "guess" $\langle \bar{Q}, \bar{Q}^*, \underset{\sim}{T}, p, \alpha^\lambda\rangle, \ldots$ Let $G_\lambda \subseteq P_\lambda$ be generic over V, $p^\lambda \in G_\lambda$. We now try to construct a sequence $\langle p_i : i < \lambda\rangle$ of conditions in $P^{*,\lambda}_{\alpha^{*,\lambda}} \cap N^\lambda[G_\lambda]$ which will induce an $N^\lambda[G_\lambda]$-generic set. If we succeed, letting $p'_i = g^*(p_i)$ in $V[G_\lambda]$ we have $p'_i \in N_\lambda^* \cap P$ is increasing and $p_\lambda = \lim_{i<\lambda}(p'_i)$ will decide all names in $N_\lambda^*[G_\lambda]$ (p_λ has domain $N_\lambda^*[G_\lambda] \cap \alpha^*$, $p_\lambda(\beta) = \bigcup_{i<\lambda} p'_i(\beta) \cup \{\langle\lambda,\lambda\rangle\}$). Moreover, $p_\lambda(\beta)$ will be an actual function (at least above $p_\lambda\restriction\beta$) rather than just a P_β^*-name, for all $\beta \in \text{Dom}(p_\lambda) \subseteq N_\lambda^*[G_\lambda]$. This will show that in V^{P_κ}, the set

$$D \stackrel{\text{def}}{=} \{p \in P_{\alpha^*}^* : \forall\beta \in \text{Dom}(p) \exists f : p\restriction\beta \Vdash \text{“}p(\beta) = f\text{”}\}$$

is dense, $P_{\alpha^*}^*$ contains a dense subset of cardinality $\leq \kappa (= \aleph_2)$, in V^{P_κ}, which is one of the demands in $(*)$, [which implies that $P_{\alpha^*}^*$ satisfies the κ^+-c.c. The usual Δ-system argument (recalling that P_κ^* used $\aleph_1$-support) then shows that also $P_{\kappa^+}^*$ satisfies the κ^+-c.c.)].

We will try to build these p_i in $\text{otp}(N^\lambda[G_\lambda] \cap (\alpha^{*,\lambda} + 1))$ many steps, by constructing initial parts $\langle p_i{\restriction}\gamma : i \le \lambda\rangle$ for $\gamma \in N^\lambda \cap (\alpha^{*,\lambda}+1)$, by induction on γ.

However, it is possible that our construction will get stuck in some stage $\gamma \in N^\lambda[G_\lambda] \cap (\alpha^{*,\lambda}+1)$. In this case we show that we will have constructed a stationary subset of λ (which guesses $\underset{\sim}{S}^*_\gamma \cap \lambda$). We will use this set as S_λ (and hence contradict the guess, since no superset of one of the sets S_λ can appear in the second iteration as some S^*_γ). However, since our guess must be correct on a stationary set of λ's, the construction will be completed stationarily often.

Before we start the construction of the p_i's, we will try to guess its outcome. In our ground model $V = L$ we have $\diamondsuit^*_\lambda$, so as $|P_\lambda| = \lambda$, we still have in V^{P_λ} i.e. $V[G_\lambda]$, that $\aleph_2$ is λ, and $\diamondsuit_{\{i<\lambda:\mathrm{cf}(i)=\aleph_1\}}$ holds. Note $N^\lambda_i[G_\lambda]$ is well defined: $G_\lambda \cap N^\lambda_i$ is a generic subset of $P^{N^\lambda_i}_\kappa$. So we can choose for each $i < \lambda$ a sequence $\langle q_{i,\xi} : \xi < i\rangle$ and a name $\underset{\sim}{\tau}_i$ such that:

(α) every initial segment of the sequence $\langle(\underset{\sim}{\tau}_i, \langle q_{i,\xi} : \xi < i\rangle) : i < \lambda\rangle$ belongs to $N^\lambda_i[G_\lambda]$.

(β) $q_{i,\xi} \in N^\lambda_i[G_\lambda] \cap P^*_\alpha$, is increasing with ξ.

(γ) if $\langle q_\xi : \xi < \lambda\rangle$ satisfies (α) + (β) [i.e., it is increasing with ξ and $q_\xi \in (N^\lambda \cap P^*_{\alpha^{*,\lambda}})$], and $\underset{\sim}{\tau} \in N^\lambda[G_\lambda]$ is a P^*_γ-name of an ordinal then $\{i : \langle q_\xi : \xi < i\rangle = \langle q_{i,\xi} : \xi < i\rangle$ and $\underset{\sim}{\tau} = \underset{\sim}{\tau}_i\}$ is stationary.

Note that $N^\lambda[G_\lambda]$ is closed under taking i-sequences (in V^{P_λ}) for $i < \lambda$ so clause (α) is not necessary.

Now at least for ordinals i such that $\mathrm{cf}(i) = \omega$ it is not clear whether $\langle q_{i,\xi} : \xi < i\rangle$ has an upper bound in $P^{*,\lambda}_{\alpha^{*,\lambda}}$, however we can find $\alpha(i) \in N^\lambda[G_\lambda] \cap (\alpha^{*,\lambda}+1)$ and $q^*_i \in N^\lambda[G_\lambda] \cap P^{*,\lambda}_{\alpha(i)}$, $q^*_i \ge q_{i,\xi}{\restriction}\alpha(i)$ for $\xi < i$, and if $\alpha(i) < \alpha^{*,\lambda}$ then $q^*_i \Vdash_{P^*_{\alpha(i)}}$ "$\langle q_{i,\xi}(\alpha(i)) : \xi < i\rangle$ has no upper bound (in $N^\lambda[G_\lambda]$!)".

[Just let $\alpha(i) \in N_i \cap (\alpha^{*,\lambda}+1)$ be maximal such that $q^\dagger \overset{\mathrm{def}}{=} (\bigcup_{\xi<i} q_{i,\xi}){\restriction}\alpha(i)$ belongs to $P_{\alpha(i)}$; hence $q^\dagger \nVdash_{P^{*,\lambda}_{\alpha(i)}}$ "$(\bigcup_{\xi<i} q_{i,\xi})(\alpha(i)) \in \underset{\sim}{Q}^{*,\lambda}_{\alpha(i)}$" hence some $r, q^\dagger \le r \in P^{*,\lambda}_{\alpha(i)}$ is as required]. Moreover, we may also assume that q^*_i decides the value of $\underset{\sim}{\tau}_i$ if $\underset{\sim}{\tau}_i$ is a $P^{*,\lambda}_{\alpha(i)}$-name.

Now we define by induction on $\gamma \in N^\lambda[G_\lambda] \cap (\alpha^{*,\lambda}+1)$ a set C_γ, a sequence $\langle C_{\beta,\gamma} : \beta < \gamma, \beta \in N^\lambda[G_\lambda] \cap (\alpha^{*,\lambda}+1)\rangle$ and sequences $\langle p_{\gamma,i} : i \in C_\gamma\rangle$ satisfying the following:

(i) Each C_γ, and each $C_{\beta,\gamma}$ is a closed unbounded subset of λ, $C_{\beta,\gamma} \subseteq C_\beta \cap C_\gamma$.

(ii) $p_{\gamma,i} \in N^\lambda[G_\lambda] \cap P^{*,\lambda}_\gamma$, and is increasing with i.

(iii) $p_{\gamma,i} \restriction \beta = p_{\beta,i}$, for $\beta < \gamma$, $i \in C_{\beta,\gamma}$.

(iv) if $\langle p_{\gamma,j} : j < i\rangle$, $\langle q_{i,j} : j < i\rangle$ (from the diamond above) satisfy $(\forall j < i)(\exists \zeta < i)(p_{\gamma,j} \leq q_{i,\zeta}\restriction\gamma)$, $(\forall \zeta < i)(\exists j < i)(q_{i,\gamma}\restriction\gamma \leq p_{\gamma_j})$ and $\alpha(i) \geq \gamma$ *then* $q^*_i \leq p_{\gamma,i}$.

We will define this sequence by induction on γ. For $\gamma = 0$ there is nothing to do. If γ is a limit of cofinality $< \lambda$, let $\gamma = \bigcup_{\xi<\mathrm{cf}(\gamma)} \gamma_\xi$ with $\zeta < \xi < \mathrm{cf}(\gamma) \Rightarrow \gamma_\zeta < \gamma_\xi < \gamma$. We let

$$C_\gamma \stackrel{\text{def}}{=} \bigcap_{\zeta<\xi<\mathrm{cf}(\gamma)} C_{\gamma_\zeta,\gamma_\xi}.$$

For $i \in C_\gamma$, we let $p_{\gamma,i} = \bigcup_{\xi<\mathrm{cf}(\gamma)} p_{\gamma_\xi,i}$. (This is a union of a sequence of at most $\aleph_1$ conditions which are end extensions of each other (in the sense that $p_{\gamma_\xi,i} = p_{\gamma_\zeta,i}\restriction\gamma_\xi$ for $\zeta < \xi < \mathrm{cf}(\gamma)$), hence this limit exists.) Finally, we let $C_{\beta,\gamma} = C_\beta \cap C_\gamma \cap \bigcap_{\xi<\mathrm{cf}(\gamma)} C_{\beta,\gamma_\xi}$ [where we let $C_{\alpha,\beta} = \lambda$ for $\alpha \geq \beta$]. If $\mathrm{cf}(\gamma) \geq \lambda$, then we can find an increasing unbounded sequence $\langle \gamma_\xi : \xi < \lambda\rangle$ in $\gamma \cap N^\lambda[G_\lambda]$ such that for all $\zeta < \lambda$ we have $\langle \gamma_\xi : \xi < \zeta\rangle \in N^\lambda[G_\lambda]$. We let C_γ be a diagonal intersection:

$$C_\gamma \stackrel{\text{def}}{=} \{i < \lambda : i \bigcap_{\zeta<\xi<i} C_{\gamma_\zeta,\gamma_\xi}\}$$

and for $i \in C_\gamma$ we let $p_{\gamma,i} \stackrel{\text{def}}{=} \bigcup\{p_{\xi,i} : \xi < i\}$. We let

$$C_{\beta,\gamma} \stackrel{\text{def}}{=} \{i < \lambda : (\forall j < i)\, i \in C_{\beta,\gamma_j}\}.$$

An easy calculation shows that (ii) will be satisfied.

Successor step: Let $\gamma = \beta + 1$. If the set $\underset{\sim}{S}^*_\beta$ as computed by $\langle p_{\beta,i} : i \in C_\beta\rangle$ (i.e. the set $\{\varepsilon < \lambda :$ for some $i \in C_\beta$ we have $p_{\beta,i} \Vdash$ "$\varepsilon \in \underset{\sim}{S}^*_\beta$"$\})$ does not include a stationary subset of λ, let $C_\gamma \subseteq C_\beta$ be a club set disjoint from this set. We let

$C_{\beta,\gamma} \stackrel{\text{def}}{=} C_\gamma$. For $\beta' < \beta$ we let $C_{\beta',\gamma} = C_{\beta',\beta} \cap C_\gamma$. So for $i \in C_\gamma$, we will have $p_{\gamma,i} \restriction \beta = p_{\beta,i}$, so we only have to define $p_{\gamma,i}(\beta)$. We will do this by induction on i: If i is (in C_γ) the successor of j, then we let $p_{\gamma,i}(\beta)$ be a condition extending $p_{\gamma,j}(\beta)$ such that there is $\varepsilon \in C_\gamma$ with $\text{ht}(p_{\gamma,j}(\beta)) < \varepsilon < \text{ht}(p_{\gamma,i}(\beta))$ (where $\text{ht}(r) \stackrel{\text{def}}{=} \sup(\text{range}(r))$. This will ensure that in limit steps the supremum of the conditions constructed so far always exists. For limit i, we first take the supremum of the conditions constructed so far, and, if possible, increase the condition again to make it stronger than $q_i^*(\beta)$.

Finally if $\underset{\sim}{S}^*_\beta$ as computed by $\langle p_{\beta_i} : i \in C_\beta \rangle$ does contain a stationary set, we will choose this as S_λ when defining the first iteration $\bar{Q}$. Note that this choice of S_λ does not depend on $\underset{\sim}{N}^*_\lambda, \ldots$ but only on $N^\lambda, \ldots$ so all is O.K. As remarked above, this will not happen if $\diamondsuit$ has guessed correctly. $\square_{7.1}$

7.2 Theorem. The following are equi-consistent

(a) ZFC + there is a weakly compact cardinal.

(b) ZFC + G.C.H. + if $S', S'' \subseteq S^2_0$ are stationary, then for some $\delta \in S^2_1$, $S' \cap \delta$, $S'' \cap \delta$ are stationary.

(c)ZFC + G.C.H.+"if $S_i \subseteq S^2_0$ are stationary sets for $i < \omega_1$ *then* there is an increasing continuous sequence of ordinals $< \omega_2$, $\langle \alpha_i : i < \omega_1 \rangle$ such that $\alpha_i \in S_i$.

Remark. In (b), (c) we can also contradict G.C.H. (use XV §3 for Con(a) $\Rightarrow$ Con(c)).

Proof. The implication Con(a) $\Rightarrow$ Con(b) was proved by Baumgartner [Ba], the inverse by [Mg5]. Now Con(c)$\Rightarrow$Con(b) is trivial, and so the point is to prove Con(a)$\Rightarrow$Con(c) which is done just like the proof of theorem 1.4, using the forcing notion $P[\langle S_i : i < \omega_1 \rangle]$ from 4.6A. $\square_{7.2}$

Before we prove the next theorem, we recall the forcing notion for "shooting a club through S":

A known forcing is

7.3 Definition. For any set $S \subseteq \omega_2$, define the forcing notion Club(S) by

$$\text{Club}(S) \overset{\text{def}}{=} \{h : \text{for some } \alpha < \omega_2 \text{ we have:}$$
$$\text{Dom}(h) = \alpha + 1, \text{Rang}(h) \subseteq S, h \text{ continuous increasing}\}$$

This forcing notion "shoots a club through S". See on it in [BHK] and more in [AbSh:146].

For $h \in \text{Club}(S)$ let

$$\alpha(h) \overset{\text{def}}{=} \max \text{Dom}(h) \qquad \delta(h) \overset{\text{def}}{=} h(\alpha(h))$$

7.4 Lemma.

(a) If $S \cap S_0^2$ is stationary, then Club(S) does not add new ω-sequences of ordinals.

(b) If the set

$$\hat{S} \overset{\text{def}}{=} \{\delta \in S_1^2 \cap S : S \cap \delta \text{ contains a club subset of } \delta\}$$

is stationary, and CH holds, *then* Club(S) does not add ω_1-sequences of ordinals to V.

7.4A Remark. Instead CH it suffice to have for some list $\{a_\alpha : \alpha < \omega_2\}$ of subsetes of ω_2 that $\{\delta \in S_1^2$: there is a club C of δ such that $C \subseteq S$ and $\alpha < \delta \Rightarrow C \cap \alpha \in \{a_\beta : \beta < \alpha\}$.

Proof: We leave (a) to the reader as it is easier and we will need only (b).
To prove (b), let $p \in \text{Club}(S)$, $\underset{\sim}{\tau}$ a Club(S)-name such that $p \Vdash$"$\underset{\sim}{\tau}$ is a function from ω_1 to the ordinals". Let $N \prec (H(\aleph_3), \in)$ be a model of size $\aleph_1$ which contains all relevant information (i.e., $\{S, p, \underset{\sim}{\tau}\} \subseteq N$), is closed under ω-sequences and satisfies $N \cap \omega_2 \in \hat{S}$. We can find such a model because we have CH and $\hat{S}$ is stationary. Let $C \subseteq \delta \cap S$ be a club set.

Now we can find a continuous increasing sequence $\langle N_\alpha : \alpha < \omega_1 \rangle$ satisfying the following for all $\alpha < \omega_1$:

(a) $N_\alpha \in N$, $N_\alpha \prec N$, N_α countable.

(b) $\langle N_\beta : \beta \le \alpha \rangle \in N_{\alpha+1}$.

(c) $\sup(N_\alpha \cap \omega_2) \in C$.

(d) $\{S, p, \underset{\sim}{\tau}\} \subseteq N_0$.

W.l.o.g. $\bigcup_{\alpha<\omega_1} N_\alpha = N$. Let $p_0 = p$, and define a sequence $\langle p_i : i < \omega_1 \rangle$ satisfying

(0) $p_i \in N$

(1) p_{i+1} decides the value of $\underset{\sim}{\tau}(i)$

(2) Letting $\alpha_i \stackrel{\text{def}}{=} \min\{\alpha : p_i \in N_\alpha \ \&\ \alpha > \alpha(p_i)\}$, we demand $\alpha(p_{i+1}) > \sup(N_{\alpha_i} \cap \omega_2)$ and $\delta(p_{i+1}) > \sup(N_{\alpha_i} \cap \omega_2)$ (see 7.3).

(3) If $j < i$, then $p_j \le p_i$.

Given p_i, it is no problem to find p_{i+1}. If i is a limit, then letting $\alpha^* \stackrel{\text{def}}{=} \sup_{j<i} \alpha(p_j)$, $\delta^* \stackrel{\text{def}}{=} \sup_{j<i} \delta(p_j)$, we have $\delta^* = N_{\alpha^*} \cap \omega_2 \in S$, so we can let $p_i = \bigcup_{j<i} p_j \cup \{(\alpha^*, \delta^*)\}$. Note that $p_i \in N$, because N was closed under ω-sequences.

Finally, $\bigcup_{i<\omega_1} p_i$ can be extended to a condition p_{ω_1} because $\delta \in S$. Now $p_{\omega_1} \Vdash \underset{\sim}{\tau} \in V$. $\square_{7.4}$

The following solves a problem of Abraham.

7.5 Theorem. The following are equi-consistent

a)ZFC + there is a 2-Mahlo cardinal.

b)ZFC + G.C.H. + $\{\delta < \aleph_2 : \delta$ inaccessible in L$\}$ contains a closed unbounded subset of ω_2.

Proof. Con(b) $\Rightarrow$ Con(a):

Let C be a closed unbounded subset of ω_2 consisting of regular cardinals of L. So each $\delta \in C \cap S^2_0$ is inaccessible in L, hence each $\delta \in C \cap S^2_1$ is Mahlo in L, hence $\aleph_2^V$ is 2-Mahlo in L, i.e., $\{\lambda : \lambda < \aleph_2^V, \lambda$ Mahlo in $L\}$ is stationary.

Con(a)⇒Con(b):

So without loss of generality $V = L$, κ is 2-Mahlo.

We define an RCS-iteration $\bar{Q} = \langle P_i, \underset{\sim}{Q}_i : i < \kappa \rangle$, where:

1)if i is not strongly inaccessible, $\underset{\sim}{Q}_i$ is Levy collapse of $2^{\aleph_1}$ to $\aleph_1$ by countable conditions.

2)if i is strongly inaccessible but not Mahlo, $\underset{\sim}{Q}_i = \mathrm{Nm}'$.

3)if i is a strongly inaccessible Mahlo cardinal, $\underset{\sim}{Q}_i = P[S_i]$ where $S_i = \{\lambda < i : \lambda$ strongly inaccessible $\}$.

It should be easy for the reader to prove that $P_\kappa = \mathrm{Rlim}\,\bar{Q}$ satisfies the κ-chain condition, and the S-condition, $S = \{\lambda : \aleph_1 < \lambda < \kappa, \lambda$ inaccessible (in L)$\}$.

Lastly in V^{P_κ} let P^* =Club($\{\lambda < \kappa : \lambda$ inaccessible in L$\}$). So our forcing is $P_\kappa * \underset{\sim}{P}^* \in V$.

Now P^* is not even $\aleph_1$-complete, but still P was constructed so that P^* does not add $\aleph_1$-sequences by 7.4(b), and $V^{P_\kappa * P^*}$ is as required. $\square_{7.5}$

7.6 Theorem. Assume κ^* is supercompact. *Then* for some forcing notion P, in V^P, for every regular $\lambda > \aleph_1$, $\mathrm{Fr}^\dagger(\lambda)$ holds (and we can ask also GCH).

Proof. W.l.o.g. $V \models$ GCH. Let κ will be the first strongly inaccessible cardinal κ which is κ^{+7}-supercompact. Let $\mathbf{j} : \kappa \to H(\kappa)$, $j(\alpha) \in H(|\alpha|^{+6})$ be a Laver diamond under this restriction. Let $\langle P_i^-, \underset{\sim}{Q}_j^- : i \leq \kappa, j < \kappa \rangle$ be an Easton support iteration. $\underset{\sim}{Q}_j^-$ is $\mathbf{j}_0(j)$ if $\mathbf{j}(j) = \langle \mathbf{j}_\ell(j) : \ell < 2 \rangle$ and $\mathbf{j}_0(j)$ is a P_i^--name of a j-directed complete forcing notion, j strong inaccessible, $(\forall \zeta < j)(|P_\zeta| < j)$, and the trivial forcing otherwise. Let $V_0 = V$, $V_1 = V_0^{P_\kappa^-}$. Clearly $V_1 \models \diamondsuit_{\{\mu<\kappa:\mu \text{ is strongly inaccessible }\}}$. Let $R = \mathrm{Levy}(\lambda, <\kappa^*)^{V_1}$, so in V^R, for every regular $\theta \geq \kappa^*$ we have:

$(*)_\theta$ if $S \subseteq \{\delta < \theta : \mathrm{cf}(\delta) = \aleph_0\}$ is stationary then for some $\delta^* < \lambda$ we have: $\mathrm{cf}(\delta^*) = \kappa$, and $S \cap \delta^*$ is a stationary subset of δ^*. (see Fact in X 7.4)

In $V_2 = V^R$ we have $\kappa^* = \kappa^+$ and define $\bar{Q}$, P_κ, $\bar{Q}^* = \langle P_i^*, \underset{\sim}{Q}_i^* : i < \kappa^+ \rangle$, $P_{\kappa^+}^*$ as in the proof of 7.1 except that for $\delta < \kappa$ strong inaccessible, $\underset{\sim}{Q}_\delta$ is suggested by $\mathbf{j}_1(\delta)$ when $\mathbf{j}(\delta) = \langle \mathbf{j}_\ell(\delta) : \ell < 2 \rangle$ as above, $\mathbf{j}_1(\delta)$ a $(P_{\delta+1}^- * P_\delta)$-

name of a forcing notion satisfying the S-condition (in the universe $V^{P_{\delta+1}\times P_\delta}$ of course). So we force with $R' \stackrel{\rm def}{=} R * \underset{\sim}{P}_\kappa * P^*_{\kappa^+}$ (really we can arrange that $P_\kappa \in V_1$, $\underset{\sim}{P}^*_\kappa$ is an $R \times P_\kappa$-name). Looking at the proof of 7.1, the only point left is to prove $(*)_\theta$ for $\theta = \mathrm{cf}(\theta) \geq \kappa^*$. If $\theta > \kappa^*$, as the density of $P_\kappa * \underset{\sim}{P}^*_{\kappa^+}$ ($\in V_2$) is κ^*, any stationary $S \subseteq \{\delta < \theta : \mathrm{cf}(\delta) = \aleph_0\}$ from $V^{R'}$ contains a stationary subset from V_2. We can use $V_2 \vDash (*)_\theta$, so we are left with the case $\theta = \kappa^*$.

If in V_2, $\underset{\sim}{S}$ is a $P_\kappa * \underset{\sim}{P}^*_{\kappa^+}$ name, $p \in P_\kappa * \underset{\sim}{P}^*_{\kappa^+}$, $p \Vdash$ "$\underset{\sim}{S}_0 \subseteq \{\delta < \kappa^+ : \mathrm{cf}(\delta) = \aleph_0\}$ stationary", let

$$S_1 = \{\delta < \kappa^+ : \mathrm{cf}(\delta) = \aleph_0 \text{ and } p \nVdash \text{“}\delta \notin \underset{\sim}{S}_0\text{”}\}.$$

For $\delta \in S_1$ choose $p_\delta \in P_\kappa * \underset{\sim}{P}^*_{\kappa^+}$, $p \leq p_\delta$, $p_\delta \Vdash$ "$\delta \in \underset{\sim}{S}$". Let $S_2 = \{\delta \in S_1 : p_\delta \in \underset{\sim}{G}_{P_\kappa * \underset{\sim}{P}^*_{\kappa^+}}\}$, so $p \Vdash$ "$\underset{\sim}{S}_2 \subseteq \underset{\sim}{S}_0$".

Let $E = \{\alpha < \kappa^+ : \alpha \text{ limit and } \delta \in S_1 \cap \alpha \Rightarrow p_\delta \in P_\kappa * P^*_\alpha\}$ is a club of κ^+ in V_2. It is enough to show that

⊗ $W \stackrel{\rm def}{=} \{\delta < \kappa^+ : \mathrm{cf}(\delta) = \kappa, \delta \in E$ and $p \Vdash_{P_\kappa * \underset{\sim}{P}^*_\delta}$ "$\underset{\sim}{S}_2 \cap \delta$ (which is a $(P_\kappa * \underset{\sim}{P}^*_{\delta^*})$-name) is a stationary subset of δ"$\}$ is a stationary subset of κ^+.

[Why? As then instead of guaranting $\underset{\sim}{S}_2 \cap \delta$ will continue to be stationary, we guess such name and related elementary submodel in some $\alpha < \kappa$ and in Q_α take care of $\underset{\sim}{S}_2 \cap \delta$ having a closed subset of order type ω_1.]
Let $G_R \subseteq R$ be the generic subset of R.

In V_1 we can find $\delta < \kappa^*$ such that $V_1 \vDash$ "$\kappa < \delta < \kappa^*$, δ is strongly inaccessible" and letting $R_\delta = \mathrm{Levy}(\kappa, <\delta) \lessdot R$, $G_{R_\delta} = G_R \cap R_\delta$, in $V[G_{R_\delta}]$ we have $\bar{Q}$, $\bar{Q}^* \restriction \delta$ hence $P_\kappa * \underset{\sim}{P}^*_\delta$, $\langle p_i : i \in S_1 \cap \delta\rangle$ and $S_1 \cap \delta$ is stationary (and of course $V[G_{R_\delta}] \vDash \delta = \kappa^+$). Also in $V_1[G_{R_\delta}]$, the forcing notion $P_\kappa * P^*_\delta$ satisfies the δ-c.c. (just as in $V_1[G_R]$, $P_\kappa * P^*_\kappa$ satisfies the κ^*-c.c.). So for a club of $i < \delta$, $i \in S_1 \cap \delta$ implies $p_i \Vdash$ "$\underset{\sim}{S}^* \stackrel{\rm def}{=} \{j \in S_1 \cap \delta : p_j \in \underset{\sim}{G}_{P_\kappa * \underset{\sim}{P}^*_\delta}\}$ is a stationary subset of δ^*" (as in Gitik, Shelah [GiSh:310]). Choose such $i(*)$. If this holds in $V_1[G_R]$ too, then we are done so assume towards contradiction that this fails, so moving back to V_0 for some $q \in P^-_\kappa$, $\underset{\sim}{r} \in R$ and $\underset{\sim}{p}$ we have $(q, \underset{\sim}{r}) \Vdash_{P_\kappa * \underset{\sim}{R}}$ "$p_{i(*)} \leq \underset{\sim}{p} \in P_\kappa * \underset{\sim}{P}^*_\delta$" and for some $P^-_\kappa * (\underset{\sim}{R} * (\underset{\sim}{P}_\kappa * \underset{\sim}{P}^*_\delta))$-name $\underset{\sim}{E}$

we have

$$(q, \underset{\sim}{r}, \underset{\sim}{p}) \Vdash_{P_\kappa^- * \underset{\sim}{R} * (\underset{\sim}{P}_\kappa * \underset{\sim}{P}_\delta^*)} \text{“}\underset{\sim}{E} \text{ is a club of } \delta \text{ disjoint to } \underset{\sim}{S}^*\text{”}.$$

Let $M \prec (H((\kappa^*)^{++}), \in)$ be such that $\|M\| = \kappa^+$, $H(\kappa^+) \subseteq M$, $x = \{\kappa, \kappa^*, \delta, \underset{\sim}{E}, q, (\underset{\sim}{r}, \underset{\sim}{p}), \underset{\sim}{R}/G_{R_\delta}, P_\kappa * \underset{\sim}{P}_\delta^*\} \subseteq M$ and ${}^\kappa M \subseteq M$.

Let (M', x') be isomorphic say by g to (M, x), M transitive, so $g : M \to M'$. We can find $\mathfrak{B} \prec (H(\kappa^{+6}), \in)$ to which x' and M' belong, such that letting $\theta = \mathfrak{B} \cap \kappa$ we have: θ is strongly inaccessible θ^{++}-supercompact, $\mathfrak{B} \approxeq (H(\theta^{+6}), \in)$ and $\mathbf{j}(\theta) = (\mathbf{j}_0(\theta), \mathbf{j}_1(\theta)) \in H(\theta^{++})$ is such that: $\mathbf{j}_0(\theta) = g(R_\delta)$ and $\mathbf{j}_1(\theta)$ is the $g(P_\kappa^- * R_\delta * \underset{\sim}{P}_\kappa)$-name i.e. $(P_\delta^- * Q_\delta^- * \underset{\sim}{P}_\delta)$-name of $g(P_\delta^*) * \text{club}(\underset{\sim}{S}^*)$. We can finish easily. $\square_{7.5}$

§8. Relativising to a Stationary Set

8.1 Definition. For a set $\mathbb{I}$ of ideals and stationary $\mathbf{W} \subseteq \omega_1$ we define when does a forcing notion P satisfies the ($\mathbb{I}$, $\mathbf{W}$)-condition (compare with 2.6). It means that there is a function F such that (letting $J_{\omega_1}^{\mathrm{bd}}$ =the bounded subsets of ω_1):

if $(T, \mathbf{I})$, f satisfies the following properties:

(∗) (a) $(T, \mathbf{I})$ is an $(\mathbb{I} \cup \{J_{\omega_1}^{\mathrm{bd}}\})$-tree.

(b) $f : T \to P$.

(c) $\nu \leq \eta$ implies $P \models f(\nu) \leq f(\eta)$.

(d) There are fronts $J_n (n < \omega)$ of T such that every member of J_{n+1} has a proper initial segment of J_n and:

(α) If $\eta \in J_n$ then $\langle \text{Suc}(\eta), \mathbf{I}_\eta, \langle f(\nu) : \nu \in \text{Suc}_T(\eta)\rangle\rangle = F(\eta, w[\eta], \langle f(\nu) : \nu \lhd \eta\rangle)$ (where $w[\eta] = \{\kappa : \eta{\restriction}\kappa \in \cup_n J_n\}$).

(β) $\cup J_n$ is the set of splitting points of $(T, \mathbf{I})$

(γ) If n is odd, $\eta \in J_n$ then $\mathbf{I}_\eta = J_{\omega_1}^{\mathrm{bd}}$.

(δ) If n is even, $\eta \in J_n$ then $\mathbf{I}_\eta \in \mathbb{I}$.

then

(**) *if* $(T, \mathbf{I}) \leq (T^*, \mathbf{I})$, $[\eta \in T^* \cap (\bigcup_{n<\omega} J_{2n}) \Rightarrow \mathrm{Suc}_{T^*}(\eta) \notin \mathbf{I}_\eta]$ and for some limit $\delta \in \mathbf{W}$ for every $\eta \in \lim T^*$, $\delta = \sup\{\eta(k) : \eta{\restriction}k \in \bigcup_{n<\omega} J_{2n+1}\}$ *then* for some $q \in P$ we have $q \Vdash$ "$(\exists\eta)[\eta \in \lim T^* \& \bigwedge_{k<\omega} f(\eta{\restriction}k) \in \underset{\sim}{G}_P]$"

8.2 Claim. 1)If P satisfies the $\mathbb{I}$-condition, e.g. P is $\aleph_2$-complete *then* P satisfies the $(\mathbb{I}, \omega_1)$ -condition.

2)If $\mathbf{W}_1 \subseteq \mathbf{W}_2 \subseteq \omega_1$, P satisfies the $(\mathbb{I}, \mathbf{W}_2)$-condition *then* p satisfies the $(\mathbb{I}, \mathbf{W}_1)$-condition.

3) If $\mathbf{W} \subseteq \omega_1$ is stationary, $\mathbb{I}$ is a family of $\aleph_2$-complete ideals, and the forcing notion P satisfies the $(\mathbb{I}, \mathbf{W})$-condition *then* forcing with P does not collapse $\aleph_1$ and preserves the stationarity of $\mathbf{W}$.

Proof. 1) If F witnesses "P satisfies the $\mathbb{I}$ -condition", define F' such that if $(T, \mathbf{I})$, f, $\langle J_n : n < \omega\rangle$ are as in Definition 8.1, then for $n = 2m$, $\eta \in J_n$:

$$\langle \mathrm{Suc}_T(\eta), \langle f(\nu) : \nu \in \mathrm{Suc}_T(\eta)\rangle\rangle = F'(\langle \eta, w[\eta], f(\nu) : \nu \trianglelefteq \eta\rangle\rangle)$$
$$= F(\langle \eta, \{\ell : \eta{\restriction}\ell \in \bigcup_{k<\omega)} J_{2k}\}, \quad \langle f(\nu) : \nu \trianglelefteq \eta\rangle\rangle)$$

This finishes the proof.

2) Trivial.

3) Suppose $p \in P$, $p \Vdash$ "$\underset{\sim}{C}$ is a club of ω_1". Choose $(T\ ,\mathbf{I})$, f, $\langle J_n : n < \omega\rangle$ as in Definition 8.1 such that:

(i) $J_n = \{\eta \in T : \ell g(\eta) = 2n\}$

(ii)If $\eta \in T$, $\ell g(\eta)$ odd then $\mathrm{Suc}_T(\eta) = \{\eta\hat{\ }\langle 0\rangle\}$ and for some α_η

(α) $\min\,(\omega_1 \cap \mathrm{Rang}(\eta)) < \alpha_\eta < \omega_1$

(β) $f(\eta\hat{\ }\langle 0\rangle) \Vdash_P$ "$\alpha_\eta \in \underset{\sim}{C}$"

(iii)$f(\langle\rangle) = p$

There is no problem in this. Let $T_0 = \{\eta \in T :$ *if* $k < \ell g(\eta)$, $k = 4m+2$, $\ell < k$, $\ell = 2n+1$ then $\alpha_{\eta{\restriction}\ell} < \eta(k)\rangle\}$. Clearly $(T, \mathbf{I}) \leq (T^0, \mathbf{I})$ and the requirement in (**) of Definition 8.1 holds. By XV 2.6 (no vicious circle! as it does not use any intermidiate material) there is a club C^* of ω_1 and for each $\delta \in C^*$, a tree T_δ such that:

(a)$(T^0, \mathbf{I}) \leq (T_\delta, \mathbf{I})$

(b)$\eta \in T_\delta$, $\eta \in \bigcup_n J_{2n} \Rightarrow \eta \in \mathrm{sp}(T_\delta, \mathbf{I})$.

(c)$\eta \in \lim T_\delta \Rightarrow \delta = \sup(\mathrm{Rang}(\eta) \cap \omega_1) \& \delta \notin \mathrm{Rang}(\eta)$.

This last statement holds also for branches of T_δ in extensions of the universe, being absolute. Choose $\delta \in W \cap C^*$, and apply Definition 8.1 to $(T_\delta, \mathbf{I})$ (standing for $T^\dagger$ there), and get q as there. Now $q \Vdash$ "$\{f(\underset{\sim}{\eta}\restriction\ell) : \ell < \omega\} \subseteq \underset{\sim}{G}_P$, $\underset{\sim}{\eta} \in \lim T$" (for some P-name $\underset{\sim}{\eta}$). In particular $q \Vdash$ "$p \in \underset{\sim}{G}_P$" (as $p_{\langle\rangle} = p$) so w.l.o.g. $p \leq q$. Also by (ii) (α), (iii) above $q \Vdash_P$ "$\sup(\mathrm{Rang}(\eta) \cap \omega_1) = \sup\{\alpha_{\eta\restriction(2n+1)} : n < \omega\}$". And so $q \Vdash$ "$\delta = \sup\{\alpha_{\eta\restriction(2n+1)} : n < \omega\}$". As $q \Vdash$ "$f(\underset{\sim}{\eta}\restriction(2n+2)) \in \underset{\sim}{G}_P$" also (see (β) of (ii)) we have $q \Vdash$ "$\alpha_{\underset{\sim}{\eta}\restriction(2n+1)} \in \underset{\sim}{C}$" hence $q \Vdash$ "$\delta \in \underset{\sim}{C}$".

As $\delta \in \mathbf{W}$ we finish. $\square_{8.2}$

8.3 Lemma. Let $\mathbf{W} \subseteq \omega_1$, be stationary. All the theorems on preservations of the $\mathbb{I}$-condition (in §5, §6) for $\aleph_2$-complete $\mathbb{I}$ hold for the $(\mathbb{I}, \mathbf{W})$- condition.

Proof. Same proof, sometimes using XV 2.6.

XII. Improper Forcing

§0. Introduction

In Chapter X we proved general theorems on semiproper forcing notions, and iterations. We apply them to iterations of several forcings. One of them, and an important one, is Namba forcing. But to show Namba forcing is semiproper, we need essentially that $\aleph_2$ was a large cardinal which has been collapsed to $\aleph_2$ (more exactly – a consequence of this on Galvin games). In XI we took great trouble to use a notion considerably more complicated than semiproperness which is satisfied by Namba forcing. However it was not clear whether all this is necessary as we do not exclude the possibility that Namba forcing is always semiproper, or at least some other forcing, fulfilling the main function of Namba forcing (i.e., changing the cofinality of $\aleph_2$ to ω without collapsing $\aleph_1$). But we prove in 2.2 here, that: there is such semiproper forcing, iff Namba forcing is semiproper, iff player II wins in an appropriate game $\eth(\{\aleph_1\}, \omega, \aleph_2)$ (a game similar to the game of choosing a decreasing sequence of positive sets (modulo appropriate filter, see X 4.10 (towards the end) and the divide and choose game, X 4.9, Galvin games) and, in 2.5, that this implies Chang's conjecture. In our game player I divide, played II choose but here it continue to choose more possibilities later. Now it is well known that Chang's conjecture implies $0^{\#}$ exists, so e.g., in ZFC we cannot prove the existence of such semiproper forcing. An amusing consequence is that if we collapse a measurable cardinal to κ by Levy-collapsing P, then Chang's conjecture holds as by X 6.13 in V^P player II wins the game mentioned above (Silver's original proof of the consistency of Chang conjecture uses a Ramsey cardinal, but he has first to force MA + $2^{\aleph_0} > \aleph_1$ and then use a more complicated collapsing).

In Sect. 1 we give the various variants of properness equivalent formations using games 1.1, 1.7; we also show how the preservation theorems work in this setting, thus getting alternative proofs of the preservation theorems for semiproper and proper, (1.8, 1.8A, 1.9). This alternative proof for properness was later and independently discovered by Gray in [Gr]. Related games have been investigated by Jech (which are like Galvin games, but using complete Boolean algebras) but his interests were different. Compared to [Sh:b] the order of the sections is inverted.

§1. Games and Properness

1.1 Theorem. A forcing notion P is proper *iff* player II has a winning strategy in the game $\mathrm{P}\eth^\omega(p, P)$, for every p, where:

1.2 Definition. In a play of the game $\mathrm{P}\eth^\alpha(p, P)$ (α a limit ordinal, $p \in P$) in the β-th move player I chooses a P-name $\underset{\sim}{\xi}_\beta$ of an ordinal, and player II chooses an ordinal ζ_β.

In the end after α moves player II wins if there is q such that $p \leq q \in P$ and $q \Vdash$ " for every $\beta < \alpha$ we have $\underset{\sim}{\xi}_\beta \in \{\zeta_{\beta+n} : n < \omega\}$", and player I wins otherwise.

1.3 Remarks.

1) Note that we can allow player II to choose countably many ordinals $\zeta_{\beta,\ell}$ ($\ell < \omega$) and demand $q \Vdash$ "$\underset{\sim}{\zeta}_\beta \in \{\xi_{\beta+n,\ell} : n, \ell < \omega\}$". Similarly player I can choose countably many P-names, and nothing is changed, i.e., the four variants of the definition, together satisfy (or do not satisfy) "player I (II) has a winning strategy".

2) Similarly for $\mathrm{P}\partial^{\omega}(p, P, \lambda)$ (see Definition 1.4(1)).

Proof of 1.1.

The "if" part.

Let λ be big enough, $N \prec (H(\lambda), \in)$ is countable, $\{P, p\} \in N, p \in P$. Then a winning strategy of $\mathrm{P}\partial^{\omega}(p, P)$ belongs to N. So there is a play of this game $\underset{\sim}{\xi}_n, \zeta_n (n < \omega)$ in which player II uses his winning strategy in choosing $\zeta_n \in N$ and every P-name $\underset{\sim}{\xi} \in N$ of an ordinal appears in $\{\underset{\sim}{\xi}_n : n < \omega\}$ and each $\underset{\sim}{\xi}_n$ belongs to N. So clearly $\zeta_n \in N$ for every n.

So there is q, witnessing the victory of II, i.e., $p \leq q \in P$, $q \Vdash$ "$\underset{\sim}{\xi}_n \in \{\zeta_m : m < \omega\}$", (for every n), but $\zeta_m \in N$, so $q \Vdash$ "$\underset{\sim}{\xi}_n \in N$ for every n". As $\{\underset{\sim}{\xi}_n : n < \omega\}$ lists all P-names of ordinals which belong to N, q is (N, P)-generic; and $q \geq p$, so we finish.

The "only" if part.

For λ big enough, expand $(H(\lambda), \in)$ by Skolem functions and get a model M^* and we shall describe a strategy for II: If player I has chosen up to now $\underset{\sim}{\xi}_0, \dots, \underset{\sim}{\xi}_n$, let N_n be the Skolem hull of $\{p, P\} \bigcup \{\underset{\sim}{\xi}_0, \dots, \underset{\sim}{\xi}_n\}$ in M^*, and $\{\zeta_{n,\ell} : \ell < \omega\}$ will be the set of ordinals which belong to N_n (remember Remark 1.3).

Suppose $\underset{\sim}{\xi}_0, \zeta_{0,\ell}(\ell < \omega), \underset{\sim}{\xi}_1, \zeta_{1,\ell}(\ell < \omega), \dots$ is a play in which player II uses his strategy (see remark 1.3(1)). Why does he win? Clearly $N = \bigcup_n N_n$, which is the Skolem hull in M^* of $\{p, P\} \bigcup \{\underset{\sim}{\xi}_\ell : \ell < \omega\}$, is an elementary submodel of M^*, (similarly for the reducts) so there is $q \geq p$ which is (N, P)-generic, so as $\underset{\sim}{\xi}_n \in N$ we have $q \Vdash$ "$\underset{\sim}{\xi}_n \in N$" but the set of ordinals of N is $\{\zeta_{n,\ell} : n, \ell < \omega\}$. So we finish. $\square_{1.1}$

1.4 Definition.

(1) $\mathrm{P}\partial^{\alpha}(p, P, \lambda)$ is defined similarly, but $\underset{\sim}{\xi}_\beta$ are P-names of ordinals $< \lambda$. For $\lambda = \infty$ (or just $\lambda > |P|$) we get $\mathrm{P}\partial^{\alpha}(p, P)$.

(2) For a set S of cardinals, the game $\mathrm{P}\partial^{\alpha}(p, P, S)$ is defined as follows: in the β-th move player I chooses $\lambda_\beta \in S$, and a P-name $\underset{\sim}{\xi}_p$, and player II

chooses for $\gamma \le \beta$ subsets $A_{\beta,\gamma}$ of λ_γ, of power $< \lambda_\gamma (A_{\beta,\gamma} \in V)$ (for λ_γ regular these are w.l.o.g. initial segments of λ_γ).

In the end player II wins if there is $q \in P$, $q \ge p$ such that $q \Vdash$ "$\underset{\sim}{\xi}_\gamma \in \bigcup_{\gamma \le \beta < \gamma+\omega} A_{\beta,\gamma}$ for $\gamma < \alpha$". The definition for a P-name $\underset{\sim}{S}$ is similar, but player I chooses λ_β (not a P-name), and in the end, q forces only that: "if $\lambda_\beta \in \underset{\sim}{S}$ then ..."

If not said otherwise we restrict ourselves to sets of regular cardinals in V.

1.4A Remark. Note that 1.4(1) is in fact a special case of 1.4(2) when we identify λ with $S = \{\mu : \aleph_1 \le \mu \le \lambda, \mu \text{ regular}\}$, i.e. a winning strategy for $\mathrm{P}\eth^\alpha(p, P, \lambda)$ can, in a canonical way, be translated to a winning strategy in $\mathrm{P}\eth^\alpha(p, P, S)$, and conversely.

1.5 Definition.

(1) $\mathrm{P}\eth^\alpha_\ell(p, P, \lambda)$ is defined similarly for $\ell = 0, 1, 2$, but:

for $\ell = 0$ it is exactly $\mathrm{P}\eth^\alpha(p, P, \lambda)$;

for $\ell = 1$ player I chooses $\underset{\sim}{\xi}_\beta$, a P-name of an ordinal $< \lambda$. Player II chooses countably many ordinals ($\zeta_{\beta,n}$ for $n < \omega$) and player II wins if $q \Vdash$ "$\underset{\sim}{\xi}_\beta \in \{\zeta_{\beta,n} : n < \omega\}$ for each $\beta < \alpha$" for some $q \ge p$

for $\ell = 2$ player II chooses $\aleph_0$ P-names ($\underset{\sim}{\xi}_{\beta,n}$ for $n < \omega$) of ordinals $< \lambda$ and player II chooses $\aleph_0$ ordinals $\zeta_{\beta,n}$ for $n < \omega$; and player II wins if $q \Vdash$ "$\underset{\sim}{\xi}_{\beta,n} \in \{\zeta_{\beta,\ell} : \ell < \omega\}$ for every β, n" for some $q \ge p$.

(2) The games $\mathrm{P}\eth^\alpha_\ell(p, P, S)$ ($\ell = 0, 1, 2$, S a set of cardinals of V) are similar to $\mathrm{P}\eth^\alpha(p, P, S)$;

$$\mathrm{P}\eth^\alpha_0(p, P, S) = \mathrm{P}\eth^\alpha(p, P, S)$$

In $\mathrm{P}\eth^\alpha_1$, (p, P, S), player I chooses (in move β) a cardinal $\lambda_\beta \in S$ and a name of an ordinal $\underset{\sim}{\xi}_\beta < \lambda_\beta$. Player II plays a set $A_\beta \subseteq \lambda_\beta, |A_\beta| < \lambda_\beta$, $A_\beta \in V$ (if $V \models$ "λ regular" w.l.o.g. $A_\beta = \underset{\sim}{\xi}_\beta$). After α moves II wins if he can find a condition $q \ge p$ such that $q \Vdash$ "$(\forall \beta < \alpha) \underset{\sim}{\xi}_\beta \in A_\beta$".

In $\mathrm{P}\partial_2^\alpha(p,P,S)$, player I chooses $\lambda_\beta \in S$ and $\aleph_0$ names $\underset{\sim}{\xi}_{\beta,n}$ of ordinals $< \lambda_\beta$, player II chooses $A_\beta \subseteq \lambda_\beta$, as above and in the end player II has to find a condition $q \geq p$ such that $q \Vdash$ "$\forall\beta < \alpha\, \forall n < \omega\, (\underset{\sim}{\xi}_{\beta,n} \in A_\beta)$".

Similarly, we can define $\mathrm{P}\partial_\ell^\alpha(p,P,\underset{\sim}{S})$, where $\underset{\sim}{S}$ is a name of a set of regular cardinals of V.

1.6 Claim.

(1) Player II wins $\mathrm{P}\partial_1^\alpha(p,P,\lambda)$ iff he wins $\mathrm{P}\partial_2^\alpha(p,P,\lambda)$ provided that $\lambda = \aleph_1$ or $\lambda^{\aleph_0} = \lambda$ or at least there is in V a family $\{A_i : i < \lambda\}$ of countable subsets of λ such that in V $(\forall A \subseteq \lambda)(\exists i)\ (|A| \leq \aleph_0 \rightarrow A \subseteq A_i)$.

(2) Player II wins $\mathrm{P}\partial_1^\alpha(p,P,\underset{\sim}{S})$ iff he wins $\mathrm{P}\partial_2^\alpha(p,P,\underset{\sim}{S})$ (when $\underset{\sim}{S}$ is a P-name of a set of regular cardinals $> \aleph_0$).

1.6A Remark. Note that if player II wins one of the games $\mathrm{P}\partial_\ell^\alpha(p,P,\lambda)$ then p forces $(\forall A \in V^P)(\exists B \in V)\ [A \subseteq \lambda \wedge |A| \leq \aleph_0 \Rightarrow A \subseteq B \wedge |B| = \aleph_0]$.

Proof. (2) The "if" part is trivial.

For the "only if" part: Note that by the assumption $\Vdash_P$ "$\lambda \in \underset{\sim}{S} \rightarrow \lambda = \mathrm{cf}^V(\lambda) > \aleph_0$" and that as player II wins $\mathrm{P}\partial_1^\alpha(p,P,S)$ we have: for any λ, $\Vdash_P$ "$\lambda \in \underset{\sim}{S} \rightarrow \mathrm{cf}(\lambda) > \aleph_0$"). Now when player I chooses λ_β and $\{\underset{\sim}{\xi}_{\beta,n} : n < \omega\}$, player II "pretends" player I has chosen $\mathrm{Sup}\{\underset{\sim}{\xi}_{\beta,n} : n < \omega\}$, and chooses a suitable initial segment of λ_β. So he translates a play of $\mathrm{P}\partial_2^\alpha(p,P,s)$ to one of "$\mathrm{P}\partial_1^\alpha(p,P,s)$".

(1) Really the same as the proof of (2): the "if" part is trivial and for the "only if" part, player II let $\underset{\sim}{i}_\beta = \mathrm{Min}\{i : \{\underset{\sim}{\xi}_{\beta,n} : n < \omega\} \subseteq A_i\}$ where $\{A_i : i < \lambda\}$ is as mentioned in 1) (such a family exists in V if $\lambda = \aleph_1$ $(A_i = i)$ or $\lambda^{\aleph_0} = \lambda$ $(\{A_i : i < \lambda\} = \{A \subseteq \lambda : |A| \leq \aleph_0\})$ and as in the proof of part (2) above, the forcing preserves this property).

Now because player II wins $\mathrm{P}\partial_1^\alpha(p,P,\lambda)$, he can find in the β-th move (of a play of $\mathrm{P}\partial_2^\alpha(p,P,\lambda)$) ordinals $i_{\beta,n} (n < \omega)$, such that at the end of the game $q \Vdash$"$\exists n(\underset{\sim}{i}_\beta = i_{\beta,n})$", and so his move in $\mathrm{P}\partial_2^\alpha(p,P,\lambda)$ will be the countable set $\bigcup_n A_{i_{\beta,n}}$. (More formally, fixing a winning strategy F_1 for player

II in $(\mathrm{P}\partial_2^\alpha(p, P, \lambda)$, we shall describe a winning strategy F_2 of player II in $\mathrm{P}\partial_2^\alpha(p, P, \lambda)$. During the play he simulate a play of $\mathrm{P}\partial_1^\alpha(p, P, \lambda)$ by playing in the β-th move $\underset{\sim}{i}_\beta$.) $\square_{1.6}$

1.7 Theorem.

(1) P is α-proper *iff* player II wins $\mathrm{P}\partial^{\omega(1+\alpha)}(p, P; \infty)$ for every p.

(2) P is $(\alpha, 1)$-proper *iff* player II wins $\mathrm{P}\partial_2^\alpha(p, P; \infty)$ for every p.

(3) A forcing notion P is semiproper iff player II has a winning strategy in $\mathrm{P}\partial_0^\omega(p, P, \aleph_1)$ for every $p \in P$.

(4) A forcing notion P is $\underset{\sim}{S}$-semiproper *iff* player II has a winning strategy in $\mathrm{P}\partial_0^\omega(p, P, \underset{\sim}{S})$ for every $p \in P$.

Proof. Similar to 2.1, for (2) use V 2.5. $\square_{1.7}$

1.7A Remark.

(1) Call a forcing notion P "$\underset{\sim}{S}$-semi-α-proper" (for $\underset{\sim}{S}$ a P-name of a set of cardinals of V), if for all large enough cardinals λ, and all $\langle N_i : i \leq \alpha \rangle \in SQS_\alpha^0(\lambda)$ (see V2.1) for which $P, \underset{\sim}{S} \in N_0$:

for all $p \in P \cap N_0$ there is $q \geq p$, such that for every $i \leq \alpha$ we have $q \Vdash_P$ " if $\mu \in \underset{\sim}{S}$ then $\sup(\mu \cap N_i[\underset{\sim}{G}]) = \sup(\mu \cap N_i)$".

Then we have: P is $\underset{\sim}{S}$-semi-α-proper iff player II wins $\mathrm{P}\partial^{\omega(1+\alpha)}(p, P, \underset{\sim}{S})$.

(2) For countable ordinals to demand $\underset{\sim}{\xi}_\beta \in \{\zeta_{\beta,n} : n < \omega\}$" is equivalent to demanding $\underset{\sim}{\xi}_{\beta,n} < \mathrm{Sup}\{\zeta_{\beta,n} : n < \omega\}$, etc.

1.8 Theorem. The property "player II wins $\mathrm{P}\partial_\ell^\alpha(p, P, \lambda)$ for every $p \in P$" is preserved by CS iterations $\langle P_i, \underset{\sim}{Q}_i : i < \alpha(*)\rangle$, for $\ell = 0, 2$, $\alpha < \omega_1$, if $|\alpha(*)| \leq \lambda$.

1.8A Remark.

(1) So we get here an alternative proof of the preservation of properness and $(\alpha, 1)$-properness: we could give such proofs for other theorems as well.

(2) The situation is similar with $\underset{\sim}{S}$ instead of λ, and easier for RCS iteration by X 2.5 (stated in 1.9).

Proof. Let $\langle P_i, \underset{\sim}{Q}_i : i < \alpha(*)\rangle$ be a countable support iteration. Let us consider a game $\mathrm{P}\eth_0^\omega(p, P_{\alpha(*)}, \lambda)$ (the others are similar). By the hypothesis for each $i < \alpha(*)$, player II has, for every $\underset{\sim}{q} \in \underset{\sim}{Q}_i$, a winning strategy $\underset{\sim}{st}_i = \underset{\sim}{st}_i(\underset{\sim}{q})$ in the game $\mathrm{P}\eth_0^\omega(\underset{\sim}{q}, \underset{\sim}{Q}_i, \lambda)$ where $\underset{\sim}{q}$, $\underset{\sim}{Q}_i$, $\underset{\sim}{st}_i$ are P_i-names.

Without loss of generality for $\underset{\sim}{Q}_i$ we use the version of the games in which player I plays a countable set (of names of countable ordinals) at each stage, and player II answers with a single ordinal. But for $\mathrm{P}\eth_0^\omega(p, P_{\alpha(*)}, \lambda)$ we use the version where both players play singletons, this is legitimate by remark 1.3(1). (The remark at the end of the proof will explain why it is more convenient to let player I choose a countable set of names in the games $\mathrm{P}\eth_0^\omega(q, \underset{\sim}{Q}_i, \lambda)$).

Now player II plays as follows: in the n-th move he will define $w_n, p_n, t_i^n (i \in w_n)$ such that:

(1) $p_n \in P_{\alpha(*)}, p \le p_0, p_n \le p_{n+1}$,

(2) w_n is a finite subset of $\mathrm{Dom}(p_n)$, $w_n \subseteq w_{n+1}$, and if $\mathrm{Dom}(p_n) = \{i_{n,k} : k < \omega\}$ and w.l.o.g. $i_{n,0} = 0$ then $w_n = \{i_{m,k} : m < n, k < n\}$, (so eventually $\bigcup_{n<\omega} w_n = \bigcup_n \mathrm{Dom}(p_n)$ and w_n depends just on $\langle p_\ell : \ell < n\rangle$)

(3) $p_{n-1}\restriction w_n = p_n \restriction w_n$ for $n > 0$.

(4) For $i \in w_n$, let $n(i) = \mathrm{Min}\{m : i \in w_m\}$, and $t_n^i = \langle\langle \Gamma_k^i, \underset{\sim}{\zeta}_k^i\rangle : n(i) \le k \le n\rangle$ is such that Γ_k^i is a countable set of P_i-names of $\underset{\sim}{Q}_i$-names of ordinals $< \lambda$ and $\underset{\sim}{\zeta}_k^i$ a P_i-name of an ordinal, and $p_n\restriction i \Vdash_{P_i}$ "t_n^i is an initial segment of a play of $\mathrm{P}\eth_\ell^\alpha(p_{n(i)}(i), \underset{\sim}{Q}_i, \lambda)$, in which player II uses the strategy $\underset{\sim}{st}_i = \underset{\sim}{st}_i(p_{n(i)}(i))$".

(5) In the zero move (as $w_0 = \emptyset$), player I chooses a $P_{\alpha(*)}$-name of an ordinal $\underset{\sim}{\xi}_0$, and player II chooses $p_0 \ge p$, $p_0 \Vdash_{P_{\alpha(*)}}$ "$\underset{\sim}{\xi}_0 = \zeta_0$", and play the ordinal ζ_0.

(6) In the n-th move, let player I play $\underset{\sim}{\xi}_n$, a $P_{\alpha(*)}$-name of an ordinal $< \lambda$. Let $\langle j_n(m) : m < l_n\rangle$ enumerate w_n in increasing order (so $j_n(0) = 0$) as $i_{n,0} = 0$. Let $j_n(l_n) = \alpha(*)$, and let $\underset{\sim}{\xi}_n^{\alpha(*)} = \underset{\sim}{\xi}_n^{j_n(l_n)} = \underset{\sim}{\xi}_n$.

By downward induction on m ($l_n > m \ge 0$), player II will define $p_{n,m}, \Gamma_n^{j_n(m)}, \underset{\sim}{\zeta}_n^{j_n(m)}$ as follows:

A) Given $\underset{\sim}{\zeta}_n^{j_n(m+1)}$ (for $m = l_n - 1$ this is $\underset{\sim}{\xi}_n$), a $P_{j_n(m+1)}$-name of an ordinal, he can find a condition $p_{n,m}$ (but see below) of the forcing notion $P_{j_n(m+1)}/P_{j_n(m)+1}$ (in the universe $V^{P_{j_n(m)+1}}$); $p_n \restriction [j_n(m)+1, j(m+1)) \leq p_{n,m}$, such that $p_{n,m}$ decides $\underset{\sim}{\zeta}_n^{j_n(m+1)}$ up to a $P_{j_n(m)+1}$-name. So we have the freedom to choose a countable set $\Gamma_n^{j_n(m)}$ of $P_{j_n(m)+1}$-names (of ordinals $< \lambda$, but see below) such that

$$p_{n,m} \Vdash_{P_{j_n(m+1)}/P_{j_n(m+1)}} \text{“}\underset{\sim}{\zeta}_\ell^{j_n(m)} \in \Gamma_n^{j_n}(m)\text{”}.$$

B) Once $\Gamma_n^{j_n(m)}$ is defined, demand (4) yields a $P_{j_n(m)}$-name $\underset{\sim}{\zeta}_n^{j_n(m)}$.

Finally, player II plays $\zeta_n = \zeta_n^0 = \zeta_n^{j_n(0)}$ (which is a P_0-name, hence a real ordinal) and define p_{n+1} by: for $j \in w_n$, $p_{n+1}(j) = p_n(j)$, and $p_{n+1}\restriction[j_n(m)+1, j(m+1)) = p_{n,m}$.

This is easily done and in the end, as for each $i \in \bigcup_{n<\omega} w_n$, player II has simulated a play of $\mathrm{P}\eth_0^\omega(p_{n(i)}(i), \underset{\sim}{Q}_i, \lambda)$ in which the second player uses his winning strategy st_i (the union of $t_n^i (n < \omega)$), there is a P_i-name $\underset{\sim}{q}(i)$, such that $\Vdash_{P_i}$ “$p_{n(i)}(i) \leq \underset{\sim}{q}(i)$”, and $\Vdash_{P_i} [\underset{\sim}{q}(i) \Vdash_{Q_i}$ “$\bigcup\{\Gamma_m : n(i) \leq n < \omega\} \subseteq \{\underset{\sim}{\xi}_m^i : m < \omega\}$”$]$.

It is clear that q (i.e. $\mathrm{Dom}(q) = \bigcup_{n<\omega} w_n, q(i)$ defined above) belongs to $P_{\alpha(*)}$. We have to prove $q \Vdash$ “$\underset{\sim}{\xi}_\ell \in \{\zeta_n^0 : n < \omega\}$”. Let $r \geq q$ be such that $r \Vdash$ “$\xi_\ell = \xi^*$”.

Note first that $q \geq p_n$ for every $n < \omega$.

We prove by induction on $i \in \bigcup_{n<\omega} w_n$ that if $\underset{\sim}{\xi} \in \cup\{\Gamma_k^j : j < i, n(j) \leq k < \omega\}$ then $q \restriction i \Vdash_{P_i}$ “$\underset{\sim}{\xi} \in \{\zeta_n^0 : n < \omega\}$”. This suffices as for each n we have $q \Vdash \underset{\sim}{\xi}_n \in \bigcup_{i \in w_n} \Gamma_i^n$. For i limit-trivial; for i successor - use the choice of $p_{n,m}$ and the winning of the second player in $\mathrm{P}\eth_0((p_{n(i)}, \underset{\sim}{Q}_i, \lambda)$.

At first glance we get it too cheaply.

But there is a delicate point – the choice of $p_{n,m}$; it is really not a condition of $P_{j(m+1)}$ or $P_{\alpha(*)}$ but a $P_{j(m)+1}$-name for it. The delicate point is that its domain is a name, whereas it is required to be a set. However, it is not fatal as long as at least q would be a real condition, for which it is not necessary to really know $\mathrm{Dom}(p_{n,m})$, just to find a countable set including it. So into the list of names player II is manipulating he has to add names of the members of $\mathrm{Dom}(p_{n,m})$. So it is enough to ask that $\bigcup_{n<\omega} \Gamma^0_n \cap \alpha^* \subseteq \bigcup_{n<\omega} w_n$ and $\mathrm{Dom}(p_{n,m}) \subseteq \Gamma^{j_n(m)}_n$ i.e., there are $P_{j_n(m)}$-names $\underset{\sim}{\varepsilon}^{j_n(m)}_{n,\ell}(\ell < \omega)$ such that $\Vdash_{P_{j(m,n)+1}}$ "$\mathrm{Dom}(p_{n,m}) = \{\underset{\sim}{\varepsilon}^{j_n(m)}_{n,\ell} : \ell < \omega\}$" and we let $\underset{\sim}{\varepsilon}^{j_n(m)}_{n,\ell} \in \Gamma^{j_n(m)}_n$. But Player II can manipulate only names of ordinals $< \lambda$; this is why $|\alpha(*)| = \lambda$ was required so in the end we know $\mathrm{Dom}(p_{n,m}) \subseteq \{\xi_k : k < \omega\}$, and there are no problems.

* * *

The cases $\ell = 0$, $\alpha > \omega$ are proved by induction on α.

For $\ell = 2$, we already know from what we prove that player II wins $\mathrm{P}\partial^\omega_0(p, P_{\alpha(*)}; \lambda)$ for every $p \in P_{\alpha(*)}$, and the proof is simple. $\square_{1.8}$

We leave to the reader the easier theorem (by now):

1.9 Theorem.

1) RCS iterations preserves the following property of forcing notions:

   for every $p \in P$, Player II wins $PG^\omega_0(p, P, \aleph_1)$.

2) Similarly for $\mathrm{P}\partial^\alpha_0(p, P, \aleph_1)$, and variants with $\underset{\sim}{S}$.

§2. When Is Namba Forcing Semiproper, Chang's Conjecture and Games

2.1 Definition. For a filter D on a set I, and a set S of regular cardinals and ordinal α we define a game $\partial(S, \alpha, D)$: the game lasts α moves, in the β-th move player I chooses a cardinal $\lambda_\beta \in S$ and a function F_β from I to λ_β.

Then player II has to choose an ordinal $i_\beta < \lambda_\beta$. In the end of the game, player II wins if

$\{t \in I$: for every $\lambda \in S$, if for some β, $\lambda = \lambda_\beta$ then $\mathrm{Sup}\{F_\beta(t) : \lambda_\beta = \lambda\}$ is $\leq \mathrm{Sup}\{i_\beta : \lambda_\beta = \lambda\}\} \not\equiv \emptyset \bmod D$; also player I has to choose each λ_β infinitely

often otherwise player II wins (though not every $\lambda \in S$ appears).
If $I = \lambda$, $D = D^{cb}_\lambda$ (the filter of cobounded subsets of λ) we replace D by λ.

Remark.

(1) This is close to the games in Definition X 4.9, X 4.10 (toward end) and reference there but here we do not choose immediately the bound, we increase it later.

(2) We do not list the obvious monotonicity properties.

2.2 Theorem.

Namba forcing is $\{\aleph_1\}$-semiproper *iff* player II wins $\partial(\{\aleph_1\}, \omega, \aleph_2)$ *iff* there is a $\{\aleph_1\}$-semiproper forcing P changing the cofinality of $\aleph_2$ to $\aleph_0$.

2.2A Remark.

It does not matter whether we use Nm or Nm' (see XI 4.1, X 4.4), so we deal with the somewhat harder case: Nm'.

Proof. We use the criterion for semiproperness from 1.7(3)

third condition implies second condition

We assume P is $\{\aleph_1\}$-semiproper; and so (by 1.7(3)) we can choose a winning strategy for player II in $\mathrm{P}\partial = \mathrm{P}\partial^\omega(p_0, P, \{\aleph_1\})$ (in the variant where both players choose countable sets) and we shall call its players $I_{\mathrm{P}\partial}, II_{\mathrm{P}\partial}$ for clarity. Now we describe a winning strategy for player II in $\partial(\{\aleph_1\}, \omega, \aleph_2)$. Player II will simulate for this a play of $\mathrm{P}\partial$, in which $II_{\mathrm{P}\partial}$ uses his winning strategy and $I_{\mathrm{P}\partial}$ is played by him (player II in $\partial(\{\aleph_1\}, \omega, \aleph_2)$). Let $\underset{\sim}{\xi}_n$ be the P-name of the n-th element in an ω-sequence which is unbounded in $\aleph_2$.

In the n-th move let player I choose $F_n : \aleph_2 \to \aleph_1$, player II will play for $I_{\mathrm{P}\partial}$ the P-names $F_n(\underset{\sim}{\xi}_\ell)$ ($\ell < \omega$), (see 1.3(2)), so he knows the ordinals $\alpha_{n,\ell} < \omega_1$ (for $\ell < \omega$) which $II_{\mathrm{P}\partial}$ plays (according to his strategy). Now player II returns to his play and makes the move $\alpha_n = \mathrm{Sup}\{\alpha_{n,\ell} + 1 : \ell < \omega\} < \omega_1$.

In the end there is a condition $q \in P$, $q \geq p_0$ such that $q \Vdash$ "$F_n(\underset{\sim}{\xi}_\ell) < \bigcup_m \alpha_m$" for n, $\ell < \omega$. Let for each $\ell < \omega$:

$$A_\ell = \{\alpha < \omega_2 : q \text{ does not force } \underset{\sim}{\xi}_\ell \neq \alpha, \text{ i.e.,}$$
$$\text{there is } r \in P \text{ such that } r \geq q \text{ and } r \Vdash \text{“}\underset{\sim}{\xi}_\ell = \alpha\text{”}\}$$

By our choice of $\underset{\sim}{\xi}_\ell (\ell < \omega)$, for some ℓ, A_ℓ is an unbounded subset of ω_2 (otherwise $\bigcup_{\ell<\omega} A_\ell$ would be bounded, as in V we have $\mathrm{cf}(\omega_2) = \aleph_2 > \aleph_0$ so $q \Vdash$ "$\{\underset{\sim}{\xi}_\ell : \ell < \omega\} \subseteq \bigcup\limits_{\ell<\omega} A_\ell$" give a contradiction to the choice of $\langle \underset{\sim}{\xi}_\ell : \ell < \omega \rangle$). Now clearly $\alpha \in A_\ell$ implies $F_n(\alpha) < \bigcup_m \alpha_m$; hence A_ℓ is as required.

First implies third condition

Trivial

Second condition implies first condition

The proof is similar to that of X 4.12(3), it is similar for Nm and Nm', and we present the one for Nm'.

We again use the games and prove player II wins $\mathrm{P}\partial_0^\omega(p, P, \aleph_1\}$ for every $p \in P$, $P = Nm' = Nm'(D^{cb}_{\aleph_2})$ (see X Definition 4.4(4) (we can in fact replace $\aleph_1$ by $\{\lambda : \lambda$ a regular cardinal $\neq \aleph_2\}$)). For notational simplicity assume p's trunk is $\langle\rangle$ i.e. $(\forall \eta \in p)(\exists^{\aleph_2} i)(\eta\hat{\ }\langle i \rangle \in p)$.

During the play, in the n-th move player I chooses a P-name of a countable ordinal $\underset{\sim}{\xi}_n$, and player II will choose $\xi_n < \omega_1$. On the side, player II in stage n also chooses a condition $p_n \in P$ and a function F_n from $p_n \cap ({}^{n\geq}(\omega_2))$ to ω_1 such that:

(a) $p \leq p_0, p_n \leq p_{n+1}, p_n \cap ({}^{n\geq}(\omega_2)) = p_{n+1} \cap ({}^{n\geq}(\omega_2))$, and the trunk of each p_n is $\langle\rangle$.

(b) For $\eta \in p_{n+1} \cap ({}^{n}(\omega_2))$ and $\ell < n$, either $(p_{n+1})_{[\eta]} \Vdash_P$ "$\underset{\sim}{\xi}_\ell \leq F_n(\eta)$" *or* there is no q, $(p_n)_{[\eta]} \leq q$, with trunk η and $\zeta < \omega_1$, such that $q \Vdash_P$ "$\underset{\sim}{\xi}_\ell \leq \zeta$" (remember $q_{[\eta]} = \{\nu \in q : \nu \trianglelefteq \eta$ or $\eta \trianglelefteq \nu\}$.

(c) For $\eta \in p_n \cap ({}^{m}(\omega_2))$ and $m < n$ let $F_{n,\eta} : \omega_2 \to \omega_1$ be defined by $F_{n,\eta}(i) = F_n(\eta\hat{\ }\langle i \rangle)$; and we demand that $F_n(\eta)$ is defined such that

$\langle\langle F_{m+\ell,\eta}, F_{m+\ell}(\eta)\rangle : \ell < n-m\rangle$ is an initial segment of a play of $\Game(\{\aleph_2\},\omega,\aleph_1)$ in which the second player uses his winning strategy (i.e. one we choose a priori).

(d) At last player II actually plays $\zeta_n = F_n(\langle\rangle)$.

There is no problem for player II to use this strategy: Stage 0, is trivial, *stage* $n+1$: for each $\eta \in p_n \cap ({}^n\omega_2)$ define $q^\ell_{\eta,n}$ by induction on $\ell \le n+1$: $q^0_{\eta,n} = (p_n)_{[\eta]}$, $q^{\ell+1}_{\eta,n} \ge q^\ell_{\eta,n}$, has trunk η and either it forces $\underset{\sim}{\xi}_\ell \le \zeta$ for some $\zeta < \omega_1$ which we call $\zeta^\ell_{\eta,n}$ or there is no such q (satisfying the two previous conditions). Then

$$p_{n+1} = \bigcup\{q^{n+1}_{\eta,n} : \eta \in p_n \cap ({}^n(\omega_2))\}.$$

and for $\eta \in p_n \cap ({}^n(\omega_2))$.

$F_{n+1}(\eta) = \text{Max}\{\zeta^\ell_{\eta,n} : \ell \le n+1, \zeta^\ell_{\eta,n} \text{ well defined}\} \bigcup \{0\}$.

Now he defines $F_{n+1}(\eta)$ for $\eta \in p_n \cap ({}^{n>}(\omega_2)) = p_{n+1} \cap ({}^{n>}(\omega_2))$ by downward induction on $lg(\eta)$, using (c) (i.e., the winning strategy of the second player in $\Game(\{\aleph_2\},\omega,\aleph_1)$.)

In the end player II has to provide the suitable condition $q \ge p$; we define $q \cap {}^n\omega_2$ by induction on n:

$\langle\rangle \in q$, and if we have decided that $\eta \in q$ then: $\eta\char`^\langle i\rangle \in q$ *iff* $\text{Sup}\{F_n(\eta\char`^\langle i\rangle) : n < \omega, n \ge lg(\eta\char`^\langle i\rangle)\} \le \text{Sup}\{F_n(\eta) : n < \omega,\ n \ge lg(\eta)\}$ (and $\eta\char`^\langle i\rangle \in p_{\ell g(\eta)+1}$ of course).

By (c), q will be a condition; and by (b) it forces $\underset{\sim}{\xi}_n < \bigcup_{n<\omega}\zeta_n$: otherwise there is $r \ge q$, $r \Vdash_P$ "$\underset{\sim}{\xi}_n \ge \bigcup_{n<\omega}\zeta_n$". Let η be the trunk of r, and then using (b) we get a contradiction. $\square_{2.2}$

2.3 Definition. More generally, if we have a family $\mathbb{D}$ of filters, a set S of regular cardinals, and a relation $K \subseteq S \times \mathbb{D}$, we define the game $\Game(K,\alpha)$ as follows: In the βth move ($\beta < \alpha$), player I chooses a pair $(\lambda_\beta, D_\beta) \in K$ and a function F'_β from $\cup D_\beta$ to λ_β. Player II chooses an ordinal $i_\beta < \lambda_\beta$. In the end, player II wins if for each $D \in \mathbb{D}$ the set $\{t \in \cup D$: for every λ with $(\lambda, D) \in K$,

if for some $\beta, (\lambda, D) = (\lambda_\beta, D_\beta)$, then Sup $\{F_\beta(t) : (\lambda, D) = (\lambda_\beta, D_\beta)\} \le$ Sup $\{i_\beta : (\lambda, D) = (\lambda_\beta, D_\beta)\}\}$ is $\ne \emptyset$ mod D. (Also, player I has to choose each

λ_β infinitely often, otherwise II wins but some $\lambda \in S$ may never be chosen). Again, we may write κ for D^{cb}_κ.

2.4 Theorem. For a countable set S of regular cardinals, and a countable set $\mathbb{D}$ of $\aleph_1$-complete filters, player II wins $\partial(S \times \mathbb{D}, \omega)$ iff there is an S-semiproper forcing notion P, such that for all $D \in \mathbb{D}$

$$\Vdash_P \text{“}\exists w \subseteq \cup D[w \text{ countable, } w \neq \emptyset \bmod D, \text{ that is, it is not disjoint to any } A \in D (\text{so } A \in V)]\text{”}.$$

(If $D = D^{cb}_\lambda$, then the above condition is clearly equivalent to $\Vdash_P$ “ $\mathrm{cf}(\lambda) = \aleph_0$.”)

Proof. The proof is similar to the proof of 2.2. To show the “only if” part, let $\mathbb{D} = \{D_n : n < \omega\}$, where each D_n occurs infinitely often. Let $P = Nm'(T, \mathfrak{D})$, where $T = \{\eta : \eta \text{ a finite sequence, } (\forall k)[k \leq \ell g\, \eta \Rightarrow \eta\restriction k \in \mathrm{Dom}(D_k)]\}$, and for each $\eta \in T$, let $\mathfrak{D}_\eta = \{\{\eta\char`\^\langle i\rangle : i \in A\} : A \in D_{\ell g(\eta)}\}$. $\square_{2.4}$

2.5 Theorem.

(1) If player II wins $\partial = \partial(\{\aleph_1\}, \omega_1, \aleph_2)$, *then* Chang's conjecture holds.

(2) Moreover if e.g. $\chi > 2^{\aleph_2}$, M^* an expansion of $((H(\chi), \in))$ by Skolem functions, $N \prec M^*$ is countable, *then* for arbitrarily large $\alpha < \aleph_2$, there is $N_\alpha, N \prec N_\alpha \prec M^*$, $\alpha \in N_\alpha$, and N_α, N have the same countable ordinals.

(3) In (2) we can find N', $N \prec N' \prec M^*$, $N' \cap \omega_1 = N \cap \omega_1$ and $|N \cap \omega_2| = \aleph_1$.

Proof. (1) Follows easily from (2). We can easily build a strictly increasing elementary chain of countable models, of length ω_1, all having the same countable ordinals.

(2) Clearly some winning strategy for player II in ∂ belongs to N. So we can construct a play of ∂, $F_0, \alpha_0, F_1, \alpha_1, \ldots$ such that player II uses his strategy, each F_n belongs to N, and every function from ω_2 to ω_1 which belong to N appears in $\{F_n : n < \omega\}$.

As the strategy and $F_0, \dots, F_n$ belong to N, also α_n belongs to N. In the end for arbitrarily large $\alpha < \aleph_2$, $F_n(\alpha) < \bigcup_m \alpha_m$ for every n. But clearly $N \cap \omega_1$ is an initial segment of ω_1, hence $\bigcup_m \alpha_m \subseteq N$, so $F_n(\alpha) \in N$ for every α satisfying $\bigwedge_n F_n(\alpha) < \bigcup_m \alpha_m$. Now we can take as N_α the Skolem hull of $N \bigcup \{\alpha\}$.

We have to show that N and N_α have the same countable ordinals. Every $\gamma \in N_\alpha \cap \omega_1$ can be written as $\gamma = \tau(\alpha, a_1, \dots, a_\alpha)$, where $a_1, \dots a_\kappa \in N$, and τ is a Skolem term.

In N, we can define a function $f : \omega_2 \to \omega_1$, by

$$f(i) = \begin{cases} \tau(i, a_1, \dots, a_k), & \text{if this is} < \omega_1 \\ 0 & \text{otherwise.} \end{cases}$$

So $f = F_n$ for some n, but clearly $\gamma = \tau(\alpha, a_n, \dots, a_n) = f(\alpha) = F_n(\alpha) \in N$.

(3) By the proof of (1). $\square_{2.5}$

We can prove similar theorems for S not necessarily $\{\aleph_1\}$, e.g.

2.6 Theorem. If player II wins $\partial(\{\lambda\}, \omega, \mu)$, $\lambda < \mu$, M a model with universe $H((2^\mu)^+)$ and countably many relations and functions, including $\in$ and Skolem functions, $N \prec M$ countable, *then* there is $N^\dagger, N \prec N^\dagger \prec M$, $N^\dagger \neq N$, $\mathrm{Sup}(N^\dagger \cap \lambda) = \mathrm{Sup}(N \cap \lambda)$, $\mathrm{Sup}(N^\dagger \cap \mu) > \mathrm{Sup}(N \cap \mu)$, and $|N^\dagger \cap \mu| = \aleph_1$.

Proof. Similar. $\square_{2.6}$

2.7 Conclusion. (1) For some regular λ, player II wins in $\partial(\{\aleph_1\}, \omega, \lambda)$ iff there is an $\{\aleph_1\}$-semiproper forcing P not preserving "$\mathrm{cf}(\alpha) > \aleph_0$".

(2) So if e.g. $0^\#$ does not exist, a forcing notion is proper *iff* it is $\underset{\sim}{S}$-semiproper, $\underset{\sim}{S} = \{\aleph_1^V\} \cup \{\kappa : \mathrm{cf}(\kappa)^{V^P} > \aleph_0 \ \kappa \text{ regular (in } V) \text{ and } \kappa > \aleph_1\}$.

Proof. (1) By 2.4, $\mathbb{D} = \{D_\lambda^{\mathrm{cb}}\}$.

(2) If P is $\underset{\sim}{S}$-semiproper and preserves "$\mathrm{cf}(\alpha) > \aleph_0$" then by Claim X 2.3(1) P is proper (see Definition X 2.2, $\underset{\sim}{S}$ is essentialy equal to the S from Claim X 2.3(1) because P preserves "$\mathrm{cf}(\alpha) > \aleph_0$").

If P is $\underset{\sim}{S}$-semiproper not preserving "cf$(\alpha) > \aleph_0$" then for some $p \in P$, and regular $\lambda > \aleph_1$ (in V) $Q = P\restriction\{q \in P : p \leq q\}$ is $\underset{\sim}{S}$-semiproper hence $\{\aleph_1\}$-semiproper and $\emptyset \Vdash$ "cf$(\lambda) = \aleph_0$". But then by 2.7(1) player II wins the game $\eth(\{\aleph_1\}, \omega_1, \lambda)$ hence the conclusion of Theorem 2.6 holds and by well known theorems such variants of Chang conjecture imply $0^\# \in V$. $\square_{2.7}$

Note that

2.8 Conclusion. If κ is measurable in V, $P = \text{Levy}(\aleph_1, < \lambda)$ (so elements of P are countable partial functions), then in V^P, Chang's conjecture holds.

Proof. By 2.2 and X 4.11.

XIII. Large Ideals on ω_1

§0. Introduction

Here we shall start with κ e.g. supercompact, use semiproper iteration to get results like ($S \subseteq \omega_1$ stationary costationary):

(a) ZFC + GCH + $\mathcal{P}(\omega_1)/(\mathcal{D}_{\omega_1} + S)$ is *layered* + suitable forcing axiom and note that by [FMSh:252] this implies the existence of a uniform ultrafilter on ω_1 such that $\aleph_0^{\omega_1}/D = \aleph_1$ (which is stronger than "D is not regular").

(b) ZFC+GCH+$\mathcal{P}(\omega_1)/(\mathcal{D}_{\omega_1} + S)$ is *Levy* + suitable forcing axiom.

(c) ZFC+GCH+$\mathcal{P}(\omega_1)/(\mathcal{D}_{\omega_1} + S)$ is *Ulam* + suitable forcing axiom.

where (a) Ulam means

$$(\mathcal{D}_{\omega_1} + S)^+ = \{A \subseteq \omega_1 : A \cap S \neq \emptyset \mod \mathcal{D}_{\omega_1}\}$$

is the union of $\aleph_1$, $\aleph_1$-complete filters, hence on $\mathbb{R}$ there are $\aleph_1$ measures such that each $A \subseteq \mathbb{R}$ is measurable for at least one measure

(b) Levy means that, as a Boolean algebra, it is isomorphic to the completion of a Boolean algebra of the Levy collapse Levy$(\aleph_0, < \aleph_2)$

(c) layered means that the Boolean algebra is $\bigcup_{\alpha<\aleph_2} B_\alpha$, where B_α are increasing, continuous, $|B_\alpha| \leq \aleph_1$, and $\mathrm{cf}(\alpha) = \aleph_1 \Rightarrow B_\alpha \lessdot \mathcal{P}(\omega_1)/(\mathcal{D}_{\omega_1} + S)$. We also deal with reflectiveness (see 4.3).

This chapter is a rerepresentation of [Sh:253], we shall give some history later, and now just remark that this work was done (and reclaimed) *after*

[FMSh:240 §1, §2] and [W83] ([W83] starts with "ZFC+DC+ADR+θ *regular*" and forces "ZFC+CH+the club filter on some stationary $S \subseteq \omega_1$ is $\aleph_1$ dense") but *before* Woodin obtained a similar result from a huge cardinal.

In this chapter we got results by semiproper iteration iterating collapses and sealing some maximal antichains of $\mathcal{P}(\omega_1)/\mathcal{D}_{\omega_1}$ up to some large κ. So it is a natural continuation of Chapter X. Our ability to do this to enough chains comes from reflection properties of κ, which is supercompact (or limit of enough supercompacts).

The first section contains preliminaries on semi-stationary sets, relevant reflection properties and what occurs to some such properties when we force. In the second section we deal more specifically with our iterations (S-suitable iterations). In the third section we deal with getting Levy algebra and layeredness, and in the fourth we deal with reflective ideals (see 4.3) and with the Ulam property. Note that for much of the chapter the iteration is of S_3-complete forcing notion, for some (fixed) stationary $S_3 \subseteq \omega_1$, and in this case the iteration is (equivalent to) a CS one; so we will stress less the names of conditions etc.

By Foreman, Magidor and Shelah [FMSh:240], CON(ZFC+κ is supercompact) implies the consistency of ZFC+"$\mathcal{D}_{\omega_1}$ is $\aleph_2$-saturated" [i.e., if $\mathfrak{B}$ is the Boolean algebra $\mathcal{P}(\omega_1)/\mathcal{D}_{\omega_1}$, "$\mathcal{D}_{\omega_1}$ is $\aleph_2$ saturated" means "$\mathfrak{B}$ satisfies the $\aleph_2$-c.c."]. This in fact was deduced from the MM$^+$(=Martin Maximum$^+$) by [FMSh:240] whose consistency was proved by RCS iteration of semiproper forcings (see Chapter X, Chapter XVII §1). Note that [FMSh:240] refutes the thesis: in order to get an elementary embedding j of V with small critical ordinal, into some transitive class M of some generic extension V^P of V, *one should start* with an elementary embedding of j of V' into some M' and then force over V'. Previously, J. Steel and Van Wesep got the same result starting from ZF+AD+AC$_R$ (see [StVW]).

This thesis was quite strongly rooted. Note that it is closely connected to the existence of normal filters D on λ which are λ^+-saturated or at least precipitous (use for P the set of nonzero members of $\mathcal{P}(\lambda)/D$ ordered by inverse inclusion, j the generic ultrapower). See [FMSh:240] for older history.

In fact, it was shortly proved directly that $\mathrm{MM}^+ \equiv \mathrm{SPFA}^+$ and much later it was proved that MM is equivalent to the Semi-Proper Forcing Axiom (in ZFC) (see XVIII §1).

The rsults of [FMSh:240, §1, §2] motivated much activity. Woodin proves from

CON(ZF+ADR+θ regular) the consistency of ZFC+"$\mathfrak{B}\restriction S$ is $\aleph_1$-dense", for some stationary $S \subseteq \omega_1$.

By Shelah and Woodin [ShWd:241], if there is a supercompact cardinal, then every projective set of reals is Lebesgue measurable (etc.). This was obtained by combining (A) and (B) below which were proved simultaneously:

(A) The conclusion holds if there is a weakly compact cardinal κ and a forcing notion P, $|P| = \kappa$, satisfying the κ-c.c., not adding reals and $\Vdash_P$"there is a normal filter D on ω_1, $\mathfrak{B} = \mathcal{P}(\omega_1)/D$ satisfying the $\aleph_2$-c.c."

(B) There is a forcing as required in (A) (see [FMSh:240, §3]).

This was improved for projective sets which are Σ_n using approximately n cardinals κ satisfying:

(*) for every forcing notion $P \in H(\kappa)$ and stationary costationary $S \subseteq \omega_1$ there is semiproper $\underset{\sim}{Q}$, not adding reals, $\Vdash_{P*\underset{\sim}{Q}}$ "$\mathcal{D}_{\omega_1}\restriction S$ is κ-saturated, $\kappa = \aleph_2$" (and $\underset{\sim}{Q}$ is not too large).

A sufficient condition for (*) is $\mathrm{Pr}_a(\kappa) \overset{\text{def}}{=} \kappa$ is strongly inaccessible, and for every $f : \kappa \to \kappa$ there is an elementary embedding $j : V \to M$ (M is a transitive class), κ the critical ordinal of j and $H(j(f)(\kappa)) \subseteq M$. Moreover it suffice (Woodin cardinals) $\mathrm{Pr}_b(\kappa)^\dagger \overset{\text{def}}{=} \kappa$ is strongly inaccessible, and for every $f : \kappa \to \kappa$ there is $\kappa_1 < \kappa$, $(\forall \alpha < \kappa_1), f(\alpha) < \kappa_1$ and for some elementary embedding $j : V \to M$ (M is a transitive class), κ_1 is the critical ordinal of j and $H((j(f))(\kappa_1)) \subseteq M$.

By [Sh:237a] "$2^{\aleph_0} < 2^{\aleph_1} \Rightarrow \mathcal{D}_{\omega_1}$ is not $\aleph_1$-dense", and by [Sh:270] if D is a layered filter on $\lambda = \lambda^{<\lambda}$ then $D^+ = \{A \subseteq \lambda : A \notin D\}$ is the union of λ filters extending D.

† Later results of Martin, Steel and Woodin clarify the connection between determinacy and large cardinals.

This chapter is a representation of [Sh:253] which was done then, but was mistakenly held as incorrect for quite some time. The main change is that we replace part of the consistency proof of the Ulam statement, ($\mathcal{P}(\omega_1)$ is the union of $\aleph_1$ $\aleph_1$-complete nontrivial measures), by a deduction from a strong variant of layerness. Later Woodin proves from a huge cardinal CON(ZFC+ GCH+$\mathcal{P}(\omega_1)/(\mathcal{D}_{\omega_1} + S)$ is $\aleph_1$-dense).

0.1. Notation and Basic Facts.

(1) $\mathcal{P}$(A) is the power set of A, $\mathcal{S}_{<\lambda}(A) = \{B : B \subseteq A, |B| < \lambda\}$, $<^*_\lambda$ is a well ordering of $H(\lambda)$ which, for simplicity only, we assume is an end extension of $<^*_\mu$ for $\mu < \lambda$.

(2) $\mathcal{D}_\lambda$ is the club filter on a regular $\lambda > \aleph_0$ and $\mathcal{D}_{<\lambda}(A)$ is the club filter on $\mathcal{S}_{<\lambda}(A)$.

(3) (a) $\mathfrak{B}$ is the Boolean Algebra $\mathcal{P}(\omega_1)/\mathcal{D}_{\omega_1}$; we do not distinguish strictly between $A \in \mathcal{P}(\omega_1)$ and $A/\mathcal{D}_{\omega_1}$ and for stationary $S \subseteq \omega_1$, $\mathfrak{B}{\restriction}S$ is defined naturally.

(b) $\mathfrak{B}$ of course depends on the universe, so we may write $\mathfrak{B}^{V^1}$ or $\mathfrak{B}[V^1]$; instead of $\mathfrak{B}[V^P]$ we may write $\mathfrak{B}^P$ or $\mathfrak{B}[P]$.

(c) If $V^1 \subseteq V^2$, $\omega_1^{V^1} = \omega_1^{V^2}$, then $\mathfrak{B}[V^1]$ is a weak subalgebra of $\mathfrak{B}[V^2]$ (i.e., distinct elements in $\mathfrak{B}[V^1]$ may be identified in $\mathfrak{B}[V^2]$).

(d) If $P \in V$ is a forcing notion preserving stationary subsets of ω_1, then $\mathfrak{B} = \mathfrak{B}[V]$ is a subalgebra of $\mathfrak{B}^P$ (identifying $(A/\mathcal{D}_{\omega_1})^V$ and $(A/\mathcal{D}_{\omega_1})^{V^P}$ for $A \in \mathcal{P}(\omega_1)^V$). If $\bar{Q} = \langle P_i, Q_i : i < \alpha \rangle$ is an iteration (with limit P_α, so $i < j < \alpha \Rightarrow P_i \lessdot P_j$), we let $\mathfrak{B}^{\bar{Q}} = \cup_{i<\alpha} \mathfrak{B}^{P_{i+1}}$.

(4) (a) Let us say, for Boolean algebras B_1 and B_2, that $B_1 \lessdot B_2$ *iff* $B_1 \subseteq B_2$ (i.e., B_1 is a subalgebra of B_2) and every maximal antichain of B_1 is a maximal antichain of B_2.

(b) Note that, for Boolean algebras B_1 and B_2, $B_1 \lessdot B_2$ *iff* $B_1 \subseteq B_2$ and $(\forall x \in B_2 \setminus \{0\})(\exists y \in B_1 \setminus \{0\})(\forall z \in B_1)[z \cap y \neq 0 \rightarrow z \cap x \neq 0]$. Hence, if $B_1 \lessdot B_3$ and $B_1 \subseteq B_2 \subseteq B_3$, then $B_1 \lessdot B_2$.

(c) Hence, the satisfaction of "$B_1 \lessdot B_2$" does not depend on the universe of set theory, i.e., if $V \models B_1 \lessdot B_2$ and $V \subseteq V^1$ then $V^1 \models B_1 \lessdot B_2$.

(d) By Solovay and Tennenbaum [ST], $\lessdot$ is transitive, and if $\langle B_i : i < \alpha \rangle$ is $\lessdot$-increasing and continuous then $B_i \lessdot \bigcup_{j<\alpha} B_j$.

(e) Also, if $\langle B_\zeta : \zeta < \xi \rangle$ is a $\subseteq$-increasing sequence of Boolean algebras and $B_0 \lessdot B_\zeta$ for $\zeta < \xi$, then $B_0 \lessdot \bigcup_{\zeta<\xi} B_\zeta$.

(f) If $\langle B_i : i \leq \delta + 1 \rangle$ is an increasing continous sequence of Boolean algebras, $\mathrm{cf}(\delta) > \aleph_0$ and $[i < \delta \Rightarrow \|B_i\| < \delta]$, and $S \stackrel{\mathrm{def}}{=} \{i < \delta : B_i \lessdot B_{\delta+1}\}$ is a stationary subset of δ then $B_\delta \lessdot B_{\delta+1}$.

[Why? If $x \in B_{\delta+1} \setminus \{0\}$ then by clause (b) for each $\alpha \in S$ for some $y_\alpha \in B_\alpha \setminus \{0\}$ we have

$$(\forall z \in B_\alpha)[z \cap y_\alpha \neq 0 \rightarrow z \cap x \neq 0].$$

So by Fodor lemma for some j,

$$S_1^* \stackrel{\mathrm{def}}{=} \{\alpha \in S : y_\alpha \in B_j\}$$

is stationary. And so for some y the set $\{\alpha \in S_1^* : y_\alpha = y\}$ is stationary and y is as required.]

(g) For a Boolean algebra B, $X_1 \lessdot X_2$ (in B) *iff* $X_1 \subseteq X_2 \subseteq B \setminus \{0_B\}$ and every predense subset of X_1 is a predense subset of X_2 *where* Y is a predense subset of X if $Y \subseteq X$ & $\forall x \in X \exists y \in Y (\exists z \in X)(z \subseteq_B x \cap y)$. If $0_B \in X_2$ we mean $X_1 \setminus \{0_B\} \lessdot X_2 \setminus \{0_B\}$.

This definition is compatible with the one in clause (a) and the iteration in clause (b) is still true; also clause (c) holds (the others are not needed here).

(5) If in V we have $P_1 \lessdot P_2 \lessdot P_3$, in V^{P_2} we have $\mathfrak{B}^{P_1} \lessdot \mathfrak{B}^{P_2}$, and in $V^{P_3}, \mathfrak{B}^{P_2} \lessdot \mathfrak{B}^{P_3}$, *then* in $V^{P_3}, \mathfrak{B}^{P_1} \lessdot \mathfrak{B}^{P_3}$, [follows by (4)(c), (4)(d)]; similarly for $\mathfrak{B}^{P_i} \restriction S$.

(6) For a set a and forcing notion P, $\underset{\sim}{G}_P$ is the P-name of the generic set and $a[\underset{\sim}{G}_P] = a \cup \{\underset{\sim}{x}[\underset{\sim}{G}_P] : \underset{\sim}{x} \in a \text{ is a } P\text{-name}\}$. So $a[\underset{\sim}{G}_P]$ is a P-name of a set, and for $G \subseteq P$ generic over V its interpretation is $a[G] = a \cup \{\underset{\sim}{x}[G] : \underset{\sim}{x} \in a$ is a P-name$\}$ ($\underset{\sim}{x}[G]$ is the interpretation of the P-name x).

(7) If $\lambda > \aleph_0$ is a cardinal, N a countable elementary submodel of $(H(\lambda), \in), P \in N$ and $G \subseteq P$ is generic over V, then $N[G] \prec (H(\lambda)^{V^P}, \in)$ (as $H(\lambda)^{V^P} = \{\tau[G] : \underset{\sim}{\tau} \in H(\lambda)$ a P-name$\}$ and if $\Vdash_P$ "$(H(\lambda)^{V^P}, \in) \models \exists x \varphi(x, \underset{\sim}{a})$" then for some P-name $\underset{\sim}{\tau} \in H(\lambda)$ we have $\Vdash_P$ "$(H(\lambda)^{V^P}, \in) \models \varphi(\underset{\sim}{\tau}, \underset{\sim}{a})$"). See III 2.11, I 5.17(1).

(8) Also, if some $p \in G$ is (N, P)-generic then $(N, G) \prec (H(\lambda)^V, \in, G)$ (i.e., G is an extra predicate, so you may write $(N, G \cap |N|)$). Also, if R is any relation (or sequence of relations) on $H(\lambda)^V$, $N \prec (H(\lambda)^V, \in, R)$ (and $P \in N, G \subseteq P$ generic over V) and some $p \in G$ is (N, P)-generic *then* $(N, G) \prec (H(\lambda)^V, \in, R, G)$ and even $(N[G], |N|, R^N, G) \prec (H(\lambda)^{V^P}, \in, H(\lambda)^V, R, G)$. Usually we use a well ordering $<^*_\lambda$ of $H(\lambda)$.

(9) Let $N <_\kappa M$ mean $N \subseteq M$ and $N \cap \kappa$ is an initial segment of $M \cap \kappa$ and $N \prec M$; if we use it for sets (rather than models), the last demand is omitted. Note that if $N \prec M \prec (H(\mu), \in)$, $\kappa < \mu$ and $N \cap \kappa = M \cap \kappa$ then $N <_{\kappa^+} M$.

§1. Semi-Stationarity

1.1. Definition.

(1) A forcing notion P is semiproper if: *for every* regular $\lambda > 2^{|P|}$, any countable $N \prec (H(\lambda), \in)$ to which P belongs, and $p \in P \cap N$ *there is* q such that: $p \leq q \in P$ and q is (N, P)-semi-generic (see below).

(2) For a set a, forcing notion P and $q \in P$, we say q is (a, P)-semi-generic *if*: for every P-name $\underset{\sim}{\alpha} \in a$ of a countable ordinal, $q \Vdash_P$ "$\underset{\sim}{\alpha} \in a$" [i.e., if: $q \Vdash$ "$a[\underset{\sim}{G}_P] \cap \omega_1 = a \cap \omega_1$" see 0.1(6); note $a[\underset{\sim}{G}_P] = \{\underset{\sim}{x}[\underset{\sim}{G}_P] : \underset{\sim}{x} \in a$ a P-name$\}$ if a is closed enough, i.e. for $x \in a$ also $\dot{x} \in a$ where $\dot{x}[G]$ is x].

(3) We call $W \subseteq \mathcal{S}_{<\aleph_1}(A)$ (where $\omega_1 \subseteq A$) *semi-stationary* in A (or in $\mathcal{S}_{<\aleph_1}(A)$ or subset of A) *if* for every model M with universe A and countably many relations and functions, there is a countable $N \prec M$, such that $(\exists a \in W)[N \cap \omega_1 \subseteq a \subseteq N]$, [equivalently, $\{a \in \mathcal{S}_{<\aleph_1}(A) : (\exists b \in W)[a \cap \omega_1 \subseteq$

$b \subseteq a]\}$ is a stationary subset of $\mathcal{S}_{<\aleph_1}(A)$ (i.e., $\neq \emptyset \bmod \mathcal{D}_{<\aleph_1}(A)$). As we allow functions in M, we can require only $N \subseteq M$].

1.2. Claim.

(1) If $W \subseteq \mathcal{S}_{<\aleph_1}(A)$ is stationary and $\omega_1 \subseteq A$ *then* W is a semi-stationary subset of A. Also if $\omega_1 \subseteq A, W \subseteq \mathcal{S}_{<\aleph_1}(A)$ is semi-stationary in A, $C \in \mathcal{D}_{<\aleph_1}(A)$ and $[a \in W \;\&\; b \in C \;\&\; b \cap \omega_1 \subseteq a \subseteq b \Rightarrow b \in W]$ *then* W is stationary (subset of $\mathcal{S}_{<\aleph_1}(A)$).

(2) If $\omega_1 \subseteq A \subseteq B$, and $W \subseteq \mathcal{S}_{<\aleph_1}(A)$ *then:* W is semi-stationary in A *iff* W is semi-stationary in B (so we can omit "in A").

(3) If $W_1 \subseteq W_2 \subseteq \mathcal{S}_{<\aleph_1}(A)$, and W_1 is semi-stationary, *then* W_2 is semi-stationary.

(4) If $|A| = \aleph_1$, $\omega_1 \subseteq A$, $A = \cup_{i<\omega_1} a_i$, a_i increasing continuous in i, with a_i countable, *then* $W \subseteq \mathcal{S}_{<\aleph_1}(A)$ is semi-stationary iff $S_W \stackrel{\text{def}}{=} \{i : (\exists b \in W)[i \subseteq b \subseteq a_i]\}$ is stationary (as a subset of ω_1).

(5) If $p \in P$ is (b, P)-semi-generic, $b \cap \omega_1 \subseteq a \subseteq b$ *then* p is (a, P)-semi-generic.

(6) If $W \subseteq \mathcal{S}_{<\aleph_1}(\lambda)$, $\mu > \lambda$, $W \in N$, $N \prec (H(\mu), \in)$ (hence $|W| < \mu$), and for some $a \in W, N \cap \omega_1 \subseteq a \subseteq N$ *then* W is semi-stationary.

(7) Assume A is an uncountable set, $W \subseteq \mathcal{S}_{<\aleph_1}(A)$, f_1, f_2, are one to one functions from ω_1 into A, and $W_\ell \stackrel{\text{def}}{=} \{a \cup \{\alpha < \omega_1 : f_\ell(\alpha) \in a\} : a \in W\} \subseteq \mathcal{S}_{<\aleph_1}(A \cup \omega_1)$. *Then* W_1 is semi-stationary iff W_2 is semi stationary, so in Definition 1.1(3) (of semi stationarity) we can replace "$\omega_1 \subseteq A$" by "A uncountable".

(8) If A_1, A_2 are uncountable sets, f is a one to one function from A_1 to A_2, $W_2 \subseteq \mathcal{S}_{<\aleph_1}(A_2)$, $W_1 \subseteq \mathcal{S}_{<\aleph_1}(A_1)$ and $[a \in W_1 \Rightarrow f''(a) \in W_2]$ and $[b \in W_2 \Rightarrow (\exists a \in W_1) b \cap f''(A_1) = f''(a)]$ *then:* W_1 is semi-stationary iff W_2 is semi-stationary. If f is onto A_2, necessarily $W_1 = \{a \in \mathcal{S}_{<\aleph_1}(A_1) : f''(a) \in W_2\}$.

Proof. (1) - (5), (7), (8) Left to the reader.

(6) If not, some $M = (\lambda, \ldots, F_n, \ldots)$ exemplifies that W is not semi-stationary, so some such M belongs to N, hence $N \cap \lambda$ is a submodel of M (even an elementary submodel of M), a contradiction. $\square_{1.2}$

1.3. Claim. A forcing notion P is semiproper *iff* the set

$$W_P = \{a \in \mathcal{S}_{<\aleph_1}(P \cup {}^P(\omega_1+1)) : \text{for every } p \in P \cap a \text{ there is } q, \text{ such that } p \leq q \in P \text{ and } q \text{ is } (a, P)\text{-semi-generic}\}$$

contains a club of $\mathcal{S}_{<\aleph_1}(P \cup {}^P(\omega_1+1))$ where each $h : P \to (\omega_1+1)$ is interpreted as a P-name $\underset{\sim}{\alpha}_h$ with the property that: if

$$\underset{\sim}{\alpha}_h^0[G] = \min\{h(r) : r \in G\},$$

then $\underset{\sim}{\alpha}_h[G]$ is $\underset{\sim}{\alpha}_h^0[G]$ if the latter is $< \omega_1$ and zero otherwise.

Proof. Immediate. $\square_{1.3}$

1.4. Claim. The following are equivalent for a forcing notion P:

(1) P is semiproper.

(2) P preserves semi-stationarity.

(3) P preserves semi-stationarity of subsets of $\mathcal{S}_{<\aleph_1}(2^{|P|})$.

Proof. (1) $\Rightarrow$ (2). Let $\omega_1 \subseteq A$, and $W \subseteq \mathcal{S}_{<\aleph_1}(A)$ be semi-stationary. Suppose $p \in P$ and $p \Vdash_P$ "W is not semi-stationary". So there are P-names of functions $\underset{\sim}{F}_n (n < \omega)$ from A to A, $\underset{\sim}{F}_n$ is n-place, and $p \Vdash$ "if $a \subseteq A$ is countable closed under $\underset{\sim}{F}_n (n < \omega)$ *then* $\neg(\exists b)[a \cap \omega_1 \subseteq b \subseteq a \;\&\; b \in W]$".

Let λ be regular large enough. Let $N \prec (H(\lambda), \in)$ be countable so that $A, \langle \underset{\sim}{F}_n : n < \omega\rangle, p, P$ belong to N and there is $b \in W$ such that $N \cap \omega_1 \subseteq b \subseteq N$ (such N, b exist as W is semi-stationary by 1.2(2)). Let q be (N, P)-semi-generic, $p \leq q \in P$. So $q \Vdash_P$ "$N[\underset{\sim}{G}] \cap \omega_1 = N \cap \omega_1$ and $N \subseteq N[\underset{\sim}{G}]$" hence, for the b above,

$$q \Vdash_P \text{"}N[\underset{\sim}{G}] \cap \omega_1 \subseteq b \subseteq N[\underset{\sim}{G}]\text{"}.$$

Also $q \Vdash_P$ "$N[G] \cap A$ is closed under the $\underset{\sim}{F}_n$'s" (as $N[G] \prec (H(\lambda)[G], \in)$ and $\underset{\sim}{F}_n[G] \in N[G]$, see Basic Fact 0.1(7) in §0), contradicting the choice of the $\underset{\sim}{F}_n$'s.

(2) $\Rightarrow$ (3). Trivial.

$\neg(1) \Rightarrow \neg(3)$. Let $W = \mathcal{S}_{<\aleph_1}(P \cup {}^P(\omega_1+1)) \setminus W_P$ (where W_P is from 1.3). As $\neg(1)$, W is stationary, so for each $a \in W$ choose $p_a \in P \cap a$ which exemplifies $a \notin W_p$, i.e. there is no $q, p \leq q \in P$ and q is (a, P)-semi-generic. By the normality of the filter $\mathcal{D}_{<\aleph_1}(P \cup {}^P(\omega_1+1))$, for some $p(*) \in P$ the set $W_1 = \{a \in W : p_a = p(*)\}$ is stationary. Hence W_1 is semi-stationary (by 1.2(1)). But by the choice of $\langle p_a : a \in W\rangle$ and W_1, easily $p(*) \Vdash$ "W_1 is not semi-stationary". Clearly $|P \cup {}^P(\omega_1+1)| = 2^{|P|}$ (as P is infinite w.l.o.g.), so let f be a one to one function from $2^{|P|}$ onto $P \cup {}^P(\omega_1+1)$ and let $W_2 = \{a \in \mathcal{S}_{<\aleph_1}(2^{|P|}) : f''(a) \in W_1\}$. By 1.3(8) we have W_2 is semi-stationary and $p(*) \Vdash$ "W_2 is not semi-stationary" so (3) fails. $\square_{1.4}$

1.5. Definition.

(1) $\mathrm{Rss}(\kappa, \lambda)$ (reflection for semi-stationarity) is the assertion that for every semi-stationary $W \subseteq \mathcal{S}_{<\aleph_1}(\lambda)$ there is $A \subseteq \lambda$, $\omega_1 \subseteq A$, $|A| < \kappa$ such that $W \cap \mathcal{S}_{<\aleph_1}(A)$ is semi-stationary (in $\mathcal{S}_{<\aleph_1}(\lambda)$).

(2) $\mathrm{Rss}(\kappa)$ is $\mathrm{Rss}(\kappa, \lambda)$ for every $\lambda \geq \kappa$.

(3) $\mathrm{Rss}^+(\kappa, \lambda)$ means that for every semiproper P of cardinality $< \kappa$ we have $\Vdash_P$ "$\mathrm{Rss}(\kappa, \lambda)$".

(4) $\mathrm{Rss}^+(\kappa)$ is $\mathrm{Rss}^+(\kappa, \lambda)$ for every $\lambda \geq \kappa$.

1.5A Remark. In 1.5(3), we could strengthen the statement by replacing "semiproper" by "not collapsing $\aleph_1$" with no change below. If we use below forcing notion from a smaller class we could weaken the statement in 1.5(3) accordingly.

1.6. Claim.

(1) In Definition 1.5(1) we can replace λ by B, when $|B| = \lambda$, $\omega_1 \subseteq B$.

(2) If $\kappa \leq \kappa_1 \leq \lambda_1 \leq \lambda$ and $\mathrm{Rss}(\kappa, \lambda)$, *then* $\mathrm{Rss}(\kappa_1, \lambda_1)$. If $\kappa \leq \lambda_1 \leq \lambda$ and $\mathrm{Rss}^+(\kappa, \lambda)$ *then* $\mathrm{Rss}^+(\kappa, \lambda_1)$. Lastly, if $\mathrm{Rss}^+(\kappa_i, \lambda)$ (for $i < \alpha$) *then* $\mathrm{Rss}^+(\sup_{i<\alpha}\kappa_i, \lambda)$.

(3) If κ is a compact cardinal, *then* $\mathrm{Rss}(\kappa)$;

(4) If κ is a compact cardinal *then* $\mathrm{Rss}^+(\kappa)$.

(5) If κ is measurable, $W_i \subseteq \mathcal{S}_{<\aleph_1}(A)$ and $\bigcup_{i<\kappa} W_i$ is semi-stationary *then* for some $\alpha < \kappa$, $\bigcup_{i<\alpha} W_i$ is semi-stationary.

(6) If κ is a limit of compact cardinals, *then* $\mathrm{Rss}^+(\kappa)$.

(7) If κ is λ-compact, $\lambda = \lambda^{\aleph_0} \geq \kappa$ *then* $\mathrm{Rss}(\kappa, \lambda)$ and even $\mathrm{Rss}^+(\kappa, \lambda)$.

Proof. (1) Trivial.

(2) Use 1.2(2).

(3) Let $\kappa \subseteq A, W \subseteq \mathcal{S}_{<\aleph_1}(A)$, and: $W \cap \mathcal{S}_{<\aleph_1}(B)$ is not semi-stationary for every $B \subseteq A, \omega_1 \subseteq B$, with $|B| < \kappa$.

Define the set of sentences Γ:

$$\Gamma = \Gamma^a \cup \Gamma^b \cup \Gamma^c$$

where (each $c \in A$ serves as an individual constant):

$$\Gamma^a = \{c_1 \neq c_2 : c_1, c_2 \text{ are distinct members of } A\},$$

$$\Gamma^b = \{R(c_0, c_1, \dots, c_l, \dots)_{l<\omega} : c_l \in A, \{c_l : l < \omega\} \in W\},$$

Γ^c is the singleton with unique member (F_n is an n-place function symbol, remember $\omega_1 \subseteq A$):

$$(\forall x_0, x_1, \dots, x_n, \dots)_{n<\omega}\Big[\text{if } \{x_0, x_1, \dots\} \text{ is closed under } F_n (n < \omega), \textit{ then } \neg(\exists y_0, y_1, \dots)\,(R(y_0, \dots, y_n, \dots)\ \& \ \{x_l : l < \omega, \textstyle\bigvee_{i<\omega_1} x_l = i\} \subseteq \{y_l : l < \omega\} \subseteq \{x_m : m < \omega\})\Big].$$

Every subset of Γ of power $< \kappa$ has a model (if it mentions only $c \in B$ where $B \subseteq A$ and $|B| < \kappa$, then use a model witnessing "$W \cap \mathcal{S}_{<\aleph_1}(B \cup \omega_1)$ is not semi-stationary"). A model M of Γ exemplifies "W is not semi-stationary" (in $|M|$, hence in A by 1.2(2)).

(4) As forcing notions of cardinality $< \kappa$ preserve the compactness of κ.

(5) Let Γ^a, Γ^c be as in the proof of 1.6(4), and:

$$\Gamma^b_i = \{R(c_0, c_1, \dots) : c_l \in A, \{c_l : l < \omega\} \in W_i\}.$$

Now $\Gamma^a \cup \Gamma^c \cup \bigcup_{i<\kappa} \Gamma^b_i$ has no model, hence (using the Łoś theorem for L_{ω_1,ω_1} and $\aleph_1$-complete ultrafilters) for some $\alpha < \kappa$, we have: $\Gamma^a \cup \Gamma^c \bigcup_{i<\alpha} \Gamma^b_i$ has no model.

(6) Easy (use last phrase of 1.6(2)).

(7) Same proof as 1.6(3), (4). $\square_{1.6}$

1.7. Claim.

(1) If $\mathrm{Rss}(\kappa, 2^{|P|})$ and P is not semiproper, *then* P destroys the semi-stationarity of some $W \subseteq \mathcal{S}_{<\aleph_1}(A), |A| < \kappa$ (i.e. some $p \in P$ forces this)
[Why? By (1) $\Leftrightarrow$ (3) from 1.4, for some $p \in P$ and semi-stationary $W \subseteq \mathcal{S}_{<\aleph_1}(2^{|P|})$, we have $p \Vdash_P$ "W is not semi-stationary". By the assumption, for some $A \subseteq 2^{|P|}$ we have: $|A| < \kappa$ and $W_1 \stackrel{\mathrm{def}}{=} W \cap \mathcal{S}_{<\aleph_1}(A)$ is semi-stationary. Clearly by 1.2(3) we have $p \Vdash_P$ "W_1 is not semi-stationary", as required].

(2) If P destroys the semi-stationarity of $W \subseteq \mathcal{S}_{<\aleph_1}(A)$, $|A| = \aleph_1$, *then* P destroys the stationarity of $S_W \subseteq \omega_1$ [with S_W as defined in 1.2(4)], which means that S_W is stationary in V but not in V^P.

(3) If $\mathrm{Rss}(\aleph_2, 2^{|P|})$ and P preserves stationarity of subsets of ω_1, *then* P is semiproper
[Why? By parts (1), (2) above].

(4) If $W \subseteq \mathcal{S}_{<\aleph_1}(A)$ exemplifies the failure of $\mathrm{Rss}(\aleph_2, |A|)$, *then* there is a forcing notion P of power $|A|^{\aleph_0}$, not semiproper but not destroying stationarity of subsets of $\aleph_1$
[Why? Let P be $\{\bar{A} : \bar{A} = \langle A_i : i \leq \alpha\rangle$ is an increasing continuous countable sequence of countable subsets of A, each A_i satisfying $\neg(\exists a \in W)(A_i \cap \omega_1 \subseteq a \subseteq A_i)\}$, ordered by being an initial segment. As forcing with P destroy the semi stationarity of W, clearly P is not semiproper; let us prove that forcing with P preserve the stationarity of subsets of ω_1. If $p \in P$ and $p \Vdash$ "S is not stationary" where S is a stationary set of limit ordinals $< \omega_1$, we can find an increasing continuous sequence $\langle N_i : i < \omega_1\rangle$ of countable elementary submodels of $(H(\beth_7^+), \in)$, with $\{W, p, A\} \in N_0$, $N_i \in N_{i+1}$. So $C = \{\delta < \omega_1 : \delta$ a limit ordinal and

$N_\delta \cap \omega_1 = \delta\}$ is a club of ω_1. By the choice of W, for some club $C_1 \subseteq C$ of ω_1, $\delta \in C_1 \Rightarrow \neg(\exists a)(a \in W \cap \delta \subseteq a \subseteq N_\delta)$, hence we can find $\delta \in C_1 \cap S$ and $q \geq p$ which is (N, P)-generic, an easy contradiction.].

(5) $\mathrm{Rss}(\aleph_2)$ is equivalent to the assertion: every forcing notion preserving stationarity of subsets of ω_1 is semiproper.
[By parts (3), (4) above]. $\square_{1.7}$

1.8. Definition. $\langle P_i, \underset{\sim}{Q}_j : i \leq \alpha, j < \alpha \rangle$ is a semiproper iteration if:

(A) it is an RCS iteration [see Ch. X, §1];

(B) if $i < j \leq \alpha$ are non-limit, then $\Vdash_{P_i}$ "P_j/P_i is semiproper";

(C) for every $i < \alpha$ we have, $\Vdash_{P_{i+1}}$ "$(2^{\aleph_1})^{V^{P_i}}$ is collapsed to $\aleph_1$" (we can use another variant instead).

We shall use not only G_{P_i} (or $\underset{\sim}{G}_{P_i}$) but also G_i (or $\underset{\sim}{G}_i$) for the (name of the) generic subset of P_i.

1.9. Theorem. *Suppose λ is measurable, $\langle P_i, \underset{\sim}{Q}_j : i \leq \lambda, j < \lambda \rangle$ is a semiproper iteration, $|P_i| < \lambda$ for $i < \lambda$, and $\{i < \lambda : \underset{\sim}{Q}_i$ is semiproper $\}$ belongs to some normal ultrafilter D on λ. Then in V^{P_λ}, Player II wins $\eth = \eth(\{\aleph_1\}, \omega, \aleph_2)$.*

1.9A. Remarks. On $\eth$ see Ch. XII, Def. 2.1. or see below.

(1) The game lasts ω moves; on the nth move Player I chooses $f_n : \aleph_2 \to \omega_1$ and Player II chooses $\xi_n < \omega_1$. In the end Player II wins if $A \stackrel{\text{def}}{=} \{i < \aleph_2 : \bigwedge_n \bigvee_m f_n(i) < \xi_m\}$ is unbounded in $\aleph_2$.

(2) We can modify the game by requiring $A \neq \emptyset \bmod E$ for a filter E on ω_2. We then denote the game by $\eth(\{\aleph_1\}, \omega, E)$. The result is true for $E = D$.

(3) By XII 2.5(2) we know the following: if Player II wins $\eth(\{\aleph_1\}, \omega, \aleph_2)$, $\lambda > 2^{\aleph_2}$, N a countable elementary submodel of $(H(\lambda), \in, <^*_\lambda)$, *then* for arbitrarily large $i < \omega_2$, there is $N' \prec (H(\lambda), \in, <^*_\lambda)$, N' countable, $N \subseteq N'$, $i \in N'$ and $N \cap \omega_1 = N' \cap \omega_1$ (hence $N <_{\omega_2} N'$; see Basic Fact 0.1(9) in §0).
If Player II wins $\eth(\{\aleph_1\}, \omega, E)$ (where E is a filter on ω_2) then the set of such i is $\neq \emptyset \bmod E$; so we have equivalence.

(4) Can we demand in (3) (on both see XII §2 when we use E) that $N' \cap i = N \cap i$? If $\{\delta < \omega_2 : \mathrm{cf}(\delta) = \aleph_0\} \in E$ the answer is No. If $\{\delta < \omega_2 : \mathrm{cf}(\delta) = \aleph_1\} \in E$ the answer is Yes provided that we can change the game to $\Game'$: Player I is also allowed to choose regressive functions $F_n : \aleph_2 \to \aleph_2$, and Player II in the nth move has to choose also $\xi'_n < \omega_2$, and in the end Player II wins if $S = \{\delta < \aleph_2 : \text{for } n < \omega, \text{ we have } \delta \geq \xi'_n, \text{ and } f_n(\delta) < \bigcup_m \xi_m, F_n(\delta) < \bigcup_m \xi'_m\} \neq \emptyset \bmod E$.

(5) If in the theorem $\Vdash_P$ "$\{\delta < \aleph_2 : Q_\delta$ is semiproper and $\mathrm{cf}(\delta)^{V^P} = \aleph_1\} \neq \emptyset \bmod D$" *then* Player II wins also in this variant (from (4) above). The proof of 1.9 still works.

(6) We can replace $\aleph_1$ by any regular $\theta, \aleph_0 < \theta < \lambda$, (as the range of f_n) and use the game $\Game(\{\theta\}, \mu, E)$, E a normal filter on $\lambda, \langle P_i, \underset{\sim}{Q}_i : i < \lambda\rangle$ is a $(< \theta)$-revised support iteration (see Chapter XIV), such that the set of $i < \lambda$ satisfying the following belongs to D: "in V^{P_i} for $p \in P_\lambda/P_i$ in the game $PG^\omega(p, P_\lambda/P_i, \lambda, \theta)$ (see below and Chapter XII, 1.7(3), 1.4), the second player has a winning strategy".

(7) We can replace in the assumption of 1.9, "D is a normal ultrafilter on κ" by "D is a normal filter on κ" and the second player wins in $\Game'$ $(\{\aleph_1\}, \omega, D)$.

(8) If we use the strong preservation version of theorems, we do not need 1.9 (a weaker version is then proved, e.g. for $\alpha < \lambda$, $(P_\kappa/P_\alpha) * \underset{\sim}{\mathrm{Nm}}$ is semi proper) and is really changed.

Proof of 1.9. Let D be a normal ultrafilter on λ (in V), $A \in D$ a set of (strongly) inaccessible cardinals such that: $(\forall \kappa \in A)[(\forall i < \kappa)(|P_i| < \kappa)$ & Q_κ is semiproper (in V^{P_κ})].

For each $\kappa \in A$ the forcing notion P_λ/P_κ (in V^{P_κ}) is a semiproper forcing, hence for each $p \in P_\lambda/P_\kappa$ in the following game, $P\Game^\omega(p, P_\lambda/P_\kappa, \aleph_1)$, Player II has a winning strategy which we call $F_p(P_\lambda/P_\kappa)$ $(\in V^{P_\kappa})$; if $p = \emptyset_{P_\lambda/P_\kappa}$ we omit p [see Chapter XII, 1.7(3), Definition 1.4]: a play of the game lasts ω-moves, in the nth move Player I chooses a P_λ/P_κ-name $\underset{\sim}{\zeta}_n$ of a countable ordinal and

Player II chooses a countable ordinal ξ_n. Player II wins a play if

$$(\exists q)(p \leq q \in P_\lambda/P_\kappa \,\&\, q \Vdash \text{“}\bigwedge_n [\underset{\sim}{\zeta}_n < \bigcup_{m<\omega} \xi_m]\text{”});$$

without loss of generality the ξ_n are strictly increasing.

Let us describe a winning strategy for Player II in $\partial(\{\aleph_1\}, \omega_1, \aleph_2)$ in $V[G_\lambda]$, where $G_\lambda \subseteq P_\lambda$ is generic over V. In the nth move Player I chooses $f_n : \omega_2 \to \omega_1$, Player II, in addition to choosing $\xi_n < \omega_1$, chooses A_n, $\underset{\sim}{f}'_n$, α_n such that:

(0) $\alpha_n < \alpha_{n+1} < \lambda$; in stage n Player II works in $V[G_{\alpha_n}]$, so D is still an ultrafilter (pedantically: generates an ultrafilter);

(1) $A_n \in D, A_{n+1} \subseteq A_n \subseteq A$ and for all $\delta \in A_n$, we have $\alpha_n < \delta$;

(2) $\Vdash_P$ “$\underset{\sim}{f}'_n : \omega_2 \to \omega_1$”;

(3) $\underset{\sim}{f}'_n[G_\lambda] = f_n$; $\underset{\sim}{f}'_n$ is the first such name so $\underset{\sim}{f}'_n$ is from V;

(4) for $\kappa \in A_n, \langle\langle \underset{\sim}{f}'_l(\kappa), \xi_l\rangle : l \leq n\rangle$ is (a P_κ-name of) an initial segment of a play of $P\partial^\omega(\emptyset_{P_\lambda}, P_\lambda/G_\kappa, \aleph_1)$ in which Player II uses his winning strategy $F(P_\lambda/G_\kappa)$, i.e. some condition in G_{α_n} forces this.

How can Player II carry out this strategy? Suppose he arrives at stage n and Player I has chosen $f_n \in V^{P_\lambda}, f_n : \lambda \to \omega_1$. Stipulate $\alpha_0 = -1$. Let $B_n = A_{n-1}$ if $n > 0$ and $B_n = A$ if $n = 0$. Player II chooses for $\underset{\sim}{f}'_n \in V$ the first (by $<^*_\chi$, $\chi = (2^\lambda)^+$) P_λ-name $\underset{\sim}{f}_n$ such that $\underset{\sim}{f}'_n[G_\lambda] = f_n$. Now for every $\kappa \in B_n$, working in $V[G_\kappa]$, he continues the play $\langle\langle \underset{\sim}{f}'_l(\kappa), \xi^0_l\rangle : l < n\rangle$ of $P\partial^\omega(\emptyset_P, P_\lambda/G_\kappa, \aleph_1)$, letting the first player play $\underset{\sim}{f}'_n(\kappa)$, and let $\xi^0_n(\kappa)$ be the choice of the second player according to the strategy $F(P_\lambda/G_\kappa)$. So $\xi^0_n(\kappa) = \underset{\sim}{\xi}^0_n(\kappa)$ is a P_κ-name. Now (in $V[G_{\alpha_{n-1}}]$) for every $p \in P_\lambda$ and $\kappa \in B_n$ there is $q_\kappa \in P_\kappa/G_{\alpha_n}$ compatible with p and forcing a value to $\xi^0_n(\kappa)$. But as $B_n \subseteq A$, and by the choice of the set A (and X 1.6) we know that $P_\kappa = \bigcup_{i<\kappa} P_i$, so we can use the normality of D; so for some $\xi < \omega_1$, $A^n_p \in D$, $A^n_p \subseteq B_n$ and q, we have q is compatible with p in $P_\lambda/G_{\alpha_{n-1}}$ and $(\forall \kappa \in A^n_p)[q_\kappa = q$ and $q \Vdash_{P_\kappa}$ “$\underset{\sim}{\xi}^0_n(\kappa) = \xi$”$]$. So there are such $q \in G_\lambda$, and ξ (which we call ξ_n) and a set which we call A_n. It is easy to choose α_n.

We should still prove that this is a winning strategy. We shall consider one play and work in V, so everything is a P_λ-name (as we are using RCS, no problems arise). I.e. we have $p^* \in P_\lambda$ such that $p^* \Vdash_{P_\lambda}$ "$\langle \underset{\sim}{f}_n, \underset{\sim}{\xi}_n : n < \omega\rangle$ is a play of the game with Player II using his strategy, choosing on the side $\langle \underset{\sim}{f}'_n, \underset{\sim}{\alpha}_n, \underset{\sim}{A}_n : n < \omega\rangle$". Now $\underset{\sim}{f}'_n$, $\underset{\sim}{A}_n$, $\underset{\sim}{\alpha}_n$ are P_λ-names of members of V ($\underset{\sim}{f}'_n$ a P_λ-name of a P_λ-name) so there is a maximal antichain $\mathcal{J}_n$ of P_λ of conditions forcing a value to each of $\underset{\sim}{f}_n$, $\underset{\sim}{\xi}_n$, $\underset{\sim}{f}'_n$, $\underset{\sim}{\alpha}_n$, $\underset{\sim}{A}_n$. But P_λ satisfies the λ-c.c., $P_\lambda = \bigcup_{\alpha<\lambda} P_\alpha$ so for some $\alpha(*) < \lambda$, $\bigwedge_{n<\omega} \mathcal{J}_n \subseteq P_{\alpha(*)}$. Also w.l.o.g. $\alpha(*)$ is bigger than every possible value $\underset{\sim}{\alpha}_n$.

Work in $V[G_{\alpha(*)}]$. Now D is (essentially) an ultrafilter (on λ) in $V[G_{\alpha(*)}]$. Each $\underset{\sim}{A}_n$ is a P_λ-name of a member of V so really there are $< \lambda$ candidates so we can find A_ω, such that for each n we have $\Vdash_{P_\lambda/G_{\alpha(*)}}$ "$A_\omega \subseteq \underset{\sim}{A}_n$," $A_\omega \in D$ (alternatively we can compute $\bigcap_{n<\omega} A_n$ in $V[G_{\alpha(*)}]$). Now for $\kappa \in A_\omega$, $\kappa > \alpha(*)$ the sequence $\langle\langle \underset{\sim}{f}_l(\kappa), \xi_l\rangle : l < \omega\rangle$ is a play of $P\eth^\omega(\emptyset_P, P_\lambda/P_\kappa, \aleph_1)$ where Player II uses his winning strategy (this is a P_κ-name, but fortunately $\langle \underset{\sim}{\xi}_l[G_\kappa] : l < \omega\rangle \in V[G_{\alpha(*)}]$). So there is $q_\kappa \in P_\lambda/P_\kappa$ so that

$$q_\kappa \Vdash_{P_\lambda/P_\kappa} \text{“}\bigwedge_l \underset{\sim}{f}_l(\kappa) < \bigcup_n \xi_n\text{”}$$

(more exactly:

$$q_\kappa \Vdash_{(P_\lambda/G_{\alpha(*)})/(P_\kappa/G_{\alpha(*)})} \text{“}\bigwedge_l \underset{\sim}{f}_l(\kappa) < \bigcup_n \xi_n\text{”},$$

actually q_κ is a $P_\kappa/G_{\alpha(*)}$-name of a P_λ/P_κ-condition).

We can consider q_κ as a P_λ-condition with $\mathrm{Dom}(q_\kappa) \subseteq [\kappa, \lambda)$, because we use RCS iteration. Now easily $\langle q_\kappa : \kappa \in A_\omega\rangle \in V[G_{\alpha(*)}]$, and

$$\Vdash_{P_\lambda/G_{\alpha(*)}} \text{“}\{\kappa \in A : q_\kappa \in \underset{\sim}{G}_\lambda\} \text{ is unbounded in } \lambda\text{”}$$

Why? As every $r \in P_\lambda/G_{\alpha(*)}$ has domain bounded in λ, we have: q_κ is compatible with it for κ large enough. This finishes the proof that the strategy works. $\square_{1.9}$

1.10. Claim. Suppose κ is measurable, $\bar{Q}$ is a semiproper iteration, $\ell g(\bar{Q}) = \kappa, |P_i| < \kappa$ for $i < \kappa$ and $\{i : Q_i \text{ semiproper }\}$ belongs to some normal ultrafilter on κ (this holds e.g. if $\{i < \kappa : \text{if } i \text{ is strongly inaccessible and } (\forall j < i)[|P_j| < i],$ then $\underset{\sim}{Q}_i$ is semi proper$\} \in \mathcal{D}_\kappa$). *Then:*

(1) $\mathrm{Rss}^+(\kappa, \lambda)$ implies $\Vdash_{P_\kappa}$ "$\mathrm{Rss}(\kappa, \lambda)$".

(2) If $\underset{\sim}{Q}$ is a P_κ-name of a forcing notion, $(P_\kappa/P_{i+1}) * \underset{\sim}{Q}$ is semiproper for each $i < \kappa$ (i.e. this is forced for P_{i+1}) *then* $\Vdash_{P_\kappa}$ "$\underset{\sim}{Q}$ is semiproper".

(3) We can replace measurability of κ by: κ is strongly inaccessible and $\Vdash_{P_\kappa}$ "Player II wins $\eth(\{\aleph_1\}, \omega_1, \aleph_2)$".

Proof. (1) Let $\underset{\sim}{W}$ be a P_κ-name and $p \in P_\kappa$ be such that $p \Vdash_{P_\kappa}$ "$\underset{\sim}{W} \subseteq \mathcal{S}_{<\aleph_1}(\lambda)$ is semi-stationary".

For $i < \kappa$, let $\underset{\sim}{W}_i = \{a : a \in V^{P_i}, a \in \mathcal{S}_{<\aleph_1}(\lambda)$, and for some $q \in \underset{\sim}{G}_{P_i}, q \Vdash_{P_\kappa}$ "$a \in \underset{\sim}{W}$"$\}$. So $\underset{\sim}{W}_i$ is a P_i-name.

Let χ be regular and large enough, and $<^*_\chi$ a well ordering of $H(\chi)^V$.

Let $p \in G = G_\kappa \subseteq P_\kappa$, G generic over V and $G_i = G \cap P_i$ for $i < \kappa$. In $V[G_\kappa]$, as $\underset{\sim}{W}[G_\kappa]$ is semi-stationary, there is a countable $(N, G_\kappa \cap N) \prec (H(\chi)^V, \in, <^*_\chi, G_\kappa)$, such that for some $a \in \underset{\sim}{W}[G_\kappa]$ we have $N \cap \omega_1 \subseteq a \subseteq N \cap \lambda$, and $p, \underset{\sim}{W}, \lambda, \kappa, \bar{Q}$ belong to N (note: G_κ is considered a relation of those models).

So there are $q \in G_\kappa$ and P_κ-names $\underset{\sim}{N}, \underset{\sim}{a}$ such that $q \Vdash_{P_\kappa}$ "$\underset{\sim}{N}, \underset{\sim}{a}$ are as above", and as $p \in G_\kappa$, without loss of generality $p \leq q$. As N and a are countable subsets of $H(\chi)^V$ and λ respectively and $P_\kappa = \bigcup_{i<\kappa} P_i$ satisfies the κ-c.c. (by X 5.3(3)), for some $i < \kappa$ we have $\underset{\sim}{N}, \underset{\sim}{a}$ are P_i-names, $\Vdash_{P_i}$ "$\underset{\sim}{N} \cap \kappa \subseteq i$" and $q \in P_i$. Now by 1.9 + 1.9A(3), in V^{P_κ}, for arbitrarily large ordinal $\theta < \kappa, N^{[\theta]} \cap \omega_1 = N \cap \omega_1$, and Q_θ is semiproper (if not, replace it by $\theta + 1$), where we let:

$$N^{[\theta]} \stackrel{\text{def}}{=} \text{Skolem Hull } (N \cup \{\theta\})$$

(in $(H(\chi)^V, \in, <^*_\chi, G_\kappa)$, working in the universe $V[G_\kappa]$ such that $q \in G_\kappa$).

Choose such a $\theta > i$. Now $\theta \in N^{[\theta]}$ and $(N^{[\theta]}, G_\theta) \prec (H(\chi)^V, \in, <^*_\chi, G_\theta)$, as $\theta > i$ clearly $\underset{\sim}{a}[G_\theta] \in \underset{\sim}{W}_\theta[G_\theta]$ and $\omega_1 \cap N^{[\theta]} \subseteq \underset{\sim}{a}[G_\theta] \subseteq N^{[\theta]}$. Let $N_{[\theta]}$ be the

Skolem Hull of $N \cup \{\theta\}$ in $(H(\chi)^V, \in, <^*_\chi, G_\theta)$; note as $\underset{\sim}{N}$ is a P_θ-name, in $V[G_\theta]$ we can compute $\underset{\sim}{N}[G_\theta] = \underset{\sim}{N}[G_\kappa] = N$ (and $\underset{\sim}{a}[G_\theta] = \underset{\sim}{a}[G_\kappa] = a$). Clearly $N^{[\theta]} \cap \omega_1 \subseteq \underset{\sim}{a}[G_\theta] \subseteq N \subseteq N_{[\theta]} \subseteq N^{[\theta]}$; hence by 1.2(6), $V[G_\theta] \models$ "$\underset{\sim}{W}_\theta[G_\theta]$ is a semi-stationary subset of $\mathcal{S}_{<\aleph_1}(\lambda)$" (remembering that in $(H(\chi)^V, \in, <^*_\chi, G_\theta)$ we can interprete $(H(\chi)^{V[G_\theta]}, \in H(\chi)^V, <^*_\chi, G))$.

As $\mathrm{Rss}^+(\kappa,\lambda)$ clearly $V[G_\theta] \models \mathrm{Rss}(\kappa,\lambda)$, hence in $V[G_\theta]$ for some $A \subseteq \lambda, |A| < \kappa$ and $\underset{\sim}{W}_\theta[G_\theta] \cap \mathcal{S}_{<\aleph_1}(A)$ is semi-stationary. As P_κ/P_θ is semiproper (by the choice of θ) it preserves the semi-stationary of $\underset{\sim}{W}_\theta[G_\theta] \cap \mathcal{S}_{<\aleph_1}(A)$ (see 1.4), hence $V[G_\kappa] \models$ "$\underset{\sim}{W}_\theta[G_\theta] \cap \mathcal{S}_{<\aleph_1}(A)$ is semi-stationary", but $\underset{\sim}{W}[G_\theta] \subseteq \underset{\sim}{W}[G_\kappa]$ hence $V[G_\kappa] \models$ "$\underset{\sim}{W}[G_\kappa] \cap \mathcal{S}_{<\aleph_0}(A)$ is semi-stationary".

(2) This is similar: suppose $p \Vdash_{P_\kappa}$ "$\underset{\sim}{N} \prec (H(\chi)^V, \in, <^*_\chi, \underset{\sim}{G}_{P_\kappa})$ and $\underset{\sim}{p}' \in \underset{\sim}{Q} \cap \underset{\sim}{N}$ are counterexample to semiproperness of Q".

Let $G_\kappa \subseteq P_\kappa$ be generic over V and $p \in G_\kappa$. Let $\theta < \kappa$, with $\theta > \sup(\underset{\sim}{N}[G_\kappa] \cap \kappa)$, be such that $\underset{\sim}{N}$ is a P_θ-name and $\sup(\underset{\sim}{N}[G] \cap \kappa) < \kappa$ and $\underset{\sim}{N}[G_\kappa]^{[\theta]} \cap \omega_1 = \underset{\sim}{N}[G_\kappa] \cap \omega_1$. Now work in $V[G_\kappa \cap P_{\theta+1}]$ and use: $\Vdash_{P_{\theta+1}}$ "$(P_\kappa/P_{\theta+1}) * \underset{\sim}{Q}$ is semiproper". (Note that if $\mathrm{Rss}^+(\kappa)$ we can get the result by 1.7(3)). Alternatively prove that forcing with $\underset{\sim}{Q}[H_\kappa]$ preserve semi stationarity of sets.

(3) In the proof of (2) we use this only. In the proof of (1) we could have chosen θ to be a successor ordinal (so $\underset{\sim}{Q}_\theta$ is semiproper). So P_κ/G_θ preserves the semi-stationarity of $\underset{\sim}{W}$, hence $V[G_\kappa] \models$ "$\underset{\sim}{W}$ is semi-stationary". $\square_{1.10}$

1.11. Claim. Suppose $\mathrm{Rss}(\kappa, 2^\kappa)$, κ regular and: $\kappa = \aleph_2$ or $(\forall \mu < \kappa)\mu^{\aleph_0} < \kappa$. *Then* for $\lambda > 2^\kappa$ for every countable $N \prec (H(\lambda), \in, <^*_\lambda)$ to which κ belongs, for arbitrarily large $i < \kappa$, letting $N^{[i]} =$ Skolem Hull $(N \cup \{i\})$, we have $N <_{\omega_2} N^{[i]}$ (note that we do not demand $N \cap \kappa \neq N^{[i]} \cap \kappa$).

1.12. Remark. (1) The "$\kappa = \aleph_2 \ldots$" can be omitted if we replace "for arbitrarily large i" by "for some $i < \kappa$ with $i > \sup(N \cap \kappa)$".

(2) We can replace "$\kappa = \aleph_2$, or $\ldots$" by

$(*)_1$ "if $\alpha < \kappa$, then there is a closed unbounded $C \subseteq \mathcal{S}_{<\aleph_1}(\alpha)$ of power $< \kappa$" (see the proof).

It even suffices to assume

$(*)_2$ "for every stationary $W \subseteq \mathcal{S}_{<\aleph_1}(\alpha), (\alpha < \kappa)$ there is a semi-stationary $W' \subseteq W$ of cardinality $< \kappa$".

(3) If in the conclusion we want to get $N <_\kappa N^{[i]}$, we have to replace "$(\exists a \in W)(N \cap \omega_1 \subseteq a \subseteq N)$" in the definition of semi-stationary (Definition 1.1) by "$(\exists a \in W)(N \cap \kappa \subseteq a <_\kappa N \cap \kappa)$".

Proof of 1.11. Let

$$W = \{|N| : N \prec (H(\kappa^+), \in, <^*_{\kappa^+}), N \text{ countable and}$$
$$\text{for some } i_N < \kappa, \text{ for no } i \in [i_N, \kappa) \text{ do we have } N <_{\omega_2} N^{[i]}\}.$$

Assume first that W is a stationary subset of $H(\kappa^+)$. So, as $\mathrm{Rss}(\kappa, 2^\kappa)$ holds (and $|H(\kappa^+)| = 2^\kappa$) there is $A \subseteq H(\kappa^+)$, $\omega_1 \subseteq A$, $|A| < \kappa$ such that: $W_A \overset{\text{def}}{=} \{a \in W : a \subseteq A\}$ is a semi-stationary subset of $\mathcal{S}_{<\aleph_1}(A)$. Without loss of generality (see 1.2(2))

$$M \overset{\text{def}}{=} (A, \in \restriction A, <^*_{\kappa^+} \restriction A) \prec (H(\kappa^+), \in, <^*_{\kappa^+})$$

and $A \cap \kappa$ is an ordinal $< \kappa$ (remember κ is regular).

Remembering that (by the definition of W) for countable elementary submodels $N_1 \subseteq N_2$ of $(H(\kappa^+), \in, <^*_{\kappa^+})$, $|N_1| \in W$, $N_1 \cap \omega_1 = N_2 \cap \omega_1$ implies $|N_2| \in W$; by 1.2(1) clearly W_A is stationary (as a subset of $\mathcal{S}_{<\aleph_1}(A)$). We know by assumption that for some closed unbounded $C \subseteq \mathcal{S}_{<\aleph_1}(A), C$ has cardinality $< \kappa$. So

$$\zeta \overset{\text{def}}{=} \sup\{i_N : |N| \in C \cap W_A\} < \kappa.$$

Now for some club $C_1 \subseteq C$, for every $a \in C_1$, the set $a^{[\zeta]}$ = Skolem Hull of $a \cup \{\zeta\}$ (inside $(H(\kappa^+), \in, <^*_{\kappa^+})$), satisfies $a^{[\zeta]} \cap A = a$, hence $a <_{\omega_2} a^{[\zeta]}$. But we can choose $a \in C_1 \cap W_A$, contradiction.

So W is not stationary and let $C^* \subseteq \mathcal{S}_{<\aleph_1}(H(\kappa^+))$ be a club disjoint to W.

Let $\lambda > 2^\kappa$, so $H(\kappa^+)$, $<^*_{\kappa^+}$, $W \in H(\lambda)$, and let N be such that $\kappa \in N \prec (H(\lambda), \in, <^*_\lambda)$ and N is countable. So $H(\kappa^+) \in N$ (and $<^*_{\kappa^+} = <^*_\lambda \restriction H(\kappa^+)$) hence $W \in N$ and without loss of generality $C^* \in N$. Hence $N \cap H(\kappa^+) \in C^*$, and so for arbitrarily large $i < \kappa$ there is N_1^i such that $N{\restriction}H(\kappa^+) \prec N_1^i \prec (H(\kappa^+), \in, <^*_{\kappa^+}), N{\restriction}H(\kappa^+) <_{\omega_2} N_1^i$ and $i \in N_1^i$. Let N^i be the Skolem Hull of $N \cup (N_1^i \cap \kappa)$. We can easily check that $N^i \cap \kappa = N_1^i \cap \kappa$, so N^i is as required.

$\square_{1.11}$

§2. *S*-Suitable Iterations and Sealing Forcing

2.1. Definition. We say $\bar{Q} = \langle P_i, \underset{\sim}{Q}_j, \mathbf{t}_j : i \le \alpha, j < \alpha\rangle$ is S-suitable (iteration), where $S \subseteq \omega_1$ is stationary, if:

(A) $\bar{Q}$ is an RCS iteration; (i.e. if we remove the $\mathbf{t}_j$'s);

(B) we denote $|\bigcup_{j<i} P_{j+1}| = \kappa_i = \kappa_i^{\bar{Q}}$ so $\kappa_0 = 1, \kappa_i$ increasing continuous. We demand that κ_i is strictly increasing;

(C) for i successor κ_i is strongly inaccessible;

(D) for $i < j \le \alpha$ non-limit, P_j/P_i is semiproper;

(E) $\underset{\sim}{Q}_i$ satisfies the κ_{i+1}-c.c., $\aleph_2^{V^{P_{i+1}}} = \kappa_{i+1}$;

(F) if $\mathbf{t}_i = 1, i < j \le \alpha$ and j is a successor, then $\mathfrak{B}^{P_i}{\restriction}S \lessdot \mathfrak{B}^{P_j}{\restriction}S$ (see 0.1(3)(a) + (b)).

Remark: We may, but do not, use $\mathbf{t}_\beta$ which are names. Also the demand "Q_i satisfies the κ_{i+1}-c.c." is just for simplicity.

2.1A. Notation. $\alpha^{\bar{Q}} = \alpha, P_i^{\bar{Q}} = P_i, \underset{\sim}{\bar{Q}}_j^{\bar{Q}} = \underset{\sim}{Q}_j, \mathbf{t}_j^{\bar{Q}} = \mathbf{t}_j$ and (remember and recall 0.1(3)(d)): $\mathfrak{B}^{\bar{Q}} = \cup\{\mathfrak{B}^{P_{i+1}} : i < \ell g(\bar{Q})\}$.

2.2. Claim.

(1) Suppose $\bar{Q} = \langle P_i, \underset{\sim}{Q}_j : i \le \alpha, j < \alpha\rangle$ is a semiproper iteration (see 1.8 for definition). *Then:*

(a) If $i < \alpha$ is non-limit or $\underset{\sim}{Q}_i$ is semiproper or $\underset{\sim}{Q}_i$ preserves stationarity of subsets of ω_1 from V^{P_i} or i is strongly inaccessible and $\bigwedge_{j<i} |P_j| < i$, *then* every stationary subset of ω_1 in V^{P_i} is also stationary in V^{P_α} (i.e., $\mathfrak{B}[P_i]$ is a subalgebra of $\mathfrak{B}[P_\alpha]$).

(b) $\aleph_1^V = \aleph_1^{V^{P_\alpha}}$.

(c) If $\alpha > \aleph_0$ is strongly inaccessible, and $|P_i| < \alpha$ for $i < \alpha$, *then* P_α satisfies the α-c.c. and so

$$\mathcal{P}(\omega_1)^{V^{P_\alpha}} = \bigcup_{i<\alpha} \mathcal{P}(\omega_1)^{V^{P_i}} \text{ and } V^{P_\alpha} \models \text{“}2^{\aleph_1} = \aleph_2\text{”}.$$

(d) If $\omega_1 \backslash S$ is stationary, each $\underset{\sim}{Q}_i$ is $(\omega_1 \setminus S)$-complete [see V §3], *then* so is P_α, hence forcing by P_α preserve the stationarity of $\omega_1 \setminus S$ and even subsets of it and does not add ω-sequences of ordinals, hence $V^{P_\alpha} \models$ “CH”.

(e) If $\bar{Q} \in N_1 \prec N_2 \prec (H(\lambda), \in), N_2$ countable, $N_1 <_\alpha N_2, \alpha$ strongly inaccessible and belongs to $N_1, \alpha > |P_i|$ for $i < \alpha$ and q is (N_1, P_α)-semi-generic and $i = \min(\alpha \cap N_2 \setminus N_1)$ is regular, *then* q is (N_2, P_i)-semi-generic.

(2) Any S-suitable iteration $\bar{Q}$ is a semiproper iteration and $\mathbf{t}_i = 1 \Rightarrow \mathfrak{B}[P_i] \restriction S \lessdot \mathfrak{B}[P_j] \restriction S]$ *when:* $j \geq i$, and j is: successor or strongly inaccessible satisfying $[\gamma < j \Rightarrow |P_\gamma| < j]$.

(3) If (in (1)) $\kappa < \alpha$ is strongly inaccessible, $|P_i| < \kappa$ for $i < \kappa$, and $\Vdash_{P_\kappa}$ “Rss$(\aleph_2)$” *then* $\underset{\sim}{Q}_\kappa$ (and P_j/P_κ when $\kappa \leq j \leq \alpha$) are semiproper.

Proof. Left to the reader. For instance:

(1)(e) Clearly i is a strong limit [as $\{j < \kappa : j \text{ strong limit }\}$ is a club of κ which belongs to N_1, hence i necessarily belongs to it]. Also we have assumed i is regular hence i is strongly inaccessible; similarly $i > \aleph_0$ and $j < i \Rightarrow |P_j| < i$. If $\mathcal{I} \in N_2$ is a maximal antichain of P_i, *then* by X 5.3(3) for some $j < i$ we have $\mathcal{I} \subseteq P_j$, so that consequently there is such j in N_2, and hence $j \in N_1$ and also the rest is easy.

(2) If j is a successor ordinal use clause (F) of Definition 2.1, if j is strong inaccessible use 2.2(1)(c) and 0.1(4)(e).

(3) By 1.7(3) it is enough to prove that forcing with $\underset{\sim}{Q}_\kappa$ does not destroy the stationarity of any $A \subseteq \omega_1$, $A \in V^{P_\kappa}$. However, by 2.2(1)(c) (and 2.2(2)) for some $\beta < \alpha$, $A \in V^{P_\beta}$. Clearly $A \in V^{P_\beta}$ and is a stationary subset of ω_1 in

$V^{P_{\beta+1}}$. As $P_{\kappa+1}/P_{\beta+1}$ is semiproper, A is also stationary in $(V^{P_{\beta+1}})^{P_{\kappa+1}/P_{\beta+1}} = V^{P_{\kappa+1}} = (V^{P_\kappa})^{\underset{\sim}{Q}_\kappa}$, as required. $\square_{2.2}$

2.2A. Remark. It follows that if κ is strongly inaccessible, and $|P_i| < \kappa$ for $i < \kappa$, and A is a stationary subset of ω_1 in V^{P_κ}, *then* A is a stationary subset of ω_1 in V^{P_α} for every large enough $\alpha < \kappa$.

2.3. Claim. Suppose $\bar{Q} = \langle P_j, \underset{\sim}{Q}_i, \mathbf{t}_i : j \le \alpha, i < \alpha \rangle$ is an RCS iteration, α a limit ordinal and $S \subseteq \omega_1$ is stationary.

(1) If $\bar{Q} \restriction \beta$ is S-suitable for $\beta < \alpha$, *then* $\bar{Q}$ is S-suitable.

(2) If for $\beta < \alpha$, $\bar{Q} \restriction \beta$ is a semiproper iteration, *then* $\bar{Q}$ is a semiproper iteration.

(3) In (2), if $i < \alpha$ and $\underset{\sim}{A}$ is a P_i-name then: $\Vdash_{P_\alpha}$ "$\underset{\sim}{A} \lessdot \mathfrak{B}^{\bar{Q}} \restriction S$" if and only if $\alpha = \sup\{j < \alpha : \Vdash_{P_{j+1}}$ "$\underset{\sim}{A} \lessdot \mathfrak{B}^{P_{j+1}} \restriction S$"$\}$ if and only if for arbitrarily large $j < \alpha$ we have $\Vdash_{P_j}$ "$\underset{\sim}{A} \lessdot \mathfrak{B}^{P_j}$".

(4) In (2), if $\alpha > |P_i|$ for $i < \alpha$, and α is strongly inaccessible, then $\mathfrak{B}^{\bar{Q}} = \mathfrak{B}^{P_\alpha}$.

Proof. (1) For clause (D) from Definition 2.1 use the semiproper iteration lemma. The other clauses are also obvious.

(2), (3), (4) are also easy. $\square_{2.3}$

2.4. Definition. Let $\bar{\mathcal{A}} = \langle \mathcal{A}_\zeta : \zeta < \xi \rangle$ be a sequence of subalgebras or just subsets of $\mathfrak{B}(= \mathfrak{B}^V)$ such that S belongs to each $\mathcal{A}_\zeta$ where $S \subseteq \omega_1$ stationary.

(1) $\mathrm{Sm}(\bar{\mathcal{A}}, S) = \{A \subseteq S :$ for some $\zeta < \xi$, $\{x \in \mathcal{A}_\zeta : x \neq 0 \bmod \mathcal{D}_{\omega_1}$ and $x \cap A = \emptyset \bmod \mathcal{D}_{\omega_1}\}$ is pre-dense in $\mathcal{A}_\zeta\}$ (we should have written $x/\mathcal{D}_{\omega_1} \in \mathcal{A}_\zeta$ for x, $x \subseteq \omega_1$; Ξ is predense in $\mathcal{A}_\zeta$ means that for every $y \in \mathcal{A}_\zeta$, such that $\mathcal{A}_\zeta \models$ "$y \neq 0$" for some $x \in \Xi$ we have $\mathcal{A}_\zeta \models$ "$x \cap y \neq 0$").

(2) For $\Xi \subseteq \mathfrak{B}^V$ let $\mathrm{seal}(\Xi) = \{\langle a_i : i < \alpha \rangle : \alpha$ is a countable ordinal, and letting $a_\alpha = \bigcup\limits_{i<\alpha} a_i$ we have $a_i \in \mathcal{S}_{<\aleph_1}(\Xi \cup \omega_1)$, a_i $(i \le \alpha)$ is increasing continuous, each a_i countable and $a_i \cap \omega_1$ is an ordinal which belongs to $\bigcup_{A \in \Xi \cap a_i} A\}$, ordered by being an initial segment.

(3) We define the sealing forcing $\mathrm{Seal}(\bar{\mathcal{A}}, S)$ as the product with countable support of $\{$ $\mathrm{seal}(\Xi)$: for some $\zeta < \xi$, Ξ is a pre-dense subset of $\mathcal{A}_\zeta$ and

$\omega_1 \backslash S \in \Xi\}$. Let $\mathrm{Seal}'(\bar{\mathcal{A}}, S) = \{\bar{c} : \bar{c}$ a partial function from $\mathrm{Sm}(\bar{\mathcal{A}}, S)$, with countable domain, and if $A \in \mathrm{Sm}(\bar{\mathcal{A}}, S) \cap \mathrm{Dom}(\bar{c})$, then $\bar{c}_A$ is a continuously increasing function from some countable $\gamma + 1$ to $\omega_1 \setminus A\}$,
the ordering is defined by:

$$\bar{c}^1 \leq \bar{c}^2 \text{ if } A \in \mathrm{Dom}(\bar{c}^1) \text{ implies } A \in \mathrm{Dom}(\bar{c}^2) \text{ and } \bar{c}^1_A \subseteq \bar{c}^2_A.$$

(4) If $\bar{\mathcal{A}} = \langle \mathcal{A} \rangle$ we write $\mathcal{A}$ instead of $\bar{\mathcal{A}}$ in (1), (2) above and (5) below.

(5) For κ ($> \aleph_0$) strongly inaccessible we define the strong sealing forcing $\mathrm{SSeal}(\bar{\mathcal{A}}, S, \kappa)$ as P_κ, where $\langle P_i, \underset{\sim}{Q}_j : i \leq \kappa, j < \kappa \rangle$ is an RCS iteration with $Q_j = \mathrm{Seal}(\bar{\mathcal{A}}, S)^{P_j} \times \mathrm{Levy}(\aleph_1, 2^{\aleph_1})^{V[P_j]}$.

(6) We call $\Xi \subseteq \mathfrak{B}^V$ semiproper *iff* $\mathrm{seal}(\Xi)$ is a semiproper forcing notion.

(7) $\mathrm{WSeal}(S)$ is the product, with countable support, of $\mathrm{seal}(\Xi)$, Ξ semiproper, $\omega_1 \setminus S \in \Xi$.

(8) For κ not strongly inaccessible, but still $\bar{\mathcal{A}}$-inaccessible, which means:
(*) $(\forall \mu < \kappa)[\mu^{\aleph_0} < \kappa], \kappa = \mathrm{cf}(\kappa), \kappa^{|\mathcal{A}_\zeta|} = \kappa$ for $\zeta < \xi$, and $\xi = \ell g(\bar{\mathcal{A}}) \leq \kappa, \kappa > \aleph_1$,
we define the strong sealing forcing $\mathrm{SSeal}^*(\bar{\mathcal{A}}, S, \kappa)$ as P_κ where $\langle P_i, \underset{\sim}{Q}_j : i \leq \kappa, j < \kappa \rangle$ is an RCS iteration; $\underset{\sim}{Q}_j = \mathrm{seal}(\underset{\sim}{\Xi}_j, S)^{V^{P_j}}, \underset{\sim}{\Xi}_j$ is a maximal antichain of $\mathcal{A}_{\zeta(j)}$ to which $\omega_1 \setminus S$ belongs for some $\zeta(j) < \xi$ (in V^{P_j}) and every maximal antichain $\underset{\sim}{\Xi}$ of some $\mathcal{A}_\zeta$ from V^{P_κ} is $\underset{\sim}{\Xi}_j$ for some $j < \kappa$. [P_κ is not neccessarily well defined].

(9) If $\mathbf{\Xi} \subseteq \{\Xi : \Xi \subseteq \mathfrak{B}\}$ then $\mathrm{seal}(\mathbf{\Xi})$ is the product, with countable support, of $\mathrm{seal}(\Xi)$ for $\Xi \in \mathbf{\Xi}$.

2.5. Remarks.

(1) We could have used CS iteration for SSeal and SSeal*.

(2) If every maximal antichain of $\mathfrak{B}^V$ is semiproper, the difference between $\mathrm{WSeal}(S) \times \mathrm{Levy}(\aleph_1, 2^{\aleph_1})$ and $\mathrm{Seal}(\mathfrak{B}^V, S)$ defined in 2.4 (7), (3) respectively, is nominal (i.e. they are equivalent, i.e. have isomorphic completions).

(3) If $\mathcal{A}_\zeta \restriction S \lessdot \mathfrak{B}^V \restriction S$ and $|\mathcal{A}_\zeta| \leq \aleph_1$ for $\zeta < \ell g(\bar{\mathcal{A}})$, *then* $\mathrm{Seal}(\bar{\mathcal{A}}, S)$ is equivalent to $\mathrm{Levy}(\aleph_1, 2^{\aleph_1})$.

(4) If $|\mathcal{A}_\zeta| \leq \aleph_1$ (for every $\zeta < \ell g(\mathcal{A})$) *then* the difference between $\mathrm{Seal}(\bar{\mathcal{A}}, S)$ and $\mathrm{Seal}'(\bar{\mathcal{A}}, S)$ is nominal (i.e. they are equivalent i.e. have isomorphic completions).

(5) We use below mainly $\mathrm{SSeal}(\bar{\mathcal{A}}, S)$, we could use $\mathrm{SSeal}^*(\bar{\mathcal{A}}, S, \beth_2)$ instead. Also instead $\mathrm{SSeal}(\bar{\mathcal{A}}, S)$ we could use $\mathrm{SSeal}'(\bar{\mathcal{A}}, S)$ by 0.1(4)(c) (and see 0.1(g)).

(6) For convenience we shall use mostly $\mathrm{SSeal}(\bar{\mathcal{A}}, S)$. So in, e.g., 2.11, 2.13 we can deal with SSeal*.

2.6. Notation. We omit κ in $\mathrm{SSeal}(\bar{\mathcal{A}}, S, \kappa)$ *when* it is the first strongly inaccessible. We omit S when $S = \omega_1$. We write $\mathcal{A}$ instead of $\langle \mathcal{A} \rangle$.

2.7. Claim. If in V,

$$\bar{\mathcal{A}}_l = \langle \mathcal{A}^l_\zeta : \zeta < \xi_l \rangle \text{ for } l = 1, 2 \text{ and}$$
$$(\forall \zeta_1 < \xi_1)(\exists \zeta_2 < \xi_2)[\mathcal{A}^1_{\zeta_1} \lessdot \mathcal{A}^2_{\zeta_2} \text{ (inside } \mathfrak{B}^V)],$$
$$(\forall \zeta_2 < \xi_2)(\exists \zeta_1 < \xi_1)[\mathcal{A}^2_{\zeta_2} \lessdot \mathcal{A}^1_{\zeta_1} \text{ (inside } \mathfrak{B}^V)]$$

then

$$\mathrm{Sm}(\bar{\mathcal{A}}^1, S) = \mathrm{Sm}(\bar{\mathcal{A}}^2, S),$$
$$\mathrm{Seal}(\bar{\mathcal{A}}^1, S) = \mathrm{Seal}(\bar{\mathcal{A}}^2, S),$$
$$\mathrm{Seal}'(\bar{\mathcal{A}}^1, S) = \mathrm{Seal}'(\bar{\mathcal{A}}^2, S) \text{ and}$$
$$\mathrm{SSeal}(\bar{\mathcal{A}}^1, S, \kappa) = \mathrm{SSeal}(\bar{\mathcal{A}}^2, S, \kappa).$$

Proof. Easy. $\square_{2.7}$

2.8. Claim.

(1) Let $\Xi \subseteq \mathfrak{B}^V$ be pre-dense. Then Ξ is semiproper *iff*: for λ regular large enough and countable $N \prec (H(\lambda), \in)$ with $\Xi \in N$, *there is* a countable $N', N \prec N' \prec (H(\lambda), \in, <^*_\lambda)$, satisfying $N \cap \omega_1 = N' \cap \omega_1 \in \bigcup_{A \in \Xi \cap N'} A$. [Why? For the implication "$\Rightarrow$" let $q \in \mathrm{seal}(\Xi)$ be $(N, \mathrm{seal}(\Xi))$-semi-generic. Let $\underset{\sim}{a}_i[G_{\mathrm{seal}(\Xi)}]$ be a_i for any $\bar{a} = \langle a_j : j \leq \alpha \rangle \in G_{\mathrm{seal}(\Xi)}$ whenever $\alpha > i$ so $\underset{\sim}{C} = \{\underset{\sim}{a}_i \cap \omega_1 : i < \omega_1\}$ is forced to be a club of ω_1. So $\underset{\sim}{C} \in N$, hence

as q is $(N, \text{seal}(\Xi))$- semi-generic, necessarily $q \Vdash$ "$\delta \stackrel{\text{def}}{=} N \cap \omega_1 \in \underset{\sim}{C}$". In fact $\delta = \underset{\sim}{a}_\delta \cap \omega_1 = \bigcup\limits_{i<\delta} \underset{\sim}{a}_i \cap \omega_1$, so possibly increasing q, for some $\langle b_i : i \leq \delta \rangle$ $q \Vdash$ "$\underset{\sim}{a}_i = b_i$ for $i \leq \delta$", so

$$q \Vdash \text{“}\delta = \omega_1 \cap (\text{Skolem hull in } (H(\lambda), \in, <^*_\lambda) \text{ of } |N| \cup b_\delta = |N| \cup \bigcup_{i<\delta} b_i)\text{”}.$$

So this Skolem hull is N' as required. For the implication "$\Leftarrow$" use 2.8(4) below.]

(2) $\Vdash_{\text{seal}(\Xi)}$ "$\Xi \subseteq \mathfrak{B}^{[\text{seal}(\Xi)]}$ is absolutely pre-dense" (absolutely means for extensions not collapsing $\aleph_1$; more specifically in this chapter, there is a list $\langle A_i : i < \omega_1 \rangle$ of members of Ξ and a club C of ω_1 such that $\delta \in C \Rightarrow \delta \in \bigcup_{i<\delta} A_i$). [Why? Let $\langle \underset{\sim}{a}_i : i < \omega_1 \rangle$ be as in the proof of 2.8(1), so let A_i be such that $\langle A_i : i < \delta \rangle$ lists the member of Ξ in a_δ for limit $\delta < \omega_1$.]

(3) WSeal(S) is semiproper and $\Vdash_{\text{WSeal}(S)}$ "*if* $\Xi \in V$ is semiproper in $\mathfrak{B}^V$ and $(\omega_1 \backslash S) \in \Xi$, *then* Ξ is absolutely pre-dense in $\mathfrak{B}^{[\text{WSeal}(S)]}$". [Why? For semiproperness use 2.8(8) below; for absoluteness use 2.8(2) above.]

(4) seal(Ξ) is A-complete (see V §3) for $A \in \Xi$; so WSeal(S) is $(\omega_1 \setminus S)$-complete. [Why? Think.]

(5) If Ξ is pre-dense in $\mathfrak{B}[V]$, *then* seal(Ξ) preserves stationarity of subsets of ω_1; if $\mathcal{A} \subseteq \mathfrak{B}^V, \Xi$ a pre-dense subset of $\mathcal{A} \setminus \{\emptyset\}$ *then* seal(Ξ) preserves stationary of subsets of ω_1 which belongs to $\mathcal{A}$ or just are not in Sm$(\mathcal{A}, S)$. [Why? Use 2.8(4) as any A-complete forcing notion surely preserve the stationarity of subsets of A.]

(6) The forcing notion seal(Ξ) forces $|\Xi| \leq \aleph_1$ and has cardinality $\leq (|\Xi| + \aleph_1)^{\aleph_0}$. The forcing notion Seal$(\bar{\mathcal{A}}, S)$ is $(\omega_1 \setminus S)$-complete; SSeal$^*(\bar{\mathcal{A}}, S, \kappa)$ and even any initial segment of such iteration of length κ is $(\omega_1 \setminus S)$-complete and if $\kappa > \aleph_0$ is $\bar{\mathcal{A}}$-inaccessible and $S \subseteq \omega_1$ is stationary *then* it satisfies the θ-c.c. if $\theta = \text{cf}(\theta) > |\mathcal{A}_\zeta|$ for $\zeta < \ell g(\bar{\mathcal{A}})$ and $\bigwedge\limits_{\alpha<\kappa} |\alpha|^{\aleph_0} < \theta$. If $\kappa > \aleph_0$ is strongly inaccessible *then* SSeal$(\bar{\mathcal{A}}, S, \kappa)$ satisfies the κ-c.c. and is $(\omega_1 \setminus S)$-complete.

(7) If $\mathrm{Rss}(\aleph_2, \beth_2(\aleph_1))$ *then* for every pre-dense $\Xi \subseteq \mathfrak{B}(V)$, $\mathrm{seal}(\Xi)$ is semiproper. [Why? By 1.7(3) and 2.8(5).]
In this case $\mathrm{Seal}(\mathfrak{B}^V, S)$, $\mathrm{SSeal}(\mathfrak{B}^V, S, \kappa)$, $\mathrm{SSeal}^*(\mathfrak{B}^V, S, \kappa)$ are semiproper.

(8) For λ regular large enough, and countable $N \prec (H(\lambda), \in, <^*_\lambda)$ *there is* a countable N', $N \prec N' \prec (H(\lambda), \in, <^*_\lambda)$ satisfying: $N \cap \omega_1 = N' \cap \omega_1$ and for every semiproper $\Xi \subseteq \mathfrak{B}^V$ we have: $[\Xi \in N \Rightarrow N' \cap \omega_1 \in \bigcup_{A \in \Xi \cap N'} A$ [use part (1) repeatedly ω-times] and even $\Xi \in N' \Rightarrow N' \cap \omega_1 \in \bigcup_{A \in \Xi \cap N'} A$ [use the previous statement repeatedly ω-times]. $\square_{2.8}$

2.9. Claim. Suppose $\bar{\mathcal{A}} = \langle \mathcal{A}_\zeta : \zeta < \xi \rangle$ is an increasing sequence of subalgebras or just subsets of $\mathfrak{B}$, $\kappa > \aleph_0$ is strongly inaccessible or just $\mathrm{SSeal}^*(\bar{\mathcal{A}}, S, \kappa)$ is well defined. Assume $\langle \bar{\mathcal{A}}, \kappa \rangle \in N \prec (H(\lambda), \in)$, N countable, $P \stackrel{\mathrm{def}}{=} \mathrm{SSeal}(\bar{\mathcal{A}}, S, \kappa)$ or $P = \mathrm{SSeal}^*(\bar{\mathcal{A}}, S, \kappa)$ respectively and

$\oplus^N_{\bar{\mathcal{A}},S}$ if $\Xi \in N$ is a pre-dense subset of $\mathcal{A}_\zeta$ for some $\zeta \in N \cap \xi$ and $\omega_1 \setminus S \in \Xi$, then $N \cap \omega_1 \in \bigcup_{A \in \Xi \cap N} A$.

Then for every $p \in P \cap N$, *there is* $q \in P$, (N, P)-generic, $p \le q$, q force a value to $\underset{\sim}{G}_P \cap N$ and $q \Vdash$ "$\oplus^{N[\underset{\sim}{G}_P]}_{\mathcal{A},S}$ holds".

Proof. We have to find q, $p \le q \in P$, which is (N, P)-generic. We first show:

$(*)$ if $\underset{\sim}{\zeta}, \underset{\sim}{\Xi} \in N$ are P-names, $\Vdash_P$ "$\underset{\sim}{\Xi}$ is a pre-dense subset of $\mathcal{A}_{\underset{\sim}{\zeta}}$", $p \in N \cap P$, *then* for some p^2, $p \le p^2 \in N \cap P$, and for some A, ζ we have $p^2 \Vdash$ "$\underset{\sim}{\zeta} = \zeta$ and $A \in \underset{\sim}{\Xi} \cap N \cap \mathcal{A}_\zeta$" (so $A \in V$, and $A \in N \cap \mathcal{A}_\zeta$ and $A \in V$) and $N \cap \omega_1 \in A$.

Proof of $(*)$. We can find p^0, $p \le p^0 \in N \cap P$, and ζ such that $p^0 \Vdash$ "$\underset{\sim}{\zeta} = \zeta$" (so necessarily $\zeta \in N$). Next define

$$\Upsilon = \{A \in \mathcal{A}_\zeta : \text{ for some } p^1,\ p \le p^1 \in P, \text{ and } p^1 \Vdash \text{“} A \in \underset{\sim}{\Xi} \text{”}\}.$$

Clearly $\Upsilon \in N$, $\Upsilon \in V$, and Υ is a pre-dense subset of $\mathcal{A}_\zeta$, $\zeta \in N$.
By $\oplus^N_{\bar{\mathcal{A}},S}$ there is $A \in \Upsilon \cap N$ such that $N \cap \omega_1 \in A$. By the definition of Υ there is p^2, $p^0 \le p^2 \in P$ and $p^2 \Vdash$ "$A \in \underset{\sim}{\Xi}$". As p^0, A and $\underset{\sim}{\Xi}$ are all in N, we

can choose such p^2 in N, thus finishing the proof of $(*)$.

Now we continue with the proof of 2.9. We define p_n for $n < \omega$ such that:

(a) $p_0 = p$, $p_{n+1} \geq p_n$;

(b) $p_n \in P_\kappa \cap N$;

(c) for every dense subset $\mathcal{J}$ of P_κ which belongs to N for some n, $p_{n+1} \in \mathcal{J}$;

(d) *if* $j \in \kappa \cap N, \underset{\sim}{\Xi}, \underset{\sim}{\zeta}$ are P_j-names from N and $\Vdash_{P_j}$ "$\underset{\sim}{\zeta} < \xi$ and $\underset{\sim}{\Xi} \subseteq \mathcal{A}_\zeta$ is pre-dense" *then* for some $n < \omega$ and $B \in \mathfrak{B}^V \cap N$, we have $N \cap \omega_1 \in B$ and

$$p_{n+1} {\restriction} j \Vdash_{P_j} \text{"}B \in \underset{\sim}{\Xi}\text{"}.$$

This clearly suffices, as (using the notation of Definition 2.4(5)):

(α) for $j \in N \cap \kappa$ we have $(\bigcup_{n<\omega} p_n)(j)$ is in Q_j by (d), and

(β) $\bigcup_{n<\omega} p_n$ is (N, P)-generic by (c).

So we can assign the tasks, and for satisfying (b) and (c) there is no problem. For (d) use $(*)$. $\square_{2.9}$

2.10 Claim. Suppose

(a) seal(Ξ) is semiproper for every maximal antichain Ξ of $\mathfrak{B}^V$ to which $\omega_1 \setminus S$ belongs, $\bar{A} = \langle \mathfrak{B}^V \rangle = \langle \mathcal{A}_0 \rangle$

or

(a)′ $\bar{A} = \langle \bar{A}_\zeta : \zeta < \xi \rangle$, $\mathcal{A}_\zeta \subseteq \mathfrak{B}^V$, and seal($\Xi$) is semiproper for any predense subset Ξ of $\mathcal{A}_\zeta$, $\zeta < \xi$

and

(b) $\kappa > \aleph_0$ is strongly inaccessible

or at least

(b)′ $\kappa > \aleph_0$ is inaccessible or just $|\mathcal{A}_\zeta|$-inaccessible for $\zeta < \xi$ (see 2.4(5)).

Then $P \stackrel{\text{def}}{=} \text{SSeal}(\bar{A}, S, \kappa)$ if (b) or $P \stackrel{\text{def}}{=} \text{SSeal}^*(\bar{A}, S, \kappa)$ if (b)′ (both well defined), is semiproper, have the κ-c.c., is $(\omega_1 \setminus S)$-complete and $\Vdash_P$ "$(\mathcal{A}_\zeta {\restriction} S) \lessdot (\mathfrak{B}^P {\restriction} S)$" (and in case (b)′ if $\theta = \text{cf}(\theta) > |\mathcal{A}_\zeta|^{\aleph_0}$, θ-c.c.).

Remark. Some points in the proof are repeated in 2.11.

Proof. The $(\omega_1 \setminus S)$-completeness is trivial by the definition of P and Ch. V, Def. 1.1 (and the preservation theorem there i.e. by 2.8(a)).

For semiproperness let λ be regular and large enough, and $N \prec (H(\lambda), \in)$ countable, $P \in N$ and $p \in P \cap N$. Applying repeatedly 2.8(1) (or directly 2.8(8)), there is N', $N \prec N' \prec (H(\lambda), \in)$, $N \cap \omega_1 = N' \cap \omega_1, N'$ countable, and for every maximal antichain $\Xi \subseteq \mathfrak{B}$ (or just pre-dense $\Xi \subseteq \mathfrak{B}^V$ if (a) or predense subset Ξ of $\mathcal{A}_\zeta$ for some $\zeta < \xi$, if (a)$'$):

$$\Xi \in N', N \cap \omega_1 \in S \Rightarrow N \cap \omega_1 = N' \cap \omega_1 \in \bigcup_{A \in \Xi \cap N'} A.$$

Now use 2.9. (with $\langle \mathfrak{B}^V \rangle, N'$ here standing for $\langle \mathcal{A}_\zeta : \zeta < \xi \rangle, N$ there).

So we have proved that P is semiproper and by the present proof and the Δ-system lemma (alternativelly if κ is strongly inacessible by 2.2(1)(c) or 2.8(6)) P has the κ-c.c., hence $\Vdash_P$ "$(\mathcal{A}_\zeta {\restriction} S) \lessdot (\mathfrak{B}^P {\restriction} S)$" follows from the definition of P as every P_κ-name of a subset of some $\mathcal{A}_\zeta$ is a P_j-name for some $j < \kappa$ (as P_κ satisfies the κ-c.c.). $\square_{2.10}$

2.11. Claim. If $\bar{\mathcal{A}} = \langle \mathcal{A}_\zeta : \zeta < \xi \rangle, \mathcal{A}_\zeta {\restriction} S \lessdot \mathfrak{B}^V {\restriction} S$ for $\zeta < \xi$, each $\mathcal{A}_\zeta {\restriction} S$ satisfies the $\aleph_2$-c.c. (e.g. has power $\leq \aleph_1$) and $\kappa > \aleph_0$ is strongly inaccessible, *then*

(1) $P_\kappa \stackrel{\text{def}}{=} \text{SSeal}(\bar{\mathcal{A}}, S, \kappa)$ is proper;

(2) $\Vdash_{P_\kappa}$ "$\mathcal{A}_\zeta {\restriction} S \lessdot \mathfrak{B}^{P_\kappa} {\restriction} S$ for $\zeta < \xi$";

(3) in fact, P_κ is $(\omega_1 \setminus S)$-complete, strongly proper and satisfies the κ-c.c. and $\Vdash_{P_\kappa}$ "$\kappa = \aleph_2 = 2^{\aleph_1}$";

(4) if $\omega_1 \setminus S$ is stationary, P_κ does not add ω-sequences of ordinals.

Proof. (1) Let λ be regular large enough and $N \prec (H(\lambda), \in)$ countable, $\bar{Q} \in N$ (hence $P_\kappa \in N$) and $p \in P_\kappa \cap N$. We want to apply 2.9, so we have (and it suffices) to verify $\oplus$ there, i.e.

$(**)$ if $\mathcal{A}_\zeta {\restriction} S \lessdot \mathfrak{B}[V] {\restriction} S$, $\mathcal{A}_\zeta$ satisfies the $\aleph_2$-c.c., $\mathcal{A}_\zeta \in N \prec (H(\chi), \in, <^*_\chi)$, N countable, $\Xi \subseteq \mathcal{A}_\zeta$ is a pre-dense subset of $\mathcal{A}_\zeta$ and $\omega_1 \setminus S \in \Xi$ *then* $N \cap \omega_1 \in \bigcup\{A : A \in \Xi \cap N\}$.

Proof of $(**)$. As $\mathcal{A}_\zeta {\restriction} S \models$ "$\aleph_2$"-c.c., clearly without loss of generality $|\Xi| \leq \aleph_1$, so let $\Xi = \{A_i : i < \omega_1\}$ (as $\Xi \neq \emptyset$ this is possible) and say $A_0 = \omega_1 \setminus S$.

Since $\mathcal{A}_\zeta \restriction S \lessdot \mathfrak{B}^V \restriction S$, clearly Ξ is pre-dense in $\mathfrak{B}^V$, hence we know $\{\delta : \delta \in \bigcup_{i<\delta} A_i\} \in \mathcal{D}_{\omega_1}$ (otherwise the complement contradicts the pre-density of Ξ in $\mathfrak{B}^V$), so there is a closed unbounded $C \subseteq \omega_1$ such that $C \subseteq \{\delta : \delta \in \bigcup_{i<\delta} A_i\}$. As $\Xi \in N$ without loss of generality $\langle A_i : i < \omega_1 \rangle \in N$ and without loss of generality $C \in N$. As $N \prec (H(\lambda), \in)$ clearly $C \cap N$ is unbounded in $N \cap \omega_1$, hence $N \cap \omega_1 = \sup(C \cap N \cap \omega_1) \in C$, so $N \cap \omega_1 \in \bigcup \{A_i : i \in N \cap \omega_1\}$, so for some $j \in N \cap \omega_1, N \cap \omega_1 \in A_j$. But $\langle A_i : i < \omega_1 \rangle \in N$ so $A_j \in N$, as required.

(2) If $A \in \mathcal{P}(\omega_1)^{V^{P_\kappa}}$ then as P_κ satisfies the κ-c.c. (by 2.10 or as by part (1), $\{p \in P_\kappa : \mathrm{Dom}(p) \text{ is countable}\}$ is dense in P_κ, clearly we can apply the Δ-system lemma) for some $\alpha < \kappa, A \in \mathcal{P}(\omega_1)^{V^{P_\alpha}}$, and so by the definition of $\mathrm{SSeal}(\bar{\mathcal{A}}, S, \kappa)$, if $A/\mathcal{D}_{\omega_1}$ is disjoint to a dense subset of $x \in \mathcal{A}_\zeta, A \subseteq S, \zeta < \xi$ then we "shoot" a club through its completion in the $(\beta+1)$-th iterand in the iteration defining $\mathrm{SSeal}(\bar{\mathcal{A}}, S, \kappa)$ for $\beta \in (\alpha, \kappa)$ large enough. Why? As $V^{P_\kappa} \models$ "$|\mathcal{A}_\zeta| \leq \aleph_1$" (as P_1 collapses $2^{\aleph_1}$ to $\aleph_1$ see 2.4(5)) there is $\beta, \alpha < \beta < \kappa$ such that for every $x \in \mathcal{A}_\zeta$, if $x \cap A$ is not stationary in V^{P_κ}, then it is not stationary in V^{P_β}.

(3) Easy (strong properness hold by the proof of 2.9 and use IX 2.7, 2.7A for preservation of strong properness or prove directly).

(4) By 2.8(4) and V §3. $\square_{2.11}$

2.12. Claim. Let $\bar{Q} = \langle P_i, \underset{\sim}{Q}_j : i \leq \alpha, j < \alpha \rangle$ be a semiproper iteration, and α be a limit ordinal. Suppose $\Vdash_{P_\alpha}$ "$\underset{\sim}{\Xi} \subseteq \mathfrak{B}^{\bar{Q}}$ is pre-dense" and $i < \alpha$. *Then* (a) $\Leftrightarrow$ (b)$^+$ $\Rightarrow$ (b), where:

(a) $(P_\alpha/P_i) * \mathrm{seal}(\Xi)$ is semiproper (in V^{P_i});

(b) If λ is regular large enough, $\bar{Q} \in N \prec (H(\lambda), \in, <^*_\lambda), N$ countable, $\underset{\sim}{\Xi} \in N$, $p \in N \cap P_\alpha$, $i \in N \cap \alpha$, $q \in P_i$ is (N, P_i)-semi-generic, $p \restriction i \leq q$ *then* there are $N^1, p^1, q^1, \underset{\sim}{A}$ and j such that:

(i) $N \prec N^1 \prec (H(\lambda), \in, <^*_\lambda)$,

(ii) N^1 is countable, $N^1 \cap \omega_1 = N \cap \omega_1$,

(iii) $p \leq p^1 \in N^1 \cap P_\alpha$,

(iv) $i < j < \alpha$, j a non-limit ordinal,

(v) $j \in N^1$,

(vi) $q \leq q^1 \in P_j$,

(vii) q^1 is (N^1, P_j)-semi-generic,

(viii) $p^1 \restriction j \leq q^1$,

(ix) $\underset{\sim}{A} \in N^1$ is a P_j-name,

(x) $q^1 \Vdash$ "$N^1 \cap \omega_1 \in \underset{\sim}{A}$",

(xi) $q^1 \cup p^1 \restriction [j, \alpha) \Vdash_{P_\alpha}$ "$\underset{\sim}{A} \in \underset{\sim}{\Xi}$";

(b)$^+$ Like (b) but N is a P_i-name and N^1 is a P_i-name.

Remark. There is not much difference if in clause (b) (or (b)$^+$) we replace clause (ix) by

(ix) $\underset{\sim}{A} \in N$ is a P_j-name

but then j is allowed to be P_i-name.

Proof. (a) $\Rightarrow$ (b)$^+$ Let $\underset{\sim}{Q} \overset{\text{def}}{=} \text{seal}(\Xi)$ and let $q \in G_i \subseteq P_i$, G_i generic over V. In $V[G_i]$, apply the definition of "$(P_\alpha/P_i) * \text{seal}(\Xi)$ is semi porper" to the model $N = \underset{\sim}{N}[G_i]$ and the condition p, and get a condition q^0, so q^0 is $(\underset{\sim}{N}, (P_\alpha/P_i) * \underset{\sim}{Q})$-semi-generic. Let G be such that $q^0 \in G \subseteq P_\alpha * \underset{\sim}{Q}$, $G_i \subseteq G$, and G is generic over V. So by the definition of $\underset{\sim}{Q}$ =seal(Ξ) for some $A \in \underset{\sim}{\Xi}[G_\alpha] \cap N[G]$ we have $N \cap \omega_1 = N[G] \cap \omega_1 \in A$. As $A \in \underset{\sim}{\Xi}[G_\alpha] \subseteq \mathfrak{B}^{\overline{Q}} = \bigcup_{j<\alpha} \mathfrak{B}[P_{j+1}]$, for some $j_0 \in \alpha \cap N$, $A \in \mathfrak{B}[P_{j_0+1}]$, and there is a P_{j_0+1}-name $\underset{\sim}{A} \in N[G]$ such that $\underset{\sim}{A}[G] = A$, and without loss of generality q^0 forces this. Now

$$\mathcal{I} = \{r : r \in P_\alpha \text{ and } r \text{ is above } p \text{ or incompatible with } p \text{ and}$$
$$r \Vdash_{P_\alpha} \text{"}\underset{\sim}{A} \in \underset{\sim}{\Xi}\text{" or } r \Vdash_{P_\alpha} \text{"}\underset{\sim}{A} \notin \underset{\sim}{\Xi}\}$$

is a dense subset of P_α and $\underset{\sim}{r}$ = the $<^*_\lambda$-least member of $\mathcal{I}$ which belongs to $\underset{\sim}{G}_\alpha$ is a P_α-name, and $\mathcal{I} \in N$, $\underset{\sim}{r} \in N$. Hence $\underset{\sim}{r}[G] \in N[G]$ and clearly $\underset{\sim}{r}[G]$ is compatible with q^0, $p \leq \underset{\sim}{r}[G]$ and $\underset{\sim}{r}[G] \vDash$ "$\underset{\sim}{A} \in \Xi$", so w.l.o.g. $\underset{\sim}{r}[G] \leq q^0$. Let N^1 be the Skolem hull of $N \cup \{j_0, \underset{\sim}{A}, \underset{\sim}{r}[G]\}$ in $(H(\lambda), \in, <^*_\lambda)$, let $j = j_0 + 1$, $q^1 = q^0 \restriction j$ and $p_1 = \underset{\sim}{r}[G]$.

(a) $\Rightarrow$ (b) Similar proof.

(b)$^+$ $\Rightarrow$ (a) Use (b)$^+$. Specifically, for $i < \alpha$ let $G_i \subseteq P_i$ be generic over V, $i < \alpha$.

Assume the desired conclusion in clause (a) fails then this is exemplified by some N, $(p, \underset{\sim}{r})$ where $N \prec (H(\lambda)^{V[G_i]}, \in)$ is countable, $(p, \underset{\sim}{r}) \in (P_\alpha/G_i) * \text{seal}(\underset{\sim}{\Xi})$ and $(p, \underset{\sim}{r}) \in N$ (where $N \in V[G_i]$). So for some $q_0 \in G_i$ and $\underset{\sim}{x}$ we have: $\underset{\sim}{x}$ is a P_i-name, $\underset{\sim}{x}[G_i] = N$ and $q_0 \Vdash_{P_i}$ "$\underset{\sim}{x}$ and $(p, \underset{\sim}{r}) \in (P_\alpha/P_i) * \text{seal}(\underset{\sim}{\Xi})$ form a counterexample to semiproperness".

Clause $(b)^+$ applied to $\underset{\sim}{x}$, q_0, p gives $\underset{\sim}{N}^1, p^1, q^1, \underset{\sim}{A}_{q(*)}$ and j as there and w.l.o.g. $q^1 \restriction i \in Q_i$ and let $N^1 = \underset{\sim}{N}^1[G_i]$.

As P_α/P_j is semiproper, there is $q^2 \in P_\alpha$ which is (N^1, P_α)-semigeneric, $p \leq q^2$ and $q^1 = q^2 \restriction j$ and let G_α be such that $q^2 \in G_\alpha \subseteq P_\alpha$ and G_α is generic over V. By the choice of q_0 (which is $\leq q_1 \leq q^1 \leq q^2 \in G_\alpha$) without loss of generality $G_i = G_\alpha \cap P_i$. So $\underset{\sim}{\Xi}, p, \underset{\sim}{r}, \underset{\sim}{A} \in N[G_\alpha]$, where on $\underset{\sim}{A}$ see clauses (ix), (x), (xi) and as $\text{seal}(\underset{\sim}{\Xi}[G_\alpha])$ is $\underset{\sim}{A}[G_\alpha]$-complete there is $r^2 \in \text{seal}(\underset{\sim}{\Xi}[G_\alpha])$ which is $(N^1[G_\alpha], \text{seal}(\underset{\sim}{\Xi}[G_\alpha])$-semiproper and above $\underset{\sim}{r}[G_\alpha]$. So for some $\underset{\sim}{r}^*$, $q^2 \Vdash_{P_\alpha * \text{seal}(\Xi)}$ " $\underset{\sim}{r}^*$ is above $\underset{\sim}{r}$ and is $(N[\underset{\sim}{G}_\alpha], (P_\alpha/G_i) * \text{seal}(\underset{\sim}{\Xi}))$-semi generic". So $(q^2, \underset{\sim}{r}^*)$ contradict the choice of q_0 and we are done. $\square_{2.12}$

2.13 Claim. Let $\bar{Q} = \langle P_i, \underset{\sim}{Q}_j : i \leq \alpha, j < \alpha \rangle$ be a semiproper iteration and α be a limit ordinal.

(1) If we have P_α-name $\underset{\sim}{\boldsymbol{\Xi}}$ satisfying $\underset{\sim}{\boldsymbol{\Xi}} \subseteq \underset{\sim}{\boldsymbol{\Xi}}^* = \{\underset{\sim}{\Xi} \in V^{P_\alpha} : \underset{\sim}{\Xi}$ is a P_α-name of a maximal antichain or just a pre-dense subset of $\mathfrak{B}^{\bar{Q}}$, such that for every $i < \alpha$, $(P_\alpha/P_{i+1})*$ $\text{seal}(\underset{\sim}{\Xi})$ is semiproper (i.e. this is $\Vdash_{P_{i+1}}$)$\}$ *then* $(P_\alpha/P_{i+1})*$ $\text{Seal}(\underset{\sim}{\boldsymbol{\Xi}})$ is semiproper for every $i < \alpha$.

(2) If

(∗) $(P_\alpha/P_{i+1})*$ $\text{seal}(\underset{\sim}{\Xi})$ is semiproper for every $i < \alpha$ and maximal antichain (or just a pre-dense subset) $\underset{\sim}{\Xi}$ of $\mathfrak{B}^{\bar{Q}}$ (from V^{P_α}) to which $\omega_1 \setminus S$ belongs,

then for every $i < \alpha$, $(P_\alpha/P_{i+1}) * \text{Seal}(\mathfrak{B}^{\bar{Q}}, S)$ is semiproper and for $\kappa > |P_\alpha|$ strongly inaccessible $(P_\alpha/P_{i+1}) * \text{SSeal}(\mathfrak{B}^{\bar{Q}}, S, \kappa)$ is semiproper with κ-c.c.

(3) The hypothesis (∗) of (2) holds if for arbitrarily large $i < \alpha$:

$\underset{\sim}{Q}_i$ is semiproper and $\Vdash_{P_i}$ "$\text{Rss}(\aleph_2)$".

(4) If ($\underset{\sim}{\mathcal{A}}$ is a P_α-name and it is forced for P_α that) $\underset{\sim}{\Xi}$ is a predense subset of $\underset{\sim}{\mathcal{A}}$, $\bigvee_{i<\alpha} \underset{\sim}{\mathcal{A}} \subseteq \mathfrak{B}^{P_{i+1}}$, and $\underset{\sim}{\mathcal{A}} \lessdot \mathfrak{B}^{\bar{Q}}$ (for this $\alpha = \sup\{i : \underset{\sim}{\mathcal{A}} \lessdot \mathfrak{B}^{P_{i+1}}\}$ suffice), *then* $\underset{\sim}{\Xi} \in \underset{\sim}{\Xi}^*$ ($\underset{\sim}{\Xi}^*$ from part (1)).

(5) Assume

(**) $\bar{\mathcal{A}} = \langle \bar{\mathcal{A}}_\beta : \beta < \beta^* \rangle$ and for $\beta < \beta^*$ we have: $\Vdash_{P_\alpha}$ "$\mathcal{A}_\beta \subseteq \mathfrak{B}^{P_{i+1}}$ for some $i < \alpha$" and if $i < \alpha$ and $\underset{\sim}{\Xi}$ is a P_α-name of a pre-dense subset of $\mathcal{A}_\beta$ to which $\omega_1 \setminus S$ belongs *then* $\Vdash_{P_{i+1}}$ " if $\underset{\sim}{\mathcal{A}}_\beta \subseteq \mathfrak{B}^{P_{i+1}}$ then $(P_\alpha/P_{i+1}) * \mathrm{seal}(\underset{\sim}{\Xi})$ is semiproper".

Then for every $i < \alpha$, $(P_\alpha/P_{i+1}) *$ Seal $(\bar{\mathcal{A}}, S)$ is semiproper and if $\kappa > |P_\alpha|$ is strongly inaccessible then $(P_\alpha/P_{i+1}) * \mathrm{SSeal}(\bar{\mathcal{A}}, S, \kappa)$ is also semiproper, satisfies the κ-c.c., has cardinality κ, forces $\kappa = \aleph_2$ and forces $\underset{\sim}{\mathcal{A}}_\beta \lessdot \mathfrak{B}^{V[P_\alpha * \mathrm{SSeal}(\bar{\mathcal{A}}, S, \kappa)]}$.

Proof. (1) Use Claim 2.12 ω times and the definition of RCS (note that in 2.12(b) we do not get $q^1 \restriction i = q$, but we can replace q by any q', $q \leq q' \in P_i$).

(2) For the first phrase use 2.13(1). For the SSeal case, use also 2.9 with $\bar{\mathcal{A}} = \langle \mathfrak{B}^{\bar{Q}} \rangle$ (so $\xi = 1$), where the assumption of 2.9 can be gotten by the first phrase; the κ-c.c. is proved as in 2.11(3) using models N as in 2.9.

(3) By 1.7(5) the statement $\mathrm{Rss}(\aleph_2)$ implies that semiproperness and preserving stationarity of subsets of ω_1 are equivalent. Suppose $i < \alpha, Q_i$ is semiproper and $\Vdash_{P_i}$ "$\mathrm{Rss}(\aleph_2)$". As by 2.8(5), $\mathrm{seal}(\underset{\sim}{\Xi})$ (for $\Xi \subseteq \mathfrak{B}^{\bar{Q}}$ a maximal antichain) preserves stationarity of subsets of ω_1 from V^{P_i} which are stationary in V^{P_α} (and this property is preserved by composition (though not by limit)) and $P_\alpha/P_i = Q_i * (P_\alpha/P_{i+1})$ is semiproper hence preserve stationarity of subsets of ω_1, we get that $(P_\alpha/P_i)*$ $\mathrm{seal}(\underset{\sim}{\Xi})$ preserves stationarity of subsets of ω_1 hence is semiproper (in V^{P_i} of course). This holds for arbitrarily large $i < \alpha$, hence (by the composition of semiproperness) for every non-limit i, which is the demand (*) of (2).

4) As in the proof of (**) from the proof of 2.11(1), it suffices to prove clause (b)$^+$ of 2.12 for successor $i < \alpha$, so let $\underset{\sim}{\mathcal{A}}$, $\underset{\sim}{\Xi}$ be as in the assumption of 2.13(4), $q \Vdash$ "$\{\underset{\sim}{\mathcal{A}}, \underset{\sim}{\Xi}\} \in \underset{\sim}{N}$" and $\underset{\sim}{N}, i, p, q$ be as in the assumption of 2.12(b)$^+$. We know that for some $i_0 \geq i$ we have $\underset{\sim}{\mathcal{A}} \subseteq \mathfrak{B}^{P_{i_0+1}}$, so without loss of generality

(possibly increasing p and q) for some i_0, $p \Vdash$ "$\underset{\sim}{\mathcal{A}} \subseteq \mathfrak{B}_{i_0+1}$", by the preservation of semiproperness by composition without loss of generality $i_0 = i+1$. Let G_i be such that $q \in G_i \subseteq P_i$, G_i generaic over V and $N = \underset{\sim}{N}[G_i]$; in $V[G_i]$ we define $\Upsilon \stackrel{\text{def}}{=} \{A \in \mathfrak{B}^{P_{i_0+1}} : p \nVdash_{P_\alpha/G_i}$ "$A \notin \Xi$"$\}$. So $\Upsilon \in N[G_i]$ and $V[G_i] \models$ "$|\Upsilon| \leq \aleph_1$", $\Upsilon \neq \emptyset$ so let, in $V[G_i]$, $\Upsilon = \{A_\zeta : \zeta < \omega_1\}$ and without loss of generality $\langle A_\zeta : \zeta < \omega_1\rangle \in N[G_i]$. Let $B = \{\delta < \omega_1$: δ limit and $\delta \notin \bigcup_{i<\delta} A_i\}$, so $B \subseteq \omega_1$, $B \in V[G_i]$, and (in $V[G_i]$) we have: $B \cap A_\zeta = \emptyset \bmod \mathcal{D}_{\omega_1}$. So in V^{P_α}, B cannot be stationary (as $B \in \mathfrak{B}^{\bar{Q}}$, $\underset{\sim}{\mathcal{A}} \lessdot \mathfrak{B}^{\bar{Q}}$) so as P_α/G_i is semiproper also in $V[G_i]$ we know that B is not stationary, and we finish as in the proof of (**) from the proof of 2.11(1).

5) The proof of 2.13(2) (and see 2.16). $\square_{2.13}$

2.14 Claim. Suppose $\bar{Q} = \langle P_i, \underset{\sim}{Q}_j, \mathbf{t}_j : i \leq \alpha+1, j < \alpha+1\rangle$ is an RCS iteration, $\bar{Q}\restriction\alpha$ is S-suitable, and $\kappa > |P_\alpha|$ is strongly inaccessible.

(1) If $\mathbf{t}_\alpha = 0, \underset{\sim}{Q}_\alpha = \mathrm{SSeal}(\langle\mathfrak{B}[P_j] : j < \alpha, \mathbf{t}_j = 1\rangle, S, \kappa)$ *then* $\bar{Q}$ is S-suitable and also: for α successor or $\alpha = \mathrm{cf}(\alpha) > |P_i|$ for $i < \alpha$ even Q_α is proper.

(2) If α is a limit ordinal, $\bar{A} = \langle\underset{\sim}{\mathcal{A}}_\zeta : \zeta < \xi\rangle$ is a sequence of (P_α-names of) subalgebras of $\mathfrak{B}^{\bar{Q}\restriction\alpha}$ with $\bigwedge_\zeta \bigvee_{i<\alpha} \underset{\sim}{\mathcal{A}}_\zeta \subseteq \mathfrak{B}^{P_{i+1}}$, and for every $\zeta < \xi, \Vdash_{P_\alpha}$ "for $\zeta < \xi$ the set $\{i < \alpha : \underset{\sim}{\mathcal{A}}_\zeta\restriction S \lessdot \mathfrak{B}[P_{i+1}]\restriction S\}$ is unbounded below α", and $\Vdash_{P_\alpha}$ "for every $j < \alpha$ satisfying $\mathbf{t}_j = 1$ for some $\zeta, \mathfrak{B}^{P_j}\restriction S \lessdot \underset{\sim}{\mathcal{A}}_\zeta\restriction S$" and $\mathbf{t}_\alpha = 0$, and $Q_\alpha = \mathrm{SSeal}(\bar{A}, S, \kappa)$ *then* $\bar{Q}$ is S-suitable.

Proof. (1) First assume α is non-limit or $\alpha = \mathrm{cf}(\alpha) > |P_i|$ for $i < \alpha$. We have to check clauses (A) – (F) of Definition 2.1. Clause (D) holds by Claim 2.11(1); clause (E) holds by Claim 2.11(3); clause (F) holds by 2.11(2); the other parts of Definition 2.1 hold trivially. Lastly the conclusion concerning "Q_α is proper" holds by 2.11(3).

If α is limit, then this follows from 2.14(2) which is proved below.

2) Let $\underset{\sim}{\Xi} = \{\Xi : \underset{\sim}{\Xi}$ is a P_α-name of a pre-dense subset of $\mathfrak{B}[P_{i+1}]$ to which $\omega_1 \setminus S$ belongs for some $i < \alpha$ and $(P_\alpha/P_{j+1}) * \mathrm{seal}(\Xi)$ is semiproper for every $j < \alpha\}$. By 2.13(4) above: if $\underset{\sim}{\Xi}$ is a P_α-name of a maximal antichain of $\underset{\sim}{\mathcal{A}}_\zeta(\zeta < \xi)$ *then*

$\underset{\sim}{\Xi} \in \underset{\sim}{\Xi}$. So by 2.13(5) clauses (D),(E),(F) of Definition 2.1 hold (the others are trivial). $\square_{2.14}$

2.15 Claim. 1) Suppose $\bar{\mathcal{A}} = \langle \mathcal{A}_\zeta : \zeta < \xi \rangle$ is an increasing sequence of subalgebras (or just subsets) of $\mathfrak{B}, \chi$ regular, N a countable elementary submodel of $(H(\chi), \in, <^*_\chi)$ and $\oplus^N_{\bar{\mathcal{A}},S}$ from 2.9(1) holds, i.e.

$\oplus^N_{\bar{\mathcal{A}},S}$ if $\zeta \in \xi \cap N$ and $\Xi \in N$ is a pre-dense subset of $\mathcal{A}_\zeta$ and $\omega_1 \setminus S \in \Xi$ then $N \cap \omega_1 \in \bigcup_{A \in N \cap \Xi} A$.

If $Q \in N$ is a strongly proper forcing notion, $p \in Q \cap N$ *then* there is $q \in Q, p \leq q, q$ is (N, Q)-generic and $q \Vdash$ "$\oplus^{N[G_P]}_{\bar{\mathcal{A}},S}$".

2) In 2.10 we can conclude also that for a strongly proper Q which is $(\omega_1 \setminus S)$-complite and satisfies $|Q| < \kappa$, the forcing notion $Q*$ $\mathrm{S\underset{\sim}{S}eal}(\mathfrak{B}^V, S, \kappa)^{V^Q}$ is semiproper $(\omega_1 \setminus S)$-complete.

3) Parallel strengthenings of 2.11, 2.13 (see mainly 2.13(1)) and 2.14 hold.

2.15A Remark. This claim can be used in §3, §4 to get appropriate axioms: it gives a comprehensive family of forcing notions which we can use quite freely in the iterations, without making problems for what is already accomplished there.

For a more general property: see 4.6.

Proof. Straightforward (reread the proof of 2.11). $\square_{2.15}$

2.16 Claim. Assume $\bar{\mathcal{A}} = \langle \mathcal{A}_\zeta : \zeta < \xi \rangle$, $\mathcal{A}_\zeta \restriction \zeta \subseteq \mathfrak{B}^V \restriction S$,

$$W = W_{\bar{\mathcal{A}}} = \Big\{ a : a \subseteq H(\beth_2(\aleph_1)), a \text{ is countable, } a \cap \omega_1 \text{ is an ordinal and: if } \zeta < \xi, \Xi \subseteq \mathcal{A}_\zeta, \Xi \text{ is a pre-dense subset of } \mathcal{A}_\zeta, \ \omega_1 \setminus S \text{ belongs to } \Xi \text{ and } \{\zeta, \Xi\} \in a \text{ then } a \cap \omega_1 \in \bigcup \{A : A \in \Xi \cap a\} \Big\}$$

is a stationary subset of $H(\beth_2(\aleph_1))$ and $\kappa > \aleph_0$ is strongly inaccessible. *Then*

(1) $P_\kappa \stackrel{\text{def}}{=} \mathrm{SSeal}(\bar{A}, S, \kappa)$ is W-proper.

(2) $\Vdash_{P_\kappa}$ "$\mathcal{A}_\zeta \restriction S \lessdot B^V \restriction S$" for $\zeta < \xi$.

(3) In fact, P_κ is $(\omega_1 \setminus S)$-complete strongly W-proper and satisfies the κ-c.c.

(4) If $\omega_1 \setminus S$ is stationary *then* P_κ does not add ω-sequences of ordinals.

(5) If $\xi = \zeta + 1$, $\mathcal{A}_\zeta = \mathfrak{B}^V$ and $\mathrm{Rss}(\aleph_2)$ *then* P_κ is semiproper.

(6) If $\lambda > \kappa$, $\mathrm{Rss}^+(\kappa, \lambda)$ *then* $V^{P_\kappa} \models \mathrm{Rss}(\aleph_2, \lambda)$.

Proof. 1) W-properness is proved as in the proof of 2.11(1) (and 2.9) restricting ourselves to models N such that $N \cap H(\beth_2(\aleph_1)) \in W$.

2), 3), 4) As in the proof of 2.11(2), (3), (4).

5) W-properness implies semiproperness by 2.8(7), (8), (note: we can ignore $\mathcal{A}_\varepsilon$ when $\varepsilon + 1 < \xi$ as $W_{\langle \mathcal{A}_{\xi-1} \rangle} = W_{\bar{\mathcal{A}}}$).

6) Should be clear. $\square_{2.16}$

2.17 Claim.

Assume $\bar{Q} = \langle P_i, \underset{\sim}{Q}_j : i \leq \kappa, j < \kappa \rangle$ is an RCS-iteration, κ is strongly inaccessible ($i < \kappa \Rightarrow |P_i| < \kappa$) and, for stationarily many $i < \kappa$, for arbitrarily large $j \in (i, \kappa)$, $\mathfrak{B}^{\bar{Q} \restriction i} \lessdot \mathfrak{B}^{P_j}$. *Then* in V^{P_κ}, for $\bar{\mathcal{A}} = \langle B[P_\kappa] \rangle$, $W = W_{\bar{\mathcal{A}}}$ contains a club of $\mathcal{S}_{<\aleph_1}(H(\beth_2(\aleph_1)))$.

Proof. By Fodor's Lemma, $\mathfrak{B}^{P_\kappa}$ satisfies the $\aleph_2$-c.c., hence we can apply 2.11. $\square_{2.17}$

§3. On $\mathcal{P}(\omega_1)/\mathcal{D}_{\omega_1}$ Being Layered or the Levy Algebra

On layered ideals see [Sh:237a], Foreman Magidor Shelah [FMSh:252] and [Sh:270]. A reader can read sepeately 3.1 – 3.3, 3.4 – 3.8, 3.4 – 3.10. Here in 3.1, 3.2, 3.3 we deal with "$\mathfrak{B} \restriction S$ being layered"; in 3.4, 3.5, 3.6 we prepared the ground for "$\mathfrak{B} \restriction S$ being the Levy algebra" and in 3.7, 3.9 we deal with "$\mathfrak{B} \restriction S$ being the Levy algebra". We deal also with getting forcing axioms and try to present some approaches (rather than saving in consistency strength around "ZFC+ there is a supercompact cardinal").

3.1. Theorem. Suppose κ is supercompact. *Then* for some forcing notion P:

(i) P satisfies the κ-c.c., has cardinality κ, does not collapse $\aleph_1$, but collapses every $\lambda \in (\aleph_1, \kappa)$ and $\Vdash_{P_\kappa}$ "$\kappa = \aleph_2$, and $2^{\aleph_0} = 2^{\aleph_1} = \aleph_2$",

(ii) $\mathfrak{B}[P]$ is S^*-layered (see 3.1A(4) below), for some stationary $S^* \subseteq \{\delta < \kappa : \mathrm{cf}(\delta) = \aleph_1 \text{ (in } V^P)\}$,

(iii) in V^P, $Ax^+\Big[Q$ semiproper collapsing $\aleph_2$ and $\mathfrak{B}[(V^P)] \lessdot \mathfrak{B}[(V^P)^Q]\Big]$.

3.1A Remark. 1) In (iii), of course if we have Ax rather than Ax^+, we can replace the condition on the forcing Q by:

for some $R, Q \lessdot R$, R is semiproper, $\mathfrak{B}[V^P] \lessdot \mathfrak{B}[(V^P)^R]$ and R collapses $\aleph_2$ (of V^P).

2) Note for 3.1(iii) that, in V^P, we have $|\mathfrak{B}| = 2^{\aleph_1} = \aleph_2$.

3) In (iii) of 3.1 we can replace Ax^+ by Ax_{ω_1}; similarly in 3.2, 3.3(1)(iii).

4) A Boolean algebra B of regular cardinality λ is S^*-layered (for $S^* \subseteq \lambda$) if: letting $B = \bigcup_{i<\lambda} B_i$, B_i increasing continuous in i, $|B_i| < \lambda$, we have $\{\delta < \lambda : \delta \in S^* \Rightarrow B_\delta \lessdot B\} \in \mathcal{D}_\lambda$.

5) We say that a filter $\mathcal{D}$ on a set A is S-layered if $\mathcal{P}(A)/D$ is S-layered.

Proof. Let $S = \omega_1$ and let $h : \kappa \to H(\kappa)$ be a Laver diamond (see Definition VII2.8; later we may say: repeat this proof for other stationary $S \subseteq \omega_1$ and $h : \kappa \to H(\kappa)$). By induction on $i < \kappa$ we define $P_i, Q_i, \mathbf{t}_i$ such that:

(A) $\bar{Q}^\alpha = \langle P_i, \underset{\sim}{Q}_j, \mathbf{t}_j : i \leq \alpha, j < \alpha \rangle$ is an S-suitable iteration.

(B) Q_α is defined by cases:

CASE a: Assume $(*)_1$ + $(*)_2$ where

$(*)_1$ α is measurable and $\bigwedge_{i<\alpha}[|P_i| < \alpha]$ and $\big[i < \alpha \;\&\; \mathbf{t}_i = 1 \Rightarrow \mathfrak{B}[P_i] \lessdot \mathfrak{B}[P_\kappa]\big]$, and $\mathrm{Rss}^+(\alpha, 2^\alpha)$ and

$(*)_2$ $h(\alpha)$ is a P_α-name of a semiproper forcing notion, and $\Vdash_{P_\alpha * h(\alpha)}$ "$\mathfrak{B}[P_\alpha] \lessdot \mathfrak{B}[P_\alpha * h(\alpha)]$ and $\alpha = \aleph_2^{V[P_\alpha]}$ is collapsed".

Then $\mathbf{t}_\alpha = 1$ and $\underset{\sim}{Q}_\alpha = h(\alpha) * \mathrm{S\underset{\sim}{S}eal}^{V[P_\alpha * h(\alpha)]}(\langle \mathfrak{B}[P_i] : i \leq \alpha,\ \mathbf{t}_i = 1 \rangle, S)$.

CASE b: Assume $(*)_1$ but not $(*)_2$,

then $\mathbf{t}_\alpha = 1$ and $Q_\alpha = \mathrm{SSeal}\ (\langle \mathfrak{B}[P_i] : i \leq \alpha,\ \mathbf{t}_i = 1 \rangle, S)$.

CASE c: Assume not $(*)_1$.

Then $\mathbf{t}_\alpha = 0$ and $Q_\alpha = \mathrm{SSeal}^{V[P_\alpha]}((\langle \mathfrak{B}[P_i] : i < \alpha, \mathbf{t}_i = 1\rangle, S)$.

3.1B Observation. $\bar{Q}$ is S-suitable and $\beta < \kappa \Rightarrow \bar{Q}\restriction\beta \in H(\kappa)$ and: $\underset{\sim}{Q}_\beta$ is semiproper when $\beta = \mathrm{cf}(\beta) > |P_i|$ for $i < \beta$ (or β successor).

Proof of 3.1B. We prove by induction on $\beta \leq \kappa$ that $\bar{Q}\restriction\beta$ is S-suitable and when $\beta < \kappa$, then $\bar{Q}\restriction\beta$ belongs to $H(\kappa)$ and if $\beta = \alpha + 1$, $\alpha = \mathrm{cf}(\beta) > |P_i|$ for $i < \alpha$ then $\underset{\sim}{Q}_\alpha$ is semiproper.

For $\beta = 0$: trivial.

For β limit: by 2.3(1).

For $\beta = \alpha + 1$ and for α, $(*)_1$ above fails: By the induction hypotheses $\bar{Q}\restriction\alpha = \langle P_i, \underset{\sim}{Q}_j, \mathbf{t}_j : i \leq \alpha, j < \alpha\rangle$ is S-suitable, hence it is a semiproper iteration and by our choice $\bar{Q}\restriction\beta = \bar{Q}\restriction(\alpha + 1)$ is an RCS iteration and letting κ_α be the first strongly inaccessible $> |P_\alpha|$, we have $Q_\alpha = \mathrm{SSeal}\ (\langle \mathfrak{B}[P_i] : i < \alpha, \mathbf{t}_i = 1\rangle, S, \kappa_\alpha)$.

Now by 2.14(1) we are done (in particular Q_α is semiproper if: α is a successor or $\alpha = \mathrm{cf}(\alpha) > |P_i|$ for $i < \alpha$).

For $\beta = \alpha + 1$ and for α, $(*)_1$ above holds but $(*)_2$ fails

By the induction hypothesis $\bar{Q}\restriction\alpha = \langle P_i, \underset{\sim}{Q}_j : i \leq \alpha,\ j < \alpha\rangle$ is S-suitable, hence a semiproper iteration and by our choice $\bar{Q}\restriction\beta = \bar{Q}\restriction(\alpha + 1)$ is an RCS-iteration and letting κ_α be the first strongly inaccessible $> |P_\alpha|$, we have:

$\mathbf{t}_\alpha = 1$ and in V^{P_α} we have Q_α =SSeal($\langle \mathfrak{B}[P_i] : i \leq \alpha, \mathbf{t}_i = 1\rangle, S, \kappa_\alpha)$. Note that, as $(*)_1$ holds, α is measurable so $\{\gamma < \alpha$: case (c) applies and $\gamma = \mathrm{cf}(\gamma) > |P_i|$ for $i < \gamma\}$ includes all strongly inaccessible non-measurable cardinals in C, for some club C of α. It is well known that there is a normal ultrafilter on α to which this set belongs so 1.10 applies.

By 1.10(1) and, as $V \models$ "$\mathrm{Rss}^+(\alpha, 2^\alpha)$" holds by $(*)_1$, we know that in V^{P_α}, $\mathrm{Rss}(\aleph_2, 2^{\aleph_2})$ holds. So by 2.8(7) every maximal antichain Ξ of $\mathfrak{B}[P_\alpha]$ (in V^{P_α}) is semiproper. Hence by 2.10 SSeal($\mathfrak{B}, S, \kappa_\alpha$) is semiproper. Now $\mathfrak{B}^{\bar{Q}\restriction\alpha} = \mathfrak{B}^{P_\alpha}$ [as α is (by $(*)_1$) strongly inaccessible, $\bigwedge_{i<\alpha} |P_i| < \kappa$, now use

2.2(1)(c)], and $\mathrm{SSeal}(\mathfrak{B}^{\bar{Q}\restriction\alpha}, S, \kappa_\alpha) = \mathrm{SSeal}(\langle \mathfrak{B}[P_i] : i \leq \alpha, \mathbf{t}_i = 1\rangle, S, \kappa_\alpha)$ by claim 2.7, as $\mathbf{t}_\alpha = 1$ and $[i < \alpha \;\&\; \mathbf{t}_i = 1 \Rightarrow \mathfrak{B}[P_i]\restriction S \lessdot \mathfrak{B}[P_\alpha]]$. Together, Q_α is semiproper and we can check that $\bar{Q}\restriction\beta$ is S-suitable.

<u>For $\beta = \alpha + 1$, and for α, $(*)_1 + (*)_2$ above holds.</u>

Similar to the previous case, but now we use the statement in $(*)_2$ to note that $h(\alpha)$ is (in V^{P_α}) a semiproper forcing. Now by $(*)_2$ we know that $\mathfrak{B}[P_\alpha] \lessdot \mathfrak{B}[P_\alpha * h(\alpha)]$ and $V^{P_\alpha * h(\alpha)} \models$ "$\mathfrak{B}[P_\alpha]$ has cardinality $\aleph_1$" hence we can use 2.11 to show that $\mathrm{SSeal}^{V[P_\alpha * h(\alpha)]}(\mathfrak{B}^{P_\alpha}, S)$ is semiproper. $\square_{3.1B}$

Remark. Note that we could use only semiproper Q_α's (so demand in $(*)_2$ that $h(\alpha)$ is semiproper).

3.1C. Observation. (1) If $\alpha < \kappa$, and $\mathbf{t}_\alpha = 1$ (equivalently $(*)_1$ holds) *then* in V^{P_κ} we have $\mathfrak{B}[P_\alpha]\restriction S \lessdot \mathfrak{B}[P_\kappa]\restriction S$.
2) If $\mathcal{D}$ is a normal ultrafilter on $\mathcal{S}_{<\kappa}(H(\beth_8(\kappa)))$, then $\{a : a \in \mathcal{S}_{<\kappa}(H(\beth_8(\kappa)))$ and $(*)_1$ is satisfied by $a \cap \kappa\} \in \mathcal{D}$.

Proof of 3.1C. Should be clear. $\square_{3.1C}$

Letting $P = P_\kappa$ and $S^* = \{\alpha < \kappa : (*)_1 + \neg(*)_2$ holds for α or at least $(*)_1 + V^{P_\kappa} \models$ "$\mathrm{cf}(\alpha) = \aleph_1$"$\}$, we easily finish, note that for $\alpha \in S^* \cup \{\kappa\}$: $\mathfrak{B}[P_\alpha] = \bigcup_{i<\alpha} \mathfrak{B}[P_i]$ and for $\alpha \in S^*$: $\mathbf{t}_\alpha = 1$. As κ is supercompact S^* is a stationary subset of κ (by 3.1C) and forcing with P_κ preserves it (as P_κ satisfies the κ-c.c.) and $\alpha \in S^* \Rightarrow \Vdash_{P_\kappa}$ "$\mathrm{cf}(\alpha) = \aleph_1$" (check Q_α). Also the other requirement causes no problems. $\square_{3.1}$

3.2 Theorem. 1) In 3.1 we can weaken "P satisfies the κ-c.c" to "P does not collapse $\aleph_2$ and has cardinality κ" but add that we have $\mathfrak{B}[P]$ is layered, which means it is S^*-layered for $S^* \stackrel{\mathrm{def}}{=} \{\delta < \aleph_2 : \mathrm{cf}(\delta) = \aleph_1$ (in V^P)$\}$.
2) In 3.1 we can add to the conclusion ($P = P_\kappa$, $\bar{Q} = \langle P_i, \underset{\sim}{Q}_j, \mathbf{t}_j : i \leq \kappa, j < \kappa, \rangle$ is S suitable and):

(iv) In V^P, Ax [Q is semiproper and $i < \kappa \;\&\; \mathbf{t}_i = 1 \Rightarrow \mathfrak{B}^{P_i} \lessdot \mathfrak{B}[(V^P)^Q]$].

3) In 3.2(1) we can add to the conclusion ($P = P_\kappa$, $\bar{Q}$ as above and):

(iv)$^-$ In V^P we have $\mathrm{Ax}^+\Big[Q$ is semiproper changing the cofinality of $\aleph_2$ to $\aleph_0$, and $i < \kappa$ & $\mathbf{t}_i = 1 \Rightarrow \mathfrak{B}^{P_i} \lessdot \mathfrak{B}[(V^P)^Q]\Big]$.

Proof. 1) Force as in 3.1, and then let $P = P_\kappa * \underset{\sim}{Q}_\kappa$ where in $V^{P_\kappa}, Q_\kappa =$ $\mathrm{club}(S^* \cup \{\delta : \mathrm{cf}^{V^P}(\delta) = \aleph_0\})$, where for S, $\mathrm{club}(S) = \{h : h$ a strictly increasing continuous function h from some $\gamma + 1 < \sup(S)$ to $S\}$.

As, in V^{P_κ}, the set $S^* = \{\alpha < \kappa : (*)_1 + \neg(*)_2$ from the proof of 3.1 hold$\} \subseteq \{\delta < \aleph_2 : \mathrm{cf}(\delta) = \aleph_1\}$ is stationary, moreover $\alpha \in S^*$ implies: there is, in V^{P_κ}, a subset b_α of α of order type ω_1 such that $\gamma < \alpha \Rightarrow b_\alpha \cap \gamma \in \bigcup_{\beta<\alpha} V^{P_\beta}$. As $\langle \bigcup_{\beta<\alpha} (\mathcal{P}(\beta) \cap V^{P_\beta}) : \alpha < \kappa\rangle$ is increasing and continuous and $V^{P_\kappa} \models$ "$|\mathcal{P}(\beta) \cap V^{P_\beta}| = \aleph_1$", clearly Q_κ adds no bounded subsets to κ and $\kappa = \aleph_2^{V[P_\kappa]}$, so $\mathfrak{B}[P_\kappa] = \mathfrak{B}[P_\kappa * \underset{\sim}{Q}_\kappa]$ and $\Vdash_{Q_\kappa}$ "$\{\delta < \kappa : \mathrm{cf}(\delta) = \aleph_1$ but not $\mathfrak{B}[P_\delta] \restriction S \lessdot \mathfrak{B}[P_\kappa] \restriction S\}$ is not stationary.

Why does (iii) of 3.1 continue to hold? Suppose, in $V^{P * \underset{\sim}{Q}_\kappa}$, R is a semiproper forcing collapsing $\aleph_2$ such that $(V^P)^{Q_\kappa} \models [\Vdash_R$ "$\mathfrak{B} \lessdot \mathfrak{B}[R]$"$]$. Let $\underset{\sim}{R}$ be a $P_\kappa * \underset{\sim}{Q}_\kappa$-name of such a forcing notion and $(p, \underset{\sim}{q}) \in P_\kappa * Q_\kappa$. Apply (iii) to $Q_\kappa * \underset{\sim}{R}$ in $V[P_\kappa]$ (strictly speaking, its proof). I.e. by the properties of the Laver diamond, for some χ, $2^{|Q_\kappa * \underset{\sim}{R}|} < \chi$, and $M \prec (H(\chi), \in, <^*_\chi)$ to which $\bar{Q}$, Q_κ, $\underset{\sim}{R}$, and $(p, \underset{\sim}{q})$ belong and M isomorphic to some $(H(\chi_1), \in, <^*_{\chi_1})$, by the Mostowski collapsing isomorphism g, taking P_κ to P_{κ_1} where $\kappa_1 = M \cap \kappa$, and $h(\kappa_1) = g(Q_\kappa * \underset{\sim}{R})$. Clearly κ_1 satisfies $(*)_1$ and without loss of generality also $(*)_2$, hence $\mathbf{t}_{\kappa_1} = 1$. So we could have increased $(p, \underset{\sim}{q})$ to guarantee the existence of the generic enough subset of R (i.e. we use the generic subset of $g(Q_\kappa)$ to increase $\underset{\sim}{q}$).

(2) In the proof of 3.1, case b is now divided into subcases b_1 and b_2;

<u>case b_1</u>: $(*)_1$, not $(*)_2$ but

$(*)_{1.5}$ $h(\alpha)$ is a P_α-name of a semiproper forcing notion such that $i < \alpha$, $\mathbf{t}_i = 1 \Rightarrow \mathfrak{B}^{P_\alpha} \lessdot \mathfrak{B}^{P_\alpha * h(\alpha)}$.

Then we let $\mathbf{t}_\alpha = 0, \underset{\sim}{Q}_\alpha = h(\alpha) *$ $\mathrm{SSeal}^{V[P_\alpha * h(\alpha)]}(\langle \mathfrak{B}[P_i] : i \leq \alpha, \mathbf{t}_i = 1\rangle, S, \kappa_\alpha)$, where κ_α is the first strongly inaccessible $> |P_\alpha * h(\alpha)|$.

<u>case b_2</u>: $(*)_1$, not $(*)_2$ and not $(*)_{1.5}$.

Then (as in the old case b) $\mathbf{t}_\alpha = 1, Q_\alpha = \mathrm{SSeal}^{V[P_\alpha]}((\mathfrak{B}[P_i] : i \le \alpha, \mathbf{t}_i = 1), S)$.

3) Should be clear. $\square_{3.2}$

3.3 Theorem. In 3.1, 3.2 we can add, as a parameter (from V), $\bar{S} = \langle S_1, S_2, S_3 \rangle$ a partition of ω_1, $S = S_1$ is a stationary and restrict ourselves to pseudo $(*, S_3)$-complete forcing, see X 3.10, (so if S_3 is not stationary this is not a restriction) so if S_3 is stationary the forcing notions will not be adding reals; i.e.

1) There is a forcing notion P such that:

(i) P satisfies the κ-c.c., does not collapse $\aleph_1$, but collapses every $\lambda \in (\aleph_1, \kappa), \Vdash_P$ "$\kappa = \aleph_2$ and $2^{\aleph_0} \le \aleph_2, 2^{\aleph_1} = \aleph_2$ and if $S_3 = \emptyset \mod \mathcal{D}_{\omega_1}$ then $2^{\aleph_0} = \aleph_2$" and P is pseudo $(*, S_3)$-complete,

(ii) $\mathfrak{B}[P] \restriction S_1$ is S^*-layered, for some stationary $S^* \subseteq \{\delta < \kappa : \mathrm{cf}(\delta) = \aleph_1$ (in V^{P_κ})$\}$,

(iii) in $V^P, Ax^+[Q$ semiproper, pseudo $(*, S_3)$-complete collapsing $\aleph_2$ and $\mathfrak{B}[(V^{P_\kappa})] \lessdot \mathfrak{B}[(V^{P_\kappa})^Q]]$,

(iv) if S_3 is stationary, the forcing P adds no new reals (so $V^P \models CH$).

2) In 3.3(1) we can replace "P satisfies the κ-c.c." by "P does not collapse κ" and have $\mathfrak{B}[P_1]$ is layered, i.e. $S^* = \{\delta < \aleph_2 : \mathrm{cf}(\delta) = \aleph_1\}$ (in V^P).

3) We can add in 3.3(1): ($P = P_\kappa, \bar{Q} = \langle P_i, \underset{\sim}{Q}_j, \mathbf{t}_j : i \le \kappa, j < \kappa \rangle$ is S_1-suitable and)

(v) in V^P, $Ax[Q$ semiproper, pseudo $(*, S_3)$-complete and $i < \kappa$ & $\mathbf{t}_i = 1 \Rightarrow \mathfrak{B}[V^{P_i}] \lessdot \mathfrak{B}[(V^P)^Q]]$.

4) Actually in (3) it suffices "for $i < \kappa$, $(P_\kappa / P_{i+1}) * \underset{\sim}{Q}$ is semiproper, pseudo $(*, S_3)$-complete and: $j \le i$ & $\mathbf{t}_j = 1 \Rightarrow \mathfrak{B}^{P_j} \lessdot V^{P_\kappa \times \underset{\sim}{Q}}$".

3.3A. Remark. 1) In 3.2(2)(iv) and in 3.3(3)(v), if we deal with Ax (Ax$^+$) it is enough that $Q \lessdot Q'$, Q' as there, or more directly, for each $i < \kappa$, there are enough models N as in 2.9.

2) The "solution" of x/3.3(3),(4) = 3.2(3)/3.2(2) holds.

Proof. 1) Like the proof of 3.1 but we seal only $\mathfrak{B}^{V[P_i]} \restriction S_1$ when $\mathbf{t}_i = 1$ and in $(*)_2$ we add "$h(\alpha)$ is pseudo $(*, S_3)$-complete", but we have to check that all forcing notions Q_α are pseudo $(*, S_3)$-complete (and use the iteration lemma X 3.11). Now all the sealing forcing notions which we use satisfies this trivially.

2), 3), 4) Similar. $\square_{3.3}$

3.4. Claim. Suppose $\bar{Q} = \langle P_i, \underset{\sim}{Q}_j : i \leq \kappa,\ j < \kappa \rangle$ is a semiproper iteration, κ strongly inaccessible with $\kappa > |P_i|$ for $i < \kappa$, and lastly $S \subseteq \omega_1$ is stationary. Suppose further

(*) (a) for $i < \kappa$, in $V^{P_{i+1}}$, Player II wins $\partial(\{\aleph_1\}, \omega, \mathcal{D}_\kappa + E_i^+)$ where $E_i^+ = \{\delta < \kappa : \delta > i, \delta$ strongly inaccessible, $(\forall \alpha < \delta)[|P_\alpha| < \delta]$ and $\Vdash_{P_\delta / P_{i+1}}$ "Q_δ is semiproper"$\}$; (for a definition of the game see 1.9A(2)) so we are assuming $E_i^+ \neq \emptyset \mod \mathcal{D}_\kappa$ in $V^{P_{i+1}}$ for each $i < \kappa$; and let $E^+ = E_0^+$.

(b) $E^* = \{i < \kappa : \Vdash_{P_i}$ "Rss$(\aleph_2)$ and $\underset{\sim}{Q}_i$ semiproper"$\}$ is unbounded in κ.

Then $R_{i+1} \stackrel{\text{def}}{=} (P_\kappa / P_{i+1}) * \underset{\sim}{\mathrm{Nm}} * \mathrm{SS}\underset{\sim}{\mathrm{e}}\mathrm{al}(\mathfrak{B}[P_\kappa], S)$ is (in the universe $V^{P_{i+1}}$, Nm in V^{P_κ}, SSeal in $V^{P_\kappa * Nm}$ of course) is semiproper for every $i < \kappa$.

3.4A. Remark. (1) Remember that Nm $= \{T : T \subseteq {}^{\omega>}(\aleph_2)$ is closed under initial segments, is nonempty, and for every $\eta \in T$ we have $|\{\nu : \eta \trianglelefteq \nu \in T\}| = \aleph_2\}$; ordered by the inverse of inclusion. Clearly $\{T :$ for $\eta \in T$, $\mathrm{Suc}_T(\eta)$ is a singleton or has power $\aleph_2\}$ is a dense subset, so usually we restrict ourselves to it. For such T_1 the trunk is the $\eta \in T$ of minimal length such that $|\mathrm{Suc}_T(\eta')| > 1$.

(2) We can use Nm(D) instead of Nm and even Nm$'$, Nm$'(D)$.

(3) We can replace Nm by any forcing notion satisfying, e.g. pseudo $(*, S)$-completeness (see X 3.9, 10) or the $\mathbb{I}$-condition (see Chapter XI) where $\mathbb{I} \in V$ is a family of κ-complete normal ideals or even $UP(\mathbb{I})$, see Chapter XV.

(4) Instead of (*)(b) we can have "largeness" demands on κ. We need it to make $(P_\kappa / P_j)*$ seal$(\underset{\sim}{\Xi})$ semiproper for $j \in E^+, \underset{\sim}{\Xi}$ a maximal antichain of $\mathfrak{B}$ from V^{P_κ}.

(5) Note that $\Vdash_{P_\kappa}$ "cf$(\delta) = \aleph_0$" is not forbidden in the definition of E_i^+; we can in clause (a) of (*) of 3.4 in the game allow pressing down functions (see 1.9A(4)), add $\Vdash_{P_{\delta+1}}$ "cf$(\delta) = \aleph_1$"; in the proof below we strengthen the

definition of $j \in E^0_\eta$ by $j = \min(N_{\eta,j} \cap \kappa \setminus N_\eta)$ and demand E^0_η to be stationary and this somewhat simplify the proof.

Proof. We work in $V^{P_{i+1}}$ so let $G_{i+1} \subseteq P_{i+1}$ be generic over V. Let λ be regular and large enough, $N \prec (H(\lambda)[G_{i+1}], \in, <^*_\lambda)$ countable, $i \in N$, $\kappa \in N$, $\kappa \in N$, $\bar{Q} \in N$ and $(p^a, \underset{\sim}{p}^b, \underset{\sim}{p}^c) \in R_{i+1} \cap N$.

We shall choose below $q_{\langle\ \rangle} \in P_\kappa/P_{i+1}$ which is $(N, P_\kappa/P_{i+1})$-semi-generic, $p^a \leq q_{\langle\ \rangle}$ and $G_\kappa \subseteq P_\kappa$ generic over V containing $G_{i+1} \cup \{q_{\langle\rangle}\}$.

We now, in $V[G_\kappa]$ (but G_κ is defined only during the definition for $n = 0$) define by induction on n, T_n, N_η $(\eta \in T_n)$ such that:

(A) $T_n \subseteq {}^{n\geq}\kappa$,

(B) $T_0 = \{\langle\ \rangle\}$,

(C) $(\forall \nu \in T_{n+1})[\nu \restriction n \in T_n]$ and $T_{n+1} \cap {}^{n\geq}\kappa = T_n$,

(D) $(\forall \eta \in T_n)[\{i : \eta^\frown\langle i\rangle \in T_{n+1}\}$ has power $\kappa]$,

(E) $N[G_\kappa] \cap H(\lambda)[G_{i+1}] <_{\omega_2} N_{\langle\rangle} \prec (H(\lambda)[G_{i+1}], \in, <^*_\lambda, G_\kappa)$ and $N_{\langle\rangle}$ is countable and $(p^a, \underset{\sim}{p}^b, \underset{\sim}{p}^c) \in N_{\langle\rangle}$ and $\bar{Q} \in N_{\langle\rangle}$, (note, abusing notation we do not distinguish strictly between $N_{\langle\rangle}$ and $(N_{\langle\rangle}, G_\kappa \cap N_{\langle\rangle})$ and similarly for N_η)

(F) for $\eta \in T_{n+1}$ the model $N_\eta \prec \left(H(\lambda)[G_{i+1}], \in, <^*_\lambda, G_\kappa\right)$ is countable, extends $N_{\eta \restriction n}$, and $N_{\eta\restriction n} <_\kappa N_\eta$,

(G) $\eta \in N_\eta$,

(H) If $\underset{\sim}{\Xi}$ is a P_κ/P_{i+1}-name of a dense subset of $\mathfrak{B}(P_\kappa)$, $\underset{\sim}{\Xi} \in N_\eta$ and $\eta \in T_n$, *then* for some natural number $k = k(\underset{\sim}{\Xi}, \eta)$ and every ν : if $\eta \trianglelefteq \nu \in T_{n+k}$ then:

$$(\exists \underset{\sim}{A} \in N_\nu)\Big[\underset{\sim}{A} \in \underset{\sim}{\Xi} \;\&\; \underset{\sim}{A} \text{ a } (P_\kappa/P_{i+1}) - \text{name} \;\&\; N \cap \omega_1 \in \underset{\sim}{A}[G_\kappa]\Big],$$

(I) E^0_η is a stationary subset of κ, where

$$E^0_\eta \stackrel{\text{def}}{=} \{j < \kappa : N_\eta <_\kappa N_{\eta,j} \text{ where } N_{\eta,j} \text{ is the Skolem Hull of } N_\eta \cup \{j\} \text{ in } \left(H(\lambda)[G_{i+1}], \in, <^*_\lambda, G_\kappa\right) \text{ and } j \text{ is strongly inaccessible in } V \text{ and } (\forall i < j)[|P_i| < j] \text{ and } \Vdash_{P_j} \text{“}\underset{\sim}{Q}_j[G_j] \text{ is semiproper}\}\text{”}.$$

Now in carrying out the definition, (H) involves standard bookkeeping.

For $n = 0$ (we start to work in $V[G_{i+1}]$) our main problem is satisfying (I). We shall now define $q_{\langle\rangle}$. For $j < \kappa$, let N_j be the Skolem Hull of $N \cap \{j\}$ in $\left(H(\lambda)[G_{i+1}], \in, <^*_\lambda\right)$. By (*)(a) and XII 2.6.

$$E^1 = \{j < \kappa : N <_{\omega_2} N_j, j \text{ strongly inaccessible, } |P_i| < j \text{ for every } i < j \text{ and } \Vdash_{P_j/P_{i+1}} \text{“}\underset{\sim}{Q}_j \text{ is semiproper”}\}$$

is a stationary subset of κ. So by the Fodor lemma [as $\delta \in E^1 \Rightarrow \mathrm{cf}(\delta) > \aleph_0$ in $V[G_{i+1}]$ and $\mu < \kappa \Rightarrow \mu^{\aleph_0} < \kappa$] we know that for some stationary $E^2 \subseteq E^1$, $\langle N_j : j \in E^2\rangle$ form a Δ-system; let $\cap\{N_j : j \in E^2\}$ be $N'_{\langle\,\rangle}$. For $j \in E^2$ let $q_j \in P_\kappa/P_{i+1}$ be $(N_j, P_\kappa/P_{i+1})$-semi-generic and above p^a. Now we know that $P_j = \bigcup_{\zeta<j} P_\zeta$, hence by the Fodor Lemma w.l.o.g. $q_j \restriction j$ is constant, so let this constant value be called $q_{\langle\,\rangle}$. Clearly $q_{\langle\,\rangle}$ is $(N'_{\langle\rangle}, P_\kappa/P_{i+1})$-semi-generic and it is the $q_{\langle\,\rangle}$ which we promised. Now we actually choose G_κ i.e. a subset of P_κ generic over V and including $G_{i+1} \cup \{q_{\langle\rangle}\}$. Let $N_{\langle\,\rangle} = N'_{\langle\,\rangle}[G_\kappa] \cap H(\lambda)[G_{i+1}]$. So $N_{\langle\,\rangle} \prec \left(H(\lambda)[G_{i+1}], \in, <^*_\lambda\right)$, moreover $(N_{\langle\rangle}, G_\kappa \cap N_{\langle\rangle}) \prec (H(\lambda)[G_{i+1}], \in, <^*_\lambda, G_\kappa)$ so $N_{\langle\rangle})$ is as required in clause (E). As for clause (I), by the genericity of G_κ we have $\{j \in E^2 : q_j \in G_\kappa\}$ is unbounded in κ (even stationary) and it include E^0_η (think).

For $n > 0$ assume N_η are defined, $\ell g(\eta) = n - 1$. Clearly, as P_κ satisfies the κ-c.c., for some $\varepsilon_\eta < \kappa$ we have $\langle N_{\eta\restriction\ell} : \ell \le \ell g(\eta)\rangle$ belongs to $V[G_{\varepsilon_\eta}]$ and ε_η is a successor ordinal $> \sup(N_\eta \cap \kappa)$. By (I), E^0_η is a stationary subset of κ, and we shall define $E^1_\eta \supseteq E^0_\eta$ stationary and will let

$$T_{\ell g(\eta)+1} \cap \{\nu : \eta \triangleleft \nu \in {}^{(n+1)}\kappa\} = \{\eta\hat{\ }\langle j\rangle : j \in E^1_\eta\}.$$

So $T_{\ell g(\eta)+1}$ will really be constructed as required.

Actually E^0_η is the interpretation of some $P_\kappa/G_{\varepsilon_\eta}$-name $\underset{\sim}{E}^0_\eta$ forced to be as above: just read the definition in clause (I). W.l.o.g. some member of G_{ε_η} force ($\Vdash_{P_\kappa}$) that N, $\underset{\sim}{E}^0_\eta$ are as above.

In $V[G_\kappa]$ for each $\gamma \in E^0_\eta = \underset{\sim}{E}^0_\eta[G_\kappa/G_{\varepsilon_\eta}]$ there is $q^1_{\eta,\gamma} \in G_\kappa/G_{\varepsilon_\eta}$ such that $q^1_{\eta,\gamma} \Vdash$ "$\gamma \in \underset{\sim}{E}^0_\eta$". So in $V[G_\kappa]$, for some $q^2_\eta \in G_\kappa$ we have $\{\gamma \in E^0_\eta : q^1_{\eta,\gamma}\restriction\gamma = q^2_n\}$ is stationary. As we can increase ε_η w.l.o.g. $q^2_\eta \in G_{\varepsilon_\eta}$. In $V[G_{\varepsilon_\eta}]$ we define

$$E^1_\eta = \{\gamma : \text{there is } q = q^3_{\eta,\gamma} \text{ such that } q^3_{\eta,\gamma}\restriction\gamma = q^1_\eta \text{ and}$$
$$q^3_{\eta,\gamma} \Vdash_{P_\kappa/G_{\varepsilon_\eta}} \text{“}\gamma \in \underset{\sim}{E}^0_\eta\text{”}\},$$

so $E^1_\eta \in V[G_{\varepsilon_\eta}]$, $E^1_\eta \supseteq E^0_\eta$ hence E^1_η is stationary.

So, in $V[G_\kappa]$, $N^0_{\eta,\gamma}$ = the Skolem Hull of $N_\eta \cup \{\gamma\}$ in $(H(\lambda)[G_{i+1}], \in, <^*_\lambda, G_\kappa)$, clearly $N^0_{\eta,\gamma} \subseteq N_{\eta,\gamma}$ (as G_γ is definable from G_κ and γ) hence $N_\eta <_\kappa N^0_{\eta,\gamma}$. Also for every $x \in N^0_{\eta,\gamma}$ for some function $f \in N_\eta$, $\mathrm{Dom}(f) = \kappa$, $f(\gamma)$ is a P_γ-name of a member of $H(\lambda)$ and $x = f(\gamma)[G_\gamma]$.

But P_γ satisfies the γ-c.c., hence $f(\gamma)$ is a P_β-name for some $\beta < \gamma$ and let $h_f(\gamma) < \gamma$ be minimal such β so $h_f(\gamma) \in N^0_{\eta,\gamma}$, but as $N_\eta <_\kappa N^0_{\eta,\gamma}$, it follows that $h_f(\gamma) \in N_\eta$, so $\sup_{f\in N_\eta}(h_f(\gamma)) < \gamma$, hence, increasing ε_η and decreasing E^1_η (preserving their properties) w.l.o.g. we have $N^0_{\eta,\gamma} \in V[G_{\varepsilon_\eta}]$.

For $\gamma \in E^1_\eta$, choose any $G'_{\gamma+1}$ such that $q^3_{\eta,\gamma}\restriction(\gamma+1) \in G'_{\gamma+1}$, $G_\gamma \subseteq G'_{\gamma+1}$ and $G'_{\gamma+1}$ is generic over V. Let our bookkeeping give us $\underset{\sim}{\Xi}_\eta \in N_\eta \subseteq N^0_{\eta,\gamma}$, a P_κ-name of a pre-dense subset of $\mathfrak{B}[P_\kappa]$.

We shall now prove that condition (a) of 2.12 holds for the iteration

$$\langle P_j/G'_{\gamma+1}, \underset{\sim}{Q}_j : \gamma+1 \le j < \kappa\rangle$$

and any non-limit ordinal (denoted by i in 2.12) in the universe $V[G'_{\gamma+1}]$.

Let $\xi \in [\gamma+1,\kappa)$ be a non-limit (or just $\underset{\sim}{Q}_\xi$ semiproper). By $(*)$(b) from the assumptions of 3.4 we can find $\gamma(\xi) \in E^*, \xi < \gamma(\xi) < \kappa$, such that:

$$\Vdash_{P_{\gamma(\xi)}} \text{“}\mathrm{Rss}(\aleph_2^{V^{P_{\gamma(\xi)}}}) \text{ and } Q_{\gamma(\xi)} \text{ is semiproper”}.$$

Now $(P_\kappa/P_{\gamma(\xi)})*$ seal($\underset{\sim}{\Xi}$) does not destroy stationary subsets of ω_1 (as $P_\kappa/P_{\gamma(\xi)}$ is semiproper and $\underset{\sim}{\Xi}$ is pre-dense so that seal($\underset{\sim}{\Xi}$) preserves stationary subsets of ω_1); so because $\gamma(\xi) \in E^*$ this forcing is semiproper by 1.7(3). As Q_ξ is semiproper, $P_{\gamma(\xi)}/P_\xi$ is semiproper. Hence $(P_\kappa/P_\xi)*$seal($\underset{\sim}{\Xi}$) is semiproper. So condition (a) of 2.12 holds, hence condition (b) of 2.12 holds. Let $N^1_{\eta,\gamma}$ be the Skolem hull of $N^0_{\eta,\gamma}$ in $(H(\lambda)^{V[G_{i+1}]}, \in, <^*_\lambda, G'_{\gamma+1})$. Note that $q^3_{\gamma,\eta} \Vdash$ "$N^0_{\eta,\gamma} \prec N^1_{\eta,\gamma} \subseteq N_{\eta,\gamma}$", hence $q^3_{\eta,\gamma} \Vdash$ "$N_\eta \prec N^0_{\eta,\gamma} \prec N^1_{\eta,\gamma} \prec N_{\eta,\gamma}$ and $N_\eta \cap \kappa = N^0_{\eta,\gamma} \cap \gamma = N^1_{\eta,\gamma} \cap \sup(N_\eta \cap \kappa) \subseteq \kappa$".

Now by 2.12(b) applied in $V[G'_{\gamma+1}]$, there is countable model $N^2_{\eta,\gamma}$ satisfying $N^2_{\eta,\gamma} \prec \left(H(\lambda)^{V[G_{i+1}]}, \in, <^*_\lambda, G'_{\gamma+1}\right)$ such that $N^1_{\eta,\gamma} <_{\gamma+1} N^2_{\eta,\gamma}$ (remember 0.1(9), and $\Vdash_{P_{\gamma+1}}$ "$\gamma < \aleph_2$") and $q^4_{\eta,\gamma} \in P_\kappa/G_{\gamma+1}$ and $j_{\eta,\gamma} < \kappa$ successor such that:

(i) $q^4_{\eta,\gamma} \in P_{j_{\eta,\gamma}}/G'_{\gamma+1}$, $\gamma < j_{\eta,\gamma} \in N^2_{\eta,\gamma}$,

(ii) $q^4_{\eta,\gamma} \geq q^3_{\eta,\gamma}$,

(iii) $q^4_{\eta,\gamma}$ is $(N^2_{\eta,\gamma}, P_{j_{\eta,\gamma}})$-semi-generic, and

(iv) $q^4_{\eta,\gamma} \Vdash_{P_{j_{\eta,\gamma}}/P_{\gamma+1}}$ "for some $\underset{\sim}{A} \in N^2_{\eta,\gamma}$ we have: $\underset{\sim}{A} \in \underset{\sim}{\Xi}_\eta$ and $N \cap \omega_1 \in \underset{\sim}{A}$" and $\underset{\sim}{A}$ is a $P_{j_{\eta,\gamma}}$-name.

Also by $(*)$(a) of the assumption of 3.4, there is $\xi_{\eta,\gamma} > \sup(N^2_{\eta,\gamma} \cap \kappa) > \gamma$ strongly inaccessible, such that $\bigwedge_{\xi<\xi_{\eta,\gamma}} |P_\xi| < \xi_{\eta,\gamma}$, and $N^3_{\eta,\gamma}$ = Skolem Hull of $N^2_{\eta,\gamma} \cup \{\xi_{\eta,\gamma}\}$ in $\left(H(\lambda)[G_{\gamma+1}], \in, G'_{\gamma+1} <^*_\lambda\right)$ satisfyies $N^2_{\eta,\gamma}[G'_{\gamma+1}] <_\kappa N^3_{\eta,\gamma}$ and $q^4_{\eta,\gamma} \in P_{\xi_{\eta,\gamma}}$. Back to $V[G_\gamma]$, let $N^4_{\eta,\gamma}$ be the Skolem Hull of $N^1_{\eta,\gamma} \cup \{\gamma, \xi_{\eta,\gamma}\}$ in $(H(\lambda)[G_{i+1}], \in, <^*_\lambda, G_{\varepsilon_\eta})$, and $q^5_{\eta,\gamma} \in P_{\xi_{\eta,\gamma}}/G_{\varepsilon_\eta}$ forces all the above and in particular is above q^3_η and $q^4_{\eta,\gamma}$. In addition $q^5_{\eta,\gamma} \restriction [\gamma+1, \kappa) = q^4_{\eta,\gamma} \restriction [\gamma+1, \kappa) = q^4_{\eta,\gamma} \restriction [\gamma+1, \xi_{\eta,\gamma})$, and $q^5_{\eta,\gamma} \restriction (\gamma+1) \in G'_{\gamma+1}$, so $q^4_{\eta,\gamma} \restriction \varepsilon_\eta \in G_{\varepsilon_\eta}$ and $N_\eta \prec N^4_{\eta,\gamma}$, $\gamma \in N^4_{\eta,\gamma}$, $N_\eta \cap \kappa = N_\eta \cap \gamma = N^4_{\eta,\gamma} \cap \sup(N_\eta \cap \kappa)$, and

$$q^4_{\eta,\gamma} \Vdash \text{"the Skolem Hull } N^5_{\eta,\gamma} \text{ of } N^4_{\eta,\gamma} \text{ in } (H(\lambda)^{V[G_{i+1}]}, \in, <^*_\lambda, \underset{\sim}{G}_{\gamma+1})$$
$$\text{satisfies } N^5_{\eta,\gamma} \cap \sup(N_\eta \cap \kappa) = N_\eta \cap \kappa\text{"},$$

hence

$$q^5_{\eta,\gamma} \Vdash \text{"the Skolem Hull } N^6_{\eta,\gamma} \text{ of } N^4_{\gamma,\gamma} \text{ in } (H(\lambda)^{V[G_{i+1}]}, \in, <^*_\lambda, \underset{\sim}{G}_{\gamma+1})$$
$$\text{satisfies } N^6_{\eta,\gamma} \cap \sup(N_\eta \cap \kappa) = N_\eta \cap \kappa \text{ (as } V[G_{\gamma+1}] \vDash |\gamma| = \aleph_1\text{)".}$$

(looking at the definition of $\underset{\sim}{E}^0_\eta$ in clause (I) above). As we can increase ε_η and decrease E^2, w.l.o.g. $q^5_\eta \restriction \gamma \in G_{\varepsilon_\eta}$ and $q^5_\eta \restriction \gamma$ is the same for all $\gamma \in E^2$.

Now as $q^5_{\eta,\gamma} \in P_\kappa / G_{\varepsilon_\eta}$ and $q^5_{\eta,\gamma} \restriction \gamma \in G_{\varepsilon_\eta}$, easily $\Vdash_{P_\kappa/G_{\varepsilon_\eta}}$ "$\underset{\sim}{E}^1_\eta \overset{\text{def}}{=} \{\gamma : q^5_{\eta,\gamma} \in \underset{\sim}{G}_\kappa\}$ is a stationary subset of κ", so we have defined at least $\underset{\sim}{E}^1_\eta$. Now in $V[G_\kappa]$, if $\gamma \in \underset{\sim}{E}^1_\eta[G_\kappa]$ then $\gamma \in E^0_\eta$ (see above). We still have to define $N_{\eta\char`^\langle\gamma\rangle}$ and $E^0_{\eta\char`^\langle\gamma\rangle}$ (for $\gamma \in \underset{\sim}{E}^1_\eta[G_\kappa]$). For each such γ we repeat the proof in the case $n=0$ with universe $V[G_{\xi_{\eta,\gamma}}]$ and Skolem Hull of $N^4_{\eta,\gamma}$ in $(H(\lambda)^{V[G_{i+1}]}, \in, <^*_\lambda, G_{\xi_{\eta,\eta}})$ here standing for $V[G_{i+1}]$, N there.

We have carried out the construction.

We now define by induction on n, for every $\eta \in T \cap {}^n\kappa$, a condition $p^b_\eta \in \mathrm{Nm}$ and $m_\eta < \omega$ such that (note $N_\eta[G_\kappa] \prec (H(\lambda)[G_\kappa], \in, <^*_\kappa)$, $N_\eta[G_\kappa] \cap H(\lambda)[G_{i+1}] = N_\eta$):

(a) $p^b_\eta \in N_\eta[G_\kappa]$, $m_\eta < \omega$ and $p^b_{\langle\rangle} = \underset{\sim}{p}^b[G_\kappa]$,

(b) $p^b_\eta \in \mathrm{Nm}$, and $\mathrm{tr}(p^b_\eta)$ (the trunk of p^b_η) has length $\geq \ell g(\eta)$ (and has κ immediate successors in p^b_η),

(c) $p^b_{\eta\restriction\ell} \leq p^b_\eta$ and $m_{\eta\restriction\ell} \leq m_\eta$ when $\ell \leq \ell g(\eta)$; and if $p^b_{\eta\restriction\ell}$ has a trunk of length $> \ell g(\eta)$ or $m_{\eta\restriction\ell} > \ell g(\eta)$ *then*: $p^b_\eta = p^b_{\eta\restriction\ell}$ & $m_\eta = m_{\eta\restriction\ell}$,

(d) if $\eta \in T_n$, $\underset{\sim}{\alpha}$ is a Nm-name for a countable ordinal, $\underset{\sim}{\alpha} \in N_\eta[G]$, *then* for some $k = k^1(\underset{\sim}{\alpha}, \eta)$, and every $\nu \in \bigcup_{m<\omega} T_m$, for some ordinal $\beta = \beta(\underset{\sim}{\alpha}, \nu) \in N_\nu$ we have

$$k+1 = |\{\ell < \ell g(\nu) : m_{\nu\restriction\ell} < m_{\nu\restriction(\ell+1)}\}| \;\&\; \eta \trianglelefteq \nu \Rightarrow p^b_\nu \Vdash_{\mathrm{Nm}} \text{“}\underset{\sim}{\alpha} = \beta(\underset{\sim}{\alpha}, \nu)\text{”},$$

(e) if $\eta \in T_n$ and $\underset{\sim}{\Xi}$ is a Nm-name for a pre-dense subset of $\mathfrak{B}[P_\kappa]$ and $\underset{\sim}{\Xi} \in N_\eta[G_\kappa]$, *then* for some $k = k^2(\underset{\sim}{\Xi}, \eta)$, for every $\nu \in \bigcup_{m<\omega} T_m$ we have:

$$[k+1 \leq |\{\ell < \ell g(\nu) : m_{\nu\restriction\ell} < m_{\nu\restriction(\ell+1)}\}| \;\&\; \eta \trianglelefteq \nu]$$
$$\Rightarrow [\text{for some } A \in N_\nu[\underset{\sim}{G}_{N_m}] \text{ we have } N \cap \omega_1 \in A \;\&\; p^b_\nu \Vdash_{\mathrm{Nm}} \text{“}A \in \underset{\sim}{\Xi}\text{”}].$$

(f) if p^b_η has a trunk of length $\leq \ell g(\eta)$, say ν_η, and $m_\eta \leq \ell g(\eta)$ and if h_η is a one-to-one function from κ onto $\{j < \kappa : \nu_\eta\char`^\langle j\rangle \in p^b_\eta\}$, $h_\eta \in N_\eta[G_\kappa]$, then for $\eta^\wedge\langle i\rangle \in \bigcup_n T_n$ we have:

$$(\forall \rho \in p^b_{\eta^\frown\langle i\rangle})[\ell g(\rho) > \ell g(\eta)) \Rightarrow \rho(\ell g(\eta)) = h_\eta(i)],$$

(g) for $\eta \in T_n$ we have: the sequences $\langle k^1(\underset{\sim}{\alpha}, \eta) : \underset{\sim}{\alpha} \in N_\eta[G_\kappa]$ is a Nm-name of a countable ordinal$\rangle$ and $\langle k^2(\underset{\sim}{\Xi}, \eta) : \underset{\sim}{\Xi} \in N_\eta[G_\kappa]$ is a Nm-name of a predense subset of $\mathfrak{B}[P_\kappa]\rangle$ are with no repetitions, with disjoint ranges whose union is a co-infinite subset of ω [Why the m_η's? just as below Υ depend on p^b_η].

There is no problem to do this. [For (e), when we come to deal with $\underset{\sim}{\Xi}$, say at η, where p^b_η has trunk of length $\leq \ell g(\eta)$ and $m_\eta \leq \ell g(\eta)$, we let

$$\Upsilon = \{A : (\exists p)(p^b_\eta \leq p \in \text{Nm} \;\&\; p \Vdash_{\text{Nm}} \text{“}A \in \underset{\sim}{\Xi}\text{”})\}.$$

So $\Upsilon \in N_\eta[G_\kappa]$ is a pre-dense subset of $\mathfrak{B}(P_\kappa)$, and by (H) above there is $k(\underset{\sim}{\Upsilon}, \rho)$ as there, choose it as $k^2(\underset{\sim}{\Xi}, \eta)$, so we shall have $p^b_\nu = p^b_\rho$ if $\nu \lhd \rho \in T'_{\ell g(\rho)}, \ell g(\rho) \leq \ell g(\nu) + k^2(\Xi, \eta)$.]

Now in V^{P_κ} let:

$$q^b = \{\rho \in {}^{\omega>}\kappa : \rho \in \underset{\sim}{p}^b[G_\kappa] \text{ and for some } \eta \in \bigcup_n T_n, \rho \text{ is an initial segment of the trunk of } p^b_\eta\}.$$

We can easily see that $p^b \leq q^b \in \text{Nm}$ (in $V[G_\kappa]$). Also (in $V[G_\kappa]$) q^b is $(N[G_\kappa],\text{Nm})$-semi-generic and moreover

$$q^b \Vdash_{\text{Nm}} \text{“}\kappa \cap N[G_\kappa][\underset{\sim}{G}_{\text{Nm}}] = \kappa \cap \bigcup_{l<\omega} N_{\underset{\sim}{\nu}\restriction l}[G_\kappa]\text{”},$$

where $\underset{\sim}{G}_{\text{Nm}}$ is the (canonical name of the) generic subset of Nm and $\underset{\sim}{\eta}$ is the Nm-name of the ω-sequence in ${}^\omega\kappa$ which it defines naturally and $\underset{\sim}{\nu}$ is the Nm-name of the ω-sequence in ${}^\omega\kappa$ such that $\nu\restriction\eta \in T_n$ and the trunk of $p^b_{\nu\restriction\eta}$ is $\lhd\eta$. [Remember that if $N_1, N_2 \prec (H(\lambda), \in), N_1 \cap \omega_1 = N_2 \cap \omega_1$, and $i \in N_1 \cap N_2$, $i < \aleph_2$, then $N_1 \cap i = N_2 \cap i$.] Hence $q^b \Vdash_{N_m}$ “$\mathcal{P}(\omega_1)^{V[G_\kappa]} \cap N[G_\kappa][\underset{\sim}{G}_{\text{Nm}}] = \mathcal{P}(\omega_1)^{V[G_\kappa]} \cap \bigcup_{l<\omega} N_{\underset{\sim}{\nu}\restriction l}[G_\kappa]$”.

Now clearly by the above and (e) we have

$$q^b \Vdash_{\mathrm{Nm}} \text{“for every pre-dense subset } \Xi \text{ of } \mathfrak{B}[P_\kappa] \text{ in } N[G_\kappa][\underset{\sim}{G}_{\mathrm{Nm}}],$$

$$N \cap \omega_1 \in \bigcup_{A \in \Xi} \{A : A \in \Xi \cap N[G_\kappa][\underset{\sim}{G}_{\mathrm{Nm}}]\}\text{”}.$$

So we can apply Claim 2.9 to get q^c, which is $(N[G_\kappa][G_{\mathrm{Nm}}], \mathrm{SSeal}(\mathfrak{B}[P_\kappa], S))$-semi-generic $\geq \underset{\sim}{p}^c[G_\kappa][\underset{\sim}{G}_{\mathrm{Nm}}]$. Let $q^a = q_{\langle\,\rangle}$ so we are assuming just $q^a \in G_\kappa \subseteq P_\kappa, G_\kappa$ generic over V and so for some P_κ-name $\underset{\sim}{q}^b$, we have: $q^a \Vdash_{P_\kappa}$ “$\underset{\sim}{q}^b$ is as above ”. Similarly for some $\underset{\sim}{q}^c$, $(q^a, \underset{\sim}{q}^b) \Vdash_{(P_\kappa/P_{i+1})*\underset{\sim}{\mathrm{N}}\mathrm{m}}$ “$\underset{\sim}{q}^c$ is as above”. Now $(q^a, \underset{\sim}{q}^b, \underset{\sim}{q}^c)$ is as required (i.e., (R_{i+1}, N)-semi-generic). $\square_{3.4}$

3.4B Remark. It seems that we can weaken clause (a) of $(*)$ of 3.1 to
(a)$'$ for $i < \kappa$ in $V^{P_{i+1}}$, player II wins in the game $\eth(\{\aleph_1\}, \omega, \kappa)$.

See [Sh:311]

3.5 Claim. Suppose $\bar{Q} = \langle P_i, \underset{\sim}{Q}_j : i \leq \kappa,\ j < \kappa\rangle$ is a semiproper iteration, $\kappa > |P_i|$ for $i < \kappa$ and $S \subseteq \omega_1$ is stationary. Suppose further that

$(*)$ (a) for $i < \kappa$, in V^{P_i}, Player II wins in $\eth(\{\aleph_1\}, \omega, \mathcal{D}_\kappa + E_i^+)$ where $E_i^+ = \{\delta < \kappa : \delta > i,\ \delta$ strongly inaccessible, $\Vdash_{P_\delta/P_i}$ “$\underset{\sim}{Q}_\delta$ is semiproper”$\}$,

(b) $E = \{i < \kappa : \Vdash_{P_i}$ “Rss$(\aleph_2)$ and $\underset{\sim}{Q}_i$-semiproper”$\}$ is unbounded,

(c) It is forced ($\Vdash_{P_\kappa}$) that $\underset{\sim}{W} \subseteq \{\delta < \kappa : V^{P_\kappa} \models \text{“cf}(\delta) = \aleph_0\text{”}\}$ is stationary ($\underset{\sim}{W}$ a P_κ-name).

Then $(P_\kappa/P_{i+1})*\mathrm{club}_{\aleph_1}(\underset{\sim}{W})*\mathrm{SSeal}\ (\mathfrak{B}(P_\kappa), S)$ is semiproper for $i < \kappa$ where $\mathrm{club}_\mu(W) \overset{\mathrm{def}}{=} \{f :$ for some non-limit $\gamma < \mu, f$ is an increasing continuous function from γ into $W\}$.

Proof. Like the previous claim, only after defining N_η for a set $G_\kappa \subseteq P_\kappa$ generic over $V, q_{\langle\,\rangle} \in G_\kappa$, in $V[G_\kappa]$ there is $\eta \in {}^\omega\kappa$, $\bigwedge_n(\eta\restriction n \in T_n)$ such that $\eta(\ell) > \sup(N_{\eta\restriction\ell} \cap \kappa)$ and $\sup\{\eta(\ell) : \ell < \omega\}$ belong to $\underset{\sim}{W}[G_\kappa]$ and then in $V[G_\kappa]$ continue with $\bigcup_\ell N_{\eta\restriction\ell}[G]$. $\square_{3.5}$

3.5A Remark. 3.5, 3.4 are cases of a more general theorem, see XV.

3.6 Claim. In 3.4, 3.5, if we add to the hypothesis:

(∗) player II wins in $V^{P_i}(i < \kappa)$, for $\mathcal{D}_\kappa$ in the game of "divide and choose" i.e. X 4.9 for $S = \{2, \aleph_0, \aleph_1\}, \alpha = \omega$,

(∗)′ for $i < j < \kappa$ non-limit, P_j/P_i is pseudo $(*, \omega_1 \setminus S^*)$-complete,

then $(P_\kappa/P_{i+1}) * \underset{\sim}{\mathrm{Nm}}$ and $(P_\kappa/P_{i+1}) * \underset{\sim}{\mathrm{club}}_{\aleph_1}(\underset{\sim}{W})$ are pseudo $(*, \omega_1 \setminus S^*)$-complete.

Proof. Left to the reader.

3.7 Theorem. 1) Suppose $\{\mu < \kappa : \mu \text{ supercompact}\}$ is unbounded below κ and κ is 3-Mahlo.

If $\langle S_1, S_2, S_3\rangle$ is a partition of ω_1 with S_1 stationary, *then* for some semiproper pseudo $(*, S_3)$-complete forcing notion P satisfying the κ-c.c., we have:

$\Vdash_P$ "$\mathfrak{B}[P_\kappa]\restriction S_1$ has a dense subset which is (up to isomorphism) Levy$(\aleph_0, < \aleph_2)$".

Proof. We define by induction on i, P_i, $\underset{\sim}{Q}_i$, $\mathbf{t}_i$ such that

(A) $\bar{Q}^\alpha = \langle P_i, \underset{\sim}{Q}_j, \mathbf{t}_j : i \le \alpha,\ j < \alpha\rangle$ is S_1-suitable,

(B) there is no strongly inaccessible Mahlo λ, $i < \lambda \le |P_i|$,

(C) if i is a singular ordinal or $(\exists j < i)[|P_j| > i]$ or i inaccessible not a limit of supercompacts or i inaccessible not Mahlo *then* $\mathbf{t}_i = 0$, $\underset{\sim}{Q}_i = \mathrm{SSeal}(\langle \mathfrak{B}[P_j] : j \le i,\ \mathbf{t}_j = 1\rangle, S_1)$ (as defined in V^{P_i}, of course),

(D) if i is supercompact, not limit of supercompacts *then* $\mathbf{t}_i = 1, \underset{\sim}{Q}_i = \mathrm{SSeal}(\mathfrak{B}[P_i], S_1)$,

(E) if $(\forall j < i)[|P_j| < i], i$ limit of supercompacts and i is inaccessible 1-Mahlo but not 2-Mahlo, we let $\mathbf{t}_i = 1$, $\underset{\sim}{Q}_i = \underset{\sim}{\mathrm{Nm}} * \underset{\sim}{\mathrm{SSeal}}(\mathfrak{B}[P_i])$ (the SSeal in $V^{P_i * \underset{\sim}{\mathrm{Nm}}}$ of course),

(F) if $(\forall j < i)[|P_j| < i], i$ is 2-Mahlo and a limit of supercompacts then $W_i \stackrel{\text{def}}{=} \{\delta < i : \delta = \text{cf}(\delta)$ is Mahlo and a limit of supercompacts and $(\forall j < \delta)[|P_j| < \delta]\}$ is a stationary subset of i, then we let:

$$\mathbf{t}_i = 1, \quad \underset{\sim}{Q}_i = \text{club}_{\aleph_1}(W_i) * \text{S}\underset{\sim}{\text{S}}\text{eal}(\mathfrak{B}[P_i], S_1).$$

Why is $\bar{Q}$ S_1-suitable? We shall prove by induction on i that $\bar{Q}\restriction i$ is S_1-suitable.

Note that the use of SSeal guarantees (F) of Definition 2.1, as well as (E) (see 2.11(3), 2.13(2)). Remembering 2.3(2), it suffices to show by induction on i that $j < i \Rightarrow (P_i/P_{j+1}) * \underset{\sim}{Q}_i$ is semiproper (actually the only problematic case is when i is inaccessible limit of supercompacts, but then for arbitrarily large $j < i$ we have $\text{Rss}^+(j)$ (by 1.10, 1.6(2), 1.6(4)), so in V^{P_j}, every forcing notion preserving stationary sets is semiproper, but we check by cases:

For $i = 0$: trivial.

For $i+1$, and i satisfies clause (C) above (in the definition of $\bar{Q}$) the result follows by Claim 2.14.

For $i+1$, and i satisfies (D) above: first note that $|P_j| < i$ for $j < i$, hence $j < i\ \&\ \mathbf{t}_j = 1 \Rightarrow \mathfrak{B}[P_j] \lessdot \mathfrak{B}[P_i]$, hence by Claim 2.7 we have $Q_i = \text{SSeal}(\mathfrak{B}^{P_i}, S_1) = \text{SSeal}(\langle \mathfrak{B}^{P_j} : j \leq i, \mathbf{t}_j = 1\rangle, S_1)$. Now for a club C of i, $j \in C\ \&\ j = \text{cf}(j) \Rightarrow Q_j$ is semi proper (see the previous case), so by 1.6(4)+1.10 we have $\Vdash_{P_i}$ "$\text{Rss}(\kappa)$ i.e. $\text{Rss}(\aleph_2)$". Hence by claim 2.10, $\text{SSeal}(\mathfrak{B}[P_\kappa], S_1)$ is semiproper in V^{P_κ}.

For $i+1$, and i satisfying clause (E) above: we shall apply 3.4 with i here standing for κ there. Note that condition $(*)$(a) (of 3.4) holds for $E^+_{j,i} \stackrel{\text{def}}{=} \{\delta < i : \delta > j, \delta$ strongly inaccessible, not Mahlo, $\delta > |P_\zeta|$ for $\zeta < \delta$, Q_δ semiproper$\}$. Why does the second player win $\Game(\{\aleph_1\}, \omega, \mathcal{D}_i + E^+_{j,i})$ in the universe $V^{P_{j+1}}$? By 1.6(6) clearly for $j < i, V^{P_j} \models$ "$\text{Rss}^+(i)$" and use 1.11, (and 1.9A(3), i.e. XII, 2.5(2)) but this give just winning in $\Game(\{\aleph_1\}, \omega, \kappa)$. However for $\mu < i$, there is a μ-complete filter on i containing the clubs of i and $E^+_{j,i}$, so winning the game is easy, and lastly if $j < i$ is strongly inaccessible not Mahlo and $(\forall \varepsilon < j)(|P_\varepsilon| < j)$ then Q_j is even proper by 2.11. Condition $(*)$(b) of 3.4

holds by the definition of case (E): if $\lambda < i$ is supercompact then $\mathrm{Rss}^+(\lambda), Q_\lambda$ semiproper by the induction hypothesis (see previous case) so any $\lambda < i$ which is supercompact satisfies the requirement on E^*.

For $i+1$, and i satisfying clause (F) above: similar to the previous case by replacing 3.4 by Claim 3.5 (and remember 0.1(5) of the Notation).

Also each Q_i is pseudo $(*, S_3)$-complete (by 3.6), hence P_κ is pseudo $(*, S_3)$-complete so when S_3 is stationary,

$$\Vdash_{P_\kappa} \text{“}2^{\aleph_0} = \aleph_1, 2^{\aleph_1} = \aleph_2\text{”}$$

and in any case $\Vdash_{P_\kappa}$ “$2^{\aleph_1} = \kappa = \aleph_2^{(V^{P_\kappa})}$”.

Let $\mathfrak{B}_i = \mathfrak{B}[P_i]$, so $\mathbf{t}_i = 1 \Rightarrow \mathfrak{B}_i \restriction S_1 \lessdot \mathfrak{B}[P_\kappa] \restriction S_1$. Let

$$W^* \overset{\text{def}}{=} \{i < \kappa : \mathfrak{B}_i \restriction S_1 \lessdot \mathfrak{B}[P_\kappa] \restriction S_1\}.$$

So in V^{P_κ} (as case (F) occurs stationarily often),

$$W^{**} \overset{\text{def}}{=} \{\delta \in W^* : \mathrm{cf}(\delta) = \aleph_1 \text{ and } W^* \text{ contains a club of } \delta\}$$

is stationary. Hence it is well known that in V^{P_κ},

$$\begin{aligned} \mathrm{club}_\kappa(W^*) = \{h : & h \text{ an increasing continuous function} \\ & \text{from some } \alpha + 1 < \kappa \text{ to } W^*\} \end{aligned}$$

does not add bounded subsets to $\kappa (= \aleph_2)$. (More exactly, if CH holds this is straightforward. If CH fails, this holds if we can find $\bar{\mathcal{P}} = \langle \mathcal{P}_\alpha : \alpha < \kappa \rangle$, $\mathcal{P}_\alpha \subseteq \mathcal{S}_{<\aleph_1}(\alpha)$, $|\mathcal{P}_\alpha| \le \aleph_1$ ($\bar{P} \in V^{P_\kappa}$ of course) such that $\{\delta \in W^{**}$: for some unbounded C of δ we have that $C \subseteq W^*$, $\mathrm{otp}(C) = \omega_1$ and $\alpha \in C \Rightarrow C \cap \alpha \in \bigcup_{\beta < \alpha} \mathcal{P}_\beta\}$ and this holds (with $\mathcal{P}_\alpha = (\mathcal{S}_{<\aleph_1}(\alpha))^{V^{P_\alpha}}$ in fact $\alpha \in C \Rightarrow C \cap \alpha \in P_\alpha$).)

So forcing will give us a universe as required. $\square_{3.7}$

3.8 Remarks. The proof of 3.1, 3.7 exemplifies two constructions which we may interchange. Another variation is 3.9 below.

3.9 Theorem. Suppose $\{\mu < \kappa : \mu \text{ supercompact}\}$ is unbounded below κ, κ is strongly inaccessible, $h : \kappa \to H(\kappa)$, and $\langle S_1, S_2, S_3 \rangle$ is a partition of ω_1, and S_1 is stationary. *Then* for some forcing notion P:

(i) P satisfies the κ-c.c., is pseudo $(*, S_3)$-complete, has cardinality κ, does not collapse $\aleph_1$ and κ but collapses every $\lambda \in (\aleph_1, \kappa)$ and in $V^{P_\kappa}, \kappa = \aleph_2, 2^{\aleph_1} = \aleph_2$, and $2^{\aleph_0} = \aleph_1 \iff S_3$ stationary,

(ii) $\mathfrak{B}[P] \restriction S_1$ has a dense subset isomorphic to Levy $(\aleph_0, < \aleph_2)$,

(iii) in V^P, an axiom holds as strong as h is a diamond, i.e.

(a) If h is a Laver diamond for $x \in H(2^\lambda)$ *then* in $V^P, Ax[Q$ is a pseudo $(*, S_3)$-complete, semiproper$^{*[S_1]}$ (see definition below), $Q \in H(\lambda)]$ (see 3.9A below) and $Ax^+[Q$ is pseudo $(*, S_3)$-complete, semiproper$^{*[S_1]}, Q \in H(\lambda)$ and $\mathfrak{B}[V^P] \lessdot \mathfrak{B}[(V^P)^Q]$.

(b) When $\lambda = \kappa$, then we can weaken the demand on h to: for every $x \subseteq \kappa$ satisfying a Σ^1_1-sentence ψ (i.e. $(\exists z \subseteq \mathcal{P}(\kappa)$ such that $\ldots))$ *then* $\{i < \kappa : h(i) = x \cap i, (H(i), \in, x \cap i) \models \psi\}$ is stationary. *Then* a conclusion similar to the one in clause (a) holds for $Q \subseteq H(\kappa)$

where

3.9A Definition. Let $\bar{A} = \langle A_\zeta : \zeta < \xi \rangle, A_\zeta \lessdot \mathfrak{B}[V]$.

1) A forcing notion Q is semiproper$^{*[\bar{A}]}$ *if* χ regular large enough, $N \prec (H(\chi), \in, <^*_\chi)$ is countable, $Q \in N, \bar{A} \in N, p \in Q \cap N$ satisfies "$(\forall \Xi, \zeta)[\Xi \in N$ a pre-dense subset of A_ζ & $\zeta \in N \cap \xi \Rightarrow N \cap \omega_1 \in \bigcup_{A \in \Xi} A]$" (if A_ζ satisfies the $\aleph_2$-c.c. this always holds) *then* there is $q \in Q$ which is (N, Q)-semi-generic and $q \Vdash$ "if $\zeta \in \xi \cap N[G_Q]$ and $\Xi \in N[G_Q]$ is a pre-dense subset of A_ζ, *then* $N \cap \omega_1 \in \cup\{A : A \in N[G_Q]\}$".

2) If $\xi = 1, A_\zeta = \{A \subseteq \omega_1 : A \cap (\omega_1 \setminus S) \in \{\emptyset, \omega_1 \setminus S\}\}$, write $*[S]$ instead $*[\bar{A}]$. We do not strictly distinguish between $\mathfrak{B}[V] \restriction S$ and $\{A \in \mathfrak{B}[V] : A \cap (\omega_1 \setminus S) \in \{\emptyset, \omega_1 \setminus S\}\}$.

Proof. We define by induction on $\alpha < \kappa, P_i, \underset{\sim}{Q}_i, \mathbf{t}_i$ for $i < \alpha$ such that:

(A) $\bar{Q}^\alpha = \langle P_i, \underset{\sim}{Q}_j, \mathbf{t}_j : i \le \alpha, j < \alpha \rangle$ is S_1-suitable, $|P_i| < \kappa$ for $i < \kappa$, and for $\alpha < \kappa, \bar{Q}^\alpha \in H(\kappa)$ and $\mathbf{t}_i = 1 \Leftrightarrow (i$ successor *or* i strongly inaccessible

& $\bigwedge_{j<i} |P_j| < i$), (note that for i limit we are trying to get $\mathfrak{B}^{\bar{Q}\restriction i} \lessdot \mathfrak{B}^{P_\kappa}$, not $\mathfrak{B}^{P_i} \lessdot \mathfrak{B}^{P_\kappa}$). Let $\underset{\sim}{\mathcal{A}}_j$ be the following P_j-name: if $j = 0$ we let $\underset{\sim}{\mathcal{A}}_j$ be trivial, if $j > 0$ we let it be $\mathfrak{B}^{\bar{Q}\restriction j} = \bigcup_{\beta<j} \mathfrak{B}[P_{\beta+1}]$.

(B) For i non-limit, let κ_i be the first supercompact $> |P_i|$,

$$\text{if } i = 0, \text{ let } Q_i = \mathrm{Levy}(\aleph_1, < \kappa_0),$$

$$\text{if } i > 0, \text{ let } Q_i = \mathrm{SSeal}(\langle \underset{\sim}{\mathcal{A}}_j : j \le i\rangle, S_1, \kappa_i).$$

(C) For i limit $< \kappa$ such that $h(i)$ is a P_i-name of a pseudo $(*, S_3)$-complete semiproper$^{*[\bar{\mathcal{A}}^i]}$, where $\bar{\mathcal{A}}^i \overset{\text{def}}{=} \langle \mathcal{A}_j : j \le i\rangle$, remember $\mathcal{A}_i = \mathfrak{B}^{\bar{Q}\restriction i}$.
Let $\kappa_{i+1} < \kappa$ be such that $h(i) \in H(\kappa_{i+1})$, κ_{i+1} supercompact and $Q_i = h(i) * \mathrm{S\underset{\sim}{S}eal}(\langle \bar{\mathcal{A}}^i, S_1, \kappa_{i+1})$.

(D) For i limit, but (C) does not hold, let $Q_i = \mathrm{SSeal}(\langle \underset{\sim}{\mathcal{A}}_j : j \le i\rangle, S_1, \kappa_{i+1})$, κ_{i+1} as before.

We can prove by induction on α that $\bar{Q}^\alpha$ is S_1-suitable and Q_α is semiproper, and if $i < \kappa$ is successor, then $\Vdash_{P_i}$ "$\mathrm{Rss}(\aleph_2)$". If α is limit ordinal use 2.3(1) and for $\alpha = 0$ this should be clear. If $\alpha = \beta + 1$, β not limit by 2.11 we can see that $\bar{Q}^\alpha$ is S_1-suitable, i.e. the first phrase holds. For the second, clearly by 2.7 we have $\underset{\sim}{Q}_\beta = \mathrm{SSeal}(\mathfrak{B}[P_\beta], S_1, \kappa_{\beta+1})$, and by the induction hypothesis $V^{P_\beta} \models$ "$\mathrm{Rss}(\aleph_2), \aleph_2 = \kappa_{\beta+1}$", hence by 2.8(7), $\Vdash_{P_\beta}$ "$\underset{\sim}{Q}_\beta$ is semiproper". Moreover in V^{P_β}, Q_β is an iteration (see Definition 2.4(5)) $\langle P^\beta_i, \underset{\sim}{Q}^\beta_j : i \le \kappa_{\beta+1}, j < \kappa_{\beta+1}\rangle$ and for every strongly inaccessible $j < \kappa_\beta$, Q^β_j and even $P^\beta_{\kappa_\beta}/P^\beta_j$ are proper by 2.11. So by 1.10 we have $\Vdash_{P_\alpha}$ "$\mathrm{Rss}(\kappa_\beta)$". For $\alpha = \beta + 1$, β limit use 4.9 from the next section and 2.11 for the first phrase (if clause (D) apply then use 2.13), the second is proved as in the previous case. Remembering strong preservation of pseudo $(*, S_3)$-completeness we have no problems. $\square_{3.9}$

3.9B Remark. We can wave in the proof some $\mathbf{t}_i = 1$, more acurately some $\underset{\sim}{\mathcal{A}}_\zeta$'s and then get stronger forcing axioms.

§4. $\mathcal{P}(\omega_1)/(\mathcal{D}_{\omega_1} + S)$ is Reflective or Ulam

In 4.3 we deal with reflectiveness: if $A_i \subseteq S \subseteq \omega_1$ is stationary for $i < \aleph_2$ then for some $W \subseteq \aleph_2$ of cardinality $\aleph_2$, $[w \subseteq W$ & $|w| \leq \aleph_0 \Rightarrow \bigcap_{i \in w} A_i$ is stationary$]$. Claims 4.1, 4.2 prepare the ground. In 4.4 we deal with the Ulam property, for this we prove in ZFC a sufficient condition for a filter to satisfy the Ulam property (see 4.5A – 4.5F, Definition 4.6 and the proof of the consistency of the Ulam property (i.e. 4.4) in 4.7). The rest of the section deal with the forcing.

4.1 Claim. Suppose $S \subseteq \omega_1$ is stationary $\bar{Q} = \langle P_i, \underset{\sim}{Q}_j : i \leq \alpha,\ j < \alpha \rangle$ is a semiproper iteration, $\mu < \alpha$ ($\mu = 0$ is allowed), and $\Vdash_{P_\mu}$ "Rss$(\aleph_2[V^{P_\mu}])$" (e.g., if μ is supercompact, $[i < \mu \Rightarrow |P_i| < \mu]$ and $\{i < \mu : \underset{\sim}{Q}_i$ is semiproper (i.e. $\Vdash_{P_i}$ "$\underset{\sim}{Q}_i$ is semiproper")$\}$ belongs to some normal ultrafilter on μ); note that $\Vdash_{P_\mu}$ "$\mu = \aleph_2$" if μ is strongly inaccessible, $|P_i| < \mu$ for $i < \mu$.

Let $\underset{\sim}{A}$ be a P_α-name for a subset of S and $\underset{\sim}{B}$ a P_α-name for a member of $\mathfrak{B}[P_\mu]$ such that:

$$\Vdash_{P_\alpha} \text{“}(\forall X \in \mathfrak{B}^{P_\mu})[0 < X \leq \underset{\sim}{B} \Rightarrow X \cap \underset{\sim}{A} \neq 0 \text{ (in } \mathfrak{B}^{P_\alpha})]\text{”}.$$

Then

⊗ *if* λ is regular and large enough, $N \prec (H(\lambda), \in, <^*_\lambda)$ is countable, and $\bar{Q}$, λ, p, $\underset{\sim}{A}$, $\underset{\sim}{B}$ and μ belong to N, $p \in P_\alpha \cap N$ and $q \in P_\mu$ is (N, P_μ)-generic, $p{\restriction}\mu \leq q$ and $q \cup p{\restriction}[\mu, \alpha) \Vdash_{P_\alpha}$ "$N \cap \omega_1 \in \underset{\sim}{B}$" (if $\underset{\sim}{B}$ is a P_μ-name this means $q \Vdash_{P_\mu}$ "$N \cap \omega_1 \in \underset{\sim}{B}$"), *then* there is a (N, P_α)-semi generic condition $q' \in P_\alpha$ satisfying $q'{\restriction}\mu = q$ such that $q' \Vdash_{P_\alpha}$ "$N \cap \omega_1 \in \underset{\sim}{A}$".

4.1A Remark. (1) If $\bar{Q}$ is S-suitable, $\mathbf{t}_\mu = 1$, and $\underset{\sim}{A} \neq \emptyset \bmod \mathcal{D}_{\omega_1}$, $\underset{\sim}{A}$ is a P_β-name for some $\beta < \alpha$, then we know that such $\underset{\sim}{B}$ exists as $\mathbf{t}_\mu = 1$ (by definition 2.1).

(2) Note, e.g., for S-suitable $\bar{Q}$, $\ell g(\bar{Q}) = \alpha = \bigcup_{n<\omega} \alpha_n, \alpha_n < \alpha_{n+1}, \mathbf{t}_{\alpha_n} = 1$, we can use $Q_\alpha = \text{SSeal}(\mathfrak{B}^{\bar{Q}}, S)$ and not only $\text{SSeal}(\mathfrak{B}^{P_\alpha}, S)$ [because in 2.13 we had demanded "$(P_\alpha/P_{i+1})*$ seal (Ξ) is semiproper"].

Proof. As we can increase p, without loss of generality p forces $\underset{\sim}{B}$ to be equal to some P_μ-name, so without loss of generality $\underset{\sim}{B}$ is a P_μ-name.

Let us fix p, $\underset{\sim}{A}$, $\underset{\sim}{B}$, μ and work in $V[G_\mu]$, $G_\mu \subseteq P_\mu$ generic over V such that $q \in G_\mu$. Let

$$\begin{aligned} W_\lambda \stackrel{\text{def}}{=} \{N \prec (H(\lambda)^{V[G_\mu]}, \in, <^*_\lambda) : \ & N \text{ is countable and } N \cap \omega_1 \in \underset{\sim}{B}[G_\mu], \text{ but} \\ & \text{there is no } r \in P_\alpha/G_\mu \text{ such that:} \\ & r \text{ is } (N, P_\alpha/G_\mu)\text{-semi-generic}, p{\restriction}[\mu, \alpha) \le r \text{ and} \\ & \ r \Vdash_{P_\alpha/G_\mu} \text{“}N \cap \omega_1 \in \underset{\sim}{A}\text{”}\}. \end{aligned}$$

If $W_\lambda = \emptyset \bmod \mathcal{D}_{<\aleph_1}(H(\lambda)^{V[P_\mu]})$, we can easily get the desired result (as in the proof of 1.11)): let λ_1 be such minimal that $2^{\lambda_1} < \lambda$, and $\bar{Q} \in H(\lambda_1)$. Clearly also $W_{\lambda_1} = \emptyset \bmod \mathcal{D}_{<\aleph_1}(H(\lambda)^{V[P_\mu]})$ and let $W'_{\lambda_1} \subseteq \mathcal{S}_{<\aleph_1}(H(\lambda_1)^{V[P_\mu]}$ be closed unbound disjoint to it. So if N is as in the assumption of $\otimes$, then necessarily $\lambda_1 \in N$ hence $W_{\lambda_1} \in N$ and w.l.o.g. $W'_{\lambda_1} \in N$. Then clearly $N \cap H(\lambda_1) \in W'_{\lambda_1}$, hence $N \cap H(\lambda_1) \notin W_{\lambda_1}$, hence $N \notin W_\lambda$, which suffices. So (in $V[G_\mu]$) the set W is a stationary subset of $\mathcal{S}_{<\aleph_1}(H(\lambda))$, hence semi-stationary. As $V[G_\mu] \models$ “Rss($\aleph_2$)” there is $u \subseteq H(\lambda)$ such that $\omega_1 \subseteq u$, $|u| < \aleph_2$ (in $V[G_\mu]$) and $W \cap \mathcal{S}_{<\aleph_1}(u)$ is semi-stationary; now by 1.2(2) without loss of generality $(u, \in, <^*_\lambda {\restriction} u) \prec (H(\lambda), \in, <^*_\lambda)$. Let $u = \bigcup_{\zeta<\omega_1} u_\zeta$, with each u_ζ countable and u_ζ is increasing and continuous. So

$$B_1 = \{\zeta < \omega_1 : (\exists N \in W)(\omega_1 \cap u_\zeta \subseteq N \subseteq u_\zeta)\}$$

is a stationary subset of ω_1 (see 1.2(4)) which belongs to $\mathfrak{B}[P_\mu]$, and obviously:

$$(*) \qquad p \Vdash_{P_\alpha/G_\mu} \text{“}\underset{\sim}{A} \cap B_1 \text{ is not stationary”}.$$

[Why? For $\zeta \in B_1$ let $\omega_1 \cap u_\zeta \subseteq N_\zeta \subseteq u_\zeta$, $N_\zeta \in W$ and for $\xi < \omega_1$ let N'_ξ be the Skolem Hull in $(H(\lambda)^{V[G_\mu]}, \in, <^*_\lambda)$ of $\{\zeta : \zeta < \xi\} \cup \{p, \langle u_\zeta, N_\zeta : \zeta \in B_1, \zeta < \xi\rangle\}$, and

$$\underset{\sim}{C} = \{\xi < \omega_1 : N'_\xi[\underset{\sim}{G}_{P_\alpha}] \cap \omega_1 = \xi \text{ and } N'_\xi[\underset{\sim}{G}_{P_\alpha}] \cap u = u_\xi\}.$$

As $\langle N'_\xi[\underset{\sim}{G}_{P_\alpha}] : \xi < \omega_1\rangle$ is increasing continuous, clearly $\underset{\sim}{C}$ is a P_α/G_μ-name of a club of ω_1. Now $\underset{\sim}{C} \cap \underset{\sim}{A}$ is necessarily disjoint to B_1 by the definition of W: if $\zeta < \omega_1, q \in P_\alpha/G_\mu$, and $q \Vdash_{P_\alpha/G_\mu}$ "$\zeta \in \underset{\sim}{C} \cap \underset{\sim}{A} \cap B_1$", then $N_\zeta \in W$ is defined (because $\zeta \in B_1$) and q_α is $(N_\zeta, P_\alpha/G_\mu)$-semi-generic, and $q_\alpha \Vdash_{P_\alpha/G_\mu}$ "$N_\zeta \cap \omega_1 \in \underset{\sim}{A}$", contradicting "$N_\zeta \in W$" so $(*)$ holds]. Also

$$(**) \qquad B_1 \subseteq B$$

by the clause "$N \cap \omega_1 \in \underset{\sim}{B}[G_\mu]$" in the definition of W.

Of course $B_1 \in V^{P_\mu}$ and as said after the definition of B_1, it is stationary so we get a contradiction to an assumption on $\underset{\sim}{A}, \underset{\sim}{B}$.

$\square_{4.1}$

4.2 Claim. (1) Suppose $\overline{Q} = \langle P_i, \underset{\sim}{Q}_j : i \leq \alpha, j < \alpha\rangle$ is a semiproper iteration, $\langle \mu_\zeta : \zeta < \xi\rangle$ an increasing sequence of strongly inaccessible cardinals $\leq \alpha$, $\bigwedge_{\zeta<\xi} \left[(\forall i < \mu_\zeta)(|P_i| < \mu_\zeta) \text{ and } \Vdash_{P_{\mu_\zeta}} \text{"Rss}(\mu_\zeta)\text{"}\right]$ and
$(*)$ every countable set of ordinals from V^{P_α} is included in a countable set of ordinals from V.

Suppose further that $\underset{\sim}{B}$ is a P_{μ_0}-name of a subset of ω_1, $\underset{\sim}{A}_\zeta$ is a $P_{\mu_{\zeta+1}}$-name of a subset of ω_1 (if $\zeta + 1 = \xi$ we stipulate $\mu_{\zeta+1} = \alpha$) and $p \in P$ satisfies:

$p{\restriction}\mu_0 \Vdash_{P_{\mu_0}}$ "$\underset{\sim}{B}$ is stationary",

$p{\restriction}\mu_{\zeta+1} \Vdash_{P_{\mu_{\zeta+1}}}$ " for every $X \in \mathfrak{B}[P_{\mu_\zeta}] \setminus \{0\}$, if $X \subseteq \underset{\sim}{B}$ *then* $\underset{\sim}{A}_\zeta \cap X$ is stationary ".

Then $p \Vdash_{P_\alpha}$ "the intersection of any countable subset of $\{A_\zeta : \zeta < \xi\}$ is stationary".

(2) In 4.2(1) we can replace the assumption $(*)$ by:

$(*)^-$ if $\delta \in (\mu_0, \alpha)$ is strongly inaccessible and $[i < \delta \Rightarrow |P_i| < \delta]$, *then* $\Vdash_{P_\alpha}$ "$\text{cf}(\delta) > \aleph_0$".

Proof. 1) Let $\underset{\sim}{w}$ be a P_α-name for a countable subset of ξ. So without loss of generality $\underset{\sim}{w} = w$ and let $w = \{\zeta(n) : n < \omega\}$. Let Y be the closure of $\{\mu_\zeta : \zeta < \xi\} \cup \{\alpha\}$ (in the order topology on the ordinals). If the conclusion of 4.2

fails then (as we can increase p) without loss of generality $p \Vdash_{P_\alpha}$ "$\bigcap_{n<\omega} \underset{\sim}{A}_{\zeta(n)}$ is disjoint to $\underset{\sim}{C}$ where $\underset{\sim}{C}$ is a club of ω_1".

We now prove by induction on $j \in Y$:

$\otimes_j$ *if* $\mu_0 \leq i < j$, both in Y, λ regular and large enough, $N \prec (H(\lambda), \in, <^*_\lambda)$ countable, $\underset{\sim}{C} \in N$, $\underset{\sim}{B} \in N, \langle \mu_\zeta, \underset{\sim}{A}_\zeta : \zeta < \xi \rangle \in N$ and $\{i,\ j,\ \bar{Q}\} \in N$, $p \leq p' \in N \cap P_\alpha$ and $q \in P_i$ is (N, P_i)-semi-generic, $p' \restriction i \leq q$, and $q \Vdash_{P_i}$ "$N \cap \omega_1 \in \underset{\sim}{B}$ and for $n < \omega$ we have $[\mu_{\zeta(n)} \leq i \Rightarrow N \cap \omega_1 \in \underset{\sim}{A}_{\zeta(n)}]$", *then* there is $q' \in P_j, (N, P_j)$-semi-generic, $p' \restriction j \leq q', q' \restriction i = q$ and $q' \Vdash_{P_j}$ "$N \cap \omega_1 \in \underset{\sim}{B}$ and for $n < \omega$ we have $[\mu_{\zeta(n)} \leq j \Rightarrow N \cap \omega_1 \in \underset{\sim}{A}_{\zeta(n)}]$".

Clearly this is enough (apply it with $p' = p,\ i = \mu_0,\ j = \alpha$, and there are $N,\ q$ as required and $\underset{\sim}{B}$ is a P_{μ_0}-name of a stationary subset of $\subseteq \omega_1$).

Case 1. $j = \mu_0$. Trivial.

Case 2. j is an accumulation point of Y (hence is of countable cofinality). As in the proof of the iteration lemma for semiproperness.

Case 3. $j = \mu_{\zeta+1}$.

Apply the previous claim 4.1 (for $\bar{Q} \restriction \mu_{\zeta+1}$ and μ_ζ).

2) The proof is similar but $\underset{\sim}{w}$ is P_α-name of a countable subset of ζ, and for $j \in [\mu_0, \alpha]$ the statement $\otimes_j$ is now for every $\underset{\sim}{w}$ which is a P_j-name (not P_α-name) of a countable subset of $\{\zeta : \mu_\zeta < j\}$. So proving it we increase $p \restriction [i,j]$ also for this purpose and $i \in [\mu_0, j)$. Cases 1, 3 remain as before. Note that we can replace $\underset{\sim}{w}$ by a larger set

Case 2A. $j > \sup[j \cap \{\mu_\zeta : \zeta < \xi\}]$

Trivial

Case 2B. $j = \sup[j \cap \{\mu_\zeta : \zeta < \xi\}]$.

W.l.o.g. p force a value to $\sup \underset{\sim}{w} \cap \{\mu_\zeta : \zeta < \xi\}$, call it ξ^*.

Subcase α. $\xi^* < j$: the proof is as in case 2A, as increasing $\underset{\sim}{w}$ w.l.o.g. it is P_{ξ^*+1}-name.

Subcase β. $\xi^* = j$: for some $i_1 < j$, $p \restriction i_1 \nVdash_{P_{i_1}}$ "$\mathrm{cf}(j) = j$" is easy too. W.l.o.g. $i_1 < i$ (by the induction hypothesis), $p \restriction i_1 \Vdash$ "$\mathrm{cf}(j) < j$". So in V^{P_i} we know $\mathrm{cf}(j)$, and it is $\aleph_0$ or $\aleph_1$. Now $\aleph_1$ is impossible (as $\xi^* = j$) and if it is $\aleph_0$ act as in the old case 2.

But by $(*)^-$ of 4.2(2), one of the subcases occurs. $\square_{4.2}$

4.3 Theorem. Suppose $\{\mu < \kappa : \mu \text{ supercompact }\}$ is a stationary subset of κ, $\langle S_1, S_2, S_3 \rangle$ is a partition of ω_1 with each S_i stationary. Let $h : \kappa \to H(\chi)$ and assume $\{\mu < \kappa : \mu \text{ supercompact}, h(\mu) = 0\}$ is stationary. *Then* for some forcing notion P:

(i) $P = P_\kappa$ ($= R\mathrm{Lim}\bar{Q}$ for some S_1-suitable $\bar{Q}$) is S_3-complete, or at least S_3-proper and satisfies the κ-c.c.

(ii) In V^{P_κ}, from any $\aleph_2$ stationary subsets of $S_1 \subseteq \omega_1$, there are $\aleph_2$ of them such that the intersection of any countably many of them is stationary (and $\mathfrak{B}^{P_\kappa}$ is layered, of course). We then call $\mathfrak{B}[V^{P_\kappa}]$ reflective.

(iii) A forcing axiom as strong as h holds (see the proof and 3.9).

4.3A Remark. 1) We really use a weaker assumption

(a) $\{\mu < \kappa : \mu \text{ measurable}\}$ is stationary;

(b) $\{\mu < \kappa\text{: for } \chi < \kappa, \mu \text{ is } \chi\text{-compact}\}$ is unbounded; use 1.6(2), 1.6(3), 1.10(1). See more in XVI§2.

2) The situation is similar in 4.4, where we get better bound (i.e. using smaller large cardinals) for a stronger result (but lose in forcing axiom.)

3) We can demand only "S_1 is stationary" etc. if we use 4.1(2) instead of 4.1(1), but then we should satisfy $(*)^-$ of 4.1(2).

Proof. We define by induction on $\alpha \leq \kappa$ the iteration $\bar{Q}^\alpha = \langle P_i, \underset{\sim}{Q}_i, \mathbf{t}_j : i \leq \alpha, j < \alpha \rangle$ such that:

(A) $\bar{Q}^\alpha$ is S_1-suitable.

(B) Each Q_i is S_3-proper.

(C) $\bar{Q}^\alpha$, $P_\alpha \in H(\kappa)$ when $\alpha < \kappa$.

(D) If $h(i) = \langle \mathbf{t}, \underset{\sim}{R} \rangle$, i measurable, $\mathbf{t}$ a truth value, $\underset{\sim}{R}$ a P_i-name and $\Big[\mathbf{t} = 1 \Rightarrow \mathfrak{B}[P_i] \restriction S_1 \lessdot \mathfrak{B}[P_i * \underset{\sim}{R}] \restriction S_1\Big]$ and $\Big[j < i \ \&\ \mathbf{t}_j = 1 \Rightarrow \mathfrak{B}[P_j] \restriction S_1 \lessdot \mathfrak{B}[P_i * \underset{\sim}{R}] \restriction S_1\Big]$ and $(P_i/P_{j+1}) * \underset{\sim}{R}$ is semiproper and S_3-proper for $j < i$ *then* $\mathbf{t}_i = \mathbf{t}, \underset{\sim}{Q}_i = \underset{\sim}{R} *$ SSeal $(\langle \mathfrak{B}[P_j] : j \leq i, \mathbf{t}_j = 1 \rangle)$.

(E) If not (D), i is inaccessible, $|P_j| < i$ for $j < i$, $h(i) = 0$, and $\Vdash_{P_i}$ "Rss($\aleph_2$)," then $\mathbf{t}_i = 1$ and $\underset{\sim}{Q}_i = \mathrm{SSeal}(\langle \mathfrak{B}[P_j] : j \leq i,\ \mathbf{t}_j = 1 \rangle, S)$.

(F) If neither (D) nor (E) then $\mathbf{t}_i = 0$, and

$$Q_i = \text{SSeal}(\langle \mathfrak{B}[P_j] : j \leq i, \mathbf{t}_j = 1 \rangle, S).$$

We can carry out the construction and prove by induction on α that $\bar{Q}^\alpha$ is S_1-suitable.

$\alpha = 0$. Trivial.

α *limit.* By Claim 2.3(1).

$\alpha = \beta + 1$, (F) *applies to* β. By 2.14(1).

$\alpha = \beta + 1$, (D) *applies to* β. By Claim 2.11.

$\alpha = \beta + 1$, (E) applied to β. By 2.16 note:

(*) if $i \in B = \{i : i \text{ inaccessible}, j < i \Rightarrow |P_j| < i\}$ then: if for i clause (E) or clause (F) occur then Q_i is semiproper. [Why? If $\Vdash_{P_i}$ "$\text{Rss}(\aleph_2)$" by 1.7(5), otherwise clause (F) applies, $\mathbf{t}_i = 0$ and we can use 2.11.]

But clause (D) does not apply to i non-measurable so

(**) for i non measurable $\Vdash_{P_i}$ "$\underset{\sim}{Q}_i$ is semiproper".

Now suppose $p \in P_\kappa$, $\langle \underset{\sim}{A}_i : i < \kappa \rangle$ a P_κ-name and $p \Vdash$ "$\underset{\sim}{A}_i \subseteq S_1$ is stationary".

Let $Y = \{\mu < \kappa : \mu \text{ strongly inaccessible}, \bigwedge_{i<\mu} |P_i| < \mu$ and $\Vdash_{P_i}$ "$\text{Rss}(\aleph_2)$" and $\mathbf{t}_\mu = 1\}$. Note that in V^{P_κ}, $Y \subseteq \{\delta < \kappa : \text{cf}^{V[P_\kappa]}(\delta) = \aleph_1\}$ is stationary because if $\mu < \kappa$ is measurable, limit of supercompacts, $\bigwedge_{j<\mu} |P_j| < \mu$ and D is a normal filter on μ, concentrating on non-measurable so $X_\mu = \{i < \mu : i$ inaccessible, not measurable, $\bigwedge_{j<i} |P_j| < \mu\} \in D$. We use 1.10(1) (noting $\text{Rss}^+(\mu)$ holds, by 1.6(6)) to get $\mu \in Y$. Now for each $\mu \in Y$ choose $p_\mu \in P_\kappa$ and a P_μ-name $\underset{\sim}{B}_\mu$ such that:

$$p \leq p_\mu,$$
$$p_\mu{\restriction}\mu \Vdash \text{"}\underset{\sim}{B}_\mu \subseteq S_1 \text{ is stationary, } \underset{\sim}{B}_\mu \in \mathfrak{B}[P_\mu]\text{"},$$
$$p_\mu \Vdash_{P_\kappa} \text{"for every nonzero } X \in \mathfrak{B}[P_\mu],$$
$$\text{if } X \leq \underset{\sim}{B}_\mu \text{ then } X \cap \underset{\sim}{A}_\mu \text{ is stationary".}$$

Why does such a p_μ exist? As $\mathfrak{B}[P_\mu] \lessdot \mathfrak{B}[P_\kappa]$ (and see 0.1(4)(b)). Remember that P_μ satisfies μ-c.c. so $\underset{\sim}{B}_\mu \in H(\chi_\mu)$ for some $\chi_\mu < \mu$ and without loss of generality $\underset{\sim}{B}$ is a P_{χ_μ}-name and $\mathbf{t}_{\chi_\mu} = 1$ (i.e. by increasing χ_μ; also, Y is stationary by a hypothesis).

By Fodor's lemma, for some stationary $Y_1 \subseteq Y$, there are p and $\underset{\sim}{B}$ such that for $\mu \in Y_1 : p_\mu \restriction \mu = p$, and $\underset{\sim}{B}_\mu = \underset{\sim}{B}$.
As each $\underset{\sim}{A}_\zeta$ is a P_λ-name for some $\lambda > \zeta, \lambda \in Y$, without loss of generality $[\mu_1 < \mu_2$ in $Y_1 \Rightarrow \underset{\sim}{A}_{\mu_1}$ is a P_{μ_2}-name$]$. Now, for $\mu \in Y_1$ let $\underset{\sim}{A}'_\mu$ be $\underset{\sim}{A}_\mu$ if $p_\mu \in \underset{\sim}{G}_{P_\kappa}$ and S otherwise.

Note that $Y_1 \in V$ and every countable subset of Y_1 is contained in a countable set from V [Why? Remembering S_3 is stationary, by S_3-properness.] Now we apply the previous Claim 4.2 to $\underset{\sim}{B}, \langle \underset{\sim}{A}'_\mu : \mu \in Y_1 \rangle$. $\square_{4.3}$

4.4. Theorem. Suppose $\kappa = \sup\{\lambda < \kappa : \lambda$ a compact cardinal$\}$ and $\langle S_1, S_2, S_3 \rangle$ is a partition of ω_1 to stationary sets. *Then* for some forcing notion $P \in V$:

(i) V^P is a model of: ZFC $+ 2^{\aleph_0} = \aleph_1 + 2^{\aleph_1} = \aleph_2$,

(ii) in V^P, the statement Ulam$(\mathcal{D}_{\omega_1} + S_1)$ holds, where, for a uniform filter D on λ, Ulam(D) means: there are many λ λ-complete filters extending D, such that every D-positive set belongs to at least one of them (A is D-positive if $A \subseteq \lambda$, and $(\lambda \setminus A) \notin D$).

Remark. So in V^P, Ulam's problem has a positive solution: there are $\aleph_1$ measures on $[0,1]_{\mathbb{R}}$, each countably additive, such that every $A \subseteq [0,1]_{\mathbb{R}}$ is measurable with respect to at least one of them.

Proof. Before we do the forcing, we work out some combinatorics, which will tell us what will suffice.

4.5A. Context and Notation.

(1) $\lambda = \lambda^{<\lambda}$ is a fixed regular uncountable cardinal.

(2) W denotes a fixed class of ordinals (in the actual case $W \subseteq \lambda^+$), $0 \in W$, for every i, $i+1 \in W$, and

$$\aleph_0 \leq \mathrm{cf}(i) < \lambda \Rightarrow i \notin W.$$

(3) B will denote a Boolean algebra.

(4) For a Boolean Algebra B, let $B^+ = B \setminus \{0\}$.

(5) $\Pr(a_1, a_2, B_1, B_2)$ means: B_1, B_2 are Boolean algebras, $B_1 \subseteq B_2$, $a_1 \in B_1^+$, $a_2 \in B_2^+$, and $(\forall x)[x \in B_1^+ \ \&\ x \le a_1 \to x \cap a_2 \ne 0]$.

(6) If the identity of B_2 is clear (when dealing with one Boolean Algebra and its subalgebras) we just write $Pr(a_1, a_2, B_1)$.

4.5B. Observation.

(a) $\Pr(1, x, B_1, B_2)$ for $x \in B_2^+, |B_1| = 2$;

(b) if $B_a \subseteq B_b \subseteq B$, $x \in B_a^+$, $y \in B_b^+$, $z \in B$ and $\Pr(x, y, B_a)$, $\Pr(y, z, B_b)$ then $\Pr(x, z, B_a)$;

(c) if $\Pr(x, y, B_1, B_2), 0 < x' \le x, x' \in B_1, y \le y' \in B_2$ *then* $\Pr(x', y', B_1)$ and $\Pr(x', y \cap x, B_1)$.

4.5C. Notation and Definition.

(1) We call $\bar{B}$ 1-o.k. (for W) if $\bar{B} = \langle B_i : i < \alpha \rangle$ is an increasing continuous sequence, each B_i a Boolean Algebra of cardinality $\le \lambda$, $[i, j \in \alpha \cap W$ and $i < j \Rightarrow B_i \lessdot B_j,]$ and $[i \in W \cap \alpha \Rightarrow B_i$ is λ-complete$]$.

(2) We call $w \subseteq W \cap \alpha$ closed (subset of $W \cap \alpha$) *if*

(i) for every accumulation point $\delta < \alpha$ of the closure of w (that is $\delta = \sup(w \cap \delta)$ & $\delta < \alpha$) we have

(a) $\delta \notin W$ & $\delta + 1 < \alpha \Rightarrow \delta + 1 \in w$

(b) $\delta \in W$ & $\delta + 1 < \alpha \Rightarrow \delta \in w$,

(ii) for every $\delta < \alpha$ we have: $\mathrm{Min}(w) < (\delta + 1) \in w$ & $\aleph_0 \le \mathrm{cf}(\delta) < \lambda \Rightarrow \delta = \sup(\delta \cap w)$,

(iii) if $\mathrm{Min}(w) < \beta \in W$, $\beta + 1 \in w$ then $\beta \in w$.

(3) Let $\mathrm{CSb}(\alpha) = \{w : w$ a closed subset of $W \cap \alpha$ of power $< \lambda\}$,
$\mathrm{CSb}_u(\alpha) = \{w \in \mathrm{CSb}(\alpha) : w$ unbounded below $\alpha\}$.
(Clearly $\mathrm{CSb}_u(\alpha) \ne \emptyset \Rightarrow \aleph_0 \le \mathrm{cf}(\alpha) < \lambda$).

(4) For $w \in \mathrm{CSb}(\alpha)$ and $\bar{B} = \langle B_i : i < \beta \rangle$ which is 1-o.k. such that $\beta > \alpha$, let

(i) $\mathrm{Seq}_w(\bar{B}) = \{\langle a_i : i \in w \rangle : a_i \in B_i^+, a_i$ is decreasing;

if $i \in w, i = \delta + 1$, δ limit of course, $\delta \notin W, i > \mathrm{Min}(w)$ *then* $a_i = \bigcap_{j \in w \cap i} a_j$;

if $i \in w$, $i > \mathrm{Min}(w), \mathrm{cf}(i) \geq \lambda$, *then* $a_i = a_j \in B_j$ for some $j \in i \cap w$;

and if $i < j$ are in $w (\subseteq W)$ then $\mathrm{Pr}(a_i, a_j, B_i, B_j)\}$.

(ii) Let $\mathrm{Seq}(\bar{B}) \overset{\mathrm{def}}{=} \bigcup\{\mathrm{Seq}_w(\bar{B})$: for some α $(\leq \ell g(\bar{B}))$, $\alpha = \ell g(\bar{B})$ or α is successor of a member of W, we have $w \in \mathrm{CSb}(\alpha)\}$.

Let $\mathrm{Seq}_u(\bar{B}) \overset{\mathrm{def}}{=} \bigcup\{\mathrm{Seq}_w(\bar{B}) : w \in \mathrm{CSb}_u(\ell g(\bar{B}))\}$.

It is naturally ordered by $\bar{a}^1 \leq \bar{a}^2$ if letting $\bar{a}^\ell = \langle a_i^\ell : i \in w_\ell \rangle$ then $w^1 \subseteq w^2$ and $[\zeta \in w^1 \Rightarrow a_\zeta^1 \geq a_\zeta^2]$.

(iii) *When* $\alpha = \delta + 1 \leq \ell g(\bar{B}), w \in \mathrm{CSb}(\delta)$ let

$$Z_w(\bar{B}) \overset{\mathrm{def}}{=} \left\{ \bigcap_{i \in w} a_i : \langle a_i : i \in w \rangle \in \mathrm{Seq}_w(\bar{B} \restriction (\delta + 1)) \right\},$$

$Z^\delta(\bar{B}) \overset{\mathrm{def}}{=} \bigcup\{Z_w(\bar{B}) : w \in \mathrm{CSb}_u(\delta)\}$.

If $\ell g(\bar{B}) = \delta + 2$, we may omit δ.

(5) We call $\bar{B}$ 2-o.k. if for every limit $\delta < \ell g(\bar{B})$, $0 \notin Z^\delta(\bar{B})$ and $\bar{B}$ is 1-o.k.

(6) We call $\bar{B}$ 3-o.k. if it is 2-o.k. and for limit $\delta < \ell g(\bar{B})$ of cofinality $< \lambda$ we have: $Z^\delta(\bar{B})$ is a dense subset of $B_{\delta+1}$.

(7) If $\bar{B}$ is not continuous, we identify it with the obvious correction for the purpose of our definitions.

(8) We call $\Upsilon \subseteq \mathrm{Seq}_u(\bar{B})$ dense if for every $\bar{a} \in \mathrm{Seq}_u(\bar{B})$ for some $\bar{a}' \in \Upsilon$ we have $\bar{a} \leq \bar{a}'$. We say Υ' refines Υ if $(\forall \bar{a} \in \Upsilon)(\exists \bar{a}' \in \Upsilon')$ $[\bar{a} \leq \bar{a}']$. We say $\Upsilon \subseteq \mathrm{Seq}_u(\bar{B})$ is open if $\bar{a} \leq \bar{a}'$ (in $\mathrm{Seq}_u(\bar{B})$), $\bar{a} \in \Upsilon$ implies $\bar{a}' \in \Upsilon$.

4.5D. Fact. Suppose $\bar{B}$ is 2-o.k., $\bar{B} = \langle B_i : i \leq \delta + 1 \rangle, \aleph_0 \leq \mathrm{cf}(\delta) < \lambda$. Then:

(0) (i) $\mathrm{CSb}_u(\delta) \neq \emptyset$, moreover for every $\alpha \leq \delta + 1$ and $v \subseteq \alpha$ of cardinality $< \lambda$ we have:

$\aleph_0 \leq \mathrm{cf}(\alpha) < \lambda \Rightarrow$ there is $w \in \mathrm{CSb}_u(\alpha)$ such that $v \subseteq w$ and

$\mathrm{cf}(\alpha) \geq \lambda \Rightarrow$ there is $w \in \mathrm{CSb}(\alpha)$, such that $v \subseteq w$ and

$\alpha = i + 1 \,\&\, i \in W \Rightarrow$ there is $w \in \mathrm{CSb}(\alpha)$ such that $v \subseteq w$ and $i \in w$.

(ii) If $w \in \mathrm{CSb}_u(\delta), \alpha < \delta$, *then* $w \setminus \alpha \in \mathrm{CSb}_u(\delta)$; similarly for $\mathrm{CSb}(\alpha)$.

(iii) If $\alpha < \delta$ and $w \in \mathrm{CSb}_u(\alpha)$ and $\alpha = \varepsilon + 1$ and $\aleph_0 \le \mathrm{cf}(\varepsilon) < \lambda$ *then* $w \cup \{\alpha\} \in \mathrm{CSb}(\delta)$.

(iv) If $w \in \mathrm{CSb}_u(\delta), \alpha \in w$ *then* there is $\beta \in w \setminus \alpha$ such that $\beta \notin \{\varepsilon + 1 : \aleph_0 \le \mathrm{cf}(\varepsilon) < \lambda\}$; in fact $\beta = \min(w \setminus (\alpha + 1))$ is as required.

(v) If $w \in \mathrm{CSb}(\alpha)$ and $\beta < \alpha$, *then* $w \cap \beta \in \mathrm{CSb}(\beta)$.

(1) If $w \in \mathrm{CSb}(\delta + 1)$ *then* $Z_w(\bar{B})$ includes $B_{\mathrm{Min}(w)}$, hence: $\aleph_0 \le \mathrm{cf}(\delta) < \lambda \Rightarrow B_\delta \subseteq Z^\delta(\bar{B})$.

(2) (i) If $w_1,\ w_2 \in \mathrm{CSb}(\delta)$ and $\mathrm{Min}(w_\ell) < \mathrm{Min}(w_{3-\ell}) \Rightarrow \mathrm{Min}(w_{3-\ell}) \in w_\ell$ *then* $w_1 \cup w_2 \in \mathrm{CSb}(\delta)$.

(ii) Similarly for $\mathrm{CSb}_u(\delta)$.

(3) (i) If $w_1 \subseteq w_2$ are both in $\mathrm{CSb}(\alpha), \min(w_1) = \min(w_2)$, $\langle a_i : i \in w_1\rangle \in \mathrm{Seq}_{w_1}(\bar{B})$ *then* $\langle a_i : i \in w_2\rangle \in \mathrm{Seq}_{w_2}(\bar{B})$ provided that for $i \in w_2 \backslash w_1$ we define $a_i = a_{\max(i \cap w_1)}$ which is well defined.

(ii) If $\alpha < \sup(w)$, and $w \in \mathrm{CSb}(\delta)$, and $\langle a_i : i \in w\rangle \in \mathrm{Seq}_w(\bar{B})$ *then* $\langle a_i : i \in w \setminus \alpha\rangle \in \mathrm{Seq}_{w\setminus\alpha}(\bar{B})$ and $\langle a_i : i \in w \cap \alpha\rangle \in \mathrm{Seq}_{w\cap\alpha}(\bar{B})$.

(iii) If $w_1 \subseteq w_2$ are both in $CSb(\delta)$ and $\langle a_i : i \in w_2\rangle \in \mathrm{Seq}_{w_2}(\delta)$ and $[i \in w_2 \,\&\, \mathrm{cf}(i) \ge \lambda \,\&\, \mathrm{Min}(w_1) < i \Rightarrow (\exists j)(j \in w_1 \cap i \,\&\, a_j = a_i)]$ *then* $\langle a_i : i \in w_1\rangle \in \mathrm{Seq}_{w_1}(\delta)$.

(iv) If $\beta < \delta$, β is a successor of a limit ordinal, $w \in \mathrm{CSb}_u(\beta - 1)$, $\mathrm{Min}(w) < \beta \le \varepsilon + 1$, $w_1 = w \cup \{\beta\}$ and $\langle a_i : i \in w\rangle \in Z_w(\bar{B})$ *then* we can find a_β such that $\langle a_i : i \in w_1\rangle \in Z_{w_1}(\bar{B})$.

(4) If $w_1 \subseteq w_2$ are both in $\mathrm{CSb}_u(\delta)$ and $\min(w_1) = \min(w_2)$ *then* $Z_{w_1}(\bar{B}) \subseteq Z_{w_2}(\bar{B})$.

(5) If $\langle a_i : i \in w_1\rangle, \langle b_j : j \in w_2\rangle$ are in $\mathrm{Seq}_{w_1}(\bar{B})$, $\mathrm{Seq}_{w_2}(\bar{B})$ respectively, and $(\forall i \in w_1)(\exists j \in w_2) a_i \le b_j$ *then* $\bigcap_{i\in w_1} a_i \le \bigcap_{j\in w_2} b_j$.

(6) If $\langle a_i : i \in w\rangle \in \mathrm{Seq}_w(\bar{B}), 0 < b < a_{\min(w)}$ and $b \in B_{\min(w)}$ *then* $\langle a_i \cap b : i \in w\rangle \in \mathrm{Seq}_w(\bar{B})$.

(7) If $\bar{B} = \langle B_i : i \le \alpha\rangle$ is l-o.k. ($l = 1, 2, 3$), $\gamma_i \le \alpha$ (for $i \le i(*)$) is strictly increasing continuous and $[i \in W \Leftrightarrow \gamma_i \in W]$ and $[\aleph_0 \le \mathrm{cf}(i) < \lambda \Rightarrow \gamma_{i+1} = \gamma_i + 1]$ *then* $\langle B_{\gamma_i} : i \le i(*)\rangle$ is l-o.k.

(8) Assume $\bar{B} = \langle B_i : i \le \alpha\rangle$ is 1-o.k.

(i) if $\beta < \gamma < \alpha$, $\beta \in W$, $\gamma \in W$, $b \in B_\gamma$ *then* for at most one a we have:

(∗) $a \in B_\beta$ and $B_\gamma \models$ "$a \geq b$" and $\Pr(a, b, B_\beta, B_\gamma)$,

(ii) if $\bar{a}^\ell \in \mathrm{Seq}_{w_\ell}(\bar{B})$ for $\ell = 1, 2$, *then* $\{\beta : \beta \in w_1 \cap w_2$ and $a^1_\beta = a^2_\beta\}$ is an initial segment of $w_1 \cap w_2$.

(9) Assume $\bar{B} = \langle B_i : i \leq \alpha \rangle$ is 3-o.k., $[\delta < \alpha \,\&\, \mathrm{cf}(\delta) \geq \lambda \Rightarrow \delta \in W]$.

(i) If $v \in \mathrm{CSb}(\beta)$, $\beta \leq \alpha + 1$ is a successor of a member of W, and $\gamma \in v$ and $d \in B_\gamma$ *then* the set $\Gamma = \Gamma_{\alpha,v,\gamma,d} = \{\bar{a} \in \mathrm{Seq}\,(\bar{B}) : v \subseteq \mathrm{Dom}(\bar{a})$ and $[a_\gamma \cap d = 0$ *or* $a_\gamma \leq d]\}$ is a dense and open subset of $\mathrm{Seq}\,(\bar{B})$.

(ii) If $\aleph_0 \leq \mathrm{cf}(\alpha) < \lambda$, $v \in \mathrm{CSb}_u(\alpha)$, $\gamma \in v$ and $d \in B_\gamma$ *then* the set $\Gamma = \Gamma_\alpha = \{\bar{a} \in \mathrm{Seq}_u(\bar{B}) : v \subseteq \mathrm{Dom}(\bar{a})$ and $[a_\gamma \cap d = 0$ or $a_\gamma \leq d]\}$ is a dense open subset of $\mathrm{Seq}\,(\bar{B})$.

(iii) If $\beta \leq \alpha$, $d \in B_\beta \setminus \{0\}$ and $v \subseteq \beta + 1$, $|v| < \lambda$ *then* there is w satisfying $v \subseteq w \in \mathrm{CSb}(\beta + 1)$, $\beta \in w$ and $\bar{a} \in \mathrm{Seq}_w(\bar{B})$ such that $a_\beta \leq d$.

Proof. Easy, e.g.,

0)(i) We prove it by induction on α. For α non-limit the result is trivial so assume α is a limit. So for every $j < \alpha$ there is w_j such that: $v \cap j \subseteq w_j$ and $[\aleph_0 \leq \mathrm{cf}(j) < \lambda \Rightarrow w_j \in \mathrm{CSb}_u(j)]$ and $[\mathrm{cf}(j) \notin [\aleph_0, \lambda) \Rightarrow [\mathrm{cf}(j) \geq \lambda \vee (\exists i)(j = i + 1 \,\&\, i \in W) \Rightarrow w_j \in \mathrm{CSb}(j)]]$. Let $\langle j_\varepsilon : \varepsilon < \mathrm{cf}(\alpha) \rangle$ be an increasing continuous sequence of ordinals $< \alpha$ with limit α. If $\mathrm{cf}(\alpha) = \aleph_0$ *then* w.l.o.g. $j_n + 2 \in v$ for $n < \omega$ and then $w \stackrel{\mathrm{def}}{=} \bigcup\{w_{j_{n+1}+3} \setminus (j_n + 3) : n < \omega\} \cup w_{j_0+3}$ is as required (remember $i + 1 \in W$ for any ordinal i by 4.5A(2)). If $\mathrm{cf}(\alpha) \geq \lambda$ then for some $j < \alpha$, we have $v \subseteq j$ and we can use the induction hypothesis. If $\mathrm{cf}(\alpha) > \aleph_0$ but still it is $< \lambda$, without loss of generality each j_ε is a limit ordinal with cofinality $< \mathrm{cf}(\alpha) < \lambda$. Let $w = \{j_\varepsilon + 1 : \varepsilon < \mathrm{cf}(\alpha)\} \cup \bigcup_{\varepsilon < \mathrm{cf}(\alpha)} (w_{j_{\varepsilon+1}+3} \setminus (j_\varepsilon + 3)) \cup w_{j_0+3}$ and note that it belongs to $\mathrm{CSb}_u(\alpha)$ and includes v, as required.

(0)(iv) See the last phrase of 4.5C(2).

(1) For the first phrase note that for $a \in B_{\min(w)}$, $\bar{b}_a = \langle a : i \in w \rangle \in \mathrm{Seq}_w(\bar{B})$ (see Definition 4.5C(4)(i), (iii)).

The second phrase follows by the definition of $Z^\delta(\bar{B})$ and 4.5D(0)(i).

(3)(i) Why is $\max(i \cap w_1)$ well defined?

First note: $i \cap w_1 \neq \emptyset$ as $i \notin w_1$ implies $i \neq \min(w_1)$, but $\min(w_1) = \min(w_2)$

so $i > \min(w_1)$ hence $\min(w_1) \in i \cap w_1$.

Second note: if $i \cap w_1$ has no last element, let $\beta = \sup(i \cap w_1)$, so $\beta \leq i$ and $\aleph_0 \leq \mathrm{cf}(\beta) \leq |w_1| < \lambda$, hence $\beta \notin W$, so $\beta \notin w_2$ and $\beta < i$. Also $\beta + 1 \in w_1$ (as w_1 is closed and $\beta < i < \alpha$ so $\beta + 1 < \delta$), so $\beta + 1$ cannot be in $i \cap w_1$, hence $i = \beta + 1 \in w_1$, contradicting the assumption on i (i.e. $i \in w_2 \setminus w_1$).

(3)(iv) Note that, as $w \in \mathrm{CSb}_u(\beta - 1)$, necessarily $\mathrm{cf}(\beta - 1) < \lambda$. Also $w_1 \in \mathrm{CSb}(\delta + 2)$, so $Z_w(\bar{B})$ is well defined. Also $a_\beta \stackrel{\mathrm{def}}{=} \bigcap_{i \in w} a_i$ is well defined as B_β is λ-complete, and $|w| < \lambda$ as $w \in\subseteq \mathrm{CSb}_u(\beta) \subseteq \mathrm{CSb}(\beta)$ (see Definition 4.5C(3)). As $\bar{B}$ is 2-o.k. (see Definition 4.5C(5)), $0_{B_{\delta+1}} \notin Z^\delta(\bar{B})$, but clearly $a_\beta \in Z_w(\bar{B}) \subseteq Z^\beta(\bar{B})$ hence $a_\beta \neq 0_{B_{\delta+1}}$. The order requirements for $\langle a_i : i \in w_1 \rangle \in \mathrm{Seq}_{w_1}(\bar{B})$ are easy too.

(4) Use (3)(i).

(6) Let for $i \in w$, $c_i \stackrel{\mathrm{def}}{=} a_i \cap b$ and $\beta = \min(w)$. So

(i) $c_i = a_i \cap b \in B_i$ [as $a_i \in B_i, b \in B_\beta \subseteq B_i$];

(ii) for $i < j$ from w, $c_j \leq c_i$ [as $a_j \leq a_i$, clearly $a_j \cap b \leq a_i \cap b$];

(iii) for $i < j$ from w, $\Pr(c_i, c_j, B_i)$.

[Why? Let $0 < d \leq c_i$, $d \in B_i$ then $0 < d \leq a_i$, $d \in B$, hence (by $\Pr(a_i, a_j, B_i)$) $d \cap a_j \neq 0$ and $d \leq c_i = a_j \cap b \leq b$ so $d \cap b = d$, hence

$$d \cap c_j = d \cap (a_j \cap b) = (d \cap b) \cap a_j = d \cap a_j$$

so $d \cap c_j \neq 0$ as required.]

The other conditions are easy too.

So $\langle c_j : j \in w \rangle \in \mathrm{Seq}_w(\bar{B})$.

(9) We prove this by induction on α ((i), (ii) and (iii) together). In parts (i) and (ii), Γ being open is immediate, so let us prove density. So assume $\bar{c} = \langle c_i : i \in v_0 \rangle$ belongs to $\mathrm{Seq}(\bar{B})$ (for 9(i)) or $\mathrm{Seq}_u(\bar{B})$ (for 9(ii)), $v \subseteq \alpha$, $\gamma \in v$, $d \in B_\gamma$ as there and we shall find $\bar{b}$, $\bar{c} \leq \bar{b} \in \Gamma$ (see end of 4.5C(4)(ii)). In the cases below for 9(iii) only the assumption on α is relevant.

Case 1: $\alpha = 0$

Trivial.

Case 2: $\alpha = \varepsilon + 1$, $\varepsilon \in W$.

For 9(iii) note that by the induction hypothesis we have to prove only the case $\beta = \alpha = \varepsilon + 1$ and $d \in B_\beta = B_\alpha$ is given. Let $d_1 \in B^+_\beta$ be such that $\mathrm{Pr}_1(d_1, d, B_\varepsilon, B_\alpha)$. By the induction hypothesis we can find $w_1 \in \mathrm{CSb}(\varepsilon + 1)$, such that $\varepsilon \in w_1$, and $\bar{a} \in \mathrm{Seq}_{w_1}(\bar{B})$ such that $a_\varepsilon \leq d_1$.

Let $w \stackrel{\mathrm{def}}{=} w_1 \cup \{\alpha\}$, $a_\alpha \stackrel{\mathrm{def}}{=} a_\beta \cap d \in B^+$ (not zero as $a_\beta \leq d_1$ and the choice of d_1). So $\langle a_i : i \in w\rangle \in \mathrm{Seq}_w(\bar{B})$ is as required.

Now as α is a successor only 9(i) is left, by the induction hypothesis $\beta = \alpha + 1$ and by the assumptions of 9(i), β is a successor of a member of W so $\alpha \in W$ hence v_0 has a last element. Let d_0 be: $c_{\max(v_0)} \cap d$ if not zero and $c_{\max(v_0)}$ otherwise and as we have proved 9(iii), there is $\langle a_i : i \in w\rangle \in \mathrm{Seq}_w(\bar{B})$ satisfying $w \in \mathrm{CSb}(\alpha + 1)$ such that: $v_0 \cup v \cup \{\varepsilon, \alpha\} \subseteq w$, and $a_\alpha \leq d_0$. So $a_\alpha \leq d$ or $a_\alpha \cap d = 0$; by 4.5D(8)(ii) we are done.

Case 3: $\alpha = \varepsilon + 1$, $\varepsilon \notin W$ (so only 9(i)+(iii) apply and ε is a limit ordinal) (as $\beta \notin W$, $\mathrm{cf}(\varepsilon) < \lambda$).

For 9(i) as in case 2 it follows from 9(iii), so let us prove 9(iii), by the induction hypothesis w.l.o.g. $\beta = \alpha$. As $\bar{B}$ is 3-o.k. by Definition 4.5C(6) there are $w_0 \in \mathrm{CSb}_u(\varepsilon)$ and $\langle b^0_i : i \in w_0\rangle \in \mathrm{Seq}_{w_0}(\bar{B})$ such that $\bigcap_{i \in w_0} b^0_i$ is not zero and is $d_0 \leq d$.

By the induction hypothesis we can apply 4.5D(9)(ii) to ε, $\mathrm{CSb}_u(\varepsilon)$, $\langle b^0_i : i \in w_0\rangle$ and so we can find $\langle b_i : i \in w_1\rangle$ such that $\langle b^0_i : i \in w_0\rangle \leq \langle b_i : i \in w_1\rangle \in \mathrm{Seq}\,(\bar{B}{\restriction}\alpha)$ and $v \subseteq w_1$. As $\bar{B}$ is 2-o.k., $b_\alpha \stackrel{\mathrm{def}}{=} \bigcap_{i \in w_1} b_i \in B_\alpha$ is not zero. Let $w = w_1 \cup \{\alpha\}$, so $w \in \mathrm{CSb}(\alpha + 1)$ and $\langle b_i : i \in w\rangle \in \mathrm{Seq}_u(\bar{B})$ is as required.

Case 4: α is a limit ordinal, $\mathrm{cf}(\alpha) \geq \lambda$ (so $\alpha \in W$ by an assumption of 4.5D(9)). As $B_\alpha = \bigcup_{\beta < \alpha} B_\beta$ and the third requirement in the definition of $\bar{a} \in \mathrm{Seq}$ (see 4.5C(4)(i)) it is easy.

Case 5: α is a limit ordinal, $\mathrm{cf}(\alpha) < \lambda$.

So 9(i), 9(iii) does not apply. First as for 9(ii) we can assume $v \setminus \min(v_0) = v_0$.

[Why? By 4.5D(0)(i) w.l.o.g. $0 \in v$ & $v_0 \subseteq v$. But $\sup(v_0 \cup v) \geq \sup(v_0) = \alpha$, so $v = v_0 \cup v \in \mathrm{CSb}_u(\alpha)$, and lastly apply 4.5D(3)(i) to replace v_0 by $v \setminus \min(v_0)$.] Second we are given $\gamma \in v$, $d \in B_\gamma$ (so, as we can increase γ w.l.o.g. $\gamma \in v_0$). Now $v_0 \cap (\gamma+1) \in \mathrm{CSb}(\gamma+1)$ and so $\langle c_i : i \in v_0 \cap (\beta+1)\rangle \in \mathrm{Seq}\,(\bar{B})$, and by the induction hypothesis (on 9(i)) we can find $\langle b_i^0 : i \in w_0\rangle \in \mathrm{Seq}\,(B\restriction(\beta+1))$ such that $\langle c_i : i \in v_0 \cap (\beta+1)\rangle \leq \langle b_i^0 : i \in w_0\rangle$ and $b_\beta^0 \leq d \vee b_\beta^0 \cap d = 0$. Define $w = w_0 \cup v_0$,

$$b_i = \begin{cases} b_i^0 & \text{if } i \leq \beta \text{ (so } i \in w_0); \\ b_\beta^0 \cap c_i & \text{if } i > \beta \text{ (so } i \in v_0). \end{cases}$$

Now $\langle b_i : i \in w\rangle$ is as required. $\square_{4.5D}$

4.5E. Claim. If $\bar{B} = \langle B_i : i \leq \lambda^+\rangle$ is 3-o.k. and $[i < \lambda^+$ & $\mathrm{cf}(i) \geq \lambda \Rightarrow i \in W]$ *then* $B^+_{\lambda^+}$ is the union of λ many λ-complete filters.

Proof. Note that by 4.5D(9) we have:

($*$) for every $\alpha \in W$ and $x \in B^+_\alpha$ for some $w \in \mathrm{CSb}(\alpha+1)$ and $\langle a_i : i \in w\rangle \in \mathrm{Seq}_w(\langle B_i : i < \alpha+1\rangle)$ we have $0 < \bigcap_{i\in w} a_i \leq x$, $0 \in w$, and w is closed and has a last element α.

Now remember that $\mathrm{Seq}\,(\bar{B}) = \bigcup\{\mathrm{Seq}\,(\langle B_i : i \leq \alpha\rangle) : \ \alpha < \lambda^+$ is a successor of a member of $W\}$.

It is well known that there is $H : \{w \subseteq \lambda^+ : |w| < \lambda\} \to \lambda$ such that: $H(w) = H(u), \alpha \in w \cap u$ implies $\alpha \cap w = \alpha \cap u$; also $H(w) = H(u)$ implies that w, u have the same order type (let $f_\alpha : \alpha \to \lambda$ be one to one, $H^0(w) = \{\langle \mathrm{otp}(w \cap \alpha), \mathrm{otp}(w \cap \beta), f_\beta(\alpha)\rangle : \alpha < \beta \text{ in } w\}$. Now H^0 is as required except that $\mathrm{Rang}(H^0) \not\subseteq \lambda$, but $|\mathrm{Rang}(H^0)| = \lambda$, so we can correct this).

Let F_i be a one-to-one function from B_{i+1} into λ. We say $\langle a_i^1 : i \in w_1\rangle, \langle a_i^2 : i \in w_2\rangle \in \mathrm{Seq}\,(\bar{B})$ (hence w_1, w_2 have last element) are *equivalent if*:

(a) $H(w_1) = H(w_2)$ and

(b) *if* $\alpha_1 \in w_1$ and $\alpha_2 \in w_2$, and $w_1 \cap \alpha_1, w_2 \cap \alpha_2$ have the same order type and $\alpha_1 = \gamma_1 + 1$, $\alpha_2 = \gamma_2 + 1$, *then*

$$F_{\gamma_1}(a^1_{\alpha_1}) = F_{\gamma_2}(a^2_{\alpha_2}).$$

Now the number of equivalence classes is $\leq \lambda^{<\lambda} = \lambda$. So it is enough to show that if $\langle a_i^\zeta : i \in w_\zeta\rangle \in \mathrm{Seq}(\bar{B})$ are equivalent for $\zeta < \zeta(*) < \lambda$, $0 \in w_\zeta$, $\max(w_\zeta) \in w_\zeta$, then $\bigcap_{\zeta<\zeta(*)} a^\zeta_{\max(w_\zeta)} \neq 0$ (see $(*)$). Note that if $\alpha \in w_{\zeta_1} \cap w_{\zeta_2}$, then $a_\alpha^{\zeta_1} = a_\alpha^{\zeta_2}$.

Toward this end we prove by induction on $\alpha \in W$:

$(*)$ (1) $x_\alpha \stackrel{\text{def}}{=} \bigcap_{\zeta<\zeta(*)} a^\zeta_{\max(w_\zeta\cap(\alpha+1))}$ is not zero (and belongs to B_α);

(2) if $\gamma < \alpha$ (and $\gamma \in W$) then $\Pr(x_\gamma, x_\alpha, B_\beta)$;

(3) if $\gamma \leq \alpha$ is a limit ordinal then:

(a) $\mathrm{cf}(\gamma) < \lambda \Rightarrow x_{\gamma+1} = \cap\{x_\varepsilon : \varepsilon \in \gamma \cap W\}$,

(b) $\mathrm{cf}(\gamma) \geq \lambda \Rightarrow x_\gamma = x_\varepsilon$ for every large enough $\varepsilon < \gamma$.

Clearly x_α is decreasing (as a_α^ζ is decreasing in α for each ζ) and well defined as $\max(w_\zeta \cap (\alpha+1))$ belongs to w_ζ when $\alpha \in W$ (remembering $0 \in w_\zeta$).

Case 1. $\alpha = 0$

Then $\max(w_\zeta \cap (\alpha+1)) = 0$ and $a_0^\zeta = a_0^0 \in B_0^+$ for every $\zeta < \zeta(*)$. So $(*)(1)$ holds and $(*)(2)$, (3) do not apply.

Case 2. $\alpha = \beta+1, \beta \in W$

Note that if ($\zeta < \zeta(*)$ and) $\alpha = \beta + 1 \notin w_\zeta$ then $a^\zeta_{\max(w_\zeta\cap(\alpha+1))} = a^\zeta_{\max(w_\zeta\cap(\beta+1))}$.

So if $\alpha \notin w_\zeta$ for every $\zeta < \zeta(*)$ then $x_\alpha = x_\beta$, so $(*)(1)$ holds. As for $(*)(2)$: for $\gamma < \beta$ use the induction hypothesis; for $\gamma = \beta$ this is easy. Similarly for $(*)(3)$.

If for some $\zeta < \zeta(*)$ we have $\alpha \in w_\zeta$, let $v = \{\zeta < \zeta(*) : \alpha \in w_\zeta\}$. So $x_\alpha = \bigcap_{\zeta\notin v} a^\zeta_{\max(w_\zeta\cap(\beta+1))} \cap \bigcap_{\zeta\in v} a_\alpha^\zeta$. By the definition of the equivalence relation and the F_i's, for some a we have $[\zeta \in v \Rightarrow a_\alpha^\zeta = a \leq a^\zeta_{\max(w_\zeta\cap(\beta+1))}]$ and $[\zeta, \xi \in v \Rightarrow w_\zeta \cap (\alpha+1) = w_\xi \cap (\alpha+1)]$. Clearly

$$\begin{aligned} x_\alpha &= \bigcap_{\zeta\notin v} a^\zeta_{\max(w_\zeta\cap(\beta+1))} \cap \bigcap_{\zeta\in v} a_\alpha^\zeta \\ &= \bigcap_{\zeta<\zeta(*)} a^\zeta_{\max(w_\zeta\cap(\beta+1))} \cap \bigcap_{\zeta\in v} a_\alpha^\zeta \\ &= x_\beta \cap a. \end{aligned}$$

Now as $\beta \in W$, B_β is λ-complete, hence $x_\beta \in B_\beta$. Now $a \in B_\alpha$ and let $\zeta(0) = \min(v)$, $\gamma(0) = \max(w_{\zeta(0)} \cap (\beta+1))$, the maximum exists as said above. Clearly $\gamma(0) = \beta$ (see 4.5C(2)(iii)), $a \le a^{\zeta(0)}_{\gamma(0)}$ and $\Pr(a^{\zeta(0)}_{\gamma(0)}, a, B_\beta)$ by the last clause in the definition of $\langle a^{\zeta(0)}_i : i \in w_{\zeta(0)}\rangle \in \mathrm{Seq}_{w_{\zeta(0)}}(\bar{B})$ (see 4.5D(4)(i)). As $x_\beta \in B_\beta$, and easily $a^{\zeta(0)}_{\gamma(0)} \ge x_\beta > 0$, clearly $x_\beta \cap a \ne 0$. So $(*)(1)$ holds. As for $(*)(2)$, by 4.5B(b) as there is a maximal $\gamma \in w \cap \alpha$, i.e. $\beta = \gamma(0)$ (see above) it is enough to prove $(*)(2)$ for $\gamma = \beta = \gamma(0)$. So let $d \in B_\beta$, $0 < d \le x_\beta$. Then $d \le a^{\zeta(0)}_{\gamma(0)}$, hence by $\Pr(a^{\zeta(0)}_{\gamma(0)}, a, B_\beta), a \cap d \ne 0$, but $a \cap d = d \cap x_\beta \cap a = d \cap x_\alpha$, so we are done. Lastly $(*)(3)$ holds by the induction hypothesis.

Case 3. $\alpha = \beta + 1, \beta \notin W$

By an assumption of 4.5E, $\aleph_0 \le \mathrm{cf}(\beta) < \lambda$ so by 4.5D(0)(i) there is $w \in \mathrm{CSb}_u(\beta)$ such that $\zeta < \zeta(*) \Rightarrow w_\zeta \subseteq w$ and $i \in w \,\&\, \mathrm{cf}(i) = \lambda \Rightarrow (\exists j)(\sup(\bigcup_\zeta w_\zeta \cap i) < j < i \,\&\, j \in w)$. Note that

$$a^\zeta_{\max(w_\zeta \cap (\alpha+1))} = \bigcap_{\gamma<\beta} a^\zeta_{\max(w_\zeta \cap (\gamma+1))}.$$

[Why? If $\alpha \notin w_\zeta$, as $\langle a^\zeta_{\max(w_\zeta \cap (\gamma+1))} : \gamma < \beta\rangle$ is nonincreasing and eventually constant (because $\langle \max(w_\zeta \cap (\gamma+1)) : \gamma < \beta\rangle$ is eventually constant), it is equal to

$$a^\zeta_{\max(w_\zeta \cap (\alpha+1))} = \bigcap_{\gamma<\beta} a^\zeta_{\max(w_\zeta \cap (\gamma+1))}.$$

If $\alpha \in w_\zeta$, as $\langle a^\zeta_\gamma : \gamma \in w_\zeta\rangle \in \mathrm{Seq}(B)$ (see in Definition 4.5C(4)(i) the second clause in the definition of $\bar{a}^\zeta \in \mathrm{Seq}(\bar{B})$).] Now:

$$x_\alpha = \bigcap_{\zeta<\zeta(*)} a^\zeta_{\max(w_\zeta \cap (\alpha+1))} = \bigcap_{\zeta<\zeta(*)} \bigcap_{\gamma<\beta} a^\zeta_{\max(w_\zeta \cap (\gamma+1))}$$
$$= \bigcap_{\gamma<\beta} \Big(\bigcap_{\zeta<\zeta(*)} a^\zeta_{\max(w_\zeta \cap (\gamma+1))} \Big) = \bigcap_{\gamma<\beta} x_\gamma.$$

So $(*)(3)$ holds (as $\gamma = \beta$ is the only new case). Also it can be checked that $\langle x_\varepsilon : \varepsilon \in w\rangle \in \mathrm{Seq}_w(\bar{B})$ (in Definition 4.5C(4)(i), the first clause by the definition of x_α and $(*)(1)$, the second clause (x_ε decreasing) is shown above, the third clause (continuity) by $(*)(3)$, the fourth clause by the choice of w and the definition of x_ε, the fifth clause by $(*)(2)$). As $\bar{B}$ is 2-o.k. (see 4.5C (5)) (as $(*)(2)$ holds below β) we get that $x_\alpha = \bigcap\{x_\varepsilon : \varepsilon \in u \cap \beta\} \neq 0$. Similarly, using 4.5D(6), we can check $(*)(2)$.

Case 4. α limit

As $\alpha \in W$, necessarily $\mathrm{cf}(\alpha) = \lambda$. But then, by the definition of $\mathrm{Seq}_w(\bar{B})$, if $\alpha \in w_\zeta$ though necessarily $\max(w_\zeta \cap (\alpha+1)) \neq \max(w_\zeta \cap (\gamma+1))$ for $\gamma < \alpha$, still for $\gamma < \alpha$ large enough $a^\zeta_\alpha = a^\zeta{}_{\max(w_\zeta \cap (\gamma+1))}$, hence $a^\zeta{}_{\max(w_\zeta\cap(\alpha+1))} = \bigcap_{\gamma<\beta} a^\zeta{}_{\max(w_\zeta\cap(\gamma+1))}$ for every large enough $\beta < \alpha$. If $\alpha \notin w_\zeta$ this holds on simpler grounds. But $\zeta(*) < \lambda = \mathrm{cf}(\alpha)$. So $x_\alpha = x_\gamma$ for every large enough $\gamma < \alpha$, and we can finish easily. $\square_{4.5E}$

4.5F Remark. The proof is written such that it will be easy to change it for $\bar{B} = \langle B_i : i < \gamma\rangle$, $\gamma < (2^\lambda)^+$, so $|B_i| = |i| + \lambda$, B_{i+1} is generated by $B_i \cup B'_i$, $|B'_i| = \lambda$, B'_i is λ-complete and in the definition of $\mathrm{Seq}_w(\bar{B})$ add: if $i = \beta+1, \beta \in w$ then $(\exists x \in B'_\beta)[a_i = a_\beta \cap x]$.

Just use $H : \{a : a \subseteq 2^\lambda, |a| < \lambda\} \to \lambda$ such that $H(a) = H(b)\ \&\ \alpha \in a \cap b \Rightarrow \mathrm{otp}\,(a\cap\alpha) = \mathrm{otp}(b \cap \alpha)$ which exists by Engelking and Karlowic [EK]. But it is not clear whether there is interest in this.

4.6 Definition. 1) We say $\bar{Q} = \langle P_i, \underset{\sim}{Q}_j, \mathbf{t}_j, \underset{\sim}{A}_i : i \leq \alpha, j < \alpha\rangle$ is an $\bar{S}$-o.k. sequence for $\underset{\sim}{W}$ (where $\bar{S} = \langle S_1, S_2, S_3\rangle$, a partition of ω_1) if:

(A) $\bar{Q}$ is a S_1-suitable iteration (forgetting the $\underset{\sim}{A}_i$'s).

(B) Each Q_i is S_3-complete.

(C) $\underset{\sim}{A}_i$ is a P_i-name of a subalgebra (or just subset) of $\mathfrak{B}^{P_i}$.

(D) $\underset{\sim}{A}_i$ is increasing continuous.

(E) $\mathbf{t}_j \in \{0,1\}$ and: if $\mathbf{t}_i = 1$, then $\Vdash_{P_i}$ "$\underset{\sim}{A}_i \lessdot \mathfrak{B}^{P_i}\restriction S_1$".

(F) $\mathbf{t}_i = 1$ for every successor ordinal i.

(G) $\Vdash_{P_i}$ "$\langle \underset{\sim}{A}_j : j \leq i, j \in \underset{\sim}{W}\rangle$ is 3-o.k. for $\underset{\sim}{W}$" where on $\underset{\sim}{W}$ see clause (H) below and λ from 4.5A(1) is chosen as $\aleph_1$ (see below and 4.5C(1), (5), (6)).

(H) If i is successor or zero then $i \in \underset{\sim}{W}$. If, in $V^{P_{i+1}}$, i is a limit of cofinality $\aleph_0$ then $i \notin \underset{\sim}{W}$. Also "$i \in \underset{\sim}{W}$" is a P_{i+1}-name and $\underset{\sim}{W} \subseteq \alpha$.

(I) $\Vdash_{P_{i+1}}$ "Rss($\aleph_2$)".

(J) If i is neither a limit nor a successor of a limit ordinal, then $\mathcal{A}_i = \mathfrak{B}^{P_i}\restriction S_1$.

2) If $\underset{\sim}{W}$ is not given we mean $\{i < \alpha : \text{if } i \text{ is limit then (in } V^{P_{i+1}}) \operatorname{cf}(\alpha) \geq \aleph_1\}$.

4.6A Remark. Note that $\underset{\sim}{W}$ determines $\langle \mathbf{t}_\alpha : \alpha < \kappa \rangle$ in Definition 4.6, so we could in 4.7 below forget it.

4.7 Proof of 4.4. Let $h : \kappa \to H(\kappa)$. We define $\bar{Q}^\alpha = \langle P_i, \underset{\sim}{Q}_j, \mathbf{t}_j, \underset{\sim}{\mathcal{A}}_i : i \leq \alpha, j < \alpha \rangle$ by induction on $\alpha \leq \kappa$ such that in stage α, the objects $\underset{\sim}{Q}_j (j < \alpha), P_j (j \leq \alpha)$, $\mathbf{t}_j$ $(j+1 \leq \alpha+1)$ and $\underset{\sim}{\mathcal{A}}_j (j \leq \alpha)$ (and the truth value of "$j \in \underset{\sim}{W}$" is as in 4.6(2)) have already been defined and for successor i, $\mathfrak{B}^{P_i} \lessdot \underset{\sim}{\mathcal{A}}_{i+1}$ and:

(A) $\bar{Q}^\alpha$ is $\bar{S}$-o.k. (and increases with α).

(B) $\bar{Q}^\alpha \in H(\kappa)$ for $\alpha < \kappa$.

(C) If α is non-limit, let $\kappa_{\alpha+1}$ be the first compact cardinal $> |P_\alpha|$, and $\underset{\sim}{Q}_\alpha = \text{SSeal}(\mathfrak{B}^{P_\alpha}, S_1, \kappa_{\alpha+1})$ if α is successor and Levy$(\aleph_1, <\kappa_{\alpha+1})$ if α is zero and $\mathcal{A}_{\alpha+1} = \mathfrak{B}^{P_{\alpha+1}}\restriction S_1$ (and $\underset{\sim}{\mathcal{A}}_0$ the trivial algebra). Lastly of course $\mathbf{t}_{\alpha+1} = 1$ and "$\alpha \in \underset{\sim}{W}$" is true also $\mathbf{t}_0 = 0$ and "$0 \in \underset{\sim}{W}$".

(D) *If* α is a limit ordinal, $h(\alpha) = \langle \mathbf{t}, \underset{\sim}{Q}, \underset{\sim}{\mathcal{A}} \rangle$, $\underset{\sim}{Q}$ a P_α-name of a forcing notion, $\underset{\sim}{\mathcal{A}}$ a $P_\alpha * \underset{\sim}{Q}$-name, and for some $\underset{\sim}{R} \in H(\kappa)$ we have $\Vdash_{P_i}$ "$\underset{\sim}{Q} \lessdot \underset{\sim}{R}$" and by the following choices for $\bar{Q}^{\alpha+1}$ we get a $\bar{S}$-o.k., *then* so we choose $\bar{Q}^{\alpha+1}$; where the choices are: $\mathbf{t}_\alpha = \mathbf{t}, \underset{\sim}{Q}_\alpha = \underset{\sim}{R}$, and $\mathcal{A}_\alpha = \mathfrak{B}^{\bar{Q}\restriction\alpha}\restriction S_1$, $P_{\alpha+1} = P_\alpha * \underset{\sim}{R}$ and $\mathcal{A}_{\alpha+1} = \mathfrak{B}[P_{\alpha+1}]$. If possible we choose $\underset{\sim}{R} = \underset{\sim}{Q}$.

(E) If clauses (C), (D) do not produce a definition of $\bar{Q}^{\alpha+1}$, let $\kappa_{\alpha+1}$ be the first compact cardinal $> |P_\alpha|$, and then:

first case if in V^{P_α}, cf$(\alpha) > \aleph_0$ *then*

$$\begin{aligned}
&\underset{\sim}{\mathcal{A}}_\alpha \overset{\text{def}}{=} \mathfrak{B}^{\bar{Q}\restriction\alpha}\restriction S_1 \text{ i.e. } \underset{\sim}{\mathcal{A}}_\alpha \overset{\text{def}}{=} \bigcup_{j<\alpha} \mathfrak{B}^{P_{j+2}}\restriction S_1 = \bigcup_{j<\alpha} \underset{\sim}{\mathcal{A}}_j, \\
&\underset{\sim}{Q}_\alpha = \text{SSeal}(\langle \mathcal{A}_j : j \leq \alpha\rangle, S_1, \kappa_{\alpha+1}) = \\
&\quad = \text{SSeal}(\langle \mathfrak{B}^{P_{j+2}} : j < \alpha\rangle \hat{}\, \langle \mathcal{A}^{P_\alpha}\rangle, S_1, \kappa_{\alpha+1}) \qquad : \\
&\underset{\sim}{\mathcal{A}}_{\alpha+1}, = \underset{\sim}{\mathfrak{B}}^{P_{\alpha+1}}, \\
&\mathbf{t}_\alpha = \mathbf{t}_{\alpha+1} = 1
\end{aligned}$$

second case if in V^{P_α}, cf$(\alpha) = \aleph_0$ (i.e. $\alpha \notin W$) *then*

$$\underset{\sim}{\mathcal{A}}_\alpha = \mathfrak{B}^{\bar{Q}\restriction\alpha}\restriction S_1,$$

(in V^{P_α}) let $\underset{\sim}{\mathcal{A}}'_{\alpha+1}$ be $Z^\alpha = Z^\alpha(\langle \mathcal{A}_j : j \leq \alpha\rangle \hat{}\, \langle \mathfrak{B}^{P_\alpha}\restriction S_1\rangle)$ (a subset of $\mathfrak{B}^{P_\alpha}\restriction S_1$, see Definition 2.4(2)) and $\underset{\sim}{\mathcal{A}}_{\alpha+1}$ be the subalgebra of $\mathfrak{B}^{P_{\alpha+1}}\restriction S_1$ which $\underset{\sim}{\mathcal{A}}'_{\alpha+1}$ generates. We let

$$\underset{\sim}{Q}_\alpha = \text{SSeal}(\langle \mathfrak{B}^{P_{j+2}} : j < \alpha\rangle \hat{}\, \langle \underset{\sim}{\mathcal{A}}_{\alpha+1}\rangle, S_1, \kappa_{\alpha+1}),$$

$$\mathbf{t}_\alpha = 0, \quad \mathbf{t}_{\alpha+1} = 1.$$

If we succeed to carry the induction, then letting $G \subseteq P_\kappa$ be generic over V we know:

(a) $\aleph_1^{V[G]} = \aleph_1^V$ and $\langle S_1, S_2, S_3\rangle$ is a partition of ω_1 to stationary subsets (as P_κ is semiproper by clause (A) of Definition 4.6).

(b) $\aleph_2^{V[G]} = \kappa$ (similarly, noting that P_κ satisfies the κ-c.c.).

(c) Every countable set of ordinals from V^{P_κ} is included in one from V (see (e) below).

(d) $\langle \underset{\sim}{\mathcal{A}}_i[G] : i < \kappa\rangle$ is 3-o.k. (by clause (G) of Definition 4.6).

(e) P_κ is S_3-complete (see clause (B) of Definition 4.6) hence, as S_3 is sationary, P_κ adds no reals so $V[G_\kappa] \models$ "$2^{\aleph_0} = \aleph_1$, so $\lambda = \lambda^{<\lambda}$".

(f) $\mathfrak{B}^{V[G]}\restriction S_1$ is Ulam i.e., omitting 0, it is the union of $\lambda = \aleph_1$ many λ-complete filters. [Why? By 4.5(E) and (d) above (as $W = \{\alpha < \kappa : \alpha$ zero, successor or has cofinality $\geq \aleph_1$ (in V, equivalently in $V[G_\kappa]$)$\}$).]

To carry the induction it is enough to show that when clause (E) in the construction is applied, we get an $\bar{S}$-o.k. iteration; this is dealt by 4.9 below $+2.13+2.16$ for $\langle \bar{\mathfrak{B}}^{\bar{Q}}\restriction S\rangle$ for the first case, and by 4.10 below $+2.13+2.16$ for W defined by 4.10 for the second case. $\square_{4.7}$

4.8 Remarks. 1) We could have allowed in clause (D) during the proof 4.7 (of 4.4) to decide if $i \in \underset{\sim}{W}$, i.e. decide $\underset{\sim}{W}_i = \underset{\sim}{W} \cap i$, i.e. demand $h(\alpha) = \langle \mathbf{t}, \underset{\sim}{\mathbf{s}}, \underset{\sim}{Q}, \underset{\sim}{A}\rangle$, and try to define $\bar{Q}^{\alpha+1}$ as there with the following addition: the truth value of "$\alpha \in \underset{\sim}{W}$" is $\underset{\sim}{\mathbf{s}}$, a $P_\alpha * \underset{\sim}{Q}$-name, and at the end shoot a suitable club of κ through the "good" places.

2) We could have gotten a forcing axiom, as before.

3) In fact we can weaken the large cardinals demand to "$\kappa = \sup\{\lambda < \kappa : \lambda$ strongly inaccessible and $\mathrm{Rss}^+(\lambda)$ or at least $\bigwedge_{\mu<\kappa} \mathrm{Rss}^+(\lambda,\mu)\}$".

4.9 Claim. Suppose $S \subseteq \omega_1$ is stationary, $\bar{Q} = \langle P_i, \underset{\sim}{Q}_j : i \leq \alpha,\ j < \alpha\rangle$ is a semiproper iteration, α a limit ordinal, and, for simplicity, $\Vdash_{P_{i+1}}$ "$\mathrm{Rss}(\aleph_2[V^{P_{i+1}}])$" for $i < \alpha$. Let $\underset{\sim}{\Upsilon}$ be a P_α-name of a dense subset of $\bar{\mathfrak{B}}^{\bar{Q}}\restriction S = \bigcup_{i<\alpha} \mathfrak{B}^{P_i}\restriction S$ for $i \in W^*$.

Then

⊗ *if* λ is regular and large enough, $N \prec (H(\lambda), \in, <^*_\lambda)$ is countable, and $\bar{Q}$, λ, p, $\underset{\sim}{\Upsilon}$, belong to N, $p \in P_\alpha \cap N$, $\beta \in \alpha \cap N$ a successor ordinal and $q \in P_\beta$ is (N, P_β)-semi-generic, $p\restriction\beta \leq q$ and $N \cap \omega_1 \in S$, *then* there is a countable N', $N \leq_\beta N' \prec (H(\lambda), \in, <^*_\lambda)$, $N \cap \omega_1 = N' \cap \omega_1$, successor $\gamma \in [\beta, \alpha)$, P_γ-name $\underset{\sim}{A} \in N'$ and q', p' satisfying $p \leq p' \in P_\alpha \cap N'$, $q' \in P_\gamma$, $p'\restriction\gamma \leq q'$, $q'\restriction\beta = q$ and $p'\restriction\beta = p\restriction\beta$ such that $q' \Vdash_{P_\alpha}$ "$N \cap \omega_1 \in \underset{\sim}{A}$" and $p' \Vdash$ "$\underset{\sim}{A} \in \underset{\sim}{\Upsilon}$".

4.9A Remark. 1) Note that $\mathfrak{B}^{P_{i+1}}\restriction S \lessdot \mathfrak{B}^{\bar{Q}}\restriction S$ for $i < \alpha$.

2) The claim gives more chains than used in 4.7.

3) This is naturally used together with 2.13.

Proof. Let us fix p, $\underset{\sim}{\Upsilon}$, β and work in $V[G_\beta]$ where $G_\beta \subseteq P_\beta$ generic over V and $q \in G_\beta$. Let λ be large enough and

$$\mathbf{W} \stackrel{\text{def}}{=} \{M \prec (H(\lambda), \in, <^*_\lambda) : M \text{ is countable, } M \cap \omega_1 \in S, \text{ but}$$
$$\text{there are no successor } \gamma \in M \cap [\beta, \alpha), r \in P_\gamma/G_\beta \text{ and}$$
$$\underset{\sim}{A} \in M \text{ (a } P_\gamma\text{-name) and } p' \in P_\alpha/G_\beta \cap N \text{ such that:}$$
$$r \text{ is } (M, P_\gamma/G_\beta)\text{-semi-generic, } p \leq p', p' \restriction \gamma \leq r \text{ and}$$
$$r \Vdash_{P_\alpha/G_\beta} \text{“} M \cap \omega_1 \in \underset{\sim}{A}\text{”}, p' \Vdash \text{“}\underset{\sim}{A} \in \underset{\sim}{\Upsilon}\text{”}\}.$$

If $\mathbf{W} = \emptyset \bmod \mathcal{D}_{<\aleph_1}(H(\lambda))$, we can easily get the desired result (as in the proof of 1.11)).

So (in $V[G_\beta]$) the set $\mathbf{W}$ is a stationary subset of $\mathcal{S}_{<\aleph_1}(H(\lambda))$, hence semi-stationary. As $V[G_\beta] \models$ “Rss($\aleph_2$)” (remember β is a successor ordinal and clause (I) of Definition 4.6) there is $u \subseteq H(\lambda)$ such that $\omega_1 \subseteq u$ and $|u| < \aleph_2$ (in $V[G_\beta]$) and $W \cap \mathcal{S}_{<\aleph_1}(u)$ is semi-stationary. Now without loss of generality $(u, \in, <^*_\lambda \restriction u) \prec (H(\lambda), \in, <^*_\lambda)$. Let $u = \bigcup_{\zeta<\omega_1} u_\zeta$, u_ζ is countable, increasing and continuous. So

$$B_1 \stackrel{\text{def}}{=} \{\zeta \in S : (\exists M \in \mathbf{W})(\omega_1 \cap u_\zeta \subseteq M \subseteq u_\zeta)\}$$

is a stationary subset of $S \subseteq \omega_1$ (see 1.2(4)), it belongs to $\mathfrak{B}[P_\beta]$, and we shall prove:

(*) $\quad p \Vdash_{P_\alpha/G_\beta}$ “ for every $X \in \underset{\sim}{\Upsilon}$ the set $X \cap \underset{\sim}{A} \cap B_1$ is not stationary”.

[Why (*)? If not then for some p' and P_α-name $\underset{\sim}{A}$, $p \leq p' \in P_\alpha/G_\beta$ and $p' \Vdash_{P_\alpha/G_\beta}$ “$\underset{\sim}{A} \in \underset{\sim}{\Upsilon}$ and $\underset{\sim}{A} \cap B_1$ is stationary”. As $\underset{\sim}{\Upsilon} \subseteq \mathfrak{B}^{\bar{Q}}$, for some $\gamma < \alpha$, $\underset{\sim}{A}[G_{P_\alpha}]$ is in $\mathfrak{B}^{P_\gamma}$, so (possibly increasing p) without loss of generality for some successor $\gamma \in [\beta, \alpha)$, $\underset{\sim}{A}$ is a P_γ-name of a member of $\mathfrak{B}^{P_\gamma}$. For $\zeta \in B_1$, let the model M_ζ be any member of $\mathbf{W}$ which satisfies $\omega_1 \cap u_\zeta \subseteq M_\zeta \subseteq u_\zeta$ (see the definition of B_1). For $\xi < \omega_1$, let N'_ξ be the Skolem Hull (in $(H(\lambda), \in, <^*_\lambda)$) of $\{\zeta : \zeta < \xi\} \cup \{p, p', \underset{\sim}{A}, W, \langle u_\zeta, M_\zeta : \zeta \in B_1\rangle\}$, and

$$\underset{\sim}{C}=\{\xi<\omega_1 : N'_\xi[G_{P_\alpha}]\cap\omega_1=\xi \text{ and } N'_\xi[G_{P_\alpha}]\cap u=u_\xi\}.$$

As $\langle N'_\xi[G_{P_\alpha}]:\xi<\omega_1\rangle$ is increasing continuous, $\underset{\sim}{C}$ is a P_α/G_β-name of a club of ω_1. Clearly $\underset{\sim}{C}\cap\underset{\sim}{A}$ is necessarily disjoint to B_1 by the definition of W: if $\zeta<\omega_1, q\in P_\alpha/G_\beta$, and $q\Vdash_{P_\alpha/G_\beta}$ "$\zeta\in\underset{\sim}{C}\cap\underset{\sim}{A}\cap\underset{\sim}{B}_1$", then $N_\zeta\in W$ is defined, q_α is $(N_\zeta,P_\alpha/G_\beta)$-semi-generic, and $q_\alpha\Vdash_{P_\alpha/G_\beta}$ "$N_\zeta\cap\omega_1\in\underset{\sim}{A}$", contradicting "$N_\zeta\in W$" so $(*)$ holds.]

But $(*)$ contradicts $p\Vdash_{P_\alpha/G_\beta}$ "$\underset{\sim}{\Upsilon}\subseteq\mathfrak{B}^{\bar{Q}}$ is dense " as $B_1\in\mathfrak{B}^{P_\beta}\subseteq\mathfrak{B}^{\bar{Q}}$.

$\square_{4.9}$

4.10 Claim. Suppose $S\subseteq\omega_1$ is stationary, $\bar{Q}=\langle P_i,\underset{\sim}{Q}_j,\mathbf{t}_j,\underset{\sim}{\mathcal{A}}_i : i\le j, j<\alpha\rangle$ 3-o.k. sequence for W, S-suitable iteration, α limit ordinal and (for simplicity) $\mathrm{cf}(\alpha)=\aleph_0$ and let $\underset{\sim}{\mathcal{A}}_\alpha=\mathfrak{B}^{\bar{Q}}$ (a P_α-name). Let $\bar{\mathcal{A}}=\langle\underset{\sim}{\mathcal{A}}_i : i\le\alpha \text{ and } i\in\underset{\sim}{W}\rangle$. *Then:*

- $\otimes$ *if* λ is regular large enough, $N\prec(H(\lambda),\in,<^*_\lambda)$ countable and $\bar{Q},\lambda,p$ belong to N, $p\in P_\alpha\cap N$, $\beta\in\alpha\cap N$ a successor ordinal and $q\in P_\beta$ is (N,P_β)-semi generic, $p{\restriction}\beta\le q$ and $N\cap\omega_1\in S$ *then* there is an (N,P_α)-semi-generic $q'\in P_\alpha, q'{\restriction}\beta\ge q$ and $q'\Vdash_{P_\alpha}$ " for every dense open $\Upsilon\subseteq\mathrm{Seq}_u(\bar{\mathcal{A}})$ (see 4.5A(4)(ii), (9)) which belongs to $N[G_\alpha]$, for some $\langle A_i : i\in w\rangle\in N[G_\alpha]\cap\Upsilon$ we have $N\cap\omega_1\in\bigcap_{i\in w}A_i$".

4.10A Remark. 1) If "$\mathrm{cf}(\alpha)\ne\aleph_0$" we can still assume p forces $\mathrm{cf}(\alpha)=\aleph_0$ or p forces $\mathrm{cf}(\alpha)=\aleph_1$ or "α is inaccessible, $\bigwedge_{i<\alpha}|P_i|<\alpha$" and in the first case prove 4.10 with minimal changes.

2) Note that $Z^\alpha[\langle\underset{\sim}{\mathcal{A}}_i : i\le\alpha\rangle]$ is a subset of $\mathfrak{B}^{P_\alpha}$ extending $\mathfrak{B}^{\bar{Q}}$, 0 is not in it, *but* there is no reason for it to be closed under differences.

Proof. Standard, by now. Let $\langle\beta_\ell:\ell<\omega\rangle\in N$ be an increasing sequence of successor ordinals with $\beta_0=\beta$, $\bigcup_{\ell<\omega}\beta_\ell=\alpha$. Let $\bar{\underset{\sim}{\Upsilon}}=\langle\underset{\sim}{\Upsilon}_n : n<\omega\rangle$ list the sets $\underset{\sim}{\Upsilon}\in N$ which are P_α-names (forced to be) pre-dense subsets of $\mathrm{Seq}_u(\bar{\mathcal{A}})$. We choose by induction $p_n,q_n,N_n,\bar{a}^n,G_{\beta_n}$ such that:

(a) $G_{\beta_n}\subseteq P_{\beta_n}$ generic over V, $G_{\beta_n}\subseteq G_{\beta_{n+1}}$,

(b) $N_0=N, p_0=p, q_0=q$,

(c) $N_n\le_{\omega_2}N_{n+1}, N_n\prec(H(\lambda),\in,<^*_\lambda)$ and $N_n\in V[G_{\beta_n}]$,

(d) $p_n \leq p_{n+1}, p_n \in N \cap P_\alpha/G_{\beta_n}$,

(e) $q_n \in G_{\beta_n} \subseteq P_{\beta_n}$ is (N_n, P_{β_n})-semi-generic,

(f) $p_n \restriction \beta_n \leq q_n$ (in P_{β_n}),

(g) $\bar{a}^n = \langle \underset{\sim}{a}^n_\zeta : \zeta \in w_n \rangle \in \mathrm{Seq}\,(\langle \mathcal{A}_i : i < \alpha \rangle) \cap \Upsilon_n$,

(h) $w_n \subseteq w_{n+1}$ and $a^{n+1}_\zeta \leq a^n_\zeta$ for $\zeta \in w_n$,

(i) $\{\beta_\ell : \ell < n\} \subseteq w_n$,

(j) $N_n \cap \omega_1 \in \underset{\sim}{a}^n_{\beta_n}$ that is $q_n \cup p_n \restriction [\beta_n, \alpha)$ forces this.

The induction step is by 4.1 (and 4.5D(9)). As we are using RCS iteration, this suffices (i.e. we can make the G_{β_n} disappear).
The details are left to the reader. This induction suffices as we can use RCS iteration, so we can find q' as required. $\square_{4.10}$

XIV. Iterated Forcing with Uncountable Support

§0. Introduction

This chapter is [Sh:250], revised. Here we consider revised support for the not necessarily countable case. In §1 we define and present the basic properties of κ-RS iterated. This includes the case $\kappa = \aleph_1$ and so it can serve as a substitute to X §1. The main difference is that here we have to use names which sometimes have no value as we cannot use rank as there.

Unlike Chapter X, we do not have a useful properness to generalize, so naturally the generalizations of completeness are in the center. In 2.1 we introduce, and in 2.4 we show it does not matter much if we use the version with games of length $\kappa = \text{cf}\kappa$ or the version with a side order $\leq_0$, the "pure" extension which is κ-complete. Then we define iterations of such forcing notions and prove the basic properties (2.5–2.8). This repeats §1, so against dullness this time we waive the associativity law and simplify somewhat the definition of the iteration. In the definition of the order except finitely many places (which are names) the extensions are pure (i.e. $\leq_0$) in the old places. The first use of "pure" extensions is Prikry [Pr], and the first use of iterations with the distinction between old and new places (in normal support of course) is Gitik [Gi] which uses Easton support iteration $\bar{Q}$'s for high inaccessibles, each Q_i is $(\{2\}, \kappa_i)$-complete where for the important i's $\kappa_i = i$; a subsequent proof more similar to our case is [Sh:276, §1]. The application we have in mind is

$\kappa = \mu^+$, $\{\theta : \aleph_0 < \theta = \mathrm{cf}(\theta) \leq \kappa\} \subseteq S \subseteq \{2, \aleph_0\} \cup \{\theta : \theta = \mathrm{cf}(\theta) \leq \kappa\}$, and we shall iterate the forcing notion Q_i which has $(S, < \kappa)$-Pr, (so necessarily the cardinals $\theta \leq \kappa$ remain cardinals), the iteration being κ-Sp$_2$-iteration, and characteristically the length of iteration is some quite large cardinal λ, $i < \lambda \Rightarrow |P_i| < \lambda$, and we collapse all cardinals $> \kappa$, $< \lambda$ (so μ, κ play the role of $\aleph_0$, $\aleph_1$ in Chapter X). So we need to know of such iterations of forcing notions having $(S, < \kappa)$-Pr$_1$, which is done in 2.9. We could also deal similarly with iterations $\bar{Q}$ of length λ, λ strongly inaccessible $[i < \lambda \Rightarrow |P_i| < \lambda]$ and $S \subseteq \kappa$ unbounded in λ. In 2.18, 2.19 we look at the case essentially cofinalities are preserved (i.e. no $\theta = \mathrm{cf}(\theta) \geq \kappa$ becomes of cardinality $< \kappa$).

In the third section we indicate what forcing axioms we can get (3.4), and show how e.g. Mathias forcing fits in assuming $\mathrm{MA}_{<\kappa}$ (in 3.3). We then give a solution of the first Abraham problem (3.5).

In the fourth section we show how to fit Sacks forcing. The last section is a real application- to the second Abraham problem. In it we consider a forcing e.g. preserving $\theta \leq \kappa$, making the cofinality of κ^+ to $\aleph_0$, assuming only a weak form of "on κ^+ there is a large ideal" in which there the ideal disappears.

§1. κ-Revised Support Iteration

$\mathcal{D}_\kappa$ is the closed unbounded filter on κ.

A work of Groszek and Jech (see [J86] deal with making the continuum large (in a different way and effect, done about the same time independently).

1.1 Definition. Here κ is an infinite cardinal, but when it is an infinite ordinal which is not a cardinal we mean $|\kappa|^+$ (this is intended just for the case κ is collapsed during the iteration). We define the following notions and those in 1.2 and prove 1.4 by simultaneous induction on α:

(A) $\bar{Q} = \langle P_i, \underset{\sim}{Q}_i : i < \alpha \rangle$ is a κ-RS interaction (RS stands for revised support)

(B) a $\bar{Q}$-named ordinal

(C) a $\bar{Q}$-named atomic condition $\underset{\sim}{q}$, and we define $\underset{\sim}{q}\restriction\xi, \underset{\sim}{q}\restriction\{\xi\}$ for a $\bar{Q}$-named atomic condition $\underset{\sim}{q}$ and ordinal ξ when $\xi \le \alpha$, $\xi < \alpha$ respectively and $\underset{\sim}{q}\restriction[\zeta,\xi)$ when $\zeta < \xi \le \alpha$.

(D) the κ-RS limit of $\bar{Q}$, $\text{Rlim}_\kappa\bar{Q}$ which satisfies $P_i \lessdot \text{Rlim}_\kappa\bar{Q}$ for every $i < \alpha$ and we define $p\restriction\beta \in P_\beta$ for $p \in \text{Rlim}_\kappa\bar{Q}$ and $\beta \le \alpha$ (We may omit κ if clear from the context).

Let us define and prove

(A) We define "$\bar{Q}$ is a κ-RS iteration"

$\alpha = 0$: no condition.

α is limit: $\bar{Q} = \langle P_i, \underset{\sim}{Q}_i : i < \alpha\rangle$ is a κ-RS iteration iff for every $\beta < \alpha, \bar{Q}\restriction\beta$ is one.

$\alpha + 1 : \bar{Q}$ is an κ-RS iteration iff $\bar{Q}\restriction\beta$ is one, $P_\beta = \text{Rlim}_\kappa(\bar{Q}\restriction\beta)$, and $\underset{\sim}{Q}_\beta$ is a P_β-name of a forcing notion.

(B) We define "$\underset{\sim}{\xi}$ is a $\bar{Q}$-named ordinal". It means:

(1) $\underset{\sim}{\xi}$ is a function, $\text{Rang}(\underset{\sim}{\xi}) \subseteq \text{Ord}$.

(2) for $r \in \text{Dom}(\underset{\sim}{\xi})$, letting $\beta = \underset{\sim}{\xi}(r)$, we have $\beta < \alpha$, and $r \in P_\beta * \underset{\sim}{Q}_\beta$. (see an identification later).

(3) for every $r_1, r_2 \in \text{Dom}(\underset{\sim}{\xi})$, if r_1, r_2 are compatible, then $\underset{\sim}{\xi}(r_1) = \underset{\sim}{\xi}(r_2)$.

[What do we mean by "r_1, r_2 are compatible"? Let $r_1 \in P_{\beta_1} * \underset{\sim}{Q}_{\beta_1}$ and $r_2 \in P_{\beta_2} * \underset{\sim}{Q}_{\beta_2}$. If $\beta_1 = \beta_2$, there is no problem in defining compatibility. Otherwise, without loss of generality $\beta_1 < \beta_2$. Then, as noted in 1.4, $P_{\beta_1} * \underset{\sim}{Q}_{\beta_1}$ is essentially the same as P_{β_1+1} and $P_{\beta_1+1} \lessdot P_{\beta_2} \lessdot P_{\beta_2} * \underset{\sim}{Q}_{\beta_2}$, so we can test compatibility in $P_{\beta_2} * \underset{\sim}{Q}_{\beta_2}$].

1.1A Remark. For α a limit ordinal, $\text{Dom}(\underset{\sim}{\xi})$ is essentially a subset of $\bigcup_{\beta<\alpha} P_\beta$, so $\underset{\sim}{\xi}$ is a "partial name" for an ordinal. Note that $\text{Dom}(\underset{\sim}{\xi})$ is not necessarily pre-dense (there is no point in requiring it to be pre-dense in $\bigcup_{\beta<\alpha} P_\beta$, since this will not imply pre-density in $P_\alpha = \text{Rlim}_\kappa\bar{Q}$, which is the forcing we are interested in).

Continuation of 1.1.

(C) We say "$\underset{\sim}{q}$ is a $\bar{Q}$-named atomic condition" if:

(1) $\underset{\sim}{q}$ is a pair of functions $(\underset{\sim}{\zeta}_q, \mathrm{cnd}_q)$ with common domain $D = D_q$.

(2) $\underset{\sim}{\zeta}_q$ is a $\bar{Q}$-named ordinal.

(3) if $r_1, r_2 \in D_q$ and r_1, r_2 are compatible (see above) then $\mathrm{cnd}_q(r_1) = \mathrm{cnd}_q(r_2)$.

(4) if $r \in D_q$, letting $\beta = \underset{\sim}{\zeta}_q(r)$, we have:

$$r \restriction \beta \Vdash \text{"}\mathrm{cnd}_q(r) \in \underset{\sim}{Q}_\beta\text{"}$$

(note: we can add: it is forced ($\Vdash_{P_\beta}$) that $Q_\beta \models$ "$r(\beta) \leq \mathrm{cnd}_q(r)$" with little subsequent change). We define $\underset{\sim}{q} \restriction \xi$ as $(\underset{\sim}{\zeta}_q \restriction D_1, \mathrm{cnd}_q \restriction D_1)$ where $D_1 = \{p \in D_q : \underset{\sim}{\zeta}_q(p) < \xi\}$. We define $\underset{\sim}{q} \restriction \{\xi\}$ as $(\underset{\sim}{\zeta}_q \restriction D_2, \mathrm{cnd}_q \restriction D_2)$ where $D_2 = \{p \in D_q : \underset{\sim}{\zeta}_q(p) = \xi\}$, and $\underset{\sim}{q} \restriction [\zeta, \xi)$ similarly.

1.1B Remark. The definition would become simpler if we demand $r \in P_\beta$ instead of $r \in P_\beta * Q_\beta$ in (B2). (e.g. we could then drop the clause "$r(\beta) \leq \mathrm{cnd}_q(r)$" in (4)). However, we need this more complicated definition if we want associativity i.e. 1.5(3):

Consider a $\kappa - RS$ iteration $\bar{Q} = \langle P_\alpha, Q_\alpha : \alpha < \alpha^* \rangle$. Then a condition in P_{α^*} could be of the form $p = (\{(r, \beta)\}, \{(r, \underset{\sim}{q})\})$ with $r \in P_\beta$, $\underset{\sim}{q} \in \underset{\sim}{Q}_\beta$. Now assume that $\langle \alpha(\xi) : \xi \leq \xi^* \rangle$ is an increasing continuous sequence with $\alpha(0) = 0, \alpha(\xi^*) = \alpha^*$, and $\alpha(\xi) < \beta < \beta + 1 < \alpha(\xi + 1)$. Then in the natural "isomorphic copy" of p in $P'_{\xi^*} = \mathrm{Rlim}_\kappa \bar{Q}'$, where

$$\bar{Q}' = \langle P_{\alpha(\xi)}, P_{\alpha(\xi+1)}/P_{\alpha(\xi)} : \xi < \xi^* \rangle = \langle P'_\xi, \underset{\sim}{Q}'_\xi : \xi < \xi^* \rangle$$

β would become ξ (as $\underset{\sim}{q}$ correspond to an element of $\underset{\sim}{Q}'_\xi$). However, r may not be in P'_ξ but can only be found in $P'_\xi * \underset{\sim}{Q}'_\xi$. However this is mainly an aesthetic problem– saving here costs us some cumbersomeness in application, but no real damage: when we prove statements on iteration $\bar{Q}$ we cannot restrict ourselves to length $\alpha = 0, 1, 2$, or $\alpha = \mathrm{cf}(\alpha)$ etc. For diversity, we do use this way in §2.

(D) We define $\mathrm{Rlim}_\kappa \bar{Q}$ as follows:

if $\alpha = 0$: $\mathrm{Rlim}_\kappa \bar{Q}$ is a trivial forcing with just two compatible conditions i.e. $\mathrm{Rlim}_\kappa \bar{Q} = \{\emptyset, \{(\emptyset, \emptyset)\}\}$ say with $\emptyset \le \{(\emptyset, \emptyset)\}$

if $\alpha > 0$: we call q an atomic condition of $\mathrm{Rlim}_\kappa \bar{Q}$, if it is a $\bar{Q}$-named atomic condition.

The set of conditions in $\mathrm{Rlim}_\kappa \bar{Q}$ is

$$\begin{aligned}&\{p : p \text{ a set of } \lambda \text{ atomic conditions for some } \lambda < \kappa;\\ &\quad \text{and for every } \beta < \alpha, p{\restriction}\beta \stackrel{\text{def}}{=} \{q{\restriction}\beta : q \in p\} \in P_\beta,\\ &\quad \text{and } p{\restriction}\beta \Vdash_{P_\beta} \text{“the set } \{q{\restriction}\{\beta\} : q \in p\} \text{ has an upper bound in } \underset{\sim}{Q}_\beta\text{”}\}\end{aligned}$$

More precisely, the last condition in the previous paragraph means

$$\begin{aligned}p{\restriction}\beta \Vdash_{P_\beta} \text{“}&\exists q_0 \in \underset{\sim}{Q}_\beta \forall \underset{\sim}{q} \in p \forall r \in D_q :\\ &\text{if } \zeta_q(r) = \beta \text{ and } r{\restriction}\beta \in \underset{\sim}{G}_{P_\beta} \text{ then}\\ &q_0 \Vdash_{\underset{\sim}{Q}_\beta} [\text{if } r(\beta) \in \underset{\sim}{G}_{Q_\beta} \text{then } \mathrm{cnd}_{\underset{\sim}{q}}(r) \in \underset{\sim}{G}_{Q_\beta}]\text{”}\end{aligned}$$

(where $\underset{\sim}{G}_P$ is the canonical name for the generic set for P).

Remember that we have defined $p{\restriction}\beta = \{q{\restriction}\beta : q \in p\}$ and $p{\restriction}[\beta, \gamma)$ for $\beta < \gamma \le \alpha$, similarly.

The order: $p_0 \le p_1$ iff $p_0 \subseteq p_1$ or just $p_0 \subseteq \{\underset{\sim}{q}{\restriction}\beta : \underset{\sim}{q} \in p_1 \text{ and } \beta \le \alpha\}$.

The identification. Clearly for $\beta < \alpha$, we have $P_\beta \subseteq P_\alpha$. We can identify $P_\beta * \underset{\sim}{Q}_\beta$ with a subset of P_α when $\beta + 1 = \alpha$: $(p, \underset{\sim}{q})$ is identified with $p \cup \{[\underset{\sim}{q}]\}$ where $[\underset{\sim}{q}] = (\zeta, \mathrm{cnd})$, $\mathrm{Dom}(\underset{\sim}{\zeta}) = \{\emptyset\}$($\emptyset$ the empty condition of P_β), $\underset{\sim}{\zeta}(\emptyset) = \beta$ and $\mathrm{cnd}(\emptyset) = \underset{\sim}{q}$.

It is easy to check the demands, e.g. under this identification $P_\beta * \underset{\sim}{Q}_\beta$ is a dense subset of $P_{\beta+1}$.

Now we have to show $P_\beta \lessdot \mathrm{Rlim}_\kappa \bar{Q}$ (for $\beta < \alpha$). Note that any $\bar{Q}{\restriction}\beta$-named ordinal (or condition) is a $\bar{Q}$-named ordinal (or condition), and see Claim 1.4(1) below.

1.1C Remark. Note that for the sake of 1.5(3) we allow κ to be not a cardinal and then we really use $|\kappa|^+$.

1.1D Remark. We can obviously define $\bar{Q}$-named sets; but for condition (and ordinals for them) we want to avoid the vicious circle of using names which are interpreted only after forcing with them below.

1.2 Definition.

(1) Suppose $\bar{Q}$ is a κ-RS iteration, $\underset{\sim}{\zeta}$ is a $\bar{Q}$-named ordinal, $\alpha = \ell g(\bar{Q})$, $G \in \mathrm{Gen}(\bar{Q})$ (see part (2) of the Definition below). We define $\underset{\sim}{\zeta}[G]$ by:

(i) $\underset{\sim}{\zeta}[G] = \gamma$ if ($\gamma < \alpha \overset{\text{def}}{=} \ell g(\bar{Q})$ and) for some $p \in \mathrm{Dom}(\underset{\sim}{\zeta}) \cap G_{\gamma+1}$ which is in $P_\gamma * \underset{\sim}{Q}_\gamma$ we have $\underset{\sim}{\zeta}(p) = \gamma$.

(ii) otherwise (i.e., $G \cap \mathrm{Dom}(\underset{\sim}{\zeta}) = \emptyset$) $\underset{\sim}{\zeta}[G]$ is not defined.

(1A) For a $\bar{Q}$-named condition $\underset{\sim}{q}$, we defined $\underset{\sim}{q}[G]$ similarly.

(2) We denote the set of $G \subseteq \bigcup_{i<\alpha} P_{i+1}$ such that $G \cap P_{i+1}$ is generic over V for each $i < \alpha$ by $\mathrm{Gen}(\bar{Q})$. We let $G_i = G \cap P_i$.

(3) For $\underset{\sim}{\zeta}$ a $\bar{Q}$-named ordinal and $q \in \bigcup_{i<\alpha} P_i$ let $q \Vdash_{\bar{Q}}$ "$\underset{\sim}{\zeta} = \xi$" if for every $G \in \mathrm{Gen}(\bar{Q})$ we have: $q \in G \Rightarrow \underset{\sim}{\zeta}[G] = \xi$, i.e. if $q \Vdash$ "$\underset{\sim}{\xi}[G]$ is defined and equal to ξ". Similarly for $p \Vdash_{\bar{Q}}$ "$\underset{\sim}{q} = r$" and for $p \Vdash_{\bar{Q}}$ "$q \in \underset{\sim}{G}$".

1.3 Remark. 1) From where is G taken in (2), (3)? E.g., V is a countable model of set theory, G taken from the "true" universe.
2) If $p, p' \in P_\alpha, p \subseteq p'$ and for all $(\underset{\sim}{\xi}, \mathrm{cnd}) \in p' \setminus p$ there is a $\beta < \alpha$ such that $p{\restriction}\beta \Vdash$ "$\underset{\sim}{\xi}[G]$ undefined", then p and p' are essentially equivalent, i.e. for all $q \supseteq p$ we have: q and p' are compatible; or equivalently, $p \Vdash$ "$p' \in G$".

Now we point out some properties of κ-RS iteration.

1.4 Claim. Let $\bar{Q} = \langle P_i, \underset{\sim}{Q}_i : i < \alpha \rangle$ be a κ-RS iteration, $P_\alpha = \mathrm{Rlim}_\kappa \bar{Q}$.

(1) If $\beta < \alpha$ then: $P_\beta \subseteq P_\alpha$ and $p \in P_\beta \Rightarrow p{\restriction}\beta = p$ & $P_\alpha \models p{\restriction}\beta \leq p$; for $p_1, p_2 \in P_\beta$ we have $[P_\beta \models p_1 \leq p_2$ iff $P_\alpha \models p_1 \leq p_2]$; and $P_\beta \lessdot P_\alpha$.

Moreover, if $q \in P_\beta, p \in P_\alpha$, then q, p are compatible in P_α iff $q, p\restriction\beta$ are compatible in P_β; if $p\restriction\beta \le q$, a least upper bound of $q \in P_\beta$, $p \in P_\alpha$ is $q \cup p\restriction[\beta, \alpha)$. Also for $\beta_1 \le \beta_2 < \alpha$ and $p \in P_\alpha$ then $p\restriction\beta_1 = (p\restriction\beta_2)\restriction\beta_1$.

(2) If $\underset{\sim}{\zeta}$ is a $\bar{Q}$-named ordinal and $G, G' \in \mathrm{Gen}(\bar{Q})$ and $G \cap P_{\xi+1} = G' \cap P_{\xi+1}$ and $\underset{\sim}{\zeta}[G] = \xi$ then $\underset{\sim}{\zeta}[G'] = \xi$; hence we write $\underset{\sim}{\zeta}[G \cap P_{\xi+1}] = \xi$.

(3) If $\underset{\sim}{\beta}, \underset{\sim}{\gamma}$ are $\bar{Q}$-named ordinals, then Max $\{\underset{\sim}{\beta}, \underset{\sim}{\gamma}\}$ (for a generic $G \in \mathrm{Gen}(\bar{Q})$, this name is defined if both are defined and its value is the maximum) is a $\bar{Q}$-named ordinal. Also Min$\{\underset{\sim}{\beta}, \underset{\sim}{\gamma}\}$ (defined if at least one of them is defined, if only one is defined the value is its value, if both are defined the value is the minimum).

(4) If $\alpha = \beta_0 + 1$, in Definition 1.1(D), in defining the set of elements of P_α, in the demand "$\beta < \alpha \Rightarrow p\restriction\beta \in P_\beta$", we can restrict ourselves to $\beta = \beta_0$.

(5) The following set is dense in P_α : $\{p \in P_\alpha$; for every $\beta < \alpha$, if $r_1, r_2 \in p$, then $\Vdash_{P_\beta}$ "if $r_1\restriction\{\beta\} \ne \emptyset, r_2\restriction\{\beta\} \ne \emptyset$ then they are equal" $\}$ where $Y \subseteq P_\alpha$ is dense iff for every $p \in P$ there is q, $p \le q$ and q is equivalent to some $q' \in Y$ (i.e. $q \Vdash$ "$q' \in \underset{\sim}{G}$" and $q' \Vdash$ "$q \in \underset{\sim}{G}$") (can even we $\Vdash_{\bar{Q}\restriction\beta}$).

(6) $|P_\alpha| \le (\prod_{i<\alpha} 2^{|P_i|})^{<\kappa}$ for limit α (where $|P|$ is the number of elements of P up to equivalence). Also if $\beta < \alpha \Rightarrow \mathrm{density}(P_\alpha) \le \lambda = \mathrm{cf}(\lambda)$ and $\alpha \le \lambda$ (or just $\alpha < \lambda^+$) *then* $\mathrm{density}(P_\alpha) \le 2^\lambda$.

(7) If $\Vdash_{P_i}$ "$|\underset{\sim}{Q}_i| \le \lambda$ & $\underset{\sim}{Q}_i \subseteq V$" (and $\lambda \ge 2$), then (essentially) $|P_{i+1}| \le \lambda^{|P_i|}$. (Why "essentially"? We have to identify P_i-names of members of Q_i which $\Vdash_{P_i}$ "they are equal".) We can replace $|P_i|$ by $\mathrm{density}(P_i)$ and get $\mathrm{density}(P_{i+1}) \le \mathrm{density}(P_i) + \lambda + \aleph_0$. Instead of "$\underset{\sim}{Q}_i \subseteq V$" it suffices that: $\lambda^{<\mu} = \lambda$ and: Q (i.e. set of members) is included in the closure of V under taking subsets of power $< \mu$.

(8) Suppose $\bar{Q}$ is an κ-RS iteration, $\varphi(x, y)$ is a formula (possibly with parameters from V) such that:

 (a) for every $G \in \mathrm{Gen}(\bar{Q})$ there is at most one x such that $(V[G], V, G) \models$ "$\varphi(x, G)$", this x is called $\underset{\sim}{x}[G]$ if there is such x, and $\underset{\sim}{x}[G]$ is not defined otherwise.

(b) if $G \in \text{Gen}(\bar{Q})$, $\underset{\sim}{x}[G]$ defined then it is an ordinal $< \ell g(\bar{Q})$, call it β, moreover for some $p \in P_{\beta+1}$, such that if $\beta + 1 = \ell g(\bar{Q})$ then $p \in P_\beta * \underset{\sim}{Q}_\beta$, we have:

$[p \in G' \in \text{Gen}(\bar{Q}) \Rightarrow \underset{\sim}{x}[G'] = \beta]$.

Then there is a $\bar{Q}$-named ordinal $\underset{\sim}{\zeta}$ such that: for every $G \in \text{Gen}(\bar{Q})$, $\underset{\sim}{x}[G] = \underset{\sim}{\zeta}[G]$ (i.e. they are both defined with the same value, or they are both undefined).

(9) Suppose $\bar{Q}, \varphi(x,y), \underset{\sim}{x}[G]($ for $G \in \text{Gen}(\bar{Q}))$ are as in (8) except that clause (b) is replaced by:

(b)′ if $G \in \text{Gen}(\bar{Q})$ and $\underset{\sim}{x}[G]$ is defined, then it has the form $(\zeta, p), \zeta < \ell g(\bar{Q}), p \in \underset{\sim}{Q}_\zeta[G_\zeta]$ and for some q we have: $q \in G \cap P_{\zeta+1}$, and $[\zeta + 1 = \ell g(\bar{Q}) \Rightarrow q \in G \cap P_\zeta]$ (and if we make the addition in 1.1(C) clause (4) then $\underset{\sim}{Q}_\zeta[G \cap P_\zeta] \models$ "$q{\restriction}\{\zeta\} \leq p$") and $[q \in G' \in \text{Gen}(\bar{Q}) \Rightarrow \underset{\sim}{x}[G] = \underset{\sim}{x}[G']]$.

Then there is a $\bar{Q}$-named condition $\underset{\sim}{q}$ such that:

for every $G \in \text{Gen}(\bar{Q})$, $\underset{\sim}{x}[G] = \langle \underset{\sim}{\zeta}_q[G], \underset{\sim}{q}[G] \rangle$ (so both are defined and equal or both are not defined).

Proof: By induction on α.

1.4A Remark. The inverse of 1.4(8) and of 1.4(9) hold, of course.

1.5 Lemma. *The Iteration Lemma*

(1) Suppose F is a function, *then* for every ordinal α there is one and only one κ-RS iteration $\bar{Q} = \langle P_i, \underset{\sim}{Q}_i : i < \alpha^\dagger \rangle$ such that:

(a) for every $i, \underset{\sim}{Q}_i = F(\bar{Q}{\restriction}i)$,

(b) $\alpha^\dagger \leq \alpha$,

(c) either $\alpha^\dagger = \alpha$ or $F(\bar{Q})$ is not an $(\text{Rlim}_\kappa \bar{Q})$-name of a forcing notion.

(2) Suppose $\bar{Q}$ is a κ-RS-iteration, $\alpha = \ell g(\bar{Q}), \beta < \alpha, G_\beta \subseteq P_\beta$ is generic over V. *Then* in $V[G_\beta], \bar{Q}/G_\beta = \langle P_i/G_\beta, \bar{Q}_i : \beta \leq i < \kappa \rangle$ is a κ-RS-iteration and $\text{Rlim}_\kappa \bar{Q} = P_\beta * (\text{Rlim}_\kappa \bar{Q}/G_\beta)$ (essentially).

(3) The Association Law: If $\alpha_\xi = \alpha(\xi)$ $(\xi \leq \xi^*)$ is increasing and continuous, $\alpha_0 = 0; \bar{Q} = \langle P_i, \underset{\sim}{Q}_i : i < \alpha^* \rangle$ is a κ-RS-iteration, $P_{\alpha^*} = \mathrm{Rlim}_\kappa \bar{Q}$ and $\alpha_{\xi^*} = \alpha^*$ and κ is a successor cardinal (for κ inaccessible we need to assume more); *then* so are $\langle P_{\alpha(\xi)}, P_{\alpha(\xi+1)}/P_{\alpha(\xi)} : \xi < \xi^* \rangle$ and $\langle P_i/P_{\alpha(\xi)}, \underset{\sim}{Q}_i : \alpha(\xi) \leq i < \alpha(\xi+1) \rangle$ (with κ-RS-Limits $P_{\alpha_{\xi^*}}$ and $P_{\alpha(\xi+1)}/P_{\alpha(\xi)}$ respectively) and vice versa.

1.5A Remark. In (3) we can use α_ξ's which are names.

Proof: (1) Easy.

(2) Pedantically, we should formalize the assertion as follows:

(*) There are function F_0, F_1(= definable classes) such that for every κ-RS-iteration $\bar{Q}$ with $\ell g(\bar{Q}) = \alpha$, and $\beta < \alpha$, $F_0(\bar{Q}, \beta)$ is a P_β-name $\underset{\sim}{\bar{Q}}^\dagger$ such that:

(a) $\Vdash_{P_\beta}$ "$\bar{Q}^\dagger$ is a κ-RS-iteration of length $\alpha - \beta$".

(b) $P_\beta * (\mathrm{Rlim}_\kappa \underset{\sim}{\bar{Q}}^\dagger)$ is equivalent to $P_\alpha = \mathrm{Rlim}_\kappa \bar{Q}$, by $F_1(\bar{Q}, \beta)$ (i.e., $F_1(\bar{q}, \beta)$ is an isomorphism between the corresponding completions to Boolean algebras)

(c) if $\beta \leq \gamma \leq \alpha$ then $\Vdash_{P_\beta}$ "$F_0(\bar{Q}\restriction\gamma, \beta) = F_0(\bar{Q}, \beta)\restriction(\gamma - \beta)$" and $F_1(\bar{Q}, \beta)$ extends $F_1(\bar{Q}\restriction\gamma, \beta)$ and $F_1(\bar{Q}\restriction\gamma, \beta)$ transfers the P_γ-name $\underset{\sim}{Q}_\gamma$ to a P_β-name of $(\mathrm{Rlim}_\kappa(\underset{\sim}{\bar{Q}}^\dagger\restriction(\gamma-\beta))$-name of $\underset{\sim}{Q}^\dagger_{\gamma-\beta}$ (where $\underset{\sim}{Q}^\dagger_{\gamma-\beta} = \mathrm{Rlim}_\kappa \langle \underset{\sim}{Q}^\dagger_{\beta+i} : i < \gamma - \beta \rangle$).

The proof is induction on α, and there are no special problems.

(3) Again, pedantically the formulation is: There are functions F_3, F_4 such that

(*) For $\bar{Q}$-iteration, $\ell g(\bar{Q}) = \alpha_{\xi^*}, \bar{\alpha} = \langle \alpha_\xi : \xi \leq \xi^* \rangle$ increasing continuous, $F_3(\bar{Q}, \bar{\alpha})$ is a κ-RS-iteration $\bar{Q}^\dagger$ of length α_{ξ^*} such that

(a) $F_4(\bar{Q}, \bar{\alpha})$ is an equivalence of the forcing notions $\mathrm{Rlim}_\kappa \bar{Q}, \mathrm{Rlim}_\kappa \bar{Q}^\dagger$.

(b) $F_3(\bar{Q}\restriction\alpha_\xi, \alpha\restriction(\xi+1)) = F_3(\bar{Q}, \bar{\alpha})\restriction\xi$

(c) $\underset{\sim}{Q}^\dagger_\xi$ is the image by $F_4(\bar{Q}\restriction\alpha_\xi, \bar{\alpha}\restriction(\xi+1))$ of the $P_{\alpha_\xi} = \mathrm{Rlim}_\kappa(\bar{Q}\restriction\alpha_\xi)$-name $F_0(\bar{Q}\restriction\alpha_{\xi+1}, \alpha_\xi)$.

The proof is tedious but straightforward. $\square_{1.5}$

1.6 Claim. Suppose we add in Definition 1.1(B) also:

1.1(B)(4) if α is inaccessible $\geq \kappa$, and for some $\beta < \alpha$ for every γ satisfying $\beta \leq \gamma < \alpha$ we have $\Vdash_{P_\beta}$ "$|P_\gamma/P_\beta| < \alpha$" *then* $(\exists \beta < \alpha)[\mathrm{Dom}(\underset{\sim}{\zeta}) \subseteq P_\beta]$.

Then nothing changes in the above, and if λ is an inaccessible cardinal $> \kappa$ and $|P_i| < \lambda$ for every $i < \lambda$ and $\bar{Q} = \langle P_i, \underset{\sim}{Q}_i : i < \lambda \rangle$ is an RS_κ-iteration, then

(1) every $\bar{Q}$-named ordinal is in fact a $(\bar{Q} \restriction i)$-named ordinal for some $i < \lambda$,

(2) like (1) for $\bar{Q}$-named conditions.

(3) $P_\lambda = \bigcup_{i<\lambda} P_i$.

(4) if λ is a Mahlo cardinal *then* P_λ satisfies the λ-c.c. (in a strong way).

1.6A Remark. As in XI §1, actually if "$\theta = \mathrm{cf}(\theta) \geq \kappa$" is preserved by every P_α for $\alpha < \alpha^*$, then: $\alpha < \alpha^* \;\&\; \mathrm{cf}(\alpha) = \theta$ implies $\bigcup_{\beta<\alpha} P_\beta$ is dense in P_α. In this case, if α^* is strongly inaccessible $> \theta$ and $[\alpha < \alpha^* \Rightarrow \mathrm{density}(P_\alpha) < \alpha^*]$ *then* P_{α^*} satisfies the α^*-c.c.

§2. Pseudo-Completeness

We think here of replacing $\aleph_1$ by, say, κ^+. So we want to deal with forcing notions not collapsing any cardinal $\leq \kappa^+$, but possibly collapsing κ^{++}, and possibly adding reals and changing the cofinality of κ^{++} to say $\aleph_0$. So on the one hand we want to have support $\leq \kappa$, and even a κ^+-RS; and on the other hand some amount of pseudo completeness (expressed in Definition 2.1 below). Further consideration lead to finite pure.

We deal with forcing notions Q satisfying:

2.1 Definition. Let γ be an ordinal, $S \subseteq \{2\} \cup \{\lambda : \lambda \text{ a regular cardinal }\}$.

1) Now Q satisfies (S, γ)-Pr_1 if:

(i) $Q = (|Q|, \leq, \leq_0)$ (here $|Q|$ is the set of elements of Q)

(ii) as a forcing notion, Q is $(|Q|, \leq)$, with a least element $\emptyset_Q$

(iii) $\leq_0$ is a partial order (of $|Q|$).

(iv) $[p \leq_0 q \Rightarrow p \leq q]$

(v) *pure decidability: for every* cardinal $\theta \in S$ and Q-name $\underset{\sim}{\tau}$, such that $\Vdash_Q$ "$\underset{\sim}{\tau} \in \theta$" and $p \in Q$ *for some* $q \in Q$ and $\beta \in \theta$ we have: $p \leq_0 q$ and $q \Vdash_Q$ "if $\theta = 2$ then $\underset{\sim}{\tau} = \beta$ and if $\theta \geq \aleph_0$ then $\underset{\sim}{\tau} < \beta$"

(vi) for each $p \in Q$ in the following game player I has a winning strategy: for $i < \gamma$ player I chooses $p_{2i} \in Q$ such that $p \leq_0 p_{2i} \wedge \bigwedge_{j<2i} p_j \leq_0 p_{2i}$ and then player II chooses $p_{2i+1} \in Q$ such that $p_{2i} \leq_0 p_{2i+1}$.

Player I loses if he has some time no legal move, which can occur in limit stages only.

2) $Q = (|Q|, \leq, \leq_0)$ satisfies (S,γ)-Pr_1^+, if (i)-(v) hold and $(Q, \leq_0)$ is γ-complete (i.e. if $p_i \in Q$ for $i < \beta < \gamma$, and $i < j < \beta \Rightarrow p_i \leq_0 p_j$ *then* for some $p \in Q$ we have: $i < \beta \Rightarrow p_i \leq_0 p$).

3) A forcing notion $(Q, \leq)$ satisfies (S,γ)-Pr_1 (or (S,γ)-Pr_1^+), if there is a relation $\leq_0$ such that $(Q, \leq, \leq_0)$ satisfies (S,γ)-Pr_1 (or (S,γ)-Pr_1^+)).

4) Q satisfies $(S,\gamma) - \mathrm{Pr}_1^-$ or $S - \mathrm{Pr}_1^-$ if it satisfies (i) - (v) of part (1) (note: the ordinal γ does not appear in conditions (i)-(v) of 2.1(1)).

5) If a member of S is an infinite ordinal δ which is not a regular cardinal, we mean $\mathrm{cf}(\delta)$ (occurs e.g. when $Q \in V^P$ and $S \in V$).

6) If $Q = (|Q|, \leq, \leq_0)$ then $\hat{Q}$ is defined as follows:

the set of elements is $\{u : u \subseteq Q$, and if $u \neq \emptyset$ then for some $q \in u$, $(\forall p \in u)(\exists r \in u)(p \leq r \;\&\; q \leq_0 r)$ and there is $r^* \in Q$ such that for every such q, r^* is a $\leq_0$-upper bound of $\{r \in u : q \leq_0 r\}\}$,

the order $u_1 \leq u_2$ iff $u_1 = u_2$ or for some $q_2 \in u_2$ for every $q_1 \in u_1$, $q_1 \leq q_2$,

the pure order $u_1 \leq_0 u_2$ iff $u_1 = u_2$ or for some $q_2 \in u_2$ witnessing $u_2 \in \hat{Q}$ for every $q_1 \in u_1$ witnessing $u_1 \in \hat{Q}$ we have $(\forall p \in u_1)(\exists r \in u_1)[p \leq r \,\&\, q_1 \leq_0 r \leq_0 q_2]$ (this is naturally used in 2.7; we usually identify $p \in Q$ with $\{p\} \in \hat{Q}$).

2.2 Fact.

(1) If $\kappa < \gamma_1, \gamma_2 < \kappa^+$ *then* (S,γ_1)-Pr_1 is equivalent to (S,γ_2)-Pr_1.

(2) If $\kappa + 1 \leq \gamma < \kappa^+$ and $\square_\kappa$ (which can be stated as, i.e. an equivalent formulation is: there is a sequence $\langle C_\alpha : \alpha < \kappa^+\rangle, C_\alpha \subseteq \alpha$ closed, for

limit α the set C_α is unbounded in α and $[\alpha_1 \in C_\alpha \Rightarrow C_{\alpha_1} = C_\alpha \cap \alpha_1]$, and $\mathrm{cf}(\alpha) < \kappa \Rightarrow |C_\alpha| < \kappa)$ and Q satisfies (S,γ)-Pr_1 *then* Q satisfies (S,κ^+)-Pr_1.

(3) Assume Q satisfies (S,γ)-Pr_1. If $\lambda \leq \gamma$, and $\lambda \in S$ *then* in V^Q still λ is a regular cardinal (or at least $\Vdash_Q$ "$\mathrm{cf}(\lambda) = \mathrm{cf}^V(\lambda)$"). If $2 \in S$, *then* Q does not add bounded subsets to γ.

(4) If Q satisfies (S, γ)-Pr_1, $\lambda \in S$, λ regular, and for every regular μ, $\gamma \leq \mu < \lambda \Rightarrow \Vdash_Q$ "μ is not regular" (e.g., $[\gamma, \lambda)$ contains no regular cardinal) *then* λ is regular in V^Q.

(5) If Q satisfies (S, γ)-Pr_1, $\gamma \geq \omega + 1$, *then* Q is S-semiproper.

(6) Similar assertions (to 1-5) holds for (S, γ)-Pr_1^+ (but in (2) we do not need $\square_\kappa$) and (S,γ)-Pr_1^+ implies (S,γ)-Pr_1.

(7) In 2.1(6), $\leq^{\hat{Q}}$, $\leq_0^{\hat{Q}}$ are quasi orders of the set of elements of $\hat{Q}$ and for p, $q \in Q$ we have

(i) $Q \models p \leq q \Leftrightarrow \hat{Q} \models \{p\} \leq \{q\}$,

(ii) $Q \models p \leq_0 q \Leftrightarrow \hat{Q} \models \{p\} \leq_0 \{q\}$

(on incompleteness see inside 2.7(D)).

(8) Assume $Q = (|Q|, \leq, \leq_0)$ satisfies:

$$(*) \qquad p \leq q \leq r \;\&\; p \leq_0 r \Rightarrow p \leq_0 q$$

then in Definition 2.1(6):

if q', $q'' \in u_1$ witness $u_1 \in \hat{Q}$ *then* there is $q \in u_1$ such that $q' \leq_0 q \;\&\; q'' \leq_0 q$.

Also: if $q' \in u_1$ witness $u_1 \in \hat{Q}$ and $q' \leq q \in u_1$ then q witness $u_1 \in \hat{Q}$, provided that

$$(*)' \qquad p \leq q \leq r \;\&\; p \leq_0 r \Rightarrow q \leq_0 r.$$

(9) In definition 2.1(6), if $u_1 \neq u_2$ and q_2 witness $u_1 \leq_0 u_2$ then $u_1 \leq_0 q_2 \leq_0 u_2$ (or more formally $u_1 \leq_0 \{q_2\} \leq_0 u_2$) so $\{\{p\} : p \in Q\} \subseteq Q$ is dense.

Proof: Straightforward. E.g. for (2), note that $\mathrm{otp}(C_\delta) \le \kappa$ for all $\delta < \kappa^+$. Without loss of generality $\alpha \in C_\delta \Rightarrow \alpha$ is even (so $C_{2\alpha+1} = \emptyset$). So in stages α, player I can apply his strategy to the play $\langle p_\gamma, p_{\gamma+1} : \gamma \in C_\alpha \rangle$). $\square_{2.2}$

2.2A Remark. Concerning 2.2(2) note that $\square_{\aleph_0}$ always holds trivially (see [Sh:351, §4]). The equivalence of this formulation of square to the standard one is similar to the proof in [Sh:351, §4].

2.3 Definition. $(S, < \kappa)$-Pr_1 will mean $(S, \gamma) - \mathrm{Pr}_1$ holds for every $\gamma < \kappa$.

2.4 Fact. The following three conditions on a forcing notion Q, a set $S \subseteq \{2\} \cup \{\lambda : \lambda$ a regular cardinal $\}$ and regular κ are equivalent:

(a) there is $Q' = (Q', \le, \le_0)$ such that $(Q', \le)$, $(Q, \le)$ are equivalent forcing notions and Q' satisfies (S, κ)-Pr_1.

(b) for each $p \in Q$, in the following game (which lasts κ moves) player II has a winning strategy:

in the ith move player I chooses $\lambda_i \in S$ and a Q-name $\underset{\sim}{\tau}_i$ of an ordinal $< \lambda_i$, then player II chooses an ordinal $\alpha_i < \lambda_i$. In the end player II wins if for every $\alpha < \kappa$ there is $p_\alpha \in Q$, $p \le p_\alpha$ such that for every $i < \alpha$ we have $p_\alpha \Vdash$ "either $\lambda_i = 2$ & $\underset{\sim}{\tau}_i = \alpha_i$ or $\lambda_i \ge \alpha_0$ & $\underset{\sim}{\tau}_i < \alpha_i$".

(c) like (a) but moreover $(Q', \le_0)$ is κ-complete (i.e. Q' satisfies $(S, \kappa) - Pr_1^+$).

(d) like (a) but moreover $(Q' \le_0)$ is κ-directed complete, i.e.

if $B \subseteq Q$, $|B| < \kappa$ and for each finite $B' \subseteq B$ there is a $\le_0$-upper bound to B', *then* B has a $\le_0$-least upper bound.

Proof. $(d) \Rightarrow (c) \Rightarrow (a)$: trivial.

$(a) \Rightarrow (b)$: As Q, Q' are equivalent, there is a forcing notion P and $f : Q \to P$, $f' : Q' \to P$ both preserving $\le$ and incompatibility and with dense ranges. Choose $q \in Q'$ which essentially is above p i.e. $f'(q) \Vdash_P$ "$f(p) \in \underset{\sim}{G}_P$" . We describe a winning strategy (in the game from (b) of 2.4) for player II: he plays on the side a play (for q) of the game from 2.1(vi) for Q' where he uses a winning strategy (whose existence in guaranteed by (a)). In step i of the play (for 2.4(b)) he already has the initial segment $\langle p_j : j < 2i \rangle$ of the simulated play for 2.1(vi). If player I plays λ_i, $\underset{\sim}{\tau}_i$ in the actual game, player II defines $p_{2i} \in Q'$ (for player I) in the simulated play by the winning strategy of player I there and then he chooses $p_{2i+1}, p_{2i} \leq_0 p_{2i+1} \in Q'$, which force for some α_i: $\underset{\sim}{\tau}_i = \alpha_i$ if $\lambda_i = 2$, $\underset{\sim}{\tau}_i < \alpha_i$ if $\lambda_i \geq \aleph_0$ (exists by 2.1(v)) (more formally, for some $r_i \in Q$, we have $f'(p_{2i+1}) \Vdash_P$ "$f(r_i) \in \underset{\sim}{G}_P$" and r_i forces $\underset{\sim}{\tau}_i = \alpha_i$ & $\lambda_i = 2$ or $\underset{\sim}{\tau}_i < \alpha_i$ & $\lambda_i \geq \aleph_0$, alternatively we can interpret $\underset{\sim}{\tau}_i$ as a Q'-name using f, f') and then plays α_i in the actual play. In the end for $\alpha < \kappa$, there is $p^*_\alpha \in Q$ such that $f(p^*_{2\alpha}) \Vdash_P$ "$f'(p_{2\alpha}) \in \underset{\sim}{G}_P$", now p^*_α exists and is as required.

$(b) \Rightarrow (d)$: Fix a winning strategy St_p for player II in the game from 2.4(b) for each $p \in Q$. We define Q' as follows

$$Q' = \{(p, \langle \lambda_i, \underset{\sim}{\tau}_i, \alpha_i : i < \xi \rangle) : p \in Q, \text{ and } \langle \lambda_i, \underset{\sim}{\tau}_i, \alpha_i : i < \xi \rangle \text{ is an initial segment of a play of the game from 2.4(b) for } p \text{ in which player II uses his winning strategy } \mathrm{St}_p\}.$$

The order $\leq_0$ is:

$$(p, \langle \lambda_i, \underset{\sim}{\tau}_i, \alpha_i : i < \xi \rangle) \leq_0 (p', \langle \lambda'_i, \underset{\sim}{\tau}'_i, \alpha'_i : i < \xi' \rangle)$$

iff (both are in Q') and

$$p = p', \ \xi \leq \xi', \text{ and for } i < \xi$$
$$\lambda_i = \lambda'_i, \underset{\sim}{\tau}_i = \underset{\sim}{\tau}'_i, \ \alpha_i = \alpha'_i$$

and the order $\leq$ on Q' is

$$\mathbf{p} = (p, \langle \lambda_i, \underset{\sim}{\tau}_i, \alpha_i : i < \xi \rangle) \leq \mathbf{p}' = (p', \langle \lambda'_i, \underset{\sim}{\tau}'_i, \alpha'_i : i < \xi' \rangle)$$

iff (both are in Q' and) $\mathbf{p} \leq_0 \mathbf{p}'$ or $Q \models p \leq p'$, and $p' \Vdash_Q$ "$\lambda_i = 2$ & $\underset{\sim}{\tau}_i = \alpha_i$ or $\lambda_i \geq \aleph_0$ & $\underset{\sim}{\tau}_i < \alpha_i$" for every $i < \xi$.

The checking is easy. Note that

(α) the map $p \mapsto (p, \langle\rangle)$ is a dense embedding of $(Q, \leq)$ into $(Q', \leq)$.

(β) hence Q'-names are essentially Q-names,

(γ) $(p, \langle \lambda_i, \underset{\sim}{\tau}_i, \alpha_i : i < \xi \rangle) \Vdash_{Q'}$ "$(\forall i < \xi)[(\lambda_i = 2 \to \underset{\sim}{\tau}_i = \alpha_i) \,\&\, (\lambda_i \geq \aleph_0 \to \underset{\sim}{\tau}_i < \alpha_i)]$".

(δ) for (d) note that every $\leq_0$-directed set is linearly ordered by $\leq_0$ and if its cardinality is $< \kappa$ then it has a $\leq_0$-lub. $\square_{2.4}$

2.4A Remark . So $(S, \kappa) - \mathrm{Pr}_1$ and $(S, \kappa) - \mathrm{Pr}_1^+$ are "essentially" the same (for κ regular).

2.5 Definition.

(1) Assume $\bar{P}$ is a $\lessdot$-increasing sequence of forcing notions.

(a) Let

$$\mathrm{Gen}^r(\bar{P}) \stackrel{\mathrm{def}}{=} \{G : \text{for some (set) forcing notion } P^* : \bigwedge_{i<\alpha} P_i \lessdot P^* \text{ and } G^* \subseteq P^* \text{ generic over } V \text{ and } G = G^* \cap \bigcup_{i<\alpha} P_i\}.$$

(b) For a set E of regular cardinals we say that $\bar{P}$ obeys E if for $\gamma \in E$ we have: $P_\alpha = \bigcup_{\beta<\gamma} P_\beta$ and P_β satisfies the γ-c.c. for $\beta < \gamma$. We say that $\bar{P}$ strongly obey E if in addition $\beta < \gamma \in E \Rightarrow |P_\beta| < \gamma$.

(c) Let $E(\bar{P}) = \{\gamma \leq \ell g(\bar{P}) : \gamma$ is strongly inaccessible , uncountable and $\beta < \gamma \Rightarrow P_\gamma$ satisfies the γ-c.c.$\}$, $E_s(\bar{P}) = \{\gamma \in E(\bar{P}) : \beta < \gamma \Rightarrow |P_\beta| < \gamma\}$

(2) If $\bar{Q} = \langle P_i : i < \alpha \rangle$ or $\bar{Q} = \langle P_i, \underset{\sim}{Q}_i : i < \alpha \rangle$ where P_i is $\lessdot$-increasing, obeying E (so here we ignore the $\underset{\sim}{Q}_i$'s) we define a $\bar{Q}$-E-name $\underset{\sim}{\tau}$ almost as we define $\left(\bigcup_{i<\alpha} P_i\right)$-names, but we do not use *maximal* antichains of $\bigcup_{i<\alpha} P_i$:

($*$) $\underset{\sim}{\tau}$ is a function, $\mathrm{Dom}(\underset{\sim}{\tau}) \subseteq \bigcup_{i<\alpha} P_i$ and for every directed $G \in \mathrm{Gen}^r(\bar{Q})$, $\underset{\sim}{\tau}[G]$ is defined iff $\mathrm{Dom}(\underset{\sim}{\tau}) \cap G \neq \emptyset$ and then $\underset{\sim}{\tau}[G] \in V[G]$ [from where "every $G \ldots$" is taken? e.g., V is countable, G any set from the true universe] and $\underset{\sim}{\tau}$ is definable with parameters from $V \cup \{G\}$ (so

$\underset{\sim}{\tau}$ is really a first-order formula with the variable G and parameters from V) and $(\forall \beta \in E \cap E(\bar{Q}))(\exists \gamma < \beta)[\mathrm{Dom}(\underset{\sim}{\tau}) \cap P_\beta \subseteq P_\gamma]$.

Now $\Vdash_{\bar{Q}}$ has a natural meaning. If E is not mentioned we mean: any fixed $E \cap (\ell g(\bar{Q}) + 1) = E(\bar{Q})$ understood from the context, normally just $E(\langle P_i : i < \alpha \rangle)$, below we fix $E = \mathbf{E}^*$.

(3) For $\bar{Q}$-names $\underset{\sim}{\tau}_0, \ldots, \underset{\sim}{\tau}_{n-1}$ we let $\{\underset{\sim}{\tau}_0, \ldots, \underset{\sim}{\tau}_{n-1}\}$ be the name for the set that contains exactly those $\underset{\sim}{\tau}_i[\bar{Q}]$ that are defined. For $p \in \bar{Q}$ (i.e., $p \in \bigcup_{i<\alpha} P_i$) we let $p \Vdash$ "$\underset{\sim}{\tau} = x$" if for every G such that $p \in G \in \mathrm{Gen}^r(\bar{Q})$ we have $\underset{\sim}{\tau}[G] = x$. (But see 2.6(2).)

(4) A $\bar{Q}$-E-named $[j, \beta)$-ordinal $\underset{\sim}{\zeta}$ is a $\bar{Q}$-E-name $\underset{\sim}{\zeta}$ such that if $\underset{\sim}{\zeta}[G] = \xi$ then $j \leq \xi < \beta$ and $(\exists p \in G \cap P_{\xi \cap \alpha}) p \Vdash_{\bar{Q}}$ "$\underset{\sim}{\zeta} = \xi$" (where $\alpha = \ell g(\bar{Q})$). If we omit "$[j, \beta)$" we mean $[0, \ell g((\bar{Q})) = [0, \alpha)$.

2.5A Remark. 1) We can restrict in the definition of $\mathrm{Gen}^r(Q)$ to P^* in some class K, and get a K-variant of our notions.

2) Note: even if in 2.5(1) we ask $\mathrm{Dom}(\underset{\sim}{\tau})$ to be a maximal antichain it will not be meaningful as in the appropriate P_δ, we have $\bigwedge_{i<\delta} P_i \lessdot P_\delta$ but it will not in general be a maximal antichain.

2.5B Remark. Note that we wrote $P_{\xi \cap \alpha}$ not $P_{(\xi+1) \cap \alpha}$. Compare this to the remark 1.1B. We will not have a general associativity law, but the definition of $Sp_2 - \mathrm{Lim}_\kappa \bar{Q}$ will be slightly simplified. As said earlier we can interchange decisions on this matter (this does not mean this is the same iteration, just that it has the same relevant properties). Of course also Ch.X can be represented with this iteration.

2.5C Remark. Note that a $\bar{Q} - \emptyset$-named ordinal $\underset{\sim}{\zeta}$ is $\bar{Q} - E^*$-named ordinal *iff* for every $\beta \in E^* \cap E(\bar{Q})$ for some $\gamma < \beta$ we have $\Vdash_{\bar{Q}}$ "$\underset{\sim}{\zeta} \notin [\gamma, \beta)$".

2.6 Fact.

(1) For $\bar{P} = \langle P_i : i < \ell g\bar{P}\rangle$, a $\lessdot$ increasing sequence of forcing notions and $\bar{P}$-named $[j,\beta)$-ordinal $\underset{\sim}{\zeta}$ and $p \in \bigcup_{i<\alpha} P_i$ there are ξ, q and q_1 such that $p \leq q \in \bigcup_{i<\ell g\bar{P}} P_i$ and: either $q \Vdash_{\bar{P}}$ "$q_1 \in \underset{\sim}{G}$", $q_1 \in P_\xi$, $\xi < \alpha$, $[p \in P_\xi \Rightarrow q = q_1]$ and $q_1 \Vdash_{\bar{P}}$ "$\underset{\sim}{\zeta} = \xi$" or $q \Vdash_{\bar{P}}$ "$\underset{\sim}{\zeta}$ is not defined" (and even $p \Vdash_{\bar{P}}$ "$\underset{\sim}{\zeta}$ is not defined").

(2) For $\bar{P}$ as above, and $\bar{P}$-named $[j,\beta)$-ordinals $\underset{\sim}{\zeta}, \underset{\sim}{\xi}$, also $\mathrm{Min}\{\underset{\sim}{\zeta}, \underset{\sim}{\xi}\}$, $\mathrm{Max}\{\underset{\sim}{\zeta}, \underset{\sim}{\xi}\}$ (naturally defined, so $\mathrm{Max}\{\underset{\sim}{\zeta}, \underset{\sim}{\xi}\}[G]$ is defined iff a $\underset{\sim}{\zeta}[G]$, $\underset{\sim}{\xi}[G]$ are defined, and $\mathrm{Min}\{\underset{\sim}{\zeta}, \underset{\sim}{\xi}\}[G]$ is defined iff $\underset{\sim}{\zeta}[G]$ is defined or $\underset{\sim}{\xi}[G]$ is defined); both are $\bar{P}$-named $[j,\beta)$-ordinals.
Similarly for $\mathrm{Min}\{\underset{\sim}{\xi}_0, \dots, \underset{\sim}{\xi}_{n-1}\}$, $\mathrm{Max}\{\underset{\sim}{\xi}_0, \dots, \underset{\sim}{\xi}_{n-1}\}$ for $\bar{P} - E$-named ordinals.

(3) For $\bar{P}$ as above, $n < \omega$ and $\bar{P}$-named ordinals $\underset{\sim}{\xi}_1, \dots, \underset{\sim}{\xi}_n$ and $p \in \bigcup_{i<\ell g(\bar{P})} P_i$ there are $\zeta < \alpha$ and $q \in P_\zeta$ such that, first: $p \leq q$ or at least $q \Vdash_{P_\zeta}$ "$p \in P_i/\underset{\sim}{G}_{P_\zeta}$ for some $i < \ell g(\bar{P})$" (actually $i = \min\{i : p \in P_i\}$) and second: for some $\ell \in \{1, \dots, n\}$ we have $q{\restriction}\zeta \Vdash_{\bar{P}}$ "$\zeta = \underset{\sim}{\xi}_\ell = \mathrm{Max}\{\underset{\sim}{\xi}_1, \dots, \underset{\sim}{\xi}_n\}$" or $p \Vdash_{\bar{P}}$ "$\mathrm{Max}\{\underset{\sim}{\xi}_1, \dots, \underset{\sim}{\xi}_n\}$ not defined", in the second case we can add $q \Vdash_{\bar{P}}$ "$\underset{\sim}{\xi}_\ell$ not defined". Similarly for Min.

(4) *Convention:* If $\bar{Q} = \langle P_i, \underset{\sim}{Q}_i : i < \alpha\rangle$, P_i is $\lessdot$-increasing, we may write $\bar{Q}$ instead of $\langle P_i : i < \alpha\rangle$.

2.6A Convention. $\mathbf{E}^*$ is a class of strongly inaccessible cardinals $> \kappa$ fixed for this section, not mentioned usually. So a $\bar{P}$-named (e.g. ordinal) mean a $\bar{P}$-$(\mathbf{E} \cap E(\bar{P}))$-named (e.g. ordinal). Outside this section the default value is the class of strongly inaccessible $> \kappa$.
[The reader can simplify life using $\mathbf{E}^* = \emptyset$, he will lose only 2.7(4), hence case II of 3.4, so this is a reasonable choice.]

2.7 Definition and Claim. Let $e \in \{1, 2\}$. We define and prove by induction on α the following simultaneously (all forcing notions satisfying 2.1 (i)- (iv)):

(A) $\bar{Q} = \langle P_i, \underset{\sim}{Q}_i : i < \alpha \rangle$ is a $\kappa - \mathrm{Sp}_e$-iteration or really $\kappa - \mathrm{Sp}_e - \mathbf{E}^*$-iteration (the $\bar{Q}$'s below will have this form).

(B) A $\bar{Q}$-named (that is $\bar{Q}$-$\mathbf{E}^*$-named) atomic condition $\underset{\sim}{q}$ (or atomic $[j, \beta)$-condition, $\beta \leq \alpha$) and we define $\underset{\sim}{q} \restriction \xi$, $\underset{\sim}{q} \restriction \{\xi\}$, $\underset{\sim}{q} \restriction [\xi, \zeta)$ for a $\bar{Q}$-named atomic condition $\underset{\sim}{q}$ and ordinal $\xi \leq \zeta \leq \alpha$ (or $\bar{Q}$-named ordinals $\underset{\sim}{\xi}, \underset{\sim}{\zeta}$ instead ξ, ζ).

(C) If $\underset{\sim}{q}$ is a $\bar{Q}$-named (or really $\bar{Q}$-$\mathbf{E}^*$-named) atomic $[j, \beta)$-condition, $\xi < \alpha$, *then* $\underset{\sim}{q} \restriction \xi$ is a $(\bar{Q} \restriction \xi)$-named atomic $[j, \mathrm{Min}\,\{\beta, \xi\})$-condition and $\underset{\sim}{q} \restriction \{\xi\}$ is a P_ξ-name of a member of $\underset{\sim}{Q}_\xi$ or undefined (and then it may be assigned the value $\emptyset_{\underset{\sim}{Q}_\xi}$, the minimal member of $\underset{\sim}{Q}_\xi$).

(D) The $\kappa - \mathrm{Sp}_e$-limit of $\bar{Q}$, $\mathrm{Sp}_e - \mathrm{Lim}_\kappa \bar{Q}$, (really $\mathrm{Sp}_e - \mathbf{E}^* - \mathrm{Lim}_\kappa \bar{Q}$) denoted by P_α for $\bar{Q}$ as in clause (A), and $p \restriction \xi$ and $\mathrm{Dom}(p)$ for $p \in \mathrm{Sp}_e - \mathrm{Lim}_\kappa \bar{Q}$, ξ an ordinal $\leq \alpha$ (or $\bar{Q}$-named ordinal $\underset{\sim}{\xi}$ etc.).

(E) $\mathrm{Sp}_e - \mathrm{Lim}_\kappa \bar{Q}$ satisfies (i)-(iv) of Definition 1.2 and it obeys $\mathbf{E}^* \cap E(\bar{Q})$ (so if $\ell g(\bar{Q}) \in \mathbf{E}^* \cap E(\bar{Q})$ then $\mathrm{Sp}_e - \mathrm{Lim}_\kappa(\bar{Q}) = \bigcup_{\beta < lg(\bar{Q})} P_\beta$). Also if $\beta \leq \alpha$, $\beta \in \mathbf{E}^* \cap E(Q)$ and $\underset{\sim}{\zeta}$ is a $(\bar{Q} \restriction \beta)$-named ordinal then it is a $(\bar{Q} \restriction \gamma)$-named ordinal for some $\gamma < \beta$; similarly for atomic condition.

(F) If $\beta < \alpha = \ell g(\bar{Q})$ then $P_\beta \subseteq \mathrm{Sp}_e - \mathrm{Lim}_\kappa \bar{Q}$ (as models with two partial orders, even compatibility is preserved) and $[p \in P_\beta \Rightarrow p \restriction \beta = p]$ and $[P_\alpha \vDash \text{“}p \leq q\text{”} \Rightarrow P_\beta \vDash \text{“}p \restriction \beta \leq q \restriction \beta\text{”}]$ and $[P_\alpha \vDash \text{“}p \leq_0 q\text{”} \Rightarrow P_\beta \vDash \text{“}p \restriction \beta \leq_0 q \restriction \beta\text{”}]$ and $P_\alpha \vDash \text{“}p \restriction \beta \leq p\text{”}$. Also $q \in P_\beta, p \in \mathrm{Sp}_e - \mathrm{Lim}_\kappa \bar{Q}$ are compatible iff $q, p \restriction \beta$ are compatible in P_β. In fact if $q \in P_\beta$, $P_\beta \models \text{“}p \restriction \beta \leq q\text{”}$ then $q \cup (p \restriction [\beta, \alpha)$ is a least upper bound of p, q, and if $P_\beta \models \text{“}p \restriction \beta \leq_0 q\text{”}$ even a $\leq_0$-least upper bound of q. Hence $P_\beta \lessdot (\kappa - \mathrm{Sp}_e - \mathrm{Lim}_\kappa(\bar{Q}))$ and so $\beta < \gamma < \ell g(\bar{Q}) \Rightarrow P_\beta \lessdot P_\gamma$.

(G) The set of $p \in P_\alpha$ such that for every $\beta < \alpha$ we have $\Vdash_{P_\beta}$ “$p \restriction \{\beta\}$ is a singleton or empty”, is a dense subset of P_α. Also we can replace $\underset{\sim}{Q}_\beta$ by $\hat{Q}_\beta$ (see Definition 2.1(6)) and the set of “old” $p \in P_\alpha$ is a dense subset of the new (but actually do not use this).

Proof and Definition.

(A) $\bar{Q} = \langle P_i, \underset{\sim}{Q}_i : i < \alpha \rangle$ is a $\kappa - \mathrm{Sp}_e$-iteration if $\bar{Q} \restriction \beta$ is a $\kappa - \mathrm{Sp}_e$-iteration for $\beta < \alpha$, and if $\alpha = \beta + 1$ then $P_\beta = \mathrm{Sp}_e\text{-}\mathrm{Lim}_\kappa(\bar{Q} \restriction \beta)$ and $\underset{\sim}{Q}_\beta$ is a P_β-name of a forcing notion as in Definition 2.1(1)(i)-(iv).

(B) We say $\underset{\sim}{q}$ is a $\bar{Q}$-named atomic $[j, \beta)$-condition when : $\underset{\sim}{q}$ is a $\bar{Q}$-name (i.e. a $\bar{Q} - \mathbf{E}^*$-name), and for some $\underset{\sim}{\zeta} = \underset{\sim}{\zeta}_{\underset{\sim}{q}}$, a $\bar{Q}$-named $[j, \beta)$-ordinal (i.e. a $\bar{Q} - \mathbf{E}^*$-named $[j, \beta)$-ordinal), we have $\Vdash_{\bar{Q}}$ "$\underset{\sim}{\zeta}$ has a value iff $\underset{\sim}{q}$ has, and if they have then $j \leq \underset{\sim}{\zeta} < \mathrm{Min}\{\beta, \ell g(\bar{Q})\}$ and $\underset{\sim}{q} \in \underset{\sim}{Q}_{\underset{\sim}{\zeta}}$". Now $\underset{\sim}{q} \restriction \xi$ will have a value iff $\underset{\sim}{\zeta}_{\underset{\sim}{q}}$ has a value $< \xi$ and then its value is the value of $\underset{\sim}{q}$. Lastly, $\underset{\sim}{q} \restriction \{\xi\}$ will have a value iff $\underset{\sim}{\zeta}_{\underset{\sim}{q}}$ has the value ξ and then its value is the value of $\underset{\sim}{q}$ (similarly for $\underset{\sim}{\xi}$ and $\underset{\sim}{q} \restriction [\zeta, \xi)$ and $\underset{\sim}{q} \restriction [\underset{\sim}{\zeta}, \underset{\sim}{\xi})$).

(C) Left to the reader.

(D) We are defining $\mathrm{Sp}_e - \mathrm{Lim}_\kappa \bar{Q}$ (where $\bar{Q} = \langle P_i, \underset{\sim}{Q}_i : i < \alpha \rangle$ of course). It is a triple $P_\alpha = (|P_\alpha|, \leq, \leq_0)$ where

(a) $|P_\alpha|$ is the set of $p = \{\underset{\sim}{q}_i : i < i^*\}$ satisfying:

(i) $i^* < \kappa$,

(ii) if $e = 1$, $\emptyset \leq p$ (see below)

(iii) each $\underset{\sim}{q}_i$ is a $\bar{Q}$-named atomic condition, and for every $\xi < \alpha$, $\Vdash_{P_\xi}$ "$p \restriction \{\xi\} \stackrel{\mathrm{def}}{=} \{q_i \restriction \{\xi\} : i < i^*\}$ if not empty, has a $\leq_0$-upper bound in $\underset{\sim}{Q}_\xi$ or at least a weak $\leq_0$-upper bound i.e. for some nonempty $u \subseteq i^*$ and $r \in \underset{\sim}{Q}_\xi$ we have $\bigwedge_{i<i^*} \bigvee_{j \in u} q_i \restriction \{\xi\} \leq q_j \restriction \{\xi\}$ and $\bigwedge_{j \in u} q_i \restriction \{\xi\} \leq_0 r$ and $\bigvee_{i \in u} \bigwedge_{j \in u} q_i \restriction \{\xi\} \leq_0 q_j \restriction \{\xi\}$ (i.e. $q \restriction \{\xi\} \in \underset{\sim}{\hat{Q}}_\xi$)".

(b) for $p \in \mathrm{Sp}_e - \mathrm{Lim}_\kappa(\bar{Q})$ and $\xi < \ell g(\bar{Q})$ we let:
$p \restriction \xi \stackrel{\mathrm{def}}{=} \{r \restriction \xi : r \in p\}$
$p \restriction \{\xi\} \stackrel{\mathrm{def}}{=} \{r \restriction \{\xi\} : r \in p\}$,
we define similarly $p \restriction [\zeta, \xi)$, $p \restriction \{\underset{\sim}{\zeta}\}$, $p \restriction [\underset{\sim}{\zeta}, \underset{\sim}{\xi})$.

(c) $P_\alpha \models$ "$p^1 \leq_0 p^2$" iff for every $\xi < \alpha$ we have (letting $p^\ell = \{q_i^\ell : i < i^\ell(*)\}$ for $\ell = 1, 2$):
$\{q_i^2 \restriction \xi : i < i^2(*)\} \Vdash_{P_\xi}$ "$p^2 \restriction \{\xi\} = \emptyset \Rightarrow p^1 \restriction \{\xi\} = \emptyset$ and one of the following holds:

(i) $\{q_i^\ell \restriction \{\xi\}: i < i^\ell(*)\}$ are equal for $\ell = 1, 2$,

(ii) letting u^1, u^2 be as in clause (a)(iii) above for $q^1 \restriction \{\xi\}$, $q^2 \restriction \{\xi\}$ respectively for some $j_2 \in u^2$ for all $j_1 < i^1(*)$ if $\underset{\sim}{\zeta}_{q^1_{j_1}} = \xi$ and $j_1 \in u^1$ then $\hat{Q}_\zeta \models [q^1_{j_1} \restriction \{\xi\} \leq_0 q^2_{j_2} \restriction \{\xi\}]$
(note (i)$\vee$(ii) means $\hat{\underset{\sim}{Q}}_\xi \models p^1 \restriction \{\xi\} \leq_0 p^2 \restriction \{\xi\}$)

(iii) $e = 2$ and $p^1 \restriction \{\xi\} = \emptyset$".

(d) $P_\alpha \models p^1 \leq p^2$ iff

(i) for every $\xi < \ell g(\bar{Q})$ we have (letting $p^\ell = \{q_i^\ell : i < i^\ell(*)\}$, $\ell = 1, 2$): $\{q_i^2 \restriction \xi : i < i^2(*)\} \Vdash_{P_\xi}$ "$p^2 \restriction \{\xi\} = \emptyset \Rightarrow p^1 \restriction \{\xi\} = \emptyset$ and one of the following occurs: $p^1 \restriction \{\xi\}, p^2 \restriction \{\xi\}$ are equal as subsets of $\underset{\sim}{Q}_\xi$, or for some $j_2 < i^2(*)$ for all $j_1 < i^1(*)$ we have $Q_\xi \models$ "$[q^1_{j_1} \leq q^2_{j_2}]$" (i.e. the order of $\hat{\underset{\sim}{Q}}_\xi$).

(ii) for some $n < \omega$ and $\bar{Q}$-named ordinals $\underset{\sim}{\xi}_1, \ldots, \underset{\sim}{\xi}_n$ we have:
for each $\zeta < \ell g(\bar{Q})$, $p_2 \restriction \zeta \Vdash_{P_\zeta}$ "if $\zeta \notin \{\underset{\sim}{\xi}_1, \ldots, \underset{\sim}{\xi}_n\}$ then: $p^1 \restriction \{\zeta\} = \emptyset$ and $e = 2$ or $\hat{\underset{\sim}{Q}}_\xi \models$ "$[\{r[\underset{\sim}{G}_{P_\zeta}] : r \in p_1, \underset{\sim}{\zeta}_r = \zeta\} \leq_0 \{r[\underset{\sim}{G}_{P_\zeta}] : r \in p_2, \underset{\sim}{\zeta}_r = \zeta\}]$", note that the truth value of $\zeta = \underset{\sim}{\xi}_\ell$ is a P_ζ-name so this is well defined. Note: $p^1 \restriction \{\zeta\} = \emptyset$ not just $= \emptyset_{Q_i}$ but $\underset{\sim}{q} \in p^1 \Rightarrow \neg[\xi = \underset{\sim}{\zeta}_q[G_\zeta]]$.

We then (i.e. if (i)+(ii)) say: $p_1 \leq p_2$ over $\{\underset{\sim}{\xi}_1, \ldots, \underset{\sim}{\xi}_n\}$.

Lastly (as said above) if $p \in \kappa - \mathrm{Sp}_e - \mathrm{Lim}_\kappa \bar{Q}$ then we let $p \restriction \xi \overset{\mathrm{def}}{=} \{r \restriction \xi : r \in P\}$ and $\mathrm{Dom}(p) = \{\underset{\sim}{\zeta}_q : \underset{\sim}{q} \in p\}$ and similarly $p \restriction \underset{\sim}{\zeta}$, $p \restriction [\zeta, \xi)$, $p \restriction [\underset{\sim}{\zeta}, \underset{\sim}{\xi})$.

(E) : Let us check Definition 2.1 (1)(i)-(iv) for $P_\alpha \overset{\mathrm{def}}{=} \mathrm{Sp}_e - \mathrm{Lim}_\kappa \bar{Q}$:

$\leq^{P_\alpha}$ is a partial order: Suppose $p_0 \leq p_1 \leq p_2$. Let n^ℓ, $\underset{\sim}{\xi}_1^\ell, \ldots, \underset{\sim}{\xi}_{n^\ell}^\ell$ appear in the definition of $p_\ell \leq p_{\ell+1}$. Let $n = n^0 + n^1$, and

$$\underset{\sim}{\xi}_i = \begin{cases} \underset{\sim}{\xi}_i^0 & \text{if } 1 \leq i \leq n^1 \\ \underset{\sim}{\xi}_{i-n}^1 & \text{if } n^1 < i \leq n^1 + n^2. \end{cases}$$

Now for $\ell = 0, 1$ and $\xi < \alpha$ we have $\Vdash_{P_\xi}$ "if $p_{\ell+1} \restriction \xi$ is in the set $\underset{\sim}{G}_{\bar{Q}}$ then $p_\ell \restriction \{\xi\} \leq p_{\ell+1} \restriction \{\xi\}$ in $\hat{\underset{\sim}{Q}}_\xi$ ", hence $\Vdash_{P_\xi}$ "if $p_2 \restriction \xi$ is in the set $\underset{\sim}{G}_\xi$ then $p_0 \restriction \{\xi\} \leq p_2 \restriction \{\xi\}$ in $\hat{\underset{\sim}{Q}}_\xi$ ".

Also for $\zeta < \alpha$ we have $p_2{\restriction}\zeta \Vdash_{P_\zeta}$ "if $\zeta \notin \{\underset{\sim}{\xi}_1, \ldots, \underset{\sim}{\xi}_n\}$ then (in $\underset{\sim}{\hat{Q}}_\zeta$) $p_0{\restriction}\{\zeta\} \leq_0 p_1{\restriction}\{\zeta\} \leq_0 p_2{\restriction}\{\zeta\}$" or $e = 2$, $p_0{\restriction}\{\zeta\} = \emptyset$". So we finish.

$\leq_0$ is a partial order: check.

$p \leq_0 q \Rightarrow p \leq q$: By the definition; easy.

So in Definition 2.1, (i), (ii), (iii), and (iv) hold.

We still have to check that $\bar{Q}$ obeys $\mathbf{E}^* \cap E(\bar{Q})$, now by the induction hypothesis the only thing to check is: if $\alpha = \ell g(\bar{Q}) = \beta + 1$, $\beta \in \mathbf{E}^* \cap E(Q)$ then $P_\beta = \bigcup_{\gamma<\beta} P_\gamma$. This follows as $\beta \geq \kappa$, and each $(\bar{Q}{\restriction}\beta)$-named ordinal is a $(\bar{Q}{\restriction}\gamma)$-named ordinal for some γ. This is true by the definition of a $\bar{Q}-\mathbf{E}$-named ordinal.

(F) , (G) We leave the checking to the reader (for the first sentence of (G) see 2.10(1) below). $\square_{2.7}$

2.8 Claim. Suppose $\bar{Q} = \langle P_i, \underset{\sim}{Q}_i : i < \alpha \rangle$ is a $\kappa - \mathrm{Sp}_e$-iteration (so $P_\alpha = \mathrm{Sp}_e{-}\mathrm{Lim}_\kappa(\bar{Q})$).

1) If $p \leq q$ in P_α *then* there are $r, n, \xi_1 < \ldots < \xi_n < \alpha$ such that:

(a) $r \in P_\alpha$

(b) $q \leq r$

(c) $p \leq r$ above $\{\xi_1, \ldots, \xi_n\}$.

2) We can find such r simultaneously for finitely many $p_k \leq q$.

Remark. In fact we can have $r{\restriction}[\xi_n, \alpha) = q{\restriction}[\xi_n, \alpha)$.

Proof. 1) We prove this by induction on α

Case 1: $\alpha = 0$. Trivial.

Case 2: $\alpha = \beta + 1$.

Apply the induction hypothesis to $\bar{Q}{\restriction}\beta, p{\restriction}\beta, q{\restriction}\beta$ (clearly $\bar{Q}{\restriction}\beta$ is an $\kappa{-}Sp_e$-iteration, $p{\restriction}\beta = P_\beta$, $q{\restriction}\beta \in P_\beta$ and $P_\beta \models$ "$p \leq q$", by 2.7).

So we can find $r', m, \{\xi'_1, \ldots, \xi'_m\}$ such that:

(a)$'$ $r' \in P_\beta$

(b)$'$ $P_\beta \models q{\restriction}\beta \leq r'$

(c)′ $p \leq r'$ (in P_β) above $\{\xi_1, \ldots, \xi_m\}$.

Let $n \stackrel{\text{def}}{=} m + 1$, and

$$\xi_\ell = \begin{cases} \xi'_\ell & \textit{if } \ell \in \{1 \ldots, m\} \\ \beta & \textit{if } \ell = n \end{cases}$$

and lastly $r = r \cup (q \restriction \{\beta\})$.

Case 3: α is a limit ordinal.

Let $p \leq q$ (in P_α) above $\{\underset{\sim}{\xi}_1, \ldots, \underset{\sim}{\xi}_n\}$. We choose by induction on $\ell \leq n$, $r_\ell, \beta_\ell, \xi^*_\ell$ such that

(α) $r_\ell \in P_{\beta_\ell}$,

(β) $r_\ell \leq r_{\ell+1}$

(γ) $q \restriction \beta_\ell \leq r_\ell$

(δ) $\beta_\ell \leq \beta_{\ell+1} < \alpha$

(ε) $\beta_0 = 0$, $r_0 = \emptyset_{P_0}$

(ζ) for $\ell \in \{0, \ldots, n-1\}$ we have: either $r_{\ell+1} \Vdash$ "$\underset{\sim}{\xi}_{\ell+1} = \xi^*_{\ell+1}$" and $\xi^*_{\ell+1} \leq \beta_{\ell+1}$ *or* $\beta_{\ell+1} = \beta_\ell \,\&\, r_{\ell+1} = r_\ell$ and $r_\ell \cup (q \restriction [\beta, \alpha)) \Vdash_{P_\alpha}$ "$\underset{\sim}{\xi}_{\ell+1}$ is not defined".

Carrying the definition is straight: for $i = 0$ use clause (ε). For $\ell + 1 \leq n$ when the second possibility of clause (ζ) fails there is r', such that $r_\ell \cup (q \restriction [\beta, \alpha)) \leq r' \in P_\alpha$, and $r' \Vdash_{P_\alpha}$ "$\underset{\sim}{\xi}_{\ell+1}$ is defined", so there are r'', $\xi^*_{\ell+1}$ such that $r' \leq r'' \in P_\alpha$ and $r'' \Vdash$ "$\underset{\sim}{\xi}_{\ell+1} = \xi^*_{\ell+1}$" so as "$\xi^*_{\ell+1}$ is a $\bar{Q}$-named ordinal" we know that $\xi^*_{\ell+1} < \alpha$ and $r'' \restriction \xi^*_{\ell+1} \Vdash_{P_{\xi^*_{\ell+1}}}$ "$\underset{\sim}{\xi}_{\ell+1} = \xi^*_{\ell+1}$". Let $\beta_{\ell+1} \stackrel{\text{def}}{=} \max\{\beta_\ell, \xi^*_{\ell+1}\}$, and $r_{\ell+1} \stackrel{\text{def}}{=} r'' \restriction \beta_{\ell+1}$. So we have carried the induction.

Apply the induction hypothesis to $\bar{Q} \restriction \beta_n\ p \restriction \beta_n$, r_n; it is applicable as $\beta_n < \alpha$, and $P_{\beta_n} \models$ "$p \restriction \beta_n \leq q \restriction \beta_n \leq r_n$". So there are $m < \omega$, $\xi_1 < \ldots < \xi_m < \beta_n$ and r^* such that $P_{\beta_n} \models r_n \leq r^*$" and $p \leq r^*$ (in P_{β_n}) above $\{\xi_1, \ldots, \xi_m\}$. Now let $r \stackrel{\text{def}}{=} r^* \cup (q \restriction [\beta_n, \alpha))$, clearly $q \leq r$ and $p \leq r$ above $\{\xi_1, \ldots, \xi_m, \beta_n\}$.

2) Should be clear. $\square_{2.8}$

2.8A Claim. Let $\bar{Q}$ be a $\kappa - \mathrm{Sp}_2$-iteration (of length α).

(1) If $\beta < \alpha$ and $\underset{\sim}{\zeta}$ is a P_β-name of a $\bar{Q}$-named $[\beta, \alpha)$-ordinal *then* for some $\bar{Q}$-named $[\beta, \alpha)$-ordinal $\underset{\sim}{\xi}$

$$\Vdash_{\bar{Q}} \text{“}\underset{\sim}{\zeta} = \underset{\sim}{\xi}\text{”}$$

(2) The same holds if we replace "ordinal" by "atomic condition".

(3) If $\underset{\sim}{\beta}$ is a $\bar{Q}$-named ordinal, and for each $\beta < \alpha$, $\underset{\sim}{\zeta}_\beta$ is a $\bar{Q}$-named $[\beta, \alpha)$-ordinal then for some $\bar{Q}$-named ordinal $\underset{\sim}{\xi}$

$$\Vdash_{\bar{Q}} \text{“if } \underset{\sim}{\beta}[\underset{\sim}{G}] = \beta \text{ then } \underset{\sim}{\xi}[\underset{\sim}{G}] = \zeta_\beta[\underset{\sim}{G}]\text{”}$$

(4) Similarly for atomic conditions.

Proof. Easy. $\square_{2.8A}$

2.8B Discussion. Why do we use iteration of kind Sp_2 when Sp_1 may seem simpler? Think that we want say κ, κ^+ to play the roles of $\aleph_0$, $\aleph_1$ in Ch.X. Suppose $\langle P_i, \underset{\sim}{Q}_i : i < \kappa^+\rangle$ is an $\kappa^+ - \mathrm{Sp}_1$-iteration which is nice enough such that $\bigcup_{i<\kappa^+} P_i$ is a dense subset of P_{κ^+}. Suppose further that for $i < \kappa^+$, we have $\{p_i\} \in P_\alpha$ such that: $\Vdash_{P_i}$ "$p_i \in \underset{\sim}{Q}_i$, and for every q such that $q \in \underset{\sim}{Q}_i$, $\emptyset_{\underset{\sim}{Q}_i} \leq_0 q$ there is r such that $q \leq_0 r \in \underset{\sim}{Q}_i$ and r is incompatible with p_i (in $\underset{\sim}{Q}_i$)". These are reasonable assumptions for the iterations we have in mind.

Let $\underset{\sim}{u} = \{i < \kappa^+ : \{p_i\} \in \underset{\sim}{G}_{P_\alpha}\}$, so this is a P_α-name of a subset of κ^+. As $\bigcup_{i<\kappa^+} P_i$ is a dense subset of κ^+ clearly $\Vdash_{P_{\kappa^+}}$ "$\underset{\sim}{u}$ is an unbounded subset of κ^+". But for each $p \in P_{\kappa^+}$ and $\alpha < \kappa^+$ w.l.o.g. we have some n and $\xi_1 < \ldots < \xi_n < \kappa^+$ such that $i \in \kappa^+ \setminus \{\xi_1, \ldots, \xi_n\} \Rightarrow p{\restriction}i \Vdash$ "$\emptyset_{\underset{\sim}{Q}_i} \leq_0 p{\restriction}\{i\}$".

Now there is $q \in P_{\kappa^+}$, such that $p \leq_0 q$ (i.e. for every i we have $q{\restriction}i \Vdash_{P_i}$ "$p{\restriction}\{i\} \leq_0 q{\restriction}\{i\}$") and for every $i \in \alpha \setminus \{\xi_1, \ldots, \xi_n\}$ we have $\Vdash_{P_i}$ "$q(i)$, p_i are incompatible in $\underset{\sim}{Q}_i$". So $q \Vdash$ "$\underset{\sim}{u} \cap \alpha \subseteq \{\xi_1, \ldots, \xi_n\}$". As α was an arbitrary ordinal $< \kappa^+$, necessarily $\Vdash_{P_{\kappa^+}}$ "$\underset{\sim}{u}$ has order type $\leq \omega$", but as indicated earlier $\Vdash_{P_\alpha}$ "$\sup(\underset{\sim}{u}) = \kappa^+$". Together $\Vdash_{P_{\kappa^+}}$ "$\mathrm{cf}(\kappa^+) = \aleph_0$", certainly contrary to our desires.

So the use of our choice is $e = 2$. Where is this used? In the proof see end of the proof of 2.13 (hence also in 2.14, 2.15).

2.9 Claim. Let $\bar{Q} = \langle P_i, \underset{\sim}{Q}_i : i < \alpha \rangle$ be an κ-Sp$_2$-iteration, $P_\alpha = \mathrm{Sp}_2 - \mathrm{Lim}_\kappa \bar{Q}$ (as usual).

(1) If $\underset{\sim}{\beta}, \underset{\sim}{\gamma}$ are $\bar{Q}$-named $[j, \ell g(\bar{Q}))$-ordinals, then $\mathrm{Max}\{\underset{\sim}{\beta}, \underset{\sim}{\gamma}\}$ (defined naturally) is a $\bar{Q}$-named $[j, \ell g(\bar{Q}))$-ordinal.

(2) If $\alpha = \beta_0 + 1$, in Definition 2.7, part (D), in defining the set of elements of P_α we can restrict ourselves to $\beta = \beta_0$. Also in such a case, $P_\alpha = P_{\beta_0} * \underset{\sim}{Q}_{\beta_0}$ (essentially). More exactly, $\{p \bigcup \{\underset{\sim}{q}\} : p \in P_{\beta_0}, \underset{\sim}{q}$ a P_{β_0}-name of a member of $\underset{\sim}{Q}_{\beta_0}\}$ is a dense subset of P_α, and the order $p_1 \bigcup \{\underset{\sim}{q}_1\} \leq p_2 \bigcup \{\underset{\sim}{q}_2\}$ iff $[p_1 \leq p_2$ (in P_{β_0}) and $p_2 \Vdash_{P_{P_0}}$ "$\underset{\sim}{q}_1 \leq \underset{\sim}{q}_2$ in $\underset{\sim}{Q}_{\beta_0}$"$]$ is equivalent to that of P_α, in fact is the restriction of $<^{P_\alpha}$ to this set, so we get the same completion to a Boolean Algebra.

(3) $|P_\alpha| \leq (\sum_{i<\alpha} 2^{|P_i|})^{|\alpha|}$, for limit α (where of course $|P| = |\{p/\approx : p \in P\}|$ $p_0 \approx p_1$ iff $p_\ell \Vdash$ "$p_{1-\ell} \in \underset{\sim}{G}_P$" for $\ell = 0, 1$).

(4) If $\Vdash_{P_i}$ "$|\underset{\sim}{Q}_i| \leq \mu$", μ a cardinal, then $|P_{i+1}| \leq 2^{|P_i|} + \mu$.

(5) If $\Vdash_{P_i}$ "$d(Q_i) \leq \mu$" then $d(P_{i+1}) \leq d(P_i) + \mu$, where $d(P)$ is the density of P.

(6) For α limit $d(P_\alpha) \leq 2^{\Sigma_{i<\alpha} d(P_i)}$.

(7) If $P = \hat{Q}$ *then* P is essentially complete, i.e. for every maximal antichain $\mathcal{I}_0 \cup \mathcal{I}_1$ of P with $\mathcal{I}_0 \cap \mathcal{I}_1 = \emptyset$, for some $q \in P$, for every $p \in \mathcal{I}_0 \cup \mathcal{I}_1$, q is compatible with p iff $p \in \mathcal{I}_0$.

Proof. Check. $\square_{2.9}$

2.10 Claim. Suppose (κ is regular and):

(a) $\bar{Q} = \langle P_i, \underset{\sim}{Q}_i : i < \alpha \rangle$ is a κ – Sp$_2$-iteration (and of course $P_\alpha = \mathrm{Sp}_2 - \mathrm{Lim}_\kappa(\bar{Q})$)

(b) $\Vdash_{P_i}$ "$(Q_i, \leq_0)$ is θ-complete" for $i < \alpha$

(c) $\theta \leq \kappa$

Then

(1) $(P_\alpha, \leq_0)$ is θ-complete, in fact: if $\delta < \theta$, $\langle p_i : i < \delta\rangle$ is $\leq_0$-increasing then it has an $\leq_0$-upper bound (in fact, as in 2.7(G))

(2) for $\beta < \alpha$, P_α/P_β is θ-complete.

(3) In fact we can get $\leq_0$ $-$lub if this holds for each Q_i.

Remark. We deal with Pr^+ and not Pr_1 (here and later) just for simplicity presentation, as it does not matter much by 2.4.

Proof: Straightforward.

(1) So assume $\delta < \theta$ and $p_i \in P_\alpha$ for $i < \delta$ and $[i < j < \delta \Rightarrow p_i \leq_0 p_j]$. Now it is enough to find $p \in P_0$ such that

$$i < \delta \Rightarrow p_i \leq_0 p$$

$$\Vdash_{\bar{Q}} \text{“}\mathrm{Dom}(p) = \bigcup_{i<\delta} \mathrm{Dom}(p_i)\text{”}$$

$$\Vdash_{P_\zeta} \text{“}p{\restriction}\{\zeta\} \text{ is a singleton or } \emptyset\text{”}.$$

Let $p_i = \{\underset{\sim}{q}^i_\gamma : \gamma < \gamma_i\}$ where $\gamma_i < \kappa$ and for each $\zeta < \zeta_i, \underset{\sim}{q}^i_\zeta$ is $\bar{Q}$-named atomic condition, say $\Vdash_{\bar{Q}}$ “$\underset{\sim}{q}^i_\gamma \in Q_{\underset{\sim}{\zeta}^i_\gamma}$”, where $\underset{\sim}{\zeta}^i_\gamma$ is a $\bar{Q}$-named ordinal which is $\zeta_{\underset{\sim}{q}^i_\gamma}$. Now for each $\beta < \alpha$ let $\underset{\sim}{\leq}^*_\beta$ be a P_β-name of a well ordering of $\underset{\sim}{Q}_\beta$. For each $i(*) < \delta, \gamma(*) < \gamma_i$ let $r^{i(*)}_{\gamma(*)}$ be the following $\bar{Q}$-named atomic condition:

Let $\zeta < \alpha$, $G_\zeta \subseteq P_\zeta$ generic over V and $\underset{\sim}{\zeta}^{i(*)}_{\gamma(*)}[G_\zeta] = \zeta$, now work in $V[G_\zeta]$, let $w_\zeta = \{i < \delta : \text{for some } \gamma \text{ we have } \underset{\sim}{\zeta}^i_\gamma[G_\zeta] = \zeta\}$, and for each $i \in w_\zeta$ let $u^\zeta_i = \{\gamma < \gamma_i : \underset{\sim}{\zeta}^i_\gamma[G_\zeta] = \zeta\}$, clearly not empty; moreover for some $\beta_i \in u^\zeta_i$ we have

$$(\forall \xi \in u^\zeta_i)(\exists \gamma \in u^\zeta_i)(\underset{\sim}{q}^i_{\beta_i}[G_\zeta] \leq_0 \underset{\sim}{q}^i_\gamma[G_\zeta] \;\&\; \underset{\sim}{q}^i_\xi[G_\zeta] \leq \underset{\sim}{q}^i_\gamma[G_\zeta])$$

and let $v^\zeta_i = \{\gamma \in u^\zeta_i : \underset{\sim}{q}^i_{\beta_i}[G_\zeta] \leq_0 \underset{\sim}{q}^i_\gamma[G_\zeta]\}$, clearly also v^ζ_i is not empty. (As p_i is $\leq_0$-increasing, w_ζ is an end segment of δ and $i(*) \in w_\zeta$.) We define $\underset{\sim}{r}^i_\gamma[G_\zeta] \in \underset{\sim}{Q}_\zeta[G_\zeta]$.

Case 1: For some $i \in w_\zeta$ we have:

$$(\forall j)\,[i \leq j \in w_\zeta \to p_i{\restriction}\{\zeta\}[G_\zeta] = p_j{\restriction}\{\zeta\}[G_\zeta]]\,.$$

By the definition there is $r \in \underset{\sim}{Q}_i$ such that $\gamma \in v_i^\zeta \Rightarrow \underset{\sim}{q}_\gamma^i[G_\zeta] \leq_0 r$.

We let $\underset{\sim}{r}_{\gamma(*)}^{i(*)}$ be the $\leq_\zeta[G_\zeta]$ - first such r.

Case 2: Not case 1.

Let $w'_\zeta = \{i \in w_\zeta :$ for no $j \in w_\zeta \cap i$ do we have $(p_j \restriction \{\zeta\})[G_\zeta] = (p_i \restriction \{\zeta\})[G_\zeta]\}$.

Note. w'_ζ has no last member and moreover is unbounded in δ; let $j(i) = \min(w'_\zeta \setminus (i+1))$ for $i \in w'_\zeta$.

For each $i \in w'_\zeta$, by 2.2(9) we know there is $\beta_i \in v_{j(i)}^\zeta$ such that

$$[\gamma \in v_i^\zeta \Rightarrow \underset{\sim}{Q}_\zeta[G_\zeta] \models \text{“}q_\gamma^i[G_\zeta] \leq_0 q_{\beta_i}^{j(i)}[G_\zeta]\text{”}].$$

Hence by 2.2(9) we know $\langle q_{\beta_i}^{j(i)}[G_\zeta] : i \in w'_\zeta\rangle$ is a $\leq_0$-increasing sequence in $\underset{\sim}{Q}_\zeta[G_\zeta]$, hence it has a $\leq_0$-upper bound; so let $r_{\gamma(*)}^{i(*)}$ be the $\underset{\sim}{\leq}_\zeta^*[G]$-least $\leq_0$-upper bound of such a sequence in $\underset{\sim}{Q}_\zeta[G_\zeta]$. Because of the “such” the choice depend on ζ, G_ζ but not on $i(*)$, $\gamma(*)$. Now

$$p \overset{\text{def}}{=} \{\underset{\sim}{r}_{\gamma(*)}^{i(*)} : i(*) < \delta \text{ and } \gamma(*) < \gamma_{i(*)}\}$$

is as required.

2), 3) Similar proof and will not be used (or use the associativity law, see 2.21(3) (which could be proved before 2.10)). $\square_{2.10}$

2.11 Definition. Let $\bar{Q} = \langle P_i, \underset{\sim}{Q}_i : i < \alpha\rangle$ be an $\kappa - \mathrm{Sp}_2$-iteration

(1) We say $\underset{\sim}{y}$ is a $(\bar{Q}, \underset{\sim}{\zeta})-\mathbf{E}^*$-name (again we usually omit $\mathbf{E}^*$) *if:* $\underset{\sim}{y}$ is a P_α-name, $\underset{\sim}{\zeta}$ is a $\bar{Q} - \mathbf{E}^*$-named $[0, \alpha)$-ordinal, and: if $\beta < \alpha$, $G_{P_\alpha} \subseteq P_\alpha$ is generic over V and for some $r \in G_{P_\alpha} \cap P_\beta$, $r \Vdash_{\bar{Q}}$ “$\underset{\sim}{\zeta} = \beta$”, *then* $\underset{\sim}{y}[G_{P_\alpha}] \in V[G_{P_\beta}]$ is well defined and depends only on $G_{P_\alpha} \cap P_\beta$ so we write $\underset{\sim}{y}[G_{P_\alpha} \cap P_\beta]$; and if $G_{P_\alpha} \subseteq P_\alpha$ is generic over V, $\underset{\sim}{\zeta}[G_{P_\alpha}]$ not well defined then $\underset{\sim}{y}[G_{P_\alpha}]$ is not well defined.

(2) If $p \in P_\alpha$, $G_{P_\alpha} \subseteq P_\alpha$ generic over V, (or just in $\mathrm{Gen}^r(\bar{Q})$), *then* $p[G_{P_\alpha}]$ is a function, $\mathrm{Dom}(p[G_{P_\alpha}]) = \{\zeta_{\underset{\sim}{q}}[G_{P_\alpha}] : \underset{\sim}{q} \in p\}$ and $(p[G_{P_\alpha}])(\varepsilon) = \{\underset{\sim}{q}[G_{P_\alpha}] : \underset{\sim}{q} \in p \text{ and } \zeta_{\underset{\sim}{q}}[G_{P_\alpha}] = \varepsilon\}$.

2.12 Claim. Suppose

(a) $\bar{Q} = \langle P_i, \underset{\sim}{Q}_i : i < \alpha \rangle$ is a $\kappa - \mathrm{Sp}_2$-iteration

(1) Assume that q is an atomic $\bar{Q}$-named condition and $\underset{\sim}{\xi}$ is a $\bar{Q}$-named ordinal and $\underset{\sim}{q}{\restriction}\underset{\sim}{\xi}$ is defined naturally (i.e. if $G \in \mathrm{Gen}(\bar{Q})$, and $\xi_0 = \underset{\sim}{\xi}[G]$ and $\zeta = \zeta_{\underset{\sim}{q}}[G]$ then $\zeta < \xi \Rightarrow (\underset{\sim}{q}{\restriction}\underset{\sim}{\zeta})[G] = \underset{\sim}{q}[G]$, and $\zeta \geq \xi \Rightarrow (\underset{\sim}{q}{\restriction}\underset{\sim}{\xi})[G]$ not defined). Then $\underset{\sim}{q}{\restriction}\underset{\sim}{\xi}$ is an atomic $\bar{Q}$-condition with $\zeta_{(\underset{\sim}{q}{\restriction}\underset{\sim}{\xi})} = \min\{\zeta_{\underset{\sim}{q}}, \underset{\sim}{\xi}\}$ (see 2.9(1)). Similarly for $\underset{\sim}{q}{\restriction}[\xi, \alpha)$. we can let $P_{\underset{\sim}{\xi}} \overset{\mathrm{def}}{=} \{p \in P_\alpha : p{\restriction}\underset{\sim}{\xi} = p\}$ and it has the natural properties.

(2) Assume in addition

(b) $p \in P_\alpha$, $\underset{\sim}{\zeta}$ is a $\bar{Q}$-named $[0, \alpha)$-ordinals

(c) $\underset{\sim}{r}$ is a $(\bar{Q}, \underset{\sim}{\zeta})$-named member of $P_\alpha / P_{\underset{\sim}{\zeta}}$

(d) κ a successor (or just not a limit cardinal)

Then: there is $q \in P_\alpha$ such that:

(*) *if* $\xi < \alpha$, $G_\xi \subseteq P_\xi$ generic over V, and $\underset{\sim}{\zeta}[G_\xi] = \xi$ *then* $(p{\restriction}\xi)[G_\xi] = (q{\restriction}\xi)[G_\xi]$ and $(q{\restriction}[\xi, \alpha))[G_\xi] = \underset{\sim}{r}{\restriction}[\xi, \alpha)[G_\xi]$

In fact $q = (p{\restriction}\underset{\sim}{\zeta}) \cup [\underset{\sim}{r}{\restriction}[\underset{\sim}{\zeta}, \alpha)]$ will do where $p{\restriction}\underset{\sim}{\zeta} = \{\underset{\sim}{p}'{\restriction}\underset{\sim}{\zeta} : \underset{\sim}{p}' \in p\}$, $\underset{\sim}{r}{\restriction}[\underset{\sim}{\zeta}, \alpha) = \{\underset{\sim}{r}'{\restriction}[\underset{\sim}{\zeta}, \alpha) : \underset{\sim}{r}' \in r\}$.

(3) If in (2) in addition

(c)′ $\underset{\sim}{r}$ is a $(\bar{Q}, \underset{\sim}{\zeta})$-name of a member of $P_\alpha / P_{\underset{\sim}{\zeta}}$ above $p{\restriction}[\underset{\sim}{\zeta}, \alpha)$

then we can add $p \leq q$ (but now $q = p \cup (r{\restriction}[\underset{\sim}{\zeta}, \alpha))$.

(4) If in (2) in addition

(c)$^+$ $\underset{\sim}{r}$ is a $(\bar{Q}, \underset{\sim}{\zeta})$-named of a member of $P_\alpha / P_{\underset{\sim}{\zeta}}$ purely above $p{\restriction}[\underset{\sim}{\zeta}, \alpha)$

then we can add $p \leq_0 q$ (and now $q = p \cup (r{\restriction}[\underset{\sim}{\zeta}, \alpha))$.

Proof. Straightforward (think particularly on the case $\underset{\sim}{\zeta}[G] \in \mathbf{E}^* \cap E(\bar{Q})$).

2.13 Claim. Suppose

(a) $\bar{Q} = \langle P_i, \underset{\sim}{Q}_i : i < \alpha \rangle$ is a κ – Sp_2-iteration

(b) each Q_i satisfies $(\{2, \theta_1\}, \aleph_0)$ – Pr_1

(c) each $\underset{\sim}{Q}_i$ satisfies 2.1(1)(v) for $\theta \in \{2, \theta_1\}$

(d) κ is successor

Then

1) P_α satisfies 2.1(1)(v) for $\theta \in \{2, \theta_1\}$

2) for $\beta < \alpha$, P_α / P_β satisfies 2.1(1)(v) for $\theta \in \{2, \theta_1\}$

Proof. 1) Let $p \in P_\alpha$ and $\underset{\sim}{\tau}$ be a P_α-name of an ordinal $< \theta$, $\theta \in \{2, \theta_1\}$. We define a $\bar{Q}$-named $[0, \alpha)$-ordinal $\underset{\sim}{\zeta}$: for $r \in P_\beta$, $r \Vdash$ "$\underset{\sim}{\zeta} = \beta$" *iff*

(a) there are q, γ such that $r \cup (p \restriction [\beta, \alpha)) \leq_0 q \in P_\alpha$ and $q \restriction \beta = r \restriction \beta$ $(= r)$ and $\gamma < \theta$ and $q \Vdash$ "if $\theta = 2$, then $\underset{\sim}{\tau} = \gamma$ and: if $\theta > 2$ then $\underset{\sim}{\tau} < \gamma$"

(b) for no $\beta' < \beta$ and r', $r \restriction \beta' \leq r' \in P_{\beta'}$ does (r', β') satisfies (a).

Note that: if β is a limit cardinal we can get (by 2.8) a contradiction to clause (b).

However we would like to apply 2.12(2) and for this we need to prove that $\underset{\sim}{\zeta}$ is a $\underset{\sim}{Q}$-named ordinal, i.e. a $\underset{\sim}{Q}$ – $\mathbf{E}^*$-named ordinal. So (by 2.5C) let $\beta \leq \lambda$, $\beta \in \mathbf{E}^* \cap E(\bar{Q})$, and it suffice to find $\gamma < \beta$ such that $\Vdash_{\bar{Q}}$ "$\underset{\sim}{\zeta} \notin [\gamma, \beta)$". But $p \restriction \beta \in P_\beta = \bigcup_{\gamma < \beta} P_\gamma$, so for some $\gamma < \beta$, $p \restriction \beta \in P_\gamma$ and this γ is as required because we are using Sp_2 (and not Sp_1), that is; because if $p \restriction \beta \leq q \in P_\beta$ then $(q \restriction \beta) \cup p \restriction [\beta, \gamma) = q \restriction \beta \leq_0 q$.

Let $\underset{\sim}{q}^*$ be a $(\bar{Q}, \underset{\sim}{\zeta})$-named member of P_α as in clause (a). Let $p_0 \overset{\text{def}}{=} p$ and $p_1 = p_0 \cup (\underset{\sim}{q}^* \restriction [\underset{\sim}{\zeta}, \alpha))$, now $p_0 \leq_0 p_1 \in P_\alpha$ by 2.12(4). We now define $p_2 = p_1 \cup \{\underset{\sim}{r}_q : \underset{\sim}{q} \in p_1\}$ where $\underset{\sim}{r}_q$ is defined as follows

(*) if $\beta < \alpha$, $G_{P_\beta} \subseteq P_\beta$ generic over V and $\underset{\approx}{\zeta}_q[G_{P_\beta}] = \beta$ and in $V[G_{P_\beta}]$ there is r such that

  (i) $\hat{Q}_\beta[G_{P_\beta}] \models$ "$p \restriction \{\beta\} \leq_0 \{r\} \in \hat{Q}_\beta[G_{P_\beta}]$"

  (ii) for some $r_1 \in P_{\beta+1}$ we have: $r_1 \restriction \beta \in G_{P_\beta}$ and $r_1 \restriction \{\beta\} = r$ and: r_1 forces $(\Vdash_{P_{\beta+1}})$ $\underset{\sim}{\zeta} = \beta + 1$ or r_1 forces $\underset{\sim}{\zeta} \neq \beta + 1$ and in the former case if $\theta = 2$ the condition $r_1 \cup (p_1 \restriction [\beta+1, \alpha))$ forces a value to $\underset{\sim}{\tau}$, if $\theta \neq 2$ forces a bound to $\underset{\sim}{\tau}$

then $\underset{\sim}{r}_q[G_\beta]$ is the $\underset{\sim}{\leq}^*_\beta[G_\beta]$-first such a member otherwise it is $\emptyset_{\underset{\sim}{Q}_\beta}$.

Let us choose now $\beta \leq \alpha$ minimal such that

$\otimes$ there are $r_1 \in P_\beta$ and $\gamma < \theta$ such that $p_2 \restriction \beta \leq r_1$ and $r_1 \cup (p_2 \restriction [\beta, \alpha)) \Vdash$ "$\theta = 2, \underset{\sim}{\tau} < \gamma$ or $\theta \neq 2$, $\underset{\sim}{\tau} \neq \gamma$".

There is such β as for some $\beta < \alpha$ and r we have $p_2 \restriction \beta \leq r \in P_\beta$, and $r \Vdash$ "$\underset{\sim}{\zeta} = \beta$" and see the choice of $\underset{\sim}{\zeta}$ and p_1, p_2; actually we can use $\beta = \alpha$. If $\beta = 0$ we are done. If β is limit without loss of generality, by 2.8 for some $n < \omega$ and $\xi_1 < \ldots < \xi_n < \beta$ we have: $p_2 \restriction \beta \leq r_1$ above $\{\xi_1, \ldots, \xi_n\}$.

By the choice of r_1 there are $q \in P_\alpha$ and $\gamma < \theta$ such that: $q \restriction \beta = r_1$ and $r_1 \cup (p_2 \restriction [\beta, \alpha)) \leq_0 q$ and $q \Vdash_{P_\alpha}$ "$\theta = 2$, $\underset{\sim}{\tau} = \gamma$ or $\theta \neq 2$, $\underset{\sim}{\tau} < \gamma$", just use $q = r_1 \cup (p_2 \restriction [\beta, \alpha))$. Hence $\beta' = \xi_n + 1$, $r' = r_1 \restriction (\xi_n + 1)$ satisfy: $r' \in P_{\xi_n+1}$, $p_2 \restriction \beta' \leq r'$, $q \restriction \beta' = q \restriction (\xi_n + 1) = r_1 \restriction (\xi_n + 1) = r'$, and $q \Vdash_{P_\alpha}$ "$\theta = 2$, $\underset{\sim}{\tau} = \gamma$ or $\theta \neq 2$, $\underset{\sim}{\tau} < \gamma$" and $r' \cup (p \restriction [\beta', \alpha)) \leq_0 r' \cup (q \restriction [\beta', \alpha))$. So by the definition of $\underset{\sim}{\zeta}$ we have $r' \Vdash$ "$\underset{\sim}{\zeta} \leq \beta'$" and of course $\beta' < \beta$. So we get a contradiction to the choice of β. Lastly assume $\beta = j + 1$, and let $G_{P_j} \subseteq P_j$ be generic over V such that $r_1 \restriction j \in G_{P_j}$. If for some $\underset{\sim}{q} \in p_2$ we have $\underset{\sim}{\zeta}_q[G_{P_j}] = j$ (so we could use $\underset{\sim}{r}_q \in p_1$) the contradiction is gotten similarly using the definition of p_2 (note that for $\theta \neq 2$ we use the result for $\theta = 2$!). In the remaining case we can decrease β by the definition of $\leq_0^{P_\alpha}$ (as we use Sp_2 rather than Sp_1!).

2) Same proof (or 2.21(2)). $\square_{2.13}$

Now 2.10 + 2.13 suffice to show that no bounded subset of κ is added by the $\kappa - \mathrm{Sp}_2$- iteration (if say each $\underset{\sim}{Q}_i$ has $(\{\theta : \aleph_0 \leq \theta = \mathrm{cf}(\theta) < \kappa\} \cup \{2\}, \gamma) - \mathrm{Pr}_1^+$ for $\gamma < \kappa$. But we may like to deal with iterations which e.g. add reals. The next claim does better.

2.14 Claim. Assume

(a) $\bar{Q} = \langle P_i, \underset{\sim}{Q}_i : i < \alpha \rangle$ is a $\kappa - \mathrm{Sp}_2$-iteration

(b) each Q_i satisfies the $(\{\theta\}, \aleph_1) - \mathrm{Pr}_1^+$, and $\theta = \mathrm{cf}(\theta) > \aleph_0$

(c) κ is a successor cardinal.

Then

(1) P_α satisfies $(\{\theta\}, \aleph_1) - \mathrm{Pr}_1^+$

(2) also for $\beta < \alpha$, P_α / P_β satisfies $(\{\theta\}, \aleph_1) - \mathrm{Pr}_1^+$.

Proof of 2.14. Before proving, (in 2.14E) we define and prove in 2.14A – 2.14D, retaining our θ, κ, $\kappa - \mathrm{Sp}_2$-iteration $\bar{Q}$, and $\alpha = \ell g(\bar{Q})$. We can assume that 2.14 holds for any case with $\alpha' < \alpha$ instead of α

2.14A Definition.

(1) $\Gamma_1 = \{(p, \underset{\sim}{\zeta}, \underset{\sim}{\tau}) : p \in P_\alpha$ and $\underset{\sim}{\zeta}$ is a $\bar{Q}$-named $[0, \alpha)$-ordinal and $\underset{\sim}{\tau}$ is a P_α-name of an ordinal $< \theta$ such that: if $p \in G_\alpha \subseteq P_\alpha$, G_α generic over V and $\underset{\sim}{\zeta}[G_\alpha] = \zeta$ then for some $r \in G_\alpha \cap P_\zeta$ and $\varepsilon < \theta$ we have $r \Vdash_{P_\alpha}$ "$\underset{\sim}{\tau} < \varepsilon$"$\}$

(2) $\Gamma_2 = \{(p, \underset{\sim}{\zeta}, \underset{\sim}{\tau}) \in \Gamma_1$: $\Vdash_{\bar{Q}}$ "$\underset{\sim}{\zeta}$ is a non limit ordinal", and for every $\beta < \alpha$ we have: if there are r and $\underset{\sim}{q}$ such that $P_\beta \models$ "$p{\restriction}\beta \le r$", $r \in P_\beta$, $\Vdash_{P_\beta}$ "$\underset{\sim}{q} \in \underset{\sim}{Q}_\beta$ & $p{\restriction}\{\beta\} \le_0 \underset{\sim}{q}$" and $r \cup \{\underset{\sim}{q}\} \Vdash_{\bar{Q}}$ "$\underset{\sim}{\zeta} = \beta + 1$" or $r \cup \{\underset{\sim}{q}\} \Vdash_{\bar{Q}}$ "$\underset{\sim}{\zeta} \neq \beta + 1$" *then* we can use $q = p{\restriction}\{\beta\}\}$

(3) For $y = (p, \underset{\sim}{\zeta}, \underset{\sim}{\tau})$ let us define a P_α-name $\underset{\sim}{w}_y$: for $G_\alpha \subseteq P_\alpha$ generic over V

$$\begin{aligned} \underset{\sim}{w}_y[G_\alpha] = \{\beta < \alpha : \text{ for some } r \in G_\alpha \cap P_\beta \text{ and } \underset{\sim}{q} \in p \text{ we have} \\ r{\restriction}\beta \Vdash_{\bar{Q}} \text{“}\underset{\sim}{\zeta}_{\underset{\sim}{q}}[G_{P_\beta}] = \beta\text{” and} \\ p{\restriction}\beta \le r \text{ and } \neg(\exists r' \in P_{\beta+1})(r \le (r'{\restriction}\beta) \\ \& \; r'{\restriction}\beta \Vdash_{P_\beta} \text{“}p{\restriction}\{\beta\} \le_0 r'{\restriction}\{\beta\}\text{”} \\ \& \; r' \Vdash_{\bar{Q}} \text{“}\underset{\sim}{\zeta} \text{ is not } = \beta + 1\text{”})\}. \end{aligned}$$

2.14B Subclaim.

(1) If $\ell \in \{1, 2\}$ and $(p, \underset{\sim}{\zeta}, \underset{\sim}{\tau}) \in \Gamma_\ell$ and $p \le_0 p_1 \in P_\alpha$ *then* $(p_1, \underset{\sim}{\zeta}, \underset{\sim}{\tau}) \in \Gamma_\ell$

(2) If $p \in P_\alpha$ and $\underset{\sim}{\tau}$ is a P_α-name of an ordinal $< \theta$ *then* for some $(p', \underset{\sim}{\zeta}', \underset{\sim}{\tau}') \in \Gamma_1$ we have $p \le_0 p'$ and $p' \Vdash_{P_\alpha}$ "$\underset{\sim}{\tau}' = \underset{\sim}{\tau}$"

(3) If $(p, \underset{\sim}{\zeta}, \underset{\sim}{\tau}) \in \Gamma_1$ *then* for some $(p', \underset{\sim}{\zeta}', \underset{\sim}{\tau}') \in \Gamma_2$ we have $p \leq_0 p'$, $\Vdash_{P_\alpha}$ "$\underset{\sim}{\zeta}' \leq \underset{\sim}{\zeta}$ and $\underset{\sim}{\tau}' \geq \underset{\sim}{\tau}$".

Proof. 1) Read the definitions.
2), 3) Like the proof of 2.13. $\square_{2.14B}$

2.14C Subclaim. If $y = (p, \underset{\sim}{\zeta}, \underset{\sim}{\tau}) \in \Gamma_2$ then the (P_α-name of a) set $\underset{\sim}{w}_y$ satisfies:

(a) $\Vdash_{P_\alpha}$ "$\beta \in \underset{\sim}{w}_y \Rightarrow \beta < \underset{\sim}{\zeta}$"

(b) if $G_\alpha \subseteq P_\alpha$ is generic over V and $\beta \in \underset{\sim}{w}_y[G_\alpha]$ *then* some $r \in G_\alpha \cap P_\beta$ forces this; in fact if $r \Vdash_{P_\alpha}$ "$\beta \in \underset{\sim}{w}_y$" then $r{\restriction}\beta \Vdash_{P_\alpha}$ "$\beta \in \underset{\sim}{w}_y$".

(c) $\Vdash_{P_\alpha}$ "$\underset{\sim}{w}_y$ is a finite subset of α"

Proof. For clauses (a), (b) read 2.14A(3), so let us prove clause (c).

If not, for some $G_{P_\alpha} \subseteq P_\alpha$ generic over V, and $\underset{\sim}{w}_y[G_{P_\alpha}]$ is infinite, and let $\zeta_0 < \zeta_1 \ldots$ be the first ω members. Let $\delta \stackrel{\text{def}}{=} \bigcup_{n<\omega} \zeta_n$, so $V[G_{P_\alpha}] \models$ "$\text{cf}(\delta) = \aleph_0$".

Let $\underset{\sim}{\delta}$ and $\langle \underset{\sim}{\zeta}_n : n < \omega \rangle$ be the corresponding P_α-names, so there are $r \in G_{P_\alpha}$ and β and $\delta \leq \alpha$ such that $r \Vdash_{P_\alpha}$ "$\underset{\sim}{\zeta} = \beta$ and $\underset{\sim}{\delta} = \delta$ (and $\underset{\sim}{w}_y$ is infinite)". Now as $\underset{\sim}{\zeta}_n[G_\alpha] \in \underset{\sim}{w}_y$, by Definition 2.14A(3) (the clause $p{\restriction}\beta \leq r$) we have $p{\restriction}\underset{\sim}{\zeta}_n[G_\alpha] \in G_\alpha$; and as this hold for every n we have $p{\restriction}\delta \in G_\alpha$, so as we can increase r w.l.o.g. $p{\restriction}\delta \leq r$. Hence by 2.8 without loss of generality for some $n < \omega$, and $\xi_1 < \ldots, < \xi_n < \alpha$, we have $p{\restriction}\delta \leq r$ above $\{\xi_1, \ldots, \xi_n\}$ so letting ξ_0 be: β if $\beta < \delta$, 0 if $\beta \geq \delta$ and letting $\xi = \sup[\delta \cap \{\xi_0, \xi_1, \ldots, \xi_n\}]$ we know that $\xi < \delta$ and $r \Vdash_{P_\alpha}$ "$\underset{\sim}{w}_y \cap (\xi, \delta) \neq \emptyset$" hence for some ε and r_1, we have $r \leq r_1 \in P_\alpha$ and $\varepsilon \in (\xi, \delta)$ and $r_1 \Vdash_{P_\alpha}$ "$\varepsilon \in \underset{\sim}{w}_y$". But by the definition of $\underset{\sim}{w}_y$, i.e. by 2.14C(b) we have: $r_1{\restriction}\varepsilon \Vdash_{P_\varepsilon}$ "$\varepsilon \in \underset{\sim}{w}_y$" and clearly $r_1{\restriction}\varepsilon \Vdash_{P_\varepsilon}$ "$p{\restriction}\{\varepsilon\} \leq_0 r{\restriction}\{\varepsilon\}$" hence by the definition of $\underset{\sim}{w}_y$ we have $(r_1{\restriction}\varepsilon) \cup (r{\restriction}\{\varepsilon\}) \nVdash_{P_{\varepsilon+1}}$ "$\underset{\sim}{\zeta}$ is not $= \varepsilon + 1$" hence (as $\underset{\sim}{\zeta}$ is a $\bar{Q}$-named ordinal) there is r_2 such that $(r_1{\restriction}\varepsilon) \cup (r{\restriction}\{\varepsilon\}) \leq r_2 \in P_{\varepsilon+1}$ and $r_2 \Vdash_{\bar{Q}}$ "$\underset{\sim}{\zeta} = \varepsilon + 1$". But $\varepsilon + 1 \neq \beta$ by the choice of ξ_0 and ξ and ε, and $r \Vdash_{P_\alpha}$ "$\underset{\sim}{\zeta} = \beta$" so r_2, r should be incompatible in P. But

$$r{\restriction}(\varepsilon + 1) \leq (r_1{\restriction}\varepsilon) \cup (r{\restriction}\{\varepsilon\}) \leq r_2 \in P_{\varepsilon+1}.$$

[Why? As $r \leq r_1$, and choice of r_2 (twice).]
Hence by 2.7 clause (F) we know r, r_2 are compatible, contradiction. $\square_{2.14C}$

2.14D Subclaim. If $y = (p, \underset{\sim}{\zeta}, \underset{\sim}{\tau}) \in \Gamma_2$ *then* we can find p^+, $\underset{\sim}{\zeta}_n$, $\underset{\sim}{\tau}_n$ for $n < \omega$ such that:

(a) $p \leq_0 p^+ \in P_\alpha$

(b) $(p^+, \underset{\sim}{\zeta}_n, \underset{\sim}{\tau}_n) \in \Gamma_1$

(c) if $G_\alpha \subseteq P_\alpha$ is generic over V, then $\underset{\sim}{\zeta}_n[G_\alpha]$ is the n-th member of $\underset{\sim}{w}_y[G_\alpha]$ if there is one, if so then for some $r \in G_{\underset{\sim}{\zeta}_n[G_\alpha]}$ we have $p^+ \restriction \underset{\sim}{\zeta}_n[G_\alpha] \leq r$ and $r \cup (p^+ \restriction \{\underset{\sim}{\zeta}_n[G_\alpha]\}) \Vdash_{P_\alpha}$ " if $\underset{\sim}{\zeta}[\underset{\sim}{G}_{P_\alpha}] = \underset{\sim}{\zeta}_n[G_\alpha] + 1$ then $\underset{\sim}{\tau}[\underset{\sim}{G}_\alpha] \leq \underset{\sim}{\tau}_n[\underset{\sim}{G}_\alpha]$".

Proof. Straight using 2.10 (for $\theta = \aleph_1$) to have a $\leq_0$-upper bound, and taking care of n work as in 2.14B(2) i.e. as in 2.13. $\square_{2.14D}$

2.14E Completion of the proof of 2.14. We concentrate on part (1) (the proof of part (2) being similar or use 2.21(3), and it is not used). By 2.7, clause (E) we know that (i) - (iv) of Definition 2.1(1) holds, and by 2.10(1) not only clause (vi) of Definition 2.1(1) and the extra demand from Definition 2.1(2) hold, so the problem is to verify clause (v) in Definition 2.1(1), i.e. the pure decidability. So let $p \in P_\alpha$ and $\underset{\sim}{\tau}$ is a P_α-name and $p \Vdash$ "$\underset{\sim}{\tau} < \theta$", we have to find q, j such that $p \leq_0 q \in P_\alpha$ and $j < \theta$ and $q \Vdash_{P_\alpha}$ "$\underset{\sim}{\tau} \leq j$". So we can replace p, $\underset{\sim}{\tau}$ by p', $\underset{\sim}{\tau}'$ if $p \leq_0 p'$ and $p' \Vdash$ "$\underset{\sim}{\tau} \leq \underset{\sim}{\tau}' < \theta$". By subclaim 2.14B(2)+(3) w.l.o.g. for some $\underset{\sim}{\zeta}$ the triple $(p, \underset{\sim}{\zeta}, \underset{\sim}{\tau})$ belongs to Γ_2. We choose by induction on $n < \omega$, p_n, p_n^+ and $\langle (\underset{\sim}{\zeta}_\eta, \underset{\sim}{\tau}_\eta, \underset{\sim}{j}_\eta, \underset{\sim}{\tau}'_\eta) : \eta \in {}^n\omega \rangle$ such that:

(a) $p_0 = p$, $\tau_{\langle\rangle} = \underset{\sim}{\tau}$, $\underset{\sim}{\zeta}_{\langle\rangle} = \underset{\sim}{\zeta} = \underset{\sim}{j}_{\langle\rangle}$

(b) $(p_n, \underset{\sim}{\zeta}_\eta, \underset{\sim}{\tau}_\eta) \in \Gamma_2$ for each $n < \omega$, $\eta \in {}^n\omega$

(c) $p_n \leq_0 p_n^+ \leq_0 p_{n+1}$

(d) for each n and $\eta \in {}^n\omega$ we have:

  (i) $\{\underset{\sim}{j}_{\eta\hat{}\langle k\rangle} : k < \omega\}$ list $\underset{\sim}{w}_{(p_n, \underset{\sim}{\zeta}_\eta, \underset{\sim}{\tau}_\eta)} \cup \{0\}$

  (ii) $\underset{\sim}{j}_{\eta\hat{}\langle k\rangle}$ is a $\bar{Q}$-named $[0, \alpha)$-ordinal

  (iii) $\underset{\sim}{\tau}'_{\eta\hat{}\langle k\rangle}$ is a $P_{\underset{\sim}{j}_{\eta\hat{}\langle k\rangle}}$-name of an ordinal $< \theta$

(iv) if $G_\alpha \subseteq P_\alpha$ is generic over V, and $\beta = \underset{\sim}{j}_\eta[G_\alpha]$, and $p^+_n \restriction (\beta+1) \in G_\alpha$ and $\beta = \underset{\sim}{\zeta}_\eta[G_\alpha]$ *then* $\underset{\sim}{\tau}'_{\eta\hat{}\langle k\rangle}[G_\alpha] \geq \underset{\sim}{\tau}_\eta[G_\alpha]$

(v) $(p^+_n, \underset{\sim}{j}_{\eta\hat{}\langle k\rangle}, \underset{\sim}{\tau}'_{\eta\hat{}\langle k\rangle}) \in \Gamma_1$

(e) $p_{n+1} \Vdash_{P_\alpha}$ "$\underset{\sim}{\zeta}_{\eta\hat{}\langle k\rangle} \leq \underset{\sim}{j}_{\eta\hat{}\langle k\rangle}$, and $\underset{\sim}{\zeta}_{\eta\hat{}\langle k\rangle}$ is non limit and $\underset{\sim}{\tau}_{\eta\hat{}\langle k\rangle} \geq \underset{\sim}{\tau}'_{\eta\hat{}\langle k\rangle}$" for $n < \omega$, $\eta \in {}^n\omega$, and $k < \omega$.

The case $n = 0$ is straight. Having arrived to stage n, i.e. p_n and $\underset{\sim}{\zeta}_\eta$, $\underset{\sim}{\tau}_\eta$ for $\eta \in {}^{n\geq}\omega$ are defined and as required, list $({}^n\omega)$ as $\langle \eta_\ell : \ell < \omega\rangle$ and choose by induction on $\ell < \omega$, $p^+_{n,\ell}$, $\underset{\sim}{j}_{\eta_\ell\hat{}\langle k\rangle}$, $\tau'_{\eta_\ell\hat{}\langle k\rangle}$ for $k < \omega$ such that $p^+_{n,0} = p_n$, $p^+_{n,\ell} \leq_0 p^+_{n,\ell+1}$ and $p^+_{n,\ell+1}$ satisfies the requirements of p^+_n for $\eta = \eta_\ell$ which is possible by Subclaim 2.14D. Then let $p^+_n = \bigcup_\ell p^+_{n,\ell}$ (it is a $\leq_0$-upper bound of $\{p^+_{n,\ell} : \ell < \omega\}$; by 2.10 + assumption (b) of 2.14 it exists, but why it still satisfies the demands? By Subclaim 2.14B(1)). Now the choice of p_{n+1}, $\underset{\sim}{\zeta}_{\eta\hat{}\langle k\rangle}$, $\underset{\sim}{\tau}_{\eta\hat{}\langle k\rangle}$ for $\eta \in {}^n\omega$, $k < \omega$ is by Subclaim 2.14B(3) again using 2.10(1).

Again there is p^+, a $\leq_0$-upper bound of $\{p_n : n < \omega\}$, it satisfies $p \leq_0 p_n \leq_0 p^+ \in P_\alpha$ and letting $\gamma = \sup\{\underset{\sim}{\tau}_\eta : \underset{\sim}{j}_\eta = 0\} < \theta$ we can prove $p^+ \Vdash$ "$\underset{\sim}{\tau}_\eta \leq \gamma$ when $\underset{\sim}{\tau}_\eta$ is defined". For this we prove by induction on $j < \alpha$ that $p\restriction j \Vdash_{P_j}$ "if $\eta \in {}^{\omega>}\omega$ and $\underset{\sim}{j}_\eta \leq j$ then $\underset{\sim}{\tau}_\eta \leq \gamma$"(similarly to the proof of 2.13). As $\underset{\sim}{\tau} = \underset{\sim}{\tau}_{\langle\rangle}$ we are done. $\square_{2.14}$

Remark. Actually the tree we use is of finite splitting.

2.15 Conclusion. Assume

(a) $\bar{Q} = \langle P_i, \underset{\sim}{Q}_i : i < \alpha\rangle$ is a $\kappa - \mathrm{Sp}_2 - \mathbf{E}^*$-iteration

(b) each $\underset{\sim}{Q}_i$ satisfies $(S, <\kappa) - \mathrm{Pr}^+_1$ and $(\aleph_0 \in S \Rightarrow 2 \in S)$

(c) κ successor

Then

(1) P_α satisfies $(S, <\kappa) - \mathrm{Pr}^+_1$

(2) for $\beta < \alpha$, P_α/P_β satisfies $(S, <\kappa) - \mathrm{Pr}^+_1$

Remark. 1) Note: if κ is not a cardinal we can replace it by $|\kappa|^+$; *but* during the iteration $|\kappa|^+$ may increase.

2) We can replace Pr_1^+ by Pr_1 with minor changes but because of 2.6 the gain is doubtful.

Proof. By 2.10, 2.14 (and 2.7 of course). $\square_{2.15}$

2.16 Discussion. Suppose α is an ordinal and $\bar{Q} = \langle P_i, \underset{\sim}{Q}_i : i < \alpha\rangle$ is a $\kappa - \mathrm{Sp}_2$-iteration (so $P_\alpha = \mathrm{Sp}_2 - \mathrm{Lim}_\kappa(\bar{Q})$). We may wonder whether:

(a) If α is strong inaccessible and density$(P_i) < \alpha$ for $i < \alpha$ *then* density$(P_\alpha) = \alpha$.

(b) *If* α is a Mahlo and $\forall i < \alpha[|P_i| < \alpha]$, *then* P_α satisfies the α-c.c.

As unlike X §1 we use antichains of $\bigcup_{i<\alpha} P_i$ (rather than antichains which are maximal in P whenever $\bigwedge_{i<\alpha} P_i \lessdot P$) this is not clear. Note that in 2.17 below, we can weaken the Pr_1 demand to $\Vdash_{P_\alpha}$ "θ remains regular".

2.17 Lemma. Suppose $\bar{Q} = \langle P_i, \underset{\sim}{Q}_i : i < \alpha\rangle$ is a κ-$\mathrm{Sp}_2 - \mathbf{E}^*$-iteration, $\kappa > \aleph_0$ a successor cardinal, $S \subseteq \{2\} \cup \{\mu : \aleph_0 \le \mu, \mu \text{ regular }\}$, $\aleph_0 \in S = 2 \in S$, and each Q_i (in V^{P_i}), has $(S, < \kappa) - \mathrm{Pr}_1^+$ (see 2.1(4)). Then:

(1) If $\kappa \le \theta \in S$ and $\mathrm{cf}(\alpha) = \theta$ *then* $\bigcup_{i<\alpha} P_i$ is dense in P_α.

(2) If α is strongly inaccessible $> \min(S \setminus \kappa)$, $\alpha > |P_i| + \kappa$ for $i < \alpha$ (or just P_i satisfies the α-c.c.) and $\alpha \in \mathbf{E}^*$ *then* P_α satisfies the α-chain condition (in a strong sense).

(3) If each Q_i satisfies $(\mathrm{RCar} \setminus \kappa, \kappa) - \mathrm{Pr}_1^+$ and has power $\le \chi$, *then* P_α has a dense subset (even a $\le_0$-dense subset) of power $(2^{|\alpha|+\chi})^{<\kappa}$ and satisfies $(\chi^{<\kappa})^+$-c.c.

(3) If α is strongly inaccessible and $\mathbf{E}^* \cap \alpha$ is a stationary subset of α and $[i < \alpha \Rightarrow |P_i| < \alpha]$ or at least $[i < \alpha \Rightarrow P_i$ satisfies the λ_i-c.c. for some $\lambda_i < \alpha]$ *then* P_α satisfies the α-chain condition (a strong version indeed).

Proof: 1) Left to the reader.

2) Choose $\theta \in S \setminus \kappa$. Let $\langle p_j : j < \alpha\rangle$ be a sequence of elements of P_α now as $\alpha \in \mathbf{E}^*$ we have $p_j \in \bigcup_{i<\alpha} P_i$. Let $A = \{j < \alpha : \mathrm{cf}(j) = \theta\}$, this is a stationary set. For $j \in A$ choose $r_j \in \bigcup_{i<j} P_i$, $r_j \ge p_j \restriction j$ (why such r_j exist? by part (1)),

say $r_j \in P_{i(j)}$, $i(j) < j$. Let $C = \{i : j \text{ limit}, \forall i < j \exists \gamma < j[p_i \in P_\gamma]\}$. This is a club subset of α.

By Fodor's lemma we can find $B \subseteq C \cap A$ stationary such that for all $j_1, j_2 \in B$ we have $i(j_1) = i(j_2)$ and $r_{j_1} = r_{j_2} = r$ or at least r_{j_1}, r_{j_2} are compatible in $\bigcup_{i<i(j_1)} P_i$ and let $r_{j_1}, r_{j_2} \leq r \in \bigcup_{i<i(j_1)} P_i$.

(Remember that $|P_i| < \alpha$ or at least P_i satisfies the α-c.c. for $i < \alpha$)

But for any such $j_1 < j_2$ the condition $r \cup p_{j_1} \restriction [j_1, j_2) \cup p_{j_2} \restriction [j_2, \alpha)$ is a common upper bound for p_{j_1}, p_{j_2}.

(3) Like III 4.1 use only names which are hereditarily $< \kappa$ (see below).

(4) Like part (2) using 2.7(E) (so 2.5(1)(b)) instead using part (1). $\square_{2.17}$

We may wonder about $\kappa - \mathrm{Sp}_e$-iterations which essentially do not change cofinality.

2.18 Definition. We define for an Sp_e-iteration $\bar{Q}$, and cardinal μ (μ regular), what is a $\bar{Q}$-name hereditarily $< \mu$, and in particular a $\bar{Q}$-named $[j, \alpha)$-ordinal hereditarily $< \mu$ and a $\bar{Q}$-named $[j, \alpha)$-atomic condition hereditarily $< \mu$, and which conditions of $\mathrm{Sp}_e\text{-}\mathrm{Lim}_\kappa \bar{Q}$ are hereditarily $< \mu$ (formally they are not special cases of the corresponding notions without the "hereditarily $< \mu$"). For simplicity we are assuming that the set of members of Q_i is in V. This is done by induction on $\alpha = \ell g(\bar{Q})$.

First case. $\alpha = 0$

trivial

Second case. $\alpha > 0$

(A) A $\bar{Q}$-named $[j, \alpha)$-ordinal $\underset{\sim}{\xi}$ hereditarily $< \mu$ is a $\bar{Q}$-named $[j, \alpha)$-ordinal which can be represented as follows: there is $\langle (p_i, \xi_i) : i < i^* \rangle, i^* < \mu$, each ξ_i an ordinal in $[j, \alpha), p_i \in P_{\xi_i}$ is a member of P_{ξ_i} hereditarily $< \mu$ and for any $G \in \mathrm{Gen}^r(\bar{Q}), \underset{\sim}{\zeta}[G]$ is ζ iff for some i we have

(a) $p_i \in G, \zeta_i = \zeta$

(b) if $p_j \in G$ then $\zeta_i < \zeta_j \vee (\zeta_i = \zeta_j \& i < j)$

(B) A $\bar{Q}$-named $[j, \alpha)$-atomic condition $\underset{\sim}{q}$ hereditarily $< \mu$, is a $\bar{Q}$-named $[j, \alpha)$-atomic condition which can be represented as follows: there is $\langle (p_i, \zeta_i, q_i) : i < i^* \rangle, i^* < \mu, \zeta_i \in [j, \alpha), p_i \in P_{\zeta_i}, q_i \in V$, and for any $G \in \mathrm{Gen}^r(\bar{Q}), \underset{\sim}{q}[G]$ is q iff for some i:

(a) $p_i \in G, q = q_i$, and $p_i \Vdash_{P_{\zeta_i}}$ "$q \in \underset{\sim}{Q}_{\zeta_i}$"

(b) if $p_j \in G$ then $\zeta_i < \zeta_j \vee (\zeta_i = \zeta_j \ \&\ i < j)$

(C) A member p of $P_\alpha = \mathrm{Sp}_e - \mathrm{Lim}_\kappa(\bar{Q})$ is hereditarily $< \mu$ if each member of r is a $\bar{Q}$-named atomic condition hereditarily $< \mu$.

(D) A $\bar{Q}$-name of a member of V hereditarily $< \mu$ is defined as in clause (B), similarly for member $x \in V^{P_\alpha}$ such that $y \in$ transitive closure of $x \ \&\ y \notin V \Rightarrow |y| < \mu$.

2.19 Claim. Suppose $\langle P_i, Q_i : i < \alpha \rangle$ is an $\kappa - \mathrm{Sp}_e$-iteration, $P_\alpha = \mathrm{Sp}_e\text{-}\mathrm{Lim}_\kappa \bar{Q}$, κ a successor cardinal, each Q_i (in V^{P_i}) satisfies $(\mathrm{RCar}^{V^{P_i}} \setminus \kappa, \kappa) - \mathrm{Pr}_1^+$. *Then*

(1) $\{p \in P_\alpha : p$ hereditarily $< \kappa\}$ is a dense subset of P_α, even a $\leq_0$-dense subset

(2) P_α preserve "$\mathrm{cf}(\delta) \geq \kappa$"

(3) for every $p \in P_\alpha$ and $\underset{\sim}{\tau}$ a P_α-name of an ordinal there are $p^*, p \leq_0 p^* \in P_\alpha$, and $A \in V$, a set of $< \kappa$ ordinals such that $q \Vdash_{p_\alpha}$ "$\underset{\sim}{\tau} \in A$".

Proof. Should be clear. $\square_{2.19}$

2.20 Remark. We can also get a similar theorem for forcing notions $(Q, \leq, \leq_0)$ as in 2.1 where instead of $\leq_0$ is κ-directed complete (see 2.4(d)) we demand that (vi) ("strategical completeness" of $\leq_0$).

2.21 Claim. (1) Suppose F is a function and $e = \{1, 2\}$, *then* for every ordinal α there is Sp_e-iteration $\bar{Q} = \langle P_i, \underset{\sim}{Q}_i : i < \alpha^\dagger \rangle$, such that:

(a) for every $i, \underset{\sim}{Q}_i = F(\underset{\sim}{Q} \restriction i)$,

(b) $\alpha^\dagger \leq \alpha$,

(c) either $\alpha^\dagger = \alpha$ or $F(\bar{Q})$ is not an $(\mathrm{Rlim}\bar{Q})$-name of a forcing notion.

(2) Suppose $\bar{Q}$ is an $\kappa-\mathrm{Sp}_e$-iteration of length α and $\beta < \alpha$, $G_\beta \subseteq P_\beta$ is generic over V, *then* in $V[G_\beta]$, $\bar{Q}/G_\beta = \langle P_i/G_\beta, \underset{\sim}{Q}_i : \beta \le i < \alpha\rangle$ is an Sp_e-iteration and $\mathrm{Sp}_e - \mathrm{Lim}(\bar{Q}) = P_\beta * (\mathrm{Sp}_e - \mathrm{Lim}\bar{Q}/\underset{\sim}{G}_\beta)$ (essentially).

(3) If $\bar{Q}$ is an $\kappa - \mathrm{Sp}_2$-iteration, $p \in \mathrm{Sp}_2 - \mathrm{Lim}(\bar{Q})$, $P'_i = \{q \in P_i : q \ge p\restriction i\}, \underset{\sim}{Q}'_i = \{p \in \underset{\sim}{Q}_i : p \ge p\restriction\{i\}\}$ *then* $\bar{Q} = \langle P'_i, Q'_i : i < \ell g\bar{Q}\rangle$ is (essentially) an Sp_e iteration (and $\mathrm{Sp}_2\mathrm{Lim}(\bar{Q}')$ is $P'_{\ell g\bar{Q}}$).

Proof. Should be clear. $\square_{2.12}$

§3. Axioms

We can get from the lemma of preservation of forcing with $(S,\gamma) - Pr_1^+$ by $\kappa - Sp_2$ iteration (and on the λ-c.c. for then) forcing axioms. We list below some variations.

3.1 Notation. 1) Reasonable choices for S are

(A) $S^0_\kappa = \mathrm{RUCar}_{\le\kappa} = \{\mu : \mu \text{ a regular cardinal}, \aleph_0 < \mu \le \kappa\}$

(B) $S^2_\kappa = \{2\} \cup \mathrm{RCar}_{\le\kappa} = \{2\} \cup \{\mu : \mu \text{ a regular cardinal}, \aleph_0 \le \mu \le \kappa\}$

(C) If we write "$< \kappa$" instead $\le \kappa$ (and $S^\ell_{<\kappa}$ instead S^ℓ_κ) the meaning should be clear.

2) [Convention] $\mathbf{E}^*$ is the class of strongly inaccessible cardinal $> \kappa$.

3.2 Fact. Suppose the forcing notion P satisfies $(S,\gamma) - \mathrm{Pr}_1$

(1) If $2 \in S$ *then* P does not add any bounded subset of γ.

(2) If μ is regular, and $\lambda_i (i < \mu)$ are regular, and $\{\mu\} \cup \{\lambda_i : i < \mu\} \subseteq S$, D here is a uniform ultrafilter on $\mu, \theta = \mathrm{cf}(\prod_{i<\mu} \lambda_i/D)$ (λ_i-as an ordered set) *then* P satisfies $(S \cup \{\theta\}, \gamma') - \mathrm{Pr}_1$ whenever $\mu\gamma' \le \gamma$, ($\mu\gamma'$ is ordinal multiplication). We can do this for all such θ simultaneously.

(3) If $\lambda \in S$ is regular, $\mu < \gamma$ *then* for every $f : \mu \to \lambda$ from V^P, for some $g : \mu \to \lambda$ from V for every $\alpha < \mu$, $f(\alpha) < g(\alpha)$.

Proof. (1) and (3) are clear.
For (2), fix an increasing cofinal (modulo D) sequence $\langle f^\theta_\alpha : \alpha < \theta \rangle$ in $\prod_{i<\mu_\theta} \lambda^\theta_i$ if $\theta = \mathrm{cf}(\prod_{i<\mu} \lambda_i / D_\theta)$ with $\lambda^\theta_i \in S$, $\mu = \mu_\theta \in S$.
To play the game ∂_1 for $(S \cup \{\theta\}, \gamma')-\mathrm{Pr}_1$ from 2.4(b), player II on the side plays the ∂_2 game for (S, γ) but for move β in ∂_1, he uses moves $\langle \zeta : \bigcup_{\gamma<\beta} \zeta_\gamma \leq \zeta < \zeta_\beta \rangle$ in ∂_2, he also chooses ζ_α's during the play. If player I chooses a $\lambda_\beta \in S, \underset{\sim}{I}_\beta$, in ∂_1, player II copies I's move to ∂_2 and plays his answer from there and let $\zeta_\beta = \bigcup_{\gamma<\beta} \zeta_\gamma + 1$. If player I plays in the β-th move $\lambda_\beta = \mathrm{cf}(\prod_{j<\mu} \lambda_{\beta,j} / D_\beta)$, $\underset{\sim}{I}_\beta$, player II simulates μ moves of ∂_2:

$$\langle \lambda_{\beta,j},\ f^{\lambda_\beta}_{\underset{\sim}{I}_\beta}(j),\ \alpha^\beta_j : j < \mu \rangle.$$

Then player II finds α_i such that $\{j : \alpha^\beta_j < f_{\alpha_i}(j)\} \in D$ and plays this α_i and let $\zeta_\beta = \bigcup_{\gamma<\beta} \zeta_\gamma + \mu$.
It is clear that α_i is as required, and as $\mu\gamma' \leq \gamma$, ∂_2 does not end before ∂_1.
$\square_{3.2}$

3.3 Claim. Suppose $\mathrm{MA}_{<\kappa}$ holds (i.e., for every P satisfying the $\aleph_1$-c.c. and dense $\mathcal{I}_i \subseteq P$ (for $i < \alpha < \kappa$) there is a directed $G \subseteq Q$ such that $\bigwedge_{i<\alpha} G \cap \mathcal{I}_i \neq \emptyset$). *Then* the following forcing notions are equivalent to forcing notions having the $(\mathrm{RUCar}, \kappa) - \mathrm{Pr}_1$.

(1) Mathias forcing; $\{(w, A) : w \subseteq \omega \text{ finite}, A \subseteq \omega \text{ infinite}\}$ with the order

$$(w_1, A_1) \leq (w_2, A_2) \text{ iff } w_1 \subseteq w_2 \subseteq w_1 \cup A_1,\ A_2 \subseteq A_1.$$

(2) The forcing from VI §6(=[Sh:207], Sect. 2).

Proof. (1) Let P' be the set of (w, A, B) satisfying: $w \subseteq \omega$ finite, $B \subseteq \omega$ infinite, $B \subseteq A \subseteq \omega$, with the order

$$(w_1, A_1, B_1) \leq (w_2, A_2, B_2) \text{ iff } (w_1, A_1) \leq (w_2, A_2)$$
$$\text{and } B_2 \subseteq^* B_1 (\text{i.e. }, B_2 \setminus B_1 \text{ finite})$$

$$(w_1, A_1, B_1) \leq_0 (w_2, A_2, B_2) \text{ iff: } w_1 = w_2$$
$$A_1 = A_2$$
$$B_2 \subseteq^* B_1.$$

Let us check Definition 2.1: (i) – (iv) easy.

Note that $\{(w, A, A) : (w, A, A) \in P'\}$ is dense in P', and isomorphic to P.

Proof of (v). Let $\mu > \aleph_0$ be a regular cardinal, $\underset{\sim}{\tau}$ a P'-name, $\Vdash_P$ "$\underset{\sim}{\tau} < \mu$. Let $p = (w, A, B)$ be given. Choose by induction on $i < \omega, n_i, B_i$ such that

(a) $B_0 = B(\subseteq A)$

(b) $n_i = \text{Min}(B_i)$

(c) $B_{i+1} \subseteq B_i \setminus \{n_i\}$

(d) for every $u \subseteq \{0, 1, 2, \ldots, n_i\}$ (not just $\subseteq \{n_0, n_1, \ldots, n_i\}$!) one of the following occurs:

for some $\alpha_{i,u} < \mu$, we have $(u, B_{i+1}, B_{i+1}) \Vdash_P$ "$\underset{\sim}{\tau} = \alpha_{i,u}$"

or for no infinite $C \subseteq B_{i+1}$ and $\alpha < \mu$ do we have $(u, C, C) \Vdash$ "$\underset{\sim}{\tau} = \alpha$"

There is no problem to do this, now $q \overset{\text{def}}{=} (w, A, \{n_i : i < \omega\})$ satisfies:

(e) $p \leq q \in P'$ and even $p \leq_0 q$.

(f) $q \Vdash_{P'}$ "$\underset{\sim}{\tau} \in \{\alpha_{i,u} : i < \omega \text{ and } u \subseteq \{0, 1, 2, \ldots, n_i\}\}$".

[Why? If not, then for some $\alpha \in \mu \setminus \{\alpha_{i,u} : i < \omega \text{ and } u \subseteq \{0, 1, \ldots, n_i\}\}$ and r we have $q \leq r \in P'$ and $r \Vdash_{P'}$ "$\underset{\sim}{\tau} = \alpha$". Let $r = (v, A', B')$ so $B' \subseteq A'$, B' is infinite, $B' \subseteq^* \{n_i : i < \omega\}$ and $A' \subseteq A$. As v is finite and by the definition of $\subseteq^*$ there is $i < \omega$ such that: $v \subseteq \{0, \ldots, n_i\}$ and $B' \setminus \{0, \ldots, n_i\} \subseteq \{n_j : j < \omega\}$. So without loss of generality $\text{Min}(B') > n_i$, and $A' = B'$; so by the choice of B_{i+1}, $(v, B_{i+1}, B_{i+1}) \Vdash$ "$\underset{\sim}{\tau} = \alpha_{i,u}$", but $\alpha \neq \alpha_{i,v}$ so (v, B_{i+1}, B_{i+1}), $r = (v, A', B')$ are incompatible, contradiction]. So q is as required.

Proof of (vi). Suppose $p_i (i < \gamma)$ is $\leq_0$-increasing so $p_i = (w, A, B_i)$ and $B_i \subseteq A$, and B_i is $\subseteq^*$ decreasing. It is well known that for $\gamma < \kappa$, $MA_{<\kappa}$ implies the existence of an infinite $B \subseteq \omega$ such that $(\forall i < \gamma) B \subseteq^* B_i$. Now $(w, A, B) \in P'$ and $i < \gamma \Rightarrow p_i \leq_0 (w, A, B)$, as required.

(2) Left to the reader (similar to the proof of (1)). $\square_{3.3}$

3.4 Discussion - Proofs. Let $\kappa < \lambda$, λ regular. Each of the following gives rise naturally to a forcing axiom, stronger as λ is demanded to be a larger cardinal (so if λ is supercompact we get parallels to PFA).

If φ is a property of forcing notions, let $Ax_{<\alpha}(\varphi, \lambda, \mu)$ be the following statement: For every forcing notion P of size $< \mu$ if P satisfies φ and $\bar{\mathcal{I}} = \langle \mathcal{I}_i : i < i^* < \lambda \rangle$ is a sequence of dense subsets of P and $\langle (\kappa_j, \underset{\sim}{S}_j) : j < j^* < \alpha \rangle$ a sequence of pairs, with $\underset{\sim}{S}_j$ a P-name of stationary subset of κ_j, where κ_j is a regular uncountable cardinal $< \lambda$ *then* there is a $\bar{\mathcal{I}}$-generic subset G of P such that (as say $i < i^* \Rightarrow \underset{\sim}{\mathcal{I}}_i \cap G \neq \emptyset$ and) $j < j^* \Rightarrow \underset{\sim}{S}_j[G]$ a stationary subset of κ_j.

Case I. We assume κ is successor cardinal and use $\bar{Q}$ of length λ, a $\kappa - \mathrm{Sp}_2$-iteration, $\Vdash_{P_i}$ "$|Q_i| < \lambda$", each Q_i having (S^ℓ_κ, κ)- Pr^+_1 and $\ell \in \{0, 2\}$ and (usually) $\diamondsuit_{\{\mu : \mu < \lambda \text{ is strongly inaccessible}\}}$.

Now $P_\lambda = \kappa - \mathrm{Sp}_2\text{-}\mathrm{Lim}_\kappa \bar{Q}$ have the $(S^\ell_\kappa, \kappa) - \mathrm{Pr}^+_1$ by 2.15, so all regular $\mu \le \kappa$ remain regular and every $\lambda' \in (\kappa, \lambda)$ is collapsed (in the general case i.e. if $\bar{Q}$ "generic" enough). But λ is not collapsed if it is strongly inaccessible (by 2.17(2)). If $2 \in S^\ell_\kappa$, no bounded subset of κ is added. We can get $Ax_{<\kappa}\Big((S^\ell_\kappa, \kappa) - \mathrm{Pr}_1, \lambda, \lambda^+\Big)$. Note: if λ is in V, supercompact with Laver diamond, we get $Ax\Big((S^\ell_\kappa, < \kappa) - \mathrm{Pr}_1, \lambda, \infty\Big)$ (see VII).

So (even if we assume $\mathbf{E}^* = \emptyset$) the theorems of §2 are strong enough to deal with such iterations get forcing axioms etc. Of course you may then look for forcing notions which can serve as iterant, of course κ-complete and θ-complete θ^+-c.c. forcing notions can serve. For some more see §4, §5 below.

Case II. Like Case I (but κ may be limit $> \aleph_0$) with $(\kappa + 1) - \mathrm{Sp}_2$-iteration each $\underset{\sim}{Q}_i$-having $(S^\ell_\kappa, \kappa + 1) - \mathrm{Pr}^+_1$ and every $\lambda' \in (\kappa, \lambda)$ is collapsed. Here we can get $Ax_{<\lambda}\Big((S^\ell_\kappa, \kappa + 1) - \mathrm{Pr}^+_1, \lambda, \lambda^+\Big)$. Here λ is not collapsed (even P_λ satisfies the λ-c.c.) if it is strongly inaccessible Mahlo (by 2.17(4)). If λ is supercompact with Laver diamond we get $Ax_{<\lambda}\Big((S^\ell_\kappa, \kappa + 1) - \mathrm{Pr}, \lambda, \infty\Big)$.

The situation is similar to that of case 1: this time better using a non empty $\mathbf{E}^*$ e.g. the one of 3.1(2).

Case III: Like case 1 but $\underset{\sim}{Q}_i$ satisfies $(S^\ell_{\kappa_i}, \kappa_i) - \mathrm{Pr}^+_1$, κ_{i+1} >density(P_i) and κ_i strictly increasing with i. So in V^{P_λ}, λ is still inaccessible (though not strongly inaccessible).

Here we better do a variant of §2 (i.e. 2.6A -2.21) without κ. Let $\mathbf{E}^*$ be the class of strongly inaccessible $> \aleph_0$. In Definition 2.7 there, the restriction of $|p|$ for $p \in \kappa - \mathrm{Sp}_2 - \mathrm{Lim}\bar{Q}$ is only: $\beta \in \mathbf{E}^* \cap E(\bar{Q}) \Rightarrow$ for some $\gamma < \beta$, $p{\restriction}\beta \in P_\gamma$ (this change 2.7(D), the above statement replaces (A)(i)). For any $\bar{Q}$ and $\beta < \alpha$ $(= \ell g(\bar{Q})$ we define a partial order $\leq_{0,\beta}$ on P_α: $p \leq_{0,\beta} q$ iff $p{\restriction}\beta = q{\restriction}\beta$ and $p \leq_0 q$. Now 2.10 is changed to

(∗) if $i \in (\beta, \alpha) \Rightarrow \Vdash_{P_i} (Q_i, \leq_0^{Q_i})$ is θ-complete then $(P_\alpha, \leq_{0,\beta})$ is θ-complete.

In Claim 2.12 we can omit clause 2.12(2)(d).

In Claim 2.12 becomes

(∗) for (our kind of) $\bar{Q}$, and $\beta < \alpha = \ell g(\bar{Q})$, and regular θ assume $i \in \beta(\alpha) \Rightarrow \Vdash_{P_i}$ "Q_i satisfies $(\{\theta\}; \aleph_1) - \mathrm{Pr}^+_1$, and P_β satisfies the θ-c.c. and $p \Vdash_{P_\alpha}$ "$\underset{\sim}{\tau} < \theta$" then for some q and ζ, we have $p \leq_{0,\beta} q \in P_\alpha$, and $q \Vdash_{P_\alpha}$ "$\underset{\sim}{\tau} < \zeta < \theta$".

Case VI: Like case I but κ is an uncountable inaccessible (possibly weakly) cardinal.

The problem with applying §2 is rooted in assumption (d) in 2.12(2), which is needed for the iteration as presented. We should change 2.7 as follows in 2.7(D) allow $i^* \leq \kappa$, but demand $\Vdash_{\bar{Q}}$ "$\{\zeta_{q_i}[\underset{\sim}{G}] : i < i^*\}$ has order type $< \kappa$".

Of course we should assume each P_i has at least $(\{\kappa\}, \aleph_1) - \mathrm{Pr}^+_1$. However, does this really add compared to Case II?

3.5 Conclusion. Suppose λ is strongly inaccessible, limit of measurables, $\lambda > \kappa$, κ successor. Then for some λ-c.c. forcing P not adding bounded subsets of κ, $|P| = \lambda$, and $\Vdash_P$ "$2^\kappa = \lambda = \kappa^+$, and for every $A \subseteq \kappa$ there is a countable subset of λ not in $[L(A)]$".

Proof: Use case I of 3.4. We use $\kappa - \mathrm{Sp}_2$-iteration $\langle P_i, \underset{\sim}{Q}_i : i < \lambda\rangle$, $|P_i| < \lambda$. For i even: let κ_i be the first measurable $> |P_i|$, (but necessarily $< \lambda$). Then Q_i is

Prikry forcing on κ_i and Q_{i+1} is Levy collapse of κ_i^+ to κ. (Compare X 5.5.)

$\square_{3.5}$

§4. On Sacks Forcing

We continue 3.3, 3.4. Assume for simplicity λ is strongly inaccessible, $\lambda > \kappa$. We want to show that we can find an $\kappa-\mathrm{Sp}_2$-iteration which force some $Ax[\ldots]$. A natural way is to use a preliminary forcing notion R.

A natural candidate is: $R = \{\bar{Q} : \bar{Q} \in H(\lambda), \bar{Q}$ an $\kappa - \mathrm{Sp}_2$-iteration of forcing notions satisfying $(S,\kappa) - \mathrm{Pr}_1^+\}$. As an example, we will prove this for Sacks forcing.

4.1 Lemma. Suppose

(i) R is an $\aleph_1$-complete forcing notion.

(ii) For $r \in R$, $\bar{Q}^r = \langle P_i^r : i \leq \alpha^r\rangle$, P_i^r is $\lessdot$-increasing in i and if $j \leq \alpha^r$ has cofinality ω_1, *then* every countable subset of ω, which belongs to $V^{P_j^r}$ belongs to $V^{P_i^r}$ for some $i < j$. We write P^r for $P_{\alpha^r}^r$

(iii) If $r^1 \leq r^2$ then $\bar{Q}^{r^1} \leq \bar{Q}^{r^2}$ (end extension), so $P^{r_1} \lessdot P^{r_2}$

(iv) If $r \in R$ and $\underset{\sim}{Q}$ is a $P_{\alpha^r}^r$-name of a forcing notion, *then* for some $r^1 \geq r$, $\alpha^{r_1} > \alpha^r$ and $P_{\alpha^r+1}^{r_1} = P_{\alpha^r} * \underset{\sim}{Q}'$ and $\Vdash_{P_{\alpha^{r_1}}^{r_1}}$ "if $\underset{\sim}{Q}$ does satisfy the c.c.c. then $\underset{\sim}{Q}' = \underset{\sim}{Q}$"

(v) If r^ζ for $(\zeta < \delta)$ is increasing, $\delta \leq \omega_1$, *then* for some r

$$\bigwedge_{\zeta<\delta} r^\zeta \leq r \text{ and } \alpha_r = \bigcup_{\zeta<\delta} \alpha_{r^\zeta}.$$

Let $P[\underset{\sim}{G}_R]$ be $\bigcup\{P_i^r : r \in \underset{\sim}{G}_R,\, i \leq \alpha_r\}$, so it is an R-name of a forcing notion. *Then* $\Vdash_R [\Vdash_{P[\underset{\sim}{G}_R]}$ "for any $\aleph_1$ dense subsets of Sacks forcing, there is a directed subset of Sacks forcing not disjoint to any of them"].

Remark. Remember $Q_{Sacks} = \{\tau : \tau \subseteq {}^{\omega>}2$ is closed under initial segments nonempty and $(\forall \eta \in \tau)(\exists \nu)(\eta \lhd \nu \;\&\; \nu{}^\frown\langle 0\rangle \in \tau \;\&\; \nu{}^\frown\langle 1\rangle \in \tau)\}$ and $\tau_1 \leq \tau_2$ if $\tau_2 \subseteq \tau_1$.

Proof: Let for $\xi < \omega_1$, $\underset{\sim}{\mathcal{I}}_\xi$ be $R * P[\underset{\sim}{G}_R]$-name of dense subset of $Q_{Sacks}^{R*P[\underset{\sim}{G}_R]}$ for $\xi < \omega_1$ (Q^V_{Sacks} is Sacks forcing in the universe V). W.l.o.g. the $\underset{\sim}{\mathcal{I}}_\xi$ are open. We will find a c.c.c. subset $\underset{\sim}{Q}'$ of $Q_{Sacks}^{R*P[G_R]}$ such that $\underset{\sim}{\mathcal{I}}_\xi \cap \underset{\sim}{Q}'$ is dense in $\underset{\sim}{Q}'$ for each $\xi < \omega_1$. Then any generic subset of $\underset{\sim}{Q}'$ intersects all $\underset{\sim}{\mathcal{I}}_\xi$'s.

For a subset E of Sacks forcing let var(E) be $\{(n, \tau) : \tau \in E, n < \omega\}$ ordered by $(n_1, \tau_1) \leq (n_2, \tau_2)$ iff $n_1 \leq n_2$, $\tau_2 \subseteq \tau_1$ and $\tau_1 \cap^{n_1 \geq} 2 = \tau_2 \cap^{n_1 \geq} 2$. (If $D \subseteq$ var(Q_{Sacks}) is sufficiently generic, then $r_D = \cup\{\tau \restriction n : (n, \tau) \in d\}$ is a condition in Q_{Sacks}). We now define by induction on $\zeta \leq \omega_1$, $r(\zeta)$, and D_ζ such that (the order as the one on Q_{Sacks}):

(a) $r(\zeta) \in R$ is increasing, $\alpha_{r(\zeta)}$-increasing continuous.

(b) D_ζ is a $P^{r(\zeta+1)}$-name of a countable subset of Q_{Sacks}.

(c) If $\tau \in D_\zeta$, $\eta \in \tau$ then $\tau_{(\eta)} \overset{\text{def}}{=} \{\nu : \eta \,\hat{}\, \nu \in \tau\}$ belongs to D_ζ. (We use round parentheses to distinguish it from $\tau_{[\eta]}$, see clause (f)).

(d) If $\tau_1, \tau_2 \in D_\zeta$ then $\{\langle\ \rangle, \langle 0\rangle \hat{}\, \eta : \eta \in \tau_1\}$, $\{\langle\rangle, \langle 1\rangle \hat{}\, \eta : \eta \in \tau_2\}$ and their union belongs to D_ζ.

(e) Let $\xi < \zeta$, then for every $\tau_1 \in D_\xi$ there is $\tau_2 \in D_\zeta$ such that $\tau_1 \leq \tau_2$ and for every $\tau_2 \in D_\zeta$ there is $\tau_1 \in D_\xi$ such that $\tau_1 \leq \tau_2$.

(f) If $\tau \in D_\zeta$ then for some n for every $\eta \in {}^n2 \cap \tau$ we have $\tau_{[\eta]} \overset{\text{def}}{=} \{\nu \in \tau : \nu \trianglelefteq \eta$ or $\eta \trianglelefteq \nu\}$ belongs to $\underset{\sim}{\mathcal{I}}_\zeta$ (i.e. is forced ($\Vdash_R$) to belong to it).

(g) Suppose ζ is limit, then $P^{r(\zeta+1)}_{\alpha_r(\zeta)+1} = P^{r(\zeta)}_{\alpha_r(\zeta)} * Q_\zeta$, Q_ζ is $\left[\text{var} \bigcup_{\xi<\zeta} D_\xi\right]^\omega$ (the ω-th power, with finite support).

(h) the generic subset of $\underset{\sim}{Q}_\zeta$ gives a sequence of length ω of Sacks conditions; closing the set of those conditions by (c) + (d) + (f) we get D_ζ.

We have to prove that Q_ζ satisfies the $\aleph_1$-c.c. in $V^{R*P_{\alpha_r(\zeta)}}$: (to get a generic subset by (iv)). If $\zeta < \omega_1$ this follows by countability. Let $\zeta = \omega_1$. It suffices to prove that $\left[\text{var} \bigcup_{\xi<\zeta} D_\xi\right]^n$ satisfies the $\aleph_1$-c.c. where $n < \omega$. So let $\underset{\sim}{\mathcal{J}}$ be a $R * \underset{\sim}{P}^{r(\zeta)}_{\alpha_r(\zeta)}$-name of a maximal antichain of $\left[\text{var} \bigcup_{<\zeta} D_\xi\right]^n$. We can find a $\xi < \zeta$, cf$(\xi) = \aleph_0$ such that $\mathcal{I}^*_\xi \overset{\text{def}}{=} \{x : x \in V_{\alpha_r(\zeta)}^{R*\underset{\sim}{P}^{r(\zeta)}}$ and every

$p \in (R{*}\underset{\sim}{P}^{r(\zeta)}_{\alpha_r(\zeta)})/(R{*}\underset{\sim}{P}^{r(\xi)}_{\alpha_r(\xi)})$ force x to be in $\mathcal{J}\}$ is pre-dense in $\left[\text{var} \bigcup_{\gamma<\xi} D_\gamma\right]^n$ (exists by (e) and assumption (i)). Check the rest.

Notice that we have used:

(a) *if* $Y \subseteq \bigcup_{\xi<\xi_0} D_\xi$, $\xi_0 \le \zeta$, $Y \in V^{P_{\alpha(\xi_0)}}$ and Y is a pre-dense subset of $\bigcup_{\xi<\xi_0} D_\xi$ (it does not matter where but e.g., in V^P) *then* Y is a pre-dense subset of $\bigcup_{\xi<\zeta} D_\xi$; *because*

(a1) every $\tau \in D_{\xi_0}$ is included in a finite union of members of Y.

(a2) every $\tau \in \bigcup_{\xi<\varepsilon<\zeta} D_\varepsilon$ is included in some member of D_{ξ_0}.

4.2 Remark. 1) This argument works for many other forcing notions, e.g., Laver forcing.

2) The *var*(*Sacks*) was introduced by author to show Sacks forcing may not collapsed $\aleph_2$ (see Baumgartner and Laver [BL]).

3) In later work Velickovic get results for $> \aleph_1$ dense sets.

§5. Abraham's Second Problem – Iterating Changing Cofinality to ω

5.1 Definition. Let S be a subset of $\{2\} \cup \{\kappa : \kappa \text{ is regular cardinal}\}$, D a filter on a cardinal λ (or any other set). For any ordinal γ, we define a game $\partial^*(S, \gamma, D)$. It lasts γ moves. In the i-th move player I choose a cardinal $\kappa_i \in S$ and function F_i from λ to κ_i and then player II chooses $\alpha_i < \kappa_i$.

Player II wins a play if for every $i < \gamma$,

$$d(\langle \kappa_j, F_j, \alpha_j : j < i\rangle) \stackrel{\text{def}}{=} \{\zeta < \lambda : \text{ for every } j < i \text{ we have}$$

$$[\kappa_i = 2 \Rightarrow F_j(\zeta) = \alpha_j] \text{ and } [\kappa_j > 2 \Rightarrow F_j(\zeta) < \alpha_j]\} \ne \emptyset \bmod D.$$

5.1A Remark.

(1) This is similar to the game of X4.9, but there we also demand $d(\langle \kappa_j, F_j, \alpha_j : j < \gamma) \ne \emptyset \bmod D$.

(2) If not said otherwise, we assume that $\lambda \setminus \{\zeta\} \in D$ for $\zeta < \lambda$.

(3) If D is an ultrafilter on λ which is κ^+-complete for each $\kappa \in S$ and $|\gamma|^+$-complete (if γ a cardinal γ-complete) *then* player II has a winning strategy (if γ is a cardinal, γ-completeness suffices).

(4) Of course only $F_i \restriction d(\langle \kappa_j, F_j, \alpha_j : j < i \rangle)$ matters so player I can choose only it.

5.2 Definition. For $\mathbf{F}$ a winning strategy for player II in $\eth^*(S, \gamma, D)$, D a filter on λ (we write $\lambda = \lambda(D)$), we define $Q = Q_{\mathbf{F},\lambda} = Q_{\mathbf{F},S,\gamma,D}$, with $Q = (|Q|, \leq, \leq_0)$ as follows.

Part A. Let $(T, H) \in Q$ iff

(i) T is a nonempty set of finite sequences of ordinals $< \lambda$.

(ii) $\eta \in T \Rightarrow \eta \restriction \ell \in T$, and for some (unique) n and η of length n we have: $T \cap {}^{n\geq}\lambda = \{\eta \restriction \ell : \ell \leq n\}$, $|T \cap {}^{n+1}\lambda| \geq 2$; we call η the trunk of T, $\eta = \mathrm{tr}(T) = \mathrm{tr}(T, H)$ (it is unique).

(iii) H is a function, $T \setminus \{\mathrm{tr}(T) \restriction \ell : \ell < \ell g(\mathrm{tr}(T))\} \subseteq \mathrm{Dom}(H) \subseteq {}^{\omega>}\lambda$.

(iv) for each $\eta \in \mathrm{Dom}(H)$, $H(\eta)$ is a proper initial segment of a play of the game $\eth^*(S, \gamma, D)$ in which player II use his strategy $\mathbf{F}$ so $H(\eta) = \langle \lambda_i^{H(\eta)}, F_i^{H(\eta)}, \alpha_i^{H(\eta)} : i < i^{H(\eta)} \rangle$ and $i^{H(\eta)} < \gamma$.

(v) if $\mathrm{tr}(T) \trianglelefteq \eta \in \mathrm{Dom}(H) \cap T$ we have $\{\zeta < \lambda : \eta\hat{\ }\langle \zeta \rangle \in T\} = d(H(\eta))$ (see Definition 5.1).

(vi) convention: if $p = (T, H)$ we may write $\eta \in p$ for $\eta \in T$.

Part B. $(T_1, H_1) \leq (T_2, H_2)$ (where both belong to Q) iff $T_2 \subseteq T_1$ and for each $\eta \in T_2$, if $\mathrm{tr}(T_2) \trianglelefteq \eta$ then $H_1(\eta)$ is an initial segment of $H_2(\eta)$.

Part C. $(T_1, H_1) \leq_0 (T_2, H_2)$ (where both belong to Q) if $(T_1, H_1) \leq (T_2, H_2)$ and $\mathrm{tr}(T_1) = \mathrm{tr}(T_2)$.

5.2A Remark . (1) So if $(T, H) \in Q_{\mathbf{F},\lambda}$ and $\mathbf{F}, \lambda, D, \gamma, S$ are as above, $\eta \in T$, $\eta \trianglerighteq \mathrm{tr}(T)$ *then* $d(H(\eta)) \neq \emptyset \bmod D$. (So this forcing is similar to Namba forcing, but here we have better control of the sets $Suc_T(\eta)$.)

(2) We can of course generalize this to cases where we have different strategies (and even different λ's and D's) in different nodes.

(3) If $(T, H_\ell) \in Q_{F,\lambda}$ for $\ell = 1, 2$ $H_1 \restriction T = H_2 \restriction T$ then $(T, H_1), (T, H_2)$ are equivalent (see Chapter II).

5.2B Notation . For $p = (T,\ H) \in Q_{\mathbf{F},\,\lambda}$ and $\eta \in T$ let $p^{[\eta]} = (T^{[\eta]},\ H)$, where $T^{[\eta]} = \{\nu \in T : \nu \trianglelefteq \eta \text{ or } \eta \trianglelefteq \nu\}$. Clearly $p \leq p^{[\eta]} \in Q_{\mathbf{F},\,\lambda}$.

5.3 Lemma. If $Q = Q_{\mathbf{F},S,\,\gamma,D}$ and D is a uniform filter on $\lambda(D)$ *then*

$$\Vdash_Q \text{“cf}[\lambda(D)] = \aleph_0\text{”}.$$

Proof. Let $\underset{\sim}{\eta}_Q = \bigcup\{\mathrm{tr}(p) : p \in \underset{\sim}{G}_Q\}$.

Clearly if $(T_\ell,\ H_\ell) \in \underset{\sim}{G}_Q$ for $\ell = 1, 2$ then for some $(T,\ H) \in \underset{\sim}{G}_Q$, $(T_\ell,\ H_\ell) \leq (T,\ H)$; hence $\mathrm{tr}(T_\ell) \trianglelefteq \mathrm{tr}(T)$, hence $\mathrm{tr}(T_1,\ H_1) \cup \mathrm{tr}(T_2, H_2)$ is in ${}^{\omega>}\lambda$. Hence $\underset{\sim}{\eta}_Q$ is a sequence of ordinals of length $\leq \omega$.

For every $p = (T,\ H) \in Q$, and n, there is $\eta \in T \cap {}^n\lambda$, hence $p \leq p^{[\eta]} \in Q$ (see 5.2B), and $p^{[\eta]} \Vdash$ “$\ell g(\underset{\sim}{\eta}_Q) \geq n$” because $\eta \trianglelefteq \mathrm{tr}(p^{[\eta]})$ and for every $q \in Q$ we have $q \Vdash_Q$ “$\mathrm{tr}(q) \trianglelefteq \underset{\sim}{\eta}_Q$”. So $\Vdash_Q$ “$\underset{\sim}{\eta}_Q$ has length $\geq n$” hence $\Vdash_Q$ “$\underset{\sim}{\eta}_Q$ has length ω”.

Obviously, $\Vdash_Q$ “$\mathrm{Rang}(\underset{\sim}{\eta}_Q) \subseteq \lambda$”. Why $\Vdash_Q$ “$\sup \mathrm{Rang}(\underset{\sim}{\eta}_Q) = \lambda$”? Because for every $(T,\ H) \in Q$ and $\alpha < \lambda$, letting $\eta \overset{\mathrm{def}}{=} \mathrm{tr}(T)$, clearly $d(H(\eta)) \neq \emptyset \bmod D$ (see Definition 5.2) but D is uniform, hence there is $\beta \in d(H(\eta))$, $\beta > \alpha$, so $\eta\,\hat{}\,\langle\beta\rangle \in T$, and $(T,\ H) \leq (T,\ H)^{[\eta\,\hat{}\,\langle\beta\rangle]} \in Q$ and $(T,\ H)^{[\eta\,\hat{}\,\langle\beta\rangle]} \Vdash_Q$ “$\eta\,\hat{}\,\langle\beta\rangle \leq \underset{\sim}{\eta}_Q$” hence $(T,\ H)^{[\eta\,\hat{}\,\langle\beta\rangle]} \Vdash$ “$\mathrm{supRang}(\underset{\sim}{\eta}_Q) \geq \beta$”, as $\alpha < \beta$ we finish. $\square_{5.3}$

5.4 Lemma. If λ, S, γ, D are as in Definition 5.1, $\aleph_0 \notin S, \mathbf{F}$ a winning strategy of player II in $\eth^*(S, \gamma, D)$ and $\mathrm{cf}(\gamma) > \aleph_0$, *then* $Q_{F,\lambda}$ satisfies $(S, \mathrm{cf}(\gamma)) - \mathrm{Pr}_1^+$ (see Definition 2.1(2)). [So if $2 \in S$, then forcing by $Q_{F,\lambda}$ add no bounded subsets of γ].

Proof. In Definition 2.1, parts (i), (ii), (iii), (iv) and part (vi) are clear. So let us check part (v). Let $\kappa \in S$, $\underset{\sim}{\tau}$ be a Q-name, $\Vdash_Q$ “$\underset{\sim}{\tau} \in \kappa$” and $p = (T,\ H) \in Q$. We define by induction on n, $p_n = (T_n,\ H_n)$ and $\langle \alpha_\eta : \eta \in T_n \cap {}^n\lambda\rangle$ such that:

(α) $p_0 = p$, $p_n \leq_0 p_{n+1}$, $T_n \cap {}^{n\geq}\lambda = T_{n+1} \cap {}^{n\geq}\lambda$

(β) if $\eta \in T_n \cap {}^{n}\lambda$, and there are $q \in Q$ and $\alpha < \kappa$ satisfying

($*$) If $p_n^{[\eta]} \leq_0 q \in Q$, $\alpha < \kappa$ and $q \Vdash$ "if $\kappa = 2$, $\underset{\sim}{\tau} = \alpha$, if $\kappa > \aleph_0$, $\underset{\sim}{\tau} < \alpha$"

then $p_{n+1}^{[\eta]}$, α_η satisfy this.

Let p_ω be the limit of $\langle p_n : n < \omega \rangle$, i. e., $p_\omega = (T_\omega, H_\omega)$ where $T_\omega \stackrel{\text{def}}{=} \bigcap_{n<\omega} T_n$ and $H_\omega(\eta)$ is the limit of the sequences $H_n(\eta)$ (for $\eta \in T_\omega \setminus \{\text{tr}(T){\restriction}\ell : \ell < \ell g(\text{tr}T)\}$). It is well defined as $\text{cf}(\gamma) > \aleph_0$ and $p_n \leq_0 p_\omega \in Q$. We now prove two facts:

5.4A Fact . If $p = (T, H) \in Q$ and $f : T \cap {}^{n+1}\lambda \to \kappa$ and $\kappa \in S$, *then* there is $p' = (T', H') \in Q$ and $\langle \beta_\eta : \eta \in {}^{n}\lambda \cap T \rangle$ with $\beta_\eta < \kappa$, such that:

(a) $p \leq_0 p'$

(b) $T_n \cap {}^{n\geq}\lambda = T' \cap {}^{n\geq}\lambda$

(c) for every $\eta \in T' \cap {}^{n}\lambda$ we have: $\kappa = 2$ and $f{\restriction}\text{Suc}_{T'}(\eta)$ is constantly $= \beta_\eta$ *or* $\kappa > \aleph_0$ and $\text{Rang}(f{\restriction}\text{Suc}_{T'}(\eta)) \subseteq \beta_\eta$.

Note that we may allow f to be a partial function; now if $\kappa = 2$ then $f{\restriction}\text{Suc}_{T'}(\eta)$ is defined on all or undefined on all. If $\kappa \geq \aleph_0$, $f{\restriction}\text{Suc}_{T'}(\eta)$ may be a partial function. Similarly in 5.4B.

Proof. For each $\eta \in T \cap {}^{n}\lambda$ we have: $H(\eta)$ is a proper initial segment of a play of the game $\partial^*(S, \gamma, D)$, and it lasts $i^{H(\eta)}$ moves. Player I could choose in his $i^{H(\eta)}$-th move the cardinal κ and the function $f_\eta : \lambda \to \kappa$,

$$f_\eta(\zeta) = f(\eta\hat{\ }\langle\zeta\rangle) \quad (\text{which is } < \kappa) \text{ if } \eta\hat{\ }\langle\zeta\rangle \in T$$
$$f_\eta(\zeta) = 0 \quad \text{if otherwise.}$$

So, for some β_η, $H(\eta)\hat{\ }\langle\alpha, f_\eta, \beta_\eta\rangle$ is also a proper initial segment of a play of $\partial^*(S, \gamma, D)$ in which player II uses the strategy $\mathbf{F}$. So there is $p' = (T', H')$ such that $H'{\restriction}(T' \setminus {}^{n}\lambda) = H{\restriction}(T' \setminus {}^{n}\lambda)$ and $H'(\eta) = H(\eta)^\wedge\langle(\kappa, f_\eta, \beta_\eta)\rangle$ for $\eta \in T \cap {}^{n}\lambda$. (If is partial for $\kappa = 2$ we should do this twice: for definability and for value.)

$\square_{5.4A}$

We can easily show

5.4B Fact . If $p = (T, H) \in Q$, $m < \omega$, and $\kappa \in S$ and $f : T \to \kappa$, *then* for some $p_1 = (T_1, H_1) \in Q$, $p \leq_0 p_1$, and for every $k \leq m$ we have $[\kappa = 2$ & $f\restriction(T_1 \cap {}^k\lambda)$ is constant$]$ or $[\kappa \geq \aleph_1$ & $f\restriction(T_1 \cap {}^k\lambda)$ is bounded below $\kappa]$.

Proof. W.l.o.g. $m > \ell g(\mathrm{tr} p)$. We define by downward induction on $n \in [\ell g(\mathrm{tr} T), m]$ the condition r^n, $p \leq_0 r^{n+1} \leq_0 r^n \in Q$, r^n satisfying the conclusion of 5.4B for $p^{[\eta]}$ for every $\eta \in p$ of length n. For $n = m$ this is trivial. For $n < m$, use Fact 5.4A $m - n$ times, for $k \in (n, m]$ for the function $f_k^{n+1} : T^{r_{n+1}} \cap {}^{n+1}\lambda \to \kappa$ defined by: $f_k^{n+1}(\eta)$ is γ if

$$(\forall \nu)[\nu \in T^{r_{n+1}} \cap {}^k\lambda \to (\kappa = 2 \ \&\ f(\nu) = \gamma) \ \vee \ (\kappa > \aleph_0 \ \&\ f(\nu) < \gamma)];$$

now $r^{\ell g(\mathrm{tr} p)}$ is as required]. $\square_{5.4B}$

Continuation of the proof of 5.4. By repeated application of 5.4B we can define by induction on n, $q_n \in Q$ such that $q_0 = p_\omega$ (see before 5.4A) and $q_n \leq_0 q_{n+1}$ and β_η^n for $\eta \in T^{q_n}$, $\ell g(\eta) \leq n$ such that:

(a) $\beta_\eta^0 = \alpha_\eta$ if this is well-defined, β_η^0 =-1 otherwise (on α_η see (β) above).

(b) when $\kappa > \aleph_0$: $\ell g(\eta) \leq n$ & $\eta\hat{\ }\langle\zeta\rangle \in T^{q_{n+1}} \Rightarrow \beta_\eta^{n+1} \geq \beta^n_{\eta\hat{\ }<\zeta>}$ whenever the later is well defined.

(c) when $\kappa = 2$: $\ell g(\eta) \leq n$ & $\eta\hat{\ }\langle\zeta\rangle \in T^{q_{n+1}} \Rightarrow \beta_\eta^{n+1} = \beta^n_{\eta\hat{\ }<\zeta>}$ (so both are defined or both not defined).

Lastly let $q_\omega \in Q$ be such that $q_n \leq_0 q_\omega$ for $n < \omega$.

Now if $\kappa > \aleph_0$ (is regular), we claim

$$q_\omega \Vdash_Q \text{“}\underset{\sim}{\tau} \leq \bigcup_{n<\omega} \beta^n_{<>}\text{”}$$

Clearly $p \leq_0 q_\omega \in Q$ and $\bigcup_{n<\omega} \beta^n_{<>} < \kappa$ so this suffices. Why does this hold? If not, then for some q' and β, $q_\omega \leq q' \in Q$, $q' \Vdash_Q$ “$\underset{\sim}{\tau} = \beta$” and $\kappa > \beta > \bigcup_n \beta^n_{<>}$. Let $\eta = \mathrm{tr}(q')$, so $\eta \in T^{q_\omega}$, and α_η is well defined, and $> \beta$. But as $\eta \in \bigcap_{n<\omega} T^{q_n}$ and $\beta^{\ell g(\eta)}_{<>} \geq \beta^0_\eta = \alpha_\eta$, and we get a contradiction.

If $\kappa = 2$, we note just that for some $\eta \in T^{q_\omega}$, the number α_η is well defined, hence $\beta_\eta^{\ell g(\eta)}$ is defined hence β_η^ℓ is defined for $\ell \in [\mathrm{tr}(q_\omega), \ell g(\eta)]$. $\square_{5.4}$

Remark. We can rephrase much of this lemma as a partition theorem on trees as in [RuSh:117].

5.5 Lemma . Suppose $\bar{Q} = \langle P_i, Q_i : i < \lambda \rangle$ is a $\kappa - \mathrm{Sp}_2$-iteration, $|P_i| < \lambda$ for $i < \lambda, \gamma \leq \kappa$, each Q_i has $(S, \gamma) - Pr_1^+$ and κ regular and even successor, $S \subseteq \{2\} \cup \{\theta : \theta \text{ regular uncountable} \leq \kappa\}$ and in V, D is a normal ultrafilter on λ (so λ is a measurable cardinal). *Then* $\Vdash_{P_\lambda}$ "player II wins $\Game^*(S, \gamma, D)$".

Proof: Let $A = \{\mu < \lambda : (\forall i < \mu)[|P_i| < \mu]$, μ strongly inaccessible Mahlo cardinal $> \kappa\}$.

Let $G_\lambda \subseteq P_\lambda$ be generic over V and for $\alpha < \lambda$ let $G_\alpha = G \cap P_\alpha$.

W.l.o.g. player I choose P_λ-names of functions and cardinals in S. Now we work in V and describe player II's strategy there (see proof of XIII 1.9). For each $\mu \in A$ the forcing notion P_λ / P_μ has $(S, \gamma) - \mathrm{Pr}_1^+$; hence, player II has a winning strategy $\underset{\sim}{\mathbf{F}}(P_\lambda / \underset{\sim}{G}_\mu) \in V[\underset{\sim}{G}_\mu]$ for the game from 2.1(1)(vi), so $\underset{\sim}{\mathbf{F}}(P_\lambda / \underset{\sim}{G}_\mu)$ is a P_μ-name, $\langle \underset{\sim}{\mathbf{F}}(P_\lambda / \underset{\sim}{G}_\mu) : \mu \rangle$ a P_λ-name. Let us describe a winning strategy for player II (for the game $\Game^*(S, \gamma, D)$).

So in the i-th move player I chooses $\underset{\sim}{\theta}_i \in S$ and $\underset{\sim}{f}_i : \lambda \to \underset{\sim}{\theta}_i$. Player II chooses in his i-th move not only $\underset{\sim}{\alpha}_i < \underset{\sim}{\theta}_i$ but also A_i, $\underset{\sim}{f}_i$, $\underset{\sim}{\gamma}_i$, $\langle \langle \underset{\sim}{p}_j^\mu : j \leq i \rangle : \mu \in A_i \rangle$ such that

(0) γ_i is an ordinal $< \lambda$,

(1) $j < i \Rightarrow \gamma_j < \gamma_i$

(2) $A_i \in D$, $A_i \in V$, $A_i \subseteq \bigcap_{j<i} A_j$ and $A_\delta = \bigcap_{j<\delta} A_j$

(3) $\Vdash$ "$\underset{\sim}{f}_i : \lambda \to \underset{\sim}{\theta}_i$, $\underset{\sim}{\theta}_i \in S$"

(4) for $\mu \in A_i$, $\langle \underset{\sim}{p}_j^\mu : j \leq 2i + 2 \rangle$ is a P_μ-name of an initial segment of a play as in (vi) of 2.1(1) for the forcing P_λ / G_μ, $p_{2j+1}^\mu \Vdash_{P_\lambda / G_\mu}$ "$\underset{\sim}{f}_j(\mu) = \underset{\sim}{\alpha}_j^\mu$ if $\theta_j = 2$, $f_j(\mu) < \alpha_j^\mu$ if $\theta_j \geq \aleph_0$", $\underset{\sim}{\alpha}_j^\mu$ a P_μ-name.

In the i-th stage clearly $A_i^0 \overset{\text{def}}{=} \bigcap_{j<i} A_j \cap A$ is in D, and let $\gamma_i^0 = \sup_{j<i}\gamma_i$, so $\gamma_i^0 < \lambda$ and choose $\gamma_i' \in (\gamma_i^0, \lambda)$ such that $\underset{\sim}{\theta}_i$ is a $P_{\gamma_i'}$-name. For every $\mu \in A_i^0$, $\mu > \gamma_i'$, we can define P_μ-names $\underset{\sim}{p}_{2i}^\mu$, $\underset{\sim}{p}_{2i+1}^\mu$, α_i^μ such that:

(a) $\Vdash_{P_\mu}$ "$\langle \underset{\sim}{p}_j^\mu : j < 2i+2\rangle$ is an initial segment of a part as in (v) of 1.1(1) for P_λ/P_μ in which player II uses his winning strategy $\underset{\sim}{\mathbf{F}}(P_\lambda/\underset{\sim}{G}_\mu)$.

(b) $p_{2i+1}^\mu \Vdash_{P_\lambda/P_\mu}$ "$\underset{\sim}{f}_i(\mu) = \underset{\sim}{\alpha}_i^\mu$ if $\underset{\sim}{\theta} = 2$, $\underset{\sim}{f}_i(\mu) < \alpha_i^\mu$ if $\underset{\sim}{\theta}_i \geq \aleph_0$".

Now as $\underset{\sim}{\alpha}_i^\mu$ is a P_μ-name of an ordinal $< \kappa \leq \mu$, it is $P_{\beta(\mu)}$-name for some $\beta[\mu] < \mu$ (as P_μ satisfies the μ-c.c. see 2.17(2)). By the normality of the ultrafilter D, on some $A_i^1 \subseteq A_i^0$, $\beta[\mu] = \beta_i$ for every $\mu \in A_i^1$. Let $\gamma_i = \gamma_i^1 + \beta_i$.

Easily for each $i < \sigma$, $\Vdash_{P_\lambda}$ "$\{\mu \in A_i : p_{2i+1}^\mu \in \underset{\sim}{G}_\lambda\} \neq \emptyset \bmod D$", so we finish.

$\square_{5.5}$

5.5A Comment.

We can present it (and the proof of XIII 1.4) slightly differently.

In V let

$$\mathbf{W}_i = \{\langle \bar{p}^\mu : \mu \in A\rangle : A \in D, \bar{p}^\mu = \langle \underset{\sim}{p}_j^\mu : j \leq i\rangle,$$
$$\Vdash_{P_\mu} \text{"}\langle \underset{\sim}{p}_j^\mu : j < i\rangle \text{ is a } \leq_0 \text{-increasing sequence in } P_\lambda/P_\mu\text{"}\}$$

and let

$$\mathbf{W} = \bigcup_i \mathbf{W}_i$$

We define on $\mathbf{W}$ a relation $\leq$ by:

$$\langle \bar{p}^{\mu,1} : \mu \in A_1\rangle \leq \langle \bar{p}^{\mu,2} : \mu \in A_2\rangle \text{ iff}$$
$$A_1 \supseteq A_2 \text{ and } \mu \in A_2 \Rightarrow \bar{p}^{\mu,1} \text{ is an initial segment of } \bar{p}^{\mu,2}$$

Clearly $\leq$ is a partial order on $\mathbf{W}$, and for $\mathbf{p} = \langle \bar{p}_j^\mu : j < i\rangle$ let

$$\underset{\sim}{B}(\mathbf{p}) = \{\mu \in A : \{p_j^\mu : j < i\} \subseteq \underset{\sim}{G}_\lambda\},$$

so clearly

(a) $\Vdash_{P_\lambda}$ "$\mathbf{p}^1 \leq \mathbf{p}^2$ implies $\underset{\sim}{B}(\mathbf{p}^1) \supseteq B(\mathbf{p}^2)$"

(b) $\Vdash_{P_\lambda}$ "$\underset{\sim}{B}(\mathbf{p}) \neq \emptyset \mod D$"

(c) For every $\alpha < \lambda$, $\mathbf{p} \in \mathbf{W}_i$ and P_λ-names $\underset{\sim}{\theta} \in S$ and $\underset{\sim}{f} : \lambda \to \underset{\sim}{\theta}$ *there is* $\mathbf{q}$, $\mathbf{p} \leq \mathbf{q} \in \mathbf{W}_{i+1}$ and a P_λ name $\underset{\sim}{\tau}$ of an ordinal $< \theta$ such that:

(∗) if $\mathbf{q} = \langle \bar{q}^\mu : \mu \in A^q \rangle$, and $\mu \in A^q$, then $\underset{\sim}{q}^\mu_{i+1}$ forces

$$(\underset{\sim}{\theta} = 2 \ \& \ \underset{\sim}{f}(\mu) = \underset{\sim}{\tau}) \ \vee \ (\underset{\sim}{\theta} \geq \aleph_0 \ \& \ \underset{\sim}{f}(\mu) < \underset{\sim}{\tau})$$

Now we can solve the second Abraham problem. (See also X 5.5.)

5.6 Conclusion . Suppose λ is strongly inaccessible $\{\mu < \lambda : \mu \text{ measurable }\}$ is stationary, $\kappa = \mathrm{cf}(\kappa) < \lambda$ a successor cardinal, $\kappa^+ \cap \mathrm{RUCar} \subseteq S \subseteq \{2\} \cup \{\theta : \theta \leq \kappa \text{ regular uncountable}\}$. *Then* for some forcing notion P we have: $|P| = \lambda, P$ satisfies λ-c.c. and $(S, \kappa) - Pr_1^+$ and $\Vdash_P$ "$\lambda = |\kappa|^+$" (so $\Vdash_{P_\mu} 2^{|\kappa|} = \lambda$): and for every $A \subseteq \lambda$, for some $\delta < \lambda$, there is a countable set $a \subseteq \delta$, which is not in $V[A \cap \delta]$. We can also get suitable axiom (see 3.5).

Proof. Should be clear (see 3.4 Case I (and 5.4)). $\square_{5.6}$

5.6A Remark . 1) We can also prove (by the same forcing) the consistency of "there is a normal filter on λ to which $\{\delta < \lambda : \mathrm{cf}(\delta) = \aleph_0\}$ belongs which is precipitous" if in addition there is a normal ultrafilter on λ concentrating on measurables.

2) We can use $(S, < \kappa) - \mathrm{Pr}_1-$ forcing notions.

5.7 Discussion. Can we weaken the assumption $\mathrm{cf}(\gamma) > \aleph_0$ in 5.4 to $\mathrm{cf}(\gamma) \geq \aleph_0$ and/or allow $\kappa = \aleph_0$? The answer is yes if $\{2, \kappa\} \subseteq S$.

As in 5.4A, 5.4B we can assume $p = (T, H)$ satisfies

(∗) for $\eta \in p$, $\ell g(\eta) \geq \mathrm{tr}(p)$ and there are q and $\alpha < \kappa$ such that $p^{[\eta]} \leq_0 q \in Q$ and $q \Vdash$ "$\kappa = 2 \ \& \ \underset{\sim}{\tau} = \alpha$ or $\kappa > \aleph_0 \ \& \ \underset{\sim}{\tau} < \alpha$" then $p^{[\eta]}$, α_η satisfies this.

Let for $\eta \in T^p$ of length $\geq \ell g(\mathrm{tr}(p))$:

$$\mathcal{P}_p(\eta) \stackrel{\mathrm{def}}{=} \{A \subseteq \lambda : \text{for some initial segment } y \text{ of the game continuing } H^p(\eta), \text{ we have } d(y) \subseteq A\}$$

(where d is from Definition 5.1).

Note that as $2 \in S$ we have

(**) $A \subseteq \lambda \Rightarrow A \in \mathcal{P}_p(\eta) \ \vee \ (\lambda \setminus A) \in \mathcal{P}_p(\eta)$

(* * *) if $A \in \mathcal{P}_p(\eta)$, $A \subseteq B \in \lambda$ then $B \in \mathcal{P}_p(\eta)$.

We define a function $\mathrm{rk}_p : T^* = \{\eta : \mathrm{tr}(p) \trianglelefteq \eta \in T\} \to \mathrm{Ord} \cup \{\infty\}$ by defining by induction on the ordinal ζ when $\mathrm{rk}_p(\eta) \geq \zeta$, the definition is splited to cases.

Case A. ζ limit

$\mathrm{rk}_p(\eta) \geq \zeta$ *iff* $(\forall \xi < \zeta)[\mathrm{rk}_p(\eta) \geq \xi]$.

Case B. $\zeta = 1$

$\mathrm{rk}_p(\eta) \geq 1$ iff α_η is not well defined (and $\eta \in T^*$)

Case C. $\zeta = \varepsilon + 1 > 1$

so $\varepsilon > 0$; let $\mathrm{rk}_p(\eta) \geq \zeta$ *iff*: $\mathrm{tr}(p) \trianglelefteq \eta \in T^p$ and the set $\{\beta < \lambda : \mathrm{rk}_p(\eta\hat{\ }\langle\beta\rangle) \geq \varepsilon\}$ belongs to $\mathcal{P}_p(\eta)$.

So $\mathrm{rk}(\eta) = 0$ if α_η is well defined and $\mathrm{rk}(\eta) = \zeta > 0$ if $\neg(\mathrm{rk}(\eta) \geq \zeta + 1)$, ζ minimal, and $\mathrm{rk}(\eta) = \infty$ if $\mathrm{rk}(\eta) \geq \zeta$ for every $\zeta \geq 1$. Now the proof is splited:

Subcase C1. $\mathrm{rk}_p(\mathrm{tr}(p)) = \infty$.

Clearly for $\eta \in T^*$, if $\mathrm{rk}_p(\eta) = \infty$ then $\{\beta < \lambda : \mathrm{rk}_p(\eta\hat{\ }\langle\beta\rangle) = \infty\} \in \mathcal{P}_p(\eta)$. Hence we can find q such that $p \leq_0 q \in Q$ such that:

$$\mathrm{tr}(p) \trianglelefteq \eta \in T^q \Rightarrow \mathrm{rk}_p(\eta) = \infty.$$

There is r such that $q \leq r \in Q$ and r forces a value to $\underset{\sim}{\tau}$, so $\alpha_{\mathrm{tr}(r)}$ is well defined but $\mathrm{tr}(p) \trianglelefteq \mathrm{tr}(r) \in T^r \subseteq T^q$ hence $\mathrm{rk}_p(\mathrm{tr}(r)) = \infty$ hence $\alpha_{\mathrm{tr}(r)}$ is not well defined, contradiction.

Subcase C2. $\mathrm{rk}_p(\mathrm{tr}(p)) < \infty$.

So choose $\eta \in T^q$, $\mathrm{tr}(p) \trianglelefteq \eta \in T^p$ such that α_η is not defined and, under those restrictions, $\mathrm{rk}_p(\eta)$ is minimal.

Let

$$A = \{\gamma < \lambda : \eta\char`^\langle\gamma\rangle \in p \text{ and } \alpha_{\eta\char`^\langle\gamma\rangle} \text{ is not defined}\}.$$

We can find q, $p^{[\eta]} \leq_0 q$ and $d(H^q(\eta))$ is included in A or disjoint to it. In the second case we can easily get "α_η well defined", contradiction. So assume $d(H^q(\eta)) \subseteq A$, and neccessarily there is $\nu \in d(H^Q(\eta))$ such that $\mathrm{rk}_p(\nu) < \mathrm{rk}_p(\eta)$ by the definition of rank. We get easy contradiction.

XV. A More General Iterable Condition Ensuring $\aleph_1$ Is Not Collapsed

§0. Introduction

Chapter XI was restricted to forcing notions not adding reals in a specific way so that under CH, Nm is always permissible. This was used to show that various combinatorical principles of $\aleph_2$ were equiconsistent with the existence of (small) large cardinals. We constructed our models starting from CH without adding reals, so that CH also holds in the final model. But what if we want CH to fail in the final model? Can we phrase a condition preserved by iterations, implying $\aleph_1$ does not collapse and include semiproper forcing and Nm? This, promised in the first version of this book, is carried out here. We start with notions similar to the one in Chapter X, and then move in the direction of semiproperness. Further theorems (which shed light on preservation of not adding reals) will appear elsewhere (see [Sh:311]). The preservation theorems from this chapter are sufficient to prove analogue of some theorems from Chapter XI with the negation of CH. For example adding Cohen reals to the construction of XI 1.4 we can show: If "ZFC+ $\exists$ weakly compact cardinal" is consistent, then so is "ZFC+ $2^{\aleph_0} = \aleph_2$ + for every stationary $S \subseteq S_0^2$ there is a closed copy of ω_1 included in it". Generally the preservation proofs generalize those of Chapter XI, except in the case of "iterating up to a strongly inaccessible and doing one more step (in this case 3.6). We generalize Gitik and Shelah [GiSh:191] which improve the relevant theorem in XI (i.e. [Sh:b, XI]).

Of course we can also add reals to the construction in XI 1.2 and get an extension V^P where $\aleph_1^V = \aleph_1^{V^P}$, $\kappa = \aleph_2^{V^P}$ and the filter generated by the measure on κ in V will include S_0^2, and it is not clear that it will be precipitous, see X §6. As in Chapter XI, we use for proving the preservation, partition theorems and Δ-system theorems on trees: mainly 2.6 and 2.6A, 2.6B, 2.6C. Some of them are from Rubin and Shelah [RuSh:117], see detailed history there, on pages 47, 48.

§1. Preliminaries

The replacement of RCS (revised countable support) by GRCS (defined below) is not essential - it is intended to simplify the preservation theorems (one of the cases in Chapter VI refers to GRCS).

1.1 Conventions. A forcing notion here, P, is a nonempty set (denoted by P too) and two partial orders $\leq_{pr}, \leq$ and a minimal element $\emptyset_P \in P$, $[p \leq_{pr} q \rightarrow p \leq q]$. We call $p \in P$ pure if $\emptyset_P \leq_{pr} p$ and we call q a pure extension of p if $p \leq_{pr} q$. (In Chapter XIV=[Sh:250] this was written $\leq_0$).

We denote forcing notions by P, Q, R. (The forcing relation of course refers to the partial order $\leq$).

1.2 Definition. Let $MAC(P)$ be the set of maximal antichains of the forcing notion P.

1.2A Remark. 1) Note: $|MAC(P)| \leq 2^{|P|}$, P satisfies the $|P|^+$-c.c. and if P satisfies the λ-c.c. then $|MAC(P)| \leq |P|^{<\lambda}$.

2) Note

(∗) if Q is a forcing notion, $\lambda = \lambda^{<\lambda} > |Q| + \aleph_0, \Vdash_Q$ "$(\forall \mu < \lambda)\mu^{\aleph_0} < \lambda$" and $Q' = Q * \underset{\sim}{\mathrm{Levy}}(\aleph_1, < \lambda)$ *then* $|MAC(Q')| = |Q'| = \lambda$

1.3 Notation. Car is the class of cardinals.

IRCar is the class of infinite regular cardinals.

RCar $=$ IRCar $\cup \{0\}$.

RUCar is the class of uncountable regular cardinals.

$\mathcal{D}^{\rm cb}_\lambda$ is the filter of co-bounded subsets of λ.

$\eta^- = \eta\restriction(\ell g(\eta) - 1)$ for $\eta \in V^{<\omega}$ and $\ell g(\eta) > 0$

1.4 Notation. $H(\chi)$ is the family of sets with transitive closure of power $< \chi$; let $<^*_\chi$ be a well ordering of $H(\chi)$.

1.5 Definition. $GRCS$ iteration is as defined in Ch X, except that, for each condition all but finitely many of the atomic conditions in it are pure (or as in Chapter XIV §1 for $\kappa = \aleph_1$, $e = 1$).

1.6 Fact. $(*)_1$ if $\bar{Q}$ is a $GRCS$ iteration, and for each i $\leq^{Q_i}_{pr} = \leq^{Q_i}$ *then* $\bar{Q}$ is an RCS iteration.

$(*)_2$ if $\bar{Q}$ is an $GRCS$ iteration, and for each i the order $\leq^{Q_i}_{pr}$ is equality *then* $\bar{Q}$ is essentially a finite support iteration.

$(*)_3$ the distributivity law, etc. (Chapter X 1.5, and §1 generally) holds for GRCS (by Chapter XIV §1).

1.7 Claim. Suppose we want to prove for all generic extensions V^Q of V, that for iteration $\langle P_i, \underset{\sim}{Q}_j : i \leq \alpha, j < \alpha\rangle$ as in 3.1 below, for a property φ that:

$(*)$ if Q and each Q_i has the property φ (of course Q in V, $\underset{\sim}{Q}_i$ in $V^{Q * \underset{\sim}{P}_i}$)

then P_α has the property φ (in V^Q).

Then it is enough to prove (*) when (a) and (b) below hold:

(a) for $i < j < \alpha, P_j/P_i$ has the property φ (in $V^{Q * \underset{\sim}{P}_i}$)

(b) $\ell g(\bar{Q})$ is: 2, or ω, or ω_1, or strongly inaccessible $> |P_i|$ for each $i < \ell g(\bar{Q})$.

Remark. You may add:

(c) $(*)$ holds for all $\bar{Q}'$, $\langle P'_i, Q'_j : i \leq \alpha', j < \alpha'\rangle$ for which $\alpha' < \alpha$ (not just in V, but in every generic extension of it).

Proof. By induction on α, using 3.7 (later in this chapter) and X §1 (or XIV §1).

§2. Trees of Models and UP

2.1 Definition.

1) A tagged tree is a pair $\langle T, \mathbf{I}\rangle$ such that:

   (a) T is a ω-tree, which here means a nonempty set of finite sequences of ordinals such that if $\eta \in T$ then any initial segment of η belongs to T. T is ordered by initial segments, i.e., $\eta \trianglelefteq \nu$ iff η is an initial segment of ν.

   (b) $\mathbf{I}$ is a partial function from T such that for every $\eta \in T$: if $\mathbf{I}(\eta) = \mathbf{I}_\eta$ is defined then $\mathbf{I}(\eta)$ is an ideal of subsets of some set called the domain of $\mathbf{I}_\eta, \mathrm{Dom}(\mathbf{I}_\eta)$, and

$$\mathrm{Suc}_T(\eta) \stackrel{\mathrm{def}}{=} \{\nu : \nu \text{ is an immediate successor of } \eta \text{ in } T\} \subseteq \mathrm{Dom}(\mathbf{I}_\eta),$$

   and if not said otherwise $\mathrm{Suc}_T(\eta) \notin \mathbf{I}_\eta$. Usually $\mathbf{I}_\eta$ is $\aleph_2$-complete.

   (c) For every $\eta \in T$ we have $\mathrm{Suc}_T(\eta) \neq \emptyset$.

2) We call $(T, \mathbf{I})$ normal if $\eta \in \mathrm{Dom}(\mathbf{I}_\eta) \Rightarrow \mathrm{Dom}(\mathbf{I}_\eta) = \mathrm{Suc}_T(\eta)$.

2.1A Convention. For any tagged tree $(T, \mathbf{I})$ we can define $\mathbf{I}^\dagger$, by:

$$\mathrm{Dom}(\mathbf{I}') = \{\eta : \mathrm{Suc}_T(\eta) \subseteq \mathrm{Dom}(\mathbf{I}_\eta), \text{ and } \mathrm{Suc}_T(\eta) \notin \mathbf{I}_\eta\} \text{ and}$$

$$\mathbf{I}^\dagger_\eta = \{\{\alpha : \eta\hat{\ }\langle\alpha\rangle \in A\} : A \in \mathbf{I}_\eta\};$$

we sometimes, in an abuse of notation, do not distinguish between $\mathbf{I}$ and $\mathbf{I}^\dagger$ e.g. if $\mathbf{I}^\dagger_\eta$ is constantly I^*, we write I^* instead of $\mathbf{I}$.

2.2 Definition. 1) η will be called a splitting point of $(T, \mathbf{I})$ if $\mathbf{I}_\eta$ is defined and $\mathrm{Suc}_T(\eta) \notin \mathbf{I}_\eta$ (normally this follows but we may forget to decrease the domain of $\mathbf{I}$). Let split$(T, \mathbf{I})$ be the set of splitting points. We will only consider trees where each branch meets split$(T, \mathbf{I})$ infinitely often.

2) For $\eta \in T$, $T^{[\eta]} \stackrel{\text{def}}{=} \{\nu \in T : \nu = \eta \text{ or } \nu \triangleleft \eta \text{ or } \eta \triangleleft \nu\}$.

2.3 Definition. We now define orders between tagged trees:

a) $(T_1, \mathbf{I}_1) \le (T_2, \mathbf{I}_2)$ if $T_2 \subseteq T_1$, and split$(T_2, \mathbf{I}_2) \subseteq$ split$(T_1, \mathbf{I}_1)$, and $\forall \eta \in$ split$(T_2, \mathbf{I}_2) : \mathbf{I}_2(\eta) \restriction \mathrm{Suc}_{T_2}(\eta) = \mathbf{I}_1(\eta) \restriction \mathrm{Suc}_{T_2}(\eta)$.
(where $I \restriction A = \{B : B \subseteq A \text{ and } B \in I\}$). (So every splitting point of T_2 is a splitting point of T_1, and $\mathbf{I}_2$ is completely determined by $\mathbf{I}_1$ and split$(T_2, \mathbf{I}_2)$ provided that $\mathbf{I}_2$ is normal.)

b) $(T_1, \mathbf{I}_1) \le^* (T_2, \mathbf{I}_2)$ *iff* $(T_1, \mathbf{I}_1) \le (T_2, \mathbf{I}_2)$ and split$(T_2, \mathbf{I}_2) =$ split$(T_1, \mathbf{I}_1) \cap T_2$.

c) $(T_1, \mathbf{I}_1) \le^{\otimes} (T_2, \mathbf{I}_2)$ if $(T_1, \mathbf{I}_1) \le^* (T_2, \mathbf{I}_2)$ and $\eta \in T_2 \setminus$ split$(T_1, \mathbf{I}_1) \Rightarrow$ $\mathrm{Suc}_{T_2}(\eta) = \mathrm{Suc}_{T_1}(\eta)$

(d) $(T_1, \mathbf{I}_1) \le^{\otimes}_{\mu} (T_2, \mathbf{I}_2)$ *if* $(T_1, \mathbf{I}_1) \le (T_2, \mathbf{I}_2)$ and $\eta \in T_2$ & $|\mathrm{Suc}_{T_1}(\eta)| < \mu \Rightarrow$ $\mathrm{Suc}_{T_2}(\eta) = \mathrm{Suc}_{T_1}(\eta)$ and $\eta \in T_2$ & $|\mathrm{Suc}_{T_1}(\eta)| \ge \mu$ & $\eta \in \mathrm{Sp}(T_1, \mathbf{I}_1) \Rightarrow \eta \in$ $\mathrm{Sp}(T_2, \mathbf{I}_2)$

2.4 Definition. 1) For a set $\mathbb{I}$ of ideals, a tagged tree $(T, \mathbf{I})$ is an $\mathbb{I}$-tree if for every splitting point $\eta \in T$ we have $\mathbf{I}_\eta \in \mathbb{I}$ (up to an isomorphism).

2) For a set $\mathbf{S}$ of regular cardinals, an $\mathbf{S}$-tree T is a tree such that for any point $\eta \in T$ we have: $|\mathrm{Suc}_T(\eta)| \in \mathbf{S}$ or $|\mathrm{Suc}_T(\eta)| = 1$.

3) We omit $\mathbf{I}$ and denote a tagged tree $(T, \mathbf{I})$ by T whenever $\mathbf{I}_\eta = \{A \subseteq \mathrm{Suc}_T(\eta) : |A| < |\mathrm{Suc}_T(\eta)|\}$ and $|\mathrm{Suc}_T(\eta)| \in \mathrm{IRCar} \cup \{1\}$ for every $\eta \in T$.

4) For a tree T, $\lim T$ is the set of *branches* of T, i.e. all ω-sequences of ordinals, such that every finite initial segment of them is a member of T: $\lim T = \{s \in {}^{\omega}\mathrm{Ord} : (\forall n)\, s \restriction n \in T\}$.

5) A subset J of a tree T is a *front* if: $\eta \ne \nu \in J$ implies none of them is an initial segment of the other, and every $\eta \in \lim T$ has an initial segment which is a member of J.

6) $(T, \mathbf{I})$ is standard if for every nonsplitting point $\eta \in T$, $|\mathrm{Suc}_T(\eta)| = 1$.

7) $(T, \mathbf{I})$ is full if every $\eta \in T$ is a splitting point.

2.4A Remark. (1) The set lim$T$ is not absolute, i.e., if $V_1 \subseteq V_2$ are two universes of set theory then in general $(\lim T)^{V_1}$ will be a proper subset of $(\lim T)^{V_2}$.

(2) However, the notion of being a front is absolute: if $V_1 \models$ "F is a front in T", then there is a depth function $\rho : T \to \mathrm{Ord}$ satisfying $\eta \lhd \nu \,\&\, \forall k \leq \ell g(\eta)[\eta \restriction k \notin F] \to \rho(\eta) > \rho(\nu)$. This function will also witness in V_2 that F is a front.

(3) $F \subseteq T$ contains a front iff F meets every branch of T. So if $F \subseteq T$ contains a front of T and $T' \subseteq T$, then $F \cap T'$ contains a front of T'. Also this notion is absolute.

2.4B Notation. In several places in this chapter we will have an occasion to use the following notation: Assume that $(T, \mathbf{I})$ is a tagged tree, and for all $\eta \in T$ there is a family a_η of subsets of $T^{[\eta]}$ such that $\eta \lhd \nu \Rightarrow \forall A \in a_\eta \, \exists B \in a_\nu$ $[B \subseteq A]$. Then we can define for all $\alpha \in \mathrm{Ord} \cup \{\infty\}$

$\mathrm{Dp}_\alpha(\eta)$ iff $\forall \beta < \alpha \forall A \in a_\eta \exists \nu \in A \cap \mathrm{split}(T)[\{\rho : \rho \in \mathrm{Suc}_T(\nu) \,\&\, \mathrm{Dp}_\beta(\rho)\} \notin \mathbf{I}_\nu]$.

Then it is easy to see that

$\mathrm{Dp}(\eta) \stackrel{\mathrm{def}}{=} \max\{\alpha \in \mathrm{Ord} \cup \{\infty\} : \mathrm{Dp}_\alpha(\eta)\}$

is well defined, and $\mathrm{Dp}_\alpha(\eta) \Leftrightarrow \mathrm{Dp}(\eta) \geq \alpha$. We call $\mathrm{Dp}(\eta)$ the "depth" of η (with respect to the family $\mathbf{a} = \langle a_\eta : \eta \in T \rangle$ and the tagged tree $(T, \mathbf{I})$). It is easy to check that $\eta \lhd \nu \Rightarrow \mathrm{Dp}(\eta) \geq \mathrm{Dp}(\nu)$.

2.5 Definition. 1) An ideal I is λ-complete if any union of less than λ members of I is still a member of I.

2) A tagged tree $(T, \mathbf{I})$ is λ-complete if for each $\eta \in T \cap \mathrm{Dom}(\mathbf{I})$ the ideal $\mathbf{I}_\eta$ is λ-complete.

3) A family $\mathbb{I}$ of ideals is λ-complete if each $I \in \mathbb{I}$ is λ-complete. We will only consider $\aleph_2$-complete families $\mathbb{I}$.

4) A family $\mathbb{I}$ is restriction-closed *if* $I \in \mathbb{I}$, $A \subseteq \mathrm{Dom}(I)$, $A \notin I$ implies $I \restriction A = \{B \in I : B \subseteq A\}$ belongs to $\mathbb{I}$.

5) The restriction closure of $\mathbb{I}$, res-cl$(\mathbb{I})$ is $\{I \restriction A : I \in \mathbb{I}, A \subseteq \mathrm{Dom}(I), A \notin I\}$.

6) I is λ-indecomposable *if* for every $A \subseteq \mathrm{Dom}(I), A \notin I$, and $h : A \to \lambda$ there is $Y \subseteq \lambda, |Y| < \lambda$ such that $h^{-1}(Y) \notin I$. We say $\mathbf{I}$ or $\mathbb{I}$, is λ-indecomposable if each $\mathbf{I}_\eta$ (or $I \in \mathbb{I}$) is λ-indecomposable.
7) I is strongly λ-indecomposable if for $A_i \in I (i < \lambda)$ and $A \subseteq \mathrm{Dom}(I), A \notin I$ we can find $B \subseteq A$ of cardinality $< \lambda$ such that for no $i < \lambda$ does A_i include B.

2.5A Remark. As indicated by the names, if I is strongly λ-indecomposable then I is λ-indecomposable at least when λ is regular. [Why? Given A, h as in 2.5(6), let $A_i = h^{-1}(\{j : j < i\})$; if for some i, $A_i \notin I$ we are done, otherwise by 2.5(7) there is $Y \subseteq A$, $|Y| < \lambda \bigwedge_i Y \not\subseteq A_i$. But as λ is regular $> |Y|$, $i(*) = \sup\{h(x) + 1 : x \in Y\} < \lambda$ hence $Y \subseteq A_{i(*)}$, contradiction.]

2.6 Lemma. Let θ be an uncountable regular cardinal (the main case here is $\theta = \aleph_1$). Let $\mathbb{I}$ be a family of θ^+-complete ideals, $(T_0, \mathbf{I})$ a tagged tree, $A = \{\eta \in T : 0 < |\mathrm{Suc}_{T_0}(\eta)| \leq \theta\}$, $[\eta \in T_0 \setminus A \Rightarrow \mathbf{I}_\eta \in \mathbb{I} \,\&\, \mathrm{Suc}_{T_0}(\eta) \notin \mathbf{I}_\eta]$ and $[\eta \in A \Rightarrow \mathrm{Suc}_{T_0}(\eta) \subseteq \{\eta\hat{\ }\langle i\rangle : i < \theta\}]$ and $H : T_0 \to \theta$ and $\bar{e} = \langle e_\eta : \eta \in A\rangle$, is such that e_η is a club of θ. *Then* there is a club C of θ such that: for each $\delta \in C$ there is $T_\delta \subseteq T_0$ satisfying:

(a) T_δ a tree.
(b) If $\eta \in T_\delta$, $|\mathrm{Suc}_{T_0}(\eta)| < \theta$, then $\mathrm{Suc}_{T_\delta}(\eta) = \mathrm{Suc}_{T_0}(\eta)$, and if $|\mathrm{Suc}(\eta)| = \theta$, then $\mathrm{Suc}_{T_\delta}(\eta) = \{\eta\hat{\ }\langle i\rangle : i < \delta\} \cap \mathrm{Suc}_{T_0}(\eta)$ and $\delta \in e_\eta$.
(c) $\eta \in T_\delta \setminus A$ implies $\mathrm{Suc}_{T_\delta}(\eta) \notin \mathbf{I}_\eta$.
(d) for every $\eta \in T_\delta$: $H(\eta) < \delta$.

Proof. For each $\zeta < \theta$ we define a game $\eth_\zeta$. The game lasts ω moves, in the nth move $\eta_n \in T_0$ of length n is chosen.

For $n = 0$ necessarily $\eta_0 = \langle\,\rangle$.

For $n = m + 1$: If $|\mathrm{Suc}_{T_0}(\eta_m)| = \theta$, then the *first* player chooses $\eta_{m+1} \in \mathrm{Suc}_{T_0}(\eta_n)$, $\eta_{m+1}(m) < \zeta$.

If $|\mathrm{Suc}_{T_0}(\eta_m)| < \theta$, then the *first* player chooses any $\eta_{m+1} \in \mathrm{Suc}_{T_0}(\eta_m)$.

If $\eta_m \notin A$, then the first player chooses $A_m \in \mathbf{I}_{\eta_m}$, and then the *second* player chooses $\eta_{m+1} \in \mathrm{Suc}_{T_0}(\eta_m) \setminus A_m$.

At the end, the second player wins if for all n, $H(\eta_n) < \zeta$ and $|\mathrm{Suc}_{T_0}(\eta_n)| = \theta \Rightarrow \zeta \in e_{\eta_n}$. Now clearly

(*) if for a club of $\zeta < \theta$ the second player has a winning strategy for the game ∂_ζ, *then* there are trees T_δ (as required).

Let $S = \{\delta < \theta :$ second player does not have a winning strategy for the game $\partial_\delta\}$; we assume that S is stationary, and get a contradiction.

Let for $\delta \in S, F_\delta$ be a winning strategy for first player in ∂_δ (he has a winning strategy as the game is determined being closed for the second player). So F_δ gives for the first $(n-1)$-moves of the second player, the nth move of the first player.

Let χ be regular large enough, and let $(N_0, \in) \prec (H(\chi), \in)$ be such that $\theta + 1 \subseteq N_0$, $|N_0| = \theta$, $(T_0, \mathbf{I}) \in N_0$, $\bar{e} \in N_0$, and $\langle F_\delta : \delta \in S\rangle \in N_0$. We can find $N_1 \prec N_0$ such that $|N_1| < \theta$, $N_1 \cap \theta$ is an ordinal and $(T_0, \mathbf{I}) \in N_1$, $\langle F_\delta : \delta \in S\rangle \in N_1$ and $\bar{e} \in N_1$. Let $\delta \stackrel{\mathrm{def}}{=} N_1 \cap \theta$. Since S was assumed to be stationary, we may assume $\delta \in S$.

Now we shall define by induction on $n, \eta_n \in T_0 \cap N_1$ of length n, such that $\langle \eta_\ell : \ell \leq n\rangle$ is an initial segment of a play of the game ∂_δ in which the first player uses his winning strategy F_δ.

Case 1. $n = 0$:
We let $\eta_0 = \langle\rangle$. (The $A_\ell \in \mathbf{I}_{\eta_\ell}$ are not mentioned as they are not arguments of F_δ).

Case 2. For $n = m+1$, $\eta_m \in A$: the first player has a winning strategy F_δ for the game ∂_δ. So F_δ gives us η_n. Now if $|\mathrm{Suc}_{T_0}(\eta_m)| < \theta$ then $\mathrm{Suc}_{T_0}(\eta_m) \subseteq N_1$ (because T_0, η_m belongs and $N_1 \cap \theta$ is an ordinal), hence $\eta_n \in N_1$ as required. If $|\mathrm{Suc}_{T_0}(\eta_m)| = \theta$ then necessarily $\mathrm{Suc}_{T_0}(\eta_m) \subseteq \{\eta_m{}^\frown\langle i\rangle : i < \theta\}$, $\eta_n = \eta_m{}^\frown\langle i\rangle, i < \delta$ (as the play is of the game ∂_δ), so necessarily $i \in N_1$ hence (as $\eta_m \in N_1$) also $\eta_n \in N_1$.

Case 3. Lastly if $n = m+1, \eta_m \notin A$: so F_δ gives us $A^\delta_m \in \mathbf{I}_{\eta_m}$ which is not necessarily in N_1, however let $A^* = \bigcup\{A^\zeta_m : \zeta \in S$, and there is a play of ∂_ζ in which $\langle \eta_\ell : \ell \leq m\rangle$ were played and the first player plays according to F_ζ (this play is unique) and the strategy F_ζ dictates to the first player to choose $A^\zeta_m\}$.

Now A^* is in N_1 (as $\bar{F} \in N_1$) and as the union of $\leq \theta$ members of $\mathbf{I}_{\eta_m}$ it belongs to $\mathbf{I}_{\eta_m}$ hence $A^* \cap \mathrm{Suc}_{T_0}(\eta_m)$ is a proper subset of $\mathrm{Suc}_{T_0}(\eta_m)$, so there is $\eta_m{}^\frown\langle i\rangle \in \mathrm{Suc}_{T_0}(\eta_m) \setminus A^*$, so there is such $i \in N_1$ (so necessarily $i < \delta$). Let the second player choose $\eta_n = \eta_m{}^\frown\langle i\rangle$.

So we have played a sequence $\langle \eta_n : n \in \omega\rangle$ of elements of N_1, always obeying F_δ so this sequence was produced by a play of ∂_δ in which the first player plays according to the strategy F_δ. But then for all n : $\eta_n \in N_1 \Rightarrow H(\eta_n) \in N_1$, so $H(\eta_n) < \delta$, and

$$\eta_n \in N_1 \Rightarrow e_{\eta_n} \in N_1 \Rightarrow \delta = \sup(e_{\eta_n} \cap \delta) \Rightarrow \delta \in e_{\eta_n};$$

hence second player wins in this play. So F_δ cannot be a winning strategy. Contradiction, so S is not stationary. $\square_{2.6}$

2.6A Lemma. Suppose $(T, \mathbf{I})$ is an $\mathbb{I}$-tree, θ regular uncountable, $\langle A_\eta : \eta \in T\rangle$ is such that: A_η is a set of ordinals, $[\eta \triangleleft \nu \Rightarrow A_\eta \subseteq A_\nu]$ and

(*) (a) $\mathbf{S} \subseteq$ RUCar,

(b) $\mathbb{I}' \overset{\mathrm{def}}{=} \mathbb{I} \setminus \{I : |\mathrm{Dom}(I)| < \mu\}$ is μ^+-complete or at least strongly μ-indecomposable for every μ such that $\mu \in \mathbf{S}$ or $\mu \in \mathrm{pcf}\,(\mathbf{S} \cap A_\eta)$ for some $\eta \in T$ and

(c) $\mathbb{I}$ is θ-complete and $|\mathrm{pcf}\,(\mathbf{S} \cap A_\eta)| < \theta$ for $\eta \in T$ and $\theta \leq \min(\mathbf{S})$

(d) $|A_\eta| < \min(\mathbf{S})$ for $\eta \in T$

Then there is $T^\dagger$, $(T, \mathbf{I}) \leq^* (T^\dagger, \mathbf{I})$, such that:

if $\lambda \in A_\nu \cap \mathbf{S}$ and $\nu \in T^\dagger$ then for some $\alpha_\nu(\lambda) < \lambda$ for every ρ such that $\nu \triangleleft \rho \in \lim T^\dagger$ we have $\alpha_\nu(\lambda) \geq \sup(\lambda \cap \bigcup_{n<\omega} A_{\rho\restriction n})$.

Proof. It is enough to prove the existence of a $T^\dagger$ as required just for $\nu = \langle\ \rangle$ (as we can repeat the proof going up in the tree). This can be proved by induction on max(pcf $(\mathbf{S} \cap A_{\langle\ \rangle})$) (exist see [Sh:g, I 1.9]). Let $\alpha_\lambda(\eta) = \sup(A_\eta \cap \lambda)$.

As this lemma (2.6A) is not used in this book we assume knowledge of [Sh:g].

Let $\mathfrak{a} \stackrel{\text{def}}{=} \mathbf{S} \cap A_{\langle\rangle}$ (if $\mathfrak{a}$ is empty we have nothing to do), $\mu =$ max pcf $(\mathfrak{a})$, and $\langle f_\alpha : \alpha < \mu\rangle$ be $<_{J_{<\mu}[\mathfrak{a}]}$-increasing cofinal. Let $\{\mathfrak{b}_\varepsilon : \varepsilon < \varepsilon(*)\}$ be cofinal in $J_{<\mu}[\mathfrak{a}]$ e.g. this set is $\{\bigcup_{\theta\in\mathfrak{c}} b_\theta[\mathfrak{a}] : \mathfrak{c} \subseteq \text{pcf}\,\mathfrak{a} \setminus \{\mu\} \text{ finite}\}$ so $\varepsilon(*) < \theta$ hence by an assumption $\mathbb{I}'$ is $|\varepsilon(*)|^+$-complete.

For $\varepsilon < \varepsilon(*)$, $\zeta < \mu$ we define:

$(*)^\varepsilon_\zeta$ there is a subtree T' of T, $(T, \mathbf{I}) \leq^* (T', \mathbf{I})$ such that for every $\eta \in \lim(T')$ and $\lambda_n \in \mathfrak{a} \setminus \mathfrak{b}_\varepsilon$ we have $\alpha_{\lambda_n}(\eta\restriction n) \leq f_\zeta(\lambda_n)$.

It suffices to find such T' (for some ε, ζ) as: max pcf $(\mathfrak{b}_\varepsilon) <$ max pcf $(\mathfrak{a})$, so we can apply the induction hypothesis on T'.

In V define for $\zeta < \mu$ and $\varepsilon < \varepsilon(*)$.

$$B_\zeta \stackrel{\text{def}}{=} \{\eta \in \lim(T) : \text{ for some } \varepsilon < \varepsilon(*) \text{ for every } \lambda \in \mathfrak{a} \setminus \mathfrak{b}_\varepsilon \text{ we have } n < \omega \Rightarrow \alpha_\lambda(\eta\restriction n) \leq f_\zeta(\lambda)\}$$

$$B_{\zeta,\varepsilon} \stackrel{\text{def}}{=} \{\eta \in \lim(T) : \text{ for every } \lambda \in \mathfrak{a} \setminus \mathfrak{b}_\varepsilon, n < \omega \Rightarrow \alpha_\lambda(\eta\restriction n) \leq f_\zeta(\lambda)\}$$

Clearly $B_{\zeta,\varepsilon}$ is closed and $B_\zeta = \bigcup\limits_{\varepsilon<\varepsilon(*)} B_{\zeta,\varepsilon}$. Now $\zeta < \xi < \mu \Rightarrow B_\zeta \subseteq B_\xi$, (as $f_\zeta <_{J_{<\mu}[\mathfrak{a}]} f_\xi$) and $\lim(T) = \bigcup_{\zeta<\mu} B_\zeta$ (as $\langle f_\zeta : \zeta < \mu\rangle$ is cofinal in $\prod_{n<\omega} \lambda_n$), hence using 2.6B(3) below (with μ, $\varepsilon(*)$ here standing for θ, ε_i there) for some $\zeta(*) < \mu$ and $\varepsilon < \varepsilon(*)$ and T' we have $(T, \mathbf{I}) \leq^* (T', \mathbf{I})$ and $\lim(T') \subseteq B_{\zeta,\varepsilon}$. So $(*)^\varepsilon_\zeta$ holds, but as said above this suffices. $\square_{2.6A}$

Question. If $\mathbb{I} \in H(\chi)$ is there a countable $N \prec (H(\chi), \in, <^*_\chi)$ such that: $\mathbb{I} \in N$ and for every $\lambda \in \text{RCar} \cap N$, letting $\mathbb{I}^{[\lambda]} \stackrel{\text{def}}{=} \{J \in \mathbb{I} : J \text{ is } \lambda^+\text{-complete}\}$, there is $\langle N^\lambda_\eta : \eta \in (T, \mathbf{I})\rangle$ an $\mathbb{I}^{[\lambda]}$-suitable tree (see Definition 2.10) such that $N <_\lambda N^\lambda_{\langle\rangle}$? (Or replace RCar by a thinner set.)

2.6B Lemma. Let $(T, \mathbf{I})$ be an $\mathbb{I}$-tree, $\mathbb{I}$ a family of ideals,

1) If $H : T \to \mu$ and $\mu^{\aleph_0} < \lambda$ and $\mathbb{I}$ is λ - complete *then* there is T' such that

$$(T, \mathbf{I}) \leq^* (T', \mathbf{I})$$
$$\eta, \nu \in T' \ \& \ \ell g(\eta) = \ell g(\nu) \Rightarrow H(\eta) = H(\nu)$$

2) If $\lim(T) = \bigcup_{i<\theta} B_i, \theta < \lambda$, $\mathbb{I}$ is λ - complete and each B_i is a Borel set *then* there is T' such that

$$(T, \mathbf{I}) \leq^* (T', \mathbf{I})$$
$$\text{for some } i : [\lim T \subseteq B_i]$$

3) If θ is regular uncountable, $\lim(T) = \bigcup\limits_{i<\theta} B_i$, and B_i is a Borel subset of $\lim(T)$, increasing with i, and $(*)$ below holds then

(a) for some $i < \theta$ and T' we have $(T, \mathbf{I}) \leq^* (T', \mathbf{I})$ and $\lim(T') \subseteq B_i$

(b) if in addition $\eta \in T \setminus \mathrm{split}((T, \mathbf{I})) \Rightarrow |\mathrm{Suc}_T(\eta)| < \theta$ then in (a) we can demand $(T, \mathbf{I}) \leq^{\otimes} (T', \mathbf{I})$

where

$(*)$ every $I \in \mathbb{I}$ is θ^+-complete or at least strongly θ-indecomposable (see 2.5(7)).

4) Assume $\lim(T) = \bigcup_{i<\theta} \bigcup_{\varepsilon<\varepsilon_i} B_{i,\varepsilon}$, each $B_{i,\varepsilon}$ is a Borel set, $[i < \theta \Rightarrow \varepsilon_i < \sigma]$, $\mathbb{I}$ is σ-complete, and each $I \in \mathbb{I}$ is strongly θ-indecomposable, and $B_i \stackrel{\text{def}}{=} \bigcup_{\varepsilon<\varepsilon_i} B_{i,\varepsilon}$ is increasing in i and

$$[\eta \in T \setminus \mathrm{split}(T, \mathbf{I}) \Rightarrow |\mathrm{Suc}_T(\eta)| < \sigma].$$

Then for some $i < \theta$ and $\varepsilon < \varepsilon_i$ and T' we have $(T, \mathbf{I}) \leq^{\otimes} (T', \mathbf{I})$, and $\lim(T') \subseteq B_{i,\varepsilon}$.

2.6C Remark. 1) We can combine 2.6B(3), (4) with 2.6A.

2) To what can we weaken "strongly θ-indecomposable"? A sufficient condition is the existence of a precipitous normal filter E on θ such that for every $I \in \mathbb{I}$ and $A_i \in I$ for $i < \theta$ and $A^* \in I^+$ there are $x_i \in A^*$ for $i < \theta$ such that $\{i \in A^* : \{x_j : j < i\} \nsubseteq A_i\} \neq \emptyset \mod E$

3) We can elaborate 2.6B(4). We can have $t \subseteq {}^{\omega>}\mathrm{Ord}$ be a tree with no ω-branch, $\langle B_\eta : \eta \in t\rangle$ a sequence of subsets of $\lim(T)$ such that:

(α) if $\eta \in \max(t)$ then B_η is Borel

(β) if $\eta \in t$ is not maximal, then (a) or (b)

(a) $\mathbb{I}$ is $|\mathrm{Suc}_t(\eta)|^+$-complete, $B_\eta = \bigcup\{B_\nu : \nu \in \mathrm{Suc}_t(\eta)\}$

(b) $\langle B_{\eta\hat{}\langle i\rangle} : \eta\hat{}\langle i\rangle \in \mathrm{Suc}_t(\eta)\rangle$ is increasing and letting $\theta_\eta = \mathrm{cf}(\mathrm{otp}(\{i : \eta\hat{}\langle i\rangle \in t\}))$, $\mathbb{I}$ is strongly θ_η-indecomposable.

Now we prove by downward induction on $\eta \in t$ that

$(*)_\eta$ there are $(T', \mathbf{I})$ and ν such that: $\eta \trianglelefteq \nu \in \max(t)$, $(T, \mathbf{I}) \leq^* (T', \mathbf{I})$ and $\lim(T') \subseteq B_i$ *or* in the game corresponding to $\bigcup\{B_\rho : \eta \trianglelefteq \rho \in \max(t)\}$ the first player wins

4) We can combine 2.6C(3) with 2.6B(3).

Proof. 1), 2) By [RuSh:117] or see here XI 3.5, 3.5A.

3) Similar to the proof of 2.6. First we prove clause (a). Without loss of generality $(T, \mathbf{I})$ is standard, so for notational simplicity it is full (see Definition 2.4(6), (7)). For each $\zeta < \theta$ let $\eth_\zeta$ be the following game with ω moves, letting $\eta_0 = \langle\rangle$ and in the n'th move $\eta_n \in T$ is chosen; the first player chooses $A_n \in \mathbf{I}_{\eta_n}$ and the second player $\eta_{n+1} \in \mathrm{Suc}_T(\eta_n) \setminus A_n$. In the end $\bigcup_{n<\omega} \eta_n \in \lim(T)$, and the second player wins the play if $\bigcup_n \eta_n \in B_\zeta$. It suffices to prove for some $\zeta < \theta$, the second player has a winning strategy. So otherwise for each ζ the first player has a winning strategy F_ζ. Let χ be large enough, $N_1 \prec (H(\chi), \in, <^*_\chi)$, $\|N_1\| < \theta$, $\delta \overset{\text{def}}{=} N_1 \cap \theta < \theta$ such that $(T, \mathbf{I}), \langle B_\zeta : \zeta < \theta\rangle$ and $\langle F_\zeta : \zeta < \theta\rangle$ belongs to N_1. We shall simulate a play $\langle A_m, \eta_{m+1} : m < \omega\rangle$ of $\eth_\delta$ such that $\eta_{m+1} \in N_0$. Assume $\langle A_\ell, \eta_{\ell+1} : \ell < m\rangle$ is already defined. Let $S'_m = \{\zeta < \theta$: there is an initial segment of a play of $\eth_\zeta$ in which the first player uses the strategy F_ζ and the second player plays $\langle \eta_\ell : \ell \leq m\rangle\}$, note that such initial segment is unique, for a given ζ. For $\zeta \in S'_m$ let A^ζ_m be the $(m+1)$'th move of the first player, for such a play with the second player using the strategy F_ζ, so $\langle A^\zeta_m : \zeta \in S'_m\rangle \in N_1$, also clearly $\delta \in S'_m$, hence $|S'_m| = \theta$ and by the assumption $(*)$ for some $B \in N_1$, $B \subseteq \mathrm{Suc}_T(\eta_m)$, $|B| < \theta$ and $\bigwedge_{\zeta \in S'_m} B \not\subseteq A^\zeta_m$. As $B \in N_1$ and $N_1 \cap \theta = \delta$, clearly $B \subseteq N_1$ and choose $i \in B \setminus A^\delta_m$ and let $\eta_{n+1} = \eta_m\hat{}\langle i\rangle$.

For proving clause (b), defining ∂_ζ if $|\mathrm{Suc}_T(\eta_n)| < \theta$ we change the rule and let player I choose $\eta_{n+1} \in \mathrm{Suc}_T(\eta_n)$.
4) We define, for $\zeta < \theta$ and $\varepsilon < \varepsilon_\zeta$ a game $\partial_{\zeta,\varepsilon}$ as in the proof of 2.6B(3) clause (a) for $B_{\zeta,\varepsilon}$. If for some $\zeta < \theta$, $\varepsilon < \varepsilon_\zeta$ the second player wins then we get the desired conclusion. Otherwise as each such game is determind (as $B_{\zeta,\varepsilon}$ is a Borel set) there is a winning strategy $F_{\zeta,\varepsilon}$ for the first player. As $\mathbf{I}$ is $|\varepsilon_\zeta|^+$-complete, there is one strategy F_ζ good in all the games $F_{\zeta,\varepsilon}(\varepsilon < \varepsilon_\zeta)$ simultaniously (take the union of the sets suggested by all those strategies). So F_ζ is a winning strategy in ∂_ζ, and we can proceed as in the proof of 2.6B(3).

$\square_{2.6B}$

2.7 Definition. Let $\mathbb{I}$ be a set of $\aleph_2$-complete ideals, $\mathbf{S}$ a set of regular cardinals, $\aleph_1 = \mathrm{Min}\,(\mathbf{S})$ and P a forcing notion.

1) We say that $(T, \mathbf{I}, \bar\lambda, \bar{\underset{\sim}{\xi}}, \bar\zeta)$ is a $(\mathbb{I}, P, \mathbf{S})$-tree if:

a) $(T, \mathbf{I})$ is a $\mathbb{I}$-tree (see Definition 2.4(2))

b) $\bar\lambda$ is a function from T to $\mathbf{S}$

c) $\bar{\underset{\sim}{\xi}}$ is a function with domain T such that for every $\eta \in T, \bar{\underset{\sim}{\xi}}(\eta)$ is a P-name of an ordinal $< \bar\lambda(\eta)$

d) $\bar\zeta$ is a function from $T \setminus \{\langle\rangle\}$ such that each $\bar\zeta(\eta)$ is an ordinal.

2) We say that the $(\mathbb{I}, P, \mathbf{S})$-tree $(T, \mathbf{I}, \bar\lambda, \bar{\underset{\sim}{\xi}}, \bar\zeta)$ *obeys* a function F if there are fronts $J_n \subseteq T$ for $n < \omega$ (see Definition 2.5 (2)) such that every member of J_{n+1} has a strict initial segment in J_n and $\eta \in J_n$ implies

$$\Big\langle \mathrm{Suc}_T(\eta), \mathbf{I}_\eta, \langle \bar\zeta(\nu) : \nu \in \mathrm{Suc}_T(\eta)\rangle \Big\rangle = F\Big\langle \eta, w[\eta], \langle\langle \bar\lambda(\eta{\restriction}\ell), \bar{\underset{\sim}{\xi}}(\eta{\restriction}\ell),\ \bar\zeta(\eta{\restriction}\ell)\rangle\ : \ell \le \ell g(\eta)\rangle\Big\rangle$$

where $w[\eta]$ is $\{k : \eta{\restriction}k \in \bigcup_{\ell<\omega} J_\ell\}$.

2.7A Definition. We say that the forcing notion P satisfies $UP(\mathbb{I}, \underset{\sim}{\mathbf{S}}, \mathbf{W})$ (the "universal property"), where $\mathbf{W} \subseteq \omega_1$ is stationary, $\underset{\sim}{\mathbf{S}}$ a P-name of a set of uncountable regular cardinals (in V) which contains $\aleph_1^V$, provided that: letting $\mathbf{S}^* = \mathbf{S}^*[\underset{\sim}{\mathbf{S}}] = \{\kappa : \kappa \text{ regular} \le |P|, \nVdash \text{"}\kappa \notin \underset{\sim}{\mathbf{S}}\text{"}\}$, for every $p \in P$ there is a function F_p (with domain and range as implied implicitly in (2)) such that: for any $(\mathbb{I}, P, \mathbf{S}^*)$-tree $(T, \mathbf{I}, \bar{\lambda}, \bar{\xi}, \bar{\zeta})$ obeying F_p and any $T^\dagger$, $(T, \mathbf{I}) \leq^* (T^\dagger, \mathbf{I})$ there is $q \in P$, $p \leq_{pr} q$ such that:

$q \Vdash_P$ "there is $\eta \in \lim T^\dagger$ such that: *if* $\sup\{\bar{\zeta}(\eta{\restriction}\ell) : \ell < \omega$ and $\bar{\zeta}(\eta{\restriction}\ell) < \omega_1\}$ is not obtained and belongs to $\mathbf{W}$ *then*: for every $m < \omega$ satisfying $\bar{\lambda}(\eta{\restriction}m) \in \underset{\sim}{\mathbf{S}}$, for some $\ell < \omega$ we have $\underset{\sim}{\bar{\xi}}(\eta{\restriction}m)[\underset{\sim}{G}_P] < \bar{\zeta}(\eta{\restriction}\ell) < \bar{\lambda}(\eta{\restriction}m)$"

2.7B Notation. 1) If $\mathbb{I}$ is the set $\{J^{\rm bd}_\lambda : \lambda \geq \aleph_2, \lambda$ is regular$\}$ (where $J^{\rm bd}_A = \{B \subseteq A : \sup(B) < \sup(A)\}$) then we may omit it. We let $\bar{\lambda}_\eta \overset{\rm def}{=} \bar{\lambda}(\eta)$, $\underset{\sim}{\bar{\xi}}_\eta \overset{\rm def}{=} \underset{\sim}{\bar{\xi}}(\eta)$, $\bar{\zeta}_\eta \overset{\rm def}{=} \bar{\zeta}(\eta)$. If $\underset{\sim}{\mathbf{S}} = \{\aleph_1\}$ we may omit it and omit $\bar{\lambda}$. If $\underset{\sim}{\mathbf{S}} = \mathrm{RUCar}^V$ we may write $*$ instead of $\underset{\sim}{\mathbf{S}}$. If $\mathbf{W} = \omega_1$ we may omit it (note: no object can serve as two among $\mathbb{I}, \mathbf{S}$ and $\mathbf{W}$, so no confusion should arise).

It is always understood that the trivial I is in $\mathbb{I}$ (even if we write $\mathbb{I} = \emptyset$), a trivial I is the empty set with domain a singleton.

2) If not said otherwise, we shall ignore the non-$\aleph_2$-complete members of $\mathbb{I}$, i.e. $UP(\mathbb{I}, \mathbf{S}, \mathbf{W})$ means $UP(\mathbb{I}', \mathbf{S}, \mathbf{W})$ where $\mathbb{I}' = \{I \in \mathbb{I} : I$ is $\aleph_2$-complete$\}$.

2.7C Remark. 1)Why do we use $\mathbf{S}^*$ and why can we require $\mathbf{S}^* \subseteq |P|^+$?

(a) $\underset{\sim}{\mathbf{S}}$ is only a name (if $\mathbf{S}$ was a set $\in V$, $\mathbf{S}^* = \mathbf{S}$ is o.k.) and

(b) P-names of an ordinal $< \lambda$, $\lambda = \mathrm{cf}\lambda \geq |P|^+$ have an apriori bound.

2) A reader may use $\mathbf{S} = \{\aleph_1\}$ all the time.

2.7D Claim. 1) In Definition 2.7, if $\underset{\sim}{\mathbf{S}} = \mathrm{RUCar}^V$ we can replace in 2.7(1)(c) "a P-name of an ordinal $< \bar{\lambda}(\eta)$" by "a P-name of a member of V", in 2.7A demand $\underset{\sim}{\bar{\xi}}(\eta{\restriction}m) = \bar{\zeta}(\eta{\restriction}(\ell))$ and omit $\bar{\lambda}$ and get an equivalent definition (we can also replace $\leq |P|$ by $< \mathrm{Min}\{\kappa : P$ satisfies the κ-c.c.$\}$).

2) The forcing notion P satisfies $UP(\mathbb{I}, \mathbf{S}, \mathbf{W})$ iff its completion (to a complete Boolean algebra) satisfies it (assuming $\leq_{\rm pr} = \leq$).

3) If Q satisfies $UP(\mathbb{I}, *, \mathbf{W})$ (i.e. as in part (1)) and $\mathbb{I}$ is μ^+-complete (e.g. $\mathbb{I} = \emptyset$) *then* any "new" countable set of ordinals $< \mu$ is included in an "old" countable set of ordinals i.e. one from V.

4) Q satisfies $UP(\emptyset, *)$ iff Q is proper

5) Q satisfies $UP(\emptyset, \{\aleph_1\})$ iff Q is semiproper.
6) If Q satisfies $UP(\mathbb{I}, \underset{\sim}{\mathbf{S}}, \mathbf{W})$ and $\mathbb{I} \subseteq \mathbb{I}_1, \underset{\sim}{\mathbf{S}}_1 \subseteq \underset{\sim}{\mathbf{S}}$ and $\mathbf{W}_1 \subseteq \mathbf{W}$ *then* Q satisfies $\mathrm{UP}(\mathbb{I}_1, \underset{\sim}{\mathbf{S}}_1, \mathbf{W}_1)$.
7) In Def. 2.7A, we can replace $\underset{\sim}{\mathbf{S}}$ by any set $\underset{\sim}{\mathbf{S}}'$ of uncountable regular cardinals of V, such that $\Vdash_P$ "$\underset{\sim}{\mathbf{S}} \cap |P|^+ = \underset{\sim}{\mathbf{S}}' \cap |P|^+$".

Proof. (sketch) (1) is easy.

(2) Note that F is defined on sequences of names, and it is well known that P-names can be canonically translated to Q-names, if P is a dense subset of Q.

(3) Use 2.6B(2) repeatedly.

(4), (5): If $\mathbb{I} = \emptyset$, then each branch of an $\mathbb{I}$-tree is itself an $\mathbb{I}$-tree, so a strategy from XII 1.1 (or 1.7(3)) easily yields a function F.

(6) Easy.

(7) By 2.7C(1)(b). $\square_{2.7D}$

2.7E Convention. 1) We write $F_w(\eta, \langle \lambda_\ell, \underset{\sim}{\xi}_\ell, \zeta_\ell : \ell \leq lg(\eta) \rangle)$ for

$$F\Big(\eta, w, \Big\langle \langle \lambda_\ell, \underset{\sim}{\xi}_\ell, \zeta_\ell \rangle : \ell \leq lg(\eta) \Big\rangle\Big);$$

we omit λ_ℓ when $\underset{\sim}{\mathbf{S}} = \{\aleph_1\}$.

In Definition 2.7, the value F gives to $\mathrm{Suc}_T(\eta)$ is w.l.o.g. $\{\eta\hat{\ }\langle\alpha\rangle : \alpha < \lambda\}$ for some λ, and we do not strictly distinguish between λ and $\mathrm{Suc}_T(\eta)$.

2.8 Definition. 1) For an ideal collection $\mathbb{I}$, a set $\mathbf{S}$ of uncountable regular cardinals, (where $\aleph_1 = \min(\mathbf{S})$, and $\mathbb{I}$ is $\aleph_2$-complete) and χ regular large enough, we say a countable model $N \prec (H(\chi), \in, <^*_\chi)$ is strictly $(\mathbb{I}, \mathbf{S}, \mathbf{W})$-suitable for χ *if*: $N \cap \omega_1 \in \mathbf{W}$ and in the following game the second player has a winning strategy (letting $N_0 = N$).

in the nth move: the first player chooses $I_n \in \mathbb{I} \cap N_n$ and set A_n (not necessarily in N_n), $A_n \subseteq \mathrm{Dom}(I_n), A_n \in I_n$,
then the second player chooses $x_n \in (\mathrm{Dom}(I)) \setminus A_1$ and let $N_{n+1} \supseteq$ Skolem

Hull of $N_n \cup \{x_n\}$ such that for each $\lambda \in \mathbf{S} \cap N_n$:

$$\sup(N_{n+1} \cap \lambda) = \sup(N_n \cap \lambda)$$

2) If $\mathbf{W}$ is omitted we mean $\mathbf{W} = \omega_1$, if $\mathbf{S}$ is omitted we mean $\{\aleph_1\}$, if both are omitted we write strictly $\mathbb{I}$-suitable.

2.9 Claim. A model $N \prec (H(\chi), \in, <^*_\chi)$ is strictly $(\mathbb{I}, \mathbf{S}, \mathbf{W})$-suitable for χ *iff* there is an $\mathbb{I}$-tagged tree $(T, \mathbf{I})$ and $\langle N_\eta : \eta \in T\rangle$ such that:

a) $N = N_{\langle\,\rangle}, \{\mathbb{I}, \mathbf{S}, \mathbf{W}\} \in N$

b) $N_\eta \prec (H(\chi), \in, <^*_\chi)$ is countable

c) $N_{\eta\restriction k} \prec N_\eta$

d) for $\lambda \in \mathbf{S} \cap N_{\eta\restriction k}$, $k < lg(\eta)$ we have: $\sup(N_\eta \cap \lambda) = \sup(N_{\eta\restriction k} \cap \lambda)$

e) for every $\eta \in T$ and $I \in \mathbb{I} \cap N_\eta$

$\{\nu : \eta \trianglelefteq \nu, \nu \text{ a splitting of } (T, \mathbf{I}) \text{ and } \mathbf{I}_\nu = I\}$ contains a front of $T^{[\eta]}$

f) $\eta \in N_\eta$.

g) $N \cap \omega_1 \in \mathbf{W}$.

Proof Easy: from a winning strategy we can build a tree, and for any such tree $\langle N_\eta : \eta \in T\rangle$ a winning strategy of player II is to choose some $\eta_{n+1} \in T, \eta_n \trianglelefteq \eta_{n+1}$ preserving $\bigcup_{\ell \leq n} N_\ell \cup N \subseteq N_n = N_{\eta_n}$. $\square_{2.9}$

2.10 Definition. Fix $\mathbb{I}, \mathbf{S}, \mathbf{W}$.

1) An $\mathbb{I}$-tagged tree of models is an $\mathbb{I}$-tagged tree $(T, \mathbf{I})$ whose nodes η are used to label countable models N_η (we write this as $\bar{N} = \langle N_\eta : \eta \in (T^*, \mathbf{I})\rangle$) satisfying the following:

(a) for $\eta \in T$ we have $N_\eta \prec (H(\chi), \in, <^*_\chi)$ is a countable model.

(b) $N_{\langle\rangle}$ contains all necessary information, in particular $\mathbb{I}, \mathbf{S}, \mathbf{W}$.

(c) $\eta \triangleleft \nu \in T$ implies $N_\eta \prec N_\nu$

(d) for $\eta \in T$ we have $\eta \in N_\eta$ and $\mathbf{I}_\eta \in N_\eta$.

Whenever we have such an $\mathbb{I}$-tagged tree $\bar{N}$ of models, we write $N_\eta = \bigcup_{k<\omega} N_{\eta\restriction k}$ for all $\eta \in \lim(T)$.

2) We call such a tree $\mathbb{I}$-*suitable* if

(e) $\forall \eta \in T \forall I \in \mathbb{I} \cap N_\eta \{\nu \in T^{[\eta]} : \nu \in \text{split}(T), \mathbf{I}_\nu = I$ (or just they are isomorphic)$\}$ contains a front of $T^{[\eta]}$.

3) We call $\bar{N}$ *suitable** if instead of (e) we only have

(e)* $\forall \eta \in T \forall I \in \mathbb{I} \cap N \{\nu \in T^{[\nu]} : \nu \in \text{split}(T), \mathbf{I}_\nu \leq_{\text{RK}} I\}$ (see 2.10A below) contains a front of T.

4) We call $\bar{N}$ $\aleph_1$-strictly $(\mathbb{I}, \mathbf{S}, \mathbf{W})$-suitable if $\bar{N}$ is suitable and in addition

(f) for some $\delta \in \mathbf{W}$, for all $\eta \in T$ we have: $N_\eta \cap \omega_1 = \delta$

5) we call $\bar{N}$ strictly $(\mathbb{I}, \mathbf{S}, \mathbf{W})$-suitable, if in addition to clauses (a) – (e) we have:

(g) for all $\nu \in T$, $\lambda \in \mathbf{S} \cap N_\delta$ there is $\delta_\lambda < \lambda$ such that $\forall \eta \in T$ $[\nu \trianglelefteq \eta \Rightarrow \sup(N_\eta \cap \lambda) = \delta_\lambda]$.

6) We call $\bar{N}$ uniformly suitable or $\aleph_1$-uniformly suitable if (g) or (f) respectively hold only for all $\eta \in \lim(T)$.

Remark. Note: for suitable trees, $\mathbf{S}$ is essentially redundant so we may omit it or allow names. Similarly so for $\mathbf{W}$. In 2.9 and 2.10 we omit $\mathbf{W}$ when it is ω_1, and omit $\mathbf{S}$ when $\mathbf{S} = \{\aleph_1\}$, so $\mathbb{I}$-suitable means $(\mathbb{I}, \{\aleph_1\}, \omega_1)$-suitable. Let $\eta \in (T, \mathbf{I})$ means $\eta \in T$ and we write T when $\mathbf{I}$ is clear.

2.10A Definition. 1) For ideals J_1, J_2 we say

$$J_1 \leq_{RK} J_2$$

if there is a function h witnessing it, i.e. $h : \text{Dom}(J_2) \to \text{Dom}(J_1)$ is such that

$$\text{for every } A \subseteq \text{Dom}(J_2) : A \neq \emptyset \text{ mod } J_2 \Rightarrow h''(A) \neq \emptyset \text{ mod } J_1$$

or equivalently, $J_2 \supseteq \{h^{-1}(A) : A \in J_1\}$.

2) For families $\mathbb{I}_1, \mathbb{I}_2$ of ideals we say $\mathbb{I}_1 \leq_{RK} \mathbb{I}_2$ if there is a function H witnessing it i.e.

(i) H is a function from $\mathbb{I}_1$ into $\mathbb{I}_2$

(ii) for every $J \in \mathbb{I}_1$ we have $J \leq_{RK} H(J)$

3) For families $\mathbb{I}_1, \mathbb{I}_2$ of ideals, $\mathbb{I}_1 \equiv_{RK} \mathbb{I}_2$ *if* $\mathbb{I}_1 \leq_{RK} \mathbb{I}_2 \& \mathbb{I}_2 \leq_{RK} \mathbb{I}_1$.

2.10B Fact. Assume $\mathbb{I} \leq_{RK} \mathbb{I}'$, where $\mathbb{I}, \mathbb{I}'$ are families of ideals.

1) If $\langle N_\eta : \eta \in (T, \mathbf{I})\rangle$ is a $\mathbb{I}'$-suitable* tree and $\mathbb{I} \in N_{\langle\rangle}$, *then* $\langle N_\eta : \eta \in (T, \mathbf{I})\rangle$ is also $\mathbb{I}$-suitable*.
2) If $\langle N_\eta : \eta \in (T, \mathbf{I})\rangle$ is $\mathbb{I}$-suitable*, *then* there is a tree $(T', \mathbf{I}')$ satisfying the following:
   (a) $T' \subseteq T$ (but in general not $T \leq T'$, as the function $\mathbf{I}'$ will be different)
   (b) $\mathrm{split}(T', \mathbf{I}') = T' \cap \mathrm{split}(T, \mathbf{I})$
   (c) $\langle N_\eta : \eta \in (T', \mathbf{I}')\rangle$ is $\mathbb{I}$-suitable.

Proof. (1) Should be clear, as $\leq_{\mathrm{RK}}$ is transitive (as a relation among ideals and also among families of ideals).

(2) For each $\eta \in \mathrm{split}(T, \mathbf{I})$ pick an ideal $\mathbf{I}'_\eta \in \mathbb{I} \cap N_\eta$, $\mathbf{I}'_\eta \leq_{\mathrm{RK}} \mathbf{I}_\eta$ such that: for all $\nu \in T$, for all $I' \in \mathbb{I} \cap N_\nu : \{\eta \in T^{[\nu]} : I' = \mathbf{I}'_\eta\}$ contains a front of $T^{[\nu]}$. This can be done using a bookkeeping argument.

Now define T' as follows: If $\eta \in T' \setminus \mathrm{split}(T, \mathbf{I})$, then $\mathrm{Suc}_{T'}(\eta) = \mathrm{Suc}_T(\eta)$. If $\eta \in T' \cap \mathrm{split}(T, \mathbf{I})$, then $\mathbf{I}'_\eta$ is already defined and it belongs to N_η. Let g_η be a witness for $\mathbf{I}'_\eta \leq_{\mathrm{RK}} \mathbf{I}_\eta$, so g_η introduces an equivalence relation on $\mathrm{Suc}_T(\eta)$. Let A_η be a selector set for this equivalence relation, i.e. $g_\eta \restriction A_\eta$ is 1-1 and has the same range as g_η. Note that we can choose g_η and A_η in N_η. So without loss of generality we may assume that $g_\eta \restriction A_\eta$ is the identity, and let $\mathrm{Suc}_T(\eta) = A_\eta$.

$\square_{2.10B}$

2.11 Claim. Assume $\mathbb{I}$ is a restriction closed family of ideals, $\underset{\sim}{\mathbf{S}}$ a P-name of a set of regular uncountable cardinals, P a forcing notion, $\mathbb{I}$ is $\aleph_2$-complete and $\mathbf{W} \subseteq \omega_1$. Then TFAE:

(A) P satisfies $UP(\mathbb{I}, \underset{\sim}{\mathbf{S}}, \mathbf{W})$.

(B) for large enough regular χ, *if* $\mathbf{S}^* = \{\lambda :\nVdash_P \text{“}\lambda \notin \mathbf{S}\text{”} \text{ and } \lambda \leq |P|\}$ and $\overline{N} = \langle N_\eta : \eta \in (T, \mathbf{I})\rangle$ is a $(\mathbb{I}, \mathbf{S}^*, \mathbf{W})$-suitable tree of models for χ (see Definition 2.10(2)) and $p \in N_{\langle\,\rangle} \cap P$, *then* there is a $q \in P$, $p \leq_{pr} q$, such that:

$$q \Vdash_P \text{" for some } \eta \in \lim T \text{ (in } (V^P)) \bigcup_{k<\omega} N_{\eta \restriction k} \cap \omega_1 \notin \mathbf{W} \text{ or :}$$

$$\text{for every } k < \omega, \text{ and } \lambda \in \underset{\sim}{\mathbf{S}} \cap N_{\eta \restriction k} \text{ and}$$

$$\underset{\sim}{\alpha} \in N_{\eta \restriction k}, \text{ a } P\text{-name of an ordinal } < \lambda$$

$$\text{we have } \underset{\sim}{\alpha}[G_P] < \sup[\bigcup_n (N_{\eta \restriction n} \cap \lambda)]\text{"}$$

i.e. $q \Vdash_P$ "for some $\eta \in \lim(T)$: if $\sup(\bigcup_{k<\omega} N_{\eta \restriction k} \cap \omega_1) \in \mathbf{W}$ then $\sup(\bigcup_k N_{\eta \restriction k}[G_P] \cap \lambda) = \sup(\bigcup_k N_{\eta \restriction k} \cap \lambda)$ for $\lambda \in \underset{\sim}{\mathbf{S}} \cap (\bigcup_k N_{\eta \restriction k})$"

(B)* Like (B) replacing suitable by suitable*.

2.11A Remark. We can use, in (B), "$\lambda \in \underset{\sim}{\mathbf{S}} \cap N_{\eta \restriction k}[G_P]$" instead of "$\lambda \in \underset{\sim}{\mathbf{S}} \cap N_{\eta \restriction k}$" if in 2.7(1) we change all $\bar{\lambda}(\eta)$ to be P-names. Such a change would not hurt the rest of this chapter.

Proof. (A) ⇒ (B)

So let $\langle N_\eta : \eta \in (T, \mathbf{I})\rangle$ be $(\mathbb{I}, \underset{\sim}{\mathbf{S}}, \mathbf{W})$-suitable tree of models for χ and $p \in N_{\langle\rangle} \cap P$. We should find q as in (B). There are $F \in N_{\langle\,\rangle}$ witnessing UP$(\mathbb{I}, \underset{\sim}{\mathbf{S}}, \mathbf{W})$ for p and $\chi_0 \in N_{\langle\,\rangle}$ (such that $<^*_{\chi_0} \in N_{\langle\rangle}$) where $\{F, P, 2^{|P|}\} \in N_{\langle\,\rangle} \cap H(\chi_0)$.

Now we form an $(\mathbb{I}, P, \underset{\sim}{\mathbf{S}})$-tree $(T^\dagger, \mathbf{I}^\dagger, \bar{\lambda}, \bar{\underset{\sim}{\xi}}, \bar{\zeta})$ which obey F, and a function $h : T^\dagger \to T$ satisfying $[\eta \trianglelefteq \nu \Rightarrow h(\eta) \trianglelefteq h(\nu)]$ and $[\eta \in T^\dagger \Rightarrow \{\eta, \mathbf{I}^\dagger, \bar{\lambda}(\eta), \bar{\underset{\sim}{\xi}}(\eta), \bar{\zeta}(\eta)\} \in N_{h(\eta)}]$, and:

$(*)_1$ for every $\eta \in T^\dagger, \lambda \in N_\eta \cap \mathbf{S}^*, I \in \mathbb{I} \cap N_\eta$ and $\underset{\sim}{\xi} \in N_\eta$ a P-name of an ordinal $< \lambda$, for some front J of $T^{[\eta]}$ consisting of splitting nodes of $(T^\dagger, \mathbf{I})$ above η,

$$[\nu \in J \Rightarrow \langle \bar{\lambda}(\nu), \bar{\underset{\sim}{\xi}}(\nu)\rangle = \langle \lambda, \underset{\sim}{\xi}\rangle]$$

$$[\nu \in J \Rightarrow \mathbf{I}^\dagger_\nu = I]$$

Note that as $F \in N_{\langle\rangle} \prec N_\nu$ necessarily

$$[\nu \in J \;\&\; \rho \in \mathrm{Suc}_T(\nu) \Rightarrow \zeta_\rho \in N_\rho].$$

Now apply Def 2.7A to $(T^\dagger, \mathbf{I}^\dagger, \bar{\lambda}, \bar{\xi}, \bar{\zeta})$ so in V^P we get $q \in P$ and $\underset{\sim}{\eta} \in \lim(T^\dagger)$ (a P-name) as required there (i.e. forced by q to be so), now there is a P-name $\underset{\sim}{\nu} \in \lim(T)$ such that $\bigwedge_{k<\omega} h(\underset{\sim}{\eta}\restriction k) \lhd \underset{\sim}{\nu}$; so q, $\underset{\sim}{\nu}$ are as required.

$\underline{(B) \Rightarrow (A)}$

Easy. Choose χ large enough, and let us define a function F which will exemplify (A). Let $\langle A_n : n < \omega\rangle$ be pairwise disjoint infinite subsets of ω, with $\mathrm{Min}(A_n) \geq n$ and $\omega = \bigcup_{n<\omega} A_n$.

Now

$$F(\eta, w, \langle\langle \lambda_l, \underset{\sim}{\xi}_l, \zeta_l\rangle : l \leq \ell g(\eta)\rangle) = \Big\langle Y^*, I^*, \langle x(\nu) : \nu \in Y^*\rangle\Big\rangle$$

is defined as follows: let n be the unique $n < \omega$ such that $|w| \in A_n$, so $n \leq \ell g(\eta)$, and let $\nu = \nu_\eta = \eta\restriction n$, we let $N_\eta \stackrel{\text{def}}{=}$ the Skolem Hull of $\{\mathbf{S}, \mathbb{I}, \eta\} \cup \langle\langle \lambda_l, \underset{\sim}{\xi}_l, \zeta_l\rangle : l \leq n\rangle$ in $(H(\chi), \in, <^*_\chi)$; and let $\langle (I^\nu_m, \lambda^\nu_m, \underset{\sim}{\xi}^\nu_m, \zeta^\nu_m) : m \in A_n\rangle$ be the $<^*_\chi$-first list of this form of all tuples $(I,\ \lambda,\ \underset{\sim}{\xi}, \zeta)$ such that $I \in N_\nu \cap \mathbb{I}$, $\lambda \in \mathrm{RUCar} \cap N_\nu, \underset{\sim}{\xi} \in N_\nu$ a P-name of an ordinal $< \lambda$ and $\zeta \in N_\nu$ an ordinal.

Lastly,

$$\begin{aligned}
Y^* &= \{\eta\hat{\ }\langle x\rangle : x \in \mathrm{Dom}(I^\nu_{|w|})\} \\
I^* &= \{\{\eta\hat{\ }\langle x\rangle : x \in B\} : B \in I^\nu_{|w|}\} \\
\lambda(\eta\hat{\ }\langle x\rangle) &= \lambda^\nu_{|w|} \\
\underset{\sim}{\xi}(\eta\hat{\ }\langle x\rangle) &= \underset{\sim}{\xi}^x_{|w|} \\
\zeta(\eta\hat{\ }\langle x\rangle) &= \ \sup(\lambda_{|w|} \cap N_\nu)
\end{aligned}$$

So let $\langle T, \mathbf{I}, \bar{\lambda}, \bar{\xi}, \bar{\zeta}\rangle$ be an $(\mathbb{I}, P, \mathbf{S})$-tree obeying F.

Now apply (B) to $\langle N_\nu : \nu \in T\rangle$ and get $q, \underset{\sim}{\eta}$ as required in (A), i.e. they are as required in 2.7(3).

$\underline{(B)^* \Rightarrow (B)}$ Easy as a suitable tree is a suitable* tree.

$\underline{(B) \Rightarrow (B)^*}$ By 2.10B(2). $\square_{2.11}$

2.12 Claim. If $\underset{\sim}{\mathbf{S}}, P, \mathbb{I}, \mathbf{W}, x \in H(\chi)$ and $\mathbf{S}^*$ is as in 2.7A (i.e. $\mathbf{S}^* = \{\theta : \theta = \mathrm{cf}(\theta) \leq |P|$ and $\nVdash_P$ "$\theta \notin \underset{\sim}{\mathbf{S}}$"$\}$, e.g. $\mathbf{S} = \{\aleph_1\}$), $\mathbb{I}$ is $\aleph_2$-complete *or* for each $I \in \mathbb{I}$, $\kappa \in \mathbf{S}$ we have I is κ-indecomposable, *then* there is a $\aleph_1$-strictly, uniformly $(\mathbb{I}, \underset{\sim}{\mathbf{S}}, \mathbf{W})$-suitable tree $\bar{N}$ with $x \in N_{\langle\rangle}$.

Proof. We will construct this tree in three steps: first we find a suitable tree, then we thin it out to be a uniformly suitable tree, then we blow up the models to make it $\aleph_1$-strict. For notational simplicity let $\underset{\sim}{\mathbf{S}} = \{\aleph_1\}$ so $\mathbf{S}^* = \{\aleph_1\}$.

First Step: An easy bookkeeping argument (to ensure 2.10(e)) yields an $(\mathbb{I} \cup \{J^{\mathrm{bd}}_{\omega_1}\})$-suitable tree $\langle N_\eta : \eta \in (T, \mathbf{I})\rangle$; so for $\eta \in \lim(T)$ we let $N_\eta = \bigcup\limits_{\ell<\omega} N_{\eta\restriction\ell}$. Hence we get that for all $\eta \in \lim(T)$, for all $I \in (\mathbb{I} \cap N_\eta) \cup \{J^{\mathrm{bd}}_{\omega_1}\}$, there are infinitely may k such that $\eta\restriction k \in \mathrm{split}(T, \mathbf{I})$ and $\mathrm{Suc}_T(\eta\restriction k) = \{\eta\hat{\ }\langle x\rangle : x \in \mathrm{Dom}(I)\}$.

Second Step: Define $H : T \to \omega_1$ by $H(\eta) = \sup(N_\eta \cap \omega_1) < \omega_1$. Apply 2.6 to get a subtree T', and a limit ordinal $\delta \in \mathbf{W} \subseteq \omega_1$ such that clauses (a) – (d) of 2.6 hold. By clause (d) of 2.1, for all $\eta \in T'$, $N_\eta \cap \omega_1 \subseteq \delta$. Let $\delta_0 < \delta_1 < \dots$, $\bigcup_n \delta_n = \delta$, and let

$$T_2 \stackrel{\mathrm{def}}{=} \{\eta \in T' : \forall k < \ell g(\eta),\ \text{if } \mathrm{Suc}_T(\eta\restriction k) = \{\eta\restriction k\hat{\ }\langle\alpha\rangle : \alpha < \omega_1\}$$
$$(\text{so } \mathrm{Suc}_{T'}(\eta\restriction k) = \{\eta\restriction k\hat{\ }\langle\alpha\rangle :\ \alpha < \delta\}) \text{ then } \eta(k) = \delta_k\}.$$

Clearly T_2 will be $\aleph_1$-uniformly suitable.

Third Step: For $\eta \in T_2$, let N'_η =the Skolem hull of $N_\eta \cup \delta$. So $N'_\eta \cap \omega_1 \supseteq \delta$. Conversely, let $\nu \in \lim(T_2)$, $\eta \lhd \nu$, then $N_\eta \cup \delta \subseteq N_\nu$, so $N'_\eta \subseteq N_\nu$ hence $N'_\eta \cap \omega \subseteq \delta$. So $N'_\eta \cap \omega_1 = \delta$, i.e. $\langle N'_\eta : \eta \in T_2\rangle$ is an $\aleph_1$-strictly by $(\mathbb{I}, \underset{\sim}{\mathbf{S}}, \mathbf{W})$-tree of models (see Definition 2.10(4)).

We claim that this tree is still suitable. Indeed, let $\eta \in T_2$, $\nu \in \lim(T_2)$, $\eta \trianglelefteq \nu$ and $I \in \mathbb{I} \cap N'_\eta$. Then for some $\alpha < \delta$, I is in the Skolem hull of $N_\eta \cup \alpha$. Let $k < \omega$ be such that $\alpha \in N_{\nu\restriction k} \cap \omega_1$, $k \leq \ell g(\eta)$. Then since $\langle N_\eta : \eta \in T_2\rangle$ was suitable, there is $\ell \geq k$ such that $\mathbf{I}_{\nu\restriction\ell} = I$. So $\langle N'_\eta : \eta \in T_2\rangle$ is also suitable. $\square_{2.12}$

2.12A Conclusion. If P satisfies $UP(\mathbb{I}, \underset{\sim}{\mathbf{S}}, \mathbf{W})$ and $\underset{\sim}{\mathbf{S}}$ is as in 2.7A (or $\underset{\sim}{\mathbf{S}} = \{\aleph_1\}$) (recall that this notation implies $\mathbb{I}$ is $\aleph_2$-complete, $\aleph_1 \in \underset{\sim}{\mathbf{S}}$, $\mathbf{W} \subseteq \omega_1$ stationary) then $\Vdash_P$ "$\mathbf{W}$ is stationary". Moreover, if $\mathbf{W}' \subseteq \mathbf{W}$ is stationary then also $\Vdash_P$ "$\mathbf{W}$ is a stationary subset of ω_1".

Proof. The "moreover" fact is by 2.7D(6) (i.e. monotonicity in $\mathbf{W}$).

Assume that $p \Vdash$ "$\underset{\sim}{C}$ is a club of ω_1 and $\underset{\sim}{C} \cap \mathbf{W} = \emptyset$". By 2.12 we can find an $\aleph_1$-strictly $(\mathbb{I}, \underset{\sim}{\mathbf{S}}, \mathbf{W})$-suitable tree of models $\langle N_\eta : \eta \in (T, \mathbf{I})\rangle$ with $\underset{\sim}{C}, p \in N_{\langle\rangle}$. Let $\delta = N \cap \omega_1$, so $\delta \in \mathbf{W}$. By $UP(\mathbb{I}, \underset{\sim}{\mathbf{S}}, \mathbf{W})$ we can find a condition q as in 2.11(B) in particular $p \leq_{\rm pr} q$. Clearly $q \Vdash$ "$N_{\langle\rangle}[G] \cap \omega_1 = \delta$" and, trivially $p \Vdash_P$ "$\underset{\sim}{C}$ is unbounded in $N_{\langle\rangle}[G] \cap \omega_1$" hence $p \Vdash$ "$N_{\langle\rangle}[G] \cap \omega_1 \in C$". So $q \Vdash$ "$\delta \in \underset{\sim}{C} \cap \mathbf{W}$". $\square_{2.12A}$

2.12B Remark. From now we shall use 2.11+2.12 freely. Usually we assume $\mathbb{I}, \mathbf{S}$ satisfies 2.6A($*$)(a)+(b), $\mathbf{S} = \{\aleph_1\}$ is the main case. We could have started with 2.11(B) as a definition of UP but did not as the definition 2.7 was closer to Chapter XI.

2.13 Remark. From the proof of 2.12 we can conclude that in 2.11; in clause (B) we can replace "$(\mathbb{I}, \mathbf{S}, \mathbf{W})$-suitable" by "$\aleph_1$-strictly $(\mathbb{I}, \mathbf{S}, \mathbf{W})$-suitable, $N_\eta \cap \omega_1 = \delta \in \mathbf{W}$", and then the condition q will be $N_{\langle\rangle}$-semi generic.

2.14 Conclusion. 1) If P satisfies $UP(\mathbb{I}, \underset{\sim}{\mathbf{S}}, \mathbf{W}), \underset{\sim}{Q}$ a P-name of a purely proper forcing *then* $P * \underset{\sim}{Q}$ satisfies $UP(\mathbb{I}, \underset{\sim}{\mathbf{S}}, \mathbf{W})$.

2) If $\underset{\sim}{\mathbf{S}} = \{\aleph_1\}, \underset{\sim}{Q}$ purely semiproper is enough.

3) Generally $\underset{\sim}{Q}$ is purely $(\underset{\sim}{\mathbf{S}}, \mathbf{W})$-semiproper is enough where:

Q is $(\underset{\sim}{\mathbf{S}}, \mathbf{W})$-semiproper *when*: if χ regular large enough, $Q \in N \prec (H(\chi), \in, <^*_\chi)$, $\|N\| = \aleph_0$, $p \in Q \cap N$ and $N \cap \omega_1 \in \mathbf{W}$ *then* there is q, $p \leq_{\rm pr} q \in Q$, such that:

$q \Vdash$ "for every $\lambda \in N \cap \underset{\sim}{\mathbf{S}}$, if $\underset{\sim}{\alpha} \in N$ is a Q-name of an ordinal $< \lambda$ then $\underset{\sim}{\alpha}[\underset{\sim}{G}_Q] <$ sup $(N \cap \lambda)$".

(Note that Q is $(\underset{\sim}{\mathbf{S}}, \mathbf{W})$-semiproper iff Q satisfies the $UP(\emptyset, \underset{\sim}{\mathbf{S}}, \mathbf{W})$).

4) Suppose Q_0 is a proper forcing (in V), $\lambda \geq |Q_0|^{\aleph_0}$, (of course $\lambda = \lambda^{\aleph_0} \geq$ density of Q_0 suffices), $\mathbb{I} \in V$ is a λ^+-complete family of ideals which is λ^+-directed under $\leq_{\rm RK}$ and $\Vdash_{Q_0}$ "$\underset{\sim}{Q}_1$ is a forcing notion satisfying $UP(\mathbb{I})$".

Then $Q_0 * \underset{\sim}{Q}_1$ satisfies $UP(\mathbb{I})$ (in V).

5) In 4) we can add $\mathbf{W}$.

6) If Q satisfies the λ-c.c., satisfaction of "Q satisfies $UP(\mathbb{I}, \underset{\sim}{\mathbf{S}}, \mathbf{W})$" depend on $\underset{\sim}{\mathbf{S}} \cap \lambda$ only so we shall ignore $\underset{\sim}{\mathbf{S}} \setminus \lambda$. For notational convenience we will demand URCard$\setminus\lambda \subseteq \underset{\sim}{\mathbf{S}}$.

Proof. 1), 2), 3), 5), 6). Left to the reader.

4) Let χ be regular large enough and let $\langle N_\eta : \eta \in (T, \mathbf{I})\rangle$ be an $\mathbb{I}$-suitable tree of models for χ, $(p_0, \underset{\sim}{p}_1) \in Q_0 * \underset{\sim}{Q}_1$ and $\{(p_0, \underset{\sim}{p}_1), Q_0 * \underset{\sim}{Q}_1, \mathbb{I}\} \in N_{\langle\rangle}$. For $\eta \in \lim(T)$ we let $N_\eta \overset{\rm def}{=} \bigcup_{k<\omega} N_{\eta\restriction k}$. As $\lambda \geq |Q_0|^{\aleph_0}$, as $\mathbb{I}$ is λ^+-complete, by 2.6B(1) w.l.o.g. for $\eta \in T$ we have: $N_\eta \cap Q_0$ depends only on $\ell g(\eta)$ and hence $N_\eta \cap Q_0$ is the same for all branches $\eta \in \lim(T)$. Now for each $\eta \in \lim(T)$, in V N_η is a countable elementary submodel of $(H(\chi), \in, <^*_\chi)$ hence there is $q^\circ \in Q_0, p_0 \leq_{\rm pr} q^\circ$, and q° is (N_η, Q_0)-generic.

Now for each $q, p_0 \leq_{\rm pr} q \in Q_0$, let

$$B_q = \{\eta \in \lim(T) : q \text{ is } (N_\eta, Q_0)\text{-generic}\}.$$

So $\lim(T) = \cup\{B_q : p_0 \leq_{\rm pr} q \in Q\}$

Note

(∗) for $\eta \in \lim(T)$, $p_0 \leq_{\rm pr} q \in Q_0$, we have: $\eta \in B_q$ *iff* for any maximal antichain $\mathcal{J} \in N_\eta$ of Q_0, we have: $[r \in \mathcal{J} \setminus N_\eta \Rightarrow r, q$ incompatible$]$.

Hence, B_q is a closed subset of $\lim(T)$, (as if $\eta \in \lim(T) \setminus B_q$ then for some $\mathcal{J} \in MAC(Q_0) \cap N_\eta$ and $r \in \mathcal{J} \setminus N_\eta$ we have r, q are compatible; then for some $m < \omega$, $\mathcal{J} \in N_{\eta\restriction m}$, and $\eta\restriction m \triangleleft \nu \in \lim(T)$ still implies $r \in \mathcal{J} \setminus N_\nu$ (because $N_\eta \cap Q_o = N_\nu \cap Q_0$) but r, q compatible. So $\lim(\underset{\sim}{T}) \setminus B_q$ contains the neighborhood determined by $\eta\restriction m$).

So by 2.6B(2) if $\lambda \geq |Q_0|^{\aleph_0}$, for some $q \in Q_0$ and T' we have: $p^0 \leq_{\rm pr} q$, $(T, \mathbf{I}) \leq^* (T', \mathbf{I})$ and $\lim(\underset{\sim}{T}') \subseteq B_q$. So $q \Vdash_{Q_0}$ "$N_\eta[\underset{\sim}{G}_{Q_0}] \cap \omega_1 = N_\eta \cap \omega_1$ for every $\eta \in T'$" and clearly $q \Vdash$ "$\langle N_\eta[\underset{\sim}{G}_{Q_0}] : \eta \in T'\rangle$ is an $\mathbb{I}$ -suitable* tree of models for χ". Why the suitable* not suitable? There may be $\eta \in T'$, $I \in \mathbb{I} \cap N_\eta[\underset{\sim}{G}_{Q_0}] \setminus N_\eta$; we get the $\mathbb{I}$-suitable* by 2.14A below.

So we can finish easily. $\square_{2.14}$

2.14A. Assume that $\langle N_\eta : \eta \in (T, \mathbf{I})\rangle$ is a $(\mathbb{I}, \underset{\sim}{\mathbf{S}}, \mathbf{W})$-suitable* tree, Q is a forcing notion satisfying κ-c.c. and $(\mathbb{I}, \leq_{RK})$ is κ-directed. *Then* $\Vdash_Q$ "$\langle N_\eta[G] : \eta \in (T, \mathbf{I})\rangle$ is $(\mathbb{I}, \underset{\sim}{\mathbf{S}}, \mathbf{W})$-suitable*".

Proof. First we claim that for each name $\underset{\sim}{I}$, if $p \Vdash$ "$\underset{\sim}{I} \in \mathbb{I}$" then there is $J \in \mathbb{I}$ such that $p \Vdash$ "$\underset{\sim}{I} \leq_{RK} J$". Indeed, since Q satisfies the κ-c.c. we can find a set $Y \subseteq \mathbb{I}$, $|Y| < \kappa$ such that $p \Vdash$ "$\underset{\sim}{I} \in Y$". Now let J be a $\leq_{RK}$-upper bound for Y. So for all $I' \in Y$ we have $I' \leq_{RK} J$. The function witnessing this relation will also witness it in V^Q, hence $p \Vdash$ "$\underset{\sim}{I} \leq_{RK} J$".

Now work in $V[G]$. Let $I \in N_\eta[G] \cap \mathbb{I}$. Applying the claim we have just proved, in N_η we can find $J \in N_\eta \cap \mathbb{I}$ such that $I \leq_{RK} J$. In $V[G]$ the set $\{x \in T^{[\eta]} : \eta \trianglelefteq \nu, J \leq_{RK} \mathbf{I}_\nu\}$ contains a front F of $T^{[\eta]}$. F is also a front in $V[G]$, so by transitivity of $\leq_{RK}$ we are done. $\square_{2.14A}$

2.15 Theorem. Suppose

a) Q_0 is a forcing notion, satisfying $UP(\mathbb{I}_0, \mathbf{S}_0, \mathbf{W})$

b) $\Vdash_{Q_0}$ "$\underset{\sim}{Q}_1$ is a forcing notion satisfying $UP(\underset{\sim}{\mathbb{I}}_1, \underset{\sim}{\mathbf{S}}_1, \mathbf{W})$". So: $\underset{\sim}{\mathbf{S}}_0, \underset{\sim}{\mathbb{I}}_1$ are $\underset{\sim}{Q}_0$-names and $\underset{\sim}{\mathbf{S}}_1$ is a $Q_0 * \underset{\sim}{Q}_1$-name.

c) $\lambda = \lambda^{\aleph_0} \geq |MAC(Q_0)|$, and $[I \in \mathbb{I}_0 \Rightarrow \lambda^{|{\rm Dom}(I)|} = \lambda]$

d) $\underset{\sim}{\mathbb{I}}_1$ is λ^+-complete. (i.e. $\Vdash_{Q_0}$ "each $I \in \underset{\sim}{\mathbb{I}}_1$ is λ^+-complete ").

e) $\{\aleph_1\} \subseteq \underset{\sim}{\mathbf{S}}_0 \subseteq \{\mu : \aleph_1 \leq \mu = {\rm cf}(\mu) \leq \lambda\}$,

f) $\mathbb{I}_0 \subseteq \mathbb{I}$ and $\mathbb{I} \setminus \mathbb{I}_0$ is λ^+-complete and $(\mathbb{I} \setminus \mathbb{I}_0, <_{\rm RK})$ is λ^+-directed (or just κ-directed where Q_0 satisfies the κ-c.c).

g) $\Vdash_{Q_0}$ "for every $I \in \underset{\sim}{\mathbb{I}}_1$ for some $I' \in \mathbb{I}, I \leq_{\rm RK} I'$" ($\leq_{RK}$ - Rudin Keisler order, see 2.10A), moreover $I' \in \mathbb{I} \setminus \mathbb{I}_0$ (hard to fail this addition).

h) $\mathbf{S} = \mathbf{S}_0 \cap \mathbf{S}_1$ i.e. $\mathbf{S}_1 \cap \left(\mathbf{S}_0 \cup (|Q_0|, |Q_0 * Q_1|]\right)$.

Then $Q_0 * Q_1$ satisfies $UP(\mathbb{I}, \mathbf{S}, \mathbf{W})$

2.15A Remark. 1) Comparing with Ch XI, 5.1 we lose a little: we demand $\lambda \geq |MAC(Q_0)|$ instead demanding $\lambda \geq |Q_0|$ but this seems marginal, (see (*) of 1.2A(2)).
2) More on $\leq_{RK}$ is this context see §4.

2.15B Example. Let Q_0 = Nm, Q_1 = Levy$(\aleph_1, \lambda_1)$ (for some large enough λ_1 (in V^{Q_0})) Q_2 = Nm (in $V^{Q_0 \times Q_1}$), Q_3 = Levy$(\aleph_1, \lambda_3)$ for some even larger λ_3, etc., then Q_0, $Q_0 * (Q_1 * Q_2)$, $(Q_0 * (Q_1 * Q_2)) * (Q_3 * Q_4), \ldots$ satisfy $UP(\mathbb{I})$ for appropriate $\mathbb{I}$, by 2.15.

Before we prove 2.15 we will remind the reader of a definition and a combinatorial lemma.

2.16 Definition. For a subset A of (an ω-tree) T we define by induction on the length of a sequence η, $\mathrm{res}_T(\eta, A)$ for each $\eta \in T$. Let $\mathrm{res}_T(\langle\ \rangle, A) = \langle\ \rangle$. Assume $\mathrm{res}_T(\eta, A)$ is already defined and we define $\mathrm{res}_T(\eta\hat{\ }\langle\alpha\rangle, A)$ for all members $\eta\hat{\ }\langle\alpha\rangle$ of $\mathrm{Suc}_T(\eta)$. If $\eta \in A$ then $\mathrm{res}_T(\eta\hat{\ }\langle\alpha\rangle, A) = \mathrm{res}_T(\eta, A)\hat{\ }\langle\alpha\rangle$, and if $\eta \notin A$ then $\mathrm{res}_T(\eta\hat{\ }\langle\alpha\rangle, A) = \mathrm{res}_T(\eta, A)\hat{\ }\langle 0\rangle$. If $\eta \in \lim(T)$, we let $\mathrm{res}(\eta, A) = \bigcup_{k\in\omega} \mathrm{res}(\eta\restriction k, A)$.

Explanation. Thus $\mathrm{res}(T, A) \stackrel{\mathrm{def}}{=} \{\mathrm{res}_T(\eta, A) : \eta \in T\}$ is a tree obtained by projecting, i.e., gluing together all members of $\mathrm{Suc}_T(\nu)$ whenever $\nu \notin A$.

We state now (see Chapter XI, 5.3):

2.17 Lemma. Let λ, μ be uncountable cardinals satisfying $\lambda^{<\mu} = \lambda$ and let $(T, \mathbf{I})$ be a tagged tree in which for each $\eta \in T$ either $|\mathrm{Suc}_T(\eta)| < \mu$ or $\mathbf{I}(\eta)$ is λ^+-complete. *Then* for every function $H : T \to \lambda$ there exist T' satisfying $(T, \mathbf{I}) \leq^* (T', \mathbf{I})$ such that for $\eta^1, \eta^2 \in T'$ we have: (letting $A = \{\mu \in T : |\mathrm{Suc}_T(\mu)| < \mu\}$):
$\mathrm{res}_T(\eta^1, A) = \mathrm{res}_T(\eta^2, A)$ implies:

$H(\eta^1) = H(\eta^2)$ and $\eta^1 \in A \Leftrightarrow \eta^2 \in A$, and: if $\eta \in T' \cap A$, then $\mathrm{Suc}_T(\eta) = \mathrm{Suc}_{T'}(\eta)$

Proof of Theorem 2.15. Let χ be large enough. Let $\langle N_\eta : \eta \in (T, \mathbf{I})\rangle$ be an $\aleph_1$-strictly $(\mathbb{I}, \underset{\sim}{\mathbf{S}}, \mathbf{W})$-suitable tree of models for χ such that

$$\{Q_0, \underset{\sim}{Q}_1, \underset{\sim}{\mathbf{S}}_0, \underset{\sim}{\mathbf{S}}_1, \mathbb{I}_0, \underset{\sim}{\mathbb{I}}_1, \mathbf{W}\} \in N_{\langle\,\rangle}, \text{ and } (p_0, \underset{\sim}{p}_1) \in (Q_0 * \underset{\sim}{Q}_1) \cap N_{\langle\,\rangle},$$

let $\mu = \mathrm{Min}\{\mu : \lambda^\mu > \lambda\}$, so $\mu > \aleph_0, \mu > |\mathrm{Dom}(I)|$ for $I \in \mathbb{I}_0$, $\mu = \mathrm{cf}(\mu)$, and $\lambda = \lambda^{<\mu}$. Let us define a function H with domain T: $H(\eta)$ is the pair

$$\Big(N_\eta \cap MAC(Q_0), \text{ isomorphism type of } (N_\eta, N_{\eta\restriction 0} \ldots, N_{\eta\restriction(lg(\eta)-1)}, \eta, c)_{c \in N_{\langle\rangle}}\Big)$$

so $|\mathrm{Rang}(H)| \le \lambda$. By the lemma above there is T^1 satisfying $(T, \mathbf{I}) \le^* (T^1, \mathbf{I})$ such that for $\eta, \nu \in T^1$:

$$\mathrm{res}_T(\eta, A) = \mathrm{res}_T(\nu, A) \Rightarrow H(\eta) = H(\nu) \;\&\; [\eta \in A \Longleftrightarrow \nu \in A]$$
$$\text{where } A = \{\eta \in T : \ |\mathrm{Suc}_T(\eta)| < \mu\}$$

let $T^* = \{\mathrm{res}_T(\eta, A) : \eta \in T^1\}$.

We can find T^2 satisfying $(T^1, \mathbf{I}) \le (T^2, \mathbf{I})$ such that the mapping $\eta \mapsto \mathrm{res}_T(\eta, A)$ on T^2, is one to one onto T^*. By 2.6A (for $\mathbf{S} = \{\aleph_1\}$) without loss of generality for some $\delta < \omega_1$, $\eta \in \lim(T^1) \Rightarrow \delta = \bigcup_{\ell<\omega} N_{\eta\restriction\ell} \cap \omega_1$, by the proof of 2.12 without loss of generality $\eta \in T^1 \Rightarrow N_\eta \cap \omega_1 = \delta$; and looking at the definition without loss of generality $\delta \in \mathbf{W}$. Let $N'_{\mathrm{res}_T(\eta,A)} = N_\eta$ for $\eta \in T^2$.

By assumption (f) we have $\langle N'_\nu : \nu \in T^*\rangle$ is an $(\mathbb{I}_0, \underset{\sim}{\mathbf{S}}_0, \mathbf{W})$-suitable tree for χ, $p_0 \in Q_0 \cap N'_{\langle\,\rangle}$. So there are $q_0, p_0 \le_{\mathrm{pr}} q_0 \in Q_0$, and Q_0-name $\underset{\sim}{\nu} \in \lim(T^*)$ such that $q_0 \Vdash_{Q_0}$ "$\bigcup_k N'_{\underset{\sim}{\nu}\restriction k}[\underset{\sim}{G}_{Q_0}] \cap \lambda$ and $\bigcup_k N'_{\underset{\sim}{\nu}\restriction k} \cap \lambda$ has the same supremum for $\lambda \in \underset{\sim}{\mathbf{S}}_0 \cap N_{\underset{\sim}{\nu}\restriction k}$".

Let $\underset{\sim}{T}^+ = \{\eta \in T^1 : \ \mathrm{res}_T(\eta, A) = \underset{\sim}{\nu}\restriction \ell g(\eta)\}$, this is a Q_0-name.

Let $q_0 \in G \subseteq Q_0, G$ generic over V, and let $\nu = \underset{\sim}{\nu}[G]$.

Now we need:

2.18 Fact. 1) $\langle N_\eta[G] : \eta \in (\underset{\sim}{T}^+[G], \mathbf{I})\rangle$ is an $\underset{\sim}{\mathbb{I}}_1[G]$- suitable* tree for χ.

2) For $\eta \in \lim(\underset{\sim}{T}^+[G])$ and $\kappa \in (\mathbf{S}_1^* \cup \underset{\sim}{\mathbf{S}}_0[G]) \cap \bigcup_{k<\omega} N_{\eta\restriction k}$ we have:

$$\sup(\bigcup_k N_{\eta\restriction k}[G] \cap \kappa) = \sup(\bigcup_k N_{\eta\restriction k} \cap \kappa).$$

3) Moreover if in (2) we choose η in some further generic extension, it still holds.

Proof of 2.18. 1) One point is $N_\eta[G] \cap \omega_1 = N_\eta \cap \omega_1 (\in \omega_1)$ which follows by the choice of q_0, ν as $\aleph_1 \in \mathbf{S}_0$ (in fact $N_\eta[G] \cap \omega_1 = \delta$, for every $\eta \in T$, as $\bar{N}$ was $\aleph_1$-strict). Another point is that for $I \in \underset{\sim}{\mathbb{I}}_1[G] \cap N_\eta[G]$, $\eta \in \underset{\sim}{T}^+[G]$ we have

$$\{\nu : \eta \trianglelefteq \nu \in \underset{\sim}{T}^+[G] \text{ and } I = \mathbf{I}_\eta \text{ or at least } I \leq_{RK} \mathbf{I}_\eta\}$$

is a front of $(\underset{\sim}{T}^+[G])^{[\eta]}$, this follows as : if $I \in N_\eta[G] \cap \underset{\sim}{\mathbb{I}}_1[G]$ then there is $Y \in N_\eta$, $|Y| \leq |Q_0|$ (even $|Y| < \kappa$ if Q_0 satisfies the κ-c.c) such that

$$(\exists I')(I \leq_{RK} I' \in Y \ \&\ I' \in \mathbb{I} \setminus \mathbb{I}_0),$$

note $I \notin \mathbb{I}_0$. Now $Y \cap \mathbb{I} \setminus \mathbb{I}_0$ has a $\leq_{\rm RK}$-upper bound in $\mathbb{I}$ hence in $\mathbb{I} \cap N_\eta$ by the assumption on $\mathbb{I} \setminus \mathbb{I}_0$ being λ^+-directed, $\lambda \geq |Q_0|$ (or κ-directed, Q_0 satisfying the κ-c.c.).

2) If $\kappa > \lambda$ then $\kappa > |Q_0|$ hence this is immediate; so assume $\kappa \leq \lambda$. Let $\nu = \underset{\sim}{\nu}[G]$ be the branch we obtained by applying $UP(\mathbb{I}, \mathbf{S}, \mathbf{W})$ to Q_0, and let $\eta \in \lim(T^+)$ be any branch. Now there is an isomorphism $g = g_\eta$ from $\bigcup_k N_{\eta\restriction k}$ onto $\bigcup_k N'_{\nu\restriction k}$ such that $g(\eta\restriction l) = \nu\restriction l$ for $\ell < \omega$, $g\restriction N_{\langle\rangle}$ = the identity, $g''(N_{\eta\restriction\ell}) = N'_{\nu\restriction\ell}$ for $\ell < \omega$, and necessarily $g\restriction(MAC(Q_0) \cap N_{\eta\restriction k})$ = the identity (as $N_\eta \cap MAC(Q_0) = N'_\nu \cap MAC(Q_0)$). Now for every $\underset{\sim}{\alpha} \in \bigcup_k N'_{\nu\restriction k}$ a Q_0-name of an ordinal $< \kappa, \underset{\sim}{\alpha}$ is just a maximal antichain of Q_0 with a function from it to ordinals. So $g(\underset{\sim}{\alpha}) = \underset{\sim}{\alpha}$ and of course $g(\kappa) = \kappa$. So as the isomorphism g is onto $\bigcup_k N_{\nu\restriction k}$ we see that $q_0 \Vdash_{Q_0}$ "if $\eta \in \lim(\underset{\sim}{T}^+[\underset{\sim}{G}_{Q_0}])$ then $\underset{\sim}{\alpha} = g_\eta(\underset{\sim}{\alpha}) < \sup(\bigcup_k N_{\eta\restriction k} \cap \kappa)$" as required.

3) Same proof as g can still be defined. $\square_{2.18}$

Continuation of the proof of 2.15: By the fact 2.18 in $V[G]$ there are $\underset{\sim}{\nu}_1$ and q_1 such that $\underset{\sim}{Q}_1[G] \models [p_1 \leq_{\mathrm{pr}} q_1;\quad q_1 \Vdash$ "$\underset{\sim}{\nu}_1$ a branch of $\underset{\sim}{T}^+[G]$, and for $\kappa \in \underset{\sim}{\mathbf{S}}_1[G] \cap N_{\nu_1}$ we have $\sup(\bigcup_{\ell<\omega} N_{\nu_1 \restriction \ell} \cap \kappa) = \sup \bigcup_{\ell<\omega}(N_{\nu_1 \restriction \ell}[G] \cap \kappa)$"$]$.

Now, back in V, there are $\underset{\sim}{\nu}_1, \underset{\sim}{q}_1$ and $q_0' \in G$ such that $q_0' \Vdash_{Q_0}$ "$\underset{\sim}{q}_1, \underset{\sim}{\nu}_1$ are as above" (actually $\underset{\sim}{\nu}_1$ is a Q_0-name of a $\underset{\sim}{Q}_1$-name); w.l.o.g. $q_0 = q_0' (\in Q_0)$ as the existence proof works for any G such that $q_0 \in G$. Note $(\underset{\sim}{q}_0', \underset{\sim}{q}_1) \in Q$ and $\underset{\sim}{\nu}_1 \in \lim(\underset{\sim}{T}^+) \subseteq \lim(T)$ are as required (remembering 2.15A(3)).

$\square_{2.15}$

Now clearly

2.19 Claim. 1) If (a forcing notion) P satisfies the $(\mathbb{I}, \mathbf{W})$ - condition (see Ch XI) *then* P satisfies $UP(\mathbb{I}, \mathbf{W})$ [look at Definition 2.7, 2.7A]

2) if $P = \mathrm{Nm}'(D)$ (see chapter X), D an $\aleph_2$ - complete filter, $\mathbb{I} = \{I :$ for some $A \subseteq \mathrm{Dom}(D)$ satisfying $A \neq \emptyset \mod D$ we have $I = \{X \subseteq A : X = \emptyset \mod D\}\}$ *then* P satisfies $UP(\mathbb{I})$

3) Let $(T^*, \mathbf{I}^*)$ be an $\mathbb{I}$-tagged full tree, and

$$P = \{(T, \mathbf{I}) : (T^*, \mathbf{I}^*) \leq (T, \mathbf{I}), \text{ and for every } \eta \in \lim(T) \text{ we have } (\exists^\infty n)[\eta \restriction n \text{ is a splitting point of } (T, \mathbf{I})\}$$

ordered by inverse inclusion.

$$P' = \{(T, \mathbf{I}) : (T^*, \mathbf{I}^*)^{[\eta]} \leq^* (T, \mathbf{I}) \text{ for some } \eta \in T^*\}$$

ordered by inverse inclusion.

Then

(a) P, P' satisfies $\mathrm{UP}(\mathbb{I})$

(b) if for λ regular

$$\forall \eta \in \lim(T^*)\ \exists^\infty n\ \forall A \in (\mathbf{I}^*_{\eta \restriction n})^+ [\mathbf{I}_\eta \restriction A \text{ is not } \lambda\text{-indecomposable}]$$

then $\Vdash_{P'}$ "$\mathrm{cf}(\lambda) = \aleph_0$".

(c) if $(\forall \eta \in \lim(T^*)) \exists n \bigwedge_{m \geq n} \forall A \in (\mathbf{I}^*_{\eta \restriction n})^+$ $[\mathbf{I}_\eta \restriction A$ is not λ-indecomposable$]$ then $\Vdash_P$ "$\mathrm{cf}(\lambda) = \aleph_0$".

4) Let $\lambda = \mathrm{cf}(\lambda) > \aleph_1$, $S \subseteq \{\delta < \lambda : \mathrm{cf}(\delta) = \aleph_0\}$ stationary, $\mathrm{club}\,(S) = \{h : h$ an increasing continuous function from some $i+1 < \omega_1$ into $S\}$ ordered by inclusion. *Then* $\mathrm{club}\,(S)$ satisfies $\mathrm{UP}(\{I\})$ if I is a uniform filter on λ.

Proof. We will only give a sketch of (2), leaving the other claims to the reader. We will use the following fact about $\mathrm{Nm}'(D)$:

(*) If $p \in \mathrm{Nm}'(D)$, $\underset{\sim}{\alpha}$ is a $\mathrm{Nm}'(D)$-name of an ordinal, then there is q, $p \le^* q$ such that the set $\{\eta \in q :$ for some β we have $q^{[\eta]} \Vdash$ "$\underset{\sim}{\alpha} = \beta$"$\}$ contains a front.

This fact follows easily from 2.6B(2) (let $H : P \to \{0,1\}$ be defined by $H(\eta) = 1$ iff $p^{[\eta]}$ decides $\underset{\sim}{\alpha}$, define $H(\eta) = \lim_{n\in\omega}(H(\eta\restriction n))$ for $\eta \in \lim(p)$, and find q such that H is constant on $\lim(q)$). Let $\mathbb{I}$ be such that the ideal dual to D is in it.

Now let $\langle N_\eta : \eta \in (T, \mathbf{I})\rangle$ be an $\aleph_1$-strictly $\mathbb{I}$-suitable tree, $\{p, D\} \in N_{\langle\rangle}$ a condition. We can now find a condition q, $p \le^* q$, an index set $\langle p_\eta : \eta \in p\rangle$ of conditions and a function $f : q \to T$ satisfying the following:

1. If $\eta \triangleleft \nu$ in q, then $f(\eta) \triangleleft f(\nu)$
2. For all η in q, $\mathrm{Suc}_T(f(\eta)) \ne 0 \bmod D$ and $\mathbf{I}_\eta$ is the ideal dual to D
3. For all η in q, $\mathrm{Suc}_q(\eta) \subseteq \mathrm{Suc}_T(f(\eta))$
4. For all η in q, $p_\eta \in N_{f(\eta)}, \mathrm{tr}(p_\eta) = \eta, p^{[\eta]} \le^* p_\eta$.
5. For all η in q, $p_\eta \le q^{[\eta]}$.
6. For all η in q, all names $\underset{\sim}{\alpha}$ in $N_{f(\eta)}$, the set $\{\nu \in q : p_\nu$ decides $\alpha\}$ contains a front of p_η.

We can do this as follows: by induction on $n < \omega$ we choose $q \cap {}^n(\mathrm{Dom}(D))$ and $\langle (f(\eta), p_\eta, \mathrm{Suc}_q(\eta)) : \eta \in q \cap {}^n(\mathrm{Dom}(D))\rangle$ satisfying the relevant demands. If $n \le \ell g(\mathrm{tr}(p))$ this is trivial. If we have defined for n, for each $\nu \in q \cap {}^n(\mathrm{Dom}(D))$ and $\eta \in \mathrm{Suc}_q(\nu)$, we do the following. We can find $f(\eta)$ satisfying (1)+(2) because $\langle N_\eta : \eta \in (T, \mathbf{I})\rangle$ is $\mathbb{I}$-suitable. We choose p_η using a bookkeeping argument to take care of (4)+(6), using (*). Then we choose $\mathrm{Suc}_q(\eta)$ such that (3) and (5) are satisfied.

Now let G be $\mathrm{Nm}'(D)$-generic, $q \in G$. Now G defines a generic branch η through q. This induces a branch ν through T by: $\nu = \bigcup_{n\in\omega} f(\eta\restriction n)$. Let $\underset{\sim}{\alpha} \in N_{\nu\restriction k}$, then there is ℓ such that $p_{\eta\restriction\ell} \Vdash$ "$\underset{\sim}{\alpha} = \beta$ and $\beta \in N_{f(\eta\restriction\ell)} \subseteq N_\nu$".

$\square_{2.19}$

2.19A Remark. 1) Note: 2.19(1) tells us that various specific forcing notions satisfy $UP(\mathbb{I},\mathbf{W})$ via Chapter XI 4.4, 4.4A, 4.5, 4.6.
2) We leave to the reader to compute the natural $\mathbf{S}$'s.

§3. Preservation of the $UP(\mathbb{I},\mathbf{S},\mathbf{W})$ by Iteration

3.1 Definition. We say that $\bar{Q} = \langle P_i, \underset{\sim}{Q}_i : i < \alpha\rangle$ satisfies $\langle\, \mathbb{I}_{i,j}, \lambda_{i,j}, \mu_{i,j}, \underset{\sim}{\mathbf{S}}_{i,j} : \langle i,j\rangle \in W^*\rangle$ for $\mathbf{W}$ provided that the following hold:

(0) $W^* \subseteq \{\langle i,j\rangle : \ i < j \leq \alpha, i$ is not strongly inaccessible$\}$, $W^* \supseteq \{\langle i+1,j\rangle : i < j < \alpha\}$ (we can use some variants, but there is no need),

(1) $\bar{Q}$ is a GRCS iteration.

(2) $P_{i,j} = P_j/P_i$ satisfies $UP(\mathbb{I}_{i,j}, \underset{\sim}{\mathbf{S}}_{i,j}, \mathbf{W})$ for $\langle i,j\rangle \in W^*$ (in V^{P_i}).

(3) for every $I \in \mathbb{I}_{i,j}$, the set $\mathrm{Dom}(I)$ is a cardinal, I is $\lambda^+_{i,j}$-complete (in V), $\lambda_{i,j} < |\mathrm{Dom}(I)| < \mu_{i,j}$ and $|MAC(P_i)| \leq \lambda_{i,j}$, and $\lambda_{i,j} \geq \aleph_2$ and $(\mathbb{I}_{i,j}, \leq_{RK})$ is $\lambda^+_{i,j}$ - directed (note that $\mathbb{I}_{i,j}$ is from V and not V^{P_i}, and $i \leq \lambda_{i,j} < \mu_{i,j}$).

(4) if $i(0) < i(1) < i(2) < \alpha, \langle i(0), i(1)\rangle \in W^*, \langle i(1), i(2)\rangle \in W^*$ *then* $(\lambda_{i(1),i(2)})^{<\mu_{i(0),i(1)}} = \lambda_{i(1),i(2)}$. (Hence $\lambda_{i(0),i(1)} < \mu_{i(0),i(1)} \leq \lambda_{i(1),i(2)}$.)

(5) for every $I \in \mathbb{I}_{i(2),i(3)}$ and $i(0) < i(1) \leq i(2) \leq i(3)$ such that $\langle i(0), i(1)\rangle \in W^*$ and $\langle i(2), i(3)\rangle \in W^*$ we have: I is $\lambda^+_{i(0),i(1)}$-complete.

(6) if $i(0) < i(1) < i(2)$,

$$\langle i(0), i(1)\rangle \in W^*,\ \langle i(1), i(2)\rangle \in W^*,\ \langle i(0), i(2)\rangle \in W^*$$

and then $\underset{\sim}{\mathbf{S}}_{i(0),i(2)}$ is $\underset{\sim}{\mathbf{S}}_{i(0),i(1)} \cap \underset{\sim}{\mathbf{S}}_{i(1),i(2)}$ [this holds if (always) $\mathbf{S}_{i,j} = \{\aleph_1\}$, and essentially if (always) $\mathbf{S}_{i,j}$ = RUCar]. (Remember - by 2.13(6) every $\chi \in$ RCar $\setminus |P_{i(1)}|^+ \subseteq$ RCar $\setminus \lambda_{i(1),i(2)}$ is considered to be in $\mathbf{S}_{i(0),i(1)}$.)

Note:

3.1A Remark. If $\mathbb{I}$ is λ^+ - complete, $\kappa \leq \lambda^+$,

$$\bigwedge_{I \in \mathbb{I}} |\mathrm{Dom}(I)| < \mu, \quad [\bigwedge_{i<\alpha<\kappa} \mu_i < \mu \Rightarrow \prod_{i<\alpha} \mu_i < \mu] \quad \text{and}$$

$$\mathbb{I}^{[\kappa]} \stackrel{\text{def}}{=} \mathbb{I} \cup \{\Pi_{i<\alpha} I_i : \quad \alpha < \kappa, \quad \kappa < \mu, I_i \in \mathbb{I}\}$$

(product $\prod_{i<\alpha} I_i$ defined in 4.11(1)) *then* $\mathbb{I}^{[\kappa]}$ is λ^+ - complete, $I \in \mathbb{I}^{[\kappa]} \Rightarrow |\mathrm{Dom}(I)| < \mu$ and $(\mathbb{I}^{[\kappa]}, \leq_{\mathrm{RK}})$ is κ - directed (see more in 4.11).

We shall use 3.1A freely.

3.2 Lemma. If $\bar{Q} = \langle P_n, \underset{\sim}{Q}_n : n < \omega \rangle$ satisfies $\langle \mathbb{I}_{i,j}, \lambda_{i,j}, \mu_{i,j}, \underset{\sim}{\mathbf{S}}_{i,j} : i < j < \omega \rangle$ for $\mathbf{W}$ and $\mathbb{I} = \bigcup_{n<\omega} \mathbb{I}_{n,n+1}$ *then* $P_\omega = \mathrm{Rlim}\bar{Q}$ satisfies $UP(\mathbb{I}, \underset{\sim}{\mathbf{S}}, \mathbf{W})$ where $\underset{\sim}{\mathbf{S}} = \{\lambda : \lambda$ is regular $> \aleph_0$ and for every n, $\mathrm{cf}(\lambda) \in \underset{\sim}{\mathbf{S}}_n\}$ (a P_ω-name.)

3.2A Remark. For the case $\leq_{pr} \neq \leq$ use VI 1.10.

Proof. Let $\mathbb{I}_n = \mathbb{I}_{n,n+1}$ and $\lambda_i = \lambda_{i,i+1}$ and $\mu_i = \mu_{i,i+1}$, note that $P_{i,i+1} = \underset{\sim}{Q}_i$ [see 3.1(2)], so $\underset{\sim}{Q}_n$ satisfies the $\mathbb{I}_n$-condition, $|P_i| \leq \lambda_i$, and $\mu_i \leq \lambda_{i+1} = (\lambda_{i+1})^{<\mu_i} < \mu_{i+1}$.

Let $\langle N_\eta : \eta \in (T, \mathbf{I}) \rangle$ be an $\aleph_1$-strict $(\mathbb{I}, \underset{\sim}{\mathbf{S}}, \mathbf{W})$-suitable tree of models for $\chi, N_{\langle\rangle} \cap \omega_1 \in \mathbf{W}, p \in N_{\langle\rangle} \cap P_\omega$ and $\{\mathbf{W}, \bar{Q}, \langle\langle \mathbb{I}_n, \lambda_n, \mu_n \rangle : n < \omega \rangle\} \in N_{\langle\rangle}$.

The proof will combine the proof of Ch XI 6.2 and the argument in preservation of (semi) properness.

We now define by induction (similarly to Ch XI§1) on $n < \omega$, a tree T_n such that (letting $A_n = \{\eta \in T : |Suc_T(\eta)| < \mu_n\}$):

(i) $T_0 = T$

(ii) $(T_n, \mathbf{I})$ is an $\mathbb{I}$-tree

(iii) $(T_n, \mathbf{I}) \leq^* (T_{n+1}, \mathbf{I})$

(iv) if $\eta, \nu \in T_{n+1}, \mathrm{res}(\eta, A_n) = \mathrm{res}(\nu, A_n)$ *then*

(a) $N_\eta \cap \mathrm{MAC}(P_{n+1}) = N_\nu \cap \mathrm{MAC}(P_{n+1})$

(b) the structures

$$\langle N_\eta, N_{\eta\restriction(lg(\eta)-1)}, \ldots, N_{\langle\,\rangle}, \eta, c\rangle_{c \in N_{\langle\rangle}} \text{ and}$$

$$\langle N_\nu, N_{\nu\restriction lg(\nu)-1)}, \ldots, N_{\langle\,\rangle}, \nu, c\rangle_{c \in N_{\langle\rangle}}$$

are isomorphic

(v) if $\eta \in T_{n+1}, |\mathrm{Suc}_{T_n}(\eta)| < \mu_n$ *then* $\mathrm{Suc}_{T_{n+1}}(\eta) = \mathrm{Suc}_{T_n}(\eta)$.

This is done by applying to $(T_n, \mathbf{I})$ Lemma 2.17 (for the function H implicit in (iv), and (λ, μ) there correspond to (λ_{n+1}, μ_n) here).

In the end let $T^* \stackrel{\mathrm{def}}{=} \bigcap_{n<\omega} T_n$; now for every n we have $(T_n, \mathbf{I}) \leq^* (T^*, \mathbf{I})$; why? if $\eta \in T^*$, then for some n $(\forall k \leq \ell g(\eta))[\mathbf{I}_{\eta\restriction k} \in \bigcup_{\ell \leq n} \mathbb{I}_\ell]$, hence for $k > n$, $\mathrm{Suc}_{T_k}(\eta) = \mathrm{Suc}_{T_{k+1}}(\eta)$.

We let $T_n^- \stackrel{\mathrm{def}}{=} \{\mathrm{res}(\nu, A_n) : \nu \in T^*\}$.

We now define by induction on $n < \omega$, p_n, q_n, η_n and P_n-name $\underset{\sim}{\nu}_n$ such that:

(a) $q_n \in P_n$

(b) $q_{n+1}\restriction n = q_n$

(c) $p\restriction n \leq_{pr} q_n$

(d) $q_n \Vdash$ "$\underset{\sim}{\nu}_n \in \lim(T_n^-)$" (lim is taken in V^{P_n})

(e) $q_{n+1} \Vdash$ "$\mathrm{res}(\underset{\sim}{\nu}_{n+1}, A_n) = \underset{\sim}{\nu}_n$" (more exactly $\mathrm{res}(\underset{\sim}{\nu}_{n+1}, \{\mathrm{res}(\rho, A_{n+1}) : \rho \in A_n\}) = \underset{\sim}{\nu}_n)$

(f) $q_{n+1} \Vdash$ "*if* $\rho \in \lim T^*, \mathrm{res}(\rho, A_n) = \underset{\sim}{\nu}_n$ *then*

$$(\bigcup_{n<\omega} N_{\rho\restriction n})[G_{P_n}] \cap \omega_1 = N_{\langle\,\rangle} \cap \omega_1$$

moreover for every $\kappa \in \bigcap_{\ell<n}(\mathbf{S}_{\ell,\ell+1} \cup [|P_\ell|^+, |P_\omega|])$ which belongs to $\bigcup_{n<\omega} N_{\rho\restriction n}$ we have

$$\sup\left[(\bigcup_{n<\omega} N_{\rho\restriction n})[G_{P_n}] \cap \kappa\right] = \sup\left[(\bigcup_{n<\omega} N_{\eta\restriction n} \cap \kappa\right]\text{''}$$

(g) $\eta_n \in T^*, \eta_n \triangleleft \eta_{n+1}, lg(\eta_n) = n$

(h) $\mathrm{res}(\eta_n, A_n) = \underset{\sim}{\nu}_n\restriction n$

(i) $p_0 = p, p_n \leq_{pr} p_{n+1} \in N_{\eta_{n+1}}$ and $p_n\restriction n = p_{n+1}\restriction n$ (we can get this from chapter VI).

(j) $p_n\restriction n \leq_{pr} q_n$

(k) if $\kappa \in N_{\eta_n}$ a regular cardinal, $\underset{\sim}{\alpha} \in N_{\eta_n}$ is a P_ω-name of an ordinal $< \kappa$ *then* for some m, and $\underset{\sim}{\beta} \in N_{\eta_m}$ we have $\underset{\sim}{\beta}$ is a P_m-name of an ordinal $< \kappa$ and $p_{m+1} \Vdash$ "$\underset{\sim}{\alpha} \leq \underset{\sim}{\beta}$ or $\kappa \notin \underset{\sim}{S}$".

The induction step is done as in the proof of 2.15 (remember that by 3.1(3) $(\mathbb{I}_{i,j}, \leq_{\mathrm{RK}})$ is $\lambda^+_{i,j}$-directed), (plus bookkeeping for (k) if $\leq_{\mathrm{pr}} \neq \leq$ we use VI 1.10). $\square_{3.2}$

3.3 Lemma.

1) If $\bar{Q} = \langle P_\alpha, \underset{\sim}{Q}_\alpha : \alpha < \omega_1\rangle$ satisfies $\langle \mathbb{I}_{\alpha,\beta}, \lambda_{\alpha,\beta}, \mu_{\alpha,\beta}, \underset{\sim}{\mathbf{S}}_{\alpha,\beta} : \alpha < \beta < \omega_1$ $\&\, \alpha$ non-limit $\rangle$ for $\mathbf{W}$, and $\mathbb{I} = \bigcup\{\mathbb{I}_{i,j} : i < j < \omega_1, i$ non-limit $\}$ *then* $P = P_{\omega_1} = \mathrm{Rlim}\bar{Q}$ satisfies $UP(\mathbb{I}, \underset{\sim}{\mathbf{S}}, \mathbf{W})$ where

$$\underset{\sim}{\mathbf{S}} = \bigcap_{\langle\alpha,\beta\rangle} \left(\underset{\sim}{\mathbf{S}}_{\alpha,\beta} \cup [|P_\beta|^+, ||P_{\omega_1}||]\right).$$

2) For $P = R\lim \bar{Q}$ as above, $\bigcup_{\alpha<\omega_1} P_\alpha$ is a dense subset of P, moreover for every $p \in P$ there is q such that $p \leq_{pr} q \in \bigcup_{\alpha<\omega_1} P_\alpha$.

3) We can replace ω_1 by a δ such that $\mathrm{cf}^V\delta < \lambda_{i,j}$ and $\Vdash_{P_i}$ "$\mathrm{cf}\delta \in \underset{\sim}{\mathbf{S}}_{i,j}$" for any $\langle i,j\rangle \in W^*$.

Proof. 1) Let $\langle N_\eta : \eta \in (T, \mathbf{I})\rangle$ be an $\aleph_1$-strict $(\mathbb{I}, \underset{\sim}{\mathbf{S}}, \mathbf{W})$-suitable tree of models for $\chi, N_{\langle\,\rangle} \cap \omega_1 \in \mathbf{W}$, $\{\mathbf{W}, \bar{Q}, \langle\langle \mathbb{I}_{\alpha,\beta}, \lambda_{\alpha,\beta}, \mu_{\alpha,\beta}, \underset{\sim}{\mathbf{S}}_{\alpha,\beta}\rangle : \alpha < \beta <$

ω_1, α non-limit $\rangle\}$ belongs to $N_{\langle\,\rangle}$ and $p \in P_{\omega_1} \cap N_{\langle\,\rangle}$. So $\mathbb{I} \in N_{\langle\,\rangle}$, hence (as $N_\eta \cap \omega_1 = N_{\langle\rangle} \cap \omega_1$):

(*) for every $\eta \in T, \mathbf{I}_\eta \in \mathbb{I}_{i,j}$ for some $i < j < N_{\langle\,\rangle} \cap \omega_1$.

Let $0 = \gamma_0 < \gamma_1 < \gamma_2 < \ldots, \bigcup_{n<\omega} \gamma_n = \delta = N_{\langle\,\rangle} \cap \omega_1$ and each γ_{1+n} is a successor ordinal. We repeat the proof of 3.2 (again remembering 3.1(3)) using P_{γ_n} instead P_n, and get $q = q_\omega$, $\underset{\sim}{\nu}_\omega$ as there.

The new point is why $p \leq_{pr} q$, and not only $p \restriction \delta = p \restriction (\bigcup_{n<\omega} \gamma_n) \leq_{pr} q$. The answer (as in Chapter XI, proof of 2.6) is: Let $\underset{\sim}{\zeta}^n(p_i)$ be prompt names as in XI 1.9. By (k) above q forces $\underset{\sim}{\zeta}^n(p_i)$ to be bounded by $\delta = N_{\langle\,\rangle} \cap \omega_1$, so we can finish.

2) We have proved this: for every $p \in P_{\omega_1}$ by 2.12 we can find $\langle N_\eta : \eta \in (T, \mathbf{I}) \rangle$ and $q \in P_{\omega_1}, p \leq_{pr} q$, as above; q is as required.

3) Almost the same proof. $\square_{3.3}$

3.4 Conclusion. 1) For $\bar{Q}$ an iteration as in 3.1, and limit $\delta \leq \ell g(\bar{Q})$ such that $\mathrm{cf}(\delta) = \omega_1$ and $[i < j < \delta,\ i$ non-limit $\Rightarrow \langle i, j \rangle \in W^*]$ *then* $\bigcup_{i<\delta} P_i$ is a dense subset of P_δ.

2) Instead $\mathrm{cf}(\delta) = \omega_1$, it is enough that for some $i < \delta, \Vdash_{P_i}$ "$\mathrm{cf}(\delta) = \omega_1$",

3) Also if δ is strongly inaccessible, $|P_i| < \delta$ for $i < \delta$ *then*

(a) conclusion of (1) holds.

(b) P_κ satisfies the κ-c.c. (in a strong sense: Δ-system lemma!)

4) In (1) we can weaken the demand on W^*, it is enough: for some unbounded $A \subseteq \delta$ we have $[i < j \ \&\ i \in A \ \&\ j \in A \Rightarrow \langle i, j \rangle \in W^*]$.

5) Moreover in (4) we can replace A by a set of strictly increasing sequences of ordinals $< \delta$, such that $[\eta \in t \Rightarrow \eta(0) = 0], [m < k < \ell g\eta \ \&\ \eta \in t \Rightarrow \langle \eta(n), \eta(k) \rangle \in W^*]$ and $[\eta \in t \ \&\ \alpha < \delta \Rightarrow \bigvee_{\beta \in (\alpha,\delta)} \eta\,\hat{}\,\langle \beta \rangle \in t]$. Of course this is because we can use a sequence from t as $\langle \gamma_0, \gamma_1, \ldots \rangle$ in 3.3. Similar claims holds for 3.5, 3.6.

Proof. Easy. $\square_{3.4}$

3.5 Lemma. Suppose $\bar{Q} = \langle P_i, \underset{\sim}{Q}_i : i < \kappa \rangle$ satisfies $\langle \mathbb{I}_{i,j}, \lambda_{i,j}, \mu_{i,j}, \underset{\sim}{\mathbf{S}}_{i,j} : i < j < \kappa,\ i$ non-limit $\rangle$ for $\mathbf{W}, \kappa$ is strongly inaccessible $|P_i| + \lambda_{i,j} + \mu_{i,j} + |\mathrm{Dom}(I)| < \kappa$

for every $\langle i,j\rangle \in W^*, I \in \mathbb{I}_{i,j}$ and $\mathbb{I} = \bigcup_{i,j} \mathbb{I}_{i,j}$. *Then* $P_\kappa = \lim(\bar{Q})$ satisfies the condition $UP(\mathbb{I}, \underset{\sim}{\mathbf{S}}, \mathbf{W})$ where $\underset{\sim}{\mathbf{S}} = \{\lambda :$ for every $(i,j) \in W^*$, in V^{P_i}, $\mathrm{cf}(\lambda) \in \underset{\sim}{\mathbf{S}}_i$ (or $\lambda = \mathrm{cf}\,\lambda > |P_j|)\}$.

Proof. Let $\langle N_\eta : \eta \in (T, \mathbf{I})\rangle$ be an $\aleph_1$-strict $(\mathbb{I}, \underset{\sim}{\mathbf{S}}, \mathbf{W})$-suitable tree of models for χ. Choose for each $\eta \in T$ a strictly increasing sequence $\langle \gamma^n_\eta : n < \omega\rangle$ of non-limit ordinals from $N_\eta \cap \kappa$ such that $0 = \gamma^0_\eta$, $\sup(N_\eta \cap \kappa) = \bigcup_{n<\omega} \gamma^n_\eta$, and $\gamma^n_{\eta \restriction k} < \gamma^n_\eta$ for $k < \ell g(\eta)$. Let $A_{\eta,n} = \{\rho \in T : |\mathrm{Suc}_T(\eta)| < \mu_{\gamma^n_\eta, \gamma^{n+1}_\eta}\}$.

We define by induction on n, T_n such that

(i) $T_0 = T$

(ii) $(T_n, \mathbf{I})$ is an $\mathbb{I}$-tree

(iii) $(T_n, \mathbf{I}) \le^* (T_{n+1}, \mathbf{I})$

(iv) if $\eta \in T_n$, and $\ell g(\eta) \le n$ then $\eta \in T_{n+1}$

(v) if $\eta \in T_n$, $\ell g(\eta) = n$, $\eta \trianglelefteq \nu_1 \in T_{n+1}, \eta \trianglelefteq \nu_2 \in T_{n+1}$ and $\mathrm{res}(\nu_1, A_{\eta,n}) = \mathrm{res}(\nu_2, A_{\eta,n})$ then

(a) $N_{\nu_1} \cap MAC(P_{\gamma^n_\eta}) = N_{\nu_2} \cap MAC(P_{\gamma^n_\eta})$.

(b) the structures

$$\langle N_{\nu_1}, N_{\nu_1 \restriction (\ell g(\nu_1)-1)}, \ldots, N_{\langle\,\rangle}, \nu_1, c\rangle_{c \in N_\eta} \quad \text{and}$$

$$\langle N_{\nu_2}, N_{\nu_2 \restriction (\ell g(\nu_2)-1)}, \ldots, N_\eta \ldots, N_{\langle\,\rangle}, \nu_2, c\rangle_{c \in N_\eta}$$

are isomorphic.

There is no problem in this: Let $T^* = \bigcap_{n<\omega} T_n$, easily $(T_n, \mathbf{I}) \le^* (T^*, \mathbf{I})$ for every n (by (iii) and (iv)). The rest is like the proof of 3.3. The only difference is that instead of actual ordinals γ_n we will have prompt names: $\underset{\sim}{\gamma}_n = \gamma^n_{\eta_n}$. We can use XI 1.10 to get the conditions $p_n \in P_{\underset{\sim}{\gamma}_n}$. Also remember that every $\bar{Q}$-named ordinal $\underset{\sim}{\zeta}(<\kappa)$ is bounded below κ (as for $\delta < \kappa$ of cofinality $\aleph_1, \bigcup_{\alpha<\delta} P_\alpha$ is a dense subset of P_δ). $\square_{3.5}$

3.6 Theorem. Suppose

(a) κ is strongly inaccessible,

(b) $\bar{Q} = \langle P_i, \underset{\sim}{Q}_j : i < \kappa\rangle$ satisfies $\langle \mathbb{I}_{i,j}, \lambda_{i,j}, \mu_{i,j}, \underset{\sim}{\mathbf{S}}_{i,j} : i < j < \kappa, i$ non-limit$\rangle$ for $\chi, \mathbf{W}$, $|P_i| < \kappa$ for $i < \kappa$, $P_\kappa = \bigcup_{i<\kappa} P_i$.

(c) $\Vdash_{P_\kappa}$ "$\underset{\sim}{Q}_\kappa$ satisfies $UP(\mathbb{I}_\kappa, \underset{\sim}{\mathbf{S}}_\kappa, \mathbf{W})$, $\mathbb{I}_\kappa$ is κ-complete" and $(\mathbb{I}_\kappa, \leq_{RK})$ is $(< \kappa)$ - directed.

(d) $\mathbb{I} = \mathbb{I}_\kappa \cup \bigcup\{\mathbb{I}_{i,j} : \langle i, j\rangle \in W^*\}$.

(e) $\underset{\sim}{\mathbf{S}} = \bigcap_{\langle i,j\rangle \in W^*} \underset{\sim}{\mathbf{S}}_{i,j} \cap \underset{\sim}{\mathbf{S}}_\kappa$.

Then $P_\kappa * \underset{\sim}{Q}_\kappa$ satisfies $UP(\mathbb{I}, \underset{\sim}{\mathbf{S}}, \mathbf{W})$.

Remark. This generalizes Gitik, Shelah [GiSh:191] which improves the relevant theorem in XI §6.

Proof. Let $\langle N_\eta : \eta \in (T, \mathbf{I})\rangle$ be an $\aleph_1$-strict $(\mathbb{I}, \underset{\sim}{\mathbf{S}}, \mathbf{W})$-suitable tree of models for $\chi, N_{\langle\rangle} \cap \omega_1 \in \mathbf{W}, (p_a, p_b) \in (P_\kappa * \underset{\sim}{Q}_\kappa) \cap N_{\langle\rangle}$. You may assume, for simplicity that $\underset{\sim}{\mathbf{S}}_{i,j} = \{\aleph_1\}, \underset{\sim}{\mathbf{S}}_\kappa = \{\aleph_1\} = \underset{\sim}{\mathbf{S}}$. Let T^* be as in the proof of the previous theorem. Let $G^* \subseteq \text{Levy}(\aleph_0, 2^\chi)$ be generic over V. Let $\kappa = \bigcup_{n<\omega} \alpha_n$, each α_{n+1} a successor, $\alpha_0 = 0$, $\alpha_n < \alpha_{n+1}$, $\bigcup_{n<\omega} \alpha_n = \kappa$ (in $V[G^*]$!).

We choose by induction on $n, \beta_n, G_n \in V[G^*], T_n, \nu_n$ such that

(a) $G_n \subseteq P_{\beta_n}$, G_n generic over V, $G_{n+1} \cap P_{\beta_n} = G_n$, $\alpha_n \leq \beta_n < \beta_{n+1} < \kappa$,

(b) $T^0 = T^*$, $T^{n+1} \subseteq T^n$,

(c) $\langle N_\eta[G_n] : \eta \in (T^n, \mathbf{I})\rangle$ is an $\mathbb{I}_n$-suitable* tree of models for χ, $\mathbb{I}_n = \bigcup\{\mathbb{I}_{i,j} : \langle i, j\rangle \in W^*,\ i \geq \beta_n\} \cup \mathbb{I}_\kappa$,

(d) $N_\eta[G_n] \cap \omega_1 = N_{\langle\rangle} \cap \omega_1$ for all $\eta \in T_n$

(e) T^n has a unique member of length k_n, ν_n,

(f) if H is a function (from V), $\text{Dom}(H) = T$, $H(\eta) \in \mathbf{I}_\eta$, for $n < \omega$, J_n is a front of T and $[\eta \in J_{n+1} \Rightarrow \bigvee_{\ell<\ell g\eta} \eta\restriction\ell \in J_n]$ and $\eta \in J_n \Rightarrow \mathbf{I}_\eta \in \mathbb{I}_\kappa$ *then* for infinitely many n, there exists m_n, $k_n \geq n$ such that $\nu_n\restriction k_n \in J_{m_n}$ and $\nu_{n+1}(k_n) \in H(\nu_n\restriction k_n)$.

Now $G = \bigcup\limits_{n<\omega} G_n$ is a generic subset of P_κ over V (as P_κ satisfies the κ-c.c., every maximal antichain in P_κ is contained in some P_{α_n}, hence meets some G_n).

We define $T^a = \{\eta \in T : N_\eta[G] \cap \omega_1 = N_{\langle\rangle} \cap \omega_1\}$.

We define a depth function on T^a:

$Dp(\eta) \geq 0$ iff $\eta \in T^a$.

$Dp(\eta) \geq \alpha(> 0)$ iff for every $\beta < \alpha$

for every $I \in \mathbb{I}_\kappa \cap N_\eta[G]$, there is ν, $I \leq_{\mathrm{RK}} I_\nu$, $\eta \triangleleft \nu \in T^a$ such that $\{i : \nu\char`^\langle i\rangle \in T^a, Dp(\nu\char`^\langle i\rangle) \geq \beta\} \neq \emptyset \bmod I_\nu$.

Easily $Dp \in V[G]$ and its definition is absolute. $[\eta \triangleleft \nu \in T^a \Rightarrow Dp(\nu) \leq Dp(\eta))]$ and $Dp(\langle\,\rangle) = \infty$ as $\bigcup_{n<\omega} \nu_n$ witnesses (in $V[G^*]$).

So in $V[G]$, $T^b \stackrel{\text{def}}{=} \{\nu \in T^a : \ Dp(\nu) = \infty\}$ is the desired tree (i.e. we can continue as in 2.15 with $P_\kappa, \underset{\sim}{Q}_\kappa$ here corresponding to Q_0, $\underset{\sim}{Q}_1$ there). $\square_{3.6}$

3.7 Lemma. Suppose $\bar{Q} = \langle P_i, \underset{\sim}{Q}_j : i \leq \alpha, j < \alpha\rangle$ satisfies $\langle \mathbb{I}_{i,j}, \lambda_{i,j}, \mu_{i,j}, \underset{\sim}{\mathbf{S}}_{i,j} : i < j \leq \alpha, i \text{ non-limit}\rangle$ for $\mathbf{W}$; $i(*) < \alpha$ is a non-limit, $G_{i(*)} \subseteq P_{i(*)}$ generic over V, and $\langle i_\zeta : \zeta \leq \beta\rangle$ is an increasing continuous sequence of ordinals in $V[G_{i(*)}]$, $i_0 = i(*)$, $i_\beta = \alpha$, each $\alpha_{\zeta+1}$ a successor ordinal.

In $V[G_{i(*)}]$, we define $P'_\zeta = P_{i_\zeta}/G_{i(*)}$, $\underset{\sim}{Q}'_\zeta = \underset{\sim}{Q}_{\alpha_\zeta}/G_{i(*)}$, $\bar{Q}' = \langle P'_\zeta, \underset{\sim}{Q}'_\xi : \zeta \leq \beta, \xi < \beta\rangle$, *then,* in $V[G_{i(*)}]$, $\bar{Q}'$ satisfies $\langle \mathbb{I}_{\alpha_\zeta, i_\xi}, \lambda_{i_\zeta, i_\xi}, \mu_{i_\zeta, i_\xi}, \mathbf{S}_{i_\zeta, i_\xi} : \zeta < \zeta \leq \beta, \zeta$ a non-limit$\rangle$ for $\mathbf{W}$.

Proof. Straightforward. $\square_{3.7}$

3.8 Conclusion. For every function F, stationary $\mathbf{W} \subseteq \omega_1$ and ordinal α^* there are $\alpha \leq \alpha^*$ and a GRCS iteration $\bar{Q}$ of length α satisfying $\langle \mathbb{I}_{i,j}, \lambda_{i,j}, \mu_{i,j}, \underset{\sim}{\mathbf{S}}_{i,j}; i < j \leq \alpha$, i non-limit $\rangle$ for $\mathbf{W}$ with $(\underset{\sim}{Q}_i, \mathbb{I}_{i,i+1}, \underset{\sim}{\mathbf{S}}_{i,i+1}) = F(\bar{Q}\restriction i)$ and $\alpha = \alpha^*$ *or* $\alpha < \alpha^*$, and $F(\bar{Q})$ does not satisfy $(*)$ below or there is β, $\beta + \omega \leq \alpha$ and $\bigwedge_n \nVdash_{P_{\beta+n}}$ "$|\operatorname{MAC}(P_\beta)| = \aleph_1$"

$(*) F(\bar{Q})$ has the form $(\underset{\sim}{Q}, \mathbb{I}, \underset{\sim}{\mathbf{S}})$, $\underset{\sim}{Q}$ a P_α-name of a forcing notion satisfying $\mathrm{UP}(\mathbb{I}, \underset{\sim}{\mathbf{S}}, \mathbf{W})$, and (a) or (b) below holds:

(a) $\Vdash_{P_\alpha}$ "$\alpha \neq \aleph_2$" (i.e. $\Vdash_\alpha$ "$|\alpha| < \aleph_2$") and for some λ the family $\mathbb{I}$ is λ^+-complete, where $\lambda = \lambda^{|\mathrm{Dom}(I)|}$ whenever $I \in \mathbb{I}_{i,j}, i < j \leq \alpha$, and $|MAC(P_\alpha)| \leq \lambda$ and $(\mathbb{I}, \leq_{RK})$-is λ^+-directed.

(b) $\Vdash_{P_\alpha}$ "$\alpha = \aleph_2$" (i.e. α is strongly inaccessible, $|P_i| < \alpha$ for $i < \alpha$), $\mathbb{I}$ is α-complete and $I \in \mathbb{I}_{i,j} \& i < j < \alpha \Rightarrow |\mathrm{Dom}(I)| < \alpha$ and $(\mathbb{I}, \leq_{RK})$ is α-directed.

Proof. Straightforward. $\square_{3.8}$

§4. Families of Ideals and Families of Partial Orders

4.1 Definition. 1) We call an ideal J fine if $\{x\} \in J$ for every $x \in \mathrm{Dom}(J)$.
2) We call the ideal with domain $\{0\}$, which is $\{\emptyset\}$, the trivial ideal.

4.2 Claim. 1) If an ideal J is not fine then $J \leq_{RK}$ "the trivial ideal". (See 2.10A for the definition of $\leq_{RK}$).
2) In 2.10B we can weaken the hypothesis to $\mathbb{I}_1 \leq_{RK} \mathbb{I}'_2$ where $\mathbb{I}'_2 \stackrel{\text{def}}{=} \mathbb{I}_2 \cup$ {the trivial ideal}. The same holds in similar situations.
3) $\leq_{RK}$ is a partial quasiorder (among ideals and also among families of ideals).

Proof. Easy. $\square_{4.2}$

4.3 Definition.

1) For an (upward) directed partial or just quasi order $L = (B, \leq)$ we define an ideal id_L:

$$\mathrm{id}_L = \{A \subseteq B : \text{ for some } y \in L \quad \text{we have} \quad A \subseteq \{x \in B : \neg y \leq x\}\}.$$

(Equivalently the dual filter fil_L is generated by the "cones" $L_y \stackrel{\text{def}}{=} \{x \in L : y \leq x\}$.) We call such an ideal a partial order ideal or a quasi order ideal. We let $\mathrm{Dom}(L) = \mathrm{Dom}(\mathrm{id}_L)(= B)$, but we may use L instead of $\mathrm{Dom}(L)$ (like $\forall x \in L$) abusing notation as usual.

2) For a partial order L let $\text{dens}(L) = \text{Min}\{|\Theta| : \Theta \subseteq \text{Dom}(L)$ is dense i.e. $(\forall a \in \text{Dom}(L))(\exists b \in \Theta)[a \leq b]\}$ (this applies also to ideals considered as the quasi order $(I, \subseteq)$).
3) For a family $\mathcal{L}$ of directed quasi orders let $\text{id}_{\mathcal{L}} = \{\text{id}_L : L \in \mathcal{L}\}$.

4.4 Fact.

1) id_L is λ-complete *iff* L is λ-directed.
2) $\text{dens}(L) = \text{dens}(\text{id}_{(L,<)}, \subseteq)$
3) If $h : L_1 \to L_2$ preserves order (i.e. $\forall x, y \in L$, $(x \leq y \Rightarrow h(x) \leq h(y))$) and has cofinal range (i.e. $\forall x \in L_2 \exists y \in L_1 (x \leq h(y))$) *then* $\text{id}_{L_2} \leq_{RK} \text{id}_{L_1}$.
4) $h : L_1 \to L_2$ exemplifies $\text{id}_{L_2} \leq_{RK} \text{id}_{L_1}$ *iff* for every $x_2 \in L_2$ there is $x_1 \in L_1$ such that: $y \in L_1$ & $x_1 \leq_{L_1} y \Rightarrow x_2 \leq_{L_2} h(y)$ (i.e. for $y \in L_1 : \neg x_2 \leq_{L_2} h(y) \Rightarrow \neg x_1 \leq_{L_1} y$ but h is not necessarily order preserving).
5) the ideal $\text{id}_{(L,<)}$ is fine iff $(L, <)$ has no maximal element.

Proof. Straight. E.g.

4) Note: h exemplifies $\text{id}_{L_2} \leq \text{id}_{L_1}$ *iff*

$$(\forall A \subseteq L_1)(A \neq \emptyset \mod \text{id}_{L_1} \to (\forall x_2 \in L_2)[h''(A) \cap \{y \in L_2 : x_2 \leq_{L_2} y\} \neq \emptyset])$$

iff

$$(\forall x_2 \in L_2)(\forall A \subseteq L_1)[A \neq \emptyset \mod \text{id}_{L_1} \to h''(A) \cap \{y \in L_2 : x_2 \leq_{L_2} y\} \neq \emptyset]$$

iff

$$(\forall x_2 \in L_2)[\{y \in L_1 : \neg x_2 \leq_{L_2} h(y)\} = \emptyset \mod \text{id}_{L_1}]$$

iff

$$(\forall x_2 \in L_2)(\exists x_1 \in L_1)(\forall y \in L_1)(\neg x_2 \leq_{L_2} h(y) \to \neg x_1 \leq_{L_1} y)$$

iff

$$(\forall x_2 \in L_2)(\exists x_1 \in L_1)(\forall y \in L_1)(x_1 \leq_{L_1} y \to x_2 \leq_{L_2} h(y))$$

$\square_{4.4}$

4.5 Fact. 1) For every ideal J (such that $(\mathrm{Dom}(J)) \notin J$), let $J_1 = \mathrm{id}_{(J,\subseteq)}$, *then*

(i) J_1 is a partial order ideal

(ii) $|\mathrm{Dom}(J_1)| = |J| \leq 2^{|\mathrm{Dom}(J)|}$

(iii) $J \leq_{RK} J_1$

(iv) if J is λ-complete *then* $(J, \subseteq)$ is λ-directed hence J_1 is λ-complete

(v) $\mathrm{dens}(J, \subseteq) = \mathrm{dens}(J_1, \subseteq)$

2) For every dense $\Theta \subseteq J$ we can use $\mathrm{id}_{(\Theta,\subseteq)}$ and get the same conclusions.

3) For every ideal J there is a directed order L such that:

$$J \leq_{RK} \mathrm{id}_L, \qquad \mathrm{dens}(J) = \mathrm{dens}(L) \text{ and:}$$

$$\text{for every } \lambda \text{ if } J \text{ is } \lambda\text{-complete then so is } \mathrm{id}_L.$$

Proof. Least trivial is (1)(iii), let $h : J \to \mathrm{Dom}(J)$ be such that $h(A) \in (\mathrm{Dom}(J)) \setminus A$ (exists as $(\mathrm{Dom}(J)) \notin J$). Let $J_1 = \mathrm{id}_{(J,\subseteq)}$.

If $X \subseteq \mathrm{Dom}(J_1) = J, X \notin J_1$ and $A \overset{\mathrm{def}}{=} h''(X)$ belongs to J, then $\{B \in J : \neg A \subseteq B\} \in \mathrm{id}_{(J,\subseteq)} = J_1$ (by the definition of $\mathrm{id}_{(J,\subseteq)}$) hence (as $X \notin J_1$) for some $B \in X, A \subseteq B$, so $h(B) \in h''(X) = A$ contradicting the choice of $h(B)$ (as $A \subseteq B$). $\square_{4.5}$

4.5A Remark. So we can replace the ideals by partial orders without changing much the relevant invariants such as completeness or density.

4.6 Conclusion. For any family of ideals $\mathbb{I}$ there is a family of $\mathcal{L}$ of directed partial order such that

(i) $\mathbb{I} \leq_{RK} \{\mathrm{id}_{(L,<)} : (L,<) \in \mathcal{L}\}$

(ii) $|\mathcal{L}| \leq |\mathbb{I}|$

(iii) $\sup\{|L| : (L,<) \in \mathcal{L}\} \leq \sup\{|J| : J \in \mathbb{I}\}) \quad \leq (\sup\{2^{|\mathrm{Dom}(J)|} : J \in \mathbb{I}\})$

(iv) $\sup\{\mathrm{dens}(L,<) : (L,<) \in \mathcal{L}\} = \sup\{\mathrm{dens}(J,\subseteq) : J \in \mathbb{I}\}$

(v) if $\mathbb{I}$ is λ-complete then every $(L,<) \in \mathcal{L}$ is λ-directed.

Proof. Easy. $\square_{4.6}$

4.7 Definition. For a forcing notion Q, satisfying the κ-c.c., a Q-name $\underset{\sim}{L}$ of a directed partial (or just quasi) order with (for notational simplicity) $\mathrm{Dom}(\underset{\sim}{L}) \in V$; let $L^* = ap_\kappa(\underset{\sim}{L})$ be the following partial order

$$\mathrm{Dom}(L^*) = \{a : \ a \subseteq \mathrm{Dom}(\underset{\sim}{L}) \text{ and } |a| < \kappa\}$$

$$a \leq^* b \text{ iff } \Vdash_Q \text{"}(\forall y \in a)(\exists x \in b)[\underset{\sim}{L} \models y < x]\text{"}$$

(this is a quasi order only, e.g. maybe $a \leq^* b \leq^* a$ but $a \neq b$).

4.8 Claim. For a forcing notion Q satisfying the κ-c.c. and a Q-name $\underset{\sim}{L}$ of a λ-directed partial order (with $\mathrm{Dom}(\underset{\sim}{L}) \in V$ for simplicity) such that $\lambda \geq \kappa$ we have:

(i) $ap_\kappa(\underset{\sim}{L})$ is λ-directed partial order (in V and also in V^Q).

(ii) $|ap_\kappa(\underset{\sim}{L})| \leq |\mathrm{Dom}(\underset{\sim}{L})|^{<\kappa}$

(iii) $\Vdash_Q$ "$\mathrm{id}_{\underset{\sim}{L}[G]} \leq_{RK} \mathrm{id}_{ap_\kappa(\underset{\sim}{L})}$"

Proof. We leave (i), (ii) to the reader. We check (iii). Let $G \subseteq Q$ be generic over V, and in $V[G]$ we define a function h from $ap_\kappa(\underset{\sim}{L})$ to $\mathrm{Dom}(\underset{\sim}{L}[G])$:

$h(a)$ will be an element of $\mathrm{Dom}(\underset{\sim}{L}[G])$ such that

$$(\forall x \in a)\underset{\sim}{L}[G] \models \text{"}x < h(a)\text{"}.$$

We can now easily verify the condition in 4.4(4). $\square_{4.8}$

4.9 Conclusion. 1) Suppose Q is a forcing notion satisfying the κ-c.c., $\underset{\sim}{\mathbb{I}}_1$ a Q-name of a family of λ-complete filters and $\lambda \geq \kappa$. *Then* there is, (in V), a family $\mathbb{I}_2$ of λ-complete filters such that:

(i) $\Vdash_Q$ "$\underset{\sim}{\mathbb{I}}_1 \leq_{RK} \mathbb{I}_2$"

(ii) $|\mathbb{I}_2| = |\mathbb{I}_1|$

(iii) $\sup\{|\mathrm{Dom}(J)| : J \in \mathbb{I}_2\} = \sup\{(2^\mu)^{<\kappa}$: some $q \in Q$ forces that some $J \in \underset{\sim}{\mathbb{I}}_1$ has domain of power $\mu\}$.

2) If $\mathbb{I}_1$ has the form $\{\mathrm{id}_{(L,<)} : (L,<) \in \mathcal{L}\}$ then in (iii) we can have

(iii)′ $\sup\{|\mathrm{Dom}(J)| : J \in \mathbb{I}_2\} = \sup\{\mu^{<\kappa}$: some $q \in Q$ force some $(L,<) \in \mathcal{L}$ has power $\mu\}$.

Proof. Easy. $\square_{4.9}$

4.9A Remark. The aim of 4.8, 4.9 is the following: We will consider iterations $\langle P_i, \underset{\sim}{Q}_i : i < \alpha\rangle$ where $\Vdash_{P_i}$ "$\underset{\sim}{Q}_i$ satisfies $UP(\underset{\sim}{\mathbb{I}}_i)$", but $\underset{\sim}{\mathbb{I}}_i$ may not be a subset of the ground model V. Now 4.9 gives us a good $\leq_{RK}$-bound $\mathbb{I}_i$ in V, and we can prove (under suitable assumptions) that P_α will satisfy the $UP(\bigcup_{i<\alpha} \mathbb{I}_i)$.

4.10 Definition. 1) We say a family $\mathbb{I}$ of ideals is κ-closed *if: for every* $\alpha < \kappa$ and $J_i \in \mathbb{I}$ for $i < \alpha$ *there is* $J \in \mathbb{I}$, $\bigwedge_{i<\alpha} J_i \leq_{RK} J$. It is strongly κ-closed it is κ-closed, and it is closed under restriction.
2) We say a family $\mathcal{L}$ of partial orders is κ-closed if $\{\mathrm{id}_L : L \in \mathcal{L}\}$ is.

4.11 Fact. 1) Let $\langle J_i : i < \alpha\rangle$ be a sequence of ideals; we define $J = \prod_{i<\alpha} J_i$ as the ideal on $\prod_{i<\alpha}(\mathrm{Dom}(J_i))$ generated by $\{\bigcup_{j<\alpha} \prod_{i<\alpha} A_i^j$: for $i, j < \alpha$ we have $A_i^j \subseteq \mathrm{Dom}(J_i)$ and for each $i < \alpha$ we have $A_i^i \in J_i\}$, *then*

(i) J is an ideal

(ii) $|\mathrm{Dom}(J)| = \prod_{i<\alpha} |\mathrm{Dom}(J_i)|$

(iii) $\mathrm{dens}(J) \leq \prod_{i<\alpha} \mathrm{dens}(J_i)$

(iv) if each J_i is λ-complete then J is λ-complete

(v) $J_i \leq_{RK} J$ for each $i < \alpha$

(vi) if for each i, $(\mathrm{Dom}(J_i)) \notin J_i$ then $(\mathrm{Dom}(J)) \notin J$

(vii) if $J_i = \mathrm{id}_{(L_i,<_i)}$ then J is naturally isomorphic to $\mathrm{id}_{(L,<)}$ where $(L,<) = \prod_{i<\alpha}(L_i,<_i)$

2) This product is associative.

4.12 Definition. 1) For κ a regular cardinal the κ-closure of a family $\mathbb{I}$ of ideals is

$$\mathbb{I} \cup \{\prod_{i<\alpha} J_i :\ \alpha < \kappa, J_i \in \mathbb{I}\}$$

2) Similarly for a family of partial orders

4.13 Fact. For a family $\mathbb{I}$ of ideals let $\mathbb{I}'$ be the κ-closure of $\mathbb{I}$, then:

(i) $|\mathbb{I}'| \leq |\mathbb{I}|^{<\kappa}$

(ii) $\mathbb{I}'$ is κ-closed

(iii) $\sup_{J\in\mathbb{I}'} |\mathrm{Dom}(J)| \leq (\ \sup_{J\in\mathbb{I}}(|\mathrm{Dom}(J)|)^{<\kappa}$

(iv) if $\mathbb{I}$ is λ-complete so is $\mathbb{I}'$

(v) $\sup_{J\in\mathbb{I}}(\operatorname{dens}(J)) \le (\sup_{J\in\mathbb{I}} \operatorname{dens}(J))^{<\kappa}$

(vi) if $\mathbb{I} = \{\mathrm{id}_L : L \in \mathcal{L}\}$ then $\mathbb{I}' \equiv_{RK} \mathrm{id}_{\mathcal{L}'}$ where $\mathcal{L}'$ is the κ-closure of $\mathcal{L}$ (in fact, $\mathbb{I}'$, $\mathrm{id}_{\mathcal{L}}$, are isomorphic)

Proof: Easy. $\square_{4.13}$

The following claim gives better cardinality restrictions in §3 (and 2.17) and not having to use "not too large $\mathbb{I}$ for P_i in the iteration for the sake of Q_i" (also alternative proofs). Here $\underset{\sim}{\mathbf{S}}$ is just $\{\aleph_1\}$.

4.14 Claim. Suppose Q satisfies $UP(\mathbb{I}, \mathbf{W})$, Q satisfies the κ-c.c. and $\langle N_\eta : \eta \in (T, \mathbf{I})\rangle$ is an $\aleph_1$-strict $(\mathbb{I}, \mathbf{W})$ - suitable tree of models (for χ). Let $\mathbb{I}' = \{I \in \mathbb{I} : \quad I$ is κ-complete $\}$ and assume $\mathbb{I}'$ is κ-closed, $N_\eta \cap \omega_1 = \delta \in \mathbf{W}$.

Then for every $p \in N_{\langle\rangle} \cap Q$ there is an $(N_{\langle\rangle}, Q)$ - semi generic $q, p \le_{\mathrm{pr}} q \in Q$ such that

$$q \Vdash_Q \text{" there is } T' \subseteq T \text{ such that } \langle N_\eta[\underset{\sim}{G}_Q] : \eta \in (T', \mathbf{I})\rangle$$
$$\text{is a } \aleph_1\text{-strictly } (\mathbb{I}', \mathbf{W}) \text{ - suitable* tree of models"}$$

Proof. Let $G \subseteq Q$ be generic over V. Let $\delta = N_{\langle\rangle} \cap \omega_1$. By 2.14A we know that $\langle N_\eta[G] : \eta \in (T, \mathbf{I})\rangle$ is $(\mathbb{I}', \mathbf{W})$-suitable*, but it is not necessarily $\aleph_1$-strict. So let (in $V[G]$) :

$$T^* = \underset{\sim}{T}^*[G] \stackrel{\text{def}}{=} \{\eta \in T : N_\eta[G] \cap \omega_1 = \delta\}.$$

$UP(\mathbb{I}, \mathbf{W})$ implies that we can find q such that $p \le_{\mathrm{pr}} q$ and q forces that $\underset{\sim}{T}^*$ contains a branch, but we want $\underset{\sim}{T}^*$ to contain even an $(\mathbb{I}', \mathbf{W})$-suitable* tree.

Define (in $V[G]$) a depth function Dp_T as follows:

$$\mathrm{Dp}_T(\eta) \ge \alpha \text{ iff} : \eta \in T^* \text{ and } \forall \beta < \alpha \forall I \in \mathbb{I}' \cap N_\eta[G]\, \exists \nu_\eta \in T^*$$
$$[\eta \trianglelefteq \nu_\eta \;\&\; I \le_{RK} \mathbf{I}_{\nu_\eta} \;\&\; \{\rho : \rho \in \mathrm{Suc}_{T^*}(\nu_\eta),\ \mathrm{Dp}_T(\rho) \ge \beta\} \notin \mathbf{I}_{\nu_\eta}].$$

Clearly $\mathrm{Dp}_T : \ T^* \to \mathrm{Ord} \cup \{\infty\}$ is well-defined, and if $\eta \trianglelefteq \nu$, then $\mathrm{Dp}_T(\eta) \ge \mathrm{Dp}_T(\nu)$.

For each $\eta \in T$, define A_η as follows:

$$\text{if } \eta \notin \mathrm{split}(T^*, \mathbf{I}) \text{ or } \mathbf{I}_\eta \notin \mathbb{I}', \text{ then } A_\eta = \emptyset$$
$$\text{otherwise } A_\eta = \{\rho \in \mathrm{Suc}_{T^*}(\eta) : \mathrm{Dp}_T(\eta) = \mathrm{Dp}_T(\rho)\}.$$

If $A_\eta \in \mathbf{I}_\eta$ let $B_\eta = A_\eta$, otherwise let $B_\eta = \emptyset$. Now we return to V. So for each η we have a name $\underset{\sim}{B}_\eta$ such that $\Vdash_Q$ "$\underset{\sim}{B}_\eta \in \mathbf{I}_\eta \in \mathbb{I}'$". As $\mathbb{I}'$ is κ-complete and Q satisfies the κ-c.c., there is $B^*_\eta \in \mathbf{I}_\eta$ such that $\Vdash_Q$ "$\underset{\sim}{B}_\eta \subseteq B^*_\eta$". Now define T^0 as follows:

$$T^0 = \{\eta: \text{ for all } \ell < \ell g(\eta), \text{ if } \mathbf{I}_{\eta\restriction\ell} \in \mathbb{I}' \text{ and } \eta\restriction\ell \in \text{split}(T, \mathbf{I}), \text{ then } \eta(\ell) \notin B^*_\eta\}.$$

So we have $(T, \mathbf{I}) \leq^* (T^0, \mathbf{I})$, and $\langle N_\eta : \eta \in (T^0, \mathbf{I})\rangle$ is still an $\aleph_1$-strictly $(\mathbb{I}, \mathbf{W})$-suitable* tree of models. So we can find a condition q and a name $\underset{\sim}{\eta}$ such that $p \leq_{\text{pr}} q$ and $q \Vdash$ "$\underset{\sim}{\eta} \in \lim(T^0)$ and for all $\ell < \omega$: $N_{\eta\restriction\ell}[G] \cap \omega_1 = \delta$". We now claim

$(*)$ $q \Vdash$ " for all $\ell < \omega$, $\text{Dp}_T(\underset{\sim}{\eta}\restriction\ell) = \infty$".

So work in $V[G]$, where $q \in G$. Clearly $\eta\restriction\ell \in T^*$ for $\ell < \omega$ and assume toward condition $\bigvee_\ell \text{Dp}_T(\eta\restriction\ell) < \infty$. As $\eta \trianglelefteq \nu (\in T^*[G]) \Rightarrow \text{Dp}(\eta) \geq Dp(\nu)$ for some $\ell_0 < \omega$, $(\forall \ell \geq \ell_0)[\text{Dp}_T(\eta\restriction\ell) = \alpha_0 < \infty]$. Let $\eta_0 = \eta\restriction\ell_0$. By definition of Dp_T, there are $I \in N_\eta[G] \cap \mathbb{I}'$ and $\beta < \alpha_0 + 1$ such that for all $\nu \in T^*$: if $\eta \trianglelefteq \nu$, and $I \leq_{RK} \mathbf{I}_\nu$ and $\nu \in \text{split}(T, \mathbf{I})$ then $\{\rho \in \text{Suc}_{T^*}(\nu) : \text{Dp}_T(\rho) \geq \beta\} \in \mathbf{I}_\eta$. W.l.o.g. $\beta = \alpha_0$. Since $\langle N_\nu[G] : \nu \in T^0\rangle$ is suitable*, and $\underset{\sim}{\eta}$ is a branch, we can find $\ell_1 > \ell_0$, such that (letting $\eta_1 = \eta\restriction\ell_1$): $I \leq_{RK} \mathbf{I}_{\eta_1}$ and $\text{Suc}_T(\eta_1) \notin \mathbf{I}_{\eta_1}$; now as $\ell_1 > \ell_0$ clearly $\eta \trianglelefteq \eta_1$ and (by the choice of ℓ_0) $\text{Dp}_T(\eta_1) = \alpha_0$; by those things and by the previous sentence $\{\rho \in \text{Suc}_{T^*}(\eta_1) : \text{Dp}_T(\rho) = \alpha_0\} \in \mathbf{I}_{\eta_1}$. But then we must have $\eta\restriction(\ell_1 + 1) \in \{\rho \in \text{Suc}_{T^*}(\eta_1)\text{: } \text{Dp}_T(\rho) = \alpha_0\} \subseteq A_{\eta_1} = B_{\eta_1} \subseteq B_\eta$. This is impossible as $\eta\restriction(\ell_1 + 1) \in T^0$. So we have proved $(*)$. Now it is easy to see that $\underset{\sim}{T}'[G] = \{\eta \in \underset{\sim}{T}^*[G] : \text{Dp}_T(\eta) = \infty\}$ satisfies all requirements. $\square_{4.14}$

We can conclude (and it should be easy for a reader who has arrived here):

4.15 Iteration Lemma. Suppose:

(a) $\langle P_i, \underset{\sim}{Q}_j : i \leq \alpha, j < \alpha\rangle$ is an *RCS* iteration

(b) for every i for some n we have $\Vdash_{P_{i+n}}$ "$|P_i| \leq \aleph_1$"

(c) $\mathbf{W} \subseteq \omega_1$ stationary

(d) for each i for some P_i-name of regular cardinal $\kappa_i \geq \aleph_1$ (in V) and P_i-name $\mathbb{I}_i$:

(α) P_i satisfies the κ_i-c.c. (i.e. if $p \Vdash_{P_i}$ "$\kappa_i = \kappa$" then $P_i \restriction \{q \in P_i, q \geq p\}$ satisfies the κ-c.c.) and

(β) $\Vdash_{P_i}$ "Q_i satisfies $UP(\mathbb{I}_i, \mathbf{W})$ and $\mathbb{I}_i$ is κ_i - complete."

Then

(1) P_α satisfies $UP(\mathbb{I}, \mathbf{W})$ for some $(\mathrm{Min}_{i<\alpha}\kappa_i)$-complete $\mathbb{I}(\in V)$ (i.e. $\mathbb{I}$ is κ-complete where $\kappa \stackrel{\text{def}}{=} \mathrm{Min}\{\kappa : \text{for some } i \text{ and } p \in P_i, p \Vdash_{P_i} \kappa_i = \kappa)$.

(2) $\bigcup_{i<\delta} P_i$ is a dense subset of P_δ (δ limit ordinal $\leq \alpha$) if: $\mathrm{cf}(\delta) = \aleph_1$ or $\Vdash_{P_\delta}$ "$\mathrm{cf}(\delta) = \aleph_1$" or δ strongly inaccessible and $\bigwedge_{i<\delta} |P_i| < \delta$.

(3) also the existence lemma holds, (like 3.8).

Proof. Should be clear. $\square_{4.15}$

We note:

4.15A Claim. 1) In 4.15, we can use the "strong preservation" version (and it works).

4.16 Lemma. The following property, $UP_{con}(\mathbb{I}, \mathbf{W})$, is preserved (even strongly preserved) by iterations as in 4.15, and implies that forcing by Q add no real, where:

$UP_{con}(\mathbb{I}, \mathbf{W})$ is satisfied by the forcing notion Q, if: *for any* $\langle N_\eta : \eta \in (T, \mathbf{I})\rangle$ an $\aleph_1$-strict $(\mathbb{I}, \mathbf{W})$-suitable tree of models for χ, such that for every $\eta, \nu \in T$, of the same length $h_{\eta,\nu}$ is an isomorphism from N_η onto N_ν, $h_{\eta,\nu}(Q) = Q$, $h_{\eta\restriction\ell,\nu\restriction\ell} \subseteq h_{\eta,\nu}$ and: *if* $\eta^* \in \lim(T)$ and G_{η^*} is a directed subset of $\bigcup_{\ell<\omega} N_{\eta^*\restriction\ell} \cap Q$, not disjoint to any dense subset of $\bigcup_{\ell<\omega} N_{\eta^*\restriction\ell} \cap Q$ defined in $(\bigcup_{m<\omega} N_{\eta^*\restriction m},\ N_{\eta^*\restriction\ell}, Q, \mathbf{I}_{\eta^*\restriction\ell})_{\ell<\omega}$ *then there is* $q \in Q$ such that $q \Vdash_Q$ "there is $\nu \in \lim(T)$ (in V^Q) such that $\bigcup_{\ell<\omega} h_{\eta^*\restriction\ell,\nu\restriction\ell}(G \cap N_{\nu\restriction\ell})$ is a subset of G_Q".

4.16A Remark. 1) This property relates to the $\mathrm{UP}(\mathbb{I}, \mathbf{W})$ as E-complete relate to E-proper (see V §1).

2) Who satisfies this condition? $\mathbf{W}$-complete forcing notions, $\mathrm{Nm}'(D)$, $\mathrm{Nm}(D)$

(D is $\aleph_2$-complete) $\mathrm{Nm}^{(\prime)}(T,\mathfrak{D})$ ($\mathfrak{D}$ is $\aleph_2$-complete), and shooting a club through a stationary subset of some $\lambda = \mathrm{cf}(\lambda) > \aleph_1$ consisting of ordinals of cofinality ω (and generally those satisfying the $\mathbb{I}$-condition from Chapter XI).

Proof: Should be clear (and will be elaborated elsewhere, see [Sh:311]). $\square_{4.16}$

XVI. Large Ideals on $\aleph_1$ from Smaller Cardinals

§0. Introduction

We give here better consistency strength than in XIII for having some large ideal on ω_1; possibly without adding a real using e.g. a Woodin cardinal. By this we keep old promises from 84 – 85, mentioned in [Sh:253], Shelah and Woodin [ShWd:241], (part of the delay was because it was originally intended to be part of [ShWd:241] which later was splitted to three). This will be continued elsewhere - getting suitable axioms in 2.4, 2.5, 2.6+2.10. Woodin told the author that the results (in 2.4 – 2.6(+2.7)) threw some light on the structure of universes of set theory satisfying AD. In §2 we use from §1 only 1.2(1),(2), 1.3(1), 1.8 for 2.1; weakening somewhat the results in §2, we can use 2.8, 2.9 instead of 2.1 (so replace $(*)^a_{ab}[\lambda]$ by "λ is a Woodin cardinal" in 2.4, 2.4A, 2.5, 2.6 thus using only 1.14, 1.15, 2.2 – 2.10).

The large cardinals from [ShWd:241] are defined in 1.14, 1.15.

§1. Bigness of Stationary $T \subseteq \mathcal{S}_{\leq\aleph_0}(\lambda)$

1.1 Notation. 1) λ a fixed regular cardinal $> \aleph_0$.

2) For sets a, b let $a \leq_\kappa b$ mean: $a \cap \kappa = b \cap \kappa$ and $a \subseteq b$ and let $a <_\kappa b$ means:

$a \subseteq b$ and $a \cap \kappa = b \cap \sup(a \cap \kappa)$ (i.e. $\alpha \in a \cap \kappa \Rightarrow a \cap (\alpha+1) = b \cap (\alpha+1)$), so $a <_\kappa b \not\Rightarrow a \leq_\kappa b$! And $a <_\kappa a$ holds!

3) $H(\alpha)$ is the family of sets x whose transitive closure has cardinality $< \alpha$, and if α is not cardinal we add: x of rank $< \alpha$. Let $<^*_\alpha$ be some well ordering of $H(\alpha)$ increasing with α. We let N denote a model (usually $N \prec (H(\chi), \in, <^*_\chi)$, N countable), $|N|$ its universe, and $\|N\|$ its cardinality. We write $N_1 \leq_\kappa N_2$ instead $|N_1| \leq_\kappa |N_2|$, similarly for $<_\kappa$.

4) $\mathcal{S}_{\leq\mu}(A) = \{b : \ b \subseteq B, \ |b| \leq \mu\}$

$\mathcal{D}_{\leq\mu}(B)$ is the filter on $\mathcal{S}_{\leq\mu}(B)$ generated by the closed unbounded subsets of $\mathcal{S}_{\leq\mu}(B)$ (similarly $\mathcal{D}_{<\mu}(A)$ for μ regular uncountable).

5) S, T denote subsets of some $\mathcal{S}_{\leq\mu}(A)$. We concentrate on $\mu = \aleph_0$.

1.2 Definition. 1) $T \subseteq \mathcal{S}_{\leq\aleph_0}(\lambda)$ is (θ, C^*)-big (where $\aleph_0 < \theta = \mathrm{cf}(\theta) \leq \lambda$ and $C^* \subseteq \lambda$ closed unbounded) *if*:

for every $\alpha < \lambda$ there is β, $\alpha \leq \beta < \mathrm{Min}(C^* \setminus (\alpha+1))$ such that for every $C \in \mathcal{D}_{\leq\aleph_0}(\beta)$ the set $\{a \in \mathcal{S}_{\leq\aleph_0}(\alpha) : (\exists b \in C \cap T)[a <_\theta b]\}$ belongs to $\mathcal{D}_{\leq\aleph_0}(\alpha)$.

We say T is $(<\sigma, C^*)$-big if for each $\theta < \sigma$ we have T is (θ^+, C^*)-big.

We define T is (θ, f)-big where $f : \lambda \to \lambda$ similarly only "$\beta < f(\alpha)$" replace "$\beta < \mathrm{Min}(C^* \setminus (\alpha+1))$". If $\neg(\aleph_0 < \theta = \mathrm{cf}\theta)$ we mean the first such $\theta^1 > \theta$.

2) $T \subseteq \mathcal{S}_{\leq\aleph_0}(B)$, is $\theta-{}^*$big (where $\theta \subseteq B$) *if*: for every χ regular large enough and countable $N \prec (H(\chi), \in, <^*_\chi)$ to which T, B, θ belong there is $N', N <_\theta N' \prec (H(\chi), \in, <^*_\chi)$ such that $N' \cap B \in T$.

3) Let $\lambda \subseteq B$ and $\theta \subseteq B$. We say $T \subseteq \mathcal{S}_{\leq\aleph_0}(B)$ is θ-big (for B; if the identity of λ not clear we add "in λ") *if*: for every $C \in \mathcal{D}_{\leq\aleph_0}(B)$ and $\alpha < \lambda$ such that $[\theta < \lambda \Rightarrow \alpha \geq \theta]$ we have: $\{a$: $a \in \mathcal{S}_{\leq\aleph_0}(\alpha)$, and for some $b \in T \cap C$ we have $a <_\theta b\} \in \mathcal{D}_{\leq\aleph_0}(\alpha)$.

If $\theta = \lambda$ we may omit θ (remember that λ is fixed (see 1.1(1))). If $B = \lambda$ we may omit it.

4) We say T is $(\theta, *)$-big if $T \subseteq \mathcal{S}_{\leq\aleph_0}(\lambda)$ is (θ, f)-big for some $f : \lambda \to \lambda$; equivalently (θ, C^*)-big for some club C^* of λ.

If not said otherwise, and λ is strongly inaccessible, then we assume: (here as well as in 1.11, 1.13) for $\beta \in C^*$, we have $f(\beta)$ (or $\mathrm{Min}(C^* \setminus (\beta+1))$) is a strong limit of cofinality $> \beta$.

1.3 Definition.

(1) For cardinals $\mu \geq \lambda \geq \theta$ we say $T \subseteq \mathcal{S}_{\leq\aleph_0}(\lambda)$ is (μ,θ)-big or big for (μ,θ) *if*: for every $C^1 \in \mathcal{D}_{\leq\aleph_0}(\mu)$ for some $C \in \mathcal{D}_{\leq\aleph_0}(\mu)$:

$$(\forall a \in C)(\exists b)[a \subseteq b \in C^1 \ \&\ a <_\theta b \ \&\ b \cap \lambda \in T]$$

(2) We say T is $(<\mu,\theta)$-big if it is (μ_1,θ)-big for every $\mu_1, \lambda \leq \mu_1 < \mu$.

1.4 Definition. Suppose $\lambda \subseteq B$. We say $T \subseteq \mathcal{S}_{\leq\aleph_0}(B)$ is θ-essentially end extension closed set (for B) *if* for some $E \in \mathcal{D}_{\leq\aleph_0}(B)$:

$$[a \in E \ \&\ b \in E \ \&\ a <_\theta b \ \&\ a \in T \Rightarrow b \in T]$$

In short we write θ-EEEC and we call E a *witness* for T. If $E = \mathcal{S}_{\leq\aleph_1}(B)$ then we say T is θ-end extension closed set (for B), in short θ-EEC.

1.5 Definition. 1) $\mathrm{Pr}^0_\theta(\lambda)$ means: every θ-EEEC θ-big (see Definition 1.2(3)) set $T \subseteq \mathcal{S}_{\leq\aleph_0}(\lambda)$ is also $(2^\lambda,\theta)$-big (see Definition 1.3(1)).

2) $\mathrm{Pr}^1_\theta(\lambda)$ means:

for every semiproper forcing notion P of cardinality $< \lambda$,

$$\Vdash_P \text{“}\mathrm{Pr}^0_\theta(\lambda)\text{''}$$

3) $\mathrm{Pr}_\ell(\lambda)$ means $\mathrm{Pr}^\ell_\lambda(\lambda)$.

1.6 Fact. 1) In Definition 1.3(1) we can replace μ by any set A satisfying $\lambda \subseteq A$, $|A| = \mu$.

2) If $\theta_1 \leq \theta_2 \leq \lambda \leq \mu_1 \leq \mu_2$ and $T \subseteq \mathcal{S}_{\leq\aleph_0}(\lambda)$ is (μ_2,θ_2)-big (see Definition 1.3(1)) *then* T is (μ_1,θ_1)-big.

3) If $\lambda \subseteq B, \theta_1 \leq \theta_2$ and $T \subseteq \mathcal{S}_{\leq\aleph_0}(B)$ is θ_2-big (see Definition 1.2(3) so for B, in λ) *then* it is θ_1-big.

1.7 Fact. 1) If λ is weakly compact, $T \subseteq \mathcal{S}_{\leq\aleph_0}(\lambda)$ is θ-big (see Definition 1.2(3)) and $\theta < \lambda$ *then* T is (λ,θ)-big (see Definition 1.3(1)).
2) If λ is weakly compact, $T \subseteq \mathcal{S}_{\leq\aleph_0}(\lambda)$ is big (i.e. λ-big) *then* T is $(\lambda,*)$-big.

Proof. 1) Let $C^1 \in \mathcal{D}_{\aleph_0}(\lambda)$ be given. Let

$$E = \{a \in C^1 : \neg(\exists a' \in C^1)[a <_\theta a' \in C^1 \& a' \cap \lambda \in T]\}$$

If $E = \emptyset \bmod \mathcal{D}_{\leq\aleph_0}(\lambda)$ we finish.

Otherwise by weak compactness, for some $\lambda^* < \lambda$ (inaccessible, $\theta < \lambda \Rightarrow \lambda^* \Rightarrow \theta$) we have $E \cap \mathcal{S}_{\leq\aleph_0}(\lambda^*) \neq \emptyset \bmod \mathcal{D}_{\leq\aleph_0}(\lambda^*)$. As T is θ-big we get a contradiction.

2) Easy too. $\square_{1.7}$

1.8 Fact. 1) If $T \subseteq \mathcal{S}_{\leq\aleph_0}(\lambda)$ is big for $(2^\lambda,\theta)$ (see definition 1.3(1)) *then* T is $\theta-{}^*$big (see Definition 1.2(2)).
2) If $T \subseteq \mathcal{S}_{\leq\aleph_0}(\lambda)$ is $\theta-{}^*$big, and $\mu \geq \lambda$ then T is (μ,θ)-big.

1.8A Remark. So the two conditions in 1.8(1) are equivalent.

Proof. 1) We check definition 1.2(2), so say $\chi > 2^\lambda$. Clearly $H(\lambda^+) \in N$, $|H(\lambda^+)| = 2^\lambda$, and $Sb \stackrel{\text{def}}{=} \{M \prec (H(\lambda^+),\in,<^*_{\lambda^+}) : \|M\| = \aleph_0$ and T,λ,θ belong to $M\} \in N$ and $Sb \in \mathcal{D}_{\leq\aleph_0}(H(\lambda^+))$. By the assumption of 1.8(1) for some $C \in \mathcal{D}_{\aleph_0}(H(\lambda^+))$:

$$(\forall a \in C)(\exists a')[a <_\theta a' \in Sb]$$

As all the parameters in the requirements on C belong to N, without loss of generality $C \in N$. As $C \in N$ is a club of $\mathcal{S}_{\leq\aleph_0}(H(\lambda^+))$, clearly $N \cap H(\lambda^+) \in C$. So there is $N' \in Sb, N \cap H(\lambda^+) <_\theta N'$.

Now $\lambda \cap$ Skolem Hull $[N \cup (N' \cap \lambda)] = N \cap \lambda$ (Skolem Hull - in $(H(\chi), \in, <^*_\chi)$) and this implies the conclusion. $\Big[$Why the equality holds? Enough to look at $\tau(x,y)$ for τ a term, $x \in N, y \in N' \cap \lambda$ such that $\forall xy[\tau(x,y) \in \lambda)]$. In N there are $x' \in N \cap H(\lambda^+)$ and a term τ' such that $(\forall y \in \lambda)[\tau(x,y) = \tau'(x',y)]$. Now $x' \in N'$ and we finish.$\Big]$

2) Easy. $\square_{1.8}$

1.9 Fact. Suppose T is big for $(\lambda, \theta), 2^\lambda = \lambda^+$ and

(*) *for every* $u \subseteq \mathcal{S}_{\leq\aleph_0}(\lambda^+)$ such that $u \neq \emptyset \bmod \mathcal{D}_{\leq\aleph_0}(\lambda^+)$ *for some* $B \subseteq \lambda^+$, $u \cap \mathcal{S}_{\leq\aleph_0}(B) \neq \emptyset \bmod \mathcal{D}_{\leq\aleph_0}(B)$ and $|B| < \lambda$.

Then T is θ–*big.

Proof. Like the proof of 1.7 (remembering 1.8).

Similarly we can prove

1.9A Fact. Suppose for every μ, such that $\lambda \leq \mu \leq 2^\lambda$ we have:

$(*)_1$ $(\forall \text{ stat } E \subseteq S_{\leq\aleph_0}(\mu))(\exists A \subseteq \mu)\Big[|A| < \mu \ \& \ [\theta < \lambda \Rightarrow \theta \subseteq A] \ \& \ E \cap S_{\leq\aleph_0}(A) \neq \emptyset \bmod \mathcal{D}_{\leq\aleph_0}(A)\Big]$.

Then every θ-big T is $(2^\lambda, \theta)$-big (equivalently, θ–*big.)

1.10 Fact. If $Pr_0(\lambda)$ (see Definition 1.5(3) and 1.5(1)) and $\lambda = \kappa^+ = 2^\kappa$ *then* $\mathcal{D}_{\leq\aleph_0}(\kappa)$ is precipitous; moreover, semiproper (see below).

Proof. Let χ be regular large enough, $N \prec (H(\chi), \in, <^*_\chi)$ countable.

It suffices to prove $\mathcal{D}_{\leq\aleph_0}(\kappa)$ is semiproper; i.e.:

1.10A Definition. $\mathcal{D}_{\leq\aleph_0}(\kappa)$ is semiproper provided that the following holds. If $\langle B_i : i < \lambda \rangle$ is a maximal antichain of stationary subsets of $\mathcal{S}_{\leq\aleph_0}(\kappa)$ which belongs to N where $N \prec (H(\chi), \in, <^*_\chi)$ and χ large enough *then* there is a countable $M, M \prec (H(\chi), \in <^*_\chi), N \prec M, N <_{\kappa^+} M$ and $M \cap \kappa \in \bigcup_{i \in M} B_i$ [i.e. sealing forcing is semiproper].

(Hence by repeating one such M works for every such $\langle B_i : i < \lambda\rangle$ which belongs to it; this definition is what we need; from this precipitousness follows).

Continuation of the proof of 1.10: So let $N \prec (H(\chi), \in, <^*_\chi)$ and $\langle B_i : i < \lambda\rangle \in N$ be as in Definition 1.10A. Let $T = \{a \in \mathcal{S}_{\leq\aleph_0}(\lambda) : a \cap \kappa \in \bigcup_{i\in a} B_i\}$. We shall first prove that for α in the interval $[\kappa, \lambda)$ the set $E_\alpha = \{N \cap \alpha : N \prec (H(\chi), \in, <^*_\chi), N \cap \kappa \in \bigcap_{i \in N} B_i\}$ belongs to $\mathcal{D}_{\leq\aleph_0}(\alpha)$.

If $E_\alpha \notin \mathcal{D}_{\leq\aleph_0}(\alpha)$ let $f : \alpha \to \kappa$ be one to one onto, let

$C' = \{a \subseteq \alpha : f''(a) = a \cap \alpha \text{ and } a = f^{-1}{}''(a \cap \alpha)\}$.

Clearly $C' \in \mathcal{D}_{\leq\aleph_0}(\alpha)$ and

$[(\mathcal{S}_{\leq\aleph_0}(\alpha) \setminus E_\alpha) \cap C')]{\restriction}\kappa$ is stationary (i.e. $\neq \emptyset \bmod \mathcal{D}_{\leq\aleph_0}(\kappa)$)

(where $E^*{\restriction}\kappa \stackrel{\text{def}}{=} \{a \cap \kappa : a \in E^*\}$), hence this set is not disjoint to some $B_{i(*)}$ and then we get an easy contradiction. So $E_\alpha \in \mathcal{D}_{\leq\aleph_0}(\alpha)$ for $\alpha \in [\kappa, \kappa^+)$. Let $\theta \stackrel{\text{def}}{=} \lambda (= \kappa^+)$, clearly T is $\theta - EEEC$ (see Definition 1.4, use as witness $C = \mathcal{S}_{\leq\aleph_0}(\lambda)$, noting that $B = \lambda$ here). Also as $E_\alpha \in \mathcal{D}_{\leq\aleph_0}(\alpha)$ for $\alpha \in [\kappa, \kappa^+)$, clearly T is θ-big (see Definition 1.2(3)). But by an assumption $\text{Pr}_0(\lambda) = \text{Pr}^0_\theta(\lambda)$ (as $\theta = \lambda$) hence we can deduce T is $(2^\lambda, \theta)$-big (Definition 1.3(1)), hence by Fact 1.8(1), T is $\theta-{}^*$big. So by Definition 1.2(2) there is N', $N <_\theta N' \prec (H(\chi), \in, <^*_\chi)$ such that $N' \cap \lambda \in T$. By the choice of T there is $i \in N' \cap \lambda$, such that $N' \cap \kappa \in B_i$, as required in Definition 1.10A.

So we have proved semiproperness. $\square_{1.10}$

1.11 Definition. 1) $Pr^2_\theta(\lambda, D, C^*)$ means: C^* a club of λ, D is a normal filter on λ concentrating on regular cardinals and for every θ-EEEC $(\theta,\ C^*)$-big $T \subseteq \mathcal{S}_{\leq\aleph_0}(\lambda)$ we have:

$$\{\kappa < \lambda : T \cap \mathcal{S}_{\leq\aleph_0}(\kappa) \text{ is } (2^\kappa, \theta \cap \kappa)\text{−big}\} \in D$$

(so here we use Def. 1.3(1) with λ replaced by κ.)
We may replace C^* by a function $f : \lambda \to \lambda$ as in Definition 1.2(1).

1.11A Remark. for "θ-EEEC" see Definition 1.4.

1.12 Definition. 1) $\mathrm{Pr}^3_\theta(\lambda, D, C^*)$ means that: for every semiproper forcing P of power $< \lambda$, we have $\Vdash$ "$\mathrm{Pr}^2_\theta(\lambda, D, C^*)$".

(D generates a normal filter in V^P and we do not strictly distinguish between the two).

1.13 Definition. 1) $\mathrm{Pr}^2_\theta(\lambda)$ means $(\exists D)(\forall C^*)\mathrm{Pr}^2_\theta(\lambda, D, C^*)$.

2) $Pr_3(\lambda, C)$ means: for some fixed D, for every semiproper P of power $< \lambda$, $\Vdash_P \mathrm{Pr}^2_\theta(\lambda, D, C)$.

From Shelah and Woodin [ShWd:241]:

1.14 Definition. (Shelah) 1) $\mathrm{Pr}_a(\kappa)$ means: $\mathrm{Pr}_a(\kappa, f)$ for every $f : \kappa \to \kappa$, where

2) $\mathrm{Pr}_a(\kappa, f)$ means: $f : \kappa \to \kappa$ and there is $\mathbf{j} : V \to M$ (elementary embedding into a transitive class) with critical point κ (i.e. $\mathbf{j}$ is the identity on κ hence on $H(\kappa)$) such that $H\big(\mathbf{j}(f)(\kappa)\big) \subseteq M$ and $M^{<\kappa} \subseteq M$. Let $\mathrm{Pr}_a(\kappa, f, D)$ means $\mathrm{Pr}_a(\kappa, f)$ is witnessed by $\mathbf{j}$ and $D = \{A \subseteq \kappa : \kappa \in \mathbf{j}(A)\}$. Note κ is necessarily measurable in all those cases.

1.15 Definition. (Woodin) $\mathrm{Pr}_b(\kappa)$, now called "κ is a Woodin cardinal" means:

for every $f : \kappa \to \kappa$ there is $\lambda < \kappa$ such that $\mathrm{Pr}_a(\lambda, f{\restriction}\lambda)$; equivalently

for every $f : \kappa \to \kappa$, there is an elementary embedding $\mathbf{j} : V \to M$ with critical point $\lambda < \kappa$, such that $H\big(\mathbf{j}(f)(\lambda)\big) \subseteq M$ and $M^{<\kappa} = M$.

So κ is a Mahlo cardinal, but not necessarily a weakly compact cardinal.

We can add

1.16 Definition. For $W \subseteq \kappa$, we can add:

1) $\mathrm{Pr}_a(\kappa, W)$ means $\mathrm{Pr}_a(\kappa, W, f)$ for every $f : \kappa \to \kappa$, which means $\mathrm{Pr}_a(\kappa, f, D)$ for some D to which W belongs.

2) Let $\mathrm{Pr}_b(\kappa, W)$ mean for every $f : \kappa \to \kappa$ there is $\lambda < \kappa$ such that $\mathrm{Pr}_a(\lambda, W \cap \lambda, f{\restriction}\lambda)$ (so in particular $\mathrm{Rang}(f{\restriction}\lambda) \subseteq \lambda$)

§2. Getting Large Ideals on $\aleph_1$

Note Ξ is a maximal antichain of $\mathcal{D}_{\omega_1}$ if $\Xi \subseteq \mathcal{P}(\omega_1)$ and for no stationary $S \subseteq \omega_1$ do we have $(\forall A \in \Xi)(A \cap S = \emptyset \mod \mathcal{D}_{\omega_1})$; we do not strictly distinguish $A \in \Xi$ and $A/\mathcal{D}_{\omega_1}$ or Ξ and $\{A/\mathcal{D}_{\omega_1} : A \in \Xi\}$.
Remember $\mathfrak{B} = \mathcal{P}(\omega_1)/\mathcal{D}_{\omega_1}$; on seal($\Xi$) and variations see XIII 2.4(2).

2.1 Lemma. *A Sealing is a Semiproper Criterion:* Let λ be strongly inaccessible, $C \subseteq \lambda$ closed unbounded, $[\delta \in C \Rightarrow (H(\delta), \in, <^*_\delta) \prec (H(\lambda), \in, <^*_\lambda)]$ (so each $\delta \in C$ is a strong limit cardinal). The following conditions satisfy (B)$^+$ $\Rightarrow$ (C) $\Rightarrow$ (A) $\Rightarrow$ (B)$^-$.

(A) Let C be an end segment of C^*. For every Levy($\aleph_1, < \lambda$)- name $\underset{\sim}{\Xi} = \{\underset{\sim}{A}_i : i < \lambda\}$ of a maximal antichain of $\mathcal{D}_{\omega_1}^{V[\text{Levy}(\aleph_1, <\lambda)]}$, the forcing notion Levy($\aleph_1, < \lambda$)$*$ seal($\underset{\sim}{\Xi}$) is semiproper, provided that:

$$\oplus_A \text{ for } \delta \in C, \Big(H(\delta), \in, <^*_\delta, \underset{\sim}{\Xi} \cap H(\delta)\Big) \prec \Big(H(\lambda), \in, <^*_\lambda, \underset{\sim}{\Xi}\Big).$$

(B)$^-$ Let C be an end segment of C^*. If $T \subseteq \mathcal{S}_{\leq\aleph_0}(\lambda)$ is $(< \lambda, C)$-big (see Definition 1.2(1)) and λ - EEEC (see Definition 1.4) *then* T is $(2^\lambda, \omega_2)$-big (see Definition 1.3(1)) provided that:

$$\oplus_B \text{ for } \delta \in C, \Big(H(\delta), \in, <^*_\delta, T \cap (\bigcup_{\alpha<\delta} \mathcal{S}_{\leq\aleph_0}(\alpha))\Big) \prec \Big(H(\lambda), \in, <^*_\lambda, T\Big).$$

(B)$^+$ Let C be an end segment of C^*. If $T \subseteq \mathcal{S}_{\leq\aleph_0}(\lambda)$ is $(\aleph_2, C)$-big (a weaker assumption see Definition 1.2(1)) λ-EEEC and *then* T is $(2^\lambda, \omega_2)$-big provided that $\oplus_B$ holds.

(C) Let C be an end segment of C^*. Suppose $\bar{P} = \langle P_i : i < \lambda\rangle$ is $\lessdot$-increasing, for $i < \lambda$, $P_i \in H(\lambda), P_i \lessdot P_\lambda$ where $P_\lambda \overset{\text{def}}{=} \bigcup_{\alpha<\lambda} P_\alpha$, and the forcing notions $P_i, P_\lambda/P_i$ are semiproper, P_λ satisfies the λ - c.c., $\Vdash_{P_\lambda}$ "$\lambda = \aleph_2$", and $\underset{\sim}{\Xi} = \{\underset{\sim}{A}_i/\mathcal{D}_{\omega_1} : i < \lambda\}$ a P_λ - name of a maximal antichain of $\mathcal{D}_{\omega_1}$.

Then $P_\lambda *$ seal(Ξ) is semiproper provided that:

$$\oplus_C \text{ for } \delta \in C,\ \Big(H(\delta), \in, <^*_\delta, \bar{P}\restriction\delta, \underset{\sim}{\Xi} \cap H(\delta)\Big) \prec \Big(H(\lambda), \in, <^*_\lambda, \bar{P}, \underset{\sim}{\Xi}\Big)$$

2.1A Definition. Assume λ is strongly inacessible, $\bar{P} = \langle P_i : i < \lambda\rangle$, P are as in clause (C) of 2.1 or $P_i = \text{Levy}(\aleph_1, < i)$, $P = \bigcup_{i<\lambda} P_i$ (for some closed unbounded $C \subseteq \lambda$). If $\underset{\sim}{A}$ is a P-name of a subset of ω_1 let

$$i(\underset{\sim}{A}/\mathcal{D}_{\omega_1}) = \min\{i : \text{ for some } P_i\text{-name } \underset{\sim}{A}', \Vdash_P \text{“}\underset{\sim}{A} = \underset{\sim}{A}'_i\text{” mod } \mathcal{D}_{\omega_1}\}$$

(note $i(A) < \lambda$ as P satisfies the λ-c.c.). Let us redefine

$$\underset{\sim}{A}/\mathcal{D}_{\omega_1} = \{\underset{\sim}{B} : \underset{\sim}{B} \text{ is a } P_{i(\underset{\sim}{A})}\text{-name of a subset of } \omega_1 \text{ such that } \Vdash_{P_\lambda} \text{“}\underset{\sim}{B} = \underset{\sim}{A}\text{”}\}.$$

Proof. Clearly $(B)^+ \Rightarrow (B)^-$, just read Definition 1.2(1).

$\neg(A) \Rightarrow \neg(C)$

Immediate: use $P = \text{Levy}(\aleph_1, < \lambda)$, and $P_i = \text{Levy}(\aleph_1, < i)$.

$\neg(B)^- \rightarrow \neg(A)$

Let T, C be a counterexample to $(\text{B})^-$, in particular $\oplus_B$ holds and we can choose a club $E \subseteq \mathcal{S}_{<\aleph_1}(\lambda)$ witnessing T is EEEC i.e. $a \in E$ & $b \in E$ & $a <_\lambda b$ & $a \in T \Rightarrow b \in T$.

Let

$$W \stackrel{\text{def}}{=} \Big\{\delta < \lambda : (\forall \alpha < \delta)(\forall a \in \mathcal{S}_{\leq\aleph_0}(\alpha)) \\ \big[(\exists b)[b \in T \ \&\ a <_\lambda b] \Rightarrow (\exists b)[b \in T \ \&\ \sup(b) < \delta \ \&\ a <_\lambda b]\big]\Big\}.$$

So W is a club of λ, definable in $(H(\lambda), \in, <^*_\lambda, T)$ hence by $\oplus_B$ for $\delta \in C$ we have $\sup(W \cap \delta) = \delta$, and $W \supseteq C$.

For $\delta < \lambda$ after forcing with $\text{Levy}(\aleph_1, < |\delta|^+)$ we have $\langle a^\delta_\zeta : \zeta < \omega_1\rangle$ increasing continuous, each a^δ_ζ countable, $\bigcup_{\zeta<\omega_1} a^\delta_\zeta = \delta$. Let $\langle \underset{\sim}{a}^\delta_\zeta : \zeta < \omega_1\rangle$ be a $\text{Levy}(\aleph_1, < |\delta|^+)$-name for such a sequence, and $\underset{\sim}{B}_\delta \stackrel{\text{def}}{=} \{\zeta : \underset{\sim}{a}^\delta_\zeta \in T\}$, this is a

Levy($\aleph_1, < |\delta|^+$)-name; and then let (again a Levy($\aleph_1, < |\delta|^+$)-name):
$\underset{\sim}{A}_\delta \stackrel{\text{def}}{=} \underset{\sim}{B}_\delta - \underset{\alpha<\delta}{\nabla} \underset{\sim}{B}_\alpha$ (∇-diagonal union, actually well defined only mod $\mathcal{D}_{\omega_1}$).

As T is λ-EEEC, clearly in $V^{\text{Levy}(\aleph_1,<\lambda)}$, $B_\delta/\mathcal{D}_{\omega_1}$ ($\delta \in W$) is increasing and is the least upper bound of $\{A_\alpha/\mathcal{D}_{\omega_1} : \alpha \in (\delta+1) \cap W\}$ (in $(\mathcal{P}(\omega_1)/\mathcal{D}_{\omega_1})^{\text{Levy}(\aleph_1,<\lambda)}$). Let $\underset{\sim}{W}^* = \{\alpha : \alpha \in W, \text{ and } \underset{\sim}{A}_\alpha \neq \emptyset \mod \mathcal{D}_{\omega_1}\}$ (it is a Levy($\aleph_1, < \lambda$)-name)

Clearly $\underset{\sim}{\Xi} = \{\underset{\sim}{A}_\alpha/\mathcal{D}_{\omega_1} : \alpha \in \underset{\sim}{W}^*\}$ is an antichain (we should not mind the $\emptyset/\mathcal{D}_{\omega_1}$'s, i.e. some A_α's are not stationary).

Clearly $\underset{\sim}{\Xi}$ is a Levy($\aleph_1, < \lambda$)-name satisfying $\oplus_A$.

2.1B Fact. $\underset{\sim}{\Xi}$ is a maximal antichain.

Suppose toward contradiction that $\underset{\sim}{A}$ is a Levy($\aleph_1, < \lambda$)-name of a stationary subset of ω_1, but $p \in$ Levy($\aleph_1, < \lambda$) force it is a counterexample. So for some $\theta < \lambda$, $\underset{\sim}{A}$ is a Levy($\aleph_1, < \theta$)-name, and $p \in$ Levy($\aleph_1, < \theta$). Let $\theta_1 = (2^\theta)^+$, $\mu = 2^{\theta_1}$, and

$$Y^\mu_p \stackrel{\text{def}}{=} \{a \in \mathcal{S}_{\leq\aleph_0}(H(\mu)) : \text{there is } q \in \text{Levy}(\aleph_1, <\theta) \text{ such that: } p \leq q,$$
$$q \text{ is an } (a, \text{Levy}(\aleph_1, <\theta))\text{-generic condition},$$
$$\text{and } q \Vdash \text{“}a \cap \omega_1 \in \underset{\sim}{A}\text{”}.\}$$

Clearly $Y^\mu_q \neq \emptyset \mod \mathcal{D}_{\leq\aleph_0}(H(\mu))$.

Now, as λ is strongly inaccessible, $2^\mu < \lambda$, and as T is (θ_1, C)-big (as T exemplifies $\neg(B)$), there is β satisfying $2^\mu < \beta < \lambda$ (and moreover $2^\mu < \beta < \min(C \setminus (\beta+1))$), such that: for every $E \in \mathcal{D}_{\leq\aleph_0}(\beta)$ we have:

$$\{a \in \mathcal{S}_{\leq\aleph_0}(2^\mu) : \text{ there is } b \text{ such that } a <_{\theta_1} b \text{ and } b \in E \cap T\} \in \mathcal{D}_{\leq\aleph_0}(2^\mu).$$

Hence, as $|H(\mu)| \leq 2^\mu$, for every $E \in \mathcal{D}_{\leq\aleph_0}(\beta)$

$$\{a \in \mathcal{S}(H(\mu)) : \text{there is } b \text{ such that } a <_{\theta_1} b \text{ and } b \in E,$$
$$\text{and } b \cap \beta \in T\} \in \mathcal{D}_{\leq\aleph_0}(H(\mu)).$$

Let

$$E_1 \stackrel{\text{def}}{=} \{N : N \text{ is a countable elementary submodel of } (H(\beth_7(\lambda)^+), \in, <^*)$$
$$\text{to which } p,\ \lambda,\ \theta,\ \mu,\ \beta,\ \underset{\sim}{A},\ \langle(\underset{\sim}{B}_\alpha, \underset{\sim}{A}_\alpha) : \alpha < \lambda\rangle$$
$$\text{and } \langle\langle \underset{\sim}{a}^\alpha_\zeta : \zeta < \omega_1\rangle : \alpha < \lambda\rangle \text{ belong}\}.$$

Clearly it is a club of $\mathcal{S}_{<\aleph_1}(H(\beth_7(\lambda)^+))$, and let

$$E_2 \stackrel{\text{def}}{=} \{N \cap \beta :\ N \in E_1\},$$

clearly it belongs to $\mathcal{D}_{<\aleph_1}(H(\beta))$. So we can use E_2 as E above hence

$$E_3 \stackrel{\text{def}}{=} \{N : N \text{ is a countable, elementary submodel of } (H(\mu), \in),$$
$$\text{such that } p, \theta, \underset{\sim}{A}, \beta \text{ belong to it and for some } M_N \in E_1 \text{ we have}$$
$$M_N \cap \beta \in T \text{ and } N <_{\theta_1} M_N \in T\}$$

belongs to $\mathcal{D}_{<\aleph_1}(H(\mu))$. Hence we can find $N \in E_3 \cap Y^\mu_p$, hence by the definition of Y^μ_p there is a condition $q \in \text{Levy}(\aleph_1, <\theta)$ such that $p \le q \in \text{Levy}(\aleph_1, <\theta)$ and q is $(N, \text{Levy}(\aleph_1, <\theta))$-generic and $q \Vdash$ "$N \cap \omega_1 \in \underset{\sim}{A}$". As $N \in E_3$ clearly M_N is well defined (see the definition of E_3), so $M_N \in E_1$, $N \prec M_N \in E_1$ and $N <_{\theta_1} M_N$, hence $N \cap 2^\theta = M_N \cap 2^\theta$, hence $N \cap \text{Levy}(\aleph_1, <\theta) = M_N \cap \text{Levy}(\aleph_1, <\theta)$ and moreover $N \cap \mathcal{P}(\text{Levy}(\aleph_1, <\theta)) = M_N \cap \mathcal{P}(\text{Levy}(\aleph_1, <\theta))$; hence as q is $(N, \text{Levy}(\aleph_1, <\theta))$-generic we know that q is $(M_N, \text{Levy}(\aleph_1, <\theta))$-generic. As $\text{Levy}(\aleph_1, <\lambda)/\text{Levy}(\aleph_1, <\theta)$ is $\aleph_1$-complete there is $q_1 \in \text{Levy}(\aleph_1, <\lambda)$ such that $q_1 \restriction \theta = q$ and q_1 is $(M_N, \text{Levy}(\aleph_1, <\mu))$-generic, hence clearly $q_1 \Vdash$ "$M_N \cap \beta = \underset{\sim}{a}^\beta_{M_N \cap \omega_1} = a^\beta_{N \cap \omega_1}$" but $M_N \cap \beta \in T$ (see the definition of E_3) hence $q_1 \Vdash$ "$N \cap \omega_1 \in \underset{\sim}{B}_\mu$".

There is $\underset{\sim}{C}'$ such that $\Vdash_{\text{Levy}(\aleph_1, <\lambda)}$ "if $\underset{\sim}{B}_\beta \cap \underset{\sim}{A}$ is not stationary then $\underset{\sim}{B}_\beta \cap \underset{\sim}{A} \cap \underset{\sim}{C}' = \emptyset$, and $\underset{\sim}{C}'$ is a club of ω_1". So as $\underset{\sim}{B}_\beta$, $\underset{\sim}{A} \in N \subseteq M_N$, clearly w.l.o.g. $\underset{\sim}{C}' \in M_N$ hence $q_1 \Vdash$ "$N \cap \omega_1 \in \underset{\sim}{C}'$" hence $q_1 \Vdash$ "$\underset{\sim}{B}_\beta \cap \underset{\sim}{A} \cap \underset{\sim}{C}' \neq \emptyset$ hence $q_1 \Vdash_{\text{Levy}(\aleph_1, <\lambda)}$ "$\underset{\sim}{B}_\beta \cap \underset{\sim}{A}$ is stationary" which is enough for the fact 2.1B as $B_\beta = \underset{\alpha \le \delta}{\nabla} A_\alpha \mod \mathcal{D}_{\omega_1}$. $\square_{2.1B}$

Continuation of the proof of 2.1.

Lastly to show that $\neg(A)$ holds, we still have to show that:
the forcing notion $Q \stackrel{\text{def}}{=} \text{Levy}(\aleph_1, < \lambda) * \text{seal}(\underset{\sim}{\Xi})$ is not semi proper.

Suppose it is semiproper, χ large enough. Let $N \prec (H(\chi), \in, <^*_\chi)$ be countable, $Q, T, \underset{\sim}{\Xi}, C \in N$. Let $\delta = N \cap \omega_1$.

So there is $p \in Q$ which is (N, Q)-semi generic. So for some q satisfying $p \leq q \in Q$, and α, we have $q \Vdash_Q$ "$\alpha \in \underset{\sim}{W}^*$, $\delta \in \underset{\sim}{A}_\alpha$, $\alpha \in N[\underset{\sim}{G}_Q]$", ($\underset{\sim}{W}^*$ was defined just before $\underset{\sim}{\Xi}$); so $\underset{\sim}{A}_\alpha$, $\underset{\sim}{B}_\alpha \in N[\underset{\sim}{G}_Q]$ and clearly $q \Vdash_Q$ "$\underset{\sim}{a}^\alpha_\zeta = N[\underset{\sim}{G}_Q] \cap \alpha$", hence necessarily also $q \Vdash_Q$ "$\delta \in \underset{\sim}{B}_\alpha$".

Hence $\quad q \Vdash_Q$ "$N[\underset{\sim}{G}_Q] \cap \alpha \in T$" (read the definition of $\underset{\sim}{B}_\alpha$).

Hence w.l.o.g. for some $b \in T$ we have $\quad q \Vdash$ "$N[\underset{\sim}{G}_Q] \cap \alpha = b$".

Let N_1 be the Skolem Hull of $|N| \cup b$ in $(H(\chi), \in, <^*_\chi)$. Clearly $N_1 \cap \alpha = b \in T$ and $N_1 \cap \omega_1 = b \cap \omega_1 = \delta$ so $N <_{\aleph_2} N_1$. This shows T is a $\aleph_2 -$ *big (see Definition 1.2(2)), which by 1.8 is equivalent to "T is $(2^\lambda, \aleph_2)$-big"; but this is a contradiction to our assumption "T exemplifies $\neg(B)^-$".

$\underline{\neg(C) \Rightarrow \neg(B)^+}$

We also prove $\neg(A) \rightarrow \neg(B)^+$

Let $\bar{P}, \underset{\sim}{\Xi} = \{\underset{\sim}{A}_i : i < \lambda\}$ and C contradict (C) or (A) (in the later case $P_i = \text{Levy}(\aleph_1, < i)$). Let, for each $p \in P_\lambda$:

$$\begin{aligned} T_p \stackrel{\text{def}}{=} \{N \cap \lambda : \; & p \in N, \text{ for some strong limit cardinal } \sigma < \lambda, \\ & N \prec (H(\sigma), \in, <^*_\lambda, \bar{P}{\restriction}\sigma, \underset{\sim}{\Xi}{\restriction}\sigma, \bar{\underset{\sim}{A}}{\restriction}\sigma) \text{ so we consider} \\ & \bar{P}, \underset{\sim}{\Xi}, \bar{\underset{\sim}{A}} \text{ as predicates, and } N \text{ is countable,} \\ & \{\bar{P}, \underset{\sim}{\Xi}, \langle \underset{\sim}{A}_i : i < \lambda\rangle\} \text{ belongs to } N, \\ & \text{and there are } j, i \in N \cap \sigma \text{ and } q \in P_i, \text{ such that} \\ & p \leq q, q \text{ is } (N, P_i)\text{-semi-generic, } \underset{\sim}{A}_j \text{ is a } P_i\text{-name,} \\ & \text{and } q \Vdash_{P_i} \text{"} N \cap \omega_1 \in \underset{\sim}{A}_j \text{ and } j \in N[\underset{\sim}{G}_{P_i}]\text{"}\}. \end{aligned}$$

$$T^+_p \stackrel{\text{def}}{=} \{b \in \mathcal{S}_{\leq\aleph_1}(\lambda) : \text{ for some } a \in T_p, a <_\lambda b\}.$$

Assume first that every T_p^+ is $(2^\lambda, \aleph_2)$-big. So for every $\chi > 2^\lambda$ and countable $N \prec (H(\chi), \in, <^*_\chi)$, to which $\bar{P}$, Ξ, $\langle \underset{\sim}{A}_i : i < \lambda\rangle$ belong, and $p \in N \cap P_\lambda$, we know $\lambda \in N$ hence $H(\lambda) \in N$ and $<^*_\lambda \in N$ hence $T_p \in N$. By 1.8(1) we know T_p hence T_p^+ is $\aleph_2-{}^*$big, hence (Definition 1.2(2)) we can find M, $N <_{\aleph_2} M \prec (H(\chi), \in, <^*_\chi)$, M countable, $M \cap \lambda \in T_p^+$, hence for some $M_1 \in T_p$ we have $M_1 \cap \lambda <_\lambda M \cap \lambda$. Clearly for α, |beta $\in M_1 \cap \lambda$ we have

$$M_1 \models \text{“}\alpha \text{ is cardinal”} \Leftrightarrow M \models \text{“}\alpha \text{ is a cardinal,”}$$

$$M_1 \models \text{“}2^\alpha = \beta\text{”} \Leftrightarrow M \models \text{“}2^\alpha = \beta\text{“};$$

and so $(\forall \sigma \in M_1 \cap \lambda)(2^\sigma \in M_1 \cap \lambda)$; as $[\sigma \in M_1 \cap \lambda \Rightarrow H(\sigma)$ is an initial segment by $<^*_\lambda$ of $H(\lambda)]$, easily $\sigma \in M_1 \cap \lambda \Rightarrow M_1 \cap H(\sigma) = M \cap H(\sigma)$, hence $M - 1 <_\lambda M$. Let q, σ, i, j witness $M_1 \in T_p$ (see the definition of T_p), and easily we can deduce what semiproperness would have required. But $P_\lambda * \text{seal}(\underset{\sim}{\Xi})$ is not semiproper (as $\bar{P}, \underset{\sim}{\Xi}$ contradict (C)). So the assumption above was wrong, i.e., for some $p \in P_\lambda$, T_p^+ is not $(2^\lambda, \aleph_2)$-big. Let $j(*) = \min\{j \in C : p \in H(j)\}$, and let $C' = C \setminus j(*)$, we shall prove that T_p^+, C' exemplify $\neg(B)^+$, renaming $C' = C$ i.e. $p \in H(\min(C))$. Also $\oplus_B$ holds for T_p^+ easily. Let $j(*) = \min\{j \in C : p \in H(j)\}$, and let $C^* = C \setminus j(*)$, we shall prove that T_p^+, C^* exemplify $\neg(B)^+$, renaming $C^* = C$ i.e. $p \in H(\min(C))$. Also T_p^+ is λ-EEEC by its definition. To complete the proof of "T_p^+ exemplifies $\neg(B)^+$" we need only to prove "T_p^+ is $(<\lambda, C)$-big" (see Definition 1.2(1)). So let $\theta < \lambda$, $\theta \geq \aleph_2$, and we shall prove that T_p is (θ, C)-big; this suffices. We can find $i(*)$ such that $\Vdash_{P_{i(*)}}$ "$|\theta| = \aleph_1$". So let $\alpha < \lambda$ be given such that $\alpha > i(*)$, θ. We define in $(H(\lambda), \in, <^*_\lambda, \bar{P}, \underset{\sim}{\Xi})$ a function g from $\mathcal{P}(\mathcal{S}_{\leq\aleph_0}(\alpha))$ to λ, $g(X)$ is: the first strong limit cardinal of uncountable cofinality $\beta < \lambda$ such that $\beta > i(*)$, $\beta > \alpha$, $\beta > \theta$ and:

$(*)^{\alpha,\beta}_X$ for every $\mathcal{U} \in \mathcal{D}_{\leq\aleph_0}(\beta)$, the set $\{a \in X : (\exists b \in \mathcal{U} \cap T_p)[a <_\theta b]\}$ is $\neq \emptyset \bmod \mathcal{D}_{\leq\aleph_0}(\alpha)$ *if* there is such β, and $\alpha + 1$ *otherwise.*

So g is definable in $(H(\lambda), \in, <^*_\lambda, \bar{P}, \underset{\sim}{\Xi})$ with the parameters θ, α, $i(*)$ hence $\beta^* = \text{supRang}(g) < \text{Min}(C \setminus (\alpha + 1))$ (remember $\oplus_C$ is assumed). If β^* is not as required in Definition 1.2(1), then there is $\mathcal{U}^* \in \mathcal{D}_{\leq\aleph_0}(\beta^*)$, such that $X \overset{\text{def}}{=} \{a \in \mathcal{S}_{\leq\aleph_0}(\alpha) : \neg(\exists b \in \mathcal{U}^* \cap T_p)[a <_\theta b]\}$ is $\neq \emptyset \bmod \mathcal{D}_{\leq\aleph_0}(\alpha)$. Now we know $\Vdash_P$ "$|\alpha| \leq \aleph_1$", and as P satisfies the λ-c.c., there is a name for a function exemplifying this mentioning only members of some $P_i (i < \lambda)$, but $P_i \lessdot P$, so $\Vdash_{P_i}$ "$|\alpha| \leq \aleph_1$", say $\underset{\sim}{h}$ is P_i-name of a function from ω_1 onto α. As P_i is semiproper, by the assumption on X we have $\Vdash_{P_i}$ "$\underset{\sim}{Y} \overset{\text{def}}{=} \{\varepsilon < \omega_1 :$ there is $a \in X$, $\varepsilon \subseteq a \subseteq h''(\varepsilon), \varepsilon = a \cap \omega_1\}$ is a stationary subset of ω_1". Hence $\Vdash_P$ "$\underset{\sim}{Y} \subseteq \omega_1$ is stationary" hence $\Vdash_P$ " for some $\xi < \lambda$, $\underset{\sim}{Y} \cap \underset{\sim}{A}_\xi \subseteq \omega_1$ is stationary". Hence for some $j \in (i, \lambda)$ and P_j-name $\underset{\sim}{\xi}$ of an ordinal $< j$ we have: $\underset{\sim}{Y}$, $\underset{\sim}{\xi}$ and $\underset{\sim}{A}_{\underset{\sim}{\xi}}$ are P_j-names and $\Vdash_{P_j}$ "$\underset{\sim}{Y} \cap \underset{\sim}{A}_{\underset{\sim}{\xi}} \subseteq \omega_1$ is stationary".

Hence there is a strong limit $j_1 \in (j, \lambda)$ such that

$$(H(j_1), \in, <^*_\lambda \restriction H(j_1), \bar{P}\restriction j_1, \Xi\restriction H(j_1)) \prec (H(\lambda), \in, <^*_\lambda, \lessdot \bar{P}, \underset{\sim}{\Xi}),$$

$\text{cf}(j_1) > \aleph_0$, and $\underset{\sim}{Y}$, $\underset{\sim}{\xi}$, i, j, α, β^*, $\mathcal{U}^* \in H(j_1)$. Now there are $\delta < \omega_1$ and countable $N \prec (H(\chi), \in, <^*_\chi, \bar{P}, \underset{\sim}{\Xi})$ and q such that: $\{\underset{\sim}{Y}, \underset{\sim}{\xi}, i, j, \alpha, j_1\} \in N$, $p \leq q \in P_j$, q is (N, P_j)-semi generic, $N \cap \omega_1 = \delta$, $q \Vdash_{P_j}$ "$\delta \in \underset{\sim}{Y} \cap \underset{\sim}{A}_{\underset{\sim}{\xi}}$", and (remember the definition of $\underset{\sim}{Y}$) there is $a^* \in X$, $\delta \subseteq a^* \subseteq N$. Clearly $j_1 \in N$, $N \restriction H(j_1) \prec (H(\chi), \in, <^*_\chi, \bar{P}, \underset{\sim}{\Xi})$ and $N\restriction H(j_1) \in T_p$ (see the definition of T_p). As $j_1 \in N$, $X \in N$ this implies that for every $\mathcal{U} \in \mathcal{D}_{\leq\aleph_0}(j_1)$ we have $\{a \in X : (\exists b \in \mathcal{U} \cap T_P)[a <_{\aleph_2} b]\} \neq \emptyset \bmod \mathcal{D}_{\leq\aleph_0}(\alpha)$. So $(*)^{\alpha, j_1}_X$ holds; hence by the definition of g and β^* without loss of generality $j_1 \leq \beta^*$, hence (check definition) $(*)^{\alpha,\beta^*}_X$ hold, but this contradicts the choice of of X. So together we have gotten a counterexample to $(\text{B})^+$. $\square_{2.1}$

2.2 Definition. 1) $(*)^a[\lambda, C]$ means condition (C) of 2.1 holds for $\bar{P}$ and C (so C satisfies $\otimes_C$) such that

$$\{\delta < \lambda : \text{ if } \delta \text{ is strongly inaccessible then } P_\delta = \bigcup_{i<\delta} P_i\}$$

contains a club of λ (so for many C's this is empty demand).

2) $(*)^a_{ab}[\lambda, C]$ means that for every semiproper forcing Q from $H(\text{Min}C)$ we have $\Vdash_Q$ "$(*)^a[\lambda, C]$".

3) We omit C if this holds for every club C of λ.

2.3 Conclusion. Suppose $(*)^a_{ab}[\lambda, C]$, λ strongly inaccessible. If $\bar{P}$, C and $\underset{\sim}{\Xi}$ are as in 2.1(C) and $i < \lambda$, *then* in V^{P_i} the forcing notion $(P_\lambda/P_i)*$ seal $(\underset{\sim}{\Xi})$ is semiproper.

2.3A Remark. So if $\bar{Q}$ is a semiproper iteration, $\langle P_{i+1} : i < \lambda\rangle, C$, $\underset{\sim}{\Xi}$ as in 2.1(C) *then* $\bar{Q}\,\hat{}\,\langle \mathrm{Rlim}\bar{Q}, \mathrm{seal}(\underset{\sim}{\Xi})\rangle$ is a semi proper iteration.

2.4 Theorem. Suppose κ is strongly inaccessible, and:

$(*)^b_{ab}[\kappa]$ for every closed unbounded $C \subseteq \kappa$, for some $\lambda \in C$ (strongly inaccessible) we have $\lambda = \sup(C \cap \lambda)$, and $(*)^a_{ab}[\lambda,\ C \cap \lambda]$.

Let $S \subseteq \omega_1$ be stationary.

Then for some semiproper forcing P of cardinality λ satisfying the λ-c.c., we have $\Vdash_P$ "$\mathcal{D}_{\omega_1} + S$ is $\aleph_2$-saturated".

Also P is $(\omega_1 \setminus S)$-complete hence if $\omega_1 \setminus S$ is stationary it does not add ω-sequences of ordinals.

Moreover

2.4A Lemma. 1) The following homogeneous forcing can serve in 2.4. We define by induction on α a semiproper iteration $\bar{Q}^\alpha = \langle P_i, \underset{\sim}{Q}_i : i < \alpha\rangle$ with $|P_i| < \lambda$ (and, for simplicity, $\bar{Q}^i \in H(\lambda)$) for $i < \alpha$ (see XIII 1.8) as follows. If P_i is defined, i strongly inaccessible and $j < i \Rightarrow |P_j|$, then let, in V^{P_i}, Q_i be the product with countable support of $\{\mathrm{seal}(\Xi) : \Xi \in \Xi_i\} \cup \mathrm{Levy}(\aleph_1, 2^{\aleph_2})^{V^{P_i}}$ where Ξ_i is $\{\Xi : \Xi$ (in V^{P_i}) is a maximal antichain of $\mathcal{D}_{\omega_1}$ and for every $j < i$, $\Vdash_{P_{j+1}}$ "$P_i/P_{j+1} * \mathrm{seal}(\underset{\sim}{\Xi})$ is semiproper", such that $\omega_1 \setminus S \in \Xi$ if it is stationary $\}$. Otherwise Q_i is $\mathrm{Levy}(\aleph_1, 2^{\aleph_2})^{V^{P_i}}$.

2) Moreover we can replace Ξ_i by

$$\Xi'_i = \{\Xi : \Xi \text{ (in } V^{P_i}) \text{ is a maximal antichain of } \mathcal{D}_{\omega_1}$$
$$\text{which is semi proper (that is seal}(\Xi) \text{ is semi proper)}$$
$$\text{such that } \omega_1 \setminus S \in \Xi \text{ if it is stationary}\}$$

provided that λ is Woodin.

2.4B Remark. 1) We can e.g. use Levy$(\aleph_1, 2^{\aleph_2})^{V^{P_i}}$ when i is not strongly inaccessible and the CS product of $\{\text{seal}(\Xi) : \Xi \in \Xi_i\}$ otherwise.
2) By 2.7(3) below if κ is Woodin then it satisfies the assumption of Theorem 2.4. Similarly in 2.5 and 2.6 concerning the μ in the definition of W^*.
3) If $\omega_1 \setminus S$ is stationary, the iteration is essentially CS (as the condition with a "real" support are dense).
4) Homogeneity is actually gotten also in the other proofs, in particular 2.5, 2.6 (and results in Chapter XIII).

Proof of 2.4. Follow by 2.4A.

Proof of 2.4A. By XIII 2.13(1) clearly $\bar{Q}^i = \langle P_j, \underset{\sim}{Q}_j \ : j < i\rangle$ is a semiproper iteration ($P_i = \text{Rlim}\bar{Q}^i$) and if $j < i$ then $\Vdash_{P_{j+1}}$ "$(P_i/P_{j+1}) * \underset{\sim}{Q}_i$ is semiproper". Also the $(\omega_1 \setminus S)$-completeness and λ-c.c. are clear. Why $\Vdash_{P_\lambda}$ "$\mathcal{D}_{\omega_1} + S$ is $\aleph_2$-saturated"? Let $\underset{\sim}{\Xi}$ be a P_λ-name of a maximal antichain of $\mathcal{D}_{\omega_1}$ (to which $\omega_1 \setminus S$ belongs if stationary), so let $\underset{\sim}{\Xi} = \{\underset{\sim}{A}_i : i < \lambda\}$. Let

$$C = \{\mu < \lambda : (H(\mu), \in, <^*_\mu, \{(\bar{Q}^i, \underset{\sim}{A}_i/\mathcal{D}_{\omega_1}) : \ i < \mu\})$$
$$\prec (H(\lambda), \in, <^*_\lambda, \ \{(\bar{Q}^i, \underset{\sim}{A}_i/\mathcal{D}_{\omega_1}) : \ i < \lambda\})$$
$$\text{and } \mu \text{ is strong limit}\}$$

So by the assumption of 2.4 for some regular (hence strongly inaccessible) $\mu \in C$ we have $\mu = \sup(\mu \cap C)$, and $(*)^a_{ab}[\mu, C \cap \mu]$. For part (1), by 2.3, $\{\underset{\sim}{A}_i : i \in I\} \in \Xi_\mu$, and the rest is easy. For part (2) similarly using 2.8. $\square_{2.4}$

2.5 Theorem. Suppose λ is strongly inaccessible, $\bar{S} = \langle S_1, S_2, S_3\rangle$ a partition of ω_1, S_1 stationary and

$W^* \stackrel{\text{def}}{=} \{\mu < \lambda : \mu$ strongly inaccessible and $(*)^b_{ab}[\mu]\}$ is a stationary subset of λ.

1) *Then* for some forcing notion P:

(a) $|P| = \lambda$, P satisfies the λ-c.c.

(b) P is semiproper.

(c) $\Vdash_P$ "$\mathcal{P}(\omega_1)/(\mathcal{D}_{\omega_1} + S_1)$ is W^*-layered" (see XIII 3.1A(4), (5)).

(d) P is S_3-complete hence if S_3 is stationary, then P adds no new ω-sequences of ordinals.

(e) $\Vdash_P$ "W^* is a stationary subset of $\{\delta < \aleph_2 = \lambda : \mathrm{cf}(\delta) = \aleph_1\}$".

2) Hence, if Q is the forcing notion of shooting a club through $\{\delta < \aleph_2 : \mathrm{cf}(\delta) = \aleph_0\} \cup W^*$ in the universe V^P, then in $V^{P * \underset{\sim}{Q}}$ we have: $\mathcal{P}(\omega_1)/(\mathcal{D}_{\omega_1} + S_1)$ is layered (see XIII 3.1A(4),(5)) (and hence e.g. there is a uniform ultrafilter E on ω_1 such that $\aleph_0^{\omega_1}/E = \aleph_1$ so E not regular; by [FMSh:252]).

Proof. 1) Similar to XIII 3.1 (see on history there).

We define by induction on $i < \kappa, P_i, \underset{\sim}{Q}_i, \mathbf{t}_i$ such that:

(A) $\bar{Q}^\alpha = \langle P_i, \underset{\sim}{Q}_j, \mathbf{t}_j : \mathbf{t}_j : i \leq \alpha, j < \alpha \rangle$ is an S_1-suitable iteration (see Definition XIII 2.1).

(B) $\mathbf{t}_\alpha$ is 1 iff: α is strongly inaccessible, $[i < \alpha \Rightarrow |P_i| < \alpha]$ and $\Vdash_{P_\alpha}$ "$\mathfrak{B}^{V^{P_\alpha}} \restriction S_1$ satisfies the α-c.c. i.e. $\aleph_2$-c.c.".

(C) $\underset{\sim}{Q}_\alpha$ is defined , in V^{P_α}, as (where $\kappa_{\alpha+1}$ is the first strongly inaccessible $> |P_\alpha|$) $Q^0_\alpha *$ SSeal$(\langle \mathfrak{B}^{P_i} : i \leq \alpha, \mathbf{t}_i = 1 \rangle, S_1, \kappa_{\alpha+1})$ (see Definition XIII 2.4(5)) where Q^0_α is the product with countable support of $\{\mathrm{seal}(\Xi) : \Xi \in \Xi_\alpha\}$ (defined as in the proof of 2.4(1); or use Ξ'' from 2.4A(2)).

We should prove by induction on α that $\bar{Q}^\alpha$ is an S_1-suitable iteration.

first case for $\alpha = 0$ - this is trivial.

second case for α limit - this holds by XIII 2.3(1).

third case for $\alpha = \beta + 1, \mathbf{t}_\beta = 0$.

We should repeat the proof of XIII 2.14(1); we do this case in details.

Let χ be regular large enough, $i < \beta, G_{i+1} \subseteq P_{i+1}$ generic over V, in $V[G_{i+1}]$, N is a countable elementary submodel of $(H(\chi)[G_{i+1}], \in, <^*_\chi)$ such that $\bar{Q}^\alpha \in N$, $p \in P_\alpha/G_{i+1}, p \in N$.

We should find $q, p \le q \in P_\alpha/G_{i+1}$, and q is $(N[G_{i+1}], P_\alpha/G_{i+1})$-generic. By repeating the use XIII 2.12 ω times, we can find $q_0 \in P_\beta/G_{i+1}, p{\restriction}\beta \le q_0$ such that if $G_\beta \subseteq P_\beta$ is generic over $V, G_{i+1} \cup \{q_0\} \subseteq G_\beta$ then:

(*) in $V[G_\beta]$, there is $N', N \subseteq N' \prec (H(\chi)[G_\beta], \in, <^*_\chi), N'$ countable, $N' \cap \omega_1 = N \cap \omega_1$, and: for every $\Xi \in N' \cap H(\kappa)$ a dense subset of $\mathfrak{B}_\gamma$ for some $\gamma \in N' \cap \beta$, such that $\mathbf{t}_\gamma = 1$ we have $N' \cap \omega_1 \in \bigcup_{A \in N' \cap \Xi} A$.

In $V[G_\beta]$ we can find $p_n \in \underset{\sim}{Q}^0_\beta[G_\beta], p_n \in N', p_n \le p_{n+1}, p_0$ the $\underset{\sim}{Q}^0_\beta[G]$-component of $p(\beta)$, such that

(a) *if* $\mathcal{I} \in N'$ is a dense subset of $\underset{\sim}{Q}^0_\beta[G_\beta]$ *then* for some $n, p_n \in \mathcal{I}$.

(b) if Ξ is a $\underset{\sim}{Q}^0_\beta[G_\beta]$-name of a pre-dense subset of $\mathfrak{B}^{P_\gamma}, \gamma \in \beta \cap N, \mathbf{t}_\gamma = 1$, *then* for some n and A, $p_n \Vdash_{\underset{\sim}{Q}^0_\beta[G_\beta]}$ "$A \in \underset{\sim}{\Xi}$" and $N' \cap \omega_1 \in A$.

By standard bookkeeping there are no problem; taking care of an instance of (b) is just like the proof of XIII 2.9, as

(**) *if* $\gamma \in \beta \cap N', \mathbf{t}_\gamma = 1, \Xi \in N'$ is a pre-dense subset of $\mathfrak{B}^{P_\gamma}, \omega_1 \setminus S \in \Xi$ *then* $N' \cap \omega_1 \in \bigcup_{A \in \Xi \cap N'} A$.

Why does this hold? As β is strongly inaccessible $\bigwedge_{\gamma<\beta} |P_\gamma| < \beta$, we know $\mathfrak{B}^{P_\beta} = \bigcup_{\gamma<\beta} \mathfrak{B}^{P_{\gamma+1}}$, hence $[\mathbf{t}_\gamma = 1 \Rightarrow \mathfrak{B}^{P_\gamma} \lessdot \mathfrak{B}^{P_\beta}]$ and $|\mathfrak{B}^{P_\gamma}| = \aleph_1$ in V^{P_β}.

<u>fourth case</u> $\alpha = \beta + 1, \mathbf{t}_\beta = 1$.

$\underset{\sim}{Q}^0_\beta$ is semiproper by XIII 2.8(3) and SSeal$(\langle \mathfrak{B}^{P_\gamma} : \gamma \le \beta, \mathbf{t}_\gamma = 1\rangle, S_1, \kappa_{\beta+1})$ is the same as SSeal$(\mathfrak{B}^{P_\beta}, S_1, \kappa_{\beta+1})$ which is semiproper by XIII 2.14(1).

* * *

Now if $\lambda \in W^*$, $\bigwedge_{\gamma<\lambda} [|P_\gamma| < \lambda]$, then exactly as in the proof of Theorem 2.4, $\Vdash_{P_\lambda}$ "$\mathfrak{B}^{P_\lambda}$ satisfies the λ-c.c.", hence $\mathbf{t}_\lambda = 1$, hence $\mathfrak{B}^{P_\lambda} \lessdot \mathfrak{B}^{P_\kappa}$.

As $\mathfrak{B}^{P_\lambda} = \mathfrak{B}^{\bar{Q}{\restriction}\lambda}$ and $\langle \bar{\mathfrak{B}}^{\bar{Q}{\restriction}\alpha} : \alpha < \kappa\rangle$ is increasing continuous with limit $\mathcal{B}^{P_\kappa}$, clearly P_κ is as required.

2) No problem (or see proof of XIII 3.1). $\square_{2.5}$

2.5A Remark. Of course, we know $|P_i| \leq$ first strongly inaccessible $\geq |P_i|$ (by a variant could have gotten $|P_i| \leq \beth_{i+1}$).

2.6 Theorem. Suppose λ strongly inaccessible and the set
$W^* = \{\mu < \lambda : \quad \mu$ measurable and $(*)^b_{ab}[\mu].\}$
is not only stationary, but for stationarity many $\kappa < \lambda$, $W^* \cap \kappa$ is stationary.

Let $\langle S_1, S_2, S_3 \rangle$ be a partition of ω_1, S_1 is stationary.
Then for some forcing notion P

(a) $|P| = \lambda$, P satisfies the λ- cc.,

(b) P is semiproper.

(c) $\Vdash_P$ "$\mathcal{P}(\omega_1)/(\mathcal{D}_{\omega_1} + S_1)$ is the Levy algebra" (i.e. as isomorphic to the complete Boolean algebra which Levy $(\aleph_0, < \aleph_2)$ generate).

(d) P is pseudo $(*, S_3)$-complete hence if S_3 is stationary then P adds no reals.

Proof. Similar to XIII 3.7.

* * *

Of course we can translate our assumptions to a standard large cardinal hierarchy, essentially by Shelah and Woodin [ShWd:241], i.e. we note:

2.7 Fact. 1) Suppose $\mathrm{Pr}_a(\lambda, f)$ (see Definition 1.14), C a club of λ, $[\delta \in C \Rightarrow (H(\mu), \in, <^*_\mu) \prec (H(\lambda), \in, <^*_\lambda)]$ and $f(i) \leq \mathrm{Min}(C \setminus (i+1))$. *Then* $(*)^a[\lambda, C]$ (see Definition 2.2(1)).
2) As $\mathrm{Pr}_a(\lambda, f)$ is preserved by forcing of cardinality $< \lambda$, we can deduce in (1) also $(*)^a_{ab}[\lambda, C]$ (see Definition 2.2(2)).
3) If λ is a Woodin cardinal i.e. $\mathrm{Pr}_b(\lambda)$ (see Definition 1.15) *then* $(*)^b_{ab}[\lambda]$ (see definition in Theorem 2.4).

Proof. 1) By 2.8 below, condition (C) of 2.1 holds in the cases refered to in Definition 2.2, hence (see Definition 2.2(1)) we get $(*)^a[\lambda, C]$.
2) Easy.

3) See Definition 1.15 and part (2) of 2.7. $\square_{2.7}$

2.7A Remark. If you want to get versions of 2.4, 2.5, 2.6 without §1 + 2.1, you can use 2.8 below (+2.9).

2.8 Claim. Sealing is Semiproper Criterion.

Suppose

(i) $\bar{P} = \langle P_i : i < \lambda \rangle$ is $\lessdot$-increasing sequence of forcing notion, $P_i \in H(\lambda)$ and $\Vdash_{P_i}$ "$\aleph_1^V$ is a cardinal", and for any $j < \lambda$ for some i, $j < i < \lambda$ and $2^{\aleph_2}$ of V^{P_j} is collapsed to $\aleph_1$ in V^{P_i}.

(ii) $\mathrm{Pr}_a(\lambda, f, D)$ (defined in Definition 1.14).

(iii) $\{\delta < \lambda : P_\delta = \bigcup_{i<\delta} P_i\} \in D$ [hence $P_i \lessdot P_\lambda$ where $P_\lambda \stackrel{\text{def}}{=} \bigcup_{i<\lambda} P_i$ and P_λ satisfies the λ-c.c.]

Hence

$$B_0 \stackrel{\text{def}}{=} \{\mu < \lambda : \text{(a) } \mu \text{ is a strong limit}$$
$$\text{(b) } P_\mu = \bigcup_{i<\mu} P_i$$
$$\text{(c) } P_\mu \text{ satisfies the } \mu\text{-c.c.}$$
$$\text{(d) } \Vdash_{P_\mu} \text{“}\mu = \aleph_2\text{” and}$$
$$\text{(e) for } A \in \mathcal{P}(\omega_1)^{V^{P_\mu}} \text{ the statement}$$
$$\text{“}A \subseteq \omega_1 \text{ is stationary”}$$
$$\text{is preserved by } P/P_\mu\} \in D$$

hence

$$B_1 \stackrel{\text{def}}{=} \{\delta < \lambda : P/P_\delta \text{ preserves the stationarity of } A \in \mathcal{P}(\omega_1)^{V^{P_\delta}}\}$$

is unbounded in λ.

(iv) $B = \{\alpha < \lambda : P_\lambda / P_\alpha \text{ is semiproper}\}$ is unbounded in λ.

(v) $\{\delta < \lambda : P_\lambda / P_\delta$ does not destroy semi stationarity (see Definition XIII 1.1(3)) of subsets of $\mathcal{S}_{\leq\aleph_0}(2^{\aleph_2})$) (where $2^{\aleph_2}$ is computed in V^{P_δ})$\} \in D$. By Claim XIII 1.4, (P_λ/P_δ) being semiproper is enough.

(vi) $\Vdash_{P_\lambda}$ "$\bar{\underset{\sim}{A}} = \langle \underset{\sim}{A}_i : i < \lambda \rangle$ is a maximal antichain of $\mathcal{D}_{\omega_1}$" and

(vii) The following set belongs to D:

$\{\delta < \lambda : f(\delta)$ is a strong limit and for some β satisfying $\delta < \beta < f(\delta)$ we have $\Vdash_{P_\beta}$ "$(2^{\aleph_2})^{V^{P_\delta}}$ is collapsed to $\aleph_1$", $P_\beta \in H(f(\delta))$ and for every P_β-name $\underset{\sim}{A}$ of a subset of ω_1 stationary in V^{P_λ}, for some $\alpha(*) \in B$ (see clause (iv)) and $\underset{\sim}{i}$ we have: $\underset{\sim}{A}_{\underset{\sim}{i}} \cap \underset{\sim}{A}$ is forced to be stationary, $\underset{\sim}{i}$ and $\underset{\sim}{A}_{\underset{\sim}{i}}$ are $P_{\alpha(*)}$-names, $\mu < \alpha(*) < f(\delta)$, and $\{\underset{\sim}{A}, P_{\alpha(*)}, \alpha(*), \underset{\sim}{i}, \underset{\sim}{A}_{\underset{\sim}{i}}\} \in H(f(\delta))\}$.

Then $\Vdash_{P_\lambda}$ "$\bar{\underset{\sim}{A}}$ is semiproper" (see XIII 2.4(6)).

Proof. Assume the conclusion fails. Let $\mathbf{j} : V \to M$ be an elementary embedding, M a transitive class, and

$\left[H\big((\mathbf{j}(f))(\lambda)\big)\right]^V \subseteq M$ and $D = \{A \subseteq \lambda : \ \lambda \in \mathbf{j}(A)\}$ and $M^{<\lambda} \subseteq M$

(exists as $\mathrm{Pr}_a(\lambda, f, D)$ holds by assumption (ii)).

By assumption (iii) we have $M \models$ "$P_\lambda = (\mathbf{j}(\bar{P}))\ (\lambda)$", let $\mathbf{j}(\bar{P}_\lambda) = \langle P_i : i < \mathbf{j}(\lambda)\rangle$ and $P_{\mathbf{j}(\lambda)} = \mathbf{j}(P_\lambda) = \bigcup_{i<\mathbf{j}(\lambda)} P_i$ (Note: the two definitions of P_i for $i \le \lambda$ are compatible by the beginning of this sentence). Similarly let $\mathbf{j}(\bar{A}) = \langle \underset{\sim}{A}_i : i < \mathbf{j}(\lambda)\rangle$.

By (vii) we have $M \models$ "$(\mathbf{j}(f))(\lambda)$ is strong limit", so as $[H((\mathbf{j}(f))(\lambda))]^V \subseteq M$, really $\mathbf{j}(f)(\lambda)$ is strong limit in V so for statements in $H(\mathbf{j}(f)(\lambda))$ we can move freely between V and M. Let $G_{\mathbf{j}(\lambda)} \subseteq P_{\mathbf{j}(\lambda)}$ be generic over M, so we let $G_i \stackrel{\text{def}}{=} G_{\mathbf{j}(\lambda)} \cap P_i$.

Clearly $G_{\lambda+1} \subseteq P_{\lambda+1}$ is generic over V because, generally $P_i \in H\big(\mathbf{j}(f)(\lambda)\big)$ implies G_i is generic over V and $P_{\lambda+1} \in H((\mathbf{j}(f))(\lambda))$ by (vii). Until almost the end we shall use G_λ only. Note: in $V[G_\lambda]$ we have $M[G_\lambda]^{<\lambda} \subseteq M[G_\lambda]$ because P_λ satisfies the λ-c.c. (see (iii)).

Remembering (vi), in $V[G_\lambda]$, $\bar{\underset{\sim}{A}}[G_\lambda] = \langle \underset{\sim}{A}_i[G_\lambda] : i < \lambda\rangle$ is a maximal antichain of $\mathcal{D}_{\omega_1}$ and seal$(\bar{A})$ has cardinality $(2^{\aleph_1})^{V[G_\lambda]} = \lambda = \aleph_2^{V[G_\lambda]}$ (remember (i)). Let

$S \stackrel{\text{def}}{=} \{N : N \prec (H(\lambda^+)^{V[G_\lambda]}, \in, <^*),\quad N$ countable, and there is no N_1,
$N \prec N_1 \prec (H(\lambda^+)^{V[G_\lambda]}, \in, <^*), N_1$ countable and $N_1 \cap \omega_1 = N \cap \omega_1 \in \bigcup_{i \in N_1} \underset{\sim}{A}_i[G_\lambda]\}$

In $V[G_\lambda]$ the set S is semi-stationary (subset of $\mathcal{S}_{\leq\aleph_0}(H(\lambda^+)^{V[G_\lambda]})$, as we are assuming that the conclusion failed — by XIII 1.3; we note: $H(\lambda^+)^{V[G_\lambda]} = H(\lambda^+)^{M[G_\lambda]}$. Clearly λ belongs to the set defined in assumption (vii), so in M there is β as there, so $\lambda < \beta < f(\lambda)$, $\Vdash_{P_\beta}$ "$(2^{\aleph_2})^{V^{P_\lambda}}$ is collapsed to $\aleph_1$", $P_\beta \in H((\mathbf{j}(f))(\lambda))$ and the last condition there hold.

So there is a P_β-name $\langle \underset{\sim}{a}_\zeta : \zeta < \omega_1\rangle$ such that:

$\Vdash_{P_\beta}$ "$\langle \underset{\sim}{a}_\zeta : \zeta < \omega_1\rangle$ is increasing continuous, each $\underset{\sim}{a}_\zeta$ countable,

$\bigcup_{\zeta<\omega_1} \underset{\sim}{a}_\zeta = H(\lambda^+)^{V[G_\lambda]}$".

Let $\underset{\sim}{A} = \{\zeta : \quad (\exists N \in S)[\omega_1 \cap \underset{\sim}{a}_\zeta \subseteq |N| \subseteq \underset{\sim}{a}_\zeta]\}$, clearly it is a P_β-name.

By assumption (v), $\Vdash_{P_\beta}$ "$\underset{\sim}{A}$ is a stationary subset of ω_1 " hence by the last condition in (vii) for some $\underset{\sim}{i}$, $\alpha(*)$ we have: $\alpha(*) \in \mathbf{j}(B)$, $\underset{\sim}{i}$ and $\underset{\sim}{A}_{\underset{\sim}{i}}$ are $P_{\alpha(*)}$-names, $\beta < \alpha(*) < \mathbf{j}(f)(\lambda)$ and $\{\underset{\sim}{i}, \alpha(*), \underset{\sim}{A}, \underset{\sim}{A}_{\underset{\sim}{i}}, \quad P_{\alpha(*)}\} \in H\Big((\mathbf{j}(f))(\lambda)\Big)$. So for some regular μ (in M and in V) we have $\mu < (\mathbf{j}(f))(\lambda)$ and this set $\in H(\mu) = H(\mu)^M$ moreover $\mathcal{P}(P_{\alpha(*)}) \in H(\mu)$. So in $M[G_{\alpha(*)}]$, we have $\underset{\sim}{A}_{\underset{\sim}{i}}[G_{\alpha(*)}] \cap \underset{\sim}{A}[G_\beta]$ is a stationary subset of ω_1. Again this holds in $V[G_{\alpha(*)}]$ too, (and of course in $V[G_{\alpha(*)}]$ $\aleph_1$ is not collapsed). Let, in M, $w = \{i < \alpha(*) : \underset{\sim}{A}_i$ is a $P_{\alpha(*)}$-name$\}$, so $w \in H(\mu)^M = H(\mu)$.

So in $M[G_\lambda]$

$$S_1 \stackrel{\text{def}}{=} \{N \prec \left(H(\mu)^{M[G_\lambda]}, \in, <^*, M, G_\lambda\right) : N \text{ is countable} \\ \{\mathbf{j}(\bar{Q}){\restriction}\alpha(*), \bar{\underset{\sim}{A}}{\restriction}w, \underset{\sim}{A}_{\underset{\sim}{i}}\} \in N \\ \text{and for some } p \in P_{\alpha(*)}/G_\lambda, \\ p \text{ is } (N, P_{\alpha(*)}/G_\lambda)\text{-semi-generic} \\ \text{and } p \Vdash \text{“} N \cap \omega_1 \in \underset{\sim}{A} \cap \underset{\sim}{A}_{\underset{\sim}{i}} \text{”}\}$$

is stationary subset of $\Big[\mathcal{S}_{\leq\aleph_0}(H(\mu))\Big]^{M[G_\lambda]}$ in $M[G_\lambda]$, hence in $V[G_\lambda]$ too. Note that $\mathbf{j}$ induces a unique elementary embedding $\mathbf{j}^+$ from $V[G_\lambda]$ into $M[G_{\mathbf{j}(\lambda)}]$, $\mathbf{j}^+$ is really $\underset{\sim}{\mathbf{j}}^+$, a $P_{\mathbf{j}(\lambda)}$-name, and if $x \in H(\lambda^+)^{M[G_\lambda]}$ then $\underset{\sim}{\mathbf{j}}^+(x) \in M[G_\lambda]$, that is the name belong, and it can be considered a $P_{\mathbf{j}(\lambda)}/G_\lambda$-name (but $\underset{\sim}{\mathbf{j}}^+ \notin M[G_\lambda]$).

In $V[G_\lambda]$,

$$S_2 \stackrel{\text{def}}{=} \{N \prec \left(H[\beth_3(\mathbf{j}(\lambda^+))^+]^{M[G_\lambda]}, \in, <^*, M, G_\lambda\right) : N \text{ countable, and}$$
$$\langle \underset{\sim}{a}_\zeta : \zeta < \omega_1\rangle,\ \mathbf{j}(\bar{P}),\ \alpha(*),$$
$$\underset{\sim}{A}, \underset{\sim}{A}_{\underset{\sim}{i}}, \text{ and } \underset{\sim}{\mathbf{j}}^+{\restriction}H(\lambda^+), H[\beth_2(\mathbf{j}(\lambda^+))], \bar{P}, G_\lambda$$
$$\text{belong to } N\}.$$

is in $\left[\mathcal{D}_{\leq\aleph_0}\left(H(\beth_3(\mathbf{j}(\lambda^+)))^+)^{M[G_\lambda]}\right)\right]^{V[G_\lambda]}$ and is a subset of $M[G_\lambda]$ (though not a member) as $V[G_\lambda] \models$ "$M[G_\lambda]^{<\lambda} \subseteq M[G_\lambda]$".

So there are $N_1 \in S_1$, $N_2 \in S_2$, such that $N_2{\restriction}H(\mu)^{M[G_\lambda]} = N_1$ and $p \in P_{\alpha(*)}/G_\lambda$ witnessing $N_1 \in S_1$ (see the definition of S_1). Let $\delta \stackrel{\text{def}}{=} N_1 \cap \omega_1$. Note: $N_1, N_2 \in M[G_\lambda]$.

Now as $p \Vdash_{P_{\alpha(*)}/G_\lambda}$ "$\delta \in \underset{\sim}{A}$", by the definition of $\underset{\sim}{A}$ there are q and b satisfying $p \leq q \in P_{\alpha(*)}/G_\lambda$, and $N \in S$, such that letting $b \stackrel{\text{def}}{=} |N|$, we have $q \Vdash$ "$\delta \subseteq b \subseteq \underset{\sim}{a}_\delta$" so $b \in M[G_\lambda]$ as $b \in S$.

Also as q is $(N_1, \ P_{\alpha(*)}/G_\lambda)$-semi-generic (being above p, as p witness $N_1 \in S_1$) and $\langle \underset{\sim}{a}_\zeta : \zeta < \omega_1\rangle \in N_1$ (as it belongs to N_2 and to $H(\mu)^{M[G_\lambda]}$) clearly

$$q \Vdash \text{"}\underset{\sim}{a}_\delta = N_1[\underset{\sim}{G}_{P_{\alpha(*)}/G_\lambda}] \cap H(\lambda^+)^{M[G_\lambda]}\text{"}.$$

[Why? As $\langle \underset{\sim}{a}_\zeta : \zeta < \omega\rangle \in N_1$ and $H(\lambda^+)^{M[G_\lambda]} \in N_1$, clearly the function $h_1 : H(\lambda^+)^{M[G_\lambda]} \to \omega_1$, $h_1(x) = \min\{\zeta < \omega_1 : x \in \underset{\sim}{a}_\zeta\}$ belongs to $N_1[\underset{\sim}{G}_{P_{\alpha(*)}/G_\lambda}]$ and also some function $h_2 : \omega_1 \times \omega \to H(\lambda^+)^{M[G_\lambda]}$ such that $\underset{\sim}{a}_\zeta = \{h_2(\zeta, n) : n < \omega\}$ belongs to $N_1[G_{P_{\alpha(*)}/G_\lambda}]$.]

Hence

$$q \Vdash_{P_{\alpha(*)}/G_\lambda} \text{"}b \subseteq N_1[\underset{\sim}{G}_{P_{\alpha(*)}/G_\lambda}] \cap H(\lambda^+)^{M[G_\lambda]}\text{"}.$$

As $N_1 \cap \mathcal{P}(P_{\alpha(*)}/G_\lambda) = N_2 \cap \mathcal{P}(P_{\alpha(*)}/G_\lambda)$ (power set in $M[G_\lambda]$), we can also replace in those statements N_1 by N_2. As $P_{\mathbf{j}(\lambda)}/P_{\alpha(*)}$ is semiproper in $M[G_\lambda]$ ($\alpha(*)$ being in $\mathbf{j}(B)$) there is q', $q \leq q' \in P_{\mathbf{j}(\lambda)}$ such that q' is $(N_2, P_{\mathbf{j}(\lambda)}/G_\lambda)$ semi-generic in $M[G_\lambda]$.

W.l.o.g. $q' \in G_{\mathbf{j}(\lambda)}$ as only G_λ was used. Work in $M[G_{\mathbf{j}(\lambda)}]$, remember $N_2 \in M[G_\lambda]$. So really $b \subseteq N_2[G_{\mathbf{j}(\lambda)}]$, now as $\mathbf{j}^+$ maps $N_2[G_{P_{\mathbf{j}(\lambda)}/G_\lambda}] \cap H(\lambda^+)^{M[G_\lambda]}$

into $N_2[G_{P_{j(\lambda)}/G_\lambda}]$ (see the definition of S_2) clearly $b_1 = j^{+\prime\prime}(b) = \{j^+(x) : x \in b\} \subseteq N_2[G_{j(\lambda)}]$. Now as $M[G_\lambda] \models$ "b is countable", necessarily $b_1 = j^+(b)$. By the properties of j^+, $b_1 = j^+(b) \in M[G_{j(\lambda)}]$; remember j^+ is the elementary embedding j induces from $V[G_\lambda]$ into $M[G_{j(\lambda)}]$, so as $b \in S$ we have: $M[G_{j(\lambda)}] \models$ "$b_1 \in j^+(S)$",

But as $q \leq q' \in G_{j(\lambda)}$, $N_2[G_{j(\lambda)}/G_\lambda]$, $\underset{\sim}{i}[G_{j(\lambda)}/G_\lambda]$ contradicts this. So we have finished proving 2.8. $\square_{2.8}$

When you want to accomplish other things by forcing remember XIII 1.10 (2):

2.9 Conclusion. 1) Assume

(i) $\bar{P} = \langle P_i : i < \lambda \rangle$ is $\lessdot$-increasing sequence of forcing notion, $P_i \in H(\lambda)$ and $\Vdash_{P_i}$ "$\aleph_1^V$ is a cardinal", and for any $j < \lambda$ for some i, $j < i < \lambda$ and $2^{\aleph_2}$ of V^{P_j} is collapsed to $\aleph_1$ in V^{P_i}, let $P_\lambda = \bigcup_{i<\lambda} P_i$,

(ii) $\mathrm{Pr}_b(\lambda)$, i.e. λ is a Woodin cardinal,

(iii) for a club of cardinals $\mu < \lambda$, if μ is strongly inaccessible then $P_\mu = \bigcup_{i<\mu} P_i$, P_λ / P_μ is semiproper.

Then in V^{P_λ}, every maximal antichain Ξ of $\mathcal{D}_{\omega_1}$ is semiproper i.e. seal(Ξ) is a semiproper forcing.

2) We can above replace (ii), (iii) by

(ii)′ $\mathrm{Pr}_b(\lambda, W)$,

(iii)′ $W = \{\delta < \lambda : P_\delta = \bigcup_{i<\delta} P_i$ and P_λ / P_δ is semiproper$\}$.

2.10 Concluding Remarks. Can we improve 2.6?

Note: we do not know imitate XIII 3.9 (on the Ulam property) as the supercompactness was used more deeply. But even trying to imitate XIII 3.7 (getting the Levy algebra, that is weakening the assumption of 2.6 to "for stationary many $\mu_0 < \lambda$, for stationary many $\mu_1 < \mu_0$ we have $(*)^a_{ab}[\mu_1]$) we have a problem: Is Nm semi proper? In 2.6 the measurability demand in the definition of W^* solves the problem. But it is natural and better to use $W^* = \{\mu < \lambda : \mu$ strongly inaccessible and $(*)^b_{[ab]}[\mu]\}$ or $W^{**} = \{\mu < \lambda : \mathrm{Pr}_b(\mu)\}$

To get such a theorem it is natural to use XV §3 to prove that the forcing does not collapse $\aleph_1$ and does not destroy stationary subsets of ω_1. If $S_3 = \emptyset$ we finish. To prove (d) - relativize Chapter XI to S_1 (as done in XI §8, or see XV). Still we have to check the parallel of 2.8. We intend to continue in [Sh:311].

XVII. Forcing Axioms

§0. Introduction

This chapter reports various researches done at different times in the later eighties. In Sect. 1, 2 we represent [Sh:263] which deals with the relationship of various forcing axioms, mainly SPFA = MM, SPFA $\nvdash$ PFA$^+$ (=Ax_1[proper]) but SPFA implies some weaker such axioms ($\mathrm{Ax}_1[\aleph_1$-complete], see 2.14, and more in 2.15, 2.16). See references in each section.

In sections 3, 4 we deal with the canonical functions (from ω_1 to ω_1) modulo normal filters on ω_1. We show in §3 that even PFA$^+$ does not imply Chang's conjecture [even is consistent with the existence of $g \in {}^{\omega_1}\omega_1$ such that for no $\alpha < \aleph_2$ is g smaller (modulo $\mathcal{D}_{\omega_1}$) than the α-th function]. Then we present a proof that Ax[α-proper] $\nvdash$ Ax [β-proper] where $\alpha < \beta < \omega_1$, β is additively indecomposable (and state that any CS iteration of c.c.c. and $\aleph_1$-complete forcing notions is α-proper for every α).

In the fourth section we get models of CH + "ω_1 is a canonical function" without $0^\#$, using iteration not adding reals, and some variation (say ω_1 is the α-th function, CH + $2^{\aleph_1} = \aleph_3$ $|\alpha| = \aleph_2$ (see 4.7(3)). The proof is in line of the various iteration theorems in this book, so here we deal with using large cardinals consistent with $V = L$.

Historical comments are introduced in each section as they are not so strongly related.

We recall definition VII 2.10: If φ is a property of forcing notions, $\alpha \leq \omega_1$ then we write $\mathrm{Ax}_\alpha[\varphi]$ for the statement:
whenever P is a forcing notion satisfying φ, $\langle \mathcal{I}_i : i < \omega_1 \rangle$ are pre-dense subsets of P, $\langle \underset{\sim}{S}_\beta : \beta < \alpha \rangle$ are P-names of stationary subsets of ω_1,
then there is a directed, downward closed set $G \subseteq P$ such that for all $i < \omega_1$, $\mathcal{I}_i \cap G \neq \emptyset$ and for all $\beta < \alpha$ the set $\underset{\sim}{S}_\beta[G]$ is stationary.
We write $\mathrm{Ax}[\varphi]$ for $\mathrm{Ax}_0[\varphi]$ and $\mathrm{Ax}^+[\varphi]$ for $\mathrm{Ax}_1[\varphi]$, PFA for Ax[proper], SPFA for Ax[semiproper], similarly PFA^+ and SPFA^+.

§1. Semiproper Forcing Axiom Implies Martin's Maximum

We prove that Ax[preserving every stationarity of $S \subseteq \omega_1$] = MM (= Martin maximum) is equivalent (in ZFC) to the older axiom Ax[semiproper] = SPFA (= semiproper forcing axiom).

1.1 Lemma. If $\mathrm{Ax}_1[\aleph_1\text{-complete}]$, P is a forcing notion satisfying $(*)_1$ (below) *then* P is semiproper, where
$(*)_1 \stackrel{\mathrm{def}}{=}$ "the forcing notion P preserves stationary subsets of ω_1".

1.1A Remark. 1) This is from Foreman, Magidor and Shelah [FMSh:240].
2) It follows that $\mathrm{SPFA}^+ = \mathrm{Ax}_1$[semiproper] is equivalent to MM^+ (compare [FMSh:240]). The conclusion is superseded by 1.2, but not the lemma.
3) The proof is very similar to III 4.2.
4) Of course every semiproper forcing preserves stationarity of subsets of ω_1 (see X 2.3(8)).

Proof. Clearly $\mathrm{Ax}_1[\aleph_1\text{-complete}]$ implies $\mathrm{Rss}(\aleph_1, \kappa)$ for any κ (see Defefinition XIII 1.5(1).). By XIII 1.7(3) "forcing with P does not destroy semi-stationarity of subsets of $\mathcal{S}_{<\aleph_1}(2^{|P|})$" implies P is semiproper. (So by 1.1A(4) these two properties are equivalent). $\square_{1.1}$

1.2 Theorem.
Ax [not destroying stationarity of subsets of ω_1] $\equiv$Ax [semiproper], i.e. MM (= Martin Maximum) $\equiv$ SPFA (i.e., proved in ZFC).

Proof. As every semiproper forcing preserves stationary subsets of ω_1 (X 2.3(8)), clearly MM $\Rightarrow$ SPFA. So it suffices to prove:

1.3 Lemma. [SPFA.]
Every forcing notion P satisfying $(*)_1$ is semiproper, where
$(*)_1 \stackrel{\text{def}}{=}$ "the forcing notion P preserves stationarity of subsets of ω_1".

Proof. We assume $(*)_1$. Without loss of generality the set of members (= conditions) of P is a cardinal $\lambda_0 = \lambda(0)$. Too generously, for $\ell = 0, 1, 2, 3$, let $\lambda_{\ell+1} = \lambda(\ell+1) = (2^{|H(\lambda_\ell)|})^+$. Let $<^*_{\lambda_\ell}$ be a well ordering of $H(\lambda_\ell)$, end extending $<^*_{\lambda_m}$ for $m < \ell$. Let

$$K_P^{\text{neg}} \stackrel{\text{def}}{=} \{N : N \prec (H(\lambda_2), \in, <^*_{\lambda_2}), \|N\| = \aleph_0, P \in N \text{ (hence } \lambda_0, \lambda_1 \in N) \text{ and}$$
$$\neg(\forall p \in P \cap N)(\exists q)[p \leq q \in P \text{ and } q \text{ is semi generic for } (N, P)]\}$$

and

$$K_P^{\text{pos}} \stackrel{\text{def}}{=} \{N : N \prec (H(\lambda_2), \in, <^*_{\lambda_2})), \|N\| = \aleph_0, P \in N \text{ (hence } \lambda_0, \lambda_1 \in N)$$
$$\text{and } \neg(\exists N')[N \prec N' \in K_P^{\text{neg}} \text{ and } N \cap \omega_1 = N' \cap \omega_1]\}.$$

We now define a forcing notion Q

$$Q \stackrel{\text{def}}{=} \{\langle N_i : i \leq \alpha\rangle : \alpha < \omega_1, N_i \in K_P^{\text{neg}} \cup K_P^{\text{pos}},$$
$$N_i \in N_{i+1}, \text{ and } N_i \text{ is increasing continuous in } i\}.$$

The order on Q is being an initial segment.
The rest of the proof of Lemma 1.3 is broken to facts 1.4 — 1.11.

1.4 Fact. If $P \in M_0 \prec (H(\lambda_3), \in, <^*_{\lambda_3}), \|M_0\| = \aleph_0$, *then* there is M_1 such that $M_0 \prec M_1 \prec (H(\lambda_3), \in, <^*_{\lambda_3}), \|M_1\| = \aleph_0, M_0 \cap \omega_1 = M_1 \cap \omega_1$ and $M_1 \restriction H(\lambda_2) \in K_P^{\text{neg}} \cup K_P^{\text{pos}}$.

Proof. As $P \in M_0$, clearly $\lambda_0 \in M_0$; hence $\lambda_1, \lambda_2 \in M_0$ hence $(H(\lambda_\ell), \in, <^*_{\lambda_\ell})$ belong to M_0 for $\ell = 0, 1, 2$, so $K_P^{\rm pos} \in M_0$ and $K_P^{\rm neg} \in M_0$. We can assume $M_0 \restriction H(\lambda_2) \notin K_P^{\rm pos}$, so by the definition of $K_P^{\rm pos}$ there is N' such that (abusing our notation) $M_0 \cap H(\lambda_2) = M_0 \restriction H(\lambda_2) \prec N' \in K_P^{\rm neg}$, $\|N'\| = \aleph_0$ and $N' \cap \omega_1 = (M_0 \restriction H(\lambda_0)) \cap \omega_1$; hence $N' \cap \omega_1 = M_0 \cap \omega_1$.
Let M_1 be the Skolem Hull of $M_0 \cup (N' \cap H(\lambda_1))$ in $(H(\lambda_3), \in, <^*_{\lambda_3})$. So

$$\begin{array}{llcl} H(\lambda_3): & M_0 & \xrightarrow{\hspace{4em}} & M_1 \\ H(\lambda_2): & M_0 \cap H(\lambda_2) & \prec\ N' & \uparrow \\ H(\lambda_1): & & & N' \cap H(\lambda_1) \end{array}$$

We claim that $M_1 \cap H(\lambda_1) = N' \cap H(\lambda_1)$. To prove this claim, let x be an arbitrary element of $M_1 \cap H(\lambda_1)$. Now x must be of the form $f(y)$, where f is a Skolem function of $(H(\lambda_3), \in, <^*_{\lambda_3})$ with parameters in M_0, and $y \in N' \cap H(\lambda_1)$ (note that $N' \cap H(\lambda_1)$ is closed under taking finite sequences). Note that f's definition may use parameters outside $H(\lambda_2)$, but $f' \stackrel{\rm def}{=} f \cap (H(\lambda_1) \times H(\lambda_1))$ belongs to $H(\lambda_2)$, so $f' \in M_0 \cap H(\lambda_2) \subseteq N'$, so also $x = f(y) = f'(y) \in N'$.
So we have

$$M_1 \cap \omega_1 = N' \cap \omega_1 = M_0 \cap \omega_1,$$

$$M_0 \prec M_1 \prec (H(\lambda_3), \in, <^*_{\lambda_3}),$$

$$||M_1|| = \aleph_0 \text{ (as } ||M_0||, ||N'|| = \aleph_0).$$

$$M_1 \cap H(\lambda_1) = N' \cap H(\lambda_1)$$

We can conclude by 1.5(1) below that $M_1 \restriction H(\lambda_2) \in K_P^{\rm neg}$, thus finishing the proof of Fact 1.4, as:

1.5 Subfact. 1) Suppose for $\ell = 0, 1$, N^ℓ is countable, $P \in N^\ell \prec (H(\lambda_2), \in, <^*_{\lambda_2})$ and $N^0 \cap H(\lambda_1) = N^1 \cap H(\lambda_1)$, *then* $N^1 \in K_P^{\rm neg} \Leftrightarrow N^2 \in K_P^{\rm neg}$.
2) Really, even $N^1 \cap \omega_1 \subseteq N^0 \subseteq N^1 \prec (H(\lambda_2), \in, <^*_{\lambda_2})$, $N^0 \in K_P^{\rm neg}$ implies $N^1 \in K_P^{\rm neg}$ (we can also fix the P in the definition of "$N \in K_P^{\rm neg}$").

Proof. 1) Because in "q is (N, P)-semi generic", not "whole N" is meaningful, just $N \cap \omega_1$, the set $N \cap P$ and the set of P-names of countable ordinals which

belong to N, hence (for "reasonably closed N") this depends only on $N \cap 2^{|P|}$ (even $|P|^{<\kappa}$, when $P \models \kappa$-c.c.).
2) Assume $N^1 \notin K_P^{\text{neg}}$. If $p \in P \cap N^0$ then $p \in P \cap N^1$, hence there is $q \in P$ which is (N^1, P)-semi generic, $q \geq p$. But as $N^0 \prec N^1$ have the same countable ordinals, q is also (N^0, P)-semi generic. $\square_{1.5,1.4}$

1.6 Fact. Q is a semiproper forcing.

Proof. Let $Q, P \in M \prec (H(\lambda_3), \in, <^*_{\lambda_3}), M$ countable. Let $p \in Q \cap M$. It is enough to prove that there is a q such that $p \leq q \in Q$ and q is semi generic for (M, Q).
Let $\delta = M \cap \omega_1$. By Fact 1.4 there is M_1, with $M \prec M_1 \prec (H(\lambda_3), \in, <^*_{\lambda_3})$, $||M_1|| = \aleph_0, M_1 \cap \omega_1 = \delta$ and $M_1 \restriction H(\lambda_2) \in K_P^{\text{neg}} \cup K_P^{\text{pos}}$. We can find by induction on n a condition $q_n = \langle N_i : i \leq \delta_n \rangle \in Q \cap M_1, q_n \leq q_{n+1}, q_0 = p$, such that: *for every* Q-name $\underset{\sim}{\gamma}$ of an ordinal which belongs to M_1 for some natural number $n = n(\underset{\sim}{\gamma})$ and ordinal $\alpha(\underset{\sim}{\gamma}) \in M_1$ we have $q_n \Vdash_Q$ "$\underset{\sim}{\gamma} = \alpha(\underset{\sim}{\gamma})$" and for every dense subset $\mathcal{I}$ of Q which belongs to M_1, for some n, $q_n \in \mathcal{J}$.
Now $q \stackrel{\text{def}}{=} \langle N_i : i \leq \delta^* \rangle$ with $\delta^* = \bigcup_{n<\omega} \delta_n$ and $N_{\delta^*} \stackrel{\text{def}}{=} \bigcup_{i<\delta^*} N_i$ will be (M_1, Q)-generic if it is a condition in Q at all, as for this the least obvious part is $N_{\delta^*} \in K_P^{\text{neg}} \cup K_P^{\text{pos}}$. Clearly (by 1.4) for each $x \in H(\lambda_2)$, $\mathcal{I}_x = \{\langle M'_i : i \leq j \rangle \in Q : x \in \bigcup_{i \leq j} M'_i\}$ is a dense subset of Q and $[x \in M_1 \cap H(\lambda_2) \Rightarrow \mathcal{I}_x \in M_1]$ and $\langle M'_i : i \leq j \rangle \in Q \cap M_1 \Rightarrow \bigcup_{i \leq j} M'_i \subseteq M_1$ (as M'_i, j are countable), and so $\bigcup_{i<\delta^*} N_i = M_1 \restriction H(\lambda_2)$, which belongs to $K_P^{\text{neg}} \cup K_P^{\text{pos}}$ by the choice of M_1.
Now $q \geq q_0 = p$; and, as q is (M_1, Q)-generic it is (M_1, Q)-semi generic hence as in the proof of 1.5 (or see X2.3(9)), as $M \prec M_1, M \cap \omega_1 = M_1 \cap \omega_1$, we know q is also (M, Q)-semi generic, as required. By the way, necessarily $\delta^* = \delta$. $\square_{1.6}$

1.7 Conclusion. [SPFA] There is a sequence $\langle N^*_i : i < \omega_1 \rangle$ such that

$$(\forall \alpha < \omega_1)[\langle N^*_i : i \leq \alpha \rangle \in Q].$$

Proof. By Fact 1.6 and SPFA (and as $\mathcal{I}_{\alpha_0} = \{\langle N_i : i \leq \alpha \rangle : \alpha \geq \alpha_0\}$ is a dense subset of Q for every $\alpha_0 < \omega_1$; which can be proved by induction on α_0 : for

$\alpha_0 = 0$ or $\alpha_0 = \beta + 1$ by Fact 1.4, for limit α_0 by the proof of Fact 1.6 or simpler).

$\square_{1.7}$

1.8 Observation. $i \subseteq N_i^*$ for $i < \omega_1$.

Proof. As $[i < j \Rightarrow N_i \subseteq N_j]$ and as $N_i^* \in N_{i+1}^*$ (see the definition of Q), we can prove this statement by induction on i. $\square_{1.8}$

1.9 Definition. $S \stackrel{\text{def}}{=} \{i < \omega_1 : N_i^* \in K_P^{\text{neg}}\}$.

1.10 Fact. S is not stationary.

Proof. Suppose it is; then for every $i \in S$ for some $p_i \in N_i^* \cap P$ there is no (N_i^*, P)-semi-generic q such that $p_i \leq q \in P$. By Fodor's lemma (as N_i^* is increasing continuous and each N_i^* is countable), for some $p \in \bigcup_{i<\omega_1} N_i^* \cap P$ the set $S_p \stackrel{\text{def}}{=} \{i \in S : p_i = p\}$ is stationary.

If $p \in G \subseteq P$ and G generic over V, then in $V[G]$ we can find an increasing continuous sequence $\langle N_i : i < \omega_1\rangle$ of countable elementary submodels of $(H^V(\lambda_2), \in, <^*_{\lambda_2}, G)$ (with G as a predicate), $N_i^* \subseteq N_i$. As P preserves stationarity of subsets of ω_1, and $E = \{i : N_i^* \cap \omega_1 = N_i \cap \omega_1 = i\}$ is a club of ω_1 (in $V[G]$), and $S_p \subseteq \omega_1$ is stationary (in V, hence in $V[G]$), it follows that there is $\delta \in S_p$ with $N_\delta^* \cap \omega_1 = N_\delta \cap \omega_1 = \delta$. As this holds in $V[G], p \in G$, clearly there is $q \in G, q \geq p$, such that $q \Vdash$ "δ and $\langle \underset{\sim}{N}_i : i < \omega_1\rangle$ are as above". As $q \Vdash$ "$N_\delta^* \subseteq N_\delta^*[\underset{\sim}{G}] \subseteq \underset{\sim}{N}_\delta$ and $\delta \in E$", also $q \Vdash$ "$N_\delta^* \cap \omega_1 = N_\delta^*[\underset{\sim}{G}] \cap \omega_1$", so q is (N_δ^*, P)-semi generic, contradiction to the definition of S and K_P^{neg} and the choice of $p_\delta = p$. $\square_{1.10}$

1.11 Fact. P is semiproper.

Proof. As S is not stationary, for some club $C \subseteq \omega_1, (\forall \delta \in C) N_\delta^* \in K_P^{\text{pos}}$. Now if $M \prec (H(\lambda_3), \in, <^*_{\lambda_3})$ is countable, and $P, \langle N_i^* : i < \omega_1\rangle, C$ belong to M, then $M \cap \bigcup_{i<\omega_1} N_i^* = N_\delta^*$ for some $\delta \in C$; hence $N_\delta^* \subseteq M \restriction H(\lambda_2)$; as both N_δ^*

and $M{\restriction}H(\lambda_2)$ are elementary submodels of $\big(H(\lambda_2), \in, <^*_{\lambda_2}\big)$ we get

$$N^*_\delta \prec M{\restriction}H(\lambda_2) \prec \big(H(\lambda_2), \in, <^*_{\lambda_2}\big).$$

Clearly $N^*_\delta \cap \omega_1 = \delta = M \cap \omega_1$. As $M{\restriction}H(\lambda_2)$ is countable and by the meaning of "$N^*_\delta \in K^{\rm pos}_P$" we have $M{\restriction}H(\lambda_2) \notin K^{\rm neg}_P$, i.e., for every $p \in P \cap M (= P \cap (M{\restriction}H(\lambda_2))))$ there is an $(M{\restriction}H(\lambda_2), P)$-semi-generic $q, p \le q \in P$. Necessarily q is (M, P)-semi-generic (as in the proof of 1.5(1)); this is enough. $\square_{1.11,1.3,1.2}$

1.12 Conclusion. SPFA implies $\mathcal{P}(\omega_1)/\mathcal{D}_{\omega_1}$ is $\aleph_2$-saturated i.e. satisfies the $\aleph_2$-c.c.

Proof. Actually it follows by Foreman Magidor Shelah [FMSh:240], and Theorem 1.2, but as this is a book we give a proof.

Let $\Xi \subsetneq \mathcal{P}(\omega_1)$ be a maximal antichain modulo $\mathcal{D}_{\omega_1}$. Remember seal(Ξ) = $\{\langle(\gamma_i, a_i) : i \le \alpha\rangle : \alpha < \omega_1,\ a_i$ is a countable subset of Ξ, non empty for simplicity, $\gamma_i < \omega_1, a_i$ and γ_i are strictly increasing continuous in i, and for limit $\delta \le \alpha$ we have $\gamma_\delta \in \bigcup_{i<\delta} \bigcup_{A \in a_i} A\}$. This forcing is S-complete for every $S \in \Xi$ (see XIII 2.8) hence does not destroy the stationarity of subsets of ω_1. Hence by 1.3 seal(Ξ) is semiproper.

Now $\mathcal{I}_i = \{\bar a \in {\rm seal}(\Xi) : \ell g(\bar a) \ge i\}$ is a dense subset of seal(Ξ). So by SPFA there is a directed $G \subsetneq {\rm seal}(\Xi)$ satisfying $\bigwedge_{i<\omega_1} G \cap \mathcal{I}_i \neq \emptyset$. Let $\bigcup G$ be $\langle(\gamma_i, a_i) : i < \omega_1\rangle$. We claim $\Xi = \bigcup\{a_i : i < \omega_1\}$. Let $C \stackrel{\rm def}{=} \{\gamma_i : i = \gamma_i = \omega i$ is a limit$\}$, $a_i = \{A_\alpha : \alpha < \omega i\}$, $A \stackrel{\rm def}{=} \{\delta < \omega_1 : (\exists i < \delta)(\delta \in A_i)\}$. Now if $S \in \Xi \setminus \{A_i : i < \omega_1\}$, then for all $i < \omega_1$, $S \cap A_i$ is nonstationary, so also $S \cap A$ is nonstationary, which is impossible as $C \subseteq A$ and C is a club. $\square_{1.12}$

§2. SPFA Does Not Imply PFA$^+$

It is folklore that in the usual forcing for PFA(=Ax[proper]) (or SPFA= Ax[semiproper]) any subsequent reasonably closed forcing preserves PFA (or

SPFA). Magidor and Beaudoin refine this, showing that starting from a model of PFA, forcing a stationary subset of $\{\delta < \omega_2 : \mathrm{cf}(\delta) = \aleph_0\}$ by

$$P = \{h : h \text{ a function from some } \alpha < \omega_2 \text{ to } \{0,1\} \text{ such that :}$$
$$\text{for all } \delta \in S^2_1 \text{ we have : } h^{-1}(\{1\}) \cap \delta \text{ is not stationary in } \delta\}$$

(ordered by inclusion) produces a stationary subset of $\{\delta < \omega_2 : \mathrm{cf}(\delta) = \aleph_0\}$ which does not reflect, and this still preserves PFA but easily makes PFA$^+$ (and SPFA) fail.

We can also start with $V \models$PFA, and force $h : \omega_2 \to \omega_1$ such that no $h^{-1}(\{\alpha\}) \cap \delta$ is stationary in δ, where $\alpha < \omega_1, \delta < \omega_2$, and $\mathrm{cf}(\delta) = \aleph_1$.

It had remained open whether SPFA $\vdash$ SPFA$^+$ and we present here the solution, first starting with a supercompact limit of supercompacts and then only from one supercompact. I thank Todorcevic and Magidor for asking me this question.

2.1 Theorem. Suppose κ is a supercompact limit of supercompacts. *Then,* in some generic extension, SPFA holds but PFA$^+$ fails.

The proof is presented in 2.3 - 2.9.

Overview of the Proof. Let f^* be a Laver diamond for κ (see Definition VII 2.8, as Laver shows w.l.o.g. it exists). Our proof will unfold as follows. We shall first define a semiproper iteration $\bar{Q}^\kappa$. Now $\Vdash_{P_\kappa}$ "SPFA" is as in the proof of X 2.8. We then define in V^{P_κ} a proper forcing notion R and an R-name $\underset{\sim}{S}$, $\Vdash_R$ "$\underset{\sim}{S} \subseteq \omega_1$ is stationary". We then show, that for no directed $G \subseteq R$ in V^{P_κ} is $\underset{\sim}{S}[G]$ well defined (i.e., $(\forall i < \omega_1)(\exists p \in G)[p \Vdash_R$ "$i \in \underset{\sim}{S}$" or $p \Vdash_R$ "$i \notin \underset{\sim}{S}$"$])$, and stationary (i.e., $\{i < \omega_1 : (\exists p \in G) p \Vdash$ "$i \in \underset{\sim}{S}$"$\}$ is stationary).

Before we start our iteration, we will define several forcing notions (which we will use later when we construct R, and also during the iteration), and we will explain some basic properties of these forcing notions.

Convention. Trees T will be such that members are sequences with the order being $\triangleleft$ (initial segments) and T closed under initial segments so $\ell g(\eta)$ is the level of η in T. But later we will use trees T whose members are sets of ordinal ordered by initial segments, so we can identify a name η if η is strictly increasing sequence of ordinals, $a = \mathrm{Rang}(\eta)$.

2.2 Fact. Let T be a tree of height ω_1, $\kappa \geq \aleph_1$ with $\kappa = 2^{\aleph_2}$ if not said otherwise. Let $P = R_1 * \underset{\sim}{R}_2$, where R_1 is Cohen forcing and R_2 is Levy$(\aleph_1, \kappa)$ (computed in V^{R_1}). *Then* every ω_1-branch of T in V^P is already in V.

Proof. Well-known and included essentially in the proof of III 6.1.

2.3 Definition. Let T be a tree of height ω_1 with $\aleph_1$ nodes and $\leq \aleph_1$ many ω_1-branches $\{B_i : i < i^* \leq \omega_1\}$ and let $\{y_i : i < \omega_1 \text{ and } [i < 2i^* \Rightarrow i \text{ odd}]\}$ list the members of T such that: $[y_j <_T y_i \Rightarrow j < i]$. Let B^*_i be: B_j if $i = 2j$, $j < i^*$ or $\{y_j\}$ if y_j is defined. Let $B'_j = B^*_j \setminus \bigcup_{i<j} B^*_i$, $x_j = \min(B'_j)$ if $B'_j \neq \emptyset$ so that the sets B'_j are disjoint nonempty end segments of some branch $B_{j'}$, or the singletons $\{y_j\}$ or $\emptyset$; let $B'_j \neq \emptyset \Leftrightarrow i \in w$ and so $\langle B'_j : j \in w\rangle$ form a partition of T. Let $A = \{x_i : i \in w\}$ (so A does not include any linearly ordered uncountable set). The forcing "sealing the branches of T" is defined as (see proof of 2.4(3)):

$$P_T = \{f : f \text{ a finite function from } A \text{ to } \omega, \text{ and}$$
$$\text{if } x < y \text{ are in } \mathrm{Dom}(f), \text{ then } f(x) \neq f(y)\}.$$

See its history in VII 3.23.

2.4 Lemma. For T, P_T as in Definition 2.3:

(1) P_T satisfies the c.c.c.

(2) Moreover: If $\langle p_i : i < \omega_1\rangle$ are conditions in P, then there are disjoint uncountable sets S_1, $S_2 \subseteq \omega_1$ such that: whenever $i < j$, $i \in S_1$, $j \in S_2$, *then* p_i and p_j are compatible.

(3) If $G \subseteq P_T$ is generic over V, $V[G] \subseteq V^*$, and $\aleph_1^{V^*} = \aleph_1^V$, then all ω_1-branches of T in V^* are already in V.

Proof. (1) Follows by (2).

(2) Recall that p and q are incompatible if:

either $p \cup q$ is not a function *or* there are $\eta \in \mathrm{Dom}(p)$, $\nu \in \mathrm{Dom}(q)$ such that $p(\eta) = q(\nu)$, and η and ν are distinct but comparable, i.e. $\eta <_T \nu$ or $\nu <_T \eta$.

Let $\langle p_i : i < \omega_1 \rangle$ be a sequence of conditions in P_T. By the usual Δ-system argument we may assume that for all $i, j < \omega_1$ $p_i \cup p_j$ is a function, and we may also assume that $|\mathrm{Dom}(p_i)| = n$ for all $i < \omega_1$. We will now get the desired result by applying the following subclaim n^2 times:

2.4A Subclaim. If $\langle \eta^1_\alpha : \alpha \in S_1 \rangle$, $\langle \eta^2_\alpha : \alpha \in S_2 \rangle$ are lists of members of A without repetitions, S_1, S_2 are uncountable, *then* there are uncountable sets $S'_1 \subseteq S_1$, $S'_2 \subseteq S_2$ such that: $\alpha \in S'_1$, $\beta \in S'_2 \Rightarrow \eta^1_\alpha, \eta^2_\beta$ are incomparable.

Proof of the subclaim. for $\ell = 1, 2$ and $\zeta < \omega_1$, let:

$$L_\ell(\zeta) = \{\eta^\ell_\alpha \restriction \zeta : \alpha < \omega_1,\ \ell g(\eta^\ell_\alpha) \geq \zeta\}.$$

Let $\zeta_\ell = \min\{\zeta : L_\ell(\zeta) \text{ is uncountable}\}$, and if all $L_\ell(\zeta)$ are countable, let $\zeta_\ell = \omega_1$.

We now distinguish 4 cases:

Case 1: $\zeta_1 < \zeta_2$: Since $L_2(\zeta_1)$ is countable, for some η the set $S'_1 \stackrel{\text{def}}{=} \{\alpha < \omega_1 : \ell g(\eta^2_\alpha) > \zeta_1 \text{ and } \eta <_T \eta^2_\alpha\}$ is uncountable (as $\aleph_1 = \mathrm{cf}(\aleph_1) > \aleph_0$), and as $L_1(\zeta_1)$ is uncountable, $S'_2 \stackrel{\text{def}}{=} \{\alpha < \omega_1 : \ell g(\eta^1_\alpha) \geq \zeta \text{ and } \neg\eta <_T \eta^1_\alpha\}$ is uncountable. So S'_1, S'_2 as required. We are done.

Case 2: $\zeta_2 < \zeta_1$: Similar.

Case 3: $\zeta_1 = \zeta_2 < \omega_1$: By induction on $\gamma < \omega_1$ choose $\beta(1,\gamma)$ and $\beta(2,\gamma)$ such that:

$$\ell g(\eta^1_{\beta(1,\gamma)}) \geq \zeta_1 \text{ and } \eta^1_{\beta(1,\gamma)} \restriction \zeta_1 \notin \{\eta^\ell_{\beta(\ell,\gamma')} \restriction \zeta_1 : \gamma' < \gamma,\ \ell = 1, 2\}$$

$$\ell g(\eta^2_{\beta(2,\gamma)}) \geq \zeta_2 \text{ and } \eta^2_{\beta(2,\gamma)} \restriction \zeta_2 \notin \{\eta^\ell_{\beta(\ell,\gamma')} \restriction \zeta_2 : \gamma' < \gamma, \ell = 1, 2\} \cup \{\eta^1_{\beta(1,\gamma)} \restriction \zeta_1\}$$

and let $S'_\ell = \{\beta(\ell,\gamma) : \gamma < \omega_1\}$, $\ell = 1, 2$.

Case 4: $\zeta_1 = \zeta_2 = \omega_1$ and no earlier case. For $\ell = 1, 2$, $\zeta < \omega_1$ let $A^\ell_\zeta = \{\eta \in T : \ell g(\eta) = \zeta$ and there are $\aleph_1$ many α with $\eta^\ell_\alpha \restriction \zeta = \eta\}$, clearly $A^\ell_\zeta \neq \emptyset$.

So $T^\ell \stackrel{\text{def}}{=} \bigcup_{\zeta<\omega_1} A^\ell_\zeta$ is a downward closed subtree of T, possibly only a single branch.

Subcase 4a: For some ℓ and ζ, $|A^\ell_\zeta| > 1$. Without loss of generality $|A^1_\zeta| > 1$. Let $\nu_2 \in A^2_\zeta$, $\nu_1 \in A^1_\zeta \setminus \{\nu_2\}$, for $\ell = 1, 2$ we let $S_\ell = \{\alpha < \omega_1 : \nu_\ell <_T \eta^\ell_\alpha\}$.

Subcase 4b: For each $\ell = 1, 2$ the set $T^\ell = \bigcup_{\zeta<\omega_1} A^\ell_\zeta$ is a branch, say $B_{i(\ell)}$. If $i(1) \neq i(2)$ then we can again find ν_1 and ν_2 as in case 4a. So let $i = i(1) = i(2)$. It is impossible that uncountably many η^ℓ_α are on B_i (by the choice of A in Definition 2.3), so we may assume that no η^ℓ_α is on B_i. By induction we can find uncountable sets $S_1' \subseteq S_1$, $S_2' \subseteq S_1$ and sequences $\langle \nu^1_\alpha : \alpha \in S_1'\rangle$, $\langle \nu^2_\alpha : \alpha \in S_2'\rangle$ such that: $\nu^\ell_\alpha \in B_i$, $\nu^\ell_\alpha <_T \eta^\ell_\alpha$, $\eta^\ell_\alpha{\restriction}(\ell g(\nu^\ell_\alpha) + 1) \notin B_i$, and $\{\nu^1_\alpha : \alpha \in S_1'\} \cap \{\nu^2_\alpha : \alpha \in S_2'\} = \emptyset$. This shows that for $\alpha \in S_1'$, $\beta \in S_2'$ the nodes η^1_α and η^2_α are incomparable. So we have proved the subclaim and hence 2.4(2).

Proof of 2.4(3). Since $T = \bigcup_{j<\omega_1} B_j'$ is a partition of T, we can for each $y \in T$ find a unique $j = j(y)$ with $y \in B_j'$. Let $h(y) = \min B'_{j(y)} \in A$. In V^{P_T} we have a generic function $g : A \to \omega$, and we can extend it to a function $g : T \to \omega$ by demanding $g(y) = g(h(y))$. Now let B^* be an ω_1-branch of T in some $\aleph_1$-preserving extension of V^{P_T}. Clearly $g{\restriction}B^*$ takes some value uncountably many times, but $g(y_1) = g(y_2) \,\&\, y_1 <_T y_2$ implies $j(y_1) = j(y_2)$, so $B^* \subseteq B_j$ for some j. $\square_{2.4}$

2.5 Fact. There is a family $\langle \eta_\delta : \delta < \omega_1, \delta \text{ limit}\rangle$ such that:

(A) $\eta_\delta : \omega \to \delta$, and $\sup\{\eta_\delta(n) : n < \omega\} = \delta$

(B) For all limit $\delta_1, \delta_2 < \omega_1$ and $n_1, n_2 < \omega$ we have: if $\eta_{\delta_1}(n_1) = \eta_{\delta_2}(n_2)$, then $n_1 = n_2$ and $\eta_{\delta_1}{\restriction}n_1 = \eta_{\delta_2}{\restriction}n_2$.

(C) if $m < \ell < \omega$ and $\delta < \omega_1$ is limit, then $\eta_\delta(m) + \omega \leq \eta_\delta(\ell) + \omega$.

Proof. Easy. Let $H : {}^{\omega>}\omega_1 \to \omega_1$ be a 1-1 map such that for all $\eta \in {}^{\omega>}\omega_1$ we have $H(\eta) \in [\,\mathrm{maxRang}(\eta), \mathrm{maxRang}(\eta) + \omega)$ (and can add $\nu \triangleleft \eta \Rightarrow H(\nu) < H(\eta)$).

Now for any limit ordinal δ, let $\alpha_0 < \alpha_1 < \cdots$ be cofinal in δ, and define η_δ

inductively by

$$\eta_\delta(n) = H(\eta_\delta{\restriction}n\hat{\ }\langle\alpha_n\rangle).$$

$\square_{2.5}$

2.6 Definition. Assume that $\langle\eta_\delta : \delta < \omega_1, \delta \text{ limit}\rangle$ is as above.

(1) For $\eta \in {}^{\omega>}\omega_1$, let $E_\eta = \{\delta : \eta \trianglelefteq \eta_\delta\}$.

(2) Let $\mathbf{Z} = \{\eta \in {}^{\omega>}\omega_1 : E_\eta \text{ is stationary}\}$, $C_0 = \{\delta < \omega_1 : (\forall n < \omega)\, \eta_\delta{\restriction}n \in \mathbf{Z}\}$.

(3) Let $\mathbf{Z}^* = \{\eta \in \mathbf{Z} : (\exists^{\aleph_1} i < \omega_1)\, \eta\hat{\ }\langle i\rangle \in \mathbf{Z}\}$.

(4) Let $C^* = \{\delta \in C_0 : (\exists^\infty n)\, \eta_\delta{\restriction}n \in \mathbf{Z}^*\}$.

(5) Let $\mathbf{Z}_0 = \{\eta \in \mathbf{Z} : (\forall k < \ell g(\eta))\, \eta{\restriction}k \notin \mathbf{Z}^*\}$

2.6A Fact.

(1) $\mathbf{Z}$ is closed under initial segments, so $\mathbf{Z}$ is a tree (of height ω). $\mathbf{Z}^*$ is the set of those nodes of $\mathbf{Z}$ which have uncountably many successors.

(2) $\mathbf{Z}$ defines a natural topology on C_0, if we take the sets E_η as basic neighborhoods.

(3) C_0 and even C^* contains a club of ω_1.

(4) For every finite $u \subseteq \mathbf{Z} \setminus \mathbf{Z}_0$ there is $\rho \in \mathbf{Z}$ which is $\triangleleft$-incomparable with every $\eta \in u$ moreover $\rho \in \mathbf{Z} \setminus \mathbf{Z}_0$.

Proof. (1) and (2) should be clear.

For (3), let χ be some large enough regular cardinal. If $\omega_1 \setminus C^*$ as stationary, we could find a countable elementary submodel $N \prec (H(\chi), \in)$ such that $\delta \stackrel{\text{def}}{=} N \cap \omega_1 \notin C^*$ and $\langle\eta_\delta : \delta < \omega_1 \text{ limit}\rangle$ belongs to N (hence $\langle E_\eta : \eta \in {}^{\omega>}(\omega_1)\rangle$, $\mathbf{Z}$, C_0, $\mathbf{Z}^*$, C^*, $\mathbf{Z}_0$ belong to N). Assume that for some $n_0 < \omega$ for all $n \in (n_0, \omega)$ we have $\eta_\delta{\restriction}n \notin \mathbf{Z}^*$. So the set

$$Y \stackrel{\text{def}}{=} \{\nu \in \mathbf{Z} : \nu \trianglelefteq \eta_\delta{\restriction}n_0 \textit{ or: } \eta_\delta{\restriction}n_0 \trianglelefteq \nu \text{ and } (\forall k \in (n_0, \ell g(\nu)))\, \nu{\restriction}k \notin \mathbf{Z}^*\}$$

is a subtree of $\mathbf{Z}$ with countable splitting, hence is countable. Let $\delta' = \sup\{\nu(k) : \nu \in Y, k \in \mathrm{Dom}(\nu)\}$. Since $Y \in N$, also $\delta' \in N$, but $(\forall k)\, [\eta_\delta{\restriction}k \in Y]$, so $\eta_\delta(k) \leq \delta' < \delta$, contradicting $\delta = \sup\{\eta_\delta(k) : k < \omega\}$.

(4) So if $u \subseteq \mathbf{Z} \setminus \mathbf{Z}_0$ is finite, let $\eta \in u$ be of minimal length and as $\eta \notin \mathbf{Z}_0$ there is $\nu \lhd \eta$, such that $\nu \in \mathbf{Z}^*$, so for some $i < \omega_1$, $\rho \stackrel{\text{def}}{=} \nu\char`^\langle i \rangle \in \mathbf{Z}$ and ρ is $\lhd$-incomparable with every $\eta' \in u$ and $\rho \notin \mathbf{Z}_0$ as $\nu \lhd \rho$, $\nu \in \mathbf{Z}^*$. $\square_{2.6A}$

From $\mathbf{Z}$ we can now define the forcing notion R_4, to be used below:

2.6B Definition.

$$R_4 = \{(u, w) : w \text{ a finite set of limit ordinals} < \omega_1, u \text{ a finite subset of } \mathbf{Z} \setminus \mathbf{Z}_0, \text{ and } w \cap E_\eta = \emptyset \text{ for } \eta \in u\}.$$

with the natural order: $(u_1, w_1) \leq (u_2, w_2)$ *iff* $u_1 \subseteq u_2$ & $w_1 \subseteq w_2$.

Note that $w \cap E_\eta = \emptyset$ just means that for all $\delta \in w$, $\eta \ntrianglelefteq \eta_\delta$. Actually $\eta = \eta_\delta$ never occurs as $[\eta \in w \Rightarrow \ell g(\eta) < \omega]$ and $[\delta \in u \Rightarrow \ell g(\eta_\delta) = \omega]$.

So we have that (u, w) and (u', w') are incompatible iff $(u \cup u', w \cup w')$ is not in R_4, i.e., either there is $\eta \in u$, $\delta \in w'$ such that $\eta \trianglelefteq \eta_\delta$, or there are such $\eta \in u'$, $\delta \in w$.

R_4 produces a generic set $\underset{\sim}{S}^4 = \bigcup\{w : (\exists u)[(u, w) \in \underset{\sim}{G}_{R_4}]\}$ (i.e. this is an R_4-name), which can easily be shown to be a stationary subset of ω_1(in V^{R_4}, see 2.6E(1))(actually $V[\underset{\sim}{S}^4] = V[\underset{\sim}{G}_{R_4}]$).

2.6C Fact. R_4 satisfies the $\aleph_1$-c.c.; in fact for every $\aleph_1$ conditions there are $\aleph_1$ pairwise compatible (and more).

Proof. Let $(u_i, w_i) \in R_4$ for $i < \omega_1$. Let $v_i \stackrel{\text{def}}{=} \bigcup\{\text{Rang}(\eta) : \eta \in u_i\}$.
Thinning out to a Δ-system we may assume that there are $\alpha < \omega_1$, $w^* \subseteq \alpha$, $v^* \subseteq \alpha$, $u^* \subseteq {}^{\omega>}\alpha$ such that for all $i < \omega_1 \setminus \alpha$,

$$w_i \cap \alpha = w^*, \quad v_i \cap \alpha = v^*, \quad u_i \cap {}^{\omega>}\alpha = u^*$$

and for all $i \neq j$: $w_i \cap w_j = w^*$, $v_i \cap v_j = v^*$ and $u_i \cap u_j = u^*$. So $\eta \in u_j \setminus u^* \Rightarrow$ $\max\text{Rang}(\eta) > \alpha$. We may also assume that none of the v_i or w_i is a subset of α, and thinning out further we may also assume that for all $i < j$ we have $\alpha < \max(w_i) < \min(v_j \setminus \alpha)$.

Now if $i < j$ and (u_i, w_i) and (u_j, w_j) are incompatible, then we must have one of the following:

(a) $(\exists \eta \in u_i \setminus u^*)\,(\exists \delta \in w_j)\,\eta \trianglelefteq \eta_\delta$

(b) $(\exists \eta \in u_j \setminus u^*)\,(\exists \delta \in w_i)\,\eta \trianglelefteq \eta_\delta$

Now if if clause (b) holds for $\eta \in u_j \setminus u^*$ and $\delta \in w_i$, this implies $\delta \leq \max(w_i) < \min(v_j \setminus \alpha) \leq \max(\mathrm{Rang}(\eta)) < \delta$. [Why? As $\delta \in w_i$; by assumption above; as $\eta \in u_j \setminus u^*$; as $\eta \trianglelefteq \eta_\delta$ and the choice of η_δ (see 2.5(1)) respectively.] A contradiction, so clause (a) must hold. Now we claim that: for each $j < \omega_1$ the set $s_j \stackrel{\text{def}}{=} \{i < j : p_i \text{ and } p_j \text{ are incompatible}\}$ is finite.

Why? Assume not; by the above for $i \in s_j$ necessarily there are $\eta^i \in u_i \setminus u^*$ and $\delta_i \in w_j$ such that $\eta^i \triangleleft \eta_{\delta_i}$. But for $i(0) < i(1)$, both in s_j, we get that $\eta^{i(0)}$ and $\eta^{i(1)}$ must be incomparable, since neither of $\mathrm{Rang}(\eta^{i(0)})$ and $\mathrm{Rang}(\eta^{i(1)})$ can be a subset of the other. Hence all the $\delta_i (i \in s_j)$ are distinct — a contradiction as w_j is finite. $\square_{2.6C}$

2.6D Fact.

(1) If $A \subseteq \omega_1$ is stationary, $n < \omega$, *then* there is $\delta \in A$ such that $E_{\eta_\delta \restriction n} \cap A$ is stationary.

(2) If $B \subseteq \omega_1$ is stationary, then also the set

$$B' \stackrel{\text{def}}{=} \{\delta \in B : (\forall n < \omega)\ [E_{\eta_\delta \restriction n} \cap B \text{ is stationary}]\}$$

is stationary, and in fact $B \setminus B'$ is nonstationary.

Proof. (1) Using Fodor's lemma we can find a stationary set $A' \subseteq A$ and a finite sequence η^* such that for all $\delta \in A'$ we have $\eta_\delta \restriction n = \eta^*$. So $A' \subseteq A \cap E_{\eta^*} = A \cap E_{\eta_\delta \restriction n}$ for all $\delta \in A'$.

(2) Let $A \stackrel{\text{def}}{=} B \setminus B'$, $A_n \stackrel{\text{def}}{=} \{\delta \in B : E_{\eta_\delta \restriction n} \cap B \text{ is nonstationary}\}$. By (1), each A_n must be nonstationary, so also $A = \bigcup_n A_n$ is nonstationary. $\square_{2.6D}$

2.6E Fact. Let $\underset{\sim}{S}^4$ be the R_4-name of a subset of ω_1 defined in 2.6B. *Then* we have

(1) $\underset{\sim}{S}^4$ is stationary in V^{R_4}.

(2) If $A \subseteq \omega_1$ is stationary in V, then in V^{R_4} there is $\eta \in \mathbf{Z}$ such that $A \cap E_\eta$ is stationary and $E_\eta \cap \underset{\sim}{S}^4 = \emptyset$.

(3) Every stationary subset of ω_1 from V has (in V^{R_4}) a stationary intersection with $\omega_1 \setminus \underset{\sim}{S}^4$.

Proof. (1) Easy; for each $p = (u, w) \in R_4$ and club $E \in V$ of ω_1, as $u \subseteq \mathbf{Z} \setminus \mathbf{Z}_0$ is finite there is $\eta \in \mathbf{Z} \setminus \mathbf{Z}_0$ which is $\triangleleft$-incomparable with every $\nu \in u$ (see 2.6A(4)) so E_η is stationary hence we can find $\delta \in E \cap E_\eta \setminus (\sup(w) + 1)$, so $q = (u, w \cup \{\delta\}) \in R_4$, $p \leq q$ and $q \Vdash_{R_4}$ "$\underset{\sim}{S}^4 \cap E \neq \emptyset$". As R_4 satisfies the c.c.c. this suffice.

(2) Let A be stationary. By 2.6A(3) w.l.o.g. $A \subseteq C^*$ and by 2.6D(2) we may w.l.o.g. assume that $(\forall \delta \in A)\, (\forall n < \omega)[E_{\eta_\delta \restriction n} \cap A$ is stationary$]$. Fix a condition $(u, w) \in R_4$. Choose $\delta \in A \setminus w$, then for some large enough n, $E_{\eta_\delta \restriction n} \cap w = \emptyset$ and $\eta_\delta \restriction n \notin \mathbf{Z}_0$, so $(u \cup \{\eta_\delta \restriction n\}, w)$ is a condition in R_4 above $(u, v) \in R_4$ and it clearly forces $A \cap E_{\eta_\delta \restriction n} \cap \underset{\sim}{S}^4 = \emptyset$.

(3) Follows from (2). $\square_{2.6E}$

2.7 Definition of the iteration. We define by induction on $\zeta \leq \kappa$ an RCS iteration (see X, §1) $\bar{Q}^\zeta = \langle P_i, \underset{\sim}{Q}_j : i \leq \zeta, j < \zeta \rangle$, and if $\zeta < \kappa$, $\bar{Q}^\zeta \in H(\kappa)$, which is a semiproper iteration (i.e. for $i < j \leq \zeta$, i non-limit P_j/P_i is semiproper but for a limit ordinal j the forcing notion $\underset{\sim}{Q}_j$ is not necessarily semiproper) and, if $\zeta = \delta$, δ a limit ordinal, also P_ζ-names, A_ζ, T_ζ (of a tree), and $P_{\zeta+1}$-name $W_\zeta = \langle H^\zeta_\alpha(a) : \alpha \in a \in A_\zeta \rangle$, as follows:

(a) Suppose ζ is non-limit, let $\kappa_\zeta < \kappa$ be the first supercompact $> |P_\zeta|$, so κ_ζ is a supercompact cardinal even in V^{P_ζ}, and let Q_ζ be a semiproper forcing notion of power κ_ζ collapsing κ_ζ to $\aleph_2$ such that $\Vdash_{P_\zeta * \underset{\sim}{Q}_\zeta}$ "any forcing notion not destroying stationary subsets of ω_1 is semiproper",
[it exists e.g. by Lemma 1.3 and X 2.8 but really $Q_\zeta = \text{Levy}(\aleph_1, < \kappa_\zeta)$ (in V^{P_ζ}) is okay, as
$$\Vdash_{P_\zeta * \underset{\sim}{Q}_\zeta} \text{“}Ax_{\omega_1}[\aleph_1 - \text{complete}]\text{”}$$
and even $\text{Ax}_1[\aleph_1\text{-complete}]$ implies (by 1.1) the required statement.]

(b) Suppose ζ is limit, $\underset{\sim}{Q}_\zeta$ will be of the form $\underset{\sim}{Q}^a * \underset{\sim}{Q}^b * \underset{\sim}{Q}^c$. Remember that $f^* : \kappa \to H(\kappa)$ is a Laver Diamond (see Definition VII 2.8).

If $f^*(\zeta)$ is a P_ζ-name, $\Vdash_{P_\zeta}$ "$f^*(\zeta)$ is a semiproper forcing notion", *then* let $\underset{\sim}{Q}^a_\zeta = f^*(\zeta)$. If $f^*(\zeta)$ is not like that, let $\underset{\sim}{Q}^a_\zeta$ =the trivial forcing.

$\underset{\sim}{Q}^b_\zeta$ will satisfy the following property:

(∗) If $\xi < \zeta$, ξ is non-limit, $A \in V^{P_\xi}$, $A \subseteq \omega_1$, and A is stationary in V^{P_ξ} (equivalently in V^{P_ζ}) *then* A is stationary in $V^{P_\zeta * \underset{\sim}{Q}^a_\zeta * \underset{\sim}{Q}^b_\zeta}$.

(This property (∗) will follow from 2.6E, it will assure that the iteration remains semiproper)

If ζ is divisible by ω^2, we will let $\underset{\sim}{Q}^b_\zeta = \underset{\sim}{Q}^1_\zeta * \underset{\sim}{Q}^2_\zeta * \underset{\sim}{Q}^3_\zeta$. First in V^{P_ζ} choose (see 2.1, 2.3) $\underset{\sim}{Q}^1_\zeta = R_1 * \underset{\sim}{R}_2 * \underset{\sim}{P}_{T_\zeta}$, where $T_\zeta = \{b : b$ an initial segment of some $a \in \bigcup_{\xi<\zeta} A_\xi\}$ ordered by being initial segment (for the definition of A_ξ see the definition of W_ξ below). From the generic subset of $\underset{\sim}{Q}^1_\zeta$ (and $P_\zeta * \underset{\sim}{Q}^a_\zeta$) we can define, for each ω_1-branch B of T_ζ, a 2-coloring $H_\alpha(B)$ of $\omega_1 : H_\alpha(B) = \bigcup\{H^\zeta_\alpha(a) : \xi \in a \in B$ and $\zeta > \xi \geq \alpha$ and $H^\xi_\alpha(a)$ is well defined$\}$. (See the definition of W_ζ below, we can say that if $H_\alpha(B)$ is not a 2-coloring of ω_1 we use trivial forcing). Remember 2.4(3).

To define $\underset{\sim}{Q}^2_\zeta$, we need the following concept:

We will say that a function $h : [\omega_1]^2 \to 2$ is almost homogeneous if there is a partition $\omega_1 = \bigcup_{n<\omega} A_n$ and an $\ell \in \{0,1\}$ such that for all n the function $h{\restriction}[A_n]^2$ is constantly $= \ell$. We may say h is almost homogeneous with value ℓ.

We choose $\underset{\sim}{Q}^2_\zeta \in H(\kappa)$ such that

⊗ *if* there is $\underset{\sim}{Q} \in H(\kappa)$ such that

(i) $\underset{\sim}{Q}$ is a $P_\zeta * \underset{\sim}{Q}^a_\zeta * \underset{\sim}{Q}^1_\zeta$-name of a forcing notion

(ii) For every $\xi < \zeta$ the forcing notion $(P_\zeta * \underset{\sim}{Q}^a_\zeta * \underset{\sim}{Q}^1_\zeta * \underset{\sim}{Q})/P_{\xi+1}$ is semi proper, (equivalently, preserves stationarity of subsets of ω_1⟩

(iii) if, in $V^{P_\zeta * \underset{\sim}{Q}^a_\zeta * \underset{\sim}{Q}^1_\zeta}$, B is a branch of T_ζ cofinal[†] in ζ, $\alpha < \omega_1$, then the coloring $H_\alpha(B)$ of ω_1, is almost homogeneous in $V^{P_\zeta * \underset{\sim}{Q}^a_\zeta * \underset{\sim}{Q}^1_\zeta * \underset{\sim}{Q}}$

then $\underset{\sim}{Q}^2_\zeta$ satisfies this.

Otherwise $\underset{\sim}{Q}^2_\zeta$ is trivial.

[†] Note: members of B are subsets of ζ with last element, so $\{\max(a) : a \in B\}$ is a subset of ζ.

In $V^{P_\zeta * \underset{\sim}{Q}^a_\zeta * \underset{\sim}{Q}^1_\zeta * \underset{\sim}{Q}^2_\zeta}$ we now define a set S_ζ, which is supposed to guess the set $\underset{\sim}{S}[G]$. More on $\underset{\sim}{S}$ will be said below (and see "overview").
We let $\alpha \in S_\zeta$ if for all the ω_1-branches B of T_ζ cofinal in ζ (i.e. such that $\bigcup\{a : a \in B, \text{otp}(a) \text{ a successor ordinal}\}$ is unbounded in ζ) the function $H_\alpha(B)$ is almost homogeneous with value 1.
Now we let $\underset{\sim}{Q}^3_\zeta$ be the forcing notion which shots a club through the complement of S_ζ, unless S_ζ includes modulo $\mathcal{D}_{\aleph_1}$ some stationary set from $\bigcup_{\xi<\zeta} V^{P_\xi}$, in which case Q^3_ζ will be trivial. This completes the definition of Q^b_ζ when ζ is divisible by ω^2, otherwise Q^b_ζ is trivial.

We let $\underset{\sim}{Q}_\zeta = \underset{\sim}{Q}^a_\zeta * \underset{\sim}{Q}^b_\zeta * \underset{\sim}{Q}^c_\zeta$ where $\underset{\sim}{Q}^c_\zeta$ is the addition of $(\aleph_1 + 2^{\aleph_0})^{V^{P_\zeta}}$ Cohen reals with finite support. Clearly for $\xi < \zeta$, $(P_\zeta/P_{\xi+1}) * \underset{\sim}{Q}_\zeta$ preserves stationarity of subsets of ω_1, hence it is semiproper (see (a)), so Q_ζ is o.k. An alternative to (b): we can demand $\underset{\sim}{Q}^a_\zeta$ forces SPFA. If ζ is not divisible by ω^2 let $\underset{\sim}{Q}_\zeta$ be $\underset{\sim}{Q}^a_\zeta * \underset{\sim}{Q}^b_\zeta * \underset{\sim}{Q}^c_\zeta$, with $\underset{\sim}{Q}^a_\zeta, \underset{\sim}{Q}^b_\zeta$ trivial, $\underset{\sim}{Q}^c_\zeta$ as above.

(c) For ζ limit we also have to define W_ζ (in $V^{P_{\zeta+1}}$).
  (i) W_ζ is a function whose domain is $A_\zeta = \{a : a \subseteq \zeta{+}1, \zeta \in a \in V^{P_\zeta}$, and a is a countable set of limit ordinals and $\xi \in a \Rightarrow a \cap (\xi+1) \in V^{P_\xi}\}$.
  (ii) For $a \in A_\zeta$, $W_\zeta(a) = \langle H^\zeta_\alpha(a) : \alpha < \text{otp}(a)\rangle$, where $H^\zeta_\alpha(a)$ is a function from $[\text{otp}(a)]^2 = \{\{j_1, j_2\} : j_1 < j_2 < \text{otp}(a)\}$ to $\{0,1\}$ (where $\text{otp}(a)$ is the order type of a).
  (iii) For every $\xi \in a \in A_\zeta$ (check definition of A_ζ), $a \cap (\xi+1) \in A_\xi$, and for $\alpha < \text{otp}(a \cap (\xi+1))$, $H^\xi_\alpha(a \cap (\xi+1))$ is $H^\zeta_\alpha(a)$ restricted to $[\text{otp}(A \cap (\xi+1))]^2$.
  (iv) If $a \in A_\zeta$, we use the Cohen reals from $\underset{\sim}{Q}^c_\zeta$ to choose the values of $H^\zeta_\alpha(a)(\{j_1, j_2\})$ when $\alpha = \text{otp}(a \cap \zeta)$ or $j_1 = \text{otp}(a \cap \zeta)$ or $j_2 = \text{otp}(a \cap \zeta)$ that is when not defined implicitly by condition (iii), i.e. by H^ξ_α (not using the same digit twice (digit from the Cohen reals from $\underset{\sim}{Q}^c_\zeta$)).
  (v) $T_\zeta (\in V^{P_\zeta})$ is the tree $(\bigcup\{A_\delta : \delta < \zeta \text{ a limit ordinal}\}, <_{T_\zeta})$, ($<_{T_\zeta}$ is being an initial segment i.e. $a < b$ iff $a = b \cap (\max(a) + 1)$).

There is no problem to carry the inductive definition.

Note that we can separate according to whether the cofinality of ζ in V^{P_ζ} is $\aleph_0$ or $\geq \aleph_1$ (so for a club of $\zeta < \kappa$ we can ask this in V) and in each case some parts of the definition trivialize.

2.7A Toward the proof: Clearly P_κ is semiproper, satisfies the κ-c.c., and $|P_\kappa| = \kappa$. In $V_0 = V^{P_\kappa}$ let $T^* = \bigcup\{A_\delta : \delta < \kappa \text{ (limit)}\}$, and let $<_{T^*}$ be the order: being initial segment. Let $T = \{a : a \text{ an initial segment of some } b \in T^*\}$.
So T is a tree, and the $(\alpha+1)$'th level of T is $\{\alpha \in T : \text{otp}(a) = \alpha + 1\}$. The height of T is ω_1 (since all elements of T are countable) and all elements of T have $\kappa = \aleph_2$ many successors and every member of T belongs to some ω_1-branch.

For every ω_1-branch B of T we get a family of ω_1 many coloring functions $H_\alpha(B) : [\omega_1]^2 \to 2$, by letting $H_\alpha(B)(\{j_1, j_2\}) = H_\alpha{}^{\max(a)}(a)(j_1, j_2)$ for any $a \in B$ with $\text{otp}(a) > \max(j_1, j_2, \alpha)$ successor ordinal. Now we want to show that PFA$^+$ fails in V^{P_κ}. To this end, we will define a proper forcing notion R and R-name $\underset{\sim}{S}$ of a stationary set of ω_1. R will be obtained by composition. The components of R and of the proof are not new.

2.8 Definition of R. Let $V_0 = V^{P_\kappa}$. Let R_0 be Levy$(\aleph_1, \aleph_2)$ (in V_0). In $V_1 = V_0^{R_0}$, let R_1 be the Cohen forcing; in $V_2 \overset{\text{def}}{=} V_1^{R_1}$ let R_2 be Levy$(\aleph_1, 2^{\aleph_2})$. Let $V_3 = V_2^{R_2}$. Let $\langle B_i : i < i^* \rangle \in V_1$ list the ω_1-branches of T in V_1 and $i_0^* < i^*$ be such that $i < i_0^* \Leftrightarrow \kappa > \sup[\bigcup\{a : a \in B_i\}]$. Easily in V_1, T has ω_1-branches with supremum κ (just build by hand) so really $i_0^* < i^*$. Forcing with $R_1 * \underset{\sim}{R}_2$ over V_1 does not add ω_1-branches to T (by 2.2), hence in V_3 it has $\leq \aleph_1$ ω_1-branches, so let us essentially specialize it (see 2.4(3)), using the forcing notion $R_3 = P_T$ from 2.3. Let $V_4 = V_3^{R_3}$. Let R_4 be the forcing defined in 2.6B, and let $V_5 = V_4^{R_4}$. In V^5 we now define R_5: it is the product with finite support of $R^5_{\alpha,i}(\alpha < \omega_1, i_0^* \leq i < i^*)$, where the aim of $R^5_{\alpha,i}$ is making ω_1 the union of $\aleph_0$ sets, on each of which $H_\alpha^{[i]} \overset{\text{def}}{=} H_\alpha(B_i)$ is constantly 0 if $\alpha \in S^4$, constantly 1 *if* $\alpha \notin S^4$ (remember $H_\alpha(B_i)$ was defined just before 2.8 and S^4 was defined from G_{R_4}), see definition below. See definition 2.6B and Fact 2.6E. Let $V_6 = V_5^{R_5}$. So the decision does not depend on i.

Now $R^5_{\alpha,i}$ is just the set of finite functions h from ω_1 to ω so that on each $h^{-1}(\{n\})$ the coloring $H^{[i]}_\alpha$ is constantly 0 or constantly 1, as required above (so some case for all $n < \omega$).

Lastly, let $R = R_0 * \underset{\sim}{R}_1 * \underset{\sim}{R}_2 * \underset{\sim}{R}_3 * \underset{\sim}{R}_4 * \underset{\sim}{R}_5$. We define $\underset{\sim}{S}$ such that $\underset{\sim}{S}^4 \subseteq \underset{\sim}{S} \subseteq \underset{\sim}{S}^4 \cup \{\gamma + 1 : \gamma < \omega_1\}$ and, if $G \subseteq R$ is directed and $\underset{\sim}{S}[G]$ well defined, then all relevant information is decided; specifically: for the model N of cardinality $\aleph_1$ chosen below, for every R-name $\underset{\sim}{\alpha}$ of an ordinal which belongs to N we have $(\exists p \in G)$ [p forces a value to $\underset{\sim}{\alpha}$] (i.e., what is needed below including a well ordering of ω_1 of order type $\underset{\sim}{\zeta}_\alpha$ for $\alpha < \omega_2$).

2.9 Fact. The forcing R is proper (in V_0).

As properness is preserved by composition, we just have to check each R_i in V_i. The only nontrivial one (from earlier facts) is R_5. For this it suffices to show that the product of any finitely many $R^5_{\alpha,i}$ satisfies the $\aleph_1$-c.c. Let $m < \omega$, and let the pairs (α_l, i_l) for $l < m$ be distinct (so $\alpha_l < \omega_1, i^*_0 \leq i_l < i^*$). Note that each B_{i_ℓ} (an ω_1-branch of T) is from V_1. So for some $\beta^* < \omega_1$, $i_{\ell_1} \neq i_{\ell_2} \Rightarrow B_{i_{\ell_1}}, B_{i_{\ell_2}}$ have no common member of level $\geq \beta^*$. Now we claim that in V_5 (on $H^{[i]}_\alpha$ see in 2.8):

(∗) If for each $\ell < m$, $\langle w^\ell_\gamma : \gamma < \omega_1\rangle$ is a sequence of pairwise disjoint finite subsets of $\omega_1 \setminus \beta^*$, then for some $\gamma(1), \gamma(2) < \omega_1$, for each even $\ell < m$

$$[x \in w^\ell_{\gamma(1)} \ \& \ y \in w^\ell_{\gamma(2)} \Rightarrow H^{[i_\ell]}_{\alpha_\ell}(\{x, y\}) = 0]$$

and for each odd $\ell < m$

$$[x \in w^\ell_{\gamma(1)} \ \& \ y \in w^\ell_{\gamma(2)} \Rightarrow H^{[i_\ell]}_{\alpha_\ell}(\{x, y\}) = 1].$$

Why? First we show that this holds in V_1 (note: $R_5 \in V_1$!). Because R_0 is $\aleph_1$-complete, it adds no new ω-sequence of members of V_0, hence for some $\zeta < \kappa, \{\langle \ell, w^\ell_\gamma\rangle : \gamma < \omega, \ell < m\}$ belongs to V^{P_ζ} and to $H(\zeta)$. Note that for each $\ell < m$, the sequence $\langle w^\ell_\gamma : \ell < m, \gamma < \omega\rangle$ is a sequence of pairwise disjoint subsets of $\omega_1 \setminus \beta^*$ and remember the way we use the Cohen reals to define the $H^\xi_i(a)$'s. We can show that for any possible candidate $\langle w^\ell : \ell < m\rangle$ for $\langle w^\ell_\varepsilon : \ell < m\rangle$ or even just for a sequence $\langle w^\ell : \ell < m\rangle$, $w^\ell \subseteq w^\ell_\varepsilon$ (for any $\varepsilon < \omega_1$ large enough) for infinitely many $\gamma < \omega$, the conclusion of $(*)$ holds for $(\gamma(1), \gamma(2)) = (\gamma, \varepsilon)$.

Clearly $(*)$ implies that any finite product of $R^5_{\alpha,i}$ satisfies the $\aleph_1$-c.c if it holds in the right universe (V_5). So for proving the fact we need to show that the subsequent forcing by R_1, R_2, R_3, R_4 preserves the satisfaction of $(*)$.

The least trivial is why R_3 preserves it (as $\underset{\sim}{R}_2$ is $\aleph_1$-complete and as R_1 and R_4 satisfy: among $\aleph_1$ conditions $\aleph_1$ are pairwise compatible (see 2.6(C)).

Recall from 2.4 that for any sequence $\langle p_i : i < \omega_1\rangle$ of conditions we can find disjoint uncountable sets S_1, S_2 such that for $i \in S_1$, $j \in S_2$ the conditions p_i and p_j are compatible. (This is also true for R_1 and R_4). We will work in V_3. So assume that $\langle \underset{\sim}{w}^\ell_\gamma : \gamma < \omega_1, \ell < m\rangle$ is an R_3-name of a sequence contradicting property $(*)$ in $V_3^{R_3}$. For $\gamma < \omega_1$ let p_γ be a condition deciding $\langle w^\ell_\gamma : \ell < m\rangle$, say $p_\gamma \Vdash \underset{\sim}{w}^\ell_\gamma = {}^*w^\ell_\gamma$. Let S_1, S_2 be as above, $S_k = \{\gamma^k_\alpha : \alpha < \omega_1\}$. Let $u^\ell_\alpha = {}^*w^\ell_{\gamma^1_\alpha} \cup {}^*w^\ell_{\gamma^2_\alpha}$ for $\ell < m$. By thinnings out we may without loss of generality assume that the sets $\bigcup_{\ell<m} w^\ell_\alpha$ for $\alpha < \omega_1$ are pairwise disjoint, so we can apply $(*)$ in V_3. This gives us $\alpha(1)$, $\alpha(2)$ such that for all even ℓ, $x \in u^\ell_{\alpha(1)}$, $y \in u^\ell_{\alpha(2)} \Rightarrow H^{[i_\ell]}_{\alpha_\ell}(\{x, y\}) = 0$ and similarly for odd ℓ we have $x \in u^\ell_{\alpha(1)}$ & $y \in u^\ell_{\alpha(2)} \Rightarrow H^{[i_\ell]}_{\alpha_\ell}(\{x, y\}) = 1$. Let q be a condition extending $p_{\gamma^1_{\alpha(1)}}$ and $p_{\gamma^2_{\alpha(2)}}$, then $q \Vdash$ "$\gamma^1_{\alpha(1)}$ and $\gamma^2_{\alpha(2)}$ are as required". $\square_{2.9}$

So R is proper in V_0; as in V_5, S^4 is stationary and R_5 satisfies the $\aleph_1$-c.c, clearly S^4 is a stationary subset of ω_1 in V_6 too; hence, by the choice of $\underset{\sim}{S}$ (just before 2.9) we have $\Vdash_R$ "$\underset{\sim}{S} \subseteq \omega_1$ is stationary".

2.9A Claim. In V^{P_κ}, PFA$^+$ fail as exemplified by R, $\underset{\sim}{S}$.

Proof. In V^{P_κ}, let χ be e.g. $\beth_3(\kappa)^+$ and let $N \prec (H(\chi), \in, <^*_\lambda)$ be a model of cardinality $\aleph_1$ containing all necessary information. i.e. the following belongs to N: i (if $i \leq \omega_1$), $\langle R_0, \underset{\sim}{R}_1, \underset{\sim}{R}_2, \underset{\sim}{R}_3, \underset{\sim}{R}_4, \underset{\sim}{R}_5\rangle$, $\langle P_i, \underset{\sim}{Q}_j : i \leq \kappa, j < \kappa\rangle$, G_{P_κ}, $\underset{\sim}{S}^4$ (but not $\underset{\sim}{S}$!), $\underset{\sim}{f}$ (see below), $\langle \underset{\sim}{B}_i : i < i^*\rangle, i^*_0$. Suppose that $G \in V^{P_\kappa}$, $G \subseteq R$ is directed and meets all dense sets of R which are in N. It suffices to show that $\underset{\sim}{S}[G]$ is not stationary. Note that N is a model of ZFC$^-$ etc.

Let $\underset{\sim}{f} \in N$ be the R_0-name of the function from ω_1 onto κ, then easily $\underset{\sim}{f}[G]$ is a function from ω_1 onto some $\delta < \kappa$, $\mathrm{cf}(\delta) = \aleph_1$, in V^{P_κ}. Note that $\underset{\sim}{T}[G] \in N[G]$ is just T_δ, and if $N[G] \models$ "$\underset{\sim}{B}[G]$ is an ω_1-branch of T cofinal in κ", then $\underset{\sim}{B}[G]$ is as ω_1-branch of T_δ cofinal in δ, and similarly with the coloring. We will now show how we could have predicted this situation in V^{P_δ}: Let $\underset{\sim}{h} : \omega_1 \times \omega_1 \to T$ be an R-name (belonging to N) which enumerates all ω_1-branches of T (we use the essential specialization by R_3) i.e.

$$\Vdash_R \text{“}\{\{\underset{\sim}{h}(i,j) : j < \omega_1\} : i < \omega_1\} = \{\underset{\sim}{B}_i : i < i^*\}\text{”}.$$

Then each set $\{\underset{\sim}{h}(i,j)[G] : j < \omega_1\}$ (for $i < \omega_1$) will be an ω_1-branch of T_δ (remember $T_\delta = \bigcup\{A_\zeta : \zeta < \delta \text{ limit}\}$), some of them cofinal in δ, and these ω_1-branches will be in $V^{P_\delta * \underset{\sim}{Q}^a_\delta}$, as $\underset{\sim}{Q}^b_\delta$ (more exactly $\underset{\sim}{Q}^1_\delta$, see 2.7) was chosen in such a way that no ω_1-branch can be added to T_δ without collapsing $\aleph_1$. Also all the ω_1-branches of $\underset{\sim}{T}[G] = T_\zeta$ will appear in this list.

Now we can recall how the set S_δ was defined: For each ω_1-branch B of T_δ (in $V^{P_\delta * \underset{\sim}{Q}^a_\delta * \underset{\sim}{Q}^1_\delta * \underset{\sim}{Q}^2_\delta}$ equivalently in $V^{P_\delta * \underset{\sim}{Q}^a_\delta}$) which is cofinal in δ, we have $\aleph_1$ many coloring functions $H_\alpha(B)$, and there are such ω_1-branches. We let $\alpha \in S_\delta$ if for all these ω_1-branches B the function $H_\alpha(B)$ is almost homogeneous with value 1.

Now note that the set G also interprets the names for the homogeneous sets for the colorings $H^{[i]}_\alpha$. These homogeneous sets exist in V^{P_κ} hence in V^{P_ξ} for $\xi < \kappa$ large enough, so in $V^{P_\delta * \underset{\sim}{Q}^a_\delta * \underset{\sim}{Q}^1_\delta}$ there is a forcing producing such sets, which, for every $\xi < \delta$ preserves stationarity of sets A, which are stationary subsets of ω_1 in $V^{P_{\xi+1}}$ (the forcing is $\underset{\sim}{Q}^3_\delta * \underset{\sim}{Q}^c_\delta * (P_\kappa / P_{\delta+1})$). Using the supercompactness of κ we can get such a forcing in $H(\kappa)$. But this implies that these sets are already almost homogeneous in $V^{P_\delta * \underset{\sim}{Q}^a_\delta * \underset{\sim}{Q}^1_\delta * \underset{\sim}{Q}^2_\delta}$ (see clause (b) in 2.7), so also $\underset{\sim}{S}[G]$ is in $V^{P_\delta * \underset{\sim}{Q}^a_\delta * \underset{\sim}{Q}^1_\delta * \underset{\sim}{Q}^2_\delta}$ (see the choice of R_5 in 2.8) and $\underset{\sim}{S}[G] = S_\delta$. But the forcing $\underset{\sim}{Q}^3_\delta$ ensures that S_δ is not stationary. $\square_{2.1}$

2.10 Lemma. We can reduce the assumption in 2.1 to "κ is supercompact"

Proof. We repeat the proof of 2.1 with some changes indicated below. We demand that every Q_δ is semiproper. We need some changes also in clause (b) of 2.7 (in the inductive definition of $\underset{\sim}{Q}_i$), we let $\underset{\sim}{Q}^a_\zeta = f^*(\zeta)$ only if: $f^*(\zeta)$ is a P_ζ-name, $\Vdash_{P_\zeta}$ "$f^*(\zeta)$ is semiproper" and let $\underset{\sim}{Q}^a_\zeta$ be trivial othervise. Let $\underset{\sim}{Q}^b_\zeta$ be trivial except when for some $\lambda_\zeta < \kappa, f^*(\zeta) \in H(\lambda_\zeta)$, and ζ is $\beth_8(\lambda_\zeta)$-supercompact. In this case we let (in $V^{P_\zeta * \underset{\sim}{Q}^a_\zeta}$), $\underset{\sim}{Q}^1_\zeta$ be defined as in the proof of 2.1 except that the $R^5_{\alpha,j}$ are now as defined below, $\underset{\sim}{Q}^2_\zeta$ is a forcing notion of cardinality $(2^{\aleph_1})^{V^{P_\zeta * \underset{\sim}{Q}^a_\zeta * \underset{\sim}{Q}^1_\zeta}}$ which forces MA. Now let $\underset{\sim}{S}_\delta \in V^{P_\zeta * \underset{\sim}{Q}^a_\zeta * \underset{\sim}{Q}^1_\zeta * \underset{\sim}{Q}^2_\zeta}$ be as described below, and $\underset{\sim}{Q}^3_\zeta$ is shooting a club through $\omega_1 \setminus \underset{\sim}{S}_\delta$ if $\underset{\sim}{Q}^a_\zeta * \underset{\sim}{Q}^1_\zeta * \underset{\sim}{Q}^2_\zeta * \underset{\sim}{Q}^3_\zeta$ is semiproper, and trivial otherwise. Now $\underset{\sim}{Q}^b_\zeta = \underset{\sim}{Q}^1_\zeta * \underset{\sim}{Q}^2_\zeta * \underset{\sim}{Q}^3_\zeta$. Lastly $\underset{\sim}{Q}^c_\zeta$ is as in the proof of 2.1 and $\underset{\sim}{Q}_\zeta = \underset{\sim}{Q}^a_\zeta * \underset{\sim}{Q}^b_\zeta * \underset{\sim}{Q}^c_\zeta$, now clearly $\underset{\sim}{Q}_\zeta \in H(\beth_8(\lambda_\zeta))$. This does not change the proof of 2.1. Now we let Q_κ = shooting a club called $\underset{\sim}{E}$ (of order type κ) through $\{i < \kappa : V \models$ "cf$(i) = \aleph_0$" or $V \models$ "i is strongly inaccessible in V, λ_ζ well defined and i is $\beth_8(\lambda_\zeta)$-supercompact"$\}$ (ordered by being an initial segment). Now it is easy and folklore that, for such Q_κ, we have $V^{P_\kappa * \underset{\sim}{Q}_\kappa} \models$SPFA, and show as before $V^{P_\kappa * \underset{\sim}{Q}_\kappa} \models \neg$PFA$^+$.

* * *

Why the need to change $\underset{\sim}{Q}^2_\zeta$? As the result of an iteration we ask "is there $\underset{\sim}{Q}$ such that (i), (ii), (iii) of $\otimes$", and this may well defeat our desire that $\underset{\sim}{Q}_\zeta$ hence $\underset{\sim}{Q}^1_\delta$ belongs to $H(\beth_8(\lambda_\zeta))$. We want to be able to "decipher" the possible "codings" fast, i.e., by a forcing notion of small cardinality, so we change $R^5_{\alpha,i}$'s inside the definition of R, in Definition 2.8).

We let $\gamma_{\alpha,j}$ be 0 if $\alpha \in S^4$ and 1 otherwise, and let $R^5_{\alpha,j}$ be defined by:

$$R^5_{\alpha,j} = \{(w,h) : w \text{ is a finite subset of } \omega_1 \text{ and } h \text{ is a finite function from the family of nonempty subsets of } w \text{ to } \omega \text{ such that: } \textit{if } u_1, u_2 \in \text{Dom}(h) \text{ and } h(u_1) = h(u_2) \ \textit{then } |u_1| = |u_2| \text{ and } [\zeta \in u_1 \setminus u_2 \ \&\ \xi \in u_2 \setminus u_1 \ \&\ \zeta < \xi \Rightarrow H^{[j]}_\alpha\{\{\zeta,\xi\}\} = \gamma_{\alpha,j}]\}.$$

(actually coloring pairs suffice).

2.10A Definition. 1) A function $H : [\omega_1]^2 \to \{0,1\}$ is called *ℓ-colored* (where $[A]^\kappa = \{a \subseteq A : |a| = \kappa\}$) if $\ell \in \{0,1\}$ and there is a function $h : S_{<\aleph_0}(\omega_1) \to \omega$ such that: if u_1, u_2 are finite subsets of ω_1 and $h(u_1) = h(u_2)$ then $|u_1| = |u_2|$ and $[\zeta \in u_1 \setminus u_2 \ \&\ \xi \in u_2 \setminus u_1 \ \&\ \zeta < \xi \Rightarrow H(\{\zeta,\xi\}) = \ell]$.
2) Called H (as above) explicitly non-ℓ-colored if there is a sequence $\langle u_\gamma : \gamma < \omega_1\rangle$ of pairwise disjoint finite subsets of ω_1 such that: for any $\alpha < \beta < \omega_1$ there are $\zeta \in u_\alpha$, $\xi \in u_\beta$ such that $H(\{\zeta,\xi\}) \neq \ell$.

2.10B Claim. 1) 1-colored, 0-colored are contradictory.
2) If H is explicitly non-ℓ-colored then it is not ℓ-colored.
3) If $MA + 2^{\aleph_0} > \aleph_1$, $\ell < 2$ and $H : [\omega_1]^2 \to \{0,1\}$ *then* H is ℓ-colored or explicitly non ℓ-colored.

Proof. 1) Clearly H cannot be both 0-colored and 1-colored.
2) Note also that if H is ℓ-colored, and u_ζ ($\zeta < \omega_1$) are pairwise disjoint non empty finite subsets of ω_1 such that $\zeta < \xi \Rightarrow \sup(u_\zeta) < \min(u_\zeta)$ *then* for some $\zeta < \xi$, $H(u_\zeta) = H(u_\xi)$ hence $H{\restriction}\{\{\alpha,\beta\} : \alpha \in u_\zeta, \beta \in u_\xi\}$ is constantly ℓ.
3) Use R defined like $R^5_{\alpha,j}$ from above.

If it satisfies the c.c.c., from a generic enough subset of R_H we can define a "witness" h to H being ℓ-colored. If R_H is not c.c.c. a failure is exemplified say by $\langle u_\zeta : \zeta < \omega_1\rangle$; without loss of generality it is a Δ-system i.e. $\zeta < \xi < \omega_1 \Rightarrow u_\zeta \cap u_\xi = u^*$. Reflection shows that $\langle u_\zeta \setminus u^* : \zeta < \omega_1\rangle$ exemplifies "explicitly non-ℓ-colored". $\square_{2.10B}$

The needed forcing $\underset{\sim}{Q}^2_\zeta$ is not too large ($\leq \lambda_\zeta$), and by 2.10B it essentially determines the $\gamma_{\alpha,j}$ (i.e., we can find $\gamma^0_{\alpha,j}$ so that if we have an appropriate G, the values of the $\gamma_{\alpha,j}$ will be $\gamma^0_{\alpha,j}$). So we have at most one candidate for $\underset{\sim}{S}[G]$, namely S_δ, and if $\omega_1 \setminus S_\delta$ is not disjoint to any stationary subset of ω_1 from V^{P_δ} modulo $\mathcal{D}_{\aleph_1}$, we end the finite iteration defining Q_δ by shooting a club through $\omega_1 \setminus S_\delta$.

Why is Q_δ still semiproper? Clearly $\underset{\sim}{Q}^a_\zeta$, $\underset{\sim}{Q}^1_\zeta$, $\underset{\sim}{Q}^2_\zeta$ are semiproper and so preserve stationarity of subsets of ω_1, and also $\underset{\sim}{Q}^3_\zeta$ do this and $\underset{\sim}{Q}^c_\zeta$ satisfies the c.c.c. So it is enough to prove that. Now use Rss (see chapter XIII §1 but assume on δ (remember we should shoot a club through $\underset{\sim}{E}$) that we have enough supercompactness for δ) to show that we still have semiproper $\equiv$ not destroying the stationarity of subsets of ω_1 for the relevant forcing.

This finish the proof that we can define the iteration $\bar{Q}$ as required. Lastly in the proof of the parallel of 2.9A we use also $E \in N$ hence $\delta \in E$. $\square_{2.10}$

2.11 Claim. If $\alpha(0), \alpha(1) \leq \omega_1$ and $|\alpha(0)| < |\alpha(1)|$, *then* $Ax_{\alpha(0)}$ [semiproper] $\nvdash Ax_{\alpha(1)}$ [proper] (assuming the consistency of ZFC+$\exists$ a supercompact).

Proof: Similar. [Now the Laver Diamond is used to guess triples of the form $(\bar{Q}{\restriction}\delta, \underset{\sim}{Q}_\delta, \langle \underset{\sim}{S}_i : i < \alpha(1)\rangle)$, $\underset{\sim}{Q}_\delta$ is a P_δ-name of a semiproper forcing, $\Vdash_{P_\delta + \underset{\sim}{Q}_\delta}$ "$\underset{\sim}{S}_i$ is a stationary subset of ω_1". In (b) from the colourings corresponding to the branches we decode a sequence $\langle \underset{\sim}{S}^*_\alpha : \alpha < \alpha(2)\rangle$ of stationary sets and try to shoot a club through $\omega_1 \setminus \underset{\sim}{S}^*_\alpha$ for one of them such that $\underset{\sim}{S}^\delta_i \setminus \underset{\sim}{S}_\delta$ is stationary for every $i < \alpha(1)$ (in addition to the earlier demands.] $\square_{2.11}$

2.12 Observation. Properness is not productive, i.e. (provably in ZFC) there are two proper forcings whose product is not proper.

Proof: Let T be the tree $({}^{\omega_1 >}(\omega_2), \triangleleft)$; now one forcing, P, adds a generic branch with supremum ω_2, e.g., $P = T$ (it is $\aleph_1$-complete). The second forcing, Q, guarantees that in any extension of V^Q, as long as $\aleph_1$ is not collapsed, T

will have no ω_1-branch with supremum ω_2^V. Use $Q = Q_1 * \underset{\sim}{Q}_2 * \underset{\sim}{Q}_3$, where Q_1 is Cohen forcing, $Q_2 = \text{Levy}(\aleph_1, 2^{\aleph_1})$ in V^{Q_1} (so it is well known that in $V^{Q_1 * Q_2}$, $\text{cf}(\omega_2^V) = \omega_1$, and T has no branch with supremum ω_2 and has no new ω_1-branch so has $\leq \aleph_1$ ω_1-branchs), and Q_3 is the appropriate specialization of T (like R_3 in the proof of 2.1, see Definition 2.3). Since in $V^{P\times Q}$ there is a branch of T cofinal in ${\omega_2}^V$ not from V and $V^{P\times Q}$ is an extension of V^Q, $\aleph_1$ must have been collapsed (see 2.4(3)).

We could also have used the tree ${}^{\omega_1 >}2$, but then we should replace "no ω_1-branch with supremum ${\omega_2}^V$" by "no branch of T which is not in V". $\square_{2.12}$

2.13 Discussion. Beaudoin asks whether SPFA $\vdash Ax_1$[finite iteration of $\aleph_1$-complete and c.c.c. forcing notions]. So the proofs of 2.1 (and 2.2) show the implications fail (whereas it is well known that already Ax(c.c.c.)$\Rightarrow$Ax$_1$(c.c.c.)).

But $\aleph_1$-complete forcing would be a somewhat better counterexample. We have

2.14 Fact. SPFA $\vdash Ax_1[\aleph_1\text{-complete}]$.

2.14A Reminder. We recall the following facts and definitions (see XIII):

(1) If P and Q are $\aleph_1$-complete, then $\Vdash_P$ "Q is $\aleph_1$-complete".

(2) For $\langle A_i : i < \omega_1\rangle$ such that $A_i \subseteq \omega_1$ we define the diagonal union of these sets as $\nabla_{i<\omega_1} A_i = \{\delta < \omega_1 : (\exists i < \delta)(\delta \in A_i)\}$.

If $A_i \subseteq \omega_1$ is nonstationary for all $i < \omega_1$, then $\nabla_{i<\omega_1} A_i$ is nonstationary (and if A_i is stationary for some i, then $\nabla_{i<\omega_1} A_i \supseteq A_i \setminus (i+1)$ is stationary).

(3) If $S \subseteq \omega_1$ is stationary, then the forcing of "shooting a club through S" is defined as $\text{club}(S) = \{h : h$ an increasing continuous function from some non-limit $\alpha < \omega_1$ into $S\}$. We have $\Vdash_{\text{club}(S)}$ "$\omega_1 \setminus S$ is nonstationary", and for every stationary $A \subseteq S$ we have $\Vdash_{\text{club}(S)}$ "A is stationary".

Proof of 2.14. Suppose $V \models$SPFA, and P is an $\aleph_1$-complete forcing, $\underset{\sim}{S}$ is a P-name, and $\Vdash_P$ "$\underset{\sim}{S} \subseteq \omega_1$ is stationary". For $i < \omega_1$ let $(P_i, \underset{\sim}{S}_i)$ be isomorphic to $(P, \underset{\sim}{S})$, and let P^* be the product of $P_i (i < \omega_1)$ with countable support; so

$P_i \lessdot P^*, P^*$ is $\aleph_1$-complete, and $\underset{\sim}{S}_i$ is a P^*-name and $\Vdash_{P_i}$ "P^*/P_i does not destroy stationarity of subsets of ω_1".

Let $\Xi = \{A \in V : A \subseteq \omega_1, A$ is stationary and $\Vdash_P$ "$\underset{\sim}{S} \cap A$ is not stationary"$\}$. Clearly if $A \in \Xi$ and $B \subseteq A$ is stationary then $B \in \Xi$. Let $\{A_i : i < i^*\} \subseteq \Xi$ be a maximal antichain $\subseteq \Xi$ (i.e., the intersection of any two elements is not stationary).

So, by 1.12 $|i^*| \leq \omega_1$, so without loss of generality $i^* \leq \omega_1$ and define $A_i = \emptyset$ for $i \in [i^*, \omega_1)$. Let $A = \nabla_{i<\omega_1} A_i$. Then also $\Vdash_P$ "$A = \nabla_{i<\omega_1} A_i$", so we have:

(i) $\Vdash_P$ "$\underset{\sim}{S} \cap A$ is not stationary", and

(ii) for every stationary $B \subseteq \omega_1 \setminus A$, for some $p \in P$, we have $p \Vdash_{P^*}$ "$\underset{\sim}{S} \cap B$ is stationary".

Let $\hat{S} \overset{\text{def}}{=} \omega_1 \setminus A$. So $\hat{S}$ is stationary (as $\Vdash_P$ "$\underset{\sim}{S}$ is stationary"). Also, clearly,

(iii) for each $i < \omega_1$, and stationary $B \subseteq \hat{S}$ for some $p \in P_i \lessdot P^*$, we have $p \Vdash_{P^*}$ "$\underset{\sim}{S}_i \cap B$ is stationary".

As P^* is the product of the P_i with countable support, P^*/P_i does not destroy stationarity of subsets of ω_1, so we have

(iv) for every stationary $B \subseteq \hat{S}, \Vdash_{P^*}$ "for some $i, \underset{\sim}{S}_i \cap B$ is stationary".

Let $\underset{\sim}{S}^*$ be the P^*-name: $\nabla_{i<\omega_1} \underset{\sim}{S}_i \overset{\text{def}}{=} \{\alpha < \omega_1 : (\exists i < \alpha)\alpha \in \underset{\sim}{S}_i\}$. So $\Vdash_{P^*}$ "for every stationary $B \subseteq \hat{S}$ (from V), we have $B \cap \underset{\sim}{S}^*$ is stationary".

In V^{P^*} let Q^* be shooting a club $\underset{\sim}{C}$ through $A \cup S^*$ (i.e., $Q^* = \{h : h$ an increasing continuous function from some non-limit $\alpha < \omega_1$ into $A \cup S\}$ ordered naturally). Now Q^* does not destroy any stationary subset of ω_1 from V (though it destroys some from V^{P^*}). So $P^* * \underset{\sim}{Q}^*$ does not destroy any stationary subsets of ω_1 from V; hence by Lemma 1.3 it is semiproper. Now if $G \subseteq P^* * \underset{\sim}{Q}^*$ is generic enough, for each $i < \omega_1, G \cap P_i$ is generic enough such that $\underset{\sim}{S}_i[G]$ is well-defined, and since $C^* = \underset{\sim}{C}[G]$ is a club set and $C^* \subseteq A \cup \nabla_{i<\omega_1} \underset{\sim}{S}_i[G]$, we have $\hat{S} \cap C^* \subseteq \nabla_{i<\omega_1} \underset{\sim}{S}_i[G]$. As $\hat{S}$ is stationary, for some $i, \underset{\sim}{S}_i[G]$ is stationary so the projection of G to $G_i \subseteq P_i$ is as required, and we have finished. $\square_{2.14}$

2.15 Remark. A similar proof works if $P = P^a * \underset{\sim}{P}^b$, where P^a satisfies the $\aleph_1$-c.c. and $\underset{\sim}{P}^b$ is $\aleph_1$-complete in V^{P_a}, if we use $P^* = \{f : f$ a function from ω_1 to $\underset{\sim}{P}, f(i) = (p_i, q_i) \in P^a * \underset{\sim}{P}^b, |\{i : p_i \neq \emptyset\}| < \aleph_0, |\{i : q_i \neq \emptyset\}| < \aleph_1\}$. Note that necessarily even any finite power of P^a satisfies the $\aleph_1$-c.c. In short, we need that some product of copies of P is semiproper, i.e:

2.16 Fact. [SPFA] Suppose Q is a semi proper forcing notion, and there is a forcing notion P and a family of complete embeddings f_i $(i < i^*)$ of P into Q such that:

(a) for any $p \in P$ and $q \in Q$ for some i, the conditions $f_i(p), q$ are compatible with Q.

(b) the forcing $Q/f_i(P)$ does not destroy the stationarity of subsets of ω_1.

Then for any dense subsets $\mathcal{I}_\alpha$ of P for $\alpha < \omega_1$, and $\underset{\sim}{S}$ a P-name of a subset of ω_1, $\Vdash_P$ "$\underset{\sim}{S} \subseteq \omega_1$ is stationary" *there is* a directed $G \subseteq P$, not disjoint to any $\mathcal{I}_\alpha$ (for $\alpha < \omega_1$) such that $\underset{\sim}{S}[G]$ is a well defined stationary subset of ω_1.

Proof. Like 2.14. We define $A \subseteq \omega_1$ satisfying for $\underset{\sim}{S}$ and P the following conditions (from the proof of 2.14): (i), (ii), hence (iii), (iv) (with $P_i = f_i(P)$ and $\underset{\sim}{S}_i = f_i(\underset{\sim}{S})$). $\square_{2.16}$

§3. Canonical Functions for ω_1

3.1 Definition. 1) We define by induction on α, when a function $f : \omega_1 \to$ ordinals is an α-th canonical function:

f is an α-th canonical function (sometimes abbreviated "f is an α-th function" *iff*

(a) for every $\beta < \alpha$ there is a β-th function, $f_\beta < f$ mod $\mathcal{D}_{\omega_1}$

(b) f is a function from ω_1 to the ordinals, and for every $f^1 : \omega_1 \to$ Ord, if $A^1 = \{i < \omega_1 : f^1(i) < f(i)\}$ is stationary then for some $\beta < \alpha$ and β-th function $f^2 : \omega_1 \to$ Ord the set $A^2 \overset{\text{def}}{=} \{i \in A^1 : f^2(i) = f^1(i)\}$ is stationary,

2) If we replace a "stationary subset of ω_1" by "$\neq \emptyset \bmod \mathcal{D}$" ($\mathcal{D}$ any filter on ω_1); we write "f is a $(\mathcal{D}, \alpha)$-th function". Of course we can replace ω_1 by higher cardinals.

Remember

3.2 Claim. 1) If $\alpha < \omega_2, \alpha = \bigcup_{i<\omega_1} a_i$, $\langle a_i : i < \omega_1 \rangle$ is increasing continuous, each a_i is countable, and $f_\alpha(i) \overset{\text{def}}{=} \text{otp}(a_i)$ *then* f_α is an α-th function.
2) If for every α there is an α-th function, *then* $\mathcal{D}_{\omega_1}$ is precipitous; really "for every $\alpha < (2^{\aleph_1})^+$ there is α-th function" suffices, in fact those three statements are equivalent.
3) If f is an α-th function; $Q = \mathcal{D}^+_{\omega_1} = \{A \subseteq \omega_1 : A \text{ is stationary}\}$ (ordered by inverse inclusion) then $\Vdash_Q$ "in $V^{\omega_1}/\underset{\sim}{G}_Q$, we have: $\{x : V^{\omega_1}/\underset{\sim}{G}_Q \models$ "x is an ordinal $< f_\alpha/\underset{\sim}{G}_Q$"$\}$ is well ordered of order type α" (remember $V^{\omega_1}/\underset{\sim}{G}_Q$ is the "generic ultrapower" with universe $\{f/\underset{\sim}{G}_Q : f \in V \text{ and } f : \omega_1 \to V\}$ and $\underset{\sim}{G}_Q$ is an ultrafilter on the Boolean algebra $\mathcal{P}(\omega_1)^V$).
4) Any two α-th functions are equal modulo $\mathcal{D}_{\omega_1}$.
5) Similarly for the other filters (we have to require them to be $\aleph_1$-complete, and for (1) - also normal).

Proof. Well known, see [J]. We will only show (1): Let $A^1 = \{i : f(i) < f_\alpha(i)\}$ be stationary. So there is a countable elementary model $N \prec H(\chi)$ (for some large χ) containing $\alpha, f, \langle a_i : i < \omega_1 \rangle$ such that $\delta \overset{\text{def}}{=} N \cap \omega_1 \in A^1$. We have $f(\delta) < f_\alpha(\delta) = \text{otp}(a_\delta)$, and $a_\delta = \bigcup_{i \in N} a_i \subseteq N$, so there is $\beta \in N$ such that $f(\delta) = \text{otp}(a_\delta \cap \beta)$. Let $A^2 = \{i \in A^1 : f(i) = \text{otp}(a_i \cap \beta)\}$. Since $A^2 \in N$, $f \in N$, $\beta \in N$, $\langle a_i : i < \omega_1 \rangle \in N$ and $\delta \in A^2$, we can deduce A^2 is stationary. $\square_{3.2}$

The following answers a question of Velickovic:

3.3 Theorem. Let κ be a supercompact. For some κ-c.c. forcing notion P not collapsing $\aleph_1$ we have that V^P satisfies:

(a) there is $f \in {}^{\omega_1}\omega_1$ bigger (mod $\mathcal{D}_{\omega_1}$) than the first ω_2 function hence the Chang conjecture fails.

(b) PFA (so $\mathcal{D}_{\omega_1}$ is semiproper hence precipitous).

(c) not PFA^+

Outline of the proof: In 3.4 we define a statement $(*)_g$, which we may assume to hold in the ground model (3.5). We define a set $S^g_\chi \subseteq \mathcal{S}_{<\aleph_1}(\chi)$ and we show that if $(*)_g$ holds, then S^g_χ is stationary (3.8). In 3.9 we recall that the class of S^g_χ-proper forcing notions is closed under CS iterations, so assuming a supercompact cardinal we can, in the usual way, force $Ax[S^g_\chi\text{-proper}]$. Finally we find, for each $\alpha < \omega_2$, an S^g_χ-proper forcing notion R_α such that $Ax[R_\alpha] \Rightarrow f_\alpha <_{\mathcal{D}_{\omega_2}} g$.

3.3A Remark. Remember that the first clause of 3.3(a) implies that Chang's conjecture fails, so the negation of 3.3(a) is sometimes called the "weak Chang conjecture".

Proof of 3.3*A*. Let $M = (M, E, \omega_1, \ldots)$ be a model with universe ω_2 which codes enough set theory. Assume that there exists an elementary submodel $N \prec M$ with $\|N\| = \aleph_1$, $|\omega_1^N| = \aleph_0$. Let $\delta = \omega_1^N = \omega_1 \cap N$. In M we have the function f from 3.3(a) and also a family $\langle f_\alpha, E_\alpha : \alpha < \omega_2\rangle$, ($f_\alpha$ is an α-th canonical function, $E_\alpha \subseteq \omega_1$ is a club set, $f_\alpha \restriction E_\alpha < f \restriction E_\alpha$) as well as a family $\langle E_{\alpha,\beta} : \alpha < \beta < \omega_2\rangle$ of clubs of ω_1 satisfying $f_\alpha \restriction E_{\alpha\beta} < f_\beta \restriction E_{\alpha\beta}$. For $\alpha < \beta$, $\alpha, \beta \in N$ we have $\delta \in E_{\alpha,\beta} \cap E_\beta$, so

(A) $(\forall \alpha, \beta \in N)\ [\alpha < \beta \Rightarrow f_\alpha(\delta) < f_\beta(\delta)]$

(B) $(\forall \alpha \in N)$: $[f_\alpha(\delta) < f(\delta)]$

So the set $\{f_\alpha(\delta) : \alpha \in N\}$ is uncountable (by (A)) and bounded in ω_1 (by (B)), a contradiction. $\square_{3.3A}$

3.4 Definition. Let f_α be the α'th canonical function for every $\alpha < \omega_2$ (so without loss of generality the f_α are of the form described in 3.2(1)). Let

$g : \omega_1 \to \mathrm{Ord}$. We let $(*)_g$ be the statement:

$$(*)_g \qquad \text{for all } \alpha < \omega_2 \text{ we have} \quad \neg(g <_{\mathcal{D}_{\omega_1}} f_\alpha).$$

By 3.2(4) this definition does not depend on the choice of $\langle f_\alpha : \alpha < \omega_2 \rangle$.

3.5 Remark. It is easy to force a function $g : \omega_1 \to \omega_1$ for which $(*)_g$ holds: let $P = \{h : \text{for some } i < \omega_1,\ h : i \longrightarrow \omega_1\}$ ordered by inclusion. P is $\aleph_1$-complete and $(2^{\aleph_0})^+$-c.c., so assuming CH we get $\aleph_1^{V^P} = \aleph_1^V$ and $\aleph_2^{V^P} = \aleph_2^V$. Let $\langle f_\alpha : \alpha < \omega_2 \rangle$ be the first ω_2 canonical function in V, then they are still canonical in V^P, and it is easy to see that for any $f : \omega_1 \to \omega_1$ in V we have $V^P \models \neg(g <_{\mathcal{D}_{\omega_1}} f)$ where g is the generic function for P.

3.6 Definition. 1) We call $N \prec (H(\chi), \in, <^*_\lambda)$ g-small (in short $g - sm$ or more precisely (g, χ)-small) if N is countable and $\mathrm{otp}(N \cap \chi) < g(N \cap \omega_1)$.
2) We let $S^g_\chi \overset{\mathrm{def}}{=} \{a : a \in \mathcal{S}_{\leq \aleph_0}(\chi), a \cap \omega_1 \text{ is an ordinal and } \mathrm{otp}(a) < g(a \cap \omega_1)\}$

3.7 Definition. We call a forcing notion Q g-small proper *if*: for any large enough χ and $N \prec (H(\chi), \in, <^*_\chi)$, satisfying $||N|| = \aleph_0$, $Q \in N$, $p \in N \cap Q$ such that N is g-small *there is* $q \geq p$ which is (N, Q)-generic. We write g-sm for g-small.

3.7A Observation. 1) Any proper forcing is g-sm proper.
2) Without loss of generality g is nondecreasing.

Proof. 1) Trivial.
2) Let $E = \{\alpha < \omega_1 : \alpha \text{ is a limit ordinal such that } \beta < \alpha \Rightarrow g(\beta) < \alpha$ and $(\forall \beta < \alpha)(\exists \gamma)(\beta < \gamma < \alpha \,\&\, g(\gamma) > \beta)\}$, and let

$$g'(\alpha) = \begin{cases} g(\alpha) & \text{if } \alpha \in E, g(\alpha) \geq \alpha \\ \sup\{g(\beta) : \beta < \alpha\} & \text{otherwise.} \end{cases}$$

Now, for our definition g', g are equivalent but g' is not decreasing. $\square_{3.7A}$

3.8 Claim. 1) $(*)_g$ holds
iff for every $\chi \geq \aleph_2$ the set S^g_χ is a stationary subset of $\mathcal{S}_{<\aleph_1}(\chi)$
iff $S^g_{\aleph_2}$ is a stationary subset of $\mathcal{S}_{<\aleph_1}(\aleph_2)$
iff for some $\chi \geq \aleph_2$, S^g_χ is a stationary subset of $\mathcal{S}_{<\aleph_1}(\chi)$.
2) For a forcing notion Q and $\chi > 2^{|Q|}$ we have: Q is g-sm proper *iff* Q is S^g_χ-proper (see V1.1(2)).
3) If $(*)_g$ holds and Q is g-sm proper *then*

$$\Vdash_Q \text{“}(*)_g\text{”}$$

Proof. 1) first implies second

Assume $(*)_g$ holds, $\chi \geq \aleph_2$ is given, and we shall prove that S^g_χ is a stationary subset of $\mathcal{S}_{<\aleph_1}(\chi)$. Let $x \in H(\chi_1)$ and $\chi_1 = \beth_3(\chi)^+$(e.g $x = S^g_\chi$).

We can choose by induction on $i < \omega_1, N_i \prec (H(\chi_1), \in, <^*_{\chi_1})$ increasing continuous, countable, $x \in N_i \in N_{i+1}$. Clearly for each i we have $\delta_i \stackrel{\text{def}}{=} N_i \cap \omega_1$ is a countable ordinal, and the sequence $\langle \delta_i : i < \omega_1 \rangle$ is strictly increasing continuous. Now letting $N = \bigcup_{i<\omega_1} N_i$, then $\omega_1 + 1 \subseteq N \prec (H(\chi_1), \in, <^*_{\chi_1})$ and N has cardinality $\aleph_1$, so $\text{otp}(N \cap \chi) = \alpha$ for some $\alpha < \omega_2$; let $h : N \cap \chi \to \alpha$ be order preserving from $N \cap \chi$ onto α.
Note: letting $a^1_i \stackrel{\text{def}}{=} N_i \cap \chi, a_i = \text{rang}(h \restriction a^1_i)$ we have: α is $\bigcup_{i<\omega_1} a_i$ where a_i is countable increasing continuous in i and $f_{\alpha+1}(i) \stackrel{\text{def}}{=} \text{otp}(a_i) + 1$ is an $(\alpha + 1)$-th function (see 3.2(1)). Also $C = \{i : \delta_i = i\}$ is a club of ω_1 so by $(*)_g$ we can find $i \in C$ such that $f_{\alpha+1}(i) \leq g(i)$, so $\text{otp}(N_i \cap \chi) = \text{otp}(a^1_i) = \text{otp}(a_i) < f_{\alpha+1}(i) \leq g(i) = g(\delta_i) = g(N_i \cap \omega_1)$. I.e. for this i, N_i is g-sm; easily $N_i \cap \chi \in S^g_\chi$ and it exemplifies that S^g_χ is stationary.

second implies fourth. Trivial

fourth implies third. Check. (note: for $\chi \geq \aleph_2$, $\text{otp}(\chi \cap N) \geq \text{otp}(\omega_2 \cap N)$).

third implies first. Let $\alpha < \omega_2, \alpha = \bigcup_{i<\omega_1} a_i$, where a_i are increasing continuous each a_i countable, so $f_\alpha(i) \stackrel{\text{def}}{=} \text{otp}(a_i)$ is an α-th function and let C be a club of ω_1. Let $\bar{a} = \langle a_i : i < \omega_1 \rangle$. Let χ be regular large enough (e.g.

$\geq \beth_3^+$). Clearly

$$\{N \cap \aleph_2 : N \text{ is countable, } N \prec (H(\chi), \in, <^*_\chi)\}$$

is a club of $\mathcal{S}_{\aleph_0}(\aleph_2)$. So by assumption for some countable $N \prec (H(\chi), \in, <^*_\chi)$ we have $C, \bar{a} \in N$ and

(i) $\mathrm{otp}(N \cap \aleph_2) < g(N \cap \omega_1)$.

But as $\bar{a} \in N$ also $f_\alpha \in N$ and we have $[j < N \cap \omega_1 \Rightarrow a_j \in N \Rightarrow a_j \subseteq N]$ hence $\bigcup\{a_j : j < N \cap \omega_1\} \subseteq N \cap \alpha$ but this union is equal to $a_{N \cap \omega_1}$($\bar{a}$ is increasing continuous:) so, as $\alpha \in N$,

(ii) $\mathrm{otp}(a_{N\cap\omega_1}) < \mathrm{otp}(a_{N\cap\omega_1} \cup \{\alpha\}) \leq \mathrm{otp}(N \cap \omega_2)$.

But

(iii) $f_\alpha(N \cap \omega_1) = \mathrm{otp}(a_{N\cap\omega_1})$.

By (i) + (ii) + (iii) we get $f_\alpha(N \cap \omega_1) < g(N \cap \omega_1)$ and trivially $N \cap \omega_1 \in C$, but C was any club of ω_1, hence $\{j < \omega_1 : f_\alpha(j) < g(j)\}$ is stationary. As α was any ordinal $< \omega_2$ we get the desired conclusion.

(2) This is almost trivial, the only point is that to check S^g_χ-properness it is enough to consider models $N \prec (H(\chi), \in, <^*_\chi)$, but for sm-g properness we should consider $N \prec (H(\chi_0), \in, <^*_{\chi_0})$ for all large enough χ_0. First assume Q is g-sm proper, and we shall prove that Q is S^g_χ-proper; and let χ_0 be large enough (say $> \beth_2(\chi)$). Let M be the Skolem Hull of $\{\alpha : \alpha \leq 2^{|Q|}\} \cup \{Q, \chi\}$ in $(H(\chi_0), \in, <^*_{\chi_0})$. Note $\|M\| = 2^{|Q|} < \chi$ hence $\mathrm{otp}(M \cap \chi_0) < \chi$ and there is an order-preserving $h : M \cap \chi \to (2^{|Q|})^+ \leq \chi$ onto an ordinal belonging to N. Let N be a countable elementary submodel of $(H(\chi_0) \in, <^*_{\chi_0})$ to which $x = \langle Q, \chi, M, h\rangle$ belongs, and $(N \cap \chi) \in S^g_\chi$. Let $N' \overset{\mathrm{def}}{=} N \cap M$, so $N' \cap \omega_1 = N \cap \omega_1$, N' is a countable elementary submodel of $(H(\chi_0), \in, <^*_{\chi_0})$ and

$$\begin{aligned}\mathrm{otp}(N' \cap \chi_0) = \mathrm{otp}(h''(N' \cap \chi_0)) &\leq \mathrm{otp}(N \cap \mathrm{Rang}(h)) \\ &\leq \mathrm{otp}(N \cap \chi) < g(N \cap \omega_1) = g(N' \cap \omega_1).\end{aligned}$$

[Why? as h is order presserving; as N is closed under h, h^{-1} and $N' \prec N$; as $\mathrm{rang}(h) \subseteq \chi$; as $N \cap \chi \in S^g_\chi$; as $N' = N \cap M$ respectively.]

Applying "Q is g-sm proper" to N', for every $p \in Q \cap N'$ there is q such that

$p \le q \in Q$ and q is (N', p)-generic. But $Q \cap N = Q \cap N'$ and [q is (N', p)-generic $\Leftrightarrow q$ is (N, p)-generic] as $N \cap 2^{|Q|} = N' \cap 2^{|Q|}$. As we can eliminate "$x \in N$" (as some such x for some χ', $H(\chi') \in H(\chi_0)$ and χ' belongs to N) we have proved Q is S^g_χ-proper.

The other direction should be clear too.

3) Let $\chi = (2^{|Q|})^+$.

By part (2) we know Q is S^g_χ-proper; by V 1.3 - 1.4(2) as Q is S^g_χ-proper, we have that $\Vdash_Q$ "$(S^g_\chi)^V \subseteq \mathcal{S}_{<\aleph_1}(\chi)^{V^Q}$ is stationary". Clearly $\Vdash_Q$ "$(S^g_\chi)^V \subseteq (S^g_\chi)^{V^Q}$" hence $\Vdash_Q$ "$(S^g_\chi)^{V^Q}$ is a stationary subset of $\mathcal{S}_{<\aleph_1}(\chi)$". So by part (1) (fourth implies first), we have $\Vdash_Q$ "$(*)_g$". $\square_{3.8}$

3.9 Claim.

Assume $(*)_g$ (where $g \in {}^{\omega_1}\omega_1$). *Then* the property "(a forcing notion is) g-sm proper" is preserved by countable support iteration (and even strongly preserved).

Proof. Immediate by V 2.3 and by 3.8(2) above. $\square_{3.9}$

3.10 Claim. Suppose, $g \in {}^{\omega_1}\omega_1$, and $(*)_g$ holds, κ supercompact, $L^* : \kappa \to H(\kappa)$ is a Laver diamond (see VII 2.8) and we define $\bar{Q} = \langle P_i, \underset{\sim}{Q}_j : i \le \kappa, j < \kappa \rangle$ as follows:

(i) it is a countable support iteration

(ii) for each i, *if* $L^*(i)$ is a P_i-name of a g-sm proper forcing and i is limit then $\underset{\sim}{Q}_i = L^*(i)$, *otherwise* $\underset{\sim}{Q}_i = \mathrm{Levy}(\aleph_1, 2^{\aleph_2})$, (in V^{P_i}, i.e. a P_i-name).

Then

(a) P_κ is g-sm proper, κ-c.c. forcing notion of cardinality κ, and $\aleph_2^{V[P_\kappa]} = \kappa$

(b) Ax_{ω_1}[g-sm proper] holds in V^{P_κ}

(c) PFA holds in V^{P_κ}

(d) in V^{P_κ} for every $\alpha < \kappa, g$ is above the α-th function (by $<_{\mathcal{D}_{\omega_1}}$).

Proof. $\bar{Q}$ is well defined by III 3.1B.

Clearly $\Vdash_{P_i}$ "Q_i is g-sm proper" - by choice or as Levy$(\aleph_1, 2^{\aleph_2})^{V[P_i]}$ is $\aleph_1$-complete hence proper hence (by 3.7) g-sm proper. So by 3.9 the forcing P_κ is g-sm proper; P satisfies κ-c.c. by III 4.1 hence $\Vdash_{P_\kappa}$ "κ regular, $\aleph_1^V$ regular".

The use of Levy $(\aleph_1, 2^{\aleph_2})^{V[P_i]}$ for i non-limit will guarantee $\kappa = \aleph_2$ in V^{P_κ}. Also $|P_\kappa| = \kappa$ is trivial, so (a) holds.

The proof of (b) is like the consistency of $\Vdash_P$ "Ax_{ω_1}[proper]", in VII 2.8 hence (by 3.7A(1)) we have $\Vdash_{P_\kappa}$ "PFA" i.e. (c) hold.

So it remains to prove (d), so let $\alpha < \aleph_2^{V[P_\kappa]} = \kappa$. This will follow from 3.10A, 3.10B, 3.10C below together with (b) above. Let us define a forcing notion R_α:

3.10A Definition. $R_\alpha = \{\langle a_i : i \leq j\rangle : j$ is a countable ordinal, each a_i is a countable subset of α and $\langle a_i : i \leq j\rangle$ is increasing continuous, and for i a limit ordinal $\text{otp}(a_i) < g(i)\}$. The order is: $p \leq q$ iff p is an initial segment of q. We can assume g is nondecreasing (see 3.7A(2)).

3.10B Observation. R_α is g-sm proper.

Proof. Left to the reader.

3.10C Observation. If $G \subseteq R_\alpha$ is sufficiently generic, *then* G defines an increasing continuous sequence $\langle a_i : i < \omega_1\rangle$ with $\bigcup_{i<\omega_1} a_i = \alpha$ and hence defines an α-th canonical function below g. $\square_{3.10,3.3}$

* * *

Answering a question of Judah:

Question. Does Ax[Countably Complete $*$ c.c.c.] imply PFA?

3.11 Claim. The answer is no.

Proof. Countably complete forcings and c.c.c. forcings and also their composition are ω-proper. So we have

$PFA \Rightarrow$Ax[ω-proper]$\Rightarrow$Ax[countably complete $*$ c.c.c.].

We will show that the first implication cannot be reversed:

3.12 Definition. $\bar{c} = \langle c(i) : i < \omega_1 \rangle$ is a ω-club guessing for ω_1 means that $c(i)$ is an unbounded subset of i of order type ω for each limit ordinal i less than ω_1, such that every closed unbounded subset c of ω_1 includes $c(i)$ for some limit ordinal $i < \omega_1$.

3.13 Claim. (1) If $\bar{c}$ is a ω-club guessing for ω_1, and P is ω-proper, then $\Vdash_P$ "$\bar{c}$ is a ω-club guessing for ω_1".
(2) $\diamondsuit_{\omega_1}$ implies that there is a ω-club guessing for ω_1 (so a ω-club guessing can be obtained by a small forcing notion).

Proof. (1): Let $\underset{\sim}{C}$ be a name for a closed unbounded subset of ω_1, $p \in P$. We need to find a condition $q \geq p$ and some $i < \omega_1$ such that $q \Vdash_P$ "$c(i) \subseteq \underset{\sim}{C}$". Let $\langle N_i : i < \omega_1 \rangle$ be an increasing continuous sequence of countable models $N_i \prec (H(\chi), \in <^*_\chi)$, χ large enough, $\{p, \underset{\sim}{C}, P\} \in N_0$. Let $\delta_i = N_i \cap \omega_1$. Let $C^* = \{i < \omega_1 : \delta_i = i\}$. Now C^* is closed unbounded, so there is some i such that $c(i) \subseteq C^*$, say $c(i) = \{i_0, i_1, \ldots\}$, $i_0 < i_1 < \ldots$. Let $q \geq p$ be N_{i_ℓ}-generic for all $n < \omega$. So $q \Vdash$ "$i_\ell = N_{i_\ell}[G] \cap \omega_1 = N_{i_\ell} \cap \omega_1$", and clearly $\Vdash$ "$N_{i_\ell}[G] \cap \omega_1 \in \underset{\sim}{C}$", so $q \Vdash$ "$c(i) \subseteq \underset{\sim}{C}$".
(2) Should be clear. $\square_{3.13}$

3.14 Claim. Suppose $\bar{c} = \langle c_\delta : \delta < \omega_1 \rangle$ is such that: c_δ is a closed subset of δ of order type $\leq \alpha^*$. Let
$R_{\bar{c}} \overset{\text{def}}{=} \{(i, C) : i < \omega_1, C$ is a closed subset of $i+1$, such that for every $\delta \leq i$, $\sup(c_\delta \cap C) < \delta\}$,
order is natural. Let
$\mathcal{I}_\gamma \overset{\text{def}}{=} \{(i, C) \in R_{\bar{c}} : \gamma < \max(C)\}$.

Then: $R_{\bar{c}}$ is proper, each $\mathcal{I}_\gamma$ is a dense subset of $R_{\bar{c}}$, and if $G \subseteq R_c$ is directed not disjoint to each $\mathcal{I}_\gamma$, then $C^* = \cup\{C : (i, C) \in G\}$ is a club of $\aleph_1$ such that: $\delta < \omega_1 \Rightarrow \sup(C \cap c_\delta) < \delta$.

Proof. Straight.

For proving "$R_{\bar{c}}$ is proper" denote $q = (i^q, C^q)$, $i^q = \mathrm{Dom}(q)$, let $N \prec (H(\chi), \in, <^*_\chi)$, N countable, $p \in N \cap R_{\bar{c}}$, and $\{\bar{c}, R_{\bar{b}}, \alpha\} \in N$. W.l.o.g. $\beth_7^+ < \chi$. Let $\delta = N \cap \omega_1$, and so we can find $\langle N_i : i < \delta\rangle$, an increasing continuous sequence of elementary submodels of $(H(\beth_7^+), \in)$, $N_i \subseteq N$, $N \cap H(\beth_7^+) = \bigcup_{i<\delta} N_i$ and $p \in N_0$. So we can find $i_0 < i_1 < \ldots$, $\delta = \bigcup_{\ell<\omega} i_\ell$ such that $\omega_1 \cap N_{i_\ell+1} \setminus N_{i_\ell}$ is disjoint to c_δ. Let $\langle \underset{\sim}{\tau}_n : n < \omega\rangle$ list the $R_{\bar{c}}$-names of ordinals from N, and we can choose by induction on n a condition p_n, q_n such that: $p \leq p_0 \in N_{i_0+1}$, i^{p_0} is $N_{i_0} \cap \omega_1$, and $[i^p, N_{i_0+1} \cap \omega_1)$ is disjoint to C^{p_0}, $p_n \leq q_n \in R_{\bar{c}} \cap N_{i_n+1}$, q_n force a value to $\underset{\sim}{\tau}_\ell$ if $\ell \leq n$ & $\underset{\sim}{\tau}_\ell \in N_{i_n+1}$, and $q_n \leq p_{n+1}$, $i^{p_{n+1}} = N_{i_{n+1}} \cap \omega_1$, and $[i^{p_{n+1}}, N_{i_{n+1}} \cap \omega_1)$ is disjoint to $c^{p_{n+1}}$. Now $\langle p_n : n < \omega\rangle$ has a limit as required.

Another presentation is noting:

(*) for each $p^* = (i^*, C^*) \in R_{\bar{c}}$ and dense subset $\mathcal{I}$ of P, there is a club $E = E_{q,\mathcal{I}}$ of ω_1 such that:

for every $\alpha \in E$, $\alpha > i^*$, and there is $(i^\alpha, C^\alpha) \in R_{\bar{c}}$, $(i^\alpha, C^\alpha) \geq (\alpha, C^*) \geq (i^*, C^*)$, (i^α, C^α) is in $\mathcal{I}$ and $i^\alpha < \min(E \setminus (\alpha+1))$.

(**) *if* $p \in N \prec (H(\chi), \in, <^*_\chi)$, N countable, $\{\bar{c}, R_{\bar{c}}, \alpha^*\} \in N$, and $\mathcal{I} \in N$ a dense subset of $R_{\bar{c}}$, *then* $E_{p,\mathcal{I}} \cap N$ has order type $N \cap \omega_1$ hence for unbounded many $\alpha \in N \cap E_{p,\mathcal{I}}$, the interval $[\alpha, \min(E \setminus (\alpha+1)))$ is disjoint to $c_{N\cap\omega_1}$. $\square_{3.14}$

3.14A Conclusion. PFA$\Rightarrow$ there is no ω-club guessing on ω_1. On the other hand "Ax[ω-proper]+ there is a ω-club guessing" is consistent, since starting from a supercompact we can force Ax[ω-proper] with an ω-proper iteration (see V3.5). $\square_{3.11}$

3.15 Remark. The generalization to higher properness should be clear: for α additively indecomposable, Ax[α-proper] is consistent with existence of $\langle c(i) : i < \omega_1$ and α divides $i\rangle$ as in 3.12 only the order type of $c(i)$ is α (for a club of i's), for it to be preserved we use $\bar{c} = \langle c(i) : i < \omega_1$, and α devides $i\rangle$ such that for every γ the set $\{c(i) \cap \gamma : i < \omega_1$ divisible by α and $\gamma \in C(i)\}$ is countable.

On the other hand Ax[α-proper] implies there is no $\langle c(i) : i < \omega_1, \alpha\omega$ divides $i\rangle$ such that: $c(i)$ is a club of i of order type $\alpha\omega$ and for every club C of ω_1 for some i, $c(i) \subseteq C$.

§4. A Largeness of $\mathcal{D}_{\omega_1}$ in Forcing Extensions of L and Canonical Functions

The existence of canonical functions is a "large cardinal property" of ω_1, or more precisely, of the filter $\mathcal{D}_{\omega_1}$. For example, the statement "the α-th canonical function exists for any α" will hold if $\mathcal{D}_{\omega_1}$ is $\aleph_2$-saturated, and it implies that the generic ultrapower V^{ω_1}/G_Q (see 3.2(3)) is well-founded. If we know only that ω_1 is a canonical function, we can conclude that the generic ultrapower is well-founded at least below ω_1^V.

It was shown by Jech and Powell [JePo] that the statement "ω_1 is a canonical function" implies the consistency of various mildly large cardinals. Jech and Shelah [JeSh:378] showed how to force the $\aleph_2$-th (or the θth, for any θ) canonical function to exist (this is weaker than "ω_1 is a cannonical function"). After this paper Jech reasked me a question from [JePo]: "if the function ω_1 is a canonical function, does $0^\#$ exist?" We give here a negative answer. Our proof which uses large cardinals whose existence is compatible with the axiom $V = L$, is in the general style of this book: quite flexible iterations, quite specific to preserving $\aleph_1$. We thank Menachem Magidor for many stimulating discussions on the subject. Subsequently Magidor and Woodin find an equiconsistency results with different method.

This section consists of two parts: First we define a large cardinal property $(*)^1_\lambda$ and show (in 4.3)

$$\mathrm{Con}\Big(\ (\exists G)\Big[V = L[G] + G \subseteq \omega_1 \text{ is generic for a forcing in } L + (\exists\lambda)(*)^1_\lambda\Big]\Big),$$

assuming the existence of $0^\#$ or some suitable strong partition relation. Then we show (in 4.6, 4.7) that $(*)^1_\lambda$ implies that there is a generic extension of the universe in which ω_1 is a λ-function, and make some remarks about possible cardinal arithmetic in this extension.

We think that the proof of 4.6 is also interesting for its own sake, as it gives a method for proving large cardinal properties of $\mathcal{D}_{\omega_1}$ from consistency assumptions below $0^\#$.

4.1 Definition. $\lambda \rightarrow^+ (\kappa)^{<\omega}_\mu$ means that for every club C of λ and function F : $[\lambda]^{<\omega} \rightarrow \mu$ there is $X \subseteq C, \mathrm{otp}(X) = \kappa$ such that: $u_1, u_2 \subseteq X \cup \min(X), |u_1| = |u_2| < \aleph_0, u_1 \cap \min(X) = u_2 \cap \min(X)$ implies $F(u_1) = F(u_2)$. Let $\lambda \rightarrow (\kappa)^{<\omega}_{<\lambda}$ mean: if $F : [\lambda]^{<\omega} \rightarrow \lambda$, $F(u) < \min(u \cup \{\lambda\})$, then for some $X \subseteq \lambda$, $\mathrm{otp}(X) = \kappa$ and $F \restriction [X]^n$ constant for each n.

By the known analysis

4.2 Remark. 1) If λ is minimal such that $\lambda \rightarrow (\kappa)^{<\omega}_\mu$ then $\lambda \rightarrow (\kappa)^{<\omega}_{<\lambda}$ and λ is regular and $2^\theta < \lambda$ for $\theta < \lambda$, from which it is easy to see $\lambda \rightarrow^+ (\kappa)^{<\omega}_\mu$. Such λ's are Erdős cardinals, which for $\kappa \geq \omega_1$ implies the existence of $0^\#$ so implies $V \neq L$. But of course it has consequences in L.

2) Remember $A^{[n]} = \{b : b \subseteq A, |b| = n\}$.

3) Of course $\mu \geq 2$ is assumed.

4) $\lambda \rightarrow^+ (\kappa)^{<\omega}_\mu$ implies λ is regular, $\mu < \lambda$, and $\lambda \rightarrow^+ (\kappa)^{<\omega}_{\mu_1}$ for any $\mu_1 < \lambda$.

4.3 Claim. If in V: $\lambda \rightarrow^+ (\kappa)^{<\omega}_\kappa$ and κ is regular uncountable, (hence $\lambda > 2^\kappa$) *then* in $V^{\mathrm{Levy}(\aleph_0, <\kappa)}$ and even in $L^{\mathrm{Levy}(\aleph_0, <\kappa)}$ (the constructible universe after we force with the Levy collapse) $(*)^1_\lambda$ is satisfied, where:

4.4 Definition. For λ an ordinal, $(*)^1_\lambda$ is the following postulate: for any $\chi > 2^\lambda$, and $x \in H(\chi)$, there are N_0, N_1 such that:

(a) N_0, N_1 are countable elementary submodels of $(H(\chi), \in, <^*_\chi)$

(b) $x \in N_0 \prec N_1$

(c) $\mathrm{otp}(N_0 \cap \lambda) = \mathrm{otp}(N_1 \cap \omega_1)$

(d) in N_1 there is a subset of $\mathrm{Levy}(\aleph_0, N_0 \cap \omega_1)$ generic over N_0.

(e) The collapsing map $f : N_0 \cap \lambda \to \omega_1$ defined by $f(\alpha) = \text{otp}(N_0 \cap \alpha)$ satisfies:

whenever $u \in N_0$, $u \subseteq \lambda$, $|u| \leq \aleph_1$, then $f{\restriction}u \in N_1$ (note $f{\restriction}u$ is $f{\restriction}(u \cap N_0)$).

Proof of 4.3. Straightforward: let $G \subseteq \text{Levy}(\aleph_0, <\kappa)$ be generic over V hence it is also generic over L (note: $\text{Levy}(\aleph_0, <\kappa)^V = \text{Levy}(\aleph_0, <\kappa)^L$). It is also easy to check that $V[G] \models$ "$\lambda \to^+ (\kappa)^{<\omega}_\kappa$ and even $\lambda \to^+ (\kappa)^{<\omega}_{(2^\kappa)}$" because $|\text{Levy}(\aleph_0, <\kappa)| < \lambda$, see 4.2.

Let $\chi > 2^\lambda$, in $L[G]$ and we shall find N_0, N_1, f as required for $L[G], x \in H(\chi)^{L[G]}$ (because $L[G]$ is the case we shall use, $V[G]$ we leave to the reader). In V we can find a strictly increasing sequence $\langle \alpha_i : i < \kappa \rangle$ of ordinals $< \lambda$, indiscernible in $(H(\chi)^{L[G]}, \in, \lambda, G)$, each $\alpha_i \in C^* \stackrel{\text{def}}{=} \{\alpha < \lambda : \alpha$ belongs to any club of λ definable in $(H(\chi)^{L[G]}, \in, \lambda, G)\}$ (so each α_i is a cardinal in $L[G]$). We define, by induction on $n, i_n, N_{0,n}, N_{1,n}$ such that

(α) $\omega \leq i_n < i_{n+1} < \omega_1, i_n$ is limit, $i_0 = \omega$

(β) $N_{0,n}$ is the Skolem Hull of $\{x\} \cup \{\alpha_i : i < i_n\}$ in $(H(\chi)^{L[G]}, \in, \lambda, G)$

(γ) $N_{1,n}$ is the Skolem Hull of $N_{0,n} \cup \bigcup\{\text{otp}(N_{0,n} \cap \lambda) + 1\} \cup \{f_u : u \in N_{0,n}$ is a set of at most $\aleph_1$ of ordinals $< \lambda\}$ where $f_u : u \cap N_{0,n} \to \omega_2$ is defined by $f_u(\alpha) = \text{otp}(N_{0,n} \cap \alpha)$ in the model $(H(\chi)^{L[G]}, \in, \lambda, G)$.

(δ) $i_{n+1} = \text{otp}(N_{1,n} \cap \omega_1)$.

There is no problem to do this. Let $i_\infty \stackrel{\text{def}}{=} \sup\{i_n : n < \omega\}$.

Finally let $N_0 = \bigcup_{n<\omega} N_{0,n}$ and $N_1 = \bigcup_{n<\omega} N_{1,n}$. Now N_0, N_1, f are not necessarily in $L[G]$ but we now proceed to show that they satisfy requirements (a)–(e) from $(*)^1_\lambda$. Clauses (a) and (b) are clear, since the models N_0 and N_1 are unions of elementary chains and $N^0_n \prec N^1_n$ and $x \in N_{0,n}$.

Clearly $N_{1,n} \cap \kappa$ is an initial segment of κ (as $V[G] \models \kappa = \aleph_1$), so $N_{1,n} \cap \kappa$ is an initial segment of $N_{1,n+1} \cap \kappa$. Hence $\text{otp}(N_1 \cap \kappa) = \sup\{\text{otp}(N_{1,n} \cap \kappa) : n < \omega\} = \sup\{i_n : n < \omega\} = i_\infty$. Since $\{\alpha_i : i < i_\infty\} \subseteq N_0$ and the α_i are strictly increasing, we have $\text{otp}(N_0 \cap \lambda) \geq \text{otp}\{i_\alpha : \alpha < \bigcup_{n<\omega} i_n\} = i_\infty$. So $\text{otp}(N_0 \cap \lambda) \geq \text{otp}(N_1 \cap \kappa)$.

For the converse inequality, note that $N_{0,n} \cap \lambda$ is an initial segment of $N_{0,n+1} \cap \lambda$ (as the α_i are indiscernible and in C^* and see Definition 4.1) so $\text{otp}(N_0 \cap \lambda) =$

$\sup\{\text{otp}(N_{0,n} \cap \lambda) : n < \omega\} \leq \sup\{\text{otp}(N_{1,n+1} \cap \omega_1) : n < \omega_1\} \leq \text{otp}(N_1 \cap \omega_1)$. So (c) holds.

Next we have to check (d). Note that N_0 is the Skolem Hull of $\{\alpha_i : i < i_\infty\}$. Let $\delta = N_0 \cap \kappa$; by the previous sentence also $\delta = N_{0,n} \cap \kappa$, and even $N_0 \cap L_\kappa = N_{0,n} \cap L_\kappa$. Let $G = \langle G_\alpha : \alpha < \kappa \rangle$, so $\bigcup G_\alpha$ is a function from ω onto α. Define $Q = \text{Levy}(\aleph_0, \aleph_1)^{N_0}$, $\mathcal{P} = \{\mathcal{I} \cap Q : N_0 \models$ "$\mathcal{I}$ is a dense subset of $Q\}$". Now in $V[G]$, we see that Q is $\text{Levy}(\aleph_0, \delta)$ and $\mathcal{P}$ is a countable family of subsets of Q. Hence for some $\alpha < \kappa$, Q and $\mathcal{P}$ belongs to $V[\langle G_\beta : \beta < \alpha \rangle]$. Without loss of generality $\alpha > \delta$, and α is divisible by $\delta \times \delta$ and without loss of generality $\alpha \in N_{1,1}$ (this is a minor change in the choice of the $N_{0,n}, N_{1,n}$'s). Define $f : \alpha \to \delta$ by $f(\delta i + j) = j$ when $j < \delta$, now $f \circ (\bigcup G_\alpha)$ is a function from ω onto δ, is generic over $V[\langle G_\beta : \beta < \alpha \rangle]$ (for $\text{Levy}(\aleph_0, \alpha)$) hence is generic over N_0 and it belongs to N_1, so demand (d) holds (alternatively we can demand $\langle \alpha_i : i < \kappa \rangle \in V$ and proceed from this.)

Finally clause (e) follows as $N_{0,n} \cap \lambda$ is an initial segment of $N_0 \cap \lambda$ hence defining $f : N_0 \cap \lambda \to \kappa$ by $f(\alpha) \stackrel{\text{def}}{=} \text{otp}(N_0 \cap \alpha)$, used in clause (e) we have: for $u \in N_{0,n}$, $|u| \leq \aleph_1$, $u \subseteq \lambda$, we have $u \cap N_{0,n} = u \cap N_{0,n+1} = u \cap N_0$ (by the choice of the α_i's) and f_u (defined is clause (γ) above) is $f{\restriction}u$ (i.e. $f{\restriction}(u \cap N_0)$) which we have put in $N_{1,n+1}$.

So N_0, N_1, f are as required except possibly not being in $L[G]$. But the statement that such models N_0, N_1 exist is absolute between $L[G]$ and $V[G]$. $\square_{4.3}$

4.5 Claim. $0^\#$ implies that if $\aleph_0 < \kappa < \lambda$ (in V) *then* $L^{\text{Levy}[\aleph_0, <\kappa]}$ satisfies $(*)^1_\lambda$.

Proof. Left to the reader as it is similar to the proof of 4.3. $\square_{4.5}$

4.6 Main Lemma. If $(*)^1_\lambda$, $\lambda = \text{cf}(\lambda) > \aleph_1$, and $2^{\aleph_0} = \aleph_1$ *then* for some forcing notion P:

(i) P satisfies the $\aleph_2$-c.c and has cardinality $(\lambda^{\aleph_1})^+$.

(ii) P does not add new ω-sequences of ordinals.

(iii) $\Vdash_P$ "ω_1 (i.e. the function $\langle \omega_1 : \alpha < \omega_1 \rangle$) is a λ-function".

(iv) $\Vdash_P$ "$2^{\aleph_1} = |P| = [(\lambda^{\aleph_1})^+]^V$" (so for $\mu \geq \aleph_1$ we have $(2^\mu)^{[V^P]} = (2^\mu)^V + \lambda^{\aleph_1}$).

(v) in V^P, for large enough χ and $x \in H(\chi)$ and stationary $S \subseteq \omega_1$ there is a countable $N \prec (H(\chi), \in)$, $x \in N$ such that $N \cap \omega_1 \in S$ and $(\forall f \in N)\big[f \in N \& f \in {}^{\omega_1}\omega_1 \Rightarrow (\exists \alpha \in \lambda \cap N)[N \cap \omega_1 \in \mathrm{eq}(f_\alpha, f)]\big]$, where $\mathrm{eq}(f_\alpha, f) \stackrel{\mathrm{def}}{=} \{i < \omega_1 : f_\alpha(i) = f(i)\}$, and f_α is an α-th function (and $\langle f_\alpha : \alpha < \lambda \rangle \in N$).

4.6A Remark. (a) Let us call a model $N \prec (H(\chi), \in, <^*_\chi)$ "good" if $(\forall f \in N \cap {}^{\omega_1}\omega_1)(\exists \alpha \in \lambda \cap N)\,[N \cap \omega_1 \in \mathrm{eq}(f_\alpha, f)]$ (where $\bar{f} = \langle f_\alpha : \alpha < \lambda \rangle$ is as above); note that this implies $\mathrm{eq}(f_\alpha, f) \subseteq \omega_1$ is stationary.

Let, for $x \in H(\chi)$,

$$\mathcal{M}_x \stackrel{\mathrm{def}}{=} \{N \cap 2^{\aleph_1} : N \text{ is good and, } x \in N\}$$

Note $\mathcal{M}_x \cap \mathcal{M}_y = \mathcal{M}_{\{x,y\}}$. So (v) can be rephrased as:

(v)$'$ The family $\langle \mathcal{M}_x : x \in H(\chi) \rangle$ is a base for a nontrivial filter on $\mathcal{S}_{<\aleph_1}(2^{\aleph_1})$ (i.e. on the Boolean algebra $(\mathcal{S}_{<\aleph_1}(2^{\aleph_1}))$.)

(b) Note that 4.6(ii) implies $\Vdash_P$CH, and (i) and (ii) together imply that P does not change any cofinalities.

(c) 4.6(v) implies almost 4.6(iii): for some $\beta \leq \lambda$, $\langle \omega_1 : \alpha < \omega_1 \rangle$ is a β-th function.

Proof of (c). Let $f : \omega_1 \to \mathrm{Ord}$, $S \stackrel{\mathrm{def}}{=} \{i : f(i) < \omega_1\}$ is stationary, and assume that for all $\alpha < \lambda$ and α-th function f_α the set $\mathrm{eq}(f, f_\alpha) \cap S$ is nonstationary (if there is such a f_α) say disjoint to the club set C_α. Let N be a model as in (v) containing all relevant information. Let $\delta = N \cap \omega_1$ so $\delta \in S$. Then for some $\alpha \in N$ we have $\delta \in \mathrm{eq}(f, f_\alpha) \cap S$ where $f_\alpha \in N$ is an α-th function. But as $\alpha \in N$ we also have $\delta \in C_\alpha$, a contradiction.

4.7 Conclusion. 1) If in V we have $\lambda \to^+ (\kappa)^{<\omega}_\kappa$ (or just $0^\# \in V, \aleph_0 < \kappa < \lambda$ are cardinals in V or just $V = L^{\mathrm{Levy}(\aleph_0, <\kappa)}$ and $V \models (*)^1_\lambda$), *then* in some generic extension V^P of L, $2^{\aleph_0} = \kappa = \aleph_1^V$ and $2^\mu = \lambda^+$ when $\kappa \leq \mu \leq \lambda, 2^\mu = \mu^+$ when $\mu > \lambda$ and ω_1 is a λ-th function (and (v) of 4.6).

2) We can, in the proof of 4.6 below, have $\alpha^* = \gamma$ if $\mathrm{cf}(\gamma) > \lambda$, γ divisible by $|\gamma|$ and $|\gamma| = |\gamma|^{\aleph_1}$ (just more care in bookkeeping) so $\Vdash_P$ "$2^{\aleph_1} = |\gamma|$" is also possible.

3) If e.g. (1) above, and we let $Q = \mathrm{Levy}(\aleph_2, \lambda^+)^{V^P}$ *then* in V^{P*Q} we have $2^{\aleph_0} = \aleph_1$, $2^{\aleph_1} = \aleph_2$ (and conditions (iii)+(v) from 4.6 hold but λ is no longer a cardinal) and V^P, V^{P*Q} has the same functions from ω_1 to the ordinals.

4) We can have in 4.6(1), that V^P satisfies $2^\mu = \lambda$ for $\mu \in [\kappa, \lambda)$ and $2^{\aleph_1} = \lambda$ (and $2^\mu = \mu^+$ when $\mu \geq \lambda$ and ω_1 is a λ-th function).

We shall prove 4.7 later.

Proof of Lemma 4.6. We use a countable support iteration $\bar{Q} = \langle P_\alpha, \underset{\sim}{Q}_\beta : \alpha \leq \alpha^*, \beta < \alpha^* \rangle$, such that:

(1) $\alpha^* = (\lambda^{\aleph_1})^+$

(2) if $\beta < \lambda$, then $\underset{\sim}{Q}_\beta$ is adding a function $f^*_\beta : \omega_1 \to \omega_1$:

$$Q_\beta = \{f : \text{for some non-limit countable ordinal } i < \omega_1,$$
$$f \text{ is a function from } i \text{ to } \omega_1\},$$

order: inclusion.

(3) if $\beta = \lambda + \lambda\beta_1 + \beta_2$ where $\beta_1 < \beta_2 < \lambda$ then $\underset{\sim}{Q}_\beta$ is shooting a club to ω_1 on which $f^*_{\beta_1}$ is smaller than $f^*_{\beta_2}$:

$$\underset{\sim}{Q}_\beta = \big\{a : \text{ for some } i < \omega_1, a \text{ is a function from } \{j : j \leq i\} \text{ to } \{0,1\}$$
$$\text{such that: } \{j \leq i : a(j) = 1\} \text{ is a closed subset of } \mathrm{sm}(\underset{\sim}{f}^*_{\beta_1}, \underset{\sim}{f}^*_{\beta_2})\big\}$$

where $\mathrm{sm}(f,g) \overset{\mathrm{def}}{=} \{i < \omega_1 : f(i) < g(i)\}$,

order: inclusion.

(4) *if* $\beta < (\lambda^{\aleph_1})^+, \beta \geq \lambda^2$ and for some $\underset{\sim}{g}, \underset{\sim}{A}$ and $\gamma \leq \beta$ and p we have

$\otimes^{\beta}_{\underset{\sim}{g},\underset{\sim}{A},\gamma,p}$ $\underset{\sim}{g}$ is a P_γ-name of a function from ω_1 to ω_1, $\underset{\sim}{A}$ is a P_γ-name of a subset of ω_1 and $p \in P_\beta$:

$$p \Vdash_{P_\beta} \text{“}\underset{\sim}{A} \text{ is a stationary subset of } \omega_1, \text{ but for no } \alpha < \lambda, \\ \text{is } \mathrm{eq}[\underset{\sim}{g}, \underset{\sim}{f}_\alpha] \cap \underset{\sim}{A} \text{ stationary”}$$

then for some such $(\underset{\sim}{g}^*_\beta, \underset{\sim}{A}^*_\beta, \gamma^*_\beta, p^*_\beta)$, with minimal γ_β, the forcing notion $\underset{\sim}{Q}_\beta$ is killing the stationarity of $\underset{\sim}{A}^*_\beta$, that is: $\underset{\sim}{Q}_\beta = \{a :$ for some $i < \omega_1, a$ is a function from $\{j : j \le i\}$ to $\{0,1\}$ and $\{j : j \le i$ and $a(j) = 1\}$ is closed and if $p^*_\beta \in \underset{\sim}{G}_{P_\beta}$ then a is disjoint to $\underset{\sim}{A}^*_\beta\}$
order: inclusion

(5) if no previous case applies let $\underset{\sim}{A}_\beta = \emptyset, \gamma_\beta = 0, \underset{\sim}{g}_\beta = 0_{\omega_1}$, and define Q_β as in (4).

There are no problems in defining $\bar{Q}$. Let $P = P_{(\lambda^{\aleph_1})^+}$.

Explanation. We start by forcing the f_α's, which are the witnesses for the desired conclusion and then forcing the easy condition: $f_\alpha < f_\beta \bmod \mathcal{D}_{\omega_1}$ for $\alpha < \beta < \lambda$. Then we start killing undesirable stationary sets. Note that given $f \in V^{P_i}$, maybe in V^{P_i} we have $S = \{\alpha < \lambda : \mathrm{eq}[f, f_\alpha]$ is stationary in $V^{P_i}\}$ has cardinality λ, and increasing i it decreases slowly until it becomes empty, so it is natural to use iteration of length of cofinality $> \lambda$ e.g. $\lambda^{\aleph_1} \times \lambda^+$ (ordinal multiplication) is O.K. The problem is proving e.g. that $\aleph_1$ is not collapsed.

Continuation of the proof of 4.6.

The main point is to prove by simultaneous induction that for $\alpha \le (\lambda^{\aleph_1})^+$ the conditions $(a)_\alpha - (e)_\alpha$ listed below hold:

$(a)_\alpha$ forcing with P_α adds no new ω-sequences of ordinals.

$(b)_\alpha$ P_α satisfies $\aleph_2$-c.c.

$(c)_\alpha$ the set P'_α of $p \in P_\alpha$ such that each $p(\beta)$ is an actual function (not just a P_β-name) is dense.

Before we proceed to define $(d)_\alpha$, note that for each $\beta < \alpha$ (using the induction hypothesis),

$$\Vdash_{P_\beta} \text{“}CH \text{ and } |\underset{\sim}{Q}_\beta| = \aleph_1 \text{ and } \underset{\sim}{Q}_\beta \text{ is a subset of}$$

$$H \overset{\text{def}}{=} \{h : h \in V \text{ is a function from some } i < \omega_1 \text{ to } \omega_1\} \in V$$

$$\text{ordered by inclusion”}.$$

So (as P_β satisfies the $\aleph_2$-c.c.), the name $\underset{\sim}{Q}_\beta$ can be represented by $\aleph_1$ maximal antichains of $P_\beta : \langle\langle p^\beta_{\zeta,h} : \zeta < \omega_1\rangle : h \in H\rangle$, i.e. for each $\zeta < \omega_1, p^\beta_{\zeta,h}$ forces $h \in \underset{\sim}{Q}_\beta$ or forces $h \notin \underset{\sim}{Q}_\beta$. So, $u^*_\beta \overset{\text{def}}{=} \bigcup_{\zeta,\ell} \text{Dom}(p^\beta_{\zeta,\ell})$ is a subset of β of cardinality $\leq \aleph_1$ (all done in V). We may increase u^*_β as long as it is a subset of β of cardinality $\leq \aleph_1$. W.l.o.g. $p^\beta_{\zeta,h} \in P'_\beta$.

Call $u \subseteq \alpha$ closed (more exactly $\bar{Q}$-closed) if $\beta \in u$ implies: $u^*_\beta \subseteq u$ and $\underset{\sim}{g}^*_\beta, \underset{\sim}{A}^*_\beta$ are names represented by $\aleph_1$ maximal antichains $\subseteq P'_\beta$ with union of domains $\subseteq u^*_\beta$ and $\text{Dom}(p^*_\beta) \subseteq u^*_\beta$. W.l.o.g. each u^*_β is closed. For a closed $u \subseteq \alpha$ we define P_u by induction on $\sup(u)$: let $P_u = \{p \in P_\alpha\colon \text{Dom}(p) \subseteq u$ and for each $\beta \in \text{Dom}(p), p(\beta)$ is a $P_{u\cap\beta}$-name$\}$. Let $P'_u = P_u \cap P'_\alpha$. Lastly let

$(d)_\alpha$ $P_u \lessdot P_\alpha$ for every closed $u \subseteq \alpha$; moreover

$(e)_\alpha$ if $u \subseteq \alpha$ is closed, $p \in P'_\alpha$ then:

(1) $p{\restriction}u \in P'_u \subseteq P'_\alpha$ and

(2) $p{\restriction}u \leq q \in P'_u$ implies $q \cup [p{\restriction} (\text{Dom}(p) \setminus u)]$ is a least upper bound of p, q (in P'_α).

Of course the induction is divided to cases (but $(a)_\alpha$ is proved separately). Note that $(\text{e})_\alpha \Rightarrow (\text{d})_\alpha$.

Case A: $\alpha = 0$ Trivial

Case B: $\alpha = \beta + 1$, proof of $(b)_\alpha, (c)_\alpha, (d)_\alpha, (e)_\alpha$.

So we know that $(a)_\beta - (e)_\beta$ holds. By $(a)_\beta$ (as noted above), Q_β has power $\aleph_1$. So we know P_β satisfies $\aleph_2$-c.c., and $\Vdash_{P_\beta}$ “Q_β satisfies the $\aleph_2$-c.c.” hence P_α satisfies the $\aleph_2$-c.c., i.e. $(b)_\alpha$ holds.

If $p \in P_\alpha$, then $p(\beta)$ is a countable subset of $\omega_1 \times \omega_1$ from V^{P_β}, hence by $(a)_\beta$ for some $f \in V$ and q we have $p{\restriction}\beta \leq q \in P_\beta$ and $q \Vdash_{P_\beta}$ “$p(\beta) = f$”. By

$(c)_\beta$ w.l.o.g. q is in P'_β. So $q \cup \{\langle \beta, f \rangle\}$ is in P_α, is $\geq p$ and is in P'_α; so $(c)_\alpha$ holds.

As for $(d)_\alpha$ and $(e)_\alpha$, if $p \in P'_\alpha$, we can observe $(e)_\alpha(1)$ which says: "$p{\restriction}u \in P_u \subseteq P_\alpha$". [Why? If $\beta \notin u$, it is easy, so assume $\beta \in u$; now just note that $p{\restriction}(\beta \cap u) \in P_{\beta\cap u} \lessdot P_\alpha$ by the induction hypothesis, now $p{\restriction}\beta \Vdash_{P_\beta}$ "$p(\beta) \in \underset{\sim}{Q}_\beta$", but $\underset{\sim}{Q}_\beta$ is a $P_{\beta\cap u}$-name, $P_{\beta\cap u} \lessdot P_\beta$ (as u is closed and the induction hypothesis), so by $(d)_\beta$ we have $(p{\restriction}u){\restriction}\beta \Vdash_{P_{u\cap\beta}}$ "$p(\beta) \in Q_\beta$"; so $p{\restriction}u \in P_\alpha$ and as $\mathrm{Dom}(p{\restriction}u) \subseteq u$ we have $p{\restriction}u \in P_u$.]

Next $(e)_\alpha(2)$ follows (check) and then $(d)_\alpha$, $(e)_\alpha$ follows.

Case C: α limit $\mathrm{cf}(\alpha) > \aleph_0$, proof of $(b)_\alpha, (c)_\alpha, (d)_\alpha, (e)_\alpha$.

Clearly $P_\alpha = \bigcup_{\beta<\alpha} P_\beta$ (as the iteration is with countable support), hence $(c)_\alpha$ follows immediately; from $(c)_\alpha$ clearly $(b)_\alpha$ is very easy [use a Δ-system argument, and CH], and clause $(\mathrm{e})_\alpha$ also follows hence $(\mathrm{d})_\alpha$.

Case D: α is limit $\mathrm{cf}(\alpha) = \aleph_0$, proof of $(b)_\alpha, (c)_\alpha, (d)_\alpha, (e)_\alpha$.

As in Case (C), it is enough to prove $(c)_\alpha$. So let $p \in P_\alpha$. Let χ be regular large enough; $N_0 \prec N_1$ be a pair of countable elementary submodels of $(H(\chi), \in, <^*_\chi)$ to which $\bar{Q}, \alpha, \lambda, p$ belongs, satisfying (a)–(e) of $(*)^1_\lambda$ in Def 4.4.

We can find an ω-sequence $\langle u_m : m < \omega \rangle$ such that:

(i) each u_m is a member of N_0, and is a bounded subset of α of power $\leq \aleph_1$ which is closed for $\bar{Q}{\restriction}\alpha$

(ii) $u_m \subseteq u_{m+1}$

(iii) if $u \in N_0$ is a bounded subset of α of power $\leq \aleph_1$ closed for $\bar{Q}{\restriction}\alpha$ *then* for some m we have $u \subseteq u_m$.

There is no problem to choose such a sequence as the family of such u's is directed and countable. Let $\langle \mathcal{I}_m : m < \omega \rangle$ be a list of the dense open subsets of P_α which belong to N_0.

Note that in general, neither $\langle u_m : m < \omega \rangle$ nor $\langle \mathcal{I}_m : m < \omega \rangle$ are in N_1.

Let $\delta \overset{\mathrm{def}}{=} N_0 \cap \omega_1$ and note that $\delta \in N_1$. Let R be $\mathrm{Levy}(\aleph_0, \delta)^\omega$, the ω-th power of $\mathrm{Levy}(\aleph_0, \delta)$ with finite support, so R is isomorphic to $\mathrm{Levy}(\aleph_0, \delta)$ and it (and such isomorphisms) belongs to N_1 so there is $G^* \in N_1$, a (directed) subset of R, generic over N_0. Note that from the point of view of N_0, $\mathrm{Levy}(\aleph_0, \delta)$ is $\mathrm{Levy}(\aleph_0, \aleph_1)$ hence $((\mathrm{Levy}(\aleph_0, \aleph_1))^\omega)^{N_0} = (\mathrm{Levy}(\aleph_0, \delta))^\omega$, so G^* is an N_0-

generic subset of $(\mathrm{Levy}(\aleph_0,\aleph_1)^\omega)^{N_0}$. Let $G^* = \langle G^*_\ell : \ell < \omega\rangle$. Note that $N_0[G^*] \models ZFC^-$ and $N_0[G^*] \subseteq N_1$.

By the induction hypothesis $P_{u_m} \lessdot P_{u_{m+1}} \lessdot P_{(\sup u_{m+1})+1} \lessdot P_\alpha$ for every m. Now we choose by induction on $m < \omega, p_m$ and $G_m \subseteq P_\alpha \cap N_0$ such that:

$$p \le p_m \le p_{m+1},$$

$$p_{m+1} \in \mathcal{I}_m \cap N_0$$

$$p_m \restriction u_m \in G_m$$

$$G_m \subseteq N_0 \cap P'_{u_m} \text{ is generic over } N_0$$

$$\bigcup_{\ell<m} G_\ell \subseteq G_m,$$

$$G_m \in N_1, \text{ moreover } G_m \in N_0[\langle G^*_\ell : \ell \le m\rangle].$$

Why is this possible? Arriving to $m(>0)$ we have $P'_{u_{m-1}} \lessdot P_\alpha$, $G_{m-1} \subseteq P'_{u_{m-1}} \cap N_0$ is generic for N_0, we can choose p_m as required ($p_m \in \mathcal{I}_m \cap N_0$ and $p_{m-1} \le p_m$ and $p_m \restriction u_{m-1} \in G_{m-1}$). Also $P'_{u_m} = P_{u_m} \cap P'_\alpha$ belongs to N_0, (as $\bar{Q}, P'_\alpha$, and u_m belongs), now it has cardinality $\aleph_1$ (and of course all its members are in V as well as itself), so some list $\langle r^{u_m}_\zeta : \zeta < \omega_1\rangle$ of the members of P'_{u_m} of length ω_1 belongs to N_0. So as $\delta = N_0 \cap \omega_1 \in N_1$, clearly $P'_{u_m} \cap N_0 = \{r^{u_m}_\zeta : \zeta < \delta\}$ belongs to N_1 and N_1 "know" that it is countable.

As G^*_m is a subset of $\mathrm{Levy}(\aleph_0,\aleph_1)^{N_0} = \mathrm{Levy}(\aleph_0,\aleph_1^{N_0})^{N_0[\langle G^*_\ell:\ell<m\rangle]}$, generic over $N_0[\langle G^*_\ell : \ell < m\rangle]$ there is in $N[\langle G^*_\ell : \ell \le m\rangle]$ a subset of $P'_{u_m} \cap N_0$ generic for $\{\mathcal{I} : \mathcal{I} \in N_0[G_{m-1}]$ and $\mathcal{I} \subseteq P'_{u_m}$ and $\mathcal{I}$ is dense in $P_u\}$ extending G_{m-1}. So in N_1 and even $N_0[\langle G^*_\ell : \ell \le m\rangle]$ we can find $G_m \subseteq P_{u_m} \cap N_0$ generic over N_0 with $p_m \restriction u_m \in G_m$ and $G_{m-1} \subseteq G_m$.

Note: as $P_{u_m} \lessdot P_{u_{m+1}}$ we succeeded to take care of "$G_m \subseteq G_{m+1}$". Let $G = \bigcup_m G_m$, $\delta = N_0 \cap \omega_1$. We define $q = q_G$, a function with domain $\alpha \cap N_0$: for $\beta \in u_m \cap N_0$ let

$q'_G(\beta) = \bigcup\{r(\beta)$: for some $m < \omega$ we have $r \in G_m$ and $r(\beta)$ is an actual (function not just a P_β-name) $\}$

$q_G(\beta)$ is: $q'_G(\beta) \cup \{\langle \delta, \mathrm{otp}(N_0 \cap \beta)\rangle\}$ if $\beta < \lambda$, and $q'_G(\beta) \cup \{\langle \delta, 1\rangle\}$ if $\beta \ge \lambda$.

Clearly q is a function with domain $\alpha \cap N_0$, each $q(\beta)$ a function from $\delta+1$ to ω_1. (Here we use the induction hypothesis $(\mathrm{c})_\beta$.)

If $q \in P_\alpha$ then we will have $q \in P'_\alpha$ and q is a least upper bound of $\bigcup_{m<\omega} G_m$ and of $\{p_m : m < \omega\}$. Hence in particular $q \geq p$ thus finishing the proof of $(c)_\alpha$, hence (as said above) of the present case (Case D). Now we shall show:

$\otimes$ $q \restriction u_m \in N_1$ for each $m < \omega$

Clearly $q'_G \restriction u_m \in N_1$ as $G_m \in N_1$ (and $P'_{u_m} \in N_1$), hence to prove $\otimes$ we have to show that $\{\langle \beta, (q_G(\beta))(\delta)\rangle : \beta \in u_m\}$ belongs to N_1. Now $\{\langle \beta, (q(\beta))(\delta)\rangle : \beta \in u_m \cap N_0 \backslash \lambda\}$ is $\{\langle \beta, 1\rangle : \beta \in u_m \cap N_0 \backslash \lambda\} = (u_m \cap N_0 \backslash \lambda) \times \{1\}$ belongs to N_1 as $u_m \in N_0 \prec N_1$ and as said earlier, as $N_0 \cap \omega_1 \in N_1$, $N_0 \models |u_0| \leq \aleph_1$ we have $u_m \cap N_0 \in N_1$ and $\lambda \in N_0 \prec N_1$. Next the set $\{\langle \beta, q(\beta)(\delta)\rangle : \beta \in u_m \cap N_0 \cap \lambda\}$ is exactly $f \restriction u_m$, where f is the function from 4.4(e).

So by Claim 4.8 below we finish.

<u>Case E</u>: α nonzero, proof of $(a)_\alpha$.

So by cases $(B), (C), (D)$ we know that $(b)_\alpha, (c)_\alpha, (d)_\alpha, (e)_\alpha$ holds.

Now we imitate the proof of Case (D) except that in (i) and (iii) we omit the "bounded in α". So now $P_{u_m} \lessdot P_\alpha$" is justified not by "$(c)_\beta$ for $\beta < \alpha$" but by $(c)_\alpha + (d)_\alpha$. We can finish now, by using again 4.8.

4.8 Claim. If

(a) $N_0 \prec N_1 \prec (H(\chi), \in, <^*_\chi)$ are countable, $\bar{Q}$ is as in the proof of 4.6, $\bar{Q} \in N_0, \alpha = \ell g(\bar{Q}) \in N_0$, $\delta = N_0 \cap \omega_1$, $\mathrm{otp}(\lambda \cap N_0) = \mathrm{otp}(N_1 \cap \omega_1)$, and part (d) of $(*)^1_\lambda$ of Definition 4.4 holds.

(b) $G \subseteq P_\alpha \cap N_0, G$ is directed,

(c) there is a family U such that:

 (α) if $u \in U$ then $u \in N_0, u \subseteq \alpha$ is closed (for $\bar{Q}$ i.e. $\alpha \in u \Rightarrow u^*_\alpha \subseteq u$) of power $\leq \aleph_1$,

 (β) $\bigcup\{u : u \in U\} = N_0 \cap \alpha$, U is directed (by $\subseteq$) and if $u \in N_0$ is closed (for $\bar{Q}$) bounded subset of α of cardinality $\leq \aleph_1$ then $u \in U$.

 (γ) if $u \in U$ then $G \cap P_u$ is generic over N_0

 (δ) if $u \in U$ then $G \cap P_u \in N_1$

(d) $q = q_G$ is defined as in case D of the proof of 4.6 above, i.e. $\mathrm{Dom}(q) = \alpha \cap N_0$ and

$q'(\beta) = \bigcup\{r(\beta)$ for some $u \in U$, $r \in G_m$, $r(\beta)$ an actual function$\}$.

$q(\beta)$ is: $q'(\beta) \cup \{\langle \delta, \mathrm{otp}(N_0 \cap \beta)\rangle\}$ if $\beta < \lambda$, $q'(\beta) \cup \{\langle \delta, 1\rangle\}$ otherwise.

Then

(i) q is in P_α (and even in P'_α)

(ii) $q \in P'_\alpha$ is a least upper bound of G.

Proof. We prove by induction on $\beta \in N_0 \cap \alpha$ that $q{\restriction}\beta \in P_\alpha$ (hence $\in P'_\alpha$).

This easily suffices.

Note. if $u \in N_0$ is closed and $\subseteq u' \in U$ then we can add it to U.

Case 1: $\beta = 0$, or β is limit. Trivial.

Case 2: $\beta = \gamma + 1, \gamma < \lambda$. Check.

Case 3: $\beta = \gamma + 1, \beta \geq \lambda$.

We should prove $q{\restriction}\gamma \Vdash_{P_\gamma}$ "$q(\gamma) \in \underset{\sim}{Q}_\gamma$". Recall that u^*_γ is the subset of γ (of size $\aleph_1$) which was needed for the antichains defining Q_γ, and $\delta = N_0 \cap \omega_1$. Clearly u^*_γ and $u^*_\gamma \cup \{\gamma\}$ belongs to U (being closed bounded and in N_0). As $G \cap P_{u^*_\gamma \cup \{\gamma\}}$ is generic over N_0, clearly

$$q{\restriction}\gamma \Vdash_{P_\gamma} \text{"}q(\gamma) \text{ is a function from } \delta + 1 \text{ to } \omega_1, \text{ such that}$$
$$\text{for every non limit } \zeta < \delta \text{ we have } q(\gamma){\restriction}\zeta \in \underset{\sim}{Q}_\gamma\text{"}.$$

Noting $(q(\gamma)){\restriction}\zeta$, where $\zeta \leq \delta$, is of the right form; and $\gamma \geq \lambda \Rightarrow (q(\gamma))^{-1}(\{1\})$ is closed and by the choice of $q(\gamma)(\delta)$, clearly it is enough to prove that:

$\otimes_a$ if $\lambda \leq \beta < \lambda^2$ and $\beta = \lambda + \lambda\beta_1 + \beta_2, \beta_1 < \beta_2 < \lambda$ *then* $q{\restriction}\beta \Vdash_{P_\beta}$ "$f^*_{\beta_1}(\delta) < f^*_{\beta_2}(\delta)$"

$\otimes_b$ if $\lambda^2 \leq \beta < \ell g(\bar{Q})$ *then* $q{\restriction}\beta \Vdash$ "$p^*_\beta \in \underset{\sim}{G}_{Q_\beta} \Rightarrow \delta \notin \underset{\sim}{A}^*_\beta$".

Now $\otimes_a$ holds as $q \Vdash_{P_\alpha}$ "$\langle f^*_\gamma(\delta) : \gamma \in N_0 \cap \alpha\rangle$ is strictly increasing" (just see how we have defined $q_G(\gamma)$ in clause (d) of 4.8 above).

So let us prove $\otimes_b$; remember $\underset{\sim}{Q}_\beta$ is a $P_{u^*_\beta}$-name and (u^*_β being closed) $\underset{\sim}{A}_\beta, g^*_\beta$ are $P_{u^*_\beta}$-names, $p^*_\beta \in N_0 \cap P'_{u_\beta}$. If $q{\restriction}u^*_\beta \Vdash$ "$\delta \notin \underset{\sim}{A}_\beta$ or $p^*_\beta \notin G_{P_{u^*_\beta}}$" we finish. Otherwise there is $r, q\restriction u^*_\beta \le r \in P_{u^*_\beta}$ and $r \Vdash$ "$\delta \in \underset{\sim}{A}_\beta$ & $p^*_\beta \in \underset{\sim}{G}_{P_\beta}$"; w.l.o.g. $r \in P'_{u^*_\beta}$. As $G\restriction P_{u^*_\beta} \in N_1$ by the proof of $\otimes$ in 4.6, case D (near the end), also $q\restriction u^*_\beta \in N_1$, and remembering $\beta \in N_0 \Rightarrow P_\beta \in N_0$ and $\delta \in N_1$, and $P_{u^*_\beta}$, $P'_{u^*_\beta} \in N_1$ and $\underset{\sim}{A}_\beta, p^*_\beta \in N_1$, clearly w.l.o.g. $r \in N_1$. As $\beta \in N_0, \underset{\sim}{g}^*_\beta \in N_0 \subseteq N_1$ is a $P_{u^*_\beta}$-name and $\delta \in N_1$, w.l.o.g. r forces a value to $\underset{\sim}{g}^*_\beta(\delta)$, say $\Vdash$ "$\underset{\sim}{g}^*_\beta(\delta) = \xi(*)$".

Now $\xi(*) \in N_1$ hence $\xi(*) < \text{otp}(N_1 \cap \omega_1) \le \text{otp}\ (N_0 \cap \lambda)$ (here we are finally using 4.4(c)), hence there is $\gamma \in \lambda \cap N_0$ such that $\xi(*) = \text{otp}(N_0 \cap \gamma)$.

But now (see definition of $\underset{\sim}{Q}_\beta$) we have $r \Vdash_{P_\beta}$ "$eq[\underset{\sim}{g}^*_\beta, \underset{\sim}{f}^*_\gamma] \cap \underset{\sim}{A}_\beta$ is not stationary, so it is disjoint to some club $\underset{\sim}{C}^*_\beta$ of ω_1" where $\underset{\sim}{C}^*_\beta$ is a P_β-name and w.l.o.g. $\underset{\sim}{C}^*_\beta \in N_0$.

[Why? As $g^*_\beta, f^*_\gamma, \underset{\sim}{A}_\beta \in N_0$ there is a P_β-name $\underset{\sim}{C}^*_\beta$ such that $\Vdash_{P_\beta}$ " if eq$[g^*_\xi, f^*_\gamma] \cap \underset{\sim}{A}_\beta$ is not a stationary subset of ω_1 then $\underset{\sim}{C}^*_\beta$ is a club of ω_1 disjoint to this intersection, otherwise $\underset{\sim}{C}^*_\beta = \omega_1$"].

So $\Vdash$ "$\underset{\sim}{C}^*_\beta$ is a club of ω_1". By the induction hypothesis for β (in particular $(\text{b})_\beta$ from the proof of 4.6 which says that P_β satisfies the $\aleph_2$-c.c.), for some $\bar{Q}$-closed bounded $u \subseteq \beta, |u| \le \aleph_1, u \in N_0$ and $\underset{\sim}{C}^*_\beta$ is a P_u-name.

By the induction hypothesis $q\restriction\beta \in P'_\beta$; now by the construction of $q, q\restriction\beta \Vdash_{P_\beta}$ "$\underset{\sim}{C}^*_\beta \cap \delta$ is unbounded in δ" hence $(q\restriction\beta) \cup r$ [i.e. $r \cup (q\restriction(\beta \cap \text{Dom}(q) \setminus u^*_\beta))$] is in P'_α, is an upper bound of $q\restriction\beta$ and r and it forces $\delta \in \underset{\sim}{C}^*_\beta$, hence $\delta \in eq[g^*_\beta, f^*_\gamma] \Rightarrow \delta \notin A^*_\beta$. But the antecedent holds by the choice of r, γ and $\xi(*)$. So we finish the proof. $\square_{4.8}$

Continuation of the proof of 4.6: So we have to check if conditions (i)-(v) of 4.6 hold for $P = P_{\alpha^*}$. Now (i) holds by $(b)_{\alpha^*} + (c)_{\alpha^*}$ (α^* is the length of the iteration- $(\lambda^{\aleph_1})^+$); condition (ii) holds by $(a)_{\alpha^*}$. Condition (iii) should be clear from the way $Q_\alpha(\lambda \le \alpha < \alpha^*)$ were defined (see the explanation after the definition of Q_α). Prove by induction on $\gamma < \lambda^+$ that

$(*)_\gamma$ if $\underset{\sim}{g}$ is a P_γ-name of a function from ω_1 to ω_1, $\underset{\sim}{A}$ is a P_γ-name of a subset of ω_1 and $p^* \in P_\gamma$ then:

if $p^* \Vdash$ " for every $\alpha < \lambda$ the set $\underset{\sim}{A} \cap eq(\underset{\sim}{g}, \underset{\sim}{f}_\alpha)$ is not stationary subset of ω_1"

then $p^* \Vdash$ "$\underset{\sim}{A} \subseteq \omega_1$ is not stationary".

Arriving to γ let $\langle(\underset{\sim}{g}_\zeta, \underset{\sim}{A}_\zeta, p^*_\zeta) : \zeta < \lambda\rangle$ list the set of such triples (their number is $\leq \lambda$ as $|P_\gamma| \leq \lambda = \lambda^{\aleph_1}$ and P_γ satisfies $\aleph_2$-c.c. and the list includes such triples for smaller γ's). For each ζ we can find a club E_ζ of λ^+ such that: if $\alpha < \beta \in E_\zeta$, then for some P_β-name $\underset{\sim}{C}_{\alpha,\underset{\sim}{A}_\zeta,\underset{\sim}{g}_\zeta}$ we have

$$\Vdash_{P_{\lambda^+}} \text{"}\mathit{if}\ \underset{\sim}{A}_\zeta \cap eq(\underset{\sim}{g}_\zeta, \underset{\sim}{f}_\alpha) \text{ is not stationary}$$

$$\mathit{then} \text{ it is disjoint to } \underset{\sim}{C}_{\alpha,\underset{\sim}{A}_\zeta,\underset{\sim}{g}_\zeta}\text{"}$$

$$\Vdash_{P_{\lambda^+}} \text{"}\underset{\sim}{C}_{\alpha,\underset{\sim}{A}_\zeta,\underset{\sim}{g}_\zeta} \text{ is a club of } \omega_1\text{"}.$$

For any $\delta \in \bigcap_{\zeta<\lambda} E_\zeta$ which has cofinality $> \aleph_1$, we ask whether when choosing $(g^*_\beta, \underset{\sim}{A}_\beta, \gamma_\beta, p^*_\beta)$ do we have a candidate $(\underset{\sim}{g}, \underset{\sim}{A}, \gamma', p)$ as in $\otimes^\delta_{g,A,\gamma'}$, $\gamma' \leq \gamma$.

If for every such δ the answer is no, we have proved $(*)$; if yes, we get easy contradiction.

For finishing the proof of condition (iii) note that we can let $f_\lambda(i) = \omega_1$, and prove by induction on $\alpha \leq \lambda$ that $\underset{\sim}{f}_\alpha$, is an α'th function as follows: $\beta < \alpha < \lambda \Rightarrow f_\beta <_{\mathcal{D}_{\omega_1}} f_\alpha$ (see $Q_{\lambda+\lambda\beta+\alpha}$'s definition) and if $S \subseteq \omega_1$, $f \in {}^{\omega_1}\omega_1$, $S \cap \mathrm{eq}(f, \underset{\sim}{f}_\alpha)$ not stationary for every $\alpha < \lambda$ we get S is not stationary by the definition of Q_β (for $\beta \in [\lambda^2, \alpha^*)$) so if $g <_{\mathcal{D}_{\omega_1}} f_\alpha$ then for every $\beta \in [\alpha, \lambda)$ the set $\mathrm{eq}[g, f_\beta]$ is not stationary and compare the definition of the α'th function and the definition of the forcing condition).

Lastly clause (iv) of 4.6 holds as $\alpha^* = (\lambda^{\aleph_1})^+$, each Q_α has cardinality $\aleph_1$, and P'_{α^*} is a dense subset of P_{α^*}. Finally, condition (v) follows from 4.8.

$\square_{4.6}$

4.9. Proof of 4.7. 1)By 4.3, $(*)^1_\lambda$ holds in $L^{\mathrm{Levy}(\aleph_0,<\kappa)}$ and λ is regular hence $\lambda^{\aleph_1} = \lambda$. By 4.6 we can define a forcing notion P in $L^{\mathrm{Levy}(\aleph_1<\kappa)}$, $|P| = [\lambda^+]^{L[\mathrm{Levy}(\aleph_0,<\kappa)]} = \lambda^+$ as required.

2) Iterate as above for α^* with careful bookkeeping.

3) Left to the reader.

4) Lastly over V^P force with $\mathrm{Levy}(\lambda, \lambda^+)$ such that $2^{\aleph_1} = \lambda$. $\square_{4.7}$

4.10 Discussion. 1) Can we omit the Levy collapse of λ^+ in the proof of 4.7(4) and still get $2^{\aleph_1} = \lambda$ (and $\langle \omega_1 : i < \omega_1 \rangle$ is the λ-th function)? Yes, if we strengthen suitably $(*)^1_\lambda$. (e.g. saying a little more than there is a stationary set of such $\lambda' < \lambda, (*)^1_{\lambda'}$).

2) In 4.6 we can add e.g. that in V^P, Ax[proper of cardinality $\aleph_1$ not adding reals as in XVIII §2]. We have to combine the two proofs.

3) Suppose $V \models$ "$(*)^1_\lambda$", and for simplicity, $V \models$ "G.C.H., λ is regular $\neg(\exists \mu)[\lambda = \mu^+ \; \& \; \mu > \mathrm{cf}\mu \leq \aleph_1]$". (E.g. $L^{\mathrm{Levy}(\aleph_0, <\kappa)}$ when $0^\#$ exists, κ is a cardinal of V.) For some forcing notion P, $|P| = \lambda^+$, and in V^P we have: ω_1 is an ω_3-th function, $\Vdash_P$ "$\aleph_1 = \aleph_1^V, \aleph_2 = (\aleph_2)^V, \aleph_3 = \lambda, \aleph_4 = (\lambda^+)^V$ and CH and $2^{\aleph_1} = \aleph_4$", (so we can then force by Levy$(\aleph_3, \aleph_4)$ and get $2^{\aleph_1} = \aleph_3$).

Proof. 3) Let $R = \mathrm{Levy}(\aleph_2, < \lambda)$, R is $\aleph_2$-complete and satisfies the λ-c.c. and $|R| = \lambda$, so forcing by R adds no new ω_1-sequences of ordinals, make λ to $\aleph_3$. Let P'_{α^*} be the one from 4.6 (or 4.7(2)). As R is $\aleph_2$-complete, also in V^R we have: P'_{α^*} satisfies the $\aleph_2$-c.c., and P'_{α^*} has the same set of maximal antichains as in V. So the family of P'_{α^*}-name of a subset of ω_1 (or a function from ω_1 to ω_1) is the same in V and V^R. So clearly $P'_{\alpha^*} \times R$ is as required. $\square_{4.10}$

Problem. Is ZFC + "θ is an α-th function for some α (for $\mathcal{D}_{\omega_1}$)" + $\neg 0^\#$ consistent? For $\theta \in \{\aleph_1, \aleph_{\omega_1}\}$ or any preassumed θ? (Which will be $< 2^{\aleph_1}$.)

XVIII. More on Proper Forcing

§0. Introduction

From the last eight chapters you may have gotten the impression that we are done with properness, but this is not so. First, we turn to the problem of not adding reals; remember that by V §7, VIII §4, for CS iterations of proper forcing notions not adding reals, the limit does not add reals, provided that two additional conditions hold: one is $\mathbb{D}$-completeness (for, say, a simple 2-completeness system) and the second is $(< \omega_1)$-properness (see V §2). Now, the first restriction is justified by the weak diamond (see V 5.1, 5.1A and AP §1); that is not to say that we have to demand exactly $\mathbb{D}$-completeness, but certainly we have to demand something in this direction. However, there was nothing there to justify the second demand: α-proper for every α. In the first section, (following [Sh:177]), we show that we cannot just omit it, even if we use an $\aleph_1$-completeness system. It is natural to hope that this counterexample will lead to a principle like the weak diamond (so provable from CH). Thus the construction of this counterexample leads to questions like: Assuming CH, can we find $\langle C_\delta : \delta < \omega_1 \rangle$, C_δ an unbounded subset of δ, say of order type ω, such that for every club E of ω_1, for stationarily many limit $\delta < \omega_1$, $C_\delta \subseteq E$ or $\delta = \sup(C_\delta \cap E)$ or $(\forall \alpha \in \delta)[\min(E \setminus \alpha) < \min(C_\delta \setminus (\alpha + 1))]$? (They are kin to "the guessing clubs", the existence of which for, e.g. $\aleph_2$, follows from ZFC, see [Sh:g].) It interests us as the theorems (and proofs) from V, VIII §4 do not give us the consistency of their negation with CH. However, those statements do not follow from ZFC + CH; for this we prove in the second section a preservation theorem for CS iterations of proper forcing not adding reals. Again we have two conditions (called there $(4)_2$ or $(4)_{\aleph_0}$ and (5)). The first, $(4)_2$, is done "against" the weak diamond, and is weaker than the older $\mathbb{D}$-completeness, but this is just a side gain. The second condition says our forcing remains proper even if we force with "forcing notion from our family, in particular not adding reals". Note that for forcing notions of cardinality $\aleph_1$, this is a very mild condition. So, the results of §1 remain the only restrictions on theorems on preservation by CS iteration, and there is a gap between them and the results of §2.

Then, in the third section we turn to other preservation theorems, giving an alternative to the theorem from VI §1 – §3, and dealing with some examples. (For a simplification of possibility A in 3.3, see [Go]).

Finally, in the fourth section we turn to the problem of a unique P-point. In VI §4 we have proved that there may be no P-point; remember, a P-point F is a nonprincipal ultrafilter on ω such that if $A_n \in F$ for $n < \omega$ then for some $A \in F$ we have $\bigwedge_n A \subseteq_{\mathrm{ae}} A_n$ ($A \subseteq_{\mathrm{ae}} A_n$ means $A \setminus A_n$ is finite). Now, to prove the consistency of "there is a unique object" is typically harder then proving there is no one. Unique here means up to permutations of ω. In VI §5 we have proved a weaker result: there may be a unique Ramsey ultrafilter, but there could have been many P-points above it which were not isomorphic. We continue this and prove the consistency of "there is a unique P-point".

§1. No New Reals: A Counterexample and New Questions

1.1 Lemma. Suppose V satisfies $2^{\aleph_0} = \aleph_1, 2^{\aleph_1} = \aleph_2$, and for some $A \subseteq \omega_1$, every $B \subseteq \omega_1$ belongs to $L[A]$ and for limit $\delta < \omega_1$,

$$L_\delta[A \cap \delta] \models \text{"}\delta \text{ is countable"}.$$

Then we can define a countable support iteration $\bar{Q} = \langle P_i, \underset{\sim}{Q}_i : i < i^* \rangle$ such that the following conditions hold:

(a) Each $\underset{\sim}{Q}_i$ is proper and $\Vdash_{P_i}$ "$\underset{\sim}{Q}_i$ has power $\aleph_1$".

(b) Each $\underset{\sim}{Q}_i$ is $\mathbb{D}$-complete for some simple $\aleph_1$-completeness system $\mathbb{D}$ (hence does not add reals).

(c) Forcing with $P_{i^*} = \mathrm{Lim}\bar{Q}$ adds reals.

Proof. We shall define $\underset{\sim}{Q}_i$ by induction on $i < i^*$, $i^* \leq \omega^2$, so that conditions (a) and (b) are satisfied, and $\underset{\sim}{C}_i$ is a $\underset{\sim}{Q}_i$-name of a closed unbounded subset of ω_1. Let $\langle f^*_\xi : \xi < \omega_1 \rangle \in L[A]$ be a list of all functions f which are from δ to δ for some limit $\delta < \omega_1$, and let $h : \omega_1 \to \omega_1$, $h \in L[A]$, be defined by $h(\alpha) = \mathrm{Min}\{\beta : \beta > \alpha \text{ and } L_\beta[A \cap \alpha] \models \text{“}|\alpha| = \aleph_0\text{”}\}$.

Suppose we have defined $\underset{\sim}{Q}_j$ for every $j < i$; then P_i is defined, is proper (as each $\underset{\sim}{Q}_j$, $j < i$, is proper, and III 3.2) and has a dense subset of power $\aleph_1$ (by III 4.1). Let $G_i \subseteq P_i$ be generic, so, clearly, there is a $B_i \subseteq \omega_1$ such that in $V[G_i]$, every subset of ω_1 belongs to $L[A, B_i]$. The following now follows:

1.1A Fact. In $V[G_i]$, every countable $N \prec (H(\aleph_2), \in, A, B_i)$ is isomorphic to $L_\beta[A \cap \delta, B_i \cap \delta]$ for some $\beta < h(\delta)$, where $\delta = \delta(N) \overset{\text{def}}{=} \omega_1 \cap N$.

We shall assume also that $V[G_i]$ has the same reals as V (otherwise we already have an example).

We now define by induction on $\alpha < \omega_1$, a set $T_\alpha = T^i_\alpha$ such that the following conditions are satisfied:

(i) Each $f \in T_\alpha$ is the characteristic function of a closed subset of some successor ordinal $\beta < \alpha$, i.e., $\mathrm{Dom}(f) = \beta$, and $f^{-1}(\{1\})$ is a closed subset of β and is included in the set of accumulation points of $\bigcap_{j<i} C_j$. If $\gamma < \alpha$, then $T_\gamma \subseteq T_\alpha$.

(ii) If $f \in T_\alpha, \gamma + 1 \leq \mathrm{Dom}(f)$, then $f{\restriction}(\gamma + 1) \in T_\alpha$, and even $f{\restriction}(\gamma + 1) \in T_\beta$ for $\gamma + 1 < \beta \leq \alpha$.

(iii) If $f \in T_\alpha$, $\mathrm{Dom}(f) = \beta$, $\beta < \gamma < \alpha$, γ a successor, then $f' = f \cup 0_{[\beta,\gamma)} \in T_\alpha$, i.e., $\mathrm{Dom}(f') = \gamma$, and

$$f'(\xi) = \begin{cases} f(\xi), & \textit{if } i < \beta, \\ 0, & \textit{if } \beta \le \xi < \gamma. \end{cases}$$

(iv) If $f, g \in T_\alpha$, $f(\beta) \neq g(\beta)$, then $f^{-1}(\{1\}) \cap g^{-1}(\{1\}) \setminus \beta$ is finite.

(v) If $f \in T_\alpha$, $\gamma \ge \beta = \mathrm{Dom}(f)$, $\gamma + 1 < \alpha$, γ is an accumulation point of $\bigcap_{j<i} C_j$ and the order type of $f^{-1}(\{1\})$ has the form $\xi + 2$ or is > 0 and $< \omega$, then $f' = f \cup 0_{[\beta,\gamma)} \cup \{\langle \gamma, 1\rangle\} \in T_\alpha$.

(vi) If $f \in T_\alpha$, $\delta + 1 = \mathrm{Dom}(f)$, δ limit, and $f(\beta) = 1$ for arbitrarily large $\beta < \delta$, then $\mathrm{Min}\{\xi : f{\restriction}\delta = f^*_\xi\}$ is larger than $\mathrm{Min}\{\xi : \delta \le \xi \in C_j\}$ (for $j < i$).

(vii) If $\delta + 1 < \alpha$, δ is an accumulation point of $\bigcap_{j<i} C_j$, $\xi^* < \omega_1$, and $f \in T_\delta \cap L_\delta[A \cap \delta]$, *then* there is a $g \in T_\alpha$, $\delta + 1 = \mathrm{Dom}(g)$, such that for every $\mathcal{J} \in L_{h(\delta)}[A \cap \delta, B_i \cap \delta]$ (an open dense subset of $T_\delta \cap L_\delta[A \cap \delta]$ (ordered by inclusion)), for some $\gamma < \delta$ we have $g{\restriction}\gamma \in \mathcal{J}$ and $g{\restriction}\delta \notin \{f^*_\xi : \xi < \xi^*\}$ and $f = g{\restriction}\mathrm{Dom}(f)$.

(viii) For $f \in T_\alpha$, if $\delta = \sup(\delta \cap f^{-1}(\{1\}))$ (hence $f(\delta) = 1$), $\delta < \beta$, and $f(\beta) = 1$, then for every $j < i$, for some $\gamma < \beta$, the characteristic function of C_j restricted to δ is f^*_γ; and if δ, $f{\restriction}\delta$ and β satisfy this then $f{\restriction}(\delta + 1) \cup 0_{[\delta+1,\beta)} \cup 1_{[\beta,\beta+1)}$ belongs to $T_{\beta+2}$.

Let us carry the induction.

Case A. α is limit, or $\alpha = \gamma + 1$, γ limit. Let $T_\alpha = \bigcup_{\beta<\alpha} T_\beta$ or $T_\alpha = \bigcup_{\beta<\gamma} T_\beta$.

Case B. $\alpha < \omega$. Let $T_\alpha = \{f : f \text{ a function from } \beta < \alpha \text{ to } \{0\}\}$.

Case C. $\alpha = \beta + 3 > \omega$. Let $T_\alpha = T_{\beta+2} \cup \{f : \mathrm{Dom}(f) = \beta + 2, f{\restriction}(\beta+1) \in T_{\beta+2},$ provided that (viii) is satisfied $\}$.

Case D. $\alpha = \delta + 2$, δ limit, $\delta \in \mathrm{acc}\left(\bigcap_{j<i} C_j\right)$ (acc- denotes the set of accumulation points). This is the main case. Let $\{f^\delta_\ell : \ell < \omega\}$ be a list of $T_\delta \cap L_\delta[A \cap \delta]$, each appearing $\aleph_0$ times, and $\{\mathcal{J}^{\delta,i}_\ell : \ell < \omega\}$ be a list of all open dense subsets $\mathcal{J}$ of $(T_\delta \cap L_\delta[A \cap \delta], \subseteq)$ which satisfy: $\mathcal{J}$ belongs to

$L_{h(\delta)}[A\cap\delta, B_i\cap\delta]$ or $\mathcal{J} = \{f \in T_\delta \cap L_\delta[A\cap\delta] : f \not\subset f^*_\xi\}$ for some $\xi < h(\delta)$. We now define by induction on $n<\omega$, an ordinal $\beta_n = \beta_n^{\delta,\alpha} < \delta$ and a finite set $F_n = F_n^{\delta,\alpha} \subseteq \{f \in T_\delta \cap L_\delta[A\cap\delta] : \beta_n = \mathrm{Dom}(f)\}$ such that

$$(*) \qquad (\forall f \in F_n)(\exists g \in F_{n+1})(f \subseteq g) \text{ and}$$

$$\text{if } n \geq 1,\ (\forall f, g \in F_n)\left(f{\restriction}\beta_{n-1} \neq g{\restriction}\beta_{n-1} \Rightarrow f^{-1}(\{1\}) \cap g^{-1}(\{1\}) \subseteq \beta_{n-1}\right).$$

Subcase α. If $n = 0 \bmod 3$, then $\beta_{n+1} = \beta_n + 1$ and $F_{n+1} = \{f \cup \{\langle\beta_n, 0\rangle\} : f \in F_n\}$; and if $n=0$, then $F_n = \emptyset$ and $\beta_n = 0$.

Subcase β. If $n = 1 \bmod 3$, then $\beta_{n+1} = \beta_n + 1$; let $f'_n = f^\delta_{(n-1)/3}$ and $\beta^*_n = \mathrm{Dom}(f^\delta_{(n-1)/3})$ if $\mathrm{Dom}(f^\delta_{(n-1)/3})$ is $< \beta_n$, and let $f'_n = \emptyset, \beta^*_n = 0$ otherwise; now let

$$F_{n+1} = \{f \cup 0_{[\beta_n,\beta_{n+1})} : f \in F_n\} \cup \{f'_n \cup 0_{[\beta^*_n, \beta_{n+1})}\}.$$

Subcase γ. If $n = 2 \bmod 3$, $(n-2)/3 = m^2 + k$, $k \leq 2m$, then every $f \in F_{n+1}$ belongs to $\mathcal{J}_k = \mathcal{J}^{\delta,i}_k$. Note that we have to take care to satisfy $(*)$; hence let† $F_n = \{f^n_\ell : \ell < |F_n|\}$, and define β^n_ℓ for $\ell \leq |F_n|$ and g^n_ℓ for $\ell < |F_n|$ by induction on ℓ : $\beta^n_0 = \beta_n$; if β^n_ℓ is defined, choose g^n_ℓ, $f^n_\ell \cup 0_{[\beta_n, \beta^n_\ell]} \subseteq g^n_\ell \in \mathcal{J}_k$, and $\beta^n_{\ell+1} = \mathrm{Dom}(g^n_\ell)$. Now let

$$\beta_{n+1} = \beta^n_{|F_n|} \text{ and } F_{n+1} = \{g^n_\ell \cup 0_{[\beta^n_{\ell+1}, \beta_{n+1})} : \ell < |F_n|\}.$$

Note that only in Subcase γ, do we have a free choice, and we eliminate it by choosing the first candidate for F_{n+1} by the canonical well-ordering of $L[A, B_i]$, and we also require that $\langle \mathcal{J}^{\delta,i}_\ell : \ell < \omega\rangle$ be the first such sequence in the canonical well ordering of $L_\delta[A\cap\delta, B_i\cap\delta]$. So we have finished defining the F_n's and we let

$$\begin{aligned} T_{\delta+2} = T_\delta \cup \{f :& \mathrm{Dom}(f) = \delta + 1 \text{ and : either } f = f' \cup 0_{[\gamma,\delta+1)},\\ & \text{where } f' \in T_\delta,\ \gamma = \mathrm{Dom}(f') \textit{ or } \text{for some } k < \omega,\\ & (\forall n > k)[f{\restriction}\beta_n \in F_n] \text{ and}\\ & f(\delta) = 1 \Leftrightarrow \delta = \sup f^{-1}(\{1\})\}. \end{aligned}$$

† Of course, we suppress the dependency of $\beta_n, f^n_\ell, \alpha^n_\ell, g^n_\ell$ on δ and i.

It is easy to check that that $T_{\delta+2}$ is as required. (Case β in the definition of F_n enables us to satisfy demand (vii)).

Case E. $\alpha = \delta + 2$, δ limit, $\delta \notin \text{acc}\left(\bigcap_{j<i} C_j\right)$. Let $T_\alpha = T_\delta \cup \{f : \text{Dom}(f) = \delta + 1, (\exists g \in T_\delta)[g \subseteq f \ \&\ f{\restriction}((\delta+1) \setminus \text{Dom}(g)) \text{ is zero }]\}$.

So we have defined $T_\alpha = T^i_\alpha$ for $\alpha < \omega_1$, and let $Q_i \in V[G_i]$ be $\bigcup_{\alpha<\omega_1} T^i_\alpha$ ordered by inclusion; and it is easy to see that $\underset{\sim}{Q}_i$ is as required (in (a) and (b) of 1.1). Let $\underset{\sim}{C}_i = \bigcup\{f^{-1}(\{1\}) : f \in \underset{\sim}{G}_{Q_i}\}$, so $\Vdash_{Q_i}$ "$\underset{\sim}{C}_i$ is a club of ω_1".

So $\bar{Q} = \langle P_i, \underset{\sim}{Q}_i : i < \omega^2\rangle$ is defined, and it is easy to see that we can replace (in $V[G_i]$) B_i by $\bar{C}^i = \langle C_j : j < i\rangle$. Let $G \subseteq P_{\omega^2}$ be generic, and C_i the interpretation of $\underset{\sim}{C}_i$. Let f_i be the characteristic function of C_i, and $C \overset{\text{def}}{=} \bigcap_{i<\omega^2} C_i$, and $\{\alpha_\zeta : \zeta < \omega_1\}$ an enumeration of C (in increasing order). We shall suppose that forcing by P_{ω^2} does not add reals, and shall deduce that $\langle f_i : i < \omega^2\rangle \in V$, which is clearly false, as $\Vdash_{Q_0}$ "$C_0 \notin V$".

By the assumption the sequence $\langle f_i{\restriction}\alpha_0 : i < \omega^2\rangle$ belongs to V, and we shall show how to compute $\langle f_i{\restriction}\alpha_\zeta : i < \omega^2\rangle$ for every ζ, by induction on ζ; as the computation is done in V we get the desired contradiction. More formally, there is a function F in V such that

$$\langle f_i{\restriction}\alpha_{\zeta+1} : i < \omega^2\rangle = F\big[\langle f_i{\restriction}\alpha_\zeta : i < \omega^2\rangle\big].$$

So suppose $\langle f_i{\restriction}\alpha_\zeta : i < \omega^2\rangle$ is given, and let, for $i < \omega^2$:

$$\beta_i \overset{\text{def}}{=} \text{Min}\,(C_i \setminus (\alpha_\zeta + 1)), \quad \xi_i \overset{\text{def}}{=} \text{Min}\{\xi : f_i{\restriction}\alpha_\zeta = f^*_\xi\}.$$

By demand (i) in the definition of the T^i_α's $C_i \subseteq \text{acc}\left(\bigcap_{j<i} C_j\right)$. So clearly $\beta_j < \beta_i$, for $j < i$ and $\beta_i \in C_j$ for $j \leq i$. Also by demand (vi) on the T^i_α's, $\beta_j < \xi_i$ for $j < i$, and by demand (viii) on the T^i_α's $\xi_j < \beta_i$ for $j < i$. We can conclude that $\text{Sup}\{\beta_i : i < \omega n\} = \text{Sup}\{\xi_i : i < \omega n\}$ for all $n \in \omega$; but from $\langle f_i{\restriction}\alpha_\zeta : i < \omega^2\rangle$ we can compute $\gamma_n \overset{\text{def}}{=} \text{Sup}\{\xi_i : i < \omega n\}$. As $\beta_i \in C_j$ for $j < i$, $\gamma_n \in C_j$ when $j < \omega n$, and clearly $\gamma_n < \gamma_{n+1}$, so we have $\gamma \overset{\text{def}}{=} \bigcup_{n<\omega} \gamma_n \in \bigcap_{j<\omega^2} C_j$. By the definition of the α_ζ's, $\gamma = \alpha_{\zeta+1}$. As we know $T^0_\gamma \cap L_\delta[A]$, and we know $\{\gamma_n : n < \omega\} \subseteq C_0$; $f_0{\restriction}\gamma$ is uniquely determined (by demand (iv)). Similarly we continue to reconstruct $f_i \restriction \gamma$ by induction on $i < \omega^2$ (see end of Case D in the construction - the canonical choice), thus finishing the proof. $\square_{1.1}$

1.2 Remark. The ω^2 in 1.1. is best possible.

1.3 Lemma. (1) Fixing $\langle f^*_\alpha : \alpha < \omega_1 \rangle$, a list of $F = \{f : f : (\beta + 1) \to \{0,1\}, f^{-1}(\{1\})$, closed, $\beta < \omega_1 (f \in V$, of course) $\}$ (so we are assuming CH) and $h : \omega_1 \to \omega_1$, we can repeat the construction in the proof of 1.1 (omitting the assumption on A), and its conclusion holds provided that

$(*)_1$ (α) if χ is large enough, $T \subseteq \{f^*_\alpha \restriction \gamma \colon \gamma < \omega_1, \alpha < \omega_1\}$, $T \in N \prec (H(\chi), \in, <^*_\chi)$, N countable, $\mathcal{J} \in N$ a dense subset of $(T, \subseteq)$, *then* $\mathcal{J} \cap N \in \{\mathcal{J}^\delta_\ell : \ell < h(\delta)\}$ where $\delta = \omega_1 \cap N$, and $\{J^\delta_\ell : \ell < \omega_1\}$ is a list of all subsets of $\{\gamma^*_\alpha \restriction \gamma : \gamma, \alpha < \delta\}$.

(β) moreover, after CS iteration of length $i < \omega^2$ of forcing notions of this form $((T, \subseteq))$, giving generic sets $G_j (j < i)$, (α) continues to hold with $\{\mathcal{J}^\delta_\ell : \ell < h(\delta)\}$ replaced with the family of subsets of $\{f^*_\alpha \restriction \gamma : \gamma, \alpha < \delta\}$ definable in $(\delta \cup \{f^*_\alpha \restriction \gamma : \gamma, \alpha < \delta\} \cup \{\langle \gamma, \alpha, \mathcal{J}^\gamma_\alpha \rangle \colon \gamma \le \delta, \alpha \le h(\gamma)\}, G_j)_{j<i}$,

or, at least

$(*)_2$ for each $\delta < \omega_1$, we have $\bar{g}^\delta = \langle g^\delta_\eta : \eta$ a sequence of successor length $< \omega^2$, each $\eta(i)$ is in ${}^\omega 2\rangle$ such that:

(α) $g^\delta_{\eta(i)} : \delta \to \{0,1\}, (g^\delta_{\eta(i)})^{-1}(\{1\})$ is an unbounded subset of δ (and if η, ν have length $i+1$, $\eta \restriction i \ne \nu \restriction i$, then $(g^\delta_{\eta(i)})^{-1}(\{1\}) \cap (g^\delta_{\nu(i)})^{-1}(\{1\})$ is bounded in δ).

(β) Suppose $i < \omega^2, \chi$ large enough, $N \prec (H(\chi), \in, <^*_\chi)$, N countable, $\delta = N \cap \omega_1, \bar{Q}^i = \langle P_j, \underset{\sim}{Q}_j : j \le i \rangle \in N$ defined as in 1.1, is a CS iteration, each $\underset{\sim}{Q}_j$ proper satisfying (i) - (viii) from the proof of 1.1 (with (vii) rephrased in terms of $\{J^\delta_\ell : \ell < \kappa(\delta)\}$), P_i adds no reals, and $\langle g^\delta_{\eta(j)} : j < i \rangle$ is generic for $P_i \cap N$, then for at least one $\nu \in {}^\omega 2, g^\delta_{\eta \frown \nu}$ is $(N[g^\delta_{\eta(j)} : j < i], Q_i)$-generic.

(2) If Q^* is adding $\aleph_1$ Cohen reals, and $V \models$CH then $V^{Q^*} \models (*)_2$.

Proof of 1.3 (1) Same proof as 1.1.

(2) Left to the reader. Note: let $Q = \{f : f$ a finite function from ω_1 to $\{0,1\}\}$,

let f^* be the generic function; now in $V[f^*]$, if

$$N \prec (H(\chi)^{V[f^*]}, \in, V \cap H(\chi), f^*, <^*_\chi) \text{ countable,}$$

then $N \in V[f^* \restriction (N \cap \omega_1)]$. So for δ, defining $\bar{g}^\delta$, we have to consider only $N \in V[f^* \restriction \delta]$, so $f^* \restriction [\delta, \omega_1)$ is "free" to be used for defining $\bar{g}^\delta$.

1.4 Lemma. (1) We could weaken the demands on V (in 1.1) to $V \models$CH, provided that we also waive the requirement $\Vdash_{P_i}$ "$|\underset{\sim}{Q}_i| = \aleph_1$".

(2) Assume CH and

$(*)_3$ there are $\bar{C} = \langle C_\delta : \delta < \omega_1 \text{ limit}\rangle$ and $h : \omega \to \omega$ such that:

(α) C_δ is an unbounded subset of δ of order type ω

(β) for every club E of ω_1, $S_h(E, \bar{C}) \stackrel{\text{def}}{=} \{\delta < \omega_1 : C_\delta \cap E$ is unbounded in δ, and, moreover, for arbitrarily large $\alpha \in C_\delta$, $|C_\delta \cap \text{Min}(C_\delta \cap E \setminus \alpha)| \geq h(|\alpha \cap C_\delta|)\}$ contains a club of ω_1†,

(γ) h diverges to infinity.

Then the conclusion of 1.1 holds except that we weaken condition (a) to: Q_i satisfies the $\aleph_2$-pic and is proper.

(3) Assume CH (for clarity). There is a forcing notion Q, $|Q| = 2^{\aleph_1}$, Q is $\aleph_1$-complete satisfying $\aleph_2$-pic, and $\Vdash_Q$ "$(*)_3$"

Proof of 1.4(1): By (2) and (3).

Proof of 1.4(2): The proof is similar to the proof of 1.1. The main difference is that defining $T^i_{\delta+2}$ for $\delta \in \text{acc}\left(\bigcap_{j<i} C_j\right)$, we do not choose the members f of $T_{\delta+1}$ such that $\delta = \sup f^{-1}(\{1\})$ by "inverse limit construction" i.e., by constructing the F_n's, but by induction on $\zeta < \omega_1$. W.l.o.g. h is non decreasing Choose $\langle h_i : i < \omega_1\rangle$, $h_i : \omega \to \omega$ diverging to ∞, non decreasing, [$i < j \Rightarrow$ for every k large enough, $h_j(k) < h_i(k) < h(k)$] and [$i, \omega_1, k < \omega \Rightarrow h_{i+1}(\ell)/h_i(\ell + k)$ goes to infinity]; why can we? choose h_i by induction on $i < \omega_1$, for each i we diagonalize. Defining Q_i, we shall assume that in

† So if $\bar{C} \in N \prec (H(\chi), E, <^*_\chi)$, $\delta = N \cap \omega_1 < \omega_1$, and $\delta \in S_h(E, \bar{C})$, *then*: if $E \in N$ is a club of ω_1 then $C_\delta \cap E$ is unbounded below δ.

$V[G_{P_i}], (\bar{C}, h_i)$ still exemplify $(*)_3$. So, for limit i, we have to repeat the proof of preservation of properness and preserve $(*)_3$.

We now define the Q_i's. First, we define Q_i^0: initiating the construction in the proof of 1.1, in Case D we have to change somewhat (to guarantee that forcing with Q_i preserves "$\bar{C}$ exemplifies $(*)_3$"). We choose by induction on $\zeta < \omega_1$, a function $f_\zeta^{\delta,i} : \delta \to \delta$ such that: (letting $C_\delta = \{\beta_n^\delta : n < \omega\}$, increasing in n)

(α) for each $\gamma < \delta$ we have

$$f_\zeta^{\delta,i} \restriction \gamma \in T_\delta^i \cap \{f_\xi^* \restriction \gamma : \xi < \delta\}.$$

(β) The set $Y_\zeta^{\delta,i} = \{n : f_\zeta^{\delta,i} \restriction [\beta_n^\delta, \beta_{n+1}^\delta) \neq 0_{[\beta_n^\delta, \beta_{n+1}^\delta)}\}$ satisfies: for n large enough if $n < m$ are successive members of $Y_\zeta^{\delta,i}$ then $n + h_{2i}(n) < m < n + h_{2i+1}(n)$.

(γ) if $\xi < \zeta$ then $Y_\zeta^{\delta,i} \cap Y_\xi^{\delta,i}$ is finite.

(δ) if $\langle \mathcal{J}_\ell^* : \ell < \omega \rangle$ is a list of dense subsets of $T_\delta^i \cap \{f_\xi^* \restriction \gamma : \xi < \delta, \gamma < \delta\}$, each satisfying $\otimes_{\mathcal{J}_\ell^*}$ (see below), then for some ζ, for every $n \in Y_\zeta^{\delta,i}$: if $m \leq n$ and there is g, $f_\zeta^{\delta,i} \restriction \beta_m^\delta \subseteq g \in T_\delta^i \cap \{f_\xi^* \restriction \gamma : \xi, \gamma < \delta\}$, $g \in \bigcap_{\ell < n} \mathcal{J}_\ell^*$, then $f_\zeta^{\delta,i}$ satisfies this for such maximal $m (\leq n)$:

$$|(f_\xi^{\delta,i})^{-1}(\{1\}) \cap (\beta_{n+1}^\delta \setminus \beta_n^\delta)|.$$

Where

$\otimes_{\mathcal{J}}$ if $f \in T_\delta^i \cap \{f_\xi^* \restriction \gamma : \xi < \delta, \gamma < \delta\}$ and

$$\begin{aligned} A_{f,\mathcal{J}} = \{\alpha \in C_\delta :\ & \text{if } f \subseteq f' \in T_\delta^i \cap \{f_\xi^* \restriction \gamma : \xi < \alpha, \gamma < \alpha\} \text{ then for} \\ & \text{some } f'' \in \mathcal{J} \text{ we have } f' \subseteq f'' \in T_\delta^i \cap \{f_\xi^* \restriction \gamma : \xi < \alpha, \gamma < \delta\} \end{aligned}$$

then for infinitely many $\alpha \in C_\delta$ we have

$$|\{\beta \in C_\delta : \alpha \leq \beta \text{ and } (\alpha, \beta) \subseteq A_{f,\mathcal{J}}\}| \geq h(|C_\delta \cap n|)$$

How? First choose inductively m_i such that: $m_i + i + 1 < m_{i+1} < \omega$, and for i large enough $m_i + i + 1 + h_{2i}(m_i + i + 1) < m_{i+1}$, $m_{i+1} + i + 2 < m_i + h_{2i+1}(m_i)$ (this is possible as $k < \omega \Rightarrow \langle h_{2i+1}(m)/h_{2i}(m + k) : m < \omega \rangle$ goes to infinity). Second list $\{j : j < i\}$ as $\langle j_k : k < \omega \rangle$, and diagonally choose $Y^{\delta,i}_\zeta \cap [m_i, m_i + i + 1)$, a singleton. Now, for $\alpha \in Y^{\delta,i}_\zeta$ we deal with the $\mathcal{J}^*_{g_\zeta(i)}$ where: for each ℓ, for some k no two successive members of $g^{-1}_\zeta\{\ell\}$ are of difference $> k$.

Now, Q^0_i is defined analogously to the Case D in the proof of Lemma 1.1. Then Q_i is the result of CS iteration starting with Q^0_i, and continuing with shooting club through $S_{h_{i+1}}[E, \bar{C}]$ for every club $E \subseteq \omega_1$, (by initial segments).
(3) Q is forcing $\bar{C}$ by initial segments and then CS iteration (of length $2^{\aleph_1}$) of shooting club through $S_h(E, \bar{C})$ for every club $E \subseteq \omega_1$ (by initial segments).

$\square_{1.4}$

1.5 Claim. Under the assumptions of 1.1 for $\varepsilon < \omega_1$ additively indecomposable, we can add to the conclusion: for $\zeta < \varepsilon$ and $i < i^*$, the forcing notion Q_i is ζ-proper.

Proof. We again assume $G_i \subseteq P_i$ generic is given; hence $\langle C_j : j < i \rangle$ (which serve as B_i too) is also given and by induction on α we define T^i_α, so that in the definition of T^i_α we use A and $\langle C_j \cap \alpha : j < i \rangle$ only (and the list $\{f^*_\xi : \xi < \omega_1\} \in L[A]$), so that a variant of (i) - (viii) holds. The changes are:

(iv)′ if $f, g \in T^i_\alpha$, $f(i) \neq g(i)$, then $f^{-1}(\{1\}) \cap g^{-1}(\{1\}) \setminus i$ has order-type $< \varepsilon$.

(vii)′ In addition to (vii), if $\langle \delta_\zeta : \zeta \leq \zeta^* \rangle$ is an increasing sequence of accumulation points of $\bigcap_{j<i} C_j$, $\langle \delta_\xi : \xi \leq \zeta \rangle \in L_{\delta_{\zeta+1}}[A \cap \delta_{\zeta+1}]$, for $\zeta < \zeta^*$, $f \in T^i_{\delta_0} \cap L_{\delta_0}[A \cap \delta_0]$, $f_m \in T^i_{\zeta^*+1}$ for $m < m^*$ and $m^* < \omega$, $\zeta^* < \varepsilon$, then there is $g \in T^i_{\zeta^*+2}$, $f \subseteq g$, $\mathrm{Dom}(g) = \zeta^* + 1$, such that the following conditions hold:

(α) For every $\mathcal{J} \in L_{h(\delta)}[A \cap \delta, B_i \cap \delta]$ (an open dense subset of $T^i_\delta \cap L_\delta[A \cap \delta]$ (ordered by inclusion)) for some $\gamma < \delta$, $g \restriction \gamma \in \mathcal{J}$, where $\delta \in \{\delta_\zeta : \zeta \leq \zeta^*\}$.

(β) $g^{-1}(\{1\}) \setminus \{\delta_\zeta : \zeta \leq \zeta^*\}$ is a bounded subset of δ_{ζ^*}.

(γ) For every $m < m^*, g^{-1}(\{1\}) \cap f_m^{-1}(\{1\}) \setminus \{\delta_\zeta : \zeta \leq \zeta^*\} \subseteq \mathrm{Dom}(f)$.

In the proof of Case D, we use the canonical well-ordering of $H(\aleph_1)^{L[A]}$ on our assignments (for the existence of $g \in T^i_{\delta+2}$, $\mathrm{Dom}(g) = \delta + 1$), and construct a witness, preserving and using (vii)$'$. $\square_{1.5}$

1.6 Discussion. 1) Also 1.3, 1.4 can be generalized to this context.
2) We have shown that just excluding the forcing notions like the one from Example V.5.1 (by demanding $\mathbb{D}$-completeness for a simple 2-completeness system) is not enough to ensure that CS iteration of proper forcing does not add reals. In VIII §4, on the other hand, we have quite weak restrictions on such $\underset{\sim}{Q}_i$ ensuring $\mathrm{Lim}\langle P_i, \underset{\sim}{Q}_i : i < \alpha\rangle$ does not add reals. However, the examples above (1.1-1.4) lead naturally to forcing notions which fall in between (and the corresponding consistency problems), which we now proceed to represent.

1.7 Problem. Let $f_\delta : \delta \to \delta$ for any limit $\delta < \omega_1$. Is there $f : \omega_1 \to \omega_1$ such that for every $\delta < \omega_1$, for arbitrarily large $\alpha < \delta$, $f_\delta(\alpha) < f(\alpha)$? I.e., the problem is, assuming CH, whether it is possible that for every such $\langle f_\delta : \delta < \omega_1\rangle$ there is a suitable f [negative answer follows from $\diamondsuit_{\aleph_1}$, and c.c.c. forcing preserves a negative answer].

1.7A Definition. For any sequence $\bar{f} = \langle f_\delta : \delta < \omega_1\rangle$, $f_\delta : \delta \to \delta$, let $P^0_{\bar{f}} = \{g : g$ a function from some $\alpha < \omega_1$ into ω_1, such that for every (limit) $\delta \leq \alpha$, for arbitrarily large $\beta < \delta$, $f_\delta(\beta) < g(\beta)\}$;
ordered by inclusion.

1.7B Discussion. Now if CH + Ax[forcing notions of the form $P^0_{\bar{f}}$] holds in some universe, the answer to 1.7 is yes (in that universe). So it is enough to show that if we iterate, with countable support, such forcing notions, then no real is added. A positive answer follows by 1.8 below and next section. A negative answer could have been viewed as proving a very weak form of diamond. The situation is similar for the other problems here.

1.8 Problem. Let $C_\delta \subseteq \delta$ be an unbounded subset of δ, for $\delta < \omega_1$. Is there a closed unbounded $C \subseteq \omega_1$ such that for no δ, $C_\delta \subseteq C$? Consider in particular the cases when we restrict ourselves to:

(a) C_δ has order-type ω, $\delta = \text{Sup } C_\delta$,

(b)$_\xi$ C_δ has order-type ξ, $\delta = \text{Sup } C_\delta$ and ξ limit,

(c) C_δ has order-type $< \delta$, $\delta = \text{Sup } C_\delta$,

(d) $C_\delta \equiv \emptyset \bmod D_\delta$, D_δ a filter on δ, $\delta = \text{Sup } C_\delta$, for a given $\bar{D} = \langle D_\delta : \delta < \omega_1 \rangle$.

1.8A Definition. For $\bar{C} = \langle C_\delta : \delta < \omega_1 \rangle, C_\delta \subseteq \delta$, let $P^1_{\bar{C}} = \{f : f$ a function from some $\alpha + 1 < \omega_1$ to $\{0,1\}, f^{-1}(\{1\})$ is closed and for no $\delta \le \alpha$, $C_\delta \subseteq f^{-1}(\{1\})\}$.

Order: inclusion.

We may consider also

1.8B Definition. For $\bar{D} = \langle D_\delta : \delta < \omega_1 \rangle$, D_δ a filter on δ, let

$$P^1_{\bar{D}} = \{f : f \text{ a function from some } \alpha + 1 < \omega_1 \text{ to } \{0,1\} \text{ such that } f^{-1}(\{1\}) \text{ is closed and for no } \delta \le \alpha,\ f^{-1}(\{1\}) \cap \delta = \delta \bmod D_\delta\}$$

Order: inclusion.

1.9 Problem. Let C_δ be an unbounded subset of δ, for $\delta < \omega_1$. Is there a closed unbounded $C \subseteq \omega_1$ such that for every δ, $C \cap C_\delta$ is a bounded subset of δ, when we restrict ourselves as in 1.8?

1.9A Definition. For a sequence $\bar{C} = \langle C_\delta : \delta < \omega_1 \rangle$, C_δ an unbounded subset of δ, let

$$P^2_{\bar{C}} = \{g : g \text{ a function from some } \alpha + 1 < \omega_1 \text{ to } \{0,1\}, \text{ such that } g^{-1}(\{1\}) \text{ is closed and for every } \delta \le \alpha, \text{Sup}[C_\delta \cap g^{-1}(\{1\})] < \delta\}.$$

Order: inclusion.

1.9B Definition. For a sequence $\bar{D} = \langle D_\delta : \delta < \omega_1 \rangle$, D_δ a filter on δ, let

$$P^2_{\bar{D}} = \{g : g \text{ a function from some } \alpha + 1 < \omega_1 \text{ to } \{0,1\} \text{ such that } g^{-1}(\{1\}) \text{ is closed and for every } \delta \leq \alpha,\ g^{-1}(\{1\}) \cap \delta \equiv \emptyset \bmod D_\delta\}$$

1.10 Claim. 1) $P^0_{\bar{f}}, P^1_{\bar{C}}$ and $P^2_{\bar{C}}$ (when one of the Cases (a)-(d) from 1.8 holds) are proper and $\mathbb{D}$-complete for some simple $\aleph_1$-completeness systems and
2) $P^0_{\bar{f}}, P^1_{\bar{C}}$ is strongly proper.
3) $P^2_{\bar{C}}$ is proper (and does not add reals) even in V^Q if forcing by Q adds no reals (for $P^0_{\bar{f}}, P^1_{\bar{C}}$ this follows by part 2).

Proof. Left to the reader.

1.11 Definition. For each $\delta < \omega_1$, let F_δ be a function, from $\mathrm{Dom}(F_\delta) = \{f : f$ a function from some $\alpha+1 < \delta$ to $\{0,1\}$ such that $f^{-1}(\{1\})$ is closed $\}$ to ω. Let $C_\delta \subseteq \delta$ be an unbounded subset of δ of order type ω and $\bar{C} = \langle C_\delta : \delta < \omega_1 \rangle$. Let

$$\begin{aligned} P_{\bar{C},F} = \{g : \; & g \text{ a function from some } \alpha + 1 < \omega_1 \text{ to } \{0,1\} \\ & g^{-1}(\{1\}) \text{ is closed and for every } \delta \leq \alpha \text{ for} \\ & \text{some } n_\delta : \text{if } \beta \in C_\delta, |C_\delta \cap \beta| > n_\delta, \text{ and} \\ & \mathrm{Min}(C_\delta \setminus (\beta+1)) > \mathrm{Min}\,(g^{-1}(\{1\}) \setminus (\beta+1)) \text{ and} \\ & \beta < \gamma \in C_\delta, \mathrm{Min}\,(C_\delta \setminus (\gamma+1)) > \mathrm{Min}\,(g^{-1}(\{1\}) \setminus (\gamma+1)), \textit{ then} \\ & |\gamma \cap C_\delta| > F_\delta(g{\restriction}(\mathrm{Min}\,(C_\delta \setminus (\beta+1))))\}. \end{aligned}$$

1.11A Claim. $P_{F,\bar{C}}$ (for $F, \bar{C}$ as in definition 1.11) is proper, $\mathbb{D}$-complete for some simple $\aleph_1$-completeness system and

$\otimes$ it is proper not adding reals even after forcing by any proper forcing notion not adding reals.
(see 2.13(2)).

Proof. Left to the reader.

Remark. In 1.11 (and 1.11A) demand

$$|\gamma \cap C_\delta| > F_\delta(|C_\delta \cap (\beta+1)|, g{\restriction}(\min(C_\delta \setminus (\beta+1)))).$$

§2. Not Adding Reals

We prove here that we can iterate (CS iterations) the forcing notions introduced at the end of the previous section, and not add reals. The real work is in Definition 2.2 and Lemma 2.8, but the reader may look at Conclusion 2.12 (or at 2.16). For our aim , naturally, we phrased a condition NNR_2, on CS iterations of proper forcing, saying we add no reals (condition (3)), a quite weak condition for avoiding "collision" with the weak diamond, and another condition, (5), intended to avoid collision with the counterexample of §1. It says each Q_i stays proper even if we force with forcing notions of the kind we iterate. Having phrased the condition, the main point is proving it is preserved by CS iteration, mainly in limit stages.

So, suppose $\bar{Q}$ has length δ, and for each $\alpha < \delta$, $\bar{Q}{\restriction}\alpha$ is as required. Assume for simplicity we do not try to kill $\Phi^{\aleph_0}_{\aleph_1}$; say, using $\aleph_1$-completeness systems. As seems natural, we start with a countable $N \prec (H(\chi), \in, <^*_\chi)$ and $p \in P_\delta \cap N$ and try to find q, $p \le q \in P_\delta$, q is (N, P_δ)-generic and determine $\underset{\sim}{G}_{P_\delta} \cap N$, which should be an old set if we succeed. So, if $\sup(\delta \cap N) = \bigcup_{n<\omega} \alpha_n$, $\alpha_n < \alpha_{n+1}$, $\alpha_n \in N$ we should try to choose approximations $q_n \in P_{\alpha_n}$, $q_{n+1}{\restriction}\alpha_n = q_n$. But, as we do not have $\aleph_1$-completeness we cannot do this per se. In V§7 a major point in the proof is that we have "above" N a sequence $\bar{N} = \langle N_i : 1 \le i \le \zeta \rangle$, ζ and each N_i countable (letting $N_0 = N$), $\bar{N}$ quite "long" in suitable sense, and we demand q_n is (N_i, P_{α_n})-generic for "many" i's. So if e.g. $\alpha_{n+1} = \alpha_n + 1$, $i < \zeta$ such that q_n is (N_i, P_{α_n})-generic, we have only $\aleph_0$ candidates for members of the relevant completeness system. However "using N_i is destroying it", "it is consumed ", as q_{n+1} is not $(N_i, P_{\alpha_{n+1}})$-generic. Why is this so? In the first step, say choosing $G_{Q_0} \cap N_0$, we have no problem; for $\underset{\sim}{G}_{Q_1} \cap N_0[\underset{\sim}{G}_{Q_0}]$ we have to choose a common member from all the candidates A to be "a subset of $Q_1 \cap N_0[\underset{\sim}{G}_{Q_0}]$ in the appropriate family $\mathbb{D}_x$". Now the common member is naturally not in N_1. We can use stronger induction hypothesis, then use only $\aleph_0$-completeness system or even 2-completeness system, so we have for $\underset{\sim}{G}_{Q_0} \cap N_1$ only finitely many (or two) candidates, so a common member exists in N_1. But after ω steps it is not clear how to guarantee $G_{P_\omega} \cap N_0 \in N_1$.

One approach, suggested in [Sh:177], is to weaken "Q is α-proper" to "for $p \in Q \cap N_0$, $\bar{N} = \langle N_i : i \le \alpha \rangle$ as usual, there is $q, p \le q \in Q$, q is (N_i, Q)-generic for "many" $i \le \alpha$"; this work for "easy" cases like interpreting "many" as "having the same order type". While this work for e.g. $P^1_{\underset{\sim}{C}}$ (from 1.8(a), 1.8A), this does not seem strong enough , but it covers the forcing notion of V for specializing Aronszajn tree, which the present condition do not. Here we rather say: having two candidates for $\underset{\sim}{G} \cap N_1$, we demand they are a subset of $(Q_0 \cap N) \times (Q_0 \cap N)$ generic over N_0; for making this work we are carried to the following.

Here we have N_1, $N_0 \prec N_1$, q_n is demanded to be (N_1, P_{α_n})-generic too; We have several possibilities, we actually have a finite tree of possibilities for $\underset{\sim}{G}_{P_{\alpha_n}} \cap N_1$ which is generic for an appropriate product of finitely many copies of $P_i's$, $i \le \alpha_n$. But to proceed with this we have N_2, where again we have a finite tree of possibilities for $\underset{\sim}{G}_{P_{\alpha_n}} \cap N_2$. But only each one is generic over N_2. Above this we imitate V 4.4. Now q_n is (N_ℓ, P_{α_n})-generic for $\ell = 3, 4, 5$ and for each P_{α_n}-name $\underset{\sim}{\tau} \in N_4$ of an ordinal it allows only finitely many possibilities, but unlike V 4.4 we do not use ω-properness. So we have for each n a "tower" of six models. For higher $\ell \le 5$, q_n "knows" less on N_ℓ, but our knowledge goes down "slowly" so moving from n to $n+1$, taking care of ℓ, the knowledge on $\underset{\sim}{G}_{P_{\alpha_n}} \cap N_{\ell+1}$ is enough to move ahead. Probably this explanation is meaningless for many readers, but will be helpful if read in the end or middle of reading the proof.

Note that §1 (particularly 1.5) show the impossibility of too good iteration theorems (say CS, of proper forcing not adding reals) but do not block consistency of appropriate forcing axiom with CH as we may instead of forcing with candidates, spoil their being candidates (as in III, V).

2.1 Definition. 1) For a finite tree t (i.e. $t = (|t|, <^t), |t|$ a finite set, $<^t$ a partial order on $|t|$ such that for $x \in t$, $\{y : y <^t x\}$ is linearly ordered), let

$$\begin{aligned} \text{trind}_{\alpha}(t) = \{\bar{\alpha} : \bar{\alpha} = \langle \alpha_\eta : \eta \in t\rangle,\ &\text{each } \alpha_\eta \text{ is an ordinal } \leq \alpha \\ &\text{and } \eta < \nu \text{ in } t \text{ implies } \alpha_\eta \leq \alpha_\nu\} \\ \text{trind}_{<\alpha}(t) = \bigcup_{\beta<\alpha} \text{trind}_{\beta}(t).& \end{aligned}$$

2) For a given iteration $\bar{Q} = \langle P_i, \underset{\sim}{Q}_j : i \leq \alpha, j < \alpha\rangle$, a finite tree t and $\bar{\alpha} \in \text{trind}_\alpha(t)$ let

$P_{\bar{\alpha}} = \{\bar{p} : \bar{p} = \langle p_\eta : \eta \in t\rangle$, and for $\eta \in t$ we have $p_\eta \in P_{\alpha_\eta}$ and if $t \models \eta < \nu$ then $p_\eta = p_\nu \restriction \alpha_\eta\}$

ordered by

$\bar{p} \leq \bar{q}$ *iff* for every $\eta \in t$, $P_{\alpha_\eta} \models p_\eta \leq q_\eta$.

3) $t \subseteq_{\text{end}} s$ *if* t is s restricted to the set of members of t and: $s \models$ ["$\eta < \nu$", $\nu \in t$] implies [$\eta \in t$].

4) We write $\langle i\rangle$ for $\bar{\alpha} = \langle \alpha_\eta : \eta \in t\rangle$, when t has one element, say $<>$ and $\alpha_{<>} = i$.

5) For $\bar{Q}$ an iteration of length α, t a finite tree, $\bar{\alpha} \in \text{trind}_\alpha(t)$, and model N:

(a) $a\text{Gen}^{\bar{\alpha}}_{\bar{Q}}(N) = \{G : G$ a subset of $P_{\bar{\alpha}} \cap N$ generic over N such that for each $\eta \in t, G_\eta \overset{\text{def}}{=} \{p_\eta : \bar{p} \in G\}$ has an upper bound in $P_{\alpha_\eta}\}$

[Note: G can essentially be identified with $\langle G_\eta : \eta \in t\rangle$].

(b) $s\text{Gen}^{\bar{\alpha}}_{\bar{Q}}(N) = \{G : G$ a subset of $P_{\bar{\alpha}} \cap N$ generic over N which has an upper bound in $P_{\bar{\alpha}}\}$.

(c) $\text{Gen}^{\bar{\alpha}}_{\bar{Q}}(N) = \{G : G$ a subset of $P_{\bar{\alpha}} \cap N$ generic over $N\}$

6) If $t_1 \subseteq t_2$, $\bar{\alpha}^\ell \in \text{trind}_{\alpha_\ell}(t_\ell)$, and $\bar{p}^\ell \in P_{\bar{\alpha}^\ell}$ for $\ell \in \{1,2\}$, and $\bigwedge_{\eta \in t_1} \alpha^1_\eta \leq \alpha^2_\eta$ then $\bar{p}^1 \leq \bar{p}^2$ means $\bigwedge_{\eta \in t_1} p^1_\eta \leq p^2_\eta \restriction \alpha^1_\eta$.

2.2 Definition. $\bar{Q}$ is an NNR_2-iteration for $(\mathcal{E}_0, \mathcal{E}_1, \mathcal{E}_2)$ means:

(1) $\bar{Q}$ is a CS iteration of $\mathcal{E}_0$-proper forcing (see V2.2).

(2) For $\ell = 0, 1, 2, \mathcal{E}_\ell$ is a stationary subset of $\mathcal{S}_{<\aleph_1}(\lambda_\ell)$ for some uncountable λ_ℓ.

(3) Forcing with P_α adds no reals for $\alpha \leq \ell g(\bar{Q})$.

$(4)_2$ *If*:

(a) $N \prec (H(\chi), \in, <^*_\chi)$ is countable, (χ regular large enough),

(b) $N \cap \lambda_1 \in \mathcal{E}_1$,

(c) $\langle \bar{Q}, \mathcal{E}_0, \mathcal{E}_1, \mathcal{E}_2 \rangle$ belongs to N,

(d) $i^* \leq i \leq j \leq \alpha \leq \ell g(\bar{Q})$, $i^* \in N$, $i \in N$, $j \in N$ and i^*, i are non-limit,

(e) $G^a \in \mathrm{Gen}^{<i>}_{\bar{Q}}(N)$,

(f) $p \in N \cap P_j$, $p{\restriction}i \in G^a$, and

(g) q_0, q_1 are upper bounds of G^a in P_i, and $q_0{\restriction}i^* = q_1{\restriction}i^*$,

then there is $G' \in \mathrm{Gen}^{<j>}_{\bar{Q}}(N)$, extending G^a, $p{\restriction}j \in G'$ and q_0^+, $q_1^+ \in P_j$ such that: $q_0 \leq q_0^+{\restriction}i$, $q_1 \leq q_1^+{\restriction}i$, and for $\ell = 0, 1$, q_ℓ^+ is an upper bound to G'. Let G' extends G^a, mean that $[p' \in G' \Rightarrow p'{\restriction}i \in G^a]$ and $q_0^+{\restriction}i^* = q_0^+{\restriction}i^*$.

(5) Assume i are not limit ordinals, $i < j \leq \alpha$, $\bar{Q}'$ a CS iteration of length β satisfying (1) - (4), $\alpha \leq \beta, \bar{Q} = \bar{Q}'{\restriction}\alpha$, $\bar{Q}'$ satisfies (1)-(4), t a finite tree, $\eta^* \in t$, $\bar{\alpha} \in \mathrm{trind}_\beta(t)$, $\alpha_{\eta^*} = i$, $s = t{\restriction}\{\eta : \eta \leq \eta^*\}$, (so $P'_{\bar{\alpha}{\restriction}s}$ is essentially P_i), and let $\underset{\sim}{R} \stackrel{\text{def}}{=} P'_{\bar{\alpha}}/P_i$ (a P_i-name).

If $\underset{\sim}{R}$ is an $\mathcal{E}_2$ - proper forcing not adding reals, *then*

$\Vdash_{P_i * \underset{\sim}{R}}$ "P_j/P_i is a $\mathcal{E}_2$−proper forcing not adding reals"†.

2.2A Remark. Note that for $i < j \leq \ell g(\bar{Q})$, i non-limit, we have: P_j/P_i is $(\mathcal{E}_0 \cup \mathcal{E}_1 \cup \mathcal{E}_2)$-proper.

2.3 Definition. $\bar{Q}$ is an $NNR_{\aleph_0}$ - iteration for $(\mathcal{E}_0, \mathcal{E}_1, \mathcal{E}_2)$ is defined similarly, replacing $(4)_2$ by $(4)_{\aleph_0}$ below (so, in clause (5) now we mean this (4)) where:

† Actually, e.g. if Q_0, Q_1 are proper forcings not adding reals and $Q_0 \times Q_1$ is proper, then $Q_0 \times Q_1$ does not add reals; in fact by 2.5 the "not adding reals" is redundant.

$(4)_{\aleph_0}$ *if:*

(a) $N \prec (H(\chi), \in, <^*_\chi)$ is countable, (χ regular large enough),

(b) $N \cap \lambda_1 \in \mathcal{E}_1$,

(c) $\langle \bar{Q}, \mathcal{E}_0, \mathcal{E}_1, \mathcal{E}_2 \rangle$ belongs to N,

(d) $i < j \leq \alpha$, i non-limit, $i \in N$ and $j \in N$,

(e) $G^a \in \mathrm{Gen}^{<i>}_{\bar{Q}}(N)$,

(f) $p \in N \cap P_j$, $p{\restriction}i \in G^a$,

(g) $t \in N$ a finite tree, $\bar{\alpha} \in \mathrm{trind}_i(t)$, each α_η non-limit,

(h) $\bar{q} \in P_{\bar{\alpha}}$, and if $\eta \in t, \alpha_\eta = i$, then q_η is a bound to G^a, and

(i) $\bar{\beta} = \langle \beta_\eta : \eta \in t \rangle$ where for $\eta \in t$:

$$\beta_\eta = \begin{cases} \alpha_\eta & \text{if } \alpha_\eta < i \\ j & \text{if } \alpha_\eta = i; \end{cases}$$

Then there are $G' \in \mathrm{Gen}^{<j>}_{\bar{Q}}(N)$ extending G^a and $\bar{r} \in P_{\bar{\beta}}$, such that:

(α) each $r_\eta (\eta \in t, \alpha_\eta = i)$ is a bound of G'

(β) $\bar{q} \leq \bar{r}$.

Before we state and prove the main lemma, we prove a few claims.

2.4 Claim. 1) Suppose $x \in \{2, \aleph_0\}$, $\bar{Q}'$ is an iteration satisfying (1) - (3),$(4)_x$ of Definition 2.2 or 2.3, $\bar{Q} = \bar{Q}'{\restriction}\alpha$, $\beta = \ell g(\bar{Q}')$, $\bar{Q}$ is an NNR_x-iteration for $(\mathcal{E}_0, \mathcal{E}_1, \mathcal{E}_2)$, χ is regular large enough, $N \prec (H(\chi), \in, <^*_\chi)$ is countable, $\langle \bar{Q}', \mathcal{E}_0, \mathcal{E}_1, \mathcal{E}_2, \alpha \rangle$ belongs to N, $N \cap \lambda_2 \in \mathcal{E}_2$, $t \subseteq_{\mathrm{end}} s$ are finite trees, $\bar{\beta} \in N$ is in $\mathrm{trind}_\beta(s)$, $\bar{\alpha} = \bar{\beta}{\restriction}t$, $\mathrm{Rang}[\bar{\beta}{\restriction}(s \setminus t)] \subseteq \alpha$, $G_{\bar{\alpha}} \in \mathrm{Gen}^{\bar{\alpha}}_{\bar{Q}}(N)$ and $\bar{q} = \langle q_\eta : \eta \in t \rangle \in P_{\bar{\alpha}}$ is above $G_{\bar{\alpha}}$, $\bar{p} \in P_{\bar{\beta}} \cap N$, and $\bar{p}{\restriction}t \in G_{\bar{\alpha}}$.

Assume in addition:

$(*)_t$ for each $\eta \in t$, the forcing notion $P_{\bar{\alpha}}/P_{\bar{\alpha}{\restriction}\{\nu:\nu \leq \eta\}}$ is $\mathcal{E}_2$-proper not adding reals.

Then there is a $G_{\bar{\beta}} \in \mathrm{Gen}^{\bar{\beta}}_{\bar{Q}}(N)$ extending $G_{\bar{\alpha}}$ (recall, $G_{\bar{\beta}}$ extending $G_{\bar{\alpha}}$ means $[\bar{\sigma} \in G_{\bar{\beta}} \Rightarrow \bar{\sigma}{\restriction}t \in G_{\bar{\alpha}}]$) to which $\bar{p}$ belongs and $\bar{r} \in P_{\bar{\beta}}$ above $G_{\bar{\beta}}$, $\bar{q} \leq \bar{r}{\restriction}t$, (and it follows $\bar{p} \leq \bar{r}$).

2) Moreover, if $\eta \in t$, and $\nu \leq^t \eta$ is maximal such that $(\exists \rho \in s \setminus t)(\nu < \rho)$, *then* $r_\eta{\restriction}[\alpha_\nu, \alpha_\eta) = q_\eta{\restriction}[\alpha_\nu, \alpha_\eta)$.

Proof. We prove it by induction on the number of elements of s.

By the induction hypothesis, we can show that if $t \subseteq t_1 \subseteq s \,\&\, t_1 \neq s$, then $(*)_{t_1}$ holds. Hence, it is easy to reduce the claim to the case $s \setminus t$ has a unique element, say η. Assume first that there is a maximal $\nu \in t$ with $\nu <^s \eta$ and let $t^* = t\restriction\{\rho : \rho \in t,\ \rho \leq \nu\}$; so by $(*)_t$ we know $P_{\bar\alpha}/P_{\bar\alpha\restriction t^*}$ is a $\mathcal{E}_2$-proper forcing not adding reals. Let R be the $P_{\bar\alpha\restriction t^*}$-name $P_{\bar\alpha}/P_{\bar\alpha\restriction t^*}$. Note that $P_{\bar\alpha\restriction t^*}$ is isomorphic to P_{α_ν}. Let $i = \alpha_\nu$, $j = \alpha_\eta$ and apply (5) of Definition 2.2 (or of 2.3, of course). We obtain $r = \langle r_\rho : \rho \in s\rangle$.

But why is r_ρ a condition in P_{α_η} and not a $P_{\bar\alpha}$-name of such a condition? As all the influence of $P_{\bar\alpha}/P_{\bar\alpha\restriction t^*}$ is on the set of dense subsets of P_{α_η} in $N[G_{P_{\bar\alpha}}]$, which we know (and it is in V) so as we know there is r_ρ we can inspect each candidate (not i), we use our demanding $q \leq \bar r\restriction t$ rather than $\bar q = \bar r\restriction t$. $\square_{2.4}$

2.5 Claim. If $\bar Q$ satisfies (1), (2), (3) of Definition 2.2, $\alpha = \ell g(\bar Q)$, t is a finite tree, $\bar\beta \in \mathrm{trind}_\alpha(t)$ and $P_{\bar\beta}$ is $\mathcal{E}_2$-proper, *then* it does not add reals. Also, if $G_{\bar\alpha} \subseteq P_{\bar\alpha}$ is generic over V, we let $G_\eta = \{q_\eta : \bar q \in G_{\bar\alpha}\}$, then $\langle G_\eta : \eta \in \mathrm{Dom}(\bar\alpha)\rangle$ determines $G_{\bar\alpha}$ (hence we do not distinguish strictly). Similarly, for "$G \subseteq P_{\bar\alpha} \cap N$ generic over N".

Proof. Immediate.

2.6 Claim. Let (i) $\bar Q$ be a CS iteration of ${}^\omega\omega$-bounding proper forcing notions.

(ii) $i < j \leq \ell g(\bar Q), N_0 \prec N_1$ are countable elementary submodels of $(H(\chi), \in, <^*_\chi)$, $\langle \bar Q, i, j\rangle \in N_0$, χ regular large enough, and $N_0 \in N_1$.

(iii) $p \in P_j \cap N_0$, $q \in P_i$, $p\restriction i \leq q$, q is (N_1, P_i)-generic, and (N_0, P_i)-generic.

(iv) for every pre-dense $\mathcal{I} \subseteq P_i$ from N_0, some finite $\mathcal{J} \subseteq \mathcal{I} \cap N_0$ is pre-dense above q.

Then there is $r \in P_j$, $r\restriction i = q$, $p \leq r$, r is (N_1, P_j)-generic and (N_0, P_j)-generic such that for every pre-dense $\mathcal{I} \subseteq P_j$ from N_0 some finite $\mathcal{J} \subseteq \mathcal{I} \cap N_0$ is pre-dense above r.

2.6A Remark. This claim is from [Sh:177] and is implicit in VI §1.

Proof of 2.6. By VI Theorem 0.A, P_i, P_j/P_i are ${}^\omega\omega$-bounding. Let $\langle \underset{\sim}{\tau}_n : n < \omega\rangle \in N_1$ list all P_j-names of ordinals which belong to N_0.

We can find a functions $F, H \in N_0$ such that: for every $\underset{\sim}{p}$, a P_i-name of a member of $P_j/\underset{\sim}{G}_{P_i}$, and $\underset{\sim}{\tau}$, a P_j-name of a ordinal, we have $\underset{\sim}{p}' = F(\underset{\sim}{p}, \underset{\sim}{\tau})$ is a P_i-name of a member of $P_j/\underset{\sim}{G}_{P_i}$ satisfying $\underset{\sim}{p}'[G_{P_i}]\restriction i = \underset{\sim}{p}[G_{P_i}]\restriction i$, $\underset{\sim}{p}[G_{P_i}] \leq^{P_j} \underset{\sim}{p}'[G_{P_i}]$ in P and $\underset{\sim}{\sigma} = H(\underset{\sim}{p}, \underset{\sim}{\tau})$ is a P_i-name of an ordinal such that

$$\underset{\sim}{p}'[G_{P_i}] \Vdash_{P_j/G_{P_i}} \text{“}\underset{\sim}{\tau} = \underset{\sim}{\sigma}[G_{P_i}]\text{”}.$$

Let $\underset{\sim}{p}_0 = p\restriction[i,j)$, $\underset{\sim}{p}_{n+1} = F(\underset{\sim}{p}_n, \underset{\sim}{\tau}_n)$, $\underset{\sim}{\sigma}_n = H(\underset{\sim}{p}_n, \underset{\sim}{\tau}_n)$, so $\underset{\sim}{p}_n$, $\underset{\sim}{\sigma}_n \in N_0$, and $\underset{\sim}{\sigma}_n$ is a P_i-name of an ordinal and $\langle(\underset{\sim}{p}_n, \underset{\sim}{\sigma}_n) : n < \omega\rangle \in N_1$. For each n we can find $(\underset{\sim}{p}_n^+, \underset{\sim}{\bar{v}}^n) \in N_1$, moreover $\langle(\underset{\sim}{p}_n^+, \underset{\sim}{\bar{v}}^n) : n < \omega\rangle$ belongs to N_1 such that $\underset{\sim}{\bar{v}}^n$ is a P_i-name of a sequence of finite sets of ordinals which belongs to N_1 (so $\Vdash_{P_i}$ “$\underset{\sim}{\bar{v}} \in V$”) $\underset{\sim}{p}_n^+$ a P_i-name of a member of $P_j/\underset{\sim}{G}_{P_i}$, and in $V[G_{P_i}]$ $\underset{\sim}{p}_n^+[G_{P_i}^+]$ is above $\underset{\sim}{p}_n[G_{P_i}]$ (in $P_j/\underset{\sim}{G}_{P_n}$) and is $(N[G_{P_i}], P_j/G_{P_i})$-generic and $\underset{\sim}{\bar{v}}[G_{P_i}] \in N_1[G_{P_i}] \cap V$ (for any $G_{P_i} \subseteq P_i$ generic over V). Let $\underset{\sim}{\bar{v}} = \langle \underset{\sim}{v}_m : m < \omega\rangle$, $\underset{\sim}{v}_m = \bigcup_{n \leq m} v_m^n$ so $\underset{\sim}{\bar{v}} \in N_1$ is a P_i-name, $\Vdash_{P_i}$ “$\underset{\sim}{\bar{v}}$ is an ω-sequence of finite sets of ordinals” (we can make $\Vdash_{P_i}$ “$\underset{\sim}{\bar{v}} \in V$” but we not use).

Let $\langle \bar{u}^n : n < \omega\rangle$ list all sequences, $\bar{u} = \langle u_m : m < \omega\rangle \in N_1$, u_m a finite set of ordinals. Let $\bar{u}^n = \langle u_m^n : m < \omega\rangle$. Choose $\langle u_n^* : n < \omega\rangle \in V$, a sequence of finite sets of ordinals, such that:

$(*)$ for each $n < \omega$ and for every large enough $m, u_m^n \subseteq u_m^*$.

As $\underset{\sim}{\sigma}_n \in N_0$ are P_i-names, by assumption (iv), we can find $\langle v_n^* : n < \omega\rangle \in V$ a sequence of finite sets of ordinals such that

$$q \Vdash_{P_i} \text{“}\underset{\sim}{\sigma}_n \in v_n^*\text{”}$$

Now, clearly it suffices to prove:

$\otimes$ $q \Vdash_{P_i}$ “there is a condition $r \in P_j, r\restriction i = q, [i,j) \cap \mathrm{Dom}(r) = N_1 \cap [i,j), r$ is (N_1, P_j)-generic, r is above some p_n, and $r \Vdash_{P_j} [\bigwedge_n \underset{\sim}{\tau}_n \in v_n^* \cup u_n^*]$”

[as there is a P_i-name of such a condition, and we know the domain, there exists an actual such $r \in P_j$].

Why $\otimes$ holds? Let $G_i \subseteq P_i$ be generic over V, $q \in G_i$. Then $\underset{\sim}{\bar{v}}$ being a P_i-name from N_1, satisfies $\underset{\sim}{\bar{v}}[G_i] = \langle v_n : n < \omega \rangle \in N_1[G_i]$ so $v_n \subseteq N_1 \cap \mathrm{Ord}$, hence for some $\langle v'_n : n < \omega \rangle \in V$, and ω-sequence of finite sets of ordinals, $\bigwedge_{n<\omega} v_n \subseteq v'_n$, so w.l.o.g. $\langle v'_n : n < \omega \rangle \in N_1$; so for some $n(*)$, $\bigwedge_{m \geq n(*)} v_n \subseteq v'_n \subseteq u^*_n$. Choose r as $(N_1[G_1], P_j/P_i)$-generic such that $\mathrm{Dom}(r) = N_1[G_1] \cap [i,j) = N_1 \cap [i,j)$, and $p^+_{n(*)}[\underset{\sim}{G}_i], p_{n(*)} \leq r$, such r exists by the theorem of preservation of properness. Now r is as required in $\otimes$. As we have done it in any $V[G_i], G_i \subseteq P_i$ generic over $V, q \in G_i$, clearly q forces ($\Vdash_{P_i}$) there is such r. $\square_{2.6}$

2.7 Claim. Let

(i) $\bar{Q}$ be a semiproper iteration of ${}^\omega\omega$-bounding forcing notions

(ii) $i < j \leq \ell g(\bar{Q}), N_0 \prec N_1 \prec (H(\chi), \in, <^*_\chi), \langle \bar{Q}, i, j \rangle \in N_0$ and $N_0 \in N_1$ both countable, and χ regular large enough.

(iii) $p \in P_j \cap N_0, q \in P_i, p \restriction i \leq q, q$ is (N_1, P_i)-semi generic and (N_0, P_i)-generic,

(iv) for every $\underset{\sim}{\tau} \in N_0$ a P_i-name of a countable ordinal for some finite u, $q \Vdash$ "$\underset{\sim}{\tau} \in u$".

Then, there is an $r \in P_j, r \restriction i = q,\ p \leq q, r$ is (N_j, P_j)-semi generic and (N_0, P_j)-semi generic such that for every P_j-name $\underset{\sim}{\tau} \in N_0$ of a countable ordinal, for some finite u, we have $r \Vdash_{P_j}$ "$\underset{\sim}{\tau} \in u$".

Proof. Same as 2.6 except that: using RCS, the issue of the domain of r disappears, and the names we deal with are names of countable ordinals. $\square_{2.7}$

2.8 Main Lemma. If $x \in \{2, \aleph_0\}$, $\bar{Q}$ is a CS iteration of (limit) length δ and for every $\alpha < \delta$, $\bar{Q} \restriction \alpha$ is an NNR_x-iteration for $(\mathcal{E}_0, \mathcal{E}_1, \mathcal{E}_2)$, *then* $\bar{Q}$ is an NNR_x-iteration for $(\mathcal{E}_0, \mathcal{E}_1, \mathcal{E}_2)$.

2.9 Remark. 1) Our main object is usually to preserve clause (3) of Definition 2.2: adding no real.

2) Comparing this with the result is V §7 and in VIII §4, we gain in replacing the completeness system by condition $(4)_x$, which is weaker; but "$(<\omega_1)$-proper" seems incomparable with condition (5) which replaces it.

2.10 Proof of 2.8 . We have to prove the five conditions from Definition 2.2 (or 2.3).

Conditions (1) and (2) are easy (part (1) follows from part (2), for part (2) see 2.2A and V).

Condition (3) follows from condition (4) (use $i = 0, j = \delta$, $\underset{\sim}{\tau}$ a name of a real).

So, it is enough to prove:

(a) condition (4) and

(b) condition (5) assuming (3), (4) hold.

2.10A Proof of Condition (5), Assuming Conditions (3), (4). So, forcing with P_δ adds no reals and $\bar{Q}$ satisfies (1) — (4).

Let i, j be non limit ordinals, $i < j \leq \delta$, $\underset{\sim}{R}$, $\bar{Q}'$, s, t, $\bar{\alpha}$, η^* be as in the assumption of (5). So let χ be regular large enough, N a countable elementary submodel of $(H(\chi), \in, <^*_\chi)$, $N \cap \lambda_2 \in \mathcal{E}_2, i \in N, j \in N, \underset{\sim}{R} \in N, \bar{Q} \in N, (q_0, q_1) \in P_i * \underset{\sim}{R}$ is $(N, P_i * \underset{\sim}{R})$-generic and force $\underset{\sim}{G}_{P_i * \underset{\sim}{R}} \cap N$ to be $G^a, p \in P_j \cap N, p{\restriction}i \in G^a \cap P_i$ (equivalently $p{\restriction}i \leq q_0$). It suffices to find $r \in P_j$ above q_0 and above p, and r is $(N[G^a], P_j)$-generic, and r forces a value to $\underset{\sim}{G}_{P_j} \cap N$.

By the assumption, $\alpha < \delta \Rightarrow \bar{Q}{\restriction}\alpha$ is NNR_x-iteration for $(\mathcal{E}_0, \mathcal{E}_1, \mathcal{E}_2)$;hence if $j < \delta$ the conclusion holds. So, assume $j = \delta$.

Let $N_0 = N$ and choose N_1 satisfying:

(α) N_1 is countable

(β) $N_0 \prec N_1 \prec (H(\chi), \in, <^*_\chi)$

(γ) $N_1 \cap \lambda_0 \in \mathcal{E}_0$

(δ) $N_0 \in N_1$

(ε) G^a, q_0, $q_1 \in N_1$.

Let $i = i_0 < i_1 < i_2 < \cdots < i_n < \cdots (n < \omega)$ be such that:

$$i_n \text{ is not limit, } i_n \in N_0 \cap \delta \text{ and } \sup[\delta \cap N_0] = \sup\{i_n : n < \omega\}.$$

Let $\langle \underset{\sim}{\mathcal{I}}_n : n < \omega \rangle \in N_1$ be a list of the $P_i * \underset{\sim}{R}$-names of dense subsets of P_j/P_i in N_0.

Now we choose by induction on n, p^n, q^n such that:

(A) (1) $q^n \in P_{i_n}$,

(2) q^n is (N_ℓ, P_{i_n})-generic for $\ell = 0, 1$,

(3) $\mathrm{Dom}(q_n) = i_n \cap N_1$

(4) $q_0 \leq q^0$

(5) $q^{n+1} \restriction i_n = q^n$

(B) (1) p^n is (a P_{i_n} - name of) a member of $P_\delta \cap N_0$

(2) $p^n \restriction i_n \leq q^n$

(3) $p^n \leq p^{n+1}$, $p^0 = p$

(4) $p^{n+1} \in \underset{\sim}{\mathcal{I}}_n$ (more exactly: $p^{n+1} \in \underset{\sim}{\mathcal{I}}_n[G^a]$ i.e. $q^{n+1} \Vdash_{P_{i_{n+1}}}$ "$p^{n+1} \in \underset{\sim}{\mathcal{I}}_n[G^a]$" where $\underset{\sim}{\mathcal{I}}_n[G^a] = \{r \in N : \text{for some } p' \in G^a \ p' \Vdash_{P_i} \text{"} r \in \underset{\sim}{\mathcal{I}}_n \text{"}\}$).

(C) $q^n \Vdash$ "$\underset{\sim}{G}_{i_n} \cap N_0$ is generic for $\Big(N_0[G^a], (P_{i_n}/P_i) \cap N_0\Big)$".

Note in (B)(1) that p^n should not depend on $\underset{\sim}{G}_{\underset{\sim}{R}}$.

For $n = 0$ - easy.

For the induction step, defining for $n+1$, first note that

$$(*) \qquad \Vdash_{P_i * [\underset{\sim}{R} \times (P_{i_n}/P_i)]} \text{"} P_{i_{n+1}}/P_{i_n} \text{ is } \mathcal{E}_2 - \text{proper not adding reals"}.$$

We get $(*)$ by applying (5) of the Definition with $\bar{Q} \restriction i_{n+1}$ (which is NNR_x-iteration for $(\mathcal{E}_0, \mathcal{E}_1, \mathcal{E}_2)$), $\bar{Q}$, $t^*, \bar{\alpha}^*$, η^{n+1}, i_n, i_{n+1} here standing for $\bar{Q}, \bar{Q}'$, t, $\bar{\alpha}$, η^*, i, j there, where: t^* is t when we add η^n just above η^* and η^{n+1} just above η^n and let $\bar{\alpha}^* \restriction t = \bar{\alpha}$, $\alpha^*_{\eta^n} = i_n$ and $\alpha^*_{\eta^{n+1}} = i_{n+1}$.

To apply condition (5) we have, however, to know that $P_{\bar{\alpha}^* \restriction (t \cup \{\eta^n\})}$ is $\mathcal{E}_2$-proper not adding reals; but this is guaranteed by Claim 2.4.

So, $(*)$ above holds; so, after forcing with $P_i * [\underset{\sim}{R} \times (P_{i_n}/P_i)]$ (with q^n, (q_0, q_1) in the generic set) we shall find a $q \in P_{i_{n+1}}/P_{i_n}$ generic for $\big(N_0\big[G^a, \underset{\sim}{G}_{P_{i_n}/P_i}\big], \big[P_{i_{n+1}}/P_{i_n}\big]\big)$.

(Note: $N_0[G^a]$ had no new members of V, so no new members of $P_{i_{n+1}}/P_{i_n}$). Now, the forcing with $\underset{\sim}{R}$ is irrelevant (except for the information in G^a and $G^a \in V$!) So, there is a P_{i_n}-name of such a q, and there is a P_{i_n}-name of a q', $q \leq q' \in P_{i_{n+1}}/P_{i_n}$, q' forcing a value to $\underset{\sim}{G}_{P_{i_{n+1}}} \cap N_0$.
So, there is a $\underset{\sim}{q}' \in N_1$, a P_{i_n}-name of a condition from $P_{i_{n+1}}/P_{i_n}$ satisfying the above if there is such q' at all.

Now, as in the proof on the preservation of properness, we can choose q_{n+1}.

In the end, let $r' = q^0 \cup \bigcup_n q^{n+1} \restriction [i_n, i_{n+1})$. Now $\langle \underset{\sim}{\mathcal{I}}_n \cap \underset{\sim}{G}_{P_j} : n < \omega \rangle$ is clearly a P_j-name, so, by condition (3), there is some r, $r' \leq r \in P_j$, and r forces an (old) value to it. Now, this r finishes the proof. $\square_{2.10A}$

2.10B Proof of condition $(4)_{\aleph_0}$ when we deal with $NNR_{\aleph_0}$:

So let N, i, j, $\bar{\alpha}$, $\bar{\beta}$, G^a, p, t, $\langle q_\eta : \eta \in t \rangle$ are as there. By the assumption w.l.o.g. $j = \delta$.

Choose N_ℓ (for $\ell = 0, 1, 2, 3, 4, 5$) such that:

(α) every N_ℓ is countable, $N_0 = N$

(β) $N_\ell \prec N_{\ell+1} \prec (H(\chi), \in, <^*_\chi)$ for $\ell = 0, 1, 2, 3, 4$

(γ) $N_1 \cap \lambda_1 \in \mathcal{E}_1$, $N_2 \cap \lambda_2 \in \mathcal{E}_2$, $N_3 \cap \lambda_0 \in \mathcal{E}_0$, $N_4 \cap \lambda_0 \in \mathcal{E}_0$, $N_5 \cap \lambda_1 \in \mathcal{E}_0$, (remember $N_0 \cap \lambda_1 \in \mathcal{E}_1$)

(δ) $N_\ell \in N_{\ell+1}$ for $\ell = 0, 1, 2, 3, 4$

(ε) $\bar{q} \in N_1$, $G^a \in N_1$.

Let $i = i_0 < i_1 < i_2 < \ldots < i_n < \ldots (n < \omega)$ be such that: each i_n is non-limit, belongs to $N_0 \cap \delta$, and

$$\sup(\delta \cap N_0) = \sup\{i_n : n < \omega\}.$$

Let $\langle \mathcal{I}_n : n < \omega \rangle \in N_1$ be a list of the dense subsets of P_δ which belong to N_0. For simplicity, w.l.o.g. we can assume t is a subset of ${}^{\omega>}\omega$ ordered by $\lhd$ (being an initial segment), $\alpha_\eta = i \Leftrightarrow \ell g(\eta) = m^*$ (remember only $\eta <^t \nu \Rightarrow \alpha_\eta \leq \alpha_\nu$ was required). Let $t^* = t \cap {}^{m^*}\omega$ and stipulate $t_{-1} = t \setminus t^*$.

Now we define by induction on $n < \omega$, p_n, $q^n_\eta (\eta \in t^*)$, G^a_n, t_n, $\bar{\alpha}^n$, G^b_η, G^c_η (for $\eta \in t_n$) such that:

(A) (1) $q^n_\eta \in P_{i_n}$ (for $\eta \in t^*$)

(2) q^n_η is (N_ℓ, P_{i_n})-generic for $\ell = 0,1,2,3,4,5$

(3) For every pre-dense subset $\mathcal{I}$ of P_{i_n} from N_4 for some finite $\mathcal{J} \subseteq \mathcal{I} \cap N_4$, $\mathcal{J}$ is pre-dense in P_{i_n} above q^n_η (hence this holds for $\ell \le 4$)

(4) $q_\eta \le q^0_\eta$ for $\eta \in t^*$

(5) if $\nu \in t \setminus t^*, \nu \lhd \eta_1 \in t^*$, $\nu \lhd \eta_2 \in t^*$ then $q^0_{\eta_1} \restriction \alpha_\nu = q^0_{\eta_2} \restriction \alpha_\nu$

(6) $q^{n+1}_\eta \restriction i_n = q^n_\eta$ for $\eta \in t^*$

(7) $\mathrm{Dom}(q^n_\eta)$ is $i_n \cap N_5$

(B) (1) G^a_n is a generic subset of $P_{i_n} \cap N_0$ over N_0

(2) $G^a_{n+1} \cap P_{i_n} = G^a_n$

(3) $G^a_0 = G^a$

(4) $q^n_\eta \Vdash_{P_{i_n}}$ "$\underset{\sim}{G}_{P_{i_n}} \cap N_0 = G^a_n$" for $\eta \in t^*$

(C) (1) $p_n \in N_0 \cap P_\delta$

(2) $p \le p_n \le p_{n+1}$

(3) $p_{n+1} \in \mathcal{I}_n$

(4) $p_n \restriction i_n \in G^a_n$

(D) (1) t_n is a nonempty finite tree, $t_0 = t, t_n \subseteq_{\mathrm{end}} t_{n+1}$,

(2) $\bar\alpha^n = \langle \alpha^n_\eta : \eta \in t_n \rangle$,

(3) $\bar\alpha^n = \bar\alpha^{n+1} \restriction t_n, \bar\alpha^0 = \bar\alpha$, so we may write α_η for α^n_η when $\eta \in t_n$

(4) if $\eta \in t_{n+1} \setminus t_n$ then there is a $\nu_\eta \in t_n$ such that: η is an immediate successor of ν_η, $\alpha_\eta = i_{n+1}$, $\alpha_{\nu_\eta} = i_n$.

(E) (1) $\langle G^b_\eta : \eta \in t_n \rangle$ belongs to $s\mathrm{Gen}^{\bar\alpha^n}_{\bar Q \restriction i_n}(N_1)$

(2) $G^b_\eta \in N_2$

(3) $G^c_\eta \in \mathrm{Gen}^{\langle \alpha_\eta \rangle}_{\bar Q \restriction i_n}(N_2)$

(4) $t_n \models \eta < \nu$ implies $G^c_\eta \subseteq G^c_\nu$

(5) $G^c_\eta \in N_3$

(6) $G^a_{\ell g(\eta) - m^*} \subseteq G^b_\eta \subseteq G^c_\eta$ for $\eta \in (t_n \setminus t) \cup t^*$ so $\ell g(\eta) \ge m^*$, of course

(7) if $\eta \in t^*$ then $q^n_\eta \Vdash$ "for some $\rho \in t_n \setminus \bigcup_{m<n} t_m$ we have: $\alpha^n_\rho = i_n, \eta \trianglelefteq \rho$ and $G^c_\rho \subseteq \underset{\sim}{G}_{P_{i_n}}$".

If we succeed, then let $r_\eta = q^0_\eta \cup \bigcup_{n<\omega} q^{n+1}_\eta \restriction [i_n, i_{n+1})$ (for $\eta \in t^*$, so $\alpha_\eta = i$ and $\beta_\eta = j = \delta$) and let $r_\eta = q_\eta$ for $\eta \in t \setminus t^*$, all are members of P_δ.

For $\eta \in t^*, r_\eta$ is (N_0, P_δ)-generic and forces $\underset{\sim}{G}_{P_\delta} \cap N = \underset{\sim}{G}_{P_\delta} \cap N_0 = \bigcup_{n<\omega} G^a_n$ and $\langle r_\eta : \eta \in t \rangle \in P_{\bar{\beta}}$. So, it is enough to carry the definition.

The case $n = 0$ is easy (better to define $\langle q^0_\eta : \eta \in t_0 \rangle \in P_{\bar{\alpha}}$ by steps i.e. choose q^0_η by induction on $\ell g(\eta)$; remember $P_{\bar{\alpha}}$ is proper not adding reals as $\bar{Q} \restriction i_0$ is an $\mathrm{NNR}_{\aleph_0}$-iteration).

Let us do the induction step: defining for $n+1$.

First Step. Choose $p_{n+1} \in \mathcal{I}_n \cap N_0$ such that $p_n \leq p_{n+1}$ and $p_{n+1} \restriction i_n \in G^a_n$. Straightforward.

Second Step.
First note:

$(*)_1$ the following set is a dense subset of $P_{\bar{\alpha}^n}$:

$$\begin{aligned} \mathcal{J} = \{\bar{q}' : & \bar{q}' \in P_{\bar{\alpha}^n}, \text{ and } \textit{either} \text{ for some } \eta \in t_n, \alpha_\eta = i_n \text{ and} \\ & q'_\eta \Vdash_{P_{i_n}} \text{“}\underset{\sim}{G}_{P_{i_n}} \cap N_0 \neq G^a_n\text{”} \\ & \textit{or} \text{ there is a } G' \in \mathrm{Gen}_{P_{i_{n+1}}}(N_0) \text{ such that} \\ & p_{n+1} \restriction i_{n+1} \in G', G' \cap P_{i_n} = G^a_n \text{ and:} \\ & \eta \in t_n \,\&\, \alpha_\eta = i' \Rightarrow q'_\eta \Vdash_{P_{i_n}} \text{“in } P_{i_{n+1}}/P_{i_n} \text{ the set } G' \\ & \qquad \text{has an upper bound”}\}. \end{aligned}$$

This follows by $(4)_{\aleph_0}$ for $\bar{Q} \restriction i_{n+1}$ (which is an $NNR_{\aleph_0}$-iteration).
Also

$(*)_2$ there is a $\bar{q}' \in \mathcal{J}$ which belongs to $\langle G^b_\eta : \eta \in t_n \rangle$ (i.e. $\eta \in t_n \Rightarrow q'_\eta \in G^b_\eta$ and $\bar{q}' \in P_{\bar{\alpha}^n}$) such that $[\eta \in t_n \,\&\, \alpha_\eta = i_n \Rightarrow q'_\eta$ is above $G^a_n]$.
(this is as there is $\bar{q}^* \in \langle G^b_\eta : \eta \in t_n \rangle$ which belongs to $\mathcal{J}$ (as $\mathcal{J} \in N_1$ and is a dense subset of $P_{\bar{\alpha}^n}$ and $\langle G^b_\eta : \eta \in t_n \rangle$ is in $\mathrm{sGen}_{P_{\bar{\alpha}^n}}(N_1)$ and the first possibility in the definition of $\mathcal{J}$ cannot hold as $G^a_n \subseteq G^b_\eta$ whenever $\alpha_\eta = i_n$).
Now choose G^a_{n+1} satisfying: (B)(1), (B)(2) and for every $\eta \in t_n$ of length $n+m^*$ for some $q^*_\eta \in P_{i_{n+1}} \cap N_1 : q^*_\eta \restriction i_n \in G^b_\eta$ and $q^*_\eta \Vdash$ “$\underset{\sim}{G}_{P_{i_{n+1}}} \cap N_0 = G^a_{n+1}$”.
This is possible by $(*)_1 + (*)_2$.

Third Step. Let for each $\nu \in t_n$ with $\alpha_\nu = i_n$, $\langle G_{\nu,m} : m < m_\nu \rangle$ be such that: $q^n_{\nu \restriction m^*} \Vdash_{P_{i_n}}$ "*if* $G^c_\nu \subseteq \underset{\sim}{G}_{P_{i_n}}$ *then* $\underset{\sim}{G}_{P_{i_n}} \cap N_3 \in \{G_{\nu,m} : m < m_\nu\}$.
(This sequence exists and is finite by (A)(3), and as P_{i_n} adds no new reals). Now we let

$$t_{n+1} = t_n \cup \{\nu\hat{\ }\langle m \rangle : m < m_\nu, \nu \in t_n, \alpha_\nu = i_n\}.$$

and choose $\bar{\alpha}^{n+1}$ by $(D)(3)$ and $(D)(4)$.

Fourth Step. Repeating the proof of 2.4 (but choosing the appropriate forcing conditions from $G^c_\eta(\eta \in t_n, \alpha_\eta = i_n)$), we choose $\langle G^b_\eta : \eta \in t_{n+1} \setminus t_n \rangle$ and $\langle r^b_\eta : \eta \in t_{n+1} \setminus t_n \rangle$ such that: $\langle G^b_\eta : \eta \in t_{n+1} \rangle \in \mathrm{sGen}^{\bar{\alpha}^{n+1}}_{\bar{Q} \restriction i_{n+1}}(N_1)$ and $q^*_{\eta \restriction (m^*+n)} \in G^b_\eta$ ($q^*_{\eta \restriction (m^*+n)}$ is from the end of the second step) and $r^b_\eta \in P_{i_{n+1}} \cap N_2$, which is an upper bound to G^b_η and $r^b_n \restriction i_n \in G^c_{\eta \restriction (m^*+n)}$ (just order $t_{n+1} \setminus t_n$, and then choose (G^b_η, r^b_η) by induction on η, see 2.4(2)).

Fifth Step. We choose $\langle G^c_\eta : \eta \in t_{n+1} \setminus t_n \rangle$, $\langle r^c_\nu : \nu \in t_{n+1} \setminus t_n \rangle$ satisfying (E) and $\left[\nu \in t_{n+1} \,\&\, \alpha_\nu = i_{n+1} \Rightarrow r^b_\nu \in G^c_\nu\right]$ and for $\eta \in t_{n+1} \setminus t_n$, $\eta = \nu\hat{\ }\langle m \rangle$, we have $r^c_\eta \in P_{i_{n+1}} \cap N_3$, $r^c_\eta \restriction i_n \in G_{\nu,m}$, r^c_η a bound of G^c_η; this is possible as in the proof of the preservation of properness.

Sixth Step. We choose $\langle q^{n+1}_\eta : \eta \in t \setminus t_{-1}$ so $\alpha_\eta = i \rangle$ by Claim 2.6 making sure that $\{r^c_\nu : \nu \in t_{n+1} \setminus t_n, \eta \triangleleft \nu\}$ is pre-dense above q^{n+1}_η, this to guarantee (E)(7); do it for each such η separately.

So, we have finished the induction step, hence the proof of $(4)_{\aleph_0}$. Hence, the proof of the Main Lemma. $\square_{2.10B}$

2.10C Proof of Condition (4)$_2$ When we are Dealing with NNR_2

We mix the proof of VIII, §4 and the previous proof [†].

So let $\chi, N, i^*, i, j, G^a, p, q_0, q_1$ are as there. By the assumption w.l.o.g. $j = \delta$. Let $\chi_1 = (2^\chi)^+, t = \{\langle\rangle, \langle 0 \rangle, \langle 1 \rangle\} \subseteq {}^{\omega>}\omega, \alpha_{<>} = i^*, \alpha_{<0>} = \alpha_{<1>} =$

[†] The readers who are happy to have the details should thank Lee Stanley for his advice.

$i, q_{<>} = q_0 \restriction i^* (= q_1 \restriction i^*)$ and $q_{<0>} = q_0, q_{<1>} = q_1$, and $\bar{q} = \langle q_\eta : \eta \in t \rangle$, stipulate $t_{-1} = \{\langle\rangle\}$.

Choose $N_\ell (\ell = 0,1,2,3,4,5)$ such that:

(α) every N_ℓ is countable, $N_0 = N$,

(β) $N_\ell \subseteq N_{\ell+1} \prec (H(\chi_1), \in, <^*_{\chi_1})$ for $\ell = 0,1,2,3,4$,

(γ) $N_1 \cap \lambda_1 \in \mathcal{E}_2$, $N_2 \cap \lambda_2 \in \mathcal{E}_2$, $N_3 \cap \lambda_0 \in \mathcal{E}_0$, $N_4 \cap \lambda_0 \in \mathcal{E}_0$, $N_5 \cap \lambda_0 \in \mathcal{E}_0$,

(δ) $N_\ell \in N_{\ell+1}$ for $\ell = 0,1,2,3,4$,

(ε) $\bar{q} \in N_1$, $G^a \in N_1$.

Let $i = i_0 < i_1 < i_2 < \ldots < i_n < \ldots (n < \omega)$ be such that: each i_n belong to $N_0 \cap \delta$, is a non-limit ordinal and

$$\sup(\delta \cap N_0) = \sup\{i_n : n < \omega\}.$$

Let $\langle \mathcal{I}_n : n < \omega \rangle \in N_1$ be a list of the dense subsets of P_δ which belong to N_0. Let $t^* = t \cap {}^1\omega$.

Now we define by induction on $n < \omega$, $k_n \in \omega, \langle M_k : k \leq k_n \rangle$, p_n, $q^n_\eta (\eta \in t^*)$, $G^a_n, t_n, \bar{\alpha}^n, s_k, \bar{\beta}^{n,k}, h_k, h^n_k (k \leq k_n), G^b_\eta, G^c_\eta$ (for $\eta \in t_n$) such that:

(A) (1) $q^n_\eta \in P_{i_n} (\eta \in t^*)$

(2) q^n_η is (N_ℓ, P_{i_n})-generic for $\ell = 0,1,2,3,4,5$

(3) For every pre-dense subset $\mathcal{I}$ of P_{i_n} from N_4, for some finite $\mathcal{J} \subseteq \mathcal{I} \cap N_4$, $\mathcal{J}$ is pre-dense in P_{i_n} over q^n_η (hence this holds for $\ell \leq 4$)

(4) $q_\eta \leq q^0_\eta$ for $\eta \in t^*$

(5) $q^0_{\langle 0 \rangle} \restriction \alpha_{\langle\rangle} = q^0_{\langle 1 \rangle} \restriction \alpha_{\langle\rangle}$

(6) $q^{n+1}_\eta \restriction i_n = q^n_\eta$

(7) $\mathrm{Dom}(q^n_\eta)$ is $i_n \cap N_5$

(B) (1) G^a_n is a generic subset of $P_{i_n} \cap N_0$ over N_0

(2) $G^a_{n+1} \cap P_{i_n} = G^a_n$

(3) $G^a_0 = G^a$

(4) $q^n_\eta \Vdash_{P_{i_n}}$ "$\underset{\sim}{G}_{P_{i_n}} \cap N_0 = G^a_n$" (for $\eta \in t^*$).

(C) (1) $p_n \in N_0 \cap P_\delta$

(2) $p \leq p_n \leq p_{n+1}$

(3) $p_{n+1} \in \mathcal{I}_n$ for $\eta \in t^*$.

(4) $p_n \restriction i_n \in G^a_n$

(D) (1) t_n is a nonempty finite tree, $t_0 = t, t_n \subseteq_{\text{end}} t_{n+1}$,

(2) $\bar{\alpha}^n = \langle \alpha^n_\eta : \eta \in t_n \rangle$,

(3) $\bar{\alpha}^n = \bar{\alpha}^{n+1} \restriction t_n, \bar{\alpha}^0 = \bar{\alpha}$, so we may write α_η for α^n_η

(4) if $\eta \in t_{n+1} \setminus t_n$ then there is $\nu_\eta \in t_n$ such that: η is an immediate successor of ν_η, $\alpha_\eta = i_{n+1}, \alpha_{\nu_\eta} = i_n$

(E) (1) $\langle G^b_\eta : \eta \in t_n \rangle$ belongs to $\text{sGen}^{\bar{\alpha}^n}_{\bar{Q} \restriction i_n}(N_1)$

(2) $G^b_\eta \in N_2$

(3) $G^c_\eta \in \text{Gen}^{<\alpha_\eta>}_{\bar{Q} \restriction i_n}(N_2)$ for $\eta \in t^*$

(4) $t_n \models \eta < \nu$ implies $G^c_\eta \subseteq G^c_\nu$

(5) $G^c_\eta \in N_3$

(6) $G^a_{\ell g(\eta)-1} \subseteq G^b_\eta \subseteq G^c_\eta$ for $\eta \in t_n (\ell g(\eta) \geq 1$, of course)

(7) if $\eta \in t^*$ then

$$q^n_\eta \Vdash_{P_{i_n}} \text{"for some } \rho \in t_n \setminus \{\langle\rangle\} \text{ we have: } \alpha^n_\rho = i_n \text{ and } G^c_\rho \subseteq \underset{\sim}{G}_{P_{i_n}}\text{"}$$

(we can demand it is a P_{i_n}-name $\underset{\sim}{\rho}_n$ and $\underset{\sim}{\rho}_n \lhd \underset{\sim}{\rho}_{n+1}$).

(F) (1) $M_0 = N_0$

(2) $M_k \prec M_{k+1} \prec (H(\chi), \in, <^*_\chi)$ for $k < k_n$

(3) M_k is countable,

(4) $M_k \in M_{k+1}$

(5) $M_k \in N_1$

(6) $k_n < k_{n+1} < \omega, k_0 = 1$ (stipulate $k_{-1} = -1$)

(G)(1) $s_0 = \{<>\}, s_1 = t$

(2) if $k_n < k \leq k_{n+1}$ then $s_k = s_{k_n} \cup \{\nu\hat{\ }\langle \ell \rangle : \ell < 2^{k-k_n}, \nu \in s_{k_n}, \ell g(\nu) = n+1\}$

(3) for $k_n \leq k < k_{n+1}$, we define h_k, a function with domain s_{k+1} and range s_k: $h_k \restriction s_{k_n} =$ identity, and for $\nu\hat{\ }\langle \ell \rangle \in s_{k+1} \setminus s_{k_n}$ $h_k(\nu\hat{\ }\langle \ell \rangle) = \nu\hat{\ }\langle [\ell/2] \rangle$

(4) $\bar{\beta}^{n,k} = \langle \beta^{n,k}_\nu : \nu \in s_{k_n}, k \leq k_n \rangle$ is defined as follows: $\beta^{n,0}_{<>} = i_n$ (remember $s_0 = \{\langle\rangle\}$) and if $k > 0, \beta^{n,k}_\nu$ is i^* if $\ell g(\nu) = 0, i_{\ell g(\nu)-1}$

if $0 < \ell g(\nu)$ but ν is not maximal in s_k, and finally i_n if ν is maximal in s_{k_n}

(5) $s_{k_n} = t_n$

(6) For k such that $k_n \le k \le k_{n+1}$, h^n_k is the function with domain $s_{k_{n+1}} = t_{n+1}$ to s_k, $h^n_k(\eta) = h_k \circ h_{k+1} \circ h_{k_{n+1}-1}(\eta)$ (and $h^n_{k_{n+1}} = \mathrm{id}_{t_{n+1}}$), also if $k \le k_n$, h^n_k is defined by the downward induction on m as $h^m_k \circ h^n_{k_{m+1}}$ where $k_m \le k \le k_{m+1}$ (no incompatibility).

(H)(1) if $\nu, \eta \in t_n = s_{k_n}$, $k \le k_{n+1}$ and $h^n_k(\eta) = h^n_k(\nu)$ (both well defined) then $G^b_\nu \cap M_k = G^b_\eta \cap M_k$, and we denote this value by $G^{b,n,k}_{h^n_k(\nu)}$,

(2) $\langle G^{b,n,k}_\rho : \rho \in s_k \rangle \in s\mathrm{Gen}^{\bar\beta^{n,k}}_{\bar Q}(M_k)$ and it belongs to M_{k+1} (and to N_1).

If we succeed, then let $r_\eta = q^0_\eta \cup \bigcup_{n<\omega} q^{n+1}_\eta \restriction [i_n, i_{n+1})$ (for $\eta \in t^*$, so $\alpha_\eta = i$) and let $r_\eta = q_\eta$ for $\eta \in t \setminus t^*$, they are members of P_δ. For $\eta \in t^*$, r_η is (N_0, P_δ)-generic and forces $\underset{\sim}{G}_{P_\delta} \cap N = \underset{\sim}{G}_{P_\delta} \cap N_0 = \bigcup_{n<\omega} G^a_n$ (remember (C)(1)–(4)), and $\langle r_\eta : \eta \in t \rangle \in P_{\bar\beta}$. Here, $\bar\beta$ is as in Definition 2.3(i). So, it is enough to carry the definition.

The case $n = 0$ is easy (better to define $\langle q^n_\eta : \eta \in t_0 \rangle \in P_{\bar\alpha}$ by steps).

Let us do the induction step: defining for $n+1$.

First Step. Choose $p_{n+1} \in \mathcal{I}_n \cap N_0$ such that $p_n \le p_{n+1}$ and $p_{n+1} \restriction i_n \in G^a_n$.

Second Step.

First Note:

$(*)_1$ the following set is a dense subset of $P_{\bar\beta^{n,1}}$:

$$\begin{aligned}\mathcal{J} = \{\bar q' : \; & \bar q' \in P_{\bar\beta^{n,1}} \text{ and } \textit{either} \text{ for some } \eta \in s_1, \beta^{n,1}_\eta = i_n \text{ and} \\ & q'_\eta \Vdash_{P_{i_n}} \text{“}\underset{\sim}{G}_{P_{i_n}} \cap N_0 \ne G^a_n\text{”} \\ & \textit{or there} \text{ is a } G' \in \mathrm{Gen}_{P_{i_{n+1}}}(N_0) \text{ such that:} \\ & p_{n+1} \restriction i_{n+1} \in G' \cap P_{i_n} = G^a_n \text{ and:} \\ & \eta \in s_1 \,\&\, \beta^{n,1}_\eta = i_n \Rightarrow q'_\eta \Vdash_{P_{i_n}} \text{“ in } P_{i_{n+1}}/P_{i_n} \text{ the set } G' \\ & \text{has an upper bound”}\}.\end{aligned}$$

This follows by $(4)_2$ for $\bar{Q}\restriction i_{n+1}$ (which is an NNR_2-iteration).

Also

$(*)_2$ there is a $\bar{q}' \in \langle G^b_\eta : \eta \in s_1 \rangle$ (i.e. $\bigwedge_{\eta \in s_1} q'_\eta \in G^{b,n,1}_\eta$) such that $[\eta \in s_1 \,\&\, \beta^{n,1}_\eta = i_n \Rightarrow q'_\eta$ is above $G^a_n]$ and $\bar{q}' \in \mathcal{J}$.
(This is as $\langle G^{b,n,1}_\eta : \eta \in s_1 \rangle$ is in $\mathrm{sGen}_{P_{\beta^{n,1}}}(M_1)$ and $G^a_n \subseteq G^{b,n,1}_\eta$ whenever $\beta^{n,1}_\eta = i_n$).

Now choose G^a_{n+1} satisfying: (B)(1), (B)(2) and for every $\eta \in s_1$ with $\beta^{n,1}_\eta = i_n$ for some $q^{n,0}_\eta \in P_{i_{n+1}} \cap M_1$ we have: $q^{n,0}_\eta \restriction i_n \in G^{b,n,1}_\eta$ and $q^{n,0}_\eta \Vdash$ "$\underset{\sim}{G}_{P_{i_{n+1}}} \cap N_0 = G^a_{n+1}$".

This is possible by $(*)_1 + (*)_2$.

Third Step. Let for each $\nu \in t_n (= s_{k_n})$ with $\beta^{n,k_n}_\nu = i_n$, $\langle G_{\nu,m} : m < m_\nu \rangle$ be such that (on q^n_ν see (A)(1), (2), (3)):
$q^n_{\nu\restriction 1} \Vdash_{P_{i_n}}$ "*if* $G^c_\nu \subseteq \underset{\sim}{G}_{P_{i_n}}$ then $\underset{\sim}{G}_{P_{i_n}} \cap N_3 \in \{G_{\nu,m} : m < m_\nu\}$".
(This sequence exists and is finite by (A)(3) and as P_{i_n} adds no new reals). W.l.o.g. m_ν is a power of 2, $m_\nu = 2^{n_\nu}$, and does not depend on ν, and let k_{n+1} be such that $k_{n+1} - k_n = 2^{m_\nu}$ for any such ν. So $s_k, k_n < k \le k_{n+1}$ and t_{n+1} are well defined. Now we can choose appropriate $M_k (k_n < k \le k_{n+1})$ such that†: $M_k \prec N_1 \restriction H(\chi), M_k \in N_1$, $M_{k-1} \prec M_k$, $M_{k-1} \in M_k$, $M_k[\langle G^b_\eta : \eta \in t_n \rangle] \prec (N_1 \restriction H(\chi))[\langle G^b_\eta : \eta \in t_n \rangle]$. Why can we choose such M_k's? By (E)(1), $\langle G^b_\eta : \eta \in t_n \rangle \in \mathrm{sGen}^{\bar{\alpha}^n}_{\bar{Q}\restriction i_n}(N_1)$, and $P_{\bar{\alpha}^n}$ is $\mathcal{E}_0$-proper. Let $G^{b,n,k}_\eta = G^b_\eta \cap M_k$ for $\eta \in t_n$. Also $\beta^{n+1,k} (k \le k_{n+1})$ and $s_k (k \le k_{n+1})$ are well defined now. Now we define by induction on $k = 0, \dots, k_{n+1}$, a condition $q^{n,k}_\eta (\eta \in s_k \,\&\, \beta^{n+1,k}_\eta = i_{n+1})$ and $\langle G^{b,n+1,k}_\eta : \eta \in s_k \rangle \in \mathrm{sGen}^{\bar{\beta}^{n+1,k}}_{\bar{Q}}(M_k)$ such

† Remember $\chi_1 = (2^\chi)^+$ and $N_\ell \prec (H(\chi_1), \in, <^*_{\chi_1})$ for $\ell = 1, 2, 3, 4, 5$.

that

$(a)\ k \le k_{n+1} \& \eta \in s_k \& \beta^{n,k}_\eta < i_{n+1} \Rightarrow G^{b,n+1,k}_\eta = G^{b,n,k}_\eta$

$(b)\ q^{n,k}_\eta \in M_{k+1}$ for $\eta \in s_{k_{n+1}}$(if $k = k_{n+1}$ then $q^{n,k}_n \in N_1$)

$(c)\ q^{n,k}_\eta \restriction i_n \in G^{b,n,k}_\eta$ when $\eta \in s_{k_{n+1}} \& \beta^{n,k}_\eta = i.$

$(d)\ q^{n,k}_\eta \in G^{b,n+1,k+1}_\eta$

$(e)\ G^{b,n+1,k}_{h^n_k(\eta)} \subseteq G^{b,n+1,k}_\eta$ for $\eta \in s_{n_{k+1}}$.

For $k = 0$ $q^{n,0}_\eta$ was already defined and let $G^{b,n+1,0}_{\langle\rangle} = G^a_{n+1}$-see second step. For $k+1$ we repeat the proof.

Fourth Step. Repeating the proof of 2.4 (but, choosing the appropriate forcing conditions from $G^c_\eta(\eta \in t_n \setminus \{\langle\rangle\}, \alpha_\eta = i_n)$), we choose $\langle G^b_\eta : \eta \in t_{n+1} \setminus t_n\rangle$ and $\langle r^b_\eta : \eta \in t_{n+1} \setminus t_n\rangle$ such that: $q^{n,k_{n+1}}_{\eta\restriction(1+n)} \in G^b_\eta$ and $r^b_\eta \in P_{i_{n+1}} \cap N_2$, which is an upper bound to G^b_η and $r^b_n \restriction i_n \in G^c_{\eta\restriction(1+n)}$ (just order $t_{n+1} \setminus t_n$, and then choose (G^b_η, t^b_η) by induction on η see 2.4(2)).

Fifth Step. We choose $\langle G^c_\eta : \eta \in t_{n+1}\rangle$, $\langle r^c_{r_\nu} : \nu \in t_{n+1} \setminus t_n\rangle$ satisfying (E) and $[\nu \in t_{n+1} \& \alpha_\nu = i_{n+1} \Rightarrow r^b_\nu \in G^c_\nu]$ and $r^c_\eta \in P_{i_{n+1}} \cap N_3$, $r_\eta \restriction i_n \in G_{\eta\restriction(1+n),\eta(1+n)}$, r^c_η a bound of G^c_η; this is possible as in the proof of the preservation of properness.

Sixth Step. We choose $\langle q_\eta : \eta \in t \setminus \{0\}, \alpha_\eta = i\rangle$ by Claim 2.6 for each such η separately taking care that $\{r^c_\nu : \nu \in t_{n+1} \setminus t_n, \eta \triangleleft \nu\}$ is pre-dense above q^{n+1}_η (this will guarantee (E)(7)).

So, we have finished the induction step hence the proof of $(4)_2$. Hence, the proof of the Main Lemma also for $x = 2$. $\square_{2.10C}$

2.11 Claim. If $\bar{Q}$ has length $\alpha + 1$, $\bar{Q}\restriction\alpha$ is an NNR_x-iteration for $(\mathcal{E}_0, \mathcal{E}_1, \mathcal{E}_2)$, $\Vdash_{P_\alpha}$ "Q_α is strongly proper, and condition $(4)_x$ holds for $i = \alpha$, $j = \alpha + 1$" *then* $\bar{Q}$ is an NNR_x-iteration for $(\mathcal{E}_0, \mathcal{E}_1, \mathcal{E}_2)$.

Proof. Straight.

Now we can phrase various conclusions on sufficient conditions for the limit of a CS iteration not to add reals.

2.12 Conclusion. Suppose $\bar{Q} = \langle P_i, \underset{\sim}{Q}_j : i \leq \alpha, j < \alpha \rangle$ is a countable support iteration of strongly proper forcing satisfying $(*)$ defined below. *Then* we can conclude that forcing with P_α adds no reals (hence, being proper, no new ω-sequences of ordinals, and in fact $\bar{Q}$ is an NNR_2-iteration) where

$(*)$ If $i_0 < i_1 < \alpha$ then $(*)^{i_0,i_1,i_1+1}_{\bar{Q}}$ holds, where we let

$(*)^{i_0,i_1,i_2}_{\bar{Q}}$ $i_0 < i_1 < i_2 \leq \alpha = \ell g(\bar{Q})$ and in $V^{P_{i_0}}$: if $N \prec (H(\chi), \in, <^*_\chi)$ is countable, $\langle \bar{Q}, i_0, i_1, i_2 \rangle \in N$, $p \in [P_{i_2}/P_{i_0}] \cap N$, $q', q'' \in P_{i_1}/P_{i_0}$ are $(N[G_{P_{i_0}}], P_{i_1}/P_{i_0})$-generic, $p{\restriction}i_1 \leq q'_1$, $p{\restriction}i_1 \leq q''$ and q', q'' force $\underset{\sim}{G}_{P_{i_1}/P_{i_0}} \cap N = G^1$, *then* for some $(N[G_{P_{i_0}}], P_{i_2}/P_{i_0})$-generic r', $r'' \in P_{i_2}/P_{i_0}$ we have: $p \leq r', p \leq r''$, $q' \leq r', q'' \leq r''$ and r', r'' force $(\underset{\sim}{G}_{P_{i_2}}/P_{i_0}) \cap N = G^r$ for some G^r.

Proof. Straight.

2.13 Claim. 1) A sufficient condition for $(*)$ from 2.12 is that each $\underset{\sim}{Q}_i$ is $(\mathbb{D}, \mathcal{E})$-complete for some simple 2-completeness system (see VIII, 4.2, 4.4).

2) We can in 2.11, 2.12 replace strongly proper by:

$\otimes$ "proper not adding reals even after forcing by any proper forcing notion not adding reals."

(3) If $V \models$ CH, κ supercompact with Laver diamond *then* for some proper forcing P not adding reals, of cardinality κ, satisfying the κ-c.c., in V^{P_κ} we have $\aleph_1 = \aleph_1^V$, $\aleph_2 = \kappa$, $2^{\aleph_0} = \aleph_1$, $2^{\aleph_1} = \aleph_2$ of course and:

$\mathrm{Ax}_{\omega_1}[\Pr(Q)]$ where $\Pr(Q)$ means:

(A) forcing with Q does not add reals

(B) part (A) holds even in a larger universe which has the same reals gotten by a proper forcing

(C) the forcing notion Q is proper and for some simple 2-completeness system $\mathbb{D}$ (or, even a $\aleph_1$-completeness system) Q is $\mathbb{D}$-complete.

2.14 Remark. 1) Part 3 is a specific case, of course.

We can now conclude the consistency of appropriate other axioms (see Ch. VIII).

2) We can now solve the problems from the end of §1.

2.15 Definition. 1) A finite tree t is simple if it has a root(= a minimal member) and all maximal $\eta \in t$ are from the same level (the level of η in t is $\ell g\eta \stackrel{\text{def}}{=} |\{\nu : \nu < \eta\}|$). t is called standard if $t \subseteq {}^{\omega>}\omega$ is closed under initial segments, the order being $\triangleleft$. Let $\max(t)$ be the set of maximal members of t.
2) If $\bar{\varepsilon}$ is a finite non-decreasing sequence of ordinals, $n = \ell g\bar{\varepsilon}$, t a simple finite tree with n levels then $\bar{\alpha}_{t,\bar{\varepsilon}} = \langle \alpha_\eta : \eta \in t\rangle$ where $\alpha_\eta = \varepsilon_{\ell g\eta}$.

2.16 Theorem. Suppose $\mathcal{E} \subseteq \mathcal{S}_{\leq\aleph_0}(\lambda)$ is stationary, $\bar{Q} = \langle P_i, \underset{\sim}{Q}_j : i \leq \alpha^*, j < \alpha^*\rangle$ a CS iteration, and for each $\alpha < \alpha^*$, $(*)^{\alpha,\alpha+1}_{\bar{Q},\mathcal{E}}$ holds (see below), *then* forcing with P_{α^*} adds no reals, where for $\beta < \gamma \leq \alpha^*$ we define:

$(*)^{\beta,\gamma}_{\bar{Q},\mathcal{E}}$ Assume

(a) $k < \omega, n < \omega, \bar{\varepsilon} = \langle \varepsilon_0, \ldots, \varepsilon_{n-1}\rangle, \varepsilon_0 < \ldots < \varepsilon_{n-1} \leq \beta$,
$m_i < \omega$ for $i < n$,
t a standard simple tree with n levels,
$t^*_\ell = t \cup \{\eta\hat{\ }\langle i\rangle : i < 2^\ell, \eta \in \max(t)\}$

$$h_\ell : t_{\ell+1} \to t_\ell \text{ is } h(\nu) = \begin{cases} \nu & \text{if } \nu \in t \\ \eta\hat{\ }\langle [i/2]\rangle & \nu = \eta\hat{\ }\langle i\rangle, \eta \in \max(t) \end{cases}$$

and let $h = h_k$, $t_0 = t^*_k$, $t_1 = t^*_{k+1}$.
If $\bar{q} = \langle q_\eta : \eta \in t_0\rangle$, let $\bar{q}^h = \langle q_{h(\eta)} : \eta \in t_1\rangle$

(b) $N \prec (H(\chi), \in, <^*_\chi)$ is countable, $\bar{Q}, \lambda_0, \bar{\varepsilon}, \beta, \gamma \in N$ and $N \cap \lambda_0 \in \mathcal{E}_0$, while $\beta \leq \gamma \leq \alpha^*$.

(c) $G_0 \subseteq P_{\bar{\alpha}_{t_0,\bar{\varepsilon}\hat{\ }\langle\beta\rangle}} \cap N$ is generic over N, (so we may write $G_0 = \langle G^0_\eta : \eta \in t_0\rangle$).

(d) $\bar{p} \in N \cap P_{\bar{\alpha}_{t_0,\bar{\varepsilon}\hat{\ }\langle\gamma\rangle}}$ is compatible with G^0 (note $P_{\bar{\alpha}_{t_0,\bar{\varepsilon}\hat{\ }\langle\beta\rangle}} \subseteq P_{\alpha_{t_0,\bar{\varepsilon}\hat{\ }\langle\gamma\rangle}}$, so this means $\bigwedge_{\eta\in t_0} p_\eta{\restriction}\beta \in G^0_\eta$).

(e) $\bar{q} \in P_{\bar{\alpha}_{t_1,\bar{\varepsilon}\hat{\ }\langle\beta\rangle}}$ such that it is above G^h_0 i.e., $\bar{r} \in G_0 \Rightarrow \bar{r}^h \leq q$.

Then we can find $G_1, \bar{r}$, such that

(α) $G_1 \subseteq P_{\bar{\alpha}_{t_0},\varepsilon \frown \langle\gamma\rangle} \cap N$ is generic over N

(β) $\bar{p} \in G_1$.

(γ) $G_0 \subseteq G_1$ (see remark in (d)).

(δ) $\bar{r} \in P_{\bar{\alpha}_{t_1},\bar{\varepsilon} \frown \langle\gamma\rangle}$, $q \le \bar{r}$.

(ε) $\bar{r}$ is above G_1^h.

Proof. We prove by induction on $\alpha \le \ell g(\bar{Q})$ that for every $\beta < \gamma \le \alpha$, $(*)^{\beta,\gamma}_{\bar{Q},\mathcal{E}}$
(in particular that $\bar{Q}\restriction\alpha$ is a CS iteration of $\mathcal{E}$-proper forcing). The main point is the case $\gamma = \alpha$ is a limit ordinal whose proof is similar to the proof in 2.10C.
$\square_{2.16}$

§3. Other Preservations

A central theme in this book is that it is worthwhile to have general preservation theorems on iterated forcing. While it seems that this is reasonably accepted in the community for properness, this seemingly is not so for preservation theorems like "proper+${}^\omega\omega$-bounding" and even less for a general framework for them. So here we try another way to materialize the theme (in 3.1-3.6). We then present several applications (but, generally, we do not repeat VI). A simple case of our framework is [Sh:326, A 2.6(3), pp.397-9]

This section passed through several versions, e.g. in most of them the proof of 3.6 was left to the reader. Goldstern [Go] starts from an earlier one, he cuts the generality for the sake of completeness. Relative to the present version he restricts himself to the case A and $\alpha^* = \omega$, in Definition 3.4 omit demand (xi) ((x) irrelevant) and demand it adds reals. Also $R_n \subseteq R_{n+1}$ and he omits S and $\mathbf{g}$ (so uses $({}^\omega\omega)^V$ as a cover: a $\mathbf{g}_{a\cap V}$ is chosen in the proof.).
Lately we added the proof of 3.6 (and added 3.4B, 3.13) and in some of the cases (i.e. when $d[a] \in a$, $\alpha^* > 1$ and we are not in Possibility $C(C^*)$) we added the condition $\oplus_k$ (or $\oplus_1$).

3.1 Context. $S \subseteq \mathcal{S}_{<\aleph_1}(A)$ for some $A = \bigcup S$ (usually S is stationary). For $a \in S$ $c[a], d[a]$ are subsets of a, and there are $c'[a], d'[a]$ defined such that:

Case a: if $d[a] \in a$ then $c'[a] = c[a]$, $d'[a] = d[a]$.

Case b: if $d[a] \notin a$ then $c[a] \notin a$, $c[a] = c'[a] \cap a$, $d[a] = d'[a] \cap a$.

Advise to the reader: At first reading the reader may think of a typical case: $\chi_0 << \chi$, $A = H(\chi_0)$, and elements of S are of the form $N \cap H(\chi_0)$, for some $N \prec (H(\chi), \in, <^*_\chi)$ such that $\chi_0 \in N$, all in the original universe V_0. A typical case for $d[a] \notin a$ would be $d[a] = a$, or $d[a] = a \cap \omega_1$, and below (in Definition 3.2) choose one possibility, say (B).

In addition we have $\mathbf{g} = \langle \mathbf{g}_a : a \in S \rangle$ where $\mathbf{g}_a$ is a function from $d[a]$ to $c[a]$ and α^* is an ordinal > 0.

The set $\bigcup S$ is, for simplicity, transitive, $\bar{R}$ is a three place relation, (more exactly a definition of one) written as $fR_\alpha g$, and whenever $fR_\alpha g$, for some $a \in S$ we have: $\alpha \in \alpha^* \cap a$ and f, g are functions from $d[a]$ to $c[a]$; for notational simplicity $\left[d[a] \in a \Leftrightarrow c[a] \in a\right]$ and $(\forall a \in S)[d[a] \in a]$ or $(\forall a \in S)[d[a] \notin a]$; and $d'[a], c'[a] \in a$ (of course $d'[a] \cap a = d[a]$, $c'[a] \cap a = c[a]$), and $\pm R_\alpha$ is absolute (enough to restrict to extension by forcings e.g. by proper forcing). Generally, saying absolutely or in any generic extension V^Q, we may mean for generic extensions by proper forcing, or any other property preserved by the iterations to which we want to apply this section.

3.2 Definition. 1) We say $(\bar{R}, S, \mathbf{g})$ covers (in V) *if* for χ large enough, for every $x \in H(\chi)^V$ there is a countable $N \prec (H(\chi), \in, <^*_\chi)$ to which $(\bar{R}, S, \mathbf{g})$ and x belong, and N is $(\bar{R}, S, \mathbf{g})$-good, which means:
$a \stackrel{\text{def}}{=} N \cap (\bigcup S)$ belongs to S, (so $\{d'[a], c'[a]\} \in N$) and: *for every* function $f \in N$ such that f maps $d[a]$ into $c[a]$, (so $d[a] \subseteq \mathrm{Dom}(f)$ but not necessarily $\mathrm{Dom}(f) \subseteq d[a]$) *for some* $\beta \in \alpha^* \cap a$, we have $(f \restriction d[a]) R_\beta \mathbf{g}_a$, the most natural case is: f a function from $d'[a]$ to $c'[a]$.
2) We say $(\bar{R}, S, \mathbf{g})$ fully covers (in V) if: the above holds for every countable $N \prec (H(\chi), \in, <^*_\chi)$ to which $(\bar{R}, S, \mathbf{g})$ and x belong and $N \cap (\bigcup S) \in S$ and in addition S is stationary.

3) We say $(\bar{R}, S, \mathbf{g})$ weakly covers if $d'[a] = d$, $c'[a] = c$ for every $a \in S$ (so c, d are constants, for example ω) and for every $f \in {}^d c$ for some a, α we have $f R_\alpha \mathbf{g}_a$.

3.2A Remark.

1) Actually, if the function $a \mapsto \mathbf{g}_a$ is one to one, then we can omit α and write $f R \mathbf{g}_a$ where R is defined by $fR\mathbf{g}$ iff $(\exists a \in S)[\mathbf{g} = \mathbf{g}_a \ \& \ \bigvee_{\alpha \in a \cap \alpha^*} f R_\alpha \mathbf{g}_a]$; the notation above is just more natural in the applications we have in mind.
2) Of course, in Definition 3.2, x is not necessary.
3) If $V^1 \subseteq V^2 \subseteq V^3$ are universes, $(\bar{R}^1, S^1, \mathbf{g}^1) \in V^1$ weakly covers in V^2 and $(\bar{R}^2, S^2, \mathbf{g}^2) \in V^2$ weakly covers in V^3, $\ell g \bar{R}^1, \ell g \bar{R}^2 < \omega_1$ and $\bigvee_{\alpha \in a} R^\ell_\alpha$ have the same definition for all $\ell = 1, 2$ and $a \in S^\ell$ (which is absolute for the cases of extension) and are partial orders and S^1 is a stationary subset of $\mathcal{S}_{<\aleph_1}(\bigcup S^1)$ even in V^3 *then* $(\bar{R}, S^1, \mathbf{g}^1)$ weakly covers in V^3.
4) We can translate an instance of Case a (in 3.1) to an instance of Case b, by replacing $d[a]$ by a and replacing $f \in {}^{d[a]}c[a]$ by a function $f^{[a]}$ where the function $f^{[a]}$ is $f \cup 0_{a \setminus d[a]}$, for example. This may help to apply e.g. 3.3. Possibility A, the case $a \in S \Rightarrow d[a] \notin a$ *but* has a price: $d[a] \notin a$ makes Definition 3.4 stronger, as the assumption becomes weaker (see clauses (vii)+(ix)), though we add the assumption in clause (x) so really there is no clear order.

3.3 Definition.; We say $(\bar{R}, S, \mathbf{g})$ strongly covers if (it is as in 3.1 and) it covers (in V, see Definition 3.2(1)) and one of the following possibilities holds:
Possibility A: Each R_α is closed (2-place relation on ${}^{d[a]}c[a]$)† (note that if R_α is open then $R = \bigcup_{n<\omega} R_{\alpha,n}$ where each $R_{\alpha,n}$ is closed, hence this possibility applies replacing α^* by $\omega\alpha^*$, using $R'_{\omega\alpha+n} = R_{\alpha,n}$) and: $[a \in S \Rightarrow d[a] \notin a]$ or $\alpha^* = 1$ or $\oplus_k$ for every $k < \omega$, which means††

† It is enough that each $\{f : \ f R_\alpha \mathbf{g}_a\}$ is closed.
†† Instead of the forcing notion P we can just demand that this holds absolutely.

$\oplus_k$ *if*

(a) P is a proper forcing notion preserving "$(\bar{R}, S, \mathbf{g})$-covers" and in V^P, $\underset{\sim}{Q}$ is a proper forcing in V^P [or just P, $\underset{\sim}{Q}$ are P_i, $\underset{\sim}{Q}_i$ as we get in our iterations]

(b) in V^P, $N \prec (H(\chi), \in, <^*_\chi)^{V^P}$ is countable, and $(\bar{R}, S, \mathbf{g})$-good (so in particular $(\bar{R}, S, \mathbf{g}) \in N$, $a = N \cap \bigcup S \in S$) and $\underset{\sim}{Q} \in N$, $p \in \underset{\sim}{Q} \cap N$

(c) for each $\ell < k$ we have: $\underset{\sim}{f}_\ell \in N$ is a Q-name of a member of ${}^{d'[a]}c'[a]$,

(d) $\chi_1 < \chi$ (χ_1 large enough e.g. $(P, \underset{\sim}{Q}) \in H(\chi_1)$ but $2^{\chi_1} < \chi$), $N_1 \prec (H(\chi_1), \in, <^*_{\chi_1})$ is countable, $N_1 \in N$, $\{Q, p, \bar{R}, S, \mathbf{g}, \underset{\sim}{f}_\ell\} \in N_1$, $p \in G^1 \in \mathrm{Gen}(N_1, Q) \cap N$.

(e) $\beta_\ell \in a \cap \alpha^*$ and $\underset{\sim}{f}_\ell \restriction d[a_1][G^1] R_{\beta_\ell} \mathbf{g}_a$

then for any $y \in N \cap H(\chi_1)$ there are N_2, G_2 satisfying (the parallel of) clause (d), such that $y \in N_2$ and: for some $\gamma_\ell \in a$, $\gamma_\ell \leq \beta_\ell$ (for $\ell < k$) we have $\underset{\sim}{f}_\ell[G_2] R_{\gamma_\ell} \mathbf{g}_a$ (for $\ell < k$).

Also instead of $\oplus_k$ we can require:

$\oplus'_k$ *if* in some (e.g. proper) forcing extension, N is $(\bar{R}, S, \mathbf{g})$-good, $N \cap \bigcup S = a \in S$, $k < \omega$, for $\ell < k$ we have $f^*_\ell R_{\beta_\ell} \mathbf{g}_a$ (where $\beta_\ell \in a \cap \alpha^*$), $\langle f^*_{\ell,n} : n < \omega \rangle$ converge to f^*_ℓ (i.e. $f^*_{\ell,n} \in {}^{d[a]}c[a]$, $\forall x \in d[a] \exists m \forall n > m [f^*_{\ell,n}(x) = f^*_\ell(x)]$) and $\langle f^*_{\ell,n} : n < \omega \rangle$, $f^*_\ell \in N$ *then* for some $\gamma_\ell \leq \beta_\ell, \gamma_\ell \in a$ we have $\bigvee_{n<\omega} \bigwedge_{\ell<\kappa} f^*_{\ell,n} R_{\gamma_\ell} \mathbf{g}_a$

Remark:

1) we can specify how $f^*_\ell, f^*_{\ell,n}$ come from N_1, (see the proof of 3.7E) (possibly in some V^Q, Q $(\bar{R}, S, \mathbf{g})$-preserving). This is close to VI §1 (if $\eta \in {}^\omega\omega$, $\eta_n \in {}^\omega\omega$ for $n < \omega$ and $\eta_n \restriction n = \eta \restriction n$ and $x \in \mathrm{Dom}(R)$ then for some T, xRT and $\eta \in \lim T$, $(\exists^\infty n)(\eta_n \in \lim T)$). The original $\oplus_k$ is better when not all $\langle \langle p^n_m : m < \omega \rangle : n < \omega \rangle$ work, but some do.

2) So possibility A splits to four cases: $[a \in S \Rightarrow d[a] \notin S]$, $\alpha^* = 1$, $\bigwedge_k \oplus_k$ and $\bigwedge_k \oplus'_k$.

Possibility B: Here we assume $d[a] \notin a$ for $a \in S$ or $\alpha^* = 1$ or at least $\oplus_k$ for every $k < \omega$. Let χ be large enough. For each $a \in S$ if (Skolem hull of a in

$(H(\chi), \in, <^*_\chi)\big) \cap \bigcup S = a$, then player II has an absolute winning strategy (i.e. an absolute definition of it) which works in any generic extension V^Q of V by a (proper) forcing notion $Q \in H(\chi)$; during the play, stipulating $b_{-1} = \emptyset$, in the n'th move player I chooses $f^n_0, \dots, f^n_n$ satisfying $f^n_\ell \restriction d[a] \in {}^{d[a]}c[a]$ (see clause (b) below) and $\alpha^n_0, \dots, \alpha^n_{n-1}, \alpha^n_n$,

such that:

(α) for $\ell < n$ $\alpha^n_\ell \in a \cap \alpha^*$, and $\alpha^n_\ell \leq \alpha^{n-1}_\ell$

(β) if $\ell < n, \alpha^n_\ell = \alpha^{n-1}_\ell$ then $f^n_\ell \restriction b_{n-1} = f^{n-1}_\ell \restriction b_{n-1}$

(γ) $f^n_\ell R_{\alpha^n_\ell} \mathbf{g}_a$ for $\ell \leq n$ (hence $\alpha^n_\ell \in a$)

Player II chooses finite b_n, $b_{n-1} \subseteq b_n \subseteq a$.

In the end player II wins if:

(a) letting $\alpha_\ell = \min\{\alpha^n_\ell : \ell < n < \omega\}$ and $n(\ell) = \min\{n : \alpha^n_\ell = \alpha_\ell\}$ and $f_\ell = \bigcup_{\substack{n<\omega \\ n \geq n(\ell)}} f^n_\ell \restriction b_n$, we have $f_\ell R_{\alpha_\ell} \mathbf{g}_a$

or

(b) $a \neq (\bigcup S) \cap$ (Skolem hull of $a \cup \{f^n_\ell : \ell \leq n < \omega\}$).

Possibility C: Let χ be large enough. For each $a \in S$ in any forcing extension of V (of our family) player II has a winning strategy in the following game. In the n'th move: player I chooses N_n, H_n such that:

(a) N_n is a countable model of ZFC$^-$ (so $\in^{N_n}$ is $\in \restriction N_n$ but N_n is not necessarily transitive), $N_n \cap (\bigcup S) = a, S \in N_n$, $\mathbf{g} \in N_n$, $\bar{R} \in N_n$ (and $d'[a] \in N_n, c'[a] \in N_n$) and $[\ell < n \Rightarrow N_\ell \subseteq N_n]$ and $N_n \models$ "$(\bar{R}, S, \mathbf{g})$ covers" and

$$[f \in {}^{d'[a]}c'[a] \;\&\; f \in N_n \Rightarrow (f \restriction d[a]) R_a \mathbf{g}_a]$$

(where $R_a = \bigvee_{\alpha \in a \cap \alpha^*} R_\alpha$)

(b) $H_n \subseteq \{\langle f_0, \dots, f_{n-1}\rangle$: for some finite $d \subseteq d'[a]$, each f_ℓ is a function from d to $c'[a]\}$ and $H_n \in N_n$ is not empty.

(c) if $\langle f_0, \dots, f_{n-1}\rangle \in H_n$ and $d \subseteq \mathrm{Dom}(f_0)$ is finite *then* $\langle f_0 \restriction d, \dots, f_{n-1} \restriction d\rangle \in H_n$.

(d) if $\langle f_0, \cdots, f_{n-1}\rangle \in H_n, \mathrm{Dom}(f_0) \subseteq d$, d finite $\subseteq d[a]$ *then* for some $\langle f'_0, \dots, f'_{n-1}\rangle \in H_n$ we have $\mathrm{Dom}(f'_\ell) = d$, and $f_\ell \subseteq f'_\ell$

(e) $m < n \;\&\; \langle f_0, \dots, f_{n-1}\rangle \in H_n \Rightarrow \langle f_0, \dots, f_{m-1}\rangle \in H^*_m$ (see below).

Player II chooses $\langle f_0^n, \ldots, f_{n-1}^n \rangle \in H_n \cap N_n$ and let[†]
$H_n^* = \{\langle f_0, \ldots, f_{n-1}\rangle$: for each ℓ the functions f_ℓ, f_ℓ^n are compatible[††] $\}$

In the end player II wins if: for every $m < \omega$, $\bigcup_{n \geq m} f_m^n$ is a function which has domain $d[a]$ and $(\bigcup_{n \geq m} f_m^n) R_a \mathbf{g}_a$ [note: if e.g. $\{f : f R_a \mathbf{g}_a\}$ is a Borel set, then the game is determined and a winning strategy is absolute].

Possibility A^*: Each R_α is closed and

$\otimes$ if $a_1, a_2 \in S$, $a_1 \in a_2$, *then* $(c'[a_1], d'[a_1]) = (c'[a_2], d'[a_2])$ and absolutely for every $f \in {}^{d'[a_2]}c'[a_2]$ we have: $(f{\restriction}d[a_1]) R_{a_1} \mathbf{g}_{a_1} \Rightarrow (f{\restriction}d[a_2]) R_{a_2} \mathbf{g}_{a_2}$

and : $(\forall a \in S)(d[a] \notin a)$ or $\alpha^* = 1$ or $\oplus_1$. Note that in cases A^*, B^*, C^*, for some (c', d') we have $(c'[a], d'[a]) = (c', d')$ for every $a \in S$ (as S is directed).

Possibility B^*: We assume

$\otimes$ if $a_1, a_2 \in S$, $a_1 \in a_2$ *then* $(c'[a_1], d'[a_1]) = (c'[a_2], d'[a_2])$ and absolutely for every $f \in {}^{d'[a_2]}c'[a_2]$ we have $(f{\restriction}d[a_1]) R_{a_1} \mathbf{g}_{a_1} \Rightarrow (f{\restriction}d[a_2]) R_{a_2} \mathbf{g}_{a_2}$,

and player II has an absolute winning strategy in a game similar to the one in Possibility B except that only f_0^n, α_0^n, b_n are chosen. And: $(\forall a \in S)(d[a] \notin a)$ or $\alpha^* = 1$ or $\oplus_1$.

Possibility C^*: We assume

$\otimes$ if $a_1, a_2 \in S$, $a_1 \in a_2$ *then* $(c'[a_1], d'[a_1]) = (c'[a_2], d'[a_2])$ and absolutely for every $f \in {}^{d'[a_2]}c'[a_2]$ we have $(f{\restriction}d[a_1]) R_{a_1} \mathbf{g}_{a_1} \Rightarrow (f{\restriction}d[a_2]) R_{a_2} \mathbf{g}_{a_2}$,

and player II has an absolute winning strategy in a game similar to the one in Possibility C

(a) as before

(b)* $H_n \subset \{f$: for some finite $d \subseteq d'[a]$, f_0 is a function from d to $c'[a]\}$

(c)* if $f \in H_n$, $d \subseteq \mathrm{Dom}(f)$ is finite *then* $f{\restriction}d \in H_n$

(d)* if $f \in H_n$, $d \subseteq \mathrm{Dom}(f)$, $d \subseteq d'[a]$ *then* for some $f' \in H_n$, we have $\mathrm{Dom}(f') = d$ and $f \subseteq f'$

(e)* $H_n \subseteq H_{n+1}$

[†] We could give the second player more influence, see proof of 3.6.
[††] We could add $\ell < m < n \Rightarrow f_\ell^m \subseteq f_\ell^n$, no real difference.

3.3A Remark. 1) In Possibility B, we can restrict the forcing to a suitable family.

2) Below in the cases $d[a] \notin a$ we use (see Possibility C) $d'[a] = c'[a] = \bigcup S$. This is essentially a notational change.

3) In Possibility C^* we can weaken $\otimes_1$ to the weaker version

$\otimes_1^-$ *if* for some forcing notion P, in V^P, $(\bar{R}, S, \bar{\mathbf{g}})$ still covers, N is a countable elementary submodel of $(H(\chi)^{V^P}, \in)$ to which $(\bar{R}, S, \mathbf{g})$ belongs, and so is a model of ZFC$^-$, and $a \stackrel{\text{def}}{=} N \cap (\bigcup S) \in S$ and if $a_1 \in S \cap N$ and $f \in N \cap ({}^{d[a]}c[a])$ then for some $a_2, a_1 \in a_2 \in S \cap N$ and $f{\restriction}d[a_2]R_{a_2}\mathbf{g}_{a_2}$ *then* $fR_a\mathbf{g}_a$.

3.3B Observation. 1) In[†] Definition 3.3

(a) $(\forall a \in S)[d[a] \notin a]$ & Possibility B* implies Possibility B.

(b) $(\forall a \in S)[d[a] \notin a]$ & Possibility C* implies Possibility C.

(c) $(\forall a \in S)[d[a] \notin a]$ & Possibility A implies Possibility B.

(d) Possibility A^* implies Possibility B^*

2) If Possibilities A* or B* or C* of Definition 3.3 hold, (or just $\otimes$ from there), Q is a proper forcing and $\Vdash_Q$ "for every $f \in {}^{d'[a]}c'[a]$, for every $a_1 \in S \cap N$ for some a_2 satisfying $a_1 \in a_2 \in S \cap N$ we have $(f{\restriction}d[a_2])R_{a_2}\mathbf{g}_{a_2}$" and $Q \in N \prec (H(\chi), \in, <^*_\chi)$, $N \cap (\bigcup S) \in S$, N countable and $q \in Q$ is (N, Q)-generic *then* $q \Vdash$ "$N[\underset{\sim}{G}_Q]$ is $(\bar{R}, S, \mathbf{g})$-good".

3) A sufficient condition for $\oplus_k$ of Definition 3.3 is[††]

$\oplus_k^*$ if (a),(b),(c), (d), (e) are as in $\oplus_k$ of Definition 3.3, then for some $p' \in G_1$, $\gamma_\ell \in (\beta_\ell + 1) \cap a$ and Borel set (even Σ_1 set over $\bigcup S$ i.e. quantifying over ${}^\omega(\bigcup S)$, with $\bar{R}$, S, $\mathbf{g}$ as parameters will do), $A_\ell \in N$ (for $\ell < k$) we have

(α) $p' \Vdash_Q$ "$\underset{\sim}{f}_\ell \in A_\ell$ for $\ell < k$"

(β) $(\forall f \in A_\ell)(\exists \gamma \leq \beta_\ell)(fR_\gamma\mathbf{g}_a)$

Proof. (1) Easy, For clause (a) note that:

---

† We can replace $(\forall a \in S)[d(a) \notin a]$ by $\alpha^* = 1$, and/or add $X \in \{A, B, C\}$ & $(\forall a \in S)[d(a) \notin a]$ implies Possibility $X \Leftrightarrow$ Possibility X^*. Note that for possibility C and C^*, w.l.o.g. $\alpha^* = 1$.

†† Many times this is easy.

(i) increasing b_n may only help player II as it just strengthen the restrictions on player II,

(ii) having more f^n_ℓ may only help player II as it make the satisfaction of clause (b) of possibility B (or B*) more probable. So for player II, having a winning strategy in the two games are equivalent (but not so for the 'player I has no winning strategy; see hopefully in [Sh:311]). Similarly for clause (b).

(c) We should give a winning strategy for player II. Let $a = \{x_i : i < \omega\}$ and his strategy is to choose $b_n = \{x_\ell : \ell < n\}$.

2), 3) Left to the reader. $\square_{3.3B}$

3.4 Definition. We say that a forcing notion Q is $(\bar{R}, S, \mathbf{g})$-preserving for possibility X if (where $X \in \{A, B, C, A^*, B^*, C^*\}$, for Possibilities C, C* (in Def 3.3) we can omit (iv)-(xi) and conclusion (a) as they hold vacuously; if we omit "for possibility X" we mean $X = C$):

$(*)$ Assume (i) χ_1 is large enough, $\chi > 2^{\chi_1}$

(ii) $N \prec (H(\chi), \in, <^*_\chi)$, N countable, $N \cap (\bigcup S) = a \in S$ and $\langle Q, S, \mathbf{g}, \chi_1 \rangle \in N$

(iii) N is $(\bar{R}, S, \mathbf{g})$− good (see Definition 3.2(1)) and $p \in Q \cap N$.

(iv) In Possibilities A, B we have $k < \omega$ and for $\ell < k$ we have $\underset{\sim}{f}_\ell \in N$ is a Q-name of a function, $\Vdash_Q$ "$\mathrm{Dom}(\underset{\sim}{f}_\ell) = d'[a]$"; for Possibilities A*, B* the situation is similar but $k = 1$. For Possibilities C, C* we can let $k = 0$.

(v) if $\ell < k$, then f^*_ℓ is a function and $\mathrm{Dom}(f^*_\ell) = d[a]$

(vi) for $n < \omega$ we have: $p, p_n \in Q \cap N$, $p \le p_n \le p_{n+1}$

(vii) if $d[a] \in a$ then $\langle p_n : n < \omega \rangle \in N$ and $\langle f^*_\ell : \ell < k \rangle \in N$

(viii) for each $x \in \mathrm{Dom}(f^*_\ell)$ and $\ell < k$, for every n large enough $p_n \Vdash_Q$ "$\underset{\sim}{f}_\ell(x) = f^*_\ell(x)$"

(ix) for $\ell < k$ we have $f^*_\ell R_{\beta_\ell} \mathbf{g}_a$ where $\beta_\ell \in a \cap \alpha^*$.

(x) if $d[a] \notin a$, $\mathcal{I} \in N$ a dense open subset of Q then for some n, $p_n \in \mathcal{I}$

(xi) if $d[a] \in a$, then for some N_1 a countable elementary submodel of $(H(\chi_1), \in, <^*_{\chi_1})$ which belong to N and include

$d[a] \cup c[a] \cup \{d[a], c[a]\} \cup \{Q, S, \mathbf{g}\} \cup \{\underset{\sim}{f}_\ell : \ell < k\}$ we have*:

$$\bigwedge_n p_n \in N_1, \bigvee_n p_n \in \mathcal{I} \text{ for any } \mathcal{I} \in N_1, \text{ a dense subset of } Q.$$

Then there is a q, $p \leq q \in Q$ such that: q is (N, Q)-generic and

(a) $q \Vdash_Q$ "$(\underset{\sim}{f}_\ell \restriction d[a]) R_{\gamma_\ell} \mathbf{g}_a$ for some γ_ℓ, $\gamma_\ell \leq \beta_\ell \,\&\, \gamma_\ell \in a \cap \alpha^*$" for each $\ell < k$

(b) $q \Vdash_Q$ "$N[\underset{\sim}{G}_Q]$ is $(\bar{R}, S, \mathbf{g})$-good"

3.4A Claim. 1) If $\alpha^* = 1$ *then* "Q is $(\bar{R}, S, \mathbf{g})$-preserving" (see 3.4 above) is equivalent to : if $N \prec (H(\chi), \in, <^*_\chi)$, N countable, N is $(\bar{R}, S, \mathbf{g})$-good, $Q \in N$, $p \in N \cap Q$ *then* for some (N, Q)-generic $q \in Q$, $q \geq p$ we have $q \Vdash$ "$N[\underset{\sim}{G}_Q]$ is $(\bar{R}, S, \mathbf{g})$-good"†.

2) If $\otimes$ (of possibilities A^*, B^*, C^* of Definition 3.3) hold, Q proper and $\alpha^* = 1$ *then:* "Q is $(\bar{R}, S, \mathbf{g})$-preserving" is equivalent to : for every $f \in {}^{d'[a]}c'[a]$ from V^Q for some a_2 we have $a \in a_2 \in S$, $(f \restriction d[a_2]) R_{a_2} \mathbf{g}_{a_2}$.

3) If $(\bar{R}, S, \mathbf{g})$ is as in Possibility A* (of Definition 3.3) and $(\forall a \in S)([d[a] \in a])$ and $\otimes^+$ below holds then: for any proper forcing notion Q, if $\Vdash_Q$ "$(\bar{R}, S, \mathbf{g})$ covers" *then* Q is $(\bar{R}, S, \mathbf{g})$-preserving for possibility A* where

$\otimes^+$ Assume†† we have a countable $N \prec (H(\chi), \in, <^*_\chi)$ such that $(\bar{R}, S, \mathbf{g}) \in N$, $a_1 \in a_2 \cap S$, $a_2 = N \cap (\bigcup S) \in S$, $(c[a_1], d[a_1]) = (c[a_2], d[a_2])$ and $\{f, \langle f_n : n < \omega \rangle\} \in N$ and $f R_\alpha \mathbf{g}_{a_2}$, and $\{f, f_n : n < \omega\} \subseteq {}^{d[a_1]}c[a_1]$, $f_n R_{\alpha_n} \mathbf{g}_{a_1}$ and $(\forall x \in d[a])(\forall^* n)(f_n(x) = f(x))$ and α, $\alpha_n \in \alpha^* \cap a_1$. Then for some $n < \omega$ and finite $d \subseteq d[a_1]$ we have

---

* We may add $a_1 \subseteq N_1$

$$N_1 \cap \bigcup S = a_1 \text{ and}$$

$$(c[a_1], d[a_1]) = (c[a], d[a])$$

and similarly add in $\oplus_k$ of Definition 3.2. Then in the proof of 3.5, 3.6 change somewhat (as in the proof of 3.4A), using some absoluteness for $xR\mathbf{g}_a$.

† This gives the results of VI §3.

†† We can add $N_1 \in N$, $N_1 \prec N$, $N_1 \cap (\bigcup S) = a_1$ and even more in this direction.

($*$) *if* $f' \in {}^{d[a_2]}c[a_2]$, $f' R_{\alpha_n} \mathbf{g}_{a_1}$ and $f' \restriction d = f_n \restriction d$ then $f' R_\alpha \mathbf{g}_{a_2}$

(we can look for $f' \in V$, or in $N_1[G_Q]$ for every $G_Q \subseteq Q$ generic over V where $Q \in N_1$ is proper, $N_1[G] \cap V = N_1$, $N[G] \cap V = N$, the second is more restrictive)

4) If Q is $(\bar{R}, S, \mathbf{g})$-preserving for possibility X for some X *then* Q is $(\bar{R}, S, \mathbf{g})$-preserving.

Proof. 1) Left to the reader.

2) Remember that (by $\otimes$ of case A*, B*) there is a pair (c', d') such that: $a \in S \Rightarrow (c'[a], d'[a]) = (c', d')$. Also note

$$a_1 \in S \;\&\; a_1 \in a_2 \in S \;\&\; a_2 = N \cap \bigcup S \;\&\; S \in N \prec (H(\chi), \in) \Rightarrow a_1 \subseteq a_2.$$

First we assume "Q is $(\bar{R}, S, \mathbf{g})$-preserving" and let $p \in Q$, $a \in S$ and $\underset{\sim}{f}$ be such that $p \Vdash_Q$ "$\underset{\sim}{f} \in {}^{d'[a]}c'[a]$" i.e. $p \Vdash_Q$ "$\underset{\sim}{f} \in {}^{d'}c'$". Take $N \prec (H(\chi), \in, <^*_\chi)$ such that a, $(\bar{R}, S, \mathbf{g})$, p, $f \in N$, and N is $(\bar{R}, S, \mathbf{g})$-good. So by the assumption, for some (N, Q)-generic q we have $p \leq q \in Q$ and $q \Vdash_Q$ "$N[\underset{\sim}{G}_Q]$ is $(\bar{R}, S, \mathbf{g})$-good". Let a_2 be $N \cap (\bigcup S)$, so q"$\Vdash \underset{\sim}{f} \restriction d[a_2] \in {}^{d[a_2]}c[a_2]$ satisfies $\underset{\sim}{f} \restriction d[a_2] R_{a_2} \mathbf{g}_{a_2}$", as required.

Second, to prove $\Rightarrow$ i.e. the "if" direction, assume that in V^Q for every $f \in {}^{d'}c'$ from V^Q for some a_1 we have $a_1 \in S$ and $f \restriction d[a_1] R_{a_1} \mathbf{g}_{a_1}$. This means: for every $G_Q \subseteq Q$ generic over V the statement above holds. Now let, in V, $N \prec (H(\chi), \in, <^*_\chi)$ be $(\bar{R}, S, \mathbf{g})$-good and assume $q \in Q$ is (N, Q)-generic. Let $q \in G_Q \subseteq Q$, G_Q generic over V, so it suffices to prove $V[G_Q] \Vdash$ "$N[G_Q]$ is $(\bar{R}, S, \mathbf{g})$-good". So let $a_2 = N \cap (\bigcup S)$, and let $f \in N[G_Q]$, $f \in {}^{d'[a_2]}c'[a_2] = {}^{d'}c'$. So for some $a_1 \in S$ we have $f \restriction d[a_1] R_{a_1} \mathbf{g}_{a_1}$, but $N[G_Q] \prec (H(\chi)[G_Q], \in)$ hence w.l.o.g. $a_1 \in N[G_Q] \cap S = N \cap S$. Now apply $\otimes$ of Definition 3.3 possibility A* (or B*, or C*), which we are assuming, to deduce $f \restriction d[a_2] R_{a_2} \mathbf{g}_{a_2}$. As this holds for every such f really $V[G_Q] \models N[G_Q]$ is $(\bar{R}, S, \mathbf{g})$-good.

3) Let N, N_1, $\underset{\sim}{f}^*_0$, β_0, $\bar{p} = \langle p_n : n < \omega \rangle$ be as in Definition 3.4 for possibility A*. Let $a = N \cap (\bigcup S)$. See in particular clause (xi) there. We can find $M_2 \prec N_2 \prec (H(\chi_1), \in, <^*_{\chi_1})$, $\{N_1, \langle p_n : n < \omega \rangle, \underset{\sim}{f}_0\} \in M_2 \in N_2 \in N$ and $N_2 \cap \bigcup S = a_2 \in S$,

$M_2 \cap (|bigcupS) = b_2$ and $(c[b_2], d[b_2]) = (c[a_2], d[a_2]) = (c[a], d[a])$. Also we can find $\langle p_{n,m} : n, m < \omega \rangle$, $\langle f_n : n < \omega \rangle$ such that: $p_n \leq p_{n,m} \leq p_{n,m+1}$, $p_{n,0}$ is (M_2, Q)-generic, and $p_{n,0} \Vdash$ "$\underset{\sim}{f}_0 R_{\alpha_n} \mathbf{g}_{b_2}$" for some $\alpha_n \in b_2 \cap \alpha^*$ and $\langle p_{n,m} : m < \omega \rangle$ is a generic sequence for N_2 (i.e. if $\mathcal{I} \subseteq Q$ is dense, $\mathcal{I} \in N_2$ then $(\exists m)(\exists r \in \mathcal{I} \cap N_2)(r \leq p_{n,m})$), $f_n \in {}^{d[a]}c[a]$, and

$$\forall x \in d[a] \forall n < \omega \forall^* m (p_{n,m} \Vdash \underset{\sim}{f}_0(x) = f_n(x)).$$

W.l.o.g. $\langle p_{n,m} : n, m < \omega \rangle$, $\langle \alpha_n : n < \omega \rangle$ and $\langle f_n : n < \omega \rangle$ belongs to N. Clearly $f_n R_{\alpha_n} \mathbf{g}_{b_2}$. (Here we used $\{f : f R_{\alpha_n} \mathbf{g}_{b_2}\}$ is closed and $\langle p_{n,m} : m < \omega \rangle$ is generic enough; Borel suffices. Why? Let $G_n = \{p \in Q \cap N_2 : (\exists m) p \leq p_{n,m}\}$ be a subset of $Q \cap N_2$ generic over N_2, so $N_2[G_n] \models$ "$\underset{\sim}{f}_0[G_n] R_{\alpha_n} \mathbf{g}_{a_1}$" but $f_n = \underset{\sim}{f}_0[G_n]$.)

Now apply $\otimes^+$ with b_2, a, f_0^*, β_0, $\langle f_n : n < \omega \rangle$, $\langle \alpha_n : n < \omega \rangle$ here standing for a_1, a_2, f, α, $\langle f_n : n < \omega \rangle$, $\langle \alpha_n : n < \omega \rangle$ there, and get n and d_n as there. Let m be such that $p_{n,m}$ force a value to $\underset{\sim}{f}_0 \restriction d_n$, so it is $f_n \restriction d_n$. Let $q \in Q$ be (N, Q)-generic such that $p_{n,m} \leq q$. Now suppose $q \in G_Q \subseteq Q$, G_Q generic over V; by the conclusion $(*)$ of $\otimes^+$ (i.e. the choice of n, d_n) we get $\underset{\sim}{f}_0[G_Q] R_{\beta_0} \mathbf{g}_a$. We still have to prove "$N[G_Q]$ is $(\bar{R}, S, \mathbf{g})$-good". But this holds by the proof of 3.4(2) above.

4) Easy. $\square_{3.4A}$

3.4B Claim. 1) Assume

(a) $(\bar{R}, S, \mathbf{g})$ is as in 3.1, $(\forall a \in S)(d[a] \in a)$,

(b) $(\bar{R}, S, \mathbf{g})$ covers,

(c) we have

$\oplus_1^+$ Assume we have a countable $N \prec (H(\chi), \in, <^*_\chi)$ such that $(\bar{R}, S, \mathbf{g}) \in N$, $a_1 \in a_2 \cap S$, $a_2 = N \cap (\bigcup S) \in S$, $(c[a_1], d[a_1]) = (c[a_2], d[a_2])$ and $\{f, \langle f_n : n < \omega \rangle\} \in N$ and $f R_\alpha \mathbf{g}_{a_2}$, and $\{f, f_n : n < \omega\} \subseteq {}^{d[a_1]}c[a_1]$, $f_n R_{\alpha_n} \mathbf{g}_{a_1}$ and $\forall x \in d[a_1] (\forall^* n)(f_n(x) = f(x))$ and α, $\alpha_n \in \alpha^* \cap a_1$. Then for some $n < \omega$ and finite $d \subseteq d[a_1]$ we have

$(*)$ *if* $f' \in {}^{d[a_2]}c[a_2]$, $f' R_{\alpha_n} \mathbf{g}_{a_1}$ and $f' \restriction d = f_n \restriction d$ then $f' R_\alpha \mathbf{g}_{a_2}$, moreover

(c)$^+$ for every proper forcing P preserving "$(\bar{R}, S, \mathbf{g})$-covers" we have $\oplus_1^+$ in V^P.

Then in the definition of "$(\bar{R}, S, \mathbf{g})$ strongly covers for possibility X" $X =$ A*, B^* we can omit $\oplus_1$ of Definition 3.2.

2) Assume (a), (b) as in (1) above and

(c) for each $k < \omega$ we have

$\oplus_k^{++}$ as in $\oplus_1^+$ but in the conclusion we replace "some n" by "for every n large enough"

or at least

$\oplus_k^+$ Assume we have a countable $N \prec (H(\chi), \in, <^*_\chi)$ such that $(\bar{R}, S, \mathbf{g}) \in N$, $a_1 \in a_2 \cap S$, $a_2 = N \cap (\bigcup S) \in S$ and $(c[a_1], d[a_1]) = (c[a_2], d[a_2])$ and $\{f_\ell : \ell < k\} \cup \{\langle f^\ell_n : n < \omega\rangle : \ell < k\} \in N$ and $f_\ell R_{\alpha(\ell)} \mathbf{g}_{a_1}$, and $\{f_\ell, f^\ell_n : \ell < k, n < \omega\} \subseteq {}^{d[a_1]}c[a_1]$ $f^\ell_n R_{\alpha_n(\ell)} \mathbf{g}_{a_1}$ and $\alpha(\ell)$, $\alpha_n(\ell) \in a_1 \cap \alpha^*$. *Then* for some $n < \omega$ and finite $d \subseteq d[a_1]$ we have

(∗) if $\ell < k$, $f'_\ell \in {}^{d[a_1]}c[a_1]$, $f'_\ell R_{\alpha_n(\ell)} \mathbf{g}_{a_1}$ and $f'_\ell \restriction d = f^\ell_n \restriction d$ *then* $f'_\ell R_\alpha \mathbf{g}_{a_2}$.

(c)′ Moreover (c) is preserved by proper forcing preserving "$(\bar{R}, S, \mathbf{g})$- covers". *Then* in the definition of "$(\bar{R}, S, \mathbf{g})$ strongly cover for possibility X", when $\forall a \in S$ $(c[a], d[a]) = (c, d)$ $X =$ A, B we can omit $(\forall k)\oplus_k$.

Proof. Like the proof of 3.4A(3). $\square_{3.4B}$

3.5 Claim. 1) If $(\bar{R}, S, \mathbf{g})$ covers in V and Q is an $(\bar{R}, S, \mathbf{g})$-preserving forcing notion *then* in V^G, $(\bar{R}, S, \mathbf{g})$ still covers.

2) Assume $(\bar{R}, S, \mathbf{g})$ covers. The property "$(\bar{R}, S, \mathbf{g})$-preserving for possibility X" is preserved by composition (of forcing notions).

Proof. 1) Just read the definitions.

2) Each part has some versions, according to whether in Definition 3.4 we choose Possibility A, A*, B, B* or Possibility C, C* and whether $d[a] \in a$ or not.

Let $\underset{\sim}{Q} = Q_0 * \underset{\sim}{Q}_1$; let χ_1, χ, N, N_1, a, k, $\underset{\sim}{f}_\ell$, β_ℓ, f^*_ℓ (for $\ell < k$), p, p_n ($n < \omega$) be as in Definition 3.4. Let $p = (q^0, \underset{\sim}{q}^1)$ and $p_\ell = (q^0_\ell, \underset{\sim}{q}^1_\ell)$. By condition (vi) of

Definition 3.4, for each $n < m < \omega$ we have $q^0_m \Vdash_{Q_0}$ "$\underset{\sim}{Q}_1 \models \underset{\sim}{q}^1 \le \underset{\sim}{q}^1_n \le \underset{\sim}{q}^1_m$", hence without loss of generality:

$(*)_1$ $\Vdash_{Q_0}$ "$\underset{\sim}{Q}_1 \models \underset{\sim}{q}^1 \le \underset{\sim}{q}^1_n \le \underset{\sim}{q}^1_m$ for $n < m < \omega$".

$(*)_2$ for every $x \in d[a]$ for every $n < \omega$ large enough, $(\emptyset, q^1_n)$ forces $\underset{\sim}{f}_\ell(x)$ to be equal to some (specific) Q_0-name, for each $\ell < k$.

[Why? By clause (x) or (xi) of Definition 3.4.]

Now we define $\underset{\sim}{f}'_\ell$, a Q_0-name of a member of ${}^{d[a]}c[a]$, such that $\Vdash_{Q_0}$ " for each $x \in d[a]$, for every n large enough $\underset{\sim}{q}^1_n \Vdash_{\underset{\sim}{Q}_1}$ '$[\underset{\sim}{f}'_\ell(x) = \underset{\sim}{f}_\ell(x)]$'". Easily: $d[a] \in a \Rightarrow \underset{\sim}{f}'_\ell \in N$.

By Definition 3.4 (and the assumption) there is $q_0 \in Q_0$ which is (N, Q_0)–generic, is above q^0 (in Q_0) and forces $N[\underset{\sim}{G}_{Q_0}]$ to be $(\bar{R}, S, \mathbf{g})$-good and for some $\gamma'_\ell \le \gamma_\ell$, $\gamma'_\ell \in N$ we have $q_0 \Vdash_{Q_0}$ "$\underset{\sim}{f}_\ell R_{\gamma'_\ell} \mathbf{g}_a$ for $\ell < k$".

Let $G_0 \subseteq Q_0$ be generic over V such that $q_0 \in G_0$. We want to apply Definition 3.4 with $N[G_0]$, $\underset{\sim}{q}^1[G_0]$, $\langle \underset{\sim}{q}^1_\ell[G_0] : \ell < \omega\rangle$, $\langle \underset{\sim}{f}_\ell[G_0] : \ell < k\rangle$, $\langle \underset{\sim}{f}'_\ell[G_0] : \ell < k\rangle$, $\langle \gamma'_\ell : \ell < k\rangle$, $\underset{\sim}{Q}_1[G_0]$ (and sometimes $N_1[G_0]$) here standing for N, p, $\langle p_\ell : \ell < \omega\rangle$, $\langle \underset{\sim}{f}_\ell : \ell < k\rangle$, $\langle f^*_\ell : \ell < k\rangle$, $\langle \beta_\ell : \ell < k\rangle$, Q there (and sometimes N_1) (and same $(\bar{R}, S, \mathbf{g})$).

So we have to check the assumptions of Definition 3.4; now we check all clauses of Definition 3.4.

clause (i): clear by the "old" (i).

clause (ii): holds as $q_0 \in G_0$ is (N, Q_0)-generic so $N[G_0] \cap (\bigcup S) = N \cap (\bigcup S)$ and the "old" (ii).

clause (iii): holds by the choice of $q_0 \in G_0$ that is $q_0 \Vdash$ "$N[G_0]$ is $(\bar{R}, S, \mathbf{g})$-good" by the choice of q_0 and clause (b) in the conclusion in Definition 3.4.

clause (iv): clear by the "old" (iv).

clause (v): If $x \in d[a]$, then ($x \in N$ or $\in N_1$ and) for some ℓ and Q_0-name $\underset{\sim}{\tau} \in N$ or $\in N_1$ we have $\Vdash_{Q_0}$ "$\left[q^1_\ell \Vdash_{\underset{\sim}{Q}_1} \text{“}\underset{\sim}{f}_\ell(x) = \underset{\sim}{\tau} \in c[a]\text{”}\right]$" (as the set of $(r_0, \underset{\sim}{r}_1) \in Q_0 \times \underset{\sim}{Q}_1$ such that

$$\Vdash_{Q_0} \text{“}\left[\underset{\sim}{r}_1 \Vdash_{\underset{\sim}{Q}_1} \text{“}\underset{\sim}{f}_\ell(x) = \underset{\sim}{\tau}\text{”}\right]\text{”}$$

for some Q_0-name $\underset{\sim}{\tau}$, is dense open subset of $(Q_0, \underset{\sim}{Q}_1)$ so some $(q_\ell^0, \underset{\sim}{q}_\ell^1)$ is in it, and there is such $\underset{\sim}{\tau}$ so w.l.o.g. $\underset{\sim}{\tau} \in N$ or $\in N_1$). So $\underset{\sim}{f}'_\ell[G_0](x) = \underset{\sim}{\tau}[G_0] \in c[a]$.

clause (vi): by $(*)_1$ above (in our proof) this holds.

clause (vii): Check (see $(*)_2$).

clause (viii): This is by the choice of $\underset{\sim}{f}'_\ell(x)$ and $\langle \underset{\sim}{q}_\ell^1 : \ell < \omega \rangle$.

clause (ix): by the choice of q_0 (and as $q_0 \in G_0$) and the choice of γ'_ℓ (for $\ell < k$).

clause (x): by the "old" clause (x) and as in the proof of clause (v) above. In details, if $N[G_0] \models$ "$\mathcal{I} \subseteq \underset{\sim}{Q}_1[G_0]$ is dense open" so $\mathcal{I} \in N[G_0]$ then for some $\underset{\sim}{\mathcal{I}}' \in N$ we have $\Vdash_{Q_0}$ "$\underset{\sim}{\mathcal{I}}'$ is a dense open subset of $\underset{\sim}{Q}_1$" and $\mathcal{I} = \underset{\sim}{\mathcal{I}}'[G_0]$; let

$$\mathcal{J} = \{(r_0, \underset{\sim}{r}_1) \in Q_0 * \underset{\sim}{Q}_1 : \Vdash_{Q_0} \text{“}\underset{\sim}{r}_1 \in \underset{\sim}{\mathcal{I}}'\text{”}\},$$

clearly $\mathcal{J} \in N_1$ is a dense open subset of $Q_0 * \underset{\sim}{Q}_1$ hence for every large enough ℓ,

$$(q_\ell^0, \underset{\sim}{q}_\ell^1) \in \mathcal{J} \text{ hence } \underset{\sim}{q}_\ell^1[G_0] \in \underset{\sim}{\mathcal{I}}'[G_0] = \mathcal{I},$$

hence we finish.

clause (xi): Use a_1, $N_1[G_0]$. Note that we do not require $N_1[G_0] \cap V = N_1$, still $N_1[G_0] \prec N[G_0]$, $N_1[G_0] \in N[G_0]$ and $\langle \underset{\sim}{q}_\ell^1[G_0] : \ell < \omega \rangle$ is as required there.

So really we can apply 3.4 and get $q_1 \in \underset{\sim}{Q}_1[G_0]$ which is $(N[G_0], \underset{\sim}{Q}_1[G_0])$-generic, and $\underset{\sim}{Q}_1[G_0] \models$ "$\underset{\sim}{q}^1[G_0] \leq q_1$" and $\langle \gamma_\ell : \ell < k \rangle$, $\gamma_\ell \leq \gamma'_\ell$ such that $q_1 \Vdash_{\underset{\sim}{Q}_1[G_0]}$ "$\underset{\sim}{f}_\ell R_{\gamma_\ell} \mathbf{g}_a$". As G_0 was any generic subset of Q_0 to which q_0 belongs, for some Q_0-name $\underset{\sim}{q}_1$ we have $q_0 \Vdash_{Q_0}$ "$\underset{\sim}{q}_1$ is as above". Now $(q_0, \underset{\sim}{q}_1)$, $\langle \gamma_\ell : \ell < k \rangle$ are as required. If we do have the demands on a_1 in Definition 3.4, clause (xi) we should replace N_1 ny another model in the intermediate stage as done in the proof of 3.4A (but we use absoluteness of $x R \mathbf{g}_a$). $\square_{3.5}$

3.6 Theorem. 1) Suppose $X \in \{A, B, C, A^*, B^*, C^*\}$ and in V we have $(\bar{R}, S, \mathbf{g})$ strongly covers, $\langle P_i, \underset{\sim}{Q}_j : i \leq \alpha, j < \alpha \rangle$ is a CS iteration of proper, $(\bar{R}, S, \mathbf{g})$-preserving for possibility X forcing notions, *then* P_α is a proper, $(\bar{R}, S, \mathbf{g})$-preserving for possibility X forcing notion.

2) This is true also for more general iterations, as in XV when $|\alpha^*| \le \aleph_1$ (in fact all cases in VI 0.1 apply) †.

Proof. 1) We prove by induction on $\zeta \le \alpha$, that: for every $\xi \le \zeta$, P_ζ/P_ξ is $(\bar{R}, S, \mathbf{g})$-preserving for possibility X (in V^{P_ξ}), moreover in Definition 3.4 we can get $\mathrm{Dom}(q) = (\zeta \setminus \xi) \cap N$. For ζ zero, there is nothing to prove, for ζ successor - use 3.5(2), so let ζ be limit, $\xi < \zeta$. Let $G_{P_\xi} \subseteq P_\xi$ be generic over V and $\chi, N, p, k, \underset{\sim}{f}_\ell, \underset{\sim}{f}^*_\ell, \beta_\ell$ (for $\ell < k$), and possibly p_n, χ_1, N_1 be as in (*) of Definition 3.4 (with $P_\zeta/P_\xi, V[G_{P_\xi}]$ here standing for Q, V there); for $X =$ C, C* we have $k = 0$ so $\underset{\sim}{f}_\ell$, $\underset{\sim}{f}^*_\ell$, β_ℓ disappear and for cases $d[a] \notin a$ we have no N_1 and for $X = A^*$, B^* we have $k = 1$. Let $G_0 = \{p \in P_\zeta/G_{P_\xi} : p \in N_1$ when well defined and $p \in N$ otherwise and for some n, $p \le p_n\}$ (used in the proof of possibility B, $d[a] \in a$).) We can choose $\zeta_n, \zeta_0 = \xi, \zeta_n < \zeta_{n+1} \in N \cap \zeta$ and $\sup(N \cap \zeta) = \bigcup_{n<\omega} \zeta_n$. Let $q_0 \in G_{P_\xi}$ force all this (so we can work in V, so we have $\underset{\sim}{G}_0$).

The proofs are built after the proofs of preservation of properness and the proofs in VI §1, VI §3 (particularly the proof of Possibilities A, $d[a] \in a$).

The case when $\mathrm{cf}(\zeta) > \aleph_0$ is elaborated when possibility B, $d[a] \in a$, is considered (note that the arguments there apply to all Possibilities).

Possibility C: Let $\langle \underset{\sim}{f}_\ell : \ell < \omega\rangle$ list the P_ζ-names $\underset{\sim}{f} \in {}^{d'[a]}c'[d]$ satisfying $\underset{\sim}{f} \in N$. Let $\langle \underset{\sim}{\tau}_n : n < \omega\rangle$ list the P_ζ-names of ordinals which belong to N. We choose by induction on n, $q_n, f_n, \underset{\sim}{H}_n$ $\langle \underset{\sim}{f}^n_\ell : \ell \le n\rangle$ such that:

(a) $q_n \in P_{\zeta_n}$, $\mathrm{Dom}(q_n) \setminus \xi = N \cap \zeta_n$, $q_{n+1} \restriction \zeta_n = q_n$ (of course q_0 is given)

(b) q_n is $(N[G_{P_\xi}], P_{\zeta_n})$-generic

(c) $q_n \Vdash$ "$N[G_{P_{\zeta_n}}]$ is $(\bar{R}, S, \mathbf{g})$-good".

† But in the applications presented here we "forget" this. Of course if we consider forcing notions with an additional order $\le_{\mathrm{pr}}$ on them, and the corresponding iteration (see XV), then "pure $(\theta, 2)$-decidability" has to be added for appropriate θ (mainly $d[a] \in N, \theta = \aleph_0$).

(d) $\underset{\sim}{p}_n$ is a P_{ζ_n}-name of a member of $P_\zeta \cap N$ such that

$$q_n \Vdash_{P_{\zeta_n}} \text{``}\underset{\sim}{p}_n \restriction \zeta_n \in \underset{\sim}{G}_{P_{\zeta_n}}\text{''}$$

(e) $\underset{\sim}{H}_n$ is a P_{ζ_n}-name, $\underset{\sim}{H}_n = \{\langle f_0, \ldots, f_{n-1}\rangle : d \subseteq d[a]$ is finite and $\underset{\sim}{p}_n \nVdash_{P_\zeta/\underset{\sim}{G}_{P_{\zeta_n}}} \text{``}\langle \underset{\sim}{f}_0 \restriction d, \ldots, \underset{\sim}{f}_{n-1} \restriction d\rangle \neq \langle f_0, \ldots, f_n\rangle\text{''}\}$

(f) $\underset{\sim}{f}^n_\ell$ is a P_{ζ_n}-name such that

$$q_n \Vdash_{P_{\zeta_n}} \text{``}\langle \underset{\sim}{f}^n_\ell : \ell < n\rangle \in \underset{\sim}{H}_n \text{ and for every } m \leq n \text{ we have}$$

$$\underset{\sim}{p}_{n+1}[\underset{\sim}{G}_{P_{\zeta_n}}] \nVdash_{P/P_{\zeta_n}} \text{``}\neg \bigwedge_{\ell < m} \underset{\sim}{f}_\ell \supseteq \underset{\sim}{f}^m_\ell\text{''''}.$$

(We can demand that q_n forces that $\underset{\sim}{p}_{n+1}[\underset{\sim}{G}_{P_{\zeta_n}}]$ forces $\ell < n \Rightarrow \underset{\sim}{f}^n_\ell[\underset{\sim}{G}_{P_{\zeta_n}}] \subseteq \underset{\sim}{f}_\ell$, minor difference.)

(g) $q_n \Vdash$ "$\underset{\sim}{p}_{n+1}$ forces a value to $\underset{\sim}{\tau}_n$".

Now there is no problem to carry out the definition but still we have freedom to choose $\langle \underset{\sim}{f}^n_\ell : \ell < n\rangle$. For this we use the winning strategy from Possibility C of Definition 3.3; choosing there the nth move of player I as:

$$N_n \stackrel{\text{def}}{=} N[\underset{\sim}{G}_{P_{\zeta_n}}]$$

$$\underset{\sim}{H}_n[\underset{\sim}{G}_{P_{\zeta_n}}] \stackrel{\text{def}}{=} \{\langle g_0, \ldots, g_{n-1}\rangle : \text{ for some finite } d \subseteq d[a] \text{ with have:}$$
$$g_\ell \in {}^d c[a] \text{ for } \ell < n \text{ and}$$
$$\underset{\sim}{p}_n[\underset{\sim}{G}_{P_{\zeta_n}}] \nVdash \text{``}\langle \underset{\sim}{f}_0 \restriction d, \ldots, \underset{\sim}{f}_{n-1} \restriction d\rangle \neq \langle g_0, \ldots, g_{n-1}\rangle\text{''}\}$$

(so the nth move is defined in $V^{P_{\zeta_n}}$; we can work in $V^{\text{Levy}(\aleph_0, (2^{|P_\alpha|})^+)}$). Now of course while playing, the universe changes but as the winning strategy is absolute there is no problem.

*Possibility C** By 3.3B(2) it is enough to show that for every P_ζ-name $\underset{\sim}{f}_0$ of a function from $d'[a]$ to $c'[a]$ for some $b \in S$, $((c'[b], d'[b]) = (c'[a], d'[a]))$ and $\underset{\sim}{f}_0 R_b \mathbf{g}_b$. This is proved as in the proof of Possibility C, dealing only with $\underset{\sim}{f}_0$ (and using Possibility C* of Definition 3.3 of course.)

Possibility A, $d[a] \notin a$: Let $\{\underset{\sim}{\tau}_j : j < \omega\}$ list the P_ζ-names of ordinals which belong to N. We shall choose by induction on $j < \omega$, $n_j < \omega$ such that :

(A) $n_j < n_{j+1} < \omega$

(B) for some sequence $\langle \underset{\sim}{\tau}_{j,\ell} : \ell \le j\rangle \in N$, with $\underset{\sim}{\tau}_{j,\ell}$ a P_{ζ_ℓ}-name we have:

  (α) $p_{n_j}\restriction[\zeta_j,\zeta) \Vdash_{P_\zeta}$ "$\underset{\sim}{\tau}_j = \underset{\sim}{\tau}_{j,j}$"

  (β) for $\ell < j$ we have $p_{n_j}\restriction[\zeta_\ell,\zeta_{\ell+1}) \Vdash_{P_{\zeta_{\ell+1}}}$ "$\underset{\sim}{\tau}_{j,\ell+1} = \underset{\sim}{\tau}_{j,\ell}$"

(C) if $j = i+1$, $\ell \le i$ then $\Vdash_{P_{\zeta_{\ell+1}}}$ "$p_{n_i}\restriction[\zeta_\ell,\zeta_{\ell+1}) \le p_{n_j}\restriction[\zeta_\ell,\zeta_{\ell+1})$".

(D) if $j = i+1$ then $\Vdash_{P_\zeta}$ "$p_{n_i}\restriction[\zeta_i,\zeta) \le p_{n_j}\restriction[\zeta_i,\zeta)$"

[Why can we carry the induction? It is enough to prove for each j that, given $\langle n_\ell : \ell < j\rangle$ as required, the set of candidates for $p \in P_\zeta$ satisfying the requirements on p_{n_j} is dense which is easy by clause (x) of (*) of Definition 3.4.]

Let $\{\underset{\sim}{f}_j : j < \omega\}$ list the P_ζ-names of members of ${}^{d'[a]}c'[a]$ which belong to N (for $\ell < k$ we let $\underset{\sim}{f}_j$ be as given). Note also that we can replace $\langle p_n : n < \omega\rangle$ by $\langle p_{n_j} : j < \omega\rangle$.

Hence without loss of generality we have $\underset{\sim}{\tau}_{\ell,j} \in N$ for $j \le \ell < \omega$, $\underset{\sim}{\tau}_{\ell,j}$ a P_{ζ_j}-name such that $p_\ell\restriction[\zeta_\ell,\zeta) \Vdash_{P_{\zeta_\ell}}$ "$\underset{\sim}{\tau}_\ell = \underset{\sim}{\tau}_{\ell,\ell}$", $p_\ell\restriction[\zeta_j,\zeta_{j+1}) \Vdash_{P_{\zeta_{j+1}}}$ "$\underset{\sim}{\tau}_{\ell,j+1} = \underset{\sim}{\tau}_{\ell,j}$". Let $h(j,x) < \omega$ be such that $\underset{\sim}{\tau}_{h(j,x)} = \underset{\sim}{f}_j(x)$. We can now define for $n < \omega$, $j < \omega$, $\underset{\sim}{f}^*_{n,j}$ a P_{ζ_n}-name of a function from $d[a]$ to $c[a]$. Let $\underset{\sim}{f}^*_{n,j}(x)$ be $\underset{\sim}{\tau}_{h(j,x),n}$ if $h(j,x) \ge n$ and $\underset{\sim}{\tau}_{h(j,x),h(j,x)}$ if $h(j,x) < n$ so $\underset{\sim}{f}^*_{0,j} = f^*_j$ for $j < k$.

We choose by induction on n, $q_n, \underset{\sim}{k}_n$, $\underset{\sim}{\alpha}^n_\ell$ (for $\ell < k+n$) such that:

(a) $q_n \in P_{\zeta_n}$, $\mathrm{Dom}(q_n) \setminus \xi = N \cap \zeta_n$, $q_{n+1}\restriction\zeta_n = q_n$, ($q_0$ is given).

(b) q_n is (N, P_{ζ_n})-generic

(c) $q_n \Vdash_{P_{\zeta_n}}$ "$N[\underset{\sim}{G}_{P_{\zeta_n}}]$ is $(\bar{R}, S, \mathbf{g})$-good"

(d) $\underset{\sim}{k}_n$ is a P_{ζ_n}/G_{P_ξ}-name of a natural number, $\underset{\sim}{k}_n < \underset{\sim}{k}_{n+1}$ (for Possibility (A), with which we are dealing) $\underset{\sim}{k}_n = n+1$ is O.K).

(e) $p_{k_0}\restriction\zeta_0 \le q_0$ (in P_{ζ_0}).

(f) $q_n\restriction\zeta_n \Vdash_{P_{\zeta_n}}$ "$p_{\underset{\sim}{k}_{n+1}}\restriction[\zeta_n,\zeta_{n+1}) \le q_{n+1}\restriction[\zeta_n,\zeta_{n+1}]$ (in $P_{\zeta_{n+1}}/P_{\zeta_n}$)".

(g) for $\ell < k+n$, $\underset{\sim}{\alpha}^n_\ell$ is a P_{ζ_n}-name of an ordinal in $a \cap \alpha^*$, $\alpha^{n+1}_\ell \le \alpha^n_\ell$, $\alpha^0_\ell = \beta_\ell$

(h) for $\ell < k+n$ we have $q_n \Vdash_{P_{\zeta_n}}$ "$f^*_{n,\ell} R_{\alpha^n_\ell} \mathbf{g}_a$".

The induction step is by the induction hypothesis (and Definition of "$(\bar{R}, S, \mathbf{g})$-preserving" (see Definition 3.4)). In the end let $q = \bigcup_{n<\omega} q_n$. Now why q is (N, P_ζ)-generic? Clearly $q \in P_\zeta$ (by condition (a)); let $q \in G_\zeta \subseteq P_\zeta$ be generic over V, $G_\xi \subseteq G_\zeta$, and $G_{\zeta_n} \stackrel{\text{def}}{=} G_\zeta \cap P_{\zeta_n}$. Now for each P_ζ-name $\underset{\sim}{\tau}$ of an ordinal, for some $j<\omega$, $\underset{\sim}{\tau} = \underset{\sim}{\tau}_j$ necessarily $k(*) \stackrel{\text{def}}{=} \underset{\sim}{k}_j[\underset{\sim}{G}_{P_{\zeta_{j+1}}}] > j$ (see condition (d)) hence: q_j forces that $\underset{\sim}{\tau}_{j,j}[G_{\zeta_j}] \in N$. But for $\ell \geq j$ and $j_1 \in [j,\omega)$ we have $p_j \restriction [\zeta_\ell, \zeta_{\ell+1}) \leq p_{j_1} \restriction [\zeta_\ell, \zeta_{\ell+1})$ hence by (e)+(f), $p_j \restriction [\zeta_\ell, \zeta_{\ell+1}] \leq q_{\ell+1}$, so together $p_j \restriction [\zeta_j, \bigcup_{i<\omega} \zeta_i) \leq q$ so $p_j \restriction [\zeta_j, \zeta) \leq q$; hence also q forces $\underset{\sim}{\tau}_j = \underset{\sim}{\tau}_{j,j}$. By the last two sentences $q \Vdash_{P_\zeta}$ "$\underset{\sim}{\tau}_j[\underset{\sim}{G}_{P_\zeta}] \in N \cap \mathrm{Ord}$" so q is really (N, P_ζ)-generic. Now for each ℓ the sequence $\langle \alpha^n_\ell[G_\zeta] : \ell \leq n < \omega \rangle$ is non increasing (see condition (g)) hence eventually constant; say for $n \in [n_\ell, \omega)$ has value α^*_ℓ. Now if $x \in d[a]$, $j<\omega$ then for $n > h(j,x)$ clearly $k_n > h(j,x)$ so $\underset{\sim}{f}_j(x) = \underset{\sim}{f}^*_{j,n}(x)$, so for every finite $b \subseteq d[a]$, $\langle (\underset{\sim}{f}^*_{j,n} \restriction b)[\underset{\sim}{G}_{P_{\zeta_n}}] : n < \omega \rangle$ is eventually constant, equal to $(\underset{\sim}{f}_j \restriction b)[G_\zeta]$. So for n large enough, $(\underset{\sim}{f}_j \restriction b)[G_{P_\zeta}] = (\underset{\sim}{f}^*_{j,n} \restriction b)[\underset{\sim}{G}_{P_{\zeta_n}}]$ and $\underset{\sim}{f}^*_{j,n}[\underset{\sim}{G}_{P_{\zeta_n}}] R_{\alpha^*_j} \mathbf{g}_a$.

So in $V[G_\zeta]$, $[\underset{\sim}{f}_j][G_\zeta]$ satisfies

⊗ for every finite $b \subseteq d[a]$ for some f', $\underset{\sim}{f}_j[G] \restriction b = f' \restriction b$ and $f' R_{\alpha^*_j} \mathbf{g}_a$.

But we are in Possibility A of Definition 3.3, so $R_{\alpha^*_j}$ is closed, so $\underset{\sim}{f}_j R_{\alpha^*_j} \mathbf{g}_a$. This finishes the proof that $q \Vdash$ "$N[G_\zeta]$ is $(\bar{R}, S, \mathbf{g})$-good". The last point is noting $\alpha^*_\ell \leq \underset{\sim}{\alpha}_{\ell,n}[G_\zeta] \leq \underset{\sim}{\alpha}_{\ell,0} = \beta_\ell$ for $\ell < k$, so we finish.

Possibility B, $\underline{d[a] \notin a}$: The proof is similar to the previous case, only the winning strategy in the game is described in Definition 3.3 (Possibility B) to make the $\underset{\sim}{k}_n$ large enough such that the part of the proof concerning $\underset{\sim}{f}_\ell[\underset{\sim}{G}_\zeta] R_{\alpha^*_\ell} \mathbf{g}_a$ works.

*Possibility A** $\underline{d[a] \notin a}$: By 3.2B(1), the next case implies it.

*Possibility B**, $\underline{d[a] \notin a}$: By 3.3B(2), we have to take care of $\underset{\sim}{f}_0$ only, and this is done as in Possibility B, $d[a] \notin a$, not increasing the set of f_ℓ's we consider.

Possibility B, $\underline{d[a] \in a}$: We shall reason as in the proof of Possibility A, $d[a] \notin a$, for N_1 (which vary).

If $\mathrm{cf}(\zeta) > \aleph_0$ then the set $\mathcal{I} = \{p \in P_\zeta :$ for some $\zeta' < \zeta$, and P_ζ-names $\underset{\sim}{g}_\ell$ we have $p \Vdash$ "$\underset{\sim}{f}_\ell = \underset{\sim}{g}_\ell$ for $\ell < k$"$\}$ is a dense subset of P_ζ and belongs to N_1. So for some n, $p_n \in \mathcal{I}$, by renaming without loss of generality $p \in \mathcal{I}$; w.l.o.g. $\zeta' \leq \zeta_1$. We can easily find $\langle q_n : n < \omega\rangle$, $\langle \underset{\sim}{p}'_n : n < \omega\rangle$ such that $q_{n+1}\restriction\zeta_n = q_n$, $q_n \in P_{\zeta_n}$, $q_n \Vdash$ "$N[\underset{\sim}{G}_{P_{\zeta_n}}]$ is $(\bar{R}, S, \mathbf{g})$-good", $q_1 \Vdash$ "$\underset{\sim}{f}_\ell R_{\gamma_\ell} \mathbf{g}_a$" for some $\gamma_\ell \in a$, $\gamma_\ell \leq \beta_\ell$ and $q_n \Vdash_{P_{\zeta_n}}$ "$\underset{\sim}{p}'_n \in P_\zeta \cap N$, $\underset{\sim}{p}'_n\restriction\zeta_n \in \underset{\sim}{G}_{P_{\zeta_n}}$ and $m < n \Rightarrow \underset{\sim}{p}'_m \leq \underset{\sim}{p}'_n$", and $\underset{\sim}{p}'_0 = p$, and for every P_ζ-name of ordinal $\underset{\sim}{\tau} \in N$ for some n, $q_n \Vdash_{P_{\zeta_n}} [\underset{\sim}{p}'_n \Vdash_{P_\zeta}$ "$\underset{\sim}{\tau} = \underset{\sim}{\alpha}_{\underset{\sim}{\tau}}$", $\underset{\sim}{\alpha}_\tau \in N]$ where $\underset{\sim}{\alpha}_\tau$ is a P_{ζ_n}-name of an ordinal. Now $q_\omega = \bigcup_{n<\omega} q_n$ is (N, P_ζ)-generic, and $p \leq q_\omega$; so q_ω is as required, so in the case $\mathrm{cf}(\zeta) > \aleph_0$ we are done[†].

So we are left with the case $\mathrm{cf}(\zeta) = \alpha_0$. We have $\aleph^* = 1$ or $\bigwedge_k \oplus_k$; as the later case is harder we speak on it. This time we use the full version of clause (xi) of Definition 3.4. Let $\{\underset{\sim}{\tau}'_j : j < \omega\}$ list the P_ζ-names of ordinals from N and $\{\underset{\sim}{f}'_j : j < \omega\}$ list the P_ζ-names of functions $f \in {}^{d[a]}c[a]$ which belong to N with $\underset{\sim}{f}'_j = \underset{\sim}{f}_j$ for $j < k$ and $\{x_j : j < \omega\}$ list $d[a]$. We now define by induction on $n < \omega$, $\underset{\sim}{M}_n$, $\underset{\sim}{G}^n$, q_n, $\underset{\sim}{p}'_n$, $\underset{\sim}{b}_n$, $\underset{\sim}{\alpha}^n_\ell(\ell < k+n)$ (note that q_0 and also G^0 are already given):

(a) $q_n \in P_{\zeta_n}$, $\mathrm{Dom}(q_n) \setminus \xi = N \cap \zeta_n \setminus \xi$, $q_{n+1}\restriction\zeta_n = q_n$ (q_0 is given).

(b) q_n is (N, P_{ζ_n})-generic

(c) $q_n \Vdash_{P_{\zeta_n}}$ "$N[\underset{\sim}{G}_{P_{\zeta_n}}]$ is $(\bar{R}, S, \mathbf{g})$-good"

(d) $\underset{\sim}{M}_n$, $\underset{\sim}{G}^n$, $\underset{\sim}{p}'_n$, $\underset{\sim}{b}_n$, $\underset{\sim}{\alpha}^n_\ell$ (for $\ell < k+n$) are P_{ζ_n}-names

(e) $q_n \Vdash_{P_{\zeta_n}}$ "$\underset{\sim}{b}_n$ is a finite subset of $d[a]$, $\underset{\sim}{M}_n$ a countable elementary submodel of $(H(\chi_1)[\underset{\sim}{G}_{P_{\zeta_n}}], \in, <^*_{\chi_1})$ which belongs to $N[\underset{\sim}{G}_{P_{\zeta_n}}]$[†] and $\underset{\sim}{b}_n \subseteq \underset{\sim}{M}_n$"

[†] This applies to all possibilities.

[†] And if we adopt the demand on a_1 in clause (xi) of Definition 3.4, we should add $\underset{\sim}{M}_n \cap (\bigcup S) \in S$

(f) $q_n \Vdash_{P_{\zeta_n}}$ "$\underset{\sim}{G}^n \subseteq P_\zeta/\underset{\sim}{G}_{\zeta_n} \cap \underset{\sim}{M}_n[\underset{\sim}{G}_{P_{\zeta_n}}]$ is generic over $\underset{\sim}{M}_n[\underset{\sim}{G}_{P_{\zeta_n}}]$, $\underset{\sim}{p}_n[\underset{\sim}{G}_{P_{\zeta_n}}] \in \underset{\sim}{G}^n$ so $\underset{\sim}{p}'_n \in P_{\zeta_n} \cap N$ (i.e. $\underset{\sim}{p}'_n[\underset{\sim}{G}_{P_{\zeta_n}}] \in P_{\zeta_n} \cap N$) and $\underset{\sim}{p}'_n \restriction \zeta_n \in \underset{\sim}{G}_{P_{\zeta_n}}$ and $\underset{\sim}{f}_j \in \underset{\sim}{M}_n$, $\underset{\sim}{f}_j[\underset{\sim}{G}^n] R_{\alpha_j^n} \mathbf{g}_\alpha$"

(g) $q_{n+1} \Vdash$ "$\underset{\sim}{p}'_n \leq \underset{\sim}{p}'_{n+1}$, $\underset{\sim}{f}_j[\underset{\sim}{G}^n] \restriction b_n \subseteq \underset{\sim}{f}_j[\underset{\sim}{G}^{n+1}] \restriction \underset{\sim}{b}_n$ for $j < k+n$".

(h) $q_n \Vdash$ "$\underset{\sim}{p}'_{n+1}[\underset{\sim}{G}_{P_{\zeta_n}}]$ forces a value to $\underset{\sim}{\tau}'_n$ (in $P_\zeta/\underset{\sim}{G}_{P_{\zeta_n}}$)" and to $\underset{\sim}{f}_j \restriction \underset{\sim}{b}_n$ for $j < k+n$.

There is no problem to carry the definition using $\bigwedge_k \oplus_k$. Now we have some freedom: choosing the $\underset{\sim}{b}_n$. So actually this is a play of the game, where the choices made above are fixing the moves of player I (with some extras). It will suffice to have player II winning, which is O.K. (so less than "I wins the game" is used).

In the end we let $q_\omega = \bigcup_{n<\omega} q_n$ and continue as in Possibility A, $d[a] \notin a$.

Possibility B^*, <u>$d[a] \in a$:</u> Combine the proofs for possibility B^* when $d[a] \notin a$ (i.e. use 3.3B(2)) but $\underset{\sim}{M}_n = N_1$ and the proof of possibility B when $d[a] \in a$.

2) Left to the reader $\square_{3.6}$

3.7 Application. Open dense subsets.

3.7A. Context and Definition. Let $\langle \eta^*_\ell : \ell < \omega \rangle$ enumerate ${}^{\omega>}\omega$ such that $\eta^*_m \restriction n \in \{\eta^*_i : i \leq m\}$, let fR_ng mean

$f, g : {}^{\omega>}\omega \to {}^{\omega>}\omega$ and $\eta \in {}^{\omega>}\omega \setminus \{\eta^*_\ell : \ell < n\}$ implies that there is ν such that $\eta \trianglelefteq \nu \triangleleft \nu\hat{\ } f(\nu) \trianglelefteq \eta\hat{\ } g(\eta)$.

Note that if fR_ng and $g' : {}^{\omega>}\omega \to {}^{\omega>}\omega$ and $(\forall \eta)(g(\eta) \trianglelefteq g'(\eta))$ then fR_ng'.

Let, for some subuniverse V', $S \subseteq \mathcal{S}_{<\aleph_1}(H(\aleph_1)^{V'})$, and for $a \in S$, $\mathbf{g}_a \in \bigcup S$ be such that $(\forall f)(f \in a \,\&\, f$ is a function from ${}^{\omega>}\omega$ to ${}^{\omega>}\omega \Rightarrow \bigvee_n fR_n\mathbf{g}_a)$. Clearly such $\mathbf{g} = \langle \mathbf{g}_a : a \in S \rangle \in V'$ exists.

Let $R = \bigvee_{n<\omega} R_n$, and let F^* be the family of functions from ${}^{\omega>}\omega$ to ${}^{\omega>}\omega$.

3.7B Claim. 1) $(\bar{R}, S, \mathbf{g})$ covers *iff* S is stationary and

$$(\forall f \in V)[f \in F^* \to (\exists g \in \bigcup S)[fRg]]$$

2) If $(\bar{R}, S, \mathbf{g})$ covers *then* it strongly covers (for possibility A*) and

$$(\forall f \in {}^\omega\omega)(\exists g \in \bigcup S)(\bigwedge_n f(n) < g(n)).$$

3) If $N \prec (H(\chi), \in, <^*_\chi)$ is countable, $(\bar{R}, S, \mathbf{g})$ covers and $N \cap (\bigcup S) = a \in S$ *then* N is $(\bar{R}, S, \mathbf{g})$-good.

4) Each R_n and $R = \bigvee_{m<\omega} R_m$ are transitive.

Proof. Straightforward. E.g.

(2) First let us show that $\oplus_1^+$ of 3.4B(1) hold. So suppose that N, a_1, a_2, f, $\langle f_n : n < \omega \rangle$ and α, α_n are as assumptions of $\oplus_1^+$. For $n < \omega$ we define

$$d_n^0 = \{\eta_\ell^* : \ell < n \text{ and } (\forall m \le \ell)(\forall \eta \trianglelefteq \eta_m^* \hat{\ } \mathbf{g}_{a_1}(\eta_m^*))(f_n(\eta) = f(\eta))\},$$

$$d_n^1 = \{\eta_\ell^* : (\exists \nu \in d_n^0)(\eta_\ell^* \trianglelefteq \nu \hat{\ } \mathbf{g}_{a_1}(\nu))\},$$

$$d_n^2 = d_n^1 \cup \{\eta_\ell^* : \ell < \alpha_n\}.$$

Note that

$(*)_1$ $d_n^0 \subseteq d_n^1 \subseteq d_n^2$ are finite subsets of ${}^{<\omega}\omega$,

$(*)_2$ each d_n^i $(i < 3)$ is closed under initial segments,

$(*)_3$ $(\forall \nu \in {}^{\omega>}\omega)(\forall^* n)(\nu \in d_n^0)$.

Using $(*)_3$ one easily constructs a function $f^* \in F^* \cap N$ such that

$(*)_4$ $(\forall k < \omega)(\eta_k^* \ne d_n^0 \Rightarrow f_n R_k f^*)$

(note that the sequence $\langle d_n^0 : n < \omega \rangle$ is in N). Then for some $\beta < \omega$ we have

$$f^* R_\beta \mathbf{g}_{a_2} \qquad \text{and} \qquad \mathbf{g}_{a_1} R_\beta \mathbf{g}_{a_2}.$$

Take n such that

$(*)_5$ $(\forall m \le \beta)(\eta_m^* \in d_n^)$)

and put $d = \{\nu : (\exists \eta \in d_n^2)(\nu \trianglelefteq \eta \hat{\ } \mathbf{g}_{a_2}(\eta)\}$.

Suppose that $f' \in F^*$ is such that $f' R_{\alpha_n} \mathbf{g}_{a_1}$ and $f' \restriction d = f_n \restriction d$. We are going to show that $f_n R_\alpha \mathbf{g}_{a_2}$. To this end suppose that $\ell \le \alpha$ and consider the following three cases.

Case 1: $\eta^*_\ell \notin d^2_n$

Then $\eta^*_\ell \notin d^0_n$ and hence (by $(*)_5$) $\ell > \beta$. Since $\mathbf{g}_{a_1} R_\beta \mathbf{g}_{a_2}$ we find $\nu \in {}^{\omega>}\omega$ such that

$$\eta^*_\ell \trianglelefteq \nu \triangleleft \nu\hat{\ }\mathbf{g}_{a_1}(\nu) \trianglelefteq \eta^*_\ell\hat{\ }\mathbf{g}_{a_2}(\eta^*_\ell).$$

It follows from $(*)_2$ that $\nu \notin d^2_n$, so $\nu = \eta^*_k$ for some $k \geq \alpha_n$. Since $f' R_{\alpha_n} \mathbf{g}_{a_1}$ we find η such that

$$\eta^*_\ell \trianglelefteq \nu \trianglelefteq \eta \triangleleft \eta\hat{\ }f'(\eta) \trianglelefteq \nu^*\hat{\ }\mathbf{g}_{a_1}(\nu) \trianglelefteq \eta^*_\ell\hat{\ }\mathbf{g}_{a_2}(\eta^*_\ell),$$

as required.

Case 2: $\eta^*_\ell \in d^2_n \setminus d^0_n$

Since $\eta^*_\ell \notin d^0_n$ we know $\ell > \beta$. As $f^* R_\beta \mathbf{g}_{a_2}$, we find k such that

$$\eta^*_\ell \trianglelefteq \eta^*_k \triangleleft \eta^*_k\hat{\ }f^*(\eta^*_k) \trianglelefteq \eta^*_\ell\hat{\ }\mathbf{g}_{a_2}(\eta^*_\ell).$$

Necessarily $\eta^*_k \notin d^0_n$ and therefore $f_n R_k f^*$. Consequently we find ν such that

$$\eta^*_\ell \trianglelefteq \eta^*_k \trianglelefteq \nu \triangleleft \nu\hat{\ }f_n(\nu) \trianglelefteq \eta^*_k\hat{\ }f^*(\eta^*_k) \trianglelefteq \eta^*_\ell\hat{\ }\mathbf{g}_{a_2}(\eta^*_\ell).$$

Plainly $\nu \in d$ (as $\eta^*_\ell \in d^2_n$) and therefore $f_n(\nu) = f'(\nu)$, so we get what is required.

Case 3: $\eta^*_\ell \in d^0_n$

Since $f R_\alpha \mathbf{g}_{a_2}$ we find ν such that

$$\eta^*_\ell \trianglelefteq \nu \triangleleft \nu\hat{\ }f(\nu) \trianglelefteq \eta^*_\ell\hat{\ }\mathbf{g}_{a_2}(\eta^*_\ell).$$

As $\eta^*_\ell \in d^0_n$ we know that $f_n(\nu) = f(\nu)$ so we conclude

$$\eta^*_\ell \trianglelefteq \nu \triangleleft \nu\hat{\ }f_n(\nu) \trianglelefteq \eta^*_\ell\hat{\ }\mathbf{g}_{a_2}(\eta^*_\ell).$$

This finishes verifying the clause $\oplus^+_1$. Now we may apply 3.4B(1) and easily check that $(\bar{R}, S, \mathbf{g})$ strongly cover for possibility A* (i.e. this claim gives $\oplus_1$ of Definition 3.3).

The other parts should be clear. $\square_{3.7B}$

3.7C Claim. Suppose in V, $(\bar{R}, S, \mathbf{g})$ covers, Q is a proper forcing notion, *then*: Q is $(\bar{R}, S, \mathbf{g})$-preserving for possibility A* *iff*

$$V^Q \models (\forall f \in F^*)(\exists g \in \bigcup S)[fRg]$$

(so also $\Vdash_Q$ "$(\bar{R}, S, \mathbf{g})$ covers" is equivalent to them).

Proof. The "only if" part is straightforward.

The converse implication follows from 3.4A(3) (note that the demand $\otimes^+$ was proved in the proof of 3.7B(2)). $\square_{3.7C}$

3.7D Claim. If $(\bar{R}, S, \mathbf{g})$ covers then "proper+ $(\bar{R}, S, \mathbf{g})$ - preserving" is preserved by composition, and more generally by CS iteration.

3.7E Claim. 1) Suppose $(\bar{R}, S, \mathbf{g})$ covers, *then* for every dense open $A \subseteq {}^{\omega>}\omega$ there is a dense open $B \subseteq {}^{\omega>}\omega, B \in \bigcup S$ and $B \subseteq A$.
2) If F^V is the family of functions from ${}^{\omega>}\omega$ to ${}^{\omega>}\omega$ and $F \subseteq F^V$ is such that $\forall g \exists f\ [gRf]$ and $S \subseteq \mathcal{S}_{<\aleph_1}(H(\chi_1))$ is stationary then we can find $\mathbf{g} = \langle \mathbf{g}_a : a \in S\rangle$, $\mathbf{g}_a \in F$ such that $(\bar{R}, S, \mathbf{g})$ covers.

Proof. 1) For a dense open set $A \subseteq {}^{\omega>}\omega$ define $f_A \in F$ by

$$f_A(\eta) \text{ is such that } \eta\,\hat{}\, f_A(\eta) \in A.$$

Let $n < \omega$, $g \in \bigcup S$ be such that $f_A R_n g$ and define

$$B \stackrel{\text{def}}{=} \{\eta \in {}^{\omega>}\omega : \text{ for some } \nu \in {}^{\omega>}\omega \setminus \{\eta^*_\ell : \ell < n\} \text{ we have } \nu\,\hat{}\, g(\nu) \trianglelefteq \eta\},$$

Clearly B is open dense, $B \in \bigcup S$, and $B \subseteq A$.

2) Straightforward. $\square_{3.7E}$

3.7F Remark. 1) In 3.7A we could have weakened fR_ng to: $\eta \notin \{\eta^*_\ell : \ell < n\}$ implies that for some $\nu, \nu \triangleleft \nu\,\hat{}\, f(\nu) \trianglelefteq \eta\,\hat{}\, g(\eta)$, call it R^w_n (and $\bar{R}^w, R^w, (S, \bar{R}^w, \mathbf{g})$ are defined accordingly). So we can demand $\langle\rangle \notin \text{Rang}(g)$. Then 3.7 B-E holds for this version too. (For 3.7B(2) second clause: for every

$f \in {}^{\omega}\omega$ let $f' : {}^{\omega>}\omega \to {}^{\omega>}\omega$ be such that $f'(\langle n \rangle) = \langle n, n+1, \dots, n+f(n) \rangle$, and $f'(\langle\rangle) = f'(\langle 0 \rangle)$. So there are $g' \in \bigcup S$ and n such that $f' R_n g'$. Let $g(n) = \min\{\ell g(\nu\,\hat{}\,g'(\nu)) : \nu\,\hat{}\,g'(\nu) \trianglelefteq \langle n \rangle\,\hat{}\,f'(\langle n \rangle)\}$; easily $f^* <^* g \in \bigcup S$.)
2) Assume $\bar{R}$ is as in 3.7A and S, $\mathbf{g}$ as in 3.1. Then also the inverse of 3.7E(1) holds, see 3.7H.

3.7H Claim. Suppose $(\bar{R}, S^1, \mathbf{g}^1), (\bar{R}^w, S^2, \mathbf{g}^2)$ is as in 3.7A for the same V' (for $\bar{R}^w$ defined in 3.7F) and S^1, $S^2 \subseteq S_{\leq \aleph_0}(\aleph_1)^{V'}$ are stationary even in V, *then*: $(\bar{R}, S^1, \mathbf{g}^1)$ covers
iff for every dense open $A \subseteq {}^{\omega>}\omega$ there is a dense open $B \subseteq {}^{\omega>}\omega$ such that $B \in \bigcup S^2 = \bigcup S^1$ and $B \subseteq A$
iff $(\bar{R}^w, S^2, \mathbf{g}^2)$ covers.

Proof. first $\Rightarrow$ *second*: this is 3.7E(1).

second $\Rightarrow$ *third*:

Let $f (\in V)$ be a function from ${}^{\omega>}\omega$ to ${}^{\omega>}\omega$; we define $A_f = \{\rho : \rho \in {}^{\omega>}\omega$ and $(\exists \nu)(\nu\,\hat{}\,f(\nu) \trianglelefteq \rho)\}$. Clearly $A_f \in V$ is a dense open subset of ${}^{\omega>}\omega$. So by the assumption there is a dense open $B \subseteq {}^{\omega>}\omega$ which belongs to $\bigcup S^2$ and $B \subseteq A_f$. So, working in V' there is $g \in \bigcup S^2$ such that: g is a function from ${}^{\omega>}\omega$ to ${}^{\omega>}\omega$ and for every $\eta \in {}^{\omega>}\omega$ we have $\eta\,\hat{}\,g(\eta) \in B$. It suffices to prove that $f R_0 g$ (as $f R_0 g \Rightarrow f R g$ and R is a partial order). Now for every $\eta \in {}^{\omega>}\omega$, we know $\eta\,\hat{}\,g(\eta) \in B$ hence $\eta\,\hat{}\,g(\eta) \in A_f$, but by its definition this implies the existence of $\nu \in {}^{\omega>}\omega$ such that $\nu\,\hat{}\,f(\nu) \trianglelefteq \eta\,\hat{}\,g(\eta)$. So ν is as required.

third $\Rightarrow$ *first:*

Let f be a function from ${}^{\omega>}\omega$ to ${}^{\omega>}\omega$. Let us define a function f' from ${}^{\omega>}\omega$ to ${}^{\omega>}\omega$ as follows. For $\eta \in {}^{\omega>}\omega$, let $\langle \rho^k_\eta : k < k_\eta \rangle$ be a list of $\{\rho : \rho \in {}^{\omega>}\omega, \ell g(\rho) = \ell g(\eta)$ and $\bigwedge_{\ell < \ell g(\eta)} \rho(\ell) \leq \eta(\ell)\}$, so η appears in it and $1 \leq k_\eta < \omega$. W.l.o.g. $\eta = \rho_\eta^{(k_\eta)-1}$. We now choose by induction on $k \leq k_\eta$, a sequence $\nu^k_\eta \in {}^{\omega>}\omega$. Let $\nu^0_\eta = \eta$, and ν^{k+1}_η be:

$$(\rho^k_\eta \cup \nu^k_\eta {\restriction} [\ell g \eta, \ell g \nu^k_\eta))\,\hat{}\,f[\rho^k_\eta \cup (\nu^k_\eta {\restriction} [\ell g \eta, \ell g \nu^k_n))]\,\hat{}\,\langle 0 \rangle.$$

Finally $f'(\eta)$ is defined by $\eta\,\hat{}\,f'(\eta) = \nu^{(k_\eta)}_\eta$, remember $\eta = \rho^{k_\eta - 1}_\eta$.

So by the assumption there is $g' \in \bigcup S^2$ such that g' is a function from ${}^{\omega>}\omega$ to ${}^{\omega>}\omega$ and $f' R^w g'$. As $\bigcup S^2$ includes the set of functions from ${}^{\omega>}\omega$ to ${}^{\omega>}\omega$ in V', without loss of generality $f' R^w_0 g'$, and as $\langle\rangle \notin \mathrm{Rang}(f')$, by 3.7F we know $\forall \eta[\ell g(f'(\eta)) > 0)$. We now define a function g from ${}^{\omega>}\omega$ to ${}^{\omega>}\omega$; we define $g(\eta)$ by induction on $k = \ell g(\eta)$; given η of length k, we choose by induction on $\ell < k$ natural numbers $i_\ell \in \{\eta(\ell), \eta(\ell)+1\}$ such that for $m \leq k$ we have i_ℓ is *not* the first element of $f'(\langle i_0, \ldots, i_{\ell-1}\rangle)$ (possible as $f'(\langle i_0, \ldots, i_{\ell-1}\rangle)$ has length > 0).
Let $\eta' = \langle i_0, \ldots, i_{k-1}\rangle$ and $g(\eta) = g'(\eta')$. Note: η' is well defined and for every $\ell < k$ the sequence η' (and even $\eta' \restriction (\ell+1)$) is not an initial segment of $(\eta' \restriction \ell)\hat{}\, f'(\eta' \restriction \ell)$. By the choice of g' and definition of R^w_0 we know that there is $\nu^0 \in {}^{\omega>}\omega$ such that $\nu^0 \trianglelefteq \nu^0\hat{}\, f'(\nu^0) \trianglelefteq \eta'\hat{}\, g'(\eta')$. By the choice of $\eta', \neg(\nu^0 \triangleleft \eta')$ so necessarily $\eta' \trianglelefteq \nu^0$. Let $\nu^1 = \eta \cup (\nu^0 \restriction [k, \ell g \nu^0))$, so $\eta \trianglelefteq \nu^1$, $\ell g(\nu^1) = \ell g(\nu^0)$ and $(\forall \ell)[\nu^1(\ell) \leq \nu^0(\ell)]$. Hence by the choice of $f'(\nu^0)$ there is $\nu^2, \nu^1 \trianglelefteq \nu^2 \trianglelefteq \nu^2\hat{}\, f(\nu^2) \triangleleft \nu^2\hat{}\, f(\nu^2)\hat{}\,\langle 0\rangle \trianglelefteq \nu^1\hat{}\, f'(\nu^0)$, just choose m such that $\nu^1 = \rho^m_{\nu^0}$ and put $\nu^2 \stackrel{\mathrm{def}}{=} \rho^m_{\nu^0}\hat{}\,(\nu^m_{\nu^0} \restriction [\ell g \nu^0, \ell g \nu^m_{\nu^0}))$. Note that $\nu^1\hat{}\, f'(\nu^0) \trianglelefteq \eta\hat{}\, g'(\eta') = \eta\hat{}\, g(\eta)$ and hence So $f R_0 g$. As g was defined from f' alone; and R is a partial order so we may easily finish. $\square_{3.7H}$

3.8 Application. Old reals of positive measure:

This is closely related to Judah Shelah [JdSh:308, §1].

3.8A Context and Definition.

Let $S \subseteq \mathcal{S}_{<\aleph_1}(H(\aleph_1))^{V^1}$, $A \stackrel{\mathrm{def}}{=} \bigcup S$ transitive model of ZFC$^-$ and S a stationary subset of $\mathcal{S}_{<\aleph_1}(\bigcup S)$. For $a \in S$ let $\mathbf{g}_a \in {}^{\omega}2$ be random over a, for simplicity: $\mathbf{g}_a \in \bigcup S$ and $\alpha^* = \omega$. For $n < \alpha^*$ we define relation R_n by $f R_n g$ iff: $g \in {}^{\omega}2$, f a sequence of nonempty rational intervals (in our context means $I_\rho = \{\eta \in {}^{\omega}2 : \rho \triangleleft \eta\}$ for some $\rho \in {}^{\omega>}2$) and† $\sum_{\ell<\omega} \mathrm{Lb}\,(f(\ell)) \leq 1$ (where $\mathrm{Lb}(\{\eta \in {}^{\omega}2 : \rho \triangleleft \eta\}) \stackrel{\mathrm{def}}{=} 2^{-\ell g(\rho)}$), and $m \geq n \Rightarrow g \notin f(m)$.

† Lb stands for Lebesgue measure.

3.8B. Claim.

1) If $(\bar{R}, S, \mathbf{g})$ covers *then* it strongly covers (for possibility A).

2) If $(\bar{R}, S, \mathbf{g})$ covers *then* $S \cap {}^{\omega}2$ does not have measure zero (equivalently, it has positive outer measure).

3) If $(\bar{R}, S, \mathbf{g})$ covers *then* "proper + $(\bar{R}, S, \mathbf{g})$ - preserving" is preserved by composition and more generally by CS iteration.

4) If in V, $A \subseteq {}^{\omega}2$ is not null (i.e. does not have Lb measure zero) and $S \subseteq S_{<\aleph_1}(H(\aleph_1))$ is stationary *then* for some $\mathbf{g} = \langle \mathbf{g}_a : a \in S\rangle$, we have: $(\bar{R}, S, \mathbf{g})$-covers and $a \in S \Rightarrow \mathbf{g}_a \in A$.

Proof. 1) We check that Possibility A holds, so we have to check $\oplus_k$. So in V^P let $Q, N, a, N_1, a_1, G^1, p, k, \underset{\sim}{f}_\ell, \beta_\ell, f^*_\ell$ $(\ell < k)$, x, y be given as there (so by 3.8A we have $d[a] \in a$). Let $\langle p_n : n < \omega\rangle$ be such that $p \le p_n \le p_{n+1} \in G^1$, (so p, $p_n \in Q \cap N_1$) and $\bigwedge_{q \in G^1} \bigvee_{n<\omega} q \le p_n$. Let $N_2 \prec (H(\chi_1), \in, <^*_{\chi_1})$ be countable such that

$$\{N_1, \langle p_n : n < \omega\rangle, \langle \underset{\sim}{f}_\ell, f^*_\ell : \ell < k\rangle, x, y\} \in N_2$$

and $a_2 \overset{\text{def}}{=} N_2 \cap \bigcup S \in S$ and $N_2 \in N$. Let $\langle p^n_m : m < \omega\rangle$, $f^*_{\ell,n}$ be such that: $p^n_0 = p_n$, $p^n_m \le p^n_{m+1}$, $\langle p^n_m : m < \omega\rangle$ is a generic sequence for (N_2, Q) and $p^n_m \Vdash$ "$\underset{\sim}{f}_\ell \restriction m = f^*_{\ell,n} \restriction m$"; without loss of generality $\langle f^*_{\ell,n}, p^n_m : \ell < k, n < \omega, m < \omega\rangle \in N$. Clearly for some $m^*_{\ell,n} < \omega$ we have $f^*_{\ell,n} R_{m^*_{\ell,n}} \mathbf{g}_a$. As we can thin the sequence $\langle\langle p_n, p^n_m : n < \omega\rangle : n < \omega\rangle$ as long as it belongs to N without loss of generality for some rational $u_n \in \mathbb{Q}$, $0 \le u < 1$, and $p_n \Vdash \sum\limits_i \mathrm{Lb}(\underset{\sim}{f}_\ell(i)) \in (u^\ell_n, u^\ell_n + 1/k2^{2^n}]$ and $\langle u^\ell_n : n < \omega\rangle \in N$ is strictly increasing and $\langle u^\ell_n + 1/k2^{2^n} : n < \omega\rangle$ is strictly decreasing, and p_n forces a value to $\underset{\sim}{f}_\ell \restriction m_{\ell,n}$ such that $\sum\limits_{i<m_{\ell,n}} \underset{\sim}{f}_\ell(i) > u^\ell_n$ and $\langle m_{\ell,n} : n < \omega\rangle \in N$. So it is forced by p_n that $\underset{\sim}{f}_\ell \restriction m_{\ell,m}$ has the value above, and $\sum\limits_{i \ge m_{\ell,n}} \underset{\sim}{f}(i) < 1/k2^{2^n}$, so $f^*_{\ell,n}$ satisfy this too. For every n we have $\sum_{\ell<k} \sum_{i \ge m_{\ell,n}} \mathrm{Lb}\,(f^*_\ell(i)) < 1/2^{2^n}$ and $p^n_0 \Vdash$ "$\underset{\sim}{f}_\ell \restriction m_{\ell,n} = f^*_\ell \restriction m_{\ell,n}$", hence $\sum\limits_i \sum\{\mathrm{Lb}\,(I_\rho)$: for some $\ell < k$, $n < \omega$ we have $I_\rho = f^*_{\ell,n}(i)\} \le \sum\limits_{\ell<k} \sum\limits_i \mathrm{Lb} f^*_\ell(i) + \sum\limits_\ell \sum\limits_n \sum\limits_{i \ge m_{\ell,n}} \mathrm{Lb} f^*_{\ell,n}(i) \le \sum\limits_\ell \sum\limits_i \mathrm{Lb} f^*_\ell(i) + \sum\limits_\ell \sum_n 1/2^{2^n}$ so this is a sum of two reals $< \infty$ (note that in first sum for each

i it is on the set double appearances not counted). Hence $\mathbf{g}_a$ belongs to only finitely many of the sets $\bigcup\{I_\rho : \text{ for some } \ell < k, n < \omega, I_\rho = \underset{\sim}{f}_{\ell,n}(i)\}$, so the rest is easy.

2), 3) are left to the reader.

4) Straightforward. $\square_{3.8B}$

3.8C Claim.

(1) Assume $S \subseteq \mathcal{S}_{<\aleph_1}(H(\aleph_1))$, and S is stationary as a subset of $\mathcal{S}_{<\aleph_1}(\bigcup S)$, and $\mathbf{g} : S \to {}^\omega 2$ is such that:

$(*)_{S,\mathbf{g}}$ if x, $S \in H(\chi)$ *then* for some countable $N \prec (H(\chi), \in <^*_\chi)$ we have $\{x, S, \mathbf{g}\} \in N$ and $N \cap (\bigcup S) = a \in S$ and $\mathbf{g}_a$ belongs to no measure zero set from N.

Then: if $\langle P_i, \underset{\sim}{Q}_j : i \leq \alpha, j < \alpha\rangle$ is a CS iteration of proper forcing proper notions, each Q_i preserving $(*)_{S^1,\mathbf{g}^1}$ whenever $\bigcup S \in \bigcup S^1$, $(\forall a \in S^1)(a \cap (\bigcup S) \in S)$, $\mathbf{g}^1_a = \mathbf{g}_{a\cap(\bigcup S)}$, this means $V^{P_i} \models$ " if $(*)_{S^1,\mathbf{g}^1}$ then $\Vdash_{Q_i}$ $(*)_{S^1,\mathbf{g}^1}$") then P_α preserves $(*)_{S,\mathbf{g}}$.

(2) Assume $X \subseteq {}^\omega 2$ has positive (outer) Lebesgue measure. If $\langle P_i, \underset{\sim}{Q}_j : i \leq \alpha, j < \alpha\rangle$, is CS iteration of proper forcing, each Q_i preserve the property $(*)_{S,\mathbf{g}}$ whenever $\mathbf{g} : S \to X$, *then* P_α preserves the property of "being of positive outer measure" for $X' \subseteq X$.

Proof. 1) As we can replace $\bar{Q}$ by $\langle P_{\beta+i}/P_\beta, \underset{\sim}{Q}_j : i \leq \alpha-\beta, j < \alpha-\beta\rangle$, and S by $S_1 \subseteq S$ as long as $(*)_{S_1,\mathbf{g}\restriction S_1}$ holds in V^{P_β}, it is enough to prove $\Vdash_{P_\alpha}$ "$(*)_{S,\mathbf{g}}$". Now letting $S^* = \{a \in S : \mathbf{g}_a \text{ is random over } a\}$, clearly $S^* \subseteq S$ is stationary and S^*, $\mathbf{g}\restriction S^*$ fit 3.8A.

We prove by induction on i that

(a) P_i is $(\bar{R}, S^*, \mathbf{g}\restriction S^*)$-preserving (for possibility A) and

(b) $\Vdash_{P_i}$ "$\underset{\sim}{Q}_i$ is $(\bar{R}, S^*, \mathbf{g}\restriction S^*)$-preserving".

Arriving to i, clause (a) holds by 3.8B(3). To prove clause (b) first we deal only with clause (b) of definition 3.4 and, for χ large enough in V^{P_i} we let

$$W = \{N : \text{(i) } N \text{ is a countable elementary submodel of } (H(\chi), \in, <^*_\chi), \text{ to which } S^*, \mathbf{g}, \underset{\sim}{Q}_i \text{ belong and } a = N \cap (\bigcup S) \in S, \text{ and } \mathbf{g}_a \text{ is random over } N$$
$$\text{(ii) for some } p \in \underset{\sim}{Q}_i \cap N \text{ there is no } q \text{ such that } p \leq q \in \underset{\sim}{Q}_i, q \text{ is } (N, \underset{\sim}{Q}_i)\text{-generic and } q \Vdash_{Q_i} \text{“}N[\underset{\sim}{G}_{Q_i}] \text{ is } (\bar{R}, S^*, \mathbf{g}{\restriction}S^*)\text{-good}\}.$$

If (b) fails then W is stationary (otherwise if $\chi' = (2^\chi)^+$, $\{W, \chi\} \in N \prec (H(\chi'), \in, <^*_\chi)$ then for N the required conclusion holds and we clearly finish). For $p \in \underset{\sim}{Q}_i$ let W_p be defined like W with p (in clause (ii)) fixed. So by normality for some p, W_p is stationary. But defining $\mathbf{g}^1 \overset{\text{def}}{=} \langle \mathbf{g}^1_N = \mathbf{g}_{N \cap \bigcup S} : N \in W_p \rangle$, clearly $V^{P_i} \Vdash$ “$(*)_{W_p, \mathbf{g}^1}$” but $V^{P_i} \models$ “ $\Vdash_{Q_i} \neg(*)_{W_p, \mathbf{g}^1}$” contradicting the assumption.

But we have to deal also with clause (b) in the conclusion of Definition 3.4, so define

$$W' = \{N : (i) \text{ as before}$$
$$(ii) \text{ for some } p \in \underset{\sim}{Q}_i \cap N \text{ and } k < \omega \text{ and } \underset{\sim}{f}_\ell \in N (\ell < k)$$
$$N_1, a_1, \langle p_n : n < \omega \rangle, f^*_\ell$$
$$\text{as in } (*) \text{ of Definition 3.4}$$
$$\text{there is no } q \text{ satisfying } (a) + (b) \text{ of Definition 3.4 } \}.$$

Assume toward contradiction that clause (b) here fails, hence $W \subset \mathcal{S}_{\leq \aleph_0}(H(\chi))$ is stationary and w.l.o.g. let $\langle m_{\ell,n} : \ell < k, n < \omega \rangle \in N$ be as in the proof of 3.8B. So for some $x = \langle p, h, \langle p_n : n < \omega \rangle, N_1, a_1, \langle f^*_\ell : \ell < k \rangle, \langle m_{\ell,n} : \ell, n \rangle \rangle$ we have

$$W'_x \overset{\text{def}}{=} \{N \in W' : x \in N \text{ gives a counterexample in (ii) of the Definition of } W'\}$$

is stationary. Let $\chi_1 >> \chi$. In $(V^{P_i})^{\underset{\sim}{Q}_i}$, clearly $\{\mathbf{g}_a : a = N \cap \bigcup S, N \in W'_x\}$ is not null. Also for every club E of $\mathcal{S}_{\leq\aleph_0}(H(\aleph_1)^{V^{P_i}})$ we have: the set $\{\mathbf{g}_a : a \in U_e\}$ is not null where $U_E = \{N \cap H(\chi_1) : N \in W''_x\} \cap E$.

So for some club E^*, the outer measure of $\{\mathbf{g}_a : a \in U_E\}$ is minimal. So really in V^{P_i} we have a $\underset{\sim}{Q}_i$-name $\underset{\sim}{E}^* \in H(\chi_1)$. We can find $\chi_0 < \chi$ large enough such that letting

$$\underset{\sim}{E}' = \{a \cap H(\chi_0) : a \in \underset{\sim}{E}^*\} \text{ and } W''_x = \{N \cap H(\chi_0) : N \in W'_x\}$$

we have all those properties and there are $<^*$-first hence belong to N_1. Replacing N_1, $\langle p_n : n < \omega\rangle$ by N_2, $\langle p'_n : n < \omega\rangle$ by 3.8B, we have $\{W''_x, \chi_1, \underset{\sim}{E}', x\} \in N_1$. So choose $N \in W'_x$.

Now for some n, p_n force outer Lebesgue measure of $\{\mathbf{g}_a : a \in U''_{\underset{\sim}{E}'}\}$ is $> 1/n^*$, $n^* > 0$, and if n is large enough, it forces value to $\underset{\sim}{f}_\ell{\restriction}m$, and force $\sum\limits_{i \geq m} \mathrm{Lb}(\underset{\sim}{f}(i)) < 1/n^*(k+1)$. Let $p_n \in G_{Q_i} \subseteq Q_i$, G_{Q_i} generic over V^{P_i}.

So $E^\otimes = \{N \prec H(\chi) : N[G_{Q_i}] \cap {}^i\mathrm{Ord} \subseteq N\}$ is a club, so restricting ourselves to it does not change the outer measure. Let $N \in U_{\underset{\sim}{E}'} \cap E^\otimes$, then $\bigvee\limits_{\ell} \neg \underset{\sim}{f}_\ell R_{\beta_\ell} g_{N \cap \bigcup S}$. There are $2k$ possibilities: which i, and if bad i is $\geq km_{\ell,n}$ or $< m_{\ell,n}$, later is impossible.

The outer measure of former is $< k1/n^*(k+1) < 1/n^*$, but by the choice of the club $\underset{\sim}{E}^*$ contradiction.

Remark. Really this is part of a quite general theorem. We shall return to it elsewhere.

2) Should be clear. $\square_{3.8C}$

3.9 Application. Souslinity of an ω_1-tree.

Here we return to the issue of IX §4.

3.9A Context and Definition. Let T be an ω_1-tree, say with $[\omega\alpha, \omega\alpha + \omega) \setminus \{0\}$ being the $(1+\alpha)$-th level. Let $W \subseteq \omega_1$ be the set of limit ordinals $\delta = \omega\delta$ (for clarity). Let for $t \in T_\gamma, \beta \leq \gamma, t{\restriction}\beta$ be the unique $s \in T_\beta$, such that $s \leq_T t$. Let for $\delta \in W$, $a_\delta = \delta \cup {}^{\omega>}\delta$, $S = \{a_\delta : \delta \in W\}$, $d[a] = a$, $c[a] = a$ (so $d' = \omega_1$,

$c' = {}^{\omega>}\omega_1$) and $\{t^\delta_n : n < \gamma_\delta \leq \omega\}$ be a subset of T_δ for some non zero $\gamma_\delta \leq \omega$. Let $\alpha^* = \omega_1$ and lastly we choose† $\mathbf{g}_a$ such that: for $\alpha \in a \cap \alpha^*$, we let $f R_\alpha \mathbf{g}_a$ *iff* one of the following holds:

(α) $\alpha = 0$ and $f^{-1}(\{1\}) \cap \{s \in T_{<\delta} : 0 < s <_T t^\delta_{f(0)}\} \neq \emptyset$ or

(β) $0 < \alpha < \delta$ and $f^{-1}(\{1\}) \cap \{s \in T_{<\delta} : t^\delta_{f(0)} \restriction \alpha \leq s$ and $s \neq 0\} = \emptyset$ or

(γ) $\neg(f(0) \in \gamma_\delta)$.

Let $Y = \{t^\delta_n : n < \gamma_\delta, \delta \in W\}$; we say the tree T is Y-Souslin if: for χ large enough, for every $x \in H(\chi)$ for some N we have: x, $T \in N \prec (H(\chi), \in, <^*_\chi)$, N countable, $\delta \stackrel{\text{def}}{=} N \cap \omega_1$, and for $n < \gamma_\delta$, $\{s : s <_T t^\delta_n\}$ is (N, T)-generic. For $W' \subseteq W$ let

$$Y \restriction W' = \{t^\delta_n : n < \gamma_\delta \text{ and } \delta \in W'\}.$$

3.9B Claim. 1) If $\bigwedge_\delta T_\delta = \{t^\delta_n : n < \gamma_\delta\}$ *then* Y-Souslin means Souslin. If T is Y-Souslin then T is not special, even not W-special.

2) If T is a Y-Souslin tree then $(\bar{R}, S, \mathbf{g})$ fully covers (so for any forcing notion Q, if in V^Q the tree T is still Y-Souslin, then $(\bar{R}, S, \mathbf{g})$ still fully covers),

3) If $(\bar{R}, S, \mathbf{g})$ covers then $(\bar{R}, S, \mathbf{g})$ strongly covers for possibility A.

Proof. 1), 2) Straightforward.

3) Clearly each R_α is closed and as $[a \in S \Rightarrow d[a] \notin a]$ we are done.

3.9C Claim. A forcing notion Q is $(\bar{R}, S, \mathbf{g})$-preserving *iff* Q is $(\bar{R}, S, \mathbf{g})$-preserving for possibility A.

Proof.

The "only if" direction.

Let $N \prec (H(\chi), \in, <^*_\chi)$ be countable $(\bar{R}, S, \mathbf{g}) \in N$, and p, $\langle p_n : n < \omega\rangle$, $\langle f^*_\ell : \ell < k\rangle$, $\langle \underset{\sim}{f}_\ell : \ell < k\rangle$, $\langle \beta_\ell : \ell < k\rangle$ be as in Definition 3.4.

We can assume $2^{\aleph_1} < \chi_1 = \mathrm{cf}(\chi_1)$, $2^{\chi_1} < \chi$.

---

† As commented earlier, actually the identity of $\mathbf{g}_a$ does not matter only the sets $R_{\alpha,a} = \{f : f R_\alpha \mathbf{g}_a\}$

Let $w = \{\ell < k : f_\ell(0) < \gamma_\delta \text{ and } \beta_\ell \neq 0\}$. For $\ell \in k \setminus w$ choose $x_\ell \in T \cap N$ such that $x_\ell <_T t^\delta_{f_\ell(0)}$ and $\bigvee_{n<\omega} p_n \Vdash$ "$\underset{\sim}{f}_\ell(x_\ell) = 1$ or $\underset{\sim}{f}_\ell(0) \geq \gamma_\delta$". So for some $n(*) < \omega$,

$$p_{n(*)} \Vdash \text{``} \bigwedge_{\ell \in k \setminus w} [\underset{\sim}{f}_\ell(x_\ell) = 1 \text{ or } \underset{\sim}{f}_\ell(0) \geq \gamma_\delta]\text{''}.$$

Let

$$\mathcal{I} = \{q \in Q : \text{for each } \ell \in w, q \text{ forces a value to } \underset{\sim}{f}_\ell(0), \text{ say } m_\ell, \text{ and it forces a truth value to } (\exists x)(t^\delta_{m_\ell} \restriction \beta_\ell <_T x \ \&\ \underset{\sim}{f}_\ell(x) = 1)\}.$$

So for some $n > n(*)$, we have $p_n \in \mathcal{I}$, so those truth values which it forces are all false (as if $p_n \Vdash (\exists x)(t^\delta_{m_\ell} \restriction \beta_\ell <_T x \ \&\ \underset{\sim}{f}_\ell(x) = 1)$ then for some $n' > n$, $p_{n'}$ forces a specific such x so $F^*_\ell(x) = 1$, contradiction). So any (N, Q)-generic $q \in Q$ which is $\geq p_n$ and satisfies $(*)_q$ below is as required, where

$(*)_q$ for $n < \gamma_{N \cap \omega_1}$, the branch $\{t : t <_T t^\delta_n\}$ of $T \cap N$ is (N, T)-generic.

Its existence follows from "Q is $(\bar{R}, S, \mathbf{g})$-preserving."

The "if" direction.

It is trivial (reread Definition 3.4). $\square_{3.9C}$

3.9 D Definition. We say Q is Y-preserving when: if $N \prec (H(\chi), \in, <^*_\chi)$ countable, $\delta = N \cap W_1$, $\{Y, T\} \in N$, and $p \in Q$ such that $n < \gamma \Rightarrow \{t : t <_T t^\delta_n\}$ is (N, T)- generic, *then* there is q, $p \leq q \in Q$, $q \Vdash$ "$n < \gamma_\delta \Rightarrow \{t : t < t^\delta_n\}$ is $(N[\underset{\sim}{G}_Q], T)$-generic."

3.9 E Fact. Q is Y-preserving iff Q is $(\bar{R}, S, \mathbf{g})$-preserving.

3.9 F Conclusion.

If T is an ω_1 - tree, $Y \subseteq T$ *then* the property "Q is Y-preserving and is proper" is preserved by CS iterations (and composition).

3.10 Application. Being a nonmeager set

3.10A Context and Definition. Let $S \subseteq \mathcal{S}_{<\aleph_1}(H(\chi))$, for $a \in S$, $d[a] = c[a] = {}^{\omega>}\omega$. Let fRg iff (f is a function from ${}^{\omega>}\omega$ to ${}^{\omega>}\omega$ and g a function from ω to ω) & $(\exists^\infty m)[(g{\restriction}m)\char`^ f(g{\restriction}m) \trianglelefteq g]$. Let $\mathbf{g} = \langle \mathbf{g}_a : a \in S\rangle$ where $\mathbf{g}_a \in {}^\omega\omega$, (so $\alpha^* = 1, R = R_0$).

Remark. Note that if N is a model of ZFC$^-$, then: "g is Cohen over N" *iff*

$$(\forall f \in N)(f : {}^{\omega>}\omega \to {}^{\omega>}\omega \Rightarrow (\exists^\infty m)[(g{\restriction}m)\char`^ f(g{\restriction}m) \trianglelefteq g])$$

iff

$$\text{“}(\forall f \in N)(f : {}^{\omega>}\omega \to {}^{\omega>}\omega \Rightarrow (\exists m)[(g{\restriction}m)\char`^ f(g{\restriction}m) \trianglelefteq g])$$

(as N is closed enough).

3.10B Claim. 1) If $(\bar{R}, S, \mathbf{g})$ covers in V *then* it strongly covers in V (by Possibility B, C).

2) In V, if $A \subseteq {}^\omega\omega$ is not meager and S stationary subset of e.g. $\mathcal{S}_{<\aleph_1}(H(\aleph_1))$ *then* for some $\mathbf{g}$ we have $(\bar{R}, S, \mathbf{g})$ covers in V and $\mathbf{g}(a) \in A$ for $a \in S$.

Proof. 1) We can show that in Definition 3.3 Possibility B holds. The winning strategy is in stage n, to choose b_n so large that for $\ell \leq n$, there are at least n members in solutions of $\{m : (\mathbf{g}_a{\restriction}m)\char`^ f^n_\ell(\mathbf{g}_a{\restriction}m) \triangleleft g\}$ are guaranteed (similar to VI §3, because the property has the form $(\exists^\infty m)$) (i.e. G_δ Borel set) (remember $\alpha^* = 1$ so $\oplus_k$ is not needed). The proof for possibility C is similar.

2) Straightforward. $\square_{3.10B}$

3.10C Claim. If $(\bar{R}, S, \mathbf{g})$ covers, then "proper $+(\bar{R}, S, \mathbf{g})$-preserving" is preserved by composition and more generally by CS iterations.

Proof. Remember that $(\bar{R}, S, \mathbf{g})$-preserving means "for possibility C" (the case where Definition 3.4 is more transparent). Now use 3.6. $\square_{3.10C}$

3.10D Claim.

(1) Assume $S \subseteq \mathcal{S}_{<\aleph_1}(H(\aleph_1))$, and S is stationary as a subset of $\mathcal{S}_{<\aleph_1}(\bigcup S)$, and $\mathbf{g} : S \to {}^\omega 2$ is such that:

$(*)_{S,\mathbf{g}}$ if x, $S \in H(\chi)$ *then* for some countable $N \prec (H(\chi), \in <^*_\chi)$ we have $\{x, S, \mathbf{g}\} \in N \cap (\bigcup S) \in S$ and $\mathbf{g}_a$ belongs to no meagre set from N.

Then: if $\langle P_i, \underset{\sim}{Q}_j : i \leq \alpha, j < \alpha \rangle$ is a CS iteration of proper forcing proper notions, each Q_i preserving $(*)_{S^1,\mathbf{g}^1}$ whenever $\bigcup S \in \bigcup S^1$, $(\forall a \in S^1)(a \cap (\bigcup S) \in S)$, $\mathbf{g}^1_a = \mathbf{g}_{a \cap (\bigcup S)}$ (and $(*)_{S^1,\mathbf{g}^1}$ is defined as in part (1); this means $V^{P_1} \models$ " if $(*)_{S^1,\mathbf{g}^1}$ then $\Vdash_{Q_i} (*)_{S^1,\mathbf{g}^1}$") then P_α preserve $(*)_{S,\mathbf{g}}$.

(2) Assume $X \subseteq {}^\omega 2$ is not meagre. If $\langle P_i, \underset{\sim}{Q}_j : i \leq \alpha, j < \alpha \rangle$, is CS iteration of proper forcing, each Q_i preserves the property $(*)_{s,\mathbf{g}}$ whenever $\mathbf{g} : S \to X$, *then* P_α preserves the property of "being of not meagre" for $X' \subseteq X$.

Proof. Like 3.8C.

3.10E Claim. If $(\bar{R}, S, \mathbf{g})$ covers in V and Q is a forcing notion which is Souslin-proper in any extension (i.e., we have a Souslin definition which in any generic extension is Souslin-proper) and $\Vdash_Q$ "$V \cap {}^\omega 2$ is not meager" (in every extension) *then* in V^Q we have: $(\bar{R}, S, \mathbf{g})$ still covers and Q is $(\bar{R}, S, \mathbf{g})$-preserving.

Proof. It follows from Lemma 3.11 below.

3.11 Lemma. [Goldstern and Shelah] Assume that Q is a Souslin proper forcing, say definable with a real parameter r^*, with the property

$$\Vdash_Q \text{"}V \cap {}^\omega 2 \text{ is not meager"}$$

and continues to have these properties in any extension of V (by set forcing). If $N \prec (H(\chi), \in, <^*_\chi)$ countable, x_0 a Cohen real over N and $p \in N \cap Q$, then there exists a condition $q \geq p$, q is (N, Q)-generic (i.e. (N, Q^V)-generic), and $q \Vdash$ "x_0 is Cohen over $N[\underset{\sim}{G}_Q]$."

We will prove this through a sequence of lemmatas. We always assume that Q is a forcing notion satisfying the assumptions of our lemma, N is a countable elementary submodel of some $(H(\chi), \in)$ (χ big enough, regular), M is a countable transitive model satisfying a large enough fragment of ZFC. We let $\lambda \in N$ be a regular cardinal that is reasonably big (say $\lambda > \beth_2$) but still small compared to χ, say $2^{2^\lambda} < \chi$.)

3.11A. Fact. Assume B is a complete Boolean algebra, $B_0 \subseteq B$ a complete subalgebra and $\{B, B_0\} \in N$. If $G_0 \subseteq B_0$ is N-generic, *then* there exists an N-generic filter $G \subseteq B$ extending G_0.

Proof. Easy.

3.11B. Fact. Assume $B \in N$ is a forcing notion, $x_0 \in {}^\omega 2$ a Cohen real over N. Assume $\underset{\sim}{c}$ is a B-name such that

$$\Vdash_B \text{“}\underset{\sim}{c} \text{ is a Cohen real over } V\text{”}$$

Then there is a N-generic filter $G_B \subseteq B$ such that $\underset{\sim}{c}[G_B]$ is almost equal to x_0.

Proof. Without loss of generality we assume that B is a complete boolean algebra. For any formula φ in the forcing language of B we write $[\![\varphi]\!]$ for the Boolean value of φ. We write $[\![\varphi]\!] = 0$ if $\Vdash_B \neg\varphi$. Assume that $x_0 \in {}^\omega 2$ is Cohen-generic over N, and $\underset{\sim}{c} \in {}^\omega 2$ is forced to be Cohen-generic over V. Let

$$T := \{\eta \in {}^{\omega>}2 : [\![\eta \subseteq \underset{\sim}{c}]\!] \neq 0\}$$

Then T is a tree, and $\Vdash_B$ “$\underset{\sim}{c} \in \lim T$”. So $\mathrm{Lim} T$ cannot be nowhere dense, so for some $\eta_0 \in T$ we must have $(\forall \eta)(\eta_0 \triangleleft \eta \in {}^{\omega>}2 \Rightarrow [\eta \in T])$. For notational simplicity only we assume $\eta_0 = \emptyset$ (otherwise we have to consider $\underset{\sim}{c}\restriction[\ell g(\eta_0), \omega)$ and $x_0 \restriction [\ell g(\eta_0), \omega)$ instead of $\underset{\sim}{c}$ and x_0).
Let $B_0 \subseteq B$ be the complete Boolean algebra generated by the elements $[\![\eta \subseteq \underset{\sim}{c}]\!]$, where η ranges over ${}^{\omega>}2$. Then B_0 is a complete subalgebra of B, and the map that sends $\eta \in {}^{\omega>}2$ to $[\![\eta \subseteq \underset{\sim}{c}]\!]$ is a dense embedding of ${}^{\omega>}2$ into B_0. Thus x_0 induces an (N, Q^N)-generic filter $G_0 \subseteq B_0$. By 3.11A, G_0 can be extended to an N-generic filter $G \subseteq B$. Clearly $\underset{\sim}{c}[G] = x_0$, as for every $n \in \omega$, letting $\eta := x_0 \restriction n$, we have $[\![\eta \subseteq \underset{\sim}{c}]\!] \in G_0 \subseteq G$. $\square_{3.11B}$

3.11C. Lemma. The formula "$M \subseteq \omega\times\omega$ codes a well founded model of ZFC$^-$ with universe ω, q is (M,Q)-generic, and $q \Vdash_Q$ x is Cohen over $M[G_Q]$" is equivalent to a Π^1_1-formula (about x, q, and M as parameters). (M-generic means M'-generic, where M' is the transitive collapse of M. We will not notationally distinguish between M and M'.)

Proof. First we note that "q is (M,Q)-generic" is a Π^1_1-statement, as it is equivalent to

for every $A \in M$ such that $M \models$ "A is pre-dense in Q", and

for every $r \geq q$ there is $a \in M, M \models$ "$a \in A$", and a, r are compatible.

(Recall that in a Souslin forcing notion the compatibility relation is Σ^1_1 and Π^1_1.)

If q is (M,Q)-generic, then we have: $q \Vdash$ "x is Cohen over $M[G]$" iff for all $\underset{\sim}{T} \in M$ such that $M \models$ "$\underset{\sim}{T}$ is a Q-name of a nowhere dense tree $\subseteq {}^{\omega>}2$", and *for all* $r \geq q$ there exists a condition $p' \in M \cap Q$ and a natural number n such that p', r are compatible, and $M \models$ "$p' \Vdash x{\restriction}n \notin \underset{\sim}{T}$". Again it is easy to see that this can be written as a Π^1_1-statement. $\square_{3.11C}$

3.11D. Lemma. Assume M is as above, $p \in Q \cap M$, A a comeager Borel set. Then there exist a real $x \in A$ and a condition $q \geq p$ such that q is (M,Q)-generic, and $q \Vdash$ "x is Cohen over $M[\underset{\sim}{G}_Q]$."

Proof. Let $q_0 \geq p$ be (M,Q)-generic. Work in $V[G]$, where $q_0 \in G \subseteq Q$, G is generic over V. Since $V \cap 2^\omega$ is not meager (in $V[G]$), $A^{V[G]}$ is comeager, and the union of all meager sets coded in $M[G]$ is meager, we can find $x \in V \cap A^{V[G]}$ which is Cohen over $M[G]$, by absoluteness $x \in A$. Now let $q \geq q_0$ be a condition which forces this. $\square_{3.11D}$

3.11E. *Proof of the Lemma 3.11:* Recall that $\lambda \in N$ is much bigger than ω, but much smaller than χ. Let $M \overset{\text{def}}{=} (H(\lambda), \in)$. So $M \in N$.
Let B be the algebra that collapses $H(\lambda)$ to a countable set (using finite conditions) i.e. Levy$(\aleph_0, |H(\lambda)|)$. Clearly $\Vdash_B$"M is a countable model of ZFC$^-$." We assert that

$(*) \Vdash_B$ "There exists x (Cohen over V) and $q \in Q$, q is (M, Q)-generic, and $\Vdash_Q$ x is Cohen over $M[G_Q]$".

To prove this assertion, work in V^B. The set of all closed nowhere dense sets coded in V is now countable, so the set of Cohen reals over V is comeager, and hence contains some comeager Borel set A. Now apply the previous lemma 3.11D. This finishes the proof of the assertion $(*)$.

From the assertion we can get names $\underset{\sim}{x}$ and $\underset{\sim}{q}$ such that all the above is forced by the trivial condition of B. Clearly we can assume that $\underset{\sim}{x}$ and $\underset{\sim}{q}$ are in N.

Now apply Fact 3.11B to get an N-generic filter $G \subseteq B$ (in V!) such that $x \stackrel{\text{def}}{=} \underset{\sim}{x}[G]$ is almost equal to x_0. Let $q \stackrel{\text{def}}{=} \underset{\sim}{q}[G]$. Then

$$N[G] \models \text{"}q \text{ is } (M, Q)\text{-generic, and } q \Vdash_Q c \text{ is Cohen over } M[\underset{\sim}{G}_Q]\text{"}$$

and $N[G] \cap \mathrm{Ord} = N \cap \mathrm{Ord}$.

Since Π^1_1-formulas are absolute, we can replace $N[G]$ by V (remember $N[G] \subseteq V$). We can also replace x by x_0, since modifying a Cohen real in finitely many places still leaves it a Cohen real. Thus,

$$V \models \text{"}q \text{ is } (M \cap N, Q)\text{-generic, and } q \Vdash_Q x_0 \text{ is Cohen over } M[\underset{\sim}{G}_Q].\text{"}$$

(Why $M \cap N$ and not M? As we should look at M as interpreted in $N[G]$, note $N[G] \models$ "M is countable"). As $M \cap N$ and N have the same dense sets of Q, q is (M, Q)-generic iff it is N-generic. Similarly, x_0 is Cohen over $M[G]$ iff it is Cohen over $N[G]$, so we are done. $\square_{3.11}$

Remark. We shall deal with more general theorems in [Sh:630].

3.12 Concluding Remarks. 1) We may consider the following variant of this section's framework concentrating on $d[a] \in a$ (of course if $R_{\alpha,s} = R_\alpha$ we get back the previous version).

(A) We replace R_α by $R_{\alpha,t}$ for $t \in \mathbb{Q}$ such that $s < t \Rightarrow R_{\alpha,s} \subseteq R_{\alpha,t}$; we may use $R_\alpha = \bigvee_{s \in \mathbb{Q}} R_{\alpha,s}$.

(B) N is $(\bar{R}, S, \mathbf{g})$-good *iff* $a \stackrel{\text{def}}{=} N \cap (\bigcup S) \in S$ and for every $f \in N$ satisfying $f \in {}^{d[a]}c[a]$ for some $\alpha < \alpha^*$ and $t \in \mathbb{Q}$ we have $f R_{\alpha,t} g$.

(C) "Strongly covers" is defined as before except that $\oplus_k$ is changed parallely to the change in (D) below, i.e. I.e.

$\oplus_k$ if (a) – (d) of $(*)$ of $\oplus_k$ from Definition 3.3 and

(e) $\beta_\ell \in a \cap \alpha^*$, $t_\ell < s_\ell$ are rationals and $\underset{\sim}{f}_\ell[G^1] R_{\alpha,t} \mathbf{g}_a$

then for any $y \in N \cap H(\chi)$ there are N_2, G_2 satisfying (the parallel of) clause (d) such that $y \in N_2$ and: for some $\gamma_\ell \in a$, $\gamma_\ell \le \beta_\ell$, $s'_\ell \in Q$, $s'_\ell \le s_\ell$ (for $\ell < k$) we have $\underset{\sim}{f}_\ell[G_2] R_{\gamma_\ell, s'_\ell} \mathbf{g}_a$ (for $\ell < k$).

(D) Q is $(\bar{R}, S, \mathbf{g})$-preserving means: if $(*)$ of Definition 3.4 holds (having now $f^*_\ell R_{\beta_\ell, t_\ell} \mathbf{g}_a$ and $s_\ell > t_\ell$, $s_\ell \in \mathbb{Q}$) *then* there is an (N, Q)-generic q, $p \le q \in Q$ such that $q \Vdash_Q$ "for $\ell < k$ there are $\gamma_\ell \le \beta_\ell$ and $s'_\ell \in \mathbb{Q}$ such that $[\gamma_\ell = \beta_\ell \Rightarrow s'_\ell \le s_\ell]$ and $\underset{\sim}{f}_\ell R_{\gamma_\ell, s'_\ell} \mathbf{g}_a$.

2) If we have $\oplus^*_k$ of 3.3B(3) *then* (b) of the conclusion in Definition 3.4 can be omitted (as it follows from "q is (N, Q)-generic" under the circumstances).

3) Another variant of our framework is as follows.

(α) Let R be a definition of a forcing notion, i.e. partial order, $\{\mathcal{I}_y : y \in Y\}$ be a definition of a family of dense subsets of it (e.g. all), all absolute enough, K be a definition of a family of forcing notions closed under CS iterations (so e.g. if $Q_\ell \in K^{V_\ell}$, $V_{\ell+1} = V_\ell^{Q_\ell}$ then $Q_1 * \underset{\sim}{Q}_2 \in K^{V_1}$, similarly for limit. We have: if in $V_0^{Q_0}$, $p \le q \in R$, $y \in Y^{(V^{Q_0})}$, $p \in \mathcal{I}_y$ *then* this holds in $V_0^{Q_0 * \underset{\sim}{Q}_1}$).

(β) $S \in V_0$, S a stationary subset of $\mathcal{S}_{<\aleph_1} H(\chi^*)^V$

(γ) for $a \in S$, $\mathbf{g}_a$ is a directed subset of $R \cap a$ not disjoint to $a \cap \mathcal{I}_y$ for $y \in Y \cap a$ (absolute as in (α)).

(δ) in $V_0^{Q_0}$, N is $(R, S, \mathbf{g})$-good if: $N \prec (H(\chi), \in, <^*_\chi)$ is countable, $(\bar{R}, S, \mathbf{g}) \in N$, $a \stackrel{\text{def}}{=} N \cap H(\chi^*)^{V_0} \in S$ and $[y \in N \cap Y^{V^{Q_0}} \Rightarrow \exists p \in \mathcal{I}_y \exists q \in \mathbf{g}_a (p \le q)]$ (of course $\mathbf{g}_a$ is still a directed subset of $R^{V^{Q_0}} \cap a$).

3.13 Preservations connected to Norms.

3.13A Context and Definitions.

1) Assume we have $(\bar{n}^*, \text{n\=or}, \bar{w})$ where

   (α) $\bar{n}^* = \langle n_i^* : i < \omega \rangle$ is strictly increasing.

   (β) $\text{n\=or} = \langle \text{nor}_i : i < \omega \rangle$, where $\text{nor}_i : \mathcal{P}([n_i^*, n_{i+1}^*)) \to \omega$ satisfies:

$$u_1 \subseteq u_2 \subseteq [n_i^*, n_{i+1}^*) \Rightarrow \text{nor}_i(u_1) \leq \text{nor}_i(u_2)$$

$$\text{nor}_i([n_i^*, n_{i+1}^*)) > 0$$

$$\langle \text{nor}_i([n_i^*, n_{i+1}^*)) : i < \omega \rangle \text{ converge to infinity}$$

   (γ) $\bar{w} = \langle w_i : i < \omega \rangle$ where $\text{Dom}(w_i) = \omega$, $\text{Rang}(w_i) \subseteq \{U : U \subseteq \mathcal{P}([n_i^*, n_{i+1}^*))$ is downward closed, $U \neq \emptyset\}$ or even† $\text{Dom}(w_i) = {}^{i+1}\omega$ and for every $u \subseteq [n_i^*, n_{i+1}^*)$ with $\text{nor}_i(u) > 0$ and $x \in \omega$ (or $\bar{x} \in {}^{i+1}\omega$ otherwise) for some $u' \subseteq u$ we have $u' \in w_n(x)$ and $\text{nor}_i(u') \geq \text{nor}_i(u) - 1$ (if $\bar{w}$ is omitted then $w_i(x)$ is: let $x(i)$ code $w_{x(i)} \subseteq \mathcal{P}([n_i^*, n_{i+1}^*))$, now let $w_i(x) = w_{x(i)}$ if $\langle w_i(x) : i < \omega \rangle$ is O.K. and let $w_i(x) = \mathcal{P}([n_i^*, n_{i+1}^*))$ otherwise.

2) Let

$$\prod_{i<\omega}^{*} \mathcal{P}([n_i^*, n_{i+1}^*)) = \{\bar{t} : t_i \subseteq [n_i^*, n_{i+1}^*) \text{ and } \langle \text{nor}_i(t_i) : i < \omega \rangle \text{ converge to infinity } \text{nor}_i(t_i) > 0\}$$

3) We define two partial orders on $\prod_{i<\omega}^{*} \mathcal{P}([n_i^*, n_{i+1}^*))$:

$$\bar{t} \leq \bar{s} \text{ iff } t_i \supseteq s_i \text{ for every } i < \omega$$

$$\bar{t} \leq^* \bar{s} \text{ iff } t_i \supseteq s_i \text{ for every } i < \omega \text{ large enough}$$

   (note that $\leq$ is a partial order, $\leq^*$ is a partial order such that every increasing ω-chain has an upper bound)

---

† Instead of $\omega = \text{Dom}(w_i)$ or ${}^i\omega = \text{Dom}(w_i)$ we can use other finite or countable set.

4) We call $\Gamma \subseteq \prod^*_{i<\omega} \mathcal{P}([n^*_i, n^*_{i+1}))$ a nice set if:

(α) Γ is $\leq^*$-directed

(β) every $\leq^*$-increasing ω-chain in Γ has an upper bound in Γ

(γ) for every $\bar{x} \in {}^\omega\omega$, for some $\bar{t} \in \Gamma$ we have $t_i \in w_i(x_i)$ (or $t_i \in w_i(\bar{x}\restriction(i+1))$) for every $i < \omega$ large enough.

3.13B Fact.

1) $(\prod^*_{i<\omega} \mathcal{P}([n^*_i, n^*_{i+1})), \leq)$ is a partial order.

2) $(\prod^*_{i<\omega} \mathcal{P}([n^*_i, n^*_{n+1})), \leq^*)$ is a partial order with any $\leq^*$-increasing ω-chain having an upper bound.

3) $\bar{t} \leq \bar{s} \Rightarrow \bar{t} \leq^* \bar{s}$.

4) If CH (or just MA) *then* there exists a nice Γ and hence S, $\mathbf{g}$ as in 3.13C (2), (3) below exist.

Proof. Straightforward.

We want to show that niceness of Γ is preserved under limit of CS proper iteration

3.13C Context and Definition.

1) Let Γ be nice in a universe V_0,

2) $S \subseteq \mathcal{S}_{<\aleph_1}(H(\aleph_1))$ be stationary,

3) $\mathbf{g} : S \to \Gamma$ be such that: $\mathbf{g}_a = \langle \mathbf{g}_{a,i} : i < \omega \rangle \in \prod^*_{i<\omega} \mathcal{P}([n^*_i, n^*_{i+1}))$ and

(α) for every $x \in ({}^\omega\omega) \cap a$ we have $(\forall^* i < \omega)[\mathbf{g}_{a,i} \in w_i(x_i)]$

(β) for $a_1 \in a_2$ from S we have $\mathbf{g}_{a_1} <^* \mathbf{g}_{a_2}$ (can ask that moreover $\mathbf{g}_{a_1} \in a_2$)

4) $d[a] = c[a] = \omega$

5) $\bar{R} = \langle R_n : n < \omega \rangle$ and $x R_n \mathbf{g}_a$ mean $(\forall i < \omega)[i \geq n \to \mathbf{g}_{a,i} \in W_i(x_i)]$

3.13D Claim.

1) $(\bar{R}, S, \mathbf{g})$ is as in 3.1, it covers (in V, see 3.2, we are assuming 3.13C of course)

2) If in V^P, we have "$(\bar{R}, S, \mathbf{g})$ covers" then it also strongly covers by Possibility A*

3) If $(\bar{R}, S, \mathbf{g})$ cover in V^P, Q is a proper forcing notion preserving "Γ is nice" *then* Q is $(\bar{R}, S, \mathbf{g})$-preserving.

Proof: 1) Read definitions.

2) Check (for $\oplus_1$, can apply 3.4B and the proof of $\oplus^+$ from there in the proof of part (3) below).

3) We use 3.4A(3), so the least trivial condition is $\otimes^+$ from there. Let V^P, N, a_1, a_2, f, $\langle f_n : n < \omega \rangle$, α, $\langle \alpha_n : n < \omega \rangle$ be as there. We can find a finite $d \subseteq \omega$ such that:

$(*)$ if $\ell \in \omega \setminus d$ then $\mathbf{g}_{a_1,\ell} \supseteq \mathbf{g}_{a_2,\ell}$,

w.l.o.g. also $\{0, \dots, \alpha - 1\} \subseteq d$, $\{0, \dots, \alpha_0 - 1\} \subseteq d$ (remember $\alpha < \alpha^* = \omega$). Let $k_i \geq i$ be maximal such that $f_i \restriction k_i = f \restriction k_i$, so $\lim\limits_{i \to \infty} k_i = \infty$.

Also w.l.o.g. $\alpha_i > k_i > \sup(d)$ and we can find an infinite $A \subseteq \omega$ such that $\langle [k_i, \alpha_i) : i \in A \rangle$ are pairwise disjoint, and w.l.o.g. $A \in N$ and $n \in [\min A, \omega) \Rightarrow f_n(0) = f(0)$. Now define $g \in {}^\omega\omega$ by:

$$g \restriction [k_i, \alpha_i) = f_i \restriction [k_i, \alpha_i) \text{ for } i \in A \text{ and } g \restriction (\omega \setminus \bigcup_{i \in A} [k_i, \alpha_i)) \subseteq f,$$

so clearly $g \in {}^\omega\omega \cap N$, hence $\bigvee\limits_k g R_k \mathbf{g}_{a_2}$ so w.l.o.g. $\ell \in \omega \setminus d \Rightarrow \mathbf{g}_{a_2,\ell} \in w_\ell(g(\ell))$. Omitting finitely many members of A we can assume $i \in A \Rightarrow d \subseteq k_i$ and hence $f_i \restriction d = f \restriction d$. We will show that any $i \in A$, $d_i = \{0, \dots, \alpha_i - 1\}$ are as required in $\oplus^+$, so assume $f' \in {}^{d[a_2]}c[a_2] = {}^\omega\omega$, $f' \restriction d_i = f_n \restriction d_i$ and $f' R_{\alpha_i} \mathbf{g}_{a_1}$. So let $\ell \in \omega \setminus \alpha$, and we should prove $\mathbf{g}_{a_2,\ell} \in w_\ell(f'(\ell))$, thus proving $f' R_\alpha \mathbf{g}_{a_2}$ and finishing the proof of $\oplus^+$; we divide this to cases.

case 1: $\ell \notin d_i$

So $\ell \geq \alpha_i$ and we know $f' R_{\alpha_i} \mathbf{g}_{a_1}$ hence $\mathbf{g}_{a_1,\ell} \in w_\ell(f'(\ell))$; but also $\ell \notin d$ hence (see $(*)$ above) $\mathbf{g}_{a_1,\ell} \supseteq \mathbf{g}_{a_2,\ell}$, and $w_\ell(f'(\ell))$ is downward closed so $\mathbf{g}_{a_2,\ell} \in w_\ell(f'(\ell))$ as required.

case 2: $\ell \in d_i \setminus \{0, \dots, k_i - 1\}$

So $k_i \leq \ell < \alpha_i$. As $\ell \in d_i$ we know $f'(\ell) = f_i(\ell)$ and $f_i(\ell) = g(\ell)$, but as $\ell \notin d$ we have $\mathbf{g}_{a_2,\ell} \in w_\ell(g(\ell))$, together we finish.

case 3: $\ell < k_i$

But $f{\restriction}k_i = f_i{\restriction}k_i = f'{\restriction}k_i$, hence $f'(\ell) = f(\ell)$, and as $f R_\alpha \mathbf{g}_{a_2}$ we are done. So we have finished checking the condition $\otimes^+$ of 3.4A(3), thus proving 3.13D(3). $\square_{3.13D}$

3.13E Conclusion. If Γ is nice, $\bar{Q} = \langle P_j, \underset{\sim}{Q}_i : j \le \delta, i < \delta\rangle$ is a CS iteration of proper forcing, each Q_i preserves the niceness of Γ *then* P_δ preserves the niceness of Γ.

3.13F Remark. Similarly for the other variants in VI 0.1, for pure preserving.

3.14 Example (of 3.13).

3.14A Context. We work inside the subcontext of 3.13.

Let $\bar{n}^* = \langle n^*_i : i < \omega\rangle$, $n^*_i << m^*_i << k^*_i << n^*_{i+1}$.

By renaming we replace $[n^*_i, n^*_{i+1})$ by $c^*_i = c^{t_i} = \{(\ell_1, \ell_2) : \ell_1, \ell_2 \in [n^*_i, k^*_i)\}$, so we consider subsets of C^*_i only, but actually can consider instead $e \in E_i$ only where:

$$E_i \overset{\text{def}}{=} \{e : e \text{ an equivalence relation on } [n^*_i, k^*_i) \text{ and each equivalence class has exactly } n^*_i + 1 \text{ elements, except possibly one.}\}$$

For $e \in E_i$ we let $\mathrm{Dom}(e) = \bigcup\{x/e : |x/e| = n^*_i + 1\}$. To make it fit we identify e with

$$s_e = \{(\ell_1, \ell_2) : \ell_1 \in [n^*_i, k^*_i) \text{ and } \ell_2 \in [n^*_i, k^*_i) \text{ and } \neg(\ell_1 e \ell_2)\},$$

we will not continue to mention the minor changes; now we let

$$\mathrm{nor}^{t_i}(e) = \log_2\log_2(k^*_i - n^*_i - |\mathrm{Dom}(e)|)/m^*_i$$

rounded (to maximal natural number $\le$ than this or zero if it is negative). For $\bar{x} = \langle x_j : j \le i\rangle$ we define $w_i(\bar{x})$: we consider x_i as (being or just coding) a pair (f_{x_i}, A_{x_i}), where $A_{x_i} \subsetneq \omega$ finite non empty and $f_{x_i} : [n^*_i, n^*_{i+1}) \to$

$({}^{A_{x_i}}\{0,\dots,n_i^*\})$ (so $a \in A_{x_i} \Rightarrow f_{x_i}[a] : [n_i^*, n_{i+1}^*) \to \{0,\dots,n_i^*\}$)

$$w_i(\bar{x}) = \Big\{ e \in E_i : \text{ for } \geq (n_i^*)^{n_i^*+2} \text{ equivalence classes } u \text{ of } e, \\ 1 - \frac{1}{(\log_2(n_i^*))^{x_0}} \leq \frac{|\{a \in A_{x_i} : (f^i_{x_i}[a]){\restriction}u \text{ is not one to one }\}|}{|A_{x_i}|} \Big\}.$$

We should check

$(*)_1$ if (i is large enough and) $e_0 \in E_i$ and $\mathrm{nor}^{t_i}(e_1) \geq \ell + 1$ and $\bar{x} = \langle x_j : j \leq i+1\rangle$ as above *then* for some $e_2 \in E_i$, $s_{e_2} \subseteq s_{e_0}$, $\mathrm{nor}^{t_i}(e_2) \geq \mathrm{nor}^{t_i}(e_0) - 1$ and $e_2 \in w_i(\bar{x})$.

[Why $(*)_1$ holds? Choose by induction on $m < n^* \overset{\text{def}}{=} (n_i^*)^{(n_i^*+2)}$ a set $u_m \subseteq [n_i^*, n_{i+1}^*)$ satisfying $|u_m| = n_{i+1}^*$ and u_m disjoint to $\bigcup\limits_{m'<m} u_{m'} \cup \bigcup\{x/e_1 : |x/e_1| = n_i^* + 1\}$ and:

$$1 - \frac{2}{n_i^*+2} \leq |\{a \in A_{x_i} : (f_{x_i}[a]){\restriction}u_n \text{ is not one to one}\}|/|A_{x_i}|.$$

(Why? If $v \subseteq [n_i^*, n_{i+1}^*)$, $|v| = n_i^* + 2$ is disjoint to the set above then for each $a \in A_{x_i}$,

$$|\{u \in [v]^{n_i^*+1} : (f_{x_i}[a]){\restriction}u \text{ not one to one}\}| \geq (1 - \frac{2}{n_i^*+2}) \times |[v]^{n_i^*+1}|,$$

so by the "finitary Fubini", some $u \in [v]^{n_i^*+1}$ is (much more than) as required, increasing v we get better estimates.)

Let $e_2 \in E_i$ be such that: the set of e_2-equivalence classes of cardinality $n_i^* + 1$ is

$$\{x/e_1 : |x/e_1| = n_i^* + 1\} \cup \{u_m : m < (n_i^*)^{n_i^*} + 2\}.$$

Now

$$\begin{aligned}\mathrm{nor}^{t_i}(e_2) &= \log_2\log_2(k_i^* - n_i^* - |\mathrm{Dom}(e_2)|)/m_i^* \\ &= \log_2\log_2(k_i^* - n_i^* - |\mathrm{Dom}(e_1)| - n_i^*(n_i^*)^{n_i^*+2})/m_i^* \\ &= \log_2\log_2((k_i^* - n_i^* - |\mathrm{Dom}(e_1)|) \times (1 - \frac{n_i^*(n_i^*)^{n_i^*+2}}{k_i^* - n_i^* - |\mathrm{Dom}(e_1)|}))/m_i^* \\ &= \log_2[\log_2(k_i^* - n_i^* - |\mathrm{Dom}(e_1)|) + \log_2(1 - \frac{n_i^*(n_i^*)^{n_i^*+2}}{k_i^* - n_i^* - |\mathrm{Dom}(e_1)|})]/m_i^*\end{aligned}$$

but as $\mathrm{nor}^{t_i}(e_0) > 0$, necessarily

$$k_i^* - n_i^* - |\mathrm{Dom}(e_1)| \geq 2^{2^{m_i^*}} >> n_i^*(n_i^*)^{n_i^*+1},$$

hence

$$\log_2(1 - \frac{n_i^*(n_i^*)^{n_i^*+2}}{k_i^* - n_i^* - |\mathrm{Dom}(e_1)|}) \geq -1/n_i^*.$$

Hence $\mathrm{nor}^{t_i}(e_2) \geq \mathrm{nor}^{t_i}(e_0) - 1$.]

Moreover, the proof gives

$(*)_2$ if $e_0 \in E_i$, $\mathrm{nor}^{t_i}(e_0) \geq \ell + 1$ and X is a set of n_i^* (or less) $(i+1)$-tuples $\bar{x} = \langle x_j : j \leq i+1 \rangle$ as above then for some $e_2 \in E_i$, $s_{e_2} \subseteq s_{e_1}$ and $\mathrm{nor}^{t_i}(e_2) \geq \mathrm{nor}^{t_i}(e_1) - 1$ and $\bigwedge_{\bar{x} \in X} e_2 \in w_i(\bar{x})$

[Why? We define above u_m for $m < ((n_i^*)^{n_i^*+2}) \times |X|$ dealing with each $x \in X$ by u_m, $(n_i^*)^{n_i^*+2}$ times. As $|X| \leq n_i^*$ there is no problem.] $\square_{3.14A}$

Remark. 1) Think first for the case A'_x a singleton.
2) $(\log(n_i^*))^{x_0+2}$ serves as the $f(-,-)$ in [RoSh:470]

3.14B Claim. If the forcing P preserve "Γ is nice" then there is no Cohen real over V in V^P.

Proof. For this the case A_{x_i} is a singleton suffices. If $\eta \in {}^\omega\omega$ is Cohen over V then

$$(\forall \bar{s} \in \Gamma)(\exists^\infty i)(\eta \text{ is not 1-to-1 on any equivalence class from } s_i)$$

(better look at $\{\eta \in {}^\omega\omega : \ell < n_{i+1} \Rightarrow \eta(\ell) \leq n_i^*\}$) $\square_{3.14B}$

3.14C Claim. Random real forcing preserves a nice Γ.

Proof. Let $p \in$Random be such that $p \Vdash$ "$\underset{\sim}{\bar{x}} = \langle \underset{\sim}{x}_\ell : \ell < \omega \rangle \in {}^\omega\omega$". W.l.o.g. $p \Vdash$ "$\underset{\sim}{x}_0 = x_0$, and $\underset{\sim}{x}_i = (\underset{\sim}{f}_{x_i}, \underset{\sim}{A}_{x_i})$, $\emptyset \neq \underset{\sim}{A}_{x_i} \subseteq \omega$ finite, $\underset{\sim}{f}_{x_i} \in {}^{\underset{\sim}{A}_{x_i}}\{0, \dots, n_i^*\}$".

As Random forcing is ${}^\omega\omega$-bounding w.l.o.g. $p \Vdash_Q$ "$|\underset{\sim}{A}^i_x| \le \ell_i$", where $\langle \ell_i : i < \omega \rangle \in V \cap {}^\omega(\omega \setminus \{0\})$ and as we can replace A by any $A \times B$ w.l.o.g. $\underset{\sim}{A}_{x_i} = A^*_i$ (not name). Now define g_i, $\mathrm{Dom}(g_i) = [n^*_i, n^*_{i+1}) \times A^*_i \times \{0, \dots, n^*_i\}$, as follows: if $m \in [n^*_i, n^*_{i+1})$, $a \in A^*_i$ and $\ell \in \{0, \dots, n^*_i\}$ then

$$g_i(m, a, \ell) \stackrel{\mathrm{def}}{=} \mathrm{Lb}\Big(\text{maximal } q \ge p \text{ forcing } (\underset{\sim}{f}^i_{\underset{\sim}{x}_i}[a](m)) = \ell)\Big) / \mathrm{Lb}(\lim p)$$

W.l.o.g. $p = \lim(T)$ where T is a closed subtree of ${}^{\omega>}2$ and we can choose for each $i < \omega$, a natural number t_i large enough so that from $\eta \in p \cap T \cap {}^{t_i}2$ we can read $\underset{\sim}{f}_{\underset{\sim}{x}_i}$ that is for any $\eta \in T \cap {}^{t_i}2$ we have $\lim(T^{[\eta]})$ force a value to $\underset{\sim}{f}_{x_i}$ (where $T^{[\eta]} = \{\nu \in T : \nu \trianglelefteq \eta \ \vee \ \eta \trianglelefteq \eta\}$ of course). For $i < \omega$ we let $A'_i = A^*_i \times ({}^{(t_i)}2 \cap p)$, and we let g'_i be the function from $[n^*_i, n^*_{i+1})$ to ${}^{(A'_i)}\{0, \dots, n^*_i\}$ defined as follows: for $(a, \eta) \in A'_i$ and $m \in [n^*_i, n^*_{i+1})$ we let $(g'_i[(a, \eta)])(m) = \ell$ iff $\lim(T^{[\eta]}) \Vdash (\underset{\sim}{f}^i_{\underset{\sim}{x}_i}[a])(m) = \ell$. So apply "$\Gamma$ nice" to $\bar{x}'' = \langle x_0 + 1, g_1, g_2, \dots \rangle$. $\square_{3.14C}$

3.14D Claim. If Q has the Laver property or just is (f, g)-bounding with $f(i) = 2^{2^{k_i k_i}}$, $g(i) = n^*_i$, *then* Q preserves any nice Γ.

Proof. Assume $p \in Q$, $p \Vdash_Q$ "$\underset{\sim}{\bar{x}} = \langle \underset{\sim}{x}_n : n < \omega \rangle$, $x_0 < \omega$, and x_n codes $\underset{\sim}{f}_{\underset{\sim}{x}_i} : [n^*_i, n^*_{i+1}) \to {}^{(\underset{\sim}{A}_{x_i})}\{0, \dots, n^*_i\}$" and we shall find p', a such that $p \le p' \in Q$, $a \in S$ and $p' \Vdash_Q \le \bar{x} R \mathbf{g}_a$", this is enough. So w.l.o.g. $p \Vdash$ "$\underset{\sim}{x}_0 = x_0$". For each $u \subseteq [n^*_i, k^*_i)$, $|u| = n^*_i + 1$, we let $\underset{\sim}{\mathbf{t}}_{i,u}$ be the truth value of the statement

$$1 - 1/(\log(n^*_i))^{x_0+2} \le |\{a \in \underset{\sim}{A}_{x_i} : (\underset{\sim}{f}_{\underset{\sim}{x}_i}[a]) \restriction u \text{ is not one to one}\}| / |\underset{\sim}{A}_{x_i}|.$$

Let $\underset{\sim}{\bar{\mathbf{t}}}_i = \langle \mathbf{t}_{i,u} : u \subseteq [n^*_i, k^*_i), |u| = n^*_i + 1 \rangle$. The number of possible u is $\le 2^{k_i k_i}$, hence the number of possible interpretation of $\underset{\sim}{\bar{\mathbf{t}}}$ is $\le 2^{2^{k_i k_i}}$. By the assumption w.l.o.g. for each i we have $\langle \bar{\mathbf{t}}^{i,\ell} : \ell < n^*_i \rangle$ (all in V not names) such that $p \Vdash$ "$\bigvee_{\ell < n^*_i} \bar{\mathbf{t}}_i = \bar{\mathbf{t}}^{i,\ell}$".

So we can find, in V, $\langle (A^i_\ell, \underset{\sim}{f}^i_\ell) : \ell < n^*_i, i < \omega \rangle$ such that $(A^i_\ell, \underset{\sim}{f}^i_\ell)$ is a possible case of $(\underset{\sim}{A}^i_{\underset{\sim}{x}_i}, \underset{\sim}{f}^i_{\underset{\sim}{x}_i})$. By the way the norm was defined (for i large enough) by

dropping the norm by 1 we can deal not just with one case (i.e. one possible $\bar{\mathbf{t}}^{i,\ell}$ i.e one (A^i_ℓ, f^i_ℓ)) but even with n^* of them. This is $(*)_2$ of 3.15A.
Note: if $p \in G \subseteq Q$, G generic over V then for some $\ell < n^*_\ell$, $(\underset{\sim}{A}'_{x_i}, \underset{\sim}{f}^i_{x_i}) = (A^i_\ell, f^i_\ell)$, and they have the same $w_i(-)$.

3.15E Claim. The forcing as in [FrSh:406] is like that.

Proof. W.l.o.g. the i-th splittings are included in $(2^{2^{k^*_i}}, \log_2\log_2(n^*_{i+1}))$, so follows by 3.15D the $\langle (2^{(k^*_i)^{n*_i+1}}, n^*_i) : i < \omega \rangle$-bounding version.

§4. There May Be a Unique P-Point

This section continues VI §5.

4.1 Theorem. Assume V satisfies $2^{\aleph_0} = \aleph_1$ and $2^{\aleph_1} = \aleph_2$, F_0 is a Ramsey ultrafilter on ω. Then for some $\aleph_2$-c.c. proper, ${}^\omega\omega$-bounding forcing notion P of cardinality $\aleph_2$, in V^P there is a unique P-point, and it is F_0 (i.e. the filter it generates in V^P).

4.1A Remark. In fact, in V^P, F_0 is a Ramsey ultrafilter (actually this follows).

Proof. By the proof of VI 5.13, it suffices to prove the following lemma:

4.2 Lemma. Suppose

$(*)_0$ F_0, F_1 are ultrafilters on ω, F_0 is a Ramsey ultrafilter, F_1 is a P-point, $F_0 \leq_{\mathrm{RK}} F_1$ but not $F_1 \leq_{\mathrm{RK}} F_0$.

Then there is a forcing notion Q such that:

(a) Q has the PP-property, (hence is ${}^\omega\omega$ bounding) and is of cardinality $2^{\aleph_0}$ and

(b) $\Vdash_Q$ "F_0 is an ultrafilter", but

(c) if $Q \lessdot Q'$, Q' has the PP-property *then* in $V^{Q'}$ we have: F_1 cannot be extended with to a P-point (ultrafilter).

4.2A Remark. During the proof of 4.1 we use the forcing notions SP*(F) from Definition VI 5.4 to kill the P-points F with $F_0 \not\leq_{\mathrm{RK}} F$.

The rest of this section is dedicated to the proof of this Lemma.

Proof. Since $F_0 \leq_{\mathrm{RK}} F_1$ and F_1 is a P-point, there is a function $h : \omega \to \omega$ such that

$(*)_1$ $h(F_1) = F_0$ and for each $\ell < \omega$ the set $I(\ell) = I_\ell \overset{\mathrm{def}}{=} h^{-1}(\{\ell\})$ is finite.

Note that then $[A \subseteq \omega \ \& \ \bigwedge_\ell 1 \geq |I_\ell \cap A| \Rightarrow A \notin F_1]$ because $F_1 \not\leq_{\mathrm{RK}} F_0$.

Now in Definition 4.4 below we define a forcing notion $Q = SP^*(F_0, F_1, h)$ and then prove in 4.3 — 4.9 that it has all the required properties thus finishing the proof of 4.2 and 4.1.

4.3 Claim. In the following game player I has no winning strategy: In the n'th move player I chooses $A_n \in F_0$ and $B_n \in F_1$; player II chooses $k_n \in A_n$ ($k_n > k_\ell$ for $\ell < n$) and $w_n \subseteq B_n \cap I_{k_n}$. In the end player II wins the play if $\{k_n : n < \omega\} \in F_0$ and $\bigcup\{w_n : n < \omega\} \in F_1$ (the first demand follows from the second).

4.3A Remark. Clearly player II has no better choice than $w_n = B_n \cap I_{k_n}$. Remember $I_{k_n} = h^{-1}(\{k_n\})$ is finite.

Proof. Suppose H is a wining strategy of player I. Let λ be big enough, $N \prec (H(\lambda), \in, <^*_\lambda)$ be such that $\{F_0, F_1, h, H\} \in N$ and N is countable. As F_ℓ is a P-point there are, for $\ell \in \{0, 1\}$ sets $A^*_\ell \in F_\ell$ such that $A^*_\ell \subseteq_{ae} B$ (i.e. $A^*_\ell \setminus B$ finite) for every $B \in F_\ell \cap N$.

Now we can find an increasing sequence $\langle M_n : n < \omega\rangle$ of finite subsets of N, $N = \bigcup_{n<\omega} M_n$ such that it increases rapidly enough; more exactly:

α) $H, F_0, F_1, h \in M_0$ and $M_n \in M_{n+1}$,

β) if $\varphi(x, a_0, \ldots)$ is a formula of length $\leq 1000 + |M_n|$ with parameters from $M_n \cup \{M_n\}$ satisfied by some $x \in N$, *then* it is satisfied by some $x \in M_{n+1}$,

γ) if $\ell \in \{0, 1\}$, $B \in F_\ell \cap N$, $B \in M_n$ then $B \cup M_{n+1} \supseteq A^*_\ell$, δ) $M_0 \cap \omega = \emptyset$,

ε) if $\ell \in M_n$ then $I(\ell) \subseteq M_{n+1}$ and M_n is closed under h (we can demand $m \in M_n \Leftrightarrow h(m) \in M_n$ if we make the domains of F_0, F_1 disjoint).

Let $u_{n+1} = (M_{n+1} \setminus M_n) \cap \omega$. So $\langle u_n : n < \omega \rangle$ forms a partition of ω into finite sets. As F_0 is Ramsey, we can find $A \in F_0$ such that $\bigwedge_n |u_n \cap A| \leq 1$ and $A \subseteq A_0^*$ and

$$u_n \cap A \neq \emptyset \;\&\; u_m \cap A \neq \emptyset \;\&\; n < m \;\Rightarrow\; m - n \geq 10.$$

Let $A = \{i_\zeta : \zeta < \omega\}$ (increasing), $i_\zeta \in u_{n_\zeta}$. Now we define by induction on ζ, A_ζ, B_ζ, k_ζ, w_ζ such that

(a) $\langle A_\xi, B_\xi, k_\xi, w_\xi : \xi < \zeta \rangle$ is an initial segment of a play of the game in which Player I uses his winning strategy.

(b) $\langle A_\xi, B_\xi, k_\xi, w_\xi : \xi \leq \zeta \rangle$ belongs to $M_{n_\zeta + 3}$.

(c) $k_\zeta = i_\zeta$ and $w_\zeta = B_\zeta \cap I(k_\zeta) \cap A_1^*$.

There is no problem to carry out the definition, and clearly Player II wins because not only $\{k_\zeta : \zeta < \omega\} = \{i_\zeta : \zeta < \omega\} = A \subseteq A_0^*$ but also

$$\bigcup_{\zeta<\omega} w_\zeta = A_1^* \cap \bigcup_{\zeta<\omega} w_\zeta = A_1^* \cap \{j < \omega : h(j) = i_\zeta \text{ for some } \zeta < \omega\}$$

$$= A_1^* \cap \{j : h(j) \in A\} \in F_1.$$

[Why? As respectively: $w_\zeta \subseteq A_1^*$; as $A_1^* \setminus A_\xi \subseteq \bigcup\{w_\zeta : \zeta \leq i_\xi + 4\}$ by clause (γ) above; as $A = \{i_\zeta : \zeta < \omega\}$; as $A_1^* \in F_1$ and $A \in F_0$ hence $\{j : h(j) \in A\} \in F_1$.] Contradiction. $\square_{4.3}$

4.4 Definition. Let $T_n^h = \prod_{\ell<n} {}^{I(\ell)\times\ell}2$ and let $T^h = \bigcup_{n<\omega} T_n^h$. Note that T^h is a perfect tree with finite branching ordered by $\triangleleft$ (being initial segment). Let $Q = \mathrm{SP}^*(F_0, F_1, h) = \{T :\ T$ is a perfect subtree of T^h and for each $k < \omega$ for some $A_k \in F_0$ and $B_k \in F_1$ we have: if $\ell \in A_k$ and $\eta \in T^{[\ell]} \overset{\text{def}}{=} T \cap T_\ell^h$ and $\rho \in {}^{(B_k \cap I(\ell))\times k}2$ then for some $\nu \in {}^{I(\ell)\times\ell}2$ we have $\rho \subseteq \nu$ and $\eta\hat{}\langle\nu\rangle \in T\}$. The order: inverse inclusion.

4.5 Claim. 1) If $T \in Q$, $T^{[n]} = \{\eta_1, \dots, \eta_k\}$ (with no repetition) $T_\ell = T_{[\eta_\ell]} \stackrel{\text{def}}{=} \{\nu \in T : \eta_\ell \trianglelefteq \nu \text{ or } \nu \trianglelefteq \eta_\ell\}$, $T^\dagger_\ell \in Q$, $T_\ell \leq T^\dagger_\ell$ (i.e. $T^\dagger_\ell \subseteq T_\ell$) *then* $T \leq T^\dagger \stackrel{\text{def}}{=} \bigcup_{\ell=1}^{k} T_\ell \in Q$.

2) If $\underset{\sim}{\tau}$ is a Q-name of an ordinal and $n < \omega$ *then* there is $T^\dagger$, $T \leq T^\dagger \in Q$ such that $T^\dagger \Vdash_Q$ "$\underset{\sim}{\tau} \in A$" for some A satisfying $|A| \leq |T^{[n]}|$, and $T \cap \bigcup_{\ell \leq n} T^{[\ell]} = T^\dagger \cap \bigcup_{\ell \leq n} T^{[\ell]}$. Moreover for each $\eta \in T^{[n]}, T^\dagger_{[\eta]}$ determines $\underset{\sim}{\tau}$.

Proof. Same as in the proof of VI 5.5. $\square_{4.5}$

4.6 Claim.

Q is proper, in fact α-proper for every $\alpha < \omega_1$, and has the strong PP-property (see VI 2.12E(3)).

Proof. First we prove properness. Let λ be regular $> 2^{\aleph_1}$, $N \prec (H(\lambda), \in, <^*_\lambda)$ be countable, $\{Q, F_0, F_1, h\} \in N$ and $T \in N \cap Q$.

Let $\{\underset{\sim}{\tau}_\ell : \ell < \omega\}$ list the Q-names of ordinals from N. We now define a strategy for player I in the game from Claim 4.3. In the n'th move player I chooses $A_n \in F_0 \cap N$, $B_n \in F_1 \cap N$ and player II chooses $k_n \in A_n$ and $w_n \stackrel{\text{def}}{=} B_n \cap I_{k_n}$ (remember 4.3A); on the side player I chooses $T_n \in N \cap Q$ and m_n such that $T_0 = T$, $T_n \leq T_{n+1}$, $T_n^{[m_{n+1}]} = T_{n+1}^{[m_{n+1}]}$ and $m_n > \max\{m_{n-1}, k_{n'} : n' < n\}$ and $m_0 = 1$.

In the $(n+1)$'th move, player I first chooses m_{n+1} as above then he chooses $T_{n+1} \in Q$, $T_n \leq T_{n+1}$, $T_{n+1}^{[m_{n+1}]} = T_n^{[m_{n+1}]}$ such that for every $\eta \in T_n^{[m_{n+1}]}$, $(T_{n+1})_{[\eta]}$ forces a value to $\underset{\sim}{\tau}_\ell$ for $\ell \leq m_{n+1}$. This is possible by 4.5. Then as $T_{n+1} \in Q \cap N$ there are sets $A_{n+1} \in F_0 \cap N$, $B_{n+1} \in F_1 \cap N$ such that for every $k \in A_{n+1}$, $\eta \in (T_{n+1})^{[k]}$ and $\rho \in {}^{(B_{n+1} \cap I(k)) \times n}2$ for some $\nu \in {}^{I(k) \times k}2$, we have: $\rho \subseteq \nu$ and $\eta\hat{\ }\langle\nu\rangle \in T_{n+1}$ and for simplicity $A_{n+1} \cap m_n = A_n \cap m_n$. Note that the amount of free choice player II retains is in N.

So by 4.3 for some such play, player II wins. Now $T^* \stackrel{\text{def}}{=} \bigcap_{n<\omega} T_n \in Q$ as $\{k_n : n < \omega\} \in F_0$ and $\bigcup\limits_{n<\omega} B_n \cap I(k_n) \in F_1$ witness; of course $T_n \leq T^*$ for each n hence $T = T_0 \leq T^*$ and $T^* \Vdash$ "$\underset{\sim}{\tau}_\ell[\underset{\sim}{G}_Q] \in N \cap Q_n$" (as $T_{\ell+1} \leq T^*$, see its choice).

So Q is proper. The proof also shows that Q has the strong PP-property (see VI 2.12: for more details see the proof of VI 4.4.). The proof of α-properness is as in VI 4.4 (and anyhow it is not used). $\square_{4.6}$

4.7 Lemma. Suppose $((*)_0$ of 4.2, $Q = \mathrm{SP}^*(F_0, F_1, h)$ as defined in 4.4 of course and) $Q \lessdot P$ and P has the PP-property.
Then in V^P, F_1 cannot be extended to a P-point.

Proof. Suppose $p \in P$ forces that $\underset{\sim}{E}$ is an extension of F_1 to a P-point (in V^P). Let $\langle \underset{\sim}{r}_n : n < \omega \rangle$ be the sequence of reals which Q introduces, i.e. $r_n(i) = \ell \in \{0,1\}$ is defined as follows: clearly for a unique $k < \omega$, $i \in I_k$; now $\underset{\sim}{r}_n(i) = \ell$ iff: $n \geq k$, $\ell = 0$ *or* for some $T \in \underset{\sim}{G}_Q$, $T^{[k+1]} = \{\eta\}$ and $(\eta(k))(i,n) = \ell$ (remember that $\eta(k)$ is a function from $I(k) \times k$ to $\{0,1\}$). Define a P-name $\underset{\sim}{h}$:

$\underset{\sim}{h}(n)$ is 1 if $\{i < \omega : \underset{\sim}{r}_n(i) = 1\} \in \underset{\sim}{E}$ and

$\underset{\sim}{h}(n)$ is 0 if $\{i < \omega : \underset{\sim}{r}_n(i) = 0\} \in \underset{\sim}{E}$

So $p \Vdash$ "$\underset{\sim}{h} \in {}^{\omega}2$". Now as P has the PP-property, by VI 2.12D, there are $p_1 \geq p$, $(p_1 \in P)$, and $\langle \langle \langle k(n), \langle i_n(\ell), j_n(\ell) \rangle : \ell \leq k(n) \rangle : n < \omega \rangle$ in V such that $k(n) < \omega$, $i_n(0) < j_n(0) < i_n(1) < j_n(1) < \ldots < i_n(k(n)) < j_n(k(n))$, and $j_n(k(n)) < i_{n+1}(0)$ such that:

$p_1 \Vdash_P$ " for every $n < \omega$ for some $\ell \leq k(n)$ we have $\underset{\sim}{h}(i_n(\ell)) = \underset{\sim}{h}(j_n(\ell))$"

Now define the following P-names:

$$\underset{\sim}{A}_n = \{m < \omega : \text{ for some } \ell \leq \underset{\sim}{k}(n), \underset{\sim}{r}_{i_n(\ell)}(m) = \underset{\sim}{r}_{j_n(\ell)}(m)\}.$$

We can conclude as in the proofs of VI 4.7, VI 5.8. $\square_{4.7}$

4.8 Claim. In V^Q, F_0 still generates an ultrafilter.

Proof. If not, then for some $T_0 \in Q$, and Q-name $\underset{\sim}{A}$ we have $T_0 \Vdash_Q$ "$\underset{\sim}{A} \subseteq \omega$ and $\underset{\sim}{A}, \omega \setminus \underset{\sim}{A}$ are $\neq \emptyset \bmod F_0$".

By the proof of 4.6 without loss of generality, for some $A_0 \in F_0$ we have: for $k \in A_0$ and $\eta \in T_0^{[k+1]}$, $(T_0)_{[\eta]}$ forces a truth value to "$k \in \underset{\sim}{A}$" which we call $\mathbf{t}(T_0, \eta)$; without loss of generality for $\eta \in T_0^{[k]}, k \notin A_0 \Rightarrow |\mathrm{suc}_{T_0}(\eta)| = 1$.

Now for every $T \geq T_0$ and $\ell < \omega$ there are $A(T,\ell), B(T,\ell)$ as in Definition 4.4. For every $\ell < \omega$, $T \geq T_0$ and $k \in A(T,\ell)$ fix an arbitrary $\eta(T,\ell,k) \in T^{[k]}$.

Then, by observation 4.9 below, there are $m_{T,\ell,k} \in I(k) \cap B(T,\ell)$ and a partition $\langle u_i(T,\ell,k) : i < 3\rangle$ of $I(k) \cap B(T,\ell)$ and a triple $\langle \mathbf{t}_i(T,\ell,k) : i < 3\rangle$ of truth values and $j_k(T,\ell) \in \{0,1\}$ and truth value $\mathbf{s}_k(T,\ell)$ such that:

(∗) (a) if $j_k(T,\ell) = 0$ then for $i < 3$, for every $\rho \in {}^{u_i(T,\ell,k)\times\ell}2$ there is $\nu \in {}^{I(k)\times k}2$ such that $\rho \subseteq \nu$ and $\eta(T,\ell,k)\hat{}\langle\nu\rangle \in T$ and

$$T_{[\eta\hat{}\langle\nu\rangle]} \Vdash_Q \text{“}k \in \underset{\sim}{A} \text{ iff } \mathbf{t}_i(T,\ell,k)\text{”}.$$

(Clearly $\mathbf{t}_i(T,\ell,k) = \mathbf{t}(T_0, \eta\hat{}\langle\nu\rangle)$).

(b) if $j_k(T,\ell) = 1$ then for every $\rho \in {}^{(I(k)\cap B(T,\ell)\setminus\{m_{T,\ell,k}\})\times\ell}2$ there is $\nu \in {}^{(I(k)\times k)}2$ such that: $\rho \subseteq \nu$ and $(\eta(T,\ell,k))\hat{}\langle\nu\rangle \in T$ and $T_{[\eta\hat{}\langle\nu\rangle]} \Vdash_Q$ "$k \in \underset{\sim}{A}$ iff $\mathbf{s}_k(T,\ell)$".

So for some $j(T,\ell) < 2$ and $i(T,\ell) < 3$ and truth value $\mathbf{t}(T,\ell)$ we have

(α) if $j(T,\ell) = 0$, then

$$\bigcup\{u_{i(T,\ell)}(T,\ell,k) : \; j_k(T,\ell) = 0, \; k \in A(T,\ell), \mathbf{t}_{i(T,\ell)}(T,\ell,k) = \mathbf{t}(T,\ell) \in F_1$$

(β) if $j(T,\ell) = 1$ then $\{k \in A(T,\ell) : j_k(T,\ell) = 1, \mathbf{s}_k(T,\ell) = \mathbf{t}(T,\ell)\} \in F_0$.

Note:

⊗ for (T,ℓ) as above there are $A = A^*(T,\ell) \in F_0$, $B = B^*(T,\ell) \in F_1$ satisfying: for every $k \in A$ there is $\eta \in T$, $\ell g(\eta) = k$ such that: every $\rho \in {}^{((I(k)\cap B)\times\ell)}2$ can be extended to $\nu \in {}^{I(k)\times k}2$ satisfying: $\eta\hat{}\langle\nu\rangle \in T$, $T_{[\eta\hat{}\langle\nu\rangle]} \Vdash_Q$ "$k \in \underset{\sim}{A}$ iff $\mathbf{t}(T,\ell)$"

[Why? If $j(T,\ell) = 0$ let

$$B = \bigcup\{u_{i(T,\ell)}(T,\ell,k) : j_k(T,\ell) = 0, \; k \in A(T,\ell), \mathbf{t}_{i(T,\ell)}(T,\ell,k) = \mathbf{t}(T,\ell)\}$$

and $A = \{k : I(k) \cap B \neq \emptyset\}$. Check the demand by clauses $(*)(a)$ and (α) above. So assume $j(T,\ell) = 1$ and let

$$B = \bigcup\{I(k) \cap B(T,\ell) \setminus \{m_{T,k,\ell}\} : k \in A(T,\ell) \text{ and } j_k(T,\ell) = 1, \\ \text{and } \mathbf{s}_k(T,\ell) = \mathbf{t}(T,\ell)\};$$

(why $B \in F_1$? because $F_1 \not\leq_{\mathrm{RK}} F_0$!). Put $A = \{k : I_k \cap B \neq \emptyset\}$ and check the demand by clauses $(*)$(b) and (β) above].

Note that we have been dealing with fixed T, ℓ.

As we can increase T_0 without loss of generality: for some truth value $\mathbf{t}^*$ for a dense set of $T' \geq T_0$ for the F_0-majority of $\ell < \omega$ we have and $\mathbf{t}(T',\ell) = \mathbf{t}^*$.

Now we can define a strategy for player I in the game from 4.3. So in the n'th move player I chooses A_n, B_n and player II chooses k_n, w_n; but we let player I play "on the side" also T_n, ℓ_n (chosen in the n'th move) such that:

(A) $T \leq T_n \leq T_{n+1}$, $T_n^{[k_n+1]} = T_{n+1}^{[k_n+1]}$, $\omega > \ell_{n+1} > \ell_n$, and $\mathbf{t}^* = \mathbf{t}((T_n)_{[\eta]}, \ell_n)$ for $n > 0$ and $\eta \in T_n^{[k_n+1]}$.

(B) For every $k \in A_{n+1}$ and $\eta \in T_n^{[k_n+1]}$ there is η_1, $\eta \trianglelefteq \eta_1 \in T_n^{[k]}$ such that for every $\rho \in {}^{(B_{n+1} \cap I(k)) \times \ell_{n+1}}2$ there is ν, $\rho \subseteq \nu$, $\eta_1{}^\frown\langle\nu\rangle \in T_n$, $\mathbf{t}(T_n, \ell_n, k_n) = \mathbf{t}(T_n, \ell_n) = \mathbf{t}^*$, (note T_{n+1} is chosen only after k_{n+1}, w_{n+1} were chosen).

We should prove that player I can carry out his strategy. For stage $n+1$ let $\{\eta_0^n, \ldots, \eta_{m(n)}^n\}$ list $T_n^{[k_n+1]}$, so for some $\ell_{n+1} > \ell_n$, for each $\zeta \leq m(n)$ there is $T_{n,\zeta} \geq (T_n)_{[\eta_\zeta^n]}$ such that $\mathbf{t}(T_{n,\zeta}, \ell_{n+1}) = \mathbf{t}^*$. Let $B_{n+1} = \bigcap_{\zeta \leq m(n)} B^*(T_{n,\zeta}, \ell_{n+1})$ and $A_{n+1} = \{k \in A_n : k > k_n \text{ and } I(k) \cap B_{n+1} \neq \emptyset\}$.

By clause (B) above, after player II moves, we can choose T_{n+1} as required. As this is a strategy, by Claim 4.3 for some play in which player I uses it he looses. For this play $\{k_n : n < \omega\} \in F_0$, $\bigcup_{n<\omega} w_n \in F_1$, so $T \overset{\text{def}}{=} \bigcap_{n<\omega} T_n \in Q$. By tracing the demands on the $\mathbf{t}$'s:

$\oplus$ for $n < \omega$, $\eta \in T$, $\ell g(\eta) = k_n + 1$ we have $T_{[\eta]} \Vdash$ "$k_n \in \underset{\sim}{A}$ iff $\mathbf{t}^*$".

We conclude: $T \Vdash$ "$\{k_n : n < \omega\} \cap \underset{\sim}{A}$ is $\emptyset$ or is $\underset{\sim}{A}$" as $\{k_n : n < \omega\} \in F_0$ we get the desired conclusion. $\square_{4.8}$

4.9 Observation. Suppose $\mathbf{t}$ is a function from $X^* = \prod_{t\in u} A_t$ to $\{0,1\}$, u finite.

Then at least one of the following holds:

(α) we can find u_i, X_i ($i < 3$) such that :

(a) $\langle u_i : i < 3\rangle$ is a partition of u,

(b) $X_i \subseteq X^*$,

(c) $\mathbf{t}{\restriction}X_i$ is constant,

(d) for every $i < 3$ and $\rho \in \prod_{t\in u_i} A_t$ there is $\nu \in X_i$, $\rho \subseteq \nu$,

(β) for some $x \in u$, there is $X \subseteq X^*$ such that $\mathbf{t}{\restriction}X$ is constant and for every $\rho \in \prod_{t\in u\setminus\{x\}} A_t$ there is $\nu \in X$, $\rho \subseteq \nu$.

Proof. Let for $j \in \{0,1\}, P_j = \{v : v \subseteq u$ and there is $X \subseteq X^*$ such that $\mathbf{t}{\restriction}X$ is constantly j and for every $\rho \in \prod_{t\in v} A_t$ there is $\nu \in X$, $\rho \subseteq \nu\}$. Clearly

(A) $u_1 \in P_j, u_0 \subseteq u_1$ implies $u_0 \in P_j$. [Why? Same X witnesses this.]

(B) $u_1 \subseteq u$ & $u_1 \notin P_j$ implies $u \setminus u_1 \in P_{1-j}$ [Why? As $u_1 \notin P_j$, for some $\rho \in \prod_{t\in u_1} A_t$ for no $\nu \in \prod_{t\in u\setminus u_1} A_t$ does $\mathbf{t}(\rho\cup\nu) = j$; let $X \stackrel{\text{def}}{=} \{\nu \in \prod_{t\in u} A_t : \rho \subseteq \nu\}$, it is as required for $u \setminus u_1$.]

(C) $\emptyset \in P_0 \cup P_1$. [Why? Trivially.]

Case (i): $P_0 \cup P_1$ is not an ideal.

So there are $u_0, u_1 \in P_0 \cup P_1$ with $v \stackrel{\text{def}}{=} u_0 \cup u_1 \notin P_0 \cup P_1$. By (A) without loss of generality $u_0 \cap u_1 = \emptyset$. Let $u_2 = u \setminus v$, so $\langle u_0, u_1, u_2\rangle$ is a partition of u. Now by clause (B) we know that $u_2 \in P_0$ (and to P_1) as $v = u \setminus u_2$ does not belong to P_1 (and to P_0). Now we know $u_0, u_1, u_2 \in P_0 \cup P_1$, so for some $\langle j_\ell : \ell < 3\rangle$ we have $u_\ell \in P_{j_\ell}$ for $\ell < 3$, and let X_ℓ be a witness. Now check that clause (α) in the conclusion holds.

Case (ii): $P_0 \cup P_1$ is an ideal.

If $u \in P_0 \cup P_1$, then $\mathbf{t}$ is constant so conclusion (α) is trivial, so assume not. By (B) above the ideal is a maximal ideal so it is principal (because u is finite), i.e. for some $x \in u$, $u \setminus \{x\} \in P_0 \cup P_1$, $\{x\} \notin P_0 \cup P_1$ so we have finished. (Reflection shows we get more than required in (β): reread the proof of (B)).

$\square_{4.2,4.1}$

Appendix. On Weak Diamonds and the Power of Ext

§0. Introduction

In [DvSh:65] K. Devlin and S. Shelah introduced a combinatorial principle Φ which they called the weak diamond. It explains some of the restrictions in theorems of the form "the limit of iteration does not add reals". See more on this in [Sh:186] and Mekler and Shelah [MkSh:274] (on consistency of uniformization properties) [Sh:208] (consistency of "ZFC+$2^{\aleph_0} < 2^{\aleph_1} < 2^{\aleph_2} + \neg\Phi_{\{\delta<\aleph_2:\mathrm{cf}(\delta)=\aleph_1\}}$") and very lately [Sh:587].

Explanation. Jensen's diamond for $\aleph_1$, denoted $\diamondsuit_{\aleph_1}$, see [Jn], can be formulated as: There exists a sequence of functions $\{g_\alpha : g_\alpha$ a function from α to α where $\alpha < \omega_1\}$ such that for every $f : \omega_1 \to \omega_1$ we have $\{\alpha < \omega_1 : f\restriction\alpha = g_\alpha\} \not\equiv 0 \bmod \mathcal{D}_{\aleph_1}$ (recall that $\mathcal{D}_{\aleph_1}$ is the filter on λ generated by the family of closed unbounded subsets of λ). Clearly $\diamondsuit_{\aleph_1} \to 2^{\aleph_0} = \aleph_1$. Jensen (see [DeJo]) also proved that $2^{\aleph_0} = \aleph_1 \not\Rightarrow \diamondsuit_{\aleph_1}$ (see Chapters V and VII remembering that $\diamondsuit_{\aleph_1}$ implies existence of an Aronszajn tree which is not special (even a Souslin tree)). You may ask, is there a diamond like principle which follows from $2^{\aleph_0} = \aleph_1$?

K. Devlin and S. Shelah [DvSh:65] answered this question positively, formulating a principle Φ which says:

$(*)_1$ $(\forall F : {}^{\omega_1>}2 \to 2)(\exists h : \omega_1 \to 2)(\forall \eta : \omega_1 \to 2)$
$\{\alpha < \omega_1 : F(\eta\restriction\alpha) = h(\alpha)\} \not\equiv 0 \bmod \mathcal{D}_{\aleph_1}$.

The author had hoped that $2^{\aleph_0} < 2^{\aleph_1} < 2^{\aleph_2}$ would imply that S^2_0 is not small, i.e. for all $F : {}^{\aleph_2 >}\aleph_2 \to 2$ there exists $\eta \in {}^{\aleph_2}2$ such that for all $g : \aleph_2 \to \aleph_2$, for all C club of $\aleph_2$ there is $S \in S^2_0 \cap C$ with $\eta(\delta) \neq F(g{\restriction}\delta)$. In [Sh:208] a consistency result contradicting this was proved.

In fact $2^{\aleph_0} < 2^{\aleph_1} \iff \Phi$. If the statement above holds for F, h we say that h is a weak diamond say for (the colouring) F. The principle Φ was used as a successful substitute for $\diamondsuit_{\aleph_1}$ in [Sh:88], [AbSh:114], [Sh:140] and [Sh:192].

An equivalent form of Φ is (just replace h by $1 - h$)

$(*)_2$ $(\forall F : {}^{\omega_1 >}2 \to 2)(\exists h \in {}^{\omega_1}2)(\forall \eta \in {}^{\omega_1}2)$
$[\{\alpha < \omega_1 : F(\eta{\restriction}\alpha) = h(\alpha)\} \not\equiv \lambda \mod \mathcal{D}_{\aleph_1}]$.

Φ can easily be generalized to higher cardinals than $\aleph_1$, for example define for uncountable regular λ and $\kappa \leq \lambda$:

$$\Phi^\kappa_\lambda \iff (\forall F : {}^{\lambda >}2 \to \kappa)(\exists h : \lambda \to \kappa)(\forall \eta : \lambda \to 2)$$
$$[\{\alpha < \lambda : F(h{\restriction}\alpha) = \eta(\alpha)\} \not\equiv 0 \operatorname{mod} \mathcal{D}_\lambda].$$

So $\Phi \iff \Phi^2_{\aleph_1}$.

We thank Grossberg for reminding us that because of a flaw in [DvSh:65] he and Magidor saw conclusion 1.15 after which this section was written.

There is natural generalization. Instead of quantifying over $\eta \in {}^\lambda 2 = \bigtimes_{i<\lambda} 2$ consider quantifying over $\eta \in \bigtimes_{i<\lambda} \bar{\mu}_i$ (and change the domain of F accordingly).

These generalizations are our goal in the first section but instead of generalizing $\Phi^2_{\aleph_1}$ we generalize its negation. Another possible generalization is $\Phi^\kappa_{\aleph_1}$ for $2 < \kappa \leq \aleph_0$ which by VIII §4 is stronger (its negation is consistent with G.C.H.). We do not assume the reader is familiar with [DvSh:65], for example the hard direction of $\Phi^2_{\aleph_1} \iff 2^{\aleph_0} < 2^{\aleph_1}$ follows from Theorem 1.10 substituting $\lambda = \aleph_1$ and $\mu = 2$. This generalization of $\Phi^2_{\aleph_1}$ was used in [Sh:88 §6] and mentioned there in a remark; since we were asked to explain it, we present it here.

In Sect. 2 we present applications of the principle from §1 to the Whitehead problem, we shall use it for two theorems. The first, Theorem 2.2, evaluates the cardinality of Ext(G, H), and the second one is Theorem 2.4 where we give information on the torsion free rank of Ext (G, H). We shall define here all the group theoretical terminology and shall use only one easy lemma which we quote from somewhere else. But this section is not an introduction to the subject of the Whitehead problem; the interested reader is referred to the book of P. Eklof and A. Mekler [EM], to the exposition [E] or to the original papers where the corresponding theorems were proved (from stronger set theoretical hypotheses) [Sh:44],[HHSh:91].

In [Sh:64] another combinatorial principle was introduced:
For a limit ordinal δ less than ω_1, an increasing ω-sequence η_δ of ordinals cofinal in δ is called a *ladder* on δ. A ladder system $\bar{\eta}$ is $\{\eta_\delta : \delta \in S\}$, where $S \subseteq \omega_1$; we say that such a ladder system $\bar{\eta}$ has *the uniformization property* if for every $\{c_\delta \in {}^{\omega}2 : \delta \in S\}$ there exists $h \in {}^{\omega_1}2$ such that $(\forall \delta \in S)(\exists n < \omega)(\forall k < \omega)[k > n \rightarrow c_\delta(k) = h(\eta_\delta(k))]$. In §3 we define the uniformization property for a ladder system $\bar{\eta} = \langle \eta_\delta : \delta \in S \rangle$, where S a set of ordinals with each member of cofinality $\aleph_0$, in particular $S = \{\delta < \aleph_2 : \mathrm{cf}(\delta) = \aleph_0\}$. We try to prove an analogous result to the one in Sect. 1, and we shall prove it assuming $2^{\aleph_0} = \aleph_1$; for more details see the introduction to Sect. 3. Sect. 3 does not depend on sections 1 and 2.

§1. Unif: a Strong Negation of the Weak Diamond

1.0 Notation. We will write $\bigtimes_{i<\lambda} \mu_i$ for the cartesian product of the ordinals μ_i (that is for $\{f : f$ a function with domain λ such that $f(i) < \mu_i\}$), and will write $\prod_{i<\lambda} \mu_i$ for the cardinality of this product.

Let's recall that (see $(*)_2$ in the introduction) the negation of $\Phi^2_{\aleph_1}$ is:

$$(\exists F : {}^{\omega_1 >}2 \rightarrow 2)(\forall h : \omega_1 \rightarrow 2)(\exists \eta : \omega_1 \rightarrow 2)$$
$$[\{\alpha < \omega_1 : F(\eta \restriction \alpha) = h(\alpha)\} \in \mathcal{D}_{\aleph_1}].$$

This is the motivation for the following definition (we replace sometimes functions by sequences, when sequences are easier to handle).

1.1 Definition. For a regular uncountable λ and sequences $\bar{\mu} = \langle \bar{\mu}(i) : i < \lambda \rangle$, $\bar{\chi} = \langle \bar{\chi}(i) : i < \lambda \rangle$ of cardinals ≥ 1 let Unif $(\lambda, \bar{\mu}, \bar{\chi})$ mean: There is a function F with domain $D(\bar{\mu}) \stackrel{\text{def}}{=} \bigcup_{\alpha<\lambda} \bigtimes_{i<\alpha} \bar{\mu}(i)$ such that:

(a) for every $\alpha < \lambda$ and $\eta \in D_\alpha(\bar{\mu}) \stackrel{\text{def}}{=} \bigtimes_{i<\alpha} \bar{\mu}(i)$ we have $F(\eta) < \bar{\chi}(\alpha)$.

(b) for every $h \in \bigtimes_{\alpha<\lambda} \bar{\chi}(\alpha)$ there exists $\eta \in \bigtimes_{\alpha<\lambda} \bar{\mu}(\alpha)$ such that the set $\{\alpha < \lambda : F(\eta \restriction \alpha) = h(\alpha)\}$ belongs to $\mathcal{D}_\lambda$.

1.1A Notation. (1) If $\bar{\mu}$ is constant, i.e., $\bar{\mu} = \langle \mu : i < \lambda \rangle$ we may write μ; similarly for $\bar{\chi}$.

(2) If $(\forall \alpha < \lambda)[\bar{\mu}(1 + \alpha) = \bar{\mu}(1)]$ we may write $\langle \mu(0), \mu(1) \rangle$ instead of $\bar{\mu}$ and Unif $(\lambda, \mu(0), \mu(1), \bar{\chi})$ instead of Unif $(\lambda, \bar{\mu}, \bar{\chi})$. We let

$$D_\alpha(\mu_0, \mu_1) \stackrel{\text{def}}{=} \{\eta : \eta \in {}^{\alpha}\text{Ord}, \eta(0) < \mu_0 \text{ and } \eta(1 + i) < \mu_1\}$$

and

$$D(\langle \mu_0, \mu_1 \rangle) = D(\mu_0, \mu_1) \stackrel{\text{def}}{=} D_{<\lambda}(\mu_0, \mu_1) = \bigcup_{\alpha<\lambda} D_\alpha(\mu_0, \mu_1).$$

Similarly we define $D_\alpha(\langle \mu \rangle) = D_\alpha(\mu)$, so $D(\mu) = {}^{\lambda>}\mu$.

(3) From now on we assume that λ is an uncountable regular cardinal.

(4) Remember that we use δ always as limit ordinal; so for $S \subseteq \lambda$ the set $\{\delta < \lambda : \delta \in S\}$ is the set of limit ordinals which belong to S.

1.1B Remark. (1) Unif $(\aleph_1, 2, 2)$ is the negation of $\Phi^2_{\aleph_1}$ i.e., it is the negation of the weak diamond.

(2) We shall say (concerning Definition 1.1) that the function F exemplifies Unif $(\lambda, \bar{\mu}, \bar{\chi})$.

(3) If $2^{\aleph_0} = 2^{\aleph_1}$, then Unif $(\aleph_1, 2, 2)$ holds. (Noted by Abraham: the converse is a theorem: see 1.10.)

Proof of (3). Let $H : {}^{\omega}2 \to {}^{\omega_1}2$ be onto. Define $F : {}^{\omega_1>}2 \to 2$ as follows:

If $\eta \in {}^n 2$, $n < \omega$, then $F(\eta) = 0$

If $\eta \in {}^\alpha 2$, $\alpha \geq \omega$, then $F(\eta) = H(\eta{\restriction}\omega)(\alpha)$.

Now check that F witnesses Unif $(\aleph_1, 2, 2)$.

Recall that we can strengthen the statement in $\diamondsuit$ by working only on a stationary set $S \subseteq \lambda$. Similarly we can consider stronger forms of the weak diamond, i.e. weaker forms of Unif by relativizing to a stationary set S.

1.2 Definition. Let $\lambda, \bar{\mu}, \bar{\chi}$ be as in Definition 1.1 and let $S \subseteq \lambda$.

(1) Unif $(\lambda, S, \bar{\mu}, \bar{\chi})$ is defined similarly to the definition of Unif $(\lambda, \bar{\mu}, \bar{\chi})$: just replace (b) there by

(b′) for every $h \in \bigtimes_{\alpha<\lambda} \bar{\chi}(\alpha)$ there exists $\eta \in \bigtimes_{\alpha<\lambda} \bar{\mu}(\alpha)$ such that the set $\{\delta \in S : F(\eta{\restriction}\delta) = h(\delta)\}$ belongs to $\mathcal{D}_\lambda + S$.

(2) Let Id $-$ Unif $(\lambda, \bar{\mu}, \bar{\chi}) \stackrel{\text{def}}{=} \{S \in \lambda :$ Unif $(\lambda, S, \bar{\mu}, \bar{\chi})$ holds $\}$.

(3) If $(\forall \alpha)$ $(\bar{\mu}(1 + \alpha) = \mu_1)$ we may write Unif$(\lambda, S, \mu(0), \mu_1, \bar{\chi})$ and Id $-$ Unif $(\lambda, \mu(0), \mu_1, \bar{\chi})$ in parts (1) and (2) respectively. So Unif $(\lambda, \mu_0, \mu_1, \bar{\chi})$ mean Unif $(\lambda, \lambda, \mu_0, \mu_1, \bar{\chi})$.

(4) If $\bar{\chi}$ is constantly χ we may write χ (in Definitions 1.1, 1.2(1), (2), (3)).

1.2A Remark. The notation of Definition 1.2(2) will be justified in Lemma 1.9 where we shall prove that if Unif $(\lambda, \bar{\mu}, \bar{\chi})$ fails, then Id $-$ Unif $(\lambda, \bar{\mu}, \bar{\chi})$ is an ideal. Note also that Unif $(\lambda, \bar{\mu}, \bar{\chi})$ is trivially equivalent to Id $-$ Unif $(\lambda, \bar{\mu}, \bar{\chi}) \neq \mathcal{P}(\lambda)$.

1.3 Remark. The diamond $\diamondsuit_\lambda$ implies the weak diamond Φ^2_λ, and more generally $\diamondsuit_\lambda(S)$ implies the failure of Unif $(\lambda, S, 2, 2, 2)$.

Proof. Let $\langle \eta_\alpha : \alpha \in S \rangle$ be such that for every $\eta : \lambda \to 2$ the set $\{\alpha \in S : \eta{\restriction}\alpha = \eta_\alpha\}$ is stationary. Now if $F : {}^{\lambda>}2 \to 2$, then we let $h : \lambda \to 2$ be defined by $h(\alpha) = F(\eta_\alpha)$, so clearly for any $\eta : \lambda \to 2$ the set $\{\alpha \in S : F(\eta{\restriction}\alpha) = h(\alpha)\}$ will be stationary. $\square_{1.3}$

1.4 Lemma. Let $\lambda, S, \bar{\mu}, \bar{\chi}$ be as in Definition 1.2.

(1) If $\{i < \lambda : \bar{\chi}(i) = 1\} \in \mathcal{D}_\lambda$ then Unif $(\lambda, S, \bar{\mu}, \bar{\chi})$ holds.

(2) Let $\bar{\chi}^1, \bar{\chi}^2$ satisfy the requirements for $\bar{\chi}$ in Definition 1.1. *then*

$$\{i \in S : \bar{\chi}^1(i) = \bar{\chi}^2(i)\} \in \mathcal{D}_\lambda + S \text{ imply that}$$

$$\text{Unif}\,(\lambda, S, \bar{\mu}, \bar{\chi}^1) \iff \text{Unif}\,(\lambda, S, \bar{\mu}, \bar{\chi}^2)$$

(3) Unif $(\lambda, S, \bar{\mu}, \bar{\chi})$ implies that $|\bigtimes_{\alpha<\lambda} \bar{\chi}(\alpha)/(\mathcal{D}_\lambda + S)| \le \prod_{\alpha<\lambda} \bar{\mu}(\alpha)$ (notice that the left hand side of the inequality is the cardinality of a reduced product).

(4) If there exists a $\beta < \lambda$ such that $|\bigtimes_{\alpha<\lambda} \bar{\chi}(\alpha)/(\mathcal{D}_\lambda + S)| \le \prod_{\alpha<\beta} \bar{\mu}(\alpha)$, *then* Unif $(\lambda, S, \bar{\mu}, \bar{\chi})$ holds.

(5) Let $\bar{\mu}$, $\bar{\chi}, \bar{\mu}^*, \bar{\chi}^*$ be sequences of cardinals ≥ 1 of length λ such that for every $\alpha < \lambda$ we have $\bar{\chi}^*(\alpha) \le \bar{\chi}(\alpha)$ and $\bar{\mu}(\alpha) \le \bar{\mu}^*(\alpha)$. *Then* Unif $(\lambda, S, \bar{\mu}, \bar{\chi}) \Rightarrow$ Unif $(\lambda, S, \bar{\mu}^*, \bar{\chi}^*)$.

(6) If $S^* = \{\delta \in S : \bar{\chi}(\delta) > 1\}$ then Unif $(\lambda, S, \bar{\mu}, \bar{\chi}) \Leftrightarrow$ Unif $(\lambda, S^*, \bar{\mu}, \bar{\chi})$.

Proof. Easy (note that part (4) can be proved just like 1.1B(3)).

1.5 Lemma. Let $\lambda, S, \bar{\mu}, \bar{\chi}$ be as in Definition 1.2. Let us define the following cardinals $\mu_0 \overset{\text{def}}{=} \sum_{\alpha<\lambda} \prod_{i<\alpha} \bar{\mu}(i)$, and $\mu_1 \overset{\text{def}}{=} \text{Min}_{\alpha<\lambda} \sum_{\beta<\lambda} \prod_{i<\beta} \bar{\mu}(\alpha+i)$; *then* the following are equivalent.

(A) Unif $(\lambda, S, \bar{\mu}, \bar{\chi})$

(B) Unif $(\lambda, S, \mu_0, \mu_1, \bar{\chi})$ (see 1.2(3)).

The proof will use the following easy fact.

1.5A Fact. Assume that $D(\bar{\mu})$ can be embedded into $D(\bar{\mu}^*)$, i.e., there is a partial function $g : D(\bar{\mu}) \to D(\bar{\mu}^*)$ such that:

(a) If $\eta \lhd \nu$ are both in Dom(g), then $g(\eta) \lhd g(\nu)$

(b) g is one - to - one

(c) g is continuous, i.e., whenever $\langle \eta_\alpha : \alpha < \delta \rangle$ is a sequence of elements of Dom(g) satisfying $\alpha_1 < \alpha_2 \Rightarrow \eta_{\alpha_1} \lhd \eta_{\alpha_2}$, then also $\eta_\delta \overset{\text{def}}{=} \bigcup_{\alpha<\delta} \eta_\alpha$ is in Dom(g), and $g(\eta_\delta) = \bigcup_{\alpha<\delta} \eta_\alpha$.

(d) For every $\eta \in \bigtimes_{i<\lambda} \bar{\mu}(i)$, the set $\{i < \lambda : \eta{\restriction}i \in \mathrm{Dom}(g)\}$ is unbounded in λ (by (c), this set will also be closed).

Then Unif $(\lambda, S, \bar{\mu}, \bar{\chi})$ implies Unif $(\lambda, S, \bar{\mu}^*, \bar{\chi})$.

Proof. Assume Unif $(\lambda, S, \bar{\mu}, \bar{\chi})$ holds. Let $g' \stackrel{\mathrm{def}}{=} g{\restriction}\{\eta \in \mathrm{Dom}(g) : \ell g(\eta) = \ell g(g(\eta))\}$. The function g' will also satisfy (a) — (d). Choose F which witnesses Unif $(\lambda, S, \bar{\mu}, \bar{\chi})$, and define F^* on $D(\bar{\mu}^*)$ as follows:

$$F^*(\nu) = \begin{cases} F(\eta), & \text{if } g'(\eta) = \nu \text{ for some } \eta \in \mathrm{Dom}(g') \\ 0 & \text{otherwise.} \end{cases}$$

Note that $F^*(\nu)$ is well defined as there is at most one $\eta \in \mathrm{Dom}(g')$ such that $g'(\eta) = \nu$ as g' is a one to one function. Let $h \in \prod_{i<\lambda} \bar{\chi}(i)$, so as F witnesses Unif $(\lambda, S, \bar{\mu}, \bar{\chi})$, necessarily there is $\eta \in \bigtimes_{i<\lambda} \bar{\mu}(i)$ such that $S_0 \stackrel{\mathrm{def}}{=} \{\delta < \lambda : F(\eta{\restriction}\delta) = h(\delta)\}$ belongs to $D_\lambda + S$.

By clause (d) of the assumption, the set $C \stackrel{\mathrm{def}}{=} \{\delta < \lambda : \eta{\restriction}\delta \in \mathrm{Dom}(g')\}$ is a closed unbounded subset of λ. So $\delta \in C \Rightarrow \ell g(g'(\eta{\restriction}\delta)) = \delta$. Let $\nu = \bigcup_{i \in C} g(\eta{\restriction}i)$, clearly $\nu \in \prod_{i<\lambda} \bar{\mu}^*(i)$ and $\delta \in S_0 \cap C \Rightarrow F^*(\nu{\restriction}\delta) = F^*(g'(\eta{\restriction}\delta)) = F(\eta{\restriction}\delta) = h(\delta)$. So it is easy to see that F^* witnesses Unif $(\lambda, S, \bar{\mu}^*, \bar{\chi})$. $\square_{1.5A}$

Proof of 1.5.

(A) $\Rightarrow$ (B)

Let $\alpha^* < \lambda$ be such that for all i we have: $\alpha^* \le i < \lambda \Rightarrow \bar{\mu}(i) \le \mu_1$, and let $\{\nu_\xi : \xi < \mu_0'\}$ be a $1-1$ enumeration of $\bigtimes_{i<\alpha^*} \bar{\mu}(i)$, where $\mu_0' \stackrel{\mathrm{def}}{=} \prod_{i<\alpha^*} \bar{\mu}(i) \le \mu_0$ by the definition of μ_0. Now define a partial function $g : D(\bar{\mu}) \to D(\mu_0, \mu_1)$ by the following conditions:

$\mathrm{Dom}(g) = \{\eta \in D(\bar{\mu}) : \ell g(\eta) \ge \alpha^*\}$

$g(\nu_\xi {}^\frown \eta) = \langle \xi \rangle {}^\frown \eta$, whenever $\xi < \mu_0'$, $\nu_\xi {}^\frown \eta \in D(\bar{\mu})$

Clearly g satisfies clauses (a) - (d) of fact 1.5A, so Unif $(\lambda, S, \bar{\mu}, \bar{\chi})$ implies Unif $(\lambda, S, \mu_0, \mu_1, \bar{\chi})$.

(B) $\Rightarrow$ (A)

This time we will construct an embedding $g : D(\mu_0, \mu_1) \to D(\bar{\mu})$ and again use 1.5A. For simplicity, let us first assume

$\otimes$ $1 \leq i < j < \lambda \Rightarrow \bar{\mu}(i) \leq \bar{\mu}(j)$.

Let $\alpha^* < \lambda$ be such that for all $\beta \in [\alpha^*, \lambda)$ we have $|D(\bar{\mu}\restriction[\beta, \lambda))| = \mu_1$, i.e.

$$\beta \geq \alpha^* \Rightarrow \sum_{\gamma<\lambda} \prod_{i<\gamma} \bar{\mu}(\beta + i) = \mu_1.$$

W.l.o.g. $\alpha^* > 2$. We claim that:

(a) There exists an antichain $\langle \nu^0_\xi : \xi < \mu_0 \rangle$ in $D(\bar{\mu})$ (and w.l.o.g. $\xi < \mu_0 \Rightarrow \ell g(\nu^0_\xi) \geq \alpha^*$)

(b) For each $\eta \in D(\bar{\mu})$ there exists an antichain $\langle \nu^\eta_\xi : \xi < \mu_1 \rangle$ in $D(\bar{\mu})$ satisfying $\xi < \mu_1 \Rightarrow \eta \lhd \nu^\eta_\xi$.

("Antichain" means that for $\xi \neq \zeta$ we have neither $\nu^\eta_\xi \trianglelefteq \nu^\eta_\zeta$ nor $\nu^\eta_\zeta \trianglelefteq \nu^\eta_\xi$).

We will prove only (a), as the proof for (b) is similar. For each $\nu \in D(\bar{\mu})$, define $g^*(\nu)$ as follows:

$$\ell g(g^*(\nu)) = \alpha^* + 2\ell g(\nu) + 2, \text{ and}$$

$$g^*(\nu)(i) = \begin{cases} \nu(0) & \text{if } i = 0 \\ 0 & \text{if } i < \alpha^*,\ i \geq 1 \\ \nu(j) & \text{if } i = \alpha^* + 2j,\ j < \ell g(\nu) \\ 0 & \text{if } i = \alpha^* + 2j + 1,\ j < \ell g(\nu) \\ 1 & \text{if } i = \alpha^* + 2\ell g(\nu) \text{ or } i = \alpha^* + 2\ell g(\nu) + 1. \end{cases}$$

Then $\{g^*(\nu) : \nu \in D(\bar{\nu})\}$ is an antichain of size μ_0. This ends the proof of (a). (We needed $\otimes$ to ensure $\nu(i) < \bar{\mu}(i)$.)

Now we define $g : D(\mu_0, \mu_1) \to D(\bar{\mu})$ inductively as follows:

$$g(\emptyset) = \emptyset,$$

$$g(\langle \xi \rangle) = \nu^0_\xi \text{ when } \xi < \mu_0,$$

$$g(\eta\hat{\ }\langle\xi\rangle) = \nu_\xi^{g(\eta)} \text{ when } \ell g(\eta) \geq 1, \xi < \mu_1,$$

$$g(\eta) = \bigcup_{\alpha<\ell g(\eta)} g(\eta{\restriction}\alpha), \text{ when } \ell g(\eta) \text{ is a limit ordinal.}$$

Again g satisfies clauses (a) - (d) of 1.5A, so we are done.

We have only one problem left: what occurs if $\otimes$ fails? Really this is not serious, e.g. by the following claim 1.6 (if $\bar\mu^*(j+i) = 1$ for every i, then $\mu_0 = \prod_{i<j}\mu(i)$, $\mu_1 = 1$, so the lemma becomes trivial, by 1.4(3), (4), as $\mu_1 = 1$). $\square_{1.5}$

1.6 Claim. Let $\lambda, S, \bar\mu, \bar\chi$ be as in Definition 1.1.

(1) For every $\{\alpha_i : i < \lambda\} \subseteq \lambda$ increasing and continuous such that $\alpha_0 = 0$, and $\bigcup_{i<\lambda}\alpha_i = \lambda$; for every $i < \lambda$ define $\bar\mu^*(i) \stackrel{\text{def}}{=} \prod_{\alpha_i \leq j < \alpha_{i+1}} \bar\mu(j)$. We have that Unif $(\lambda, S, \bar\mu, \bar\chi)$, and Unif $(\lambda, S, \bar\mu^*, \bar\chi)$ are equivalent.

(2) For any $\bar\mu$ there exist $\{\alpha_i : i < \lambda\} \subseteq \lambda$ as in (1) such that letting $\bar\mu^*$ be defined using α_i's as in (1) we have $\bar\mu^*(1+i) \leq \mu^*(1+j)$ for $i \leq j$ and $\bar\mu^*(i) \geq 1$.

Proof. (1) Similar to the proof of 1.5.

(2) Let κ^* be minimal such that $\{i < \lambda : \bar\mu(i) \geq \kappa^*\}$ is bounded in λ, so for some $\alpha_1 < \lambda$ we have $[\alpha_1 \leq i < \lambda \Rightarrow \bar\mu(i) < \kappa^*]$. If $\kappa^* = \kappa^+$, it is enough to choose inductively α_i (when $1 \leq i < \lambda$, increasing continuous) such that: $\{j : \alpha_i < j < \alpha_{i+1}, \bar\mu(j) = \kappa\}$ has the same order type (hence the same cardinality) as α_{i+1}, hence $\prod\limits_{j\in[\alpha_i,\alpha_{i+1})} \bar\mu(i) = \kappa^{|\alpha_{i+1}|}$ will be non decreasing for $i \in [1, \lambda)$.

If κ^* is limit, necessarily $\mathrm{cf}(\kappa^*) \leq \lambda$.

If $\mathrm{cf}(\kappa^*) = \lambda$ choose α_i (when $1 < i < \lambda$, increasing continuous) such that for $i > 0$, $\{\beta : \alpha_i \leq \beta < \alpha_{i+1}, \bar\mu(\beta) > \sup\{\bar\mu(\gamma) : \alpha_1 \leq \gamma < \alpha_i\}\}$ has cardinality $\geq |\alpha_i|$.

If $\mathrm{cf}(\kappa^*) = \theta < \lambda$ let $\langle\kappa_\varepsilon : \varepsilon < \theta\rangle$ be a strictly increasing sequence of cardinals $< \kappa^*$ with limit κ^* and choose α_i (when $1 < i < \lambda$, increasing continuous) such that for every $i \geq 1$ we have the order type of $\{\beta : \alpha_i \leq \beta < \alpha_{i+1}$ and $\bar\mu(\beta) \geq \kappa_\varepsilon\}$ is α_{i+1} for each $\varepsilon < \theta$. $\square_{1.6}$

1.7 Claim. (1) If Unif $(\lambda, S, \bar{\mu}, \bar{\chi})$, $\kappa < \lambda$ and $\bar{\mu}^*(i) = \bar{\mu}(i)^\kappa$, $\bar{\chi}^*(i) = \bar{\chi}(i)^\kappa$ for $i < \lambda$ *then* Unif $(\lambda, S, \bar{\mu}^*, \bar{\chi}^*)$

(2) If Unif $(\lambda, S, \bar{\mu}_\xi, \bar{\chi}_\xi)$ for $\xi < \kappa$, $\kappa < \lambda$ and $\bar{\mu}(i) = \prod_{\xi<\kappa} \bar{\mu}_\xi(i)$ and $\bar{\chi}(i) = \prod_{\xi<\kappa} \bar{\chi}_\xi(i)$ *then* Unif $(\lambda, S, \bar{\mu}, \bar{\chi})$

(3) If $\bar{\mu}$ is a nondecreasing sequence of infinite cardinals and Unif $(\lambda, S, \bar{\mu}, \bar{\chi})$ and $\bar{\chi}^*(i) \le (\bar{\chi}(i))^{|i|}$ *then* Unif $(\lambda, S, \bar{\mu}, \bar{\chi}^*)$.

Proof. (1) Easy. Let $G^i_\xi : \bar{\mu}^*(i) \to \bar{\mu}(i)$ (for $\xi < \kappa$) be such that for every $\langle \alpha_\xi : \xi < \kappa \rangle \in {}^\kappa\bar{\mu}(i)$ there is a unique $\gamma < \bar{\mu}^*(i)$ such that $(\forall \xi < \kappa) G^i_\xi(\gamma) = \alpha_\xi$ that is, identifying $\bar{\mu}(i)^\kappa$ with the cartesian product ${}^\kappa\bar{\mu}(i)$, the function G^i_ξ is the projection onto the ξ-th coordinate. Similarly $H^i_\xi : \bar{\chi}^*(i) \to \bar{\chi}(i)$ for $\xi < \kappa$. If F exemplifies Unif $(\lambda, S, \bar{\mu}, \bar{\chi})$ let us define F^*:

For $\eta \in D(\bar{\mu}^*)$ let $F^*(\eta)$ be the unique $\gamma < \chi^*(\ell g(\eta))$ such that

$$(\forall \xi < \kappa)[F(\langle G^i_\xi(\eta(i)) : i < \ell g(\eta) \rangle) = H^i_\xi(\gamma)]$$

So given $h \in \bigtimes_{i<\lambda} \bar{\chi}^*(i)$ we have to find appropriate η. Let $h_\xi \in \bigtimes_{i<\lambda} \bar{\chi}(i)$ be such that $h_\xi(i) = H^i_\xi(h(i))$. By the choice of F, for each $\xi < \kappa$ there is $\eta_\xi \in \bigtimes_{i<\lambda} \bar{\mu}(i)$ such that $C_\xi \overset{\text{def}}{=} \{\delta \in S : F(\eta_\xi \restriction \delta) = h_\xi(\delta)\} \in \mathcal{D}_\lambda + S$. Define $\eta(i)$ as the unique $\gamma < \bar{\mu}^*(i)$ such that $\langle \eta_\xi(i) : \xi < \kappa \rangle = \langle G^i_\xi(\gamma) : \xi < \kappa \rangle$. Now $\bigcap_{\xi<\kappa} C_\xi \in \mathcal{D}_\lambda + S$ and for every $\delta \in \bigcap_{\xi<\kappa} C_\xi$ we have $F^*(\eta) = h(\delta)$ so we finish.

(2) Similarly.

(3) Without loss of generality $\bar{\chi}^*(i) = |\bar{\chi}(i)|^{|i|}$ (by 1.4(5)). Let $\langle h_\zeta : \zeta < \lambda \rangle$ be such that: h_ζ is a strictly increasing function from λ to λ and $\langle \mathrm{Rang}(h_\zeta) : \zeta < \lambda \rangle$ are pairwise disjoint and $\bar{\mu}(i) \le \bar{\mu}(h_\zeta(i))$ (for $\zeta < \lambda, i < \lambda$). Let $H^i_\xi : \bar{\chi}^*(i) \to \bar{\chi}(i)$ for $\xi < i < \lambda$ be as in the proof of part (1). Let

$$C^* = \{\delta < \lambda : \delta \text{ a limit ordinal such that for every } \zeta < \delta, \text{ the order type of } \delta \cap \mathrm{Rang}(h_\zeta) \text{ is } \delta \text{ so } h_\zeta \text{ maps } \delta \text{ to } \delta\}.$$

Lastly define F^* by: if $\delta \in C^*$ and $\eta \in D_\delta(\bar{\mu})$, let $\eta^{[\zeta]} \in D_\delta(\bar{\mu})$ be defined by $\eta^{[\zeta]}(i) = \eta(h_\zeta(i))$, and $F^*(\eta)$ is defined such that $H^\delta_\xi(F^*(\eta)) = F(\eta^{[\xi]})$; $F^*(\eta) = 0$ otherwise. The checking is as above. $\square_{1.7}$

1.8 Conclusion. If $\mathrm{Unif}\,(\lambda, \mu(0), 2, \chi), 1 < \kappa < \lambda$ and $\mu(0)^\kappa = \mu(0)$ *then* $\mathrm{Unif}\,(\lambda, \mu(0), 2, \chi^\kappa)$.

Proof. By the previous lemma 1.7(1) we have $\mathrm{Unif}\,(\lambda, \mu(0)^\kappa, 2^\kappa, \chi^\kappa)$ and as $\mu(0) = \mu(0)^\kappa$ by applying 1.5 twice this is equivalent to $\mathrm{Unif}\,(\lambda, \mu(0), 2, \chi^\kappa)$.

1.9 Lemma . 1) $\mathrm{Id} - \mathrm{Unif}\,(\lambda, \bar{\mu}, \bar{\chi})$ is either $\mathcal{P}(\lambda)$ or an ideal on λ.
2) If $\bar{\mu}$ is non decreasing *then* $\mathrm{Id} - \mathrm{Unif}\,(\lambda, \mu, \bar{\chi})$ is either $\mathcal{P}(\lambda)$ or a normal ideal on λ (i.e., on $\mathcal{P}(\lambda)$) containing all nonstationary sets.

1.9A Remark. Note that $\mathrm{Id} - \mathrm{Unif}\,(\lambda, \bar{\mu}, \bar{\chi})$ is equal to $\mathrm{Id} - \mathrm{Unif}\,(\lambda, \mu_0, \mu_1, \bar{\chi})$ when μ_0, μ_1 are defined as in 1.5. Also $\mathrm{Id} - \mathrm{Unif}\,(\lambda, \mu_0, \mu_1, \bar{\chi})$ is equal to $\mathrm{Id} - \mathrm{Unif}\,(\lambda, \mu_0, \mu_0, \lambda)$ if $\mathrm{cov}(\mu_0, \lambda) = \mu_0$ (see Definition 1.12 below) by 1.14(5), (6) below (applied twice), so of course the normality holds in such cases.

Proof. 1) Trivial.
2) Call the ideal I. Trivially any nonstationary $S \subseteq \lambda$ belongs to I. So it is enough to prove that *if* $S \subseteq \lambda$ and f is a function from λ to λ such that $(\forall \alpha \in S) f(\alpha) < 1+\alpha$, and for every $i < \lambda$ we have $S_i \overset{\mathrm{def}}{=} \{\alpha \in S : f(\alpha) = i\} \in I$ *then* $S \in I$. Let F_i exemplify that $S_i \in I$ and $\langle h_\zeta : \zeta < \lambda \rangle$, C^* be as in the proof of 1.7(3). Let us define F: if $\eta \in D(\mu_0, \mu_1)$, $\ell g(\eta) \in S_i \cap C^*$, we let $F(\eta)$ be $F_i(\langle \eta(h_i(j)) : j < \ell g(\eta) \rangle)$, otherwise $F(\eta) = 0$, and we can finish as in the proof of 1.7(3). $\square_{1.9}$

1.10 Theorem. 1) Assume the following conditions hold:
(A) λ regular and $2^{<\lambda} < 2^\lambda$.
(B) $\mu^{\aleph_0} < 2^\lambda$.
Then $\mathrm{Unif}\,(\lambda, \mu, 2^{<\lambda}, 2^{<\lambda})$ fails.

2) Moreover in part (1) instead of (B) it suffices to assume
(B′) The following property does not hold:
(∗) There is a family $\{S_i : i < 2^\lambda\}$, $S_i \subseteq \mu$, $|S_i| = \lambda$ and $i \neq j < 2^\lambda$ implies $|S_i \cap S_j| < \aleph_0$.

1.10A Conclusion. If for some $\theta < \lambda$, $2^\theta = 2^{<\lambda} < 2^\lambda$ (hence λ regular uncountable) *then* Unif $(\lambda, 2^{<\lambda}, 2^{<\lambda}, 2)$ fails.

[Why? This holds as by 1.10 applied to $\mu = 2^{<\lambda}$ we get $\neg$ Unif $(\lambda, 2^{<\lambda}, 2^{<\lambda}, 2^{<\lambda})$ now apply 1.7(1) for $\kappa = \theta$.]

Proof. First notice that (B) $\Longrightarrow$ (B′). [Why? Assume by contradiction that (*) holds, choose $T_i \subsetneq S_i$ countable for every $i < 2^\lambda$. So necessarily $i \neq j \Rightarrow T_i \neq T_j$, and we got $\{T_i : i < 2^\lambda\} \subseteq \{S \subseteq \mu : |S| = \aleph_0\}$, i.e., $2^\lambda \le \mu^{\aleph_0}$ contradiction to $\mu^{\aleph_0} < 2^\lambda$.]

Therefore from now till the end of the proof of 1.11 we assume that (*) fails. This implies $\mu < 2^\lambda$ as if $\mu = 2^\lambda$ then the family $\{S_i : i < 2^\lambda\}$ where $S_i \overset{\text{def}}{=} \{\alpha : \lambda i \le \alpha < \lambda i + \lambda\}$ for $i < 2^\lambda$ would show that (*) holds trivially . We also assume the conclusion of the theorem fails (i.e., Unif $(\lambda, \mu, 2^{<\lambda}, 2^{<\lambda})$ holds) and eventually get a contradiction. Let F exemplify Unif $(\lambda, \mu, 2^{<\lambda}, 2^{<\lambda})$. Let us define:

$$\begin{aligned}\text{Mod} = \{\langle \alpha, C_0, g_0, C_1, g_1, \dots C_\beta, g_\beta, \dots\rangle_{\beta<\beta(0)} :\ & \beta(0), \alpha < \lambda, \\ & g_\beta \text{ a function from } \alpha \setminus \{0\} \text{ to } {}^{\lambda>}2,\ C_\beta \text{ a closed subset of } \alpha\}\end{aligned}$$

Clearly $|\text{Mod}| = 2^{<\lambda}$ hence we can fix a one-to-one function $H :$ Mod $\to {}^{\lambda>}2$. Now for every function $f : \lambda \to \{0,1\}$ we shall define by induction on $\beta < \lambda$, functions $h_{f,\beta} : \lambda \to {}^{\lambda>}2$ and $g_{f,\beta} \in D_\lambda(\mu, {}^{\lambda>}2)$ and a closed unbounded subset $C_{f,\beta}$ of λ. If we have defined for every $\beta < \gamma, \gamma < \lambda$ let us define $h_{f,\gamma}$, $g_{f,\gamma}, C_{f,\gamma}$ as follows.

If $\gamma = 0$, let $h_{f,\gamma} = g_{f,\gamma} = f$ and $C_{f,\gamma} = \lambda \setminus \{0\}$.

If $\gamma > 0$, let:

A) $h_{f,\gamma}(i)$ is $H(\langle \alpha, C_{f,0} \cap \alpha, g_{f,0} \restriction (\alpha \setminus \{0\}), \dots, C_{f,\beta} \cap \alpha, g_{f,\beta} \restriction (\alpha \setminus \{0\}), \dots\rangle_{\beta<\gamma})$ where $\alpha = \alpha(i, f, \gamma) = \text{Min}(\bigcap_{\beta<\gamma} C_{f,\beta} \setminus (i+1))$

B) As $h_{f,\gamma} : \lambda \to {}^{\lambda>}2$ is defined, and as we are assuming Unif $(\lambda, \mu, 2^{<\lambda}, 2^{<\lambda})$ is exemplified by F, there are a function $g \in D_\lambda(\mu, 2^{<\lambda})$, and a closed unbounded subset C of λ such that: $C \subseteq \{\delta < \lambda : F(g \restriction \delta) = h_{\beta,\gamma}(\delta)\}$. Now let $g_{f,\gamma} = g$ and $C_{f,\gamma}$ be the set of accumulation points of $\bigcap_{\beta<\gamma} C_{f,\beta} \cap C$. In order to finish the proof we need (proved later):

1.11 Fact. If $f_1, f_2 \in {}^{\lambda}2$, and $j_n < \lambda$ for $n < \omega$, $[n \neq m \to j_n \neq j_m]$, and $\delta_n \stackrel{\text{def}}{=} \text{Min} C_{f_1,j_n} = \text{Min} C_{f_2,j_n}$ and $g_{f_1,j_n} \restriction \delta_n = g_{f_2,j_n} \restriction \delta_n$ and $f_1(0) = f_2(0)$ *then* $f_1 = f_2$.

Continuation of the proof of 1.10.

For every $f : \lambda \to \{0,1\}$, define
$A_f = \{\langle j, g_{f,j}(0), g_{f,j} \restriction (\delta \setminus \{0\}), f(0)\rangle : j < \lambda, \delta = \text{Min} C_j\}$. Clearly $|A_f| = \lambda$. If $A_{f_1} \cap A_{f_2}$ is infinite, we can easily get the hypothesis of Fact 1.11 hence $f_1 = f_2$. So $A_{f_1} \cap A_{f_2}$ is finite for $f_1 \neq f_2$. The A_f's are not subsets of μ but of $A^* = \lambda \times \mu \times {}^{\lambda>}({}^{<\lambda}2) \times 2$, which is a set of cardinality $\mu + 2^{<\lambda}$ so $\mathcal{P} = \{A_f : f$ a function from λ to $\{0,1\}\}$ is a family of 2^λ subsets of A^*, each of power λ, the intersection of any two is finite. If $|A^*| = \mu$ we finish (having contradicted $(*)$ of $(\text{B})'$), otherwise $|A^*| = 2^{<\lambda}$ and $2^\lambda = |\{A_f : f$ a function from λ to $2\}| \leq |A^*|^{\aleph_0} \leq (2^{<\lambda})^{\aleph_0} = 2^{<\lambda} < 2^\lambda$ (second inequality-as in the proof of $(\text{B}) \Rightarrow (\text{B}')$ above), contradiction.

Proof of Fact 1.11. By Ramsey theorem, and as the ordinals are well ordered, w.l.o.g. $j_0 < j_1 < \ldots < j_n < j_{n+1} < \cdots$, and let $j \stackrel{\text{def}}{=} \bigcup_{n<\omega} j_n$

Let $C^\ell = \bigcap_{n<\omega} C_{f_\ell,j_n}$ for $\ell = 1,2$, and let $C^\ell = \{\gamma_i^\ell : i < \lambda\}$, γ_i^ℓ increasing continuous, and let $\gamma_\lambda^\ell = \lambda$.

Now we shall prove by induction on $i \leq \lambda$ that:

$$\otimes \quad \left\{ \begin{array}{l} \text{a) } \gamma_i^1 = \gamma_i^2 \\ \text{b) for every } \zeta < j,\ g_{f_1,\zeta} \restriction (\gamma_i^1 \setminus \{0\}) = g_{f_2,\zeta} \restriction (\gamma_i^2 \setminus \{0\}) \\ \qquad \text{and } C_{f_1,\zeta} \bigcap \gamma_i^1 = C_{f_2,\zeta} \bigcap \gamma_i^2. \end{array} \right.$$

This is enough, as in particular it says, for $i = \lambda, \zeta = 0$ that $g_{f_1,0} \restriction (\lambda \setminus \{0\} = g_{f_2,0} \restriction (\lambda \setminus \{0\})$, but by its definition $g_{f_\ell,0} = f_\ell$, so $f_1 \restriction (\lambda \setminus \{0\}) = f_2 \restriction (\lambda \setminus \{0\})$. But in fact we have assumed $f_1(0) = f_2(0)$, so $f_1 = f_2$, which is the desired conclusion of the fact. So for proving the fact, it suffices to prove $\otimes$.

Case I. $i = 0$

We first prove clause a) of $\otimes$. Now for $\ell = 1, 2$ clearly $\gamma_0^\ell = \text{Min}C^\ell \geq \delta_n \stackrel{\text{def}}{=} \text{Min}C_{f_\ell, j_n}$, hence $\gamma_0^\ell \geq \text{Sup}_{n<\omega}\delta_n$. On the other hand for $n < m$, $C_{f_\ell, j_m} \subseteq C_{f_\ell, j_n}$ (as $j_n < j_m$) hence $\langle \delta_m : m < \omega \rangle$ is non decreasing and $\{\delta_m : n \leq m < \omega\} \subseteq C_{f_\ell, j_n}$, hence $\text{Sup}_{m<\omega}\delta_m = \text{Sup}_{m\in[n,\omega)}\, \delta_m \in C_{f_\ell, j_n}$, hence $\text{Sup}_{m<\omega}\delta_m \in \bigcap_{n<\omega} C_{f_\ell, j_n} = C^\ell$, so $\gamma_0^\ell = \text{Min}C^\ell \leq \text{Sup}_{m<\omega}\delta_m$. Clearly we got $\gamma_0^\ell = \text{Min}C^\ell = \text{Sup}_{m<\omega}\delta_m$, so $\gamma_0^1 = \gamma_0^2$.

For clause b) of $\otimes$ we can choose large enough n, such that $\zeta < j_n (< j)$ and

$(*)_0$ $g_{f_1,\zeta}\restriction(\gamma_0^1 \setminus \{0\}) \neq g_{f_2,\zeta}\restriction(\gamma_0^2 \setminus \{0\})$ implies $g_{f_1,\zeta}\restriction(\delta_n \setminus \{0\}) \neq g_{f_2,\zeta}\restriction(\delta_n \setminus \{0\})$
and $C_{f_1,\zeta} \cap \gamma_0^1 \neq C_{f_2,\zeta} \cap \gamma_0^2$ implies $C_{f_1,\zeta} \cap \delta_n \neq C_{f_2,\zeta} \cap \delta_n$

Now we have assumed in the statement of the fact that:

$(*)_1$ $g_{f_1,j_n}\restriction\delta_n = g_{f_2,j_n}\restriction\delta_n$

hence

$(*)_2$ $F(g_{f_\ell,j_n}\restriction\delta_n) = h_{f_\ell,j_n}(\delta_n) = H(\langle \alpha, \ldots, C_{f_\ell,\beta} \bigcap \alpha, g_{f_\ell,\beta}\restriction(\alpha \setminus \{0\}), \ldots \rangle_{\beta<j_n})$
where $\alpha = \alpha(\gamma_i^\ell, f_\ell, j_n) = \text{Min}[\bigcap_{\beta<j_n} C_{f_\ell,\beta} \setminus (\delta_n + 1)]$.

We can conclude, as the left side in $(*)_2$ does not depend on ℓ, (by $(*)_1$) and as H is one-to-one, that $\beta < j_n \Rightarrow g_{f_1,\beta}\restriction(\gamma_0^1 \setminus \{0\}) = g_{f_2,\beta}\restriction(\gamma_0^2 \setminus \{0\})$ and $\beta < j_n \Rightarrow C_{f_1,\beta} \cap \gamma_0^1 = C_{f_2,\beta} \cap \gamma_0^2$. But in particular $\zeta < j_n$ hence $g_{f_1,\zeta}\restriction(\delta_n \setminus \{0\}) = g_{f_2,\zeta}\restriction(\delta_n \setminus \{0\})$ and $C_{f_1,\zeta} \cap \delta_n = C_{f_2,\zeta} \cap \delta_n$ so we have gotten $g_{f_1,\zeta}\restriction(\gamma_0^1 \setminus \{0\}) = g_{f_2,\zeta}\restriction(\gamma_0^2 \setminus \{0\})$ by $(*)_0$. So we have proved clause b) of $\otimes$ (for the case $i = 0$).

Case II. i limit

This is easy: clause a) holds as γ_ξ^ℓ (for $\xi \leq i$) is increasing continuous and $(\forall \xi < i)\ \gamma_\xi^1 = \gamma_\xi^2$ by the induction hypothesis, and similarly clause b) holds.

Case III. Prove for $i + 1$, assuming truth for i.

For any $n < \omega$, $g_{f_1,j_n}\restriction\gamma_0^1 = g_{f_2,j_n}\restriction\gamma_0^2$ by the assumption in the fact. By the induction hypothesis $g_{f_1,j_n}\restriction(\gamma_i^1 \setminus \{0\}) = g_{f_2,j_n}\restriction(\gamma_i^2 \setminus \{0\})$. Together we can conclude

(α) $g_{f_1,j_n}\restriction\gamma_i^1 = g_{f_2,j_n}\restriction\gamma_i^2$ for $n < \omega$

By the definition of g_{f_ℓ,j_n}, for $\ell = 1, 2$ we have

(β) $F(g_{f_\ell,j_n}\restriction\gamma_i^1) = h_{f_\ell,j_n}(\gamma_i^1) =$

$H(\langle\alpha_n^\ell, \ldots, C_{f_\ell,\beta} \cap \alpha_n^\ell, g_{f_\ell,\beta}\restriction(\alpha_n^\ell \setminus \{0\}), \ldots\rangle_{\beta<j_n})$

[where $\alpha_n^\ell = \alpha(\gamma_i^\ell, f_\ell, j_n) = \text{Min}[\bigcap_{\beta<j_n} C_{f_\ell,\beta} \setminus (\gamma_i^\ell + 1)]$

As H is one-to-one, by (α) and (β) we can conclude

(γ) $\langle\alpha_n^1, \ldots, C_{f_1,\beta} \cap \alpha_n^1, g_{f_1,\beta}\restriction(\alpha_n^1 \setminus \{0\}), \ldots\rangle_{\beta<j_n} =$

$= \langle\alpha_n^2, \ldots, C_{f_2,\beta} \cap \alpha_n^2, g_{f_2,\beta}\restriction(\alpha_n^2 \setminus \{0\}), \ldots\rangle_{\beta<j_n}$

So $\alpha_n^1 = \alpha_n^2$; it is also clear that, for $\ell = 1, 2$ $\alpha_0^\ell < \ldots < \alpha_n^\ell < \alpha_{n+1}^\ell < \ldots$ and $\bigcup_{n<\omega} \alpha_n^\ell = \text{Min}[\bigcap_{\beta<j} C_{f_\ell,\beta} \setminus (\gamma_i^\ell + 1)]$ is γ_{i+1}^ℓ, so we can conclude $\gamma_{i+1}^1 = \gamma_{i+1}^2$ (i.e. clause a) of $\otimes$). Also, by (γ), for every $\zeta < j$ for every n large enough, $\zeta < j_n$ and $C_{f_1,\zeta} \cap \alpha_n^1 = C_{f_2,\zeta} \cap \alpha_n^2$, and as this holds for every n and $\alpha_n^1 = \alpha_n^2$ and $\gamma_{i+1}^\ell = \bigcup_{n<\omega} \alpha_n^\ell$ clearly:

(δ) $C_{f_1,\zeta} \cap \gamma_{i+1}^1 = C_{f_2,\zeta} \cap \gamma_{i+1}^2$.

Similarly $g_{f_1,\zeta}\restriction(\gamma_{i+1}^1 \setminus \{0\}) = g_{f_2,\zeta}\restriction(\gamma_{i+1}^2 \setminus \{0\})$, and so we finish proving clause b) of $\otimes$ for $i+1$. So we have finished proving $\otimes$ for all i. As stated earlier by this we prove Fact 1.11. $\square_{1.11}$

$\square_{1.10}$

1.12 Definition. Let X be a set and λ a cardinal.

(1) A family $\mathcal{F}$ of subsets of X is an (X, λ)- cover if for all $S \subseteq X$, $|S| = \lambda$, there is $T \in \mathcal{F}$ such that $S \subseteq T$, and all the members of $\mathcal{F}$ are of cardinality $\leq \lambda$. In other words, $\mathcal{F}$ is cofinal in the directed partial order $(\mathcal{S}_{\leq\lambda}(X), \subseteq)$.

(2) The covering number of (X, λ) which is denoted by $\text{cov}(X, \lambda)$ is :

$$\text{cov}(X, \lambda) = \text{Min}\{|\mathcal{F}| : \mathcal{F} \text{ is a } (X, \lambda)\text{-cover}\}.$$

Clearly $\text{cov}(X, \lambda)$ depends just on $|X|$ and λ (see 1.13(1) below) so we usually use cardinals for X.

1.13 Lemma.

(1) $X \subseteq Y \Rightarrow \text{cov}(X, \lambda) \leq \text{cov}(Y, \lambda)$, and $|X| \leq |Y| \Rightarrow \text{cov}(X, \lambda) \leq \text{cov}(Y, \lambda)$ hence if $|X| = |Y|$ then $\text{cov}(X, \lambda) = \text{cov}(Y, \lambda)$

(2) if $\lambda < \mu$ then $\text{cov}(\mu, \lambda) \geq \mu$

(3) i. $\text{cov}(\lambda, \lambda) = 1$

ii. for $\lambda \leq \mu$ we have $\text{cov}(\mu^+, \lambda) = \text{cov}(\mu, \lambda) + \mu^+$

iii. If μ is a limit cardinal, $\lambda < \mu$ and let $\{\mu_i : i < \text{cf}\mu\}$ be an increasing sequence with limit μ and $\mu_0 > \lambda$; then $\text{cov}(\mu, \lambda) \leq \prod_{i<\text{cf}\mu} \text{cov}(\mu_i, \lambda)$.

(4) $\text{cov}(\lambda^{+\alpha}, \lambda) \leq (\lambda^{+\alpha})^{|\alpha|}$

Remark. See more in [Sh:g], [Sh:400a].

Proof. (1) E.g., if $X \subseteq Y$ and if $\mathcal{F}$ is a (Y, λ)-cover, then $F^\dagger = \{A \cap X : A \in \mathcal{F}\}$ is a (X, λ)-cover and $|\mathcal{F}^\dagger| \leq |\mathcal{F}|$.

(2) Because if $\mathcal{F}$ is a (μ, λ)-cover then $\bigcup\{A : A \in \mathcal{F}\}$ is necessarily μ hence $\mu \leq |\bigcup\{A : A \in \mathcal{F}\}| \leq \sum_{A \in \mathcal{F}} |A| \leq |\mathcal{F}|\lambda$ so $|\mathcal{F}| \geq \mu$.

(3) i. Take $F = \{\lambda\}$. It is obvious that this is a cover as required.

(3) ii. Clearly $\text{cov}(\mu^+, \lambda) \geq \text{cov}(\mu, \lambda)$ by part (1) and $\text{cov}(\mu^+, \lambda) \geq \mu^+$ by part (2). So it suffices to show that $\text{cov}(\mu^+, \lambda) \leq \text{cov}(\mu, \lambda) + \mu^+$. We do this by finding a (μ^+, λ)-cover $\mathcal{F}$ of cardinality $\text{cov}(\mu, \lambda) + \mu^+$. For every ordinal α, $\lambda \leq \alpha < \mu^+$ let $\mathcal{F}_\alpha$ be an (α, λ)-cover such that $|\mathcal{F}_\alpha| = \text{cov}(|\alpha|, \lambda) \leq \text{cov}(\mu, \lambda)$ (we use part (1)). Define $\mathcal{F} = \bigcup_{\alpha<\mu} \mathcal{F}_\alpha$, we shall prove that it is (μ^+, λ)-cover. Let $S \subseteq \mu^+$ be of cardinality λ, from the regularity of μ^+ follows the existence of α, $\mu \leq \alpha < \mu^+$ such that $S \subseteq \alpha$, since $\mathcal{F}_\alpha$ is a (α, λ)-cover there is $T \in \mathcal{F}_\alpha$ ($T \in \mathcal{F}$ since $\mathcal{F}_\alpha \subseteq \mathcal{F}$) such that $S \subseteq T$, $|T| \leq \lambda$, so we have proved one inequality. The other was done before.

(3) iii. We shall find a (μ, λ)-cover $\mathcal{F}$ of the appropriate cardinality. For $i < \text{cf}(\mu)$ let $\mathcal{F}_i$ be a (μ_i, λ)-cover exemplifying $\text{cov}(\mu_i, \lambda)$, define $\mathcal{F}_i^\dagger = \mathcal{F}_i \cup \{\emptyset\}$, $\mathcal{F} = \{\bigcup_{i \in S} s_i : s_i \in \mathcal{F}_i^\dagger, S \subseteq \text{cf}(\mu) \text{ and } |S| \leq \lambda\}$. It is easy to verify that $\mathcal{F}$ is a (μ, λ)-cover and $|\mathcal{F}| \leq \prod_{i<\text{cf}\mu} \text{cov}(\mu_i, \lambda)$.

(4) Prove by induction on $\alpha < \lambda$.

For $\alpha + 0$, we have $\text{cov}(\lambda^{+0}, \lambda) = \text{cov}(\lambda, \lambda) = 1 \leq (\lambda)^{+0}$ (by 3 (i)).

For $\alpha = \beta + 1$ we have $\text{cov}(\lambda^{+\alpha}, \lambda) = \text{cov}(\lambda^{+(\beta+1)}, \lambda) = \text{cov}((\lambda^{+\beta})^+, \lambda) =$

$\text{cov}(\lambda^{+\beta}, \lambda) + (\lambda^{+\beta})^+$ where the last equality holds by clause (ii) of part (3); now, using the induction hypothesis, $\text{cov}(\lambda^{+(\beta+1)}, \lambda) \leq (\lambda^{+\beta})^{|\beta|} + (\lambda^{+\beta})^+ \leq (\lambda^{+\alpha})^{|\alpha|}$.

For α a limit ordinal; let $\{\alpha_i : i < \text{cf}(\alpha)\}$ be a cofinal sequence in α; then by 3(iii) $\text{cov}(\lambda^{+\alpha}, \lambda) \leq \prod_{i<\text{cf}\mu} \text{cov}(\lambda^{+\alpha_i}, \lambda) \leq (\lambda^{+\alpha})^{\text{cf}\alpha} \leq (\lambda^{+\alpha})^{|\alpha|}$. $\square_{1.13}$

1.14 Lemma. 1) Let $\lambda \leq \mu < 2^\lambda$, χ_1, χ be cardinals, $\bar{\chi} = \langle \chi_i : i < \lambda \rangle$, $\chi = \sup\{\chi_i : i < \lambda\}$, λ regular uncountable, *then*

$$\text{Unif}\,(\lambda, \mu, \mu, \bar{\chi}) \text{ implies } \text{Unif}\,(\lambda, \text{cov}(\mu, \lambda), \lambda, \bar{\chi}).$$

2) In part (1) assume $\mu_0 + \mu_1 + \chi < 2^\lambda$, $\lambda \leq \chi$ and $\text{cov}(\chi, \lambda) \leq \mu_0$ (and $\mu_1 \geq 2$). *Then* $\text{Unif}\,(\lambda, \mu_0, \mu_1, \chi) \iff \text{Unif}\,(\lambda, \mu_0, \mu_1, \lambda)$.

3) In part (2) if in addition $\mu_0 \leq \mu_1$, λ is not strong limit and only $2 \leq \chi$ is required then $\text{Unif}\,(\lambda, \mu_0, \mu_1, \chi) \Leftrightarrow \text{Unif}\,(\lambda, \mu_0, \mu_1, 2)$.

4) If $\lambda \leq \mu_0 \leq \mu_1 < 2^\lambda$ and $\bar{\chi} = \langle \chi_i : i < \lambda \rangle$ is a sequence of cardinals, λ is regular uncountable *then*

$$\text{Unif}\,(\lambda, \mu_0, \mu_1, \bar{\chi}) \Rightarrow \text{Unif}\,(\lambda, \mu_0 + \text{cov}(\mu_1, \lambda), \lambda, \bar{\chi}).$$

5) In part (4) if $\mu_0 \geq \text{cov}(\mu_1, \lambda) \geq \mu_1 \geq 2$ *then*

$$\text{Unif}\,(\lambda, \mu_0, \mu_1, \bar{\chi}) \Leftrightarrow \text{Unif}\,(\lambda, \mu_0, 2, \bar{\chi}).$$

6) We get similar results if we add $S \subseteq \lambda$ is a parameter (in parts 1) - 5)).

Proof. 1) We do it by translating every $g \in D(\mu, \mu)$ to $g^* \in D(\text{cov}(\mu, \lambda), \lambda)$ where the first coordinate $g^*(0)$ codes a subset of μ of cardinality λ which covers $\text{Rang}(g)$, and $g^*(1+i)$ tells us where $g(i)$ appears in it (e.g. the place in some well ordering) of order type λ. More formally let $\kappa \overset{\text{def}}{=} \text{cov}(\mu, \lambda)$, and let $\mathcal{F} = \{A_i : i < \kappa\}$ exemplify this, where w.l.o.g. $A_i \neq \emptyset$ and let $A_i = \{\alpha_{i,j} : j < \lambda\}$ possibly with repetition. We define a function H from ${}^\lambda\mu$ to $D_\lambda(\kappa, \lambda)$. For a given $g : \lambda \to \mu$ let $h = H(g)$ be defined by:

$h(0) = \min\{i < \kappa : \{g(\alpha) : \alpha < \lambda\} \subseteq A_i\}$ and $h(1+i)$ is the first $j < \lambda$ such that $g(i) = \alpha_{h(0),j}$. Let F exemplify $\mathrm{Unif}(\lambda, \mu, \mu, \bar{\chi})$, and we shall define F^* which will exemplify $\mathrm{Unif}(\lambda, \kappa, \lambda, \bar{\chi})$: for $\eta \in D(\kappa, \lambda)$ let $F^*(\eta) = F(\langle \alpha_{\eta(0),\eta(1+i)} : 1+i < \ell g \eta \rangle)$ if $\eta \neq \langle\rangle$, and $F^*(\eta) = 0$ if $\eta = \langle\rangle$.

2) By 1.4(5) the implication $\Rightarrow$ holds.

For the other direction, assume that $\mathrm{Unif}(\lambda, \mu_0, \mu_1, \lambda)$ holds. Let $\mathcal{F} = \{A_i : i < \mu_0\}$ exemplify $\mathrm{cov}(\chi, \lambda) \leq \mu_0$ with $A_\zeta = \{\alpha_{\zeta,j} : j < \lambda\}$ and let F exemplify $\mathrm{Unif}(\lambda, \mu_0, \mu_1, \lambda)$, and let $\mathrm{pr}(-,-)$ be a pairing function on μ_0 (so it is onto μ_0). Now we define F^* as follows: $F^*(\langle\rangle) = 0$ and for $\eta \in D(\mu_0, \mu_1) \backslash \{\langle\rangle\}$, let $\eta(0) = \mathrm{pr}(\beta_0, \beta_1)$, $\nu_\eta = \langle \beta_1 \rangle {}^\frown \eta \restriction [1, \ell g(\eta))$, and we choose $F^*(\eta) \stackrel{\mathrm{def}}{=} \alpha_{\beta_0, F(\nu_\eta)}$. Now check that F^* exemplifies $\mathrm{Unif}(\lambda, \mu_0, \mu_1, \chi)$; for any $g \in {}^\lambda\chi$, let $\mathrm{Rang}(g) \subseteq A_\zeta$, $g(i) = \alpha_{\zeta,h(i)}$ where $h \in {}^\lambda\lambda$; let $\eta^* \in D_\lambda(\mu_0, \mu_1)$ be such that for some club C of λ, $\delta \in C \Rightarrow F(\eta^* \restriction \delta) = h(\delta)$. Now define $\nu^* \in D_\lambda(\mu_0, \mu_1)$ as follows: $\nu^* \restriction [1, \lambda) = \eta^* \restriction [1, \lambda)$ and $\nu^*(0) = \mathrm{pr}(\zeta, \eta^*(0))$. Easily $\delta \in C \,\&\, \delta > 0 \Rightarrow F^*(\nu^* \restriction \delta) = \alpha_{\zeta,h(\delta)} = g(\delta)$, as required

3) W.l.o.g. $2 \leq \mu_1$ (otherwise the statements are trivially false). By monotonicity (=1.4(5)) and part (2) without loss of generality $\chi = \lambda$, and we have to prove the $\Leftarrow$ direction. Now apply 1.7(3) and 1.5.

4) Repeat the proof of part (1).

5) The implication $\Rightarrow$ holds by monotonicity (that is by 1.4(5)). The implication $\Leftarrow$ holds by part (4) above and 1.5.

6) Same proofs. $\square_{1.14}$

1.15 Conclusion. Let $\mu < \aleph_{\omega_1}$ and assume $\mu^{\aleph_0} < 2^{\aleph_1}$, *then* $\mathrm{Unif}(\aleph_1, \mu, \mu, 2)$ fails.

Proof. Assume toward contradiction $\mathrm{Unif}(\aleph_1, \mu, \mu, 2)$; from Claim 1.7(3) we obtain $\mathrm{Unif}(\aleph_1, \mu, \mu, 2^{<\aleph_1})$ is true, apply Lemma 1.14(1) and we have

$$\mathrm{Unif}(\aleph_1, \mathrm{cov}(\mu, \aleph_1), \aleph_1, 2^{<\aleph_1}).$$

Now by Lemma 1.13(4) (let $\aleph_\alpha = \mu, \alpha < \omega_1$) $\mathrm{cov}(\mu, \aleph_1) \leq \aleph_\alpha^{|\alpha|} \leq \mu^{\aleph_0} < 2^{\aleph_1}$. This is a contradiction to theorem 1.10. $\square_{1.15}$

We can strengthen theorem 1.10 to

1.16 Theorem. Suppose λ is regular uncountable, $2^{<\lambda} < 2^\lambda$, and $\mu > \lambda$. If Unif $(\lambda, \mu, 2^{<\lambda}, 2^{<\lambda})$ holds *then:*

$(*)_{2^\lambda,\mu,\lambda^+}$ There is a family $\{S_i : i < 2^\lambda\}$, $S_i \subseteq \mu, |S_i| = \lambda^+$ and $|S_i \cap S_j| < \aleph_0$ for $i \neq j$.

Proof. Similar to 1.10; we may assume $\mu < 2^\lambda$, otherwise the conclusion is trivial. From the proof of 1.10 we get $2^\lambda \leq \mu^{\aleph_0}$. Hence we may assume $\mu \geq (2^{<\lambda})^{+\omega}$ (otherwise we have $\mu = (2^{<\lambda})^{+n}$ for some n, so by the Hausdorff formula we get $\mu^{\aleph_0} = \mu + (2^{<\lambda})^{\aleph_0} = \mu + 2^{<\lambda} = \mu < 2^\lambda \leq \mu^{\aleph_0}$, a contradiction). Let for every $\alpha < \lambda^+$, $\alpha = \bigcup_{i<\lambda} B_i^\alpha$, $|B_i^\alpha| < \lambda$, B_i^α increasing continuous in i, and we can assume: $B_i^\alpha \cap \lambda = i$, and $\beta, j \in B_i^\alpha \Rightarrow B_j^\beta \subseteq B_i^\alpha$. For notational convenience let $B(\alpha, i) = B_i^\alpha$. We follow the proof of 1.10 and mention only the differences. We let

Mod $= \{\langle \alpha, \ldots, C_\beta, g_\beta, \ldots \rangle_{\beta \in B(\alpha,i)} : \beta < \lambda^+, i < \lambda, g_\beta$ a function from $B_i^\alpha \setminus \{0\}$ to ${}^{\lambda>}2$, C_β a closed subset of $i\}$, so from $x \in$ Mod we can reconstruct α and $B(\alpha, i)$ hence i. Now for every $f : \lambda \to \{0, 1\}$ we define by induction on $\beta < \lambda^+$ functions $h_{f,\beta} : \lambda \to {}^{\lambda>}2$, $g_{f,\beta} \in D_\lambda(\mu, {}^{\lambda>}2)$ and a closed unbounded subset $C_{f,\beta}$ of λ.

If we have defined for every $\beta < \gamma$ and $\gamma > 0$, let

$$h_{f,\gamma}(i) = H(\langle \alpha, \ldots, C_{f,\beta} \cap \alpha, g_{f,\beta} \restriction (\alpha \setminus \{0\}), \ldots \rangle_{\beta \in B(\gamma,i)})$$

where $\alpha = \alpha(i, f, \gamma)$ is the minimal $\alpha > i, \alpha \in \bigcap \{C_{f,\beta} : \beta \in B(\gamma, i)\}$ and we let

$$C_{f,\gamma} \stackrel{\text{def}}{=} \{\delta : \text{ if } \beta \in B(\gamma, \delta) \text{ then } \delta \text{ is an accumulation point of } C_{f,\beta}\}.$$

We modify Fact 1.11 to : there are no distinct $j_n < \lambda^+$ for $n < \omega$ and $f_0 \in {}^\lambda 2$ such that the set $Y \stackrel{\text{def}}{=} \{f \in {}^\lambda 2 : g_{f,j_n}(0) = g_{f_0,j_n}(0)$ for each $n < \omega\}$ has power $> 2^{<\lambda}$ (the number is just to give us two distinct f's as required for starting the induction there).

How do we prove this new version of 1.11? Assume $\langle j_n : n < \omega \rangle$ and f_0 form a counterexample. Without loss of generality $\bigwedge_n j_n < j_{n+1}$ and choose

$i < \lambda$ large enough such that $j_n \in B(j_m, i)$ for $n < m$, and for each $f \in Y$ let $\alpha(f) = \mathrm{Min}\{\alpha : \alpha > i,\ \alpha \in \bigcap_{n<\omega} C_{f,j_n}\}$; we define a relation E on Y:

$$\begin{aligned} f_1 E f_2 \quad \textit{iff} \quad & \alpha(f_1) = \alpha(f_2) \text{ and for } n < \omega, \\ & f_1 {\restriction} \alpha(f) = f_2 {\restriction} \alpha(f) \text{ and } g_{f_1,j_n} {\restriction} \alpha(f) = g_{f_2,j_n} {\restriction} \alpha(f) \\ & \text{and } C_{f_1,j_n} \cap \alpha(f) = C_{f_2,j_n} = \alpha(f). \end{aligned}$$

Now E is an equivalence relation with $\leq \lambda \times (2^{<\lambda})^{\aleph_0} \times (2^{<\lambda})^{\aleph_0} = 2^{<\lambda}$ equivalence classes. So we can find $f_1 \neq f_2 \in Y$ which are equivalent.

Now $g_{f_1,j_n}(0) = g_{f_0,j_n}(0) = g_{f_2,j_n}(0)$ by the definition of Y. Now we can apply the proof of 1.11 to f_1, f_2, contradicting the choice of $f_1 \neq f_2$.

Why is this new version of 1.11 enough? For $f_0 \in {}^{\lambda}2$ let $Y'_{f_0} \stackrel{\text{def}}{=} \{f : f \in {}^{\lambda}2$ and for infinitely many $j < \lambda^+$ we have $g_{f,j}(0) = g_{f_0,j}(0)\}$, now the number of possible $\langle j_n : n < \omega\rangle$ is $\leq (\lambda^+)^{\aleph_0} \leq 2^{<\lambda} + \lambda^+$ which is $< 2^{\lambda}$. Moreover $\sup\{|Y'_f| : f \in {}^{\lambda}2\} \leq \lambda^+ + 2^{<\lambda} < 2^{\lambda}$. As $f_0 \in Y'_{f_1} \Leftrightarrow f_1 \in Y'_{f_0}$ we can find $F^* \subseteq {}^{\lambda}2$ such that $|F^*| = 2^{\lambda}$ and $f_0 \in F^* \;\&\; f_1 \in F^* \;\&\; f_0 \neq f_1 \Rightarrow f_0 \notin Y'_{f_1}$. So $\{\{\langle g_{f,j}(0), j\rangle : j < \lambda^+\} : f \in F^*\}$ is a family of 2^{λ} subsets of $\mu \times \lambda^+$; which by the choice of F^* satisfies: the intersection of any two is finite, confirming $(*)$ of 1.16 (note that without this symmetry we could have used Hajnal's free subset theorem [Ha61]). $\square_{1.16}$

1.17 Conclusion. If $\lambda = \mathrm{cf}(\lambda) > \aleph_0$, $2^{\lambda} > \mu \geq 2^{<\lambda} = 2^{\kappa}$, $\mathrm{cov}(\mu, \lambda) < 2^{\lambda}$ *then* Unif $(\lambda, \mu, \mu, 2)$ fails.

Proof. Let $\sigma \stackrel{\text{def}}{=} \mathrm{cov}(\mu, \lambda)$ and let us assume toward contradiction that Unif $(\lambda, \mu, \mu, 2)$. Now by Claim 1.7(3) we have Unif $(\lambda, \mu, \mu, 2^{<\lambda})$, and by Lemma 1.14(1) we have Unif $(\lambda, \mathrm{cov}(\mu, \lambda), \lambda, 2^{<\lambda})$ i.e. Unif $(\lambda, \sigma, \lambda, 2^{<\lambda})$ hence by monotonicity (i.e. 1.4(5)) we have Unif $(\lambda, \sigma, 2^{<\lambda}, 2^{<\lambda})$, so by 1.16 we know that $(*)_{2^{\lambda},\sigma,\lambda^+}$ holds. Now we would like to apply [Sh:430, 2.1(2)], with κ^+, λ, μ here standing for κ, λ, μ there, but we have to check the assumptions there: "$\mu > \lambda \geq \kappa$" is obvious, as $\mu > \lambda \geq \kappa^+$; as for "$\mathrm{cov}(\lambda, \kappa, \kappa, 2) \leq \mu$" trivially $|\mathcal{S}_{<\kappa^+}(\lambda)| \leq \mu$ suffices but $|\mathcal{S}_{<\kappa^+}(\lambda)| = \lambda^{\kappa} \leq 2^{<\lambda} \leq \mu$. Now "$\mathrm{cov}(\mu, \lambda^+, \lambda^+, 2)$"

there means $\text{cov}(\mu, \lambda)$ here, so we get $\sigma^{<\kappa^+} = \sigma$. Hence $\sigma = \sigma^{\aleph_0}$ hence $(*)_{2^\lambda,\sigma,\lambda^+}$ is impossible. $\square_{1.17}$

1.18 Conclusion. 1) If $\theta < \lambda$ are regular cardinals, $2^\theta = 2^{<\lambda} < 2^\lambda$, and $\lambda \leq \mu < 2^\lambda$ and $\neg(*)_{2^\lambda,\mu,\lambda^+}$ (this is the statement from 1.16) then $\neg\,\text{Unif}\,(\lambda, \mu, 2^{<\lambda}, 2^\theta)$.

2) Under the assumptions of 1) if $\text{cov}(\mu, \lambda) \leq \mu$ or just $\text{cov}(2^\theta, \lambda) \leq \mu$ then $\neg\,\text{Unif}\,(\lambda, \mu, \mu, \lambda)$.

Proof. 1) By 1.16 we have $\neg\,\text{Unif}\,(\lambda, \mu, 2^{<\lambda}, 2^{<\lambda})$ i.e. $\neg\,\text{Unif}\,(\lambda, \mu, 2^{<\lambda}, 2^\theta)$.
2) By part (1) and 1.14(2). $\square_{1.18}$

1.19 Conclusion. 1) If $\theta < \lambda$ are regular cardinals $2^\theta = 2^{<\lambda} < 2^\lambda$ (e.g. $\lambda = \theta^+$, $2^\theta < 2^\lambda$) and $\theta \geq \beth_\omega$ *then* for every $\mu < \lambda$, we have $\neg\,\text{Unif}\,(\lambda, \mu, 2^\theta, 2^\theta)$.
2) Moreover if $\text{cov}(\mu, \lambda) < 2^\lambda$ then $\neg\,\text{Unif}\,(\lambda, \mu, 2^\theta, \lambda)$.

Proof.
1) By 1.18 it suffices to prove $\neg(*)_{2^\lambda,\mu,\lambda^+}$ which is proved in [Sh:460]. For the reader's benefit we derive it from the main theorem of [Sh:460]. As $\mu \geq \theta \geq 1$ main theorem of [Sh:460] says that for every regular large enough $\kappa < \beth_\omega$, the κ-revised power of μ, $\mu^{[\kappa]}$, is μ where

$$\mu^{[\kappa]} = \min\{|\mathcal{P}| : \mathcal{P} \subseteq \mathcal{S}_{\leq\kappa}(\mu) \text{ and every } a \subseteq \mathcal{S}_{\leq\kappa}(\mu) \text{ is included in a union of } < \kappa \text{ members of } \mathcal{P}\}$$

Let $\mathcal{P} \subseteq \mathcal{S}_{\leq\kappa}(\mu)$ exemplified $\mu^{[\kappa]} = \mu$, and let $\mathcal{P}_1 = \{b : |b| = \kappa \text{ and } (\exists a)(b \subseteq a \in \mathcal{P})\}$, so $\mathcal{P}_1 \subseteq \mathcal{S}_{\leq\kappa}(\mu)$, $|\mathcal{P}_1| \leq \mu \times 2^\kappa \leq \mu + \beth_\omega = \mu$.
Now if $\{S_i : i < 2^\lambda\} \subseteq \mathcal{S}_{\leq\lambda^+}(\mu)$ is as required in $(*)_{2^\lambda,\mu,\lambda^+}$, each S_i contains some a_i of cardinality κ, hence for some $\zeta_i^* < \kappa$, $b_{i,\zeta} \in \mathcal{P}$ for $\zeta < \zeta_i^*$ we have $a_i \subseteq \bigcup_{\zeta<\zeta_i^*} b_{i,\zeta}$, hence for some $\zeta(i)$ we have $c_i \stackrel{\text{def}}{=} a_i \cap b_{i,\zeta(i)}$ has cardinality κ. Clearly $c_i \in \mathcal{P}_1$, but $|\mathcal{P}_1| \leq \mu < 2^\lambda$ hence for some $i < j < 2^\lambda$, $c_j = c_i$ so

$c_j = c_i$ is a subset of $S_i \cap S_j$ of cardinality κ contradiction to the choice of $\{S_i : i < 2^\lambda\}$.

2) By part (1) and 1.18(2). $\square_{1.19}$

Remark. Even for smaller λ, $(*)_{2^\lambda,\mu,\lambda^+}$ is a very strong requirement, and it is not clear if it is consistent with ZFC. By [Sh:420, §6] it implies that there are regular cardinals $\theta_i \in (2^{<\lambda}, \mu)$ for $i < \lambda$ such that $\prod\limits_{i<\lambda} \theta_i / \mathcal{S}_{<\aleph_0}(\lambda)$ is μ^+-directed and even has true cofinality which is $> \mu$.

1.20 Question. 1) Does $\lambda = \mathrm{cf}(\lambda) > \aleph_0$, $\lambda \leq 2^{<\lambda} < 2^\lambda$ imply that Unif $(\lambda, 2, 2, 2)$ fails?

2) Is it consistent with ZFC that for some strongly inaccessible λ we have Unif $(\lambda, 2, 2, 2)$ fails?

3) Can we prove in 1.14(2) equality? can we omit the "λ not strong limit" in 1.14(3)?

4) How complete is Id $-$ Unif $(\lambda, \mu_0, \mu_1, \bar{\chi})$?

§2. On the Power of Ext and Whitehead's Problem

Let the word group stand here for abelian group, for notational simplicity. A comprehensive book of set-theoretic methods in Abelian group theory is [EM].

By [Sh:44], [Sh:52] if G is a non-free group and $V = L$ then $\mathrm{Ext}(G, \mathbb{Z}) \neq \{0\}$. In Hiller, Huber and Shelah [HHSh:91], it is proved that if $V = L$, the torsion free rank of $\mathrm{Ext}(G, \mathbb{Z})$ is the immediate upper bound: $\mathrm{Min}\{2^{|K|} : K$ a subgroup of G such that G/K free $\}$.

Now in fact not the full power of the axiom $V = L$ is used, just the satisfaction of the diamond principle for every stationary subset of a regular uncountable cardinal. Devlin and Shelah [DvSh:65] introduced a weakening of this principle, and in [HHSh:91] we stated that for the result mentioned above it is enough that the weak diamond holds for every stationary subset of any regular uncountable cardinal. Here we prove a somewhat stronger result, using failure of cases of Unif, (e.g., $\chi = 2^{\aleph_0}$ suffice). Meanwhile Eklof and Huber [EkHu] found an alternative proof, more group-theoretic, for the result with weak diamond (really a slight weakening)

On the difference between weak diamond and failure of Unif, and between variants of Unif, see [Sh:98] also §1 of this chapter and VIII §4 (where we show that it is consistent that only one of them holds). On the torsion part of Ext $(G, \mathbb{Z})$ see Sageev and Shelah [SgSh:138] [SgSh:148]; an alternative proof to [SgSh:138], more group theoretic, Eklof and Huber [EkHu]; on other cardinals Grossberg and Shelah [GrSh:302] and Mekler, Roslanowski and Shelah [MRSh:314].

2.0 Definition.

(1) A group $(G, +)$ is called torsion free, if for all $g \in G \setminus \{0\}$, for all $n > 0$ we have $ng \neq 0$.

(2) The torsion free rank of an (abelian) group G, $r_0(G)$ is the maximal size of a set $\{a_i : i < \lambda\} \subseteq G$ such that for every finite non empty $S \subseteq \lambda$, for all $\langle u_i : i \in S\rangle$ $(u_i \in \mathbb{Z} \setminus \{0\})$, we have $\sum_{i \in S} u_i a_i \neq 0$.

(3) For $g \in G$ and n such that $0 < n < \omega$, we say that "n divides g in G" $(G \models n|g)$ if there is $g' \in G$ such that $ng' = g$. A subgroup $A \subseteq G$ is called a "pure" subgroup if for all $a \in A$, all n, $0 < n < \omega$ we have: $G \models n|a$ implies $A \models n|a$.

(4) If $A \subseteq G$ is a subgroup, we write G/A for the quotient group, and for $a \in G$ we let $a + A$ or a/A be the equivalence class of a.

(5) G is called divisible, if for all $a \in G$, all $n > 0$ we have $G \models n|a$.

(6) G is called free if it has a free basis, where $\langle x_i : i \in T\rangle$ is a free basis of G *iff* every element of G has a representation $\sum_{i \in S} u_i x_i$ where $S \subseteq T$ is finite and $u_i \in \mathbb{Z}$, and $\sum_{i \in S} u_i x_i = 0 \Rightarrow \bigwedge_{i \in S} u_i = 0$.

2.0A Fact.

(1) If G is torsion free, $A \subseteq G$ a pure subgroup, then G/A is torsion free.

(2) If G is not a torsion group (i.e. $\exists a \in G\ \forall n > 0\ [na \neq 0]$), then the two cardinals

$$\max\{|A| : A \subseteq G, \text{ for all } a_1 \neq a_2 \text{ in } A, \text{ all } n > 0 : na_1 \neq na_2\}$$

and

$$\min\{|A| : A \subseteq G, \text{ for all } a \in G \text{ there is } u \in \mathbb{Z}, \text{ such that } ua \in A\}$$

are equal to $\max\{r_0(G), \aleph_0\}$.

(3) If G is torsion free non zero, then $|G| = \max\{r_0(G), \aleph_0\}$.

(4) If G is an abelian group and $0 < n < \omega$ then we can find $a_i \in G$ for $i < |nG|$ such that $i \neq j \Rightarrow n(a_i - a_j) \neq 0_G$, where $nG = \{na : a \in G\}$. Note that if G is divisible then $|nG| = |G|$ as $nG = G$.

Recall (see [Fu])

2.0B Fact.

(a) If H and G/H are free (so $H \subseteq G$), then G is free.

(b) If $G = \bigcup_{i<\lambda} G_i$ where $\langle G_i : i < \lambda\rangle$ is an increasing continuous sequence of groups, G_0 is free and for all $i < \lambda$ the group G_{i+1}/G_i is free, *then* G is free.

(c) If $G = \bigcup_{i<\lambda} G_i$ where $\langle G_i : i < \lambda\rangle$ is an increasing continuous sequence of group, each G_i is free and for a closed unbounded set of $i < \lambda$ we have $(\forall j)(i \leq j < \lambda \Rightarrow G_j/G_i$ is free) *then* G is free.

After Fuchs [Fu] pp. 209-211:

2.1 Definition. For abelian groups A, H let

(1) Fact (A, H) is the family of functions $f : A \times A \longrightarrow H$ such that
$f(a, -a) = f(a, 0) = f(0, a) = 0$ and
$f(a, b + c) + f(b, c) = f(b, a + c) + f(a, c) = f(c, a + b) + f(a, b)$

(2) We make Fact (A, H) into an abelian group by coordinatewise addition.

(3) For each function $g : A \to H$ satisfying $g(0) = 0$, $g(-a) = -g(a)$ (we call such g normal) let $(\partial g) \in \text{Fact}(A, H)$ be defined by $(\partial g)(a, b) = g(a) - g(a + b) + g(b)$.

(4) $\text{Trans}(A, H)$ is $\{\partial g : g$ a normal function from A to $H\}$, and it is a subgroup of $\text{Fact}(A, H)$ and we make it to an abelian group by coordinatewise addition.

(5) $\text{Ext}(A, G) = \text{Fact}(A, G)/\text{Trans}(A, G)$ (quotient as an abelian group).

2.1A Fact.

(1) If $h : A \to B$ is a group homomorphism, *then* h induces naturally a homomorphism

$$\check{h} : \text{Fact}(B, H) \to \text{Fact}(A, H)$$

(namely, $\check{h}(f)((a_1, a_2)) \mapsto f(h(a_1), h(a_2)))$ for a_1, $a_2 \in A$ which satisfies $\hat{h}(\partial g) = \partial(g \cdot h)$ so maps $\text{Trans}(B, H)$ into $\text{Trans}(A, H)$ and so naturally induces a homomorphism

$$\hat{h} : \text{Ext}(B, H) \to \text{Ext}(A, H)$$

(satisfying $\hat{h}(f + \text{Trans}(B, H)) = \check{h}(f) + \text{Trans}(A, H)$).

(2) If h is $1 - 1$, then $\check{h}$ and $\hat{h}$ are onto.

(See [Fu, 51.3] for $\check{h}$ and [HHSh, Lemma 1] for $\hat{h}$.)

2.1B Remark. (See [Fu])

(1) If G is free, then $\text{Ext}(G, H) = \{0\}$ i.e. $\text{Trans}(G, H) = \text{Fact}(G, H)$.

(2) G is called a *Whitehead group* if $\text{Ext}(G, \mathbb{Z}) = \{0\}$.

(3) If G is divisible, then $\text{Ext}(G, H)$ is torsion free (see [Fu, 52.1 I]).

2.2 Theorem. Suppose λ is a regular uncountable cardinal, H, $G = G_\lambda$ are abelian groups, G_i (for $i < \lambda$) torsion free abelian subgroups of G, $|G| = \lambda > |G_i|$, $G = \bigcup_{i<\lambda} G_i$, $G_i (i < \lambda)$ increasing and continuous, and let $\chi(i)$ be the cardinality of $\text{Ext}(G_{i+1}/G_i, H)$. If $\text{Unif}(\lambda, |H|, \bar{\chi})$ fails *then* $|\text{Ext}(G, H)| > 1$.

Proof. First we remark that we may w.l.o.g. assume that each G_i is a pure subgroup of G (and hence of G_{i+1}): the set $C = \{i < \lambda : G_i$ is a pure subgroup of $G\}$ is a closed unbounded subset of λ, say $C = \{\xi_i : i < \lambda\}$ an increasing continuous enumeration. Let $G'_i = G_{\xi_i}$, $\bar{\chi}'(i) = |\operatorname{Ext}(G'_{i+1}/G'_i, H)|$, $E = \{i : \xi_i = i\}$, then E is closed unbounded and for $i \in E$ we have $G'_i = G_i \subseteq G_{i+1} \subseteq G'_{i+1}$ so $\bar{\chi}'(i) \geq \bar{\chi}(i)$ (by 2.1A(2)), so the failure of $\operatorname{Unif}(\lambda, |H|, \bar{\chi})$ implies the failure of $\operatorname{Unif}(\lambda, |H|, \bar{\chi}')$ by 1.4(2) and 1.4(5), and we can continue the proof with G'_i, $\bar{\chi}'(i)$ instead of G_i, $\bar{\chi}(i)$ renaming them as G_i, $\bar{\chi}(i)$. Let $\bar{\mu} = \langle \bar{\mu}(i) : i < \lambda \rangle$ be defined by $\bar{\mu}(i) = |H|^{|G_i|}$, so by 1.6(1) we get that $\operatorname{Unif}(\lambda, \bar{\mu}, \bar{\chi})$ too fails and we can assume $|G_\alpha| \geq |\alpha| + \aleph_0$.
Next we prove

2.3 Claim. Let H, A, B be abelian groups, B a pure subgroup of A, $f \in \operatorname{Fact}(B, H)$. *Then* there are $f_t \in \operatorname{Fact}(A, H)$ (for $t \in \operatorname{Ext}(A/B, H)$) extending f, such that

(∗) there are no distinct $t, s \in \operatorname{Ext}(A/B, H)$ and normal functions g_t, g_s from A to H such that $\partial g_t = f_t$, $\partial g_s = f_s$ and $g_t \restriction B = g_s \restriction B$.
In other words, for any normal function $g_0 : B \to H$ there is at most one $t \in \operatorname{Ext}(A/B, H)$ such that for some normal $g : A \to H$ extending g_0 we have $f_t = \partial g$.

Proof of the Claim 2.3 By 2.1A(2) there is $f_0 \in \operatorname{Fact}(A, H)$ extending f. Let for each $t \in \operatorname{Ext}(A/B, H), h_t \in \operatorname{Fact}(A/B, H)$ represent t, i.e., $t = h_t / \operatorname{Trans}(A/B, H)$, and w.l.o.g. h_0 is the zero function. For $t \in \operatorname{Ext}(A/B, H)$ let $f_t \in \operatorname{Fact}(A, H)$ be defined by:

⊗ for $a, b \in A$, $f_t(a, b) = f_0(a, b) + h_t(a/B, b/B)$ (where $a/B, b/B \in A/B$ are defined naturally).

Clearly each f_t is well defined and belongs to $\operatorname{Fact}(A, H)$, (and the two definitions of f_0 agree).

Suppose t, s are members of $\operatorname{Ext}(A/B, H)$, and there are normal functions g_t, g_s from A to H, $\partial g_t = f_t, \partial g_s = f_s$ and $g_t \restriction B = g_s \restriction B$. Let $f^* \stackrel{\text{def}}{=} f_t - f_s \in$

Fact (A, H), $g^* \stackrel{\text{def}}{=} g_t - g_s$ (a normal function from A to H), so clearly $\partial g^* = f^*$ and $f^* \restriction B = 0_B$, moreover, $f^*(a,b) = h^*(a/B, b/B)$ where $h^* = h_t - h_s$ (see $\otimes$ above). It is also clear that $h^* \in \text{Fact}\,(A/B, H)$ and $h^*/\,\text{Trans}\,(A/B, H) = t - s \neq 0$.

Now if in A, $c - a = b \in B$ then $h^*(a/B, b/B) = f^*(a,b) = (\partial g^*)(a,b) = g^*(a) - g^*(a+b) + g^*(b) = g^*(a) - g^*(c) + g^*(b)$. As $b \in B, g^*(b) = 0_H$ by "$g^* = g_t - g_s$ and $g_t \restriction B = g_s \restriction B$" and $b/B = 0_A/B$ hence $h^*(a/B, b/B) = h^*(a/B, 0_A/B) = 0_H$ (as $h^* \in \text{Fact}\,(A/B, G)$). So putting together the last two sentences $0_H = g^*(a) - g^*(c) + 0_H$, hence $g^*(a) = g^*(c)$.

We can conclude that $c/B = a/B$ implies $a - c \in B$ hence $g^*(a) = g^*(c)$. So there is $g^\dagger : A/B \longrightarrow H$ such that $g^*(a) = g^\dagger(a/B)$. We can check $h^* = \partial g^\dagger$, but $h^*/\,\text{Trans}\,(a/B, H) = t - s \neq 0$, contradiction. $\square_{2.3}$

Continuation of the proof of the Theorem 2.2.

Recall that we assumed that each G_i is a pure subgroup of G. We define by induction on $\alpha < \lambda$ for every $\eta \in \bigtimes_{i<\alpha} \bar{\chi}(i)$ a function f_η (note that $\bar{\chi}(i) \geq 1$ for every i) such that:

a) $f_\eta \in \text{Fact}\,(G_\alpha, H)$ (when $\ell g(\eta) = \alpha$)

b) if $\nu = \eta \restriction \beta$, and $\beta \leq \ell g(\nu)$ then $f_\nu \subseteq f_\eta$

c) if $\xi < \zeta < \bar{\chi}(\alpha)$, then there are no normal functions g_0, g_1 from $G_{\alpha+1}$ into H, such that $\partial g_0 = f_{\eta \hat{} <\xi>}$, $\partial g_1 = f_{\eta \hat{} <\zeta>}$ and $g_0 \restriction G_\alpha = g_1 \restriction G_\alpha$.

Hence

c)$'$ for any function $g_0 : G_\alpha \to H$ there is at most one $\xi < \bar{\chi}(\alpha)$ such that there exists a normal function $g : G_{\alpha+1} \to H$ extending g_0 with $f_{\eta \hat{} \langle \xi \rangle} = \partial g$.

There is no problem in the induction, as the induction step is done by the Claim 2.3.

In the end, it is enough to prove that: for some $\eta \in \bigtimes_{i<\lambda} \bar{\chi}(i)$ for no normal function g from G into H do we have $f_\eta = \partial g$. So assume that for each $\eta \in \bigtimes_{\alpha<\lambda} \bar{\chi}(\alpha)$ there is a normal function $g_\eta : G \to H$ such that $f_\eta = \partial g_\eta$. So also $f_{\eta \restriction \alpha} = f_\eta \restriction (G_\alpha \times G_\alpha) = \partial(g_\eta \restriction G_\alpha)$, if $\ell g(\eta) = \alpha$. Hence $\eta(\alpha)$ can be computed from $\langle \eta \restriction \alpha, g_\eta \restriction G_\alpha \rangle$, since it is the unique (by (c)) $\xi < \bar{\chi}(\alpha)$

such that there is a normal $g : G_{\alpha+1} \to H$ extending $g_\eta \restriction G_\alpha$ and satisfying $f_{(\eta\restriction\alpha)^\frown\langle\xi\rangle} = \partial g$. What is the cardinality of $\{(\eta\restriction\alpha, g_\eta\restriction G_\alpha) : \eta \in {}^\lambda 2\}$? Clearly at most $(\prod_{\beta<\alpha} \bar{\chi}(\beta)) \times |H|^{|G_\alpha|} \leq (\prod_{\beta<\alpha} (|H|^{|G_\beta \times G_\beta|}) \times |H|^{|G_\alpha|} = |H|^{|G_\alpha|} = \bar{\mu}(\alpha)$ as $|G_\alpha| \geq |\alpha| + \aleph_0$, we thus easily get that Unif $(\lambda, \bar{\mu}, \bar{\chi})$ holds, which is equivalent to Unif $(\lambda, |H|, \bar{\chi})$. This contradicts our assumption. $\square_{2.2}$

2.4 Theorem. Assume λ is regular uncountable, H, G are abelian groups, G_i a torsion free abelian subgroup of G increasing continuous with i for $i < \lambda$ such that $G = \bigcup_{i<\lambda} G_i$. Let $\bar{\chi}^0 = \langle \chi^0(i) : i < \lambda \rangle$ be defined by $\bar{\chi}^0(i) = |\operatorname{Ext}(G_{i+1}/G_i, H)|$ and let $\bar{\chi}^1 = \langle \bar{\chi}^1(i) : i < \lambda \rangle$ be defined by $\bar{\chi}^1(i) = |\operatorname{Ext}(G_{i+1}/G_i)/\text{Torsion}(\operatorname{Ext}(G_{i+1}/G_i, H))|$. Let $\ell(*) < 2$ and assume that Unif $(\lambda, |H|, \bar{\chi}^{\ell(*)})$ fails (note: $\bar{\chi}^1(i)$ is $\aleph_0 \times r_0(\operatorname{Ext}(G_{i+1}/G_i, H))$).

(1) Ext (G, H) is not a torsion group provided that

(∗) (a) $\ell(*) = 1$ or

(b) $\ell(*) = 0$ and the Boolean algebra $\mathcal{P}(\lambda)/\operatorname{Id} - \operatorname{Unif}(\lambda, \bar{\mu}, \bar{\chi}^1)$ is infinite.

(2) If Unif $(\lambda, \mu_0, |H|, \bar{\chi}^0)$ fails and (∗) of part (1) *then* the torsion free rank of Ext (G, H) is $> \mu_0$.

(3) Suppose Id $-$ Unif $(\lambda, |H|, \bar{\chi}^0)$ is not κ-saturated, $\aleph_0 \leq \kappa \leq \lambda$ *then* the torsion free rank of Ext (G, H) is at least 2^κ.

Remark. 1) An ideal I on λ is called κ-saturated if there are no κ pairwise disjoint non zero elements in the Boolean algebra $\mathcal{P}(\lambda)/I$.

2) An ideal I on λ is called weakly λ-saturated if there are no λ pairwise disjoint sets in $\mathcal{P}(\lambda) \setminus I$.

3) As is well known; if I is κ-complete the two notions are equivalent.

4) It is well known that the extra hypothesis in 2.4(3) is very weak (i.e. the assumption that there is a normal κ-saturated ideal on λ has high consistency strength and put other restrictions on λ e.g. λ not successor).

Proof. (1) As in the proof of 2.4 w.l.o.g. each G_i is a pure subgroup of G, hence $G_\alpha, G_{\alpha+1}/G_\alpha$ are torsion free. Also $|G_i| \geq \aleph_0$, and letting $\bar{\mu}(i) = |H|^{|G_i|}$ also $\text{Unif}\,(\lambda, \bar{\mu}, \bar{\chi}^{\ell(*)})$ fails. Now we prove two claims:

2.5 Observation. There are pairwise disjoint $S_n \subseteq \lambda, n < \omega$ such that $\text{Unif}\,(\lambda, S_n, \bar{\mu}, \bar{\chi})$ fails, provided that one of the following holds:

(α) $\bar{\mu}(i) \leq 2^{<\lambda}$

(β) $\mathcal{P}(\lambda)/\text{Id} - \text{Unif}\,(\lambda, \bar{\mu}, \bar{\chi})$ is an infinite Boolean algebra

(γ) $\bar{\mu}(i)$ non decreasing, λ not measurable.

(δ) $\bar{\chi}(\alpha) \geq 2$ for every α or just for every normal ultrafilter D on λ, $\{\alpha : \bar{\chi}(\alpha) \geq 2\} \in D$.

Proof. If clause (β) holds, this is very trivial. If the Boolean algebra $\mathcal{P}(\lambda)/\text{Id} - \text{Unif}\,(\lambda, \bar{\mu}, \bar{\chi})$ is atomless below some element, say $S/\text{Id} - \text{Unif}\,(\lambda, \bar{\mu}, \bar{\chi})$ we choose by induction on n a set $S_n \subseteq S$ such that $S_n/\text{Id} - \text{Unif}\,(\lambda, \bar{\mu}, \bar{\chi})$ is not zero and $S_n \subseteq S \setminus \bigcup_{\ell<n} S_\ell$, and $S \setminus \bigcup_{\ell \leq n} S_\ell \notin \text{Id} - \text{Unif}\,(\lambda, \bar{\mu}, \bar{\chi})$, so $\langle S_\ell : \ell < \omega \rangle$ is as required. If $\mathcal{P}(\lambda)/\text{Id} - \text{Unif}\,(\lambda, \bar{\mu}, \bar{\chi})$ is an atomic Boolean algebra, it has infinitely many atoms say $\langle S'_n/\text{Id} - \text{Unif}\,(\lambda, \bar{\mu}, \bar{\chi}) : n < \omega \rangle$ are disjoint atoms, so $S_n \stackrel{\text{def}}{=} S'_n \setminus \bigcup_{\ell \leq n} S'_\ell$ are as required. So assume clause (α), so by 1.7 w.l.o.g. $\mu_i = 2^{<\lambda}$. By induction on n try to choose pairwise disjoint sets $S_n \in \mathcal{P}(\lambda) \setminus \text{Id} - \text{Unif}\,(\lambda, \bar{\mu}, \bar{\chi})$ such that $\lambda \setminus \bigcup_{k \leq n} S_k \notin \text{Id} - \text{Unif}\,(\lambda, \bar{\mu}, \bar{\chi})$. Assume that we cannot continue the induction in stage n, then clearly $S' \stackrel{\text{def}}{=} \{\alpha < \lambda : \bar{\chi}(\alpha) = 1\}$ belongs to the ideal (by 1.4(1)), hence the restriction of $\text{Id} - \text{Unif}\,(\lambda, \bar{\mu}, \bar{\chi})$ to $S \stackrel{\text{def}}{=} \lambda \setminus \bigcup_{k<n} S_k$ is a maximal ideal. Since it is also normal by 1.9(2), λ must be measurable and the dual filter is a normal ultrafilter to which S belongs. So $\diamondsuit_S$ holds. Now it is easy to find disjoint stationary sets $S_n \subseteq S$, $n < \omega$ such that for all n the statement $\diamondsuit_\lambda(S_n)$ holds (e.g. let $\langle X_\alpha : \alpha \in S \rangle$ be a diamond sequence, and let $S_n = \{\alpha : \text{Min}(X_\alpha) = n\}$). Since $\diamondsuit_\lambda(S_n)$ implies the weak diamond on S_n i.e. $\neg\,\text{Unif}\,(\lambda, 2, 2, 2)$ (by 1.3) by 1.7 also $\neg\,\text{Unif}\,(\lambda, 2^{<\lambda}, 2^{<\lambda}, 2)$ hence by monotonicity (1.4(5)), we are done.

The proof when clause (γ) holds is included in the proof above. $\square_{2.5}$

2.6 Claim. Let H, A, B be abelian groups, B a subgroup of A, A/B torsion free and $f \in \mathrm{Fact}(B, H)$ and n, $0 < n < \omega$.

(1) Then there are $f_i \in \mathrm{Fact}(A, H)$ for $i < \chi = |\,\mathrm{Ext}(A/B, H)|$ such that:

(∗) there are no $i < j < \chi$ and normal functions g_i, g_j from A to H such that $\partial g_i = nf_i, \partial g_j = nf_j$ and $g_i {\restriction} B = g_j {\restriction} B$. This means that for every normal $g_0 : B \to H$ there is at most one $i < \chi$ such that for some normal $g : A \to h$ extending g_0 we have $nf_i = \partial g$.

(2) Then there are $f_i \in \mathrm{Fact}(A, H)$ for $i <$ (the torsion free rank of $\mathrm{Ext}(A/B, H)$ multiplied by $\aleph_0$) such that

(∗∗) There are no $i \neq j$ and functions g_i, g_j from A to H and $0 < m < \omega$ such that $mf_i = \partial g_i, mf_j = \partial g_j$ and $g_i {\restriction} B = g_j {\restriction} B$.

Proof. (1) As A/B is torsion free, $\mathrm{Ext}(A/B, H)$ is a divisible abelian group (see [Fu]), hence we can inductively find $t_i \in \mathrm{Ext}(A/B, H)$ for $i < \chi$ such that $i < j$ implies $n(t_j - t_i) \neq 0$ (see 2.0A(4)). Now repeat the proof of Claim 2.3.
(2) We can choose a sequence $\langle t_i : i < r_0(\mathrm{Ext}(A/B, H)) \times \aleph_0\rangle$ such that for $m < \omega$ if $mt_i = mt_j$ and $m \neq 0$ then $i = j$ (this is possible by 2.0A(2)), and continue as in 2.3. $\square_{2.6}$

Continuation of the Proof of 2.4(1). Let us first assume $\ell(*) = 0$ and the Boolean algebra $\mathcal{P}(\lambda)/\,\mathrm{Id} - \mathrm{Unif}(\lambda, \bar{\mu}, \bar{\chi}^0)$ is infinite (i.e. possibility (b) holds).

Let $S_n \subseteq \lambda$ (for $n < \omega$) be as in Fact 2.5, and w.l.o.g. $\lambda = \bigcup_{n<\omega} S_n$.
Let us define by induction on $\alpha \leq \lambda$ for every $\eta \in \bigtimes_{i<\alpha} \bar{\chi}(i)$ a function f_η such that

a) $f_\eta \in \mathrm{Fact}(G_\alpha, H)$ (when $\ell g(\eta) = \alpha$)

b) if $\nu = \eta{\restriction}\beta, \beta \leq \ell g(\nu)$ then $f_\nu \subseteq f_\eta$.

c) if $\alpha \in S_n, \xi < \zeta < \bar{\chi}(\alpha)$ and $\eta \in \bigtimes_{i<\alpha} \bar{\chi}(i)$ *then* there are no normal functions g_0, g_1 from $G_{\alpha+1}$ into H such that $\partial g_0 = (n+1) f_{\eta\char`^<\xi>}$, $\partial g_1 = (n+1) f_{\eta\char`^<\zeta>}$ and $g_0{\restriction}G_\alpha = g_1{\restriction}G_\alpha$.

Hence

c)′ if $\alpha \in S_n$ and $\eta \in \bigtimes_{i<\alpha} \bar{\chi}^0(i)$ *then* for every normal $g_0 : G_\alpha \to H$ there is at most one $\xi < \bar{\chi}(\alpha)$ such that for some normal function $g : G_{\alpha+1} \to H$ extending g_0 we have $(n+1) f_{\eta\char`^\langle\xi\rangle} = \partial(g)$

There is no problem in the induction as the induction step is by Claim 2.6(1), and we finish as in the proof of 2.2.

If we do not assume $(*)$(a) but rather $(*)$ (b), we have to use 2.6(2) instead of 2.6(1) and let $S_n = \lambda$ for $n < \omega$ (so in clause (c) above, the demand is for every $\alpha < \lambda$, $n < \omega$).

Proof of 2.4(2). As in the proof of part (1) w.l.o.g. G_i is a pure subgroup of G, infinite. Let $\bar{\mu} = \langle \bar{\mu}(i) : i < \lambda \rangle$ be defined as: $\bar{\mu}(0) = \mu_0 \times |H|^{|G_0|}$, and $\bar{\mu}(i) = |H|^{|G_i|}$, and again as in the proof of part (1) also $\mathrm{Unif}\,(\lambda, \bar{\mu}, \bar{\chi}^0)$ fails. Note that $\bar{\mu}(0) \geq \aleph_0$.

We define f_η as in the proof of 2.4(1). If the torsion free rank of $\mathrm{Ext}\,(G, H)$ is $\leq \mu_0$, then there are $t^\alpha \in \mathrm{Ext}\,(G, H)$ $(\alpha < \mu_0)$ such that for any $t \in \mathrm{Ext}\,(G, H)$, for some $n > 0$ and for some α we have $nt = t^\alpha$ (note that w.l.o.g. $\mu_0 \geq \aleph_0$ by the proof of 2.4(1)). So there are $g^\alpha \in \mathrm{Fact}\,(G, H)$ for $\alpha < \mu_0$, so that for every $f \in \mathrm{Fact}\,(G, H)$ there are $n_f > 0$ and a function g_f from G to H and $\alpha(f) < \mu_0$ such that

$$n_f f = \partial g_f + g^{\alpha(f)}.$$

In particular this holds for every $f_\eta, \eta \in \bigtimes_{i<\lambda} \bar{\chi}(i)$. First assume $(*)$(b), so we have defined the S_n's. So for each $\eta \in \bigtimes_{i<\lambda} \chi(i)$, for each $i \in S_{n_f}$ and $g' : G_i \to H$ satisfying $\partial g' = f_\eta \restriction G_i$ we have $\eta(i)$ can be computed from $\langle \alpha(f_\eta), \eta \restriction i \rangle$ as

$(*)$ "the unique $\xi < \bar{\chi}^{\ell(*)}(i)$ such that for some normal $g : G_{i+1} \to H$, and we have $n_f \times f_{\eta \restriction \beta^\frown \langle \xi \rangle} = \partial g + g^{\alpha(f_{\eta \restriction \beta})}$."

This contradicts $\neg\,\mathrm{Unif}\,(\lambda, \bar{\mu}, \bar{\chi})$ (which was deduced above). If we assume $(*)$(a) holds just replace "$i \in S_{n_f}$" by "$i < \lambda$" and use 2.6(1) rather than 2.6(2) and in $(*)$ replace "and we have $n_f\times$" by "and for some n we have $n\times$"

Proof of 2.4(3).

As in the prove of part (1) w.l.o.g. G_i is infinite pure subgroup of G and $\mathrm{Unif}\,(\lambda, \bar{\mu}, \bar{\chi}^0)$ fail with $\bar{\mu}(i) = |H|^{|G_i|}$. Let $\langle S^n_i : i < \kappa, n < \omega \rangle$ be pairwise disjoint subsets of λ which are positive modulo $\mathrm{Id} - \mathrm{Unif}\,(\lambda, |H|, \bar{\chi})$, and let $S_i \stackrel{\text{def}}{=} \bigcup\limits_{0<n<\omega} S_i^n$. Using a lemma similar to 2.6(1) we can define a family $\langle h_\eta : \eta \in {}^{\lambda>}2 \rangle$, $h_\eta \in \operatorname{Fact}(G_{\ell g(\eta)}, H)$ such that:

> whenever $\alpha \in S_i^n$ (so $n > 0$), $\bar{\chi}(\alpha) > 1$, $\eta \in {}^\alpha 2$, $g : G_{\alpha+1} \to H$ is normal and $n(h_{\eta \frown \langle 0 \rangle} - h_{\eta \frown \langle 1 \rangle}) = \partial g$, *then* $g \restriction G_\alpha \neq 0$.

For each $\eta \in {}^\lambda 2$ we thus get a function $h_\eta = \bigcup\limits_{\alpha<\lambda} h_{\eta\restriction\alpha} \in \operatorname{Fact}(G, H)$. Below we will select 2^κ many $\eta \in {}^\lambda 2$ such that the corresponding h_η witness $r_0(\operatorname{Ext}(G, H)) \geq 2^\kappa$.

Let $\langle A_\varepsilon : \varepsilon < 2^\kappa \rangle$ be subsets of κ such that for any $\varepsilon_1 \neq \varepsilon_2$ the set $A_{\varepsilon_1} \setminus A_{\varepsilon_2}$ is nonempty.

For each $i < \kappa$, $n < \omega$ define F_i^n on $\bigcup\limits_{\alpha \in S_i^n} {}^\alpha 2 \times {}^\alpha 2 \times {}^{G_\alpha} H$ as follows: if η_1, $\eta_2 \in {}^\alpha 2$, $g_0 : G_\alpha \to H$ is normal, $\alpha \in S_i^n$ and there is a normal $g^+ : G_{\alpha+1} \to H$ extending g_0 such that

$$n(h_{\eta_1 \frown \langle 0 \rangle} - h_{\eta_2 \frown \langle 0 \rangle}) = \partial g^+$$

then $F_i^n(\eta_1, \eta_2, g_0) = 1$, otherwise $F_i^n(\eta_1, \eta_2, g_0) = 0$.

Since $S_i^n \notin \operatorname{Id} - \operatorname{Unif}(\lambda, \bar{\mu}, 2)$ we can find a weak diamond f_i^n for F_i^n and S_i^n (so only $f_i^n \restriction S_i^n$ matters).

Now for $\varepsilon < 2^\kappa$ define $\eta(\varepsilon) \in {}^\lambda 2$ by

$$\eta(\varepsilon)(\gamma) = \begin{cases} f_i^n(\gamma) & \text{if } \gamma \in S_i^n, i \in A_\varepsilon \\ 0 & \text{otherwise.} \end{cases}$$

We now claim that for all $\varepsilon_1 \neq \varepsilon_2$, for all $n > 0$

$$nh_{\eta(\varepsilon_1)} \not\equiv nh_{\eta(\varepsilon_2)} \mod \operatorname{Trans}(G, H).$$

(This claim will finish the proof of 2.4(3).)

So assume that $nh_{\eta(\varepsilon_1)} - nh_{\eta(\varepsilon_2)} = \partial g$ for some normal $g : G \to H$. Let $\eta_1 \stackrel{\text{def}}{=} \eta(\varepsilon_1)$, $\eta_2 \stackrel{\text{def}}{=} \eta(\varepsilon_2)$. Choose $i \in A_{\varepsilon_1} \setminus A_{\varepsilon_2}$. Since f_i^n was a weak diamond for F_i^n on S_i^n, the set

$$\{\alpha \in S_i^n : F_i^n(\eta_1 \restriction \alpha, \eta_2 \restriction \alpha, g \restriction \alpha) = f_i^n(\alpha)\}$$

is nonempty, so let α be an element of this set.

Case 1. $f^n_i(\alpha) = 1$.

So there is normal $g' : G_{\alpha+1} \to H$ such that

$$n(h_{\eta_1 \frown \langle 0 \rangle} - h_{\eta_2 \frown \langle 0 \rangle}) = \partial g' \text{ and } g' \text{ extends } g{\restriction}G_\alpha.$$

But we also have

$$n(h_{\eta_1{\restriction}(\alpha+1)} - h_{\eta_2{\restriction}(\alpha+1)}) = \partial(g{\restriction}G_{\alpha+1}).$$

Note that $\eta_1(\alpha) = f^n_i(\alpha) = 1$, $\eta_2(\alpha) = 0$, since $\alpha \in S^n_i$, $i \in A_{\varepsilon_1} \setminus A_{\varepsilon_2}$. So subtracting the two equations above, we get

$$n(h_{\eta_1 \frown \langle 0 \rangle} - h_{\eta_1 \frown \langle 1 \rangle}) = \partial((g' - g){\restriction}G_{\alpha+1}).$$

Since $((g' - g){\restriction}G_{\alpha+1}){\restriction}G_\alpha = 0$, this contradicts our choice of $\langle h_\eta : \eta \in {}^{\lambda>}2\rangle$.

Case 2. $f^n_i(\alpha) = 0$.

So there is no normal $g' : G_{\alpha+1} \to H$ satisfying

$$n(h_{\eta_1 \frown \langle 0 \rangle} - h_{\eta_2 \frown \langle 0 \rangle}) = \partial g',\ g' \text{ extends } g{\restriction}\alpha.$$

This is a contradiction, since $g' \stackrel{\text{def}}{=} g{\restriction}G_{\alpha+1}$ satisfies the requirements (as $\eta_1(\alpha) = \eta_2(\alpha) = 0$). $\square_{2.4}$

2.7 Conclusion. Assume that

⊕ for every regular uncountable λ, for all stationary subsets $S \subseteq \lambda$, the weak diamond holds on S, or just $\text{Unif}\,(\lambda, S, 2, \langle 2^{|i|} : i < \lambda\rangle)$ fails.

Then

(a) Every Whitehead group is free.

(b) If G is torsion free but not free, uncountable and for all subgroups H of cardinality $|H| < |G|$ the quotient group G/H is not free, *then* the torsion free rank of $\text{Ext}\,(G, \mathbb{Z})$ is $2^{|G|}$.

Remark. 1) If there is no inaccessible cardinal then ⊕ is equivalent, by 1.7(2), to

$\oplus'$ for every regular uncountable λ, for every stationary subset S of λ, the weak diamond holds for S, that is $\mathrm{Unif}(\lambda, S, 2, 2)$.

2) We can get a weaker version, still sufficient for our theorem, if we restrict F in the definition of Unif to the particular kind of functions implicit in the proof. See generally on such version of the weak diamond in [Sh:576, §2].

Proof. First note that (b) implies (a). Indeed, let G be a nonfree Whitehead group of minimal size. The countable case is well known (see below) so assume $|G| > \aleph_0$. Then G is almost free, so all subgroups H such that $|H| < |G|$ are free, hence G/H is not free (by 2.0B) so G satisfies the assumption of (b), hence its conclusion so $|\mathrm{Ext}(G, \mathbb{Z})| > 1$.

Proof of (b). We prove by induction on λ.

The case $|G| = \aleph_0$ is well known (see e.g. [HHSh:91]) and the case $|G|$ is singular is just like [HHSh:91]. So assume $\lambda = |G|$ is regular $> \aleph_0$.

Let $G = \bigcup_{\gamma<\lambda} G_\gamma$ with G_γ a continuous increasing in γ, each G_γ a pure subgroup of G of size $< \lambda$ such that:

(*) If G/G_γ is not almost free i.e. if $(\exists\beta)(\gamma < \beta < \lambda \;\&\; G_\beta/G_\gamma$ not free$)$, then $G_{\gamma+1}/G_\gamma$ is not free.

Let

$$S = \{\gamma : G_{\gamma+1}/G_\gamma \text{ is not free}\}.$$

$$\bar{\chi} = \langle \bar{\chi}(i) : i < \lambda \rangle$$

$$\bar{\chi}(i) = r_0(\mathrm{Ext}\,(G_{\gamma+1}/G_\gamma, \mathbb{Z}) \times \aleph_0$$

Note that by induction hypothesis for all $\gamma \in S$ we have $\chi(\gamma) \geq 2$ (in fact $\geq 2^{\aleph_0}$).

If S is stationary, then S can be divided into λ many stationary sets $\langle S_i : i < \lambda \rangle$. By our assumption, all the sets S_i will be $\not\equiv 0 \mod \mathrm{Id} - \mathrm{Unif}(\lambda, 2, 2, 2) = \mathrm{Id} - \mathrm{Unif}(\lambda, \aleph_0, \aleph_0, 2)$, so by 2.4(3) we know that $\mathrm{Ext}\,(G, \mathbb{Z})$ has torsion free rank 2^λ.

If S is not stationary then by (*) we have a continuous increasing sequence $\langle \gamma_i : i < \lambda \rangle$, $\bigcup_{i<\lambda} \gamma_i = \lambda$ with $i < \lambda \Rightarrow G_{\gamma_{i+1}}/G_{\gamma_i}$ is free. Then it is easy to see that G/G_{γ_0} is free (see 2.0B, clause (c)), contradicting an assumption (of clause (b) of 2.7). $\square_{2.7}$

A more detailed analysis of the situation shows that for a given group G of cardinality λ (regular uncountable), we do not need the full strength of 2.7$\oplus$ (assuming the induction hypothesis of 2.7(b)).

2.7A Theorem. Assume G satisfies the assumption on G of clause (b) from 2.7, $|G| = \lambda$, λ regular uncountable and that all groups of size $< \lambda$ satisfy 2.7(b) or just: $|H| < \lambda$, H not free $\Rightarrow$ $\mathrm{Ext}\,(H, \mathbb{Z}) \neq 0$. Let $G = \bigcup_{i<\lambda} G_i$ be an increasing union of (w.l.o.g. pure) subgroups of G, and let

$$S^* \subseteq \{i < \lambda : G_{i+1}/G_i \text{ is not free}\}.$$

(Note that S^* is stationary since G is not free.)
Now assume that S^* is not in $\mathrm{Id} - \mathrm{Unif}\,(\lambda, 2, 2, \bar{\chi})$, $\chi_i = \mathrm{Ext}\,(G_{i+1}/G_i, \mathbb{Z})$ (so $i \in S^* \Rightarrow \chi_i \geq 2$) and $i \in S^*$, i inaccessible $\Rightarrow$ $\mathrm{Ext}\,(G_{i+1}/G_i, \mathbb{Z}) = 2^i$. Then $r_0(\mathrm{Ext}\,(G, \mathbb{Z})) = 2^\lambda$.

Proof. As remarked in 2.0 ([Fu], or see essentially [HHSh:91, Lemma 1 p.41]) if $G^\dagger$ is a subgroup of G, then $\mathrm{Ext}\,(G^\dagger, H)$ is a homomorphic image of $\mathrm{Ext}\,(G, H)$, hence the torsion free rank of $\mathrm{Ext}\,(G, H)$ is not smaller than the torsion free rank of $\mathrm{Ext}\,(G^\dagger, H)$, so we shall freely replace G by some subgroups during the proof.

We split the proof to cases.

Case I: G has subgroups $G^*, G_\alpha (\alpha < \lambda)$ such that:
$|G_\alpha| < \lambda, G^* \subseteq G_\alpha, G_\alpha/G^*$ is not free and $\{G_\alpha : \alpha < \lambda\}$ is independent over G^*, (i.e., if $n \in (0, \omega)$ and $x_m \in G_{\alpha_m} \setminus G^*$ for $m < n$, the α_m's distinct *then* $\sum_{m<n} x_m \notin G^*$). W.l.o.g. $G = \sum_\alpha G_\alpha$.

We choose, for any $n < \omega, \alpha < \lambda$, a function $f^n_\alpha \in \mathrm{Fact}\,(G_\alpha/G^*, \mathbb{Z})$ such that $f^0_\alpha = 0$, and for $n \neq 0$ we have $nf^n_\alpha/\,\mathrm{Trans}\,(G_\alpha/G^*, \mathbb{Z}) \neq \{0\}$. Let $F : \omega \times \lambda \to \lambda$, be one to one onto. Let $\{A_i : i < 2^\lambda\}$ be a family of distinct subsets of λ, and define, for $i < 2^\lambda$, a function $\xi_i : \lambda \to \omega$ by: $\xi_i(\alpha) = n$ if for some $\zeta \in A_i$, $\alpha = F(n, 2\zeta)$ or for some $\zeta \in \lambda \setminus A_i$, $\alpha = F(n, 2\zeta + 1)$, and $\xi_i(\alpha) = 0$ otherwise.

So we have defined functions ξ_i (for $i < 2^\lambda$), from λ to ω, such that for every $n < \omega$ and $i \neq j < 2^\lambda$ for some $\alpha < \lambda$ we have $\xi_i(\alpha) = 0, \xi_j(\alpha) = n$.

For every $i < 2^\lambda$ we define $h_i \in \mathrm{Fact}\,(\sum_{\alpha<\lambda} G_\alpha, \mathbb{Z})$: if $x = \sum_\alpha x_\alpha, y = \sum_\alpha y_\alpha$ and $x_\alpha, y_\alpha \in G_\alpha$ (so $x_i = y_i = 0$ for all but finitely many i's) then $h_i(x, y) = \sum_\alpha f_\alpha^{\xi_i(\alpha)}(x_\alpha/G^*, y_\alpha/G^*)$ (the representation $x = \sum_\alpha x_\alpha$ is not unique, but for any two representations $x = \sum_\alpha x_\alpha = \sum_\alpha x^\dagger_\alpha$ we get $x_\alpha/G^* = x^\dagger_\alpha/G^*$, so h_i is well defined).
It is easy to check $h_i \in \mathrm{Fact}\,(\sum_\alpha G_\alpha, \mathbb{Z})$.

Now if the torsion free rank of G $(= \sum_\alpha G_\alpha)$ is $< 2^\lambda$, there is an n, $0 < n < \omega$ such that $\{nh_i/\mathrm{Trans}\,(G^\dagger, \mathbb{Z}) : i < 2^\lambda\}$ has power $< 2^\lambda$. We know that $2^{|G^*|} < 2^\lambda$ (if $2^{|G^*|} = 2^\lambda$, then letting $\bar{\chi}(\alpha) = \bar{\mu}(\alpha) = 2$ we get $\prod\limits_{\alpha<\lambda} \bar{\chi}(\alpha) = \prod\limits_{\alpha<|G^*|} \bar{\mu}(\alpha)$, so $\mathrm{Unif}\,(\lambda, \bar{\mu}, \bar{\chi})$ holds by 1.4(4)) so without loss of generality (by renaming) $nh_i/\mathrm{Trans}\,(G, \mathbb{Z})$ are equal, for $i < (2^{|G^*|})^+$. Hence there are normal functions $g_i : G \to \mathbb{Z}$ such that $nh_i - nh_0 = \partial g_i$ for $i < (2^{|G^*|})^+$. Now the number of $g_i{\restriction}G^*$ is $\leq (2^{|G^*|})$, hence without loss of generality for every i such that $0 < i < (2^{|G^*|})^+$ we have $g_i{\restriction}G^* = g^*$.

We can choose $\alpha < \lambda$ such that $\xi_1(\alpha) = 0$, $\xi_2(\alpha) = n$. Now restricting ourselves to G_α, note for some k (namely $k = \xi_0(\alpha)$), $h_0{\restriction}(G_\alpha \times G_\alpha) = f^k_\alpha$ and $(h_1 - h_0){\restriction}(G_\alpha \times G_\alpha) = f^0_\alpha - f^k_\alpha$, $(h_2 - h_0){\restriction}(G_\alpha \times G_\alpha) = f^n_\alpha - f^\kappa_\alpha$ and now we can apply the proof of Claim 2.3, and get a contradiction.

So we have finished Case I.

* * *

Let from now on, $G = \bigcup_{i<\lambda} G_i, G_i$ increasing and continuous, $|G_i| < \lambda$, all G_i are pure subgroups of G, hence all the quotients G/G_i are torsion free.

2.8 Subclaim. If Case I does not hold, we can assume that:

(a) for every $\gamma < \lambda$, there is no $G^\dagger$, $G_\gamma \subseteq G^\dagger \subseteq G$, $|G^\dagger| < \lambda$, $G^\dagger \cap G_{\gamma+1} = G_\gamma$ and $G^\dagger/G_\gamma$ is not free.

(b) for every limit δ, G/G_δ is $(\mathrm{cf}\delta)$-free, except, maybe, when $\mathrm{cf}(\delta) = \aleph_0$.

Proof of the Subclaim. We define by induction on $i < \lambda$, $\alpha_i < \lambda$, increasing and continuous.

Let $\alpha_0 = 0$, and for a limit i let $\alpha_i = \bigcup_{j<i} \alpha_j$. If α_i is defined let $\{G^i_\zeta : \zeta < \zeta_i\}$ be a maximal family of subgroups of G, satisfying: $G_{\alpha_i} \subseteq G^i_\zeta, |G^i_\zeta| < \lambda, G^i_\zeta/G_{\alpha_i}$ not free and $\{G^i_\zeta : \zeta < \zeta_i\}$ is independent over G_{α_i}; such a family exists by Zorn's Lemma and $\zeta_0 < \lambda$ as Case I does not hold.

Let $\alpha_{i+1} = \mathrm{Min}\{\alpha : \alpha_i < \alpha \text{ and } G^i_\zeta \subseteq G_\alpha \text{ for every } \zeta < \zeta_i\}$.

We know that α_{i+1} exists as λ is regular, $|G^i_\zeta| < \lambda$, $\zeta_i < \lambda$. Also there is no $G^\dagger, G_{\alpha_i} \subseteq G^\dagger \subseteq G$, $|G^\dagger| < \lambda, G^\dagger \cap G_{\alpha_{i+1}} = G_{\alpha_i}$ and $G^\dagger/G_{\alpha_i}$ not free, as this would contradict the choice of $\{G^i_\zeta : \zeta < \zeta_i\}$ as a maximal family.

Now we can replace $\langle G_\alpha : \alpha < \lambda\rangle$ by $\langle G_{\alpha_i} : i < \lambda\rangle$ and clause (a) of the subclaim will hold, so without loss of generality (a) holds, i.e., $\alpha_i = i$. What about (b)? Now we will show that (a) implies (b). So assume that G/G_δ is not $\mathrm{cf}(\delta)$-free, where $\mathrm{cf}(\delta) > \aleph_0$. Let G^*/G_δ be a non-free subgroup of cardinality $\kappa < \mathrm{cf}(\delta)$. Let $\{x_j : j < \kappa\}$ be a set of representatives, and let K be the group generated by this set. Clearly $|K| = \kappa$ ($\kappa \geq \aleph_0$, as G/G_δ and hence G^*/G_δ are torsion free). So there is an ordinal $\gamma < \delta$ such that $K_\delta \cap G_\delta \subseteq G_\gamma$. Hence $(K_\delta + G_\gamma) \cap G_{\gamma+1} = G_\gamma$, and

$$(K_\delta + G_\gamma)/G_\gamma \cong K_\delta/K_\delta \cap G_\gamma = K_\delta/K_\delta \cap G_\delta \cong (K_\delta + G_\delta)/G_\delta = G^*/G_\delta$$

is not free. This contradicts condition (a) for γ. $\square_{2.8}$

Continuation of the proof of 2.7A Recall $S^* \subseteq \{\gamma < \lambda : G_{\gamma+1}/G_\gamma \text{ is not free}\}$ and let $S \overset{\text{def}}{=} \{\gamma \in S^* : \gamma \text{ is a regular limit (i.e. inaccessible) cardinal}\}$. Let $\bar{\chi} = \langle \bar{\chi}(\gamma) : \gamma < \lambda\rangle$, $\bar{\chi}(\gamma) = |\,\mathrm{Ext}\,(G_{\gamma+1}/G_\gamma, \mathbb{Z})|$

Case II: not Case I and $S^* \setminus S \notin \mathrm{Id} - \mathrm{Unif}\,(\lambda, \aleph_0, \bar{\chi})$.

We can use 2.4(3), because of the following well-known theorem:

Theorem. Assume λ is regular, D a normal filter on λ, $S^0 \not\equiv \emptyset \bmod D$ and $\delta \in S^0 \Rightarrow \mathrm{cf}(\delta) < \delta$ (i.e. δ not a regular cardinal). *Then* there are pairwise disjoint $S_\alpha \subseteq S^0(\alpha < \lambda), S_\alpha \not\equiv 0 \bmod D$.

Proof.

Clearly $\mathrm{cf}(-)$ is a regressive function on $S^0 \setminus \{0\}$, hence for some κ and $S^1 \subseteq S$, $S^1 \not\equiv \emptyset \bmod D$, $(\forall \delta \in S^*)[\mathrm{cf}(\delta) = \kappa]$. For each $\delta \in S^1$, choose $\langle \alpha(\delta, \xi) : \xi < \kappa \rangle$ an increasing continuous converging to δ, and let $A_{\xi,j} = \{\delta \in S^1 : \alpha(\delta, \xi) = j\}$. Now we can prove that for some ξ for λ ordinals j we have $A_{\xi,j} \not\equiv 0 \bmod D$, and as $A_{\xi,j} \cap A_{\xi,i} = \emptyset$ for $i \neq j$ we will finish.

So we have finished Case II.

Continuation of the proof of 2.7A

Case III: $S^* \setminus S \in \mathrm{Id} - \mathrm{Unif}\,(\lambda, \aleph_0, \bar{\chi})$.

So by our assumption $S \notin \mathrm{Id} - \mathrm{Unif}\,(\lambda, \aleph_0, \bar{\chi})$. Note that by an assumption we have $S^* \setminus \delta \in S \Rightarrow \chi(\delta) = 2^\delta$.

We first state (and prove later).

2.9 Subclaim. Assume G^0, G^1 are torsion free, G^0 a pure subgroup of G^1, $f_i \in \mathrm{Fact}\,(G^0, \mathbb{Z})$, for $i < \chi$ and the torsion free rank of $\mathrm{Ext}\,(G^1/G^0, \mathbb{Z})$ is $\geq \lambda \geq \chi$ and $\lambda > \aleph_0$. *Then* we can define $f_{i,\alpha} \in \mathrm{Fact}\,(G^1, \mathbb{Z})$, $f_i \subseteq f_{i,\alpha}$ for $\alpha < \lambda$ such that:

$(*)$ if $\beta \neq \gamma < \chi$ and $0 < n < \omega$ and $g : G^0 \to \mathbb{Z}$ is a normal function *then* for at most one α there is a normal function $g^\dagger : G^1 \to \mathbb{Z}$ extending g, such that $nf_{\beta,\alpha} - nf_{\gamma,\alpha} = \partial g^\dagger$.

Continuation of the proof of 2.7A

So let us prove the theorem in Case III. We define by induction on $i \leq \lambda$, for every $\eta \in \bigtimes_{j<i} \chi_j$ and $A \subseteq i$, a function $f_{\eta,A} \in \mathrm{Fact}\,(G_i, \mathbb{Z})$ such that

a) if $j < \ell g(\eta), \eta \in \bigtimes_{j<i} \bar{\chi}(j), A \subseteq i$ *then* $f_{\eta \restriction j, A \cap j} = f_{\eta,A} \restriction (G_j \times G_j)$.

b) if $\eta \in \bigtimes_{j<i+1} \bar{\chi}(j)$, $A, B \subseteq i+1$, $A \cap i = B \cap i$ *then* $f_{\eta,A} = f_{\eta,B}$.

c) if $\delta \in S$ (so $\bar{\chi}(\delta) = 2^{|\delta|}$), $\eta \in \bigtimes_{j<\delta} \bar{\chi}(j), A \subseteq \delta, B \subseteq \delta, g : G_\delta \to \mathbb{Z}$ is normal and $0 < n < \omega$ *then* for at most one $j < \bar{\chi}(\delta)$ there is a normal $g^\dagger : G_{\delta+1} \to \mathbb{Z}$ extending g such that $nf_{\eta\char"5E\langle j\rangle,A} - nf_{\eta\char"5E\langle j\rangle,B} = \partial g^\dagger$.

There is no problem in the definition: for c) use the subclaim 2.9, remembering $\delta \in S \Rightarrow \bar{\chi}(\delta) = 2^\delta$. Now for at least one $\eta \in \bigtimes_{i<\lambda} \bar{\chi}(i)$, for every distinct $A, B \subseteq \lambda$ and $0 < n < \omega$, $nf_{\eta,A} - nf_{\eta,B} \notin \mathrm{Trans}(G, \mathbb{Z})$. Otherwise for every $\eta \in \bigtimes_{i<\lambda} \bar{\xi}(i)$ there are $A_\eta \neq B_\eta \subseteq \lambda$ and $0 < n_\eta < \omega$ and $g_\eta : G \to \mathbb{Z}$ such that

$$n_\eta f_{\eta,A} - n_\eta f_{\eta,B} = \partial g_\eta$$

By condition c) above, for every $\delta \in S$, from n_η, $f_{\eta,A}\restriction(G_\delta \times G_\delta) = f_{\eta\restriction\delta,A\cap\delta}$, $f_{\eta,B}\restriction(G_\delta \times G_\delta) = f_{\eta\restriction\delta,B\cap\delta}$ and $g_\eta\restriction G_\delta$ we can compute $\eta(\delta)$, so this contradicts $S \not\equiv \emptyset \mod \mathrm{Id} - \mathrm{Unif}(\lambda, \aleph_0, \bar{\chi})$. Now for such an η, $\{f_{\eta,A} : A \subseteq \lambda\}$ exemplify that the torsion free rank of $\mathrm{Ext}(G, \mathbb{Z})$ is $\geq 2^\lambda$. $\square_{2.7A}$

Proof of the subclaim 2.9.

Let $\{(i_\zeta, \alpha_\zeta) : \zeta < \lambda\}$ be a list of all pairs $(i, \alpha), i < \chi, \alpha < \lambda$, and we define f_{i_ζ,α_ζ} by induction on ζ. Suppose we have defined f_{i_ξ,α_ξ} for every $\xi < \zeta, \zeta < \lambda$ and they are required, and let us define f_{i_ζ,α_ζ}.

Let $\{t(j) : j < \lambda\}$ be members of $\mathrm{Ext}(G^1/G^0, \mathbb{Z})$ such that $nt(j_1) - nt(j_2) \neq 0$ for $n > 0$, $j_1 \neq j_2$. By Claim 2.6(2) there are $f^j (j < \lambda)$ such that: $f^j \in \mathrm{Fact}(G^1, \mathbb{Z})$ extend f_{i_ζ}, and there are no $n > 0$, $j(1) \neq j(2) < \lambda$ and normal $g : G^1 \to \mathbb{Z}$ such that $nf^{j(1)} - nf^{j(2)} = \partial g$ and $g\restriction G^0 = 0$.

We can try to let $f_{i_\zeta,\alpha_\zeta} = f^j$ for any $j < \lambda$ and assume toward contradiction that it always fail. The only thing that can go wrong is $(*)$ from the subclaim. So for every j there are $\beta_j, \gamma_j, n_j > 0$ and normal $g_j : G^0 \to \mathbb{Z}$ and $\alpha_j^1 \neq \alpha_j^2$ and normal $g_j^1 : G^1 \to \mathbb{Z}, g_j^2 : G^1 \to \mathbb{Z}$ extending g_j such that $\{(\beta_j, \alpha_j^1), (\gamma_j, \alpha_j^1), (\beta_j, \alpha_j^2), (\gamma_j, \alpha_j^2)\} \subseteq \{(i_\zeta, \alpha_\zeta) : \zeta \leq j\}$ and letting $f_{i_\zeta,\alpha_\zeta} = f^j$ we have:

$(**)$ $n_j f_{\beta_j,\alpha_j^1} - n_j f_{\gamma_j,\alpha_j^1} = \partial g_j^1$ and $n_j f_{\beta_j,\alpha_j^2} - n_j f_{\gamma_j,\alpha_j^2} = \partial g_j^2$.

Now there are λ ordinals j and only $\aleph_0 \times |\zeta+1| \times |\zeta+1| \times |\zeta+1| \times |\zeta+1| < \lambda$ possible 5-tuples $\langle n_j, \beta_j, \gamma_j, \alpha_j^1, \alpha_j^2 \rangle$: so without loss of generality for $j < \omega$ we have the same $n, \beta, \gamma, \alpha_1, \alpha_2$. Also by the induction hypothesis, at least one of $\{(\beta, \alpha_1), (\gamma, \alpha_1), (\beta, \alpha_2), (\gamma, \alpha_2)\}$ is not in $\{(i_\xi, \alpha_\xi) : \xi < \zeta\}$ hence is (i_ζ, α_ζ), so by symmetry without loss of generality $(\beta, \alpha_1) = (i_\xi, \alpha_\xi)$. As $\beta \neq \gamma, \alpha_1 \neq \alpha_2$ clearly $\{(\gamma, \alpha_1), (\beta, \alpha_2), (\gamma, \alpha_2)\} \subseteq \{(i_\xi, \alpha_\xi) : \xi < \zeta\}$. So for each $j < \omega$ (subtracting the equations in (**)) we have:

$$nf^j - nf_{\gamma,\alpha_1} - nf_{\beta,\alpha_2} + nf_{\gamma,\alpha_2} = \partial g_j^1 - \partial g_j^2 = \partial(g_j^1 - g_j^2)$$

Subtracting the equations for $j = 0, 1$

$$nf^1 - nf^0 = \partial(g_1^1 - g_1^2) - \partial(g_0^1 - g_0^2) = \partial(g_1^1 - g_1^2 - g_0^1 + g_0^2)$$

clearly $(g_1^1 - g_1^2)\restriction G^0 = 0$ and $(g_0^1 - g_1^2)\restriction G^0 = 0$ so we get a contradiction to the choice of the f^j's. $\square_{2.9}$

Now similarly to [HHSh:91] by our proof:

2.10 Conclusions. If $\oplus$ of 2.10 holds, G a torsion free group, $\lambda = \mathrm{Min}\{|G^\dagger| : G/G^\dagger$ is free $\}$, *then* $\mathrm{Ext}\,(G, \mathbb{Z})$ has torsion free rank 2^λ.

Remark. The use of $\mathbb{Z}$ instead H in 2.13, 2.10 is just for simplicity.

How strong are the assumptions of theorem 2.7?

Unlike the full diamond, the weak diamond has only little influence on the behavior of the exponentiation function $\kappa \mapsto 2^\kappa$, as the following theorem shows:

2.11 Theorem. Assume $V \models GCH$, F is a function defined on the regular cardinals, $F(\lambda)$ a cardinal, $(\forall\lambda)\mathrm{cf}(F(\lambda)) > \lambda$,

$\otimes$ $\forall\lambda[\sum_{\mu \in \mathbf{Reg} \cap \lambda} F(\mu) < F(\lambda)]$ (so in particular F is strictly increasing).

Let P_f be Easton forcing for F. (So $(\forall\lambda \in \mathrm{RCar})[V^{P_F} \models 2^\lambda = F(\lambda)]$.) Then $V^{P_F} \models \forall\lambda$ regular, $\forall S \subseteq \lambda$ stationary, $\neg\,\mathrm{Unif}\,(\lambda, S, 2, 2, 2)$ holds. (Note that for inaccessible λ, $2^{<\lambda} = 2^\lambda$ implies the failure of the weak diamond, so $\otimes$ is a reasonable hypothesis.)

Proof. Recall that Easton forcing $P_F = \prod_{\lambda\in\mathrm{RCar}} P_\lambda$ with Easton support (i.e. bounded below inaccessibles, full support below non-inaccessibles), where

$$P_\lambda = \{F : \mathrm{Dom}(f) \in [F(\lambda)]^{<\lambda} \text{ and } \mathrm{Rang}(f) \subseteq \{0,1\}\}.$$

So fix λ and a name $\underset{\sim}{S}$ for a stationary subset of λ. We will work in $V_1 \overset{\text{def}}{=} V^{\prod_{\kappa>\lambda} P_\kappa}$. Note that V_1 satisfies GCH up to λ, as $\prod_{\kappa>\lambda} P_\kappa$ is λ^+-closed, hence does not add any subsets of λ. So we have to deal with the forcing $P^0 \times P_\lambda$, where $P^0 \overset{\text{def}}{=} \prod_{\mu<\lambda} P_\mu$. Let $\underset{\sim}{F}$ be the name of a function, $\Vdash_{P^0\times P_\lambda}$ "$\underset{\sim}{F} : {}^{\lambda>}2 \to 2$". $\mathrm{Dom}(\underset{\sim}{F})$ is in $V_1^{P^0}$, as P_λ adds no bounded subsets to λ. Since $P^0 \times P_\lambda$ satisfies the λ^+-c.c., we can find a set $A \subseteq F(\lambda)$, satisfying $|F(\lambda) \setminus A| = \lambda$ such that $\underset{\sim}{F}$ and $\underset{\sim}{S}$ are $P^0 \times (P_\lambda{\restriction}A)$-names, where $P_\lambda{\restriction}A \overset{\text{def}}{=} \{f \in P_\lambda : \mathrm{Dom}(f) \subseteq A\}$. (We can even find such A of size λ.)

Assume that $p \Vdash$ " there is no weak diamond on $\underset{\sim}{S}$ for $\underset{\sim}{F}$".

We may also assume $p \in P^0 \times (P_\lambda{\restriction}A)$, and for notational convenience assume $A = [\lambda, F(\lambda))$.

Let $\underset{\sim}{f}_\lambda : F(\lambda) \to 2$ be the name for the generic function for P_λ. We claim that $\underset{\sim}{d} \overset{\text{def}}{=} \underset{\sim}{f}_\lambda{\restriction}\lambda$ is a weak diamond for F on $\underset{\sim}{S}$. So assume that $\underset{\sim}{\eta}$ is a $P^0 \times P_\lambda$-name such that

$$p \Vdash \text{“}\underset{\sim}{\eta} \in {}^\lambda 2, \underset{\sim}{C}_1 \overset{\text{def}}{=} \{\alpha : \underset{\sim}{F}(\underset{\sim}{\eta}{\restriction}\alpha) \neq \underset{\sim}{f}_\lambda(\alpha)\} \in \mathcal{D}_\lambda\text{”}.$$

Let $\bar{N} = \langle N_i : i < \lambda\rangle$ be a continuous increasing sequence of elementary submodels of $H(\chi)$ (for some large enough χ) satisfying

$$(\forall i < \lambda)[\bar{N}{\restriction}(i+1) \in N_{i+1}]$$

$$\underset{\sim}{C}_1,\ p,\ \underset{\sim}{\eta},\ \underset{\sim}{S},\ \underset{\sim}{F}, \ldots \in N_0.$$

Define a name $\underset{\sim}{C}_2$ by

$$\Vdash \text{“}\underset{\sim}{C}_2 = \{\alpha : N_\alpha[G_{P^0 \times P_\lambda}] \cap \lambda = N_\alpha \cap \lambda = \alpha\}\text{”}.$$

Since $\underset{\sim}{C}_2$ is the name of a club set, we can find an ordinal δ and a condition $q \geq p$ in $P^0 \times P_\lambda$ such that

$$q \Vdash \text{“}\delta \in \underset{\sim}{S} \cap \underset{\sim}{C}_2\text{”}.$$

As $q \Vdash$ “$\delta \in \underset{\sim}{S}$” clearly the set $\mathcal{J}_\delta \subseteq P^0 \times P_\lambda{\restriction}A$ is predense above q where

$$\mathcal{J}_\delta = \{r \in P^0 \times (P_\lambda{\restriction}A) : r \text{ forces that } \delta \in \underset{\sim}{S}\}.$$

As $q \Vdash$ “$\delta \in \underset{\sim}{C}_2$”, clearly for every $\alpha < \delta$ the set $\mathcal{I}_{\delta,\alpha} \subseteq P^0 \times (P_\lambda{\restriction}(\delta \cup A))$ is predense above q, where

$$\mathcal{I}_{\delta,\alpha} = \{r \in P^0 \times (P_\lambda{\restriction}(\delta \cup A)) : \text{ for some } \beta \in (\alpha, \delta)\ r \text{ forces that } \beta \in \underset{\sim}{C}_1\}.$$

Why? Let $G \subseteq P^0 \times P_\lambda$ be generic over V, and $g \in G$, so $\delta \in \underset{\sim}{C}_2[G]$ hence $N_\delta[G] \cap \lambda = \delta \subseteq N$, so there is $\beta \in (\alpha, \delta) \cap \underset{\sim}{C}_1[G]$ hence for some $p \in N_\delta[G] \cap G$ we have $p \Vdash$ “$\beta \in \underset{\sim}{C}_1$”, so $p \in \mathcal{I}_{\delta,\alpha} \cap G$.

Define $q' \in P^0 \times P_\lambda$ by $q'{\restriction}P^0 = q{\restriction}P^0$, $q'{\restriction}P_\lambda = q{\restriction}(P_\lambda{\restriction}(\delta \cup A))$. It is clear that also $\mathcal{J}_\delta$ and $\mathcal{I}_{\delta,\alpha}$ (for $\alpha < \delta$) are predense above q' hence

$$q' \Vdash \text{“}\delta \in \underset{\sim}{S} \text{ and } \delta = \sup(\underset{\sim}{C}_1 \cap \delta) \text{ hence } \delta \in \underset{\sim}{C}_1\text{”}.$$

(Alternatively for every $(P^0 \times P_\lambda)$-name $\underset{\sim}{\tau} \in N_\delta$ of an ordinal $< \lambda$ the set

$$\mathcal{I}_{\underset{\sim}{\tau}} = \{p : p \in P^0 \times P_\lambda \text{ and } p{\restriction}P_\lambda \in P_\lambda{\restriction}(\delta \cup A) \text{ and } p \text{ forces a value to } \underset{\sim}{\tau} \text{ which is } \gamma_{\underset{\sim}{\tau},p} \text{ and is } < \delta\}$$

is predense above q', hence

$$q' \Vdash \text{“}\delta \in \underset{\sim}{S} \cap \underset{\sim}{C}_2\text{”}.$$

So since $\Vdash$ "$\underset{\sim}{C}_1 \in N_0[G] \subseteq N_\delta[G]$", we also have $q' \Vdash$ "$\delta \in \underset{\sim}{C}_1$".)
But now we can extend q' to a condition q'' forcing a value to $\underset{\sim}{F}(\underset{\sim}{\eta}\restriction\alpha)$, say ℓ^*, again by the choice of A w.l.o.g. $q'' \in P_0 \times P\restriction(\delta \cup A)$. Now we can extend q'' to a condition forcing $\underset{\sim}{f}_\lambda(\alpha) = \ell^*$, a contradiction. $\square_{2.11}$

The following variation of the weak diamond is also sufficient for our purposes (see more in [Sh:576, §1, §3]).

2.12 Definition. 1) We say $F : {}^{\lambda>}2 \to 2$ is "μ-definable" if for some $Y \subseteq \lambda$, for every $\delta < \lambda$, $\eta \in {}^\delta 2$ we can compute $F(\eta)$ in $L[\eta, Y]$. If $\mu = \lambda$ we may omit it.
2) We say F is "weakly definable" if it is μ-definable for some $\mu < 2^\lambda$.

2.13 Remark. 1) For the proof of 2.7A it is enough to have the weak diamond for all *weakly definable* F. (We let the set Y code G, $\langle G_\alpha : \alpha < \lambda\rangle$, H, and for each α where $G_{\alpha+1}/G_\alpha$ is not free, Y computes a function $f \in \text{Fact}\,(G_{\alpha+1}, H)$, $f\restriction(G_\alpha \times G_\alpha = 0$, and in V there is no $g \in \text{Trans}\,(G_{\alpha+1}, H)$, $g\restriction G_\alpha = 0$, $f = \delta g$. See [MkSh:313] for a related argument.)
2) Now all Easton forcings P_f (not just the ones satisfying $\otimes$ from Theorem 2.11 stating with universe satisfying GCH) satisfies: in V^{P_f} the definable weak diamonds hold for $S \subseteq \lambda$ whenever λ is regular uncountable, S stationary.

§3. Weak Diamond for $\aleph_2$ Assuming CH

3.1 Definition. Let λ be a cardinal and $S \subseteq \lambda$. The sequence $\bar{\eta} = \langle \eta_\delta : \delta \in S\rangle$ is called a *ladder system* if for all $\delta \in S$, $\eta_\delta = \langle \eta_\delta(i) : i < \ell g(\eta_\delta)\rangle$ is increasing and cofinal in δ. We say that $\bar{\eta}$ is *continuous* if each η_δ is continuous.
$\bar{\eta}$ has the *uniformization* [alternatively: *club uniformization*] *property* if:
Whenever $\bar{c} = \langle c_\delta : \delta \in S\rangle$ is a sequence of functions $c_\delta : \ell g(\eta_\delta) \to 2$, then we can find a function $h : \lambda \to 2$ such that for each $\delta \in S$ the set

$$\{i < \ell g(\eta_\delta) : c_\delta(i) = h(\eta_\delta(i))\}$$

is cobounded [alternatively: contains a closed unbounded set]. (In this case we say that h "uniformizes" $\bar{c}$.)

3.2 Remark.

(1) If $\bar{\eta}$ is a ladder system on S then we can thin out $\bar{\eta}$ to a ladder system $\bar{\eta}'$ on S satisfying $\ell g(\eta'_\delta) = \mathrm{cf}(\delta)$ for all $\delta \in S$. Moreover, if $\bar{\eta}$ was continuous, and if $\bar{\eta}$ had the uniformization property, then also $\bar{\eta}'$ will have it.

(2) If $2^{\aleph_0} < 2^{\aleph_1}$, then no ladder system on $S^1_0 \stackrel{\text{def}}{=} \{\delta < \aleph_1 : \mathrm{cf}(\delta) = \aleph_0\}$ has the uniformization property.

Proof of (2). ¿From $2^{\aleph_0} < 2^{\aleph_1}$ we conclude that Unif $(\aleph_1, 2, 2^{\aleph_0})$ fails (by 1.10). Let $=^*$ be the equivalence relation on ${}^\omega 2$ defined by $f =^* g$ iff $\forall k\, \exists n \geq k$ such that $f(n) = g(n)$. Let $A \stackrel{\text{def}}{=} {}^\omega 2/ =^*$ be the set of equivalence classes. By the failure of Unif $(\aleph_1, 2, 2^{\aleph_0})$ we know that

(*) $\forall F : {}^{\omega_1 >}2 \to A \exists h : \omega_1 \to A \forall g : \omega_1 \to 2\ [\{\alpha : F(g{\restriction}\alpha) \neq h(\alpha)\}$ stationary].

Fix a ladder system $\bar{\eta} = \langle \eta_\delta : \delta \in S^1_0 \rangle$. We will show that η does not have the uniformization property. Let

$$F(s) = (s \circ \eta_\delta / =^*) \in A \text{ for } s \in {}^{\omega_1 >}2.$$

Let $h : \omega_1 \to A$ be as in (*), and let $\bar{h} : \omega_1 \to 2^\omega$ be such that

$$h(\alpha) = (\bar{h}(\alpha)/ =^*).$$

Define $c_\delta : \omega \to 2$ by $c_\delta(n) = \bar{h}(\delta)(n)$. Now check that $\bar{c} = \langle c_\delta : \delta \in S^1_0 \rangle$ witnesses the failure of the uniformization property of $\bar{\eta}$. $\square_{3.2}$

Recall that $S^2_1 \stackrel{\text{def}}{=} \{i < \aleph_2 : \mathrm{cf}(i) = \aleph_1\}$.
In this section we will consider continuous ladder systems on S^2_1, and we ask the following

3.3 Question. Can $\bar{\eta} = \langle \eta_\delta : \delta \in S^2_1 \rangle$ have the (club) uniformization property (with η_δ increasing continuous with limit δ, of length $\mathrm{cf}(\delta)$)?

We shall answer this question negatively even for club uniformization property in Conclusion 3.7 assuming $2^{\aleph_0} = \aleph_1$.

3.4 Why only for continuous η_δ?. The reader may ask what happens if we waive the restriction that η_δ be a continuous sequence and require just η_δ which is cofinal in δ? By works of the author (see in [Sh:80], Steinhorn and King [SK] and [Sh:186] and very lately [Sh:587]) even assuming GCH a sequence $\langle \eta_\delta : \delta \in S^2_1 \rangle$ may have the uniformization property. But if we require e.g. each c_δ to be eventually constant, for every η_δ which enumerates a club of δ, we have consistency. Also if we restrict ourselves to $\langle \eta_\delta : \delta \in S \rangle$ where $S \subseteq S^2_1$, $S^2_1 \setminus S$ stationary we have consistency results.

3.4A Discussion. This shows the impossibility of some generalizations of MA to $\aleph_1$-complete forcing notions. Why? Suppose $\bar\eta = \langle \eta_\delta : \delta \in S^2_1 \rangle$, η_δ is increasing continuous with limit δ, and $\bar c = \langle c_\delta : \delta \in S^2_1 \rangle$, $c_\delta \in {}^{\omega_1}2$. We define $P_{\bar\eta,\bar c} = \{p : p = (u, i, \bar d, f) = (u^p, i^p, \bar d^p, f^p)$ where u is a countable subset of S^2_1, i a successor ordinal $< \omega_1$, $\bar d = \langle d_\delta : \delta \in u \rangle$, d^p_δ a closed subset of i, f is a function from $\mathrm{Dom}(f) = \{\eta_\delta(j) : \delta \in u, j < i\}$ to $\{0,1\}$ such that $j \in d_\delta$ & $\delta \in u \Rightarrow f(\eta_\delta(j)) = c_\delta(j)\}$ ordered by $p \le q$ iff $u^p \subseteq u^q$, $i^p \le i^q$, $[\delta \in u^p \Rightarrow d^p_\delta = d^q_\delta \cap i^p]$, $f^p \subseteq f^q$, and $i^p < i^q$ & $\delta_1 \in u^p$ & $\delta_2 \in u^p$ &$\delta_1 \ne \delta_2 \Rightarrow$ $\{\eta_{\delta_1}(j) : j \in [i^q, \omega_1)\} \cap \{\eta_{\delta_2(j)} : j \in [i^q, \omega_1)\} = \emptyset$.

So:

(∗) if the answer to 3.3 is no as exemplified by $\bar c$, *then* there is no directed $G \subseteq P_{\bar\eta,\bar c}$ which intersect each $\mathcal{I}_{\delta,i} = \{p \in P_{\bar\eta,\bar c} : \delta \in u^p$ and $i \le i^p$ and $d^p_\delta \setminus i \ne \emptyset\}$ which is dense.

So any generalization of MA as above necessarily does not include $P_{\bar\eta,\bar c}$, which is a quite nice forcing notion: it is $\aleph_1$-complete, and can be divided to $\aleph_1$ formulas, each $\aleph_1$-directed.

3.5 Convention. Let F denote a function from
$\{h : h$ a function, $\mathrm{Dom}(h) \subseteq \omega_2$ is countable, $\mathrm{Rang}(h) \subseteq 2\}$ into $2 = \{0,1\}$.

3.6 Theorem. 1) $(2^{\aleph_0} = \aleph_1)$: *For any* function F and $\bar\eta = \langle \eta_\delta : \delta \in S^2_1 \rangle$ as in 3.1 *there is* $\langle d_\delta : \delta \in S^2_1 \rangle, d_\delta \in {}^{\omega_1}2$, (we can call it a weak diamond sequence) such that *for any* $h : \omega_2 \to 2$, for stationarily many $\delta \in S^2_1$, for stationarily many $i < \omega_1$,

$$d_\delta(i) = F(h\restriction\{\eta_\delta(j) : j \leq i\}).$$

2) Suppose

(a) $\theta < \kappa = \mathrm{cf}(\kappa)$, $2^\theta = 2^{<\kappa} = \kappa$ (so $\kappa = \kappa^{<\kappa}$).

(b) $S = \{\delta < \kappa^+ : \mathrm{cf}(\delta) = \kappa\}$.

(c) for each $\delta \in S, \eta_\delta$ is a strictly increasing continuous function from κ to δ with limit δ.

(d) F is a function with domain $\{h : h$ a partial function from κ^+ to $\{0, 1\}$ of cardinality $< \kappa\}$ with range $\{0, 1\}$.

Then we can find $\langle d_\delta : \delta \in S\rangle$, $d_\delta \in {}^\kappa 2$ such that for any $h : \kappa^+ \to \{0, 1\}$ for stationarily many $\delta \in S$ for stationarily many $i < \kappa$ we have

$$d_\delta(i) = F(h\restriction\mathrm{Rang}(\{\eta_\delta(j) : j \leq i\})).$$

3.6A Remark. Note the "$j \leq i$" rather than "$j < i$" in part (1).

3.7 Conclusion. (CH) $\bar\eta = \langle\eta_\delta : \delta \in S^2_1\rangle$ does not have the club uniformization property.

Proof of 3.7. Let $F(h)$ be $h(\mathrm{MaxDom}(h))$ if defined, zero otherwise. By 3.6 there are for F, η_δ a sequence $\langle d_\delta : \delta \in S^2_1\rangle$ as there; let $c_\delta(i) = 1 - d_\delta(i)$.

Proof of 3.6. We prove part (1) as (2) has essentially the same proof. Let λ be big enough (e.g., $(2^{\aleph_2})^+$), and M^* be an expansion of $(H(\lambda), \in)$ by Skolem functions (if it has a definable well ordering it suffices).

Suppose $\bar\eta, F$ form a counterexample. It is known that there is a function G from $\{A : A \subseteq \omega_2, |A| \leq \aleph_0\}$ to ω_1 such that $G(A) = G(B)$ implies A, B have the same order type and their intersection is an initial segment of both (e.g. if $h_\alpha : \alpha \to \omega_1$ is one-to-one for $\alpha < \omega_1$, we let $G_0(A) \stackrel{\mathrm{def}}{=} \{(\mathrm{otp}(A \cap \alpha), \mathrm{otp}(A \cap \beta), h_\beta(\alpha)) : \alpha \in A$ and $\beta \in A\}$. Now G_0 is as required except that $\mathrm{Rang}(G_0) \not\subseteq \omega_1$ but $|\mathrm{Rang}(G_0)| \leq \aleph_1$ so we can correct this).

We now define a procedure for defining for any $p \in H(\lambda)$, $\langle c^p_\delta : \delta \in S^2_1\rangle$ where $c^p_\delta : \omega_1 \to H(\omega_1)$, which we shall use later.

For every $\delta \in S^2_1$, $i < \omega_1$, let $N^p_{\delta,i}$ be the Skolem hull of $\{\delta, i, p\}$ in M^*, and let

$$\oplus \quad c^p_\delta(i) \stackrel{\text{def}}{=} \langle \text{ isomorphism type } (N^p_{\delta,i}, p, \delta, i), G(N^p_{\delta,i} \cap \aleph_2)\rangle.$$

Remarks. 1) The model of $(N^p_{\delta,i}, p, \delta, i)$ is not in $H(\aleph_1)$, but since $N^p_{\delta,i}$ is countable we can assume its isomorphism type does belong.
2)$(N^p_{\delta,i}, p, i, \delta)$ is $N^p_{\delta,i}$ expanded by three individual constants.

Now remember we have assumed $F, \bar{\eta}$ form a counterexample. So for every $c_\delta \in {}^{\omega_1}2$ ($\delta \in S^2_1$) there is $h_\delta : \omega_2 \to 2$ such that for a closed unbounded set of $\delta \in S^2_1$, for a closed unbounded set of $i < \omega_1$, $c_\delta(i) = F(h_\delta \restriction \{\eta_\delta(j) : j \le i\})$.

Now we can easily replace 2 by the set ${}^\omega 2$ as follows.

For h a function into ${}^\omega 2$, let $h^{[n]}$ be $h^{[n]}(i) = (h(i))(n)$ for $i \in \mathrm{Dom}(h)$. Define F^* by: $F^*(h) = \langle F(h^{[n]}) : n < \omega\rangle$; now if we are given $\langle c_\delta : \delta \in S^2_1\rangle$ where $c_\delta \in {}^{\omega_1}({}^\omega 2)$, i.e., $c_\delta : \omega_1 \to {}^\omega 2$, so $c^{[n]}_\delta \in {}^{\omega_1}2$ is well defined for each $\delta \in S^2_1$ and let $h^{[n]} : \aleph_2 \to 2$ be such that for a club of $\delta \in S^2_1$ for a club of $i < \omega_1$ we have

$$c^{[n]}(i) = F(h^{[n]}(i) \restriction \{\eta_\delta(j) : j \le i\}).$$

Define $h : \aleph_2 \to {}^\omega 2$ by $h(i) = \langle h^{[n]}(i) : n < \omega\rangle$, it is as required.

Now as $|{}^\omega 2| = 2^{\aleph_0} = |H(\aleph_1)|$, we conclude:

(∗) for every $c_\delta \in {}^{\omega_1}H(\aleph_1)$ ($\delta \in S^2_1$) there is $h : \omega_2 \to H(\aleph_1)$ such that for a club of $\delta \in S^2_1$ for a club of $i < \omega_1$, $c_\delta(i) = F^*(h \restriction \{\eta_\delta(j) : j \le i\})$.

Now we define by induction on $n < \omega$, $p(n) \in H(\lambda)$, and $h_n : \omega_2 \to H(\aleph_1)$.

Let $p(0) = \langle \bar{\eta}\rangle$. If we have defined $p(n)$, let $c^{p(n)}_\delta : \omega_1 \to H(\aleph_1)$ be as we have defined before (in $\oplus$), so by (∗) there is a suitable $h_n : \aleph_2 \to H(\aleph_1)$; i.e., there is a closed unbounded $W^n \subseteq \aleph_2$ such that for every $\delta \in W^n \cap S^2_1$, there is a closed unbounded $W^n_\delta \subseteq \omega_1$ such that for $i \in W^n_\delta, \delta \in W^n \cap S^2_1$ we have: $c^{p(n)}_\delta(i) = F^*(h_n \restriction \{\eta_\delta(j) : j \leq i\})$.

Let $p(n+1) \stackrel{\text{def}}{=}$

$\langle p(n), h_n, W^n, \langle W^n_\delta : \delta \in W^n \cap S^2_1 \rangle, \langle \langle N^{p(n)}_{\delta,i} : i < \omega_1 \rangle : \delta \in S^2_1 \rangle \rangle$.

Now let $W = \bigcap_{n<\omega} W^n$, and for $\delta \in W$ let $W_\delta = \bigcap_{n<\omega} W^n_\delta$. Clearly W is a closed unbounded subset of $\aleph_2$, and W_δ is a closed unbounded subset of ω_1. So for every $\delta \in W \cap S^2_1$, there is $i(\delta) \in W_\delta$; so as $\eta_\delta(i(\delta)) < \delta$ by Fodor lemma, for some $i < \aleph_2$ and $i^* < \aleph_1$ the set $\{\delta \in W \cap S^2_1 : \eta_\delta(i(\delta)) = i \text{ and } i(\delta) = i^*\}$ is stationary. As CH holds there are δ_1, δ_2 in $W \cap S^2_1$ and $\xi < \omega_1$ such that

A) $\eta_{\delta_1}(\xi) = \eta_{\delta_2}(\xi)$ moreover $\eta_{\delta_1} \restriction (\xi+1) = \eta_{\delta_2} \restriction (\xi+1)$
B) $\delta_1 < \delta_2$
C) $\xi \in W_{\delta_\ell}$ for $\ell = 1, 2$.

So clearly we can assume

D) there are no $\delta^\dagger_1, \delta^\dagger_2$ satisfying (A), (B) and (C) such that $\delta^\dagger_1 \leq \delta_1$, $\delta^\dagger_2 \leq \delta_2$ and $(\delta^\dagger_1, \delta^\dagger_2) \neq (\delta_1, \delta_2)$.

Now as $\delta_1 < \delta_2$, for some $i > \xi, \eta_{\delta_1}(i) \neq \eta_{\delta_2}(i)$, and there is a minimal such i; but as $\eta_{\delta_1}, \eta_{\delta_2}$ are increasing and continuous, such minimal i should be a succesor ordinal, so there is a maximal ζ among those satisfying $\zeta < \omega_1, \eta_{\delta_1} \restriction \zeta = \eta_{\delta_2} \restriction \zeta$, $\eta_{\delta_1}(\zeta) = \eta_{\delta_2}(\zeta)$ and $\zeta \in W_{\delta_1} \cap W_{\delta_2}$. So $\omega_1 > \zeta^\dagger > \zeta, \bigwedge_{\ell=1,2} \zeta^\dagger \in W_{\delta_\ell}$ implies $\eta_{\delta_1}(\zeta^\dagger) \neq \eta_{\delta_2}(\zeta^\dagger)$ or at least $\eta_{\delta_1} \restriction (\zeta^\dagger + 1) \neq \eta_{\delta_2} \restriction (\zeta^\dagger + 1)$.

So for every n

$$(\alpha) \quad c^{p(n)}_{\delta_1}(\zeta) = c^{p(n)}_{\delta_2}(\zeta)$$

as both are equal to $F^*(h_n \restriction \{\eta_{\delta_\ell}(j) : j \leq \zeta\})$. Looking at the definition of $c^{p(n)}_\delta(\zeta)$ (see $\oplus$) we see that $N^{p(n)}_{\delta_1,\zeta}$ is isomorphic to $N^{p(n)}_{\delta_2,\zeta}$, and let the isomorphism be called g_n. Note that the isomorphism is unique (as $\in$ in those models is transitive well founded).

By the definition of $c^{p(n)}_\delta(\zeta)$, clearly without loss of generality

$$g_n[p(n)] = p(n), g_n(\delta_1) = \delta_2, g_n(\zeta) = \zeta$$

Looking at $p(n)$'s definition we see that $g_n(\eta_{\delta_1}) = \eta_{\delta_2}$ and for $n > 0$ $g_n(W^{n-1}) = W^{n-1}$ and $g_n(W^{n-1}_{\delta_1}) = W^{n-1}_{\delta_2}$ and $g_n(N^{p(n-1)}_{\delta_1,\zeta}) = N^{p(n-1)}_{\delta_2,\zeta} \in N^{p(n)}_{\delta_2,\zeta}$.

As $N^{p(n-1)}_{\delta_\ell,\zeta}$ is countable and belongs to $N^{p(n)}_{\delta_\ell,\zeta}$, it is also included in it, hence $g_n \restriction N^{p(n-1)}_{\delta_1,\zeta}$ is an isomorphism from $N^{p(n-1)}_{\delta_1,\zeta}$ onto $N^{p(n-1)}_{\delta_2,\zeta}$ hence (by the uniqueness of g_n)

(β) $\quad g_n \supseteq g_{n-1}$.

For $\ell = 1,2$ let $N_\ell = \bigcup_{n<\omega} N^{p(n)}_{\delta_\ell,\zeta}$ and $g = \bigcup_{n<\omega} g_n$; so g is an isomorphism from N_1 to N_2.
By the definition of $c^{p(n)}_{\delta_\ell}(\zeta)$, clearly the second coordinates are the same, thus:

(γ) $\quad G(N^{p(n)}_{\delta_1,\zeta} \cap \omega_2) = G(N^{p(n)}_{\delta_2,\zeta} \cap \omega_2)$,

hence those sets have their intersection an initial segment of both hence also $N_1 \cap \omega_2, N_2 \cap \omega_2$ have their intersection an initial segment of both (as usually, we are not strictly distinguishing between a model and its universe), hence g is the identity on $N_1 \cap N_2 \cap \omega_2$.

Note that clearly $\delta_1 \notin N_2$ as $g(\delta_1) = \delta_2 \neq \delta_1$, hence $\delta_2 \notin N_1$.

Let $\delta^*_\ell \overset{\text{def}}{=} \text{Min}(\omega_2 \cap N_\ell \setminus (N_1 \cap N_2))$, so clearly $\delta^*_\ell \leq \delta_\ell$, $g(\delta^*_1) = \delta^*_2$ and so $\text{cf}(\delta^*_1) = \text{cf}(\delta^*_2)$.

(δ) $\quad \text{cf}(\delta^*_\ell) = \aleph_1$.

Why? Otherwise $\text{cf}(\delta^*_1) = \aleph_0$, and as $\delta^*_1 \in N_1$ for some n, $\delta_1 \in N^{p(n)}_{\delta_1,\zeta}$, hence there is $\{\beta_m : m < \omega\} \subseteq \delta^*_1 \cap N^{p(n)}_{\delta_1,\zeta}$ cofinal in δ^*_1. By the choice of $\delta^*_1, \beta_m \in N_1 \cap N_2$, hence $g(\beta_m) = \beta_m$; let $\beta^* = \min(N^{p(n)}_{\delta_2,\zeta} \setminus \bigcup_m \beta_m)$, so $\beta^* \in N^{p(n)}_{\delta_2,\zeta} \subseteq N^{p(n+1)}_{\delta_2,\zeta}$, so $\delta^*_1 = \text{Sup}\{\beta_m : m < \omega\} = \sup(\beta^* \cap N^{p(n)}_{\delta_2,\zeta}) \in N_2$, contradiction.

So we have proved (δ).

Now let for $\ell = 1, 2, \alpha_\ell \stackrel{\text{def}}{=} N_\ell \cap \omega_1$, (it is an initial segment) and $\beta_\ell \stackrel{\text{def}}{=} \sup(N_\ell \cap \delta^*_\ell)$ hence $\beta_1 = \beta_2$ (by δ^*_ℓ definition) and call it β. As $\mathrm{cf}(\delta^*_\ell) \geq \aleph_1$ clearly $\delta^*_\ell \geq \omega_1$, and so clearly by g's existence $\alpha_1 = \alpha_2$ and call it α (also as $\omega_1 \in N_1 \cap N_2 \cap \omega_2$, necessarily $N_1 \cap \omega_1 = N_2 \cap \omega_1$).

As $\eta_{\delta^*_1}$ is a one to one function (being increasing) from ω_1, clearly

$$\eta_{\delta^*_1}(i) \in N_1 \text{ iff } i < \alpha.$$

Also $N_1 \models$ "$\langle \eta_{\delta^*_1}(i) : i < \omega_1 \rangle$ is unbounded below δ^*_1" (remember $N_1 \prec M^*$ as $N^{p(n)}_{\delta_1,\zeta} \prec M^*$ for each n).

So clearly $\beta = \mathrm{Sup}\{\eta_{\delta^*_1}(i) : i < \alpha\}$; but $\eta_{\delta^*_1}$ is increasing continuous and α is a limit ordinal (being $N_\ell \cap \omega_1$), hence $\beta = \eta_{\delta^*_1}(\alpha)$.

For the same reasons $\beta = \eta_{\delta^*_2}(\alpha)$.

Now $\eta_{\delta^*_1} \restriction \alpha = \eta_{\delta^*_2} \restriction \alpha$ because $g(\eta_{\delta^*_1}) = \eta_{\delta^*_2}$, and $\alpha \in W^n_{\delta^*_\ell}$ for each $n < \omega (\ell = 1, 2)$ as $N_\ell \models$ "$W^n_{\delta^*_\ell}$ is a closed unbounded subset of ω_1". For similar reasons $\delta^*_\ell \in W_n$ for each n: as $W_n \in N^{p(n+1)}_{\delta_\ell,\zeta}$ hence $W_n \in N_\ell$ hence $W_n \in N_1 \cap N_2$, and as $N_1, N_2 \prec M^*, M^*$ has Skolem functions, clearly $N_1 \cap N_2 \prec M^*$, so W_n is an unbounded subset of $N_1 \cap N_2 \cap \omega_2$. So in N_ℓ, W_n is unbounded in $\delta^*_\ell = \mathrm{Min}[(\omega_2 \cap N_\ell) \setminus (N_1 \cap N_2)]$, hence $N_\ell \models$ "$\delta^*_\ell \in W_n$" hence $\delta^*_\ell \in W_n$.

We can conclude that $\delta^*_1, \delta^*_2, \beta$ satisfy the requirements (A), (B), (C) on δ_1, δ_2, ξ. Hence by requirement (D) on them, $\delta_1 = \delta^*_1$, $\delta_2 = \delta^*_2$. But, $\zeta \in N^{p(n)}_{\delta_\ell,\zeta} \subseteq N_\ell$ hence $\zeta < \omega_1 \cap N_1 \cap N_2$ hence $\zeta < \alpha$, so clause (α) contradicts the choice of ζ, so we get a contradiction, thus finishing the proof of the theorem (3.6). $\square_{3.6}$

3.8 Concluding Remarks. 1) If $\lambda = \kappa^+$, κ is strongly inaccessible then the conclusion of 3.6(2) may fail (see [Sh:186], we repeat the proof in [Sh:64], see more in [Sh:587]).

2) If $2^{\aleph_0} = 2^{\aleph_2}$, then it follows that for some F and $\bar{\eta}$ we have uniformization. Just choose $\bar{\eta} = \langle \eta_\delta : \delta \in S^2_1 \rangle$ such that $\langle \eta_\delta \restriction \omega : \delta \in S^2_1 \rangle$ are pairwise distinct and for every $\delta \in S^2_1$ and non successor $i < \omega_1$ and $n < \omega$ for some non successor $j < \omega_2$ we have $\eta_\delta(i + n) = j + n$. Now let $\langle \langle c^\gamma_\delta : \delta \in S^2_1 \rangle : \gamma < 2^{\aleph_2} \rangle$ list the set

of sequences $\langle c_\delta : \delta \in S^2_1\rangle, c_\delta \in {}^{\omega_1}2$. Let $\langle r_\alpha : \alpha < 2^{\aleph_0}\rangle$ list distinct reals, and we let $h^\gamma \in {}^{\omega_2}2$ be: $h^\gamma(i+n) = r_\gamma(n)$ for any non-successor ordinal $i < \omega_1$. Now define F by: $F(h) = c^\gamma_\delta(i)$ if $\mathrm{Dom}(h) = \{\alpha_j : j \le i\}$ with α_j increasing, $i \ge \omega$, $\langle h(\alpha_n) : n < \omega\rangle = r_\gamma$.

3) In 3.6(2) we may demand that (e) $F(h\restriction \mathrm{Rang}(\eta_\delta\restriction(i+1)))$ only depend on $h(\eta_\delta(i))$ and i. Then we can weaken clause (a) there as follows.

3.9 Theorem. Suppose

(a) $\aleph_0 < \mathrm{cf}(\theta) = \theta < \kappa = 2^\theta$,

(b) $S = \{\delta < \kappa^+ : \mathrm{cf}(\delta) \ge \theta^+\}$.

(c) for each $\delta \in S, \eta_\delta$ is a strictly increasing continuous function from $\mathrm{cf}(\delta)$ to δ with limit δ.

(d) F is a function with domain $\{h : h$ a partial function from κ^+ to κ such that $|\mathrm{Dom}(h)| \le \theta\}$ with range $\{0,1\}$.

(e) $\bar{a} = \langle a^\delta_i : \delta \in S, i < \mathrm{cf}(\delta)\rangle$, $a^\delta_i \subseteq \eta_\delta(i)+1$ and $|a_\alpha| \le \theta$,

Then we can find $\langle d_\delta : \delta \in S\rangle$, $d_\delta \in {}^\kappa 2$ such that for any $h : \kappa^+ \to \kappa$ for stationarily many $\delta \in S$ for stationarily many $i < \mathrm{cf}(\delta)$, $d_\delta(i) = F(h\restriction a^\delta_i)$.

3.10 Conclusion. If $\theta, \kappa, \bar{\eta}$ as above *then* $\bar{\eta} = \langle \eta_\delta : \delta \in S\rangle$ does not have the club uniformization property.

Proof of 3.10. Let $F(h) = h(\mathrm{MaxDom}(h))$ if defined, zero otherwise. By 3.6 there are for F, η_δ a sequence $\langle d_\delta : \delta \in S^2_1\rangle$; let $c_\delta(i) = 1 - d_\delta(i)$.

The proof of 3.10 is very similar to that of 3.6.

Proof of 3.9. Let λ be big enough (e.g., $(\beth_3(\kappa))^+$), and M^* be an expansion of $(H(\lambda), \in, \bar{\eta}, \bar{a}, i)_{i \le \theta}$ be Skolem functions (if it has a definable well ordering it suffices).

Suppose $\bar{\eta}, F$ form a counterexample. It is known that there is a function G from $\{A : A \subseteq \kappa^+, |A| \le \theta\}$ to κ such that $G(A) = G(B)$ implies $A \cap \kappa = B \cap \kappa$, A, B have the same order type and their intersection is an initial segment of both (e.g. if $h_\alpha : \alpha \to \kappa$ is one-to-one for $\alpha < \kappa^+$, we let $G_0(A) \stackrel{\mathrm{def}}{=}$

$\{(\mathrm{otp}(A \cap \alpha), \mathrm{otp}(A \cap \beta), h_\beta(\alpha)) : \alpha \in A$ and $\beta \in A\}$. Now G_0 is as required except that $\mathrm{Rang}(G_0) \not\subseteq \kappa$ but $|\mathrm{Rang}(G_0)| \leq \kappa^\theta = \kappa$ so we can correct this).

We now define a procedure for defining for any $p \in H(\lambda)$, $\langle c^p_\delta : \delta \in S\rangle$, $c^p_\delta : \mathrm{cf}(\delta) \to H(\theta^+)$, which we shall use later.

For every $\delta \in S$, $i < \omega_1$, let $N^p_{\delta,i}$ be the Skolem hull of $\{\delta, i, p\} \cup \{\alpha : \alpha \leq \theta\}$ in M^*, and let

$$\oplus \quad c^p_\delta(i) \stackrel{\text{def}}{=} \langle \text{ isomorphism type } (N^p_{\delta,i}, p, \delta, i), G(N^p_{\delta,i} \cap \theta)\rangle.$$

Remarks. 1) The model of $\langle N^p_{\delta,i}, p, \delta, i\rangle$ is not in $H(\theta^+)$, but since $N^p_{\delta,i}$ has cardinality $\leq \theta$ we can assume its isomorphism type does belong.
2)$(N^p_{\delta,i}, p, i, \delta)$ is $N^p_{\delta,i}$ expanded by three individual constants.

Now remember we have assumed

$$\otimes \quad F,\ \bar{a},\ \bar{\eta} \text{ form a counterexample.}$$

So for every $c_\delta \in {}^{\mathrm{cf}(\delta)}2$ (for $\delta \in S$) there is $h_\delta : \kappa^+ \to \kappa$ such that for a closed unbounded set of $\delta \in S$, for a closed unbounded set of $i < \mathrm{cf}(\delta)$, $c_\delta(i) = F(h_\delta \restriction \{\eta_\delta(j) : j \leq i\})$.

Now we can easily replace 2 by the set ${}^\theta 2$ as follows.

For $\varepsilon < \theta$ and h a function into ${}^\theta 2$, let $h^{[\varepsilon]}$ be $h^{[\varepsilon]}(i) = (h(i))(\varepsilon)$ for $i \in \mathrm{Dom}(h)$. Define F^* by: $F^*(h) = \langle F(h^{[\varepsilon]}) : \varepsilon < \theta\rangle$; now if $c_\delta \in {}^{\mathrm{cf}(\delta)}({}^\theta 2)$ for $\delta \in S$, i.e., $c_\delta : \mathrm{cf}(\delta) \to {}^\theta 2$ (so $c^{[\varepsilon]}_\delta$ are well defined for $\varepsilon < \theta$). So by the assumption "F, $\bar{a}$ and $\bar{\eta}$ form a counterexample" for each $\varepsilon < \theta$ there is $h^{[\varepsilon]} : \kappa^+ \to 2$ be such that for a club of $\delta \in S$ for a club of $i < \mathrm{cf}(\delta)$

$$c^{[\varepsilon]}(i) = F(h^{[\varepsilon]}(i) \restriction \{a^\delta_i\}).$$

Define the function $h : \kappa^+ \to {}^\theta 2$ by $h(i) = \langle h^{[\varepsilon]}(i) : \varepsilon < \omega\rangle$.

Now as $|{}^{\theta}2| = \kappa = |H(\theta^+)|$, we conclude:

$(*)$ for every $c_\delta \in {}^{\mathrm{cf}(\delta)}H(\theta^+)$ $(\delta \in S)$ there is $h : \kappa^+ \to H(\theta^+)$ such that for a club of $\delta \in S$ for a club of $i < \mathrm{cf}(\delta)$ we have $c_\delta(i) = F^*(h{\restriction}a_i^\delta)$.

Now we define by induction on $n < \omega$, $p(n) \in H(\lambda)$, and $h_n : \kappa^+ \to H(\theta^+)$.

Let $p(0) = \langle \bar{\eta}, \bar{a}, F\rangle$. If we have defined $p(n)$, let $c_\delta^{p(n)} : \mathrm{cf}(\delta) \to H(\theta^+)$ be as we have defined before (in $\oplus$), so by $(*)$ there is a suitable $h_n : \kappa^+ \to H(\theta^+)$; i.e., there is a closed unbounded $W^n \subseteq \kappa^+$ such that for every $\delta \in W^n \cap S$, there is a closed unbounded $W_\delta^n \subseteq \mathrm{cf}(\delta)$ such that for $i \in W_\delta^n, \delta \in W^n \cap S$ we have: $c_{\delta,i}^{p(n)}(i) = F^*(h_n{\restriction}a_i^\delta)$.

Let

$$p(n+1) \stackrel{\mathrm{def}}{=} \langle p(n), h_n, W^n, \langle W_\delta^n : \delta \in W^n \cap S\rangle, \langle\langle N_{\delta,i}^{p(n)} : i < \mathrm{cf}(\delta)\rangle : \delta \in S\rangle\rangle.$$

Now let $W = \bigcap_{n<\omega} W^n$, and for $\delta \in W$, $W_\delta = \bigcap_{n<\omega} W_\delta^n$. Clearly W is a closed unbounded subset of κ^+, and if $\delta \in W \cap S$ then W_δ is a closed unbounded subset of $cf(\delta)$. So for every $\delta \in W \cap S$, there is $i(\delta) \in W_\delta$; so as $\eta_\delta(i(\delta)) < \delta$ for some $i < \kappa^+$ and $i^* < \kappa$ and $\delta = \mathrm{cf}(\delta) \le \kappa$ the set $\{\delta \in W \cap S : \eta_\delta(i(\delta)) = i, i(\delta) = i^* \text{ and } \mathrm{cf}(\delta) = \delta\}$ is stationary. As $\kappa = \kappa^\theta$ holds there are δ_1, δ_2 in $W \cap S$ and $\xi < \mathrm{cf}(\delta_1)$ such that

A) $\eta_{\delta_1}(\xi) = \eta_{\delta_2}(\xi)$ and $\mathrm{cf}(\delta_1) = \mathrm{cf}(\delta_2)$

B) $\delta_1 < \delta_2$ (so both in $W \cap S$)

C) $\xi \in W_{\delta_\ell}$ for $\ell = 1, 2$ (so $\xi < \mathrm{cf}(\delta)$).

So clearly we can assume

D) there are no $\delta_1^\dagger, \delta_2^\dagger$ satisfying (A), (B) and (C) such that $\delta_1^\dagger \le \delta_1$, $\delta_2^\dagger \le \delta_2$ and $(\delta_1^\dagger, \delta_2^\dagger) \ne (\delta_1, \delta_2)$.

Now as $\delta_1 < \delta_2$ for every large enough $i < \mathrm{cf}(\delta_1)$, $\eta_{\delta_2}(i) > \delta_1$, hence $\{\zeta < \mathrm{cf}(\delta) : \zeta \in W_{\delta_1}, \zeta \in W_{\delta_2} \text{ and } \eta_{\delta_1}(\zeta) = \eta_{\delta_2}(\zeta)\}$ is a bounded subset of $\mathrm{cf}(\delta_1)$. As $W_{\delta_1}, W_{\delta_2}$ are clubs of $\mathrm{cf}(\delta_1)$ and $\eta_{\delta_1}, \eta_{\delta_2}$ are increasing continuous, the set above is closed hence it has a last element. So there is $\zeta < \mathrm{cf}(\delta_1)$ such that $\eta_{\delta_1}(\zeta) = \eta_{\delta_2}(\zeta)$ and $\zeta \in W_{\delta_1} \cap W_{\delta_2}$, but $\zeta^\dagger > \zeta, \bigwedge_{\ell=1,2} \zeta^\dagger \in W_{\delta_\ell}$ implies $\eta_{\delta_1}(\zeta^\dagger) \neq \eta_{\delta_2}(\zeta^\dagger)$.

So for every n

$$(\alpha)\ c^{p(n)}_{\delta_1}(\zeta) = c^{p(n)}_{\delta_2}(\zeta)$$

as both are equal to $F^*(h_n \restriction a^{\delta_\ell}_{\eta_{\delta_\ell}}(\zeta))$, which do not depend on ℓ as $\eta_{\delta_1}(\zeta) = \eta_{\delta_2}(\zeta))$ and they are equal to $h_{n+1}(\eta_{\delta_\ell}(\zeta))$. Looking at the definition of $c^{p(n)}_\delta(\zeta)$ (see $\oplus$ above) we see that $N^{p(n)}_{\delta_1,\zeta}$ is isomorphic to $N^{p(n)}_{\delta_2,\zeta}$, and let the isomorphism be g_n. Note that the isomorphism is unique (as $\in$ in those models is transitive well founded).

By the definition of $c^{p(n)}_\delta(\zeta)$, clearly without loss of generality

$$g_n(p(n)) = p(n), g_n(\delta_1) = \delta_2, g_n(\zeta) = \zeta$$

Looking at the definition of M^* and $p(n)$, $p(0)$ we see that $g_n(\eta_{\delta_1}) = \eta_{\delta_2}$ and for $n > 0$ we have $g_n(W^{n-1}) = W^{n-1}$ and $g_n(W^{n-1}_{\delta_1}) = W^{n-1}_{\delta_2}$ and $g_n(N^{p(n-1)}_{\delta_1,\zeta}) = N^{p(n-1)}_{\delta_2,\zeta} \in N^{p(n)}_{\delta_2,\zeta}$.

As $N^{p(n-1)}_{\delta_\ell,\zeta}$ is of cardinality θ and belongs to $N^{p(n)}_{\delta_\ell,\zeta}$, and $\theta + 1 \subseteq N^{p(n)}_{\delta_i}$ clearly $N^{p(n-1)}_{\delta_\ell,\zeta}$ is also included in it, hence $g_n \restriction N^{p(n-1)}_{\delta_1,\zeta}$ is an isomorphism from $N^{p(n-1)}_{\delta_1,\zeta}$ onto $N^{p(n-1)}_{\delta_2,\zeta}$ hence (by the uniqueness of g_n and the previous sentence)

$$(\beta)\ g_n \supseteq g_{n-1}.$$

For $\ell = 1,2$ let $N_\ell = \bigcup_{n<\omega} N^{p(n)}_{\delta_\ell,\zeta}$ and $g = \bigcup_{n<\omega} g_n$; so g is an isomorphism from N_1 to N_2.

By the definition of $c^{p(n)}_{\delta_\ell}(\zeta)$, clearly:

$$(\gamma)\ G(N^{p(n)}_{\delta_1,\zeta} \cap \kappa^+) = G(N^{p(n)}_{\delta_2,\zeta} \cap \kappa^+),$$

hence sets $N_1 \cap \kappa^+$, $N_2 \cap \kappa^+$ have the same intersection with κ and have their intersection an initial segment of both (as usually, we are not strictly distinguishing between a model and its universe), hence g is the identity on $N_1 \cap N_2 \cap \kappa^+$.

Note that clearly $\delta_1 \notin N_2$ as $g(\delta_1) = \delta_2 \neq \delta_1$, hence $\delta_2 \notin N_1$.

Let $\delta^*_\ell \stackrel{\text{def}}{=} \mathrm{Min}(\kappa^+ \cap N_\ell \setminus (N_1 \cap N_2))$, so clearly $\delta^*_\ell \leq \delta_\ell$, $g(\delta^*_1) = \delta^*_2$. Note $\mathrm{cf}(\delta^*_\ell) \leq \kappa$ (as $\delta^*_\ell < \kappa^+$) so $\mathrm{cf}(\delta^*_\ell) \in N^*_\ell \cap (\kappa+1) \subseteq N_1 \cap N_2 \cap \kappa^+$ and so $\mathrm{cf}(\delta^*_1) = \mathrm{cf}(\delta^*_2)$. Call it σ, so $\sigma \in N_1 \cap N_2 \cap (\kappa+1)$ is regular.

(δ) $\mathrm{cf}(\delta^*_\ell) > \theta$.

[Why? Otherwise $\mathrm{cf}(\delta^*_1) \leq \theta$, and as $\delta^*_1 \in N_1$ for some n, $\delta_1 \in N^{p(n)}_{\delta_1,\zeta}$, hence there is $b \in N^{p(n)}_{\delta_1,\zeta}$, $b = \{\beta_\varepsilon : \varepsilon < \sigma\} \subseteq \delta^*_1$ cofinal in δ^*_1. As $|b| = \sigma \leq \theta$, $b \in N^{p(n)}_{\delta_1,\zeta}$ and $\theta + 1 \subseteq N^{p(n)}_{\delta_1,\zeta}$ necessarily $b = \{\beta_\varepsilon : \varepsilon < \sigma\} \subseteq N^{p(n)}_{\delta_1,\zeta}$. By the choice of $\delta^*_1, \beta_\varepsilon \in N_1 \cap N_2 \cap \kappa^+$, hence $g(\beta_\varepsilon) = \beta_\varepsilon$. Easily $g(b) = \{g(\beta_\varepsilon) : \varepsilon < \sigma\} = \{\beta_\varepsilon : \varepsilon < \sigma\} = b$ (as $\theta + 1 \subseteq N_1 \cap N_2$) and $N_1 \models$ "$\delta^*_1 = \sup(b)$" hence $N_2 \models$ "$g(\delta^*_1) = \sup(g(b))$" that is $N_2 \models$ "$\delta^*_2 = \sup(b)$" so $\delta^*_1 = \delta^*_2$, contradiction.]

So we have proved (δ).

Now for $\ell = 1, 2$ let $\alpha_\ell \stackrel{\text{def}}{=} \sup[N_\ell \cap \mathrm{cf}(\delta_1)]$, so as $N_1 \cap \kappa = N_2 \cap \kappa$ clearly $\alpha_1 = \alpha_2$ call it α. Let $\beta_\ell \stackrel{\text{def}}{=} \sup(N_\ell \cap \delta^*_\ell)$ hence $\beta_1 = \beta_2$ (by δ^*_ℓ's definition) and call it β.

As $\eta_{\delta^*_1}$ is a one to one function (being increasing) from σ, clearly

$$\eta_{\delta^*_1}(i) \in N_1 \text{ iff } i \in \sigma \cap N_1.$$

Also $N_1 \models$ "$\langle \eta_{\delta^*_1}(i) : i < \sigma \rangle$ is unbounded below δ^*_1" (remember $N_1 \prec M^*$ as $N^{p(n)}_{\delta_1,\zeta} \prec M^*$ for each n).

So clearly $\beta = \mathrm{Sup}\{\eta_{\delta^*_1}(i) : i < \alpha\}$; but $\eta_{\delta^*_1}$ is increasing continuous and α is a limit ordinal (being $\sup(N_\ell \cap \sigma)$), hence $\beta = \eta_{\delta^*_1}(\alpha)$.

For the same reasons $\beta = \eta_{\delta^*_2}(\alpha)$.

So $\eta_{\delta_1^*}(\alpha) = \eta_{\delta_2^*}(\alpha)$ and $\alpha \in W^n_{\delta^*_\ell}$ for each $n < \omega (\ell = 1, 2)$ as $N_\ell \models$ "$W^n_{\delta^*_\ell}$ is a closed unbounded subset of σ". For similar reasons $\delta^*_\ell \in W_n$ for each n: as $W_n \in N^{p(n+1)}_{\delta_\ell,\zeta}$ hence $W_n \in N_\ell$, hence $W_n \in N_1 \cap N_2$, and as $N_1, N_2 \prec M^*, M^*$ has Skolem functions, clearly $N_1 \cap N_2 \prec M^*$, so W_n is an unbounded subset of $N_1 \cap N_2 \cap \kappa^+$. So in N_ℓ, W_n is unbounded in $\delta^*_\ell = \mathrm{Min}[(\kappa^+ \cap N_\ell) \setminus (N_1 \cap N_2)]$, hence $N_\ell \models$ "$\delta^*_\ell \in W_n$" hence $\delta^*_\ell \in W_n$.

We can conclude that δ^*_1, δ^*_2, β satisfy the requirements (A), (B), (C) on δ_1, δ_2, ξ. Hence by require-mint (D) on them, $\delta_1 = \delta^*_1$, $\delta_2 = \delta^*_2$. But, $\zeta \in N^{p(n)}_{\delta_\ell,\zeta} \subseteq N_\ell$ hence $\zeta \in \kappa \cap N_1 \cap N_2$ hence $\zeta < \alpha$, so clause (α) contradicts the choice of ζ, so we get a contradiction, thus finishing the proof of the theorem (3.9). $\square_{3.9}$

3.11 Remark. We can replace in the conclusion of 3.9, $F(h \restriction a^\delta_i)$ by $F^\delta_i(h)$, so F is replaced by $\langle F^\delta_i : \delta \in S, i < \mathrm{cf}(\delta) \rangle$, where F^δ_i is a function from ${}^{\kappa^+}\kappa$ to $\{0, 1\}$. Also we may weaken $a^\delta_i \subseteq \eta_0(i) + 1$ to $a^\delta_i \subseteq \lambda^+$.

References

[Ab] Uri Abraham. Lectures on proper forcing. In M. Foreman A. Kanamori and M. Magidor, editors, *Handbook of Set Theory.*

[A] Uri Abraham. *Isomorphisms of Aronszajn trees and forcing without the generalized continuum hypothesis* (in Hebrew). PhD thesis, The Hebrew University, Jerusalem, 1979.

[AbSh 146] Uri Abraham and Saharon Shelah. Forcing closed unbounded sets. *The Journal of Symbolic Logic*, **48**:643–657, 1983.

[AbSh 114] Uri Abraham and Saharon Shelah. Isomorphism types of Aronszajn trees. *Israel Journal of Mathematics*, **50**:75–113, 1985.

[AbSh 403] Uri Abraham and Saharon Shelah. A Δ^2_2 well-order of the reals and incompactness of $L(Q^{MM})$. *Annals of Pure and Applied Logic*, **59**:1–32, 1993.

[ADSh 81] Uri Avraham (Abraham), Keith J. Devlin, and Saharon Shelah. The consistency with CH of some consequences of Martin's axiom plus $2^{\aleph_0} > \aleph_1$. *Israel Journal of Mathematics*, **31**:19–33, 1978.

[AbSh 106] Uri Avraham (Abraham) and Saharon Shelah. Martin's axiom does not imply that every two $\aleph_1$-dense sets of reals are isomorphic. *Israel Journal of Mathematics*, **38**:161–176, 1981.

[AbSh 102] Uri Avraham (Abraham) and Saharon Shelah. Forcing with stable posets. *The Journal of Symbolic Logic*, **47**:37–42, 1982.

[BPS] Bohuslav Balcar, Jan Pelant, and Petr Simon. The space of ultrafilters on N covered by nowhere dense sets. *Fundamenta Mathematicae*, **CX**:11–24, 1980.

[BS] Bohuslav Balcar and Petr Simon. Cardinal invariants in Boolean spaces. In J. Novak, editor, *General Topology and its Relation to Modern Analysis and Algebra, Proceedings of the fifth Prague Topology Symposion 1981*, volume V, pages 39–47, Berlin, 1982. Helderman Verlag.

[BaJu95] Tomek Bartoszyński and Haim Judah. *Set Theory: On the Structure of the Real Line.* A K Peters, Wellesley, Massachusetts, 1995.

[B] James E. Baumgartner. Decomposition of embedding of trees. *Notices Amer. Math. Soc.*, **17**:967, 1970.

[B1] James E. Baumgartner. *Results and independence proofs in combinatorial set theory.* PhD thesis, Univ. of Calif. Berkeley, 1970.

[B4] James E. Baumgartner. All $\aleph_1$-dense sets of reals can be isomorphic. *Fund. Math.*, **79**:101–106, 1973.

[B2] James E. Baumgartner. A new class of order types. *Annals of Math. Logic*, **9**:187–222, 1976.

[B5] James E. Baumgartner. Almost disjoint sets, the dense set problem and partition calculus. *Annals of Math Logic*, **9**:401–439, 1976.

[Ba] James E. Baumgartner. Almost disjoint sets, the dense set problem and partition calculus. *Annals of Math Logic*, **9**:401–439, 1976.

[B3] James E. Baumgartner. Iterated forcing. In A. Mathias, editor, *Surveys in Set Theory*, volume 87 of *London Mathematical Society Lecture Notes*, pages 1–59, Cambridge, Britain, 1978.

[B6] James E. Baumgartner. Ultrafilters on ω. *The Journal of Symbolic Logic*, **60**:624–639, 1995.

[BHK] James E. Baumgartner, Leo Harrington, and Eugene M. Kleinberg. Adding a closed unbounded set. *The Journal of Symbolic Logic*, **41**:481–487, 1976.

[BL79] James E. Baumgartner and Richard Laver. Iterated perfect-set forcing. *Annals of Mathematical Logic*, **17**:271–288, 1979.

[BMR] James E. Baumgartner, Jerome Malitz, and William Reinhart. Embedding trees in the rationals. *Proc. Nat. Acad. Sci. U.S.A.*, **67**:1748–1753, 1970.

[BD] Shai Ben David. On Shelah's compactness of cardinals. *Israel J. of Math.*, **31**:34–56 and 394, 1978.

[BsSh 242] Andreas Blass and Saharon Shelah. There may be simple $P_{\aleph_1}$- and $P_{\aleph_2}$-points and the Rudin-Keisler ordering may be downward directed. *Annals of Pure and Applied Logic*, **33**:213–243, 1987.

[BuSh 437] Max R. Burke and Saharon Shelah. Linear liftings for non complete probability space. *Israel Journal of Mathematics*, **79**:289–296, 1992.

[C] Paul Cohen. *Set Theory and the Continuum Hypothesis*. Benjamin, New York, 1966.

[De1] Keith J. Devlin. $\aleph_1$ trees. *Ann. Math. Logic*, pages 267–330, 1978.

[De2] Keith J. Devlin. Concerning the consistency of the Souslin Hypothesis with the continuum hypothesis. *Ann. Math. Logic*, **19**:115–125, 1980.

[DeJo] Keith J. Devlin and Havard Johnsbraten. *The Souslin problem*, volume 405 of *Lecture Notes in Mathematics*. Springer, Berlin, 1974.

[DvSh 65] Keith J. Devlin and Saharon Shelah. A weak version of $\diamondsuit$ which follows from $2^{\aleph_0} < 2^{\aleph_1}$. *Israel Journal of Mathematics*, **29**:239–247, 1978.

[DvSh 85] Keith J. Devlin and Saharon Shelah. A note on the normal Moore space conjecture. *Canadian Journal of Mathematics. Journal Canadien de Mathematiques*, **31**:241–251, 1979.

[DM] Ben Dushnik and E. W. Miller. Partially ordered sets. *American Journal of Mathematics*, **63**:600–610, 1941.

[DjSh 604] Mirna Džamonja and Saharon Shelah. ♣ does not imply the existence of a Suslin tree. *Israel Journal of Mathematics* (submitted).

[EkHu1] Paul C. Eklof and Martin Huber. On the rank of Ext. *Math. Zeitschrift*, **174**:159–185, 1980.

[EkHu] Paul C. Eklof and Martin Huber. On the p-ranks of $Ext(A, G)$, assuming CH. In *Abelian Group Theory*, volume 874 of *Lecture Notes in Math.*, pages 93–108. Springer, 1981. (Oberwolfach 1981).

[EM] Paul C. Eklof and Alan Mekler. *Almost free modules; Set theoretic methods.* North Holland Library, 1990.

[EMSh 441] Paul C. Eklof, Alan H. Mekler, and Saharon Shelah. Uniformization and the diversity of Whitehead groups. *Israel Journal of Mathematics*, **80**:301–321, 1992.

[EMSh 442] Paul C. Eklof, Alan H. Mekler, and Saharon Shelah. Hereditarily separable groups and monochromatic uniformization. *Israel Journal of Mathematics*, **88**:213–235, 1994.

[EkSh 505] Paul C. Eklof and Saharon Shelah. A Combinatorial Principle Equivalent to the Existence of Non-free Whitehead Groups. In *Abelian group theory and related topics*, volume 171 of *Contemporary Mathematics*, pages 79–98. American Mathematical Society, Providence, RI, 1994, edited by R. Goebel, P. Hill and W. Liebert, Oberwolfach proceedings.

[EK] Ryszard Engelking and Monika Karłowicz. Some theorems of set theory and their topological consequences. *Fundamenta Math.*, **57**:275–285, 1965.

[FMSh 240] Matthew Foreman, Menachem Magidor, and Saharon Shelah. Martin's maximum, saturated ideals, and nonregular ultrafilters. I. *Annals of Mathematics. Second Series*, **127**:1–47, 1988.

[FMSh 252] Matthew Foreman, Menachem Magidor, and Saharon Shelah. Martin's maximum, saturated ideals and nonregular ultrafilters. II. *Annals of Mathematics. Second Series*, **127**:521–545, 1988. will reapeare (in a revised form) in chapter XIII of [Sh f].

[FrSh 406] David H. Fremlin and Saharon Shelah. Pointwise compact and stable sets of measurable functions. *Journal of Symbolic Logic*, **58**:435–455, 1993.

[Fr] Harvey Friedman. One hundred and two problems in mathematical logic. *J. of Symb. Logic*, **40**:113–124, 1975.

[FShS 544] Sakaé Fuchino, Saharon Shelah, and Lajos Soukup. Sticks and clubs. *preprint.*

[Fu] Laszlo Fuchs. *Infinite Abelian Groups*, volume I, II. Academic Press, New York, 1970, 1973.

[GJM] Fred Galvin, Thomas Jech, and Menachem Magidor. An ideal game. *J. of Symb. Logic.*, **43**:284–292, 1978.

[Gi] Moti Gitik. The negation of SCH from $o(\kappa) = \kappa^{++}$. *Annals of Pure and Applied Logic*, **43**:209–234, 1989.

[GiSh 310] Moti Gitik and Saharon Shelah. Cardinal preserving ideals. *Journal of Symbolic Logic* (submitted).

[GiSh 191] Moti Gitik and Saharon Shelah. On the $\mathbb{I}$-condition. *Israel Journal of Mathematics*, **48**:148–158, 1984.

[Go] Martin Goldstern. Tools for your forcing construction. In *Set Theory of the Reals*, volume 6 of *Israel Mathematical Conference Proceedings*, pages 305–360.

[Gr] C. Gray. *Iterated Forcing from the Strategic Point of View*. PhD thesis, UC Berkeley, 1982.

[GrSh 302] Rami Grossberg and Saharon Shelah. On the structure of $\mathrm{Ext}_p(G, \mathbf{Z})$. *Journal of Algebra*, **121**:117–128, 1989.

[Ha61] Andras Hajnal. Proof of a conjecture of S.Ruziewicz. *Fundamenta Mathematicae*, **50**:123–128, 1961.

[HrSh:99] Leo Harrington and Saharon Shelah. Some exact equiconsistency results in set theory. *Notre Dame Journal of Formal Logic*, **26**:178–188, 1985. Proceedings of the 1980/1 Jerusalem Model Theory year.

[HLSh 162] Bradd Hart, Claude Laflamme, and Saharon Shelah. Models with second order properties, V: A General principle. *Annals of Pure and Applied Logic*, **64**:169–194, 1993.

[HHSh:91] Howard L. Hiller, Martin Huber, and Saharon Shelah. The structure of $\mathrm{Ext}(A, \mathbf{Z})$ and $V = L$. *Mathematische Zeitschrift*, **162**:39–50, 1978.

[J] Thomas Jech. *Set theory.* Academic Press, New York, 1978.

[J86] Thomas Jech. *Multiple forcing*, volume 88 of *Cambridge Tracts in Mathematics.* Cambridge University Press, 1986.

[JMMP] Thomas Jech, Menachem Magidor, William Mitchell, and Karel Prikry. On precipitous ideals. *J. of Symb. Logic*, **45**:1–8, 1980.

[JePo] Thomas Jech and William C. Powell. Standard models of set theory with predication. *Bulletin of the AMS*, **77**(5):808–813, Sep 1971.

[JP1] Thomas Jech and Karel Prikry. On ideals of sets and power set operation. *Bull. Amer. Math. Soc.*, **82**:593–595, 1976.

[JP2] Thomas Jech and Karel Prikry. *Ideals over uncountable sets: application of almost disjoint functions and generic ultrapowers*, volume 214, 18(2) of *Amer. Math. Soc. Memoir.* AMS, 1979.

[JeSh 378] Thomas Jech and Saharon Shelah. A note on canonical functions. *Israel Journal of Mathematics*, **68**:376–380, 1989.

[Jn] Ronald B. Jensen. The fine structure of the constructible hierarchy. *Annals of Math. Logic*, **4**:229–308, 1972.

[JuRe] Haim Judah and Miroslav Repicky. No random reals in countable support iterations. *Israel Journal of Mathematics*, **92**:349–359, 1995.

[JdSh 308] Haim Judah and Saharon Shelah. The Kunen-Miller chart (Lebesgue measure, the Baire property, Laver reals and preservation theorems for forcing). *The Journal of Symbolic Logic*, **55**:909–927, 1990.

[JShW 339] Haim Judah, Saharon Shelah, and Hugh Woodin. The Borel conjecture. *Annals of Pure and Applied Logic*, **50**:255–269, 1990. Correction of third section has appeared in the book by Bartoszynski and Judah.

[Ju92] Winfried Just. A modification of Shelah's oracle-cc with applications. *Transactions of the AMS*, **329**:325–356, 1992.

[Ko1] Peter Komjath. Set systems with finite chromatic number. *European Journal of Combinatorics*, **10**:543–549, 1989.

[Ku83] Kenneth Kunen. *Set Theory: An introduction to independence proofs*, volume 102 of *Studies in Logic and the Foundations of Mathematics*. North–Holland Publishing Co, 1983.

[Ku35] C. Kurepa. Ensembles ordonnes et ramifies. *Publ. Math. Univ. Belgrade*, **9**:1–38, 1935.

[K] C. Kurepa. Transformations monotones des ensembles partiellement ordonne. *C.R. Acad. Sci. Paris*, **205**:1033–1035, 1937.

[L1] Richard Laver. On the consistency of Borel's conjecture. *Acta Math.*, **137**:151–169, 1976.

[L] Richard Laver. Making supercompact indestructible under κ-directed forcing. *Israel J. of Math.*, **29**:385–388, 1978.

[Lv82] Richard Laver. Saturated ideals and nonregular ultrafilters. In *Procedings of the Bernays Conference 1980, Pabras, Greece.* North Holland, 1982.

[LvSh 104] Richard Laver and Saharon Shelah. The $\aleph_2$-Souslin hypothesis. *Transactions of the American Mathematical Society*, **264**:411–417, 1981.

[Mg] Menachem Magidor. On the singular cardinals problem I. *Israel J. Math.*, **28**:1–31, 1977.

[Mg4] Menachem Magidor. Changing cofinality of cardinals. *Fund. Math.*, **XCIX**:61–71, 1978.

[Mg80] Menachem Magidor. Precipitous ideals and Σ^1_4 sets. *Israel J. of Math.*, **35**:109–134, 1980.

[MS] Donald Martin and Robert M. Solovay. Internal Cohen extensions. *Annals of Math Logic*, **2**:143–178, 1970.

[Mr75] Donald A. Martin. Borel Determinacy. *Annals of Mathematics*, **102**:363–371, 1975.

[Mt3] A. Mathias. Happy families. *Annals of Mathematical Logic*, **12**:59–111, 1977.

[Mk84] Alan H. Mekler. Finitely additive measures on **N** and the additive property. *Proceedings of the American Mathematical Society*, **92**:439–444, 1984.

[MRSh 314] Alan H. Mekler, Andrzej Rosłanowski, and Saharon Shelah. On the p-rank of Ext. *Israel Journal of Mathematics* (submitted).

[MkSh 313] Alan H. Mekler and Saharon Shelah. Diamond and λ-systems. *Fundamenta Mathematicae*, **131**:45–51, 1988.

[MkSh 274] Alan H. Mekler and Saharon Shelah. Uniformization principles. *The Journal of Symbolic Logic*, **54**:441–459, 1989.

[MiRa] Eric Milner and Richard Rado. The pigeon-hole principle for ordinal numbers. *Proc. London Math. Soc.*, **15**:750–768, 1965.

[Mi1] William Mitchell. Aronszajn trees and the independence of the transfer property. *Annals of Mathematical Logic*, **5**:21–46, 1972/73.

[Nm] Kanji Namba. Independence proof of (ω, ω_α)-distributive law in complete Boolean algebras. *Comment Math. Univ. St. Pauli*, **19**:1–12, 1970.

[Os] A. Ostaszewski. On countably compact perfectly normal spaces. *Journal of London Mathematical Society*, **14**:505–516, 1975.

[Ox] John C. Oxtoby. *Measure and Category.* Springer, 1980.

[Pr] Karel Prikry. Changing measurable to accessible cardinals. *Rozprawy Matematyczne*, **LXVIII**:5–52, 1970.

[RoSh 470] Andrzej Rosłanowski and Saharon Shelah. Norms on possibilities I: forcing with trees and creatures. *Memoirs of the AMS* (accepted).

[RuSh 117] Matatyahu Rubin and Saharon Shelah. Combinatorial problems on trees: partitions, Δ-systems and large free subtrees. *Annals of Pure and Applied Logic*, **33**:43–81, 1987.

[SgSh 148] G. Sageev and Saharon Shelah. Weak compactness and the structure of Ext$(A, \mathbf{Z})$. In *Abelian group theory (Oberwolfach, 1981)*, volume 874 of *Lecture Notes in Mathematics*, pages 87–92. Springer, Berlin New York, 1981, ed. Goebel, R. and Walker, A.E.

[SgSh 138] G. Sageev and Saharon Shelah. On the structure of Ext$(A, \mathbf{Z})$ in ZFC$^+$. *The Journal of Symbolic Logic*, **50**:302–315, 1985.

[Sh 311] Saharon Shelah. A more general iterable condition ensuring $\aleph_1$ is not collapsed, II. *preprint.*

[Sh 576] Saharon Shelah. Categoricity of an abstract elementary class in two successive cardinals. *Israel Journal of Mathematics* (submitted).

[Sh 630] Saharon Shelah. Non-elementary proper forcing notions. *In preparation.*

[Sh 587] Saharon Shelah. Not collapsing cardinals $\leq \kappa$ in $(< \kappa)$–support iterations. *In preparation.*

[Sh 656] Saharon Shelah. On CS iterations not adding reals. *In preparation.*

[Sh 460] Saharon Shelah. The Generalized Continuum Hypothesis revisited. *Israel Journal of Mathematics* (submitted).

[Sh 594] Saharon Shelah. There may be no nowhere dense ultrafilter. In *Proceedings of the Logic Colloquium Haifa'95* (accepted). Springer.

[Sh 486] Saharon Shelah. Uniformization. *Preprint.*

[Sh 44] Saharon Shelah. Infinite abelian groups, Whitehead problem and some constructions. *Israel Journal of Mathematics*, **18**:243–256, 1974.

[Sh 52] Saharon Shelah. A compactness theorem for singular cardinals, free algebras, Whitehead problem and transversals. *Israel Journal of Mathematics*, **21**:319–349, 1975.

[Sh 64] Saharon Shelah. Whitehead groups may be not free, even assuming CH. I. *Israel Journal of Mathematics*, **28**:193–204, 1977.

[Sh 80] Saharon Shelah. A weak generalization of MA to higher cardinals. *Israel Journal of Mathematics*, **30**:297–306, 1978.

[Sh 73] Saharon Shelah. Models with second-order properties. II. Trees with no undefined branches. *Annals of Mathematical Logic*, **14**:73–87, 1978.

[Sh 100] Saharon Shelah. Independence results. *The Journal of Symbolic Logic*, **45**:563–573, 1980.

[Sh:98] Saharon Shelah. Whitehead groups may not be free, even assuming CH. II. *Israel Journal of Mathematics*, **35**:257–285, 1980.

[Sh 120] Saharon Shelah. Free limits of forcing and more on Aronszajn trees. *Israel Journal of Mathematics*, **38**:315–334, 1981.

[Sh 119] Saharon Shelah. Iterated forcing and changing cofinalities. *Israel Journal of Mathematics*, **40**:1–32, 1981.

[Sh 82] Saharon Shelah. Models with second order properties. III. Omitting types for $L(Q)$. *Archiv fur Mathematische Logik und Grundlagenforschung*, **21**:1–11, 1981.

[Sh 140] Saharon Shelah. On endo-rigid, strongly $\aleph_1$-free abelian groups in $\aleph_1$. *Israel Journal of Mathematics*, **40**:291–295, 1981.

[Sh 122] Saharon Shelah. On Fleissner's diamond. *Notre Dame Journal of Formal Logic*, **22**:29–35, 1981.

[Sh 125] Saharon Shelah. The consistency of $\mathrm{Ext}(G, \mathbf{Z}) = \mathbf{Q}$. *Israel Journal of Mathematics*, **39**:74–82, 1981.

[Sh:b] Saharon Shelah. *Proper forcing*, volume 940 of *Lecture Notes in Mathematics*. Springer-Verlag, Berlin-New York, xxix+496 pp, 1982.

[Sh 87a] Saharon Shelah. Classification theory for nonelementary classes, I. The number of uncountable models of $\psi \in L_{\omega_1,\omega}$. Part A. *Israel Journal of Mathematics*, **46**:212–240, 1983.

[Sh 87b] Saharon Shelah. Classification theory for nonelementary classes, I. The number of uncountable models of $\psi \in L_{\omega_1,\omega}$. Part B. *Israel Journal of Mathematics*, **46**:241–273, 1983.

[Sh 136] Saharon Shelah. Constructions of many complicated uncountable structures and Boolean algebras. *Israel Journal of Mathematics*, **45**:100–146, 1983.

[Sh 185] Saharon Shelah. Lifting problem of the measure algebra. *Israel Journal of Mathematics*, **45**:90–96, 1983.

[Sh 107] Saharon Shelah. Models with second order properties. IV. A general method and eliminating diamonds. *Annals of Pure and Applied Logic*, **25**:183–212, 1983.

[Sh 176] Saharon Shelah. Can you take Solovay's inaccessible away? *Israel Journal of Mathematics*, **48**:1–47, 1984.

[Sh 186] Saharon Shelah. Diamonds, uniformization. *The Journal of Symbolic Logic*, **49**:1022–1033, 1984.

[Sh 177] Saharon Shelah. More on proper forcing. *The Journal of Symbolic Logic*, **49**:1034–1038, 1984.

[Sh 207] Saharon Shelah. On cardinal invariants of the continuum. In *Axiomatic set theory (Boulder, Colo., 1983)*, volume 31 of *Contemp. Mathematics*, pages 183–207. Amer. Math. Soc., Providence, RI, 1984. Proceedings of the Conference in Set Theory, Boulder, June 1983; ed. Baumgartner J., Martin, D. and Shelah, S.

[Sh 208] Saharon Shelah. More on the weak diamond. *Annals of Pure and Applied Logic*, **28**:315–318, 1985.

[Sh 237a] Saharon Shelah. On normal ideals and Boolean algebras. In *Around classification theory of models*, volume 1182 of *Lecture Notes in Mathematics*, pages 247–259. Springer, Berlin, 1986.

[Sh 88] Saharon Shelah. Classification of nonelementary classes. II. Abstract elementary classes. In *Classification theory (Chicago, IL, 1985)*, volume 1292 of *Lecture Notes in Mathematics*, pages 419–497. Springer, Berlin, 1987. Proceedings of the USA–Israel Conference on Classification Theory, Chicago, December 1985; ed. Baldwin, J.T.

[Sh 253] Saharon Shelah. Iterated forcing and normal ideals on ω_1. *Israel Journal of Mathematics*, **60**:345–380, 1987.

[Sh 263] Saharon Shelah. Semiproper forcing axiom implies Martin maximum but not PFA^+. *The Journal of Symbolic Logic*, **52**:360–367, 1987.

[Sh 192] Saharon Shelah. Uncountable groups have many nonconjugate subgroups. *Annals of Pure and Applied Logic*, **36**:153–206, 1987.

[Sh 250] Saharon Shelah. Some notes on iterated forcing with $2^{\aleph_0} > \aleph_2$. *Notre Dame Journal of Formal Logic*, **29**:1–17, 1988.

[Sh 276] Saharon Shelah. Was Sierpiński right? I. *Israel Journal of Mathematics*, **62**:355–380, 1988.

[Sh 270] Saharon Shelah. Baire irresolvable spaces and lifting for a layered ideal. *Topology and its Applications*, **33**:217–221, 1989.

[Sh 351] Saharon Shelah. Reflecting stationary sets and successors of singular cardinals. *Archive for Mathematical Logic*, **31**:25–53, 1991.

[Sh 400a] Saharon Shelah. Cardinal arithmetic for skeptics. *American Mathematical Society. Bulletin. New Series*, **26**:197–210, 1992.

[Sh 407] Saharon Shelah. CON($\mathbf{u} > \mathbf{i}$). *Archive for Mathematical Logic*, **31**:433–443, 1992.

[Sh 326] Saharon Shelah. Viva la difference I: Nonisomorphism of ultrapowers of countable models. In *Set Theory of the Continuum*, volume 26 of *Mathematical Sciences Research Institute Publications*, pages 357–405. Springer Verlag, 1992.

[Sh 420] Saharon Shelah. Advances in Cardinal Arithmetic. In *Finite and Infinite Combinatorics in Sets and Logic*, pages 355–383. Kluwer Academic Publishers, 1993. N.W. Sauer et al (eds.).

[Sh:g] Saharon Shelah. *Cardinal Arithmetic*, volume 29 of *Oxford Logic Guides*. Oxford University Press, 1994.

[Sh 430] Saharon Shelah. Further cardinal arithmetic. *Israel Journal of Mathematics*, **95**:61–114, 1996.

[ShSt 154] Saharon Shelah and Lee Stanley. Generalized Martin's axiom and Souslin's hypothesis for higher cardinals. *Israel Journal of Mathematics*, **43**:225–236, 1982.

[ShSt 154a] Saharon Shelah and Lee Stanley. Corrigendum to: "Generalized Martin's axiom and Souslin's hypothesis for higher cardinals" [Israel Journal of Mathematics 43 (1982), no. 3, 225–236; MR 84h:03120]. *Israel Journal of Mathematics*, **53**:304–314, 1986.

[ShSr 315] Saharon Shelah and Juris Steprans. PFA implies all automorphisms are trivial. *Proceedings of the American Mathematical Society*, **104**:1220–1225, 1988.

[ShSr 296] Saharon Shelah and Juris Steprans. Nontrivial homeomorphisms of $\beta\mathbf{N}\setminus\mathbf{N}$ without the continuum hypothesis. *Fundamenta Mathematicae*, **132**:135–141, 1989.

[ShSr 427] Saharon Shelah and Juris Steprans. Somewhere trivial automorphisms. *Journal of the London Mathematical Society*, **49**:569–580, 1994.

[ShWd 241] Saharon Shelah and Hugh Woodin. Large cardinals imply that every reasonably definable set of reals is Lebesgue measurable. *Israel Journal of Mathematics*, **70**:381–394, 1990.

[Si67] Jack Silver. The independence of Kurepa's conjecture and two-cardinal conjectures in model theory. In *Axiomatic Set Theory*, volume XIII of *Proc. Symp in Pure Math.*, pages 383–390, 1967.

[So70] Robert M. Solovay. A model of set theory in which every set of reals is Lebesgue measurable. *Annals of Math.*, **92**:1–56, 1970.

[ST] Robert M. Solovay and S. Tennenbaum. Iterated Cohen extensions and Souslin's problem. *Annals of Math.*, **94**:201–245, 1971.

[StVW] John Steel and Robert van Wesep. Two consequences of determinacy consistent with choice. *Transactions of the AMS*, **272**:67–85, 1982.

[SK] Charles L. Steinhorn and James H. King. The uniformization property for $\aleph_2$. *Israel J. Math.*, **36**:248–256, 1980.

[D] Eric K. van Douwen. The integers and topology. In K. Kunen and J. E. Vaughan, editors, *Handbook of Set-Theoretic Topology*, pages 111–167. North-Holland, 1984.

[Ve86] Boban Velickovic. Definable automorphisms of $P(\omega)/fin$. *Proceedings of the AMS*, **96**:130–135, 1986.

[Ve93] Boban Velickovic. OCA and automorphisms of $P(\omega/fin)$. *Topology and Applications*, **49**:1–13, 1993.

[Wi] Edward L. Wimmers. The Shelah P-point independence theorem. *Israel Journal of Mathematics*, **43**:28–48, 1982.

[W83] Hugh Woodin. Some consistency results in ZF using AD. In *Cabal Seminar*, volume 1019 of *Lecture Notes in Mathematics*, pages 172–199. 1983.

More References

For the reader's convenience we list here again all references to papers of the author, this time sorted by number.

[Sh:b] Saharon Shelah. *Proper forcing*, volume 940 of *Lecture Notes in Mathematics*. Springer-Verlag, Berlin-New York, xxix+496 pp, 1982.

[Sh:g] Saharon Shelah. *Cardinal Arithmetic*, volume 29 of *Oxford Logic Guides*. Oxford University Press, 1994.

[Sh 44] Saharon Shelah. Infinite abelian groups, Whitehead problem and some constructions. *Israel Journal of Mathematics*, **18**:243–256, 1974.

[Sh 52] Saharon Shelah. A compactness theorem for singular cardinals, free algebras, Whitehead problem and transversals. *Israel Journal of Mathematics*, **21**:319–349, 1975.

[Sh 64] Saharon Shelah. Whitehead groups may be not free, even assuming CH. I. *Israel Journal of Mathematics*, **28**:193–204, 1977.

[DvSh 65] Keith J. Devlin and Saharon Shelah. A weak version of $\diamondsuit$ which follows from $2^{\aleph_0} < 2^{\aleph_1}$. *Israel Journal of Mathematics*, **29**:239–247, 1978.

[Sh 73] Saharon Shelah. Models with second-order properties. II. Trees with no undefined branches. *Annals of Mathematical Logic*, **14**:73–87, 1978.

[Sh 80] Saharon Shelah. A weak generalization of MA to higher cardinals. *Israel Journal of Mathematics*, **30**:297–306, 1978.

[ADSh 81] Uri Avraham (Abraham), Keith J. Devlin, and Saharon Shelah. The consistency with CH of some consequences of Martin's axiom plus $2^{\aleph_0} > \aleph_1$. *Israel Journal of Mathematics*, **31**:19–33, 1978.

[Sh 82] Saharon Shelah. Models with second order properties. III. Omitting types for $L(Q)$. *Archiv fur Mathematische Logik und Grundlagenforschung*, **21**:1–11, 1981.

[DvSh 85] Keith J. Devlin and Saharon Shelah. A note on the normal Moore space conjecture. *Canadian Journal of Mathematics. Journal Canadien de Mathematiques*, **31**:241–251, 1979.

[Sh 87a] Saharon Shelah. Classification theory for nonelementary classes, I. The number of uncountable models of $\psi \in L_{\omega_1,\omega}$. Part A. *Israel Journal of Mathematics*, **46**:212–240, 1983.

[Sh 87b] Saharon Shelah. Classification theory for nonelementary classes, I. The number of uncountable models of $\psi \in L_{\omega_1,\omega}$. Part B. *Israel Journal of Mathematics*, **46**:241–273, 1983.

[Sh 88] Saharon Shelah. Classification of nonelementary classes. II. Abstract elementary classes. In *Classification theory (Chicago, IL, 1985)*, volume 1292 of *Lecture Notes in Mathematics*, pages 419–497. Springer, Berlin, 1987. Proceedings of the USA–Israel Conference on Classification Theory, Chicago, December 1985; ed. Baldwin, J.T.

[HHSh:91] Howard L. Hiller, Martin Huber, and Saharon Shelah. The structure of $\mathrm{Ext}(A, \mathbf{Z})$ and $V = L$. *Mathematische Zeitschrift*, **162**:39–50, 1978.

[Sh:98] Saharon Shelah. Whitehead groups may not be free, even assuming CH. II. *Israel Journal of Mathematics*, **35**:257–285, 1980.

[HrSh:99] Leo Harrington and Saharon Shelah. Some exact equiconsistency results in set theory. *Notre Dame Journal of Formal Logic*, **26**:178–188, 1985. Proceedings of the 1980/1 Jerusalem Model Theory year.

[Sh 100] Saharon Shelah. Independence results. *The Journal of Symbolic Logic*, **45**:563–573, 1980.

[AbSh 102] Uri Avraham (Abraham) and Saharon Shelah. Forcing with stable posets. *The Journal of Symbolic Logic*, **47**:37–42, 1982.

[LvSh 104] Richard Laver and Saharon Shelah. The $\aleph_2$-Souslin hypothesis. *Transactions of the American Mathematical Society*, **264**:411–417, 1981.

[AbSh 106] Uri Avraham (Abraham) and Saharon Shelah. Martin's axiom does not imply that every two $\aleph_1$-dense sets of reals are isomorphic. *Israel Journal of Mathematics*, **38**:161–176, 1981.

[Sh 107] Saharon Shelah. Models with second order properties. IV. A general method and eliminating diamonds. *Annals of Pure and Applied Logic*, **25**:183–212, 1983.

[AbSh 114] Uri Abraham and Saharon Shelah. Isomorphism types of Aronszajn trees. *Israel Journal of Mathematics*, **50**:75–113, 1985.

[RuSh 117] Matatyahu Rubin and Saharon Shelah. Combinatorial problems on trees: partitions, Δ-systems and large free subtrees. *Annals of Pure and Applied Logic*, **33**:43–81, 1987.

[Sh 119] Saharon Shelah. Iterated forcing and changing cofinalities. *Israel Journal of Mathematics*, **40**:1–32, 1981.

[Sh 120] Saharon Shelah. Free limits of forcing and more on Aronszajn trees. *Israel Journal of Mathematics*, **38**:315–334, 1981.

[Sh 122] Saharon Shelah. On Fleissner's diamond. *Notre Dame Journal of Formal Logic*, **22**:29–35, 1981.

[Sh 125] Saharon Shelah. The consistency of $\mathrm{Ext}(G, \mathbf{Z}) = \mathbf{Q}$. *Israel Journal of Mathematics*, **39**:74–82, 1981.

[Sh 136] Saharon Shelah. Constructions of many complicated uncountable structures and Boolean algebras. *Israel Journal of Mathematics*, **45**:100–146, 1983.

[SgSh 138] G. Sageev and Saharon Shelah. On the structure of $\mathrm{Ext}(A, \mathbf{Z})$ in ZFC^+. *The Journal of Symbolic Logic*, **50**:302–315, 1985.

[Sh 140] Saharon Shelah. On endo-rigid, strongly $\aleph_1$-free abelian groups in $\aleph_1$. *Israel Journal of Mathematics*, **40**:291–295, 1981.

[AbSh 146] Uri Abraham and Saharon Shelah. Forcing closed unbounded sets. *The Journal of Symbolic Logic*, **48**:643–657, 1983.

[SgSh 148] G. Sageev and Saharon Shelah. Weak compactness and the structure of Ext(A, $\mathbf{Z}$). In *Abelian group theory (Oberwolfach, 1981)*, volume 874 of *Lecture Notes in Mathematics*, pages 87–92. Springer, Berlin-New York, 1981. ed. Goebel, R. and Walker, A.E.

[ShSt 154] Saharon Shelah and Lee Stanley. Generalized Martin's axiom and Souslin's hypothesis for higher cardinals. *Israel Journal of Mathematics*, **43**:225–236, 1982.

[ShSt 154a] Saharon Shelah and Lee Stanley. Corrigendum to: "Generalized Martin's axiom and Souslin's hypothesis for higher cardinals" [Israel Journal of Mathematics 43 (1982), no. 3, 225–236; MR 84h:03120]. *Israel Journal of Mathematics*, **53**:304–314, 1986.

[HLSh 162] Bradd Hart, Claude Laflamme, and Saharon Shelah. Models with second order properties, V: A General principle. *Annals of Pure and Applied Logic*, **64**:169–194, 1993.

[Sh 176] Saharon Shelah. Can you take Solovay's inaccessible away? *Israel Journal of Mathematics*, **48**:1–47, 1984.

[Sh 177] Saharon Shelah. More on proper forcing. *The Journal of Symbolic Logic*, **49**:1034–1038, 1984.

[Sh 185] Saharon Shelah. Lifting problem of the measure algebra. *Israel Journal of Mathematics*, **45**:90–96, 1983.

[Sh 186] Saharon Shelah. Diamonds, uniformization. *The Journal of Symbolic Logic*, **49**:1022–1033, 1984.

[GiSh 191] Moti Gitik and Saharon Shelah. On the $\mathbb{I}$-condition. *Israel Journal of Mathematics*, **48**:148–158, 1984.

[Sh 192] Saharon Shelah. Uncountable groups have many nonconjugate subgroups. *Annals of Pure and Applied Logic*, **36**:153–206, 1987.

[Sh 207] Saharon Shelah. On cardinal invariants of the continuum. In *Axiomatic set theory (Boulder, Colo., 1983)*, volume 31 of *Contemp. Mathematics*, pages 183–207. Amer. Math. Soc., Providence, RI, 1984. Proceedings of the Conference in Set Theory, Boulder, June 1983; ed. Baumgartner J., Martin, D. and Shelah, S.

[Sh 208] Saharon Shelah. More on the weak diamond. *Annals of Pure and Applied Logic*, **28**:315–318, 1985.

[Sh 237a] Saharon Shelah. On normal ideals and Boolean algebras. In *Around classification theory of models*, volume 1182 of *Lecture Notes in Mathematics*, pages 247–259. Springer, Berlin, 1986.

[FMSh 240] Matthew Foreman, Menachem Magidor, and Saharon Shelah. Martin's maximum, saturated ideals, and nonregular ultrafilters. I. *Annals of Mathematics. Second Series*, **127**:1–47, 1988.

[ShWd 241] Saharon Shelah and Hugh Woodin. Large cardinals imply that every reasonably definable set of reals is Lebesgue measurable. *Israel Journal of Mathematics*, **70**:381–394, 1990.

[BsSh 242] Andreas Blass and Saharon Shelah. There may be simple $P_{\aleph_1}$- and $P_{\aleph_2}$-points and the Rudin-Keisler ordering may be downward directed. *Annals of Pure and Applied Logic*, **33**:213–243, 1987.

[Sh 250] Saharon Shelah. Some notes on iterated forcing with $2^{\aleph_0} > \aleph_2$. *Notre Dame Journal of Formal Logic*, **29**:1–17, 1988.

[FMSh 252] Matthew Foreman, Menachem Magidor, and Saharon Shelah. Martin's maximum, saturated ideals and nonregular ultrafilters. II. *Annals of Mathematics. Second Series*, **127**:521–545, 1988. will reapeare (in a revised form) in chapter XIII of [Sh f].

[Sh 253] Saharon Shelah. Iterated forcing and normal ideals on ω_1. *Israel Journal of Mathematics*, **60**:345–380, 1987.

[Sh 263] Saharon Shelah. Semiproper forcing axiom implies Martin maximum but not PFA$^+$. *The Journal of Symbolic Logic*, **52**:360–367, 1987.

[Sh 270] Saharon Shelah. Baire irresolvable spaces and lifting for a layered ideal. *Topology and its Applications*, **33**:217–221, 1989.

[MkSh 274] Alan H. Mekler and Saharon Shelah. Uniformization principles. *The Journal of Symbolic Logic*, **54**:441–459, 1989.

[Sh 276] Saharon Shelah. Was Sierpiński right? I. *Israel Journal of Mathematics*, **62**:355–380, 1988.

[ShSr 296] Saharon Shelah and Juris Steprans. Nontrivial homeomorphisms of $\beta\mathbf{N} \setminus \mathbf{N}$ without the continuum hypothesis. *Fundamenta Mathematicae*, **132**:135–141, 1989.

[GrSh 302] Rami Grossberg and Saharon Shelah. On the structure of $\mathrm{Ext}_p(G, \mathbf{Z})$. *Journal of Algebra*, **121**:117–128, 1989.

[JdSh 308] Haim Judah and Saharon Shelah. The Kunen-Miller chart (Lebesgue measure, the Baire property, Laver reals and preservation theorems for forcing). *The Journal of Symbolic Logic*, **55**:909–927, 1990.

[GiSh 310] Moti Gitik and Saharon Shelah. Cardinal preserving ideals. *Journal of Symbolic Logic*, **submitted**.

[Sh 311] Saharon Shelah. A more general iterable condition ensuring $\aleph_1$ is not collapsed, II. *preprint*.

[MkSh 313] Alan H. Mekler and Saharon Shelah. Diamond and λ-systems. *Fundamenta Mathematicae*, **131**:45–51, 1988.

[MRSh 314] Alan H. Mekler, Andrzej Rosłanowski, and Saharon Shelah. On the p-rank of Ext. *Israel Journal of Mathematics*, **submitted**.

[ShSr 315] Saharon Shelah and Juris Steprans. PFA implies all automorphisms are trivial. *Proceedings of the American Mathematical Society*, **104**:1220–1225, 1988.

[Sh 326] Saharon Shelah. Viva la difference I: Nonisomorphism of ultrapowers of countable models. In *Set Theory of the Continuum*, volume 26 of *Mathematical Sciences Research Institute Publications*, pages 357–405. Springer Verlag, 1992.

[JShW 339] Haim Judah, Saharon Shelah, and Hugh Woodin. The Borel conjecture. *Annals of Pure and Applied Logic*, **50**:255–269, 1990. Correction of third section has appeared in the book by Bartoszynski and Judah.

[Sh 351] Saharon Shelah. Reflecting stationary sets and successors of singular cardinals. *Archive for Mathematical Logic*, **31**:25–53, 1991.

[JeSh 378] Thomas Jech and Saharon Shelah. A note on canonical functions. *Israel Journal of Mathematics*, **68**:376–380, 1989.

[Sh 400a] Saharon Shelah. Cardinal arithmetic for skeptics. *American Mathematical Society. Bulletin. New Series*, **26**:197–210, 1992.

[AbSh 403] Uri Abraham and Saharon Shelah. A Δ^2_2 well-order of the reals and incompactness of $L(Q^{MM})$. *Annals of Pure and Applied Logic*, **59**:1–32, 1993.

[FrSh 406] David H. Fremlin and Saharon Shelah. Pointwise compact and stable sets of measurable functions. *Journal of Symbolic Logic*, **58**:435–455, 1993.

[Sh 407] Saharon Shelah. CON($\mathfrak{u} > \mathfrak{i}$). *Archive for Mathematical Logic*, **31**:433–443, 1992.

[Sh 420] Saharon Shelah. Advances in Cardinal Arithmetic. In *Finite and Infinite Combinatorics in Sets and Logic*, pages 355–383. Kluwer Academic Publishers, 1993. N.W. Sauer et al (eds.).

[ShSr 427] Saharon Shelah and Juris Steprans. Somewhere trivial automorphisms. *Journal of the London Mathematical Society*, **49**:569–580, 1994.

[Sh 430] Saharon Shelah. Further cardinal arithmetic. *Israel Journal of Mathematics*, **95**:61–114, 1996.

[BuSh 437] Max R. Burke and Saharon Shelah. Linear liftings for non complete probability space. *Israel Journal of Mathematics*, **79**:289–296, 1992.

[EMSh 441] Paul C. Eklof, Alan H. Mekler, and Saharon Shelah. Uniformization and the diversity of Whitehead groups. *Israel Journal of Mathematics*, **80**:301–321, 1992.

[EMSh 442] Paul C. Eklof, Alan H. Mekler, and Saharon Shelah. Hereditarily separable groups and monochromatic uniformization. *Israel Journal of Mathematics*, **88**:213–235, 1994.

[Sh 460] Saharon Shelah. The Generalized Continuum Hypothesis revisited. *Israel Journal of Mathematics*, **submitted**.

[RoSh 470] Andrzej Rosłanowski and Saharon Shelah. Norms on possibilities I: forcing with trees and creatures. *Memoirs of the AMS*, **accepted**.

[Sh 486] Saharon Shelah. Uniformization. *preprint*.

[EkSh 505] Paul C. Eklof and Saharon Shelah. A Combinatorial Principle Equivalent to the Existence of Non-free Whitehead Groups. In *Abelian group theory and related topics*, volume 171 of *Contemporary Mathematics*, pages 79–98. American Mathematical Society, Providence, RI, 1994. edited by R. Goebel, P. Hill and W. Liebert, Oberwolfach proceedings.

[FShS 544] Sakaé Fuchino, Saharon Shelah, and Lajos Soukup. Sticks and clubs. *preprint*.

[Sh 576] Saharon Shelah. Categoricity of an abstract elementary class in two successive cardinals. *Israel Journal of Mathematics*, **submitted**.

[Sh 587] Saharon Shelah. Not collapsing cardinals $\leq \kappa$ in $(< \kappa)$–support iterations. *in preparation.*

[Sh 594] Saharon Shelah. There may be no nowhere dense ultrafilter. In *Proceedings of the Logic Colloquium Haifa'95*, accepted. Springer.

[DjSh 604] Mirna Džamonja and Saharon Shelah. $\clubsuit$ does not imply the existence of a Suslin tree. *Israel Journal of Mathematics*, submitted.

[Sh 630] Saharon Shelah. Non-elementary proper forcing notions. *in preparation.*

[Sh 656] Saharon Shelah. On CS iterations not adding reals. *in preparation.*